HANDBOOK OF
PHOTOMEDICINE

HANDBOOK OF PHOTOMEDICINE

Edited by

Michael R. Hamblin, PhD
Ying-Ying Huang, MD

CRC Press
Taylor & Francis Group
Boca Raton London New York

CRC Press is an imprint of the
Taylor & Francis Group, an **informa** business

Taylor & Francis
Taylor & Francis Group
6000 Broken Sound Parkway NW, Suite 300
Boca Raton, FL 33487-2742

First issued in paperback 2020

© 2014 by Taylor & Francis Group, LLC
Taylor & Francis is an Informa business

No claim to original U.S. Government works

Version Date: 20130318

ISBN 13 : 978-0-367-57629-5 (pbk)
ISBN 13 : 978-1-4398-8469-0 (hbk)

Library of Congress Cataloging-in-Publication Data

Handbook of photomedicine / editors, Michael R. Hamblin and Ying-Ying Huang.
    p. ; cm.
  Includes bibliographical references and index.
  ISBN 978-1-4398-8469-0 (alk. paper)
  I. Hamblin, Michael R. II. Huang, Ying-Ying.
  [DNLM: 1. Phototherapy--methods. 2. Ultraviolet Rays--adverse effects. WB 480]

RM840
615.8'31--dc23
                                             2013010111

Visit the Taylor & Francis Web site at
http://www.taylorandfrancis.com

and the CRC Press Web site at
http://www.crcpress.com

To my dearest wife Angela, without whom this book would not have been possible.

M. R. H.

To my father Guangwu Huang, my mother Xinxin Zheng, and my dearest husband Longde Lin, for their understanding and everlasting support.

Y.-Y. H.

Contents

SECTION III   Ultraviolet Phototherapy

SECTION IV   Photodynamic Therapy (PDT)

SECTION V Low-Level Laser (Light) Therapy (LLLT)

SECTION VI   Surgical Laser Therapy

SECTION VII Other Phototherapies and Future Outlook

Preface

This handbook is designed to be of interest to clinicians and researchers in the broad field of photomedicine including dermatology, oncology, dentistry, ophthalmology, infectious disease, neurology, orthopedics, and even psychiatry. Basic researchers in chemistry, physics, cell biology, and photobiology will also find much of interest.

Photomedicine has developed to become one of the most exciting fields in biomedical research in the past 50 years. Although all life has evolved to depend on sunlight as its basic energy source, light can also be harmful and can cause diseases in certain subjects. Studies on the effects of ultraviolet phototherapy and photodynamic therapy on skin and cancers have led to a fruitful collaboration between basic scientists and clinicians. Applications of lasers and intense pulsed light sources play a critical role in clinical practice with a growing and demanding market. Low-level light therapy is routinely practiced as part of physical therapy and cosmetic procedures and is under intense investigation for many serious conditions and even life-threatening diseases, with coherent light sources (lasers) or noncoherent light sources consisting of filtered lamps or light-emitting diodes (LEDs) being used throughout the world.

The 70 chapters are grouped into 7 sections, progressing from the history and fundamentals of photomedicine to a diverse collection of therapeutic applications of light known collectively as phototherapy. The reader is thus provided with information in a logical sequence that facilitates the understanding of diseases in humans caused by light, the rationale for photoprotection, and the major applications of phototherapy in clinical practice. Section I contains an Introduction consisting of a series of historical vignettes of some of the pioneers in photomedicine over the last two centuries and chapters focused on the history and fundamentals of photomedicine, followed by the diseases caused by light covered in Section II. Sections III to VII cover the most important clinical applications of the different kinds of light that vary widely in both wavelength and intensity. Section III covers ultraviolet phototherapy for skin diseases and infection. Section IV covers the basic science of photodynamic therapy and its use in cancer therapy, antimicrobial therapy, and its application in different medical specialties. Section V discusses the mechanistic studies and clinical applications of low-level laser (light) therapy. Section VI covers the use of high power or surgical laser therapy in different specialties. Section VII is a collection of miscellaneous types of phototherapy that do not fit easily into other parts of the handbook.

Our purpose was to create a book that is comprehensive and up-to-date, yet is user-friendly to its intended readers. Each chapter is written by different authors from different parts of the world, chosen for their expertise in their respective fields. We hope that the topics covered in this handbook will provide a worldwide view of photomedicine and will serve as an authoritative reference work for many years to come.

Acknowledgments

The editors express their gratitude to the following for their valuable editorial assistance: Pinar Avci, Tyler G. St. Denis, Magesh Sadasivam, Rakkiyappan Chandran, Tianhong Dai, Rui Yin, Wanessa CMA de Melo, Asheesh Gupta, Daniela Vecchio, Fatma Vatansever, Shih-Fong Huang, Ana Elisa S. Jorge, Alex Lee, and Vida de Arce.

Acknowledgements

Editors

 **Michael R. Hamblin, PhD,** is a principal investigator at the Wellman Center for Photomedicine at Massachusetts General Hospital, an associate professor of dermatology at Harvard Medical School, and a member of the affiliated faculty of the Harvard-MIT Division of Health Science and Technology. He was trained as a synthetic organic chemist and earned his PhD from Trent University in England. Dr. Hamblin's research interests lie in the area of photomedicine, concentrating on photodynamic therapy (PDT) for infections, cancer, and heart disease and on low-level light therapy (LLLT) for wound healing, arthritis, traumatic brain injury, and hair regrowth. He directs a laboratory of approximately 16 postdoctoral fellows, visiting scientists, and graduate students. His research program is supported by the National Institutes of Health (NIH), Congressionally Directed Medical Research Programs (CDMRP), United States Air Force Office of Scientific Research (USAFOSR), and Center for Integration of Medicine and Innovative Technology (CIMIT), among other funding agencies. He has published over 190 peer-reviewed articles, more than 150 conference proceedings, book chapters, and international abstracts and holds 8 patents. He is an associate editor for 7 journals and serves on NIH Study Sections. For the past 7 years, Dr. Hamblin has chaired an annual conference at SPIE Photonics West entitled "Mechanisms for low level light therapy," and he has edited the seven proceedings volumes together with two other major textbooks on PDT. In 2011, Dr. Hamblin was honored by election as a fellow of SPIE for services to laser biomedicine.

 **Ying-Ying Huang, MD,** has been a postdoctoral fellow in Dr. Hamblin's laboratory for 4 years. She was promoted to instructor at Harvard Medical School in 2013. She received her MD from China in 2004, earned her MMed in dermatology in China, and was trained as a dermatologist at Guangxi Medical University. Dr. Huang's research interests lie in photodynamic therapy (PDT) for infections, cancer and mechanisms of low-level light therapy (LLLT). She has published 40 peer-reviewed articles and 15 conference proceedings and book chapters.

Contributors

Jan-Bonne Aans
Center for Optical Diagnostics and
  Therapy
Erasmus Medical Center
Rotterdam, the Netherlands

Heidi Abrahamse
Laser Research Centre
Faculty of Health Sciences
University of Johannesburg
Johannesburg, South Africa

Orn Adalsteinsson
International Strategic Cancer Alliance
Philadelphia, Pennsylvania

Vaqar Adhami
Department of Dermatology
University of Wisconsin-Madison
Madison, Wisconsin

Patrizia Agostinis
Laboratory of Cell Death Research and
  Therapy
University of Leuven (KU Leuven)
Leuven, Belgium

Luciano Alleruzzo
Immunophotonics Inc.
Columbia, Missouri

Arjen Amelink
Center for Optical Diagnostics and
  Therapy
Erasmus Medical Center
Rotterdam, the Netherlands

Karl E. Anderson
Department of Preventive Medicine and
  Community Health
The University of Texas Medical Branch
Galveston, Texas

R. Rox Anderson
Wellman Center for Photomedicine
Massachusetts General Hospital
and
Department of Dermatology
Harvard Medical School
Boston, Massachusetts

Takahiro Ando
Wellman Center for Photomedicine
Massachusetts General Hospital
Boston, Massachusetts

and

Department of Electronics and Electrical
  Engineering
Keio University
Tokyo, Japan

Pinar Avci
Wellman Center for Photomedicine
Massachusetts General Hospital
and
Department of Dermatology
Harvard Medical School
Boston, Massachusetts

Muriel Barberi-Heyob
Centre de Recherche en Automatique de
  Nancy (CRAN)
Université de Lorraine
Nancy, France

and

Médicaments Photoactivables–
  Photochimiothérapie (PHOTOMED)
Limoges, France

Anca M. Barbu
Center for Laryngeal Surgery and Voice
  Rehabilitation
Department of Surgery
and
Massachusetts General Hospital
Harvard Medical School
Boston, Massachusetts

Juliana Basko-Plluska
Section of Dermatology
University of Chicago Medical Center
Chicago, Illinois

Rene-Jean Bensadoun
Radiation Oncology Department
University Hospital and Faculty of
  Medicine of Poitiers
Poitiers, France

Kristian Berg
Department of Radiation Biology
Institute for Cancer Research
and
The Norwegian Radium Hospital
Oslo University Hospital
Oslo, Norway

Naeem Bhojani
Endourology Fellow
Indiana University School of Medicine
Indianapolis, Indiana

Jan Magnus Bjordal
Section of Physiotherapy Science
University of Bergen
Bergen, Norway

James A. Burns
Center for Laryngeal Surgery and Voice
  Rehabilitation
Department of Surgery
Massachusetts General Hospital
and
Harvard Medical School
Boston, Massachusetts

Theresa M. Busch
Department of Radiation Oncology
Perelman School of Medicine
University of Pennsylvania
Philadelphia, Pennsylvania

Scott N. Byrne
Discipline of Infectious Diseases and
   Immunology
and
Discipline of Dermatology
Bosch Institute
University of Sydney
and
Sydney Cancer Centre
Sydney, New South Wales, Australia

PierGiacomo Calzavara-Pinton
Department of Dermatology
University of Brescia
Brescia, Italy

James D. Carroll
THOR Photomedicine Ltd
Chesham, United Kingdom

Keith A. Cengel
Department of Radiation Oncology
University of Pennsylvania
Philadelphia, Pennsylvania

Wei R. Chen
Department of Engineering and Physics
University of Central Oklahoma
Edmond, Oklahoma

Michael A. Choma
Departments of Diagnostic Radiology,
   Biomedical Engineering, and
   Pediatrics
Yale University
New Haven, Connecticut

Roberta Chow
Nerve Research Foundation
Brain and Mind Institute
The University of Sydney
Camperdown, New South Wales, Australia

Denise Collangettes
Unité de Laserthérapie et Soins Dentaires
Centre Jean-Perrin
Clermont-Ferrand, France

Melissa Culligan
Department of Radiation Oncology
University of Pennsylvania
Philadelphia, Pennsylvania

Tianhong Dai
Wellman Center for Photomedicine
Massachusetts General Hospital
and
Department of Dermatology
Harvard Medical School
Boston, Massachusetts

Diona L. Damian
Discipline of Dermatology
Bosch Institute
University of Sydney
and
Sydney Cancer Centre
Sydney, Australia

Vida de Arce
Wellman Center for Photomedicine
Massachusetts General Hospital
Harvard Medical School
Boston, Massachusetts

Elodie Debefve
Medical Photonics Group
Swiss Federal Institute of Technology
   (EPFL)
Lausanne, Switzerland

Fiona Dignan
Clinical Research Fellow in Haematology
St John's Institute of Dermatology
St Thomas' Hospital
and
Section of Haemato-oncology
Royal Marsden Hospital
London, United Kingdom

Ryan F. Donnelly
School of Pharmacy
Queen's University Belfast
Medical Biology Centre
Belfast, United Kingdom

Anja Eichner
Department of Dermatology
Regensburg University Hospital
Regensburg, Germany

Sam Eljamel
The Scottish PDT Centre
Ninewells Hospital and Medical School
Scotland, United Kingdom

Cleber Ferraresi
Laboratory of
   Electro-Thermo-Phototherapy
Department of Physical Therapy
and
Department of Biotechnology
Federal University of São Carlos
and
Center for Research in Optics and Photonics
Physics Institute of São Carlos
University of São Paulo
São Carlos, Brazil

Leila Soares Ferreira
School of Dentistry
University of São Paulo
São Paulo, Brazil

Gabriela Ferrel
Hospital Nacional Edgardo Rebagliati
   Martins
Lima, Peru

Jarod C. Finlay
Department of Radiation Oncology
University of Pennsylvania
Philadelphia, Pennsylvania

Cristina Flors
Madrid Institute for Advanced Studies in
   Nanoscience (IMDEA Nanociencia)
Madrid, Spain

Joseph S. Friedberg
Department of Radiation Oncology
University of Pennsylvania
Philadelphia, Pennsylvania

Céline Frochot
Laboratoire Réactions et Génie des
   Procédés (LRGP)
Université de Lorraine
Nancy, France

Anne-Marie Gagné
Centre de Recherche de l'Institut
Universitaire en Santé Mentale de Québec
Quebec, Quebec, Canada

Shannon M. Gallagher-Colombo
Department of Radiation Oncology
Perelman School of Medicine
University of Pennsylvania
Philadelphia, Pennsylvania

Abhishek D. Garg
Laboratory of Cell Death Research and
    Therapy
University of Leuven (KU Leuven)
Leuven, Belgium

Lilach Gavish
Faculty of Medicine of
Institute for Medical Research-IMRIC
The Hebrew University Hadassah
    Medical School
Jerusalem, Israel

Fernanda Pereira Gonzales
Department of Dermatology
Regensburg University Hospital
Regensburg, Germany

Eric Gou
Department of Preventive Medicine and
    Community Health
The University of Texas Medical Branch
Galveston, Texas

Asheesh Gupta
Wellman Center for Photomedicine
Massachusetts General Hospital
and
Department of Dermatology
Harvard Medical School
Boston, Massachusetts

and

Defence Institute of Physiology and
    Allied Sciences
Delhi, India

Pradeep Kumar Gupta
Laser Biomedical Applications and
    Instrumentations Division
Raja Ramanna Centre for Advanced
    Technology
Indore, India

Gary M. Halliday
Discipline of Dermatology
Bosch Institute
University of Sydney
and
Sydney Cancer Centre
Sydney, Australia

Michael R. Hamblin
Wellman Center for Photomedicine
Massachusetts General Hospital
and
Department of Dermatology
Harvard Medical School
Boston, Massachusetts

Prue H. Hart
Telethon Institute for Child Health
    Research
and
Centre for Child Health Research
University of Western Australia
Perth, Australia

Tayyaba Hasan
Wellman Center for Photomedicine
Massachusetts General Hospital
Harvard Medical School
Boston, Massachusetts

John L. M. Hawk
Dermatological Photobiology
Photodermatology Unit
St John's Institute of Dermatology
and
St Thomas' Hospital
London, United Kingdom

Marc Hébert
Centre de Recherche de l'Institut
Universitaire en Santé Mentale de Québec
Quebec, Quebec, Canada

Lars Hode
Swedish Laser-Medical Society
Lidingö, Sweden

Tomas Hode
Immunophotonics Inc.
Columbia, Missouri

Anders Høgset
PCI Biotech AS
Oslo, Norway

Michael F. Holick
Department of Medicine
Section of Endocrinology, Nutrition, and
    Diabetes
Vitamin D, Skin and Bone Research
    Laboratory
Boston University Medical Center
Boston, Massachusetts

Herbert Hönigsmann
Department of Dermatology
Medical University of Vienna
Vienna, Austria

Yih-Chih Hsu
Department of Bioscience Technology
Chung Yuan Christian University
Zongli, Taiwan

Brendan K. Huang
Departments of Diagnostic Radiology,
    Biomedical Engineering, and Pediatrics
Yale University
New Haven, Connecticut

Huang-Chiao Huang
Wellman Center for Photomedicine
Massachusetts General Hospital
Boston, Massachusetts

Ying-Ying Huang
Wellman Center for Photomedicine
Massachusetts General Hospital
and
Department of Dermatology
Harvard Medical School
Boston, Massachusetts

and

Guangxi Medical University
Nanning, China

Zheng Huang
Department of Bioscience Technology
University of Colorado
Zongli, Taiwan

Mohammad Ilyas
School of Environmental Engineering
University Malaysia Perlis
Kangar, Malaysia

Eric Jadaud
Service de Radiothérapie
Institut de Cancérologie de la Loire
Angers, France

Peter Jenkins
Irradia LLC
Lawndale, North Carolina

Patrice Jichlinski
Department of Urology
CHUV University Hospital
Lausanne, Switzerland

Kirpa Johar
Laser Dentistry Research and Review
Johar's Laser Dental Clinic
Bangalor, Kamataka, India

Penny Joshi
Photodynamic Therapy Center
Department of Cell Stress Biology
Roswell Park Cancer Institute
Buffalo, New York

Asta Juzeniene
Department of Radiation Biology
Institute for Cancer Research
and
The Norwegian Radium Hospital
Oslo University Hospital
Oslo, Norway

Edidiong Ntuen Kaminska
Section of Dermatology
University of Chicago Medical Center
Chicago, Illinois

Tiina I. Karu
Institute of Laser and Information
   Technologies
Russian Academy of Sciences
Moscow Region, Russian Federation

Stanley Kimani
Wellman Center for Photomedicine
Massachusetts General Hospital
Harvard Medical School
Boston, Massachusetts

Irene E. Kochevar
Wellman Center for Photomedicine
Massachusetts General Hospital
Boston, Massachusetts

Garuna Kositratna
Wellman Center for Photomedicine
Massachusetts General Hospital
Boston, Massachusetts

Thorsten Krueger
Thoracic and Vascular Department
Centre Hospitalier Universitaire Vaudois
   (CHUV)
Lausanne, Switzerland

Pavel Kucera
Department of Biomedical Engineering
Czech Technical University
Prague, Czech Republic

Norbert Lange
Laboratory of Pharmaceutical
   Technology
University of Geneva
Geneva, Switzerland

Xiaosong Li
Department of Oncology
The First Affiliated Hospital of Chinese
   PLA General Hospital
Beijing, China

Henry W. Lim
Department of Dermatology
Henry Ford Hospital
Detroit, Michigan

James Lingeman
Indiana University School of Medicine
Indianapolis, Indiana

Rodrigo A. B. Lopes-Martins
Department of Pharmacology
Institute of Biomedical Sciences
University of São Paulo
São Paulo, Brazil

John Lunn
Commonwealth Medical Research
   Institute
Nassau, The Bahamas

J. Guy Lyons
Discipline of Dermatology
Bosch Institute
University of Sydney
and
Sydney Cancer Centre
and
Sydney Head and Neck Cancer Institute
Sydney Cancer Centre
Royal Prince Alfred Hospital
Sydney, New South Wales, Australia

Tim Maisch
Department of Dermatology
Regensburg University Hospital
Regensburg, Germany

Iqbal Massodi
Wellman Center for Photomedicine
Massachusetts General Hospital
Harvard Medical School
Boston, Massachusetts

Daiane Thais Meneguzzo
School of Dentistry
São Leopoldo Mandic
Campinas, Brazil

Patrycja Mikolajewska
Department of Radiation Biology
Institute for Cancer Research
and
The Norwegian Radium Hospital
Oslo University Hospital
Oslo, Norway

Lee Miller
Scripps Clinic
San Diego, California

Joseph R. Missert
Photodynamic Therapy Center
Department of Cell Stress Biology
Roswell Park Cancer Institute
Buffalo, New York

Eiman Mukhtar
Department of Dermatology
University of Wisconsin-Madison
Madison, Wisconsin

Hasan Mukhtar
Department of Dermatology
University of Wisconsin-Madison
Madison, Wisconsin

Raj G. Nair
Centre for Medicine and Oral Health
School of Dentistry and Oral Health
Griffith University
Southport, Australia

Mark Naylor
Dermatology Associates of San Antonio
San Antonio, Texas

Santi Nonell
IQS School of Engineering
Universitat Ramon Llull
Barcelona, Spain

Robert E. Nordquist
Immunophotonics Inc.
Columbia, Missouri

Mary Norval
Biomedical Sciences
University of Edinburgh Medical School
Edinburgh, Scotland

Patrycja Nowak-Sliwinska
Medical Photonics Group
Swiss Federal Institute of Technology
  (EPFL)
Lausanne, Switzerland

Theodore Nyame
Division of Plastic and Reconstructive
  Surgery
Harvard Medical School
Boston, Massachusetts

Kris Nys
Laboratory of Cell Death Research and
  Therapy
University of Leuven
Leuven, Belgium

Girgis Obaid
School of Chemistry
University of East Anglia
Norwich, United Kingdom

Bernhard Ortel
Section of Dermatology
University of Chicago Medical Center
Chicago, Illinois

Ravindra K. Pandey
Photodynamic Therapy Center
Department of Cell Stress Biology
Roswell Park Cancer Institute
Buffalo, New York

Nivaldo A. Parizotto
Department of Physical Therapy
and
Graduate Program in Bioengineering
  Interunits
Federal University of São Carlos
São Carlos, Brazil

and

Wellman Center for Photomedicine
Massachusetts General Hospital
Harvard Medical School
Boston, Massachusetts

Harishankar Patel
Laser Biomedical Applications and
  Instrumentations Division
Raja Ramanna Centre for Advanced
  Technology
Indore, India

Thierry J. Patrice
Laboratoire de Photobiologie des Cancers
Département Laser
Nantes, France

Marlène Pernot
Centre de Recherche en Automatique de
  Nancy (CRAN)
Université de Lorraine
Nancy, France

James Ramos
School of Biological and Health Systems
  Engineering
Arizona State University
Tempe, Arizona

Robert W. Redmond
Wellman Center for Photomedicine
Massachusetts General Hospital
Boston, Massachusetts

Kaushal Rege
Chemical Engineering
Arizona State University
Tempe, Arizona

Dominic J. Robinson
Center for Optical Diagnostics and
  Therapy
Erasmus Medical Center
Rotterdam, the Netherlands

Edward Victor Ross
Laser & Cosmetic Dermatology Center
Scripps Clinic
San Diego, California

David A. Russell
School of Chemistry
University of East Anglia
Norwich, United Kingdom

Magesh Sadisivam
Wellman Center for Photomedicine
Massachusetts General Hospital
Boston, Massachusetts

Courtney Saenz
Photodynamic Therapy Center
Department of Cell Stress Biology
Roswell Park Cancer Institute
Buffalo, New York

Raffaella Sala
Department of Dermatology
University Hospital Spedali Civili
Brescia, Italy

Yoram Salomon
Department of Biological Regulation
The Weizmann Institute of Science
Rehovot, Israel

Julia Scarisbrick
University Hospital Birmingham
Birmingham, United Kingdom

Drew Schembre
Swedish Gastroenterology
Seattle, Washington

Avigdor Scherz
Department of Plant Sciences
The Weizmann Institute of Science
Rehovot, Israel

R. Bryan Sears
Wellman Center for Photomedicine
Massachusetts General Hospital
Harvard Medical School
Boston, Massachusetts

Pål K. Selbo
Department of Radiation Biology
Institute for Cancer Research
The Norwegian Radium Hospital
and
Oslo University Hospital
and
PCI Biotech AS
Oslo, Norway

Michel Sickenberg
Medical Photonics Group
Swiss Federal Institute of Technology
  (EPFL)
Lausanne, Switzerland

Alexis Sidoroff
Department of Dermatology and
  Venereology
Medical University of Innsbruck
Innsbruck, Austria

Charles B. Simone
Department of Radiation Oncology
University of Pennsylvania
Philadelphia, Pennsylvania

Tyler G. St. Denis
Department of Chemistry
Columbia University
New York City

and

Wellman Center for Photomedicine
Massachusetts General Hospital
Boston, Massachusetts

Henricus J. C. M. Sterenborg
Center for Optical Diagnostics and
    Therapy
Erasmus Medical Center
Rotterdam, the Netherlands

Mahesh Kumar Swami
Laser Biomedical Applications and
    Instrumentations Division
Raja Ramanna Centre for Advanced
    Technology
Indore, India

Elena G. Vakulovskaya
N.N. Blochin
Russian Cancer Research Center
Russian Academy of Medical Sciences
Moscow, Russian Federation

Hubert van den Bergh
Medical Photonics Group
Swiss Federal Institute of Technology
    (EPFL)
Lausanne, Switzerland

Robert L. P. van Veen
Center for Optical Diagnostics and
    Therapy
Erasmus Medical Center
Rotterdam, the Netherlands

Régis Vanderesse
Laboratoire de Chimie Physique
    Macromoléculaire (LCPM)
Université de Lorraine
Nancy, France

and

Médicaments Photoactivables-
    Photochimiothérapie (PHOTOMED)
Limoges, France

Fatma Vatansever
Wellman Center for Photomedicine
Massachusetts General Hospital
Harvard Medical School
Boston, Massachusetts

Georges Wagnières
Medical Photonics Group
Swiss Federal Institute of Technology
    (EPFL)
Lausanne, Switzerland

Yabo Wang
Thoracic and Vascular Department
Centre Hospitalier Universitaire Vaudois
    (CHUV)
Lausanne, Switzerland

Robert Weersink
Radiation Medicine Program
University Health Network
Ontario, Toronto, Canada

Andrea Weiss
Medical Photonics Group
Swiss Federal Institute of Technology
    (EPFL)
Lausanne, Switzerland

Anette Weyergang
Department of Radiation Biology
Institute for Cancer Research
The Norwegian Radium Hospital
Oslo University Hospital
Oslo, Norway

Brian C. Wilson
Department of Medical Biophysics
Faculty of Medicine
University of Toronto
and
Ontario Cancer Institute
University Health Network
Ontario, Toronto, Canada

Peter Wolf
Research Unit for Photodermatology
Department of Dermatology
Medical University of Graz
Graz, Austria

Chanisada Wongpraparut
Department of Dermatology
Faculty of Medicine Siriraj Hospital
Bangkok, Thailand

Charles Wormington
Salus University
Elkins Park, Pennsylvania

Shengnan Wu
MOE Key Laboratory of Laser Life
    Science
Institute of Laser Life Science
College of Biophotonics
South China Normal University
Guangzhou, China

Da Xing
MOE Key Laboratory of Laser Life
    Science
Institute of Laser Life Science
College of Biophotonics
South China Normal University
Guangzhou, China

Rui Yin
Department of Dermatology
Southwest Hospital
Third Military Medical University
Chongqing, China

and

Wellman Center for Photomedicine
Massachusetts General Hospital
and
Department of Dermatology
Harvard Medical School
Boston, Massachusetts

Lei Zak Zheng
Wellman Center for Photomedicine
Massachusetts General Hospital
Harvard Medical School
Boston, Massachusetts

History and Fundamentals

I

1

Michael R. Hamblin
Massachusetts General Hospital

Ying-Ying Huang
Massachusetts General Hospital
Guangxi Medical University

Introduction: Historical Vignettes from the Field of Photomedicine

The present handbook aims to cover the entire area of medical knowledge known as "photomedicine." This field is concerned with two broad areas of medicine: (1) diseases caused by light and (2) the therapeutic applications of light known as phototherapy. The latter broad area can be further subdivided into five more-specialized applications of phototherapy: (i) ultraviolet phototherapy, (ii) photodynamic therapy (PDT), (iii) low-level laser (light) therapy, (iv) laser surgery and medicine, and (v) phototherapy for miscellaneous applications.

In this introductory chapter, we will attempt to draw biographical sketches of some great historical figures who have made seminal contributions to the field of photomedicine.

The history of photomedicine goes back more than 3000 years to India, where sunlight was employed for therapeutic purposes as recorded in the sacred Hindu text *Atharva Veda* dating from 1400 B.C. Persons with vitiligo (a patchy depigmentation of the skin then thought to be a form of leprosy) were given certain plant extracts (*Eclipta prostrata*, *Citrullus colocynthis*, and *Curcuma longa*) and then exposed to the sun (Fitzpatrick and Pathak 1959). In Ayurvedic and Chinese traditional medicine, the plant *Psoralea corylifolia*, which contains the furanocoumarin compound called "psoralen," was used for similar purposes. In ancient Egypt, *Ammi majus*, another furanocoumarin-containing plant, was employed.

Starting in the 18th century, reports began to appear in the medical literature indicating that sunlight ameliorated different diseases. In 1735, Fiennius (Giese 1964) described a case in which he cured a cancerous growth on the lip using a sunbath. In 1774, Faure (Russell and Russell 1927) reported that he successfully treated skin ulcers with sunlight, and in 1776, LePeyre and LeConte (Rollier 1923a) found that sunlight concentrated through a lens accelerated wound healing and destroyed tumors. There were also reports that sunlight had beneficial effects on internal maladies. In 1782, Harris (Giese 1964) used irradiated mollusk shells to improve a case of rickets (fragile bones resulting from vitamin D deficiency).

In 1801, Johann Wilhelm Ritter, a Polish physicist working at the University of Jena in Germany, discovered a form of light beyond the violet end of the spectrum that he called "chemical rays" and that later became known as "ultraviolet" light (Frercksa, Weberb, and Wiesenfeldt 2009). In 1845, Bonnet (1845) first reported that sunlight could be used to treat tuberculosis arthritis (a bacterial infection of the joints).

In the second half of the 19th century, the therapeutic application of sunlight known as heliotherapy gradually became popular. In 1855, Rikli from Switzerland opened a thermal station in Veldes in Slovenia for the provision of heliotherapy (Barth and Kohler 1992). In 1877, Downes and Blunt (1877) discovered, by chance, that sunlight could kill bacteria. They noted that sugar water placed on a windowsill turned cloudy in the shade but remained clear while in the sun. Upon microscopic examination of the two solutions, they realized that bacteria were growing in the shaded solution but not in the one exposed to sunlight.

Theobald Adrian Palm (1848–1928) discovered the role of sunlight in the prevention of rickets. His biography has been recorded by Russell Chesney (Chesney 2012).

Palm graduated in medicine from Edinburgh University and became a missionary for the Edinburgh Medical Mission. He spent 10 years in Japan and noted that rickets was essentially absent, in contrast to the situation in the United Kingdom where the urban prevalence figures ranged from 60 to 90% in the late 19th and early 20th centuries (Zappert 1910). He first wrote on the 'want of light' in a letter (Palm 1888) to the *British Medical Journal* in 1888, when he was living in Birkenhead, near Liverpool, and saw children with rickets. Palm speculated that the therapy of rickets should include 'the systematic use of sun-baths'. Palm wrote in more detail about his observations after making three investigations. First, he wrote to and assembled the replies from medical missionaries in the southeast region of Asia and North Africa. Second, he analyzed the topography of rickets in the UK based upon a medical research report (Owens 1889). Third, he analyzed the rates of rickets in other parts of Europe. Other medical missionaries from China, Mongolia, India, Morocco, Ceylon, and other parts of Japan rarely or never encountered rickets (Palm 1890). The geography of rickets appeared to involve the northern latitudes of Europe: Germany, England, Holland, Belgium, France and northern Italy. More southerly regions such as southern Italy, southern Spain, Turkey and Greece 'enjoy a notable immunity from it' (Palm 1890). Rickets abounded in the UK in large towns and industrialized regions. In Scotland there were Glasgow and Edinburgh and the coal-mining regions of the Clyde-Forth region. In England and Wales there were Tyneside, Lancaster and Yorkshire, Birmingham and Manchester, Cardiff and Swansea, and the whole of London except for the prosperous areas. These, apart from London, were the coal-mining districts and industrial areas of Britain (Clark and Jacks 2007). They were regions of iron and steel works and of the manufacture of bridges, steam engines and ships. Edinburgh was particularly hazy and smoggy, and the air was filled with soot leading to it being referred to as 'auld reeky'. It was also an area with the highest prevalence of rickets (Palm 1890). Palm focused upon the absence of sunlight. He recommended the use of sunbaths and the relocation of rachitic children to areas where sunshine is common (Findlay 1908). There had only been one previous mention of sunlight's role in the cure of rickets, by Jedrzej Sniadecki in 1822, who noted less rickets in children from rural districts of western Poland (Findlay 1908) but this was a local finding.

Nils Ryberg Finsen (1860–1904) (Figure 1.1) was born in the Faroe Islands and studied medicine at the University of Copenhagen, qualifying in 1890. He suffered from an illness that turned out to be Niemann-Pick's disease and is characterized by progressive thickening of the connective tissue of the liver, the heart, and the spleen. He received the Nobel Prize for physiology or medicine in 1903 and the following is taken from his Nobel lecture (Finsen 1967).

My disease has played a very great role for my whole development. The disease was responsible for my starting investigations on light: I suffered from anemia and tiredness, and since I lived in a house facing the north, I began to believe

that I might be helped if I received more sun. I therefore spent as much time as possible in its rays. As an enthusiastic medical man I was of course interested to know what benefit the sun really gave. I considered it from the physiological point of view but got no answer. I drew the conclusion that I was right and the physiology wrong. From this time (about 1888) I collected all possible observations about animals seeking the sun, and my conviction that the sun had a useful and important effect on the organism (especially the blood?) that became stronger and stronger. What this useful effect really was, I could not find; I have been working for this goal ever since but have not been able to find exactly what I have been seeking, though we have gone somewhat forward. My intention was even then (about 15 years ago) to use the beneficial effects of the sun in the form of sun bathing or artificial light baths; but I understood that it would be inappropriate to bring it into practical use if the theory was not built upon scientific investigations and definite facts. During my work towards this goal I encountered several effects of light. I then devised the treatment of small-pox with red light (1893) and further the treatment of lupus (1895). Both these things are therefore in a sense side-issues, but they completely occupied my time for several years and have partly drawn me away from my main goal.

Finsen demonstrated that "chemical rays" refracted from sunlight or from an electric arc lamp may have a stimulating effect on the tissues. If the irradiation is too strong, however, it may give rise to tissue damage. With smallpox, Finsen thought that the multiple scars might be avoided if the patient was protected from the chemical rays and used red light to improve healing. The experiments with such patients were successful. On the other hand, chemical rays free from heat rays might be used to obtain a useful effect on lupus vulgaris (cutaneous tuberculosis) or other skin diseases or employed as general sunbaths, which, on Finsen's suggestion, was tried in cases of nonpulmonary or "surgical" tuberculosis. The Finsen Institute was formed in Copenhagen in 1896 and was extended some years later as a result of the generosity of two Danish donors, Mr. Hageman and Mr. Jörgensen. Finsen became a knight of the Order of Dannebrog in 1899 to which, a few years later, the Silver Cross was added. Sadly, he died at the age of 44.

Finsen realized that the northern climate was not well suited for such therapy. The treatment of tuberculosis by sunlight was championed in Switzerland by O. Bernhard at St. Moritz and A. Rollier at Leysin, Switzerland. Solar therapy, as practiced by these workers, included increasing graduated exposures of parts of the body to sunlight and was thought to be accentuated by the fresh mountain air. The descriptions of the contributions of Bernhard and Rollier are taken from a paper by R.A. Hobday entitled "Sunlight Therapy and Solar Architecture" (Hobday 1997).

Oskar Bernhard (1861–1939) (Figure 1.2) was born in Samaden, Switzerland, and studied medicine at Zurich, Heidelberg, and Berne, developing a particular interest in surgery. He returned

FIGURE 1.1 Nils Ryberg Finsen.

FIGURE 1.2 Oskar Bernhard.

to the Upper Engadine, where he built up a large surgical practice and helped to establish the district hospital at Samaden, which opened in 1898. He began to use sunlight to heal first wounds and then tuberculosis. On February 2, 1902, an Italian patient was brought to the hospital with severe knife wounds. As the injured man was in danger of bleeding to death, Bernhard had to remove the spleen, which had been ruptured. Eight days after the laparotomy, the operation wound burst open and gaped widely. Re-stitching failed, and none of the treatments used to dry it up had any effect, so Bernhard took the unusual step of exposing the wound to sunlight. He gave the following account of the discovery (Bernhard 1926):

> As I entered the hospital one beautiful morning, and the sun shone warmly through the open window, while a refreshing and stimulating atmosphere filled the whole ward, the thought suddenly occurred to me of exposing this large wound to the sun and air; for the mountain peasant of the Grisons also exposes fresh pieces of flesh to the sun and dry air to preserve them, and in this way makes a nourishing and tasty food, the well-known 'Bindenfleisch.' So I resolved to try this antiseptic and drying effect of the sunlight and air on the living tissues. Then, to the great astonishment of the staff, I had the bed put to the open window and laid the large wound open. By the end of the first hour and a half of exposure a marked improvement was noticeable, and the wound presented quite a different appearance. The granulations became visibly more normal and healthy, and the enormous wound skinned over quickly under the treatment.

This success led Bernhard to treat all granulating and infected wounds with sunlight. He then began to treat open tuberculous cavities and, soon after, closed tuberculous foci of the bones, joints, and glands with sunlight. In 1905, Bernhard had established his own small private clinic for sunlight therapy at St. Moritz. It could accommodate some 33 patients and had south-facing balconies on two of the upper floors. This clinic is described in Bernhard's book of 1917, *Sonnenlichtbehandlung in der Chirurgie*, (Bernhard 1917) in which he put forward general design recommendations for heliotherapy wards and balconies and also outlined the plans for a more ambitious institution: a large sunlight-treatment clinic in the form of an amphitheater with a large solarium on the roof and open-air balconies built into the terraces that formed each story. During the early years of the First World War, Bernhard served in German military hospitals. In the summer of 1915, he began to use heliotherapy at Bad Durrheim in the Black Forest. At the request of the health department of the 14th German Army Corps, he started a sun clinic in the Association Hospital at Kindersolbad for soldiers with indolent wounds and external tuberculosis.

Auguste Rollier (1874–1954) was born at St. Aubin in the Swiss Canton of Neuchatel and graduated in medicine from Zurich and Berne. He became deeply disillusioned with the poor results obtained by surgery in the treatment of skeletal tuberculosis, which led him to abandon a promising career in surgery and take up general practice. Rollier left Berne and went into a rural general practice at Leysin in the Alpes Vaudoises and began to treat nonpulmonary tuberculosis with sunshine and fresh air. Although he consulted Oskar Bernhard on the practice of open-air methods, the degree to which Bernhard influenced him in other respects is not clear. Over the next 40 years, the technique Rollier devised for irradiating the body with sunlight (Rollier's Sunlight Therapy or Heliotherapy) came to be generally accepted (Rollier 1923b). The method of slow tanning practiced by Rollier was developed for debilitated patients who were often very seriously ill with tuberculosis and could not respond as well to sunlight as a healthy adult. Gradual exposure to cold air was important because it raised and maintained high metabolic activity in patients. This, in turn, was considered to improve their general health and to enable them to resist and overcome infection. Cool conditions and winter or low-angle summer sun also reduced the risk of overexposure to the sun's rays. Rollier regarded exposure to the sun at temperatures greater than 18°C to be a "hot-air bath" and not a sunbath. On arrival at Leysin, his patients underwent a thorough medical examination. Then, after they had completed a period of acclimatization, they were carefully exposed to cold air. After 1 to 2 weeks of this open-air treatment, sunlight treatment could begin in the solaria and on the balconies of his clinics (Figure 1.3a). His first clinic, called "Le Chalet" and opened in 1903, was in a converted boarding-house. His second clinic, called "Les Chamois" and opened in 1909, was converted from a hotel, and his biggest clinic, called "Les Frênes," was the first large, purpose-built sunlight-therapy clinic to be constructed in Europe. It was built in 1911 and consisted of a central, south-facing block and two large wings (Figure 1.3b).

FIGURE 1.3 (a) Sunlight treatment could begin on the solaria and balconies. (b) "Les Frênes," the first large, purpose-built sunlight-therapy clinic to be constructed in Europe.

PDT was discovered in 1900 in Munich, Germany, by Oscar Raab, a medical student working with Professor Herman von Tappeiner in Munich (Figure 1.4). They made the chance discovery during a thunderstorm that the combination of acridine red and light killed Infusoria, a species of paramecium (Raab 1900). He went on to show that this cytotoxic effect was greater than that of either acridine red alone, light alone, or acridine red exposed to light and then added to the paramecium. Raab associated this property of dyes (light-mediated cytotoxicity) with the optical property of fluorescence. He postulated that the effect was caused by the transfer of energy from light to the chemical, similar to the process of photosynthesis seen in plants after the absorption of light by chlorophyll. In a second paper, von Tappeiner (1900) discussed the potential future application of fluorescent substances in medicine. The first report of systemic administration of a photosensitizer (PS) in humans was in 1900 by Prime, a French neurologist, who used the dye eosin (a brominated

derivative of fluoroscein) orally in the treatment of epilepsy but observed that this treatment induced dermatitis in sun-exposed areas of skin (Prime 1900). This observation led to the first therapeutic clinical application of an interaction between a PS and light in which von Tappeiner, together with a dermatologist named Jesionek, used a combination of topically applied eosin and white light to treat skin tumors (von Tappeiner and Jesionek 1903). Together with Jodlbauer, von Tappeiner went on to demonstrate the requirement of oxygen in these photosensitization reactions (von Tappeiner and Jodlbauer 1904), and in 1907, they introduced the term "photodynamic action" to describe this phenomenon (von Tappeiner and Jodlbauer 1907).

The real application of PDT in medicine, however, had to wait more than 60 years and took place in the United States. Thomas J. Dougherty (Figure 1.5) is often credited as being the "father of PDT" and describes his contribution as follows (personal communication):

FIGURE 1.4 Herman von Tappeiner.

My role in developing Photodynamic Therapy (PDT) came about as a result of a series of events I could not have expected. For example my knowledge of photochemistry was not from my formal education as this was a new and undeveloped field at that time, but directly from George Hammond, the man known as the 'father of photochemistry', who was a consultant at DuPont where I had landed directly after graduate school and where I was assigned a project involving photo-degradation of one of their products. When I decided that industry was not for me I thought I might be able to do more fulfilling research at Roswell Park Cancer Institute which was in the same city (Buffalo) as was the DuPont Lab where I worked (and also my hometown). I managed to get a starting position at Roswell (working on someone's grant) with the help of George Hammond who wrote a letter of recommendation for me

to a friend he knew there (who happened to have worked with Linus Pauling). When I finished with the grant project, I obtained a more secure position in the Department of Radiation Oncology where I was able to work on my own ideas. I started by studying compounds that would produce oxygen when exposed to ionizing radiation in order to get around the low efficacy of radiation therapy in low oxygenated areas of tumors and can result in tumor regrowth. In testing the cellular toxicity of the compound I synthesized I was warned by a technician that I should do the cell culture test in the dark or the light would kill all the cells (the test was based on fluorescence of fluorescein developed only in viable cells!) I wondered if anyone had tried this approach to cancer treatment, so I exposed cancer cells to sunlight after adding fluorescein. They all died! We then went on to optimization of parameters first in experimental animals and then in patients. I think the readers of this book will know where this is going.

I learned how to initiate and treat tumors in animals, design, write and assist in human clinical trials, deal with the FDA, negotiate with pharmaceutical companies (quite an experience) and of course how to write grant applications—which thanks to many members of our group have supported our research since 1974.

As time went on I also learned that I was not the only one who thought of using light-activated drugs therapeutically—the first one was in 1799! Most of this came to me through contacts since I did not have access to journals going that far back. However these were mostly single patient or anecdotal reports, some of which were negative mainly to lack of optimized PDT variables. If I had known this at the time I may not have continued. Sometimes 'ignorance is bliss'.

The discovery of low-level laser therapy can be attributed to Endre Mester (1903–1984) (Figure 1.6) in Hungary. His contribution is described in a tribute by Professor Lajos Gáspár, his friend and colleague from Hungary (Gaspar 2009).

FIGURE 1.5 Thomas J. Dougherty.

FIGURE 1.6 Endre Mester.

As a schoolboy, he excelled as a violin player and developed an interest in medicine. With such diverse interests, he decided to sit university entrance examinations in both medicine and music. He succeeded in gaining a place to read medicine at the University of Budapest. His continued interest in music was relegated to social enjoyment with friends on weekends. Having successfully completed his undergraduate medical study and graduation, he was invited to receive training in surgery as a resident staff member of the 3rd Department of Surgery at the University of Budapest. He subsequently obtained his first position outside the university as head of children's surgery in St. Steven's Hospital in Budapest. From there, greater recognition beckoned and his next position was head of a major hospital in Budapest, the Bajcsy Zsilinszky. There he combined clinical work with administration, being appointed director of the hospital. During this period, Mester undertook research and scientific work in the field of abdominal surgery, with a special focus on bile duct surgery. His interest in the latter led to his gaining his PhD. Promotion led to his full professorship and director of the 2nd Department of Surgery of Semmelweis University in Budapest and election as President of the Society of Hungarian Surgeons. At Semmelweis, Mester started laser research in 1965; this was five years after the development of the first laser by Theodore H. Maiman (Maiman 1960) and only two years after surgical laser applications had been pioneered by notable workers such as Paul McGuff (McGuff et al. 1963) and Leon Goldman (Goldman et al. 1963). Like others of his era, Mester attempted to use a "high-power" laser to destroy malignant tumors. Early in his experiments, he implanted tumor cells beneath the skin of laboratory rats and exposed them to a customized ruby laser, based on Maiman's earlier model. To his surprise, the tumor cells were not destroyed by doses of what was presumed to be high power laser energy. Instead, it was observed that in many cases the skin incisions made to implant the abnormal cells appeared to heal faster in treated animals, compared to incisions of control animals that were not treated with light (Mester et al. 1968). Mester was baffled as to how a device that was intended to destroy tumor cells had instead promoted tissue repair. His custom-designed ruby laser was weak and certainly not as powerful as he thought it to be. Instead of being photo-ablative, the low-power light had no effect on the tumor. Indeed, it stimulated the skin to heal faster. This fortuitous encounter opened the field of monochromatic light treatment. This casual observation led him to design an experiment to ascertain his suspicion that treatment with red light accelerated healing of the surgical skin incisions he made to implant the cells. The experiment was successful as it showed that treatment with red light indeed produced faster healing of the skin wounds (Mester et al. 1971). Following his earlier observations on accelerated healing in skin, he developed studies into hair growth in mice, since regrowth of hair on depilated

skin was faster after exposure to laser energies as low as 1 Joule/cm2 per day (Mester, Szende, and Gartner 1968). Experimental research into laser biostimulation continued most actively during a period of 6 years and Mester was awarded a Scientific Doctorate by the Hungarian Academy of Sciences in recognition of his work.

John A. Parrish, MD, (Figure 1.7) graduated from Duke University and Yale University School of Medicine. He served in the US Marine Corps and was a battlefield doctor in Vietnam. On returning to civilian life, Parrish began dermatology research working with Professor Thomas B. Fitzpatrick at Massachusetts General Hospital in Boston. Becoming intrigued with the historical reports of psoralens being used for treatment of vitiligo, he experimented with making this a reliable medical treatment. His discovery is outlined in his recent autobiography *Autopsy of War* (Parrish 2012).

I set about to produce an artificial light source for activating orally ingested psoralen that reached the skin by way of blood flow. Using lasers, monochromators, and other instruments and lenses, I determined that the activating portion of the sun was a portion of the invisible ultraviolet radiation, referred to as UV-A, or "black light" because it activated fluorescent materials and made them glow in the dark. By ingesting the drug myself, and subjecting postage-stamp-sized areas of my skin to many different durations of UV-A exposure, I determined the ideal timing. After doing this twice a week for months, my entire body, except for my genitals, hands, and face, was covered in pigmented squares labeled with ink. The pigmentation lasted for months.

Working with physicists from GTE Sylvania, we developed a UV-A source capable of activating psoralen over large areas of skin. The new UV-A source made the

FIGURE 1.7   John A. Parrish.

treatment of vitiligo safer and more convenient (Parrish et al. 1976) but it still required months of twice-weekly trips to the UV-A source and was not always effective. Its contributions to dermatology proved to be limited, but several subsequent observations and events led to a huge impact.

A colleague at the University of Michigan observed improvement in a small plaque of psoriasis after application of psoralen to the affected skin and subsequent exposure to UV-A from a small handheld device. My physicist collaborators improved the UV-A source, making it possible to safely deliver uniform amounts of radiation to the entire skin surface. Then Mr. Gregory walked into our clinic.

A forty-five-year-old homeless alcoholic, Mr. Gregory was covered head to toe with red, raw, scaling disfiguring psoriasis. He looked like a red monster and constantly shed sheets of scales, which clung like dirty snow to his ragged clothing. Because of the appearance of his skin, he was shunned by shelters and even by the other homeless men and women who slept on the streets or under the bridges. He agreed to have one half of his body treated with our recently developed method for treating vitiligo. We hospitalized him in order to observe him carefully and because he had no place else to go.

After Mr. Gregory ingested the correct amount of psoralen, we carefully shielded the left side of his body with sheets, waited the right amount of time, and then exposed his right side to our new UV-A source for the ideal duration. After eight exposures over two weeks, Mr. Gregory's right side looked entirely normal. There was no evidence of psoriasis, and he even sported a nice tan. We then treated the left side with the same astonishing result. Gregory was ecstatic, left the hospital without permission, got drunk, and that night slept in Boston's best shelter for the homeless, a very happy man.

After further testing confirmed the results in another twelve patients, Dr. Fitzpatrick arranged for the new treatment to be announced simultaneously on the front page of the New York Times, as the lead article of the most prestigious medical journal (Parrish et al. 1974), and to thousands of dermatologists from the speaker's platform at the annual meeting of the American Academy of Dermatology. Initial reaction was total disbelief. Then, as we accumulated more patients, dermatologists around the world were dazzled by our accomplishment. For the next two years we were the buzz of the dermatology world. Subsequent studies by me and others showed that this treatment was effective in many other skin disorders, including severe eczema. We called the treatment PUVA (psoralen plus UV-A) (LeVine, Parrish, and Fitzpatrick 1981).

One volunteer for clinical testing was Gullan Wellman, a beautiful blond woman whose self-esteem was dependent on her appearance. Her second multimillionaire husband was Arthur O. Wellman, who had made a fortune in the wool industry. He had then turned his business over to his sons and entertained himself and made another fortune as a "wildcat" oil driller. His brother, "Wild Bill" Wellman, was no less flamboyant and was a famous Hollywood movie producer. As an adult Mrs. Wellman developed generalized disfiguring psoriasis, primarily on her face, arms, and hands, which for her was a psychosocial disaster. She had failed to respond to multiple standard treatment regimes. Yet after a course of PUVA, she looked normal—which for her was beautiful. In gratitude Mr. Wellman became a benefactor and eventually contributed more than $1 million to help me create the MGH Wellman Laboratories of Photomedicine.

Wellman Labs later grew to be the Wellman Center for Photomedicine (the world leader of its kind), which, at present, has 250 employees and is the editors' home institution. Parrish also went on to establish the Cutaneous Biology Research Center, the Center for Integration of Medicine and Innovative Technology (CIMIT), and the Home Base Program for traumatic brain injury and post-traumatic stress disorder.

The editors hope that this highlighting of the contributions of some of the great figures in the field of photomedicine has whetted the readers' appetite for the "meat" of the subject that will be found in the following 69 chapters.

References

Barth, J., and U. Kohler. 1992. Photodermatologie in Dresden—ein historischer Abriss. Festschrift anlasslich des 75. Geburtstages von Prof. Dr. Dr. Dr. h.c. H.-E. Kleine-Natrop (1917–1985), Dresden.

Bernhard, O. 1917. *Sonnenlichtbehandlung in der Chirurgie*. Enke, Stuttgart.

Bernhard, O. 1926. *Light Treatment in Surgery*. Edward Arnold, London.

Bonnet, A. 1845. *Traite des maladies des articulations*. Bailliere, Paris.

Chesney, R. W. 2012. Theobald Palm and his remarkable observation: How the sunshine vitamin came to be recognized. *Nutrients* 4:42–51.

Clark, G., and D. Jacks. 2007. Coal and the industrial revolution, 1700–1869. *Eur Rev Econ Hist* 11:39–72.

Downes, A., and T. P. Blunt. 1877. Researches on the effect of light upon bacteria and other organisms. *Proc Royal Soc London* 26:488–500.

Findlay, L. 1908. The etiology of rickets: A clinical and experimental study. *Br Med J* 2:13–17.

Finsen, N. R. 1967. *Nobel Lectures, Physiology or Medicine 1901–1921*. Elsevier Publishing Company, Amsterdam.

Fitzpatrick, T. B., and M. A. Pathak. 1959. Historical aspects of methoxsalen and other furocoumarins. *J Invest Dermatol* 32:229–231.

Frercksa, J., H. Weberb, and G. Wiesenfeldt. 2009. Reception and discovery: The nature of Johann Wilhelm Ritter's invisible rays. *Stud Hist Philos Sci A* 40:143–156.

Gaspar, L. 2009. Professor Endre Mester, the father of photobiomodulation. *J Laser Dentistry* 17:146–148.

Giese, A. C. 1964. Historical introduction. In *Photophysiology*. A. C. Giese, editor. Academic Press, New York, 1–18.

Goldman, L., D. J. Blaney, D. J. Kindel, Jr., and E. K. Franke. 1963. Effect of the laser beam on the skin. Preliminary report. *J Invest Dermatol* 40:121–122.

Hobday, R. A. 1997. Sunlight therapy and solar architecture. *Med Hist* 41:455–472.

LeVine, M. J., J. A. Parrish, and T. B. Fitzpatrick. 1981. Oral methoxsalen photochemotherapy (PUVA) of dyshidrotic eczema. *Acta Derm Venereol* 61:570–571.

Maiman, T. H. 1960. Stimulated optical radiation in ruby. *Nature* 187:493–494.

McGuff, P. E., D. Bushnell, H. S. Soroff, and R. A. Deterling, Jr. 1963. Studies of the surgical applications of laser (light amplification by stimulated emission of radiation). *Surg Forum* 14:143–145.

Mester, E., G. Ludány, M. Sellyei, B. Szende, and J. Tota. 1968. The simulating effect of low power laser rays on biological systems. *Laser Rev* 1:3.

Mester, E., T. Spiry, B. Szende, and J. G. Tota. 1971. Effect of laser rays on wound healing. *Am J Surg* 122:532–535.

Mester, E., B. Szende, and P. Gartner. 1968. The effect of laser beams on the growth of hair in mice. *Radiobiol Radiother (Berl)* 9:621–626.

Owens, I. 1889. Reports of the Collective Investigation Committee of the British Medical Association. Geographical distribution of rickets, acute and subacute rheumatism, chorea and urinary calculus in the British Islands. *Br Med J* 1:113–116.

Palm, T. A. 1888. Letter to the editor. *Br Med J* 2:1247.

Palm, T. A. 1890. The geographic distribution and etiology of rickets. *Practitioner* 45:270–279.

Parrish, J. A. 2012. *Autopsy of War*. St Martins Press, New York.

Parrish, J. A., T. B. Fitzpatrick, C. Shea, and M. A. Pathak. 1976. Photochemotherapy of vitiligo. Use of orally administered psoralens and a high-intensity long-wave ultraviolet light system. *Arch Dermatol* 112:1531–1534.

Parrish, J. A., T. B. Fitzpatrick, L. Tanenbaum, and M. A. Pathak. 1974. Photochemotherapy of psoriasis with oral methoxsalen and longwave ultraviolet light. *N Engl J Med* 291:1207–1211.

Prime, J. 1900. Les Accidentes Toxiques par L'eosinate de Sodium. Jouve & Boyer, Paris.

Raab, O. 1900. The effect of fluorescent agents on infusoria (in German). *Z Biol* 39:524–526.

Rollier, A. 1923a. *Heliotherapy*. Oxford Medical Publishers, London.

Rollier, A. 1923b. *Heliotherapy: With Special Consideration of Surgical Tuberculosis*. Frowde and Hodder & Stoughton, London.

Russell, E. H., and W. K. Russell. 1927. *Ultraviolet Radiation and Actinotherapy*. William Wood, New York.

von Tappeiner, H., and A. Jesionek. 1903. Therapeutische versuche mit fluoreszierenden stoffen. *Münch Med Wochenschr* 47:2042–2044.

von Tappeiner, H., and A. Jodlbauer. 1904. Über die Wirkung der photodynamischen (fluorescierenden) Stoffe auf Protozoen und Enzyme. *Dtsch Arch Klin Med* 80:427–487.

von Tappeiner, H., and A. Jodlbauer. 1907. Die Sensibilisierende Wirkung Fluorescierender Substanzer. Untersuchungen Uber die Photodynamische Erscheinung. FCW Vogel, Leipzig.

Zappert, J. 1910. *Rickets (Rachitis)*. D. Appleton and Co., New York, 236–284.

History and Fundamentals of Lasers and Light Sources in Photomedicine

Edward Victor Ross
Scripps Clinic

Lee Miller
Scripps Clinic

2.1 Introduction

The race to create the laser is full of intrigue, and several books have addressed the topic. In short, Charles Townes created the MASER (microwave amplification by stimulated emission of radiation), which was a nonvisible light predecessor for the LASER (light amplification by stimulated emission of radiation). Townes and Arthur Schawlow at Bell Labs published their ideas in a scientific journal (Schawlow 1965) about the same time that Gordon Gould (who had coined the term "laser") filed a patent application (Hecht 1992). There is still debate about the real inventor of the laser. Townes and Schawlow, along with Soviet pioneers, received the Nobel Prize for their work with masers and lasers. Gould, after many legal struggles, was recognized for his four patents in 1988. Theodore Maiman was the eventual winner of the great laser race (at least in terms of building the first working device) when, on May 16, 1960, despite being discouraged by his employer (Hughes Labs), he was able to put mirrors at the end of a ruby rod and create the first working laser. The rod was surrounded by a helical flash lamp (Figure 2.1). Javan later produced the first gas laser (Helium neon). In 1962, the first semiconductor laser was created. Q-switched lasers were first introduced in 1962 and mode-locking lasers in 1964 (*vide infra*).

Light represents one portion of a much broader electromagnetic spectrum. It can be divided into the UV (200–400 nm), VIS (400–700 nm), NIR "I" (755–940 nm), NIR "II" (940–1300 nm), MIR (1.3–3 μm), and Far IR (3 μm and beyond). Light is normally thought of as existing as waves and photons (the dual nature of light). When propagating through space, the wave characteristics of light are most useful to describe its behavior, whereas in characterizing interactions between atoms and energy transfer between atoms, the particle and quantum natures of light are the center stage. The energy of a single photon is described by

$$E_{photon} = hc/\lambda \qquad (2.1)$$

where h is Plank's constant (6.6×10^{-34} J s), and c is the speed of light (3×10^{10} cm/s) (Hillenkamp 1980).

Light sources are a fundamental part of photomedicine. The range of applications, from diagnostics to therapeutics, continues to expand, and the demands on light sources have increased in response. Fortunately, engineering advances have allowed for a range of reasonably priced laser and nonlaser light sources. The first "artificial" light sources used in medicine were related to photochemotherapy, which now celebrates its 65th anniversary. In this review of phototechnology, the physics and basic principles of light sources are examined (Grossweiner 1994). For both diagnostics and some therapeutics, conventional light sources have played a major role. In spectroscopy, various lamp sources are used with integrating spheres. UV light sources and visible (VIS) light as well as NIR sources are all important. The EM spectrum includes a large range of wavelengths; however, in photomedicine, we are only concerned with UV, VIS, and infrared light. The colors of the visible light portion of spectrum are violet (400–455 nm), blue (455–492 nm), green (492–577 nm), yellow (577–597 nm), orange (597–622 nm), and red (622–780 nm).

Conventional light sources include incandescent lamps, fluorescent lamps, and electric arcs. Incandescent lamps consist of a tungsten filament heated to approximately 2500 K. The

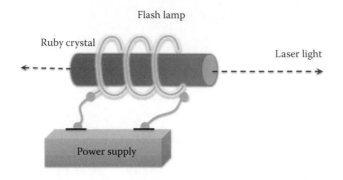

Flash lamp

Ruby crystal

Laser light

Power supply

FIGURE 2.1 Schematic drawing of laser.

output of the lamp is described by black body radiation physics and depends on the temperature. Wien's displacement law further describes the spectrum of light. Arc lamps are created by passing current through a vapor. At low pressures, a Hg lamp transmits primarily 254 nm light through a quartz envelope. High-pressure Hg and Xe lamps are very bright, the brightest of nonlaser light sources. Xenon flash lamps are popular in intense pulsed-light sources. They emit a bright flash, the output determined by the intensity of the lamp pumping. These lamps are typically encased in glass and cooled by circulating water jackets. The lamps can be spectrally modulated by filters. Both dielectric and absorption filters are used (see Figure 2.2). Halogen lamps are used in skin tightening. They are deployed over longer times (usually seconds versus milliseconds) than conventional intense pulsed-light devices. Fluorescent lamps are low-pressure Hg lamps with phosphorus on the inner coating.

As Rox Anderson (1994), MD, has noted, lasers represent a convenient source of photons. Its invention had been predicted by Einstein in 1917, but it was much later that the MASER (1954) and even later than the first visible light ruby laser was created. This author had the good fortune of meeting the only surviving laser inventor (Dr. Townes) who related (ASLMS Annual Conference, Phoenix, AZ, 2010) that his graduate student reported the much-anticipated amplification to him. He

confided that they did not realize the potential medical uses of lasers.

Laser light enjoys a great amount of attention in movies and sci-fi dramas, but the principle components of any laser are a lasing medium, a pumping mechanism, and a delivery device (Figure 2.3). An electronic power supply enables the pumping entity (whether that is another laser, a flash lamp, or an electrical discharge). Lasers really are the product of fluorescence of the active medium. In normal fluorescent conditions (e.g., a typical lamp source), photons are emitted in many directions. With a laser, "pumping" creates a scenario in which the number of atoms in the upper energy state is greater than those in the lower energy state in a condition known as *population inversion*. Most solid-state and dye lasers are optically pumped, whereas gas lasers must be electrically pumped because of the active medium not being sufficiently absorbing to create the population inversion with visible light. The major differentiating properties of most lasers (versus conventional light sources) are monochromaticity, collimation (and the associated directionality), high power, and spatial/temporal coherence (see Figure 2.4). With respect to lasing media, there are diode, solid-state, dye, and gas lasers. The feedback mechanism consists of mirrors of which one mirror reflects 100% of the light, and the other transmits a small fraction of light (Reinisch 1996). An example of a solid-state laser is the alexandrite laser. A solid-state laser consists of a rod that is pumped by a flash lamp. The lamp pumps the rod for stimulated emission. The rod and lamp assembly (the laser head) must be designed for adequate cooling. High-powered lasers typically are fragile because all of the components are driven near their damage thresholds. As an example of this concept, consider the pulsed dye laser (PDL). As the dye degrades, the lamps must work harder to generate higher pulse energies from the dye. Also, mirrors become contaminated over time such that the lamps must work harder and harder. These demands stress the power supply. Thus, eventually, the dye kit, power supply, lamps, and dye are all working at their maximal output. To make matters worse, medical lasers often are manufactured on mobile platforms where

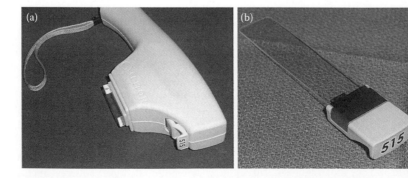

FIGURE 2.2 (a) Plug-in filter: 515 nm indicates the cutoff wavelength (the filter allows very little light below 515 nm); (b) handpiece with filter inserted. The sapphire window is applied to the skin surface.

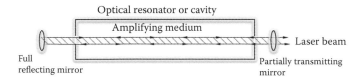

FIGURE 2.3 Simplified schematic of typical solid-state laser.

fine optical components are jostled as the units are pushed and pulled from one area of the clinic to another. All of the components must be aligned properly to optimize the beam propagation from the laser head to the delivery system. Figure 2.5 depicts popular laser wavelengths.

The scientific principle on which lasers are based is *stimulated emission*. Einstein proposed that, in addition to spontaneous emission, there could be stimulated emission, in which an excited atom is illuminated with photons with the same energy that correlated to the transition to the lower state. The light could be emitted from the re-excited atom. With spontaneous emission, electrons transition to the lower level in a random process. With stimulated emission, the emission occurs only in the presence of photons of certain energy (the transition energy from one state to another) given by

$$\Delta E = E_2 - E_1 \qquad (2.2)$$

The critical point is maintaining a condition in which the population of photons in a higher energy state is larger than that in the lower state. However, stimulated emission is a challenge because the natural state for atoms is the lower state, so incoming photons tend to excite the lower-state atoms that exist in much greater numbers. To achieve a population inversion and stimulated emission, one must populate the upper level and

depopulate the lower levels simultaneously. Quantization had been shown by the Bohr model and considered an assembly of identical atoms excited by EMR and in thermal equilibrium with the environment. As in Figure 2.6 in which two atomic levels are shown, N_1 atoms occupy the lower state and N_2 the upper state. The number in each state is noted by

$$N_2/N_1 = \exp[-(E_2-E_1)]/kT \qquad (2.3)$$

where k is the Boltzmann constant (1.38×10^{-23} J/K).

Stimulated emission is hard to accomplish. Under normal conditions, N_2/N_1 is very small, as small as 1.5×10^{-17}. To reach a situation in which $N_2 > N_1$, one requires a metastable state with a longer lifetime to have a chance of keeping a large percentage of the atoms in the excited state. The population of the upper state N_2 cannot exceed N_1 in thermal equilibrium. Solely increasing the energy or heating the medium cannot excite a significant number of atoms (or T would have to be negative in Equation 2.3). However, one can achieve the inversion as long as the energy levels are not in thermal equilibrium. To increase the likelihood of stimulated emission, metastable states are required (vida supra). This is the case with the original ruby laser (the three-level laser, Maiman's ruby laser) (Figure 2.7).

Once the population inversion is established, light will be emitted in all directions. One needs a resonant cavity to amplify the beam. By adding a mirror on one end of the laser, the gain can be increased. Most lasers have a full mirror and a partial reflecting mirror.

Amplification and gain: Stimulated emission can amplify light because one photon can stimulate a cascade of other photons at the same wavelength. Then we have the potential for gain or the degree of amplification given by

$$\text{Output/input} = \text{amplification} = (1 + \text{gain})^{\text{length}} \qquad (2.4)$$

For example, a gain of 3 cm^{-1} implies that one photon generates three additional photons over each 3 cm that it travels (Hecht 1992). There is no real extra energy—the energy originally derives from the pumping source so that conservation of energy is always enforced. Gain and power are increased by placing mirrors at the end of the rod. The mirrors extend the optical path length so that power increases. For optimal resonance, twice the length of the cavity equals an integral number of wavelengths, or $N\lambda = 2L$, where L is the cavity length.

Population inversion can be achieved if the pumping radiation frequency is $h/(E_2 - E_1)$, where E_1 and E_2 represent the energies of the two respective atomic states. Then the number of atoms excited per second is proportional to the light intensity $I(v)$, the number of atoms in the lower state N_1, and an efficiency parameter, $B(v)$, that depends on the absorption efficiency at that particular pumping wavelength. The atoms in the upper state are unstable and spontaneously decay at a rate A, so that the decay rate is AN_2; then the total rate of downward energy transmissions is given by

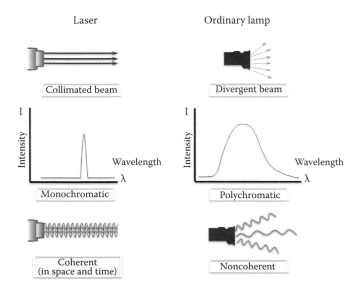

FIGURE 2.4 Comparison of laser and ordinary light sources.

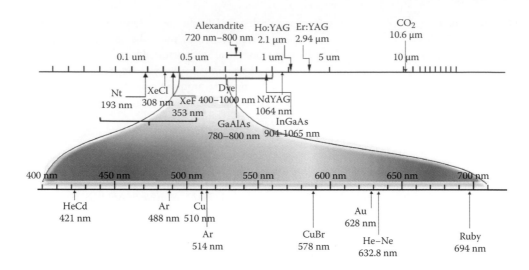

FIGURE 2.5 Wavelengths of popular lasers.

$$[A + B(v)' \, I(v)]N_1 \qquad (2.5)$$

where B' is the Einstein coefficient for stimulated emission. Then the light intensity distribution $I(v)$ is

$$\frac{I(v) = (A/B)}{[\exp(-hv/kT)-1]} \qquad (2.6)$$

To achieve stimulated emission, one needs a three-level energy system because in any two-level system, even the most efficient processes typically lead to $N_1 = N_2$ (Diels and Arissian 2011).

Light device terminology: Basic parameters for light sources are power, time, and spot size for continuous wave (CW) lasers and, for pulsed sources, the energy per pulse, pulse duration, spot size, fluence, repetition rate, and the total number of pulses (Welch and van Gemert 1995). Energy is measured in joules (J). The amount of energy delivered per unit area is the fluence, sometimes called

the dose or radiant exposure, given in joules per square centimeter (J/cm^2). The rate of energy delivery is called power, measured in watts (W). One watt is 1 J/s (W = J/s). The power delivered per unit area is called the irradiance or power density, usually given in watts per square centimeter (W/cm^2). Laser exposure duration (called pulse width for pulsed lasers) is the time over which energy is delivered. Fluence is equal to the irradiance times the exposure duration (Anderson and Ross 2000). Power density is a critical parameter, for it often determines the action mechanism in cutaneous applications. For example, a very low irradiance emission (in a typical range of 2–10 mW/cm^2) does not heat tissue and is associated with diagnostic applications, photochemical processes, and biostimulation. On the other extreme, a very short nanosecond pulse can generate high peak power densities associated with shock waves and even plasma formation (Fisher 1996). Plasma is a "spark" resulting from the ionization of matter.

Another factor is the laser exposure spot size, particularly important for visible light and NIR lasers in which the ratios of scattering to absorption are high. Other characteristics of importance are whether the incident light is convergent, divergent, or diffuse and the uniformity of irradiance over the exposure area (spatial beam profile). The pulse profile, that is, the character of the pulse shapes in time (instantaneous power versus time), also affects the tissue response (Tanghetti et al. 2002).

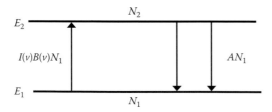

FIGURE 2.6 In Einsteinian radiation theory, identical atoms are in thermal equilibrium with a heat reservoir. N_1 atoms occupy the lower energy state at E_1, and N_2 atoms occupy the upper energy state at E_2. The imposition of light of special intensity distribution $l(v)$ induces absorption by atoms in state 1 and stimulated emission by atoms in state 2 at rates proportional to coefficients B and B', respectively. Atoms in state 2 also undergo spontaneous emission at a rate proportional to coefficient A. Assuming the Bohr frequency condition, $v = (E_2 - E_2/h)$, $B' = B$, and that thermal equilibrium is maintained, this leads to the Planck spectral distribution of thermal radiation.

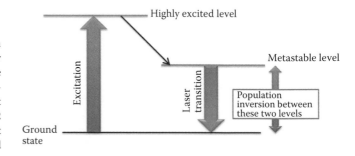

FIGURE 2.7 Energy levels in a three-level laser.

Many lasers in medicine are pulsed, and the user interface shows pulse duration, fluence, spot size, and, in certain devices, cooling system settings. Some multiwavelength lasers also allow for wavelength selection. Some older lasers, for example, a popular CO_2 laser, showed only the pulse energy on the instrument panel or, in CW mode, the number of watts. In these cases, one uses the exposure area and exposure time to calculate the total light dose (fluence) (Grossweiner 1994):

$$fluence = \left(\frac{Power \times time}{area} \right) \quad (2.7)$$

With the exception of some photodynamic therapy (PDT) sources and CW CO_2 lasers, most lasers create pulsed light. In many CW applications (e.g., wart treatment with a CO_2 laser), the fluence is not of great importance in characterizing the overall tissue effect. A more important parameter is power density (in which higher power densities achieve ablation and lower power densities cause charring). In these cases, the physician stops the procedure when an appropriate end point is reached. On the other hand, in photodynamic therapy (PDT) applications with CW light in which the clinical end point might be delayed, the total fluence *and* power density are important predictors of the tissue response.

A thorough knowledge of a specific laser's operation and quirks is imperative for optimal and safe lasering. Vendors are creating lasers that are more intuitive to operate. Increasingly, manufacturers have added touch-screen interfaces with application-driven menus and skin type-specific preset parameters. Some devices record an individual patient's laser parameters to be stored for future reference. Most lasers are designed such that the handpiece and instrument panels are electronically interfaced. It follows that the laser control module "knows" what spot size and/or scanner is being used at the operator's "side" of the action. Typically this "handshake" occurs when one inserts the handpiece into the calibration port or through a control cable from the handpiece to the laser "main frame." With others, one selects the spot size on the display, and the laser calculates the fluence accordingly.

Most lasers calibrate through a system in which the end of the handpiece is placed in a portal on the base unit. This configuration allows for interrogation of the entire system, from the pumping lamps to the fiber/articulated arm to the handpiece optics. For example, if a mirror is damaged, the laser will fail calibration, and an error message appears. Other systems measure the output within the distal end of the handpiece using a small calibration module that "picks off" a portion of the beam. Some systems only calibrate inside the laser case so that any compromised delivery apparatus is not assessed. This configuration demands that the user pay particular attention to the tissue effect and beam profile as the laser will not account for any damage to components beyond the point in the internal energy assessment.

There are some simple ways to interrogate for system integrity. One can examine the aiming beam as it illuminates a piece

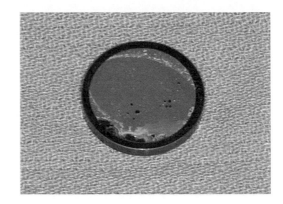

FIGURE 2.8 Damaged mirror from the arm of a CO_2 laser.

of white paper, checking that the beam edges are sharp; this suggests that the treatment beam is also sharp and the profile is according to the manufacturer's specifications. Also, burn paper or a tongue depressor can be used. Here, the laser is used with low energy, and the spot is checked for uniformity from beam edge to edge. By checking the impact pattern, one can diagnose damaged mirrors in the knuckle of the articulated arm or a damaged focusing lens that renders the laser unstable or unsafe (Figure 2.8). Likewise, for scanners, one can ensure that skin coverage will be uniform.

Waste heat is a major factor in laser design and a practical issue in a small room with poor AC. Much of the electrical energy used to create laser emissions is wasted as heat, which is why water is used for cooling most solid-state lasers. On the other hand, air cooling is used for some high-powered flash lamps and many diode lasers. For example, some Nd:YAG lasers might waste as much as 98% of energy as heat. On the other hand, diode lasers tend to have a higher wall-plug efficiency and operate with less wasted heat.

2.2 Types of Light Devices

In principle, many nonlaser devices could be used for medicine in which precise tissue heating or ablation is required (Hillenkamp 1980). Most properties of laser light (e.g., coherence) are unimportant insofar as the way light interacts with tissue in therapeutic applications. And, although collimation (lack of divergence) of the incident beam might increase the percentage of transmitted light in the skin with laser versus intense pulsed light (IPL), the increasing use of filtered flash lamps in dermatology suggests that losses from IPL beam divergence are not critical in most biomedical applications.

The intensity, directionality, and monochromaticity of laser light allow the beam to be expanded or focused quite easily. Beam divergence is a key feature of lasers versus conventional light sources. It is measured as half of the full angle at which the beam spreads. The beam divergence is measured in milliradians in which 1 radian = 57.3°. A laser beam can only be focused to its diffraction limited spot given by the following:

$$S = f\lambda/D \qquad (2.8)$$

where S is the focused spot size, D is the lens diameter, and f is the focal length of the lens (Hecht 1992).

With nonlaser sources, such as flash lamps or light emitting diodes (LEDs), the light intensity at the skin surface cannot exceed the brightness of the source lamp. With many lasers, a lamp similar to the IPL flash lamp pumps the laser cavity (Ross 2006). The *amplification of light* within the laser cavity sets laser light apart from other sources (Katzir 1993). Focusing is one characteristic that distinguishes a laser source from nonlaser light sources. Although nonlaser light sources can be focused, the intensity cannot exceed the source brightness. For example, with an IPL, the fluence cannot exceed the emitted fluence at the lamp surface.

For most visible light applications, a laser represents a conversion from lamplight or nonvisible light diode lasers to an amplified monochromatic form (Anderson 1994). The high power possible with lasers (especially *peak power*) is achieved through *resonance* in the laser cavity. For many applications requiring millisecond or longer pulses delivered to large areas of tissue, IPLs are either adequate or preferable to lasers.

Laser systems differ with regard to the duration and power of the emitted laser radiation. In continuous-wave lasers (CW mode) with power outputs of up to 10^3 W, the lasing medium is excited continuously. With pulsed lasers, excitation is effected in a single pulse or in online pulses (free-running mode). Peak powers of 10^5 W can be developed for a duration of 100 µs to 10 ms. A Q-switcher is an optical shutter that blocks the optical path until population inversion is achieved. This results in a single high energy pulse with a very short pulse duration (usually in the nanosecond range). Storing the excitation energy and releasing it suddenly (Q-switch mode or mode locking) leads to a peak power increase of up to 10^{10} to 10^{12} W and a pulse duration of 10 ps to 100 ns (Anderson 1994). Mode locking results in even shorter pulses: as short as pulses in the picosecond and femtosecond range. Duty cycle refers to the percentage of time the laser is on for a pulsed laser in which

$$\text{duty cycle} = \text{rep rate} \times \text{pulse width.} \qquad (2.9)$$

There are two basic types of pulsed lasers in medicine. In one scenario, lasers use multiple, very low energy pulses with very high repetition rates that are quasi-CW with a high duty cycle. In the other, very high energy pulses are delivered with low duty cycles. Typical peak powers are on the order of kilowatts for millisecond outputs.

A summary of wavelength ranges useful in medicine are as follows:

1. UV laser and light sources have been used primarily for treatment of inflammatory skin diseases and/or vitiligo, as well as striae and corneal refraction. The presumed action in skin at lower power densities is immunomodulatory.

The XeCl excimer laser emits at 308 nm, near the peak action spectrum for psoriasis. Other UV nonlaser sources have also been used for hypopigmentation, striae, and various inflammatory diseases (Alexiades-Armenakas et al. 2004; Raulin et al. 2004). Excimer lasers at 193 nm have been used for corneal ablation.

2. Violet IPL emissions, low-power 410 nm LED, and fluorescent lamps are used either alone or with aminolevulinic acid (ALA). Alone, the devices take advantage of endogenous porphyrins and kill *P acnes* (Gold and Goldman 2004). After application of ALA, this wavelength range is highly effective in creating singlet O_2 after absorption by PpIX. Uses include treatment of actinic keratoses, actinic cheilitis, and basal cell carcinomas (Itkin and Gilchrest 2004).

3. VIS (GY): These wavelengths are highly absorbed by HgB and melanin and are especially useful in treating epidermal pigmented lesions and superficial vessels (Anderson and Parrish 1981, 1982, 1983).

4. Red and near IR (I) (630, 694, 755, and 810 nm): Deeply penetrating red light (630 nm) CW devices are efficient activators of PpIX after topical application of ALA. The pulsed 694 nm (ruby) laser is optimized for pigment reduction and hair reduction in lighter skin types. The 810 nm diode and 755 nm alexandrite laser, depending on spot size, cooling, pulse duration, and fluence, can be configured to optimize outcomes for hair reduction, lentigines, or blood vessels (Trafeli et al. 2007). By decreasing the pulse width into the nanosecond range, the alexandrite laser is a first line treatment for many tattoo colors.

5. Near IR (II) 940 and Nd:YAG (1064 nm): These two wavelengths have been used for a broad range of vessel sizes on the leg and face (Passeron et al. 2003). They occupy a unique place in the absorption spectrum of our *big 3* chromophores, that is, *blood, melanin, and water*. Because of the depth of penetration (on the order of millimeters), they are especially useful for hair reduction and coagulation of deeper blood vessels. By varying fluence and spot size, reticular ectatic veins, as well as those associated with nodular port wine stains or hemangiomas, can be safely targeted. The Q-switched YAG laser plays an expanding role in the treatment of tattoos, nevus of Ota, and even melasma.

6. MIR-lasers and deeply penetrating halogen lamps: These lasers and lamps heat tissue water. With macrowounding (>1 mm) spot sizes, depending on where we want to heat, we can "choreograph" our laser and/or cooling settings to maximize the skin temperature in certain skin layers. In general, with more deeply penetrating wavelengths, larger volumes are heated. On the other hand, achieving temperature elevations in the volume will require higher fluences than with highly absorbing wavelengths. Without surface cooling, unless very small fluences are applied, a top-to-bottom thermal injury occurs.

7. Far infrared systems. The major lasers are the CO_2, erbium:YAG, and erbium YSGG (chromium: yttrium–scandium–gallium–garnet) lasers. Overall, the ratio of ablation to heating is much higher with the erbium:YAG laser. However, one can enhance the thermal effects of the Er:YAG laser by extending the pulse or increasing the repetition rate, and, likewise, one can decrease residual thermal damage (RTD) of the CO_2 laser by decreasing pw (Majaron et al. 2001a,b). Where precision is required in ablation, Er:YAG is preferred. On the other hand, depending on settings, the CO_2 laser enjoys a desirable blend of ablation and heating.

2.3 Beam Profiles: Top Hat versus Gaussian

As we noted earlier, the laser is an optical resonator in which EM waves move back and forth. Depending on the interference pattern inside the cavity, various stable modes are created. When a beam is referred to as "multimode," this description usually applies to the transverse mode and not the longitudinal mode. Transverse modes are the most important and determine the intensity of the beam as a function of the distance from the beam axis (Katzir 1993). The TEM 00 mode is the most desirable for many laser applications and tends to be the mode out of the cavity. However, the beam profile for many lasers may not be constant across the breadth of the beam. Creation of a top-hat profile can be achieved by a fiber delivery system in which modes are mixed. Laser beam profiles vary based on intercavity design, lasing medium, and the delivery system. A common profile is Gaussian or bell-shaped (Figure 2.9). For many lasers, this profile represents the fundamental optimized mode of the laser. This shape is usually observed when the beam has been delivered through an articulated arm. For some wavelengths, this is an effective way to deliver energy (CO_2 and erbium). The disadvantage of the

rigid arm is limited flexibility, the typically short arm length, the possibility of misalignment from even minor impact, and a tendency for nonflat beam profiles (Reinisch and Ossoff 1996). The Gaussian profile can be modified outside the cavity, which is desirable in many applications. With a fiber-equipped delivery system, the beam is mixed within the fiber and can be shaped to be more flat-topped. Although fiber delivery systems are usually preferred by physicians, some laser beams are difficult to deliver through a fiber. Examples include far IR wavelengths and very short pulses (i.e., 5–10 ns with typical Q-switched Nd:YAG lasers whose high peak power exceeds the damage threshold of most fibers).

2.4 Pulse Profiles: Square versus Spiky

Power pulse profiles vary among different lasers and can have a profound effect on tissue effects. The total energy of a laser pulse or train of pulses is given by the following:

$$E = \int P \, dt \qquad (2.10)$$

The average power is also important and can be defined as

$$\sum \frac{E}{t} \qquad (2.11)$$

The pulse profile is the temporal shape of the laser pulse (Figure 2.10) (Shafirstein et al. 2004). In many pulsed laser applications, the macro pulse is composed of several shorter micropulses (Mulholland 2009).

The laser head and the power supply are two main laser components. The laser head is composed of two mirrors and the medium. The power supply is required either to charge the lamps or, in the case of gas lasers, to create a high-voltage electrical

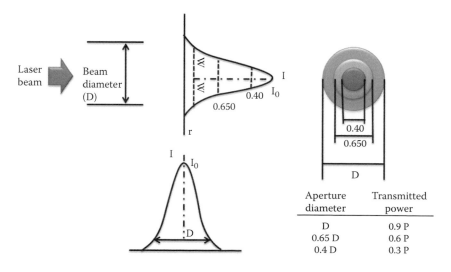

FIGURE 2.9 Gaussian laser beam profile: intensity distribution.

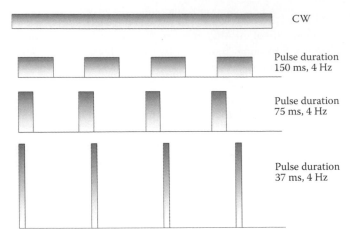

CW

Pulse duration
150 ms, 4 Hz

Pulse duration
75 ms, 4 Hz

Pulse duration
37 ms, 4 Hz

FIGURE 2.10   CW and pulsed laser beams.

discharge. Ergonomics plays a great role in the design of beam-delivery systems for lasers in medicine. Ideally, the delivery system is flexible (fibers "trump" articulated arms), and the distal end of any system is light and easy to manipulate for the operator. One concern with any delivery system is the alignment of the beam. With an articulated arm, mirrors direct the beam down the arm at segments known as knuckles. Fiber delivery systems typically use a focusing lens at both the laser and the distal end of the fiber. The distal handpiece usually uses a lens system to create a flat-topped beam of various spot sizes. Some devices have a zoom system that allows for a quick rotation of the handpiece to form various spots (Figure 2.11).

Lasers are also classified by their power level, where a class 1 device is incapable of producing damage. Class 2 is low powered enough that the eye aversion response is sufficient. Class 3 devices may be a hazard with a direct shot but not a reflection. A class 4 laser presents an eye, skin, and fire hazard. The permissible dose of light for these different devices is determined by the American National Standards Institute (ANSI) instruction.

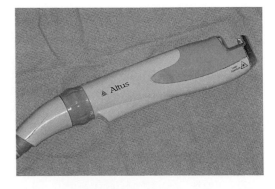

FIGURE 2.11   Typical handpiece with adjustable spot size: a zoom feature allows for quick conversion from 3 to 10 mm spot in 2–3 mm intervals.

2.5   Tour of Medically Useful Lasers

Excimer lasers use rare halide gases that are stable only in their excited state. They are used at 308 nm for skin applications and often used in corneal shaping at 193 nm in procedures such as laser-assisted *in situ* keratomileusis (LASIK) and photorefractive keratectomy (PRK).

The Nd:YAG laser is commonly used in surgery and dermatology applications. It has a low efficiency (from a wall plug of about 0.1%–1%). The Nd:YAG laser can be flash-lamp pumped or diode pumped. Wavelengths include 1.06 μm, 1320 nm, and 1440 nm, depending on the cavity design and pumping details. They can be arc-lamp driven or flash-lamp driven. Increasingly, a crystal (potassium titanyl phosphate or KTP) is often used to double the frequency 1064 to 532 nm.

The Er:YAG laser is popular in dermatology and dentistry. At 2.94 μm, there is very strong absorption by tissue water. Er:YAG lasers are pulsed, are flash-lamp driven, and normally deliver pulse widths of approximately 300 μs. Some erbium:YAG lasers have been designed to develop pulse trains (up to 32 ms) that can extend the range of thermal damage. Unlike the CO_2 laser, all erbium laser resurfacing systems are pulsed, and, overall, the available systems are similar in basic operation. Energy is delivered through an articulated arm (Ross and Anderson 1999).

Holmium lasers emit "light" at 2.1 μm. The laser beam can be delivered via a fiber and has been used in urology for destroying kidney stones. Thulium lasers (1.97 μm) have been used in dermatology as both fiber lasers and as laser rods pumped by other lasers.

Laser rods can be drawn into a thin fiber in which the fiber becomes the lasing medium. One company (Solta Medical, Hayward, CA) deploys a diode-pumped thulium fiber laser. Solta also manufactures a 1550 nm fiber laser. Advantages of the fiber laser are the single mode nature of the beam. The Solta device uses an air-cooled diode laser to pump the erbium glass fiber. One company has embraced the concept of laser-pumped lasers. In one scenario, an alexandrite laser end-pumps the thulium rod to produce 1.94 μm.

The pulsed-dye laser was the first device designed to exploit the concept of selective photothermolysis (SPT). One dye, rhodamine 6G, is frequently used and optically pumped. Dye lasers can be operated in pulsed or CW mode and can be tuned to emit various wavelengths in a range.

Some solid-state lasers are tunable. One commonly used laser is the alexandrite laser. They are used in skin applications, such as tattoo removal, in Q-switched mode and, in the longer pulsed modes, are applied to blood vessels and pigmented lesions.

LEDs are becoming commonplace in dermatology (Figure 2.12). Primarily used as a PDT and biostimulation light source, they are similar to semiconductor lasers in that they use electrical current placed between two types of semiconductors. However, they lack an amplification process (no mirrors). LEDs do not produce coherent beams but can produce monochromatic light. A semiconductor is characterized by having two energy bands rather than discrete energy levels as is the case for atoms

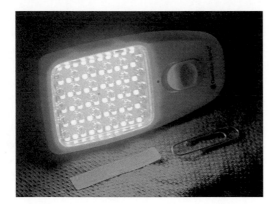

FIGURE 2.12 Small LED panel for photodynamic therapy and biostimulation.

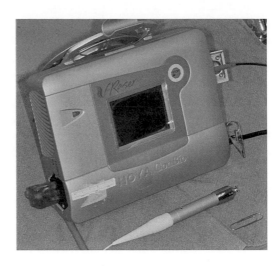

FIGURE 2.14 Small diode laser for vascular lesions: note paper clip as reference for size.

and molecules (Diels and Arissian 2011). Photons are emitted by the transition of an electron from the higher band (conduction band) to the lower energy band (valence band). As voltage is applied to the diode—positive on the p side, negative on the n side—electrons will cascade down from the high-energy conduction band into the valence band at the location of the junction so that a population inversion is created (Figure 2.13). Semiconductor lasers are notable for their small size (see little 980 nm laser, Figure 2.14) in which individual lasers are as small as 0.1 mm (less than the size of a grain of salt). Low laser thresholds are better because semiconductor lasers are very susceptible to heat. Most solid-state lasers emit round beams because of the nature of the rod. However, diode lasers emit a beam that depends largely on the emitting area. One needs a lens system to collimate the typical diode beam that would otherwise diverge more than that of a flashlight (Milonni and Eberly 2010). Semiconductor lasers enjoy more than 30% efficiency, among the highest of all lasers (Katzir 1993). Newer laser diodes, such as the 800 nm laser used in hair removal, are capable of high powers (>1000 W). Most semiconductor (diode) lasers are operated

in CW mode but can be pulsed. New visible light semiconductor lasers are available; also, laser diode arrays are available in which scientists have created large numbers of semiconductor lasers on one substrate. Some diode lasers are housed separately from the handpiece and delivered by fiber optics. Others are configured with the laser diodes in the handpiece as arrays (Figure 2.15) and another in which the array is focused into a fiber.

CO_2 lasers are pumped electrically; that is, the laser cavity is excited by electrical discharges. CO_2 lasers use a mixture of CO_2, N_2, and helium as the lasing medium. Doping of the CO_2 gas with other elements, particularly nitrogen, increases the laser's efficiency. There are multiple energy transitions for the CO_2 laser, so 10.6 μm is not the only possible emitted wavelength. The 10.6 μm photons are created when CO_2 molecules drop from a higher-energy asymmetric stretching mode to a lower-energy stretching or bending mode. The wavelengths are longer than

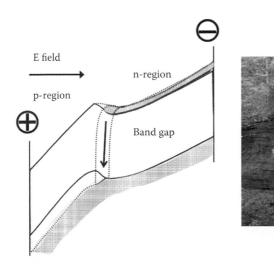

FIGURE 2.13 Note schematic of diode laser.

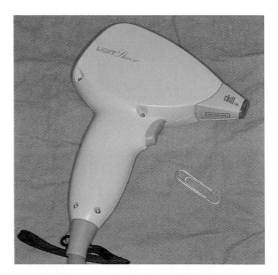

FIGURE 2.15 Diode array laser handpiece (810 nm): note size of handpiece.

visible-light lasers because the transitions are of the rotational/vibrational type, whereas the ruby laser, for example, involves moving electrons into higher and lower states (electronic transitions). The overall efficiency (5%–20% from the wall plug) of CO_2 lasers is among the highest of all lasers.

Solid fiber delivery systems presently are impractical for IR radiation as the materials are hygroscopic and not durable enough for pulsed laser systems. However, hollow waveguide delivery devices are used in one commercially available laser. The waveguide allows for less encumbered movement of the handpiece versus more typical articulated arms.

IPL devices are becoming increasingly comparable to lasers that emit millisecond domain pulses (Ross et al. 2005). The absorption spectra of skin chromophores show multiple peaks (HgB) or can be broad (melanin) (Boulnois 1986), and therefore a broadband light source is a logical alternative to lasers. Proper filtration of a xenon lamp tailors the output spectrum for a particular application. Some concessions are made with direct use of lamplight. For example, rapid beam divergence obliges that the lamp source be near the skin surface. This requirement makes for a typically heavier handpiece compared with most lasers (Figure 2.2), the exception being some diode arrays in which the light source is also housed in the handpiece. IPL cannot be adapted to fibers for subsurface delivery. High-energy short pulses (Q-switched nanosecond pulses) are not possible with flash lamps. They can, however, be used to pump a laser, and some modern IPLs feature a laser attachment in which the flash lamp and laser rod are in the handpiece. In this case, lasers and the IPL itself share the base power supply and other components to save costs. The lasers are attached to the base power supply (like accessories on a vacuum cleaner). In general, the size, weight, and cost of both laser and flash lamp technology are steadily decreasing. Optical filters are combined with flash lamps for some skin applications. Newer IPL systems use partial discharge technology, which ensures a more even energy flow (Cartier 2005). The lamp is cooled by circulating water around a quartz envelope. The quartz envelope filters out harmful far-ultraviolet (UV) output. The lamp output is "focused" toward the distal end of the handpiece and is usually coupled into the skin surface via a sapphire or quartz block. Cooling of the skin surface may be achieved by cryogen spray, forced refrigerated air, or contact cooling on the distal end of the handpiece depending on the particular IPL device.

A major advantage of IPL devices is their therapeutic versatility. By varying filtration, lamp type, or current density, an IPL can be configured for different emission spectra (Ross 2006). Overall, IPL devices have a good safety profile and are intrinsically more eye-safe than lasers because of the beam divergence. Although IPLs typically have a slow repetition rate (0.3 to 2 Hz), they have large spot sizes, which allows for greater skin coverage per pulse and rapid treatment (Ross et al. 2005). In general, for most devices, as the fluence increases, more time is required for lamp recovery and recharging, and the pulse repetition decreases accordingly. Larger handpieces tend to be more bulky, making maneuverability over irregular skin surfaces more difficult.

Treatment settings between various manufacturers or even successive device models are generally not interchangeable because of variances in fluences and filtering. For example, changes in lamp pumping affect the pulse profile and spectral emission, creating a domino effect with simple changes of fluence impacting other treatment parameters. Various devices emit different wavelength ranges and spectral shapes, allowing for different depths of light penetration and absorption by human skin with the same macro-pulse duration, nominal filtration, and fluence. Different companies use various techniques for measuring fluences. Some display an effective fluence (a fluence to tissue based on tissue optics and the "raw" energy available at the crystal surface) on the graphical user interface (GUI), whereas others report only the energy at the crystal surface. Thus, the fluence that is given on the instrument panel might not be equivalent to the energy density at the tip.

Unlike lasers, older IPL outputs tended to vary from pulse to pulse or even during a pulse as a result of the instantaneous pumping voltage of the capacitors. During the course of a pulse, the beginning and end portions are less energetic while the middle of the pulse is more energetic, termed "spectral jitter." Newer devices use sophisticated computer control systems to minimize this phenomenon.

Another concern with IPLs is that the spots are so large that background normal skin is within the treatment field. In cases where there is low contrast in the chromophore between the lesion and the surrounding skin, this configuration exposes the innocent bystander skin to light doses that might exceed its damage threshold. To address this issue, manufacturers have designed smaller spots. For example, the Acutip (Cutera, Burlingame, CA) has a smaller 6 mm spot size that allows for treatment of discrete brown and red dyschromic lesions. This small spot size is achieved by using a long cylindrical sapphire waveguide and high-performance reflectors. Another IPL device has the option of a 4 mm adaptor tip (Figure 2.16) for their MaxG

FIGURE 2.16 CO_2 laser with hollow waveguide delivery system.

handpiece (Palomar Medical Technologies Inc., Burlington, MA). The BBL (Sciton Inc., Palo Alto, CA) also has an adapter tip in three different sizes for harder-to-reach areas. If smaller spot sizes are not available, a white plastic mask with small holes (Palomar Medical Technologies Inc., Burlington, MA, and Alma Lasers, Buffalo Grove, IL) can be added to achieve a smaller area of irradiation.

As IPLs continue to gain popularity, newer technologies make these devices more user friendly. The Skintel Reader (Palomar Medical Technologies Inc., Burlington, MA) (Figure 2.17) is an accessory to an IPL device designed to objectively measure melanin content in the skin. The melanin reader guides selection of appropriate treatment settings to improve safety by minimizing operator error and optimizing treatment settings. It also provides guidance in tracking changes in an individual's sun exposure. Rather than subjective "eyeball" assessment of background epidermal melanin content, the operator takes three sample readings with the Skintel Reader on clear skin, which is then averaged into a melanin index value. This value is communicated wirelessly to the IPL system. The operator then chooses a suitable pulse width using the Skintel Test Spot Tables and proceeds to enter a test spot fluence based on assessment of the client's skin and the Skintel value.

Also, there is a new appreciation for modulation of lamp pumping to optimize outcomes for vascular disorders. For example, by elongating the pulse duration (up to a single pulse of 100 ms), a typical spectrum red shifts so that deeper, larger vessels are preferentially heated. In this way, a larger venous lake or larger 1 mm nasal vessel tends to close more completely than if a shorter pulse were delivered. Also, some devices allow for a train of two to three pulses over a similar or longer time period. This multiple pulse exposure has been shown to enhance vessel coagulation with smaller peak fluences (Jia et al. 2012).

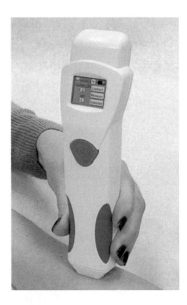

FIGURE 2.17 Skin pigment meter.

Many laser companies now manufacture cases in which different laser heads can be added. For example, Sciton makes a laser in which 1320, 1064, 755, and 2940 nm wavelengths can be added in a modular fashion. This type of design decreases the risk of early obsolescence.

2.5.1 Laser Accessories

Scanners allow for more uniform and rapid tissue removal than freehand movements, and accordingly, most companies have incorporated the option of a scanner from a third-party source (Sahar Technologies, San Diego, CA); other companies have integrated their own proprietary models. Most available scanners restrict the operator to a set working distance. An exception is the computer pattern generator (Lumenis, Santa Clara, CA). Like the scanner on the Ultrapulse CO_2 laser, it delivers a scanned collimated beam that preserves the spatial quality of the scan over a liberal working distance. All available scanners permit a range of overlap, shapes, and sizes (Ross 2005).

Fibers are an important way to deliver laser energy into cavities as well as to the skin surface. Depending on the fiber design, they can be optimized for cellulite and tumor ablation, or, with diffusers, can be used for PDT in larger cavities such as the uterus or esophagus. Optical fibers have the appearance and size of a hair. There are optical fibers in nature, such as with polar bears for whom light is conducted to the skin by surface hairs. The principle of light concentration in fibers is total internal reflection. If a light ray is incident on the interface between a medium of a high index for refraction (n) and a lower index between the core and a lower n for the cladding, there will be a critical angle at which no light is transmitted and all light remains in the fiber. If the diameter of the core is large, a large number of rays with different angles can be trapped by total internal reflection (Katzir 1993).

The increasing miniaturization of devices has allowed for home use in a range of applications. Multiple home hair-removal devices are available as are LED panels for acne. Any home-based therapy must overcome eye-safety issues. Typically, this is achieved by not allowing the laser to fire unless it is in contact with the skin or by limiting output to below the eye-damage threshold (often the case for LEDs). With "home" fractional lasers, overtreatment risks are decreased by time limiting the number of total pulses in any given time period. Several studies show significant hair reduction with home-use IPL devices (Alster and Tanzi 2009; Emerson and Town 2009; Mulholland 2009). However, these devices require extended treatment times and increased frequency of use compared to in-office devices. A concern with home-use devices is that additional safety measures with skin-sensor technologies are required to prevent use on darker skin types. The results of home devices are impressive but still inferior to office-based devices.

Optical clearing, either through specially designed compression handpieces or application of topical solutions, has been used to enhance light penetration in the turbid dermis. These approaches have been shown to enhance clearing of tattoos and vascular lesions. A recent article examined optical clearing with both glycerol (one optical clearing agent) and compression. In porcine skin, they examined images through a range of skin thicknesses. They found that both optical clearing agents and compression enhance light transmission but that the compression methods outperformed chemical agents as far as image resolution. The compression method resulted in tissue thickness reductions. They used very small compressive windows. Another study examined optical compression pins to heat the dermis and hypodermis. Zelickson et al. studied such a device in which 1540 and 1208 nm laser microbeams (mb) are coaligned with optical pins.

2.5.2 Fractional Laser Systems

With water as the chromophore for these devices, several skin components are targeted. In general, the affected structures are keratinocytes, collagen, and blood vessels. Unlike nonfractional light sources in which HgB and melanin are specifically targeted, the spatial confinement in fractional lasers depends on the geometry of the microbeams, the pulse duration, and the wavelength. Fractional lasers typically damage less than 50% of the surface area and micro-injuries are no greater than 500 μm in diameter.

2.5.3 Total Reflection Amplification of Spontaneous Emission of Radiation

The total reflection amplification of spontaneous emission of radiation (TRASER) contains an internally reflecting body hosting a fluorescent substance, a dye cell (center); flash lamps (above and below); a rear mirror (end left); an optional output waveguide (right) (Figure 2.18); an optional reflector cavity housing

the light source; and the internally reflecting body (Zachary and Gustavsson 2012). Photons from the flash lamps excite the fluorescent material within the dye cells, which spontaneously emit a different narrow spectrum of light. Typically, the dye cell traps 45%–62% of the spontaneous emission by internal reflection. The internally reflected photons propagate axially along the length of the dye cell or crystal in both directions. At the proximal end of the cell, a mirror is placed, redirecting the light forward. The light is passively coupled out at the distal end of the cell. A waveguide is used to target the treatment area. The TRASER (Zachary and Gustavsson 2012) uses total internal reflection (TIR) to amplify light versus a resonator and a laser rod. The potential advantages are lower manufacturing costs, increased versatility, and enhanced power levels versus similarly configured lasers (Zachary and Gustavsson 2012). Like all physical processes, LASER, despite the acronym, only amplifies radiation within the confines of conservation of energy. The TRASER has no optical resonator nor an output coupler, and the emitted light is noncoherent and noncollimated. The amplification of the light is achieved by capturing and "herding" the photons with a liquid jet or solid body of high refractive index using TIR. The lack of population inversion distinguishes the TRASER versus the laser.

2.6 Conclusions

The rapid advances in tissue optics, electronics, and engineering should increase the role of energy-based technologies in the medical arena. Noninvasive imaging devices will be deployed alongside our high-energy lasers in which, presumably, the feedback from the former will optimize treatment algorithms for the latter. One day, for example, a dermatologist might perform an optical biopsy to diagnose a basal cell carcinoma (BCC) followed by same-day PDT treatment. In another room, a provider will be guided by a skin assessment meter to select optimal settings for the laser treatment of a photodamaged chest. All of this will occur as technology costs decrease for noninvasive imaging and as we witness a progressive miniaturization in device dimensions.

References

Alexiades-Armenakas, M. R., L. J. Bernstein, P. M. Friedman, and R. G. Geronemus. 2004. The safety and efficacy of the 308-nm excimer laser for pigment correction of hypopigmented scars and striae alba. *Arch Dermatol* 140:955–960.

Alster, T. S., and E. L. Tanzi. 2009. Effect of a novel low-energy pulsed-light device for home-use hair removal. *Dermatol Surg* 35:483–489.

Anderson, R. R. 1994. Laser tissue interactions. In *Cutaneous Laser Surgery—The Art and Science of Selective Photothermolysis*. M. Goldman, and R. Fitzparick, editors. Mosby, St Louis. 3–5.

Anderson, R. R., and J. A. Parrish. 1981. Microvasculature can be selectively damaged using dye lasers: a basic theory and experimental evidence in human skin. *Lasers Surg Med* 1:263–276.

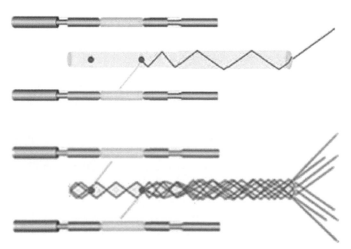

FIGURE 2.18 TRASER contains an internally reflecting body hosting a fluorescent substance, "dye cell" (center); flash lamps (above and below); an optional reflector cavity housing the light source; and the internally reflecting body (not shown).

Anderson, R. R., and J. A. Parrish. 1982. Lasers in dermatology provide a model for exploring new applications in surgical oncology. *Int Adv Surg Oncol* 5:341–358.

Anderson, R. R., and J. A. Parrish. 1983. Selective photothermolysis: precise microsurgery by selective absorption of pulsed radiation. *Science* 220:524–527.

Anderson, R. R., and E. Ross. 2000. Laser–tissue interactions. In *Cosmetic Laser Surgery*. R. Fitzpatrick, and M. Goldman, editors. Mosby, St. Louis. 1–30.

Boulnois, J. 1986. Photophysical processes in recent medical laser developments- a review. *Lasers Med Sci* 1:47–66.

Cartier, H. 2005. Use of intense pulsed light in the treatment of scars. *J Cosmet Dermatol* 4:34–40.

Diels, J.-C., and L. Arissian. 2011. *Lasers: The Power and Precision of Light*. Wiley-VCH Verlag GmbH & Co. KGaA, Weinheim, Germany.

Emerson, R., and G. Town. 2009. Hair removal with a novel, low fluence, home-use intense pulsed light device. *J Cosmet Laser Ther* 11:98–105.

Fisher, J. C. 1996. Basic biophysical principles of resurfacing of human skin by means of the carbon dioxide laser. *J Clin Laser Med Surg* 14:193–210.

Gold, M. H., and M. P. Goldman. 2004. 5-Aminolevulinic acid photodynamic therapy: Where we have been and where we are going. *Dermatol Surg* 30:1077–1083.

Grossweiner, L. 1994. *The Science of Phototherapy*. CRC, Boca Raton.

Hecht, J. 1992. *Understanding Lasers*. IEEE Press, New York.

Hillenkamp, F. 1980. Interaction between laser radiation and biological systems. In *Lasers in Medicine and Biology*. F. Hillenkamp, P. R, and C. Sacchi, editors. Plenum, New York. 37–68.

Itkin, A., and B. A. Gilchrest. 2004. delta-Aminolevulinic acid and blue light photodynamic therapy for treatment of multiple basal cell carcinomas in two patients with nevoid basal cell carcinoma syndrome. *Dermatol Surg* 30:1054–1061.

Jia, W., N. Tran, V. Sun et al. 2012. Photocoagulation of dermal blood vessels with multiple laser pulses in an in vivo microvascular model. *Lasers Surg Med* 44:144–151.

Katzir, A. 1993. *Lasers and Optical Fibers in Medicine*. Academic Press, San Diego.

Majaron, B., K. M. Kelly, H. B. Park, W. Verkruysse, and J. S. Nelson. 2001a. Er:YAG laser skin resurfacing using repetitive long-pulse exposure and cryogen spray cooling: I. Histological study. *Lasers Surg Med* 28:121–131.

Majaron, B., W. Verkruysse, K. M. Kelly, and J. S. Nelson. 2001b. Er:YAG laser skin resurfacing using repetitive long-pulse exposure and cryogen spray cooling: II. Theoretical analysis. *Lasers Surg Med* 28:131–138.

Milonni, P. W., and J. H. Eberly. 2010. Some specific lasers and amplifiers, in Laser Physics. In *Laser Physics*. John Wiley & Sons, Hoboken, NJ. 497–560.

Mulholland, R. S. 2009. Silk'n—A novel device using home pulsed light for hair removal at home. *J Cosmet Laser Ther* 11:106–109.

Passeron, T., V. Olivier, L. Duteil et al. 2003. The new 940-nanometer diode laser: an effective treatment for leg venulectasia. *J Am Acad Dermatol* 48:768–774.

Raulin, C., B. Greve, S. H. Warncke, and C. Gundogan. 2004. [Excimer laser. Treatment of iatrogenic hypopigmentation following skin resurfacing]. *Hautarzt* 55:746–748.

Reinisch, L. 1996. Laser physics and tissue interactions. *Otolaryngol Clin North Am* 29:893–914.

Reinisch, L., and R. H. Ossoff. 1996. Laser applications in otolaryngology. *Otolaryngol Clin North Am* 29:891–892.

Ross, E. V. 2005. CW and pulsed CO_2 lasers. In *Principles and Practice in Cutaneous Laser Surgery*. A. Kauvar, editor. Taylor & Francis, Boca Raton, FL.

Ross, E. V. 2006. Laser versus intense pulsed light: Competing technologies in dermatology. *Lasers Surg Med* 38:261–272.

Ross, E. V., and R. R. Anderson. 1999. Erbium laser resurfacing. In *Cosmetic Laser Surgery*. T. S. Alster and D. B. Apfelberg, editor. Wiley-Liss, Hoboken, NJ.

Ross, E. V., M. Smirnov, M. Pankratov, and G. Altshuler. 2005. Intense pulsed light and laser treatment of facial telangiectasias and dyspigmentation: some theoretical and practical comparisons. *Dermatol Surg* 31:1188–1198.

Schawlow, A. L. 1965. Lasers. *Science* 149:13 22.

Shafirstein, G., W. Baumler, M. Lapidoth et al. 2004. A new mathematical approach to the diffusion approximation theory for selective photothermolysis modeling and its implication in laser treatment of port-wine stains. *Lasers Surg Med* 34:335–347.

Tanghetti, E., R. A. Sierra, E. A. Sherr, and M. Mirkov. 2002. Evaluation of pulse-duration on purpuric threshold using extended pulse pulsed dye laser (cynosure V-star). *Lasers Surg Med* 31:363–366.

Trafeli, J. P., J. M. Kwan, K. J. Meehan et al. 2007. Use of a long-pulse alexandrite laser in the treatment of superficial pigmented lesions. *Dermatol Surg* 33:1477–1482.

Welch, A. J., and M. J. van Gemert. 1995. Overview of optical and thermal interaction and nomenclature. In *Optical Thermal Response of Laser-Irradiated Tissue*. A. J. Welch, and M. J. van Gemert, editors. Plenum, New York. 1–14.

Zachary, C. B., and M. Gustavsson. 2012. TRASER—Total reflection amplification of spontaneous emission of radiation. *PLoS ONE* 7:e35899.

3

Light–Tissue Interactions

Pradeep Kumar Gupta
Raja Ramanna Centre for Advanced Technology

Mahesh Kumar Swami
Raja Ramanna Centre for Advanced Technology

Harishankar Patel
Raja Ramanna Centre for Advanced Technology

"ādityasya namaskaraṁ ye kurvanti dincdinc |
janmāṁ tarasahasre ṣudridhryaṁ nopajāyate ||
akālamṛtyuharaṇm sarvavyādhivinaśanam |
sūryapādodakaṁ tīrtham jaṭharedhārayāmyaham ||"

—Rigveda

("Those who pay obeisance to Sun every day do not face poverty of health in life, early death or suffer from diseases. One should drink the water kept before the Sun.")

In this chapter, we provide an overview of the fundamentals of the interaction of light with tissue and how this interaction is utilized in photomedicine.

3.1 Introduction

The interaction of light with biological matter plays an important role in our life; photosynthesis and vision are good examples. Therefore, studies on the use of this interaction to address an issue of utmost importance to mankind, the quality of its health care, have always been an important scientific pursuit. Use of sunlight for therapy has been explored since time immemorial by the ancient civilizations. For example, in India, the therapeutic potential of sunlight was well appreciated, and, for good health, exposure to light from the rising sun and consumption of water kept in sunlight were recommended. Indian medical literature dating back to 1400 B.C. documents the combined use of *Psoralea corylifolia* L. and sunlight for the treatment of nonpigmented skin lesions (vitiligo) (Pathak and Fitzpatrick 1992). In his book *A History of Medicine: Medieval Medicine*, Plinio Prioreschi (2003) notes the medieval practices of the use of red light for treatment of smallpox. A resurgence of the use of light in therapy happened in the second half of the 19th century when

the role of sunlight for the treatment of rickets, tuberculosis, etc. was explored and the ability of the invisible ultraviolet (UV) radiation to kill microorganisms was scientifically established (McDonagh 2001). Perhaps the biggest thrust toward phototherapy was a result of the efforts of Niels Finsen, a Danish physician, who carried out several interesting experiments on the therapeutic effects of light and established the role of UV light in curing skin tuberculosis (lupus vulgaris) (Bie 1899) for which he was awarded the Nobel Prize in physiology or medicine in 1903. The use of light for the treatment of tuberculosis remained popular until 1946 when a more effective treatment and cure became possible with the development of the antibiotic streptomycin. Another important discovery made in the early 20th century was by Oscar Rabb, a student working in the laboratory of von Tappeiner in Munich, who found that a low concentration of acridine, which had no effect in the dark, led to rapid killing of the protozoan paramecium on illumination (Moan and Peng 2003). This discovery of photodynamic action also spurred a great deal of interest in investigating the effect of light on living systems and has led to the development of photodynamic therapy (PDT). Presently, PDT is an accepted modality for the treatment of several forms of cancer and some other diseases (Moan and Peng 2003). Other notable uses of light in medicine developed in the 20th century include treatment of neonatal jaundice using UV light (Cremer, Perryman, and Richards 1958), use of the UV-A spectrum of light to suppress the immune system and reduce inflammatory responses in psoriasis (Parrish 1977), and treatment of seasonal affective disorders (Lam et al. 2006).

The use of light in therapy received a major boost with the invention of lasers in 1960. Laser applications in medicine exploit one or more of the several rather remarkable properties of lasers, viz., high directionality, monochromaticity, brightness, the ability to generate short-duration pulses, etc. Because of its high directionality, a laser beam can be focused to very small spot sizes, a few tenths of a micrometer for visible lasers. This micro-irradiation capability of lasers, in conjunction with the control of the laser pulse duration, energy, and intensity, helps the surgeon to elicit the desired response in the tissue, making possible an ultraprecise surgery. Further, the laser light can be efficiently coupled to thin optical fibers and thus guided endoscopically to internal organs for therapeutic applications without any major incision, considerably reducing the patient trauma and hospitalization time. The use of lasers in surgery that started in 1961 (within a year of its invention) most often exploits the heating of the target tissue. It is pertinent to note that although tissue heating can be achieved by several other means, none can provide the selectivity made possible by the exquisite control on laser parameters. By use of laser pulses of duration shorter than the thermal relaxation time of the target tissue, heat can be confined within the target tissue so that it can be vaporized or coagulated without significant thermal effect on surrounding tissue. Laser's monochromaticity provides further control for selective processing of a constituent of a multicomponent tissue. For example, before lasers, no satisfactory treatment existed for port-wine stains (purple birthmark), a cutaneous vascular disorder. These are now effectively managed by the use of pulsed lasers with a wavelength tuned to the hemoglobin absorption peaks to selectively destroy the vasculature without affecting the overlying or nearby structures. More recently, there has also been interest in the use of lasers as well as noncoherent light for promoting the healing of wounds and for providing relief in pains of different etiology, neuralgia, arthritis, etc. (Fulop et al. 2010; Woodruff et al. 2004), presumably exploiting the photodynamic effect in endogenous photosensitizers present in the tissue.

Apart from therapy, the observant ancient civilizations had also used visual inspection of the patient through the light scattered from the patient's body as a tool to diagnose diseases. The invention of the microscope in the 17th century helped evolve histology, and in the early 19th century, light penetration in tissue was utilized for diagnosis of hydrocephalus [an increase in the volume of cerebrospinal fluid (CSF) in the head] in children and intraventricular hemorrhage (Gibson and Dehghani 2009). Other important breakthroughs made in this period include the development of endoscopes, which allowed noninvasive examination of the internals of a hollow organ or body cavity and the invention of the ophthalmoscope by Hermann von Helmholtz (Keeler 2002), which made possible *in vivo* investigations of retinal disorders. During the early 20th century, the use of transillumination for imaging the human breast was also initiated (Cutler 1931). As a result of the significant difference in the transmission characteristics of the constituents of the breast tissue (fat being highly translucent; fibrous tissue less translucent; solid epithelial masses, fibro-epithelial masses, and epithelial debris being moderately opaque; and blood being intensely opaque), the transillumination of the breast was found to be a valuable aid in the interpretation of pathological conditions in the mammary gland.

Since the early 20th century, with the advent of quantum theory, the use of optical spectroscopic methods for disease diagnosis also started getting attention. Differences in the light absorption properties of oxyhemoglobin and deoxyhemoglobin were used by several groups (Millikan 1942) for monitoring the oxygen content of arterial blood. These measurements led to the development of the earlobe-based oxymeter and formed the basis for the present-day pulse oxymeter. Jobsis (1977) demonstrated the use of near-infrared (NIR) spectroscopy for monitoring of myocardial oxygen saturation in the animal model as well as in the neonatal head. Over the last few decades, there has been considerable growth in the use of different optical spectroscopic techniques, such as fluorescence, Raman, and nonlinear spectroscopy, to probe the biochemical composition or the morphology of the tissue (Tuchin 2011). Presently, the advances in optics and instrumentation and the large image processing capability of computers are making possible a much more comprehensive use of the information content of the light coming from the tissue for quantitative; sensitive; and higher-resolution, noninvasive diagnosis.

In this chapter, we first provide a brief overview of light propagation in tissue and then discuss the use of light for biomedical imaging, diagnosis, and therapeutic applications.

3.2 Light Propagation in Tissue

When light falls on tissue, a part of it is reflected, transmitted, or scattered from the tissue. Some of the incident light may also be absorbed by the tissue constituents. The energy absorbed can be reemitted as fluorescence or dissipated as heat in the tissue. The light emerging from the tissue (as a result of reflection, transmission, scattering, or reemission) depends on its optical properties and therefore can be used for tissue diagnostics.

The therapeutic effects arise as a result of the energy absorbed in the tissue. The major contributors for absorption in the UV spectral range are DNA and proteins. In the visible and NIR wavelength range, the absorption in tissue is dominated by hemoglobin and melanin. Although cytochromes, the terminal member of the respiratory chain, have a very large molar extinction coefficient (higher than that for oxyhemoglobin and deoxyhemoglobin in the NIR region), these do not have significant influence on tissue absorption characteristics because their relative abundance is very small. Absorption by water, the main constituent of all tissues, becomes significant beyond ~1000 nm and becomes dominant for wavelengths higher than ~2000 nm. It should be emphasized that for a multicomponent tissue, the absorption at a given wavelength is a weighted average of the absorption by its constituents at the wavelength.

For wavelengths greater than ~650 nm and smaller than 2000 nm, the tissue absorption is weak, so the light can penetrate more deeply. For biomedical imaging and diagnostic applications, it is desirable to have minimal absorption in the tissue for two reasons: First, it would allow probing larger depths of the tissue, and second, deposition of energy in the tissue may result in irreversible changes. Therefore, generally, light in the so-called diagnostic window (700 to, say, 1500 nm) in which tissue absorption is minimal is used for biomedical imaging and diagnosis. It should also be noted here that for some diagnostic techniques, which utilize fluorescence of native tissue or a suitable biomarker for diagnosis, use of light that is absorbed by the fluorescing component(s) will be required.

Attenuation of light propagating in a nonscattering medium is completely described by Beer–Lambert's law $I = I_0 \exp(-\mu_a z)$, where μ_a is the absorption coefficient. Because scattering removes photons from the direction of propagation, it also leads to attenuation of the light. One may think that replacing μ_a with $\mu_a + \mu_s$, where μ_s is the scattering coefficient, in Beer–Lambert's law can describe the variation of irradiance of the collimated light with depth. However, it should be noted that while photons may get scattered out of the beam path, multiple scattering events might also bring photons back into the beam path. These photons, although not part of the collimated beam, also add to the irradiance at a given point along the direction of propagation of the beam limiting the validity of Beer–Lambert's law. The degree of attenuation arising as a result of the scattering depends on the angular distribution of the scattered photons, which, in turn, has a strong dependence on the size of the scatterer. For scatterers with size ≪ wavelength, the phase of the electromagnetic field across the scatterer can be treated as constant. Therefore, the light scattered by all the induced dipoles in the scatterer adds up in phase, resulting in dipole-like scattering. Here, the angular distribution of the scattered light, often referred to as the "phase function," shows no dependence on the angle of scattering in the plane transverse to the electric field of the incident light, but in the plane containing the electric field, it shows a cosine square intensity pattern with minima along the dipole axis resulting from the transverse nature of the electromagnetic wave (Figure 3.1a). Rayleigh was the first to show that for such small scatterers, the scattered intensity is inversely proportional to the fourth power of the wavelength and that this is responsible for the blue color of the sky (Bohren and Huffman 1983). For larger-sized scatterers (>λ), light scattered by all the induced dipoles in the scatterer do not add up in phase except only in the forward direction, making the angular distribution of the scattered light peak in the forward direction (Figure 3.1b and c). An exact mathematical description of scattering from spherical particles of size > wavelength was provided by Ludwig Valentine Lorenz and Gustav Mie (Bohren and Huffman 1983). This is, therefore, referred to as Lorenz–Mie scattering or often just Mie scattering. In the Mie regime, the wavelength dependence of scattering coefficients for different tissue is given as λ^{-k}, where k typically varies from 1 to 2 (Tuchin 2007).

The first moment of the phase function is the average cosine of the scattering angle, denoted by g. It is also referred to as the anisotropy parameter. The value of g' ranges from –1 to +1, where $g = 0$ corresponds to isotropic scattering (Rayleigh scattering), $g = +1$ corresponds to ideal forward scattering, and $g = -1$ corresponds to ideal backward scattering. A photon acquires random direction after approximately $1/(1-g)$ scattering events,

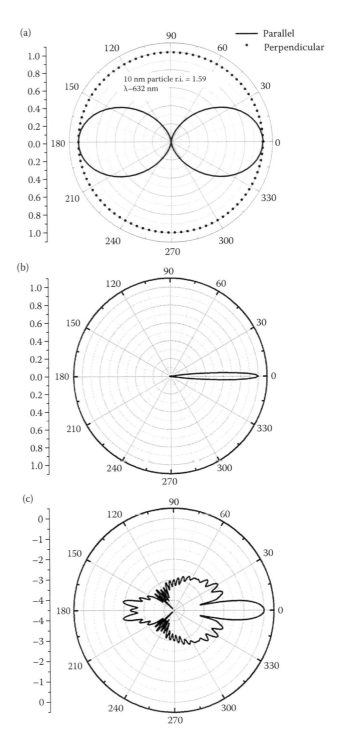

FIGURE 3.1 (a) Angular distribution of scattering from a dipole in the plane perpendicular to the incident electric field (dotted line) and in the plane containing the electric field (black line). (b) Angular distribution of scattering for a Mie scatterer (diameter = 2 μm, r.i. = 1.59 wavelength 632 nm) and (c) the same in log scale to highlight the oscillation in angular scattering distribution.

which is only 5 for $g = 0.8$. Typical values of g for biological tissues vary from 0.7 to 0.99. Another parameter frequently used to describe the scattering properties of the tissue is $\mu_s'(-\mu_s(1-g))$.

This is referred to as the reduced scattering coefficient. It defines the path length over which the incident light loses its directional information; that is, the angular distribution of the scattered light becomes isotropic.

The dependence of the angular distribution of the scattered light on the size and shape of the scatterer is widely utilized in flow cytometers for cell identification and characterization. The forward scattering is correlated with the total volume of the cell while the side scattering is related to the inner complexity of the cell, such as granularities in cytoplasm, shape of nucleolus, and membrane roughness. In addition to elastically scattered light, a very small fraction of the incident light can also be scattered inelastically by processes such as Raman scattering, which involve energy transfer to or from internal excitations of the medium. This inelastically scattered light is a very sensitive probe for the biochemical composition or morphology of the tissue and is therefore being utilized for biomedical diagnosis.

A rigorous, electromagnetic theory-based approach for analyzing light propagation in tissue will need to identify and incorporate the spatial/temporal distribution and the size distribution of tissue structures and their absorption and scattering properties. Quite evidently, this is not straightforward. Therefore, heuristic approaches with different levels of approximations have been developed to model light transport in tissues (Ishimaru 1978). The radiative transfer theory has been the most successful in modeling light transport in tissues. In this approach, originally developed to explain light propagation in the stellar atmosphere, light propagation in tissue is described by the transport of energy by the motion of photons through a medium containing discrete scattering and absorption centers. Because there are no exact general solutions to the transport equations in a tissue, for practical applications, it is convenient to work with some approximate solutions. Further, for tissues, the tenuous scattering approximation (volume density of scattering particles < 10^{-3}) rarely applies. Tissues are therefore usually modeled either as intermediate scattering media or dense media (volume density > 10^{-2}). The intermediate scattering case is the most difficult to handle rigorously. Several approaches, such as the Kubelka–Munk model, have been developed (Kubelka 1948); however, the range of validity of these models is not well established, and the formulation cannot be generalized to most cases of practical interest. For weakly absorbing, dense media, in which scattering predominates ($\mu'_s \gg \mu_a$), the equation of radiation transfer reduces to a simpler photon diffusion equation (Ishimaru 1978), which, for an isotropic point source placed on an optically dense slab far away from the boundary, leads to the following expression for the diffuse flux density $\varphi_d(z)$:

$$\varphi_d(z) = k\,\varphi_o \exp(-\mu_{eff} z) \qquad (3.1)$$

where $\varphi_d(0) = k\,\varphi_o$, φ_o being the incident fluence, and μ_{eff} is $(3(\mu_a + \mu'_s)\mu_a)^{1/2}$. The factor k accounts for the enhancement of fluence just below the boundary arising as a result of back scattering. It follows from Equation 3.1 that the penetration depth, the depth at which the fluence reduces to $1/e$ of the incident

fluence, is $[1 + \ln(k)]\mu_{eff}^{-1}$. For typical tissue parameters, the depth of penetration can be 1.5 to 3 times $(1/\mu_{eff})$. The typical values for the absorption coefficient, reduced scattering coefficient, anisotropy parameter, and effective attenuation coefficient for some human tissues at different wavelengths are listed in Table 3.1.

The diffusion approximation provides a fair representation for light transport in soft tissue in the visible and NIR spectral region. It is, however, only valid away from the source and boundaries. The other widely used method for modeling light transport in tissues is Monte Carlo simulation (Wang, Jacques, and Zheng 1995). In this statistical approach to radiative transfer, the multiple scattering trajectories of individual photons are determined using random numbers to predict the probability of each microscopic event. The superposition of many photon paths approaches the actual photon distribution in time and space. Although Monte Carlo simulation may require lengthy computation time, it can be performed for any experimental geometry and is considered the gold standard of tissue optics calculations. For a tissue with optical transport parameters $\mu'_s = 2$ mm^{-1}, $\mu_a = 0.05$ mm^{-1}, and $g = 0.9$ (typical of tissue in the visible wavelength range), the calculated depth distribution of fluence using one-dimensional diffusion approximation and exact Monte Carlo simulation is shown in Figure 3.2.

One can see that away from the boundary, the predictions from diffusion theory are in good agreement with the Monte Carlo simulations. One should also note that, contrary to normal expectations, the fluence just below the boundary ($z = 0$) is significantly larger than the incident fluence. Scattering also leads to another interesting consequence. For finite beams, for a given value for irradiance, the depth of penetration increases with an increase in the size of the area of illumination (Star 1997). Figure 3.3 shows the variation in penetration depth with increasing size of beam diameter for fixed-incident fluence. These results, which may appear counterintuitive, further emphasize the need to have a clear understanding of light propagation in tissue because, for most practical applications, the spatial distribution of light in tissue is expected to play an important role.

TABLE 3.1 Typical Values for Absorption Coefficient (μ_a), Reduced Scattering Coefficient (μ'_s), Anisotropy Parameter (g), and Effective Attenuation Coefficient (μ_{eff}) for Some Human Tissues

Tissue Type	λ (nm)	μ_a(mm^{-1})	μ'_s(mm^{-1})	g
Breast (normal)[a]	530	0.11	1.85	~0.88
Breast (malignant)[a]	530	0.21	2.87	~0.96
Lung[b]	635	0.81	8.1	0.75
Liver[b]	635	0.23	10.0	0.68
Myocardium[b]	1064	0.14	1.3	0.96
Skin (dermis)[b]	630	0.27	3.7	0.8
Skin (epidermis)[b]	630	3.5	8.8	0.8

[a] Ghosh, Mohanty, Majumder, and Gupta (2001).

[b] Tuchin, V. *Tissue optics: Light scattering methods and instruments for medical diagnosis*, 2nd edition, Bellingham, SPIE Press, 2007.

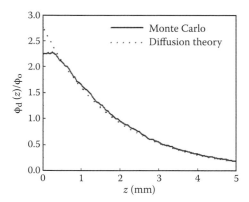

FIGURE 3.2 Theoretically computed depth distribution of fluence for a turbid medium with $\mu_s' = 2$ mm^{-1}, $\mu_a = 0.05$ mm^{-1}, and $g = 0.9$. Dotted line shows the results obtained using one-dimensional diffusion theory, and the solid line represents the results obtained by Monte Carlo simulation.

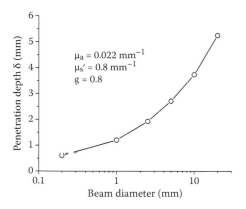

FIGURE 3.3 Dependence of the penetration depth on the diameter of illuminating beam.

Polarization of light should also be considered while modeling the propagation of light in tissue because for a number of tissue constituents, such as collagen, elastin, and other structural proteins, the refractive index is polarization dependent. Similarly, account should be taken of the fact that the presence of chiral molecules, such as glucose and proteins, can also rotate the plane of polarization of the incident light.

3.3 Measurements of Tissue Optical Parameters

In order to be able to model light transport in tissue, reasonable estimates for tissue optical parameters are required. Multiple scattering in tissue scrambles the information content and makes direct measurement of these parameters difficult. An obvious approach to eliminate multiple scattering effects is to use a tissue sample with thickness $d \ll 1/\mu_s$. Using such thin tissue sections, the refractive index of tissue is usually determined by applying the traditional refractive index measurement techniques, such as white light interferometry or prismatic

dispersion. More recently, optical coherence tomography (OCT) (Ding et al. 2006), digital holography (Bhaduri et al. 2012), and light interference microscopy (Wang et al. 2011) have been used to determine the refractive index of tissue. These studies suggest that refractive index distribution of tissue can provide valuable diagnostic information and may serve as a basis for label-free histopathology (Wang et al. 2011). Measurements at the cellular and subcellular level are usually carried out using light-scattering and phase microscopy (Beuthany et al. 1996). Although the refractive index of tissue constituents varies over a range of 1.3 to 1.6 (Table 3.2), the average refractive index of tissue is close to that of water because water is the major constituent of tissue (typically ~75% by weight).

For a tissue sample with thickness $d \ll 1/\mu_s$, the absorption coefficient can be determined by using an integrating sphere to measure all the photons transmitted or scattered by the sample, so that the loss is only a result of the absorption. Having determined μ_a, μ_s can be determined from a measurement of the collimated fraction of transmitted light as this will provide an estimate for the sum total of both the absorption and scattering coefficients. This approach, however, has a fundamental limitation. Because the typical value of μ_s for soft tissues is in the range of 100–1000 cm^{-1}, the sample thickness must be less than approximately 10 μm. The methods used for the preparation of these thin sections may alter the tissue optical properties. Further, the signals are weak. For example, for soft tissues in the visible wavelength range, $\mu_a \sim \mu_s/100$. Therefore, the loss of photons as a result of absorption in sample thickness $d \ll 1/\mu_s$ is very small. The weak signal may easily be swamped by several artifacts, such as unavoidable fluctuations in incident light flux, nonuniformity of response of the integrating spheres, etc. Although methods have been developed to minimize these effects, this approach may not provide very accurate measurements. Estimation of the phase function and the anisotropy parameter requires measurement of angular distribution using a goniometer. A reasonable estimate of the anisotropy parameter of tissue can be obtained by fitting the experimentally measured scattering phase function to the Henyey-Greenstein phase function (Ghosh et al. 2001).

A more established method, which can also be used for *in situ* determination of tissue optical parameters, involves either spatial or temporal distribution of the diffuse radiance as both of these are strongly influenced by μ_a and μ_s'. The lower the value of μ_s', the larger the spatial extent over which photons will be distributed, and consequently, the steady-state relative radiance curve $R(r)$ will be less steep. Similarly, with increasing absorption, the radiance at larger distances will be affected more than at shorter distances, leading to a steeper $R(r)$ curve with lower intensity. Similar considerations apply to temporal broadening of an incident light pulse. The slope of the tail of the local time-dependent reflectance curve gives μ_a directly (Patterson, Chance, and Wilson 1989). However, for finite geometries, the slope also depends on the tissue geometry and source–detector separation. The time-resolved measurements can be done in both time and frequency domain. The

TABLE 3.2 Refractive Index of Some of the Important Tissue Constituents

Sr. No.	Tissue Constituents	Size (μm)[a]	Refractive Index
1.	Nucleus (Choi et al. 2007)	3–12	Nucleoli ~1.39 Nucleoplasm ~1.35
2.	Mitochondria (Wax and Backman 2010)	0.5–1	~1.4
3.	Lysosomes (Wax and Backman 2010)	0.25–0.8	~1.6
5.	Cell membrane (Wax and Backman 2010)	–	~1.46
6.	Cytoplasm (Wax and Backman 2010)	–	~1.37
7.	Collagen fiber (Choi et al. 2007; Leonard and Meek 1997)	0.5–3	1.32 to 1.45 (along axis) 1.40 to 1.61 (radial)
8.	Red blood cells (Tuchin 2007)	7.1–9.2	1.39–1.41

[a] Largest dimension.

time domain measurements have the potential to yield μ_a and μ'_s directly. Frequency domain measurements by themselves and in combination with steady-state measurements have also been explored for determination of μ_a and μ'_s (Gurfinkel, Pan, and Sevick-Muraca 2004).

In vivo measurements of tissue parameters using the steady-state spatial radiance profile have received more attention. The parameters μ_a and μ'_s can be estimated by iterative fitting of the data either with an analytical solution of the diffusion theory when applicable or by Monte Carlo simulations when necessary. It has also been shown that the slope of the ln $R(r)$ versus r curve is strongly correlated to the reduced transport coefficient $\mu'_t = \mu_a + \mu'_s$ for smaller values of r and to the effective attenuation coefficient μ_{eff} for larger values of r. Although the technique has been successfully used for *in vivo* measurements, its sensitivity to local variations in tissue optical parameters can affect the accuracy of the estimates. Indeed, a large variation was observed in the estimated values of μ_{eff} and μ'_s for the human esophagus from *in vivo* measurements (Bays et al. 1996). While this may represent real variance in tissue optical properties, measurement errors, in particular, those induced by tissue inhomogeneities, are also expected to contribute.

3.4 Use of Light–Tissue Interaction for Biomedical Imaging and Diagnosis

The motivation for the use of light for biomedical imaging and diagnosis arises because optical techniques not only can image tissue with a resolution down to a few micrometers but also can provide valuable biochemical and morphological information on the tissue. It is pertinent to note here that only for thin tissue sections can transmitted light be used for imaging or diagnosis. For *in situ* imaging or diagnosis, the light scattered or reemitted in the backward direction (with respect to the direction of the illuminating beam) is utilized.

The use of light for biomedical imaging has to contend with the problem that, in contrast to x-ray photons, visible light photons undergo multiple scattering in the tissue leading to a blurring of the image. This can be best seen if one shines a torch at one's palm. One can see a pinkish glow but not the outline of the bones in the path of the beam. The pinkish glow arises because the red component of the light is least attenuated and is therefore dominant in the light that emerges from the palm. The bones are not visible because of the multiple scattering of light in the tissue.

For optical imaging of objects embedded in a turbid medium, basically two schemes have been used. One scheme is to filter out the multiply scattered light, and the other, referred to as the inverse approach, is to use the multiply scattered light emerging from various positions around the object for imaging. For filtering out the multiply scattered light, one may exploit the depolarization or the loss of coherence of the scattered light or the fact that the scattered light emerges from the tissue in all directions and also takes a longer time to emerge as compared to the unscattered (ballistic) or predominantly forward scattered (snake-like) components. The latter essentially travel in a forward direction and so arrive earlier. Coherence gating filters out the ballistic photons having the highest image information and hence can provide images with the best resolution (down to a few micrometers, limited by the coherence length of the source). However, the number of ballistic photons decreases exponentially on propagation through a turbid medium and will be of the order of e^{-100} of the incident photons on propagation through 1-cm-thick tissue with a scattering coefficient of ~100 cm^{-1}. Therefore, coherence gating can only be used for imaging the full depth of transparent objects (such as ocular structure) or a few millimeter's depth of turbid tissue. OCT, the approach that exploits coherence gating for optical imaging, has emerged as a rapid, noncontact, noninvasive, high-resolution imaging technique and is finding clinical applications in ophthalmology, dermatology, etc. (Zysk et al. 2007). The contrast in OCT images can be further augmented by incorporating polarization sensitive detection, which would provide information about the birefringent properties of the tissue. This helps monitor changes in the morphology of the birefringent constituents (collagen, tendon, etc.) of the tissue and thus can be used for noninvasive monitoring of the healing of wounds (Sahu et al. 2010). Similarly, OCT can be extended to incorporate other functional imaging parameters, such as absorption or flow velocities in the spectral and Doppler OCT, respectively (Bouma and Tearney 2002).

A larger depth of imaging, but with a poorer resolution, is possible by the use of polarization gating or time gating because in these approaches both ballistic and forward scattered components are utilized for imaging. The combined magnitude of these on propagation through 1-cm-thick tissue is proportional to $\exp(-\mu_s')$. This, for an anisotropy parameter of 0.9 and μ_s of 100 cm^{-1}, implies that even after a depth of 1 cm, the signal strength is reasonable: $\sim e^{-10}$ as compared to e^{-100} for the ballistic component. It may be noted that the size distribution of the scatterer also affects the temporal profile of the transmitted laser pulse and thus the contrast in time-gated optical imaging (Rao et al. 2005). Further, with the use of nonlinear optical time-gating techniques, such as stimulated Raman scattering, the gated light can be amplified to generate reasonable signal levels even after propagating through a few centimeters thick tissue. For polarization-gated imaging, the rate of depolarization and hence the depth of imaging as well as the contrast depends both on the size distribution of the scatterer and the incident polarization state (Ghosh, Patel, and Gupta 2003).

While the polarization-gated optical imaging normally utilizes the polarized fraction of the light, the depolarized fraction has also been effectively utilized for the imaging of tissue vasculature. If the tissue is illuminated with a linearly polarized light and the backscattered light is detected in a cross-polarization channel, not only is the specular reflection from samples surface eliminated but also the depolarized light coming from deeper layers effectively back illuminates the tissue for the imaging of microvasculature (Groner et al. 1999). Such orthogonal polarization spectral (OPS) imaging has been used for visualization and quantitative imaging of microcirculation in a number of diagnostic applications (Cerny, Turek, and Parizkova 2007).

Imaging through still larger depths as required for imaging of the human brain or female breast necessitates the use of diffuse photons with a concomitant decrease in the spatial resolution (at best, a few millimeters). In a typical diffuse optical tomography (DOT) system, data are acquired by placing the source and detectors at various locations around the object. Both time-domain and frequency-domain measurements are used for DOT. In the time-domain approach, the temporal profile of the detected signal following illumination by ultrafast pulse is used to reconstruct the three-dimensional map of the optical parameter distribution, whereas in the frequency-domain systems, demodulation and changes in phase of the intensity-modulated light are used for estimation of the tissue parameter distribution. Although frequency-domain DOT setups have a limitation in that measurements are made at only few discrete frequencies, it is still more widely used being less expensive and portable compared to time-domain setups (Gibson and Dehghani 2009). DOT provides useful information about the blood dynamics, blood volume, blood oxygen saturation, and water and lipid content of the tissues because the absorption spectra of major tissue chromophores, such as oxyhemoglobin and deoxyhemoglobin, cytochrome, and water, differ significantly in the NIR region.

While for imaging one exploits the intensity, coherence, or polarization of the scattered light, other parameters of the scattered light, such as its angular distribution and the spectral content, also contain significant diagnostic information. As noted earlier, the angular distribution of the scattered light can provide information about the size of the scatterers. Further, measurements on the polarized component of back-scattered light can also be used to filter out the multiply scattered light from deeper layers of the tissue and thus look at the light scattered from the superficial tissue layer. This approach has been used to extract the size distribution and the density of the nuclei in the superficial epithelial cell layer, which can be used to monitor neoplastic changes in biological tissues (Gurjar et al. 2001). Changes in the polarization parameters of the tissue arising as a result of its birefringent (collagen, tendon, etc.) and chiral (glucose) constituents can be separated from the change in polarization arising as a result of scattering (Manhas et al. 2006) and exploited for diagnostics. Indeed, significant differences were observed in the polarization parameters (retardance, diattenuation, and depolarization) of the normal and malignant sites of oral and breast tissue, and these were shown to correlate well with the changes expected in the structure of the collagen present in these tissues (Manhas et al. 2009). Similarly, there is interest in the estimation of the concentration of chiral molecules, such as glucose in tissue, by measuring the rotation of the plane of polarization of the back-scattered light (Manhas et al. 2006). The differences in the structural and functional properties of normal and malignant tissue lead to differences in their wavelength-dependent absorption and scattering characteristics (Ghosh et al. 2001), which can also be used for diagnostics by itself or as a result of the effect these changes have on the elastically and inelastically scattered light.

The inelastically scattered light is much weaker compared to the elastically scattered light and therefore requires use of high brightness sources, for example, lasers and appropriate light delivery and collection systems, for its use for practical applications. Both fluorescence and Raman spectroscopic (Vo-Dinh 2003; Tuchin 2011) approaches are being actively pursued for their diagnostic potential. The fluorescence of native tissue originates from a number of endogenous fluorophores that are present in tissue. The prominent fluorophores include aromatic amino acids, such as tryptophan; structural proteins, such as collagen and elastin; coenzymes, such as NADH and flavins; and the porphyrins. Their excitation maxima typically lie in the range 280–500 nm, whereas their emission maxima lie in the range 300–700 nm. The observed fluorescence emission of native tissue is essentially a convolution of the emission spectra of the endogenous fluorophores of tissue and therefore strongly depends on the wavelength of the light of excitation. Only those endogenous fluorophores are excited and emit fluorescence whose absorption bands have an overlap with the wavelength of the excitation light. Because the excitation light and the emitted fluorescence have to propagate through the turbid tissue, the measured tissue fluorescence is also influenced by the absorption and scattering at both the excitation and the emission wavelengths. This makes it difficult to extract, from the measured fluorescence, the intrinsic fluorescence of the tissue, which may have valuable biochemical information about the tissue. Changes in the absorption and scattering characteristics of tissue can also lead to subtle changes in its fluorescence characteristics. For example, it has been shown that malignant breast tissue sites have a larger scattering coefficient as compared to normal. This has an interesting consequence. Whereas for thin tissue sections (thickness < optical transport length), the depolarization of fluorescence was observed to be smaller in malignant tissues compared to normal (as a result of the changes in the biochemical environment of the fluorophores), the reverse was observed for a thicker tissue section because there scattering is the major cause of depolarization (Mohanty et al. 2001). As a consequence, the fluorescence from the superficial layer of tissue is the least depolarized, and that originating from deeper layers becomes increasingly more depolarized. Therefore, measurements on the polarized component of fluorescence could be used for depth-resolved measurements on fluorescence (Ghosh et al. 2005).

While, because of significantly larger strength, the fluorescence-based technique is presently better developed for clinical applications, the utilization of Raman spectroscopy for diagnostic applications is also receiving a lot of attention. Because of its molecular specificity, the Raman technique makes obtaining specific biochemical information about the tissue so much easier. For further details on the use of optical spectroscopy for biomedical diagnosis, the reader is referred to Vo-Dinh (2003) and Tuchin (2011).

3.5 Surgical and Therapeutic Applications

Surgical and therapeutic applications depend on absorption of light. The absorbed laser energy can broadly lead to three effects (Gupta, Ghosh, and Patel 2007). The most common effect is a rise in tissue temperature (the photothermal effect). At high intensities associated with lasers operating in short-pulse durations (nanosecond, picosecond), absorption of laser radiation may lead to the generation of pressure waves or shockwaves (photomechanical effects). Short-wavelength lasers can cause electronic excitation of chromophores in the tissue and thus initiate a photochemical reaction (photochemical effect). The relative role played by the three depends primarily on the laser wavelength, irradiance, and pulse duration.

3.5.1 Photothermal Effects

Most of the surgical applications of light exploit a photothermal effect, that is, a rise in tissue temperature following absorption of light. The biological effect depends on the level of rise in tissue temperature, which is determined by two factors: the tissue volume in which a given energy is deposited and the time in which the energy is deposited *vis à vis* the thermal relaxation time (the inverse of which determines the rate of flow of heat from heated tissue to the surrounding cold tissue). A small rise in temperature (5°C–10°C) can influence the activity of enzymes and lead to changes in blood flow and vessel permeability. Tissues heated to a temperature of 45°C–80°C may get denatured as a result of breakage of van der Wall bonds, which stabilize the conformation of proteins and other macromolecules. Thermal denaturation is exploited for therapy in several ways. For example, hemostasis occurs because of increased blood viscosity caused by denaturation of plasma proteins, hemoglobin, and perivascular tissue. When the temperature exceeds 100°C, boiling of water in the tissue takes place. Because of the large latent heat of water, the main constituent of tissue, energy added to tissue at 100°C first results in generation of steam without further increase in temperature. A volume expansion by ~1670-fold occurs when water is vaporized isobarically. When this large and rapid expansion occurs within tissue, physical separation or "cutting" occurs. Tissue surrounding the region being vaporized will also be heated, resulting in coagulation of the tissue at the wound edges and thus preventing blood loss.

If the rate of deposition of energy is faster than that required for the boiling of water, the tissue is superheated and can be thermally ablated. Thermal ablation or explosive boiling is similar to what happens when cold water is sprinkled on a very hot iron. In contrast to the vaporization in which the tissue temperature is ~100°C, for ablation, the tissue temperature is much higher (500°C or more), and the kinetics involved are considerably faster. In ablation, practically all the energy deposited in the tissue is converted into the kinetic energy of the ablation products with the result of minimal thermal damage to the adjoining tissues.

It is pertinent to emphasize that by appropriate choice of laser wavelength, it is possible to selectively deposit energy in a target site if it has greater absorption than the surrounding tissue. Further, by use of laser pulses of duration shorter than the thermal relaxation time, heat can be confined within the target tissue so that it can be vaporized without significant effect on surrounding tissue. Such selective photothermolysis has been exploited for several therapeutic applications, such as laser treatment of port-wine stains. Another approach that is receiving attention for controlled localized heating involves the use of near infrared light tuned to surface plasmon resonance of metallic nanoparticles. Such heating of metallic nanoparticles that have been selectively deposited in target cells can be used for applications such as hyperthermia for cancer treatment.

3.5.2 Photomechanical Effects

Photomechanical effects are usually important only at high intensities typical of short-duration (10^{-9}–10^{-12} s) laser pulses. The localized absorption of intense laser radiation can lead to very large temperature gradients, resulting in enormous pressure waves and localized photomechanical disruption. Such disruption is useful, for example, in the laser removal of tattoo marks. Tattoo ink has pigmented molecular particles too large for the body's immune system to eliminate. Photodisruption of these into smaller particles enables the body's lymphatic system to dispose them, resulting in removal of the tattoo mark.

At high intensities, the electric field strength of radiation is also very large (about 3×10^7 V/cm at an intensity of 10^{12} W/cm²) and can cause dielectric breakdown in the tissue. The resulting plasma absorbs energy and expands, creating shockwaves, which can shear off the tissue. These plasma-mediated shockwaves are used for breaking stones in the kidney or urethra (lithotripsy) and in posterior capsulotomy for the removal of an opacified posterior capsule of the eye lens.

3.5.3 Photochemical Effects

For laser irradiation at power levels in which there is no significant rise in temperature of the tissue, the photothermal and photomechanical effects are not possible. In such a situation, only photochemical effects can take place provided the energy of the laser photon is adequate to cause electronic excitation of biomolecules, which can be either endogenous or externally injected. The photoexcitation of molecules and the resulting biochemical reactions can lead to either bioactivation, exploited in various phototherapies (Karu 2010), or generation of some free radicals or toxins, which are harmful for the host tissue. The latter process is used for PDT of cancer (Agostinis et al. 2011). Short-wavelength radiation of sufficient energy can also break molecular bonds and impart kinetic energy to the fragments by which they get ejected from the tissue. This photochemical removal (photoablation) can occur only from the area of tissue exposed to the light and will have no effect on the surrounding unexposed tissue. In photoablation, there is no rise in the tissue temperature, and the tissue removal can be achieved in an extremely precise manner because of the small penetration depth of light at these short wavelengths. It is for this reason that UV excimer lasers are being widely used for reprofiling the cornea to correct vision disorders. Though conventional light sources are also useful for phototherapy, a better control of light characteristics can often make phototherapy more convenient with the use of laser. PDT, because of its high selectivity, has received a lot of attention for the treatment of cancer. Its use for inactivation of microbes, referred to as antimicrobial PDT (APDT), is also attracting attention. The advantage of PDT over conventional antimicrobials is that the treatment is restricted to the light-irradiated regions of the drug-treated area and therefore reduces the risk of adverse systemic effects. Also, reactive oxygen species generated in the photochemical reactions cause nonspecific damage to cellular components, and therefore, it is highly unlikely for bacteria to develop resistance (Maisch 2007).

3.6 Concluding Remarks

Light has been used in photomedicine since time immemorial. This field has seen remarkable growth in the past few decades fueled by the invention of lasers that provide a light source with much better control on its parameters and other technological developments, such as sensitive detection systems, the large information-processing capability of present-day computers, etc. Optical techniques are contributing significantly toward the realization of noninvasive, near real-time biomedical imaging and diagnosis, and also therapeutic modalities providing higher selectivity than conventional modalities. Effective utilization of these techniques and further improvement in their performance requires a good understanding of light propagation in tissue. In spite of the complexities of the problem, significant advancements have been made. It is expected that continued interest in photomedicine will provide further impetus to this field and should lead to more innovative methods for furthering photomedicine.

References

Agostinis, P., K. Berg, K. A. Cengel, T. H. Foster, A. W. Girotti, S. O. Gollnick, S. M. Hahn, M. R. Hamblin, A. Juzeniene, D. Kessel, M. Korbelik, J. Moan, P. Mroz, D. Nowis, J. Piette, B. C. Wilson, and J. Golab. 2011. Photodynamic therapy of cancer: An update. *CA Cancer J Clin* 61:250–281.

Bays, R., G. Wagnieres, D. Robert et al. 1996. Clinical determination of tissue optical properties by endoscopic spatially resolved reflectometry. *Appl Opt* 35:1756–1766.

Beuthany, J., O. Minety, J. Helfmannz, M. Herrigz, and G. Muller. 1996. The spatial variation of the refractive index in biological cells. *Phys Med Biol* 41:369–382.

Bhaduri, B., H. Pham, M. Mir, and G. Popescu. 2012. Diffraction phase microscopy with white light. *Opt Lett* 37:1094–1096.

Bie, V. 1899. Remarks on Finsen's phototherapy. *Br Med J* 30:825–830.

Bohren, C. F., and D. Huffman. 1983. *Absorption and Scattering of Light by Small Particles*. John Wiley, New York.

Cerny, V., Z. Turek, and R. Parizkova. 2007. Orthogonal polarization spectral imaging. *Physiol Res* 56:141–147.

Choi, W., C. Fang-Yen, K. Badizadegan et al. 2007. Tomographic phase microscopy. *Nat Methods* 4:717–719.

Cremer, R. J., P. W. Perryman, and D. H. Richards. 1958. Influence of light on the hyperbilirubinaemia of infants. *Lancet* 271:1094–1097.

Cutler, M. 1931. Transillumination of the breast. *Ann Surg* 93:223–234.

Ding, H., J. Q. Lu, W. A. Wooden, P. J. Kragel, and X. Hu. 2006. Refractive indices of human skin tissues at eight wavelengths and estimated dispersion relations between 300 and 1600 nm. *Phys Med Biol* 51:1479–1489.

Fulop, A. M., S. Dhimmer, J. R. Deluca et al. 2010. A meta-analysis of the efficacy of laser phototherapy on pain relief. *Clin J Pain* 26:729–736.

Ghosh, N., S. K. Mohanty, S. K. Majumder, and P. K. Gupta. 2001. Measurement of optical transport properties of normal and malignant human breast tissue. *Appl Opt* 40:176–184.

Ghosh, N., H. S. Patel, and P. K. Gupta. 2003. Depolarization of light in tissue phantoms – effect of a distribution in the size of scatterers. *Opt Express* 11:2198–2205.

Ghosh, N., S. K. Majumder, H. S. Patel, and P. K. Gupta. 2005. Depth-resolved fluorescence measurement in a layered turbid medium by polarized fluorescence spectroscopy. *Opt Lett* 30:162–164.

Gibson, A., and H. Dehghani 2009. Diffuse optical imaging. *Phil Trans R Soc A* 367:3055–3072.

Groner, W., J. W. Winkelman, A. Harris et al. 1999. Orthogonal polarization spectral imaging: a new method for study of the microcirculation. *Nat Med* 5:1209–1212.

Gupta, P. K., N. Ghosh, and H. S. Patel. 2007. *Lasers and Laser Tissue Interaction. Fundamentals & Applications of Biophotonics in Dentistry.* A. Kishen and A. Asundi, editors. Imperial College Press, London. 123–148.

Gurfinkel, M., T. Pan, and E. Sevick-Muraca. 2004. Determination of optical properties in semi-infinite turbid media using imaging measurements of frequency-domain photon migration obtained with an intensified charge-coupled device. *J Biomed Opt* 9:1336–1346.

Gurjar, R. S., V. Backman, L. T. Perelman et al. 2001. Imaging human epithelial properties with polarized light-scattering spectroscopy. *Nat Med* 7:1245–1248.

Ishimaru, A. 1978. *Wave Propagation and Scattering in Random Media. Vol 1: Single Scattering and Transport Theory.* Academic Press, Waltham, MA.

Jobsis, F. 1977. Noninvasive, infrared monitoring of cerebral and myocardial oxygen sufficiency and circulatory parameters. *Science* 198:1264–1267.

Karu, T. I. 2010. Multiple roles of cytochrome c oxidase in mammalian cells under action of red and IR-A radiation. *IUBMB Life* 62:607–610.

Keeler, C. R. 2002. The ophthalmoscope in the lifetime of Hermann von Helmholtz. *Arch Ophthalmol* 120:194–201.

Kubelka, P. 1948. New contributions to the optics of intensely light-scattering materials. Part I. *JOSA* 38:448–457.

Lam, R. W., A. J. Levitt, R. D. Levitan, M. Enns, R. Morehouse, E. E. Michalak, and E. M. Tam. 2006. The Can-SAD Study: A randomized controlled trial of the effectiveness of light therapy and fluoxetine in patients with winter seasonal affective disorder. *Am J Psychiatry* 163:805–812.

Leonard, D. W., and K. M. Meek. 1997. Refractive indices of the collagen fibrils and extrafibrillar material of the corneal stroma. *Biophys J* 72:1382–1387.

Maisch, T. 2007. Anti-microbial photodynamic therapy: useful in the future? *Lasers Med Sci* 22:83–91.

Manhas, S., M. K. Swami, P. Buddhiwant, N. Ghosh, P. K. Gupta, and K. Singh. 2006. Mueller matrix approach for determination of optical rotation in chiral turbid media in backscattering geometry. *Opt Express* 14:190–202.

Manhas, S., M. K. Swami, H. S. Patel, A. Uppal, N. Ghosh, and P. K. Gupta. 2009. Polarized diffuse reflectance measurements on cancerous and noncancerous tissues. *J Biophotonics* 2:581–587.

McDonagh, A. F. 2001. Phototherapy: from ancient Egypt to the new millennium. *J Perinatol* 21:S7–S12.

Millikan, G. A. 1942. The oximeter: An instrument for measuring continuously oxygen-saturation of arterial blood in man. *Rev Sci Instrum* 13:434–444.

Moan, J., and Q. Peng. 2003. An outline of the history of PDT. In *Photodynamic Therapy.* T. Patrice, editor. The Royal Society of Chemistry, London. 1–18.

Mohanty, S. K., N. Ghosh, S. K. Majumder, and P. K. Gupta. 2001. Depolarization of autofluorescence from malignant and normal human breast tissues. *Appl Opt* 40:1147–1154.

Parrish, J. A. 1977. Treatment of psoriasis with long-wave ultraviolet light. *Arch Dermatol* 113:1525–1528.

Pathak, M. A., and T. B. Fitzpatrick. 1992. The evolution of photochemotherapy with psoralens and UVA (PUVA): 2000 BC to 1992 AD. *J Photochem Photobiol B* 30:3–22.

Patterson, M. S., B. Chance, and B. C. Wilson. 1989. Time resolved reflectance and transmittance for the non-invasive measurement of tissue optical properties. *Appl Opt* 28:2331–2336.

Prioreschi, P. A. 2003. *A History of Medicine: Medieval Medicine.* Horatius Press, Omaha.

Rao, D., H. S. Patel, B. Jain, and P. K. Gupta. 2005. Time-gated optical imaging through turbid media using stimulated Raman scattering: Studies on image contrast. Pramana. *Journal of Physics* 64:229–238.

Sahu, K., Y. Verma, M. Sharma, K. D. Rao, and P. K. Gupta. 2010. Non-invasive assessment of healing of bacteria infected and uninfected wounds using optical coherence tomography. *Skin Res Technol* 16:428–437.

Star, W. M. 1997. Light dosimetry in vivo. *Phys Med Biol* 42:763–787.

Tuchin, V. V. (editor). 2011 *Handbook of Photonics for Biomedical Sciences.* CRC Press, Boca Raton, FL.

Tuchin, V. V. 2007. *Tissue Optics: Light Scattering Methods and Instruments for Medical Diagnosis.* 2nd edition. SPIE Press, Bellingham, WA.

Vo-Dinh, T. (editor). 2003. *Biomedical Photonics Handbook.* CRC Press, Boca Raton, FL.

Wang, L., S. L. Jacques, and L. Zheng. 1995. MCML—Monte Carlo modeling of light transport in multi-layered tissues. *Comput Methods Programs Biomed* 47:131–146.

Wang, Z., K. Tangella, A. Balla, and G. Popescu. 2011. Tissue refractive index as marker of disease. *J Biomed Opt* 16:116017.

Wax, A. and V. Backman (editors). 2010. *Biomedical Applications of Light Scattering.* McGraw-Hill, New York.

Welch, A. J., and M. J. C. van Gemert. 1995. *Optical-Thermal Response of Laser-Irradiated Tissue.* Plenum Press, New York.

Woodruff, L. D., J. M. Bounkeo, W. M. Brannon et al. 2004. The efficacy of laser therapy in wound repair: A Meta-analysis of the literature. *Photomed Laser Surg* 22:241–247.

Zysk, A. M., F. T. Nguyen, A. L. Oldenburg, D. L. Marks, and S. A. Boppart. 2007. Optical coherence tomography: a review of clinical development from bench to bedside. *J Biomed Opt* 12:051403.

History and Fundamentals of Photodynamic Therapy

Tyler G. St. Denis
Columbia University,
Massachusetts General Hospital

Michael R. Hamblin
Massachusetts General Hospital

4.1 Photochemical and Photophysical Fundamentals of Photodynamic Therapy

Before explaining the historical, and chiefly biological, developments of photodynamic therapy (PDT) research, we would like to provide a survey of the chemical and physical properties of PDT. We would like to preface this section by saying that more detailed photophysical and photochemical explanations—as aided by quantum mechanical explanations—may be found elsewhere (Hamblin and Mróz 2008).

PDT consists of the application of light in the presence of a photosensitizer (PS) and oxygen. When combined, these three components produce reactive oxygen species (ROS) as illustrated in Figure 4.1. The PS exists in a ground singlet state in which all electrons are spin paired in low-energy orbitals. Upon illumination at the wavelength corresponding to absorption peaks of the PS, an electron in the highest occupied molecular orbital (HOMO) of the PS is excited to the lowest unoccupied molecular orbital (LUMO) of the PS. Because of the aromatic character of many PSs, the energy difference between the HOMO and LUMO is fairly low, and light used is typically in the visible and near-infrared portions of the electromagnetic spectrum. Because of the optical properties of tissue, red light is desirable for PDT as a result of its deeper tissue penetration while still maintaining sufficient energy for molecular excitations. Moreover, red light is not absorbed by collagens, melanin, water, hemoglobin, and proteins and is thus in the ideal optical therapeutic window (Hamblin and Mróz 2008).

Once electronic excitation from the HOMO to the LUMO occurs, the PS is in an unstable and transient excited singlet state. In this state, several distinct processes may occur (Foote 1991). In the case of PDT, the most critical process is the reversal of the spin of the excited electron, known as intersystem crossing to the triplet state of the PS. Intersystem crossing is favored in that the PS triplet state is of lower energy compared to the singlet state, but the triplet state has a considerably longer lifetime. This is because the excited electron, now with a spin parallel to its former paired electron, may not immediately relax to the singlet state (as it would then have identical quantum numbers to that of its paired electron, thus violating the Pauli exclusion principle). Accordingly, the excited triplet-state electron may first obtain correct spin orientation (a slow process) and then relax to the ground singlet state (phosphorescence), or the PS excited triplet state electron may interact with molecules abundant in its immediate environment.

Because of selection rules that specify that triplet–triplet interactions are spin-allowed while triplet–singlet interactions are spin-forbidden, the excited PS may then only interact with a molecule that is also triplet in electron spin. One of the few available molecules in the environment, molecular oxygen (O_2) is a triplet in its ground state as evident from its diradical nature. Therefore, energy transfer to oxygen from a triplet-state PS is a remarkably favored process.

The photodestructive capabilities of PDT may be attributed to the generation of free radical and electronically excited oxygen species. As mentioned above, the ground state of O_2 is a triplet state in which the two outermost orbitals are unpaired but spin

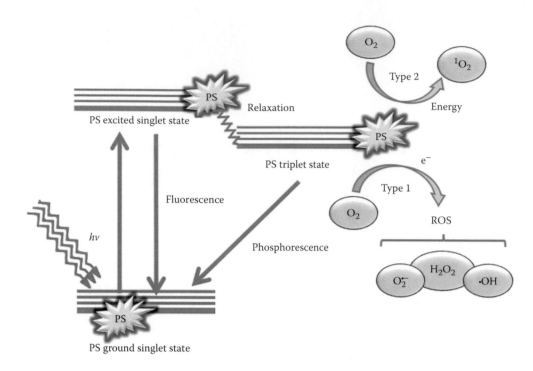

FIGURE 4.1 Jablonski diagram illustrating PDT mechanisms, including the Type 1 and Type 2 mechanisms. A PS absorbs light of the appropriate wavelength (usually in the visible range), which causes excitation to the excited singlet state. The singlet excited PS can decay back to the ground state upon release of energy (fluorescence) or relax down to the long-lived triplet state. In the triplet state, reactions between the PS and oxygen (with a ground triplet) are favorable and permitted by selectivity rules.

parallel (responsible for the paramagnetic qualities of O_2). When the PS is in the long-lived excited triplet state, it may interact with O_2 in two distinctly different ways (Foote 1991; Tanielian et al. 2000). The so-called Type 1 process occurs when the PS directly transfers an electron, first yielding the superoxide anion ($O_2^{\bullet-}$), which may then form other radical species, including the hydroxyl radical (•OH) and hydrogen peroxide (H_2O_2).

Alternatively, the Type 2 photochemical process occurs when the energy used to excite the PS to the triplet state is simply transferred to O_2. This has the effect of flipping the spin of an outermost O_2 electron and shifting it into the orbital containing the other, previously unpaired, electron of (now) opposite spin. This leaves one molecular orbital entirely unoccupied (a violation of Hund's rule). This species of O_2 is said to be in the excited singlet state and is accordingly termed singlet oxygen ($^1O_2$). This form of oxygen is not a radical as all electrons are spin paired; however, it is extremely reactive and short lived because of its instability in electron configuration.

The formation of ROS results in the extensive oxidation of biological macromolecules of importance to life, including lipids, proteins, and nucleic acids (Ochsner 1997). •OH is arguably the most reactive of all radicals formed and will abstract electrons to become a hydroxide ion, which is considerably more stable than its radical analogue, such that subsequent protonation yields a water molecule. $O_2^{\bullet-}$ may also abstract an electron, forming a peroxide ion, O_2^{2-}, that is immediately protonated to form H_2O_2. That being said, $O_2^{\bullet-}$ is relatively unreactive in biological systems because of the antioxidant enzyme superoxide

dismutase, which converts $O_2^{\bullet-}$ to H_2O_2 and O_2. H_2O_2 may then be converted to H_2O and O_2 by catalase. Nonetheless, H_2O_2 may react with extremely small concentrations of one-electron reducing agents, such as ferrous iron (Valko, Morris, and Cronin 2005), resulting in the heterolytic fission of the O–O bond in H_2O_2, yielding a hydroxide ion and •OH via oxidation of ferrous iron to ferric iron. While •OH is not degraded by any enzymes, it may be quenched by antioxidant peptides (e.g., glutathione) or by antioxidant vitamins (e.g., ascorbic acid) (Buettner 1993).

The Type I pathway may also lead indirectly to the formation of reactive nitrogen species, because $O_2^{\bullet-}$ may react with nitric oxide (NO•) to yield peroxynitrite anions ($ONOO^-$). $ONOO^-$ is highly reactive and short lived and will rapidly undergo homolytic fission, forming •OH and nitrogen dioxide (•NO_2). $ONOO^-$ may also react with CO_2 to form carbonate radical anions ($CO_3^{\bullet-}$) and •NO_2. All of these generated radical species are dangerous as they will readily undergo radical propagation until a radical–radical termination process may occur.

Though $^1O_2$ is not a radical, it still reacts with macromolecules and may do so in several distinct manners. $^1O_2$ may serve as a dienophile for Diels–Alder cycloaddition reactions. Aromatic as well as conjugated systems may serve as dienes, resulting in the degradation of many lipids and proteins. Double bonds, sulfur moieties (both of which have high electron density and are abundant in biological molecules), and other electron-rich species may also attack $^1O_2$ (Clennan and Pace 2005). Unlike radicals, $^1O_2$ cannot be broken down by enzymes but can be quenched by antioxidants, including carotenoids.

4.2 Photochemotherapy and the Pre-PDT Era

Many would be quick to say that PDT and photodynamic processes were discovered in the last century by Oskar Raab and his advisor Professor Herman von Tappeiner; however, this is not technically true. The use of light and light-sensitive dyes (PS) as a medical modality can be traced from antiquity. That said, the rediscovery, coinage, and mechanism elucidation of PDT and photodynamic processes occurred over the last 100 years. This photomedical historical gap from the ancients to 20th-century Europe may be explained by the fact that the only photonic source for the ancients' photochemotherapeutics was the sun. With the advent of Christianity in Europe, heliotherapy was considered sun worship, an unacceptable paganism; this stigma, in turn, retarded the development and advancement of all phototherapies (Daniell and Hill 1991).

Photochemotherapy's history is grounded in the pigment psoralen (Figure 4.2a) of the furanocoumarin class of natural products and psoralen + ultraviolet A (PUVA) therapy. The first documented photochemotherapy comes from the ancient Hindu scripture *Artharva Veda* (Fitzpatrick and Pathak 1959). Dating back to 1400 B.C. or earlier, *Artharva Veda* describes the application of a dark, black plant for the repigmentation of leprous and vitiliginous skin. While the exact plant used is unspecified, many scholars speculate that the photoactive natural product used was psoralen, found in the seeds of *Psoralea corylifolia,* which was ubiquitous in ancient Indian medicine. *Ammi majus,* commonly known as Bishop's weed, grows along the south bank of the Nile River and is a natural source of psoralens. Consequently, around the 12th century C.E., the Egyptians discovered Bishop's weed's photoactive properties and applied it in salve form to combat vitiligo (Fitzpatrick and Pathak 1959).

Unbeknownst to the ancients, their concomitant application of psoralens and sunlight (a source of ultraviolet A light) may not only induce photocyclizations between nucleic bases but also generate $^1O_2$ and represents a quasi-PDT (Jones, Young, and Truscott 1993). However, it would take almost 1800 years after the Egyptians before the potential of photochemotherapy, and specifically PDT, would be realized. Before continuing, it is important to contextualize the PDT discoveries of the 20th century by noting that the 19th century consisted of a plethora of dye research mainly fueled by the interests of the textile industry. Consequently, PDT was not discovered with natural product phytoalexin pigments but rather with synthetic, coal tar-derived dyes.

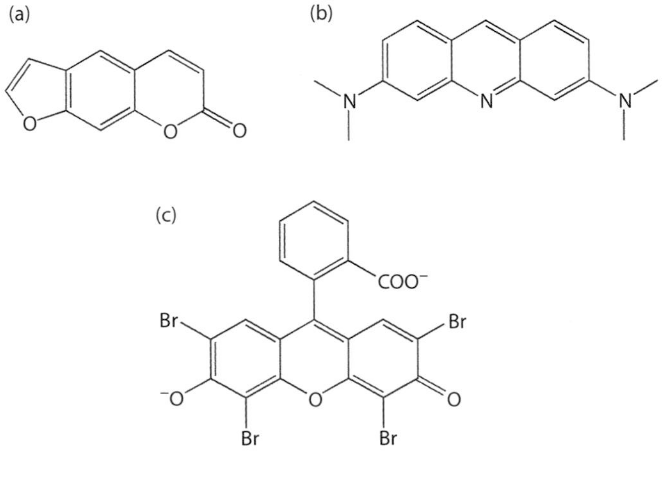

4.3 PDT: An Accidental Discovery

In 1900, more than 100 years ago, Oskar Raab, a German medical student working under the guidance of Professor Herman von Tappeiner at the Pharmacological Institute of Munich, Germany, first observed an *in vitro* photodynamic effect (Raab 1900). At the time, von Tappeiner was interested in the potential antimalarial properties of acridine orange (Figure 4.2b) and used *Paramecium spp.* protozoans as a model system for *Plasmodium spp.* parasites. Some time before 1900, Raab noticed an inconsistency in his data: The lowest concentration of acridine orange serial dilutions killed paramecia within 60–100 min of application, and earlier experimentation demonstrated that the same dilution enabled paramecia to live 800–1000 min. This greater than tenfold decrease in life expectancy initially confused both Raab and von Tappeiner, causing them to thoroughly and scrupulously analyze the protocols of the two experiments. The only difference between the two trials was that one was carried out during a thunderstorm with a good deal of lightning that modified ambient lighting conditions. This sparked their scientific curiosity, and the two went on to prove that light combined with acridine orange is more effective at killing paramecia than light or acridine orange alone. Raab postulated that the concomitant application of light and acridine orange resulted in some sort of energy transfer but did not yet understand the importance of oxygen in the destruction of paramecia by photodynamic mechanisms. Nevertheless, Raab and Tappeiner understood the importance of this finding, and, as early as 1900, the two realized the potential dermatological applications of this dye excitation (von Tappeiner 1900).

In February of 1903, von Tappeiner began collaborating with Albert Jesionek, an assistant professor in the department of dermatology at the University of Munich. The two investigated the application of eosin (Figure 4.2c) and light for the treatment of skin cancer, lupus vulgaris, and condylomata and, in November of 1903, published the first data on the topic (von Tappeiner and Jesionek 1903). In 1904, von Tappeiner coined the term *photodynamische,* translating to "photodynamic effect" when noting that an oxygen-consuming reaction occurred in protozoa upon adjunct application of aniline-based dyes and light (von Tappeiner and Joblauer 1904). Remarkably, 4 years after PDT's discovery, the importance of the light, PS, and oxygen—the holy trinity of PDT—had already been realized without the aid of the aforementioned photophysical and photochemical explanations.

FIGURE 4.2 PS associated with the pre-PDT era and the early 1900s (the discovery of PDT). (a) Psoralen; (b) acridine orange; (c) eosin.

Around 1905, von Tappeiner and Jesionek continued their clinical investigations on the PDT-mediated eradication of skin tumors. Three patients' tumors were covered and/or injected with 5% eosin solution and either irradiated with sunlight or artificial arch lamp light for several weeks, and some early improvement was noted (Jesionek and von Tappeiner 1905). This pair extended their studies to six patients with skin cancer, modifying the concentration of eosin used as well as combining eosin with fluorescein (Figure 4.2c) or sodium dichloroanthracene disulfonate (Jesionek and von Tappeiner 1905). One patient was incubated with Grubler's Magdalene red, and in all instances, favorable results were observed. While these studies were the first to methodically employ PDT for the treatment of skin disorders, they were not the first clinical administrations of dyes. Five years earlier, in 1900, Jean Prime, a French neurologist, administered eosin orally in an attempt to treat epilepsy and observed the blistering, swelling, and nail loss of treated epileptics when exposed to light (Prime 1900). Unknown to Prime, he had observed perhaps the greatest drawback of PDT: photosensivity. Nevertheless, this phenomenon was hardly new seeing as in 1892, Charles Darwin and Karl Dammann had reported exanthema of animals fed with buckwheat (Letner 1990). This was later attributed to the photoreactivity of fagopyrin in the buckwheat plant (Hinneburg and Neubert 2005).

4.4 Intermediate Years

While coal tar-derived dyes initiated the discovery of PDT, subsequent PDT research was predominately fueled by the biological class of dyes known as porphyrins. Porphyrins are aromatic cyclic tetrapyrrole compounds that possess a red-purple color (hence the name's derivation from the Greek *porphuros*, meaning purple) and are found in all forms of life, serving myriad functions (Figure 4.3a). W. Haussmann of Vienna, Austria, is unequivocally responsible for spearheading porphyrin-based PDT with hematoporphyrin (Figure 4.3b). Hematoporphyrin is derived from the acid hydrolysis of hemoglobin and was first produced by the German chemist Johann Joseph von Scherer in 1841 when he treated dried blood with sulfuric acid, thus removing the iron from the heme in hemoglobin (von Scherer 1841). J.J. von Scherer noted that the iron-free adduct still maintained a red color, and this was attributed to the presence of the remaining dechelated heme. In 1908, Haussmann employed hematoporphyrin in the successful destruction of paramecia and erythrocytes and, soon after in 1911, injected mice with hematoporphyrin, observing murine skin photosensitivity (Hausman 1908, 1911).

Two years later, in 1913, the curious German physician Friedrich Meyer-Betz sought to understand the physiological effects of intravenous application of hematoporphyrin in mammals—specifically humans—and, consequently, injected himself with 200 mg of hematoporphyrin (Meyer-Betz 1913). Within minutes after exposure to the sun, Meyer-Betz experienced swelling and extreme pain on his face and hands, the only

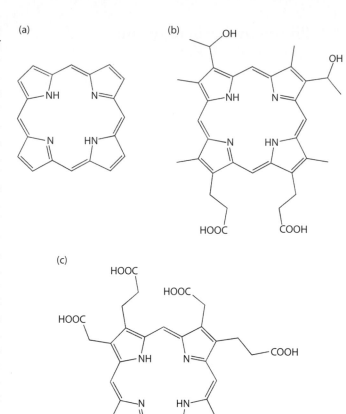

FIGURE 4.3 Porphyrin-based PS. (a) Porphyrin macrocycle; (b) hematoporphyrin; (c) uroporphyrin.

parts of his body exposed to light (Figure 4.4). This acute photosensitivity lasted for more than 2 months and highlighted the potential biomedical applications of targeted porphyrin therapeutics. While the bulk of Germany's chemical research was relocated for the creation of chemical weaponry during World

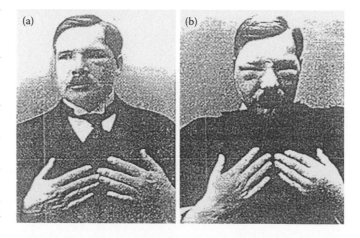

FIGURE 4.4 The famous photosensitization of Friedrich Meyer-Betz. Note the before and after injection of 200 mg hematoporphyrin.

War I, pyrrole chemist and Nobel laureate Hans Fischer continued studying porphyrins. In 1925, he noticed that uroporphyrin (Figure 4.3c) was comparable to hematoporphyrin in toxicity (Fischer et al. 1925).

In 1924, the French scientist Policard observed red fluorescence in rat sarcoma following ultraviolet illumination (via a Woofs lamp) of rats subjected to porphyrin injections (Policard 1924). This was the first demonstration of selective accumulation of porphyrins by malignant cells. Policard's findings were corroborated in 1942 by Auler and Banzer in Berlin (Auler and Banzer 1942). During World War II, Auler and Banzer studied hematoporphyrin's interaction with neoplastic tissue and demonstrated increased uptake and retention of hematoporphyrin by malignant cells. Almost 30 years before PDT would be rigorously investigated for clinical applications, one of the principal features of PDT—hyperproliferating tissue's enhanced susceptibility to photodynamic damage—had been noted. Unfortunately, the world wars intervened and derailed PDT-related research. It would be more than 20 years before the "rediscovery" of PDT would occur, and PDT research would be on a charted course.

4.5 Porphyrin and Rediscovery of PDT: Hematoporphyrin and Hematoporphyrin Derivatives and the Move into Clinical Setting

No thorough chronicle of PDT can be told without discussing hematoporphyrin and its derivatives. Several years after World War II, scientists were once again free to pursue porphyrin chemistry. In 1951, Manganiello and Figge explored the effects of injecting hematoporphyrin in patients with head and neck malignancies (Manganiello and Figge 1951). Curiously, tumor fluorescence was not noticed, and this was attributed to hematoporphyrin doses that were too low. This underscores the principle issue of dosimetry in PDT; that is, one must effectively determine the ideal concentration of PS *and* the amount of light needed to achieve the ideal therapeutic response. In 1955, David Samuel Rasmussen-Taxdal, an American neurosurgeon, noted the fluorescence of tumors prior to hematoporphyrin injection, confirming selective accumulation of the PS in neoplastic cells (Rasmussen-Taxdal, Ward and Figge 1955). In the same year, Samuel Schwartz and colleagues at the University of Minnesota published their analysis of hematoporphyrin preparations and discovered that the hematoporphyrin typically used was not exclusively a hematoporphyrin solution but rather an impure mixture (Schwartz, Absolon, and Vermund 1955). Curiously, hematoporphyrin alone proved to be a worse PS than the impure hematoporphyrin. In attempts to further purify the active component of the hematoporphyrin mixture, Schwartz isolated various fractions of the hematoporphyrin mixture and discovered that samples treated with acetic–sulfuric acid mixtures had the best tumor localization. This sulfuric and acetic acid-treated hematoporphyrin was actually a mixture of hematoporphyrin stereoisomers, hematoporphyrin vinyl deuteroporphyrin

isomers, and protoporphyrin and later became known as hematoporphyrin derivative (HpD). Lipson and colleagues at the Mayo Clinic later conducted tumor localization studies of HpD throughout the 1960s and began clinical work with HpD (Lipson and Baldes 1960; Lipson, Baldes, and Gray 1967; Lipson, Baldes, and Olsen 1961a,b, 1964; Lipson et al. 1964). Specifically, in 1966, Lipson employed HpD for the elimination of recurrent carcinoma of the breast, and while the lesion did return, there was substantial evidence that the HpD-mediated PDT employed did lead to neoplastic tissue death. It is worth noting here that HpD, now marketed as Photofrin, is an anomaly for intravenous pharmaceuticals because of its being a mixture.

In 1972, Ivan Diamond of the University of California, San Francisco, published a study in *The Lancet* demonstrating the first *in vivo* application of HpD and light for the destruction of glioma cells (Diamond et al. 1972). Implanted flank glioma tumors treated with HpD and light suffered growth arrest for up to 20 days, and photosensitization drastically reduced tumor volumes. At about the same time as Diamond was conducting his work, the PDT pioneer Thomas J. Dougherty III at the Roswell Park Cancer Institute of Buffalo, New York, was conducting PDT research seminal to the field (Dougherty et al. 1975). In 1978, Dougherty published a clinical trial on hematoporphyrin and red light for the treatment of 113 neoplastic legions on 10 patients (Dougherty et al. 1978). The lesions studied include metastatic malignant melanoma, recurrent colon carcinoma, metastatic breast carcinoma, mycocis fungoides, metastatic chondrosarcoma, metastatic prostatic carcinoma, metastatic squamous cell carcinoma of the skin, metastatic endometrial carcinoma, and metastatic angiosarcoma. After administration of intravenous 2.5 or 5.0 mg/kg HpD, patients were exposed to red light from a xenon arc lamp at times from 24 to 169 h postinjections. In 111 out of 113 tumors, there were complete or partial responses to HpD-mediated PDT. This study demonstrated for the first time that PDT could serve as an alternative anticancer therapy when other modalities had failed. Dougherty's work launched a series of clinical trials eventually leading to an explosion of PDT interest. Ohi and Tsuchiya (1983) explored PDT for the treatment of superficial bladder tumors in 11 patients, and Hayata et al. reported the effects of PDT for superficial esophageal lesions as well as early gastric cancer.

In the 1980s, the success of the exogenous applications of PS—by topical application or intravenous injection—inspired researchers to investigate the possibility of endogenous PS generation. This idea was inspired by the porphyrias, a series of hereditary and acute diseases that are a consequence of dysfunctional heme biosynthesis (Layer et al. 2010). Heme biosynthesis starts with a small building block common to biological porphyrins and their derivatives—5-aminolevulinic acid (5-ALA). Biosynthesis of heme from 5-ALA is a calculated process that evades host photosensitivity by the inclusion of a central iron atom (Teng et al. 2011). Nevertheless, increased activity of 5-ALA-synthase and porphobilinogen was noted to increase host photosensitivity (Sassa and Kappas 2000). Consequently, in 1987, Malik and Lugaci demonstrated that exogenous application of 5-ALA with red light resulted in enhanced destruction of erythroleukemic cells by the endogenous synthesis of uroporphyrin and protoporphyrin IX (Malik and Lugaci 1987). This represents the first targeted PDT approach as heme biosynthesis occurs near mitochondria. Nevertheless, the true brilliance of this approach lies in its bypassing of the biological feedback loops in place to ensure that autophotosensitization is avoided. Malik's work inspired a plethora of studies, including clinical trials, on 5-ALA and eventually led to its acceptance as a viable, targeted medical technology.

As a result of the success of HpD and 5-ALA in the 1980s, several medical regulatory agencies approved the use of PS for the clinical setting. In 1993, HpD, marketed as Photofrin, was approved for the treatment of bladder cancer in Canada. In the United States, Photofrin was approved by the US Food and Drug Administration (FDA) in 1998 for the treatment of completely or partially obstructive esophageal cancer, and in 1998, the FDA approved Photofrin for complete or completely or partially obstructive endobronchial non-small cell lung cancer. In 1999, the FDA approved of Levulan Kerastick (a topical 5-ALA salve) for the treatment of precancerous and cancerous skin legions.

In addition to PDT's approval for the treatment of neoplastic conditions, the FDA has approved several PS, including Visudyne—a verteporfin preparation—for the treatment of ophthalmological disease, including age-related macular degeneration (AMD). By 2005, Visudyne had been approved for the treatment of AMD for various diseases characterized by hyperproliferation in more than 50 countries. In 2001, Foscan was approved with obligations by the European Medical Agency (EMEA) for the treatment of head and neck cancers. In 2003, Photofrin was approved for the treatment of high-grade dysplasia associated with Barrett's esophagus (Hamblin and Mróz 2008).

From 1900 to 1990, a few papers were published on the microbiological potential of PDT (Macmillan, Maxwell, and Chichester 1966). In the 1970s, a few reports published the antiviral potential of PDT (Horvath 1977; Morison 1975). But antimicrobial PDT really took off in the 1990s and 2000s as a consequence of the success of anticancer PDT. With the emergence of antimicrobial resistance in the latter half of the last century, a resurgence of interest in novel antimicrobials occurred. In the 1990s, susceptibility differences between Gram-positive and Gram-negative bacteria to PDT was noted, but by enhancing the permeability of the Gram-negative outer membrane of Gram-negative bacteria or by using cationic PS, these differences could be evaded (Hamblin and Giulio 2011). Since then, antimicrobial PDT has emerged as a flourishing field, attracting the attention of many chemists, microbiologists, and clinicians. A more comprehensive review of antimicrobial PDT is provided in Chapter 34.

4.6 Second-Generation PSs

While the PSs approved by regulatory agencies are capable of removing neoplastic tissue, they are not without adverse side effects, namely, photosensitivity. In an effort to improve targeted PDT, many new PSs have been and are being generated. Among these promising second-generation PSs is silicon (v) phthalocyanine 4 (PC4). PC4 was developed by Olenick and her colleagues at Case Western Reserve University and has been intensely explored for the treatment of retinoblastoma and epidermoid carcinoma (Ahmad et al. 2000; Ahmad, Gupta, and Mukhtar 1999). Moreover, Lee, Baron, and Foster (2008) explored PC4 for clinical trials in the elimination of cutaneous T-cell lymphoma.

Other second-generation PSs possess longer absorption wavelengths than their first-generation counterparts. These second-generation PSs include benzoporphyrins, chlorins, and porphycenes as well as metalated derivatives. Several of these have already been approved for clinical applications, including the benzoporphyrin Visudyne and the chlorin Temoporfin (Josefsen and Boyle 2008). The creation of PSs that have absorption profiles well into the red portion of the electromagnetic spectrum and are also highly selective in their localization represent the third generation of PSs. The future of PDT undoubtedly lies in the design of novel PSs that are highly selective and possess deep red-absorption profiles.

4.7 Conclusion and Future Outlook

The field of photodynamic therapeutics has emerged as a flourishing field since its humble beginning in 1900. Figure 4.5 details the rapid increase in PDT-related publications since 1990. To date, PDT has been hybridized with nanotechnological techniques (e.g., quantum dots, fullerenes) and antibody technologies and is actively being explored to treat ophthalmological, oncological, neurological, dermatological, and gastroenterological disease, infections, and atherosclerotic plaques. In the past, pharmaceutical companies have not been attracted to PDT because of the low cost of PSs and the relatively high cost of illumination devices. Nevertheless, the prospect of these new PDT applications will not only hopefully inspire pharmaceutical companies to design, produce, and market PSs but also inspire clinicians to utilize PDT to cure a plethora of ailments, therefore increasing the demand for these technologies.

Aside from PS design, visible-light dosimetry is preventing PDT's widespread clinical acceptance. As evident from a brief survey of the literature, this is likely because there are no set standards for what amount of light is appropriate for effective PDT treatment and how that amount of light is dependent on the amount of PS administered to a patient. Moreover, current illumination technologies are expensive investments; however, with the advent of light-emitting diodes, these issues of expensive illumination devices may be avoided.

In spite of these challenges, we are optimistic that in the next 100 years, PDT will be greatly improved. Target cell selectivity will drastically increase, photosensitivity side effects will decrease, and many lives will be saved in the process. Thousands of researchers are actively pursuing PDT research in the hopes of having the global medical community recognize PDT's tremendous potential. It is our strong conviction that PDT will make

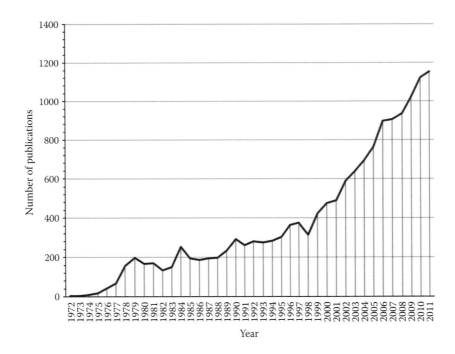

FIGURE 4.5 PDT publications every year from 1972. Note the exponential increase in PDT-related publications. Values were acquired from PubMed searches restricted by year and do not include papers and correspondences that are not catalogued by the NCBI-PubMed database.

its way into the clinical setting as a first-line defense against disease. And perhaps one day PDT will be as common a term to the laity as is chemotherapy or radiation therapy.

Acknowledgments

Research conducted by Tyler G. St. Denis is supported by the Columbia University I. I. Rabi Science Fellowship. Research conducted in the laboratory of Michael R. Hamblin was supported by NIH (RO1 AI050875 to MRH) and the US Air Force MFEL Program (FA9550-04-1-0079).

References

Ahmad, N., S. Gupta, D. K. Feyes, and H. Mukhtar. 2000. Involvement of Fas (APO-1/CD-95) during photodynamic-therapy-mediated apoptosis in human epidermoid carcinoma A431 cells. *J Invest Dermatol* 115:1041–1046.

Ahmad, N., S. Gupta, and H. Mukhtar. 1999. Involvement of retinoblastoma (Rb) and E2F transcription factors during photodynamic therapy of human epidermoid carcinoma cells A431. *Oncogene* 18:1891–1896.

Auler, H., and G. Banzer. 1942. Untersuchungen über die Rolle der Porphyrine bei geschwulstkranken Menschen und Tieren. *Z Krebsforsch* 53:65–68.

Buettner, G. R. 1993. The pecking order of free radicals and antioxidants: lipid peroxidation, alpha-tocopherol, and ascorbate. *Arch Biochem Biophys* 300:535–543.

Clennan, E. L., and A. Pace. 2005. Advances in singlet oxygen chemistry. *Tetrahedron* 61:6665–6691.

Daniell, M. D., and J. S. Hill. 1991. A history of photodynamic therapy. *Aust N Z J Surg* 61:340–348.

Diamond, I., S. G. Granelli, A. F. McDonagh et al. 1972. Photodynamic therapy of malignant tumours. *Lancet* 2:1175–1177.

Dougherty, T. J., G. B. Grindey, R. Fiel, K. R. Weishaupt, and D. G. Boyle. 1975. Photoradiation therapy. II. Cure of animal tumors with hematoporphyrin and light. *J Natl Cancer Inst* 55:115–121.

Dougherty, T. J., J. E. Kaufman, A. Goldfarb et al. 1978. Photoradiation therapy for the treatment of malignant tumors. *Cancer Res* 38:2628–2635.

Fischer, H., H. Hilmer, F. Linder, and B. Putzer. 1925. Chemische Befunde bei einem Fall von Porphyrie (Petry). *Z Physiol Chem* 150:44.

Fitzpatrick, T. B., and M. A. Pathak. 1959. Historical aspects of methoxsalen and other furocoumarins. *J Invest Dermatol* 32:229–231.

Foote, C. S. 1991. Definition of type I and type II photosensitized oxidation. *Photochem Photobiol* 54:659.

Hamblin, M. R., and J. Giulio. 2011. *Photodynamic Inactivation of Microbial Pathogens: Medical and Environmental Applications*. Royal Society of Chemistry, Cambridge, UK.

Hamblin, M. R., and P. Mróz. 2008. *Advances in Photodynamic Therapy: Basic, Translational, and Clinical*. Artech House, Norwood, MA.

Hausman, W. 1908. Die sensibilisierende Wirkung tierscher Farbstoffe und ihre physiologische Bedeutung. *Wien Klin Wochnschr* 21:1527–1529.

Hausman, W. 1911. Die sensibilisierende Wirkung des Hamatoporphyrins. *Biochem Z* 30:276–316.

Hinneburg, I., and R. H. Neubert. 2005. Influence of extraction parameters on the phytochemical characteristics of extracts from buckwheat (Fagopyrum esculentum) herb. *J Agric Food Chem* 53:3–7.

Horvath, P. 1977. [Recurrent herpes simplex. The situation in the United States]. *Z Hautkr* 52:529–532.

Jesionek, A., and H. von Tappeiner. 1905. Zur Behandlung der Hautcarcinome mit fluorescierenden Stoffen. *Arch Klin Med* 82:223.

Jones, S. G., A. R. Young, and T. G. Truscott. 1993. Singlet oxygen yields of furocoumarins and related molecules—The effect of excitation wavelength. *J Photochem Photobiol B* 21:223–227.

Josefsen, L. B., and R. W. Boyle. 2008. Photodynamic therapy: novel third-generation photosensitizers one step closer? *Br J Pharmacol* 154:1–3.

Layer, G., J. Reichelt, D. Jahn, and D. W. Heinz. 2010. Structure and function of enzymes in heme biosynthesis. *Protein Sci* 19:1137–1161.

Lee, T. K., E. D. Baron, and T. H. Foster. 2008. Monitoring Pc 4 photodynamic therapy in clinical trials of cutaneous T-cell lymphoma using noninvasive spectroscopy. *J Biomed Opt* 13:030507.

Letner, A. 1990. *Zur Geschichte der Lichttherapie: Von der Heliotherapie der Antike zur modernen ultravioletten Phototherapie.* Thesis, Düsseldorf University, Düsseldorf.

Lipson, R. L., and E. J. Baldes. 1960. The photodynamic properties of a particular hematoporphyrin derivative. *Arch Dermatol* 82:508–516.

Lipson, R. L., E. J. Baldes, and M. J. Gray. 1967. Hematoporphyrin derivative for detection and management of cancer. *Cancer* 20:2255–2257.

Lipson, R. L., E. J. Baldes, and A. M. Olsen. 1961a. Hematoporphyrin derivative: a new aid for endoscopic detection of malignant disease. *J Thorac Cardiovasc Surg* 42:623–629.

Lipson, R. L., E. J. Baldes, and A. M. Olsen. 1961b. The use of a derivative of hematoporphyrin in tumor detection. *J Natl Cancer Inst* 26:1–11.

Lipson, R. L., E. J. Baldes, and A. M. Olsen. 1964a. Further evaluation of the use of hematoporphyrin derivative as a new aid for the endoscopic detection of malignant disease. *Dis Chest* 46:676–679.

Lipson, R. L., J. H. Pratt, E. J. Baldes, and M. B. Dockerty. 1964b. Hematoporphyrine derivative for detection of cervical cancer. *Obstet Gynecol* 24:78–84.

Macmillan, J. D., W. A. Maxwell, and C. O. Chichester. 1966. Lethal photosensitization of microorganisms with light from a continuous-wave gas laser. *Photochem Photobiol* 5:555–565.

Malik, Z., and H. Lugaci. 1987. Destruction of erythroleukaemic cells by photoactivation of endogenous porphyrins. *Br J Cancer* 56:589–595.

Manganiello, L. O., and F. H. Figge. 1951. Cancer detection and therapy II. Methods of preparation and biological effects of metalloporphyrins. *Bull. School. Med. Univ. Maryland* 36:3–7.

Meyer-Betz, F. 1913. Untersuchungen über die biologische (photodynamische) Wirkung des Hamatoporphyrins und anderer Derivate des Blut- und Galenfarbstoffs. *Dtch Arch Klin Med* 112:476–503.

Morison, W. L. 1975. Anti-viral treatment of warts. *Br J Dermatol* 92:97–99.

Ochsner, M. 1997. Photophysical and photobiological processes in the photodynamic therapy of tumours. *J Photochem Photobiol B* 39:1–18.

Ohi, T., and T. Tsuchiya. 1983. Superficial bladder tumors. In *Laser Photoradiation for Tumor Detection and Treatment.* Y. Hayata, and T. J. Dougherty, editors. Igaku-Shoin, Tokyo. 79.

Policard, A. 1924. Etude sur les aspects offerts par des tumeurs experimentales examinés a la lumière de Wood. *C R Soc Biol* 91:1423–1428.

Prime, J. 1900. *Les accidentes toxiques par l'eosinate de sodium.* Jouve & Boyer, Paris.

Raab, O. 1900. Über die Wirkung fluoreszierender Stoffe auf Infusori. *Z Biol* 39:524–536.

Rasmussen-Taxdal, D. S., G. E. Ward, and F. H. Figge. 1955. Fluorescence of human lymphatic and cancer tissues following high doses of intravenous hematoporphyrin. *Surg Forum* 5:619–624.

Sassa, S., and A. Kappas. 2000. Molecular aspects of the inherited porphyrias. *J Intern Med* 247:169–178.

Schwartz, S. K., K. Absolon, and H. Vermund. 1955. Some relationships of porphyrins, x-rays and tumours. *Univ Min Med Bull* 27:7–8.

Tanielian, C., R. Mechin, R. Seghrouchni, and C. Schweitzer. 2000. Mechanistic and kinetic aspects of photosensitization in the presence of oxygen. *Photochem Photobiol* 71:12–19.

Teng, L., M. Nakada, S. G. Zhao et al. 2011. Silencing of ferrochelatase enhances 5-aminolevulinic acid-based fluorescence and photodynamic therapy efficacy. *Br J Cancer* 104:798–807.

Valko, M., H. Morris, and M. T. Cronin. 2005. Metals, toxicity and oxidative stress. *Curr Med Chem* 12:1161–1208.

von Scherer, J. J. 1841. Chemisch-physiologische Untersuchungen. *Liebs Ann Chem Pharm* 40:1–64.

von Tappeiner, H. 1900. Über die Wirkung fluorescierenden Stoffe auf Infusorien nach Versuchen von O. Raab. *Munch Med Wochenschr* 47:5.

von Tappeiner, H., and A. Jesionek. 1903. Therapeutische Versuche mit fluorescierenden Stoffen. *Munch Med Wochenschr* 47:2042–2044.

von Tappeiner, H., and A. Joblauer. 1904. Über die Wirkung der photodynamischen (fluorescierenden) Stoffe auf Protozoan und Enzyme. *Dtsch Arch Klin Med* 80:427–487.

5

History and Fundamentals of Low-Level Laser (Light) Therapy

Asheesh Gupta
Massachusetts General Hospital

Michael R. Hamblin
Massachusetts General Hospital

5.1 Introduction and History

The use of laser or other light sources for reducing pain and inflammation, augmenting tissue repair and regeneration, healing deeper tissues and nerves, and preventing tissue damage (among other medical applications) is known as low-level laser (or light) therapy (LLLT), phototherapy, or photobiomodulation. Over many centuries, light therapy has been one of the oldest therapeutic modalities used to treat various health conditions (Daniell and Hill 1991). Ancient Egyptian, Indian, and Chinese civilizations used sunlight or heliotherapy to treat various diseases, including psoriasis, rickets, vitiligo, and skin cancer. Modern phototherapy was later rediscovered by Niels Ryberg Finsen, a Danish physician and scientist who won the 1903 Nobel Prize in physiology or medicine in recognition of his contribution to the treatment of diseases, notably lupus vulgaris. Finsen developed a "chemical rays" lamp with which he treated many patients with skin tuberculosis. He also found that red-light exposure prevents the formation and discharge of smallpox pustules and can be used to treat this disease. This was the beginning of modern phototherapy using an artificial irradiation source (Honigsmann 2013; Roelandts 2005).

It was only many years later that the therapeutic benefits of light were uncovered again using other segments of the electromagnetic spectrum with visible and near-infrared (NIR) wavelengths. In the late 1960s, LLLT as a therapeutic modality was first introduced by Endre Mester and colleagues who noted improvement in wound healing with application of a low-energy (1 J/cm²) ruby laser (Mester, Mester, and Mester 1985). In 1967, a few years after the first working laser was invented, E. Mester of Semmelweis Medical University, Budapest, Hungary, wanted to test if laser radiation might cause cancer in mice (Mester, Szende, and Gartner 1968). He shaved the dorsal hair, divided the mice into two groups, and gave a laser treatment with a low-powered ruby laser (694 nm) to one group. The mice did not get cancer, and to his surprise, the hair on the treated group grew back more quickly than the untreated group. This was the first demonstration of photobiostimulation, which generated an increased interest in understanding and developing further low-energy laser technologies and applications. This casual observation prompted him to conduct other studies that provided support for the efficacy of red light on wound healing. Mester also demonstrated that the HeNe laser could stimulate wound healing in mice (Mester et al. 1971). He soon applied his findings to human patients, using lasers to treat patients with nonhealing skin ulcers (Mester et al. 1972, 1976). Since then, large numbers of reports demonstrating the positive effects of LLLT in various *in vitro*, *in vivo*, and clinical studies have been published. Recently, medical treatment with coherent light sources (lasers) or noncoherent light sources consisting of filtered lamps or light-emitting diodes (LEDs) has spread throughout the world (Barolet 2008).

The research area of LED photobiomodulation was mainly developed by the National Aeronautics and Space Administration (NASA). NASA research came about as a result of the effects noted when light of a specific wavelength was shown to accelerate plant growth. Because of the deficient level of wound healing experienced by astronauts in zero-gravity space conditions and Navy Seals in submarines under high atmospheric pressure, NASA investigated the use of LED therapy in wound healing and obtained positive results (Barolet 2008; Sobanko and Alster 2008).

Currently, LLLT is routinely practiced as part of physical therapy and is under investigation for many serious conditions and even life-threatening diseases (see Figure 5.1 for a graphical timeline of phototherapy and LLLT). LLLT involves exposing cells or tissue to low levels of red and NIR light and is referred to as "low level" because of the use of light at energy or power densities that are low compared to other forms of laser therapy that are used for ablation, cutting, and thermally coagulating tissue. LLLT is also known as "cold-laser" or "soft-laser" therapy as the power densities used are lower than those needed to produce heating of tissue. Although solar and UV light have been popular for many decades, they can damage tissues with long-time exposure. Other artificial light sources, including halogen lights, can cause thermal damage to the irradiated tissues because they also include a broad wavelength spectrum light. The use of lasers and LEDs as nonthermal and narrow wavelength spectrum light sources, nonablative laser therapy, and LLLT light sources is currently being explored as the next step in the technological development of photomedicine and phototherapy.

Although LLLT is now used to treat a wide variety of ailments, it remains somewhat controversial as a therapy for two principle reasons: First, there are uncertainties about the fundamental molecular and cellular mechanisms responsible for transducing signals from the photons incident on the cells to the biological effects that take place in the irradiated tissue. Second, there are a large number of dosimetry parameters, which are mainly categorized in two ways, that is, the irradiation or the "medicine" (wavelength, irradiance or power density, pulse structure, coherence, polarization), and the delivered fluence or "dose" (energy, fluence, irradiation time, repetition regimen). A less-than-optimal choice of parameters can result in reduced effectiveness of the treatment or even a negative therapeutic outcome (Chung et al. 2012). As a result, many of the published results on LLLT include negative results simply because of an inappropriate choice of light source and dosage (Posten et al. 2005). It is important to consider that there is an optimal dose of light for any particular application, and doses higher or lower than this optimal value may have no therapeutic effect. In fact, one important point that has been demonstrated by multiple studies is the concept of a biphasic dose response depicted by LLLT (Chung et al. 2012).

It is well established that if the light applied is not of sufficient irradiance or the irradiation time is too short, then there is no response. If the irradiance is too high or irradiation time is too long, then the response may be inhibited. Somewhere in between is the optimal combination of irradiance and time for stimulation. This dose response is often likened to the biphasic response known as the Arndt–Schulz law (Sommer et al. 2001), which dates back to 1887 when Hugo Schulz published a paper showing that various poisons have a stimulatory effect on yeast metabolism when given in low doses; then, later with Rudolph Arndt, they developed their principle, claiming that a weak stimulus slightly accelerates activity, a stronger stimulus raises it further, but a peak is reached, and a stronger stimulus will suppress activity. A biphasic dose response has been frequently observed in which low levels of light have a much better effect on stimulating and repairing tissues than higher levels of light. The so-called "Arndt–Schulz curve" is frequently used to describe this biphasic dose response (Huang et al. 2009). Evidence suggests that both energy density and power density are key biological parameters for the effectiveness of LLLT, and they may both operate with thresholds (i.e., a lower and an upper threshold for both parameters between which laser therapy is effective and outside of which laser therapy is too weak to have any effect or so intense that the tissue is inhibited) (Sommer et al. 2001).

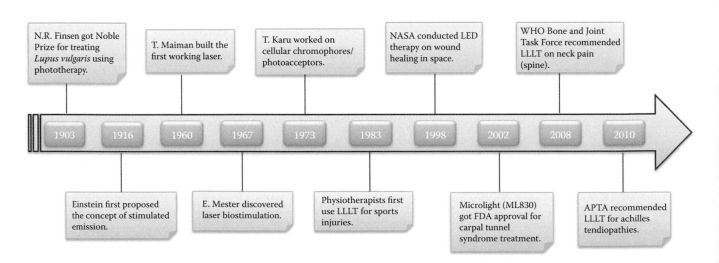

FIGURE 5.1 Historical timeline of phototherapy and LLLT. APTA, American Physical Therapy Association; FDA, Food and Drug Administration; LED, light-emitting diodes; NASA, National Aeronautics and Space Administration; UK, United Kingdom; US, United States; WHO, World Health Organization.

5.2 Biophysical and Biochemical Mechanisms of Laser–Tissue Interactions

5.2.1 Light and Laser

Light is part of the spectrum of electromagnetic radiation (ER), which ranges from radio waves to gamma rays. ER has a dual nature as both particles and waves. As a wave, light has amplitude, which is the brightness of the light, wavelength (λ), which determines the color of the light, and an angle at which it is vibrating, called polarization. In terms of modern quantum theory, ER consists of particles called photons, which are packets ("quanta") of energy that move at the speed of light.

The laser was first developed in 1960 by Theodore Maiman in California, USA, and quickly found a huge demand in the biomedical field (Ohshiro and Calderhead 1991). A laser is a device that emits light through a process of optical amplification based on the stimulated emission of photons. The emitted laser light is notable for its high degree of spatial and temporal coherence (Chung et al. 2012; McGuff et al. 1963). For a detailed overview on laser (light)–tissue interactions, readers are referred to Chapter 3 in this book by P. K. Gupta et al. Currently, one of the biggest sources of debate in the choice of light sources for LLLT is the choice between lasers and LEDs. LEDs have become widespread in LLLT devices. Early studies have shown promising results with LED-induced increases in fibroblasts and decreases in wound sizes in rat models (Whelan et al. 2001). Small, poorly controlled human studies have demonstrated that LED can also reverse signs of photoaging and improve the outcome of other thermal-based rejuvenation treatments (Trelles and Allones 2006; Weiss et al. 2005). Most initial work in LLLT used the HeNe laser, which emits light of wavelength 632.8 nm, while, nowadays, semiconductor diode lasers, such as gallium arsenide (GaAs) lasers, have increased in popularity. It was originally believed that the coherence of laser light was crucial to achieving the therapeutic effects of LLLT, but recently this notion has been challenged by the use of LEDs, which emit noncoherent light over a wider range of wavelengths than lasers. It has yet to be determined whether there is a real difference between laser and LED and, if it indeed exists, whether the difference results from the coherence or the monochromaticity of laser light as opposed to the noncoherence and wider bandwidth of LED light.

A future development in LLLT devices will be the use of organic LEDs (OLEDs). These are LEDs in which the emissive electroluminescent layer is a film of organic compounds that emit light in response to an electric current (Xiao et al. 2011). They operate in a similar manner to traditional semiconductor material whereby electrons and the holes recombine forming an exciton. The decay of this excited state results in a relaxation of the energy levels of the electron, accompanied by emission of radiation whose frequency is in the visible region.

Further, there is a considerable debate in the literature about the differences between the cellular effects of monochromatic laser light and polychromatic light from nonlaser light sources

(Flemming, Cullum, and Nelson 1999; Pontinen, Aaltokallio, and Kolari 1996). In this context, it has been demonstrated that LLLT stimulated healing of dermal wounds in mice and reported no difference between noncoherent, 635 ± 15 nm light, and coherent, 632.8 nm, laser (Demidova-Rice et al. 2007) irradiations performed using the same spot size and same fluence (2 J/cm²).

5.2.2 Tissue Photobiology and Optics

In LLLT, the question is no longer whether it has biological effects but rather what the optimal light parameters are for different use. Different wavelengths have different chromophores and can have various effects on tissue (Barolet 2008). Wavelengths are often referred to using their associated color and include blue (400–470 nm), green (470–550 nm), red (630–700 nm), and NIR (700–1200) lights. In general, the longer the wavelength, the deeper the penetration into tissues (Simpson et al. 1998). Depending on the type of tissue, the penetration depth is less than 1 mm at 400 nm, 0.5–2 mm at 514 nm, 1–6 mm at 630 nm, and maximal at 700 to 900 nm (Simpson et al. 1998).

An important consideration in LLLT involves the optical properties of tissue. Both the absorption and scattering of light in tissue are wavelength dependent (both much higher in the blue region of the spectrum than the red), and the principle tissue chromophores (hemoglobin and melanin) have high absorption bands at wavelengths shorter than 600 nm. Water begins to absorb significantly at wavelengths greater than 1150 nm. For these reasons, there is a so-called "optical window" in tissue covering the red and NIR wavelengths in which the effective tissue penetration of light is maximized (Figure 5.2) (Barolet 2008; Chung et al. 2012). The transmission of light through tissue is highly wavelength-specific. An optical window exists in tissue in the approximate range 600–1100 nm. Therefore, although blue, green, and yellow light may have significant effects on cells growing in optically transparent culture medium, the use

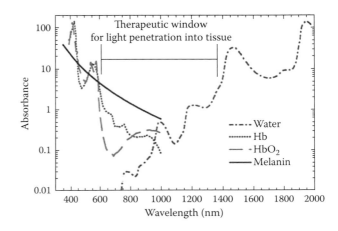

FIGURE 5.2 Absorption spectra of the important tissue chromophores showing the optical window in which visible and NIR light can penetrate deepest into tissue.

of LLLT in animals and patients almost exclusively involves red and NIR light in the range of 600–1100 nm.

Laser radiation or noncoherent (LED) light has a wavelength-dependent capability to alter cellular behavior in the absence of significant heating (Basford 1995). Phototherapy includes wavelengths of between 600 and 1100 nm and typically involves radiations as a continuous wave or pulsed light that consist of a constant beam of relatively low fluence delivery (0.04–50 J/cm²) at the target tissue or monolayer of cells using powers measured in milliwatts (AlGhamdi, Kumar, and Moussa 2012). In general, the power densities used for LLLT are lower than those needed to produce heating of tissue, that is, less than 100 mW/cm², depending on wavelength and tissue type. Wavelengths in the range 600–700 nm are used to treat superficial tissue, and longer wavelengths in the range 780–950 nm, which penetrate further, are used to treat deeper-seated tissues. Wavelengths in the range 700–770 nm have been found to have limited biochemical activity and are therefore not used (Chung et al. 2012). Various substrates used in LLLT include such inert gases and semiconductor laser diodes as helium neon (HeNe, 632.8 nm), ruby (694 nm), argon (488 and 514 nm), krypton (521, 530, 568, and 647 nm), gallium arsenide (GaAs, 904 nm), and gallium aluminum arsenide (GaAlAs, 820 and 830 nm) (Posten et al. 2005).

5.2.3 Photoacceptors and Cellular Chromophores

The first law of photobiology states that for low-power visible light to have any effect on a living biological system, the photons must be absorbed by electronic absorption bands belonging to some molecular chromophore or photoacceptor (Sutherland 2002). There may be one (or more than one) chromophore that

leads to the multitude of biological effects of LLLT, as shown in Figure 5.3. One approach to finding the identity of this chromophore(s) is to carry out action spectra. This is a graph representing biological photoresponse as a function of wavelength and should resemble the absorption spectrum of the photoacceptor molecule. The fact that a structured action spectrum can be constructed supports the hypothesis of the existence of cellular photoacceptors and signaling pathways stimulated by light. Chapter 46 in this book by T.I. Karu discusses the photoacceptors and cellular chromophores in detail.

5.2.4 Cellular and Tissular Mechanisms of LLLT

The mechanism associated with the photobiostimulation by LLLT is not yet fully understood. From observation, it appears that LLLT has a wide range of effects at the molecular, cellular, and tissular levels. The basic biological mechanism behind the effects of LLLT is thought to be through absorption of red and NIR light by chromophores, in particular, cytochrome c oxidase (CCO), which is contained in the respiratory chain located within the mitochondria (Greco et al. 1989; Karu and Kolyakov 2005), and perhaps also the plasma membrane in cells functions as a photoacceptor of photons, and thereafter, a cascade of events occurs in the mitochondria, leading to biostimulation of various processes (Oron 2011). Absorption spectra obtained for CCO in different oxidation states were recorded and found to be very similar to the action spectra for biological responses to the light (Karu and Kolyakov 2005). It is assumed that this absorption of light energy may cause photodissociation of inhibitory nitric oxide from CCO (Lane 2006), leading to increased enzyme activity and electron transport (Pastore et al. 1994). Furthermore, it

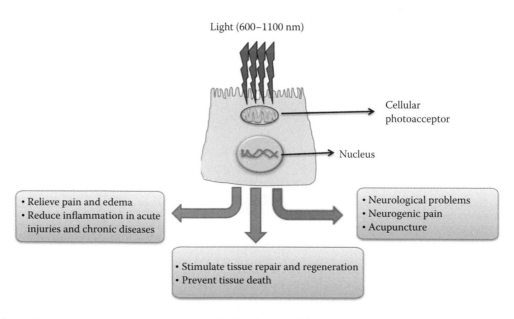

FIGURE 5.3 Schematic representation showing absorption of red and NIR light by specific chromophores or photoacceptors, which lead to a wide range of beneficial biomedical effects.

has been demonstrated that LLLT enhances mitochondrial respiration and ATP production and increases proton gradient, leading to increased activity of the Na+/H+ and Ca²⁺/Na+ antiporters and of all the ATP-driven carriers for ions, such as Na+/K+ ATPase and Ca²⁺ pumps (Harris 1991; Karu 1999). A change in ion balance leads to a short-term increase in the intracellular pH (ΔpH), which is one of the required components involved in transmission of mitogenic signals in cells (Pouyssegur et al. 1985). ATP is the substrate for adenyl cyclase, and therefore, the ATP levels control the amount of cAMP. Ca²⁺, K⁺, and cAMP are very important second messengers, influencing cell proliferation, and Ca²⁺ ions, in particular, have a significant role in regulating gene expression and triggering mitosis (Friedmann et al. 1991). Several transcription factors are regulated by changes in the cellular redox state, such as redox factor-1-dependent activator protein-1 (AP-1) (a heterodimer of c-Fos and c-Jun), nuclear factor kappa B (NFκB) (Chen et al. 2011), p53, activating transcription factor/cAMP–response element-binding protein (ATF/CREB), and hypoxia-inducible factor (HIF)-1α, an HIF-like factor (Mucaj, Shay, and Simon 2012). In turn, LLLT by altering the redox state can induce the activation of numerous intracellular signaling pathways, alter the affinity of transcription factors for DNA, increase RNA and DNA synthesis, and regulate nucleic acid and protein synthesis, enzyme activation, and cell-cycle progression (Karu and Kolyakov 2005; Liu et al. 2005). Figure 5.4 shows that the proposed photosignal transduction and amplification chain induced by light leads to an increase in cell proliferation (stimulation of DNA and RNA synthesis), survival, and tissue repair and regeneration. It is suggested that LLLT produces a shift in overall cell redox potential in the direction of greater oxidation (Karu 1999). Different cells at a range of growth conditions have distinct redox states. Therefore, the effects of LLLT can vary considerably. Cells being initially at a more reduced state (low intracellular pH) have a high potential to respond to LLLT while cells at the optimal redox state respond weakly or do not respond to treatment with light. For a detailed overview on the cellular and tissular mechanisms of LLLT, readers are referred to Chapter 47 by Da Xing and Shengnan Wu.

5.3 Biomedical Applications of LLLT

Figure 5.5 depicts that LLLT acts on multiple diseases and conditions and has gained considerable recognition and importance among the available treatment modalities. There are perhaps three main areas of medicine and veterinary practice where LLLT has a major role to play (Figure 5.3). These are (1) to reduce pain, edema, and inflammation in acute injuries and chronic diseases (Bjordal et al. 2003; Castano et al. 2007; Chow et al. 2009); (2) to promote tissue repair and regeneration and prevent tissue death (Bisht et al. 1994; Demidova-Rice et al. 2007; Fushimi et al. 2012; Gigo-Benato, Geuna, and Rochkind 2005); and (3) to relieve neurogenic pain and some neurological problems (Chow et al. 2009; Christie et al. 2007). These applications appear in a wide range of preclinical and clinical settings, ranging from dermatology to dentistry to rheumatology and physiotherapy.

5.3.1 Effects of LLLT on Immune System and Pain Relief

It has been shown that LLLT affects lymphocyte metabolism and immune system function (Harris 1991). Mast cells, which play a crucial role in the movement of leukocytes, are of considerable importance in inflammation. Specific wavelengths of light are able to trigger mast cell degranulation, which results in the release of the proinflammatory cytokine TNF-α from the cells (el Sayed and Dyson 1996). This leads to increased infiltration of

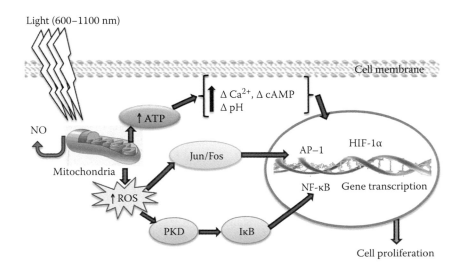

FIGURE 5.4 Schematic representation showing possible mechanism of action of LLLT-mediated repair and regeneration. Light is initially absorbed by mitochondrial chromophore or photoacceptor (CCO) and causes increased production of ATP and reactive oxygen species (ROS) and release of nitric oxide (NO), which, in turn, cause changes in cellular redox potential, Ca²⁺, K+, cAMP, and pH levels and induce several transcription factors (AP–1, NFkB, HIF-1α) concerned with cell proliferation, survival, and tissue repair and regeneration.

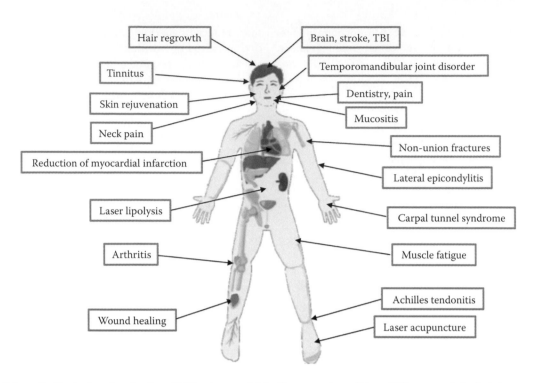

FIGURE 5.5 Schematic illustration showing how LLLT acts on multiple diseases and conditions. TBI, traumatic brain injury.

the tissues by leukocytes followed by its proliferation. The ability of macrophages to act as phagocytes is also enhanced under the application of LLLT. Infrared LLLT has been shown to increase both the phagocytic and chemotactic activity of human leukocytes *in vitro*. The other immune system-related diseases, such as atopic dermatitis, some forms of eczema, and asthma, have responded well to LLLT (Tadakuma 1993). The studies have suggested that laser biostimulation activates immune system components that help to attenuate the undesired inflammation responses (release of histamine, prostaglandins, kinins, etc.) (Harris 1991).

In recent years, there is growing interest in the use of laser biostimulation and phototherapy as a therapeutic modality for pain management. Many published reports documented the positive results of laser biostimulation in pain management. A review of 16 randomized clinical trials, including a total of 820 patients, found that LLLT reduced acute neck pain immediately after treatment and for up to 22 weeks after completion of treatment in patients with chronic neck pain (Chow et al. 2009). LLLT has also been shown to relieve pain because of cervical dentinal hypersensitivity or from periodontal pain during orthodontic tooth movement (Bicakci et al. 2012). A study of 88 randomized controlled trials indicated that LLLT can significantly reduce pain and improve health in chronic joint disorders, such as osteoarthritis, patellofemoral pain syndrome, and mechanical spine disorders (Bjordal et al. 2003). However, the authors of the study urge caution in interpreting the results because of heterogeneity in patients, treatments, and trial designs involved. Another study reported that LLLT induced reduction of joint swelling in rats with inflammatory arthritis

caused by intra-articular injection of zymosan, and this reduction in swelling correlated with reduction in the inflammatory marker serum prostaglandin E2 (PGE2) with illumination of 810 nm laser (Castano et al. 2007).

5.3.2 Effects of LLLT on Repair and Regeneration

Wound healing was one of the first applications of LLLT in which HeNe lasers were used by Mester et al. (1971, 1972, 1976) to treat skin ulcers. Light in the red to NIR range (600–1100 nm) generated by using low-energy laser or LED arrays has been reported to have beneficial biological effects in many injury models by eliciting significant biological effects at the cellular and tissular levels. LLLT can prevent cell apoptosis and improve cell proliferation, migration, and adhesion at low levels of red or NIR light illumination (Huang et al. 2009). The fact that red and NIR light can penetrate into a deep tissue injury allows noninvasive treatment to be carried out for augmented healing processes. Many *in vitro* (Hawkins and Abrahamse 2006; Zhang et al. 2003) and *in vivo* (Bisht et al. 1994; Demidova-Rice et al. 2007; Chung et al. 2012; Gigo-Benato, Geuna, and Rochkind 2005; Meirelles et al. 2008) studies have demonstrated that LLLT acts on multiple events of healing. LLLT is therefore believed to affect all three phases of wound healing, that is, inflammatory, new tissue formation, and remodeling phases (Chung et al. 2012). It has been shown that LLLT induces cell proliferation (Fushimi et al. 2012; Hawkins and Abrahamse 2006; Mvula, Moore, and Abrahamse 2010; Saygun et al. 2012); promotes angiogenesis (Chen, Hung, and Hsu 2008), proliferation, and migration of keratinocytes (Fushimi et al. 2012); and allows the wound site to close more quickly (Demidova-Rice et al. 2007; Fushimi et al. 2012). Furthermore, photobiomodulation mediated by LLLT has been observed to increase mitochondrial metabolism (Hu et al. 2007) and ATP synthesis (Demidova-Rice et al. 2007; Karu, Pyatibrat, and Kalendo 1995), induce cell proliferation (Hu et al. 2007; Pereira et al. 2002), accelerate collagen synthesis, increase tensile strength (Pereira et al. 2002; Prabhu et al. 2012), and influence the concentration of prostaglandins (pain control) (Chow et al. 2009). LLLT was also shown to stimulate the expression of multiple genes related to cellular migration, proliferation, anti-apoptosis, and pro-survival elements responsive to NFkB (Chen et al. 2011) besides modulating the production of growth factors (bFGF, VEGF, PDGF, TGF-β, IGF-1) and cytokines (IL-α, IL-8) (Fushimi et al. 2012; Hawkins and Abrahamse 2006; Kipshidze et al. 2001; Mvula, Moore, and Abrahamse 2010; Peplow et al. 2011; Saygun et al. 2012; Yu et al. 1996; Zhang et al. 2003). Readers are referred to Chapter 50 by L. Gavish in this book discussing LLLT for wound healing in detail.

Stem or progenitor cell-based therapy has been suggested as a potential solution to life-threatening injuries, such as myocardial infarction and brain strokes, because these cells have the ability to continuously self-renew, proliferate, and give rise to one or more differentiated cell types and hence are being considered as a therapeutic option in reparative and regenerative medicine. Resulting from the fact that stem cells grow and proliferate at very slow rates and their harvest yields are low, in this context, the usage of light stimulation (LLLT) has been found to be beneficial to overcome these issues (Oron 2011, Tuby, Maltz, and Oron 2007). Studies on LLLT and stem cells have demonstrated that LLLT can change the metabolism of stem cells; promote their migration, proliferation, and differentiation; and increase their viability. LLLT was also found to activate early cell cycle regulatory genes, mitogen-activated protein kinase (MAPK), and extracellular signal regulated kinase (ERK) cascades in mesenchymal and cardiac stem cells (Mvula, Moore, and Abrahamse 2010; Oron 2011; Tuby, Maltz, and Oron 2007). For a detailed overview on the effects of LLLT on stem cells, readers are referred to Chapter 57 by H. Abrahamse.

Further, a growing body of evidence suggests that LLLT promotes skeletal muscle repair and regeneration, which offers a new preventive approach to muscular fibrosis (Luo et al. 2013; Oron 2011). Studies have also demonstrated that LLLT improved liver function and increased the number of Kupffer and hepatic stellate cells in chemically induced liver cirrhosis (Oliveira-Junior et al. 2013) and a biostimulate regeneration process conducive to both the formation of new hepatocytes and mesenchymal stem cells and angiogenesis followed by hepatectomy (Oron et al. 2010). Moreover, LLLT also improved renal function and antioxidant defense capabilities in the kidneys of Type I diabetic rats (Lim et al. 2010).

On the other hand, one of the most commercially successful applications of LLLT is the stimulation of hair regrowth in balding individuals. The photobiomodulation activity of LLLT can cause more hair follicles to move from the telogen phase into the anagen phase. The FDA has given approval for Hairmax Lasercomb. Recently, a different LLLT device received FDA clearance in women suffering from androgenetic alopecia (Chung et al. 2012).

5.3.3 Effects of LLLT on Neurological Problems

Evidence that LLLT is a beneficial treatment for serious neurological conditions, such as traumatic brain injury (TBI), stroke, spinal cord injury, and degenerative central nervous system disease, is rapidly accumulating (Huang et al. 2012). The benefits of transcranial LLLT appear to be based on many mechanisms (Chung et al. 2012). Several studies have shown that LLLT is effective in increasing neurological performance, memory, and learning in mouse models of TBI (Ando et al. 2011; Wu et al. 2012). The anti-inflammatory, antiedema, and proangiogenic effects of LLLT may also have a role in the beneficial effects (Huang et al. 2012). Transcranial LLLT has also been shown to have a noticeable effect on acute stroke patients (Chung et al. 2012). It has been postulated that the most exciting possible beneficial mechanisms of LLLT may be a result of stimulation of neurogenesis or increasing the ability of the brain to repair itself and further encourage brain cells to form new synaptic connections (synaptogenesis or synaptic plasticity) (Huang et al. 2012). Moreover, LLLT has also been considered as a candidate for treating neurodegenerative diseases, such as Alzheimer's disease, Parkinson's disease, familial amyotrophic lateral sclerosis (FALS), and many diverse psychiatric disorders (Chung et al. 2012).

5.3.4 Effects of LLLT on Cardiovascular Problems

Because of the known property of LLLT to prevent apoptosis and reduce tissue death and its anti-inflammatory properties, it was plausible that delivering light to the heart or intravascularly could lessen the problems associated with cardiovascular disease. Oron et al. (2001) demonstrated the cardioprotective effect of LLLT evidenced by attenuation of infarct size and reduced formation of scar tissue after myocardial infarction in rats and dogs. Additionally, it was demonstrated that long-term LLLT irradiation resulted in elevation of antioxidants in the blood and expression of heat shock proteins on infarction and reperfusion myocardiac injury in rats (Yaakobi et al. 2001).

It is believed that restenosis following coronary interventions is the result of endothelial denudation that leads to thrombus formation, vascular remodeling, and smooth muscle cell proliferation. Kipshidze and coworkers demonstrated that LLLT stimulated rapid endothelial regeneration and reduced restenosis in rabbits and pigs (De Scheerder et al. 1998) and suggested that intravascular low-level red light caused induction of nitric oxide synthase expression and increased cyclic GMP in the arterial wall. Clinical trials were also carried out that demonstrated safety and showed significant reductions in restenosis at both medium- and long-term follow-ups (Kaul et al. 1998). Further, it has been hypothesized that the cardioprotective effect of LLLT may be attributed to the photodissociation of NO, not only from CCO but also from intracellular stores, such as nitrosylated forms of both hemoglobin and myoglobin, leading to vasodilation (Lohr et al. 2009).

Intravascular laser irradiation for therapy of cardiocirculatory diseases was first presented in the *American Heart Journal* in 1982 (Lee et al. 1982). Intravenous or intravascular blood irradiation involves the *in vivo* illumination of the blood by feeding low-level laser light generated by a 1–3 mW low-power laser at a variety of wavelengths through a fiber optic inserted in a vascular channel, usually a vein in the forearm, under the assumption that any therapeutic effect will be circulated through the circulatory system (Chung et al. 2012). Although it is, at present, uncertain what the mechanisms of intravascular laser actually are and why it differs from traditional laser therapy, it has been hypothesized to affect particular components of the blood. Blood lipids (low-density lipoprotein, high-density lipoprotein, and cholesterol) are said to be "normalized"; platelets are thought to be rendered less likely to aggregate, thus lessening the likelihood of clot formation; and the immune system (dendritic cells, macrophages, and lymphocytes) may be activated (Chung et al. 2012).

5.4 Conclusion and Future Perspective

The biomedical applications of LLLT are tremendously diverse. They include stimulation of healing in many situations of damage and trauma to multiple tissue types. LLLT is able to prevent cell apoptosis and tissue death in diverse diseases and situations where tissue is otherwise doomed to die. LLLT can reduce pain and inflammation in many traumatic, acute, and chronic injuries and diseases. It is only in relatively recent years that the basic molecular and cellular mechanisms of LLLT have begun to be understood. The realization that LLLT seems to have preferential action on various types of stem cells causing them to proliferate, migrate, and differentiate underlies many (if not all) of its beneficial effects. We believe that LLLT will steadily progress to be better accepted by the medical profession, physical therapists, and the general public at large. The number of published negative reports will continue to decline as the optimum LLLT parameters become better understood. There will need to be a large number of clinical and experimental studies carried out before the effects of changing all these variables are completely understood. Nevertheless, as LLLT grows in acceptance in the future, it is likely that more and more serious diseases will become amenable to phototherapy approaches, including widespread killer diseases, such as strokes and heart attacks.

References

AlGhamdi, K. M., A. Kumar, and N. A. Moussa. 2012. Low-level laser therapy: a useful technique for enhancing the proliferation of various cultured cells. *Lasers Med Sci* 27:237–249.

Ando, T., W. Xuan, T. Xu et al. 2011. Comparison of therapeutic effects between pulsed and continuous wave 810-nm wavelength laser irradiation for traumatic brain injury in mice. *PLoS One* 6:e26212.

Barolet, D. 2008. Light-emitting diodes (LEDs) in dermatology. *Semin Cutan Med Surg* 27:227–238.

Basford, J. R. 1995. Low intensity laser therapy: still not an established clinical tool. *Lasers Surg Med* 16:331–342.

Bicakci, A. A., B. Kocoglu-Altan, H. Toker, I. Mutaf, and Z. Sumer. 2012. Efficiency of low-level laser therapy in reducing pain induced by orthodontic forces. *Photomed Laser Surg* 30:460–465.

Bisht, D., S. C. Gupta, V. Misra, V. P. Mital, and P. Sharma. 1994. Effect of low intensity laser radiation on healing of open skin wounds in rats. *Indian J Med Res* 100:43–46.

Bjordal, J. M., C. Couppe, R. T. Chow, J. Tuner, and E. A. Ljunggren. 2003. A systematic review of low level laser therapy with location-specific doses for pain from chronic joint disorders. *Aust J Physiother* 49:107–116.

Castano, A. P., T. Dai, I. Yaroslavsky et al. 2007. Low-level laser therapy for zymosan-induced arthritis in rats: Importance of illumination time. *Lasers Surg Med* 39:543–550.

Chen, A. C., P. R. Arany, Y. Y. Huang et al. 2011. Low-level laser therapy activates NF-kB via generation of reactive oxygen species in mouse embryonic fibroblasts. *PloS one* 6:e22453.

Chen, C. H., H. S. Hung, and S. H. Hsu. 2008. Low-energy laser irradiation increases endothelial cell proliferation, migration, and eNOS gene expression possibly via PI3K signal pathway. *Lasers Surg Med* 40:46–54.

Chow, R. T., M. I. Johnson, R. A. Lopes-Martins, and J. M. Bjordal. 2009. Efficacy of low-level laser therapy in the management of neck pain: a systematic review and meta-analysis of randomised placebo or active-treatment controlled trials. *Lancet* 374:1897–1908.

Christie, A., G. Jamtvedt, K. T. Dahm et al. 2007. Effectiveness of nonpharmacological and nonsurgical interventions for patients with rheumatoid arthritis: an overview of systematic reviews. *Phys Ther* 87:1697–1715.

Chung, H., T. Dai, S. K. Sharma et al. 2012. The nuts and bolts of low-level laser (light) therapy. *Annals of Biomedical Engineering* 40:516–533.

Daniell, M. D., and J. S. Hill. 1991. A history of photodynamic therapy. *Aust N Z J Surg* 61:340–348.

De Scheerder, I. K., K. Wang, X. R. Zhou et al. 1998. Intravascular low power red laser light as an adjunct to coronary stent implantation evaluated in a porcine coronary model. *J Invasive Cardiol* 10:263–268.

Demidova-Rice, T. N., E. V. Salomatina, A. N. Yaroslavsky, I. M. Herman, and M. R. Hamblin. 2007. Low-level light stimulates excisional wound healing in mice. *Lasers Surg Med* 39:706–715.

el Sayed, S. O., and M. Dyson. 1996. Effect of laser pulse repetition rate and pulse duration on mast cell number and degranulation. *Lasers Surg Med* 19:433–437.

Flemming, K. A., N. A. Cullum, and E. A. Nelson. 1999. A systematic review of laser therapy for venous leg ulcers. *J Wound Care* 8:111–114.

Friedmann, H., R. Lubart, I. Laulicht, and S. Rochkind. 1991. A possible explanation of laser-induced stimulation and damage of cell cultures. *J Photochem Photobiol B* 11:87–91.

Fushimi, T., S. Inui, T. Nakajima et al. 2012. Green light emitting diodes accelerate wound healing: characterization of the effect and its molecular basis in vitro and in vivo. *Wound Repair Regen* 20:226–235.

Gigo-Benato, D., S. Geuna, and S. Rochkind. 2005. Phototherapy for enhancing peripheral nerve repair: a review of the literature. *Muscle Nerve* 31:694–701.

Greco, M., G. Guida, E. Perlino, E. Marra, and E. Quagliariello. 1989. Increase in RNA and protein synthesis by mitochondria irradiated with helium-neon laser. *Biochem Biophys Res Commun* 163:1428–1434.

Harris, D. M. 1991. Editorial comment: Biomolecular mechanisms of laser biostimulation. *J Clin Laser Med Surg* 9:277–280.

Hawkins, D. H., and H. Abrahamse. 2006. The role of laser fluence in cell viability, proliferation, and membrane integrity of wounded human skin fibroblasts following helium-neon laser irradiation. *Lasers Surg Med* 38:74–83.

Honigsmann, H. 2013. History of phototherapy in dermatology. *Photochem Photobiol Sci* 12:16–21.

Hu, W. P., J. J. Wang, C. L. Yu et al. 2007. Helium-neon laser irradiation stimulates cell proliferation through photostimulatory effects in mitochondria. *J Invest Dermatol* 127:2048–2057.

Huang, Y. Y., A. C. Chen, J. D. Carroll, and M. R. Hamblin. 2009. Biphasic dose response in low level light therapy. *Dose Response* 7:358–383.

Huang, Y. Y., A. Gupta, D. Vecchio et al. 2012. Transcranial low level laser (light) therapy for traumatic brain injury. *J Biophotonics* 5:827–837.

Karu, T. I. 1999. Primary and secondary mechanisms of action of visible to near-IR radiation on cells. *J Photochem Photobiol B* 49:1–17.

Karu, T. I., and S. F. Kolyakov. 2005. Exact action spectra for cellular responses relevant to phototherapy. *Photomed Laser Surg* 23:355–361.

Karu, T., L. Pyatibrat, and G. Kalendo. 1995. Irradiation with He-Ne laser increases ATP level in cells cultivated in vitro. *J Photochem Photobiol B* 27:219–223.

Kaul, U., B. Singh, D. Sudan, T. Ghose, and N. Kipshidze. 1998. Intravascular red light therapy after coronary stenting N angiographic and clinical follow-up study in humans. *J Invasive Cardiol* 10:534–538.

Kipshidze, N., V. Nikolaychik, M. H. Keelan et al. 2001. Low-power helium: neon laser irradiation enhances production of vascular endothelial growth factor and promotes growth of endothelial cells in vitro. *Lasers Surg Med* 28:355–364.

Lane, N. 2006. Cell biology: power games. *Nature* 443:901–903.

Lee, G., R. M. Ikeda, R. M. Dwyer et al. 1982. Feasibility of intravascular laser irradiation for in vivo visualization and therapy of cardiocirculatory diseases. *Am Heart J* 103:1076–1077.

Lim, J., R. A. Sanders, A. C. Snyder et al. 2010. Effects of low-level light therapy on streptozotocin-induced diabetic kidney. *J Photochem Photobiol B* 99:105–110.

Liu, H., R. Colavitti, Rovira, II, and T. Finkel. 2005. Redox-dependent transcriptional regulation. *Circ Res* 97:967–974.

Lohr, N. L., A. Keszler, P. Pratt et al. 2009. Enhancement of nitric oxide release from nitrosyl hemoglobin and nitrosyl myoglobin by red/near infrared radiation: potential role in cardioprotection. *J Mol Cell Cardiol* 47:256–263.

Luo, L., Z. Sun, L. Zhang et al. 2013. Effects of low-level laser therapy on ROS homeostasis and expression of IGF-1 and TGF-beta1 in skeletal muscle during the repair process. *Lasers Med Sci* 28:725–734.

McGuff, P. E., D. Bushnell, H. S. Soroff, and R. A. Deterling, Jr. 1963. Studies of the surgical applications of laser (light amplification by stimulated emission of radiation). *Surg Forum* 14:143–145.

Meirelles, G. C., J. N. Santos, P. O. Chagas, A. P. Moura, and A. L. Pinheiro. 2008. A comparative study of the effects of laser photobiomodulation on the healing of third-degree burns: a histological study in rats. *Photomed Laser Surg* 26:159–166.

Mester, E., A. F. Mester, and A. Mester. 1985. The biomedical effects of laser application. *Lasers Surg Med* 5:31–39.

Mester, E., S. Nagylucskay, A. Doklen, and S. Tisza. 1976. Laser stimulation of wound healing. *Acta Chir Acad Sci Hung* 17:49–55.

Mester, E., T. Spiry, B. Szende, and J. G. Tota. 1971. Effect of laser rays on wound healing. *Am J Surg* 122:532–535.

Mester, E., B. Szende, and P. Gartner. 1968. [The effect of laser beams on the growth of hair in mice]. *Radiobiol Radiother (Berl)* 9:621–626.

Mester, E., B. Szende, T. Spiry, and A. Scher. 1972. Stimulation of wound healing by laser rays. *Acta Chir Acad Sci Hung* 13:315–324.

Mucaj, V., J. E. Shay, and M. C. Simon. 2012. Effects of hypoxia and HIFs on cancer metabolism. *Int J Hematol* 95:464–470.

Mvula, B., T. J. Moore, and H. Abrahamse. 2010. Effect of low-level laser irradiation and epidermal growth factor on adult human adipose-derived stem cells. *Lasers Med Sci* 25:33–39.

Ohshiro, T., and R. G. Calderhead. 1991. Development of low reactive-level laser therapy and its present status. *J Clin Laser Med Surg* 9:267–275.

Oliviera-Junior, M. C., A. S. Monteiro, E. C. Junior et al. 2013. Low-level laser therapy ameliorates CCl(4)-induced liver cirrhosis in rats. *Photochem Photobiol* 89:173–178.

Oron, U. 2011. Light therapy and stem cells: a therapeutic intervention of the future? *Interv Cardiol* 3:627–629.

Oron, U., L. Maltz, H. Tuby, V. Sorin, and A. Czerniak. 2010. Enhanced liver regeneration following acute hepatectomy by low-level laser therapy. *Photomed Laser Surg* 28:675–678.

Oron, U., T. Yaakobi, A. Oron et al. 2001. Low-energy laser irradiation reduces formation of scar tissue after myocardial infarction in rats and dogs. *Circulation* 103:296–301.

Pastore, D., M. Greco, V. A. Petragallo, and S. Passarella. 1994. Increase in <— H+/e- ratio of the cytochrome c oxidase reaction in mitochondria irradiated with helium-neon laser. *Biochem Mol Biol Int* 34:817–826.

Peplow, P. V., T. Y. Chung, B. Ryan, and G. D. Baxter. 2011. Laser photobiomodulation of gene expression and release of growth factors and cytokines from cells in culture: a review of human and animal studies. *Photomed Laser Surg* 29:285–304.

Pereira, A. N., C. de P. Eduardo, E. Matson, and M. M. Marques. 2002. Effect of low-power laser irradiation on cell growth and procollagen synthesis of cultured fibroblasts. *Lasers Surg Med* 31:263–267.

Pontinen, P. J., T. Aaltokallio, and P. J. Kolari. 1996. Comparative effects of exposure to different light sources (He-Ne laser, InGaAl diode laser, a specific type of noncoherent LED) on skin blood flow for the head. *Acupunct Electrother Res* 21:105–118.

Posten, W., D. A. Wrone, J. S. Dover et al. 2005. Low-level laser therapy for wound healing: mechanism and efficacy. *Dermatol Surg* 31:334–340.

Pouyssegur, J., A. Franchi, G. L'Allemain, and S. Paris. 1985. Cytoplasmic pH, a key determinant of growth factor-induced DNA synthesis in quiescent fibroblasts. *FEBS Lett* 190:115–119.

Prabhu, V., S. B. Rao, S. Chandra et al. 2012. Spectroscopic and histological evaluation of wound healing progression following low level laser therapy (LLLT). *J Biophotonics* 5:168–184.

Roelandts, R. 2005. A new light on Niels Finsen, a century after his Nobel Prize. *Photodermatol Photoimmunol Photomed* 21:115–117.

Saygun, I., N. Nizam, A. U. Ural et al. 2012. Low-level laser irradiation affects the release of basic fibroblast growth factor (bFGF), insulin-like growth factor-I (IGF-I), and receptor of IGF-I (IGFBP3) from osteoblasts. *Photomed Laser Surg* 30:149–154.

Simpson, C. R., M. Kohl, M. Essenpreis, and M. Cope. 1998. Near-infrared optical properties of ex vivo human skin and subcutaneous tissues measured using the Monte Carlo inversion technique. *Phys Med Biol* 43:2465–2478.

Sobanko, J. F., and T. S. Alster. 2008. Efficacy of low-level laser therapy for chronic cutaneous ulceration in humans: a review and discussion. *Dermatol Surg* 34:991–1000.

Sommer, A. P., A. L. Pinheiro, A. R. Mester, R. P. Franke, and H. T. Whelan. 2001. Biostimulatory windows in low-intensity laser activation: Lasers, scanners, and NASA's light-emitting diode array system. *J Clin Laser Med Surg* 19:29–33.

Sutherland, J. C. 2002. Biological effects of polychromatic light. *Photochemistry and Photobiology* 76:164–170.

Tadakuma, T. 1993. Possible application of the laser in immunobiology. *Keio J Med* 42:180–182.

Trelles, M. A., and I. Allones. 2006. Red light-emitting diode (LED) therapy accelerates wound healing post-blepharoplasty and periocular laser ablative resurfacing. *J Cosmet Laser Ther* 8:39–42.

Tuby, H., L. Maltz, and U. Oron. 2007. Low-level laser irradiation (LLLI) promotes proliferation of mesenchymal and cardiac stem cells in culture. *Lasers Surg Med* 39:373–378.

Weiss, R. A., D. H. McDaniel, R. G. Geronemus et al. 2005. Clinical experience with light-emitting diode (LED) photomodulation. *Dermatol Surg* 31:1199–1205.

Whelan, H. T., R. L. Smits, Jr., E. V. Buchman et al. 2001. Effect of NASA light-emitting diode irradiation on wound healing. *J Clin Laser Med Surg* 19:305–314.

Wu, Q., W. Xuan, T. Ando et al. 2012. Low-level laser therapy for closed-head traumatic brain injury in mice: effect of different wavelengths. *Lasers in Surgery and Medicine* 44:218–226.

Xiao, L., Z. Chen, B. Qu et al. 2011. Recent progresses on materials for electrophosphorescent organic light-emitting devices. *Adv Mater* 23:926–952.

Yaakobi, T., Y. Shoshany, S. Levkovitz et al. 2001. Long-term effect of low energy laser irradiation on infarction and reperfusion injury in the rat heart. *J Appl Physiol* 90:2411–2419.

Yu, H. S., K. L. Chang, C. L. Yu, J. W. Chen, and G. S. Chen. 1996. Low-energy helium-neon laser irradiation stimulates interleukin-1 alpha and interleukin-8 release from cultured human keratinocytes. *J Invest Dermatol* 107:593–596.

Zhang, Y., S. Song, C. C. Fong et al. 2003. cDNA microarray analysis of gene expression profiles in human fibroblast cells irradiated with red light. *J Invest Dermatol* 120:849–857.

II

Diseases Caused by Light

6

UV Effects on the Skin

Kris Nys
University of Leuven

Patrizia Agostinis
University of Leuven (KU Leuven)

6.1 The Skin

The skin is the largest organ of the adult human body, constituting approximately 12%–14% of our total body weight. It provides a physical barrier at the interface with the external environment, thereby protecting against dehydration and defending internal tissues against a large variety of chemical and environmental insults. Moreover, it plays a role in thermoregulation and as a sensory organ.

The skin consists of an outer squamous epithelium, the epidermis, and an inner connective tissue, the dermis (also containing pilosebaceous units, nails, and sweat glands). The epidermis fulfills the crucial barrier function of the skin and undergoes continuous self-renewal as a result of mitotic activity of the stem cells in the basal layer that provide new keratinocytes. The keratinocytes (the major cellular skin component constituting approximately 90%–95% of the epidermis) complete a differentiation-induced cell death program (called cornification) while moving upward through the different epidermal layers to become the corneocytes in the outer layers of the epidermis before they are shed from the skin (Figure 6.1).

A subpopulation of the basal keratinocytes (in the *stratum basale*), the epidermal stem cells, is responsible for maintaining the epidermal turnover and replenishing damaged cells. Typically they display a slow cell-cycling state *in vivo*, while, when released from quiescence, they exhibit a high proliferative potential and a long life span. Populational asymmetry is the most commonly accepted model for the proliferative organization of the epidermis, meaning that one stem cell divides to replace itself and form a daughter cell that is capable of dividing a few more times (limited proliferative capacity), depending on the need of the epidermis. Thereafter, the progeny of this intermediate population enters the suprabasal layer, withdraws from the cell cycle, and starts to undergo terminal differentiation. Typically the differentiating keratinocytes in the successive

epidermal layers (in the *stratum spinosum* and *granulosum*) will start forming desmosomes that extend from their surface and serve to create interactions with their neighbors, consecutively leading to an accumulation of keratohyalin granules (containing profilaggrin and loricrin) and a flattened, polyhedral morphology (triggered by the intracellular rise in Ca^{2+} in the differentiating cell). Finally, after losing their nucleus and other organelles, these dead yet functionally important corneocytes (in the *stratum corneum*) consist mostly of bundled keratin filaments enclosed within the cornified outer epidermal envelope (a rigid structure of highly cross-linked insoluble proteins). This envelope, by binding lipids and organizing them in orderly lamellae, will carry out the physical and water-barrier functions of the skin (Houben, De Paepe, and Rogiers 2007; Lippens et al. 2005).

Imbalances in the delicate physiological turnover of proliferating or differentiating keratinocytes can result in the disturbance of the skin barrier function and are reflected in many skin disorders (Lippens et al. 2005).

In the epidermal layer, the keratinocytes reside in close contact and interact through homotypic E-cadherin binding with the melanocytes, specialized pigment-producing cells, mainly derived from the neural crest. Melanocytes are able to survive considerable genotoxic stress, forming the so-called "epidermal melanin unit," whereby a constant, skin type-independent ratio of keratinocytes to melanocytes is maintained (approximately 35:1).

Melanocytes have long been thought only to arise directly from neural crest cells (NCCs), which, during development, delaminate from the neural tube and, shortly thereafter, at the location of the dorsal neural tube, commit to a melanoblast fate as instructed by secreted factors released from the dorsal neural tube. These committed melanoblasts migrate under the epidermis to cover the skin of the body and, during this process, are exposed to signals, which lead to a massive increase in population

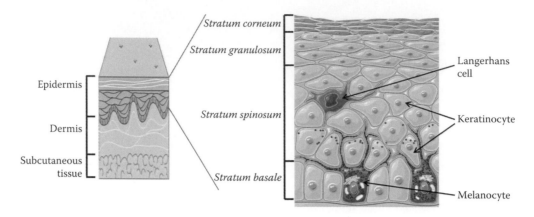

FIGURE 6.1 Schematic overview of the human skin. The subcutaneous tissue is mainly composed of adipose tissue and supports the dermis, which contains connective tissue, fibroblasts, blood and lymphatic vessels, nerves, and appendages. The epithelial epidermis is built up out of several layers (*stratum basale, spinosum, granulosum,* and *corneum*) of terminally differentiating keratinocytes. The *stratum basale* also contains the pigment-producing melanocytes.

size (i.e., mitogens) as well as "homing factors" determining the final position of the melanocytes in the skin. New results show that melanocytes also arise from immature glial cells in nerves innervating the skin. Melanocytes can therefore also be generated in a process likely involving an inductive recruitment of cells from nerves innervating the body surface (Ernfors 2010).

The interactions between keratinocytes and melanocytes are of fundamental relevance for skin homeostasis and a defense function against major environmental insults, such as ultraviolet (UV) radiation. On one hand, the keratinocytes appear to be capable of regulating the genetic and phenotypic profiles of the melanocytes as well as their proliferation and melanogenesis. On the other hand, the melanocytes, through the production of melanin and stimulation of the tanning response, further protect the skin keratinocytes from the adverse genotoxic effects of UV light (discussed further in Section 6.2).

Though far less in number, Langerhans cells (epidermal antigen-presenting cells), Merkel cells (thought to be the mechanoreceptors), and lymphocytes are also components of the epidermis.

The basement membrane, a thin sheet of fibers comprised out of several laminae, forms the contact site between the epidermis and the underlying dermis. Hemidesmosomes anchor the basal epidermal cells to the basement membrane, and anchoring fibrils extend in the dermis, thereby intertwining the tissue ridges. The dermis itself consists mainly of connective tissue, produced by resident fibroblasts, which harbors the local skin nerves and vasculature to provide structural support and nutrients, thereby safeguarding the multiple epidermal functions. Underneath the dermis, the adipose tissue of the hypodermis provides additional insulation and protection. The different skin layers are graphically represented in Figure 6.1.

6.2 Ultraviolet

Sunlight is a continuous spectrum of electromagnetic radiation that can be divided into several major ranges of wavelength: UV, visible, and infrared. The UV spectrum, accounting for 3% of

the total solar radiation, is the most significant part in terms of the induction of photodamage and skin carcinogenesis. UV (200–400 nm) itself can be further subdivided into three wavelength ranges: UVC (200–290 nm), UVB (290–320 nm), and UVA (320–400 nm) (Figure 6.2).

The portion of the UV radiation that reaches the Earth's surface can be influenced by several factors. The ozone layer in the stratospheric atmosphere, protecting life on Earth from the sun's UV rays, contributes importantly to this filtering process. Additionally, other factors, such as the time of day and the year (mainly depending on the angle of the sun's rays, e.g., between 10 a.m. and 4 p.m. or during summer, solar irradiation is more intense), latitude (i.e., the further away from the equator, the greater the distance rays have to travel through the ozone layer), altitude (i.e., radiation increases 10%–12% every 1000 m), ground reflection (e.g., snow reflects and increases exposure), fog, clouds, and pollutants (block and decrease exposure), also play a role.

UVC is probably the most harmful radiation, being a strong, direct genotoxic stress, and has been extensively used during the pioneering years of research on UV-induced signaling pathways (typically using germicidal lamps maximally emitting at 254 nm). However, UVC is not physiologically relevant for human health because this UV fraction is completely absorbed by the stratospheric ozone layer before reaching the Earth's surface. Most of the UVB and a small fraction of the UVA radiation are also filtered out by the ozone layer, allowing 1%–10% and 90%–99% of these UV radiations, respectively, to pass through (Narayanan, Saladi, and Fox 2010) (Figure 6.2).

Every effect of UV radiation on the skin includes a consecutive series of events, starting with the absorption of the radiation by chromophores in the skin (photoexcitation), followed by photochemical reactions, which induce changes in cell and tissue biology. Depending on the wavelength and energy level of the UV photons hitting the skin, specific chromophores absorb the photon's energy (closely matching their absorption maxima as determined by the energy amount needed to transition their electrons from the ground state into an excited state). As a result, UVA and

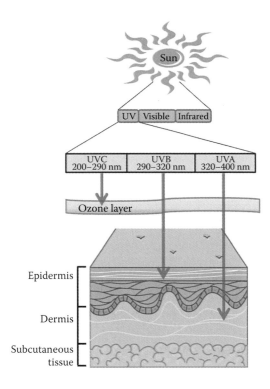

FIGURE 6.2 UV spectrum and wavelength-dependent penetration in the skin. UV light (200–400 nm) is an electromagnetic radiation covering wavelength/energy status between x-rays and visible light. It can be subdivided into UVC, UVB, and UVA. Highly energetic UVC is nearly completely blocked by the ozone layer. Whereas UVB is mainly absorbed by the epidermis, UVA can reach the dermis.

UVB affect different chromophores, resulting in chemically modified photoproducts, which can be very unstable and show altered characteristics. But, also, the depth of the penetration through the epidermal layers increases with wavelength because the highly energetic shorter wavelengths are scattered and absorbed to a greater extent. Therefore, UVB mainly reaches the epidermis while the less energetic UVA rays also affect the dermal skin layers.

UVA and UVB will therefore elicit different biological effects culminating in what is generally called the UVA and UVB response.

The major cellular chromophores that absorb in the UVB range are nucleic acids (DNA, RNA) and proteins (mainly tryptophan and tyrosine amino acids in these proteins). But also other biomolecules, such as NADH, quinones, flavins, porphyrins, 7-dehydrocholesterol, and urocanic acid, absorb UVB light. Of all these molecules, the photodamage and lack of repair of DNA is linked to most carcinogenic effects of UVB (1000–10,000 times stronger than those of UVA) (Trautinger 2001).

When going toward the longer wavelengths (UVA and UVB), direct cellular damage by absorption of UV photons becomes less prominent while indirect damage resulting from energy transfer to other molecules, such as oxygen, leading to the generation of reactive oxygen species (ROS), increases (Masaki 2010).

By now, it is well established that, as a result of these UV-induced molecular changes, a signaling network is activated that will decide on the eventual fate of the UV-exposed cell. Depending on the severity of the UV radiation, a cell will first try to survive by undergoing growth arrest and repairing the damage, but when the induced damage is irreparable, it will initiate the apoptotic program. In the case of keratinocytes, this process is called sunburn cell (SBC) formation. Simultaneously, intercellular communication is affected, following UV irradiation to regulate inflammatory and proliferative responses. The final skin phenotype will depend on the outcome of these (inter) cellular processes (Assefa et al. 2005).

In Sections 6.3 and 6.4, the main UV-induced signal transduction cascades and responses will be addressed.

6.3 Molecular UV Response

6.3.1 DNA Damage and Repair

Exposure to solar UV is a major etiological factor for skin cancer. It results in the direct generation of DNA photoproducts, mainly in the form of cyclobutane pyrimidine dimers (CPDs), in addition to pyrimidine 6,4-pyrimidone photoproducts (6,4-PP, or its Dewar form; the frequency of 6,4-PP is only one third of that of CPDs). These pyrimidine dimers are the principal DNA photoproducts that are formed following absorption of short-wavelength UVB by DNA (leaving a typical UVB fingerprint) and are characterized by the covalent link between carbons of adjacent pyrimidines. Surprisingly, adjacent thymines do not appear to be associated with mutations, possibly by default insertion of an adenine in a synthesized DNA strand opposite a "noninstructive" (i.e., damaged) base in the template strand, a typical action of DNA polymerase-η (this is referred to as the "A rule").

Although these lesions are the main and best-studied types of UV-induced DNA damage, UV radiation also induces a wider range of DNA damage. Oxidative DNA damage (mainly through UVA but, to a lesser extent, also through UVB), a consequence of the generation of ROS, mainly results in 7,8-dihydro-8-oxoguanine (8-oxodG) that originates from the oxidation of guanine, the base in DNA that is highly susceptible to this type of modification. Moreover, protein–DNA cross-links and single-strand breaks are possible. Although the nonpyrimidine dimer types of DNA damage can also be induced by other agents and hence such damage would not necessarily be recognized as caused by UV radiation, they may contribute to UV-induced carcinogenesis (de Gruijl, van Kranen, and Mullenders 2001).

Bulky DNA damage, such as UV-induced DNA photoproducts, needs to be repaired, but in order to have adequate time for DNA repair, cells are arrested in the G1 phase as well as in the G2 phase of the cell cycle, and both types of arrest can be regulated by the tumor-suppressor protein p53 (Decraene et al. 2001). This sequence-specific transcription factor (TF) is one of the most extensively studied proteins involved in sensing and integrating extracellular stress stimuli in general and UV radiation in particular. p53 is being referred to as "the guardian of the genome" for its unique role in the induction of cell-cycle arrest, followed by the well-orchestrated DNA repair process to prevent replication of damaged DNA and accumulation of genetic mutations. In addition, p53 has a crucial function in cellular proofreading, whereby it induces apoptosis or senescence to prevent the cells with severely damaged DNA to survive and potentially become cancerous (Latonen and Laiho 2005). It is generally thought that low doses of UV radiation mediate a transient induction of p53 accompanied with a transient cell-cycle arrest, whereas higher UV radiation doses instigate apoptosis correlating with a pronounced and more sustained induction of p53.

When UV-induced DNA damage occurs, these lesions are detected by intracellular sensors [e.g., ataxia telangiectasia-mutated (ATM) and Rad3-related protein (ATR)]. Subsequently, p53 is stabilized, mainly through phosphorylation-mediated interruption of the human double minute 2 (HDM2)-p53 complex, which prevents p53 from ubiquitin-mediated proteolysis and promotes its translocation and retention in the nucleus. p53 can trigger cell-cycle arrest in the G1 phase, thereby facilitating DNA repair by the induction of the potent inhibitor of cyclin-dependent kinases (Cdks) p21[WAF1/CIP1], whereas the G2/M transition is prevented by the p53-dependent induction of 14-3-3σ and growth arrest and DNA damage (GADD45) inducible gene products (Latonen and Laiho 2005; Matsumura and Ananthaswamy 2004). Interestingly, p38[MAPK] has also been identified as an important inducer of cell-cycle arrest (G1/S and G2/M) via several mechanisms, acting in both p53-dependent and -independent fashions. ROS-mediated p38[MAPK] activation has been shown *in vivo* to stabilize p21[WAF1/CIP1] to induce G1/S or G2/M arrest via either direct phosphorylation of p21[WAF1/CIP1] (Kim et al. 2002) or activation of p53 through direct phosphorylation (Gong et al. 2010). UV-induced double-strand breaks have also been shown to activate p38[MAPK] through the DNA damage sensors ATM and ATR in a largely p53-independent manner (Reinhardt et al. 2007). Additionally, p38[MAPK] has been reported to reduce levels of cyclin D1 and Cdc25A, a Cdk phosphatase (Casanovas et al. 2000; Goloudina et al. 2003).

After the initial growth arrest, the NER will attempt to repair the UV-induced damage. The elucidation of the mechanisms underlying DNA repair was made possible by identifying the mutations in various NER genes that underlie the different complementation groups of xeroderma pigmentosum (XP) (Cleaver 1968). The XP paradigm provided unequivocal evidence for the significance of NER in photocarcinogenesis because patients with XP have 100-fold higher risk for skin cancer than the general population (Kraemer et al. 1994). A sub-pathway of NER is the transcription-coupled repair (TCR), which targets DNA alterations that interfere with the translocation of RNA polymerase to genes that are to be transcribed. The CPDs present in the transcribed strand are more rapidly repaired than those in the nontranscribed strand because of the targeted recruitment of lesion-recognition factors to the site of the arrested RNA polymerase (Mellon and Hanawalt 1989). The second NER sub-pathway is the global genomic repair (GGR) that deals with damage in both transcribed and nontranscribed genes. Lesions such as 6,4-PP are efficiently repaired by GGR. In general, the mechanism for NER includes the following sequence of events:

lesion recognition, excision of damaged DNA, synthesis of a DNA fragment, and ligation of the repair patch.

Recently, both p53 and its homolog p63, which plays an important role in epidermal development and homeostasis, have been shown to contribute to the GGR of CPDs in human epidermal keratinocytes (Ferguson-Yates et al. 2008). This suggests that in human keratinocytes exposed to UV(B), p63 could maintain genomic stability in the absence of p53, thereby preventing skin cancer (Ferguson-Yates et al. 2008). Interestingly, also certain cytokines were shown to affect DNA repair after UVB exposure. IL-12 could, for example, suppress UVB-induced apoptosis in keratinocytes by inducing NER-mediated DNA repair both *in vitro* and *in vivo* (Schwarz et al. 2002).

From the above, it is clear that pyrimidine dimers are the main form of solar UV-induced DNA damage and that NER is the main line of defense against the genetic alterations that these dimers may cause. However, it is also clear that other forms of DNA damage and repair [e.g., removal of 8-oxodG by base excision repair (BER)] play a role and may be important under specific circumstances (D'Errico, Parlanti, and Dogliotti 2008).

6.3.2 ROS Production: A Major UV-Induced Mediator

ROS are short-lived entities that include, among others, singlet oxygen ($^1O_2$), superoxide (O_2^-), and H_2O_2. While $^1O_2$ is a very strong oxidant itself, both O_2^- and H_2O_2 can be converted, through iron (Fe^{2+})-catalyzed Fenton reactions, into highly reactive hydroxyl radicals ($OH^•$), but alternatively, O_2^- can also react with nitric oxide (NO), producing peroxynitrite ($ONOO^-$). Normal skin cells continuously generate ROS at low levels during the course of aerobic metabolism. Various environmental insults, such as UV radiation, have been shown to incite the production of ROS in excessive amounts, which quickly overwhelm the endogenous antioxidant defense systems (see the following). ROS will eventually react with and chemically modify cellular biomolecules, thereby inflicting damage and leading to cellular demise (Masaki 2010).

The type of ROS produced after UV irradiation depends on the wavelength. Whereas UVA mainly induces the formation of $^1O_2$ (through a photosensitizing reaction with internal chromophores, such as riboflavin), UVB rather leads to the production of O_2^- and H_2O_2 through the activation of NADPH oxidase and respiratory chain reactions (Masaki 2010).

Although ROS in high amounts are very dangerous and deleterious for a cell, they are also endowed with signaling properties when produced in a regulated and localized way.

Interestingly, two waves of ROS can be distinguished after UVB exposure as illustrated by measurement of UVB-induced ROS in keratinocytes showing a biphasic production (Rezvani et al. 2007). Whereas, at later time points (several hours after insult), a sustained amount of ROS is produced by damaged and malfunctioning mitochondria, a transient ROS wave is generated immediately after UVB irradiation mainly at the plasma membrane (Beak, Lee, and Kim 2004; Rezvani et al. 2007; Van Laethem et al. 2006). This apical ROS peak is thought to derive from the UVB-induced ligand-independent activation of the epidermal growth factor receptor (EGFR) (Beak, Lee, and Kim 2004; Van Laethem et al. 2006; Yao et al. 2009), which subsequently activates the ROS-generating plasma membrane–bound NADPH oxidase. In agreement with this, in human keratinocytes, inhibition of either EGFR or NADPH oxidase blunted ROS production after UVB (Wang and Kochevar 2005).

Moreover, ROS generated after UVA/B irradiation can enhance or sustain receptor-mediated signaling through the oxidation-mediated inhibition of different members of the protein tyrosine phosphatases (PTP) family (Gross et al. 1999; Gulati et al. 2004).

Several studies have shown that UVB-induced signaling in the presence of ROS scavengers is drastically hampered in epidermal cells (Di Domenico et al. 2009; Rezvani et al. 2007; Van Laethem et al. 2006). Moreover, although UVB-induced DNA damage (CPDs) has been shown to be capable of initiating a UVB stress response (Stege et al. 2000 and as illustrated previously), cell enucleation experiments have clearly demonstrated that the UVB response can occur independent of nuclear events (Devary et al. 1993). This suggests that apical cytosolic and ROS-mediated signaling pathways are crucial mediators of the UVB response.

As mentioned before, ROS can cause major cellular damage and potentially induce mutations. Therefore, several antioxidant defense systems have been evolutionarily designed in order to protect cells against a ROS overload. It should, however, be mentioned that the efficiency and importance of these separate mechanisms depend drastically on the cell type and severity and type of induced ROS.

Antioxidant enzymes transform ROS into less-toxic molecules; for example, superoxide dismutase (SOD) assists in the dismutation of superoxide (O_2^-) into H_2O_2, which, in turn, will be neutralized into O_2 and H_2O by catalase. Under basal conditions, their combined action scavenges most superoxide-based ROS. Also other redox-sensitive enzymes, for example, the ubiquitous oxidoreductase thioredoxin (Trx) and glutathione peroxidases (GPx), in the presence of glutathione (GSH), can break down H_2O_2. Alternatively GPx decomposes lipid hydroperoxides into their respective alcohols. Next to GSH, epidermal cells have a variety of nonenzymatic antioxidants at their disposal (either newly synthetized or provided by the cellular environment), such as the antioxidant vitamins ascorbic acid (vitamin C) and α-tocopherol (vitamin E) and the ROS-scavenging, heavy metal ion-induced, cysteine-rich peptide metallothionein (Masaki 2010).

In response to excessive oxidative stress, the main redox defense switch, the TF nuclear factor erythroid 2-related factor 2 (Nrf2), is activated, generally leading to a coordinated induction of a battery of targets that function to reinforce the intracellular antioxidant capacity. Nrf2, which is stabilized by the dissociation of Kelch-like ECH-associated protein 1 (Keap1) only when oxidative stress occurs, binds to an antioxidant response element in the promoter region of a vast number of antioxidant genes, thereby upregulating their transcription. These Nrf2 downstream genes include most classes of antioxidant defense molecules, detoxifying enzymes, and intracellular redox-balancing proteins, including, besides those described above, γ-glutamylcysteine synthetase (a rate-limiting enzyme of GSH synthesis), thioredoxin reductase, peroxiredoxin, heme oxygenase-1, glutathione S transferase, NAD(P)H quinone oxidoreductase-1, and UDP-glucuronosyltransferase (Schafer et al. 2010).

6.3.3 Apoptosis

When a UV-exposed epidermal cell is too severely damaged and all repair systems have failed, this cell will undergo apoptosis (a highly regulated cell death mechanism elaborately explained in Box 6.1), following the idea that it is better to die than to risk malignant transformation.

The importance of epidermal cell death after genotoxic/cytotoxic UV exposure is to limit the survival of irreparably damaged cells, which often carry mutations in tumor-suppressor genes or proto-oncogenes that could be transmitted to their offspring. This notion becomes clear in chronically irradiated skin where growing resistance to apoptosis results in increased skin carcinogenesis (Narayanan, Saladi, and Fox 2010).

The key role of apoptosis as a tumor-suppressor mechanism is well illustrated by the relevance of a regulated SBC formation in preventing skin carcinogenesis. UVB-mediated apoptosis has been thoroughly studied and shown to entail molecular events triggered at the plasma membrane and cytoplasm as well the nucleus (Assefa et al. 2005; Herrlich, Karin, and Weiss 2008).

However, it should be kept in mind that apoptosis following UV irradiation is a complex process that depends on the cell's genomic background, the UVB dose, the microenvironment, and the balanced presence of survival/death factors. In the following paragraphs, the major molecular players in UVB-induced apoptosis will be described.

While p53-mediated growth arrest usually requires the sequence-specific transactivation of genes involved in cell-cycle regulation (as discussed in Section 6.3.1), it remains an unclear question whether p53 is essential for UVB-induced apoptosis (Assefa et al. 2005; McKay et al. 2000). Although functional aberrations of p53 signaling have been linked with skin cancer (Brash et al. 1996), keratinocytes are still able to activate the apoptotic machinery in response to UVB damage in the absence of p53 (Assefa et al. 2005; Van Laethem et al. 2005). Moreover, loss of Noxa has a more pronounced protective effect than the loss of p53 in UV-irradiated skin, thus indicating that p53-dependent pathways of apoptosis cannot solely explain SBC formation (Naik et al. 2007; Nys et al. 2010).

Next, it is generally accepted that p38[MAPK] has a proapoptotic function in UVB-exposed keratinocytes. For example, the inhibition of p38[MAPK] was shown to almost completely block UVB-induced apoptosis in both normal keratinocytes and HaCaT cells (Shimizu et al. 1999; Van Laethem et al. 2004, 2006) as well

Box 6.1   Apoptotic Pathways and Their Regulators: A Snapshot

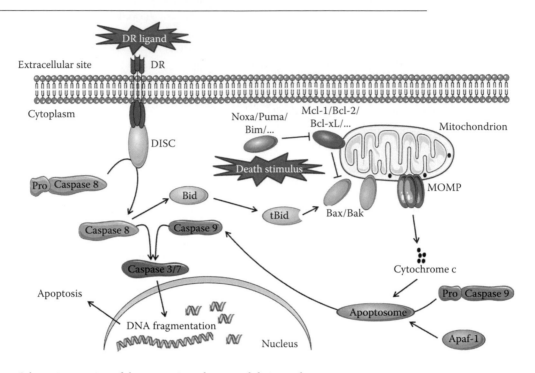

FIGURE 6.3   Schematic overview of the apoptotic pathways and their regulators.

Apoptosis can be initiated by a wide variety of intracellular and extracellular stimuli, but its execution occurs in a remarkably morphologically uniform and evolutionarily conserved way because of the activation of several members of the cysteinyl aspartate-specific proteases (caspases). This family of proteases, of which, to date, approximately 15 members have been identified in mammals, comprises proteins involved in apoptosis (i.e., caspase 2, 3, 6, 7, 8, 9, and 10), inflammation (i.e., caspase 1, 4, 5, and 12), as well as differentiation (i.e., caspase 14, a mediator of keratinocyte differentiation). Apoptotic caspases are further subdivided into initiators (i.e., caspase 2, 8, 9, and 10) and effectors (i.e., caspase 3, 6, and 7), according to their presumptive function and location in signaling pathways (Garrido et al. 2006; Stennicke and Salvesen 2000). Two distinct signaling pathways lead to the activation of the apoptotic caspases (Figure 6.3).

The extrinsic or death receptor (DR) pathway is mediated by ligand-dependent activation of DRs belonging to the tumor necrosis factor (TNF) receptor superfamily. DR engagement leads to the formation of a death-inducing signaling complex (DISC) that recruits initiator procaspase 8(/10) and triggers their dimerization-induced activation. Caspase 8 then activates effector caspases (e.g., caspase 3 and 7), thereby committing the cells to apoptosis (Lavrik, Golks, and Krammer 2005).

The intrinsic or mitochondrial pathway for caspase activation is engaged by the permeabilization of the outer mitochondrial membrane (MOMP) and the release of intermembrane proteins, such as cytochrome c, into the cytosol following a variety of cellular insults, such as ROS and DNA damage. In the cytosol, the apoptotic protease–activating factor-1 (Apaf-1), together with cytochrome c and (d)ATP, forms the apoptosome, a molecular platform for the recruitment and activation of procaspase 9. Subsequently, initiator caspase 9 directly cleaves and activates the effector caspases, resulting in the orchestration of the biochemical execution of the cell (Green 2005). In addition, caspase 8 has been shown to cleave Bid, a member of the Bcl-2 family of proteins, generating tBid, a C-terminal fragment that engages the mitochondrial pathway (illustrating that mitochondria can function both as initiators or amplifiers of caspase activation). At the level of integrity/permeability of the mitochondrial membrane, intrinsic apoptosis is crucially regulated by the interplay or balance of proapoptotic and antiapoptotic Bcl-2 family members (Figure 6.3). Consequently, the Bcl-2 family proteins play a pivotal role in determining whether a cell will live or die. They are found in the cytosol or localized in membranes of the mitochondria, the endoplasmic reticulum

(ER), and nucleus. Although their overall amino acid homology is relatively low, all Bcl-2 family members possess at least one of the four highly conserved motifs known as Bcl-2 homology (BH) domains (BH1–BH4), which correspond to the α-helical segments that confer their specific structure and function.

The Bcl-2 family can be classically grouped into three classes. One class inhibits apoptosis (e.g., Bcl-2, Bcl-xL, Mcl-1), whereas a second class promotes apoptosis (e.g., Bax, Bak). A third class of BH3-only proteins (e.g., Bid, Bim, Noxa, Puma, BNIP3) has a conserved BH3 domain that can bind and regulate the antiapoptotic Bcl-2 proteins to promote apoptosis. It appears that the proapoptotic family members Bax and Bak, either directly or indirectly, are crucial for inducing MOMP and the subsequent release of apoptotic molecules (such as cytochrome c), which leads to caspase activation. The antiapoptotic family members, such as Bcl-2 and Bcl-xL, inhibit Bax and Bak. Recent evidence indicates that BH3-only proteins de-repress Bax and Bak by direct binding to and inhibiting Bcl-2 and other antiapoptotic family members. By contrast, an opposing model postulates direct activation of Bax and Bak by some BH3-only proteins (specifically Bim, tBid, and Puma).

Although it is commonly thought that Bax and Bak form pores in membranes, the biochemical nature of such pores and how antiapoptotic Bcl-2 family proteins might regulate them remains a key and controversial issue in the field of cell death. At the same time as the cytochrome c release (or immediately before), Bax and Bak contribute to the fragmentation of mitochondria into more numerous and smaller units, which suggests connections between mitochondrial fission processes and the functions of the Bcl-2 family (Lippens et al. 2009; Lomonosova and Chinnadurai 2008).

as to reduce SBC formation in the epidermis of UVB-irradiated mice (Hildesheim, Awwad, and Fornace 2004). A functional cross talk between the p38MAPK pathway, hypoxia-inducible factor-1α (HIF-1α), and certain proapoptotic members of the Bcl-2 family has been suggested (Nys et al. 2010; Rezvani et al. 2007), whereas UVB-induced p38-mediated apoptosis is mainly p53-independent (Nys et al. 2010). However, the protective (proapoptotic) effects of p38MAPK after acute UVB damage may be overruled by chronic epidermal UVB exposure, during which the contribution of p38MAPK to a persistent proinflammatory, tumor-promoting microenvironment may become more important (Bowden 2004; Cooper and Bowden 2007).

In normal melanocytes, UVB-induced JNK activation has been shown to be a major contributor of the apoptotic pathway by promoting lysosomal membrane permeabilization with subsequent release of cathepsins and by regulating Bim functionality (Bivik and Ollinger 2008). Irrespective of its emerging, cell-specific function in the epidermis, UVB-mediated action of the stress-activated MAPKs happens in a mainly ROS-dependent manner (Nys et al. 2012; Van Laethem et al. 2006).

UVB is capable of activating both extrinsic and intrinsic apoptotic pathways at doses that fall within the physiological range of exposure of the human skin. UVB can induce ligand-independent clustering and activation of membrane DRs (TNF-R1, Fas, and TRAIL, probably as a result of irradiation-induced cross-linking or deactivation of PTPs) (Gulati et al. 2004; Wehrli et al. 2000). Recent studies, however, indicate that the extrinsic cascade may not have a dominant role in overall UVB-induced cell death because blocking TRAIL- or Fas-mediated signals does not substantially influence UVB-induced apoptosis (Eckert et al. 2002). In contrast to FasL-induced apoptosis, UVB induces procaspase 8 cleavage through a cytosolic and Bcl-2 inhibitable mechanism, which involves lysosomal proteases suggesting that caspase 8 activation is a consequence rather than the cause of cell death after UVB exposure (Assefa et al. 2003).

The dependence of UVB-induced apoptosis on the mitochondrial intrinsic pathway, on the other hand, has been clearly shown. Bcl-2 or Bcl-xL overexpression could completely abrogate UV-induced apoptosis both *in vitro* and *in vivo* (Assefa et al. 2003; Knezevic et al. 2007; Naik et al. 2007; Takahashi et al. 2001). Interestingly, Naik et al. (2007) could identify Noxa as the dominant BH3-only protein during UVB-induced apoptosis *in vivo* because loss of Noxa drastically suppressed the formation of apoptotic keratinocytes. Concomitantly, Mcl-1, reported to be the main degradation-target for Noxa's proapoptotic function, has recently been shown to function as the major epidermal survival protein (Sitailo, Jerome-Morais, and Denning 2009), and its elimination was proven to be required for the initiation of apoptosis following UV irradiation in epithelial cells (Nijhawan et al. 2003).

All together, these studies illustrate that UVB exerts its apoptotic effect mainly by signals affecting the balance of proapoptotic versus antiapoptotic Bcl-2 family members converging on MOMP, implicating mitochondria as the central coordinators of UVB-induced cell death.

However, next to the typical proapoptotic caspases activated during intrinsic and/or extrinsic apoptosis (see Box 6.1), caspase 2 has been linked to UV-induced apoptosis (Paroni et al. 2001). In response to a genotoxic stress, caspase 2 has been recognized as an important mediator of apoptosis, cell-cycle, and DNA repair (Vakifahmetoglu-Norberg and Zhivotovsky 2010) through a mechanism depending on the PIDDosome, a large protein complex involving p53-induced protein with a death domain (PIDD) (Tinel and Tschopp 2004). Although the proapoptotic function of this apical caspase in response to DNA damage has been clearly shown (Krumschnabel et al. 2009), PIDD-activated caspase 2 rather signals toward DNA repair and survival in UVB-irradiated skin because *in vivo* PIDD deficiency has been recently associated with cellular sensitization to UVB-induced apoptosis (Logette et al. 2011).

6.3.4 UVB-Activated Survival Pathways

As mentioned before, UVB-induced signaling involves many different pathways, and as a consequence, survival pathways are activated, possibly to avoid premature apoptosis and to allow the damaged cell more time for repair.

A major and well-known survival cascade (especially in the keratinocytes) is the phosphatidylinositol 3 kinase (PI3K)/Akt pathway. The Ser/Thr protein kinase B (PKB)/Akt can be activated through UVB-mediated IGFR signaling. This kinase functions as an important antagonist of UVB-induced apoptosis. For example Akt can directly phosphorylate and inactivate several apoptotic molecules, such as Bad, a BH3-only protein that is subsequently released from Bcl-xL, thereby allowing this prosurvival protein to more efficiently perform its antiapoptotic function (Datta et al. 1997, 2000). Also ASK-1 (Kim et al. 2001) and procaspase 9 (Cardone et al. 1998) have been identified as phosphorylation targets of Akt, thereby blocking their essential apoptotic function following UVB.

Next, UVB can trigger NF-κB activation and nuclear translocation (via ROS-, membrane receptor-, and/or DNA damage-dependent mechanisms) through the degradation of inhibitor of κB (IκB) (Cooper and Bowden 2007; Herrlich, Karin, and Weiss 2008). Concomitantly, AP-1-mediated signaling has been shown to be activated following UVB (Cooper and Bowden 2007; Herrlich, Karin, and Weiss 2008). Activation of both NF-κB and AP-1 pathways has been linked to decreased susceptibility to UVB-induced apoptosis and increased tumorigenesis (Cooper and Bowden 2007; Herrlich, Karin, and Weiss 2008), although some publications also report that NF-κB blockade, in combination with oncogenic Ras, is associated with epidermal neoplasia (Dajee et al. 2003).

Next to the prosurvival pathways, UVB is known to activate several inflammatory mechanisms in keratinocytes, supporting the concept that these are important immune-competent cells. Both NF-κB and AP-1 are known inflammatory mediators, inducing an array of cytokines/chemokines (Cooper and Bowden 2007). Moreover, after UVB-induced inflammasome-mediated activation, caspase 1 leads to maturation and secretion of IL-1β, a major proinflammatory cytokine (Feldmeyer et al. 2007).

As a consequence of the antiapoptotic, prosurvival, and inflammatory role of these pathways, their deregulation is directly linked to the induction of cancer.

6.3.5 Eventual Outcome (Survival or Apoptosis)

It is essential to realize that both cell survival and cell death mechanisms are often concomitantly activated after UV and share common molecular mediators. So, depending on the severity of the insult (i.e., the UV dose), the cellular background, and additional microenvironmental factors, the balance between cell survival and death signals will eventually decide on the fate of the irradiated cell (Figure 6.4).

6.3.6 Mild Hypoxia as a Physiological Skin Microenvironment

It is important to mention that these studies were all performed in a 21% O_2-containing environment, whereas normal human epidermal cells are constantly exposed to a mild hypoxic microenvironment, which could affect UV-induced cellular signaling.

A distinguished characteristic of rodent and human epidermis is the absence of vasculature that results in a constitutive low level of tissue oxygenation. A recent evaluation of the O_2 tension (pO_2) in human skin showed that while the dermis is well oxygenated and vascularized, displaying a pO_2 of 10%, in the epidermis, the pO_2 gradient ranges from mildly hypoxic (e.g., 5%) to severely hypoxic (e.g., 0.5%) in some skin appendages, such as the sebaceous glands and hair follicles (Evans et al. 2006). Consistent with this, murine and human skin exhibits an extensive binding of hypoxia-sensitive agents (such as nitroimidazole EF5) particularly in the basal epidermal compartment along with increased levels of the hypoxia inducible factor-1α (HIF-1α, a typical hypoxia-regulated protein) (Bedogni and Powell 2006; Boutin et al. 2008; Evans et al. 2006). Interestingly, recent research has shown that a mild hypoxic environment could sensitize keratinocytes to UVB-induced apoptosis through a sustained p38^{MAPK}/JNK activation, whereas melanocytes were slightly protected (Nys et al. 2012).

6.4 Cellular/Tissue Response

6.4.1 Acute Effects

The main effects of acute UV irradiation on human skin include sunburn, tanning, and local and systemic immunosuppression (Matsumura and Ananthaswamy 2004; Verschooten et al. 2006).

Sunburn is a skin reaction (its susceptibility is skin type–dependent with fair skin being more sensitive than dark skin) to solar damage that ranges from mild erythema (redness as a consequence of vasodilatation) to inflammation with invasion of inflammatory cells in the epidermal and dermal compartment to more severe cases with edema, pain, blistering, and subsequent desquamation of the skin. In very severe cases, sunburn is accompanied by systemic symptoms, such as nausea and fever (Honigsmann 2002).

Tanning is another (skin-type specific) response to UV. Upon irradiation, augmented pigmentation is caused by enhanced melanin synthesis in and release by melanocytes (in the epidermal melanin unit). There are two main pigment classes, the brown/black eumelanin and the reddish/brown pheomelanin, which determine the differences in human skin color, a critical factor in the development of skin cancer (Costin and Hearing 2007; Ernfors 2010).

Melanin is packaged and delivered to keratinocytes by melanosomes, lysosome-like structures, where this UV absorbing and ROS neutralizing molecule is positioned over the sun-exposed side of the nuclei to form cap-like structures that protect against DNA damage. The production of melanin is

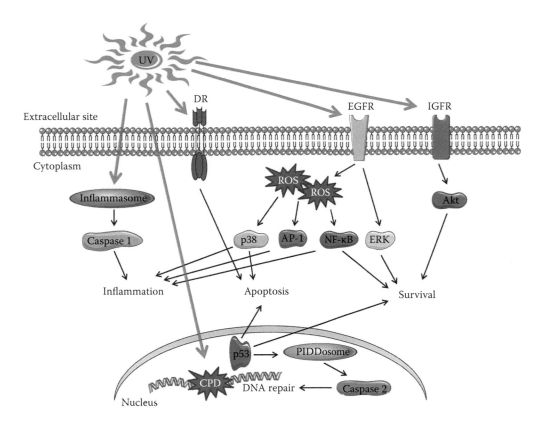

FIGURE 6.4 Overview of the UVB response of a keratinocyte. In response to UVB exposure of a keratinocyte, several signaling pathways are induced, leading to survival, apoptosis, and/or inflammation. DRs contribute to apoptosis, whereas growth factor receptors (e.g., IGFR, EGFR) can activate survival pathways (e.g., PI3K/Akt, ERK). EGFR also results in ROS-mediated NF-κB, AP-1, and p38^MAPK signaling. Next, several intracellular protein complexes, such as the PIDDosome and the inflammasome, can be formed, leading to the activation of caspase 2 and 1, respectively. Direct UVB-induced DNA damage can activate p53 resulting in DNA repair and consecutive survival/apoptosis.

therefore a major line of defense against the deleterious effects of UV light. When this defense mechanism is inefficient, as it occurs in fair-skinned people, cancer-promoting genotoxic damage is more likely to emerge (Costin and Hearing 2007). In addition to this, certain genetic variants of the melanocortin 1 receptor (MC1R; activation stimulates the release and synthesis of melanin in melanocytes) gene predispose to cutaneous malignant melanoma (Kennedy et al. 2001) and even nonmelanoma skin cancer (Bastiaens et al. 2001; Box et al. 2001) in a largely skin color–independent way. Melanogenesis is triggered by UVB-induced DNA damage in melanocytes and keratinocytes (Gilchrest and Eller 1999). A role for p53 in the upregulation of melanogenic genes Tyrosinase and Tyrosinase-related protein 1 (TRP-1) in melanocytes was shown (Khlgatian et al. 2002), and recently also α-melanocyte-stimulating hormone (α-MSH, a ligand for MC1R) after UVB exposure of keratinocytes has been reported to be upregulated in a p53-dependent fashion (Cui et al. 2007). It was shown that mice lacking p53 fail to mount a tanning response to UVB, which may account for their increased propensity to develop UVB-induced skin cancers (Cui et al. 2007). So it seems that the role of p53 in photocarcinogenesis is more complex than originally anticipated, and its function as a protector of the genomes of keratinocytes and melanocytes

against genotoxic damage (and thus the risk of malignant transformation) may partially rely on its ability to promote the tanning response.

Next to p53, p38^MAPK is also known to mediate the tanning response. As mentioned before, p38^MAPK after UVB is a known p53 regulator and could consequently, in that way, affect melanogenesis. Furthermore, several studies have identified p38^MAPK as a mediator of the upstream stimulatory factor-1 (USF-1), a key component of the tanning response. Following UVB irradiation of melanocytes, USF1 is phosphorylated by p38^MAPK and directly upregulates the expression of α-MSH, MC1R, Tyrosinase, and TRP-1 (Corre et al. 2004, 2009).

UVB provides the skin with local and systemic immunosuppression, which is mediated by different mechanisms involving Langerhans cell depletion and the production of immunosuppressant cytokines, such as IL-10, predominantly by irradiated keratinocytes (Schwarz 2005). DNA damage has been proposed as the initial trigger for UVB-induced immunosuppression because the suppressive effects can be prevented by pretreatment with T4 endonuclease V. But also urocanic acid, produced by keratinocytes in the superficial cornifying epidermis, could be involved (Schade, Esser, and Krutmann 2005).

The contribution of immunosuppression to an increased risk of malignancies and more—in particular, skin cancer—is supported both by studies in mice and by the increased risk for developing skin cancer in patients undergoing immunosuppressive therapies (Matsumura and Ananthaswamy 2004).

6.4.2 Chronic Effects

Skin photoaging is the result of cumulative, chronic sun exposure, which contributes to the gradual deterioration of cutaneous structures and functions in addition to the intrinsic aging process of the skin. It is predominantly observed in fair-skinned Caucasians and mainly involves the face, neck, or extensor surface of the upper extremities. Photoaged skin appears dry (rough), irregularly pigmented, and wrinkled and shows elastosis and telangiectasia. Whereas the more energetic UVB photons are mostly responsible for sunburn, tanning, and photocarcinogenesis, UVA has been shown to contribute substantially to photoaged skin through the degradation of the collagen matrix, either directly or via the activation of, for example, metalloproteinases, downregulation of new collagen synthesis, and induction of vascular changes. As skin ages, apoptotic mechanisms are found to decline, which could be associated with a decrease in the induction of SBCs by UVB (Matsumura and Ananthaswamy 2004).

Next to the induction of photoaging, long-term and recurrent exposure to UV irradiation leads to the accumulation of DNA damage, resulting in photocarcinogenesis. The balance between cellular survival on the one hand and induction of apoptosis on the other hand has an important role in many physiological processes, such as epidermal homeostasis of the skin. A disturbance of this fragile equilibrium can be associated with the development and progression of skin cancer.

Squamous cell carcinoma (SCC) and cutaneous malignant melanoma (CMM) are the most important malignant tumors of the skin. SCC, together with basal cell carcinoma (BCC) (also called nonmelanoma skin cancer), is the most frequent cancer within the Caucasian population, and CMM is one of the deadliest human diseases showing the highest increase in incidence. The steep rising incidence of nonmelanoma (SCC, BCC) and melanoma (CMM) skin cancer during the last decade is considered to be a direct consequence of the increased exposure to genotoxic and mutagenic UV radiation. Although there is no doubt that long-wave UVA (320–400 nm) can also contribute to skin cancer, most of the mutagenic and carcinogenic properties of sunlight have been attributed to UVB (290–320 nm) (Afaq, Adhami, and Mukhtar 2005). While BCC and CMM are related to intermittent sun overexposure during childhood, SCC is linked with cumulative UV exposure (Armstrong and Kricker 2001; de Gruijl, van Kranen, and Mullenders 2001).

Skin cancer progression is invariably seen as a multistep process in which different mutations accumulate to initiate tumor formation and progression. UVB is accepted as a complete carcinogen because this UV fraction is capable of inducing skin cancers in mice in the absence of any other carcinogenic agents (Bowden 2004). UVB can act as an initiator event by inducing DNA damage that, if unrepaired, may lead to mutations. Chronic exposure to UVB and, as such, accumulating mutations may result in reduced sensitivity to apoptosis and selection of cells with autonomous growth capability, which may give rise to precursor lesions. Additional genetic alterations may subsequently lead to further proliferation of neoplastic cells, genomic instability, acquisition of invasive capacity, and finally, metastatic capacity (Bowden 2004).

Because skin cancer is associated with the acquisition of defects in the apoptotic machinery (a major line of defense against the induction of cancer) and because therapeutic strategies designed to overcome these defects remain unsatisfactory, it is extremely important to further investigate, define, and understand apoptotic signaling.

References

Afaq, F., V. M. Adhami, and H. Mukhtar. 2005. Photochemoprevention of ultraviolet B signaling and photocarcinogenesis. *Mutat Res* 571(1–2):153–173.

Armstrong, B. K., and A. Kricker. 2001. The epidemiology of UV induced skin cancer. *J Photochem Photobiol B* 63(1–3):8–18.

Assefa, Z., M. Garmyn, A. Vantieghem et al. 2003. Ultraviolet B radiation-induced apoptosis in human keratinocytes: Cytosolic activation of procaspase-8 and the role of Bcl-2. *FEBS Lett* 540(1–3):125–132.

Assefa, Z., A. Van Laethem, M. Garmyn, and P. Agostinis. 2005. Ultraviolet radiation-induced apoptosis in keratinocytes: on the role of cytosolic factors. *Biochim Biophys Acta* 1755(2):90–106.

Bastiaens, M. T., J. A. ter Huurne, C. Kielich et al. 2001. Melanocortin-1 receptor gene variants determine the risk of nonmelanoma skin cancer independently of fair skin and red hair. *Am J Hum Genet* 68(4):884–894.

Beak, S. M., Y. S. Lee, and J. A. Kim. 2004. NADPH oxidase and cyclooxygenase mediate the ultraviolet B-induced generation of reactive oxygen species and activation of nuclear factor-kappaB in HaCaT human keratinocytes. *Biochimie* 86(7):425–429.

Bedogni, B., and M. B. Powell. 2006. Skin hypoxia: A promoting environmental factor in melanomagenesis. *Cell Cycle* 5(12):1258–1261.

Bivik, C. and K. Ollinger. 2008. JNK mediates UVB-induced apoptosis upstream lysosomal membrane permeabilization and Bcl-2 family proteins. *Apoptosis* 13(9):1111–1120.

Boutin, A. T., A. Weidemann, Z. Fu et al. 2008. Epidermal sensing of oxygen is essential for systemic hypoxic response. *Cell* 133(2):223–234.

Bowden, G. T. 2004. Prevention of non-melanoma skin cancer by targeting ultraviolet-B-light signalling. *Nat Rev Cancer* 4(1):23–35.

Box, N. F., D. L. Duffy, R. E. Irving et al. 2001. Melanocortin-1 receptor genotype is a risk factor for basal and squamous cell carcinoma. *J Invest Dermatol* 116(2):224–229.

Brash, D. E., A. Ziegler, A. S. Jonason et al. 1996. Sunlight and sunburn in human skin cancer: p53, apoptosis, and tumor promotion. *J Investig Dermatol Symp Proc* 1(2):136–142.

Cardone, M. H., N. Roy, H. R. Stennicke et al. 1998. Regulation of cell death protease caspase-9 by phosphorylation. *Science* 282(5392):1318–1321.

Casanovas, O., F. Miro, J. M. Estanyol et al. 2000. Osmotic stress regulates the stability of cyclin D1 in a p38SAPK2-dependent manner. *J Biol Chem* 275(45):35091–35097.

Cleaver, J. E. 1968. Defective repair replication of DNA in xeroderma pigmentosum. *Nature* 218(5142):652–656.

Cooper, S. J., and G. T. Bowden. 2007. Ultraviolet B regulation of transcription factor families: Roles of nuclear factor-kappa B (NF-kappaB) and activator protein-1 (AP-1) in UVB-induced skin carcinogenesis. *Curr Cancer Drug Targets* 7(4):325–334.

Corre, S., A. Primot, Y. Baron et al. 2009. Target gene specificity of USF-1 is directed via p38-mediated phosphorylation-dependent acetylation. *J Biol Chem* 284(28):18851–18862.

Corre, S., A. Primot, E. Sviderskaya et al. 2004. UV-induced expression of key component of the tanning process, the POMC and MC1R genes, is dependent on the p-38-activated upstream stimulating factor-1 (USF-1). *J Biol Chem* 279(49):51226–51233.

Costin, G. E., and V. J. Hearing. 2007. Human skin pigmentation: melanocytes modulate skin color in response to stress. *FASEB J* 21(4):976–994.

Cui, R., H. R. Widlund, E. Feige et al. 2007. Central role of p53 in the suntan response and pathologic hyperpigmentation. *Cell* 128(5):853–864.

D'Errico, M., E. Parlanti, and E. Dogliotti. 2008. Mechanism of oxidative DNA damage repair and relevance to human pathology. *Mutat Res* 659(1–2):4–14.

Dajee, M., M. Lazarov, J. Y. Zhang et al. 2003. NF-kappaB blockade and oncogenic Ras trigger invasive human epidermal neoplasia. *Nature* 421(6923):639–643.

Datta, S. R., H. Dudek, X. Tao et al. 1997. Akt phosphorylation of BAD couples survival signals to the cell-intrinsic death machinery. *Cell* 91(2):231–241.

Datta, S. R., A. Katsov, L. Hu et al. 2000. 14-3-3 proteins and survival kinases cooperate to inactivate BAD by BH3 domain phosphorylation. *Mol Cell* 6(1):41–51.

de Gruijl, F. R., H. J. van Kranen, and L. H. Mullenders. 2001. UV-induced DNA damage, repair, mutations and oncogenic pathways in skin cancer. *J Photochem Photobiol B* 63(1–3):19–27.

Decraene, D., P. Agostinis, A. Pupe, P. de Haes, and M. Garmyn. 2001. Acute response of human skin to solar radiation: regulation and function of the p53 protein. *J Photochem Photobiol B* 63(1–3):78–83.

Devary, Y., C. Rosette, J. A. DiDonato, and M. Karin. 1993. NF-kappa B activation by ultraviolet light not dependent on a nuclear signal. *Science* 261(5127):1442–1445.

Di Domenico, F., M. Perluigi, C. Foppoli et al. 2009. Protective effect of ferulic acid ethyl ester against oxidative stress mediated by UVB irradiation in human epidermal melanocytes. *Free Radic Res* 43(4):365–375.

Eckert, R. L., T. Efimova, S. R. Dashti et al. 2002. Keratinocyte survival, differentiation, and death: many roads lead to mitogen-activated protein kinase. *J Investig Dermatol Symp Proc* 7(1):36–40.

Ernfors, P. 2010. Cellular origin and developmental mechanisms during the formation of skin melanocytes. *Exp Cell Res* 316(8):1397–1407.

Evans, S. M., A. E. Schrlau, A. A. Chalian, P. Zhang, and C. J. Koch. 2006. Oxygen levels in normal and previously irradiated human skin as assessed by EF5 binding. *J Invest Dermatol* 126(12):2596–2606.

Feldmeyer, L., M. Keller, G. Niklaus et al. 2007. The inflammasome mediates UVB-induced activation and secretion of interleukin-1beta by keratinocytes. *Curr Biol* 17(13):1140–1145.

Ferguson-Yates, B. E., H. Li, T. K. Dong, J. L. Hsiao, and D. H. Oh. 2008. Impaired repair of cyclobutane pyrimidine dimers in human keratinocytes deficient in p53 and p63. *Carcinogenesis* 29(1):70–75.

Garrido, C., L. Galluzzi, M. Brunet et al. 2006. Mechanisms of cytochrome c release from mitochondria. *Cell Death Differ* 13(9):1423–1433.

Gilchrest, B. A., and M. S. Eller. 1999. DNA photodamage stimulates melanogenesis and other photoprotective responses. *J Investig Dermatol Symp Proc* 4(1):35–40.

Goloudina, A., H. Yamaguchi, D. B. Chervyakova et al. 2003. Regulation of human Cdc25A stability by Serine 75 phosphorylation is not sufficient to activate a S phase checkpoint. *Cell Cycle* 2(5):473–478.

Gong, X., A. Liu, X. Ming, P. Deng, and Y. Jiang. 2010. UV-induced interaction between p38 MAPK and p53 serves as a molecular switch in determining cell fate. *FEBS Lett* 584(23):4711–4716.

Green, D. R. 2005. Apoptotic pathways: Ten minutes to dead. *Cell* 121(5):671–674.

Gross, S., A. Knebel, T. Tenev et al. 1999. Inactivation of protein-tyrosine phosphatases as mechanism of UV-induced signal transduction. *J Biol Chem* 274(37):26378–26386.

Gulati, P., B. Markova, M. Gottlicher, F. D. Bohmer, and P. A. Herrlich. 2004. UVA inactivates protein tyrosine phosphatases by calpain-mediated degradation. *EMBO Rep* 5(8):812–817.

Herrlich, P., M. Karin, and C. Weiss. 2008. Supreme EnLIGHTenment: Damage recognition and signaling in the mammalian UV response. *Mol Cell* 29(3):279–290.

Hildesheim, J., R. T. Awwad, and A. J. Fornace, Jr. 2004. p38 Mitogen-activated protein kinase inhibitor protects the epidermis against the acute damaging effects of ultraviolet irradiation by blocking apoptosis and inflammatory responses. *J Invest Dermatol* 122(2):497–502.

Honigsmann, H. 2002. Erythema and pigmentation. *Photodermatol Photoimmunol Photomed* 18(2):75–81.

Houben, E., K. De Paepe, and V. Rogiers. 2007. A keratinocyte's course of life. *Skin Pharmacol Physiol* 20(3):122–132.

Kennedy, C., J. ter Huurne, M. Berkhout et al. 2001. Melanocortin 1 receptor (MC1R) gene variants are associated with an increased risk for cutaneous melanoma which is largely independent of skin type and hair color. *J Invest Dermatol* 117(2):294–300.

Khlgatian, M. K., I. M. Hadshiew, P. Asawanonda et al. 2002. Tyrosinase gene expression is regulated by p53. *J Invest Dermatol* 118(1):126–132.

Kim, A. H., G. Khursigara, X. Sun, T. F. Franke, and M. V. Chao. 2001. Akt phosphorylates and negatively regulates apoptosis signal-regulating kinase 1. *Mol Cell Biol* 21(3):893–901.

Kim, G. Y., S. E. Mercer, D. Z. Ewton et al. 2002. The stress-activated protein kinases p38 alpha and JNK1 stabilize p21(Cip1) by phosphorylation. *J Biol Chem* 277(33):29792–29802.

Knezevic, D., W. Zhang, P. J. Rochette, and D. E. Brash. 2007. Bcl-2 is the target of a UV-inducible apoptosis switch and a node for UV signaling. *Proc Natl Acad Sci U S A* 104(27):11286–11291.

Kraemer, K. H., M. M. Lee, A. D. Andrews, and W. C. Lambert. 1994. The role of sunlight and DNA repair in melanoma and nonmelanoma skin cancer. The xeroderma pigmentosum paradigm. *Arch Dermatol* 130(8):1018–1021.

Krumschnabel, G., B. Sohm, F. Bock, C. Manzl, and A. Villunger. 2009. The enigma of caspase-2: The laymen's view. *Cell Death Differ* 16(2):195–207.

Latonen, L., and M. Laiho. 2005. Cellular UV damage responses—Functions of tumor suppressor p53. *Biochim Biophys Acta* 1755(2):71–89.

Lavrik, I., A. Golks, and P. H. Krammer. 2005. Death receptor signaling. *J Cell Sci* 118(Pt 2):265–267.

Lippens, S., G. Denecker, P. Ovaere, P. Vandenabeele, and W. Declercq. 2005. Death penalty for keratinocytes: Apoptosis versus cornification. *Cell Death Differ* 12(Suppl 2):1497–1508.

Lippens, S., E. Hoste, P. Vandenabeele, P. Agostinis, and W. Declercq. 2009. Cell death in the skin. *Apoptosis* (4):549–569.

Logette, E., S. Schuepbach-Mallepell, M. J. Eckert et al. 2011. PIDD orchestrates translesion DNA synthesis in response to UV irradiation. *Cell Death Differ* 18(6):1036–1045.

Lomonosova, E., and G. Chinnadurai. 2008. BH3-only proteins in apoptosis and beyond: An overview. *Oncogene* 27(Suppl 1):S2–S19.

Masaki, H. 2010. Role of antioxidants in the skin: Anti-aging effects. *J Dermatol Sci* 58(2):85–90.

Matsumura, Y., and H. N. Ananthaswamy. 2004. Toxic effects of ultraviolet radiation on the skin. *Toxicol Appl Pharmacol* 195(3):298–308.

McKay, B. C., F. Chen, C. R. Perumalswami, F. Zhang, and M. Ljungman. 2000. The tumor suppressor p53 can both stimulate and inhibit ultraviolet light-induced apoptosis. *Mol Biol Cell* 11(8):2543–2551.

Mellon, I., and P. C. Hanawalt. 1989. Induction of the Escherichia coli lactose operon selectively increases repair of its transcribed DNA strand. *Nature* 342(6245):95–98.

Naik, E., E. M. Michalak, A. Villunger, J. M. Adams, and A. Strasser. 2007. Ultraviolet radiation triggers apoptosis of fibroblasts and skin keratinocytes mainly via the BH3-only protein Noxa. *J Cell Biol* 176(4):415–424.

Narayanan, D. L., R. N. Saladi, and J. L. Fox. 2010. Ultraviolet radiation and skin cancer. *Int J Dermatol* 49(9):978–986.

Nijhawan, D., M. Fang, E. Traer et al. 2003. Elimination of Mcl-1 is required for the initiation of apoptosis following ultraviolet irradiation. *Genes Dev* 17(12):1475–1486.

Nys, K., H. Maes, G. Andrei et al. 2012. Skin mild hypoxia enhances killing of UVB-damaged keratinocytes through reactive oxygen species-mediated apoptosis requiring Noxa and Bim. *Free Radic Biol Med* 52(6):1111–1120.

Nys, K., A. Van Laethem, C. Michiels et al. 2010. A p38(MAPK)/HIF-1 pathway initiated by UVB irradiation is required to induce Noxa and apoptosis of human keratinocytes. *J Invest Dermatol* 130(9):2269–2276.

Paroni, G., C. Henderson, C. Schneider, and C. Brancolini. 2001. Caspase-2-induced apoptosis is dependent on caspase-9, but its processing during UV- or tumor necrosis factor-dependent cell death requires caspase-3. *J Biol Chem* 276(24):21907–21915.

Reinhardt, H. C., A. S. Aslanian, J. A. Lees, and M. B. Yaffe. 2007. p53-deficient cells rely on ATM- and ATR-mediated checkpoint signaling through the p38MAPK/MK2 pathway for survival after DNA damage. *Cancer Cell* 11(2):175–189.

Rezvani, H. R., S. Dedieu, S. North et al. 2007. Hypoxia-inducible factor-1alpha, a key factor in the keratinocyte response to UVB exposure. *J Biol Chem* 282(22):16413–16422.

Schade, N., C. Esser, and J. Krutmann. 2005. Ultraviolet B radiation-induced immunosuppression: Molecular mechanisms and cellular alterations. *Photochem Photobiol Sci* 4(9):699–708.

Schafer, M., S. Dutsch, Keller U. auf dem, and S. Werner. 2010. Nrf2: A central regulator of UV protection in the epidermis. *Cell Cycle* 9(15):2917–2918.

Schwarz, A., S. Stander, M. Berneburg et al. 2002. Interleukin-12 suppresses ultraviolet radiation-induced apoptosis by inducing DNA repair. *Nat Cell Biol* 4(1):26–31.

Schwarz, T. 2005. Mechanisms of UV-induced immunosuppression. *Keio J Med* 54(4):165–171.

Shimizu, H., Y. Banno, N. Sumi et al. 1999. Activation of p38 mitogen-activated protein kinase and caspases in UVB-induced apoptosis of human keratinocyte HaCaT cells. *J Invest Dermatol* 112(5):769–774.

Sitailo, L. A., A. Jerome-Morais, and M. F. Denning. 2009. Mcl-1 functions as major epidermal survival protein required for proper keratinocyte differentiation. *J Invest Dermatol* 129(6):1351–1360.

Stege, H., L. Roza, A. A. Vink et al. 2000. Enzyme plus light therapy to repair DNA damage in ultraviolet-B-irradiated human skin. *Proc Natl Acad Sci U S A* 97(4):1790–1795.

Stennicke, H. R., and G. S. Salvesen. 2000. Caspases—Controlling intracellular signals by protease zymogen activation. *Biochim Biophys Acta* 1477(1–2):299–306.

Takahashi, H., M. Honma, A. Ishida-Yamamoto et al. 2001. In vitro and in vivo transfer of bcl-2 gene into keratinocytes suppresses UVB-induced apoptosis. *Photochem Photobiol* 74(4):579–586.

Tinel, A., and J. Tschopp. 2004. The PIDDosome, a protein complex implicated in activation of caspase-2 in response to genotoxic stress. *Science* 304(5672):843–846.

Trautinger, F. 2001. Mechanisms of photodamage of the skin and its functional consequences for skin ageing. *Clin Exp Dermatol* 26(7):573–577.

Vakifahmetoglu-Norberg, H. and B. Zhivotovsky. 2010. The unpredictable caspase-2: what can it do? *Trends Cell Biol* 20(3):150–159.

Van Laethem, A., K. Nys, S. Van Kelst et al. 2006. Apoptosis signal regulating kinase-1 connects reactive oxygen species to p38 MAPK-induced mitochondrial apoptosis in UVB-irradiated human keratinocytes. *Free Radic Biol Med* 41(9):1361–1371.

Van Laethem, A., S. Claerhout, M. Garmyn, and P. Agostinis. 2005. The sunburn cell: Regulation of death and survival of the keratinocyte. *Int J Biochem Cell Biol* 37(8):1547–1553.

Van Laethem, A., S. Van Kelst, S. Lippens et al. 2004. Activation of p38 MAPK is required for Bax translocation to mitochondria, cytochrome c release and apoptosis induced by UVB irradiation in human keratinocyte. *FASEB J* 18(15):1946–1948.

Verschooten, L., S. Claerhout, A. Van Laethem, P. Agostinis, and M. Garmyn. 2006. New strategies of photoprotection. *Photochem Photobiol* 82(4):1016–1023.

Wang, H., and I. E. Kochevar. 2005. Involvement of UVB-induced reactive oxygen species in TGF-beta biosynthesis and activation in keratinocytes. *Free Radic Biol Med* 38(7):890–897.

Wehrli, P., I. Viard, R. Bullani, J. Tschopp, and L. E. French. 2000. Death receptors in cutaneous biology and disease. *J Invest Dermatol* 115(2):141–148.

Yao, Y., J. E. Wolverton, Q. Zhang et al. 2009. Ultraviolet B radiation generated platelet-activating factor receptor agonist formation involves EGF-R-mediated reactive oxygen species. *J Immunol* 182(5):2842–2848.

7

Photocarcinogenesis Nonmelanoma Skin Cancer

Gary M. Halliday
University of Sydney

Scott N. Byrne
University of Sydney

J. Guy Lyons
University of Sydney and Royal Prince Alfred Hospital

Diona L. Damian
University of Sydney

7.1 Introduction

Nonmelanoma skin cancers (NMSCs) originate from cells other than melanocytes. Most of these arise from keratinocytes, which make up the vast majority of cells in the epidermis, or outer layer, of the skin. This group of cancers is very frequent in humans and is mostly caused by exposure to ultraviolet (UV) radiation. The study of UV-induced skin cancers is referred to as photocarcinogenesis, and this can be investigated in small animal models. Mice are the most frequently used animal model for photocarcinogenesis as they provide a good mimic of human skin cancer. This chapter describes human skin cancers and their treatment. It also provides a practical guide for setting up a photocarcinogenesis laboratory. Molecular responses to UV can also be used as surrogate endpoints for photocarcinogenesis in animal models and humans. These can be used to develop preventative strategies. These molecular responses, which include genetic damage, disruptions to cell cycling and differentiation, oncogenic signaling pathways, and immunosuppression, are also described in this chapter.

7.2 Nonmelanoma Skin Cancer

NMSC, comprising mainly basal cell carcinoma (BCC) and squamous cell carcinoma (SCC), is the most common malignancy in Caucasian populations (Staples et al. 2006). In fair-skinned populations living in areas of high solar irradiance, such as Australia, NMSC is four times as common as all other cancers combined (Staples et al. 2006). UV radiation in sunlight is the main cause of skin cancer. Exposure to artificial UV sources, such as tanning beds, also increases the risk of both melanoma and SCC (IARC 2007). The chemical carcinogen arsenic, which contaminates groundwater in various geographic areas, most notably throughout Bangladesh and West Bengal, also contributes to the global burden of skin cancer. Arsenic and UV radiation can act as cocarcinogens to produce large numbers of NMSC in darker-skinned individuals who would otherwise be at low skin-cancer risk (Rahman et al. 2001).

7.2.1 Basal Cell Carcinoma

BCC (Figure 7.1) is the most common form of skin cancer, classically presenting as a slowly progressing red scaling patch (superficial BCC) or as a pearly plaque or papule (nodular BCC). Some BCC variants, such as infiltrating or morphoeic BCC, present as subtle, scar-like lesions that may be difficult to diagnose and which may have large amounts of subclinical extension. While metastasis to lymph nodes or distant spread of BCC is rare, these cancers will cause local tissue destruction if left untreated. Treatment options for superficial BCC include excision, topical immunotherapy with imiquimod, freezing, curettage, superficial radiotherapy, and photodynamic therapy (Galiczynski and Vidimos 2011). Thicker nodular BCCs and infiltrating lesions are most effectively treated surgically.

7.2.2 Squamous Cell Carcinoma

SCCs generally present as more rapidly growing lesions than BCC (Figure 7.2) and have metastatic potential. Treatment is surgical. Approximately 5% of SCCs at higher risk sites, such as the head and neck, will metastasize, usually to the lymph nodes, although distant metastases may also occur. Other risk factors for metastasis include host immunosuppression, tumor thickness, presence of perineural invasion, and tumor differentiation. Keratoacanthomas are spontaneously regressing SCC variants, which present as rapidly growing, crateriform lesions that begin to involute after a 4–6 week growth phase. They can be difficult to histologically distinguish from SCC and are generally excised.

7.2.3 Actinic Keratoses

Actinic keratoses (AKs) (Figure 7.3) are premalignant, scaling skin lesions that commonly occur on chronically sun-exposed sites, such as the face, arms, and hands. AKs sometimes regress spontaneously but can also progress to invasive SCC (Thompson, Jolley,

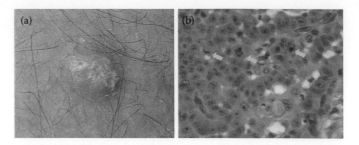

FIGURE 7.2 SCC is frequent on chronically sun-exposed sites, such as the scalp in elderly men (a). SCC is both more common and more aggressive in immunocompromised individuals. Histopathology (b) from a 2.2-mm-thick, moderately differentiated SCC on the face of an immunosuppressed renal transplant recipient shows frequent mitoses (examples highlighted with arrows) and high-grade nuclear pleomorphism and prominent nucleoli. This lesion later metastasized to lymph nodes and lungs.

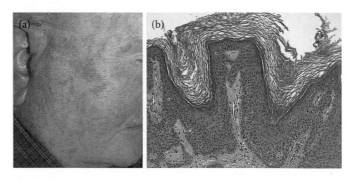

FIGURE 7.3 AKs are scaling erythematous lesions (a); histopathology shows mild squamous atypia in the lower epidermis and a few superficial mitoses (b).

and Marks 1993). Treatment options for AKs include cryotherapy, imiquimod, topical 5-fluorouracil cream, diclofenac cream, and photodynamic therapy (Galiczynski and Vidimos 2011). Daily use of sunscreen on the face can also assist regression of these lesions and can, on average, cause a 40% reduction in AK counts within just a few months (Thompson, Jolley, and Marks 1993).

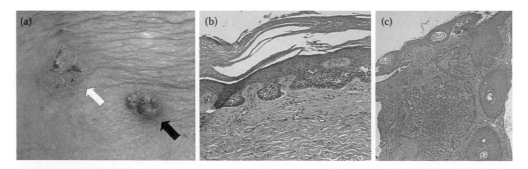

FIGURE 7.1 BCC is the most common form of skin cancer. Superficial BCC presents as scaling patches (a, white arrow) with histopathology showing basaloid tumor cells budding from the base of the epidermis (b). Nodular BCC (a, black arrow) shows peripheral tumor cell palisading on histopathology (c). These BCCs were among ~100 concurrent lesions found on an elderly, sun-damaged patient who had been exposed in childhood to sodium arsenite in Fowler's solution. Fowler's solution was used from the 1700s until the 1950s to treat a range of diseases including psoriasis and asthma.

7.3 Experimental Models for NMSC

7.3.1 Human Skin Experimental Models

The carcinogenic effects of UV can be investigated on tissue obtained from human subjects. Explants of human skin can be kept alive in organ culture for several days to weeks to investigate the biological effects of UV on skin. The source is most commonly neonatal foreskins or adult skin removed during reductive plastic surgery. The skin can be grafted onto immunodeficient mice, enabling longer-term investigations to be conducted and the effects of systemic parameters to be tested, at least in the absence of a competent immune system.

Engineered human skin (EHS), consisting of an extracellular matrix, epidermal keratinocytes, and other cells from the skin, was developed to overcome the limitations of monolayer cultures while providing an *in vitro* model in which cells and their microenvironment can be readily manipulated (Breitkreutz et al. 1984). Typically, EHSs are grown on a suspended porous membrane onto which a type I collagen gel is formed. Fibroblasts and sometimes other stromal cells are included in the collagen prior to gelling. After a period of consolidation, epidermal cells are added, and the surface of the culture is raised to the air–liquid interface, stimulating the keratinocytes to stratify and differentiate in a manner similar to that of normal skin. The EHS can be irradiated with UV of defined wavelengths and the genetic and biological effects determined (Huang, Bernerd, and Halliday 2009). The keratinocytes can be genetically manipulated using lentiviral vectors to investigate the roles of particular genes or signaling pathways in UV responses.

7.3.2 NMSC Induction in Experimental Animals

The ability of UV to cause skin cancer in animal models, combined with epidemiological studies, provides conclusive evidence that the UV wavebands within sunlight are the prime cause of skin cancer in humans (Halliday and Lyons 2008). For practical reasons, mice are the most commonly used animal model. They are convenient laboratory animals to handle and house, and their responses to UV are similar to humans. Skin cancers induced in mice are a good model of AK formation and development into SCC. They resemble human skin cancers clinically and histologically. However, it is rare for UV-induced BCC to be reported in murine studies. Different mouse strains have been used for photocarcinogenesis studies, including Skh:hr-1 hairless mice and conventional haired strains, such as C57BL/6. Unlike hairless athymic mice, the Skh:hr-1 strain has a normal functioning immune system, which is essential for photocarcinogenesis studies. This mouse strain is very sensitive to UV carcinogenesis and has the practical advantage of being hairless and therefore not needing to be shaved before UV exposure. This also makes it easier to monitor tumor growth (Benavides et al. 2009). However, these mice are not well characterized genetically. It is uncertain whether mutations in skin appendages, such as hair, and other skin abnormalities, such as dermal cysts and enlargement of sebaceous glands, in these mice influence photocarcinogenesis. C57BL/6 mice have the disadvantages of needing to be shaved to remove hair prior to UV exposure and being less sensitive than Skh:hr-1 mice to skin-cancer induction. However, they are better characterized, and a large number of genetic knockout or transgenic mice are available on this background strain, enabling the use of transgenic technology for studying the role of particular genes or molecules during photocarcinogenesis.

7.4 Experimental Equipment Required for Photocarcinogenesis Experiments

If haired mice are used, then the hair needs to be removed by shaving as it stops UV from reaching the skin. Small animal clippers, such as those made for small dogs, can be used to remove most hair. A closer shave is then achieved with a commercial men's electric dry shaver. Mice will require shaving about once per week, sometimes with a touch-up midweek.

A number of different UV sources can be used. Custom-built banks of fluorescent tubes are frequently used, which are relatively inexpensive (Figure 7.4). For statistical reasons, an experiment is likely to consist of 20–30 (or more) mice per group with multiple groups being irradiated most days per week for 20 or more weeks, sometimes for more than a year. Irradiation of large numbers of mice for many weeks can be achieved using an irradiation table consisting of a bank of 2- to 3-m-long fluorescent tubes. The tubes can be mounted on the underside of the top bench with the mice in boxes on a shelf below the fluorescent lamps. The fluorescent tubes should be individually switched to enable versatility in spectrum. A layer of cellulose triacetate film can be used to filter any contaminating UVC and attenuate low wavelength UVB emitted from fluorescent tubes in order

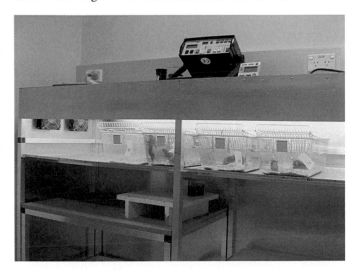

FIGURE 7.4 Sunbed for UV irradiation of mice. UV fluorescent tubes are mounted under the top shelf of a table to irradiate mice in their cages placed on the second shelf. Mice have been shaved on their back. Fans forcing air along the tubes help keep them at a constant temperature to reduce fluctuations in irradiance. The broadband radiometer can be seen on top of the sunbed.

to create an irradiation spectrum more closely matching that of sunlight. The intensity of the UV source is temperature dependent. Hence, fans should be built into the irradiation source for cooling and the tubes switched on for at least 20 min prior to use to attain a constant temperature. A ratio of approximately 2 UVB- to 4 UVA-emitting tubes can be used to produce a spectrum that approximates sunlight. We aim for a UVB/UVA ratio of 1:10 while sunlight has about half this UVB content. For unknown reasons, a higher UVA ratio often results in chronic scratching by the mice, causing skin damage, which needs to be avoided. Nail clipping the mice can also reduce scratching. Examples of tubes include FS72T12-UVB-HO UVB-emitting tubes and Cosmolux RA plus A1-14-100W UVA-emitting tubes.

Monitoring of the spectrum, intensity, and uniformity of UV over the irradiation area is essential. This needs to be done with a scanning spectroradiometer calibrated against traceable standard lamps. The probe of the spectroradiometer needs to be placed at the same distance from the lamps as the surface to be irradiated. The intensity of the UV source can be measured at short intervals, usually 1–2 nm, generating the spectrum of the source. Integration of the area under the curve between any two wavelengths gives the intensity of that waveband. Intensity needs to be measured before and after each irradiation as it is dependent on a number of conditions, including temperature and fluctuations in the electricity current. Lamp output changes with time. Measurement with the scanning spectroradiometer takes several hours, which is not practical for daily measurement of fluctuations in intensity. For this purpose, a broadband radiometer needs to be used. As its broadband spectral responsiveness will not match the waveband of the UV source being monitored, it determines relative fluctuations in intensity, rather than absolute intensity of the UV source. Therefore, it needs to be calibrated against the UV source by taking a reading immediately after assessment of intensity with the scanning spectroradiometer. A mathematical calibration factor, or ratio of intensity measured with the scanning spectroradiometer compared to the broadband radiometer, can be used on a routine basis to convert the broadband reading to intensity. The broadband radiometer measurement only takes a few minutes and therefore is convenient to routinely measure intensity. The spectrum of the source and calibration factor for the broadband radiometer needs to be determined at regular intervals as optics of the broadband probe can be altered (solarized) by continuous exposure to UV and therefore lose accuracy. Similarly, the optics of the lamps and of any UV filters used to modify the UV spectrum will also alter with age and UV exposure. Hence, regular spectral measurements are essential.

SCCs can be induced in C57BL/6 mice irradiated about 4 days per week for 25 weeks with 250 mJ/cm² UVB and accompanying UVA from the source. This dose induces barely detectable sunburn in mice and is therefore the minimum erythemal dose (MED). Cumulative exposure to one MED 4 days per week would cause sunburn, including blistering and pain in mice that have not been acclimatized to UV. The development of adaptive responses, such as skin thickening, needs to occur before they can receive one MED each day. This can be achieved by using lower incremental doses, such as 50 mJ/cm² UVB for the first week, increasing gradually over a 4-week period to the final dose, or the number of days the mice are irradiated can start at 2 days per week and increase. Using such a protocol, the irradiation can cease after approximately 25 weeks, and skin cancer monitoring can then continue for a longer period, depending on the experiment, perhaps for up to week 40.

To monitor skin cancer development, we use a body map to record the position of skin lesions. The diameter of each skin lesion in two vertical directions can then be measured weekly with engineers' calipers to determine the average diameter of each tumor. They grow slowly, and therefore, more frequent measurement is usually not required. It is not unusual for small lesions to regress after a few weeks. It is also advisable to photograph each lesion on a regular basis. Each skin lesion should be removed and formalin-fixed for routine histopathological analysis. In an experiment, some mice may need to be euthanized before the time at which the experiment is to be concluded if the tumors become too large, if the tumors appear as if they are close to ulcerating, or if the mice become unhealthy.

Survival analysis, such as the log-rank (Mantel–Cox) test, is usually used to assess incidence. The mouse is regarded as not surviving at the time of appearance of its first tumor. Tumor growth can be assessed by comparing diameter at different times by repeated measures ANOVA (with Tukey's multiple comparison *post hoc* test). For this analysis, tumor diameter on mice culled before the end time point for ethical reasons needs to remain at its final diameter as any further change in growth cannot be assumed. Tumor multiplicity takes into account whether more than one tumor arises on mice with time as this measure is not included in the incidence data. This can also be assessed by repeated measures ANOVA. Histopathological analysis is usually descriptive.

7.5 UV-Induced Genetic Mutations That Cause NMSC

Photocarcinogenesis takes considerable time and cannot always be achieved experimentally, particularly when studying human material. Therefore, mechanistic endpoints, such as mutations, are often used. UV-induced mutations in key genes that control keratinocyte growth or differentiation are fundamental for carcinogenesis. UV causes different types of damage to DNA. Dimerization of adjacent pyrimidine residues upon direct absorption of UV by DNA forms cyclobutane pyrimidine dimers (CPDs). These photoproducts distort the DNA structure and can lead to mutations. UV also causes oxidation products, the most frequent being oxidation of guanine to 8-oxo-7,8-dihydro-2′-deoxyguanosine (8-oxodGuo) (Halliday 2005). DNA photodamage is repaired by complex processes. These include cell-cycle arrest to give the cell sufficient time to repair damaged DNA prior to cell division. Chromatin remodeling enables access to damaged DNA by repair enzymes, which then use different pathways, including nucleotide and base excision repair to mend the damage (Farrell, Halliday, and Lyons 2011). Failure to repair the damage before DNA replication can result in an incorrect nucleotide (or mutation) being incorporated into the newly replicated DNA. Thus, DNA repair prior to cell division is a key step in preventing UV induced photocarcinogenesis. In genetic conditions where DNA repair capacity is markedly impaired, such as xeroderma pigmentosum, the risk of melanoma and NMSC increases by 1000-fold (Kraemer, Lee, and Scotto 1987).

Mutations could occur in a number of genes that contribute to photocarcinogenesis. The most well studied is p53, which is frequently mutated in human skin cancer (Agar et al. 2004). This gene is a master regulator of the cell cycle, DNA repair, and apoptosis and, therefore, is important in many cellular responses during photocarcinogenesis. In mouse models, mutated p53 can be found in UV-irradiated skin prior to the appearance of clinically detectable tumors (Ananthaswamy et al. 1997). Many of these mutations occur in patches of keratinocyte clones that appear to be potential precursors of SCC (de Gruijl and Rebel 2008). Thus, mutation of this gene is likely to be an early event in photocarcinogenesis. p53 knockout mice have a highly enhanced susceptibility to photocarcinogenesis (Jiang et al. 1999), confirming the critical role of this gene in protection from UV damage. Gorlin's syndrome (basal cell nevus syndrome) results from inactivation of the patched (PTCH1) gene, which may be inherited in an autosomal dominant manner or may arise from a spontaneous mutation. Second-hit mutations in p53 then cause multiple BCCs with the first lesions arising in late childhood or adolescence (Leger et al. 2011). Recently, we have described a hotspot mutation in the Brahma (BRM) gene in human skin cancers (Moloney et al. 2009). This gene codes for one of the two catalytic subunits of the SWI/SNF chromatin remodeling complex. Like p53, SWI/SNF is a master regulator of many cellular processes as it controls access to DNA. This mutation is consistent with being caused by UV-induced oxidative damage, but whether it is a tumor suppressor gene involved in photocarcinogenesis needs to be determined by further studies.

7.6 Cellular Changes that Accompany UV-Induced Malignant Transformation of Keratinocytes

7.6.1 Enhanced Proliferation

Keratinocytes normally proliferate only in the basal layer of the epidermis, undergoing cell-cycle arrest as they lose contact with the basement membrane. UV causes a short-term hyperplastic response in the epidermis, resulting in a thickening of the precornified layers (Ananthaswamy et al. 1999). However, in epidermis that has undergone pro-oncogenic mutagenesis, hyperplasia is sustained, and proliferation also occurs in suprabasal cells.

7.6.2 Impaired Cell Death

Contributing to the higher proliferation of NMSC compared with normal keratinocytes is an impairment of programmed cell death. The most obvious form of programmed cell death is terminal differentiation of keratinocytes from the basal, proliferative layer through intermediate spinous and granular layers to a fully cornified corneocyte that is shed into the environment. In NMSC, this process is partially or completely inhibited, there by contributing to the thickening of the epidermis in early lesions *in situ*. A partial retention of the ability to terminally differentiate can be retained, even in large SCCs, giving rise to characteristic keratin whorls within the tumors. UV can induce G2 cell-cycle arrest in keratinocytes. These cells can transit the epidermis and be shed without undergoing programmed cell death (Stout et al. 2005).

Another important form of programmed cell death that is impaired in NMSC is apoptosis. UV causes DNA damage in keratinocytes, which can initiate apoptosis if the damage is too severe to be repaired. The apoptotic bodies can be seen histologically as morphologically distinct "sunburn cells" and through methods such as terminal deoxynucleotidyl transferase-meditated dUTP nick end labeling (TUNEL). NMSC and their precursor cells often have inactivating mutations in genes, such as p53, whose activity is critical to apoptosis.

7.6.3 Plasticity and Microenvironmental Changes

Epidermal keratinocytes are potentially plastic, able to undergo a change resembling an epithelial–mesenchymal transition (EMT) under conditions of UV-induced carcinogenesis and wound healing. UV upregulates EMT-inducing transcription factors, such as SNAILs, through the activation of signaling pathways, such as the epidermal growth factor receptor (EGFR), resulting in reduced cell–cell affinity and enhanced motility (Sou et al. 2010). The same plasticity regulators can increase proinflammatory cytokines, attracting leukocytes that themselves can have complex pro-oncogenic effects (Halliday and Lyons 2008).

7.7 Oncogenic Signaling Pathways That Are Regulated by UV

Many pathways can contribute to photocarcinogenesis by being mutated as discussed above or by being deregulated by UV. These are also useful intermediates for the study of photocarcinogenesis.

7.7.1 p53

p53 is a tumor-suppressing, DNA-binding transcription factor that has many biological roles, including the promotion of apoptosis and cell-cycle arrest. Genes whose transcription is activated by p53 include those involved in DNA repair (GADD45, PCNA, XPCC, DDB2), cell-cycle arrest (p21), and apoptosis (BAX, PUMA, NOXA, FAS). Its activity is subjected to many mechanisms of regulation (Decraene et al. 2001; Kruse and Gu 2009). Exposure of keratinocytes to UV stabilizes and therefore increases p53 protein levels (Liu et al. 1994). This is regulated by a ubiquitin ligase, MDM2, which inhibits p53 transactivation and targets it for proteasomal degradation by ubiquitination. UV-mediated DNA damage activates the protein kinases ATM, ATR, and DNA-PK. These kinases and others that they, in turn, activate, including Chk1, Chk2, and p38, phosphorylate amino terminal serine sites in p53, which weakens

MDM2 binding and consequently stabilizes p53. p14ARF disrupts the p53–MDM2 interaction, thereby increasing p53 activity.

7.7.2 EGFR

The hyperplastic response of epidermis to UV is driven by the EGFR (El-Abaseri, Putta, and Hansen 2006). This is activated through autocrine or paracrine stimulation by EGFR ligands, released from cell surface precursors by UV activation of ADAM family proteinases (He et al. 2008). EGFR promotes cell-cycle progression via MAPKs and cyclin D1 and, in keratinocytes, activates transcription of matrix metalloproteinases that can promote invasion (Lyons et al. 1993). EGFR also inhibits terminal differentiation of keratinocytes by downregulating Notch signaling (Kolev et al. 2008). The EGFR gene rarely acquires an activating mutation in SCCs but is often amplified and overexpressed.

7.7.3 Notch

Efficient signaling via the Notch pathway is essential for the terminal differentiation of epidermal keratinocytes (Rangarajan et al. 2001). A mouse model in which the epidermis is Notch-deficient has demonstrated that Notch has a tumor-suppressor role. These mice acquire spontaneous BCC-like tumors and cooperate with oncogenic mutant forms of Hras to produce SCCs (Nicolas et

al. 2003). Notch pathway genes, including Notch1, Notch2, and Reln, are frequently mutated in SCCs of both the head and neck region and the skin (Agrawal et al. 2011; Durinck et al. 2011).

7.7.4 Hedgehog

Sonic Hedgehog (SHH) and related Hedgehog-family proteins are secreted molecules that signal through Patched cell surface receptor proteins, PTCH1 and PTCH2 (Daya-Grosjean and Couve-Privat 2005). PTCH inhibits the signaling activity of another membrane-bound protein of the same cell, Smoothened (SMO), which is relieved upon binding of SHH. SMO then activates GLI transcription factors, which deregulate cell-cycle control. GLI synergizes with EGFR signaling through c-Jun AP-1 transcription factor complexes that bind to a subset of GLI-activated genes, leading to increased *in vitro* transformation and tumorigenicity *in vivo* (Schnidar et al. 2009).

7.8 Immune System Control of UV-Induced NMSC

The adaptive immune system, made up of antigen-specific T and B lymphocytes, has the capacity to "survey" our bodies for tumors. MacFarlane Burnet pioneered the "tumor immunosurveillance"

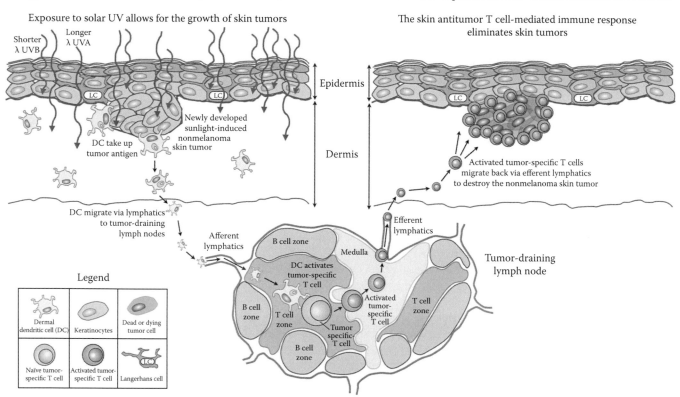

FIGURE 7.5 Anti-skin tumor T cell-mediated adaptive immune response. Cutaneous DCs, including Langerhans cells (LCs), can detect and phagocytose UV-induced NMSC. Activated DC loaded with tumor antigens then migrate via afferent lymphatic vessels to local skin-draining lymph nodes where they home toward the T cell zones. There, DC will present tumor antigens to naïve CD4+ T cells. DCs also have the capacity to cross-present tumor antigens to naïve CD8+ CTLs. The newly activated tumor-specific T cells will undergo clonal expansion and differentiation into effector T cells before migrating via efferent lymphatic vessels back to the skin to destroy the developing tumor.

model more than 50 years ago. Cutaneous dendritic cells (DCs) phagocytose tumor cells and then migrate to skin-draining lymph nodes where they present processed tumor antigens to naïve T lymphocytes (Figure 7.5). Both CD4⁺ helper and CD8⁺ cytotoxic T lymphocytes (CTLs) are required for an effective antitumor immune response. The DC-activated tumor-specific effector T cells then migrate back to the tumor site where they mediate tumor destruction. Hence, events occurring in both the skin and skin-draining lymph nodes are important for an effective antiskin tumor immune response.

The importance of the skin's immune system for defense against skin cancer is illustrated by organ transplant–recipient studies. These patients are deliberately and chronically immunosuppressed by antirejection medication. They have rates of SCC 50- to 80-fold higher and rates of BCC approximately five times higher than those in the immunocompetent population (Carroll et al. 2003). This highlights the greater responsiveness of squamous lesions (SCCs and AKs) than of BCCs to changes in host immunity. Individuals immunosuppressed as a result of HIV infection or chronic lymphoid malignancy are also at hugely elevated risk of skin cancer development and metastasis (Wilkins et al. 2006; Otley 2006).

7.9 UV-Induced NMSC Evade Antitumor Immunity

In humans, NMSCs frequently undergo spontaneous regression when the host adaptive immune response successfully detects and then destroys them (Halliday et al. 1995). Cell lines established from UV-induced murine tumors provide a good model for this aspect of human tumors. Many regress when transplanted into unirradiated, genetically identical recipients. They initially grow before being immunologically destroyed. Other UV tumor cell lines evade immunological destruction and instead grow progressively upon transplantation into naïve immunocompetent hosts. Mechanisms by which progressor NMSC lines evade immunity

include expression of CD95 Ligand (CD95L or FasL) (Byrne and Halliday 2003). CD95L on the surface of T cells binds to its receptor CD95 on tumor cells, inducing apoptosis of the tumor. To evade immunity, skin tumor cells themselves express CD95L, enabling them to kill the T cells (Hahne et al. 1996). The progressor UV-induced tumor cells also express elevated levels of surface major histocompatibility complex (MHC) class II molecules, but none of the costimulatory molecules required for T cell activation. MHC II molecules present tumor peptides to CD4+ T cells, the main lymphocyte population increased in spontaneously regressing human NMSC (Halliday et al. 1995). Tumors that express MHC II in the absence of additional costimulatory molecules are unable to activate immunity (Baskar et al. 1993).

UV-induced progressor tumors are infiltrated by significantly higher numbers of MHC IIʰⁱᵍʰ macrophages than regressors. This recruitment of macrophages is facilitated by the progressing tumors secreting large amounts of transforming growth factor β (TGF β) (Byrne, Knox, and Halliday 2008). Therefore, immunomodulating, tumor-derived cytokines, such as TGF β, can have a profound effect on the antitumor immune response and hence tumor growth (Figure 7.6).

7.10 Solar UV Radiation Suppresses the Antitumor Immune Response

UV radiation in sunlight can markedly suppress skin immune responses in otherwise immunocompetent individuals, even at very low doses well below the sunburn threshold (Norval and Halliday 2011). Because UV immunosuppression is a critical step in photocarcinogenesis, it can be used as an endpoint to develop therapeutic and preventative strategies. Both UVB (290–320 nm) and also long-wave UVA at 365–385 nm are immunosuppressive after exposures equivalent to ~4–5 and 10–15 min of summer sunlight, respectively (Damian et al. 2011). UV can suppress skin immunity when the antigen contact is local

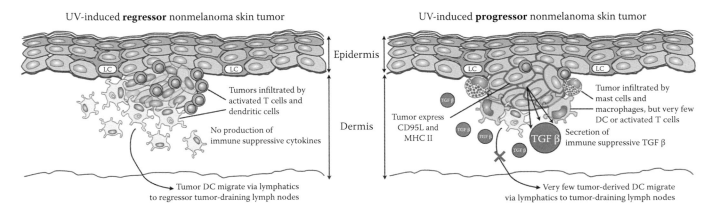

FIGURE 7.6 Cellular and molecular mechanisms involved in UV-induced NMSC evasion of antitumor immunity. Many UV-induced NMSC will be detected and destroyed by the host antitumor immune response (left-hand panel and outlined in Figure 7.5). Some UV-induced NMSCs acquire the ability to evade immunological destruction. They grow progressively (right-hand panel). Mechanisms include the production of immunosuppressive cytokines (e.g., TGF β), the expression of immunomodulating surface molecules (e.g., FasL and MHC II), and the recruitment of tumor-promoting cells (e.g., tumor-associated macrophages and mast cells).

at the site of irradiation or when it is systemic to the irradiation site. Otherwise healthy people with a history of skin cancer are more susceptible to this UV-induced immunosuppression than age-matched controls (Norval and Halliday 2011). Sunscreens, especially broad-spectrum products filtering long-wave UVA as well as UVB, can reduce UV immunosuppression in humans (Norval and Halliday 2011). Daily use of sunscreen can reduce SCC incidence by 40% within 2 years (van der Pols et al. 2006) and can reduce numbers of AKs by 30%–40% within a few months (Thompson, Jolley, and Marks 1993). Protection from UV immunosuppression is thought to be a central mechanism of this rapid regression.

Men have a higher incidence of and mortality from skin cancer (Staples et al. 2006). This may, in part, reflect gender differences in susceptibility to UV-induced immunosuppression. Paralleling findings in murine models, men may have greater susceptibility to UV immunosuppression than women (Damian et al. 2008).

Antitumor immunity is a complex, tightly regulated process. Solar UV targets multiple points in the pathway for suppression. Within minutes of UV exposure, a cascade of immunosuppressive cellular and molecular events are set in motion that culminate in inhibition of effector T cell activation and the generation of antigen-specific regulatory cells (Table 7.1) (Ullrich and Byrne 2012). A number of skin-associated inflammatory events occur almost immediately following UV exposure, including isomerization of *trans*-urocanic acid (UCA) to the immunosuppressive *cis* isoform. Oxidation of membrane phospholipids and the generation of reactive oxygen species also occur very early after UV and all contribute to the developing immunosuppressive milieu.

These early events can either cause immunosuppression directly by binding to specific receptors on the surface of cells or lead to downstream production and release of immunosuppressive cytokines, chemokines, and complement proteins. These soluble mediators direct the migration of cells into and away from the exposed skin, ultimately resulting in the activation of long-lived, antigen-specific regulatory cells (Table 7.1).

Much of our understanding of how UV suppresses immunity has come from animal models. Contact as well as delayed type hypersensitivity and, more recently, *in vivo* cytotoxicity (CTL) assays (Rana, Rogers, and Halliday 2011) have been used to model the antitumor immune response (Figure 7.7). These models have been highly useful in deciphering the cellular and molecular mechanisms of how UV suppresses immunity. They have exposed and cemented the relationship that exists between UV-induced DNA damage, immunosuppression, and skin cancer. Consequently, these animal models are now facilitating translation of preclinical findings into novel therapeutic options for the prevention and treatment of NMSC. For example, the demonstration that *cis*-UCA and PAF receptor antagonists block UV immunosuppression and accelerate DNA repair led to studies showing that they protect mice from UV induced NMSC (Sreevidya et al. 2010). In humans, nicotinamide (vitamin B3) prevents UV from suppressing immunity (Damian et al. 2008), probably because it preserves ATP levels in UV-irradiated keratinocytes, enabling optimal immune function and repair of damaged DNA (Park 2010). This has led to a clinical trial showing that oral nicotinamide reduces the incidence of AKs and NMSC in humans (Surjana et al. 2012).

TABLE 7.1 Immunosuppressive Cellular and Molecular Events Caused by UV

	UV Induced Event	Mechanism and Immunological Effect
Very early[a]	*trans*-Urocanic acid converted to *cis*-urocanic acid (*cis*-**UCA**)	Binds to serotonin receptors ($5HT_{2a}$) on immune cells
	Production of oxidized lipids, e.g., platelet activating factor (**PAF**)	Binds to PAF receptors on immune cells causing immunosuppression
	Generation of reactive oxygen and nitrogen species (**ROS** and **NOS**)	Damages DNA and promotes inflammation
	DNA damage	Immunosuppressive
Early[b]	**Neutrophil** recruitment into skin	Produce immunosuppressive IL-4
	Mast cell recruitment into skin	Produce anti-inflammatory IL-10
	Dendritic cells destroyed or induced to migrate damaged to lymph nodes	Failure to activate T cells or activation of regulatory cells in lymph nodes
	Production of cytokines (**IL-1** α and β, **IL-4, IL-10, TNF, PGE**₂)	Anti-inflammatory and immunosuppressive properties
Intermediate[c]	**Mast cell** migration to skin draining lymph nodes	Produce IL-10, inhibit antibody production, may activate T and B regs
	Classical and **alternative** complement pathways activated	Binds to complement receptors on mast cells and lymphocytes
	Macrophages infiltrate skin	Produce immunosuppressive IL-10
	Regulatory B cells (UV-B-regs) activated	Suppress immune induction via IL-10
Late[d]	**Effector** and **memory** T cells (CD4+ helper and CD8+ CTL) not activated	Absence of T cell immunity as well as immunological memory
	Regulatory T cells activated (IL-4 + NK-T cells; CD4+FoxP3+CTLA-4+)	Long-term, antigen-specific immunosuppression

[a] Minutes to hours post UV.
[b] Less than 24 h post UV.
[c] 24 to 96 h post UV.
[d] Days to weeks post UV.

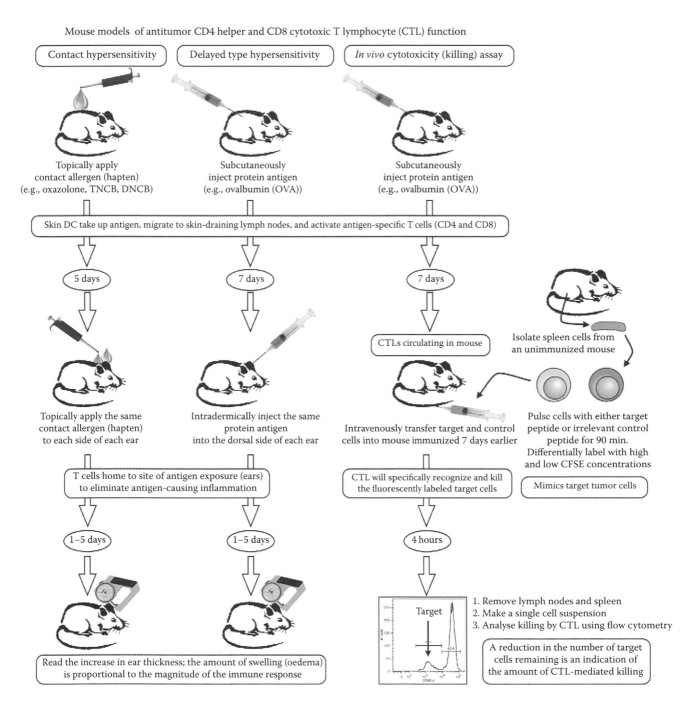

FIGURE 7.7 Models used to explore UV suppression of T cell-mediated antitumor immunity. Three models have been used to decipher the cellular and molecular mechanisms of UV immunosuppression: contact hypersensitivity (left column), delayed type hypersensitivity (middle column), and, more recently, the *in vivo* CTL assay (right column).

7.11 Conclusions

NMSC is the most common cancer in Caucasians. While the majority of these cancers can be treated, as a result of their high incidence, they cause a sizeable number of deaths in addition to excessive morbidity and use a disproportionately large amount of the health budget. The major cause of NMSC is exposure to UV radiation from sunlight. As they have an environmental cause, they are theoretically preventable. While the dose of UV required to cause skin cancer in humans is unknown, suberythemal doses are sufficient to cause the molecular damage that leads to skin cancer, including DNA damage and immunosuppression. Public health campaigns urging sun protection are failing to reduce skin cancer incidence to manageable levels, suggesting that protection from suberythemal doses of UV is required to substantially reduce skin cancer incidence. This may not be possible as these levels of UV are required for vitamin D production, which is essential for good health, and humans cannot avoid sun exposure during normal daily and work-related activities. The development of better preventative strategies is dependent upon photobiology research determining how UV causes genetic damage, activates oncogenic signaling pathways in the skin, and suppresses immunity.

Photocarcinogenesis studies in mice have contributed greatly to our understanding of skin cancer and development of preventative and therapeutic strategies as many aspects of human NMSC are well modeled by mice. A large amount of expertise and specialized equipment is required for photocarcinogenesis experiments with the monitoring of UV dose and spectra being particularly expensive and requiring exacting attention. The use of transgenic mice and specific inhibitors enables key mechanisms to be unraveled.

Genetic damage, disturbances in cell-cycle control, activation of oncogenic signaling processes, and immunosuppression are essential effects of UV on the skin for photocarcinogenesis. All of these are required, and blocking any of these mechanisms would be expected to reduce the incidence of skin cancer. UV causes many different types of genetic damage and activates a large number of molecular mechanisms that disrupt cell cycle, chromatin remodeling, and DNA repair processes. These influence whether genetic damage will proceed to a mutation that may contribute to carcinogenesis if it occurs in a gene that controls keratinocyte proliferation or differentiation. Impairment of cell death, induction of EMT, and activation of oncogenic signaling via molecular mechanisms, such as the EGFR, hedgehog, Notch, and the p53 pathway, all contribute to photocarcinogenesis. UV initiates a cascade of immunosuppressive signals. These include reactive oxygen mediated signaling, *cis*-UCA, and immunosuppressive cytokines, such as IL-10. These inhibit activation of effector and memory T cell responses, whereas they activate B and T regulatory cells that suppress antitumor immunity. Inhibiting these mechanisms by preventative measures, such as blockade of specific receptors, inhibition of reactive oxygen signaling, and maintaining ATP levels with nicotinamide to optimize the immune response, all appear promising for prevention of photocarcinogenesis.

Acknowledgments

We thank the National Health and Medical Research Council of Australia, the Cancer Council of NSW, the Cancer Institute NSW, Cure Cancer Australia, and Epiderm for financial support.

References

Agar, N. S., G. M. Halliday, R. S. Barnetson et al. 2004. The basal layer in human squamous tumors harbors more UVA than UVB fingerprint mutations: A role for UVA in human skin carcinogenesis. *Proc Natl Acad Sci U S A* 101:4954–4959.

Agrawal, N., M. J. Frederick, C. R. Pickering et al. 2011. Exome sequencing of head and neck squamous cell carcinoma reveals inactivating mutations in NOTCH1. *Science* 333:1154–1157.

Ananthaswamy, H. N., S. M. Loughlin, P. Cox et al. 1997. Sunlight and skin cancer: inhibition of p53 mutations in UV-irradiated mouse skin by sunscreens. *Nat Med* 3:510–514.

Ananthaswamy, H. N., A. Ouhtit, R. L. Evans et al. 1999. Persistence of p53 mutations and resistance of keratinocytes to apoptosis are associated with the increased susceptibility of mice lacking the XPC gene to UV carcinogenesis. *Oncogene* 18:7395–7398.

Baskar, S., S. Ostrandrosenberg, N. Nabavi et al. 1993. Constitutive expression of B7 restores immunogenicity of tumor cells expressing truncated major histocompatibility complex class-II molecules. *Proc Natl Acad Sci U S A* 90:5687–5690.

Benavides, F., T. M. Oberyszyn, A. M. VanBuskirk, V. E. Reeve, and D. F. Kusewitt. 2009. The hairless mouse in skin research. *J Dermatol Sci* 53:10–18.

Breitkreutz, D., A. Bohnert, E. Herzmann et al. 1984. Differentiation specific functions in cultured and transplanted mouse keratinocytes: environmental influences on ultrastructure and keratin expression. *Differentiation* 26:154–169.

Byrne, S. N., and G. M. Halliday. 2003. High levels of Fas ligand and MHC class II in the absence of CD80 or CD86 expression and a decreased CD4(+) T cell infiltration, enables murine skin tumours to progress. *Cancer Immunol Immunother* 52:396–402.

Byrne, S. N., M. C. Knox, and G. M. Halliday. 2008. TGF beta is responsible for skin tumour infiltration by macrophages enabling the tumours to escape immune destruction. *Immunol Cell Biol* 86:92–97.

Carroll, R. P., H. M. Ramsay, A. A. Fryer et al. 2003. Incidence and prediction of nonmelanoma skin cancer post-renal transplantation: a prospective study in Queensland, Australia. *Am J Kidney Dis* 41:676–683.

Damian, D. L., Y. J. Matthews, T. A. Phan, and G. M. Halliday. 2011. An action spectrum for ultraviolet radiation-induced immunosuppression in humans. *Br J Dermatol* 164:657–659.

Damian, D. L., C. R. S. Patterson, M. Stapelberg et al. 2008. UV radiation-induced immunosuppression is greater in men and prevented by topical nicotinamide. *J Invest Dermatol* 128:447–454.

Daya-Grosjean, L., and S. Couve-Privat. 2005. Sonic hedgehog signaling in basal cell carcinomas. *Cancer Lett* 225:181–192.

de Gruijl, F. R., and H. Rebel. 2008. Early events in UV carcinogenesis—DNA damage, target cells and mutant p53 foci. *Photochem Photobiol* 84:382–387.

Decraene, D., P. Agostinis, A. Pupe, P. de Haes, and M. Garmyn. 2001. Acute response of human skin to solar radiation: regulation and function of the p53 protein. *J Photochem Photobiol B Biol* 63:78–83.

Durinck, S., C. Ho, N. J. Wang et al. 2011. Temporal dissection of tumorigenesis in primary cancers. *Cancer Discov* 1:137–143.

El-Abaseri, T. B., S. Putta, and L. A. Hansen. 2006. Ultraviolet irradiation induces keratinocyte proliferation and epidermal hyperplasia through the activation of the epidermal growth factor receptor. *Carcinogenesis* 27:225–231.

Farrell, A. W., G. M. Halliday, and J. G. Lyons. 2011. Chromatin structure following UV-induced DNA damage-repair or death? *Int J Mol Sci* 12:8063–8085.

Galiczynski, E. M., and A. T. Vidimos. 2011. Nonsurgical treatment of nonmelanoma skin cancer. *Dermatol Clin* 29:297–309, x.

Hahne, M., D. Rimoldi, M. Schroter et al. 1996. Melanoma cell expression of Fas(Apo-1/Cd95) ligand—Implications for tumor immune escape. *Science* 274:1363–1366.

Halliday, G. M. 2005. Inflammation, gene mutation and photo-immunosuppression in response to UVR-induced oxidative damage contributes to photocarcinogenesis. *Mutat Res* 571:107–120.

Halliday, G. M., and J. G. Lyons. 2008. Inflammatory doses of UV may not be necessary for skin carcinogenesis. *Photochem Photobiol* 84:272–283.

Halliday, G. M., A. Patel, M. J. Hunt, F. J. Tefany, and R. S. C. Barnetson. 1995. Spontaneous regression of human melanoma/non-melanoma skin cancer: Association with infiltrating CD4+ T cells. *World J Surg* 19:352–358.

He, Y. Y., S. E. Council, L. Feng, and C. F. Chignell. 2008. UVA-induced cell cycle progression is mediated by a disintegrin and metalloprotease/epidermal growth factor receptor/AKT/Cyclin D1 pathways in keratinocytes. *Cancer Res* 68:3752–3758.

Huang, X. X., F. Bernerd, and G. M. Halliday. 2009. Ultraviolet A within sunlight induces mutations in the epidermal basal layer of engineered human skin. *Am J Pathol* 174:1534–1543.

IARC. 2007. International Agency for Research on Cancer working group on artificial ultraviolet (UV) light and skin cancer. The association of use of sunbeds with cutaneous malignant melanoma and other skin cancers: A systematic review. *Int J Cancer* 120:1116–1122.

Jiang, W., H. N. Ananthaswamy, H. K. Muller, and M. L. Kripke. 1999. p53 protects against skin cancer induction by UV-B radiation. *Oncogene* 18:4247–4253.

Kolev, V., A. Mandinova, J. Guinea-Viniegra et al. 2008. EGFR signalling as a negative regulator of Notch1 gene transcription and function in proliferating keratinocytes and cancer. *Nat Cell Biol* 10:902–911.

Kraemer, K. H., M. M. Lee, and J. Scotto. 1987. Xeroderma pigmentosum. Cutaneous, ocular, and neurologic abnormalities in 830 published cases. *Arch Dermatol* 123:241–250.

Kruse, J. P., and W. Gu. 2009. Modes of p53 regulation. *Cell* 137:609–622.

Leger, M., A. Quintana, J. Tzu et al. 2011. Nevoid basal cell carcinoma syndrome. *Dermatol Online J* 17:23.

Liu, M., K. R. Dhanwada, D. F. Birt, S. Hecht, and J. C. Pelling. 1994. Increase in p53 protein half-life in mouse keratinocytes following UV-B irradiation. *Carcinogenesis* 15:1089–1092.

Lyons, J. G., B. Birkedal-Hansen, M. C. Pierson, J. M. Whitelock, and H. Birkedal-Hansen. 1993. Interleukin-1 beta and transforming growth factor-alpha/epidermal growth factor induce expression of M(r) 95,000 type IV collagenase/gelatinase and interstitial fibroblast-type collagenase by rat mucosal keratinocytes. *J Biol Chem* 268:19143–19151.

Moloney, F. J., J. G. Lyons, V. L. Bock et al. 2009. Hotspot mutation of Brahma in non-melanoma skin cancer. *J Invest Dermatol* 129:1012–1015.

Nicolas, M., A. Wolfer, K. Raj et al. 2003. Notch1 functions as a tumor suppressor in mouse skin. *Nat Genet* 33:416–421.

Norval, M., and G. M. Halliday. 2011. The consequences of UV-induced immunosuppression for human health. *Photochem Photobiol* 87:965–977.

Otley, C. C. 2006. Non-Hodgkin lymphoma and skin cancer: A dangerous combination. *Australas J Dermatol* 47:231–236.

Park, J., G. M. Halliday, D. Surjana, and D. L. Damian. 2010. Nicotinamide prevents ultraviolet radiation-induced cellular energy loss. *Photochemistry and Photobiology* 86(4):942–948.

Rahman, M. M., U. K. Chowdhury, S. C. Mukherjee et al. 2001. Chronic arsenic toxicity in Bangladesh and West Bengal, India—A review and commentary. *J Toxicol Clin Toxicol* 39:683–700.

Rana, S., L. J. Rogers, and G. M. Halliday. 2011. Systemic low-dose UVB inhibits CD8 T cells and skin inflammation by alternative and novel mechanisms. *Am J Pathol* 178:2783–2791.

Rangarajan, A., C. Talora, R. Okuyama et al. 2001. Notch signaling is a direct determinant of keratinocyte growth arrest and entry into differentiation. *EMBO J* 20:3427–3436.

Schnidar, H., M. Eberl, S. Klingler et al. 2009. Epidermal growth factor receptor signaling synergizes with Hedgehog/GLI in oncogenic transformation via activation of the MEK/ERK/JUN pathway. *Cancer Res* 69:1284–1292.

Sou, P. W., N. C. Delic, G. M. Halliday, and J. G. Lyons. 2010. Snail transcription factors in keratinocytes: Enough to make your skin crawl. *Int J Biochem Cell Biol* 42:1940–1944.

Sreevidya, C. S., A. Fukunaga, N. M. Khaskhely et al. 2010. Agents that reverse UV-Induced immune suppression and photocarcinogenesis affect DNA repair. *J Invest Dermatol* 130:1428–1437.

Staples, M. P., M. Elwood, R. C. Burton et al. 2006. Non-melanoma skin cancer in Australia: the 2002 national survey and trends since 1985. *Med J Aust* 184:6–10.

Stout, G. J., D. Westdijk, D. M. Calkhoven et al. 2005. Epidermal transit of replication-arrested, undifferentiated keratinocytes in UV-exposed XPC mice: an alternative to in situ apoptosis. *Proc Natl Acad Sci U S A* 102:18980–18985.

Surjana, D., G. M. Halliday, A. J. Martin, F. J. Moloney, and D. L. Damian. 2012. Oral nicotinamide reduces actinic keratoses in phase II double-blinded randomized controlled trials. *J Invest Dermatol* 132:1497–1500.

Thompson, S. C., D. Jolley, and R. Marks. 1993. Reduction of solar keratoses by regular sunscreen use. *N Engl J Med* 329:1147–1151.

Ullrich, S. E., and S. N. Byrne. 2012. The immunologic revolution: photoimmunology. *J Invest Dermatol* 132:896–905.

van der Pols, J. C., G. M. Williams, N. Pandeya, V. Logan, and A. C. Green. 2006. Prolonged prevention of squamous cell carcinoma of the skin by regular sunscreen use. *Cancer Epidemiol Biomarkers Prev* 15:2546–2548.

Wilkins, K., R. Turner, J. C. Dolev et al. 2006. Cutaneous malignancy and human immunodeficiency virus disease. *J Am Acad Dermatol* 54:189–206.

Autoimmune Photodermatoses

John L. M. Hawk
St Thomas' Hospital

8.1 Acquired Autoimmune Photodermatoses

8.1.1 Polymorphic (Polymorphous) Light Eruption

Polymorphic light eruption (PMLE) is very common world wide, affecting some 20% of southern Scandinavians, 10%–15% of northern Americans and southern British, and 5% of southern Australians (Pao et al. 1994), but almost no equatorial Singaporeans. Its prevalence therefore decreases steadily with decreasing latitude, although this has been challenged in a later study, but purely of indoor workers with no regular sun exposure (Rhodes et al. 2010).

A delayed-type hypersensitivity (DTH) response to sunlight-induced cutaneous photoantigen was first suggested as the cause of PMLE as early as 1942, based on the delay between ultraviolet radiation (UVR) exposure and the eruption onset as well as its histology. This has now proven almost certainly to be the case, although the major evidence in favor does remain circumstantial. Determination of the nature of the inflammatory infiltrate in sunlight-induced PMLE of uncertain and varying age was originally inconclusive until timed biopsies from lesions induced by low-dose solar-simulated irradiation eventually demonstrated the consistent appearance of a T cell-dominated perivascular infiltrate within a few hours, peaking by 72 h. CD4+ T cells were initially most numerous, but by 72 h, CD8+ T cells were predominant, perhaps ablating the response (Norris et al. 1989a). More dermal and epidermal Langerhans cells, as well as increased dermal macrophages, were also present. This picture strongly supported a DTH response, as already recognized in the similar allergic-contact dermatitis and tuberculin reactions, as the basis for PMLE.

In further support of this, intercellular adhesion molecule-1 (ICAM-1) expression, particularly on keratinocytes, as already observed in DTH reactions but not irritant contact dermatitis or after UVB irradiation of normal skin, was also noted on keratinocytes above the perivascular infiltrate in PMLE (Norris et al. 1992). Finally, high-dose UVB and UVA irradiation of PMLE epidermal cells in one study increased their attraction of autologous peripheral blood monocytes (Gonzales-Amaro et al. 1991), a response not seen in nonirradiated cells, again supporting an immunological response against an endogenous UVR-modified antigen.

As a likely cause of the PMLE response, it has more recently been shown that PMLE patients do not immunosuppress in the precipitating induction phase of the disorder as easily as normal subjects, thereby leaving them capable of recognizing a putative UVR-induced cutaneous antigen. Thus, contact sensitization to dinitrochlorobenzene (DNCB) following solar-simulated irradiation of the sensitization site occurred significantly more easily in PMLE patients than in normals (van de Pas et al. 2004). On the other hand, elicitation of the allergic contact response to DNCB in already sensitized PMLE patients and normals was gradually and equally suppressed in both groups by irradiation (Palmer et al. 2005), apparently also explaining the frequent development of immunologic tolerance, often imprecisely called "hardening," in PMLE as summer progresses, as well as the efficacy of prophylactic phototherapy in the condition.

The radiation-absorbing molecules initiating PMLE have not been identified, and a number of different molecules may well be involved in both the same and separate patients. Once UVR absorption and molecular distortion occur, it appears likely that the altered molecule itself may often become antigenic, while, on other occasions, secondarily produced free radicals may interact with and alter other nearby molecules to produce an antigen, or both.

The radiation wavelengths responsible for initiating PMLE appear to be very variable between patients as might be expected if multiple inducing antigens are possible. However, the task of defining them in any given patient with monochromatic or even broadband irradiation is also often difficult even in patients who react easily in sunlight. This is probably because the skin areas exposed artificially and the irradiation doses or dose rates used are too small to produce a sufficient rash-causing antigen, while any immunologic skin tolerance induced by recent sunlight exposure will make a response even more unlikely (Palmer et al. 2005). Nevertheless, UVA (315 to 400 nm) has usually been more effective than UVB (280 to 315 nm) (Ortel et al. 1986) at initiating the eruption, with UVA being successful in 56% of patients exposed to UVA or UVB daily for 4 to 8 days in one study (Ortel et al. 1986), UVB in 17%, and both in 27%. Another report, however, has suggested that UVB may sometimes be effective in up to 57% of patients (Miyamoto 1989). In very broad terms, though, it may well be that approximately 50% of PMLE patients are sensitive to UVB and 75% to UVA with approximately 25% of each of these groups being sensitive to both, while visible light may also be responsible on very rare occasions. As a result, paradoxically, any use by PMLE patients of mostly UVB-protective sunscreens, which tend preferentially to remove the more immunosuppressive UVB while simultaneously permitting the passage of UVA, may have a significant PMLE-enhancing effect as often reported by patients. However, most modern sunscreens protect against both wavebands.

The major predisposing factor to the tendency to PMLE seems very probably genetic, one study even suggesting approximately 70% of all subjects have a tendency to the condition (McGregor et al. 2000). Its actual expression then appears to depend on gene penetrance, although sufficient initial UVR exposure may also be important in predisposed subjects to induce enough putative antigen for initial rash development.

PMLE usually has its onset before age 30 years and affects females two to three times more frequently than males with a positive family history present in approximately a fifth of patients (McGregor et al. 2000).

It may affect all skin types and racial groups. The eruption typically recurs each spring, on sunny vacations or after sunbed use, often improving with continuing exposure. It may occur through window glass and also rarely in winter after UVR reflection from snow. Individual susceptibility varies, the period of continuing exposure needed to initiate the eruption usually being from about one half to several hours or often up to several days on vacations. Lesions may then appear hours to, rarely, days after sufficient exposure but usually in not less than approximately half an hour, although itching may develop sooner. Once exposure ceases, the lesions gradually disappear fully without scarring over one to several days, occasionally a week or two, rarely longer. In any given patient, eruption outbreaks tend always to affect the same exposed sites, generally symmetrically, although perhaps gradually spreading or receding overall over time. Only some exposed skin is affected in most patients, particularly that normally covered, with sparing of widespread areas possible. Juvenile spring eruption, affecting mainly boys in spring, is an apparent extreme example of this, manifesting itchy papules and vesicles of just the ear helices, although typical PMLE may sometimes coexist (Hawk 1996). Systemic PMLE symptoms are very rare but include malaise, nausea, and other variable sensations. A very few patients with PMLE may also develop lupus, and there is a higher chance than normal of prior PMLE in patients with lupus (Murphy and Hawk 1991). Once established, PMLE may often gradually improve over years and even clear fully in approximately 10% of patients (Jansen and Darvonen 1984).

PMLE has many clinical variants, all with a presumed similar pathogenesis and prognosis. Papular, papulovesicular, plaque, vesiculobullous, insect bite-like, and erythema multiforme-like forms have been described, while pruritus alone is rarely possible though not, apparently, a macular variant. The papular variant of large or small, separate or confluent lesions, generally in clusters, is most common, followed by the papulovesicular and plaque variants; the others are rare. An eczematous form has been claimed, but this is almost certainly either mild chronic actinic dermatitis (CAD) or photoexacerbated seborrheic or atopic eczema. Differing lesional variants may also occur at different skin sites in the same patient; diffuse facial erythema with swelling, for example, sometimes occurs with typical papular lesions at other sites. A last variant, a small papular PMLE generally sparing the face and occurring after several days' continuing exposure in summer, has been called benign summer light eruption in continental Europe (Thomas and Amblard 1988).

The histologic features (Hawk and Calonje 2005) of PMLE are characteristic but not pathognomonic and vary with the clinical presentation. There is normally a moderate to dense perivascular infiltrate of the upper and mid-dermis in all variants; this consists predominantly of T cells with occasional neutrophils and infrequent eosinophils. Other features may often include upper dermal and perivascular edema with endothelial cell swelling, while frequent epidermal changes include variable spongiosis and occasional dyskeratosis, exocytosis, and basal cell vacuolization.

Assessment of the circulating antinuclear and extractable nuclear antibody titers is advisable in all patients to exclude subacute cutaneous or other lupus and, if necessary, the red blood cell protoporphyrin concentration to exclude erythropoietic protoporphyria.

Cutaneous phototesting with an irradiation monochromator confirms photosensitivity in up to about half of cases but often does not discriminate from other photodermatoses. Photoprovocation testing with the solar simulator or other broadband sources, however, sometimes necessary over several successive days, often induces the typical eruption (Ortel et al. 1986) for biopsy if necessary if the clinical features themselves are not diagnostic. Figure 8.1 shows a photoprovocation test for PMLE.

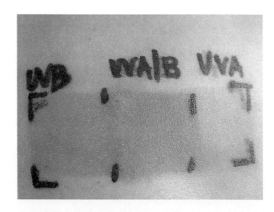

FIGURE 8.1 **(See color insert.)** Photoprovocation test for PMLE.

Mild PMLE is often controlled by reducing summer sun exposure; wearing appropriate clothing; and regularly applying broad-spectrum, high-protection sunscreens; mostly UVB sunscreens are often ineffective. Patients who develop PMLE only infrequently, such as on vacations, usually respond well to oral corticosteroids prescribed to be taken only if their eruption develops (Patel et al. 2000), approximately 25 mg prednis(ol)one to be taken at the very first sign of itching and then each morning until clear. This is regularly effective after, at most, several days, following which recurrence is relatively rare on the same vacation. Rare adverse effects of nausea, depression, or abdominal upset only occasionally necessitate stopping the drug. This treatment, if well tolerated, may be safely repeated every few months if required. More severely affected subjects suffering repeated PMLE throughout the summer may need courses of preventive low-dose photo(chemo)therapy (PUVA) in spring (Bilsland et al. 1993). This appears to be more effective than broadband UVB, controlling symptoms in up to 90% as compared with 60% of cases. However, narrowband 312 nm UVB phototherapy is now the usual treatment of choice, being simpler to administer and safer than PUVA, although the UVA doses are low anyway, with, at most, slightly reduced efficacy. Phototherapy may, on occasion, trigger the eruption, particularly in severely affected subjects, but brief administration of oral corticosteroid therapy usually ablates this. Multiple other therapies have also been tried but appear to be largely ineffective, including the traditional hydroxychloroquine (Murphy, Hawk, and Magnus 1987), which is perhaps occasionally useful; beta-carotene (Corbett et al. 1982), which is probably never effective; nicotinamide, which is also ineffective; and omega-3 polyunsaturated fatty acids, which are perhaps of, at most, moderate assistance in some patients. In the small proportion of patients then remaining who are unsuitable for, unable to tolerate, or not helped by any of these approaches, oral immunosuppressive therapy, usually intermittent, with azathioprine (Norris and Hawk 1989) or cyclosporine may be considered, if appropriate, and is generally helpful.

8.1.2 Actinic Prurigo

Actinic prurigo (AP) (Norris and Hawk 1999) appears to occur worldwide, but particularly affects native Americans at all latitudes. It appears sunlight-induced, being more severe in spring and summer, and, reasonably often, demonstrates abnormal skin phototest responses to the UVB or UVA wavelengths or both. In addition, sunlight exposure and solar-simulated radiation may sometimes induce a rash resembling PMLE, while a dermal, perivascular mononuclear cell infiltrate similar to that of PMLE may occur in early lesions. AP may therefore be a slowly evolving, excoriated form of PMLE and, thus, also a DTH reaction, a suggestion further supported by the fact that many AP patients have close relatives with PMLE (McGregor et al. 2000). In addition, human leukocyte antigen (HLA) DR4B1*0401 (DR4), present in some 30% of normal subjects, occurs in approximately 80%–90% of those with AP, while HLA DRB1*0407, present in some 6% of normal subjects and not infrequently in native Americans,

occurs in approximately 60% (Grabczynska et al. 1999), such that this inherited feature may well be responsible for converting PMLE into AP. In addition, some patients with the AP tissue type demonstrate clinical PMLE but also have persistent lesions, while some with clinical AP convert to clinical PMLE and some with clinical PMLE change to clinical AP (Grabczynska et al. 1999), all further suggesting a relationship between the two disorders. The cutaneous molecular UVR absorbers responsible for initiating the eruption are not known but may well be diverse as suggested for PMLE.

AP is more common in females and usually begins by age 10 years, often improving or resolving in adolescence, although persistence into adult life is possible (Norris and Hawk 1999). A positive family history of either AP or PMLE is present in approximately one fifth of patients (McGregor et al. 2000). The eruption is often present all year round, but it is generally worse in summer although, very rarely, in winter or both spring and fall; immunologic tolerance in these instances presumably develops during summer. Exacerbations tend to have onset gradually during fine weather rather than following specific sun exposure, although PMLE-like outbreaks are also possible.

Lesions are typically very pruritic, usually excoriated papules or nodules, sometimes associated with variable eczematization, lichenification, or crusting. All exposed areas are usually affected, particularly consistently uncovered sites, with gradual fading toward normally covered skin; the latter is also often mildly affected, particularly over the sacral area and buttocks. Cheilitis, particularly of the lower lip, and conjunctivitis are also possible, particularly in native Americans, while healed facial lesions may leave small, pitted, or linear shallow scars.

Early papular lesions show changes similar to those of PMLE, namely, mild acanthosis, exocytosis, and spongiosis in the epidermis and moderate lymphohistiocytic, dermal perivascular infiltration (Hawk and Calonje 2005), even very rarely suggesting lymphoma. In persistent lesions, however, excoriations, increasing acanthosis, variable lichenification, and dense mononuclear cell infiltration lead to a nonspecific appearance.

Assessment of the circulating antinuclear and extractable nuclear antibody titers should be undertaken to exclude subacute cutaneous or other lupus and particularly also the HLA type, HLA DRB1*0401 (DR4) or DRB1*0407, especially the latter, both supporting AP.

Cutaneous phototesting with the monochromator may confirm light sensitivity in up to approximately half of cases (Norris and Hawk 1999), but as in PMLE, it does not discriminate from other photodermatoses, and provocation testing with the solar simulator or other broadband sources may, on occasion, induce a PMLE-like eruption.

Restricted sun exposure and the use of broad-spectrum, high-protection factor sunscreens may help milder AP, assisted by intermittent topical and, perhaps rarely, oral steroids, while oral thalidomide, generally in low doses (50 to 200 mg at night) and preferably intermittently, is almost always effective for more resistant disease within weeks (Lovell et al. 1983). Generally mild adverse effects may include drowsiness, headache, constipation, and weight gain, while careful nerve conduction studies every few months are important to avoid a moderate, probably dose-related, risk of slowly progressive peripheral neuropathy. Pregnancy must also be rigorously avoided because of the high risk of teratogenicity. If thalidomide is unavailable or unsuitable, phototherapy with narrowband UVB or PUVA may occasionally help (Farr and Diffey 1989), perhaps more reliably if the skin has been cleared first with oral steroids. Topical tacrolimus or pimecrolimus may also, perhaps, help, on occasion, if the skin is again cleared first, and oral immunosuppressive therapy with azathioprine or cyclosporine may well also be useful if the other therapies are ineffective, unsuitable, or not tolerated.

8.1.3 Hydroa Vacciniforme

Hydroa vacciniforme (HV) apparently occurs worldwide as a very rare sunlight-induced blistering, often severely scarring condition (Norris and Hawk 1999). Its definitive pathogenesis is unknown, but some patients have reduced UVA minimal erythema doses, although the UVB reaction is normal in most patients (Sonnex and Hawk 1988). Blood, urine, and stool porphyrin concentrations are normal as are all other laboratory parameters, including circulating lupus titers. Nevertheless, the relationship of the eruption to sunlight exposure, its distribution, and its early clinical appearances are all very similar to those of PMLE, strongly suggesting a possible relationship with that disorder, although recent work has shown that the condition is often associated with Epstein–Barr virus (EBV) particles in skin, which may conceivably somehow become antigenic with sun exposure (Verneuil et al. 2010). This association may, in some way, make the fully developed HV eruption much more severe than that of PMLE, but definitive evidence is difficult to acquire because of its rarity and occurrence mainly in children. HV in Asian and Mexican patients is also associated with EBV infection, and many of these are even more severe and also linked with natural killer cell or T cell lymphoma, which is not, for some reason, the case in Europe and North America (Iwatsuki et al. 2006), although the reasons for this are under consideration (Cohen et al. 2009).

Most HV arises in early childhood and resolves spontaneously by puberty, although some patients suffer lifelong. Familial incidence is exceptional. The eruption typically occurs in summer (Creamer et al. 1998); a burning or stinging sensation is followed by the development of individual or confluent papules and then vesicles within hours of sun exposure, followed by their umbilication, crusting, and progression to permanent pock scarring over weeks. The cheeks and, to a lesser extent, other areas of the face are mostly affected, but somewhat also the backs of the hands and outer arms, generally symmetrically. It rarely affects other sites. In more detail, initial erythema, sometimes with swelling, followed by symmetrically scattered, usually tender papules within 24 h; vesiculation; occasionally confluent and hemorrhagic umbilication; then crusting and detachment of the lesions with permanent, depressed, hypopigmented scarring within weeks are usual. Oral ulcers and eye signs also rarely occur.

The histological changes are essentially pathognomonic and include intraepidermal vesicle formation with later focal epidermal keratinocyte necrosis and spongiosis in association with dermal perivascular neutrophil and lymphocyte infiltration (Hawk and Calonje 2005). Older lesions show necrosis, ulceration, and scarring. Blood, urine, and stool porphyrin concentrations should also be assessed to exclude cutaneous porphyria as should the circulating antinuclear factor and extractable nuclear antibody titers to exclude the unlikely possibility of cutaneous lupus. Phototesting may show reduced minimal erythema doses in the short-wavelength UVA in some cases, but it is not discriminatory from other photodermatoses. Solar simulated irradiation may also induce erythema at reduced irradiation doses or occasionally typical HV vesiculation. Viral studies to check for herpes or other viral disorders should be undertaken if clinically appropriate.

The treatment of HV consists of reduction of sun exposure and the use of high-protection broad-spectrum sunscreens. Antimalarials have been reported to help but have not proven effective in practice as is also the case for beta-carotene and omega-3 fatty acids. As for PMLE, prophylactic phototherapy with narrowband UVB or PUVA may be helpful, particularly the latter, but must be administered with care to avoid disease exacerbation (Jaschke and Honigsmann 1981). In addition, topical or intermittent oral steroids, topical calcineurin inhibitors, and, perhaps, oral immunosuppressive medication might be tried if clinically appropriate, but no treatment at all, apart from UVR avoidance and sometimes sunscreen use, has proven reliably effective.

8.1.4 Chronic Actinic Dermatitis

CAD, synonymous with the outdated terms actinic reticuloid, photosensitive eczema, photosensitivity dermatitis and actinic reticuloid syndrome and persistent light reaction (Figure 8.2), has been noted worldwide, although perhaps more commonly in temperate regions, and may affect all skin types (Menagé and Hawk 1999). It is likely that pathogenesis has gradually been clarified, although not fully, and detailed studies of its clinical, histologic, and immunohistochemical features all show it, precisely, to resemble the DTH reaction, allergic contact dermatitis (Norris et al. 1989b; Menagé et al. 1996), even in its severe pseudolymphomatous form, formerly called actinic reticuloid, in which the clinical and histological features are the same as those seen in long-standing allergic contact dermatitis (Orbaneja et al. 1976). CAD therefore appears similar, but in that it occurs just following irradiation in the absence of any contact allergen in place, the condition is presumably a reaction against a photoinduced endogenous skin antigen. If CAD is such a response, it must follow either direct absorptive or secondary oxidative molecular distortion to form an antigen, for which important support comes from the fact that albumin can become antigenic *in vitro* through photooxidation of its contained histidine (Kochevar and Harber 1977). There is no evidence for a genetic predisposition to CAD, but one stimulus for its development

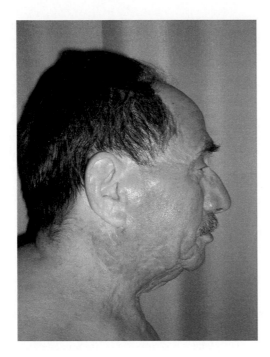

FIGURE 8.2 Chronic actinic dermatitis.

may also be the frequent presence of true allergic contact dermatitis, often airborne, to widespread exogenous sensitizers or photosensitizers (Menagé et al. 1995b), which may act as a predisposing factor through increasing skin immune activity to permit endogenous photoantigen recognition. Long-standing, prior endogenous eczema (Creamer et al. 1998), drug-induced photosensitivity, human immunodeficiency virus infection, or possibly PMLE may apparently have the same effect. On the other hand, chronic photodamage in constantly sun-exposed elderly outdoor enthusiasts, who most often develop CAD (Menagé and Hawk 1999), may, in addition or instead, decrease normal UVR-induced skin immunosuppression sufficiently for an endogenous UVR-induced photoantigen to be recognized as occurs genetically in PMLE. Assessment of the inducing action spectrum for CAD not possible with the other photodermatoses because of too much imprecision, can theoretically identify the postulated antigen, and this has been shown to resemble, in shape, that for sunburn in many patients (Menagé et al. 1995a), and photobiological theory suggests that the UVR absorber in such cases is the same as, or a substance associated with, that causing sunburn, namely, DNA (Menagé et al. 1995a) but, in this instance, acting as an antigen leading to eczema. In other CAD, however, the antigen must be different, since some patients reacting to just UVA and, perhaps, a very few to just 600 nm visible light. In spite of this strong circumstantial evidence, however, final proof that CAD is an UVR-induced endogenous antigenic process leading to a contact dermatitis-like reaction remains lacking.

CAD usually affects middle-aged or elderly men, or women slightly less often (Menagé et Hawk 1999), with occurrence under 50 years of age being unusual except in patients with prior atopic eczema (Creamer et al. 1998). The condition usually persists for years before not uncommon gradual resolution (Dawe et al. 2000). An association with cutaneous T cell lymphoma (CTCL) has been suggested on a few occasions, but T cell receptor, immunoglobulin gene rearrangement, and other similar studies in CAD have not suggested malignancy, and life expectancy is normal (Dawe et al. 2000). However, CTCL itself although CTCL itself may very rarely present with severe CAD-like light sensitivity, and careful investigation to exclude this is necessary if it is suspected (Agar et al. 2009). CAD is worse in summer and often also within minutes to hours of specific sun exposure with itching and a confluent, erythematous rash occasionally settling over days with scaling if exposure ceases and the condition is mild. The rash is usually patchy or confluent, and eczematous, whether chronic, subacute, or acute, with lichenification in more severe cases and, more rarely, scattered or widespread, erythematous, shiny, infiltrated, pseudolymphomatous papules or plaques arising on erythematous, eczematous, or normal skin. Exposed areas are most often affected, particularly the face, scalp, back and sides of neck, upper chest, outer aspects of forearms, and backs of hands, often with sharp cutoff at lines of clothing and sparing, particularly in the depths of skin creases, on the upper eyelids, in the finger webs, and on the skin behind the ear lobes. In severe disease, eczema of the palms and soles is also possible; eyebrow, eyelash, and scalp hair may be stubbly or lost from constant rubbing and scratching; and erythroderma, usually but not always, accentuated on exposed sites, may rarely supervene. Variable, sometimes geographical, sparing of exposed areas of the face or elsewhere, as well as irregular hyperpigmentation and hypopigmentation, sometimes vitiligo-like, may also occasionally occur.

Histologic features include epidermal spongiosis, acanthosis, and sometimes hyperplasia, along with predominantly perivascular lymphocytic cellular infiltration confined to the upper dermis with milder cases just demonstrating chronic eczema (Hawk and Calonje 2005). Severe CAD, however, may histologically mimic CTCL, on occasion being virtually or totally indistinguishable, with epidermal Pautrier-like microabscesses and deep, dense epidermotropic mononuclear cell infiltration, sometimes with hyperchromatic convoluted nuclei and giant cells but no marked increase in mitoses. T cell receptor gene rearrangement studies should be undertaken if there is any suspicion of lymphoma. Assessment of the circulating antinuclear and extractable nuclear antibody titers is advisable in all patients to exclude the unlikely possibility of subacute cutaneous or other lupus, and in severe or erythrodermic CAD, there may be large numbers of circulating CD8+ Sézary cells without other suggestion of malignancy (Chu et al. 1986). Patient HIV status should also be checked if there is suspicion that this may be predisposing to the CAD.

Phototesting is essential to confirm CAD and is virtually always characterized by low erythemal thresholds and eczematous or pseudolymphomatous responses characteristic of the disorder following irradiation with UVB, usually UVA, and rarely also the visible wavelengths, ideally from an irradiation monochromator or other monochromatic source (Menagé and Hawk 1999). A small number of patients react just to UVA, and a very few react to just 600 nm visible light. The uninvolved skin of the back free of topical and systemic steroid effects for, at least, the preceding few days should be used to avoid false negative results. Broad-spectrum studies are also generally positive, normally demonstrating early acute eczema. Patch and photopatch testing is also essential in suspected CAD; contact sensitivity to often airborne allergens, such as Compositae oleoresins, colophony, perfume mix, and, more recently, paraphenylenediamine (Chew et al. 2010) often closely resemble CAD or, more frequently, coexist with the condition. In addition, secondary contact or photocontact sensitivity to sunscreens or other applications may worsen the clinical picture.

The treatment of CAD is often difficult and not fully effective. Rigorous avoidance of UVR and exacerbating contact allergen exposure is essential, along with the regular application of high protection factor broad-spectrum topical sunscreens. Strong topical steroids are also often needed and frequently produce marked symptomatic relief without adverse effects even if continued, provided their use is confined to affected skin. Intermittent oral steroid use is often helpful, too, for CAD flares, while more resistant disease has responded on occasion to the topical calcineurin inhibitors tacrolimus, and pimecrolimus (Evans, Palmer, and Hawk 2004). However, for severe CAD, oral immunosuppressive therapy is almost always necessary and generally helpful if tolerated; azathioprine 1.5–2.5 mg/kg/day often achieves remission in months (Murphy et al. 1987), when it may be reduced in dose or discontinued, often for the winter, and cyclosporine 3.5–5 mg/kg/day also is usually effective. Mycophenolate mofetil is less often useful. Finally, long-term, low-dose phototherapy with PUVA, usually several times weekly initially followed by maintenance exposures every 3 weeks or so, may help (Hindson et al. 1990), generally under initial oral and topical steroid cover to avoid disease flare.

8.1.5 Solar Urticaria

Solar urticaria (SU) probably occurs worldwide in perhaps 3 per 100,000 of the population (Beattie et al. 2003). Primary SU is an immediate type 1 hypersensitivity response against a cutaneous or circulating photoallergen generated from a precursor following UVR or visible light absorption. Both circulating photoallergen and reaginic antibodies have been demonstrated. Very rarely, secondary SU may occur in association with drug photosensitivity, cutaneous porphyria, or lupus. There appears to be no genetic basis for the condition. Two forms of primary SU have been proposed: type 1 being an IgE-mediated hypersensitivity against specific photoallergens generated only in SU patients, and type 2 being an IgE-mediated hypersensitivity against a nonspecific photoallergen generated in both SU patients and normal subjects (Leenutaphong et al. 1989). A wide range of possible inducing wavelengths is presumably attributable to different photoallergens; patients with type 1 appear to have photoallergens of molecular mass 25 to 34 kDa and an action spectrum in the visible region, and those with type 2 photoallergens of molecular mass 25 to 1000 kDa and a variable action spectrum.

Primary SU is slightly more common in females and may arise at any age; the first attack is sometimes following marked sunlight or occasionally sunbed exposure (Beattie et al. 2003). Typically, 5 to 10 or, rarely, 20 to 30 minutes' exposure leads to itching erythema followed by patchy or confluent urticarial whealing with gradual resolution over a maximum of 1 to 2 h. Very rarely, just itching, perhaps, occurs; or symptom onset may be delayed for up to several hours; or just scattered fixed sites may be affected. With extensive whealing, patients may describe headache, nausea, bronchospasm, fainting, and syncope, rarely life-threatening. Secondary SU should be excluded by ruling out drug photosensitivity, cutaneous porphyria, and lupus. SU often persists indefinitely, sometimes with deterioration but also sometimes with improvement, there being a probability of clinical resolution at 5 and 10 years of 12% and 26%, respectively.

SU usually affects all exposed skin with an initial macular erythema, generally, but not always, followed rapidly by separate or confluent whealing, often with clear demarcation at lines of clothing. Rarely, there is sparing of the face and hands as a result of UVR-induced tolerance, or just fixed sites may be affected.

Histologically, there is dermal vasodilation, edema, and, predominantly, perivascular neutrophil and eosinophil infiltration at 5 min and 2 but not 24 h (Hawk and Calonje 2005). Endothelial cell swelling occurs early on with mononuclear cell infiltration occurring later at higher irradiation doses. Extensive eosinophil granule major basic protein deposition is also present in the dermis at 2 and 24 h, suggesting eosinophil degranulation. The circulating antinuclear factor and extractable nuclear antibody titers should be assessed to exclude cutaneous lupus as should the blood, urine, and stool porphyrin concentrations to exclude cutaneous porphyria, both very rare accompaniments of SU. Irradiation monochromator, broad-spectrum or sunlight phototesting generally allows confirmation of the diagnosis of SU and definition of the inducing wavelengths. However, negative phototests do not necessarily exclude the disorder as the ease of SU induction, particularly in mild cases, may vary. Broadband sources, including sunlight, may be used instead, and minimal urticarial dose estimation may help the assessment of treatment efficacy. Figure 8.3 shows minimal urticarial dose assessment.

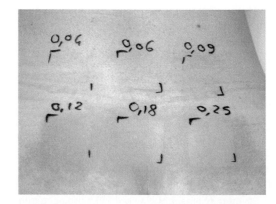

FIGURE 8.3   (See color insert.) Minimal urticarial dose assessment with UVB.

Restriction of sun exposure; high-protection, broad-spectrum sunscreen use; and appropriate clothing cover may be helpful for UVA-sensitive but, generally, not visible light-induced SU in which dark clothing may be helpful. Nonsedating, often higher-than-normal dose H1 antihistamines, best taken an hour or so before exposure, are very effective in approximately one third of patients and partially in a further third. In patients who develop SU tolerance as summer advances, prophylactic phototherapy may be helpful and also sometimes in persistent disease, although the therapy then generally needs to be continued to maintain efficacy with a consequent risk of long-term adverse effects. Phototherapy should be undertaken with extreme care early on to avoid the risk of anaphylaxis, particularly in severely affected subjects. Refractory patients may respond to the rather unpleasant plasma exchange, or plasmapheresis, particularly if shown to have a circulating SU-associated serum factor by its intradermal injection after irradiation beforehand; remissions in some cases are long-lived (Bissonnette et al. 1999). Intravenous immunoglobulin has been reported as helpful by several research groups (Darras et al. 2004), although it is not always so, and oral cyclosporine may also, rarely, be of use. Some patients, however, respond poorly to all therapy.

References

Agar, N., S. Morris, R. Russell-Jones, J. Hawk, and S. Whittaker. 2009. Case report of four patients with erythrodermic cutaneous T-cell lymphoma and severe photosensitivity mimicking chronic actinic dermatitis. *Br J Dermatol* 160:698.

Beattie, P., R. S. Dawe, S. H. Ibbotson, and J. Ferguson. 2003. Characteristics and prognosis of idiopathic solar urticaria: A cohort of 87 cases. *Arch Dermatol* 139:1149.

Bilsland, D., S. A. George, N. K. Gibbs et al. 1993. A comparison of narrow band phototherapy (TL-01) and photochemotherapy (PUVA) in the management of polymorphic light eruption. *Br J Dermatol* 129:708.

Bissonnette, R., N. Buskard, D. I. McLean, and H. Lui. 1999. Treatment of refractory solar urticaria with plasma exchange. *J Cutan Med Surg* 3:236.

Chew, A. L., S. J. Bashir, J. L. Hawk et al. 2010. Contact and photocontact sensitization in chronic actinic dermatitis: A changing picture. *Contact Dermatitis* 62:42.

Chu, A. C., D. Robinson, J. L. Hawk et al. 1986. Immunologic differentiation of the Sézary syndrome due to cutaneous T-cell lymphoma and chronic actinic dermatitis. *J Invest Dermatol* 86:134.

Cohen, J., H. Kimura, S. Nakamura, Y. H. Ko, and E. S. Jaffe. 2009. Epstein–Barr virus-associated lymphoproliferative disease in non-immunocompromised hosts: A status report and summary of an international meeting, 8–9 September 2008. *Ann Oncol* 20:1472.

Corbett, M. F., J. L. Hawk, A. Herxheimer, and I. A. Magnus. 1982. Controlled therapeutic trials in polymorphous light eruption. *Br J Dermatol* 107:571.

Creamer, D., J. M. McGregor, and J. L. Hawk. 1998. Chronic actinic dermatitis occurring in young patients with atopic dermatitis. *Br J Dermatol* 139:1112.

Darras, S., M. Ségard, L. Mortier, A. Bonnevalle, and P. Thomas. 2004. Treatment of solar urticaria by intravenous immunoglobulins and PUVA therapy. *Ann Dermatol Venereol* 131:65.

Dawe, R. S., I. K. Crombie, and J. Ferguson. 2000. The natural history of chronic actinic dermatitis. *Arch Derm* 136:1215.

Evans, A. V., R. A. Palmer, and J. L. M. Hawk. 2004. Erythrodermic chronic actinic dermatitis responding only to topical tacrolimus. *Photodermatol Photoimmunol Photomed* 20:59.

Farr, P. M., and B. L. Diffey. 1989. Treatment of actinic prurigo with PUVA: Mechanism of action. *Br J Dermatol* 120:411.

Gonzales-Amaro, R., L. Baranda, J. F. Salazar-Gonzalez, C. Abud-Mendoza, and B. Moncada. 1991. Immune sensitization against epidermal antigen in polymorphous light eruption. *J Am Acad Dermatol* 24:70.

Grabczynska, S. A., J. M. McGregor, E. Kondeatis, R. W. Vaughan, and J. L. M. Hawk. 1999. Actinic prurigo and polymorphic light eruption: Common pathogenesis and the importance of HLA-DR4/DRB1*0407. *Br J Dermatol* 140:232.

Hawk, J. 1996. Juvenile spring eruption is a variant of polymorphic light eruption. *N Z Med J* 109:389.

Hawk, J. L. M., and E. Calonje. 2005. The photosensitivity disorders. In *Lever's Histopathology of the Skin*. 9th edition. D. E. Elder, R. Elenitsas, B. L. Johnson Jr., and G. F. Murphy, editors. Lippincott, Williams and Wilkins, Philadelphia, 345.

Hindson, C., A. Downey, S. Sinclair, and B. Cominos. 1990. PUVA therapy of chronic actinic dermatitis: A 5 year follow-up. *Br J Dermatol* 123:273.

Iwatsuki, K., M. Satoh, T. Yamamoto et al. 2006. Pathogenic link between hydroa vacciniforme and Epstein–Barr virus associated haematologic disorders. *Arch Dermatol* 142:587.

Jansen, C. T., and J. Darvonen. 1984. Polymorphous light eruption. A seven-year follow-up evaluation of 114 patients. *Arch Dermatol* 120:862.

Jaschke, E., and H. Honigsmann. 1981. Hydroa vacciniforme—Aktionsspektrum, UV-Toleranz nach Photochemotherapie. *Hautarzt* 32:350.

Kochevar, I. E., and L. C. Harber. 1977. Photoreactions of 3,3′,4′,5-tetrachlorosalicylanilide with proteins. *J Invest Dermatol* 68:151.

Leenutaphong, V., E. Hölzle, and G. Plewig. 1989. Pathogenesis and classification of solar urticaria: A new concept. *J Am Acad Dermatol* 21:237.

Lovell, C., J. L. Hawk, C. D. Calnan, and I. A. Magnus. 1983. Thalidomide in actinic prurigo. *Br J Dermatol* 108:467.

McGregor, J. M., S. Grabczynska, R. Vaughan, J. L. Hawk, and C. M. Lewis. 2000. Genetic modeling of abnormal photosensitivity in families with polymorphic light eruption and actinic prurigo. *J Invest Dermatol* 115:471.

Menagé H. duP., and J. L. M. Hawk. 1999. The idiopathic photodermatoses: Chronic actinic dermatitis (photosensitivity dermatitis/actinic reticuloid syndrome). In *Photodermatology*. J. L. M. Hawk, editor. Arnold, London, 127.

Menagé H. duP., G. I. Harrison, C. S. Potten, A. R. Young, and J. L. Hawk. 1995a. The action spectrum for induction of chronic actinic dermatitis is similar to that for sunburn inflammation. *Photochem Photobiol* 62:976.

Menagé H. duP., J. S. Ross, P. G. Norris, J. L. Hawk, and I. R. White. 1995b. Contact and photocontact sensitization in chronic actinic dermatitis: Sesquiterpene lactone mix is an important allergen. *Br J Dermatol* 132:543.

Menagé H. duP., N. K. Sattar, D. O. Haskard, J. L. Hawk, and S. M. Breathnach. 1996. A study of the kinetics and pattern of adhesion molecule expression in induced lesions of chronic actinic dermatitis. *Br J Dermatol* 134:262.

Miyamoto, C. 1989. Polymorphous light eruption: Successful reproduction of skin lesions, including papulovesicular light eruption, with ultraviolet B. *Photodermatology* 6:69.

Murphy, G. M., and J. L. M. Hawk. 1991. The prevalence of antinuclear antibodies in patients with apparent polymorphic light eruption. *Br J Dermatol* 125:448.

Murphy, G. M., J. L. Hawk, and I. A. Magnus. 1987. Hydroxychloroquine in polymorphic light eruption: A controlled trial with drug and visual sensitivity monitoring. *Br J Dermatol* 116:379.

Murphy, G. M., P. D. Maurice, P. G. Norris, R. W. Morris, and J. L. Hawk. 1987. A double-blind controlled trial of azathioprine in chronic actinic dermatitis. *Br J Dermatol* 117:16.

Norris, P. G., J. N. Barker, M. H. Allen et al. 1992. Adhesion molecule expression in polymorphic light eruption. *J Invest Dermatol* 99:104.

Norris, P. G., and J. L. M. Hawk. 1989. Successful treatment of severe polymorphic light eruption with azathioprine. *Arch Dermatol* 125:1377.

Norris, P., and J. Hawk. 1999. The idiopathic photodermatoses: Polymorphic light eruption, actinic prurigo and hydroa vacciniforme. In *Photodermatology*. J. Hawk, editor. Arnold, London.

Norris, P., J. Morris, D. M. McGibbon, A. C. Chu, and J. L. Hawk. 1989a. Polymorphic light eruption: An immunopathological study of evolving lesions. *Br J Dermatol* 120:173.

Norris, P. G., J. Morris, N. P. Smith, A. C. Chu, and J. L. Hawk. 1989b. Chronic actinic dermatitis: An immunohistological and photobiological study. *J Am Acad Dermatol* 21:966.

Orbaneja, J. G., L. I. Diez, J. L. Lozano, and L. C. Salazar. 1976. Lymphomatoid contact dermatitis. A syndrome produced by epicutaneous hypersensitivity with clinical features and a histopathologic picture similar to that of mycosis fungoides. *Contact Dermatitis* 2:139.

Ortel, B., A. Tanew, K. Wolff, and H. Hönigsmann. 1986. Polymorphous light eruption: Action spectrum and photoprotection. *J Am Acad Dermatol* 14:748.

Palmer, R. A., J. L. M. Hawk, A. R. Young, and S. L. Walker. 2005. The effect of solar-simulated radiation on the elicitation phase of contact hypersensitivity does not differ between controls and patients with polymorphic light eruption. *J Invest Dermatol* 124:1308.

Pao, C., P. G. Norris, M. Corbett, and J. L. Hawk. 1994. Polymorphic light eruption: Prevalence in Australia and England. *Br J Dermatol* 130:62.

Patel, D. C., G. J. Bellaney, P. T. Seed, J. M. McGregor, and J. L. Hawk. 2000. Efficacy of short-course oral prednisolone in polymorphic light eruption: A randomized controlled trial. *Br J Dermatol* 143:828.

Rhodes, L. E., M. Bock, A. S. Janssens et al. 2010. Polymorphic light eruption occurs in 18% of Europeans and does not show higher prevalence with increasing latitude: Multicenter survey of 6,895 individuals residing from the Mediterranean to Scandinavia. *J Invest Dermatol* 130:626.

Sonnex, T. S., and J. L. M. Hawk. 1988. Hydroa vacciniforme: A review of ten cases. *Br J Dermatol* 118:101.

Thomas, P., and P. Amblard. 1988. Lucite estivale bénigne. In *Photodermatologie et Photothérapie*. Masson, Paris, 49.

van de Pas, C. B., D. A. Kelly, P. T. Seed et al. 2004. Ultraviolet-radiation-induced erythema and suppression of contact hypersensitivity responses in patients with polymorphic light eruption. *J Invest Dermatol* 122:295.

Verneuil, L., S. Gouarin, F. Comoz et al. 2010. Epstein–Barr virus involvement in the pathogenesis of hydroa vacciniforme: An assessment of seven adult patients with long-term follow-up. *Br J Dermatol* 163:174.

9

Photoaggravated Dermatoses

PierGiacomo
Calzavara-Pinton
University of Brescia

Raffaella Sala
University Hospital Spedali Civili

9.1 Introduction

Many diseases that are affected by exposure to ultraviolet radiation (UVR) are not actually caused by UVR. Many inflammatory diseases improve or clear following UVR exposure. This chapter will encompass diseases not primarily caused by UVR but usually or occasionally exacerbated by it. Aggravation or worsening of diseases by UVR may occur for a variety of different mechanisms. The disease may be an inflammatory disease, and in some individuals, UV exposure may just add to inflammation. The disease may be complicated by coincidence with a sun-induced inflammatory disorder and appear to be aggravated by UV exposure. For the disease to be truly photoaggravated, it should show exacerbation with primary lesions of the disease itself. In this situation, the disorder may actually be induced by UV exposure. A moderate number of dermatoses that are not actually induced by UVR may be worsened by it (Table 9.1). The immunological activity involved in their disease mechanism may be increased by UVR, or (more rarely) any already-present inflammation can be increased by UVR.

9.2 Autoimmune Photoaggravated Dermatoses

9.2.1 Systemic Lupus Erythematosus

Skin lesions are found in up to 90% of patients with lupus erythematosus (LE). Two clinical variants, acute cutaneous LE (ACLE) and subacute cutaneous LE (SCLE), are almost always associated with systemic involvement. Generalized ACLE presents with a macular and papular eruption on the whole skin surface, and it is frequently referred to as the systemic LE (SLE) rash. Localized ACLE is a rash that extends across the cheeks of the face and the bridge of the nose and looks like a butterfly. It can be flat or raised, and the color ranges from bright red to light pink. SCLE presents with scaly erythematous papules that evolve as polycyclic annular lesions or psoriasiform plaques. ACLE and SCLE are asymptomatic and nonscarring.

Other cutaneous LE manifestations can exist with or without skin systemic manifestations. Discoid LE (DLE), also known as chronic cutaneous LE (CCLE), is coin-shaped or oval in shape, like a disk, on the face, scalp, and upper trunk. It is erythematous, infiltrated, and scaly, and the skin around the lesion is either lighter or darker in color. When DLE heals, it can leave behind a scar. DLE lesions are usually painless and typically do not itch. Lupus tumidus (LT) presents with elevated areas of red skin with no scale or scarring, and Lupus profundus presents with discoid skin lesions in conjunction with panniculitis. Chilblains lupus is associated with itchy, cold sensations in the extremities and toes as well as painful, dark red swelling.

It has been known for a long time that there is a clear relationship between sunlight exposure and cutaneous LE manifestations. Lesions tend to occur in sun-exposed skin, and sunlight can induce new skin eruptions, exacerbate existing lesions, cause progression of the disease to non-UV-exposed areas, and, in rare cases, even induce systemic symptoms, such as weakness, fatigue, and joint pain.

Photosensitivity by the definition of the American College of Rheumatology (ACR)—"a rash as a result of an unusual reaction to sunlight by patient history or physician observation" (Hochberg 1997)—occurs in up to 80%–90% of patients

TABLE 9.1 Diseases Sometimes Aggravated by Ultraviolet Irradiation (the Commonest in Bold Type)

Acne

Atopic eczema

Carcinoid syndrome

Cutaneous T-cell lymphoma

Dermatomyositis

Disseminated superficial actinic porokeratosis

Erythema multiforme

Familial benign chronic pemphigus (Hailey–Hailey disease)

Keratosis follicularis (Darier's disease)

Lichen planus

Lupus erythematosus

Pellagra

Pemphigus foliaceus (erythematosus)

Pityriasis rubra pilaris

Psoriasis

Reticulate erythematous mucinosis syndrome

Rosacea

Seborrheic eczema

Transient acantholytic dermatosis (Grover's disease)

Viral infections

with certain cutaneous subsets, such as ACLE, SCLE, and LT. Photosensitivity is less common in other cutaneous LE subsets, such as DLE, in which it approaches only 50%, and lupus panniculitis.

However, the incidence of photosensitivity in LE could be overestimated because the ACR definition is extremely broad, and it can be fulfilled by a variety of other diseases, such as PMLE, photoallergic contact dermatitis, and dermatomyositis (Doria et al. 1996).

At the same time, the incidence of photoaggravated LE could be underestimated as many patients are not aware of the link with sun exposure because their disease develops after a delay of several days or even weeks and may persist for months.

A complete understanding of the diverse pathophysiological mechanisms of photoinduced LE lesions does not exist.

UV light is a potent inducer of apoptosis of keratinocytes, the so-called "sunburn cells" and lymphocytes. A variety of signaling pathways are involved with a waveband dependency. Apoptosis by UVA is mediated by death receptor activation and formation of reactive oxygen species. The same mechanisms, together with apoptotic pathways by direct DNA damage, are activated by UVB.

Apoptosis is characterized by a well-defined orderly process of nuclear condensation, surface blebbing, cytoplasmic contraction, and packaging of cellular components within membranes before their budding from the apoptotic cell as apoptotic bodies. Apoptotic cells express nuclear and cytoplasmic antigens on their outer surface, which are targeted by a wide array of autoantibodies. Serum autoantibodies against SSA/Ro and SSB/La antigens are very frequently correlated with photosensitivity (Sontheimer 1996).

Once present, autoantibodies may impair or delay the removal of apoptotic cells, and their accumulation is considered one of the major factors that may break tolerance leading to overt inflammatory skin lesions in patients with established SLE.

Beside autoantibodies, other pathogenetic factors may have a key pathogenetic role in the UV-mediated induction of cutaneous inflammatory lesions in SLE patients.

Release of RNA and DNA from apoptotic cells induces the production of interferon-α (IFNα) by plasmocytoid dendritic cells (PDCs) (Vermi et al. 2009).

IFNα is a potent and rapid inducer of CXCR3 ligands on keratinocytes and dermal fibroblasts. PDCs abundantly express the CXCR3 receptor on their surface and subsequently accumulate in LE lesions, and their activation results in subsequent release of effector cytokines, which amplifies and perpetuates the recruitment of leukocytes and the release of chemokines (Vermi et al. 2009).

The decrease in the suppressive function of peripheral blood CD4(+) CD25 (high) Tregs may play a key role in the pathogenesis of LE l photosensitive skin lesions E (Wolf, Byrne, and Gruber-Wackernagel 2009).

Specific phototest protocols with multiple (most commonly three) daily exposures to UVA (60–100 J/cm²) and UVB (1.5 MED) wavebands can reproduce skin lesions in a variable ratio of patients according to the various subtypes of LE: 45% DLE, 76% LET, 63% SCLE, 41% hypertrophic LE, lupus panniculitis, and chilblain lupus, and 60% SLE (Kuhn et al. 2001). Interestingly, the development of UV light–induced skin lesions is considerably slower, up to 2–3 weeks, and the persistence is longer than in other photodermatoses, such as PLE (Kuhn et al. 2001). In addition to lesion induction, a decreased threshold for the induction of erythema has been described. Thus, patients with either systemic or cutaneous lupus develop prolonged skin redness at lower doses of UV light than normal controls (Lehmann et al. 1990).

Of interest, results of phototests often correlate poorly with a personal history of photosensitivity (Kuhn et al. 2001).

Some authors have suggested that PLE and LE share a common pathogenesis (Millard et al. 2001; Nyberg et al. 1997). Millard et al. (2001) reported a high prevalence of PLE in LE patients as well as clustering of PLE among first-degree relatives of patients with SCLE and DLE. Nyberg et al. (1997) studied 337 lupus patients and observed that the prevalence of PLE in this group greatly exceeded the prevalence of PLE in the general population at that geographical latitude. PLE has been described as a presenting symptom in patients with LE and thus might be considered a predisposing factor for LE in some PLE patients. Additionally, several studies have demonstrated the presence of elevated antinuclear antibody (ANA) titers in the absence of other apparent LE symptoms in some PLE patients (Jansen and Karvonen 1984).

Treatment of SLE is guided by the individual patient's manifestations. Avoiding sunlight is the primary change to the lifestyle of SLE sufferers as sunlight is known to exacerbate the disease, as is the debilitating effect of intense fatigue. Fever, rash, musculoskeletal manifestations, and serositis generally respond to treatment with hydroxychloroquine, nonsteroidal anti-inflammatory drugs (NSAIDs), and low- to moderate-dose steroids, as necessary, for acute flares. Medications, such as methotrexate, may be useful in chronic lupus arthritis, and azathioprine and mycophenolate have been widely used in moderate-severity lupus. The central nervous system (CNS) involvement and renal disease constitute more serious disease and often require high-dose steroids and other immunosuppression agents, such as cyclophosphamide, azathioprine, or mycophenolate. Monitoring of levels of anti-dsDNA and treatment with steroids as soon as there is a significant rise in this marker prevents relapse in most cases without increasing the cumulative dose of steroid given.

9.2.2 Dermatomyositis

Dermatomyositis (DM) is an idiopathic inflammatory myopathy with a pathognomonic persistent violaceous to dusky erythematous rash with a variable degree of edema and scaling in a symmetrical distribution involving the periorbital skin (heliotrope rash) (Dourmishev, Meffert, and Piazena 2004), malar erythema, and poikiloderma of the dorsal skin of the arms, neck, upper back (shawl sign), and/or the upper-lateral thighs (holster sign). All these lesions are localized in sun-exposed skin, and sun exposure is also related to the exacerbation of the muscle disease and other systemic manifestations. Photosensitivity in DM is directly related to the presence of anti-Mi-2 autoantibodies (Callen and Wortmann 2006). However, the development or the worsening of skin lesions may be delayed following solar exposure, and therefore, patients with DM rarely complain of photosensitivity (Cheong et al. 1994). Phototesting has found a decreased erythematous threshold to UVB in approximately half of the patients (Dourmishev, Meffert, and Piazena 2004). However, a standardized provocative phototest is not available so far.

Indeed, other clinical features, for example, Gottron' papules, periungual and cuticular changes, and alopecia, do not seem related to sun exposure. Unfortunately, the action spectrum of the disease has not been assessed with phototesting so far.

Treatment commonly involves a steroid drug called prednisone. For patients in whom prednisone is not effective, other immunosuppressants such as azathioprine and methotrexate may be prescribed. Recently, a drug called intravenous immunoglobulin was shown to be effective and safe in the treatment of the disease. Physical therapy is usually recommended to preserve muscle function and avoid muscle atrophy. Most cases of dermatomyositis respond to therapy. The disease is usually more severe and resistant to therapy in patients with cardiac or pulmonary problems. Removal of an underlying tumor may produce some improvement of the muscle weakness in adult dermatomyositis, although this is often only temporary. Systemic steroids may provide relief—prednisolone is the drug of choice—initially at 20–60 mg/day and then reduced to maintenance doses. Systemic azathioprine at 150 mg/day may be used alone or as a steroid-sparing agent. Physical rest during active disease should be considered because bed rest produces a significant fall in enzyme levels. Methotrexate may be tried if corticosteroids are unsuccessful.

9.3 Seborrheic Dermatitis

Seborrheic dermatitis is a recurrent, chronic inflammation of the skin that occurs on sebum-rich areas, such as the face, scalp, and chest, and is characterized by red, scaly lesions. It often occurs together with rosacea and may represent a consequence of failure to adequately police epidermal flora because of an abnormal stratum corneum (in atopy) or impaired immune responsiveness in the immunosuppressed. In an increasing proportion of individuals, it is now recognized to flare with sun exposure, failing to clear up as is the norm. Light-testing these individuals shows some with normal responses, as in rosacea, but a proportion with abnormal responses. Treatment of the basic disorder eradicates the UV-induced flare (Palmer and Hawk 2004). Photosensitive seborrheic dermatitis may occur in patients taking immunosuppressive medication and with HIV disease; abnormal phototest responses were found in one patient with monochromator test results showing UVB and UVA photosensitivity.

Topical therapies are the main treatments as the condition is recurrent, usually mild, and responds well to these agents. Antifungal agents (Faergemann, Borgers, and Degreef 2007), topical steroids, and topical tacrolimus prevent the photoaggravation.

9.4 Pellagra

Pellagra is a nutritional disease caused by the deficiency of niacin characterized by a photodistributed rash, gastrointestinal symptoms, and neuropsychiatric disturbances (Karthikeyan and Thappa 2002). Apart from nutritional deficiency, chronic alcoholism, drugs, and carcinoid syndrome can cause pellagra. Niacin can be obtained directly from diet or is synthesized from tryptophan. Dietary niacin is mainly in the form of nicotinamide adenine nucleotide (NAD) and nicotinamide adenine dinucleotide phosphate (NADPH). These molecules undergo hydrolysis in the intestinal lumen to form nicotinamide, which can be converted to nicotinic acid by intestinal bacteria or absorbed directly into the bloodstream (Nogueira et al. 2009). Nicotinamide and nicotinic acid are then reincorporated as a component of coenzymes NAD and NADP, which, in turn, intervene in essential oxidation–reduction reactions. Tissues with high energy requirements, such as the brain, or high turnover rates, such as the gut or skin, are primarily affected. The exquisite photosensivity seen in pellagra may result from a deficiency of urocanic acid and/or cutaneous accumulation of kynurenic acid, which may induce a phototoxic reaction (Wan, Moat, and Anstey 2011). Clinical studies on phototesting in pellagra have yet to establish the action spectrum for this photosensitivity disorder, but preliminary data in a single patient suggest that this will be within the UVA spectrum (Wan, Moat, and Anstey 2011). The skin eruption is characteristically a photosensitive rash affecting the dorsal surfaces of the hands, face, neck, arms, and feet. In the acute phase, it resembles sunburn with erythema and bullae (wet pellagra) but progresses to a chronic, symmetric, scaly rash that exacerbates following re-exposure to sunlight. The Casal necklace extends as a fairly broadband or collar around the entire neck (cervical dermatome with C3 and C4 innervation). The other characteristic features are diarrhea and progressive dementia (Wan, Moat, and Anstey 2011).

The treatment of pellagra is with oral nicotinamide supplementation, 100–300 mg daily in three to four separate doses, until resolution of the major acute symptoms occurs. The dosage can then be reduced to 50 mg every 8–12 h until the skin lesions heal. Resolution of dermatitis occurs in 3–4 weeks (Srinivas, Sekar, and Jayashree 2012).

9.5 Disseminated Superficial Actinic Porokeratosis

Disseminated superficial actinic porokeratosis (DSAP) is an inherited skin condition causing dry patches mainly on the arms and legs.

DSAP is a special type of inherited sunspot and is sometimes confused with solar keratoses, but solar keratoses are more likely to arise on the face and hands.

The tendency to DSAP is inherited as an autosomal dominant characteristic, which means, on average, half of the children of an affected parent will also have the tendency. However a certain amount of accumulated sun exposure and perhaps other factors, such as immunosuppression, are needed to bring this tendency out.

DSAP is an uncommon skin condition that leads to reddish-brown, scaly spots. The spots are mostly seen on the arms and legs but sometimes will show up on other sun-damaged skin. It is a result of an abnormal sun sensitivity leading to precancerous skin cells. It is not a serious condition.

Most cases are inherited, but some occur in people whose immune systems are not working well. DSAP will only show up after sun damage has already occurred, so it is usually seen only in fair-skinned people midlife and beyond. For some reason, mostly women are affected. Once a spot of DSAP forms, it may slowly enlarge to form a ring or circle. The spots seem to grow or itch after sun exposure.

Once DSAP has been diagnosed, the best thing one can do is avoid further sun damage by wearing long sleeves and using strong sunscreens. Unfortunately, treatment of DSAP is not very satisfactory. Creams such as Retin-A, Tazorac, Efudex, and Aldara offer some slight help. Cryosurgery may be used but can lead to areas of hypopigmentation. Photodynamic therapy has been used with mixed results.

9.6 Rosacea

Rosacea is a chronic condition characterized by facial erythema (redness). Pimples are sometimes included as part of the definition. Unless it affects the eyes, it is typically a harmless cosmetic condition. Rosacea affects both sexes, but is almost three times more common in women. It has a peak age of onset between 30 and 60. Rosacea typically begins as redness on the central face across the cheeks, nose, or forehead but can also, less commonly, affect the neck, chest, ears, and scalp. In some cases, additional symptoms, such as semipermanent redness; telangiectasia (dilation of superficial blood vessels on the face); red-domed papules (small bumps) and pustules; red, gritty eyes; burning and stinging sensations; and, in some advanced cases, a red, lobulated nose (rhinophyma), may develop.

Triggers that cause episodes of flushing and blushing play a part in the development of rosacea. Exposure to temperature extremes can cause the face to become flushed as well as strenuous exercise, heat from sunlight, severe sunburn, stress, anxiety, cold wind, and moving to a warm or hot environment from a cold one, such as heated shops and offices during the winter.

Treatment in the form of topical steroids can aggravate the condition. Treating rosacea varies depending on severity and subtypes. A subtype-directed approach to treating rosacea patients is recommended to dermatologists. Mild cases are often not treated at all or are simply covered up with normal cosmetics. Therapy for the treatment of rosacea is not curative and is best measured in terms of a reduction in the amount of erythema and inflammatory lesions; a decrease in the number, duration, and intensity of flares; and concomitant symptoms of itching, burning, and tenderness. The two primary modalities of rosacea treatment are topical and oral antibiotic agents. While medications often produce a temporary remission of redness within a few weeks, the redness typically returns shortly after treatment is suspended. Long-term treatment, usually 1 to 2 years, may result in permanent control of the condition for some patients. Lifelong treatment is often necessary, although some cases resolve after a while and go into a permanent remission.

9.7 Darier Disease

Darier disease is also known as "keratosis follicularis." It is a rare genetic disorder that is manifested predominantly by skin changes. Onset of skin changes is usually in adolescence, and the disease is usually chronic. It is inherited in an autosomal dominant pattern, which means that a single gene passed from one parent causes the condition. The chance of a child inheriting the abnormal gene if one parent is affected is 1 in 2 (50%), but not all people with the abnormal gene will develop symptoms of the disease.

The abnormal gene in Darier disease has been identified as *ATP2A2*, found on chromosome 12q23-24.1 (Craddock et al. 1993). This gene codes for the sarcoendoplasmic reticulum calcium–ATPase (SERCA) enzyme or pump, which is required to transport calcium within the cell. The exact mechanism by which this abnormal gene causes the disease is still under investigation, but it appears that the way in which skin cells join together may be disrupted. The skin cells (keratinocytes) stick together via structures called desmosomes, and it seems the desmosomes do not assemble properly if there is insufficient calcium.

Treatment is required only if there are troublesome symptoms. For patients with mild disease, simple moisturizers, sun protection, and selection of the right clothing to avoid heat and sweating are usually sufficient, but avoidance of excessive heat, humidity, stress, and tight-fitting clothes (and general cleanliness) is advised. Treatment of choice for severe cases is oral retinoids. During flares, topical or oral antibiotics may be administered. Cyclosporin and prescription-only topical corticosteroids, for example, betamethasone, have been used during acute flares. Some patients are able to prevent flares with use of topical sunscreens and oral vitamin C.

9.8 Hailey–Hailey Disease

Hailey–Hailey disease, sometimes called "familial benign chronic pemphigus,'" is a rare hereditary blistering skin disease. The Hailey brothers first described it in 1939 (Hailey 1939).

Hailey–Hailey disease usually appears in the third or fourth decade, although it can occur at any age. It typically begins as a painful erosive skin rash in the skin folds. Common sites include the armpits, groins, neck, under the breasts, and between the buttocks. The lesions tend to come and go and leave no scars. If the lesions are present for some time, they may become thickened. The skin then tends to macerate leaving quite painful cracks. Secondary bacterial infection, which is not uncommon, can give rise to an unpleasant smell. White bands on the fingernails and pits in the palms can also occur. Sunburn, sweating, and friction often exacerbate the disease, and most patients have worse symptoms during the summer months.

Most patients with familial benign pemphigus respond well to anti-infective therapy and short courses of corticosteroids. Topical tacrolimus ointment has been a valuable addition to the treatment regimen and has been able to control familial benign pemphigus well even without the adjunctive use of topical corticosteroids (Rabeni and Cunningham 2002), and photodynamic therapy with 5-aminolevulinic acid has been used for recalcitrant cases (Ruiz-Rodriguez et al. 2002). Soothing compresses (aluminum acetate in a 1:40 dilution) followed by intermittent use of mild corticosteroid preparations and topical antibiotics result in transient improvement. More widespread flares of familial benign pemphigus may require systemic antibiotics to suppress protease activation and acantholysis. Bacterial culture and sensitivity can help guide appropriate therapy. Reports indicate that low-dose botulinum toxin type A injection may be of benefit for familial benign pemphigus (Lapiere et al. 2000).

9.9 Conclusion

Exact details of photoaggravated dermatoses have only occasionally been investigated. The basic condition may commonly be severely exacerbated even if it originally was only minor or often subclinical, especially in seborrheic (Palmer and Hawk 2004) and, less often, atopic eczema and acne (Norris and Hawk 1989), and these disorders may form a significant proportion of all photodermatoses in some countries. Such conditions, particularly the eczemas, psoriasis, and acne, improve with UVR exposure in most patients; normal skin immune reactivity generally is reduced as a result, but in a small proportion of subjects, it is instead significantly increased. If photoaggravation does occur, the new eruption generally develops or worsens initially at sites typical of the basic disorder, followed sometimes by spread to all exposed areas. In photoaggravated seborrheic eczema, however, an unpleasant sensation of the exposed sites may be the first or only feature. Treatment is by minimizing UVR exposure; protection by suitable clothing; application of high-protection, broad-spectrum sunscreens; and particularly very careful therapy of the basic disorder, even if mild or subclinical, especially to include daily anti-eczema shampoo use in seborrheic eczema, all of which, frequently if perhaps surprisingly, totally abort the photosensitivity (Palmer and Hawk 2004). If this is inadequate, prophylactic low-dose phototherapy as for PMLE can sometimes help in conditions usually responding to such treatment, as for example seborrheic or atopic eczema and psoriasis, although not cutaneous lupus or dermatomyositis, in which the possible aggravation of systemic disease is a significant risk. Photoaggravated acne generally requires oral isotretinoin (Norris and Hawk 1989).

References

Callen, J. P., and R. L. Wortmann. 2006. Dermatomyositis. *Clin Dermatol* 24:363–373.

Cheong, W. K., G. R. Hughes, P. G. Norris, and J. L. Hawk. 1994. Cutaneous photosensitivity in dermatomyositis. *Br J Dermatol* 131:205–208.

Craddock, N., E. Dawson, S. Burge et al. 1993. The gene for Darier's disease maps to chromosome 12q23-q24.1. *Hum Mol Genet* 2:1941–1943.

Doria, A., C. Biasinutto, A. Ghirardello et al. 1996. Photosensitivity in systemic lupus erythematosus: Laboratory testing of ARA/ACR definition. *Lupus* 5:263–268.

Dourmishev, L., H. Meffert, and H. Piazena. 2004. Dermatomyositis: Comparative studies of cutaneous photosensitivity in lupus erythematosus and normal subjects. *Photodermatol Photoimmunol Photomed* 20:230–234.

Faergemann, J., M. Borgers, and H. Degreef. 2007. A new ketoconazole topical gel formulation in seborrhoeic dermatitis: An updated review of the mechanism. *Expert Opin Pharmacother* 8:1365–1371.

Hailey, H., and H. Hailey. 1939. Familial benign chronic pemphigus. *Arch Dermatol* 39:679–685.

Hochberg, M. C. 1997. Updating the American College of Rheumatology revised criteria for the classification of systemic lupus erythematosus. *Arthritis Rheum* 40:1725.

Jansen, C. T., and J. Karvonen. 1984. Polymorphous light eruption. A seven-year follow-up evaluation of 114 patients. *Arch Dermatol* 120:862–865.

Karthikeyan, K., and D. M. Thappa. 2002. Pellagra and skin. *Int J Dermatol* 41:476–481.

Kuhn, A., M. Sonntag, D. Richter-Hintz et al. 2001. Phototesting in lupus erythematosus: A 15-year experience. *J Am Acad Dermatol* 45:86–95.

Lapiere, J. C., A. Hirsh, K. B. Gordon, B. Cook, and A. Montalvo. 2000. Botulinum toxin type A for the treatment of axillary Hailey–Hailey disease. *Dermatol Surg* 26:371–374.

Lehmann, P., E. Holzle, P. Kind, G. Goerz, and G. Plewig. 1990. Experimental reproduction of skin lesions in lupus erythematosus by UVA and UVB radiation. *J Am Acad Dermatol* 22:181–187.

Millard, T. P., C. M. Lewis, M. A. Khamashta et al. 2001. Familial clustering of polymorphic light eruption in relatives of patients with lupus erythematosus: Evidence of a shared pathogenesis. *Br J Dermatol* 144:334–338.

Nogueira, A., A. F. Duarte, S. Magina, and F. Azevedo. 2009. Pellagra associated with esophageal carcinoma and alcoholism. *Dermatol Online J* 15:8.

Norris, P. G., and J. L. Hawk. 1989. Actinic folliculitis—Response to isotretinoin. *Clin Exp Dermatol* 14:69–71.

Nyberg, F., T. Hasan, P. Puska et al. 1997. Occurrence of polymorphous light eruption in lupus erythematosus. *Br J Dermatol* 136:217–221.

Palmer, R. A., and J. L. Hawk. 2004. Light-induced seborrhoeic eczema: severe photoprovocation from subclinical disease. *Photodermatol Photoimmunol Photomed* 20:62–63.

Rabeni, E. J., and N. M. Cunningham. 2002. Effective treatment of Hailey–Hailey disease with topical tacrolimus. *J Am Acad Dermatol* 47:797–798.

Ruiz-Rodriguez, R., J. G. Alvarez, P. Jaen, A. Acevedo, and S. Cordoba. 2002. Photodynamic therapy with 5-aminolevulinic acid for recalcitrant familial benign pemphigus (Hailey–Hailey disease). *J Am Acad Dermatol* 47:740–742.

Sontheimer, R. D. 1996. Photoimmunology of lupus erythematosus and dermatomyositis: A speculative review. *Photochem Photobiol* 63:583–594.

Srinivas, C. R., C. S. Sekar, and R. Jayashree. 2012. Photodermatoses in India. *Indian J Dermatol Venereol Leprol* 78 Suppl 1:S1–8.

Vermi, W., S. Lonardi, M. Morassi et al. 2009. Cutaneous distribution of plasmacytoid dendritic cells in lupus erythematosus. Selective tropism at the site of epithelial apoptotic damage. *Immunobiology* 214:877–886.

Wan, P., S. Moat, and A. Anstey. 2011. Pellagra: A review with emphasis on photosensitivity. *Br J Dermatol* 164:1188–1200.

Wolf, P., S. N. Byrne, and A. Gruber-Wackernagel. 2009. New insights into the mechanisms of polymorphic light eruption: Resistance to ultraviolet radiation-induced immune suppression as an aetiological factor. *Exp Dermatol* 18:350–356.

10

Photoaging

Rui Yin
Third Military Medical University
Massachusetts General Hospital

Michael R. Hamblin
Massachusetts General Hospital

10.1 Definition

Skin aging is a complex process and can be divided into two basic processes: intrinsic aging and photoaging. The term "photoaging" was first coined in 1986 and described the effects of chronic ultraviolet (UV) light exposure on skin (Kligman and Kligman 1986). "Photo" is derived from the Greek word "phos," which means "light." Photoaging is premature aging of the skin caused by cumulative exposure to UV radiation (UVR) from the sun and artificial UV sources.

The basic biologic processes involved in aging lead to a reduction in the function and ability to tolerate injury. There are two general theories of aging (Wolff et al. 2008). The first one states that aging is a genetically determined process. Telomere lengths support this theory, in which the terminal portions of chromosomes can be shortened at every cell cycle. Once the telomeres reach a critical length, either cell-cycle arrest or apoptosis occurs (Vaziri and Benchimol 1996). Another theory suggests that aging is mainly a result of the accumulation of environmental damage (Wolff et al. 2008; Yasui and Sakurai 2003). For example, free radicals can be generated from oxygen during normal metabolism and are likely to contribute to this process (Harman 1956). Organisms have evolved cellular defense systems against the toxicity of free radicals, particularly oxygen-based free radicals or reactive oxygen species (ROS). Long-lived species were found to have higher degrees of enzymatic protection against ROS (Tolmasoff, Ono,

and Cutler 1980). The activity of antioxidant enzymes and the levels of nonenzymatic antioxidants decline with age (Hoppe et al. 1999; Yasui and Sakurai 2003), allowing oxidative damage to occur. In the skin, both genetic and environmental mechanisms likely promote the aging process and may share a common final pathway to destroy the skin (Kosmadaki and Gilchrest 2004).

10.2 Clinical and Histological Manifestation

Typical photoaged skin is characterized by dryness, laxity, dyspigmentation, telangiectasia, yellowish color, plaque-like thickening, deep creases and wrinkles, a leathery appearance, and cutaneous malignancies (Figure 10.1) (Gilchrest 1989; Gordon and Brieva 2012; Helfrich, Sachs, and Voorhees 2008; Zhang et al. 2011). Histological changes manifest these clinical changes in both the epidermis and the dermis. The epidermis may be relatively normal or present alterations, such as epidermal hyperplasia or atrophy, disappearance of dermal papillae, thickening of the basement membrane, focally increased numbers and irregular distribution of melanocytes and melanosomes, atypical keratinocytes, parakeratosis, and thickening of the stratum corneum. The most obvious dermal histological defect, which presents the character of photoaged skin, is the accumulation of cell population, such as numerous and hyperplastic fibroblasts, abundant

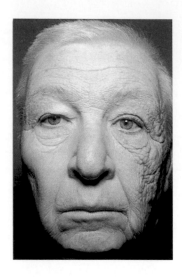

FIGURE 10.1 Unilateral dermatoheliosis. A 69-year-old driver was exposed to UVA for 25 years. He presented with asymptomatic thickening and wrinkling of the skin on the left side of his face. It confirms that chronic UVA exposure can result in thickening of the epidermis and stratum corneum as well as destruction of elastic fibers. (From Gordon and Brieva, *N Engl Med* 366:e25, 2012)

inflammatory cell infiltration, etc. (Chen et al. 2009). Other observations in the upper and middle dermis of photoaged skin include the presence of deformed collagen fibers, a decrease in the total amount of collagen, and an increase in the amounts of ground substance (Oikarinen 1990). Fibrolasts in photoaging skin are elongated and collapsed (Mermut et al. 2009). Microvasculature is dilated, and vessel walls are thickened with deposition of a basement membrane-like material (Schastak et al. 2008). Because of the effects of chronic inflammation, elastin is accumulated and seems to occupy the areas previously held by collagen (Staneloudi et al. 2007). The abnormal accumulated elastin associated with repeated sun exposure is called solar elastosis (Ichinose et al. 2006) and appears in the upper and middle layers of dermis. By electron microscopy, fully developed photoaged skin shows alternating areas of fibrous, granular, and homogeneous elastotic material (Heinonen et al. 1999; Pulkkinen, Ringpfeil, and Uitto 2002; Uitto, Pulkkinen, and Ringpfeil 2002). An increased number of thickened and tangled elastic fibers are observed around the fibrous areas, and the granular and homogeneous areas are thought to be the result of fragmentation of these thickened and tangled fibers (Oikarinen 1990).

10.3 Pathogenesis of Photoaging

10.3.1 UV-Induced ROS Formation

Photoaging is a process of aging of the skin resulting from continuous, long-term exposure to natural or synthetic UVR, specifically UVA and UVB, with a wavelength range of 245–290 nm. UVR has numerous direct and indirect effects on the skin. It is estimated that approximately 50% of UV-induced damage is from the formation of free radicals (Bernstein et al. 2004).

When skin is exposed to sunlight, UVR is absorbed by skin molecules that can generate ROS, resulting in direct and indirect damage to cellular components, such as cell walls, lipid membranes, mitochondria, DNA (Rosenthal 1991), and the extracellular matrix (ECM) proteins (Morgan et al. 1989). Furthermore, oxidative stress can result in an influx of inflammatory cells, including neutrophils (Oxford, Pooler, and Narahashi 1977; Spikes 1975). Clinically, infiltration of neutrophils is capable of causing significant tissue damage, which is accompanied by erythema of the skin. They are packed with proteolytic enzymes, including neutrophil elastase and metalloproteinases. Meanwhile, activated neutrophils also can generate and release ROS and thus damage collagen fibers, specifically the elastic fibers (Spikes 1975).

10.3.2 UV-Induced Molecular and Genetic Changes

DNA maximally absorbs UVR within the range of 245 to 290 nm, which is the wavelength of UVB (Tornaletti and Pfeifer 1996). Therefore, UVB rays are considered as a primary potent mutagen that can penetrate through the epidermal or outermost layer of the skin, resulting in DNA mutations (Linge 1996). Although specific mechanisms of DNA mutations induced by UV photodamage have not been elucidated, these DNA mutations arise attributed to chemical changes, through the formation of cyclobutane pyrimidine dimers and photoproducts formed between adjacent pyrimidine bases (Baugh et al. 2001). These mutations may be clinically related to specific signs of photoaging, such as wrinkling, increased elastin, and collagen damage, which are observed in animals exposed to UVB (Zeisser-Laboube et al. 2006).

UVA can also damage DNA indirectly through the generation of ROS, which include superoxide anions, peroxide, and singlet oxygen. As stated previously, these ROS damage cellular DNA as well as lipids and proteins (Agostinis et al. 2002; Helfrich, Sachs, and Voorhees 2008). UVA-induced mutagenesis may involve trans-urocanic acid and result in DNA nicks (Croce et al. 2011; Muller and Wilson 2006). 8-Hydroxy guanine is also a product of UVA mutagenesis via ROS induction (Figure 10.2) (Kochevar 1995).

10.3.3 UV-Induced Pigmentation and Vascular Alteration

The epidermal layer contains melanocytes as well as basal cells, which are both embedded in this layer. Upon exposure to UVB rays, melanocytes will produce melanin, the pigment that gives the skin its color tone. Tanning occurs in two steps: immediate pigment darkening, occurring in individuals with Fitzpatrick skin types III–IV, followed by delayed formation of new melanin called "delayed tanning." Immediate pigment changes and, therefore, changes in skin color are observed after UV exposure and result from structural changes in melanocytes and keratinocytes and chemical modification of preexisting melanin (Soter 1995). Delayed tanning is associated with an increase

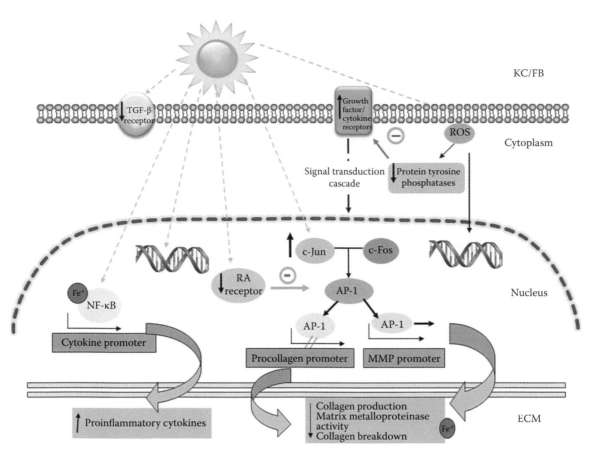

FIGURE 10.2 Effects of UV light on the keratinocyte (KC) and fibroblast (FB). Effects of UVR on skin. UV induces ROS, which can damage DNA and influence a number of transcription factors within the nucleus. Downstream effects lead to decreased production and extensive destruction of collagen. These effects have been related to collagen production and breakdown as well as to inflammatory cytokine production. AP, activator protein; MMP, matrix metalloproteinase; NF, nuclear factor; TGF-ß, transforming growth factor-ß; RA, retinoic acid.

in the activity and number of melanocytes, whose function is photoprotection (Halaban 2003). Hence, UV can cause the formation of freckles and dark spots, both of which are symptoms of photoaging. With constant exposure to UVB rays, signs of photoaging might appear, and precancerous lesions or skin cancer may develop. A recent study shows that UVB is more effective at stimulating skin pigmentation with appropriate doses, although the mechanism is not clear. UVA has no effect on melanin content, activation of melanocyte differentiation, or other parameters measured, yet UVA exposure stimulates visible skin pigmentation significantly (Ana et al. 2012; Coelho et al. 2009).

UV exposure can also lead to inflammation and vasodilation, which are clinically manifested as sunburn (Soter 1995). UVR activates the transcription factor, NF-κB, which is the first step in inflammation. NF-κB activation can result in the increase in proinflammatory cytokines, for example, interleukin 1 (IL-1), IL-6, vascular endothelial growth factor (VEGF), and tumor necrosis factor (TNF)-α. This then attracts neutrophils leading to an increase in oxidative damage through the generation of free radicals. Additionally, UVR causes the downregulation of an angiogenesis inhibitor, thrombospondin-1, and the upregulation of an angiogenesis activator, platelet-derived endothelial

cell growth factor, in keratinocytes (Henderson and Dougherty 1992). These enhance angiogenesis and trigger growth of UV-induced neoplasms (Tegos et al. 2005). UVA rays are capable of penetrating deeper into the skin as compared to UVB rays. Because of the presence of blood vessels in the dermis, UVA rays can lead to dilated or broken blood vessels most commonly visible on the nose and cheeks.

10.3.4 UV-Induced Immunosuppression

It has also been reported that UVR leads to local and systemic immunosuppression (Geraldo-Martins, Lepri, and Palma-Dibb 2012) resulting from DNA damage and altered cytokine expression as stated previously. It benefits for the surveillance of a cutaneous tumor (Arbabzadah et al. 2011). Following UV exposure, the Langerhans cells undergo changes in terms of quantity, morphology, and function and eventually become depleted. The underlying cause for the immunosuppression is thought to be the need for suppression or prevention of a possible immune response to inflammatory products resulting from UV-mediated damage (Gonzalez-Mosquera et al. 2011; Jia, Wang, and Liu 2011; Ryu, Lee, and Yoon 2011).

10.3.5 UV-Induced Degradation of Collagen

Increased breakdown and decreased production of collagen are the cornerstones of photoaging. Collagen in skin undergoes continuous remodeling and turnover, which is controlled by TGF-beta and activator protein (AP)-1. While TGF-beta promotes collagen formation, AP-1 promotes collagen breakdown by upregulating enzymes called matrix metalloproteinases (MMPs). Following UV irradiation, AP-1 becomes elevated at least for 24 h (Maxwell and Chichester 1971) and upregulates the transcription of MMPs (MMP-1 being the major metalloproteinase for collagen degradation) with a subsequent increased breakdown of collagen. This upregulation of MMPs can occur even after minimal exposure to UV. The entire process is aided by the presence of ROS that inhibit protein-tyrosine phosphatases via oxidation, thereby resulting in upregulation of the above-mentioned receptors. Another transcription factor, NF-κB, activated by UV light, is known to increase the expression of MMP-9. It is important to note that even exposure to UVR that is inadequate to cause sunburn can still facilitate the degradation of skin collagen and lead to photoaging. Through impaired spreading and attachment of fibroblasts on the degraded collagen, the presence of damaged collagen also seems to downregulate new collagen synthesis. In addition to this, UV irradiation leads to decreased expression of TGF-beta, which promotes collagen formation as mentioned previously. Therefore, it can cause decreased collagen production (Koch, Neumueller, and Schenck 1961). UV exposure also leads to the activation of receptors for epidermal growth factor, IL-1 and TNF-α in keratinocytes and fibroblasts, which then activates signaling kinases throughout the skin via an unknown mechanism to increase degradation of collagen. Each UV induces a wound response with a subsequent imperfect repair, leaving an invisible "solar scar." Repetitive UV insults over a lifetime eventually lead to development of a visible solar scar, manifesting as a visible wrinkle (Olivo, Du, and Bay 2006).

10.3.6 Retinoic Acids for Photodamaged Skin

Retinoic acid (RA), a vitamin A derivative, is essential for normal epithelial growth and differentiation as well as for maintenance of normal skin homeostasis (Tsurumaki Jdo et al. 2011). UVR decreases the expression of both RA receptors (RARs) and retinoid X receptors (RXRs) in human skin, thereby resulting in a complete loss of induction of RA-responsive genes (Aranha and de Paula Eduardo 2012). This, in turn, leads to an increase in AP-1 activity, increasing MMPs, and results in functional deficiency of vitamin A in the skin, which can contribute to collagen breakdown (Arslan et al. 2012; Cvikl et al. 2012).

10.4 Endogenous Defense Mechanisms against UVR

After UVR, numerous endogenous defense mechanisms provide protection for the skin against the damages induced by UV. Increased epidermal and dermal mitotic activity can be observed 24 to 48 h after acute UV exposure (Kato et al. 2011). Consequently, it can lead to an increased epidermal thickness to protect skin from further UV damage (Ishida et al. 2011). The endogenous protection from UV includes the redistribution of melanin, DNA repair, and formation of antioxidants.

10.4.1 Pigment

The protective role of melanin pigment should not be underestimated. Black skin differs from white skin with respect to the size and number of melanosomes as well as the aggregation pattern within melanocytes and keratinocytes (Szabo 1959). The distribution of melanin is thought to provide protection from sunburn, photoaging, and carcinogenesis by absorbing and scattering detrimental UV rays (Kadekaro et al. 2006; Kaidbey et al. 1979). A study also showed that white skin when compared to dark skin is more prone to DNA photodamage and has increased infiltration of neutrophils, keratinocyte activation, and IL-10 expression following UV exposure (Rijken et al. 2004).

10.4.2 Repair of DNA Mutation and Apoptosis

DNA damage following UV exposure leads to increased expression of p53, thereby leading to eventual arrest in the G1 phase of the cell cycle. This allows DNA repair mediated by the nucleotide excision repair system, which belongs to the endogenous mechanisms (Helfrich, Sachs, and Voorhees 2008; Perussi et al. 2012). However, when the damage is too severe, apoptosis occurs. On the other hand, apoptotic mechanisms decline with age, and if neither the DNA repair mechanism nor apoptosis occurs, cutaneous tumorigenesis may result (Young et al. 2003).

10.4.3 Antioxidants

The skin consists of several endogenous antioxidants, which include vitamin E, coenzyme Q10 (CoQ10), ascorbate, carotenoids (Shindo et al. 1994), and enzymatic antioxidants, which include superoxide dismutase, catalase, and glutathione peroxidase (Leccia et al. 2001). These antioxidants are important to the skin's defense mechanism against UVR and photocarcinogenesis by providing protection from ROS produced during normal cellular metabolism. However, in case of too much UV exposure, the antioxidant produced may not be sufficient to reduce ROS and therefore cause oxidative stress.

10.5 Skin Damage Classification Scales

10.5.1 Fitzpatrick Classification Scale

The landmark Fitzpatrick Classification Scale was developed in 1975 (Table 10.1). The scale offers a useful method of classifying a person's skin complexion and his or her tolerance to sunlight (de Oliveira, Sampaio, and Marcantonio 2010). Although this system remains the gold standard of classification systems, it does little to predict the skin's response to trauma from certain

TABLE 10.1 Fitzpatrick Classification Scale

Skin Type	Skin Color	Characteristics
I	White, very fair, red or blond hair, blue eyes, freckles	Always burns, never tans
II	White; fair; red or blond hair; blue, hazel, or green eyes	Usually burns, tans with difficulty
III	Cream white, fair with any eye or hair color, very common	Sometimes mild burn, gradually tans
IV	Brown, typical Mediterranean Caucasian skin	Rarely burns, tans with ease
V	Dark brown, middle-eastern skin types	Very rarely burns, tans very easily
VI	Black	Never burns, tans very easily

procedures, such as laser and surgery. In addition, the Fitzpatrick Classification System is potentially misleading because all skin types, even those classified as Fitzpatrick types V and VI, are susceptible to some elements of burning from UVR (Onay et al. 2010).

10.5.2 Glogau Classification of Photoaging

The Glogau classification scale was developed in 1994 (Table 10.2) (Glogau 1994). It classifies the degree of photoaging and categorizes the amount of wrinkling and discoloration in patients' skin. It helps practitioners pick the best procedures to treat photoaging. While undeniably useful, the Glogau system does not emphasize the signs of photoaging in mixed ethnic-racial skin types, which may include various pigmentary variations, midfacial decent, and periorbital darkening (Dundar and Guzel 2011).

There are some other skin classification systems, including the Kawada Skin Classification System (1986) (Kelbauskiene et al. 2011), Lancer Ethnicity Scale (1998) (Yavari et al. 2010), Goldman World Classification type (2002) (Goldman 2002), Fanous Classification (2002) (Olivi et al. 2010), Willis and Earles Scale (2005) (Willis 2005), Taylor Hyperpigmentation Scale (2006) (Gomez-Santos et al. 2010), Baumann Skin Type Solution (2006), and the Roberts Skin Type Classification System, which includes a hyperpigmentation scale and a scarring scale (2008) (Haedersdal et al. 2009). The Roberts system can be a predictor of an impending complication, such as hyperpigmentation and scarring, which can then be avoided. In addition, it includes the skin phototype and photoaging propensity (Haedersdal et al. 2009).

In summary, classification of skin types is rather a new practice, and there has been considerable progress made with continuing awareness. The Fitzpatrick Skin Phototype Classification remains the gold standard. While the Glogau classification scale classifies the degree of photoaging, the Roberts Skin Type Classification System is a good tool to predict the skin response to injury and insult from dermatologic and cosmetic procedures and to identify the propensity of sequelae from inflammatory skin disorders. However, there is still not a perfect classification yet that covers whole skin types.

10.6 Animal Model Study for Photoaging

Photoaging is a complicated and lengthy process. Choosing a suitable study model is critical for investigating cellular and molecular mechanisms of photodamage and photoaging. However, it is difficult to design animal models that can mimic all the physiological and pathological processes to study the exact mechanisms. At present, only a few animal models have been successfully used to investigate the aging process; however, these models were only able to evaluate acute photodamage. The three major mouse models used for photoaging studies are C57BL/6J, SKH1, and BALB/c mice (Jantschitsch et al. 2012; Sharma, Werth, and Werth 2011; Singh et al. 2012). Adult females from three strains of mice were exposed to UVB and then investigated 3 or 20 h after the last irradiation. Skin from UVB-exposed C57BL/6J mice showed pathological features resembling human photodamage, which include epidermal thickness; infiltration of inflammatory cells in the dermis; induction of tumor necrosis factor-a (TNF-a) mRNA; accumulation of glycosaminoglycans, particularly hyaluronan in the epidermis; and loss of collagen (Sharma, Werth, and Werth 2011). Hairless SKH1 mouse skin responded similarly to C57BL/6J mice, except that no production of TNF-a mRNA nor chondroitin sulfate was observed (Singh et al. 2012). Irradiated BALB/c mice were found to be the least similar to humans. The results in C57BL/6J mice and, to a lesser extent, in SKH1 mice show cutaneous responses to a course of UVB irradiation that mirror those seen in human skin (Jantschitsch et al. 2012). Another animal model is albino hairless mice (Skh: HR-1), which are used to study the UVB-induced DNA mutations that occurred by chemical change (Bissett, Hannon, and Orr 1989; Kligman and Sayre 1991).

TABLE 10.2 Glogau Classification of Photoaging

Group	Classification	Typical Age	Description	Skin Characteristics
I	Mild	28–35	No wrinkles	Early photoaging: mild pigment changes, no keratosis, minimal wrinkles, minimal or no makeup
II	Moderate	35–50	Wrinkles in motion	Early to moderate photoaging: early brown spots visible, keratosis palpable but not visible, parallel smile lines begin to appear, wears some foundation
III	Advanced	50–65	Wrinkles at rest	Advanced photoaging: obvious discolorations, visible capillaries (telangiectasias), visible keratosis, always wears heavier foundation
IV	Severe	60–75	Only wrinkles	Severe photoaging: yellow-gray skin color, prior skin malignancies, wrinkles throughout, no normal skin, cannot wear makeup because it cakes and cracks

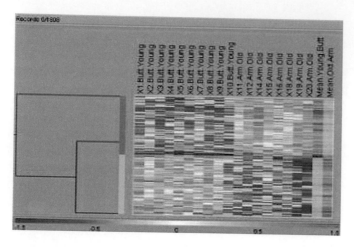

FIGURE 10.3 Photodamage at gene expression. This figure explains the combined effects of aging and photodamage at the gene expression level. The black color indicates genes that are downregulated, the grey color is genes that are upregulated, and white indicates no change.

The next level of photoaging studies will be possible through gene chip technology, which provides an understanding of skin aging that deals by analysis of various metabolic pathways at the gene-expression level (Figure 10.3).

10.7 Treatment of Photoaging

Rejuvenation or slowing of the aging process has always been one of the major goals of the area of dermatology. Strategies for medical treatment and intervention of photoaging included (1) measures for prevention against UV damage and (2) medications and procedures to reverse existing damage.

10.7.1 Photoprotection

Photoprotection refers to different practices for protecting the skin from UV damage and is achieved by sunscreens, sun-protective clothing, and sun avoidance. Sunscreens are broadly defined as agents that protect against UV damage and therefore against sunburn, wrinkles, and pigmentary changes (Tierney et al. 2009). Sun-protection factor (SPF) refers to the degree of protection from UVB. People should be advised to choose a sunscreen with a minimum SPF of 15 or higher and to apply it liberally and frequently to all exposed body sites, especially around the face and neck. Sunscreens should be applied every 2 to 3 h, especially if the person is engaged in outdoor activities. Aside from protection from UVB, protection from UVA is also very important. Chemical blockers of UVA include oxybenzone and avobenzone. Recently, some newer UVA blockers have become available, including ecamsule, which is the most photostable UVA blocker available and sold under the trade name Mexoryl (Anthelios, La Roche Posay/L'Oreal), and Helioplex, a stabilized form of avobenzone (Neutrogena). Sunscreens that contain physical and chemical blockers, such as titanium dioxide and zinc oxide, confer protection against both UVB and UVA.

Newer technologies, such as micronization, have been developed over recent years to make these sunscreens cosmetically more acceptable.

Sun-protective behavior can be achieved through general education, and avoidance of midday sun exposure when UVR is most intense, engaging in outdoor activities early or late in the day, avoidance of sunbathing (even with sunscreens), and seeking shady, covered areas rather than direct sunlight are all advised. In addition to this, avoiding tanning beds is also recommended.

10.7.2 Topical Retinoids

Topical tretinoin was first observed to ameliorate the clinical signs of photoaging by Cordero (1983) and Kligman and Kligman (1986) (Goldberg 2005). Retinoids are derivatives of vitamin A that have antiaging properties. A large-number clinical trial proved that skin wrinkles, surface roughness, and dyspigmentation have been improved with topical tretinoin therapy (Goldberg and Russell 2006; Hamaliia et al. 2005; Hongcharu et al. 2000; Trelles et al. 2005; Wiegell and Wulf 2006). Tretinoin (all-trans-retinoic acid), a nonselective agent that activates all RARs directly and RXRs indirectly (Itoh et al. 2000), has been shown to improve the clinical signs of photoaging in controlled clinical trials (Casas et al. 1999; Harth, Hirshowitz, and Kaplan 1998; Hurlimann, Hanggi, and Panizzon 1998; Orenstein et al. 2000).

The benefits of retinoids are thought to be mediated, at least in section, by their effects on collagenase induction. Pretreatment with all-trans-retinoic acid inhibits the formation of c-Jun protein, AP-1, and MMPs, which is mediated by UV exposure (Jeffes et al. 1997; Wolf et al. 1997). Pretreatment also reduces loss and accelerates recovery of RAR-g and RXRea following UV exposure (Fritsch et al. 1996). Partial restoration of markedly reduced collagen appears to be responsible for the observed clinical improvement (Stringer et al. 1996).

Tazarotene is a second-generation retinoid that selectively binds RAR-g and RAR-b (Szeimies et al. 1996b). Like tretinoin, tazarotene is effective in the treatment of photodamage, which can reduce atypia and restoration of keratinocyte polarity (Ammann and Hunziker 1995). In a 24-week randomized, controlled, double-blind study, treatment with 0.1% tazarotene resulted in significant improvement in numerous clinical assessments of photodamage. However, in two other studies, clinical improvement occurred during an open-label extension but did not reach a plateau within weeks (Lui et al. 1995; Orenstein et al. 1995). When compared with a standard dose of tretinoin, a high-dose tazarotene regimen produced faster improvements in fine wrinkling and mottled pigmentation (Lang et al. 1995). Tazarotene is also a strong irritant and is thought to inhibit AP-1e–dependent gene expression as similar to tretinoin (Karrer et al. 1995).

The main complaints about retinoid are irritation in the form of erythema, peeling, stinging, and itching at the first using stage (Hoerauf et al. 1994; Szeimies, Sassy, and Landthaler 1994).

Some patients are not able to tolerate the side effects. The greatest obstacle of treatment with retinoid is the long time period required until the clinical effects are observed.

An active area of research is the development of receptor-selective retinoids to optimize therapy and minimize side effects (Wolf, Rieger, and Kerl 1993). Upregulation of RA response elements and the antagonizing actions of AP-1 are not linked (Goff et al. 1992), which suggests that receptor-selective retinoids hold promise.

10.7.3 Cosmeceuticals

Cosmeceuticals are agents that are marketed as cosmetic products and that contain biologically active ingredients, including peptides, antioxidants, and botanicals. Peptides are amino acid chains that are fragments of large proteins, such as collagen, that is, Pal-KTTS. Pal-KTTS, by penetrating into the dermis and thereby stimulating collagen production, were shown to aid in wound healing (Hage et al. 2004). Antioxidants are very popular ingredients in cosmeceuticals. They are molecules that can prevent or reverse clinical signs associated with photoaging secondary to ROS, which are generated by UV damage and which lead to breakdown of collagen (Sakamoto, Torezan, and Anderson 2010). However, there are only a few published studies on the efficacy of these agents.

10.7.4 Coenzyme Q10

Coenzyme Q10 is a component of the mitochondrial electron transport chain, which is a potent antioxidant for the skin. It can reduce skin roughness, increase skin hydration, and reduce skin wrinkles and is associated with an improvement in overall global assessment of photoaged skin (Tsai et al. 2004). In a vehicle-controlled 6-month pilot study conducted by Steimer et al., use of topical CoQ10 has been shown to be effective against UVA-mediated oxidative stress in human keratinocytes in terms of thiol depletion, activation of specific phosphotyrosine kinases, and prevention of oxidative DNA damage. It also can significantly suppress the expression of collagenase in human dermal fibroblasts following UVA irradiation and reduction in wrinkle measurements assessed by optical profilometry (Hoppe et al. 1999).

10.7.5 Topical Vitamin C

Vitamin C, a potent antioxidant, has been shown to prevent erythema and sunburn cell formation after UV exposure (Kimura et al. 2004; Pass 1993; Yin et al. 2010). Vitamin C can also upregulate collagen and tissue inhibitors of MMP (TIMP) synthesis in human skin (Itoh et al. 2001). A 5% cream applied for 6 months led to clinical improvement in the appearance of photoaged skin with regard to firmness, smoothness, and dryness compared to vehicle (Al-Watban and Zhang 2005). Topical vitamin C stimulates the collagen-producing activity of the dermis (Cerburkovas et al. 2001). Because of the short half-life of vitamin C and its hydrophilic character (Bissonette, Bergeron, and Liu 2004; Stables et al. 1997), skin-care formulations commonly include its derivatives, which do not penetrate the skin as readily (Bissonnette and Lui 1997). Recently, various technologies, such as liposome and amphiphile, have been developed to improve the penetration and the stability of vitamin C.

10.7.6 α-Lipoic Acid

α-Lipoic acid is an antioxidant and anti-inflammatory agent that has been previously shown to reduce the production of transcription factors, such as NF-kB, and indirectly affect the gene expression of inflammatory cytokines. Topical or systemic treatment with α-lipoic acid has led to significant decreases in oxidative stress and, consequently, lipid peroxidation from 30% to 40% ($P < 0.005$) in blood serum (Morganti et al. 2002). It also can cause significant improvement in clinical and objective measurements of photoaging, including laser profilometry (Calzavara-Pinton et al. 1996). As we know, psoralen and UVA (PUVA) treatment disrupts the integrity of cellular membranes and impairs homeostasis and function of the cellular antioxidant system with a significant decrease in glutathione and hydrogen peroxide-detoxifying enzyme activities (Frippiat et al. 2001). Recently, one study *in vitro* showed that supplementation of PUVA-treated fibroblasts with α-lipoic acid abrogated the increased ROS generation and rescued fibroblasts from the ROS-dependent changes into the cellular senescence phenotype, such as cytoplasmic enlargement and enhanced expression of senescence-associated β-galactosidase and matrix metalloproteinase-1, hallmarks of photoaging and intrinsic aging (Briganti et al. 2008).

10.7.7 Estrogens

In a cross-sectional analysis, oral estrogen use was associated with a statistically significant improvement for dry skin and wrinkling, but no effect on skin atrophy has been observed (Lui and Anderson 1993). These clinical changes may be a result of an increase in collagen content (Szeimies et al. 1996a). However, oral estrogen is a potential risk factor for breast cancer. On the other hand, topical estrogen therapy can also lead to significant improvement in the amounts of collagen (Frazier 1996), firmness, and elasticity as well as reduction in wrinkle depth measured by optical profilometry (Lui 1994).

10.7.8 Fucose-Rich Polysaccharide

The fucose-rich polysaccharide FROP-3 increases glycosaminoglycan biosynthesis in fibroblast cultures; downregulates skin matrix degrading enzymes, such as MMP-2 and MMP-9; promote free-radical scavenging; increases skin elastin biosynthesis; and improves collagen fibrillogenesis (Dijkstra et al. 2001; Lui and Anderson 1992; Robert, Robert, and Robert 2005; Stranadko et al. 2001). In a pilot study examining skin-surface microrelief on the pattern of fine wrinkling found in skin of any age, a cream containing FROP-3 showed a 10- to 15-year decrease in apparent age after 4 weeks of treatment in a majority of patients (Stranadko et al. 2001). In a 5-week pilot study, an extract of date palm kernel was shown to reduce wrinkles by optical profilometry and visual assessment compared with the placebo (Kulapaditharom and Boonkitticharoen 1999; Morton et al. 1998).

10.7.9 Soy Isoflavones

Soy isoflavones can improve the activity of endogenous antioxidant enzymes (Rook et al. 2010), inhibit UVB-induced apoptosis and inflammation (Chiu et al. 2009), and protect against UV-induced aging. Mice fed with a solution containing isoflavones (primarily genistein and daidzein) and chronically exposed to UV for 4 weeks exhibited significant decreases in skin roughness measured by optical profilometry. In addition, epidermal thickness was significantly lower, and the level of procollagen was higher in the isoflavone-treated group. Dose-dependent decreases in MMP induction by UVR were also noted in an *in vitro* study of human fibroblasts treated with isoflavones (Lee, Baron, and Foster 2008). Similar research showed that an isoflavone extract decreased UVB-induced HaCaT cell death and the phosphorylation of p38, JNK, and ERK1/2 *in vitro*, reduced the expressions of cyclooxygenase-2 (COX-2) and proliferating cell nuclear antigen (PCNA), and increased catalase concentration *in vivo* (Chiu et al. 2009).

10.7.10 Topical Genistein and N-Acetylcysteines

Topical genistein was shown to prevent c-Jun and collagenase upregulation after UV exposure in human skin *in vivo* (Coors and von den Driesch 2004). Beyond its antioxidant activity, genistein is an inhibitor of tyrosine kinase activity (Umegaki et al. 2004) and may inhibit signal transduction induced by UV light. Some reports got similar effects with the antioxidant N-acetyl cysteine, a precursor to glutathione, which is also a benefit for improving photoaging (De Vries and De Flora 1993; Kang et al. 2003).

10.7.11 Gluconolactone

Gluconolactone is a polyhydroxy acid, related to α-hydroxy acid (AHA) such as glycolic acid (GA). Gluconolactone has antioxidant properties while sharing some of the effects of AHAs. Pretreatment with gluconolactone was shown to reduce UV induction of elastin by 50% in murine fibroblast cultures, potentially through its free-radical scavenger activity. Gluconolactone has already been incorporated into numerous cosmetic preparations, apparently serving as a preventive treatment for solar elastosis (Bernstein et al. 2004; Bucko et al. 2010; Lindsay et al. 1997).

10.7.12 Green Tea Polyphenols

Green tea polyphenols (GTPs) are potent antioxidants found in numerous skin-care products (Katiyar and Elmets 2001).

Oral administration of GTPs markedly inhibited UV-induced expression of MMP in mouse skin, which suggests that GTP has a potential antiphotoaging effect (Vayalil et al. 2004). Even in the absence of UV light, (e)-epigallocatechin-3-gallate, a component of green tea, was shown to inhibit the expression of various MMPs (Dell'Aica et al. 2002). GTP can also protect skin against the UVB-induced stress both via interacting with UVB-induced ROS and attenuating mitochondrion-mediated apoptosis (Wu et al. 2009).

10.7.13 N(6)-Furfuryladenine

N(6)-furfuryladenine (kinetin) is a synthetic plant growth hormone with antioxidant properties and several antiaging effects reported for human cells and fruit flies (Berge, Kristensen, and Rattan 2006). It has been shown to decrease or delay some of the age-related changes that occur in human fibroblasts during serial passage in cell culture (Rattan and Clark 1994). It also reduces ROS-mediated damage to DNA (Olsen et al. 1999). Although there are only a few published clinical studies for this compound (Chiu et al. 2007), it has been introduced into cosmeceuticals and may be useful in people who are unable to tolerate retinoids (Glaser 2004). Recent research has shown that the cosmetic formulation containing a dispersion of liposome with kinetin has photoprotective effects in skin barrier function as well as a pronounced hydration effect on human skin, which suggests that this dispersion has potential antiaging effects (Campos et al. 2012).

10.7.14 Iron Chelators

Because MMP activation is dependent on iron (Polte and Tyrrell 2004), the iron chelator kojic acid was investigated to determine its potential preventive effects on photoaging. Kojic acid, which is produced by the fungus *Aspergillus oryzae*, is found in Japanese soy-based products with antioxidant properties (Niwa and Akamatsu 1991). It is a tyrosinase inhibitor that has been used in the treatment of hyperpigmentation disorders, such as melasma (Garcia and Fulton 1996). Pretreatment of mice with kojic acid before long-term UV exposure was found to reduce wrinkling. Furthermore, histologically, it reduced UV-induced epidermal hyperplasia, dermal fibrosis, and increased amounts of the dermal dermatan sulfate chondroitin. Kojic acid has now been widely used in many Japanese cosmetic products (Mitani et al. 2001).

10.7.15 Aesthetic Surgery

10.7.15.1 Botulinum Toxin and Soft Tissue Fillers

Botulinum toxin A is a naturally occurring exotoxin produced by *Clostridium botulinum* that prevents local neuromuscular transmission. It was approved for cosmetic use by the Food and Drug Administration in 2002 in the United States. The toxin facilitates cleavage of synaptosomal-associated membrane protein (SNAP)-25, which is required for exocytosis of acetylcholine (Montecucco and Schiavo 1993), thereby, through neuromuscular inhibition, preventing muscle contraction. Although botulinum toxin A does not directly reverse changes in the ECM caused by photodamage, it gives the appearance of rejuvenation by relaxation of the underlying musculature (Rabe et al. 2006). Purified botulinum toxin type A (Botox, Allergan Inc., Irvine, California) is the most popular product used to paralyze various muscle groups of the face for cosmetic improvement of wrinkles in current clinical cosmetic treatments. It is most commonly used to treat wrinkles of the glabella, forehead, and periocular regions. Paralysis of these small muscle groups of the face results in a more youthful appearance with effects lasting from 3 to 6 months. With time and repeated injections, many patients will note the softening or disappearance of particular facial lines. Side effects of Botox injections include pain, bruising, and paralysis of the nerves that control eyelid function (Helfrich, Sachs, and Voorhees 2008).

Soft tissue augmentation or "fillers" are designed to address the subcutaneous atrophy that accompanies senescence. Fillers have been used to treat fine lines and sallowness in photoaging but have a greater market in intrinsically aged skin and for other cosmetic purposes. First approved in 1981, bovine collagen was the gold standard for soft tissue augmentation, but the drawbacks, such as immunogenicity (Trentham 1986) and potential hypersensitivity reactions, restrict its clinical use. In recent years, non-animal stabilized hyaluronic acid (HA) gel has gained tremendous popularity among patients and physicians and is currently the most widely used filler worldwide (Coleman and Carruthers 2006). The HA derivatives, derived from rooster combs or through bacterial fermentation (Manna et al. 1999), are less immunogenic compared with bovine collagen preparations because HA is chemically identical across species (Larsen et al. 1993). Recently, to assess the efficacy, patient satisfaction, and safety of the HA filler in periorbital rejuvenation, a multicenter, 6-month, open-label study showed that HA is suitable for rejuvenation of the periorbital region, which leads to safe results, long-lasting efficacy, and high levels of patient satisfaction (Rzany et al. 2012). At present, available fillers include Restylane (Medicis Pharmaceuticals, Scottsdale, Arizona), calcium hydroxylapatite (Radiesse, BioForm Medical Inc., San Mateo, California), poly-L-lactic acid (Sculptra, Dermik Laboratories, Bridgewater, New Jersey), and human-based collagen (Cosmoderm and Cosmoplast, both made by Allergan Inc., Irvine, California). They are most commonly used to improve the appearance of the nasolabial folds, which become more pronounced as a result of photoaging and chronological aging. They are also injected into cheeks, periocular areas, and glabellar lines and are often used in combination with Botox for maximal effect because they address different aspects of aging skin (Coleman and Carruthers 2006). Soft tissue fillers have been thought to exert their effect by volume expansion, but recent work investigating the mechanism of action of Restylane suggests that the filler stretches fibroblasts, leading to new collagen formation (Wang et al. 2007).

10.7.15.2 Chemical Peels

A variety of chemical peels, including AHAs, salicylic acid, trichloroacetic acid, and phenol, are used to treat acne, acne scars, photodamage, and mottled hyperpigmentation (Ghersetich et al. 2004; Kauvar and Dover 2001). They are classified as superficial, medium, and deep, which correlate with the depth of injury induced (Kauvar and Dover 2001). Portions of the epidermis and dermis are damaged with subsequent regeneration, resulting in a controlled wound and re-epithelialization with rejuvenation of the skin (Brody 1999). GA is an AHA superficial peel that improves skin texture and reduces fine wrinkling and the number of actinic keratoses. It can also reduce the stratum corneum and epidermis thickness as well as increase dermal collagen (Newman et al. 1996; Bertin et al. 2008). GA is found in many skin creams and has been shown to modestly improve photodamage when used in this fashion (Stiller et al. 1996). On the other hand, GA can also increase sunburn cell formation and sensitivity to UV-induced erythema, DNA damage, and sunburn cell formation (Kornhauser et al. 2009). Therefore, it may paradoxically enhance short-term sensitivity to the damaging effects of UV light (Kaidbey et al. 2003; Kornhauser et al. 2009). It is important to combine GA and sunscreen to reduce its drawbacks maximally.

10.7.15.3 Laser and Light Treatment

There are numerous applications for cutaneous laser surgery, including destruction of vascular and pigmented lesions, striae, and verrucae, as well as dermal remodeling for treatment of photodamage (Tanzi, Lupton, and Alster 2003). Most lasers work through selective photothermolysis, in which controlled destruction of a chromophore occurs without damage to surrounding normal tissue (Anderson and Parrish 1983). Ablative and nonablative laser systems have been successfully used in the treatment of photodamage and wrinkles and for increasing collagen production; however, the exact mechanism is yet unknown (Tanzi, Lupton, and Alster 2003).

10.7.15.3.1 Ablative Laser Systems

Ablative systems include the carbon dioxide (CO_2) and erbium: yttrium–aluminum–garnet (YAG) lasers. The CO_2 laser is considered the gold standard for skin rejuvenation. Facial resurfacing with the CO_2 laser typically produces at least a 50% improvement in overall skin tone, wrinkle depth, and atrophic scar depth (Apfelberg and Smoller 1997; Tanzi, Lupton, and Alster 2003; West and Alster 1998). Changes seen after CO_2 laser resurfacing include increased mRNA of several cytokines (IL-1b, TNF-a, and TGF-b1), type I and type III procollagen, and MMPs (Orringer et al. 2004). The erbium:YAG laser, developed to reduce the morbidity associated with CO_2 laser resurfacing (Tanzi, Lupton, and Alster 2003), has demonstrated comparable results with fewer side effects (Munker 2001; Weiss et al. 1999). Undesired effects of ablative systems include the possibility of hypertrophic scar formation and pigmentary alterations, especially in skin of colored races. The main downside of these lasers is that they induce significant morbidity until re-epithelialization occurs, which lasts at least 1 week, and the full recovery period can be a month or more. Therefore, the long downtime period has limited its clinical use (Tajirian and Goldberg 2011). Fractional ablative technology is a relatively recent development, which combines a lesser amount of downtime with similar efficacy. Recently, fractional laser therapy (FLT) has become a widely accepted modality for skin rejuvenation and has also been used for various other skin diseases, including pigment disorders, scars, and striae (Alexiades-Armenaka et al. 2011; Tajirian and Goldberg 2011). Ablative FLT may induce fibrosis formation (Wind et al. 2012). A multicenter clinical study using the fractional carbon dioxide (CO_2) laser (SmartXide DOT, DEKA) for the treatment of rhytides, photoaging, scars, and striae distensae revealed good results and safety (Alexiades-Armenaka et al. 2011). Other studies have proven to be effective and well tolerated for the correction of superficial brown epidermal dyschromia, mild to moderate winkles, and even moderately deep acne scarring employed with the 2790 nm wavelength YSGG laser in clinical practice (Smith and Schachter 2011). In summary, fractional ablative laser systems can shorten the downtime, have fewer side effects, and show improvement in clinical satisfaction compared with ablative laser systems. They have gained popularity among both physicians and patients for skin rejuvenation as a cosmetic laser treatment.

10.7.15.3.2 Nonablative Laser Systems

Nonablative fractional resurfacing (NFR) uses a mode of delivery with preservation of the stratum corneum, a true resurfacing with epidermal extrusion, and creation of microscopic thermal zones of injury that ultimately lead to neocollagenesis (Narurkar 2009). NFRs include the long-pulsed neodymium YAG (1064 nm), 1320 nm, radiofrequency, etc. Each of these treatment modalities is performed multiple times, usually several weeks apart. They are popular among patients who are unwilling or unable to undergo a postoperative recovery period associated with ablative procedures (Tanzi and Alster, 2004). They are much less effective than ablative lasers for the treatment of photoaging (Tanzi, Lupton, and Alster 2003) but can reduce hyperpigmentation and telangiectasia. Therefore, it has relatively lower risks and fewer side effects, especially in colored races. A controlled half-face study of the 1450 nm diode laser demonstrated significant clinical improvement in periorbital rhytides as well as increases in dermal collagen assessed histologically. Recently, a clinical study described the safety and efficacy results of treatment with a fractional nonablative 1540 nm erbium:glass laser in 51 patients with Fitzpatrick skin types II to IV for striae (de Angelis et al. 2011). However, in a separate study, 25 dermatologists clinically evaluated patients after 1450 nm diode treatment, and although all patients reported mild to moderate improvement, only 2 of the 25 dermatologists recorded a significant positive treatment effect, suggesting that modest changes induced by the laser may not be clinically meaningful (Kopera et al. 2004). So far, none of the nonablative system lasers can replace the ablative procedures.

10.7.15.3.3 Photodynamic Therapy

Topical photodynamic therapy (PDT) has been shown to be effective in the treatment and prevention of nonmelanoma skin cancer. PDT photorejuvenation, as a procedure for aesthetic and cosmetic laser surgeons, was adopted when Bitter (2000) published the first clinical article on the topic of photorejuvenation. In Bitter's study, more than 90% of the patients studied had a greater than 75% improvement in rosacea symptoms (facial erythema and flushing), 84% had an improvement in their fine wrinkles, 78% had significant changes in their facial pigmentation, and 49% noted an improvement in their pore size. Each patient in this study received five full-face, intense, pulsed-light treatments at a 1-month interval. Until now, the aesthetic effects of PDT for photoaged skin have been well documented (Goldberg and Samady 2001; Kohl and Karrer 2011; Sadick 2003; Weiss, Weiss, and Beasley 2002). The most commonly used photosensitizer is 20% 5-aminolevulinic acid (ALA). After application of the photosensitizer and an appropriate incubation time, ALA is converted to protoporphyrin IX and can be targeted by light. Blue light has classically been used in PDT because it targets the largest absorption band of protoporphyrin IX, known as the Soret band; however, when used for photorejuvenation, a variety of other light sources, including red light, intense pulsed light (IPL), or 585 nm pulsed dye laser, are used. Multiple studies have demonstrated significant photorejuvenation with PDT, including appreciable decreases in erythema, dyspigmentation, and fine wrinkles (Goldberg 2008), although how much is a result of the laser or IPL device versus the PDT effect is debated (Doherty et al. 2009; Goldberg 2008). Methyl 5-aminolevulinate (MAL) is an ester derivative of ALA. It is more lipophilic and has higher penetration when compared to ALA. While one study performed by Kujipers et al. (2006) demonstrated no difference in terms of short-term efficacy and side effects between ALA and MAL and suggested that both can be equally recommended as topical photosensitizers for PDT, another clinical trial performed among 69 patients with AKs demonstrated that PDT with MAL induced less pain than ALA and was better tolerated (Kasche et al. 2006). The clinical outcome is based on the different light sources, different photosensitizer concentrations, and incubation time on the skin. However, optimal levels for each of these parameters for photorejuvenation with PDT are still under debate. The formulation of standard guidelines still seems to be in the future.

References

Agostinis, P., A. Vantieghem, W. Merlevede, and P. A. de Witte. 2002. Hypericin in cancer treatment: More light on the way. *Int J Biochem Cell Biol* 34:221–241.

Al-Watban, F. A., and X. Y. Zhang. 2005. Photodynamic therapy of human undifferentiated thyroid carcinoma-bearing nude mice using topical 5-aminolevulinic acid. *Photomed Laser Surg* 23:206–211.

Alexiades-Armenaka, M., D. Sarnoff, R. Gotkin, and N. Sadick. 2011. Multi-center clinical study and review of fractional ablative CO2 laser resurfacing for the treatment of rhytides, photoaging, scars and striae. *J Drugs Dermatol* 10:352–362.

Ammann, R., and T. Hunziker. 1995. Photodynamic therapy for mycosis fungoides after topical photosensitization with 5-aminolevulinic acid. *J Am Acad Dermatol* 33:541.

Ana, P. A., C. P. Tabchoury, J. A. Cury, and D. M. Zezell. 2012. Effect of Er,Cr:YSGG laser and professional fluoride application on enamel demineralization and on fluoride retention. *Caries Res* 46:441–451.

Anderson, R. R., and J. A. Parrish. 1983. Selective photothermolysis: Precise microsurgery by selective absorption of pulsed radiation. *Science* 220:524–527.

Apfelberg, D. B., and B. Smoller. 1997. UltraPulse carbon dioxide laser with CPG scanner for deepithelialization: Clinical and histologic study. *Plast Reconstr Surg* 99:2089–2094.

Aranha, A. C., and C. de Paula Eduardo. 2012. Effects of Er:YAG and Er,Cr:YSGG lasers on dentine hypersensitivity. Short-term clinical evaluation. *Lasers Med Sci* 27:813–818.

Arbabzadah, E., S. Chard, H. Amrania, C. Phillips, and M. Damzen. 2011. Comparison of a diode pumped Er:YSGG and Er:YAG laser in the bounce geometry at the 3 mum transition. *Opt Express* 19:25860–25865.

Arslan, S., A. R. Yazici, J. Gorucu et al. 2012. Comparison of the effects of Er,Cr:YSGG laser and different cavity disinfection agents on microleakage of current adhesives. *Lasers Med Sci* 27:805–811.

Baugh, S. D., Z. Yang, D. K. Leung, D. M. Wilson, and R. Breslow. 2001. Cyclodextrin dimers as cleavable carriers of photodynamic sensitizers. *J Am Chem Soc* 123:12488–12494.

Berge, U., P. Kristensen, and S. I. Rattan. 2006. Kinetin-induced differentiation of normal human keratinocytes undergoing aging in vitro. *Ann N Y Acad Sci* 1067:332–336.

Bernstein, E. F., D. B. Brown, M. D. Schwartz, K. Kaidbey, and S. M. Ksenzenko. 2004. The polyhydroxy acid gluconolactone protects against ultraviolet radiation in an in vitro model of cutaneous photoaging. *Dermatol Surg* 30:189–195; discussion 196.

Bertin, C., H. Zunino, M. Lanctin et al. 2008. Combined retinol-lactose-glycolic acid effects on photoaged skin: A double-blind placebo-controlled study. *Int J Cosmet Sci* 30:175–182.

Bissett, D. L., D. P. Hannon, and T. V. Orr. 1989. Wavelength dependence of histological, physical, and visible changes in chronically UV-irradiated hairless mouse skin. *Photochem Photobiol* 50:763–769.

Bissonette, R., A. Bergeron, and Y. Liu. 2004. Large surface photodynamic therapy with aminolevulinic acid: Treatment of actinic keratoses and beyond. *J Drugs Dermatol* 3:S26–S31.

Bissonnette, R., and H. Lui. 1997. Current status of photodynamic therapy in dermatology. *Dermatol Clin* 15:507–519.

Bitter, P. H. 2000. Noninvasive rejuvenation of photodamaged skin using serial, full-face intense pulsed light treatments. *Dermatol Surg* 26:835–842; discussion 843.

Briganti, S., M. Wlaschek, C. Hinrichs et al. 2008. Small molecular antioxidants effectively protect from PUVA-induced oxidative stress responses underlying fibroblast senescence and photoaging. *Free Radic Biol Med* 45:636–644.

Brody, H. J. 1999. Chemical peeling: An updated review. *J Cutan Med Surg* 3 Suppl 4:S14–S20.

Bucko, M., P. Gemeiner, A. Vikartovska et al. 2010. Coencapsulation of oxygen carriers and glucose oxidase in polyelectrolyte complex capsules for the enhancement of D-gluconic acid and delta-gluconolactone production. *Artif Cells Blood Substit Immobil Biotechnol* 38:90–98.

Calzavara-Pinton, P. G., R. M. Szeimies, B. Ortel, and C. Zane. 1996. Photodynamic therapy with systemic administration of photosensitizers in dermatology. *J Photochem Photobiol B* 36:225–231.

Campos, P. M., F. B. de Camargo Junior, J. P. de Andrade, and L. R. Gaspar. 2012. Efficacy of cosmetic formulations containing dispersion of liposome with magnesium ascorbyl phosphate, alpha-lipoic acid and kinetin. *Photochem Photobiol* 88:748–752.

Casas, A., H. Fukuda, R. Meiss, and A. M. Batlle. 1999. Topical and intratumoral photodynamic therapy with 5-aminolevulinic acid in a subcutaneous murine mammary adenocarcinoma. *Cancer Lett* 141:29–38.

Cerburkovas, O., M. Krause, J. Ulrich, B. Bonnekoh, and H. Gollnick. 2001. [Disseminated actinic keratoses. Comparison of topical photodynamic therapy with 5-aminolevulinic acid and topical 5% imiquimod cream]. *Hautarzt* 52:942–946.

Chen, K., A. Preuss, S. Hackbarth et al. 2009. Novel photosensitizer-protein nanoparticles for photodynamic therapy: Photophysical characterization and in vitro investigations. *J Photochem Photobiol B* 96:66–74.

Chiu, P. C., C. C. Chan, H. M. Lin, and H. C. Chiu. 2007. The clinical anti-aging effects of topical kinetin and niacinamide in Asians: A randomized, double-blind, placebo-controlled, split-face comparative trial. *J Cosmet Dermatol* 6:243–249.

Chiu, T. M., C. C. Huang, T. J. Lin et al. 2009. In vitro and in vivo anti-photoaging effects of an isoflavone extract from soybean cake. *J Ethnopharmacol* 126:108–113.

Coelho, S. G., W. Choi, M. Brenner et al. 2009. Short- and long-term effects of ultraviolet radiation on the pigmentation of human skin. *J Invest Dermatol Symp Proc* 14:32–35.

Coleman, K. R., and J. Carruthers. 2006. Combination therapy with BOTOX and fillers: The new rejuvenation paradigm. *Dermatol Ther* 19:177–188.

Coors, E. A., and P. von den Driesch. 2004. Topical photodynamic therapy for patients with therapy-resistant lesions of cutaneous T-cell lymphoma. *J Am Acad Dermatol* 50:363–367.

Cordero, A. J. 1983. La vitamina a acida en la piel senile. *Actualiz Terapeut Dermatologica* 6:49–54.

Croce, A. C., E. Fasani, M. G. Bottone et al. 2011. Hypocrellin-B acetate as a fluorogenic substrate for enzyme-assisted cell photosensitization. *Photochem Photobiol Sci* 10:1783–1790.

Cvikl, B., G. Moser, J. Wernisch et al. 2012. The impact of Er,Cr:YSGG laser on the shear strength of the bond between dentin and ceramic is dependent on the adhesive material. *Lasers Med Sci* 27:717–722.

de Angelis, F., L. Kolesnikova, F. Renato, and G. Liguori. 2011. Fractional nonablative 1540-nm laser treatment of striae distensae in Fitzpatrick skin types II to IV: Clinical and histological results. *Aesthet Surg J* 31:411–419.

de Oliveira, G. J., J. E. Sampaio, and R. A. Marcantonio. 2010. Effects of Er,Cr:YSGG laser irradiation on root surfaces for adhesion of blood components and morphology. *Photomed Laser Surg* 28:751–756.

De Vries, N., and S. De Flora. 1993. N-acetyl-l-cysteine. *J Cell Biochem Suppl* 17F:270–277.

Dell'Aica, I., M. Dona, L. Sartor, E. Pezzato, and S. Garbisa. 2002. (-)Epigallocatechin-3-gallate directly inhibits MT1-MMP activity, leading to accumulation of nonactivated MMP-2 at the cell surface. *Lab Invest* 82:1685–1693.

Dijkstra, A. T., I. M. Majoie, J. W. van Dongen, H. van Weelden, and W. A. van Vloten. 2001. Photodynamic therapy with violet light and topical 6-aminolaevulinic acid in the treatment of actinic keratosis, Bowen's disease and basal cell carcinoma. *J Eur Acad Dermatol Venereol* 15:550–554.

Doherty, S. D., C. B. Doherty, J. S. Markus, and R. F. Markus. 2009. A paradigm for facial skin rejuvenation. *Facial Plast Surg* 25:245–251.

Dundar, B., and K. G. Guzel. 2011. An analysis of the shear strength of the bond between enamel and porcelain laminate veneers with different etching systems: Acid and Er,Cr:YSGG laser separately and combined. *Lasers Med Sci* 26:777–782.

Frazier, C. C. 1996. Photodynamic therapy in dermatology. *Int J Dermatol* 35:312–316.

Frippiat, C., Q. M. Chen, S. Zdanov et al. 2001. Sublethal H_2O_2 stress triggers a release of TGF-b1 which induce biomarkers of cellular senescence of human diploid fibroblasts. *J Biol Chem* 276:2531–2537.

Fritsch, C., B. Verwohlt, K. Bolsen, T. Ruzicka, and G. Goerz. 1996. Influence of topical photodynamic therapy with 5-aminolevulinic acid on porphyrin metabolism. *Arch Dermatol Res* 288:517–521.

Garcia, A., and J. E. Fulton, Jr. 1996. The combination of glycolic acid and hydroquinone or kojic acid for the treatment of melasma and related conditions. *Dermatol Surg* 22:443–447.

Geraldo-Martins, V. R., C. P. Lepri, and R. G. Palma-Dibb. 2012. Influence of Er,Cr:YSGG laser irradiation on enamel caries prevention. *Lasers Med Sci* 28:33–39.

Ghersetich, I., B. Brazzini, K. Peris et al. 2004. Pyruvic acid peels for the treatment of photoaging. *Dermatol Surg* 30:32–36; discussion 36.

Gilchrest, B. A. 1989. Skin aging and photoaging: An overview. *J Am Acad Dermatol* 21:610–613.

Glaser, D. A. 2004. Anti-aging products and cosmeceuticals. *Facial Plast Surg Clin North Am* 12:363–372, vii.

Glogau, R. 1994. Chemical peeling and aging skin. *J Geriatr Dermatol* 1:31.

Goff, B. A., R. Bachor, N. Kollias, and T. Hasan. 1992. Effects of photodynamic therapy with topical application of 5-aminolevulinic acid on normal skin of hairless guinea pigs. *J Photochem Photobiol B* 15:239–251.

Goldberg, D. J. 2005. Nonablative laser surgery for pigmented skin. *Dermatol Surg* 31:1263–1267.

Goldberg, D. J. 2008. Photodynamic therapy in skin rejuvenation. *Clin Dermatol* 26:608–613.

Goldberg, D. J., and B. A. Russell. 2006. Combination blue (415 nm) and red (633 nm) LED phototherapy in the treatment of mild to severe acne vulgaris. *J Cosmet Laser Ther* 8:71–75.

Goldberg, D. J., and J. A. Samady. 2001. Intense pulsed light and Nd:YAG laser non-ablative treatment of facial rhytids. *Lasers Surg Med* 28:141–144.

Goldman, M. 2002. Universal classification of skin type. *J Cosmet Dermatol* 15:53–54.

Gomez-Santos, L., J. Arnabat-Dominguez, A. Sierra-Rebolledo, and C. Gay-Escoda. 2010. Thermal increment due to ErCr:YSGG and CO2 laser irradiation of different implant surfaces. A pilot study. *Med Oral Patol Oral Cir Bucal* 15:e782–787.

Gonzalez-Mosquera, A., J. Seoane, L. Garcia-Caballero et al. 2011. Er,CR:YSGG lasers induce fewer dysplastic-like epithelial artefacts than CO(2) lasers: an in vivo experimental study on oral mucosa. *Br J Oral Maxillofac Surg* 50:508–512.

Gordon, J. R., and J. C. Brieva. 2012. Images in clinical medicine. Unilateral dermatoheliosis. *N Engl J Med* 366:e25.

Haedersdal, M., K. E. Moreau, D. M. Beyer, P. Nymann, and B. Alsbjorn. 2009. Fractional nonablative 1540 nm laser resurfacing for thermal burn scars: a randomized controlled trial. *Lasers Surg Med* 41:189–195.

Hage, M., P. D. Siersema, H. van Dekken et al. 2004. 5-Aminolevulinic acid photodynamic therapy versus argon plasma coagulation for ablation of Barrett's oesophagus: A randomised trial. *Gut* 53:785–790.

Halaban, R., D. Hebert, and G. J. Fisher. 2003. Biology of melanocytes. In *Fitzpatrick's Dermatology in General Medicine*, 6th edition. 127–148. Freedberg, I. M., A. Z. Eisen, K. Wolff, F. Austen, L. A. Goldsmith, and S. I. Katz., eds. New York: McGraw Hill.

Hamaliia, M. F., V. V. Kutsenok, O. B. Horobets et al. 2005. [Photodynamic therapy of experimental tumors using 5-aminolevulinic acid]. *Fiziol Zh* 51:65–70.

Harman, D. 1956. Aging: A theory based on free radical and radiation chemistry. *J Gerontol* 11:298–300.

Harth, Y., B. Hirshowitz, and B. Kaplan. 1998. Modified topical photodynamic therapy of superficial skin tumors, utilizing aminolevulinic acid, penetration enhancers, red light, and hyperthermia. *Dermatol Surg* 24:723–726.

Heinonen, S., M. Mannikko, J. F. Klement et al. 1999. Targeted inactivation of the type VII collagen gene (Col7a1) in mice results in severe blistering phenotype: A model for recessive dystrophic epidermolysis bullosa. *J Cell Sci* 112 (Pt 21):3641–3648.

Helfrich, Y. R., D. L. Sachs, and J. J. Voorhees. 2008. Overview of skin aging and photoaging. *Dermatol Nurs* 20:177–183; quiz 184.

Henderson, B. W., and T. J. Dougherty. 1992. How does photo-dynamic therapy work? *Photochem Photobiol* 55:145–157.

Hoerauf, H., G. Huttmann, H. Diddens, B. Thiele, and H. Laqua. 1994. [Photodynamic therapy of eyelid basalioma after topical administration of delta-aminolevulinic acid]. *Ophthalmologe* 91:824–829.

Hongcharu, W., C. R. Taylor, Y. Chang et al. 2000. Topical ALA-photodynamic therapy for the treatment of acne vulgaris. *J Invest Dermatol* 115:183–192.

Hoppe, U., J. Bergemann, W. Diembeck et al. 1999. Coenzyme Q10, a cutaneous antioxidant and energizer. *Biofactors* 9:371–378.

Hurlimann, A. F., G. Hanggi, and R. G. Panizzon. 1998. Photo-dynamic therapy of superficial basal cell carcinomas using topical 5-aminolevulinic acid in a nanocolloid lotion. *Dermatology* 197:248–254.

Ichinose, S., J. Usuda, T. Hirata et al. 2006. Lysosomal cathepsin initiates apoptosis, which is regulated by photodamage to Bcl-2 at mitochondria in photodynamic therapy using a novel photosensitizer, ATX-s10 (Na). *Int J Oncol* 29:349–355.

Ishida, K., T. Endo, K. Shinkai, and Y. Katoh. 2011. Shear bond strength of rebonded brackets after removal of adhesives with Er,Cr:YSGG laser. *Odontology* 99:129–134.

Itoh, Y., Y. Ninomiya, T. Henta, S. Tajima, and A. Ishibashi. 2000. Topical delta-aminolevulinic acid-based photody-namic therapy for Japanese actinic keratoses. *J Dermatol* 27:513–518.

Itoh, Y., Y. Ninomiya, S. Tajima, and A. Ishibashi. 2001. Photo-dynamic therapy of acne vulgaris with topical delta-amino-laevulinic acid and incoherent light in Japanese patients. *Br J Dermatol* 144:575–579.

Jantschitsch, C., M. Weichenthal, E. Proksch, T. Schwarz, and A. Schwarz. 2012. IL-12 and IL-23 affect photocarcinogenesis differently. *J Invest Dermatol* 132:1479–1486.

Jeffes, E. W., J. L. McCullough, G. D. Weinstein et al. 1997. Photodynamic therapy of actinic keratosis with topical 5-aminolevulinic acid. A pilot dose-ranging study. *Arch Dermatol* 133:727–732.

Jia, X. Y., F. Y. Wang, and C. Liu. 2011. [Study on the microle-akage between the composite resin and cavity wall after Er,Cr:YSGG laser preparation]. *Shanghai Kou Qiang Yi Xue* 20:577–583.

Kadekaro, A. L., K. Wakamatsu, S. Ito, and Z. A. Abdel-Malek. 2006. Cutaneous photoprotection and melanoma suscepti-bility: Reaching beyond melanin content to the frontiers of DNA repair. *Front Biosci* 11:2157–2173.

Kaidbey, K. H., P. P. Agin, R. M. Sayre, and A. M. Kligman. 1979. Photoprotection by melanin—A comparison of black and Caucasian skin. *J Am Acad Dermatol* 1:249–260.

Kaidbey, K., B. Sutherland, P. Bennett et al. 2003. Topical gly-colic acid enhances photodamage by ultraviolet light. *Photodermatol Photoimmunol Photomed* 19:21–27.

Kang, S., J. H. Chung, J. H. Lee et al. 2003. Topical N-acetyl cyste-ine and genistein prevent ultraviolet-light-induced signal-ing that leads to photoaging in human skin in vivo. *J Invest Dermatol* 120:835–841.

Karrer, S., R. M. Szeimies, U. Hohenleutner, A. Heine, and M. Landthaler. 1995. Unilateral localized basaliomatosis: Treatment with topical photodynamic therapy after appli-cation of 5-aminolevulinic acid. *Dermatology* 190:218–222.

Kasche, A., S. Luderschmidt, J. Ring, and R. Hein. 2006. Photo-dynamic therapy induces less pain in patients treated with methyl aminolevulinate compared to aminolevulinic acid. *J Drugs Dermatol* 5:353–356.

Katiyar, S. K., and C. A. Elmets. 2001. Green tea polyphenolic antioxidants and skin photoprotection (review). *Int J Oncol* 18:1307–1313.

Kato, C., Y. Taira, M. Suzuki, K. Shinkai, and Y. Katoh. 2011. Conditioning effects of cavities prepared with an Er,Cr:YSGG laser and an air-turbine. *Odontology* 100:164–171.

Kauvar, A. N., and J. S. Dover. 2001. Facial skin rejuvenation: Laser resurfacing or chemical peel: Choose your weapon. *Dermatol Surg* 27:209–212.

Kelbauskiene, S., N. Baseviciene, K. Goharkhay, A. Moritz, and V. Machiulskiene. 2011. One-year clinical results of Er,Cr:YSGG laser application in addition to scaling and root planing in patients with early to moderate periodonti-tis. *Lasers Med Sci* 26:445–452.

Kimura, M., Y. Itoh, Y. Tokuoka, and N. Kawashima. 2004. Delta-aminolevulinic acid-based photodynamic therapy for acne on the body. *J Dermatol* 31:956–960.

Kligman, L. H., and A. M. Kligman. 1986. The nature of photo-aging: Its prevention and repair. *Photodermatol* 3:215–227.

Kligman, L. H., and R. M. Sayre. 1991. An action spectrum for ultra-violet induced elastosis in hairless mice: Quantification of elastosis by image analysis. *Photochem Photobiol* 53:237–242.

Koch, R., A. Neumueller, and G. O. Schenck. 1961. [Possibilities of the use of sensitizing and desensitizing supplements in radiation chemistry and radiation biology. II. The effect of photodynamic sensitizers on tumor formation and acute lethality of roentgen-irradiated animals]. *Strahlentherapie* 114:508–524.

Kochevar, I. 1995. *Molecular and Cellular Effects of UV Radiation Relevant to Chronic Photodamage*. Blackwell Science, Cambridge, MA.

Kohl, E., and S. Karrer. 2011. Photodynamic therapy for photore-juvenation and non-oncologic indications: Overview and update. *G Ital Dermatol Venereol* 146:473–485.

Kopera, D., J. Smolle, S. Kaddu, and H. Kerl. 2004. Nonablative laser treatment of wrinkles: Meeting the objective? Assessment by 25 dermatologists. *Br J Dermatol* 150:936–939.

Kornhauser, A., R. R. Wei, Y. Yamaguchi et al. 2009. The effects of topically applied glycolic acid and salicylic acid on ultravio-let radiation-induced erythema, DNA damage and sunburn cell formation in human skin. *J Dermatol Sci* 55:10–17.

Kosmadaki, M. G., and B. A. Gilchrest. 2004. The role of telo-meres in skin aging/photoaging. *Micron* 35:155–159.

Kuijpers, D. I., M. R. Thissen, C. A. Thissen, and M. H. Neumann. 2006. Similar effectiveness of methyl aminolevulinate and 5-aminolevulinate in topical photodynamic therapy for nodular basal cell carcinoma. *J Drugs Dermatol* 5:642–645.

Kulapaditharom, B., and V. Boonkitticharoen. 1999. Photodynamic therapy for residual or recurrent cancer of the nasopharynx. *J Med Assoc Thai* 82:1111–1117.

Lang, S., R. Baumgartner, R. Struck et al. 1995. [Photodynamic diagnosis and therapy of neoplasms of the facial skin after topical administration of 5-aminolevulinic acid]. *Laryngorhinootologie* 74:85–89.

Larsen, N. E., C. T. Pollak, K. Reiner, E. Leshchiner, and E. A. Balazs. 1993. Hylan gel biomaterial: Dermal and immunologic compatibility. *J Biomed Mater Res* 27:1129–1134.

Leccia, M. T., M. Yaar, N. Allen, M. Gleason, and B. A. Gilchrest. 2001. Solar simulated irradiation modulates gene expression and activity of antioxidant enzymes in cultured human dermal fibroblasts. *Exp Dermatol* 10:272–279.

Lee, T. K., E. D. Baron, and T. H. Foster. 2008. Monitoring Pc 4 photodynamic therapy in clinical trials of cutaneous T-cell lymphoma using noninvasive spectroscopy. *J Biomed Opt* 13:030507.

Lindsay, R. M., W. Smith, W. K. Lee, M. H. Dominiczak, and J. D. Baird. 1997. The effect of delta-gluconolactone, an oxidised analogue of glucose, on the nonenzymatic glycation of human and rat haemoglobin. *Clin Chim Acta* 263:239–247.

Linge, C. 1996. Relevance of in vitro melanocytic cell studies to the understanding of melanoma. *Cancer Surv* 26:71–87.

Lui, H. 1994. Photodynamic therapy in dermatology with porfimer sodium and benzoporphyrin derivative: An update. *Semin Oncol* 21:11–14.

Lui, H., and R. R. Anderson. 1992. Photodynamic therapy in dermatology. Shedding a different light on skin disease. *Arch Dermatol* 128:1631–1636.

Lui, H., and R. R. Anderson. 1993. Photodynamic therapy in dermatology: recent developments. *Dermatol Clin* 11:1–13.

Lui, H., S. Salasche, N. Kollias et al. 1995. Photodynamic therapy of nonmelanoma skin cancer with topical aminolevulinic acid: A clinical and histologic study. *Arch Dermatol* 131:737–738.

Manna, F., M. Dentini, P. Desideri et al. 1999. Comparative chemical evaluation of two commercially available derivatives of hyaluronic acid (hylaform from rooster combs and restylane from streptococcus) used for soft tissue augmentation. *J Eur Acad Dermatol Venereol* 13:183–192.

Maxwell, W. A., and C. O. Chichester. 1971. Photodynamic responses in *Rhodotorula glutinis* in the absence of added sensitizers. *Photochem Photobiol* 13:259–273.

Mermut, O., K. R. Diamond, J. F. Cormier et al. 2009. The use of magnetic field effects on photosensitizer luminescence as a novel probe for optical monitoring of oxygen in photodynamic therapy. *Phys Med Biol* 54:1–16.

Mitani, H., I. Koshiishi, T. Sumita, and T. Imanari. 2001. Prevention of the photodamage in the hairless mouse dorsal skin by kojic acid as an iron chelator. *Eur J Pharmacol* 411:169–174.

Montecucco, C., and G. Schiavo. 1993. Tetanus and botulism neurotoxins: A new group of zinc proteases. *Trends Biochem Sci* 18:324–327.

Morgan, A. R., A. Rampersaud, G. M. Garbo, R. W. Keck, and S. H. Selman. 1989. New sensitizers for photodynamic therapy: Controlled synthesis of purpurins and their effect on normal tissue. *J Med Chem* 32:904–908.

Morganti, P., C. Bruno, F. Guarneri et al. 2002. Role of topical and nutritional supplement to modify the oxidative stress. *Int J Cosmet Sci* 24:331–339.

Morton, C. A., R. M. MacKie, C. Whitehurst, J. V. Moore, and J. H. McColl. 1998. Photodynamic therapy for basal cell carcinoma: Effect of tumor thickness and duration of photosensitizer application on response. *Arch Dermatol* 134:248–249.

Muller, P. J., and B. C. Wilson. 2006. Photodynamic therapy of brain tumors—A work in progress. *Lasers Surg Med* 38:384–389.

Munker, R. 2001. Laser blepharoplasty and periorbital laser skin resurfacing. *Facial Plast Surg* 17:209–217.

Narurkar, V. A. 2009. Nonablative fractional laser resurfacing. *Dermatol Clin* 27:473–478, vi.

Newman, N., A. Newman, L. S. Moy et al. 1996. Clinical improvement of photoaged skin with 50% glycolic acid. A double-blind vehicle-controlled study. *Dermatol Surg* 22:455–460.

Niwa, Y., and H. Akamatsu. 1991. Kojic acid scavenges free radicals while potentiating leukocyte functions including free radical generation. *Inflammation* 15:303–315.

Oikarinen, A. 1990. The aging of skin: Chronoaging versus photoaging. *Photodermatol Photoimmunol Photomed* 7:3–4.

Olivi, G., G. Chaumanet, M. D. Genovese, C. Beneduce, and S. Andreana. 2010. Er,Cr:YSGG laser labial frenectomy: A clinical retrospective evaluation of 156 consecutive cases. *Gen Dent* 58:e126–133.

Olivo, M., H. Y. Du, and B. H. Bay. 2006. Hypericin lights up the way for the potential treatment of nasopharyngeal cancer by photodynamic therapy. *Curr Clin Pharmacol* 1:217–222.

Olsen, A., G. E. Siboska, B. F. Clark, and S. I. Rattan. 1999. N(6)-Furfuryladenine, kinetin, protects against Fenton reaction-mediated oxidative damage to DNA. *Biochem Biophys Res Commun* 265:499–502.

Onay, E. O., H. Orucoglu, A. Kiremitci, Y. Korkmaz, and G. Berk. 2010. Effect of Er,Cr:YSGG laser irradiation on the apical sealing ability of AH Plus/gutta-percha and hybrid root seal/resilon combinations. *Oral Surg Oral Med Oral Pathol Oral Radiol Endod* 110:657–664.

Orenstein, A., J. Haik, J. Tamir et al. 2000. Photodynamic therapy of cutaneous lymphoma using 5-aminolevulinic acid topical application. *Dermatol Surg* 26:765–769; discussion 769–770.

Orenstein, A., G. Kostenich, H. Tsur, L. Kogan, and Z. Malik. 1995. Temperature monitoring during photodynamic therapy of skin tumors with topical 5-aminolevulinic acid application. *Cancer Lett* 93:227–232.

Orringer, J. S., S. Kang, T. M. Johnson et al. 2004. Connective tissue remodeling induced by carbon dioxide laser resurfacing of photodamaged human skin. *Archives of dermatology* 140:1326–1332.

Oxford, G. S., J. P. Pooler, and T. Narahashi. 1977. Internal and external application of photodynamic sensitizers on squid giant axons. *J Membr Biol* 36:159–173.

Pass, H. I. 1993. Photodynamic therapy in oncology: Mechanisms and clinical use. *J Natl Cancer Inst* 85:443–456.

Perussi, L. R., C. Pavone, G. J. de Oliveira, P. S. Cerri, and R. A. Marcantonio. 2012. Effects of the Er,Cr:YSGG laser on bone and soft tissue in a rat model. *Lasers Med Sci* 27:95–102.

Polte, T., and R. M. Tyrrell. 2004. Involvement of lipid peroxidation and organic peroxides in UVA-induced matrix metalloproteinase-1 expression. *Free Radic Biol Med* 36:1566–1574.

Pulkkinen, L., F. Ringpfeil, and J. Uitto. 2002. Progress in heritable skin diseases: Molecular bases and clinical implications. *J Am Acad Dermatol* 47:91–104.

Rabe, J. H., A. J. Mamelak, P. J. McElgunn, W. L. Morison, and D. N. Sauder. 2006. Photoaging: Mechanisms and repair. *J Am Acad Dermatol* 55:1–19.

Rattan, S. I., and B. F. Clark. 1994. Kinetin delays the onset of ageing characteristics in human fibroblasts. *Biochem Biophys Res Commun* 201:665–672.

Rijken, F., P. L. Bruijnzeel, H. van Weelden, and R. C. Kiekens. 2004. Responses of black and white skin to solar-simulating radiation: Differences in DNA photodamage, infiltrating neutrophils, proteolytic enzymes induced, keratinocyte activation, and IL-10 expression. *J Invest Dermatol* 122:1448–1455.

Robert, C., A. M. Robert, and L. Robert. 2005. Effect of a preparation containing a fucose-rich polysaccharide on periorbital wrinkles of human voluntaries. *Skin Res Technol* 11:47–52.

Rook, A. H., G. S. Wood, M. Duvic et al. 2010. A phase II placebo-controlled study of photodynamic therapy with topical hypericin and visible light irradiation in the treatment of cutaneous T-cell lymphoma and psoriasis. *J Am Acad Dermatol* 63:984–990.

Rosenthal, I. 1991. Phthalocyanines as photodynamic sensitizers. *Photochem Photobiol* 53:859–870.

Ryu, S. W., S. H. Lee, and H. J. Yoon. 2011. A comparative histological and immunohistochemical study of wound healing following incision with a scalpel, CO(2) laser or Er,Cr:YSGG laser in the Guinea pig oral mucosa. *Acta Odontol Scand* 70:448–454.

Rzany, B., H. Cartier, P. Kestemont et al. 2012. Correction of tear troughs and periorbital lines with a range of customized hyaluronic acid fillers. *J Drugs Dermatol* 11:27–34.

Sadick, N. S. 2003. Update on non-ablative light therapy for rejuvenation: A review. *Lasers Surg Med* 32:120–128.

Sakamoto, F. H., L. Torezan, and R. R. Anderson. 2010. Photodynamic therapy for acne vulgaris: A critical review from basics to clinical practice: Section II. Understanding parameters for acne treatment with photodynamic therapy. *J Am Acad Dermatol* 63:195–211; quiz 211–192.

Schastak, S., Y. Yafai, W. Geyer et al. 2008. Initiation of apoptosis by photodynamic therapy using a novel positively charged and water-soluble near infra-red photosensitizer and white light irradiation. *Methods Find Exp Clin Pharmacol* 30:17–23.

Sharma, M. R., B. Werth, and V. P. Werth. 2011. Animal models of acute photodamage: Comparisons of anatomic, cellular and molecular responses in C57BL/6J, SKH1 and Balb/c mice. *Photochem Photobiol* 87:690–698.

Shindo, Y., E. Witt, D. Han, W. Epstein, and L. Packer. 1994. Enzymic and non-enzymic antioxidants in epidermis and dermis of human skin. *J Invest Dermatol* 102:122–124.

Singh, T., S. C. Chaudhary, P. Kapur et al. 2012. Nitric oxide donor exisulind is an effective inhibitor of murine photocarcinogenesis (dagger). *Photochem Photobiol* 88:1141–1148.

Smith, K. C., and G. D. Schachter. 2011. YSGG 2790-nm superficial ablative and fractional ablative laser treatment. *Facial Plast Surg Clin North Am* 19:253–260.

Soter, N. 1995. *Sunburn and Suntan: Immediate Manifestations of Photodamage*. Blackwell Science, Cambridge, MA.

Spikes, J. D. 1975. Porphyrins and related compounds as photodynamic sensitizers. *Ann N Y Acad Sci* 244:496–508.

Stables, G. I., M. R. Stringer, D. J. Robinson, and D. V. Ash. 1997. Large patches of Bowen's disease treated by topical aminolaevulinic acid photodynamic therapy. *Br J Dermatol* 136:957–960.

Staneloudi, C., K. A. Smith, R. Hudson et al. 2007. Development and characterization of novel photosensitizer:scFv conjugates for use in photodynamic therapy of cancer. *Immunology* 120:512–517.

Stiller, M. J., J. Bartolone, R. Stern et al. 1996. Topical 8% glycolic acid and 8% L-lactic acid creams for the treatment of photodamaged skin. A double-blind vehicle-controlled clinical trial. *Arch Dermatol* 132:631–636.

Stranadko, E. F., M. I. Garbuzov, V. G. Zenger et al. 2001. [Photodynamic therapy of recurrent and residual oropharyngeal and laryngeal tumors]. *Vestn Otorinolaringol*: 36–39.

Stringer, M. R., P. Collins, D. J. Robinson, G. I. Stables, and R. A. Sheehan-Dare. 1996. The accumulation of protoporphyrin IX in plaque psoriasis after topical application of 5-aminolevulinic acid indicates a potential for superficial photodynamic therapy. *J Invest Dermatol* 107:76–81.

Szabo, G. 1959. *Pigment Cell Biology*. Academic Press, New York.

Szeimies, R. M., P. Calzavara-Pinton, S. Karrer, B. Ortel, and M. Landthaler. 1996a. Topical photodynamic therapy in dermatology. *J Photochem Photobiol B* 36:213–219.

Szeimies, R. M., S. Karrer, A. Sauerwald, and M. Landthaler. 1996b. Photodynamic therapy with topical application of 5-aminolevulinic acid in the treatment of actinic keratoses: An initial clinical study. *Dermatology* 192:246–251.

Szeimies, R. M., T. Sassy, and M. Landthaler. 1994. Penetration potency of topical applied delta-aminolevulinic acid for photodynamic therapy of basal cell carcinoma. *Photochem Photobiol* 59:73–76.

Tajirian, A. L., and D. J. Goldberg. 2011. Fractional ablative laser skin resurfacing: A review. *J Cosmet Laser Ther* 13:262–264.

Tanzi, E. L., J. R. Lupton, and T. S. Alster. 2003. Lasers in dermatology: four decades of progress. *J Am Acad Dermatol* 49:1–31; quiz 31–34.

Tanzi, E. L., and Alster T. S. 2004. Comparison of a 1450-nm diode laser and a 1320-nm Nd:YAG laser in the treatment of atrophic facial scars: a prospective clinical and histologic study. *Dermatol Surg,* 30(2 Pt 1): 152–157.

Tegos, G. P., T. N. Demidova, D. Arcila-Lopez et al. 2005. Cationic fullerenes are effective and selective antimicrobial photosensitizers. *Chem Biol* 12:1127–1135.

Tierney, E., B. H. Mahmoud, D. Srivastava, D. Ozog, and D. J. Kouba. 2009. Treatment of surgical scars with nonablative fractional laser versus pulsed dye laser: A randomized controlled trial. *Dermatol Surg* 35:1172–1180.

Tolmasoff, J. M., T. Ono, and R. G. Cutler. 1980. Superoxide dismutase: Correlation with life-span and specific metabolic rate in primate species. *Proc Natl Acad Sci U S A* 77:2777–2781.

Tornaletti, S., and G. P. Pfeifer. 1996. UV damage and repair mechanisms in mammalian cells. *Bioessays* 18:221–228.

Trelles, M. A., X. Alvarez, M. J. Martin-Vazquez et al. 2005. Assessment of the efficacy of nonablative long-pulsed 1064-nm Nd:YAG laser treatment of wrinkles compared at 2, 4, and 6 months. *Facial Plast Surg* 21:145–153.

Trentham, D. E. 1986. Adverse reactions to bovine collagen implants. Additional evidence for immune response gene control of collagen reactivity in humans. *Arch Dermatol* 122:643–644.

Tsai, J. C., C. P. Chiang, H. M. Chen et al. 2004. Photodynamic Therapy of oral dysplasia with topical 5-aminolevulinic acid and light-emitting diode array. *Lasers Surg Med* 34:18–24.

Tsurumaki Jdo, N., B. H. Souto, G. J. Oliveira et al. 2011. Effect of instrumentation using curettes, piezoelectric ultrasonic scaler and Er,Cr:YSGG laser on the morphology and adhesion of blood components on root surfaces: A SEM study. *Braz Dent J* 22:185–192.

Uitto, J., L. Pulkkinen, and F. Ringpfeil. 2002. Progress in molecular genetics of heritable skin diseases: The paradigms of epidermolysis bullosa and pseudoxanthoma elasticum. *J Invest Dermatol Symp Proc* 7:6–16.

Umegaki, N., R. Moritsugu, S. Katoh et al. 2004. Photodynamic therapy may be useful in debulking cutaneous lymphoma prior to radiotherapy. *Clin Exp Dermatol* 29:42–45.

Vayalil, P. K., A. Mittal, Y. Hara, C. A. Elmets, and S. K. Katiyar. 2004. Green tea polyphenols prevent ultraviolet light-induced oxidative damage and matrix metalloproteinases expression in mouse skin. *J Invest Dermatol* 122:1480–1487.

Vaziri, H., and S. Benchimol. 1996. From telomere loss to p53 induction and activation of a DNA-damage pathway at senescence: The telomere loss/DNA damage model of cell aging. *Exp Gerontol* 31:295–301.

Wang, F., L. A. Garza, S. Kang et al. 2007. In vivo stimulation of de novo collagen production caused by cross-linked hyaluronic acid dermal filler injections in photodamaged human skin. *Arch Dermatol* 143:155–163.

Weiss, R. A., A. C. Harrington, R. C. Pfau, M. A. Weiss, and S. Marwaha. 1999. Periorbital skin resurfacing using high energy erbium:YAG laser: Results in 50 patients. *Lasers Surg Med* 24:81–86.

Weiss, R. A., M. A. Weiss, and K. L. Beasley. 2002. Rejuvenation of photoaged skin: 5 years results with intense pulsed light of the face, neck, and chest. *Dermatol Surg* 28:1115–1119.

West, T. B., and T. S. Alster. 1998. Improvement of infraorbital hyperpigmentation following carbon dioxide laser resurfacing. *Dermatol Surg* 24:615–616.

Wiegell, S. R., and H. C. Wulf. 2006. Photodynamic therapy of acne vulgaris using 5-aminolevulinic acid versus methyl aminolevulinate. *J Am Acad Dermatol* 54:647–651.

Willis, I., and R. M. Earles. 2005. A new skin classification system relevant to people of African descent. *J Cosmet Dermatol* 18:209–216.

Wind, B. S., A. A. Meesters, M. W. Kroon et al. 2012. Formation of fibrosis after nonablative and ablative fractional laser therapy. *Dermatol Surg* 38:437–442.

Wolf, P., R. Fink-Puches, A. Reimann-Weber, and H. Kerl. 1997. Development of malignant melanoma after repeated topical photodynamic therapy with 5-aminolevulinic acid at the exposed site. *Dermatology* 194:53–54.

Wolf, P., E. Rieger, and H. Kerl. 1993. Topical photodynamic therapy with endogenous porphyrins after application of 5-aminolevulinic acid. An alternative treatment modality for solar keratoses, superficial squamous cell carcinomas, and basal cell carcinomas? *J Am Acad Dermatol* 28:17–21.

Wolff, K., L. Goldsmith, S. Katz, B. Gilchrest, A. Paller, and D. Leffell. 2008. *Fitzpatrick's Dermatology in General Medicine*, 7th edition. 286. New York: McGraw Hill.

Wu, L. Y., X. Q. Zheng, J. L. Lu, and Y. R. Liang. 2009. Protective effect of green tea polyphenols against ultraviolet B-induced damage to HaCaT cells. *Hum Cell* 22:18–24.

Yasui, H., and H. Sakurai. 2003. Age-dependent generation of reactive oxygen species in the skin of live hairless rats exposed to UVA light. *Exp Dermatol* 12:655–661.

Yavari, H. R., S. Rahimi, S. Shahi et al. 2010. Effect of Er, Cr: YSGG laser irradiation on *Enterococcus faecalis* in infected root canals. *Photomed Laser Surg* 28 Suppl 1:S91–S96.

Yin, R., F. Hao, J. Deng, X. C. Yang, and H. Yan. 2010. Investigation of optimal aminolaevulinic acid concentration applied in topical aminolaevulinic acid-photodynamic therapy for treatment of moderate to severe acne: A pilot study in Chinese subjects. *Br J Dermatol* 163:1064–1071.

Young, L. C., J. B. Hays, V. A. Tron, and S. E. Andrew. 2003. DNA mismatch repair proteins: Potential guardians against genomic instability and tumorigenesis induced by ultraviolet photoproducts. *J Invest Dermatol* 121:435–440.

Zeisser-Labouebe, M., N. Lange, R. Gurny, and F. Delie. 2006. Hypericin-loaded nanoparticles for the photodynamic treatment of ovarian cancer. *Int J Pharm* 326:174–181.

Zhang, J., L. Deng, J. Yao et al. 2011. Synthesis and photobiological study of a novel chlorin photosensitizer BCPD-18MA for photodynamic therapy. *Bioorg Med Chem* 19:5520–5528.

11

UVR-Induced Immunosuppression

Mary Norval
*University of Edinburgh
Medical School*

Prue H. Hart
University of Western Australia

11.1 Introduction

The suppression of immune responses to an antigen encountered within a short period after exposure to ultraviolet radiation (UVR) was first described by Margaret Kripke and her colleagues almost 40 years ago. They revealed that when highly antigenic tumor cells were implanted into mice that had been previously irradiated, the cells were not rejected, as might be expected, but grew into tumors. The suppression was shown subsequently to be immunologically mediated, antigen-specific, and long-lasting (Kripke 1981). These results stimulated new ideas and discoveries of great interest and relevance to skin biologists and immunologists. Despite considerable advances, the exact mechanism whereby UVR can lead to changes in immune responses is not entirely clear at present, and no unifying scenario has been proposed.

Several chromophores located at or near the skin surface are involved as initiators of the process following their absorption of UV photons, and then several pathways are possible, ending with the generation of a variety of antigen-specific immune regulatory cells and mediators that suppress immunity. Which of these chromophores or pathways is of major importance may be influenced by various factors that have rarely been defined, but considerable redundancy in the responses is indicated. These variables include the UV source itself (spectrum, dose, intensity, frequency of exposure, area of body irradiated, length of time between exposure, and antigen application), the antigen (whether novel or previously experienced, type, quantity, route of administration, host species,

and strain), and the immune response generated (innate, T cell–mediated, B cell–mediated, in skin, or internal organ). Much of the information relating to UV-induced immunosuppression has been obtained in mouse models, frequently using highly manipulated or transgenic strains, or in *in vitro* systems. It has not been possible to test many of these findings in human subjects for obvious ethical and practical reasons, but, in the instances where this has been achieved, similar results have been obtained.

In this chapter, the chromophores and the production of immune mediators will be described first, followed by sections on the mechanisms involved when the antigen is applied directly to the irradiated site (termed local UV-induced immunosuppression) and when the antigen is applied to a site distant from the site of UVR (termed systemic UV-induced immunosuppression) exposure. The beneficial and harmful consequences of the downregulation in immunity for human health are then considered before various conclusions are reached.

11.2 Chromophores and Induction of Immune Mediators

The UVR in terrestrial sunlight covers the wavelengths between 290 and 400 nm and is divided into the UVB (290–320 nm) and UVA (320–440 nm) wavebands, with the latter being subdivided into UVAI (340–400 nm) and UVAII (320–340 nm). On a clear day in summer, UVB constitutes approximately 6% of terrestrial UVR. This percentage varies considerably with latitude, altitude, season of the year, time of day, cloud cover, aerial pollution, and surface reflection. UVB has poor penetrating power into the skin layers with almost all being absorbed in the epidermis. On the other hand, UVA can penetrate further and reach the dermis or even the subcutis layer. Almost all the evidence to date indicates that chromophores present in the epidermis, and perhaps in the dermis, are required for the initiation of the immunomodulation that follows the exposure. It is thought very unlikely that the UVR can directly affect immune cells in the blood or lymph circulating through the skin.

The production of an action spectrum for UV-induced immunosuppression could provide an indication for which cutaneous chromophore might be involved. Such an action spectrum was constructed in 1983, using mice irradiated with different wavelengths of UVR followed by application of a contact sensitizer on a distant unirradiated site and then challenged with the contact sensitizer several days later with subsequent measurement of the contact hypersensitivity (CHS) response. The peak effectiveness for the systemic suppression in this response was at 260–280 nm (De Fabo and Noonan 1983). However, more recent studies, also in mice, have shown a more complicated picture. In addition to UVB, UVAI caused systemic suppression of the induction of CHS and of delayed hypersensitivity (DTH), but, unlike the case with UVB where the dose response was linear, the dose response to UVAI was bell-shaped with no suppressive effect at either low or high doses (Halliday et al. 2012). An action spectrum for the local suppression of the memory CHS response to nickel in human subjects has also been established (Halliday et al. 2012). Two peaks of effectiveness were found: one in the UVB waveband at 300 nm and another in the UVAI waveband at 370 nm.

Thus, the situation at present is far from clear, but, after exposure to the range of UVB and UVA wavelengths present in natural sunlight, the involvement of more than one chromophore is likely. Details of the chromophores for which there is most evidence are described below.

11.2.1 DNA

DNA is a major absorber of UVR in the skin. The exposure results in the production of specific DNA lesions. Following UVB, cyclobutane pyrimidine dimers (CPDs) and pyrimidine (6-4) pyrimidone photoproducts (6-4 PPs) are formed, which, if unrepaired, represent the "UVB signature" mutations found in many skin tumors. Following UVA, there is excitation of non-DNA chromophores with the formation of products, such as 8-oxo-7,8-dihydroguanine, induced by reactive oxygen species (ROS) base oxidation. Very recently, UVAI irradiation was shown to cause the formation of CPDs (but not 6-4 PPs) and that their number increased with increasing epidermal depth, whereas it decreased following UVB (Tewari, Sarkay, and Young 2012).

A series of elegant experiments in mice in the 1990s demonstrated that UV-induced DNA damage was critical to the induction of immunosuppression. Mice were irradiated followed by topical application of a contact sensitizer to an unirradiated site, and then challenged with the contact sensitizer on the ears a few

days later. Suppression in the CHS (as measured by ear swelling) was observed. However, if the irradiated mice were treated with liposomes containing a dimer-specific excision-repair enzyme (T4 endonuclease V) or a dimer-specific photolyase to repair the CPDs before the sensitization step, the subsequent CHS was not suppressed (Vink et al. 1998).

The DNA damage leads to the production of a range of cytokines, such as interleukin-6 (IL-6), IL-1, and tumor necrosis factor-α (TNF-α), and other immune mediators, such as platelet activating factor (PAF), prostaglandin E_2 (PGE_2), and histamine, which have significant effects on several cell populations within the skin (see Section 11.3). Interestingly, IL-12, IL-18, and IL-23 have all been shown to reduce cutaneous DNA damage and hence to inhibit the immunosuppression that follows UVR (Schwarz and Schwarz 2011). They are thought to do this by the induction or activation of nucleotide excision repair. Another role of IL-12 and IL-23 is to inhibit the activity of T regulatory cells (Tregs) (see Section 11.3.1) (Schwarz and Schwarz 2011).

11.2.2 Urocanic Acid

Urocanic acid (UCA) is produced as the *trans*-isomer from histidine in the upper layers of the epidermis in which filaggrin, a histidine-rich protein, is cleaved. *Trans*-UCA accumulates in a high concentration (at least 6 nmol/cm² of surface skin in humans) as the enzyme urocanase, which catabolizes it, is not present in this site. *Trans*-UCA is crucial for efficient epidermal barrier function and is also a major epidermal chromophore. Through absorption of UVR, it converts to the *cis*-isomer in a dose-dependent manner until the photostationary state is reached at approximately 60% *cis*-UCA. The action spectrum for the *trans* to *cis* isomerization peaks at approximately 300–315 nm in mouse skin versus 280–310 nm in human skin. Some photoisomerization occurs after exposure to both UVAI and UVAII radiation. *Cis*-UCA remains in the epidermis of human subjects for at least 2 weeks following UVR, gradually returning to the background level during this time period. After a single exposure, it is also present in the dermis for several weeks, is in the blood for 1 to 2 days, and is excreted in the urine for approximately 2 weeks. These properties are reviewed in Gibbs, Tye, and Norval (2008).

Cis-UCA was first suggested by de Fabo and Noonan (1983) to act as an initiator for UV-induced immunosuppression. This was based on its superficial location in the epidermis, the similarity of its absorption spectrum to the action spectrum of the suppression in CHS in irradiated mice, and its photochemical properties (de Fabo and Noonan 1983). Since these early days, *cis*-UCA, administered as a "pure" molecule, has been shown to mimic many of the down-regulatory effects of UVR on immunity in mice (reviewed in Norval and El-Ghorr 2002). In addition, if mice are treated with a *cis*-UCA monoclonal antibody just prior to exposure to UVR and administration of a complex protein, on challenge with the protein, the expected suppression in DTH is abrogated (Gibbs, Tye, and Norval 2008).

How, or even where, *cis*-UCA acts as a suppressor of immune responses is not clear currently. The involvement of a range of mediators and cell populations is probable. These possibilities include the generation of ROS with oxidative DNA damage and release of PAF and PGE_2, binding to the serotonin (5HT-2A) receptor, thus activating it with effects on mast cells and release of anti-inflammatory cytokines, induction of keratinocyte apoptosis and cell growth arrest, and stimulation of the production of neuropeptides that cause mast cell activation and perhaps their migration (Gibbs, Tye, and Norval 2008; Sreevidya et al. 2010; Kaneko et al. 2011).

11.2.3 Membrane Phospholipids

Clustering and subsequent internalization of cell membrane–bound growth factor receptors have been identified as initiating events in UVB-induced immunosuppression (Rosette and Karin 1996). This change occurs independently of the cell nucleus and therefore does not involve DNA. Tryptophan found in the cytoplasm is thought to be involved as the chromophore in this stress response. On UVR, tryptophan forms a number of photoproducts that serve as arylhydrocarbon (AhR) ligands (reviewed in Krutmann, Morita, and Chung 2012). Activation of the AhR signaling pathway increases the transcription of a number of genes, including cyclooxygenase-2, which, in turn, can upregulate PAF and PGE_2 production and cytokine release.

11.2.4 7-Dehydrocholesterol

Upon absorption of UVB photons, 7-dehydrocholesterol in the skin is converted to pre-vitamin D_3, which then isomerizes to vitamin D_3 (cholecalciferol) through a thermal reaction. Much of the vitamin D_3 is stepwise hydroxylated in the liver to 25-hydroxyvitamin D_3 [25(OH)D] and then in the kidney to 1,25-dihydroxyvitamin D_3 [1,25(OH)$_2$D], the active form of vitamin D. In addition to this route, some evidence indicates that 1,25(OH)$_2$D can be formed within the skin itself within 16 h of UVR exposure (Bouillon et al. 2008).

All immune cells express the vitamin D receptor (VDR), and ligand binding signals alter transcription of more than 200 genes. 1,25(OH)$_2$D in the skin activates innate immune responses, such as the production of antimicrobial peptides, which can enhance microbial killing, and the stimulation of macrophage differentiation and phagocytosis. In the skin and draining lymph nodes, 1,25(OH)$_2$D suppresses adaptive immune responses, largely by blocking the maturation of myeloid dendritic cells through reducing the expression of costimulatory molecules, by inhibiting IL-12 secretion, and by increasing IL-10 production (Hart, Gorman, and Finlay-Jones 2011). Direct effects on T cells have been reported, including inhibiting the secretion of IFN-γ by T helper 1 (Th1) cells; stimulating Th2 cell development with greater production of IL-4, IL-5, and IL-10; and increasing the number of IL-10-producing T regulatory cells (Tregs) (Hart, Gorman, and Finlay-Jones 2011). Topical 1,25(OH)$_2$D can enhance the regulatory activity of CD4+CD25+ cells in the draining lymph nodes.

The contribution of UVR-induced vitamin D to the immunosuppression that follows UV exposure is not clear. Mice deficient in the VDR or CYP27B1, the enzyme that converts 25(OH)D to 1,25(OH)$_2$D, should help to answer this question. However, these mice have serious developmental problems and abnormal skin physiology, and therefore cannot reliably help to determine vitamin D's role. The most important test will be whether vitamin D supplementation can reduce the risk of diseases associated with the lack of UVR exposure (see Section 11.5.5 and Hart, Gorman, and Finlay-Jones 2011).

11.3 Mechanisms of Local Immunosuppression

In irradiated sites following absorption of UV photons by the cutaneous chromophores, the production of immune mediators, such as PAF, histamine, PGE_2, interleukins, and other cytokines, has profound effects on several cell populations locally and on their ability to respond to antigens. Three of these populations are described in the following sections, and the process is depicted in Figure 11.1. The end result is an altered interaction with T and B cells in the draining lymph nodes with generation of antigen-specific regulatory cells capable of producing immunosuppressive factors themselves or stimulating the release of immunosuppressive cytokines from other cell types.

11.3.1 Role of Langerhans Cells

As early as 1980, Toews, Bergstresser, and Streilein revealed that the Langerhans cells (LCs) were dramatically modified by UVR (Toews, Bergstresser, and Streilein 1980). These cells normally form a dendritic network in the epidermis and migrate to the draining lymph node on contact with antigens. They mature functionally and phenotypically as they move. After UV exposure, the network is lost and, in a process stimulated largely by TNF-α, a proportion of the LCs migrates to the lymph nodes, draining the irradiated site. Evidence has been published to indicate that if an antigen is applied to the irradiated site within a short period of the exposure, there is an imbalance in the T cell response in the draining lymph nodes with inhibition of T helper 1 cytokine production and promotion of the immunosuppressive cytokines IL-10 and IL-4 (Simon et al. 1992). In contrast to the increased expression of IL-12 by CD11c+ cells (predominantly dendritic cells) in the draining lymph nodes following administration of a contact sensitizer, there is no upregulation of this cytokine when the sensitizer is applied to the UV-irradiated site (Toichi et al. 2008).

Until recently, the LCs were assumed to represent the major antigen presenting cells of the skin, responsible for initiating immunostimulatory responses to exogenous insults. Recently, this view has changed, and their main function is proposed to be the activation of regulatory cells in the draining lymph nodes (Romani, Brunner, and Stingl 2012). Fukunaga et al. (2010) showed that the LCs that migrate to the lymph nodes in response to UVR and antigens can activate natural killer T (NKT) cells to produce IL-4. Although expressing surface markers of both T cells (CD4, ab-TCR) and NK cells (NK1.1, DX5), NKT cells do

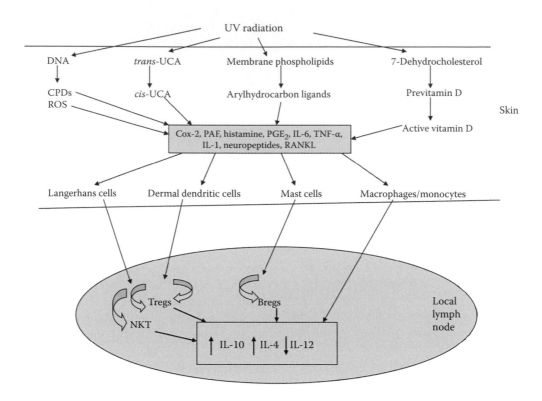

FIGURE 11.1 Outline of the steps leading to local UV-induced immunosuppression. Abbreviations: Bregs: B regulatory cells; Cox: cyclooxygenase; CPD: cyclobutane pyrimidine dimer; IL: interleukin; NKT: natural killer T cells; PAF: platelet activating factor; PGE_2: prostaglandin E_2; RANKL: receptor activator of the NF-κB ligand; ROS: reactive oxygen species; TNF: tumor necrosis factor; Tregs: T regulatory cells; UCA: urocanic acid.

not recognize antigen–MHC complexes but instead are activated by lipid antigens presented by CD1. Other studies demonstrate that UV-damaged but still viable LCs can induce Tregs in the draining lymph nodes (Maeda et al. 2008). Such UVR-Tregs have the phenotype CD4+CD25+CTLA-4+CD62L+FoxP3+neuropilin+ and are generally considered antigen-specific. They release IL-10 on activation, are cytotoxic for antigen-presenting cells, and can suppress the activation, cytokine production, and proliferation of other types of immunostimulatory T cells (reviewed in Schwarz and Schwarz 2010). Once generated, these Tregs are long-lasting, leading to tolerance so that, if the same antigen is encountered at a future date, the immune response continues to be suppressed.

To add further to this already complex picture, some doubt has been expressed about whether LCs play any role at all in UV-induced immunosuppression. Indeed, the suggestion is that, rather than the LCs, the Langerin+ dermal dendritic cells might be key in this respect (Wang, Jameson, and Hogquist 2009), at least in the mouse, although such cells have not been described as yet in human skin.

11.3.2 Role of Mast Cells

Mast cells are located in all organs close to nerves and blood vessels and provide a conduit for integrating inflammatory responses between immune cells and the soluble mediators they produce, nerves, and changes in blood flow. Mast cells are proinflammatory upon IgE binding. However, depending on the environment, mast cells can be both negative and positive regulators of immunity. They have also been described as versatile and "tunable" as they can control the intensity of immune responses.

There are dynamic changes in the number of dermal mast cells in response to erythemal UV exposure. Mast cell density quickly increases and peaks in skin 6 h after irradiation of mice; this is followed by their migration to skin-draining lymph nodes after 24 h by a CXCR4-CXCL12-dependent pathway (Ullrich and Byrne 2011). There may also be several signals necessary for activation of mast cells in UV-irradiated skin, such as neuropeptides, *cis*-UCA, and PAF; on the other hand, for mast cell activation in skin, upon chronic exposure to low-dose UVR, 1,25(OH)₂D is necessary. The immunomodulatory products of activated mast cells include TNF-α, histamine, and IL-10 (Ullrich and Byrne 2011; Hart, Gorman, and Finlay-Jones 2011), each of which may contribute to immunosuppression by exerting their effects on antigen-presenting cells and T lymphocytes. CXCR4+IL-10+ mast cells draining from UV-irradiated skin locate to the lymph node B cell areas where they suppress T-dependent antibody responses by an IL-10-dependent process involving reduced T follicular helper cell function and suppressed germinal center formation (Chacon-Salinas et al. 2011).

It is plausible that different products are released from the mast cells according to the intensity of the activating stimulus (high- or low-dose UVR) and their location. In consideration of their more traditional role, mast cells may also help to regulate the vascular and cellular responses to UVR-induced skin inflammation.

11.3.3 Role of Macrophages/Monocytes

A special population of macrophages/monocytes (F4/80⁺CD11c⁻) infiltrate the skin after UVR. They first appear in the dermis and then in the epidermis. Three hours after application of a contact sensitizer to the irradiated site, they migrate to the draining lymph nodes (Toichi et al. 2008). Such cells are capable of releasing IL-10 and are therefore thought to be involved in UV-induced immunosuppression.

11.4 Mechanisms of Systemic Immunosuppression

As indicated above, local immunosuppression by UVR may reflect inefficient handling of antigens by antigen-presenting cells, which are initially located at the UV-irradiated site, later migrating to the draining nodes. However, in terms of systemic immunosuppression by UVR whereby responses to antigens applied to distant nonirradiated sites are reduced, the immunological pathways are less clear. The possibilities are outlined in the following sections and illustrated in Figure 11.2.

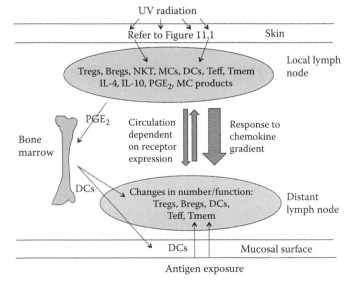

FIGURE 11.2 Cells and mediators associated with systemic UV-induced immunosuppression. Abbreviations: Bregs: B regulatory cells; DCs: dendritic cells; IL: interleukin; MCs: mast cells; NKT: natural killer cells; PGE₂: prostaglandin E₂; Teff: T effector cells; Tmem: T memory cells; Tregs: T regulatory cells.

11.4.1 Changes in Dendritic Cells in Bone Marrow and Lymph Nodes Draining and Not Draining the UV-Irradiated Skin

It has been proposed that dendritic cells at nonirradiated sites are changed such that they have altered antigen-presenting abilities and thus, in a manner similar to that reported for dendritic cells from UV-irradiated sites, they stimulate reduced effector T cell responses and induce antigen-specific Tregs. In studies of lymph nodes that drain nonirradiated sites in mice, the results have varied, with some indicating dendritic cell-enriched populations that produce lower levels of IL-12 (or inactive IL-12 p40 homodimers) and are therefore inefficient at antigen presentation (Ullrich and Byrne 2012). Although contributory in some models and possibly of greater importance with increased time after UV exposure, the primary defect in UVR-induced systemic immunosuppression may not always relate to immature or functionally impaired antigen-presenting cells in lymph nodes distant to UV-irradiated skin. It is likely that soluble mediators produced in the UV-irradiated site or regulatory cells induced in nodes draining UV-irradiated sites have long-range immune effects. UVR-induced Tregs can "switch" antigen-presenting cells from a stimulatory to a regulatory phenotype (Schwarz and Schwarz 2011). Similarly, UVR-induced Tregs and their migration responses can be reprogrammed by tissue dendritic cells *in vivo* to move from lymph nodes to peripheral tissues (Schwarz et al. 2011).

Erythemal UVR also stimulates the production of poorly immunogenic dendritic cells from bone marrow (Ng et al. 2010). In initial studies, bone marrow cells, harvested 3 days after irradiation of mice and cultured for 7 days in a medium containing growth factors that stimulate the differentiation of dendritic cells, were assessed for their priming ability. The generation of poorly priming dendritic cells was still apparent when bone marrow cells were harvested up to 10 days after UV exposure. The effect of UVR on the bone marrow dendritic cell progenitors was not detected if the mice were treated with indomethacin, a cyclooxygenase inhibitor, prior to UV exposure (Ng et al. 2010). Further, poorly immunogenic dendritic cells were generated from the bone marrow of PGE₂ administered mice. Thus, PGE₂, directly or indirectly, regulates dendritic cell progenitors in the bone marrow. Indeed, many studies have shown an important role for PGE₂ in UVR-induced systemic immunosuppression, with keratinocyte-derived PGE₂ implicated in UVR-suppressed CHS responses (Hart, Gorman, and Finlay-Jones 2011). Recently, PGE₂ binding to the prostaglandin E receptor subtype 4 (EP4) on keratinocytes was shown to regulate the expression of RANKL, the epidermal receptor activator of the NF-κB ligand (Soontrapa et al. 2011). This leads to the activation of epidermal dendritic cells, which express RANK, the receptor for RANKL, and subsequently the migration of these cells to the draining lymph nodes (Loser et al. 2006). Once there, they stimulate the proliferation of Tregs.

The UV-induced change to the dendritic cell progenitors in bone marrow may provide a more sustained response compared with other immune alterations as the bone marrow is the site of development of hematopoietic cells for subsequent migration to the periphery and for the replenishment of dying cells during an inflammatory response.

11.4.2 Role of Dermal Mast Cells

The downregulatory role for dermal mast cells in UVR-induced systemic immunosuppression was first reported by Hart et al. (1998). This group demonstrated that UVR-susceptible mice (defined as requiring only a low UVR dose for suppression of CHS) had a significantly increased prevalence of dermal mast cells in dorsal skin, while a relatively UVR-resistant mouse strain (defined as requiring a higher UVR dose for suppression of CHS) had a reduced prevalence of dermal mast cells. Further, the ability of UVR to reduce CHS responses to haptens was reduced in mice depleted in skin mast cells. The suppressive ability of UVR was restored if the mice were experimentally reconstituted in their skin with bone marrow-derived mast cells prior to UV exposure. Several groups have now confirmed a controlling influence of mast cells in responses to both UVB and UVA irradiation (Hart, Gorman, and Finlay-Jones 2011; Ullrich and Byrne 2012). In the initial experiments of Hart et al. (1998), UVR suppression of local CHS responses to haptens was not altered by depletion of dermal mast cells. Thus, it is possible that mast cells may play a different and perhaps more significant role in systems compared to local UV-induced immunosuppression (see Section 11.3.2).

11.4.3 Role of NKT Cells

As mentioned in Section 11.3.1, recent experimentation suggests that LCs from UV-irradiated skin colocalize with NKT cells in the lymph nodes draining the UV-irradiated skin, leading to increased production of IL-4, a cytokine that can suppress Th1 immunity (Fukunaga et al. 2010). IL-4 has previously been implicated in UVR-induced systemic immunosuppression (Ullrich and Byrne 2012). The importance of this pathway was highlighted by the absence of UVR-induced systemic immunosuppression in mice deficient in LCs, or if LCs from UV-irradiated mice were transferred into mice without NKT cells (CD1$^{-/-}$ or Ja18$^{-/-}$), or if the mice were treated with an anti-IL-4 antibody (Fukunaga et al. 2010). These results support a previous report in which transferring NKT cells from the spleens of mice exposed to chronic UVR suppressed the immune response in the recipient mice (Ullrich and Byrne 2012). It is possible that UVR damage to skin may alter glycolipids that bind to CD1$^+$ LCs, which, upon migration to the lymph nodes, stimulate the NKT cells to produce immunoregulatory IL-4.

11.4.4 Role of T and B Regulatory Cells

As discussed in Section 11.4.1, Tregs may be induced by UVR-damaged dendritic cells (possibly consequent to oxidative damage to dendritic cell DNA, membrane lipids, and proteins) that

suboptimally present antigens to cells in the draining nodes. By expression of CD62L, these Tregs accumulate in the lymph nodes and not in skin; they are efficient at preventing the induction of immune responses but poor at controlling inflammatory responses (such as the elicitation of CHS responses). However, UVR-induced Tregs can be reprogrammed by dendritic cells from skin (Schwarz et al. 2011) and may have greater potential *in vivo* in controlling disease than originally thought. Not all UV-induced CD4$^+$ Tregs in lymph nodes draining UV-irradiated skin are antigen-specific. Gorman et al. (2007) demonstrated that CD4$^+$CD25$^+$ cells from the nodes of UV-irradiated skin could suppress immune responses to new antigens both *in vitro* and *in vivo*. As topically applied 1,25(OH)$_2$D led to the same outcome (Hart, Gorman, and Finlay-Jones 2011), the increased regulatory activity of CD4$^+$CD25$^+$ cells in the skin-draining lymph nodes of UVR-treated skin is probably a result of UVR-induced vitamin D$_3$.

IL-10 secreting Bregs have also been described in the lymph nodes draining sites of skin exposed to erythemal UVR (Byrne and Halliday 2005). These cells suppress dendritic cell function. Evidence suggests that PAF and serotonin are involved in activation of the Bregs (Ullrich and Byrne 2012).

Not all responses suppressed by UVR have been linked with the induction of Tregs or Bregs. Fewer effector CD4$^+$CD25$^+$ cells have been recorded in the trachea and airway draining lymph nodes of mice with reduced allergic airway disease resulting from prior UV irradiation of skin, but no UVR-induced cells with regulatory properties could be identified (McGlade et al. 2010). Similarly, in mice irradiated with multiple low doses of UV, in the absence of Tregs, CHS responses were reduced (Rana et al. 2008). In these models, it is proposed that fewer memory T cells develop in response to smaller pools of activated T cells in the UV-irradiated mice. It seems likely that there are other UVR-induced regulatory cell types that have not yet been identified.

11.5 Consequences of UV-Induced Immunosuppression for Human Diseases

The immunosuppression that follows UV and antigen exposure has some harmful as well as some beneficial outcomes on human health. These are depicted in outline in Figure 11.3 and are described in the following sections.

11.5.1 For Photosensitivity

One beneficial outcome of the immunosuppression that follows UVR might be the prevention of an inflammatory allergic response to any neoantigens produced in the skin as a result of the mutagenic activity of UVR. However, there are a number of photosensitivity disorders, all associated with an abnormal cutaneous response following UVR or visible light. The commonest of these is polymorphic light eruption (PLE). It is classified as an idiopathic (immunological) disease and is present in

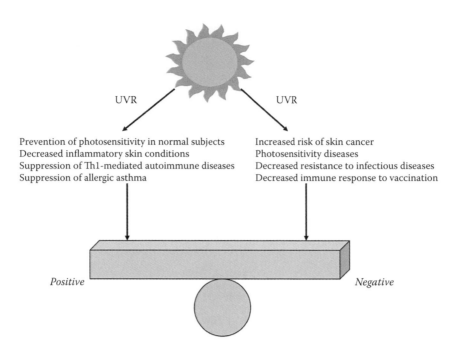

FIGURE 11.3 Potential positive and negative consequences of solar UV radiation for human health.

approximately 5%–20% of the population of Europe and North America. The disease is characterized by red, itchy eruptions that develop on body sites exposed to the sun. The most effective wavelengths for provoking the symptoms of PLE are not clear and may vary from person to person and also depend on the dose and spectrum of sunlight. Even though the amount of UV exposure varies in different latitudes, a recent European survey demonstrated no latitude gradient in the prevalence of PLE.

The evidence published to date suggests that individuals with PLE respond to any sunlight-induced neoantigens by an allergic reaction, a form of DTH; that is, they lack the ability to be immunosuppressed. The mechanism involved has not been defined but has been attributed to fewer neutrophils and suppressor macrophages migrating into, and fewer LCs migrating away from, the exposed skin site compared with normal subjects (reviewed in Wolf, Byrne, and Wackernagel 2009). An enhanced proinflammatory cytokine response may also occur as might altered neuroendocrine signaling locally in the exposed site. A further change in PLE in comparison with normal people is that the ability to be tolerized (defined as long-term immune unresponsiveness) by UVR is impaired in PLE.

However one advantage that the PLE subjects might be predicted to have is a reduction in the incidence of skin cancer because of the lack of immunosuppression that follows exposure to sunlight, the major environmental risk factor for skin cancer. This was found to be true in a prospective case-control study in which the prevalence of PLE was 7.5% in subjects with skin cancer compared with 21.4% in gender and age-matched controls without skin cancer (Lembo et al. 2008). The harmful consequence of UV-induced immunosuppression relating to skin cancer is discussed in Section 11.5.3.

Another example of photosensitivity is found as part of the inflammatory autoimmune disease lupus erythematosus (LE). The skin lesions appear mainly on sun-exposed body sites and are provoked by UV wavelengths between 280 and 340 nm (Lehmann et al. 1990). They consist of red, scaly plaques, persisting for months and then resolving with or without scarring, or as a more diffuse photosensitivity. Although many of the symptoms of LE reflect the systemic nature of this disorder, the effect of UVR in the skin of patients with cutaneous LE is different from that in healthy subjects. On exposure of LE patients to UVR, apoptosis of epidermal keratinocytes occurs with resulting clustering of autoantigens at the cell surface. There is probably delayed clearance of these cells, possibly as a result of impaired macrophage function (Chen et al. 2011). The accumulation of apoptotic cells leads to the local release of proinflammatory factors, such as TNF-α, with development of the inflammatory plaques. Further immune changes reflect the properties of Tregs and mast cells. TNF-α inhibits the function of Tregs, and the number of Tregs is reduced in the LE skin lesions when compared to other chronic inflammatory diseases. The number of mast cells, on the other hand, is doubled in cutaneous LE lesions when compared to normal skin. The mast cells may be induced to infiltrate by the release of IL-15 and chemokine (CV-C motif) ligand 5 from keratinocytes, both recognized chemoattractants for mast cells (van Nguyen et al. 2011).

11.5.2 For Inflammatory Skin Conditions

Psoriasis and atopic dermatitis are two of the most common inflammatory skin diseases. Each is characterized by the proliferation and abnormal differentiation of keratinocytes, by epidermal barrier defects, and by the infiltration of inflammatory dendritic cells together with Th1/Th17 cells in psoriasis and Th2/Th22 cells in atopic dermatitis (Guttman-Yassky, Nograles, and Krueger 2011). To complement the improvement of symptoms in summer, both topical application of vitamin D and exposure to narrowband UVB (311–313 nm) have been used successfully to treat these conditions. UVB phototherapy can increase vitamin D levels, which, in turn, may be immunomodulatory, as well as promoting the expression of antimicrobial peptides (Vahavihu et al. 2010). Vitamin D-independent benefits of UVR exposure have been reported, with suppression of the IL-23/IL-17 pathway in psoriatic plaques and a reduced number of inflammatory dendritic cells in skin and in their products. The mechanisms of suppression of inflammatory responses by UVR and UVR-induced vitamin D are discussed previously in Section 11.2 and may vary from patient to patient.

11.5.3 For Skin Cancer

Exposure to sunlight is the major environmental risk factor for the three common types of skin cancer, namely, malignant melanoma (MM), basal cell carcinoma (BCC), and squamous cell carcinoma (SCC). The effects of UVR in causing DNA damage and in suppressing immune responses are both crucial in cutaneous carcinogenesis. The skin tumors arise most frequently on areas of the body most exposed to the sun, such as the face, neck, and backs of the hands. There is a higher prevalence of these tumors in immunosupppressed individuals than in the general population, estimated at 65 times higher for SCCs, 10 times higher for BCCs, and 7 times higher for MMs. This situation is particularly evident for SCCs in transplant recipients who are treated therapeutically with drugs that suppress mainly T cell activity to prevent rejection of the organ. The production of the Th1 cytokines is downregulated by UVR, and such mediators have been demonstrated to protect against SCCs in mice. Furthermore, Tregs, known to be induced by UVR, as outlined in Section 11.3.1, have been shown to infiltrate SCCs in mice and to surround BCCs in humans. These Tregs are likely to be immunosuppressive by the release of the immunosuppressive cytokine IL-10.

The minimal UV dose required to induce local suppression of CHS has enabled strains of mice to be divided into UV-susceptible and UV-resistant. One report published more than 20 years ago suggests that people can be divided similarly. Yoshikawa et al. (1990) revealed that 92% of individuals with a past history of skin cancer were UV-susceptible, whereas only 35% of individuals without such a history fell into this category. Thus, the ability to be suppressed following low UVR doses could be a factor in determining the propensity of a subject to skin cancer development. It has long been recognized that people with fair skin who sunburn readily and do not tan (phototype I/II) have a higher risk of skin cancer than people with darker skins who burn rarely and tan easily (phototype III/IV). Kelly et al. (2000) suggested that the explanation lay in the susceptibility of the two phototypes to local UV-induced immunosuppression. Thus, for a particular dose of solar UVR, the phototype I/II subjects were more susceptible to the local suppression of CHS than the phototype III/IV subjects. This difference was most prominent when suberythemal UV doses were used.

In contrast to the CHS studies, Grimbaldeston, Finlay-Jones, and Hart (2006) concentrated on mast cells. As a high density of mast cells in MMs and SCCs had been associated with poor prognosis, they counted the number of dermal mast cells in subjects with skin cancer and controls. They found that the higher the density of mast cells, the higher the risk of BCC and MM (Grimbaldeston, Finlay-Jones, and Hart 2006). As explained in Section 11.2.3, the mast cells contribute to UV-induced immunosuppression in various ways, such as by producing a range of immune mediators upon activation by UVR and by migrating to the draining lymph nodes to induce the production of Bregs.

After the initial findings of Kripke (1981), many experiments in mice have confirmed the contribution to UV-induced immunosuppression in skin cancer development. For example, if mice were treated with a *cis*-UCA monoclonal antibody, a PAF receptor antagonist, or a serotonin receptor antagonist during the chronic UV exposures required to induce carcinogenesis, the number of skin tumors that developed and their rate of growth were reduced. A role for UV-induced IFN-γ in melanomagenesis has been discovered recently (Zaidi et al. 2011). Neonates of a novel transgenic mouse strain were UVB irradiated on a single occasion, which led to the infiltration of macrophages into the skin within 24 h and production of IFN-γ by these macrophages. This cytokine then activated melanocytes to proliferate and migrate into the epidermis. In addition, it has been shown that exposing the skin of neonatal mice to UVR results in generation of Tregs, which persist into adulthood. Such cells could represent a predisposing factor for the development of MM later in life.

11.5.4 For Infectious Diseases and Vaccination

As outlined previously, UVR can lead to less efficient antigen presentation, downregulation of Th1 cytokine responses, and generation of antigen-specific Tregs and Bregs. All of these responses are required for the effective control of microbial infections, particularly those that are intracellular. Thus there is the potential for increased microbial load, higher risk of symptomatic infections, reactivation from latency, and enhanced microbial tumorigenesis in irradiated animals compared with the unirradiated controls. To test this hypothesis, at least 20 animal models of infection have been examined in which UVR has been administered either before or shortly after the microbial infection. The animals ranged from mice and rats to guinea pigs; the organisms comprised viruses, bacteria, yeasts, protozoa, and nematodes; and the protocols included examples of systemic infections as well as skin infection. In almost all cases, there was significant suppression of microbial-specific T cell responses and some B cell responses in the irradiated animals with higher microbial load and more severe symptoms or even death on occasion (Norval 2006).

However, when humans are considered, the evidence that UVR can affect the course of infections is less convincing. It is limited to lesions occurring on the skin on UV-exposed body sites caused by the herpes simplex virus (reactivation from latency with development of a cold sore), some human papillomavirus types (enhanced conversion of infection to SCC), varicella zoster virus (reactivation from latency with development of shingles), Merkel cell polyomavirus (Merkel cell carcinoma), *Mycobacterium leprae* (leprous lesions), and *Leishmania donovani* (post-kala-azar lesions). Reasons for the discrepancy between the animal models and human infections could be a result of differences in gene regulation and innate defenses, particularly the generation of antimicrobial peptides in irradiated skin. In addition, the amount of the infecting dose and the route of inoculation may be critical, as well as the quantity, spectrum, and intensity of the UVR. Furthermore, epidemiological studies relating to the prevalence and severity of human infections rarely include data about solar UVR or personal UV exposure with potential confounders including ambient temperature and humidity that can affect the transmission of many microbes. It is noteworthy that, although UVR seems to have little impact on the replication of microorganisms in humans, it has been demonstrated to reduce memory (recall) immune responses (as assessed by DTH) to several microbial antigens. These include tuberculin purified protein derivative (Mantoux reaction), lepromin, and the standard multitest antigens. The last is perhaps of most interest as the suppression in DTH occurred after the subjects had been exposed to natural sunlight during a holiday and was found if the antigens were applied to both the irradiated skin and to unexposed skin.

It is very important to consider whether any adverse effect of UVR on immune responses to vaccines in human subjects occurs. There are at least four animal models that have shown that UVR suppresses the efficacy of a vaccination so that, on challenge, the resistance is significantly reduced (reviewed in Norval and Woods 2011). The epidemiological surveys and other studies relating to ambient UVR and vaccination in humans are outlined in Table 11.1. The only experimental study has involved the hepatitis B subunit vaccine. This was administered to one group of volunteers who had been UV-irradiated beforehand and to another unirradiated group. Various immune responses were then compared. While the natural killer and CHS responses were suppressed in the irradiated group compared with the unirradiated, no difference was found in the hepatitis-specific T cell or antibody responses between the two groups. This vaccine contains an alum adjuvant to promote Th2 cytokines and is given at a high dose to induce immunity in poor responders. Both of these factors could mask any downregulatory effect of UVR. However, further analysis revealed that a subset of the irradiated subjects with a minor IL-1β polymorphism did have suppressed hepatitis-specific antibody responses, and those irradiated subjects with high cutaneous *cis*-UCA concentration showed suppressed hepatitis-specific T cell responses. Thus, while UVR may not affect the immune response to this vaccine in everyone, it can do so in some individuals.

In summary, the evidence is not definitive that UVR can significantly downregulate the primary immune response to vaccines or the resistance to reinfection following vaccination. On the other hand, it can be argued that there is sufficient published information to make further studies in this area a matter of priority, particularly for those commonly used vaccines, such as measles, mumps, rubella, varicella, tuberculosis, and *Salmonella typhi*, that promote predominantly Th1 cytokine responses and that might be predicted to be most affected by UVR.

11.5.5 For Several Systemic Immune Diseases, Including Multiple Sclerosis, Allergic Asthma, and Types 1 and 2 Diabetes

Multiple sclerosis (MS) is a debilitating autoimmune disease of the central nervous system characterized by a bias toward Th1 and Th17 cell responses. A seasonal variation in disease expression has been reported with a peak in winter months, and a positive latitude gradient in prevalence has been observed (Hart, Gorman, and Finlay-Jones 2011). Sun exposure during all phases of life may benefit patients with MS, although increased sun exposure in childhood may have significantly enhanced benefits. The season of birth can influence the risk of developing the disease with an inverse association between low UVR exposure in the first trimester and increased risk of MS in the offspring. Vitamin D has been proposed as the molecule responsible for

TABLE 11.1 Information that UVR Might Adversely Affect the Efficacy of Human Vaccination (in Chronological Order)

Vaccine	UVR Effect
Poliovirus	Antibody response higher in temperate vs. tropical areas
Poliovirus	Antibody response higher if administered in winter vs. summer months
Influenza virus	Immunogenicity higher if administered in winter vs. summer months
BCG (tuberculosis)	More protective with increasing distance from equator (less solar UVB)
Hepatitis B virus	Antibody response initially higher if administered in winter vs. summer months
Hepatitis B virus	No effect overall on hepatitis-specific T cell or antibody responses, but suppressed T cell response if high epidermal *cis*-UCA concentration, and suppressed antibody response in those with minor IL-1β polymorphism
Measles virus	Immunity declines with high solar UVR exposure
Measles virus and poliovirus	Promotion of Th2 cytokines
Rubella virus	Antibody response higher if administered in winter vs. summer

Source: Norval M., and G. Woods, *Photochem Photobiol Sci*, 10, 1267–1274, 2011.

the regulatory effects of sun exposure; deficient and insufficient vitamin D levels have been linked with increased risk of MS. However, supplementation with vitamin D has not proved causality between vitamin D deficiency and disease pathogenesis in clinical trials thus far (Plum and DeLuca 2010). In addition, recent epidemiological findings suggest that vitamin D and sun exposure are independent risk factors for MS development. Further, studies using an experimental model of MS in mice have suggested that hypercalcemia may be responsible for some of the reported reduction in disease severity induced by vitamin D (Hart, Gorman, and Finlay-Jones 2011).

For allergic asthma, a Th2-driven disease, a positive latitude gradient has been reported, suggesting that UV exposure can modulate the prevalence of the disease. However, this effect has not been detected in all studies, which may reflect the complexities of the disease definition and its pathogenesis (Hart, Gorman, and Finlay-Jones 2011; Norval and Halliday 2011). In both pediatric and adult patients, many correlations have been found between low vitamin D status and higher incidence of asthma, increased allergen sensitivity (high IgE levels), bronchial hyperresponsiveness, poor lung function, and reduced responses to steroids. Insufficient UV exposure and vitamin D insufficiency in children may lead to increased atopic sensitization and early infections of the lower respiratory tract; these represent the two greatest risk factors for asthma. The effects of UV irradiation on skin integrity may provide a prelude to the "atopic march" and the development of atopy and asthma.

Type 1 diabetes is an autoimmune disease in which pancreatic B cells are destroyed by Th1 cells and immune processes, causing insulin deficiency. A positive latitude gradient has been demonstrated and may account for up to 40% of the variation in the incidence of this disease. A seasonal variation in the diagnosis of type I diabetes has been reported with peaks in winter and troughs in summer; less consistently, an association has been claimed between spring births and an increased likelihood of type I diabetes. These effects have been generally ascribed to the influence of UVR-induced vitamin D with epidemiological studies reporting an association between higher vitamin D status and a significantly reduced risk of diabetes in both children and adults (Baeke et al. 2010). Models using the nonobese diabetic (NOD) mouse support vitamin D as an important immunomodulator of disease progression. However, the success of supplementation of children early in life with vitamin D to reduce the risk of type I diabetes has varied (Baeke et al. 2010). Perhaps insufficient amounts of vitamin D were administered in some studies, or it is difficult to reverse genetically determined autoimmune processes once initiated.

For MS, allergic asthma, and types I and II diabetes, the outcomes of well-controlled cohort studies of vitamin D supplementation are eagerly awaited. It is necessary to demonstrate that vitamin D insufficiency is the cause, rather than the consequence, of the disease and that vitamin D supplementation can ameliorate the symptoms. Whether UVR-induced mediators other than vitamin D (Sections 11.2 and 11.4) may influence the risk and severity of these immune diseases is unknown.

11.6 Conclusions

There is no doubt that suppression of the immune response to antigens, encountered for the first time within a short period of UVR, occurs. The pathways initiated by the absorption of UVR in the skin are astonishingly complex and demonstrate remarkable redundancy with the involvement of several chromophores and many interacting cell phenotypes, together with an array of immune mediators. The mechanisms leading to local UV-induced immunosuppression differ from those leading to systemic immunosuppression, although there are also many common elements. Details of the steps resulting in downregulation of already-established immune responses remain scanty. The consequences for human diseases of the changes in immunity due to UVR are multiple, and many remain to be further examined. It is hoped that the elucidation of the pathways altered by UVR that affect such diseases may allow them to be boosted and targeted in novel therapeutic approaches. Further research in this intriguing area is required not only to further understand the skin immune system but also to inform the public about best practices regarding the sun exposure that is likely to result in optimal health outcomes.

Acknowledgment

PHH is supported by a principal research fellowship from the National Health and Medical Research Council of Australia.

References

Baeke, F., T. Takiishi, H. Korf, C. Gysemans, and C. Mathieu. 2010. Vitamin D: Modulator of the immune system. *Curr Opin Pharmacol* 10:482–496.

Bouillon, R., G. Carmeliet, L. Verlinden et al. 2008. Vitamin D and human health: Lessons from vitamin D receptor null mice. *Endocr Rev* 29:726–776.

Byrne, S. N., and G. M. Halliday. 2005. B cells activated in lymph nodes in response to ultraviolet irradiation by interleukin-10 inhibit dendritic cell induction of immunity. *J Invest Dermatol* 124:570–578.

Chacon-Salinas, R., A. Y. Limon-Flores, A. D. Chavez-Blanco, A. Gonzalez-Estrada, and S. E. Ullrich. 2011. Mast cell-derived IL-10 suppresses germinal center formation by affecting T follicular helper cells function. *J Immunol* 186:25–31.

Chen, X. W., Y. Shen, C. Y. Sun, F. X. Wu, Y. Chen, and C. D. Yang. 2011. Anti-class A scavenger receptor autoantibodies from systemic lupus erythematosus patients impair phagocytic clearance of apoptotic cells by macrophages in vitro. *Arthritis Res Ther* 13:R9.

De Fabo, E. C., and F. P. Noonan. 1983. Mechanism of immune suppression by ultraviolet irradiation in vivo. I. Evidence for the existence of an unique photoreceptor in skin and its role in photoimmunology. *J Exp Med* 158:84–98.

Fukunaga, A., N. M. Khaskhely, Y. Ma et al. 2010. Langerhans cells serve as immunoregulatory cells by activating NKT cells. *J Immunol* 185:4633–4640.

Gibbs, N. K., J. Tye, and M. Norval. 2008. Recent advances in urocanic acid photochemistry, photobiology and photoimmunology. *Photochem Photobiol Sci* 7:655–667.

Gorman, S., J. W.-Y. Tan, S. T. Yerkovich, J. J. Finlay-Jones, and P. H. Hart. 2007. CD4+ T cells in lymph nodes of UVB-irradiated mice suppress immune responses to new antigens both in vitro and in vivo. *J Invest Dermatol* 127:915–924.

Grimbaldeston, M. A., J. J. Finlay-Jones, and P. H. Hart. 2006. Mast cells in photodamaged skin: What is their role in skin cancer? *Photochem Photobiol Sci* 5:177–183.

Guttman-Yassky, E., K. E. Nograles, and J. G. Krueger. 2011. Contrasting pathogenesis of atopic dermatitis and psoriasis—Part I: Clinical and pathologic concepts. *J Allergy Clin Immunol* 127:1110–1118.

Halliday, G. M., D. L. Damian, S. Rana, and S. N. Byrne. 2012. The suppressive effects of ultraviolet radiation on immunity in the skin and internal organs: Implications for autoimmunity. *J Dermatol Sci* 66:176–182.

Hart, P. H., S. Gorman, and J. J. Finlay-Jones. 2011. Modulation of the immune system by UV radiation: More than just the effects of vitamin D? *Nat Rev Immunol* 11:584–596.

Hart, P. H., M. A. Grimbaldeston, G. J. Swift, A. Jaksic, F. P. Noonan, and J. J. Finlay-Jones. 1998. Dermal mast cells determine susceptibility to ultraviolet B-induced systemic suppression of contact hypersensitivity responses in mice. *J Exp Med* 187:2045–2053.

Kaneko, K., S. L. Walker, J. Lai-Cheong, M. S. Matsui, M. Norval, and A. R. Young. 2011. *Cis*-urocanic acid enhances prostaglandin E2 release and apoptotic cell death via reactive oxygen species in human keratinocytes. *J Invest Dermatol* 131:1262–1271.

Kelly, D. A., A. R. Young, J. M. McGregor, P. T. Seed, C. S. Potten, and S. L. Walker. 2000. Sensitivity to sunburn is associated with susceptibility to ultraviolet radiation-induced suppression of cutaneous cell-mediated immunity. *J Exp Med* 191:561–566.

Kripke, M. L. 1981. Immunologic mechanisms in UV radiation carcinogenesis. *Adv Cancer Res* 34:69–106.

Krutmann, J., A. Morita, and J. H. Chung. 2012. Sun exposure: What molecular photodermatology tells us about its good and bad sides. *J Invest Dermatol* 132:976–984.

Lehmann, P., E. Holzle, P. Kind, G. Goerz, and G. Plewig. 1990. Experimental reproduction of skin lesions in lupus erythematosus by UVA and UVB radiation. *J Am Acad Dermatol* 22:181–187.

Lembo, S., J. Fallon, P. O'Kelly, and G. M. Murphy. 2008. Polymorphic light eruption and skin cancer prevalence; is one protective against the other? *Br J Dermatol* 159:1342–1347.

Loser, K., A. Mehling, S. Loser et al. 2006. Epidermal RANKL controls regulatory T-cell numbers via activation of dendritic cells. *Nat Med* 12:1372–1379.

Maeda, A., S. Beissert, T. Schwarz, and A. Schwarz. 2008. Phenotypic and functional characterization of ultraviolet radiation-induced regulatory T cells. *J Immunol* 180:3065–3071.

McGlade, J. P., D. H. Strickland, M. J. M. Lambert et al. 2010. UV inhibits allergic airways disease in mice by reducing effector CD4+ T cells. *Clin Exp Allergy* 40:772–785.

Ng, R. L. X., J. L. Bisley, S. Gorman, M. Norval, and P. H. Hart. 2010. UV-irradiation of mice reduces the competency of bone marrow-derived CD11c+ cells via an indomethacin-inhibitable pathway. *J Immunol* 185:7207–7215.

Norval, M. 2006. The effects of ultraviolet radiation on human viral infections. *Photochem Photobiol* 86:1495–1504.

Norval, M., and A. A. El-Ghorr. 2002. Studies to determine the immunomodulating effects of *cis*-urocanic acid. *Methods* 28:63–70.

Norval, M., and G. M. Halliday. 2011. The consequences of UV-induced immunosuppression for human health. *Photochem Photobiol* 87:965–977.

Norval, M., and G. M. Woods. 2011. UV-*induced* immunosuppression and the efficacy of vaccination. *Photochem Photobiol Sci* 10:1267–1274.

Plum, L. A., and H. F. DeLuca. 2010. Vitamin D, disease and therapeutic opportunities. *Nat Rev Drug Discov* 9:941–955.

Rana, S., S. N. Byrne, L. J. MacDonald, C. Y. Chan, and G. M. Halliday. 2008. Ultraviolet B suppresses immunity by inhibiting effector and memory T cells. *Am J Pathol* 172:993–1004.

Romani, N., P. M. Brunner, and G. Stingl. 2012. Changing views of the role of Langerhans cells. *J Invest Dermatol* 132:872–881.

Rosette, C., and M. Karin. 1996. Ultraviolet light and osmotic stress: Activation of the JNK cascade through multiple growth factor and cytokine receptors. *Science* 274:1194–1197.

Schwarz, A., F. Navid, T. Sparwasser, B. E. Clausen, and T. Schwarz. 2011. In vivo reprogramming of UV radiation-induced regulatory T-cell migration to inhibit the elicitation of contact hypersensitivity. *J Allergy Clin Immunol* 128: 823–833.

Schwarz, A., and T. Schwarz. 2010. UVR-induced regulatory T cells switch antigen-presenting cells from a stimulatory to a regulatory phenotype. *J Invest Dermatol* 130:1914–1921.

Schwarz, T., and A. Schwarz. 2011. Molecular mechanisms of ultraviolet radiation-induced immunosuppression. *Eur J Cell Biol* 90:560–564.

Simon, J. C., J. Krutmann, C. A. Elmets, P. R. Bergstresser, and P. D. Cruz, Jr. 1992. Ultraviolet B-irradiated antigen-presenting cells display altered accessory signalling for T-cell activation: Relevance to immune responses initiated in skin. *J Invest Dermatol* 98(6 Suppl):S66–S69.

Soontrapa, K., T. Honda, D. Sakata et al. 2011. Prostaglandin E2-prostaglandin E receptor subtype 4 (EP4) signalling mediates UV irradiation-induced systemic immunosuppression. *Proc Natl Acad Sci U S A* 108:6668–6673.

Sreevidya, C. S., A. Fukunaga, N. M. Khaskhely et al. 2010. Agents that reverse UV-induced immune suppression and photocarcinogenesis affect DNA repair. *J Invest Dermatol* 130:1428–1437.

Tewari, A., R. P. Sarkay, and A. R. Young. 2012. UVA1 induces cyclobutane pyrimidine dimers but not 6-4 photoproducts in human skin in vivo. *J Invest Dermatol* 132:394–400.

Toews, G. B., P. R. Bergstresser, and J. W. Streilein. 1980. Epidermal Langerhans cell density determines whether contact hypersensitivity or unresponsiveness follows skin painting with DNFB. *J Immunol* 124:445–449.

Toichi, E., K. Q. Lu, A. R. Swick, T. S. McCormick, and K. D. Cooper. 2008. Skin-infiltrating monocytes/macrophages migrate to draining lymph nodes and produce IL-10 after contact sensitizer exposure to UV-irradiated skin. *J Invest Dermatol* 128:2705–2715.

Ullrich, S. E., and S. N. Byrne. 2012. The immunologic revolution: Photoimmunology. *J Invest Dermatol* 132:896–905.

Vahavihu, K., M. Ala-Houhala, M. Peric et al. 2010. Narrowband ultraviolet B treatment improves vitamin D balance and alters antimicrobial peptide expression in skin lesions of psoriasis and atopic dermatitis. *Br J Dermatol* 163:321–328.

Van Nguyen, H., N. Di Girolamo, N. Jackson et al. 2011. Ultraviolet radiation-induced cytokines promote mast cell accumulation and matrix metalloproteinase production: Potential role in lupus erythematosus. *Scand J Rheumatol* 40:197–204.

Vink, A. A., V. Shreedhar, L. Roza, J. Krutmann, and M. L. Kripke. 1998. Cellular target of UVB-induced DNA damage resulting in local suppression of contact hypersensitivity. *J Photochem Photobiol B* 44:107–111.

Wang, L., S. C. Jameson, and K. A. Hogquist. 2009. Epidermal Langerhans cells are not required for UV-induced immunosuppression. *J Immunol* 183:5548–5553.

Wolf, P., S. N. Byrne, and A. Gruber-Wackernagel. 2009. New insights into the mechanisms of polymorphic light eruption: Resistance to ultraviolet radiation-induced immune expression as an aetiological factor. *Exp Dermatol* 18:350–356.

Yoshikawa, T., V. Rae, W. Bruins-Slot et al. 1990. Susceptibility to effects of UVB radiation on induction of contact hypersensitivity as a risk factor for skin cancer in humans. *J Invest Dermatol* 95:530–536.

Zaidi, M. R., S. David, F. P. Noonan et al. 2011. Interferon-γ links ultraviolet radiation to melanomagenesis in mice. *Nature* 469:548–553.

12

The Porphyrias

Eric Gou
The University of Texas Medical Branch

Karl E. Anderson
The University of Texas Medical Branch

12.1 Introduction

Photosensitivity is a prominent feature of many of the porphyrias (Phillips and Anderson 2010). Chronic blistering lesions affecting sun-exposed skin occur in several types of porphyria and is the only clinical manifestation in porphyria cutanea tarda (PCT), the most common porphyria. In contrast, an acute, painful nonblistering photosensitivity occurs in erythropoietic protoporphyria (EPP), the third most common porphyria (Cox 2003; Sarkany 2008). These manifestations occur primarily from sunlight exposure. But especially in EPP, patients may be sensitive to fluorescent lighting and require accommodation at the workplace and at home. Also in EPP, operating-room lights during liver transplant surgery can cause marked phototoxic injury (Wahlin et al. 2008). Thus, knowledge of photomedicine is applicable to the understanding and management of these diseases.

The porphyrias are distinct metabolic disorders, each resulting from an alteration of a specific enzyme of the heme biosynthetic pathway (Figure 12.1). Heme serves multiple important roles in the human body thanks to the inclusion of iron within its structure. Iron has two ionic forms, ferrous (Fe^{2+}) and ferric (Fe^{3+}), depending on whether it takes up or donates an electron. Hemoproteins are particularly important in transporting and storing molecular oxygen (O_2, e.g., hemoglobin and myoglobin, respectively), transporting electrons (e.g., respiratory cytochromes), and oxidation–reduction reactions (e.g., cytochrome P450 enzymes, nitric oxide synthase, and catalase).

The heme biosynthetic pathway is a sequence of eight steps, each catalyzed by a different enzyme (Table 12.1). The first and last three of these enzymes are mitochondrial and the other four are cytosolic. Heme synthesis begins with glycine and succinyl coenzyme A, which are substrates for the first enzyme δ-aminolevulinic acid synthase (ALAS). The product of this enzyme is an amino acid, δ-aminolevulinic acid (ALA), which is committed only to the synthesis of heme as well as other intermediates in the pathway, including porphyrins. Because ALA is an effective precursor of porphyrins, it was developed as a photosensitizer that is applied topically to actinic keratosis to generate protoporphyrin that is activated with phototherapy, causing necrosis of the *in situ* neoplasm (Lang et al. 2001). Two molecules of ALA are combined to form porphobilinogen (PBG), a pyrrole. Four molecules of PBG are then assembled to form a linear tetrapyrrole (hydroxymethylbilane or HMB) and then a series of cyclic tetrapyrroles (i.e., porphyrins). These porphyrins are further modified before insertion of iron into the porphyrin macrocycle to form heme. With one exception, the porphyrin intermediates in this pathway are in their reduced forms (i.e., porphyrinogens). Protoporphyrin IX, the last intermediate, is an oxidized porphyrin. A specific enzyme alteration in this pathway can lead to accumulation of intermediates that determines the clinical features of a particular type of porphyria (Phillips and Anderson 2010).

An oxidized porphyrin has a heterocyclic aromatic structure, which allows some of its electrons to delocalize, providing greater stability for the molecule. However, exposure to violet light (at a wavelength of ~410 nm) causes these delocalized electrons to increase in energy level. This energy may be released as red fluorescent light or transferred to molecular oxygen, forming reactive singlet-state oxygen. This and other reactive oxygen

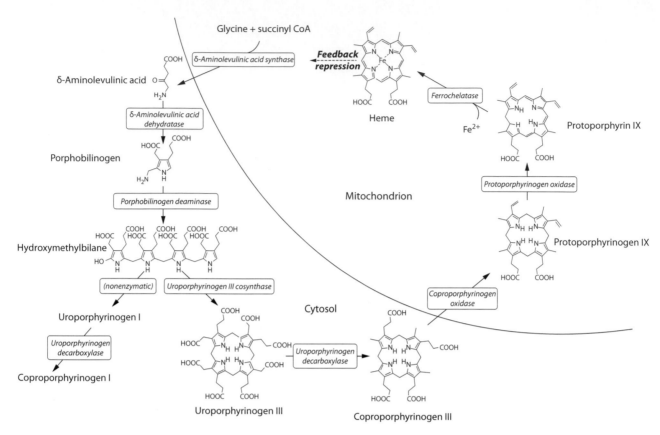

FIGURE 12.1 Enzymes and intermediates of the heme biosynthetic pathway. (From Anderson, K. E. 2006. In *Zakim and Boyer's Hepatology: A Textbook of Liver Disease.* T. D. Boyer, T. L. Wright, and M. P. Manns, editors. 1391–1432. With permission.)

TABLE 12.1 Enzymes Altered in the Different Types of Porphyria, Their Patterns of Inheritance, and Classifications

Altered Enzyme (abbreviation)	Porphyria	Inheritance	Classifications
ALA synthase–erythroid (ALAS2)	XLP	X-linked	Erythropoietic/cutaneous
ALA dehydratase (ALAD)	ADP	Autosomal recessive	Hepatic/acute
Porphobilinogen deaminase (PBGD)	AIP	Autosomal dominant	Hepatic/acute
Uroporphyrinogen III (co)synthase (UROS)	CEP	Autosomal recessive	Erythropoietic/cutaneous
Uroporphyrinogen decarboxylase (UROD)	PCT type 1	Acquired	Hepatic/cutaneous
	PCT type 2	Autosomal dominant	
	HEP	Autosomal recessive	
Coproporphyrinogen oxidase (CPOX)	HCP	Autosomal dominant	Hepatic/acute and cutaneous
Protoporphyrinogen oxidase (PPOX)	VP	Autosomal dominant	Hepatic/acute and cutaneous
Ferrochelatase (FECH)	EPP	Autosomal recessive	Erythropoietic/cutaneous

Note: ADP, ALA dehydratase porphyria; AIP, Acute intermittent porphyria; ALA, δ-aminolevulinic acid; CEP, Congenital erythropoietic porphyria; EPP, erythropoietic protoporphyria; HCP, Hereditary coproporphyria; HEP, Hepatoerythropoietic porphyria; PCT, Porphyria cutanea tarda; VP, Variegate porphyria; XLP, X-linked protoporphyria.

species may directly damage cellular constituents, including proteins and lipids, or cause mast cell degranulation and complement activation. Differences in water solubility may affect tissue and subcellular localization of various types of porphyrins and determine development of blistering or nonblistering phototoxic effects (Sarkany 2008).

Porphyrias are often classified based on clinical manifestations. *Cutaneous porphyrias*, which are the emphasis of this chapter, cause cutaneous photosensitivity, whereas *acute porphyrias* present with a variety of neurological symptoms and signs usually occurring as acute attacks. However, some porphyrias can cause both types of manifestations. These diseases are also classified based on pathophysiology as either *hepatic* or *erythropoietic*. In hepatic porphyrias, the site of initial accumulation of excess pathway intermediates is the liver, whereas the marrow is the site of accumulation of porphyrins in the erythropoietic forms (Phillips and Anderson 2010).

12.2 PCT and Hepatoerythropoietic Porphyria

12.2.1 Overview

PCT is the most common porphyria and is prototypic for the porphyrias that cause chronic blistering phototoxicity. This hepatic porphyria is caused by reduced activity of hepatic uroporphyrinogen decarboxylase (UROD), the fifth enzyme in the heme biosynthetic pathway, and is characterized by chronic blistering skin lesions most prominently seen on the dorsa of the hands and also on other sun-exposed areas, such as the forearms, face, ears, neck, and sometimes the feet. PCT is primarily an acquired iron-related disorder and is the only porphyria that can develop in the absence of a mutation of the affected enzyme (Elder 2003). A UROD inhibitor has been isolated from the liver of a mouse model of PCT (Phillips et al. 2011).

A multiplicity of susceptibility factors is associated with PCT in individual patients, although none is essential for its development. These include alcohol use, smoking, estrogen use, hepatitis C, HIV infection, hemochromatosis gene (*HFE*) mutations, and *UROD* mutations. *UROD* mutations are found in only ~20% of patients, and these cases are classified as having familial (type 2) PCT. Those without *UROD* mutations are classified as having sporadic PCT (type 1) and represent ~80% of patients. Rare families with more than one affected individual but without *UROD* mutations are classified as type 3. These three types are generally not distinguishable clinically and are all associated with multiple susceptibility factors (Elder 2003; Jalil et al. 2010).

At least 100 different disease-related *UROD* mutations have been identified (Balwani and Desnick 2012). An inherited heterozygous *UROD* mutation, which reduces UROD activity to ~50% of normal, is inherited as an autosomal dominant trait with low penetrance. In those with *UROD* mutations, other factors must be present to reduce hepatic UROD activity to approximately 20% of normal before PCT becomes manifest.

Hepatoerythropoietic porphyria (HEP) is the homozygous form of type 2 PCT and is usually more severe, presents in childhood, and resembles congenital erythropoietic porphyria (CEP) (Elder 2003).

12.2.2 Pathophysiology

UROD converts uroporphyrinogen (an octacarboxyl porphyrinogen) to coproporphyrinogen in a stepwise fashion by the removal of four carboxyl groups. Decreased activity of hepatic UROD results in accumulation of uroporphyrinogen as well as the intermediate heptacarboxyl, hexacarboxyl, and pentacarboxyl porphyrinogens, which are then autoxidized to the corresponding porphyrins. These porphyrins, usually mostly uroporphyrin and heptacarboxylporphyrin, are measurably increased in plasma and urine. Skin lesions are caused by activation of porphyrins by long-wave ultraviolet or violet light after their transport in plasma from the liver to the skin (Elder 2003; Phillips and Anderson 2010).

12.2.3 Diagnosis

A diagnosis of PCT is suggested by typical skin lesions and elevations of highly carboxylated porphyrins in urine or plasma. Levels of PBG are normal, and ALA is normal or slightly elevated. Erythrocyte porphyrins are normal or only slightly elevated.

12.2.4 Treatment

Patients should be questioned for known susceptibility factors and advised to stop drinking and smoking and discontinue estrogens if applicable. Other susceptibility factors, such as hepatitis C, HIV, *HFE*, and *UROD* mutations, should be identified or excluded by specific testing. Symptoms may improve after removal of one or more susceptibility factors, but the response without further treatment is not dependable. PCT is the most treatable of the porphyrias and responds to iron reduction by repeated phlebotomy or, alternatively, to a low-dose regimen of hydroxychloroquine (or chloroquine). Patients with substantial iron overload (assessed by serum ferritin) should be treated by phlebotomy. Other considerations in choice of treatment are discussed elsewhere (Singal et al. 2012).

Repeated phlebotomy consists of removing ~450 mL of blood at 2-week intervals. Hemoglobin or hematocrit levels should be monitored to prevent symptomatic anemia. The therapeutic target is a serum ferritin below ~15 ng/mL, which allows tissue iron depletion but usually does not cause iron deficiency anemia (Ratnaike et al. 1988). Usually six to eight phlebotomies are sufficient to reach the target ferritin, but more are needed in patients with substantial iron overload. Plasma or serum porphyrin levels continue to decline after phlebotomy is completed. Low-dose hydroxychloroquine (100 mg twice weekly) appears to be as effective as phlebotomies and is especially appropriate when phlebotomy is contraindicated by anemia or is poorly tolerated. Full therapeutic doses of these medications should be avoided as they initially cause hepatocellular damage and exacerbate PCT symptoms (Singal et al. 2012). Rarely, hydroxychloroquine causes damage to the retina, and patients should be evaluated before treatment by an ophthalmologist for retinopathy by dilated fundus exam and either automated threshold visual field testing or spectral domain optical coherence tomography (Marmor et al. 2011). Treatment is continued until plasma or urine porphyrins are normalized for at least a month.

Treatment of hepatitis C should usually be postponed until PCT has been treated and is in remission. However, treatment of HIV should not be interrupted for treatment of PCT.

12.3 Hereditary Coproporphyria and Variegate Porphyria

12.3.1 Overview

These hepatic porphyrias can present with blistering skin lesions that are identical to those seen in PCT. However, they also cause neurological symptoms that are identical to those that occur in acute intermittent porphyria (AIP). Skin manifestations are common in variegate porphyria (VP) but uncommon in hereditary coproporphyria (HCP).

HCP and VP are caused by deficiencies of coproporphyrinogen oxidase (CPOX) and protoporphyrinogen oxidase (PPOX), the sixth and seventh enzymes of the heme biosynthetic pathway, respectively. CPOX catalyzes the two-step decarboxylation of coproporphyrinogen III to form protoporphyrinogen IX with the intermediate formation of harderoporphyrinogen (a tricarboxylate porphyrinogen) (Nordmann et al. 1983). PPOX removes six protons from protoporphyrinogen IX to form protoporphyrin IX, the only oxidized porphyrin that serves as an intermediate in the pathway (Phillips and Anderson 2010).

Inheritance is autosomal dominant with variable penetrance in both of these disorders. More than 60 mutations have been identified in HCP (Balwani and Desnick 2012; Lamoril et al. 2001). A rare variant form of HCP, termed harderoporphyria, results from certain *CPOX* mutations that cause harderoporphyrinogen to be released prematurely from the enzyme. At least 165 mutations have been identified to date in VP (Balwani and Desnick 2012; Whatley et al. 1999). VP is especially common in South Africans of Dutch descent, as a result of a founder effect, with an incidence of 3 per 1000 population (Meissner, Hift, and Corrigall 2003). Homozygous cases of HCP and VP have been described with development early in life of severe photosensitivity, blistering of sun-exposed areas, brachydactyly, and short stature (Grandchamp, Phung, and Nordmann 1977; Poblete-Gutierrez et al. 2006).

12.3.2 Pathophysiology

In active cases of HCP, the deficiency of CPOX causes marked accumulation of coproporphyrinogen III, which undergoes autoxidation to coproporphyrin III, which is excreted in urine and feces. The deficiency of PPOX in VP leads to accumulation of protoporphyrinogen IX as well as coproporphyrinogen III, which appear in urine and feces as protoporphyrin IX and coproporphyrin III. Inhibition of PBGD by protoporphyrinogen IX and coproporphyrinogen III may explain the increases in PBG that are found in these porphyrias, particularly during acute attacks of neurological symptoms (Meissner, Adams and Kirsch 1993). Clinical expression is influenced by the same factors seen in AIP.

12.3.3 Diagnosis

As in AIP, measurement of urinary PBG is important for diagnosis of HCP and VP when they present with acute neurovisceral symptoms. When cutaneous blisters are the presenting manifestation, differentiation from PCT is essential. Plasma porphyrin concentrations are commonly increased in VP even in the absence of skin lesions. Moreover, fluorescence scanning of plasma diluted at neutral pH displays an emission peak at ~626 nm, which dependably differentiates VP from all other porphyrias (Hift et al. 2004; Poh-Fitzpatrick and Lamola 1976). This distinctive feature in VP may result from covalent binding of protoporphyrinogen IX to plasma proteins (Longas and Poh-Fitzpatrick 1982). Plasma porphyrin concentrations are seldom increased in HCP unless there are cutaneous manifestations.

Fecal porphyrins are markedly elevated in symptomatic cases of HCP and VP and in some asymptomatic individuals as well. In HCP, the increased fecal porphyrins are almost entirely coproporphyrin III, and in asymptomatic cases, an increase in the fecal coproporphyrin III:I ratio is a sensitive diagnostic test (Blake et al. 1992). In VP, fecal porphyrins consist mostly of coproporphyrin III and protoporphyrin IX in roughly equal amounts.

After the diagnosis is established biochemically, DNA studies should follow to identify the disease-causing mutation. This is important for confirmation of the diagnosis and to serve as the basis for directed mutation analysis to identify asymptomatic carriers of specific mutations in families.

12.3.4 Treatment

Treatment of acute attacks is the same as in AIP. Chronic cutaneous symptoms are difficult to treat but may respond to removal of factors that worsen acute hepatic porphyrias. Otherwise, avoidance of sunlight and use of protective clothing is important. Unfortunately, HCP and VP do not respond to phlebotomies or low-dose hydroxychloroquine because the underlying pathophysiology is fundamentally different from PCT.

12.4 Congenital Erythropoietic Porphyria

12.4.1 Overview

CEP, also known as Günther disease, results from an inherited deficiency of uroporphyrinogen III cosynthase (UROS), the fourth enzyme in the heme biosynthetic pathway. Inheritance of this very rare disease is autosomal recessive, and affected individuals are homozygous (or compound heterozygous) for two severe *UROS* mutations. However, at least one mutation must allow for production of some UROS enzyme. CEP is characterized by chronic, severe photosensitivity and hemolytic anemia. Erythrodontia, the brown staining and red fluorescence of the teeth under long-wave ultraviolet light, is characteristic of CEP and arises from deposition of porphyrins in the teeth *in utero*. The disease becomes manifest *in utero*, most commonly soon after birth, and rarely in adult life (Phillips and Anderson 2010).

The cutaneous lesions in CEP resemble those in PCT but are typically much more severe and are often complicated by infection, scarring, and disfiguring mutilation of light-exposed areas—especially the face and hands. Porphyrin levels in plasma in CEP are typically an order of magnitude higher than in PCT and are also accompanied by marked elevations in erythrocyte porphyrins, which can explain the more severe cutaneous manifestations.

Rarely, the disease develops in adults with myelodysplastic or myeloproliferative disorders resulting from expansion of a clone of erythrocyte precursors that carry a *UROS* mutation, which may be somatic or inherited (Sassa et al. 2002). Mild adult cases of CEP may occur in the absence of a bone marrow disorder and can be mistakenly diagnosed as PCT unless erythrocyte porphyrins are measured and found to be elevated.

12.4.2 Pathophysiology

The profound deficiency of UROS activity leads to accumulation of the substrate hydroxymethylbilane (HMB), a linear tetrapyrrole. UROS catalyzes inversion of one of the pyrroles of HMB and closes this linear molecule to form an asymmetric macrocycle, uroporphyrinogen III. When UROS is deficient, HMB spontaneously forms the symmetrical molecule uroporphyrinogen I, which can be further metabolized (by UROD) to coproporphyrinogen I but not to heme. These porphyrinogens accumulate in CEP and are autoxidized to uroporphyrinogen I and coproporphyrin I. Porphyrin concentrations are increased in the marrow, circulating erythrocytes and plasma, urine, and feces. Hemolysis results from excess erythrocyte porphyrins, with exposure to light in the dermal capillaries possibly contributing.

At least 39 different mutations of the *UROS* gene have been identified. More severe mutations are associated with more severe clinical manifestations (Balwani and Desnick 2012).

12.4.3 Diagnosis

CEP is usually suspected after birth when pink to dark brown staining of diapers with red fluorescence under long wave ultraviolet light is noted. Severe cases may be recognized *in utero* as a cause of nonimmune hydrops (Verstraeten et al. 1993). Urine, plasma, and erythrocyte porphyrin levels are markedly increased and are predominantly uroporphyrin I and coproporphyrin I. Protoporphyrin IX may predominate in erythrocytes in milder cases. In all cases, the diagnosis should be confirmed by DNA studies to identify causative mutations.

12.4.4 Therapy

Hematopoietic stem cell transplantation in early childhood is the most effective treatment for severe cases of CEP (Dupuis-Girod et al. 2005). This effectively lowers porphyrin levels and corrects anemia and transfusion dependence. Otherwise, patients are advised to avoid sunlight, cutaneous trauma, and infection to prevent severe scarring and loss of facial features and digits. Occlusive sunscreens are marginally beneficial.

12.5 Erythropoietic Protoporphyria

12.5.1 Overview

EPP is characterized by nonblistering cutaneous photosensitivity with onset in early childhood in most cases. Symptoms typically consist of pain, redness, itching, and swelling that develop immediately with sunlight exposure and resolve gradually over hours or days. These skin symptoms are acute with few blisters or other lasting manifestations.

EPP is the third most common porphyria and the most common in children. It often remains undiagnosed for years. The disease can greatly affect quality of life and limit lifestyle and choice of occupation (Holme et al. 2006).

Photosensitivity in this disease is a result of increased circulating protoporphyrin IX, the last intermediate in the pathway for heme synthesis. This dicarboxylate porphyrin is insoluble in water and is excreted from the body by uptake in hepatocytes and excretion in bile. The risk of gallstones (containing protoporphyrin) is increased in EPP. Protoporphyrin is potentially hepatotoxic and cholestatic. Less than 5% of patients develop protoporphyric hepatopathy, which can be rapidly progressive and life threatening and require liver transplantation. As liver dysfunction progresses, plasma protoporphyrin levels and photosensitivity increase as well (Anstey and Hift 2007; Cox 2003). At surgery, protective filters that screen wavelengths responsible for porphyrin excitation are used to prevent extensive burns of the skin and peritoneal surfaces exposed to operating-room lights (Wahlin et al. 2008).

12.5.2 Pathophysiology

EPP usually results from a partial deficiency of ferrochelatase (FECH), the last enzyme of eight in the heme biosynthetic pathway, leading to accumulation of protoporphyrin IX. FECH activity is reduced to 15%–25% of normal in most EPP patients. In >90% of such individuals, this is a result of a severe loss-of-function *FECH* mutation inherited from one parent (who is usually not affected) and a common, low expression or "hypomorphic" variant *FECH* allele (IVS3-48T > C) from the other parent (Gouya et al. 2002). The hypomorphic variant *FECH* allele is found in ~10% of Caucasians and does not itself cause disease. Inheritance of two severe loss-of-function *FECH* mutations, one of which must produce some enzyme activity, in the absence of the common hypomorphic allele is found in <5% of EPP patients (Cox 2003).

Inheritance of EPP was previously described as autosomal dominant with low penetrance. But because it is now understood that inheritance of two variant loss-of-function *FECH* alleles is required to cause EPP, this disease may best be described as autosomal recessive. Rare adult onset cases have been associated with myelodysplastic or myeloproliferative disorders (Goodwin et al. 2006).

In a variant form, termed X-linked protoporphyria (XLP), there are no *FECH* mutations, and the disease results from one of several recently described *ALAS2* gain-of-function C terminal deletions. *ALAS2* mutations were suspected in these families after a pattern of X-linked inheritance was noticed (Whatley et al. 2008). The clinical phenotype is the same except that erythrocytes contain more zinc protoporphyrin in XLP than in EPP.

In EPP and XLP, excess protoporphyrin IX accumulates primarily in bone marrow reticulocytes (Bottomley, Tanaka, and Everett 1975; Clark and Nicholson 1971; Piomelli et al. 1975). Protoporphyrin IX accumulates in EPP as a result of deficient

FECH activity. In XLP, it accumulates because with increased ALAS2 activity, protoporphyrin IX production exceeds the amounts needed for heme synthesis. In both types of protoporphyria, protoporphyrin IX is increased also in circulating erythrocytes and plasma and is excreted exclusively in bile and feces. Other heme pathway intermediates do not accumulate. Therefore, urinary porphyrin and porphyrin precursor concentrations are not increased.

The excess protoporphyrin found in erythrocytes in EPP and XLP is predominantly metal-free, whereas zinc protoporphyrin predominates in other conditions, including iron deficiency, anemia of chronic disease (Hastka et al. 1993), hemolytic anemias (Anderson et al. 1977), lead poisoning, and homozygous cases of autosomal dominant porphyrias. Formation of zinc protoporphyrin is dependent on FECH activity, which is deficient in EPP but not in XLP. Therefore, in XLP, zinc protoporphyrin accounts for a greater proportion of erythrocyte protoporphyrin than in EPP because of FECH deficiency (Whatley et al. 2008).

12.5.3 Diagnosis

This diagnosis should be prompted in children or adults who complain of painful, nonblistering photosensitivity. The screening test for EPP is measurement of total erythrocyte porphyrins (or protoporphyrin).

It is important to select a laboratory that measures total erythrocyte protoporphyrin and also accurately reports the amounts of metal-free and zinc protoporphyrin. Laboratories in the United States that currently do this are ARUP Laboratories, Mayo Medical Laboratories, and the Porphyria Laboratory at the University of Texas Medical Branch at Galveston. It is also important not to request a measurement of free erythrocyte protoporphyrin (FEP) because this terminology is misleading. The FEP test is an obsolete term previously applied to testing for lead poisoning before it was known that zinc protoporphyrin is the predominant porphyrin in erythrocytes after excess lead exposure and all other conditions except EPP (Labbe 1992). Unfortunately, some major laboratories still measure erythrocyte protoporphyrin using a hematofluorometer, an instrument that is tuned to measure only zinc protoporphyrin, and was developed in the 1970s for screening for lead poisoning. These laboratories are not suitable for diagnosis of EPP because they do not measure total or free protoporphyrin and may incorrectly report the amount of zinc protoporphyrin as erythrocyte protoporphyrin or FEP.

An elevation of erythrocyte protoporphyrin is a nonspecific finding and occurs in many conditions, such as iron deficiency, anemia of chronic disease, thalassemias, and lead poisoning (Labbe 1992). A diagnosis of EPP is established by demonstrating a predominance of free protoporphyrin relative to zinc protoporphyrin. FECH deficiency impairs formation of both heme (iron protoporphyrin) as well as zinc protoporphyrin, reflecting the fact that FECH will accept metals other than iron, including zinc, for chelation with protoporphyrin IX. Free and zinc-chelated protoporphyrin is measured by ethanol extraction or high-performance liquid chromatography (Deacon and Elder

2001). In EPP resulting from FECH deficiency, the proportion of erythrocyte protoporphyrin that is metal-free is more than ~90%. In XLP, the amount of metal-free protoporphyrin ranges from approximately 50% to 85%.

Plasma porphyrins are less elevated in EPP than in other cutaneous porphyrias and are very light sensitive during processing and laboratory analysis (Poh-Fitzpatrick and DeLeo 1977). Therefore, measurement of plasma porphyrins is less useful than measuring erythrocyte porphyrins for diagnosis of EPP. However, fluorescence scanning and finding an emission peak at neutral pH at ~634 nm can help in confirming the diagnosis (Poh-Fitzpatrick and Lamola 1976).

Other heme pathway intermediates do not accumulate in EPP. Therefore, urinary porphyrin and porphyrin precursor concentrations are not increased.

12.5.4 Therapy

Avoidance of sunlight and other sources of ultraviolet light is essential for minimizing symptoms. Acute symptoms usually resolve within hours or days; symptomatic treatment with cold compresses, anti-inflammatory medications, antihistamines, and glucocorticoids may be helpful.

Patients typically adjust their daily activities to avoid sunlight, but children especially may suffer embarrassment that is accentuated if they are undiagnosed and have no explanation for their limitation. Topical sunscreens that absorb UVA and sunblocks containing zinc oxide or titanium dioxide may be somewhat helpful to prevent symptoms. Oral pharmaceutical grade β-carotene (Lumitine, Tishcon) may be useful for improving tolerance to sunlight. Its proposed protective mechanism is quenching of oxygen free radicals. Daily doses of 30 to 300 mg (1 to 10 capsules) in adults and maintaining serum carotene levels in the range of 600 to 800 mcg/dL have been recommended (Mathews-Roth et al. 1977). An effective dose level will cause mild skin discoloration (carotoderma), especially on the palms of the hands, which is often embarrassing for children. Oral cysteine at doses of 500 mg twice daily is reported to improve sunlight tolerance in EPP (Mathews-Roth and Rosner 2002), whereas N-acetylcysteine was ineffective in two double-blind, crossover, placebo-controlled trials (Bijlmer-Iest et al. 1992). A marginal benefit of vitamin C did not achieve statistical significance in a double-blind, randomized trial (Boffa et al. 1996).

Narrow-wave UV-B therapy can increase skin melanin and improve sunlight tolerance (Warren and George 1998). Afamelanotide, a synthetic analogue of α-melanocyte stimulating hormone, increases skin pigmentation and is undergoing development for treatment of EPP (Harms et al. 2009).

12.6 Acute Intermittent Porphyria
12.6.1 Overview

AIP is the most common of the four acute porphyrias and the prototype for these disorders that are characterized by acute neurovisceral manifestations. HCP and VP were discussed above because they can also cause chronic, blistering photosensitivity. Skin manifestations never occur in AIP, except rarely in patients with concurrent renal failure (Sardh et al. 2009).

AIP is caused by the deficient activity of porphobilinogen deaminase (PBGD), the third enzyme in the heme pathway, also known as HMB synthase (HMBS). Inheritance is autosomal dominant with variable penetrance. AIP manifests clinically as acute neurovisceral attacks with effects on the enteric, autonomic, peripheral, and central nervous systems. Symptoms develop after puberty and more commonly in women than men.

These attacks are usually severe and potentially life-threatening and often require hospitalization. Abdominal pain is the most common symptom and is frequently accompanied by nausea; vomiting; constipation (less commonly diarrhea); distension; urinary retention; pain in the back, chest, or extremities; agitation; and seizures. The most common physical finding is tachycardia often occurring with hypertension and decreased bowel sounds but little fever or abdominal tenderness. Peripheral neuropathy that is primarily motor can lead to quadriplegia and respiratory weakness with sensory involvement manifested by extremity pain. Hyponatremia is common with severe attacks and is sometimes a result of hypothalamic involvement and the syndrome of inappropriate antidiuretic hormone secretion (Anderson et al. 2005).

Some patients have only one or a few attacks. Others have repeated attacks over many years, especially during the luteal phase of the menstrual cycle in women. Chronic pain and depression may develop and increase the risk of suicide. There is an increased risk of hepatocellular carcinoma in AIP as well as HCP and VP (Andant et al. 2000). The risk for chronic hypertension and renal disease is also increased and may necessitate long-term dialysis or renal transplantation (Barone et al. 2001). Some patients with AIP and end-stage renal disease have developed cutaneous lesions resembling those in PCT (Sardh et al. 2009). In rare homozygous cases, severe neurological symptoms develop in early childhood, but acute attacks are not prominent (Solis et al. 2004).

12.6.2 Pathophysiology

More than 300 *PBGD* mutations causing AIP have been described (Balwani and Desnick 2012). Most heterozygotes do not develop symptoms because the normal *PBGD* allele produces sufficient enzyme for heme synthesis under normal conditions. However, certain drugs, hormones, and other factors can induce ALAS1, the first enzyme in the heme biosynthetic pathway, which is rate-limiting for heme synthesis in the liver. With ALAS1 induction and increased production of ALA and PBG, the deficient activity of PBGD may then become rate-limiting. This can impair heme synthesis and cause accumulation of ALA and PBG in the liver. It is suspected that ALA is neurotoxic, but this is not proven (Anderson et al. 2005).

Drugs that induce hepatic cytochrome P450 enzymes also induce ALAS1 and are especially likely to exacerbate AIP and other acute porphyrias. Websites of the American Porphyria Foundation (www.porphyriafoundation.com) and the European Porphyria Network (www.porphyria-europe.com) provide updated information on drugs that are safe and unsafe in these disorders. Other factors that can precipitate attacks include use of progesterone, synthetic progestins, alcohol, and reduced intake of calories and carbohydrates (Anderson et al. 2005).

12.6.3 Diagnosis

Acute porphyrias should be suspected in patients presenting with abdominal pain or other symptoms after an initial workup for more common conditions does not reveal the cause. Although the symptoms of acute porphyrias are very nonspecific, the diagnosis is readily ruled in or out by measuring urinary porphobilinogen (PBG), which is markedly elevated especially during acute attacks of AIP, HCP, or VP. The same spot urine specimen should be saved to measure ALA, PBG, and total porphyrins to confirm the qualitative PBG result. This will detect isolated elevations of ALA and coproporphyrin III in rare patients with ALA dehydratase porphyria (ADP) and some cases of HCP and VP, in which porphyrin elevations may be more persistent than ALA and PBG. In AIP patients with renal dysfunction, PBG elevation can be demonstrated in serum.

A specific diagnosis of AIP is confirmed by finding an elevation in PBG with little or no elevation in total fecal porphyrins, which are markedly increased in both HCP and VP. Plasma porphyrins also are normal or slightly elevated in AIP but are increased in VP. Erythrocyte PBGD activity is approximately half-normal in ~90% of AIP patients. Identification of the disease-causing mutation provides further confirmation and enables other gene carriers to be identified by DNA testing.

12.6.4 Treatment

Most attacks require hospitalization for treatment of severe symptoms and monitoring of respiration, electrolytes, and nutritional status. Offending drugs should be discontinued if possible. Intravenous hemin is the most effective treatment for acute attacks. Carbohydrate loading by oral or intravenous glucose is less effective but may suffice for milder attacks (Anderson et al. 2005).

Avoiding harmful drugs and maintaining a well-balanced diet can help prevent further attacks. Frequent cyclic attacks can be prevented with use of a gonadotropin release hormone analogue (Anderson et al. 1990). Recurrent noncyclic attacks may be prevented by prophylactic infusions of hemin (Lamon et al. 1978). Liver transplantation can be curative in patients with frequent exacerbations that are unresponsive to other therapies and is evidence that in AIP the liver is the source for the neurological manifestations (Dar et al. 2010; Soonawalla et al. 2004).

12.7 ALA Dehydratase Porphyria

12.7.1 Overview

ADP is an autosomal recessive disorder caused by severe deficiency of δ-ALA dehydratase (ALAD) activity, the second enzyme in the heme biosynthetic pathway. ADP is the rarest of the porphyrias, with only six documented cases reported (Akagi et al. 2000). Symptoms occurring as acute attacks are the same as in the other acute porphyrias. All reported cases have been males. One severely affected infant did not improve after liver transplantation (Thunell et al. 1992). A patient with a heterozygous *ALAD* mutation developed late onset ADP associated with polycythemia vera and expansion of a clone of erythroid cells with the defective *ALAD* allele (Akagi et al. 2000).

12.7.2 Pathophysiology

ADP is associated with marked elevations in ALA and coproporphyrin III. Coproporphyrin III is also increased in normal subjects loaded with ALA (Shimizu et al. 1978). Therefore, excess coproporphyrin III in ADP may originate from reentry of ALA into the heme pathway in a tissue other than the site of origin of the excess ALA. ALA is suspected to be neurotoxic. Erythrocyte zinc protoporphyrin is also markedly elevated. ADP is often classified as a hepatic porphyria, but this is uncertain.

12.7.3 Diagnosis

Characteristic biochemical findings in ADP include marked elevations in urinary ALA and coproporphyrin III and erythrocyte zinc protoporphyrin with normal or only slightly increased levels of urinary PBG. Erythrocyte ALAD activity is approximately half-normal in both parents and markedly decreased in the patient. It is essential to exclude other causes of ALAD deficiency, including lead poisoning and hereditary tyrosinemia type I, and to confirm the diagnosis of ADP by DNA studies.

12.7.4 Treatment

Treatment data are limited. Most patients benefited with hemin infusions but not from glucose. A severely affected infant did not respond to glucose, hemin, or liver transplantation (Thunell et al. 1992).

References

Akagi, R., C. Nishitani, H. Harigae et al. 2000. Molecular analysis of delta-aminolevulinate dehydratase deficiency in a patient with an unusual late-onset porphyria. *Blood* 96:3618–3623.

Andant, C., H. Puy, C. Bogard et al. 2000. Hepatocellular carcinoma in patients with acute hepatic porphyria: Frequency of occurrence and related factors. *J Hepatol* 32:933–939.

Anderson, K. E. 2006. The porphyrias. In *Zakim and Boyer's Hepatology: A Textbook of Liver Disease*. T. D. Boyer, T. L. Wright, and M. P. Manns, editors. Saunders, Philadelphia, PA, 1391–1432.

Anderson, K. E., J. R. Bloomer, H. L. Bonkovsky et al. 2005. Recommendations for the diagnosis and treatment of the acute porphyrias. *Ann Intern Med* 142:439–450.

Anderson, K. E., S. Sassa, C. M. Peterson, and A. Kappas. 1977. Increased erythrocyte uroporphyrinogen-l-synthetase, delta-aminolevulinic acid dehydratase and protoporphyrin in hemolytic anemias. *Am J Med* 63:359–364.

Anderson, K. E., I. M. Spitz, C. W. Bardin, and A. Kappas. 1990. A gonadotropin releasing hormone analogue prevents cyclical attacks of porphyria. *Arch Intern Med* 150:1469–1474.

Anstey, A. V., and R. J. Hift. 2007. Liver disease in erythropoietic protoporphyria: Insights and implications for management. *Gut* 56:1009–1018.

Balwani, M., and R. J. Desnick. 2012. The porphyrias: Advances in diagnosis and treatment. *Blood* 19–27. Prepublished online July 12, 2012; doi: 10.1182/blood-2012-05-423186.

Barone, G. W., B. J. Gurley, K. E. Anderson, B. L. Ketel, and S. R. Abul-Ezz. 2001. The tolerability of newer immunosuppressive medications in a patient with acute intermittent porphyria. *J Clin Pharmacol* 41:113–115.

Bijlmer-Iest, J. C., H. Baart de la Faille, B. S. van Asbeck et al. 1992. Protoporphyrin photosensitivity cannot be attenuated by oral N-acetylcysteine. *Photodermatol Photoimmunol Photomed* 9:245–249.

Blake, D., J. McManus, V. Cronin, and S. Ratnaike. 1992. Fecal coproporphyrin isomers in hereditary coproporphyria. *Clin Chem* 38:96–100.

Boffa, M. J., R. D. Ead, P. Reed, and C. Weinkove. 1996. A double-blind, placebo-controlled, crossover trial of oral vitamin C in erythropoietic protoporphyria. *Photodermatol Photoimmunol Photomed* 12:27–30.

Bottomley, S. S., M. Tanaka, and M. A. Everett. 1975. Diminished erythroid ferrochelatase activity in protoporphyria. *J Lab Clin Med* 86:126–131.

Clark, K. G., and D. C. Nicholson. 1971. Erythrocyte protoporphyrin and iron uptake in erythropoietic protoporphyria. *Clin Sci* 41:363–370.

Cox, T. M. 2003. Protoporphyria. In *The Porphyrin Handbook*. K. Kadish, K. Smith, and R. Guilard, editors. Elsevier Science, San Diego, CA, 121–149.

Dar, F. S., K. Asai, A. R. Haque et al. 2010. Liver transplantation for acute intermittent porphyria: A viable treatment? *Hepatobiliary Pancreat Dis Int* 9:93–96.

Deacon, A. C., and G. H. Elder. 2001. ACP Best Practice No 165: Front line tests for the investigation of suspected porphyria. *J Clin Pathol* 54:500–507.

Dupuis-Girod, S., V. Akkari, C. Ged et al. 2005. Successful match-unrelated donor bone marrow transplantation for congenital erythropoietic porphyria (Gunther disease). *Eur J Pediatr* 164:104–107.

Elder, G. H. 2003. Porphyria cutanea tarda and related disorders. In *The Porphyrin Handbook*. K. Kadish, K. Smith, and R. Guilard, editors. Elsevier Science, San Diego, CA, 67–92.

Goodwin, R. G., W. J. Kell, P. Laidler et al. 2006. Photosensitivity and acute liver injury in myeloproliferative disorder secondary to late-onset protoporphyria caused by deletion of a ferrochelatase gene in hematopoietic cells. *Blood* 107:60–62.

Gouya, L., H. Puy, A. M. Robreau et al. 2002. The penetrance of dominant erythropoietic protoporphyria is modulated by expression of wildtype FECH. *Nat Genet* 30:27–28.

Grandchamp, B., N. Phung, and Y. Nordmann. 1977. Homozygous case of hereditary coproporphyria. *Lancet* 2:1348–1349.

Harms, J., S. Lautenschlager, C. E. Minder, and E. I. Minder. 2009. An alpha-melanocyte-stimulating hormone analogue in erythropoietic protoporphyria. *N Engl J Med* 360:306–307.

Hastka, J., J. J. Lasserre, A. Schwarzbeck, M. Strauch, and R. Hehlmann. 1993. Zinc protoporphyrin in anemia of chronic disorders. *Blood* 81:1200–1204.

Hift, R. J., B. P. Davidson, C. van der Hooft, D. M. Meissner, and P. N. Meissner. 2004. Plasma fluorescence scanning and fecal porphyrin analysis for the diagnosis of variegate porphyria: Precise determination of sensitivity and specificity with detection of protoporphyrinogen oxidase mutations as a reference standard. *Clin Chem* 50:915–923.

Holme, S. A., A. V. Anstey, A. Y. Finlay, G. H. Elder, and M. N. Badminton. 2006. Erythropoietic protoporphyria in the U.K.: Clinical features and effect on quality of life. *Br J Dermatol* 155:574–581.

Jalil, S., J. J. Grady, C. Lee, and K. E. Anderson. 2010. Associations among behavior-related susceptibility factors in porphyria cutanea tarda. *Clin Gastroenterol Hepatol* 8:297–302, 302 e291.

Labbe, R. F. 1992. Clinical utility of zinc protoporphyrin. *Clin Chem* 38:2167–2168.

Lamon, J. M., B. C. Frykholm, M. Bennett, and D. P. Tschudy. 1978. Prevention of acute porphyric attacks by intravenous haematin. *Lancet* 2:492–494.

Lamoril, J., H. Puy, S. D. Whatley et al. 2001. Characterization of mutations in the CPO gene in British patients demonstrates absence of genotype-phenotype correlation and identifies relationship between hereditary coproporphyria and harderoporphyria. *Am J Hum Genet* 68:1130–1138.

Lang, K., K. W. Schulte, T. Ruzicka, and C. Fritsch. 2001. Aminolevulinic acid (Levulan) in photodynamic therapy of actinic keratoses. *Skin Therapy Lett* 6:1–2, 5.

Longas, M. O., and M. B. Poh-Fitzpatrick. 1982. A tightly bound protein-porphyrin complex isolated from the plasma of a patient with variegate porphyria. *Clin Chim Acta* 118:219–228.

Marmor, M. F., U. Kellner, T. Y. Lai, J. S. Lyons, and W. F. Mieler. 2011. Revised recommendations on screening for chloroquine and hydroxychloroquine retinopathy. *Ophthalmology* 118:415–422.

Mathews-Roth, M. M., M. A. Pathak, T. B. Fitzpatrick, L. H. Harber, and E. H. Kass. 1977. Beta carotene therapy for erythropoietic protoporphyria and other photosensitivity diseases. *Arch Dermatol* 113:1229–1232.

Mathews-Roth, M. M., and B. Rosner. 2002. Long-term treatment of erythropoietic protoporphyria with cysteine. *Photodermatol Photoimmunol Photomed* 18:307–309.

Meissner, P., P. Adams, and R. Kirsch. 1993. Allosteric inhibition of human lymphoblast and purified porphobilinogen deaminase by protoporphyrinogen and coproporphyrinogen. A possible mechanism for the acute attack of variegate porphyria. *J Clin Invest* 91:1436–1444.

Meissner, P., R. Hift, and A. Corrigall. 2003. Variegate porphyria. In *The Porphyrin Handbook*. K. Kadish, K. Smith, and R. Guilard, editors. Elsevier Science, San Diego, CA, 93–120.

Nordmann, Y., B. Grandchamp, H. de Verneuil et al. 1983. Harderoporphyria: A variant hereditary coproporphyria. *J Clin Invest* 72:1139–1149.

Phillips, J. D., and K. E. Anderson. 2010. The porphyrias. In *Williams Hematology*. McGraw-Hill Professional, New York, 839–863.

Phillips, J. D., J. P. Kushner, H. A. Bergonia, and M. R. Franklin. 2011. Uroporphyria in the Cyp1a2-/- mouse. *Blood Cells Mol Dis* 47:249–254.

Piomelli, S., A. A. Lamola, M. F. Poh-Fitzpatrick, C. Seaman, and L. C. Harber. 1975. Erythropoietic protoporphyria and lead intoxication: the molecular basis for difference in cutaneous photosensitivity. I. Different rates of disappearance of protoporphyrin from the erythrocytes, both in vivo and in vitro. *J Clin Invest* 56:1519–1527.

Poblete-Gutierrez, P., C. Wolff, R. Farias, and J. Frank. 2006. A Chilean boy with severe photosensitivity and finger shortening: The first case of homozygous variegate porphyria in South America. *Br J Dermatol* 154:368–371.

Poh-Fitzpatrick, M. B., and V. A. DeLeo. 1977. Rates of plasma porphyrin disappearance in fluorescent vs. red incandescent light exposure. *J Invest Dermatol* 69:510–512.

Poh-Fitzpatrick, M. B., and A. A. Lamola. 1976. Direct spectrofluorometry of diluted erythrocytes and plasma: A rapid diagnostic method in primary and secondary porphyrinemias. *J Lab Clin Med* 87:362–370.

Ratnaike, S., D. Blake, D. Campbell, P. Cowen, and G. Varigos. 1988. Plasma ferritin levels as a guide to the treatment of porphyria cutanea tarda by venesection. *Australas J Dermatol* 29:3–8.

Sardh, E., D. E. Andersson, A. Henrichson, and P. Harper. 2009. Porphyrin precursors and porphyrins in three patients with acute intermittent porphyria and end-stage renal disease under different therapy regimes. *Cell Mol Biol (Noisy-le-grand)* 55:66–71.

Sarkany, R. P. E. 2008. Making sense of the porphyrias. *Photodermatol Photoimmunol Photomed* 24:102–108.

Sassa, S., R. Akagi, C. Nishitani, H. Harigae, and K. Furuyama. 2002. Late-onset porphyrias: What are they? *Cell Mol Biol (Noisy-le-grand)* 48:97–101.

Shimizu, Y., S. Ida, H. Naruto, and G. Urata. 1978. Excretion of porphyrins in urine and bile after the administration of delta-aminolevulinic acid. *J Lab Clin Med* 92:795–802.

Singal, A. K., C. Kormos-Hallberg, C. Lee et al. 2012. Low-dose hydroxychloroquine is as effective as phlebotomy in treatment of patients with porphyria cutanea tarda. *Clin Gastroenterol Hepatol* 14:038.

Solis, C., A. Martinez-Bermejo, T. P. Naidich et al. 2004. Acute intermittent porphyria: Studies of the severe homozygous dominant disease provides insights into the neurologic attacks in acute porphyrias. *Arch Neurol* 61:1764–1770.

Soonawalla, Z. F., T. Orug, M. N. Badminton et al. 2004. Liver transplantation as a cure for acute intermittent porphyria. *Lancet* 363:705–706.

Thunell, S., A. Henrichson, Y. Floderus et al. 1992. Liver transplantation in a boy with acute porphyria due to amino-laevulinate dehydratase deficiency. *Eur J Clin Chem Clin Biochem* 30:599–606.

Verstraeten, L., N. Van Regemorter, A. Pardou et al. 1993. Biochemical diagnosis of a fatal case of Gunther's disease in a newborn with hydrops foetalis. *Eur J Clin Chem Clin Biochem* 31:121–128.

Wahlin, S., N. Srikanthan, B. Hamre, P. Harper, and A. Brun. 2008. Protection from phototoxic injury during surgery and endoscopy in erythropoietic protoporphyria. *Liver Transpl* 14:1340–1346.

Warren, L. J., and S. George. 1998. Erythropoietic protoporphyria treated with narrow-band (TL-01) UVB phototherapy. *Australas J Dermatol* 39:179–182.

Whatley, S. D., S. Ducamp, L. Gouya et al. 2008. C-terminal deletions in the ALAS2 gene lead to gain of function and cause X-linked dominant protoporphyria without anemia or iron overload. *Am J Hum Genet* 83:408–414.

Whatley, S. D., H. Puy, R. R. Morgan et al. 1999. Variegate porphyria in Western Europe: Identification of PPOX gene mutations in 104 families, extent of allelic heterogeneity, and absence of correlation between phenotype and type of mutation. *Am J Hum Genet* 65:984–994.

13

Photoprotection

Chanisada Wongpraparut
Faculty of Medicine
Siriraj Hospital

Henry W. Lim
Henry Ford Hospital

13.1 Introduction

Ultraviolet (UV) radiation emitted by the sun is divided into three bands: UVA, UVB, and UVC. UVA is further subdivided into UVA2 (320–340 nm) and UVA1 (340–400 nm). The nitrogen and oxygen molecules in the ozone layer absorb UVC, so only UVA and UVB reach the Earth's surface. Side effects of UV exposure have been well studied and can be divided into acute and chronic effects. The most well-recognized, acute effect is erythema, which is the predominant biologic effect of UVB and, to a lesser extent, UVA2. UVB exposure also induces cutaneous vitamin D synthesis. Immediate and persistent pigment darkening (PPD) and delayed tanning are caused predominantly following UVA exposure, although in dark-skinned individuals, it can be induced by exposure to visible light (Mahmoud et al. 2010). Epidermal hyperplasia occurs within days of UV exposure. Chronic effects are photoaging and photocarcinogenesis.

With the increasing rates of melanomas and nonmelanoma skin cancers, photoprotection is advocated by physicians and public health organizations around the world. Proper photoprotection includes seeking shade during the peak UV hours of 10 a.m. to 2 p.m.; wearing photoprotective clothing, wide-brimmed hats, and sunglasses; and using broad-spectrum sunscreens on exposed skin. In this chapter, photoprotection by sunscreens, the emerging field of systemic photoprotection, and photoprotection by clothing and glass will be discussed.

13.2 Sunscreens

The use of sunscreen was reported more than a hundred years ago. Tannin was used as a photoprotective agent in 1887; this was followed by the use of zinc oxide, magnesium salt, and bismuth in the early 1900s. In 1928, the first commercial sunscreen became available in the United States with benzyl salicylate and benzyl cinnamate as active ingredients. Since then, many UV filters were developed and widely used globally. Regular use of sunscreen has been shown to reduce the incidence of actinic keratosis and squamous cell carcinoma (SCC) (Naylor et al. 1995; Thompson, Jolley, and Marks 1993; van der Pols et al. 2006). However, the photoprotective effect of sunscreens on basal cell carcinoma (BCC) and malignant melanoma (MM) is still inconclusive. A randomized, controlled trial on the effect of SPF 16 broad-spectrum sunscreen on the development of skin cancers in 1621 individuals demonstrated that after 4.5 years of study, the rate of SCC was reduced in sunscreen users (Green et al. 1999). An 8-year follow-up showed that SCC tumor rates were decreased significantly by almost 40%. BCC tumor rates showed a decrease of 25%; however, this did not achieve statistical significance (van der Pols et al. 2006). A 10-year follow-up showed that the sunscreen group developed fewer melanomas and had fewer invasive melanomas (Green et al. 2011).

13.2.1 UV Filters

UV filters are classified into three groups according to the mechanism of action: organic UV absorbers, inorganic particulates, and organic particulates. The properties of these filters are summarized in Table 13.1; selected filters are further discussed here.

13.2.1.1 Organic UV Absorbers

Organic UV absorbers or organic filters, previously termed "chemical sunscreens," are generally aromatic compounds conjugated with a carbonyl group. These chemicals absorb UV rays resulting in excitation to a higher energy state. The

TABLE 13.1   Properties of UV Filters

INCI Name	USAN (Brand Name)	Peak Absorption λ max (nm)	UV Action Spectrum	Comment
			Organic Filters	
3-Benzylidene camphor	–	294	UVB	
Benzylidene camphor sulfonic acid	–	294	UVB	
Diethylhexyl butamido triazone	Iscotrizinol	311	UVB	Pending FDA approval via TEA process
Ethylhexyl methoxycinnamate	Octinoxate	311	UVB	Most common UV filter used worldwide
Ethylhexyl salicylate	Octisalate	305	UVB	
Ethylhexyl triazone	–	314	UVB	Pending FDA approval via TEA process
Homosalate	Homosalate	306	UVB	
Isoamyl P-methoxycinnamate	Amiloxate	308	UVB	Pending FDA approval via TEA process
4-Methylbenzylidene camphor	Enzacamene	300	UVB	Pending FDA approval via TEA process
Octocrylene	Octocrylene	303	UVB	
Para-amino benzoic acid	PABA	283	UVB	
Phenylbenzimidazole sulfonic acid	Ensulizole	302	UVB	
Benzophenone-3	Oxybenzone	286, 324	UVB, UVA 2	
Benzophenone-4	Sulisobenzone	286, 324	UVB, UVA 2	
Benzophenone-8	Dioxybenzone	284, 327	UVB, UVA 2	
Butyl methoxydibenzoylmethane	Avobenzone	357	UVA 2, UVA 1	
Diethylamino hydroxy benzoyl hexyl benzoate	–	354	UVA 2, UVA 1	
Disodium phenyl dibenzimidazole tetrasulfonate	Bisdisulizole disodium	335	UVA 2, UVA 1	
Methyl anthranilate	Meradimate	336	UVA 2	
Bis-ethylhexyloxyphenol methoxyphenyl triazine	Bemotrizinol (Tinosorb S)	310, 343	UVB, UVA 2, UVA 1	Pending FDA approval via TEA process
Drometrizole trisiloxane	Silatriazole (Mexoryl XL)	303, 341	UVB, UVA 2, UVA 1	Pending FDA approval via TEA process
Terephthalylidene dicamphor sulfonic acid	Ecamsule (Mexoryl SX)	345	UVB, UVA 2, UVA 1	Pending FDA approval via TEA process (US: approved in certain formulations up to 3% via New Drug Application (NDA) Route
			Organic Particulates	
Methylene bis-benzotriazolyl tetramethylbutylphenol	Bisoctrizole (Tinosorb M)	305, 360	UVB, UVA 2, UVA 1	Pending FDA approval via TEA process
			Inorganic Filters	
Titanium dioxide	Titanium dioxide	280–350	Depend on particle size	
Zinc oxide	Zinc oxide	280–390	Depend on particle size	

Note: FDA, Food and Drug Administration; INCI, International Nomenclature of Cosmetic Ingredients; TEA, Time and Extent Application; USAN, United States Adopted Name.

molecule returns to the ground state by releasing the excess energy as phosphorescence or as heat (Antoniou et al. 2008). In this process, some undergo photodegradation, rendering them ineffective as further functioning effective filters; octinoxate and avobenzone are examples of photounstable filters. Organic UV absorbers are classified as either UVA or UVB filters.

13.2.1.1.1 UVB Filters

13.2.1.1.1.1 Cinnamates   Ethylhexyl methoxycinnamate (octinoxate) is the most common cinnamate and offers protection against UVB with peak absorption at 311 nm. It is probably the most common UV filter used globally because it rarely causes irritation (Kullavanijaya and Lim 2005). Upon sunlight exposure, octinoxate can be degraded, resulting in a decrease in efficacy. However, encapsulating the filter in an inert shell minimizes the photodegradation.

13.2.1.1.1.2 Salicylates   Salicylates include ethylhexyl salicylate (octisalate), homosalate, and trolamine salicylate. They are weak organic UVB absorbers; however, they are photostable and have favorable safety profiles (Sambandan and Ratner 2011).

13.2.1.1.2 UVA Filters

13.2.1.1.2.1 Benzophenones
Benzophenones are commonly used UVA filters. The US Food and Drug Administration (FDA) has approved three benzophenones: benzophenone-3 (oxybenzone), benzophenone-4 (sulisobenzone), and benzophenone-8 (dioxybenzone). Among these, oxybenzone is the most commonly used. Among all UV filters, it is the most common cause of photoallergy (Darvay et al. 2001).

13.2.1.1.2.2 Butyl Methoxydibenzoylmethane (Avobenzone)
Butyl methoxydibenzoylmethane (avobenzone) is a widely used UVA1 absorber with the peak absorption at 357 nm. It was approved by the US FDA in 1988 and is now being used worldwide. Currently, it is the best UVA1 filter available in the United States; however, it undergoes significant photodegradation upon exposure to sunlight. The presence of avobenzone together with octinoxate accelerate the photodegradation of both; hence, this combination should not be used in a final product.

The photodegradation of avobenzone can be overcome by combining it with other photostable UV filters, such as octocrylene, oxybenzone, salicylates, 4-methylbenzylidene camphor, or bis-ethylhexyloxyphenol methoxyphenyl triazine (Tinosorb S), and non-UV filters, such as diethylhexyl 2,6 naphthalate (DEHN), Oxynex ST, or caprylyl glycol (Tuchinda et al. 2006). These photostabilizers absorb energy for excited-state avobenzone, hence minimizing the photodegradation of avobenzone.

13.2.1.1.2.3 Terephthalylidene Dicamphor Sulfonic Acid (TDSA, Ecamsule, Mexoryl SX)
This is an efficient UVA filter, which absorbs UV radiation between 290 and 390 nm with peak absorption at 345 nm. Mexoryl SX was developed and patented by L'Oréal (Paris, France) in 1982 and approved by the European Economic Community (EEC) in 1991. In July 2006, Mexoryl SX was approved in the United States as an active ingredient in a sunscreen formulation. Several studies showed promising results of sunscreen products containing Mexoryl SX in the prevention of photoaging, UV-induced skin pigmentation, UV-induced immunosuppression and carcinogenesis, and UVA-induced phototoxicity from medication (Duteil et al. 2002; Fourtanier 1996; Moyal 2004).

This UVA filter was also shown to be useful for patients with photodermatoses. Upon exposure to sunlight, sunscreen with SPF 50+ and UVA-PF 28 (containing octocrylene, Mexoryl SX, Mexoryl XL, avobenzone, or TiO_2) was demonstrated to decrease the development of polymorphous light eruption (PLE) and lupus erythematosus (LE) lesions in all subjects, and sunscreen with high SPF but low UVA-PF can only partially protect against them (Stege et al. 2000, 2002).

13.2.1.1.2.4 Disodium Phenyl Dibenzimidazole Tetrasulfonate (DPDT, Neo Heliopan AP)
This is a water-soluble UVA filter with a peak absorption spectrum at 335 nm. It shows a synergistic effect when combined with filters in the oil phase. It has been approved in the European Union since 2000 but has not yet been approved in the United States. It is also available in China, South Korea, Australia, New Zealand, and South Africa and the ASEAN and MERCOSUR (Argentina, Brazil, Paraguay, Uruguay) countries.

13.2.1.1.2.5 Diethylamino Hydroxybenzoyl Hexyl Benzoate (DHHB, Uvinul A Plus)
DHHB provides the UV-spectral properties in the UVA-1 range, similar to avobenzone, but with superior photostability. Currently, it is not yet approved by the US FDA.

13.2.1.1.3 Broad-Spectrum UVB and UVA Filters

13.2.1.1.3.1 Drometriazole Trisiloxane (DTS, Silatriazole, Mexoryl XL)
This is the first photostable broad UVA and UVB filter. It is now available in Europe, China, Japan, ASEAN countries, Australia, and New Zealand. At the time of this writing, it is not yet listed in the US monograph.

Mexoryl SX combined with Mexoryl XL was shown to be beneficial in different studies. A broad-spectrum sunscreen (SPF > 60, UVA-PF 28, Anthelios XL containing octocrylene, Mexoryl SX, Mexoryl XL, avobenzone, TiO_2) was demonstrated to prevent lesions of photodermatoses; suppress UV-induced, delayed-type hypersensitivity response (Moyal et al. 2002); as well as prevent against drug-induced phototoxicity (Duteil et al. 2002).

13.2.1.1.3.2 Bis-Ethylhexyloxyphenol Methoxyphenyl Triazine (BEMT, Bemotrizinol, Tinosorb S)
Tinosorb S is a highly efficient, broad-spectrum, photostable UV filter, which has strong absorption peaks at 310 and 343 nm. It has been shown to prevent photodegradation of avobenzone in a concentration-dependent way and to improve the photostability of product that contained two photolabile filters, avobenzone and octinoxate (Chatelain and Gabard 2001).

13.2.1.2 Organic Particulates

This new group contains a new UV filter that has the properties of an organic UV absorber and inorganic particulates (both UV absorption and reflection/scattering). Methylene-bis-benzotriazolyl tetramethylbutylphenol (MBBT, Tinosorb M) is an example agent in this group.

13.2.1.2.1 Methylene-Bis-Benzotriazolyl Tetramethylbutylphenol (MBBT, Bisoctrizole; Tinosorb M)
Tinosorb M is a new class of UV absorber developed by Ciba Chemicals (now part of BASF, Ludwigshafen, Germany). It is a photostable, broad-spectrum UV filter, which has strong absorption peaks in UVB and UVA at 305 and 360 nm. It is now available in Europe, China, Japan, South Korea, ASEAN countries, Australia, and New Zealand. Because the Tinosorb M has a high molecular weight (> 500 Da), the potential for its systemic absorption following topical application is minimal.

13.2.1.3 Inorganic Particulates

This group is also known as inorganic filters and was previously termed "physical sunscreens." The major mechanism is to reflect or scatter the UV, visible, and infrared radiation, although, at small particle size, some absorption does occur (Sambandan and Ratner 2011). The major inorganic agents are zinc oxide and titanium dioxide. The photoprotective effect of inorganic particulates is influenced by their particle size. When the particle size is large (200–500 nm), it scatters visible light, resulting in the visual perception of whitening of the skin. With developments in sunscreen technology, the particle size can now be decreased into micronized form (10–50 nm), resulting in less scattering of visible light, thus improving the cosmetic acceptability and allowing easier incorporation into formulations. However, by reducing the particle size of inorganic sunscreen, especially titanium dioxide, the peak of the absorption spectrum is shifted to the shorter wavelength (Kullavanijaya and Lim 2005). Compared to titanium dioxide, zinc oxide provides superior protection against long-wave UVA (UVA1) and reflects less visible light, hence appearing less white (Beasley and Meyer 2010; Pinnell et al. 2000).

13.2.2 Sun Protection Factor

Sun protection factor (SPF) is a reflection of the efficacy of the sunscreen against sunburn. It is defined as a ratio of the time of exposure to the solar-simulated radiation necessary to produce minimally detectable erythema in sunscreen-protected skin to that time for unprotected skin (Schalka and Reis 2011). Because the effect of UV on inducing skin erythema is mainly from the UVB (290–320 nm) and UVA2 wavelength (320–340 nm), a sunscreen with high SPF provides good UVB protection but may not necessarily protect against long-wave UVA.

The amount of sunscreen used for the determination of SPF values is 2 mg/cm^2. However, in actual use, most individuals apply only around a quarter or half the concentration used for testing (0.5–1 mg/cm^2) (Bech-Thomsen and Wulf 1992; Reich et al. 2009). It is known that there is an exponential relationship between the amount of sunscreen applied

and the in-use SPF values (Faurschou and Wulf 2007; Kim et al. 2010; Schalka, dos Reis, and Cuce 2009); at 0.5 mg/cm^2, the in-use SPF of SPF 15 sunscreen is 4, and the in-use SPF of SPF 30 is 7 (Schalka, dos Reis, and Cuce 2009). Therefore, in general, the in-use SPF is several magnitudes lower than the labeled SPF.

13.2.3 Standard for UVA Protection

To date, there is still no worldwide agreement on a UVA assessment method. Several methods for the assessment of UVA protection by sunscreen are currently available and in use in different parts of the world (Table 13.2).

The *in vitro* methods include critical wavelength (CW), which is defined as the wavelength in which 90% of UV absorbance occurs as measured from 290 to 400 nm; therefore, it is an assessment of the breath of absorption of the product. In the United States and the European Union, a broad-spectrum sunscreen must have a CW of ≥370 nm. In the United Kingdom, there is a system called the Boots Star Rating, which evaluated the ratio of UVA to UVB absorbed by the sunscreen *in vitro* (Diffey 1994). In Australia, the UVA test method uses *in vitro* absorbance of 320–360 nm of UV through a layer of the sunscreen product.

The *in vivo* UVA test methods include immediate pigment darkening (IPD), persistent pigment darkening (PPD), and protection factor UVA (PFA), which, respectively, assess the development of IPD, PPD, or UVA erythema in sunscreen-protected skin compared with that observed in unprotected skin (Kaidbey and Barnes 1991; Moyal, Chardon, and Kollias 2000; Nash, Tanner, and Matts 2006). The most widely used method is the PPD method. In Japan, the PPD method is used

TABLE 13.2 Worldwide Standard for UVA Protection

Country/Region	Method	Definition	Rating
Australia	*In vitro* transmittance measure	Measure the UV transmission within the range 320–360 nm	Only products with SPF > 15 and pass the UVA test can be labeled as "broad spectrum"
European Union	Recommendation, September 2006 – Ratio UVA-PF to SPF – CW	– COLIPA *in vitro* method (ratio of UVA-PF to SPF) – CW reflects wavelength below which 90% of absorbance curve resides	– UVA-PF/SPF > 1/3 – CW ≥ 370 nm defines "broad-spectrum" protection
Japan	PPD	UV protection to prevent persistent darkening of the skin	Protection grade of UVA (PA) PA+ = PPD 2–4 PA++ = PPD 4–8 PA+++ = PPD > 8
United Kingdom	Boots Star Rating system, revised in 2008	Ratio of UVA absorbance to mean UVB absorbance	UVA/UVB limits 0–0.2 no claim 0.21–0.4 minimal (★) 0.41–0.6 moderate (★★) 0.61–0.8 good (★★★) 0.81–0.9 superior (★★★★) > 0.91 ultra (★★★★★)
United States	FDA, June 2011 *In vitro* CW	CW reflects wavelength below which 90% of absorbance curve resides	Only product with CW ≥ 370 nm are allowed to use "broad-spectrum" label

Note: COLIPA, European Cosmetic, Toiletry and Perfumery Association; CW, critical wavelength; PPD, persistent pigment darkening; UVA-PF, UVA protection factor.

with a classification of PA+, PA++, and PA+++. This system is also widely used in many countries in Asia.

On September 22, 2006, the European Commission published a recommendation on the efficacy of sunscreen stating that the value of the UVA protection factor (UVA-PF) should be at least one third of the SPF; hence, a product with an SPF of 30 must have a UVA-PF of 10. Products that fulfill this requirement will then be eligible to have the acronym UVA enclosed within a circle on the package (Wang, Stanfield, and Osterwalder 2008). The UVA-PF is measured by the PPD method.

On June 17, 2011, the US FDA issued the final ruling on labeling and effectiveness testing of sunscreen products in the United States (Wang and Lim 2011). The 2011 final rule was to be implemented on June 17, 2012. The FDA has adopted the *in vitro* CW as the method for assessing UVA protection. It is a pass/fail system; those with CW ≥ 370 nm can then put on the package the label "Broad Spectrum," which should be of the same font size as the SPF value (Wang and Lim 2011).

13.2.4 Water Resistance

The terms "water resistant" and "very water resistant" represent the photoprotective properties of sunscreen after two 20-min periods of water exposure and four 20-min periods of water exposure, respectively. However, the 2011 FDA final rule of sunscreen regulation indicated that the term "very water resistant" will no longer be allowed. The terms "sunblock," "waterproof," or "sweat proof" also will be abandoned. Instead, either the statement "Water Resistant (40 minutes)" or "Water Resistant (80 minutes)" will be used (Wang and Lim 2011). However, it should be noted that for the previously mentioned water-resistance testing, subjects did not towel-dry themselves in between water immersions. Therefore, even for water-resistant sunscreens, consumers should be instructed to reapply sunscreen after towel drying and at least every 2 h when outdoors.

13.2.5 Concerns on the Safety Issue of Sunscreens

13.2.5.1 Systemic Toxicity of Topical Sunscreens

There is a growing concern about endocrine disruption or other systemic toxicities from UV filters because there have been reports of systemic absorption of some organic UV filters. There was a study that demonstrated that following repeated whole-body topical application of 2 mg/cm^2 of sunscreen containing benzophenone-3, octyl methoxycinnamate, and 4-methylbenzilidene camphor, all these UV filters were detected in human plasma and urine (Janjua et al. 2008). However, the plasma level of these UV filters was too low to cause detectable systemic toxic effects in humans. In a recent analysis, it was estimated that it would take 34.5 years of daily application at 2 mg/cm^2 of oxybenzone-containing sunscreen to achieve levels that resulted in estrogenic effects in a study in rats (Wang, Burnett, and Lim 2011).

13.2.5.2 Contact Dermatitis or Photoallergic Contact Dermatitis from Sunscreens

Increasing use of sunscreens has led to increased reporting of adverse reactions to sunscreens. However, true allergic contact dermatitis and photoallergy to sunscreens are quite rare (Darvay et al. 2001; Shaw et al. 2010). A review of 1155 individuals who underwent photopatch testing in 17 centers across the United Kingdom, Ireland, and the Netherlands showed that only 4.4% were found to have a true photoallergy (PA), 5.5% were found to have a contact allergy (CA), and 1.3% had combined PA and CA. The most common photoallergen was benzophenone-3 (Bryden et al. 2006). Because now there are many new UV filters, continued revision of the photopatch test series is necessary to detect allergic and photoallergic reactions to new UV filters (Schauder and Ippen 1997).

13.2.5.3 Other Safety Concerns

The retinyl palmate used in some sunscreen products has been cited as a safety concern resulting from the photocarcinogenicity shown in an animal model. However, careful analysis of the data did not indicate that this concern is justified (Wang, Dusza, and Lim 2010).

The safety of nanoparticles used in sunscreens has also been questioned. Currently, there is no evidence that nanoparticles could penetrate the intact epidermis. However, the trans-epidermal penetration of nanoparticles through inflamed skin where the epidermal barrier has been compromised is not clearly known (Burnett and Wang 2011).

13.2.5.4 Vitamin D and Sunscreens

Adequate vitamin D levels are well established as being important to bone health, and while vitamin D has been suggested to be beneficial for other diseases, data are still evolving, and hence, no firm conclusion on the role of vitamin D in other diseases can be drawn at this time (Vanchinathan and Lim 2012). Serum 25-hydroxyvitamin D (25 (OH) D) levels are widely used to measure the vitamin D status in the body.

There are only three sources of vitamin D: sunlight, diet, and vitamin D supplements. The synthesis of vitamin D in the skin requires UVB (300 ± 5 nm), and it is influenced by many factors, such as the intensity of sunlight, skin type, age, and photoprotection (Holick 2007). While more than 50% of individuals who practice rigorous photoprotection, such as patients with photosensitive LE or erythropoietic protoporphyria (Cusack et al. 2008; Holme et al. 2008), have been shown to have inadequate serum vitamin D levels, normal usage of sunscreens was not associated with inadequate serum vitamin D levels (Linos et al. 2012; Norval and Wulf 2009). This is most likely a result of the known findings that, in actual use, most individuals apply a lesser amount of sunscreen than the amount mandated for SPF testing (2 mg/cm^2). However, individuals with a high risk of vitamin D deficiency, such as those of older age group, individuals with dark skin, and individuals who usually avoid the sun or wear clothing that covers most parts of the body, should have

TABLE 13.3   2011 Requirement for Vitamin D from The Institute of Medicine

Age Group	Recommended Dietary Allowances (RDAs) per Day
0–1 year	400 IU/day
1–70 years	600 IU/day
71 years and older	800 IU/day

Source: Ross, A. et al., *J Clin Endocrinol Metabol*, 96, 53–58, 2011.
Note: RDAs; covering requirements of ≥97.5% of the population; IU: international unit.

vitamin D supplementation. Furthermore, because vitamin D is a fat-soluble vitamin absorbed in the gastrointestinal tract, individuals who are obese or those with fat malabsorption should also consider vitamin D supplementation.

The current Institute of Medicine recommendations for vitamin D intake are listed in Table 13.3 (Ross et al. 2011). It should be noted that these recommendations are based on data on bone health. Furthermore, because the upper intake levels (i.e., the highest daily intake likely to pose no risk) for different age groups range from 1000 to 4000 IU/day, a daily intake of 1000 IU/day should be safe, although benefits are not clear.

13.2.6 Technologies to Improve the Efficacy of Sunscreen Products

Technologies are available to improve the efficacy of UV filters, such as encapsulation of UV absorbers and the use of nonabsorbing material to bolster SPF values.

13.2.6.1 Encapsulation of Conventional UV Filters

In this technique, the organic UV filter is entrapped within a microcapsule silica shell so the active sunscreen itself does not directly contact the skin. It decreases the probability of an allergic or irritant reaction to sunscreen and overcomes the incompatibility problem of active sunscreen ingredients. An example of this technology is Eusolex UV-Pearls (encapsulated ethylhexyl methoxycinnamate), the first sunscreen product using this technique. By using this encapsulation technique, the photostability of the combined avobenzone/ethylhexyl methoxycinnamate product is improved because there is no chemical interaction between these two UV filters (Tuchinda et al. 2006).

13.2.6.2 SPF-Boosting Agents

A method to boost the efficacy of UV filters is the incorporation of nonabsorbing particles that scatter the UV radiation, thus resulting in a longer pathway than the UV photons would have to travel through the sunscreen film on the skin. This would increase the probability of photons being absorbed by the organic filters in the film.

SunSpheres is an example of a non-UV absorbing booster material, developed by Rohm and Haas Company (Philadelphia, Pennsylvania). It is a styrene/acrylates copolymer, which was designed to enhance the effectiveness of the active ingredients present in the formula. When Sun-Spheres are manufactured, they are filled with water. When the product is applied to the

skin, the internal water migrates out of the sphere, leaving microscopic hollow beads. When UV hits these hollow beads, the UV radiation is scattered and travels sideways instead of straight down onto the skin. The spheres increase the probability of UV radiation coming into contact with the UV active ingredients present in the sunscreen formulation, hence "boosting" the efficacy of the UV filter (Tuchinda et al. 2006).

13.3 Systemic Photoprotective Agents

Limitations of topical sunscreens are the application process itself, the need of applying it to all parts of the body in order to provide photoprotection, and the need for reapplication. Systemic photoprotective agents are simpler to administer, do not have to be readministered every few hours, and could provide overall cutaneous protection. Recently, several promising systemic agents have been developed and studied; however, none has been proven to provide excellent photoprotection by itself.

13.3.1 Polypodium Leucotomos

Polypodium leucotomos (PL) is a natural extract of fern leaves obtained from tropical and subtropical regions of America. It has potent antioxidant and anti-inflammatory properties. *In vitro*, PL was shown to inhibit the harmful effects of UV radiation, such as generation of reactive oxygen species, development of DNA damage and isomerization, decomposition of trans-urocanic acid, and prevention of apoptosis and necrosis from UV exposure (Gonzalez, Gilaberte, and Philips 2010). Oral administration of PL extract (7.5 mg/kg) was demonstrated to reduce the UV-induced erythema reaction in healthy participants of skin types II to III. Histologically, PL-treated biopsy specimens also showed significantly fewer sunburn cells and a lesser amount of cyclobutane pyrimidine dimers compared to untreated skin (Middelkamp-Hup et al. 2004). It has been shown to be able to downregulate UVB and PUVA-induced erythema and development of solar urticaria (Caccialanza et al. 2007). It has also been demonstrated that PL can prevent PLE lesions in patients with long-standing PLE (Tanew et al. 2012). It is possible that, in the future, PL might be an option for the prevention of various types of photodermatoses.

13.3.2 Carotenoids

Carotenoids are pigments that naturally occur in plants. Currently, there are more than 600 known carotenoids. Important carotenoids that are rich in vegetables and fruits include alpha-carotene, beta-carotene, lycopene, lutein, and zeaxanthin. Carotenoids were demonstrated to quench excited species, such as singlet oxygen and free radicals. A study in healthy volunteers demonstrated that 10–12 weeks after ingestion of lycopene, a major carotenoid in tomatoes, there was a decrease in the sensitivity of UV-induced erythema (Stahl et al. 2006). In a placebo-controlled study comparing oral beta-carotene (24 mg/day); a carotenoid mix consisting of the three main dietary carotenoids: beta-carotene, lutein, and lycopene (8 mg/d each); and a placebo, it was demonstrated that only the two groups that received carotenoids showed an inhibition of UV-induced erythema (Heinrich et al. 2003). Beta-carotene is used clinically to manage photosensitivity in erythropoietic protoporphyria, although its efficacy has been questioned (Mathews-Roth 1990).

13.3.3 Green Tea

In green tea, there are epicatechin derivatives, commonly called polyphenols. These derivatives have antioxidant, anti-inflammatory, and anticarcinogenic properties (Katiyar 2003). The *in vitro* and *in vivo* animal and human studies suggest that green tea polyphenols had a photoprotective effect and might have some benefits for the prevention of UV-induced erythema, photoaging, and skin cancers (Elmets et al. 2001; Katiyar 2003).

13.3.4 Afamelanotide

Afamelanotide (Nle^4-D-Phe^7-α-melanocyte-stimulating hormone, also known as SCENESSE, CUV1647) is an α-melanocyte–stimulating hormone analog developed at the University of Arizona. It binds to a melanocortin-1 receptor on melanocytes, resulting in stimulation of melanocyte proliferation and upregulation of tyrosinase activity (Langan, Nie, and Rhodes 2010). Currently, afamelanotide, delivered as subcutaneous implants, is in various stages of clinical trials for several dermatological conditions, including erythropoietic protoporphyria, PLE, solar urticaria, vitiligo, phototoxicity associated with systemic photodynamic therapy, actinic keratosis, and SCC in patients who have received organ transplants (Harms et al. 2009; Haylett et al. 2011).

13.4 Photoprotection by Fabric

Wearing appropriate clothing and hats is one of the photoprotection strategies. For fabrics, the term "UV protection factor (UPF)" is used as the measure of UV radiation penetration through the fabric (Morison 2003).

The UPF of a fabric depends on several factors, such as fiber content, weave, fabric color, finishing processes, and the presence of additives (Hoffmann et al. 2001). Polyester generally provides high UPF. Nylon, wool, and silk provide more UPF than cotton, viscose, and rayon. The UPF is increased with the darker fabric color (Hoffmann et al. 2001). The performance of a fabric also depends upon stretching, shrinkage, hydration, and laundering. Stanford, Georgouras, and Pailthorpe studied the UV penetration of cotton T-shirts. They found that UPF increased significantly from a mean of 19.0 to 40.6 in 10 weeks after use and wash. The increase in UPF was explained by the reduction in the hole size of the fabric resulting from shrinkage (Stanford, Georgouras, and Pailthorpe 1995). Currently, there are UV-absorbing additives that can be added to the fabric during laundering; binding of these additives to the material can enhance the UV protection of clothing (Wang et al. 2001).

Wearing a hat is helpful against solar UV for the face. Similar to clothing, hats made from dark fabric provide better UV protection than light fabric. Gies et al. studied the UV protection provided by different types of hats (broad-brimmed hats, bucket hats, caps, legionnaire hats) used at school in Australia. They found that the broad-brimmed hats and bucket hats provided the most UV protection. All types of hat provided excellent protection on the forehead area (protection factor 8.8–16). A Legionnaire hat is the hat that has a wide brim to protect the ears and posterior neck. Legionnaire hats protected well on the nose and neck area. The cap provided the least UV protection especially on the cheeks, ears, and chin area (protection factor 1.1) (Gies et al. 2006).

13.5 Photoprotection by Glass

In everyday life, individuals may not be aware that they can be exposed to a significant amount of UV while being indoors. Standard window glass filters out UVB; however, UVA, visible light, and infrared radiation are transmitted. With continuous exposure, it is possible for individuals to be exposed to a high amount of UVA that is transmitted through window glass.

New developments in the manufacture of glass resulted in the development of several types of glass that offer a good protection against UVA and infrared. Tuchinda, Srivannaboon, and Lim have done a comprehensive review of the photoprotective role of glass. They showed that the UV-blocking glass has the highest UV-protective property, followed by laminated glass, low-E glass, tinted glass, and clear glass. Only 0.1% of UV passes through the double-glazed UV-blocking glass, and only 0.5% of UV passes through double-glazed laminated glass while 57% of UV passes through double-glazed clear glass (Tuchinda, Srivannaboon, and Lim 2006).

Another source of UV exposure through glass is inside cars. For safety reasons, all windshields are made from laminated glass, which can filter most of UVA (only 2%–3% UVA transmission). However, individuals traveling by car can be exposed to a substantial amount of UVA (nearly 50% of UVA) through the side and rear windows, which are usually made from tempered glass (nonlaminated glass) (Tuchinda, Srivannaboon, and Lim 2006). Studies have shown that signs of chronic UV exposure, such as photoaging and premalignant skin cancers, are more predominant on the driver's side (Foley, Lanzer, and Marks 1986; Singer et al. 1994). It should be noted that there is now a trend of using laminated glass even for side and rear windows for safety reasons.

For tempered glass used in auto side and back windows, the best way to reduce UV penetration is to put an aftermarket film on the window. The darkness of the film determines visible light transmission. One study demonstrated that window film with 35% and 20% visible light transmittance filtered UVA below 370 and 380 nm, respectively (Johnson and Fusaro 1992). It has been reported that the side window with UV absorbing film allows only 0.4% UV transmission, while side window glass without film allows as high as 79% of UVR transmission (Bernstein et al.

2006). The use of window films needs to comply with the federal and state regulations on the minimal allowable transmission of visible light.

13.6 Conclusions and Future Perspective

To maximize the photoprotective effect, individuals should practice several methods of photoprotection, such as seeking shade during the peak period of 10 a.m. to 2 p.m.; wearing appropriated clothing, hats, and sunglasses; and the regular use of broad-spectrum sunscreens with an SPF of at least 30. The use of UV-protective glass and window films could also be helpful. During the past 10 years, novel technologies have been developed to increase the efficacy and safety of sunscreens. Nowadays, newly broad-spectrum sunscreens are becoming available worldwide. In the future, harmonization of the UVA protection assessment would benefit all.

References

Antoniou, C., M. G. Kosmadaki, A. J. Stratigos, and A. D. Katsambas. 2008. Sunscreens—What's important to know. *J Eur Acad Dermatol Venereol* 22:1110–1118.

Beasley, D. G., and T. A. Meyer. 2010. Characterization of the UVA protection provided by avobenzone, zinc oxide, and titanium dioxide in broad-spectrum sunscreen products. *Am J Clin Dermatol* 11:413–421.

Bech-Thomsen, N., and H. C. Wulf. 1992. Sunbathers' application of sunscreen is probably inadequate to obtain the sun protection factor assigned to the preparation. *Photodermatol Photoimmunol Photomed* 9:242–244.

Bernstein, E. F., M. Schwartz, R. Viehmeyer et al. 2006. Measurement of protection afforded by ultraviolet-absorbing window film using an *in vitro* model of photodamage. *Lasers Surg Med* 38:337–342.

Bryden, A. M., H. Moseley, S. H. Ibbotson et al. 2006. Photopatch testing of 1155 patients: Results of the U.K. multicentre photopatch study group. *Br J Dermatol* 155:737–747.

Burnett, M. E., and S. Q. Wang. 2011. Current sunscreen controversies: A critical review. *Photodermatol Photoimmunol Photomed* 27:58–67.

Caccialanza, M., S. Percivalle, R. Piccinno, and R. Brambilla. 2007. Photoprotective activity of oral polypodium leucotomos extract in 25 patients with idiopathic photodermatoses. *Photodermatol Photoimmunol Photomed* 23:46–47.

Chatelain, E., and B. Gabard. 2001. Photostabilization of butyl methoxydibenzoylmethane (Avobenzone) and ethylhexyl methoxycinnamate by bis-ethylhexyloxyphenol methoxyphenyl triazine (Tinosorb S), a new UV broadband filter. *Photochem Photobiol* 74:401–406.

Cusack, C., C. Danby, J. C. Fallon et al. 2008. Photoprotective behaviour and sunscreen use: impact on vitamin D levels in cutaneous lupus erythematosus. *Photodermatol Photoimmunol Photomed* 24:260–267.

Darvay, A., I. R. White, R. J. Rycroft et al. 2001. Photoallergic contact dermatitis is uncommon. *Br J Dermatol* 145:597–601.

Diffey, B. L. 1994. A method for broad spectrum classification of sunscreens. *Int J Cosmet Sci* 16:47–52.

Duteil, I., C. Queille-Roussel, A. Rougier, A. Richard, and J. P. Ortonne. 2002. High protective effect of a broad-spectrum sunscreen against tetracycline phototoxicity. *Eur J Dermatol* 12:X–XI.

Elmets, C. A., D. Singh, K. Tubesing et al. 2001. Cutaneous photoprotection from ultraviolet injury by green tea polyphenols. *J Am Acad Dermatol* 44:425–432.

Faurschou, A., and H. C. Wulf. 2007. The relation between sun protection factor and amount of sunscreen applied in vivo. *Br J Dermatol* 156:716–719.

Foley, P., D. Lanzer, and R. Marks. 1986. Are solar keratoses more common on the driver's side? *Br Med J (Clin Res Ed)* 293:18.

Fourtanier, A. 1996. Mexoryl SX protects against solar-simulated UVR-induced photocarcinogenesis in mice. *Photochem Photobiol* 64:688–693.

Gies, P., J. Javorniczky, C. Roy, and S. Henderson. 2006. Measurements of the UVR protection provided by hats used at school. *Photochem Photobiol* 82:750–754.

Gonzalez, S., Y. Gilaberte, and N. Philips. 2010. Mechanistic insights in the use of a *Polypodium leucotomos* extract as an oral and topical photoprotective agent. *Photochem Photobiol Sci* 9:559–563.

Green, A., G. Williams, R. Neale et al. 1999. Daily sunscreen application and betacarotene supplementation in prevention of basal-cell and squamous-cell carcinomas of the skin: A randomised controlled trial. *Lancet* 354:723–729.

Green, A. C., G. M. Williams, V. Logan, and G. M. Strutton. 2011. Reduced melanoma after regular sunscreen use: Randomized trial follow-up. *J Clin Oncol* 29:257–263.

Harms, J., S. Lautenschlager, C. E. Minder, and E. I. Minder. 2009. An alpha-melanocyte-stimulating hormone analogue in erythropoietic protoporphyria. *N Engl J Med* 360:306–307.

Haylett, A. K., Z. Nie, M. Brownrigg, R. Taylor, and L. E. Rhodes. 2011. Systemic photoprotection in solar urticaria with alpha-melanocyte-stimulating hormone analogue [Nle4-D-Phe7]-alpha-MSH. *Br J Dermatol* 164:407–414.

Heinrich, U., C. Gartner, M. Wiebusch et al. 2003. Supplementation with beta-carotene or a similar amount of mixed carotenoids protects humans from UV-induced erythema. *J Nutr* 133:98–101.

Hoffmann, K., J. Laperre, A. Avermaete, P. Altmeyer, and T. Gambichler. 2001. Defined UV protection by apparel textiles. *Arch Dermatol* 137:1089–1094.

Holick, M. F. 2007. Optimal vitamin D status for the prevention and treatment of osteoporosis. *Drugs Aging* 24:1017–1029.

Holme, S. A., A. V. Anstey, M. N. Badminton, and G. H. Elder. 2008. Serum 25-hydroxyvitamin D in erythropoietic protoporphyria. *Br J Dermatol* 159:211–213.

Janjua, N. R., B. Kongshoj, A. M. Andersson, and H. C. Wulf. 2008. Sunscreens in human plasma and urine after repeated whole-body topical application. *J Eur Acad Dermatol Venereol* 22:456–461.

Johnson, J. A., and R. M. Fusaro. 1992. Broad-spectrum photoprotection: The roles of tinted auto windows, sunscreens and browning agents in the diagnosis and treatment of photosensitivity. *Dermatology* 185:237–241.

Kaidbey, K. H., and A. Barnes. 1991. Determination of UVA protection factors by means of immediate pigment darkening in normal skin. *J Am Acad Dermatol* 25:262–266.

Katiyar, S. K. 2003. Skin photoprotection by green tea: antioxidant and immunomodulatory effects. *Curr Drug Targets Immune Endocr Metabol Disord* 3:234–242.

Kim, S. M., B. H. Oh, Y. W. Lee, Y. B. Choe, and K. J. Ahn. 2010. The relation between the amount of sunscreen applied and the sun protection factor in Asian skin. *J Am Acad Dermatol* 62:218–222.

Kullavanijaya, P., and H. W. Lim. 2005. Photoprotection. *J Am Acad Dermatol* 52:937–958; quiz 959–962.

Langan, E. A., Z. Nie, and L. E. Rhodes. 2010. Melanotropic peptides: More than just 'Barbie drugs' and 'sun-tan jabs'? *Br J Dermatol* 163:451–455.

Linos, E., E. Keiser, M. Kanzler et al. 2012. Sun protective behaviors and vitamin D levels in the US population: NHANES 2003–2006. *Cancer Causes Control* 23:133–140.

Mahmoud, B. H., E. Ruvolo, C. L. Hexsel et al. 2010. Impact of long-wavelength UVA and visible light on melanocompetent skin. *J Invest Dermatol* 130:2092–2097.

Mathews-Roth, M. M. 1990. Carotenoid functions in photoprotection and cancer prevention. *J Environ Pathol Toxicol Oncol* 10:181–192.

Middelkamp-Hup, M. A., M. A. Pathak, C. Parrado et al. 2004. Oral Polypodium leucotomos extract decreases ultraviolet-induced damage of human skin. *J Am Acad Dermatol* 51:910–918.

Morison, W. L. 2003. Photoprotection by clothing. *Dermatol Ther* 16:16–22.

Moyal, D. 2004. Prevention of ultraviolet-induced skin pigmentation. *Photodermatol Photoimmunol Photomed* 20:243–247.

Moyal, D., A. Chardon, and N. Kollias. 2000. UVA protection efficacy of sunscreens can be determined by the persistent pigment darkening (PPD) method. (Section 2). *Photodermatol Photoimmunol Photomed* 16:250–255.

Moyal, D., I. Duteil, C. Queille-Roussel et al. 2002. Prevention of solar-induced immunosuppression by a new highly protective broadspectrum sunscreen. *Eur J Dermatol* 12:XII–XIV.

Nash, J. F., P. R. Tanner, and P. J. Matts. 2006. Ultraviolet A radiation: testing and labeling for sunscreen products. *Dermatol Clin* 24:63–74.

Naylor, M. F., A. Boyd, D. W. Smith et al. 1995. High sun protection factor sunscreens in the suppression of actinic neoplasia. *Arch Dermatol* 131:170–175.

Norval, M., and H. C. Wulf. 2009. Does chronic sunscreen use reduce vitamin D production to insufficient levels? *Br J Dermatol* 161:732–736.

Pinnell, S. R., D. Fairhurst, R. Gillies, M. A. Mitchnick, and N. Kollias. 2000. Microfine zinc oxide is a superior sunscreen ingredient to microfine titanium dioxide. *Dermatol Surg* 26:309–314.

Reich, A., M. Harupa, M. Bury, J. Chrzaszcz, and A. Starczewska. 2009. Application of sunscreen preparations: A need to change the regulations. *Photodermatol Photoimmunol Photomed* 25:242–244.

Ross, A. C., J. E. Manson, S. A. Abrams et al. 2011. The 2011 report on dietary reference intakes for calcium and vitamin D from the Institute of Medicine: What clinicians need to know. *J Clin Endocrinol Metab* 96:53–58.

Sambandan, D. R., and D. Ratner. 2011. Sunscreens: An overview and update. *J Am Acad Dermatol* 64:748–758.

Schalka, S., V. M. dos Reis, and L. C. Cuce. 2009. The influence of the amount of sunscreen applied and its sun protection factor (SPF): Evaluation of two sunscreens including the same ingredients at different concentrations. *Photodermatol Photoimmunol Photomed* 25:175–180.

Schalka, S., and V. M. dos Reis. 2011. Sun protection factor: meaning and controversies. *An Bras Dermatol* 86:507–515.

Schauder, S., and H. Ippen. 1997. Contact and photocontact sensitivity to sunscreens. Review of a 15-year experience and of the literature. *Contact Dermatitis* 37:221–232.

Shaw, T., B. Simpson, B. Wilson et al. 2010. True photoallergy to sunscreens is rare despite popular belief. *Dermatitis* 21:185–198.

Singer, R. S., T. A. Hamilton, J. J. Voorhees, and C. E. Griffiths. 1994. Association of asymmetrical facial photodamage with automobile driving. *Arch Dermatol* 130:121–123.

Stahl, W., U. Heinrich, O. Aust, H. Tronnier, and H. Sies. 2006. Lycopene-rich products and dietary photoprotection. *Photochem Photobiol Sci* 5:238–242.

Stanford, D. G., K. E. Georgouras, and M. T. Pailthorpe. 1995. Sun protection by a summer-weight garment: The effect of washing and wearing. *The Medical Journal of Australia* 162:422–425.

Stege, H., M. Budde, S. Grether-Beck et al. 2002. Sunscreens with high SPF values are not equivalent in protection from UVA induced polymorphous light eruption. *Eur J Dermatol* 12:IV–VI.

Stege, H., M. A. Budde, S. Grether-Beck, and J. Krutmann. 2000. Evaluation of the capacity of sunscreens to photoprotect lupus erythematosus patients by employing the photoprovocation test. *Photodermatol Photoimmunol Photomed* 16:256–259.

Tanew, A., S. Radakovic, S. Gonzalez, M. Venturini, and P. Calzavara-Pinton. 2012. Oral administration of a hydrophilic extract of *Polypodium leucotomos* for the prevention of polymorphic light eruption. *J Am Acad Dermatol* 66:58–62.

Thompson, S. C., D. Jolley, and R. Marks. 1993. Reduction of solar keratoses by regular sunscreen use. *N Engl J Med* 329:1147–1151.

Tuchinda, C., H. W. Lim, U. Osterwalder, and A. Rougier. 2006a. Novel emerging sunscreen technologies. *Dermatol Clin* 24:105–117.

Tuchinda, C., S. Srivannaboon, and H. W. Lim. 2006b. Photoprotection by window glass, automobile glass, and sunglasses. *J Am Acad Dermatol* 54:845–854.

van der Pols, J. C., G. M. Williams, N. Pandeya, V. Logan, and A. C. Green. 2006. Prolonged prevention of squamous cell carcinoma of the skin by regular sunscreen use. *Cancer Epidemiol Biomarkers Prev* 15:2546–2548.

Vanchinathan, V., and H. W. Lim. 2012. A dermatologist's perspective on vitamin D. *Mayo Clin Proc* 87:372–380.

Wang, S. Q., M. E. Burnett, and H. W. Lim. 2011. Safety of oxybenzone: Putting numbers into perspective. *Archives of dermatology* 147:865–866.

Wang, S. Q., S. W. Dusza, and H. W. Lim. 2010. Safety of retinyl palmitate in sunscreens: A critical analysis. *J Am Acad Dermatol* 63:903–906.

Wang, S. Q., A. W. Kopf, J. Marx et al. 2001. Reduction of ultraviolet transmission through cotton T-shirt fabrics with low ultraviolet protection by various laundering methods and dyeing: Clinical implications. *J Am Acad Dermatol* 44:767–774.

Wang, S. Q., and H. W. Lim. 2011. Current status of the sunscreen regulation in the United States: 2011 Food and Drug Administration's final rule on labeling and effectiveness testing. *J Am Acad Dermatol* 65:863–869.

Wang, S. Q., J. W. Stanfield, and U. Osterwalder. 2008. *In vitro* assessments of UVA protection by popular sunscreens available in the United States. *J Am Acad Dermatol* 59:934–942.

14

Botanical Antioxidants for Photochemoprevention

Eiman Mukhtar
University of Wisconsin-Madison

Vaqar Adhami
University of Wisconsin-Madison

Hasan Mukhtar
University of Wisconsin-Madison

14.1 Introduction

Exposure to ultraviolet (UV) radiation is the major cause of skin disorders, such as sunburn, photodamage, and skin cancer. Skin cancer is the most common of all cancers as the number of basal cell and squamous cell skin cancers (i.e., nonmelanoma skin cancers or NMSC) and its estimation has been limited because these cases are not required to be reported to cancer registries. One report on NMSC occurrence in the United States estimated that 3.5 million cases were diagnosed and 2.2 million people were treated in 2006 (Siegel, Naishadham, and Jemal 2013). It is also the easiest to cure if diagnosed and treated early; however, once allowed to progress, skin cancer can result in disfigurement morbidity and even mortality. Similar trends are reported from many other countries with a predominantly Caucasian population. In contrast, melanoma is the most severe form of skin cancer with an estimated 76,690 new cases in the United States in 2013, according to recent statistics (Siegel, Naishadham, and Jemal 2013). This reflects a significant rise of 6460 cases from 2011 figures. The number of melanoma-related deaths has risen as well by almost 300 cases—from 9180 deaths in 2012 to a predicted 9480 deaths in 2013.

Exposure of human or mouse skin to UVB radiation is reported to induce an accelerated generation of reactive oxygen species (ROS) (Bickers and Athar 2006), which are an inherent cellular metabolism. Through a series of one-electron subtractions, molecular oxygen is converted to superoxide anion, hydrogen peroxide, hydroxyl radical, and, finally, water. Most of these reactions occur in the mitochondria and are related to energy production. Cellular enzymes and controlled metabolic processes ordinarily keep oxidative damage to cells at a minimum

(Bergendi et al. 1999). However, in times of increased oxidative stress, including high metabolic demands and external injuries, such as sunlight, smoking, and pollution, oxidative damage may occur. Previous observations suggest that a ROS can act as both an initiator as well as a promoter of tumors by damaging critical cellular macromolecules, such as DNA, proteins, and lipids, and by acting as a stimulator or inducer of cell-signaling molecules. Skin-related disorders, such as photoaging and photocarcinogenesis, are believed to be mediated, at least in part, by ROS generation. Furthermore, UV irradiation results in the induction of matrix metalloproteinases (MMPs), which degrade the collagen and connective tissue components of the skin and block the transforming growth factor-beta type II (TGF-β2) receptor/Smad signaling and also activate an activator protein (AP)-l, which, in turn, is triggered by a series of mitogen-activated protein kinases (MAPKs) (Quan et al. 2004). In addition, nuclear factor kappa B (NF-κB), a transcription factor, is also activated by UV irradiation. It is suggested that both AP-1 and NF-κB, which are activated by ROS, may provide the complex driving force that results in a series of complex biological interactions, resulting in skin diseases (Bowden 2004) (Figure 14.1).

Exposure of humans to solar radiation is unavoidable, and therefore, adverse effects associated with UVB exposure need to be addressed. While avoiding excessive exposure to the sun, wearing protective clothing, and using sunscreen lotions are popular recommendations, there is still need for the development of effective nontoxic chemical agents that are capable of decreasing the UVB adverse effects. Experimental and epidemiologic studies have suggested that some botanical polyphenols have significant photoprotective effects, and many

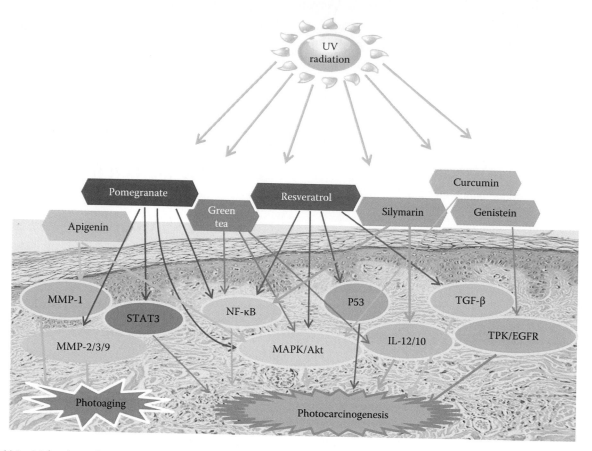

FIGURE 14.1 Molecular pathways deregulated by UV radiation and their modulation by botanical dietary agents.

plant-derived phytochemicals possess photoprotective properties. Some of these agents, popularly known as "photochemopreventive agents," are also present in the human diet. Green tea polyphenols (GTPs) have been one of the first identified agents to afford protection against UVB radiation-induced carcinogenesis (Wang et al. 1991). A series of subsequent studies have led to the identification of other natural compounds and the mechanisms associated with the inhibition of photocarcinogenesis (Afaq, Ahmad, and Mukhtar 2003). Modulations in NF-κB and MAPKs pathways have been observed to be the molecular basis for the photochemopreventive effects of the green tea constituent epigallocatechin-3-gallate (EGCG) (Afaq, Ahmad, and Mukhtar 2003) (Table 14.1). Many such agents have found a place in various skin-care products. For this reason, it is essential to understand how such agents exert their effects, helping improvement of and more effective photochemopreventive products for general human use.

14.2 Photocarcinogenesis

Photocarcinogenesis is a complex multistage phenomenon, which involves three distinct stages exemplified by initiation, promotion, and progression (Bowden 2004), and each of these stages is mediated by several cellular, biochemical, and molecular events. First, the carcinogenic initiation is an irreversible

process in which genetic alterations occur that ultimately lead to DNA damage. Second, the tumor promotion is a vital process in cancer development, involving clonal expansion of initiated cells, giving rise to premalignant and then to malignant lesions, essentially by alterations in signal transduction pathways. Third, the tumor progression involves the conversion of premalignant and malignant lesions into an invasive and potentially metastatic malignant tumor (Bowden 2004). The generation of squamous cell carcinoma is attributed mostly to the UVB spectrum, although there is a peak of activity in the UVA range (320–400 nm) (deGruijl and Van der Leun 1994). Whereas UVB is important for tumor initiation, UVA predominantly causes tumor promotion. The process of skin cancer development involves stimulation of DNA synthesis, DNA damage, cell proliferation, inflammation, epidermal hyperplasia, immunosuppression, cell-cycle deregulation, depletion of antioxidant enzymes, impairment of signal transduction pathways, and induction of ornithine decarboxylase (ODC) and cyclooxygenase-2 (COX-2) (Afaq et al. 2005).

14.3 UV Radiation-Induced Aging or Photoaging

Skin aging is a dynamic, multifactorial process consisting of two distinct components: (1) intrinsic or natural aging that is

TABLE 14.1 Summary of the Effect of Botanical Agents in Photochemoprevention

Botanicals	*In Vitro* Studies	*In Vivo* Studies	References
Green tea	Keratinocytes		Katiyar et al. (2001); Afaq, Ahmad, and Mukhtar (2003)
	Fibroblasts *XPA* –proficient cells		Katiyar et al. (2010)
		SKH-1 mice	Afaq, Adhami, and Ahmad (2003); Afaq, Ahmad, and Mukhtar (2003); Mittal et al. (2003)
		C3H/Hen mice	Katiyar et al. (1999); Katiyar and Mukhtar (2001)
		Human skin	Katiyar et al. (1999); Katiyar et al. (2001); Elmets et al. (2001)
Resveratrol	Keratinocytes		Adhami, Afaq, and Ahmad (2003)
		SKH-1 mice	Afaq, Adhami, and Ahmad (2003)
	Melanoma cells Human A431 SCC cells		Lee, Kumar, and Glickman (2012); Kim et al. (2011)
Silymarin	Fibroblasts *XPA* -proficient cells	C3H/Hen mice SKH-1 mice	Katiyar (2002); Meeran et al. (2006); Gu et al. (2005)
Genistein	A431 cells Human reconstituted skin NIH 3T3 cells	SENCAR mice	Moore et al. (2006); Okura et al. (1988)
Curcumin	HaCaT cells		Park and Lee (2007)
Apigenin	Human dermal fibroblasts HaCaT cells		Sim et al. (2007) Hwang et al. (2011)
Pomegranate	Keratinocytes	SKH-1 mice Mouse skin	Afaq et al. (2005); Afaq et al. (2010); Khan et al. (2011)

genetically determined and is unavoidable and (2) extrinsic aging that also manifests in cutaneous changes, originates from exogenous sources, and is avoidable. The process of aging, both intrinsic and extrinsic, is influenced by the formation of ROS (Quan et al. 2004). Extrinsic aging results from various factors, but exposure to solar UV radiation is the primary cause. Depending on the amount and form of the UVB radiation, as well as the skin type of the individual, UVB irradiation causes premature skin aging, referred to as photoaging (Quan et al. 2004). Solar UV radiation has a profound effect on exposed skin, producing accelerated age-related changes consisting of fine and coarse wrinkling, rough skin texture, dryness, telangiectasia, and dyspigmentation abnormalities, including lentigines as well as guttate hypermelanosis and hypomelanosis. Studies have suggested that there is an increased generation of ROS in skin upon UV exposure, and it can overwhelm antioxidant defense mechanisms, resulting in oxidative stress and oxidative photodamage of proteins and other macromolecules in the skin. DNA photodamage and UV-mediated generation of ROS are the initial molecular events that lead to most of the typical histological and clinical manifestations of chronic photodamage of the skin. These events are believed to be critical mediators of the photoaging process (Quan et al. 2004); for instance, ROS can modify proteins in tissue to form carbonyl derivatives, which accumulate in the papillary dermis of photodamaged skin (Sander et al. 2002). Mast cells and macrophages are increased in photodamaged skin and are believed to be involved in photoaging (Bosset et al. 2003). Exposure of human or mouse skin to UV irradiation induces MMPs, which have been implicated in photoaging (Fisher et al. 1996). To date, UV radiation activates AP-l by MAPKs induction and, as a result, stimulates transcription of MMP genes that encode MMP-l (collagenase), MMP-9 (gelantinase), and MMP-3 (stromelysin-l) in skin cells. Together, these MMPs are capable of degrading the collagen framework and other components of skin connective tissue (Fisher et al. 1996). Studies have suggested that solar UV reduces collagen in photoaged human skin by blocking TGF-β2 receptor/Smad signaling (Quan et al. 2004). In addition, NF-κB is also activated by UV irradiation, which stimulates neutrophil attraction, bringing neutrophil collagenase (MMP-8) into the irradiation site to further aggravate matrix degradation.

14.4 Photochemoprevention of Skin Cancer

In recent years, photochemoprevention has become an important armamentarium in the fight against UVR-induced damage to the skin. For prevention of photodamage and skin cancer, education about the harmful effects of solar UV light, the need to avoid its excessive exposure, wearing protective clothing, and the use of sunscreen have been emphasized. Unfortunately, these advices are only partially effective. Therefore, additional efforts are needed to protect skin against the deleterious effect of UVB exposure. One approach to impede the occurrence of skin cancer is through chemoprevention, which, by definition, is a means of cancer control in which the disease can be entirely prevented, slowed, or reversed by topical or oral administration of naturally occurring agents. An expanded definition of cancer chemoprevention also includes the chemotherapy of precancerous lesions, which, for skin cancer, would mean the reversal of actinic keratosis.

14.5 Sunscreens Are Not Enough

One of the cornerstones of UV radiation protection is the conservative use of potent sunscreens. However, the protection is not fully achieved even if the sunscreen is effective in blocking harmful UV rays. This is a result of several reasons that include uneven application, insufficient amount, rub-off with sweat, etc. The "gold standard" in protecting skin from photodamage involves the development of sunscreens that can absorb UV light efficiently and protect against erythema. However, in actual use, sunscreens provide much less protection than expected. Sun protection factor (SPF) is measured and tested at an application to skin of 2 mg/cm^2 (Wulf, Stender, and Lock-Andersen 1997). Controlled studies of actual sunscreen use reveal that sunscreens are applied to skin at only 0.5 mg/cm^2 or less. For instance, SPF is not linearly proportional; thus, at an application of 0.5 mg/cm^2, no sunscreen provides more than three- to fivefold protection (Wulf, Stender, and Lock-Andersen 1997). Moreover, important biological events such as DNA damage as measured by thymine dimer and 8-hydroxy-2′-deoxyguanosine formation, as well as p53 induction and UV immunosuppression, continue at sub-erythemal levels of irradiation (Liardet et al. 2001). Sunscreens may give a false sense of security as none provides full spectral protection against UV light. Some of the ingredients contained in sunscreens can result in free radical generation when activated by UV irradiation, and sunscreen chemicals may be absorbed into the skin and potentially cause harm (Liardet et al. 2001).

14.6 Photochemoprevention by Botanical Antioxidants

Plants are continuously exposed to sunlight and therefore have to protect themselves from the sun. In fact, they have an even greater struggle to avoid being oxidized to death because they are unable to move to avoid sunlight. All plants synthesize vitamin C and vitamin E to protect themselves from the sun (Smirnoff, Conklin, and Loewus 2001). Moreover, they synthesize flavonoids, the powerful polyphenolic antioxidant compounds. Plant extracts have been widely used as topical applications for wound-healing, antiaging, and disease treatments. Examples of these include green tea, ginkgo biloba, echinacea, ginseng, grape seed, lemon, lavender, rosemary, thuja, sarsaparilla, soy, prickly pear, sagebrush, jojoba, aloe vera, allantoin, feverwort, bloodroot, apache plume, papaya, and more. These plants share a common characteristic by producing flavonoid compounds with phenolic structures. These phytochemicals are highly reactive with other compounds, such as ROS and biologic macromolecules, to neutralize free radicals or initiate biological effects. A number of phenolic phytochemicals with promising properties to benefit

human health include a group of polyphenol compounds, called catechins, found in green tea.

14.6.1 Green Tea

Epigallocatechin-3-gallate (EGCG), the major and the most active polyphenolic constituent found in green tea [dried fresh leaves of the plant *Camellia sinensis* L. (Theaceae)], is one of the most extensively investigated phytochemicals for photoprotection. EGCG acts as a potent antioxidant and can scavenge ROS, such as lipid free radicals, superoxide radicals, hydroxyl radicals, hydrogen peroxide, and singlet oxygen (Afaq, Ahmad, and Mukhtar 2003). The treatment of normal human epidermal keratinocytes (NHEKs) with EGCG resulted in inhibition of UVB-mediated phosphorylation and degradation of IκBα, activation of IKKα and NF-κB, and phosphorylation of MAPK in a dose- and time-dependent manner (Afaq, Ahmad, and Mukhtar 2003). These data suggested that EGCG protects against the adverse effects of UV radiation via modulations in NF-κB and MAPK pathways. EGCG protects against UV-induced oxidative stress in humans as well. The skin of healthy volunteers treated with green tea extracts reduced the number of sunburn cells and protected epidermal Langerhans cells from UV damage. Green tea extracts also reduced the DNA damage that formed after UV radiation (Elmets et al. 2001). In another study, it was applied to the volunteer's skin just before exposure to a four times minimal erythema dose (MED) of UVB radiation; it significantly decreased the production of hydrogen peroxide and nitric oxide production as well as lipid peroxidation in the dermis and epidermis (Katiyar et al. 2001). An investigation about the effect of oral EGCG as a powerful antioxidant on MED and UV-induced skin damage has found that regular intake of EGCG strengthens the skin's tolerance by increasing MED and therefore prevents UV-induced alteration of the epidermal barrier function and skin damage. These suggest that EGCG is a potent candidate for systemic photoprotection (Jeon et al. 2009).

Topical treatment of SKH-1 hairless mice with EGCG in a hydrophilic ointment-based formulation resulted in exceptionally strong inhibition of tumor incidence (60%), tumor multiplicity (86%), and tumor growth in terms of total tumor volume (95%) per group (Mittal et al. 2003). These results indicate that the use of EGCG in topical formulation might increase the penetration or absorption capacity inside the skin layers. Topical application of EGCG to mouse skin prior to UV irradiation was shown to result in inhibition of the contact hypersensitivity response (CHS) to contact sensitizer, a reduction in the number of infiltrating macrophages (CD11b$^+$ cells) and neutrophils, downregulation in UV-induced production of IL-10, and increased production of IL-12 in the skin and draining lymph nodes (Katiyar et al. 1999). EGCG was also found to balance the alterations in the IL-10/IL-12 cytokines. This effect may be mediated by the antigen-presenting cells in the skin and draining lymph nodes or by blocking the infiltration of IL-10-secreting CD11b$^+$ macrophages into the irradiated site (Katiyar et al. 1999). Oral administration of GTPs in drinking water avoids photocarcinogenesis in mice by preventing UV-induced suppression of the contact hypersensitivity (CHS) response to a contact sensitizer in local and systemic models of CHS. The authors showed that GTPs repaired UV-induced DNA damage faster in the skin of mice as evidenced by a reduced number of cyclobutane pyrimidine dimer (CPD)-positive cells and also by reduced migration of CPD-positive cells from the skin to draining lymph nodes. GTPs also repaired UV-induced CPD in xeroderma pigmentosum complementation group A (XPA)-proficient cells obtained from a healthy person but did not repair XPA-deficient cells obtained from patients suffering from XPA. These data suggest a mechanism by which drinking GTPs prevent UV-induced immunosuppression, and this may contribute to the chemopreventive activity of GTPs in prevention of photocarcinogenesis (Katiyar et al. 2010). Furthermore, UV-induced erythema decreased significantly in the intervention group by 16% and 25% after 6 and 12 weeks, respectively. GTP beverage intake for 12 weeks increased blood flow and oxygen delivery to the skin. In conclusion, these statements suggested that GTP delivered in a beverage was shown to increase microcirculation and protect skin against harmful UV radiation to improve overall skin quality of women (Heinrich et al. 2011).

14.6.2 Resveratrol

Resveratrol (trans-3,4′,5-trihydroxystilbene, RES) is a polyphenolic phytoalexin found largely in red grapes, berries, nuts, fruits, and red wine and is considered a potent antioxidant with anti-inflammatory and antiproliferative properties (Afaq, Adhami, and Ahmad 2003; Adhami, Afaq, and Ahmad 2003). A single topical application of RES to SKH-1 hairless mice prior to UVB irradiation resulted in significant inhibition of UVB-induced skin edema and caused a significant decrease in UVB-mediated generation of H_2O_2 and infiltration of leukocytes (Afaq, Adhami, and Ahmad 2003). RES treatment of mouse skin was also found to result in significant inhibition of UVB-mediated induction of COX and ODC enzyme activities and protein expression, which are well-established markers for tumor promotion. It was also observed that RES inhibited a UVB-mediated increased level of lipid peroxidation, a marker of oxidative stress (Afaq, Adhami, and Ahmad 2003). In another study, pretreatment of *NHEK* with RES inhibited UVB-mediated activation of NF-κB pathway in a dose- and time-dependent fashion (Adhami, Afaq, and Ahmad 2003). In addition, RES treatment of NHEK also inhibited UVB-mediated phosphorylation and degradation of IκBα and activation of IKKα (Adhami, Afaq, and Ahmad 2003). Recently, modulation of phytochemical study showed that pretreatment of the cells with 1 to 2 μM of RES or ursolic acid (UA) significantly reduced the amount of phosphorylated NF-κB at 24 h post exposure (Lee, Kumar, and Glickman 2012). Pretreatment of the cells with RES reduced the burden of light-induced protein carbonyl adducts by up to 25% in exposed cells (Lee, Kumar, and Glickman 2012). The sensitivity of melanoma cells was markedly increased with UA treatment to UV radiation while conferring some photoprotection to retinal pigment epithelial cells. These

observations indicated that phytochemicals interact with a specific cell signaling pathway (Lee, Kumar, and Glickman 2012).

Oral administration of RES to highly tumor-susceptible p53(+/−)/SKH-1 mice noticeably delayed UV-induced skin tumorigenesis and reduced the malignant conversion of benign papillomas to SCC. While RES increased the level of epithelial cadherin, TGF-β2 was decreased in RES-treated SCC skin. This RES-mediated TGF-β2 downregulation led to inhibition of both TGF-β2/Smad-dependent and -independent pathways and suppressed the invasiveness of A431 cells. In addition, the authors found that TGF-β2, but not TGF-β1, rescued the RES-mediated downregulation of p-extracellular, signal-regulated kinases 1/2, p-Smad3, and α-smooth muscle actin. RES treatment decreased phosphorylation of Akt and *pCREB*. Expression of constitutively active Akt blocked RES inhibition of CREB and TGF-β2 and rescued RES inhibition of cellular invasiveness. These data suggest that RES suppresses UV-induced malignant tumor progression in p53(+/−)/SKH-1 mice. Likewise, human A431 SCC cells' invasiveness inhibition by RES appears to occur, in part, through the Akt-mediated downregulation of TGF-β2 (Kim et al. 2011). Recent studies indicate that RES induces premature senescence in human A431 SCC cells, which are associated with a blockade of autolysosome formation. Further, the authors show that RES downregulates the level of Rictor, a component of mTORC2, leading to decreased RhoA-GTPase and altered actin cytoskeleton organization. Exogenous overexpression of Rictor restores RhoA-GTPase activity and the actin cytoskeleton network and decreases RES-induced, senescence-associated β-gal activity, indicating a direct role of Rictor in senescence induction. RES decreases the overexpression of Rictor in UV-induced murine SCCs. These data suggest that RES attenuates autophagic processes via Rictor and that downregulation of Rictor may be a mechanism of tumor suppression associated with UV induction (Back et al. 2012).

14.6.3 Pomegranate

Pomegranate (*Punica granatum*) contains anthocyanins and hydrolyzable tannins and retains strong antioxidant, anti-inflammatory, and antiproliferative properties. A study demonstrated that NHEK treatment with pomegranate fruit extract (PFE) inhibited UVB-mediated phosphorylation of ERK1/2, JNK1/2, and p38 proteins in a dose- and time-dependent manner (Afaq et al. 2005). This study also found that PFE treatment of NHEK resulted in dose- and time-dependent inhibition of UVB-mediated degradation and phosphorylation of IκBα and activation of IKKα. The authors found that PFE treatment of NHEK resulted in dose- and time-dependent inhibition of UVB-mediated nuclear translocation and phosphorylation of NF-κB/p65 at Ser[536]. The authors concluded that PFE protects against the adverse effects of UVB radiation by inhibiting UVB-induced modulations of NF-κB and MAPK pathways and provides a molecular basis for the photochemopreventive effects of PFE. The same authors in another study have shown that the effect of pomegranate-derived products against UVB-mediated damage using reconstituted human skin revealed inhibition of UVB-induced MMP-2 and MMP-9 activities. Pomegranate-derived products also caused a decrease in UVB-induced protein expression of c-Fos and phosphorylation of c-Jun. Thereafter, these results suggest that pomegranate may prevent UVB-induced damage to human skin (Afaq et al. 2009).

Similar to what occurs with the UVB interaction, the treatment of NHEK with PFE before exposure to UVA results in a dose-dependent inhibition of UVA-mediated phosphorylation of signal transducers and activators of transcription-3 (STAT3) at Tyr^{705}, AKT at Ser^{473}, and ERK1/2. The same authors found that UVA radiation of NHEK resulted in the phosphorylation of mammalian target rapamycin (mTOR) at Thr^{2448} and p70S6K at Thr^{421}/Ser^{424}. PFE pretreatment resulted in a dose-dependent inhibition in the phosphorylation of mTOR at Thr^{2448} and p70S6K at Thr^{421}/Ser^{424} and was also found to induce UVA-mediated activation in the protein expression of Bax and Bad and downregulation of Bcl-xL (Syed et al. 2006). Oral feeding of PFE to SKH-1 hairless mice resulted in inhibition of a UVB-induced increase in skin edema, hyperplasia, infiltration of leukocytes, generation of hydrogen peroxide, and DNA damage in the form of CPDs and 8-hydroxy-2'-deoxyguanosine (Afaq et al. 2010). Furthermore, oral feeding of PFE resulted in inhibition of UVB-induced PCNA, ODC, and COX-2 protein expression and induction of a UVB-mediated increase in p21 and p53 protein expression (Afaq et al. 2010). Another study with SKH 1 mice proved that this intake inhibited UVB-induced epidermal hyperplasia, infiltration of leukocytes, protein oxidation, and lipid peroxidation. This was demonstrated by inhibition of nuclear translocation and phosphorylation of NF-κB/p65, phosphorylation and degradation of IκBα, activation of IKKα/IKKβ, and phosphorylation of mitogen-activated protein kinase proteins and c-Jun. PFE intake also repressed UVB-induced protein expression of COX-2 and iNOS, PCNA, and cyclin D1 and MMP-2, -3, and -9 in mouse skin. Taken together, these data show that PFE consumption afforded protection to mouse skin against the adverse effects of UVB radiation by modulating UVB-induced signaling pathways (Khan et al. 2011).

14.6.4 Silymarin

Silymarin, a polyphenolic flavonoid isolated from milk thistle plant (*Silybum marianum L. Gaertn*), is a mixture of several flavonolignans, including silybin, silibinin, silydianin, silychristin, and isosylibin. Silibinin was found to restore UVB-caused depletion of survivin, which is associated with the upregulation of NF-κB DNA-binding activity (Dhanalakshmi et al. 2004). Further, silibinin treatment upregulated UVB-induced ERK 1/2 phosphorylation and increased the duration of the S phase, possibly providing a prolonged time for efficient DNA repair (Dhanalakshmi et al. 2004).

Treatment of C3H/HeN mice with topically applied silymarin or silibinin, a major component of silymarin, markedly inhibited UVB-induced suppression of the CHS in a local model of immunosuppression and had a moderate inhibitory effect in a systemic model of contact hypersensitivity (Meeran et al. 2006). Silymarin reduced the UVB-induced enhancement of the levels in the skin of the immunosuppressive cytokine (IL-10) draining lymph nodes and enhanced the levels of the immunostimulatory cytokine IL-12 (Meeran et al. 2006). Dietary feeding of silibinin to SKH-1 hairless mice for 2 weeks before a single UVB irradiation resulted in a strong and significant decrease in UVB-induced thymine dimer-positive cells and proliferating cell nuclear antigen (PCNA), terminal deoxynucleotidyl transferase-mediated dUTP nick end labeling, and apoptotic sunburn cells together with an increase in p53 and p21/cip1-positive cell population in the epidermis (Gu et al. 2005).

Moreover, topical application of silymarin protects against photocarcinogenesis in mice. SKH-1 hairless mice were subjected to three different protocols: tumor initiation with UVB, followed by tumor promotion with tetradecanoylphorbol-13-acetate (group 1); tumor initiation with 7,12-dimethylbenz[a]anthracene (DMBA), followed by tumor promotion with UVB (group 2); and tumor initiation and promotion with UVB (group 3). In all three groups, topical application of silymarin prior to exposure to UVB or DMBA significantly reduced tumor incidence, tumor multiplicity, and average tumor volume per mouse (Katiyar et al. 1997). Furthermore, in short-term experiments, the topical application of silymarin was found to result in significant inhibition against UVB-induced skin edema, skin sunburn, and cell apoptosis; depletion of catalase activity; and induction of COX-2 and ODC activities and ODC mRNA expression. These studies suggest that topical application of silymarin provides considerable protection against UVB-mediated damage in mouse skin, conceivably by its strong antioxidant properties at different stages of UVB-induced carcinogenesis (Katiyar et al. 1997). A recent study by Roy et al. (2012) indicates that silibinin also prevents UVB radiation-induced damage in mouse epidermal JB6 cells and SKH1 hairless mouse skin. In this study, silibinin pretreatment protected JB6 cells against apoptosis and accelerated the repair of CPD induced by a moderate dose of UVB (50 mJ/cm^2), the daily risk of exposure. Moreover, silibinin reversed UVB-induced S-phase arrest, reducing both active DNA synthesizing and inactive S-phase populations. Additionally, UVB-irradiated cells showed a transient upregulation of both phosphorylated (Ser-15 and Ser-392) and total p53, while silibinin pretreatment led to a more sustained upregulation and stronger nuclear localization of p53. GADD45α, a downstream target of p53, also upregulated by silibinin, was implicated in DNA repair and cell-cycle regulation. Topical application of silibinin prior to or immediately after UVB irradiation in the epidermis of SKH1 hairless mice resulted in a sustained increase in p53 and GADD45α levels and accelerated CPD removal. This study proposed that silibinin protects against UVB-induced photodamage and reduces the risk of or prevents early onset of nonmelanoma skin cancer (Roy et al. 2012). In addition, a 2011 study demonstrated that silymarin is a promising new sunscreen agent. This determination is based on the physical demonstration of energy reduction in the UV range. Approximately 15 mg of an oil-in-water

(O/W) emulsion containing silymarin at various concentrations was applied on roughened polymethylmethacrylate (PMMA) plates. Measurements were carried out using a spectrophotometer equipped with an integrating sphere. Incorporated in O/W creams at a concentration of 10% (w/w), silymarin gives a SPF similar to that of octylmethoxicinnamate (Couteau et al. 2011).

14.6.5 Genistein

Genistein (5,7,4′-trihydroxyisoflavone) is an isoflavone that was first isolated from soybeans. It displays a very low level of toxicity in most animal species. Although soybeans contain a number of ingredients with demonstrated anticancer activities, genistein is the most important agent that has been extensively investigated for its chemopreventive and anticancer activity. Genistein specifically inhibits the growth of ras-oncogene transfected NIH 3T3 cells without affecting the normal cells' growth and diminishes the c-Fos and c-Jun expression in CH310T1/2 fibroblasts induced by platelet-derived growth factor (Okura et al. 1988).

It has been shown recently in human reconstituted skin that genistein dose-dependently preserved cutaneous proliferation and repair mechanics as evidenced by proliferating cell populations with increasing genistein concentrations and a noticeable paucity in PCNA immunoreactivity in the absence of genistein. Genistein inhibited UV-induced DNA damage, evaluated with CPD immunohistochemical expression profiles, and demonstrated an inverse relationship with increasing topical genistein concentrations. Irradiation with UVB substantially induced CPD formation in the absence of genistein, and a dose-dependent inhibition of UVB-induced CPD formation was observed relative to increasing genistein concentrations. Collectively, all genistein pretreated samples demonstrated appreciable histologic architectural preservation when compared with untreated specimens (Moore et al. 2006). Pretreatment of hairless mice with genistein prior to UVB exposure significantly inhibited UVB-induced H_2O_2 and MDA in skin and 8-hydroxy 2′-deoxyguanosine (8-OHdG) in the epidermis as well as internal organs. Suppression of 8-OHdG formation by genistein has been corroborated in purified DNA irradiated with UVA and UVB (Wei et al. 2002).

Application of genistein significantly decreased psoralen plus UVA radiation (PUVA)-induced skin thickening and greatly diminished cutaneous erythema and ulceration in a dose-dependent manner. Histological examination showed that PUVA treatment of mouse skin induced dramatic inflammatory changes throughout the epidermis; topical genistein prevented these changes without noticeable adverse effects. Cells containing cleaved poly (ADP-ribose) polymerase (PARP) and active caspase-3 were significantly increased in PUVA-treated skin as compared with unexposed control skin. Topical genistein completely inhibited cleavage of PARP and caspase-3. PCNA positive cells were observed in suprabasal areas of the epidermis and were significantly decreased in PUVA-treated skin compared with both control samples and samples treated with PUVA plus topical genistein (Shyong et al. 2002).

Furthermore, topical application of genistein before UVB radiation reduced c-Fos and c-Jun expression in SENCAR mouse skin in a dose-dependent manner. Inhibition was more pronounced in skin exposed to the low dose (5 kJ/m²) than to the high dose (15 kJ/m²) of UVB radiation. Application of genistein after UVB exposure downregulated the expressions of c-Fos and c-Jun but to a lesser extent compared with preapplication. Genistein also downregulated the UVB-mediated phosphorylation of the TPK-dependent epidermal growth factor receptor in a dose-dependent manner in A431 human epidermoid carcinoma cells (Wang et al. 1998).

Another study has investigated the photoprotective effects of different concentrations of genistein and daidzein individually or combined. Expression levels of the COX-2, growth arrest, and DNA-damage inducible (Gadd45) genes were measured in BJ-5ta human skin fibroblasts irradiated with 60 mJ/cm² UVB. The authors also have determined the cellular response to UVB-induced DNA damage. As a result, they have suggested that administration of genistein and daidzein combined at a specific concentration and ratio exerted a synergistic photoprotective effect that was greater than the effect obtained with each isoflavone alone. These may be good candidate ingredients for protective agents against UV-induced photodamage (Iovine et al. 2011).

14.6.6 Curcumin

Curcumin, a natural compound extracted from the rhizome of *Curcuma longa* L., has been widely studied for its tumor-inhibiting properties. Lately, attention has been focused on dietary phytochemicals, such as curcumin, in an effort to repair photodamage, preventing skin degeneration caused by UV radiation. Curcumin is also described as an agent capable of inducing apoptosis in numerous cellular systems. The varied biological property of curcumin and low toxicity even when administered at higher doses make it attractive to explore its use in various skin disorders. The molecular basis of the anticarcinogenic and chemopreventive effects of curcumin is attributed to its effect on several targets, including transcription factors, growth regulators, adhesion molecules, apoptotic genes, angiogenesis regulators, and cellular signaling molecules.

It has been revealed that on treatment of HaCaT cells with UVB and curcumin, there was induction of apoptosis as evidenced by DNA fragmentation. The combination of UVB irradiation with curcumin also synergistically induced apoptotic cell death in HaCaT cells through activation of caspase-8, -3, and -9 followed by release of cytochrome *c* (Park and Lee 2007). Treatment with curcumin strongly inhibited COX-2 mRNA and protein expressions in UVB-irradiated HaCaT cells. Notably, there was effective inhibition by curcumin on UVB-induced activations of p38 MAPK and JNK in HaCaT cells. The DNA-binding activity of AP-1 transcription factor was also markedly decreased with curcumin treatment in UVB-irradiated HaCaT cells. These results collectively suggest that curcumin may inhibit COX-2 expression by suppressing p38 MAPK and JNK activities in UVB-irradiated HaCaT cells (Cho et al. 2005).

14.6.7 Apigenin

Apigenin (5,7,4′-trihydroxyflavone) is a natural flavonoid present in the leaves, stems, fruits, and vegetables of vascular plants. A beneficial role of flavonoids in the treatment and prevention of skin disorders has been suggested. Initial studies showed that apigenin treatment to murine skin resulted in inhibition of UV-mediated induction of ODC activity as well as reduction in cancer incidence with an increase in tumor-free survival (Birt et al. 1997). In addition, delivering apigenin into viable epidermis seemed to be essential for an apigenin formulation to be effective in skin cancer prevention as apigenin strongly absorbs UV light with three maximum absorption wavelengths at 212, 269, and 337 nm (Li and Birt 1996).

Naturally occurring flavonoids, including apigenin, have been shown to inhibit MMP-1 activity and downregulate MMP-1 expression via an inhibition of the AP-1 activation, although the cellular inhibitory mechanisms differ depending on their chemical structures. The structure–activity relationship of the antioxidative property of flavonoids was studied by Sim et al. (2007), and it was found that the inhibitory effects of flavonoids on the collagenase in human dermal fibroblasts depend on the number of the OH group in the flavonoid structure, and those with a higher number of OH group may be more useful in protecting the skin from photoaging. A study has assessed the effects of UVA radiation on HaCaT cells that were treated with apigenin and luteolin for the indicated times. Results indicate that apigenin and luteolin inhibited UVA-induced collagenolytic MMP-1 production by interfering with Ca (2+)-dependent MAPKs and AP-1 signaling. This suggests that these flavonoids may be potentially useful in the prevention and treatment of skin photoaging (Hwang et al. 2011).

14.7 Conclusion

Nutritional contribution to systemic photoprotection is emerging as a topic of interest in public health and preventive medicine. *In vitro* experiments in cell culture and various animal studies provide further evidence that dietary components afford protection against excess UV light. Information from various studies supports the idea that photocarcinogenic damage can be a result of overexposure to sunlight and its relationship to oxidative stress. Over the past several decades, there has been a significant increase in the amount of UVB radiation that people receive, and this, consequently, has led to an increase in the incidence of skin cancer. Although the rate of skin cancer is increasing constantly, it is necessary to mention that it is considered one of the most preventable types of cancer. Healthier preventive strategies are required to reduce UV radiation-caused photodamage. Education about the harmful effects of UVB existing in sunlight, wearing protective clothing to avoid excessive exposure, and the use of sunscreen is mandatory for prevention of photodamage and cutaneous disease. While protection through individual dietary components, in terms of SPF, may be considerably lower than that achieved using topical sunscreens, an increased lifelong overall protection via dietary supplement may contribute significantly to skin health. There is evidence from *in vitro*, animal, and human studies demonstrating the actions of dietary botanical agents, with antioxidant properties existing as anti-inflammatory, cancer-preventive, antiphotoaging, and photoprotectants. Finally, the success of chemoprevention strategies will be assisted by the use of skin-care products supplemented with several effective botanical agents in combination with the use of sunscreens. This may be an effective approach for reducing UVB-generated ROS-mediated photoaging and skin cancer in humans.

References

Adhami, V. M., F. Afaq, and N. Ahmad. 2003. Suppression of ultraviolet B exposure-mediated activation of NF-kappaB in normal human keratinocytes by resveratrol. *Neoplasia* 5:74–82.

Afaq, F., V. M. Adhami, and N. Ahmad. 2003. Prevention of short-term ultraviolet B radiation-mediated damages by resveratrol in SKH-1 hairless mice. *Toxicol Appl Pharmacol* 186:28–37.

Afaq, F., N. Ahmad, and H. Mukhtar. 2003. Suppression of UVB-induced phosphorylation of mitogen-activated protein kinases and nuclear factor kappa B by green tea polyphenol in SKH-1 hairless mice. *Oncogene* 22:9254–9264.

Afaq, F., N. Khan, D. N. Syed, and H. Mukhtar. 2010. Oral feeding of pomegranate fruit extract inhibits early biomarkers of UVB radiation-induced carcinogenesis in SKH-1 hairless mouse epidermis. *Photochem Photobiol* 86:1318–1326.

Afaq, F., A. Malik, D. Syed et al. 2005. Pomegranate fruit extract modulates UV-B-mediated phosphorylation of mitogen-activated protein kinases and activation of nuclear factor kappa B in normal human epidermal keratinocytes paragraph sign. *Photochem Photobiol* 81:38–45.

Afaq, F., M. A. Zaid, N. Khan, M. Dreher, and H. Mukhtar. 2009. Protective effect of pomegranate-derived products on UVB-mediated damage in human reconstituted skin. *Exp Dermatol* 18:553–561.

Back, J. H., Y. Zhu, A. Calabro et al. 2012. Resveratrol-mediated downregulation of rictor attenuates autophagic process and suppresses UV-induced skin carcinogenesis. *Photochem Photobiol* 88:1165–1172.

Bergendi, L., L. Benes, Z. Durackova, and M. Ferencik. 1999. Chemistry, physiology and pathology of free radicals. *Life Sci* 65:1865–1874.

Bickers, D. R., and M. Athar. 2006. Oxidative stress in the pathogenesis of skin disease. *J Invest Dermatol* 126:2565–2575.

Birt, D. F., D. Mitchell, B. Gold, P. Pour, and H. C. Pinch. 1997. Inhibition of ultraviolet light induced skin carcinogenesis in SKH-1 mice by apigenin, a plant flavonoid. *Anticancer Res* 17:85–91.

Bosset, S., M. Bonnet-Duquennoy, P. Barre et al. 2003. Photoageing shows histological features of chronic skin inflammation without clinical and molecular abnormalities. *Br J Dermatol* 149:826–835.

Bowden, G. T. 2004. Prevention of non-melanoma skin cancer by targeting ultraviolet-B-light signalling. *Nat Rev Cancer* 4:23–35.

Cho, J. W., K. Park, G. R. Kweon et al. 2005. Curcumin inhibits the expression of COX-2 in UVB-irradiated human keratinocytes (HaCaT) by inhibiting activation of AP-1: p38 MAP kinase and JNK as potential upstream targets. *Exp Mol Med* 37:186–192.

Couteau, C., C. Cheignon, E. Paparis, and L. J. Coiffard. 2011. Silymarin, a molecule of interest for topical photoprotection. *Nat Prod Res* 26:2211–2214.

de Gruijl, F. R., and J. C. Van der Leun. 1994. Estimate of the wavelength dependency of ultraviolet carcinogenesis in humans and its relevance to the risk assessment of a stratospheric ozone depletion. *Health Phys* 67:319–325.

Dhanalakshmi, S., G. U. Mallikarjuna, R. P. Singh, and R. Agarwal. 2004. Dual efficacy of silibinin in protecting or enhancing ultraviolet B radiation-caused apoptosis in HaCaT human immortalized keratinocytes. *Carcinogenesis* 25:99–106.

Elmets, C. A., D. Singh, K. Tubesing et al. 2001. Cutaneous photoprotection from ultraviolet injury by green tea polyphenols. *J Am Acad Dermatol* 44:425–432.

Fisher, G. J., S. C. Datta, H. S. Talwar et al. 1996. Molecular basis of sun-induced premature skin ageing and retinoid antagonism. *Nature* 379:335–339.

Gu, M., S. Dhanalakshmi, R. P. Singh, and R. Agarwal. 2005. Dietary feeding of silibinin prevents early biomarkers of UVB radiation-induced carcinogenesis in SKH-1 hairless mouse epidermis. *Cancer Epidemiol Biomarkers Prev* 14:1344–1349.

Heinrich, U., C. E. Moore, S. De Spirt, H. Tronnier, and W. Stahl. 2011. Green tea polyphenols provide photoprotection, increase microcirculation, and modulate skin properties of women. *J Nutr* 141:1202–1208.

Hwang, Y. P., K. N. Oh, H. J. Yun, and H. G. Jeong. 2011. The flavonoids apigenin and luteolin suppress ultraviolet A-induced matrix metalloproteinase-1 expression via MAPKs and AP-1-dependent signaling in HaCaT cells. *J Dermatol Sci* 61:23–31.

Iovine, B., M. L. Iannella, F. Gasparri, G. Monfrecola, and M. A. Bevilacqua. 2011. Synergic effect of genistein and daidzein on UVB-induced DNA damage: An effective photoprotective combination. *J Biomed Biotechnol* 2011:692846.

Jeon, H. Y., J. K. Kim, W. G. Kim, and S. J. Lee. 2009. Effects of oral epigallocatechin gallate supplementation on the minimal erythema dose and UV-induced skin damage. *Skin Pharmacol Physiol* 22:137–141.

Katiyar, S. K. 2002. Treatment of silymarin, a plant flavonoid, prevents ultraviolet light-induced immune suppression and oxidative stress in mouse skin. *Int J Oncol* 6:1213–1222.

Katiyar, S. K. and H. Mukhtar. 2001. Green tea polyphenol (−)-epigallocatechin-3-gallate treatment to mouse skin prevents UVB-induced infiltration of leukocytes, depletion of antigen-presenting cells, and oxidative stress. *J Leukocyte Biol* 5:719–726.

Katiyar, S. K., F. Afaq, A. Perez, and H. Mukhtar. 2001. Green tea polyphenol (−)-epigallocatechin-3-gallate treatment of human skin inhibits ultraviolet radiation-induced oxidative stress. *Carcinogenesis* 22:287–294.

Katiyar, S. K., A. Challa, T. S. McCormick, K. D. Cooper, and H. Mukhtar. 1999. Prevention of UVB-induced immunosuppression in mice by the green tea polyphenol (−)-epigallocatechin-3-gallate may be associated with alterations in IL-10 and IL-12 production. *Carcinogenesis* 20:2117–2124.

Katiyar, S. K., N. J. Korman, H. Mukhtar, and R. Agarwal. 1997. Protective effects of silymarin against photocarcinogenesis in a mouse skin model. *J Natl Cancer Inst* 89:556–566.

Katiyar, S. K., M. Vaid, H. van Steeg, and S. M. Meeran. 2010. Green tea polyphenols prevent UV-induced immunosuppression by rapid repair of DNA damage and enhancement of nucleotide excision repair genes. *Cancer Prev Res (Phila)* 3:179–189.

Khan, N., D. N. Syed, H. C. Pal, H. Mukhtar, and F. Afaq. 2011. Pomegranate fruit extract inhibits UVB-induced inflammation and proliferation by modulating NF-kappaB and MAPK signaling pathways in mouse skin. *Photochem Photobiol* 88:1126–1134.

Kim, K. H., J. H. Back, Y. Zhu et al. 2011. Resveratrol targets transforming growth factor-beta2 signaling to block UV-induced tumor progression. *J Invest Dermatol* 131:195–202.

Lee, Y. H., N. C. Kumar, and R. D. Glickman. 2012. Modulation of photochemical damage in normal and malignant cells by naturally-occurring compounds. *Photochem Photobiol* 88:1385–1395.

Li, B., and D. F. Birt. 1996. In vivo and in vitro percutaneous absorption of cancer preventive flavonoid apigenin in different vehicles in mouse skin. *Pharm Res* 13:1710–1715.

Liardet, S., C. Scaletta, R. Panizzon, P. Hohlfeld, and L. Laurent-Applegate. 2001. Protection against pyrimidine dimers, p53, and 8-hydroxy-2′-deoxyguanosine expression in ultraviolet-irradiated human skin by sunscreens: Difference between UVB + UVA and UVB alone sunscreens. *J Invest Dermatol* 117:1437–1441.

Meeran, S. M., S. Katiyar, C. A. Elmets, and S. K. Katiyar. 2006. Silymarin inhibits UV radiation-induced immunosuppression through augmentation of interleukin-12 in mice. *Mol Cancer Ther* 5:1660–1668.

Mittal, A., C. Piyathilake, Y. Hara, and S. K. Katiyar. 2003. Exceptionally high protection of photocarcinogenesis by topical application of (−)-epigallocatechin-3-gallate in hydrophilic cream in SKH-1 hairless mouse model: Relationship to inhibition of UVB-induced global DNA hypomethylation. *Neoplasia* 5:555–565.

Moore, J. O., Y. Wang, W. G. Stebbins et al. 2006. Photoprotective effect of isoflavone genistein on ultraviolet B-induced pyrimidine dimer formation and PCNA expression in human reconstituted skin and its implications in dermatology and prevention of cutaneous carcinogenesis. *Carcinogenesis* 27:1627–1635.

Okura, A., H. Arakawa, H. Oka, T. Yoshinari, and Y. Monden. 1988. Effect of genistein on topoisomerase activity and on the growth of [Val 12]Ha-ras-transformed NIH 3T3 cells. *Biochem Biophys Res Commun* 157:183–189.

Park, K., and J. H. Lee. 2007. Photosensitizer effect of curcumin on UVB-irradiated HaCaT cells through activation of caspase pathways. *Oncol Rep* 17:537–540.

Quan, T., T. He, S. Kang, J. J. Voorhees, and G. J. Fisher. 2004. Solar ultraviolet irradiation reduces collagen in photoaged human skin by blocking transforming growth factor-beta type II receptor/Smad signaling. *Am J Pathol* 165:741–751.

Roy, S., G. Deep, C. Agarwal, and R. Agarwal. 2012. Silibinin prevents ultraviolet B radiation-induced epidermal damages in JB6 cells and mouse skin in a p53-GADD45alpha-dependent manner. *Carcinogenesis* 33:629–636.

Sander, C. S., H. Chang, S. Salzmann et al. 2002. Photoaging is associated with protein oxidation in human skin in vivo. *J Invest Dermatol* 118:618–625.

Shyong, E. Q., Y. Lu, A. Lazinsky et al. 2002. Effects of the isoflavone 4′,5,7-trihydroxyisoflavone (genistein) on psoralen plus ultraviolet A radiation (PUVA)-induced photodamage. *Carcinogenesis* 23:317–321.

Siegel, R., D. Naishadham, and A. Jemal. 2013. Cancer statistics, 2012. *CA Cancer J Clin* 62:10–29.

Sim, G. S., B. C. Lee, H. S. Cho et al. 2007. Structure activity relationship of antioxidative property of flavonoids and inhibitory effect on matrix metalloproteinase activity in UVA-irradiated human dermal fibroblast. *Arch Pharm Res* 30:290–298.

Smirnoff, N., P. L. Conklin, and F. A. Loewus. 2001. Biosynthesis of ascorbic acid in plants: A renaissance. *Annu Rev Plant Physiol Plant Mol Biol* 52:437–467.

Syed, D. N., A. Malik, N. Hadi et al. 2006. Photochemopreventive effect of pomegranate fruit extract on UVA-mediated activation of cellular pathways in normal human epidermal keratinocytes. *Photochem Photobiol* 82:398–405.

Wang, Y., X. Zhang, M. Lebwohl, V. DeLeo, and H. Wei. 1998. Inhibition of ultraviolet B (UVB)-induced c-fos and c-jun expression in vivo by a tyrosine kinase inhibitor genistein. *Carcinogenesis* 19:649–654.

Wang, Z. Y., R. Agarwal, D. R. Bickers, and H. Mukhtar. 1991. Protection against ultraviolet B radiation-induced photocarcinogenesis in hairless mice by green tea polyphenols. *Carcinogenesis* 12:1527–1530.

Wei, H., X. Zhang, Y. Wang, and M. Lebwohl. 2002. Inhibition of ultraviolet light-induced oxidative events in the skin and internal organs of hairless mice by isoflavone genistein. *Cancer Lett* 185:21–29.

Wulf, H. C., I. M. Stender, and J. Lock-Andersen. 1997. Sunscreens used at the beach do not protect against erythema: a new definition of SPF is proposed. *Photodermatol Photoimmunol Photomed* 13:129–132.

Reversal of DNA Damage to the Skin with DNA Repair Liposomes

Peter Wolf
Medical University of Graz

15.1 Significance of Functional DNA Repair

Exposure of the skin to ultraviolet (UV) radiation leads to DNA damage and, if unrepaired, to subsequent tumor initiation and skin cancer formation (Ananthaswamy and Kanjilal 1996). The importance of effective DNA repair to restrain skin carcinogenesis is best highlighted by the genetic disease xeroderma pigmentosum (XP) in which a defective nucleotide excision repair leads to the development of multiple skin cancers starting early on in life at a rate more than a thousandfold greater than in the normal population (Kraemer et al. 1994). A molecular analysis of nonmelanoma skin cancers from XP patients and individuals from the normal population reveals typical UV fingerprints (i.e., C to T or CC to TT transitions) at dipyrimidine sites of tumor suppressor genes, such as p53, INK4a-ARF, and patched or oncogenes, such as RAS (Ananthaswamy and Kanjilal 1996), as a direct consequence of unrepaired DNA photoproducts, such as cyclobutane pyrimidine dimers (CPDs). However, UV-induced DNA damage is also one of the triggers of UV-induced immune suppression (Applegate et al. 1989; Kripke et al. 1992), which is crucial in skin carcinogenesis in rodents and, most likely, also humans. For instance, XP patients have been shown to exhibit severe immunologic alterations, which may accelerate tumor growth after tumor initiation (Norris et al. 1990). However, in individuals of the normal population, excessive sunlight-caused DNA damage that overloads the endogenous DNA repair capacity and/or a simultaneous presence of a slight weakness in DNA repair may result in skin cancer as well. Munch-Peterson et al. (1985) observed a decrease in the cellular tolerance to UV radiation in lymphocytes of normal individuals with multiple skin cancers, using unscheduled DNA synthesis. Lambert, Ringborg, and Swanbeck (1976) found reduced DNA repair after UV irradiation of lymphocytes *in vitro* in patients with solar keratoses,

which are potential precursors of squamous cell carcinoma (SCC). After irradiation with various doses of UV, the lymphocytes were incubated with [3H]thymidine in the presence of hydroxyurea. A dose–response relationship for the UV-induced DNA repair synthesis was established for each individual, and the average repair capacity in patients with actinic keratoses was approximately 30% below that of controls. Sbano et al. (1978) utilized autoradiographic counting to measure the UV-induced unscheduled DNA synthesis of skin fibroblasts from patients with chronic actinic keratosis and from healthy donors of similar age. In order to study a possible regional difference of DNA repair between the parts of the body ordinarily exposed and those parts unexposed to sunlight, two cell strains were used for each examined subject: one developed from the forehead skin and the other from the abdominal or axillary skin. Unscheduled DNA synthesis analysis confirmed a depressed DNA repair in actinic keratosis patients as compared with controls. Thielmann et al. (1987) demonstrated that fibroblast strains from patients with SCC and basal cell carcinoma (BCC) had diminished DNA repair capacity (as measured by the alkaline elution technique) upon exposure to UV radiation by up to 82% compared to controls. Alcalay et al. (1990) showed that the skin of subjects with BCC had decreased repair of solar simulated radiation-induced CPD (as measured by a dimer-specific endonuclease assay) compared to the epidermis of a normal control population. Grossman and Wei (1995) and Wei et al. (1994, 1995) reported molecular epidemiologic data indicating that impaired DNA repair is involved in the etiopathogenesis of BCC in the normal population. They found that a low DNA repair capacity correlated to the number of skin tumors in subjects with one or more BCC (Wei et al. 1994). Wang et al. (2005) assessed UVB-induced chromatid breaks in a hospital-based, case-control study and found a dose–response relationship between mutagen sensitivity and risk for both BCC and SCC.

15.2 Liposomes with DNA Repair Enzymes

15.2.1 Enzyme Technology

Despite the availability of powerful sunscreens with high sun protection factors, skin cancer remains a major health problem. For instance, BCC and SCC taken together as the most common forms of skin cancer affect more than a million people each year alone in the United States (http://www.skincancer.org). Therefore, besides sunscreens, secondary approaches to reduce sun damage are desirable.

Liposomes (consisting of multilamellar phospholipid layers) containing prokaryotic DNA repair enzymes, such as T4 endonuclease V (T4N5), have been introduced as a novel concept of secondary photoprotection (Wolf et al. 1995; Wolf, Yarosh, and Kripke 1993; reviewed in Cafardi and Elmets 2008; Zahid and Brownell 2008). T4N5, a product of the denV gene of bacteriophage T4, initiates repair of CPD by cleaving the glycosylic bond of the 5′ pyrimidine and then breaking the DNA at the apyrimidinic site, a process that is thought to be a rate-limiting step of excision repair (Yarosh et al. 1996 and cited herein). The denV gene has been cloned and the complete nucleotide and amino acid coding sequence determined (Radnay et al. 1984). The expressed T4N5 protein has been purified to homogeneity in commercial quantities. T4N5 liposomes can penetrate the stratum corneum of murine skin and human skin explants and deliver T4N5 to epidermal cells (Ceccoli et al. 1989; Yarosh et al. 1994). The membrane of the liposomes employed is pH sensitive, which is what facilitates the intracellular release of the enzyme (Yarosh et al. 1994). However, liposomes are of importance not only to shuttle the enzyme into the skin but also to protect the enzyme from inactivation through bacteria and/or the acid environment of the skin surface. Moreover, liposomes have the capacity to assist in protein refolding by interaction with a secondary protein structure (Zardeneta and Horowitz 1994). This may have happened in certain studies in which a heat-denatured enzyme was used as a vehicle control and found partially active (Yarosh et al. 1999). However, particularly under daily life conditions, liposome-assisted protein refolding may be of great help to preserve the preparation and/or the activity of a preparation, for instance, on the beach when the preparation and/or the skin is transiently exposed to high temperatures that may damage the ingredients of such a preparation with DNA repair activity.

15.2.2 Preclinical Work with Topical DNA Repair Enzymes

T4N5 liposomes increased DNA repair in fibroblasts and keratinocytes cultured from XP patients and DNA repair-proficient subjects (Yarosh et al. 1991). In various other studies, DNA repair liposomes affected other UV-induced alterations. Gilchrest et al. (1993) reported that T4N5 treatment enhanced *in vitro* UV-induced melanogenesis as measured within 16 to 96 h after UV exposure by melanin content, tyrosinase activity, 14C-dopa incorporation, and visual assessment of both murine melanoma cells and human keratinocytes. The most significant increase in the T4N5-enhanced melanogenesis was seen at the late time points, consistent with the delayed tanning reaction induced by natural sunlight. Liposomes containing extracts from *Micrococcus luteus* (a widely dispersed organism found in the normal flora of the mammalian skin, unpasteurized milk, marine waters, and even soil) did also enhance DNA repair and melanogenesis in Cloudman S91 melanoma cells (Yarosh, Kibitel, and O'Connor 1997). However, the exact mechanisms of DNA repair enzyme-induced melanogenesis after UV exposure remain to be determined.

In repair-proficient mice, treatment with T4N5 liposomes after UV exposure had strong effects on a variety of biologic endpoints. T4N5 liposomes enhanced CPD repair (Yarosh et al. 1992) and diminished UV-induced sunburn cell formation. In addition, they protected against UV-induced alterations of Langerhans cells and Thy-1+ dendritic epidermal T cells (Wolf et al. 1995), impairment of antigen-presenting cell function (Vink et al. 1996), production of the immunosuppressive cytokine Interleukin-10 (Nishigori et al. 1996), and the appearance of T suppressor (i.e., nowadays regulatory) cells (Kripke et al. 1992) as well as both local and systemic functional immune suppression (Kripke et al. 1992; Wolf et al. 1995; Wolf, Yarosh, and Kripke 1993). Moreover, T4N5 treatment delayed the abnormal rise of p53 protein expression in chronically UV-irradiated mice (Bito et al. 1995) and reduced both the incidence and yield of skin cancer in mice (Bito et al. 1995; Yarosh et al. 1992). Together, the data indicated that T4N5 liposomes can enhance DNA repair in UV-irradiated mice and may prevent skin cancer by reducing tumor-initiating DNA damage and simultaneously preserving the immune response essential for tumor rejection. Notably, T4N5 liposomes only marginally affected the sunburn reaction as measured by skin edema but clearly reduced DNA damage and other UV-induced biologic alterations, including immune suppression (Wolf et al. 1995). Vink et al. (1997) used liposomes containing photolyase, which, upon absorption of photoreactivating visible light, splits UVB-induced CPD to restore the function of antigen-presenting activity of dendritic cells in the skin of mice. They used isopsoralen plus UVA (PUVA) as a control to demonstrate that the photoreactivation repair process was specific for UVB damage because it was unable to restore PUVA photoproducts. In one of the studies in mice (Wolf, Yarosh, and Kripke 1993), the effect of sunscreens containing standard UV filters (2-ethylhexyl-p-methoxycinnamate, octyl-N-dimethyl-p-aminobenzoate, or benzophenone-3) was directly compared to that of T4N5 liposomes. Notably, the sunscreens much better protected against the UV-induced inflammatory response than the liposomes (protection up to 97% vs. 39%); in contrast, the liposomes much better protected against immune suppression (as measured by systemic suppression of delayed type hypersensitivity to *C. albicans*) than the sunscreens (82% vs. 42%).

15.2.3 Clinical Studies with Topical DNA Repair Enzyme Preparations

The results of clinical studies with topical DNA repair enzyme preparation are summarized in Table 15.1. Yarosh et al. (1996) first reported on the acute and chronic safety testing of T4N5 liposomes, showing neither adverse reactions nor significant changes in serum chemistry or in skin histology of treated individuals. The skin of XP patients treated with the DNA repair liposomes had fewer CPD in DNA and showed less erythema than control sites.

Wolf et al. (2000) investigated the penetration of T4N5 liposomes and their effects on biologic endpoints in UV-exposed buttock skin of non-XP skin patients with a history of multiple skin cancers (Wolf et al. 2000). Transmission electron microscopic studies after immunogold labeling and anti-T4N5 staining indicated that T4N5 enzyme incorporated into the liposomes penetrated into skin. T4N5-labeled gold particles were found in the cytoplasm and nuclei of keratinocytes and Langerhans cells. The investigators observed no effect of the T4N5 liposomes on UV-induced erythema as measured by skin reflectance spectroscopy and only weak improvement of removal of CPD as measured by immunohistochemical antibody staining. However, UV-induced upregulation of interleukin-10 and tumor necrosis factor-alpha was nearly entirely abrogated when the test fields exposed to two minimal erythema doses were treated after UV exposure with active T4N5 liposomes compared to heat-inactivated control liposomes. Together, this suggested that even a moderate increase in overall DNA repair may lead to large effects on subsequent UV-induced biologic alterations, such as immune function, presumably by selective repair of certain sites in the genome (Kripke et al. 1992; Yarosh et al. 1996).

Halliday et al. (2004) and Kuchel, Barnetson, and Halliday (2005) used T4N5 liposomes to determine whether CPD formation is involved in UV-induced suppression of efferent immunity.

Nickel-allergic volunteers were irradiated with a range of doses of solar-simulated UV and T4N5, or empty liposomes were applied after irradiation. Nickel-induced recall immunity was assessed by reflectance spectrometry. T4N5 liposomes inhibited suppression of the nickel response, prevented solar-simulated UV radiation from reducing the number of epidermal dendritic cells, and reduced macrophage infiltration into irradiated skin. Consistent with the results of other studies (Wolf et al. 2000), T4N5 liposomes did not significantly affect the UV-induced erythema response.

Stege et al. (2000) employed the DNA-repair enzyme photolyase, derived from *Anacystis nidulans*, that specifically converts CPDs into their original DNA structure after exposure to photoreactivating light to repair UVB radiation-induced DNA damage. Topical application of photolyase-containing liposomes to UVB-irradiated skin and subsequent exposure to photoreactivating light decreased the number of UVB radiation-induced CPD by 40%–45%. Moreover, application of photolyase prevented UVB-induced suppression of intercellular adhesion molecule-1 (ICAM-1), a molecule required for immunity and inflammatory events in the epidermis. Photolyase treatment also prevented UV-induced suppression of elicitation of hypersensitivity to nickel sulfate, sunburn cell formation, and erythema. No effect on all these responses was observed if empty liposomes (containing no photolyase) were used or photoreactivating light exposure was not given after administration of photolyase-filled liposomes.

Ke et al. (2008) investigated the effect of topical formulations containing either DNA-repair enzyme extracts from *Micrococcus luteus* or RNA fragments (UVC-irradiated rabbit globin mRNA to increase the resistance of human keratinocytes to UVB damage and enhance DNA repair) on UV-induced local contact hypersensitivity (CHS) suppression in eight human volunteers as measured *in vivo* using the contact allergen dinitrochlorobenzene. Exposure to a single 0.75 minimum erythema dose (MED) simulated solar radiation resulted in 64% CHS suppression in unprotected subjects compared with unirradiated sensitized

TABLE 15.1 Summary of Clinical Studies on the Effect of Liposomes Containing DNA Repair Enzymes

Study	Enzyme(s)	Effect
Yarosh et al. 1996	T4N5	Improvement of DNA repair in XP patients
Wolf et al. 2000	T4N5	Abrogation of UV-induced IL-10 and TNF-alpha formation in skin cancer patients
Stege et al. 2000	Photolyase	Reduction of UV-induced CPDs and restoration of interferon-gamma-induced ICAM-1 expression in the skin
Yarosh et al. 2001	T4N5	Reduction of the number of actinic keratoses and BCCs in patients with XP
Halliday et al. 2004	T4N5	Prevention of UV-induced immune suppression in the nickel-allergy model
Kuchel, Barnetson, and Halliday 2005	T4N5	Reduction of UV-induced immune suppression in the nickel-allergy model and epidermal dendritic cell alteration and macrophage infiltration of the skin
Ke et al. 2008	*Micrococcus luteus* lysate	Reduction of local immune suppression in the DNCB contact allergy model
Lucas et al. 2008	*Micrococcus luteus* lysate	Prevention of immune suppression in the DNCB contact allergy model
DeBoyes et al. 2010	T4N5	Reduction in the number of actinic keratoses in normal individuals with moderate-to-severe photodamaged skin
Hofer et al. 2011	Photolyase and *Micrococcus luteus* lysate	Protection from PLE

Note: T4N5, bacteriophage T4 endoclease V; photolyase, extracted from *Anacystis nidulans*; *Micrococcus luteus* lysate, containing DNA-repair enzymes with UV-specific endoclease activity; DNCB, dinitrochlorobenzene.

controls. In contrast, UV-induced CHS suppression was reduced to 19% with DNA-repair enzymes and 7% with RNA fragments. Biopsies from an additional nine volunteers showed an 18% decrease in thymine dimers by both DNA-repair enzymes and RNA fragments in UV-irradiated skin, relative to unprotected UV-irradiated skin. The authors suggested that both topical RNA fragments as well as DNA-repair enzymes may be useful as photoprotective agents. Similar results on UV-induced suppression of the induction of CHS were reported by Lucas et al. (2008) in a study using a commercial moisturizer with DNA-repair enzymes.

Yarosh et al. (2001) reported the results of a prospective, multicenter, double-blind, placebo-controlled study in patients with XP in whom the regular daily administration (as close to midday as possible) of an after-sun lotion containing T4N5 liposomes over a period of 12 months reduced the incidence of actinic keratoses and BCC by 68% and 30%, respectively. Interestingly, an effect on the number of actinic keratoses was observed as early as 3 months after the start of treatment, suggesting that a quickly acting process, such as DNA repair enzyme-induced immune protection (rather than prevention of mutation formation), might have been responsible for the therapeutic action of the after-sun lotion. However, there were discrepancies about the statistical methods used for analysis of the data (Lachenbruch et al. 2001; Yarosh 2001), and therefore, more work is necessary to substantiate the efficacy of T4N5 liposomes in XP.

More recently, DeBoyes et al. (2010) evaluated 17 normal individuals with moderate-to-severe photodamaged skin for differences in actinic keratoses following topical application of T4N5 liposome lotion over 48 weeks. Compared to baseline, a statistically significant reduction in the number of actinic keratoses was seen following the treatment period. Wolf et al. evaluated the effect of a proprietary SPF 30 sunscreen and after-sun lotion, both containing a combination of photolyase from *Anacystis nidulans* and *Micrococcus luteus* lysate in a pilot study of intensified photoprotection in repair-deficient and -proficient patients with a history of multiple skin cancers (http://www.clinicaltrials.gov/show/NCT00555633). Thirteen patients (including five XP, one XP variant, three basal cell nevus syndrome, and four normal patients) were instructed to apply the sunscreen regularly before sun exposure and the after-sun lotion to their face and arms daily as close to midday as possible for 24 months. There was a trend for less BCC occurring in the patients during the 24-month study period compared to a 24-month pre-study period. In addition, the results of this study indicated that the intensified photoprotection strategy may have helped in reducing skin aging. The analysis of patient self-reports revealed a statistically significant improvement for several endpoints, including smoothness, wrinkles, color spots, and telangiectasia of the skin, starting as early as at 3 months of treatment with a maximum effect seen at 12 months (Figure 15.1). No adverse effects were noted during the study.

Hofer et al. (2011) led the research with liposomal DNA-repair enzymes into the field of photodermatoses (Gruber-Wackernagel, Byrne, and Wolf 2009; Wolf, Byrne, and Gruber-Wackernagel 2009). They used a proprietary after-sun lotion, containing a

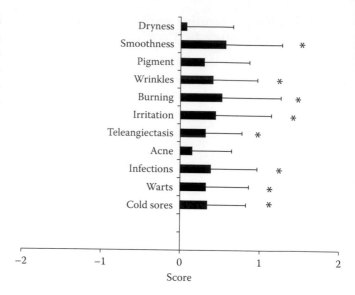

FIGURE 15.1 Effect of intensified photoprotection with a proprietary SPF 30 sunscreen and after-sun lotion, both containing a combination of photolyase from *Anacystis nidulans* and *Micrococcus luteus* lysate in a pilot study in repair-deficient (XP) or -proficient patients with a history of multiple skin cancers. Patients scored the skin status of treated areas (face, forearms, and hands) in 3-month intervals during a 24-month study period for various endpoints on a scale from –2 (maximum worsening) to +2 (maximum improvement). Data shown are from the 12-month observation time point. * $p < 0.05$ (Wilcoxon test).

combination of photolyase and *Micrococcus luteus lysate* in a randomized, double-blind, left-right body side comparison study in patients with polymorphic light eruption (PLE). Fourteen PLE patients were treated on four consecutive days with near-erythemal solar-simulated UV exposures fields on symmetrically located, individual PLE predilection sites. As shown by a newly established specific PLE test score, PLE symptoms were significantly fewer on test sites treated with active DNA repair lotion followed by blue light treatment than on untreated or placebo-treated test sites. At 144 h after first UV exposure (the time point of maximal PLE symptoms), the mean test score for the active enzyme-treated test fields were 61% lower compared to the untreated test fields (whereas the reduction by placebo treatment was only 27%). These results provided evidence that DNA damage may trigger PLE, and increased DNA repair may prevent the induction of PLE symptoms. However, the mechanism by which improved DNA repair may diminish PLE remains to be elucidated. Photoimmunoprotection is not a good mechanistic candidate because PLE patients have been shown to be more resistant to UV-induced immune suppression (as measured by induction of CHS) than normal controls (Koulu et al. 2010; Palmer and Friedmann 2004; van de Pas et al. 2004). Moreover, a lack of skin infiltration of neutrophilic granulocytes after UV exposure and subsequent failure of immune suppression to UV-induced neoantigens has been hypothesized to be responsible for PLE (Schornagel et al. 2004). Therefore, at least theoretically, photo (photo immune protection) through enhanced

DNA repair should deteriorate rather than prevent PLE. The authors speculated that enhanced DNA repair might reduce the formation of DNA photoproduct-related neoantigens as the potential starting point of the pathogenic chain in PLE. DNA-repair enzyme treatment had a highly significant effect on PLE symptoms but did not significantly alter UV-induced erythema. This is consistent with the results of previous studies in which topical DNA-repair enzymes affected the sunburn reaction only marginally or not at all but reduced DNA damage and other UV-induced biologic alterations, including immune suppression (Ke et al. 2008; Stege et al. 2000; Wolf et al. 1995, 2000).

15.3 Outlook

The topical application of liposomal-incorporated DNA repair enzymes is a novel strategy in photoprotection. Several companies are already marketing cosmetic sun-care preparations (after-sun lotions and sunscreens) containing DNA-repair enzymes (photolyase from *Anacystis nidulans* and/or endonucleases from *Micrococcus luteus* lysate). Whereas conventional sunscreens containing chemical and/or physical UV filters must be applied before UV exposure to be protective, liposomal preparations containing DNA-repair enzymes may be effective when applied after UV exposure and initiation of a sunburn reaction. However, they hardly protect against erythema (Kuchel, Barnetson, and Halliday 2005; Wolf et al. 2000). Therefore, they need to be combined in sunscreens with standard chemical and/or physical UV filters. The results of a controlled study have suggested that topical DNA-repair enzymes may reduce skin cancer, at least in DNA repair-deficient XP patients (Yarosh et al. 2001). However, so far, only cosmeceutical preparations with DNA-repair enzymes are marketed (Hofer et al. 2011; Lucas et al. 2008), and no pharmaceutical preparation is available for humans. More work and controlled clinical studies are necessary before a preparation with DNA-repair enzyme(s) may be able to receive drug approval, most likely first under orphan drug designation for patients with XP. Besides, comparative studies are necessary to determine which enzyme strategy (endonuclease-empowered excision repair vs. photoreactivating light repair) is most effective for secondary photoprotection.

References

Alcalay, J., S. E. Freeman, L. H. Goldberg, and J. E. Wolf. 1990. Excision repair of pyrimidine dimers induced by simulated solar radiation in the skin of patients with basal cell carcinoma. *J Invest Dermatol* 95:506–509.

Ananthaswamy, H. N., and S. Kanjilal. 1996. Oncogenes and tumor suppressor genes in photocarcinogenesis. *Photochem Photobiol* 63:428–432.

Applegate, L. A., R. D. Ley, J. Alcalay, and M. L. Kripke. 1989. Identification of the molecular target for the suppression of contact hypersensitivity by ultraviolet radiation. *J Exp Med* 170:1117–1131.

Bito, T., M. Ueda, T. Nagano, S. Fujii, and M. Ichihashi. 1995. Reduction of ultraviolet-induced skin cancer in mice by topical application of DNA excision repair enzymes. *Photodermatol Photoimmunol Photomed* 11:9–13.

Cafardi, J. A., and C. A. Elmets. 2008. T4 endonuclease V: Review and application to dermatology. *Expert Opin Biol Ther* 8:829–838.

Ceccoli, J., N. Rosales, J. Tsimis, and D. B. Yarosh. 1989. Encapsulation of the UV-DNA repair enzyme T4 endonuclease V in liposomes and delivery to human cells. *J Invest Dermatol* 93:190–194.

DeBoyes, T., D. Kouba, D. Ozog et al. 2010. Reduced number of actinic keratoses with topical application of DNA repair enzyme creams. *J Drugs Dermatol* 9:1519–1521.

Gilchrest, B. A., S. Zhai, M. S. Eller, D. B. Yarosh, and M. Yaar. 1993. Treatment of human melanocytes and S91 melanoma cells with the DNA repair enzyme T4 endonuclease V enhances melanogenesis after ultraviolet irradiation. *J Invest Dermatol* 101:666–672.

Grossman, L., and Q. Wei. 1995. DNA repair and epidemiology of basal cell carcinoma. *Clin Chem* 41:1854–1863.

Gruber-Wackernagel, A., S. N. Byrne, and P. Wolf. 2009. Pathogenic mechanisms of polymorphic light eruption. *Front Biosci (Elite Ed)* 1:341–354.

Halliday, G. M., S. N. Byrne, J. M. Kuchel, T. S. Poon, and R. S. Barnetson. 2004. The suppression of immunity by ultraviolet radiation: UVA, nitric oxide and DNA damage. *Photochem Photobiol Sci* 3:736–740.

Hofer, A., F. J. Legat, A. Gruber-Wackernagel, F. Quehenberger, and P. Wolf. 2011. Topical liposomal DNA-repair enzymes in polymorphic light eruption. *Photochem Photobiol Sci* 10:1118–1128.

Ke, M. S., M. M. Camouse, F. R. Swain et al. 2008. UV protective effects of DNA repair enzymes and RNA lotion. *Photochem Photobiol* 84:180–184.

Koulu, L. M., J. K. Laihia, H. H. Peltoniemi, and C. T. Jansen. 2010. UV-induced tolerance to a contact allergen is impaired in polymorphic light eruption. *J Invest Dermatol* 130:2578–2582.

Kraemer, K. H., M. M. Lee, A. D. Andrews, and W. C. Lambert. 1994. The role of sunlight and DNA repair in melanoma and nonmelanoma skin cancer. The xeroderma pigmentosum paradigm. *Arch Dermatol* 130:1018–1021.

Kripke, M. L., P. A. Cox, L. G. Alas, and D. B. Yarosh. 1992. Pyrimidine dimers in DNA initiate systemic immunosuppression in UV-irradiated mice. *Proc Natl Acad Sci U S A* 89:7516–7520.

Kuchel, J. M., R. S. Barnetson, and G. M. Halliday. 2005. Cyclobutane pyrimidine dimer formation is a molecular trigger for solar-simulated ultraviolet radiation-induced suppression of memory immunity in humans. *Photochem Photobiol Sci* 4:577–582.

Lachenbruch, P., L. Marzella, W. Schwieterman, K. Weiss, and J. Siegel. 2001. Poisson distribution to assess actinic keratoses in xeroderma pigmentosum. *Lancet* 358:925.

Lambert, B., U. Ringborg, and G. Swanbeck. 1976. Ultraviolet-induced DNA repair synthesis in lymphocytes from patients with actinic keratosis. *J Invest Dermatol* 67:594–598.

Lucas, C. R., M. S. Ke, M. S. Matsui et al. 2008. Immune protective effect of a moisturizer with DNA repair ingredients. *J Cosmet Dermatol* 7:132–135.

Munch-Petersen, B., G. Frentz, B. Squire et al. 1985. Abnormal lymphocyte response to U.V. radiation in multiple skin cancer. *Carcinogenesis* 6:843–845.

Nishigori, C., D. B. Yarosh, S. E. Ullrich et al. 1996. Evidence that DNA damage triggers interleukin 10 cytokine production in UV-irradiated murine keratinocytes. *Proc Natl Acad Sci U S A* 93:10354–10359.

Norris, P. G., G. A. Limb, A. S. Hamblin et al. 1990. Immune function, mutant frequency, and cancer risk in the DNA repair defective genodermatoses xeroderma pigmentosum, Cockayne's syndrome, and trichothiodystrophy. *J Invest Dermatol* 4:94–100.

Palmer, R. A., and P. S. Friedmann. 2004. Ultraviolet radiation causes less immunosuppression in patients with polymorphic light eruption than in controls. *J Invest Dermatol* 122:291–294.

Radnay, E., L. Naumovski, J. Love et al. 1984. Physical mapping and complete nucleotide sequence of denV gene of bacteriophage T4. *J Virol* 52:846–856.

Sbano, E., L. Andreassi, M. Fimiani, A. Valentino, and R. Baiocchi. 1978. DNA-repair after UV-irradiation in skin fibroblasts from patients with actinic keratosis. *Arch Dermatol Res* 262:55–61.

Schornagel, I. J., V. Sigurdsson, E. H. Nijhuis, C. A. Bruijnzeel-Koomen, and E. F. Knol. 2004. Decreased neutrophil skin infiltration after UVB exposure in patients with polymorphous light eruption. *J Invest Dermatol* 123:202–206.

Stege, H., L. Roza, A. A. Vink et al. 2000. Enzyme plus light therapy to repair DNA damage in ultraviolet-B-irradiated human skin. *Proc Natl Acad Sci U S A* 97:1790–1795.

Thielmann, H. W., L. Edler, M. R. Burkhardt, and E. G. Jung. 1987. DNA repair synthesis in fibroblast strains from patients with actinic keratosis, squamous cell carcinoma, basal cell carcinoma, or malignant melanoma after treatment with ultraviolet light, N-acetoxy-2-acetyl-aminofluorene, methyl methanesulfonate, and N-methyl-N-nitrosourea. *J Cancer Res Clin Oncol* 113:171–186.

van de Pas, C. B., D. A. Kelly, P. T. Seed et al. 2004. Ultraviolet-radiation-induced erythema and suppression of contact hypersensitivity responses in patients with polymorphic light eruption. *J Invest Dermatol* 122:295–299.

Vink, A. A., A. M. Moodycliffe, V. Shreedhar et al. 1997. The inhibition of antigen-presenting activity of dendritic cells resulting from UV irradiation of murine skin is restored by in vitro photorepair of cyclobutane pyrimidine dimers. *Proc Natl Acad Sci U S A* 94:5255–5260.

Vink, A. A., F. M. Strickland, C. Bucana et al. 1996. Localization of DNA damage and its role in altered antigen-presenting cell function in ultraviolet-irradiated mice. *J Exp Med* 183:1491–1500.

Wang, L. E., P. Xiong, S. S. Strom et al. 2005. In vitro sensitivity to ultraviolet B light and skin cancer risk: A case-control analysis. *J Natl Cancer Inst* 97:1822–1831.

Wei, Q., G. M. Matanoski, E. R. Farmer, M. A. Hedayati, and L. Grossman. 1994. DNA repair and susceptibility to basal cell carcinoma: a case-control study. *Am J Epidemiol* 140:598–607.

Wei, Q., G. M. Matanoski, E. R. Farmer, M. A. Hedayati, and L. Grossman. 1995. DNA repair capacity for ultraviolet light-induced damage is reduced in peripheral lymphocytes from patients with basal cell carcinoma. *J Invest Dermatol* 104:933–936.

Wolf, P., S. N. Byrne, and A. Gruber-Wackernagel. 2009. New insights into the mechanisms of polymorphic light eruption: resistance to ultraviolet radiation-induced immune suppression as an aetiological factor. *Exp Dermatol* 18:350–356.

Wolf, P., P. Cox, D. B. Yarosh, and M. L. Kripke. 1995. Sunscreens and T4N5 liposomes differ in their ability to protect against ultraviolet-induced sunburn cell formation, alterations of dendritic epidermal cells, and local suppression of contact hypersensitivity. *J Invest Dermatol* 104:287–292.

Wolf, P., H. Maier, R. R. Mullegger et al. 2000. Topical treatment with liposomes containing T4 endonuclease V protects human skin in vivo from ultraviolet-induced upregulation of interleukin-10 and tumor necrosis factor-alpha. *J Invest Dermatol* 114:149–156.

Wolf, P., D. B. Yarosh, and M. L. Kripke. 1993. Effects of sunscreens and a DNA excision repair enzyme on ultraviolet radiation-induced inflammation, immune suppression, and cyclobutane pyrimidine dimer formation in mice. *J Invest Dermatol* 101:523–527.

Yarosh, D. for the XP study group. 2001. Poisson distribution to assess actinic keratoses in xeroderma pigmentosum. *Lancet* 358:925.

Yarosh, D., L. G. Alas, V. Yee et al. 1992. Pyrimidine dimer removal enhanced by DNA repair liposomes reduces the incidence of UV skin cancer in mice. *Cancer Res* 52:4227–4231.

Yarosh, D., C. Bucana, P. Cox et al. 1994. Localization of liposomes containing a DNA repair enzyme in murine skin. *J Invest Dermatol* 103:461–468.

Yarosh, D., J. Kibitel, and A. O'Connor. 1997. DNA repair liposomes in antimutagenesis. *J Environ Pathol Toxicol Oncol* 16:287–292.

Yarosh, D., J. Klein, J. Kibitel et al. 1996. Enzyme therapy of xeroderma pigmentosum: Safety and efficacy testing of T4N5 liposome lotion containing a prokaryotic DNA repair enzyme. *Photodermatol Photoimmunol Photomed* 12:122–130.

Yarosh, D., J. Klein, A. O'Connor et al. 2001. Effect of topically applied T4 endonuclease V in liposomes on skin cancer in xeroderma pigmentosum: a randomised study. Xeroderma Pigmentosum Study Group. *Lancet* 357: 926–929.

Yarosh, D. B., J. T. Kibitel, L. A. Green, and A. Spinowitz. 1991. Enhanced unscheduled DNA synthesis in UV-irradiated human skin explants treated with T4N5 liposomes. *J Invest Dermatol* 97:147–150.

Yarosh, D. B., A. O'Connor, L. Alas, C. Potten, and P. Wolf. 1999. Photoprotection by topical DNA repair enzymes: molecular correlates of clinical studies. *Photochem Photobiol* 69:136–140.

Zahid, S., and I. Brownell. 2008. Repairing DNA damage in xeroderma pigmentosum: T4N5 lotion and gene therapy. *J Drugs Dermatol* 7:405–408.

Zardeneta, G., and P. M. Horowitz. 1994. Detergent, liposome, and micelle-assisted protein refolding. *Anal Biochem* 223:1–6.

Climate Change and Ultraviolet Radiation Exposure

Mohammad Ilyas
University Malaysia Perlis

16.1 Introduction

Solar radiation is the basic thermal energy source for the planetary surface. A small portion of the incoming solar flux lies in the short-wave ultraviolet (UV) radiation region [ultraviolet B (UVB)]. Exposure to UVB is harmful to humans in several ways, including sunburn and skin cancer (SC) as well as effects to the eye and immune system. Most of the incoming UVB radiation is absorbed by the stratospheric ozone layer, thereby providing an important protective shield by blocking most of the UVB radiation from reaching the Earth's surface.

Although ozone is dangerously toxic and fatal when inhaled in large quantities by humans, we owe our existence in part to ozone because of its presence in the atmosphere. This is a result of the strong absorption of ozone in the UV band, thereby shielding the living organisms at the Earth's surface from the very harmful solar radiation between 200 and 320 nm by absorbing most of it before it reaches the Earth's surface.

Atmospheric ozone, closely linked to the evolution of atmospheric oxygen (Ilyas 1975; Blake and Carver 1977), was discovered in the mid-19th century (Schonbein 1840; Barigney 1885). Subsequently, it was recognized that atmospheric ozone, in particular the stratospheric ozone layer, has an important protective role against the incoming solar UV radiation through the photoabsorption process, and measurements of atmospheric ozone followed (Dobson and Harrison 1926). This further helped in a greater understanding of ozone's climatic importance (Chapman

1930; Ohring and Muench 1960), and basic UV–ozone photointeractions were explored (Dutsch 1971).

16.2 Solar UV Radiation, the Ozone Hole, and Human Health

The issue of solar UV radiation climatology in the context of human health through atmospheric ozone modulation came into scientific prominence approximately 40 years ago when it was recognized that inadvertent human activities (e.g., projected supersonic aircraft travel, extensive use of chlorofluorocarbons (CFCs) and halons, and nuclear explosions) could lead to ozone depletion, which later came to be popularly known as the Antarctic Ozone Hole (Crutzen 1970; Johnston 1971; Whitten, Borucki, and Turco 1975; Rowland and Molina 1975). Ozone depletion's potential significance at lower latitudes was also duly recognized (Ilyas 1981, 1986, 1989a, 1991). Tracer experiments were initiated, and climatic effects were considered (Lovelock 1971; Dickinson 1974). Subsequently, intensive measurements of UV radiation and modeling were undertaken (Mo and Green 1974; Ilyas and Apandi 1979; Ilyas and Barton 1983; Ilyas 1993), including some in relation to SC (Scotto, Fears, and Gori 1976; Robertson 1969). These were followed by intensive studies involving measurements, modeling, effects, and other related aspects worldwide. In consideration of the potential danger to the tropics, a rare series of integrated measurements by Ilyas and colleagues was initiated with simultaneous measurements of vertical ozone (soundings), UVB, UVA, surface ozone, and climatic parameters, including cloud cover in Malaysia covering a period of 25 years (Ilyas 1994). Based on these quantitative measurements, it was demonstrated that the UVB climatology in the context of adverse human health effects, including tropical diseases, is of particular significance in the tropics because of high UVB radiation dosages coupled with hot and humid climate (Ilyas 1986, 1989b,c). Indeed, a case for important policy considerations was made on a global scale (Ilyas 1989a, 1997).

In consideration of the wide-ranging potential adverse effects from ozone depletion, intensive efforts have been made to reverse the process through the Montreal Protocol, resulting in a gradual phase-out of most ozone-depleting gases. Because of the success of gradual ozone recovery, ever-decreasing amounts of extra UVB from ozone depletion are anticipated in the coming decades (Prestre, Reid, and Morehouse 1998).

However, all is not well with the tropics. By putting together input from the previously mentioned equatorial–tropical measurements, including ozone, climate, cloud cover, UV radiation, and so forth, it was recently demonstrated that in the tropics and subtropics, we are faced with a "climate ozone hole" scenario (Ilyas 2004, 2007). This is because, in the tropics, prevailing high temperatures and humidity contribute additively to UV-induced health effects, which we find to be more serious than we thought previously. As a result, the effect of future climate change will further contribute to radiative damage on human health in a similar way as ozone depletion-induced UVB would have: deepening the climate ozone hole. A detailed account of this issue is the focus of this chapter.

16.3 UV, Ozone Photochemistry, and Climate

Although it is customary to speak of an ozone layer, the concentration of atmospheric ozone, in fact, never exceeds a few parts per million (Figure 16.1); the average total amount of ozone in the vertical column corresponds to 0.3–0.5 mm of pure ozone at standard temperature and pressure (STP) (the comparable values for total air are 8 km; for nitrogen [N_2], 1.68 km; for oxygen [O_2], 6.25 km; and for carbon dioxide [CO_2], 2.6 m).

Although present at all altitudes up to the mesosphere, almost all the ozone is in the stratosphere (Figure 16.2). Variability of the ozone column (at STP) with time and latitude is shown in Figure 16.3. Formed principally above 25 km by photochemical processes, it is carried downward to the lower stratosphere (15–25 km) and troposphere (0–15 km) by mixing processes.

Ozone is produced mainly in the upper stratosphere through the photodissociation of oxygen by UV radiation from the sun. The radiation must have a wavelength shorter than 242 nm to have sufficient energy to break the oxygen bond. Ozone itself also strongly absorbs UV radiation and can itself be photodissociated. Hence, a photochemical equilibrium is established between the production and loss mechanisms that depend on the particular region of the atmosphere being irradiated. The chemical equations describing this process, the Chapman reactions, are given below:

Formation

$$O_2 + h\nu \ (\lambda < 242 \text{ nm}) \rightarrow O + O_2 \tag{16.1}$$

$$O + O_2 + M \rightarrow O_3 + M \tag{16.2}$$

Destruction

$$O_3 + h\nu \rightarrow O_2 + O \tag{16.3}$$

$$O_3 + O \rightarrow 2O_2 \tag{16.4}$$

It can be seen that Equation 16.1 is limited by the amount of UV radiation available because there is an abundant supply

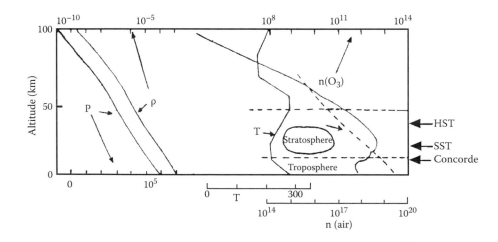

FIGURE 16.1 Vertical distribution of pressure (P), temperature (T, K), mass density, and so forth and approximate flight altitude for some proposed sonic aircraft.

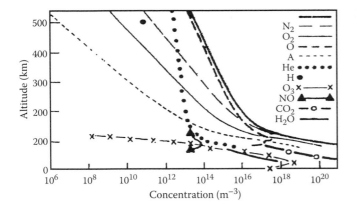

FIGURE 16.2 Distribution of various gaseous constituents in the upper atmosphere.

of oxygen, and Equation 16.2 favors reactions at higher pressure and lower altitudes because there are more molecules for three-body reactions. Hence, the production of ozone will vary with the radiation path throughout the atmosphere and the sun angle. There is, therefore, more ozone produced over tropical regions than over polar regions and also at altitudes in which there is both sufficient radiation at wavelengths < 242 nm and where there is also sufficient atmospheric density for three body collisions.

Beside photodissociation, the other significant sink for ozone is its reaction with any oxidizable material present in the atmosphere because it is such a strong oxidizing agent. Because of the turbulent mixing in the troposphere, any ozone penetrating into the troposphere has a relatively short lifetime, and hence, the concentration of ozone in the troposphere is much

lower than that in the stratosphere. Ozone is also removed by catalytic cycles in which a species X abstracts an oxygen atom from ozone and is then regenerated in a reaction with atomic oxygen:

$$X + O_3 \rightarrow XO + O_2 \tag{16.5}$$

$$XO + O \rightarrow X + O_2 \tag{16.6}$$

$$O + O_3 \rightarrow O_2 + O_2 \tag{16.5} + \text{(16.6)}$$

Such cycles have been recognized for X = H, HO, NO, Cl, and Br. They achieve the same result as Equation 16.4 but with relatively smaller activation energies.

Equations 16.1 and 16.3 together are thus responsible for the depletion of the solar UV radiation reaching the surface of vacuum UV (VUV) wavelengths over the 100 to 290 nm range.

The energy absorbed by ozone heats up the stratosphere, giving the temperature increase with height (inversion) responsible for this region's stability (Figure 16.4) and thus ozone's meteorological importance. Therefore, the vertical distribution of ozone as well as the integrated column depth is related to the circulation in the stratosphere and the weather phenomenon.

Although a minor constituent in the atmospheric gases and restricted to altitudes below approximately 80 km, it nevertheless plays an important meteorological and climatic role in the lower atmospheric phenomena because energy from the previously mentioned absorption process is the principal source of thermal energy in the stratosphere and is responsible for the increase of temperature with height in the upper stratosphere (Ilyas 1991).

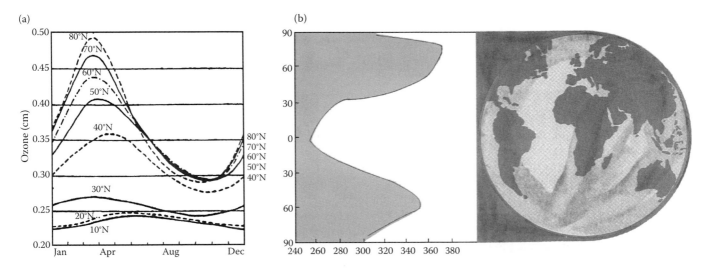

FIGURE 16.3 (a) Annual variation of total ozone for each 10° of geographic latitude. See the lowest levels of ozone at lower latitudes in the tropics and minimum variability, which result in maximum UVB radiation reaching the Earth's surface. (b) Annual variation of total ozone at different latitudes with clear indication of a minimum ozone column at the equatorial–tropical region, in which UV radiation is maximum both outside the atmosphere and at the surface (see Figure 16.7).

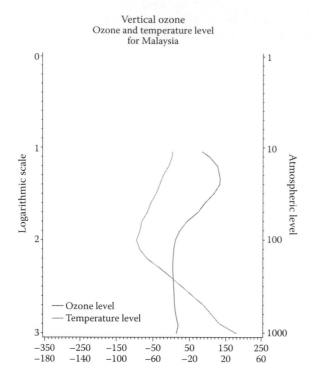

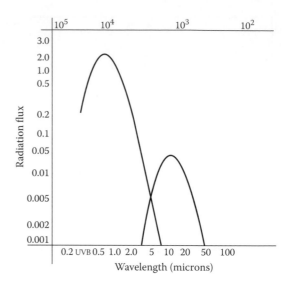

FIGURE 16.4 Average ozone layer and temperature vertical profiles near the equator (Malaysia). Notice the thermal impact of ozone absorption of UV radiation as discussed in the text.

FIGURE 16.5 Incoming solar radiation outside the atmosphere and at the Earth's surface after absorption of most short-wave radiation, including UV on the far left side. Also, the long-wave infrared and radio wave radiation outgoing from the Earth's surface is shown in the right-side envelope. The radiation band marked "UVB" is affected by the ozone layer in the lower stratosphere as discussed in the text.

16.4 Medically Important UVB and Ozone Depletion

Although ozone acts as a very strong shield for UV radiation, some harmful UV radiation in the 280–320 nm band (denoted by UVB) still reaches the Earth's surface (Figure 16.5). This causes injuries to plants and animals, but more importantly, it causes sunburn, advanced aging of the skin, and SC among humans; the UV flux also enhances photochemical air pollution (Urbach 1969; Prestre, Reid, and Morehouse 1998).

In general, there is a natural balance between ozone's production and its destruction, with a net ozone surplus sufficient to absorb most of the incoming solar UVB radiation transporting to the surface.

The solar UVB dosage reaching the surface varies with time (seasonally) and with latitude, increasing from the poles to the tropics with an equatorial maximum, primarily because of the variability of the vertical ozone content with season and latitude (Ilyas 1991).

Under the prevailing natural equilibrium conditions, there are only small dosages of UV radiation reaching the Earth's surface except in the equatorial–tropical region and the highland regions in the subtropics. Any significant ozone depletion would cause an increase in UV radiation, resulting in more adverse radiation effects to people, including SC, cataracts, and immune system deficiencies, besides affecting living organisms and the aquatic system.

For several years, we have been studying the dangers that some human activities may pose to the ozone layer. We may mention that, way back, one of the major concerns about stratospheric pollution centered around the introduction of supersonic aircraft, especially the proposed American supersonic transports (SSTs). This concern was because the SSTs can inject exhaust emissions of water vapor and, most importantly, NO*x* directly into the stratosphere, which would result in the depletion of the ozone through the NO*x* cycle as shown previously.

Fluorocarbons received much more attention than the SSTs in relation to the ozone layer. It was first pointed out by Rowland and Molina (1975) that, as no significant destruction processes for these compounds on land, in the oceans, or in the troposphere exist, these compounds would be transported upward to the stratosphere at altitudes above 25 km. Therefore, they would be decomposed by far-UV light (180–220 nm) or by reaction with singlet oxygen to yield chlorine atoms, which can destroy ozone.

The UV radiation may also enhance the production of reactive chemicals, including ozone at the surface level. Being very reactive and toxic, ozone at the surface level is harmful to humans, plants and vegetation, and animals and damaging to certain materials including rubber, timber, and paints. Near the surface, ozone also acts as a greenhouse gas and may contribute to the Earth's greenhouse warming and climate change.

As mentioned, under the prevailing natural equilibrium condition, there are only small dosages of UV radiation reaching the surface to which populations, living organisms, plants, and aquatic systems have generally become adapted. This, however, is not true for the tropics and the equatorial belt. This is clear from the seasonal distribution of vertical ozone columns for different latitudes (Figure 16.6). At the lower latitudes, not only is the ozone column thickness significantly smaller, but it also does not vary much seasonally. The solar UV radiation penetration is thus maximum at Earth's lower latitudes throughout the year. The overall effect of the low ozone content coupled with smaller seasonal changes in solar declination results in the latitudinal distribution of annual erythemal dosage (incoming radiation weighted according to skin erythemal action spectrum) (Mo and Green 1974) as shown in Figure 16.7. This diagram illustrates the manifold increase in the damaging UVB dosage from high latitudes to the equator.

The industrial injection of chemicals, such as CO_2, CH_4 (methane), and ozone-depleting CFCs, into the atmosphere also contributes to the greenhouse effect. Global warming through CFCs, CO_2, methane, and other greenhouse gases, if sufficiently realized, can contribute to a significant rise in the sea level. Recent concerns about climate change and sea level rise-related issues have come under closer scrutiny through the Intergovernmental Panel on Climate Change (IPCC).

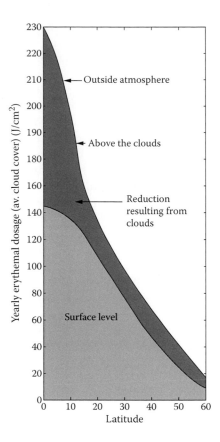

FIGURE 16.7 Yearly incoming erythemal solar ultraviolet dosage for clear sky (upper curve) and at surface level (lower curve) after accounting for modulation resulting from average cloud cover at different latitudes (see Figure 16.6). The large reduction in the radiation received at the equatorial–tropical latitude region is a result of relatively high cloud cover experienced in the region. The net surface-level radiation is responsible for skin cancer and related medical effects, such as immune suppression, as discussed in the paper. The radiation effect is further augmented by the large climate parameters in the tropics (see discussion).

16.5 Climate and Human Health Effects

Initial indications that climate elements play an important role in human health and well-being came from physiological studies on thermal comfort. From these, we recognized a temperature "comfort zone" for optimum environmental conditions, and as a result, we try to regulate environmental thermal conditions through cooling and heating in order to achieve thermal comfort for the human body (McGregor and Loh 1941). Environmental humidity also plays a discomforting role in a physiological sense for the human body (Ellis 1953). For instance, an increase in humidity can be translated into an equivalent temperature increase; in other words, the greater the air humidity level, the lower the temperature we need to maintain the same level of physiological body comfort. In this sense, an increase in temperature is equivalent to an increase in humidity and vice versa. This temperature–humidity combination is particularly potent for human health in

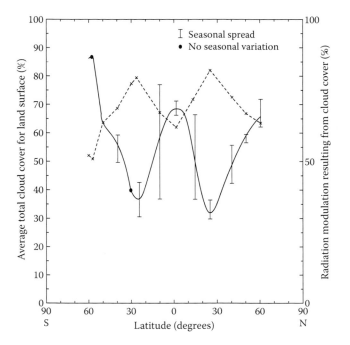

FIGURE 16.6 Yearly average total cloud cover for global land surface as a function of latitude (left scale) and radiation modulation effect resulting from cloud cover represented as percentage of incoming radiation with no cloud (dotted curve). The bars on the cloud cover line reflect seasonal spread in the data. High cloud cover in the equatorial region leads to large radiation modulation (reduction) (see Figure 16.7).

the humid and hot tropics and contributes adversely to the aging of human skin and a general decline in human health. Primarily for this reason, the British are reported to have introduced the early retirement age of 55 years in Malaysian civil service, presuming that, as a result of high air temperature and humidity conditions to which people were exposed over a long period of time, they would have done all the productive work by this age and should retire for a well-deserved rest (Ilyas 2007).

In a series of more direct experiments involving animals, it was shown clearly that both temperature and humidity contribute to SC in a way similar to the UVB effect, which can be quantified. These studies also provide us with a confirmation of the general physiological impact on the thermal comfort considerations and enable us to evaluate climatic impact at cooler temperatures in which experimental results are not applicable.

In this chapter, we discuss the climatic impact on UVB radiation dosage under prevailing conditions in the tropics and the impact of potential climate change with reference to human SC (UV carcinogenesis), in particular, to the nonmelanoma skin cancer (NMSC) problem and other damaging effects. This would help us in future evaluations of the danger that climate change can pose in the lower-latitude region directly and through climate augmentation of UVB irrespective of the ozone recovery.

16.6 Climate Ozone Hole and UVB Radiation Dose in the Tropics

In present times, no effort seems to have been made to understand the augmentative temperature effect on UVB radiation for real environmental conditions in a practical way. Bain, Rusch, and Kline (1943) and Freeman and Knox (1964) studied the temperature impact on UVB dose damage (DD) using animals. They showed that the effect of temperature on the UVB radiation dosage for SC (UV carcinogenesis) is positive and contributes to greater damage for the same radiation dosage under a higher-temperature condition above approximately 23°C. This leads to an increase in the DD, which we can define as effective dose damage (EDD). This temperature effect on the damage is not strictly linear and varied between +3% and +7% for each degree Celsius of temperature increase. We adopt a value of +5% per 1°C temperature as a reasonable estimate, and the results for animals could be applied to humans as a first approximation (van der Leun and de Gruijl 2002).

The bulk of the UVB radiation at the surface is received within a couple of hours of solar noon. Hence, the best indicator for the temperature condition prevailing during this time would be the daily average maximum temperature. Valuable data on thermal and humidity conditions at numerous places using many years of measurements are available in Pearce and Smith (1984). For our reference, we consider a surface-level location in the equatorial zone as the radiation input is nearly constant for a wide latitude zone. Typically, we find (Pearce and Smith 1984) that annual mean daily average maximum temperatures in the equatorial–tropical region are approximately 30°C–32°C or close

to 10°C higher than the threshold for positive temperature effect (Freeman and Knox 1964). Later, we will see that in our comparison with higher latitudes, even a smaller temperature difference would lead to considerable augmentation in the tropics. There would be some months with greater temperature differential and others with smaller values, but in the tropical zone, the daytime maximum temperature would be effectively above the threshold with positive augmentation. So we may say that the temperature enhancement to EDD for SC is an increase of approximately +50% (5% per degree Celsius × 10°C).

Experimental work pertaining to the humidity effect on UV carcinogenesis involving mice was undertaken by Owen and Knox (1978) and clearly showed the positive effect of humidity in augmenting the UVB injury; the greater the humidity, the more the damage. The quantitative evaluation of DD under humidity was done using albino rabbits. The measurements showed up to a 40% reduction in the energy requirement for erythema production, with a mean value of 33%. Assuming a linear relationship between damage and humidity over the range of 100% to 0% humidity values, we estimate an average effect of +4% for a 10% humidity change. Humidity values in the tropics around noontime are in the range of 70%–80% (Pearce and Smith 1984), and a linear relationship is not particularly critical as we are close to one end of the measurement range. For the humid (highly populated) tropical region, taking an annual mean daily average humidity value around midday (minimum value for the daytime) as 70%, we estimate the humidity augmentation of +30%, and again assuming that the effects in mice and rabbits can be translated to humans, we can translate this into a damage dose for human skin following this approach as a first approximation as explained previously (Ilyas 2007). Another way to evaluate the effect of humidity in a way that is more relevant to humans is to utilize the approach of the equivalent temperature for the same physiological comfort. Even though the physiological effect relationship (McGreger and Loh 1941; Ellis 1953) to erythema is not clear, we should get some indication of how this approach compares with the experimental work mentioned previously. It has the advantage that we have an intermediate data point for better interpolation. In this way, we utilize the fact that the greater the humidity level we have, the lower the temperature we need to provide comfort equilibrium for the human body. In other words, an increase in humidity is equivalent to an increase in temperature. The UVB DD may be assumed to be affected by humidity in a positive way, as is temperature as well, as mentioned previously. Based on a standard thermal comfort approach for the human body (Ilyas, Pang, and Chan 1981), we can translate a particular humidity level to an equivalent temperature to provide the same physiological comfort level under a typical wind condition (10 cm/s) as follows:

Humidity (%)	Temperature (°C)
0	38
42	32
100	27

For our discussion, for the lower tropical region, we take a typical humidity level of approximately 70%, requiring a 30°C temperature for the same comfort. In other words, a 70% humidity level (compared to 0%) amounts to an effective temperature increase by +8°C, which is equivalent to an effective UVB damage enhancement for SC or EDD increase by 40% (5% per degree Celsius × 8°C). It is indeed interesting to note that this leads us to a similar value as the one deduced earlier based on animal measurements, and therefore, our assumption that results can be translated from animal studies to humans is reasonable.

UVA radiation, by itself, does not seriously contribute to skin carcinogenesis, but in the presence of UVB, it enhances the damaging effect as well (Diffey 1982). In the tropics, we receive large amounts of UVB and UVA radiation (Ilyas, Pandy, and Jaafar 2001; Ilyas, Aziz, and Tajuddin 1988), and as a result, we estimate a small additive UVA contribution to EDD by +10% in the presence of high UVB radiation. Therefore, we can combine the previous three factors for SC damage enhancement for the tropics:

Total EDD enhancement (50% + 40% + 10%) = +100%

We may say that the effect of prevailing climate in the tropics, resulting in the doubling of the effective dose, is equivalent to a 50% ozone depletion (taking a typical ozone depletion-to-UVB amplification factor of 2). We may refer to this as a permanent climate ozone hole of very significant magnitude, leading to enhanced UVB radiative adverse effects for human health.

16.7 Enhanced UVB Radiation Damage Dose in the Tropics

Based on direct measurements of UVB radiation both outside the atmosphere and at the surface level and incorporating the global cloud cover effect (Ilyas 1987) at the surface (Figure 16.6), we estimate that we receive DD (Figures 16.7 and 16.8) as follows (dose units in J/cm^2) (Ilyas 1991):

| Midlatitudes | 27 |
| Tropics | 145 |

We find that the damaging UVB dosage received in the tropics (EDD) is close to 5.4 times that of the midlatitudes (i.e., more than approximately 500%).

In evaluating the overall EDD in the tropics, we can now incorporate the +100% (or a factor of 2) increase resulting from an augmentative effect of temperature, humidity, and UVA. The basic EDD in the tropics would be therefore enhanced from 500% (145 units) to effectively 1000% (500 × 2, i.e., 10 times) relative to the midlatitudes as humidity and temperature contributions at midlatitudes will be not significant, as explained in the following paragraphs.

It is widely understood that, generally, human populations in the tropics have developed a darker skin tan, which provides a

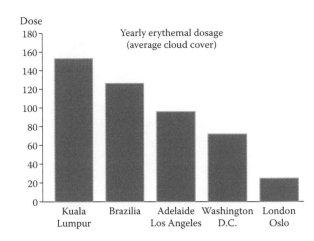

FIGURE 16.8 The surface-level erythemal ultraviolet radiation for average cloud cover (based on data in Figure 16.7) is compared for several places along the latitude belt to illustrate the high level of input received at the equatorial–tropical region (represented by Malaysia and Kuala Lumpur).

greater measure of protection against the previously mentioned intense UV radiation damage. While this (positive role) may well be true, the great disparity in the input radiation dosage between tropics and midlatitudes is generally not taken into consideration in such evaluations (Ilyas 1989a, 1991). Typically, for similar radiation conditions, the damage efficiency dose (DE) (i.e., photon damage effectiveness) varies between 100 for white skin, 10 for brown skin, and 1 for dark skin (Ilyas 2007). For large populations in the tropical regions, this factor could well be between 30 (darker populations in equatorial regions, such as Indonesia and Malaysia) and 70 (near-white populations in the Iran and Kashmir regions), especially if the damage factor is incident radiation dependent. So despite this negative demodulation resulting from skin pigmentation, the 10-times-more-effective dosages received in the tropics mentioned in the previous section (Figure 16.7) would still mean a great deal of damage to most populations in the tropical region; the damage dosages approach similar values (and perhaps even more in some cases) to those for light-skinned populations living in the midlatitudes.

To obtain an overall perspective on the relative degree of danger in the tropics, we calculate a total danger factor (TDF), that is, EDD × DE, by using a rather conservative value of 15 and a more realistic value of 30 for DE in the tropics under the present climatic conditions under a best-case scenario (not including future climate change):

TDF (tropics 1)   (290 × 15) = 4350 units

TDF (tropics 2)   (290 × 30) = 8700 units

In the best-case scenario (15), TDF is nearly twice that of a worst-case scenario at midlatitudes (see the following section).

16.8 Impact of Climate Change on Human Health at Low to High Latitudes

We first compare the tropical dose with a similar factor applicable to the higher-latitude region. The incoming radiation input at higher latitudes (say 60° latitude) is quite small (approximately 5 units) for any appreciable effect. Populations with light skin reside in midlatitudes and need to be considered. We find that places near 40° latitude, such as Akita, Cagliari, Napier, and Palma, experience annual mean temperatures and humidity of approximately 19°C and 65%, respectively, while those near 45° latitude, such as Christchurch and Venice, experience annual mean temperatures and humidity of approximately 17°C and 65%, respectively. These temperatures are generally below the threshold level of 22°C and thus would not contribute to temperature augmentation at these latitudes.

The studies by Owen and Knox (1978) were done at approximately 32°C, and we cannot use these results for cooler conditions. However, from the thermal comfort perspective, at a near 20°C environmental temperature, humidity does not contribute to physiological discomfort. Therefore, we could say that at midlatitudes, the temperature and humidity augmentation of the UVB dose will be insignificant. Even though we have significant UVA radiation input at the midlatitudes comparable to what we get at tropical latitudes, the additive effect for SC and immune deficiency effects may not be significant. This is because UVA has an augmentative effect in the presence of UVB input, and as the surface-level UVB flux at the midlatitudes is relatively rather small (27 units), the overall contribution from this factor is not expected to be large. In any case, we can compare the 27 units producing a TDF (using an upper range value for DE as 85 for a worst-case scenario):

$$\text{TDF (midlatitudes) } (27 \times 85) = 2295$$

We see that the TDF for the tropical region for a best-case scenario (see previous section) is far more significant compared to the midlatitudes for a worst-case scenario even if we were to include the greater effect resulting from UVA and so forth, as a result of high radiation input and large climate augmentation.

Among the various climatic change scenarios (IPCC 2006; Hegerl et al. 2006), there are wide variations on the amount of projected temperature increase.

However, a median plausible condition for temperature (T) and humidity (e) is the following:

(a) $\Delta T_{increase} \sim +4°C$
(b) $\Delta e_{increase}$ (70% → 90%) or an equivalent temperature effect of +2°C
(c) Hence, total effect ~ +6°C
     or ~ +30% (5% per degree Celsius × 6°C) increase to UVB radiation EDD

As a result of global warming, there would be an increase in clouds. We assume a cloud cover increase by 20% for the previous base consideration (resulting from a humidity increase from 70% to 90%); we estimate (Ilyas 1987; IPCC 2006) the decrease in UVB radiation reaching the surface to be approximately –11%. Therefore, we can say that the overall effect will be a significant +19% (30% – 11%) increase to EDD or UVB radiation SC DD.

16.9 Discussion

There are two key elements that have an exacerbating effect on the damage dosage in the tropics. First, we deal with large populations, which lag behind in education and awareness. They tend to have a carefree attitude with respect to taking protective action against exposure to UV radiation. Second, because of the widespread poverty, populations suffer from malnutrition as well as poor health and are increasingly susceptible to the UV radiation damage to their immune systems and associated diseases (e.g., herpes)—contributors to lower life expectancy. This is further enhanced as a result of the fact that considerable work-related daytime exposure to mostly high solar UV radiation (Figures 16.9 and 16.10) prevails in the tropical region with high temperature and humidity conditions, which would heat up darker skin more than whiter skin as a result of stronger visible radiation absorption as well, as discussed by De Fabo and Noonan (1983).

Given the previously mentioned information, we find that, overall, the potential for skin damage in the tropics is comparable to that in the midlatitudes. Unfortunately, there is very little

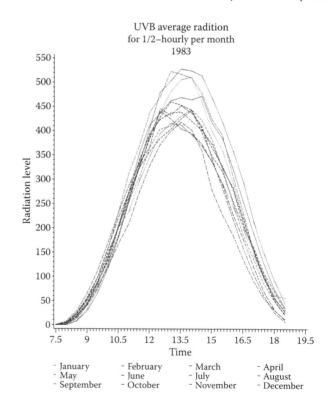

FIGURE 16.9 The diurnal UVB radiation variation at different times of year (seasons) at the equatorial location of Penang. We notice that the radiation input dose practically remains approximately the same and rather high throughout the year unlike at the high latitudes, where the radiation dose is only high in the summer and the quantity is small.

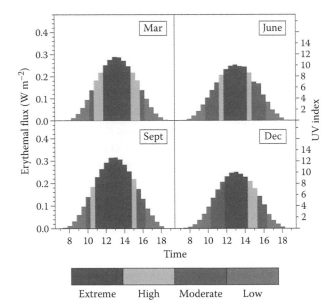

FIGURE 16.10 Measured data on UVB at near-equatorial location (Penang, Malaysia) for clear weather shows that the radiation flux remains between extreme and high levels on the UV index system through most of the day practically homogeneously during much of the year, indicating great potential for erythemal and related damaging health effects through radiation and climatic augmentation, as discussed in the text. Given this high radiation level for clear sky conditions, we can understand that the use of average cloud-modulated UVB radiation data (see text) means that the calculated numbers for DD are indicative of the lower threshold; for clearer weather (refer to Figure 16.7), the values will be revised upward.

research work on collating the available information on the incidence of SC in the tropics. This could well be because nobody has had the opportunity to make any organized study because skin ailments (except acne) are treated in a general way. Also, it is our feeling that average life expectancy in the tropics is relatively lower and many people perhaps do not live to an age when the cumulative effect becomes more easily seen. Indeed, early death could well be contributed to by an impaired immune system and frequent infections, which commonly affect the tropical populations. It is therefore of interest if dermatologists and others, such as biologists, health authorities, and so forth, could undertake organized studies in compiling the data for these regions, and studies are also undertaken for immune deficiency research in this context. In addition, specific measurements of UVB radiation in the tropics are sparse (SAP 2007), and more effort is needed for wider coverage. Just for the record, we have highlighted the special concern we have pertaining to high UVB radiation-related health effects at several scientific and policy forums (Ilyas 1989b; 1997).

16.10 Conclusions

Human activities resulting in the ozone layer depletion, such as the release of CFCs into the atmosphere, are being effectively controlled through the Montreal Protocol. As a result, the recovery process for the ozone layer is making progress, thereby restricting any UVB increase at the surface resulting from inadvertent activities. However, it is not generally recognized that, under hot and humid environmental conditions, the temperature and humidity can have a significant additive contribution to the damaging UVB effects experienced at present and that would prevail during and even after the ozone recovery phase. Additionally, future climate change may further exacerbate the cumulative contribution to the UV radiation damage dosage on humans. This effect will be particularly serious in the tropical–equatorial region, in which existing UVB levels are higher than those at midlatitudes and higher latitudes and protective mechanisms against sunlight are far less effective resulting from outdoor activities undertaken at wrong times (Figure 16.10), for example, physical education lessons in schools in the afternoon. This is important for human health in most populations in the tropics and should be taken into account when evaluating the UVB and climate impact on immune deficiency, viral infections, and overall public health considerations in the context of present and future climate scenarios. Besides, in the tropical environment, high levels of temperature and humidity are directly detrimental to human health.

Acknowledgments

Several persons have helped in the preparation of this chapter. In particular, I wish to record my appreciation for technical support by Anjung UniMAP Kulim, University Malaysia Perlis, and Ms. Norernahezreen Hamzah.

References

Bain, J., H. Rusch, and B. Kline. 1943. The effect of temperature upon ultraviolet carcinogenesis with wavelengths 2800–3400A. *Cancer Res* 3:610–612.

Berigny, A. 1855. Observations faites a l'observatoire meteorologique de Versailles avec le papierditozonemetrique de M. Schonbein (de Bale) pendant le moisd'aout1855 a 6 heures du matin, midi, 6 heures du soir et minui. *Crit Rev* 41:426.

Blake, A., and J. Carver. 1977. The evolutionary role of atmospheric ozone. *J Atmos Sci* 34:720.

Chapman, S. 1930. A theory of upper atmospheric ozone. *J R Meteorological Soc* 3:103.

Crutzen, P. 1970. The influence of nitrogen oxides on the atmospheric ozone concept. *Q J R Meteorological Soc* 96:320–325.

De Fabo, E., and F. Noonan. 1983. Mechanism of immune suppression by ultraviolet irradiation in vivo I: Evidence for the existence of a unique photoreceptor in skin and its role in photoimmunology. *J Exp Med* 157:84–98.

Dickinson, R. 1974. Climatic effects of stratospheric chemistry. *Can J Chem* 52:1616–1624.

Diffey, B. 1982. *Ultraviolet Radiation in Medicine*. Adam Hilger, Bristol.

Dobson, G., and D. Harrison. 1926. Measurements of the amount of ozone in the Earth's atmosphere and its relation to other geophysical conditions. *Proc R Soc Lond* 110:660.

Dutsch, H. 1971. Photochemistry of atmospheric ozone. *Adv Geophys* 15:219–322.

Ellis, E. 1953. Thermal comfort in warm and humid atmospheres. *J Hyg* 51:386.

Freeman, R., and J. Knox. 1964. Influence of temperature on ultraviolet injury. *Arch Dermatol* 89:858–864.

Hegerl, G., T. Crowley, W. Hyde, and D. Frame. 2006. Climate sensitivity constrained by temperature reconstructions over the past seven centuries. *Nature* 440:1029–1032.

Ilyas, M. 1975. The evolution of atmospheric oxygen. *Sci Rep* 12:251–256.

Ilyas, M. 1981. Adverse biological and climatic effects of ozone layer depleting activities—SSTs, aerosol cans, nitrogen fertilizers: An overview in Malaysian context. *Sains Malays* 8:13–37.

Ilyas, M. 1986. Ozone modification: Importance for countries in tropical/equatorial region. In *Effects of Changes in Stratospheric Ozone and Global Climate*, vol. 2. Environmental Protection Agency, Washington, DC.

Ilyas, M. 1987. Effect of cloudiness on solar ultraviolet radiation reaching the surface. *Atmos Environ* 21:1483–1484.

Ilyas, M. 1989a. Potential effects of ozone depletion with special reference to the developing countries. Keynote paper presented at the Ministerial London Conference on Saving the Ozone Layer, London, March 5–7, 1989.

Ilyas, M. 1989b. Atmospheric ozone and solar ultraviolet radiation: The tropical scene. In *Ozone in the Atmosphere*, R. Bojkov and P. Fabian, editors. Deepak Publishing, Hampton, VA.

Ilyas, M. 1989c. The danger of ozone depletion to the tropics. *Search* 20:148–149.

Ilyas, M., editor. 1991. *Ozone Depletion: Implications for the Tropics*. United Nations Environment Programme, Nairobi, Kenya.

Ilyas, M. 1993. UV radiation in the tropics (1979–1989). In *Frontiers of Photobiology*, A. Shima et al. Elsevier Science, New York.

Ilyas, M. 1994. *Proceedings of the Quadrennial International Ozone Symposium*, Charlottesville. Deepak Publishing, Hampton, Virginia.

Ilyas, M. 1997. Issues in ecosystem effects of ozone depletion. *10th Anniversary Colloquium, Montreal Protocol*, Montreal, Canada, September 13, 1997.

Ilyas, M. 2004. Impact of climate change on ozone depletion and UV radiation effects in tropics. *Proceedings of the SPARC General Assembly 3*, Canada.

Ilyas, M. 2007. Climate augmentation of erythemal UV-B radiation dose damage in the tropics and global change. *Curr Sci* 91:1605.

Ilyas, M. and A. Apandi. 1979. *Medical Journal of Malaysia* 34:181–184.

Ilyas, M., D. Aziz, and M. Tajuddin. 1988. Medically important solar ultraviolet-A radiation measurements. *Int J Dermatol* 27:315–318.

Ilyas, M., and I. Barton. 1983. Surface dosage measurements of erythemal ultraviolet radiation near the equator. *Atmos Environ* 17:2069–2073.

Ilyas, M., A. Pandy, and M. Jaafar. 2001. Changes to the surface level solar ultraviolet-B radiation due to haze perturbation. *J Atmos Chem* 40:111–121.

Ilyas, M., C. Pang, and A. Chan. 1981. Effective temperature comfort indices for some Malaysian towns. *Singap J Trop Geogr* 2:27–31.

IPCC. 2006. *Impact of Climate Change (IPCC Fourth Assessment Report)*. Cambridge University Press, Cambridge.

Johnston, H. 1971. Reduction of stratospheric ozone by nitrogen oxide catalyst from supersonic transport exhaust. *Science* 173:517–522.

Lovelock, J. 1971. Atmospheric fluorine compounds as indicators of air movements. *Nature (Lond)* 230:379.

McGregor, R., and G. Loh. 1941. The influence of a tropical environment upon the basal metabolism, pulse rate and blood pressure in Europeans. *J Physiol* 99:496–509.

Mo, T., and A. Green. 1974. A climatology of solar erythema dose. *Photochem Photobiol* 20:483–496.

Ohring, G., and H. Muench. 1960. Relationships between ozone and meteorological parameters in the lower stratosphere. *J Meteorol* 17:195–206.

Owen, D., and J. Knox. 1978. Influence of heat, wind, and humidity on ultraviolet radiation injury. In *International Conference on Ultraviolet Carcinogenesis, National Cancer Institute Monograph 50*, M. Kripke and E. Sass, editors. U.S. Department of Health, Education and Welfare Publication no. (NIH) 78-1532.

Pearce, E., and C. Smith. 1984. *The Hutchinson World Weather Guide*. Hutchinson, London.

Prestre, P., J. Reid, and E. Morehouse Jr., editors. 1998. *Protecting the Ozone Layer: Lessons, Models, and Prospects*. Kluwer Academic, Dordrecht, The Netherlands.

Robertson, D. 1969. Correlation of observed ultraviolet exposure and skin cancer, incidence in the populations in Queensland and New Guinea. In *The Biological Effects of UV Radiation*, F. Urbach, editor. Pergamon Press, Oxford.

Rowland, F., and M. Molina. 1975. Chlorofluoromethanes in the environment. *Rev Geophys Space Phys* 13:1–36.

SAP. 2007. Surface ultraviolet radiation. In *Scientific Assessment of Ozone Depletion*. WMO/UNEP, Geneva, Switzerland.

Schonbein, C. 1840. Recherchessur la nature de l'odeur qui se manifestedanscer-taines actions chimiques. *Crit Rev* 10:706.

Scotto, J., T. Fears, and G. Gori, editors. 1976. *Measurements of Ultraviolet Radiation in the United States and Comparisons with Skin Cancer Data*, DHEW NO (NEH) 76-1029. National Cancer Institute, Washington, DC.

Urbach, F., editor. 1969. *The Biologic Effects of Ultraviolet Radiation*. Pergamon Press, Oxford.

van der Leun, J., and F. de Gruijl. 2002. Climate change and skin cancer. *Photochem Photobiol Sci* 1:324–326.

Whitten, R., W. Borucki, and R. Turco. 1975. Possible ozone depletions following nuclear explosions. *Nature (Lond)* 257:38–39.

Photochemistry and Photobiology of Vitamin D

Michael F. Holick
Boston University Medical Center

17.1 Prehistorical Perspective

Vitamin D is one of the oldest hormones, if not the oldest, which has existed in life forms for more than 500 million years (Holick 1989). *Emiliania huxleyi* is phytoplankton that has existed unchanged in the Sargasso Sea for more than 500 million years. When this organism was cultured and then exposed to ultraviolet (UV) radiation, the ergosterol (previtamin D$_2$) in this organism was converted to previtamin D$_2$ (Holick 2003). Similarly, *Skeletonema menzelii*, a phytoplankton species that contains calcium carbonate for its cytoskeleton, when exposed to UV radiation, also converted its ergosterol to previtamin D$_2$. Previtamin D$_2$, in turn, isomerized to vitamin D$_2$. Remarkably, the amount of ergosterol present in *E. huxleyi* was 1 μg/g weight. An evaluation of plankton net tows from the Atlantic Ocean revealed a wide variety of provitamin Ds, including ergosterol and 7-dehydrocholesterol, which were present in the collection of organisms that included phytoplankton and zooplankton (Holick 1989, 2003).

Although the function of provitamin D is not known in either phytoplankton or zooplankton, it has been speculated that there are at least three possible functions for its presence. Based on studies in human skin, it is likely that the ergosterol is present in the plasma membrane that is sandwiched in between the fatty acid side chains and polar head group of triglycerides. Ergosterol efficiently absorbs UV radiation up to 315 nm (Figure 17.1). When ergosterol

absorbs UV radiation, it converts to previtamin D$_2$. As can be seen in Figure 17.1, the absorption spectra for previtamin D$_2$ and its thermal isomerization product vitamin D$_2$ are almost identical to DNA and RNA, and thus, they can absorb the solar UV radiation before it damages the DNA and RNA. One of the photoproducts of previtamin D$_2$ is tachysterol, which has a UV absorption maximum at 282 nm and, thus, can efficiently absorb UV radiation that is damaging to UV-sensitive proteins (Holick 2003) (Figure 17.1). Therefore, ergosterol, previtamin D$_2$ and its photoproducts, and vitamin D$_2$ are ideally suited to act as a natural sunscreen, absorbing UV radiation that is most damaging to UV radiation-sensitive macromolecules including DNA, RNA, and proteins.

Another possible function of ergosterol being photolyzed to previtamin D$_2$ is to open up the plasma membrane to permit the entrance of the plentiful calcium present in the ocean environment. Once previtamin D$_2$ is made, it rapidly converts to vitamin D$_2$, which then exits the plasma membrane, potentially opening up a channel to permit calcium to enter the cell. This could be the explanation for why, throughout evolution, sunlight has played such an important role in calcium metabolism and the evolution of vertebrates (Holick 2003).

It has also been speculated that the photoproduction of vitamin D was so important for the maintenance of the vertebrate skeleton that when the asteroid struck Earth 65 million years ago, one of the reasons for the rapid demise of the dinosaurs was their

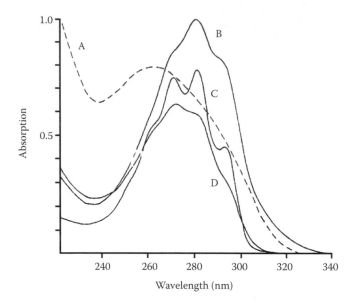

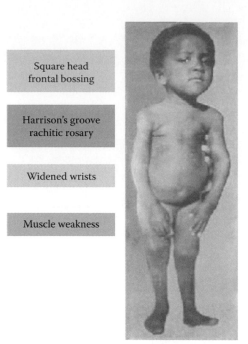

FIGURE 17.2 Child with rickets demonstrating muscle weakness, rachitic rosary, Harrison's groove, widened wrists, frontal bossing, and square head. (Reproduced from Holick, M., *N. Engl. J. Med.*, 357, 266–281, 2007. With permission.)

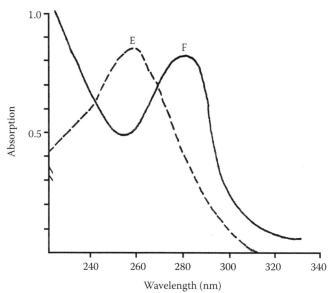

FIGURE 17.1 Ultraviolet absorption spectra for (A) previtamin D_3, (B) tachysterol, (C) provitamin D_3 (7-dehydrocholesterol), (D) lumisterol, (E) DNA, and (F) albumin. (Reproduced from Holick, M., *N. Engl. J. Med.*, 357, 266–281, 2007. With permission.)

inability to produce vitamin D because of the ash cloud that enveloped the Earth for many years. This would have led to poorly mineralized skeletons and a disruption in calcium signaling, which is important for most metabolic functions (Holick 2003).

17.2 Historical Perspective

In the mid-1600s, it was realized that children living in industrialized centers in Europe developed a devastating bone-deforming disease known as rickets. This disease caused growth retardation and skeletal deformities of the legs, wrists, rib cage, and skull (Holick 2006) (Figure 17.2). Rickets became the scourge of the industrial revolution in Northern Europe and spread to

the northeastern United States. In the early 1900s, it was estimated that more than 80% of children living in Europe and the northeastern United States showed evidence of rickets (Holick 2006). Although cod liver oil was found to be effective in the mid-1800s for preventing and treating rickets (Holick 2006), it was not until 1919 that Huldschinsky (1919) reported that children exposed to mercury arc lamps had dramatic radiologic improvement in their rachitic lesions (Figure 17.3). Two years later, Hess and Unger (1921) reported that children exposed to sunlight also had dramatic improvement in their rachitic lesions. The observation that exposure to sunlight or UV radiation could cure rickets led Hess and Weinstock (1924) and Steenbock and Black (1924) to introduce the concept of irradiating food with UV radiation to promote antirachitic activity. This ultimately led to the fortification of milk with vitamin D by adding ergosterol and then irradiating the milk (Steenbock 1924). When vitamin D could be commercially produced inexpensively, it was added to milk. This simple process essentially eradicated rickets throughout the United States and Europe in the 1930s and 1940s. However, in the 1950s, an outbreak of "vitamin D intoxication" associated with hypercalcemia in young children in Great Britain was thought to be a result of the overfortification of milk with vitamin D (British Pediatric Association 1956). This led to great hysteria, resulting in laws that forbid the fortification of any food or product with vitamin D throughout Europe that are still practiced in most European countries. It is likely, however, that these children suffered from Williams syndrome, a disease now known to cause sensitivity to vitamin D. Only recently, Finland and Sweden have permitted dairy products to be fortified with vitamin D.

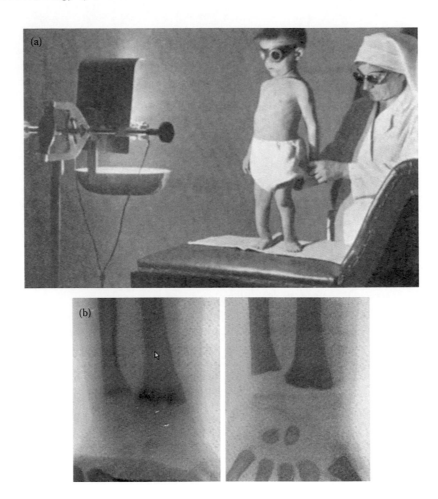

FIGURE 17.3 (a) A child with rickets being exposed to ultraviolet radiation (UVR). (b) Florid rickets of the hand and wrist (left panel) and the same wrist and hand radiograph taken after treatment with 1 h UVR two times a week for 8 weeks. Note mineralization of the carpal bones and epiphyseal plates (right panel).

17.3 Photochemistry of Provitamin D

During exposure to sunlight, the ultraviolet B (UVB) radiation with wavelengths of 290–315 nm (Figure 17.4) that are able to reach the skin's surface penetrate into the epidermis, resulting in its absorption by 7-dehydrocholesterol (MacLaughlin, Anderson, and Holick 1982). This results in 7-dehydrocholesterol opening its B ring, in turn resulting in the formation of previtamin D_3 (Holick et al. 1980; Holick, Tian, and Allen 1995) (Figure 17.5). Normally, at body temperature (37°C), it takes approximately 24 h for 50% of the previtamin D_3 to thermally isomerize to vitamin D_3. However, in human skin, this isomerization occurs at a much faster rate, and nearly 100% of previtamin D_3 is converted to vitamin D_3 within 8 h (Tian et al. 1993) (Figure 17.5). The reason for the rapid isomerization is the fact that 7-dehydrocholesterol is incorporated into the lipid bilayer and, when exposed to UVB radiation, converts to the thermodynamically less favorable *cis–cis* conformer, which, in turn, rapidly converts to vitamin D_3 (Tian et al. 1993) (Figure 17.6). The more thermodynamically stable conformer (*cis–trans*) is unable to isomerize to vitamin D_3. In solution, previtamin D_3 is

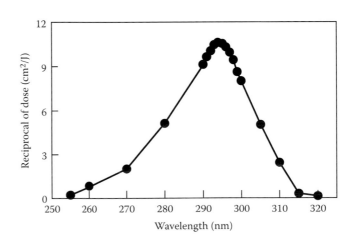

FIGURE 17.4 Action spectrum of previtamin D_3 formation from 7-DHC. (Reproduced from Holick, M., *N. Engl. J. Med.*, 357, 266–281, 2007. With permission.)

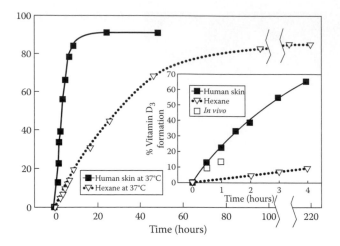

FIGURE 17.5 Thermoconversion of pre-D$_3$ to vitamin D$_3$ as a function of time in human skin and in *n*-hexane at 37°C. The inset depicts the thermoconversion of preD$_3$ to vitamin D$_3$ in human skin *in vivo* (*open squares*) and compares them with those in *n*-hexane and in human skin *in vitro* at 37°C. (Reproduced from Holick, 1993. With permission.)

mainly in the stable *cis–trans* conformer, which is in equilibrium with its *cis–cis* relative. Thus, it takes a considerable amount of time for the *cis–trans* conformer to equilibrate to the *cis–cis* conformer, which can then isomerize to vitamin D$_3$. This nonenzymatic catalytic reaction in human skin also occurs in lizard skin and even mushrooms and has likely been in existence for more than 500 million years (Holick, Tian, and Allen 1995).

Once vitamin D is made in the living skin cell, it moves into the extracellular space and then diffuses into the dermal capillary bed. It binds to the vitamin D binding protein (DBP) and is transported to the liver (Haddad et al. 1976).

17.4 Regulation of Cutaneous Vitamin D$_3$ Synthesis

There has never been a documented case of vitamin D intoxication from sun exposure. It was speculated that skin pigmentation developed to prevent humans who evolved near the equator from becoming vitamin D intoxicated from sun exposure (Loomis 1967). However, during exposure to sunlight, previtamin D$_3$ absorbs UVB photons and is converted to lumisterol and tachysterol (Holick, MacLaughlin, and Doppelt 1981) (Figure 17.7). Neither of these photoproducts have any effect on calcium metabolism (Holick, MacLaughlin, and Doppelt 1981). When vitamin D$_3$ is made in the skin, it too can absorb UVB photons to form suprasterols (Webb, Kline, and Holick 1988). Thus, sunlight itself is responsible for regulating the production of vitamin D$_3$ in the skin, and there was no need for melanin to have evolved to suppress vitamin D production in our ancestors living near the equator.

Melanin most likely evolved to protect the skin from damage resulting from exposure to high-intensity ultraviolet A (UVA) and UVB radiation. Melanin is an effective natural sunscreen that efficiently absorbs solar UVA and UVB radiation. Therefore, it absorbs UVB radiation, which is responsible for producing

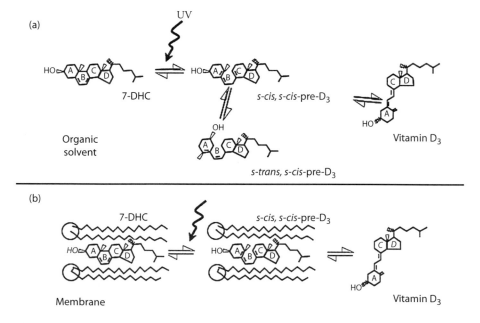

FIGURE 17.6 (a) Photolysis of provitamin D$_3$ (pro-D$_3$; 7-dehydrocholesterol) into previtamin D$_3$ (pre-D$_3$) and its thermal isomerization to vitamin D$_3$ in hexane and in skin. In hexane, pro-D$_3$ photolyzed to *s-cis,s-cis*-pre-D$_3$. Once formed, this energetically unstable conformation undergoes a conformational change to the *s-trans,s-cis*-pre-D$_3$. Only the *s-cis,s-cis*-pre-D$_3$ can undergo thermal isomerization to vitamin D$_3$. (b) The *s-cis,s-cis* conformer of pre-D$_3$ is stabilized in the phospholipid bilayer by hydrophilic interactions between the 3 β-hydroxl group and the polar head of the lipids as well as by the van der Waals interactions between the steroid ring and side-chain structure and the hydrophobic tail of the lipids. These interactions significantly decrease the conversion of the *s-cis,s-cis* conformer to the *s-trans,s-cis* conformer, thereby facilitating the thermal isomerization of *s-cis,s-cis*-pre-D$_3$ to vitamin D$_3$. (From Holick, M., *N. Engl. J. Med.*, 357, 266–281, 2007. With permission.)

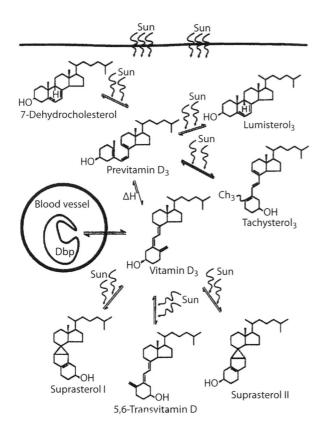

FIGURE 17.7 Photochemical events that lead to the production and regulation of vitamin D_3 in the skin. DBP, plasma vitamin D binding protein. (Reproduced from Holick, M., *N. Engl. J. Med.*, 357, 266–281, 2007. With permission.)

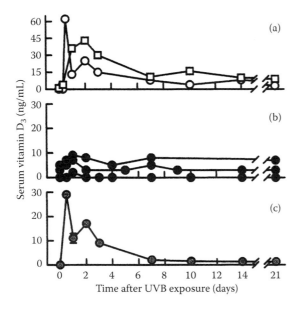

FIGURE 17.8 Change in serum concentrations of vitamin D in (a) two lightly pigmented white (skin type II) and (b) three heavily pigmented black subjects (skin type V) after total-body exposure to 54 mJ/cm² of UVB radiation. (c) Serial change in circulation vitamin D after reexposure of one black subject in panel B to a 320 mJ/cm² dose of UVB radiation. (Reproduced from Clemens, T. et al., *Lancet*, 1:74–76, 1982. With permission.)

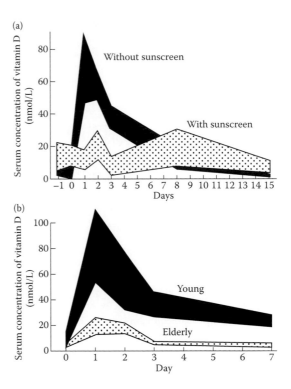

FIGURE 17.9 (a) Circulating concentrations of vitamin D_3 after a single exposure to 1 minimal erythemal dose of simulated sunlight either with a sunscreen with a sun protection factor of 8 (SPF-8) or a topical placebo cream. (b) Circulating concentrations of vitamin D in response to a whole-body exposure to 1 minimal erythemal dose in healthy young and elderly subjects. (Reproduced from Holick, 1994. With permission.)

vitamin D_3. An increase in skin melanin content reduces the efficiency of sunlight to produce vitamin D_3. When Caucasian and African American adults were exposed to the same amount of simulated sunlight in a tanning bed, the Caucasian adults raised their blood level of vitamin D_3 by more than 60-fold, whereas there was no significant change in the blood levels of vitamin D_3 in the African-American adults (Clemens et al. 1982) (Figure 17.8). When an African American adult was exposed to approximately six times the amount of simulated sunlight, he or she finally could raise the blood level of vitamin D_3 by approximately 30-fold. This is similar to a Caucasian wearing a sunscreen with a sun protection factor (SPF) of 10–15. The topical application of a sunscreen with an SPF of 15 reduces the ability of the skin to produce vitamin D_3 by more than 90%. A sunscreen with an SPF of 30 properly applied reduces the cutaneous vitamin D_3 synthesis by more than 97% (Matsuoka et al. 1987) (Figure 17.9).

17.5 Factors Affecting Cutaneous Vitamin D_3 Production

The solar zenith angle of the sun has a dramatic influence on the cutaneous production of previtamin D_3 (Webb, Kline, and Holick 1988) (Figure 17.10). Only about 0.5%–1% of UVB radiation is able to penetrate through the ozone layer and reach the Earth's surface in the summer at noontime. Thus, as the zenith angle of

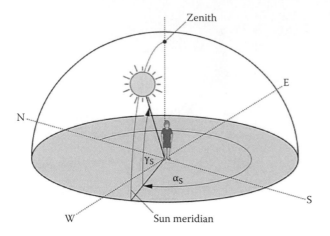

FIGURE 17.10 The solar zenith angle is the angle made by the sun's light with respect to the vertical (the sun being directly overhead). This angle is increased at higher latitudes, in the early morning, and in the late afternoon when the sun is not directly overhead and during the winter months. As the solar zenith angle increases, the amount of UVB radiation reaching the Earth's surface is reduced. Therefore, at higher latitudes, a greater distance from the equator, more of the UVB radiation is absorbed by the ozone layer, thereby reducing or eliminating the cutaneous production of vitamin D_3. (Reproduced from Holick, M., *J. Clin. Investig.*, 116, 8, 2062–2072, 2006. With permission.)

the sun increases, the UVB radiation has a longer path length to transit through the ozone layer, and thus, more of the UVB radiation is absorbed by the ozone layer. In the winter, essentially all UVB photons are absorbed by the ozone layer, and therefore, very little previtamin D_3 can be made in the skin. Living above and below ~33° latitude, children and adults can make previtamin D_3 during sun exposure in the spring, summer, and fall months but not during the winter (Holick 2003; Webb, Kline, and Holick 1988) (Figures 17.11 and 17.12). Furthermore, the time of day also influences the zenith angle of the sun, and therefore, not only season but also the time of day has had a dramatic effect on the cutaneous production of previtamin D_3 (Figure 17.11). Very little, if any, previtamin D_3 is produced in the early morning and late afternoon even in the summer. The most efficient times are between 10 a.m. and 3 p.m. (Holick 2003; Holick, Chen, and Sauter 2007) (Figures 17.11 and 17.12). Thus, what many dermatologists have suggested, that is, that exposure to early-morning and late-afternoon sun is the best time to be outside without sun protection because it is the least damaging to the skin while vitamin D_3 is produced, is incorrect. Morning and late afternoon are the worst times to be exposed to sunlight because the skin is receiving high doses of UVA radiation, which increases risk for wrinkling and nonmelanoma and melanoma skin cancer while not providing any significant amount of vitamin D_3.

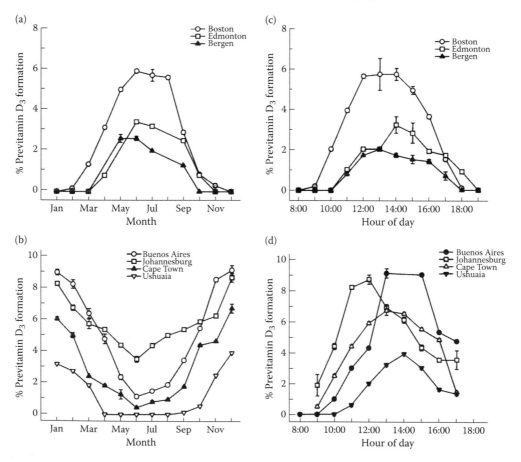

FIGURE 17.11 Influence of season, time of day, and latitude on the synthesis of previtamin D_3 in the northern (a and c) and southern hemispheres (b and d). The hour indicated in (c) and (d) is the end of the 1-hour exposure time. (Reproduced from Holick, 1998. With permission.)

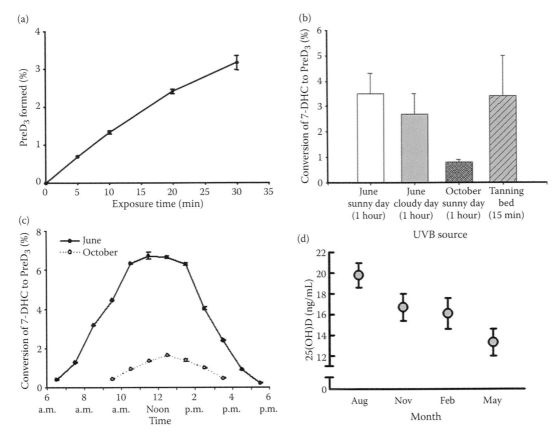

FIGURE 17.12 (a) Ampoules containing 7-dehydrocholesterol in ethanol were exposed to sunlight at noon in June in Boston. High-performance liquid chromatography analysis was performed to determine the production of previtamin D_3. (b) An ampoule of 7-dehydrocholesterol (7-DHC) was exposed between the hours of 12 p.m. and 1 p.m. in June on a sunny day, in June on a cloudy day, and in October on a sunny day in Boston. The conversion of 7-dehydrocholesterol to previtamin D_3 (preD$_3$) was determined by high-performance liquid chromatography, and the results were compared to conversion of 7-dehydrocholesterol to previtamin D_3 that occurred in a tanning bed after exposure for 15 min. (c) Conversion of 7-dehydrocholesterol (7-DHC) to previtamin D_3 (preD$_3$) at various times throughout the day in June and in October on a sunny day in Boston. Note that the data points are plotted every half hour to represent what occurred before and 30 min after that time point, that is, 6 a.m.–7 a.m. and so forth. (d) Circulating levels of 25(OH)D were measured in healthy free-living nursing home residents at various seasons of the year. (Reproduced from Holick, M. et al., *J. Bone Miner. Res.*, 22, V28–V33, 2007. With permission.)

The higher the altitude, the less ozone there is, and thus, the skin is exposed to more UVB radiation, resulting in the production of vitamin D_3. As seen in Figure 17.13, at the same latitude in November, little previtamin D_3 is produced in Agra, India, the home of the Taj Mahal; however, substantial previtamin D_3 production occurred at higher altitudes and at the base camp of Mount Everest. Similarly, clouds absorb UVB radiation and also diminish the production of vitamin D_3 in the skin (Holick, Chen, and Sauter 2007) (Figure 17.12).

Aging influences a wide variety of metabolic processes and has a dramatic effect on the composition and integrity of the skin. The amount of 7-dehydrocholesterol present in the epidermis of a 70-year-old is approximately three to four times less than that of a 20-year-old (MacLaughlin and Holick 1985) (Figure 17.14). As a result, for the same amount of sun exposure, a 70-year-old makes approximately 25% of the amount of vitamin D_3 that a 20-year-old would make (Holick 2009) (Figure

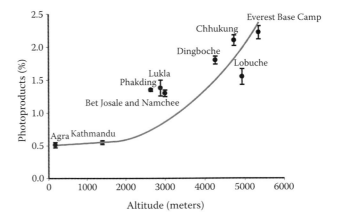

FIGURE 17.13 Ampoules containing 7-dehydrocholesterol in ethanol were exposed for 1 h between 11:30 a.m. and 12:30 p.m. at 27° N in India at various altitudes. The conversion of 7-dehydrocholesterol to previtamin D_3 and its photoproducts was determined by high-performance liquid chromatography. (Reproduced from Holick, M. et al., *J. Bone Miner. Res.*, 22, V28–V33, 2007. With permission.)

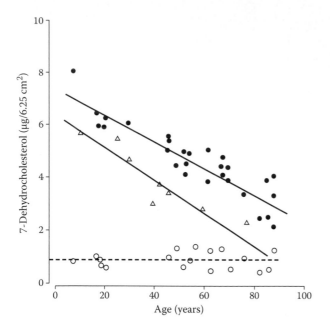

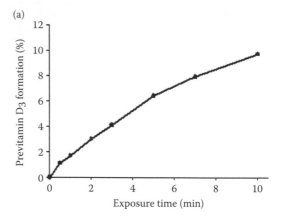

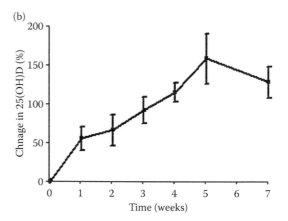

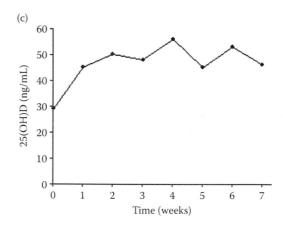

FIGURE 17.14 Effect of aging on 7-dehydrocholesterol concentrations in human epidermis and dermis. Concentrations of 7-dehydrocholesterol (provitamin D_3) per unit area of human epidermis (●), stratum basale (△), and dermis (○) obtained from surgical specimens from donors of various ages. A linear regression analysis revealed slopes of –0.05, –0.06, and –0.0005 for the epidermis ($r = -.89$), stratum basale ($r = -.92$), and dermis ($r = -.04$), respectively. The slopes of the epidermis and stratum basale are significantly different from the slope of the dermis ($P < .001$). (Reproduced from MacLaughlin, J. and Holick, M., *J. Clin. Investig.*, 76, 1536–1538, 1985. With permission.)

17.9). However, because the skin has such a large capacity to produce vitamin D_3, even elders can improve their vitamin D status when exposed to sunlight (Holick, Chen, and Sauter 2007) (Figure 17.15).

17.6 Blood Levels of Vitamin D_3 in Response to UVB Irradiation

A study was conducted to determine what the equivalency was regarding exposure to simulated sunlight in raising blood levels of vitamin D_3 when compared to taking an oral dose of vitamin D. Healthy adults in a bathing suit exposed to 1 minimal erythemal dose (1 MED) raised their blood levels of vitamin D_3 to an amount that was comparable to taking approximately 15,000–20,000 IU of vitamin D (Holick 2009) (Figure 17.16). It is also interesting to note that the blood levels of vitamin D_3 were sustained for a longer period of time when the vitamin D_3 came from the skin when compared to taking a supplement. This is not at all unexpected because previtamin D_3 requires time to convert to vitamin D_3, and then vitamin D_3 takes additional time to diffuse into the bloodstream, thereby maintaining the blood

FIGURE 17.15 (a) Ampoules containing 7-dehydrocholesterol were placed in a tanning bed at various times, and conversion of 7-dehydrocholesterol to previtamin D_3 was measured by high-performance liquid chromatography. (b) Healthy adults were exposed to 0.75 MED in a tanning bed three times a week for 7 weeks. Circulating concentrations of 25(OH)D were determined at baseline and once a week thereafter. (c) A 76-year-old healthy male was exposed to tanning bed radiation equivalent to 0.75 MED three times a week for 7 weeks. His circulating concentrations of 25(OH)D were obtained at weekly intervals. (Reproduced from Holick, M., *N. Engl. J. Med.*, 357, 266–281, 2007. With permission.)

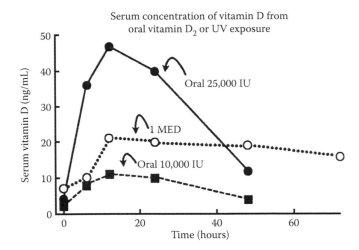

FIGURE 17.16 Comparison of serum vitamin D_3 levels after a whole-body exposure (in a bathing suit, bikini for women) to 1 minimal erythemal dose (MED) of simulated sunlight compared with a single oral dose of either 10,000 or 25,000 IU of vitamin D_2. (Reproduced from Holick, 1994. With permission.)

level of vitamin D_3 for at least two to three times longer than when taken as an oral dose.

17.7 Vitamin D Metabolism for Calcium Homeostasis

Once vitamin D_3 is made in the skin or when vitamin D_2 or vitamin D_3 (vitamin D represents D_2 or D_3) is absorbed from the diet, it travels to the liver, where it is converted to 25-hydroxyvitamin D [25(OH)D] (Figure 17.17). 25(OH)D is a major circulating form of vitamin D that is used by clinicians to determine a person's vitamin D status. However, this metabolite is biologically inactive in calcium metabolism and requires further hydroxylation in the kidneys on carbon-1 to form 1,25-dihydroxyvitamin D [1,25(OH)$_2$D]. 1,25(OH)$_2$D is considered to be the biologically active form of vitamin D. It travels to the small intestine and interacts with its vitamin D receptor (VDR), unlocking genetic information resulting in an increase in intestinal calcium absorption (Christakos et al. 2003). It also travels to the bone and interacts with its VDR in the osteoblast, which stimulates the receptor activator of nuclear factor kappa-light-chain-enhancer of activated B cells (NF-κB) (RANKL). This ligand interacts with its receptor on mononuclear cells, stimulating them to become mature bone-resorbing osteoclasts (Holick 2007; Khosla 2001) (Figure 17.17). Thus, the major function of the renal production of 1,25(OH)$_2$D is to maintain serum calcium levels within a physiologic normal range, which is important for the maintenance of the signal transduction for a wide variety of metabolic functions as well as the mineralization of the skeleton.

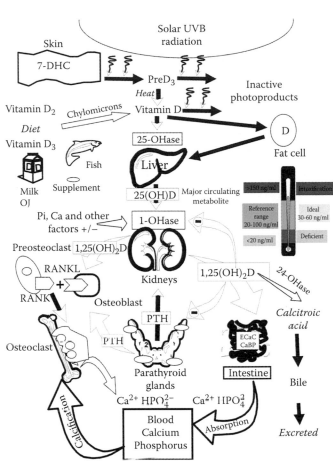

FIGURE 17.17 Schematic representation of the synthesis and metabolism of vitamin D for regulating calcium, phosphorus, and bone metabolism. During exposure to sunlight, 7-dehydrocholesterol in the skin is converted to previtamin D_3. Vitamin D (D represents D_2 or D_3) made in the skin or ingested in the diet is transported it to the liver, where vitamin D is converted by the vitamin D-25-hydroxylase to 25-hydroxyvitamin D [25(OH)D]. This is the major circulating form of vitamin D that is used by clinicians to measure vitamin D status (although most reference laboratories report the normal range to be 20–100 ng/mL, the preferred healthful range is 30–60 ng/mL) converted in the kidneys by 25-hydroxyvitamin D-1α-hydroxylase (1-OHase) to its biologically active form, 1,25-dihydroxyvitamin D [1,25(OH)$_2$D]. 1,25(OH)$_2$D enhances intestinal calcium absorption in the small intestine by stimulating the expression of the epithelial calcium channel (ECaC) and the calbindin 9K (calcium-binding protein [CaBP]). 1,25(OH)$_2$D is recognized by its receptor in osteoblasts, causing an increase in the expression of receptor activator of NF-κB ligand (RANKL). Its receptor RANK on the preosteoclast binds RANKL, which induces the preosteoclast to become a mature osteoclast. The mature osteoclast removes calcium and phosphorus from the bone to maintain blood calcium and phosphorus levels. Adequate calcium and phosphorus levels promote the mineralization of the skeleton. (Reproduced from Holick, M., *N. Engl. J. Med.*, 357, 266–281, 2007. With permission.)

17.8 Extrarenal Metabolism of 25(OH)D and Noncalcemic Functions

It is now recognized that most cells and organs, including the skin, breast, colon, brain, prostate, pancreas, and macrophages, not only have a VDR but also have the 25-hydroxyvitamin D-1-hydroxylase (1-OHase) (Adams and Hewison 2010; Bikle 2009; Merewood et al. 2010) (Figure 17.18). It has been

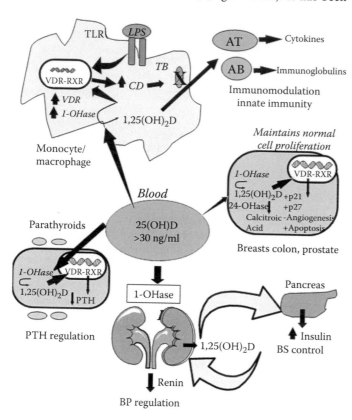

FIGURE 17.18 Metabolism of 25-hydroxyvitamin D [25(OH)D] to 1,25-dihydroxyvitamin D [1,25(OH)₂D] for nonskeletal functions. When a monocyte/macrophage is stimulated through its toll-like receptor 2/1 (TLR2/1) by an infective agent, such as *Mycobacterium tuberculosis* (TB), or its lipopolysaccharide (LPS), the signal upregulates the expression of vitamin D receptor (VDR) and the 25-hydroxyvitamin D-1-hydroxylase (1-OHase). 25(OH)D levels >30 ng/mL provide adequate substrate for the 1-OHase to convert it to 1,25(OH)₂D. 1,25(OH)₂D returns to the nucleus, where it increases the expression of cathelicidin (CD), which is a peptide capable of promoting innate immunity and inducing the destruction of infective agents, such as TB. 1,25(OH)₂D also regulates genes to control cell growth and genes to regulate blood pressure and glucose metabolism. The production of 1,25(OH)₂D in the kidney enters the circulation and is able to downregulate renin production in the kidney and to stimulate insulin secretion in the β- islet cells of the pancreas. (Reproduced from Holick, M., *N. Engl. J. Med.*, 357, 266–281, 2007. With permission.)

estimated that as many as 2000 genes are either directly or indirectly controlled by 1,25(OH)₂D. 1,25(OH)₂D is a hormone that is capable of inhibiting DNA synthesis and inducing cellular maturation. It also inhibits angiogenesis and induces apoptosis (Mantell et al. 2000; Nagpal, Na, and Rathnachalam 2005). 1,25(OH)₂D is also a potent modulator of the immune system. When a macrophage ingests an infective agent, such as tuberculosis, the toll-like receptors are activated, resulting in signal transduction to stimulate the expression of VDR and the 1-OHase (Liu et al. 2006) (Figure 17.18). This results in the conversion of 25(OH)D to 1,25(OH)₂D. 1,25(OH)₂D reenters the macrophage's nucleus to increase the expression of cathelicidin. Cathelicidin is a defensin peptide that interacts with infectious agents, resulting in their death. 1,25(OH)₂D stimulates the production of insulin and, thus, plays an important role in glucose metabolism (Pittas et al. 2006). It was also shown to downregulate renin, a hormone that controls blood pressure (Figure 17.18).

17.9 Definition of Vitamin D Deficiency and Insufficiency

It is now recognized that vitamin D deficiency, defined as 25(OH)D <20 ng/mL, and vitamin D insufficiency, defined as 25(OH)D of 21–29 ng/mL, is a worldwide epidemic (Figure 17.19). In the United States, it has been reported that 32% of the US population is vitamin D deficient (Holick et al. 2012). A survey of children revealed that 50% were vitamin D deficient or insufficient between the ages of 1–5 years, and 70% were deficient or insufficient between the ages of 6–11 years (Mansbach, Ginde, and Camargo 2009). Even in Australia, children and adults are at very high risk (Daly et al. 2012). The major cause for the vitamin D deficiency pandemic is the misbelief that children and adults can obtain an adequate amount of vitamin D from dietary sources. However, even in the United States, where dairy products, juice products, and other foods are fortified with vitamin D, neither children nor adults are able to obtain enough vitamin D to satisfy their vitamin D requirement (Figure 17.19). Another reason is the lack of appreciation that the major source of vitamin D for humans throughout evolution has been and continues to be exposure to sunlight (Holick, Chen, and Sauter 2007; Holick 2007). It is well documented that the blood levels of 25(OH)D are at their nadir in the winter and are maximum at the end of the summer (Holick, Chen, and Sauter 2007; Brot et al. 2001) (Figures 17.12 and 17.20). Both the amount of sun exposure and the season of the year had a dramatic influence on the circulating levels of 25(OH)D both in adults living in Denmark and in Aborigines in Australia (Figures 17.20 and 17.21).

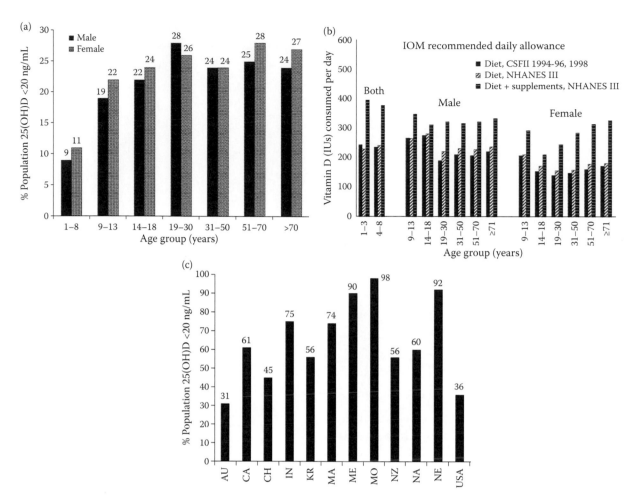

FIGURE 17.19 (a) Prevalence at risk of vitamin D deficiency defined as a 25-hydroxyvitamin D < 12–20 ng/mL by age and sex, United States, 2001–2006 (b) Mean intake of vitamin D (IU) from food and food plus dietary supplements from the Continuing Survey of Food Intakes by Individuals (CSFII), 1994–1996 and 1998, and the Third National Health and Nutrition Examination Survey (NHANES III), 1988–1994. (c) Reported incidence of vitamin D deficiency defined as a 25-hydroxyvitamin D < 20 ng/mL around the globe, including Australia (AU), Canada (CA), China (CH), India (IN), Korea (KR), Malaysia (MA), the Middle East (ME), Mongolia (MO), New Zealand (NZ), North Africa (NA), Northern Europe (NE), and the United States (USA). (Reproduced from Holick, M. et al., *J. Clin. Endocrinol. Metab.*, 97, 1153–1158, 2012. With permission.)

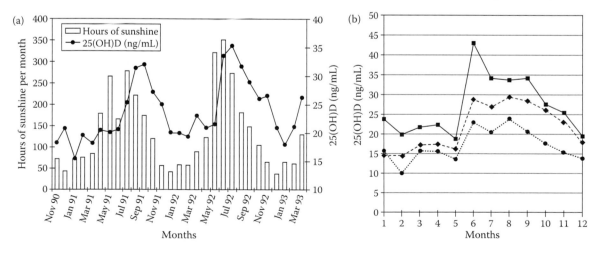

FIGURE 17.20 (a) Relationship between hours of sunshine and serum 25(OH)D. ● 25(OH)D (ng/mL). (b) Seasonal fluctuation of serum 25(OH)D according to frequency of sun exposure. ■ Regular sun exposure; ◆ occasional sun exposure; ● avoiding direct sun exposure. (Reproduced from Brot, C. et al., *Br. J. Nutr.*, 86, 1, S97–S103, 2001. With permission.)

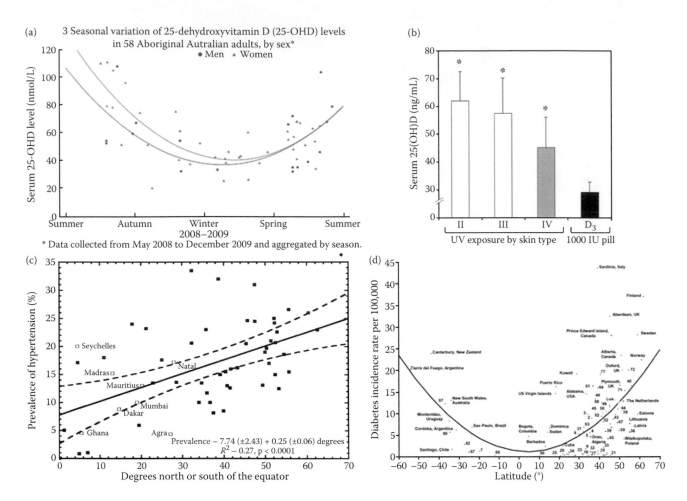

FIGURE 17.21 (a) Seasonal variation of 25-hydroxyvitamin D in 58 Aboriginal Australian men (solid circle) and women (solid triangle). (b) Comparison of serum 25(OH)D levels in either healthy adults who were in a bathing suit and exposed to suberythemal doses (0.5 MED) of ultraviolet B radiation once a week for 3 months or healthy adults who received 1000 units of vitamin D_3 daily during the winter and early spring for a period of 11 weeks. Skin type is based on the Fitzpatrick scale. The data represent mean ± standard error of the mean (SEM). (c) Relationship of prevalence of hypertension to distance north or south of the equator. Broken lines represent 95% confidence limits. Regression line and confidence limits are derived from INTERSALT centers only. (d) Age-standardized incidence rates of type 1 diabetes per 100,000 boys <14 years of age by latitude in 51 regions worldwide in 2002. $R^2 = .25$, $P < .001$. (Reproduced from Holick, M. F. *Endocrine Practice*, 17, 143–149, 2011a. With permission.)

17.10 Chronic Illnesses Related to Lack of Sun Exposure and Inadequate Vitamin D

As early as 1915, it was recognized that workers who were mainly indoors were more likely to die of cancer than workers who worked outside (Spina et al. 2006). Apperly (1941) reported that living in the northeastern United States increased the risk of dying of cancer compared to living in the southern United States. It was also reported that if you are born and live near the equator, you have a 10–15 times lower risk for developing type 1 diabetes (Mohr et al. 2008) (Figure 17.21). If you were born and lived for the first 10 years of your life at a latitude above Atlanta,

Georgia (~33°N), you have a 100% increased risk for developing multiple sclerosis for the rest of your life no matter where you live afterward (Posonby, McMichael, and van der Mei 2002). It is also documented that living at higher latitudes increases risk for having higher blood pressure (Rostand 1979) (Figure 17.21).

During the past three decades, there have been a variety of association studies linking higher circulating levels of 25(OH)D with a variety of chronic illnesses, including breast, colon, and prostate cancer; autoimmune diseases, including type 1 diabetes, rheumatoid arthritis, and multiple sclerosis; cardiovascular disease; type 2 diabetes; and infectious diseases (Holick 2011) (Figure 17.22). More recent prospective studies have also supported these association studies. Children in Japan who received

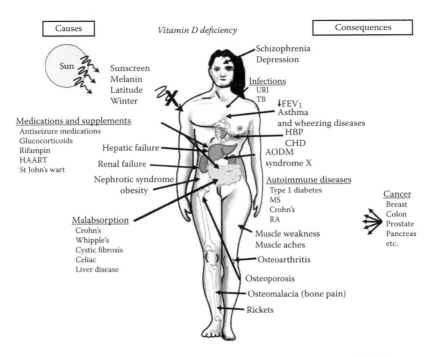

FIGURE 17.22 A schematic representation of the major causes for vitamin D deficiency and potential health consequences. (Reproduced from Holick, 2010. With permission.)

1200 IU of vitamin D_3 during the winter reduced their risk of developing influenza A infection by 42% (Urashima et al. 2010). An Australian study revealed over a period of 5 years that adults who had the highest circulating blood levels of 25(OH)D were less likely to develop pre–type 2 diabetes (Gagnon et al. 2012). A study in African American teenagers who received 2000 IU of vitamin D_3 daily for 4 months had improved vascular health (Dong et al. 2010). Even all-cause mortality was reduced when circulating concentrations of 25(OH)D were >30 ng/mL (Melamed et al. 2008; Thomas et al. 2012).

17.11 Treatment and Prevention of Vitamin D Deficiency

The Endocrine Practice Guidelines Committee recommended that to prevent and treat vitamin D deficiency, children aged 0–1 year require 400–1000 IU/d, children aged 1–18 years require 600–1000 IU/d, and adults require 1500–2000 IU/d, which is substantially higher than what is recommended by the Institute of Medicine (Holick et al. 2012; IOM 2011) (Table 17.1). Because body fat in obese people can sequester vitamin D, they often need two to three times more, (Holick et al. 2012). An effective method

to treat vitamin D deficiency is to give 50,000 IU of vitamin D_2 once a week for 8 weeks. To prevent recurrence of vitamin D deficiency and insufficiency, patients can be treated with 50,000 IU once every 2 weeks thereafter (Pietras et al. 2009). This is an effective strategy for maintaining a healthy blood level of 25(OH)D of 40–60 ng/mL (Pietras et al. 2009) (Figure 17.23). None of the patients had any side effects from this treatment for up to 6 years. Sensible sun exposure is also a good source of vitamin D. The amount of exposure depends on time of day, season of the year, latitude, altitude, and weather conditions. The rule of thumb is that if you know that you would develop a mild pinkness to your skin 24 h after being exposed to 1 h of sunlight (1 MED), then exposing as much of your skin as is reasonable, that is, arms, legs, abdomen, and back, to about 50% of that time (i.e., 30 min), will produce approximately 1,000–15,000 IU of vitamin D_3, depending upon the area of exposure (Figure 17.21). Do not expose the face because it is only 9% of the body surface and is most sun damaged and most prone to developing nonmelanoma skin cancer. If you do this two to three times a week, it will help support your vitamin D status. For those who wish to stay outside longer, I recommend properly using a sunscreen with an SPF of at least 30, thereby taking advantage of the beneficial effect of sunlight while preventing the damaging effects from excessive exposure.

TABLE 17.1 Recommended Dietary Allowances for Vitamin D

Life Stage Group	IOM Recommendations				Committee Recommendations	
	AI	EAR	RDA	UL	Daily Allowance (IU/d)	UL (IU)
Infants						
0 to 6 months	400 IU (10 µg)	— -	— -	1000 IU (25 µg)	400–1000	2000
6 to 12 months	400 IU (10 µg)	— -	— -	1500 IU (38 µg)	400–1000	2000
Children						
1–3 years	— -	400 IU (10 µg)	600 IU (15 µg)	2500 IU (63 µg)	600–1000	4000
4–8 years	— -	400 IU (10 µg)	600 IU (15 µg)	3000 IU (75 µg)	600–1000	4000
Males						
9–13 years	— -	400 IU (10 µg)	600 IU (15 µg)	4000 IU (100 µg)	600–1000	4000
14–18 years	— -	400 IU (10 µg)	600 IU (15 µg)	4000 IU (100 µg)	600–1000	4000
19–30 years	— -	400 IU (10 µg)	600 IU (15 µg)	4000 IU (100 µg)	1500–2000	10,000
31–50 years	— -	400 IU (10 µg)	600 IU (15 µg)	4000 IU (100 µg)	1500–2000	10,000
51–70 years	— -	400 IU (10 µg)	600 IU (15 µg)	4000 IU (100 µg)	1500–2000	10,000
>70 years	— -	400 IU (10 µg)	800 IU (20 µg)	4000 IU (100 µg)	1500–2000	10,000
Females						
9–13 years	— -	400 IU (10 µg)	600 IU (15 µg)	4000 IU (100 µg)	600–1000	4000
14–18 years	— -	400 IU (10 µg)	600 IU (15 µg)	4000 IU (100 µg)	600–1000	4000
19–30 years	— -	400 IU (10 µg)	600 IU (15 µg)	4000 IU (100 µg)	1500–2000	10,000
31–50 years	— -	400 IU (10 µg)	600 IU (15 µg)	4000 IU (100 µg)	1500–2000	10,000
51–70 years	— -	400 IU (10 µg)	600 IU (15 µg)	4000 IU (100 µg)	1500–2000	10,000
>70 years	— -	400 IU (10 µg)	800 IU (20 µg)	4000 IU (100 µg)	1500–2000	10,000
Pregnancy						
14–18 years	— -	400 IU (10 µg)	600 IU (15 µg)	4000 IU (100 µg)	600–1000	4000
19–30 years	— -	400 IU (10 µg)	600 IU (15 µg)	4000 IU (100 µg)	1500–2000	10,000
31–50 years	— -	400 IU (10 µg)	600 IU (15 µg)	4000 IU (100 µg)	1500–2000	10,000
Lactation[a]						
14–18 years	— -	400 IU (10 µg)	600 IU (15 µg)	4000 IU (100 µg)	600–1000	4000
19–30 years	— -	400 IU (10 µg)	600 IU (15 µg)	4000 IU (100 µg)	1500–2000	10,000
31–50 years	— -	400 IU (10 µg)	600 IU (15 µg)	4000 IU (100 µg)	1500–2000	10,000

Source: Reproduced from Hollick et al. *J. Clin. Endocrinal Metab.*, 97, 1153–1158, 2012.

Note: AI = adequate intake; EAR = estimated average requirement; IU = international units; RDA = recommended dietary allowance; UL = tolerable upper intake level.

[a] Mother's requirement, 4000–6000 IU/d (mother's intake for infant's requirement if infant is not receiving 400 IU/d).

17.12 Conclusion

There has been a lot of debate over the past four decades as to whether humans should be exposed to any direct sunlight (Wolpowitz and Gilchrest 2006). The American Academy of Dermatology continues to recommend avoidance of all direct sun exposure. Although it is true that excessive sun exposure increases the risk for nonmelanoma skin cancer, these cancers are often easy to detect and treat and, if caught early, are not usually lethal. Melanoma, which is often lethal, has been associated with sun exposure (Wolpowitz and Gilchrest 2006). However, most melanomas occur on the least sun-exposed areas, and occupational sun exposure decreases risk (Kennedy et al. 2003). The major causes are being red-headed, having multiple sunburn experiences as a child and young adult, being genetically prone to developing this deadly cancer, and having a large number of moles (Kennedy et al. 2003). Even in the "skin cancer capital" of the world, Australia, vitamin D deficiency has

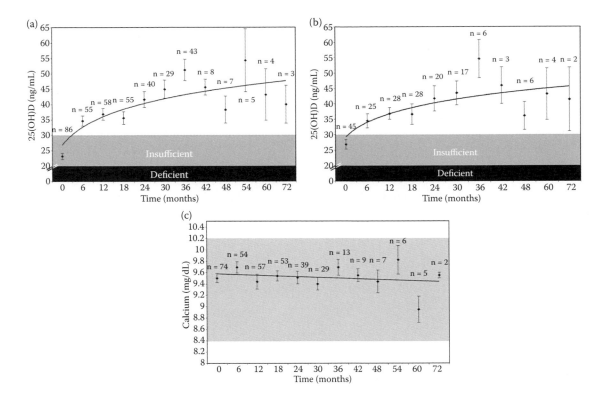

FIGURE 17.23 Mean serum 25-hydroxyvitamin D [25(OH)D] and calcium levels. Results are given as mean (SEM) values averaged over 6-month intervals. Time 0 is initiation of treatment. (a) Mean 25(OH)D levels in all patients treated with 50,000 IU of ergocalciferol (vitamin D_2) every 2 weeks (maintenance therapy, $n = 86$). Forty-one of the patients were vitamin D insufficient or deficient and first received 50,000 IU ergocalciferol weekly for 8 weeks before being placed on maintenance therapy of 50,000 IU of ergocalciferol every 2 weeks. The mean 25(OH)D level of each 6-month interval was compared with initial mean 25(OH)D level and showed a significant difference of $P < .001$ for all time points. To convert 25(OH)D to nanomoles per liter, multiply by 2.496. (b) Mean serum 25(OH)D levels in patients receiving maintenance therapy only. There were 38 patients who were vitamin D insufficient [25(OH)D levels <21–29 ng/mL] and 7 patients who were vitamin D sufficient [25(OH)D levels ≤ 30 ng/mL] who were treated only with maintenance therapy of 50,000 IU of ergocalciferol (vitamin D_2) every 2 weeks. The mean 25(OH)D levels in each 6-month interval were compared with mean initial 25(OH)D levels and showed a significant difference of $P < .001$ for all time points up to 48 months. The data for interval months 60 and 72 were pooled, and there was a significant difference of $P < .01$ compared with the baseline value. (c) Serum calcium levels. Results for all 86 patients who were treated with 50,000 IU of ergocalciferol (vitamin D_2). The reference range for serum calcium level is 8.5 to 10.2 mg/dL (to convert to millimoles per liter, multiply by 0.25). (Reproduced from Pietras et al., *Arch. Intern. Med.*, 169, 1806–1808, 2009. With permission.)

become such a significant health problem (Daly et al. 2012) that the Australian Dermatology Society now recommends some sensible sun exposure as an important source of vitamin D for children and adults. As noted by the Endocrine Society's guidelines, there is, essentially, no downside to improving a person's vitamin D status. There continues to be strong evidence that improvement in children's and adults' vitamin D status not only will maximize bone health and reduce risk for osteoporosis and fracture later in life but also could potentially reduce risk of many chronic illnesses, as outlined in Figure 17.22. It is reasonable to maintain

a blood level of 25(OH)D of >30 ng/mL, and up to 100 ng/mL is considered to be safe (unless you have a chronic granulomatous disorder, such as sarcoidosis) (Holick et al. 2012). The preferred level of 40–60 ng/mL can be achieved with sensible sun exposure along with dietary and supplemental vitamin D.

Acknowledgments

This work is supported in part by the UV Foundation and the Mushroom Council.

References

Adams, J., and M. Hewison. 2010. Update in vitamin D. *J Clin Endocrinol Metab* 95(2):471–478.

Apperly, F. 1941. The relation of solar radiation to cancer mortality in North America, *Cancer Res* 1:191–195.

Bikle, D. 2009. Nonclassic actions of vitamin D. *J Clin Endocrinol Metab* 94(1):26–34.

British Pediatric Association. 1956. Hypercalcemia in infants and vitamin D. *Br Med J* 2:149–151.

Brot, C., P. Vestergaard, N. Kolthoff, J. Gram, A. Hermann, and O. Sorensen. 2001. Vitamin D status and its adequacy in healthy Danish perimenopausal women: Relationships to dietary intake, sun exposure and serum parathyroid hormone. *Br J Nutr* 86(1):S97–S103.

Christakos, S., P. Dhawan, Y. Liu, X. Peng, and A. Porta. 2003. New insights into the mechanisms of vitamin D action. *J Cell Biochem* 88:695–705.

Clemens, T., S. Henderson, J. Adams et al. 1982. Increased skin pigment reduces the capacity of skin to synthesise vitamin D_3. *Lancet* 1:74–76.

Daly, R., C. Gagnon, Z. Lu, D. Magliano, D. Dunstan, K. Sikaris, P. Zimmet, P. Ebeling, and J. Shaw. 2012. Prevalence of vitamin D deficiency and its determinants in Australian adults aged 25 years and older: A national, population-based study. *Clin Endocrinol* 77:26–35.

Dong, Y., I. Stallmann-Jorgenson, N. Pollock, R. Harris, D. Keeton, Y. Huang, R. Bassali, D. Gao, J. Thomas, G. Pierce, J. White, M. Holick, and H. Zhu. 2010. A 16-week randomized clinical trial of 2,000 IU daily vitamin D_3 supplementation in black youth: 25-hydroxyvitamin D, adiposity, and arterial stiffness. *J Clin Endocrinol Metab* 95(10):4584–4591.

Gagnon, C., Z. Lu, D. Magliano, D. Dunstan, J. Shaw, P. Zimmet, K. Sikaris, P. Ebeling, and R. Daly. 2012. Low serum 25-hydroxyvitamin D is associated with increased risk of the development of the metabolic syndrome at five years: Results from a national, population-based prospective study (The Australian Diabetes, Obesity and Lifestyle Study: AusDiab). *J Clin Endocrinol Metab* 97(6)1953–1961.

Haddad, J., J. Walgate, C. Miyyn et al. 1976. Vitamin D metabolite-binding proteins in humantissue. *Biochim Biophys Acta* 444: 921–925.

Hess, A., and L. Unger. 1921. The cure of infantile rickets by sunlight. *J Am Med Assoc* 77:39–41.

Hess, A., and M. Weinstock. 1924. Antirachitic properties imparted to inert fluids and to green vegetables by ultraviolet irradiation. *J Biol Chem* 62:301–313.

Holick, M. 1989. Phylogenetic and evolutionary aspects of vitamin D from phytoplankton to humans. In *Vertebrate Endocrinology: Fundamentals and Biomedical Implications*, vol. 3, P. Pang, and M. Schreibman, editors. Academic Press, Orlando, FL.

Holick, M. 2003. Vitamin D: A millennium perspective. *J Cell Biochem* 88:296–307.

Holick, M. 2006. Resurrection of vitamin D deficiency and rickets. *J Clin Invest* 116(8):2062–2072.

Holick, M. 2007. Vitamin D deficiency. *N Engl J Med* 357:266–281.

Holick, M. 2009. Vitamin D and health: Evolution, biologic functions, and recommended dietary intakes for vitamin D. *Clin Rev Bone Miner Metab* 7(1):2–19.

Holick, M. 2011. Health benefits of vitamin D and sunlight: A D-bate. *Nat Rev Endocrinol* 7:73–75.

Holick M. F. 2011a. The D-Batable Institute of Medicine Report: A D-lightful perspective. *Endocrine Practice* 17(1):143–149.

Holick, M., N. Binkley, H. Bischoff-Ferrari, C. Gordon, D. Hanley, R. Heaney, M. Murad, and C. Weaver. 2012. Controversy in clinical endocrinology: Guidelines for preventing and treating vitamin D deficiency and insufficiency revisited. *J Clin Endocrinol Metab* 97:1153–1158.

Holick, M., T. Chen, and E. Sauter. 2007. Vitamin D and skin physiology: A D-lightful story. *J Bone Miner Res* 22(suppl. 2):V28–V33.

Holick, M., J. MacLaughlin, M. Clark, S. Holick, J. Potts, Jr., R. Anderson, I. Blank, J. Parrish, and P. Elias. 1980. Photosynthesis of previtamin D_3 in human skin and the physiologic consequences. *Science* 210:203–205.

Holick, M., J. MacLaughlin and S. Doppelt. 1981. Regulation of cutaneous previtamin D_3 photosynthesis in man: Skin pigment is not an essential regulator. *Science* 211:590–593.

Holick, M., X. Tian, and M. Allen. 1995. Evolutionary importance for the membrane enhancement of the production of vitamin D_3 in the skin of poikilothermic animals. *Proc. Natl Acad Sci* 92:3124–3126.

Huldschinsky, K. 1919. Heilung von Rachitis durch kunstliche Hohensonne. *Dtsch Med Wochenschr* 45:712–713.

IOM (Institute of Medicine). 2011. *Dietary Reference Intakes for Calcium and Vitamin D*. The National Academies Press, Washington, DC.

Kennedy, C., C. Bajdik, R. Willemze et al. 2003. The influence of painful sunburns and lifetime of sun exposure on the risk of actinic keratoses, seborrheic warts, melanocytic nevi, atypical nevi and skin cancer. *J Invest Dermatol* 120(6): 1087–1093.

Khosla, S. 2001. The OPG/RANKL/RANK system. *Endocrinology* 142(12):5050–5055.

Liu, P., S. Stenger, H. Li, L. Wenzel, B. Tan, S. Krutzik, M. Ochoa, J. Schauber, K. Wu, C. Meinken, D. Kamen, M. Wagner, R. Bals, A. Steinmeyer, U. Zugel, R. Gallo, D. Eisenberg, M. Hewison, B. Hollis, J. Adams, B. Bloom, and R. Modlin. 2006. Toll-like receptor triggering of a vitamin D-mediated human antimicrobial response. *Science* 3:1770–1773.

Loomis, W. 1967. Skin pigment regulation of vitamin D biosynthesis in man. *Science* 157(3788):501–506.

MacLaughlin, J., R. Anderson, and M. Holick. 1982. Spectral character of sunlight modulates the photosynthesis of previtamin D_3 and its photoisomers in human skin. *Science* 216:1001–1003.

MacLaughlin, J., and M. Holick. 1985. Aging decreases the capacity of human skin to produce vitamin D_3. *J Clin Invest* 76:1536–1538.

Mansbach, J., A. Ginde, and C. Camargo. 2009. Serum 25-hydroxyvitamin D levels among US children aged 1 to 11 years: Do children need more vitamin D? *Pediatrics* 124:1404–1410.

Mantell, D., P. Owens, N. Bundred et al. 2000. $1\alpha,25$-dihydroxyvitamin D_3 inhibits angiogenesis *in vitro* and in vivo. *Circ Res* 87:214–220.

Matsuoka, L., L. Ide, J. Wortsman et al. 1987. Sunscreens suppress cutaneous vitamin D_3 synthesis. *J Clin Endocrinol Metab* 64:1165–1168.

Melamed, M., E. Michos, W. Post, and B. Astor. 2008. 25-hydroxyvitamin D levels and the risk of mortality in the general population. *Arch Intern Med* 168(15):1629–1637.

Merewood, A., S. Mehta, X. Grossman, T. Chen, J. Mathieu, M. Holick et al. 2010. Widespread vitamin D deficiency in urban Massachusetts newborns and their mothers. *Pediatrics* 125:640–647.

Mohr, S., C. Garland, E. Gorham, and F. Garland. 2008. The association between ultraviolet B irradiance, vitamin D status and incidence rates of type 1 diabetes in 51 regions worldwide. *Diabetologia* 51:1391–1398.

Nagpal, S., S. Na, and R. Rathnachalam. 2005. Noncalcemic actions of vitamin D receptor ligands. *Endocr Rev* 26:662–687.

Pietras, S., B. Obayan, M. Cai, and M. Holick. 2009. Vitamin D_2 treatment for vitamin D deficiency and insufficiency for up to 6 years. *Arch Intern Med* 169:1806–1808.

Pittas, A., B. Dawson-Hughes, T. Li et al. 2006. Vitamin D and calcium intake in relation to type 2 diabetes in women. *Diabetes Care* 29:650–656.

Ponsonby, A., A. McMichael, and I. van der Mei. 2002. Ultraviolet radiation and autoimmune disease: Insights from epidemiological research. *Toxocology* 181–182:71–78.

Rostand, S. 1979. Ultraviolet light may contribute to geographic and racial blood pressure differences. *Hypertension* 30: 150–156.

Spina, C., V. Tangpricha, M. Uskokovic, L. Adorinic, H. Maehr, and M. Holick. 2006. Vitamin D and cancer. *Anticancer Res* 26(4a):2515–2524.

Steenbock, H. 1924. The induction of growth-prompting and calcifying properties in a ration exposed to light. *Science* 60:224–225.

Steenbock, H., and A. Black. 1924. The reduction of growth-promoting and calcifying properties in a ration by exposure to ultraviolet light. *J Biol Chem* 61:408–422.

Thomas, G., B. Hartaigh, J. Bosch, S. Pilz, A. Loerbroks, M. Kleber, J. Fischer, T. Grammer, B. Bohm, and W. Marz. 2012. Vitamin D levels predict all-cause and cardiovascular disease mortality in subjects with the metabolic syndrome: The Ludwigshafen Risk and Cardiovascular Health (LURIC) study. *Diabetes Care* 35:1158–1164.

Tian, X., T. Chen, L. Matsuoka, J. Wortsman, and M. Holick. 1993. Kinetic and thermodynamic studies of the conversion of previtamin D_3 in human skin. *J Biol Chem* 268: 14888–14892.

Urashima, M., T. Segawa, M. Okazaki, M. Kurihara, Y. Wada, and H. Ida. 2010. Randomized trial of vitamin D supplementation to prevent seasonal influenza A in schoolchildren. *Am J Clin Nutr* 91:1255–1260.

Webb, A., L. Kline, and M. Holick. 1988. Influence of season and latitude on the cutaneous synthesis of vitamin D_3: Exposure to winter sunlight in Boston and Edmonton will not promote vitamin D_3 synthesis in human skin. *J Clin Endocrinol Metab* 67:373–378.

Wolpowitz, D., and B. Gilchrest. 2006. The vitamin D questions: How much do you need and how should you get it? *J Am Acad Dermatol* 54:301–317.

III

Ultraviolet Phototherapy

18

UVB Phototherapy for Psoriasis

Edidiong Ntuen
Kaminska
*University of Chicago
Medical Center*

PierGiacomo
Calzavara-Pinton
University of Brescia

Bernhard Ortel
*University of Chicago
Medical Center*

18.1 Introduction to UVB Phototherapy

Phototherapy is a valuable therapeutic option for many cutaneous disorders and has significantly enriched the treatment options in dermatology. The goals of therapeutic photomedicine are the prevention and reversal of disease processes through the modulation or suppression of pathogenic mechanisms. Ultraviolet B (UVB) phototherapy is based on repeated exposures to nonionizing radiation with wavelengths ranging from 290 nm to 320 nm and has been successfully used for decades to treat psoriasis and other inflammatory skin disorders. This chapter will provide a general overview of UVB in regard to psoriasis because it is one of the best-studied modalities for treating this very common inflammatory skin disease.

18.1.1 Historical Perspective

Natural sunlight has been utilized empirically for thousands of years as phototherapy for skin ailments. The notion of ultraviolet (UV) rays was not discovered until the 19th century. During the latter portion of this period, artificial radiation sources became available to treat the most important cutaneous indications of that time: acne, leprosy, lupus vulgaris, pellagra, psoriasis, prurigo, and syphilis (Roelandts 2002). In 1903, the dermatologist Niels Ryberg Finsen was awarded the Nobel Prize in medicine and physiology for his groundbreaking work in treating lupus vulgaris, a form of skin tuberculosis, with concentrated radiation from a carbon arc lamp. This artificial radiation source was also applied to psoriatic patients. As physicians began to understand the physics of phototherapy more thoroughly, specific wavelengths were targeted to treat various cutaneous diseases,

including psoriasis, and the development of novel radiation sources has not ceased.

18.1.2 What Is Psoriasis?

In order to better understand the use of phototherapies in the management of psoriasis, we introduce here this skin disease for those without a medical background. Psoriasis is a common dermatosis that affects approximately 2% of the world population with regional variations. It presents as a thickening of the skin with redness and a cover of whitish, thick (micaceous) scales. Psoriasis involvement may range from a single, coin-size spot to the whole body surface being affected, the latter case being very rare. The cause of this inflammatory skin disease is multifactorial. Aggregation of cases in families and studies in twins has long suggested a genetic component. Several genes have been identified that confer increased risk for developing psoriasis. In addition, multiple triggers have been described that may precipitate psoriasis in those that carry the genetic background. Physical trauma has been implied as a factor for the location over the large joints, namely the knees and elbows. Minor injuries to the skin, such as scratches, sunburn, or rashes, may precipitate lesions at those sites. Streptococcal infections ("strep throat") are known to induce a generalized eruption of drop-size psoriasis, specifically in young people. Other infections, endocrine alterations, medications, and emotional stress also contribute. High alcohol consumption, smoking, and obesity have been associated with psoriasis. Although there are multiple therapeutic ways to induce remission in psoriasis, it tends to recur in many patients. A cure does not exist at this time because of the genetic predisposition that cannot be altered.

Figure 18.1 shows the knee of an adult with chronic psoriasis. In the affected area, the skin is thick and hard and covered with a solid layer of white scale. Where the scale is missing, the red, infiltrated underlying skin becomes visible, and this area easily bleeds. The skin is composed of three basic layers: the epidermis, which is composed of layers of keratinocytes; the dermis, which lies beneath the epidermis and contains the blood supply that nourishes the epidermis; and the subcutis, which is the fat layer of the skin. The histological image shows the thickened epidermis with compromised keratinocyte maturation (this corresponds to the visible scale), the inflammatory cell infiltrate in the dermis, and the widened blood vessels (these vessels contribute to bleeding when the scale is peeled). The small inset shows a section of normal surrounding skin, which is much thinner and matures to a thin, wispy top layer; this is very different from the thick, red-stained scale layer on top of the psoriatic epidermis.

Psoriasis used to be viewed as a primary disturbance of the epidermal cells, the keratinocytes, which proliferate at a much higher rate in psoriatic compared to normal skin. In the past 25 years, it has become clear that this hyperproliferation is driven by inflammatory cells and mediators. Nowadays, T lymphocytes are viewed as the main drivers of the pathogenic process. As the pathogenic concept of psoriasis has expanded, so has the understanding of how phototherapies work. The early mechanistic idea contended that cell division of hyperproliferative keratinocytes was inhibited by UV-induced DNA lesions. Consistent with the current improved understanding of the role of lymphocytes in pathogenesis, experimental data now indicate that apoptotic cell death of activated infiltrating lymphocytes is decisive in the response to phototherapy.

18.2 Principles and Mechanisms

UVB phototherapy refers to the use of artificial UVB radiation without the addition of an exogenous photosensitizer. It is effective for psoriasis because UVB alters the cutaneous immune function. The photobiological effects are located primarily within the epidermis and the uppermost dermis because of limited UVB penetration into the skin. Because psoriasis is a disease that is associated with hyperproliferation of epidermal keratinocytes and infiltration of activated T cells in the upper dermis, UVB is a logical treatment option. Although a complete understanding of how UVB improves psoriasis and other cutaneous diseases has not been attained, a number of photobiological effects have been identified.

- UVB radiation reduces proliferation of keratinocytes and lymphocytes by affecting DNA synthesis. Nuclear DNA is a major endogenous chromophore for UVB radiation; it absorbs UVB and generates photoproducts (Kulms et al. 1999).
- The predominant DNA lesions are pyrimidine dimers, which are the substrate for most photobiological effects; other DNA effects, such as the formation of the 6,4'-photoproduct, are much smaller in number but may have an important impact, for example, on the carcinogenic potential of UVB (Kulms et al. 1999).
- UVB-induced DNA lesions play a role in cell-cycle control via the tumor-suppressor gene p53 (Lapolla et al. 2011), affecting DNA repair, cell survival decisions, and activation of apoptotic and necrotic death pathways.
- Independently of DNA damage, UVB affects cell surface structures, including receptor molecules, such as those

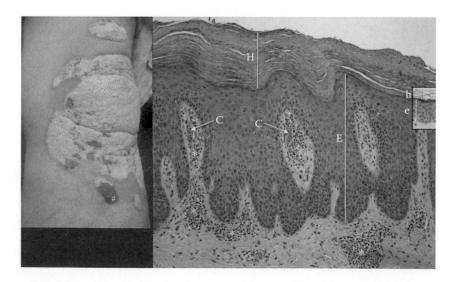

FIGURE 18.1   (See color insert.) Psoriasis. On the left side, the clinical appearance of chronic plaque psoriasis. # indicates area of psoriatic skin without scale. On the right side, a microscopic image of a vertical section of psoriatic skin; the inset on the right shows normal epidermis for comparison. E and e indicate the thickness of psoriatic and normal epidermis; H and h indicate the thickness of psoriatic stratum corneum (scale) and that of normal skin; C indicates widened capillaries with red blood cells inside; * indicates infiltrates of inflammatory cells.

for cytokines, interleukin-1, and the epidermal growth factor, which have effects on transcriptional regulation, proliferation, and survival also through the death receptor CD95 (Rosette and Karin 1996; Leverkus, Yaar, and Gilchrest 1997).

- UVB modulates inflammatory cells by diminishing inflammatory cytokines, shifting the ratio of T-helper lymphocytes, inducing regulatory T cells, and changing T-cell morphology (Weichenthal and Schwarz 2005). UVB can inhibit and reduce the number of Langerhans cells, which play a central role in cutaneous immune functions (Gruner et al. 1992; DeSilva et al. 2008).
- UVB isomerizes urocanic acid from the *cis* to the *trans* form, inducing additional immunosuppressive effects.

As a therapeutic consequence of successful UVB phototherapy, cellular and molecular changes found in a psoriatic epidermis revert to the pattern of normal epidermis. For instance, proteins that are markers of hyperproliferation, such as keratin 16, α3-integrin, and insulin-like growth factor receptors that are increased in psoriatic skin, are reduced after UVB treatments (Krueger et al. 1995).

18.3 Action Spectrum

An action spectrum is the magnitude of an effect plotted against the wavelength; it demonstrates which wavelength of radiation is most effective in producing a specific photon-based outcome. One example seen in Figure 18.2 is the erythema action spectrum, which describes the relative efficiency of UV wavelengths to induce a sunburn reaction. The determination

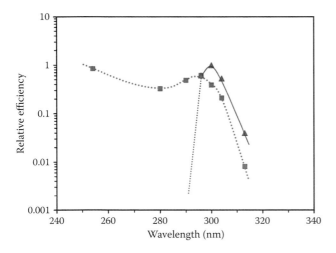

FIGURE 18.2 Phototherapy action spectrum of psoriasis. The action spectrum for clearing psoriasis in a patient is plotted as the reciprocal of the lowest effective daily dose to clearing versus wavelength (triangles). The dashed line shows the erythema action spectrum in the same patient (squares). In this patient, 300 to 313 nm was the most efficient wavelength range. The pooled data favored 313 nm. (From Parrish, J. and Jaenicke, K., *J. Investig. Dermatol.*, 76, 359–362, 1981.)

of a therapeutic action spectrum for psoriasis is more difficult because there are multiple exposures required to induce clearing in psoriasis. Nevertheless, several researchers have contributed to identifying the most effective wavelengths for treating psoriasis.

18.3.1 Experimental Data

Torkel Fischer (1976) reported that 313 nm radiation was very effective in clearing plaque psoriasis. Although he tested only a limited number of wavelengths, his work pointed at a specific wavelength range for therapeutic efficacy. Parrish and Jaenicke (1981) expanded Fischer's line of investigation by studying a larger number of wavelengths in the UVC and UVB range for their effectiveness with psoriasis. Their work showed that wavelengths up to 290 nm—although erythemogenic—were therapeutically ineffective (Figure 18.2). However, erythemogenic doses of radiation between 296 and 313 nm produced significant clearing. Although the study was performed on a small number of patients, the conclusions again favored 313 nm (Parrish and Jaenicke 1981). Van Weelden, Young, and van der Leun (1980) also studied broadband UV sources with different peak emissions in psoriasis and reported that while longer wavelengths did not contribute to the therapeutic benefits, shorter wavelengths decreased effectiveness of treatment, and that UVB was most effective at approximately 310 nm. The Phillips TL-01 fluorescent lamp that emits a narrow peak at approximately 311 nm [now commonly called narrowband UVB (NB-UVB)] has now become the most popular radiation source for the management of psoriasis, validating these experimental data.

18.4 Clinical Phototherapy for Psoriasis

Phototherapy continues to be an efficient and safe treatment option for psoriatic patients. It is generally available as broadband UVB (BB-UVB) and NB-UVB, depending on the spectral properties of the radiation source. Phototherapy can be administered in a hospital, in an outpatient clinic, or even in the patient's home. Whole-body exposures are most often administered in booth-like units that are lined with fluorescent tubes that emit UV radiation (Figure 18.3). Photochemotherapy using psoralen plus UVA (PUVA) is a different type of cutaneous phototherapy and is described in detail in another chapter of this book.

18.4.1 BB-UVB Versus NB-UVB

BB-UVB was the first phototherapy modality other than heliotherapy to be used in treating psoriasis. Conventional BB-UVB lamps emit radiation over the whole UVB spectrum, including the longer wavelengths with the optimal therapeutic efficiency for skin diseases and the shorter wavelengths that are more relevant for causing sunburn. Excessive phototoxicity counteracts the benefit of phototherapy by limiting the amount of

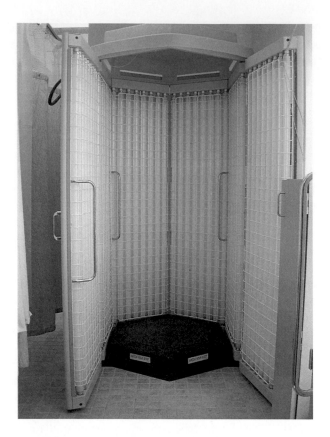

FIGURE 18.3 Phototherapy unit. This booth has a hexagonal cross-section and is lined with UVB-emitting fluorescent tubes. Other units may have a circular or a square layout. Most modern units have built-in electronic controls and integrated dosimetry to avoid hazards of unintentional overexposure. To keep the inside from overheating during operation, air is blown up from the bottom platform and escapes through the slits in the top of the booth.

therapeutic UVB that can be delivered and by compromising patient comfort. Over the years, new radiation sources have been developed to improve efficiency and to reduce potential side effects. Most commonly, these involved the reduction of output in the less-efficient shorter wavelength range of UVB (290–300 nm). Figure 18.4 shows spectra for therapeutic UVB radiation sources.

In recent years, NB-UVB has, in many places, replaced BB-UVB because it appears to be more effective for treating psoriasis. The output of NB-UVB lamps is concentrated in a narrow peak within the optimal therapeutic range, and there is minimal emission in the shorter wavelength range. Based on theoretical considerations, NB-UVB is safer and more efficient than conventional BB-UVB with respect to both clearing and more rapid remission. Regardless, there is a small proportion of psoriasis patients who do not tolerate NB-UVB but demonstrate an excellent clinical response to BB-UVB phototherapy, which thus remains a valid alternative regimen (Pugashetti, Lim, and Koo 2010).

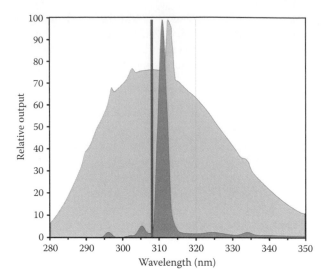

FIGURE 18.4 Emission spectra of UVB radiation sources for phototherapy. Light grey indicates the Philips TL12 broadband UVB tube; dark grey indicates the Philips TL01 narrowband UVB tube; the black line indicates the XeCl excimer laser. Wavelengths below 296 nm contribute to skin phototoxicity but not antipsoriatic activity; the narrowing of the emission spectrum by using narrowband sources thus aims at improved therapeutic efficiency.

18.4.2 Excimer Lasers and Lamps

Monochromatic excimer light (MEL) can be administered by both laser and lamp and it is useful for targeted phototherapy of localized psoriatic lesions involving less than 10% of the body surface and for recalcitrant lesions. An excimer laser uses a combination of an inert noble gas (e.g., xenon) and a reactive halide gas (e.g., chlorine) to create an excited dimer (excimer) that gives rise to emission in the UV range (in this case, 308 nm). An excimer lamp is a radiation source that emits non-coherent radiation at 308 nm by also using xenon and chlorine gases; its advantages over the excimer laser include the ability to treat larger areas and being cost effective (Lapolla et al. 2011; Mudigonda, Dabade, and Feldman 2012). In comparison to whole-body radiation, MEL for targeted NB-UVB therapy results in reduced toxicity, fewer treatments, and lower cumulative UV doses (Stein, Pearce, and Feldman 2008).

18.4.3 Treatment Regimens

Before starting phototherapy, patients should optimally be evaluated for their individual UV sensitivity. This is achieved by determining the minimal erythema dose (MED), the lowest fluence that leads to a well-circumscribed pink erythema 24 h after test exposure. The initial therapeutic UVB dose is between 50% and 70% of the MED, and treatments are given two to five times weekly. Subsequent irradiations are increased based on the skin reaction to the previous exposure. Because peak UVB erythema occurs within 24 h after exposure, incremental changes may occur with each successive treatment.

A previously published guideline for psoriasis treatment with BB-UVB recommends starting at 70% MED with increases of 10% of MED per visit at a frequency of three to four sessions each week for 15–25 visits (Zanolli 2004). Multiple treatment guidelines have been suggested for NB-UVB. Standard regimens involve starting at 50%–70% MED and three weekly exposures with increments of 10%–30% MED of previous exposure with each treatment session (Lapolla et al. 2011). UVB treatment is continued until clearance is achieved or there is no further improvement that can be obtained.

18.4.4 Adjunctive Therapies

UVB can be added to or combined with topical or systemic therapies that are used for improved efficacy and/or to reduce cumulative UV exposure in order to decrease the incidence of long-term side effects. Calcipotriene, a vitamin D derivative, can enhance phototherapy but should be applied after UVB radiation because the compound is inactivated by UVB (Kragballe 2002). Anthralin with UVB is a well-established and effective combination (Hönigsmann 2001). Bathwater psoralen and systemic psoralen or topical and systemic retinoids (ReUVB) also enhance UVB therapy (Ortel et al. 1993; Guenther 2003). Oral methotrexate and UVB have also been shown to work well (Asawanonda and Nateetongrungsak 2006), but UV radiation recall reactions remain a concerning possible adverse reaction.

18.4.5 Other Applications

Applications of UVB phototherapy for skin diseases other than psoriasis are now being studied, primarily with NB-UVB. Cutaneous disorders that can be managed by UVB therapy include atopic dermatitis, vitiligo, cutaneous T-cell lymphoma, lichen planus, certain types of chronic urticaria, pruritus, seborrheic dermatitis, graft-versus-host disease, and others. Photosensitive disorders, such as polymorphic light eruption, solar urticaria, and hydroa vacciniforme, may—seemingly paradoxically—benefit from "hardening" with UVB phototherapy (Bandow and Koo 2004; Gambichler et al. 2005; Feldstein et al. 2011). Describing the therapeutic characteristics of these diseases is outside of the scope of this chapter.

18.5 Contraindications

Absolute contraindications for any phototherapy include patients who have underlying photosensitive disorders, such as but not limited to lupus erythematosus; porphyria cutanea tarda; and genodermatoses with compromised DNA repair, such as xeroderma pigmentosum, and those suffering from autoimmune blistering disorders. Medically unfit patients, such as those with severe cardiovascular or respiratory disease that prevents standing, should not be referred for phototherapy.

Relative contraindications to UVB phototherapy include a personal history of skin cancer or atypical nevus syndrome,

previous exposure to radiotherapy, and photoinduced epilepsy or poorly controlled epilepsy. Common photosensitizing drugs include amiodarone, diltiazem, furosemide, thiazide diuretics, tetracyclines, sulfonamides, oral antifungals, and certain antidepressants, which are all relative contraindications. Notably, there is no contraindication for UVB therapy in pregnant patients.

18.6 Adverse Effects

Although UVB is generally effective for psoriasis, there are associated disadvantages. Socially, phototherapy can be time intensive and an inconvenient option. The most common short-term adverse effect is excessive phototoxicity, which may include pruritus, erythema, or phototoxic blisters; severe UVB burns can result in pain and general malaise resulting from cytokine release. These complications are managed with topical or systemic corticosteroids and nonsteroidal anti-inflammatory agents. Additional unwanted effects include dry skin and recurrence of herpes simplex; emollients, gentle skin care, or antivirals are used in their management.

The most concerning long-term risk of chronic UVB exposure is skin cancer because UV exposure is a well-known risk factor for it. To date, numerous published studies have failed to show an increased risk of melanoma or nonmelanoma skin cancers in patients treated with UVB phototherapy (Lee, Koo, and Berger 2005; Hearn et al. 2008). Longer follow-up may be needed to fully characterize the carcinogenesis risk.

18.7 Summary

UVB phototherapy is a safe, effective, and valuable tool in the treatment of psoriasis. A thorough, detailed patient medical and medication history is essential prior to initiating phototherapy. Clinicians should be aware of the advantages and disadvantages of phototherapy and adjust treatment regimens according to patients' needs. Additionally, other first-line therapeutic options, such as topical steroids, systemic medications, and biologics, should be offered as deemed appropriate. Just as importantly, patient preferences should be considered as these may determine overall compliance.

References

Asawanonda, P., and Y. Nateetongrungsak. 2006. Methotrexate plus narrowband UVB phototherapy versus narrowband UVB phototherapy alone in the treatment of plaque-type psoriasis: A randomized, placebo-controlled study. *J Am Acad Dermatol* 54(6):1013–1018.

Bandow, G., and J. Koo. 2004. Narrow-band ultraviolet B radiation: A review of the current literature. *Int J Dermatol* 43(8):555–561.

DeSilva, B., R. McKenzie, J. Hunter, and M. Norval. 2008. Local effects of TL01 phototherapy in psoriasis. *Photodermatol Photoimmunol Photomed* 24:268–269.

Dotterud, L., T. Wilsgaard, L. Vorland, and E. Falk. 2008. The effect of UVB radiation on skin microbiota in patients with atopic dermatitis and healthy controls. *Int J Circumpolar Health* 67:254–260.

Feldstein, J., J. Bolaños-Meade, V. Anders, and R. Abuav. 2011. Narrowband ultraviolet B phototherapy for the treatment of steroid-refractory and steroid-dependent acute graft-versus-host disease of the skin. *J Am Acad Dermatol* 65(4):733–738.

Fischer, T. 1976. UV-light treatment of psoriasis. *Acta Derm Venereol* 56:473–479.

Gambichler, T., F. Breuckmann, S. Boms, P. Altmeyer, and A. Kreuter. 2005. Narrowband UVB phototherapy in skin conditions beyond psoriasis. *J Am Acad Dermatol* 52(4):660–670.

Glaser, R., F. Navid, W. Schuller, C. Jantschitsch, J. Harder, J. Schroder et al. 2009. UV-B radiation induces the expression of antimicrobial peptides in human keratinocytes in vitro and in vivo. *J Allergy Clin Immunol* 123:1117–1123.

Gruner, S., A. Zwirner, D. Strunk, and N. Sonnichsen. 1992. The influence of topical dermatological treatment modalities on epidermal Langerhans cells and contact sensitization in mice. *Contact Dermatitis* 26:241–247.

Guenther, L. 2003. Optimizing treatment with topical tazarotene. *Am J Clin Dermatol* 4(3):197–202.

Hearn, R., A. Kerr, K. Rahim, J. Ferguson, and R. Dawe. 2008. Incidence of skin cancers in 3867 patients treated with narrow-band ultraviolet B phototherapy. *Br J Dermatol* 159(4):931–935.

Hönigsmann, H. 2001. Phototherapy for psoriasis. *Clin Exp Dermatol* 26(4):343–350.

Kragballe, K. 2002. Vitamin D and UVB radiation therapy. *Cutis* 70(suppl. 5):9–12.

Krueger, J., J. Wolfe, R. Nabeya, V. Vallat, P. Gilleaudeau, N. Heftler, L. Austin, and A. Gottlieb. 1995. Successful ultraviolet B treatment of psoriasis is accompanied by a reversal of keratinocyte pathology and by selective depletion of intraepidermal T cells. *J Exp Med* 182(6):2057–2068.

Kulms, D., B. Poppelmann, D. Yarosh, T. Luger, J. Krutmann, and T. Schwarz. 1999. Nuclear and cell membrane effects contribute independently to the induction of apoptosis in human cells exposed to ultraviolet B radiation. *Proc Natl Acad Sci* 96:7974–7979.

Lapolla, W., B. Yentzer, J. Bagel, C. Halvorson, and S. Feldman. 2011. A review of phototherapy protocols for psoriasis treatment. *J Am Acad Dermatol* 64(5):936–949.

Lee, E., J. Koo, and T. Berger. 2005. UVB phototherapy and skin cancer risk: A review of the literature. *Int J Dermatol* 44(5):355–360.

Leverkus, M., M. Yaar, and B. Gilchrest. 1997. Fas/Fas ligand interaction contributes to UV-induced apoptosis in human keratinocytes. *Exp Cell Res* 232:255–262.

Mudigonda, T., T. Dabade, and S. Feldman. 2012. A review of targeted ultraviolet B phototherapy for psoriasis. *J Am Acad Dermatol* 66(4):664–672.

Ortel, B., S. Perl, T. Kinaciyan, P. Calzavara-Pinton, and H. Hönigsmann. 1993. Comparison of narrow-band (311 nm) UVB and broad-band UVA after oral or bath-water 8-methoxypsoralen in the treatment of psoriasis. *J Am Acad Dermatol* 29(5, pt. 1):736–740.

Parrish, J., and K. Jaenicke. 1981. Action spectrum for phototherapy of psoriasis. *J Invest Dermatol* 76:359–362.

Pugashetti, R., H. Lim, and J. Koo. 2010. Broadband UVB revisited: Is the narrowband UVB fad limiting our therapeutic options? *J Dermatolog Treat* 21(6):326–330.

Roelandts, R. 2002. The history of phototherapy: Something new under the sun? *J Am Acad Dermatol* 46(6):926–930.

Rosette, C., and M. Karin. 1996. Ultraviolet light and osmotic stress: Activation of the JNK cascade through multiple growth factor and cytokine receptors. *Science* 274(5290):1194–1197.

Stein, K., D. Pearce, and S. Feldman. 2008. Targeted UV therapy in the treatment of psoriasis. *J Dermatolog Treat* 19:141–145.

Van Weelden, H., E. Young, and J. van der Leun. 1980. Therapy of psoriasis: Comparison of photochemotherapy and several variants of phototherapy. *Br J Dermatol* 103(1):1–9.

Weichenthal, M., and T. Schwarz. 2005. Phototherapy: How does UV work? *Photodermatol Photoimmunol Photomed* 21:260–266.

Yoshimura, M., S. Namura, H. Akamatsu, and T. Horio. 1996. Antimicrobial effects of phototherapy and photochemotherapy in vivo and in vitro. *Br J Dermatol* 135:528–532.

Zanolli, M. 2004. Phototherapy arsenal in the treatment of psoriasis. *Dermatol Clin* 22:397–406.

19

PUVA Therapy

Juliana Basko-Plluska
University of Chicago
Medical Center

Herbert Hönigsmann
Medical University of Vienna

Bernhard Ortel
University of Chicago
Medical Center

19.1 Introduction

Ultraviolet radiation (UVR) penetrates the human skin at different shallow depths depending on the wavelength and is absorbed by a number of biomolecules. UVR exerts a variety of biological effects, some of which are of therapeutic relevance in the management of skin diseases. In particular, ultraviolet (UV) phototherapies exert an immunomodulatory effect through altered expression of cell surface–associated molecules, production of soluble mediators, and induction of apoptosis in pathogenic cells.

Psoralen photochemotherapy (PUVA) is a special form of phototherapy in which a psoralen and ultraviolet A (UVA) radiation (320–400 nm) are combined. As a UVA-activated compound, psoralen's main role in PUVA is to enhance the inherent biological effect of the UVA wavelength range, which is much weaker compared to ultraviolet B (UVB). Although sunlight-based PUVA therapy using herbal preparations for treatment of vitiligo dates back thousands of years, it was not until 1974 that PUVA proved to be an effective treatment for psoriasis (Parrish et al. 1974). Following approval of PUVA by the Federal Drug Administration (FDA), PUVA became the main therapy for patients with severe psoriasis who had previously failed to respond to other treatment modalities or required hospitalization. PUVA was approved by the FDA for the treatment of psoriasis in 1982. For the last 30 years, it has been successfully used in the management of various other skin disorders, including vitiligo, mycosis fungoides, atopic dermatitis, and many others. Recently, the availability of novel alternative treatments and concerns about carcinogenicity have led to a decline in the utilization of PUVA.

19.2 Principles and Mechanisms

The principle of PUVA is the combination of the two components of photochemotherapy at the site of desired activity. Systemic psoralen is distributed throughout the body but gets activated only in the skin to the depth of UVA penetration. Neither the psoralen nor the low dose of UVA by itself has a therapeutic benefit in PUVA.

19.2.1 Dark Reaction

Psoralens are naturally occurring furocoumarins derived from plants. Furocoumarins are tricyclic, lipophilic compounds that readily penetrate into cells, where they intercalate into the apolar environment inside the DNA double helix, a process that occurs in the absence of UV radiation. For the purpose of this chapter, most references to psoralen will pertain to 8-methoxypsoralen (methoxsalen, 8-MOP), which has been most widely used. Several other furocoumarins, including unsubstituted psoralen, 5-methoxypsoralen, and the synthetic 4,5′,8-trimethylpsoralen, have also been utilized for PUVA (Figure 19.1). Psoralens also associate with other lipophilic structures, such as the cellular membranes.

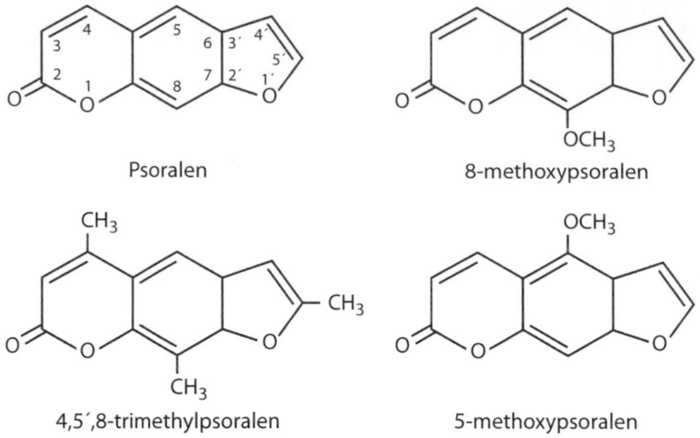

FIGURE 19.1 Structures of psoralens that are being utilized for PUVA. The most extensive experimental and clinical data exist for the use of 8-methoxypsoralen. Trimethylpsoralen (TMP) has been predominantly used for topical sensitization in bath PUVA.

19.2.2 Photochemical Reactions

Psoralens are reactive only when activated by UVR, and the activity is confined to those layers of skin that are penetrated by UVA, namely the epidermis and the papillary dermis. Upon absorption of a photon in the UVA range (320–400 nm), psoralens are activated to the excited singlet state. One portion of these excited molecules undergoes intersystem crossing to form the excited triplet state, which is responsible for two types of photochemical reactions.

Type I (direct) photochemistry results in photoaddition of the psoralen to a pyrimidine base, forming monofunctional adducts at either the reacting 3,4- or the 4′,5′ double bond of the psoralen. Only the 4′,5′ monoadduct can subsequently undergo binding of the psoralen's 3,4 double bond to a nucleic acid of the complementary DNA chain to form a bifunctional psoralen adduct (cross-link) upon absorption of a second photon (Figure 19.2). The psoralen monoadducts and cross-links inhibit DNA synthesis and, therefore, inhibit cellular proliferation of keratinocytes and lymphocytes (Stern 2007). These photobiological effects appear to form the basis for the therapeutic action.

Type II (indirect) photochemical reactions result from formation of reactive oxygen species (ROS), such as singlet oxygen (Joshi et al. 1983). ROS, photochemically produced by psoralen, have been shown to damage cellular membranes, cause mitochondrial dysfunction, and induce apoptosis of keratinocytes and lymphocytes. However, the relevance of type II reactions in the clinical setting is not known.

19.2.3 Action Spectrum

The peak action spectrum for 8-MOP-induced delayed erythema is approximately 330–335 nm (Cripps, Lowe, and Lerner 1982; Kaidbey 1985). Therapeutic relevance of the erythema action spectrum has been evaluated only for psoriasis. Studies have shown that the antipsoriatic activity parallels the erythema

action spectrum for 8-MOP (Farr et al. 1991). The psoralen action spectrum is well covered by conventional therapeutic UVA fluorescent tubes.

19.3 Biological Effects

As discussed within the photochemical reactions section, psoralen adducting with biomolecules and ROS produced through the process lead to phototoxic effects at the cellular level. Some of the cellular and molecular pathways by which PUVA exerts its effects on the treatment of psoriasis have been elucidated (McEvoy and Stern 1987; Laskin et al. 1994). Psoralen photosensitization leads to an altered expression of multiple cytokines and growth factors, including the reduced expression of tumor necrosis factor (TNF)-alpha and vascular endothelial growth factor (VEGF). It alters the expression of cell-surface molecules and also inhibits epidermal growth factor (EGF) by binding to its receptor, which further downregulates the hyperproliferation that is characteristic of psoriasis. Langerhans cells, the main antigen-presenting cells in the skin, are highly sensitive to psoralen phototoxicity (Erkin et al. 2007). Activated T lymphocytes, which have a dominant role in the pathogenesis of psoriasis, readily undergo apoptosis when exposed to PUVA (Coven 1999). These combined effects on the hyperproliferative keratinocytes and on the key cells that drive the inflammatory process account for the high efficacy of PUVA in the treatment of psoriasis.

19.4 Clinical Application

The therapeutic use of PUVA involves repeated controlled phototoxic exposures over the course of several weeks that ultimately lead to remission of cutaneous diseases (Figure 19.3). Psoriasis is the most common dermatosis to be managed by PUVA as the modern-day development of PUVA by Parrish et al. (1974) used psoriasis as the targeted disease, and most of the detailed studies have been performed on this skin condition. In addition to psoriasis, vitiligo is another skin condition for which PUVA has received approval by the FDA. Off-label uses of PUVA are numerous and include mycosis fungoides, atopic dermatitis, pityriasis lichenoides, lymphomatoid papulosis, urticaria pigmentosa, prurigo nodularis, idiopathic pruritus, polymorphous light eruption, and solar urticaria. For most of these conditions, PUVA protocols have been slightly modified before application.

19.5 Treatment Approach

19.5.1 Protocols

All treatment protocols consist of a two-step process. First, the psoralen is administered; then the skin is exposed to UVA. Treatments are repeated until the desired therapeutic effect is achieved. Two distinct treatment approaches have evolved in the treatment of psoriasis by PUVA (Lapolla et al. 2011). The protocol that is most commonly followed in the United States involves administering an initial UVA dose that is determined according

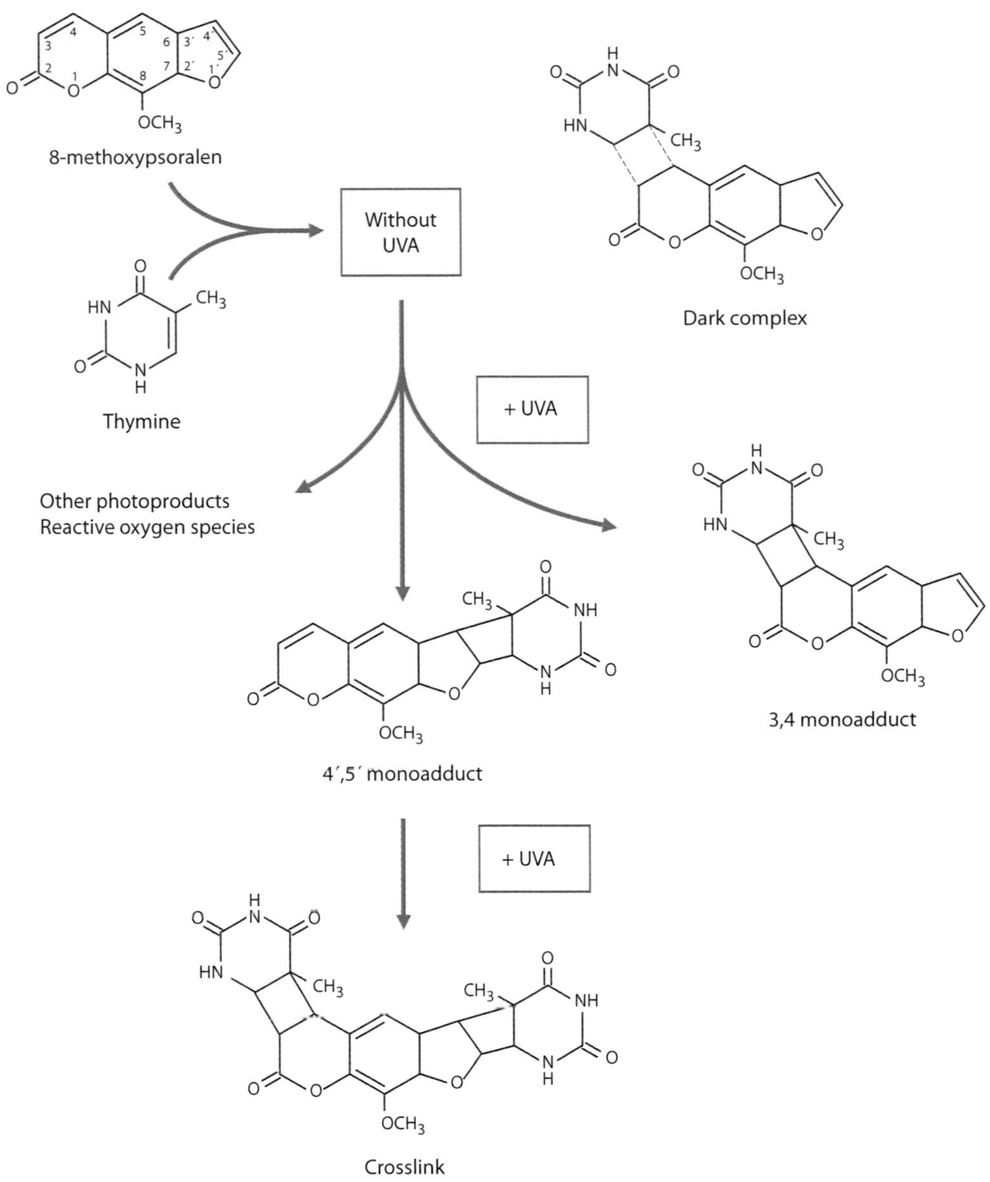

FIGURE 19.2 Scheme of UVA-induced photochemistry of 8-methoxypsoralen (8-MOP). 8-MOP intercalates into DNA in the dark. Exposure to UVA leads to a variety of photochemical reactions. The 8-MOP–DNA cross-link appears to be phototherapeutically most relevant. It requires sequential absorption by 8-MOP of two UVA photons.

to the patient's skin phototype. Radiation dose increments remain fixed, and two to three treatments per week are performed. On the other hand, the protocol used in most European centers bases the initial radiation dose on the patient's minimum phototoxic dose (MPD), which is defined as the minimal UVA irradiation that produces uniform redness at 72 h after exposure. UVA doses are increased by a fraction of the previous exposure. Four treatments per week, with an intermission on Wednesdays, are performed until clearing is achieved. Comparative studies have not shown a clinically significant difference in terms of the results achieved among patients treated per either protocol (Collins et al. 1996).

19.5.1.1 Psoralen Administration

Psoralens are lipophilic compounds, which are poorly soluble in water and can be administered systemically or topically. Once oral psoralen is absorbed from the gastrointestinal tract, it reversibly binds to serum proteins, mainly albumin, and they are metabolized in the liver by hydroxylation and glucuronide formation. As the result of a considerable first-pass effect, small changes in the oral psoralen dose may alter plasma levels considerably; this fact and inconsistent intestinal absorption account for large interindividual and intraindividual variations of 8-MOP plasma concentrations. In the United States, 8-MOP is the only form that is currently used for PUVA. 8-MOP is available in

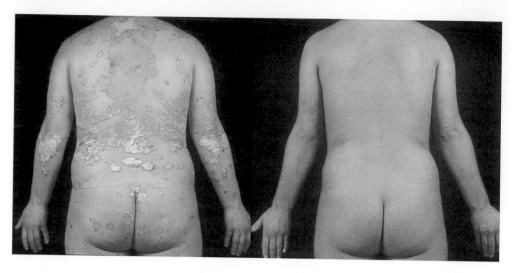

FIGURE 19.3 Therapeutic efficacy of PUVA in the clinical setting. PUVA therapy of a patient with chronic plaque psoriasis with pictures taken before (left) and after (right) a course of oral PUVA. Note on the left pretreatment photograph the erythematous circles on the buttocks from phototoxicity testing.

capsules, either in a crystalline form, known as 8-MOP capsules, or solubilized in a gel, Oxsoralen Ultra. 8-MOP is ingested 2 h before UVA irradiation at a dose of 0.6–0.8 mg/kg. Oxsoralen Ultra is taken 90 min before UVA irradiation at a dose of 0.4–0.6 mg/kg because of its more rapid, efficient, and reproducible delivery. Topical drug delivery can be achieved using creams or lotions or by immersion in water containing psoralen. As previously mentioned, all topical applications also require subsequent exposure to UVA for psoralen activation.

19.5.1.2 Bath PUVA

Bathwater delivery of psoralens is becoming increasingly popular because it provides a uniform drug distribution over the skin surface, very low psoralen plasma levels, and a quick elimination of free psoralens from the skin, thereby reducing the period of photosensitivity. Because of the absence of systemic photosensitization, bathwater delivery of 8-MOP circumvents any gastrointestinal side effects and potential ocular effects. Skin psoralen levels are highly reproducible, and photosensitivity lasts no more than 2 h. Originally, in Scandinavia, bath PUVA was performed with trimethylpsoralen (TMP), but 8-MOP is now being used as well. Bath PUVA consists of 15–20 min of whole-body (or hands and feet) immersion in solutions of 0.5–5.0 mg of 8-MOP per liter of bathwater. Irradiation has to be performed immediately thereafter as photosensitivity decreases rather rapidly. TMP is more phototoxic after topical application and thus is used at lower concentrations than 8-MOP.

19.5.1.3 UVA Exposure

Exposure to UVA is an outpatient procedure. In the most common irradiation units, the patient stands in an upright light box lined with high-output fluorescent lamps. UVA-blocking goggles are required during exposure, and UVA-opaque sunglasses need to be worn outdoors, during daylight, for up to 12 h after ingesting the psoralen. Additional sun and tanning-booth exposure should be avoided on the day of treatment and the following day because of persistent photosensitivity. Unlike sunburn caused by exposure to UVB, which peaks before 24 h, cutaneous psoralen phototoxicity peaks after 72 to 96 h. That is why the European protocol has an intermission period after two subsequent exposures to avoid cumulative excessive phototoxicity. It also illustrates why the MPD is determined after 72 h. A randomized half-body controlled study of twice versus thrice weekly PUVA showed no difference in the psoriasis area and severity index (PASI) score after 25 sessions. However, the cumulative UVA dose of the twice-weekly regimen was considerably lower (Valbuena et al. 2007). A recent randomized controlled study of darker skin gave similar results in this group of patients, demonstrating that twice-weekly phototherapy maintained PUVA efficacy and improved the risk–benefit ratio (El-Mofty et al. 2008).

19.5.1.4 Treatment Response

Treatment response is assessed after the completion of 10 to 15 exposures. After 20–30 treatments, nearly 90% of patients achieve marked improvement or clearing (Melski et al. 1977). Once most of the lesions are cleared, the frequency of therapy is gradually reduced over 4–12 weeks (maintenance therapy) then discontinued. Remissions typically last for 3–6 months (Spuls et al. 1997; Griffiths et al. 2000). Therapy ideally should be restricted to 50 or fewer treatments per year; however, patients with more severe disease may require a greater number of treatments. There are limited data regarding the risks and benefits of maintenance PUVA. Some patients are able to achieve long-term clearance even without maintenance therapy (Koo and Lebwohl 1999). Many recommend discontinuation of treatment once clearance is achieved and also recommend maintenance PUVA only in case there is rapid relapse (British Photodermatology Group 1994). One prospective, right–left comparison study assessing the effect of short-term maintenance treatment for patients with chronic relapsing plaque psoriasis did not show any significant benefit to short-term maintenance therapy (Radakovic et al. 2009).

Because of its high efficacy, PUVA is often used in comparison with other regimens. A different type of phototherapy using UVB is described in another chapter. PUVA has been shown to be more effective than conventional broadband UVB (BB-UVB) (Boer et al. 1984; Morison 1995) and at least equally as effective as narrowband UVB (NB-UVB) in the treatment of psoriasis (Spuls et al. 1997; Yones et al. 2006; Gordon et al. 1999; Markham, Rogers, and Collins 2003). Despite equal efficacy, PUVA patients usually achieve faster clearance and require fewer treatment sessions compared to patients treated with NB-UVB (Dayal, Mayanka, and Jain 2010). A randomized study of 100 patients demonstrated a significantly higher clearance rate with PUVA (84%) versus NB-UVB (63%). In addition, 12% of patients treated with NB-UVB were clear at 6 months as compared to 35% of those treated with PUVA (Gordon et al. 1999).

19.5.1.5 Combination Treatments

The therapeutic efficacy of PUVA therapy is considerably enhanced when combined with oral retinoid intake initiated about a week before the first PUVA exposure. Retinoids are vitamin A derivatives that have modulating effects on epidermal proliferation and differentiation. This combination is also known as RePUVA. RePUVA reduces the number, duration, and cumulative dose of exposure necessary for clearing psoriatic lesions. One study showed a 42% reduction in the total UVA doses required and 10 days less treatment time for patients treated with RePUVA versus those receiving placebo PUVA (Tanew, Guggenbichler, and Hönigsmann 1991). Furthermore, RePUVA has been shown to clear poor PUVA responders who cannot be brought into complete remission by PUVA alone (Fritsch et al. 1978; Saurat et al. 1988; Lebwohl et al. 2001). The combination of PUVA with oral retinoids is ideal because the treatments are not only synergistic, but they also reduce one another's side effects. The mechanism of the synergistic action of retinoids and PUVA is not completely understood. Accelerated desquamation of the psoriatic plaques does enhance the penetration of UVA by thinning or removing the highly scattering psoriatic scales. An immunomodulatory effect of the retinoid may contribute to the reduction of the inflammatory infiltrate as well. Other combination regimens have not reached general acceptance.

19.6 Contraindications

Absolute contraindications to PUVA therapy include the presence of autoimmune disorders, such as pemphigus, bullous pemphigoid, or lupus erythematosus, as these may be precipitated or worsened by PUVA. In addition to this, patients with xeroderma pigmentosum or other genetic disorders that are associated with compromised DNA repair as well as those with a history of a hypersensitivity reaction to psoralens (Table 19.1) are contraindicated. Relative contraindications include photosensitivity or intake

TABLE 19.1 Adverse Effects

Short Term	Long Term
A. Psoralen ingestion	
Nausea	
Vomiting	
B. Phototoxicity	
Pruritus	Nonmelanoma skin cancer
Xerosis	Melanoma
Erythema	Lentigenes
Fever/malaise	Photoaging/telangiectasias
Photo-onycholysis	
Subungual hemorrhage	
Friction blisters	
Hypertrichosis	

of photosensitizing medications, a personal or family history of melanoma, a personal history of skin cancer, and young age.

PUVA is also contraindicated in pregnancy and in nursing mothers. Lastly, patients with hepatic, cardiac, or renal dysfunction as well as patients with malabsorption syndromes need to use caution.

19.7 Adverse Effects

19.7.1 Short-Term Adverse Effects

The short-term side effects of PUVA are related either to the intake of psoralens or to excessive phototoxicity (Morison 1990; Morison, Marwaha, and Beck 1997). 8-MOP can induce systemic adverse effects in the absence of UVA, mainly nausea and vomiting. These can be managed by taking it with ginger ale or by dividing the 8-MOP capsules over a 30-minute period and taking them with food; antiemetics are rarely required. Where available, 5-MOP at 1.2 to 1.5 mg/kg may replace 8-MOP as 5-MOP does not cause nausea.

Phototoxic side effects may be heralded by generalized pruritus, xerosis, or a tingling sensation. Acute phototoxicity resembles sunburn; however, the time course of cutaneous phototoxicity is prolonged compared to regular sunburn. If phototoxicity is severe, it can lead to blister formation and epidermal necrosis and may be accompanied by systemic symptoms, including fever and general malaise resulting from massive cytokine release. It may be necessary to reduce subsequent UVA doses or skip a few PUVA sessions depending on the severity of the symptoms, which can be managed symptomatically. Additional short-term adverse effects of PUVA phototoxicity include photo-onycholysis, subungual hemorrhages, conjunctivitis, friction blisters, and hypertrichosis.

19.7.2 Long-Term Adverse Effects

The long-term adverse effects of PUVA therapy are related to its mutagenic and immunosuppressive effects. An increased risk of nonmelanoma skin cancer and melanoma has been reported with extensive use of PUVA. Most of the data on the carcinogenicity of PUVA therapy originate from the PUVA follow-up study in the United States. This multi-institutional prospective cohort study evaluated 1380 patients who began PUVA treatment for psoriasis between 1975 and 1976 and included nearly 30,000 person-years of follow-up of those with moderate to severe psoriasis. This cohort of patients developed 30 times more squamous cell cancers (SCCs) than what would be expected for a comparable population sample. The risk of developing SCCs was PUVA dose dependent. While exposure to fewer than 150 treatments did not lead to significantly increased risk of SCCs, more than 350 exposures were associated with a 20-fold increase. In this cohort of patients, the genitalia of male patients previously treated with tar and UVB appeared to be particularly susceptible to the carcinogenic stimuli of PUVA (Stern 1990), but the risk seemed not to be increased when PUVA was used alone (Wolff and Hönigsmann 1991). In a retrospective study from France comprising 5400 patients treated between 1978 and 1998, no case of genital skin cancer was found despite the fact that the genital area had not been protected during UVA exposure, and this raises the question of whether genital shielding is absolutely necessary (Aubin et al. 2001).

In terms of basal cell carcinoma (BCC), the risk was only four-fold higher even with more than 450 treatments (Stern 2012). These concerns have led to a more prudent use of PUVA. Patients without a history of skin cancer or photodamage are considered to be good candidates. PUVA is also considered for patients with preexisting conditions that would predispose them to elevated risks associated with alternative psoriasis therapies, such as certain immunosuppressive systemic therapies.

The PUVA follow-up study also documented the occurrence of melanoma among cohort subjects since they were first treated with PUVA in 1975 and 1976 (Stern 2001). Even in a large cohort with substantial exposure to PUVA, an increased risk of melanoma was not detectable until approximately 15 years after the first treatment (a fivefold increase in patients treated with high doses). The data also showed that the risk of melanoma was even higher among patients exposed to higher doses of PUVA. Melanomas were exclusively seen among patients with skin type I–II; no melanomas were seen in patients with skin types IV or higher. This study suggested that latency (i.e., time from exposure) is an especially important factor in determining the ultimate risk of melanoma among PUVA-treated patients. On the other hand, in European populations, this magnitude of carcinogenicity has not been confirmed. The different approach to PUVA in Europe may account for the difference.

Additional long-term side effects include photoaging manifested as dermal sclerosis, telangiectasias, epidermal atrophy, and PUVA lentigines (Stern et al. 1985; Rhodes, Harrist, and Momtaz 1983; Stern 1994).

19.8 Monitoring Guidelines

Monitoring guidelines for patients treated with PUVA have been established. A baseline skin and ophthalmologic examination as well as laboratory investigations of renal and liver function are recommended prior to initiation of therapy. Although ophthalmologic sequelae, including cataracts, have been considered, the only major clinically applicable concern is PUVA-induced conjunctivitis if eye protection is neglected (Calzavara-Pinton et al. 1994). Complete skin examinations, including biopsy of suspicious lesions, are recommended every 6–12 months while on PUVA therapy and yearly after cessation of PUVA therapy (Drake et al. 1994).

19.9 Summary and Outlook

The history of modern PUVA involved the rapid rise of a new therapeutic principle and technology in the early 1980s. An efficacy that was previously rarely seen propelled this novel regimen forward, and it became a serious and often successful consideration in a variety of skin diseases besides psoriasis. Trimethylpsoralen and 5-methoxypsoralen have also been used in multiple clinical trials and in routine clinical applications. When prospective studies documented risks of cutaneous carcinogenicity, PUVA has been progressively substituted with NB-UVB phototherapy. It has become increasingly more challenging to obtain 8-MOP; all other psoralens are very difficult to obtain for therapy in many countries. However, it is important that PUVA remains a therapeutic option for patients with, for example, psoriasis, graft-versus-host disease, eczema, vitiligo, and others because of its reliability even in severe diseases.

References

Aubin, F., E. Puzenat, P. Arveux et al. 2001. Genital squamous cell carcinoma in men treated by photochemotherapy: A cancer registry-based study from 1978 to 1998. *Br J Dermatol* 144(6):1204–1206.

Boer, J., J. Hermans, A. Schothorst et al. 1984. Comparison of phototherapy (UV-B) and photochemotherapy (PUVA) for clearing and maintenance therapy for psoriasis. *Arch Dermatol* 120:52–57.

British Photodermatology Group. 1994. British Photodermatology Group guidelines for PUVA. *Br J Dermatol* 130:246–255.

Calzavara-Pinton, P., A. Carlino, E. Manfredi et al. 1994. Ocular side-effects PUVA-treated patients refusing eye sun protection. *Acta Derm Venereol Suppl* 186:164–165.

Collins, P., N. Wainwright, I. Amorim et al. 1996. 8-MOP PUVA for psoriasis: A comparison of a minimal phototoxic dose-based regimen with a skin-type approach. *Br J Dermatol* 135:248–254.

Coven, T. R., I. B. Walters, I. Cardinale, and J. G. Krueger. 1999. PUVA-induced lymphocyte apoptosis: Mechanism of action in psoriasis. *Photodermatol Photoimmunol Photomed* 15(1):22–27.

Cripps, D., N. Lowe, and A. Lerner. 1982. Action spectra of topical psoralens: A re-evaluation. *Br J Dermatol* 107:77–82.

Dayal, S., Mayanka, and V. Jain. 2010. Comparative evaluation of NBUVB phototherapy and PUVA photochemotherapy in chronic plaque psoriasis. *Indian J Dermatol Venereol Leprol* 76(5):533–537.

Drake, L., R. Ceilley, W. Dorner et al. 1994. Guidelines of care for phototherapy and photochemotherapy. *J Am Acad Dermatol* 31:643–648.

El-Mofty, M., H. El Weshahy, R. Youssef et al. 2008. A comparative study of different treatment frequencies of psoralen and ultraviolet A in psoriatic patients with darker skin types (randomized-control study). *Photodermatol Photoimmunol Photomed* 24(1):38–42.

Erkin, G., Y. Ugur, C. Gurer et al. 2007. Effect of PUVA, narrowband UVB and cyclosporine on inflammatory cells of the psoriatic plaque. *J Cutan Pathol* 34:213–219.

Farr, P., B. Diffey, E. Higgins et al. 1991. The action spectrum between 320 nm and 400 nm for clearance of psoriasis by psoralenphotochemotherapy. *Br J Dermatol* 124:443–448.

Fritsch, P., H. Hönigsmann, E. Jaschke et al. 1978. Augmentation of oral methoxsalen-photochemotherapy with an oral retinoic acid derivative. *J Invest Dermatol* 70:178–182.

Gordon, P., B. Diffey, J. Matthews et al. 1999. A randomized comparison of narrow-band TL-01 phototherapy and PUVA photocheotherapy for psoriasis. *J Am Acad Dermatol* 41:728–732.

Griffiths, C., C. Clarck, R. Charlmers et al. 2000. A systematic review of treatments for severe psoriasis. *Health Technol Assess* 4:1–125.

Kaidbey, K. H. 1985. An action spectrum for 8-methoxypsoralen-sensitized inhibition of DNA synthesis in vivo. *J Invest Dermatol* 85(2):98–101.

Koo, J., and M. Lebwohl. 1999. Duration of remission of psoriasis therapies. *J Am Acad Dermatol* 41:51–59.

Lapolla, W., B. Yentzer, J. Bagel et al. 2011. A review of phototherapy protocols for psoriasis treatment. *J Am Acad Dermatol* 64:936–949.

Laskin, J., E. Lee, D. Laskin et al. 1986. Psoralens potentiate ultraviolet light-induced inhibition of epidermal growth factor binding. *Proc. Natl Acad Sci* 83:8211–8215.

Lebwohl, M., L. Drake, A. Menter et al. 2001. Consensus conference: Acitretin in combination with UVB or PUVA in the treatment of psoriasis. *J Am Acad Dermatol* 45:544–553.

Markham, T., S. Rogers, and P. Collins. 2003. Narrowband UV-B (TL-01) phototherapy vs oral 8-methoxypsoralen psoralen-UV-A for the treatment of chronic plaque psoriasis. *Arch Dermatol* 139:325–328.

McEvoy, M., and R. Stern. 1987. Psoralens and related compounds in the treatment of psoriasis. *Pharmacol Ther* 34:75–97.

Melski, J., L. Tanenbaum, J. Parrish et al. 1977. Oral methoxsalen-photochemotherapy for the treatment of psoriasis: A cooperative clinical trial. *J Invest Dermatol* 68:328–335.

Morison, W. 1990. *Phototherapy and Photochemotherapy of Skin Disease*, 2nd ed. Raven Press, New York.

Morison, W., S. Marwaha, and L. Beck. 1997. PUVA-induced phototoxicity: Incidence and causes. *J Am Acad Dermatol* 36:183–185.

Parrish, J., T. Fitzpatrick, L. Tanenbaum et al. 1974. Photochemotherapy of psoriasis with oral methoxsalen and long-wave ultraviolet light. *N Engl J Med* 291:1207–1211.

Pathak, M. A., and P. C. Joshi. 1983. The nature and molecular basis of cutaneous photosensitivity reactions to psoralens and coal tar. *J Invest Dermatol* 80:66–74.

Radakovic, S., A. Seeber, H. Hönigsmann et al. 2009. Failure of short-term psoralen and ultraviolet A light maintenance treatment to prevent early relapse in patients with chronic recurring plaque-type psoriasis. *Photodermatol Photoimmunol Photomed* 25(2):90–93.

Rhodes, A., T. Harrist, and T. Momtaz. 1983. The PUVA-induced pigmented macule: Alentiginous proliferation of large, sometimes cytologically atypical, melanocytes. *J Am Acad Dermatol* 9:47–58.

Saurat, J., J. Geiger, P. Amblard et al. 1988. Randomized double-blind multicenter study comparing acitretin-PUVA, etretinate-PUVA and placebo-PUVA in the treatment of severe psoriasis. *Dermatologica* 177(4):218–224.

Spuls, P., L. Witkamp, P. Bossuyt et al. 1997. A systematic review of five systemic treatments for severe psoriasis. *Br J Dermatol* 137:943–949.

Stern, R. 1990. Genital tumors among men with psoriasis exposed to psoralens and ultraviolet A radiation (PUVA) and ultraviolet B radiation. The Photochemotherapy Follow-up Study. *N Engl J Med* 322(16):1093–1097.

Stern, R. 1994. The Photochemotherapy Follow-up Study. 1994. Ocular lens findings in patients treated with PUVA. *J Invest Dermatol* 103:534–538.

Stern, R. 2001. The risk of melanoma in association with long-term exposure to PUVA. *J Am Acad Dermatol* 44:755–761.

Stern, R. 2007. Psoralen and ultraviolet A light therapy for psoriasis. *N Engl J Med* 357:682–690.

Stern, R. 2012. The risk of squamous cell and basal cell cancer associated with psoralen and ultraviolet A therapy: A 30-year prospective study. *J Am Acad Dermatol* 4:1–10.

Stern, R., J. Parrish, T. Fitzpatrick et al. 1985. Actinic degeneration in association with long-term use of PUVA. *J Invest Dermatol* 84:135–138.

Tanew, A., A. Guggenbichler, and H. Hönigsmann. 1991. Photochemotherapy for severe psoriasis without or in combination with acitretin: A randomized, double-blind comparison study. *J Am Acad Dermatol* 25:682–684.

Valbuena, M., O. Hernandez, M. Rey et al. 2007. Twice- vs. thrice-weekly MPD PUVA in psoriasis: A randomized-controlled efficacy study. *Photodermatol Photoimmunol Photomed* 23:126–129.

Wolff, K., and H. Hönigsmann. 1991. Genital carcinomas in psoriasis patients treated with photochemotherapy. *Lancet* 337(8738):439.

Yones, S., R. Palmer, T. T. Garibaldinos, and J. L. Hawk. 2006. Randomized double-blind trial of the treatment of chronic plaque psoriasis: Efficacy of psoralen-UV-A therapy vs narrowband UV-B therapy. *Arch Dermatol* 142(7):836–842.

20

Extracorporeal Photopheresis

Fiona Dignan
*St. Thomas' Hospital and
Royal Marsden Hospital*

Julia Scarisbrick
University Hospital Birmingham

20.1 Introduction

Extracorporeal photopheresis (ECP) is a cell-based immune-modulatory therapy, which was first reported by Edelson et al. (1987) for the treatment of erythrodermic cutaneous T-cell lymphoma (CTCL). This technique is now used to treat CTCL as well as other T-cell-mediated conditions, including acute and chronic graft-versus-host disease (GvHD) and rejection of solid organ transplants. This chapter will summarize the research work undertaken to elucidate the mechanism of action of ECP and outline the technical procedure and its safety profile. The role of ECP in the treatment of CTCL, GvHD, and other conditions will also be discussed.

20.2 Mechanism of Action of ECP

The mechanism of action of ECP is a topic of active research worldwide. It is likely that ECP works by a multifaceted mechanism that is only partially understood at the present time. The technique has caused particular interest as it seems to cause two opposite effects: immune system activation in CTCL and downregulation of autoallogeneic immune responses in GvHD. Patients who undergo long-term ECP treatment for CTCL or GvHD do not appear to be at increased risk of either malignancy or infection. In addition, the treatment does not seem to suppress T- or B-cell responses to novel or recall antigens. These observations suggest that ECP does not cause a generalized immunosuppressive effect but may instead exert its effect by modulation of the immune system (reviewed in Marshall 2006).

20.2.1 Cellular Apoptosis

The combination of 8-methoxypsoralen (8-MOP) and ultraviolet (UV) light is known to cause DNA damage by the formation of monoadducts and covalent cross-links of DNA, and psoralen plus ultraviolet A (PUVA) has been shown to induce apoptosis of lymphoid cells (Marks and Fox 1991). Only a small proportion of circulating T cells (approximately 10%) are exposed to PUVA during an ECP procedure, suggesting that lymphocyte apoptosis alone cannot account for the efficacy of the technique (reviewed in Fimiani, Di Renzo, and Rubegni 2004).

20.2.2 Anticlonal Immunity

In the treatment of CTCL, ECP may induce anticlonal immunity by inducing a vaccine-like response. It has been shown in murine models that reinfusion of a clone of pathogenic T cells that had been exposed to 8-MOP/UVA led to a response by which untreated T cells of the same clone were targeted (Khavari et al. 1988).

It is hypothesized that this "vaccination" effect occurs as a result of stimulation of monocytes. There have been a number of reports suggesting that monocytes are resistant to apoptosis by ECP (reviewed in Bladon and Taylor 2006). Berger et al.

(2001) reported that monocytes that were incubated overnight after ECP treatment differentiated into immature dendritic cells. This was felt to be a result of the temporary adherence of monocytes to the plastic surfaces of the ECP kit procedure (Berger et al. 2001). These activated dendritic cells engulfed apoptotic cells, presented tumor antigens, and stimulated cytotoxic T cells and natural killer cells in the equivalent of a vaccine-like response (reviewed in Marshall 2006).

This theory is supported by clinical data in CTCL as patients who respond to ECP have a detectable malignant peripheral T-cell clone and are relatively immune competent, allowing for the generation of a vaccine-like reaction (reviewed in Bladon and Taylor 2006).

20.2.3 Changes in Cytokine Production

The process of clearing of apoptotic cells by antigen-presenting cells (APCs) may lead to regulation of immune responses and tolerance-inducing APCs. The clearance of apoptotic cells has been shown to modulate cytokine production with an overall anti-inflammatory effect. In particular, there is elevation of interleukin (IL)-10 and transforming growth factor β (reviewed in Marshall 2006). ECP has been shown to reduce levels of pro-inflammatory cytokines, including tumor necrosis factor (TNF) α, IL-1A, IL-1B, IL-6, and IL-8, which may explain its effect on inflammatory disorders, including GvHD (reviewed in Bladon and Taylor 2006).

20.2.4 Regulatory T Cells

Regulatory T cells may have a role in the pathogenesis of GvHD, although this has not been confirmed in all studies (Edinger et al. 2003). It has been hypothesized that clearance of apoptotic cells following ECP may stimulate the production of regulatory T cells (reviewed in Peritt 2006). Gatza et al. (2008) have shown in a murine model that ECP reduces the allogeneic responses of donor effector T cells and increases Foxp3+ T regulatory cells.

20.3 ECP: The Technique

The treatment is a three-stage procedure: (1) leukapheresis, (2) ex vivo photoactivation with 8-MOP/UVA, and (3) reinfusion of the buffy coat. The procedure takes 3 to 4 h. One cycle of ECP comprises two treatments on two consecutive (Scarisbrick 2009). Venous access requires either peripheral access (via a Kimal black-eye access needle, size 16–18 G) or a single-lumen apheresis line (10.8 Fr) two consecutive days.

During the collection phase, whole blood enters a latham centrifuge bowl, and the leukocytes are separated at speeds greater than 4500 r.p.m., generating force fields of up to 2700 g. The buffy coat consists of leukocytes with some plasma and erythrocytes (Scarisbrick et al. 2008). Originally, systemic 8-MOP was used for photoactivation, but this has now been replaced by an extracorporeal photosensitizer (UVADEX, Therakos, Ascot, UK). UVADEX is injected directly into the buffy coat

before UVA irradiation, thereby avoiding the systemic toxicities of oral psoralens. This product is used specifically as part of a second-generation UVAR XTS (Therakos) system and is the only closed system commercially available for ECP, although separate components have been used in some centers in Europe. A third-generation system (CellEx, Therakos) is now available, which allows more consistent buffy coat harvest while reducing treatment times from 4 h to less than 1 h (Scarisbrick 2009).

20.3.1 Safety Considerations

The safety profile of ECP has been documented in several studies (reviewed in Scarisbrick 2009). The procedure is relatively safe, and severe side effects are rare. Mild transient side effects can include fatigue, headache, fever, chills, and nausea. Nausea is the most commonly reported, with the total number of adverse reactions recorded as typically less than 1%. More severe side effects can include hypotension and vasovagal syncope, but these are rare. There are no long-term side effects despite more than 5 years' follow-up in some patients, and the technique is not associated with an increase in malignancy or infection (Scarisbrick 2009). A very important aspect of this treatment is that patients do not become immunosuppressed; therefore, there is no increase in infections, and the success of the treatment allows the medical therapies to be tapered off and finally withdrawn (reviewed in Marshall 2006). Patients who require indwelling vascular access devices may develop complications associated with device insertion or line-associated infections or thrombosis.

Heparin is generally used as an anticoagulant during cytapheresis, but side effects have rarely been reported (Heshmati 2010). Patients may develop anemia or thrombocytopenia, particularly if they are treated with intensive ECP schedules. Transfusion support may be required (Heshmati 2010). Reinfusion may be associated with fluid overload in susceptible patients, but the newer CellEx system allows more flexibility and can eliminate extracorporeal volume by using streamlined blood prime methods (www.therakos.com).

20.4 The Role of ECP in CTCL

20.4.1 Cutaneous CTCL

Primary CTCL defines a group of primary cutaneous lymphomas of which mycosis fungoides (MF) and Sézary syndrome (SS) are the most common. The incidence of CTCL in the United States has been shown to be increasing. The overall incidence is 6.4 per million persons, with a higher incidence in males and older age groups.

CTCL is frequently an indolent disease, but more aggressive variants occur, such as SS, which has a median survival of just 3 years. Skin-directed therapies (SDT) are the most appropriate treatment for early-stage CTCL. Patients who are refractory to SDT or with advanced disease may require systemic therapy and, when possible, should be entered in clinical trials.

ECP is a recommended therapy for stage III or IVA erythrodermic CTCL (Kim et al. 2003; Willemze et al. 2005). This includes patients with erythrodermic MF and SS. The use of ECP in the early stages of the disease has shown some benefit but should be limited to clinical trials.

20.4.2 ECP in CTCL

20.4.2.1 Early-Stage IA–IIA MF

ECP therapy in early clinical stages (IB, IIA) and skin stages (T2 and T4) of MF and SS is efficacious (Zic 2003). The use of ECP to treat 1A disease is controversial because of the high comparative costs for ECP and the excellent prognosis in this group of patients.

A review of the current literature on ECP in CTCL by Miller et al. (2007) reports on 124 early-stage patients treated with ECP or ECP plus adjuvant therapy from 16 published studies between 1987 and 2007. Response rates with ECP and ECP plus adjuvant therapy ranged from 33% to 88%.

A recent study of 19 patients, all with early-stage MF [IA (n = 3), IB (n = 14), and IIA (n = 2)], showed an overall response rate of 42% (8 out of 19, including 7 partial responses and 1 complete response) with a median of 12 ECP sessions (range, 3–32) given over a median of 12 months (3–32 months) and with an overall duration of response of 6.5 months (range 1–48 months). Quality-of-life questionnaires also showed an improvement in emotional scores (Talpur et al. 2011). A randomized crossover study of 16 patients with MF treated with PUVA twice weekly for 3 months followed by ECP once monthly for 6 months at relapse or vice-versa, showed improved skin scores after PUVA compared to ECP in the eight patients who completed the study (Child et al. 2004).

Given the excellent safety profile of ECP compared with other systemic therapies for CTCL and its demonstrated efficacy, this treatment modality is possibly beneficial for patients with earlier stages of CTCL. Randomized trials comparing ECP to other standard therapies are needed to determine if there is a role for ECP in the early stages of CTCL.

20.4.2.2 Intermediate-Risk (Stage III) MF/SS and High-Risk (Stage IIB/IVA/IVB) MF/SS

ECP is an effective therapy for erythrodermic CTCL (Scarisbrick 2009; Crovetti et al. 2000). Response rates to ECP have been shown to vary widely between different study groups. A systematic review of nonrandomized and mostly retrospective studies of ECP in CTCL on more than 650 patients from 30 published studies of CTCL treated with ECP showed a mean response rate of 63% (range, 43%–100%) (Scarisbrick et al. 2008). Response rates were higher in those with erythrodermic CTCL. Complete responses were recorded in 27 studies involving 527 patients with a mean complete response (CR) of 20%. The differences in response rates between centers may relate to different patient selection for treatment with ECP, such as the presence of a peripheral T-cell clone, stage of disease, prior treatment, ECP

protocol, duration of ECP, and definition of response. Similar considerations need to be applied when reporting on survival in patients with erythrodermic CTCL receiving ECP.

Although these studies have been uncontrolled, the evidence suggests that ECP can be an effective and well-tolerated therapy for erythrodermic patients, including those with SS.

20.4.3 Initiation and Length of Therapy

Treatment is typically given on two consecutive days as initially performed by Edelson et al. (1987). Treatments are repeated every 2 to 4 weeks and may be given most frequently in those with a high blood tumor burden. Responses may take up to 6 months, and ECP therapy typically continues until loss of response. The median number of treatments varies between centers from 10 to 32 (Scarisbrick et al. 2008). The range is wide, with some patients continuing on ECP for more than 5 years.

Treatment may be initiated as monotherapy or with adjuvant therapy. Adjuvant therapy is most frequent with interferon-alpha and/or bexarotene but may be safely used in combination with a variety of therapies (Suchin et al. 2002; Duvic, Chiao, and Talpur 2003; Tsirigotis et al. 2007; Scarisbrick et al. 2008). Adjuvant therapy may also be added safely to improve response in those receiving ECP, and some patients may benefit from combination ECP regimens.

20.4.4 Monitoring Response to Treatment

The UK consensus statement recommends formal assessment every 3 months (Scarisbrick et al. 2008). Responses may be measured in the skin, blood, and lymph nodes. Quality-of-life questionnaires should be available and have shown a positive response to ECP. Studies should comply with the clinical end points and response criteria (Olsen et al. 2011).

20.5 The Role of ECP in GvHD

20.5.1 GvHD

Hematopoietic stem cell transplantation (HSCT) is a curative treatment modality for many patients with both malignant and nonmalignant hematological disorders. A major barrier to a successful outcome of HSCT is the development of GvHD. GvHD occurs as a result of the immunological disparity between patients and their donors and results in serious morbidity and mortality. GvHD is a relatively common complication following HSCT. The prevalence of acute GvHD ranges from 35%–45% in recipients of fully matched sibling donor grafts to 60%–80% in people receiving one-antigen HLA mismatched unrelated donor grafts (Ferrara et al. 2009). Acute GvHD classically occurs early after transplantation and targets the skin, gut, and liver. The clinical presentation of chronic GvHD is very broad and heterogenous, with the possibility of almost every organ system being involved. The National Institute for Health has recently defined diagnostic criteria for chronic GvHD (Filipovich et al. 2005).

The treatment of GvHD is targeted toward suppressing the overreactive donor immune system. The mainstay of treatment is using immunosuppressive agents, such as steroids and calcineurin inhibitors, which can be associated with an increased risk of infections and, potentially, disease relapse. Patients with steroid-refractory GvHD have historically been notoriously difficult to treat. A number of third-line therapies exist (e.g., chemotherapy agents, mycophenolate mofetil, monoclonal antibodies), but all of these have been associated with significant side effects, including infection. The success rate of such therapies varies between studies but is approximately 30% (Antin et al. 2004). ECP has been used for several years as a treatment for the chronic manifestations of chronic GvHD and, more recently, as a promising treatment for acute GvHD.

20.5.2 Chronic GvHD

Owsianowski et al. (1994) first reported the use of ECP in chronic GvHD in 1994 in one patient who had an improvement in lichenified skin changes, joint contractures, and sicca syndrome. Promising responses and a steroid sparing effect have been reported in patients with chronic skin GvHD in several nonrandomized, nonblinded studies, even in those with sclerodermatous skin.

A prospective study at St. Thomas' hospital published in 2003 reported on 28 patients with cutaneous chronic GvHD. After 6 months, median skin scores were 53% lower ($P = .003$) in sclerodermoid and lichenoid disease than at presentation (Seaton et al. 2003). Apisarnthanarax et al. (2003) retrospectively reviewed 32 patients who had received ECP for steroid-dependent or steroid-refractory disease. After a median of 36 cycles of ECP, complete or partial responses were seen in 56% of patients. In addition, 64% of patients were able to reduce their steroid dose by 50% (Apisarnthanarax et al. 2003). Couriel et al. (2006) reported a response rate of 61% in 71 patients with skin, oral, and liver disease and, 1 year after starting ECP, 22% of patients had discontinued steroids. We recently reported on 82 patients treated for mucocutaneous disease. A partial or complete response was observed in 65 out of 69 evaluable patients, and 77% of evaluable patients achieved a reduction in immunosuppression (Dignan et al. 2012).

Flowers et al. (2008) recently published the first multicenter, randomized, controlled, prospective phase 2 trial of ECP in the treatment of patients with GvHD. This study included patients who were steroid dependent, steroid refractory, and intolerant of steroids. Ninety-five patients were randomized to receive either ECP and standard therapy (corticosteroids plus other immunosuppressive agents, including cyclosporine, tacrolimus, or mycophenolate mofetil) or standard therapy. The study used percentage improvement in total skin scores after 12 weeks of ECP treatment as the primary end point. The percentage reduction in total skin score from baseline was greater in the ECP arm compared to the non-ECP arm, but this did not achieve statistical significance ($P = .48$). The proportion of patients who had at least a 50% reduction in steroid dose and at least a 25% decrease in total skin score was 8.3% in the ECP arm at week 12 and 0% in the control arm ($P = .04$) (Flowers et al. 2008). This study has several limitations because of the methodologic challenges of conducting clinical trials in patients with chronic GvHD. These include the short duration of treatment, using only skin as the primary end point to assess response, the limited time allowed for reduction in steroids (6 weeks), and the large variation in immunosuppressive regimens used.

The response reported in patients with visceral GvHD, for example, in the liver, is more variable. In 2006, Greinix et al. reported a CR rate of 68% for liver chronic GvHD (17 out of 25 patients) (Greinix et al. 2006a). Similarly, Couriel et al. (2006) reported a partial response rate of 15 out of 21 (71%) for liver chronic GvHD. These results have not been reflected in all studies. Foss et al. (2005) reported no response in six patients with liver GvHD, and Seaton et al. (2003) reported a partial response in only 8 of 25 patients. The number of patients with chronic GvHD involving the gut, lungs, joints, and eyes enrolled in studies to date is more limited, and the role of ECP in the treatment of these manifestations is less clear.

20.5.3 ECP in Acute GvHD

There are fewer reports detailing the role of ECP in acute GvHD compared to chronic GvHD. The initial reports included small patient numbers but did suggest efficacy of ECP in the acute setting (Smith et al. 1998). A retrospective series of 23 patients with acute steroid-refractory GvHD reported a complete response rate of 52%, although no patients with grade IV GvHD had a CR. A trend for improved survival was seen in grade III/IV GvHD compared to matched controls (38% vs. 16%, $P = .08$) (Perfetti et al. 2008). Greinix et al. (2006b) have published the largest series to date. This phase 2 prospective study included 59 patients with steroid-refractory or steroid-dependent GvHD treated with two consecutive ECP treatments every week. Complete responses were reported in 82% of patients with cutaneous involvement, 61% with liver involvement, and 61% with gut involvement.

20.5.4 Survival and Quality of Life

A significantly longer survival has been reported in patients who respond to ECP compared to nonresponders. Couriel et al. (2006) suggested that response to ECP and platelet count at initiation of therapy were the strongest predictors of mortality. In acute GvHD, Greinix et al. (2006b) reported a significant difference in the transplant-related mortality in patients who responded to ECP (14%) compared to those who did not respond (73%, $P < .0001$). Patients receiving ECP seem to have an improvement in their quality of life (Flowers et al. 2008). This trend may reflect a reduction in their corticosteroid dose or improvement in the symptoms of their disease, although, as patients were aware of their treatment assignment, the placebo effect may also be implicated (Flowers et al. 2008). The Karnofsky performance scores can, however, be used as a surrogate marker of quality of life, and these have been shown to increase from 50%–60% to 90% following ECP treatment (Smith et al. 1998).

20.5.5 Initiation of Treatment

In chronic GvHD, studies have focused on the use of ECP in patients who are steroid refractory, steroid dependent, or steroid intolerant. Patients have been on a variety of additional immunosuppressive medications. In a recent European consensus report, it has been suggested that ECP may be helpful as a second-line treatment option in patients with chronic GvHD (Wolff et al. 2011). In acute GvHD, Greinix et al. (2006b) commenced treatment early after the onset of GvHD, with patients starting treatment a median of 17 days after starting steroids.

20.5.6 Intensity of Treatment

In chronic GvHD, there is variation in the intensity of ECP treatments. There is also variation in the treatment schedules and duration of ECP in the literature. Some include a variety of treatment schedules (Foss et al. 2005; Apisarnthanarax et al. 2005), and others reduce the frequency, depending on response (Couriel et al. 2006). Couriel et al. (2006) reported a very intensive initial regimen of two to four treatments per week, which was reduced according to response and a median of 32 ECP procedures.

A UK consensus statement on the use of ECP in chronic GvHD suggests that patients with cutaneous, mucous membrane, and hepatic manifestations of GvHD should be given priority for treatment as it is particularly efficacious in this setting. This consensus group recommended a treatment schedule of two ECP treatments on 2 consecutive days every 2 weeks, with less-frequent monthly treatment in those who respond (Scarisbrick et al. 2008). We have recently reported on 82 patients who received ECP treatment for mucocutaneous GvHD using this schedule and found this regimen to be feasible and effective in the management of chronic GvHD. This fortnightly schedule may be more feasible for centers where patients have to travel a long distance for treatment. The median number of ECP cycles was 15 (30 treatments), and the median duration of treatment was 330 days (Dignan et al. 2012).

In acute GvHD, Greinix et al. (2006b) used a schedule of two ECP treatments each week, and the median duration of treatment was 1.2 months. Treatment was discontinued when maximal response was seen.

20.5.7 Monitoring Response to Treatment

The UK consensus group recommends formal assessment of response after 3 months of ECP. Response can be assessed in terms of reduction of immunosuppressive medications and clinical response based on NIH criteria (Pavletic et al. 2006). Formal assessment of quality of life can also be undertaken using standardized questionnaires (Pidala et al. 2011).

20.6 The Use of ECP for Other Conditions

The successful use of ECP for CTCL and its excellent safety profile prompted investigation into use of the procedure for other T-cell–mediated disorders. Successful use of ECP in the management of rejection of lung, renal, and cardiac transplants has been reported. Investigation into the role of ECP in management of autoimmune conditions, including Crohn's disease, diabetes, and multiple sclerosis, has also been undertaken.

20.6.1 Rejection of Solid Organ Transplants

ECP has been used in the setting of solid organ transplants, primarily in the setting of heart and lung transplantation. More than 200 patients have been treated for cardiac transplant rejection and 170 for lung transplant rejection (reviewed in Marques and Schwartz 2011). A multicenter, randomized, double-blind study including 60 patients showed the benefit of ECP for the prevention of acute rejection of cardiac transplants. After 6 months of follow-up, the number of acute rejection episodes was significantly lower in the group treated with ECP plus standard therapy compared to standard therapy alone (Barr et al. 1998). ECP has also been investigated as a treatment for acute rejection of cardiac transplants.

ECP was first used in the treatment of rejection of lung transplants in 1995, and initial studies suggested that ECP could lead to stabilization of lung function tests (Slovis, Loyd, and King 1995). More recently, two larger studies have been published. Benden et al. (2008) reported on 24 patients treated with ECP as a result of bronchiolitis obliterans syndrome (BOS) or acute rejection. There was no significant improvement in forced expiratory volume in 1 second (FEV1) value, although patients did have a significantly slower rate of decline in FEV1 than prior to initiation of ECP (Benden et al. 2008). Patients with early-stage BOS seemed to do better than those with more advanced disease. In the second study, Morrell et al. (2010) reported that ECP slowed down the rate of lung function in 60 patients.

The role of ECP has also been explored in liver transplantation. Urbani et al. (2008) reported on the possible benefit of ECP in allowing delayed introduction of calcineurin inhibitors in prophylaxis of acute cellular rejection in ABO-incompatible liver transplant recipients and reduction of the immunosuppressive burden in hepatitis C-positive patients. There have also been reports of promising results with using ECP to treat renal allograft rejection, although no large studies have been published to date (Dall'Amico et al. 1998).

20.6.2 Autoimmune Disease

The role of ECP in a number of autoimmune diseases has been investigated. There have been several small studies suggesting a positive effect of ECP on Crohn's disease activity index (CDAI) scores and steroid doses. The largest study to date, undertaken by Abreu et al. (2009), included 28 patients in a prospective, uncontrolled pilot study. Patients had moderate-to-severe disease as measured by the CDAI and were intolerant or refractory to immunosuppression and/or anti-TNF agents. Clinical response was seen in 50% of patients, and 25% obtained remission (Abreu et al. 2009). Randomized, controlled studies on the role of ECP in the management of type 1 diabetes, multiple sclerosis, and systemic sclerosis have been undertaken but did not show any definite clinical benefit (reviewed in Maeda 2009).

20.7 Summary

ECP is currently used in the management of CTCL and acute and chronic GvHD and in the management of rejection of solid organ transplants. The treatment has an excellent safety profile and appears to be effective in the management of these conditions. More studies are required to fully determine the optimal regimen of ECP and duration of treatment. Its mechanism of action is yet to be fully elucidated and is the work of active research worldwide.

References

Abreu, M., C. von Tirpitz, R. Hardi et al. 2009. Crohn's disease photopheresis study group: Extracorporeal photopheresis for the treatment of refractory Crohn's disease: Results of an open-label pilot study. *Inflamm Bowel Dis* 15(6):829–836.

Antin, J., A. Chen, D. Couriel et al. 2004. Novel approaches to the therapy of steroid-resistant acute graft-versus-host disease. *Biol Blood Marrow Transplant* 10(10):655–668.

Apisarnthanarax, N., M. Donato, M. Körbling et al. 2003. Extracorporeal photopheresis therapy in the management of steroid-refractory or steroid-dependent cutaneous chronic graft-versus-host disease after allogeneic stem cell transplantation: Feasibility and results. *Bone Marrow Transplant* 31:459–465.

Barr, M., B. Meiser, H. Eisen et al. 1998. Photopheresis for the prevention of rejection in cardiac transplantation: Photopheresis Transplantation Study Group. *N Engl J Med* 339(24):1744–1751.

Benden, C., R. Speich, G. Hofbauer et al. 2008. Extracorporeal photopheresis after lung transplantation: A 10-year single-center experience. *Transplantation* 86(11):1625–1627.

Berger, C., A. Xu, D. Hanlon et al. 2001. Induction of human tumor-loaded dendritic cells. *Int J Cancer* 91(4):438–447.

Bladon, J., and P. Taylor. 2006. Extracorporeal photopheresis: A focus on apoptosis and cytokines. *J Dermatol Sci* 43(2):85–94.

Child, F., T. Mitchell, S. Whittaker et al. 2004. A randomized cross-over study to compare PUVA and extracorporeal photopheresis in the treatment of plaque stage (T2) mycosis fungoides. *Clin Exp Dermatol* 29(3):231–236.

Couriel, D., C. Hosing, R. Saliba et al. 2006. Extracorporeal photochemotherapy for the treatment of steroid-resistant chronic GVHD. *Blood* 107(8):3074–3080.

Crovetti, G., A. Carabelli, E. Berti et al. 2000. Photopheresis in cutaneous T-cell lymphoma: Five-year experience. *Int J Artif Organs* 23(1):55–62.

Dall'Amico, R., L. Murer, G. Montini et al. 1998. Successful treatment of recurrent rejection in renal transplant patients with photopheresis. *J Am Soc Nephrol* 9(1):121–127.

Dignan, F., D. Greenblatt, M. Cox et al. 2012. Efficacy of bimonthly extracorporeal photopheresis in refractory chronic mucocutaneous GVHD. *Bone Marrow Transplant* 47(6):824–830.

Duvic, M., N. Chiao, and R. Talpur. 2003. Extracorporeal photopheresis for the treatment of cutaneous T-cell lymphoma. *J Cutan Med Surg* 7(4S):3–7.

Edelson, R., C. Berger, F. Gasparro et al. 1987. Treatment of cutaneous T-cell lymphoma by extracorporeal photochemotherapy: Preliminary results. *N Engl J Med* 316:297–303.

Edinger, M., P. Hoffmann, J. Ermann et al. 2003. CD4+ CD25+ regulatory T cells preserve graft-versus-tumor activity while inhibiting graft-versus-host disease after bone marrow transplantation. *Nat Med* 9(9):1144–1150.

Ferrara, J., J. Levine, P. Reddy, and E. Holler. 2009. Graft-versus-hostdisease. *Lancet* 2373(9674):1550–1561.

Filipovich, A., D. Weisdorf, S. Pavletic et al. 2005. National Institutes of Health consensus development project on criteria for clinical trials in chronic graft-versus-host disease: I. Diagnosis and staging working group report. *Biol Blood Marrow Transplant* 11(12):945–956.

Fimiani, M., M. Di Renzo, and P. Rubegni. 2004. Mechanism of action of extracorporeal photochemotherapy in chronic graft-versus-host disease. *Br J Dermatol* 150(6):1055–1060.

Flowers, M., J. Apperley, K. van Besien et al. 2008. A multicenter prospective phase 2 randomized study of extracorporeal photopheresis for treatment of chronic graft-versus-host disease. *Blood* 112(7):2667–2674.

Foss, F., G. DiVenuti, K. Chin et al. 2005. Prospective study of extracorporeal photopheres is in steroid-refractory or steroid-resistant extensive chronic graft-versus-host disease: Analysis of response and survival incorporating prognostic factors. *Bone Marrow Transplant* 35(12):1187–1193.

Gatza, E., C. Rogers, S. Clouthier et al. 2008. Extracorporeal photopheresis reverses experimental graft-versus-host disease through regulatory T cells. *Blood* 112(4):1515–1521.

Greinix, H., R. Knobler, N. Worel et al. 2006. The effect of intensified extracorporeal photochemotherapy on long-term survival in patients with severe acute graft-versus-host disease. *Haematologica* 91:405–408.

Greinix, H., G. Socié, A. Bacigalupo et al. 2006. Assessing the potential role of photo pheresis in hematopoietic stem cell transplant. *Bone Marrow Transplant* 38(4):265–273.

Heshmati, F. 2010. Extracorporeal photo chemotherapy (ECP) in acute and chronic GVHD. *Transfus Apher Sci* 43(2):211–215.

Khavari, P., R. Edelson, O. Lider et al. 1988. Specific vaccination against photoinactivated cloned T cells. *Clin Res* 36:662.

Kim, Y., H. Liu, S. Mraz-Gernhard et al. 2003. Long-term outcome of 525 patients with mycosis fungoides and Sézary syndrome. *Arch Dermatol* 139:857–866.

Maeda, A. 2009. Extracorporeal photochemotherapy. *J Dermatol Sci* 54(3):150–156.

Marks, D, and R. Fox. 1991. Mechanisms of photochemotherapy-induced apoptotic cell death in lymphoid cells. *Biochem Cell Biol* 69(10–11):754–760.

Marques, M. and J. Schwartz. 2011. Update on extracorporeal photopheresis in heart and lung transplantation. *J Clin Apher* 26(3):146–151.

Marshall, S. 2006. Technology insight: ECP for the treatment of GvHD—Can we offer selective immune control without generalized immunosuppression? *Nat Clin Pract Oncol* 3(6):302–314.

Miller, J., E. Kirkland, D. Domingo et al. 2007. Review of extracorporeal photopheresis in early-stage (IA, IB, and IIA) cutaneous T-cell lymphoma. *Photodermatol Photoimmunol Photomed* 23(5):163–171.

Morrell, M., G. Despotis, D. Lublin et al. 2010. The efficacy of photopheresis for bronchiolitis obliterans syndrome after lung transplantation. *J Heart Lung Transplant* 29(4):424–431.

Olsen, E., S. Whittaker, Y. Kim et al. 2011. Clinical endpoints and response criteria in mycosis fungoides and Sézary syndrome. *J Clin Oncol* 29(18):2598–2607.

Owsianowski, M., H. Gollnick, W. Siegert, R. Schwerdtfeger, and C. Orfanos. 1994. Successful treatment of chronic graft-versus-host disease with extracorporeal photopheresis. *Bone Marrow Transplant* 14(5):845–848.

Pavletic, S., P. Martin, S. Lee et al. 2006. Measuring therapeutic response in chronic graft-versus-host disease: National Institutes of Health Consensus Development Project on Criteria for Clinical Trials in Chronic Graft-versus-Host Disease: IV. Response Criteria Working Group report. *Biol Blood Marrow Transplant* 12(3):252–266.

Perfetti, P., P. Carlier, P. Strada et al. 2008. Extracorporeal photopheresis for the treatment of steroid refractory aGvHD. *Bone Marrow Transplant* 42:609–617.

Peritt, D. 2006. Potential mechanisms of photopheresis in hematopoietic stem cell transplantation. *Biol Blood Marrow Transplant* 12(1 suppl. 2):7–12.

Pidala, J., B. Kurland, X. Chai et al. 2011. Patient-reported quality of life is associated with severity of chronic graft-versus-host disease as measured by NIH criteria: Report on baseline data from the Chronic GVHD Consortium. *Blood* 117(17):4651–4657.

Scarisbrick, J. 2009. Extracorporealphotopheresis: What is it and when should it be used? *Clin Exp Dermatol* 34(7):757–760.

Scarisbrick, J., P. Taylor, U. Holtick et al. 2008. Photopheresis Expert Group: U.K. consensus statement on the use of extracorporeal photopheresis for treatment of cutaneous T-cell lymphoma and chronic graft-versus-host disease. *Br J Dermatol* 158(4):659–678.

Seaton, E., R. Szydlo, E. Kanfer, J. Apperley, and R. Russell-Jones. 2003. Influence of extracorporeal photopheresis on clinical and laboratory parameters in chronic graft-versus-host disease and analysis of predictors of response. *Blood* 102(4):1217–1223.

Slovis, B., J. Loyd, and L. King, Jr. 1995. Photopheresis for chronic rejection of lung allografts. *N Engl J Med* 332(14):962.

Smith, E., I. Sniecinski, A. Dagis et al. 1998. Extracorporeal photochemotherapy for treatment of drug-resistant graft-vs.-host disease. *Biol Blood Marrow Transplant* 4:27–37.

Suchin, K., A. Cucchiara, S. Gottleib et al. 2002. Treatment of cutaneous T-cell lymphoma with combined immunomodulatory therapy: A14-year experience at a single institution. *Arch Dermatol* 138(8):1054–1060.

Talpur, R., M. Demierre, L. Geskin et al. 2011. Multicenter photopheresis intervention trial in early-stage mycosis fungoides. *Clin Lymphoma Myeloma Leuk* 11(2):219–227.

Tsirigotis, P., V. Pappa, S. Papageorgiou et al. 2007. Extracorporeal photopheresis in combination with bexarotene in the treatment of mycosis fungoides and Sézary syndrome. *Br J Dermatol* 156(6):1379–1381.

Urbani, L., A. Mazzoni, P. Colombatto et al. 2008. Potential applications of extracorporeal photopheresis in liver transplantation. *Transplant Proc* 40(4):1175–1178.

Willemze, R., E. Jaffe, G. Burg et al. 2005. WHO-EORTC classification for cutaneous lymphomas. *Blood* 105:3768–3785.

Wolff, D., M. Schleuning, S. von Harsdorf et al. 2011. Consensus Conference on Clinical Practice in Chronic GvHD: Second-line treatment of chronic graft-versus-host disease. *Biol Blood Marrow Transplant* 17(1):1–17.

Zic, J. 2003. The treatment of cutaneous T-cell lymphoma with photopheresis. *Dermatol Ther* 16(4):337–346.

21

Ultraviolet C Therapy for Infections

Tianhong Dai
Massachusetts General Hospital and Harvard Medical School

Michael R. Hamblin
Massachusetts General Hospital

21.1 Introduction

The prevalence of infectious diseases caused by multi-drug-resistant microorganisms is relentlessly increasing worldwide as a result of the excessive use of antibiotics (Bell 2003). Recently, a dangerous new mutation (named NDM-1) (Park 2010) that makes some bacteria resistant to almost all antibiotics has been found in the United States in patients with urinary tract infections (Centers for Disease Control and Prevention 2010). Antibiotic resistance has led to a major research effort to find alternative antimicrobial approaches to which, it is hypothesized, microorganisms will not be easily able to develop resistance.

Ultraviolet (UV) irradiation is electromagnetic irradiation with a wavelength (100–400 nm) shorter than that of visible light (400–700 nm) but longer than x-rays (<100 nm) (Figure 21.1). UV irradiation is divided into four distinct spectral areas, including vacuum UV (100–200 nm), UVC (200–280 nm), UVB (280–315 nm), and UVA (315–400 nm) (Vázquez and Hanslmeier 2006) (Figure 21.1). The mechanism of UVC's inactivation of microorganisms is to damage the genetic material in the nucleus of a cell or nucleic acids in a virus (Chang et al. 1985). The UVC spectrum, especially the range of 250–270 nm, is strongly absorbed by the nucleic acids of a microorganism and, subsequently, is the most lethal range of wavelengths for microorganisms. This range, with 262 nm being the peak germicidal wavelength, is known as the germicidal spectrum (Gurzadyan, Gorner, and Schulte-Frohlinde 1995). The light-induced damage to the DNA and RNA of a microorganism often results from the dimerization of pyrimidine molecules. In particular, thymine (which is only found in DNA) produces cyclobutane dimers. When thymine molecules are dimerized, it becomes very difficult for the nucleic acids to replicate, and if replication does occur, it often produces a defect that prevents the organism from being viable.

Although it has been known for the last 100 years that UVC irradiation is highly germicidal, the use of UVC irradiation for the prevention and treatment of localized infections is still in the very early stages of development. Most of the studies are confined to *in vitro* and *ex vivo* levels, and *in vivo* animal studies and clinical studies are much rarer. Studies that have examined UVC inactivation of antibiotic-resistant bacteria have found them to be equally susceptible as their naïve counterparts (Conner-Kerr et al. 1998). Within the UVC range, 254 nm is easily produced from a mercury low-pressure vapor lamp and has been shown to be close to the optimal wavelength for germicidal action. Because the delivery of UVC irradiation to living tissue is almost a localized process, UVC for infectious diseases is likely to be applied exclusively to localized infections.

In this chapter, we will discuss the potential of UVC irradiation as an alternative approach to localized infections. The topics include the efficacy of UVC for localized infections, effects of UVC on wound healing, effects of UVC on mammalian tissue and cells, and the possibility of the resistance development of microorganisms to UVC. To the best of our knowledge, this is the first comprehensive discussion on UVC for localized infections.

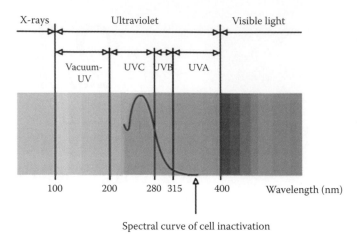

FIGURE 21.1   Spectrum of ultraviolet light.

21.2 UVC Irradiation for Infections

The mechanism of UVC inactivation of microorganisms is to cause cellular damage by inducing changes in the chemical structure of DNA chains (Chang et al. 1985). The consequence is the production of cyclobutane pyrimidine dimers (CPDs), which cause distortion of the DNA molecule, which might cause malfunctions in cell replication and lead to cell death.

21.2.1 *In Vitro/Ex Vivo* Studies

An *ex vivo* study was carried out by Taylor, Leeming, and Bannister (1993) to investigate the use of UVC irradiation (254 nm) for the prophylaxis of surgical-site infections. The authors modeled a "clean" surgical wound lightly contaminated with airborne bacteria by using agar, ovine muscle, and ovine adipose tissue, respectively. It was found that airborne bacteria were inhibited more rapidly and more completely on agar than on muscle. A coating of blood over the microorganisms on the surface of muscle substantially reduced the effectiveness of UVC. At an irradiance of 1.2 mW/cm^2 calculated at the lamp aperture, 1 min of UVC irradiation time reduced bacterial colony forming units (CFUs) by 99.1% on agar, 97.1% on muscle ($P = .046$), and 53.5% on muscle coated with blood ($P < .001$). The combination of pulsed-jet lavage and UVC was tested with the intention of removing the blood coated over the bacteria prior to UVC irradiation. The bacterial CFUs were reduced by 97.7% with the combination of pulsed-jet lavage and UVC.

Conner-Kerr et al. (1998) examined the effectiveness of UVC irradiation at 254 nm in inactivating antibiotic-resistant strains of *Staphylococcus aureus* and *Enterococcus faecalis in vitro*. Bacterial suspensions at 10^8 CFU/mL were prepared and plated on agar medium and then exposed to UVC irradiation. The calculated irradiance at the device aperture was 15.54 mW/cm^2, and the distance between the UVC lamp and agar medium was 25.4 mm. For the methicillin-resistant strain of *S. aureus* (MRSA), inactivation rates were 99.9% at 5 s and 100% at 90 s. For vancomycin-resistant *E. faecalis* (VRE), inactivation rates were 99.9% at 5 s and 100% at 45 s. These findings suggest that

UVC at 254 nm is bactericidal for antibiotic-resistant strains of *S. aureus* and *E. faecalis* at times as short as 5 s.

In a similar study, Rao et al. (2011) reported a complete (100%) eradication of the microorganisms on agar at the UVC doses ranging from a minimum of 5 s of irradiation (methicillin-resistant, coagulase-negative *Staphylococcus* and *Streptococcus pyogenes*) to a maximum of 15 s of irradiation (methicillin-susceptible *S. aureus* and *Enterococci* species). The irradiance used was 5 mW/cm^2 calculated at the lamp aperture, and the lamp–agar distance was 10 cm.

By using a prototype solid-state UVC LED device at 265 nm, Dean et al. (2011) evaluated the efficacy of UVC for treating corneal bacterial infections *in vitro*. Agar plate lawns of *S. aureus*, *Escherichia coli*, *Pseudomonas aeruginosa*, and *S. pyogenes* were exposed to UVC irradiation at an irradiance of 1.93 mW/cm^2 (calculated at the target surface) for varying lengths of time. The study demonstrated that exposure to UVC for 1 s (1.93 mJ/cm^2) was sufficient to induce 100% inhibition of growth for all the bacterial species tested. In this study, human corneal epithelial cells cultured on glass coverslips were also exposed to corresponding doses of UVC from the same device.

An idea of using UVC irradiation for disinfection of catheter biofilms was reported by Bak et al. (2009). In this study, the investigators determined the dose requirement for UVC disinfection of catheter biofilms. Contaminated urinary catheters from patients ($n = 67$) were used as test samples. The microorganisms identified from the catheter biofilms included *E. coli* ($n = 32$), coagulase-negative *Staphylococcus* ($n = 22$), *E. faecalis* ($n = 13$), *Streptococcus* ($n = 13$), *P. aeruginosa* ($n = 12$), coryneforms ($n = 7$), and so forth. Mean killing rates of the bacteria in catheter biofilms were 89.6% (11.8 mJ/cm^2), 98% (47 mJ/cm^2), and 99% (1400 mJ/cm^2). The UVC exposures were calculated at the target surface.

Mohr et al. (2009) described the use of UVC for pathogen reduction of platelet concentrates. The application of strong agitation of loosely fixed platelet concentrate bags resolved one of the problems related to the use of UVC for pathogen inactivation: namely, the quenching of the irradiation in protein-containing and turbid solutions or cell suspensions. Agitation allowed the penetration of UVC irradiation for inactivation of six bacterial species, including Gram-positive *Bacillus cereus*, *Staphylococcus aureus*, and *Staphylococcus epidermidis*, and Gram-negative *E. coli*, *Klebsiella pneumoniae*, and *P. aeruginosa*. All bacteria species tested were reduced by more than 4log$_{10}$ at 400 mJ/cm^2 (calculated at the surface of the quartz plate where platelet concentrate samples were placed). The study also proved that platelet damage by UVC irradiation was limited under the conditions used. The *in vitro* functions and the other variables measured were only moderately influenced, and the storage stability of the platelet concentrates was not impaired. Glucose consumption was slightly enhanced, and lactate accumulation was slightly increased in comparison to the untreated control samples. In contrast, irradiation of platelet concentrates with UVC irradiation leads to more enhanced platelet metabolism as evidenced by lactate accumulation and a stronger decrease in pH during storage.

Sullivan and Conner-Kerr (2000) compared the inactivation efficacies of UVC on pathogenic bacteria and fungi in both single suspensions and mixed suspensions *in vitro*. The calculated irradiance at the device aperture was 15.54 mW/cm², with the distance between the UVC lamp and suspension surface being 25.4 mm. Upon exposure to UVC, a 99.9% inactivation rate was obtained at 3 to 5 s for the bacteria (*P. aeruginosa* and *Mycobacterium abscessus*) tested. In contrast, 15 to 30 s of UVC treatment was required to obtain 99.9% inactivation of the fungi (*Candida albicans*, *Aspergillus fumigatus*) tested.

Dai et al. (2008) tested the ability of UVC to inactivate dermatophyte suspensions *in vitro* and to sterilize an *ex vivo* model of nail infection. *Trichophyton rubrum*, *Trichophyton mentagrophytes*, *Epidermophyton floccosum*, and *Microsporum canis* suspensions were irradiated with UVC (254 nm) at a dose of 120 mJ/cm² (calculated at the suspension surface) and surviving fungal cells quantified. *T. rubrum* infecting porcine hoof slices and human toenail clippings was irradiated with UVC at the doses of 36–864 J/cm². *In vitro* studies showed that 3–5log₁₀ of cell inactivation in dermatophyte suspensions were produced with 120 mJ/cm² UVC irradiation. Depending on factors such as the thickness and infectious burden of the *ex vivo* cultures, the radiant exposure of UVC needed for complete sterilization was usually on the order of tens to hundreds of J/cm².

21.2.2 Animal Studies

There has been, rather surprisingly, only one reported animal study on the use of UVC irradiation in treating infections. Dai et al. (2011a) investigated the use of UVC irradiation (254 nm) for treatment of *C. albicans* infection in mouse third-degree burns. The *C. albicans* strain was stably transformed with a version of the *Gaussia princeps* luciferase gene that allowed real time bioluminescence imaging of the progression of *C. albicans* infection. UVC treatment with a single exposure carried out on day zero (30 min postinfection) gave an average of 2.16log₁₀ (99.2%) loss of fungal luminescence when 2.92 J/cm² UVC had been

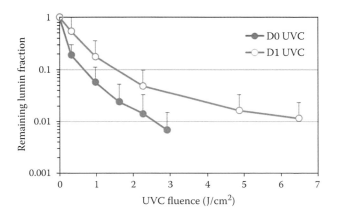

FIGURE 21.2 Dose responses of mean fungal luminescence of the mouse burns infected with *C. albicans* and treated by use of a single UVC exposure on day zero (30 min, *n* = 11) and day 1 (24 h, *n* = 12) postinfection, respectively. The data are displayed as mean ± SD. (Reprinted from Dai, T. et al., *Photochem Photobiol*, 87, 342–349, 2011. With permission.)

delivered, and UVC at 24 h postinfection gave a 1.94log₁₀ (95.8%) reduction of fungal luminescence after 6.48 J/cm² (Figure 21.2). The UVC exposures were calculated at the surfaces of mouse burns. Statistical analysis demonstrated that UVC treatment carried out on both day zero and day 1 significantly reduced the fungal burden of infected burns by 99.2% (*P* = .003) and 99.2% (*P* = .004), respectively. UVC was found to be superior to a topical antifungal drug, nystatin cream (*P* = .028).

21.2.3 Clinical Studies

The first clinical study was reported by Taylor, Bannister, and Leeming (1995), who used UVC irradiation (254 nm) for the disinfection of surgical wounds during total joint arthroplasty procedures. UVC irradiation commenced 10 min after the operation, allowing the wound to be exposed to a conventional open-air environment. Two different UVC irradiances, 0.1 mW/cm² and 0.3 mW/cm² (calculated at the lamp aperture), were used. Bacteria in wounds were measured by imprinting with 47 mm diameter 5 μm mixed cellulose acetate and nitrate membrane filters. After 10 min of UVC irradiation, the average bacterial CFU in wounds was reduced by 87% with 0.1 mW/cm² (*n* = 18, *P* < .001) and 92% with 0.3 mW/cm² (*n* = 13, *P* < .001), compared to that in the conventional environment (*n* = 13).

Shimomura et al. (1995) examined the efficacies of UVC irradiation (254 nm) on the prevention of catheter exit-site infections. First, bacterial cultures of swabbed fluid from catheter exit sites were obtained from 68 continuous ambulatory peritoneal dialysis (CAPD) outpatients six times during the 24-month observation period. Second, the bactericidal effects of UVC irradiation on the catheter exit site were examined. The authors found that the following. (1) In spite of disinfection of the catheter exit site by the strict application of povidone-iodine once or twice a day, 23–45% of the cases were found to be microorganism positive. (2) In the nasal cavity, *S. aureus* was detected in 20–25% of patients. There was a high incidence of exit-site infection among the *S. aureus* nasal carriers. (3) UVC irradiation was performed (twice per day, 30–60 s each time) in 18 cases that constantly revealed bacteria on the cultures from the catheter exit site. Ten cases (55%) became culture negative, three cases showed a microbial decrease, and five cases remained unchanged. These results suggest that UVC can eliminate bacteria and can be of prophylactic use for exit-site infections.

Thai et al. (2002) investigated the use of UVC for the treatment of cutaneous ulcer infections. In this study, three patients with chronic ulcers infected with MRSA were treated with UVC at 254 nm. UVC irradiation was applied to each wound for 180 s with an irradiance of 15.54 mW/cm² (calculated at the UVC device aperture) and a wound-to-lamp distance of 25.4 mm. In all three patients, UVC treatment reduced the relative amount of bacteria in the wounds and facilitated wound healing. Two patients had complete wound closure following 1 week of UVC treatment. In a later study performed by Thai et al. (2005), 22 patients with chronic ulcers exhibiting at least two signs of infection and critically colonized with bacteria received a single 180 s treatment of UVC. Semiquantitative swabs taken immediately before and after UVC treatment were used to assess changes in the bacterial bioburden present within the wound bed. A statistically significant ($P < .0001$) reduction in the relative amount of bacteria following a single treatment of UVC was observed. The greatest reduction in semiquantitative swab scores following UVC treatment was observed for wounds colonized with *P. aeruginosa* and wounds colonized with only one species of bacteria. Significant ($P < .05$) reductions in the relative amount of bacteria also were observed in 12 ulcers in which MRSA was present.

A study using UVC to treat toenail onychomycosis was reported by Boker et al. (2007). Thirty patients with mild-to-moderate onychomycosis involving no more than 35% of the great toenail were equally randomized to receive four weekly UVC treatments with either a low-pressure mercury lamp delivering a total UVC dose of 22 J/cm² at the surfaces of the treated toenails or via a xenon pulsed-light device delivering a total UVC dose of 2–4 J/cm² at the surface of the treated toenails. The Investigators' Global Assessment (IGA) scale was used to assess treatment efficacy. Sixty percent of patients treated with the xenon pulsed-light device showed an improvement of at least one point on their week 16 IGA scale compared to baseline ($P < .01$). An image depicting the result of a patient in the xenon pulsed-light group is presented in Figure 21.3. Of patients treated with the low-pressure mercury lamp, 26% had at least a one-point improvement in their week 16 IGA score ($P < .01$). The treatments with both devices were well tolerated. Minor and uncommon side effects included temporary mild erythema of the irradiated toe.

In summary, *in vitro* studies have reported that multi-drug-resistant pathogenic microorganisms are highly sensitive to UVC inactivation. Furthermore, microorganisms are found to be more sensitive to UVC than mammalian cells. As a result, with appropriate doses, pathogenic microorganisms may be selectively inactivated by UVC with minimum nonspecific damage to mammalian cells. This is crucial in the application of UVC irradiation for localized infections. Generally, bacterial cells are found to be more sensitive to UVC than fungal cells.

The UVC doses required to inactivate a therapeutically sufficient fraction of microorganisms *in vivo* (e.g., 180 s irradiation time in the study by Thai et al. in 2005) may be orders of magnitude higher than those for *in vitro* (e.g., 5 s irradiation time in the study by Conner-Kerr et al. in 1998). This is because the energy of UVC irradiation attenuates exponentially when penetrating into tissue.

One advantage of using UVC over antibiotics is that UVC can eradicate microorganisms in a much faster manner (2–3log₁₀ eradication of microorganism population *in vivo* could be achieved in less than 1 h), while antibiotics usually take several days to take effect, especially in burns and chronic wounds that frequently have impaired blood perfusion. UVC may also be much more cost-effective than the commonly used antibiotics.

21.3 Effects of UVC Irradiation on Wound Healing

In addition to the eradication of microorganisms that can impede wound healing, it is hypothesized that judicious UV exposure might be beneficial for wound healing and restoration of skin homeostasis. The effects of UVC on wound healing include hyperplasia and enhanced re-epithelialization or desquamation of the leading edge of peri-ulcer epidermal cells, granulation tissue formation, and sloughing of necrotic tissue (Kloth 1995). In addition, UV exposure of wounds might stimulate and restore normal melanocyte number and distribution in reepithelialized wounds while preventing hypopigmentation (Rennekampff et al. 2010). Furthermore, exposure of reepithelialized wounds to UV irradiation might exert a photoprotective effect in the skin through the production of melanin by melanocytes (Rennekampff et al. 2010). It is therefore proposed that moderate UV exposure should be commenced early in the healing process of cutaneous wounds (Kloth 1995; Rennekampff et al. 2010). Physical therapists have used UVC irradiation as a therapeutic modality for wound healing for many years; however, the physician community has been slow to adopt this technology (Ennis, Lee, and Meneses 2007).

21.3.1 *In Vitro* Studies

An *in vitro* study was reported by Morykwas and Mark (1998) on the effects of UVC irradiation at 254 nm on dermal fibroblasts. Fifteen newborn foreskin fibroblast cultures were treated with UVC light. The investigators observed that, in comparison to nonirradiated fibroblast cultures, those fibroblasts irradiated with UVC had a decreased amount of fibronectin bound to cell surfaces (mean 14%) and an increased amount of fibronectin released into the medium (mean 42%). In addition, collagen lattices constructed with irradiated fibroblasts contracted significantly faster at 7 days. The authors suggested that fibronectin release led to increased healing via wound contraction.

21.3.2 Animal Studies

Basford et al. (1986) compared the efficacy of several approaches for wound healing by using a pig model, including HeNe laser (632.8 nm), UVC irradiation (254 nm), occlusion, and air

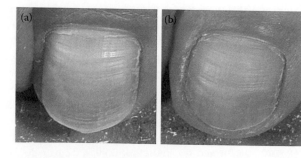

FIGURE 21.3 UVC treatment of toenail onychomycosis (a) before UVC treatment and (b) 28 weeks after UVC treatment.

exposure. UVC-treated wounds were given two minimal erythemal dose treatments, twice daily, 6 days a week. Although UVC treatment showed a tendency toward healing faster than air-exposed wounds, the tendency did not reach a clinical significance. The authors concluded that there was no advantage in using UVC treatment.

Suo, Wang, and Wang (2002) investigated the effect of UVC on the expression of transforming growth factor β (TGF-β) in rat wounds. Three full-thickness wounds were made on the dorsal surface of each rat ($n = 30$) and then treated with UVC, respectively, at 0 mJ/cm² (no UVC), 15 mJ/cm², and 60 mJ/cm² daily for 3 successive days. The UVC exposures were calculated at the wound surfaces. The expression of TGF-β was measured at both mRNA level and protein level by *in situ* hybridization and immunohistochemistry. The authors observed that on day 7 after wounding, the expression of TGF-β in the wounds treated with 15 mJ/cm² UVC was higher than that in the wounds treated with 60 mJ/cm² and the control wounds without UVC ($P < .05$). On day 21, the expression of TGF-β in the wounds treated with 60 mJ/cm² was higher than that in the wounds treated with 15 mJ/cm² as well as the control wounds without UVC treatment. The authors concluded that at the early stage of wound healing, 15 mJ/cm² UVC treatment promotes the expression of TGF-β and might be beneficial for accelerating wound healing. The level of TGF-β expression was upregulated at the later stage at the UVC dose of 60 mJ/cm².

In a later study using the same rat wound model (Suo et al. 2003), authors from the same group investigated the effects of UVC irradiation at different doses on the expression of basic fibroblast growth factor (bFGF) in rat wounds. Full-thickness wounds made on the dorsal surfaces of rats ($n = 30$) were respectively treated with UVC at 0 mJ/cm², 15 mJ/cm², and 60 mJ/cm² on a daily basis for 3 days. On day 7 after wounding, the expression of bFGF in the wounds treated with 15 mJ/cm² or 60 mJ/cm² UVC was higher than that in the control wounds without UVC ($P < .01$), and the bFGF expression in the wound treated with 60 mJ/cm² UVC was higher than that in the wounds treated with 15 mJ/cm² UVC. On day 14 after wounding, the bFGF expression in the wounds treated with 60 mJ/cm² significantly decreased and was lower than that in the wounds treated with 15 mJ/cm² and the wounds without UVC ($P < .01$). The authors concluded that in the early stage of wound healing, UVC promotes the expression of bFGF in granulation tissues. The effect of UVC with 60 mJ/cm² was acute, and the effect of UVC with 15 mJ/cm² was chronic.

21.3.3 Clinical Studies

A very early clinical study on the effect of UVC on wound healing was conducted in 1963. In this study, Freytes, Fernandez, and Fleming (1965) tested the use of UVC irradiation at 254 nm emitted from a mercury-vapor lamp for the treatment of indolent ulcers in three patients. The ulcerated area was exposed to UVC irradiation for 150 s (minimum erythema dose). Treatments were repeated once a week. The first patient had a

deeply ulcerated area 1 in. in diameter. He received four treatments, at the end of which the ulcer was approximately 6.35 mm (1/4 in.) in size, clean, and with good granulation tissue. The second patient had an ulcer of approximately 63.5 mm (2.5 in.) in diameter. She received four treatments, and complete healing was achieved. The third patient had a multiple sclerosis and a decubitus ulcer resistant to conventional treatment. The ulcer was 2 in. in diameter and 1/4 in. in depth. He received five treatments, at the end of which the ulcer was approximately 1/2 in. in diameter, clean, and with good granulation tissue.

Nussbaum, Biemann, and Mustard (1994) compared the use of UVC in combination with ultrasound (UVC/US) for wound healing of pressure ulcers with the use of low-level laser and standard nursing care alone. UVC was emitted from a cold-quartz lamp at 250 nm. Treatment parameters for UVC were based on wound appearances using erythema dosages. Ultrasound treatment was delivered at a frequency of 3 MHz and at a spatial average temporal average intensity of 0.2 W/cm² (1:4 pulse ratio). Twenty patients were randomly assigned to the three groups. Results showed that UVC/US treatment had a greater effect on wound healing than nurse care, either alone or combined with the laser. The average weekly rates of healing were 53.5%, 32.4%, and 23.7% for the UVC/US group, standard nurse care alone group, and laser group, respectively.

In summary, variable results have been reported on the effects of UVC irradiation on wound healing. The discrepancies of the results might be because of the various parameters of UVC used in different studies. For all the reported animal and clinical studies on the effects of UVC on wound healing, UVC irradiation was applied to noninfected wounds. While most of the studies reported positive results of enhanced wound healing by UVC light, no study reported any negative results that UVC delayed wound healing. While pathogenic microorganisms impede the healing of infected wounds, one can expect that the eradication of microorganisms by UVC would enhance wound healing in infected wounds.

21.4 Effects of UVC Irradiation on Mammalian Tissue and Cells

It is well known that prolonged UV irradiation is damaging to human tissue and particularly to skin. UVB irradiation of skin has been particularly well studied and is accepted as the main cause of skin cancer (Ichihashi et al. 2003). It is recommended that exposure of skin to prolonged UVB irradiation be avoided by appropriate application of sunscreen and choice of lifestyle (Albert and Ostheimer 2003; Suo, Wang, and Wang 2002). However, under appropriate risk–benefit analysis, UVB therapies are still delivered to various areas of the bodies of millions of patients in the United States every year for the treatment of a larger number of cutaneous disorders (Sage and Lim 2010). It has been documented that UVB therapy is an effective therapeutic option with an excellent safety profile and well-documented side effects (Hearn et al. 2008; Weischer et al. 2004).

UVC irradiation of human skin has been much less studied but is also known to cause the same kind of damage (Trevisan et al. 2006). In proposing to employ UVC irradiation to treat localized infections, it is clearly important to test the effect of UVC with an effective antimicrobial dose on normal mammalian cells and tissue to ensure that unacceptable damage is not inflicted.

21.4.1 Selectivity of UVC Inactivation of Microorganisms over That of Mammalian Cells

The safety issue of the UVC treatment requires that the pathogenic microorganisms be selectively inactivated while the cells in normal tissue cells are spared. Sosnin et al. (2004) compared the *in vitro* sensitivities to UVC irradiation between living mammalian cells and bacteria. The light source was a narrow-band UVC lamp with the emission peak at 206 nm. Chinese hamster ovary (CHO-K1) cells (fibroblasts) and *E. coli* (which is considered to be one of the most resistant species to UV irradiation within the enterobacteria group) were used in the study. The fibroblasts were in confluent monolayer cultures, and the *E. coli* cultures were colonies on agar plates. The authors found that the UVC dose that led to necrosis in fibroblasts was more than 10 times higher than that needed for inactivation of *E. coli*. A $2\log_{10}$ inactivation of *E. coli* was achieved at approximately 25 mJ/cm^2, and this dose did not cause any adverse effects on the fibroblasts. The authors concluded that UVC irradiation may become a method of selective bacterial decontamination of wounds without killing the host cells that strive to repair the wound. The authors also pointed out that the possible DNA damage in mammalian cells that survived the UVC treatment should be investigated in a long-term study.

Another *in vitro* study was conducted by Dai et al. (2011b) on the selectivity of UVC inactivation of bacteria in suspensions over keratinocytes in confluent monolayer cultures. On average, when 11 mJ/cm^2 UVC irradiation had been delivered, as shown in Figure 21.4, the viability loss of the keratinocytes was only approximately $0.24\log_{10}$ ($\approx 57\%$) while a more than $6\log_{10}$ inactivation of bacteria (average value of *P. aeruginosa* and *S. aureus*) was achieved under similar conditions, resulting in a more than $5\log_{10}$ selective inactivation of bacteria over keratinocytes. If we consider a $2\log_{10}$ (99%) inactivation of bacteria as a sufficiently therapeutic fraction, the viability loss of the keratinocytes would be only approximately 6% at the same UVC dose.

In another study carried out by Dai et al. (2011a), the authors compared the *in vitro* susceptibilities to UVC irradiation between keratinocytes and *C. albicans*. It was found that UVC selectively inactivated *C. albicans* compared to keratinocytes in a UVC dose-dependent manner. On average, when 19.2 mJ/cm^2 UVC irradiation had been delivered, the viability loss of keratinocytes was approximately $1.22\log_{10}$ while a $3.02\log_{10}$ inactivation of *C. albicans* was achieved at the same irradiation dose ($P < .001$), resulting in a nearly $2\log_{10}$ selective inactivation of *C. albicans* over keratinocytes. If we consider a $2\log_{10}$ (99%) inactivation of *C. albicans* as the therapeutically effective fraction, the viability loss of keratinocytes is approximately $0.77\log_{10}$ (18.9%) at the comparable UVC dose.

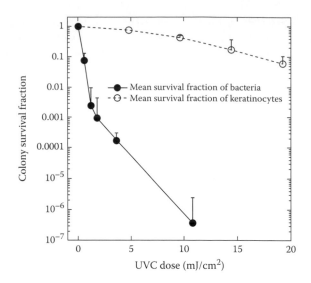

FIGURE 21.4 Comparison of averaged fluence-dependent survival fractions in response to UVC irradiation of bacteria with that of keratinocytes using identical irradiation conditions and colony-forming assays. Bars: SD. (Reprinted from Dai, T. et al., *Photochem. Photobiol.*, 87, 250–255, 2011. With permission.)

To assess the safety of using UVC irradiation to treat corneal bacterial infections, Dean et al. (2011) compared the *in vitro* sensitivity to UVC (265 nm) of human primary corneal epithelial cells with that of corneal pathogenic bacteria. The authors found that an exposure of confluent monolayer human corneal epithelial cells to 57.95 mJ/cm^2 UVC (calculated at the culture surface) gave no statistically significant decrease ($P = .877$) in the ratio of live to dead cells when compared to the nonirradiated control cultures, and an exposure to only 1.93 mJ/cm^2 UVC was sufficient to induce 100% inhibition of growth of all the bacterial species tested on agar plates. The authors suggested that UVC at appropriate doses could potentially be beneficial in treating corneal surface infections without causing significant adverse effects.

21.4.2 Effects of UVC on Host Tissue

In an animal study using BALB/c mice, Dai et al. (2011a) investigated whether mouse skin could tolerate UVC irradiation at the antifungal dose. Figure 21.5a through c shows representative morphologies of BALB/c mouse skin before, immediately after, and 24 h after being exposed to UVC at a radiant exposure of 6.48 J/cm^2, which is the effective antifungal dose for treating infections on day 1 postinfection. It can be observed that there is mild wrinkling of the skin evident immediately after UVC treatment, and this is slightly more pronounced at 24 h after UVC exposure, but the wrinkling disappeared in succeeding days. A small lesion (top of square in Figure 21.5c) was also observed, which occurred at 24 h after UVC exposure, but the lesion eventually healed without problem. Cyclobutane pyrimidine dimer–positive nuclei were observed in the immunofluorescence micrograph of the biopsy taken immediately after UVC exposure (Figure 21.5e). However, the damage was extensively repaired at 24 h after UVC exposure where only traces of fluorescence remained (Figure 21.5f). Mild epidermal shrinkage was seen at 24 h after UVC exposure (Figure 21.5l).

In another animal study using hairless mice, Sterenborg, van der Putte, and van der Leun (1988) compared the carcinogenic effect of UVC (254 nm) on a mouse with that of UVB (313 nm). UVC or UVB irradiation was applied daily at a dose of 57.5–460 mJ/cm² (calculated at the lamp aperture) per day. In all dose groups, most of the animals developed large numbers of tumors at some stage of the experiment. The large majority were classified as squamous cell carcinomas. For both UVC and UVB, the tumor induction time was proportional to a power function of the daily dose (the power was −0.2 and −0.58 for UVC and UVB, respectively). Throughout the whole range of daily doses used in the experiment, UVC was less carcinogenic than UVB. An intriguing difference between the two types of radiation was that the tumors induced by UVC appeared much more scattered over the irradiated parts of the animals than the UVB-induced tumors.

In summary, studies did find that UVC at the effective antimicrobial doses can cause DNA damage to mammalian cells to some extent. However, at that same time, it has been found that the UVC-induced DNA damages can be rapidly repaired by the DNA-repairing enzymes of the host. To further minimize the UVC-induced DNA damage, one can combine the

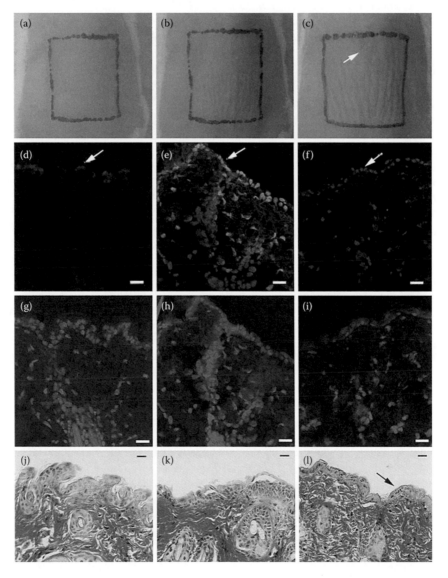

FIGURE 21.5 **(See color insert.)** (a–c) Morphologies of a representative mouse skin before, immediately after, and 24 h after being exposed to UVC at a dose of 6.48 J/cm², respectively. (d–f) Representative immunofluorescence micrographs of cyclobutane pyrimidine dimers in skin cell nuclei. (g–i) Micrographs of diamidino-2-phenylindole counterstaining of cell nuclei. (j–l) Micrographs of Masson's trichrome stained sections. Biopsies were taken before (d, g, j), immediately after (e, h, k), and 24 h after (f, i, l) being exposed to UVC at a dose of 6.48 J/cm² from the same mouse. Arrow in (c): UVC-induced lesion on mouse skin. Arrows in (d–f): mouse skin surface. Arrow in panel l shows epidermal shrinkage. Scale bars: 20 μm. (Reprinted from Dai, T. et al., *Photochem. Photobiol.*, 87, 342–349, 2011. With permission.)

use of protective agents (e.g., green tea) (Camouse et al. 2009) and DNA repair agents (e.g., DNA repair liposomes) (Wolf et al. 2000) with UVC irradiation. Green tea could be applied to the UVC-irradiated area during the UVC treatment, and liposomes could be applied after the UVC treatment. In addition, the intact skin surrounding the area to be treated could be screened from UVC irradiation (Dai et al. 2011a; Thai et al. 2005).

21.5 Will Microorganisms Develop Resistance to UVC?

The rapid rate of replication of microorganisms allows them to adapt to environmental stresses with some facility. Favorable mutations in DNA can arise, and if these mutations lead to a competitive advantage, they will spread to the whole microbial population. This consideration is, in fact, largely responsible for the emergence of drug resistance in clinical therapy. Because one of the primary functions of UVC is to damage the DNA of microorganisms, is it possible to generate mutants with increased resistance to UVC?

In a study on the use of UVC to treat onychomycosis, Dai et al. (2008) tested whether *T. rubrum* can develop resistance to UVC irradiation at 254 nm by carrying out five consecutive sublethal UVC inactivations *in vitro*. No significant difference was found in cell inactivation rates among the five consecutive cycles when 120 mJ/cm² UVC had been delivered (Figure 21.6, $P = .66$). This indicates that resistance to UVC irradiation is not rapidly acquired by *T. rubrum* cells that are repeatedly exposed to sublethal UVC irradiation.

Alcantara-Diaz, Brena-Valle, and Serment-Guerrero (2004) investigated the divergent adaptation of *E. coli* to repeated cycles

of UVC irradiation at 254 nm. Five cultures of wild-type *E. coli* PQ30 were exposed to 80 consecutive growth-irradiation cycles of UVC light. Each growth-irradiation cycle was carried out as follows: An aliquot of 1 mL from an early stationary culture (~5 × 10⁸–10⁹ cells) was irradiated with a UVC germicidal lamp at an irradiance of 0.1 mW/cm². Following irradiation, 0.1 mL of this suspension was inoculated into 1 mL of fresh LB broth and incubated at 37°C for approximately 6 h to the early stationary phase. Starting with a dose of 1 mJ/cm² for each cycle, the dose was increased twofold every 10 growth-irradiation cycles. Quantitative dose–response curves were obtained by exposing exponentially growing cells, suspended by phosphate buffer, to several UV doses. Immediately after irradiation, cells were diluted in phosphate buffer and spread on LB agar. At the end, according to the quantitative dose–response curves, all the five cultures were found to give rise to different degrees of UVC resistance. The authors found that the adaptation to cyclic UVC irradiation was a consequence of selection of advantageous mutations arising in different genes related to repair and replication of DNA.

In summary, excessive repetition of UVC irradiation may induce resistance of microorganisms to UVC. No resistance of fungal cells to UVC irradiation was observed in the study by Dai et al. (2008), which may be a result of the limited cycles of UVC inactivation carried out. While permissible UVC dose-exposure limits for human tissue do not exist, it is expected that there would be an acceptable maximum number of repetition times of UVC irradiation and possibly a total lifetime cumulative exposure. For wound infections, we expect that only limited numbers of repeated doses of UVC irradiation would be required, while the UV-induced carcinogenic mutation is a long-term effect.

21.6 Expert Commentary

The use of UVC irradiation for the prevention and treatment of localized infections is still in the very early stages of development. Most of the studies are confined to *in vitro* and *ex vivo* levels, while *in vivo* animal studies and clinical studies are much rarer.

Several *in vitro* studies have reported that multi-drug-resistant pathogenic microorganisms are highly sensitive to UVC inactivation. Bacterial cells are found to be more sensitive to UVC than fungal cells. Microorganisms are found to be more sensitive to UVC than mammalian cells.

One advantage of using UVC over antibiotics is that UVC can eradicate microorganisms in a much faster manner. UVC may also be much more cost-effective than the commonly used antibiotics.

UVC is generally much more effective and also safer for prophylaxis of wound infections than for treatment of already infected wounds. This is because when the infections get established, pathogenic microorganisms penetrate deep into the tissue, and biofilms can often be formed. Higher doses of irradiation are needed to treat established infections (e.g., as reported by Dai et al. in 2011a) because irradiation is quickly

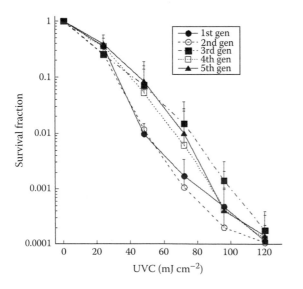

FIGURE 21.6 Dose responses of *Trichophyton rubrum in vitro* to five consecutive subtotal ultraviolet UVC inactivation. The data are representative experiments performed in triplicate and are displayed as mean ± SD. (Reprinted from Dai, T. et al., *Br. J. Dermatol.*, 158, 1239–1246, 2008. With permission.)

attenuated when penetrating into tissue. Similarly, higher irradiation doses are needed when inactivating microorganisms within biofilm than are needed for their planktonic counterparts. It is likely that the use of higher irradiation doses is accompanied by more serious side effects, caused by compromising the inactivation selectivity and increasing the UVC-induced tissue damage.

The effects of UVC irradiation on wound healing have also been investigated, and variable results have been reported. While pathogenic microorganisms impede the healing of infected wounds, one can expect that the eradication of microorganisms by UVC would enhance wound healing in infected wounds.

Studies did find that UVC at the effective antimicrobial doses can cause DNA damage to mammalian cells to some extent. However, at the same time, it has been found that the UVC-induced DNA damages can be rapidly repaired by the DNA-repairing enzymes.

In contrast to the large amount of studies regarding the chronic effects of UVB on human skin and tissue, there has not been any similar report on the chronic effects of UVC. However, it has been suggested in an animal study that UVC is less carcinogenic than UVB (Sterenborg, van der Putte, and van der Leun 1988) because of its more superficial penetration depth. "Abnormal differentiation of a layer of cells that is committed to being sloughed off anyway (UVC) is not harmful, whereas mutation of the basal cells (UVA or UVB) may result in skin cancer" (Sterenborg, van der Putte, and van der Leun 1988). On the other hand, it has been reported that UVB treatment is an effective option for a large number of cutaneous disorders in humans with excellent safety profiles. A retrospective study of 195 psoriasis patients treated with UVB did not provide evidence for increased skin cancer risk with up to 9 years of follow-up (Weischer et al. 2004). An analysis of 3867 patients receiving UVB over an 18-year period, with a median number of 29 treatments and 352 patients receiving 100 or more treatments with more than 6 months of follow-up for each patient, showed no increase in skin cancers of any kind (Hearn et al. 2008).

It has been found that resistance of microorganisms to UVC may develop after excessive repetition (e.g., 80 cycles) of UVC irradiation (Alcantara-Diaz, Brena-Valle, and Serment-Guerrero 2004). Therefore, similar to traditional antibiotics, excessive or long-term use of UVC should also be avoided.

It is worth noting that the use of UVC for sun-sensitive patients (or lupus erythematosus patients) should be cautious. It is well known that solar irradiation, mainly the part of UVA and UVB, causes photosensitivity in lupus erythematosus patients (Bijl and Kallenberg 2006), which is an abnormal reaction of human skin to solar irradiation. It was found that similar effects can also be induced by UVC (Cripps and Rankin 1973).

It is also worth noting that the penetration of UVC irradiation in human skin and tissue is limited; as a result, topical irradiation of UVC may not be sufficient to reach deeply located infections and subsequently inactivate pathogenic microorganisms.

However, with the advancement of modern optical fiber technologies (Oto et al. 2001; Parpura and Haydon 1999), this limitation could be overcome by delivering the UVC irradiation interstitially to the infected sites. In addition, optical clearing techniques (Khan et al. 2004; Tuchin 2006), which have attracted extensive attention recently, have provided another potential technique for improving the UVC penetration in human skin and tissue.

In conclusion, we believe there are situations where the risk–benefit ratio is favorable for the use of UVC for treating localized infections, particularly when the microorganisms responsible are antibiotic resistant. While permissible UVC dose-exposure limits for human tissue do not exist, it is expected that there would be an acceptable maximum number of repetition times of UVC irradiation and possibly a total lifetime cumulative exposure. For wound infections, we expect that only limited numbers of repeated doses of UVC irradiation would be required, while the UV-induced carcinogenic mutation is a long-term effect of prolonged use of UVC.

21.7 Five-Year View

The increasing emergence of antibiotic resistance of pathogens presents a serious clinical challenge for the future. UVC should be investigated as an alternative approach for prophylaxis and treatment of localized infectious diseases, especially those caused by antibiotic-resistant pathogens. UVC should be used in a way where the side effects would be minimized and the resistance of microorganisms to UVC would be avoided. As a result, more extensive animal studies and clinical studies need to be carried out to investigate and optimize UVC treatment, for example, the minimal effective antimicrobial doses of UVC irradiation for localized infections at various stages with different levels of severity. Light-delivery technologies, for example, optical fibers and optical clearing techniques, should be investigated to improve the penetration of UVC irradiation in human skin and tissue. Technologies that help reduce the side effects (e.g., enhanced repair of UVC-induced DNA damage to human cells, selective protection of human tissue and cells from UVC irradiation during UVC treatment) of UVC treatment are also worthy of being further investigated. In addition to epidermal cells, effects of UVC on human bone, vessels, nerves (which may be exposed to UVC irradiation in open wounds), and leukocytes (which are important in the local defense mechanism against dissemination of infections) also should be studied.

Financial and Competing Interests Disclosure

This study was supported in part by an Airlift Research Foundation Extremity Trauma Research Grant (grant #109421 to TD), an Orthopedic Trauma Association Basic Research Grant (2012 cycle grant # 16 to TD), and the National Institutes of Health (NIH, grant RO1AI050875 to MRH).

References

Albert, M. and K. Ostheimer. 2003. The evolution of current medical and popular attitudes toward ultraviolet light exposure: Part 2. *J Am Acad Dermatol* 48:909–918.

Alcantara-Diaz, D., M. Brena-Valle, and J. Serment-Guerrero. 2004. Divergent adaptation of *Escherichia coli* to cyclic ultraviolet light exposures. *Mutagenesis* 19:349–354.

Bak, J., S. Ladefoged, M. Tvede, T. Begovic, and A. Gregersen. 2009. Dose requirements for UVC disinfection of catheter biofilms. *Biofouling* 25:289–296.

Basford, J., H. Hallman, C. Sheffield, and G. Mackey. 1986. Comparison of cold-quartz ultraviolet, low-energy laser, and occlusion in wound healing in a swine model. *Arch Phys Med Rehabil* 67:151–154.

Bell, S. 2003. Antibiotic resistance: Is the end of an era near? *Neonatal Netw* 22:47–54.

Bijl, M., and C. Kallenberg. 2006. Ultraviolet light and cutaneous lupus. *Lupus* 15:724–727.

Boker, A., G. Rolz-Cruz, B. Cumbie, and A. Kimball. 2007. A single-center, prospective, open-label, pilot study of the safety, local tolerability, and efficacy of ultraviolet-C (UVC) phototherapy for the treatment of great toenail onychomycosis. *J Am Acad Dermatol* 58:AB82.

Camouse, M., D. Domingo, F. Swain et al. 2009. Topical application of green and white tea extracts provides protection from solar-simulated ultraviolet light in human skin. *Exp Dermatol* 18:522–526.

Centers for Disease Control and Prevention. 2010. Detection of enterobacteriaceae isolates carrying metallo-beta-lactamase: United States, 2010. *MMWR Morb Mortal Wkly Rep* 59:750.

Chang, J., S. Ossoff, D. Lobe et al. 1985. UV inactivation of pathogenic and indicator microorganisms. *Appl Environ Microbiol* 49:1361–1365.

Conner-Kerr, T., P. Sullivan, J. Gaillard, M. Franklin, and R. Jones. 1998. The effects of ultraviolet radiation on antibiotic-resistant bacteria in vitro. *Ostomy Wound Manage* 44:50–56.

Cripps, D., and J. Rankin. 1973. Action spectra of lupus erythematosus and experimental immunofluorescence. *Arch Dermatol* 107:563–567.

Dai, T., G. Kharkwal, J. Zhao et al. 2011a. Ultraviolet-C light for treatment of *Candida albicans* burn infection in mice. *Photochem Photobiol* 87:342–349.

Dai, T., G. Tegos, G. Rolz-Cruz, W. Cumbie, and M. Hamblin. 2008a. Ultraviolet C inactivation of dermatophytes: Implications for treatment of onychomycosis. *Br J Dermatol* 158:1239–1246.

Dai, T., G. Tegos, T. St Denis et al. 2011b. Ultraviolet-C irradiation for prevention of central venous catheter-related infections: An in vitro study. *Photochem Photobiol* 87:250–255.

Dean, S., A. Petty, S. Swift et al. 2011. Efficacy and safety assessment of a novel ultraviolet C device for treating corneal bacterial infections. *Clin Exp Ophthalmol* 39:156–163.

Ennis, W., C. Lee, and P. Meneses. 2007. A biochemical approach to wound healing through the use of modalities. *Clin Dermatol* 25:63–72.

Freytes, H., B. Fernandez, and W. Fleming. 1965. Ultraviolet light in the treatment of indolent ulcers. *South Med J* 58:223–226.

Gurzadyan, G., H. Gorner, and D. Schulte-Frohlinde. 1995. Ultraviolet (193, 216 and 254 nm) photoinactivation of Escherichia coli strains with different repair deficiencies. *Radiat Res* 141:244–251.

Hearn, R., A. Kerr, K. Rahim, J. Ferguson, and R. Dawe. 2008. Incidence of skin cancers in 3867 patients treated with narrow-band ultraviolet B phototherapy. *Br J Dermatol* 159:931–935.

Ichihashi, M., M. Ueda, A. Budiyanto et al. 2003. UV-induced skin damage. *Toxicology* 189:21–39.

Khan, M., B. Choi, S. Chess et al. 2004. Optical clearing of in vivo human skin: Implications for light-based diagnostic imaging and therapeutics. *Lasers Surg Med* 34:83–85.

Kloth, L. 1995. Physical modalities in wound management: UVC, therapeutic heating and electrical stimulation. *Ostomy Wound Manage* 41:18–20, 22–14, 26–17.

Mohr, H., L. Steil, U. Gravemann et al. 2009. A novel approach to pathogen reduction in platelet concentrates using short-wave ultraviolet light. *Transfusion* 49:2612–2624.

Morykwas, M., and M. Mark. 1998. Effects of ultraviolet light on fibroblast fibronectin production and lattice contraction. *Wounds* 10:111–117.

Nussbaum, E., I. Biemann, and B. Mustard. 1994. Comparison of ultrasound/ultraviolet-C and laser for treatment of pressure ulcers in patients with spinal cord injury. *Phys Ther* 74:812–823.

Oto, M., S. Kikugawa, N. Sarukura, M. Hirano, and H. Hosono. 2001. Optical fiber for deep ultraviolet light. *IEEE Photonics Technol Lett* 13:978–980.

Park, A. 2010. Antibiotics: NDM-1 how dangerous is the mutation? *Time* 176:20.

Parpura, V., and P. Haydon. 1999. Uncaging using optical fibers to deliver UV light directly to the sample. *Croatian Med J* 40:340–345.

Rao, B., P. Kumar, S. Rao, and B. Gurung. 2011. Bactericidal effect of ultraviolet C (UVC), direct and filtered through transparent plastic, on Gram-positive cocci: An in vitro study. *Ostomy Wound Manage* 57:46–52.

Rennekampff, H., M. Busche, K. Knobloch, and M. Tenenhaus. 2010. Is UV radiation beneficial in postburn wound healing? *Med Hypotheses* 75:436–438.

Sage, R., and H. Lim. 2010. UV-based therapy and vitamin D." *Dermatol Ther* 23:72–81.

Shimomura, A., D. Tahara, M. Tominaga et al. 1995. The effect of ultraviolet rays on the prevention of exit-site infections. *Adv Perit Dial* 11:152–156.

Sosnin, E., E. Stoffels, M. Erofeev, I. Kieft, and S. Kunts. 2004. The effects of UV irradiation and gas plasma treatment on living mammalian cells and bacteria: A comparative approach. *IEEE Trans Plasma Sci* 32:1544–1550.

Sterenborg, H., S. van der Putte, and J. van der Leun. 1988. The dose-response relationship of tumorigenesis by ultraviolet radiation of 254 nm. *Photochem Photobiol* 47:245–253.

Sullivan, P., and T. Conner-Kerr. 2000. A comparative study of the effects of UVC irradiation on select procaryotic and eucaryotic wound pathogens. *Ostomy Wound Manage* 46:28–34.

Suo, W., H. Guo, X. Wang, and D. Wang. 2003. Effect of ultraviolet C light on the expression of basic fibroblast growth factor in rat wounds. *Chin J Phys Med Rehabil* 25:651–654.

Suo, W., X. Wang, and D. Wang. 2002. Effect of ultraviolet c irradiation on expression of transforming growth factor-β in wound. *Chin J Rehabil Theor Pract* 8:5–7.

Taylor, G., G. Bannister, and J. Leeming. 1995. Wound disinfection with ultraviolet radiation. *J Hosp Infect* 30:85–93.

Taylor, G., J. Leeming, and G. Bannister. 1993. Effect of antiseptics, ultraviolet light and lavage on airborne bacteria in a model wound. *J Bone Joint Surg*ery *Br* 75:724–730.

Thai, T., P. Houghton, K. Campbell, and M. Woodbury. 2002. Ultraviolet light C in the treatment of chronic wounds with MRSA: A case study. *Ostomy Wound Manage* 48:52–60.

Thai, T., D. Keast, K. Campbell, M. Woodbury, and P. Houghton. 2005. Effect of ultraviolet light C on bacterial colonization in chronic wounds. *Ostomy Wound Manage* 51:32–45.

Trevisan, A., S. Piovesan, A. Leonardi et al. 2006. Unusual high exposure to ultraviolet-C radiation. *Photochem Photobiol* 82:1077–1079.

Tuchin, V. 2006. *Optical Clearing of Tissues and Blood*. SPIE Press, Bellingham, WA.

Vázquez, M., and A. Hanslmeier. 2006. *Ultraviolet Radiation in the Solar System*. Springer, Dordrecht.

Weischer, M., A. Blum, F. Eberhard, M. Rocken, and M. Berneburg. 2004. No evidence for increased skin cancer risk in psoriasis patients treated with broadband or narrowband UVB phototherapy: A first retrospective study. *Acta Dermato-Venereol* 84:370–374.

Wolf, P., H. Maier, R. Mullegger et al. 2000. Topical treatment with liposomes containing T4 endonuclease V protects human skin in vivo from ultraviolet-induced upregulation of interleukin-10 and tumor necrosis factor-alpha. *J. Invest Dermatol* 114:149–156.

IV

Photodynamic Therapy (PDT)

Recent Advances in Developing Improved Agents for Photodynamic Therapy

Penny Joshi
Roswell Park Cancer Institute

Courtney Saenz
Roswell Park Cancer Institute

Joseph R. Missert
Roswell Park Cancer Institute

Ravindra K. Pandey
Roswell Park Cancer Institute

22.1 Introduction to Photodynamic Therapy

Photodynamic therapy (PDT) involves local activation of a drug called a photosensitizer (PS), whereby the PS accumulates in tumor tissue, and by means of light, the PS is activated. The activated PS reacts with triplet O_2 to produce singlet oxygen (van Lier 1990), which a result of its cytotoxic behavior, believed to be responsible for tumor destruction (Ethirajan et al. 2011). PSs are activated by light of specific wavelengths, and the depth that light will penetrate into tissue is wavelength dependent (Gudgin Dickson, Goyan, and Pottier 2002). Thus, the combination of PS and wavelength of light for PDT is determined by the location of the tumor (National Cancer Institute, 2011. http://www.cancer.gov/cancertopics/factsheet/Therapy/photodynamic).

PDT has three components: a PS, light, and molecular oxygen. The PS transfers energy from light to molecular oxygen, to generate reactive oxygen species (ROS) at the intersection of all of these three components (Figure 22.1).

Upon absorption of the light, the PS is transformed from its ground state (^1PS) into an excited singlet state ($^1PS^*$) (Figures 22.1 and 22.2). From this state, the excited PS may decay by emitting fluorescence to ground state, whereby the emitted florescence can be utilized for photodetection of a tumor (Figure 22.1). However, to obtain a therapeutic photodynamic effect, the PS should undergo electron spin conversion to its triplet state ($^3PS^*$), (Figures 22.1 and 22.2) (http://www.cancer.gov/cancertopics/factsheet/Therapy/photodynamic). From the triplet excited state ($^3PS^*$) of the PS, energy can be transferred to molecular oxygen in two ways: The excited molecule can react directly with a substrate, by proton or electron transfer, to form radicals or radical ions, which can interact with oxygen to produce oxygenated products (type I reaction), or the energy of the excited PS can be directly transferred to oxygen to form singlet oxygen (type II reaction), which is extremely cytotoxic and is believed to play a major role in cell killing (Figure 22.2) (Dennis et al. 2003).

As we know, PSs accumulate in tumor tissue more readily than normal tissue. The problem with PSs is that some accumulate more than others, and in order to tackle that issue, we must develop a PS that accumulates in the tumor tissue more efficiently and selectively.

The first-generation PS Photofrin is the only drug approved by the Food and Drug Administration (FDA) for use in a variety of

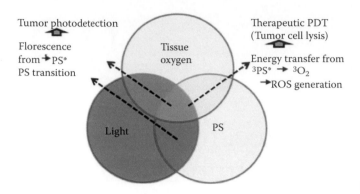

FIGURE 22.1 Venn diagram shows three components of PDT. Photodetection is achieved at the intersection of light and PS. When excited, the PS returns from its singlet excited state ($^1PS^*$) to its ground state (PS) by fluorescence; for the therapeutic outcome of PDT, molecular oxygen is an essential component, which generates singlet oxygen ($^1O_2^*$) by gaining energy from triplet state ($^3PS^*$) of PS.

cancers. However, there are several drawbacks that are observed with Photofrin, which are as follows:

- Although the PS accumulates in tumor tissue, there is an overall poor selectivity in terms of target tissue to healthy tissue ratios (Konan, Gurny, and Allemann 2002; Ethirajan et al. 2008).
- The first-generation PS has a low extinction coefficient at the irradiating wavelength, so to obtain the optimal therapeutic response, large amounts of the drug must be administered (Konan, Gurny, and Allemann 2002; Ethirajan et al. 2008).
- The first-generation PS's absorption maximum is at a comparatively short wavelength (630 nm). This short wavelength leads to poor tissue penetration when the laser light activates the PS. The poor tissue penetration ultimately prevents tumors that are deeply seated from receiving light, which is required for the photoreaction to occur to, in turn, destroy the tumor tissue (Konan, Gurny, and Allemann 2002; Ethirajan et al. 2008).

- Because the first-generation PS shows poor selectivity in regards to the target tissue to healthy tissue ratio, a high accumulation in the skin is prevalent. The high amount of the PS in surrounding skin tissue leads to photosensitivity. After PDT treatment, patients are advised by the clinicians to stay out of the sunlight up to 6 to 8 weeks to avoid the possibility of developing a severe sunburn reaction (Konan, Gurny, and Allemann 2002; Ethirajan et al. 2008).

However, the approval of Photofrin by various health organizations led researchers to develop second-generation PSs from other classes, such as chlorins, bacteriochlorins, phthalocyanines, and other porphyrin derivatives. These second-generation PSs show improvements, which include the following:

- The second-generation PSs show improved selectivity for the target tissue (Konan, Gurny, and Allemann 2002; Ethirajan et al. 2008).
- The second-generation PSs' longest absorption wavelength is longer than that of the first-generation PS, in the range of 650–800 nm, allowing improved light penetration through the target tissue (Konan, Gurny, and Allemann 2002; Ethirajan et al. 2008).
- This generation of PSs is more readily eliminated from the system, which reduces the side effects of skin phototoxicity exhibited by first-generation PSs (Konan, Gurny, and Allemann 2002; Ethirajan et al. 2008).
- Second-generation PSs have been shown to be more effective at generating singlet oxygen (Konan, Gurny, and Allemann 2002; Ethirajan et al. 2008).

22.1.1 Photophysics of the PSs

Porphyrins and other tetrapyrrolic systems are the most studied PSs. These PSs have a strong absorption band at approximately 400 nm, which is known as the Soret band (MacDonald and Dougherty 2001). However, the Soret band is not used for PDT

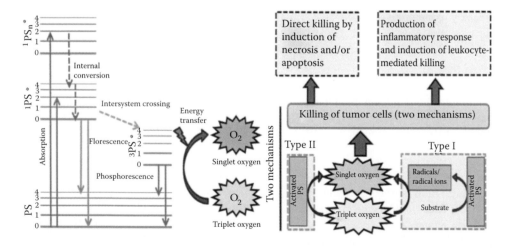

FIGURE 22.2 Various energy levels of PS are shown in a Jablonski diagram. Figure also depicts type I and type II mechanisms of generation of singlet oxygen and two mechanisms of killing of tumor cells by singlet oxygen.

because, at that wavelength, light only minimally penetrates the tissue; also, hemoglobin absorbs in this range. Based on the total number of π electrons, the main class of tetrapyrroles, porphyrin, is divided into three classes: porphyrins (1), chlorins (2), and bacteriochlorins (3) (Figure 22.3). Porphyrins are 22 π electron aromatic macrocycles, exhibiting characteristic optical spectra with a strong π–π* transition at approximately 400 nm (Soret band) and usually four Q bands. In the porphyrin system, two of the peripheral double bonds in opposite pyrrolic rings are cross-conjugated and are not required to maintain aromaticity. Reduction of one or both of the conjugated bonds maintains the aromaticity, giving rise to two systems: chlorins and bacteriochlorins, respectively (Spikes 1984).

For PDT application, the Q-bands (between 600 and 800 nm) are more beneficial to use, utilizing light that penetrates tissue more deeply (Figure 22.4). Porphyrins have a weak maximal absorption at approximately 630 nm, while chlorins have absorption maxima in the range of 650–700 nm and bacteriochlorins in the range of 720–800 nm (chemical structures shown in Figure 22.3) (MacDonald and Dougherty 2001), depending on the nature of substituents at the peripheral position. However, the bacteriochlorin chromophore is relatively less stable than chlorophyll. The energy gap between the highest occupied molecular orbital (HOMO) and the lowest unoccupied molecular orbital (LUMO) increases in the order bacteriochlorin < isobacteriochlorin <

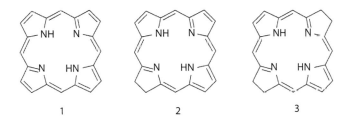

FIGURE 22.3   Basic structures of PSs.

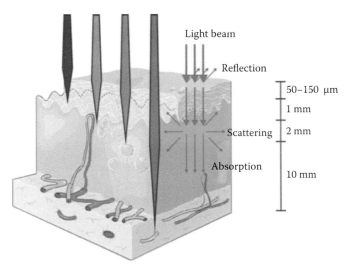

FIGURE 22.4   Light penetration through the tissues. (From Agostinis, P. et al., *CA Cancer J Clin*, 61, 2011.)

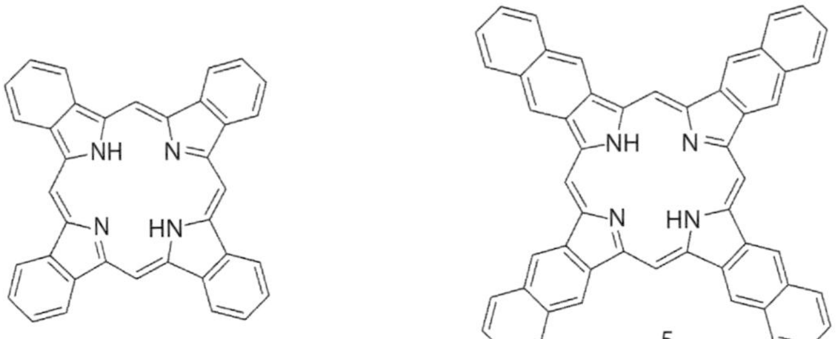

FIGURE 22.5   Basic structures of phthalocyanine and naphthalocyanine.

chlorins < porphyrins, whereas the energy required for oxidation decreases in the order porphyrins > chlorin > isobacteriochlorin > bacteriochlorin (Pandey and Zheng 2000). The loss of one double bond destabilizes the π system of a chlorin, and the HOMO rises in energy; thus, the molecule becomes easier to oxidize (Yang et al. 2011). Bacteriochlorins have the most red-shifted first absorption band, and this is the reason for most of the naturally occurring bacteriochlorin analogs being extremely sensitive to oxidation and undergoing a rapid transformation to chlorins. This feature can be modulated by the presence or absence of metals and also by various substituents.

A series of stable free-base and metallated bacteriochlorins [Zn (II) or Pd (II)] were synthesized by Pandey et al. and their photophysical and photochemical properties were studied (Fukuzumi et al. 2008). Among all the compounds, it was seen that photoexcitation of the palladium bacteriochlorin allowed the triplet excited state without fluorescence emission, resulting in formation of singlet oxygen with a high quantum yield resulting from the heavy atom effect of palladium. It was also seen, from the electrochemical studies, that the zinc bacteriochlorin has the smallest HOMO–LUMO gap, and this value was lower than the triplet excited-state energy of the compound in benzonitrile. The condition in which the HOMO–LUMO gap was smaller than the triplet energy gave rise to intermolecular, photoinduced electron transfer from the triplet excited state of the PS to its ground state to produce both the radical cation and radical anion. The radical anion thus produced can transfer an electron to molecular oxygen to produce superoxide anion. The same PS can also act as an efficient singlet oxygen generator. Thus, the same zinc bacteriochlorin can function as a sensitizer with a dual role in that it produces both singlet oxygen and superoxide anion in an aprotic solvent (benzonitrile).

Furthermore, phthalocyanines (4) and naphthalocyanines (5) (Figure 22.5) demonstrate an even greater red shift than the corresponding chlorin and bacteriochlorin, exhibiting a longer maxima absorption wavelength (Ethirajan et al. 2008).

22.2 History of PDT

Ancient Egyptian and Indian literature provides the description of use of light in Ebers papyrus (1550 B.C.) and Atharvaveda (approximately 1000 B.C.), respectively (El-Mofy 1968; Wyss 2000; Baden 1984). The ancient Egyptians employed dried powder from plants, such as parsnip, parsley, and Saint John's wort, in the treatment of depigmented skin lesions (vitiligo). These plants are now identified as containing natural PSs, psoralens. In the treatment procedure, after the application of the plant powder, patients were exposed to sunlight. Activation of PSs by sunlight produced a sunburn-like effect on the skin (El-Mofy 1968; Wyss 2000; Fahmy and Abu-Shady 1948). In another Greek book, Mofradat Al Adwiya (13th century A.D.) describes the treatment of vitiligo with a tincture of honey and powdered Aatrillal (*Ammi majus*) seeds (Fahmy and Abu-Shady 1948; Baden 1984) known to contain a number of psoralens. Similarly, Beivechi (*Psoralena corylifolia*) seeds, described for treatment of vitiligo in Atharvaveda, also contain psoralens. References for the use of *Psoralena corylifolia* in the treatment of vitiligo have also been found in Buddhist texts circa 200 B.C. and Chinese literature of the 10th century A.D. (Wyss 2000); (Magic Ray Moscow Center of Laser Medicine, http://www.magicray.ru/ENG/lecture/1.html).

In the modern world, phototherapy got attention for the first time with the work of Niels Finsen of Denmark. Finsen (1901) used red light exposures to prevent the formation and discharge of smallpox pustules and used ultraviolet light from the sun to treat cutaneous tuberculosis. He was awarded a Nobel Prize for this work in 1903 (Finsen 1901; Dennis et al. 2003). Oscar Raab (1900), a student of Professor H. Von Tappeiner at the Pharmacological Institute of Munich University, reported that an acridine solution killed paramecia upon irradiation with certain wavelengths of light. Von Tappeiner and Jesionek (1903) described the first clinical application of PDT by applying eosin topically to basal cell carcinomas prior to illumination. Later, they defined PDT as the dynamic interaction among light, a photosensitizing agent, and oxygen, resulting in tissue destruction (von Tappeiner and Jodlbauer 1907; Triesscheijna et al. 2006). Hausmann (1910) studied hematoporphyrin phototoxicity and described its effectiveness against paramecia and erythrocytes. In another important experiment, he injected mice with hematoporphyrin and showed the photosensitive and phototoxic effects of hematoporphyrin on the skin of mice (Hausmann 1908). Meyer-Betz (1913) in 1912 examined hematoporphyrin's photosensitizing effect on a human subject by injecting himself with hematoporphyrin. On exposure to sunlight, he observed edema and hyperpigmentation, which persisted for 2 months (Meyer-Betz 1913). Starting with A. Policard in 1924, a number of researchers observed that tumors accumulated hematoporphyrin more than healthy tissues, leading to the proposal of the diagnostic capabilities of hematoporphyrin fluorescence (Policard 1924; Auler and Banzer 1942). Tumor cell selectivity of porphyrins was demonstrated by Figge, Weiland, and Manganiello (1948). As hematoporphyrin itself was a mixture of active porphyrins and inert inactive porphyrins, Schwartz, Absolon, and Vermund (1955) produced the hematoporphyrin derivative (HpD) by processing hematoporphyrin with concentrated sulfuric and acetic acids. HpD was tested by Lipson and Baldes in 1960 at the Mayo Clinic and was found to produce a much better photodynamic effect (Lipson, Baldes, and Olsen 1961).

Dougherty et al. (1975) used HpD in combination with red light to eradicate mouse mammary tumor growth. Later on, clinical trials were carried out with HpD to treat patients with skin tumors and bladder cancer with great success (Dougherty et al. 1978; Kelly and Snell 1976). The Dougherty group was next able to remove the inert portion of HpD. This purified version of HpD was named Photofrin (porfimer sodium). Succeeding trials with Photofrin led to its approval as the first PS for certain cancers (Triesscheijna et al. 2006). Since then, at least three more PSs have been approved for clinical use, including 5-aminolevulinic acid (ALA, Levulan, DUSA Pharmaceuticals Inc., Wilmington, Massachusetts), the methyl ester of ALA (Metvix, Photocure ASA, Oslo, Norway), and *meso*-tetra-hydroxyphenyl-chlorin (mTHPC, temoporfin, Foscan, Biolitec Pharma Ltd., Dublin, Ireland). PDT is becoming more acceptable than ever before for the treatment of localized cancers (Triesscheijna et al. 2006).

A number of PDT agents, for example, Tookad (6) and 3-devinyl-3-(1′-hexyloxy)ethyl-pyropheophorbide (HPPH) (7), are at various stages of human clinical trials, and certain longer-wavelength PS 605 Me (8, a bacteriochlorin,) and 531 (9, iodo-chlorophyll-a analog) (Figure 22.6) are at advanced preclinical studies for fluorescence image–guided therapy and positron emission tomography (PET) imaging.

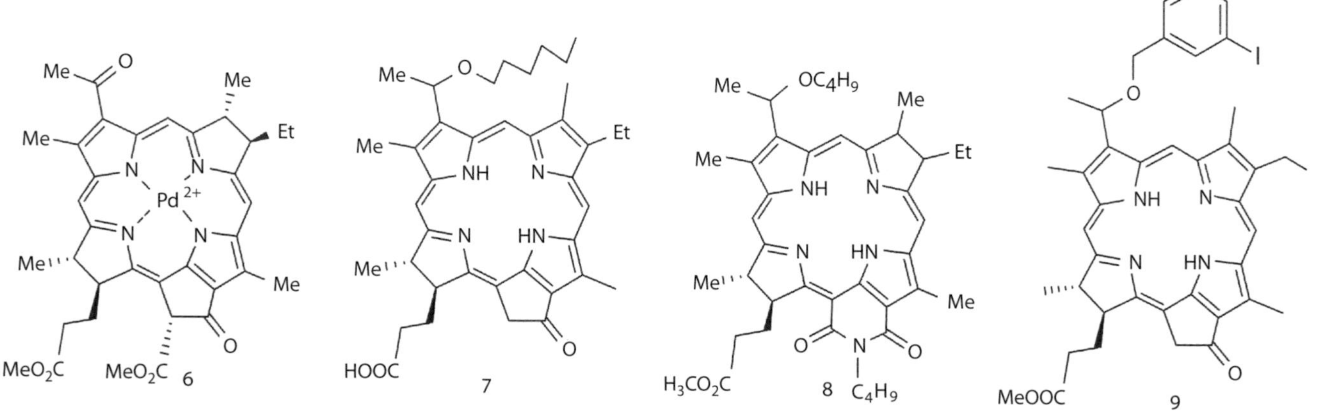

FIGURE 22.6 Structures of PDT agents.

22.3 Classes of PSs

22.3.1 Porphyrins

The activity of hematoporphyrin derivative (HpD) in tumors opened the doors for modern PDT, although the nonideal nature of such mixtures emphasized the need for the development of new PSs. Photofrin, the first FDA-approved PS, a complex mixture of oligomers derived from hematoporphyrin, is presently used in the treatment of various cancers. Though effective, it suffers from various drawbacks (Youngjae et al. 2003; Bellnier and Daugherty 1996). The drawbacks associated with Photofrin have led to the need for the search for new PSs with greater specificity and efficacy (Sharman, Allen, and van Lier 1999; Detty 2001). Various *meso*-substituents have been incorporated in the porphyrin molecule, and they provide longer-wavelength absorption maxima; several strategies have since been followed toward novel, improved porphyrin development. One such class among synthetic porphyrins is based on *meso*- 5,10,15,20-tetraarylporphyrin [e.g., H$_2$(TPP) 10]. Tetrassulfonated *meso*-tetraphenylporphyrins H$_2$(TPPS$_4$) 11 have been known for their tumor-localizing properties (Winkelaman, Slater, and Grossman 1967) (Figure 22.7). Though these compounds have a high singlet oxygen yield, a low photochemical efficiency *in vitro* and *in vivo* was reported (Evensen and Moan 1987).

In the early 1980s, tetra hydroxyphenyl porphyrin (THPP) 12 emerged as a new class of porphyrin with potential use as

PSs. Berenbaum et al. (1986) reported a comparison of three isomers of *meso*-tetra(hydroxyphenyl)porphyrin with HpD and Photofrin II (Akande et al. 1986). Songca synthesized and studied the fluorinated derivatives of tetra hydroxyphenyl porphyrins and found them to be potent PSs. The meta hydroxy compounds 5,10,15,20-tetrakis(2-fluoro-3-hydroxyphenyl)porphyrin (13) and 5,10,15,20-tetrakis(2,4-difluoro-3-hydroxyphenyl) (14) showed the highest phototoxicity (Songca 2001) (Figure 22.7).

It is known that amphiphilic molecules having both hydrophilic and hydrophobic groups show improved tumor uptake and retention (Osterloh and Vicente 2002). A series of tetraarylporphyrins and diarylporphyrins were synthesized by Banfi et al. (2006) to look into the effect of phenyl groups with different substituents at the meso position. The phototoxic effect of the series of compounds on HCT116 human colon adenocarcinoma cells was studied. It was seen that the diaryltetrapyrrole compounds were more effective than the tetrapyrrolic derivatives, whereas, within the diaryl series, the hydroxy compounds were found to be more active than the methoxy-substituted derivatives. Monoarylporphyrins, isolated as side products of the 5,15-diarylporphyrin synthesis, were found to be promising PSs with IC50 values comparable or even better than the diarylporphyrins. Compounds 15, 16, and 17 (Figure 22.8) were found to be more active than the clinically approved PSs porfimer sodium and temoporfin in the HCT116 human colon adenocarcinoma cells *in vitro* (Banfi et al. 2006).

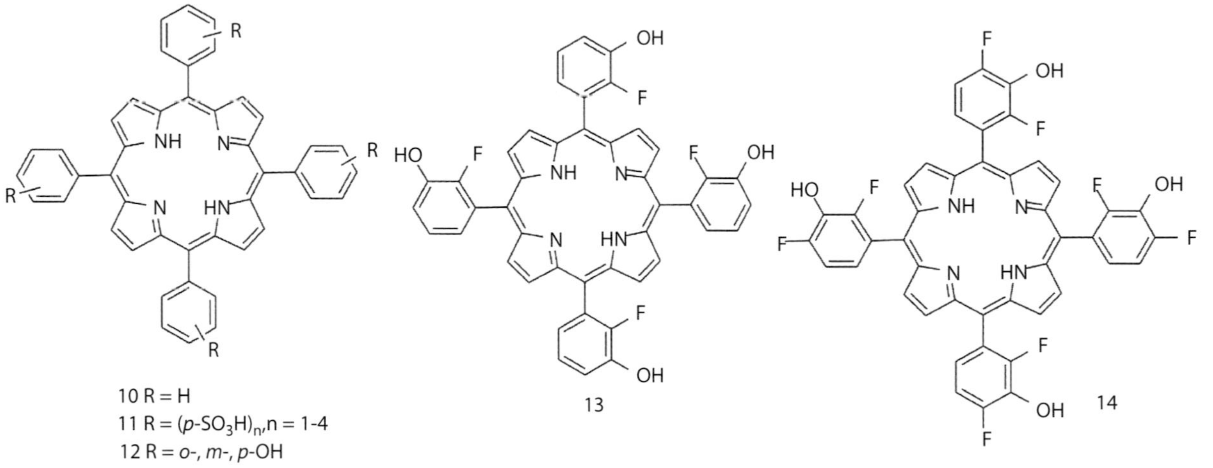

10 R = H
11 R = (*p*-SO$_3$H)$_n$, n = 1–4
12 R = *o*-, *m*-, *p*-OH

13

14

FIGURE 22.7 Tetrahydroxy porphyrins and their fluorinated derivatives.

15

16

17

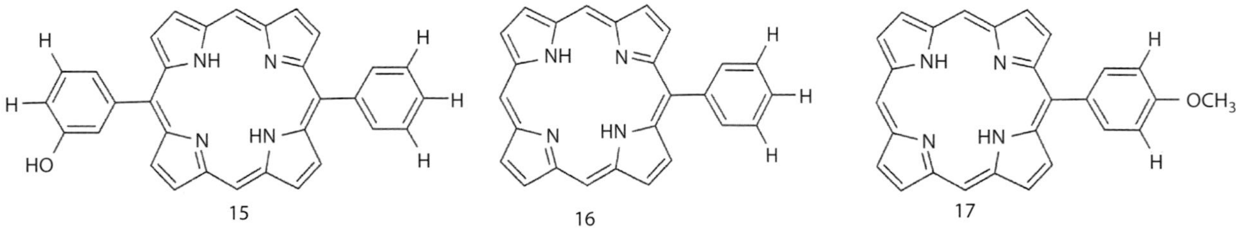

FIGURE 22.8 Structures of mono and diarylporphyrins.

Recently, glycol-functionalized porphyrins containing one to four low-molecular-weight glycol chains linked via ether bonds to either the para or meta position of meso-tetraphenyl porphyrin (18–25) were synthesized and evaluated for *in vitro* and *in vivo* PDT efficacy (Kralova et al. 2008) (Figure 22.9). The structure–activity relationship showed that fluorination enhanced the photosensitizing potential of para-phenyl derivatives as compound 22 was more efficacious than 18. Intracellular uptake and phototoxicity was exhibited by hydroxyglycol porphyrins (18, 20, 22, 24, 25) in contrast with methoxy derivatives (19, 21, 23).

Ethylene glycol in the meta position (20) exhibited better intracellular uptake and PDT efficacy both *in vitro* and *in vivo*, and it appears to be a better candidate for *in vivo* PDT. Mitra et al. (2011) evaluated the effectiveness of meso-tetra (N-methyl-4-pyridyl) porphine tetra tosylate in the PDT treatment of *Candida albicans* infection of *in vitro* and in an animal model.

22.3.2 Chlorins

The chlorins are a class of 20π electron aromatic tetrapyrroles with photophysical properties similar to those of porphyrin systems (Cubeddn et al. 1987), but they have enhanced red-shifted Q bands (approximately 670–700 nm) that make chlorin-containing systems better candidates for PDT. Chlorophyll-a is one of the prototypes of the chlorin class of natural products. It serves as a light-harvesting chromophore in photosynthetic plants, algae, and cyanobacteria (Montforts and Glasenapp-Breiling 2002). It can be extracted from certain Spirulina species without any contamination from other chlorophyll forms (Smith, Goff, and Simpson 1985). A number of lesser-known chlorins also play important biological roles. Cyclopheophorbide and related species are thought to inhibit damaging oxidative processes in certain marine species (Karuso et al. 1986; Watanabe et al. 1993), and the chlorin bonellin is a hormone responsible

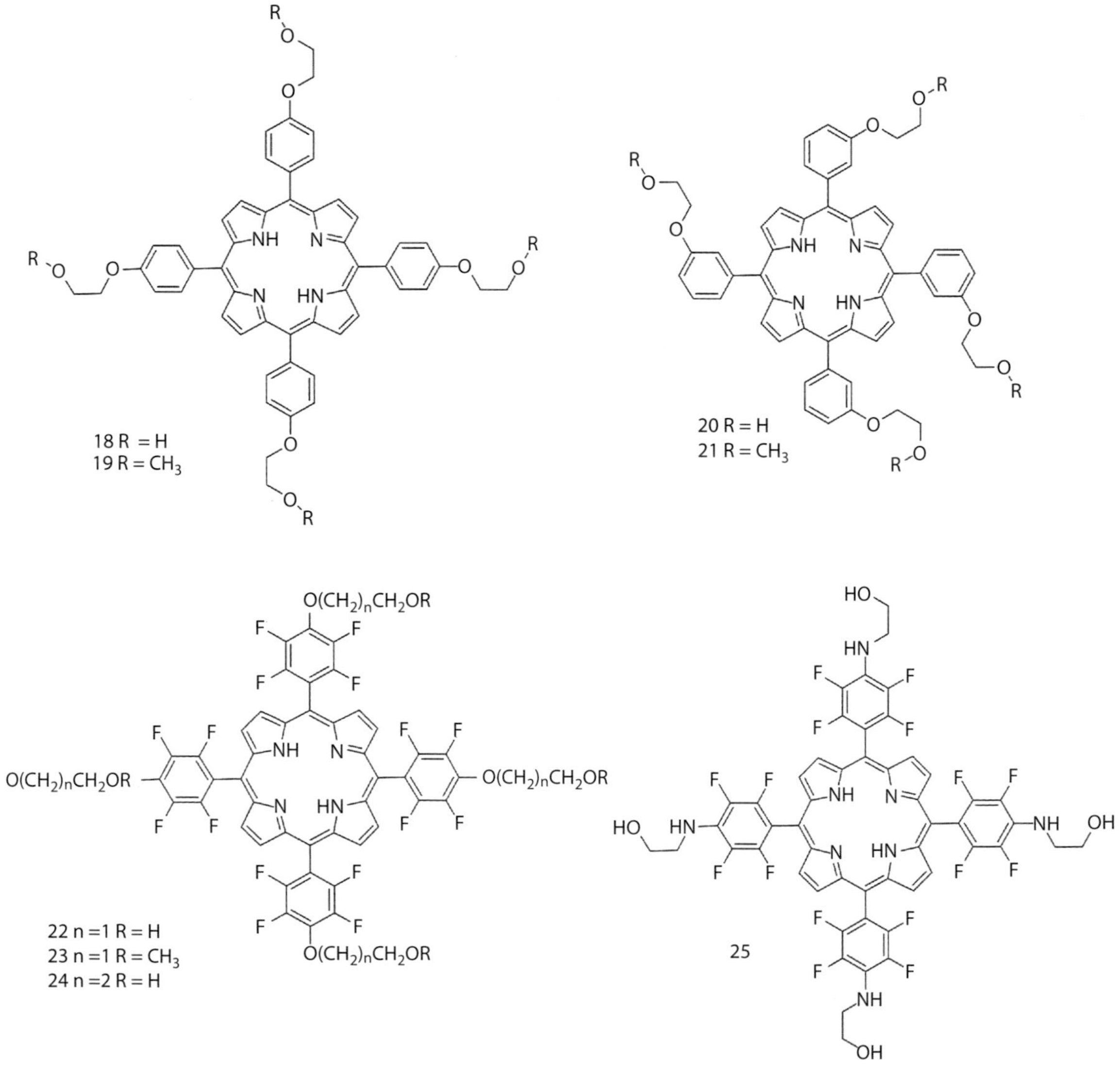

FIGURE 22.9 Structures of glycol-functionalized fluoro porphyrins.

for sexual differentiation in the marine worm *Bonellia viridis* (Agius et al. 1979). Among all the chlorophyll-a derivatives synthesized so far, pyropheophorbide *a* analogs have attracted the most attention (Pandey et al. 1991, 1996). Pandey et al. (1996) synthesized and evaluated a series of alkyl ether derivatives of pyropheophorbide with variable lipophilicity (Henderson et al. 1997). Subsequent data reported from Henderson et al. (1997) have shown structural modifications of pyropheophorbide-derived PSs, in which each PS only differed in the number of carbon atoms or the shape or flexibility of the alkyl ether side chain. The 1-hexyl derivative showed superior PDT activity over the other derivatives in the study (Henderson et al. 1997). Based on this study, 3-devinyl-3-(1′-hexyloxy)ethyl-pyropheophorbide 7 (HPPH) was selected as the best candidate from the alkyl ether derivatives of the pyropheophorbide-a system. Currently, the hexyl ether analog 7, HPPH (max = 660 nm), is under phase I/II human clinical trials for obstructive esophageal, head, and neck cancers; BCC; and Barrett's esophagus with high-grade dysplasia (Rigual et al. unpublished results). The clinical data of HPPH-PDT (Bellnier et al. 1993, 2003, 2004, 2006; Chen et al. 2005a; Anderson et al. 2003; Lobel et al. 2001; Dougherty et al. 2000; Potter et al. 1999; Magne et al. 1997; Furukawa et al. 1994a,b) show its enormous potential in treating certain types of cancers.

The SAR approach followed by Pandey et al. (1996) in developing an effective PS from a series of alkyl ether analogs of pyropheophorbide-a was extremely useful in selecting the highly effective PS HPPH (Henderson et al. 1997). They extended this approach to the purpurinimide system and synthesized a series of the corresponding alkyl ether analogs with variable carbon units at two different positions [3-(1′-O-alkyl) and 13²-N-alkyl] with long-wavelength absorption near 700 nm and efficient singlet oxygen generation (Zheng et al. 2001). In this series, PS 26 containing the N-butyl and O-butyl side chains is currently being evaluated for toxicological studies and was selected on the basis of its *in vivo* PDT efficacy and ease of synthesis (Ethirajan et al. 2011) (Figure 22.10).

The importance of fluorine in medicinal chemistry is well known (Surya Prakash and Yudin 1997). Fluorine substitutions are known to increase lipid solubility, which could result in increasing the rate of transportation of biologically active

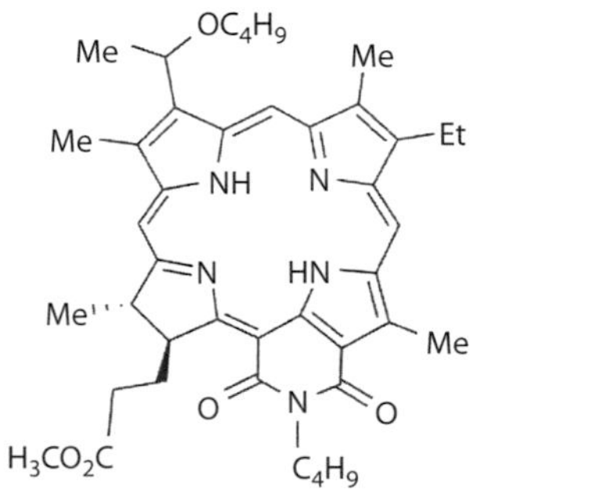

26

FIGURE 22.10 Chlorin-based PS. (From Ethirajan, M. et al., *Chem. Soc. Rev.*, 40, 2011.)

compounds across the lipid membrane (Banks, Smart, and Tatlow 1994). To look into the effect of fluorination, Gryshuk et al. utilized the earlier approach, synthesizing analogs in which the alkyl ether side chain was replaced with trifluoromethyl benzyl ether groups (Gryshuk et al. 2002). Among this series of PSs, compared to nonfluorinated analogs, the corresponding fluorinated derivatives showed improved PDT efficacy (MacDonald and Dougherty 2001).

Benzoporphyrin derivative is an example of a chlorin-like compound that has been used clinically as a PS (Richter et al. 1987).

Another class of chlorins, the benzochlorins, consists of a benzene ring fused to the tetrapyrrolic structure and was first prepared from octaethylporphyrin (OEP) by Arnold et al. (1978). Morgan et al. (1989, 1992) showed the utility of this class of compounds as PSs for PDT. A series of fluorinated and nonfluorinated octaethylporphyrin-based benzochlorins with variable lipophilicity were synthesized by Li et al. (2001), and at a concentration of 2.5 μM and a light dose of 4 J/cm², all benzochlorins produced significant *in vitro* photosensitizing efficacy. The same group later reported an *in vitro* and *in vivo* structure–activity relationship study with a new series of benzochlorins (37–46) with variable lipophilicity. The structural features evaluated in this study included the length of the alkyl or fluoroalkyl groups attached to the six-member exocyclic ring either by an ether or by a carbon–carbon bond (Graham et al. 2003) (Figure 22.11).

Among all the analogs, benzochlorins bearing methyl (38), trifluoromethyl (37), and perfluorooctanyl (42) groups on the six-membered exocyclic ring showed enhanced PDT efficacy compared to other analogs (Figure 22.12). Also, it was seen that the Zn benzochlorin (37) showed improved PDT efficacy over the related free-base drug 47.

22.3.3 Natural and Synthetic Bacteriochlorins

Earlier, we discussed that, compared to porphyrins, chlorins have a strong, more red-shifted wavelength in the range of 650–700 nm, which almost doubles the tissue penetration distance of the incident light. However, a PS that absorbs light near 800 nm should be ideal not only because that is the optimal wavelength for tissue penetration but also because of the availability of cheaper diode lasers for sensitizer excitation. Most of the naturally occurring bacteriochlorins are such ideal candidates because they exhibit long-wavelength absorption near 800 nm. However, because of their unstable nature, synthesis of stable bacteriochlorins has been a challenge for chemists worldwide.

Eisner (1957) elucidated the constitution of bacteriochlorophyll. Bacteriochlorophyll *a* 48 is a naturally occurring bacteriochlorophyll with a good singlet oxygen–producing ability and absorption at 780 nm (extinction coefficient > 70,000) near the optimum wavelength for tissue penetration. Henderson et al. (1991) found that it was an effective PS but was unstable *in vivo*. Beens et al. (1987) investigated two water-soluble derivatives of bacteriochlorophyll *a* 48, namely, bacteriochlorophyllin *a* 49 and bacteriochlorin a 50 in Fig 22.13. Bacteriochlorin 50 was proved to be effective *in vitro* and *in vivo* (Post et al. 1996).

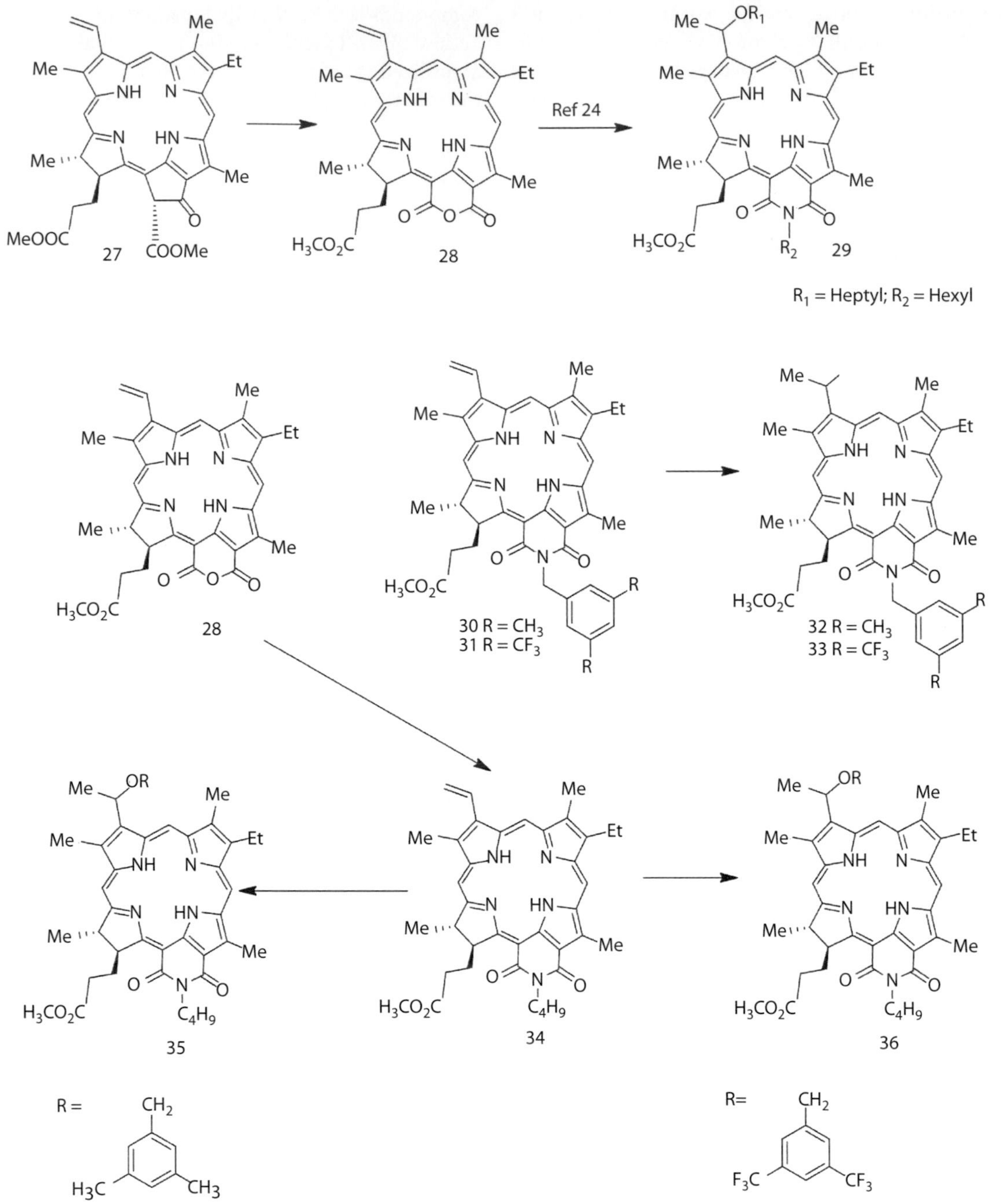

FIGURE 22.11 Synthesis of fluorinated and nonfluorinated chlorin-based PS. (From Graham, A. et al. *Photochem. Photobiol.*, 77, 5, 2003.)

Tolyporphin 51, isolated by Prinsep et al. (1992) from the blue green alga *Tolypothrix nodosa*, is the only naturally occurring bacteriochlorin that is not involved in photosynthesis. This compound was found to enhance the cytotoxicity of adriamycin or vinblastine in SK-VLB cells at doses as low as 1 μg and is characterized as a multidrug resistance (MDR) agent. Minehan and Kishi (1997) reported the synthesis of tolyporphin chromophore, but the structure of tolyporphin was revised by the same group (52) (Wang and Kishi 1999) (Figure 22.14). This naturally occurring bacteriochlorin was found to be a very potent PS of

EMT-6 tumor cells grown both *in vitro* as suspensions or monolayers and *in vivo* in tumors implanted on the backs of C.B17/Icr severe combined immunodeficient mice (Morliere et al. 1998). During PDT of EMT-6 tumor cells *in vitro*, the photokilling effectiveness of tolyporphin was approximately 5000 times higher than that of Photofrin II. *In vivo* studies done with the EMT-6 mouse tumor model showed excellent effectiveness as compared to that of Photofrin II and other second-generation PSs of the pheophorbide class, which are themselves much more potent than Photofrin.

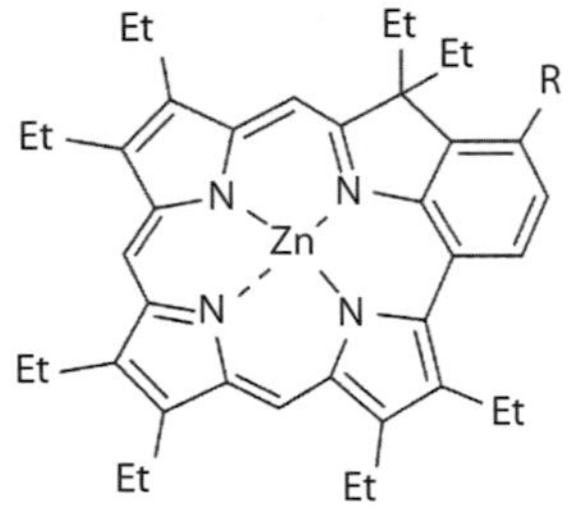

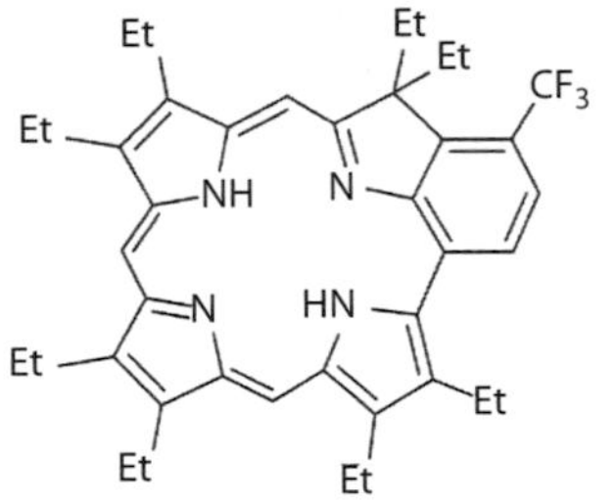

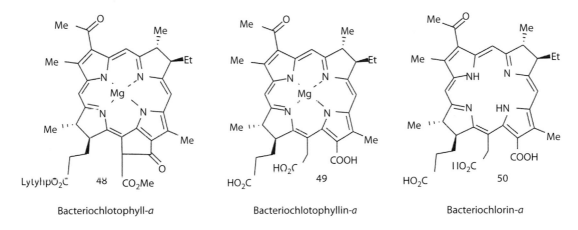

37 R = CF₃
38 R = CH₃
39 R = (CH₂)₅CH₃
40 R = (CH₂)₉CH₃
41 R = (CH₂)₁₇CH₃

42 R = (CF₂)₇CH₃
43 R = CH(OCH₃)CH₃
44 R = CH[O(CH₂)₅CH₃]CH₃
45 R = CH[O(CH₂)₉CH₃]CH₃
46 R = CH[O(CH₂)₁₇CH₃]CH₃

FIGURE 22.12 Structure of benzochlorin analogs.

Bacteriochlotophyll-*a*   Bacteriochlotophyllin-*a*   Bacteriochlorin-*a*

FIGURE 22.13 Structures of some naturally occurring bacteriochlorins. (From Beens, E. et al. *Photochem. Photobiol.*, 45, 1987.)

Naturally occurring bacteriochlorins mostly have absorptions between 760 and 780 nm and are extremely sensitive to oxidation, resulting in a rapid transformation into the chlorin state resulting in a drop of the absorption maxima to 650–660 nm. Further excitation of these bacteriochlorins during PDT *in vivo* leads to oxidation with the formation of chromophores that are outside the excitation window, reducing the photodynamic efficacy. Therefore, the synthesis of stable bacteriochlorins with appropriate photochemical properties has been an important goal to obtain long-wavelength PSs for PDT.

Chlorins containing six-membered imide ring systems (e.g., purpurinimides) were found to be more stable to oxidation than those containing anhydride and isoimide ring systems (Grahn et al. 1997). Pandey and coworkers (Chen et al. 2007) prepared a series of stable bacteriopurpurinimides from bacteriopurpurin-18, which had been isolated from *Rb. sphaeroides*. A series of

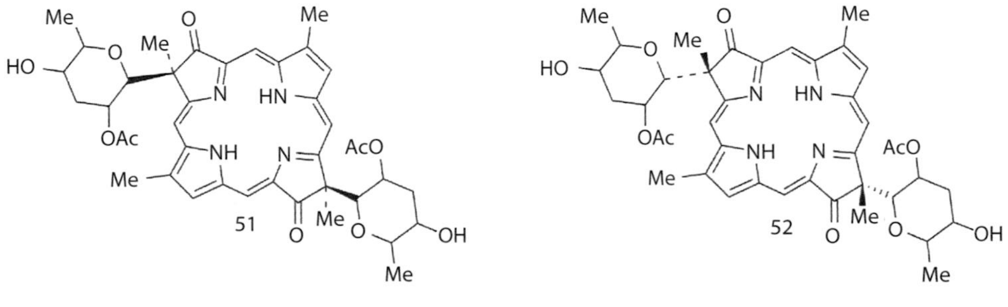

FIGURE 22.14 Original and revised structure of tolyporphin.

open-chain bacteriochlorins (bacteriochlorin p6) with variable lipophilicity were also prepared in order to look into the effect of the fused imide ring system. In their approach, bacteriochlorophyll *a* 53, a component of *Rb. sphaeroides*, was converted into bacteriopurpurin-*a* in good yield (Chen et al. 2007) (Figure 22.15).

Bacteriopurpurin-*a* in a sequence of reactions was converted to two series of bacteriochlorins: bacteriopurpurinimide with a fused imide ring system and bacteriopurpurin p_6 with an open-chain system. Between the two series, the bacteriopurpurinimides containing a fused imide ring (57–60) were found to be more effective *in vivo* in C3H mice bearing RIF tumors.

A series of N-hydroxycycloimide and N-aminocycloimide bacteriochlorin-p derivatives were developed (Mironov et al. 2003a,b, 2004; Mironov, Grin, and Tsyprovskiy 2002; Mironov, Kozyrev, and Brandis, 1993), and many of these were reported to be phototoxic to tumor cells (Mironov et al. 2003b; Sharonov et al. 2004). A detailed report on the PDT-relevant properties of this series of compounds was reported, and among the various derivatives, compounds 61–64 were found to be promising for efficacious PDT (Sharonov et al. 2006) (Figure 22.16).

To overcome the limitations of naturally occurring bacteriochlorins, a novel synthetic route was developed by Kim and Lindsey (2005) (Figure 22.17). The key feature of this synthesis was the geminal-dimethyl group in ring B and ring D (66, 67, and 68), which prevented dehydrogenation and tautomerization and thus provided chemical stability to the bacteriochlorin chromophore (Fan, Taniguchi, and Lindsey 2007; Borbas, Ruzié, and Lindsey 2008; Ruzié et al. 2008). These synthetic bacteriochlorins showed significant PDT efficacy in killing both pigmented and nonpigmented melanoma cells compared with Photofrin and Lu-Tex. Moreover, the bacteriochlorins examined (69–71), particularly bacteriochlorins 70 and 71 (Figure 22.18), were effective at significantly lower concentrations (Mroz et al. 2010). A series of synthetic bacteriochlorins (Kim and Lindsey 2005; Fan, Taniguchi, and Lindsey 2007; Borbas, Ruzié, and Lindsey 2008; Ruzié et al. 2008; Ruzié, Krayer, and Lindsey 2009) with various substituents were tested against HeLa human cervical cancer cells for their usefulness as PSs (Huang et al. 2010).

Joshi et al. (2011) synthesized a series of bacteriochlorins that were subjected to oxidation either by FeCl$_3$ or DDQ. The

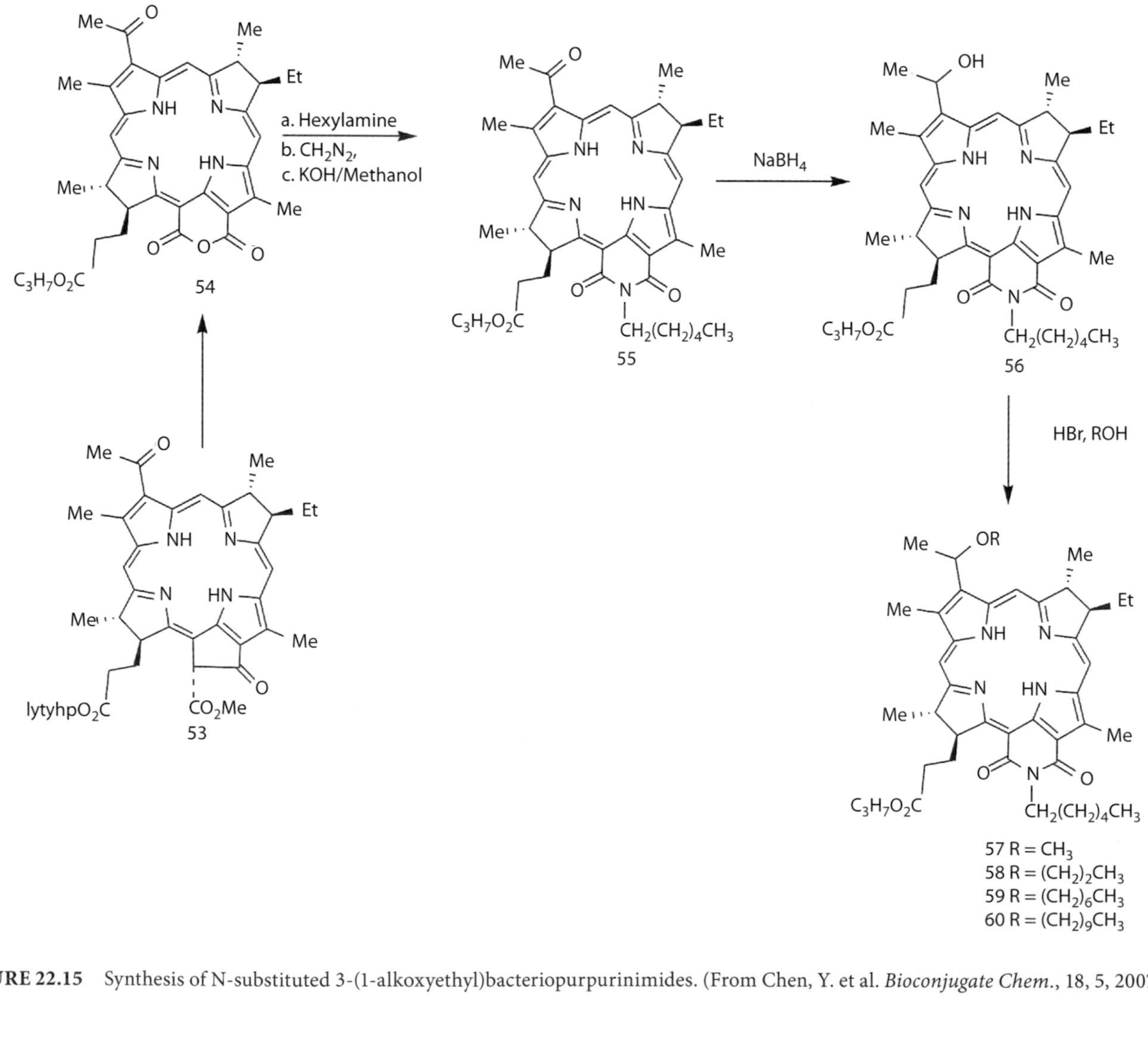

FIGURE 22.15 Synthesis of N-substituted 3-(1-alkoxyethyl)bacteriopurpurinimides. (From Chen, Y. et al. *Bioconjugate Chem.*, 18, 5, 2007.)

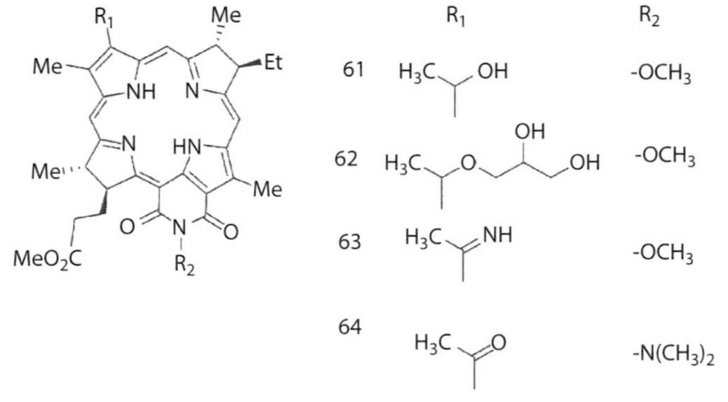

FIGURE 22.16 Structures of bacteriochlorin p$_6$ derivatives.

chlorins thus obtained were reacted under pinacol–pinacolone reaction conditions to give various keto bacteriochlroins. The newly synthesized bacteriochlorins (72–75) showed strong long-wavelength absorption and produced significant *in vitro* (Colon26 cells) photosensitizing ability (Figure 22.19). Among the compounds tested, the bacteriochlorins containing a keto-group at position 7 (compound 74) of ring-B and with the five-member isocyclic ring cleaved showed the best efficacy.

22.3.4 Isomeric Porphyrins, Porphycenes, and Inverted Porphyrins

Porphycenes (76), the first constitutional isomers of porphyrin (1), were prepared by Vogel et al. (1986) (Figure 22.20). Structurally, this class of compounds consists of two 2,2'-bipyrrole subunits linked by two double bonds resulting in a planar, aromatic macrocycle formally known as [18] porphyrin-(2.0.2.0) (Gosmann and Franck 1986). The most relevant feature of the

porphycene ring is its lower structural symmetry compared to porphyrins, resulting in 20-fold larger absorption coefficients in the red part of the spectrum (Waluk et al. 1991; Hayashi et al. 2002; Hasegawa et al. 2005; Sánchez-Ganchez and Sessler 2008). Vogel et al. (1995) found that porphycenes have stability comparable with that of porphyrins. This evident stability is attributed to the presence of strong NH–N hydrogen bonds.

Tetra-propyl-porphycene (78) was the first porphycene isomer evaluated for potential use as a PS for PDT. Guardiano et al. (1989) reported that 78 was delivered to tumor tissues with good efficiency and selectivity. Since the discovery of porphycenes, several derivatives have been synthesized, and among those, some have proven to be effective as *in vivo* antitumor agents (Richert et al. 1994; Canete et al. 1997; Luo, Chang, and Kessel 1996; Segalla et al. 1997).

Various additional structural isomers of porphycenes were constructed but out of those, only two (hemiporphycene and corrphycene) were synthesized as a result of the instability of other isomers (Wu et al. 1997) (Figure 22.21). Octaethyl-hemiporphycene 79 and octaethyl-corrporphycene 80 both generate singlet oxygen efficiently (0.58 and 0.48, respectively). Both of these compounds are potential PDT PSs. Shimakoshi et al. (2008) have reported the synthesis of a series of compounds (81–84) (Figure 22.22); wherein they have incorporated a heavy atom bromine directly into the porphycene macrocycle that can promote intersystem crossing by spin-orbit coupling (Azenha et al. 2002). Among this series of compounds, the dibrominated porphycene (82) shows the highest sensitization efficiency for singlet oxygen production in response to visible light (0.95 quantum yield) (Figure 22.23).

Recently, Ragas et al. (2010) reported the synthesis of the first aryl tricationic water-soluble porphycene and its potential use in antimicrobial PDT. The starting 2,7,12,17-tetrakis-(p-(methoxy

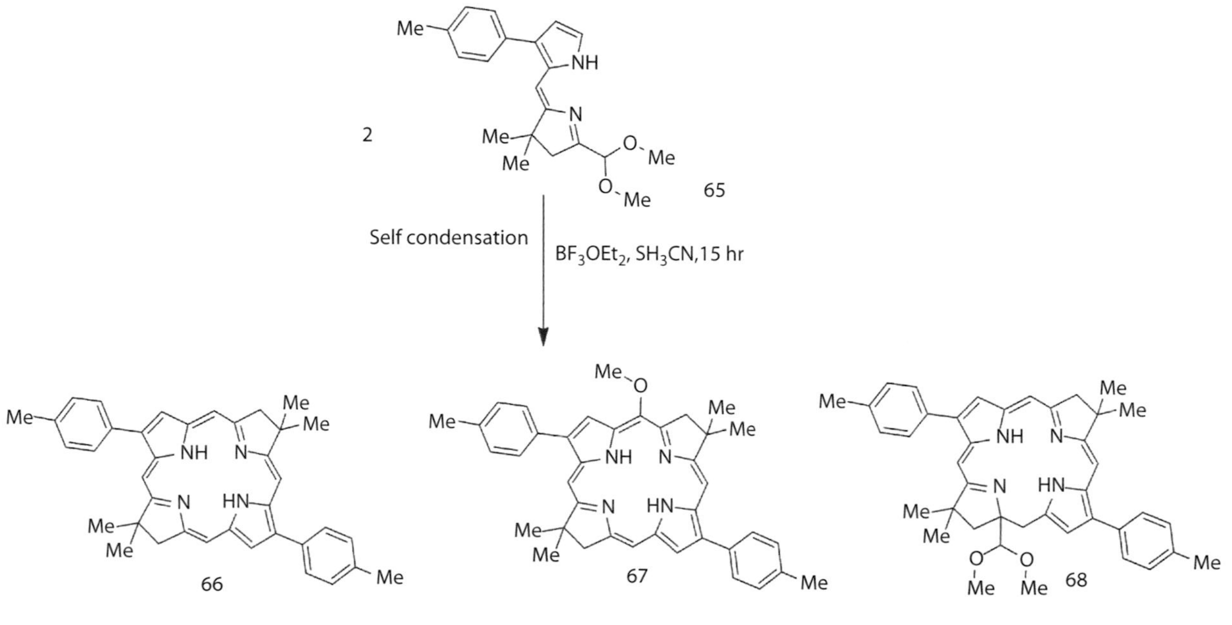

FIGURE 22.17 Synthetic route for the synthesis of bacteriochlorins.

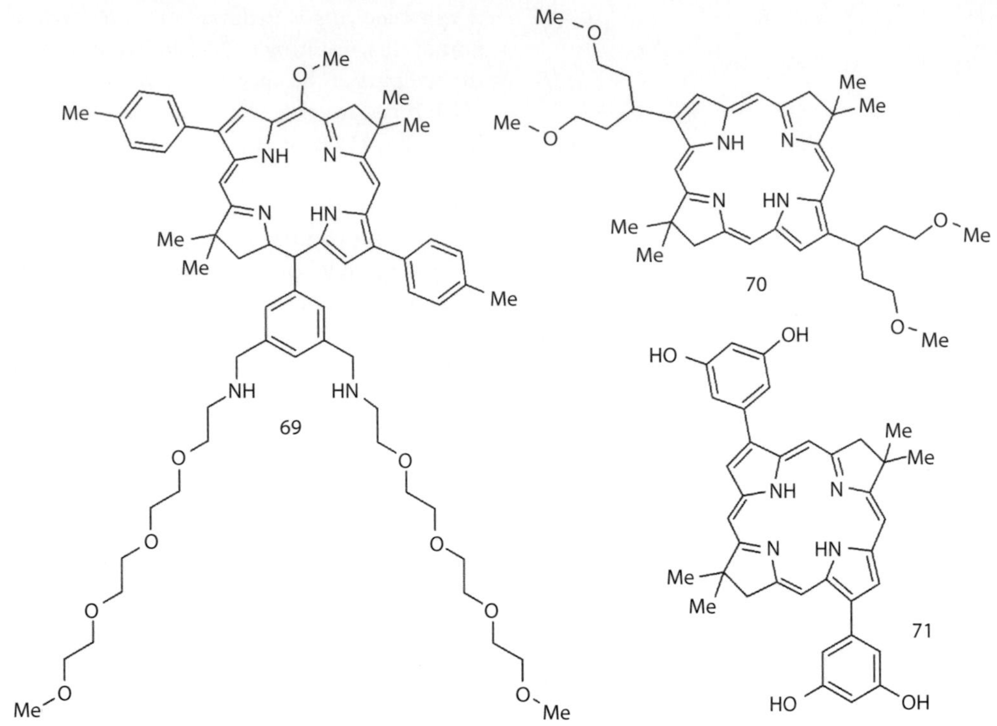

FIGURE 22.18 Highly stable synthetic bacteriochlorins.

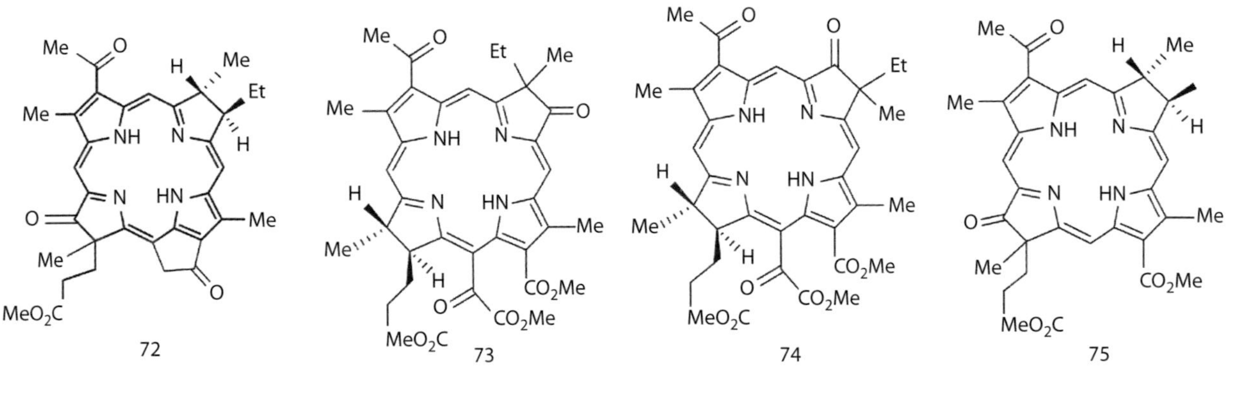

FIGURE 22.19 Structures of various ketobacteriochlorins.

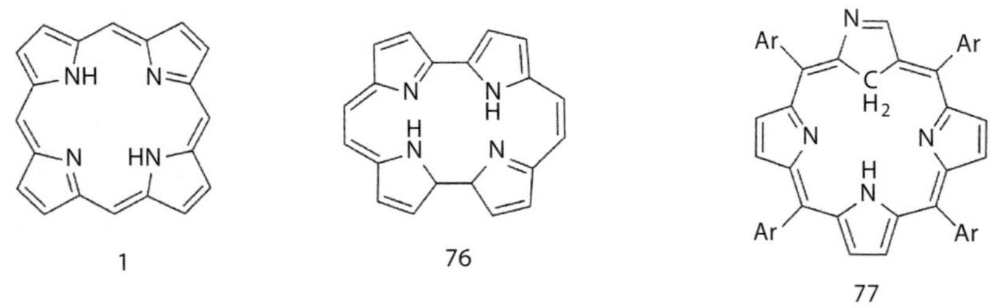

FIGURE 22.20 Constitutional isomers of porphyrin.

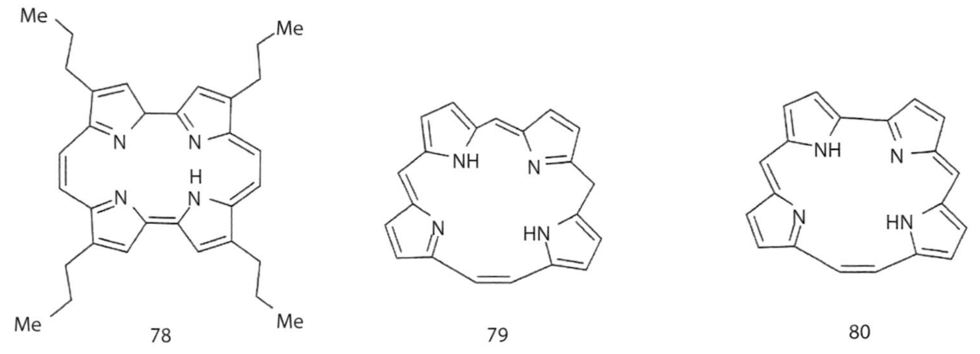

FIGURE 22.21 Structural isomers of porphyrin.

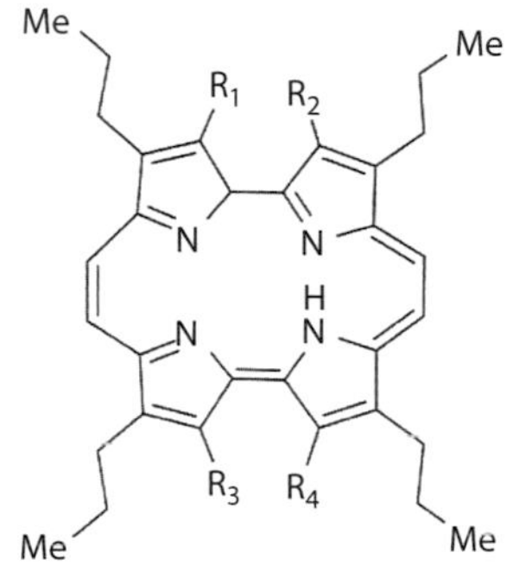

81 $R_1 = Br, R_2 = R_3 = R_4 = H$
82 $R_1 = R_4 = Br, R_2 = R_3 = H$
83 $R_1 = R_2 = R_3 = Br, R_4 = H$
84 $R_1 = R_2 = R_3 = R_4 = Br$

FIGURE 22.22 Structures of brominated porphycenes.

methyl) phenyl) porphycene (90) was synthesized using the method developed by Sanchez-Garcia, Borrell, and Nonell (2009). This aryl tricationic water-soluble porphycene (93) was successfully tested *in vitro* against different Gram-positive and Gram-negative bacteria as well as a fungal species (*Candida*) for both drug-dose and light-dose dependence (Figure 22.24). It was seen that low concentrations of porphycenes (<2 μM) as well as low light doses (<30 J/cm²) were enough to completely eliminate all the Gram-positive strains, and higher light doses (>60 J/cm²) and higher PS concentration (<10 μM) were needed for Gram-negative bacteria inactivation. They also used it *in vivo* to treat an infection model composed of mouse third-degree burns infected with a bioluminescent methicillin-resistant *Staphylococcus aureus* strain (Ragas et al. 2010).

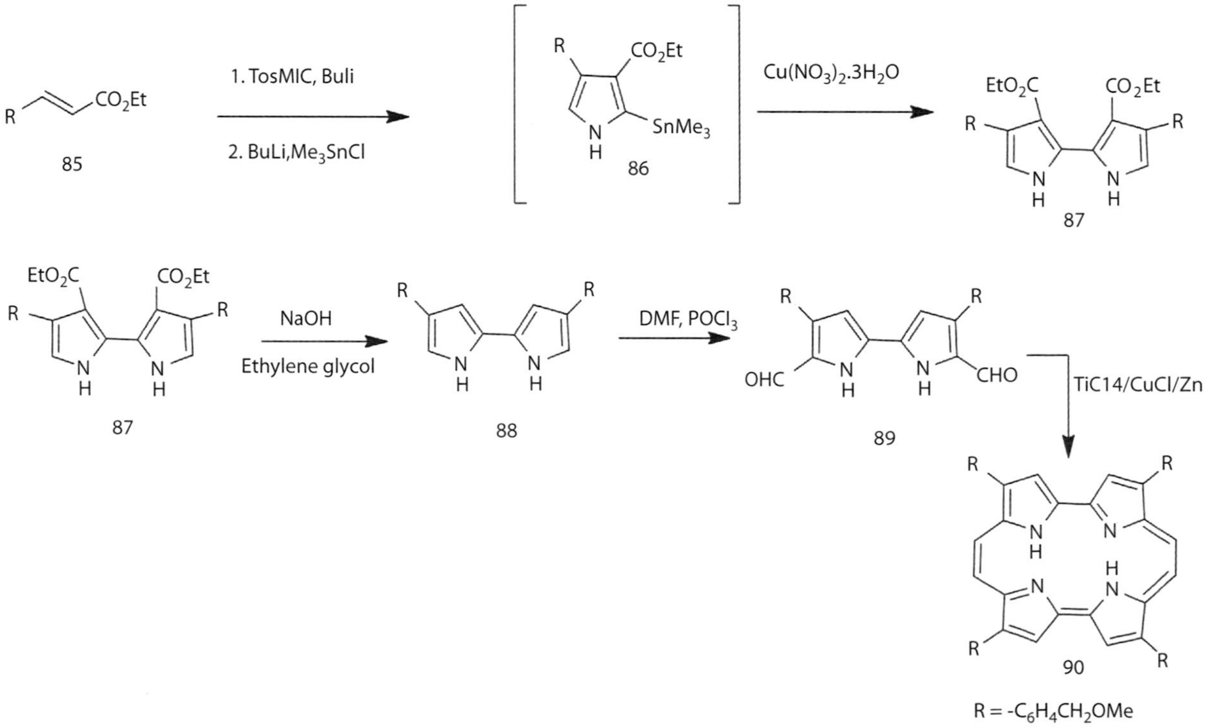

FIGURE 22.23 Synthesis of substituted porphycene. (From Azenha, E. et al. *Chem. Phys*., 280, 2002.)

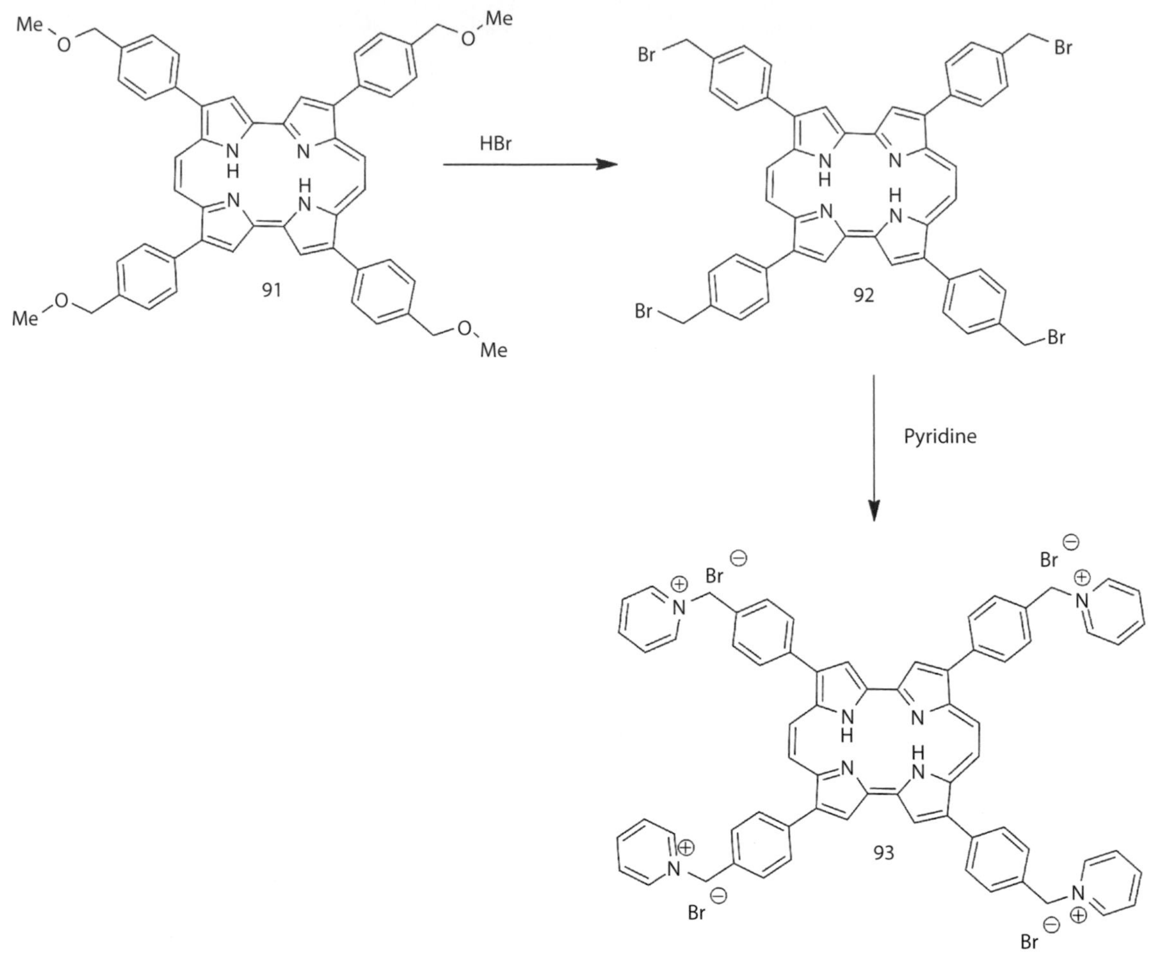

FIGURE 22.24 Synthesis of water-soluble porphycene.

The chemistry of inverted or N-confused porphyrin (NCP) or inverted porphyrins (77) and respective analogs started in 1994 when Furuta, Asano, and Ogawa (1994) and Chmielewski et al. (1994) independently and almost simultaneously reported N-confused porphyrin (NCP), in which one of the pyrrole rings is connected to meso-carbons at the α and β' positions with an otherwise identical framework of a normal porphyrin (Figure 22.20). Meso-substituted N-confused porphyrin (95, 96) were synthesized by the Rothemund type reaction (Rothemund 1939) (Figure 22.25), an acid catalyzed pyrrole–aldehyde condensation with concurrent formation of normal porphyrin. Although the isolated yield reported in 1994 was low, Geier, Haynes, and Lindsey (1999) and Geier and Lindsey (1999) have improved the reaction yield to 39% (Narayanan et al. 1998). On the other hand, meso-free type forms of NCP (99, 102, 103, and 104) were prepared in a stepwise manner using acid-catalyzed MacDonald-type [2 + 2] and [3 + 1] condensation by Liu, Brückner, and Dolphin (1996) and Lash, Richter, and Shiner (1999), respectively (Figure 22.25). NCP analogs containing hetero-atoms, such as O, S, and Se, have also been synthesized by Heo, Shin, and Lee (1996); Heo and Lee (1996); Lee

and Kim (1997); Lee, Kim, and Yoon (1999); Yoon and Lee (2000); Sprutta and Latos-Grazynski (1999); Pacholska et al. (2000); and Sprutta and Latos-Grazynski (2001) and Pushpan et al. (2001).

22.3.5 Sapphyrins and Texaphyrins

Sapphyrins (105) are the class of expanded porphyrins that were discovered accidently during the synthesis of vitamin B_{12} (Bauer et al. 1983) (Figure 22.26). Sapphyrins possess a 22 π-electron pathway and, as a consequence, have more electron affinity than the corresponding porphyrin systems (18 π-electron) (Springs et al. 1999).

These have near-infrared absorbance properties and relatively high singlet oxygen yield, which makes them potential therapeutic agents in the context of PDT (Judy et al. 1991; Maiya et al. 1990). Several novel water-soluble sapphyrins prepared (107–111) and characterized by Kral et al. (2002) (Figure 22.27) were found to localize selectively in pancreatic carcinoma tissue in a xenographic murine model. Among these, the sapphyrins bearing neutral solubilizing groups (compounds 107–109) were

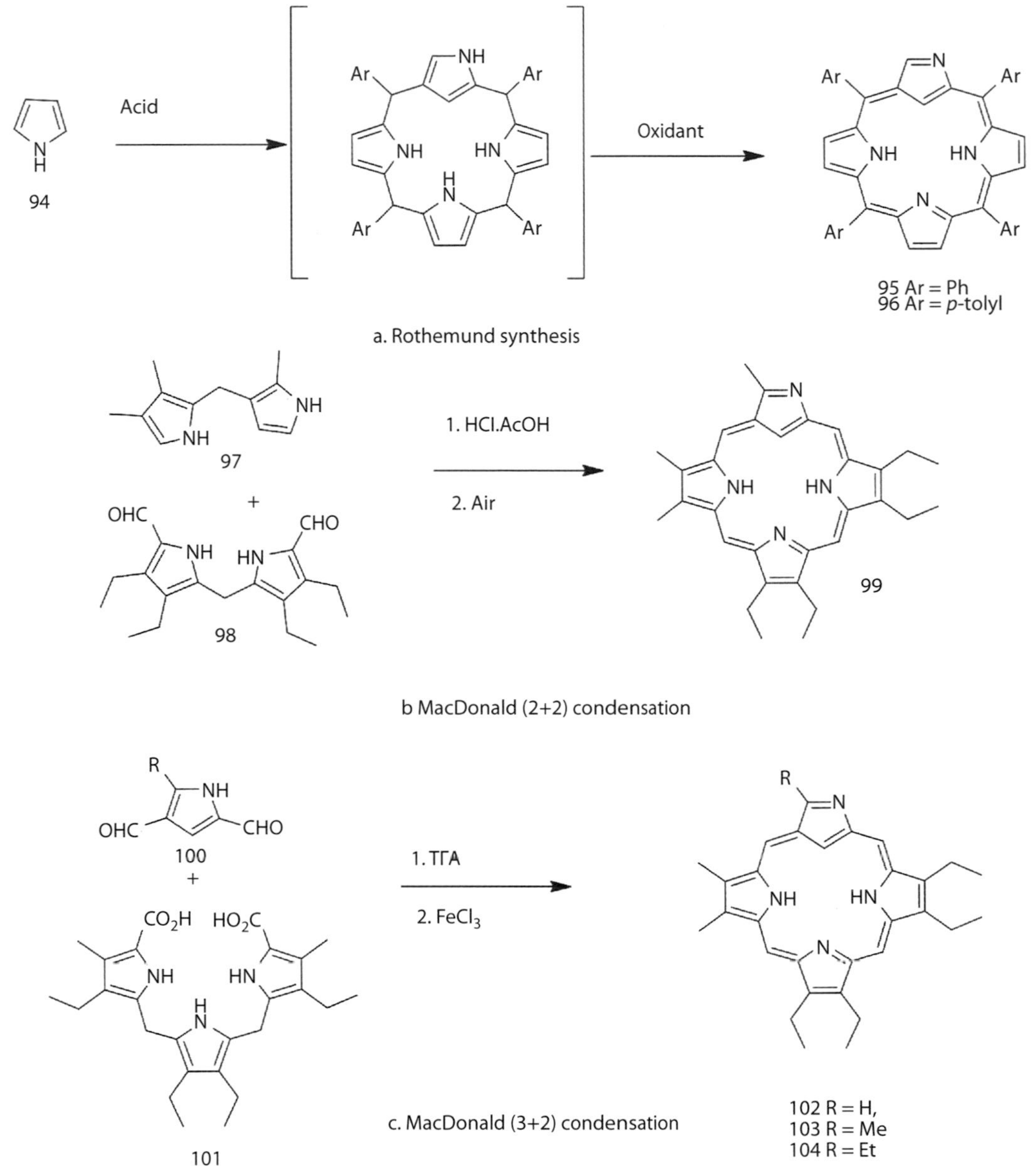

a. Rothemund synthesis

b MacDonald (2+2) condensation

c. MacDonald (3+2) condensation

102 R = H,
103 R = Me
104 R = Et

FIGURE 22.25 Synthesis of N-confused porphyrins or inverted porphyrins.

found to have selectivities for tumor tissue over surrounding tissues. The incorporation of charged moieties into the sapphyrin (compounds 110 and 111) significantly reduced the tumor localization. The tetrahydroxy sapphyrin 107 exhibited the best tumor-to-muscle ratio (280 ± 80), whereas the incorporation of glucosamide into the sapphyrin (109) cores afforded the best tumor-to-liver ratio (880 ± 120). Until now, sapphyrins have been explored solely as anticancer agents, but recently, Hooker et al. (2012) showed the activity of sapphyrins and heterosapphyrins in the presence of light against Leishmania parasites.

Another class of expanded porphyrins are the pentaazadentate macrocycles. The texaphyrins (106) first reported by Sessler et al. (1988) (Figure 22.26) were formed by the Schiff base

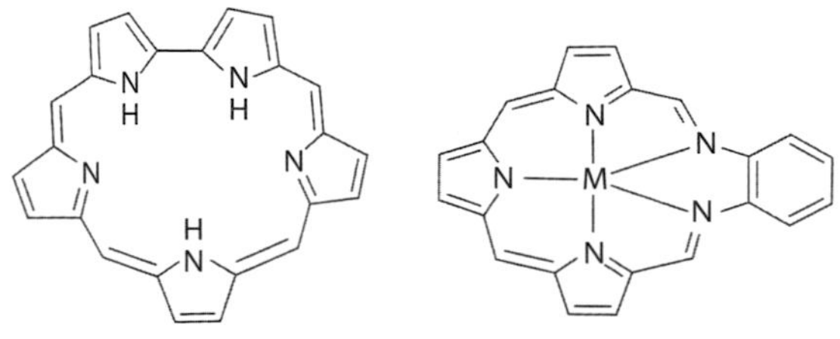

105   106

FIGURE 22.26 Structures of expanded porphyrins.

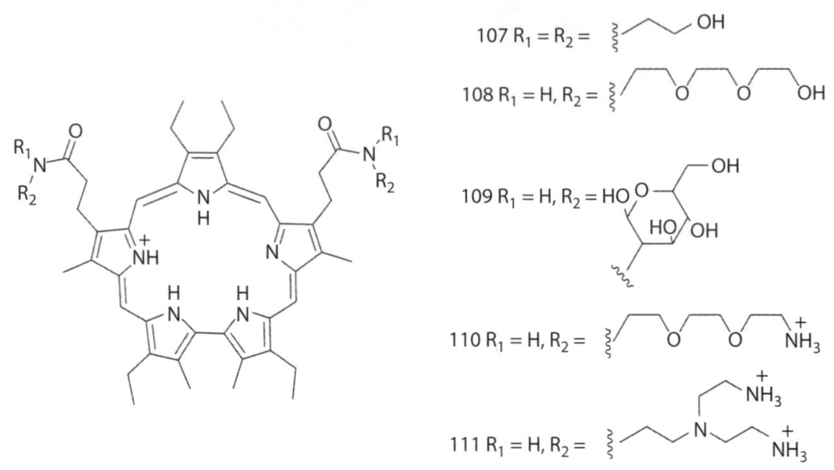

107 R$_1$ = R$_2$ =

108 R$_1$ = H, R$_2$ =

109 R$_1$ = H, R$_2$ =

110 R$_1$ = H, R$_2$ =

111 R$_1$ = H, R$_2$ =

FIGURE 22.27 Structures of water-soluble sapphyrins.

condensation between a diformyltripyrrane and a 1,2-diamine. This strategy has since been employed to obtain numerous cadmium texaphyrin and texaphyrin complexes with a wide range of metals (Sessler et al. 1994). The synthesis of a metal-free form of texaphyrin was reported by Hannah et al. (2001). Earlier texaphyrins could only be obtained in the form of metal complexes. Using a ferrocenium cation as the oxidizing agent and starting with a reduced porphyrinogen-like nonaromatic form of texaphyrin (112), the metal-free oxidized texaphyrin 113 as its HPF6 salt was isolated (Figure 22.28).

The photophysical properties of metallotexaphyrins are of particular interest in PDT. This is a result of the fact that texaphyrins absorb strongly in the 720–780 nm spectral region. The remarkable efficiency of texaphyrins as singlet oxygen-producing PSs is of great importance in PDT as well. Among all the metallotexaphyrins tested for PDT so far, lutetium (III) texaphyrin 114 appears to be the most promising (Young et al. 1996) (Figure 22.29). Compound 114 in various formulations is being tested as a PS for use in (1) the photodynamic treatment of recurrent breast cancer (Lutrin, phase II clinical trials complete), (2) photoangioplastic reduction of atherosclerosis involving peripheral arteries (Antrin, now in phase II testing), and (3) light-based treatment of age-related wet macular degeneration (Optrin, currently in phase I clinical trials), a vision-threatening disease of the retina (Sessler and Miller 2000). The magnesium texaphyrin complexes have been shown to be reliable candidates for their use as PSs in this medical therapy (Lanzo, Russo, and Sicilia 2008).

For the first time, a benzotexaphyrin 115 with an extensively delocalized π-electron system was synthesized to further shift the Q band of texaphyrins (Lu et al. 2008) (Figure 22.29).

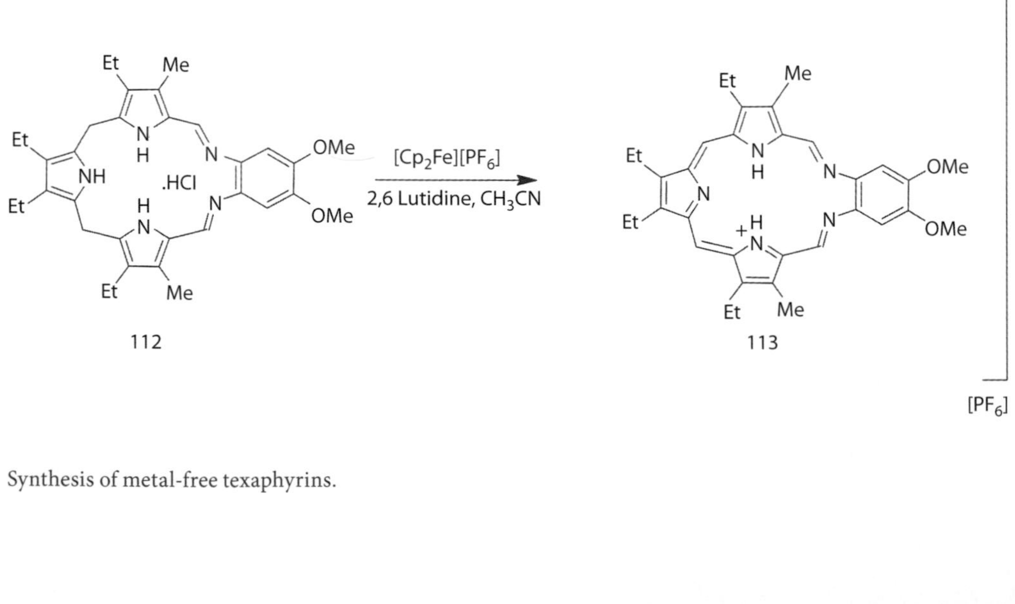

FIGURE 22.28 Synthesis of metal-free texaphyrins.

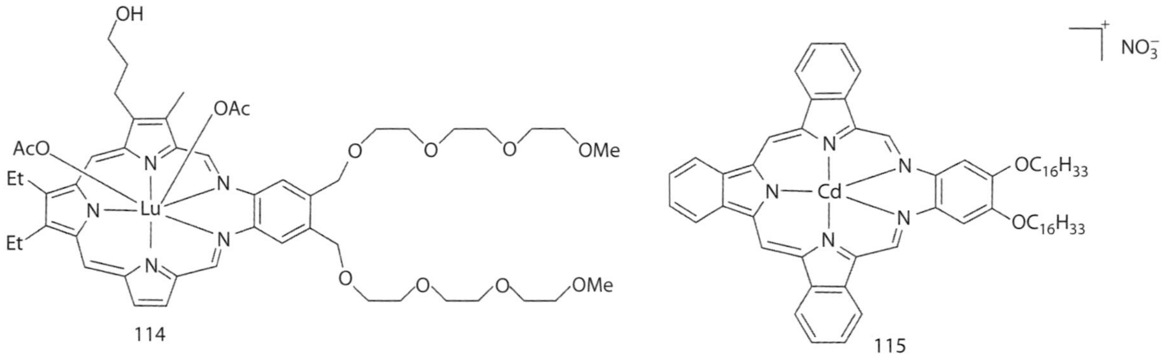

FIGURE 22.29 Structures of LUTRIN (114) and benzotexaphyrin (115).

Because of the extended π-conjugation, the Q-band of benzotexaphyrin bathochromically shifts to 810 nm, and it exhibits a high efficiency in generating singlet oxygen in methanol ($\Delta\phi = 0.65$).

Therefore, benzotexaphyrin could act as a NIR PS and emitter for PDT and have potential imaging applications.

22.3.6 Phthalocyanines and Naphthalocyanines

Phthalocyanines (Pc) 4 and naphthalocyanines (Nc) 5 can be regarded as azaporphyrins, containing four isoindole moieties linked by nitrogen atoms (van Lier and Spikes 1989) (Figure 22.5). They are being studied as PSs for PDT. Compared to Photofrin, they have high extinction coefficients (approximately 10^5), and they absorb at approximately 680 nm (Pc) and 780 nm (Nc). Both of these are excellent singlet oxygen generators, and chelation of a metal ion increases the singlet oxygen efficiency to nearly 100% (Baden 1984). For the first time, water-soluble indium(III) phthalocyanine complexes were prepared by Durmus and Nyokong (2007). 3-Hydroxypyridine tetrasubstituted indium(III) phthalocyanines (118, 121) and their quaternized derivatives (122, 123) were synthesized. Compounds 118 and 121 showed excellent water solubility and high singlet oxygen quantum yield (>0.55), making them potential PDT agents (Figure 22.30).

de Oliveira et al. (2009) reported the synthesis and photophysical evaluation of water-soluble dual PSs containing phthalocyanines-ALA (5-aminolevulinic) conjugate (124, 125) treatments (Figure 22.31).

The ΦΔ values obtained for 124 and 125 were 0.52 and 0.58, respectively, indicating that these new water-soluble phthalocyanines might be good candidates for PDT.

Jiang et al. (2010a,b) have reported silicon (IV) phthalocyanines substituted with aminomoieties (Figure 22.32). These compounds exhibit remarkable pH-dependent properties showing enhanced fluorescence and singlet oxygen-generation efficiency at lower pH.

Encouraged by the results, the same group synthesized a series of novel silicon (IV) phthalocyanines with polyamine moieties at the axial positions (126–134). These were found to be potent PSs toward the HT29 cells that have IC50 values as low as ~1 nM

(Figure 22.32). These also showed *in vivo* PDT efficacy against HT29 tumor-bearing nude mice (Jiang et al. 2011).

A new class of dual agents, the phthalocyanine–chalcone conjugates (135), has been designed by Tuncel et al. (2012) to combine the vascular disrupting effect of chalcones with the photodynamic effect of phthalocyanines (Figure 22.33). The phototoxicity, dark cytotoxicity, cellular uptake, and vascular disrupting effect of this dual agent 135 have not yet been investigated. Isosteric zinc (II) phthalocyanine 2,9(10),16(17),23(24)-tetrakis[(N-butyl-N-methylammonium) ethylsulfanyl] phthalocyaninnato zinc(II) tetraiodide (136) and 2,9(10),16(17),23(24)-tetrakis [(N-dibutyl-N-methyl ammonium) ethoxy] phthalocyaninatozinc(II) tetraiodide (137) were synthesized and assessed for their PDT action (Gauna et al. 2011) (Figure 22.34).

Earlier work by the same group had shown that a sulfur-linked cationic dye, named 2,9(10),16(17),23(24)-tetrakis[(2-trimethyl-ammonium) ethylsulfanyl] phthalocyaninatozinc(II) tetraiodide, was more active than the oxygen-linked cationic aliphatic phthalocyanine 2,9(10),16(17),23(24)-tetrakis-[(2-trimethylammonium) ethoxy] phthalocyaninatozinc(II) tetraiodide (Marino et al. 2010). The phototoxic effect of the two novel phthalocyanines 136 and 137 was evaluated on human nasopharynx KB carcinoma cells. It was seen that sulfur-linked cationic phthalocyanine 136 (IC50 = 1.45 ± 1 mM) showed a better phototoxic effect with respect to the oxygen-linked phthalocyanine 137 (IC50 = 10.5 ± 2 mM). Though the singlet oxygen yield for 137 was higher (0.67) compared to 136 (0.42), the higher cellular uptake obtained for 136, which was mainly localized within lysosomes, could explain the result (Gauna et al. 2011).

Fluorinated zinc phthalocyanines offer some advantages over nonfluorinated derivatives as PSs as shown by Oda and coworkers (Oda, Ogura, and Okura 2000; Alleman et al. 1995). Yslas, Rivarola, and Durantini (2005) showed that zinc(II) phthalocyanine derivatives bearing methoxy and trifluoromethylbenzyloxy substituents showed higher cytotoxic activity (138, 139) (Figure 22.35). Among phthalocyanine analogs, the pegylated zinc phthalocyanines (109–111, 113, 114) were reported to be highly cytotoxic toward HT29 human colorectal carcinoma and HepG2 human hepatocarcinoma cells on illumination with light with IC50 values as low as 0.02 µM (Liu et al. 2008) (Figure 22.36).

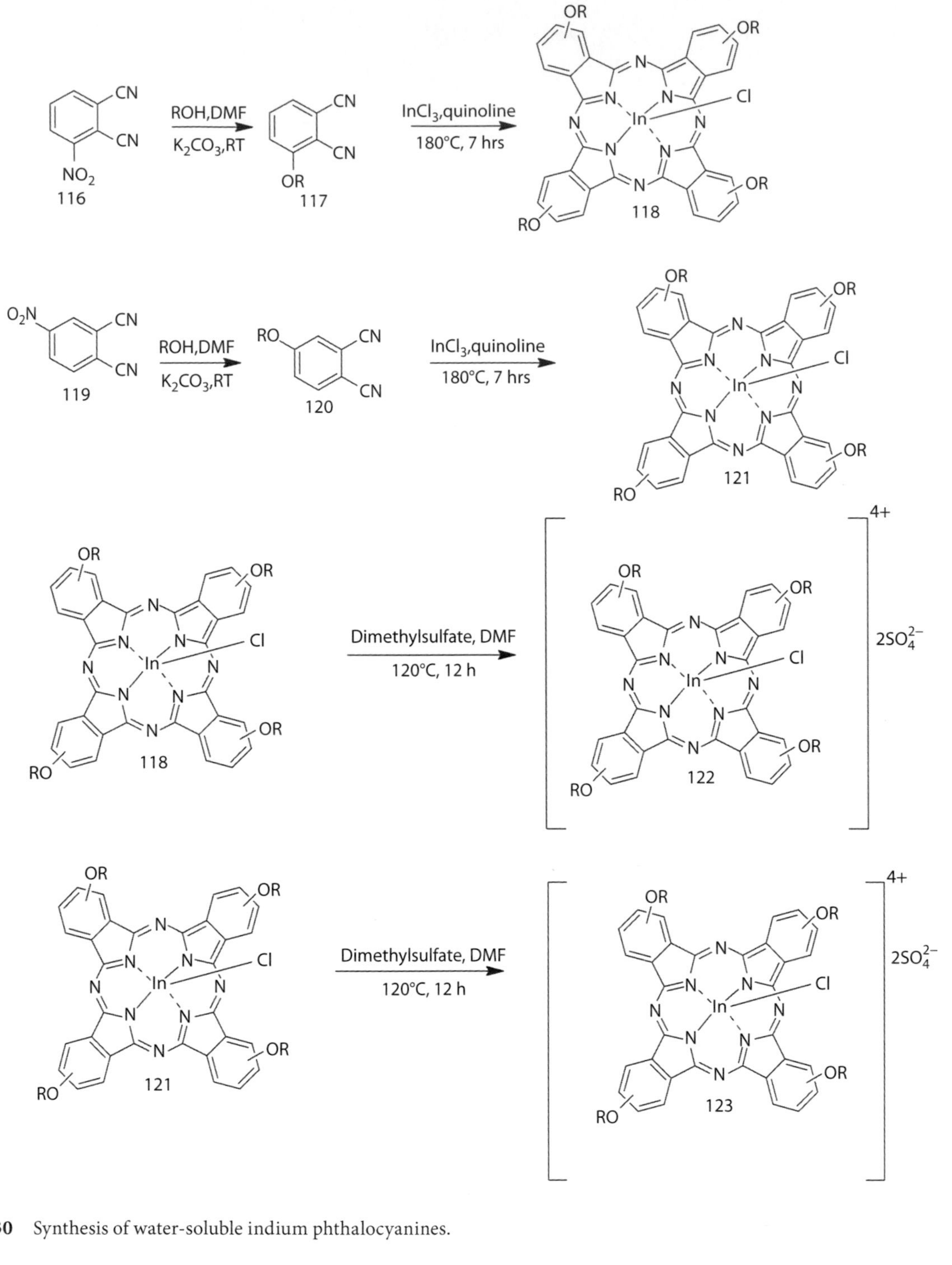

FIGURE 22.30 Synthesis of water-soluble indium phthalocyanines.

Metallo-naphthalocyanines (MNc) have an intense absorption in the 800–900 nm range and have been shown to be efficiently accumulated by solid tumors (Jori et al. 1990). They have also been shown to have successful photothermal sensitization both *in vivo* and *in vitro* (Camerin et al. 2005; Busetti et al. 1999). Recently, Camerin et al. (2009) showed affinity and photothermal sensitization activity of Pd(II)-5,9,14,18,23,27,32,36-octabutoxy-2,3-naphthalocyanine (158) and the corresponding Pt(II)-(159)

derivative toward the B78H1 amelanotic melanoma cell line (Figure 22.37).

22.4 Strategies for Developing Tumor-Specific PSs for PDT

There has been progress in developing more efficient and effective PSs, but more advancement is needed. For instance, the

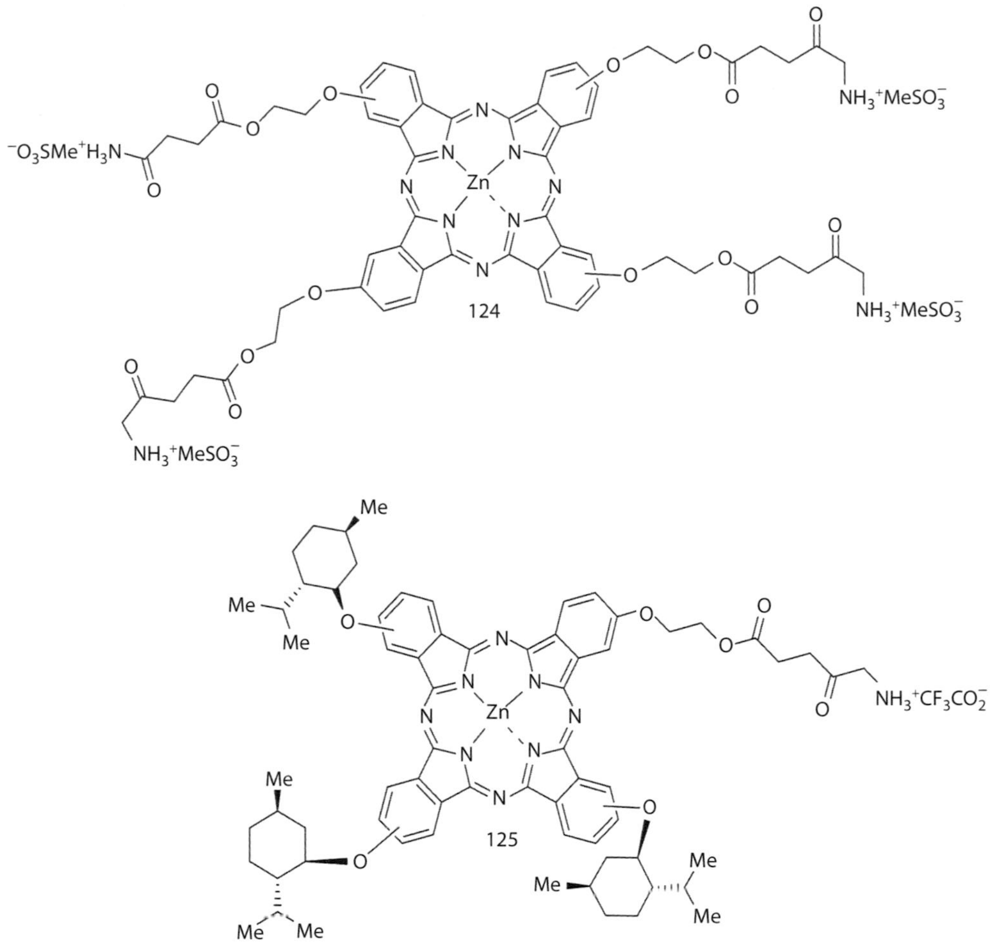

FIGURE 22.31 Dual PS: phthalocyanine–ALA conjugate.

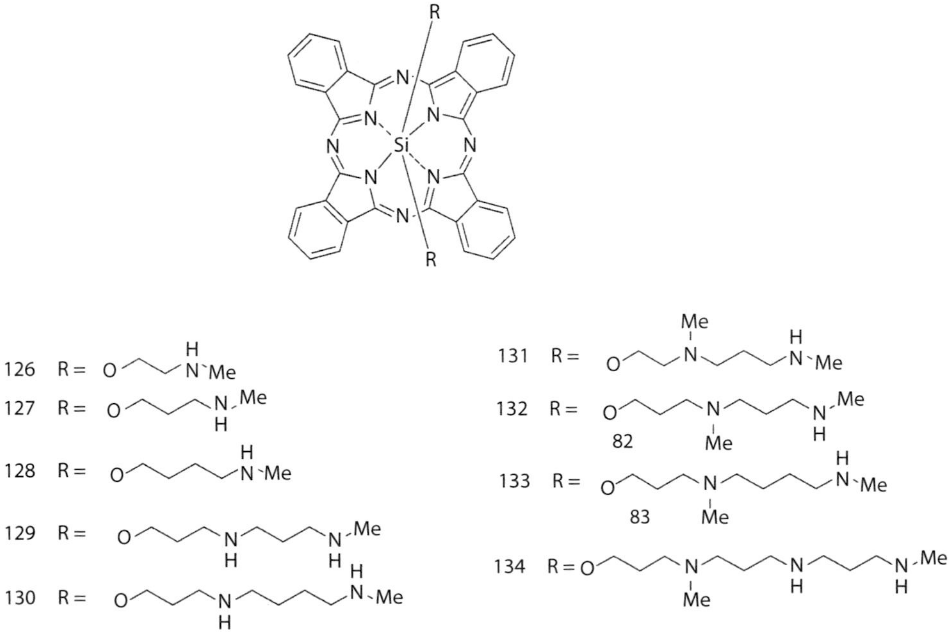

FIGURE 22.32 Structures of silicon (IV) phthalocyanines.

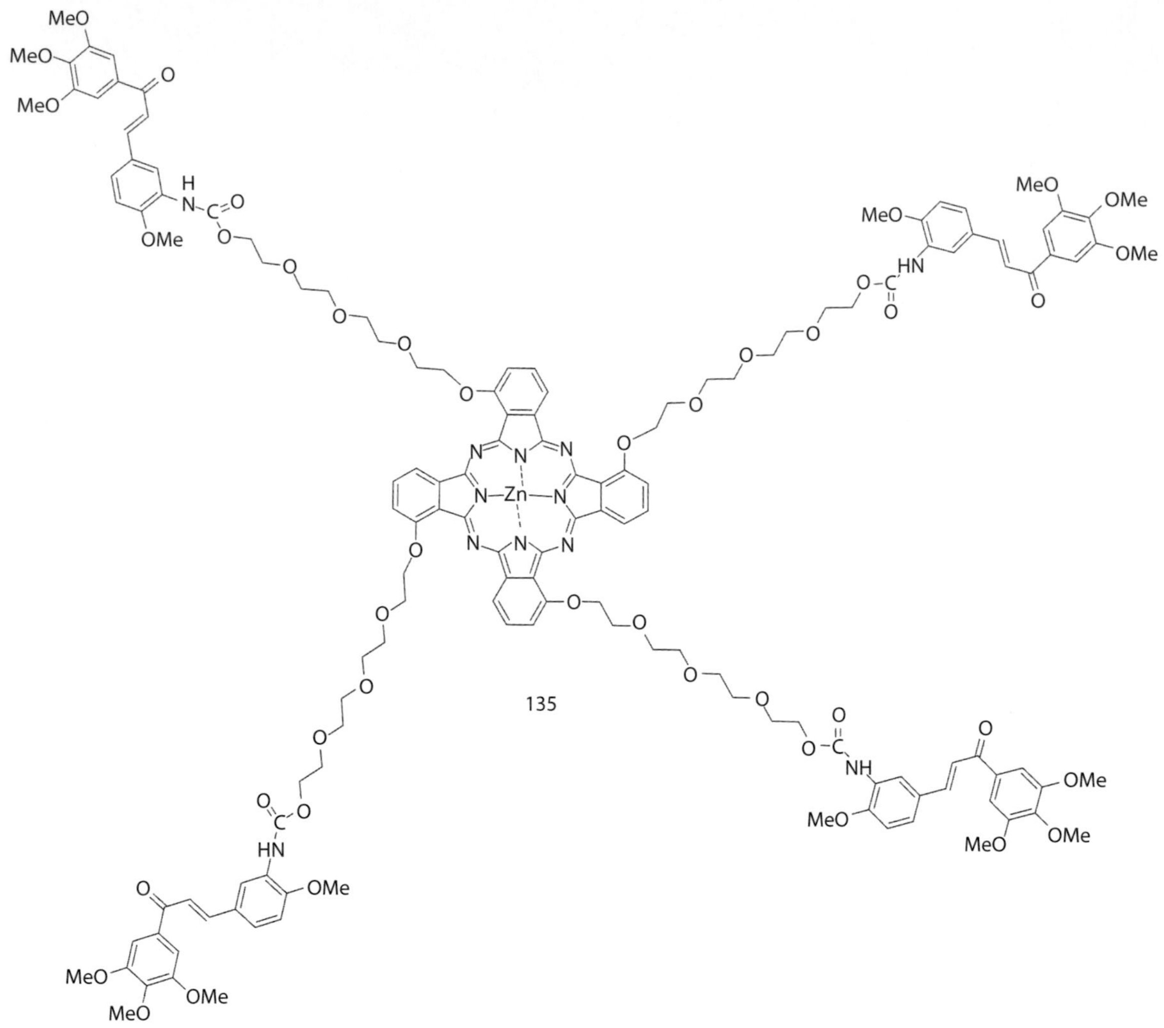

FIGURE 22.33 Structure of phthalocyanine–chalcone conjugate.

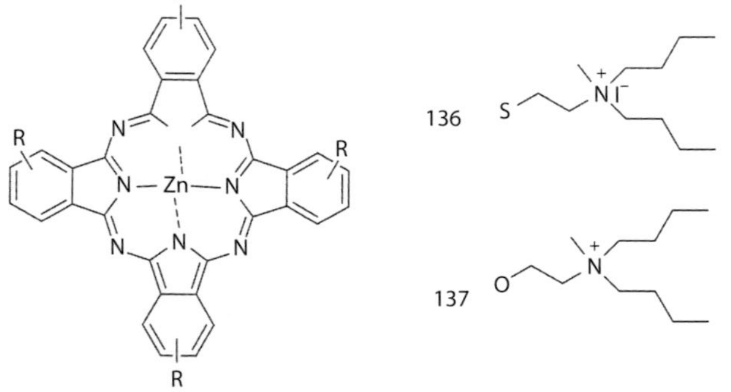

FIGURE 22.34 Structure of isosteric zinc phthalocyanines.

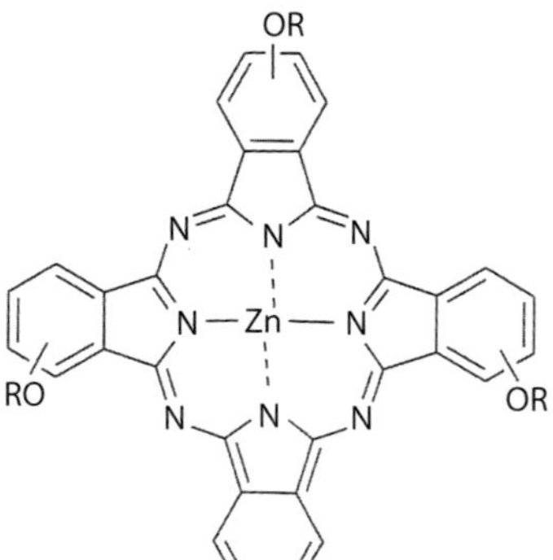

138 $R = CH_3$

139 $R = H_2C$–C$_6$H$_4$–CF_3

FIGURE 22.35 Structures of fluorinated and nonfluorinated zinc phthalocyanines.

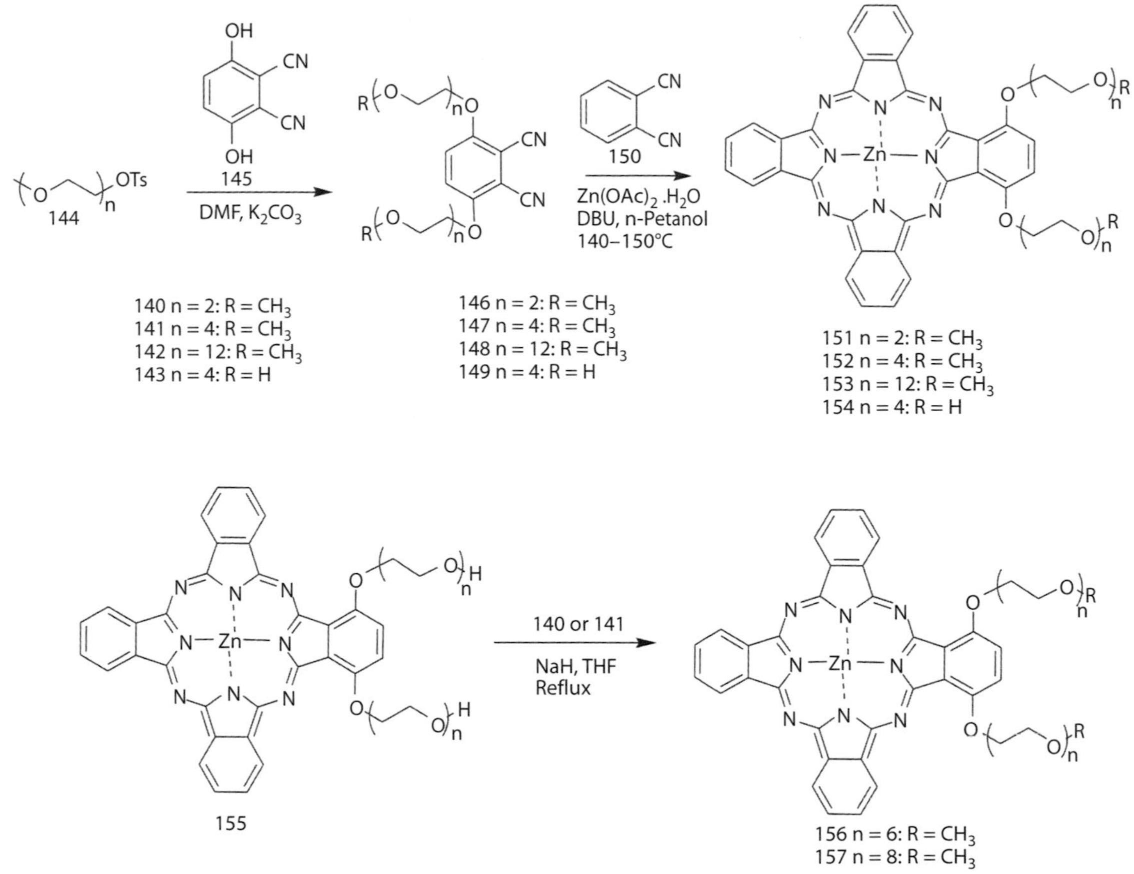

FIGURE 22.36 Synthesis of pegylated zinc phthalocyanines.

hydrophilic PSs remain in circulation longer, diffusing more slowly into cellular compartments while hydrophobic molecules diffuse comparatively faster into cellular compartments (Solban, Rizvi, and Hasan 2006). The longer time in circulation illustrated with hydrophilic molecules might lower the amount retained in the target tissue (Solban, Rizvi, and Hasan 2006). It has been reported that the destination of the PS once it encounters a cell could affect the PDT response (Mroz et al. 2009). The PSs that interfere with mitochondrial function tend to be more effective in killing cancer cells than the ones that do not affect mitochondria (Mroz et al. 2009).

Regardless of the hydrophobic PSs' ability to diffuse faster into compartments, these molecules have solubility issues. One approach to circumvent the solubility issues is by the use of a

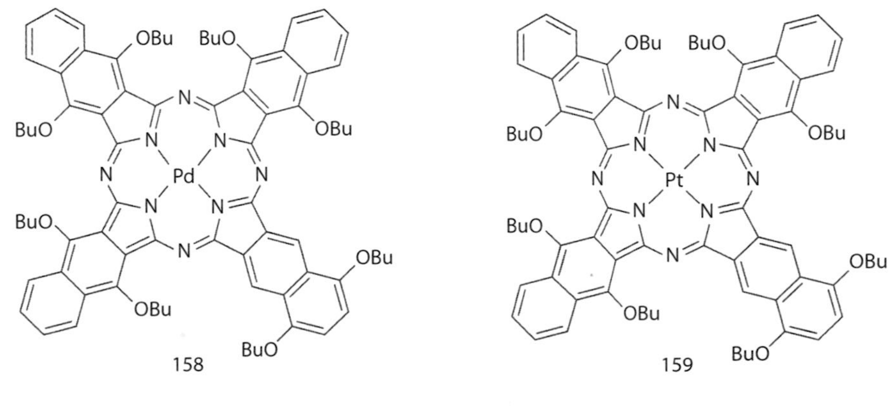

FIGURE 22.37 Structure of palladium and platinum phthalocyanines.

delivery system. The ideal delivery system may allow the selective delivery and accumulation of the desired concentration of the PS within the targeted tissue with little or no uptake by nontargeted tissue (Konan, Gurny, and Allemann 2002). When a delivery system has been established, the activity of the PS must be preserved within the carrier moiety to maintain the overall therapeutic response.

As PDT continues to evolve, a number of factors have been identified that play an important role in the therapy. These factors have guided researchers in how to change or modify the PS structural design to optimize and improve therapeutic efficacy. These factors include, but are not limited to, photophysical and photochemical characteristics, the efficiency of the light delivery, and the localization of the PS (Ethirajan et al. 2008). To optimize PDT, there are several approaches that can be utilized. One approach is to design a PS that does not cause persistent skin sensitivity. Alteration of the PS's design and its effects on the outcome of PDT has shown promising responses. There is much effort to establish the structural parameters that determine improved PS accumulation and selectivity in target tissue. The modifications to the PS are aimed to improve the utility, efficacy, and selectivity. The following strategies may be utilized: conjugating the molecule with a targeting moiety, attaching a highly fluorescent component to the PS for imaging (a bifunctional agent) (Pandey et al. 2005), applying minor modifications to the chemical structure, modifying the lipophilicity of the overall PS, or adding a metal to the core of the ring system (Chen et al. 2005b). These approaches can be addressed individually or in combination, so a more efficient PS can be administered for PDT treatment.

In the following sections, a discussion of the approaches that several researchers are utilizing will be described. First, there will be a discussion of targeting strategies that use proteins, antibodies, peptides, and carbohydrates in conjugation with a PS. Following this, there will be a brief discussion on the approaches recently used for vascular targeting and the rationale behind targeting the vasculature of a tumor. The "heavy atom effect" and the relevance of the approach to improve the photodynamic effect will be discussed. In addition, the effect of modifications to the PS's chemical structure will be addressed. Finally, there is a discussion of the recent applications of nanomaterials as delivery systems and how these nanomaterials are currently being utilized to improve the selectivity of tumor tissue.

22.4.1 Targeting Strategies

22.4.1.1 Proteins as Delivery Vehicles

In a recent review by Verma et al. (2007), strategies to enhance the PDT effect are discussed, one of which utilizes serum proteins as carrier molecules. When a drug is administered, most of the time, the drug will bind to proteins within the blood, including high- or low-density lipoproteins and albumin. However, the literature has shown a diversity of interactions of a PS with the various proteins in the blood. One class of the many proteins

in the blood that commonly bind to the PSs is low-density lipoproteins (LDLs). Allison, Pritchard, and Levy (1994) showed that many PSs may bind to LDLs when the PS is in circulation. Mishra, Patel, and Datta (2006) studied the interaction of chlorin p_6 (160) and purpurin 18 (161), structures shown in Figure 22.38, to ascertain whether these PSs bound to blood plasma, serum albumin, or lipoprotein (Mishra, Patel, and Datta 2006). Both PSs in the study bound preferably to LDL (Mishra, Patel, and Datta 2006). In addition, Misawa et al. (2005) showed that neovascular endothelial cells and tumor cells generally have a high expression of LDL receptors as a result of increased cell proliferation. The Misawa study showed the relevance of using LDL as a carrier molecule for a PS. Other groups have studied the role of LDL receptors as carrier molecules to diminish phototoxicity, which was investigated using various PSs (Verma et al. 2007; Allison, Pritchard, and Levy 1994; Polo et al. 2002; Aquaron et al. 2002; Kessel, Whitcomb, and Schulz 1992; Schmidt-Erfurth et al. 1997).

Hamblin, Miller, and Ortel (2000) discussed an approach that used a PS covalently conjugated to a ligand that was specifically recognized and internalized by a cell surface receptor. The candidate to test this approach was a class-A, type-1 scavenger receptor of macrophages, which recognizes maleylated serum albumin at a high capacity (Hamblin, Miller, and Ortel 2000). The PS used in this approach was chlorin e_6, which was covalently attached to bovine serum albumin to give different conjugates by changing the molar substitution ratio (dye to protein) (Hamblin, Miller, and Ortel 2000). The molar substitution ratios of the conjugates were 1:1 and 3:1, and the conjugates were then modified by maleylation (Hamblin, Miller, and Ortel 2000). When the conjugates were incubated using target macrophage cells, J774, they were taken up and cell killing occurred. However, when the conjugates were incubated using nonmacrophage target cells, OVCAR-5, they were minimally taken up, and cell killing was not observed (Hamblin, Miller, and Ortel 2000). Figure 22.39 illustrates the results of cellular uptake and cell killing for both purified conjugates with J774 and OVCAR-5 cells. Hamblin, Miller, and Ortel (2000) showed that scavenger receptor-targeted PDT gives a high degree of specificity toward

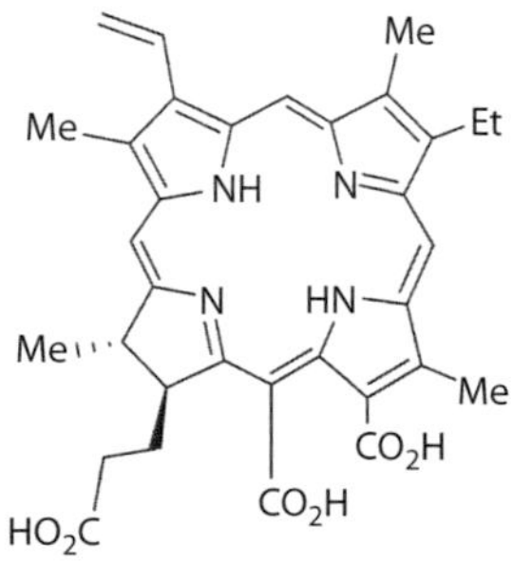

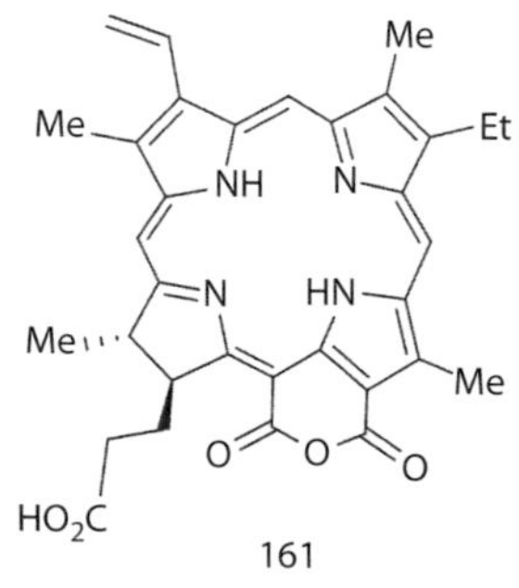

FIGURE 22.38 Structures of chlorin p_6 (160) and purpurin 18 (161). (From Mishra, P. et al., *Phys. Chem. B*, 110, 2006.)

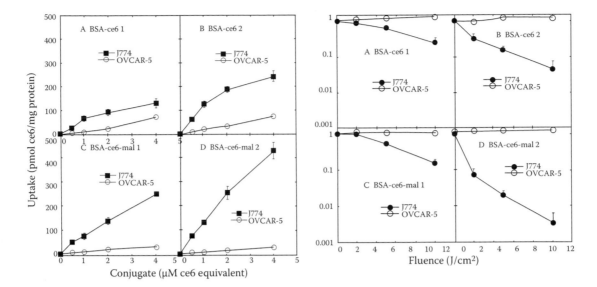

FIGURE 22.39 Cellular uptake and phototoxicity of the PS conjugates (maleylated and non-maleylated) for J774 macrophages compared to OVCAR-5 cells. (From Hamblin, M. et al., *Photochem. Photobiol.*, 72, 4, 2000.)

macrophages and may have applications in the treatment of tumors and atherosclerosis.

22.4.1.2 Monoclonal Antibodies Conjugated PS

Another approach to utilize a targeting strategy to improve selectivity is to use a monoclonal antibody (MAb). MAbs are directed toward antigens or ligands that are overexpressed on cancer cells. MAbs are not limited to only recognizing tumor-associated antigens; the MAbs can recognize proteins that bind to receptors on a certain tumor type. Kuimova et al. (2007) synthesized conjugates of pyropheophorbide-*a* (PP*a*) and verteporfin (VP) with a single-chain antibody fragment (scFv) specific to the HER2 receptor (C6.5). In this study, the group utilized this approach to target an overexpressed receptor, HER2, found on breast and ovarian cancers (Kuimova et al. 2007). The specificity of scFV conjugates toward HER2 receptors by cellular uptake of the scFv conjugates to HER2-positive (SKOV-3) and HER2-negative (KB) epithelial cell lines is illustrated in Figure 22.40 (Kuimova et al. 2007). The study indicated the usefulness of PSs conjugated to an antibody to target an overexpressed receptor found in specific breast and ovarian cancers (Kuimova et al. 2007).

22.4.1.3 Peptide-PS Conjugates

Another approach to improve selectivity is to use peptides as targeting vehicles. Specific peptides are used that recognize specific receptors that are overexpressed in a particular target tissue. Allen et al. (2002) used a peptide conjugated to aluminum disulfophthalocyanine; the peptide used in this study was Arg-Gly-Asp (RGD). The Arg-Gly-Asp (RGD) motif is known to bind to several of the 25 known integrin receptor types. Soluble RGD peptides under most circumstances induce apoptosis in a number of cell lines (Allen et al. 2002). Allen et al.

(2002) demonstrated that RGD peptide conjugated to aluminum disulfophthalocyanine was able to induce apoptosis at elevated concentrations.

In a more recent work, Srivatsan et al. (2011a) illustrated the effectiveness of a HPPH-peptide conjugate, showing the *in vitro* and *in vivo* behavior of a series of HPPH-peptide conjugates. Among the conjugates, the HPPH-cRGD conjugate, structure 165 shown in Figure 22.41, showed faster clearance, enhanced tumor imaging, and improved PDT efficacy when compared to HPPH alone. The *in vitro* photosensitizing efficacy of HPPH and the corresponding peptide conjugates are shown in Figure 22.42. Cyclic Arg-Gly-Asp (cRGD) peptide represents a selective $\alpha_v\beta_3$ integrin ligand that has been extensively used for research, therapy, and diagnosis of neoangiogenesis (Srivatsan et al. 2011a).

22.4.1.4 PS-Carbohydrate Conjugates

Carbohydrates are the most abundant class of biological molecules in nature and are fundamental to many physiological processes. The use of carbohydrates as potential drug applications for cancer provides a rationale for their use in strategies to improve PDT efficacy. McCarthy et al. (2009) developed both a glucose-modified chlorin and a bacteriochlorin using a reactive carboxylic acid linker to conjugate targeting moieties (structure 170 shown in Figure 22.43).

The glucose moieties increased the polarity of the relatively hydrophobic PS, but maintained an overall neutral charge (McCarthy et al. 2009).

A variety of porphyrins and nonporphyrin compounds have been conjugated with carbohydrates to serve as substrates for cellular lectins (e.g., galectins), which are known for their higher expression in many tumor cells (Liu and Rabinovich 2005). Chen et al. (2004a) synthesized a porphyrin–saccharide conjugate in which PDT responses varied with the conjugate structure. The

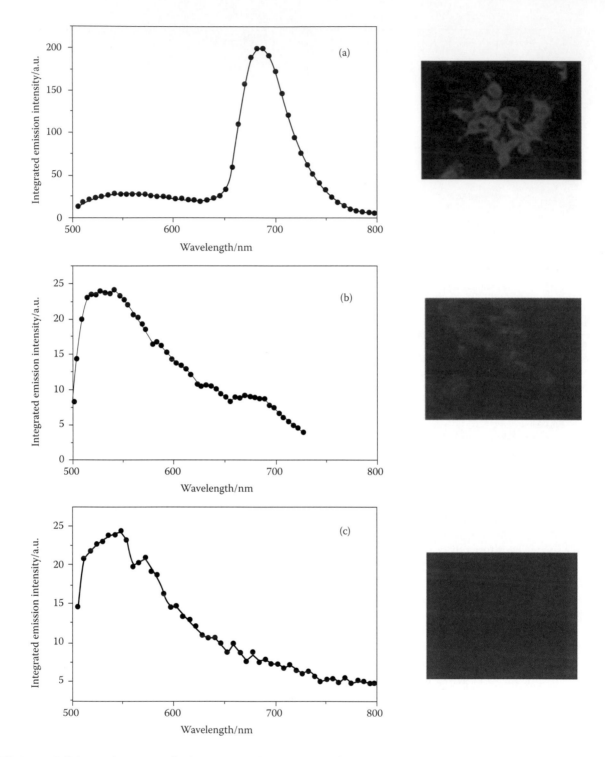

FIGURE 22.40 Cellular uptake using confocal microscopy, showing both the fluorescent spectra and the corresponding fluorescent images from the confocal microscope using fixed cells incubated with C6.5-PP*a* (a) HER2 positive SKOV-3, (b) HER2 negative KB cells, and (c) KB cells incubated with the blank medium. (From Kuimova, M. et al., *Photochem. Photobiol. Sci.*, 6, 2007.)

PDT outcome relied on the concentration of the conjugate and irradiation energy, illustrating necrosis, apoptosis, and a reduction in cell migration (Chen et al. 2004a). Zheng et al. (2009) examined the *in vitro* and *in vivo* photodynamic effects of a series of carbohydrate conjugates of HPPH, 170–178 (synthesis shown

in Figure 22.44). The PSs were conjugated to carbohydrates that were known for their high affinity to galectin-3.

Alvarez-Mico et al. (2006) reported an anomeric glycoconjugation to phthalocyanine for the first time, whereby a phthalocyanine bearing a carbohydrate was linked to the macrocycle via a

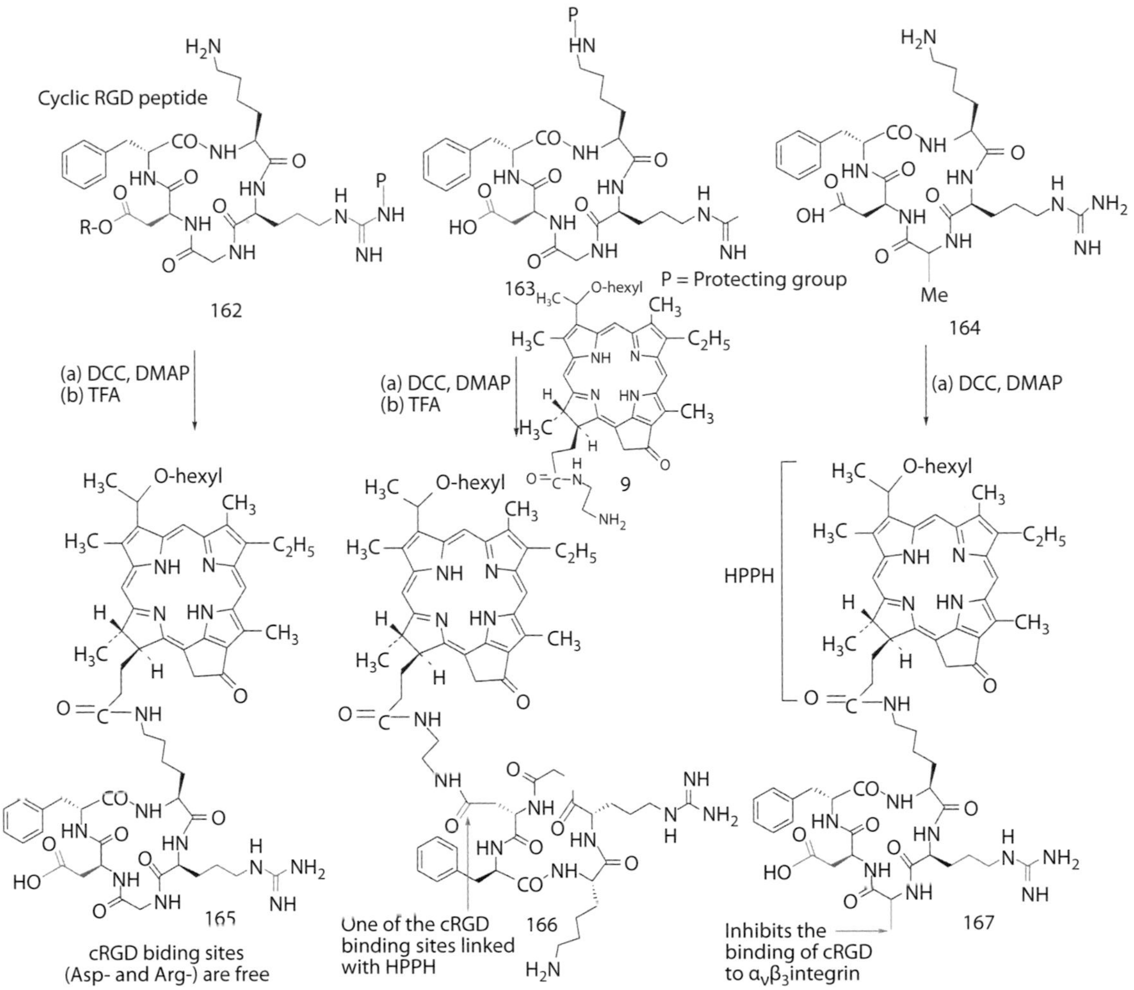

FIGURE 22.41 Scheme of HPPH-cRGD analogs. (From Srivatsan, A. et al., *Mol. Pharm*, 8, 2011.)

glycosidic bond. Ermeydan et al. (2010) synthesized two types of amphiphilic phthalocyanine–carbohydrate conjugates, shown in Figure 22.45. Conjugates, each bearing either glucose, galactose, mannose, or lactose, were synthesized and characterized.

22.4.1.5 Other Targeting Strategies: Vascular Targeting

One of the main effects of PDT is vascular damage in the tumor tissue. Vascular destruction during PDT has been reported, whereby blood stasis and hemorrhaging were found to starve tumor cells of oxygen and nutrients (MacDonald and Daugherty 2001). After light treatment, the deficiency of vital nutrients enhances tumor destruction by reducing the ability of the cancer cells to survive after treatment. Henderson and Fingar (1987) showed, by directly measuring oxygen concentration in tumor tissue, that oxygen deprivation kills tumor cells after PDT (MacDonald and Daugherty 2001). The degree of vascular shutdown or damage during irradiation influences the photodynamic effect because it will be inhibited if the oxygen supply is limited during this time (Busch 2006).

Vascular destruction seems to be an important factor for understanding the effect of a PS during and after light treatment. One way to optimize the PDT effect could be by targeting the vasculature of the tumor tissue. One of the overall goals when utilizing tumor vascular targeting is to selectively disrupt tumor vascular function for therapeutic purposes with minimal impact on normal tissue functions.

To achieve vascular targeting, the PS can be structurally modified; a targeting moiety that has a high affinity to endothelial cell markers or vessel-supporting structures is often used in the PS modification (Chen et al. 2008). The resulting PS with a targeting moiety for the vasculature is expected to be selectively accumulated in the targeted blood vessels, leading to a more precise, site-specific photosensitization upon light activation (Chen et al. 2008). The endothelial cells make up the inner layer of blood vessels, playing a key role in angiogenesis, which is one of the known causes for tumor regrowth in tumors treated with PDT (Ferrario et al. 2000). Angiogenesis can be stimulated by a dozen different proteins as well as several smaller molecules, such as those listed in Figure 22.46.

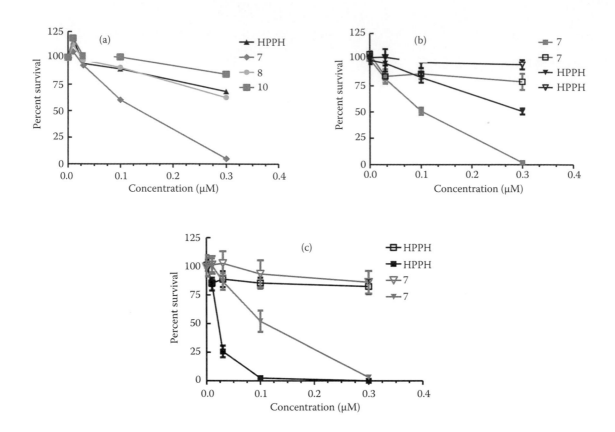

FIGURE 22.42 Comparative *in vitro* photosensitizing efficacy of the HPPH and the corresponding peptide conjugates 165, 166, and 167 at variable PS concentrations in (a) U87 ($\alpha_v\beta_3$ positive), (b) 4T1 $\alpha_v\beta_3$ positive, and (c) A431 ($\alpha_v\beta_3$ negative) tumor cells, respectively. (From Srivatsan, A. et al., *Mol. Pharm.*, 8, 2011.)

Among these molecules, two proteins appear to be the most important for sustaining tumor growth: vascular endothelial growth factor (VEGF) and basic fibroblast growth factor (bFGF). Earlier studies have shown that PDT treatment induced significant increases in VEGF expression within treated tumors, but the use of inhibitors could weaken the angiogenic action of VEGF and ultimately enhance the PDT outcome (Ethirajan et al. 2011; Ferrario et al. 2000; Bhuvaneswari et al. 2007). In addition, it is important to study how endothelial cells respond to photosensitization at cellular and molecular levels because different types of PSs have demonstrated activation of transcription factors. Volanti, Matroule, and Piette (2002) and Volanti et al. (2004) reported that photosensitization activates nuclear transcriptor NF-κB in endothelial cells through a ROS-mediated mechanism.

22.5 Optimizing the Generation of Singlet Oxygen Via Metallation

As we know, oxygen is one of the main components required for PDT. Triplet oxygen is converted to singlet oxygen, which is believed to be the main cytotoxic species. Enhancing singlet oxygen generation is one avenue being explored to improve PDT.

One approach to producing singlet oxygen efficiently is to introduce a metal at the center of the PS molecule, such as indium, palladium, or gallium. These metals have a tendency to increase the lifetime of the triplet excited state of the PS molecule and hence the triplet quantum yield of the porphyrin (Ethirajan et al. 2008; Wainwright 2008; Josefsen and Boyle 2008). The quantum yield is directly related to the efficiency of generating singlet oxygen; thus, PS molecules with a metal at the core can be more effective singlet oxygen producing PSs through the "heavy atom effect." Brun et al. (2004) determined the pharmacokinetics and tissue distribution of WST09 (6) in female EMT6 tumor-bearing BALB/c mice in order to determine if selective accumulation of this drug occurs in tumor tissue. Palladium bacteriopheophorbide, WST09, currently known as Tookad, is a pure Pd-substituted bacteriochlorophyll derivative used for the treatment of recurrent prostate cancer. Tookad is a drug that has a strong absorption in the near-infrared region (763 nm) and has the ability to efficiently generate singlet oxygen upon irradiation. Tookad has a relatively fast clearance from the body. Its effectiveness has been demonstrated in various animal models (Chen et al. 2002; Huang et al. 2004, 2005, 2007).

A Chen et al. (2005b) study involved the application of pyropheophorbides and their metal complexes. The PSs were

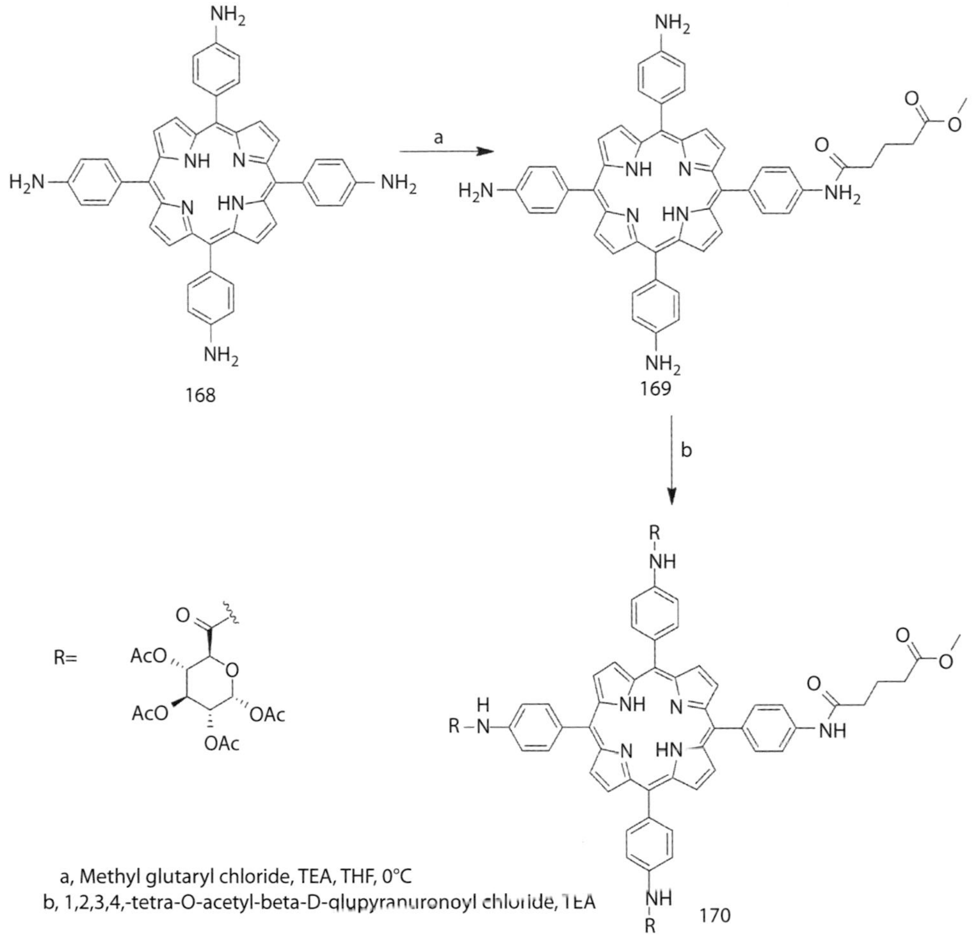

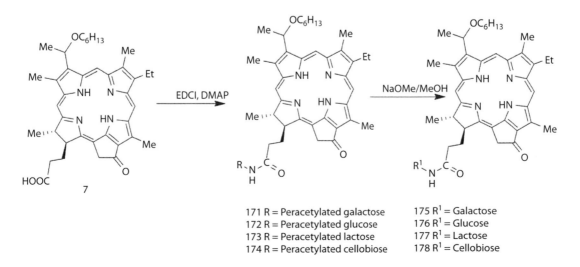

FIGURE 22.43 Synthesis of porphyrin–carbohydrate conjugates. (From McCarthy, J. et al., *Org. Biomol. Chem.*, 7, 2009.)

171 R = Peracetylated galactose
172 R = Peracetylated glucose
173 R = Peracetylated lactose
174 R = Peracetylated cellobiose

175 R^1 = Galactose
176 R^1 = Glucose
177 R^1 = Lactose
178 R^1 = Cellobiose

FIGURE 22.44 Synthesis of HPPH–carbohydrate conjugates. (From Zheng, X. et al., *J. Med. Chem.*, 52, 2009.)

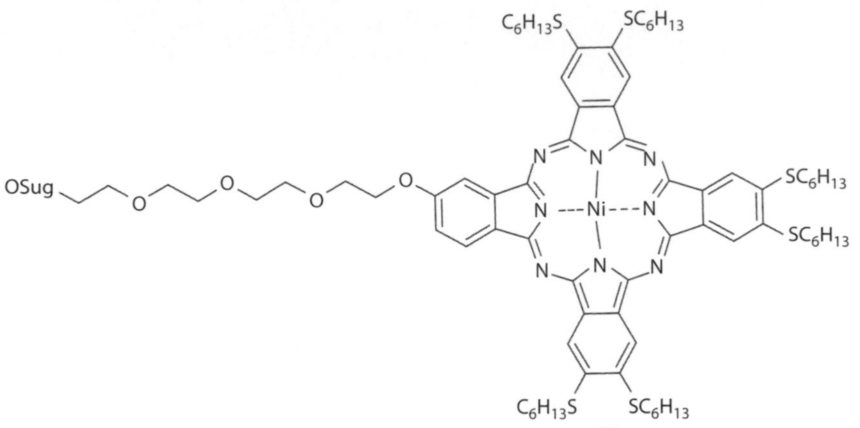

FIGURE 22.45 Structure of glycosylated phthalocyanines and clicked phthalocyanines with sugar (Sug) moiety (β-D-Glc, β-D-Gal, ά-D-Man, β-Lac). (From Ermeydan, M. et al., *N. J. Chem.*, 34, 2010.)

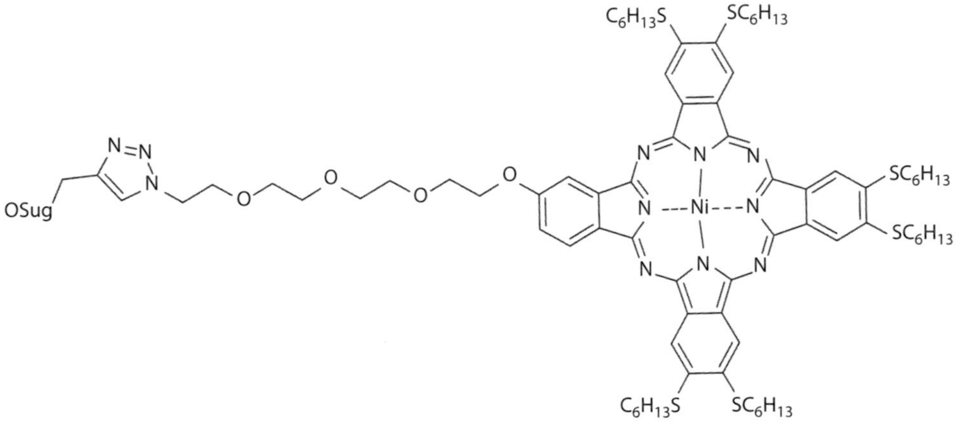

Proteins

1. Acidic fibroblast growth factor
2. Angiogenin
3. Basic fibroblast growth factor (bFGF)
4. Epidermal growth factor
5. Granulocyte colony-stimulating factor
6. Hepatocyte growth factor
7. Interleukin 8
8. Placenta growth factor
9. Platelet-derived endothelial growth factor
10. Scatter factor
11. Transforming growth factor alpha
12. Tumor necrosis factor alpha
13. Vascular endothelial growth factor (VEGF)

Small Molecules

1. Adenosine
2. 1-Butyryl glycerol
3. Nicotinamide
4. Prostaglandins E1 and E2

(National Cancer Institute website)

FIGURE 22.46 Activators of angiogenesis: some naturally occurring activators of angiogenesis.

synthesized and later investigated for their PDT efficacy. To determine the effect of the central metal, pyropheophorbide-*a* was converted into the corresponding Zn(II), In(III), and Ni(II) complexes (Figure 22.47). The presence of metals also had a significant impact on photosensitizing efficacy with the In(III) complexes proving to be the most efficacious, which could be a result of their higher singlet oxygen production. In addition, the In(III) complex of pyropheophorbide was reported as a potential candidate for the treatment of age-related wet macular degeneration (Dolmans et al. 2002; Ciulla et al. 2005), a leading cause of blindness that is most prevalent in people over the age of 60.

Saenz et al. (2011) investigated the photosensitizing efficacy of indium as a central metal in benzoporphyrin derivatives, shown in Figure 22.48. Compared to benzoporphyrin dimethyl ester and its 8-(1′-hexyloxy)ethyl analog, the corresponding In(III) complexes showed enhanced *in vivo* photosensitizing efficacy in colon 26 tumor cells, illustrated in Figure 22.49.

Rosenfield et al. (2006) investigated the effect of an additional keto group at position 13² of a series of the alkyl ether analogs of pyropheophorbide-*a* and their metal complexes. It was observed

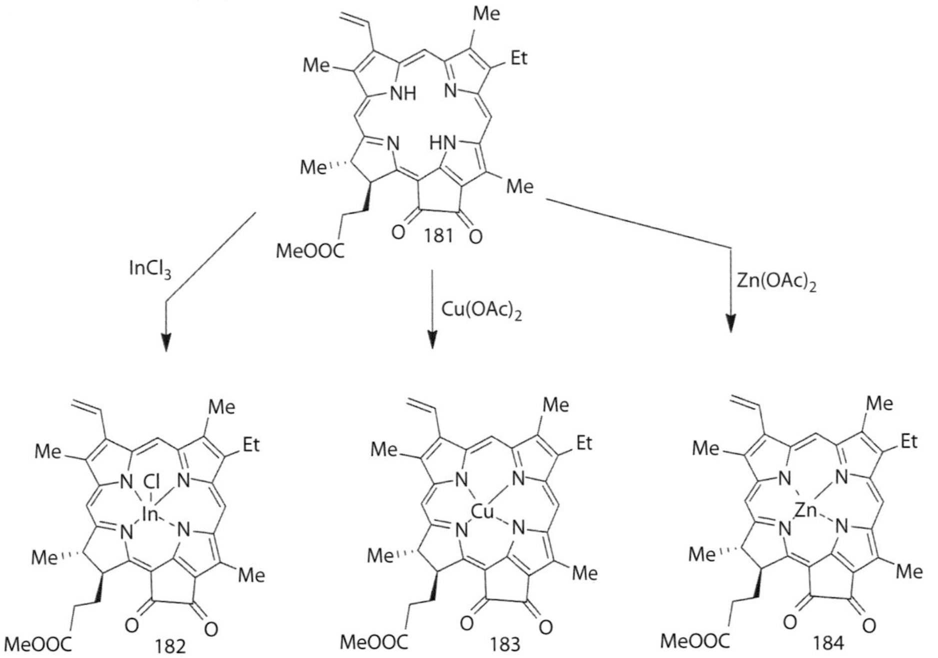

FIGURE 22.47 Scheme of pyropheophorbide-*a* and metal analogs. (From Chen, Y. et al., *J. Med. Chem.*, 48, 2005.)

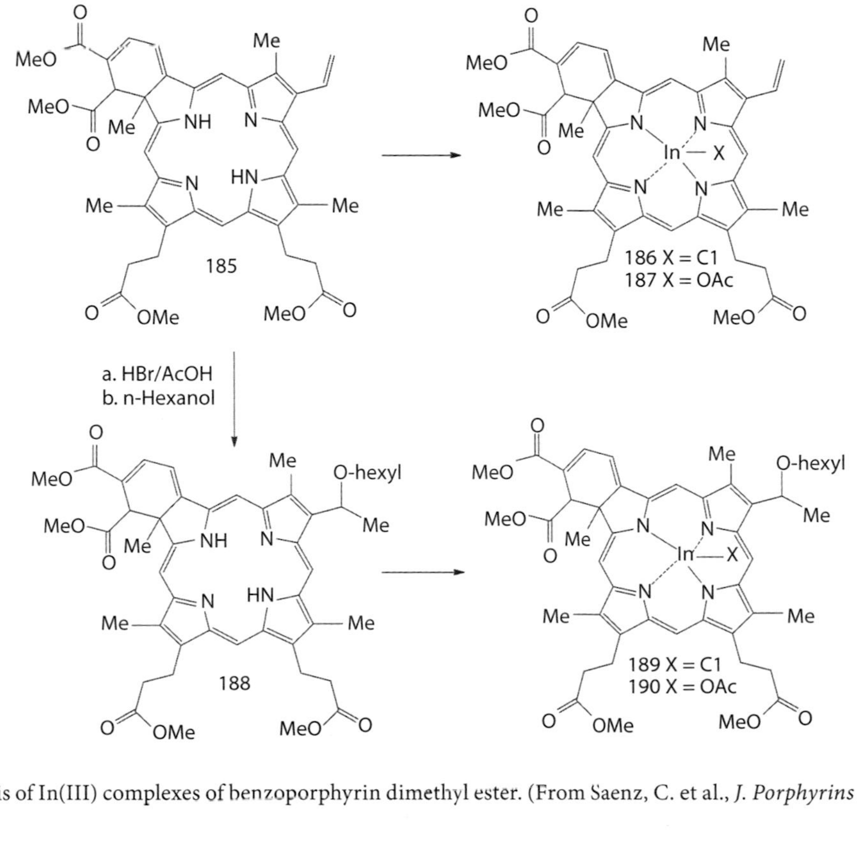

FIGURE 22.48 Synthesis of In(III) complexes of benzoporphyrin dimethyl ester. (From Saenz, C. et al., *J. Porphyrins Phthalocyanines*, 15, 2011.)

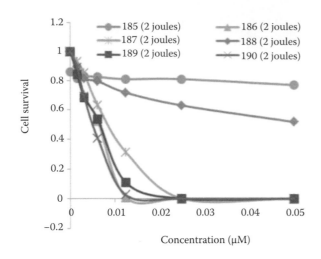

FIGURE 22.49 Dose response via MTT viability assay using Colo-26 cell line for BPD and corresponding indium(III) analogs. (From Saenz, C. et al., *J. Porphyrins Phthalocyanines*, 15, 2011.)

that the insertion of In(III) into the ring system enhanced photo-toxicity (Rosenfield et al. 2006).

With the continuous development of possible PSs that are currently described in the literature, the utilization of singlet oxygen that is generated during PDT has shown the possibilities of new applications of singlet oxygen as a cytotoxic species. In other fields, the generation and application of singlet oxygen present numerous research opportunities. To explore the use of singlet oxygen, many scientists are developing new strategies to enhance the photodynamic effect. Several recent developments have been discussed that show promising improvements for the selectivity and destruction of the targeted tissue. Of the strategies discussed, researchers are still not able to conclude that one single targeting strategy shows superiority over another. New approaches to optimize PDT will continue to be developed with the idea to improve therapeutic efficacy and specificity, thereby reducing the adverse reactions that are often seen with certain PSs.

22.6 Effect of Isomers in PDT Efficacy

The structure activity approach has been valuable in guiding PS development. A given PS may exist in different isomeric forms, for example, stereoisomers or structural isomers. How isomeric variation of a particular compound impacts PDT efficacy has been investigated for certain compounds. As pertains to stereoisomers, while most studies have found no difference in therapeutic response between *R* and *S* isomers, there are some significant studies in which one isomer is safe and effective while the other is toxic and/or ineffective (Chen et al. 2004b). The most tragic example of this situation is thalidomide [N-(2,6-dioxo-3-piperidinyl)phthalimide], a well-known drug with antinauseous activity for pregnant women. Between the two possible isomers, the S-enantiomer was found to produce the desired sedative

efficacy, whereas the R-enantiomer was teratogenic and caused severely underdeveloped limbs (Chen et al. 2004b).

In regards to PDT efficacy as it relates to *R* and *S* isomers of a given PS, there have been a few studies published. Chen et al. (2004b) showed the synthesis, separation, and identification of the isomerically pure 3-(1-heptyloxyethyl)-3-deacetylbacterio-purpurine-18-*N*-hexylimides (*R*- and *S*-isomers) and the impact of chirality in photosensitizing efficacy, which seemed to make a significant difference. Both heptyl ether derivatives (*R*- and *S*-isomers) showed similar *in vitro* photosensitizing efficacy and limited skin phototoxicity and were found to localize in mitochondria. However, in preliminary *in vivo* screening, compared to the *S*-isomer, the corresponding *R*-isomer produced enhanced *in vivo* PDT efficacy (Chen et al. 2004b). Recently, Srivatsan et al. (2011b) investigated the photosensitizing and imaging potential of isomeric PSs derived from chlorophyll-*a* and bacterio-chlorophyll-*a*. The study showed that among the PSs investigated, both ring-D and ring-B reduced chlorins containing the *m*-iodobenzyloxyethyl group at position-3 and a carboxylic acid functionality at position-17[2] showed the highest uptake by tumor cells and a light-dependent photoreaction that correlated with maximal tumor imaging and long-term PDT efficacy in BALB/c mice bearing colon 26 tumors (Srivatsan et al. 2011b).

22.7 Applications of Nanomaterials in PDT

The area of nanotechnology in medicine is rapidly developing, especially its relevance to drug delivery. Nanoparticles can potentially reduce toxicity and side effects that are shown with current FDA-approved PSs. The ideal nanoparticle delivery system would deliver the PS directly to the targeted tissue, thus avoiding damage to nearby tissue. The availability of the PS and the appropriate formulation of the nanoparticle delivery system are critical for a successful PDT outcome. The use of nanosized carriers for PSs is a promising approach, which might improve the efficiency of PDT and which can overcome side effects associated with traditional PDT (Paszko et al. 2011). As mentioned earlier, one of the main focuses in PDT research is improving the therapeutic concentration of the PS at the site of action with, ideally, no PS in normal tissue. In the review article by Paszko et al. (2011), many issues were discussed, including how to improve drug delivery by nanomaterials, how to get the maximum therapeutic efficacy by improving the solubility of poorly water-soluble drugs, how to prolong the circulation half-life in the blood, and ways of minimizing degradation of the drug after administration and to decrease side effects and increase drug bioavailability (Paszko et al. 2011).

The schematics for the formulations of nanoparticles by following various approaches (encapsulation, conjugation, and post-loading) are depicted in Figure 22.50 (Wang et al. 2011). Among the formulations tested for PDT effectiveness, the post-loaded formulation showed the best *in vitro* phototoxicity response (via MTT assay) with no visible dark toxicity; data are shown in Figure 22.51.

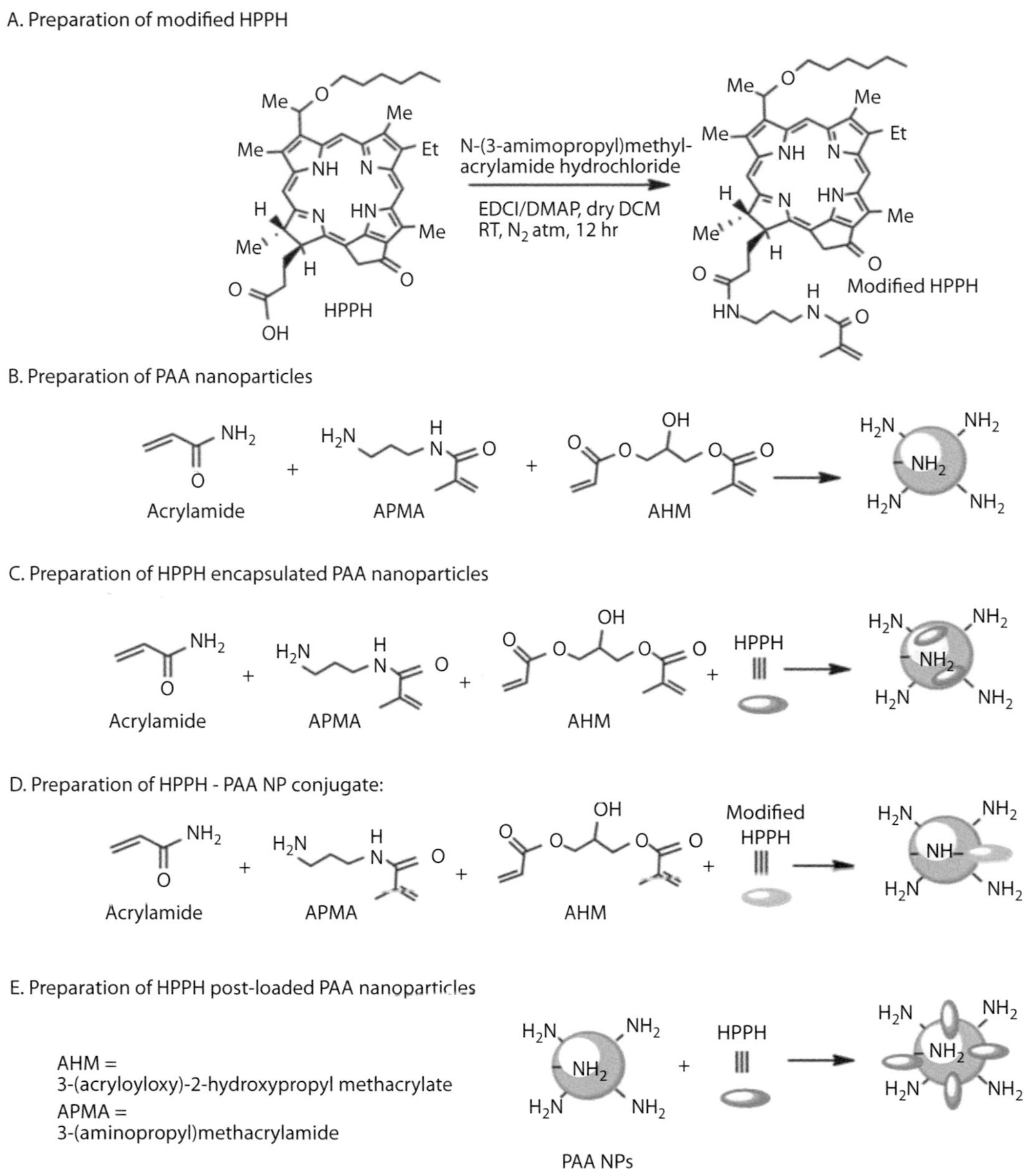

A. Preparation of modified HPPH

B. Preparation of PAA nanoparticles

C. Preparation of HPPH encapsulated PAA nanoparticles

D. Preparation of HPPH - PAA NP conjugate:

E. Preparation of HPPH post-loaded PAA nanoparticles

AHM =
3-(acryloyloxy)-2-hydroxypropyl methacrylate
APMA =
3-(aminopropyl)methacrylamide

PAA NPs

FIGURE 22.50 Preparation of modified HPPH, blank PAA NPs, and corresponding nanoformulations. (From Wang, S. et al., *Lasers Surg. Med.*, 43, 2011.)

In addition, Wang et al. (2011) showed an improved PDT response with the post-loaded HPPH, with the highest singlet oxygen yield. Cinteza et al. (2006) used a nanocarrier that was co-loaded with HPPH and magnetic Fe_3O_4 nanoparticles; this approach provided an ability to magnetically guide PDT as well as to enhance drug accumulation in the desired area.

Roy et al. (2003) showed destruction of tumor cells *in vitro* by use of PS-doped organically modified silica-based nanoparticles. The scheme of the synthesis of the ceramic-based nanoparticles is illustrated in Figure 22.52. The *in vitro* observation shows the potential of ceramic-based particles as carriers for photodynamic drugs.

There are other methods to deliver the maximum amount of PS using nanomaterials. One approach is to attach targeting moieties to the nanoparticles' surfaces. Nanoparticles that have a targeting moiety can bind to target cells through ligand–receptor interactions that induce receptor-mediated endocytosis, thus leading to drug release within the cell. Reddy et al. (2006) treated glioma-bearing rats with targeted nanoparticles followed by PDT, which showed a significant improvement in survival rate when compared with animals that received PDT after administration of nontargeted nanoparticles or systemic Photofrin. The targeting moiety used in this study was a vascular homing peptide, F3, which has been reported to have cell-penetrating

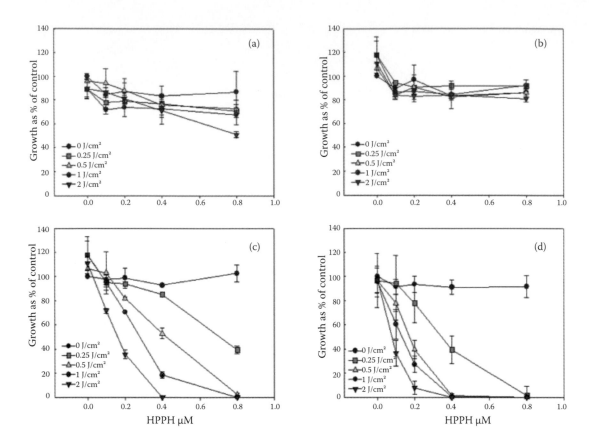

FIGURE 22.51 Phototoxicity of Colon26 cells after incubation with the different preparations of AFPAA nanoparticles for 4 h followed by light at a series of doses at a fluence rate of 3.2 mW/cm². (a) Encapsulated HPPH (PAA-E), (b) conjugated HPPH (PAA-CONJ), (c) post loaded HPPH (PAA-PL), and (d) free HPPH. (From Wang, S. et al., *Lasers Surg. Med.*, 43, 2011.)

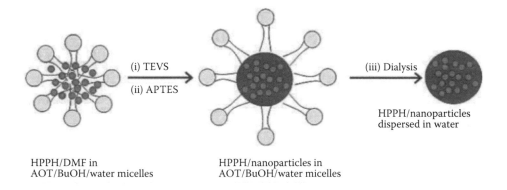

FIGURE 22.52 Scheme depicting the synthesis and purification of HPPH-doped silica-based nanoparticles in a micellar medium. (From Roy, I. et al., *J. Am. Chem. Soc.*, 125, 2003.)

properties (Reddy et al. 2006). For tumor-specific uptake of F3 peptide, cell surface nucleolin is a specific marker for angiogenic endothelial cells within the tumor vasculature (Reddy et al. 2006). Nucleolin is a shuttle protein that traffics between the membrane and nucleus (Reddy et al. 2006). Expression of it on the surface is a prerequisite for tumor-specific uptake of F3 peptide (Reddy et al. 2006). The Reddy group study demonstrates the improved PDT treatment with F3 targeted Photofrin

nanoparticles compared to nontargeted Photofrin nanoparticles and Photofrin alone.

Recently, Stuchinskaya et al. (2011) targeted breast cancer cells using an antibody PS-gold nanoparticle conjugate. Anti-HER2 MAbs were covalently bound to nanoparticles via a terminal carboxy moiety on the polyethylene glycol. Cellular experiments demonstrated that the nanoparticle conjugates selectively target breast cancer cells that overexpress the HER2 epidermal growth factor cell surface receptor and that they are effective PDT agents (Stuchinskaya et al. 2011).

22.8 Other Applications of PDT

PDT is not only used in the area of cancer. Clinical applications of antimicrobial PDT have been recently developed. The clinical application of PDT for infectious diseases is growing because of the fast bacterial killing typical of PDT and the fact that it is unlikely that bacteria will develop resistance to PDT. Gad et al. (2004) showed the potential application of PDT in drug-resistant, soft-tissue infections. In the study, Gad et al. (2004) established a soft tissue infection model using a bioluminescent *Staphylococcus aureus* in mice that were temporarily made neutropenic by cyclophosphamide administration. The infections were treated with a polycationic PS conjugate, whereby the conjugate was injected to the infected area and then exposed to a red laser light (for activation of the PS). The same group had previously shown that both a nonpathogenic strain of *Escherichia coli* and a highly pathogenic strain of *Pseudomonas aeruginosa* could effectively be killed in the wounds of living mice (Hamblin et al. 2002, 2003). Gad et al. (2004) discusses the conditions that must be met in order to treat infections effectively using PDT in a living system, which are the following:

- It is necessary for the PS to be selective for bacteria compared to host cells and tissue.
- A suitable route of administration of the PS to the infected area must exist.
- The infected area must allow effective light delivery.
- An appropriate method of monitoring the result of treatment should be employed.

In addition, antimicrobial PDT has been used successfully to control biofilms and treatments for biofilm infections of the oral cavity (Wainwright 1998; Wilson 2004; Zanin et al. 2006; Meisel and Kocher 2005). To improve antimicrobial PDT, Suci et al. (2007) coupled a PS to a targeting ligand to enhance the selectivity of antimicrobial PDT. Suci et al. reported the photodynamic inactivation of *Staphylococcus aureus* using targeted nanoplatforms conjugated to a PS. The results showed that protein cage architectures can serve as versatile templates for engineering nanoplatforms for targeted antimicrobial PDT (Suci et al. 2007).

Acknowledgment

A part of the work presented in this article was supported by the NIH (PO1 CA55791 and RO1 CA127369).

List of Abbreviations

FDA: Food and Drug Administration
HOMO: Highest occupied molecular orbital
HpD: Hematoporphyrin derivative
HPPH: 3-Devinyl-3-(1'-hexyloxy)ethylpyropheophorbide-a
LDL: Low density lipoprotein
LUMO: Lowest unoccupied molecular orbital
MDR: Multidrug resistance
MNC: Metallo-naphthalocyanine
NC: Naphthalocyanine
NCP: *N*-Confused porphyrin
NPs: Nanoparticles
PAA: Polyacrylamide
PDT: Photodynamic therapy
PET: Positron emission tomography
PL: Post-loading
PS: Photosensitizer
ROS: Reactive oxygen species
TPP: Tetraphenylporphyrin

References

Agius, L., J. Ballantine, V. Ferrito et al. 1979. The structure and physiological activity of bonellin - a unique chlorin derived from *Bonellia viridis*. *Pure Appl Chem* 51:1847.

Agostinis, P., K. Berg, K. Cengel et al. 2011. Photodynamic therapy of cancer: An update. *CA Cancer J Clin* 61:250–281.

Alleman, E., N. Brasser, O. Benrezzak et al. 1995. PEG-coated poly(lactic acid) nanoparticles for the delivery of hexadecafluoro zinc phthalocyanine to EMT-6 mouse mammary tumours. *Pharm Pharmacol* 47:382–387.

Allen, C., W. Sharman, C. La Madeleine, J. van Lier, and J. Weber. 2002. Attenuation of photodynamically induced apoptosis by an RGD containing peptide. *Photochem Photobiol Sci* 1:246–254.

Allison, B., P. Pritchard, and J. Levy. 1994. Evidence for low density protein receptor-mediated uptake of benzoporphyrin derivative. *Br J Cancer* 69:833–839.

Alvarez-Mico, X., M. Calvete, M. Hanack, and T. Ziegler. 2006. The first example of anomeric glycoconjugation to phthalocyanines. *Tetrahedron Lett* 47:3283–3286.

Aekand, M., S. Bonnett, R. Kaur et al. 1986. meso-Tetra(hydroxyphenyl)porphyrins, a new class of potent tumour photosensitisers with favourable selectivity. *Br J Cancer* 54:717–725.

Anderson, T., T. Dougherty, D. Tan et al. 2003. Photodynamic therapy for sarcoma pulmonary metastases. *Anticancer Res* 23:3713.

Aquaron, R., O. Forzano, J. Murati et al. 2002. Simple, reliable and fast spectrofluorometric method for determination of plasma verteporfin (Visudyne) levels during photodynamic therapy for choroidal neovascularization. *Cell Mol Biol* 48:925–930.

Arnold, D., R. Gaete-Holmes, A. Johnson, and G. Williams. 1978. Wittig condensation products from nickel meso-formyl – octaethyl-porphyrin aetioporphyrin I and some cyclization reactions. *J Chem Soc Perkin Trans* 1:1660.

Auler, H., and G. Banzer. 1942. Untersuchung uber die Rolle der Porphyrine bei geschwulstkranken Menschen und Tieren. *Z Krebsforsch Bd* 53:S65–S68.

Azenha, E., A. Serra, M. Pineiro et al. 2002. Heavy-atom effects on metalloporphyrins and polyhalogenated porphyrins. *Chem Phys* 280:177–190.

Baden, H., editor. 1984. *The Chemotherapy of Psoriasis*. Pergamon, New York.

Banfi, S., E. Caruso, L. Buccafurni et al. 2006. Gramatic PP. Comparison between 5,10,15,20-Tetraaryl- and 5,15-Diarylporphyrins as photosensitizers: Synthesis, photodynamic activity, and quantitative structure–activity relationship modeling. *J Med Chem* 49:3293–3330.

Banks, R., B. Smart, and J. Tatlow. 1994. *Organofluorine Chemistry: Principles and Commercial Applications*. Plenum Press, New York.

Bauer, V., D. Clive, D. Dolphin et al. Sapphyrins: Novel aromatic pentapyrrolic macrocycles. *J Am Chem Soc* 105:6429–6436.

Beens, E., T. Dubbelman, J. Lugtenberg et al. 1987. Photosensitizing properties of bacteriochlorophyllin-a and bacteriochlorin a two derivatives of bacteriochlorophyll a Photochem Photobiol., *Photochem Photobiol* 45:639.

Bellnier, D., and T. Dougherty. 1996. A preliminary pharmacokinetic study of intravenous photofrin in patients. *J Clin Laser Med Surg* 14:311–314.

Bellnier, D., W. Greco, G. Loewen et al. 2003. Population pharmacokinetics of the photodynamic therapy agent 2-[-1-hexyloxyethyl]-2-devinyl pyropheophorbide-a in cancer patients. *Cancer Res* 63:1806.

Bellnier, D., W. Greco, H. Nava, G. Loewen, A. Oseroff, and T. Dougherty. 2006. Mild skin photosensitivity in cancer patients following injection of Photochlor (2-1-[1-hexyloxyethyl]-2-devinylpyropheophorbide-a) for photodynamic therapy. *Cancer Chemother Pharmacol* 57:40.

Bellnier, D., B. Henderson, R. Pandey, W. Potter, and T. Dougherty. 1993. Murine pharmokinetics and antitumor efficacy of the photodynamic sensitizer 2-[1-hexyloxyethyl]-2-devinyl pyropheophorbide-a. *J Photochem Photobiol B* 20:55.

Berenbaum, M. C., S. L. Akande, R. Bonnett, H. Kaur, S. Ioannou, R. D. White, and U. J. Winfield. 1986. Meso-Tetra(hydroxyphenyl)porphyrins, a new class of potential tumour photosensitisers with favourable selectivity. *Br J Cancer* 54:717–25.

Bhuvaneswari, R., G. Yuen, S. Chee, and M. Olivo. 2007. Hypericin-mediated photodynamic therapy in combination with Avastin (bevacizumab) improves tumor response by downregulating angiogenic proteins. *Photochem Photobiol Sci* 6:1275–1283.

Borbas, K., C. Ruzié, and J. Lindsey. 2008. Swallowtail bacteriochlorins. Lipophilic absorbers for the near-infrared. *Org Lett* 10:1931–1934.

Brun, P., J. DeGroot, E. Dickson, M. Farahani, and R. Pottier. 2004. Determination of the in vivo pharmacokinetics of palladium-bacteriopheophorbide (WST09) in EMT6 tumour-bearing BALB/c mice using graphite furnace atomic absorption spectroscopy. *Photochem Photobiol Sci* 3:1006–1010.

Busch, T. 2006. Local physiological changes during photodynamic. *Lasers Surg Med* 38:494–499.

Busetti, A., M. Soncin, E. Reddi et al. 1999. Irradiation of amelanotic melanoma cells with 532 nm high peak power pulsed laser radiation in the presence of the photothermal sensitizer Cu(II)-hematoporphyrin: A new approach to cell photoinactivation. *Photochem Photobiol B: Biol* 53:103–109.

Camerin, M., S. Rello, A. Villanueva et al. 2005. *Eur J Cancer* 41:1203–1212.

Camerin, M., S. Rello-Varona, A. Villanueva, M. Rodgers, and G. Jori. 2009. Photothermal sensitisation as a novel therapeutic approach for tumours: Studies at the cellular and animal level. *Lasers Surg Med* 41:665–673.

Canete, M., M. Lapena, A. Juarranz et al. 1997. Uptake of tetraphenylporphycene and its photoeffects on actin and cytokeratin elements of HeLa cells. *Anticancer Drug Des* 12:543–554.

Chen, B., C. He, P. deWitte et al. 2008. Photosensitizers for photodynamic therapy. In *Advances in Photodynamic Therapy Basic, Translational and Clinical*, M. Hamblin et al., editors. Artech House, Boston.

Chen, Q., Z. Huang, D. Luck et al. 2002. Preclinical studies in normal canine prostate of a novel palladium-bacteriopheophorbide (WST09) photosensitizer for photodynamic therapy of prostate cancer. *Photochem Photobiol* 76:438–445.

Chen, X., L. Hui, D. Foster, and C. Drain. 2004a. Drain CM. Efficient synthesis and photodynamic activity of porphyrin-saccharide conjugates: Targeting and incapacitating cancer cells. *Biochemistry* 43:10918–10929.

Chen, Y., R. Miclea, T. Srikrisnan et al. 2005a. Investigation of human serum albumin (HSA) binding specificity of certain photosensitizers related to pyropheophorbide-a and bacteriopurpurinimide by circular dichroism spectroscopy and its correlation with in vivo photosensitizing efficacy. *Bioorg Med Chem Lett* 15:3189.

Chen, Y., W. Potter, J. Missert, J. Morgan, and R. Pandey. 2007. *Bioconjuga* comparative in vitro and in vivo studies on long-wavelength phototosensitizers derived from bacteriopurpurinimide and bacteriochlorin p6: Fused imide ring enhances the in vivo PDT efficacy. *Bioconjugate Chem* 18(5):1460–1473.

Chen, Y., A. Sumlin, J. Morgan et al. 2004b. Synthesis and photosensitizing efficacy of isomerically pure bacteriopurpurinimides. *J Med Chem* 47:4814–4817.

Chen, Y., X. Zheng, M. Dobhal et al. 2005b. Methyl pyropheophorbide-a analogues: Potential fluorescent probes for the peripheral-type benzodiazepine receptor. Effect of central metal in photosensitizing efficacy. *J Med Chem* 48:3692–3695.

Chmielewski, P., L. Latos-Grazynski, K. Rachlewicz, and T. Glowiak. 1994. Tetra-*p*-tolylporphyrin with an inverted pyrrole ring: A novel isomer of porphyrin. *Angew Chem Int Ed Engl* 33:779.

Cinteza, L., T. Ohulchanskyy, Y. Sahoo et al. 2006. Diacyllipid micelle-based nanocarrier for magnetically guided delivery of drugs in photodynamic therapy. *Mol Pharm* 3(4):415–423.

Ciulla, T., M. Criswell, W. Snyder, and W. Small IV. 2005. Photodynamic therapy with photopoint photosensitizer

MV6401, indium chloride methyl pyropheophorbide, achieves selective closure of rat corneal neovacularisation and rabbit choriocapillaris. *Br J Ophthalmol* 89:113–119.

Cubeddn, R., W. Keir, R. Rampson, and T. Truscott. 1987. Photophysical properties of porphyrin-chlorin systems in the presence of surfactants. *Photochem Photobiol* 46:633.

de Oliveira, K., F. de Assis, A. Ribeiro et al. 2009. Synthesis of phthalocyanines-ALA conjugates: Water-soluble compounds with low aggregation. *Org Chem* 74:7962–7965.

Dennis, E., G. Dolmans, D. Fukumura, and R. Jain. 2003. Photodynamic therapy of cancer. *Nature* 3:380.

Detty, M. 2001. Photosensitizers for the photodynamic therapy of cancer and other diseases. *Expert Opin Ther Pat* 11:1849–1860.

Dolmans, D., A. Kadambi, J. Hill et al. 2002. Vascular Accumulation of a novel photosensitizer, MV6401, causes selective thrombosis in tumor vessels after photodynamic therapy. *Cancer Res* 62:2151–2156.

Dougherty, T., G. Grindey, R. Fiel et al. 1975. Photoradiation therapy. II. Cure of animal tumors with hematoporphyrin and light. *J Natl Cancer Inst* 55:115–121.

Dougherty, T., J. Kaufman, A. Goldfarb et al. 1978. Photoradiation for the treatment of malignant tumors. *Cancer Res* 38:2628–2635.

Dougherty, T., R. Pandey, H. Nava et al. 2000. Preliminary clinicaldata on a new photodynamictherapy photosensitizer: 2-[1-hexyloxyethyl]-2-devinyl pyropheophorbide–a (HPPH) for treatment of obstructive esophageal cancer. *Proc SPIE* 3909:25.

Durmus, M., and T. Nyokong. 2007. Synthesis, photophysical and photochemical properties of novel water-soluble cationic indium (III) phthalocyanines. *Photochem Photobiol Sci* 6:659–668.

Eisner, U. 1957. Some novel hydroporphyrins. *J Chem Soc (0, Resumed)* 3461–3469.

El-Mofy, A. 1968. *Vitiligo and Psoralens*. Pergamon Press, Oxford.

Ermeydan, M., F. Dumoulin, T. Basova et al. 2010. Amphiphilic carbohydrate-phthalocyanine conjugates obtained by glycosylation or by azide-alkyne click reaction. *New J Chem* 34:1153–1162.

Ethirajan, M., Y. Chen, P. Joshi, and R. Pandey. 2011. The role of porphyrin chemistry in tumor imaging and photodynamic therapy. *Chem Soc Rev* 40:340–362.

Ethirajan, M., C. Saenz, A. Gupta, M. Dobhal, and R. Pandey. 2008. Photosensitizers for photodynamic therapy. In *Advances in Photodynamic Therapy Basic, Translational and Clinical*, Hamblin et al., editors. Artech House, Boston.

Evensen, J., and J. Moan. 1987. A test of different photosensitizers for photodynamic treatment of cancer in a murine tumor model. *Photochem Photobiol* 46:859.

Fahmy, I., and H. Abu-Shady. 1948. Ammi majus linn: The isolation and properties of ammoidin, ammidin and majudin, and their effect in the treatment of leukoderma. *J Pharm Pharmacol* 21:499–503.

Fan, D., M. Taniguchi, and J. Lindsey. 2007. Regioselective 15-bromination and functionalization of a stable synthetic bacteriohlorin. *J Org Chem* 72:5350–5357.

Ferrario, A., K. von Tiehl, N. Rucker et al. 2000. Antiangiogenic treatment enhances photodynamic therapy responsiveness in a mouse mammary carcinoma. *Cancer Res* 60:4066–4069.

Figge, F., G. Weiland, and O. Manganiello. 1948. Cancer detection and therapy: Affinity of neoplastic, embryonic, and traumatized tissues for porphyrins and metalloporphyrins. *Proc Soc Exp Biol Med* 68:640–664.

Finsen, N. 1901. *Phototherapy*. Edward Arnold, London.

Fukuzumi, S., K. Ohkuba, X. Zheng, Y. Chen, and R. Pandey. 2008. Metal bacteriochlorins which act as dual singlet oxygen and superoxide generators. *J Phys Chem B* 112(9):2738–2746.

Furukawa, K., D. Green, H. Kato, and T. Dougherty. 1994a. Point fluorescence measurements of transformed tissues using 2-[1-hexyloxyethel]-2-devinyl pyropheophorbide-a. *Proc SPIE* 2133:170.

Furukawa, K., D. Green, T. Mang, H. Kato, and T. Dougherty. 1994b. Fluorescence detection of premalignant, malignant, and micrometastatic disease using hexylpyropheophorbide. *Proc SPIE* 2371:510.

Furuta, H., T. Asano, and T. Ogawa. 1994. "N-confused porphyrin": A new isomer of tetraphenylporphyrin. *J Am Chem Soc* 116:767.

Gad, F., T. Zahra, K. Francis, T. Hasan, and M. Hamblin. 2004. Targeted photodynamic therapy of established soft-tissue infections in mice. *Photochem Photobiol Sci* 3:451–458.

Gauna, G., J. Marino, M. García, L. Roguin Vior, and J. Awruch. 2011. Synthesis and comparative photodynamic properties of two isosteric alkyl substituted zinc(II) phthalocyanines. *Eur J Med Chem* 45:5532–5539.

Geier, G., D. Haynes, and J. Lindsey. 1999. An efficient one-flask synthesis of N-confused tetraphenylporphyrin. *Org Lett* 1:1455.

Geier, G., and J. Lindsey. 1999. N-Confused tetraphenylporphyrin and tetraphenylsapphyrin formation in one-flask syntheses of tetraphenylporphyrin. *J Org Chem* 64:1596.

Gosmann, M., and B. Franck. 1986. Synthesis of a fourfold enlarged porphyrin with an extremely large, diamagnetic ring current. *Angew Chem Int Ed Engl* 25:1100.

Graham, A., G. Li, Y. Chen et al. 2003. Structure-activity relationship of new octaethylporphyrin-based benzochlorins as photosensitizers for photodynamic therapy. *Photochem Photobiol* 77(5):561–566.

Grahn, M., A. McGuinness, R. Benzie et al. 1997. Intracellular uptake, absorption spectrum and stability of the bacteriochlorin photosensitizer 5,10,15,20- tetrakis(m-hydroxyphenyl) bacteriochlorin (mTHPBC) with 5,10,15,20- tetrakis(m-hydroxyphenyl)chlorin. (mTHPC). *Photochem Photobiol B* 37(3):261–266.

Gryshuk, A., A. Graham, R. Pandey et al. 2002. A first comparative study of purpurinimide-based fluorinated vs. non-fluorinated photosensitizers for photodynamic therapy. *Photochem Photobiol* 76:555.

Guardiano, M., R. Biolo, G. Jori, and K. Schaffner. 1989. Tetra-n-propylporphycene as a tumour localizer: Pharmacokinetic and phototherapeutic studies in mice. *Cancer Lett* 44:1–6.

Gudgin Dickson, E., R. Goyan, and R. Pottier. 2002. Directions in photodynamic therapy. *Cell Mol Biol* 48(8):939–954.

Hamblin, M., J. Miller, and B. Ortel. 2000. Scavenger-receptor targeted photodynamic therapy. *Photochem Photobiol* 72(4):533–540.

Hamblin, M., D. O'Donnell, N. Murthy, C. Contag, and T. Hasan. 2002. *Photochem Photobiol* 75:51–57. *Recent Advances in Developing Improved Agents for PDT* 261.

Hamblin, M., T. Zahra, C. Contag, A. McManus, and T. Hasan. 2003. Optical monitoring and treatment of potentially lethal wound infections in vivo. *J Infect Dis* 187:1717–1725.

Hannah, S., V. Lynch, N. Gerasimchuk, D. Maqda, and J. Sessler. 2001. Synthesis of a metal-free texaphyrin. *Org Lett* 3:3911–3914.

Hasegawa, J., K. Takata, T. Miyahara et al. 2005. Excited states of porphyrin isomers and porphycene derivatives: A SAC-CI study. *J Phys Chem* 109:3187–3200.

Hausmann, W. 1908. Die sensibilisierende wirkung tierischer farbstoffe und ihre physiologische bedeutungWien. *Klin Wochenschr* 21:1527–1529.

Hausmann, W. 1910. Die sensibilisierende wirkung des hamatoporphyrins. *Biochem Z* 30:276–316.

Hayashi, T., H. Dejima, T. Matsuo et al. 2002. Blue myoglobin reconstituted with an iron porphycene shows extremely high oxygen affinity. *J Am Chem Soc* 124:11226–11227.

Henderson, B., D. Bellnier, W. Greco et al. 1997. An in vivo quantitative structure-activity relationship for a congeneric series of pyropheophorbide derivatives as photosensitizers for photodynamic therapy. *Cancer Res* 57:4000–4007.

Henderson, B. W., A. B. Sumlin, B. L. Owczarczak, T. J. Dougherty. 1991. Bacteriochlorophyll-a as a photosensitizer for photodynamic treatment of transplantable murine tumors. *J Photochem Photobiol B* 10:303–313.

Henderson, B., and V. Fingar. 1987. Relationship between tumor hypoxia and response to photodynamic therapy treatment in an experimental mouse model. *Cancer Res* 47:3110–3114.

Heo, P., and C. Lee. 1996. Rearrangement of 2,4-bisalkylpyrrole unit to 2,5-bisalkylpyrrole unit in the ligand-modified porphyrinogens. *Bull Korean Chem Soc* 17:778.

Heo, P., K. Shin, and C. Lee. 1996. Stepwise syntheses of core-modified, meso-substituted porphyrins. *Tetrahedron Lett* 37:197.

Hooker, J., V. Nguyen, V. Taylor et al. 2012. New application for expanded porphyrins: Sapphyrin and heterosapphyrins as inhibitors of Leishmania parasites. *Photochem Photobiol* 88:194–120.

Huang, Y., P. Mroz, T. Zhiyentayev et al. 2010. In vitro photodynamic therapy and quantitative structure-activity relationship studies with stable synthetic near-infrared-absorbing bacteriochlorin photosensitizers. *J Med Chem* 53:4018–4027.

Huang, Z., Q. Chen, K. Dole et al. 2007. The effect of Tookad-mediated photodynamic ablation of the prostate gland on adjacent tissues- in vivo study in a canine model. *Photochem Photobiol Sci* 6:1318–1324.

Huang, Z., Q. Chen, D. Luck et al. 2005. Studies of a vascular-acting photosensitizer, Pd-bacteriopheophorbide (TOOKAD), in normal canine prostate and spontaneous canine prostate cancer. *Lasers Surg Med* 36:390–397.

Huang, Z., Q. Chen, N. Trncic et al. 2004. Effects of Pd-bacteriopheophorbide (TOOKAD)-mediated photodynamic therapy on canine prostate pretreated with ionizing radiation. *Radiat Res* 161:723–731.

Jiang, X., P. Lo, Y. Tsang et al. 2010a. Phthalocyanine-polyamine Conjugates as pH-controlled photosensitizers for photodynamic therapy. *Chem Eur J* 16:4777–4783.

Jiang, X., P. Lo, S. Yeung, W. Fong, and D. Ng. 2010b. A pH-responsive fluorescence probe and photosensitiser based on a tetraamino silicon(IV) phthalocyanine. *Chem Commun* 46:3188–3190.

Jiang, X., S. Yeung, P. Lo, W. Fong, and D. Ng. 2011. Phthalocyanine-polyamine conjugates as highly efficient photosensitizers for photodynamic therapy. *J Med Chem* 54:320–330.

Jori, G., B. Rihter, M. Kenney, and M. Rodgers. 1990. Naphthalocyanine as a photodynamic sensitiser for experimental tumours: Pharmacokinetic and phototherapeutic studies. *Br J Cancer* 62:966–970.

Josefsen, L., and R. Boyle. 2008. Photodynamic therapy and the development of metal-based photosensitizers. *Metal-Based Drugs*, article ID 276109.

Joshi, P., M. Ethirajan, L. Goswami et al. 2011. Synthesis, spectroscopic, and in vitro photosensitizing efficacy of keto-bacteriochlorins derived from ring-B and ring-D reduced chlorins via pinacol-pinacolone rearrangement. *J Org Chem* 76:8629–8640.

Judy, M., J. Matthews, J. Newman et al. 1991. In vitro photodynamic inactivation of herpes simplex virus with sapphyrins: 22 pi-electron porphyrin-like macrocycles. *Photochem Photobiol* 53:101–107.

Karuso, P., P. Berguquist, J. Buckleton et al. 1986. 13² 17²-Cyclopheophorbide enol, the first porphyrin isolated from a sponge. *Tetrahedron Lett* 27:2177–2178.

Kelly, J., and M. Snell. 1976. Hematoporphyrin derivative: A possible aid in the diagnosis and and therapy of carcinoma of the bladder. *J Urol* 115:150–151.

Kessel, D., K. Whitcomb, and V. Schulz. 1992 Lipoprotein-mediated distribution of N-aspartylchlorin e6 in the mouse. *Photochem Photobiol* 56:51–56.

Kim, H., and J. Lindsey. 2005. De novo synthesis of stable tetrahydro-porphyrinic macrocycles: Bacteriochlorins and a tetrahydrocorrin. *J Org Chem* 70:5475–5486.

Konan, Y., R. Gurny, and E. Allemann. 2002. State of the art in the delivery of photosensitizers for photodynamic therapy. *J Photochem Photobiol B* 66:89–106.

Kral, V., J. Davis, A. Andrievsky et al. 2002. Synthesis and biolocalization of water-soluble sapphyrins. *J Med Chem* 45:1073–1078.

Kralova, J., T. Briza, I. Moserova et al. 2008. Glycol porphyrin derivatives as potent photodynamic inducers of apoptosis in tumor cells. *J Med Chem* 51:5964–5973.

Kuimova, M., M. Bhatti, M. Deonarain et al. 2007. Fluorescence characterization of multiply-loaded anti-HER2 single chain Fv-photosensitizer conjugates suitable for photodynamic therapy. *Photochem Photobiol Sci* 6:933–939.

Lash, T., D. Richter, and C. Shiner. 1999. Synthesis of hexa- and heptaalkyl-substituted inverted or N-confused porphyrins by the "3 + 1" methodology. *J Org Chem* 64:7973.

Lanzo, I., N. Russo, and E. Sicilia. 2008. *J Phys Chem B* 112:4123–4130.

Lee, C., and H. Kim. 1997. Synthesis of meso-tetraphenylthiaporphyrins bearing one inverted pyrrole. *Tetrahedron Lett* 38:3935.

Lee, C., H. Kim, and D. Yoon. 1999. Synthesis of core-modified porphyrins and studies of their temperature-dependent tautomerism. *Bull Korean Chem Soc* 20:276.

Li, G., A. Graham, W. Potter et al. 2001. A simple and efficient approach for the synthesis of fluorinated and nonfluorinated octaethylporphyrin-based benzochlorins with variable lipophilicity, their in vivo tumor uptake, and the preliminary in vitro photosensitizing efficacy. *J Org Chem* 66:1316–1332.

Lipson, R., E. Baldes, and A. Olsen. 1961. Hematoporphyrin derivative: a new aid for endoscopic detection of malignant disease. *J Thorac Cardiovasc Surg* 42:623–629.

Liu, B., C. Brückner, and D. Dolphin. 1996. A meso-unsubstituted N-confused porphyrin prepared by rational synthesis. *Chem Commun* 18:2141.

Liu, F., and G. Rabinovich. 2005. Galectins as modulators of tumour progression. *Nat Rev Cancer* 5:29–41.

Liu, J., X. Jiang, W. Fong, and D. Ng. 2008. Highly photocytotoxic 1, 4-dipegylated zinc(II) phthalocyanines. *Org Biomol Chem* 6:4560–4566.

Lobel, J., I. MacDonald, M. Ciesielski et al. 2001. 2-[1-hexyloxyethyl]-2-devinyl pyropheophorbide-a (HPPH) in a nude rat glioma. *Lasers Surg Med* 29:397.

Lu, T., P. Shao, I. Mathew, A. Sand, and W. Sun. 2008. Synthesis and photophysics of benzotexaphyrin: A near-infrared emitter and photosensitizer. *Am Chem Soc* 130(47):15782–15783.

Luo, Y., C. Chang, and D. Kessel. 1996. Rapid initiation of apoptosis by photodynamic therapy. *Photochem Photobiol* 63:528–534.

MacDonald, I., and T. Dougherty. 2001. Basic principles of photodynamic therapy. *J Porphyrins Phthalocyanines* 5:105–129.

Magic Ray Moscow Center of Laser Medicine. Retrieved from: http://www.magicray.ru/ENG/ lecture/1.html.

Magne, M., C. Rodriguez, A. Autry et al. 1997. Photodynamic therapy of facial squamous cell carcinoma in cats using a new photosensitizer. *Lasers Surg Med* 20:202.

Maiya, B., M. Cyr, A. Harriman, and J. Sessler. 1990. In vitro photodynamic inactivation of herpes simplex virus with sapphyrins: 22 pi-electron porphyrin-like macrocycles. *J Phys Chem* 94:3597–3601.

Marino, J., M. García Vior, L. Dicelio, L. Roguin, and J. Awruch. 2010. Photodynamic effects of isosteric water-soluble phthalocyanines on human nasopharynx KB carcinoma cells. *J Eur J Med Chem* 45:4129–4139.

McCarthy, J., J. Bhaumik, N. Merbouh, and R. Weissleder. 2009. High-yielding syntheses of hydrophilic conjugatable chlorins and bacteriochlorins. *Org Biomol Chem* 7:3430–3436.

Meisel, P., and T. Kocher. 2005. Photodynamic therapy for periodontal diseases: State of the art. *J Photochem Photobiol B–Biol* 79:159–170.

Meyer-Betz, F. 1913. Untersuchung uber die biologische (photodynamische) wirkung des hamatoporphyrins und anderer derivate des blutund gallenfarbstoffs. *Dtsch Arch Klin Med* 112:476–503.

Minehan, T., and Y. Kishi. 1997. Extension of the Eschenmoser sulfide contraction/iminoester cyclization method to the synthesis of tolyporphin chromophore. *Tetrahedron Lett* 39:6811.

Mironov, A., M. Grin, A. Tsyprovskiy et al. 2003a. New photosensitizers of bacteriochlorin series for photodynamic cancer therapy. *Bioorg Khim* 29:214–221.

Mironov, A., M. Grin, A. Tsyprovskiy et al. 2004. Synthesis of cationic bacteriochlorins. *Mendeleev Commun* 5:204–207.

Mironov, A., M. Grin, A. Tsyprovskiy et al. 2003b. New bacteriochlorin derivatives with a fused N-aminoimide ring. *J Porphyr Phthalocya* 7:725–730.

Mironov, A., M. Grin, and A. Tsyprovskiy. 2002. Synthesis of the first N-hydroxycycloimide in the bacteriochlorophyll-a series. *J Porphyr Phthalocya* 6:358–361.

Mironov, A., A. Kozyrev, and A. Brandis. 1993. Sensitizers of second generation for photodynamic therapy of cancer based on chlorophyll and bacteriochlotophyll derivatives. *Proc. SPIE*. 1922:204–208.

Misawa, J., S. Moriwaki, E. Kohno et al. 2005. The role of low-density lipoprotein receptors in sensitivity to killing by photofrin-mediated photodynamic therapy in cultured human tumor cell lines. *J Dermatol Sci* 40:59–61.

Mishra, P., S. Patel, and A. Datta. 2006. Effect of increased hydrophobicity on the binding of two model amphiphilic chlorin drugs for photodynamic therapy with blood plasma and its components. *J Phys Chem B* 110:21238–21244.

Mitra, S., C. Haidaris, S. Snell et al. 2011. Effective photosensitization and selectivity in vivo of Candida Albicans by meso-tetra (N-methyl-4-pyridyl) porphine tetra tosylate. *Lasers Surg Med* 43(4):324–332.

Montforts, F., and M. Glasenapp-Breiling. 2002. In *Progress in the Chemistry of Organic Natural Products*, W. Herz, H. Falk, and G. Kirby, editors. Springer, Wien, New York.

Morgan, A., G. Garbo, A. Rampersaud et al. 1989. Photodynamic action of benzochlorins. *Proc SPIE* 1065:146.

Morgan, A., S. Skalkos, G. Maguire et al. 1992. Observations on the synthesis and in vivo photodynamic activity of some benzochlorins. *Photochem Photobiol* 55:13.

Morliere, P., J. Maziere, R. Santus et al. 1998. Tolyporphin: A natural product from cyanobacteria with potent photosensitizing activity against tumor cells *in vitro* and *in vivo*. *Cancer Res* 58(16):3571–3578.

Mroz, P., J. Bhaumik, D. Dogutan et al. 2009. Imidazole metallo-porphyrins as photosensitizers for photodynamic therapy: Role of molecular charge, central metal and hydroxyl radical protection. *Cancer Lett* 282:63–76.

Mroz, P., Y. Huang, A. Szokalska et al. 2010. Stable synthetic bacteriochlorins overcome the resistance of melanoma to photodynamic therapy. *FASEB J* 24:3160–3170.

Narayanan, S., B. Sridevi, A. Srinivasan, T. Chandrashekar, and R. Roy. 1998. One step synthesis of sapphyrin and N-confused porphyrin using dipyrromethane. *Tetrahedron Lett* 39:7389.

National Cancer Institute. 2011. Retrieved from: http://www.cancer.gov/cancertopics/factsheet/Therapy/photodynamic.

Oda, K., S. Ogura, and I. Okura. 2000. Preparation of a water-soluble fluorinated zinc phthalocyanine and its effect for photodynamic therapy. *Photochem Photobiol* 59:20–25.

Osterloh, J., and M. Vicente. 2002. Mechanism of porphyrinoid localization in tumors. *J Porphyrins Phthalocyanine* 6(5):305–324.

Pacholska, E., L. Latos-Grazynski, L. Szterenberg, and Z. Ciunik. 2000. Pyrrole-inverted isomer of 5,10,15,20-tetraaryl-21-selenaporphyrin. *J Org Chem* 65:8188.

Pandey, R., D. Bellnier, K. Smith, and T. Dougherty. 1991. Chlorin and porphyrin derivatives as potential photosensitizers in photodynamic therapy. *Photochem Photobiol* 53:65.

Pandey, R., A. Sumlin, W. Potter et al. 1996. Alkyl ether analogs of chlorophyll-*a* derivatives: Part 1. Synthesis, photophysical properties and photodynamic efficacy. *Photochem Photobiol* 63:194–205.

Pandey, R., and G. Zheng. 2000. In *The Porphyrin Handbook*, K. Kadish et al., editors. Academic Press, Boston. vol. 6, 157–230.

Pandey, S., A. Gryshuk, M. Sajjad et al. 2005. Multimodality agents from tumor imaging (PET, Fluorescence) and photodynamic therapy. A possible "see and treat" approach. *J Med Chem* 48:6286–6295.

Paszko, E., C. Ehrhardt, M. Senge, D. Kelleher, and J. Reynolds. 2011. Nanodrug applications in photodynamic therapy. *Photodiagnosis Photodyn Ther* 8:14–29.

Policard, A. 1924. Etudes sur les aspects offerts par des tumeurs experimentales examines a la lumiere de wood. *CR Soc Biol* 9:1423–1424.

Polo, L., G. Valduga, G. Jori, and E. Reddi. 2002. Low-density lipoprotein receptors in the uptake of tumour photosensitizers by human and rat transformed fibroblasts. *Int J Biochem Cell Biol* 34:10–23.

Post, J., J. Poele, J. Schuitmaker, and F. Stewart. 1996. A comparison of functional bladder after intravesical photodynamic therapy with three different photosensitizers. *Photochem Photobiol* 63:314.

Potter, W., B. Henderson, D. Bellnier et al. 1999. Parabolic quantitative structure-activity relationships and photodynamic therapy: Application of a three-compartment model with clearance to the in vivo quantitative structure-activity

relationships of a congeneric series of pyropheophorbide derivatives used as photosensitizers for photodynamic therapy. *Photochem Photobiol* 70:781.

Prinsep, M., F. Caplan, R. Moore, G. Patterson, and C. Smith. 1992. Tolyporphin, a novel multidrug resistance reversing agent from the blue-green alga *Tolypothrix nodosa*. *J Am Chem Soc* 114:385.

Pushpan, S., A. Srinivasan, V. Anand et al. 2001. Inverted meso-aryl porphyrins with heteroatoms; characterization of thia, selena, and oxa N-confused porphyrins. *Org Chem* 66:15.

Raab, O. 1900. Uber die wirkung fluoreszierender stoffe auf infusorien. *Zeitung Biol* 39:524–526.

Ragas, X., D. Sanchez-Garcia, R. Ruiz-Gonzalez et al. 2010. Cationic porphycenes as potential photosensitizers for antimicrobial photodynamic therapy. *J Med Chem* 53:7796–7803.

Reddy, G., M. Bhojani, P. McConville et al. 2006. Vascular targeted nanoparticles for imaging and treatment of brain tumors. *Clin Cancer Res* 12(22):6677–6686.

Richert, C., J. Wessels, M. Muller et al. 1994. Photodynamic antitumor agents: Beta-methoxyethyl groups give access to functionalized porphycenes and enhance cellular uptake and activity. *J Med Chem* 37:2797–2807.

Richter, A., B. Kelly, J. Chow et al. 1987. Preliminary studies on a more effective phototoxic agent than hematoporphyrin. *J Natl Cancer Inst* 79:1327–1332.

Rosenfield, A., J. Morgan, L. Goswami et al. 2006. Photosensitizers derived from 13^2 – oxo-methyl pyropheophorbide-a: Enhanced effect of indium (III) as a central metal in *in vitro* and *in vivo* photosensitizing efficacy. *Photochem Photobiol* 82:626–634.

Rothemund, P. 1939. Porphyrin studies. III. The structure of the porphine ring system. *J Am Chem Soc* 61:2912.

Roy, I., T. Ohulchanskyy, H. Pudavar et al. 2003. Ceramic-based nanoparticles entrapping water-insoluble photosensitizing anticancer drugs: A novel drug-carrier system for photodynamic therapy. *J Am Chem Soc* 125:7860–7865.

Ruzié, C., M. Krayer, T. Balasubramanian, and J. Lindsey. 2008. Tailoring a bacteriochlorin building block with cationic, amphipathic, or lipophilic substituents. *J Org Chem* 73: 5806–5820.

Ruzié, C., M. Krayer, and J. Lindsey. 2009. Fast and robust route to hydroporphyrin-chalcone with extended red or near infrared absorption. *Org Lett* 11:1761–1764.

Saenz, C., M. Ethirajan, G. Iacobucci et al. 2011. Indium as a central metal enhances the photosensitizing efficacy of benzoporphyrin derivatives. *J Porphyrins Phthalocyanines* 15:1–7.

Sánchez-Ganchez, D., and J. Sessler, J. 2008. Porphycenes: Synthesis and derivatives. *Chem Soc Rev* 37:215–232.

Sanchez-Garcia, D., J. Borrell, and S. Nonell. 2009. One-pot synthesis of substituted 2,2′- bipyrroles. A straightforward route to aryl porphycenes. *Org Lett* 11:77–79.

Songca, S. P. 2001. In-vitro activity and tissue distribution of new fluorinated meso-tetrahydroxyphenylporphyrin photosensitizers. *J Pharm Pharmacol* 53:1469–1476.

Schmidt-Erfurth, U., H. Diddens, R. Birngruber, and T. Hasan. 1997. Photodynamic targeting of human retinoblastoma cells using covalently low-density lipoprotein conjugates. *Br J Cancer* 75:54–61.

Schwartz, S., K. Absolon, and H. Vermund. 1955. Some relationships of porphyrins, X-rays and tumors. *Bull Minn Univ School Med* 27:7–13.

Segalla, A., F. Fedeli, E. Redd, G. Jori, and A. Cross. 1997. Effect of chemical structure and hydrophobicity on the pharmacokinetic properties of porphycenes in tumour-bearing mice. *Int J Cancer* 72:3.

Sessler, J., G. Hemmi, T. Mody et al. 1994. Texaphyrins: Synthesis and applications. *Acc Chem Res* 27:43–50.

Sessler, J., and R. Miller. 2000. Exaphyrins: New drugs with diverse clinical applications in radiation and photodynamic therapy. *Biochem Pharmacol* 59:733–739.

Sessler, J., T. Murai, V. Lynch, and M. Cyr. 1988. An "expanded porphyrin": The synthesis and structure of a new aromatic pentadentate ligand. *J Am Chem Soc* 110:5586.

Sharman, W., C. Allen, and van Lier. 1999. Photodynamic therapeutics: Basic principles and clinical applications. *J Drug Discovery Today* 4:507–517.

Sharonov, G., T. Karmakova, R. Kassies et al. 2004. *Chem Listy* 98:s17.

Sharonov, G., T. Karmakova, R. Kassies et al. 2006. Cycloimide bacteriochlorin p derivatives: Photodynamic properties and cellular and tissue distribution. *Free Radical Biol Med* 40:407–419.

Shimakoshi, H., T. Baba, Y. Iseki et al. 2008. Photophysical and photosensitizing properties of brominated porphycenes. *Chem Commun* 2882–2884.

Smith, K., D. Goff, and D. Simpson. 1985. The meso substitution of chlorophyll derivatives: direct route for transformation of bacteriopheophorbides d into bacteriopheophorbides c. *J Am Chem Soc* 107:4946.

Solban, N., I. Rizvi, and T. Hasan. 2006. Targeted photodynamic therapy. *Lasers Surg Med* 38:522–531.

Spikes, J. 1984. In *Porphyrins Localization and Treatment of Tumors*, D. Doiron, and C. Gomer, editors. Alan R. Liss, New York.

Springs, S., D. Gosztola, M. Wasielewski et al. 1999. Picosecond dynamics of energy transfer in porphyrin-sapphyrin non-covalent assemblies. *J Am Chem Soc* 121:2281–2289.

Sprutta, N., and L. Latos-Grażyński. 1999. A tetraphenylthiaporphyrin with an inverted thiopene ring. *Tetrahedron Lett* 40:8457–8460.

Sprutta, N., and L. Latos-Grazynski. 2001. 25,27-Dithiasapphyrin and pyrrole-inverted isomer of 21, 23-dithiaporphyrin from condensation of pyrrole and 2,5-Bis(p-tolylhydroxymethyl) thiophene. *Org Lett* 3:1933.

Srivatsan, A., M. Ethirajan, S. Pandey et al. 2011a. Conjugation of cRGD peptide to chlorophyll-a based photosensitizer (HPPH) alters its pharmacokinetics with enhanced tumor-imaging and photosensitizing (PDT) efficacy. *Mol Pharm* 8:1186–1197.

Srivatsan, A., Y. Wang, P. Joshi et al. 2011b. In vitro cellular uptake and dimerization of signal transducer and activator of transcription-3 (STAT3) identify the photosensitizing and imaging-potential of isomeric photosensitizers derived from chlorophyll-a and bacteriochlorophyll-a. *J Med Chem* 54(19):6859–6873.

Stuchinskaya, T., M. Moreno, M. Cook, D. Edwards, and D. Russell. 2011. Targeted photodynamic therapy of breast cancer cells using antibody-phthalocyanine-gold nanoparticles conjugates. *Photochem Photobiol Sci* 10:822–831.

Suci, P., Z. Varpness, E. Gillitzer, T. Douglas, and M. Young. 2007. Targeting and photodynamic killing of a microbial pathogen using protein cage architectures functionalized with a photosensitizer. *Langmuir* 23:12280–12286.

Surya Prakash, G., and A. Yudin. 1997. Perfluoroalkylation with organosilicon reagents. *Chem Rev* 97:757.

Triesscheijna, M., P. Paul Baas, J. Schellensa, and F. Stewarta. 2006. Photodynamic therapy in oncology. *Oncologist* 11:1034–1044.

Tuncel, S., J. Chabert, F. Albrieux et al. 2012. Towards dual photodynamic and antiangiogenic agents: Design and synthesis of a phthalocyanine–chalcone conjugate. *Org Biol Chem* 10(6):1154–1157.

van Lier, J. 1990. Phthalocyanines as sensitizers for PDT of cancer. In *Photodynamic Therapy of Neoplastic Disease*, 1, D. Kessel editor. CRC Press, Boca Raton, FL.

van Lier, J., and J. Spikes. 1989. The chemistry, photophysics and photosensitizing properties of phythalocyanines. *Ciba Found Symp* 146:17–32.

Verma, S., G. Watt, Z. Mai, and T. Hasan. 2007. Strategies for enhanced photodynamic therapy effects. *Photochem Photobiol* 83:996–1005.

Vogel, E., M. Broring, J. Fink et al. 1995. From porphyrin isomers to octapyrrolic "Figure Eight". *Angew Chem Int Ed Eng* 34:2511.

Vogel, E., M. Kocher, H. Schmickler, and J. Lex. 1986. Porphycene-A novel porphin isomer. *Angew Chem Int Ed Eng* 25:257.

Volanti, C., G. Gloire, A. Vanderplasschen et al. 2004. Downregulation of ICAM-1 and VCAM-1 expression in endothelial cells treated by photodynamic therapy. *Oncogene* 23:8649–8658.

Volanti, C., J. Matroule, and J. Piette. 2002. Involvement of oxidative stress in NF-kB activation in endothelial cells treated by photodynamic therapy. *Photochem Photobiol* 75(1):36–45.

von Tappeiner, H., and H. Jesionek. 1903. Therapeutische versuche mit fluoreszierenden stoffen. *Munch Med Wochenschr* 47:2042–2044.

von Tappeiner, H., and A. Jodlbauer. 1907. *Die sensibilisierende wirkung fluorescierender substanzen: Gesammelte untersuchungen über die photodynamische erscheinung.* F.C.W. Vogel, Leipzig.

Wainwright, M. 1998. Photodynamic antimicrobial chemotherapy (PACT) photosensitizers. *J Antimicrob Chemother* 42:13–28.

Wainwright, M. 2008. The development of new photosensitizers. *Anti-Cancer Agents Med Chem* 8:280–291.

Waluk, J., M. Muller, P. Swiderek et al. 1991. Electronic states of porphycenes. _J Am Chem Soc_ 113:5511–5527.

Wang, S., W. Fan, G. Kim et al. 2011. Novel methods to incorporate photosensitizers into nanocarriers for cancer treatment by photodynamic therapy. _Lasers Surg Med_ 43:686–695.

Wang, W., and Y. Kishi. 1999. Synthesis and structure of tolyporphin A O,O-diacetate. _Org Lett_ 1(7):1129–1132.

Watanabe, N., K. Yamamoto, H. Ihshikawa, and A. Yagi. 1993. New chlorophyll-a related compound isolated as antioxidants from marine bivalves. _J Nat Prod_ 56:305–317.

Wilson, M. 2004. Lethal photosensitisation of oral bacteria and its potential application in the photodynamic therapy of oral infections. _Photochem Photobiol Sci_ 3:412–418.

Winkelaman, J., G. Slater, and J. Grossman. 1967. The concentration in tumor and other tissues of parenterally administered tritium- and 14-C-labeled tetraphenylporphinesulfonate. _Cancer Res_ 27:2060.

Wu, Y., K. Chan, C. Yip et al. 1997. Porphyrin isomers: Geometry, tautomerism, geometrical isomerism, and stability. _J Org Chem_ 2:9240.

Wyss, P. 2000. Photomedicine in gynecology and reproduction. In _History of Photomedicine_, P. Wyss, Y. Tadir, B. Tromberg, and U. Haller, editors. Karger, Basel.

Yang, E., C. Kirmaier, M. Krayer et al. 2011. Photophysical properties and electronic structure of stable, tunablesynthetic bacteriochlorins: Extending the features of native photosynthetic pigments. _J Phys Chem B_ 115:10801–10816.

Yoon, D., and C. Lee. 2000. Synthesis and NMR studies of core-modified, N-confused porphyrins possessing alkyl groups at the rim nitrogen. _Bull Korean Chem Soc_ 21:618.

Young, S., K. Woodburn, M. Wright et al. 1996. Lutetium texaphyrin (PCI-0123): A near-infrared, water-soluble photosensitizer. _Photochem Photobiol_ 63:892.

Youngjae, Y., S. Gibson, R. Hilf et al. 2003. Water soluble core modified porphyrins. 3. Synthesis, photophysical properties and in vitro studies of photosensitization, uptake, localization with carboxylic acid-substituted derivatives. _J Med Chem_ 46:3734–3747.

Yslas, E., V. Rivarola, and E. Durantini. 2005. Synthesis and photodynamic activity of zinc(II) phthalocyanine derivatives bearing methoxy and trifluoromethylbenzyloxy substituents in homogeneous and biological media. _Biorg Med Chem_ 13(1):39–46.

Zanin, I., M. Lobo, L. Rodrigues et al. 2006. Photosensitization of in vitro biofilms by toluidine blue O combined with a light-emitting diode. _Eur J Oral Sci_ 114:64–69.

Zheng, X., J. Morgan, S. Pandey et al. 2009. Conjugation of 2-(1′-Hexyloxyethyl)-2-devinylpyropheophorbide-a (HPPH) to carbohydrates changes its subcellular distribution and enhances photodynamic activity in vivo. _J Med Chem_ 52:4306–4318.

Zheng, G., W. Potter, S. Camacho et al. 2001. Synthesis, photophysical properties, tumor uptake, and preliminary in vivo photosensitizing efficacy, of a homologous series of, 3-(1′-alkyloxy)ethyl-3-devinylpurpurin, -18-N-alkylimides with variable lipophilicity. _J Med Chem_ 44:1540.

23

5-Aminolevulinic Acid and Its Derivatives

Asta Juzeniene
Oslo University Hospital

Patrycja Mikolajewska
Oslo University Hospital

23.1 The Early History of 5-Aminolevulinic Acid— Photodynamic Therapy

5-Aminolevulinic acid (ALA) is a naturally occurring amino acid, and it is the first compound in the biosynthetic pathway leading to the production of porphyrins and heme. In 1956, exogenous administration of ALA to humans was noted for the first time to lead to transient accumulation of porphyrins and skin photosensitization (Berlin, Neuberger, and Scott 1956). In 1979, Malik and Djaldetti demonstrated that addition of ALA increases levels of the endogenous photosensitizer protoporphyrin IX (PpIX) in Friend erythroleukemic cells. Later, it was demonstrated in mice that cells of the skin can synthesize and accumulate high concentrations of PpIX after localized administration of ALA (Pottier et al. 1986). The history of ALA–photodynamic therapy (PDT) starts in 1987, when Malik and Lugaci used ALA-induced PpIX together with light to inactivate cells *in vitro*. Three years later, Kennedy, Pottier, and Pross (1990) reported the first clinical trials using ALA-PDT for the treatment of malignant and precancerous skin tumors. Since then, ALA has been an exciting new modality for PDT (Peng et al. 1997; Klein et al. 2008). Although PDT's therapeutic usefulness has been tested for a variety of malignancies, for example, colon, genitourinary tract, brain, and so forth, the real potential of ALA-PDT is in dermatology (Peng et al. 1997; Klein et al. 2008). ALA, being a relatively small molecule (167.6 Da, Table 23.1), has the potential of penetrating the *stratum corneum,* thus increasing the prospects of effective topical PDT. Over the last decades, ALA-PDT has been shown to be successful in treating a number of dermatological disorders (Klein et al. 2008), and its use is now being extended to cosmetic dermatology for photorejuvenation,

treatment of sebaceous hyperplasia, and acne vulgaris (Zakhary and Ellis 2005; Babilas, Landthaler, and Szeimies 2006; Klein et al. 2008; Sakamoto, Lopes, and Anderson 2010).

23.2 Heme Biosynthesis

Heme biosynthesis takes place in all cells of the human body except erythrocytes (Ponka 1999). The first step takes place inside the mitochondria. ALA is formed by ALA synthase-catalyzed condensation of succinyl-coenzyme A and glycine (Figure 23.1). Two ALA molecules are transported into the cytosol and condensed into porphobilinogen (PBG) by ALA dehydratase. Four PBGs are then joined by PBG deaminase into a single molecule, hydroxymethylbilane, which then undergoes dehydratation by uroporphyrinogen III synthase to form uroporphyrinogen III. In the next step, uroporphyrinogen III decarboxylase transforms it into coproporphyrinogen III. Coproporphyrinogen III is transported back into the mitochondria, where it undergoes decarboxylation by coproporphyrinogen oxidase and forms protoporphyrinogen IX. Another oxidation process, catalyzed by protoporphyrinogen oxidase, leads to the creation of PpIX. The last step of heme biosynthesis is the incorporation of iron into PpIX by the enzyme ferrochelatase and the creation of heme. Both the first and the last steps of heme biosynthesis are rate-limiting steps. The amount of available ALA synthase is tightly regulated according to the amount of heme present in the cells by a negative-feedback control system. A high concentration of heme blocks both ALA synthase transcription and translation. On the other hand, the concentration of heme is dependent on the availability of iron in the cell. This regulation prevents accumulation of any of the intermediate products in concentrations high enough for photosensitization. Addition of

TABLE 23.1   Molecular Weight (MW), Molecular Formula and Chemical Structure of 5-Aminolevulinic Acid (ALA), Methyl Aminolevulinate (MAL), Hexyl Aminolevulinate (HAL), and Protoporphyrin IX (PpIX)

Compound	MW (Da)	Molecular Formula	Chemical Structure
ALA HCl	167.6	$C_5H_{10}ClNO_3$	
MAL HCl	181.6	$C_6H_{12}ClNO_3$	
HAL HCl	251.8	$C_{11}H_{22}ClNO_3$	
PpIX	562.7	$C_{34}H_{34}N_4O_4$	

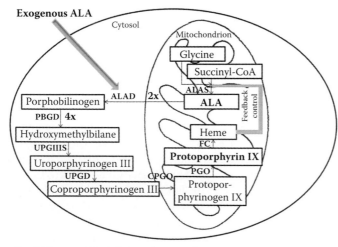

FIGURE 23.1   Heme biosynthesis pathway and mechanism of PpIX accumulation after administration of exogenous ALA. ALAD, ALA dehydratase; ALAS, ALA synthase; CPGO, coproporphyrinogen oxidase; FC, ferrochelatase; PBGD, porphobilinogen deaminase; PGO, protoporphyrinogen oxidase; UPGD, uroporphyrinogen decarboxylase; UPGIIIS, uroporphyrinogen III synthase.

exogenous ALA bypasses this feedback control and, with limited amounts of iron, allows for accumulation of PpIX in phototoxic concentrations.

23.3 Selective PpIX Formation in Neoplastic Tissues

PpIX is the main photosensitizer observed after application of ALA under most conditions. Because uroporphyrin and coproporphyrin are hydrophilic, they are rapidly removed from the body (Strauss et al. 1997). Much larger amounts of PpIX are produced in certain types of tumor tissue than in adjacent, normal tissue (Collaud et al. 2004). The reasons for the selective PpIX accumulation in neoplastic tissues might be of morphological, environmental, cell-cycle, or enzymatic origin. The selectivity is a result of a complex interactions of these factors, which depend on the type of the neoplastic disease, its localization, the stage and grade of the disease, and, finally, on the PpIX precursor that is used and its application mode. Moreover, in some malignant cells, the PBG deaminase activity is increased, ferrochelatase activity is reduced, and ALA uptake is augmented (Hinnen et al. 1998; Collaud et al. 2004; Ohgari et al. 2005). This allows for more molecules of ALA entering the neoplastic than normal cells and, combined with the poor availability of iron in those cells, leads to selective uptake of ALA into the neoplastic tissue as compared with the surrounding cells.

23.3.1 Photoactivation of PpIX

The accumulated PpIX can be activated with light at different wavelengths, according to its absorption spectrum or fluorescence excitation spectrum (Figure 23.2). Not all the wavelengths will activate PpIX equally efficiently because of the spectrum of PpIX and the penetration-depth spectra of light into the tissue (Moan, Iani, and Ma 1996; Nielsen et al. 2005). PpIX is activated most efficiently with light at 410 nm (Soret band), followed by light at 510, 540, 580, and 632 nm (Q-bands) (Figure 23.2). However, light in the blue/green part of the spectrum does not penetrate deeply into tissue (Figure 23.2). Light at 632 nm (red) penetrates deeper and, therefore, is being used in topical PDT, although it is not as efficient in photoactivation of PpIX as blue light. PpIX can be activated by blue light for the treatment of superficial lesions (down to 1 mm) while red light with greater depth of tissue penetration is the best choice for thicker lesions (Moan, Iani, and Ma 1996).

Excited PpIX emits fluorescence with a peak at 636 nm (Figure 23.2). This property, in combination with the high tumor-to-surrounding tissue ratio of PpIX accumulation, has been used for photodiagnosis (PDD) (Juzeniene, Peng, and Moan 2007). PpIX shows red fluorescence when excited with violet-blue light. Fluorescence images of the neoplastic tissue are of help both during surgery and in taking precise biopsies.

ALA-induced PpIX accumulation reaches a maximum after 1–8 h (depending on the application method, application time, and concentration) and is cleared from the body via natural mechanisms within 24–48 h (Peng et al. 1997). This makes both PDD and PDT easy and safe modalities that do not induce prolonged photosensitivity or cumulative toxicity.

23.4 Weak Points of ALA-PDT

The major weak points of topical PDT are the shallow penetration depth (<2 mm) of ALA and pain experienced during light exposure.

23.4.1 Shallow Penetration Depth

The efficiency of ALA-PDT is limited by the poor penetration of hydrophilic ALA through tissue and the limited stability of the applied formulations. ALA, even though it is a small molecule with the potential to penetrate through the *stratum corneum*, does not penetrate deeply into tissue. The penetration depth of ALA depends on its concentration, application time, formulation, and type of lesion (intact versus disrupted *stratum corneum*). Longer application times can improve the penetration depth. However, it can result in lower selectivity as compared to short (1 to 6 h) application times. In practice, there are two main approaches for improving penetration of ALA: a physical and a chemical approach (Table 23.2).

Diffusion through the *stratum corneum* is the longest and a rate-limiting process in topical ALA-PDT. Lipid bilayers efficiently prevent hydrophilic ALA molecules from rapid penetration. Mechanical removal of superficial layers by gentle curettage or debulking is an efficient method for improving ALA penetration into the lesion (Souza et al. 2007). Ultrasound, iontophoresis, electroporation, and electrophoresis also enhance ALA transport (Gerritsen et al. 2009), but the devices for these physical enhancers are more complex, and the application procedure of ALA becomes more complicated.

Temperature affects many parameters during PDT (Juzeniene, Neilsen, and Moan 2006; Juzeniene et al. 2006). An elevated skin temperature does not only improve penetration of precursors into the skin and improve cell uptake but also increases the activity of heme synthesis enzymes, leading to higher intracellular levels of PpIX. Tissue oxygenation is also increased at higher temperatures. Changes in optical penetration will occur at elevated skin temperatures and oxygen concentrations: The optical penetration will decrease in the blue region and increase in the red region. We have demonstrated *in vitro* that the photobleaching

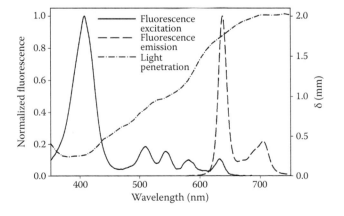

FIGURE 23.2 The normalized fluorescence excitation and emission spectra of protoporphyrin IX (PpIX) in human skin. The fluorescence excitation wavelength was set at 705 nm, and fluorescence emission wavelength was set at 405 nm. Light penetration into healthy human skin.

TABLE 23.2 Different Approaches for Improvement of ALA Penetration into Tissue

Mechanical/Physical Methods	Chemical Methods
Curettage (debulking)	Chemical penetration enhancers
Microdermabrasion	(DMSO, EDTA, oleic acid)
Ultrasound	Derivatization of ALA [methyl
Iontophoresis	aminolevulinate (MAL), butyl
Electroporation	aminolevulinate, hexyl
Electrophoresis	aminolevulinate (HAL), octyl
Microneedles	aminolevulinate]
Elevation of skin temperature	

rate of PpIX in the cells increases with increasing temperature during the light exposure (Juzeniene et al. 2006). Additionally, photoinactivation can be enhanced by heating the cells (<41°C) before light exposure or cooling them (<25°C) after light exposure. These effects are probably related to altered localization of the photosensitizer, diffusion of the singlet oxygen, and the degree of aggregation and cell repair. The PDT effect with ALA or its derivatives may be significantly enhanced by heating the tumor area before the treatment and by cooling it immediately after the treatment (Juzeniene et al. 2006). Recent clinical studies support our experimental results (Barolet and Boucher 2010; Fuchs et al. 2004).

Recently, silicon microneedles (MNs) were employed (Mikolajewska et al. 2010). The application of low concentrations [2% and 8% (w/w)] of ALA and methyl aminolevulinate (MAL) for short times (4 h), combined with the use of MNs, resulted in two to three times higher PpIX fluorescence as compared with the PpIX levels in non-pretreated skin (Mikolajewska et al. 2010). MNs provide a safe and efficient method for the improvement of skin permeability in a minimally invasive manner in comparison with, for example, tape stripping, needle puncturing, or painful curettage. Reduction of concentrations of prodrugs improves stability of the formulations. It also reduces the cost per treatment and, therefore, makes the treatment available for more patients. However efficient, physical methods (Table 23.2) can cause inhomogeneous dosage and need further investigations before being utilized in clinics.

The addition of a chemical penetration enhancer to the ALA formulation can help ALA penetration by disrupting the highly organized lipid structure of the *stratum corneum* (e.g., DMSO), interacting with intracellular proteins, or improving the partition of the drug or solvent into the *stratum corneum* (Williams 2003). Alternatively, penetration of the drug can be chemically enhanced by lowering its hydrophilicity. Derivatization of ALA can lead to increased lipophilicity and increased penetration (Gaullier et al. 1997). The longer the alkyl chain length of ALA derivatives, the more lipophilic the molecule gets and, therefore, the more efficiently it penetrates through the *stratum corneum*. Topical application of ALA esters leads to a buildup of PpIX only at or close to the site of the application, and the systemic administration of ALA results in fluorescence detectable at distant sites (Juzeniene et al. 2002). This is a result of the fact that ALA derivatives are bound within the lipophilic region of the skin, and ALA reaches the bloodstream and can be transported to other sites. This also results in a considerable lag time before PpIX production from ALA esters reaches the same levels as PpIX production from ALA in the skin. A different situation can be observed in nondermatological applications of ALA-PDT in which the permeability barriers differ from those of the *stratum corneum*, for example, the bladder (Williams 2003). The distribution of ALA-induced PpIX as well as ALA ester-induced PpIX does not entail risk for patients as PpIX from both ALA and ALA esters is cleared from the body within 24–48 h.

In addition to the problems caused by the hydrophilic nature of ALA, it tends to create dimers and other condensation products under alkaline conditions (Kaliszewski et al. 2004). This is challenging as temperature, pH, concentration, and oxygenation of the solution influence the stability of ALA formulations (Donnelly, McCarron, and Woolfson 2005). For long-term stability, solutions buffered to pH 2.00 are preferred. However, such pH values have the potential for irritation of the skin. Thus, ALA solutions buffered to physiological values (pH 5.5 or 7.4) need to be prepared fresh before each treatment.

23.4.2 Pain and Erythema

Pain during PDT with ALA or its derivatives is experienced to varying degrees by 70% of all PDT patients (Warren et al. 2009; Halldin et al. 2011; Miller et al. 2011). In 20% of cases, the treatment cannot be completed because of the pain. It is described as itching, burning, or stinging sensations that arise within seconds of the beginning of the light exposure and can last up to several hours. The severity of the pain is influenced by the localization, size, and type of the lesions. PDT of lesions localized on the face and scalp tends to be more painful than of those localized on other sites of the body (Lange et al. 1999), probably because of the density of nerve endings and/or the thickness of the skin. Treatment of multiple lesions is more painful than that of single lesions (Gadmar et al. 2002), and psoriasis patients are more troubled by the pain than patients with actinic keratosis or basal cell carcinoma (Radakovic-Fijan et al. 2005). This may be explained by different levels of keratinization leading to different patterns of PpIX accumulation in the lesions (Cottrell et al. 2008). Other factors influencing the pain sensation are wavelength, fluence, fluence rate, and mode of delivery (continuous vs. fractionated) of the light. Exposure at fluence rates below 60 mW/cm^2 delivered in small fractions separated by dark periods is less painful as compared to continuous exposure at high fluence rates (Kasche et al. 2006). The choice of wavelength has been discussed previously, and it should take into consideration the thickness and the characteristics of tissue to be destroyed. MAL has been shown to induce less pain during exposure to red light than ALA, but no difference was seen when exposed to blue light (Miller et al. 2011).

The mechanisms behind the pain remain unknown. There are some theories that can explain certain features of this sensation (peripheral nerve stimulation; transport of ALA by GABA receptors into the nerve endings; release of mediators of inflammation, e.g., histamine; or heat receptor stimuli) (Sandberg et al. 2006). Because the mechanisms behind the pain have not yet been fully elucidated, pain management is difficult to achieve. To date, only subcutaneous infiltration, nerve blocking, and cooling of the treated areas with air or water has been shown to provide effective pain relief (Clark et al. 2004). However, none of the methods are without risks. Multiple injection sites required for subcutaneous infiltration can result in swelling (especially dangerous within proximity of eyes) and present a risk for infections. Nerve blocking is the most effective of the pain relief methods but can lead to potential vessel trauma or nerve injury resulting in paresis (Halldin et al. 2009). Cooling of the treated areas does not result in any immediate risks. The effectiveness of this method is largely associated with the lowering of tissue metabolism, thus reducing the effects of injury and limiting the inflow of inflammatory mediators. Unfortunately, this also lowers the PpIX photobleaching rate and, thus, the PDT efficacy (Tyrrell, Campbell, and Curnow 2011).

Although erythema is, next to pain, the most common side effect of topical PDT, little is known about its origin or even its time course. Both its extent and duration vary significantly between patients, and it does not seem to be correlated with the treatment site nor with the type of light source.

23.5 ALA and Its Derivative-Based PDT and PDD

ALA/MAL-PDT is now the most widely practiced modality of PDT. It is used in dermatology, gynecology, urology, gastroenterology, and neurology (Krammer and Plaetzer 2008). This is for several reasons: (1) ALA or its esters alone show low dark toxicity to cells. (2) Topical delivery of ALA and its derivatives gives no prolonged photosensitivity reactions because the drugs can be selectively applied in the areas to be treated. (3) Endogenously produced PpIX is rapidly cleared from the body. (4) The short time interval needed between the administration of ALA and its esters and the maximal accumulation of PpIX in the target tissues makes ALA and its derivatives attractive for patients. Besides their usefulness in therapy, ALA derivatives can also be exploited for diagnostic (PDD) purposes.

Bacterial resistance against antibiotic treatment is an increasing problem. PDT research is being directed to achieve bacterial killing. The efficiency of photosensitization of bacterial cells is independent of the antibiotic resistance, both *in vitro* and *in vivo* (Demidova and Hamblin 2004). Induction of porphyrins by ALA can be achieved in most species of Gram-positive or Gram-negative bacteria (Nitzan et al. 2004). Possible targets of antimicrobial ALA-PDT are periodontal diseases, impetigo, atopic dermatitis, acne vulgaris, infected wounds, and superinfected psoriatic plaques (Maisch et al. 2004).

ALA-PDT is a promising new technique in a variety of cosmetic practices and is already being used for photorejuvenation and for the treatment of acne vulgaris, sebaceous gland hyperplasia, rosacea, and hirsutism (Zakhary and Ellis 2005). Photorejuvenation with ALA-PDT improves wrinkles, pore size, skin texture, rosacea, and sebaceous hyperplasia (Gold 2007).

Skin cancer incidence rates have been increasing over several decades. More than 80% of all skin cancer cases are related to sun exposure. Recent investigations have shown that traditional sunscreens probably do not afford adequate protection against skin cancer induction. Moan and Bissonnette (2001) patented a new sun cream in 2001, based on a principle that is totally different from that of traditional ones: It contains a derivative of ALA, which produces PpIX in skin cells. Experiments on mice have shown that this sun cream delays the development of UV-induced skin cancer (Sharfaei et al. 2002). Furthermore, pigmentation is induced (Monfrecola et al. 2002), and signs of photoaging and photodamage are reduced (Zakhary and Ellis 2005). In fact, this cream can also be used for photorejuvenation.

ALA-PDT has been approved for clinical application for non-hyperkeratotic actinic keratoses of the face and scalp in the United States, Canada, and many European countries since 1999 (Table 23.3). It is marketed by DUSA Pharmaceuticals Inc. as Levulan Kerastic in conjunction with Blue Light Photodynamic Therapy Illuminator BLU-U. Kerastic consists of a plastic tube containing two glass ampoules and an applicator tip. One ampoule contains ALA powder, and the other contains alcoholic solvent. The contents of the ampoules are mixed by breaking the glass inside the tube and shaking the Kerastic immediately before application. Light exposure is performed 14 to 18 h after application of the solution. The system is designed for treatment of single lesions.

PDT with ALA esters has also been recently approved. Metvix (known as Metvixia in the United States and France) is approved for the treatment of thin or non-hyperkeratotic and nonpigmented actinic keratosis on the face and scalp, superficial and/or nodular basal cell carcinomas, and Bowen's disease in the United States, Canada, European Union and European Economic Area. Additionally, it is approved for treatment of solar keratoses in New Zealand. Metvix can be used in combination with Aktilite CL 128, a LED-based, narrow-band (630 nm), red-light technology device, or CureLight Broadband (CureLight 01, red-light

TABLE 23.3 Areas of Clinical Use of PDT with ALA or Its Esters

Active Compound	Trade Name	Indication
ALA	Levulan	Treatment of **actinic keratosis**, actinic cheilitis, squamous cell carcinoma, cutaneous T-cell lymphoma, localized pagetoid reticulosis, vulval intraepithelial neoplasia, extramammary Paget's disease, acne, Hailey–Hailey disease, Darier disease, lichen planus, viral hand warts, genital warts, vulval lichen sclerosus, sarcoidosis, extragenital lichen sclerosus, cutaneous leishmaniasis, psoriasis, photorejuvenation, localized scleroderma, rosacea, perioral dermatisis, venous leg ulcer, molluscum contagiosum, epidermodysplasia verruciformis, interdigital mycoses
MAL	Metvix/Metvixia	Treatment of **basal cell carcinomas, actinic keratosis, Bowen's disease**, actinic cheilitis, cutaneous T-cell lymphoma, nephrogenic fibrosing dermopathy, extramammary Paget's disease of the vulva, acne, cutaneous leishmaniasis, photorejuvenation, necrobiosis, *Mycobacterium marinum*, lipoidica, granulation in Goltz syndrome, rosacea
HAL	Hexvix	Photodiagnosis of **bladder cancer**

Note: Bold font: Approved applications.

broadband) lamp. The cream contains 160 mg/g MAL with a shelf life of 1 week after opening the tube. The cream is applied directly on the lesion and exposed to light after 3–6 h incubation time.

Also, hexyl aminolevulinate (HAL) gained approval for PDD of non-muscle-invasive papillary cancer of the bladder in patients with known or suspected bladder cancer. It is marketed as Hexvix (Cysview outside Europe). It is sold as an 85 mg powder for intravenous delivery.

References

Babilas, P., M. Landthaler, and R. Szeimies. 2006. Photodynamic therapy in dermatology. *Eur J Dermatol* 16:340–348.

Barolet, D., and A. Boucher. 2010. Radiant near infrared light emitting diode exposure as skin preparation to enhance photodynamic therapy inflammatory type acne treatment outcome. *Lasers Surg Med* 42:171–178.

Berlin, N., A. Neuberger, and J. Scott. 1956. The metabolism of delta -aminolaevulic acid. 1. Normal pathways, studied with the aid of 15N. *Biochem J* 64:80–90.

Clark, C., R. Dawe, H. Moseley et al. 2004. The characteristics of erythema induced by topical 5-aminolaevulinic acid photodynamic therapy. *Photodermatol Photoimmunol Photomed* 20:105–107.

Collaud, S., A. Juzeniene, J. Moan et al. 2004. On the selectivity of 5-aminolevulinic acid-induced protoporphyrin IX formation. *Curr Med Chem Anticancer Agents* 4:301–316.

Cottrell, W., A. Paquette, K. Keymel et al. 2008. Irradiance-dependent photobleaching and pain in delta-aminolevulinic acid-photodynamic therapy of superficial basal cell carcinomas. *Clin Cancer Res* 14:4475–4483.

Demidova, T., and M. Hamblin. 2004. Photodynamic therapy targeted to pathogens. *Int J Immunopathol Pharmacol* 17:245–254.

Donnelly, R., P. McCarron, and A. Woolfson. 2005. Drug delivery of aminolevulinic acid from topical formulations intended for photodynamic therapy. *Photochem Photobiol* 81:750–767.

Fuchs, S., J. Fluhr, L. Bankova et al. 2004. Photodynamic therapy (PDT) and waterfiltered infrared A (wIRA) in patients with recalcitrant common hand and foot warts. *Ger Med Sci* 2:doc08.

Gadmar, O., J. Moan, E. Scheie et al. 2002. The stability of 5-aminolevulinic acid in solution. *J Photochem Photobiol B: Biol* 67:187–193.

Gaullier, J., K. Berg, Q. Peng et al. 1997. Use of 5-aminolevulinic acid esters to improve photodynamic therapy on cells in culture. *Cancer Res* 57:1481–1486.

Gerritsen, M., T. Smits, M. Kleinpenning et al. 2009. Pretreatment to enhance protoporphyrin IX accumulation in photodynamic therapy. *Dermatology* 218:193–202.

Gold, M. 2007. Photodynamic therapy with lasers and intense pulsed light. *Facial Plast Surg Clin North Am* 15:145–160.

Halldin, C., M. Gillstedt, J. Paoli et al. 2011. Predictors of pain associated with photodynamic therapy: A retrospective study of 658 treatments. *Acta Derm Venereol* 91:545–551.

Halldin, C., J. Paoli, C. Sandberg et al. 2009. Nerve blocks enable adequate pain relief during topical photodynamic therapy of field cancerization on the forehead and scalp. *Br J Dermatol* 160:795–800.

Hinnen, P., F. de Rooij, M. van Velthuysen et al. 1998. Biochemical basis of 5-aminolaevulinic acid-induced protoporphyrin IX accumulation: A study in patients with (pre)malignant lesions of the oesophagus. *Br J Cancer* 78:679–682.

Juzeniene, A., P. Juzenas, I. Bronshtein et al. 2006. The influence of temperature on photodynamic cell killing in vitro with 5-aminolevulinic acid. *J Photochem Photobiol B* 84:161–166.

Juzeniene, A., P. Juzenas, V. Iani et al. 2002. Topical application of 5-aminolevulinic acid and its methylester, hexylester and octylester derivatives: Considerations for dosimetry in mouse skin model. *Photochem Photobiol* 76:329–334.

Juzeniene, A., K. Nielsen, and J. Moan. 2006. Biophysical aspects of photodynamic therapy. *J Environ Pathol Toxicol Oncol* 25:7–28.

Juzeniene, A., Q. Peng, and J. Moan. 2007. Milestones in the development of photodynamic therapy and fluorescence diagnosis. *Photochem Photobiol Sci* 6:1234–1245.

Kaliszewski, M., M. Kwasny, J. Kaminski et al. 2004. The stability of 5-aminolevulinic acid and its ester derivatives. *Acta Pol Pharm* 61:15–19.

Kasche, A., S. Luderschmidt, J. Ring et al. 2006. Photodynamic therapy induces less pain in patients treated with methyl aminolevulinate compared to aminolevulinic acid. *J Drugs Dermatol* 5:353–356.

Kennedy, J., R. Pottier, and D. Pross. 1990. Photodynamic therapy with endogenous protoporphyrin IX: Basic principles and present clinical experience. *J Photochem Photobiol B* 6:143–148.

Klein, A., P. Babilas, S. Karrer et al. 2008. Photodynamic therapy in dermatology: An update 2008. *J Dtsch Dermatol Ges* 6:839–846.

Krammer, B., and K. Plaetzer. 2008. ALA and its clinical impact, from bench to bedside. *Photochem Photobiol Sci* 7:283–289.

Lange, N., P. Jichlinski, M. Zellweger et al. 1999. Photodetection of early human bladder cancer based on the fluorescence of 5-aminolevulinic acid hexylester-induced protoporphyrin IX. *Br J Cancer* 80:185–193.

Maisch, T., R. Szeimies, G. Jori et al. 2004. Antibacterial photodynamic therapy in dermatology. *Photochem Photobiol Sci* 3:907–917.

Malik, Z., and M. Djaldetti. 1979. 5-Aminolevulinic acid stimulation of porphyrin and hemoglobin synthesis by uninduced Friend erythroleukemic cells. *Cell Differ* 8:223–233.

Malik, Z., and H. Lugaci. 1987. Destruction of erythroleukaemic cells by photoactivation of endogenous porphyrins. *Br J Cancer* 56:589–595.

Mikolajewska, P., R. Donnelly, M. Garland et al. 2010. Microneedle pre-treatment of human skin improves 5-aminolevulininc acid (ALA)- and 5-aminolevulinic acid methyl ester (MAL)-induced PpIX production for topical photodynamic therapy without increase in pain or erythema. *Pharm Res* 27:2213–2220.

Miller, I., J. Nielsen, S. Lophaven et al. 2011. Factors related to pain during routine photodynamic therapy: A descriptive study of 301 patients. *J Eur Acad Dermatol-Venereol* 10:3083.

Moan, J., and R. Bissonnette. 2001. Skin preparation. Patent 10/275,557 US 6,911,194 B2.

Moan, J., V. Iani, and L. Ma. 1996. Choice of the proper wavelength for photochemotherapy. *Proc SPIE* 2625:544–549.

Monfrecola, G., E. Procaccini, D. D'Onofrio et al. 2002. Hyperpigmentation induced by topical 5-aminolaevulinic acid plus visible light. *J Photochem Photobiol B* 68:147–155.

Nielsen, K., A. Juzeniene, P. Juzenas et al. 2005. Choice of optimal wavelength for PDT: The significance of oxygen depletion. *Photochem Photobiol* 81:1190–1194.

Nitzan, Y., M. Salmon-Divon, E. Shporen et al. 2004. ALA induced photodynamic effects on Gram positive and negative bacteria. *Photochem Photobiol Sci* 3:430–435.

Ohgari, Y., Y. Nakayasu, S. Kitajima et al. 2005. Mechanisms involved in delta-aminolevulinic acid (ALA)-induced photosensitivity of tumor cells: Relation of ferrochelatase and uptake of ALA to the accumulation of protoporphyrin. *Biochem Pharmacol* 71:42–49.

Peng, Q., T. Warloe, K. Berg et al. 1997. 5-Aminolevulinic acid-based photodynamic therapy: Clinical research and future challenges. *Cancer* 79:2282–2308.

Ponka, P. 1999. Cell biology of heme. *Am J Med Sci* 318:241–256.

Pottier, R., Y. Chow, J. LaPlante et al. 1986. Non-invasive technique for obtaining fluorescence excitation and emission spectra in vivo. *Photochem Photobiol* 44:679–687.

Radakovic-Fijan, S., U. Blecha-Thalhammer, V. Schleyer et al. 2005. Topical aminolaevulinic acid-based photodynamic therapy as a treatment option for psoriasis? Results of a randomized, observer-blinded study. *Br J Dermatol* 152:279–283.

Sakamoto, F., J. Lopes, and R. Anderson. 2010. Photodynamic therapy for acne vulgaris: A critical review from basics to clinical practice: Part I. Acne vulgaris: When and why consider photodynamic therapy? *J Am Acad Dermatol* 63:183–193.

Sandberg, C., B. Stenquist, I. Rosdahl et al. 2006. Important factors for pain during photodynamic therapy for actinic keratosis. *Acta Derm Venereol* 86:404–408.

Sharfaei, S., P. Juzenas, J. Moan et al. 2002. Weekly topical application of methyl aminolevulinate followed by light exposure delays the appearance of UV-induced skin tumours in mice. *Arch Dermatol Res* 294:237–242.

Souza, C., A. Neves, L. Felicio et al. 2007. Optimized photodynamic therapy with systemic photosensitizer following debulking technique for nonmelanoma skin cancers. *Dermatol Surg* 33:194–198.

Strauss, W., R. Sailer, H. Schneckenburger et al. 1997. Photodynamic efficacy of naturally occurring porphyrins in endothelial cells in vitro and microvasculature in vivo. *J Photochem Photobiol B* 39:176–184.

Tyrrell, J., S. Campbell, and A. Curnow. 2011. The effect of air cooling pain relief on protoporphyrin IX photobleaching and clinical efficacy during dermatological photodynamic therapy. *J Photochem Photobiol B* 103:1–7.

Warren, C., L. Karai, A. Vidimos et al. 2009. Pain associated with aminolevulininc acid-photodynamic therapy of skin disease. *J Am Acad Dermatol* 61:1033–1043.

Williams, A. 2003. *Transdermal and Topical Drug Delivery*. Pharmaceutical Press, London.

Zakhary, K., and D. Ellis. 2005. Applications of aminolevulinic acid-based photodynamic therapy in cosmetic facial plastic practices. *Facial Plast Surg* 21:110–116.

Genetically Encoded Photosensitizers: Structure, Photosensitization Mechanisms, and Potential Application to Photodynamic Therapy

Cristina Flors
Madrid Institute for Advanced Studies in Nanoscience (IMDEA Nanociencia)

Santi Nonell
IQS School of Engineering, Universitat Ramon Llull

24.1 Introduction

The ability to control the location of a photosensitizer is crucial for inflicting photodynamic damage in a desired site. Chemical functionalization of a photosensitizer can influence its intracellular location as a consequence of solubility and/or complex formation with a specific molecule or macromolecule (Ogilby 2010). However, nonspecific binding is difficult to avoid and can lead to uncontrolled photodamage. Strategies to tag specific proteins with photosensitizers (Jing and Cornish 2011; Keppler and Ellenberg 2009) or to genetically encode photosensitizer binding sequences (Tour et al. 2003) have been realized, but these methods still need the external addition of the photosensitizer, which does not completely solve the issue of nonspecific binding. On the other hand, fully genetically encoded photosensitizers can be fused to virtually any protein, are expressed by a cell without the need to add any external cofactors, have the best possible target specificity, and thus provide absolute control of photosensitizer location. Genetically encoded photosensitizers were first based on green or red fluorescent proteins (GFPs, RFPs) derived from *Aequorea* jellyfish and other marine organisms. Some variants of these proteins, which are typically used in the context of fluorescence microscopy and biosensing, have

also been specifically evolved to efficiently generate reactive oxygen species (ROS) for their use in chromophore-assisted light inactivation (CALI) and other applications. KillerRed is a phototoxic protein derived from a GFP-like hydrozoan chromoprotein (Bulina et al. 2006a). This protein is composed of two 27 kDa units, is excited with 540–580 nm light, and was initially thought to photosensitize singlet oxygen (Bulina et al. 2006a), although it is now established that oxygen radicals, probably superoxide, and not singlet oxygen, are the main photosensitized ROS (Serebrovskaya et al. 2009; Vegh et al. 2011). Although some studies had previously investigated the issue of ROS photosensitization by fluorescent proteins (FPs), typically in the context of photobleaching in fluorescence microscopy (Bell et al. 2003; Dixit and Cyr. 2003; Greenbaum et al. 2000), the appearance of KillerRed has arguably catalyzed the study of ROS photosensitization by FPs at the molecular level. As we will discuss later, an increasing number of scientific literature is providing a more detailed view of the mechanistic aspects of ROS photosensitization and the relationship with the structure of these proteins.

More recently, another singlet oxygen photosensitizing protein, miniSOG, has been engineered from a phototropin photoreceptor (Shu et al. 2011). MiniSOG, which is a 15 kDa flavoprotein not structurally related to GFP, has much higher singlet oxygen photosensitization efficiency than GFP-like proteins and thus promises to revolutionize all applications related to genetically encoded singlet oxygen production.

So far, these applications have focused on CALI, correlative light and electron microscopy (CLEM), optogenetics, and photodynamic inactivation of bacteria (see Section 24.3). However, other potential applications have been proposed for them: importantly, photodynamic therapy (Bulina et al. 2006b; Shirmanova et al. 2013). This strategy could be achieved by appropriate gene-transfer methods, or by using fusions of the photosensitizers with antibodies to create "genetically encoded immunophotosensitizers" (Serebrovskaya et al. 2009).

In this chapter, we provide a molecular view of the photosensitization mechanisms in the different proteins, relating their photophysical behavior to their structure. We then discuss in detail the applications of genetically encoded photosensitizers mentioned above, including the prospects of their use in photomedicine.

24.2 Photosensitization Mechanisms: Superoxide Versus Singlet Oxygen (Structural Aspects)

As mentioned above, initial reports suggested that the main mechanism for KillerRed phototoxicity was the light-induced formation of singlet oxygen. However, later evidence points to a more important role of a type I mechanism in its phototoxicity (Serebrovskaya et al. 2009; Vegh et al. 2011). By using a free radical fluorescent probe and electronic paramagnetic resonance, it was shown that irradiation of KillerRed produced oxygen radicals, most probably superoxide, and hydrogen peroxide (Vegh et al. 2011). This is consistent with the observation that D_2O decreases phototoxicity compared to H_2O (Serebrovskaya et al. 2009), which is the opposite trend to that expected for a type II photosensitizer. The structural reasons for the mechanism of KilledRed's phototoxicity are still unclear, but crystallographic data have provided some interesting insight. A KillerRed monomer (it is dimeric at physiological conditions) has a typical β-barrel structure, consisting of one β-sheet with alpha helices containing the covalently bonded chromophore (autocatalytically formed from the sequence Gln65-Tyr66-Gly67) running through the center, similarly to other GFP-like proteins. A water-filled channel that connects the chromophore with the external solvent is the most remarkable feature of KillerRed's structure (Carpentier et al. 2009; Pletnev et al. 2009) (Figure 24.1a). Although this channel is also present in non-phototoxic FPs, it has been linked to an increased phototoxicity of KillerRed as it may facilitate the transit of molecular oxygen and ROS in and out of the protein beta-can (Carpentier et al. 2009; Pletnev et al. 2009; Roy et al. 2010). Electron transfer from the excited-state chromophore to molecular oxygen would produce a superoxide anion radical (Carpentier et al. 2009), which would, in turn, be (at least partially) dismutated into molecular oxygen and hydrogen peroxide (Vegh et al. 2011).

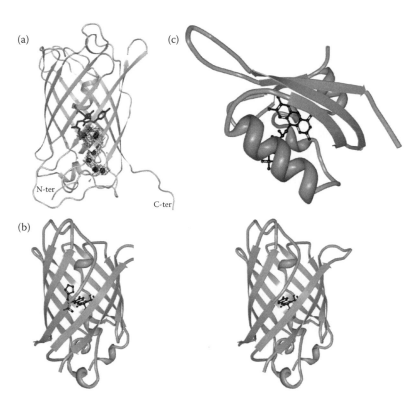

FIGURE 24.1 (a) X-ray structure of a KillerRed monomer showing the protein backbone in gray, and the chromophore and channel with water molecules in black. (Adapted from Carpentier, P. et al., *FEBS Lett.*, 583, 2839–2842, 2009. With permission.) (b) Blocked (left) and unblocked (right) channel in the beta barrel of S65T GFP resulting from the presence and absence, respectively, of the bulky amino acid His148 (c) Structure of miniSOG.

In contrast, the photosensitization by other GFP-like proteins, such as enhanced-GFP (EGFP) and TagRFP, mainly produces singlet oxygen. Self-sensitization of singlet oxygen by EGFP was suggested early on in the context of its participation in fluorescence photobleaching (Bell et al. 2003; Greenbaum et al. 2000). More recent results also point to singlet oxygen as being mainly responsible for the CALI action of EGFP as judged by the reduction of the CALI effect in the presence of sodium azide (McLean et al. 2009).

In 2008, direct spectroscopic evidence of singlet oxygen photosensitized by EGFP was provided for the first time (Jimenez-Banzo et al. 2008). Time-resolved detection of the phosphorescence of singlet oxygen at 1270 nm allowed the study of the kinetics of singlet oxygen formation and decay sensitized by EGFP and compare it with its synthetic chromophore 4-hydroxybenzylidene-1,2-dimethylimidazoline (HBDI). Singlet oxygen photosensitized by EGFP has a relatively short lifetime (4 μs compared to 20 μs for HBDI), indicating significant quenching by the surrounding protein residues. Moreover, the exceptionally long triplet lifetime of EGFP of approximately 25 μs in air-saturated solutions (compared to 3 μs for HBDI) highlights the role of the protein β-barrel in preventing oxygen access to the internally buried chromophore. By introducing specific mutations in GFP, subsequent work has highlighted the effect of β-barrel rigidity and structure on the singlet oxygen photosensitization and triplet state properties of these proteins (Jimenez-Banzo et al. 2010). For example, replacement of His148 by a less bulky residue results in enhanced singlet oxygen photosensitization efficiency resulting from better access of molecular oxygen to the chromophore (Figure 24.1b).

Because of the weak phosphorescence signals recorded at 1270 nm for GFP proteins, it was not possible to estimate a quantum yield for the photosensitization of singlet oxygen (Φ_Δ). A specific green fluorescent probe for singlet oxygen detection was used instead to estimate Φ_Δ for an RFP, TagRFP (Ragas et al. 2011). A Φ_Δ of 0.004 was measured for TagRFP. This is the first estimation of Φ_Δ for an FP and also provides a lower limit for the quantum yield of triplet state formation (Φ_T). Moreover, time-resolved spectroscopic data on the triplet state of TagRFP was used to extract structural information. A triplet lifetime of 3 μs indicates relatively high oxygen accessibility to the chromophore compared with EGFP. By putting these results in the context of other evidence obtained by x-ray crystallography (Subach et al. 2010) and molecular dynamics (Roy et al. 2010), a picture of the molecular events involved in the production and fate of singlet oxygen sensitized by FPs is starting to emerge. We have suggested that the presence of a pore in the β-barrel (as in KillerRed) is not the only mechanism for protein permeability to molecular oxygen and ROS and that transient breathing of the protein chains is also important in the overall picture of FP phototoxicity (Ragas et al. 2011).

The low efficiency of singlet oxygen production by GFP-like proteins has turned the attention of scientists to a different protein family for which a higher ability to photosensitize singlet oxygen is expected, that is, the light-oxygen-voltage 2 (LOV2)

flavin-binding domain from phototropin, a plant photoreceptor. Flavin efficiently produces singlet oxygen and exists in all cells, but LOV2 diverts light energy absorbed by flavin to create a covalent bond with a cysteine. To prevent this, cysteine was replaced by glycine to create miniSOG (Shu et al. 2011), a 15 kDa monomeric photosensitizing and green fluorescent flavoprotein (106 amino acids, half of the size of KillerRed). The nature of the chromophore and the absence of the β-barrel are the most obvious structural differences between MiniSOG and GFP-like proteins (Figure 24.1c). MiniSOG has absorption peaks at 448 and 473 nm (ε = 17,000 and 14,000 M⁻¹cm⁻¹, respectively), fluorescences in the green, and a reported Φ_Δ of 0.47, two orders of magnitude higher than TagRFP, the only GFP-like protein for which such a parameter has been reported.

24.3 Applications of Genetically Encoded Photosensitizers

24.3.1 Chromophore-Assisted Laser Inactivation

Early applications of genetically encoded photosensitizers were in chromophore-assisted light inactivation (CALI), which is a technique employed to study protein function in living cells by specifically inactivating the target protein through photochemical damage (Jacobson et al. 2008; Wang and Jay 1996). CALI is generally achieved by the irradiation of neighboring organic dyes that are able to produce ROS. However, nonspecific binding of the dyes induces uncontrolled photodamage. An alternative approach was the use of FPs, which typically produce smaller CALI effects but can be fully genetically encoded to avoid nonspecific damage. The usefulness of FPs for CALI has been demonstrated for some time using the EGFP both with one- and two-photon excitation (Monier et al. 2010; Ou et al. 2010; Raijfur et al. 2002; Surrey et al. 1998; Tanabe et al. 2005; Vitriol et al. 2007). For example, it has been shown that CALI of EGFP-tagged myosin is a useful tool to control and understand the role of this protein in asymmetric cell division (Ou et al. 2010). Examples of CALI mediated by KillerRed, which has been estimated to be sevenfold more efficient for CALI than EGFP (Bulina et al. 2006a), include the loss of membrane targeting of a lipid-interacting domain of phospholipase C δ1 (Bulina et al. 2006a), inhibition of cell-cycle progression (Serebrovskaya et al. 2011), and obstruction of water transport by aquaporin inactivation (Baumgart, Rossi, and Verkman 2012).

24.3.2 Optogenetics

Optogenetics is the combination of optical and genetic methods to manipulate and induce specific events in targeted cells in living tissue (Deisseroth 2011). Optogenetic manipulation of zebra fish expressing membrane-tagged KillerRed has been recently achieved in proof-of-principle experiments (Teh et al. 2010). It was possible to induce decreased cardiac output and subsequent pericardial edema by KillerRed-mediated ROS production in the heart of the zebra fish in a dose-dependent manner. In another study, destruction of specific KillerRed-tagged interneurons, responsible for responses to large visual stimuli, resulted in the loss of ability of zebra fish to catch prey (DelBene et al. 2010). A recent paper elegantly shows the use of miniSOG for cell ablation and suggests that it may become a valuable tool to study the function of specific cells within complex cellular networks (Qi 2012).

24.3.3 CLEM

The combination of fluorescence and electron microscopy allows the study of biological specimens at high spatial resolution and with information about molecular identity. CLEM makes use of fluorescent probes that are capable of generating contrast for electron microscopy (Giepmans 2008). Singlet oxygen generated by these probes can photooxidize 3,3′-diaminobenzidine (DAB) into an osmiophilic polymer. Addition of osmium tetroxide to a fixed sample produces an electron-opaque stain that then allows imaging by electron microscopy. Genetically encodable FPs that can generate singlet oxygen are thus highly sought tools for CLEM. EGFP has been used for CLEM in a number of studies (Grabenbauer et al. 2005; Meisslitzer-Ruppitsch et al. 2008; Monosov et al. 1996), although its low Φ_Δ has precluded more extensive use. MiniSOG, which was specifically developed with CLEM in mind, with a reported Φ_Δ of 0.47 constitutes an important breakthrough (Shu et al. 2011). It has been suggested that "miniSOG may do for EM what GFP did for fluorescence microscopy" Shu et al.

24.3.4 Photodynamic Inactivation of Microorganisms

Photodynamic inactivation of microorganisms is a promising alternative to antibiotics as it is capable of eliminating antibiotic-resistant pathogens, and, in turn, it is very unlikely to elicit the selection of resistant strains given its mode of action. Most conventional, exogenously delivered photosensitizers tend to accumulate both in the pathogens and in the surrounding healthy tissues, causing unwanted phototoxicity. Genetically encoding a photosensitizer is an attractive option as it would enhance the selectivity of photodynamic treatments. Fully genetically encoded photosensitizers have been shown to achieve photodynamic inactivation of bacteria. KillerRed was able to kill 96% of *Escherichia coli* cells after 10 min of irradiation with white light at 1 W/cm² (Bulina et al. 2006a) through a type I photodynamic mechanism (Serebrovskaya et al. 2009; Vegh et al. 2011).

More recently, singlet oxygen was shown to be the main cytotoxic species in the photodynamic inactivation of TagRFP-expressing *E. coli* (Ruiz-González et al.). Consistent with the low photosensitization efficiency of TagRFP ($\Phi_\Delta = 0.004$), a considerable light dose of 3200 J/cm² at 532 nm was needed for a $3.7\log_{10}$ population reduction in a deuterated medium ($2.4\log_{10}$ in aqueous medium) (Figure 24.2). A similar experiment was performed with *E. coli* cells expressing miniSOG in the cytosol. A population reduction of over 1-$\log_{10}$ units was achieved even after a light doses as mild as 2.5 J·cm⁻² and up to 3.5-$\log_{10}$

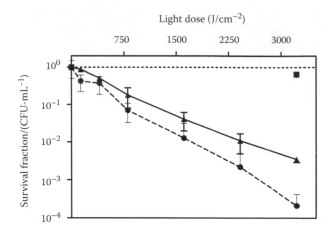

FIGURE 24.2 Photoinactivation of *E. coli* expressing TagRFP in the cytosol upon irradiation with 532-nm laser light in aqueous (solid line) and deuterated medium (broken line). Control with bacteria that do not express TagRFP (black squares). (Adapted from Ruiz-González, R. et al, *Photochem. Photobiol. Sci.*, 11, 1411–1413, 2012. With permission.)

units was observed after 12 J·cm⁻² treatment. Such a light dose is orders of magnitude lower than that used in previous bacterial photodynamic inactivation treatments with red fluorescent proteins and comparable to the dose of KillerRed. As with TagRFP, an enhancement of photoinduced cell death was observed when the experiment was performed in deuterated medium (Ruiz-González et al. 2013).

Mechanistic tests suggested that photodamage seems to occur primarily in the inner membrane, and extends to the outer membrane if the photosensitizer's efficiency is high enough (Ruiz-González et al. 2013). These observations are markedly different from those reported for an external photosensitizer (Spesia et al. 2009) and highlight that the site where singlet oxygen is primarily generated proves crucial for inflicting different types of cell damage. This work showed for the first time that intracellular generation of singlet oxygen is sufficient to kill bacteria strictly from the inside (Ruiz-González et al. 2013).

24.4 Conclusions and Outlook

Fully genetically encoded photosensitizers provide absolute control of photosensitizer location and are therefore ideal tools for any study or application that requires spatial control of ROS location. EGFP and TagRFP have been shown to sensitize singlet oxygen with low, but not insignificant, efficiency. On the other hand, it is now generally accepted that KillerRed acts by a type I mechanism, photosensitizing superoxide anion or hydrogen peroxide. The relationship between protein structure and photosensitization properties is still vague, but it appears that the access of molecular oxygen to the chromophore and the exit of ROS needs to be facilitated by at least two factors: some kind of channel in the β-barrel and/or transient breathing of the protein structure that results in permeability of the β-barrel to molecular oxygen.

Although the choice of genetically encoded photosensitizers is still scarce and the mechanistic details of ROS photosensitization are still largely unknown, their potential benefits for a wide range of applications, such as CALI, CLEM, and optogenetics, may catalyze the development of new and improved variants. From the photophysical point of view, future developments in genetically encodable photosensitizers should include a wider choice of excitation wavelength, which may enable two-color applications, for example, in CALI or optogenetics. Although TagRFP has a relatively high value compared to other orange-red FPs (Drobizhev et al. 2009), higher two-photon absorption cross sections should also be useful. Higher efficiencies of ROS photosensitization might also be of interest, although a ROS production that is too high may be counterproductive because of photosensitizer photobleaching or cell photodamage in ambient light. A better understanding is needed of the structural basis of the different photosensitization properties in terms of efficiency and type of ROS. Indeed, a more detailed knowledge of the identity of the ROS generated by each photosensitizing protein is necessary to further exploit them for the study of downstream effects of locally increased levels of ROS. Initial studies using KillerRed to induce oxidative stress in the peroxisomal matrix in a controlled spatiotemporal manner show the promise of this kind of approach (Ivashchenko et al. 2011).

Although the potential of fully genetically encodable photosensitizers was limited until recently because of the minute efficiency of ROS production, this situation has recently changed as a result of the development of KillerRed and miniSOG. The higher photosensitization efficiency allows the use of relatively lower light doses. In the context of photodynamic therapy of cancer, genetically encodable immunophotosensitizers, which avoid the challenges of chemical bioconjugation of photosensitizer drugs with antibodies, may be further explored. So far, efficient elimination of the ovarian carcinoma cell line SKOV-3 has been shown after incubation with an immunophotosensitizer that consists of a genetic fusion of KillerRed with a 4D5 single-chain Fv fragment antibody against p185 (a typical cancer marker) and subsequent irradiation with white light (Serebrovskaya et al. 2009). More recently, the phototoxic effects of KillerRed on mice tumors has been shown (Shirmanova et al. 2013). In this work, HeLa Kyoto cells stably expressing KillerRed (in the mitochondria or the nucleus) were injected into mice to form the tumor. This study represents a proof-of-principle for photodynamic therapy of cancer using genetically-encoded photosensitizers.

For photodynamic therapy of cancer or photoinactivation of bacteria, suitable gene-transfer methods will be necessary to exploit the full potential of this approach. Delivery of the photosensitizer (or immunophotosensitizer) gene into a bacterial or cancer cell using viral vectors (Yu et al. 2003, 2004) would allow very specific, light-induced cell destruction.

References

Baumgart, F., A. Rossi, and A. S. Verkman. 2012. Light inactivation of water transport and protein-protein interactions of aquaporin-Killer Red chimeras. *J Gen Physiol* 139:83–91.

Bell, A. F., D. Stoner-Ma, R. M. Wachter, and P. J. Tonge. 2003. Light-driven decarboxylation of wild-type green fluorescent protein. *J Am Chem Soc* 125:6919–6926.

Bulina, M. E., D. M. Chudakov, O. V. Britanova et al. 2006a. A genetically encoded photosensitizer. *Nat Biotechnol* 24:95–99.

Bulina, M. E., K. A. Lukyanov, O. V. Britanova et al. 2006b. Chromophore-assisted light inactivation (CALI) using the phototoxic fluorescent protein KillerRed. *Nat Protoc* 1:947–953.

Carpentier, P., S. Violot, L. Blanchoin, and D. Bourgeois. 2009. Structural basis for the phototoxicity of the fluorescent protein KillerRed. *FEBS Lett* 583:2839–2842.

Deisseroth, K. 2011. Optogenetics. *Nat Methods* 8:26–29.

Del Bene, F., C. Wyart, E. Robles et al. 2010. Filtering of visual information in the tectum by an identified neural circuit. *Science (New York)* 330:669–673.

Dixit, R., and R. Cyr. 2003. Cell damage and reactive oxygen species production induced by fluorescence microscopy: Effect on mitosis and guidelines for non-invasive fluorescence microscopy. *Plant J* 36:280–290.

Drobizhev, M., S. Tillo, N. S. Makarov, T. E. Hughes, and A. Rebane. 2009. Absolute two-photon absorption spectra and two-photon brightness of orange and red fluorescent proteins. *J Phys Chem B* 113:855–859.

Giepmans, B. N. 2008. Bridging fluorescence microscopy and electron microscopy. *Histochem Cell Biol* 130:211–217.

Grabenbauer, M., W. J. C. Geerts, J. Fernandez-Rodriguez et al. 2005. Correlative microscopy and electron tomography of GFP through photooxidation. *Nat Methods* 2:857–862.

Greenbaum, L., C. Rothmann, R. Lavie, and Z. Malik. 2000. Green fluorescent protein photobleaching: A model for protein damage by endogenous and exogenous singlet oxygen. *Biol Chem* 381:1251–1258.

Ivashchenko, O., P. P. Van Veldhoven, C. Brees et al. 2011. Intraperoxisomal redox balance in mammalian cells: Oxidative stress and interorganellar cross-talk. *Mol Biol Cell* 22:1440–1451.

Jacobson, K., Z. Rajfur, E. Vitriol, and K. Hahn. 2008. Chromophore-assisted laser inactivation in cell biology. *Trends Cell Biol* 18:443–450.

Jimenez-Banzo, A., S. Nonell, J. Hofkens, and C. Flors. 2008. Singlet oxygen photosensitization by EGFP and its chromophore HBDI. *Biophys J* 94:168–172.

Jimenez-Banzo, A., X. Ragas, S. Abbruzzetti et al. 2010. Singlet oxygen photosensitisation by GFP mutants: Oxygen accessibility to the chromophore. *Photochem Photobiol Sci* 9:1336–1341.

Jing, C. and V. W. Cornish. 2011. Chemical tags for labeling proteins inside living cells. *Acc Chem Res* 44:784–792.

Keppler, A., and J. Ellenberg. 2009. Chromophore-assisted laser inactivation of alpha- and gamma-tubulin SNAP-tag fusion proteins inside living cells. *ACS Chem Biol* 4:127–138.

McLean, M. A., Z. Rajfur, Z. Z. Chen et al. 2009. Mechanism of chromophore assisted laser inactivation employing fluorescent proteins. *Anal Chem* 81:1755–1761.

Meisslitzer-Ruppitsch, C., M. Vetterlein, H. Stangl et al. 2008. Electron microscopic visualization of fluorescent signals in cellular compartments and organelles by means of DAB-photoconversion. *Histochem Cell Biol* 130:407–419.

Monier, B., A. Pelissier-Monier, A. H. Brand, and B. Sanson. 2010. An actomyosin-based barrier inhibits cell mixing at compartmental boundaries in *Drosophila* embryos. *Nat Cell Biol* 60–65(suppl.):61–69.

Monosov, E. Z., T. J. Wenzel, G. H. Luers, J. A. Heyman, and S. Subramani. 1996. Labeling of peroxisomes with green fluorescent protein in living *P. pastoris* cells. *J Histochem Cytochem* 44:581–589.

Ogilby, P. R. 2010. Singlet oxygen: There is still something new under the sun, and it is better than ever. *Photochem Photobiol Sci* 9:1543–1560.

Ou, G., N. Stuurman, M. D'Ambrosio, and R. D. Vale. 2010. Polarized myosin produces unequal-size daughters during asymmetric cell division. *Science (New York)* 330:677–680.

Pletnev, S., N. G. Gurskaya, N. V. Pletneva et al. 2009. Structural basis for phototoxicity of the genetically encoded photosensitizer KillerRed. *J Biol Chem* 284:32028–32039.

Qi, Y. B., E. J. Garren, X. Shu, R. Y. Tsien, and Y. Jin. 2012. Photo-inducible cell ablation in *Caenorhabditis elegans* using the genetically encoded singlet oxygen generating protein miniSOG. *Proc Natl Acad Sci* 109:7499–7504.

Ragas, X., L. P. Cooper, J. H. White, S. Nonell, and C. Flors. 2011. Quantification of photosensitized singlet oxygen production by a fluorescent protein. *Chemphyschem* 12:161–165.

Raijfur, Z., P. Roy, C. Otey, L. Romer, and K. Jacobson. 2002. Dissecting the link between stress fibres and focal adhesions by CALI with EGFP fusion proteins. *Nat Cell Biol* 4:286–293.

Roy, A., P. Carpentier, D. Bourgeois, and M. Field. 2010. Diffusion pathways of oxygen species in the phototoxic fluorescent protein KillerRed. *Photochem Photobiol Sci* 9:1342–1350.

Ruiz-González, R., J. H. White, A. L. Cortajarena, M. Agut, S. Nonell and C. Flors. 2013. Fluorescent proteins as singlet oxygen photosensitizers: Mechanistic studies in photodynamic inactivation of bacteria. *Proc SPIE* 8596:859609.

Ruiz-González, R., J. H. White, M. Agut, S. Nonell, and C. Flors. 2012. A genetically-encoded photosensitiser demonstrates killing of bacteria by purely endogenous singlet oxygen. *Photochem Photobiol Sci* 11:1411–1413.

Serebrovskaya, E. O., E. F. Edelweiss, O. A. Stremovskiy et al. 2009. Targeting cancer cells by using an antireceptor antibody-photosensitizer fusion protein. *Proc Natl Acad Sci USA* 106:9221–9225.

Serebrovskaya, E. O., T. V. Gorodnicheva, G. V. Ermakova et al. 2011. Light-induced blockage of cell division with a chromatin-targeted phototoxic fluorescent protein. *Biochem J* 435:65–71.

Shirmanova, M. V., E. O. Serebrovskaya, K. A. Lukyanov, L. B. Snopova, M. A. Sirotkina, N. N. Prodanetz, M. L. Bugrova, E. A. Minakova, I. V. Turchin, V. A. Kamensky, S. A. Lukyanov and E. V. Zagaynova. 2013. Phototoxic effects of fluorescent protein KillerRed on tumor cells in mice. *J. Biophotonics* 6:283–290.

Shu, X., V. Lev-Ram, T. J. Deerinck et al. 2011. A genetically encoded tag for correlated light and electron microscopy of intact cells, tissues, and organisms. *PLoS Biol* 9:e1001041.

Spesia, M. B., D. A. Caminos, P. Pons, and E. N. Durantini. 2009. Mechanistic insight of the photodynamic inactivation of Escherichia coli by a tetracationic zinc(II) phthalocyanine derivative. *Photodiagnosis Photodyn Ther* 6:52–61.

Subach, O. M., V. N. Malashkevich, W. D. Zencheck et al. 2010. Structural characterization of acylimine-containing blue and red chromophores in mTagBFP and TagRFP fluorescent proteins. *Chem. Biol* 17:333–341.

Surrey, T., M. B. Elowitz, P.-E. Wolf et al. 1998. Chromophore-assisted light inactivation and self-organization of microtubules and motors. *Proc Natl Acad Sci USA* 95:4293–4298.

Tanabe, T., M. Oyamada, K. Fujita et al. 2005. Multiphoton excitation-evoked chromophore-assisted laser inactivation using green fluorescent protein. *Nat Methods* 2:503–505.

Teh, C., D. M. Chudakov, K. L. Poon et al. 2010. Optogenetic in vivo cell manipulation in KillerRed-expressing zebrafish transgenics. *BMC Dev Biol* 10:110.

Tour, O., R. M. Meijer, D. A. Zacharias, S. R. Adams, and R. Y. Tsien. 2003. Genetically targeted chromophore-assited light inactivation. *Nat Biotechnol* 21:1505–1508.

Vegh, R. B., K. M. Solntsev, M. K. Kuimova et al. 2011. Reactive oxygen species in photochemistry of the red fluorescent protein 'Killer Red'. *Chem Commun (Camb)* 47:4887–4889.

Vitriol, E. A., A. C. Uetrecht, F. M. Shen, K. Jacobson, and J. E. Bear. 2007. Enhanced EGFP-chromophore-assisted laser inactivation using deficient cells rescued with functional EGFP-fusion proteins. *Proc Natl Acad Sci USA* 104:6702–6707.

Wang, F. S., and D. G. Jay. 1996. Chromophore-assisted laser inactivation (CALI): Probing protein function in situ with a high degree of spatial and temporal resolution. *Trends Cell Biol* 6:442–445.

Yu, Y. A., S. Shabahang, T. M. Timiryasova et al. 2004. Visualization of tumors and metastases in live animals with bacteria and vaccinia virus encoding light-emitting proteins. *Nat Biotechnol* 22:313–320.

Yu, Y. A., T. Timiryasova, Q. Zhang, R. Beltz, and A. A. Szalay. 2003. Optical imaging: Bacteria, viruses, and mammalian cells encoding light-emitting proteins reveal the locations of primary tumors and metastases in animals. *Anal Bioanal Chem* 377:964–972.

25

Light Dosimetry for Photodynamic Therapy: Basic Concepts

Henricus J. C. M.
Sterenborg
Erasmus Medical Center

Robert L. P. van Veen
Erasmus Medical Center

Jan-Bonne Aans
Erasmus Medical Center

Arjen Amelink
Erasmus Medical Center

Dominic J. Robinson
Erasmus Medical Center

25.1 Introduction

Research on photodynamic therapy (PDT) dosimetry has been going on for more than three decades, and a lot is still going on (Wilson and Patterson 2008). Nevertheless, in clinical PDT or in preclinical experimentation, light dosimetry is still not commonly embraced. The main reason for this may be that photomedicine is a multidisciplinary field, and the high levels of physics, mathematics, and optical technology are not easily accessible to all the disciplines involved in PDT.

The present chapter aims to provide an introduction to the basic concepts of light dosimetry for PDT that is both accessible and complete.

25.2 Light Transportation in Tissue

The two dominant processes of interaction between light and tissue are absorption of light and elastic light scattering. Several other interactions can occur, such as inelastic scattering (Raman scattering) and nonlinear effects, but these will not be considered in this chapter as they do not influence the transport of light in tissue in a significant way under normal PDT conditions.

25.2.1 Light Absorption

Absorption of light takes place at the molecular level. An absorption event terminates the life of a photon and transfers its energy to the molecule that absorbs the photon. It is characterized by the absorption coefficient μ_a expressed in inverse meters and is defined by Beer's law:

$$I(\ell) = I_0 e^{(-\mu_a \ell)}$$

where I_0 stands for the initial intensity and $I(\ell)$ for the intensity at distance ℓ. The absorption coefficient μ_a is proportional to the concentration of the absorbers, C_a, and the molar absorption coefficient:

$$\mu_a = c_a \mu_{ao} \ [m^{-1}]$$

Different absorbing molecules have different absorption spectra. Common absorbers in tissue are hemoglobin, oxyhemoglobin, water, and fat (Zijlstra, Buursma, and van assendelft 2000; van Veen et al. 2005; Anderson et al. 2006; Nachabé et al. 2010) (Figure 25.1, absorption and scattering spectra). When these molecules absorb a photon, usually all energy is transferred to heat, which is rapidly dispersed into the tissue. In certain molecules, only some of the energy is transferred to heat, and a portion of the energy is reemitted in the form of a new photon with lower energy. This phenomenon is called fluorescence. Also, some of the energy can be used to initiate photochemical reactions, such as those that occur in PDT. The details of these photochemical interactions are described elsewhere in this book.

25.2.2 Elastic Light Scattering

Thinking of light and lasers, the first image that comes to mind is one of rays and beams. In PDT, the reality is very different. The dominant interaction between light and tissue is elastic scattering, an interaction that changes the direction in which an individual photon is traveling. Scattering is caused by local

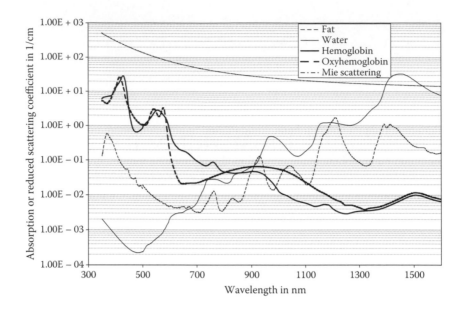

FIGURE 25.1 Illustration of Mie scattering and the absorption coefficients of the four most common absorbing constituents of tissue.

variations in the index of refraction of the tissue. On the basis of the wavelength dependence of the scattering coefficient measured in tissue, one can conclude that the dominant scattering process is Mie scattering. Mie theory applied to the observed wavelength dependence of scattering suggests that index-of-refraction variations of sizes between 100 and 200 nm are responsible for light scattering in tissue (Graaff et al. 1992). The nature of these submicroscopic variations is currently unknown. Mie theory is based on the wave aspect of light, and it derives the angular distribution of scattering probability. The actual scattering direction in a scattering event, however, is not predetermined but a random sampling of the probability density function. Hence, although Mie scattering is highly directed, it is a random phenomenon. The scattering coefficient, μ_s, is also expressed in inverse meters and defined in a similar way to μ_a according to Beer's law (but in the absence of absorption):

$$I(\ell) = I_0 e^{(-\mu_s \ell)}$$

The influence of scattering on the distribution of light is enormous. Although the behavior of individual photons can still be characterized as ballistic—they go straight forward like a bullet with the speed of light—as a group, they behave quite differently. On average, photons undergo multiple scattering events before their life terminates by absorption or by exiting the tissue. Scattering has a strong random component, and all scattering events in the group of photons will be unrelated. As a result, a parallel beam of photons, traveling through a highly scattering medium, quickly changes into a diffuse cloud of light. This phenomenon is depicted in the experiment illustrated in Figure 25.2. It shows a red laser beam traveling though water. By adding milk, a scattering component, the beam is rapidly attenuated, and the light distribution changes from a directed beam into a diffuse

cloud of light. This experiment clearly illustrates that scattering strongly influences light transport. In fact, it limits the penetration of light to a depth much smaller than without scattering.

25.2.3 Diffuse Light

The diffuse nature of light in tissue is a difficult concept to grasp as it works rather counterintuitively. When illuminating an object, we expect the illumination pattern to be determined by the illuminating beam. When illuminating a highly scattering object, the internal light distribution still depends, to some extent, on the shape of the illuminating beam but is also strongly influenced by the optical properties of the medium illuminated.

The most popular way to mathematically describe the light distribution in a highly scattering medium is the diffusion approximation (Jacques and Pogue 2008). It treats light as if it were a diffusing entity that diffuses in all directions (e.g., heat).

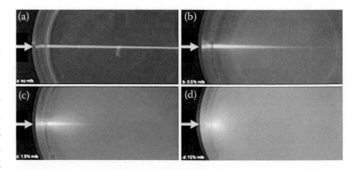

FIGURE 25.2 **(See color insert.)** Illustration of the effect of elastic light scattering on the distribution of light in a scattering medium, a mixture of water and milk in a glass. From no scattering (a) to strong scattering (d), the light distribution changes from a parallel beam to a diffuse cloud.

The diffusion equation can be derived from the Boltzmann transport equation and is based on energy conservation by accounting for the amount of photons in a specific volume at a certain position, $\vec{x}$. In the case of continuous light sources, this further simplifies to the following:

$$\mu_a \phi(\vec{x}) - D\nabla^2 \phi(\vec{x}) = S(\vec{x})$$

where ϕ stands for the fluence rate (W/m^2), μ_a for the absorption coefficient, S for the photon source term, and D for the diffusion coefficient given by the following:

$$D = \frac{1}{3(\mu_a + \mu'_s)}$$

In principle, solutions to the diffusion approximation are only simple in an infinite homogeneous medium with isotropic scattering. Tissue, however, is not homogeneous, and Mie scattering is strongly forward directed, with an average cosine of the scattering angle, g, having values in the range from 0.7 to 0.999 (van Gemert et al. 1989). To cope with this, the diffusion approach uses an effective scattering coefficient, $\mu'_s = (1 - g)\mu_s$. It assumes that $1/(1 - g)$ successive Mie scattering events have the same effect on the light distribution as a single isotropic scattering event. Other complications result from the fact that tissue is far from homogeneous:

- There is a stepwise change in index of refraction at the surface–air interface of tissue, which causes strong internal reflections. The internal reflection coefficient depends on the angle at which the photon hits the internal surface. This angular-dependent internal reflection causes the light to be not perfectly diffuse at the tissue-air boundary. Farrell, Patterson, and Wilson (1992) partially fixed this problem mathematically.
- Blood, the main absorber in tissue, is concentrated in blood vessels. These highly absorbing regions cause local deviations from the calculated diffuse light distribution. As a consequence, some of the absorbing molecules, for example, in the center of a blood vessel, are exposed to less light than others, depending on the size of the blood vessel and the absorption coefficient. This effect shields a part of the absorbers from light and decreases the overall absorption of light. (van Veen, Verkruysse and Sterenborg 2002a) suggested the use of an effective absorption coefficient to compensate for this effect.
- Inhomogeneously distributed absorbers may lead to local "cold spots" in the fluence rate. Specifically, close to blood vessels, we expect the fluence rate to be much lower than that calculated on the basis of homogeneous blood. These local "cold spots" may be of consequence to the PDT response as less light is delivered locally.
- Inhomogeneously distributed scattering may lead to both hot and cold spots in the fluence rate distribution. Such complications are expected at boundaries between tissues of different compositions (Nachabé et al. 2010).

It is technically possible to deal with all these complications when the inhomogeneities are well characterized, both in amplitude and spatial distribution. Using other mathematical approaches, such as finite element methods or Monte Carlo simulations (Wang, Jacques, and Zheng 1995; Dwivedi, Krishnan, and Suryanarayanan 2007), we can then calculate the light distribution in much more detail. The detailed distribution of the optical properties, however, will be different in every patient and for each treated tumor. In fact, it may even change during treatment. Calculation of fluence rate distributions based on detailed information on the distribution the optical properties measured in real time is presently a topic of research (see the following chapter), and currently, such methods are not available for routine use.

In spite of the limitations mentioned, diffusion theory is very useful as a first-order approximation of the light distribution in tissue. It can help to predict the general behavior of light in tissue. An often-used concept is the optical penetration depth, δ. It is defined as

$$\delta = \frac{1}{\sqrt{3\mu_a \mu_{tr}}}$$

with

$$\mu_{tr} = \mu_a + \mu'_s$$

The optical penetration depth indicates over which distance the diffuse fluence rate decreases. The decrease in fluence rate with distance is different in different geometries:

$\phi(z) \cong e^{-z/\delta}$ for a broad beam on a flat surface of a semi infinite medium

$\phi(r) \cong \dfrac{e^{-r/\delta}}{\sqrt{r/\delta}}$ for a infinitely long cylindrical source in an infinite medium

$\phi(r) \cong \dfrac{e^{-r/\delta}}{r/\delta}$ for an isotropic point source in an infinite medium

where z represents the depth below the surface and r the distance to the interstitial source. The absolute value of the fluence rate, of course, depends on the power emitted by the sources or, in the case of surface illumination, the incident irradiance. In addition, factors such as the refractive index and the scattering properties also play an important role. This is illustrated in Figure 25.3, in which we plot the fluence rate as a function of depth based on Monte Carlo simulations (Jacques and Wang 1995). Just as in Figure 25.2, we can clearly see that with scattering, the penetration of light is much less than without scattering. An additional striking feature we can observe here is the so-called subsurface

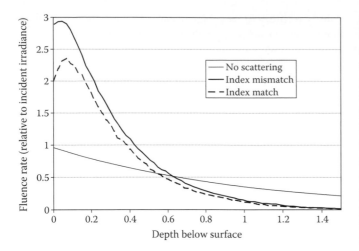

FIGURE 25.3 Fluence rates in tissue under various conditions. Note how scattering increases the fluence rate under the surface and decreases the penetration of light in tissue.

fluence peak. Scattering makes photons bounce around in a small volume, creating a higher photon density than in the incident beam. When a significant step in index of reflection is present at the surface, as in the normal tissue–air interface, the internal diffuse reflection increases the subsurface fluence even further. The fluence rate buildup resulting from scattering is also present when using a linear or spherical interstitial source, but it is less obvious. It is a very important aspect of dosimetry as it locally increases the fluence rate far above what might be expected on the basis of the delivered light.

The optical penetration depth depends on absorption and scattering and, hence, on the composition of tissue. Optical penetration depths calculated for different tissues over a wide wavelength region are given in Figure 25.4. The wavelengths most commonly used in clinical PDT, 630 nm [hematoporphyrin, protoporphyrin IX (PpIX)], 652 nm [meta-tetra(hydroxyphenyl)

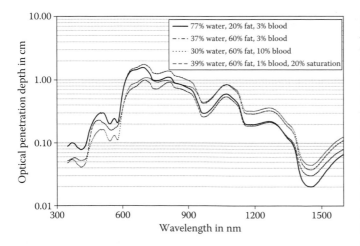

FIGURE 25.4 Optical penetration depths as a function of wavelength calculated using diffusion theory for different tissue compositions.

chlorin(mTHPC), Foscan], and 690 nm (Verteporphyrin), are situated in the region of maximum optical penetration of 1–2 cm. The therapeutic depth, that is, the depth to which the desired therapeutic effect occurs, is related but not equal to the optical penetration depth. In some cases, however, the therapeutic depth is limited by the depth distribution of the photosensitizer. For instance, in topical 5-aminolevulinic acid (ALA) PDT of skin lesions, the 632 nm light used to activate the PpIX has a penetration depth of roughly 5–10 mm. Nevertheless, ALA PDT is only effective for superficial lesions as the deposition of PpIX is limited to a depth of approximately 2 mm. As the distribution of the photosensitizer is here limited to a region less than the optical penetration depth, increasing the fluence will just increase the intensity of the effect up to a maximum possible tissue response. On the other hand, when a photosensitizer is distributed more evenly, such as in systemic administration, the volume in which a certain therapeutic effect occurs increases with increasing fluence. For instance, in interstitial PDT of pig liver, the size of the lesion increased logarithmically with the total fluence (Rovers et al. 2000). Obviously, the therapeutic depth can change with changing therapeutic effectiveness. As we describe here, both fluence rate and light fractionation strongly influence the relationship between fluence and effectiveness.

The optical properties have been documented to potentially vary during PDT. This can be caused by changes in the amount and oxygenation of blood, one of the dominant components in the optics of tissue. These may be induced either directly by PDT, such as by the oxygen consumption by PDT or by PDT-induced vascular effects, or indirectly, such as by patient movement, temperature change, or the effects of local anesthetics. Moreover, it is important to realize that the PDT-induced variations may be different, or even opposite, in different regions of the treated area, either by the different effects induced or simply by different tissue sensitivities. On the positive side, all changes in optical properties can be measured and give us real-time insight into the effect of the treatment on the tissue. These phenomena will be discussed in the following chapter.

25.3 Fluence and Fluence Rate

The choice of light treatment parameters, such as fluence and fluence rate, are clearly important considerations because they are related to the PDT dose that is delivered to tissue and the resulting effectiveness. Fluence rate has the same SI unit as irradiance, W/m^2, but is not limited to being defined over a particular surface, although it can be defined and measured in free space or within a tissue volume. Traditionally, fluence rate is expressed in units of mW/cm^2 and in clinical PDT using 630 and 800 nm typically ranging from 10 to 200 mW/cm^2. Historically, fluence rates were chosen at the upper region of this range to minimize treatment times while avoiding hyperthermia. Groundbreaking theoretical and experimental studies in the early 1990s led to a deeper understanding of the role of oxygen diffusion in PDT (Foster et al. 1991; Foster and Gao 1992; Tromberg et al. 1990) and showed that oxygen depletion is an important factor in fluence rates that are routinely used in PDT. Following studies by other investigators have shown convincing biological conformation of reduced efficacy at a high fluence rate (Chen, Chen, and Hetzel 1996; Robinson et al. 1998; Coutier et al. 2002; Henderson et al. 2004; Busch 2006).

In this section, we will consider the range of effects that come into play when light is delivered across a wide range of fluence rates from very high rates delivered in short pulses through continuous-wave illumination at a moderate fluence rate to a (very) low fluence rate and on to light-fractionated PDT.

Sterenborg and van Gemert (1996) provided a theoretical analysis that showed that short-pulsed laser sources are less effective than continuous-wave sources when the same average power is used. The reason for this effect is that the first portion of each very-high-fluence-rate (i.e., $<4 \times 10(8)$ W m^{-2}) pulse depletes the number of photosensitizer molecules in the ground state so that the remaining energy is not absorbed and does not contribute to the formation of reactive oxygen species. This means that pulsed-light sources with a high energy per pulse (>1 mJ cm^{-2}) are less effective and should be avoided.

At moderate fluence rates, photosensitizer ground state depletion is not a significant effect, but the depletion of oxygen remains critically important. Foster was the first to show that at, for example, 100 mW/cm^2, the local rate of photochemical oxygen consumption can be an order of magnitude higher than the metabolic oxygen consumption rate of any tissues containing typical concentrations of photosensitizer with a modest singlet oxygen quantum yield. It is therefore unsurprising that the vasculature of many tissues is unable to adequately supply this demand for oxygen, and it is this diffusion that limits the generation of reactive oxygen species. An elegant series of theoretical *in vitro* and *in vivo* studies incorporating more complicated effects, such as different photochemical decay modes, photosensitizer photobleaching, and spatially heterogeneous singlet oxygen generation, have resulted in further insights into the importance of oxygen depletion even at moderate fluence rates (Finlay 2004; Wang, Mitra, and Foster 2007).

Both the calculations described in the studies above and the preclinical biological evidence support the use of a lower fluence rate. Only a very small number of clinical studies have investigated the use of a lower fluence rate. This is, in part, because treatment times are substantially increased if equal fluences are to be delivered. The decision to treat for an equal amount of time at a lower fluence (rate) is a challenging concept for institutional board review committees.

The strong influence of oxygen depletion has led some investigators to study light fractionation as a method to allow reoxygenation. The concept here is that dark intervals of a few seconds every few seconds (an on–off approach) overcome oxygen depletion as tissues are reoxygenated in the short dark intervals in the illumination (Curnow, Haller, and Bown 2000). While the details of the role of reoxygenation are important, in particular, the exact fluence rate of the on cycle, these approaches are effectively the same as delivering the light at the average fluence rate.

The considerations described so far are based on the photosensitization of cells in tissues supplied by the local microvasculature. Another critical factor to consider is the situation in which cells of the vasculature are themselves photosensitized to a certain degree. This is a consideration for PDT using the majority of systemically administered, preformed photosensitizers and some topical porphyrin precursors. Here too, strong fluence rate effects have been noted (Busch 2006; Henderson et al. 2004; Cottrell et al. 2008). In this context, the role of the immune response to PDT has been shown to be different at high and low fluence rates (Henderson et al. 2004).

Reducing the fluence rate still further (or using light fractionation with dark intervals of longer than a few 10s of minutes) brings in effects that are not limited to photo physics, photochemistry, and vascular physiology. Cell repair and molecular responses to oxidative stress now become important (Oleinick and Evans 1998; Veenhuizen and Stewart 1995). The mechanism of cell death can shift from necrosis to apoptosis as shown by the groundbreaking work on metronomic PDT (Lilge, Portnoy, and Wilson 2000; Bisland et al. 2004). We have shown that the use of light fractionation with long dark intervals of the order of 2 h can lead to enhanced efficacy that is not related to reoxygenation but in which cells are rendered sensitive to a second light fraction by a small PDT dose in a first light fraction (de Bruijn et al. 2006; de Haas 2006).

In this discussion concerning the choice of and the effects of fluence rate, a critical and often overlooked factor is that the magnitude of the effects described in this section are very highly dependent on the photosensitizer; its photophysical properties; its location, which may be dependent on both the method of application and the interval between administration; and illumination. Care should be taken when generalizing between photosensitizers and the specific application. Differences in the relationship between efficacy, fluence, and fluence rate for different photosensitizers is perfectly illustrated by comparing the efficacy of Photofrin and Foscan (Mitra and Foster 2005).

25.4 Practical Dosimetry: Light Distributions in PDT Treatments

25.4.1 Devices

Two important fiber-optic devices developed for PDT light delivery and dosimetry are the linear diffusor and the isotropic diffusor (Figure 25.5). Isotropic diffusors are predominantly used for measuring diffuse light. They are equally sensitive to light from all directions. Calibration of a diffuse tip can be done in a calibrated integrating sphere, preferably in water, as this resembles the situation in tissue. Special attention has to be given to the use of diffuse tips for measuring the fluence rate on a surface of a diffuse medium (Murrer, Marijnissen, and Star 1995). Diffuse tips can also be used as a diffuse point source. More common diffuse sources are linear diffusers. They are available in various lengths. The output of a diffuse tip is also measured in a calibrated integrating sphere.

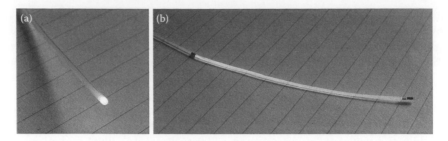

FIGURE 25.5 Isotropic diffuser (a) and linear diffuser (b) (length: 7 cm, right).

25.4.2 Surface Illumination

Illumination of a surface using a broad beam, as illustrated in Figure 25.6a, is the simplest geometry to perform PDT. It is the common approach in treatment of skin diseases and some areas in head and neck cancer. The exposure is usually characterized by how much light is delivered, the incident power per unit area. As explained previously, however, the fluence rate below the surface may be essentially higher. Measurement of the fluence rate can easily be done by putting a diffuse detector on the surface, hence collecting both the incident irradiance and the backscattered diffuse light. The use of a diffuse detector in this geometry, however, requires special calibration (Murrer, Marijnissen, and Star 1995).

In clinical practice, broad parallel beams are rarely available. A fiber-optic emitter or a filtered lamp generates a diverging beam. The resulting fluence rate strongly depends on the distance to the surface, which, depending on the location on the body, may be difficult to keep constant. A very practical approach now common in dermatology is the use of large panels of LEDs. Because of the extended size of the source at shorter distances, the resulting fluence rate is rather independent of the distance between lamp and patient.

Often, this geometry is also used in experiments with cell cultures, irradiating a Petri dish, a culture flask, or a multi-well plate with a broad beam from above. In principle, light dosimetry can be relatively straightforward here. There may be some internal reflections, especially in 96-well plates, but the contribution of this is expected to be relatively low. The problem, however, lies in the culture medium. Depending on the amount of sensitizer and fluence rates used, the oxygen in the liquid can be consumed in seconds after the start of the illumination. As a consequence, the cells at the bottom of a well depend on oxygen diffusion from the surface of the medium downward. As the diffusion length of oxygen in a watery environment is on the order of 0.1 mm, normal levels of medium on top of these cells on the order of a few millimeters will form a strong barrier against oxygen. In such geometry, PDT would take place in a low-oxygen environment, and the slightest stirring of the medium could overcome this.

25.4.3 Hollow Cavities

The internal surface of a hollow cavity behaves differently than a surface in open space. This is because elastic scattering causes a substantial amount of light to be diffusely reflected. In an open surface, this reflected light is lost, but in a hollow cavity, the reflected light from one area is incidental on the opposite surface and adds to the total dose received (Figure 25.6b). This effect is similar to that in an integrating sphere. The buildup of the fluence rate can be quite substantial. It increases with increasing diffuse reflection coefficient, which increases with increasing elastic scattering and decreasing absorption. In the bladder, Marijnissen et al. (1993) reported buildup factors up to a factor of seven. It varies between patients, and it can vary dynamically during treatment. Figure 25.7 shows an example of the fluence rate measured at the surface of the esophagus during ALA PDT (van Veen et al. 2002b). The incident irradiation was 100 mW/cm². It shows a fluence rate buildup that varies from 1.5 to 3.5, changing rapidly during treatment. We believe the variations to be caused by rapid muscular contractions in the esophageal wall, squeezing blood out of the tissue, in combination with much slower changes in the blood volume resulting from treatment response. We believe dosimetry in hollow cavities should be based on actual fluence rate measurements inside the cavity.

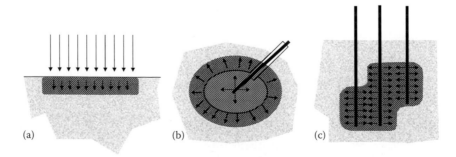

FIGURE 25.6 Treatment geometries: (a) superficial, (b) hollow cavities, (c) interstitial.

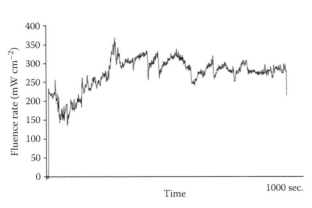
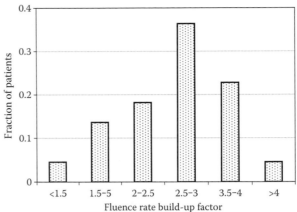

FIGURE 25.7 Fluence rate (left) measured at the surface of the esophagus during ALA PDT. Incident irradiation was 100 mW/cm². Distribution of fluence rate buildup factors (right) measured in 22 patients.

25.4.3.1 Dedicated Light Applicators

The delivery of light in a hollow cavity is technically complicated; usually, access is given through an endoscope, and light is delivered through an optical fiber. To place the fiber illuminator and measurement probes in a reproducible manner is challenging. Hence, dedicated light applicators are often used. Such applicators have been made for the bladder (Marijnissen et al. 1993), the brain (Muller and Wilson 1985), the central airways (Murrer et al. 1997), the pleural cavity (van Veen et al. 2001), the cervix (Soergel et al. 2008), the esophagus (van Veen et al. 2002b), the nasopharynx (Nyst et al. 2007), and the anal cavity (Kruijt et al. 2010). An example of a dedicated light applicator is given in Figure 25.6. It has been developed for PDT of the nasopharynx. In the design of such an applicator, we strive to generate a reproducible and, where possible, even fluence rate distribution over the surface we intend to treat and to block exposure to an area we want to protect. This, however, is challenging and often impossible because of the irregular shape of the cavity. The best compromise is usually based on limiting the highest fluence rate

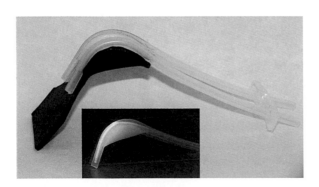

FIGURE 25.8 Applicator for PDT in the nasopharynx. The applicator is inserted through the mouth, with the two long tubes sticking out of the nose and the curved part positioned reproducibly in the nasopharynx. The inset shows the illumination by a linear diffuser. The black flap protects the soft palate from exposure.

to a maximum value to avoid oxygen depletion and extending the treatment time to ensure a sufficient dose in the low-fluence-rate areas (Figure 25.8).

25.4.4 Interstitial Illumination

Deep-seated and large tumors can be treated by inserting one or more linear diffusers into the tissue (Figure 25.6c). In theory, we can treat any size volume at any depth as long as we can reach the area with hollow needles to position the fibers and we can accurately place these needles. For this, we need spatial information on the location of the tumor as well as a method to guide the positioning. Several approaches to this have been developed. Jerjes et al. from Hopper group (Jerjes et al. 2010) pioneered interstitial PDT in the head and neck region and used bare fibers positioned using a grid and retracted them slowly during illumination. The grid approach was inspired by brachytherapy. Likewise, several investigators developed a technique for interstitial PDT (iPDT) of the prostate using the grid-based approach common in brachytherapy of prostate cancer (Haider et al. 2007; Yu et al. 2006; Swartling et al. 2010).

Current practice developed by researchers in Rotterdam in collaboration with the Netherlands Cancer Institute in Amsterdam (Karakullukcu et al. 2012) for interstitial PDT using Foscan for recurrent cancer of the base of the tongue is slightly more complicated. It consists of the following steps:

1. First, we draw the tumor boundaries in every slice of the MRI or CT scan, a standard procedure in radiotherapy (Figure 25.9, dotted red lines).
2. Pretreatment plan: Next, we determine the ideal placement of the linear diffusers and their desired amount, orientation, and required length. As we are currently using Foscan, we assume a therapeutic radius around the linear diffuser of 1 cm. We are currently not using any automated algorithm to do this. The main rules considered in placing the fibers are to reach maximum

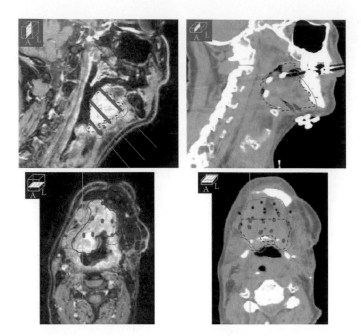

FIGURE 25.9 (See color insert.) Example of an iPDT treatment. Left: Pretreatment CT scan with tumor boundaries drawn as thick red lines. The planned position of the linear diffuser (thick red dots and lines); resulting zone of necrosis is indicated as thin red lines. Right: magnetic resonance (MR) image taken after insertion of the fibers (thick red dots and lines); the tumor volume is indicated here as the dashed line and the expected zone of necrosis as the thin red line.

coverage of the target volume, that is, the tumor, and at the same time, a minimum coverage of the risk volume. No risk volumes are visualized here. A first risk volume to consider in PDT in this region is the carotid artery. In the patient presented here, we are at sufficient distance, but in other cases, shielding is allied. Serious side effects can also be caused by damaging skin and underlying muscle. High fluences in this area will lead to holes in the skin that will have to be treated surgically. The treatment plan shown in Figure 25.9 clearly avoids high doses in these regions. Obviously, a treatment plan is more challenging when target volume and risk volumes lie closer together.

3. Placement: Usually a few days later, we bring the patient to the operating room (OR) and insert hollow needles manually under general anesthesia. Through these needles, we insert transparent catheters, and then the needles are removed. This procedure is borrowed from brachytherapy. The bore of the catheters is sufficient to allow insertion of the linear diffusers (and shielding).

4. Verification: To verify the placement, we take a CT (Figure 25.9, right). As can be seen, there is big difference. In part, this is caused by the difficulty of obtaining images of comparable orientation. The reality, however, is that it is impossible to manually position the needles exactly as planned. The pretreatment plan serves as an excellent

guideline; it makes one think carefully about the approach and possible problems before the actual procedure, but the actual positioning of the fibers is a difficult procedure, and during the procedure, there is little feedback about the actual needle positions. The verification step is therefore very important.

5. Correction: In case we find that the target volume is not completely covered, we must take action. We have several options: We can reposition a fiber, we can add an additional fiber, we can change the length of the diffuser, or we can increase the fluence to one or more of the fibers in the problem area.

6. Treatment: We deliver a standard fluence of 30 J cm^{-1} of diffuser length (output of the fiber 100 mW/cm length, wavelength 652 nm, using Foscan). For practical reasons, we do not illuminate all fibers simultaneously. Currently, we use a custom four-channel laser from Biolitec, allowing us to illuminate four fibers at the same time.

The presence of multiple fibers in the tissue would allow us to measure optical properties in real time during treatment. This could be used to modify the output power of the various fibers to correct for unexpected variations in optical properties or slight misplacement of the fibers. Johansson et al. (2007) showed that such a procedure is technically feasible. At present, however, it is not clear if this would contribute to an improvement in the clinical results. Obviously, we prefer smooth fluence and fluence rate distributions over the entire treatment volume. As described above, however, the effectiveness of the treatment also depends on other factors, such as drug and oxygen availability, and in the large tumor volumes treated, this may present an even bigger problem. Ideally, one would like to add information on oxygen and drug availability to a general dosimetry approach. Although it seems very attractive, certainly from a scientific point of view, a very fundamental problem lurks in the shadows of this approach. The point is that we have many variables we would like to control but only a few knobs to turn on. This is related to the fact that we use linear diffusers. These linear diffusers deliver the same amount of light to an entire string of tissue voxels that may have or require different fluence rates. Only if we would insert an isotropic point source in every voxel with a diameter equal to the optical penetration depth δ, then it would be possible to generate any desired fluence rate distribution. When using linear diffusers, we have to abandon the concept of a dedicated fluence rate distribution for each patient but have to be satisfied with a less demanding approach. The concept of covering a target area with a minimum fluence at a maximum fluence rate while having a maximum fluence in a risk area, borrowed from radiotherapy, is an attractive alternative as it has a direct relationship to clinical end points.

At present, we feel that the first critical step in the treatment, that is, accurate placement of the fibers, requires our attention first before adding further technology. So far, up to 24 patients have been treated with the approach described. The follow-up on these patients is still in progress.

25.5 Future Directions

25.5.1 Implicit versus Explicit Dosimetry

The concept of implicit dosimetry as introduced by Wilson, Patterson, and Lilge (1997) proposes to obtain information on the treatment effectiveness by measurement of the direct effects of the treatment. An example of this is fluorescence photobleaching of PpIX (Robinson et al. 2000). As photobleaching of PpIX is a direct effect of the photodynamic activation of the photosensitizer, the bleaching rate is a good measure of the effectiveness of the treatment (Robinson et al. 1998). Others have shown that the amount of singlet oxygen can be measured directly during therapy (Jarvi, Patterson, and Wilson 2012), but there are a range of fundamental problems with directly monitoring singlet oxygen luminescence, its spatial distribution, and its relationship to PDT dose. So far, however, implicit dosimetry has not found its way to routine clinical dosimetry. This topic will be addressed in more detail in the following chapter.

PDT started with Photofrin, a photosensitizer that has its main absorption peak at 630 nm, which has an optical penetration of approximately 5 mm. Development of photosensitizers that absorb light at longer wavelengths, leading to an improved penetration of light compared to Photofrin, has been a topic that has kept drug developers busy for several decades. The currently approved photosensitizers Foscan (652 nm) and Verteporphyrin (690 nm) have been the result of this effort. Recently, special up-conversion nanoparticles have been developed that can excite a conjugated photosensitizer by two step absorption at wavelengths of more than 900 nm. It has been suggested that at these wavelengths, an even deeper penetration could be reached. Figure 25.4, however, clearly shows that this is not the case. Moreover, the therapeutic effectiveness of a two-step excited sensitizer would drop with the square of the light intensity. Hence, the therapeutic depth would be half the optical penetration depth.

Molecular targeting of the photosensitizer has been a hot issue for nearly a decade. If successful, it can separate the thresholds for damage to normal tissue and cancer tissue substantially. At present, these thresholds are very close. Currently, selectivity is mainly based on proper treatment planning. Molecular targeting of the sensitizer may represent a major breakthrough for PDT as it will finally allow selective sensitization of the diseased tissue. It is a challenging approach, but similar targeting strategies have proven to be feasible. The biggest challenge, however, may lie in preventing the sensitizer localizing in normal tissue. With proper dosimetry, we then have better opportunities to destroy the target tissue while sparing the surrounding normal tissue.

References

Anderson, R., W. Farinelli, H. Laubach et al. 2006. Selective photo-thermolysis of lipid-rich tissues: A free electron laser study. *Lasers Surg Med* 38:913–919.

Bisland, S., L. Lilge, A. Lin, R. Rusnov, and B. Wilson. 2004. Metronomic photodynamic therapy as a new paradigm for photodynamic therapy: Rationale and preclinical evaluation of technical feasibility for treating malignant brain tumors. *Photochem Photobiol* 80:22–30.

Busch, T. 2006. Local physiological changes during photodynamic therapy. *Lasers Surg Med* 38:494–509.

Chen, Q., H. Chen, and F. Hetzel. 1996. Tumor oxygenation changes post-photodynamic therapy. *Photochem Photobiol* 63:128–131.

Cottrell, W., A. Paquette, K. Keymel, T. Foster, and A. Oseroff. 2008. Irradiance-dependent photobleaching and pain in delta-aminolevulinic acid-photodynamic therapy of superficial basal cell carcinomas. *Clin Cancer Res* 14:4475–4483.

Coutier, S., L. Bezdetnaya, T. Foster, R. Parache, and F. Guillemin. 2002. Effect of irradiation fluence rate on the efficacy of photodynamic therapy and tumor oxygenation in meta-tetra(hydroxyphenyl)chlorin mTHPC-sensitized HT29 xenografts in nude mice. *Radiat Res* 158:339–345.

Curnow, A., J. Haller, and S. Bown. 2000. Oxygen monitoring during 5-aminolaevulinic acid induced photodynamic therapy in normal rat colon: Comparison of continuous and fractionated light regimes. *J Photochem Photobiol B* 58:149–155.

de Bruijn, H., A. van der Ploeg-van den Heuvel, H. Sterenborg, and D. Robinson. 2006. Fractionated illumination after topical application of 5-aminolevulinic acid on normal skin of hairless mice: The influence of the dark interval. *J Photochem Photobiol B* 1(85):184–190.

de Haas, E., B. Kruijt, H. Sterenborg, H. Martino Neumann, and D. Robinson. 2006. Fractionated illumination significantly improves the response of superficial basal cell carcinoma to aminolevulinic acid photodynamic therapy. *J Invest Dermatol* 126:2679–2786.

Dwivedi, S., K. Krishnan, and S. Suryanarayanan. 2007. Digital mouse phantom for optical imaging. *J Biomed Opt* 12:051804.

Farrell, T., M. Patterson, and B. Wilson. 1992. A diffusion theory model of spatially resolved, steady-state diffuse reflectance for the noninvasive determination of tissue optical properties *in vivo*. *Med Phys* 19:879–888.

Finlay, J., S. Mitra, M. Patterson, and T. Foster. 2004. Photobleaching kinetics of Photofrin *in vivo* and in multi-cell tumour spheroids indicate two simultaneous bleaching mechanisms. *Phys Med Biol* 49:4837–4860.

Foster, T., and L. Gao. 1992. Dosimetry in photodynamic therapy: Oxygen and the critical importance of capillary density. *Radiat Res* 130:379–383.

Foster, T., R. Murant, R. Bryant et al. 1991. Oxygen consumption and diffusion effects in photodynamic therapy. *Radiat Res* 126:296–303.

Graaff, R., J. Aarnoudse, J. Zijp et al. 1992. Reduced light-scattering properties for mixtures of spherical particles: A simple approximation derived from Mie calculations. *Appl Opt* 31:1370–1376.

Haider, M., S. Davidson, A. Kale et al. 2007. Prostate gland: MR imaging appearance after vascular targeted photodynamic therapy with palladium-bacteriopheophorbide. *Radiology* 244:196–204.

Henderson, B., S. Gollnick, J. Snyder et al. 2004. Choice of oxygen-conserving treatment regimen determines the inflammatory response and outcome of photodynamic therapy of tumors. *Cancer Res* 64:2120–2126.

Jacques, S., and Pogue, B. 2008. Tutorial on diffuse light transport. *J Biomed Opt* 13:041302.

Jacques, S., and L. Wang. 1995. Monte Carlo modelling of light transport in tissue. In *Optical Thermal Response of Laser Irradiated Tissue*, A. Welch, and M. van Gemert, editors. Plenum Press. New York.

Jarvi, M., M. Patterson, and B. Wilson. 2012. Insights into photodynamic therapy dosimetry: Simultaneous singlet oxygen luminescence and photosensitizer photobleaching measurements. *Biophys J* 102:661–671.

Jerjes, W., T. Upile, S. Akram, and C. Hopper. 2010. The surgical palliation of advanced head and neck cancer using photodynamic therapy. *Clin Oncol (R Coll Radiol)* 22:785–791.

Johansson, A., J. Axelsson, S. Andersson-Engels, and J. Swartling. 2007. Realtime light dosimetry software tools for interstitial photodynamic therapy of the human prostate. *Med Phys* 34:4309–4321.

Karakullukcu, B., H. Nyst, R. van Veen et al. 2012. mTHPC mediated interstitial photodynamic therapy of recurrent nonmetastatic base of tongue cancers: Development of a new method. *Head Neck.* 31: doi: 10.1002/hed.21969.

Kruijt, B., E. Snoek, H. Sterenborg, A. Amelink, and D. Robinson. 2010. A dedicated light applicator for light delivery and monitoring of PDT of intra-anal intraepithelial neoplasia. *Photodiagnosis Photodyn Ther* 7:3–9.

Lilge, L., M. Portnoy, and B. Wilson, B. 2000. Apoptosis induced *in vivo* by photodynamic therapy in normal brain and intracranial tumour tissue. *Br J Cancer* 83:1110–1017.

Marijnissen, J., W. Star, H. in't Zandt, M. D'Hallewin, and L. Baert. 1993. *In situ* light dosimetry during whole bladder wall photodynamic therapy. *Phys Med Biol* 38:567–582.

Mitra, S., and T. Foster. 2005. Photophysical parameters, photosensitizer retention and tissue optical properties completely account for the higher photodynamic efficacy of meso-tetra-hydroxyphenyl-chlorin vs Photofrin. *Photochem Photobiol* 81:849–859.

Muller, P., and B. Wilson. 1985. Photodynamic therapy: Cavitary photoillumination of malignant cerebral tumours using a laser coupled inflatable balloon. *Can J Neurol Sci* 12:371–373.

Murrer, L., J. Marijnissen, P. Baas, N. van Znadwijk, and W. Star. 1997. Applicator for light delivery and *in situ* light dosimetry during endobronchial photodynamic therapy: First measurements in humans. *Lasers Med Sci* 12:253–259.

Murrer, L., J. Marijnissen, and W. Star. 1995. *Ex vivo* light dosimetry and Monte Carlo simulations for endobronchial PDT. *Phys Med Biol* 40:1807–1817.

Nachabé, R., B. Hendriks, M. van der Voort, A. Desjardins, and H. Sterenborg. 2010. Estimation of biological chromophores using diffuse optical spectroscopy: Benefit of extending the UV-VIS wavelength range to include 1000 to 1600 nm. *Biomed Opt Exp* 1:1432–1442.

Nyst, H., R. van Veen, I. Tan et al. 2007. Performance of a dedicated light delivery and dosimetry device for photodynamic therapy of nasopharyngeal carcinoma: Phantom and volunteer experiments. *Lasers Surg Med* 39:847–853.

Oleinick, N., and H. Evans. 1998. The photobiology of photodynamic therapy: Cellular targets and mechanisms. *Radiat Res* 150:S146–S156.

Robinson, D., H. de Bruijn, W. de Wolf, H. Sterenborg, and W. Star. 2000. Topical 5-aminolevulinic acid-photodynamic therapy of hairless mouse skin using two-fold illumination schemes: PpIX fluorescence kinetics, photobleaching and biological effect. *Photochem Photobiol* 72:794–802.

Robinson, D., H. de Bruijn, N. van der Veen et al. 1998. Fluorescence photobleaching of ALA-induced protoporphyrin IX during photodynamic therapy of normal hairless mouse skin: The effect of light dose and irradiance and the resulting biological effect. *Photochem Photobiol* 67:140–149.

Rovers, J., M. de Jode, Rezzoug, and M. Grahn, M. 2000. *In vivo* photodynamic characteristics of the near infreared photosensitiser 5,10,15,20-tetrakis(M-hydroxyphenyl) Bacteriochlorin. *Photochem Photobiol* 72:358–364.

Soergel, P., X. Wang, H. Stepp, H. Hertel, and P. Hillemanns. 2008. Photodynamic therapy of cervical intraepithelial neoplasia with hexaminolevulinate. *Lasers Surg Med* 40:611–615.

Sterenborg, H., and M. van Gemert. 1996. Photodynamic therapy with pulsed sources: A theoretical analysis. *Phys Med Biol* 41:835–850.

Swartling, J., J. Axelsson, G. Ahlgren et al. 2010. System for interstitial photodynamic therapy with online dosimetry: First clinical experiences of prostate cancer. *J Biomed Opt* 15:058003.

Tromberg, B., A. Orenstein, S. Kimel et al. 1990. *In vivo* tumor oxygen-tension measurements for the evaluation of the efficiency of photodynamic therapy. *Photochem Photobiol* 52:375–385.

van Gemert, M., S. Jacques, H. Sterenborg, and W. Star. 1989. Skin optics. *IEEE Trans Biomed Eng* 36:1146–1154.

van Veen, R., M. Aalders, K. Pasma et al. 2002b. *In situ* light dosimetry during photodynamic therapy of Barrett's esophagus with 5-aminolevulinic acid. *Lasers Surg Med* 31:299–304.

van Veen, R., H. Schouwink, W. Star et al. 2001. Wedge-shaped applicator for additional light delivery and dosimetry in the diaphragmal sinus during photodynamic therapy for malignant pleural mesothelioma. *Phys Med Biol* 46:1873–1883.

van Veen, R., H. Sterenborg, A. Pifferi et al. 2005. Determination of visible near-IR absorption coefficients of mammalian fat using time- and spatially resolved diffuse reflectance and transmission spectroscopy. *J. Biomed Opt* 10:054004.

van Veen, R., W. Verkruijsse, and H. Sterenborg. 2002a. Diffuse reflectance spectroscopy from 500 to 1060 nm by correction for inhomogenously distributed absorbers. *Opt Lett* 27:246–248.

Veenhuizen, R., and F. Stewart. 1995. The importance of fluence rate in photodynamic therapy: Is there a parallel with ionizing-radiation dose-rate effects. *Radiother Oncol* 37:131–135.

Wang, L., S. Jacques, and L. Zheng. 1995. MCML: Monte Carlo modeling of light transport in multi-layered tissues. *Comput Methods Programs Biomed* 47:131–146.

Wang, K., S. Mitra, and T. Foster. 2007. Comprehensive mathematical model of microscopic dose deposition in photodynamic therapy. *Med Phys* 34:282–293.

Wilson, B., and M. Patterson. 2008. The physics, biophysics and technology of photodynamic therapy. *Phys Med Biol* 2008:61–109.

Wilson, B., M. Patterson, and L. Lilge. 1997. Implicit and explicit dosimetry in photodynamic therapy: A new paradigm. *Lasers Med Sci* 12:182–199.

Yu, G., T. Durduran, C. Zhou et al. 2006. Real-time *in situ* monitoring of human prostate photodynamic therapy with diffuse light. *Photochem Photobiol* 82:1279–1284.

Zijlstra, W., A. Buursma, and O. Van Assendelft. 2000. *Visible and Near Infrared Absorption Spectra of Human and Animal Haemoglobin*. Utrecht: VSP Publishing.

26

Multimodality Dosimetry

Jarod C. Finlay
University of Pennsylvania

26.1 Introduction

The previous chapter provided an overview of light dosimetry in photodynamic therapy (PDT). This chapter extends that discussion to include other factors that impact the photodynamic dose, including sensitizer concentration, tissue oxygen supply, and physiological factors. The key to understanding how these factors play into dosimetry lies in the basic chemical and photochemical interactions among the sensitizer, oxygen, and tissue. The first section of this chapter will review these interactions and their incorporation into a dosimetric model. The remainder of the chapter will review the various applications of this model and related strategies to PDT dosimetry.

26.2 Overview of Singlet Oxygen and Sensitizer Diffusion and Reaction Modeling

The majority of work in PDT has involved type II sensitizers, and the majority of these are presumed to generate the bulk of their cytotoxic effect via reactions of singlet oxygen. The basic interactions between light, the sensitizer in its ground state and excited states, and oxygen have been modeled in various environments and conditions during the last two decades. These interactions are relatively simple at the molecular level as shown in Figure 26.1. The bold arrows indicate the reactions and interactions that constitute the intended energy-transfer process during PDT. Each of these interactions can be characterized by a reaction constant. These are indicated in the figure by the variables indicated next to each arrow. The notation used is that

proposed by Finlay, Mitra, and Foster (2004) and Georgakoudi and Foster (1998).

Light energy is absorbed by the ground state (S_0) sensitizer, promoting it to its excited singlet state (S_1). The sensitizer molecule subsequently undergoes spontaneous intersystem crossing to its excited triplet (T_1) state or spontaneous fluorescence emission, which brings it back to the ground state. The accumulation of T_1 and fluorescence emission are simultaneous processes that compete for the available S_1 population. A good sensitizer therefore requires a ratio of k_{isc} to k_f high enough to effectively populate the triplet state. The lifetime of the sensitizer singlet state is given by the following:

$$\tau_{S_1} = \frac{1}{k_{isc} + k_f} \qquad (26.1)$$

A molecule in the T_1 state can return to the ground state by emitting phosphorescence or by transferring energy to molecular oxygen ($^3O_2$). Some fraction S_Δ of the latter reactions will create singlet oxygen ($^1O_2$). As in the previous case, this is a competitive process. However, in this case, the concentration of oxygen is also important. The lifetime of the triplet state is given by the following:

$$\tau_{T_1} = \frac{1}{k_{ot}\left[^3O_2\right] + k_p} \qquad (26.2)$$

The greater the local concentration of oxygen, the shorter the triplet lifetime, and, therefore, the lower the probability of interactions involving the triplet state.

Finally, the singlet oxygen reacts with targets within the tumor cells, consuming oxygen and causing photochemical damage

293

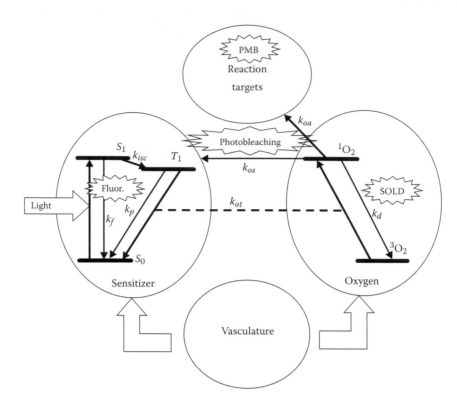

FIGURE 26.1 Schematic diagram of energy flow in PDT treatment. The steps that lead to the photodynamic effect are indicated by bold arrows. The detection technologies used for multimodality dosimetry are indicated next to the energy level transition involved in each.

that eventually results in cell death and the therapeutic effect. The details of the specific targets and mechanisms of cell death are beyond the scope of this chapter but are an area of active study. For the purposes of modeling the photodynamic kinetics, it is sufficient to assume that the concentration of singlet oxygen reacted in the tissue is predictive of the photodynamic effect.

Because the reaction of singlet oxygen is generally irreversible, the photodynamic process consumes oxygen. Oxygen is continuously replenished by diffusion from the external environment or from the vasculature. It is possible to drive the photodynamic process fast enough that oxygen consumption overwhelms resupply, leading to photochemical oxygen depletion. In the extreme case, the oxygen depletion can be significant enough to make oxygen the limiting factor in the photodynamic process. In this case, the sensitizer will continue to absorb light at the same rate, but the generation of singlet oxygen (and hence the photodynamic effect) will be significantly reduced compared to that in well-oxygenated tissue.

The sensitizer can also be consumed during PDT by reactions with singlet oxygen, an effect known as photobleaching. The probability of photobleaching is quantified by a bimolecular rate constant k_{os}. These reactions consume both oxygen and sensitizer but do not contribute to the photodynamic effect. Fortunately, these reactions are generally a negligibly small fraction of the total singlet oxygen reactions generated during a typical PDT treatment. Their contribution to the oxygen consumption can generally be ignored, but their effect on sensitizer concentration is often significant.

The dynamics of the photodynamic process can be modeled by a series of differential equations, one for each molecular state. Even for the simplest case in which the minimum number of possible reactions is considered, this yields five coupled differential equations (three for the sensitizer and two for oxygen). However, the system can be significantly simplified by the observation that the various processes involved occur over three vastly different timescales. The transport of light, the transitions among states of the individual molecules, and the reactions between molecules are fast enough that they reach equilibrium on the order of microseconds or less. Oxygen diffusion is much slower, taking seconds or tens of seconds to reach equilibrium. A typical PDT treatment takes hundreds or thousands of seconds. Because we are primarily concerned with the slower of these processes (oxygen consumption and accumulation of singlet oxygen reaction), we can make the simplifying assumption that transitions among states of sensitizer and oxygen are in equilibrium. The relative populations of the excited states of the sensitizer are given by the following:

$$[S_1] = \frac{\Phi[S_0]\frac{\sigma_s}{h\upsilon}}{(k_f + k_{isc})} \tag{26.3}$$

and

$$[T_1] = \frac{k_{isc}[S_1]}{\left(k_p + k_{ot}[^3O_2]\right)} \tag{26.4}$$

where Φ is the light fluence rate, σ_S is the absorption cross section of the sensitizer, and $h\upsilon$ converts from energy to number of photons. The corresponding relationship for singlet oxygen is given by the following:

$$\left[^1O_2 \right] = \left(\Phi [S_0] \frac{\sigma_S}{h\upsilon} S_\Delta \frac{k_{isc}}{(k_f + k_{isc})} \right) \left(\frac{k_{ot}[^3O_2]}{(k_p + k_{ot})[^3O_2]} \right) \left(\frac{1}{k_d + k_{oa}[A] + k_{os}[S_0]} \right) \quad (26.5)$$

where k_{oa} and $[A]$ represent the reaction rate constant for the reaction of singlet oxygen with targets other than the sensitizer and the concentration of those targets. The three terms in this equation can be thought of as modeling the generation of triplet sensitizer, the transfer to singlet oxygen, and the depletion of singlet oxygen, respectively.

Taken together, Equations 26.3 through 26.5 allow us to model the populations of any of the excited states of the sensitizer or oxygen relative to their respective ground states. The ground state populations can then be modeled by a combination of depletion resulting from reaction and diffusion. Additional terms can be added to these equations to account for additional effects, such as photobleaching, mediated by reactions not involving singlet oxygen (Finlay, Mitra, and Foster 2004; Georgakoudi and Foster 1998).

26.2.1 Microscopic Experimental Models for Dosimetry

The simplest experimental biological model that allows quantitative modeling of the interactions of light and sensitizer in a medium that provides oxygen and nutrients via diffusion from the outside is the multicell tumor spheroid (Dubessy et al. 2000). The spheroid provides a useful analog to the *in vivo* tumor environment in which cells receive oxygen and nutrients via diffusion from a network of microvessels as shown in Figure 26.2. Spheroids can be grown to diameters of a few hundred microns on the scale of the intervascular spacing *in vivo*. They are of biological interest because they exhibit many features of

real tumors, including gradients in oxygen and drug uptake, development of hypoxia, and, for large spheroids, spontaneous necrosis in the central, oxygen-poor region (Dubessy et al. 2000). In addition, various intercellular processes that may have significant effects *in vivo* are reproduced more closely in three-dimensional models, such as spheroids, than in monolayer cultures (Hirschhaeuser et al. 2010).

From the point of view of basic photophysics, the spheroid is an ideal model for PDT dosimetry in that it provides an environment in which the distribution of light is easily controlled, the supply and transport of oxygen is straightforward to the model, and the distribution of sensitizer can be imaged using techniques such as confocal fluorescence microscopy. Early work in this area used oxygen concentration measurements made with a microelectrode to quantify both the radial oxygen distribution and the time course of oxygen depletion induced by PDT. These measurements could be modeled by combining the basic kinetic model described previously with a spatially resolved oxygen-diffusion model (Nichols and Foster 1994). The diffusion-with-reaction model, combined with the observed oxygen depletion and recovery in the spheroid periphery, constrains the ratios of the basic reaction-rate constants described in the previous section (Georgakoudi and Foster 1998; Georgakoudi, Nichols, and Foster 1997). While the dynamics of singlet oxygen generation, the physiological response, and the light distribution are much different between the spheroids and the tissue, these basic chemical interaction constants are expected to be similar.

Furthermore, the spheroid exhibits a monotonic gradient along the radial direction in reacted singlet oxygen. The outermost layer of the spheroid, closest to the oxygen supply, receives the highest singlet oxygen dose, while the center of the spheroid becomes hypoxic and receives a lower dose. The fraction of cells surviving a treatment can be converted to a volume of cells and, because of the spherical geometry, to a limiting radius. Calculating the dose of singlet oxygen reactions at that radius allows an estimate of the threshold dose of singlet oxygen required for cell killing (Nichols and Foster 1994).

26.2.2 Extension to Macroscopic Models

The microscopic model of photodynamic deposition of singlet oxygen is quantitatively accurate at the microscopic level. The basic features of the spheroid model are preserved as shown in Figure 26.2. However, the extension of this model to bulk tissue poses several important challenges. First, the information available from measurements that can be made *in vivo* is generally limited to volume-averaged data. This is a result both of the resolution of the measurements that can practically be made clinically and the fact that the volume of tissue treated during PDT is millions of times the volume of a spheroid. Second, the supply of oxygen and nutrients is dynamic, being supplied by the vasculature, rather than static as in the case of the spheroid surrounded by an oxygen-saturated medium. The geometry of the vasculature is much more complex than that of the spheroid's oxygen supply, particularly in tumors. Third, the spheroid's small size

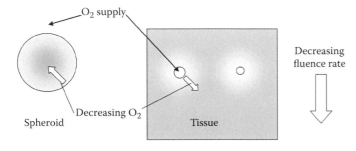

FIGURE 26.2 Illustration of the similarities and differences between the multicell spheroid environment used to determine photodynamic rate constants (left) and the *in vivo* environment (right).

and lack of hemoglobin mean that it does not significantly perturb the distribution of light incident on it, so the fluence rate is approximately uniform across the spheroid. In contrast, tissue is highly absorbing and scattering, leading to a gradient in fluence rate independent of the gradients in oxygenation and sensitizer.

Despite these challenges, the kinetics equations are useful for understanding the underlying mechanics of the photodynamic process, many of which are at least qualitatively reproduced in tissue. They predict that high fluence rates will deplete oxygen faster than it can be resupplied by the vasculature, leading to photochemically induced hypoxia. Even without explicitly considering the details of oxygen diffusion, this model qualitatively explains the effects of multiple photobleaching mechanisms at different oxygen and sensitizer concentrations (Finlay, Mitra, and Foster 2004) and the difference in effectiveness of different sensitizers (Mitra and Foster 2005).

Wang, Mitra, and Foster (2007) proposed a quantitative model that considers the microscopic reaction kinetics involved in the photodynamic process and the supply of oxygen by the vasculature simultaneously. Such a model requires simplifying assumptions about such parameters as vascular spacing, blood-flow rates, and arterial oxygenation, which are not generally accessible on the microscopic scale of the model using currently available measurements. The resulting model makes quantitative predictions about the effects of variations in these physiological parameters.

An alternative approach is to generalize the microscopic model by applying the kinetics equations not to microscopic regions of tissue but to the volume-average quantities available to clinical measurement. This approach also makes significant simplifying assumptions. For instance, if the rate of oxygen supply is modeled as a bulk supply term (Hu et al. 2005) or an approximation to diffusion (Wang et al. 2010), rather than by vessel-by-vessel diffusion, and the sensitizer and blood supply are generally assumed to be homogeneous. It is not surprising that the reaction rate constants appropriate to these models are not exactly the same as those applied to the microscopic model. However, *in vivo* experiments have shown that such a macroscopic modeling can make quantitative predictions of the depth of necrosis induced by photodynamic treatment (Wang et al. 2010).

The complexity of the reactions involved in PDT and the potential for variations in the important quantities over vastly different distance and timescales imply that the light dosimetry alone will likely not be sufficient to predict clinical outcome and that modeling the entire process from first principles will require prohibitively large amounts of input data. The remainder of this chapter will describe various approaches to addressing these challenges.

26.3 Model-Based Multimodality Dosimetry

The first set of approaches to PDT dosimetry falls into the category of dosimetric models that inform forward calculations of PDT dose or "explicit" dosimetry (Wilson, Patterson, and Lilge

1997). The common strategy in these methods is to measure the components that contribute to the PDT effect and use an empirical or theoretical model, such as those described in the previous chapter, to predict the PDT effect. Several variants of this strategy have been developed, and work in this area is ongoing.

26.3.1 Drug–Light Product

The simplest measure of explicit dosimetry is the drug–light product, referred to as the photodynamic dose (Patterson, Wilson, and Graff 1990). The rationale for using this quantity is that the rate of photon absorption by sensitizer is proportional to the product of the local sensitizer and the local light fluence rate. In the notation used previously,

$$D = \int \frac{\sigma_S}{h\upsilon} \Phi[S_0] \mathrm{d}t \qquad (26.6)$$

This definition makes the dose proportional to the total amount of energy absorbed by the photosensitizer. This quantity is analogous to the definition commonly used in ionizing radiation therapy, in which the dose is defined as the energy absorbed per unit mass of tissue. The factor of $\sigma_s/h\upsilon$ makes the definition dependent on the number of photons absorbed by the sensitizer, accounting for differences in photon energy among different light sources. In the case of the sensitizer benzoporphyrin derivative (BPD), prescribing a dose in terms of drug–light product has been shown to produce less variability in tumor response than using light dose alone (Zhou et al. 2006).

26.3.1.1 Quantification of Sensitizer Concentration

The primary challenge in using the drug–light product as a dosimetric tool is to quantitatively measure the sensitizer concentration. One solution is to measure its light absorption. This method is straightforward in clear solution but becomes more challenging in scattering media, such as cell suspensions and tissue. A number of techniques have been developed for quantitatively measuring the absorption and scattering properties of tissue using diffusely reflected or transmitted light (Farrell, Patterson, and Wilson 1992; Finlay and Foster 2004; Kienle et al. 1996; Kienle and Patterson 1997). These methods have been applied to the measurement of sensitizer absorption in a range of clinical and preclinical trials dating back two decades (Patterson et al. 1987; Yu et al. 2006). The absorption of light by the sensitizer competes with absorption by background absorbers, particularly hemoglobin. Because of this overlap, it is often necessary to use absorption spectroscopy to separate the absorption that results from the sensitizer from that of the tissue itself. Therefore, absorption measurements are best suited to sensitizers that absorb very strongly in wavelength ranges where hemoglobin is less absorbing. Many of the second-generation sensitizers were developed specifically to take advantage of this wavelength range. In the extreme case, Motexafin Lutetium (MLu) has an absorption peak at approximately 730 nm, a region in which hemoglobin absorption is near its minimum. At clinically relevant concentrations, MLu can account for the majority of the absorption at 730 nm (Zhu, Finlay, and Hahn 2005).

The majority of the photosensitizers currently in clinical use are fluorescent. Sensitizer fluorescence can often be detected in cases where absorption spectroscopy is not sufficiently sensitive. Fluorescence spectroscopy has therefore become the modality of choice for quantifying sensitizer concentration in many cases. Spectroscopy is generally more reliable than a single-wavelength measurement of fluorescence emission because of the potential for interference from tissue autofluorescence. Fluorescence spectroscopy is further complicated by the fact that both the excitation light and the emitted fluorescence are subject to absorption and scattering in the tissue, which can distort both the intensity of the fluorescence emission and its spectral shape. This distortion can be corrected by various methods, including dividing by a diffuse reflectance measurement made at wavelengths corresponding to the emission and/or excitation wavelengths (Diamond, Patterson, and Farrell 2003; Finlay et al. 2001, 2006; Muller et al. 2001; Wu, Feld, and Rava 1993), measurement of fluorescence with an optical probe designed to be insensitive to scattering (Diamond, Patterson, and Farrell 2003), or explicit modeling of the effects of absorption and scattering based on known absorbers (Finlay and Foster 2005). The methods most appropriate for a given clinical application will depend on the details of the accessibility to optical probes, the geometry of the measurement, the sensitizer and treatment wavelength being used, and the background optical properties and autofluorescence of the tissue being treated.

26.3.1.2 Effects of Photobleaching

One of the potential challenges for the drug–light product definition of dose is the fact that many sensitizers photobleach significantly over the course of a treatment. In this case, the sensitizer concentration will change dynamically during treatment, so the concentration must be measured dynamically. Measurements can be made periodically during breaks in treatment; however, interruptions in treatment are not always clinically practical. An alternative approach is to monitor the long-wavelength emission of the sensitizer excited by the treatment light itself. The treatment light is blocked by appropriate optical filters, and sensitizer concentration is quantified based on the known absorption spectrum of the sensitizer. This strategy has been successfully applied in a clinical trial of ALA-sensitized PDT of the skin (Cottrell et al. 2008) and will likely be adopted in other areas. Equation 26.6 is compatible with dynamic changes in both sensitizer concentration and fluence rate (described in the previous chapter), allowing calculation of a real-time photodynamic dose.

26.3.2 Models Incorporating Oxygenation

The third and often most dynamic component of PDT is the oxygen supplied by the tissue. The oxygen content of the tissue is related to the vascular oxygen content by a balance between diffusion and consumption. There can be significant variation in tissue oxygenation among different tissues and between tumor and normal tissue. In addition, PDT treatment can generate enough photochemical oxygen consumption to produce local hypoxia, particularly at high fluence rates (Sitnik and Henderson 1998). This relationship has been modeled in detail (Wang, Mitra, and Foster 2007); however, it is difficult to quantify tissue oxygenation on the microscopic level on which it is expected to vary.

The most common approach to the quantification of oxygenation is the measurement of hemoglobin oxygen saturation via its optical absorption. Because hemoglobin is confined to the blood, hemoglobin absorption gives a measure of the oxygen content of the vasculature. This can be related to the oxygenation of the tissue by an appropriate diffusion model (Wang, Mitra, and Foster 2007). Measurement of hemoglobin saturation has been incorporated into several PDT clinical trials (Johansson et al. 2007; Tyrrell et al. 2011; Wang et al. 2005; Zhu, Finlay, and Hahn 2005); however, the inclusion of the results into a single, unified, clinical dosimetry metric remains a work in progress.

26.4 Direct Singlet Oxygen Monitoring for Dosimetry

In biological environments, the vast majority of singlet oxygen will react with substrates while in the excited singlet state. However, a minority of singlet oxygen molecules will remain unreacted long enough to undergo spontaneous phosphorescence, returning to their triplet ground state and emitting 1270 nm light (Jimenez-Banzo et al. 2008). In principal, detection of this emission, or singlet oxygen luminescence dosimetry (SOLD), yields the dosimetric quantity most closely tied to the singlet oxygen concentration induced by PDT.

In practice, its measurement poses several significant challenges. First, the most commonly used photodetectors have low sensitivity at 1270 nm. This is predominantly a technological challenge. Recent generations of photomultipliers (PMTs) have sufficient quantum efficiency to make direct singlet oxygen detection feasible (Jimenez-Banzo et al. 2008). It is likely that future technological improvements will increase the performance and reduce the costs of SOLD-compatible detectors.

Second, the interpretation of the measured signal is nontrivial. The signal measured is proportional to the concentration of singlet oxygen; however, the proportionality depends on the singlet oxygen lifetime (Jarvi et al. 2011). The singlet oxygen lifetime depends on both the rate of spontaneous decay (k_d) and the reaction with targets ($k_{oa}[A]$). $k_{oa}[A]$ is increased, and the lifetime is consequently reduced, in environments in which singlet oxygen quenchers are prevalent and singlet oxygen reactions are likely. The singlet oxygen signal is therefore, perversely, strongest in regions in which singlet oxygen is not reacting. For this reason, simply measuring the singlet oxygen luminescence signal is not sufficient to measure the generation of singlet oxygen in cells or *in vivo* (Niedre, Patterson, and Wilson 2002). This challenge can be overcome by measuring the time-resolved singlet oxygen signal excited with a pulsed laser. The singlet oxygen luminescence exhibits a characteristic sharp (submicrosecond) rise followed by a slower decay over several microseconds. The shape of this long decay curve depends on the decay rates of both the sensitizer triplet state ($k_p + k_{ot}[^3O_2]$) and the singlet oxygen ($k_d + k_{oa}[A]$). Because the lifetime of the triplet is typically much longer than that of singlet oxygen, the decay curve can give a measure of the triplet lifetime, which implicitly indicates the local oxygen concentration. However, completely separating these two decay rates remains a technical challenge (Jarvi et al. 2011).

26.5 Chemical or Biochemical Monitoring for Dosimetry

The complexity of explicit dosimetry and the technical challenges of explicit dosimetry and direct singlet oxygen detection have motivated the development of alternative approaches in which an easily detectable surrogate for tissue effect is used to monitor PDT.

26.5.1 Sensitizer Photobleaching: Implicit Dosimetry

The simplest detectible surrogate for tissue damage is the photosensitizer itself. Because the chemical reactions that degrade the sensitizer are similar to those that damage tissue, more photobleaching can indicate more tissue damage. In the specific case of sensitizers in which photobleaching is mediated purely by singlet oxygen, it can be shown that, at the microscopic level, the fractional change in sensitizer concentration resulting from photobleaching is related to the absolute dose of reacted singlet oxygen deposited in the tissue (Georgakoudi, Nichols, and Foster 1997). This observation is the basis for "implicit dosimetry" (Wilson, Patterson, and Lilge 1997), the use of the fluorescence photobleaching measurements to evaluate PDT effect.

As predicted, less-efficient photobleaching is observed in high-irradiance, oxygen-depleting treatments, which also exhibit less-efficient therapeutic singlet oxygen reactions (Finlay 2003; Robinson et al. 1999). Implicit dosimetry automatically takes into account the effects of variable oxygen and sensitizer concentration because it measures the damage produced by the singlet oxygen, rather than the factors that lead to it. Lower oxygen concentrations make the PDT less effective but also reduce the efficiency of singlet-oxygen-mediated photobleaching. Sensitizer photobleaching has been successfully used to reduce pain in ALA-PpIX-mediated skin PDT (Cottrell et al. 2008).

One of the key advantages of implicit dosimetry is that it inherently accounts for the effects of oxygen depletion. However, this assumes that the sensitizer bleaches only via singlet oxygen reactions. The dynamics of sensitizer photobleaching vary significantly among different sensitizers. In animal models, different sensitizers, all of which exhibit a more efficient cytotoxic effect at lower irradiances, can exhibit more-efficient photobleaching (Finlay et al. 2001), indistinguishable photobleaching (Finlay, Mitra, and Foster 2004), less-efficient photobleaching, or

photobleaching with complex features not easily understood in terms of basic kinetics (Finlay, Mitra, and Foster 2002).

It has been proposed that some sensitizers may bleach via reactions of the triplet-state sensitizer. In cases in which oxygen is depleted, the collisional de-excitation of the sensitizer triplet state is reduced by the lack of available oxygen, and the triplet lifetime is consequently enhanced. This increases the population of triplet-state sensitizer as shown in Equation 26.4, making triplet reactions more likely. As a result, sensitizers with a triplet-mediated bleaching mechanism bleach more effectively in hypoxic regions despite delivering less singlet oxygen dose. This not only makes the bleaching ineffective as a dose metric but also depletes sensitizers in regions of hypoxia, which complicates any dose modeling.

Given this complexity, it is essential to understand the underlying mechanisms and dynamics of the photobleaching of a sensitizer before using implicit dosimetry clinically. It has recently been proposed that SOLD measurements could be used to validate photobleaching as a dose metric for individual sensitizers (Jarvi, Patterson, and Wilson 2012). This approach provides direct evidence of the relationship between photobleaching and singlet oxygen dose. A sensitizer thus characterized could be monitored for photobleaching in cases where SOLD is not clinically practical.

As described previously, the detection of fluorescence is complicated by absorption and scattering, which may change dynamically during PDT, and by background autofluorescence. In addition, many sensitizers react with singlet oxygen to produce one or more photoproducts, which are themselves fluorescent and, in some cases, photodynamically active. As the parent sensitizer is bleached and its signal becomes weaker, the photoproduct fluorescence may become comparable to it. Because the parent sensitizer and its photoproducts often have similar and overlapping emission spectra, sophisticated spectral fitting algorithms are required to quantitatively separate them.

The complexity of photoproduct dynamics can bring advantages; in the case of Photofrin, there are regimens of sensitizer and oxygen concentration in which photobleaching is not predictive of dose, but the accumulation of the primary photoproduct may be (Finlay, Mitra, and Foster 2004). This results from the competition among multiple bleaching mechanisms with opposite oxygen dependence (e.g., singlet oxygen- and triplet sensitizer-mediated bleaching), only one of which produces a photoproduct. In other cases, the photoproducts may provide confirmation of the photobleaching-based predictions or additional biological information (Finlay et al. 2001).

26.5.2 Conjugated or Codelivered Markers of Dose

Over the past several decades, various nanoparticles have been designed to deliver a sensitizer along with imaging contrast agents that allow its distribution to be imaged using various modalities (Celli et al. 2010). The majority of these agents are designed to target the delivery of the sensitizer to the target tissue by making it selective for a molecule overexpressed in a specific tumor type. While these developments are promising, they do not specifically address the problem of dosimetry. Recently, however, there has been interest in incorporating agents designed for dosimetry as well.

The actinometric nanoparticles proposed by Bisland et al. (2004) combine a fluorophore that can be bleached by singlet oxygen with a fluorophore that bleaches in proportion to the incident fluence via an approximately oxygen-independent mechanism. In this case, the degradation of the indicator fluorophores can be used as a metric of deposited light and singlet-oxygen doses even if the sensitizer itself does not bleach or bleaches via a mechanism that is not indicative of dose.

Photodynamic molecular beacons (PMBs) take this approach one step further. Here, the sensitizer is linked to a quencher by a peptide chain. In its administered form, the peptide holds the sensitizer in close proximity to the quencher, making it nonfluorescent. The peptide is chosen to be cleavable by a specific protease—in this case, caspase-3, which is indicative of apoptotic response to PDT. When the peptide is cleaved, the sensitizer regains its fluorescence (Stefflova et al. 2006). This approach has the advantage that dose produces positive contrast rather than the negative contrast inherent in photobleaching. Variations on this theme include nanoparticles that modulate not just fluorescence but also singlet-oxygen generation in response to specific cellular targets (Chen et al. 2008).

26.6 Physiological Monitoring for Dosimetry

All the measures of PDT dose described thus far take as their underlying premise the assumption that the concentration of singlet oxygen deposited in tissue is predictive of biological response. While this is a reasonable assumption and a rational starting point for dosimetry, developments over the last decade have called this assumption into question. Among these is the observation that the concentration of singlet oxygen required to kill cells in cell culture or spheroids appears to be an order of magnitude greater than that required to achieve a cytotoxic effect *in vivo* (Wang, Mitra, and Foster 2008).

Even if damage at the cellular level is strictly proportional to the local concentration of singlet oxygen deposited in the cell, the clinically relevant endpoints, such as local tumor control and long-term survival, are much more complicated because they involve a variety of effects beyond cell killing.

It has long been realized that PDT acts at least partially through damage to the vasculature. The balance between vascular effects and cellular effects can be influenced by the choice of sensitizer, the drug–light interval, and the irradiation protocol. Increasing the drug–light interval and, hence, the fraction of drug in the vasculature has been shown to shift the mechanism of damage away from vascular damage and toward direct cellular damage even within the same model system; however, some vascular damage remains (Chen et al. 2003, 2005). In the extreme case of Vertiporfin used to treat macular degeneration,

the intended effect is entirely vascular, so a short drug–light interval is used.

Given the propensity for PDT to cause vascular damage, it would be reasonable to assume that PDT treatment would lead to an irrecoverable decrease in blood flow in the treated region. While a long-term decrease is observed, especially in vascular-targeted treatment regimens, the overall relationship between PDT damage and vascular response is much more complicated. Yu et al. (2005) monitored the blood flow in animals during PDT with Photofrin and found a repeatable pattern of initial increase in blood flow followed by a decrease to below initial values. The specifics of the blood flow response were predictive of outcome. This result was promising in that a real-time measurement of blood flow could potentially serve as a measure of PDT dose and be used to guide treatment. Unlike the other dosimetric schemes discussed so far, this method would be sensitive to a physiological response, regardless of the underlying photochemical inputs.

One of the challenges, however, is that this response depends on the tumor microenvironment and vasculature (Maas et al. 2012), which can be expected to differ among tumors and patients. It has been observed that the details of the response differ among strains of mice with the same tumor type and sensitizer, (Mesquita et al. 2012) between mice and humans, and among humans with different tumor types with the same sensitizer (Becker et al. 2010). This does not detract from the clinical usefulness of blood flow monitoring but may limit its utility as a quantitative dose metric for guiding treatment.

26.7 Future Directions

There is increasing appreciation of the need for individualized approaches to treatment across the field of medicine. At the same time, reliance on a single modality of treatment or technology is becoming outdated. The developments in PDT treatment in general and PDT dosimetry are a part of these general trends. We can expect PDT dosimetry to increase in complexity and to incorporate more technologies and more patient-specific measurements.

The ideal dosimetry system would yield a measurement of a metric that is directly predictive of PDT outcome and, at the same time, able to provide guidance to the clinician at the time of treatment. These two goals are, to some extent, contradictory. On one hand, measurements of light fluence, drug–light product, photobleaching, or SOLD provide real-time feedback to the clinician and clear guidance as to what action to take—in this case, stopping treatment when the prescribed dose is reached. However, none of these truly measures the ultimate response of the tumor, and there are situations in which each can fail to predict response. Conversely, measurements such as the blood flow response or the accumulation of apoptotic damage reported by PMBs may directly report a response to PDT, but it may not be clear what action the physician is to take. Physiological parameters such as a given tumor's vascular structure or propensity for apoptosis are important to treatment outcome but cannot be controlled as easily as the fluence rate delivered by a laser.

Ultimately, clinical protocols will likely depend on a combination of physiological monitoring and basic dosimetry.

The other general trend in medicine that is starting to find its way into the PDT dosimetry is the move from single-data-point measurements to imaging and from point-dose calculations to three-dimensional calculations of dose distribution. The last decade has seen an explosion in optical imaging technologies, particularly in microscopy. Three-dimensional imaging of drug uptake, cellular structure, and even gene expression has become routine at the cellular level. At the opposite end of the size scale, MRI, PET, and CT technology has yielded higher resolution both in space and time and allowed imaging of a large variety of contrast agents. Because it is an optical intervention, it is likely that PDT will benefit most from imaging modalities based on optical detection, such as diffuse optical tomography and optical coherence tomography, which tend to have spatial resolution between these extremes.

References

Becker, T. L., A. D. Paquette, K. R. Keymel, B. W. Henderson, and U. Sunar. 2010. Monitoring blood flow responses during topical ALA-PDT. *Biomed Opt Express* 2:123–130.

Bisland, S. K., J. W. Austin, D. P. Hubert, and L. Lilge. 2004. Photodynamic actinometry using microspheres: Concept, development and responsivity. *Photochem Photobiol* 79:371–378.

Celli, J. P., B. Q. Spring, I. Rizvi et al. 2010. Imaging and photodynamic therapy: Mechanisms, monitoring, and optimization. *Chem Rev* 110:2795–2838.

Chen, B., B. W. Pogue, I. A. Goodwin et al. 2003. Blood flow dynamics after photodynamic therapy with verteporfin in the RIF-1 tumor. *Radiat Res* 160:452–459.

Chen, B., B. W. Pogue, P. J. Hoopes, and T. Hasan. 2005. Combining vascular and cellular targeting regimens enhances the efficacy of photodynamic therapy. *Int J Radiat Oncol Biol Phys* 61:1216–1226.

Chen, J., J. F. Lovell, P. C. Lo et al. 2008. A tumor mRNA-triggered photodynamic molecular beacon based on oligonucleotide hairpin control of singlet oxygen production. *Photochem Photobiol Sci* 7:775–781.

Cottrell, W. J., A. D. Paquette, K. R. Keymel, T. H. Foster, and A. R. Oseroff. 2008. Irradiance-dependent photobleaching and pain in delta-aminolevulinic acid-photodynamic therapy of superficial basal cell carcinomas. *Clin Cancer Res* 14:4475–4483.

Diamond, K. R., M. S. Patterson, and T. J. Farrell. 2003. Quantification of fluorophore concentration in tissue-simulating media by fluorescence measurements with a single optical fiber. *Appl. Opt.* 42:2436–2442.

Dubessy, C., J. M. Merlin, C. Marchal, and F. Guillemin. 2000. Spheroids in radiobiology and photodynamic therapy. *Crit Rev Oncol/Hematol* 36:179–192.

Farrell, T. J., M. S. Patterson, and B. Wilson. 1992. A diffusion theory model of spatially resolved, steady-state diffuse reflectance for the noninvasive determination of tissue optical properties *in vivo*. *Med Phys* 19:879–888.

Finlay, J. C. 2003. *Reflectance and Fluorescence Spectroscopies in Photodynamic Therapy*. University of Rochester, Rochester, NY.

Finlay, J. C., D. L. Conover, E. L. Hull, and T. H. Foster. 2001. Porphyrin bleaching and PDT-induced spectral changes are irradiance dependent in ALA-sensitized normal rat skin *in vivo*. *Photochem Photobiol* 73:54–63.

Finlay, J. C., and T. H. Foster. 2004. Hemoglobin oxygen saturations in phantoms and *in vivo* from measurements of steady state diffuse reflectance at a single, short source-detector separation. *Med Phys* 31:1949–1959.

Finlay, J. C., and T. H. Foster. 2005. Recovery of hemoglobin oxygen saturation and intrinsic fluorescence using a forward adjoint model of fluorescence. *Appl Opt* 44:1917–1933.

Finlay, J. C., S. Mitra, and T. H. Foster. 2002. *In vivo* mTHPC photobleaching in normal rat skin exhibits unique irradiance-dependent features. *Photochem Photobiol* 75:282–288.

Finlay, J. C., S. Mitra, and T. H. Foster. 2004. Photobleaching kinetics of Photofrin *in vivo* and in multicell tumor spheroids indicate multiple simultaneous bleaching mechanisms. *Phys Med Biol* 49:4837–4860.

Finlay, J. C., T. C. Zhu, A. Dimofte et al. 2006. Interstitial fluorescence spectroscopy in the human prostate during motexafin lutetium-mediated photodynamic therapy. *Photochem Photobiol* 82:1270–1278.

Georgakoudi, I., and T. H. Foster. 1998. Singlet oxygen- *versus* nonsinglet oxygen-mediated mechanisms of sensitizer photobleaching and their effects on photodynamic dosimetry. *Photochem Photobiol* 67:612–625.

Georgakoudi, I., M. G. Nichols, and T. H. Foster. 1997. The mechanism of Photofrin photobleaching and its consequences for photodynamic dosimetry. *Photochem Photobiol* 65:135–144.

Hirschhaeuser, F., H. Menne, C. Dittfeld et al. 2010. Multicellular tumor spheroids: An underestimated tool is catching up again. *J Biotechnol* 148:3–15.

Hu, X. H., Y. Feng, J. Q. Lu et al. 2005. Modeling of a type II photofrin-mediated photodynamic therapy process in a heterogeneous tissue phantom. *Photochem Photobiol* 81:1460–1468.

Jarvi, M. T., M. J. Niedre, M. S. Patterson, and B. C. Wilson. 2011. The influence of oxygen depletion and photosensitizer triplet-state dynamics during photodynamic therapy on accurate singlet oxygen luminescence monitoring and analysis of treatment dose response. *Photochem Photobiol* 87:223–234.

Jarvi, M. T., M. S. Patterson, and B. C. Wilson. 2012. Insights into photodynamic therapy dosimetry: Simultaneous singlet oxygen luminescence and photosensitizer photobleaching measurements. *Biophys J* 102:661–671.

Jimenez-Banzo, A., X. Ragas, P. Kapusta, and S. Nonell. 2008. Time-resolved methods in biophysics. 7. Photon counting vs. analog time-resolved singlet oxygen phosphorescence detection. *Photochem Photobiol Sci* 7:1003–1010.

Johansson, A., J. Axelsson, S. Andersson-Engels, and J. Swartling. 2007. Realtime light dosimetry software tools for interstitial photodynamic therapy of the human prostate. *Med Phys* 34:4309–4321.

Kienle, A., L. Lilge, M. S. Patterson et al. 1996. Spatially resolved absolute diffuse reflectance measurements for noninvasive determination of the optical scattering and absorption coefficients of biological tissue. *Appl Opt* 35:2304–2314.

Kienle, A., and M. S. Patterson. 1997. Determination of the optical properties of semi-infinite turbid media from frequency-domain reflectance close to the source. *Phys Med Biol* 42:1801–1819.

Maas, A. L., S. L. Carter, E. P. Wileyto et al. 2012. Tumor vascular microenvironment determines responsiveness to photodynamic therapy. *Cancer Res* 72:2079–2088.

Mesquita, R. C., S. W. Han, J. Miller et al. 2012. Tumor blood flow differs between mouse strains: Consequences for vasoresponse to photodynamic therapy. *PloS one* 7:e37322.

Mitra, S., and T. H. Foster. 2005. Photophysical parameters, photosensitizer retention and tissue optical properties completely account for the higher photodynamic efficacy of meso-tetra-hydroxyphenyl-chlorin vs Photofrin. *Photochem Photobiol* 81:849–859.

Muller, M. G., I. Georgakoudi, Q. Zhang, J. Wu, and M. S. Feld. 2001. Intrinsic fluorescence spectroscopy in turbid media: Disentangling effects of scattering and absorption. *Appl Opt* 40:4633–4646.

Nichols, M. G., and T. H. Foster. 1994. Oxygen diffusion and reaction kinetics in the photodynamic therapy of multicell tumour spheroids. *Phys Med Biol* 39:2161–2181.

Niedre, M., M. S. Patterson, and B. C. Wilson. 2002. Direct near-infrared luminescence detection of singlet oxygen generated by photodynamic therapy in cells *in vitro* and tissues *in vivo*. *Photochem Photobiol* 75:382–391.

Patterson, M. S., B. C. Wilson, J. W. Feather, D. M. Burns, and W. Pushka. 1987. The measurement of dihematoporphyrin ether concentration in tissue by reflectance spectrophotometry. *Photochem Photobiol* 46:337–343.

Patterson, M. S., B. C. Wilson, and R. Graff. 1990. *In vivo* tests of the concept of photodynamic threshold dose in normal rat liver photosensitized by aluminum chlorosulphonated phthalocyanine. *Photochem Photobiol* 51:343–349.

Robinson, D. J., H. S. de Bruijn, N. van der Veen et al. 1999. Protoporphyrin IX fluorescence photobleaching during ALA-mediated photodynamic therapy of UVB-induced tumors in hairless mouse skin. *Photochem Photobiol* 69:61–70.

Sitnik, T. M., and B. W. Henderson. 1998. Reduction of tumor oxygenation during and after photodynamic therapy *in vivo*: Effects of fluence rate. *Br J Cancer* 77:1386–1394.

Stefflova, K., J. Chen, D. Marotta, H. Li, and G. Zheng. 2006. Photodynamic therapy agent with a built-in apoptosis sensor for evaluating its own therapeutic outcome *in situ*. *J Med Chem* 49:3850–3856.

Tyrrell, J., C. Thorn, A. Shore, S. Campbell, and A. Curnow. 2011. Oxygen saturation and perfusion changes during dermatological methylaminolaevulinate photodynamic therapy. *Br J Dermatol* 165:1323–1331.

Wang, H. W., T. C. Zhu, M. E. Putt et al. 2005. Broadband reflectance measurements of light penetration, blood oxygenation, hemoglobin concentration, and drug concentration in human intraperitoneal tissues before and after photodynamic therapy. *J Biomed Opt* 10:14004.

Wang, K. K., J. C. Finlay, T. M. Busch, S. M. Hahn, and T. C. Zhu. 2010. Explicit dosimetry for photodynamic therapy: Macroscopic singlet oxygen modeling. *J Biophoton* 3:304–318.

Wang, K. K., S. Mitra, and T. H. Foster. 2007. A comprehensive mathematical model of microscopic dose deposition in photodynamic therapy. *Med Phys* 34:282–293.

Wang, K. K., S. Mitra, and T. H. Foster. 2008. Photodynamic dose does not correlate with long-term tumor response to mTHPC-PDT performed at several drug light intervals. *Med Phys* 35:3518–3526.

Wilson, B. C., M. S. Patterson, and L. Lilge. 1997. Implicit and explicit dosimetry in photodynamic therapy: A new paradigm. *Lasers Med Sci* 12:182–199.

Wu, J., M. S. Feld, and R. P. Rava. 1993. Analytical model for extracting intrinsic fluorescence in turbid media. *Appl Opt* 32:3583–3595.

Yu, G., T. Durduran, C. Zhou et al. 2005. Noninvasive monitoring of murine tumor blood flow during and after photodynamic therapy provides early assessment of therapeutic efficacy. *Clin Cancer Res* 11:3543–3552.

Yu, G., T. Durduran, C. Zhou et al. 2006. Real-time *in situ* monitoring of human prostate photodynamic therapy with diffuse light. *Photochem Photobiol* 82:1279–1284.

Zhou, X., B. W. Pogue, B. Chen et al. 2006. Pretreatment photosensitizer dosimetry reduces variation in treatment response. *Int J Radiat Oncol Biol Phys* 64:1211–1220.

Zhu, T. C., J. C. Finlay, and S. M. Hahn. 2005. Determination of the distribution of light, optical properties, drug concentration, and tissue oxygenation *in vivo* in human prostate during motexafin lutetium-mediated photodynamic therapy. *J Photochem Photobiol B Biol* 79:231–241.

Cell Death and PDT-Based Photooxidative (Phox) Stress

Abhishek D. Garg
University of Leuven (KULeuven)

Patrizia Agostinis
University of Leuven (KULeuven)

27.1 Introduction

The ability of photodynamic therapy (PDT) to cause cell death depends on a systematic process. This process starts with the accumulation of photosensitizer (PS) in cells followed by its activation (in the presence of oxygen) via light irradiation (of appropriate wavelength, which corresponds to the excitation spectra of that particular PS), all of which combined brings about the production of reactive oxygen species (ROS) within the cellular system (Agostinis et al. 2011). The ROS (or photooxidative stress) have the potential to bring about cell death by reacting with biomolecules, such as lipids or proteins, and activating cell-demise processes. Within the context of a tumor, in addition to the cancer cell-killing potential, PDT also has an ability to destroy tumor vasculature (causing tumor ischemia) as well as to activate the immunological system, thereby assisting in long-term tumor suppression (Agostinis et al. 2004; Garg et al. 2010a). Thus, there are many factors that influence the successful execution and pathways of PDT-based cell killing, for example, oxygen availability, concentration of PS, physicochemical properties and subcellular PS localization, suitable light wavelength and intensity/fluence, and cell type (Castano, Demidova, and Hamblin 2005).

In this chapter, we discuss how PDT-based photooxidative (phox)-stress leads to various cell death subroutines or photokilling and, as per current knowledge, what decides the nature and diversity of pathways that would finally seal the fate of the phox-stressed cell.

27.2 Cell Death: From Execution Biochemistry to Immunobiology

The field of PDT-induced cell death or photokilling has been constantly influenced and shaped by the trends in the broader cell death field such that any emerging cell death subroutine has (in due course of time) been associated with PDT-based phox-stress (retrospectively or prospectively) (Agostinis et al. 2011). Thus, before discussing the trends of phox-stress-induced cell death, it is important to have a bird's eye view of the broader cell death field. To this end, in the following sections, we have tried to briefly summarize the currently most prevalent cell death terminologies and cell death subroutine features, definitions, and characteristics.

27.2.1 Cell Death: Basic Biochemical Characteristics of Major Subroutines

Apoptosis and necrosis are the two main recognized cell death subroutines through which a cell may die under stress (Figure 27.1). Although the strict boundaries of these cell death

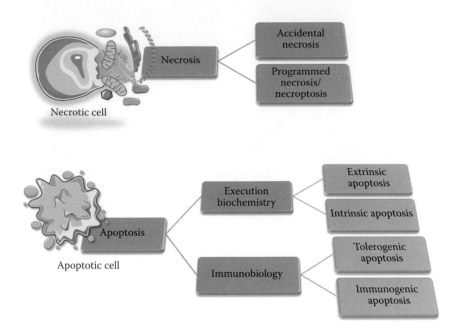

FIGURE 27.1   Cell death subroutines: from execution biochemistry to immunobiological profiles. As per the current knowledge, necrosis and apoptosis are the two main "undisputable" subroutines of cell death. As per execution biochemistry, necrosis can be divided into "accidental" or primary necrosis and "programmed" necrosis or necroptosis. Immunobiologically speaking, necrosis tends to be strongly proinflammatory irrespective of whether it is accidental or programmed. On the other hand, in terms of execution biochemistry, apoptosis is mainly divided into extrinsic and intrinsic apoptosis. However, immunobiologically considering, apoptosis can either be actively immunosuppressive or tolerogenic (tolerogenic apoptosis) or accompanied by enhanced immunogenicity (immunogenic apoptosis).

subroutines being either regulated or programmed (for apoptosis) or being unregulated or accidental (for necrosis) have started to blur in recent times, drastic biochemical and morphological differences between these subroutines still exist.

Necrosis is morphologically recognized by various cell biological signatures, including cytoplasmic swelling and breakdown of the plasma membrane, resulting in the release of the intracellular content and proinflammatory molecules (Figure 27.1). The biochemistry of necrosis is characterized mostly by the absence of caspase activation, cytochrome c release, and DNA oligonucleosomal fragmentation. Necrosis in most cases tends to be a passive, unorganized way to die and, hence, has been considered for a long time to be a kind of accidental cell-death mechanism also termed *accidental or primary necrosis*. However, recent evidence suggests that necrotic cell death can be actively propagated as part of a molecular signal transduction pathway (Vanlangenakker et al. 2008) in which the serine/threonine kinase receptor-interacting protein 1 (RIP1), caspase-8, and ROS are at the core of the molecular execution machinery (Vanlangenakker et al. 2008). This form of necrotic cell death is also termed *programmed necrosis* or *necroptosis* (Figure 27.1).

On the other hand, *apoptosis* (Figure 27.1) is a completely genetically well-controlled or programmed mode of cell death, which is distinguished by its distinctive morphological features, such as cell shrinkage, membrane blebbing, apoptotic body formations, cleavage of chromosomal DNA into internucleosomal fragments (DNA fragmentation), formation of apoptotic bodies without plasma membrane breakdown, and chromatin condensation (Plaetzer et al. 2003). Almost all of the apoptotic

pathways depend upon or converge on mitochondria, which is often regarded as the central processing organelle for apoptosis. Various apoptotic signals external or internal to the cell converge toward the mitochondrion, which then senses the stress and brings about the execution of apoptosis (Green and Reed 1998). A primary role of mitochondria in apoptosis is to release proteins, such as an apoptosis-inducing factor (AIF), certain procaspases, cytochrome c, DIABLO (direct IAP binding protein with low pI), and SMAC (second mitochondria-derived activator of caspases) into the cytoplasm in response to a variety of death stimuli (Oleinick, Morris, and Belichenko 2002). This release is usually achieved by the opening of pores in the mitochondrial outer membrane through the oligomerization of proapoptotic BAX/BAK proteins (Buytaert et al. 2006; Garrido and Kroemer 2004; Gottlieb 2000). The proteins thus released from the mitochondria activate apoptotic effectors, such as the caspases, which then bring about the signaling processes required for executing apoptosis. Caspases are a highly conserved family of cysteine-dependent, aspartate-specific proteases, of which 11 members have been identified in humans (Garrido and Kroemer 2004; Kroemer and Levine 2008). While most forms of apoptosis tend to be *caspase-dependent*, there do exist instances of *caspase-independent apoptosis* as well. Cells dying via caspase-independent apoptosis display all the main morphological characteristics of caspase-dependent apoptosis; however, rather than caspase activation, this phenomenon is regulated by the release of proteins, such as AIF, Omi/HtrA2, or endonuclease G from the mitochondria (Chipuk and Green 2005), which, at least in the case of AIF, is brought about through cleavage by calcium-dependent calpain proteases (Broker, Kruyt, and Giaccone 2005).

Where the apoptotic stimulus originates and reaches the mitochondria thereafter divides apoptosis into two types: extrinsic apoptosis and intrinsic apoptosis (Figure 27.1). *Extrinsic apoptosis*, as the name suggests, is mediated by extrinsic stimuli or ligands that activate the so-called surface death receptors (predominantly belonging to the tumor necrosis factor receptor or TNFR family), which then transmit the apoptotic signal to the mitochondria (via death-inducing signaling complex or DISC formed through the receptor's death domains) (Ashkenazi 2002; Naismith and Sprang 1998; Scaffidi et al. 1998). Caspase-8 is the main initiator caspase for extrinsic apoptosis (Scaffidi et al. 1998). On the other hand, the stimuli that lead to *intrinsic apoptosis* tend to originate inside the cells, that is, DNA damage, oxidative stress, starvation, and stress on particular organelles as well as all stress spectra that are instigated by chemotherapy, radiotherapy, or PDT (and other such approved or experimental therapies) (Kaufmann and Earnshaw 2000; Wang 2001). Here, caspase-9 tends to be the main initiator caspase. In the case of intrinsic apoptosis, the site of stress within the cell [e.g., endoplasmic reticulum (ER), lysosomes] can instigate its own unique signaling cascade that transmits the apoptotic signal to the mitochondria. This latter point is heavily relevant to PDT-induced cell death as will be discussed later.

Last, but not least, is the controversial cell death subroutine that is often referred to as *autophagic cell death*, which has recently started to emerge as a possible misnomer when compared to other cell-death terminologies (Kroemer and Levine 2008). By definition, autophagic cell death is a subroutine that occurs in absence of chromatin condensation but is accompanied by a massive amount of cytoplasmic autophagic vacuolization (Kroemer and Levine 2008). It has been proposed that this excessive autophagy in such cells may promote cell death through excessive self-digestion and degradation of essential cellular constituents or by instigating an ill-defined, caspase-independent cell-death pathway (Maiuri et al. 2007). However, it has, in recent times, strongly emerged that autophagy probably cannot execute cell death on its own, but, in fact, it accompanies cell death more as a last attempt of cells to adapt to the lethal stress and promote survival (Shen, Kepp, and Kroemer 2012). A recent large-scale screening study showed that of 1400 cytotoxic compounds tested, while a considerable number of compounds induced autophagic flux, not a single one turned out to be killing cells through autophagy induction because the knockdown of autophagy-essential genes (ATG5 or ATG7) failed to reduce cell death (Shen et al. 2011). In fact, knockdown of autophagy-essential genes in the respective treatment conditions accelerated cell death, thereby substantiating the survival-supporting role of autophagy accompanying cell death (Shen et al. 2011). Moreover, our own studies have shown that autophagy activation under conditions of PDT treatment acts more as a cellular survival pathway rather than a cell-death execution pathway (Dewaele et al. 2011). To this end, in the current chapter, we have not considered autophagic cell death as a bona fide cell-death subroutine, and the discussions have been mainly focused on the more incontrovertible cell-death subroutines, that is, necrosis, apoptosis, and their biochemical or immunobiological variants (as applicable).

27.2.2 Cell Death: Immunobiological Characteristics of Subroutines

For a better part of the last three decades, the primary focus of cell-death characterization studies has been to identify the biochemical, molecular, and morphological features of cell-death subroutines. Thus, not surprisingly (as apparent from the previous discussion), most of the classifications of cell-death subroutines have been based on biochemical parameters [e.g., the presence or absence of phosphatidylserine (PS) externalization], molecular parameters (e.g., presence or absence of caspase activation), or morphological parameters (e.g., the presence or absence of blebs or cell shrinkage) (Figure 27.1). However, in recent times, another parameter of cell-death subclassification has emerged strongly, that is, immunobiology. As per this parameter, the cell-death subroutines can be further subclassified as per their ability to mediate immunomodulation (on the level of both underlying signaling as well as the actual immunological elicitation).

In case of *necrosis*, irrespective of whether it is accidental or primary or programmed, this cell-death subroutine can be highly inflammatory, sometimes even to the extent of being harmful (Vakkila and Lotze 2004), because of the sudden release of various intracellular factors (Peter, Wesselborg, and Lauber 2009). Major immunostimulatory molecules released by necrosis have been discussed elsewhere (Garg et al. 2010b). These molecules tend to sensitize and attract various professional and nonprofessional phagocytic immune cells, which further secrete different proinflammatory cytokines, chemokines, or enzymes (e.g., IL-8, macrophage-inflammatory protein 2, and TNF) (Napirei and Mannherz 2009).

On the other hand, most forms of apoptosis tend to be immunologically silent or even tolerogenic (Tesniere et al. 2008)—hence the term *tolerogenic apoptosis* (Figure 27.1). This apparent immunological tolerogenicity toward apoptosis is considered to be a mechanism playing an important role in the host's protection (Tesniere et al. 2008). Cells undergoing tolerogenic apoptosis are usually quickly recognized by phagocytes (Savill et al. 2002) and subsequently cleaned up via phagocytosis as a result of surface exposure of various "eat me" signals [e.g., modified ICAM-3, PS, phosphatidylethanolamine, phosphatidylinositol, modified low-density lipoproteins (LDLs), PTX3 binding sites, opsonins, thrombospondin binding sites, lactoferrin, and HRG1 binding sites] (Napirei and Mannherz 2009) and suppression of "don't eat me" signals [such as plasma membrane exposed CD31 (Brown et al. 2002) and CD47 (Gardai et al. 2005)]. Phagocytic cleanup of tolerogenic apoptotic cells is carried out without eliciting any major immunological response because of the release or exposure of anti-inflammatory "eat me," "find me," or other such signals (Napirei and Mannherz 2009). In addition to these autocrine and paracrine effects of tolerogenic apoptosis, it has been observed that (Birge and Ucker 2008) these apoptotic cells can also exert their anti-inflammatory effects directly, by binding to macrophages, independent of phagocytosis process or soluble factors.

While tolerogenic apoptosis is essentially a textbook view of an apoptotic immunobiological profile, it is worth noting that the generalization of this view on all the apoptotic subroutines has come under scrutiny recently. It has been discovered that certain chemotherapeutic agents or apoptotic stimuli, which are capable of exerting ROS-based ER stress, induce an apoptotic subroutine accompanied by enhanced immunogenicity termed as *immunogenic apoptosis* (Garg et al. 2010b; Kepp et al. 2009) (Figure 27.1). Immunogenic apoptosis has all the major biochemical hallmarks of tolerogenic apoptosis except that it possesses two main properties absent in the latter: (1) the ability to expose or secrete vital immunogenic signals or damage-associated molecular patterns (DAMPs) and (2) the ability to activate the host's immune system [maturation of antigen-presenting cells (APCs) and priming of the adaptive immune cells] (Panaretakis et al. 2009; Zitvogel et al. 2010). DAMPs (also called danger signals, alarmins, or leaderless secretory proteins) are molecules that are intracellular under normal conditions but are immunostimulatory or proinflammatory when they are secreted or surface-exposed by the damaged or dying cells (Bianchi 2007; Garg et al. 2010b; Verfaillie, Garg, and Agostinis 2013). The type, diversity, and mode of emergence (exposure, secretion, or release) of DAMPs are usually intricately associated with the biochemistry of a particular cell-death pathway (Garg et al. 2011; Zitvogel et al. 2010). Here, DAMPs that are vital for this subroutine include surface-exposed calreticulin (ecto-CRT; a vital "eat me" signal) (Garg et al. 2012c,d; Obeid et al. 2007; Panaretakis et al. 2009), surface-exposed HSP90 (ecto-HSP90) (Spisek et al. 2007), secreted ATP (a "find me" signal and inflammasome activator) (Garg et al. 2012d; Ghiringhelli et al. 2009), released HMGB1 (vital for proper antigen presentation) (Green et al. 2009; Zitvogel et al. 2010), and secreted or released heat shock proteins, such as HSP70, CRT, and HSP90 (capable of causing DC maturation, vital for proper antigen processing and important tumor antigen carriers) (Garg et al. 2010b, 2011; Green et al. 2009; Kepp et al. 2009; Zitvogel, Kepp and Kroemer 2010). More information on DAMPs and their association with immunogenic apoptosis is available in other reviews (Garg et al. 2010b, 2011; Green et al. 2009; Krysko et al. 2011; Zitvogel et al. 2010).

27.3 Cell Death and PDT-Based Phox-Stress: A Bird's Eye View of the Dose and Cell-Death Subroutine Relationship

With the discovery of apoptosis in 1972, the distinctions between necrosis and apoptosis started to be elucidated (Kerr, Wyllie, and Currie 1972). It was recognized that, unlike necrosis, apoptosis was a mode of cell death that minimized overall inflammation because its products were easily recycled or removed

(as discussed previously). It was not until 1991 that apoptosis was reported to be occurring in response to PDT-based phox-stress (Agarwal et al. 1991). This discovery increased the interest of researchers in the cellular mechanisms associated with PDT, that is, apoptosis, repair and survival, and necrosis (Agostinis et al. 2011; Plaetzer et al. 2003).

Over the years, it has been observed that the severity of phox-stress-induced damage defines the particular response of cells to phox-stress (Figure 27.2). Creation of phox-stress along with some thermal stress resulting from various photochemical reactions leads to production of both lethal (pro-death) as well as sublethal (pro-death coupled with pro-survival) signals within the target cell (Moor, Ortel, and Hasan 2003). The overall decision of a cell to commit toward a particular cell-death pathway or survival pathway at this point depends upon the phox-stress or PDT dose applied such that the subcellular localization and intensity of the stress regulate the pathway of cell death that a particular cell might follow in order to die following phox-stress (Agostinis et al. 2004; Buytaert, Dewaele, and Agostinis 2007; Garg et al. 2012a; Moor, Ortel, and Hasan 2003). The escalating stress is usually responsible for severe damage to various subcellular organelles and large or small biomolecules. All of these factors together might cause as well as govern the cell's ability to die via particular mechanisms, such as apoptosis, necrosis, or cell death accompanied by autophagy.

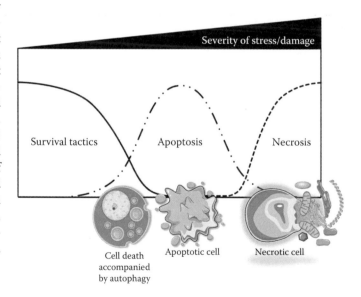

FIGURE 27.2 Relationship between stress and cellular prosurvival or pro-death responses. After cells have been exposed to stress, the cellular responses oscillate from prosurvival to apoptosis to necrosis as the amount or severity of stress goes on increasing. Low-level stress would usually cause activation of predominantly prosurvival tactics with little or no cell death. Sublethal stress would usually cause activation of mixed responses involving apoptosis as well as cells with high autophagy or other prosurvival pathways. Thereafter, as stress keeps increasing, the cellular responses would start shifting from purely apoptotic to ones with a mixture of apoptosis and necrosis, finally giving way to purely necrotic responses in case of tremendously severe stress.

To this end, *low phox-stress/PDT doses* (Table 27.1) usually lead to activation of predominantly survival tactics, for example, autophagy, p38MAPK signaling, or JNK signaling (Figure 27.2). A very low amount of apoptosis might be expected at a low phox-stress dose; however, the occurrence of (accidental) necrosis is negligible or nonexistent. Consequently, *medium phox-stress/ PDT doses* (Table 27.1) usually lead to an increase in the proportions of apoptotic cells as well as cells in which apoptosis is accompanied by autophagy (Figure 27.1). Also, at this medium dose, the presence of accidental necrosis is negligible. Furthermore, *high phox-stress/PDT doses* (Table 27.1) cause significant amounts of apoptosis in the treated cells. Such cells might still have autophagy activation within them, and accidental necrosis might be observed albeit in nonsignificant amounts. However, a further increase in dose, that is, the *highest phox-stress/PDT dose* (Table 27.1) can cause a significant amount of apoptosis possibly accompanied by a certain amount of initial accidental necrosis. Presence of cells with autophagy is still expected at this dose, however, to a lower extent than at low, medium, or high doses. Doses beyond the latter (Table 27.1) would usually lead to massive accidental necrosis of cells. However, a small amount of apoptosis and few surviving cells might still be expected. Thus, as one keeps increasing the treatment dose, the cell responses move from predominantly survival-oriented to apoptosis, from predominantly apoptosis to a considerable amount of necrosis, and finally, to massive amounts of necrosis (Garg et al. 2012a,d). In general, beyond the low phox-stress doses, a particular population of cells treated with phox-stress would be possibly composed of all three cellular response states, that is, cells showing activation of survival tactics, apoptotic cells, and necrotic cells with overlaps between the bell curves of these responses (Figure 27.2). The proportions of occurrences between these three states within a cell population at a particular given point in time would depend upon the severity of stress or damage (Garg et al. 2012a).

PDT factors, such as the type of PS, PS localization at time of irradiation, total duration and dose of treatment, light fluence rate, type of cells, and level of oxygenation modulate the cellular responses to phox-stress (Agostinis et al. 2004; Buytaert, Dewaele, and Agostinis 2007; Castano, Demidova, and Hamblin 2005; Garg et al. 2012a). In the end, the combination of PS concentrations and irradiation fluencies can produce either "pure"

or differing proportions of necrosis or apoptosis within particular treated populations of cells. The exact cell-death mechanism following phox-stress might well depend upon the site of maximum photodamage within the treated cells (especially in the case of apoptosis), which, of course, coincides with the subcellular localization site for a PS (more often than not). Thus, PS localizing in mitochondria (Garg et al. 2012a), ER (Garg et al. 2012d), or lysosomes may instigate unique organelle-associated signaling pathways culminating in apoptosis (Buytaert, Dewaele, and Agostinis 2007). In the following sections, the cell-death responses to phox-stress are discussed in detail.

27.4 Phox-Stress-Based Accidental Necrosis

Necrosis, as discussed previously, occurs as a result of extreme external physical stress or severe intracellular damage and is irreversible (Plaetzer et al. 2003). In the case of phox-stress/ PDT, high PS concentrations, production of ROS, and high fluences usually lead to necrosis (Buytaert, Dewaele, and Agostinis 2007; Garg et al. 2012a). Lipophilic dyes, anionic dyes, or PSs (especially those localizing in biomembranes) affect various unsaturated phospholipids and membrane cholesterol, leading to extensive lipid peroxidation (Zhou 1989). Peroxidation reactions of this kind during PDT treatment (usually mediated by phox-stress) led to extensive membrane changes, loss of membrane integrity and fluidity, and inactivation of membrane protein systems, leading to initiation of necrosis (Girotti 1990; Proskuryakov, Konoplyannikov, and Gabai 2003). Other than membrane enzymes, mitochondrial enzymes are also inhibited, leading to loss of cellular energy supply (a key event in phox-stress-based necrosis) (Gibson and Hilf 1985; Salet and Moreno 1990). Further events associated with phox-stress-based necrosis include DNA-repair inhibition, the stalling of membrane transport systems, and inactivation of lysosomal enzymes (Plaetzer et al. 2003). Researchers have also found that mitochondrial permeability transition (mPT), which results in loss of $\Delta\Psi_m$ (mitochondrial transmembrane potential), does not always lead to apoptosis (Yuan 2006). Persistent mPT activation might also cause necrosis, especially in an oxidative stress-induced environment (Garg et al. 2012a; Yuan 2006).

TABLE 27.1   Approximate Proportions of Cell Death at Various Phox-Stress/PDT Treatment Doses

Dose	Typical Cell Death Ranges
Low phox-stress/PDT doses	After 1–2 h, less than 5% cell death above untreated cells
	After 24 h, less than 20% cell death above untreated cells
Medium phox-stress/PDT doses	After 1–2 h, less than 5% above untreated cells
	After 24 h, approximately 50%–60% above untreated cells
High phox-stress/PDT doses	After 1–2 h, less than 10% above untreated cells
	After 24 h, approximately 60%–80% above untreated cells
Highest phox-stress/PDT doses	After 1–4 h, less than 50% above untreated cells
	After 24 h, approximately 70%–90% above untreated cells
Doses beyond the highest dose	After 1–2 h, approximately 30%–60% above untreated cells
	After 24 h, approximately 85%–95% above untreated cells

Thus, phox-stress can cause accidental necrosis either when high phox-stress doses are applied irrespective of the type of PS being used (Table 27.1) or when phox-stress is applied using a PS that predominantly localizes in the plasma membrane.

27.5 Phox-Stress-Based Programmed Necrosis or Necroptosis

The reporting of phox-stress-induced necroptosis has been rare. This is so also because this particular cell-death subroutine terminology is relatively young (introduced in 2005) as compared to accidental or primary necrosis or apoptosis (Degterev et al. 2005). Necroptosis is mainly regulated via RIP3 kinase, and it has recently been demonstrated that severe phox-mitochondrial (phox-mito) stress induced via 5-aminolevulinic acid (5-ALA)-PDT was capable of causing necroptosis (Coupienne et al. 2011). More specifically, it was found that singlet oxygen generated by 5-ALA-PDT-based phox-mito stress triggered RIP3-mediated programmed necrosis. The necrosome complex executing this mode of cell-death subroutine was found to be composed of RIP1 and RIP3 kinases but not caspase-8 and FADD (Coupienne et al. 2011).

It remains to be seen, though, how many other kinds of PSs can lead to necroptosis in a PDT setup or whether there are specific classes of PS that can cause necroptosis following PDT. Of course, it may not be ruled out that this might be simply a cell type-dependent phenomenon such that in a necroptosis-susceptible cell type (or a cell type with defects in caspase-activation pathways), any kind of phox-stress methodology might lead to necroptosis.

27.6 Phox-Stress-Based Apoptosis: Extrinsic or Intrinsic Biochemical Execution Pathways

Over the last decades, new biotechnological and imaging technologies have paved the way toward the confirmation that, indeed, apoptosis is the main cell-death pathway of PDT-mediated cell death *in vivo*, thus justifying the major efforts that have gone into studying the biochemistry of the apoptotic photo-killing process *in vitro* (Agostinis et al. 2011; Banihashemi et al. 2008; Bhuvaneswari et al. 2009). The mechanisms and molecular cascades activated by phox-stress are incredibly complex and have been the subjects of recent extensive reviews (Agostinis et al. 2011; Buytaert, Dewaele, and Agostinis 2007; Oleinick, Morris, and Belichenko 2002; Plaetzer et al. 2003).

Phox-stress is capable of instigating both the main apoptotic biochemical execution cascades, that is, *extrinsic or death-receptor (DR)* and the *intrinsic or mitochondrial* pathways (Buytaert, Dewaele, and Agostinis 2007). It has been observed that PDT-based extrinsic apoptosis is usually activated by the cytokines, which are released by the phox-stressed or dying cells (Buytaert, Dewaele, and Agostinis 2007). On the other hand, induction of intrinsic apoptotic cell death following phox-stress

tends to depend on either direct mitochondrial damage (via phox-mito stress, for instance) or is secondary to signaling pathways activated by the phox-stress-based damage to other subcellular sites, for example, lysosomes and the ER. As is the case for general apoptotic executions, for PDT as well, mitochondrial membrane permeabilization (MMP) tends to be a crucial lethal event such that it is largely p53-independent but tightly controlled by Bcl-2 family members (Buytaert, Dewaele, and Agostinis 2007). In this phox-stress-based apoptotic pathway, the multidomain BAX and BAK proapoptotic proteins tend to play a vital role (Buytaert et al. 2006).

Interestingly, it has been observed that caspase inhibition in a phox-stress setup usually only delays photo-killing but does not prevent it (as is the case for typical chemical or chemotherapeutic apoptosis inducers). This indicates that the phox-stress-mediated cell death signal can also be propagated in a *caspase-independent* manner. Moreover, caspase inhibition or genetic deficiency in key components of the apoptotic machinery has been observed to convert phox-stress-based apoptosis into necrosis (Buytaert, Dewaele, and Agostinis 2007) or necroptosis (Coupienne et al. 2011). Additionally, the phox-stress-instigated cell-death subroutines have been recently shown to be influenced by the stimulation of autophagy (Buytaert, Dewaele, and Agostinis 2007; Martinet et al. 2009). It has been observed that activation of autophagy may impede phox-stress-mediated cell killing.

On the other hand, specifically, in the case of phox-stress-based intrinsic apoptosis, subcellular localization of the PS may define what organelle-based signaling would finally guide the path toward intrinsic apoptosis. This is because the location of a PS within a cell is usually the primary cellular target for photochemical reactions because the ROS produced as a result of the photochemical reactions have a very short life (<0.05 μs), high reactivity, and a diffusion distance of <0.02 μm away from the site of production (He et al. 2008; Moan and Berg 1991). Therefore, it is most likely that these ROSs would exert their effects in and around the location of the PS's subcellular localization. Accordingly, PSs localizing near plasma membranes would favor necrosis because of their membrane-disrupting properties (as mentioned previously) rather than PSs localizing in mitochondria, cytoplasm, or other organelles, which might initiate other cell-death pathways that ultimately culminate in apoptosis (Plaetzer et al. 2003). Further, some research has also shown that after irradiation with light, the PS might relocalize itself, thereby initiating damage to subcellular components other than those in which the PS localized before irradiation (Berg et al. 1991; Kessel 2002; Marchal et al. 2007). PS subcellular localization tends to be regulated by a number of criteria. Based upon purely physico-chemical criteria, hydrophilic PSs usually localize preferentially in cytoplasm, mitochondria, and/or lysosomes while hydrophobic PSs usually localize in membranes (Plaetzer et al. 2003). However, in certain cases, cell type and PS concentration might also define the subcellular location of the PS (He et al. 2008); for example, Rose bengal (RB) has been shown to localize in the outer membrane of murine monocytes (Kochevar et al. 1994), but, in HeLa cells, RB localizes in mitochondria and Golgi apparatus (Bottone et al. 2007; Soldani et al. 2004). In case of hydrophobic PSs, PS subcellular localization may also be influenced by the affinity of a particular PS for a particular phospholipid; for example, the affinity of hypericin (Hyp) for phosphatidylcholine (Fox et al. 1998; Lajos et al. 2009; Lavie et al. 1995) might be the reason behind its preferential localization in the ER membranes (Garg et al. 2012b,d) because the cellular concentrations of phosphatidylcholine tend to increase as we go from the plasma membrane toward the ER such that the ER has the highest final concentration of phosphatidylcholine within a cell (van Meer, Voelker, and Feigenson 2008). This theory is further supported by the observation that SERCA2, the ER trans-membrane protein photodamaged by Hyp-based PDT (Buytaert et al. 2006), has been proposed to reside in a phosphatidylcholine-rich environment (Sonntag et al. 2011).

As per current knowledge, three organelle-based signaling pathways that culminate in intrinsic apoptosis have been mainly studied for phox-stress (Buytaert et al. 2007), that is, (1) ER-based apoptosis, (2) mitochondria (mito)-based apoptosis, and (3) lysosome (lyso)-based apoptosis. Of course, as discussed previously, these three signaling cascades are instigated by phox-stress involving PS associated with these respective organelles. These intrinsic apoptotic signaling cascades are discussed briefly in the following sections. It should be noted that apart from the balance between pro-death and prosurvival stress signaling (modulated by the intensity of stress) that originates as a result of phox-stress directed against the respective organelles, the usual cell-death resistance tactics downstream of mitochondrial apoptotic signaling (e.g., ablation of proapoptotic proteins and caspases, increased autophagic flux, or overexpression of antiapoptotic proteins) would be applicable to each of these signaling pathways irrespective of the site of the origin of phox-stress.

27.6.1 Photooxidative ER (Phox-ER) Stress-Based Apoptosis

Phox-stress originating as a result of PDT-based activation of ER-associated PSs (phox-ER stress) has been shown to cause the upregulation of several ER chaperones, including glucose-regulated protein or binding protein (GRP78/Bip), GRP94, and protein disulfide isomerase (PDI), along with the activation of the ER stress response, that is, the unfolded protein response (UPR) or UPR^{ER} (Agostinis et al. 2004; Buytaert, Dewaele, and Agostinis 2007; Garg et al. 2012d). Phox-ER stress is also capable of causing the release of Ca^{2+} ions from the ER into the cytosol (Hubmer et al. 1996). Moreover, (lethal) phox-ER stress can activate the CCAAT/enhancer-binding protein (C/EBP) homologous protein (CHOP), which is a key proapoptotic transcription factor in the ER stress response. Altogether, severe phox-ER stress has the ability to convey toxic signals via UPR, which are capable of bringing about MMP-based apoptosis through the expression of different BH3-only proapoptotic proteins, such as Bim (Puthalakath et al. 2007) and Noxa (Verfaillie et al. 2012).

We have recently observed that, of the three main UPR^{ER} sensors (i.e., PERK, IRE1, and ATF6) capable of conveying the pro-death signal, phox-ER stress tends to predominantly depend upon PERK for transmitting the pro-death (ROS and Ca^{2+}-based) signal from the ER to the mitochondria, which ultimately culminates in BAX/BAK-dependent apoptosis (Verfaillie et al. 2012).

It is worth mentioning here that phox-ER stress-based UPR^{ER} signaling has the ability to engage both prosurvival as well as pro-death cascades. The selection between these two cascades depends upon the intensity of phox-ER stress such that severe phox-ER stress (resulting from medium to high phox-stress doses; Table 27.1) engages more pro-death cascades (e.g., apoptosis), while mild or low phox-ER stress (resulting from low to medium phox-stress doses; Table 27.1) engages more prosurvival cascades (e.g., autophagy, p38MAPK signaling, antioxidant signaling, and JNK signaling) (Verfaillie, Garg, and Agostinis 2013).

27.6.2 Photooxidative Mitochondrial (Phox-Mito) Stress-Based Apoptosis

Phox-stress originating as a result of PDT-based activation of mitochondria-associated PSs (phox-mito stress) has been shown to directly damage mitochondria-associated antiapoptotic Bcl-2 proteins (Kessel and Castelli 2001), thereby facilitating BAX- or BAK-mediated MMP and the subsequent release of caspase activators, such as cytochrome c and SMAC/DIABLO or other pro-apoptotic molecules, including AIF.

Moreover, it has also been observed that phox-mito stress can cause considerable peroxidation of lipids present in the mitochondrial membrane, thereby leading to rapid loss of mitochondrial trans-membrane potential ($\Delta\Psi_m$) (Bernardi 1992; Lam, Oleinick, and Nieminen 2001). This reduction in $\Delta\Psi_m$ leads to the opening up of voltage-dependent anion channels (VDACs) as well as a permeability transition pore complex (PTPC) (Bernardi 1992; Oleinick, Morris and Belichenko 2002). Peroxidation of lipids surrounding an adenine nucleotide translocator (ANT) might also lead to formation of channel-like PTPCs, which might disrupt the mitochondrial membrane integrity (Plaetzer et al. 2003). Further to this, intramitochondrial Ca^{2+} ions may also amplify depolarization of membrane, finally contributing to pore opening and structural changes in mitochondrial membrane proteins (Kowaltowski and Castilho 1997; Kowaltowski, Castilho, and Vercesi 2001).

On the level of the mitochondria alone, a cell's ability to resist overall MMP or $\Delta\Psi_m$ perturbations (resulting from differing mitochondrial biochemistry, increased antiapoptotic Bcl-2 protein signaling, or mitophagy) partly underlies its ability to resist phox-mito stress-induced apoptosis (Garg et al. 2012a). For future, it would be interesting to know whether phox-mito stress is capable of activating the, as yet, not so well-characterized process of mitochondrial UPR (Mt-UPR or UPR^{mt}) (Haynes and Ron 2010). This would also explain whether, as applicable to UPR^{ER}, phox-mito stress can also engage pro-death and prosurvival UPR^{mt} signaling.

27.6.3 Photooxidative Lysosomal (Phox-Lyso) Stress-Based Apoptosis

Phox-stress originating as a result of PDT-based activation of lysosome-associated PSs (phox-lyso stress) has been found to cause the release of cathepsins from photodamaged lysosomes, resulting in the cleavage of the BH3-only proapoptotic Bid protein and consequent MMP (Buytaert, Dewaele, and Agostinis 2007; Kessel and Luo 1998). Cathepsins also have the ability to directly activate caspases (Hishita et al. 2001; Ishisaka et al. 1999) and promote intrinsic apoptosis (Stoka et al. 2001) following phox-lyso stress.

27.7 Phox-Stress-Based Apoptosis: Tolerogenic and Immunogenic Immunobiological Profiles

Most cells undergoing apoptosis following phox-stress have been shown to exhibit tolerogenic immunobiological profiles. Although, in certain cases, it has been shown that phox-stress is capable of accentuating immunogenicity (Garg et al. 2010a, 2011; Gollnick and Brackett 2010), these studies have mostly consisted of producing tumor lysates, that is, a predominantly necrotic phox-stress-based preparation of cells rather than an apoptotic one. Moreover, although phox-stress (especially phox-mito stress) has been recurrently shown to cause surface exposure (ecto-) or extracellular release (exo-) of DAMPs (Garg et al. 2011), such as ecto-CRT, ecto-HSP70, ecto-HSP60, ecto-GRP94, ecto-GRP78, and exo-HSP70 (Korbelik and Sun 2006; Korbelik, Sun, and Cecic 2005; Korbelik, Zhang, and Merchant 2011), these studies have failed to show whether the apoptosis observed after phox-stress was accompanied by enhanced immunogenicity (i.e., immunogenic apoptosis) or not.

Conventionally, immunogenic apoptosis tends to be stressor-dependent in that only selected agents are capable of inducing it and these include mitoxantrone, doxorubicin (Obeid et al. 2007; Panaretakis et al. 2009), oxaliplatin (Tesniere et al. 2010), UVC irradiation, γ-irradiation (Obeid et al. 2007; Panaretakis et al. 2009), bortezomib (Spisek et al. 2007), cyclophosphamide (Schiavoni et al. 2011), combination treatment with cisplatin and thapsigargin (Martins et al. 2011), and combination treatment with heat shock plus UVC irradiation plus γ-irradiation (Zappasodi et al. 2010). As mentioned previously, this is because only these agents are apparently capable of inducing ROS-based ER stress, although this is predominantly off target in nature (in most cases) (Garg et al. 2012b). Recently, however, we reported induction of immunogenic apoptosis for the first time via predominantly on-target ROS-based ER stress (Garg et al. 2012b) induced through Hypericin-PDT-based phox-ER stress (Garg et al. 2012b,d). Phox-ER stress-induced immunogenic apoptosis was found to be accompanied by emergence of various bona fide immunological signatures of immunogenic apoptosis, that is, ecto-CRT, secreted ATP, and ecto-HSP70, in the preapoptotic stage (the stage preceding PS exposure) (Garg et al. 2012c,d).

In fact, our observations of phox-ER stress-based accentuated immunogenicity in a prophylactic immunization mice model have been recently substantiated in a therapeutic or curative mice model (Sanovic et al. 2011). This is the first instance of any kind of phox-stress inducing immunogenic apoptosis. Moreover, we found that Photofrin-PDT, which produces predominantly phox-mito stress and a mild amount of phox-ER stress, was not capable of inducing as much preapoptotic ecto-CRT as Hyp-PDT based on on-target phox-ER stress (Garg et al. 2012d).

It is interesting to note, however, that, it has been reported recently that mitoxantrone exhibits potential as a PS and that mitoxantrone-based PDT has the ability to induce potent cell death in the treated cells (Montazerabadi et al. 2012). Thus, it would be interesting to see in the future if mitoxantrone-based PDT could further accentuate (or partly account for) the currently observed ability of mitoxantrone to induce potent immunogenic apoptosis (Garg et al. 2010b). Nevertheless, it is crucial to characterize, in the near future, whether phox-mito stress and phox-lyso stress are also capable of inducing immunogenic apoptosis apart from phox-ER stress.

References

Agarwal, M. L., M. E. Clay, E. J. Harvey et al. 1991. Photodynamic therapy induces rapid cell death by apoptosis in L5178Y mouse lymphoma cells. *Cancer Res* 51:5993–5996.

Agostinis, P., K. Berg, K. A. Cengel et al. 2011. Photodynamic therapy of cancer: an update. *CA Cancer J Clin* 61:250–281.

Agostinis, P., E. Buytaert, H. Breyssens, and N. Hendrickx. 2004. Regulatory pathways in photodynamic therapy induced apoptosis. *Photochem Photobiol Sci* 3:721–729.

Ashkenazi, A. 2002. Targeting death and decoy receptors of the tumour-necrosis factor superfamily. *Nat Rev Cancer* 2:420–430.

Banihashemi, B., R. Vlad, B. Debeljevic et al. 2008. Ultrasound imaging of apoptosis in tumor response: Novel preclinical monitoring of photodynamic therapy effects. *Cancer Res* 68:8590–8596.

Berg, K., K. Madslien, J. C. Bommer et al. 1991. Light induced relocalization of sulfonated meso-tetraphenylporphines in NHIK 3025 cells and effects of dose fractionation. *Photochem Photobiol* 53:203–210.

Bernardi, P. 1992. Modulation of the mitochondrial cyclosporin A-sensitive permeability transition pore by the proton electrochemical gradient. Evidence that the pore can be opened by membrane depolarization. *J Biol Chem* 267:8834–8839.

Bhuvaneswari, R., Y. Y. Gan, K. C. Soo, and M. Olivo. 2009. Targeting EGFR with photodynamic therapy in combination with Erbitux enhances *in vivo* bladder tumor response. *Mol Cancer* 8:94.

Bianchi, M. E. 2007. DAMPs, PAMPs and alarmins: All we need to know about danger. *J Leukoc Biol* 81:1–5.

Birge, R. B., and D. S. Ucker. 2008. Innate apoptotic immunity: The calming touch of death. *Cell Death Differ* 15:1096–1102.

Bottone, M. G., C. Soldani, A. Fraschini et al. 2007. Enzyme-assisted photosensitization with rose Bengal acetate induces structural and functional alteration of mitochondria in HeLa cells. *Histochem Cell Biol* 127:263–271.

Broker, L. E., F. A. Kruyt, and G. Giaccone. 2005. Cell death independent of caspases: A review. *Clin Cancer Res* 11:3155–3162.

Brown, S., I. Heinisch, E. Ross et al. 2002. Apoptosis disables CD31-mediated cell detachment from phagocytes promoting binding and engulfment. *Nature* 418:200–203.

Buytaert, E., G. Callewaert, N. Hendrickx et al. 2006. Role of endoplasmic reticulum depletion and multidomain proapoptotic BAX and BAK proteins in shaping cell death after hypericin-mediated photodynamic therapy. *FASEB J* 20:756–758.

Buytaert, E., M. Dewaele, and P. Agostinis. 2007. Molecular effectors of multiple cell death pathways initiated by photodynamic therapy. *Biochim Biophys Acta* 1776:86–107.

Castano, A. P., T. N. Demidova, and M. R. Hamblin. 2005. Mechanisms in photodynamic therapy: Part three—Photosensitiser pharmacokinetics, biodistribution, tumour localization and modes of tumour destruction. *Photodiagn Photodyn Ther* 2:91–106.

Chipuk, J. E., and D. R. Green. 2005. Do inducers of apoptosis trigger caspase-independent cell death? *Nat Rev Mol Cell Biol* 6:268–275.

Coupienne, I., G. Fettweis, N. Rubio, P. Agostinis, and J. Piette. 2011. 5-ALA-PDT induces RIP3-dependent necrosis in glioblastoma. *Photochem Photobiol Sci* 10:1868–1878.

Degterev, A., Z. Huang, M. Boyce et al. 2005. Chemical inhibitor of nonapoptotic cell death with therapeutic potential for ischemic brain injury. *Nat Chem Biol* 1:112–119.

Dewaele, M., W. Martinet, N. Rubio et al. 2011. Autophagy pathways activated in response to PDT contribute to cell resistance against ROS damage. *J Cell Mol Med* 15:1402–1414.

Fox, F. E., Z. Niu, A. Tobia, and A. H. Rook. 1998. Photoactivated hypericin is an anti-proliferative agent that induces a high rate of apoptotic death of normal, transformed, and malignant T lymphocytes: Implications for the treatment of cutaneous lymphoproliferative and inflammatory disorders. *J Invest Dermatol* 111:327–332.

Gardai, S. J., K. A. McPhillips, S. C. Frasch et al. 2005. Cell-surface calreticulin initiates clearance of viable or apoptotic cells through trans-activation of LRP on the phagocyte. *Cell* 123:321–334.

Garg, A. D., M. Bose, M. I. Ahmed, W. A. Bonass, and S. R. Wood. 2012a. *In vitro* studies on erythrosine-based photodynamic therapy of malignant and pre-malignant oral epithelial cells. *PLoS ONE* 7:e34475.

Garg, A. D., D. V. Krysko, P. Vandenabeele, and P. Agostinis. 2011. DAMPs and PDT-mediated photo-oxidative stress: Exploring the unknown. *Photochem Photobiol Sci* 10:670–680.

Garg, A. D., D. V. Krysko, P. Vandenabeele, and P. Agostinis. 2012b. The emergence of phox-ER stress induced immunogenic apoptosis. *Oncoimmunology* 1:787–789.

Garg, A. D., D. V. Krysko, P. Vandenabeele, and P. Agostinis. 2012c. Hypericin-based photodynamic therapy induces surface exposure of damage-associated molecular patterns like HSP70 and calreticulin. *Cancer Immunol Immunother* 61:215–221.

Garg, A. D., D. V. Krysko, T. Verfaillie et al. 2012d. A novel pathway combining calreticulin exposure and ATP secretion in immunogenic cancer cell death. *EMBO J* 31:1062–1079.

Garg, A. D., D. Nowis, J. Golab, and P. Agostinis. 2010a. Photodynamic therapy: Illuminating the road from cell death towards anti-tumour immunity. *Apoptosis* 15:1050–1071.

Garg, A. D., D. Nowis, J. Golab et al. 2010b. Immunogenic cell death, DAMPs and anticancer therapeutics: An emerging amalgamation. *Biochim Biophys Acta* 1805:53–71.

Garrido, C., and G. Kroemer. 2004. Life's smile, death's grin: Vital functions of apoptosis-executing proteins. *Curr Opin Cell Biol* 16:639–646.

Ghiringhelli, F., L. Apetoh, A. Tesniere et al. 2009. Activation of the NLRP3 inflammasome in dendritic cells induces IL-1beta-dependent adaptive immunity against tumors. *Nat Med* 15:1170–1178.

Gibson, S. L., and R. Hilf. 1985. Interdependence of fluence, drug dose and oxygen on hematoporphyrin derivative induced photosensitization of tumor mitochondria. *Photochem Photobiol* 42:367–373.

Girotti, A. W. 1990. Photodynamic lipid peroxidation in biological systems. *Photochem Photobiol* 51:497–509.

Gollnick, S. O., and C. M. Brackett. 2010. Enhancement of anti-tumor immunity by photodynamic therapy. *Immunol Res* 46:216–226.

Gottlieb, R. A. 2000. Mitochondria: execution central. *FEBS Lett* 482:6–12.

Green, D. R., T. Ferguson, L. Zitvogel, and G. Kroemer. 2009. Immunogenic and tolerogenic cell death. *Nat Rev Immunol* 9:353–363.

Green, D. R., and J. C. Reed. 1998. Mitochondria and apoptosis. *Science* 281:1309–1312.

Haynes, C. M., and D. Ron. 2010. The mitochondrial UPR—Protecting organelle protein homeostasis. *J Cell Sci* 123:3849–3855.

He, Y. Y., S. E. Council, L. Feng, M. G. Bonini, and C. F. Chignell. 2008. Spatial distribution of protein damage by singlet oxygen in keratinocytes. *Photochem Photobiol* 84:69–74.

Hishita, T., S. Tada-Oikawa, K. Tohyama et al. 2001. Caspase-3 activation by lysosomal enzymes in cytochrome c-independent apoptosis in myelodysplastic syndrome-derived cell line P39. *Cancer Res* 61:2878–2884.

Hubmer, A., A. Hermann, K. Uberriegler, and B. Krammer. 1996. Role of calcium in photodynamically induced cell damage of human fibroblasts. *Photochem Photobiol* 64:211–215.

Ishisaka, R., T. Utsumi, T. Kanno et al. 1999. Participation of a cathepsin L-type protease in the activation of caspase-3. *Cell Struct Funct* 24:465–470.

Kaufmann, S. H., and W. C. Earnshaw. 2000. Induction of apoptosis by cancer chemotherapy. *Exp Cell Res* 256:42–49.

Kepp, O., A. Tesniere, L. Zitvogel, and G. Kroemer. 2009. The immunogenicity of tumor cell death. *Curr Opin Oncol* 21:71–76.

Kerr, J. F., A. H. Wyllie, and A. R. Currie. 1972. Apoptosis: A basic biological phenomenon with wide-ranging implications in tissue kinetics. *Br J Cancer* 26:239–257.

Kessel, D. 2002. Relocalization of cationic porphyrins during photodynamic therapy. *Photochem Photobiol Sci* 1:837–840.

Kessel, D., and M. Castelli. 2001. Evidence that bcl-2 is the target of three photosensitizers that induce a rapid apoptotic response. *Photochem Photobiol* 74:318–322.

Kessel, D., and Y. Luo. 1998. Mitochondrial photodamage and PDT-induced apoptosis. *J Photochem Photobiol B* 42: 89–95.

Kochevar, I. E., J. Bouvier, M. Lynch, and C. W. Lin. 1994. Influence of dye and protein location on photosensitization of the plasma membrane. *Biochim Biophys Acta* 1196:172–180.

Korbelik, M., and J. Sun. 2006. Photodynamic therapy-generated vaccine for cancer therapy. *Cancer Immunol Immunother* 55:900–909.

Korbelik, M., J. Sun, and I. Cecic. 2005. Photodynamic therapy-induced cell surface expression and release of heat shock proteins: relevance for tumor response. *Cancer Res* 65:1018–1026.

Korbelik, M., W. Zhang, and S. Merchant. 2011. Involvement of damage-associated molecular patterns in tumor response to photodynamic therapy: Surface expression of calreticulin and high-mobility group box-1 release. *Cancer Immunol Immunother* 60:1431–1437.

Kowaltowski, A. J., and R. F. Castilho. 1997. Ca2+ acting at the external side of the inner mitochondrial membrane can stimulate mitochondrial permeability transition induced by phenylarsine oxide. *Biochim Biophys Acta* 1322: 221–229.

Kowaltowski, A. J., R. F. Castilho, and A. E. Vercesi. 2001. Mitochondrial permeability transition and oxidative stress. *FEBS Lett* 495:12–15.

Kroemer, G., and B. Levine. 2008. Autophagic cell death: The story of a misnomer. *Nat Rev Mol Cell Biol* 9:1004–1010.

Krysko, D. V., P. Agostinis, O. Krysko et al. 2011. Emerging role of damage-associated molecular patterns derived from mitochondria in inflammation. *Trends Immunol* 32:157–164.

Lajos, G., D. Jancura, P. Miskovsky, J. V. Garcìla-Ramos, and S. Sanchez-Cortes. 2009. Interaction of the photosensitizer hypericin with low-density lipoproteins and phosphatidylcholine: A surface-enhanced Raman scattering and surface-enhanced fluorescence study. *J Phys Chem C* 113: 7147–7154.

Lam, M., N. L. Oleinick, and A. L. Nieminen. 2001. Photodynamic therapy-induced apoptosis in epidermoid carcinoma cells. Reactive oxygen species and mitochondrial inner membrane permeabilization. *J Biol Chem* 276:47379–47386.

Lavie, G., Y. Mazur, D. Lavie, and D. Meruelo. 1995. The chemical and biological properties of hypericin—A compound with a broad spectrum of biological activities. *Med Res Rev* 15:111–119.

Maiuri, M. C., E. Zalckvar, A. Kimchi, and G. Kroemer. 2007. Self-eating and self-killing: Crosstalk between autophagy and apoptosis. *Nat Rev Mol Cell Biol* 8:741–752.

Marchal, S., A. Francois, D. Dumas, F. Guillemin, and L. Bezdetnaya. 2007. Relationship between subcellular localisation of Foscan and caspase activation in photosensitised MCF-7 cells. *Br J Cancer* 96:944–951.

Martinet, W., P. Agostinis, B. Vanhoecke, M. Dewaele, and G. R. De Meyer. 2009. Autophagy in disease: A double-edged sword with therapeutic potential. *Clin Sci (Lond)* 116:697–712.

Martins, I., O. Kepp, F. Schlemmer et al. 2011. Restoration of the immunogenicity of cisplatin-induced cancer cell death by endoplasmic reticulum stress. *Oncogene* 30:1147–1158.

Moan, J., and K. Berg. 1991. The photodegradation of porphyrins in cells can be used to estimate the lifetime of singlet oxygen. *Photochem Photobiol* 53:549–553.

Montazerabadi, A. R., A. Sazgarnia, M. H. Bahreyni-Toosi et al. 2012. Mitoxantrone as a prospective photosensitizer for photodynamic therapy of breast cancer. *Photodiagn Photodyn Ther* 9:46–51.

Moor, A. C. E., B. Ortel, and T. Hasan. 2003. Mechanisms of photodynamic therapy. In *Photodynamic Therapy*. T. Patrice, editor. The Royal Society of Chemistry, Cambridge, 19–57.

Naismith, J. H., and S. R. Sprang. 1998. Modularity in the TNF-receptor family. *Trends Biochem Sci* 23:74–79.

Napirei, M., and H. G. Mannherz. 2009. Molecules involved in recognition and clearance of apoptotic/necrotic cells and cell debris. In *Phagocytosis of Dying Cells*. D. V. Krysko, and P. Vandenabeele, editors. Springer Science + Business Media B.V., Berlin, 103–145.

Obeid, M., A. Tesniere, F. Ghiringhelli et al. 2007. Calreticulin exposure dictates the immunogenicity of cancer cell death. *Nat Med* 13:54–61.

Oleinick, N. L., R. L. Morris, and I. Belichenko. 2002. The role of apoptosis in response to photodynamic therapy: What, where, why, and how. *Photochem Photobiol Sci* 1:1–21.

Panaretakis, T., O. Kepp, U. Brockmeier et al. 2009. Mechanisms of pre-apoptotic calreticulin exposure in immunogenic cell death. *EMBO J* 28:578–590.

Peter, C., S. Wesselborg, and K. Lauber. 2009. Role of attraction and danger signals in the uptake of apoptotic and necrotic cells and its immunological outcome. In *Phagocytosis of Dying Cells*. D. V. Krysko, and P. Vandenabeele, editors. Springer Science + Business Media B.V., Berlin, 63–101.

Plaetzer, K., T. Kiesslich, T. Verwanger, and B. Krammer. 2003. The modes of cell death induced by PDT: An Overview. *Med Laser Appl* 18:7–19.

Proskuryakov, S. Y., A. G. Konoplyannikov, and V. L. Gabai. 2003. Necrosis: A specific form of programmed cell death? *Exp Cell Res* 283:1–16.

Puthalakath, H., L. A. O'Reilly, P. Gunn et al. 2007. ER stress triggers apoptosis by activating BH3-only protein Bim. *Cell* 129:1337–1349.

Salet, C., and G. Moreno. 1990. Photosensitization of mitochondria. Molecular and cellular aspects. *J Photochem Photobiol B* 5:133–150.

Sanovic, R., T. Verwanger, A. Hartl, and B. Krammer. 2011. Low dose hypericin-PDT induces complete tumor regression in BALB/c mice bearing CT26 colon carcinoma. *Photodiagn Photodyn Ther* 8:291–296.

Savill, J., I. Dransfield, C. Gregory, and C. Haslett. 2002. A blast from the past: clearance of apoptotic cells regulates immune responses. *Nat Rev Immunol* 2:965–975.

Scaffidi, C., S. Fulda, A. Srinivasan et al. 1998. Two CD95 (APO-1/Fas) signaling pathways. *EMBO J* 17:1675–1687.

Schiavoni, G., A. Sistigu, M. Valentini et al. 2011. Cyclophosphamide synergizes with type I interferons through systemic dendritic cell reactivation and induction of immunogenic tumor apoptosis. *Cancer Res* 71:768–778.

Shen, S., O. Kepp, and G. Kroemer. 2012. The end of autophagic cell death? *Autophagy* 8:1–3.

Shen, S., O. Kepp, M. Michaud et al. 2011. Association and dissociation of autophagy, apoptosis and necrosis by systematic chemical study. *Oncogene* 30:4544–4556.

Soldani, C., M. G. Bottone, A. C. Croce et al. 2004. The Golgi apparatus is a primary site of intracellular damage after photosensitization with Rose Bengal acetate. *Eur J Histochem* 48:443–448.

Sonntag, Y., M. Musgaard, C. Olesen et al. 2011. Mutual adaptation of a membrane protein and its lipid bilayer during conformational changes. *Nat Commun* 2:304.

Spisek, R., A. Charalambous, A. Mazumder et al. 2007. Bortezomib enhances dendritic cell (DC) mediated induction of immunity to human myeloma via exposure of cell surface heat shock protein 90 on dying tumor cells: Therapeutic implications. *Blood* 109:4839–4845.

Stoka, V., B. Turk, S. L. Schendel et al. 2001. Lysosomal protease pathways to apoptosis. Cleavage of bid, not pro-caspases, is the most likely route. *J Biol Chem* 276:3149–3157.

Tesniere, A., T. Panaretakis, O. Kepp et al. 2008. Molecular characteristics of immunogenic cancer cell death. *Cell Death Differ* 15:3–12.

Tesniere, A., F. Schlemmer, V. Boige et al. 2010. Immunogenic death of colon cancer cells treated with oxaliplatin. *Oncogene* 29:482–491.

Vakkila, J., and M. T. Lotze. 2004. Inflammation and necrosis promote tumour growth. *Nat Rev Immunol* 4:641–648.

van Meer, G., D. R. Voelker, and G. W. Feigenson. 2008. Membrane lipids: Where they are and how they behave. *Nat Rev Mol Cell Biol* 9:112–124.

Vanlangenakker, N., T. V. Berghe, D. V. Krysko, N. Festjens, and P. Vandenabeele. 2008. Molecular mechanisms and pathophysiology of necrotic cell death. *Curr Mol Med* 8:207–220.

Verfaillie, T. et al. 2012. PERK is required at the ER-mitochondrial contact sites to convey apoptosis after ROS-based ER stress. *Cell Death Differ* 19:1880–1891.

Verfaillie, T., A. D. Garg, and P. Agostinis. 2013. Targeting ER stress induced apoptosis and inflammation in cancer. *Cancer Lett* 332:249–264.

Wang, X. 2001. The expanding role of mitochondria in apoptosis. *Genes Dev* 15:2922–2933.

Yuan, J. 2006. Divergence from a dedicated cellular suicide mechanism: Exploring the evolution of cell death. *Mol Cell* 23:1–12.

Zappasodi, R., S. M. Pupa, G. C. Ghedini et al. 2010. Improved clinical outcome in indolent B-cell lymphoma patients vaccinated with autologous tumor cells experiencing immunogenic death. *Cancer Res* 70:9062–9072.

Zhou, C. N. 1989. Mechanisms of tumor necrosis induced by photodynamic therapy. *J Photochem Photobiol B* 3:299–318.

Zitvogel, L., O. Kepp, and G. Kroemer. 2010a. Decoding cell death signals in inflammation and immunity. *Cell* 140:798–804.

Zitvogel, L., O. Kepp, L. Senovilla et al. 2010b. Immunogenic tumor cell death for optimal anticancer therapy: the calreticulin exposure pathway. *Clin Cancer Res* 16:3100–3104.

28

Vascular and Cellular Targeted PDT

Shannon M.
Gallagher-Colombo
University of Pennsylvania

Theresa M. Busch
University of Pennsylvania

28.1 Introduction

Oxidative damage by photodynamic therapy (PDT) can lead to direct cytotoxicity in any tissue, malignant or stromal, that is illuminated following its accumulation of photosensitizer. In this chapter, the mechanisms and consequences of PDT-mediated cytotoxicity are initially considered, followed by a more specific emphasis on how damage to the vasculature of malignant or otherwise diseased tissue contributes to the successful application of PDT. We describe how changes in tumor blood flow report on the extent and therapeutic significance of PDT vascular damage. Lastly, tumor oxygenation, vascular composition, and other features of the tumor microenvironment are discussed in the context of how they affect the potential for PDT to create cellular and vascular damage.

28.2 Cellular Response to PDT

Cell death in response to PDT can result from multiple mechanisms, including apoptosis, macroautophagy, and necrosis; indeed, activation of multiple cell-death pathways often occurs after cytotoxic PDT doses. For example, Rose bengal acetate (RBAc)-mediated PDT is reported to set off a cascade of cell-death mechanisms in HeLa cells that begins with intrinsic apoptotic cell death, followed by extrinsic apoptotic signaling, and rapidly concluding with autophagy (Panzarini, Inguscio, and Dini 2011). In many cases, the dominant mode of cell death resulting from PDT is a factor of the site of subcellular damage, which itself is a function of photosensitizer localization within the cell. Photosensitive drugs that localize to the mitochondria, endoplasmic reticulum (ER), and the lysosomal compartment tend to induce apoptotic and/ or autophagic cell death, and photosensitizers that localize to the

plasma membrane will more commonly induce necrosis (Buytaert, Dewaele, and Agostinis 2007). Examples of drugs that localize to the mitochondria and ER include Verteporfin and mTHPC, while hypericin localizes to the lysosomes and ER. Among porphyrin-based drugs, protoporphyrin IX (PpIX) is commonly found in the mitochondria, cytosol, and cytosolic membranes, and Photofrin typically localizes to the Golgi apparatus and plasma membrane (Buytaert, Dewaele, and Agostinis 2007).

Potentially balancing the activation of cell death pathways by PDT is the induction of signal transduction along survival pathways. PDT is reported to increase expression of the epidermal growth factor receptor (EGFR), survivin, hypoxia-inducible factor (HIF)-1α, and cyclooxygenase-2 (COX-2), all of which can contribute to proangiogenic and prosurvival signaling (Edmonds et al. 2012; Ferrario et al. 2007; Gomer et al. 2006). In order to combat this consequence of light treatment, numerous reports demonstrate added benefit when combining PDT with molecular-targeting drugs against these pathways. For example, increased apoptosis, decreased expression of genes important for progression through the cell cycle (e.g., cyclin-D1, c-myc), and enhanced tumor responses in animal models have been reported when combining PDT with the EGFR-targeting antibody cetuximab (Abu-Yousif et al. 2012; Bhuvaneswari et al. 2009b; del Carmen et al. 2005). Therapeutic benefits have also been observed when combining PDT with the COX-2 inhibitor, NS-398 (Gomer et al. 2006) in RIF tumors, and *in vitro* studies using a human melanoma cell line demonstrated improved PDT responses when targeting survivin or its binding partner, heat shock protein 90 (Hsp90) (Ferrario et al. 2007).

In addition to its effects on the malignant cells themselves, PDT can also damage tumor-residing immune cells, with lymphocytes being especially vulnerable. While PDT-induced immunosuppression has also been reported and is typically associated with the suppression of the contact hypersensitivity reaction (Castano, Mroz, and Hamblin 2006), many PDT protocols stimulate the recruitment of neutrophils and macrophages to the site of injury as well as induce the secretion of cytokines that can either promote cell death or mediate cell survival (Brackett et al. 2011; Castano, Mroz, and Hamblin 2006). For example, PDT increases the production of tumor necrosis factor (TNF)-α, which contributes to the cell–cell adhesion required for leukocytes to traverse the endothelial cell barrier and enter the site of injury (e.g., tumor tissue) where they are able to assist in the destruction of the damaged tumor or endothelial cells (Bhuvaneswari et al. 2009a; Chen et al. 2006). Conversely, PDT-induced increases in the proinflammatory cytokine, interleukin (IL)-6, are reported to be cytoprotective as evidenced by the observation of increased expression of the proapoptotic protein BAX and increased cell death 8–24 h after PDT of Colo26 tumors propagated in IL-6 deficient mice as compared to wild-type mice (Brackett et al. 2011). Other cytokines produced in response to PDT include IL-1β, IL-8, IL10, and granulocyte colony stimulating factor, among others (Castano, Mroz, and Hamblin 2006); however, it should be noted that the specific panel of proinflammatory and anti-inflammatory cytokines induced by PDT can vary with the photosensitizer and treatment protocol used (Bhuvaneswari et al. 2009a).

28.3 PDT-Created Vascular Damage

The direct cytotoxic action of PDT on tissue vasculature can lead to subsequent deterioration of vessel function and further contribute to the overall therapeutic effect of PDT. Many PDT protocols lead to some form of vascular damage; however, the extent of vascular destruction can vary depending on the specific treatment conditions. One approach that is used to direct PDT damage to the vasculature of a target tissue is to perform light administration at short drug–light intervals before significant clearance of the photosensitizer from the circulation can occur (Chen et al. 2006). This so-called passive approach to PDT-induced vascular damage can be performed using photosensitizers, such as Tookad, Verteporfin, or Motexafin Lutetium. For example, illumination within 15 min of the i.v. administration of Verteporfin (Chen et al. 2006) or 3 h of the administration of Motexafin Lutetium (Busch et al. 2010) ensures the presence of high drug levels in the plasma and likely the vascular endothelium during PDT. Tookad is a hydrophobic palladium–bacteriopheophorbide compound with a half-life of ~20 min in plasma and, thus, is best applied in combination with a drug–light interval of no more than 30 min (Chen et al. 2006). Each of these drugs has been studied or used clinically in the context of vascular-damaging PDT. Verteporfin is FDA approved for PDT of choroidal neovascularization associated with age-related macular degeneration (Chen et al. 2006), Motexafin Lutetium–mediated PDT has been studied for the treatment of locally recurrent prostate cancer after radiation therapy (Patel et al. 2008), and Tookad-mediated PDT has been investigated in treatment-naïve prostate cancer (Arumainayagam et al. 2010).

Vascular damage resulting from PDT is not restricted to vascular-targeting protocols. Even treatment protocols that do not intentionally deliver light while circulating drug levels are high can induce some form of vessel damage resulting from the affinity of many photosensitizers for vascular endothelial cells (Chang et al. 1999) as well as to the basement membrane of the vessels (Maas et al. 2012). Generally speaking, vessel constriction, activation of the coagulation cascade and thrombus formation, vascular leakage, and blood-flow stasis are all potential consequences of standard PDT protocols using photosensitizers, such as Photofrin, some phthalocyanine derivatives, and chlorin-based drugs (Bhuvaneswari et al. 2009a). However, there are some photosensitizers that do not produce any acute vascular effects, including disulfonated zinc or aluminum phthalocyanine (Fingar et al. 2000).

PDT-associated vascular damage can be initiated through the loss of integrity of the endothelial cell barrier, resulting in exposure of the vascular basement membrane. This, in turn, induces platelet aggregation and a coagulation response, followed by thrombus formation and blood-flow stasis (Fingar et al. 2000). This process may be further promoted by photosensitization of the basement membrane itself, which has been shown *ex vivo* and *in vivo* to promote platelet aggregation and vascular congestion, respectively (Fungaloi et al. 2002; Maas et al. 2012). The induction of thrombosis during light delivery for PDT, while an effective aid to tumor cell kill through oxygen or nutrient deprivation, can also be detrimental to the long-term effects of PDT if it impedes light distribution and the photochemical reaction required for cytotoxicity. The addition of heparin before Photofrin-PDT was reported to enhance light distribution, oxygen supply, and blood perfusion, leading to improved tumor responses compared to treatment with PDT alone (Yang et al. 2010). Conversely, a report by Fingar et al. (1993) showed a negative effect of combining thromboxane inhibitors before and during PDT. Specifically, it was shown that the addition of inhibitors that targeted thromboxane synthesis, release, or activity reduced vascular constriction following PDT accompanied by reductions in tumor cure rates. Taken together, these reports emphasize the importance of understanding the underlying causes of PDT effect on diseased vasculature and that these causes need to be carefully considered when designing vascular-damaging protocols.

Importantly, and often unfavorably, PDT can induce the secretion of proangiogenic factors, such as vascular endothelial growth factor (VEGF), COX-2, and matrix metalloproteinases (MMPs) (Gomer et al. 2006), all of which can contribute to incomplete therapeutic response by allowing the reestablishment of the tumor vasculature. Increased VEGF secretion in response to PDT has been reported in a wide range of tumor cell lines and animal models, including those of human glioblastoma and prostate cancer (Deininger et al. 2002; Solban et al. 2006). As with survival signaling, the combination of PDT with anti-angiogenic compounds is gaining increasing interest as a mechanism of improving efficacy. For example, combination of Verteporfin-PDT with bevacizumab is reported to attenuate the PDT-induced increase in VEGF levels and leads to increased tumor cure rates in a murine model of human non-small cell lung carcinoma (H460) (Gallagher-Colombo et al.

2012). An enhanced PDT effect was also observed when combining light treatment with bevacizumab and the antiangiogenic compounds EMAPII or IM862 in murine models of human bladder carcinoma (Bhuvaneswari et al. 2010) and mouse mammary carcinoma (Ferrario et al. 2000), respectively.

Another approach toward PDT-created vascular damage is the active targeting of photosensitizers to the vasculature. These protocols require either structural modification of the photosensitizer to enhance its vascular retention or conjugation of the photosensitizer to a molecular target, such as cell-surface receptors or integrins that are preferentially expressed on endothelial cells (Chen et al. 2006). Designing specific vascular-targeting moieties has proven more difficult than is often observed for tumor-targeting protocols in that a unique marker of tumor vasculature has not yet been defined. However, some efficacy has been observed when coupling photosensitizers to peptides, which bind to proteins expressed at high levels in angiogenic endothelial cells. For example, specificity for neovasculature observed in retinas of animal models with choroidal neovascularization has been observed using Verteporfin conjugated to a VEGFR-2 binding peptide (Renno et al. 2004). Verteporfin has also been conjugated to mouse factor VII protein, which binds tissue factor, a cell-surface receptor highly expressed on angiogenic vessels and tumor cells. This conjugate has the added benefit that it preferentially targets both the tumor and endothelial cells of the tumor vasculature while sparing the normal tissue and vessels (Hu et al. 2010).

28.4 Hemodynamic Effects of PDT

An event of consequence in PDT-created vascular damage is the resulting ischemia that limits tissue access to essential oxygen and nutrients. PDT-created vascular damage leads to ischemia with many photosensitizers and photosensitizing conditions;

not surprisingly, it can result from the vasoconstrictive effects of PDT (Fingar et al. 2000) as well as from vascular congestion through PDT-induced activation of the coagulation cascade (Dolmans et al. 2002). Both vasoconstriction and thrombi formation in the form of PDT-created emboli (He, Agharkar, and Chen 2008) can be rapid and transient events and will contribute to fluctuations in the blood flow of PDT-treated tumors.

Recently, technological advances in noninvasive monitoring have facilitated the measurement of PDT effect on blood flow with increased temporal and spatial sensitivity compared to that previously possible. These measurements have revealed blood flow to be highly dynamic over the course of illumination, including both increases and decreases in flow that can be extremely rapid (only seconds or minutes in duration) (Madar-Balakirski et al. 2010). Moreover, protocol-dependent patterns in tumor hemodynamic response during illumination have emerged. For example, when Photofrin is used with a typical 24-h drug–light interval, tumor hemodynamics are characterized by an initial, acute increase in blood flow of varying magnitude that is followed by a prominent decline (Standish et al. 2008; Yu et al. 2005) (Figure 28.1). Subsequent hemodynamic fluctuations varied among animals and could contribute to a return of blood flow to pretreatment levels by the conclusion of treatment. In contrast, PDT with the photosensitizer Motexafin Lutetium and a 3 h drug–light interval produced an acute, sharp decrease in blood flow that quickly recovered before subsequently declining with a more gradual slope over the remainder of treatment (Busch et al. 2010) (Figure 28.2). Curiously, PDT with aminolevulinic acid (ALA)-PDT (topically applied with a 3 h drug–light interval) produced a similar pattern of a sharp decrease, recovery, and then gradual decline, which, in some cases, was preceded by an initial small and transient increase in blood flow (Becker et al. 2010).

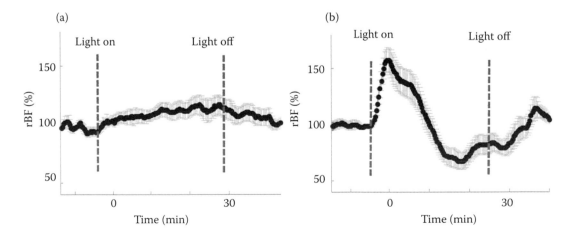

FIGURE 28.1 Hemodynamic response during Photofrin-PDT. Relative tumor blood flow (rBF) was measured by diffuse correlation spectroscopy and expressed as a percentage of baseline function prior to the initiation of light delivery. Control animals (a) received light but no photosensitizer, which led to little change in hemodynamic function. PDT-treated animals (b) received photosensitizer, leading to an increase, and then decline in blood flow when light was delivered. Trend lines plot average (± SE) relative blood flow (*n* = 10 and 15 for controls and PDT-treated, respectively). PDT was performed at 75 mW/cm², 135 J/cm² (630 nm) at 24 h after i.v. administration of 5 mg/kg Photofrin in RIF tumors. (From Yu, G. et al., *Clin Cancer Res*, 11, 2005. With permission.)

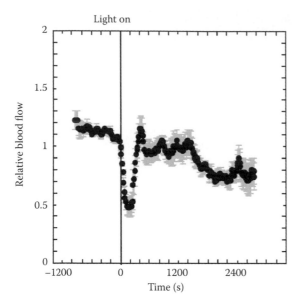

FIGURE 28.2 Hemodynamic response during PDT with Motexafin Lutetium. PDT induced an acute, but temporary decrease in blood flow upon illumination. Plot depicts a representative trace of relative tumor blood flow immediately before and during PDT of a RIF tumor with Motexafin Lutetium (10 mg/kg, 3 h incubation) and light delivery to 200 J/cm² at 75 mW/cm² (730 nm). Values are normalized to baseline function before PDT. Note that relative blood flow is expressed as a fraction instead of a percentage as used in Figure 28.1. (From Busch, T. et al., *Radiat Res*, 174, 2010. With permission.)

It is interesting to speculate how protocol-dependent differences in hemodynamic trends during PDT may relate to photosensitizer localization. For example, one could hypothesize that initial rises in blood flow during Photofrin-PDT may be secondary to treatment-created hypoxia by photochemical oxygen consumption in the tumor mass. Declines in blood flow that follow are consistent with the known vasoconstrictive and thrombotic effects of PDT with this photosensitizer (Fingar et al. 2000). The rapid, but transient, ischemic response found with Motexafin Lutetium and ALA may be indicative of a hemodynamic response dominated by strong, rapid vasoconstriction of photosensitized vasculature, followed by relaxation from consequent insult by ROS. Such a response could be facilitated by high blood levels of Motexafin Lutetium resulting from the short drug–light interval (Busch et al. 2010) or, in the case of ALA, resulting from accumulation of PpIX in the vascular endothelium (Rodriguez et al. 2009). In this regard, it is interesting to note that in studies of vascular damage via Verteporfin photosensitization at short drug–light intervals, others have reported vessel diameter to remain unchanged during PDT (He, Agharkar, and Chen 2008) or, alternatively, have identified PDT-triggered acute vasoconstriction that led to *increases* in blood flow at the level of individual vessels (Khurana et al. 2008). Although these studies assessed the PDT effect in individual vessels and the previously described studies of Motexafin Lutetium looked at tumor-averaged vessel response, taken together, the data suggest that illumination of a vascular-localized photosensitizer can produce any of several

hemodynamic effects that vary with photosensitizer type. Drug-dependent variability in hemodynamic response to vascular-targeting PDT is further emphasized in studies of Tookad as a photosensitizer, which, upon illumination, produces an extremely rapid vasodilative response in tumor-feeding arteries, followed by vasoconstriction and occlusive thrombosis (Madar-Balakirski et al. 2010). The authors of this study hypothesize that in the case of photosensitization with Tookad, which does not extravasate from the circulation, blood clots form as a result of the extreme and rapid changes in vessel diameter that are triggered by illumination as opposed to being initiated by damage to the vascular endothelium as found with other vascular-targeting photosensitizers (Madar-Balakirski et al. 2010).

One of the most significant results to be gained from studies of tumor hemodynamics during PDT is that the treatment-induced change in blood flow can be predictive of both tumor damage and long-term outcome in murine studies. For example, we have shown that the rate of decrease in tumor blood flow during Photofrin-PDT is highly correlated with tumor response, measured as the regrowth delay to a 400 mm³ tumor volume (Yu et al. 2005). This relationship was marked by an increase in the duration of the regrowth delay as a function of a decrease in the rate of blood flow reduction during PDT, which is consistent with the expectation that ischemia during illumination leads to hypoxia that limits oxidative damage by PDT. In agreement, Standish et al. (2008) have shown that more necrosis (at 24 h after PDT) develops in tumors that experienced slower rates of blood flow decline during Photofrin-PDT. Using Verteporfin with a 5 min drug–light interval, others have shown changes in blood flow during illumination with a vascular-localized drug to also report on treatment damage (Pham et al. 2001). During Verteporfin-PDT, significant decreases in tumor blood flow (measured as the change in total hemoglobin concentration) and tumor oxygenation (measured as the percent of tissue hemoglobin oxygen saturation) occurred over the course of treatment. However, in contrast to results with Photofrin-PDT, decreasing blood flow during Verteporfin-PDT was a favorable event because tumors with larger decreases in blood flow or oxygenation during PDT had more necrosis at 72 h after treatment. These results suggest that Verteporfin-PDT under these conditions led to permanent vascular occlusion, which produced the desired (necrotic) outcome while mitigating limitations to direct cytotoxicity imposed by hypoxia created during illumination.

28.5 Tumor Microenvironment as an Effector of PDT-Created Cellular and Vascular Damage

Irrespective of whether damage is mediated through a vascular or a cellular mechanism, the microenvironment of a tumor can either facilitate or impede the action of PDT. Tumor characteristics, such as its oxygenation, vascularization, and stromal content, have the potential to affect all components of the PDT reaction, including photosensitizer accumulation and distribution, light penetration within tissue, and the creation of oxygen-dependent cytotoxic species. It is generally believed that the oxygen dependence of PDT stems from type II photochemistry, which is associated with many (but not all) photosensitizers, and results in the formation of singlet oxygen—a highly reactive molecule that oxidizes its substrates, leading to PDT-created cytotoxicity. As a result of this type II photochemistry, PDT-created hypoxia can develop very quickly during illumination, when ground state oxygen is depleted faster than its replenishment through the blood supply (i.e., photochemical oxygen consumption). The effects of PDT on blood flow, as discussed above, can severely limit oxygen delivery and further contribute to a hypoxic state in the target tissue during illumination (Busch 2006). Moreover, the oxygenation status of tumors can affect light penetration at PDT-relevant wavelengths resulting from differences in the absorption spectrum of oxyhemoglobin and deoxyhemoglobin (Mitra and Foster 2004). Because of these effects, it is important to consider not only preexisting tumor hypoxia but also its development during the illumination period in studies of microenvironmental effect on PDT-mediated damage.

Minimizing PDT-created hypoxia during illumination can increase both direct cytotoxicity to tumor cells and vascular damage in the illumination field. For example, increases in cellular and/or histological damage are found after fractionation of light delivery into multiple periods or on/off intervals that may allow for tumor reoxygenation during the interruption(s) in illumination (Curnow, Haller, and Bown 2000; Iinuma et al. 1999; Sotiriou et al. 2012). Similarly, hyperbaric oxygen breathing during illumination can increase tumor oxygenation during PDT and is associated with improvements in the outcome of both preclinical and clinical investigations (Chen et al. 2002; Maier et al. 2000). In another approach, lowering the fluence rate of light delivery may facilitate maintenance of tumor oxygenation during the illumination period, which has led us to examine the effects of fluence rate on tumor oxygenation and vascular responses in a series of murine investigations (Busch et al. 2002, 2009, 2010). In this work, we found lower fluence rate to reduce the severity of PDT-induced hypoxia in tumor tissue that is distant from the vasculature (i.e., the oxygen supply) as well as in tumor cells that are immediately adjacent to the blood vessels (Busch et al. 2002). In concert with this result, a low fluence rate favored the maintenance of tumor perfusion during PDT (Busch et al. 2002) and led to significant treatment-created cytotoxicity (Busch et al. 2009). Moreover, these studies found low fluence rate to reduce intratumor heterogeneities in hypoxia during illumination, leading to the benefit that low fluence rate reduces intratumor variability in PDT-created damage (Busch et al. 2009). Yet another benefit of low fluence rate is the extended length of treatment compared to that needed to deliver the same total fluence at a higher fluence rate. Reports by ourselves (Busch et al. 2010) and others (Seshadri et al. 2008) showed that extending illumination time produces more vascular damage and better outcomes that are separate from any benefit to oxygen conservation from low fluence rate.

Although the control of treatment-created hypoxia is paramount to providing a favorable tumor microenvironment for many PDT applications, there is also a role for the preexisting tumor microenvironment in PDT responses. Early studies identified that severe, preexisting hypoxia is a limitation in PDT (Fingar et al. 1992). However, investigations of intratumor heterogeneity in tumor oxygenation found areas with low oxygen tensions (<8 mm Hg) to be more susceptible to PDT-induced decreases in oxygenation, presumably as a result of vascular damage (Pogue et al. 2001). This suggests that tumor regions with poor preexisting perfusion may be at greater risk for treatment-induced vascular shutdown. In agreement, subsequent studies have shown that vessels with a low initial flow velocity are quicker to experience a complete stoppage of blood flow after PDT than those vessels with higher preexisting flow (He, Agharkar, and Chen 2008). Most recently, it has been shown that PDT-induced ischemia during illumination varies between identical tumor models grown in different mouse strains. This could be generalized as an inherent difference in tumor hemodynamics between the strains that may result from differences in tumor vessel size, among other factors (Mesquita et al. 2012). Taken together, these data suggest that intratumor and intertumor heterogeneities in preexisting vascular structure and function are contributing factors to PDT-induced acute vascular damage.

The composition of a tumor vascular network can also alter PDT responses through an effect on photosensitizer accumulation and localization. Several reports have shown that heightened vascular permeability in tumors can contribute to their greater accumulation of the photosensitizing drug (Chen et al. 2005; Roberts and Hasan 1993). The structural composition of the vascular stroma can further determine photosensitizer distribution through its effect on drug localization to the vasculature. In this regard, we have recently shown that Photofrin localization to the collagen of the vascular basement membrane will significantly increase vascular damage in tumors that contain more preexisting collagen (Maas et al. 2012) (Figure 28.3). This was accompanied by higher cure rates after PDT of the more highly collagenated tumors. Furthermore, tumors with more collagen deposition also demonstrated more intratumor homogeneity in the distribution of blood vessels and greater intratumor homogeneity in hemodynamic response to PDT. Thus, the better PDT response of these tumors may reflect the findings of others (Zhou et al. 2006) that variability in PDT responses among tumors can be reduced by approaches that account for heterogeneities in photosensitizer uptake.

28.6 Conclusions

Cellular and vascular damage are the fundamental underlying mechanisms by which PDT acts to create a treatment effect. The extent of PDT action on each of these targets will vary as a function of the protocol under investigation and the microenvironmental characteristics of the target tissue. Short intervals between drug administration and light delivery have

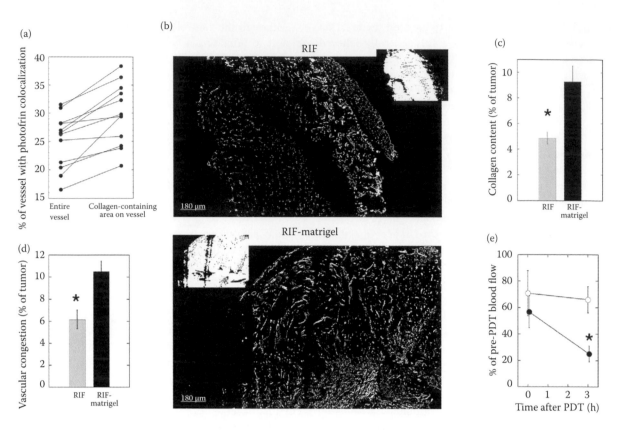

FIGURE 28.3 Photofrin localizes to collagen IV, mediating PDT-induced vascular damage. Fluorescence microscopy of Photofrin distribution, coupled with immunohistochemistry of tumor vascular structure, reveals Photofrin to localize to the collagen IV on the basement membrane of blood vessels better than to the entire vessel (a). This becomes relevant to the PDT response of tumor models that contain more collagen; for example, RIF tumors supplemented with basement membrane matrix (RIF-Matrigel) contain more collagen (image in b, quantification in c) than Matrigel-naïve RIF tumors (RIF). In association with their increased collagen content, RIF-Matrigel tumors accumulated more vascular damage after PDT, including more vascular congestion (d) and greater shutdown of the tumor blood vessels (e). PDT was performed at 75 mW/cm², 135 J/cm² (630 nm) at 24 h after i.v. administration of 5 mg/kg Photofrin. Insets in (b) depict tumor area on the section. * indicates $p < 0.05$ for the comparison between RIF and RIF-Matrigel. (From Maas, A. et al., *Cancer Res*, 72, 2012. With permission.)

been employed with great success to destroy diseased tissues via vasoconstriction and/or occlusive thrombosis of vessels in the treatment field. Importantly, however, vascular effects can accompany even those protocols designed to deliver light when photosensitizer levels are high in the tumor and its stroma, compared to the circulation. Emerging noninvasive techniques of monitoring blood flow during PDT will provide an opportunity to evaluate the real-time effect of PDT on tumor hemodynamics, which may inform on treatment outcome and potentially aid the development of better clinical protocols. Additionally, further improvements in clinical applications may come with increased knowledge of how the tissue microenvironment impacts the cellular and vascular effects of PDT, thus allowing for better choices of illumination schemes, better patient selection, and the informed investigation of approaches toward biological or molecular priming of the tumor microenvironment for PDT.

Acknowledgments

Support during the preparation of this chapter was provided by the National Institutes of Health through R01-CA085831, R01-CA129554, and PPG-CA87971 (for TMB), as well as through T32-CA009677 (for SGC).

References

Abu-Yousif, A. O., A. C. Moor, X. Zheng et al. 2012. Epidermal growth factor receptor-targeted photosensitizer selectively inhibits EGFR signaling and induces targeted phototoxicity in ovarian cancer cells. *Cancer Lett* 321(2):120–127.

Edmonds, C., S. Hagan, S. M. Gallaher-Colombo, T. M. Busch, and K. A. Cengel. 2012. Photodynamic therapy activated signaling from epidermal growth factor receptor and STAT3: Targeting survival pathways to increase PDT efficacy in ovarian and lung cancer. *Cancer Biology & Therapy* 13(14):1463–1470.

Arumainayagam, N., C. M. Moore, H. U. Ahmed, and M. Emberton. 2010. Photodynamic therapy for focal ablation of the prostate. *World J Urol* 28:571–576.

Becker, T. L., A. D. Paquette, K. R. Keymel, B. W. Henderson, and U. Sunar. 2010. Monitoring blood flow responses during topical ALA-PDT. *Biomed Opt Express* 2:123–130.

Bhuvaneswari, R., Y. Y. Gan, K. C. Soo, and M. Olivo. 2009a. The effect of photodynamic therapy on tumor angiogenesis. *Cell Molec Life Sci* 66:2275–2283.

Bhuvaneswari, R., Y. Y. Gan, K. C. Soo, and M. Olivo. 2009b. Targeting EGFR with photodynamic therapy in combination with Erbitux enhances *in vivo* bladder tumor response. *Mol Cancer* 8:94.

Bhuvaneswari, R., P. S. Thong, Y. Y. Gan, K. C. Soo, and M. Olivo. 2010. Evaluation of hypericin-mediated photodynamic therapy in combination with angiogenesis inhibitor bevacizumab using *in vivo* fluorescence confocal endomicroscopy. *J Biomed Opt* 15:011114.

Brackett, C. M., B. Owczarczak, K. Ramsey, P. G. Maier, and S. O. Gollnick. 2011. IL-6 potentiates tumor resistance to photodynamic therapy (PDT). *Lasers Surg Med* 43:676–685.

Busch, T. M. 2006. Local physiological changes during photodynamic therapy. *Lasers Surg Med* 38:494–499.

Busch, T. M., H. W. Wang, E. P. Wileyto, G. Yu, and R. M. Bunte. 2010. Increasing damage to tumor blood vessels during motexafin lutetium-PDT through use of low fluence rate. *Radiat Res* 174:331–340.

Busch, T. M., E. P. Wileyto, M. J. Emanuele et al. 2002. Photodynamic therapy creates fluence rate-dependent gradients in the intratumoral spatial distribution of oxygen. *Cancer Res* 62:7273–7279.

Busch, T. M., X. Xing, G. Yu et al. 2009. Fluence rate-dependent intratumor heterogeneity in physiologic and cytotoxic responses to Photofrin photodynamic therapy. *Photochem Photobiol Sci* 8:1683–1693.

Buytaert, E., M. Dewaele, and P. Agostinis. 2007. Molecular effectors of multiple cell death pathways initiated by photodynamic therapy. *Biochim Biophys Acta* 1776:86–107.

Castano, A. P., P. Mroz, and M. R. Hamblin. 2006. Photodynamic therapy and anti-tumour immunity. *Nat Rev Cancer* 6:535–545.

Chang, C. J., C. H. Sun, L. H. Liaw, M. W. Berns, and J. S. Nelson. 1999. *In vitro* and *in vivo* photosensitizing capabilities of 5-ALA versus photofrin in vascular endothelial cells. *Lasers Surg Med* 24:178–186.

Chen, B., B. W. Pogue, P. J. Hoopes, and T. Hasan. 2006. Vascular and cellular targeting for photodynamic therapy. *Crit Rev Eukaryot Gene Express* 16:279–305.

Chen, B., B. W. Pogue, X. Zhou et al. 2005. Effect of tumor host microenvironment on photodynamic therapy in a rat prostate tumor model. *Clin Cancer Res* 11:720–727.

Chen, Q., Z. Huang, H. Chen et al. 2002. Improvement of tumor response by manipulation of tumor oxygenation during photodynamic therapy. *Photochem Photobiol* 76:197–203.

Curnow, A., J. C. Haller, and S. G. Bown. 2000. Oxygen monitoring during 5-aminolaevulinic acid induced photodynamic therapy in normal rat colon. Comparison of continuous and fractionated light regimes. *J Photochem Photobiol B Biol* 58:149–155.

Deininger, M. H., T. Weinschenk, M. H. Morgalla, R. Meyermann, and H. J. Schluesener. 2002. Release of regulators of angiogenesis following Hypocrellin-A and -B photodynamic therapy of human brain tumor cells. *Biochem Biophys Res Commun* 298:520–530.

del Carmen, M. G., I. Rizvi, Y. Chang et al. 2005. Synergism of epidermal growth factor receptor-targeted immunotherapy with photodynamic treatment of ovarian cancer *in vivo*. *J Natl Cancer Inst* 97:1516–1524.

Dolmans, D. E., A. Kadambi, J. S. Hill et al. 2002. Vascular accumulation of a novel photosensitizer, MV6401, causes selective thrombosis in tumor vessels after photodynamic therapy. *Cancer Res* 62:2151–2156.

Ferrario, A., N. Rucker, S. Wong, M. Luna, and C. J. Gomer. 2007. Survivin, a member of the inhibitor of apoptosis family, is induced by photodynamic therapy and is a target for improving treatment response. *Cancer Res* 67:4989–4995.

Ferrario, A., K. F. von Tiehl, N. Rucker et al. 2000. Antiangiogenic treatment enhances photodynamic therapy responsiveness in a mouse mammary carcinoma. *Cancer Res* 60:4066–4069.

Fingar, V. H., K. A. Siegel, T. J. Wieman, and K. W. Doak. 1993. The effects of thromboxane inhibitors on the microvascular and tumor response to photodynamic therapy. *Photochem Photobiol* 58:393–399.

Fingar, V. H., S. W. Taber, P. S. Haydon et al. 2000. Vascular damage after photodynamic therapy of solid tumors: A view and comparison of effect in pre-clinical and clinical models at the University of Louisville. *In Vivo* 14:93–100.

Fingar, V. H., T. J. Wieman, Y. J. Park, and B. W. Henderson. 1992. Implications of a pre-existing tumor hypoxic fraction on photodynamic therapy. *J Surg Res* 53:524–528.

Fungaloi, P., R. Statius van Eps, Y. P. Wu et al. 2002. Platelet adhesion to photodynamic therapy-treated extracellular matrix proteins. *Photochem Photobiol* 75:412–417.

Gallagher-Colombo, S. M., A. L. Maas, M. Yuan, and T. M. Busch. 2012. Photodynamic therapy-induced angiogenic signaling: Consequences and solutions to improve therapeutic response. *Israel J Chem* 52:681–690.

Gomer, C. J., A. Ferrario, M. Luna, N. Rucker, and S. Wong. 2006. Photodynamic therapy: Combined modality approaches targeting the tumor microenvironment. *Lasers Surg Med* 38:516–521.

He, C., P. Agharkar, and B. Chen. 2008. Intravital microscopic analysis of vascular perfusion and macromolecule extravasation after photodynamic vascular targeting therapy. *Pharm Res* 25:1873–1880.

Hu, Z., B. Rao, S. Chen, and J. Duanmu. 2010. Targeting tissue factor on tumour cells and angiogenic vascular endothelial cells by factor VII-targeted verteporfin photodynamic therapy for breast cancer *in vitro* and *in vivo* in mice. *BMC Cancer* 10:235.

Iinuma, S., K. T. Schomacker, G. Wagnieres et al. 1999. *In vivo* fluence rate and fractionation effects on tumor response and photobleaching: Photodynamic therapy with two photosensitizers in an orthotopic rat tumor model. *Cancer Res* 59:6164–6170.

Khurana, M., E. H. Moriyama, A. Mariampillai, and B. C. Wilson. 2008. Intravital high-resolution optical imaging of individual vessel response to photodynamic treatment. *J Biomed Opt* 13:040502.

Maas, A. L., S. L. Carter, E. P. Wileyto et al. 2012. Tumor vascular microenvironment determines responsiveness to photodynamic therapy. *Cancer Res* 72:2079–2088.

Madar-Balakirski, N., C. Tempel-Brami, V. Kalchenko et al. 2010. Permanent occlusion of feeding arteries and draining veins in solid mouse tumors by vascular targeted photodynamic therapy (VTP) with Tookad. *PLoS One* 5:e10282.

Maier, A., F. Tomaselli, U. Anegg et al. 2000. Combined photodynamic therapy and hyperbaric oxygenation in carcinoma of the esophagus and the esophago-gastric junction. *Eur J Cardiothorac Surg* 18:649–654; discussion 654–655.

Mesquita, R. C., H. S. Miller, S. S. Schenkel et al. 2012. Tumor blood flow differs between mouse strains: Consequences for vasoresponse to photodynamic therapy. *PLoSONE* 7:e37322.

Mitra, S., and T. H. Foster. 2004. Carbogen breathing significantly enhances the penetration of red light in murine tumours *in vivo*. *Phys Med Biol* 49:1891–1904.

Panzarini, E., V. Inguscio, and L. Dini. 2011. Timing the multiple cell death pathways initiated by Rose Bengal acetate photodynamic therapy. *Cell Death Dis* 2:e169.

Patel, H., R. Mick, J. Finlay et al. 2008. Motexafin lutetium-photodynamic therapy of prostate cancer: Short- and long-term effects on prostate-specific antigen. *Clin Cancer Res* 14:4869–4876.

Pham, T. H., R. Hornung, M. W. Berns, Y. Tadir, and B. J. Tromberg. 2001. Monitoring tumor response during photodynamic therapy using near-infrared photon-migration spectroscopy. *Photochem Photobiol* 73:669–677.

Pogue, B. W., R. D. Braun, J. L. Lanzen, C. Erickson, and M. W. Dewhirst. 2001. Analysis of the heterogeneity of pO2 dynamics during photodynamic therapy with verteporfin. *Photochem Photobiol* 74:700–706.

Renno, R. Z., Y. Terada, M. J. Haddadin et al. 2004. Selective photodynamic therapy by targeted verteporfin delivery to experimental choroidal neovascularization mediated by a homing peptide to vascular endothelial growth factor receptor-2. *Arch Ophthalmol* 122:1002–1011.

Roberts, W. G., and T. Hasan. 1993. Tumor-secreted vascular permeability factor/vascular endothelial growth factor influences photosensitizer uptake. *Cancer Res* 53:153–157.

Rodriguez, L., H. S. de Bruijn, G. Di Venosa et al. 2009. Porphyrin synthesis from aminolevulinic acid esters in endothelial cells and its role in photodynamic therapy. *J Photochem Photobiol B* 96:249–254.

Seshadri, M., D. A. Bellnier, L. A. Vaughan et al. 2008. Light delivery over extended time periods enhances the effectiveness of photodynamic therapy. *Clin Cancer Res* 14:2796–2805.

Solban, N., P. K. Selbo, A. K. Sinha, S. K. Chang, and T. Hasan. 2006. Mechanistic investigation and implications of photodynamic therapy induction of vascular endothelial growth factor in prostate cancer. *Cancer Res* 66:5633–5640.

Sotiriou, E., Z. Apalla, E. Chovarda et al. 2012. Single vs. fractionated photodynamic therapy for face and scalp actinic keratoses: A randomized, intraindividual comparison trial with 12-month follow-up. *J Eur Acad Dermatol Venereol* 26:36–40.

Standish, B. A., K. K. Lee, X. Jin et al. 2008. Interstitial Doppler optical coherence tomography as a local tumor necrosis predictor in photodynamic therapy of prostatic carcinoma: an *in vivo* study. *Cancer Res* 68:9987–9995.

Yang, L., Y. Wei, D. Xing, and Q. Chen. 2010. Increasing the efficiency of photodynamic therapy by improved light delivery and oxygen supply using an anticoagulant in a solid tumor model. *Lasers Surg Med* 42:671–679.

Yu, G., T. Durduran, C. Zhou et al. 2005. Noninvasive monitoring of murine tumor blood flow during and after photodynamic therapy provides early assessment of therapeutic efficacy. *Clin Cancer Res* 11:3543–3552.

Zhou, X., B. W. Pogue, B. Chen et al. 2006. Pretreatment photosensitizer dosimetry reduces variation in tumor response. *Int J Radiat Oncol Biol Phys* 64:1211–1220.

29

Photodynamic Therapy for Increased Delivery of Anticancer Drugs

Elodie Debefve
Swiss Federal Institute of Technology (EPFL)

Yabo Wang
Centre Hospitalier Universitaire Vaudois (CHUV)

Thorsten Krueger
Centre Hospitalier Universitaire Vaudois (CHUV)

Hubert van den Bergh
Swiss Federal Institute of Technology (EPFL)

29.1 Introduction

29.1.1 Systemic Drug Delivery

The efficacy of any drug depends on its ability to reach its target. The drug delivery to a given tissue after systemic intravenous injection is influenced by the blood supply to the region of interest as well as transport through the vascular wall and the diffusion/convection through the interstitial space. To optimize the selective drug accumulation within the targeted area, one can play with three main parameters.

First, the drug's formulation and physicochemical properties should allow for sufficient residence time in the blood circulation to reach the intended sites of the body. This circulation time can be improved among others by nanoencapsulation, liposomal formulation, and/or PEGylation to avoid uptake in the reticuloendothelial system (RES). Second, the drug or its carrier should be designed to have efficient retention within the targeted sites, and deposition in other tissues should be minimized (i.e., selective targeting). Third, the active component of the drug should then be efficiently released to the target at a designed time at the concentration that allows for effective pharmacological activity (Bae and Park 2011; Narang and Varia 2011).

For cancer therapies, an idealized, targeted drug-delivery system delivers the drug exclusively to the target tumor. In reality, however, it is often far from that ideal situation, that is, the amount of drug delivered to a typical tumor tends to be much less than 5% of the injected dose. Thus, a large proportion of many chemotherapeutics accumulates in normal tissue, inducing side effects and limiting the drug available for the target tissue (Bae and Park 2011). Isolated organ cannulations have been explored to allow for an increased local perfusion with high concentrations of chemotherapeutics (Alexander and Butler 2010; Grootenboers et al. 2006; Han et al. 2011). This preserves many of the healthy organs from side effects but only partially increases the tumor-specific therapeutic effects. Today, many drug-delivery approaches involve pharmacological vehicles or formulations that, to some extent, will selectively target the tumor tissue and which we also hope will diminish the uneven distribution of the drug across the solid tumor's tissue. A large number of approaches have been developed to improve the access of these therapeutic agents to the tumors and their accumulation in the tissue targets as well as the reduction of cytotoxic effects.

However, no matter how great the advances are, which are made in nanoparticle design or antibody technology, no matter

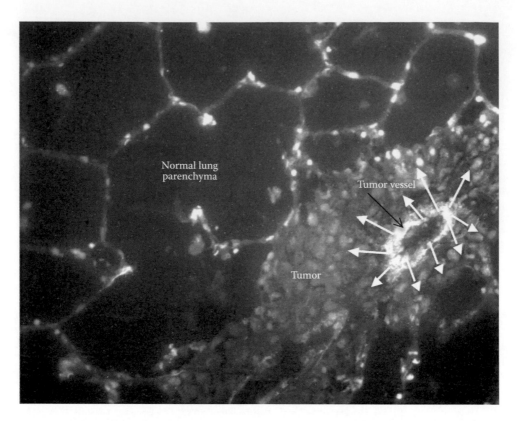

FIGURE 29.1   This picture illustrates typical poor uptake and distribution of a chemotherapeutic agent in tumor tissue.

how well the new applications of prodrugs that are locally activated at the target site only, and no matter how selective a ligand is to target a cancer cell marker, the delivery of a drug to the target and the ligand–receptor interaction will be always limited, to some extent, by the blood circulation and efficient drug extravasation from the vasculature (Bae and Park 2011). As an example, Figure 29.1 illustrates the often-typical poor uptake and distribution of a chemotherapeutic agent encountered experimentally in tumor tissue. In this image, one should note that the fluorescent signal given by the chemotherapeutic agent (doxorubicin) is much higher in the wall of the blood vessels that are feeding the tumor and in the normal lung parenchyma than in the tumor cells themselves.

Strategies can then be used, such as hyperthermia, irradiation, or the use of vasoactive factors, in order to disrupt the endothelial barrier integrity within the target site (van Nieuw Amerongen and van Hinsbergh 2002). Following the same philosophy, the novel strategy discussed in this chapter consists of improving the extravasation of a chemotherapeutic drug from the tumor-feeding blood vessel to the interstitium and the tumor cells.

29.1.2 Vascular Effect of Photodynamic Therapy

Photodynamic therapy (PDT) can have three more-or-less distinct effects on the human body: It can kill tumor cells; it can damage the vascular endothelium, causing thrombosis and stasis; and it can also modulate the immune system. Tissue localization of the damage, which is induced by PDT, is influenced by both the choice of photosensitizer (PS), its formulation and/or vehicle used, and the experimental conditions (Fingar 1996; Sharman, Allen, and van Lier 1999). For instance, shortly after intravenous injection, Visudyne, a benzoporphyrin derivative, is localized mainly in the vasculature. Accordingly, in this case, destruction of the vasculature is the main result of Visudyne PDT in which light is applied shortly (10 min) after injection. This "vascular" PDT has been clinically used since 2001 for the treatment of the classic form of exudative age-related macular degeneration (AMD). The treatment consists of the occlusion of choroidal neovessels in the retina. From the year 2000, this treatment was essentially the viable one for these patients, and it worked particularly well for small "classical" lesions, starting before the arrival of antivascular endothelial growth factor (anti-VEGF) drugs (Schmidt-Erfurth et al. 2007). Vascular PDT consumes oxygen and occludes blood vessels locally, decreasing the oxygenation of the tissue. This PDT-induced hypoxia triggers the production of HIF-1α, which, together with the PDT-induced inflammation, can result in an increase in the vascular permeability of the treated blood vessels. This kind of vascular leakage was observed during clinical trials by Sickenberg et al. following PDT with Lutetium texaphyrin (Lu-Tex) in the retina of patients treated for AMD (data unpublished). Michels and Schmidt-Erfurth (2003) documented a similar leakage at short times after Visudyne PDT in the retina of patients treated for age-related macular degeneration. This phenomenon was probably also observed by Fingar et al (1992), who were the first to perform detailed analysis of this vascular leakage induced by PDT in a preclinical model. They showed in a rat cremaster muscle model that PDT using the drug Photofrin can result, under certain drug–light conditions, in increased venule permeability and leakage.

This PDT-induced leakage was considered clinically to be a drawback of the treatment, but it was suggested that it can, in effect, be exploited as a means to increase the local drug delivery, for instance, of anti-VEGF or chemotherapeutic drugs. One can thus exploit two effects of vascular PDT sequentially to treat cancer and some other diseases. PDT can be used first as a local drug-delivery pathway, followed by the occlusion of these pathological blood vessels after the passage of the chemotherapeutics. This so-called combined PDT with an enhanced drug-delivery mechanism continued with subsequent vascular closure is schematically represented in Figure 29.2.

The mechanisms involved in this PDT-induced leakage are complex and multifactorial. The drugs intravenously injected will not simply diffuse trough the gaps induced in the vessel wall, but other biochemical and immune factors will also be involved. Several experiments to investigate this have already been conducted by other groups in this framework (Chen et al. 2006b, 2008; He, Agharkar, and Chen 2008; Hirschberg et al. 2008; Snyder et al. 2003). We now wish to confirm the preclinical models in which PDT does efficiently and selectively induce vascular leakage. Thus, we may observe and understand the different factors and mechanisms involved in this complex process and thus may be able to find the optimal treatment conditions before transposing this drug-delivery technique to clinical applications.

29.2 Experimental Data as Examples of Mechanistic Approaches

29.2.1 Proof of Concept Demonstrated in a Rodent Model

We investigated the possibility for improving the selective drug delivery of a chemotherapeutic drug (liposomal doxorubicin) to tumor tissue by pretreating the target area and its surroundings with low-dose vascular PDT. For this study, we used a Fisher rat model with a syngeneic subpleural sarcoma tumor superficially grown on the left lung. This model is illustrated in Figure 29.3a, showing the left lung and the tumor exposed by thoracotomy for the treatment. In the different treatment groups, rats received PDT using two different doses of Visudyne intravenously injected 15 min before irradiation. The irradiation is only in the lower lobe of the lung containing the tumor in the center. The experimental setup is illustrated in Figure 29.3b. Directly after irradiation, liposomal doxorubicin was intravenously injected and circulated for 1, 3, or 6 h before sacrificing the animal. The samples were then collected for high-performance liquid chromatography (HPLC) quantification of the doxorubicin concentration in the different tissues. Because the tumor and the surrounding normal lung tissue are exposed to the same PDT conditions but analyzed separately regarding drug tissue concentrations, this model enables the assessment of PDT-induced uptake of the macromolecular compound in normal and tumor tissues. The results presented in Figure 29.4 suggest that in this model, Visudyne-mediated PDT pretreatment led to a significant and selective increase in liposomal doxorubicin uptake in tumors but not in the surrounding normal lung parenchyma. Thus, PDT pretreatment resulted in a significantly better tumor to-normal lung drug ratio compared to intravenous drug application alone. This therapeutic advantage of PDT pretreatment over just simple intravenous drug application without PDT

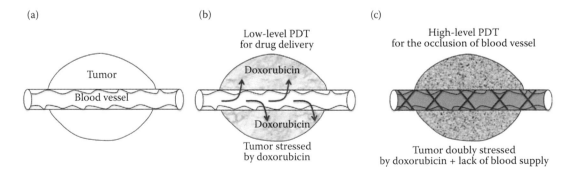

(a)　　　　　　　　　(b)　　　　　　　　　(c)

Low-level PDT for drug delivery

High-level PDT for the occlusion of blood vessel

Tumor

Blood vessel

Doxorubicin

Doxorubicin

Tumor stressed by doxorubicin

Tumor doubly stressed by doxorubicin + lack of blood supply

FIGURE 29.2 For this study, a tumor was subpleurally grown on a rodent lung. The rat was then treated with a high dose of a chemotherapeutic drug, doxorubicin, administered by isolated lung perfusion (ILP). The rat was then sacrificed for histological observations. Doxorubicin is fluorescent. Thus, it is easy to see its distribution by fluorescence microscopy in histology sections of the tumor. ILP allows for administration of a high dose of drug in the lung, avoiding the systemic circulation. Compared to intravenous administration, ILP leads to a better accumulation of the chemotherapy in the tumor. However, as observed here in histology, the drug is mainly confined within the wall of the vessel feeding the tumor but not in the tumor tissue itself. Indeed, one can thus observe that the brightest fluorescent signal comes from the vessel wall of the tumor as well as from the normal lung parenchyma. One needs to have a homogeneous fluorescent signal within the whole tumor and not only within the vessel wall. A leakage of the chemotherapy, selectively from the tumor vasculature (as illustrated by the arrows), would probably improve the therapeutic effect of the drug.

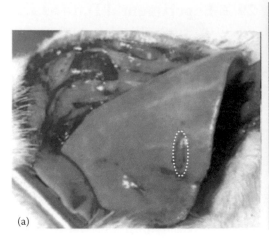

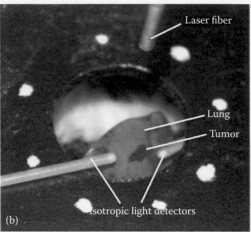

FIGURE 29.3 (a) Experimental setting with mobilized and exposed left lower lobe of the lung of the rat containing a subpleurally located superficial sarcoma tumor (encircled) through the thoracotomy. This vascularized tumor was generated 10 days previously by subpleural injection of sarcoma cell suspension. (b) Setup of the PDT pretreatment showing the shielding of surrounding tissues, the laser fiber placed above providing noncontact, nonthermal surface irradiance, and two isotropic light detectors (white fibers) positioned within the incident laser beam allowing for continuous recording of the fluence rate and the cumulative fluence.

persisted for a liposomal doxorubicin circulation time of up to 6 h. Thus, we conclude that the concept of photodynamic drug delivery in this study gave positive results (Figure 29.4). This principle was also verified by other experiments in rat models of prostate cancer (Chen et al. 2006b) and colon cancer (Snyder et al. 2003). It was further demonstrated that this pathway of PDT for selective

drug delivery to tumor tissue was effectively independent of the tumor type encountered in the lung or pleura metastasis (Cheng et al. 2010; Wang et al. 2012a) as long as the tumor was sufficiently vascularized.

However, even if this concept of PDT for selective drug delivery has now been repeatedly proven using different treatment

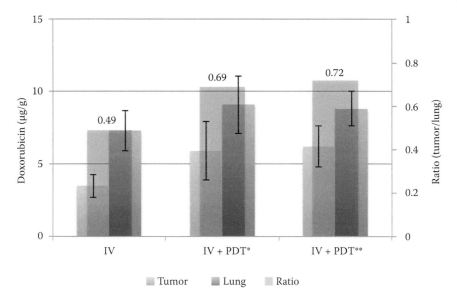

FIGURE 29.4 Impact of Visudyne-mediated PDT on liposomal doxorubicin uptake (μg/g) in the tumor (blue column) and normal lower lobe of the left lung (red column). Ratio (green column, values above the column) of tumor to normal lower lobe of the left lung Liporubicin concentrations. Comparison between IV injection alone and IV injection + PDT with two Visudyne concentrations (*0.0625 and **0.125 mg/kg) after 3 h of circulation time of 400 μg. Liporubicin: Light condition was left constant for all PDT experiments: 15 min after Visudyne IV injection, laser light (690 nm) irradiation of the tumor and surrounding lung tissue ($\varnothing$ of the treatment spot: 30 mm) with a fluence rate of 35 mW/cm^2 and a fluence of 10 J/cm^2 corresponding to a treatment time of 15 min. Photodynamic drug delivery in this study has already given good and encouraging results. In this model, Visudyne-mediated PDT pretreatment led to a significant and selective increase in liposomal doxorubicin uptake in tumors with no drug increase in the surrounding normal lung parenchyma. Thus, PDT pretreatment resulted in a better tumor-to-normal lung drug ratio as compared to IV drug application alone. The increase in tumor uptake relative to the normal parenchyma resulting from the PDT pretreatment was also found for a liposomal doxorubicin circulation time of 1 and 6 h.

conditions and animal models, the optimal conditions and parameters still had to be determined. A fundamental understanding of the photosensitization-induced vascular permeability effect is necessary for optimizing this modality to target blood vessels for the treatment of cancer, age-related macular degeneration (AMD), or other diseases in the clinic. The literature from the last decades of research on PDT-induced and other types of vascular leakage demonstrates that no single mechanism explains all vascular leakage.

To observe, understand, and optimize the parameters involved in vascular PDT used as a drug-delivery technique, noninvasive, real-time, *in vivo* observations are necessary. Optimally, we need to monitor, among various people, blood flow, the vascular response at the cellular and molecular levels, and the distribution of the macromolecular drugs in various tumor and normal tissues during and/or after PDT under various treatment conditions.

29.2.2 One Example of a Mechanistic Approach in the Chorioallantoic Membrane Model

The chorioallantoic membrane (CAM) of the chicken embryo is a useful model for simple *in vivo* monitoring of vasculature (Vargas et al. 2007). This model offers the possibility of observing, *in vivo*, the basic mechanisms involved in the PDT-induced vascular damage and/or leakage with minor intervention of the immunity and inflammatory systems, which are still immature at the time of observation, and without the problems of high tumor interstitial pressure encountered in some tumor models.

By treating the CAM vasculature with Visudyne PDT, we could observe a significant vasoconstriction of the arteries, simultaneously to platelet aggregation and blood coagulation in veins, both during and after irradiation (Debefve et al. 2008; Lange et al. 2001). This is illustrated in the sequence of pictures shown in Figure 29.5. Both phenomena contributed to the

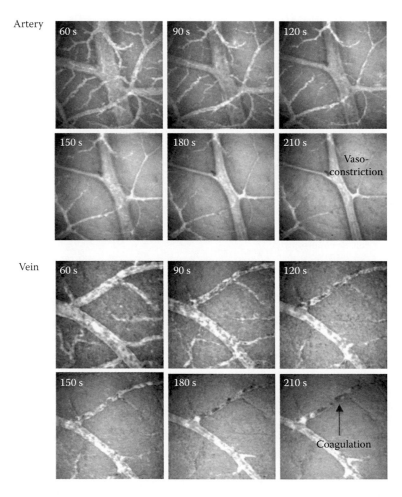

FIGURE 29.5 This figure shows two sequences of Visudyne angiography pictures taken during Visudyne PDT. The drug–light interval was 60 s, and the first picture was taken right after the start of the irradiation. The first sequence of six images shows a field comprising an artery, and the second sequence focuses on a vein with the same order of diameter (150 ± 20 μm). The whole field was irradiated with 420 ± 20 nm during 210 s to observe the PDT-induced occlusion differences in both kinds of blood vessels. In the arteries, one can observe the dramatic decrease in the vessel diameter during the time of observation, whereas the vein also presents platelet adhesion and aggregation (see the black spots appearing already after 30 s of irradiation, i.e., at time = 90 s, in the upper right branch of the blood vessel and then the thrombus progression leading to almost total occlusion of the vein after 210 s. This phenomenon was not observed in arteries during the observation time.

reduction of blood flow within the treatment area, potentially leading to the final occlusion of both types of blood vessels, thus limiting the possibility of leakage during and after PDT.

A simplified scenario of vascular occlusion by PDT is schematized in Figure 29.6 (Debefve et al. 2007). It is known that vascular PDT can cause perturbation of the cytoskeleton that leads to alteration of cell shape. Endothelial cells both becoming round and contracting have been observed. The change in shape of endothelial cells leads to the disruption of the tight junctions between the endothelial cells and a decrease in interendothelial cell communication and causes exposure of the subendothelial basement membrane, which may cause the release of certain proteins (Foster et al. 1991). This disruption of tight junctions after vascular PDT is illustrated in Figure 29.7 (Stepinac et al. 2005), which shows a histological section of retinal blood vessels in a minipig model, where the tight junctions are labeled in red by means of an antioccludin antibody. The minipig

has been injected with a PS, and one of the two eyes has been exposed to a therapeutic light dose; the other eye has not been exposed to light. One might expect to observe that, under these conditions, in the light- and drug-exposed eye, an increase in leakage through the newly fenestrated vasculature would take place. However, the photodynamic damage to the endothelial cells activated hemostasis processes, such as the release of von Willebrand factor (vWF) and other blood-clotting agents, thus limiting the possibility of PDT-induced leakage. The activated platelets stick to the gaps between the damaged rounded endothelial cells, and further aggregation of platelets takes place at these sites, starting to plug the vessel.

Indeed, to increase the permeability of the CAM vessels by PDT, limiting the hemostasis by delaying the blood clotting appears to be a pivotal process. In order to demonstrate this, we need to do a tri-therapy, using aspirin or any other antiplatelet factor together with Visudyne PDT and the cytostatic drug that

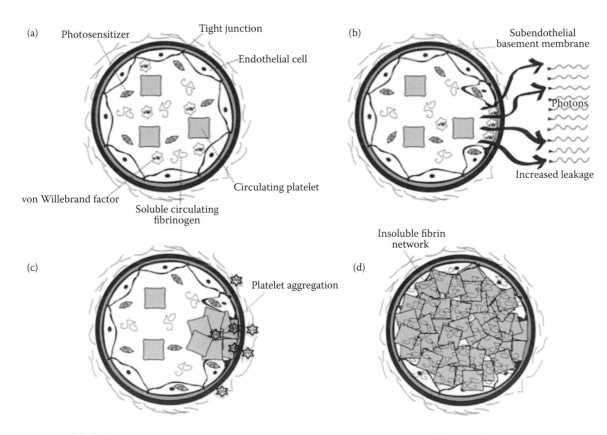

FIGURE 29.6 Simplified scheme of the PDT-induced hemostasis pathway. PS = photosensitizer; vW = von Willebrand factor. (a) A blood vessel prior to PDT. Endothelial cells lining the vessel wall are flat and linked to each other by tight junctions. After IV injection, the PS circulates in the blood flow. (b) Some PS enters the endothelial cells. PDT is then started by illuminating the area, including the blood vessel. PDT induces oxidative stress to the endothelial cells, causing modifications among others to the cytoskeleton. This damage induces a change of cell shape and opening of the tight junctions. The retraction of the endothelial cells exposes parts of the subendothelial connective tissue, inducing release of clotting factors, such as vWF, and probably also inducing increased vessel wall permeability. (c) Formation of the start of a plug. vWF and other clotting factors lead to platelet activation. PDT damage to the membrane lipids also causes the release of arachidonic acid that is transformed by cyclooxygenase into thromboxane that activates and aggregates platelets. (d) Fibrinogen is changed into fibrin, and finally, the thrombus is stabilized with fibrin by coagulation. The blood vessel is then no longer perfused.

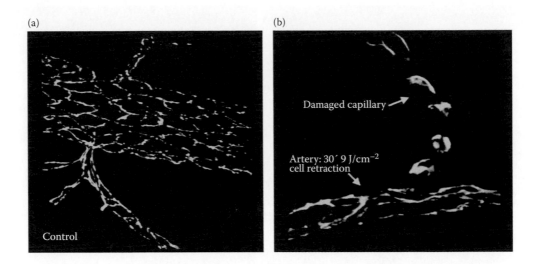

FIGURE 29.7 Confocal microscopy and immunolocalization of occludin in blood vessels of a flat-mounted minipig retina. (a) Blood vessels in this eye were not treated by PDT. One can observe the regular pattern outlining the tight junctions of the retinal circulation's endothelial cells. (b) Blood vessels in this eye were treated by PDT with a Pd-meso-tetra(4-carboxyphenyl) porphyrin (PdTCPP) and irradiation with 9 J/cm² at 532 nm. One can observe that the reticulate pattern was locally disrupted by the treatment. Local displacement, damaged capillaries, absence of occludin staining between endothelial cells, and cell retraction suggest heavy damage of the blood–retina barrier.

we need to deliver to this target area. For the sake of experimental simplicity, we replaced the cytostatic agent with a fluorescent dye in the CAM model and, after, delayed platelet aggregation with aspirin. The resulting PDT-induced leakage is illustrated in Figure 29.8 (Debefve et al. 2007). This concept of combining vascular PDT with an antiplatelet factor in order to modulate the platelet activation mechanism involved in the photo-occlusion of vascular PDT opens up new perspectives in optimizing the PDT-induced leakage for drug delivery.

29.2.3 Observations by Intravital Microscopy of Vascular PDT in the Skinfold Chamber of Nude Mice

The preparation of the skinfold chamber implanted on the back of nude mice is illustrated in Figure 29.9. This experimental model allows for real-time observation of the subcutaneous microvessels through a glass window (Lehr et al. 1993). We used this model first to assess the effect of Visudyne PDT by fluorescence

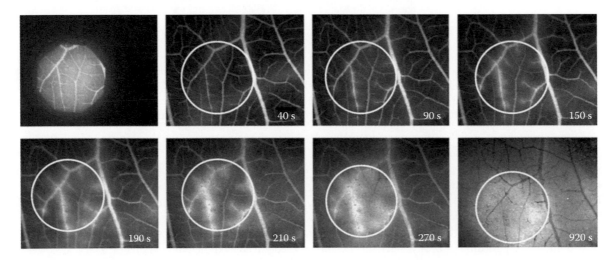

FIGURE 29.8 Typical fluorescence pharmacokinetics obtained with FITC–dextran (FITC-D) 10 kDa (25 mg/mL PBS) (λex = 470 ± 20 nm; λem = 500–550 nm) during PDT with Visudyne (2 µg/embryo) coinjected with aspirin (80 µg/embryo) from the beginning of the irradiation (λex = 420 ± 20 nm) with a light dose of 20 J/cm². The first picture shows the fluorescence of Visudyne observed at $\lambda \geq$ 610 nm, intravenously coinjected with aspirin. Irradiation was performed 1 min after injection. All the other pictures show the fluorescence pharmacokinetics of FITC-D 10 kDa during this PDT. This shows that PDT with Visudyne associated with aspirin significantly increases the FITC-D 10 kDa local leakage. The white circle shows the irradiated area.

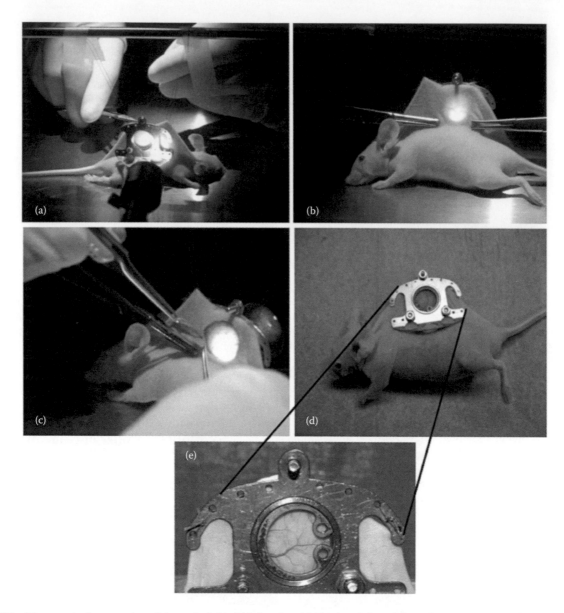

FIGURE 29.9 Microsurgical preparation of the optical skinfold chamber. (a) The dorsal skin of the nude mouse was stretched to form a double-layer skinfold. A titanium frame is placed on one side. (b) The titanium frame is fixed through the skin. The area and its vasculature are chosen by transparency of the skin. (c) A circular area 15 mm in diameter from one area of skin was carefully and completely removed under the microscope. (d) The remaining layers were covered with a glass coverslip incorporated into one of the frames. (e) Details of the observation chamber.

intravital microscopy in normal tissue with high drug and light doses, that is, 0.8 mg Visudyne/kg body weight combined with a light dose of 200 J/cm² at 420 ± 20 nm given at a fluence rate of 300 mW/cm². We believe that a better understanding of how PDT affects the normal vasculature is relevant as current PDT treatments of tumors are not only strictly limited to the cancerous tissue but also affect the peritumoral area. Under these conditions, we observed a significant PDT-induced vessel permeabilizing effect for fluorescein isothiocyanate dextran of 2000 kDa (FITC-D) in the treated area of all animals. This PDT-induced leakage is illustrated in Figure 29.10 (Debefve et al. 2010, 2011). Using this model, we also confirmed, by histological analysis, that the treated tissue was significantly inflamed after PDT.

The development of this inflammatory response following treatment was monitored before, during, and after PDT by means of the observation of the behavior of leukocytes labeled with Rhodamine-6-G in the veins. The number of "rolling leukocytes" is taken as a good indicator of inflammatory response. Inflamed endothelium of the veins expresses selectins that are recognized by the leukocytes that thus tend to slow down within the bloodstream and "roll" along the vessel wall, sticking to the endothelial cells. These leukocytes then extravasate to the interstitial tissue. We then investigated the role of inflammation in the induction of increased vascular permeability after PDT (Debefve et al. 2011). Using selective antibodies, which neutralized the selectins, we could completely inhibit the leukocyte

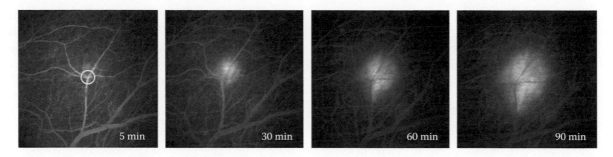

FIGURE 29.10 Typical example of FITC-D leakage (2000 kDa; 25 mg/mL PBS; λex = 470 ± 20 nm; λem = 500–550 nm) when IV injected 2 h after PDT in a tumor-free skinfold chamber implanted on nude mice. This was observed by intravital microscopy 5, 30, 60, and 90 min after IV injection of FITC-D. PDT treatment conditions: Visudyne 0.8 mg/kg; drug–light interval = 10 min; λex = 420 ± 20 nm; light dose = 200 J/cm^2; the fluence was 300 mW/cm^2. The treatment area is circled in the first picture.

recruitment by the endothelium (i.e., rolling) and thus suppress the inflammatory response after PDT. Under these conditions of inhibited inflammation, PDT did not increase the vascular permeability. This study underlines the importance of the leukocyte recruitment associated with the inflammation induced by PDT in the mechanisms of PDT-induced leakage in normal tissue. This fact has to be taken into account in the understanding of the mechanism and development of PDT for drug delivery.

We can apply active targeting strategies to the tumor vasculature in order to accumulate more PS within the tumor blood vessels as compared to normal vessels. Several molecules, such as the integrin $\alpha_v\beta_3$, have already been identified as having higher expression in the blood vessels of cancer. Thus, this cellular receptor preferentially expressed by tumor endothelium can be targeted. A good candidate for this targeting is the kistrin molecule, which is a protein extract from snake venom and which

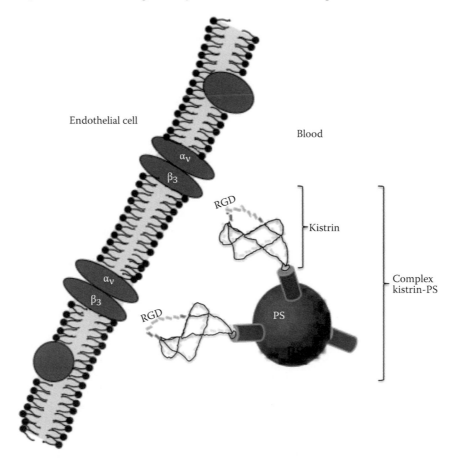

FIGURE 29.11 Scheme illustrating the possibility of targeting $\alpha_v\beta_3$ selectively expressed by neovessels, using the complex between kistrin and a photosensitizer (PS). Kistrin, a disintegrin extract from snake venom, contains RGD domains that recognize $\alpha_v\beta_3$. This protein can be coupled with photosensitive agents (PS) to vehicle them to the target vasculature, thus improving the selective accumulation of the PS in the target area and the selectivity of a PDT treatment.

comprises RGD domains that have a strong affinity for αvβ3 (Rahman et al. 2000). An ongoing study consists of coupling a photosensitive agent to kistrin to improve its selective accumulation and then confining the PDT effect to the targeted tumor vessels. This principle is illustrated in Figure 29.11. Furthermore, as kistrin is also known for its antiplatelet effect, one can expect to have potentiation of the PDT-induced leakage as a result of the delay in platelet aggregation as explained previously by our experiments in the CAM model.

It is also possible to find improved passive targeting strategies of the tumor vasculature. Thus, we can try to find differences between tumor and normal vasculature in terms of physiological responses instead of differences in protein expression. The following is an example of this. We repeated the study summarized previously, in which we observed the leukocyte activity after PDT within the veins of normal tissue, but with tumor blood vessels. We then observed significant and interesting differences between tumor and normal tissue. When anti-selectin antibodies were injected before PDT treatment of a skinfold chamber implanted with a tumor, leukocyte rolling was significantly inhibited as was the case in normal tissue (Debefve et al. 2011; Wang et al. 2012b). However, whereas this selectin neutralization by an anti-selectin antibody mixture tends to completely inhibit the PDT-induced leakage in the normal tissue (Debefve et al. 2011), it did not affect at all the induced permeability in the tumor tissue (Wang et al. 2012b). Learning from past studies of combination therapies in the CAM model summarized previously, one can thus imagine that under the influence of anti-selectins or possibly using more common molecules, such as anti-inflammatory drugs, one might be able to induce by PDT a very selective leakage in the tumor tissue without affecting the leakage in the normal surrounding tissue. It may well be possible to do this without limiting the irradiation to the tumor itself. With this combination treatment, the leakage would occur only in the tumor without affecting normal vasculature, thus clearly enlarging the treatment window. We would only need to find the treatment conditions that avoid the occlusion of the tumor vessels without caring about potential leakage in the healthy tissue, and we could be essentially independent of the selectivity of the PS for tumor vessels.

29.3 Different Parameters to Consider on the Way to Optimal Conditions of PDT for Drug Delivery

29.3.1 PDT Treatment Conditions

PDT for drug delivery necessitates a photosensitization mainly limited to the vasculature. The preference of vascular versus cellular targeting by PDT is dependent upon the relative distribution of a PS in each compartment, which is governed by the PS pharmacokinetic properties and can be effectively manipulated by the PS drug administration, by the modification of the PS's molecular structure and/or vehicle, as well as by the time of light illumination relative to the drug's injection (drug–light

interval). When the PS is intravenously administered, PDT using shorter drug–light intervals mainly targets the vasculature by confining the PS localization within the blood vessels and then the endothelium (Chen et al. 2003, 2006a; Fingar et al. 1999; Kurohane et al. 2001). The PDT-mediated vascular effects can range from transient vasoconstriction to permanent vessel occlusion by vessel wall disintegration and thrombosis. Between these two extremes, PDT can lead to vascular leakiness and, hence, to increased extravasation of therapeutics. The latter can be coadministered with the PS or added a bit later.

This timing of the therapeutic injection relative to the light application of the PDT is also dependent on the drug and light conditions. Unfavorable PDT treatment conditions may result in vasospasm or thrombosis and result in impaired or no blood perfusion and, consequently, to uneven drug distribution within tissues. This uneven distribution in that case is, in part, a result of the fact that small vessels are rather easy to occlude by PDT, whereas larger vessels may just leak and/or constrict. The relatively low fluence rate PDT regimen that proved optimal for enhancement of vascular permeability to liposomal doxorubicin did not result in vascular collapse or induction of hypoxia in the tumor during PDT treatment (Foster and Gao 1992; Sitnik, Hampton, and Henderson 1998). The timing and sequence of macromolecule injection relative to PDT can also significantly affect macromolecule delivery to the target tissue. The time during which tumor vessels are exposed to the macromolecular drug is also an important parameter influencing its retention within the target area. As an example, in one of our studies in the skinfold chamber implanted on nude mice, we observed that the leakage of a fluorescent dye observed immediately after PDT was significantly smaller than in the case where the dye was injected 2 h after the vessel-permeating PDT (Debefve et al. 2010).

29.3.2 Architecture of the Blood Vessels

Endothelial cells form the inner lining of all blood vessels and thus provide the main physical barrier that the chemotherapeutics administered intravenously need to cross to reach the target tissue. The endothelium does not only prevent unwanted extravasation of blood components but also actively and selectively regulates the efflux of blood fluid and macromolecules to the surrounding tissue (van Nieuw Amerongen and van Hinsbergh 2002). In physiological tissue, the wall of capillary and postcapillary venules consists of a layer of endothelial cells that connects to each other via closely opposed intercellular junctions. Endothelial permeability for macromolecules, such as the chemotherapeutic agent liposomal doxorubicin, depends first on the integrity of these intercellular junctions as well as the actomyosin-based cell contractility. The paracellular permeability of this barrier structure is maintained by equilibrium between the contractile force generated at the endothelial cytoskeleton and the adhesive force produced at endothelial cell–cell junctions and cell–matrix contacts. A dynamic interaction among these structural elements controls the opening and closing of paracellular pathways to macromolecular transport and thus serves as a fundamental mechanism in the regulation of blood–tissue exchange (Yuan 2002). It has been demonstrated that Visudyne PDT induces a rapid endothelial cell microtubule depolymerization, resulting in endothelial cell shrinkage and loss of interendothelial tight junctions responsible for the sealing properties between adjacent cells (illustrated in Figure 29.7 with another PS). This leaves enlarged gaps between endothelial cells and disrupts the vascular barrier (Bazzoni 2006; Chen et al. 2006b). This results in the exposure of the vascular basement membrane, thus temporarily increasing the vascular permeability to macromolecules following PDT (Fingar, Wieman, and Haydon 1997; Krammer 2001).

This leakage induced by PDT can be influenced by the calcium level within the endothelial cells. Indeed, these cells have a specific calcium handling that can be compared to that in smooth muscle cells. The elevation in the cytoplasm calcium concentration is a pivotal event in inducing endothelial hyperpermeability, which involves actin–myosin interactions. Modulation of the activity of the channels controlling the calcium levels may thus affect the integrity of junction complexes between adjacent endothelial cells (Dejana, Spagnuolo, and Bazzoni 2001; Seebach et al. 2005). Protein kinases and several protein tyrosine kinases regulate this interaction (Bogatcheva et al. 2003; Seebach et al. 2005; Yuan 2002). A modulation of their activity could modulate the endothelial permeability and so influence the efficacy of the photodynamic drug delivery. Calcium chelators and some fatty acids, phosphate esters, surfactants, cationic polymers, cyclodextrins, and nitric-oxide donors have also been identified as modulators of tight junctions, consequently increasing the endothelium permeability (Deli 2009). One can also expect to observe a variation in responsiveness of the endothelium to PDT and variation of the efficacy of the photodynamic drug delivery as a result of architectural differences between the treated blood vessels (van Nieuw Amerongen and van Hinsbergh 2002).

These vessel architectural parameters are listed as follows:

- The *diameter* of the vessel and the composition of its inner layer, *the tunica intima* (particularly the deepness of the internal elastic lamina); the composition of its *tunica media* (with or without vascular smooth muscles and their related contractility); and its external layer, the *tunica adventia* (with or without nerves and *vasa vasorum* capillaries)
- The *shape* of the blood vessel (straight/tortuous, with/without branching)
- The blood *flow* (laminar/turbulent, continuous/discontinuous)
- The *oxygenation* of the circulating blood (arteries/veins, systemic/pulmonary circulation) and the oxygen availability for PDT
- The *protein* expressions on the vessel wall and the potential interactions with blood cells (as an example, see Section 29.2.3, describing the PDT-induced inflammation and selectin expression in veins)
- The *organ* or *tissue* vascularized by the blood vessel [e.g., the blood brain barrier (BBB) in neurologic tissues,

which presents particularly important sealing properties] (Hirschberg et al. 2008; Madsen and Hirschberg 2010)
- The *maturity* of the blood vessel (neovessels versus mature vessels), expressing different proteins
- The normality of the blood vessel (*normal versus tumor*, fenestrated, tortuous, with heterogeneous blood flow)

Each difference between vascular beds can lead to a changed response in terms of PDT-induced leakage. These differences should be further studied and used as a means to improve the targeting. Indeed, most of the intervasculature differences in sensitivity to PDT treatment should not only be taken as a difficulty to finding the optimal standard treatment conditions for photodynamic drug delivery but also be exploited to improve the selectivity of our drug-delivery pathway. For example, the mature vasculature in normal (nonneoplasic) tissue is significantly less sensitive to vascular targeting PDT as compared to the vessels in tumor tissue. This difference between normal and tumor tissue allows, under low drug and light doses, for a selective photodynamic drug delivery to the tumor without affecting the drug distribution within the normal surrounding tissue.

As mentioned above, blood-flow velocity is also an important parameter in determining vascular response to PDT. Vessels with high flow velocity are more resistant to PDT-induced vascular shutdown (He, Agharkar, and Chen 2008). As the blood flow inside a tumor is generally slower than in the neighboring normal tissue, tumor blood vessels are more sensitive to vascular PDT-induced occlusion than the peripheral normal blood vessels (Fukumura and Jain 2007). This partially explains the heterogeneity of the blood vessel response to vascular-targeting PDT in the different deepness of the tumor tissue. The drug uptake increase resulting from PDT is significantly higher in the periphery than in the interior of the tumor (Chen et al. 2008). The lack of a functional blood vessel in the middle of the tumor occluded following PDT could thus limit the possibility of drug delivery (Wang et al. 2012a).

29.3.3 Physiological Responses to Vascular Injury

29.3.3.1 Hemostasis Coagulation

The opening of tight junctions by PDT, as explained in the previous paragraph, results in exposure of the vascular basement membrane, which can, in principle, temporarily increase vascular permeability and leakage, but also induces platelet binding and aggregation at the site of damage to ensure hemostasis. The activated platelets release vasoactive mediators, such as serotonin and thromboxane, that can trigger a cascade of events, including amplification of platelet activation, coagulation, and vasoconstriction. The final result is vascular collapse and blood-flow stasis, leading to the temporary or permanent occlusion of the irradiated neovascularization (Fingar 1996; Fingar et al. 1999; Fingar, Wieman, and Haydon 1997; Krammer 2001). The mechanism underlying thrombus formation induced by photodynamic vascular targeting is complicated. Reactive oxygen species generated intravascularly after PDT cause damage to multiple targets, such as red blood cells, platelets, and endothelial cells, which, in turn, leads to activation of hemostatic cascades and results in thrombus formation (Debefve et al. 2008; Fingar 1996; Krammer 2001). As an example, it was described previously in Section 29.2.2 that the use of antithrombotic or antiplatelet agents in the CAM model can extend the therapeutic window of the PDT-induced vascular changes (Debefve et al. 2007), thus potentially improving the photodynamic drug delivery.

29.3.3.2 Vasoconstriction

Following PDT, one can observe the release of different vasoactive substances that have opposing effects on vessel diameter, some leading to vessel constriction and others to dilation, at different times after PDT. As mentioned previously, depending on the architecture and the contractility of the different kinds of blood vessels treated, one can expect to observe a variety of responses in terms of diameter modifications during and after PDT. During and at short times after PDT, treatment conditions using even quite mild drug and light doses induce a vasoconstriction of the small arteries, reducing the blood flow to limit the hemorrhage in the injured area. This physiological vasoconstriction is observed after any kind of vessel injury. However, at longer times after PDT, one can observe more complex mechanisms influencing the diameter of the blood vessel. Depending on the PS used, vasoconstriction is or is not finally involved in the PDT-mediated long-term vascular occlusion (Fingar et al. 1992, 1999; He, Agharkar, and Chen 2008). This modulation in the vessel diameter can influence the PDT-induced leakage. The latter simply confirms the complexity of vessel responses to PDT (He, Agharkar, and Chen 2008) and the necessity to study the vasoactive effect of each PS at different times after irradiation under the chosen PDT treatment conditions.

Following the same philosophy as previously described in the combination therapy in which we used aspirin together with PDT, in order to avoid rapid PDT-induced vascular occlusion so as to keep the targeted blood vessels well vascularized and thus improve the photodynamic drug delivery, one could alternatively try to inhibit the vasoconstriction induced by PDT for a while.

Delaying both the phenomena of platelet aggregation and vasoconstriction might be an even better possibility to optimize photodynamic drug delivery.

29.3.3.3 Inflammation

Endothelial cells form a layer between the blood and the surrounding connective tissue. As the blood vessel, mentioned previously, the PDT-induced contraction and shrinkage of these endothelial cells lead to exposure of the tissue extracellular matrix to circulating blood. This causes white blood cell activation, principally as a result of different subtypes of selectins, expressed by endothelial cells after injury and by platelets and leukocytes themselves (Debefve et al. 2011). The leukocytes, by recognition of the expressed selectins, then tend to slow down, roll, and adhere directly to the damaged endothelial cells (deVree et al. 1996a,b; Debefve et al. 2011). Both the interactions

of leukocytes with the endothelium and the secreted products released by leukocytes acting on the endothelium contribute to the increased vascular leakage. The recruitment of leukocytes following PDT and their adherence to venules has been the subject of many studies (Debefve et al. 2011; Fingar et al. 1992; Yuan et al. 1995). This leukocyte behavior is an important part of the inflammatory response to PDT (Korbelik 2006). Depending on the severity and duration of the inflammatory reaction, different major steps contribute to initial and prolonged vascular leakage. At an early stage, inflammatory mediators induce a transient vascular leakage (van Nieuw Amerongen and van Hinsbergh 2002) that is complicated by the simultaneous action of several proteases and vasoactive agents, such as histamine, thrombin, serotonin, bradykinin, and prostaglandin E2 (PGE$_2$). Some of these molecules, released by the activated leukocyte, have the potential to further increase the leakage of macromolecules through the endothelium (Lentsch and Ward 2000). As explained in Section 29.2.3, in an experimental setup containing normal (nontumor) blood vessels, when antibodies against selectins functionally blocked the interaction between the leukocytes and the endothelium, almost no inflammatory response was observed even after high drug–light doses of PDT. In this situation, in which selectin antibodies completely suppressed leukocyte rolling, the PDT-induced microvessel permeability was significantly reduced (Debefve et al. 2011). In normal vasculature, the leukocyte–endothelial cell interaction was proven to be essential for PDT-induced drug delivery to normal tissue.

However, when the study was repeated in a mouse model of malignant mesothelioma, the leukocyte recruitment was significantly downregulated by the antiselectin antibodies as well, but in this case, the inhibition did not affect the PDT-induced extravasation of a fluorescent dye (FITC-dextran). This revealed that, unlike normal vessels, leukocyte–endothelial cell interaction is not required for PDT-induced leakage in the tumor tissue (Wang et al. 2012a). Thus, any difference between tumor and normal vasculature, for example, concerning the role of inflammation in the PDT-induced leakage, can be exploited to try to further increase the tumor selectivity of our photodynamic drug delivery.

At a slower pace, in response to the inflammatory process, the entire microvasculature undergoes dramatic remodeling initiated by angiogenic factors. Not only these angiogenic growth factors do affect the integrity of the cell junctions by enhancing endothelial cell migration but also some of them, in particular VEGF, also known as vascular permeability factor (VPF), induce a hyperpermeable status of the vasculature by themselves. VEGF has been shown to induce transendothelial pathways by formation of vesiculovacuolar organelles, which are interconnected chains of vesicles forming a kind of a pore through endothelial cells (Bates and Harper 2002; Dvorak 2002; Feng et al. 1996; van Nieuw Amerongen and van Hinsbergh 2002). However, other factors involved in vessel stabilization, such as platelet-derived growth factor (PDGF) and angiopoietins, can counteract these effects (Thurston 2002; van Nieuw Amerongen and van Hinsbergh 2002). These different stages of inflammation appearing after vascular-targeting PDT and inducing vascular leakage could be selectively modulated by different pharmacological approaches of combination therapies, thus influencing or allowing for a better control of the photodynamic drug delivery.

29.3.4 Properties of the Leaking Drug

The photodynamic drug-delivery process described in this chapter can be applied to improve the uptake of carefully selected drugs. The characteristics required for these drugs are nonexhaustively listed below. A good candidate would

- Be homogeneously distributed in circulating blood shortly after intravenous injection
- Preferentially have weak accumulation in the nontargeted tissue
- Have a relatively long circulation time and a stable structure/activity, allowing for sufficient accumulation within the target tissue before being cleared from the vasculature
- Have an appropriate pharmacological effect on the target tissue
- Have carefully chosen physicochemical properties, especially optimal molecular size for the selective leakage

Some of these physicochemical properties can be optimized by playing with the pharmaceutical formulation. A number of technologies can be used to modify the drug's circulation time. Coupling the drug with a polyethylene glycol (PEG) group would, for example, contribute to delaying its clearance, thus increasing its circulation time. One can also entrap the active molecule within vehicles, such as liposomes, nanoparticles, or microspheres, to try to influence the distribution of the drug. These vehicles can be chemically modified or labeled to make them preferentially accumulate within the target area. As mentioned previously, the size of the drug to be delivered by photodynamic leakage is probably a major factor. After intravenous administration, the spontaneous decrease in the concentration of a drug in the vasculature can be mainly attributed to its leakage from the intravascular to the extravascular compartment as well as by its clearance by the RES in the cases where the drug is chemically stable. In general, "small" molecules or particles have the tendency to leak spontaneously, whereas "big" molecules are quickly recognized and cleared by the RES (Pegaz et al. 2005) before they can reach the target. One has thus to define an optimal size and design a corresponding formulation for an appropriate circulation time for each drug to be delivered.

The transport of macromolecules across the vascular barrier occurs mainly through endothelial gaps and is thus mainly limited by the size of these gaps (Hobbs et al. 1998; Yuan et al. 1995). The characteristic of these gaps varies as a function of, among other things, the organ in which the vessel is located and the architecture of the blood vessels, their maturity, integrity, and normality (vs. neoplasy). In the case of fenestrated tumor vessels, it also depends on the tumor type and its microenvironment (Dreher et al. 2006). As mentioned previously, vascular PDT has the potential to break the tight junctions, resulting in enlarged gaps between

endothelial cells. To delineate the pore size induced by PDT, one can administer different sizes of fluorescent dyes or microspheres during and/or after PDT and observe the induced leakage from the treated area. As an example, in one of their studies, Snyder et al. showed that the increase in tumor uptake of 2000 kDa fluorescein isothiocyanate dextran (FITC-D) was more significant than that of Evans blue (Graff, Bjornaes, and Rofstad 2001). The latter strongly binds to albumin in the blood. Its behavior reflects the transport of albumin (67 kDa with a diameter of 7 nm), which is the same size as the effective pore occurring in most normal blood vessels. Thus, it probably leaks significantly out of its latter tumor vessels, which typically have larger interendothelial junctions than normal blood vessels (Roberts and Palade 1997). There might be only a little hindrance for the transvascular transport of the Evans blue–albumin complex. Therefore, a further increase in vascular permeability induced by vascular photosensitization may have little influence on the extravasation of albumin that can already cross the tumor vessel wall.

For the same reason, administration of free doxorubicin was not improved by PDT (Cowled, MacKenzie, and Forbes 1987). However, pretreating the tissue by low-dose PDT before intravenous administration of a drug/dye characterized by a larger size, such as FITC-D 2000 kDa, allowed for significantly better uptake compared to intravenous administration alone. Indeed, these larger molecules are otherwise difficult to transport across the endothelial barrier and thus need an enhanced leakage to be distributed outside the vasculature (Chen et al. 2006b).

29.3.5 Target Tissue

To bring the photodynamic drug delivery from bench to bedside, one of the first steps consists of adapting the dosimetry to the targeted organ and doing a dose-finding study to define the treatment conditions, including, in particular, the light conditions. These conditions will depend on the accessibility and the optical properties of the tissue containing the tumor and surrounding it. The vascular density and homogeneity of this vascularization are important parameters to consider as well, both determining the sensitivity of the tissue to vascular PDT. The composition of the extravascular matrix and the interstitial pressure will also influence the efficacy of the PDT-induced leakage. Endothelial gaps alone do not allow transport of water or circulating drugs without driving forces. A hydrostatic pressure gradient is required between the intravascular and interstitial space and/or dissimilar oncotic pressure in these two compartments (Debefve et al. 2011). The typical high interstitial tumor pressure can be a major barrier to the delivery and efficacy of certain chemotherapeutics. However, the leakiness of tumor vessels and the defective lymphatic drainage can enhance the accumulation of high molecular weight drugs in the neoplasic tissue as compared to the surrounding normal tissue. This phenomenon is known as the enhanced permeability and retention (EPR) effect, in which the accumulation of a drug is caused by an increased extravasation combined with a decreased clearance (Maeda et al. 2000).

Two main types of mechanisms are thought to be responsible for drug distribution in the tumor: convection and diffusion (Jain 2005). Convection follows the equation of Starling and is dependent on the hydrostatic (SHP) and oncotic pressure (OP) inside and outside tumor blood vessels:

the driving forced across the vessel wall = [SHP (in the vessel) – SHP (interstitium)] – [OP (in the vessel) – OP (interstitium)]

Note that above and beyond this equation, the loss of substance from a vessel to the interstitium will depend on the leakiness of the vessel. Furthermore, OP depends on leaking proteins, which is different in each type of tissue. Diffusion, however, is dependent on concentration differences between the intravascular and extravascular space. Because chemotherapeutics, such as liposomal doxorubicin, are macromolecules with high molecular weights and reflection coefficients, their diffusion is limited and their distribution mainly relies on convection.

It has not been generally accepted that increasing tumor vascular permeability will result in enhanced drug delivery. Jain, Tong, and Munn (2007) have even suggested that normalization of a tumor vessel may improve drug delivery by decreasing the vascular permeability, thus lowering the interstitial fluid pressure (IFP). The high interstitial hydrostatic pressure within tumors reduces the forces for convective movements.

Vascular targeting PDT has been shown to further increase tumor IFP as a result of enhancing vascular permeability (Fingar, Wieman, and Doak 1991; Leunig et al. 1994) at early time points. However, one can expect to measure diminishment of the IFP following PDT because of vasoconstriction or vascular shutdown. Much later after PDT, IFP should also reincrease through the VEGF-induced neovessel formation (Bhuvaneswari et al. 2007; Fingar, Wieman, and Doak 1991; Leunig et al. 1994). IFP changes following PDT are sequential and can thus be exploited for distribution enhancement in chemotherapy. Following the idea of combination therapies, one could also combine an anti-VEGF pretreatment with photodynamic drug delivery. An anti-VEGF pretreatment can normalize the tumor vessels, which will then decrease, at first, the IFP within the tumor. This favorable lower IFP could then facilitate the leakage that subsequently would be induced by vascular PDT and thus improve the drug delivery.

29.4 The Outlook for Bringing PDT for Drug Delivery from Bench to Bedside

29.4.1 Combination Therapies for Enhanced PDT-Induced Leakage

Based on the new and future mechanistic findings concerning the different types or stages of vascular leakage occurring during or after PDT, pharmacological intervention, which uses a number of these mechanisms, can be optimally developed. One can play with the different parameters mentioned previously to enhance the PDT-induced leakage. For example, one can delay hemostasis using antiplatelet factors to enlarge the time window of increased leakage between PDT treatment and vascular shutdown (Debefve et al. 2007). This kind of combination therapy temporarily or irreversibly inhibits one part of the PDT-induced effect, that is, the vascular occlusion, preserving the disruption of the endothelium, thus allowing an increase in drug–light doses without interrupting the bloodflow and potentially improving the drug-delivery pathway. Similarly, a triple therapy using a vasodilator together with vascular PDT and chemotherapy could counteract the PDT-induced vasoconstriction, thus becoming a good adjuvant for photodynamic drug delivery.

The differences in the photodynamic drug delivery between the tumor and normal blood vessels can also be exploited and enhanced to further improve the selectivity of PDT for tumor drug delivery, preserving the normal tissue from side effects. One can mention, furthermore, the possibility of using an anti-inflammatory drug to inhibit the leukocyte activation, which, as we have shown, is correlated with the PDT-induced leakage in normal surrounding vasculature without affecting the PDT-induced drug delivery in tumor tissue. The improved selectivity, of course, also can be attained by using PS molecules with particular physicochemical properties designed to selectively target tumor vasculature.

These steps will be combined with microcirculatory research approaches *in vivo* and clinical studies before definitive conclusions can be drawn regarding the optimal mechanisms of PDT-induced vascular leakage as well as regarding the efficacy of enhancing drugs and combined treatments for improved PDT-induced leakage. Nevertheless, these steps provide targets for new therapeutic approaches to enhance the basic idea of an effective PDT-induced leakage used as a drug delivery system in patients (van Hinsbergh and van Nieuw Amerongen 2002). Our understanding of the PDT-induced leakage is approximately at a level of maturity that will enable us to develop drugs and treatment regimens that effectively enhance the PDT-induced leakage to make photodynamic drug delivery usable to treat certain tumors clinically.

29.4.2 Examples of Possible Clinical Applications

29.4.2.1 The Pleura: Mesothelioma or Pleuropulmonary Metastasis

The treatment of superficially spreading chest malignancies, such as malignant pleural mesothelioma or pleuropulmonary metastasis, is challenging. Currently, their response rates to conventional chemotherapy range from 15% to 45% (Chojniak, Yu, and Younes 2006). This is, in part, a result of the heterogeneous cytostatic drug distribution in these tumors, leaving untreated areas behind, which could be the origin of further cancer progression. The use of low-dose PDT as a means for selective enhancement of tumor drug uptake is an attractive treatment concept for these pleural superficially spreading chemoresistant tumors, which minimize the side effects of conventional high-dose PDT on large surfaces. PDT with low drug–light conditions delivered by minimally invasive thoracoscopy before intravenous administration of a macromolecular cytostatic agent might be considered in patients referred for the thoracoscopic treatment of malignant pleural effusion. It is anticipated that low-dose PDT combined with chemotherapy could improve the outcome of these clinical cases (Wang et al. 2012a). A first clinical study, involving 30 patients anesthetized and operated on by thoracoscopy for a talc powder procedure, will be undertaken in our group in Lausanne, starting in 2012.

29.4.2.2 The Brain: Glioblastoma

PDT can also be used to open the blood–brain barrier (BBB) (Deli 2009; Hirschberg et al. 2008; Madsen and Hirschberg 2010), breaking the tight junctions. A transient barrier dysfunction has advantages to increase oxygen and nutrient delivery under ischemic conditions to penetrate the BBB for delivery of drugs into the central nervous system (CNS).

29.4.2.3 The Retina: Age-Related Macular Degeneration

In patients suffering from age-related macular degeneration, the current treatment consists of repetitive, monthly intravitreal injection of anti-VEGF drugs. Another treatment proposed to these patients is repeated PDT using Visudyne. These two treatment modalities might be combined to use PDT for drug delivery of an anti-VEGF factor, administered intravenously instead of directly into the vitreous, thus lowering the risk of nosocomial pathologies following repeated intravitreous injections and furthermore potentially decreasing the number of treatments necessary.

29.5 Conclusion

The use of PDT as a pretreatment to selectively enhance the distribution of macromolecular therapeutics to superficially spreading tumors is attractive. It depends on optimal PDT conditions with rather uniform light delivery to target tissues. One also needs to avoid vasospasm, vessel thrombosis, and tissue infarction. A careful optimization of the various parameters involved and a better understanding of how photosensitization modifies vascular function and drug distribution in tissues are of great interest to define what combination therapy should be applied to optimize the selective photodynamic drug delivery.

References

Alexander, H. R., Jr., and C. C. Butler. 2010. Development of isolated hepatic perfusion via the operative and percutaneous techniques for patients with isolated and unresectable liver metastases. *Cancer J* 16:132–141.

Bae, Y. H., and K. Park. 2011. Targeted drug delivery to tumors: Myths, reality and possibility. *J Control Release* 153:198–205.

Bates, D. O., and S. J. Harper. 2002. Regulation of vascular permeability by vascular endothelial growth factors. *Vascul Pharmacol* 39:225–237.

Bazzoni, G. 2006. Endothelial tight junctions: Permeable barriers of the vessel wall. *Thromb Haemost* 95:36–42.

Bhuvaneswari, R., Y. Y. Gan, K. K. Yee, K. C. Soo, and M. Olivo. 2007. Effect of hypericin-mediated photodynamic therapy on the expression of vascular endothelial growth factor in human nasopharyngeal carcinoma. *Int J Mol Med* 20:421–428.

Bogatcheva, N. V., S. M. Dudek, J. G. Garcia, and A. D. Verin. 2003. Mitogen-activated protein kinases in endothelial pathophysiology. *J Investig Med* 51:341–352.

Chen, B., C. Crane, C. He et al. 2008. Disparity between prostate tumor interior versus peripheral vasculature in response to verteporfin-mediated vascular-targeting therapy. *Int J Cancer* 123:695–701.

Chen, B., B. W. Pogue, I. A. Goodwin et al. 2003. Blood flow dynamics after photodynamic therapy with verteporfin in the RIF-1 tumor. *Radiat Res* 160:452–459.

Chen, B., B. W. Pogue, P. J. Hoopes, and T. Hasan. 2006a. Vascular and cellular targeting for photodynamic therapy. *Crit Rev Eukaryot Gene Expr* 16:279–305.

Chen, B., B. W. Pogue, J. M. Luna et al. 2006b. Tumor vascular permeabilization by vascular-targeting photosensitization: Effects, mechanism, and therapeutic implications. *Clin Cancer Res* 12:917–923.

Cheng, C., E. Debefve, A. Haouala et al. 2010. Photodynamic therapy selectively enhances liposomal doxorubicin uptake in sarcoma tumors to rodent lungs. *Lasers Surg Med* 42:391–399.

Chojniak, R., L. S. Yu, and R. N. Younes. 2006. Response to chemotherapy in patients with lung metastases: How many nodules should be measured? *Cancer Imaging* 6:107–112.

Cowled, P. A., L. Mackenzie, and I. J. Forbes. 1987. Pharmacological modulation of photodynamic therapy with hematoporphyrin derivative and light. *Cancer Res* 47:971–974.

de Vree, W. J., M. C. Essers, H. S. de Bruijn et al. 1996a. Evidence for an important role of neutrophils in the efficacy of photodynamic therapy *in vivo*. *Cancer Res* 56:2908–2911.

de Vree, W. J., A. N. Fontijne-Dorsman, J. F. Koster, and W. Sluiter. 1996b. Photodynamic treatment of human endothelial cells promotes the adherence of neutrophils *in vitro*. *Br J Cancer* 73:1335–1340.

Debefve, E., C. Cheng, S. C. Schaefer et al. 2010. Photodynamic therapy induces selective extravasation of macromolecules: Insights using intravital microscopy. *J Photochem Photobiol B* 98:69–76.

Debefve, E., F. Mithieux, J. Y. Perentes et al. 2011. Leukocyte-endothelial cell interaction is necessary for photodynamic therapy induced vascular permeabilization. *Lasers Surg Med* 43:696–704.

Debefve, E., B. Pegaz, J. P. Ballini, Y. N. Konan, and H. van den Bergh. 2007. Combination therapy using aspirin-enhanced photodynamic selective drug delivery. *Vascul Pharmacol* 46:171–180.

Debefve, E., B. Pegaz, H. van den Bergh et al. 2008. Video monitoring of neovessel occlusion induced by photodynamic therapy with verteporfin (Visudyne), in the CAM model. *Angiogenesis* 11:235–243.

Dejana, E., R. Spagnuolo, and G. Bazzoni. 2001. Interendothelial junctions and their role in the control of angiogenesis, vascular permeability and leukocyte transmigration. *Thromb Haemost* 86:308–315.

Deli, M. A. 2009. Potential use of tight junction modulators to reversibly open membranous barriers and improve drug delivery. Biochim Biophys Acta 1788:892–910.

Dreher, M. R., W. Liu, C. R. Michelich et al. 2006. Tumor vascular permeability, accumulation, and penetration of macromolecular drug carriers. *J Natl Cancer Inst* 98:335–344.

Dvorak, H. F. 2002. Vascular permeability factor/vascular endothelial growth factor: A critical cytokine in tumor angiogenesis and a potential target for diagnosis and therapy. *J Clin Oncol* 20:4368–4380.

Feng, D., J. A. Nagy, J. Hipp, H. F. Dvorak, and A. M. Dvorak. 1996. Vesiculo-vacuolar organelles and the regulation of venule permeability to macromolecules by vascular permeability factor, histamine, and serotonin. *J Exp Med* 183:1981–1986.

Fingar, V. H. 1996. Vascular effects of photodynamic therapy. *J Clin Laser Med Surg* 14:323–328.

Fingar, V. H., P. K. Kik, P. S. Haydon et al. 1999. Analysis of acute vascular damage after photodynamic therapy using benzoporphyrin derivative (BPD). *Br J Cancer* 79:1702–1708.

Fingar, V. H., T. J. Wieman, and K. W. Doak. 1991. Changes in tumor interstitial pressure induced by photodynamic therapy. *Photochem Photobiol* 53:763–768.

Fingar, V. H., T. J. Wieman, and P. S. Haydon. 1997. The effects of thrombocytopenia on vessel stasis and macromolecular leakage after photodynamic therapy using photofrin. *Photochem Photobiol* 66:513–517.

Fingar, V. H., T. J. Wieman, S. A. Wiehle, and P. B. Cerrito. 1992. The role of microvascular damage in photodynamic therapy: The effect of treatment on vessel constriction, permeability, and leukocyte adhesion. *Cancer Res* 52:4914–4921.

Foster, T. H., and L. Gao. 1992. Dosimetry in photodynamic therapy: Oxygen and the critical importance of capillary density. *Radiat Res* 130:379–383.

Foster, T. H., M. C. Primavera, V. J. Marder, R. Hilf, and L. A. Sporn. 1991. Photosensitized release of von Willebrand factor from cultured human endothelial cells. *Cancer Res* 51:3261–3266.

Fukumura, D., and R. K. Jain. 2007. Tumor microenvironment abnormalities: Causes, consequences, and strategies to normalize. *J Cell Biochem* 101:937–949.

Graff, B. A., I. Bjornaes, and E. K. Rofstad. 2001. Microvascular permeability of human melanoma xenografts to macromolecules: Relationships to tumor volumetric growth rate, tumor angiogenesis, and VEGF expression. *Microvasc Res* 61:187–198.

Grootenboers, M. J., J. Heeren, B. P. van Putte et al. 2006. Isolated lung perfusion for pulmonary metastases: A review and work in progress. *Perfusion* 21:267–276.

Han, D., G. M. Beasley, D. S. Tyler, and J. S. Zager. 2011. Minimally invasive intra-arterial regional therapy for metastatic melanoma: Isolated limb infusion and percutaneous hepatic perfusion. *Expert Opin Drug Metab Toxicol* 7:1383–1394.

He, C., P. Agharkar, and B. Chen. 2008. Intravital microscopic analysis of vascular perfusion and macromolecule extravasation after photodynamic vascular targeting therapy. *Pharm Res* 25:1873–1880.

Hirschberg, H., F. A. Uzal, D. Chighvinadze et al. 2008. Disruption of the blood–brain barrier following ALA-mediated photodynamic therapy. *Lasers Surg Med* 40:535–542.

Hobbs, S. K., W. L. Monsky, F. Yuan et al. 1998. Regulation of transport pathways in tumor vessels: Role of tumor type and microenvironment. *Proc Natl Acad Sci USA* 95:4607–4612.

Jain, R. K. 2005. Normalization of tumor vasculature: An emerging concept in antiangiogenic therapy. *Science* 307:58–62.

Jain, R. K., R. T. Tong, and L. L. Munn. 2007. Effect of vascular normalization by antiangiogenic therapy on interstitial hypertension, peritumor edema, and lymphatic metastasis: Insights from a mathematical model. *Cancer Res* 67:2729–2735.

Korbelik, M. 2006. PDT-associated host response and its role in the therapy outcome. *Lasers Surg Med* 38:500–508.

Krammer, B. 2001. Vascular effects of photodynamic therapy. *Anticancer Res* 21:4271–4277.

Kurohane, K., A. Tominaga, K. Sato et al. 2001. Photodynamic therapy targeted to tumor-induced angiogenic vessels. *Cancer Lett* 167:49–56.

Lange, N., J. P. Ballini, G. Wagnieres, and H. van den Bergh. 2001. A new drug-screening procedure for photosensitizing agents used in photodynamic therapy for CNV. *Invest Ophthalmol Vis Sci* 42:38–46.

Lehr, H. A., M. Leunig, M. D. Menger, D. Nolte, and K. Messmer. 1993. Dorsal skinfold chamber technique for intravital microscopy in nude mice. *Am J Pathol* 143:1055–1062.

Lentsch, A. B., and P. A. Ward. 2000. Regulation of inflammatory vascular damage. *J Pathol* 190:343–348.

Leunig, M., A. E. Goetz, F. Gamarra et al. 1994. Photodynamic therapy-induced alterations in interstitial fluid pressure, volume and water content of an amelanotic melanoma in the hamster. *Br J Cancer* 69:101–103.

Madsen, S. J., and H. Hirschberg. 2010. Site-specific opening of the blood–brain barrier. *J Biophotonics* 3:356–367.

Maeda, H., J. Wu, T. Sawa, Y. Matsumura, and K. Hori. 2000. Tumor vascular permeability and the EPR effect in macromolecular therapeutics: A review. *J Control Release* 65:271–284.

Michels, S., and U. Schmidt-Erfurth. 2003. Sequence of early vascular events after photodynamic therapy. *Invest Ophthalmol Vis Sci* 44:2147–2154.

Narang, A. S., and S. Varia. 2011. Role of tumor vascular architecture in drug delivery. *Adv Drug Deliv Rev* 63:640–658.

Pegaz, B., E. Debefve, F. Borle et al. 2005. Encapsulation of porphyrins and chlorins in biodegradable nanoparticles: The effect of dye lipophilicity on the extravasation and the photothrombic activity. A comparative study. *J Photochem Photobiol B* 80:19–27.

Rahman, S., G. Flynn, A. Aitken et al. 2000. Differential recognition of snake venom proteins expressing specific Arg-Gly-Asp (RGD) sequence motifs by wild-type and variant integrin alphaIIbbeta3: Further evidence for distinct sites of RGD ligand recognition exhibiting negative allostery. *Biochem J* 345, pt. 3:701–709.

Roberts, W. G., and G. E. Palade. 1997. Neovasculature induced by vascular endothelial growth factor is fenestrated. *Cancer Res* 57:765–772.

Schmidt-Erfurth, U. M., G. Richard, A. Augustin et al. 2007. Guidance for the treatment of neovascular age-related macular degeneration. *Acta Ophthalmol Scand* 85:486–494.

Seebach, J., H. J. Madler, B. Wojciak-Stothard, and H. J. Schnittler. 2005. Tyrosine phosphorylation and the small GTPase rac cross-talk in regulation of endothelial barrier function. *Thromb Haemost* 94:620–629.

Sharman, W. M., C. M. Allen, and J. E. van Lier. 1999. Photodynamic therapeutics: Basic principles and clinical applications. *Drug Discov Today* 4:507–517.

Sitnik, T. M., J. A. Hampton, and B. W. Henderson. 1998. Reduction of tumour oxygenation during and after photodynamic therapy in vivo: Effects of fluence rate. *Br J Cancer* 77:1386–1394.

Snyder, J. W., W. R. Greco, D. A. Bellnier, L. Vaughan, and B. W. Henderson. 2003. Photodynamic therapy: A means to enhanced drug delivery to tumors. *Cancer Res* 63:8126–8131.

Stepinac, T. K., S. R. Chamot, E. Rungger-Brandle et al. 2005. Light-induced retinal vascular damage by Pd-porphyrin luminescent oxygen probes. *Invest Ophthalmol Vis Sci* 46:956–966.

Thurston, G. 2002. Complementary actions of VEGF and angiopoietin-1 on blood vessel growth and leakage. *J Anat* 200:575–580.

van Hinsbergh, V. W., and G. P. van Nieuw Amerongen. 2002. Endothelial hyperpermeability in vascular leakage. *Vascul Pharmacol* 39:171–172.

van Nieuw Amerongen, G. P., and V. W. van Hinsbergh. 2002. Targets for pharmacological intervention of endothelial hyperpermeability and barrier function. *Vascul Pharmacol* 39:257–272.

Vargas, A., M. Zeisser-Labouebe, N. Lange, R. Gurny, and F. Delie. 2007. The chick embryo and its chorioallantoic membrane (CAM) for the *in vivo* evaluation of drug delivery systems. *Adv Drug Deliv Rev* 59:1162–1176.

Wang, Y., M. Gonzalez, C. Cheng et al. 2012a. Photodynamic induced uptake of liposomal doxorubicin to rat lung tumors parallels tumor vascular density. *Lasers Surg Med* 44:318–324.

Wang, Y., J. Y. Perentes, S. C. Schafer et al. 2012b. Photodynamic drug delivery enhancement in tumours does not depend on leukocyte-endothelial interaction in a human mesothelioma xenograft model. *Eur J Cardiothorac Surg* 42:348–354.

Yuan, F., M. Dellian, D. Fukumura et al. 1995. Vascular permeability in a human tumor xenograft: Molecular size dependence and cutoff size. *Cancer Res* 55:3752–3756.

Yuan, S. Y. 2002. Protein kinase signaling in the modulation of microvascular permeability. *Vascul Pharmacol* 39:213–223.

Targeting Strategies in Photodynamic Therapy for Cancer Treatment

Marlène Pernot
Université de Lorraine

Céline Frochot
Université de Lorraine

Régis Vanderesse
Université de Lorraine
Médicaments Photoactivables-
Photochimiothérapie (PHOTOMED)

Muriel Barberi-Heyob
Université de Lorraine
Médicaments Photoactivables-
Photochimiothérapie (PHOTOMED)

30.1 Introduction

The selective, targeted delivery of photosensitizers to diseased cells is one of the major problems in photodynamic therapy (PDT) and is still a challenge to take up. One area of importance is the elaboration of targeted photosensitizers. Targeted therapy is a promising new therapeutic strategy, created to overcome growing problems of contemporary medicine, such as drug toxicity and drug resistance. An emerging modality of this approach is targeted PDT (TPDT) with the main aim of improving delivery of the photosensitizer to cancer tissue and, at the same time, enhancing specificity and efficiency of PDT. Depending on the mechanism of targeting, we can suggest dividing the strategies of TPDT into "passive," "active," and "activatable"; in the latter case, the photosensitizer is activated only in the targeted tissue. In this review, contemporary strategies of TPDT are described, including innovative new concepts, such as targeting assisted by peptides and aptamers, multifunctional nanoplatforms with navigation by magnetic field, or "photodynamic molecular beacons" activatable by enzymes and nucleic acid. The imperative of introducing a new paradigm of PDT, focused on the concepts of heterogeneity and dynamic state of tumor, is also called for.

In this sense, to overcome the drawbacks encountered in passive photosensitizer targeting, an arsenal of targeting strategies has been recently developed, including conjugated organic molecules as well as supramolecular carrier platforms, such as dendrimers, micelles, liposomes, and nanoparticles. Numerous interesting works have clearly demonstrated that more specific drug targeting, cellular uptake, and bioavailability can be achieved by binding various ligands, known as targeting moieties, to the photosensitizers or to the surface of the carrier platforms, such as peptides, growth factors, transferrin, antibodies or antibody fragments, oligonucleotide aptamers, and small compounds, such as folate. The rationale of all these strategies is taken from the biologic and molecular characteristics of tumor tissues. For instance, alterations or increased levels of receptor expression of a specific cellular type occur in diseased tissues.

These active-targeting approaches may be particularly useful for vascular-targeted PDT (VTP). Indeed, VTP effects are mediated not only through direct killing of tumor cells but also through indirect effects, involving both initiation of an immune response against tumor cell antigens and destruction of the tumor vasculature.

This chapter will describe recent and significant advances and developments in strategies for achieving selective enhanced photosensitizer concentration in neoplastic target tissue with the emphasis on target specificity. Description is focused on the recent and significant published articles describing the different targeting strategy, and references of related reviews will be indicated.

30.2 Biodistribution and Natural Selectivity of Photosensitizers

PDT has inherent dual selectivity as a result of the control of light delivery and, to some extent, selective photosensitizer accumulation in tumors. Biodistribution and cellular incorporation of photosensitizers depend partly on the interactions between these molecules and transport molecules in plasma. Most photosensitizers, when injected into the blood circulation, behave as macromolecules either because they are bound to large protein molecules or because they have formed intermolecular aggregates. It is thought that the increased vascular permeability to macromolecules typical of tumor neovasculature is responsible for the preferential extravasation of the photosensitizer in tumors. In addition, it is thought that tumors have poorly developed lymphatic drainage, and those macromolecules, which extravasate from the hyperpermeable tumor neovasculature, are retained in the extravascular space. This means that therapeutic agents that gain interstitial access to the tumor present higher retention times than normal tissues. The combination of leaky vasculature and poor lymphatic drainage, which characterize tumor vasculature, results in the so-called enhanced permeation and retention (EPR) effect (Byrne, Betancourt, and Brannon-Peppas 2008). The effectiveness of PDT is widely determined by the efficiency of the local generation of $^1O_2$ and the degree of efficiency and selectivity to which therapeutic concentrations of the photosensitizer are delivered to the target site with minimal uptake by nontarget cells and subsequent localized light irradiation.

30.3 Tumor Neovascularization Targeting

The targeting of tumor vasculature has become a large area of focus for the development of new cancer therapeutics. It is known that when a tumor grows, its need for nutrients and oxygen increases, and thus the number and size of blood vessels increase proportionately. This phenomenon is angiogenesis. By attacking the growth of the neovasculature, the size and the metastatic capabilities of tumors can be regulated. Tumor neovasculature targeting appears as an approach of significant research interest for the development of active photosensitizer delivery systems able to enhance selectivity and efficiency of vascular PDT for cancer treatment. The main molecular targets explored in the vascular-targeting PDT for cancer include the vascular endothelial growth factor receptors [VEGFRs, such as neuropilin-1 (NRP-1)], receptor tissue factor (TF), $\alpha_v\beta_3$ integrins, and matrix metalloproteinase (MMP) activities.

30.3.1 Vascular Endothelial Growth Factor Receptors

Vascular endothelial growth factor (VEGF) is one of the most potent direct-acting angiogenic proteins known. This growth factor is upregulated in the majority of neoplastic cells, mainly as a response to hypoxia and many oncogenes, and its overproduction is correlated with high microvascular density and poor prognosis. The overexpression of VEGF results in the upregulation of VEGFR-1 [also known as fms-like tyrosine kinase (flt-1)] and VEGFR-2 [fetal liver kinase-1 (flk-1) or kinase domain region (KDR)]. VEGFR-2 is generally believed to be the main receptor that mediates VEGF biological activities and plays a major role in tumor-associated angiogenesis. It is highly expressed on endothelial cells in tumor neovasculature. Therefore, VEGFR-2 is seen as an interesting molecular target for antiangiogenic drug delivery, and specific targeting of VEGFR-2 could provide a promising approach for selective and efficient photosensitizer delivery to tumor neovasculature.

Although targeting VEGFR-2 has been widely investigated for the selective delivery of therapeutic drugs for conventional therapies, such as radiotherapy, a limited number of papers have been published for the selective delivery of photosensitizers. Renno et al. (2004) have shown that PDT using verteporfin conjugated to a peptidic motif (ATWLPPR) known to bind VEGFR-2 was more effective in causing choroidal neovascularization closure than untargeted verteporfin. As expected, targeted verteporfin also resulted in more selective treatment than an untargeted drug (Renno et al. 2004). These studies suggest that targeting VEGFR-2 can be an effective approach to confine sufficient doses of photosensitizers within the tumor neovasculature and thus potentiate the vascular photodynamic effect with minimal side effects. Our group demonstrated, in agreement with others (Perret et al. 2004; Tirand et al. 2006), that this peptide targeted NRP-1 and not KDR. Although VEGFR-2 is still a potential target to explore, the latter observations have then attracted a great interest in the potential of NRP-1 as a promising target for targeted vascular PDT.

Neuropilins (NRP) have also been identified as receptors for the $VEGF_{165}$ isoform. The transmembrane protein NRP-1 has been described as a positive modulator of VEGFR-2 and is thereby a crucial inducer of angiogenesis. Neuropilins are expressed specifically in tumor angiogenic vessels and some tumor cells and promote tumor angiogenesis and progression. Hence, targeting NRP-1 can lead to the selective vascular localization of photosensitizers and thus enhance the vascular photodynamic effects. The conjugation of a photosensitizer [5-(4-carboxyphenyl)-10,15,20-triphenyl-chlorin (TPC)] to the heptapeptide ATWLPPR has been described by our group. Our group evidenced *in vitro* by competition experiments in human umbilical vein endothelial cells (HUVECs) (Tirand et al. 2006) and *in vivo* by biodistribution studies in nude mice xenografted with U87 human malignant glioma cells that a part of the accumulation of the conjugated photosensitizer (noticed TPC-Ahx-ATWLPPR) was related to NRP-1-dependent mechanisms but also to nonspecific mechanisms. Taking advantage of RNA silencing techniques known as RNA interference, we have selectively silenced NRP-1 expression in MDA-MB-231 breast cancer cells to provide the evidence for the involvement of NRP-1 expression in the conjugate cellular uptake. We also observed *in vivo* the vascular effect by measuring the tumor blood flow during PDT using both conjugated and nonconjugated PS. The conjugate-mediated, vascular-targeted PDT produced a selective vascular effect, leading to vascular shutdown and tumor growth delay (Bechet et al. 2010; Thomas et al. 2009). From the biological mechanism point of view, the conjugate-mediated vascular effect implies the induction of TF protein expression that may lead to the thrombogenic effect within the vessel lumen. Nevertheless, affinity for NRP-1 remains low probably as a result of (1) intramolecular interactions between chlorin and peptide, steric hindrance resulting from the TPC moiety; (2) aggregation of the photosensitizer molecules; and (3) low stability of the peptide moiety that may be a result of sensitivity to circulating peptidase action. To improve our system, multifunctional nanoparticles as photosensitizer carriers have been developed (Couleaud et al. 2011).

30.3.2 TF and Coagulation Factor VII

TF is a cell membrane-bound glycoprotein and a member of the class 2 cytokine receptor family. It is composed of a hydrophilic extracellular domain, a membrane-spanning hydrophobic domain, and a cytoplasmic tail of 21 residues, including a nondisulfide-linked cysteine. TF forms a high affinity and specific complex with its endogenous ligand, coagulation factor VII (FVII), as the initial step of the blood coagulation pathway (Nemerson 1988). Accumulating evidence suggests that TF is aberrantly overexpressed by angiogenic vascular endothelial cells and various tumor cells (Chen et al. 2001), but not on endothelial cells of normal blood vessels. Several studies have generated promising results in inhibiting tumor growth in animal models by targeting TF. Moreover, it has been suggested that VEGF protein secreted by tumor cells induces endothelial cells in tumor vasculature to overexpress TF. All the aforementioned characteristics suggest that TF can serve as a specific therapeutic target for targeted drug delivery directed at tumor angiogenesis. Accordingly, several methods have been developed for selectively delivering drugs to TF-expressing tumor vasculature and tumors using FVII as a drug carrier. An immunotherapy strategy for destroying the tumor vasculature by targeting TF on tumor vascular endothelial cells with an immunoconjugate (icon) molecule has been described. More recently, this group has described an approach targeting TF on tumor cells and angiogenic vascular endothelial cells by FVII-targeted verteporfin PDT for breast cancer *in vitro* and *in vivo*. Authors demonstrated that (1) FVII protein could be conjugated with verteporfin without affecting its binding ability; (2) FVII-targeted PDT could selectively affect TF-expressing cells and VEGF-stimulated angiogenic human endothelial vein endothelial cells (HUVEC), but no side effect was observed on non-TFR-expressing unstimulated cells; (3) fVII targeting enhanced the effect of verteporfin PDT by three- to fourfold; (4) fVII-targeted PDT induced significantly stronger levels of apoptosis and necrosis than nontargeted PDT; and (5) FVII-targeted PDT had a significantly stronger effect on inhibiting breast tumor growth in mice than nontargeted PDT (Hu et al. 2010). Because receptor TF is expressed on many types of

cancer cells and selectively on angiogenic vascular endothelial cells, these findings suggest that FVII-targeted PDT could have broad therapeutic applications. Cheng et al. (2011) demonstrated that *in vivo* administration of FVII-targeted PDT significantly inhibited or eliminated subcutaneous A549 and H460 tumor xenografts in an athymic nude mouse model without any obvious side effects, concluding that this targeting strategy could be effective and safe for the treatment of human lung cancer in preclinical studies.

30.3.3 Integrin $\alpha_v\beta_3$

The $\alpha_v\beta_3$-integrin is highly expressed on neovascular endothelial cells and is important in the calcium-dependent signaling pathway, leading to endothelial cell migration. It is overexpressed in many tumor cells, and also in actively proliferating endothelial cells and around tumor tissues. Moreover, $\alpha_v\beta_3$-integrin is an endothelial cell receptor for extracellular matrix (EMC) proteins harboring the RGD peptidic sequence. Therefore, $\alpha_v\beta_3$-integrin is considered as an attractive molecular target for antivascular therapies. In accordance with this, several investigations have described the potential of such receptors in VTP. Chaleix et al. (2004) reported the solid-phase synthesis of four porphyrins bearing the $\alpha_v\beta_3$-integrin ligand RGD (H-Arg-Gly-Asp-OH) tripeptide. Three of the conjugates prepared displayed photodynamic activity on the K562 leukemia cell line to a degree comparable to that of Photofrin II. The authors later described the synthesis of a cyclic peptide containing the RGD motif and adopting conformations showing an increased affinity for integrins. To increase the selectivity of phthalocyanines, Allen and Cullis (2004) have evaluated the use of viral proteins to deliver the photosensitizer to the selected tissue. Adenoviruses cause illnesses, such as gastroenteritis, conjunctivitis, and respiratory diseases. They have received a great deal of attention as genetherapy vectors. Adenoviruses efficiently break the endosomes upon infection in the cell, and therefore, it was anticipated that adenovirus–photosensitizer dyads would target the cell nucleus more quickly than a free photosensitizer. This targeting of the nuclear cell resulted in a 2.5-fold increase in the photodynamic activity of the photosensitizer in comparison with a cytoplasmic localization. Adenovirus penton base proteins contain the RGD peptide sequence motif. Adenovirus type 2 structural proteins; the hexon, penton bases; and fiber antigen were isolated and purified. The adenovirus tetrasulfonated aluminum phthalocyanine derivatives were tested both *in vitro* and *in vivo*. The penton base conjugate was the most efficient *in vitro* as measured in two positive cell lines (A549, Hep2) expressing integrins (Allen and Cullis 2004). According to this study, it appears that adenoviral proteins can be used to target tumor cells. Despite nonoptimal photodynamic activity, *in vivo* results were encouraging. Nevertheless, tumor targeting using adenoviral protein vehicles may promote inflammation and antiprotein cellular immunity, which could limit their usefulness.

Our group conjugated a photosensitizer [5-(4-carboxyphenyl)-10,15,20-triphenylchlorin or porphyrin] to the RGD motif as a common sequence (Frochot et al. 2007). We reported an efficient solid-phase synthesis of a new family of peptidic photosensitizers with a linear or cyclic [RGDfK] RGD motif. Chlorins containing linear and constrained RGD motif were incorporated up to 98- and 80-fold more, respectively, than the unconjugated photosensitizer over a 24-h exposure in HUVEC overexpressing $\alpha_v\beta_3$ integrin. Peptidic moiety also led to a nonspecific increased cellular uptake by murine mammary carcinoma cells (EMT-6), lacking RGD binding receptors. The higher photodynamic efficiency was related to conjugates' greater cellular uptake. Survival measurements demonstrated that HUVECs were greatly sensitive to conjugate-mediated PDT. This study also showed that peptidic moiety was obviously an interesting means to introduce such a balance between hydrophilicity and hydrophobicity. More recently, we described the synthesis, characterization, fluorescence, and singlet oxygen quantum yields of tetraphenylporphyrin and tetraphenylchlorin coupled to RGD-type peptides. They reported that some of these compounds are very promising for potential PDT applications (Boisbrun et al. 2008).

30.3.4 Matrix Metalloproteinases

Tumor-associated proteases are known to function at multiple steps of tumor progression, affecting tumor establishment, growth, neovascularization, intravasation, extravasation, and metastasis. Proteolytic degradation of the EMC is crucial for cancer development, invasion, and metastasis, all of them associated with increased activities of several different protease families (Berdowska 2004; van Kempen, de Visser, and Coussens 2006). Numerous studies have documented increased expression of proteases such as MMPs in many human malignant tissue types. Thus, protease-sensitive macromolecular prodrugs have attracted interest for bioresponsive drug delivery to disease sites with upregulated proteolytic activities, and many strategies emerged from protease targeting to convert an inactive PDT prodrug into an active drug when activated *in vivo* (Law and Tung 2009). In fact, the tumor-associated proteases could act as activators of the protease-mediated PDT agent.

MMPs are tumor-promoting enzymes playing an important role in angiogenesis (Ferrario et al. 2004; Sharwani et al. 2006; Vihinen and Kahari 2002) and thus are potential interesting targets in PDT. Prodrugs activated by proteases are not recent, but applied to PDT, proteases become attractive biological triggers in drug development to control $^1O_2$ production; many strategies emerged from protease targeting to convert pharmacological inert prodrugs into active pharmacophores when activated *in vivo* (Law and Tung 2009).

As most photosensitizers used in PDT are porphyrin derivatives that absorb light energy and then transfer to oxygen or emitted fluorescence, it has been reported that $^1O_2$ generation is closely correlated with fluorescent intensity of the photosensitizing agent. So when a photosensitizer is in a fluorescent-quenched state, its ability to generate $^1O_2$ is also reduced. Molecular beacons are FRET-based, target-activatable probes, offering control of fluorescence emission in response to specific targets, such as

proteases, as a useful tool for *in vivo* cancer imaging. By combining the two principles of FRET and PDT, Zheng et al. (2007) suggested the concept of a photodynamic molecular beacon for controlling the photosensitizer's ability to generate $^1O_2$ and, thus, for controlling its PDT activity. They described the synthesis and characterization of a MMP7-triggered photodynamic molecular beacon, using (1) pyropheophorbide as the photosensitizer; (2) black hole quencher 3 (Q) as a dual fluorescence and $^1O_2$ quencher; and (3) a short peptide sequence, GPLGLARK, as the MMP7-cleavable linker. The photosensitizer and the quencher are attached to the opposite end of the MMP7-specific cleavable peptide linker to keep them in close proximity, enabling FRET and $^1O_2$ quenching to form an inactive prodrug. Thus, the photosensitizer's photoactivity is silenced until the linker interacts with the target tumor-associated MMP7.

After validating the MMP7-triggered production of $^1O_2$ in solution, the authors demonstrated the MMP7-mediated photodynamic cytotoxicity in cancer cells. *In vivo* studies also revealed the MMP7-activated PDT efficacy (Zheng et al. 2007).

30.4 PDT: Neoplastic Cell Targeting

This second part deals with photosensitizers covalently bound to different biomolecules (sugars, steroid hormones, amino acids, proteins, etc.) via a direct coupling or using linkers or spacers. These biomolecules, as targeting moieties, are used to direct the photosensitizing agents against neoplastic cell-associated specific antigens. The coupling of a vector can either (1) modulate the amphiphilicity, enhance the solubility of these compounds in biological media, and prevent self-aggregation (passive targeting); or (2) promote cellular recognition (active targeting) with the aim of increasing the biological efficiency. Several of these targeting strategies offer the advantage of transporting the photosensitizer across the cellular plasma membrane, resulting in intracellular accumulation of the photosensitizers, which may allow for targeting photosensitive intracellular sites, and thus improving photodynamic efficiency.

30.4.1 Epidermal Growth Factor Receptors

The ligands that bind the EGF receptor (EGFR HER) include EGF, transforming growth factor-α (TGFα), heparin-binding EGF-like growth factor, amphiregulin (AR), betacellulin (BTC), epiregulin (EPR), and epigen. Each of the natural peptide growth factors is characterized by a consensus sequence consisting of six spatially conserved cysteine residues. This consensus sequence is known as the EGF motif and is crucial for binding members of the HER receptor tyrosine kinase family. In addition to binding EGFR, HB-EGF, BTC, and EPR are also reported to bind HER4. Because high EGF receptor expression frequently accompanies development of several tumor types, such as squamous carcinomas, its natural ligand EGF was an attractive candidate for the conception of drug-targeting strategies. One binding to its receptor, EGF is internalized in the cell through receptor-mediated endocytosis, enabling the intracellular accumulation of photosensitizers. Based on this concept, EGFR receptors provide an interesting approach for selective photosensitizer delivery to tumor tissues.

Lutsenko et al. (1999) conjugated EGF with aluminum and cobalt disulfonated phthalocyanines, with the EGF/photosensitizer ratio being 1:1. EGF-phthalocyanines exhibited much higher photoactivity than their nonconjugated analogs. The antitumor activity was also investigated *in vivo* on the growth of solid tumors (murine melanoma cell line B16) implanted in mice. The administration of the conjugate strongly increased mean life spans and mean survival time of animals bearing tumors, whereas free phthalocyanines described practically no effect on these parameters (Lutsenko et al. 1999).

Targeting the EGFR with antibody (Ab)-delivered photoactive molecules may serve as a two-armed feedback-controlled approach: The photoactivatable compound together with illumination may target and destroy the EGFR, and the fluorescent dye may allow the progress of the treatment to be monitored. Soukos et al. (2001) studied an anti-EGFR monoclonal (m) Ab (C225) coupled to either the near-infrared fluorescent dye N,N'-di-carboxypentyl-indodicarbocyanine-5,5'-disulfonic acid (indocyanine Cy5.5) for detection or a photochemically active dye (chlorin$_{e6}$, Ce6) for therapy (1) to determine whether the intravenous injection of a conjugate between C225 and the fluorescent dye indocyanine Cy5.5 could deliver sufficient amounts of the fluorophore into clinically unrecognizable early premalignant lesions and (2) to test the photodynamic effect of a conjugate between C225 and Ce6. The authors demonstrated that immunophotodiagnosis using the C225–Cy5.5 conjugate might have potential both as a diagnostic modality for oral precancer and as a surrogate marker. As suggested, the use of a near infrared (NIR)-emitting fluorophore as a label may be an alternative to use radioisotopes with consequent increased patient safety. To extend the analogy with radioisotopes, the principle of using an anti-EGFR mAb to deliver photoactive molecules applies equally well to therapeutic photosensitizers as to fluorophores, and furthermore, immunophotodiagnosis can be used to assess response. HER2, also called human EGF receptor-2, is a highly upregulated target on tumor cell surfaces, such as breast cancer. Bhatti et al. (2008) described a complex of multiple photosensitizers (verteporfin or pyropheophorbide-a) with anti-HER2 single chain Fv (scFv). They demonstrated that the photoimmunoconjugates were more potent than either free photosensitizer, effectively killing tumor cells *in vitro* and *in vivo*. *In vivo* treatment with the photoimmunoconjugate comprising an anti-HER2 scFv with approximately 8–10 molecules of the pyropheophorbide-a led to a significant tumor regression. It is important to note that the clearance of the immunoconjugate was still more rapid than free pyropheophorbide-a, suggesting that the time from administration of scFv-pyropheophorbide-a to exposure to laser light could be therapeutically attractive with very little danger of skin photosensitivity.

Selbo et al. (2012) explored the possible cell population inactivation by photochemical internalization (PCI) of the EGFR-targeted protein toxin EGF-saponin. A wide range of anticancer drugs are substrates of the ATP-binding cassette transporter ABCG2/CD338/BCRP/MXR, which is thought to play an important role in multidrug resistance (MDR) and protection of cancer stem cells against chemotherapeutics and PDT. In this interesting study, the authors demonstrated that strong amphiphilic PCI-photosensitizers, including a chlorine, a porphyrin, and a phthalocyanine, are not substrates of ABCG2 using the MA11 cell line as a novel MDR model. As proof-of-concept, the authors also show that PCI of targeting toxins is a highly potent strategy for the site-specific elimination of EGFR-expressing MDR cells.

30.4.2 Cholesterol and Low-Density Lipoprotein

The steroid hormones are an interesting class of biomolecules to target photosensitizers to cancer cells. In particular, cholesterol is a vital component of eukaryotic cell membranes and can be taken up quickly by cancer cells. It thus appears that the covalent coupling of cholesterol to a molecule could favor its association with low-density lipoprotein (LDL) and increase its photodynamic efficiency. Furthermore, axial substitution of suitable central metals (Si, Ge, Al) by cholesterol, which is a lipophilic and bulky ligand, could decrease phthalocyanine self-aggregation (inhibition of the π–π stacking by the addition of steric hindrance on the sides of the macrocycle) and increase the crossing of the lipophilic membrane. Photodynamic activities of cholesterol–phthalocyanine derivatives were investigated on two pigmented melanoma cell lines: M3Dau and SK-MEL-2 (Barge et al. 2004). M3Dau cells were of interest as they led to melanoma tumors in nude mice, allowing further *in vivo* validations. SK-MEL-2 cells were also used in order to elucidate whether the photocytotoxicity of the phthalocyanines could vary depending on the pigmented melanoma cell line used. Segalla et al. (1994) demonstrated that cholesterol–phthalocyanine derivatives injected systemically into mice bearing an intramuscularly implanted MS-2 fibrosarcoma were quantitatively transferred to serum lipoproteins and localized into the tumor tissue. The selectivity of tumor targeting was similar to that observed for other liposome-delivered phthalocyanines (Cuomo et al. 1991). Another grafting of cholesterol was achieved by Maree and Nyokong (2001). The authors showed that, unfortunately, the conjugate was highly aggregated at low concentrations. No biological evaluation has been performed on this compound yet.

LDLs have been suggested to be able to play an important role in the transport and release of photosensitizers in tumor cells. A plethora of studies have showed that photosensitizers, such as phthalocyanine, Visudyne, and chlorin e6 mixed with LDL before administration, led to an increase in photodynamic efficiency (Konan, Gurny, and Allemann 2002). Zheng et al. (2002) prepared a delivery system comprising a pyropheophorbide cholesterol oleate photosensitizer incorporated into LDL with a modest photosensitizer payload. Interestingly, laser scanning confocal microscopy studies demonstrated that such an LDL-based photosensitizer was internalized exclusively by LDL receptors overexpressing human hepatoblastoma G2 (HepG2) tumor cells (Zheng et al. 2002). To reduce the dose for more efficient cancer detection and treatment, Li et al. (2005) designed a novel strategy to improve LDL's photosensitizer payload using a silicon–phthalocyanine as a new NIR dye. The choice of such a compound was based mainly on the fact that the central silicon atom of silicon–phthalocyanine allows axial coordination of two bulky ligands on each side of the phthalocyanine ring to prevent the stack aggregation usually encountered in the solution for the planar molecular structure because such aggregation is thought to be the major limiting factor for achieving a high-probe/LDL payload. Confocal microscopy studies demonstrated that the internalization of the conjugate by human HepG2 cells was still mediated by the LDL receptor pathway. Using clonogenic assay, the *in vitro* PDT response of HepG2 cells revealed significant enhanced efficacy of LDL receptor-targeted PDT (Li et al. 2005). More recently, Song et al. (2007) described the preparation and characterization of the naphthalocyanine reconstituted LDL nanoparticles as well as the *in vitro* and *in vivo* validation of the LDL receptor-targeting of the conjugate to cancer cells and in mice bearing human HepG2 tumors. A confocal microscopy study revealed that the photosensitizer reconstituted LDL nanoparticles retain their LDL receptor-mediated uptake by cancer cells. Its preferential uptake by tumor versus normal tissue was also confirmed *in vivo* by noninvasive optical imaging technique (Song et al. 2007).

30.4.3 Estrogen Receptors

To improve the uptake of the phthalocyanine derivatives by receptor-rich endocrine tumors, Khan et al. reported the synthesis of phthalocyanine–estradiol conjugates (Khan et al. 2003) and their photodynamic activities. Phthalocyanine–estradiol conjugates were prepared with both aliphatic and aromatic alkynyl spacers in order to modulate the overall amphiphilicity of the molecule. The highest receptor binding affinities were observed with lipophilic conjugates coupled via a relatively long spacer, while the sulfonated analogs showed little binding affinities. Surprisingly, the most hydrophilic trisulfonated phthalocyanine–estradiol described the highest photocytotoxicity (EMT6 cells) comparable to those reported for the nonconjugated phthalocyanine. The nature of the spacer did not seem to influence the biological activity. Grafting of estrone, the biosynthetic precursor of estradiol receptors overexpressed by breast cancer cells, was achieved by Maree, Phillips, and Nyokong (2002). From a photophysical point of view, aggregation was prevented because of the presence of axial ligands in all complexes. From the biological point of view, the octaestrone phthalocyanine showed the best promise for detection through fluorescence and for treatment because this complex has the longest triplet lifetime.

30.4.4 Nucleus and DNA

The cell nucleus is the most sensitive site, in contrast to cell membranes and other cytoplasmic organelles. Therefore, creation of a molecule of $^1O_2$ in close proximity to the cancer cell DNA would dramatically increase the odds of cell death whether by induction of apoptosis or by necrosis. Efforts have therefore been made to enhance delivery of the photosensitizer to the cell nucleus.

Different nuclear localization signal (NLS) sequences, contained in targeting proteins, have been tested to favor intranuclear delivery of photosensitizers into the target cells (Schneider et al. 2006). Gariépy's group has designed branched peptides that act as multitasking intracellular vehicles (Singh et al. 1999). These vehicles incorporate eight identical peptide arms coding for two functional domains, namely, a pentalysine domain acting as a cytoplasmic transport sequence and the simian virus SV40 large T antigen NLS, which guides their nuclear uptake. These squid-like intracellular vehicles are referred to as loligomers. By solid-phase synthesis, the coupling of chlorin e6 (Ce6) to a nucleus-directed linear peptide (Ce6-peptide) or a branched peptide (Ce6-loligomer) composed of eight identical arms, displaying the sequence of the Ce6-peptide, was achieved. The authors have shown that cellular distributions of the two conjugates into radiation-induced fibrosarcoma cells (RIF-1) clearly differed. Nevertheless, an enhanced (1.5-fold) nuclear localization was only significant for the Ce6-loligomer compared to the Ce6-peptide. With the goal of enhancing the photodynamic activity of Ce6 by the directed delivery of photosensitizers, Sobolev (2009) have also shown that photosensitizers can be successfully redirected within cells by using cross-linked modular polypeptide transporters possessing (1) an internalizable ligand providing a cell-specific delivery; (2) a module enabling escape from endosomes; (3) a NLS conferring interaction with importins, the cellular proteins mediating active translocation into the nucleus; and (4) a module allowing attachment of the photosensitizer (Figure 30.1).

Rosenkranz et al. (2003) describe the design, production, and characterization of bacterially expressed modular recombinant transporters (MRTs) for bacteriochlorin p_6 comprising (1) the α-melanocyte-stimulating hormone (MSH) as the internalizable ligand, (2) the optimized T-ag NLS, (3) the *Escherichia coli*

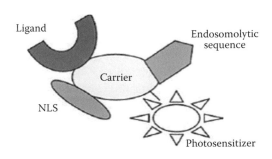

FIGURE 30.1 Scheme of distinct modules conferring cell-specific targeting for photosensitizers. The various sequence modules for cell surface binding and uptake (a ligand), an endosomolytic sequence, and a NLS as well as the photosensitizer are linked to the carrier molecule. (From Schneider, R. et al., *Anticancer Agents Med Chem* 6:469–488, 2006.)

hemoglobin-like protein HMP as a carrier, and (4) an endoso-molytic amphipathic peptide. The MRT delivered the PS into the nuclei of murine melanoma cells and provided for several orders of magnitude greater cytotoxicity than free PS.

A large number of nucleobase porphyrins, which are of interest for nucleus targeting, have also been described in the literature. Koval et al. (2001) first reported the synthesis of oligonucleotide–phthalocyanine conjugates for specific DNA modifications *in vitro* and *in vivo* for the development of sequence-specific gene targeting reagents. Because of the presence of a phthalocyanine core and adenine substituent, these compounds present strong intermolecular interactions, resulting in a poor solubility in common organic solvents and unusual spectral features (observation of fluorescence emission despite strong aggregation). No biological evaluation was performed.

30.4.5 Folic Acid Receptor

The vitamin folic acid is a ligand able to target bioactive agents quite specifically to folate receptor (FR)-positive cancer cells. The FR, which is a well-known cancer cell-associated protein, can actively internalize bound folates via endocytosis. FRs exist in three major forms, namely, FR-α, FR-β, and FR-γ. The FR-α form is overexpressed by many types of tumor cells, including ovarian, endometrial, colorectal, breast, lung, renal, and neuroendocrine carcinomas and brain metastases. Upon receptor interaction, the folic acid–receptor complex is taken up by cells and moves into many organelles involved in endocytotic trafficking, providing for cytosolic deposition. Proper synthesis procedures have been pointed out to link folic acid to molecules in order to develop targeting delivery systems. It was demonstrated that the glutamate γ-carboxyl group modification does not induce significant loss of folic acid affinity for the receptor. Moreover, because of its high stability, compatibility with both organic and aqueous solvent, low-cost, nonimmunogenic character, ability to conjugate with a wide variety of molecules, and low molecular weight, folic acid has attracted wide attention as a targeting agent for tumor detection. Numerous examples that take advantage of folic acid uptake to promote targeting and internalization include inorganic nanoparticles, polymer nanoparticles, polymeric micelles, lipoprotein nanoparticles, tumor-imaging agents, and dendrimers. For example, folic acid was recently covalently anchored on the surface of gold nanorods with a silane coupling agent. The resultant folic acid–conjugated silica-modified gold nanorods show highly selective targeting, enhanced radiation therapy, and photothermal therapy effects on MGC803 gastric cancer cells and also exhibited strong x-ray attenuation for *in vivo* x-ray and computed tomography (CT) imaging (Ding et al. 2011).

Novel amphiphilic polysaccharide/photosensitizer conjugates synthesized by chemical conjugation of heparin with pheophorbide-a and FA were investigated. The anticoagulant activity of heparin-PhA (HP) and folate-heparin-PhA (FHP) conjugates was significantly decreased compared to that of heparin, thereby potentially reducing the hemorrhagic side effects. HP and FHP

nanoparticles exhibited marked phototoxicity on HeLa cells and were minimally dark toxic without light treatment (Li et al. 2011).

Our group synthesized two new conjugates composed of folic acid coupled to 4-carboxyphenylporphyrin via two short linkers that were different in nature but similar in size. Both conjugated photosensitizers showed improved intracellular uptake in KB cells acting as a positive control because of their overexpression of FR-α. Using a short PEG, 2,2′-(ethylenedioxy)-bis-ethylamine, as a spacer arm, our group demonstrated that the uptake of this conjugate was on average sevenfold higher than tetraphenylporphyrin used as a reference and that the cells were significantly more sensitive to folic acid–conjugated porphyrin-mediated PDT. We also demonstrated that *in vivo* selectivity of meta tetra(hydroxyphenyl)chlorin (*m*-THPC)-like photosensitizer conjugated to folic acid exhibited enhanced accumulation in KB tumors *in vivo* compared to *m*-THPC, 4 h after intravenous injection (Gravier et al. 2008). Tumor-to-normal tissue ratio exhibited a very interesting selectivity for the conjugate (5:1) in KB tumors. An aspect that is important for folic acid–mediated drug delivery concerns the rate of FR recycling between the cell surface and its intracellular compartments. Accumulation of folate conjugates in KB cells will depend upon not only the number and accessibility of FRs on the cell surfaces but also the time required for unoccupied receptors to recycle back to the cell surface for additional conjugated Quantum dots (QD) uptake. Using radioactive conjugates, Paulos et al. (2004) found empty FR+ to unload their cargo and return to the cell surface in approximately 8 to 12 h. Using folic acid–linked drugs in a FR-targeting strategy could be considered as an efficient method to improve selectivity of anticancer treatment, specifically to FR+ cancer cells.

Stefflova et al. (2007) prepared a construct (pyro-peptide-folate, PPF) consisting of three compounds: (1) pyropheophorbide, (2) peptide as linker, and (3) folate as a targeting moiety. They observed a higher accumulation of PPF in KB cells compared to HT1080 used as a negative control, leading to a more effective photodynamic effect. The effect of folate was also confirmed *in vivo* with an accumulation of PPF in KB tumors (KB versus HT1080 tumors 2.5:1). It was interesting to note that the short peptide used as a spacer considerably improved the delivery efficiency with a reduction (50-fold) in PPF uptake in the liver and spleen.

30.4.6 Cholecystokinin A Receptor

A peptide hormone cholecystokinin CCK8 (H-Asp-Tyr-Met-Gly-Trp-Met-Asp-Phe-NH$_2$) is known to interact with cholecystokinin A (CCK$_A$), a receptor that is overexpressed in some colonic, gastric, and brain cancers (Reubi 2003). De Luca et al. (2001) described the synthesis of 5,10,15-tris (2,6-dichlorophenyl)-20-(4 carboxyphenyl) porphyrin (abbreviated PK) covalently bound to CCK8 using a lysine residue as a spacer. A molecular dynamics analysis indicated that the conformational characteristics of the CCK8/CCK$_A$ receptor–model complex and of the PK-CCK8/CCK$_A$ receptor–model complex were similar. This evidence supported the hypothesis that the introduction of the porphyrin-Lys moiety did not influence the mode of ligand binding to the CCK$_A$ receptor model. A nuclear magnetic resonance (NMR) conformational study of the PK–CCK8 conjugate was also carried out in dimethylsulfoxide. No nuclear overhauser effect correlation between the porphyrin ring protons and the peptide chain of the molecule was observed, indicating that the porphyrin macrocycle does not lie close to the peptide moiety of PK–CCK8. The authors concluded that the presence of the porphyrin substituent at the *N*-terminus of CCK8 did not interfere with the conformation of the peptide residues supposed to be involved in receptor binding. This study suggested that a peptide–photodynamic drug conjugate approach might be useful for targeting the tumor cells expressing the CCK$_A$ receptor.

Cawston and Miller (2010) reviewed the therapeutic potential of highly selective partial agonists and allosteric modulators of the type 1 cholecystokinin receptor. These peptides could be conjugated to PS to target cancer cells displaying overexpression of CCK receptor.

30.4.7 Gonadotropin-Releasing Hormone Receptors

Gonadotropin-releasing hormone (GnRH) is considered as a key integrator between the nervous and the endocrine systems and plays a pivotal role in the regulation of the reproductive system. Several studies revealed that GnRH and its receptors also play extra pituitary roles in numerous normal and malignant cells or tissues (for review, see Aguilar-Rojas and Huerta-Reyes 2009). Thus, GnRH analogs, both agonists and antagonists, have attracted great interest because of their potential application in the treatment of diseases, such as prostate and breast cancers. Their mechanism of action is believed to be related to gonadal steroid deprivation. This phenomenon results from GnRH analog action that leads to downregulation of the GnRH receptors (GnRH-R) and to desensitization of the pituitary gonadotropes. GnRH-R expression was identified on different tumors (breast, ovarian, endometrial, prostate, renal, brain, pancreatic, melanomas, non-Hodgkin's lymphomas, etc.), and the limited number of GnRH-R in normal tissues provided a basis for the development of diagnostic and therapeutic approaches of cancer. Rahimipour et al. (2003) described the coupling of protoporphyrin IX to peptides acting as GnRH agonists with the aim of selectively targeting GnRH-R that are overexpressed in many prostate and breast cancers. Conjugates obtained exhibited binding affinity for GnRH receptors, as assessed *in vitro* by displacement assays using [125]I[D-Lys6]GnRH as the radioligand. The affinity of the photosensitizer–peptide conjugates was found to be lower than that of the corresponding peptide. Photodynamic activity was, however, enhanced (approximately 1.5-fold) in GnRH expressing–pituitary gonadotrope aT3-1 cells, compared to unconjugated protoporphyrin IX. The photodynamic activity of the conjugates was attenuated by coincubation with the parent peptide, indicating that phototoxicity was receptor-mediated (Rahimipour et al. 2003).

30.4.8 Transferrin Receptors

As iron is an essential element for cell proliferation and metabolism and because the increased proliferative activity of malignant cells requires a higher iron supply, it is not surprising that TfR is overexpressed on a variety of malignant cells compared to normal cells. Indeed, the magnitude of TfR expression and turnover is proportional to the proliferative activity of the tumor tissue because a higher proliferation rate requires more iron. Therefore, the Tf-TfR system has been used in several formats to target photosensitizer molecules to different types of malignant cells.

Gijsens et al. (2002) prepared a Tf-conjugated polyethylene glycol (PEG)–liposome that contains the photosensitizer aluminum phthalocyanine tetrasulfonate compound. The antiproliferative activity of the targeted liposomes was evaluated and compared to the native photosensitizer and the nontargeted liposome on Hela tumor cells. After irradiation, the conjugate was 10 times more photocytotoxic than free photosensitizer, whereas the conjugated liposome displayed no photocytotoxicity at all (Gijsens et al. 2002). The high photocytotoxicity of targeted liposome was shown to be the result of a high intracellular concentration in HeLa cells, which could be lowered dramatically by incubating the conjugate with a competing transferrin concentration. Later, this group confirmed the selective uptake of the targeted liposome in human AY-27 transitional-cell carcinoma cells and in an orthotopic rat bladder tumor model (Derycke et al. 2004). Biodistribution studies revealed that among rats bearing AY-27 cell-derived bladder tumors, intravesical instillation with the targeted liposome resulted in a significant selective accumulation in tumor tissue compared to surrounding healthy tissues (normal urothelium and submucosa/muscle) (Derycke et al. 2004). More recently, Laptev et al. (2006) demonstrated that bioconjugates composed of Tf and hematoporphyrin photosensitizers significantly improve the specificity and efficiency of PDT for erythroleukemic cells.

It has been showed that transferrin is responsible for the cellular accumulation of metal-based drugs, such as platinum and ruthenium complexes via transferrin receptor-mediated endocytosis resulting from the fact that these metal-based drugs also bind to this blood transporter. Schmitt et al. (2009) have coordinated arene ruthenium moieties to pyridylporphyrins. They showed that ruthenium (II) organometallic complexes enhance uptake of the porphyrin by human melanoma cells and that these complexes remain intact following their uptake by cells. Moreover, the organometallic fragments on the pyridylporphyrin ring do not modify the photophysical properties of the photosensitizer. Using fluorescence microscopy, they studied cellular compartmentation of their complexes and noted that they accumulate in the cytoplasm of the melanoma cells. Moreover, Lottner et al. (2004) developed a hematoporphyrin (HP)–platinum(II) conjugate and demonstrated that this conjugate displays the same antiproliferative activity compared with cisplatin in J82 tumor cells.

30.4.9 Lectin Saccharide Receptors

A variety of carbohydrate-conjugated photosensitizers have been synthesized. Because oligosaccharides play important roles in cellular communication through saccharide–receptor interactions, which are usually specific and multivalent, their conjugation with a photosensitizer can confer high selectivity and specificity (DiStasio et al. 2005). Among the carbohydrate-binding proteins, galectins share a highly conserved domain with a high affinity for β-galactoside. Pandey et al. (2007) synthesized a series of carbohydrate–photosensitizer conjugates with different carbohydrate moieties, position of conjugation, and types of linkers. The comparative *in vitro*/*in vivo* studies using the RIF tumor cells and CH3 mice bearing RIF tumors of the series of positional isomers showed a substantial difference in their photosensitizing efficacy. The galectin-1 and -3 binding values for the carbohydrate conjugates obtained by the Enzyme-linked immunosorbent assay (ELISA) test showed higher affinity than the corresponding nonconjugate purpurinimides (Pandey et al. 2007). D'Auria et al. (2009) reported that human galectin-1 can be characterized as a porphyrin-binding protein based on its tight interactions with Zn/Mn and Au-porphyrins, indicating that galectin-1 may have potential as a delivery molecule to target tumor cells.

It was also reported that asialoglycoprotein (ASGP) receptors were abundantly expressed in the surface of hepatoma cells and mammalian hepatocytes. Several studies suggested that targeting could be accomplished through introduction of galactose residues, which can bind specifically to the ASGP receptors on the cells. Wu et al. (2010) described the synthesis of porphyrin and galactosyl conjugated micelles based on amphiphilic copolymer incorporating galactosyl and porphyrin, and they evaluated their targeting and PDT efficiency in HEp2 and HepG2 cells. Porphyrin and galactosyl conjugated polymer micelles displayed higher targeting and photodynamic efficacy in HepG2 cells than in receptor-negative Hep2 cells (Wu et al. 2010). Brevet et al. (2009) synthesized mesoporous silica nanoparticles combining covalent anchoring of the porphyrin to the mesoporous silica matrix and targeting of cancer cells with mannose. These nano-objects presented higher *in vitro* photoefficiency in MDA-MB-231 through mannose-dependent endocytosis than nonfunctionalized nanoparticles (Brevet et al. 2009).

Ferreira et al. (2009) reported the design of synthetic DNA oligonucleotide aptamers selected to bind specifically to O-glycan-peptide signatures of mucin glycoproteins, which are specifically expressed on the surface of a broad range of epithelial cancer cells. They showed that when conjugated to Ce6 and delivered to epithelial cancer cells, these DNA aptamers displayed a significant enhancement (>500-fold increase) in toxicity upon light activation compared to the free drug and have no cytotoxic effect on cell types lacking such O-glycan-peptide signatures (Ferreira et al. 2009). Ballut et al. (2009) reported that glycodendrimeric porphyrins interacted more significantly with a specific lectin, concavaline A, than the drug compound devoid of any sugar.

Poiroux et al. (2011) recently covalently linked a porphyrin to a plant lectin (Morniga G) known to recognize tumor-associated T and Tn antigens. The conjugation allowed a quick uptake of photosensitizer by Tn-positive Jurkat leukemia cells and efficient photosensitizer-induced phototoxicity. The conjugate strongly increased (1000×) the photosensitizer phototoxicity toward leukemic Jurkat T cells and specifically purged tumor cells from a 1:1 mixture of Jurkat leukemia (Tn-positive) and healthy (Tn-negative) lymphocytes (Poiroux et al. 2011). The same authors coupled other photosensitizers [TPPS, Al (III)-phthalocyanine AlPcS (4), and chlorin e6] to the protein. A conjugate including a single AlPcS4 per protein shows a strong phototoxicity [LD (50) = 4 nM] when irradiated in the therapeutic window; it preferentially kills cancerous lymphocytes, and the sugar-binding specificity of the lectin part of the molecule remains unaltered.

Hammerer et al. (2011) elaborated pi-conjugated porphyrin dimers and a triphenylamine centered trimer bearing monoethyleneglycol-peracetylated a-mannose targeting moieties, which were synthesized for application to two-photon absorption PDT.

30.4.10 Tumor-Specific Antigens: Targeting Antibodies

It is now widely accepted that neoplastic transformation generates new and specific antigenic biomarkers not present on normal cells. A large amount of studies has proved the effectiveness of anticancer PDT using mAb-conjugated photoactivatable molecules as drug delivery systems into tumor tissues (Solban, Rizvi, and Hasan 2006). Using mAb as antigen-specific carriers for selective delivery of the photosensitizer to the tumor is also known as photoimmunotherapy. Such an approach combines phototoxicity of the photosensitizers with the selectivity of mAb directed against tumor-associated antigens.

HER2 is a highly upregulated target receptor on tumor cell surfaces, such as breast cancer. To enhance photosensitizer immunoconjugate uptake by tumor cells, Savellano et al. (2005) suggested a multiepitope HER2 targeting strategy. Anti-HER2 photosensitizer immunoconjugates were produced via the conjugation of two mAbs, HER50 and HER66, to pyropheophorbide-a. Uptake and phototoxicity experiments using human cancer cells revealed selective uptake and a potential photodynamic effect of the immunoconjugates on the HER2-overexpressing target cells. Moreover, the multiepitope-targeted photoimmunotherapy was significantly more effective than single-epitope targeted photoimmunotherapy with a single anti-HER2 photosensitizer immunoconjugate (Savellano et al. 2005), suggesting that multitargeted photoimmunotherapy should also be useful against cancers that overexpress other antigenic receptors.

Recently, Jankun (2011) describes an antibody directed against prostate-specific membrane antigen and conjugated with HP to treat LnCAP human prostate cancer cells. His results show clearly that mAb/HP conjugates can deliver HP to tumor cells, which would result in considerably less HP in the circulation and, therefore, less delivery of HP to normal tissue, resulting in fewer side effects.

Nevertheless, it has been observed, particularly *in vivo*, that the large size of the mAb could limit the ability of the immunoconjugate to penetrate solid, deep-seated, and poorly vascularized tumors. To overcome this drawback, accumulating recent data suggest that an alternative can be the use of single-chain (sc) Fv mAb fragments (scFv), which are more efficient at penetrating tumor tissues because of their smaller size and more effectively cleared from the circulation because of the lack of the Fc domain (Staneloudi et al. 2007). Staneloudi and coworkers described the development and characterization of an immunoconjugate comprising two isothiocyanato-porphyrins attached to colorectal tumor-specific scFv. They demonstrated that these conjugates had a selective photocytotoxic effect as showed by *in vitro* assays against colorectal cell lines (Staneloudi et al. 2007). As indicated in Section 30.4.1, more recently, Bhatti et al. (2008) described a complex of multiple photosensitizer molecules (verteporfin or pyropheophorbide-a) with anti-HER2 scFv and demonstrated that the photoimmunoconjugates are more potent than either free photosensitizers, effectively killing tumor cells *in vitro* and *in vivo*. Indeed, many evidences suggest that the photoimmunoconjugates produced by coupling photosensitizers to recombinant mAb fragments scFv not only can target specifically and destroy tumor cells but also make it possible to load much more photosensitizing agents on an scFv than to the larger whole mAb (Bhatti et al. 2008) and thus increase the amount of delivered drug molecules in the tumor tissues.

30.5 Conclusion

Selective delivery of therapeutic amounts of photosensitizers in diseased tissues is recognized as an absolute requirement for efficient and safe PDT for the treatment of cancers. This is also the challenge to extend the application of PDT for the treatment of a broad range of tumor types as such modality present many advantages over conventional therapies. For this goal, development in targeted PDT continues to take advantage of the advances in the characterization of molecular mechanisms of tumor development. A large number of specific molecular targets have been identified and exploited in tumor targeting. Many photosensitizing agents have been elaborated and evaluated through *in vitro* and *in vivo* studies; however, only a very few have reached clinical evaluation phases. Each of the handful of photosensitizers has specific characteristics, but none includes all the properties of an ideal photosensitizer. Although third-generation photosensitizers have been widely described for selective targeting, very few have been evaluated for clinical applications as the *in vivo* selectivity was not sufficiently high. Nanoparticles, as multifunctional platforms, could represent emerging photosensitizer carriers that show great promise for PDT. In bionanotechnology, their development can overcome most of the shortcomings of classic photosensitizers.

References

Aguilar-Rojas, A., and M. Huerta-Reyes. 2009. Human gonadotropin-releasing hormone receptor-activated cellular functions and signaling pathways in extra-pituitary tissues and cancer cells (review). *Oncol Rep* 22:981–990.

Allen, T. M., and P. R. Cullis. 2004. Drug delivery systems: Entering the mainstream. *Science* 303:1818–1822.

Ballut, S., A. Makky, B. Loock et al. 2009. New strategy for targeting of photosensitizers: Synthesis of glycodendrimeric phenylporphyrins, incorporation into a liposome membrane and interaction with a specific lectin. *Chem Commun* 224–226.

Barge, J., R. Decreau, M. Julliard et al. 2004. Killing efficacy of a new silicon phthalocyanine in human melanoma cells treated with photodynamic therapy by early activation of mitochondrion-mediated apoptosis. *Exp Dermatol* 13:33–44.

Bechet, D., L. Tirand, B. Faivre et al. 2010. Neuropilin-1 targeting photosensitization-induced early stages of thrombosis via tissue factor release. *Pharm Res* 27:468–479.

Berdowska, I. 2004. Cysteine proteases as disease markers. *Clin Chim Acta* 342:41–69.

Bhatti, M., G. Yahioglu, L. R. Milgrom et al. 2008. Targeted photodynamic therapy with multiply-loaded recombinant antibody fragments. *Int J Cancer* 122:1155–1163.

Boisbrun, M., R. Vanderesse, P. Engrand et al. 2008. Design and photophysical properties of new RGD targeted tetraphenylchlorins and porphyrins. *Tetrahedron* 64:3494–3504.

Brevet, D., M. Gary-Bobo, L. Raehm et al. 2009. Mannose-targeted mesoporous silica nanoparticles for photodynamic therapy. *Chem Commun*:1475–1477.

Byrne, J. D., T. Betancourt, and L. Brannon-Peppas. 2008. Active targeting schemes for nanoparticle systems in cancer therapeutics. *Adv Drug Deliv Rev* 60:1615–1626.

Cawston, E. E., and L. J. Miller. 2010. Therapeutic potential for novel drugs targeting the type 1 cholecystokinin receptor. *Br J Pharmacol* 159:1009–1021.

Chaleix, V., V. Sol, M. Guilloton, R. Granet, and P. Krausz. 2004. Efficient synthesis of RGD-containing cyclic peptide-porphyrin conjugates by ring-closing metathesis on solid support. *Tetrahedron Lett* 45:5295–5299.

Chen, J., A. Bierhaus, S. Schiekofer et al. 2001. Tissue factor: A receptor involved in the control of cellular properties, including angiogenesis. *Thromb Haemost* 86:334–345.

Cheng, J., J. Xu, J. Duanmu et al. 2011. Effective treatment of human lung cancer by targeting tissue factor with a factor VII-targeted photodynamic therapy. *Curr Cancer Drug Targets* 119:1069–1081.

Couleaud, P., D. Bechet, R. Vanderesse et al. 2011. Functionalized silica-based nanoparticles for photodynamic therapy. *Nanomedicine* 6:995–1009.

Cuomo, V., G. Jori, B. Rihter, M. E. Kenney, and M. A. J. Rodgers. 1991. Tumor-localizing and tumor-photosensitizing properties of liposome-delivered Ge(Iv)-octabutoxyphthalocyanine. *Br J Cancer* 64:93–95.

D'Auria, S., L. Petrova, C. John et al. 2009. Tumor-specific protein human galectin-1 interacts with anticancer agents. *Mol Biosyst* 5:1331–1336.

De Luca, S., D. Tesauro, P. Di Lello et al. 2001. Synthesis and solution characterization of a porphyrin-CCK8 conjugate. *J Pept Sci* 7:386–394.

Derycke, A. S. L., A. Kamuhabwa, A. Gijsens et al. 2004. Transferrin-conjugated liposome targeting of photosensitizer AlPcS4 to rat bladder carcinoma cells. *J Natl Cancer Inst* 96:1620–1630.

Di Stasio, B., C. Frochot, D. Dumas et al. 2005. The 2-aminoglucosamide motif improves cellular uptake and photodynamic activity of tetraphenylporphyrin. *Eur J Med Chem* 40:1111–1122.

Ding, H., B. D. Sumer, C. W. Kessinger et al. 2011. Nanoscopic micelle delivery improves the photophysical properties and efficacy of photodynamic therapy of protoporphyrin IX. *J Control Release* 151:271–277.

Ferrario, A., C. F. Chantrain, K. von Tiehl et al. 2004. The matrix metalloproteinase inhibitor prinomastat enhances photodynamic therapy responsiveness in a mouse tumor model. *Cancer Res* 64:2328–2332.

Ferreira, C. S. M., M. C. Cheung, S. Missailidis, S. Bisland, and J. Gariepy. 2009. Phototoxic aptamers selectively enter and kill epithelial cancer cells. *Nucleic Acids Res* 37:866–876.

Frochot, C., B. D. Stasio, R. Vanderesse et al. 2007. Interest of RGD-containing linear or cyclic peptide targeted tetraphenylchlorin as novel photosensitizers for selective photodynamic activity. *Bioorg Chem* 35:205–220.

Gijsens, A., A. Derycke, L. Missiaen et al. 2002. Targeting of the photocytotoxic compound AlPcS4 to HeLa cells by transferrin conjugated PEG-liposomes. *Int J Cancer* 101:78–85.

Gravier, J., R. Schneider, C. Frochot et al. 2008. Improvement of meta-tetra(hydroxyphenyl)chlorin-like photosensitizer selectivity with folate-based targeted delivery: Synthesis and *in vivo* delivery studies. *J Med Chem* 51:3867–3877.

Hammerer, F., S. Achelle, P. Baldeck, P. Maillard, and M. P. Teulade-Fichou. 2011. Influence of carbohydrate biological vectors on the two-photon resonance of porphyrin oligomers. *J Phys Chem A* 115:6503–6508.

Hu, Z., B. Rao, S. Chen et al. 2010. Targeting tissue factor on tumour cells and angiogenic vascular endothelial cells by factor VII-targeted verteporfin photodynamic therapy for breast cancer in vitro and in vivo in mice. *BMC Cancer* 10:235.

Jankun, J. 2011. Protein-based nanotechnology: Antibody conjugated with photosensitizer in targeted anticancer photoimmunotherapy. *Int J Oncol* 39:949–953.

Khan, E. H., H. Ali, H. J. Tian et al. 2003. Synthesis and biological activities of phthalocyanine-estradiol conjugates. *Bioorg Med Chem Lett* 13:1287–1290.

Konan, Y. N., R. Gurny, and E. Allemann. 2002. State of the art in the delivery of photosensitizers for photodynamic therapy. *J Photochem Photobiol B Biol* 66:89–106.

Koval, V. V., A. A. Chernonosov, T. V. Abramova et al. 2001. Photosensitized and catalytic oxidation of DNA by metallophthalocyanine-oligonucleotide conjugates. *Nucleosides Nucleotides Nucleic Acids* 20:1259–1262.

Laptev, R., M. Nisnevitch, G. Siboni, Z. Malik, and M. A. Firer. 2006. Intracellular chemiluminescence activates targeted photodynamic destruction of leukaemic cells. *Br J Cancer* 95:189–196.

Law, B., and C. H. Tung. 2009. Proteolysis: A biological process adapted in drug delivery, therapy, and imaging. *Bioconjugate Chem* 20:1683–1695.

Li, H., D. E. Marotta, S. Kim et al. 2005. High payload delivery of optical imaging and photodynamic therapy agents to tumors using phthalocyanine-reconstituted low-density lipoprotein nanoparticles. *J Biomed Optics* 10(4):41203.

Li, L., B. C. Bae, T. H. Tran et al. 2011. Self-quenchable biofunctional nanoparticles of heparin-folate-photosensitizer conjugates for photodynamic therapy. *Carbohydr Polym* 86:708–715.

Lottner, C., R. Knuechel, G. Bernhardt, and H. Brunner. 2004. Combined chemotherapeutic and photodynamic treatment on human bladder cells by hematoporphyrin-platinum(II) conjugates. *Cancer Lett* 203:171–180.

Lutsenko, S. V., N. B. Feldman, G. V. Finakova et al. 1999. Targeting phthalocyanines to tumor cells using epidermal growth factor conjugates. *Tumor Biol* 20:218–224.

Maree, S., D. Phillips, and T. Nyokong. 2002. Synthesis, photophysical and photochemical studies of germanium and tin phthalocyanine complexes. *J Porphyrins Phthalocyanines* 6:17–25.

Maree, S. E., and T. Nyokong. 2001. Syntheses and photochemical properties of octasubstituted phthalocyaninato zinc complexes. *J Porphyrins Phthalocyanines* 5:782–792.

Nemerson, Y. 1988. Tissue factor and hemostasis. *Blood* 711:1–8.

Pandey, S. K., X. Zheng, J. Morgan et al. 2007. Purpurinimide carbohydrate conjugates: Effect of the position of the carbohydrate moiety in photosensitizing efficacy. *Mol Pharm* 4:448–464.

Paulos, C. M., J. A. Reddy, C. P. Leamon, M. J. Turk, and P. S. Low. 2004. Ligand binding and kinetics of folate receptor recycling *in vivo*: Impact on receptor-mediated drug delivery. *Mol Pharmacol* 66:1406–1414.

Perret, G. Y., A. Starzec, N. Hauet et al. 2004. *In vitro* evaluation and biodistribution of a 99mTc-labeled anti-VEGF peptide targeting neuropilin-1. *Nucl Med Biol* 31:575–581.

Poiroux, G., M. Pitie, R. Culerrier et al. 2011. Targeting of T/Tn antigens with a plant lectin to kill human leukemia cells by photochemotherapy. *PlosOne* 6:e23315.

Rahimipour, S., N. Ben-Aroya, K. Ziv et al. 2003. Receptor-mediated targeting of a photosensitizer by its conjugation to gonadotropin-releasing hormone analogues. *J Med Chem* 46:3965–3974.

Renno, R. Z., Y. Terada, M. J. Haddadin et al. 2004. Selective photodynamic therapy by targeted verteporfin delivery to experimental choroidal neovascularization mediated by a homing peptide to vascular endothelial growth factor receptor-2. *Arch Ophthalmol* 122:1002–1011.

Reubi, J. 2003. Peptide receptors as molecular targets for cancer diagnosis and therapy. *Endocr Rev* 24:389–427.

Rosenkranz, A. A., V. G. Lunin, P. V. Gulak et al. 2003. Recombinant modular transporters for cell-specific nuclear delivery of locally acting drugs enhance photosensitizer activity. *FASEB J* 17:1121.

Savellano, M. D., B. W. Pogue, P. J. Hoopes, E. S. Vitetta, and K. D. Paulsen. 2005. Multiepitope HER2 targeting enhances photoimmunotherapy of HER2-overexpressing cancer cells with pyropheophorbide-a immunoconjugates. *Cancer Res* 65:6371–6379.

Schmitt, F., P. Govindaswamy, O. Zava et al. 2009. Combined arene ruthenium porphyrins as chemotherapeutics and photosensitizers for cancer therapy. *J Biol Inorg Chem* 14:101–109.

Schneider, R., L. Tirand, C. Frochot et al. 2006. Recent improvements in the use of synthetic peptides for a selective photodynamic therapy. *Anticancer Agents Med Chem* 6:469–488.

Segalla, A., C. Milanesi, G. Jori et al. 1994. CGP 55398, a liposomal Ge(IV) phthalocyanine bearing two axially ligated cholesterol moieties: a new potential agent for photodynamic therapy of tumours. *Br J Cancer* 695: 817–825.

Selbo, P. K., A. Weyergang, M. S. Eng et al. 2012. Strongly amphiphilic photosensitizers are not substrates of the cancer stem cell marker ABCG2 and provides specific and efficient light-triggered drug delivery of an EGFR-targeted cytotoxic drug. *J Control Release* 159:197–203.

Sharwani, A., W. Jerjes, C. Hopper et al. 2006. Photodynamic therapy down-regulates the invasion promoting factors in human oral cancer. *Arch Oral Biol* 51:1104–1111.

Singh, D., S. K. Bisland, K. Kawamura, and J. Gariepy. 1999. Peptide-based intracellular shuttle able to facilitate gene transfer in mammalian cells. *Bioconjugate Chem* 10: 745–754.

Sobolev, A. S. 2009. Novel modular transporters delivering anticancer drugs and foreign DNA to the nuclei of target cancer cells. *J BUON* 14 Suppl 1: S33–42.

Solban, N., I. Rizvi, and T. Hasan. 2006. Targeted photodynamic therapy. *Lasers Surg Med* 38:522–531.

Song, L. P., H. Li, U. Sunar et al. 2007. Naphthalocyanine-reconstituted LDL nanoparticles for *in vivo* cancer imaging and treatment. *Int J Nanomed* 2:767–774.

Soukos, N. S., M. R. Hamblin, S. Keel et al. 2001. Epidermal growth factor receptor-targeted immunophotodiagnosis and photoimmunotherapy of oral precancer *in vivo*. *Cancer Res* 61:4490–4496.

Staneloudi, C., K. A. Smith, R. Hudson et al. 2007. Development and characterization of novel photosensitizer: scFv conjugates for use in photodynamic therapy of cancer. *Immunology* 120:512–517.

Stefflova, K., H. Li, J. Chen, and G. Zheng. 2007. Peptide-based pharmacomodulation of a cancer-targeted optical imaging and photodynamic therapy agent. *Bioconjugate Chem* 18:379–388.

Thomas, N., D. Bechet, P. Becuwe et al. 2009. Peptide-conjugated chlorin-type photosensitizer binds neuropilin-1 *in vitro* and *in vivo*. *J Photochem Photobiol B Biol* 96:101–108.

Tirand, L., C. Frochot, R. Vanderesse et al. 2006. A peptide competing with VEGF165 binding on neuropilin-1 mediates targeting of a chlorin-type photosensitizer and potentiates its photodynamic activity in human endothelial cells. *J Control Release* 111:153–164.

van Kempen, L. C. L., K. E. de Visser, and L. M. Coussens. 2006. Inflammation, proteases and cancer. *Eur J Cancer* 42:728–734.

Vihinen, P., and V. M. Kahari. 2002. Matrix metalloproteinases in cancer: Prognostic markers and therapeutic targets. *Int J Cancer* 99:157–166.

Wu, D. Q., Z. Y. Li, C. Li et al. 2010. Porphyrin and galactosyl conjugated micelles for targeting photodynamic therapy. *Pharm Res* 27:187–199.

Zheng, G., J. Chen, K. Stefflova et al. 2007. Photodynamic molecular beacon as an activatable photosensitizer based on protease-controlled singlet oxygen quenching and activation. *Proc Natl Acad Sci USA* 104:8989–8994.

Zheng, G., H. Li, M. Zhang et al. 2002. Low-density lipoprotein reconstituted by pyropheophorbide cholesteryl oleate as target-specific photosensitizer. *Bioconjugate Chem* 13:392–396.

31

Enhancing Photodynamic Treatment of Cancer with Mechanism-Based Combination Strategies

Lei Zak Zheng
Harvard Medical School

R. Bryan Sears
Emmanuel College

Iqbal Massodi
Harvard Medical School

Stanley Kimani
Harvard Medical School

Tayyaba Hasan
Harvard Medical School

31.1 Introduction

31.1.1 Photodynamic Effect

Photodynamic therapy (PDT) is a photochemistry-based technology that utilizes the absorption of light by a photosensitizer (PS), initiating a photochemical process and resulting in localized treatment. Current PDT regimens produce this effect in two steps in which the nontoxic chemical compound PS is administered, and after an incubation period, local light is used to generate cytotoxic effects only in the area of irradiation (Wilson and Patterson 2008) (also see Section 3.1). Recent advancements have brought PDT into the clinic for the treatment of lung, prostate, esophageal, head and neck, and skin cancers (Agostinis et al. 2011). Additionally, the first frontline therapy Visudyne (Benzoporphyrin derivative-mediated PDT) used in the treatment of the wet form of the age-related macular degeneration gained FDA approval in 2001 (Michels and Schmidt-Erfurth 2001). Currently, scientists and physicians are continuing to optimize the technology of PDT as a therapeutic, diagnostic, and monitoring modality for a diverse number of diseases.

PDT requires three components to be simultaneously present at the site of treatment: light, tissue oxygen, and PS. For the processes relevant to PDT, irradiation is achieved within the visible region of the electromagnetic spectrum (400–800 nm). A simplified Jablonski diagram is presented in Figure 31.1 showing the possible pathways of energy absorption and dissipation. Upon absorption of light, a PS can exist in either singlet (S_n) or triplet (T_n) excited states arising from the quantum spin of the excited-state molecule. For typical PS, a first singlet excited state S_1 is first observed after absorption. S_1 is short-lived within PDT-related PS because of efficient rates of intersystem crossing (k_{isc}), which allow for the interconversion of spin states. The relatively short-lived (nanoseconds) S_1 excited state results in photochemistry preferentially initiating from the longer-lived T_1 excited state (microseconds to milliseconds). In the biological milieu, the photochemical reactions of PS typically result in the production of free radicals ($\cdot OH$, $\cdot OH^-$, O_2^-) or transfer of energy to the ground state of oxygen molecules ($^3O_2$) to give rise to $^1O_2$ molecules. The collisional transfer for energy to ground-state oxygen molecules arise from the long lifetimes of the PS triplet excited state and is the predominately accepted mode of action for most PSs currently under investigation, although other competing mechanisms are also possible (Wilson and Patterson 2008).

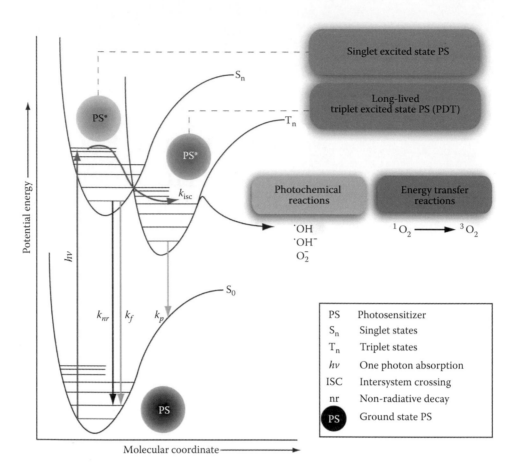

FIGURE 31.1 Simplified Jablonski diagram.

31.1.2 PDT-Induced Cell Death

Depending on the localization and type of PS photochemical reaction, different tissue and subcellular targets can be immediately damaged by the PDT photochemical products, subsequently triggering cell death. Commonly, PDT can elicit three major cell death pathways: apoptosis, necrosis, and autophagy-associated cell death (Buytaert, Dewaele, and Agostinis 2007) (also see Section 28.4). Apoptosis is a regulated cell-death process dependent on the activation of caspases that occurs following photodamage to critical organelles, such as mitochondria, endoplasmic reticulum (ER), and lysosomes. For mitochondria-associated PSs, photodamage to membrane-bound B-cell lymphoma-2 (Bcl-2) becomes a permissive signal for the mitochondria outer membrane permeabilization (MOMP) and results in the subsequent release of proapoptotic molecules, such as cytochrome c, Smac, and AIF. In the case of lysosome-localizing PS, lysosomal membrane rupture and leakage of cathepsins into the cytosol induce Bid cleavage and MOMP. These signaling events will then further execute the caspase-dependent apoptotic program (Agostinis et al. 2011; Oleinick, Morris, and Belichenko 2002).

PDT also induces cell death via necrosis, which traditionally is considered uncontrolled and without signaling components. However, recently, there is evidence showing its regulation via the activation of receptors interacting with protein kinases RIP1K and RIP3K, mitochondrial reactive oxygen species (ROS) accumulation, lysosomal damage, and intracellular Ca^{2+} overload. This type of cell death phenotypically resembles necrosis except that it is programmed like apoptosis and therefore termed necroptosis (Coupienne et al. 2011).

Photodamage to the cells can also stimulate autophagy, normally a physiology event for cells under nutrient shortage, which involves the degradation and recycling of intracellular proteins and organelles using a lysosomal pathway (Levine and Kroemer 2008; Mizushima and Komatsu 2011). Recent reports have suggested that PDT elicited autophagy serves a prosurvival function to protect cells by removing and recycling photodamaged proteins (Buytaert et al. 2006; Kessel, Vicente, and Reiners 2006). However, in cases where PDT treatment has also damaged lysosomes, excessive autophagic cargo accumulates and can potentiate the phototoxicity in apoptosis-competent cells, thereby resulting in a pro-death role of PDT-induced autophagy (Andrzejak, Price, and Kessel 2011) (also see Section 28.4).

31.1.3 PDT-Induced Tumor Destruction

Besides the direct photo-killing of cancer cells mentioned above, two additional mechanisms have been documented that can contribute to PDT tumor destruction. Damage to tumor-associated microvessels has been shown to cause thrombosis, hemorrhage, hypoxia, and nutrient shortage in tumors, which in turn, can contribute to tumor destruction (Bhuvaneswari et al. 2009a; Chen et al. 2006). Likewise, PDT treatment also induces acute inflammation and the release of cytokines or stress-response proteins. Once released, these proteins stimulate immune response and lead to the invasion of leukocytes and mobilization of lymphocytes that facilitate the destruction of both the local tumor as well as the disseminated tumor foci (Brackett and Gollnick 2011; Castano, Mroz, and Hamblin 2006) (reviewed in Section 28.4). The relative importance of each mechanism for the overall therapy outcome is yet to be defined. However, the combined PDT effects of all these components are required for optimum long-term tumor regression, especially for tumors that may have metastasized.

31.1.4 PDT-Induced Cell Signaling

PDT can effectively induce cell-death pathways in a tumor; however, post-treatment regrowth often occurs, presumably stemming from regions receiving sublethal doses of PDT (see Section 28.4). It is therefore critical to understand the changes in tumor biology and physiology induced by sublethal PDT treatment. In particular, how these treatments impact cellular signaling circuitry, such as cell death, cell-cycle progression, cellular metabolism, stromal interaction, and tumor vascularization is of critical importance. Such an understanding may help identify key molecular responses and inform the future design of effective combination strategies for optimal treatment outcomes. There have been extensive reviews detailing the subjects of cellular signaling in the context of PDT (Almeida et al. 2004; Robertson, Evans, and Abrahamse 2009). In this chapter, we will instead focus our discussion on selected families of molecular targets and the PDT combination approaches that have been reported.

31.2 Mechanism-Based PDT Combination Strategies

PDT is gaining recognition as a treatment option for many types of cancers, and the clinical results are promising. As with other therapies, for PDT to evolve as a curative modality, both our understanding of PDT mechanisms and the design of treatment strategy are necessary to provide an enhanced therapeutic outcome. Combination treatment strategies comprising PDT and other concurrent treatments increase the effectiveness through strategies that can (1) reduce detrimental molecular responses triggered by surviving tumor cells following PDT on the second treatment, (2) potentiate the antitumor immune response following PDT, and (3) increase the tumor cells' susceptibility to PDT (Castano, Mroz, and Hamblin 2006; Gomer et al. 2006; Verma et al. 2007). In this section, we retrospectively examine the key molecular pathways and therapeutic strategies that have shown enhancement to PDT. By categorizing families of emerging and existing molecular targets, we hope to bring perspectives and facilitate the creative designs of high-efficacy, PDT-based combination strategies in the future (Figure 31.2).

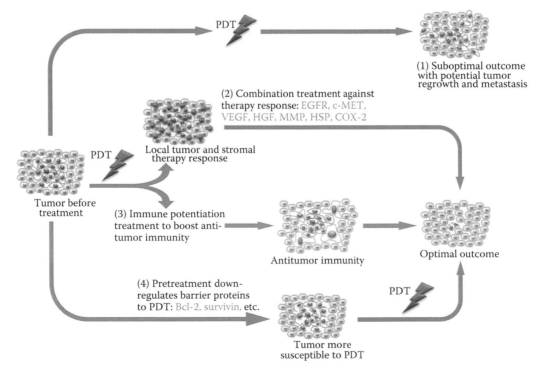

FIGURE 31.2 Combination strategies to enhance PDT. In conventional PDT (1), residual cancer cells may regrow and metastasize leading to suboptimal outcome. In (2), a secondary treatment targeting PDT molecular response may enhance the PDT outcome. In (3), antitumor immunity is boosted to enhance PDT outcome. In (4), pretreatment renders cancer cells more susceptible to PDT, therefore leading to improvement in overall outcome.

31.2.1 Enhancing Direct PDT Cell Killing

31.2.1.1 Modulating Antiapoptotic Proteins

As one of the major cell-death pathways triggered by PDT, the execution of apoptosis involves a cascade of signaling mediators and regulatory proteins that can either facilitate or negate innate cell death programs (Oleinick, Morris, and Belichenko 2002). Numerous studies support the findings that the levels and ratio of antiapoptotic and proapoptotic proteins influence the outcome of PDT treatment. Therefore, it has been believed that modulating these apoptotic protein levels may increase treatment efficacy for PDT induced cell death. Herein, we discuss several studies that have enhanced PDT efficacy through the modulation of apoptotic proteins.

The Bcl-2 is a family of apoptosis regulatory proteins that contain both proapoptotic (Bax, Bad, Bid, Bak) and antiapoptotic members (Bcl-2, Bcl-x_L, Mcl-1, Bcl-w, A/Bfl-1). Prosurvival members prevent the progression of apoptosis in response to a cytotoxic insult (Cory and Adams 2002). It has been well studied that PDT with mitochondria/ER-selective PS causes photodamage to the antiapoptotic proteins Bcl-2 and Bcl-x_L, followed by Bax activation, cytochrome c release, and caspase activation leading to apoptosis via the intrinsic pathway (Usuda et al. 2003; Vantieghem et al. 2002; Xue, Chiu, and Oleinick 2001). Conversely, Bcl-2 overexpression in tumors provides protection against PDT-induced apoptosis and cytotoxicity (Xue et al. 2007). As Bcl-2 protein level can influence PDT efficacy, Bcl-2 inhibitors have been sought to promote PDT treatment efficacy. Kessel (2008) has examined this effect using the Bcl-2 antagonist HA14-1 within murine leukemia L1210 cells. In this study, synergism has been demonstrated between a suboptimal dose of PDT and the pretreatment of cells with HA14-1. Similarly, genestein, a tyrosine kinase inhibitor, has been shown to suppress Bcl-2 expression. The use of hypericin-PDT in the presence of genestein was found to decrease the clonogenic survival of breast adenocarcinoma cells (Ferenc et al. 2010) by mediating the downregulation of Bcl-2 and upregulation of Bax, thereby enhancing the PDT efficacy.

Besides the Bcl-2 family of antiapoptotic proteins, the inhibitor of apoptosis proteins (IAPs) can be induced and phosphorylated following PDT (Ferrario et al. 2007). The IAP family member survivin that suppresses caspase activation is aberrantly overexpressed in many human cancers. Overexpressed survivin is also associated with inhibition of mitochondrial-associated apoptotic pathways. Therefore, any disruption of survivin function would change the therapeutic response of PDT. Ferrario et al. 2007 has shown that downregulation of survivin protein with the geldanamycin derivative 17-allylamino-17-demethoxygeldanamycin (17-AAG) sensitized cells to PDT-induced apoptosis resulting in increased cell death. In another study, siRNA silencing of the survivin gene before PDT treatment resulted in the enhanced cytotoxicity response and apoptosis in metastatic T47D breast cancer cells (Cogno et al. 2011). Conversely, overexpression of survivin by an expression vector diminished the effect of PDT in T47D breast cancer cells.

31.2.1.2 Inhibition of Stress Response Heat Shock Proteins

Heat shock proteins (HSPs) are evolutionarily conserved proteins produced by cells in response to various types of stresses. Acting as molecular chaperones, HSPs can protect cells from environmental stress by assisting in proper protein folding and stabilization. Additionally, these proteins help to sequester severely damaged proteins, which require degradation (Bukau, Weissman, and Horwich 2006). Because of these functionalities, HSPs are often found to be overexpressed in many cancers. A considerable number of studies have demonstrated the upregulation of various HSP family members in response to PDT both *in vitro* and *in vivo* (Gomer et al. 1996; Moor 2000). Further detailed biochemical examination indicated that oxidative stress imposed by PDT triggers direct ligation of a transcription inducer to the HSP gene-promoter region immediately following PDT stress. This result is expected as the expression of HSP is a natural cellular stress response aimed to either remove or refold excessively photodamaged proteins. Furthermore, studies have shown that the upregulation of some HSP play a role in the resistance of photooxidative stress. These results suggest that HSP upregulation following PDT may reduce PDT efficacy (Hanlon et al. 2001; Shen et al. 2005; Wang et al. 2002). A combination study in which HSP90 inhibitor 17-AAG was used with PDT in mouse mammary carcinoma cells and tumor models found that targeting HSP90 not only reduced tumor expression levels of several PDT response biomarkers (survivin, Akt, HIF-1α, MMP-2, and VEGF) but also improved long-term tumoricidal responses (Ferrario and Gomer 2010).

31.2.2 Targeting Cancer Growth Pathways

Deregulated cell-cycle progression and invasion of surrounding tissue are hallmarks of cancer growth. Molecular pathways responsible for the initiation and maintenance of these cancer phenotypes are being extensively interrogated in order to develop pathway targeted therapeutics. Several targeted drugs or biologics are in clinical use for various cancer types, either alone or in combination. Here, we present the key PDT combination studies targeting cancer growth pathways.

31.2.2.1 EGFR Pathway of Cell-Cycle Progression

Many human epithelial cancers are characterized by the upregulation of the epidermal growth factor receptor (EGFR) signaling pathway. EGFR is a member of the human EGFR family of receptor tyrosine kinases. It is a transmembrane receptor, which can be activated through ligand binding via its extracellular domain. Upon ligand binding, EGFR forms either homodimers or heterodimers with other members of the ErbB family, resulting in the autophosphorylation of the intracellular domain through intrinsic tyrosine kinase activity and subsequent activation of downstream signaling pathways. These downstream pathways include the canonical Ras/Raf, MAPK, and PI3K/AKT pathways and have been linked to cell-cycle progression in cancer. Alternatively, EGFR can be cross-activated by many other receptors, such as mesenchymal epithelial transition factor (c-MET), or can remain constitutively activated as a result of gene mutations (Takeuchi and Ito 2010; Weickhardt, Tebbutt, and Mariadason 2010).

A series of early studies have been made to understand the role of EGFR in the context of PDT. Fanuel-Barret et al. (1997) examined hematoporphyrin-PDT in glioma cell lines. In all three cell lines tested, post-PDT epidermal growth factor (EGF) treatment of the cells reduced the efficacy of PDT killing. However, pre-PDT treatment with EGF did not affect cell killing, which suggested that EGF/EGFR signaling following PDT played a role against PDT killing (Fanuel-Barret et al. 1997). Ahmad, Kalka, and Mukhtar (2001) investigated the involvement of the EGFR pathway using models of A431 cells and murine skin papillomas with Pc4-PDT. Following PDT, a time-dependent inhibition of EGFR expression and activation was observed in both *in vitro* and *in vivo* models (Ahmad, Kalka, and Mukhtar 2001). These early observations have instigated further detailed mechanistic examination of the role of EGFR in PDT.

In addition to the overall expression levels of EGFR pre- and post-PDT, Wong et al. (2003) have shown that for membrane or mitochondrial targeted PS, a window of time during which the responsiveness to EGF stimulation could be observed was lost in malignant and normal cells. Additionally, this absence of EGF responsiveness was only temporary and recovered after 72 h post-treatment (Wong et al. 2003). Indeed, a follow-up study confirmed that sublethal PDT treatment may immediately inhibit and/or degrade membrane-bound EGFR. However, the loss of EGFR from the cell membrane can be recovered at later times (Ahmad, Kalka, and Mukhtar 2001). In another case, Abu-Yousif et al. (2012) demonstrated that LD_{50} benzoporphyrine derivative (BPD)-PDT treatments result in a rapid reduction of ovarian cancer cells responsiveness to EGF. This is followed by a recovery of cell-surface EGFR and elevated responsiveness to EGF compared with the basal level, which suggests that the EGFR pathway is a therapy-associated response that favors cancer cell regrowth.

These findings not only clarified the dynamic involvement of the EGFR pathway in the PDT but also sparked preclinical studies to optimize the combination therapy using EGFR-targeted PDT. In these studies, del Carmen et al. (2005) examined the efficacy of a PDT-based treatment regimen combined with EGFR-neutralizing monoclonal antibody cetuximab in an orthotopic model of ovarian cancer, which frequently overexpresses EGFR. This combination treatment was found to be well tolerated and showed synergism between the EGFR-targeted immunotherapy and PDT, leading to significantly improved outcomes in mice (del Carmen et al. 2005). Another *in vivo* study by Bhuvaneswari et al. (2009b) confirmed that combination therapies of PDT and cetuximab potently inhibited tumor growth in a xenograft bladder cancer model.

31.2.2.2 c-MET Pathway of Invasive Growth

The hepatic growth factor (HGF)/mesenchymal epithelial transition factor (c-MET) pathway is critically involved in normal organ development but also in cancer invasive growth and metastasis. During embryo development, the c-MET pathway is required for the transient conversion of specific cells from epithelial to mesenchymal phenotype and their migration and morphogenesis into new structures. Genetic studies have strongly suggested the role of c-MET in the development of placenta, liver, lung, and postnatal tissue repair and organ regeneration (Ohnishi and Daikuhara 2003). However, when the c-MET pathway is hijacked, increased cancer cell invasion, spreading, and metastasis may result. In cancer cells, the c-MET pathway can be upregulated via multiple mechanisms, including activating gene mutations, overexpression of c-MET, or its ligand HGF. The tissue microenvironment, especially tissue hypoxia, is another direct factor that regulates the activity of the c-MET pathway. Biochemically, hypoxia triggers the transcriptional induction of c-MET in carcinoma cells via hypoxia inducible factor 1 (HIF-1α) and therefore increases the cell sensitivity to ligand stimulation. Subsequent *in vivo* studies have demonstrated that hypoxia concomitantly promotes self-renewal and a migratory phenotype in cancer cells, which prompts the cell migration toward a favorably oxygenated environment. This observation has been supported by clinical evidence that highly hypoxic tumors are much more aggressive than oxygenated tumors (Boccaccio and Comoglio 2006; Cooke et al. 2012).

It is well known that PDT treatment can lead to tumor hypoxia via both photochemical oxygen consumption and the damage to the vasculature. Therefore, it is reasonable to speculate that PDT-treated tumors can potentially mount a molecular response of the c-MET pathway driven by the hypoxia tumor microenvironment. In a study that aimed to profile the changes of angiogenesis molecular regulators following hypericin PDT, Bhuvaneswari et al. (2008) observed a 30-fold upregulation of HGF mRNA levels in bladder tumor cells. In another study, we have examined the effects of a sublethal dose of PDT treatment on various pancreatic cancer cells. A dose-dependent increase in HGF protein following a sublethal dose of PDT has been observed in PANC-1 cells both *in vitro* and in orthotopic tumor models (Figure 31.3a). In another cell line, AsPC-1, that does not express HGF, PDT treatment activated the c-MET 24 h after treatment as shown in the Western blot (Figure 31.3b). Although there is no published report on targeting c-MET or HGF in combination with PDT, the preliminary results have suggested the HGF/c-MET pathway as a potential rescuing pathway for hypoxic pancreatic cancer cells following PDT.

31.2.3 Targeting Stromal Factors

Tumor stroma is part of the solid tumor that is composed of nonmalignant, host-derived extracellular matrix, infiltrating cells, and connective tissues. Intratumoral PS localization studies have revealed that hydrophilic PS transported by albumin and globulins preferentially localized in the vascular stroma of tumor tissue. This is in contrast to hydrophobic PSs, which are commonly incorporated into lipoproteins and localized in the tumor parenchyma (Peng and Nesland 2004). Stroma not only

(a)

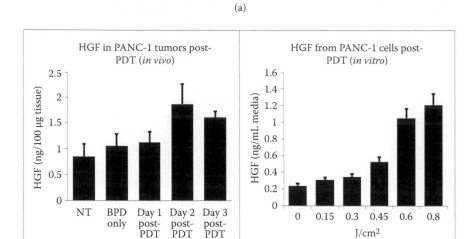

(b)

(c)

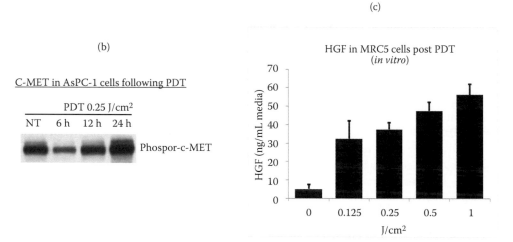

FIGURE 31.3 HGF/c-MET as molecular response to PDT in pancreatic cancer cells. (a) Orthotopic PANC-1 tumors were treated with BPD-PDT at 50 J/cm², and HGF in tumor tissue were measured using ELISA at different time following PDT (left panel). PANC-1 cells were treated with BPD-PDT at various PDT doses, and HGF levels in cell culture media were measured using ELISA (right panel). (b) AsPC-1 cells were treated with BPD-PDT at 0.25 J/cm², and phosphor-c-MET, total c-MET, and b-actin in PDT treated cells were measured using Western blotting 24 h after PDT. (c) MRC5 fibroblasts were treated with BPD-PDT at various PDT doses indicated, and secreted HGFs in culture media were measured using ELISA 24 h after PDT.

plays a passive supportive role in tumor development but also can actively promote malignant progression through production and secretion of stimulatory growth factors and cytokines. For example, tumor invasion of surrounding tissue requires extensive remodeling of the neighboring stromal environment and can affect tumor progression. Excessive stroma has also been implicated in chemotherapy resistance in pancreatic cancer; however, the clinical trial targeting a pancreatic cancer stroma-inducing factor failed to provide a survival benefit. Nevertheless, it is still interesting to consider targeting stroma and stromal factors in conjunction with PDT to be a viable combination strategy, especially given the results that PDT has been shown to target stroma and sensitize tumors to chemotherapy (Rizvi et al. 2010).

31.2.3.1 Matrix Metalloproteinases that Remodel Tumor Microenvironment

The matrix metalloproteinases (MMPs) represent a large family of proteinases mediating stromal changes within the tumor microenvironment during tumorigenesis. This family of zinc-dependent endopeptidases are initially expressed in an enzymatically inactive state and become proteolytically active after processing by other proteases. MMPs have been implicated in cancer for more than 40 years, and the MMP-mediated degradation of ECM has been linked to cancer cell invasion and metastasis. The function of MMPs *in vivo* is regulated by a balance of physiological inhibitors. The expression of MMPs and their inhibitors (TIMPs) in the tumor microenvironment is quite diverse. Although cancer cells from various tissues can express

MMPs, the major producers of these proteinases are stromal cells infiltrating the tumor (Kessenbrock, Plaks, and Werb 2010).

The relationships between PDT and MMPs have been examined in both cancer and stromal cell types. Hypericin-PDT has been found to induce MMP-1 expression in two nasopharyngeal cancer cell lines and in animal tumor models. However, MMP-9 mRNA and protein expression were downregulated following hypericin-PDT in nasopharyngeal cancer cells (Du et al. 2004, 2007). In another study, collagen-degrading MMP-1 and MMP-3 were indirectly inducible in fibroblasts following aminolevulinic acid (ALA) PDT, and the process was dependent on PDT-induced inflammatory cytokine IL-1 and IL-6 secreted by keratinocytes (Takahashi et al. 2006). A follow-up human study showed considerable improvement in photodamaged skin with a significant increase in MMP-9 expression in the dermis 3 months after treatment (Almeida Issa et al. 2009). Another PDT study using liposomal formulated ALA showed that MMP-3 was expressed in the tumor cells with the highest levels in the tissues directly adjacent to the tumors (Osiecka et al. 2010). All these observations strongly suggested that PDT treatment can directly or indirectly upregulate stromal factors, such as MMP-1, -3, and -9, that are capable of remodeling the extracellular matrix on the tumor tissue boundary, potentially paving the road for cancer cell invasion. This hypothesis was validated in a therapeutic outcome study, in which MMP-9 upregulation was clearly demonstrated, and prinomastat, a broad-spectrum MMP protease inhibitor, significantly improved PDT tumor response, suggesting the feasibility of combining PDT with inhibition of stromal-derived MMPs (Ferrario et al. 2004).

31.2.3.2 HGF/c-MET-Mediated Tumor-Stromal Crosstalk

In addition to MMPs, the paracrine signaling of stromal HGF and tumor c-MET crosstalk represents another potential targeting strategy that may enhance PDT outcomes. A series of reports in pancreatic cancer models have demonstrated that HGF upregulation in cancer stromal fibroblast, in conjunction with c-MET overexpression in cancer cells, form a tumor/stroma crosstalk that drives the aggressive growth of pancreatic cancer. PDT-induced hypoxia was demonstrated to significantly upregulate HGF following hypericin-PDT. We found that sublethal BPD-PDT treatment has significantly increased HGF in the stromal MRC5 fibroblast culture media in a dose-dependent manner (Figure 31.3c). Given that a majority of pancreatic tumors overexpress c-MET, targeting the HGF produced by stroma following PDT may prevent the stroma/tumor crosstalk mediated by the HGF/c-MET pathway and may enhance the therapeutic outcome by limiting the tumor survival effects mediated by this pathway.

31.2.4 Antagonizing Angiogenic Response

PDT has been known to cause microvasculature occlusion. As tumor growth heavily depends on functional vasculature for the supply of oxygen and nutrients, induction of damage to existing microvasculature can significantly improve treatment efficacy. Paradoxically, PDT-treated tumors are severely hypoxic because of the consumption of oxygen during PDT. In such hypoxic tissue, an induction of the cell adaptive response is observed, characterized by the production and secretion of angiogenic factors, such as vascular endothelial growth factor (VEGF). These factors play a critical role in the revascularization of tumors, and their post-PDT expression may contribute to diminished treatment efficacy. In this section, we will discuss a few efficacy-enhancing strategies to antagonize the angiogenic response switched on by PDT (Gomer et al. 2006; Kosharskyy et al. 2006; Solban et al. 2006).

31.2.4.1 Angiogenic Pathway VEGF/VEGFR

The angiogenic factor VEGF is well characterized and is expressed in response to a variety of stimuli-including cytokines, hypoxia, and oxidative stress. VEGF acts via endothelial cell-surface receptors, VEGFR-1 and VEGFR-2. The consumption of oxygen and subsequent vascular damage induced by PDT results in significant tissue hypoxia, and this further leads to VEGF expression via hypoxia-inducible factor-1 (HIF-1) transcription factor (Ferrario et al. 2000). This response mechanism protects tumor cells against PDT-mediated damage and may limit the PDT treatment efficacy. Therefore, blocking VEGF pathways with targeted therapeutic agents following PDT may result in improved PDT therapeutic outcomes.

Among the early reports, Roberts and Hasan (1993) have shown the role of VEGF in PS uptake in tumors. Additionally, Solban et al. (2006) showed that subcurative PDT using the PS BPD-MA resulted in an increase in VEGF expression in prostate cancer. Numerous studies have used different angiogenic inhibitors in combination with PDT to enhance the therapeutic outcome of PDT. These angiogenic inhibitors target different pathways in the VEGF-signaling cascade. Soluble or free VEGF can be targeted using Avastin (bevacizumab), which is an FDA-approved, humanized monoclonal antibody that binds to free VEGF and blocks VEGF receptor activation. Several studies have shown that blocking the VEGF pathway by Avastin in combination with PDT reduced tumor growth in Kaposi's sarcoma (KS) and murine and human bladder tumor models (Bhuvaneswari et al. 2007, 2011; Ferrario and Gomer 2006). Similarly, blocking VEGF receptors (VEGFR-1 and VEGFR-2) with an antibody has been shown to attenuate the VEGF signaling pathway. This was confirmed in a study by Jiang et al. (2008) in which the combination effect of PDT with antibodies against VEGFR-1 and VEGFR-2 in a U87 glioblastoma model was evaluated. Combination treatment markedly reduced the tumor growth and extended mouse survival by more than both monotherapies (Jiang et al. 2008). Apart from the use of anti-angiogenic antibodies, synthetic angiogenic inhibitors are also attractive partners for combination with PDT. In a single study of its kind, post-PDT administration of a small molecule anti-angiogenic agent TNP-470 was found to decrease the VEGF secretion and tumor growth in an orthotopic prostate cancer model (Kosharskyy et al. 2006).

31.2.4.2 Regulators of Angiogenic Factors: COX-2

The cyclooxygenase-2 (COX-2) protein belongs to the cyclooxygenase (COX) family and is the key enzyme involved in the conversion of arachidonic acid to prostaglandins. COX-2 plays an important role in inflammation, mitogenesis, and angiogenesis and apoptosis. COX-2 has also been shown to regulate the expression of PGE-2, which directly regulates VEGF expression (Chang et al. 2004). An increasing number of studies have shown the role of COX-2 in tumor development and drug resistance (Liao, Mason, and Milas 2007). PDT using porphyrin and chlorin-based PS has shown to increase the expression of COX-2 and subsequent release of PGE-2 (Hendrickx et al. 2003). Therefore, therapies that block the PDT-mediated upregulation of COX-2 are believed to improve PDT therapeutic efficacy.

Ferrario et al. (2002) have shown the COX-2 gene overexpression in mouse mammary carcinoma (RIF) cells after PDT treatment. Using a selective COX-2 inhibitor NS-398, the responsiveness to PDT treatment was enhanced in RIF tumors (Ferrario et al. 2002). Similarly, Akita et al. (2004) tested the COX-2 expression in biopsy samples from the skin and oral mucosal lesions and demonstrated the synergistic effect *in vitro* by combining 5-aminolaevulinic acid (ALA)-based PDT with selective COX-2 inhibitor nimesulide. Makowski et al. (2003) has shown that a COX-2 inhibitor administered before PDT did not result in the potentiation of treatment in C-26 colon adenocarcinoma. Conversely, the COX-2 inhibitor was found to enhance PDT efficacy post tumor illumination, therefore demonstrating the importance of post-PDT COX-2 upregulation (Makowski et al. 2003). Moreover, this study demonstrated that the sequence of drug and treatment administration plays an important role in the synergistic outcome as only post-PDT COX-2 treatments potentiated the antitumor effect. Furthermore, the COX-2 inhibition resulting in promotion of PDT efficacy has been mechanistically argued by the subsequent downregulation of angiogenic and inflammatory factors (Ferrario et al. 2005). However, the appropriate dosing of these COX-2 inhibitors still needs to be optimized to reduce undesirable cardiovascular toxicity.

31.3 Further Improving PDT Combination Therapy

31.3.1 Choosing a Combination Target

Understanding the PDT mechanism of a given therapy regimen warrants the comprehension of the molecular targets involved in treatment response in order to design high-efficacy combination strategies. Most successful PDT combination approaches comprise PDT and a secondary treatment designed to increase the PDT efficacy by either reducing potentially detrimental molecular responses following PDT or increasing the susceptibility of tumor cells to PDT. Therefore, profiling the molecular barriers to PDT in target disease will provide an unbiased entry to assess candidate targets that influence PDT efficacy. Candidate targets identified in the initial molecular profiling results can

be concentrated and critical targets can be deduced and validated in experimental models. This process of unbiased profiling and subsequent validation may optimize the combination strategy because individual disease information becomes available, which allows room for therapy design to be tailored and optimized.

Another critical issue is that cancer is known to possess multiple deregulated signaling circuitries fueled by redundant targets present in cancer cells, ensuring survival against therapeutic interventions. It is therefore justifiable to target multiple candidate targets in combination with PDT so that therapy resistance resulting from redundant targets can be overcome. Taking the angiogenic factor family for example, more than a dozen soluble factors have been identified so far that play supportive roles in neovascularization and angiogenic response, and several (i.e., VEGF, HGF, and c-MET) have been implicated in PDT response. Currently available multitargeted therapeutics, such as XL184, were designed to simultaneously inhibit multiple angiogenic pathways (including VEGFR1-3, c-MET, etc.) at nanomolar concentrations, and treatment with these high-potency agents may potentially improve upon previous PDT combinations in which only one target was inhibited along with PDT. With a growing trend of developing multitargeted small-molecule therapeutics, the arsenal against cancer becomes diverse, therefore offering exciting new opportunities for novel PDT combination strategies.

31.3.2 Strategies for Delivering Combination Therapeutics

The success of any pharmacological agent is improved by the accurate, localized delivery of the drug preferentially to the targeted area. Treatment using PDT facilitates the localized activation of the cytotoxic effect of the PS only upon irradiation with light. However, multitargeted therapies used in conjunction with PDT demand further precision in drug delivery as two agents must now be localized both spatially and temporally within the window of activation for PDT (Agostinis et al. 2011). For combination therapies, the greatest combinatorial effect occurs only when the delivery of drug and light is achieved at the right place and the right time (Buytaert, Dewaele, and Agostinis 2007). As discussed earlier in this chapter, the administration and irradiation of a PS can induce a number of angiogenic and cell-survival pathways that can sometimes result in incomplete or suboptimal treatment outcomes. Consequently, many molecular targets have been designated as critical factors in these post-treatment responses (Buytaert, Dewaele, and Agostinis 2007). Therapeutic modulation of these molecular targets with appropriate molecular-targeted chemotherapeutics before, during, or after PDT has been shown to enhance tumor destruction and decrease metastatic growth. Inherently, there is a need for precise drug delivery of these multidrug therapies, especially *in vivo*, to increase effectiveness of their use in combination. Several approaches have been developed to enhance the delivery of combination therapies in tandem with PS, a few of which are discussed here.

The pharmacology of PS varies greatly depending on their size, charge, and lipophilicity (Reddi 1997). In some of the more clinically relevant PSs, which have absorptions at longer wavelengths within the PDT activation window, extended π-conjugation contributes to largely hydrophobic, lipophilic PS moieties (Konan, Gurny, and Allemann 2002). This intermolecular property has a paradoxical role in PS pharmacology as the increased hydrophobicity increases PS localization with tumor tissue while, at the same time, resulting in poor clearance kinetics, producing prolonged periods of light sensitivity post-administration (Konan, Gurny, and Allemann 2002). For this reason, many hydrophobic PSs, including formulations of FDA-approved benzoporphyrin derivative (Visudyne), have been formulated using carrier systems (i.e., lipids, lipoproteins, dendrimers), which increase the localization of PSs into the target tissue while improving water solubility and pharmacokinetics (Jones, Vernon, and Brown 2003; Schmidt-Erfurth and Hasan 2000).

Recently, nanotechnology has begun to play an increasing role in drug formulation and represents an ideal platform for designing rational combination therapies (Al-Jamal and Kostarelos 2011; Drummond et al. 2008, 2009, 2010; McCarthy et al. 2005; Sengupta et al. 2005) (also see Sections 33.4 and 34.4). By utilizing the enhanced permeability and retention (EPR) effect, these new nanomaterials have the capability of uniting individual drug pharmacokinetic profiles while allowing for synergistic drug ratios to be obtained locally through designer packaging of the payload (Al-Jamal and Kostarelos 2011; Drummond et al. 2008, 2009, 2010; McCarthy et al. 2005; Sengupta et al. 2005). Although few published cases of formulated PS and molecular targeted therapies exist, several groups have dedicated significant effort to developing methods to use existing drug-delivery platforms to codeliver these drugs (Chatterjee, Fong, and Zhang 2008). These formulations will surely challenge the future of conventional administration of drug "cocktails" in PDT.

To this end, lipid- and polymer-based materials originally designed as monotherapy vehicles are being reassessed for their ability to facilitate the codelivery of multiple agents within a single construct (Al-Jamal and Kostarelos 2011; Boccaccio and Comoglio 2006; Cooke et al. 2012; Ohnishi and Daikuhara 2003; Sengupta et al. 2005). For example, coblock polymers of polylactic/glycolic acid (PLGA) can be used to encapsulate a broad range of both hydrophilic and hydrophobic small molecules and can be formulated into nanosized materials capable of preferentially delivering their cargo to tumor tissue (Chatterjee, Fong, and Zhang 2008; Konan et al. 2003a,b). In addition, these biodegradable polymers exhibit unique pharmacokinetics, and by changing polymer size, composition, and pegylation, tunable release kinetics can be achieved for the loaded drug cocktail (Chatterjee, Fong, and Zhang 2008; Konan et al. 2003a,b).

Several advancements in liposomal drug delivery have also proven useful to enhance the codelivery of combination therapies (Al-Jamal and Kostarelos 2011; Cinteza et al. 2006; Drummond et al. 2008, 2009, 2010; Lovell et al. 2011; Yavlovich et al. 2010). The chemical architecture of a liposome provides a unique handle for the coencapsulation of a variety of drug and PS candidates (Al-Jamal and Kostarelos 2011; Cinteza et al. 2006; Drummond et al. 2008, 2009, 2010; Lovell et al. 2011; Yavlovich et al. 2010). The liposome, composed of a hydrophilic surface, hydrophobic lipid bilayer, and aqueous core, is capable of encapsulating drugs with a range of intermolecular properties (Al-Jamal and Kostarelos 2011; Cinteza et al. 2006; Drummond et al. 2008, 2009, 2010; Lovell et al. 2011; Yavlovich et al. 2010). The polar head groups of lipids on the liposome surface can be reacted with drug moieties, forming covalent chemical conjugates for drug delivery. Lovell et al. 2011 have reported such conjugates referred to as porphysomes, which are achieved through acylation reactions of a pyropheophorbide-porphyrin PSs to lysophosphatidylcholine and spontaneous self-assembly into liposomal-like structures (Lovell et al. 2011). These conjugates were found to improve pharmacokinetics in *in vivo* models with increased PS intratumoral delivery (Lovell et al. 2011). Additionally, the covalently linked PS was released via an enzymatic cleavage of the PS-lipid linkage (Lovell et al. 2011). In addition, the lipid bilayer has been shown to be ideal for delivery of highly hydrophobic PSs (such as Verteporfin, Foscan, and Photofrin) and improves circulation and drug-dosing limitations resulting from improved solubility (Schmidt-Erfurth and Hasan 2000). Lastly, by using pH gradient titration loading techniques, liposomes can be loaded with high concentrations of drugs forming nanosized crystals in the aqueous core environment (Drummond et al. 2010). The most successful example of this is the FDA-approved drug Doxil, a liposomal formulation of the anticancer drug doxorubicin (Drummond et al. 2008).

Monoclonal antibodies (MAbs) to receptors overexpressed in tumorigenic tissues have also been highlighted as a viable and successful means of site-directed delivery of therapeutic agents. Specifically, PS antibody conjugates (commonly referred to as photoimmunoconjugates) have been successfully employed for the preferential delivery of PSs to cancers (Abu-Yousif et al. 2012; Kuimova et al. 2007; Savellano and Hasan 2003) (also see Section 31.4). In addition, these monoclonal therapies also elicit their own inherent therapeutic response, which can enhance the treatment response of PDT. Hasan et al. have shown conjugates of modified BPD-MA with the anti-EGFR conjugated to chimerical MAb C225 (Erbitux) to increase the selectivity of BPD delivery over free PS (Abu-Yousif et al. 2012; Savellano and Hasan 2003). Similarly, anti-HER2 single-chain fragments conjugated to either pyropheophorbide or vertoporfin derivatives have been shown to localize in cells overexpressing HER2 leading to preferential accumulation of the PS and increased selectivity (Kuimova et al. 2007).

31.4 Conclusions

PDT has demonstrated promise in the treatment of cancer and noncancer diseases even in cases in which conventional treatments fail. In fact, the full potential of this modality has not been explored in cancers because it is typically used in patients with advanced disease, often when the remaining cancer is resistant to other modalities such as chemotherapy and radiation therapy. It can be safely hypothesized that it would be most successful as a local therapy when the lesions are small (allowing for optimal light penetration) in nonmetastatic disease. Evidence for this comes from the noncancer and precancer applications, such as the treatment of AMD and actinic keratosis. However, even beyond this, the potential of PDT lies in mechanism-based, targeted combination therapies. As with all cancer modalities, PDT elicits molecular responses that, if not negated, could limit its full effectiveness. Studies from several groups, including ours (Gomer et al. 2006; Kosharskyy et al. 2006; Solban et al. 2006), have shown that PDT synergizes with conventional therapies and sensitizes tumor cells to several biological and chemotherapy regimens. In a study using cells from cancer patients, it was demonstrated that PDT re-sensitized chemoresistant cells to chemotherapy (Duska et al. 1999). In summary, no single therapy is likely to be effective against cancer as a curative modality and, similarly, PDT too will be most effective as part of an intelligently chosen arsenal of multimodality therapies.

Acknowledgment

Our studies have been supported by the National Cancer Institute Grants 5P01CA084203, 5RC1CA146337, and 1R01CA158415.

References

Abu-Yousif, A. O., A. C. Moor, X. Zheng et al. 2012. Epidermal growth factor receptor-targeted photosensitizer selectively inhibits EGFR signaling and induces targeted phototoxicity in ovarian cancer cells. *Cancer Lett* 321:120–127.

Agostinis, P., K. Berg, K. A. Cengel et al. 2011. Photodynamic therapy of cancer: An update. *CA Cancer J Clin* 61:250–281.

Ahmad, N., K. Kalka, and H. Mukhtar. 2001. *In vitro* and *in vivo* inhibition of epidermal growth factor receptor-tyrosine kinase pathway by photodynamic therapy. *Oncogene* 20:2314–2317.

Akita, Y., K. Kozaki, A. Nakagawa et al. 2004. Cyclooxygenase-2 is a possible target of treatment approach in conjunction with photodynamic therapy for various disorders in skin and oral cavity. *Br J Dermatol* 151:472–480.

Al-Jamal, W. T., and K. Kostarelos. 2011. Liposomes: From a clinically established drug delivery system to a nanoparticle platform for theranostic nanomedicine. *Acc Chem Res* 44:1094–1104.

Almeida Issa, M. C., J. Pineiro-Maceira, R. E. Farias et al. 2009. Immunohistochemical expression of matrix metalloproteinases in photodamaged skin by photodynamic therapy. *Br J Dermatol* 161:647–653.

Almeida, R. D., B. J. Manadas, A. P. Carvalho, and C. B. Duarte. 2004. Intracellular signaling mechanisms in photodynamic therapy. *Biochim Biophys Acta* 1704:59–86.

Andrzejak, M., M. Price, and D. H. Kessel. 2011. Apoptotic and autophagic responses to photodynamic therapy in 1c1c7 murine hepatoma cells. *Autophagy* 7:979–984.

Bhuvaneswari, R., Y. Y. Gan, S. S. Lucky et al. 2008. Molecular profiling of angiogenesis in hypericin mediated photodynamic therapy. *Mol Cancer* 7:56.

Bhuvaneswari, R., Y. Y. Gan, K. C. Soo, and M. Olivo. 2009a. The effect of photodynamic therapy on tumor angiogenesis. *Cell Mol Life Sci* 66:2275–2283.

Bhuvaneswari, R., Y. Y. Gan, K. C. Soo, and M. Olivo. 2009b. Targeting EGFR with photodynamic therapy in combination with Erbitux enhances *in vivo* bladder tumor response. *Mol Cancer* 8:94.

Bhuvaneswari, R., G. Y. Yuen, S. K. Chee, and M. Olivo. 2007. Hypericin-mediated photodynamic therapy in combination with Avastin (bevacizumab) improves tumor response by downregulating angiogenic proteins. *Photochem Photobiol Sci* 6:1275–1283.

Bhuvaneswari, R., G. Y. Yuen, S. K. Chee, and M. Olivo. 2011. Antiangiogenesis agents avastin and erbitux enhance the efficacy of photodynamic therapy in a murine bladder tumor model. *Lasers Surg Med* 43:651–662.

Boccaccio, C., and P. M. Comoglio. 2006. Invasive growth: A MET-driven genetic programme for cancer and stem cells. *Nat Rev Cancer* 6:637–645.

Brackett, C. M., and S. O. Gollnick. 2011. Photodynamic therapy enhancement of anti-tumor immunity. *Photochem Photobiol Sci* 10:649–652.

Bukau, B., J. Weissman, and A. Horwich. 2006. Molecular chaperones and protein quality control. *Cell* 125:443–451.

Buytaert, E., G. Callewaert, J. R. Vandenheede, and P. Agostinis. 2006. Deficiency in apoptotic effectors Bax and Bak reveals an autophagic cell death pathway initiated by photodamage to the endoplasmic reticulum. *Autophagy* 2:238–240.

Buytaert, E., M. Dewaele, and P. Agostinis. 2007. Molecular effectors of multiple cell death pathways initiated by photodynamic therapy. *Biochim Biophys Acta* 1776:86–107.

Castano, A. P., P. Mroz, and M. R. Hamblin. 2006. Photodynamic therapy and anti-tumour immunity. *Nat Rev Cancer* 6:535–545.

Chang, S.H., C. H. Liu, R. Conway et al. 2004. Role of prostaglandin E2-dependent angiogenic switch in cyclooxygenase 2-induced breast cancer progression. *Proc Natl Acad Sci USA* 101:591–596.

Chatterjee, D. K., L. S. Fong, and Y. Zhang. 2008. Nanoparticles in photodynamic therapy: An emerging paradigm. *Adv Drug Deliv Rev* 60:1627–1637.

Chen, B., B. W. Pogue, P. J. Hoopes, and T. Hasan. 2006. Vascular and cellular targeting for photodynamic therapy. *Crit Rev Eukaryot Gene Expr* 16:279–305.

Cinteza, L. O., T. Y. Ohulchanskyy, Y. Sahoo et al. 2006. Diacyllipid micelle-based nanocarrier for magnetically guided delivery of drugs in photodynamic therapy. *Mol Pharm* 3:415–423.

Cogno, I. S., N. B. Vittar, M. J. Lamberti, and V. A. Rivarola. 2011. Optimization of photodynamic therapy response by survivin gene knockdown in human metastatic breast cancer T47D cells. *J Photochem Photobiol B* 104:434–443.

Cooke, V. G., V. S. LeBleu, D. Keskin et al. 2012. Pericyte depletion results in hypoxia-associated epithelial-to-mesenchymal transition and metastasis mediated by met signaling pathway. *Cancer Cell* 21:66–81.

Cory, S., and J. M. Adams. 2002. The Bcl2 family: Regulators of the cellular life-or-death switch. *Nat Rev Cancer* 2:647–656.

Coupienne, I., G. Fettweis, N. Rubio, P. Agostinis, and J. Piette. 2011. 5-ALA-PDT induces RIP3-dependent necrosis in glioblastoma. *Photochem Photobiol Sci* 10:1868–1878.

del Carmen, M. G., I. Rizvi, Y. Chang et al. 2005. Synergism of epidermal growth factor receptor-targeted immunotherapy with photodynamic treatment of ovarian cancer *in vivo*. *J Natl Cancer Inst* 97:1516–1524.

Drummond, D. C., C. O. Noble, Z. Guo et al. 2009. Improved pharmacokinetics and efficacy of a highly stable nanoliposomal vinorelbine. *J Pharmacol Exp Ther* 328:321–330.

Drummond, D. C., C. O. Noble, Z. Guo et al. 2010. Development of a highly stable and targetable nanoliposomal formulation of topotecan. *J Control Release* 141:13–21.

Drummond, D., C. Noble, M. Hayes, J. Park, and D. Kirpotin. 2008. Pharmacokinetics and *in vivo* drug release rates in liposomal nanocarrier development. *J Pharm Sci* 97:4696–4740.

Du, H., M. Olivo, R. Mahendran, and B. H. Bay. 2004. Modulation of Matrix metalloproteinase-1 in nasopharyngeal cancer cells by photoactivation of hypericin. *Int J Oncol* 24:657–662.

Du, H. Y., M. Olivo, R. Mahendran et al. 2007. Hypericin photoactivation triggers down-regulation of matrix metalloproteinase-9 expression in well-differentiated human nasopharyngeal cancer cells. *Cell Mol Life Sci* 64:979–988.

Duska, L. R., M. R. Hamblin, J. L. Miller, and T. Hasan. 1999. Combination photoimmunotherapy and cisplatin: Effects on human ovarian cancer *ex vivo*. *J Natl Cancer Inst* 91:1557–1563.

Fanuel-Barret, D., T. Patrice, M. T. Foultier et al. 1997. Influence of epidermal growth factor on photodynamic therapy of glioblastoma cells *in vitro*. *Res Exp Med (Berl)* 197:219–233.

Ferenc, P., P. Solar, J. Kleban, J. Mikes, and P. Fedorocko. 2010. Down-regulation of Bcl-2 and Akt induced by combination of photoactivated hypericin and genistein in human breast cancer cells. *J Photochem Photobiol B* 98:25–34.

Ferrario, A., C. F. Chantrain, K. von Tiehl et al. 2004. The matrix metalloproteinase inhibitor prinomastat enhances photodynamic therapy responsiveness in a mouse tumor model. *Cancer Res* 64:2328–2332.

Ferrario, A., A. M. Fisher, N. Rucker, and C. J. Gomer. 2005. Celecoxib and NS-398 enhance photodynamic therapy by increasing *in vitro* apoptosis and decreasing *in vivo* inflammatory and angiogenic factors. *Cancer Res* 65:9473–9478.

Ferrario, A., and C. J. Gomer. 2006. Avastin enhances photodynamic therapy treatment of Kaposi's sarcoma in a mouse tumor model. *J Environ Pathol Toxicol Oncol* 25:251–259.

Ferrario, A., and C. J. Gomer. 2010. Targeting the 90 kDa heat shock protein improves photodynamic therapy. *Cancer Lett* 289:188–194.

Ferrario, A., N. Rucker, S. Wong, M. Luna, and C. J. Gomer. 2007. Survivin, a member of the inhibitor of apoptosis family, is induced by photodynamic therapy and is a target for improving treatment response. *Cancer Res* 67:4989–4995.

Ferrario, A., K. F. von Tiehl, N. Rucker et al. 2000. Antiangiogenic treatment enhances photodynamic therapy responsiveness in a mouse mammary carcinoma. *Cancer Res* 60:4066–4069.

Ferrario, A., K. von Tiehl, S. Wong, M. Luna, and C. J. Gomer. 2002. Cyclooxygenase-2 inhibitor treatment enhances photodynamic therapy-mediated tumor response. *Cancer Res* 62:3956–3961.

Gomer, C. J., A. Ferrario, M. Luna, N. Rucker, and S. Wong. 2006. Photodynamic therapy: Combined modality approaches targeting the tumor microenvironment. *Lasers Surg Med* 38:516–521.

Gomer, C. J., S. W. Ryter, A. Ferrario et al. 1996. Photodynamic therapy-mediated oxidative stress can induce expression of heat shock proteins. *Cancer Res* 56:2355–2360.

Hanlon, J. G., K. Adams, A. J. Rainbow, R. S. Gupta, and G. Singh. 2001. Induction of Hsp60 by photofrin-mediated photodynamic therapy. *J Photochem Photobiol B* 64:55–61.

Hendrickx, N., C. Volanti, U. Moens et al. 2003. Up-regulation of cyclooxygenase-2 and apoptosis resistance by p38 MAPK in hypericin-mediated photodynamic therapy of human cancer cells. *J Biol Chem* 278:52231–52239.

Jiang, F., X. Zhang, S. N. Kalkanis et al. 2008. Combination therapy with antiangiogenic treatment and photodynamic therapy for the nude mouse bearing U87 glioblastoma. *Photochem Photobiol* 84:128–137.

Jones, H. J., D. I. Vernon, and S. B. Brown. 2003. Photodynamic therapy effect of m-THPC (Foscan) *in vivo*: Correlation with pharmacokinetics. *Br J Cancer* 89:398–404.

Kessel, D. 2008. Promotion of PDT efficacy by a Bcl-2 antagonist. *Photochem Photobiol* 84:809–814.

Kessel, D., M. G. Vicente, and J. J. Reiners, Jr. 2006. Initiation of apoptosis and autophagy by photodynamic therapy. *Lasers Surg Med* 38:482–488.

Kessenbrock, K., V. Plaks, and Z. Werb. 2010. Matrix metalloproteinases: Regulators of the tumor microenvironment. *Cell* 141:52–67.

Konan, Y. N., R. Cerny, J. Favet et al. 2003a. Preparation and characterization of sterile sub-200 nm meso-tetra(4-hydroxylphenyl)porphyrin-loaded nanoparticles for photodynamic therapy. *Eur J Pharm Biopharm* 55:115–124.

Konan, Y. N., J. Chevallier, R. Gurny, and E. Allemann. 2003b. Encapsulation of p-THPP into nanoparticles: Cellular uptake, subcellular localization and effect of serum on photodynamic activity. *Photochem Photobiol* 77:638–644.

Konan, Y. N., R. Gurny, and E. Allemann. 2002. State of the art in the delivery of photosensitizers for photodynamic therapy. *J Photochem Photobiol B* 66:89–106.

Kosharskyy, B., N. Solban, S. K. Chang et al. 2006. A mechanism-based combination therapy reduces local tumor growth and metastasis in an orthotopic model of prostate cancer. *Cancer Res* 66:10953–10958.

Kuimova, M. K., M. Bhatti, M. Deonarain et al. 2007. Fluorescence characterisation of multiply-loaded anti-HER2 single chain Fv-photosensitizer conjugates suitable for photodynamic therapy. *Photochem Photobiol Sci* 6:933–939.

Levine, B., and G. Kroemer. 2008. Autophagy in the pathogenesis of disease. *Cell* 132:27–42.

Liao, Z., K. A. Mason, and L. Milas. 2007. Cyclo-oxygenase-2 and its inhibition in cancer: Is there a role? *Drugs* 67:821–845.

Lovell, J. F., C. S. Jin, E. Huynh et al. 2011. Porphysome nanovesicles generated by porphyrin bilayers for use as multimodal biophotonic contrast agents. *Nat Mater* 10:324–332.

Makowski, M., T. Grzela, J. Niderla et al. 2003. Inhibition of cyclo-oxygenase-2 indirectly potentiates antitumor effects of photodynamic therapy in mice. *Clin Cancer Res* 9:5417–5422.

McCarthy, J. R., J. M. Perez, C. Bruckner, and R. Weissleder. 2005. Polymeric nanoparticle preparation that eradicates tumors. *Nano Lett* 5:2552–2556.

Michels, S., and U. Schmidt-Erfurth. 2001. Photodynamic therapy with verteporfin: A new treatment in ophthalmology. *Semin Ophthalmol* 16:201–206.

Mizushima, N., and M. Komatsu. 2011. Autophagy: Renovation of cells and tissues. *Cell* 147:728–741.

Moor, A. 2000. Signaling pathways in cell death and survival after photodynamic therapy. *J Photochem Photobiol B* 57:1–13.

Ohnishi, T., and Y. Daikuhara. 2003. Hepatocyte growth factor/scatter factor in development, inflammation and carcinogenesis: Its expression and role in oral tissues. *Arch Oral Biol* 48:797–804.

Oleinick, N. L., R. L. Morris, and I. Belichenko. 2002. The role of apoptosis in response to photodynamic therapy: What, where, why, and how. *Photochem Photobiol Sci* 1:1–21.

Osiecka, B., K. Jurczyszyn, K. Symonowicz et al. 2010. *In vitro* and *in vivo* matrix metalloproteinase expression after photodynamic therapy with a liposomal formulation of aminolevulinic acid and its methyl ester. *Cell Mol Biol Lett* 15: 630–650.

Peng, Q., and J. M. Nesland. 2004. Effects of photodynamic therapy on tumor stroma. *Ultrastruct Pathol* 28:333–340.

Reddi, E. 1997. Role of delivery vehicles for photosensitizers in the photodynamic therapy of tumours. *J Photochem Photobiol B* 37:189–195.

Rizvi, I., J. P. Celli, C. L. Evans et al. 2010. Synergistic enhancement of carboplatin efficacy with photodynamic therapy in a three-dimensional model for micrometastatic ovarian cancer. *Cancer Res* 70:9319–9328.

Roberts, W. G., and T. Hasan. 1993. Tumor-secreted vascular permeability factor/vascular endothelial growth factor influences photosensitizer uptake. *Cancer Res* 53:153–157.

Robertson, C. A., D. H. Evans, and H. Abrahamse. 2009. Photodynamic therapy (PDT): A short review on cellular mechanisms and cancer research applications for PDT. *J Photochem Photobiol B* 96:1–8.

Savellano, M. D., and T. Hasan. 2003. Targeting cells that overexpress the epidermal growth factor receptor with poly-ethylene glycolated BPD verteporfin photosensitizer immunoconjugates. *Photochem Photobiol* 77:431–439.

Schmidt-Erfurth, U., and T. Hasan. 2000. Mechanisms of action of photodynamic therapy with verteporfin for the treatment of age-related macular degeneration. *Surv Ophthalmol* 45:195–214.

Sengupta, S., D. Eavarone, I. Capila et al. 2005. Temporal targeting of tumour cells and neovasculature with a nanoscale delivery system. *Nature* 436:568–572.

Shen, X. Y., N. Zacal, G. Singh, and A. J. Rainbow. 2005. Alterations in mitochondrial and apoptosis-regulating gene expression in photodynamic therapy-resistant variants of HT29 colon carcinoma cells. *Photochem Photobiol* 81:306–313.

Solban, N., P. K. Selbo, A. K. Sinha, S. K. Chang, and T. Hasan. 2006. Mechanistic investigation and implications of photodynamic therapy induction of vascular endothelial growth factor in prostate cancer. *Cancer Res* 66:5633–5640.

Takahashi, H., S. Komatsu, M. Ibe et al. 2006. ATX-S10(Na)-PDT shows more potent effect on collagen metabolism of human normal and scleroderma dermal fibroblasts than ALA-PDT. *Arch Dermatol Res* 298:257–263.

Takeuchi, K., and F. Ito. 2010. EGF receptor in relation to tumor development: Molecular basis of responsiveness of cancer cells to EGFR-targeting tyrosine kinase inhibitors. *FEBS J* 277:316–326.

Usuda, J., K. Azizuddin, S. M. Chiu, and N. L. Oleinick. 2003. Association between the photodynamic loss of Bcl-2 and the sensitivity to apoptosis caused by phthalocyanine photodynamic therapy. *Photochem Photobiol* 78:1–8.

Vantieghem, A., Y. Xu, Z. Assefa et al. 2002. Phosphorylation of Bcl-2 in G2/M phase-arrested cells following photodynamic therapy with hypericin involves a CDK1-mediated signal and delays the onset of apoptosis. *J Biol Chem* 277:37718–37731.

Verma, S., G. M. Watt, Z. Mai, and T. Hasan. 2007. Strategies for enhanced photodynamic therapy effects. *Photochem Photobiol* 83:996–1005.

Wang, H. P., J. G. Hanlon, A. J. Rainbow, M. Espiritu, and G. Singh. 2002. Up-regulation of Hsp27 plays a role in the resistance of human colon carcinoma HT29 cells to photooxidative stress. *Photochem Photobiol* 76:98–104.

Weickhardt, A. J., N. C. Tebbutt, and J. M. Mariadason. 2010. Strategies for overcoming inherent and acquired resistance to EGFR inhibitors by targeting downstream effectors in the RAS/PI3K pathway. *Curr Cancer Drug Targets* 10:824–833.

Wilson, B. C., and M. S. Patterson. 2008. The physics, biophysics and technology of photodynamic therapy. *Phys Med Biol* 53:R61–R109.

Wong, T. W., E. Tracy, A. R. Oseroff, and H. Baumann. 2003. Photodynamic therapy mediates immediate loss of cellular responsiveness to cytokines and growth factors. *Cancer Res* 63:3812–3818.

Xue, L. Y., S. M. Chiu, K. Azizuddin, S. Joseph, and N. L. Oleinick. 2007. The death of human cancer cells following photodynamic therapy: Apoptosis competence is necessary for Bcl-2 protection but not for induction of autophagy. *Photochem Photobiol* 83:1016–1023.

Xue, L. Y., S. M. Chiu, and N. L. Oleinick. 2001. Photochemical destruction of the Bcl-2 oncoprotein during photodynamic therapy with the phthalocyanine photosensitizer Pc 4. *Oncogene* 20:3420–3427.

Yavlovich, A., B. Smith, K. Gupta, R. Blumenthal, and A. Puri. 2010. Light-sensitive lipid-based nanoparticles for drug delivery: Design principles and future considerations for biological applications. *Mol Membr Biol* 27: 364–381.

Nanoparticles for Photodynamic Cancer Therapy

Girgis Obaid
University of East Anglia

David A. Russell
University of East Anglia

32.1 Introduction to Nanotechnology for Photodynamic Therapy

There has been a dramatic rise in the development of technologies using nanoparticles as a result of their intriguing physical, chemical, and optical properties. Nanoparticles are defined as particles with diameters below 100 nm. On the nanoscale, materials can possess characteristics that differ from their bulk state. From photocatalysis and photovoltaics to photodiagnosis and phototherapy, the interaction of light with nanoparticles is an area of research that is proving to be increasingly important. In addition to the advantageous nanometer size, the highly tunable size, shape, surface area-to-volume ratio, and photochemical or photophysical properties of nanoparticles make them attractive candidates for multimodal biological applications. Such applications include drug delivery, bioimaging, phototherapy, as well as various other clinical diagnostic and therapeutic applications (Burda et al. 2005). The focus of this chapter is the use of nanoparticles for the photodynamic therapy (PDT) of cancer.

Growing tumors develop an increasing demand for blood that is met by rapidly forming vascularization. The newly formed vessels are of poor quality and are more "leaky" than healthy vasculature, thus giving rise to a phenomenon referred to as the enhanced permeability and retention (EPR) effect (Maeda et al. 2003). Given sufficient systemic circulation time, nanoparticles accumulate inside tumors through the EPR effect, making them of particular interest for delivery of anticancer agents (Paciotti, Kingston, and Tamarkin 2006).

In addition to their small size, nanoparticles can serve as multifunctional platforms for therapy, allowing the combination of drug delivery and cancer specificity through the conjugation of biological targeting ligands. Nanoparticles have drawn particular interest in PDT for their ability to stabilize hydrophobic photosensitizers in aqueous environments and to preserve the sensitizers' photoactive monomeric form. Some nanoparticles also possess additional intrinsic photophysical characteristics, which include the capacity to directly produce reactive oxygen species (ROS) and the ability to enhance bioimaging (Bechet et al. 2008). In this chapter, the potential application of nanoparticles for PDT-based cancer therapy will be highlighted. The synthesis, biocompatibility, functionalization, and phototoxicity of such nanoparticle systems will be outlined, and their potential for clinical use for effective cancer treatment will also be discussed based on *in vitro* and *in vivo* experiments.

32.2 Types of Nanoparticles Used for PDT

32.2.1 Gold Nanoparticles

Gold nanostructures are of particular interest in therapies such as PDT because of their biocompatibility, stability, size control, and ease of surface functionality (Daniel and Astruc 2004). The nature of the covalent gold–sulfur bond has been exploited with the self-assembly of thiolated photosensitizer molecules onto the surface of gold nanoparticles for PDT (Hone et al. 2002). Figure 32.1 shows a gold nanoparticle with a photosensitizer supported on the surface. Excitation of the photosensitizer attached to the gold nanoparticle results in the production of singlet oxygen and/or other ROS. The gold core can be used also for plasmonic imaging or hyperthermal treatment.

The Russell group was the first to use gold nanoparticles for supporting a zinc phthalocyanine (Pc) photosensitizer via a C11 mercaptoalkyl tether (Hone et al. 2002). Reduction of gold

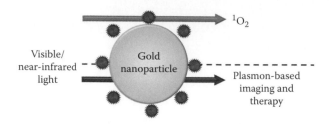

FIGURE 32.1 Schematic diagram of gold nanoparticles modified with photosensitizer. Top: Irradiation of photosensitizer with light induces production of singlet oxygen. Bottom: Irradiation of gold core with light can be used for surface plasmon-based imaging and hyperthermal therapy.

chloride by sodium borohydride in the presence of the phase transfer reagent, tetraoctylammonium bromide (TOAB), and the thiolated Pc yielded stable nanoparticles with a diameter of 2–4 nm. The TOAB not only provided solubility for the hydrophobic system in polar solvents but also enhanced the singlet oxygen quantum yield. The same Pc-gold nanoparticle system was shown to enhance the PDT effect on HeLa cervical adenocarcinoma cells as compared to an emulsion of the free photosensitizer (Wieder et al. 2006). *In vivo* studies further demonstrated the ability of these Pc gold nanoparticle conjugates to inhibit the growth of an implanted amelanotic melanoma following irradiation and showed that irradiation of the nanoparticles was effective against the tumors 3 h following intravenous administration in mice (Camerin et al. 2010). Gold nanoparticles have been synthesized carrying the same Pc photosensitizer together with polyethylene glycol (PEG) (Stuchinskaya et al. 2011). PEG provided the nanoparticle system with aqueous solubility and colloidal stability in addition to providing terminal carboxyl moieties that allowed conjugation of biomolecules. By conjugating anti-HER-2 antibodies to the gold nanoparticles through PEG, it was shown that the nanoparticles could actively target human breast cancer cells overexpressing the HER-2 receptor for enhanced cancer-specific PDT. The cancer-associated carbohydrate T antigen present at the surface of HT-29 human colon adenocarcinoma cells has been targeted using the lectin Jacalin conjugated to PEG-Pc gold nanoparticles (Obaid et al. 2012). Jacalin conjugation to the gold nanoparticles induced substantial targeted phototoxicity (95–98%) of the HT-29 colon cells.

The fate of nanoparticles of various compositions and morphologies has been the focus of much speculation in recent years. *In vivo* biodistribution, pharmacokinetics, and excretion of 5 nm PEG modified gold nanoparticles carrying a hydrophobic silicon phthalocyanine photosensitizer (Pc4) have been recently studied in a mouse tumor model by the Burda group (Cheng et al. 2011). The photosensitizer was noncovalently associated with the nanoparticle surface via hydrophobic interactions along with a covalently bound aliphatic PEG monolayer. It is known that PEG prolongs systemic nanoparticle circulation time (Paciotti, Kingston, and Tamarkin 2006). The nanoparticles caused maximal accumulation of the photosensitizer deep within the tumor 4 h following intravenous tail vein administration. Even though

the nanoparticles were not intracellularly internalized, they efficiently delivered the hydrophobic Pc4 to the cytoplasm of the cells. Clearance of the photosensitizer was monitored by fluorescence imaging and was found to involve the kidneys and the hepatobiliary system. The nanoparticles were also found to have a circulation half-life of 3 h, avoiding rapid clearance by the reticuloendothelial system (RES). Clearance of nanoparticles from the blood by the RES limits the potential of nanoparticles for accumulating at the site of the tumor via the EPR effect, leading to a rapid buildup of nanoparticles in the liver and spleen, both major organs involved in the RES (Paciotti, Kingston, and Tamarkin 2006). Although gold nanoparticles were found to accumulate in the liver and spleen throughout the 7 days following administration, the nanoparticles that were observed in the kidneys at 4 h after administration continuously decreased over the 7 days studied. This suggested that the gold nanoparticles could likely be excreted from the body through renal clearance.

Two conjugates composed of a porphyrin and brucine have been synthesized and conjugated to 3-mercaptopropionic acid decorated gold nanoparticles approximately 15 nm in diameter (Zaruba et al. 2010). These nanoparticle vehicles showed no evidence of enhanced *in vitro* PDT efficacy on PE/CA-PJ34 squamous cell carcinoma cells as compared to the free photosensitizer, possibly because of nanoparticle aggregation. Conversely, the nanoparticles were shown to completely eradicate subcutaneous tumors *in vivo* for up to 30 days post-PDT. This was thought to be a result of the interaction of the nanoparticle conjugates with serum proteins, improving their accumulation at the site of the tumor. Although this system appears to have potent antitumor properties, the *in vivo* stability of the nanoparticles should be further investigated.

The PDT efficacy of gold nanoparticles functionalized with the widely used photosensitizer hematoporphyrin (HP) was studied *in vitro* to compare particles of 15 and 45 nm in diameter (Gamaleia et al. 2010). The citrate-capped nanoparticles (15 and 45 nm) were stabilized with a layer of polyvinylpyrrolidone (PVP), and HP was then adsorbed within the polymer. It was found that upon irradiation, these gold-photosensitizer conjugates produced significantly more ROS than free HP, suggested to be a result of the catalytic effect of the gold nanoparticles. *In vitro* application of these nanoparticle conjugates on MT4 and Jurkat leukemia cells showed that the 45 nm gold nanoparticle conjugates were the most effective at photodynamic destruction.

Nanoparticles possessing optimal absorption characteristics have been used in combination with photosensitizers for dual photothermal–photodynamic therapies. An aluminum phthalocyanine photosensitizer and PEG were conjugated to gold nanorods for *in vitro* and *in vivo* photodynamic, photothermal, and combined photodynamic–photothermal therapy (Jang et al. 2011). The fluorescence of the phthalocyanine was used to monitor both the intracellular uptake and the biodistribution of the free and bound phthalocyanine. The nanorods proved to enhance the cellular uptake of the photosensitizer and to improve the physiological delivery of the photosensitizer to the tumor. *In vivo* studies of a subcutaneously implanted SCC7 mouse tumor model indicated that dual photodynamic–photo-thermal therapy was significantly more effective at inhibiting tumor growth than either therapy alone.

The surface plasmon absorption characteristics of gold nanoparticles also have been used for surface enhanced Raman scattering (SERS)-based detection of cancer cells in conjunction with PDT for theranostic (combined therapy and diagnostics) applications in cancer treatment. Silica-coated, star-shaped gold nanoparticles with a surface plasmon absorption band centered at approximately 900 nm have been synthesized for use in dual SERS-PDT theranostics (Fales, Yuan, and Vo-Dinh 2011). The silica shell was co-doped with a SERS dye to enhance the Raman scattering signal and the photosensitizer methylene blue (MB) for singlet oxygen generation. The nanoparticle conjugates were shown to induce significant cell death of BT549 human breast carcinoma cells following irradiation.

Gold has appealing physical and optical qualities that are both appropriate for chemical modification with sensitizers and effective for clinical use. Both *in vivo* and *in vitro* studies have shown the suitability of gold nanoparticles for stable delivery systems of PDT drugs. Further studies into the physiological fate of such gold nanoparticles are needed to assess the extent and duration of the nanoparticle accumulation in organs, such as the liver and spleen.

32.2.2 Silica Nanoparticles

Silica is a robust material with tunable chemical properties. Its versatile polymerization has meant that silica is frequently used for the stabilization of photosensitizers and to functionalize various types of nanoparticles for biomedical applications (Couleaud et al. 2010). A wide range of photosensitizers has been entrapped within silica nanoparticles. This method has proven to be effective in preserving the activity of hydrophobic photosensitizers in aqueous environments. Organically modified silica (ORMOSIL)-based nanoparticles have seen widespread use resulting from the flexible hydrophobic/hydrophilic properties, which can be tuned in accordance with the degree of hydrophobicity of the photosensitizer (Couleaud et al. 2010). In addition to physical entrapment of photosensitizers within a silica nanoparticle, some photosensitizing drugs have been incorporated within nanoparticles by covalently coupling to monomeric precursors of silica. Following condensation of the silica, the nanoparticles contain covalently integrated photosensitizers with a regulated sensitizer content (Couleaud et al. 2010). Figure 32.2 shows a schematic representation of silica nanoparticles doped with a photosensitizer.

The Prasad group was the first to encapsulate a photosensitizer in silica nanoparticles to yield stable nanoparticulates dispersed in an aqueous medium (Roy et al. 2003). The ORMOSIL nanoparticles were synthesized by loading the photosensitizer 2-devinyl-2-(1-hexyloxyethyl)pyropheophorbide (HPPH) into the hydrophobic core of micelles along with the silica precursor, triethoxyvinylsilane, and the aminoalkyl silica precursor, 3-aminopropyltriethoxysilane (APTES). Complete condensation

FIGURE 32.2 Schematic diagram of silica nanoparticles with photosensitizer encapsulated within the core. Irradiation of photosensitizer with light (arrows) produces singlet oxygen.

of the silica precursors yielded 30 nm nanoparticles with a hydrophobic core entrapping HPPH and the hydrophilic amine functionalized surface provided by APTES condensation. These HPPH-encapsulated nanoparticles were capable of producing singlet oxygen at levels identical to those generated by an aqueous emulsion of free HPPH. Importantly, these ORMOSIL particles caused substantial cytotoxicity at an HPPH equivalent concentration of 20 μM in both HeLa cervical adenocarcinoma cells and UCI-107 ovarian carcinoma cells. The Prasad group went on to synthesize 20 nm ORMOSIL nanoparticles by the coprecipitation of the photosensitizer iodobenzylpyropheophorbide (IP) covalently coupled to the silica precursors 4-(triethoxysilyl)-aniline and vinyltriethoxysilane (Ohulchanskyy et al. 2007). Three different types of ORMOSIL nanoparticles were synthesized by varying the ratio of the two silica precursors. All of the as-synthesized nanoparticles covalently incorporated the IP photosensitizer and preserved both the fluorescence properties of the sensitizer and its capacity to produce singlet oxygen. Fluorescence imaging confirmed that all three types of nanoparticles were readily internalized by Colon-26 mouse adenocarcinoma cells at an IP equivalent concentration of 2 μM. Phototoxicity was effectively induced with an IP equivalent concentration of 0.5 μM in a light dose-dependent manner.

ORMOSIL nanoparticles have been synthesized from the silica precursor, APTES, covalently bound to the photosensitizer protoporphyrin IX (Pp IX) (Rossi et al. 2008). Singlet oxygen production of the Pp IX derivatized, amino-functionalized silica nanoparticles (approximately 70 nm) was verified by the characteristic phosphorescence of singlet oxygen at 1270 nm and also by singlet oxygen–induced photobleaching of 1,3-diphenylisobenzofuran (DPBF) upon illumination with 532 nm light. These authors observed that the efficiency of singlet oxygen delivery of the Pp IX-nanoparticle system was higher than that of free Pp IX, further suggesting that by preserving photosensitizers in their monomeric form, nanoparticles can enhance singlet oxygen production.

Pp IX was also encapsulated in silica nanoparticles approximately 10, 25, and 60 nm in diameter (Simon et al. 2010). The silica precursor triethoxyvinylsilane was condensed in the presence of Pp IX and dioctadecyl tetramethyl indodicarbocyanine chloro benzene (DID), a fluorophore used for bioimaging experiments. Initial studies with all three nanoparticle sizes, incubated with HCT 116 human adenocarcinoma cells, showed no difference in internalization rate or cytotoxicity following irradiation.

Phototoxicity of the Pp IX doped nanoparticles (25 nm) was investigated in human colon, breast, epidermoid, and lymphoblastoid cancer cell lines. The nanoparticles were most effective on HCT-116 colon cancer cells with a half maximal effective concentration (EC$_{50}$) of 0.44 ± 0.05 μM.

Mannose-functionalized silica nanoparticles carrying a covalently bound photosensitizer have been used to target mannose-binding lectins at the surface of MDA-MB-231 breast cancer cells and have demonstrated the potency of carbohydrates as selective cytoadhesive molecules (Brevet et al. 2009). A further study has looked into the functionalization of silica nanoparticles with mannose residues to enhance bioadhesion and cellular uptake (Gary-Bobo et al. 2011). The nanoparticles contained a covalently bound porphyrin derivative photosensitizer, which was activated by multiphoton excitation using 760 nm laser light. *In vitro* PDT studies on MCF-7 and MDA-MB-231 human breast adenocarcinoma cells and on HCT-116 human colon adenocarcinoma cells concluded that the nanoparticles exhibited no significant dark toxicity; however, significant phototoxicity of the conjugates upon multiphoton excitation reduced cell viability to approximately 44%, 33%, and 27%, respectively. The authors showed that multiphoton irradiation of the silica nanoparticle conjugates intravenously administered to an HCT-116 tumor-bearing mouse model also exhibited significant *in vivo* PDT by reducing the tumor mass by approximately 70%. The silicon content of urine in mice treated with the silica nanoparticles was monitored using inductively coupled plasma mass spectrometry (ICP-MS) over the first 14 days post-PDT. It was found that the amount of silicon in urine reached a maximum approximately 6–8 days following PDT and then decreased gradually until 14 days posttreatment. These findings suggest that the silica nanoparticles could be readily excreted from the body through the renal clearance pathway within 2 weeks of intravenous administration.

Calcium phosphosilicate nanoparticles carrying the photosensitizer indocyanine green have been conjugated to anti-CD117 or anti-CD96 antibodies for targeted PDT (Barth et al. 2011). This study has shown that leukemic cells overexpressing CD117 or CD96 cell-surface receptors can be selectively targeted for cancer cell-specific PDT.

The Prasad group also investigated the *in vivo* biodistribution and clearance of 20 nm ORMOSIL nanoparticles loaded with a near-infrared fluorescent dye or radiolabeled with iodine-124 (Kumar et al. 2010). They found that almost all of the nanoparticles were cleared from the body of a mouse via the hepatobiliary system 6 days following intravenous administration. Examination of the liver, spleen, kidney, lungs, and skin 15 days post-administration showed no residual nanoparticles (as determined by the lack of fluorescence of the loaded dye) and showed no cell or tissue damage.

There have been a number of reports indicating that silica nanoparticles can interfere with cellular function and viability *in vitro* and can induce systemic toxicity and organ damage *in vivo*. The effects of different-sized silica particles (40 nm to 5 μm) on human and rat epithelial cell lines have been investigated

(Chen and von Mikecz 2005). The authors found that 40–70 nm nanoparticles were able to enter the cell nucleus, whereas larger particles (0.2–5 μm) were confined to the cytoplasm. Once inside the nucleoplasm, silica nanoparticles (40–70 nm) induced aberrant aggregates of topoisomerase I, an enzyme crucial for processes including DNA replication. Subsequently, these silica nanoparticles inhibited DNA replication, transcription, and cell proliferation, none of which were observed when the cells were incubated with 0.2–5 μm particles (Chen and von Mikecz 2005). An *in vivo* study found that intravenous administration of nonporous silica nanoparticles (70 nm) at a dose of 30 mg/kg was lethal to mice, although a higher dose (100 mg/kg) of 300 and 1000 nm silica particles had no effect (Nishimori et al. 2009). Repeated administration of a sublethal dose of 70 nm silica nanoparticles resulted in hepatic fibrosis, an indication of chronic liver injury (Nishimori et al. 2009).

Silica nanoparticles are highly versatile and robust nanoparticles useful for PDT studies. The feasibility of co-doping the nanoparticles with photosensitizers and imaging agents has suggested their potential as theranostic nanoplatforms. Both *in vitro* and *in vivo* experiments have highlighted the powerful PDT efficacy of photosensitizers entrapped within silica nanoparticles. Additionally, evidence of renal clearance of silica nanoparticles and clearance through the hepatobiliary system make them worthy of further investigation for clinical PDT. However, the reports suggesting the potential for silica nanoparticles to impair cellular function and viability and induce systemic toxicity and organ damage raise concerns over their *in vivo* use for theranostic applications, and further toxicological studies should be performed.

32.2.3 Polymeric Nanoparticles

Synthetic organic polymers have been routinely used for the solubilization, stabilization, and encapsulation of many nanoparticles in aqueous environments. Their capacity to be readily loaded with photosensitizers while providing a platform for further functionalization with biomolecules has suggested their use as nanoparticle platforms for cancer cell-specific PDT (Bechet et al. 2008). Nanoparticle size and sensitizer loading efficiency can be precisely modulated by controlling polymer composition, reaction conditions, and the chemical characteristics of the photosensitizer (Bechet et al. 2008). Some sensitizers have also been covalently tethered to monomers prior to polymerization and formation of nanoparticles to prevent release of physisorbed PDT drugs before reaching the target site (Qin et al. 2011). Many polymers used for this technology, such as the commonly used polylactic-*co*-glycolic acid (PLGA), have the added advantage of being biodegradable. As a result, polymeric nanoparticle delivery systems can be enzymatically hydrolyzed and excreted from the body, thus minimizing long-term accumulation (Bechet et al. 2008). Although polymer encapsulation techniques have been described for a wide variety of PDT drugs, the resultant particles are typically micrometer sized with diameters ranging from approximately 117 to 988 nm (Bechet et al. 2008). Figure 32.3

FIGURE 32.3 Schematic representation of polymeric nanoparticles with photosensitizer entrapped within the core. Irradiation of photosensitizer with light generates cytotoxic singlet oxygen.

demonstrates the PDT action of polymeric nanoparticles loaded with photosensitizer molecules and irradiated with a specific wavelength of visible or near-infrared light to produce cytotoxic singlet oxygen.

Early studies of polymeric nanoparticles as PDT drug carriers showed that sulfonated zinc phthalocyanine and naphthalocyanine could be efficiently incorporated into polyisobutylcyanoacrylate and polyethyl-2-butylcyanoacrylate nanoparticles with diameters ranging from 10 to 380 nm (Labib et al. 1991). The size of the nanoparticles, which were formed by polymerization in the presence of the sensitizers, was strongly dependent on pH as smaller nanoparticles were produced at alkali pHs.

The *in vitro* and *in vivo* effects of PLGA nanoparticles have been investigated for the encapsulation of the lipophilic chlorin photosensitizer meso-tetraphenylporpholactol (McCarthy et al. 2005). The nanoparticles had a hydrophobic diameter of approximately 98 nm and a meso-tetraphenylporpholactol loading efficiency of 12%. Nanoparticles (10 µM) were incubated with 9L glioblastoma cells for 16 h followed by irradiation at 650 nm, which caused approximately 95% cell cytotoxicity. Intravenous administration of the nanoparticle formulation to mice implanted with a prostate carcinoma tumor caused significant tumor shrinkage following irradiation, and tumor regrowth was not observed up to 27 days post-PDT. It was found, however, that encapsulation reduced the efficiency of singlet oxygen production as a result of photosensitizer aggregation within the nanoparticle. Also, 20% of the photosensitizer leached out of the nanoparticles when they were kept in a 0.5% lipid solution for 360 min, outlining the limited stability of polymer nanoparticles for photosensitizers.

The Kopelman group recently reported the synthesis of two variants of polyacrylamide nanoparticles covalently conjugated to two methylene blue (MB) derivatives to enhance the loading of the photosensitizer and to prevent leaching from the nanoparticles (Qin et al. 2011). The nanoparticles had an average diameter of 74.4 and 38.4 nm, depending on the MB derivative. Both types of nanoparticles were either PEGylated or conjugated to an F3 peptide, a molecule specific for the commonly overexpressed receptor nucleolin. The PDT efficacy of the nanoparticles was assessed using MDA-MB-435 human breast adenocarcinoma cells *in vitro*. A dual live/dead cell stain using Calcein-AM/propidium iodide (PI), respectively, revealed that both types of nanoparticles exhibited no phototoxic behavior; however, cytotoxicity was observed when the nanoparticles were targeted to the cells using the F3 peptide.

The Hamblin group recently reported the use of hyperbranched polyether–ester nanoparticles composed of hybrids of chlorin e6 (Ce6) and hyper-branched ether–ester monomers for the enhanced PDT of Cal-27 human tongue carcinoma cells (Li et al. 2012). The water-soluble nanoparticles had an average diameter of approximately 50 nm. Confocal fluorescence imaging confirmed the intracellular localization of the polymeric nanoparticles. The conjugates exhibited a three to four times higher PDT efficacy of Cal-27 cell kill as compared to the free sensitizer.

The effects of two different polyacrylamide nanoparticles have been explored *in vitro*: (1) nanoparticles entrapping polylysine modified aluminum phthalocyanine tetrasulfonate (approximately 45 nm) and (2) nanoparticles entrapping the same sensitizer with another surface-bound porphyrin photosensitizer, 5,10,15,20-Tetrakis(4-N-methylpyridyl)porphyrin (approximately 95 nm) (Kuruppuarachchi et al. 2011). The uptake of the nanoparticles in HT-29 human colon adenocarcinoma cells was investigated using flow cytometry and was found to be dose-dependent, reaching a maximum at 18 h. It was found that the polyacrylamide nanoparticles containing the two photosensitizers exhibited higher levels of phototoxicity. Cells irradiated in the presence of the nanoparticles without prior incubation were effectively killed, even more than those incubated for 25 h before PDT. These results suggested that the use of two photosensitizers with different absorption profiles within one nanoparticle system can be more effective than using a single photosensitizer. It was also shown that nanoparticles in close proximity to the cells were sufficient to deliver cytotoxic levels of singlet oxygen.

Experiments using nontoxic and biodegradable nanoparticles for delivering photosensitizers to cancer cells have proven to be effective. However, photosensitizer aggregation within the polymeric matrix has been reported in several instances, and leaching of the PDT drugs from the nanoparticle may also prove to be problematic as it can decrease the efficiency of photosensitizer delivery to tumors.

32.2.4 Magnetic Nanoparticles

Magnetic nanoparticles have drawn considerable attention as a result of the intrinsic properties that enable multiple functions to be performed, including enhancement of magnetic resonance imaging (MRI), magnetically guided treatment, and magnetic hyperthermal therapy (Shubayev, Pisanic, and Jin 2009). In several studies, the intrinsic characteristics of magnetic nanoparticles have been combined with the potent cytotoxicity of photosensitizers to yield multifunctional nanoparticles for combined cancer therapy and diagnostics. Colloidal suspensions of magnetic nanoparticles composed of iron oxide and manganese oxide have been utilized as theranostic agents, which act as efficient nanocarriers of photosensitizing agents for improved PDT in addition to enhancing MRI-based tumor imaging and diagnosis. Figure 32.4 is a schematic representation of a multifunctional magnetic nanoparticle modified with a photosensitizer.

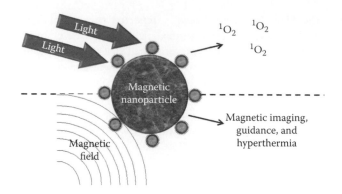

FIGURE 32.4 Schematic of multifunctional magnetic nanoparticle-based PDT system. Top: Irradiation of immobilized photosensitizer with visible or near-infrared light producing singlet oxygen. Bottom: Magnetic properties of nanoparticle core allow for imaging, magnetic guidance, and hyperthermal treatment with externally applied magnetic field.

The synthesis of biocompatible magnetic nanoparticles functionalized with a PDT agent was reported by the Tedesco group, which used magnetic nanoparticles coated with the photosensitizer zinc phthalocyanine (Oliveira et al. 2005). Spectroscopic characterization confirmed that the phthalocyanine maintained its optical characteristics following immobilization onto the magnetic nanoparticles and therefore retained its photosensitizing capacity. In this instance, the system was proposed to be a potential bimodal PDT-hyperthermia agent for cancer treatment.

The first report of *in vivo* use of nanoparticles for PDT combined with MR imaging used polyacrylamide nanoparticles co-doped with the photosensitizer Photofrin and an MRI contrast agent (Kopelman et al. 2005). The group showed that *in vivo* tumor growth was arrested following PDT using the nanoparticles that had been targeted to tumor vessels overexpressing $\alpha_v\beta_3$ integrins using an arginine-glycine-aspartate (RGD) peptide bound to the surface. The entrapped MRI contrast agent enabled *in vivo* monitoring of the destruction of the tumor.

Iron oxide nanoparticles have been synthesized and modified with a porphyrin photosensitizer (Gu et al. 2005). Intracellular uptake of the conjugates in HeLa cervical adenocarcinoma cells was observed by exploiting the fluorescence of the photosensitizer. Irradiation of the cells incubated with the nanoparticle conjugates with light ranging from 545 to 580 nm directly induced cytotoxicity of the HeLa cells. Although no attempt at hyperthermal treatment was made, their potential for dual PDT–hyperthermal treatment was proposed. The Prasad group prepared iron oxide nanoparticles encapsulated within a PEG-coated lipid micelle containing the photosensitizer 2-(1-hexyloxyethyl)-2-devinyl pyropheophorbide-a (HPPH) to treat HeLa cervical adenocarcinoma cells with the potential for magnetic guidance and enhanced delivery to cancer cells. Cellular uptake of the nanoparticle conjugates was confirmed by utilizing the fluorescence of the photosensitizer, and the ability of the system to induce cell death was demonstrated by the

3-(4,5-dimethylthiazol-2-yl)-2,5-diphenyltetrazolium bromide (MTT) viability assay.

The *in vivo* PDT efficiency of MRI-guided iron oxide nanoparticles coated with chitosan, which encapsulated the porphyrin-based photosensitizer 2,7,12,18-tetramethyl-3,8-di-(1-propoxyethyl)-13,17-bis-(3-hydroxypropyl) porphyrin (PHPP), has been investigated (Sun et al. 2009). PDT treatment of SW480 colorectal adenocarcinoma cells with the magnetic nanoparticle conjugates was demonstrated using an MTT viability assay. Importantly, the nanoparticles were successfully guided to a mouse tumor using an external magnetic field, and PDT was monitored *in vivo* using quantitative magnetic resonance-based imaging. Investigation into the biodistribution of these nanoparticles concluded that the nanoparticles accumulated at the tumor more than in the liver and skin.

Similarly, iron oxide nanoparticles have been covalently modified with the photosensitizer chlorin e6 (Ce6) (Huang et al. 2011). Using an MGC803 gastric carcinoma mouse model, the nanoparticle conjugates enabled fluorescence and MRI visualization of the tumor within 6 h of administration. As the nanoparticles accumulated at the tumor with the assistance of external magnetic guidance, the fluorescence intensity of the Ce6 bound nanoparticles increased, and the MRI contrast image darkened. Following PDT treatment, inhibition of tumor growth was the most significant for up to 28 days when the conjugates had been magnetically guided to the tumor.

An alternative material that has seen significant interest in PDT-based theranostics as a result of its paramagnetic property is manganese oxide. Hydrophobic monodispersed manganese oxide nanoparticles (approximately 14 nm in diameter) have been coated in a PEG shell with the photosensitizer Pp IX covalently attached to the polymer (Schladt et al. 2010). The study highlighted the potential use of this nanoparticle platform as an alternative to iron oxide for combined fluorescence and magnetic resonance-based imaging of cancer cells by exploiting the luminescence of the photosensitizer and the paramagnetism of the nanoparticle core, respectively. Moreover, following conjugation to the nanoparticle, the photosensitizer retained its ability to generate singlet oxygen as confirmed by its cytotoxic effects on Caki-1 human renal carcinoma cells.

These studies outline the significant potential of such magnetic nanoparticles for the improvement of PDT for cancer treatment. Such magnetic nanoparticles provide an additional dimension to *in vivo* theranostics using an external magnetic field for MRI tumor detection, nanoparticle guidance, and hyperthermal treatment using these versatile, multifunctional, composite nanosystems.

32.2.5 Photocatalytic Metal Oxide Semiconductor Nanoparticles

A large number of nanoparticles being used for studies of PDT of cancer act as stabilizers and delivery vehicles for photosensitizers. However, some nanoparticles are inherently photosensitizing and produce ROS products usually associated with type I reactions. Examples of these are titanium dioxide (TiO_2) and zinc oxide (ZnO) nanoparticles, which are semiconductor nanocrystals that catalytically generate ROS in response to photoirradiation (Chen and Mao 2007). Following excitation with photons of energies greater than or equal to their bandgap, these metallic oxide nanoparticles undergo a charge separation in their crystal lattice where electrons in the valence band move to the conduction band, leaving behind a positive hole. The electron in the conduction band becomes trapped, and recombination of this electron-hole pair (exciton) is delayed. This results in a long-lived charge separation, thus increasing the photocatalytic efficiency of ROS production at the surface of the nanoparticles. The positive hole can oxidize water molecules to produce hydroxyl radicals ($^{\bullet}OH$), and the electrons in the conduction band can reduce molecular oxygen to produce superoxide anions $\left(^{\bullet}O_2^-\right)$ and hydrogen peroxide (H_2O_2), all of which are cytotoxic oxidizing species (Chen and Mao 2007). Figure 32.5 demonstrates the photoinduced charge separation of these nanoparticles and the subsequent ROS species produced.

As a result of their high rates of photocatalysis, metallic oxide nanoparticles, in particular, TiO_2 nanoparticles, have been intensely investigated for their potential in PDT of cancer. The earliest report of using TiO_2 nanoparticles for UV-induced phototoxicity for the treatment of cancer showed that 30 nm TiO_2 nanoparticles effectively destroyed HeLa human cervical adenocarcinoma cells *in vitro* when exposed to UV light and also inhibited tumor growth *in vivo* when implanted into mice (Cai et al. 1992). Tumor inhibition was most significant when the tumor received a second cycle of PDT treatment 13 days following the initial cycle. It has been demonstrated that TiO_2 nanoparticles doped with platinum exhibited greatly improved phototoxicity against HeLa cervical adenocarcinoma cells (Liu et al. 2010). Platinum-doped TiO_2 nanoparticles caused an approximately 85% reduction in cancer cell viability compared to an approximately 35% reduction in viability induced by nondoped TiO_2 nanoparticles following UV irradiation. *In vivo* antitumor potential of TiO_2 nanoparticles was further demonstrated using intravenously administered nanoparticles with UV irradiation, which prolonged the life of U87 glioblastoma tumor-bearing mice by approximately 14 days (Wang et al. 2011a). However, to effectively excite the administered TiO_2

nanoparticles, the tumor was exposed to UV light following incision and temporary removal of the covering skin. This highlights the limitation of using UV light for the photosensitization of deeply situated tumors.

There have been several reports on the use of ZnO nanoparticles for PDT. For example, the different photoinduced toxicities of 20, 60, and 100 nm ZnO nanoparticles coated with aminopolysiloxane has been explored using SMMC-7721 hepatocellular carcinoma cells (Li et al. 2010). Significant reduction in SMMC-7721 cell activity (approximately 80%) was observed with the 20 nm nanoparticles at the lowest concentration of 2.5 µg/mL upon 245 nm UV irradiation for 180 s. At the highest concentration of 10 µg/mL, all three sizes of nanoparticles reduced cell activity by approximately 100%, although varying degrees of dark toxicity were observed. These findings clearly highlight the size and dose dependence of ZnO nanoparticles on their photodynamic activity, showing that smaller (20 nm) ZnO nanoparticles exhibit higher phototoxicity at lower concentrations; however, they also appear to exhibit higher dark toxicity at larger concentrations.

ZnO nanoparticles have also been used as photoactive delivery agents for a porphyrin photosensitizer bound to the ZnO surface through an L-cysteine linker (Liu et al. 2008b). Excitation of the 5 nm ZnO nanoparticles with 300 nm light results in two fluorescence emission bands, one of which is centered at 445 nm. The emission band at 445 nm was used to excite the attached photosensitizer through fluorescence resonance energy transfer (FRET), which was found to be 83% efficient. By incubating NIH:OVCAR-3 human ovarian adenocarcinoma cells with the nanoparticle conjugates and irradiating them for 30 min with 365 nm UV light, cell viability was reduced to approximately 10%, whereas, in the dark, viability remained at approximately 98%. Unconjugated ZnO nanoparticles alone, however, did not induce a photodynamic response under the conditions of the experiment, highlighting possible limitations of the photosensitizing capacity of ZnO.

Overall, photocatalytic metal oxide semiconductor nanoparticles provide opportunities for effective PDT, especially TiO_2 nanoparticles, which exhibit exceptional photocatalytic efficiency. However, short-wavelength excitation is a significant limitation.

32.2.6 Quantum Dots

Quantum dots (QDs) are semiconductor nanocrystals that have drawn a lot of attention for luminescence-based applications, including bioimaging, bioanalytics, and diagnostics as a result of their intrinsic photophysical characteristics. Compared to organic fluorophores, QDs exhibit remarkable photostability, higher fluorescence quantum yields, significantly longer fluorescence lifetimes, and broader absorption bands. QDs typically exhibit quantum-sized effects with finely tuned absorption and emission profiles: the bandgap of the QD widens with decreasing size causing a blue shift in absorption and photoluminescence. In addition, optical properties of QDs can be controlled

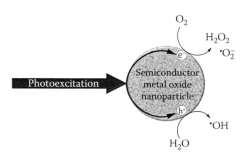

FIGURE 32.5 Schematic representation of photoinduced charge separation of semiconductor metal oxide nanoparticles resulting in the generation of a free electron (e^-) and a positive hole (h^+).

by composition and surface functionality (Alivisatos 1996). A noticeable drawback of QDs is the acute toxicity of cadmium, a common integral component of these semiconductor nanocrystals. To overcome this limitation, efforts to graft the QDs within various nonporous polymers or shells, including zinc sulfide (ZnS), have been investigated, and attempts to use nontoxic photoluminescent QDs, such as ZnS, silicon and carbon QDs are ongoing (Derfus, Chan, and Bhatia 2004). For example, Cd-free core-shell indium phosphide (InP)/ZnS QDs have been synthesized, and the photosensitizer chlorin e6 (Ce6) has been encapsulated in a thin silica shell coating the nanoparticle (Charron et al. 2012). These authors found that the rate-determining step of singlet oxygen production by Ce6 activated by the QD emission was the distance-dependent energy transfer between the nanoparticle and the sensitizer.

QDs undergo charge separation upon excitation with light of an appropriate wavelength; however, radiative exciton recombination readily occurs, resulting in strong fluorescence emission and weak ROS production. It has been reported that QDs can induce significant cytotoxicity upon UV excitation, although this is attributed to surface oxidation, crystal deterioration, and subsequent Cd^{2+} release, rather than ROS production (Derfus, Chan, and Bhatia 2004). However, in contrast, results with CdTe QDs have shown production of ROS in quantities sufficient for cancer cell kill (approximately 80% cytotoxicity); the photosensitization process was proposed to be type I, as the singlet oxygen quantum yield of these QDs was found to be only 1% (Chen et al. 2010). As a result of their unique photoluminescent properties, QDs have been exploited for nonradiative energy transfer to photosensitizers supported onto their surface. Figure 32.6 shows an example of this energy transfer process that causes activation of a photosensitizer associated with the QDs.

Samia, Chen, and Burda (2003) were the first to successfully functionalize the surface of 5 nm cadmium–selenide (CdSe) QDs with a phthalocyanine (Pc4) photosensitizer. They showed that unconjugated QDs produced singlet oxygen upon excitation in toluene but found that the singlet oxygen quantum yield was approximately 5%. By conjugating Pc4 to the QDs, a FRET efficiency of 77% was observed following direct excitation with 488 nm light, which ultimately resulted in fluorescence emission of the sensitizer at 680 nm.

Green- and red-emitting core-shell CdSe/CdS QDs coated with ZnS have been synthesized and modified with rose Bengal and chlorin e6 (Ce6), respectively (Tsay et al. 2007). Singlet oxygen production of the conjugates was confirmed by photoluminescence centered at 1270 nm and also by the photobleaching of anthracene-9, 10-dipropionic acid (ADPA). It was found that both QD systems could act as efficient FRET donors and/or photosensitizer carriers for bioimaging and PDT.

Multiphoton excitation has also been employed for photoactivation of QD-based PDT systems. CdSe/ZnS core-shell QDs (approximately 5.4 nm) have been synthesized and modified with mercaptopropionic acid (Fowley et al. 2012). These nanoparticles had an absorption band maximum at 515 nm and an emission band centered at 535 nm. Irradiation of these QDs with 400 nm light in the presence of 1,3-diphenylisobenzofuran (DPBF) indicated that singlet oxygen was not produced by the QDs. An amine-modified rose Bengal derivative with an absorption band maximum at 565 nm was covalently conjugated to the QDs, which enhanced the aqueous solubility of the sensitizer. *In vitro* cell viability was assessed using the MTT assay upon single photon excitation of the QD–Rose Bengal conjugates with 365 nm light. HeLa human cervical adenocarcinoma cells were incubated with 0, 1, 10, and 100 μM QDs with no significant dark toxicity. However, following irradiation, an approximately 33% decrease in cell viability was observed with 100 μM QDs.

CdTe QDs have also been used for targeted PDT of KB human head and neck carcinoma cells overexpressing the folate receptor-α (Morosini et al. 2011). Phototoxicity of the QDs toward KB cells was found to be significantly higher than that of the HT-29 cells, which do not express the folate receptor-α.

Because of the remarkably low photobleaching of QDs and their highly tunable photoluminescence emission, the various potential applications of QDs for photodiagnostic and phototherapeutic applications are broad. Although the excitation of QDs is usually limited to shorter wavelengths of light, it has been shown that multiphoton excitation is sufficient for PDT treatment. Nevertheless, the majority of the QD-based PDT systems investigated to date are somewhat limited by the toxicity of the Cd-based nanocrystals.

32.2.7 Scintillation Nanoparticles

Light penetration through tissue has proven to be a significant barrier in activation of PDT systems present within solid tumors. To address this deficiency in clinical PDT, scintillation nanoparticles have been investigated (Morgan et al. 2009). These self-lighting nanoparticles absorb ionizing radiation such as x-rays and γ-rays that can penetrate into human tissue deeper than the light of UV or visible wavelengths. Following absorption, the nanoparticles become excited and the radiative recombination of the electron-hole pair results in an afterglow of visible light, that is, scintillation. The scintillation nanoparticles can be utilized as FRET donors by photosensitizers supported

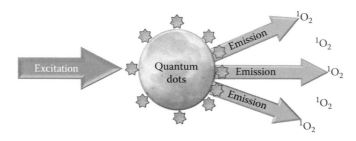

FIGURE 32.6 Schematic diagram of the FRET-based activation of photosensitizers bound to the surface of QDs following photoexcitation of the nanoparticles. Visible light emitted from photoexcited QDs is absorbed by the photosensitizer, which subsequently generates singlet oxygen.

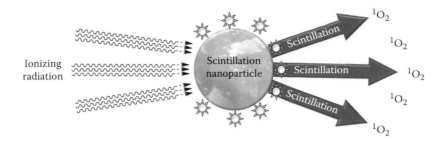

FIGURE 32.7 Schematic representation of scintillation nanoparticles with photosensitizer coating the surface. Irradiation of nanoparticles with ionizing radiation induces scintillation, thereby emitting visible light, which can activate the photosensitizer through FRET.

on the surface, assuming that sufficient spectral overlap between the scintillation and sensitizer absorption exists. This scintillation phenomenon has opened up an array of materials that can be excited by high-energy radiation for simultaneous photodynamic and radiation therapy for cancer, as well as *in vivo* imaging of solid tumors (Morgan et al. 2009). Figure 32.7 is a schematic representation of a model multifunctional scintillation nanoparticle modified with photosensitizer molecules.

Scintillating lanthanum fluoride nanoparticles doped with terbium ions ($LaF_3:Tb^{3+}$) have been synthesized and functionalized with the photosensitizer meso-tetra(4-carboxyphenyl) porphine (MTCP) (Liu et al. 2008a). Excitation of the $LaF_3:Tb^{3+}$ nanoparticles (approximately 15 nm) with 260 nm UV light resulted in multiple luminescence emission bands, centered at 489, 542, 584, and 621 nm, which exhibited significant spectral overlap with the MTCP excitation spectrum. For the emission bands at 542, 584, and 621 nm, a FRET efficiency of 68%, 52%, and 50%, respectively, was determined. The production of singlet oxygen following x-ray irradiation of the MTCP coated $LaF_3:Tb^{3+}$ nanoparticles was then measured as a function of the decrease in ADPA fluorescence emission. It was found that both free MTCP and MTCP conjugated $LaF_3:Tb^{3+}$ nanoparticles produced singlet oxygen in a time-dependent manner during x-ray irradiation, although the scintillation nanoparticle conjugates were significantly more efficient.

Amine functionalized QDs have also been used as nanoscale scintillators that emit at 520 nm upon x-ray excitation (Wang et al. 2010). The nanoparticles were modified with the photosensitizer Photofrin. By applying 6 MV of x-ray radiation, FRET at 520 nm occurred from the scintillation of the QD to the conjugated Photofrin molecules. It was shown that x-ray irradiation of these QD conjugates generated singlet oxygen in a dose-dependent manner from 6 to 30 Gy. H460 human lung carcinoma cells were then incubated with 48 nM of Photofrin-conjugated QDs for 24 h and were then exposed to a total x-ray dose of 6 Gy. Apoptosis was confirmed as the primary mechanism of PDT cytotoxicity by positive TUNEL staining.

Preliminary studies of scintillation nanoparticles have so far been mostly promising, indicating that radiotherapy of tumors could be enhanced using nanoparticles that readily absorb ionizing radiation and scintillate to activate a PDT drug. If *in vivo* studies prove successful, scintillation nanoparticles could

aid radiotherapy by reducing the dosage of high-energy rays required for tumor eradication, thereby minimizing unnecessary secondary effects of the treatment.

32.2.8 Upconverting Nanoparticles

An alternative route to overcoming the limitation of light penetration through tissue is to utilize upconversion. Upconversion is a phenomenon by which light is absorbed by a material at one wavelength, but the photons emitted following excitation are of a shorter wavelength than that used for excitation. This photophysical process has been explored for possible use in PDT, in which upconverting nanocrystals are excited by near-infrared light. These nanoparticles, which are doped with rare earth metal ions, typically lanthanide ions, emit light in the visible region of the spectrum. The visible light is then used to excite the photosensitizer supported on the upconverter surface. This process activates the photosensitizer with a wavelength of light that favors tissue penetration (Wang, Cheng, and Liu 2011). Figure 32.8 is a schematic of upconverting nanoparticles.

The PDT potential of sodium yttrium fluoride ($NaYF_4$) nanoparticles co-doped with ytterbium (Yb^{3+}) and erbium ions (Er^{3+}) has been reported (Zhang et al. 2007). The $NaYF_4:Yb^{3+},Er^{3+}$

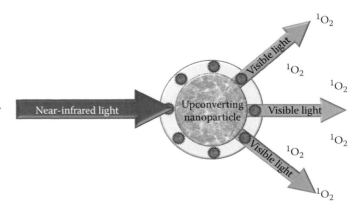

FIGURE 32.8 Schematic representation of upconverting nanoparticles coated in silica shell encapsulating photosensitizer molecules. Following irradiation of nanoparticles with near-infrared light, upconverters emit visible light, which then activates the photosensitizer to generate singlet oxygen.

upconverting nanoparticles (approximately 60–120 nm) emit green light (537 nm) following excitation with 974 nm near-infrared light. The upconverters were embedded in a silica shell doped with the photosensitizer Merocyanine 540. Following modification, the upconverting nanoparticles retained their photoluminescence properties. The nanoparticles were found to generate singlet oxygen upon near-infrared irradiation in a buffered solution. To aid cancer cell-specific uptake, the upconverting nanoparticles were further functionalized with an anti-MUC1 antibody targeted to MUC1 receptors overexpressed by MCF-7/AZ human breast adenocarcinoma cells. The cells incubated with the nanoparticle conjugates were irradiated with a 974 nm laser. Following 36 min of irradiation, cell morphology was altered, which was indicative of cytotoxicity.

In vitro and *in vivo* studies have been reported highlighting the potential of using upconverting nanoparticles for PDT of cancer (Chatterjee and Yong 2008). These authors synthesized polyethyleneimine (PEI)-coated $NaYF_4:Yb^{3+},Er^{3+}$ nanoparticles (approximately 50 nm). The nanoparticles were further modified by covalently conjugating with a cancer cell-targeting molecule, folic acid. A zinc phthalocyanine (ZnPc) photosensitizer was physisorbed onto the surface of the nanoparticles to enable PDT action of the conjugates. The PDT action of these conjugates was investigated *in vitro* using HT-29 human colon adenocarcinoma cells and the MTT assay. HT-29 cells incubated with the upconverting nanoparticle conjugates for 24 h were irradiated for 30 min with a 980 nm laser. Cell viability was significantly reduced by approximately 80–90% following PDT, indicating that the singlet oxygen production of these upconverting nanoparticle–photosensitizer conjugates upon near-infrared excitation is sufficient for effective cancer cell kill.

In vitro and *in vivo* investigations have been carried out with approximately 30 nm PEGylated $NaYF_4$ upconverting nanoparticles doped with 20% Yb and 2% Er, modified with chlorin e6 (Ce6) for PDT (Wang et al. 2011b). The Ce6 was loaded onto the nanoparticles through hydrophobic interactions with oleic acid present at the surface. In addition to singlet oxygen measurement and *in vitro* MTT viability assays, *in vivo* studies of tumor volume, mouse survival, and pharmacokinetic behavior of the upconverting nanoparticles were conducted. 4T1 mouse breast cancer cells were incubated with 5 μM Ce6 equivalent of Ce6-modified upconverting nanoparticles for 2 h and then irradiated for 10 min with a 980 nm near-infrared laser. Significant reduction of cell viability was observed in a dose-dependent fashion when the cells were incubated with the Ce6-nanoparticle conjugates and irradiated. As expected, unmodified nanoparticles and free Ce6 did not exhibit significant phototoxicity. When Ce6-nanoparticle conjugates were directly injected into the tumor of a 4T1 mouse model and irradiated with the 980 nm laser, substantial tumor reduction was observed. Of the mice treated with the nanoparticle conjugates and irradiated, approximately 70% survived up to 60 days, whereas no mice in the control groups survived longer than 23 days.

Upconverting nanoparticles have only relatively recently been considered for PDT. Preliminary *in vitro* and *in vivo* studies show huge potential. However, long-term stability and toxicity of these lanthanide-doped nanocrystals in various physiological environments must be systematically studied if they are to be used for clinical PDT.

32.3 Future Prospects for Nanoparticles in Photodynamic Cancer Therapy

An increasing number of studies have highlighted the exciting potential of the use of nanoparticles for PDT. Nanoparticles, as exemplified in this chapter, exhibit multifunctionality, which can dramatically improve PDT. The vast array of nanoparticles and their respective activities show promise to significantly enhance cancer therapy by increasing the delivery of hydrophobic photosensitizers, by radically improving photosensitizer accumulation in tumors through biofunctionality and targeting therapy, and by potentially maximizing the antitumor effect of radiotherapy. However, the majority of studies with nanoparticles for PDT have, so far, not exceeded initial stages of *in vitro* and *in vivo* experimentation. Although many of the nanoparticles reported to date demonstrate no immediate acute toxicity, and there is some evidence to suggest that some nanoparticles are cleared by the urinary and hepatobiliary systems, significant further studies are required to establish their long-term physiological effects before translation into clinical practice. However, it is clear that the future for nanoparticle technology within PDT is exceptionally promising.

References

Alivisatos, A. P. 1996. Semiconductor clusters, nanocrystals, and quantum dots. *Science* 271:933–937.

Barth, B. M., E. I. Altinoğlu, S. S. Shanmugavelandy et al. 2011. Targeted indocyanine-green-loaded calcium phosphosilicate nanoparticles for *in vivo* photodynamic therapy of leukemia. *ACS Nano* 5:5325–5337.

Bechet, D., P. Couleaud, C. Frochot et al. 2008. Nanoparticles as vehicles for delivery of photodynamic therapy agents. *Trends Biotechnol* 26:612–621.

Brevet, D., M. Gary-Bobo, L. Raehm et al. 2009. Mannose-targeted mesoporous silica nanoparticles for photodynamic therapy. *Chem Commun* 12:1475–1477.

Burda, C., X. B. Chen, R. Narayanan, and M. A. El-Sayed. 2005. Chemistry and properties of nanocrystals of different shapes. *Chem Rev* 105:1025–1102.

Cai, R., Y. Kubota, T. Shuin et al. 1992. Induction of cytotoxicity by photoexcited TiO_2 particles. *Cancer Res* 52:2346–2348.

Camerin, M., M. Magaraggia, M. Soncin et al. 2010. The *in vivo* efficacy of phthalocyanine-nanoparticle conjugates for the photodynamic therapy of amelanotic melanoma. *Eur J Cancer* 46:1910–1918.

Charron, G., T. Stuchinskaya, D. R. Edwards, D. A. Russell, and T. Nann. 2012. Insights into the mechanism of quantum dot-sensitized singlet oxygen production for photodynamic therapy. *J Phys Chem C* 116:9334–9342.

Chatterjee, D. and Z. Yong. 2008. Upconverting nanoparticles as nanotransducers for photodynamic therapy in cancer cells. *Nanomedicine* 3:73–82.

Chen, J. Y., Y. M. Lee, D. Zhao et al. 2010. Quantum dot-mediated photoproduction of reactive oxygen species for cancer cell annihilation. *Photochem Photobiol* 86:431–437.

Chen, M., and A. von Mikecz. 2005. Formation of nucleoplasmic protein aggregates impairs nuclear function in response to SiO$_2$ nanoparticles. *Exp Cell Res* 305:51–62.

Chen, X., and S. Mao. 2007. Titanium dioxide nanomaterials: Synthesis, properties, modifications and applications. *Chem Rev* 107:2891–2959.

Cheng, Y., J. D. Meyers, A. M. Broome et al. 2011. Deep penetration of a PDT drug into tumors by noncovalent drug-gold nanoparticle conjugates. *J Am Chem Soc* 133:2583–2591.

Cinteza, L. O., T. Y. Ohulchanskyy, Y. Sahoo et al. 2006. Diacyllipid micelle-based nanocarrier for magnetically guided delivery of drugs in photodynamic therapy. *Mol Pharm* 3:415–423.

Couleaud, P., V. Morosini, C. Frochot et al. 2010. Silica-based nanoparticles for photodynamic therapy applications. *Nanoscale* 2:1083–1095.

Daniel, M. C., and D. Astruc. 2004. Gold nanoparticles: Assembly, supramolecular chemistry, quantum-size-related properties, and applications toward biology, catalysis, and nanotechnology. *Chem Rev* 104:293–346.

Derfus, A. M., W. C. W. Chan, and S. N. Bhatia. 2004. Probing the cytotoxicity of semiconductor quantum dots. *Nano Letters* 4:11–18.

Fales, A. M., H. Yuan, and T. Vo-Dinh. 2011. Silica-coated gold nanostars for combined surface-enhanced Raman scattering (SERS) detection and singlet-oxygen generation: A potential nanoplatform for theranostics. *Langmuir* 27:12186–12190.

Fowley, C., N. Nomikou, A. P. McHale et al. 2012. Water soluble quantum dots as hydrophilic carriers and two-photon excited energy donors in photodynamic therapy. *J Mater Chem* 22:6456–6462.

Gamaleia, N. F., E. D. Shishko, G. A. Dolinsky et al. 2010. Photodynamic activity of hematoporphyrin conjugates with gold nanoparticles: Experiments *in vitro*. *Exp Oncol* 32:44–47.

Gary-Bobo, M., Y. Mir, C. Rouxel et al. 2011. Mannose-functionalized mesoporous silica nanoparticles for efficient two-photon photodynamic therapy of solid tumors. *Angew Chem Int Ed Engl* 50:11425–11429.

Gu, H., K. Xu, Z. Yang, C. K. Chang, and B. Xu. 2005. Synthesis and cellular uptake of porphyrin decorated iron oxide nanoparticles: A potential candidate for bimodal anticancer therapy. *Chem Commun* 4270–4272.

Hone, D. C., P. I. Walker, R. Evans-Gowing et al. 2002. Generation of cytotoxic singlet oxygen via phthalocyanine-stabilized gold nanoparticles: A potential delivery vehicle for photodynamic therapy. *Langmuir* 18:2985–2987.

Huang, P., Z. Li, J. Lin et al. 2011. Photosensitizer-conjugated magnetic nanoparticles for *in vivo* simultaneous magnetofluorescent imaging and targeting therapy. *Biomaterials* 32:3447–3458.

Jang, B., J. Y. Park, C. H. Tung, I. H. Kim, and Y. Choi. 2011. Gold nanorod-photosensitizer complex for near-infrared fluorescence imaging and photodynamic/photothermal therapy *in vivo*. *ACS Nano* 5:1086–1094.

Kopelman, R., Y.-E. L. Koo, M. Philbert et al. 2005. Multifunctional nanoparticle platforms for *in vivo* MRI enhancement and photodynamic therapy of a rat brain cancer. *J Magn Magn Mater* 293:404–410.

Kumar, R., I. Roy, T. Y. Ohulchanskky et al. 2010. *In vivo* biodistribution and clearance studies using multimodal organically modified silica nanoparticles. *ACS Nano* 4:699–708.

Kuruppuarachchi, M., H. Savoie, A. Lowry, C. Alonso, and R. Boyle. 2011. Polyacrylamide nanoparticles as a delivery system in photodynamic therapy. *Mol Pharm* 8:920–931.

Labib, A., V. Lenaerts, F. Chouinard et al. 1991. Biodegradable nanospheres containing phthalocyanines and naphthalocyanines for targeted photodynamic tumor therapy. *Pharm Res* 8:1027–1031.

Li, J., D. Guo, X. Wang et al. 2010. The photodynamic effect of different size ZnO nanoparticles on cancer cell proliferation *in vitro*. *Nanoscale Res Lett* 5:1063–1071.

Li, P., G. Zhou, X. Zhu et al. 2012. Photodynamic therapy with hyperbranched poly(ether-ester) chlorin(e6) nanoparticles on human tongue carcinoma CAL-27 cells. *Photochem Photobiol Sci* 9:76–82.

Liu, L., P. Miao, Y. Xu et al. 2010. Study of Pt/TiO$_2$ nanocomposite for cancer-cell treatment. *J Photochem Photobiol B Biol* 98:207–210.

Liu, Y., W. Chen, S. Wang, and A. Joly. 2008a. Investigation of water-soluble X-ray luminescence nanoparticles for photodynamic activation. *Appl Phys Lett* 92:43901–43903.

Liu, Y., Y. Zhang, S. Wang, C. Pope, and W. Chen. 2008b. Optical behaviors of ZnO-porphyrin conjugates and their potential applications for cancer treatment. *Appl Phys Lett* 92:143901–143903.

Maeda, H., J. Fang, T. Inutsuka, and Y. Kitamoto. 2003. Vascular permeability enhancement in solid tumor: Various factors, mechanisms involved and its implications. *Int Immunopharmacol* 3:319–328.

McCarthy, J. R., J. M. Perez, C. Bruckner, and R. Weissleder. 2005. Polymeric nanoparticle preparation that eradicates tumors. *Nano Letters* 5:2552–2556.

Morgan, N. Y., G. Kramer-Marek, P. D. Smith, K. Camphausen, and J. Capala. 2009. Nanoscintillator conjugates as photodynamic therapy-based radiosensitizers: Calculation of required physical parameters. *Radiat Res* 171:236–244.

Morosini, V., T. Bastogne, C. Frochot et al. 2011. Quantum dot-folic acid conjugates as potential photosensitizers in photodynamic therapy of cancer. *Photochem Photobiol Sci* 10:842–851.

Nishimori, H., M. Kondoh, K. Isoda et al. 2009. Silica nanoparticles as hepatotoxicants. *Eur J Pharm Biopharm* 72:496–501.

Obaid, G., I. Chambrier, M. J. Cook, and D. A. Russell. 2012. Targeting the oncofetal Thomsen–Friedenreich disaccharide using jacalin-PEG phthalocyanine gold nanoparticles for photodynamic cancer therapy. *Angew Chem Int Ed Engl* 51:6158–6162.

Ohulchanskyy, T. Y., I. Roy, L. N. Goswami et al. 2007. Organically modified silica nanoparticles with covalently incorporated photosensitizer for photodynamic therapy of cancer. *Nano Letters* 7:2835–2842.

Oliveira, D. M., P. P. Macaroff, K. F. Ribeiro et al. 2005. Studies of zinc phthalocyanine/magnetic fluid complex as a bifunctional agent for cancer treatment. *J Magn Magn Mater* 289:476–479.

Paciotti, G. F., D. G. I. Kingston, and L. Tamarkin. 2006. Colloidal gold nanoparticles: A novel nanoparticle platform for developing multifunctional tumor-targeted drug delivery vectors. *Drug Dev Res* 67:47–54.

Qin, M., H. J. Hah, G. Kim et al. 2011. Methylene blue covalently loaded polyacrylamide nanoparticles for enhanced tumor-targeted photodynamic therapy. *Photochem Photobiol Sci* 10:832–841.

Rossi, L. M., P. R. Silva, L. L. Vono et al. 2008. Protoporphyrin IX nanoparticle carrier: Preparation, optical properties, and singlet oxygen generation. *Langmuir* 24:12534–12538.

Roy, I., T. Y. Ohulchanskyy, H. E. Pudavar et al. 2003. Ceramic-based nanoparticles entrapping water-insoluble photosensitizing anticancer drugs: A novel drug–carrier system for photodynamic therapy. *J Am Chem Soc* 125:7860–7865.

Samia, A. C., X. Chen, and C. Burda. 2003. Semiconductor quantum dots for photodynamic therapy. *J Am Chem Soc* 125:15736–15737.

Schladt, T. D., K. Schneider, M. I. Shukoor et al. 2010. Highly soluble multifunctional MnO nanoparticles for simultaneous optical and MRI imaging and cancer treatment using photodynamic therapy. *J Mater Chem* 20:8297–8304.

Shubayev, V. I., T. R. Pisanic II, and S. Jin. 2009. Magnetic nanoparticles for theranostics. *Adv Drug Del Rev* 61:467–477.

Simon, V., C. Devaux, A. Darmon et al. 2010. Pp IX silica nanoparticles demonstrate differential interactions with *in vitro* tumor cell lines and *in vivo* mouse models of human cancers. *Photochem Photobiol* 86:213–222.

Stuchinskaya, T., M. Moreno, M. J. Cook, D. R. Edwards, and D. A. Russell. 2011. Targeted photodynamic therapy of breast cancer cells using antibody-phthalocyanine-gold nanoparticle conjugates. *Photochem Photobiol Sci* 10:822–831.

Sun, Y., Z. I. Chen, X.-X. Yang et al. 2009. Magnetic chitosan nanoparticles as a drug delivery system for targeting photodynamic therapy. *Nanotechnology* 20:135102.

Tsay, J. M., M. Trzoss, L. Shi et al. 2007. Singlet oxygen production by peptide-coated quantum dot-photosensitizer conjugates. *J Am Chem Soc* 129:6865–6871.

Wang, C., S. Cao, X. Tie et al. 2011a. Induction of cytotoxicity by photoexcitation of TiO_2 can prolong survival in glioma-bearing mice. *Mol Biol Rep* 38:523–530.

Wang, C., L. Cheng, and Z. Liu. 2011. Research spotlight: Upconversion nanoparticles for potential cancer theranostics. *Ther Del* 2:1235–1239.

Wang, C., H. Q. Tao, L. Cheng, and Z. Liu. 2011b. Near-infrared light induced *in vivo* photodynamic therapy of cancer based on upconversion nanoparticles. *Biomaterials* 32:6145–6154.

Wang, L., W. Yang, P. Read, J. Larner, and K. Sheng. 2010. Tumor cell apoptosis induced by nanoparticle conjugate in combination with radiation therapy. *Nanotechnology* 21:475103.

Wieder, M. E., D. C. Hone, M. J. Cook et al. 2006. Intercellular photodynamic therapy with photosensitizer-nanoparticle conjugates: Cancer therapy using a 'Trojan horse.' *Photochem Photobiol Sci* 5:727–734.

Zaruba, K., J. Kralova, P. Rezanka et al. 2010. Modified porphyrin-brucine conjugated to gold nanoparticles and their application in photodynamic therapy. *Organ Biomol Chem* 8:3202–3206.

Zhang, P., W. Steelant, M. Kumar, and M. Scholfield. 2007. Versatile photosensitizers for photodynamic therapy at infrared excitation. *J Am Chem Soc* 129:4526–4527.

33

Drug Delivery Strategies for Photodynamic Therapy

Ryan F. Donnelly
Queen's University Belfast

33.1 Introduction

In photodynamic therapy (PDT) or photodynamic antimicrobial chemotherapy (PACT), which is to be used clinically, effective delivery methods of both light and photosensitizer to the site of action are necessary. Because of limited light penetration through tissue, clinical PACT would be necessarily used or applied to areas of the body where light can be delivered relatively easily, such as the skin and body cavities. Light delivery onto the skin is simple and has been frequently employed in PDT of dysplastic and neoplastic diseases. In PDT, however, the photosensitizer or its precursor is often administered orally or intravenously. Because of disordered metabolism and blood flow peculiar to dysplastic or neoplastic tissue, photosensitizing concentrations of drug molecules accumulated in the target lesions are relatively higher. Targeting the photosensitizers to wound infections in this way is not possible, and so the drug must be applied topically. Factors such as physicochemical properties of the photosensitizer dose, percentage of photosensitizer to be delivered, barrier properties of the target site, and patient acceptability all have a major impact upon the method of drug delivery.

33.2 Topical Drug Delivery

The clinical success of topical PDT and PACT is more influenced by pharmaceutical and physicochemical considerations, such as tissue penetration, characteristics of photosensitizers, an easy way for applying it, and stability in the applied dosage form. These, in turn, are critically dependent on formulation factors, notably the design of the delivery system. Before any formulation design can be considered, it is important to appreciate the permeation process within a dosage form matrix, this being a fundamental mechanism by which most formulations control the drug-delivery process.

Molecular diffusion is a passive transport mechanism that always directs a chemical system toward a state of thermodynamic equilibrium. It is a random process, and the driving force behind the mass transfer occurs from applied topical dosage systems. If applied to skin, for example, the depth of drug penetration can be such that the drug is localized in cutaneous sites, or it may extend deeper to the underlying capillary network and bring about transdermal delivery into the systemic circulation. This would be undesirable in topical PDT and PACT. Diffusion of drug molecules occurs from areas of high concentration to those in which the concentration is lower and is described by Fick's first law of mass flow. The movement of a drug through a cross section of area, S, in unit time, t, is known as the flux, J. This is a vector that gives the direction and magnitude of the transport. Considering the one-dimensional form along the x component, the flux is shown to be proportional to the concentration gradient, as shown in the following equation:

$$J = \frac{dM}{dt} \cdot \frac{1}{S} = -D \frac{\partial C}{\partial x} \qquad (33.1)$$

where M is the amount of drug diffusing through a distance, x, and D is the diffusion coefficient.

Most of the drug release experiments use a receptor and donor phase separated by a membrane as shown in Figure 33.1. The membrane can be a biological barrier, such as excised skin, or a model membrane, such as silicone sheeting. During the initial part of the experiment, a nonsteady state exists in which Fick's second law describes how the concentration of diffusant varies at a distance, x, through the membrane with respect to time. After a period of time, a quasi-stationary state (pseudo-steady state) is established in which the concentration gradient ($\delta C / \delta x$) across the membrane remains constant and does not vary with respect

Concentration

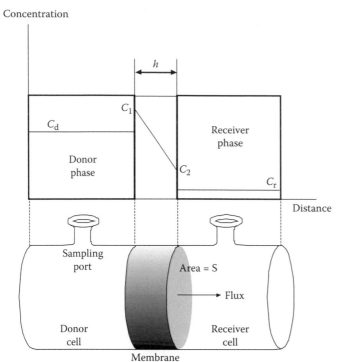

FIGURE 33.1 Schematic representation of a typical side-by-side diffusion cell apparatus. The concentrations across the cells are represented in the top part of the figure. The concentration gradient across the membrane of thickness h is shown at steady state. C_d is not expected to equal C_1 unless the partition coefficient of the drug into the membrane from the donor phase is unity.

to time. During this state, it is possible to express Equations 33.1 and 33.2, namely,

$$\frac{dM}{dt} = \frac{DSK(C_d - C_r)}{h} \tag{33.2}$$

where C_d and C_r are the concentrations in the donor and receptor phases, respectively; h is the thickness of the membrane; and K is the partition coefficient into and out of the membrane from the phases.

Drug release from a cream or ointment-type delivery system can be evaluated using a suitable diffusion cell. Diffusion of the drug through the semisolid matrix becomes an important consideration. If the donor phase is replaced by a sample of the formulation, the membrane can either be rate-limiting in terms of drug permeation to the receiver phase or can simply act as a means to separate the phases. In the latter case, the membrane poses little resistance to drug diffusion. The membrane may be dispensed with altogether in cases in which a hydrophobic formulation is poorly miscible in an aqueous receiving phase.

Incorporating a drug with limited solubility into a semisolid matrix results in a suspension-type formulation. Even moderate loading of the drug will inevitably possess a substantial fraction of the drug in the solid phase. In this case, the Higuchi equation is applicable, as shown in the following equation:

$$Q = [D(2A - C_S)\,C_S t]^{1/2} \tag{33.3}$$

where Q is the amount of drug depleted per unit area of matrix, C_S is the solubility of drug in the matrix, and A is the total drug per unit volume. When $A \gg C_S$, which is the case here, Equation 33.3 can be differentiated with respect to t to yield the following equation:

$$\frac{dQ}{dt} = \sqrt{\frac{ADC_S}{2t}}. \tag{33.4}$$

It is clear from Equation 33.4 that the rate of release of the drug can be increased by increasing the total amount of drug in the formulation, its solubility, and its diffusion coefficient. Ordinarily, drug release that is under matrix (bulk formulation) control will display linear kinetics.

Under conditions in which the drug release greatly exceeds diffusional resistance at any of the interfaces between the donor and receiver phases, the drug has significant solubility in the vehicle, and then Equation 33.3 is expected to play a major role. When the solubility of a photosensitizer in the formulation is high, then $A \gg C_S$ is no longer considered to be appropriate. However, if a biological barrier exerts a significant resistance to diffusion, then barrier resistance must be considered similarly. In these cases, when the biological barrier (e.g., the skin's *stratum corneum*) is believed to be the rate-controlling factor, Equation 33.2 can be used, providing the thickness of the barrier, h, is known and remains constant.

Many antimicrobial photosensitizers are particularly water-soluble molecules. Even when loadings as high as 20% w/w are used, no solid drug exists, and all is present in the dissolved state. In most topical systems used in PACT, it is clear that $A < C_S$ and the following solution to Fick's law can be used when a pseudo-steady state is assumed:

$$M_t \approx 2A\sqrt{\frac{Dt}{\pi}} \tag{33.5}$$

and differentiating with respect to t gives the rate of release:

$$\frac{dM_t}{dt} \approx A\sqrt{\frac{D}{\pi t}} \tag{33.6}$$

where M_t is the amount of solute released in time, t, per unit area and providing the skin acts as a perfect sink and is not a barrier to permeation. This is unlikely to be true when hydrophilic photosensitizers are applied to intact skin. However, the disordered structure associated with many superficial infected lesions and the lack of a keratinized barrier in certain mucous membranes, such as the oral cavity and vulval epithelium, may present situations in which Equation 33.5 is valid.

It is clear in Equation 33.5, as with Equation 33.3, that M_t versus D is linear. This demonstrates that the amount of a photosensitizer released from a semisolid matrix, such as a cream vehicle, does not remain constant with respect to time. Thus, the amount of photosensitizer released during the initial application phase is more than that after a period of time has elapsed. It is also clear that the drug release can be enhanced by increasing A and D. Indeed, these strategies have been manifest in the high loading characteristic of cream-based preparations and the use of alkyl esters of 5-aminolevulinic acid; the latter pro-drug approach increases the diffusivity of the parent compound through lipid-rich regions of the *stratum corneum*.

Equation 33.6 defines the relationship between the diffusivity of the drug and the rate of release. The diffusivity of a hydrophilic photosensitizer through a matrix with a less hydrophilic property or characteristic, such as an emulsion system with a high proportion of lipid, is expected to be low. This will not be so when using a more hydrophilic vehicle, such as a structured aqueous hydrogel, whereupon the water solubility of the photosensitizer will permit a more-rapid flux through the matrix. Permeation through the biological barrier arising from a crystalline photosensitizer deposition on the barrier surface is most unlikely unless dissolution occurs in advance.

Over the past 20 years, topical PDT has been the center of intense investigation. Although still widely considered to be an experimental technique, its status and value within modern clinical practice continue to grow. Aminolevulinic acid (ALA) and its derivatives have continued to be the agents of choice in topical PDT; however, a number of drawbacks limit its widespread application (Donnelly et al. 2008).

Conventional formulations such as cream- and ointment-type preparations are generally only suitable for dry, flat areas of the body, while the aqueous nature of these vehicles can affect ALA stability and the lipophilic components retard drug release. Moreover, the lack of reproducibility with regards to the applied dose in reported clinical trials makes it almost impossible to compare data. Innovative drug delivery strategies, such as novel patch-type systems, have gone a long way toward overcoming these limitations.

Whereas response rates for superficial lesions continue to impress, ALA penetration into deeper lesions, such as nodular basal cell carcinomas (BCCs), continues to be a primary concern (McLoone et al. 2008). Strategies such as prodrug esterification and use of chemical permeation enhancers have proved promising *in vitro* (Morrow et al. 2010). However, there is little clinical evidence to indicate that these approaches are significantly superior to simple ALA preparations (Donnelly, McCarron, and Woolfson 2009).

Recently a number of drug-delivery groups have made a concrete effort to develop novel techniques of maximizing ALA delivery to the target area, and *in vitro* results that are reported look encouraging if ALA penetration can be significantly enhanced, but still a number of questions exist. Most notably, will this, in turn, induce greater PpIX production and, subsequently, enhanced neoplastic cell death? Will this be accompanied with increased pain for the patient, making treatment impractical?

Several groups have begun to seek the benefits of topical delivery with the increased flexibility of preformed photosensitizers.

Physical techniques, such as the use of needle-free jet injectors and micro-fabricated microneedle arrays have the potential to facilitate delivery of relatively high molecular weight molecules across the skin (Donnelly et al. 2007, 2009; Clementoni, B-Roscher, and Munavalli 2010). Such delivery systems would not only assist penetration to deep lesions but also permit almost instant drug delivery, thus negating the time for photosensitizer production or ALA diffusion. Furthermore, preformed photosensitizers that are excited at longer wavelengths would facilitate deeper light penetration, further optimizing this treatment modality (Garland et al. 2009).

PACT is clearly a viable treatment option for localized infection. However, factors such as acceptable delivery of photosensitizers to their site of action may impede the development of PACT and reduce its clinical applicability. Aspects of drug delivery that are important in PACT include the site of action, the physicochemical characteristics of the photosensitizer, and the characteristics of the formulation (Cassidy et al. 2009).

From the oral mucosa to the skin surface, the structure and, thus, characteristics of the target site vary markedly. A drug-delivery device should be able to withstand high moisture without degradation in the case of delivery to the oral cavity. A device should provide the moist environment essential for diffusion of photosensitizer to the site of action in the case of delivery to the skin or onychomycosis-infected nails. In some cases, the temperature and pH of the site may also affect release of drugs from dosage forms.

Factors such as the size of the molecule, its pKa, and lipophilicity will affect transport of the molecule to the site of action. With respect to drug-delivery devices, the solubility and diffusivity of a drug in the dosage form and its chemical stability are vitally important.

Drug concentration within the formulation, rate of drug delivery, and required contact time with the site of action for adequate drug release are all governed by the formulation. Issues such as stability and ease of application and removal after use are important with respect to usability in a clinical situation (Cassidy et al. 2009).

To date, current research in PACT has focused on achieving significant killing *in vitro* with little emphasis on drug delivery. Therefore, simplistic methods of drug delivery have been employed as research progresses from experimental lab-based studies, through animal experiments, to the clinical environment. Preparation of photosensitizer solutions immediately prior to use is not possible in the clinical scenario, in which most patients would require treatment in one day. In this case, the scope for error in preparation of solution and dosing, in addition to the time constraint placed on clinicians, would make PACT a nonviable option. Thus, more complex drug-delivery systems that facilitate administration of photosensitizer to the site of action are required.

Significant achievements have been made to date in the field of drug delivery for PACT. The Levulan Kerastick is a device that can reduce preparation time for clinicians by removing the need for measuring photosensitizer and solvent prior to use. However, solution-based delivery systems will be almost impossible to keep in place for sufficient times to allow drug absorption. For more complex and controlled drug delivery, patch-based systems in current development provide an exciting alternative in this field. Doses may be controlled by duration of contact with these patches. Formulations adhere to the site of action and deliver highly colored photosensitizer only to the site that would be treated for the patient.

Despite the vast number of studies published in the area of PACT, a rational approach to formulation design has not taken place. This may be because this field is dominated by clinicians and basic scientists, rather than those involved in pharmaceutical formulation development. Studies published to date in PACT have used aqueous solutions, oil-in-water creams, hydrogels, organogels, cubic phase formulations, and aqueous and solvent-based patches (Cassidy et al. 2009). These dosage forms, which, in many cases, seem to have been selected at random with little regard to their nature, possess a multitude of different physicochemical properties. When formulating a drug-delivery system, the aim should be to maximize the thermodynamic activity of the drug substance in the vehicle so as to maximize the concentration drive for diffusion and the partition coefficient between the target site in the body and the vehicle (Cassidy et al. 2011). The situation is, of course, different when photosensitizers are immobilized on insoluble supports for use in water disinfection or as implantable biomaterials (Cassidy et al. 2009). In these cases, there is a real need to demonstrate broad-spectrum antimicrobial activity while ensuring damage to human tissue is minimized. From a practical point of view, systems that can eradicate high levels of both adhering microorganisms and those in close proximity in relatively short periods of time (hours rather than days) without being subject to significant photobleaching or detachment of photosensitizer from the insoluble support should be the ultimate goal. It is difficult to see the real benefit of using targeting devices for PACT photosensitizers, given their already-demonstrated affinity for microbial cells over their mammalian counterparts (Embleton et al. 2005). This is especially so as one oft-cited reason for lack of development of resistance to PACT is its multifactorial mechanism of action, which could be compromised by targeting attachment to a specific microbial structure.

In recent times, PACT has been the subject of intense investigation in the laboratory, but now it is also, to some extent, in the clinic. Although still considered as an experimental technique, the status of PACT and its value within modern clinical practice is likely to continue to grow, especially considering the increasing incidence of antibiotic-resistant microorganism-associated infections. It is hoped that small-scale clinical trials carried out by small- or medium-sized enterprises, such as Destiny Pharma and Photopharmica, will lead to more widespread use of PACT, which finally leads to the ultimate benefit of patients worldwide.

33.3 Systemic Drug Delivery

The treatment of solid tumors and angiogenic ocular diseases by PDT typically requires the injection of a photosensitizer to destroy target cells upon subsequent irradiation. Historically, injection of crude photosensitizer mixtures (e.g., hematoporphyrin derivative) has led to prolonged cutaneous photosensitivity, often lasting several weeks, as a result of poor tumor-to-normal tissue accumulation ratios (Sibani et al. 2008). There is currently great interest in the development of efficient and specific carrier-delivery platforms for systemic PDT, and much progress has been made in reduction of side effects. Much recent work in academia and the industry has focused on systemic carrier-delivery platforms for PDT with an emphasis on target specificity. Promising carrier-delivery platforms for systemic PDT, including photosensitizer conjugates (e.g., with antibodies), dendrimers, micelles, liposomes, and nanoparticles, have led to numerous clinical successes, and a number of products are now marketed (Sibani et al. 2008; Camerin et al. 2010; Bullous, Alonso, and Boyle 2011; Stuchinskaya et al. 2011). Photosensitizer conjugates and supramolecular delivery platforms can improve PDT selectivity by exploiting cellular and physiological specificities of the targeted tissue (Li et al. 2012; Schmitt and Juillerat-Jeanneret 2012). Overexpression of receptors in cancer and angiogenic endothelial cells allows their targeting by affinity-based moieties for the selective uptake of photosensitizer conjugates and encapsulating delivery carriers, while the abnormal tumor neovascularization induces a specific accumulation of heavy weighted photosensitizer carriers through the enhanced permeability and retention effect. In addition, polymeric pro-drug delivery platforms triggered by the acidic nature of the tumor microenvironment or the overexpression of proteases have been designed and extensively evaluated (Sibani et al. 2008; Schmitt and Juillerat-Jeanneret 2012).

33.4 Conclusion

It is clear that PDT has an important role to play in the treatment of neoplastic disease at body sites amenable to irradiation. While systemic delivery systems for photosensitizing drugs have reached a high level of sophistication, the same cannot be said for topical delivery vehicles, in which much work is still required in order to enhance tissue penetration and improve therapeutic outcomes for patients, especially those suffering from deeper lesions, such as nodular basal cell carcinomas. The situation with antimicrobial applications is much less developed, and this is likely to remain so without significant investment from funding agencies, which currently looks at the PACT as more of a curiosity than the viable therapy for antibiotic-resistant wound and burn infections.

References

Bullous, A. J., C. M. Alonso, and R. W. Boyle. 2011. Photosensitiser-antibody conjugates for photodynamic therapy. *Photochem Photobiol Sci* 10:721–750.

Camerin, M., M. Magaraggia, M. Soncin et al. 2010. The *in vivo* efficacy of phthalocyanine-nanoparticle conjugates for the photodynamic therapy of amelanotic melanoma. *Eur J Cancer* 46:1910–1918.

Cassidy, C. M., M. M. Tunney, P. A. McCarron, and R. F. Donnelly. 2009. Drug delivery strategies for photodynamic antimicrobial chemotherapy: From benchtop to clinical practice. *J Photochem Photobiol B* 95:71–80.

Cassidy, C. M., M. M. Tunney, D. L. Caldwell, G. P. Andrews, and R. F. Donnelly. 2011. Development of novel oral formulations prepared via hot melt extrusion for targeted delivery of photosensitizer to the colon. *Photochem Photobiol* 87:867–876.

Clementoni, M. T., M. B-Roscher, and G. S. Munavalli. 2010. Photodynamic photorejuvenation of the face with a combination of microneedling, red light, and broadband pulsed light. *Lasers Surg Med* 42:150–159.

Donnelly, R. F., P. A. McCarron, D. I. Morrow, S. A. Sibani, and A. D. Woolfson. 2008. Photosensitiser delivery for photodynamic therapy. Part 1: Topical carrier platforms. *Expert Opin Drug Deliv* 5:757–766.

Donnelly, R. F., P. A. McCarron, and D. Woolfson. 2009. Drug delivery systems for photodynamic therapy. *Recent Pat Drug Deliv Formul* 3:1–7.

Donnelly, R. F., D. I. J. Morrow, P. A. McCarron, M. J. M. Garland, and A. D. Woolfson. 2007. Influence of solution viscosity and injection protocol on distribution patterns of jet injectors: Application to photodynamic tumour targeting. *J Photochem Photobiol B* 89:98–109.

Donnelly, R. F., D. I. Morrow, P. A. McCarron et al. 2009. Microneedle arrays permit enhanced intradermal delivery of a preformed photosensitizer. *Photochem Photobiol* 85:195–204.

Embleton, M. L., S. P. Nair, W. Heywood et al. 2005. Development of a novel targeting system for lethal photosensitization of antibiotic-resistant strains of *Staphylococcus aureus*. *Antimicrob Agents Chemother* 49:3690–3696.

Garland, M. J., C. M. Cassidy, D. Woolfson, and R. F. Donnelly. 2009. Designing photosensitizers for photodynamic therapy: Strategies, challenges and promising developments. *Future Med Chem* 1:667–691.

Li, P., G. Zhou, X. Zhu et al. 2012. Photodynamic therapy with hyperbranched poly(ether-ester) chlorin(e6) nanoparticles on human tongue carcinoma CAL-27 cells. *Photodiagn Photodyn Ther* 9:76–82.

McLoone, N., R. F. Donnelly, M. Walsh et al. 2008. Aminolaevulinic acid diffusion characteristics in 'in vitro' normal human skin and actinic keratosis: Implications for topical photodynamic therapy. *Photodermatol Photoimmunol Photomed* 24:183–190.

Morrow, D. I. J., P. A. McCarron, A. D. Woolfson et al. 2010. Hexyl aminolaevulinate is a more effective topical photosensitiser precursor than methyl aminolaevulinate and 5-aminolaevulinic acids when applied in equimolar doses. *J Pharm Sci* 99:3486–3498.

Schmitt, F., and L. Juillerat-Jeanneret. 2012. Drug targeting strategies for photodynamic therapy. *Anticancer Agents Med Chem* 12:500–525.

Sibani, S. A., P. A. McCarron, A. D. Woolfson, and R. F. Donnelly. 2008. Photosensitiser delivery for photodynamic therapy. Part 2: systemic carrier platforms. *Expert Opin Drug Deliv* 5:1241–1254.

Stuchinskaya, T., M. Moreno, M. J. M. Cook, D. R. Edwards, and D. A. Russell. 2011. Targeted photodynamic therapy of breast cancer cells using antibody-phthalocyanine-gold nanoparticle conjugates. *Photochem Photobiol Sci* 10:822–831.

34

Antimicrobial PDT for Clinical Infectious Diseases

Anja Eichner
Regensburg University Hospital

Fernanda Pereira
Gonzales
Regensburg University Hospital

Tim Maisch
Regensburg University Hospital

34.1 Introduction

Photodynamic activity of chemical compounds against microorganisms was first published more than 100 years ago. Oskar Raab observed that toxicity of acridine hydrochloride against *Paramecium caudatum* was dependent on the amount of light that was incident on the experimental mixture. In addition, his teacher, Hermann von Tappeiner, reported that the toxic effects in the presence of light are not a result of heat. Afterwards, von Tappeiner termed the influence of light and oxygen in combination with a nontoxic dye as a "photodynamic reaction." Additional investigations demonstrated the involvement of oxygen in killing bacteria because the antibacterial activity of fluorescent dyes against the facultative anaerobic species *Proteus vulgaris* could not be demonstrated in the absence of oxygen.

Since the middle of the last century, antimicrobial photodynamic therapy (aPDT) passes out of mind because of the discovery of antibiotics. The first antimicrobial substance was penicillin, discovered by Alexander Fleming in 1928. This was the beginning of the "golden age of antibiotics," and almost all strains of *Staphylococcus aureus* were susceptible to penicillin. In 1944, resistance to penicillin in *S. aureus* was recognized after the first tests in patients. Therefore, methicillin was released in 1960, and only one year later, methicillin-resistant *S. aureus* strains (MRSA) appeared. Nevertheless, even new classes of antibiotics were not able to stop the worldwide rise of multiple drug–resistant strains, which cause much concern in the treatment of several diseases. The situation will be additionally exacerbated because the pipeline of new antibiotics is not fully straight. The Infectious Diseases Society of America (IDSA) highlights that, over the past several years, the number of new antibacterial drugs approved continues to decrease. Only three of the eight antibiotics approved within the last decade act toward a new mechanism of action, whereas all the others are only modifications of already known antibiotics (Spellberg 2008; Spellberg et al. 2004).

Reasons for the increasing emergence of antibiotic-resistant strains were the massive application of antibiotics against pathogenic bacteria in the last 50 years, the failure of patients to complete their therapy, and the use of antibiotics in the food and farming industries. In this context, aPDT may be an approach to substitute or act as a coadjuvant in the conventional antimicrobial therapy of the future.

In recent decades, the incidence of microbial infections has considerably increased for various reasons, for instance, the more frequent use of invasive procedures, immunosuppressive medications, and broad-spectrum antibiotics as well as the increased incidence of neutropenia and HIV infections (Donnelly, McCarron, and Tunney 2008).

Opportunistic pathogens from the genera *Candida*, *Aspergillus*, and *Cryptococcus* can cause skin and mucous membrane infections or invasive fungal infections, particularly in immunocompromised patients. In these patients, invasive fungal infections are often associated with high morbidity and mortality (Espinel-Ingroff 2009; Groll and Tragiannidis 2009). The crude mortality related to opportunistic fungal infections still exceeds 50% in most human studies (Romani 2004). Skin mycoses affect more than 20%–25% of the world's population, making them one of the most frequent forms of infection (Dismukes 2000). Only a few drugs are available for treating this kind of infection, and the emergence of antifungal resistance has decreased the efficacy of conventional therapies. Treatments are time consuming and thus demanding on health care budgets. Additionally, current drugs only have a limited spectrum of action, and toxicity and drug interactions are common.

The drug of choice for the treatment of many invasive or life-threatening mycoses is amphotericin B (Amp B), discovered in the 1950s, but it presents dose-dependent nephrotoxicity (Martino and Girmenia 1997; Neely and Ghannoum 2000). The lipid formulations of Amp B greatly reduce the side effects of the parent drug; however, the acquisition cost of these compounds is considerably more expensive than conventional Amp B, ranging from 10- to 20-fold higher in cost per dose (Dismukes 2000). Fluconazole, introduced in 1988, has been widely used for treating fungal infections; however, the recently emerged resistance to this drug may limit its use in the future. The failure of antifungal therapy in immunocompromised patients is related to the fungistatic activity of azoles (Arana, Nombela, and Pla 2010). The newest class of antifungal drugs, the echinocandins, was introduced more than 15 years ago and represents the first class of clinically useful antifungal drugs to inhibit synthesis of the fungal cell wall (Denning 2003). Clinical isolates of *C. albicans* resistant to the echinocandin drug caspofungin are slowly emerging and are linked to either point mutation (Perlin 2007) or elevation of chitin levels in the cell wall (Lee et al. 2012). Despite their broad spectrum of activity, the echinocandins do not possess *in vitro* activity against important basidiomycetes, including *Cryptococcus*, *Rhodotorula*, and *Trichosporon* (Arana et al. 2010).

Another aspect related to antifungal resistance and recurrence of infection is the ability of fungal cells to form biofilms on surfaces. Fungal biofilms not only tend to be more resistant to antimicrobial agents than their platonically grown forms but also are known to withstand host immune defenses (Douglas 2003; Kojic and Darouiche 2004).

34.2 General Aspects of aPDT

34.2.1 Mechanism of Action

The antimicrobial photodynamic process presents several positive aspects for the treatment of microbial infections, including a broad spectrum of action, the efficient inactivation of antibiotic-resistant strains, the low mutagenic potential, and the lack of selection of photoresistant microbial cells (Jori et al. 2006).

The concept of aPDT is the application of a nontoxic dye, known as a photosensitizer (PS), which can be activated by visible light (390–700 nm). Upon irradiation the PS skips from its energy ground state to a higher energy triplet state. In this setup, the activated PS can react directly with biomolecules to produce free radicals (electron transfer, type I reaction) or react with molecular oxygen to generate highly reactive singlet oxygen (energy transfer, type II reaction) (Figure 34.1). In aPDT, singlet oxygen is the major reactive species involved in the microbial cell damage, and it is suitable for killing bacteria, viruses, and fungi on localized sites (such as the skin) because of the short lifetime and the short radius of action (0.02 µm) (Rajesh et al. 2011). Therefore, the surrounding tissue is not concerned with $^1O_2$ damage, making the type II reaction an ideal candidate for therapy of topical infections. Multiple cellular targets are available for the photooxidative effect caused by singlet oxygen, including inactivation of enzymes/proteins and lipid peroxidation, leading to the lyses of cell membranes, lysosomes, and

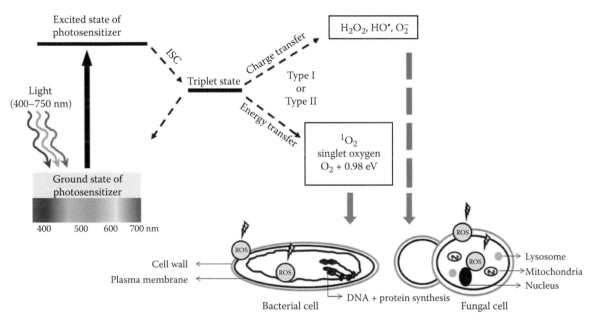

FIGURE 34.1 Schematic illustration of mechanism of action. Generation of ROS can follow two alternative pathways after light activation by PS. Upon absorption of a photon of appropriate wavelength in the visible light spectra (400–700 nm) by the ground-state PS, the excited state of the PS is formed. The excited state is short-lived and can undergo intersystem crossing to a long-lived triplet state or, alternatively, can return to the ground state by fluorescence emission or heat or both. Generally, triplet state acts as a mediator of type I/type II photosensitization processes. Type I: Generation of hydrogen peroxide (H_2O_2), hydroxyl radical (HO·), and superoxide anion $\left(O_2^-\right)$ by charge transfer from the excited PS. Type II: The triplet state of PS can directly undergo energy exchange with triplet ground state oxygen, leading to the formation of singlet oxygen, $^1O_2$. The generated ROS rapidly react with their environment depending on the localization of the excited PS: microorganism walls, lipid membranes, peptides, and nucleic acids.

mitochondria (Bertoloni et al. 1989). Thus, singlet oxygen generated by the excitation of PS is a nonspecific oxidizing agent, and there is no cellular defense against it (Donnelly, McCarron, and Tunney 2008).

The most suitable application of aPDT is the treatment of superficial and localized infected tissues and the possibility of reducing the nosocomial colonization of multiresistant bacteria of the skin. This delivery seems to be the most promising feature of aPDT because it neither harms the surrounding tissue nor disturbs the residual bacteria-flora of tissues.

34.2.2 Susceptibility of aPDT to Pathogens

The different sensitivity to aPDT between Gram-positive and Gram-negative bacteria and fungi is a result of cell wall structures and the size of these microbial cells (Figure 34.2). The cell wall of Gram-positive bacteria is composed of 40–80 nm thick outer peptidoglycan layers within lipoteichoic acid molecules. Although the cell wall consists of up to 100 peptidoglycan layers, the PS is able to cross this murein sacculus because of more or less porosity (Malik, Ladan, and Nitzan 1992). Additionally, the negatively charged lipoteichoic acids contribute to the binding of cationic agents (Lambert 2002; Minnock et al. 2000). In contrast, the cell wall of Gram-negative bacteria consists of three layers: (1) an inner cytoplasmic membrane, (2) a small layer of peptidoglycan in the periplasmic space, and (3) an outer membrane with negatively charged lipopolysaccharides (LPSs). This outer membrane provides an effective permeability barrier and limits the binding and penetration of PS (Minnock et al. 2000). Application of permeabilizing agents (e.g., Tris-EDTA or polymyxin nonapeptide) enhances the effectiveness of aPDT against Gram-negative bacteria (Malik, Ladan, and Nitzan 1992; Nitzan et al. 1992; Valduga et al. 1993) because of withdrawing divalent cations, in particular, Ca^{2+} and Mg^{2+}, which are responsible for

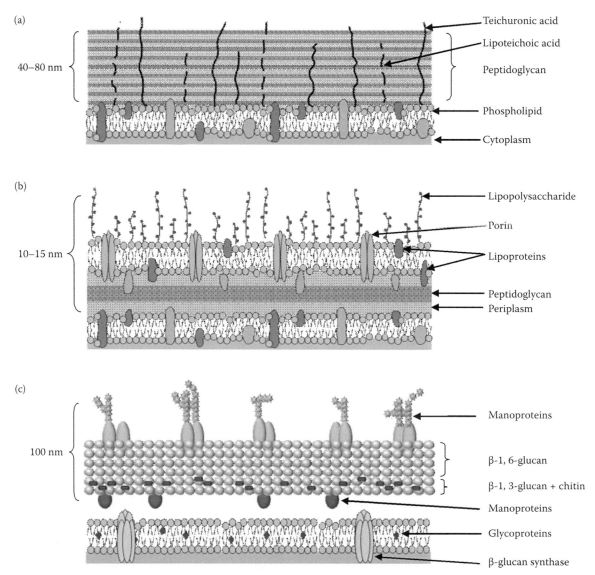

Figure 34.2 Cell walls of (a) Gram-positive bacteria, (b) Gram-negative bacteria, and (c) fungal cells. (Modified from Kharkwal, G. B. et al., *Lasers Surg Med*, 43:755–767, 2011.)

stabilization of negatively charged LPS molecules in the outer membrane.

Fungal cells are eukaryotic organisms and also present an overall negative charge. They are enveloped by a thick cell wall composed of glucan, mannan, and chitin (Figure 34.2c). They have a relatively permeable outer layer intermediate between Gram-positive and Gram-negative bacteria. Thus, even a negatively charged porphyrin derivative, such as Photofrin, is accumulated by *Candida* species, causing extensive photokilling of this microorganism (Dai, Huang, and Hamblin 2009). However, the presence of a nuclear membrane and the greater size/volume of *C. albicans* make this microorganism more resistant to aPDT than Gram-positive bacteria (Bliss et al. 2004; Demidova and Hamblin 2005).

Recent reports have shown that yeasts and dermatophytes can be efficiently killed *in vitro* by aPDT (Mang, Mikulski, and Hall 2010; Smijs and Pavel 2011). Few studies explored the effectiveness of aPDT to treat clinical infections, but these findings increased the interest in animal studies and clinical trials. One interesting aspect of *in vivo* studies is that the selectivity of aPDT may be enhanced because the diffusion distance of singlet oxygen from its site of generation is limited, and the PS and illumination are applied directly in the infected area (Donnelly, McCarron, and Tunney 2008).

PSs for antifungal PDT should present hydrophilic properties and positive charge for better penetration in the cell wall. The localization of the PS is restricted to the cytoplasm because the nuclear membrane limits molecular penetration into the nucleus. This, consequently, restricts the mutagenic potential of aPDT. The photodynamic mechanism damages fungal cells by ROS perforating cell walls and membranes, thus allowing the PS to be translocated into the cell. Once inside the cell, oxidizing species generated by light excitation induce photodamage to internal cell organelles and finally cell death (Bertoloni et al. 1989; Donnelly, McCarron, and Tunney 2008). In addition, fungal cells can be killed at photodynamic dose rates much lower than light doses that kills keratinocytes, which means that a therapeutic window is available (Zeina et al. 2002).

Irradiation parameters depend on the application of interest. Decolonization of microorganisms growing on the surface of the skin (*stratum corneum*) will be inactivated by blue light (approximately 400 nm); in deeper infections caused, for example, by dermatophytes that can colonize both the *stratum corneum* and hair follicles, red light (approximately 600 nm) is sufficient. Despite *C. albicans* infections being restricted to *stratum corneum*, extracellular enzymes may cause tissue changes in the deepest layers of the epithelium, so red light can be more useful for the treatment of such infections (Samaranayake and Samaranayake 2001). Power outputs for light sources are usually in the range of 10–100 mW/cm^2 with total light doses between 10 and 200 J/cm^2 (Donnelly, McCarron, and Tunney 2008). One crucial point in light dosimetry is to avoid thermal effects of the light itself when light doses above 100 mW cm^{-2} are needed for aPDT.

34.3 Chemical Classes of PSs and Their Application in Infectious Diseases

Different classes of PSs have shown an efficient inactivation of Gram-positive and Gram-negative bacteria as well as fungi. The most commonly used PSs in aPDT are cationic agents (Sharma et al. 2011), such as (meso-substituted) porphyrins or water-soluble cationic zinc phthalocyanines (Minnock et al. 1996). In addition, positively charged phenothiazines, such as methylene blue (MB) and toluidine blue, have been also tested successfully as PSs against Gram-positive, Gram-negative pathogens and fungal cells (Merchat et al. 1996; Wainwright 1998).

34.3.1 Phenothiazines

MB belongs to the class of phenothiazines that are characterized by a tricyclic structure (Figure 34.3). MB has a strong absorbance in the range of 550–700 nm with a quantum yield of $\Phi_\Delta = 0.52$ (Redmond and Gamlin 1999). After substituting the structure with methyl/ethyl groups [dimethyl MB (DMMB), new methylene blue (NMB); Figure 34.3] or combining with a nitro group (methylene green), the efficiency of inactivation against both Gram-positive strains *Staphylococcus aureus*, *Enterococcus faecalis*, and *Bacillus cereus* and Gram-negative strains *Escherichia coli* and *Pseudomonas aeruginosa* was increased (Wainwright et al. 1997, 1998). Pal, Ghosh, and Ghosh (1990) showed that lipophilic DMMB binds to teichoic acid molecules in the cell wall of Gram-positive bacteria. This lipophilic feature may be the reason for better inactivation of DMMB in contrast to MB, which is more lipophobic. It is also known that MB reacts with guanine bases of DNA, and therefore, MB causes irreparable DNA damage subsequently followed by cell death. Even at high concentrations of 10 mg/mL, MB has no toxicity in humans (Creagh et al. 1995; Schneider et al. 1990). For treatment of fungi cells, the presence of additional ethyl groups also yielded an enhanced phototoxicity when compared to the parental molecule MB. The lipophilic character of new MB (NMB) showed a better efficacy compared to the traditional MB for controlling of infections caused by *Candida* sp. (Dai et al. 2011).

An important clinical application is the use of MB in sterilization of human blood components. MB has no reported toxicity in humans, whereas it effectively destroys extracellular enveloped viruses existing in human plasma units (DeRosa and Crutchley 2002). Another *in vivo* application of phenothiazines was the investigation of PDT-based clinical and microbiological efficacy against the periodontopathogenic species *Fusobacterium nucleatum*. Patients with localized chronic periodontitis were treated with soft diode laser light (660 nm, 60 mW/cm^2, 6 × 10 s) in combination with a phenothiazine chloride PS solution (Sigusch et al. 2010). Patients who received PDT treatment showed a significant reduction in reddening, bleeding on probing, mean probing depth, and clinical attachment level. Furthermore, the periodontal inflammation symptoms within 3 months of observation were reduced, as well as the level of detectable *F. nucleatum* DNA in comparison to the control group (Sigusch et al. 2010).

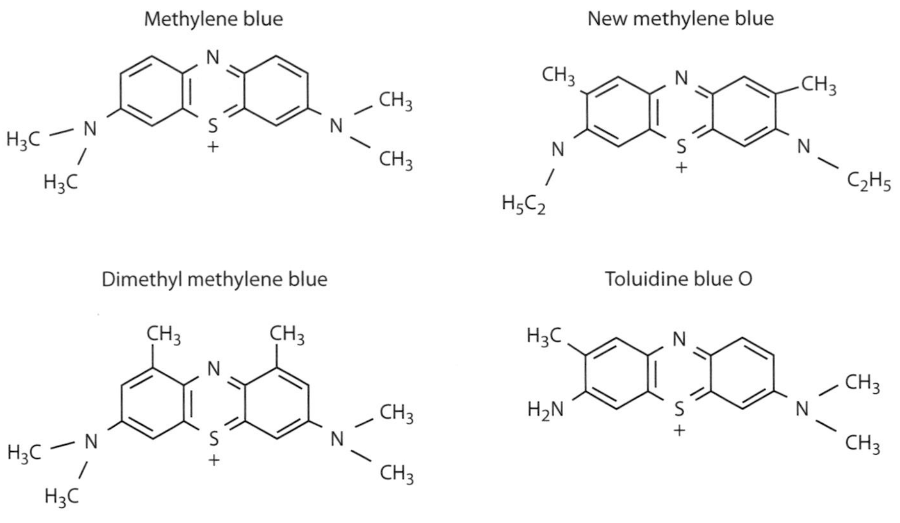

Methylene blue

New methylene blue

Dimethyl methylene blue

Toluidine blue O

FIGURE 34.3 Chemical structures of phenothiazine dyes.

Helicobacter pylori, a Gram-negative pathogen, infects the human gastric mucosa and is closely correlated with upper gastrointestinal tract diseases, such as acute/chronic gastritis, peptic ulcer, and gastric cancer (Venerito, Wex, and Malfertheiner 2011). It is supposed that *H. pylori* has a poor ability to repair DNA damage because of only few genes in its genome associated with restoration of light-induced DNA damage (Goosen and Moolenaar 2008). In addition, it was shown that *H. pylori* have an intrinsic sensitivity to photoinactivation as a result of naturally aggregating porphyrins (Hamblin et al. 2005). An *in vitro* reduction of *H. pylori* was achieved by using 0.2 mg/mL MB and an endoscopic light source. After 2 min of irradiation, the number of viable cells decreases from 2.0×10^8 to 2.3×10^6, a reduction of nearly 2 $\log_{10}$ steps. After 5 min of exposure to light (940 lx), a reduction of viable cells of 7 $\log_{10}$ steps was detected (Choi, Lee, and Chae 2010).

Toluidine blue O (TBO) is another phenothiazine dye that is predominantly used in dentistry. Pinheiro et al. (2009, 2010) showed a great reduction (81.24%–98%) of oral bacteria including *E. faecalis* and *Streptococcus mutans* in deciduous teeth with necrotic pulp and in periodontal pockets by using 50% toluidine blue at 0.005 mg/500 mL. TBO was activated by red laser light (660 nm, 4 J/cm², 40 s) or a low-intensity diode laser (632.8 nm, 4 J/cm², 60 s), respectively. Furthermore, a 3 $\log_{10}$ reduction was achieved of a suspension of *S. mutans* ($1–2 \times 10^8$ cfu mL⁻¹) in the presence of TBO (163.5 µM) and red light-emitting diode (LED) light (636 nm, 24 J/cm²) (Rolim et al. 2012).

In general the localisation of phenothiazines in yeast cells is the plasma membrane. Consequently, this structure is the first to be photodamaged. Probably, the reason for cell death is the increasing permeability from such injury (Paardekooper et al. 1995). Chromoblastomycosis, a chronic fungal skin infection caused by traumatic inoculation of dematiaceous fungi species (*Fonsecaea pedrosoi* and *Claphialophora carrionii*) (Rubin et al. 1991), is usually refractory to conventional treatment (Queiroz-Telles et al. 2009). A combination of MB and red LED light was

used in the treatment of 10 patients with chromoblastomycosis. The treatment consisted of the local application of 20% MB preparation in cream to the skin lesion for 4 h and, subsequently, illumination with a light dose of 12.1 J/cm². After six treatments at weekly intervals, all the patients showed a reduction of 80%–90% of lesion volume and tissue cicatrization, and no side effect has been noted during or after the treatment. The authors emphasize the use of aPDT as a coadjuvant treatment because the complete wound healing was not achieved, but aPDT reduced significantly the size of the compromised areas and the degree of infection.

C. albicans is not only the main species associated with oral mycoses in humans but also the most common mycosis affecting patients with HIV (Samaranayake, Keung Leung, and Jin 2009). This may be significant in the view that approximately 81% of HIV infected patients infection are estimated to be colonized with azole-resistant strains (Johnson et al. 1995), and locally administered aPDT would not be expected to cause either an increased burden on the immune system or additional side effects (Wainwright 1998). *C. albicans* strains used in the following experiment expressed luciferase, allowing real-time monitoring of the extent of infection in mice noninvasively through bioluminescence imaging. The treatment began either at 30 min or at 24 h after fungal inoculation in the surface of mouse wounds to investigate the efficacy of aPDT for both prophylaxis and treatment of infections. A combination of 400 µM of NMB and an applied dose of 78 J/cm² (for aPDT at 30 min post-infection) or 120 J/cm² (for aPDT at 24 h post-infection) was used. Almost no dark toxicity was observed, and a significant reduction in the *C. albicans* burden in both cases was achieved as demonstrated by bioluminescence imaging (Dai et al. 2011).

Even after efficient microbial killing, virulence factors may still be present and cause extensive tissue damage. An important feature of aPDT is the ability of ROS to destroy secreted virulence factors, especially those that are proteins. *C. albicans* produces extracellular enzymes that are able to cause tissue destruction and invasion during infection (Yordanov et al. 2008). Martins Jda et al. (2011) evaluated the effect of aPDT using MB and a red laser light to investigate the secretion of extracellular enzymes by *C. albicans* in a rat model of buccal candidiasis. The yeasts recovered from the oral cavity of the rats after aPDT treatment were submitted to tests of phospholipase and proteinase activity. Photodynamic treatment inhibited only proteinase cell activity (Martins Jda et al. 2011). And yet, aPDT did not cause any histological alteration in the healthy tongue dorsum, confirming that it is a safe antimicrobial treatment, without causing damage to the host's healthy tissues (Martins Jda et al. 2011). According to the author, it is probable that the yeasts, which were not killed by aPDT, exhibited lower proteinase activity, which means that aPDT affected the cell secretion pathway as a potential mechanism of action.

Furthermore, aPDT also induces inflammatory signaling that led to activation of immune competent cells, such as macrophages and neutrophil granulocytes, which are involved in the process of yeast killing (Gollnick et al. 1997). In *Candida* infections, the induction of cell immunity stimulates the proliferation of T cells with subsequent cytokine synthesis, which enhances the candidacidal functions of the neutrophils and macrophages (Ashman and Papadimitriou 1995).

34.3.2 Porphyrins and Phthalocyanines

Porphyrines and phthalocyanines are aromatic molecules and consist of four pyrrolic rings (tetrapyrrol ring system). In porphyrines, the tetrapyrrol ring system is connected via methine groups; the pyrrolic rings in phthalocyanines are linked via nitrogen bonds. Porphyrin-based PSs have a strong absorption around 400 nm (Soret band) and four satellite absorption bands (Q bands) between 600 and 650 nm. Porphyrins, such as

protoporphyrin IX (PPIX, Figure 34.4), can be used in PDT in the form of its prodrug 5-aminolevulinic acid (5-ALA). PPIX is synthesized via the heme biosynthetic pathway, which is highly conserved in nonphototrophic bacteria (Schobert and Jahn 2002).

Recently, newly synthesized phthalocyanine derivates were used to kill *E. coli*, spores of sulfite-reducing *Clostridium*, and coliphages, which are common bacteria in drinking water (Kuznetsova et al. 2009). Two octacationic oxotitanium phthalocyanines with pyridiniomethyl or cholinyl substituents (each 2.5 µM) in combination with white light (500 W halogen Osram lamp, 240 W m^{-2}) were applied to water samples of the Moscow River. After irradiation of 30 min, they determined the colony forming units per milliliter, resulting in inactivation of coliform bacteria and coliphages, and only germinates of *Clostridium* spores were observed (efficacy 95%–97%) (Kuznetsova et al. 2009).

In 2012, Maisch and colleagues showed the *in vitro* inactivation of *Bacillus atrophaeus*, *S. aureus*, methicillin-resistant *S. aureus* (MRSA), and *E. coli* using short pulses of an intense pulse light (IPL) source. TMPyP (Figure 34.4) was applied at different concentrations (1, 10, 100 µM) and activated by light flashes of 83 or 100 ms (20 to 80 J/cm^2, respectively) of a commercially available IPL. For all tested strains, they showed a inactivation efficacy of > 5 log$_{10}$ steps within a few milliseconds. *S. aureus* was inactivated by 10 µM TMPyP and 1 × 10 J/cm^2, growth of *E. coli* was reduced by 100 µM and 2 × 20 J/cm^2, MRSA viability was affected by 10 µM and 20 J/cm^2, and *B. atrophaeus* cells were killed by 10 µM and 1 × 10 J/cm^2.

Another study utilized meso-tetra (N-methyl-4-pyridyl) porphine tetra tosylate (TMPyP, Figure 34.4) for the *in vitro* treatment of planktonic and biofilm-grown *Burkholderia cepacia* complex infections (Cassidy et al. 2012). In patients with cystic fibrosis (CF), *B. cepacia* and other pathogens, such as *P.*

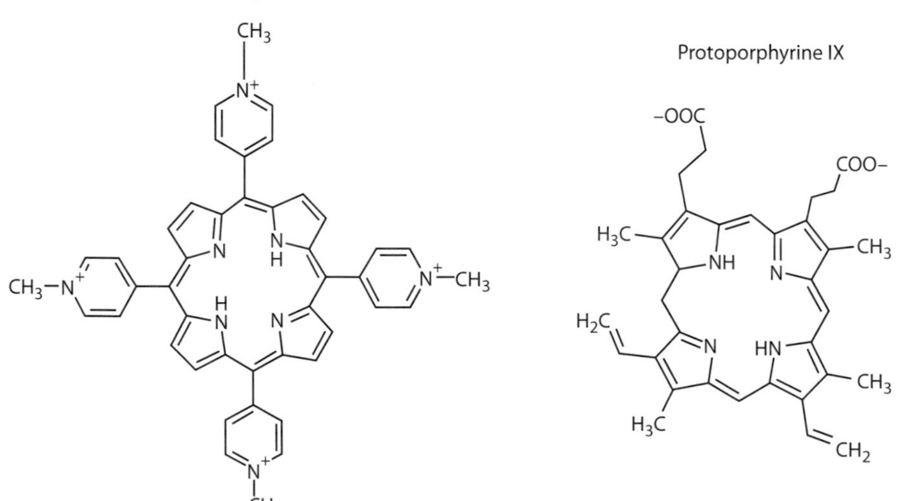

FIGURE 34.4 Chemical structures of TMPyP and PPIX.

aeruginosa, cause chronic lung infection, which leads to permanent lung inflammation and decreasing lung function, causing high morbidity and mortality. For illumination, Cassidy and colleagues used a Paterson Lamp (635 nm) at a fluency rate of 200 mW/cm^2 (total light dose of 200 J/cm^2) and different concentrations of TMPyP. They showed a 3 log$_{10}$ reduction of planktonic *B. cepacia* at 250 mg/mL TMPyP and a 3 log$_{10}$ reduction of biofilm-grown cells at 750 mg/mL TMPyP (Cassidy et al. 2012). Because treatment of *B. cepacia* complex in pulmonary infection in CF is extremely difficult as a result of their inherent resistance to a wide range of antibiotics, these results seem to be good progress for aPDT treatment in CF. In addition, the susceptibility of *P. aeruginosa* to aPDT was already shown in 2001 using a hematoporphyrin (HpD) derivate with rutin and arginine substituents (Szpakowska et al. 2001).

Because porphyrins are not taken up by yeast cells, the photodamage is caused by molecules attached or in the areas surrounding the cells (Bertoloni et al. 1987; Bliss et al. 2004). After irradiation, there is an initial limited alteration of the plasma membrane, which leads to penetration of the dye into the cell, the translocation to the inner membranes, and the consequent photodamage of intracellular targets, such as mitochondria and ribosomes (Bertoloni et al. 1987). The tetracationic porphyrin TMPyP was used in combination with green light in the treatment of an ear pinna infected with *C. albicans* in a mouse model. Infected ears displayed complete healing over time without damage to the ear pinna with a concentration of 0.3 mg/mL and an applied light dose of 90 J/cm^2 (Mitra et al. 2011). To achieve outcomes more correlated to oral candidiasis in humans, a murine model of oral candidiasis represented by white patches or pseudomembranes on the dorsal tongue was submitted to aPDT by topical administration of the hematoporphyrin derivative Photogem and illumination with LED (305 J/cm^2) at 455 nm (blue light) or 630 nm (red light). A reduction of 1.59 log$_{10}$ was achieved with 500 mg/mL of Photogem associated with red LED light, whereas blue LED light promoted reduction of 1.39 log$_{10}$ in cell viability. Although Photogem better absorbs blue light, red light achieved higher penetration rates into the tissue, increasing photodynamic efficacy. One of the most important factors concerning the choice of light source is that the emission spectrum of the used light source matches the absorption spectra of the PS used to achieve maximum treatment efficacy. Although no complete inactivation was observed, this investigation showed that aPDT was significantly effective in reducing the viability of *C. albicans* cells. Besides this, histological evaluation of the tongue showed that aPDT did not cause any significant adverse effects to the local mucosa (Mima et al. 2010).

Five patients suffering from denture stomatitis caused by *Candida* sp. were submitted to six sessions of aPDT, two times per week over 15 days. The denture and palate of these patients were sprayed with 500 mg/mL Photogem and left in the dark for 30 min. After this period, denture and palate were illuminated with LED light with a light dose of 37.5 and 122 J/cm^2, respectively. A Peltier chip and an air cooler were used to dissipate the heat and avoid thermal effects. Follow up one and two months after the treatment showed that two patients had full recovery from the infection, one demonstrated reduction in palatal inflammation, and recurrence was observed in two patients. Moreover, four patients showed no inflammation right after the treatment. Overall, these results indicate that aPDT may represent a useful alternative to treat denture stomatitis. In these cases, longer treatment may be required to eliminate *Candida* infections, and this should not represent a barrier because, during the treatment, any side effect was reported by patients, and emergence of resistance is unlikely to occur (Mima et al. 2011).

34.3.3 5-Aminolevulinic Acid

Aminolevulinic acid (ALA) and its lipophilic derivate methyl aminolevulinate (MAL, Figure 34.5), which is de-esterified into ALA by intracellular enzymes (Katsambas and Dessinioti 2008), are not photodynamically active, but they are taken up directly by bacteria, yeasts, fungi, and some parasites, which then induce the accumulation of protoporphyrin IX (PPIX). PPIX has an absorption peak at 375–405 nm and a lower peak at 630–633 nm. Light irradiation of PPIX leads to the inactivation of these organisms via photodamage to their cellular structures (Harris and Pierpoint 2012). Photodynamic therapy with ALA/MAL has proven to be efficient in the treatment of several malignant skin tumors and inflammatory diseases (Babilas, Landthaler, and Szeimies 2006). Furthermore, PDT in dermatology using either ALA or MAL is well established and effective for the treatment of acne *vulgaris*. Acne is a multifactorial pleomorphic skin disorder of the pilosebaceous follicles and characterized by chronically inflamed comedones and lesions, such as papules, pustules, and nodules. Although there is a controversial discussion of the role of *Propionibacterium acnes* in the beginning of the disease

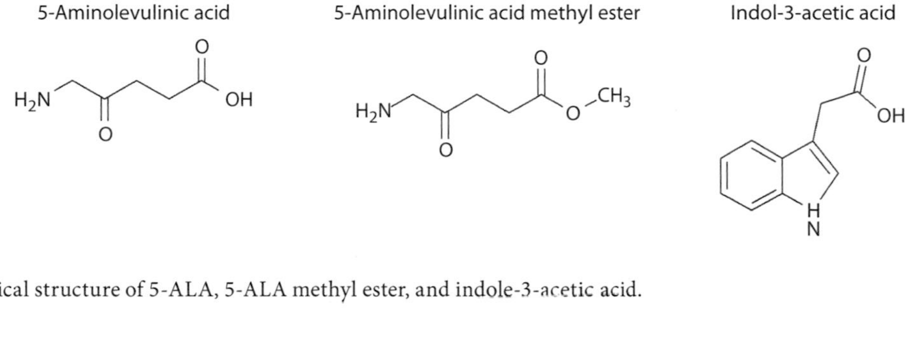

FIGURE 34.5 Chemical structure of 5-ALA, 5-ALA methyl ester, and indole-3-acetic acid.

(Shaheen and Gonzalez 2013), it is usually accepted that bacteria such as *P. acnes*, *S. aureus*, and other Gram-negative strains are involved in ongoing chronic inflammation of follicles. Several authors have been reported that topical application of ALA or MAL with a respective light source (diode laser, broadband light, LED, conventional polychromatic light) led to remarkable clinical improvement and reduction of inflammatory follicular lesions (Hongcharu et al. 2000; Horn and Wolf 2007; Pollock et al. 2004). However, adverse effects, such as pain, burning, erythema, edematous follicular reactions, and hyperpigmentation, are commonly observed in ALA-PDT.

In 2011, a new PS was introduced into aPDT treatment of acne *vulgaris*. Indole-3-acetic acid (IAA, Figure 34.5) is nontoxic by itself and can be activated by visible and ultraviolet light, whereas the most effective activation is at 520 nm (green light) (Na et al. 2011). In a pilot study, 14 acne patients were treated three times for 15 min over a 2-week interval with 0.015% IAA, which was subsequently irradiated by green light (520 nm, 9 J/cm²) after application on one side of the face (Na et al. 2011). The authors showed a significant reduction of inflammatory lesions and sebum secretion. Likewise, the treatment was not painful, and they did not observe adverse effects. In addition, they showed a remarkable decrease in viable *P. acnes* and *S. aureus* cells *in vitro* using IAA-PDT as mentioned above. The reduction of bacteria was dependent on the concentration used in the experiments. Nearly 90% of *P. acnes* and *S. aureus* were inactivated upon irradiation with green light (520 nm, 15 min, 9 J/cm²) and an IAA concentration of 10 mM (Na et al. 2011).

Onychomycosis is one of the most common infections in dermatology, and it is very difficult to treat with relapsing of the condition. This disease affects mainly immunocompromised patients, which are precluded from using systemic medications (Schalka, Nunes, and Gomes Neto 2012). There are some interesting successful cases of onychomycosis treated with topical application of ALA. A patient with onychomycosis of both toenails caused by *Trichophyton rubrum* was treated with aPDT-ALA. After surgical removal of the nail plate and nail bed hyperkeratosis, ALA was applied under an occlusive dressing and irradiated with laser light (100 J/cm²). The treatment was repeated three times at intervals of 15 days, and evaluation of

the patient after 1 year ensured a complete cure with residual mild traumatic onycholysis (Piraccini, Rech, and Tosti 2008). Another similar study used 20% solution of ALA methyl ester in aqueous cream together with laser light (630 nm, 100 J/cm²) to treat two patients with dermatophyte onychomycosis of the big toenails. The treatment, performed six to seven times at weekly intervals, led to a cure in the treated nail, and no recurrence was observed clinically at a 6-month follow-up visit. No dark toxicity and no light effect alone were observed (Watanabe et al. 2008).

Infections caused by the yeast *Malassezia* are usually recurrent, and long-term prophylaxis is often required to prevent relapse leading to the prevalence of drug-resistant strains (Levin 2009). The clinical efficacy of methyl-5-ALA (MAL) with noncoherent red light was used in the treatment of six patients with *Malassezia* folliculitis who either refused oral medication or were unable to take oral antifungal agents because of hepatotoxicity concerns. Skin lesions were treated with MAL cream and illuminated with red light (37 J/cm²). Three treatments were performed at 2-week intervals, and complete lesion and patient response rates were evaluated 1 month after the last treatment. In four patients, inflammatory lesions had decreased and improved obviously; in one patient, they had improved slightly; and one patient had not improved at all (Choi, Lee, and Chae 2010).

34.3.4 Fullerenes

A new class of PSs is named fullerenes, which are now under intensive study for potential biomedical applications. Fullerenes are arranged in a soccer ball structure and consist of 60–70 carbon atoms. They are biologically inert unless derivatized with functional hydrophilic or cationic groups (Figure 34.6) in which they become soluble and can act as a PS (Mroz et al. 2007). Fullerenes with polar diserinol and quarternary pyrrolidinium groups were reported to kill effectively *in vitro* (4–6 $\log_{10}$) both Gram-positive and Gram-negative bacteria *S. aureus*, *E. coli*, and *P. aeruginosa* as well as the fungus *C. albicans* (Tegos et al. 2005). Fullerenes were activated in 10 min with a broadband, white-light band-pass filter (400–700 nm) and a power of 200 mW cm⁻² to kill the mentioned microorganisms (Tegos et al. 2005). In addition, fullerenes were tested in an *in vivo* model

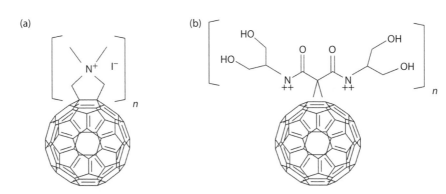

FIGURE 34.6 Chemical structures of cationic fullerenes. (a) Fullerene with quarternary pyrrolidinium groups (*n* = 1–3). (b) Fullerene with diserinol groups (*n* = 1–4).

with fatal wound infections. Bioluminescent *Proteus mirabilis* was used for an excisional wound infection of mouse backs, and after illumination with broadband white light and application of a tricationic fullerene, 82% of *P. mirabilis* infected mice survived (Lu et al. 2010). In contrast, only 8% of the mice survived without any treatment. This is the first report of a fullerene-mediated therapy *in vivo*, and therefore, fullerenes have a great potential for treatment of infectious diseases in the future.

34.4 Conclusion and Outlook

The worldwide increase in antimicrobial resistance has resulted in the search for alternative antimicrobial therapies, such as aPDT. The statements described here validate the use of aPDT for the treatment of bacterial and fungal infections susceptible to topical treatment. So far, neither the PS nor the subsequent localized reactions may cause any damage to the surrounding tissue or disturb the residual flora of the tissue. Furthermore, successful decolonization of patients colonized with multi-resistant pathogens is of interest for controlling and preventing microorganisms spread in a hospital's daily routine in order to reduce nosocomial infections. However, the clinical outcome of standard disinfection solutions for eradication of MRSA varied strongly between 6% and 75% (Krishna and Gibb 2010). The success of MRSA decolonization depends on how many body regions are colonized as well as on the compliance of the treatment protocol by the patients and the health care workers. In this case, aPDT can be an additional option to eradicate MRSA because it could be demonstrated that a porphyrin-based PS showed efficient antibacterial activity against MRSA inoculated on porcine skin upon illumination (Maisch et al. 2007). Skin toxicity of the used PS is limited by the lack of penetration beyond the *stratum corneum*. Enhanced penetration of a PS is needed when the pathogens grow within the epidermis or even when covered under overlying portions of the *stratum corneum*. Additional work is in progress on the use of aPDT containing appropriate formulations to reach interepidermal bacteria without damage to skin cells. To eradicate oral infections in a single session in patients who are more susceptible to infections and particularly likely to develop resistance, such as HIV-positive patients or patients undergoing chemotherapy, it would be attractive to both patients and health care providers.

Besides the discussed treatment options of aPDT to eradicate pathogens, the biological activities of some virulence factors produced by Gram-negative and Gram-positive bacteria have also been demonstrated to be successfully reduced. LPS molecules as well as proteases released by Gram-negative bacteria are of great importance because of their inflammatory effects and tissue damage during an infection. Recently Komerik, Wilson, and Poole (2000) demonstrated clearly the inhibitory effect of red laser light and the photosensitizer TBO on the virulence factors from *E. coli, Pseudomonas aeruginosa*. In this study, the photodynamic action of TBO could reduce the ability of LPS to stimulate the release of the proinflammatory cytokines IL-6 and IL-8 from peripheral blood mononuclear cell (PBMC),

which reduce the pathological effects of these cytokines in case of an enhanced production after LPS-induced activation. Furthermore, extracellular proteases of *Porphyromonas gingivalis*, which are thought to be the main cause of tissue damage associated with chronic periodontitis, can be also inactivated by using a combination of TBO and red light (Packer et al. 2000). Besides Gram-negative virulence factors, more than 40 different virulence factors have been identified in *S. aureus*, (Arvidson and Tegmark 2001). These may contribute to the pathology involving almost all processes from colonization of the host, thereby impairing the tissue integrity to dissemination, modulating effects on the host immune system and nutrition. Recently, Tubby, Wilson, and Nair (2009) could demonstrate for the first time a photodynamic action on V8 protease, alpha-hemolysin, and sphingomyelinase, three staphylococcal virulence factors that could reduce the harmful impact of preformed virulence factors on the host in the future.

As the mechanism of action of microbial killing is nonspecific and multiple sites are affected, it is considered unlikely that resistance will evolve, thus representing a significant advantage over conventional antibiotic treatment with which resistance is an ever-increasing problem. Whether microorganisms could develop resistance to ROS, for example, singlet oxygen, is questionable because the photodynamic process is typically multi-target, and the mechanism of action is different from that of most antimicrobial drugs.

Overall, it is important to realize that aPDT is a localized process, which may prevent a localized infection from becoming worse (systemic) in the future. There are a lot of newly developed PSs, which have demonstrated antimicrobial properties, which are discussed in this chapter, but have not been subjected to safety studies (mutagenicity and toxicology) to get approval for human use. The absence of genotoxicity and mutagenicity to human cells is a must-have of aPDT that favors its long-term safety. So far, the most promising clinical applications of aPDT are decolonization of pathogens on skin; treatments of the oral cavity, such as periodontitis, endodontitis, and mucosis; and superinfected burn wounds because these are relatively accessible for PS application and illumination.

References

Arana, D. M., C. Nombela, and J. Pla. 2010. Fluconazole at sub-inhibitory concentrations induces the oxidative- and nitrosative-responsive genes TRR1, GRE2 and YHB1, and enhances the resistance of *Candida albicans* to phagocytes. J Antimicrob Chemother 65:54–62.

Arvidson, S., and K. Tegmark. 2001. Regulation of virulence determinants in *Staphylococcus aureus*. *Int J Med Microbiol* 291:159–170.

Ashman, R. B., and J. M. Papadimitriou. 1995. Production and function of cytokines in natural and acquired immunity to *Candida albicans* infection. *Microbiol Rev* 59:646–672.

Babilas, P., M. Landthaler, and R. M. Szeimies. 2006. Photodynamic therapy in dermatology. *Eur J Dermatol* 16:340–348.

Bertoloni, G., E. Reddi, M. Gatta, C. Burlini, and G. Jori. 1989. Factors influencing the haematoporphyrin-sensitized photo in activation of *Candida albicans*. *J Gen Microbiol* 135:957–966.

Bertoloni, G., F. Zambotto, L. Conventi, E. Reddi, and G. Jori. 1987. Role of specific cellular targets in the hematoporphyrin-sensitized photoinactivation of microbial cells. *Photochem Photobiol* 46:695–698.

Bliss, J. M., C. E. Bigelow, T. H. Foster, and C. G. Haidaris. 2004. Susceptibility of *Candida* species to photodynamic effects of photofrin. *Antimicrob Agents Chemother* 48:2000–2006.

Cassidy, C. M., R. F. Donnelly, J. S. Elborn, N. D. Magee, and M. M. Tunney. 2012. Photodynamic antimicrobial chemotherapy (PACT) in combination with antibiotics for treatment of *Burkholderia cepacia* complex infection. *J Photochem Photobiol B Biol* 106:95–100.

Choi, S. S., H. K. Lee, and H. S. Chae. 2010. *In vitro* photodynamic antimicrobial activity of methylene blue and endoscopic white light against *Helicobacter pylori* 26695. *J Photochem Photobiol B Biol* 101:206–209.

Creagh, T. A., M. Gleeson, D. Travis et al. 1995. Is there a role for *in vivo* methylene blue staining in the prediction of bladder tumour recurrence? *Br J Urol* 75:477–479.

Dai, T. H., V. J. B. de Arce, G. P. Tegos, and M. R. Hamblin. 2011. Blue dye and red light, a dynamic combination for prophylaxis and treatment of cutaneous *Candida albicans* infections in mice. *Antimicrob Agents Chemother* 55:5710–5717.

Dai, T., Y. Y. Huang, and M. R. Hamblin. 2009. Photodynamic therapy for localized infections: State of the art. *Photodiagn Photodyn Ther* 6:170–188.

Demidova, T. N., and M. R. Hamblin. 2005. Effect of cell-photosensitizer binding and cell density on microbial photoinactivation. *Antimicrob Agents Chemother* 49:2329–2335.

Denning, D. W. 2003. Echinocandin antifungal drugs. *Lancet* 362:1142–1151.

DeRosa, M. C., and R. J. Crutchley. 2002. Photosensitized singlet oxygen and its applications. *Coord Chem Rev* 233–234:351–371.

Dismukes, W. E. 2000. Introduction to antifungal drugs. *Clin Infect Dis* 30:653–657.

Donnelly, R. F., P. A. McCarron, and M. M. Tunney. 2008. Antifungal photodynamic therapy. *Microbiol Res* 163:1–12.

Douglas, L. J. 2003. *Candida* biofilms and their role in infection. *Trends Microbiol* 11:30–36.

Espinel-Ingroff, A. 2009. Novel antifungal agents, targets or therapeutic strategies for the treatment of invasive fungal diseases: A review of the literature (2005–2009). *Rev Iberoam Micol* 26:15–22.

Gollnick, S. O., X. N. Liu, B. Owczarczak, D. A. Musser, and B. W. Henderson. 1997. Altered expression of interleukin 6 and interleukin 10 as a result of photodynamic therapy *in vivo*. *Cancer Res* 57:3904–3909.

Goosen, N., and G. F. Moolenaar. 2008. Repair of UV damage in bacteria. *DNA Repair* 7:353–379.

Groll, A. H., and A. Tragiannidis. 2009. Recent advances in antifungal prevention and treatment. *Semin Hematol* 46:212–229.

Hamblin, M. R., J. Viveiros, C. Yang et al. 2005. Helicobacter pylori accumulates photoactive porphyrins and is killed by visible light. *Antimicrob Agents Chemother* 49:2822–2827.

Harris, F., and L. Pierpoint. 2012. Photodynamic therapy based on 5-aminolevulinic acid and its use as an antimicrobial agent. *Med Res Rev* 32:1292–1327.

Hongcharu, W., C. R. Taylor, Y. Chang et al. 2000. Topical ALA-photodynamic therapy for the treatment of acne vulgaris. *J Invest Dermatol* 115:183–192.

Horn, M., and P. Wolf. 2007. Topical methyl aminolevulinate photodynamic therapy for the treatment of folliculitis. *Photodermatol Photoimmunol Photomed* 23:145–147.

Johnson, E. M., D. W. Warnock, J. Luker, S. R. Porter, and C. Scully. 1995. Emergence of azole drug resistance in *Candida* species from HIV-infected patients receiving prolonged fluconazole therapy for oral candidosis. *J Antimicrob Chemother* 35:103–114.

Jori, G., C. Fabris, M. Soncin et al. 2006. Photodynamic therapy in the treatment of microbial infections: Basic principles and perspective applications. *Lasers Surg Med* 38:468–481.

Katsambas, A., and C. Dessinioti. 2008. New and emerging treatments in dermatology: Acne. *Dermatol Ther* 21:86–95.

Kharkwal, G. B., S. K. Sharma, Y. Y. Huang, T. Dai, and M. R. Hamblin. 2011. Photodynamic therapy for infections: Clinical applications. *Lasers Surg Med* 43:755–767.

Kojic, E. M., and R. O. Darouiche. 2004. *Candida* infections of medical devices. *Clin Microbiol Rev* 17:255–267.

Komerik, N., M. Wilson, and S. Poole. 2000. The effect of photodynamic action on two virulence factors of Gram-negative bacteria. *Photochem Photobiol* 72:676–680.

Krishna, B. V., and A. P. Gibb. 2010. Use of octenidine dihydrochloride in methicillin-resistant Staphylococcus aureus decolonisation regimens: A literature review. *J Hosp Infect* 74:199–203.

Kuznetsova, N., D. Makarov, O. Yuzhakova et al. 2009. Photophysical properties and photodynamic activity of octacationic oxotitanium(IV) phthalocyanines. *Photochem Photobiol Sci* 8:1724–1733.

Lambert, P. A. 2002. Cellular impermeability and uptake of biocides and antibiotics in Gram-positive bacteria and mycobacteria. *J Appl Microbiol* 92:46S–54S.

Lee, K. K., D. M. Maccallum, M. D. Jacobsen et al. 2012. Elevated cell wall chitin in *Candida albicans* confers echinocandin resistance *in vivo*. *Antimicrob Agents Chemother* 56:208–217.

Levin, N. A. 2009. Beyond spaghetti and meatballs: Skin diseases associated with the Malassezia yeasts. *Dermatol Nurs* 21:7–13.

Lu, Z., T. Dai, L. Huang et al. 2010. Photodynamic therapy with a cationic functionalized fullerene rescues mice from fatal wound infections. *Nanomedicine (Lond)* 5:1525–1533.

Maisch, T., C. Bosl, R. M. Szeimies, B. Love, and C. Abels. 2007. Determination of the antibacterial efficacy of a new porphyrin-based photosensitizer against MRSA *ex vivo*. *Photochem Photobiol Sci* 6:545–551.

Maisch, T., F. Spannberger, J. Regensburger, A. Felgenträger, and W. Bäumler. 2012. Fast and effective: Intense pulse light photodynamic inactivation of bacteria. *J Ind Microbiol Biotechnol* 39:1013–1021.

Malik, Z., H. Ladan, and Y. Nitzan. 1992. Photodynamic inactivation of Gram-negative bacteria: Problems and possible solutions. *J Photochem Photobiol B Biol* 14:262–266.

Mang, T. S., L. Mikulski, and R. E. Hall. 2010. Photodynamic inactivation of normal and antifungal resistant *Candida* species. *Photodiagn Photodyn Ther* 7:98–105.

Martino, P., and C. Girmenia. 1997. Are we making progress in antifungal therapy? *Curr Opin Oncol* 9:314–320.

Martins Jda, S., J. C. Junqueira, R. Faria et al. 2011. Antimicrobial photodynamic therapy in rat experimental candidiasis: Evaluation of pathogenicity factors of *Candida albicans*. *Oral Surg Oral Med Oral Pathol Oral Radiol Endod* 111:71–77.

Merchat, M., G. Bertolini, P. Giacomini, A. Villanueva, and G. Jori. 1996. Meso-substituted cationic porphyrins as efficient photosensitizers of Gram-positive and Gram-negative bacteria. *J Photochem Photobiol B Biol* 32:153–157.

Mima, E. G., A. C. Pavarina, L. N. Dovigo et al. 2010. Susceptibility of *Candida albicans* to photodynamic therapy in a murine model of oral candidosis. *Oral Surg Oral Med Oral Pathol Oral Radiol Endod* 109:392–401.

Mima, E. G., A. C. Pavarina, M. M. Silva et al. 2011. Denture stomatitis treated with photodynamic therapy: Five cases. *Oral Surg Oral Med Oral Pathol Oral Radiol Endod* 112: 602–608.

Minnock, A., D. I. Vernon, J. Schofield, et al. 1996. Photoinactivation of bacteria: Use of a cationic water-soluble zinc phthalocyanine to photoinactivate both Gram-negative and Gram-positive bacteria. *J Photochem Photobiol B Biol* 32:159–164.

Minnock, A., D. I. Vernon, J. Schofield et al. 2000. Mechanism of uptake of a cationic water-soluble pyridinium zinc phthalocyanine across the outer membrane of *Escherichia coli*. *Antimicrob Agents Chemother* 44:522–527.

Mitra, S., C. G. Haidaris, S. B. Snell et al. 2011. Effective photosensitization and selectivity *in vivo* of *Candida Albicans* by meso-tetra (N-methyl-4-pyridyl) porphine tetra tosylate. *Lasers Surg Med* 43:324–332.

Mroz, P., A. Pawlak, M. Satti et al. 2007. Functionalized fullerenes mediate photodynamic killing of cancer cells: Type I versus type II photochemical mechanism. *Free Radical Biol Med* 43:711–719.

Na, J. I., S. Y. Kim, J. H. Kim et al. 2011. Indole-3-acetic acid: A potential new photosensitizer for photodynamic therapy of acne vulgaris. *Lasers Surg Med* 43:200–205.

Neely, M. N., and M. A. Ghannoum. 2000. The exciting future of antifungal therapy. *Eur J Clin Microbiol Infect Dis* 19:897–914.

Nitzan, Y., M. Gutterman, Z. Malik, and B. Ehrenberg. 1992. Inactivation of Gram-negative bacteria by photosensitized porphyrins. *Photochem Photobiol* 55:89–96.

Paardekooper, M., A. W. De Bruijne, J. Van Steveninck, and P. J. Van den Broek. 1995. Intracellular damage in yeast cells caused by photodynamic treatment with toluidine blue. *Photochem Photobiol* 61:84–89.

Packer, S., M. Bhatti, T. Burns, and M. Wilson. 2000. Inactivation of proteoloytic enzymes from *Porpyhromonas gingivalis* using light-activated agents. *Lasers Med Sci* 15:24–30.

Pal, M. K., T. C. Ghosh, and J. K. Ghosh. 1990. Studies on the conformation of and metal ion binding by teichoic acid of *Staphylococcus aureus*. *Biopolymers* 30:273–277.

Perlin, D. S. 2007. Resistance to echinocandin-class antifungal drugs. *Drug Resist Updat* 10:121–130.

Pinheiro, S. L., J. M. Donega, L. M. Seabra et al. 2010. Capacity of photodynamic therapy for microbial reduction in periodontal pockets. *Lasers Med Sci* 25:87–91.

Pinheiro, S. L., A. A. Schenka, A. Neto et al. 2009. Photodynamic therapy in endodontic treatment of deciduous teeth. *Lasers Med Sci* 24:521–526.

Piraccini, B. M., G. Rech, and A. Tosti. 2008. Photodynamic therapy of onychomycosis caused by *Trichophyton rubrum*. *J Am Acad Dermatol* 59:S75–S76.

Pollock, B., D. Turner, M. R. Stringer et al. 2004. Topical aminolaevulinic acid-photodynamic therapy for the treatment of acne vulgaris: A study of clinical efficacy and mechanism of action. *Br J Dermatol* 151:616–622.

Queiroz-Telles, F., P. Esterre, M. Perez-Blanco et al. 2009. Chromoblastomycosis: An overview of clinical manifestations, diagnosis and treatment. *Med Mycol* 47:3–15.

Rajesh, S., E. Koshi, K. Philip, and A. Mohan. 2011. Antimicrobial photodynamic therapy: An overview. *J Indian Soc Periodontol* 15:323–327.

Redmond, R. W., and J. N. Gamlin. 1999. A compilation of singlet oxygen yields from biologically relevant molecules. *Photochem Photobiol* 70:391–475.

Rolim, J. P., M. A. de-Melo, S. F. Guedes et al. 2012. The antimicrobial activity of photodynamic therapy against Streptococcus mutans using different photosensitizers. *J Photochem Photobiol B Biol* 106:40–46.

Romani, L. 2004. Immunity to fungal infections. *Nat Rev Immunol* 4:1–23.

Rubin, H. A., S. Bruce, T. Rosen, and M. E. McBride. 1991. Evidence for percutaneous inoculation as the mode of transmission for chromoblastomycosis. *J Am Acad Dermatol* 25:951–954.

Samaranayake, L. P., W. Keung Leung, and L. Jin. 2009. Oral mucosal fungal infections. *Periodontology* 49:39–59.

Samaranayake, Y. H., and L. P. Samaranayake. 2001. Experimental oral candidiasis in animal models. *Clin Microbiol Rev* 14:398–429.

Schalka, S., S. Nunes, and A. Gomes Neto. 2012. Comparative clinical evaluation of efficacy and safety of a formulation containing ciclopirox 8% in the form of a therapeutic nail lacquer in two different posologies for the treatment of onychomycosis of the toes. *An Bras Dermatol* 87:19–25.

Schneider, J. E., S. Price, L. Maidt, J. M. Gutteridge, and R. A. Floyd. 1990. Methylene blue plus light mediates 8-hydroxy 2′-deoxyguanosine formation in DNA preferentially over strand breakage. *Nucleic Acids Res* 18:631–635.

Schobert, M., and D. Jahn. 2002. Regulation of heme biosynthesis in non-phototrophic bacteria. *J Mol Microbiol Biotechnol* 4:287–294.

Shaheen, B., and M. Gonzalez. 2013. Acne sans P. acnes. *J Eur Acad Dermatol Venereol* 27:1–10.

Sharma, S. K., T. Dai, G. B. Kharkwal et al. 2011. Drug discovery of antimicrobial photosensitizers using animal models. *Curr Pharm Des* 17:1303–1319.

Sigusch, B. W., M. Engelbrecht, A. Volpel et al. 2010. Full-mouth antimicrobial photodynamic therapy in Fusobacterium nucleatum-infected periodontitis patients. *J Periodontol* 81:975–981.

Smijs, T. G., and S. Pavel. 2011. The susceptibility of dermatophytes to photodynamic treatment with special focus on Trichophyton rubrum. *Photochem Photobiol* 87:2–13.

Spellberg, B. 2008. Antibiotic resistance and antibiotic development. *Lancet Infect Dis* 8:211–212.

Spellberg, B., J. H. Powers, E. P. Brass, L. G. Miller, and J. E. Edwards Jr. 2004. Trends in antimicrobial drug development: Implications for the future. *Clin Infect Dis* 38:1279–1286.

Szpakowska, M., K. Lasocki, J. Grzybowski, and A. Graczyk. 2001. Photodynamic activity of the haematoporphyrin derivative with rutin and arginine substituents (HpD-Rut(2)-Arg(2)) against *Staphylococcus aureus* and *Pseudomonas aeruginosa*. *Pharmacol Res* 44:243–246.

Tegos, G. P., T. N. Demidova, D. Arcila-Lopez et al. 2005. Cationic fullerenes are effective and selective antimicrobial photosensitizers. *Chem Biol* 12:1127–1135.

Tubby, S., M. Wilson, and S. P. Nair. 2009. Inactivation of staphylococcal virulence factors using a light-activated antimicrobial agent. *BMC Microbiol* 9:211.

Valduga, G., G. Bertoloni, E. Reddi, and G. Jori. 1993. Effect of extracellularly generated singlet oxygen on Gram-positive and Gram-negative bacteria. *J Photochem Photobiol B Biol* 21:81–86.

Venerito, M., T. Wex, and P. Malfertheiner. 2011. *Helicobacter pylori* related and non-related lesions in the stomach. *Minerva Gastroenterol Dietol* 57:395–403.

Wainwright, M. 1998. Photodynamic antimicrobial chemotherapy (PACT). *J Antimicrob Chemother* 42:13–28.

Wainwright, M., D. A. Phoenix, S. L. Laycock, D. Wareing, and P. A. Wright. 1998. Photobactericidal activity of phenothiazinium dyes against methicillin-resistant strains of *Staphylococcus aureus*. *FEMS Microbiol Lett* 160:177–181.

Wainwright, M., D. A. Phoenix, J. Marland, D. R. Wareing, and F. J. Bolton. 1997. A study of photobactericidal activity in the phenothiazinium series. *FEMS Immunol Med Microbiol* 19:75–80.

Watanabe, D., C. Kawamura, Y. Masuda et al. 2008. Successful treatment of toenail onychomycosis with photodynamic therapy. *Arch Dermatol* 144:19–21.

Yordanov, M., P. Dimitrova, S. Patkar, L. Saso, and N. Ivanovska. 2008. Inhibition of *Candida albicans* extracellular enzyme activity by selected natural substances and their application in *Candida* infection. *Can J Microbiol* 54:435–440.

Zeina, B., J. Greenman, D. Corry, and W. Purcell. 2002. Cytotoxic effects of antimicrobial photodynamic therapy on keratinocytes *in vitro*. *Br J Dermatol* 146:568–573.

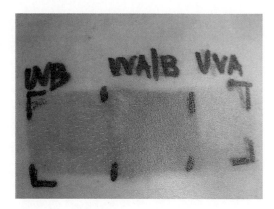

FIGURE 8.1 Photoprovocation test for PMLE.

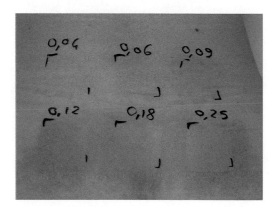

FIGURE 8.3 Minimal urticarial dose assessment with UVB.

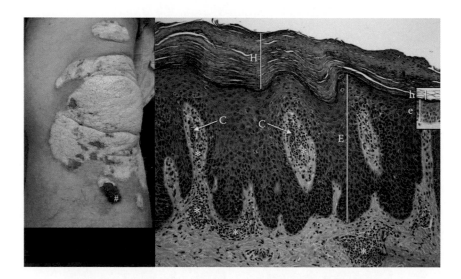

FIGURE 18.1 Psoriasis. On the left side, the clinical appearance of chronic plaque psoriasis. # indicates area of psoriatic skin without scale. On the right side, a microscopic image of a vertical section of psoriatic skin; the inset on the right shows normal epidermis for comparison. E and e indicate the thickness of psoriatic and normal epidermis; H and h indicate the thickness of psoriatic stratum corneum (scale) and that of normal skin; C indicates widened capillaries with red blood cells inside; * indicates infiltrates of inflammatory cells.

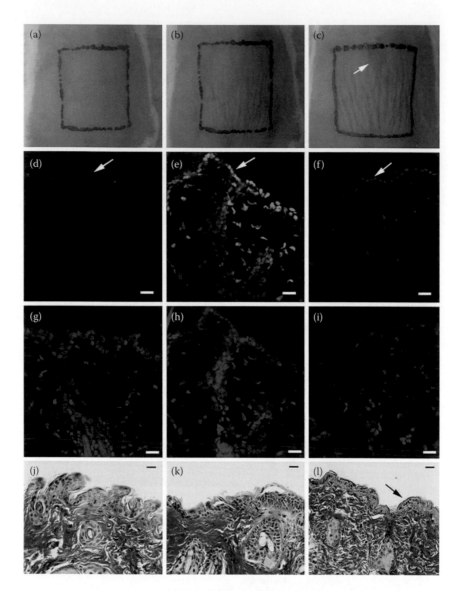

FIGURE 21.5 (a–c) Morphologies of a representative mouse skin before, immediately after, and 24 h after being exposed to UVC at a dose of 6.48 J/cm², respectively. (d–f) Representative immunofluorescence micrographs of cyclobutane pyrimidine dimers in skin cell nuclei. (g–i) Micrographs of diamidino-2-phenylindole counterstaining of cell nuclei. (j–l) Micrographs of Masson's trichrome stained sections. Biopsies were taken before (d, g, j), immediately after (e, h, k), and 24 h after (f, i, l) being exposed to UVC at a dose of 6.48 J/cm² from the same mouse. Arrow in (c): UVC-induced lesion on mouse skin. Arrows in (d–f): mouse skin surface. Arrow in panel l shows epidermal shrinkage. Scale bars: 20 μm. (Reprinted from Dai, T. et al., *Photochem. Photobiol.*, 87, 342–349, 2011. With permission.)

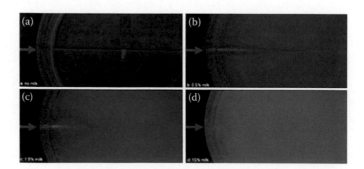

FIGURE 25.2 Illustration of the effect of elastic light scattering on the distribution of light in a scattering medium, a mixture of water and milk in a glass. From no scattering (a) to strong scattering (d), the light distribution changes from a parallel beam to a diffuse cloud.

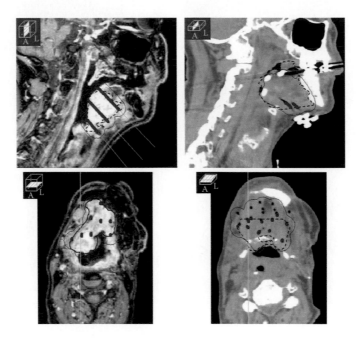

FIGURE 25.9 Example of an iPDT treatment. Left: Pretreatment CT scan with tumor boundaries drawn as thick red lines. The planned position of the linear diffuser (thick red dots and lines); resulting zone of necrosis is indicated as thin red lines. Right: magnetic resonance (MR) image taken after insertion of the fibers (thick red dots and lines); the tumor volume is indicated here as the dashed line and the expected zone of necrosis as the thin red line.

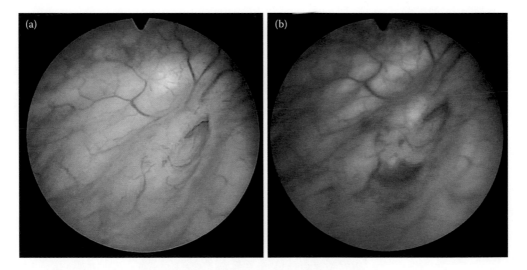

FIGURE 36.2 (a) Bladder wall observed under standard WL illumination. The early cancers are invisible. Diameter of the field of view: 15 mm. (b) Same site observed in the fluorescence mode after administration of HAL. Small superficial cancers (carcinoma in situ) are clearly glowing red due to the marker. (Courtesy of the CHUV Hospital.)

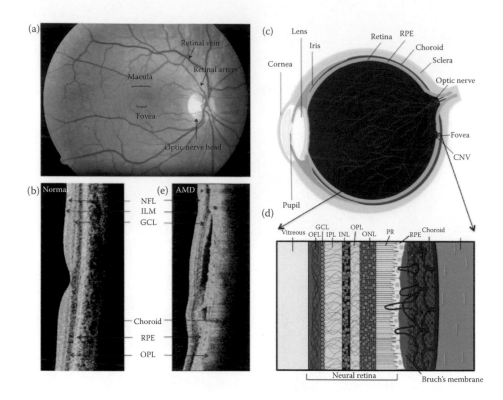

FIGURE 39.1 (a) Retinograph of healthy human eye. (b) Optical coherence tomography (OCT) scan with different layers of the healthy retina. (c) Schematic drawing of a cross section of the human eye. (d) Schematic cross section of the macula. (e) OCT image of the AMD-diagnosed retina visualizing the components of the neovascular membrane. GCL = ganglion cell layer, ILM = integral limiting membrane, IPL = inner plexiform layer, NFL = nerve fiber layer, OFL = optic fiber layer, ONL = outer nuclear layer, OPL = outer plexiform layer, RPE = retinal pigment epithelium.

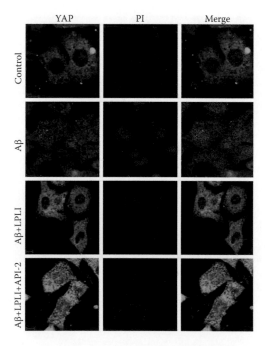

FIGURE 47.17 LLLT inhibits $A\beta_{25-35}$-induced YAP translocation from cytoplasm to nucleus through Akt activation. PC12 cells were treated with $A\beta_{25-35}$ or/and API-2, an inhibitor of Akt, and then nucleus was stained with PI to differentiate from the cytoplasm. Representative images are shown. (Adapted from Zhang, H. et al., *Cell Signal* 24: 224–232, 2012.)

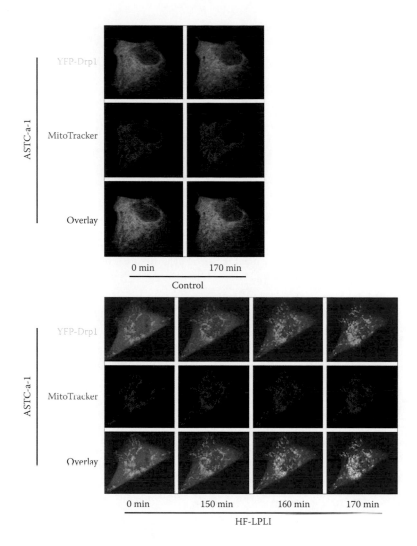

FIGURE 47.24 ASTC-a-1 cells were transiently cotransfected with pYFP-Drp1 and pDsRed-mit. Forty-eight hours after transfection, cells coexpressing YFP-Drp1 and DsRed-mit were treated with high-fluence LLLT at 120 J/cm^2. The control group received no treatment. Representative confocal microscopic images reveal increased association of YFP-Drp1 with mitochondria in response to HF-LPLI. Scale bars, 10 μm. (Adapted from Wu, S. et al., *FEBS J* 278: 941–954, 2011.)

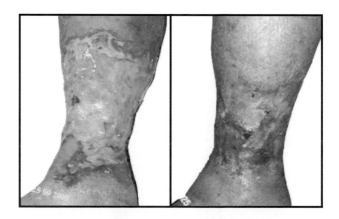

FIGURE 50.1 Ulcus cruris as a result of infected varicose vein that failed conventional treatment in a 55-year-old incapacitated male. Nearly complete cure was achieved after 5 months of LLLT. The wound remained closed and pain-free at 9 months' follow-up. The patient returned to part-time work. (Courtesy of Dr. Yosef Kleinman, Wound Care Center, Bikur Cholim Hospital, Jerusalem, Israel.)

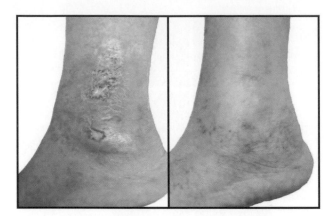

FIGURE 50.2 Recalcitrant leg ulcers caused by hydroxyurea in a 67-year-old woman with essential thrombocytosis. LLLT (see regimen in Figure 50.1) resulted in complete resolution within 3 months. The patient remained ulcer-free for 1 year of follow-up. (Courtesy of Dr. Yosef Kleinman, Wound Care Center, Bikur Cholim Hospital, Jerusalem, Israel.)

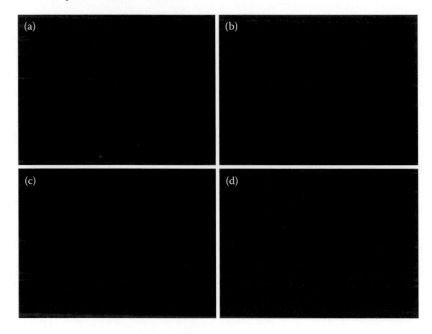

FIGURE 55.1 Micrograph of sciatic nerve submitted to LLLT treatment and control evaluable at 28 days after the lesion. The immunomarker is S-100, which stains the Schwann cells. More staining of the Schwann cells can be seen in two magnifications (20× in the left and 100× in the right) of the laser-treated (3 times a week for 4 weeks, c and d) than the control (sham, a and b).

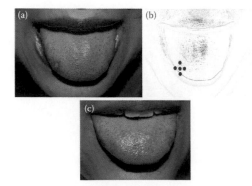

FIGURE 56.4 Treatment of an aphthous ulcer lesion with one session of LLLT. (a) Initial lesion, (b) schematic drawing of irradiation points performed with a diode laser, 660 nm, 3 J per point, and (c) clinical result after 3 days of LLLT treatment.

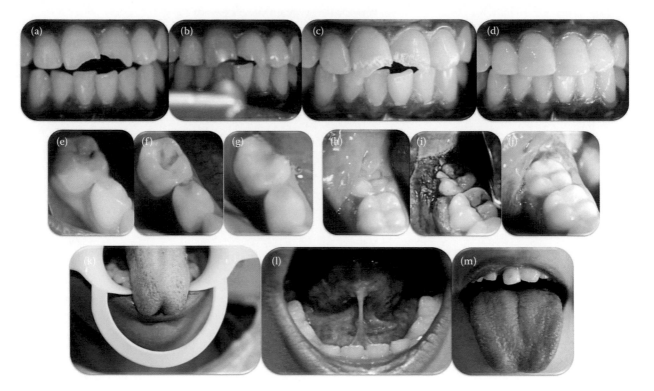

FIGURE 61.5 Clinical case of laser-assisted restoration of tooth 11 and 21. This hard tissue procedure was performed using a Fotona Fidelis laser (Slovenia). The procedure was performed with an Er:YAG laser of 2940 nm wavelength at 250 millijoule (mj) and 20 Hz (short pulse) with high-speed evacuation and with water on a 25-year-old female patient (a–d) and a 9-year-old child (e–g). (a) Ellis type 2 fracture on tooth 11 and 21. (b) Tooth surface prepared with Er:YAG (R02 handpiece, 250 mj, 20 Hz with water spray). (c) Lased tooth surface postablation. (d) Definitive restoration after Ellis type 2 fracture. (e) Occlusal caries on the first permanent molar. (f) Laser-assisted cavity preparation. (g) After composite restoration. (h, i, j) Laser-assisted operculectomy. This soft tissue procedure was performed using a 980 nm diode laser manufactured by Microscientific Instruments, New Delhi, India. The procedure was performed using a diode laser with a 400 μm optical fiber at 2.5 W power and (continuous pulse) with high-speed evacuation and with no water or air, on a 23-year-old male patient. (h) Pericoronitis with respect to tooth 48. (i) Immediately postoperative. (j) One week postoperative. (k) Preoperative view of lingual frenum showing the inability to extend the tongue with observed cleft of the tip of the tongue. (l) Er:YAG-assisted frenectomy. (m) After lingual frenectomy. Immediately postoperative, showing the ability to extend the tongue. This pediatric soft tissue procedure was performed using a Fotona Fidelis laser, an Erbium:YAG laser, with a wavelength of 2940 nm at 100 mJ and 20 Hz (very long pulse) with high-speed evacuation, and with no water or air, on a 7-year-old boy.

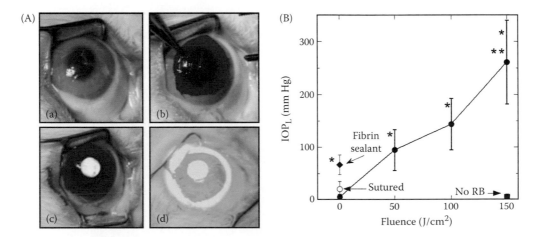

FIGURE 66.1 Sealing amniotic membrane over corneal wounds. (A) Steps in the procedure: (a) full-thickness incision in cornea, (b) RB-stained amnion placed on cornea, (c) opaque disk placed to block light from entering through pupil, and (d) irradiation with 532 nm laser. (B) The IOP_L was measured immediately after bonding treatment. *$p < 0.05$ versus control, ** $p < 0.05$ versus 100 J/cm^2. (Adapted from Verter, E. E. et al., *Invest Ophthalmol Vis Sci* 52:9470–9477, 2011. Used with permission from the Association for Research in Vision and Ophthalmology.)

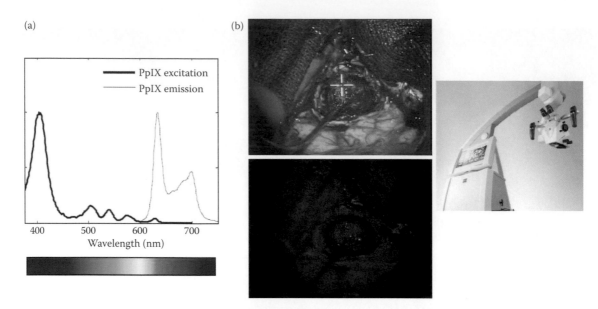

FIGURE 67.2 ALA-PpIX fluorescence imaging for guiding resection of malignant brain tumors. (a) PpIX fluorescence excitation and emission spectra and (b) white-light and PpIX fluorescence images at the end of white-light resection of high-grade glioma. (Courtesy of Dr. David Roberts and colleagues, Dartmouth-Hitchcock Medical Center, Hanover, NH, USA.) The residual tumor is clearly visible as red fluorescence on the blue scattered light background. Also shown is the working head of a fluorescence-enabled neurosurgical microscope (Pentero, Zeiss, Germany) used to take these images.

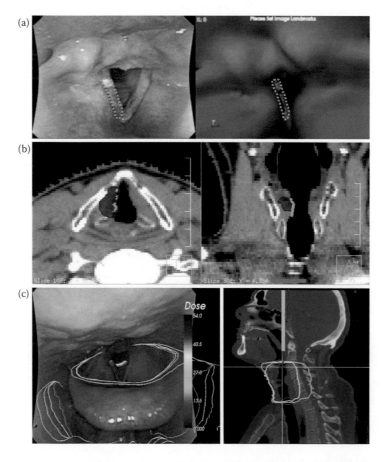

FIGURE 67.4 Example of combining real (white-light) and virtual (CT-based) endoscopy in planning radiation treatment of a glottis tumor. (a) Real (left) and virtual (right) images with the tumor margins outlined. (b) Coregistered images showing different slices of the real endoscopy-defined tumor superimposed on the virtual endoscopic mass. (c) Planned radiation dose contours superimposed on the real endoscopic image (left) and a CT volumetric image slice (right).

35

PDT and the Immune System

Fatma Vatansever
Harvard Medical School

Michael R. Hamblin
Massachusetts General Hospital

35.1 Introduction

Worldwide preclinical and clinical studies for more than two decades now have shown that photodynamic therapy (PDT) can be an integral part of a multimodal treatment approach to various cancer types and other malignancies. After the FDA's approval of Photofrin, a commercially available hematoporphyrin derivative, in 1993 for patients with early and advanced stages of lung cancer, research utilizing PDT picked up speed globally. Today, PDT is a clinically approved therapeutic modality that is being successfully used for treatment of neoplastic and nonmalignant diseases.

The PDT has three essential components: photosensitizer (PS), light, and molecular oxygen ($^3O_2$). Individually, none of these elements is toxic, but together, they initiate a cascade of photochemical reactions that culminate in generation of highly reactive singlet-oxygen ($^1O_2$) species. Because PDT is highly localized (the lifetime of the singlet oxygen is very short, approximately 10–320 ns and thus has limited diffusion to approximately 10–55 nm in cells) (Moan et al. 1989), the photodynamic damage only occurs in the vicinity of the PS location. Classical PSs, still widely in use, have a porphyrin-like nucleus, in which the energy conversion takes place. They function as energy-converting platforms in which visible light is absorbed, and during the emission process, molecular oxygen present in the media is transformed into the range of reactive oxygen species (ROS). The generated ROS then are shown to destroy tumors in multiple pathways (Castano, Mroz, and Hamblin 2006). For more detailed discussion about the fundamentals of PDT, see Chapter 4 by Tyler G. St. Denis and Michael R. Hamblin.

The most obvious pathway is PDT's direct cytotoxic effect on cancer cells via oxidative stress and cell-membrane damage, which results in necrosis and/or apoptosis (Oleinick, Morris, and Belichenko 2002). The second pathway operates by affecting tumor vasculature, in which illumination and ROS production cause a shutdown of vessels, thus depriving the tumor of oxygen and nutrients (Krammer 2001; Dolmans et al. 2002). The most tantalizing and promising PDT effects are immune system alterations that are either immunostimulatory or immunosuppressive in nature. The relative contribution of these mechanisms depends on factors that include type and dose of the PS, time span between the PS administration and light exposure, total light exposure and its fluence rate, tumor oxygen concentration, etc. (Figure 35.1). Studies looking at combined effects of these components are underway, testing under what conditions long-term tumor control is achievable with antitumor action against both primary and metastatic tumors (Castano, Mroz, and Hamblin 2006; Dolmans, Fukumura, and Jain 2003; Dougherty et al. 1998; Gollnick and Brackett 2010). For more detailed discussion about PDT-induced cell death pathways, see Chapter 27 by Patrizia Agostinis.

PDT, as an integral part of complete cancer treatment, brings to the table many advantages over conventional single-modality treatments. The most significant benefit is that clinically approved PSs do not accumulate in the cell nuclei and thus have a limiting DNA damage effect (that can be carcinogenic or can lead to development of resistant clones). Other important advantages are, to list them, localization, targeted therapy expandability, minimal invasiveness, preservation of the anatomical and functional integrity of the treated organs, minimized side effects, and reduced toxicity and drug resistance, allowing repeated treatment cycles. The list can be continued with aversion to the adverse effects of chemo and radiation therapies, improved cosmetic outcomes, no significant change

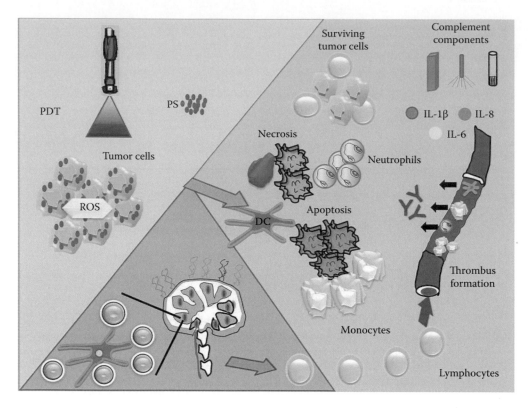

FIGURE 35.1 PDT-induced effects. In tumors, cells loaded with PS upon excitation generate ROS species. This leads to predominantly apoptotic and necrotic cell deaths. Tumor cell death is accompanied with complement cascade activation, proinflammatory cytokine activation, rapid neutrophils, DCs, and macrophage recruitment. Dying tumor cells and their debris are phagocytosed by phagocytic cells and DCs, which then migrate to the local lymph nodes and there differentiate into antigen-presenting cells. Tumor antigen presentation is then followed by clonal expansion of tumor-sensitized lymphocytes that home to the tumor site and eliminate residual tumor cells.

in tissue temperature during and after treatment, and preservation of the connective tissue at the PDT application site (thus inducing minimal fibrosis and, by doing so, alleviating excessive scarring and the need for further reconstruction procedures). Also, many of the PDT procedures can be performed in outpatient or ambulatory settings, thereby substantially lowering the cost of treatment.

PDT treatment efficacy (with its three main cell death pathways: apoptotic, necrotic, and autophagy-associated cell deaths) arises from its combination effect over tumorigenesis, its direct cytotoxic effect over cells, and an oxidative stress-initiated secondary tumoricidal effects (Figure 35.2). It turns out that secondary tumoricidal effects can trigger events with far-reaching consequences. This can happen because they have major effects over several signal transduction pathways, such as upregulation of stress proteins and early response genes, activation of the genes that regulate apoptotic cell death, and upregulation of some cytokine genes, just to name some. For instance, as a consequence of induced changes in the plasma membrane and cellular organelles' membranes, upregulation of stress proteins that play major role in cell adhesion and antigen presentation takes place. And this, in return, triggers an inflammatory or immune response affecting the whole body. Yet another PDT-induced membrane alteration is generation of photooxidative lesions of

membrane lipids. This damage prompts a rapid activation of membranous phospholipases, which causes accelerated phospholipid degradation with massive release of lipid fragments and metabolites of arachidonic acid, and these molecules are powerful inflammatory triggers.

Tumor vasculature damage can generate a cascade of signaling response as well. For instance, even minor phototoxic lesions can cause contraction in endothelial cells, thus exposing the basement membrane in the vessel wall. In response, circulating neutrophils and platelets surge in the area, attaching to the exposed sites. The result of this insult is an impairment of vasculature function followed by massive release of myriad of inflammatory mediators. This self-feeding or looping cascade of releasing immunogenic mediators is at the heart of PDT's tumor-suppressing effect.

To summarize, induction of a strong inflammatory reaction is a central paradigm in the tumorigenesis effect of PDT. A key aspect of the PDT-mediated immunogenic (both) initiation and response is the release of a wide variety of potent mediators, such as (but not limited to) radicals, cytokines, chemokines, growth factors, acute phase proteins, proteinases, peroxidases, leukocyte chemoattractants, vasoactive substances, and other immunoregulators. All these immunologic effects will be discussed in detail in this chapter.

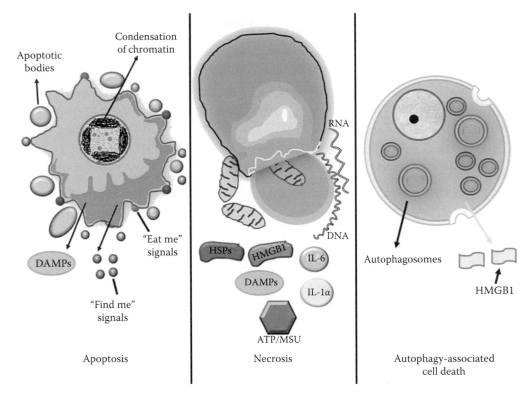

FIGURE 35.2 Major cell deaths and their immunological profiles. Apoptosis is found during normal cell turnover, and tissue homeostasis, embryogenesis, induction and maintenance of immune tolerance, and endocrine-dependent tissue atrophy affect individual cells. Characteristic morphological and biomedical features are membrane blebbing (but no loss of integrity), chromatin aggregation at the nuclear membrane, cleavage of chromosomal DNA into internucleosomal fragments, nuclear and cytoplasmic condensation or shrinking, partition of cytoplasm and nucleus, leaky mitochondria (resulting from pore formation involving proteins of the bcl-2 family), fragmentation of the cell into smaller bodies, formation of membrane-bound vesicles (apoptotic bodies) (they are rapidly recognized and phagocytized by either macrophages or/and adjacent epithelial cells—no elicited inflammatory response), release of various factors (cytochrome C, AIF) into cytoplasm by the mitochondria, activation of caspase cascade, and energy (ATP) dependence (thus, an active process and does not occur at 4°C). Necrosis occurs when cells are exposed to extreme variance from physiological conditions (hypothermia, hypoxia, ischemia, lytic viruses, etc.) and results with damage to the plasma membrane. Affects groups of contiguous cells. Necrosis begins with an impairment of the cell's ability to maintain homeostasis (leading to an influx of water and extracellular ions). Morphologically, it is characterized by swelling and rupture of intracellular organelles, mitochondria, and the entire cell (thus cells undergo lysis). As a result of the ultimate breakdown of the plasma membrane, the cytoplasmic contents, including lysosomal enzymes, are released into the extracellular fluid. Thus, necrotic cell death is associated with extensive tissue damage resulting in an extensive inflammatory response. Energetically, it is a passive process (no energy required to start and/or propagate, occurs at 4°C as well). Autophagy is characterized by a massive vacuolization of the cytoplasm. During the process, double-membrane structures called autophagosomes are formed, thus sequestering cytoplasmic components and organelles, and traffic them to the lysosomes. The fusion of autophagosomes and lysosomes results in the degradation of the cytoplasmic components by the lysosomal hydrolases. In the body, autophagy functions as a self-digestion pathway (promoting cell survival in an adverse environment) as well as a quality control mechanism (by removing damaged organelles, toxic metabolites, or intracellular pathogens).

35.2 Immunologic Effects of PDT

PDT-induced apoptotic cell death mechanisms are, in essence, highly immunogenic and capable of driving antitumor immunity (Garg et al. 2010b). Very often, PDT generates a strong, acute inflammatory reaction, the result of induced oxidative stress, which can be observed as a localized edema at the targeted site. In effect, PDT produces a chemical and physical insult in tumor tissue that the host body perceives as acute localized trauma. This identification then prompts the host to launch protective actions against this threat to tissue integrity and homeostasis (Korbelik 2006). The resulting acute inflammatory response is

the principal protective process launched in order to contain the disruption of the homeostasis, to remove the damaged cells, and, subsequently, to promote local healing and restoration of normal tissue functions. In effect, the inflammation elicited by PDT is a tumor antigen nonspecific process orchestrated by the innate immune system of the host (Korbelik 2006).

The recognition unit of this system—in particular, the pattern-recognition receptors—are the first-in-line sensors responsible for detecting the presence of PDT-inflicted and tumor-localized insult (Korbelik 2006). It has been shown that PDT is particularly effective in arousing quick and abundant alarm or danger signals, called damage-associated molecular patterns (DAMPs) or cell death-associated molecular patterns (CDAMPs), at the PDT treatment site (Korbelik 2006). After PDT-induced inflammation is launched at the tumor site, dramatic changes in the tumor vasculature take place. Vessels become leaky and therefore permeable for blood proteins as well as proadhesive for the inflammatory cells (Korbelik 2006). PDT treatment produces photooxidative damage in perivascular regions with chemotactic gradients reaching the vascular endothelium. Thereafter, inflammatory cells led by neutrophils and followed by mast cells and monocytes or macrophages rapidly and in increased numbers invade the tumor (Krosl, Korbelik, and Dougherty 1995). The primary task of these infiltrating cells is to neutralize the source of DAMPs/CDAMPs by eliminating debris containing compromised tissue and injured and dead cells. After PDT, the damaged and dysfunctional tumor vasculature walls off the damaged tissue until it is removed via phagocytosis. This vascular occlusion effect helps keep the damage localized by preventing the spread of disrupted homeostasis (Korbelik 2006). Depletion of these inflammatory cells or inhibition of their activity has shown to diminish the therapeutic effect of PDT (Korbelik and Cecic 1999, 2003; de Vree et al. 1997; Kousis et al. 2007).

Another critical factor in PDT is cytokine regulation of the inflammatory process. Studies have identified interleukins 6 and 1β (IL-6 and IL-1β) (Sun et al. 2002; Gollnick et al. 2003) as well as TNF-α, G-CSF, thromboxane, prostaglandins, leukotrienes, and histamine (Cecic and Korbelik 2002) molecules as central to tumor PDT response. Blocking the function of cell adhesion molecules has proven detrimental to PDT outcome (Sun et al. 2002; Gollnick et al. 2003). On the other hand, blocking the effect of anti-inflammatory cytokines, such as IL-10 and TGF-β, can visibly improve the cure rates after PDT (Korbelik 2006).

The population of neutrophils is also critical to launching an immune response after PDT. In groundbreaking research, the regulatory role of neutrophils in PDT and antitumor immunity has been demonstrated by Gollnick et al. of Roswell Park Cancer Institute. Their findings show that PDT treatment of murine tumors result in regimen-dependent induction of an acute inflammatory response, one outcome of which is a rapid neutrophil infiltration into the treated tumor bed. Their research showed that a PDT regimen that generates a high level of neutrophilic infiltrate also ends up generating tumor-specific primary and memory CD8[+] T-cell responses. Along the same lines, if the PDT regimen induces little or no neutrophilic infiltrate, minimal antitumor immunity has been induced. Also, mice defective in neutrophil homing to peripheral tissue or depleted of neutrophils were not able to launch strong antitumor CD8[+] T-cell responses as a result of PDT. The general conclusion of their research is that T-cell proliferation and/or survival is directly affected by the neutrophils; thus, as a result of this effect, tumor-infiltrating neutrophils play an essential role in the establishment of antitumor immunity following PDT treatment (Gollnick and Brackett 2010; Kousis et al. 2007). More detailed discussion about PDT-induced immunostimulatory and immunosuppressive effects follows.

35.3 Immunostimulatory Effect of PDT

Both preclinical and clinical studies have shown that PDT can trigger adaptive immune response (in some cases, potentiation of the adaptive immunity and, in others, immunosuppression). The exact factors and precise mechanisms that determine which immune response will predominate (potentiation versus suppression) are not entirely clear. We do know much more about these mechanisms and triggering effects than even few years back, however. It is now well established that the effect of PDT on the immune system depends strongly on variables such as treatment regimen, PS type, treated area, and so forth (Kousis et al. 2007; Hunt and Levy 1998). Decades ago, Canti et al. (1994) showed the immune potentiation inductive effect of PDT in mice by demonstrating that cells isolated from tumor-draining lymph nodes of PDT-treated mice were able to transfer tumor resistance to naïve mice. Later, it was shown that PDT application in murine tumors generates immune memory effect (Korbelik and Dougherty 1999).

Clearly, PDT efficacy depends upon the induction of antitumor immunity. *In vivo* studies have shown that long-term tumor response is diminished or even absent in immuno-compromised mice (Korbelik and Cecic 1999; Korbelik et al. 1996), but reconstitution of these mice with bone marrow or T cells from immunocompetent mice results in improved PDT efficacy. Likewise, in clinical studies, patients with vulval intraepithelial neoplasia (VIN) who did not respond to PDT with the prodrug 5-aminolevulinic acid (ALA) were more likely to have tumors that lacked class I major histocompatibility complex hetero-dimer (MHC I) than patients who responded to ALA treatment (Abdel-Hady et al. 2001). This difference appears to be a result of the fact that MHC I recognition is crucial for activation of CD8[+] T cells, and thus tumors that lack MHC I are resistant to cell-mediated antitumor immune reactions (Maeurer et al. 1996). On the other hand, patients who responded to VIN PDT treatment had increased CD8[+] T cell infiltration into the treated tumors. More recent reports have shown that clinical antitumor PDT treatment also increases the antitumor immunity. PDT performed in multifocal angiosarcoma of the head and neck increased the immune cell infiltration into distant untreated tumors with a subsequent tumor regression effect (Thong et al. 2007). PDT performed in basal cell carcinoma increased the immune cells' reactivity against a basal cell carcinoma-associated antigen (Kabingu et al. 2009).

What are the mechanistic pathways for PDT enhancement of antitumor immunity? It has been shown that PDT activates both humoral and cell-mediated antitumor immunity. For example, we know that the efficacy of PDT is reduced, in humans as well as mice, in the absence of CD8[+] T cell activation and/or tumor infiltration (Korbelik and Cecic 1999; Abdel-Hady et al. 2001; Kabingu et al. 2007). Also, it is clear now that antitumor immunity generation, as a consequence of PDT, is directly dependent on inflammation induction (Henderson et al. 2004). Therefore, it can be concluded that CD8[+] T cell induction plays a crucial role in cell-mediated antitumor immunity. It is commonly agreed upon now that PDT-induced (both acute-local and systemic) inflammation results in maturation and activation of dendritic cells (DCs); mature DCs are crucial for activation of tumor specific CD8⁺ T cell and induction of antitumor immunity (Reis and Sousa 2004). Thus, after activation, DCs migrate to tumor-draining lymph nodes, where they are thought to stimulate, in return, the T cell activation (Gollnick et al. 2003; Sur et al. 2008). PDT-triggered antitumor immunity is believed to take place as a result of dying and dead tumor cells, stimulating the DCs (Gollnick, Vaughan, and Henderson 2002). Interestingly, an acute inflammatory response, launched as a result of PDT, causes accumulation of neutrophils and other inflammatory cells in large numbers at the treated site (Krosl, Korbelik, and Dougherty 1995; Cecic, Scott, and Korbelik 2006). In addition to stimulating a local inflammation response, PDT can effectively induce a potent acute-phase response, a process that assists the maturation and activation of DCs and increases their ability to home to lymph nodes and whereby to successfully present tumor antigens and to prime the lymphocytes (Gollnick, Owczarczak, and Maier 2006), so that the latter are able to destroy distant-antigen positive tumors (Mroz et al. 2010) (Figure 35.3).

Several tumor models have been developed to demonstrate the inductive effect of PDT, both for local inflammatory response (Castano, Mroz, and Hamblin 2006; Cecic, Parkins, and Korbelik 2001) and for systemic immunological effects (Mroz et al. 2010). In a recent *in vivo* study, the ability of PDT

to elicit systemic antigen-specific immune response was demonstrated (Mroz et al. 2010) in the context of lethal BALB/c colon adenocarcinoma, CT26 wild type (CT26WT), and CT26.CL25, which expresses a tumor antigen β-galactosidase (β-gal). Study results showed that PDT treatment could elicit a systemic antigen/epitope specific antitumor immune response, which was sufficiently robust to lead to regression of distant (outside the treatment field) and well-established antigen-positive tumors. Results also showed that PDT induced a local response in both CT26WT and β-gal antigen-expressing CT26.CL25 tumors, the outcome of which was reduction of the tumor sizes. However, local tumor regrowth took place in the CT26WT cancer, whereas β-gal antigen-positive CT26.CL25 tumors were not only reduced in size completely but also, more importantly, 100% of these antigen-positive tumors stayed in remission during the course of study. Assessments of the memory immunity and the antigen specificity of the memory immunity effect were also performed with the following procedure: Some of the cured mice were inoculated with the β-gal antigen expressing CT26.CL25 tumor cells into the contralateral thigh for assessment of the memory immunity effect. And some of the cured mice (that were cured from CT26.CL25 cells) were inoculated with antigen-negative CT26WT cells in order to validate the antigen specificity of the memory immunity effect. The outcome was that more than 95% of mice rechallenged with the CT26.CL25 tumor proved immune to the tumor assault and stayed tumor free, while all antigen-negative

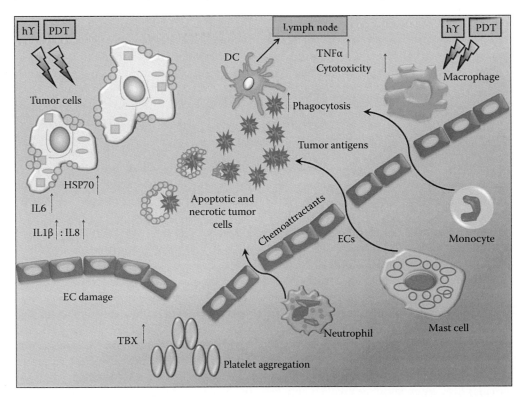

FIGURE 35.3 PDT-induced inflammation. Damaging the endothelial cells (ECs) activates a cascade of events that lead to local inflammation, vessel dilation, and platelet aggregation. Much of these effects are caused by the release of thromboxane (TBX), cytokines (such as interleukins IL1β, IL6, IL8, and tumor necrosis factor-α), and infiltration of immune system cells (necrotic and apoptotic cells provide antigens to the DCs that migrate to lymph nodes).

CT26WT tumors progressed (Mroz et al. 2010). The study also assessed the extent of local activation of the immune system by measuring secreted cytokines in the tumor (Mroz et al. 2010). It was observed that, as a result of PDT treatment, antigen-positive CT26.CL25 tumors (but not the antigen-negative CT26WT tumors) had significant elevation of TNFα and IFNγ levels, and this, in return, suggests involvement of the Th1 arm of the adaptive immune response (Mroz et al. 2010). The study showed that PDT-induced cytotoxic T cells specifically destroy antigen-positive cancer cells. The same study also showed that PDT elicits development of epitope-specific CD8+ T cells (Mroz et al. 2010). To further evaluate whether PDT treatment can generate β-gal antigen-specific systemic immune response (that is to say, one strong enough to destroy distant, established, and non-treated tumors), mice bearing two bilateral tumors were used. In this model, only one of the tumors was illuminated with PDT, while the second tumor was kept shielded. Results show that in mice bearing β-gal antigen-positive CT26.CL25 tumors (in both legs) the tumors regressed in both legs, that is, the distant and untreated tumor also shrank (Mroz et al. 2010).

In order to determine whether the observed destruction of the contralateral, established, and nontreated tumors was antigen specific, evaluations were performed with two groups of mice (each with two mismatched tumors, antigen-negative CT26WT in the left leg and antigen-positive CT26.CL25 in the right one). One group received PDT only for their CT26WT tumors and the other group only for the CT26.CL25 tumors. The PDT-treated tumors showed the expected after-PDT responses. However, because there were no effects on the size or growth rate of the contralateral untreated tumors, the experiment was cut short (the mice could not be followed for long-term outcome). To confirm the effect of the immune system in the observed PDT responses (remission of the contralateral tumors), intratumoral activated cytotoxic T cell infiltration was studied, via monitoring the LAMP-1 (CD107a) marker (Mroz et al. 2010; Betts et al. 2003; Parkinson-Lawrence et al. 2005). PDT-treated CT26.CL25 tumors revealed pronounced T cell infiltration, but more importantly, contralateral antigen-positive CT26.CL25 tumors were also heavily infiltrated by LAMP-1-positive T cells. Moreover, a direct correlation was observed between the shrinking rate of the tumor and the size of the infiltrating T cell population (Mroz et al. 2010).

The ultimate testing to confirm the observed PDT effects (in causing activation of the immune system) was done by using the same experimental protocol (as with β-gal antigen-positive CT26.CL25 tumors), but only this time, the tumor was used in immunocompromised BALB/c Nu/Nu type mice. The results of this study clearly showed that the lack of a functional adaptive immune system abrogates PDT's antitumor effects (Mroz et al. 2010). Collectively, the series of studies cited here indicate that an effective vascular PDT regimen can be used not only for local tumors but also (by inducing potent, systemic, antigen-specific antitumor immunity) for contralateral (for one reason or another, also nonoperable) tumors. The observed immunity has been shown capable of causing regression and/or curing of

distant tumors, which lie outside the illumination field as well as inducing long-term immune memory and resistance to rechallenging. This tumor-destructive effect is shown to be mediated by tumor antigen-specific cytotoxic T cells that can recognize the immunodominant epitope of β-gal antigen (Mroz et al. 2010). Taken as a whole, the *in vivo* results suggest that PDT may be a successful treatment modality for clinical trials in patients with tumors that express tumor-associated antigen (such as melanoma, renal cell carcinoma, etc.).

35.3.1 PDT-Induced Cancer Vaccines

A logical extension of the observed antitumor immunity effect induced by PDT would be antitumor vaccines created from PDT-treated tumor cells (Gollnick, Vaughan, and Henderson 2002). This hypothesis has held thus far in various studies, using a wide variety of PSs and tumor models, in both preventive and therapeutic conditions (Korbelik and Cecic 2003; Gollnick, Vaughan, and Henderson 2002; Korbelik and Sun 2006; Korbelik, Merchant, and Huang 2009). These studies have shown that incubation of immature DCs with PDT-treated tumor cells enhances the DC's maturation, activation, and ability to stimulate T cells (Gollnick, Vaughan, and Henderson 2002; Jalili et al. 2004). In a very interesting *in vivo* study set, it was observed that PDT with mono-L-aspartyl chlorine-e6 and tin etio-purpurin as PSs are causing elevations in the HSP70 mRNA levels in tumors (Gomer et al. 1996). Also, as a result of the PDT, HSP-bound tumor antigens are released and induce necrotic cell deaths, the debris of which can easily be taken up by DCs (Mroz and Hamblin 2008). The process can be summarized as follows: PDT performed on the tumor site causes both cell stress and cell death (Henderson and Gollnick 2003; Oleinick and Evans 1998). As a result, DAMPs/CAMPs are released or secreted from the dying cells; these fractions, in return, activate the DCs (Gollnick, Owczarczak, and Maier 2006; Korbelik, Sun, and Cecic 2005; Korbelik, Scott, and Sun 2007). HSP70 induced after PDT (Gomer et al. 1996) is one of the well-characterized DAMPs that interacts with the Tall-like receptors 2 and 4 (also called danger signal receptors) (Vabulas, Wagner, and Schil 2002). Research is showing that the HSP70 expression level correlates with its ability to stimulate DC maturation (Gollnick et al. 2004) and inflammation initiation (Korbelik, Sun, and Cecic 2005; Scott and Korbelik 2007).

The immunogenic profile of cancer cell death concept has been further developed in recent years, and a new (sub)class of apoptotic cancer cell death, defined as immunogenic apoptosis, has started to take shape (Garg et al. 2010b). To date, *in vivo* research has shown that agent-specific immunogenic cancer cell deaths have capabilities to induce an "anticancer vaccine effect." It has been shown that DAMPs' immunological silhouette is making the signature of these cell death pathways at molecular levels. In addition to HSP70, various intracellular molecules [such as calreticulin (CRT), heat-shock proteins (HSPs), and high-mobility group box-1 (HMGB1) protein] have been shown to be DAMPs secreted in a stress agent or factor environment and generating this type of specific cell deaths. These discoveries have motivated further research for identifying new DAMPs, new pathways for their release or secretion, and new PS agents capable of inducing immunogenic cell deaths.

Because PDT-generated vaccines are induced via cytotoxic T cell response (and are tumor-specific) because of the very nature of its mechanism, this type of vaccine does not require coadministration of an adjuvant in order to be effective (Korbelik and Sun 2006; Gollnick, Vaughan, and Henderson 2002). Korbelik and Sun (2006) incubated squamous cell carcinoma (SCC) cells with the PS benzoporphyrin derivative monoacid ring A (BPD). After Korbelik's team exposed the culture to light (killing them with the radiation exposure), they then injected the cells (now constituted as a cancer vaccine) into mice, bearing subcutaneous SCC (Korbelik and Sun 2006). Application of this PDT-generated vaccine resulted in a significant therapeutic effect (including tumor growth retardation, regression, and finally, curing). Gollnick, Vaughan, and Henderson (2002) explored the effectiveness of the PDT-generated vaccines in cancer therapy as well. In the study, they compared the PDT vaccine potential with the vaccines generated via ultraviolet and ionizing irradiations. They concluded that PDT-generated vaccines are tumor specific, induce cytotoxic T cell response, and do not require coadministration of an adjuvant in order to be effective. Their results corroborate results of the other studies mentioned above.

35.3.2 PDT-Induced Neutrophil Infiltration

The main function of the cells forming the granulocyte cell group (neutrophils, basophils, and eosinophils) is to secrete prostaglandins, leukotrienes, and other cytokines in order to stimulate inflammatory response. Several studies have looked at the effect of PDT on neutrophil activation and involvement in immunogenic responses. Gollnick et al. (2003) demonstrated that, after PDT treatment, neutrophils migrated into the treated tumor area. They also described how this occurs, namely, through transient and local elevation in the macrophage-inflammatory-protein-2 (murine equivalent of IL-8) chemokine expression. During this process, the expression of the adhesion molecule E-selectin proved to be elevated as well (Gollnick et al. 2003). In addition, their results showed that a local and systemic elevation in the expression of IL-6, otherwise critical for launching an immunological response, is not necessary for neutrophil recruitment. They also demonstrated that, after PDT, mice with defective neutrophil homing ability to the peripheral tissues along with mice depleted of neutrophils were unable to mount a strong antitumor CD8+ T cell response (Kousis et al. 2007). Following the results from PDT-neutrophil coordination studies, it seems that a lack of neutrophils critically affects the T cell proliferation and/or survival. These observations support the view that neutrophils are essential for launching antitumor immunity after PDT. Interestingly, the elevated levels of E-selectin were also accounted for; it was shown that, after PDT, neutrophils adhere to the microvascular wall (Sluiter et al. 1996) and that EC retracts, which, in turn, allows the adherence of the

neutrophils via their β2-integrin receptors to the subendothelial matrix (de Vree et al. 1996b).

The significance of neutrophil infiltration to the tumor site, in launching immune response after PDT, was proven to be so important that research was extended to study the effect of other adhesion molecules. Volanti et al. (2004) studied the coordination effect between expression levels of adhesion molecules ICAM-1 and VCAM-1 and the neutrophil migration to the tumor site (as well as the effect of lacking that ability). Their results show that expression levels of ICAM-1 and VCAM-1 on the EC were downregulated after PDT (Volanti et al. 2004) and that administration of anti-neutrophil serum, in addition to the PDT treatment, completely abrogated launching of PDT antitumor effects (de Vree et al. 1996a), thus proving the necessity of neutrophil infiltration for launching antitumor response. Blocking ICAM-1 with monoclonal antibodies is shown to reduce the number of shrunken/cured tumors. Also, anti-G-CSF antibodies decreased the neutrophil numbers as well as the efficacy of the PDT treatment.

35.4 Immunosuppressive Effect of PDT

Up to this point, we discussed the effects of PDT for inducing inflammation and immune response, as well as immunogenic apoptosis. However, the coin has another side, too. There are reports suggesting that PDT may induce immunosuppression as well (Hunt and Levy 1998; Lynch et al. 1989).

Various research studies were conducted to ascertain whether PDT-mediated immunosuppression is local or systemic (thus, adoptively transferable and antigen specific), which are the immune cells involved, and what types of cytokines are actively involved in channeling the immunosuppression process. Lynch et al. (1989) showed that PDT can lead to systemic immunosuppression. They identified the suppressor cells to be microphages and also showed that the immunosuppressive effect is adoptively transferable through viable splendocytes (Figure 35.4). As for the role of the PSs in inducing immunosuppressive effects, studies done with a series of porphyrins have shown that this process is one of the most common side effects of the PDT; PSs HpD and meso-tetra(4-sulfonatophenole)porphine were observed to generate immunosuppression, while Photofrin II and meso-tetra(4-carboxyphenyl)porphine were shown to delay the immunosuppression (Musser and Fiel 1991). Research has shown that the anatomic site of the PDT illumination is also a factor in triggering an immunosuppressive effect (Musser et al. 1999). The IL-10 cytokine is known to play a leading role in the immunologic processes. With that in mind, its role in PDT-mediated immunosuppression was extensively investigated (Simkin et al. 2000; Gollnick et al. 2001). However, data from these studies gave rise to contradictory conclusions.

Along the line of PSs' effect in inducing immunosuppression, "prophotosensitizers" were also tested in a clinical setting (Matthews and Damian 2010). In the study, healthy and PPD-positive volunteers were subjected to an ALA or MAL "prophotosensitizer" (Matthews and Damian 2010). The PPD Mantoux

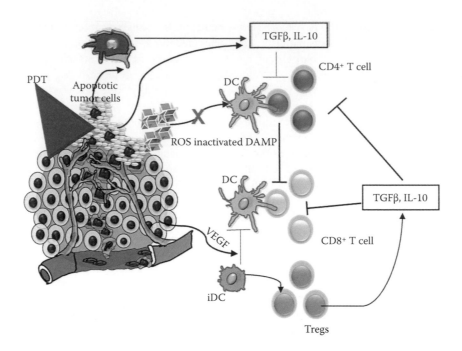

FIGURE 35.4 Potential mechanism in PDT-induced local and systemic immunosuppression. PDT-generated tissue damage leads to release of a number of antigens that cause activation of immune and regulatory responses. Local immunogenic response may cause the DCs to move to the regional lymph nodes (LNs) and subsequently end up with antigen presentation to naïve T cells. This type of PDT-mediated immune activation may trigger counteraction, which is manifested by immunosuppression. DCs may deliver the antigen to the LNs both in stimulatory and in tolerogenic manners. The T cells, if activated, may be prevented from functioning by Treg cells or immunosuppressive cytokines (locally secreted as a response to PDT). Tregs may be responsible for the observed systemic immunosuppression after PDT. Tolerance induction effect of PDT. Apoptotic cell deaths after PDT may cause DAMPs release. However, ROS produced may inactivate the immunostimulatory potential of these molecules. Apoptotic cells may also release immunosuppressive cytokines, such as IL-10 or TGFβ, or stimulate macrophages. Immunosuppressive cytokines may activate the CD4+ T cells to generate Tregs, while the lack of CD4+ T cells results in generation of anergic CD8+ T cells. Moreover, elevated levels of VEGF during PDT may effect the maturation of DCs that results in generation of highly tolerogenic and immunosuppressive media.

preparation of tuberculin protein was used to generate hypersensitivity responses at the irradiation site and at nonirradiated areas. Results revealed that both ALA-PDT and MAL-PDT significantly suppressed the Mantoux erythema as well as the diameter of the reaction. Interestingly, while red light alone significantly suppressed the reaction diameter, it did not reduce the degree of the erythema. PDTs' immunosuppressive effect was observed not only in contact hypersensitivity situations (which are T cell mediated immunity cases) but also in other immunomodulatory cases, such as skin grafting (Korbelik and Sun 2006; Qin et al. 1993; Obochi, Ratkay, and Levy 1997; Honey et al. 2000).

Along the same lines, the immunosuppressive effect of PDT in the autoimmune disease called adjuvant-enhanced arthritis with MRL/lpr mice, a model of human autoimmune arthritis and found in systemic lupus erythematosus (SLE) disease, was tested (Chowdhary et al. 1994; Ratkay et al. 1994). During these studies, it was observed that mice treated with PDT have delayed onset of the arthritis as well as reduced severity in comparison with the untreated one. Moreover, PDT prevented cartilage and bone tissue damage, which was attributed to the PDT's selective destruction of adjuvant-activated lymphocytes in both circulation and joints (Chowdhary et al. 1994; Ratkay et al. 1994).

Although the mechanism of PDT-mediated immunosuppression is not fully understood, it seems likely that the local

inflammatory reaction generated by PDT simultaneously induces a compensatory anti-inflammatory response, which, in return, limits the otherwise potentially dangerous overactive immune responses. This dual action of PDT-mediated inflammation may paradoxically cause active immunosuppression. It has also been argued that PDT treatment may lead to generation of a specific local immune privilege (Mellor and Munn 2008). Because it is localized, this "immune privilege" would be different from the systemic tolerance, and on some occasions, it may actually lead to systemic tolerance over time. Along the same lines, if the PDT treatment generates directly a local immune privilege, it would not lead to systemic antigen-specific tolerance.

In studying the factors leading to immunosuppressive effects after PTD treatment, cell death induction after PDT has to be taken into consideration. In the previous section, we discussed the immunity-inducing side of cell deaths after PDT. However, the coin has another side to it, too; it has been shown that, under some conditions, dying cells may generate immune tolerance as well (Garg et al. 2010a; Green et al. 2009). It is important to realize that the mode of cell deaths is a key determinant in generating the systemic response, whether that would be immune tolerance or an inflammatory reaction. Studies have shown that DCs that digest necrotic cell remnants can effectively activate both CD8+ and CD4+ T cells, while apoptotic cell deaths may only lead to

CD8+ T cell activation (Griffith et al. 2007). However, there is some evidence that without the help of CD4+, the CD8+ T cells may become anergic and tolerogenic (Griffith et al. 2007).

We pointed out in the previous section about the immunostimulatory effects of PDT that this type of immune response is largely based on cell deaths with subsequent release of DAMPs (Garg et al. 2010b). Studies have shown, however, that DAMP molecules can foster immunotolerance as well. One of the DAMP molecules studied is the nuclear DNA-binding HMGB1 (Scaffidi, Misteli, and Bianchi 2002), which, by binding to Toll-like receptors (TLRs), can activate the immune response (Bianchi and Manfredi 2007). Research has shown that because of generated ROS, DAMP molecules can undergo further modifications (oxidative modifications), which may hamper the immunostimulation and instead promote immunotolerance (Kazama et al. 2008).

Another factor to be taken into consideration in looking into the immunosuppressive effects of PDT is the mitochondrial changes taking place after the PDT. During apoptosis, mitochondria permeabilize and release cytochrome C, which causes activation of caspases. This cascade continues with cleaving of the NADH dehydrogenase Fe-S protein1 (NDuFS1), a component of complex 1 of the electron transport chain (Ricci et al. 2004). This inhibition of complex 1, in turn, induces production of ROS from the mitochondria that oxidize a key cysteine residue in HMGB1, thus, neutralizing the HMGB1's ability to promote immune responses (Broady, Yu, and Levings 2008). Considering the fact that, at this point, there are two different ROS in the milieu (one from the PDT-generated cell deaths and another one from the mitochondria processes), it is entirely possible that the cumulative effect may end up inducing immunotolerance. Also, there is strong evidence (Chung et al. 2007; Fadok et al. 1998; Voll et al. 1997) that cells undergoing apoptosis release immunosuppressive cytokines, such as TGFβ and IL-10, which might also contribute to developing an immunotolerance.

Another possible mechanism through which PDT may generate immunosuppression is by interrupting the maturation process of DCs. There is evidence suggesting that PDT can elevate VEGF levels, and this correlates well with the poorest PDT treatment outcomes (Solban et al. 2006). VEGF's proangiogenic effect is well known; however, a number of studies have brought to our attention its immunosuppressive effects (Solban et al. 2006; Ohm and Carbone 2001; Laxmanan et al. 2005). These studies have shown that VEGF with its inhibitory effect over DC differentiation (Ohm and Carbone 2001) acts as one of the major mediators for disrupting the DCs' differentiation process (Laxmanan et al. 2005). When this happens, accumulation of immature DCs (iDCs) could induce Treg differentiation. In other words, iDC accumulation has a two-punch effect over immune responses at the molecular level, the first one during antigen presentation and the second one during the effector phase of T cell differentiation (Johnson et al. 2007). Blocking the VEGF augments the "normal" DCs' differentiation and functioning (Ohm and Carbone 2001; Gabrilovich et al. 1999; Ohm et al. 2003). In summary, abnormal differentiation and maturation of DCs,

mediated by tumor-derived soluble factors, such as VEGF, is likely to adversely affect the T cell-mediated antitumor immune response.

In general, Tregs can be seen as a T cell population that suppresses immune responses. In addition to many pathways of immunosuppression, it has been suggested recently that Tregs control T cell activation via suppressing the DCs' activation (Veldhoen et al. 2006). Tumor-induced proliferation of Tregs has been shown to be a major obstacle to successful cancer immunotherapy because they inhibit launching of an antitumor immune response (Zou 2006). To our knowledge, Hamblin's lab was the first to realize that Tregs may have an important and suppressive effect on PDT-generated antitumor immunity as described in their seminal investigation of the interplay between the Treg population and the launching of an immune response (Mroz et al. 2011). They observed that Treg population can be efficiently depleted by a low-dose cyclophosphamide (CY) and that a combination of low-dose CY with BPD-PDT leads to significant cure of long-term reticulum-cell sarcomas as well as launching a resistance to tumor rechallenge (Castano and Hamblin 2005). To cap their conclusion, efficient Treg depletion with low-dose CY treatment and a combination of low-dose CY and BPD-PDT treatment led to resistance to tumor rechallenge, whereas neither treatment alone has this effect. Moreover, Treg depletion by CY treatment may unravel PDT-mediated immune response to mice autoantigen gp70, in colon adenocarcinoma CT26WT mouse model. Lastly, they have shown that this type of combination treatment leads to development of long-lasting immune memory effect.

All said and done, despite all our efforts thus far, there are still black holes in our knowledge about the mechanism(s) and the effects of PDT-generated immunosuppression. Understanding them may lead to designing better therapy combinations and enhancement of PDT-generated immunity response. For more details about the immunosuppressive effect of PDT, see Chapter 11 by Mary Norval and Prue H. Hart.

35.5 Novel Strategies in PDT

Research has been going on to improve the PDT treatment outcomes and to minimize its side effects. Nowadays, efforts are concentrated on developing systems that provide better targeting with better control over the response cascades and developing new generation of PSs that can be activated with lower and narrow range of energy and allowing higher penetration depth. In here we will discuss only two of these directions: two-photon absorption (TPA) concept in developing a new class of PSs and harnessing the TLR agonist effect.

35.5.1 Developing a New Class of PSs

Singlet-oxygen induction (from triplet-oxygen) lies at the heart of the PDT technique, but for that, a sufficiently high energy is required. Long-wavelength near-IR (NIR) light ($\lambda_{ex} > 750$ nm) has too low photon energy to induce triplet-to-singlet oxygen generation in a one-photon excitation process. Fortunately, with the new class of molecules/PSs, named as two-photon chromophores, this drawback is ameliorated. The photophysics of simultaneous TPA is different from the classical, single-photon absorption; it falls in the nonlinear optical properties regimen. In a way, TPA combines the energy of the two photons, and by doing that, at the emission stage, it is able to convert the triplet oxygen into a singlet one. Why should TPA be important in PDT? The answer lies in the fact that the "phototherapeutic window" wavelengths in the near IR (780–950 nm) allow deeper tissue penetration. Thus, PSs synthesized from these high cross-section TPA molecules will be capable of generating singlet oxygen in the near IR range. The use of TPA in this new class of PSs can greatly expand the PDT application range. Using their proprietary tetrapyrrole-based TPA PS on human nonsmall cell lung cancer, human breast carcinoma cells, and human pancreatic carcinoma cells, Starkey et al. (2008) demonstrated the advantages of using these novel PSs, operating in the near-IR window. For more information about the classical PSs, refer to Chapter 22 by Penny Joshi et al. and Chapter 30 by Marlène Pernot et al. Detailed discussion about genetically encoded PSs can be seen in Chapter 24 by Cristina Flors and Santi Nonell.

35.5.2 TLR Agonists in Anticancer Immunotherapy

As discussed earlier, biomolecules or segments of agents from microbial stimulators of innate immunity can be injected into the tumor or its environs before, during, or after PDT. These molecules serve as adjuvants to activate the TLR and DCs.

The field of anticancer immunotherapy has widened in the recent years. One class of agent that has been seen as very promising is TLR agonists. Once they were included in the US National Cancer Institute's list of immunotherapeutic agents with the highest potential to treat cancer, TLR agonist-related research soared. It is known that each member of the TLR family triggers the release of inflammatory cytokines of a different spectrum; thus, it is crucial to decide over which targets to agonize in order to evoke the most potent antitumor immune response. Some experts argue that TLR agonists are better used as therapeutic cancer vaccine adjuvants to activate the immune system's DCs so that DCs are highly primed to recognize specific tumor antigens. One has to keep in mind that the TLR system has evolved primarily as a defense against outside invaders rather than against cancer, and cancers take advantage of this conserved mechanism that normally prevents autoimmune response to self-antigens. This brings into play the need to tackle the tumor immunosuppression effect when using TLR agonists.

It is well known that tumors secrete immunosuppressive factors and activate the regulatory immune system. Into this process, TLR agonists appear to induce a negative feedback loop. These two factors working in synergy can limit the uncontrolled inflammation that might otherwise arise. Preclinical studies of treatment with topical imiquimod are shown to induce high

levels of the anti-inflammatory cytokine IL-10, and blockade of IL-10 enhanced the antitumor effect of imiquimod (Lu et al. 2010). The same combination strategy is planned to be used now by the same group, in a phase I trial in breast cancer patients with chest wall metastases. Similarly, TriMod Therapeutics is planning to initiate clinical trials of TLR agonist and phosphoinositide 3-kinase (PI3K) inhibitor (Guha 2012) after preclinical studies have shown that this combination is suppressing the anti-inflammatory IL-10 and TGFβ activity induced by the TLR5 agonist flagellin (Marshall et al. 2012). Yet other groups are testing TLR agonists with cytotoxic T lymphocyte antigen 4 (CTLA)-targeting drugs (Guha 2012).

35.6 Conclusion

Because of its inflammatory and immunogenesis-triggering capabilities, PDT can be used in tandem with other treatment modalities to combat and to be successful in achieving long-term tumor control. However, in doing so, one has to keep in mind the potential danger of blocking the immunoregulatory mechanisms, thus, running into the risk of generating excessive immune reactions. At the very essence, the solution to the problem is to make this therapeutic treatment more targeted. By doing so, the overwhelming inflammatory response might arise only at the tumor site and not throughout the whole body.

Abbreviations

ALA:	5-Aminolevulinic acid (prodrug that needs to be converted to protoporphyrin in order to be active as a PS)
BPD:	Benzoporphyrin derivative monoacid ring A
CDAMPs:	Cell death-associated molecular patterns
CRT:	Calreticulin
CY:	Cyclophosphamide
DAMPs:	Damage-associated molecular patterns
DCs:	Dendritic cells
HMGB1:	High-mobility group box-1 protein
HSPs:	Heat-shock proteins
iDCs:	Immature dendritic cells
IFNγ:	Interferon gamma
MAL:	Methyl aminolevulinate
MHC I:	Major histocompatibility complex class I
NDuFS1:	NADH dehydrogenase Fe-S protein1
NIR:	Near infrared (780–950 nm)
PDT:	Photodynamic therapy
PSs:	Photosensitizer(s)
ROS:	Reactive oxygen species
SCC:	Squamous cell carcinoma
SLE:	Systemic lupus erythematosus disease
TLR:	Toll-like receptor
TNFα:	Tumor necrosis factor alpha
TPA:	Two-photon absorption
Treg:	Regulatory T cells
VIN:	Vulval intraepithelial neoplasia

References

Abdel-Hady, E. S., P. Martin-Hirsch, M. Duggan-Keen et al. 2001. Immunological and viral factors associated with the response of vulval intraepithelial neoplasia to photodynamic therapy. *Cancer Res* 61:192–196.

Betts, M. R., J. M. Brenchley, D. A. Price et al. 2003. Sensitive and viable identification of antigen-specific CD8+ T cells by a flow cytometric assay for degranulation. *J Immunol Methods* 281:65–78.

Bianchi, M. E., and A. Manfredi. 2007. High-mobility group box 1 (HMGB1) protein at the crossroads between innate and adaptive immunity. *Immunol Rev* 220:35–46.

Broady, R., J. Yu, and M. K. Levings. 2008. Pro-tolerogenic effects of photodynamic therapy with TH9402 on dendritic cells. *J Clin Apheresis* 23:82–91.

Canti, G. L., D. Lattuada, A. Nicolin et al. 1994. Immunopharmacology studies on photosensitizers used in photodynamic therapy. *Proc SPIE* 2078:268–275.

Castano, A. P., and M. R. Hamblin. 2005. Anti-tumor immunity generated by photodynamic therapy in a metastatic murine tumor model. *Proc SPIE* 5695:7–16.

Castano, A. P., P. Mroz, and M. R. Hamblin. 2006. Photodynamic therapy and anti-tumor immunity. *Nat Rev Cancer* 6:535–545.

Cecic, I., and M. Korbelik. 2002. Mediators of peripheral blood neutrophilia induced by photodynamic therapy of solid tumors. *Cancer Lett* 183:43–45.

Cecic, I., C. S. Parkins, and M. Korbelik. 2001. Induction of systemic neutrophil response in mice by photodynamic therapy in solid tumors. *Photochem Photobiol* 74:712–720.

Cecic, I., B. Scott, and M. Korbelik. 2006. Acute phase response-associated systemic neutrophil mobilization in mice bearing tumors treated by photodynamic therapy. *Int Immunopharmacol* 6(8):1259–1266.

Chowdhary, R. K., L. G. Ratkay, H. C. Neyndorff et al. 1994. The use of transcutaneous photodynamic therapy in the prevention of adjuvant-enhanced arthritis in MRL/lpr mice. *Clin Immunol Immunopathol* 72(2):255–263.

Chung, E. Y., J. Liu, Y. Homma et al. 2007. Interleukin-10 expression in macrophages during phagocytosis of apoptotic cells is mediated by homeodomain proteins pbx1 and prep-1. *Immunity* 27:952–964.

de Vree, W. J., M. C. Essers, H. S. de Bruijn et al. 1996a. Evidence for an important role of neutrophils in the efficacy of photodynamic therapy *in vivo*. *Cancer Res* 56(13):2908–2911.

de Vree, W. J., M. C. Essers, J. F. Koster, and W. Sluiter. 1997. Role of interleukin 1 and granulocyte colony-stimulating factor in photofrin-based photodynamic therapy of rat rhabdomyosarcoma tumors. *Cancer Res* 57:2555–2558.

de Vree, W. J., A. N. Fontijne-Dorsman, J. F. Koster, and W. Sluiter. 1996b. Photodynamic treatment of human endothelial cells promotes the adherence of neutrophils *in vitro*. *Br J Cancer* 73(11):1335–1340.

Dolmans, D. E., A. Kadambi, J. S. Hill et al. 2002. Vascular accumulation of a novel photosensitizer, MV6401, causes selective thrombosis in tumor vessels after photodynamic therapy. *Cancer Res* 62:2151–2156.

Dolmans, T. J., D. Fukumura, and R. K. Jain. 2003. Photodynamic therapy in cancer. *Nat Rev Cancer* 3:380–387.

Dougherty, T. J., C. J. Gomer, B. W. Henderson et al. 1998. Photodynamic therapy. *J Natl Cancer Inst* 90:889–905.

Fadok, V. A., D. L. Bratton, A. Konowal et al. 1998. Macrophages that have ingested apoptotic cells *in vitro* inhibit proinflammatory cytokine production through autocrine/paracrine mechanisms involving TGF-beta, PGE2, and PAF. *J Clin Invest* 101(4):890–898.

Gabrilovich, D. G., S. Ishida, J. E. Nadaf et al. 1999. Antibodies to vascular endothelial growth factor enhance the efficacy of cancer immunotherapy by improving endogenous dendritic cell function. *Clin Cancer Res* 5:2963–2970.

Garg, A. D., D. Nowis, J. Golab, and P. Agostinis. 2010a. Photodynamic therapy: Illuminating the road from cell death towards anti-tumour immunity. *Apoptosis* 15:1050–1071.

Garg, A. D., D. Nowis, J. Golab et al. 2010b. Immunogenic cell death, DAMPs and anticancer therapeutics: An emerging amalgamation. *Biochim Biophys Acta* 1805(1):53–71.

Gollnick, S. O. and C. M. Brackett. 2010. Enhancement of anti-tumor immunity by photodynamic therapy. *Immunol Res* 46:216–226.

Gollnick, S. O., S. S. Evans, H. Baumann et al. 2003. Role of cytokines in photodynamic therapy-induced local and systemic inflammation. *Br J Cancer* 88(11):1772–1779.

Gollnick, S. O., E. Kabingu, P. C. Kousis, and B. W. Henderson. 2004. Stimulation of the host immune response by photodynamic therapy (PDT). *Proc SPIE* 5319:60–70.

Gollnick, S. O., D. A. Musser, A. R. Oseroff et al. 2001. IL-10 does not play a role in cutaneous Photofrin photodynamic therapy-induced suppression of the contact hypersensitivity response. *Photochem Photobiol* 74:811–816.

Gollnick, S. O., B. Owczarczak, and P. Maier. 2006. Photodynamic therapy and anti-tumor immunity. *Lasers Surg Med* 38:509–515.

Gollnick, S. O., L. Vaughan, and B. W. Henderson. 2002. Generation of effective antitumor vaccines using photodynamic therapy. *Cancer Res* 62(6):1604–1608.

Gomer, C. J., S. W. Ryter, A. Ferrario et al. 1996. Photodynamic therapy-mediated oxidative stress can induce expression of heat shock proteins. *Cancer Res* 56:2355–2360.

Green, D. R., T. Ferguson, L. Zitvogel, and G. Kroemer. 2009. Immunogenic and tolerogenic cell death. *Nat Rev Immunol* 9(5):353–363.

Griffith, T. S., H. Kazama, R. L. VanOosten et al. 2007. Apoptotic cells induce tolerance by generating helpless CD8+ T cells that produce TRAIL. *J Immunol* 178:2679–2687.

Gruner, S., H. Meffert, H. D. Volk, R. Grunow, and S. Jahn. 1985. The influence of haematoporphyrin derivative and visible light on murine skin graft survival, epidermal Langerhans cells and stimulation of the allogeneic mixed leukocyte reaction. *Scand J Immunol* 21:267–273.

Guha, M. 2012. Anticancer TLR agonists on the ropes. *Nat Rev Drug Discov* 11:503–505.

Henderson, B. W., and S. O. Gollnick. 2003. Mechanistic principles of photodynamic therapy. In *Biomedical Photonics Handbook*. CRC Press, Boca Raton, FL, 36.1–36.27.

Henderson, B. W., S. O. Gollnick, J. W. Snyder et al. 2004. Choice of oxygen-conserving treatment regimen determines the inflammatory response and outcome of photodynamic therapy in tumors. *Cancer Res* 64:2120–2126.

Honey, C. R., M. O. Obochi, H. Shen et al. 2000. Reduced xenograft rejection in rat striatum after pretransplant photodynamic therapy of murine neural xenografts. *J Neurosurg* 92(1):127–131.

Hunt, D. W., and J. G. Levy. 1998. Immunomodulatory aspects of photodynamic therapy. *Expert Opin Investig Drugs* 7:57–64.

Jalili, A., M. Makowski, T. Switaj et al. 2004. Effective photo-immunotherapy of murine colon carcinoma induced by the combination of photodynamic therapy and dendritic cells. *Clin Cancer Res* 10:4498–4508.

Johnson, B. F., T. M. Clay, A. C. Hobeika, H. K. Lyerly, and M. A. Morse. 2007. Vascular endothelial growth factor and immunosuppression in cancer: Current knowledge and potential for new therapy. *Expert Opin Biol Ther* 7:449–460.

Kabingu, E., A. R. Oseroff, G. E. Wilding, and S. O. Gollnick. 2009. Enhanced systemic immune reactivity to a basal cell carcinoma associated antigen following photodynamic therapy. *Clin Cancer Res* 15:4460–4466.

Kabingu, E., L. Vaughan, B. Owczarczak, K. D. Ramsey, and S. O. Gollnick. 2007. CD8+ T cell-mediated control of distant tumors following local photodynamic therapy is independent of CD4+ T cells and dependent on natural killer cells. *Br J Cancer* 96:1839–1848.

Kazama, H., J. E. Ricci, J. M. Herndon et al. 2008. Induction of immunological tolerance by apoptotic cells requires caspase-dependent oxidation of high-mobility group box-1 protein. *Immunity* 29(1):21–32.

Korbelik, M. 2006. PDT-associated host response and its role in the therapy outcome. *Lasers Surg Med* 38:500–508.

Korbelik, M., and I. Cecic. 1999. Contribution of myeloid and lymphoid host cells to the curative outcome of mouse sarcoma treatment by photodynamic therapy. *Cancer Lett* 137:91–98.

Korbelik, M. and I. Cecic. 2003. Mechanism of tumor destruction by photodynamic therapy. In *Handbook of Photochemistry and Photobiology*, H. S. Nalwa (ed), ISBN: 1-58883-004-7. American Scientific Publishers, Valencia, CA.

Korbelik, M. and G. J. Dougherty. 1999. Photodynamic therapy-mediated immune response against subcutaneous mouse tumors. *Cancer Res* 59:1941–1946.

Korbelik, M., G. Krosl, J. Krosl, and G. J. Dougherty. 1996. The role of host lymphoid populations in the response of mouse EMT6 tumor to photodynamic therapy. *Cancer Res* 56:5647–5652.

Korbelik, M., S. Merchant, and N. Huang. 2009. Exploitation of immune response-eliciting properties of hypocrellin photosensitizer SL052-based photodynamic therapy for eradication of malignant tumors. *Photochem Photobiol* 85:1418–1424.

Korbelik, M., B. Scott, and J. Sun. 2007. Photodynamic therapy-generated vaccines: Relevance of tumour cell death expression. *Br J Cancer* 97:1381–1387.

Korbelik, M., and J. Sun. 2006. Photodynamic therapy-generated vaccine for cancer therapy. *Cancer Immunol Immunother* 55:900–905.

Korbelik, M., J. Sun, and I. Cecic. 2005. Photodynamic therapy-induced cell surface expression and release of heat shock proteins: Relevance for tumor response. *Cancer Res* 65:1018–1026.

Kousis, P. C., B. W. Henderson, P. G. Maier, and S. O. Gollnick. 2007. Photodynamic therapy enhancement of antitumor immunity is regulated be neutrophils. *Cancer Res* 67:10501–10510.

Krammer, B. 2001. Vascular effects of photodynamic therapy. *Anticancer Res* 21:4271–4277.

Krosl, G., M. Korbelik, and G. J. Dougherty. 1995. Introduction of immune cell infiltration into murine SCCVII tumor by photofrin-based photodynamic therapy. *Br J Cancer* 71:549–555.

Laxmanan, S., S. W. Robertson, E. Wang et al. 2005. Vascular endothelial growth factor impairs the functional ability of dendritic cells through Id pathways. *Biochem Biophys Res Commun* 334:193–198.

Lu, H., W. Wagner, E. Gad et al. 2010. Treatment failure of a TLR-7 agonist occurs due to self-regulation of acute inflammation and can be overcome by IL-10 blockade. *J Immunol* 184:5360–5367.

Lynch, D. H., S. Haddad et al. 1989. Systemic immunosuppression induced by photodynamic therapy (PDT) is adoptively transferred by macrophages. *Photochem Photobiol* 49:453–458.

Maeurer, M. J., S. M. Gollin, W. J. Storkus et al (1996. Tumor escape from immune recognition: Loss of HLA-A2 melanoma cell surface expression is associated with a complex rearrangement of the short arm of chromosome 6. *Clin. Cancer Res* 2:641–652.

Marshall, N. A., K. C. Galvin, A. B. Corcoran et al. 2012. Immunotherapy with PI3K inhibitor and Toll-like receptor agonist induces IFN-γ+IL-17+ polyfunctional T cells that mediate rejection of murine tumors. *Cancer Res* 72:581–590.

Matthews, Y. J., and D. L. Damian. 2010. Topical photodynamic therapy is immunosuppressive in humans. *Br J Dermatol* 162:637–641.

Mellor, A. L., and D. H. Munn. 2008. Creating immune privilege: Active local suppression that benefits friends, but protects foes. *Nat Rev Immunol* 8:74–80.

Moan, J., K. Berg, E. Kvam et al. 1989. Intracellular localization of photosensitizers. *Ciba Found Symp* 146:95–107.

Mroz, P., and M. R. Hamblin. 2008. PDT and cellular immunity. In *Advances in Photodynamic Therapy: Basic, Translational, and Clinical*, M. R. Hamblin (ed), ISBN-13: 978-1-59693-277-7. Artech House Publishers, Norwood, MA.

Mroz, P., J. T. Hashmi, Y. Y. Huang, N. Lange, and M. R. Hamblin. 2011. Stimulation of anti-tumor immunity by photodynamic therapy. *Expert Rev Clin Immunol* 7(1):75–91.

Mroz, P., A. Szokalska, M. X. Wu, and M. R. Hamblin. 2010. Photodynamic therapy of tumors can lead to development of systemic antigen-specific immune response. *PLoS One* 5:e15194.

Musser, D. A., S. H. Camacho, P. A. Manderscheid, and A. R. Oseroff. 1999. The anatomic site of photodynamic therapy is a determinant for immunosuppression in a murine model. *Photochem Photobiol* 69:222–225.

Musser, D. A., and R. J. Fiel. 1991. Cutaneous photosensitizing and immunosuppressive effects of a series of tumoral localizing porphyrins. *Photochem Photobiol* 53:119–123.

Obochi, M. O., L. G. Ratkay, and J. G. Levy. 1997. Prolonged skin allograft survival after photodynamic therapy associated with modification of donor skin antigenicity. *Transplantation* 63:810–817.

Ohm, D. I., G. D. Gabrilovich, E. Sempowski et al. 2003. VEGF inhibits T-cell development and may contribute to tumor-induced immune suppression. *Blood* 101:4878–4886.

Ohm, J. E., and P. D. Carbone. 2001. VEGF as a mediator of tumor-associated immunodeficiency. *Immunol Res* 23:263–271.

Oleinick, N. L., and H. H. Evans. 1998. The photobiology of photodynamic therapy: Cellular targets and mechanisms. *Radiat Res* 150(suppl. 5):5146–5156.

Oleinick, N. L., R. L. Morris, and I. Belichenko. 2002. The role of apoptosis in response to photodynamic therapy: What, where, why, and how. *Photochem Photobiol Sci* 1:1–21.

Parkinson-Lawrence, E. J., C. J. Dean, M. Chang et al. 2005. Immunochemical analysis of CD107a (LAMP-1). *Cell Immunol* 236:161–166.

Qin, B., S. H. Selman, K. M. Payne, R. W. Keck, and D. W. Metzger. 1993. Enhanced skin allograft survival after photodynamic therapy: Association with lymphocyte inactivation and macrophage stimulation. *Transplantation* 56:1481–1486.

Ratkay, L. G., R. K. Chowdhary, H. C. Neyndorff et al. 1994. Photodynamic therapy: A comparison with other immunomodulatory treatments of adjuvant-enhanced arthritis in MRL/lpr mice. *Clin Exp Immunol* 95:373–377.

Reis, E., and C. Sousa. 2004. Activation of dendritic cells: Translating innate into adaptive immunity. *Curr Opin Immunol* 16:21–25.

Ricci, J. E., C. Munoz-Pinedo, P. Fitzgerald et al. 2004. Disruption of mitochondrial function during apoptosis is mediated by caspase cleavage of the p75 subunit of complex I of the electron transport chain. *Cell* 117:773–786.

Scaffidi, P., T. Misteli, and M. E. Bianchi. 2002. Release of chromatin protein HMGB1 by necrotic cells triggers inflammation. *Nature* 418:191–195.

Scott, B., and M. Korbelik. 2007. Activation of complement C3, C5, and C9 genes in tumors treated by photodynamic therapy. *Cancer Immunol Immunother* 56:649–658.

Simkin, G. O., J. S. Tao, J. G. Levy, and D. W. Hunt. 2000. IL-10 contributes to the inhibition of contact hypersensitivity in mice treated with photodynamic therapy. *J Immunol* 164:2457–2462.

Sluiter, W., W. J. de Vree, A. Pietrsma, and J. F. Koster. 1996. Prevention of the late lumen loss after coronary angioplasty by photodynamic therapy: Role of activated neutrophils. *Mol Cell Biochem* 157(1–2):233–238.

Solban, N., P. K. Selbo, A. K. Sinha, S. K. Chang, and T. Hasan. 2006. Mechanistic investigation and implications of photodynamic therapy induction of vascular endothelial growth factor in prostate cancer. *Cancer Res* 66(11):5633–5640.

Starkey, J., et al. 2008. New two-photon activated photodynamic therapy sensitizers induce xenograft tumor regression after near-IR laser treatment through the body of the host mouse. *Clin Cancer Res* 14(20):6564–6573.

Sun, J., I. Cecic, C. S. Parkins, and M. Korbelik. 2002. Neutrophils as inflammatory and immune effectors in photodynamic therapy-treated mouse SCCVII tumors. *Photochem Photobiol Sci* 1:690–695.

Sur, B. W., P. Nguyen, C. H. Sun, B. J. Tromberg, and E. L. Nelson. 2008. Immunotherapy using PDT combined with rapid intratumoral dendritic cell injection. *Photochem Photobiol* 84:1257–1264.

Thong, P. S., K. W. Ong, N. S. Goh et al. 2007. Photodynamic-therapy-activated immune response against distant untreated tumors in recurrent angiosarcoma. *Lancet Oncol* 8:950–952.

Vabulas, R. M., H. Wagner, and H. Schil. 2002. Heat shock proteins as ligands of toll-like receptors. *Curr Top Microbiol Immunol* 270:167–184.

Veldhoen, M., H. Moncrieffe, R. J. Hocking, C. J. Atkins, and B. Stockinger. 2006. Modulation of dendritic cell function by naïve and regulatory CD4+ T cells. *J Immunol* 176(10):6202–6210.

Volanti, C., G. Gloire, A. Vanderplasschen, N. Jacobs, Y. Habraken, and J. Piette. 2004. Downregulation of ICAM-1 and VCAM-1 expression in endothelial cells treated by photodynamic therapy. *Oncogene* 23(53):8649–8658.

Voll, R. E., M. Herrmann, E. Roth et al. 1997. Immunosuppressive effects of apoptotic cells. *Nature* 390:350–351.

Zou, W. 2006. Regulatory T cell, tumor immunity and immunotherapy. *Nat Rev Immunol* 6(4):295–307.

Detection of Bladder Cancer by Fluorescence Cystoscopy: From Bench to Bedside— the Hexvix Story

Georges Wagnières
Swiss Federal Institute of Technology (EPFL)

Patrice Jichlinski
CHUV University Hospital

Norbert Lange
University of Geneva

Pavel Kucera
Czech Technical University

Hubert van den Bergh
Swiss Federal Institute of Technology (EPFL)

36.1 Introduction

36.1.1 State of the Art and Medical Need

Bladder cancer is the fourth most common malignancy in men and the eighth most common malignancy in women, and the lifetime risk of the disease is greater than 3% (NIH 1990; Ries et al. 1997). Every year, nearly 200,000 new cases of bladder cancer are reported in Europe and the United States. Urothelial cell carcinoma comprises 90%–95% of all bladder cancers, with about 70% found initially as nonmuscle-invasive bladder cancer (NMIBC) and the remainder as invasive (Oosterlinck et al. 2002).

Carcinogenesis in bladder cancer results from urothelium mutations and epigenetic modifications of different origins partly reinforced by the influence of exogenous agents such as aromatic amines due to environmental exposure or tobacco. Distinctive aspects of urothelial carcinoma are related to its dual pathways of malignant transformation (papillary/noninvasive vs. flat/invasive) and its notorious propensity to recur in every clinical stage and setting or possibly progress to a fatal disease (Brandt 2009). It is therefore crucial to obtain accurate diagnostic information on the bladder wall during the whole process of the disease evolution.

The course of NMIBC is extremely variable. Recurrence and progression rates depend on tumor size and multiplicity, tumor stage (Ta vs. T1), histological grade, and presence of carcinoma in situ (CIS) (Oosterlink et al. 2002; Sylvester et al. 2006). Prognostic factors permit tumors to be classified into low-, intermediate-, and high-risk groups to enable the appropriate disease management. Recurrence rates range from 15% to 61% at 1 year and 31% to 78% at 5 years and tumor progression rates from <1% to 17% at 1 year and <1% to 45% at 5 years. The high prevalence of NMIBC upon recurrence makes the disease labor-intensive and costly to manage (Botteman et al. 2003). Therefore, optimization of the surgical management of NMIBC based on endoscopic techniques is essential.

NMIBC current standard of care after initial diagnosis is treatment by transurethral resection (TUR) and follow-up monitoring based on white-light cystourethroscopy (WL-CUS). This is sometimes associated with urinary cytology. In most cases, rigid or flexible WL-CUS allows for the detection of exophytic papillary tumors, even though some small lesions may be overlooked. However, flat urothelial abnormalities such as high-grade dysplasia and CIS are often missed. The accuracy of cytology depends on numerous factors such as specimen collection, transport, and available technology. Interpretation, which frequently depends on a cytopathologist's experience, is also very important. Whereas sensitivity and specificity of urinary cytology are over 80% in the presence of multiple high-grade tumors and/or large CIS, both remain extremely poor for the staging of low- and intermediate-risk disease in the absence of sophisticated labeling techniques (Bubendorf 2001; Fradet and Loskart 1997; Rathert 1993). CIS, as a surrogate marker of genetic instability and field carcinogenesis, has a high potential of malignant transformation in the whole "normal" surrounding urothelium and consequently is very important in relation to local and systemic progression (Palmeira et al. 2011). As the prevalence of concomitant CIS on normal-appearing mucosa may concern more than half of the patients with the criteria of high-grade tumors (Hara et al. 2009), it is essential to get information on the presence of CIS in real time during endoscopic inspection.

In NMIBC management, the preservation of the bladder will depend strongly on the quality of bladder wall inspection and tumor resection. Current guidelines advise a secondary TUR in the 4–6 weeks following initial treatment of high-grade disease and hold this recommendation to be mandatory if the first TUR is incomplete, even in the presence of an intermediate-risk disease. This underlines the need to optimize the initial surgical endoscopic procedure (i.e., the need for the removal of the tumor as completely as possible).

In conclusion, the availability of novel technologies to detect and/or localize NMIBCs and to carefully monitor transurethral resection of bladder tumor (TURBT) to ascertain complete resection of intravesical tumors is mandatory for progress in the management of bladder cancer. Endoscopic imaging remains the most promising approach since no very effective prognostic marker has yet been validated for routine clinical use. Therefore, the assessment of the disease is still mainly based on clinical parameters, such as tumor size and aspect, multifocality, presence/absence of urothelial flat abnormalities, residual tumor following TURBT, and recurrence rate (Brausi et al. 2002; Parmar et al. 1989). Consequently, it appears that the development of a novel approach based on fluorescence cystoscopy (FC) might well be a good strategy to respond to this medical need.

36.1.2 Basic Principles and Instrumentation for FC

The fundamental principle of FC is to image the optical contrast induced between the lesion and its surrounding normal tissue using fluorophores, often photoactive porphyrins (PAPs), including protoporphyrin IX (PpIX) (Wagnieres, Star, and Wilson 1998) (see Figures 36.1 and 36.2). These porphyrins are quite selectively induced in the lesions following the instillation in the bladder of a 50-mL solution containing one of their precursors [i.e., 5-aminolevulinic acid (5-ALA) or one of its derivatives such as its hexyl ester, known as hexaminolevulinate (HAL)]. The duration of this instillation is typically in the order of 1–3 h. Both 5-ALA and PpIX are naturally occurring intermediates in the heme biosynthesis taking place in nearly all aerobic cells in mammals (Fotinos 2006). Normally, heme inhibits the endogenous formation of excess 5-ALA by a negative feedback control mechanism, thereby avoiding natural PpIX photosensitization (Collaud et al. 2004; Peng et al. 1997). However, the presence of an excess of exogenous 5-ALA or one of its derivatives after topical application (instillation) and cellular internalization temporarily overwhelms this regulatory mechanism and results in the transient formation of excess PAPs. The latter is observed to take place preferentially in neoplastic tissues. The factors explaining this selectivity have been the subject of much discussion in the literature and have been reviewed by Collaud et al. (2004) and Fotinos et al. (2006). Apart from metabolic differences, environmental and morphological factors have been considered to affect the selective generation of PAPs in neoplastic cells. Several groups have proposed the reduced activity of the ferrochelatase enzyme, as well as a relative enhancement of the porphobilinogen deaminase activity in neoplastic tissue, as possible mechanisms for the selective production of PAPs (Fukuda, Casas, and Batlle 2005; Greenbaum et al. 2003). The latter makes NMIBC detectable by fluorescence imaging and spectroscopy (Lange, Jichlinski, and Zellweger 1999) following exposure to a PAP precursor.

Molecular fluorophores absorb light with a photon of higher energy than the reemitted photons. This energy difference, or Stokes shift, produces a change toward longer wavelengths between excitation and fluorescence light. Thus, optical filters can easily block the more intense backscattered excitation light and transmit selectively the weaker fluorescence emission of the porphyrins. Because the intensity of unspecific reflected light is much weaker, such fluorophores can be detected with a high sensitivity at concentrations of the order of 1 ppb.

Currently used instrumentation, available from several manufacturers of cystoscope systems, is designed for FC as well as classical white light (WL) endoscopy. In the fluorescence mode they display red porphyrin fluorescence together with a small amount of blue backscattered and/or autofluorescence light. This blue light helps to guide the cystoscopy since it enables the visualization of the bladder wall, including blood vessels, even in the absence of lesions. The observation with a color camera or with the naked eye of suspicious tissue by enhanced porphyrin fluorescence is by the color contrast of red fluorescence superimposed on the blue background. Normal tissue thus appears blue, whereas suspicious tissue is identified by its reddish color (Figure 36.2). To prevent practical limitations, including the photobleaching of the PAPs, the instrumentation must be highly sensitive for fluorescence detection and unnecessary light exposure is to be avoided.

Three major manufacturers supply equipment suitable for FC: Olympus Optical Co., Japan; Karl Storz GmbH, Germany; and Richard Wolf GmbH, Germany. They all allow fluorescence examination in addition to WL examination. These systems are composed of the following (see Figure 36.1a):

(1) A high-power xenon arc light source with integrated excitation filter, providing fluorescence excitation light in the range of 380–440 nm. These light sources can be switched

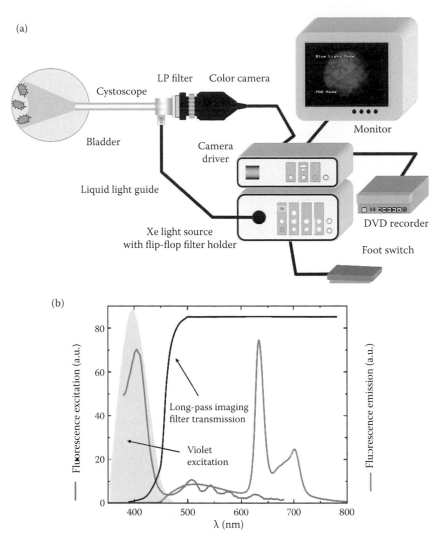

FIGURE 36.1 (a) Diagram of the experimental setup used to image the bladder wall fluorescence. The long-pass (LP) filter positioned in front of the camera enables to reject most of the blue-violet fluorescence excitation light delivered by the light source. (b) Spectral design of the experimental setup used to image the bladder wall fluorescence.

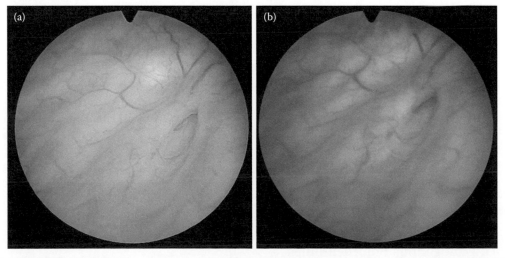

FIGURE 36.2 **(See color insert.)** (a) Bladder wall observed under standard WL illumination. The early cancers are invisible. Diameter of the field of view: 15 mm. (b) Same site observed in the fluorescence mode after administration of HAL. Small superficial cancers (carcinoma in situ) are clearly glowing red due to the marker. (Courtesy of the CHUV Hospital.)

between WL and fluorescence modes. This switch communicates with the camera control unit to adjust the gain and color balance according to the operational mode selected. Thus, by using a foot pedal or pushing a button on the camera head or the light source, one can rapidly change between either the conventional WL or fluorescence modes.

(2) A special light guide for efficient excitation light transmission and coupling to the cystoscope.

(3) A cystoscope, which is a specialized endoscope allowing direct visual inspection of the urothelium. Cystoscopes for fluorescence contain optimized fiber bundles for blue-violet illumination and an observation filter in the eyepiece. Alternatively, the filter is inserted in the camera optics. Most cystoscopes have extra tubes to guide other instruments for taking a biopsy or for the resection of superficial bladder tumors. Cystoscopes can be either rigid or flexible. TURBT can only be performed with rigid cystoscopes.

(4) A high-sensitivity color camera that is equipped with either a single-chip or triple-chip imaging detector. These cameras can be operated in a 'fluorescence mode' with preset color balance and gain values, and they communicate with the light source, as mentioned above. Images of the bladder wall are usually displayed on a standard video monitor. Some systems enable to perform two independent white or color balance calibrations for the WL and fluorescence modes, respectively.

All these systems provide standard WL cystoscopy images that are unaffected by the instillation of the precursor or the endogenous production of PAPs. Nevertheless, excessive use of WL endoscopy may contribute to the photobleaching of the PAPs. Finally, it should be noted that the use of fluorescence detection and demarcation of bladder cancer minimally changes the "standard" WL cystoscopy technique routinely used by urologists.

36.2 Historical Background of Hexvix

As proposed by Wagnières et al. (1998), the approaches to detect and characterize lesions by fluorescence imaging can be classified into three main categories: (1) endogenous fluorophores that are responsible for native tissue fluorescence (autofluorescence) (Richards-Kortum and Sevick-Muraca 1996; Wolfbeis 1993), (2) fluorophores administered as exogenous drugs, and (3) fluorophores synthesized in the tissue after external administration of a precursor molecule (specifically PAPs, including PpIX, induced by 5-ALA or its derivatives) (Fotinos et al. 2006). Each class of fluorophore outlined above has distinct advantages and limitations.

36.2.1 FC with Endogenous Fluorochromes

As outlined by Richards-Kortum and Sevick-Muraca (1996), most endogenous fluorophores are either associated with the

structural matrix of tissues or are involved in cellular metabolic processes. The most important of the former are collagen and elastin, the fluorescence of which is the result of cross-linking between amino acids. Fluorophores involved in cellular metabolism include reduced nicotinamide adenine dinucleotide (NADH) and flavins. Other fluorophores include the aromatic amino acids (e.g., tryptophan, tyrosine, phenylalanine) and various porphyrins and lipopigments (e.g., ceroids, lipofuscin), which are the end products of lipid metabolism. In addition, red fluorescence due to bacteria-induced porphyrins may be significant in certain body sites, including the bladder and/or lesions. The detection of premalignant lesions or early cancer using autofluorescence then depends on changes in one or more of (1) the fluorophore concentration or spatial distribution, (2) their metabolic status (e.g., NADH is fluorescent only in its reduced form), (3) the biochemical and/or biophysical microenvironment of the tissue, which may alter the fluorescence quantum yield, spectral peak positions, (4) the tissue architecture, such as mucosal thickening or loss of layered structure, which affects the relative contributions to the measured fluorescent signal at the tissue surface, and (5) the wavelength-dependent light attenuation due to the concentration and distribution of (nonfluorescent) chromophores, particularly hemoglobin.

Bladder autofluorescence (308 nm excitation) has been studied by Anidjar, Ettori, and Cussenot (1996). They reported that the shape of the emission spectra of malignant lesions, including CIS, differed from that of the normal urothelium. König et al. (1996) measured the ratio of the bladder autofluorescence intensities at 385 and 455 nm, upon 337 nm excitation, and reported over 95% sensitivity and specificity for (visible) tumor detection. More recently, Frimberger et al. (2001) have shown that autofluorescence may be a promising additional tool to discriminate neoplastic from benign lesions of the bladder. However, attempts to exclusively exploit endogenous fluorescence for tumor recognition in the bladder have not found their way into the routine practice of urologists (D'Hallewin, Bezdetnaya, and Guillemin 2002).

Nevertheless, it should be noted that certain commercially available instruments for FC are based on the detection of tissue autofluorescence to reinforce the tumor-to-normal (T/N) contrast produced by PAPs. The rationale for this spectral design is based on studies reported by the groups of Lausanne (Lange, Jichlinski, and Zellweger 1999) and Singapore (Zheng et al. 2003) indicating that a decrease in the blue-green autofluorescence is associated with the presence of certain pre-/early cancerous lesions of the bladder when exposed to blue-violet light. This decrease, observed for the lesions in the background blue-green image, is combined with the increase in the red fluorescence produced by the PAPs to improve the T/N chromatic contrast.

36.2.2 FC with Exogenous Fluorophores

Most exogenously administered fluorophores investigated for clinical applications have been developed primarily as sensitizers for photodynamic therapy (PDT). Among those are hematoporphyrin derivative (HpD) (Profio and Sarnaik 1984), benzoporphyrin derivative (BPD) (Kim et al. 1997), and hypericin (D'Hallewin, Bezdetnaya, and Guillemin 2002). These compounds have been designed with specific PDT characteristics, such as strong absorption at "long" red wavelengths and high triplet-state quantum yield. These properties are not optimal for fluorescence diagnostics. In addition, the selectivity of these photosensitizers for localizing premalignant and early malignant lesions has generally been rather poor. Consequently, attempts using a fluorescing photosensitizing agent such as HpD to improve endoscopic bladder cancer detection failed to translate into clinical practice due to their lack of sensitivity, specificity and presence of side effects (Jocham et al. 1989; Vicente, Chècile, and Algaba 1987).

Several groups studied or developed non- or weakly phototoxic fluorophores for *in vivo* spectroscopy and imaging. For example, in the early 1960s, researchers demonstrated that ultraviolet cystoscopy after oral administration of tetracycline was a simple and practical technique that might be useful for the recognition of malignant areas not visible with standard white-light cystoscopy. However, tumors showed a nonhomogeneous and unspecific fluorescence. In addition, benign papilloma, which are now considered as papillary urothelial neoplasm of low malignant potential, and low-grade papillary urothelial carcinoma, did not emit fluorescence (Barlow, Maurice, and Atkins 1966; Whitmore and Bush 1968). Fluorescein has also been reported to be suitable for tumor imaging (Braginskaja et al. 1993), but its efficacy for the detection of bladder cancers is unlikely because it is highly water soluble and very rapidly cleared *in vivo*. Nile blue and derivatives thereof seem to be good tumor localizers, and one of these (EtNBA) is fluorescent but nonphototoxic (Cincotta et al. 1994) and therefore a good candidate as a purely diagnostic drug. Nevertheless, none of these weakly phototoxic exogenous fluorochromes has attained extensive clinical use up to now.

36.2.3 FC with 5-ALA

The medical need for better technologies to detect, delineate, and characterize NMIBC, combined with the limited progress of the techniques based on the imaging of exogenous fluorophores or based on the tissue autofluorescence, is at the origin of the search for new fluorescence imaging strategies.

About 20 years ago, Kriegmair et al. (1993), following the early investigations performed by Malik and Lugaci (1987) and Kennedy, Pottier, and Pross (1990) on PDT in dermatology using 5-ALA, proposed the use of this prodrug for fluorescence diagnostics in the bladder. This was the first clinical implementation of 5-ALA administered intravesically. Because the fluorescence intensity obtained had much more contrast and was brighter than with previously used fluorochromes, the imaging instrumentation could be simplified (i.e., it could be used without the need of image intensification and processing). In 1993, in Lausanne, Switzerland, a collaboration was established between the Swiss Federal Institute of Technology (EPFL) and the Department of Urology of the University Hospital Center (CHUV)–Lausanne's University Hospital on photodetection

and therapeutic applications of 5-ALA in NMIBC. In the first studies, Jichlinski et al. (1997a,b) confirmed quantitatively the results reported by the Munich group (Kriegmair, Baumgartner, and Hofstetter 1992; Kriegmair et al. 1993, 1996; Kriegmair, Baumgartner, and Knuechel 1994) and underlined the clinical interest in FC. Measurement of the PpIX fluorescence intensity and spectroscopy, performed *in vivo* in human bladder cancers with an optical fiber-based spectrofluorometer, confirmed a highly selective production of PpIX in NMIBC as compared to the normal urothelium (Forrer 1995a,b). However, the PpIX fluorescence intensity was found to vary strongly between tumors. Nevertheless, 5-ALA-mediated FC quickly spread worldwide and significant amount of data has been obtained (Kriegmair, Baumgartner, and Knuechel 1994; Jichlinski et al. 1997a; König et al. 1999). The reported sensitivity obtained using 5-ALA for detecting NMIBCs in different studies is generally very high, with values ranging between 90% and 100% (D'Hallewin, Bezdetnaya, and Guillemin 2002). In addition, 5-ALA-mediated cystoscopy was shown to detect about twice as many NMIBCs than standard WL cystoscopy (D'Hallewin, Bezdetnaya, and Guillemin 2002). Despite these promising results in multiple clinical studies using 5-ALA-mediated FC for NMIBC detection, this detection procedure still lacks wide acceptance in the medical community. The failure of 5-ALA to gain marketing authorization as a diagnostic procedure for NMIBCs can be ascribed to a multitude of factors, as explained in the review of Fotinos et al. (2006). One of the main reasons is related to the physicochemical properties of 5-ALA itself. Indeed, 5-ALA is a small molecule containing five carbon atoms, with an amino group on one end and a carboxylic acid group on the other end. Under physiological conditions, more than 90% of all 5-ALA molecules are present as zwitterions and carry a positive charge at the amine terminal and a negative charge at the carboxylic terminal. Such hydrophilic compounds have limited capacities to reach and ultimately enter the target cells via passive diffusion through the cell membrane and the mitochondrial membrane. However, 5-ALA has been shown to gain access to the intracellular space via active-transport mechanisms, possibly involving dipeptide and tripeptide transporters (Doring et al. 1998; Rud et al. 2000; Whitaker et al. 2000). This molecule thus penetrates the tissue in a nonoptimal way, it is not very homogeneously distributed, and PAP production is lower and slower, so that high concentrations and quantities of this precursor are needed (Lange, Jichlinski, and Zellweger 1999). Finally, the relatively poor stability of 5-ALA formulations as well as the relatively weak patent protection/situation are also factors that can explain the failure of 5-ALA to gain marketing authorization for FC.

36.3 Development of Hexvix

36.3.1 Motivation for the Chemical Derivation of 5-ALA

For detailed explanations of the PpIX biosynthesis and the mechanisms underlying its preferential accumulation in neoplastic tissue after exogenous administration of 5-ALA or one of its derivatives, the reader is referred to Chapter 23 by Asta Juzeniene or to the reviews of Collaud et al. (2004), Fotinos et al. (2006), Fukuda, Casas, and Batlle (2005), and Peng et al. (1997).

The limitations of formulations containing 5-ALA, in particular, for the detection of NMIBCs by fluorescence imaging, are at the origin of efforts performed by several groups to improve the endogenous production of PpIX. As mentioned above, most of the disadvantages of 5-ALA can be ascribed to the physico-chemical properties of this zwitterion molecule itself. Because the lipid bilayer of biological membranes is relatively imper-meable to charged molecules, the cellular uptake of 5-ALA is limited. In addition, the low lipophilicity of 5-ALA prevents its thorough tissue penetration.

Systematic studies have shown that the modification of a drug that contains polar groups like a charged organic acid, to an ester, an amide, or a urethane, by the addition of a hydrocarbon chain improves penetration through biological barriers (Bridges, Sargent, and Upshall 1979; Jain 1987a,b). However, while apply-ing this derivation strategy to 5-ALA, one must consider that the compound must reach the cytosol to start the chain of events that finally leads to the production of PAPs. Indeed, the 5-ALA derivatives have differential interaction types with the biomem-branes and the hydrophilic cytoplasm. Thus, for instance, a long-chain ester of 5-ALA that is too lipophilic might get stuck in the cell membrane and not enter the cytosol.

Promising results have been obtained following this strategy with different alkyl esters of 5-ALA *in vivo* and *in vitro* (Gaullier et al. 1997; Kloek and Beijersbergen van Henegouwen 1996; Marti et al. 1999; Peng et al. 1996; Uehlinger et al. 2000). The Norwegian group filed a patent application in March 1995 aim-ing at protecting, in particular, esters of 5-ALA and their use in photodiagnosis and photochemotherapy (Gierskcky et al. 1995).

However, lipophilicity seems not to be the sole character-istic responsible for the improved PAP generation by 5-ALA derivatives. For example, it has been shown by Uehlinger et al. (2000) that 5-ALA hexylester and 5-ALA cyclohexylester have essentially the same log *p* values, whereas they differ widely with respect to their activity. These results were confirmed by Whitaker et al. (2000).

As mentioned above, esterases in the cells are believed to release, at least in part, 5-ALA, enabling them to enter the heme biosynthesis pathway. However, it cannot be excluded that the enzymes involved in this biosynthetic pathway also act on the ester of 5-ALA directly, finally producing esterified PpIX. For this reason, the fluorochromes synthesized following the deliv-ery of 5-ALA derivatives are usually referred to more generally as PAPs (Fotinos et al. 2006).

An important observation with respect to the bioavailabil-ity of 5-ALA derivatives is based on the initial *in vitro* experi-ments of Kloek and Beijersbergen van Henegouwen (1996). They allowed 5-ALA derivatives to penetrate the cell membrane dur-ing a short (0–30 min) incubation time. Then, after washing the cells, PAP synthesis was observed. These experiments revealed that, for minimal incubation periods, the differences between

the PAP formation induced by 5-ALA and the induction by one of the esters is maximal. These observations were attributed to the faster uptake of the 5-ALA esters as compared with 5-ALA itself. A pool of PAP precursors is thus formed within the cell that is sufficient to lead to the production of PAPs over a long period of time.

Finally, it should be noted that other strategies are of interest to improve the delivery of 5-ALA. These approaches include, in particular, the use of iron chelators (H. van den Bergh and R. Tyrell, unpublished results) or of carriers such as dimethyl sulf-oxide (DMSO) (Uehlinger et al. 2006).

36.3.2 Early Works on Hexyl-Ester of ALA during Its Development in Lausanne

The collaboration initiated in 1993 between the EPFL and the Department of Urology of the CHUV–Lausanne University Hospital was extended several years later to include the University of Lausanne's Institute of Physiology as well as the CHUV University Hospital Institute of Pathology. The com-bined efforts of *in vitro*, preclinical, and clinical studies in Lausanne, in which the drug HAL, as well as its formulation, was selected and extensively tested, eventually led to the phase II and III clinical trials undertaken together with Photocure ASA, a Norwegian company. Further development of Hexvix, the first commercial name for this formulation of HAL, was also under auspices of Photocure ASA. Today, Hexvix is sold in Europe, and in the United States, the same compound is sold under the name Cysview. Hexvix is now reimbursed in most Western countries, as described in more detail below.

It should be noted that the group of Lausanne significantly contributed not only to the development of Hexvix but also to the optimization of the spectral design of certain instruments for the imaging of the bladder wall fluorescence (see Figure 36.1b). Some of the main steps of this research are reported below.

This research program aimed at optimizing the conditions of PAP accumulation and distribution in urological cancers look-ing at different strategies. Besides 5-ALA esterification, various approaches were tried: 5-ALA was combined with the adminis-tration of DMSO and iron chelators like Desferal were admin-istered in combination with PAP precursors (H. van den Bergh and R. Tyrell, unpublished results). Desferal acts indirectly by depleting the pool of iron necessary for the transformation of PpIX into heme, thus increasing the PpIX cellular concentration (Uehlinger et al. 2006).

36.3.2.1 Studies Performed with Experimental Models

Initial optimization has been performed in living microdis-sected specimens of porcine urothelium (0.5 cm^2) obtained from freshly slaughtered animals. These samples were mounted in a thermostabilized (37°C) bicameral transparent culture cham-ber with circulating media above and below the tissue sample (Kucera and Raddatz 1980). This chamber was placed under a modified fluorescence microscope as described by Marti et al. (1999). With this model, the group of Lausanne mainly focused its attention on the study of various 5-ALA esters, mostly alkyl esters, administered at different concentrations. The analysis of the PpIX formation and distribution across the urothelium demonstrated the superiority of HAL administered at 4–8 mM (i.e., an about 25-fold lower concentration than 5-ALA). HAL was selected for FC over other n-alkyl-chain derivatives not only due to its capacity to induce the highest amounts of PAPs at much lower doses than 5-ALA but also because of its satisfactory water solubility. The formulations involving such surprising low doses of alkyl esters of ALA, including HAL, have been protected by a patent application in April 1998 (Marti et al. 1998).

Similar and more extensive results were obtained in cell cultures, including cells derived from human transition cell carcinoma, with various conditions (Uehlinger et al. 2000) (see Figure 36.3). In this dose–response study, the generation of PAPs is positively correlated to precursor concentration up to an optimum concentration, at which the highest fluorescence intensities can be observed; at concentrations higher than the optimum, PAP formation sharply decreases. This decrease in PAP formation is probably due to cytotoxic effects generated by the 5-ALA derivative itself or its metabolic products.

These results confirmed those obtained in porcine urothelia and also led to the selection of the HAL as an optimal PAP precursor. Indeed, the hexyl ester showed a significantly higher maximum PAP fluorescence than 5-ALA at concentrations lower than one to two orders of magnitude.

36.3.2.2 Pilot Clinical Measurements

The first clinical investigation of HAL involved 25 patients. The results indicated that a significant (>2×) increase in PpIX fluorescence intensity could be observed using about 20-fold lower concentrations as compared to 5-ALA (Lange, Jichlinski, and Zellweger 1999). The fluorescence intensities in papillary tumors

increased rapidly (i.e., during the first hours) as a function of time, with the highest value obtained with a HAL concentration of 8 mM. Microspectrofluorometric analysis of human bladder cancer specimens confirmed a fairly homogeneous PpIX distribution over the whole depth of the malignant urothelium with a similar increase in PpIX (Marti et al. 2003). Consequently, for obtaining the same minimal amount of fluorescence in the bladder necessary for the efficient detection of small bladder cancers and effective removal of lesions by TURBT, the HAL instillation time can be significantly reduced from more than 2 h with 5-ALA down to about 1 h (Jichlinski 2003b). A subsequent study involving 143 patients where two different protocols of HAL instillation time were compared confirmed that the reduction of the instillation time to less than 1 h does not endanger the validity of the HAL FC in terms of sensitivity and specificity. This reduction in instillation time represents a significant advantage in daily clinical practice (Jichlinski 2003a). Moreover, it was also shown that HAL, in addition to its higher lipophilicity and its better PAP production capacity at a lower dose, has a much better chemical stability in aqueous solution than 5-ALA, and hence a much better shelf life.

36.3.2.3 Pilot Clinical Study of HAL FC

The first clinical study of the sensitivity and specificity of FC using HAL (HAL-FC) was published in 2003 (Jichlinski 2003b), based on findings obtained on 52 patients with NMIBCs. Exclusion criteria for this study were patients with topical treatments (BCG) in the previous 3 months and presence of hematuria or known porphyria. The investigators concluded that not only the detection rate for the bladder tumors was higher with HAL-FC than with WL cystoscopy but also the detection of CIS alone or associated with exophytic tumors is improved. This is important, as the latter is a key prognostic factor for this pathology, as mentioned above.

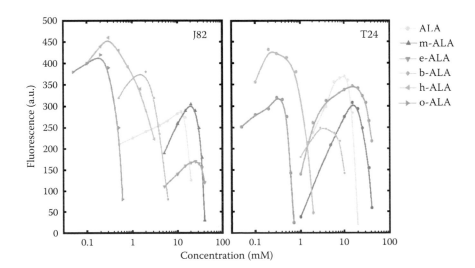

FIGURE 36.3 Concentration dependence of the PAP fluorescence intensity for two different human cell lines derived from human transitional cell carcinoma of the bladder after 3 h of incubation with different precursors. m = methyl-, e = ethyl-, b = butyl-, h = hexyl-, o = octyl ester of ALA. (Derived from Uehlinger, P. et al., *J Photochem Photobio. B: Biol* 54:72–80, 2000.)

36.3.3 Multicentric Clinical Studies Leading to Approval of HAL-FC

The preliminary results of these initial clinical studies performed in Lausanne were confirmed by a larger European trial (study: PCB301/01) involving 286 patients in 19 centers (Schmidbauer et al. 2004). The results revealed that HAL-FC detects 97% of all lesions compared to only 78% by WL cystoscopy. This study also confirmed the superiority of this fluorescence technique over the WL detection method when focusing on the detection of CIS, with detection rates of 97% for HAL-FD and only 58% for WL cystoscopy. It should be noted here that the clinicians participating in this study were generally quite experienced in WL endoscopy. This fact may well imply that, in the case of less experienced urologists, the outcome of the study might have been even more favorable for HAL. Further confirmations of the efficacy of HAL-FC have come from more recent multicentric studies in the United States (PCB302/01) (Fradet and Loskart 2007; Grossman et al. 2007), in which 311 patients underwent WL cystoscopy followed by HAL-FC.

According to the findings of the three studies outlined above, HAL-FC is easy to perform, improves the detection rate of bladder tumors, in particular CIS, and is now becoming an important tool in the management of bladder cancer. Side effects following bladder instillation or bladder-wall illumination—such as dysuria, hematuria, bladder pain, and bladder spasm—were rarely reported, nonspecific, and probably not drug-related (Jocham, Stepp, and Waidelich 2008). Indeed, these side effects are comparable to those of WL cystoscopy. They are predominantly mild, reversible, and related to the TURBT or patient comorbidity (Denzinger et al. 2007a; Fradet et al. 2007; Grossman et al. 2007; Jichlinski 2003b; Jocham et al. 2005; Schmidbauer et al. 2004). This improvement in detection efficiency of bladder cancer affects the subsequent treatment strategy in a significant proportion of the patients. This will be discussed below in more detail (Jocham et al. 2005).

36.3.4 Clinical Studies That Show the Clinical Benefit of FC with HAL and Led to Clinical Recommendations

HAL-FC has now been recommended for more than a decade by the European Association of Urology for the detection of bladder cancer in patients with known bladder cancer or a suspicion of bladder cancer based on screening cystoscopy or positive urine cytology. According to the present guidelines, HAL-FC should be used as an adjunct to standard WL cystoscopy as a guide for taking biopsies. The abundance of data that supports the use of HAL for this indication has been summarized by Witjes et al. (2010).

Briefly, a review of the literature clearly indicates that HAL-FC offers considerable benefits over WL cystoscopy in the detection of NMIBC (Fradet et al. 2007; Grossman et al. 2007; Jichlinski 2003b; Kausch et al. 2010; Jocham et al. 2005; Schmidbauer et al. 2004). The largest benefit of HAL-FC is for the detection of CIS,

as was previously suggested by the first report of Jichlinski et al. (2003b). Overall, the CIS lesion detection rate is around 20% higher with the addition of HAL-FC to WL cystoscopy versus WL cystoscopy alone (Babjuk et al. 2005; Jichlinski and Jacqmin 2008; Kriegmair et al. 2002; Riedl et al. 2001).

As mentioned above, the clinical benefit of an improved tumor detection rate for subsequent treatment or change of treatment protocol of the patients was reported by Jocham et al. (2005). Their European multicenter study found that HAL-FC resulted in improved treatment in 17% of the patients. In May 2011, Photocure ASA announced the results from a 5.5-year follow-up study of recurrence in patients with NMIBC that demonstrated the long-term benefit of the use of Hexvix as compared to patients who received WL cystoscopy only. During these 5.5 years, the number of patients with recurrence of their bladder cancer was lower, and the time it took until the recurrence occurs was longer when the patients had Hexvix-guided FC as compared to WL cystoscopy alone. This clinical study (PCB305) includes retrospective follow-up of 526 patients included in a prospective randomized phase III trial in 28 centers in the European Union and North America (Stenzl et al. 2010). Another article published by Hermann et al. (2011) reported that Hexvix-guided cystoscopy, as an adjunct to conventional WL cystoscopy, improves the detection of bladder cancer and reduces the rate of recurrence as compared to WL cystoscopy alone. This report is based on a prospective, randomized clinical trial in 233 patients, which was conducted by Photocure in two centers in Denmark. The results of the trial demonstrate that Hexvix enables the detection of lesions (including residual tumors) that were not detected with WL alone in 49% of the patients. It also demonstrated that the improved detection of lesions resulted in a significant reduction in tumor recurrence within 12 months. In this study, 31% of patients who had Ta or T1 tumors at inclusion experienced recurrence with Hexvix-guided cystoscopy compared to 47% with WL cystoscopy ($p = 0.05$), a relative reduction in recurrence of 36%. Furthermore, the recurrence-free interval was significantly longer for patients in the Hexvix arm ($p = 0.02$).

Importantly, large prospective randomized multicenter international studies (Babjuk et al. 2005; Daniltchenko et al. 2005; Denzinger 2007a; Filbeck et al. 2002) demonstrated the interest of HAL-FC–guided resections. These studies concluded that the tumor-free recurrence rates at 1 year range from 66% to 90% with HAL-FC as compared to 39% to 74% when WL only is used for the treatment. It is noteworthy that patients with multifocal or recurrent tumors appear to benefit most from HAL-FC–guided operations (Babjuk et al. 2005).

In February 2012, the data concerning the impact of Hexvix on long-term recurrence rates was presented at the European Association of Urology (EAU) meeting in Paris. These data taken from a long-term follow-up study on 551 patients clearly demonstrate that Hexvix cystoscopy significantly improves bladder cancer recurrence-free survival (Burger 2012).

A recent article reports the outcome of the discussion at the Nordic urology expert panel meeting (Malmström et al. 2012). This panel concluded that, in line with the European guidelines, HAL-FC has an important role in the initial detection of NMIBC, as well as in the case of follow-up of patients to assess tumor recurrence after initial WL cystoscopy. The panel provided practical advice, together with an algorithm on the use of the HAL-FC diagnostic procedure for urologists managing NMIBC.

FC has been associated with a relatively high rate of false positives. However, this rate is about the same as in the case of WL cystoscopy (Witjes et al. 2010). It should be noted that this rate is particularly high following recent TURBT or bacillus Calmette–Guérin (BCG) treatment, as well as after a recent or during a concomitant urinary tract infection because of possible scarring or inflammation. On the basis of recent evidence (Draga et al. 2010), it is recommended that HAL-FC should be postponed by 9–12 weeks after TURBT or BCG instillation if clinically feasible.

In parallel to studies focusing on the clinical interest in HAL-mediated FC, important resources were employed for demonstrating the safety of this methodology in humans (Klem et al. 2006).

Finally, it is of interest to note that, since the initial proposal of increasing the lipophilicity of 5-ALA to circumvent its limited local bioavailability, several different derivatives have been proposed. However, most of the clinical and preclinical data are for simple 5-ALA n-alkyl esters at present. Only two of the 5-ALA esters, methyl aminolevulinate (MAL) and HAL, have successfully finished multicenter phase III trials for treatment of different diseases. MAL gained marketing authorization for the treatment of actinic keratosis and basocellular carcinoma in Europe and Australia.

36.4 Industrial Developments

36.4.1 Business Development

The industrial development of the pharmaceutical technology based on the use of HAL has been performed by Photocure ASA, a Norwegian company founded in 1993 by a group including researchers of the Norwegian Radium Hospital located in Oslo. Photocure collaborates extensively with academic institutions worldwide. In addition to the contributions by the Norwegian Radium Hospital, Hexvix itself, as stated above, is the result of basic, preclinical, and clinical research at EPFL and CHUV University Hospital, both in Lausanne, Switzerland. At later stages of its development, approximately 60 university clinics in Europe and America were involved in clinical research. Consequently, the development of Hexvix is strongly linked to the relation between Photocure and its partners. Photocure became a public company in May 2000, during the year when the first pivotal clinical study of Hexvix was launched. Photocure received the European approval of its first patent application (no. 0820432) covering 5-ALA derivatives, including HAL, in June 2001. Due to its relatively small size, Photocure has launched several strategic partnerships with larger pharmaceutical companies to commercialize Hexvix in selected geographic areas. GE Healthcare was its licensing partner for Hexvix from 2006

to 2011. In September 2011, Photocure established a new commercial strategy that includes a partnership with Ipsen to commercialize Hexvix worldwide, excluding the United States and the European Nordic countries where Photocure itself undertakes the commercialization of this technology, which is called Cysview in the United States.

36.4.2 Regulatory Aspects and Reimbursement

The initial Hexvix documentation is based on three phase III studies performed in Europe as well as in the United States and Canada, including 759 patients with known or suspected bladder cancer (Fradet et al. 2007; Jocham et al. 2005; Schmidbauer et al. 2004). The studies were all performed with a within-patient comparison of WL and blue-light+HAL bladder inspection. In two studies, the aim was to compare the number of tumors detected with the two diagnostic modalities. In the third study (Fradet et al. 2007), the objective was to evaluate if the improved tumor detection rate found with Hexvix had a clinical benefit for the patient.

The first Marketing Authorization Application for HAL fluorescence was filed in Sweden, the reference European Union member state, in December 2002. Swedish authorities gave their approval of Hexvix in September 2004, making Sweden the first country worldwide to do so. By March 1, 2005, all European Union/European Economic Area (EU/EEA) countries had approved the Hexvix documentation through the mutual recognition procedure. Hexvix also got a number of approvals outside of Europe, including South Korea.

In the United States, Photocure submitted a new drug application (NDA) in June 2009 and achieved a priority review scheduled to be completed by December 30, 2009 (Photocure 2010). This NDA includes data from one pivotal (PCB305) and four supportive phase III studies (PCB301–PCB304). The pivotal phase III study, initiated in January 2005 and completed in September 2007, included 814 patients. Based on an intent-to-treat analysis, a significantly improved detection ($p = 0.001$) of noninvasive papillary cancer using Cysview as compared to standard WL cystoscopy in patients with noninvasive papillary bladder cancer was reported. The improved detection was followed by a significant reduction ($p = 0.026$) in recurrence at 9 months. In December 2009, Photocure received the positive response from the United States Food and Drug Administration (FDA) on this NDA. More precisely, the FDA informed Photocure that the NDA for Hexvix may be approved pending the premarket approval for the Karl Storz diagnostic system and final agreements between Photocure and FDA on labeling and postmarketing commitments. The US regulation requires that the combination of both technologies must be approved as a "combination product" (i.e., Hexvix together with the Karl Storz HAL-FC system). Thus, Photocure and Karl Storz have signed a formal collaboration for the development and marketing of Hexvix together with the Karl Storz's D-light system. The D-light system from Storz was already approved in Europe, as was the case for similar systems commercialized by Olympus Optical Co. and Richard Wolf GmbH.

The approval of Cysview was announced in May 2010, and Photocure finally announced in May 2012 that the FC system from Karl Storz got FDA approval. Following the demonstration of its cost-effectiveness, Hexvix obtained reimbursement in many Western countries including France, Spain, Denmark, Belgium, and Greece, an important milestone being its reimbursement in Germany, obtained in January 2010.

36.4.3 Cost-Effectiveness

Although patient outcome is the main focus of medical research, issues of health care funding must also be considered. Indeed, obtaining the reimbursement of a pharmaceutical technology such as Hexvix requires, in most cases, a clear demonstration of better cost-effectiveness than is found in the existing gold standard. WL cystoscopy is considered to be the gold standard for the detection, delineation, and transurethral resection of NMIBC. However, HAL-FC has been shown to improve final outcome, as mentioned in Section 36.3.4.

Bladder cancer causes higher costs than all other cancers, and thus noninvasive urothelial carcinoma of the bladder is at the origin of a large economic burden to public health systems. This is mainly due to its lifelong character and frequent recurrences.

We will now briefly review selected contributions aimed at assessing or providing useful information on the cost-effectiveness of FC.

Stenzl, Hoeltl, and Bartsch (2001) were the first to suggest that FC offers economic benefits in the management of bladder cancer. In 2001, they retrospectively analyzed 392 TURBTs performed within a 3-year period at their institution at a cost of EUR 2073 each. They calculated that if all the TURBTs had been FC-assisted, an additional cost of EUR 51,042 would have been needed to pay for the optical equipment and extra materials and working hours associated with the technique. They also estimated that 21% of the WL TURBTs would have been inefficient, resulting in the need for repeat interventions at a cost of EUR 161,710. Based on this economic assessment, Stenzl, Hoeltl, and Bartsch concluded that the cost-effectiveness of HAL-FC was quite significant.

Burger et al. (2007) reported the results obtained on a series of 301 patients with noninvasive urothelial carcinoma of the bladder. These patients were randomized prospectively to standard WL- or fluorescence-guided transurethral resections of the bladder with 5-ALA. Expenditures of subsequent procedures and FC-associated costs were assessed (median follow-up: 7.1 years). On a per-patient analysis, disease recurrence was found in 42% of WL and 18% of FC patients ($p = 0.0003$). On the average, in the WL group, 1.0 recurring urothelial carcinoma of the bladder occurred per patient versus 0.3 in the FC group, resulting in costs of EUR 1750 per WL patient versus EUR 420 per FC patient in the follow-up period, respectively. This last amount does not include the expenditure specific to FC. Nevertheless, it can be postulated that such a difference emphasizes the potential value of FC in optimizing the follow-up protocol of NMIBC in order to reduce its global cost. This is supported by the article published by Zaak et al. (2008) reporting that FC-related expenditures are "reimbursed" within the first years in their series. Moreover, although the additional cost of the light source and optics for FC is about EUR 16,000 and a specifically designed video camera is around EUR 9000 (Witjes and Douglass 2007), these costs can be easily amortized over the other applications.

Finally, it should be noted that FC may not be justifiable in all patients, and there may be a case for restricting the use of this technique to those who are likely to gain a benefit. Sylvester et al. (2006) have reported that recognized predictive factors for recurrence (e.g., multifocality, previous bladder cancer, tumor stage, histological grade) can be used to stratify risk groups. Classification of patients according to risk groups has also been advocated by the EAU guidelines on bladder cancer (Oosterlinck et al. 2006). Nevertheless, Burger et al. (2007) demonstrated that the additional FC-related cost was offset by overall savings in all risk groups, particularly in intermediate-risk patients. This finding is consistent with the low rate of recurrence seen with low-risk bladder cancer and the adverse outcome of high-risk bladder cancer despite the use of FC (Denzinger et al. 2007b; Sylvester et al. 2006).

36.5 Applications of HAL to Other Organs

Although it is well established that the optimal derivation of ALA is not the same for different types of tissues (Fotinos et al. 2006), HAL is of high interest for many detection, delineation, and characterization applications in other organs.

Because fluorescence imaging with HAL is limited to light-accessible organs that can be subject to topical administration of PAP precursors, gynecological cancers, such as ovarian or cervical cancers, and cancers of the gastrointestinal tract, such as colon and rectal (pre-)cancers and high-grade dysplasia in Barrett's esophagus, are good candidates among others.

Encouraging preclinical results were obtained by HAL-FD for Fisher rats implanted with NuTu-19 cells (rat epithelial ovarian cell line), resulting in the detection of twice as many cancer lesions compared with WL inspections. These experiments with HAL also showed a clear advantage in terms of the selective porphyrin accumulation when compared with 5-ALA (Ludicke et al. 2003). A Swiss group (Andrejevic Blant et al. 2004) examined the dependence of PAP formation in cervical intraepithelial neoplasia as a function of HAL topical application time. They determined that an application for 100 min of HAL cream was optimal to obtain the highest ratio of epithelium to underlying lamina propria PAP fluorescence. This ability of HAL to produce a significant quantity of PAPs in the cervix is at the basis of the treatment of HIV-induced cervix cancer by PDT (Soergel et al. 2011). Apart from bladder and gynecological cancers, HAL-mediated FD has been investigated for the diagnosis in the gastrointestinal (GI) tract, including rectal adenoma and cancers. For this purpose, a pilot study evaluated the efficiency of an instilled HAL solution to induce selective fluorescence of rectal tumors (adenoma and malignant tumors) (Endlicher, Gelbmann, and Knüchel 2004). Under certain conditions, PAP fluorescence appeared to emanate selectively from the tumors and a significant difference was observed between adenomas that showed homogeneous fluorescence, whereas moderately differentiated carcinomas had fluorescence only at the tumor's edge.

As mentioned above, the fact that 5-ALA esters such as HAL enter the intracellular space with less resistance as compared with 5-ALA rapidly provides a pool of PAP precursors that is sufficient to maintain PAP synthesis over relatively long periods of time, even if the topical administration is rapidly removed. Therefore, the administration time of such esters can be significantly reduced as compared to ALA. Short application times are advantageous in many clinical applications. One example is in the detection of high-grade dysplasia in Barrett's esophagus (Stepinac et al. 2003), where, after *per os* application, good contact between the drug and the mucosa can only be maintained for several minutes after administration.

Last but not least, a low-dose formulation of HAL has also been shown to be useful for improving endoscopic performance for early cancer detection in the colon. Photocure has completed a clinical proof-of-concept study that demonstrated a nearly 40% increase in detection rate when this compound was administered prior to using fluorescence colonoscopy (Mayinger et al. 2010). In 2010, Photocure licensed the compound to Salix Pharmaceuticals Inc. in the United States, who will develop and commercialize the product (Lumacan) for improving the detection of lesions in the colon. Photocure is also, together with Salix, trying to develop an oral formulation containing HAL for the diagnosis of colorectal cancer.

Consequently, the successful development of Hexvix in routine urology has opened up very promising new applications in fields that are expected to benefit from the detection, characterization, and treatment of important diseases such as colon cancer.

36.6 Conclusion and Perspectives

The historical analysis of the Hexvix story presented here illustrates the importance of each of the individual steps that lead to the development of such a technology. Resulting from academic innovations that were patented in due time before being published, the Hexvix technology was developed following an effective technology transfer to industry. This was followed by a successful development by the pharmaceutical industry that finally led to an approval by the regulatory authorities and reimbursement in many countries. The final step, which is currently underway, consists of a marketing effort targeting the appropriate community of urologists.

The development of Hexvix/Cysview has led to a cost-effective technology that, according to unanimous opinion, significantly improves the endoscopic detection and TURBT of bladder cancer. The enhanced detection and resection mediated by HAL-FC of the frequently difficult-to-see and aggressive CIS is one of the main benefits of this approach. This detection method is especially of interest for patients presenting multifocal, aggressive, and flat lesions. The high sensitivity of HAL-FC makes it feasible to refrain from taking random or useless biopsies or TUR if there is no suspicious fluorescence visible. In addition, there is no question that HAL-FC offers a more complete resection with a lower rate of residual tumor and recurrence. This is because HAL-FC facilitates a more complete ablation and is useful during the follow-up after intravesical therapy by providing a thorough understanding of the disease anatomy (Lerner et al. 2012).

Nevertheless, one weakness of HAL-FD remains its limited specificity. False positives are linked to technical factors (operator skill, tangential illumination of the bladder wall), time, and modalities of previous treatments (scar, inflammation) or often remain not clearly defined (normal mucosa, hyperplasia) (Grimbergen et al. 2003; Jocham, Stepp, and Waidelich 2008).

Thus, ongoing research aims at combining HAL-FC with tissue characterization techniques such as high magnification imaging (Lovisa et al. 2010) or optical coherence tomography (OCT) (Schmidbauer et al. 2008). Interesting results have been obtained with the former approach that make it possible, using a simple and convenient system, to image the pattern of small urothelial vessels of about 50 µm in diameter. This local high magnification (about 150×) cystoscopy allows to reject 97% of the HAL-FC false positives (Lovisa et al. 2010). OCT is relatively new in urology. The physical principle of this technique is similar to that of B-mode ultrasound but a near-infrared wavelength is employed instead of sound. It allows real-time visualization of the microarchitecture of the bladder wall by providing cross-sectional images. Therefore, this is a characterization method that intends to be a complement to HAL-FC or conventional cystoscopy. As reported by Encina et al. (2010), some authors have already studied the diagnostic capabilities of OCT, which seems to present a better specificity than WL cystoscopy.

Narrowband imaging (NBI) is also of interest to improve the detection and characterization of lesions affecting the bladder wall. NBI is based on the detection of backscattered light produced in spectral bands that correspond to the hemoglobin absorption (typically 415 and 540 nm). As the longer wavelengths penetrate deeper in the tissue, the superficial blood vessels have a bluish hue that becomes greenish for deeper-lying vessels. As reported by Encina et al. (2010), some authors have already reported on the diagnostic capabilities of NBI that seems to present a better sensitivity than WL cystoscopy, whereas its specificity is unfortunately comparable to the WL results.

Correct maintenance of the cystoscopic instrumentation is essential to the quality of information obtained with fluorescence cystoscopy, and this issue seems to be underestimated by most manufacturers. This is important because diagnosing the non-optimal functioning of the instrumentation is quite difficult for the medical staff. Therefore, a quality check option is important to avoid false diagnosis due to improperly working equipment. Such a quality check option needs to be developed urgently and integrated in the next generations of fluorescence cystoscopes. This development must integrate the fact that instrumentation for fluorescence diagnosis at present is following the general trend toward improved optical sensitivity and resolution (high-definition). In addition, other improvements will result from the integration of semiconductor-based light sources for WL and/or fluorescence excitation light at the distal end of the video-cystoscope. These improvements will lead to the availability of more user-friendly and reliable equipment, as it is the case, in particular, for flexible cystoscopes. This is important since outpatient cystoscopic follow-up is usually carried out with flexible cystoscopes. Consequently, the availability of such flexible fluorescence cystoscopes will certainly increase the interest of this cancer detection approach in private practice. Two small-scale studies of HAL-FC have already shown that fluorescence-guided flexible cystoscopy produces results comparable to rigid cystoscopy (Bertrand et al. 2012; Loidl et al. 2005; Witjes et al. 2005).

References

Andrejevic Blant, S., A. Major, F. Ludicke et al. 2004. Time-dependent hexaminolaevulinate induced protoporphyrin IX distribution after topical application in patients with cervical intraepithelial neoplasia: A fluorescence microscopy study. *Lasers Surg Med* 35:276–283.

Anidjar, M., D. Ettori, and O. Cussenot. 1996. Laser-induced autofluorescence diagnosis of bladder tumors. Dependence on the excitation wavelength. *J Urol* 156:1590–1596.

Babjuk, M., V. Soukup, R. Petrik, M. Jirsa, and J. Dvoracek. 2005. 5-aminolaevulinic acid-induced fluorescence cystoscopy during transurethral resection reduces the risk of recurrence in stage Ta/T1 bladder cancer. *BJU Int* 96:798–802.

Barlow, K. A., B. A. Maurice, and P. Atkins. 1966. Ultraviolet fluorescence of bladder tumors following oral administration of tetracycline compounds. A macroscopic, microscopic, and fluorescence spectrophotometric study. *Cancer* 19:1013–1018.

Bertrand, J., L. Soustelle, P. Grès et al. 2012. Interest of flexible videocystoscopy in blue light (+Hexvix®) in consultation for the diagnosis of vesical tumor. *Prog Urol* 22(3):172–177.

Botteman, M., C. L. Pashos, A. Redaelli, B. Laskin, and R. Hauser. 2003. The health economics of bladder cancer: A comprehensive review of the published literature. *Pharmacoeconomics* 21:1315–1330.

Braginskaja, O. V., V. V. Lazarev, V. I. Polsachev, L. B. Rubin, and V. E. Stoskii. 1993. Fluorescent diagnosis of human gastric cancer and sodium fluorescein accumulation in experimental gastric cancer in rats. *Cancer Lett* 69:117–121.

Brandt, W. D., W. Matsui, J. E. Rosenberg, X. He, S. Ling, E. M. Schaeffer, and D. M. Berman. 2009. Urothelial carcinoma: Stem cells on the edge. *Cancer and Metastasis Reviews* 28(3–4):291–304.

Brausi, M., L. Collette, K. Kurth et al. 2002. Variability in recurrence rate at first follow-up cystoscopy after TUR in stage Ta T1 transitional cell carcinoma of the bladder: A combined analysis of seven EORTC studies. *Eur Urol* 41:523–531.

Bridges, J. W., N. S. E. Sargent, and D. G. Upshall. 1979. Rapid absorption from the urinary bladder of a series of n-alkyl carbamate: A route for the recirculation of drug. *Br J Pharmacol* 66:283–289.

Bubendorf, L., B. Grilli, G. Sauter et al. 2001. Multiprobe FISH for enhanced detection of bladder cancer in voided urine specimens and bladder washings. *Am J Clin Pathol* 116:79–86.

Burger, M. 2012. Paper presented at the poster session of the 27th EAU Congress, Paris, France, February 2012. Retrieved from http://www.eauparis2012.org/.

Burger, M., D. Zaak, C. G. Stief et al. 2007. Photodynamic diagnostics and non-invasive bladder cancer: Is it cost-effective in long-term application? A Germany-based cost analysis. *Eur Urol* 52:142–147.

Cincotta, L., J. W. Foley, T. MacEachern, E. Lampros, and A. H. Cincotta. 1994. Novel photodynamic effects of a benzophenothiazine on two different murine carcinomas. *Cancer Res* 54:1249–1258.

Collaud, S., A. Juzeniene, J. Moan, and N. Lange. 2004. On the selectivity of 5-aminolevulinic acid-induced protoporphyrin IX formation. *Curr Med Chem Anti-Cancer Agents* 4:301–316.

D'Hallewin, M.-A., L. Bezdetnaya, and F. Guillemin. 2002. Fluorescence detection of bladder cancer: A review. *Eur Urol* 42:417–425.

Daniltchenko, D. I., C. R. Riedl, M. D. Sachs et al. 2005. Long-term benefit of 5- aminolevulinic acid fluorescence assisted transurethral resection of superficial bladder cancer: 5-year results of a prospective randomized study. *J Urol* 174:2129–2133.

Denzinger, S., M. Burger, B. Walter et al. 2007a. Clinically relevant reduction in risk of recurrence of superficial bladder cancer using 5-aminolevulinic acid-induced fluorescence diagnosis: 8-year results of prospective randomized study. *Urology* 69:675–679.

Denzinger, S., W. F. Wieland, W. Otto et al. 2007b. Does photodynamic transurethral resection of bladder tumour improve the outcome of initial T1 high-grade bladder cancer? A long-term follow-up of a randomized study. *BJU Int* 101:566–569.

Doring, F., J. Walter, J. Will et al. 1998. Delta-aminolevulinic acid transport by intestinal and renal peptide transporters and its physiological and clinical implications. *J Clin Invest* 101:2761–2767.

Draga, R. O. P., M. C. M. Grimbergen, E. T. Kok et al. 2010. Photodynamic diagnosis (5-aminolevulinic acid) of transitional cell carcinoma after bacillus Calmette-Guerin immunotherapy and mitomycin C intravesical therapy. *Eur Urol* 57:655–660.

Encina, J. O., A. Marco Valdenebro, J. Pelegrí Gabarró, and C. Rioja Sanz. 2010 Beyond the photodynamic diagnosis: Searching for excellence in the diagnosis of non-muscle-invasive bladder cancer. *Actas Urol Esp* 34(8):657–668.

Endlicher, E., C. M. Gelbmann, and R. Knüchel. 2004. Hexaminolevulinate-induced fluorescence endoscopy in patients with rectal adenoma and cancer: A pilot study. *Gastrointest Endosc* 60(3):449–454.

Filbeck, T., U. Pichlmeier, R. Knuechel, W. F. Wieland, and W. Roessler. 2002. Clinically relevant improvement of recurrence-free survival with 5-aminolevulinic acid induced fluorescence diagnosis in patients with superficial bladder tumors. *J Urol* 168:67–71.

Forrer, M., T. Glanzmann, and J. Mizeret. 1995a. Fluorescence excitation and emission spectra of ALA induced protoporphyrin IX in normal and tumoral tissue of the human bladder. *SPIE* 2324:84–88.

Forrer, M., J. Mizeret, D. Braichotte et al. 1995b. Fluorescence imaging photodetection of early cancer in the bronchi with mTHPC and in the bladder with ALA-induced protoporphyrin IX: Preliminary clinical results. 5th International Photodynamic Assoc. Biennial Meeting, D. A. Cortese, editor. *SPIE* 2371:109–114.

Fotinos, N., M. A. Campo, F. Popowycz, R. Gurny, and N. Lange. 2006. 5-aminolevulinic acid derivatives in photomedicine: Characteristics, application and perspectives. *Photochem Photobiol* 82:994–1015.

Fradet, Y., H. B. Grossman, L. Gomella et al. 2007. A comparison of hexaminolevulinate fluorescence cystoscopy and white-light cystoscopy for the detection of carcinoma in situ in patients with bladder cancer: A phase III, multicenter study. *J Urol* 178:68–73.

Fradet, Y., and C. Loskart. 1997. The ImmunoCyt trialists. Performance characteristics of a new monoclonal antibody test for bladder cancer: ImmunoCyt. *Can J Urol* 3:400–405.

Frimberger, D., D. Zaak, H. Stepp et al. 2001. Autofluorescence imaging to optimize 5-ALA-induced fluorescence endoscopy of bladder carcinoma. *Urology* 58:372–375.

Fukuda, H., A. Casas, and A. Batlle. 2005. Aminolevulinic acid: From its unique biological function to its star role in photodynamic therapy. *Int J Biochem Cell Biol* 37:272–276.

Gaullier, J. M., K. Berg, Q. Peng et al. 1997. Use of 5-aminolaevulinic acid esters to improve photodynamic therapy on cells in culture. *Cancer Res* 57:1481–1486.

Gierskcky, K. E., J. Moan, Q. Peng et al., Patent GB9504948.2, 1995.

Greenbaum, L., D. J. Katcoff, H. Dou et al. 2003. A porphobilinogen deaminase (PBGD) Ran-binding protein interaction is implicated in nuclear trafficking of PBGD in differentiating glioma cells. *Oncogene* 22:5221–5228.

Grimbergen, M. C. M., C. F. P. van Swol, T. G. M. Jonges, T. A. Boon, R. J. A. van Moorselaar. 2003. Reduced specificity of 5-ALA induced fluorescence in photodynamic diagnosis of transitional cell carcinoma after previous intravesical therapy. *Eur Urol* 44:51–56.

Grossman, H. B., L. Gomella, Y. Fradet et al. 2007. A phase III, multicenter comparison of hexaminolevulinate fluorescence cystoscopy and white light cystoscopy for the detection of superficial papillary lesions in patients with bladder cancer. *J Urol* 178:62–67.

Hara, T., M. Takahashi, T. Gondo et al. 2009. Risk of concomitant carcinoma in situ determining biopsy candidates among primary non-muscle-invasive bladder cancer patients: Retrospective analysis of 173 Japanese cases. *Int J Urol* 16(3):293–298.

Hermann, G. G., K. Mogensen, S. Carlsson, N. Marcussen, and S. Duun. 2011. Fluorescence-guided transurethral resection of bladder tumors reduces bladder tumour recurrence due to less residual tumor tissue in Ta/T1 patients: A randomized two-centre study. *BJU Int* 108(8b):E297–E303.

Jain, R. K. 1987a. Transport of molecules in the tumor interstitium: A review. *Cancer Res* 47:3039–3305.

Jain, R. K. 1987b. Transport of molecules across tumor vasculature. *Cancer Metastasis Rev* 6:559–593.

Jichlinski, P., M. Forrer, J. Mizeret et al. 1997a. Clinical evaluation of a method for detecting superficial transitional cell carcinoma of the bladder by light induced fluorescence of protoporphyrin IX following topical application of 5-aminolevulinic acid. Preliminary results. *Lasers Surg Med* 20:402–408.

Jichlinski, P., G. Wagnières, M. Forrer et al. 1997b. Clinical assessment of fluorescence cystoscopy during transurethral bladder resection in superficial bladder cancer. *Urol Res* 25:3–6.

Jichlinski, P., D. Aymon, G. Wagnières et al. 2003a. On the influence of the instillation time on the results of HAL (Hexvix®) fluorescence detection of superficial bladder cancer. *Proc SPIE* 5141:272–277.

Jichlinski, P., L. Guillou, S. J. Karlsen et al. 2003b. Hexyl aminolevulinate fluorescence cystoscopy: A new diagnostic tool for the photodiagnosis of superficial bladder cancer—A multicenter study. *J Urol* 170:226–229.

Jichlinski, P., and D. Jacqmin. 2008. Photodynamic diagnosis in non–muscle invasive bladder cancer. *Eur Urol* 7:529–535.

Jocham, D., R. Baumgartner, N. Fuchs et al. 1989. Die fluoreszenzdiagnose porphyrin-markierter urothelialer tumoren. *Urologe A* 28:59–64.

Jocham, D., H. Stepp, and R. Waidelich. 2008. Photodynamic diagnosis in urology: State-of-the-art. *Eur Urol* 53:1138–1150.

Jocham, D., F. Witjes, S. Wagner et al. 2005. Improved detection and treatment of bladder cancer using hexaminolevulinate imaging: A prospective, phase III multicenter study. *J Urol* 174:862–866.

Kausch, I., M. Sommerauer, F. Montorsi et al. 2010. Photodynamic diagnosis in non-muscle-invasive bladder cancer: A systematic review and cumulative analysis of prospective studies. *Eur Urol* 57:595–606.

Kennedy, J. C., R. H. Pottier, and D. C. Pross. 1990. Photodynamic therapy with endogenous protoporphyrin IX: Basic principles and present clinical experience. *J Photochem Photobiol* B6:143–148.

Kim, R. Y., L. K. Hu, T. J. Flotte, E. S. Gragoudas, and L. H. Y. Young. 1997. Digital angiography of experimental choroidal melanomas using benzoporphyrin derivative. *Am J Ophthalmol* 123:810–816.

Klem, B., G. Lappin, S. Nicholson et al. 2006. Determination of the bioavailability of [14C]-hexaminolevulinate using accelerator mass spectrometry after intravesical administration to human volunteers. *J Clin Pharmacol* 46(4):456–460.

Kloek, J., and G. M. J. Beijersbergen van Henegouwen. 1996. Prodrugs of 5-aminolaevulinic acid for photodynamic therapy. *Photochem Photobiol* 64:994–1000.

König, F., F. J. McGovern, A. F. Althausen, T. F. Deutsch, and K. T. Schomacker. 1996. Laser-induced autofluorescence diagnosis of bladder cancer. *J Urol* 156:1597–1601.

König, F., F. J. McGovern, R. Larne et al. 1999. Diagnosis of bladder carcinoma using protoporphyrin IX fluorescence induced by 5-aminolaevulinic acid. *BJU Int* 83:129–135.

Kriegmair, M., R. Baumgartner, and A. Hofstetter. 1992. Intravesikale Instillation von Delta-Aminolävulinsäure (ALA)—Eine neue Methode zur photodynamischen Diagnostik und Therapie. *Lasermedizin* 8S:83.

Kriegmair, M., R. Baumgartner, R. Knuechel et al. 1994. Fluorescence photodetection of neoplastic urothelial lesions following intravesical instillation of 5-aminolevulinic acid. *Urology* 44(6):836–841.

Kriegmair, M., R. Baumgartner, R. Knuechel et al. 1996. Detection of early bladder cancer by 5-aminolevulinic acid induced porphyrin fluorescence. *J Urol* 155:105–110.

Kriegmair, M., R. Baumgartner, W. Lumper et al. 1993. Fluorescence cystoscopy following intravesical instillation of 5-aminolevulinic acid (ALA). *J Urol* 149:24OA.

Kriegmair, M., D. Zaak, K. H. Rothenberger et al. 2002. Transurethral resection for bladder cancer using 5-aminolevulinic acid induced fluorescence endoscopy versus white light endoscopy. *J Urol* 168:475–478.

Kucera, P., and E. Raddatz. 1980. Spatio-temporal micromeasurements of the oxygen uptake in the developing chick embryo. *Resp Physiol* 39:199–215.

Lange, N., P. Jichlinski, and M. Zellweger. 1999. Photodetection of early human bladder cancer based on the fluorescence of 5-aminolaevulinic acid hexylester induced protoporphyrin IX: A pilot study. *Br J Cancer* 80:185–193.

Lerner, S. P., H. Liu, M. F. Wu, Y. K. Thomas, and J. A. Witjes. 2011. Fluorescence and white light cystoscopy for detection of carcinoma in situ of the urinary bladder. *Urol Oncol* 30(3):285–289.

Loidl, W., J. Schmidbauer, M. Susani, and M. Marberger. 2005. Flexible cystoscopy assisted by hexaminolevulinate induced fluorescence: A new approach for bladder cancer detection and surveillance? *Eur Urol* 47:323–326.

Lovisa, B., P. Jichlinski, B.-C. Weber et al. 2010. High-magnification vascular imaging to reject false positive sites in situ during Hexvix® fluorescence cystoscopy. *J Biomed Optics* 15(5):051606.

Ludicke, F., T. Gabrecht, N. Lange et al. 2003. Photodynamic diagnosis of ovarian cancer using hexaminolaevulinate: A preclinical study. *Br J Cancer* 88:1780–1784.

Malik, Z., and H. Lugaci. 1987. Destruction of erythroleukaemic cells by photoinactivation of endogenous porphyrins. *Br J Cancer* 56:589–595.

Malmström, P.-U., M. Grabe, E. S. Haug et al. 2012. Role of hexaminolevulinate-guided fluorescence cystoscopy in bladder cancer: Critical analysis of the latest data and European guidance. *Scand J Urol Nephrol* 46(2):108–116.

Marti, A., N. Lange, M. Zellweger et al., Patent FR2777782, 1998.

Marti, A., P. Jichlinski, N. Lange et al. 2003. Comparison of aminolevulinic acid and hexylester aminolevulinate induced protoporphyrin IX distribution in human bladder cancer. *J Urol* 170:428–432.

Marti, A., N. Lange, H. van den Bergh et al. 1999. Optimalisation of the formation and distribution of protoporphyrin IX in the urothelium: An *in vitro* approach. *J Urol* 162:546.

Mayinger, B., F. Neumann, C. Kastner, T. Haider, and D. Schwab. 2010. Hexaminolevulinate-induced fluorescence colonoscopy versus white light endoscopy for diagnosis of neoplastic lesions in the colon. *Endoscopy* 42(1):28–33.

National Institutes of Health (NIH). 1990. *Cancer of the Bladder*, Publication No. 90–722. National Cancer Institute, Bethesda, MD.

Oosterlinck, W., A. van der Meijden, R. Sylvester et al. 2006. *Guidelines on TaT1 (Non-muscle Invasive) Bladder Cancer*. European Association of Urology, Arnhem.

Palmeira, C., C. Lameiras, T. Amaro et al. 2011. CIS is a surrogate marker of genetic instability and field carcinogenesis in the urothelial mucosa. *Urol Oncol* 29(2):205–211.

Parmar, M. K. B., L. S. Friedman, T. B. Hargreave et al. 1989. Prognostic factors for recurrence. Follow-up policies in the treatment of superficial bladder cancer. Report from the British Medical Research Council Subgroup on Superficial Bladder Cancer. *J Urol* 142:284–288.

Peng, Q., K. Berg, J. Moan, M. Kongshaug and J. M. Nesland. 1997. 5-Aminolevulinic acid-based photodynamic therapy: Principles and experimental research. *Photochem Photobiol* 65:235–251.

Peng, Q., J. Moan, T. Warloe et al. 1996. Build-up of esterified aminolaevulinic-acid-derivative-induced porphyrin fluorescence in normal mouse skin. *J Photochem Photobiol B* 34:96–96.

Photocure press release. Retrieved June 5, 2012 from http://www.photocure.com/Pressmedia/News/US-approval-of-Cysview-/.

Profio, A. E., and J. Sarnaik. 1984. Fluorescence of Hpd for tumor detection and dosimetry in photoradiation therapy. In *Porphyrin Localization and Treatment of Tumors*, Vol. 170, D. R. Doiron, and C. J. Gomer, editors, Alan R. Liss, Inc., New York, 163–175.

Rathert, P., and S. Roth. 1993. Indications for urinary cytology. In *Urinary Cytology, Manual and Atlas*, ed. 2. P. Rather, S. Roth, and M. S. Soloway, editors. Springer-Verlag, New York, 9–13.

Richards-Kortum, R., and E. Sevick-Muraca. 1996. Quantitative optical spectroscopy for tissue diagnosis. *Annu Rev Phys Chem* 47:555–606.

Riedl, C. R., D. Daniltchenko, F. Koenig et al. 2001. Fluorescence endoscopy with 5-aminolevulinic acid reduces early recurrence rate in superficial bladder cancer. *J Urol* 165:1121–1123.

Ries, L. A. G., C. L. Kosary, B. F. Hankey et al. 1997. SEER Cancer Statistics Review 1973–1994. National Institutes of Health Publication No. 94–2789. National Cancer Institute, Bethesda, MD.

Rud, E., O. Gederaas, A. Hogset, and K. Berg. 2000. 5-aminolevulinic acid, but not 5-aminolevulinic acid esters, is transported into adenocarcinoma cells by system BETA transporters. *Photochem Photobiol* 71:640–647.

Schmidbauer, J., M. Remzi, G. Lindenau, M. Susani, and M. Marberger. 2008. Optical coherence tomography and hexaminolevulinate fluorescence cystoscopy in detecting urothelia carcinoma of the bladder. *Eur Urol Suppl* 7:77 (abstract no. 28).

Schmidbauer, J., F. Witjes, N. Schmeller et al. 2004. Improved detection of urothelial carcinoma in situ with hexaminolevulinate fluorescence cystoscopy. *J Urol* 171:135–138.

Soergel, P., L. Makowski, E. Makowski et al. 2011. Treatment of high grade cervical intraepithelial neoplasia by photodynamic therapy using hexylaminolevulinate may be cost effective compared to conisation procedures due to decreased pregnancy-related morbidity. *Lasers Surg Med* 43(7):713–720.

Stenzl, A., M. Burger, Y. Fradet et al. 2010. Hexaminolevulinate guided fluorescence cystoscopy reduces recurrence in patients with nonmuscle invasive bladder cancer. *J Urol* 184:1907–1913.

Stenzl, A., L. Hoeltl, and G. Bartsch. 2001. Fluorescence assisted transurethral resection of bladder tumours: Is it cost effective? *Eur Urol* 39:31.

Stepinac, T., C. Felley, P. Jomod et al. 2003. Endoscopic fluorescence detection of intraepithelial neoplasia in Barrett's esophagus after oral administration of aminolevulinic acid. *Endoscopy* 35:663–668.

Sylvester, R., A. P. van der Meijden, W. Oosterlinck et al. 2006. Predicting recurrence and progression in individual patients with stage Ta T1 bladder cancer using EORTC risk tables: A combined analysis of 2596 patients from seven EORTC trials. *Eur Urol* 49:466–477.

Uehlinger, P., J.-P. Ballini, H. van den Bergh, and G. Wagnières. 2006. On the role of iron and one of its chelating agents in the production of protoporphyrin IX generated by 5-aminolevulinic acid and its hexyl ester derivative tested on an epidermal equivalent of human skin. *Photochem Photobiol* 82:1069–1076.

Uehlinger, P., M. Zellweger, G. Wagnières et al. 2000. 5-aminolevulinic acid and its derivatives: Physical chemical properties and protoporphyrin IX formation in cultured cells. *J Photochem Photobio B Biol* 54:72–80.

Vicente, J., G. Chècile, and F. Algaba. 1987. Value of *in vivo* mucosa-staining test with methylene blue in the diagnosis of pretumoral and tumoral lesions of the bladder. *Eur Urol* 13:15–16.

Wagnieres, G. A., W. M. Star, and B. C. Wilson. 1998. *In vivo* fluorescence spectroscopy and imaging for oncological applications. *Photochem Photobiol* 68:603–632.

Whitaker, C. J., S. H. Battah, M. J. Forsyth et al. 2000. Photosensitization of pancreatic tumour cells by delta-aminolaevulinic acid esters. *Anticancer Drug Des* 15:161–170.

Whitmore, W. F. Jr., and I. M. Bush. 1968. Ultraviolet cystoscopy. *JAMA* 203:1057–1059.

Wolfbeis, O. S. 1993. Fluorescence of organic natural products. In *Molecular Luminescence Spectroscopy*, Vol. 1. S. G. Schulman, editor. John Wiley and Sons, New York, 167–370.

Witjes, J. A., and J. Douglass. 2007. The role of hexaminolevulinate fluorescence cystoscopy in bladder cancer. *Nat Clin Pract Urol* 4:542–549.

Witjes, J., P. M. Moonen, and A. G. van der Heijden. 2005. Comparison of hexaminolevulinate based flexible and rigid fluorescence cystoscopy with rigid white light cystoscopy in bladder cancer: Results of a prospective phase II study. *Eur Urol* 47:319–322.

Witjes, J. A., J. Palou Redorta, D. Jacqmin et al. 2010. Hexaminolevulinate-guided Fluorescence cystoscopy in the diagnosis and follow-up of patients with non-muscle-invasive bladder cancer: Review of the evidence and recommendations. *Eur Urol* 57:607–614.

Zaak, D., W. F. Wieland, C. G. Stief, and M. Burger. 2008. Routine use of photodynamic diagnosis of bladder cancer: Practical and economic issues. *Eur Urol Suppl* 7:536–541.

Zheng, W., L. Weber, Ch. Cheng, K. So, and M. Olivo. 2003. Optimal excitation-emission wavelengths for autofluorescence diagnosis of bladder tumors. *Int J Cancer* 104:477–481.

37

Photochemical Internalization: From Bench to Bedside with a Novel Technology for Targeted Macromolecule Therapy

Kristian Berg
Oslo University Hospital

Anette Weyergang
Oslo University Hospital

Anders Høgset
PCI Biotech AS

Pål K. Selbo
Oslo University Hospital

37.1 Introduction

The utilization of macromolecules in the therapy of cancer and other diseases is becoming increasingly relevant to successful tumor therapy. Recent advances in molecular biology and biotechnology have made it possible to improve the targeting and design of cytotoxic agents, DNA complexes, and other macromolecules for clinical applications. To achieve the expected biological effect of these macromolecules, in many cases, internalization to the cell cytosol is crucial. Macromolecules are frequently prohibited from penetrating cell membranes, but are instead endocytosed and without any active intervention become degraded by hydrolytic enzymes in the lysosomes. Thus, at an intracellular level, the most fundamental obstruction for cytosolic localization of therapeutic macromolecules is the membrane barrier of the endocytic vesicles (Varkouhi et al. 2010). Photochemical internalization (PCI) is a novel technology for release of endocytosed macromolecules into the cytosol. PCI has been shown to potentiate the biological activity of a large variety of macromolecules and other molecules that do not readily penetrate the plasma membrane (Berg et al. 1999; Selbo et al. 2010). PCI may therefore be utilized for treatment of most solid tumors, and the first phase I/II clinical trial with PCI of bleomycin (Berg et al. 2005) for several cancer indications has recently been completed. This review will describe the mechanistic basis for the PCI technology, the present status, the potential clinical advantages, and future perspectives.

37.2 Background

37.2.1 Photodynamic Therapy versus PCI

Photodynamic therapy (PDT) is a relatively new and promising treatment modality (Agostinis 2011) approved for multiple diseases, including several cancer indications such as basal cell carcinoma (BCC); head and neck squamous cell carcinoma; cancer of the lung, esophagus, and bladder; as well actinic keratosis and age-related macular degeneration. PDT is based on light activation of a compound called a photosensitizer (PS) that exerts a preferential retention in hyperproliferating (e.g., tumor) tissues. Light exposure of the photosensitizers with visible light generates reactive oxygen species (ROS), where singlet oxygen is considered as the most important, and thereby induces cytotoxic effects in the light-exposed area (Figure 37.1). The therapeutic effect of PDT may be exerted through killing cells exposed to both PS and light as well as vascular shutdown and inflammatory responses. The PCI technology utilizes the same ROS-generating effects of photosensitizers as in PDT, but the oxidizing effect of the ROS generated in PCI is utilized to deliver therapeutics from endocytic vesicles into the cytosol of cells instead of inducing a direct therapeutic effect by the photochemical reactions. In a PCI treatment session, the therapeutic outcome will be partly due to a PDT effect where sufficient levels of ROS are formed to exert a therapeutic effect alone and a PCI effect where the ROS formed are able to activate therapeutics entrapped in endocytic vesicles without exerting a therapeutic

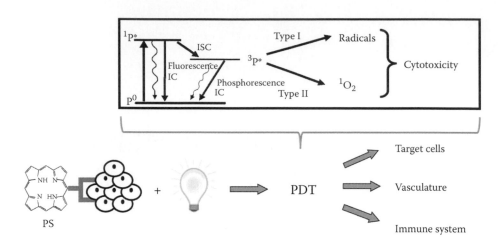

FIGURE 37.1 Schematic description of PDT and the main therapeutic effects. The PS associates with the target tissue (e.g., tumor) and, upon exposure to light absorbed by the PS, may induce therapeutic effects on the target cells and the vasculature and activate the immune system. The photophysical mechanisms involved in the photodynamic process are illustrated by the simplified Jablonski diagram (upper figure). IC = internal conversion.

effect alone. The therapeutic effect of PCI is most easily seen as necrosis in deeper tissue layers and in the tumor rim (Figure 37.2) (Dietze 2005; Norum et al. 2009b; Norum, Giercksky, and Berg 2009). In clinical PDT practice, PSs used for systemic administration are injected shortly prior to light administration to several days in advance depending on the PS properties and the treatment target (Figure 37.3). In PCI, the PS is administered

4 days prior to light exposure while the second drug is usually clinically injected 3 h prior to light exposure. However, this drug–light interval may be influenced by the pharmacokinetics of the drug to be activated by PCI.

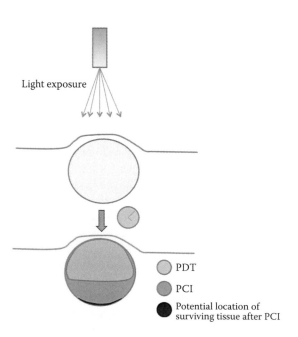

FIGURE 37.2 Illustration of the difference in therapeutic effect on a suboptimal PDT session and the addition therapeutic effect of PCI utilizing the same photosensitizer in both cases. The indicated PCI effect shows the additional effect of using PDT for intracellular delivery of another drug accumulated in endocytic vesicles.

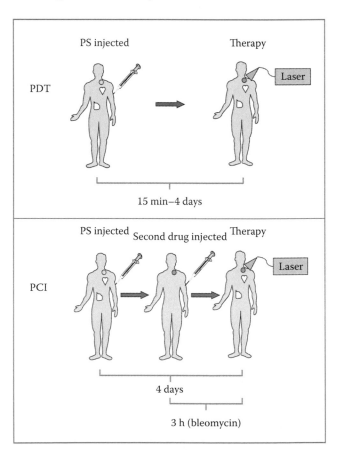

FIGURE 37.3 Schematic description of the clinical treatment procedure for systemic PDT and PCI.

37.2.2 The Mechanistic Basis for PCI Technology

The cellular uptake mechanisms as well as intracellular localization of PSs are dependent on their physicochemical properties. PSs with two or fewer carboxylate groups as the only ionic side groups of the PS core porphyrin-type structures usually penetrate the plasma membrane by passive diffusion while a higher number of carboxylate groups or PSs with side groups always completely ionized under physiological conditions (e.g., low pKa) appear unable to penetrate cellular membranes (Figure 37.4) (Berg and Moan 1997). Such PSs will instead be taken up into cells by endocytosis, either through pinocytosis for the most hydrophilic PSs and/or adsorptive/LDL-mediated endocytosis for the most amphiphilic PSs. Some highly hydrophobic PSs may sometimes be found in endocytic vesicles due to cellular uptake of aggregated PSs (Berg et al. 1993). Additionally, lysosomotropic weak bases such as acridine orange and Nile blue derivatives may penetrate cellular membranes and become trapped in endocytic vesicles due to their low intravesicular pH (Berg and Moan 1997; Lin et al. 1991). Light exposure of some of the amphiphilic PSs [such as disulfonated tetraphenylporphin (TPPS$_{2a}$) and disulfonated aluminum phthalocyanine (AlPcS$_{2a}$)] (Figure 37.4) has been shown to damage the membrane of the endocytic vesicles, leading to relocation of the PSs to other intracellular compartments during light exposure (Berg et al. 1991; Moan et al. 1994). The damage induced to the membranes of the endocytic vesicles has not been fully investigated but does not only cause translocation of small molecules such as PSs since the hydrolytic enzymes located in lysosomes (and late endosomes) have be found in the cytosol after photochemical treatment (Berg and Moan 1994). Accordingly, externally added enzymes or other molecules not penetrating cellular membranes have also be found in the cytosol in a functionally active form after such photochemical treatment (Berg et al. 1999). In contrast, hydrophilic PSs located in the matrix of the endocytic vesicles photooxidize and inactivate biomolecules located in these vesicles and may be relocated to the cytosol in an inactive form (Berg and Moan 1994). Recently, we documented that PCI photosensitizers that are highly amphiphilic are not substrates of the efflux pump ABCG2 (Selbo et al. 2012), a putative cancer stem cell marker that transports a panel of other photosensitizers out of cancer cells (Robey et al. 2005).

Lysosomes were named "suicide bags" by de Duve in the early 1960s because the rupture of lysosomes resulted in cell death (de Duve 1963). Later it was shown that some cysteine cathepsins exert enzymatic activity at physiologic pH 7.2–7.4 in contrast to most lysosomal enzymes with optimal enzymatic activity around pH 5.5 and no activity above pH 7 (Bohley and Seglen 1992; Repnik et al. 2012). These cathepsins have since been regarded as the cause of the suicidal effect of lysosomal rupture, resulting in apoptosis or necrosis depending on the extent of lysosomal rupture (Turk and Turk 2009; Wilson, Firestone, and Lenard 1987). However, the lysosomal rupture seen after TPPS$_{2a}$ or AlPcS$_{2a}$-mediated photochemical treatment induces only moderate cytotoxicity at doses sufficient to induce substantial relocation of the PS (Berg and Moan 1994). The PS TPPS$_{2a}$ has been found to relocate during light exposure and inactivate the endoplasmic reticulum (ER) marker enzyme NADPH cytochrome c reductase (Rodal et al. 1998). One may therefore envision that the cytotoxic effect of TPPS$_{2a}$-based PDT is at least partly due to

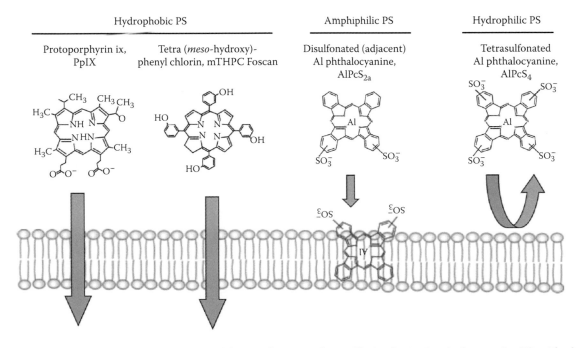

FIGURE 37.4 Photosensitizer penetration through a cellular membrane is influenced by its physicochemical properties. PSs with a low number of side groups that may be protonated at physiological pH (down to approximately pH 5 in lysosomes) are able to penetrate cellular membranes, while strongly amphphilic and hydrophilic PSs will instead be taken up into cells through endocytosis.

photooxidation of ER. Only very low levels of cathepsin activity (<10% of total activity) have been found in the cytosol after photochemical treatment (Berg and Moan 1994). This is partly due to a relatively high sensitivity of cysteine cathepsins (as reflected in cathepsin E+L activity) to PDT and a differential sensitivity of lysosomes to PDT-induced rupture leading to a fraction of intact lysosomes after treatment. In addition, a group of cathepsin inhibitors named stefins (type 1 cystatins) located in the cytosol are able to inhibit the enzymatic activity of a partial cathepsin release from lysosomes (Repnik et al. 2012). However, it should be pointed out that the levels of cathepsins and stefins are less controlled in cancer cells than in normal cells and may influence on the toxic effect of photochemically induced lysosomal rupture and PCI efficacy (Kolwijck et al. 2010; Strojan et al. 2011).

It was initially thought that the macromolecule to be released from the endocytic compartments and the photosensitizer had to be located in the same compartments at the time of light exposure. However, it has recently been found that the photochemical treatment can be performed up to 6–8 h prior to the delivery of the macromolecule without reducing the synergistic effect of the combined treatment (the "light first" procedure) (Prasmickaite et al. 2002). The light first principle was also confirmed in vivo, resulting in a strong antitumor effect in mice. Light exposure of subcutaneous WiDr colon adenocarcinoma prior to an intratumoral injection of the ribosomal inactivating protein toxin gelonin induced complete response (CR) in >80% of animals treated with PCI of gelonin, compared to 60% CR when the PCI "light after" principle was performed (Berg et al. 2006). The assumed mechanistic basis for such an observation is that newly formed vesicles are able to fuse with photochemically damaged vesicles and that the macromolecules are released into the cytosol after fusion of such vesicles. It has recently been found that PCI of targeted therapeutics, such as cetuximab–saporin, is less efficient with the light first procedure, most likely due to photochemical damage to the targeting receptor, although the amount of PS located on the plasma membrane is barely detectable (Weyergang, Kaalhus, and Berg 2008; Weyergang, Selbo, and Berg 2007; Yip et al. 2007).

37.3 Potential Therapeutics That May Benefit from PCI Technology

The major treatment modalities for cancer are still surgery, ionizing radiation, and chemotherapy. Despite substantial efforts to optimize these treatment regimens, a large number of cancer patients still die each year (approximately 550,000 per year in the United States alone). The main cause of treatment failures is related to the insufficient specificity of current treatment modalities resulting in dose-limiting toxicity and leading to local recurrences, metastasis, multidrug resistance, and death. On the basis of new knowledge on cancer characteristics [hallmarks of cancer (Hanahan and Weinberg 2011)] and in-depth understanding of expression and mutation profiles of subclasses of cancers, novel therapeutics have been developed that may be mainly grouped into macromolecular therapeutics and specific small molecular inhibitors. Compared with more traditional small molecular

chemotherapeutics, macromolecular drugs have the potential advantage of exerting higher therapeutic specificity. These macromolecules include proteins such as molecular antibodies (mAbs), ribosome-inactivating protein toxins linked to mAbs or growth factors for cell surface targeting, peptides and messenger ribonucleic acid (mRNA) for therapeutic vaccines, deoxyribonucleic acid (DNA) for gene therapy, and oligonucleotides such as ribozymes, peptide nucleic acids (PNAs), and short interfering RNA (siRNA) for gene silencing as well as nanomedicine (Leonetti and Zupi 2007; Yang et al. 2007).

Macromolecular therapeutics may include drugs where the macromolecule itself exerts a therapeutic effect as well as drugs where the macromolecule serves as a targeting moiety. Some macromolecules, such as mAbs, exert their therapeutic effect on soluble factors such as growth factors or at the surface of the target cells through inhibition of cell signaling pathways and activation of the immune system. mAbs can also be conjugated to moieties with extracellular activities such as radionuclides, immunomodulators, and prodrug-activating enzymes (Lu, Yang, and Sega 2006). In these cases where the therapeutic target is located extracellularly, there are many approved therapeutics both for cancer and noncancerous indications.

Macromolecular therapeutics with intracellular targets usually exert low efficacy due to the limited penetration through the plasma membrane, and very few drugs approved for clinical use are found in this category. Even though a molecule is unable to penetrate the plasma membrane, it is still taken up into the cells by means of endocytosis, either by receptor-mediated or adsorptive endocytosis or by pinocytosis. Endocytosed molecules are further transported to late endosomes and lysosomes, where they are degraded. Drugs with intracellular targets that are not able to penetrate the plasma membrane will be degraded in the endocytic vesicles and will not exert any therapeutic effects (Figure 37.5). As described above, PDT with selected PSs has been shown

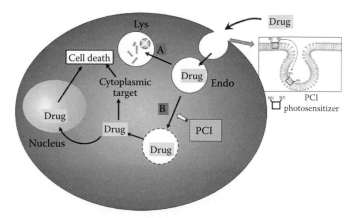

FIGURE 37.5 Schematic description of PCI technology. Drugs that are endocytosed may accumulate in lysosomes where they will be subjected to enzymatic hydrolyses and unable to exert any therapeutic effect (pathway A). Alternatively, drugs may be released into the cytosol by means of PCI before enzymatic degradation and reach the target of their therapeutic effect in the nucleus or in the cytoplasm (pathway B). Lys = lysosomes, Endo = endosomes.

to damage the endolysosomal membrane without inducing severe cytotoxicity and may therefore be utilized for cytosolic release of endocytosed drugs before they are degraded. PCI has been shown to increase the effect of ribosome-inactivating protein toxins from plants, immunotoxins, genes (in both plasmids and viruses), PNAs, siRNAs, peptides (for vaccination), doxorubicin, and bleomycin (Selbo et al. 2010). Experiments have been performed in many cell lines (>80) and 10 tumor models derived from all major cancers and shown to exert a synergistic effect in all cases. The results show that PCI can induce efficient light-directed delivery of macromolecules into the cytosol.

37.4 Clinical Development and Utilization of PCI

Clinical utilization of PCI requires a clinically developed PS as well as a suitable drug to be delivered by PCI technology. The experimental and preclinical development of PCI has been based on the two photosensitizers $TPPS_{2a}$ and $AlPcS_{2a}$ (Figure 37.6).

$TPPS_{2a}$, with its porphyrin structure and thereby low absorption in the optical window optimal for tissue light penetration (600–800 nm), has been utilized for in vitro studies, while $AlPcS_{2a}$, with strong absorption around 670 nm, has been used for in vivo documentation. The photosensitizer $AlPcS_{2a}$ is useful for preclinical studies but contains a large number of isomers potentially with batch-to-batch ratio variations (Ali et al. 1988; Berg, Bommer, and Moan 1989). The biological effect of the various isomers may vary substantially upon exposure to light and may result in batch-to-batch variations in clinical response (Brasseur et al. 1987). For clinical development of PCI, a new chemical entity for clinical use is needed that may fulfill the requirements with respect to impurities, controllable and reproducible synthesis, and so forth. On this basis, the photosensitizer disulfonate tetraphenyl chlorin ($TPCS_{2a}$, Amphinex) has been developed (Berg et al. 2011). $TPCS_{2a}$ is a chlorin-type PS and, due to its asymmetric structure, results in three isomers in reproducible ratios upon reduction of the pyrrol rings of the porphyrin $TPPS_{2a}$, exerting strong absorption at 652 nm (Figure 37.6). As with $TPPS_{2a}$ and $AlPcS_{2a}$, $TPCS_{2a}$ has been found to induce a PCI

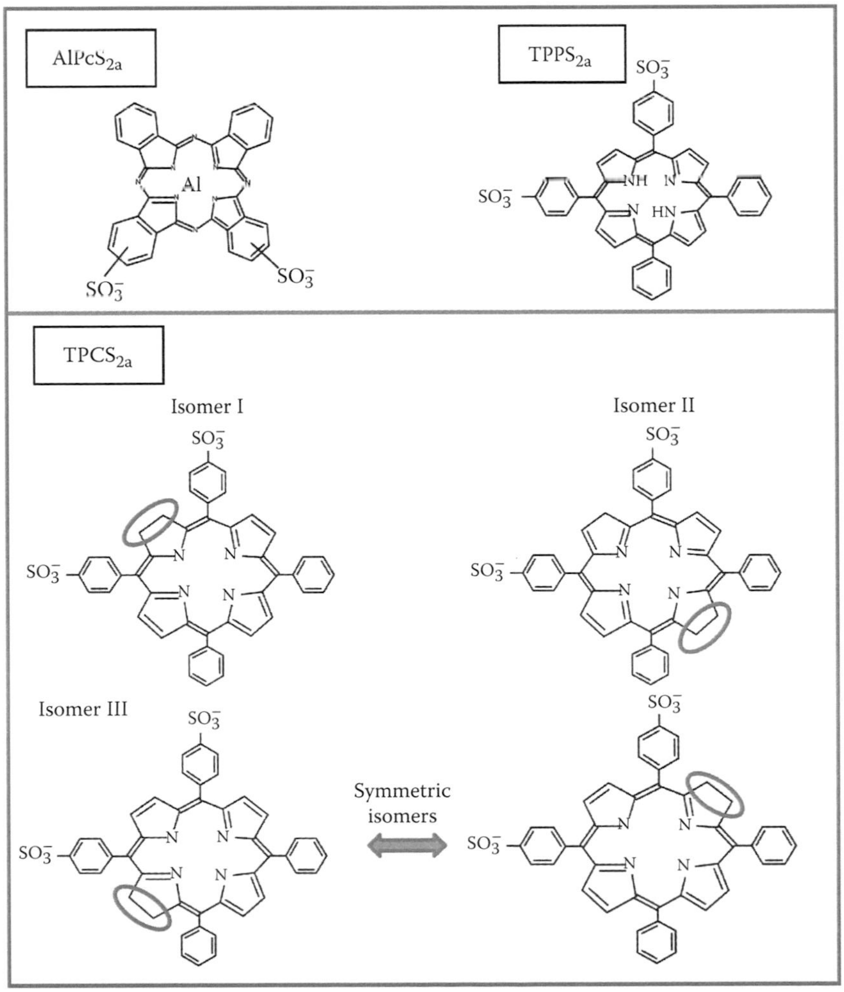

FIGURE 37.6 Photosensitizers used in PCI.

effect comparable to that of TPPS$_{2a}$ and AIPcS$_{2a}$. TPCS$_{2a}$ has also been found to be more efficient in photoinducing permeation of liposomes than meso-tetra(3-hydroxyphenyl)chlorin (mTHPC) and chlorin e$_6$ (Mojzisova et al. 2009).

Although the PCI technology has been developed for intracellular delivery of therapeutic macromolecules, some chemotherapeutic agents accumulate in endocytic vesicles probably due to their size, charge, and ability to form hydrogen bonds. Bleomycin is an approved chemotherapeutic agent for several cancer indications but is also known for accumulating in endocytic vesicles (Pron et al. 1999). The therapeutic effect of bleomycin is limited, and lung fibrosis is a severe side effect frequently observed after bleomycin treatment (Chen and Stubbe 2005). However, after entering the cytosol, bleomycin has been found to be highly cytotoxic, and it has been estimated that only approximately 500 molecules are sufficient to kill a cell (Poddevin et al. 1991). PCI of bleomycin has been shown to improve the cytotoxic effect of bleomycin both in vitro and in vivo (Berg et al. 2005; Berg et al. 2011; Hirschberg et al. 2009; Norum et al. 2009a), indicating that not only large macromolecules like proteins and DNA may profit from a PCI-based combination treatment, but also smaller molecules unable to efficiently penetrate biological membranes. A benefit of PCI might be reduced adverse effects of such chemotherapeutic agents if lower doses of the chemotherapeutics may be sufficient to exert chemotherapeutic effects. The side effects of bleomycin treatment are related to an accumulated dose, but in the clinical trial, a standard dose of bleomycin was injected only once.

On the basis of development of a clinical-grade formulation of TPCS$_{2a}$ and preclinical documentation, a phase I/II clinical study has been performed with TPCS$_{2a}$ (Amphinex) activated by a 652 nm laser light (external beam radiation) for intracellular delivery of bleomycin (Principle Investigator: Colin Hopper, UCL Hospital, London). The dose-escalation study was performed with a fixed and single dose of bleomycin, a fixed dose of light, and increasing doses of TPCS$_{2a}$ until unacceptable side effects were observed. Nineteen patients, mainly diagnosed with head and neck squamous cell carcinoma, were enrolled. Strong response to treatment was seen in all patients. Amphinex seems to be well tolerated, and no serious product-related adverse events have been reported other than photosensitivity (see www.pcibiotech.no for further details).

37.5 Future Perspectives and Clinical Advantages of the PCI Technology

Despite a large number of new treatment modalities developed for improving therapeutic efficacy and specificity, adverse effects and development of resistance mechanisms are still limiting factors for the therapeutic outcome. Photochemical activation of drugs and genes is an attractive alternative or addition to other delivery methods that may improve their therapeutic efficacy and specificity. Substantial site direction of the treatment can be obtained by the need for light activation, and the

therapeutic efficacy of drugs and genes may even be improved. The specificity is further improved by the preferential retention of the photosensitizer in tumor tissue. Photochemical delivery of therapeutic compounds may therefore increase the therapeutic window and thereby reduce the adverse reactions of the compounds. In most targeted therapeutics with an intracellular effect, a translocation-to-the-cytosol function has been built into the therapeutics (e.g., the vectors used for gene delivery and membrane translocation function of pseudomonas exotoxin). However, due to expression of targeting receptors and nonspecific uptake in normal cells, some cytotoxic effect on normal tissues is difficult to avoid. In PCI, drugs may preferentially be designed to accumulate and degrade in endocytic vesicles if not subjected to PCI. These drugs alone may therefore not exert therapeutic or adverse effects. In addition, PCI may benefit from the same properties as PDT, such as the single-treatment option that reduces the risk of developing treatment-resistant cancers and improving patient quality of life. As well, similar procedures may be utilized for several cancer indications, and the treatment effect may be due to direct parenchyma cell killing, vascular shutdown, and inflammatory responses including development of antitumor immunity.

Clinically, PCI has so far only been used to deliver bleomycin. Although this treatment regimen has shown promising results, it is to be expected that utilizing PCI of targeted therapeutics will further improve the specificity. A future goal may therefore be to clinically evaluate PCI of targeted therapeutics.

Finally, in our opinion, photochemical drug and gene activation and delivery could be considered for a large variety of clinical indications, including cancer, cardiovascular diseases, autoimmune diseases, rheumatoid arthritis, and as a delivery system for DNA vaccines (Hogset et al. 2003).

References

Agostinis, P., K. Berg, K. A. Cengel et al. 2011. Photodynamic therapy of cancer: An update. *CA Cancer J Clin* 61:250–281.

Ali, H., R. Langlois, J. R. Wagner et al. 1988. Biological activities of phthalocyanines—X. Syntheses and analyses of sulfonated phthalocyanines. *Photochem Photobiol* 47:713–717.

Berg, K., H. Anholt, J. Moan, A. Ronnestad, and C. Rimington. 1993. Photobiological properties of hematoporphyrin diesters: Evaluation for possible application in photochemotherapy of cancer. *J Photochem Photobiol B* 20:37–45.

Berg, K., J. C. Bommer, and J. Moan. 1989. Evaluation of sulfonated aluminum phthalocyanines for use in photochemotherapy. Cellular uptake studies. *Cancer Lett* 44:7–15.

Berg, K., A. Dietze, O. Kaalhus, and A. Hogset. 2005. Sitespecific drug delivery by photochemical internalization enhances the antitumor effect of bleomycin. *Clin Cancer Res* 11:8476–8485.

Berg, K., A. Høgset, L. Prasmickaite et al. 2006. Photochemical internalization (PCI): A novel technology for activation of endocytosed therapeutic agents. *Med Laser Appl* 21:239–250.

Berg, K., K. Madslien, J. C. Bommer et al. 1991. Light induced relocalization of sulfonated meso-tetraphenylporphines in NHIK 3025 cells and effects of dose fractionation. *Photochem Photobiol* 53:203–210.

Berg, K., and J. Moan. 1994. Lysosomes as photochemical targets. *Int J Cancer* 59:814–822.

Berg, K., and J. Moan. 1997. Lysosomes and microtubules as targets for photochemotherapy of cancer. *Photochem Photobiol* 65:403–409.

Berg, K., S. Nordstrand, P. K. Selbo et al. 2011. Disulfonated tetraphenyl chlorin (TPCS$_{2a}$), a novel photosensitizer developed for clinical utilization of photochemical internalization. *Photochem Photobiol Sci* 10:1637–1651.

Berg, K., P. K. Selbo, L. Prasmickaite et al. 1999. Photochemical internalization: A novel technology for delivery of macromolecules into cytosol. *Cancer Res* 59:1180–1183.

Bohley, P., and P. O. Seglen. 1992. Proteases and proteolysis in the lysosome. *Experientia* 48:151–157.

Brasseur, N., H. Ali, R. Langlois, and J. E. van Lier. 1987. Biological activities of phthalocyanines—VII. Photoinactivation of V-79 Chinese hamster cells by selectively sulfonated gallium phthalocyanines. *Photochem Photobiol* 46:739–744.

Chen, J., and J. Stubbe. 2005. Bleomycins: Towards better therapeutics. *Nat Rev Cancer* 5:102–112.

de Duve, C. 1963. The lysosome. *Sci Am* 208:64–72.

Dietze, A., Q. Peng, P. K. Selbo et al. 2005. Enhanced photodynamic destruction of a transplantable fibrosarcoma using photochemical internalisation of gelonin. *Br J Cancer* 92:2004–2009.

Hanahan, D., and R. A. Weinberg. 2011. Hallmarks of cancer: The next generation. *Cell* 144:646–674.

Hirschberg, H., M. J. Zhang, H. M. Gach et al. 2009. Targeted delivery of bleomycin to the brain using photo-chemical internalization of *Clostridium perfringens* epsilon protoxin. *J Neurooncol* 95:317–329.

Hogset, A., L. Prasmickaite, B. O. Engesaeter et al. 2003. Light directed gene transfer by photochemical internalisation. *Curr Gene Ther* 3:89–112.

Kolwijck, E., J. Kos, N. Obermajer et al. 2010. The balance between extracellular cathepsins and cystatin C is of importance for ovarian cancer. *Eur J Clin Invest* 40:591–599.

Leonetti, C., and G. Zupi. 2007. Targeting different signaling pathways with antisense oligonucleotides combination for cancer therapy. *Curr Pharm Des* 13:463–470.

Lin, C. W., J. R. Shulok, S. D. Kirley, L. Cincotta, and J. W. Foley. 1991. Lysosomal localization and mechanism of uptake of Nile blue photosensitizers in tumor cells. *Cancer Res* 51:2710–2719.

Lu, Y., J. Yang, and E. Sega. 2006. Issues related to targeted delivery of proteins and peptides. *AAPS J* 8:E466–E478.

Moan, J., K. Berg, H. Anholt, and K. Madslien. 1994. Sulfonated aluminium phthalocyanines as sensitizers for photochemotherapy. Effects of small light doses on localization, dye fluorescence and photosensitivity in V79 cells. *Int J Cancer* 58:865–870.

Mojzisova, H., S. Bonneau, P. Maillard, K. Berg, and D. Brault. 2009. Photosensitizing properties of chlorins in solution and in membrane-mimicking systems. *Photochem Photobiol Sci* 8:778–787.

Norum, O. J., O. S. Bruland, L. Gorunova, and K. Berg. 2009a. Photochemical internalization of bleomycin before external-beam radiotherapy improves locoregional control in a human sarcoma model. *Int J Radiat Oncol Biol Phys* 75:878–885.

Norum, O. J., J. V. Gaustad, E. Angell-Petersen et al. 2009b. Photochemical internalization of bleomycin is superior to photodynamic therapy due to the therapeutic effect in the tumor periphery. *Photochem Photobiol* 85:740–749.

Norum, O. J., K. E. Giercksky, and K. Berg. 2009. Photochemical internalization as an adjunct to marginal surgery in a human sarcoma model. *Photochem Photobiol Sci* 8:758–762.

Poddevin, B., S. Orlowski, J. Belehradek, Jr., and L. M. Mir. 1991. Very high cytotoxicity of bleomycin introduced into the cytosol of cells in culture. *Biochem Pharmacol* 42 Suppl:S67–S75.

Prasmickaite, L., A. Hogset, P. K. Selbo et al. 2002. Photochemical disruption of endocytic vesicles before delivery of drugs: A new strategy for cancer therapy. *Br J Cancer* 86:652–657.

Pron, G., N. Mahrour, S. Orlowski et al. 1999. Internalisation of the bleomycin molecules responsible for bleomycin toxicity: A receptor-mediated endocytosis mechanism. *Biochem Pharmacol* 57:45–56.

Repnik, U., V. Stoka, V. Turk, and B. Turk. 2012. Lysosomes and lysosomal cathepsins in cell death. *Biochim Biophys Acta* 1824:22–33.

Robey, R. W., K. Steadman, O. Polgar, and S. E. Bates. 2005. ABCG2-mediated transport of photosensitizers: Potential impact on photodynamic therapy. *Cancer Biol Ther* 4:187–194.

Rodal, G. H., S. K. Rodal, J. Moan, and K. Berg. 1998. Liposome-bound Zn (II)-phthalocyanine. Mechanisms for cellular uptake and photosensitization. *J Photochem Photobiol B* 45:150–159.

Selbo, P. K., A. Weyergang, M. S. Eng et al. 2012. Strongly amphiphilic photosensitizers are not substrates of the cancer stem cell marker ABCG2 and provides specific and efficient light-triggered drug delivery of an EGFR-targeted cytotoxic drug. *J Control Release* 159:197–203.

Selbo, P. K., A. Weyergang, A. Hogset et al. 2010. Photochemical internalization provides time- and space-controlled endolysosomal escape of therapeutic molecules. *J Control Release* 148:2–12.

Strojan, P., A. Anicin, B. Svetic, L. Smid, and J. Kos. 2011. Proteolytic profile of cysteine proteases and inhibitors determines tumor cell phenotype in squamous cell carcinoma of the head and neck. *Int J Biol Markers* 26:247–254.

Turk, B. and V. Turk. 2009. Lysosomes as "suicide bags" in cell death: Myth or reality? *J Biol Chem* 284: 21783–21787.

Varkouhi, A. K., R. M. Schiffelers, M. J. van Steenbergen et al. 2010. Photochemical internalization (PCI)-mediated enhancement of gene silencing efficiency of polymethacrylates and N,N,N-trimethylated chitosan (TMC) based siRNA polyplexes. *J Control Release* 148:e98–e99.

Weyergang, A., O. Kaalhus, and K. Berg. 2008. Photodynamic targeting of EGFR does not predict the treatment outcome in combination with the EGFR tyrosine kinase inhibitor Tyrphostin AG1478. *Photochem Photobiol Sci* 7: 1032–1040.

Weyergang, A., P. K. Selbo, and K. Berg. 2007. Y1068 phosphorylation is the most sensitive target of disulfonated tetraphenylporphyrin-based photodynamic therapy on epidermal growth factor receptor. *Biochem Pharmacol* 74: 226–235.

Wilson, P. D., R. A. Firestone, and J. Lenard. 1987. The role of lysosomal enzymes in killing of mammalian cells by the lysosomotropic detergent N-dodecylimidazole. *J Cell Biol* 104:1223–1229.

Yang, Z. R., H. F. Wang, J. Zhao et al. 2007. Recent developments in the use of adenoviruses and immunotoxins in cancer gene therapy. *Cancer Gene Ther* 14:599–615.

Yip, W. L., A. Weyergang, K. Berg, H. H. Tonnesen, and P. K. Selbo. 2007. Targeted delivery and enhanced cytotoxicity of cetuximab-saporin by photochemical internalization in EGFR-positive cancer cells. *Mol Pharm* 4:241–251.

The Story of Tookad: From Bench to Bedside

Avigdor Scherz
The Weizmann Institute of Science

Yoram Salomon
The Weizmann Institute of Science

38.1 Introduction

Tookad (TOOKAD®, WST09) and Tookad Soluble (TS), two derivatives of the photosynthetic pigment bacteriochlorophyll (Bchl-D), represent a new photosensitizer generation and a new paradigm in photodynamic therapy (PDT) of cancer. The novel approach known as vascular targeted PDT with Tookad or TS vascular targeted photodynamic therapy (TS VTP) has attained advanced clinical trials in the treatment of localized prostate cancer (Arumainayagam et al. 2011; Azzouzi et al. 2011; Moore et al. 2009, 2011). In essence, following intravenous (IV) administration, the new Bchl-Ds spontaneously form complexes with plasma proteins and stay in the circulation until cleared from the body with minimal or no extravasation. Near-infrared (NIR) illumination of these circulating Bchl-Ds generates reactive oxygen species (ROS) in the form of superoxide, hydrogen peroxide, and hydroxyl radicals. Importantly, these differ from singlet oxygen generated solely by most currently used PDT photosensitizers. In addition to acute ROS generation, secondary reactive nitrogen species (RNS) are formed. The cogeneration of these radicals in the vascular lumen leads to a plantlike hypersensitive response also seen in animals during ischemia reperfusion injury, resulting in the instantaneous tumor vascular occlusion with permanent dysfunction followed by rapid necrosis/apoptosis of the tumor tissue. Tumor ablation through radical-based processes known to trigger localized organ collapse in plants and animals in response to various pathological threats suggests a novel paradigm for cancer therapy. The progression of this treatment approach from the bench to the bedside has involved the crossing of traditional scientific disciplines and extensive interdisciplinary collaboration between research groups in academia, medical practice, and the pharmaceutical industry. This chapter describes milestones on the road to reaching that goal.

38.2 From Porphyrins to (Bacterio)Chlorins

The term PDT describes a process whereby molecules delivered into diseased tissue mediate a therapeutic process through photogeneration of ROS. Most of the applied photosensitizers do so by transferring energy (Type II) from their triplet excited state into a colliding oxygen molecule. However, in some cases, the colliding oxygen undergoes reduction through electron transfer from the photoexcited sensitizer (Type I) (Foote 1991). Modern PDT started in the first quarter of the 20th century using nonporphyrin dyes and visible light for ablation of skin carcinomas and superficial infections, for example (Moan and Peng 2003; von Tappeiner and Jodlbauer 1904). Half a century later, hematoporphyrins and derivatives thereof entered PDT and became the most commonly used sensitizers in the field (Diamond et al. 1972; Dougherty et al. 1975; Lipson and Baldes 1960; Schwartz et al. 1955). However, at a fairly early stage of development, serious drawbacks undermined the translation of preclinical results into the clinical mainstay practice. Of the major reservations to this choice of drug were (1) the spectral overlap between the applied porphyrin-based sensitizers and bloodborne porphyrins that limited treatment depth to a few millimeters (Moan and Peng 2003), (2) their very slow clearance rate that subjected patients to severe and prolonged skin phototoxicity (Dougherty et al. 1975), and (3) the differential accumulation of the therapeutic porphyrins in the normal and diseased tissues, which was insufficient to assure selectivity beyond the one restricted by the targeted illumination (Vrouenraets et al. 2003).

Reduction of porphyrins at two or four β positions of the tetrapyrrol macrocycle (Figure 38.1) converts them into chlorins and bacteriochlorins, respectively (Scheer 1991, 2003).

The reduction breaks the compound's symmetry, shifts its maximum absorption to the red/NIR domains, and increases the respective cross section for their optical absorption by almost two orders of magnitudes (Hanson 1991).

These improved properties led several research groups to direct efforts toward the synthesis and utilization of reduced porphyrins in the frame of Type II PDT. Two representatives of these efforts are Foscan, a chlorin derivative that entered advanced clinical trials but presented prolonged skin phototoxicity and collateral damages (Senge and Brandt 2011), and verteporfin, which has been approved for treatment of age-related macular degeneration (AMD) (Celli et al. 2010; Cruess et al. 2009; Verma et al. 2007).

In parallel to the introduction of sensitizers activated at longer wavelengths, significant research efforts have been devoted toward reduction of the undesirable skin phototoxicity as well as increasing their specific accumulation in the target tissue, the latter not only in malignant cells but even in particular subcellular organelles. These attempts were examined under the umbrella of achieving better Type II PDT. Consequently, a flow of new agents were developed, many of which are composite nanomaterials including combinations of chemo, contrast, and photodynamic agents bound to the delivery platforms that aimed at homing specific tumor compartments (Chen et al. 2005; Verma et al. 2007).

38.3 From Photosynthesis to Photodynamic Therapy

Confronting cancer in his family in the late 1980s/early 1990s drove Avigdor Scherz to explore alternative approaches to the commonly used chemotherapy. Photodynamic therapy appeared attractive, but the selection of porphyrins as major sensitizers and the efforts to turn them into chlorinlike molecules seemed quite ironic. In both prokaryotes and eukaryotes, porphyrins function as oxygen carriers and catalytic cofactors of reactions that involve ground state ligation of molecular oxygen. Cytochromes and hemoglobin are two representative examples (Voet and Voet 2011). Fortunately, the porphyrin cofactors in the biological redox centers have been optimized through evolution to avoid light-induced energy or electron transfer to molecular oxygen by having a relatively low extinction coefficient at the ultraviolet–visible spectroscopy (UV–VIS) and high oxidation potential (Hanson 1991; Katz et al. 1978). In contrast, providing the driving force for photosynthesis, chlorophylls (Chls) and Bchls have been optimized for energy transfer, photooxidation, and photoreduction by harvesting solar energy at 680–1000 nm (~1.5–1.0 eV) illumination (Watanabe and Kobayashi 1991). In this wavelength range, more specifically at 750–800 nm, animal tissue presents maximal transparency. Therefore, it could be anticipated that Chl- and Bchl-based

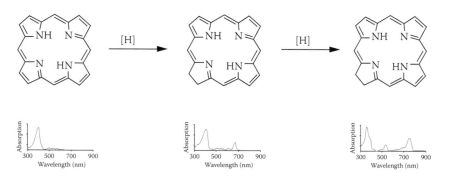

FIGURE 38.1 The conversion of porphyrin (left) to chlorin (middle) and bacteriochlorin (right) involves stepwise reduction of the molecular periphery at two b positions each time. The conversion breaks the molecule's symmetry as reflected by a redshift and intensification of the longest wavelength transition (Q band).

sensitizers would be superior to porphyrin-based sensitizers in photodynamic therapy and tumor imaging. In fact, generation of singlet oxygen by energy transfer from excited (B)Chl is a well-known deleterious phenomenon that results in the irreversible impairment of Type II photosynthetic reaction centers (Vass and Cser 2009).

Based on the experience of Scherz's lab in photosynthetic electron and energy transfer (Braun and Scherz 1991; Kolbasov and Scherz 2000; Shlyk-Kerner et al. 2006), we could appreciate that among the different abundant (B)Chls that exist in nature, those used by nonoxygenic bacteria—namely Bchla and b—may fit better for energy and electron transfer to molecular oxygen. This is due to their longer wavelengths and higher extinction for photoactivation along with their lower oxidation potential (Watanabe and Kobayashi 1991). The possibility of electron transfer from excited Bchl to molecular oxygen seemed to be an attractive property of these molecules in the context of PDT, potentially augmenting their cellular phototoxicity. Indeed, photophysical and photochemical studies conducted by our labs at that early stage and later on demonstrated high quantum yield for both singlet oxygen and oxygen radical generation that was strongly dependent on the sensitizer environment (Ashur et al. 2009; Tregub et al. 1992; Vakrat-Haglili et al. 2005). Following these observations, we started to look at cancer-related targets that could test the expected virtues of Bchl-D.

At that time, the laboratory of Yoram Salomon (the older brother of Ilan, a classmate of Avigdor Scherz in high school) studied melanoma, a lethal and incurable type of cancer. The lab focused on the machinery of hormone-regulated signal transduction aiming at harnessing the melanocyte-stimulating hormone (MSH) receptor expressed on these tumor cells as a target for tumor therapy, for example (Salomon et al. 1993). During an unintentional meeting at the Weizmann campus, we exchanged the latest information from both labs and the possibility of complementary collaboration. As a result, a therapeutic strategy to test targeted PDT action on melanoma cells was devised. In this context, the strategy aimed to examine the photocytotoxic potential of Chls and Bchl and the selective targeting ability of the 13-amino-acid alpha-MSH peptide (Tregub et al. 1992). While being fairly opaque to light in the UV-VIS, we hypothesized that *in situ* melanin would allow for significant penetration of NIR radiation. We therefore opted for the melanotic melanoma tumor to be a good challenge for NIR/Bchl-D PDT. The MSH peptide was conjugated with native Bchla, and *in vitro* studies with cultured M2R melanoma cells (Gerst et al. 1986) showed very promising results with LD_{50} (lethal dose, 50%) values in the nanomolar range (Fiedor et al. 1993). The very large amounts of MSH anticipated for *in vivo* studies in the mouse tumor model led us first to test the native sensitizer in a rather standard PDT setting, namely a 24 h drug/light time interval (DLI) and different light intensities at 740–780 nm using a white light source and filters. Despite the encouraging results in cell cultures, these first animal PDT studies failed to show any effect on the animal tumors. The first clue to account for this discrepancy was found in the pharmacokinetics of the

used Bchl-D, which indicated rapid clearance from the animal circulation within less than 1 h and minimal extravagation to surrounding tissues including the tumor (Rosenbach-Belkin et al. 1996). Once the DLI was shortened to the range of minutes, relatively large tumors of different cancer cell lines were successfully ablated despite their high opacity (Gross et al. 2003; Kelleher et al. 1999, 2003, 2004). These experiments provided proof of concept that Bchl-Ds are efficient PDT photosensitizers with best results attained at a DLI of 0–5 min. Moreover, the rapid clearance of sensitizers from the body dramatically reduced and practically eliminated skin phototoxicity (no skin toxicity 2–3 h posttreatment) (Brandis et al. 2005; Mazor et al. 2005; Rosenbach-Belkin et al. 1996). Furthermore, excitation in the NIR enabled light penetration deep into animal tissues and successful eradication of large pigmented melanoma tumors (in the gram range) (Gross et al. 2003; Mazor et al. 2005; Zilberstein et al. 1997). The use of a novel oxygen microsensor for continuous and direct *in situ* real-time measurements of the oxygen consumption during the PDT protocol pointed at instantaneous hypoxia as light was turned on (Zilberstein et al. 1997).

38.4 *In Vivo* Online Follow-Up of VTP by Blood Oxygen Level-Dependent Magnetic Resonance Imaging

To further understand the physicochemical basis of VTP, we were interested in noninvasively monitoring the photochemically driven hemodynamic process that leads to blood stasis. We hypothesized that photogeneration of ROS during VTP is tightly coupled to oxygen consumption and local elevation of circulating deoxyhemoglobin (DexHeb). The latter is a natural contrast agent that enables blood oxygen level-dependent magnetic resonance imaging (BOLD-MRI). DexHeb is the central player that permits follow-up of neuronal-stimulated hemodynamic changes in the course of functional imaging of brain activity. Applying BOLD-MRI to VTP required the manipulation of the test animal (anesthesia, Tookad injection and illumination) to be conducted within the magnet bore (Gross et al. 2003; Tempel-Brami et al. 2007). The results obtained showed that (1) light-dependent BOLD contrast MRI can indeed be generated upon local illumination of circulating Tookad in a treated animal, (2) BOLD contrast is proportional to DexHeb concentrations, (3) changes in BOLD contrast reflect on both photochemical and hemodynamic effects, and (4) laser beam pulses synchronously trigger BOLD contrast transients in the target tissue, allowing representation of the luminous spatiotemporal profile as a contrast map on the magnetic resonance monitor. The new concept of photosensitized BOLD-MRI can become a useful tool for intraoperative online monitoring of clinical VTP and other clinical PDT applications.

It took a few more years to decipher the key role played by this hypoxia in enhancing tumor immunogenicity (Preise et al. 2011) and collapse (Ashur et al. 2009; Gal et al. 2012; Gross et al. 2003; Madar-Balakirski et al. 2010).

38.5 From Photochemotherapy into Vascular Targeted Tumor Ablation

Photodynamic therapy is frequently described as photochemotherapy for cancer cells in which the photosensitizer selectively accumulates. According to this paradigm, many nontoxic light-activated sensitizers were designed to home at specific tumor cell compartments expected to generate cytotoxicity via singlet oxygen following peroxidation of vital lipids and proteins (Dougherty et al. 1975; Zhu and Finlay 2008). This paradigm has defined steps and strategies of the "good agent" development and selection. First, a compound that presents the photophysical and photochemical profiles needed for singlet oxygen generation at high yield is selected or synthesized. Second, the photocytotoxicicty of the compound is examined in cancer cell culture models. Third, additional chemical manipulations are performed to grant the molecules high efficacy by optimizing low intrinsic toxicity with enhanced avidity and uptake by the target cancer cells. Fourth, the molecule is intravenously administered to the animal tumor model and tested for PDT activity.

Most PDT agents developed along these lines are hydrophobic to enhance cell membrane penetration (Chen et al. 2004; Dougherty et al. 1975; Ethirajan et al. 2011) and require specific carriers such as liposomes, lipoproteins, micelles, or dendrimeric scaffolds for topical or systemic delivery in the above cases (Ben-Dror et al. 2006; Chen et al. 2005; Derycke and de Witte 2004; Donnelly et al. 2008; Dragicevic-Curic and Fahr 2012; Jori and Reddi 1993; Sibani et al. 2008). Under these constrains, the bioavailability of the test compounds is significantly modified compared to *in vitro* conditions, possibly altering the PDT efficacy dramatically. In most cases, the cellular uptake and PDT activity are tested in culture medium supplemented with 10% serum. In reality, the intravenously administered sensitizers confront much higher circulating serum concentration and frequently engage in noncovalent complexing with serum albumin (SA) as well as high- and low-density lipoprotein (HDL and LDL, respectively) that further alters the cellular uptake even in the tumor interstitium (Ashur et al. 2009; Celli et al. 2010; Goldshaid et al. 2010). As a result, many anticancer drugs that were found effective *in vitro* showed poor *in vivo* efficacy. The main obstacle in moving from *in vitro* to *in vivo* milieus is likely the difficulty to enter target cells at sufficiently high concentrations upon complexing with serum proteins or even aggregating in the circulation (Brandis et al. 2005).

Following these and other considerations, two pivotal observations were made during our PDT studies: (1) Under physiological (equivalent to *in vivo*) serum concentrations, Bchl does not accumulate in the target tumor cells (Brandis et al. 2005; Mazor et al. 2005), and (2) illumination immediately after IV administration of Bchl (DLI = 0–5 min) results in the highest treatment efficacy (Zilberstein et al. 2001). Consequently, we hypothesized that Bchl-Ds are photosensitized in the circulation and that PDT

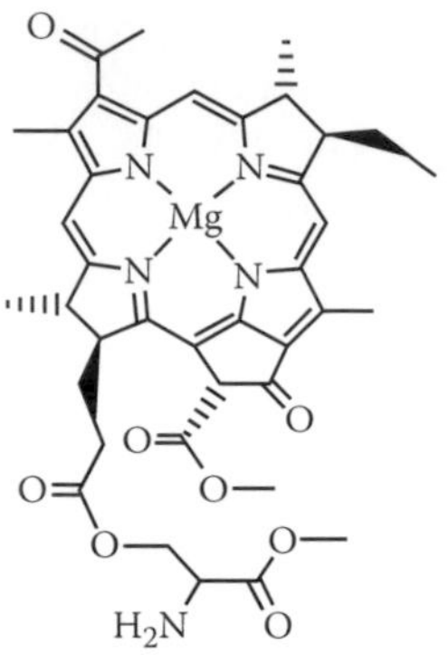

FIGURE 38.2 Catalytic transesterification of the propionyl residue at C-17 or catalytic condensation generates Bchl-Ser.

with these sensitizers aims primarily at the tumor vasculature rather than the cancer cells.

This paradigm shift coincided with the increased recognition of the tumor vasculature as a major therapeutic target in cancer management (Folkman 1971; Jain 2005). An increasing number of studies showed that tumor dormancy and regression occurs upon antiangiogenic treatment (Jain 2005; Kerbel and Folkman 2002; Rak and Kerbel 1996). In fact, during the mid-1980s, hematoporphyrin-based PDT was already shown to strongly impact the tumor vasculature (Berns and Wile 1986; Rosenbach-Belkin et al. 1998). However, at the same time, numerous studies suggested that an early antivascular effect interferes with the continuous supply of oxygen to the treated tissue and therefore may impair the success of PDT. In contrast, the instantaneous hypoxia induced by *in situ* excitation of Bchl-D (Zilberstein et al. 1997) appeared later on to be a major player in the rate of treatment success.

The understanding that Bchl-D PDT primarily targets the tumor vasculature motivated us to enhance the sensitizer's hydrophilicity, first by enzymatic transesterification of the C-17 propionic residue with different charged residues and then by catalytic condensation (Fiedor et al. 1992; Rosenbach-Belkin et al. 1996; Scherz et al. 2000). The resulting Bchl conjugates with serine (Bchl-Ser) presented significantly higher hydrophilicity than the nonconjugated Bchl derivatives even after hydrolysis of the hydrophobic esterified alcohol at C-17 position (Figure 38.2) and therefore fitted the newly suggested paradigm.

The encouraging results in animal model studies (Eichwurzel et al. 1998; Katz et al. 1998; Moser et al. 1998; Rosenbach-Belkin et al. 1998; Scherz et al. 1998; Zilberstein et al. 2001) drove toward the establishment of a collaboration with Steba N.V., currently Steba Biotech, that was licensed to develop the novel clinical PDT approach with the Bchl class of sensitizers.

38.6 From Research to Development and Backward

Signing an agreement with an industrial partner and the subsequent visions of clinical applications provide the ground for euphoric moments. Soon after comes the reality of agents' shelf life, impurities, relevance of animal studies to the clinical arena, and the treatment safety and regulation—aspects belonging to the road not taken by most academic labs. Although the early Bchl-D compounds, all containing the native central Mg^{+2} atom (Figure 38.2), presented relatively high efficacy in PDT of different tumors in animal models (Katz et al. 1998; Kelleher et al. 1999; Rosenbach-Belkin et al. 1996; Tregub et al. 1992), their shelf life appeared too short for pharmaceutical use. Furthermore, these sensitizers all contained small but significant amounts of oxidation products characterized by a new absorption band at approximately 650 nm, typical to the chlorin derivatives that are frequently observed as by-products of the Bchl chemical manipulations (Chen et al. 2004; Pandey et al. 1991). Such products can be formed by illumination of native and synthetically modified Bchl under aerobic conditions (Ashur et al. 2009; Brandis et al. 2005; Vakrat-Haglili et al. 2005). These impurities often impair further development toward clinical trials. Furthermore, the inherited instability of such Bchl-Ds interferes on additional chemical manipulations required in the course of development of future pharmaceutical applications. Following intensive exploration in collaboration with the group of Hugo Scheer (Munich), we substituted the central Mg^{+2} of Bchl-D with Pd^{+2} and thereby significantly enhanced the sensitizer stability (Hartwich et al. 1998; Noy et al. 1998; Teuchner et al. 1997). Moreover, the Pd^{+2} complexes presented a significantly higher rate of intersystem crossing (ISC) to their excited triplet state, resulting in a higher yield of ROS (Teuchner et al. 1997). The transmetalation procedure primarily used to incorporate Pd^{+2} was soon replaced by direct Pd^{+2} incorporation that avoided the use of toxic cadmium (Scherz et al. 2003). At that stage, the road to clinical applications appeared secured. The resulting Ser-Pd-Bchl, a first version of Tookad, was fairly soluble in aqueous solutions and presented good efficacy in animal studies (Kelleher et al. 2003; Zilberstein et al. 2001), which motivated the initiation of larger-scale preclinical safety trials.

Unfortunately, prolonged tests pointed at insufficient stability in the biological milieu and sent us back to the bench, where we finally decided to go for the simplest version of Pd-substituted Bchl, namely, to synthesize and use Pd-bacteriopheophorbide (Figure 38.3), later termed Tookad, which means "to burn" in Hebrew [with multiple citations in the bible: Leviticus 6:2,5,6 (http://www.mechon-mamre.org/p/pt/pt0306.htm) and Jeremiah 15:14, 17:4].

Following the proven efficacy in animal studies, Tookad (WST09) was elected to be the first lead compound for Bchl-based clinical VTP. Tookad is hydrophobic (although significantly less so than native Bchla) and needs to be in a micellar or liposomal formulation for IV administration. Tookad was found to be highly effective in animal studies (Borle et al. 2003a,b; Schreiber et al. 2002) and rapidly cleared from the animal body (Brun et al. 2004). Following the experience by others with hydrophobic drugs like paclitaxel (Taxol) (Fu et al. 2009; Liebmann et al. 1993), it was decided to use Cremophor as the Tookad excipient for injection and, based on the proven safety in preclinical studies, to advance Tookad into clinical trials.

Importantly, throughout the aforementioned research and development (R&D) cycles, our labs were kept involved in solving the emerging difficulties and providing the synthetic and biological know-how. This collaboration has enabled us to bridge the gap between the industrial ("D") and academic ("R") philosophies, as described below, that was pivotal for subsequent progress toward the clinical translation.

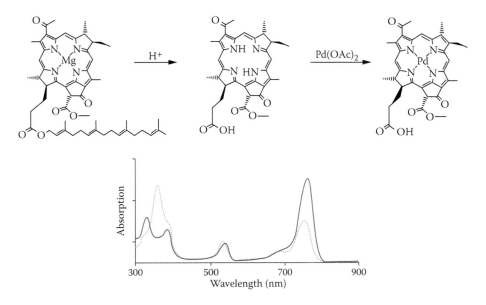

FIGURE 38.3 Acidic hydrolysis of bacteriochlorophyll a (left) followed by heterogeneous interaction with PdAc results in Tookad (right). Notably, the Q band intensity is markedly increased.

38.7 Selecting Localized Prostate Cancer as the First Clinical Indication

The choice of indication is a critical issue in advancing a bench-stage therapy to the clinic. Most frequently, the industrial partner is looking for the "blockbuster" indications. However, the less common orphan diseases (Joppi et al. 2006, 2009) with low-profit or even nonprofitable markets provide a faster clinical track that may open the door toward larger markets in the future. Regulatory affairs such as clear definition of treatment end points to be accomplished only years after initiating the clinical trials contribute toward selecting the first target indication. The academic labs that are focused on scientific and medicinal challenges are not always aware of such considerations during basic preclinical research. Furthermore, the bench research is often centered on household cancer models that may frequently be only marginally relevant to the clinical setting. Lastly, beyond the vision, translation will always require tedious and prolonged adaptation. In the settings of PDT, new sensitizer synthesis frequently involves multiple steps that are difficult to adjust and upscale to the final selected clinical targets. For example, systemic and regional toxicities are only marginally tested in the lab, and modifications required by additional thorough preclinical studies are challenging to foresee and implement.

Both the synthesis in the research laboratory and the industrial upscaling of Tookad turned out to be relatively simple, involving a small number of steps (Brandis et al. 2005; Scherz et al. 2004, 2005, 2008, 2009). The compound's stability allows for further chemical manipulation as might be required for different physiological settings and targeting approaches.

Keeping these (not necessarily purely academic aspects) in the back of our minds, a choice had to be made for the first clinical indication. The finding that Tookad VTP shows high efficacy against different types of solid tumor models (Gross et al. 2003; Kelleher et al. 1999, 2003; Koudinova et al. 2003; Plaks et al. 2004; Preise et al. 2003; Vilensky et al. 2005) agreed with the notion that the tumor vasculature is of comparable abnormal nature and a proper therapeutic target in a variety of solid tumor types. Moreover, the VTP protocol developed in the mouse and rat models were predicted and retrospectively found to be highly relevant to the clinical arena, irrespective of the selected malignancy. While these considerations allowed applications in multiple oncological settings, it was decided to first engage in the development of Tookad VTP for localized prostate cancer (PC) (Chen et al. 2002a,b; Hetzel et al. 2002; Huang et al. 2003, 2005).

The selection of prostate cancer was not trivial. On one hand, it is the second most frequent cancer in men, with approximately 700,000 new cases annually observed in the developed countries (Jemal et al. 2009). In most of these patients, the cancer is found at an early stage, localized in the prostate, and the more significant tumor(s) can be identified by combining different imaging techniques with biopsies. Administration of Tookad VTP in such patients was anticipated to be fairly straightforward.

On the other hand, prostate cancer is a slow-progressing disease until reaching the metastatic stage. As a result, it is hard to demonstrate a significant increased survival rate following application of new relative to traditional treatment approaches—an end point usually required by the regulating authorities in approving such new treatments (Guidance for Industry Clinical Trial Endpoints for the Approval of Cancer Drugs and Biologics. Retrieved 6/15, 2013, from http://www.fda.gov/downloads/Drugs/GuidanceComplianceRegulatoryInformation/Guidances/ucm071590.pdf).

During early 2000, mainstay approaches for early-stage treatments were considered established and effective, and most experimental approaches referred to the late, metastatic stage. Thus, it seemed impossible to enter the clinical arena with the intention to become a first-line therapy. Consequently, it was decided to first try Tookad VTP as a salvage treatment in prostate patients that presented failed radiation therapy. Following this strategic decision, several studies with encouraging results were conducted on canine models (Chen et al. 2002a; Hetzel et al. 2002, 2006; Vilensky et al. 2005). Not only were appropriate drug and light dose combinations effective in eradication of significant volumes of prostate tissues, but the treatment also appeared to be safe for the urethra and nerve bundles (Dole et al. 2005; Huang et al. 2007). The clinical pharmacokinetics of the drug were found to agree with the rodent data (Hetzel et al. 2002), with less than 1 h clearance half-life time suggesting insignificant skin toxicity for treated patients.

New technologies that may overcome the morbidity involved in current PC radical management usually attract urologists, and so was the case for John Trachtenberg [Princess Margaret Hospital (PMH), Toronto]. In the summer of 2002 during a short sabbatical with Brian Wilson at PMH, Avigdor Scherz together with Steba representatives met with John Trachtenberg to discuss the possible application of Tookad VTP to prostate cancer patients. It took a mere 1 h meeting to convince Trachtenberg to do clinical trials with Tookad. Based on the above considerations, Trachtenberg suggested applying Tookad VTP first to patients who presented local recurrence after failure of radiation therapy. These patients usually underwent hormonal treatment but rapidly progressed to the hormone refractory stage followed by tumor dissemination and metastases (Schellhammer et al. 1993). Hence, treatment efficacy in compliance with commonly accepted end points could be rapidly evaluated.

The stage was finally set with Mustafa El Hilali from Victoria Hospital and McGill University in Montreal and several other Canadian groups, enabling us to advance to clinical trials in the context of a multicenter team. The clinical protocols were designed by combining the input from results of our labs and the experience in brachytherapy within the clinical centers. In short, following IV infusion of Tookad to the patient (10–20 min), the selected target region in the prostate was interstitially illuminated (755 nm, 100–200 mW/cm² for 20–30 min). Using a transperineal approach and a standard brachytherapy template, illumination was delivered by ultrasound-guided optic fiber insertion (Trachtenberg et al. 2007). The resulting treatment success is monitored by contrast-enhanced MRI (day 7) and biopsies (6 months). Treatment planning based on NIR light distribution in the prostate tissue was provided by the group of Brian Wilson (Davidson et al. 2009; Weersink et al. 2005a,b). Following animal data and in consideration of the rapid clearance of Tookad, it was decided to apply light at the second half of drug infusion, a protocol that was never applied before in cancer PDT. Notably, AMD with verteporfin also benefits from short DLI (Cruess et al. 2009), but this is to avoid collateral damage upon longer illumination times where the drug extravagates into surrounding tissue. In contrast Tookad, TS, and other Bchl-Ds of this category rapidly clear from the body with no significant extravagation in periphery (Mazor et al. 2005).

The results of these clinical trials corresponded well with the preclinical predictions. When applied under optimal conditions Tookad VTP resulted in tumor ablation that was reflected in negative biopsies for 9 of 16 (>60%) patients treated with optimal Tookad VTP, with no incontinence or enhanced impotency (Trachtenberg et al. 2008). In these patients, the prostate-specific antigen (PSA) value dropped to <0.5 ng/L. Follow-up of the successfully treated patients (out of protocol) showed that 6 years, later eight of the nine successfully treated patients remained tumor-free with practically insignificant PSA values (John Trachtenberg, 2012, personal communication). These results also happened to suggest an attractive treatment alternative to such radiation-failed patients that was not previously available.

The encouraging results provided proof for the efficacy of Tookad VTP and its higher safety potential for the treatment of the entire gland as well as when compared with other local treatment modalities [e.g., cryotherapy and high-intensity focused ultrasound (HIFU)] (Mouraviev and Polascik 2006). On the other hand, it was realized that at the early developmental stage treatment, planning for this group of patients might not be safe enough, mainly because of vascular modification caused by the radiation around the diseased prostate at the rectal wall. Following these considerations and the demonstrated ability to safely eradicate large prostatic volumes with minimal damage to surrounding normal tissue, Steba Biotech decided to move toward treatment of naïve patients that presented localized, early-stage prostate cancer. This strategic decision coincided with major ongoing changes in the treatment paradigms of prostate cancer.

38.8 Tookad VTP and the Emerging Concept of Focal Therapy of Early-Stage Prostate Cancer

An increasing number of urology advocates have recently claimed that the management of prostate cancer should undergo a paradigm shift. Several large-scale studies published during the last few years pointed at the small differences in mortality between PC patients that were treated or untreated by radical approaches at the early stages of the disease (Wilt et al. 2012). These studies cast doubt on the need to treat such patients with currently approved radical methods, let alone when considering

the associated side effects and reduced quality of life. However, closer inspection of these and other recent studies points at the higher rate of disease progression in PC patients that have not been treated at the early stage. Thus, patients presented with early-stage PC have started to face the dilemma of being overtreated by the morbid mainstay approaches or undergoing active surveillance with the risk of significant disease progression.

Several less morbid treatment approaches aiming at minimally invasive whole-gland ablation have already been experimentally available in the clinic. These include HIFU, cryotherapy, and brachytherapy (Mouraviev and Polascik 2006). Unfortunately, none of these treatments could provide efficient ablation of the entire prostate while successfully sparing impotence and incontinence at relatively high proportions (Mouraviev and Polascik 2006). Tookad VTP appeared regionally safer than HIFU and cryotherapy, and initial phase II clinical trials conducted in Toronto (Trachtenberg et al. 2008), London [Mark Emberton group (Nathan et al. 2002)], and France (Azzouzi et al. 2013) showed high tissue volume ablation in patients who selected treatment over active surveillance (Moore et al. 2011). Meanwhile, a growing body of evidence suggested that although localized PC is often multifocal, comprising several tumor foci, only one, termed the index tumor, seems to drive cancer proliferation and dissemination (Ahmed and Emberton 2010; Lindner et al. 2010; Nguyen and Jones 2011; Nomura and Mimata 2012; Scardino 2009). This conclusion resulted in the launching of the focal therapy approach for clinical PC therapy. Focal prostate therapy aims at safe ablation/eradication of the index tumor in early-stage patients while sparing surrounding organs and functionalities. Application of HIFU or cryotherapy to the suspected lesion and even to an entire lobe of the prostate appears to limit collateral damage to one nerve bundle and results in less morbidity to the urethral function (Moore et al. 2011). Still, treatment of peripheral tumors is not expected to be safe because of the nonselective (thermal) damage induced by the two available methods, and treatment time increases with the intended treatment volume. In contrast, the aforementioned results with Tookad VTP suggested that this new approach may overcome the challenges presented by other focal treatment modalities, thus providing the potential of being the method of choice for minimally invasive treatment of localized prostate cancer. Yet, while good appreciable efficacy and regional safety were observed with Tookad VTP, a nonnegligible number of treated patients presented transient hypotension following drug injection, and a few presented transient cardiac side effects.

38.9 The Shift from Tookad VTP to TS (WST11) VTP

At this critical point of development, the collaboration established by the physicians, the company pharmaceutical experts, and our research teams proved critical. The explanations proposed by the medical teams as well as new formulations suggested by the company's developmental teams were rapidly studied in

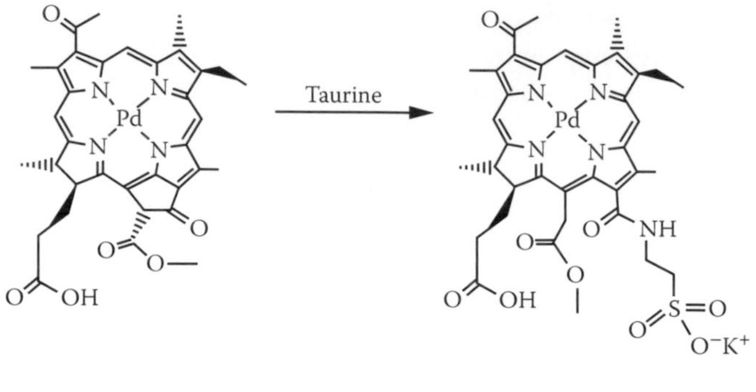

FIGURE 38.4 Aminolysis by taurine opens the isocyclic ring and generates the hydrophilic Tookad Soluble (WST11).

our labs. Based on these experiments along with literature data, we all agreed that Tookad formulation with cremophore-18 significantly contributed to the observed side effects. Correlating our experimental findings in animals with the medical data of all treated patients, we concluded that modifying Tookad at the bench side into a more hydrophilic compound is probably the optimal path for the long run.

It took us about a year to demonstrate that WST11 (Figure 38.4), now called Tookad-Soluble, provides a good solution for the above-mentioned problems.

This compound was first synthesized a few years earlier following our hypothesis that increased hydrophilicity should improve the application of VTP (Brandis et al. 2005; Brandis et al. 2006; Mazor et al. 2005; Plaks et al. 2004). Based on its lower efficacy tested in cultured endothelial cells, WST11 was first thought to be less suitable for clinical cancer therapy and was further developed, under the name Stakel, for treatment AMD (Berdugo et al. 2008). Interestingly, new experiments conducted in our labs showed that when properly applied, TS VTP is highly effective for cancer ablation (Fleshker et al. 2008; Mazor et al. 2005; Preise et al. 2003, 2009; Tempel-Brami et al. 2007; Vilensky et al. 2005). The noncovalent complexing of WST11 with SA in all animal models studied resulted in complete solubility in the blood, catalytic photogeneration of oxygen radicals (Ashur et al. 2009), and rapid clearance from the circulation with none of the adverse effect observed with Tookad (Steba Biotech, unpublished data; Chevalier et al. 2011; Fabre et al. 2007). Our lab's preclinical data obtained in rodents were further approved in larger animal (porcine) studies, assuring the absence of any myocardial- or circulation-related adverse effects under clinical conditions. At that point, Steba Biotech management and consulting teams decided to continue the clinical development with TS. This decision enforced a bridging phase I/II clinical trial that was not simple and reflected to a large extent increased faith and collaboration between the bench, pharmaceutical, and clinical teams.

38.10 TS, a Pure Type I Sensitizer in Aqueous Solutions

Previous studies conducted with micellar Tookad in organic and aqueous solutions clearly showed a transition from pure

Type II to a mixture of Type I and II mechanisms. However, this observation could be attributed to the mixed environment of the Bchls in these conditions (Vakrat-Haglili et al. 2005). The water solubility of TS enabled us for the first time to explore the photochemistry and physics of (bactrio)chlorophylls in aqueous solutions without the involvement of lipophilic encapsulation. Continuing our extensive collaboration with Tadeusz Sarna's lab in Krakow and using a host of experimental approaches, we showed generation of superoxide and hydroxyl radicals with no significant traces of singlet oxygen, an experiment conducted when WST11 in aqueous solutions was illuminated at its peak by a 755 nm laser (Ashur et al. 2009). The presence of SA stabilized the compound and significantly increased the yield of oxygen radical generation. Concomitantly, it showed rapid photoconsumption of the available oxygen. The overall reaction appeared to be a catalytic photogeneration of oxygen radicals and catalytic consumption of molecular oxygen (Ashur et al. 2009). These findings were linked to the *in vivo* observations of instantaneous vasodilation followed by immediate vasoconstriction of the blood vessels in tumors exposed to Tookad and TS VTP (Madar-Balakirski 2011).

Tumor eradication by Tookad and TS VTP has been found to be very efficient in animal models, although, as mentioned above, it does not involve direct intoxication of the tumor cells by the formed radicals (Brandis et al. 2005). These observations are in line with the finding that the extensive oxygen consumption during illumination does not interfere with the treatment efficacy. Our recently published data show that the tumor vascular dysfunction starts with the irreversible occlusion of the feeding arteries and draining veins (Madar-Balakirski et al. 2010), representing a completely new tool in antivascular cancer therapy (Nagy and Dvorak 2012). Indeed, the antivascular effect of Tookad and TS VTP does not seem to involve the classical fibrin thrombi formation, as found for verteporfin-mediated VTP, nor does it appear to primarily target the microcapillaries. The occlusion and dysfunction of the tumor's feeding arteries and draining veins by TS VTP (Madar-Balakirski et al. 2010) rapidly translates into tumor ablation. Recent studies conducted in collaboration with J. Coleman, S. Kimm and others (Coleman and Scherz 2012) at Memorial Sloan Kettering Cancer (New York) indicate that, as found in rodents, cell death propagates from the illuminated vessels to the tumor tissues, resulting in occlusive necrosis and an apoptotic front near the vessels. While at moderate fluency (<200 mW/cm²), tissue necrosis was purely conveyed by the photodynamic effect, at 500 mW/cm², a significant thermal effect could be observed, opening new avenues for the use of light-activated TS (Tarin et al. 2012).

One of the more striking phenomena related to TS is the aforementioned catalytic generation of oxygen radicals when noncovalently complexed to SA (Ashur et al. 2009). Since following IV administration, this is the circulating form of the drug, it is anticipated that a similar catalytic generation takes place *in vivo*. This catalytic generation accounts for the rapidly conveyed hypoxia in and around the illuminated space as already observed in early studies of the Bchl-Ds (Zilberstein et al. 1997).

In contrast to singlet oxygen-dependent PDT, this hypoxia appears to play a key role in the mechanism of action of TS. Apparently, as physiologically expected, a rapid local release of nitric oxide that augments the toxic effect of the oxygen radicals takes place with the de novo production of deleterious peroxynitrite (Gal et al., submitted; Madar-Balakirski 2011). The positive impact of RNS in TS VTP is another reflection of the novel treatment paradigm offered by this new drug.

38.11 Systemic Impact of TS VTP

Using CT-26, a heterotropic model of a mouse colorectal carcinoma, and 4T1 mouse mammary carcinoma as models, we recently showed that tumors in immune-competent mice respond significantly better to TS VTP than tumors in immunocompromised animals (Preise et al. 2009, 2011). We further demonstrated that the TS VTP treatment preliminary annihilated resident T cells in the tumor (CD3+) followed by a fairly massive recruitment of new T cells that peak at approximately 20 h posttreatment and slowly diminish.

TS VTP is followed by a rapid neutrophil recruitment to the tumor rim that peaks at 1 h post-VTP and lasts for less than 4 h, for example, in CT-26 colorectal carcinoma models (Preise et al. 2009, 2011). Although high neutrophil density was observed at the tumor rim, there was significant infiltration to the tumor tissue. Recruitment of F4/80 macrophages followed, with a peak at approximately 20 h posttreatment (Preise et al. 2009).

Cumulatively, the onset of TS VTP appears to intoxicate primarily resident macrophages (F4/80+), possibly granulocytes and resident T cells [hypothetically T-regulatory cells (Treg)], followed by fairly massive recruitment of nonresident immune cells that accomplish primary tumor eradication and were shown by us to initiate prolonged tumor immunity. Prolonged immunity following PDT was reported by others (for a leading review, see Mroz et al. 2011) and was found to be strongly mediated by dendritic cells (Preise et al. 2009). The participating dendritic cells appeared to undergo activation through the PDT effect and the subsequent priming and presentation of the cancer cell antigens. Nonetheless, the impact of TS VTP was sufficient to provoke such response with no addition of dendritic cells as reported by others (Jalili et al. 2004). Moreover, immunity appeared to be cross-reactive among various cells types, suggesting overlapping antigenic epitopes.

38.12 Treatment of Patients Presenting with Low-Risk, Localized Prostate Cancer with TS VTP

The replacement of Tookad by TS appeared to be a game changer in the clinical application of Tookad VTP. Following a short bridging study led by Emberton in the United Kingdom (Nathan et al. 2002) and Azzuzi in France (Azzouzi et al. 2013), extensive multicenter phase II studies involving over 200 patients were conducted in Europe and in the United States

(Coleman and Scherz 2012). Apparently, the increased water solubility circumvented the need for prior solubilization with Cremophor. Consequently, none of the adverse side effects previously reported for Tookad were observed. The first study (PCM201) aimed at defining the optimal treatment parameters to achieve negative biopsies in the treated lobes. Based on the collected data, the group of Mordon (France) was able to define the threshold density of light energy (DI) required to obtain prostatic tissue ablation at 4 mg/kg TS infused for 10 min to the treated patients (Betrouni et al. 2011).

The objectives of the second study (PCM203) were to assess the efficacy, safety, and quality of life of TS-mediated VTP administered at optimal dosing of drug and light and maintaining the above energy density by adjusting the overall fiber length to the treated volume of the prostate. This study was a multicenter, phase II, 6-month open-label clinical trial in prostate cancer patients eligible for active surveillance. A total of 85 patients were treated between September 2009 and July 2010. TS at doses of 4 and 6 mg/kg were used for small and larger prostates, respectively. Depending on the tumor location, either one lobe or both lobes were treated. The criteria for a successful score were negative biopsy in the treated lobe at 6 months (primary end point); treatment effect after 1 week monitored by Gd-DTPA (gadolinium-diethylenetriaminepentaacetic acid, a gadolinium-based MRI contrast agent) contrast-enhanced MRI; and serum PSA level, International Prostate Symptom Score (IPSS), and International Index of Erectile Function (IIEF) scored at 1, 3, and 6 months (secondary end points). Patients who exhibited a positive biopsy at month 6 were offered retreatment with TS VTP. Several combinations of targets, doses, and light energies were evaluated. Six related adverse events were reported: prostatitis (2), hematuria (1), orchiepidymitis (1), optic neuropathy (1), and one patient had a prostatic urethral stricture treated by transurethral resection of the prostate (TURP). No rectourethral fistula and no incontinence were observed. The mean percentage of ablated tissue in the treated lobes (as defined by the absence of gadolinium uptake at day 7 MRI) was 77% for all the patients and 87% in the group who received a unilateral treatment with 4 mg/kg. In conclusion, TS VTP was found to be tolerated and reproducible in agreement with optimal conditions found by the PCM201 study (Azzouzi et al. 2011). The 6 months' biopsy showed >80% tumor control in patients treated with optimal conditions (to be published soon) and provided the ground for the ongoing phase III clinical trial.

TS VTP has recently entered a multicenter phase III clinical study in >40 European centers for patients presenting with low-risk prostate cancer. The study comprises two arms, each including approximately 200 patients, where the control arm consists of men elected to undergo active surveillance (Clinicaltrials.gov 2011) ("Clinical Trial: Efficacy and Safety Study of TOOKAD Soluble for Localized Prostate Cancer Compared to Active Surveillance." Principal Investigator: Mark Emberton, Professor University College of London Hospital, United Kingdom. ClinicalTrials.gov Identifier, NCT01310894).

38.13 Moving TS and New Bchl-Ds beyond VTP

While TS VTP appears to spare the collagen scaffold of the treated tissues, direct application of TS to connective tissues followed by NIR illumination is expected to initiate collagen cross-linking by the generated oxygen radicals. This assumption led us to develop a new protocol for treatment of keratoconus and degenerative myopia—two important ophthalmic diseases. Keratoconus is an eye disease where the cornea loses its stiff caplike shape due to various processes that weaken the collagenous scaffold. The cap bulges into a conelike shape, causing image distortion. The vision of the diseased individual is worsened with time to a degree that it cannot be corrected by regular glasses and may eventually require corneal transplantation from a healthy donor. Keratoconus is presently treated by topical application of riboflavin followed by illumination of the cornea with UV light. Although UV–riboflavin treatment prevents disease progression, the use of deleterious UV illumination is hazardous, underscoring the need for alternative treatments. Degenerative myopia (severe shortsightedness) is an excessive axial enlargement of the eye causing stretching and thinning of the chorioretinal tissues that result in retinal cell atrophy, bleeding, and eventually irreversible visual loss. Degenerative myopia is a leading cause of untreatable blindness in China, Taiwan, and Japan and is ranked as the seventh leading cause of blindness in the United States. Current treatment approaches for all three diseases are noneffective to a significant portion of the population and, no less important, are not safe. Following the proven capacity of TS to generate oxygen radicals at high yield under NIR illumination and preclinical studies that confirmed its safety, we hypothesized that topical application of WST11 with NIR illumination can be safer and possibly more effective compared to UVA–riboflavin. In a recent study, we showed that corneal stiffness significantly increased in WST11/NIR treated corneas *ex vivo* and *in vivo* compared to nontreated contralateral eyes (Marcovich et al. 2012). The incremental ultimate stress of treated corneas increased by more than twofold depending on time length of drug application and illumination. To avoid damage to the eye endothelium, the WST11 was mixed with high-molecular-weight dextran (resulting in a WST-D complex) before application. This treatment restricted the stiffening to the anterior half of the cornea and markedly reduced posttreatment edema and time of epithelial healing. Histology post-WST-D/NIR showed reduction in keratocyte population in the anterior cornea with no damage to the endothelium (Marcovich et al. 2012).

As mentioned above, the relative stability and water solubility of TS markedly help in performing chemical manipulations and opened the way to a variety of new applications in which Bchl-D sensitizers form a general novel platform for medical utilization that goes beyond VTP therapy. Two examples with potentially important outcomes are the RGD conjugates and cationic compounds. The coupling of WST11, its nonmetalated analogue or compounds substituted by manganese (Mn) or copper (Cu), to the integrin activator peptide Arg-Gly-Asp (RGD) provide a new powerful means for homing different cancers that present necrotic domains, their imaging, and possibly their ablation (Goldshaid et al. 2010). Conjugation with cationic residues resulted in new molecules that were found useful in the ablation of selected embryos in pregnant rats used as a model for treatment of extrauterine pregnancy in women (Glinert et al. 2008).

38.14 Application of TS for Modeling Oxidative Stress

ROS comprise a group of noxious by-products of oxidative processes involved in many common diseases. Understanding their role in the regulation of normal physiological redox signaling is of great interest. Detailed study of the dynamic functions of ROS within the biological milieu is difficult because of their high chemical reactivity, minute concentrations, and short lifetime. TS was found to be useful in modeling acute and confined oxidative stress in cell cultures and animal models alike. Oxidative insult expansion from a few injured endothelial cells to distal sites underlies the initiation of major circulation pathologies. This possibly involves mechanisms that are important for understanding circulation physiology and for designing therapeutic management of myocardial pathologies.

Using TS and other Bchl-Ds, we first demonstrated that controlled *in situ* photogeneration of ROS at subtoxic levels can be used as a photoswitch for initiation and manipulation of respective physiological responses in cultured melanoma cells. Among these were ROS-induced responses of specific enzymes [p38, mitogen-activated protein kinase (MAPK), extracellular signal–regulated kinase (ERK), c-Jun N-terminal kinase (JNK), and serine-threonine kinase (Akt)] as well as effects on the subcellular distribution of phosphorylated p38. Furthermore, alterations in cell morphology and motility and effects on cell viability as a function of time and photosensitizer concentrations were observed (Gerst et al. 1986; Plaks et al. 2004; Posen et al. 2005).

Recently, we tested the hypothesis that a localized oxidative insult of endothelial cells propagates through gap junction intercellular communication (GJIC). To that end, we generated a pulse of oxygen radicals in endothelial cell cultures by their incubation with TS, wash, and NIR illumination at a resolution of approximately 6 μm. The propagation of the localized oxidative insult to remote cells was shown to comprise of de novo generation of oxygen and nitroxide radicals triggered by signal that propagates from the illuminated to remote cells through GJIC (Feine et al. 2012).

38.15 Concluding Remarks

The introduction of Bchl-D to the field of photodynamic therapy has opened new and unforeseen avenues for research and applications in different medicinal arrays. At the same time, it instructed us on how to bridge gaps between a variety of research and developmental cultures. The resulting collaborative efforts were essential in promoting the envisioned applications to benefit humans, which we expect to fully realize in the coming years.

Acknowledgments

A.S. is the incumbent of the Robert and Yaddele Sklare Professorial Chair in Biochemistry. Y.S. is the incumbent of the Tillie and Charles Lubin Professorial Chair in Biochemical Endocrinology. We thank Ilan Samish for critical reading and reviewing of the manuscript.

References

Ahmed, H. U., and M. Emberton. 2010. Benchmarks for success in focal therapy of prostate cancer: cure or control? *World J Urol* 28:577–582.

Arumainayagam, N., C. M. Moore, C. A. Mosse et al. 2011. Tookad soluble (WST-11) second generation vascular targeted photodynamic therapy (VTP) for prostate cancer: Safety and feasibility. *Br J Surg* 98:E10.

Ashur, I., R. Goldschmidt, I. Pinkas et al. 2009. Photocatalytic generation of oxygen radicals by the water-soluble bacteriochlorophyll derivative WST11, noncovalently bound to serum albumin. *J Phys Chem A* 113:8027–8037.

Azzouzi, A.-R., E. Barret, C. M. Moore et al. 2013. TOOKAD Soluble Vascular Targeted Photodynamic (VTP) therapy: determination of optimal treatment conditions and assessment of effects in patients with localised prostate cancer. *BJU Int* Accepted Manuscript online: 2013 Jun 5, DOI: 10.1111/bju.12265.

Azzouzi, A. R., E. Barret, A. Villers et al. 2011. Results of Tookad soluble vascular targeted photodynamic therapy (VTP) for low risk localized prostate cancer (PCM201/PCM203). *Eur Urol Suppl* 10:54.

Ben-Dror, S., I. Bronshtein, A. Wiehe et al. 2006. On the correlation between hydrophobicity, liposome binding and cellular uptake of porphyrin sensitizers. *Photochem Photobiol* 82:695–701.

Berdugo, M., R. A. Bejjani, F. Valamanesh et al. 2008. Evaluation of the new photosensitizer Stakel (WST-11) for photodynamic choroidal vessel occlusion in rabbit and rat eyes. *Invest Ophthalmol Vis Sci* 49:1633–1644.

Berns, M. W. and A. G. Wile. 1986. Hematoporphyrin phototherapy of cancer. *Radiother Oncol* 7:233–240.

Betrouni, N., R. Lopes, P. Puech, P. Colin, and S. Mordon. 2011. A model to estimate the outcome of prostate cancer photodynamic therapy with TOOKAD Soluble WST11. *Phys Med Biol* 56:4771–4783.

Borle, F., A. Radu, C. Fontolliet et al. 2003a. Selectivity of the photosensitiser Tookad for photodynamic therapy evaluated in the Syrian golden hamster cheek pouch tumour model. *Br J Cancer* 89:2320–2326.

Borle, F., A. Radu, P. Monnier, H. van den Bergh, and G. Wagnieres. 2003b. Evaluation of the photosensitizer Tookad for photodynamic therapy on the Syrian golden hamster cheek pouch model: Light dose, drug dose and drug-light interval effects. *Photochem Photobiol* 78:377–383.

Brandis, A., O. Mazor, E. Neumark et al. 2005. Novel water-soluble bacteriochlorophyll derivatives for vascular-targeted photodynamic therapy: Synthesis, solubility, phototoxicity and the effect of serum proteins. *Photochem Photobiol* 81:983–993.

Brandis, A., Y. Salomon, and A. Scherz. 2006. Bacteriochlorophyll sensitizers in photodynamic therapy. In *Chlorophylls and Bacteriochlorophylls: Biochemistry, Biophysics, Functions and Applications*. Advances in Photosynthesis and Respiration. B. Grimm, R. J. Porra, W. Rudiger, and H. Scheer, editors. Springer, Dordrecht. 485–494.

Braun, P., and A. Scherz. 1991. Polypeptides and bacteriochlorophyll organization in the light-harvesting complex B850 of *Rhodobacter sphaeroides* R-26.1. *Biochemistry* 30:5177–5184.

Brun, P. H., J. L. DeGroot, E. F. Dickson, M. Farahani, and R. H. Pottier. 2004. Determination of the *in vivo* pharmacokinetics of palladium-bacteriopheophorbide (WST09) in EMT6 tumour-bearing Balb/c mice using graphite furnace atomic absorption spectroscopy. *Photochem Photobiol Sci* 3:1006–1010.

Celli, J. P., B. Q. Spring, I. Rizvi et al. 2010. Imaging and photodynamic therapy: Mechanisms, monitoring, and optimization. *Chem Rev* 110:2795–2838.

Chen, B., B. W. Pogue, and T. Hasan. 2005. Liposomal delivery of photosensitising agents. *Expert Opin Drug Deliv* 2:477–487.

Chen, Q., Z. Huang, D. Luck et al. 2002a. Preclinical studies in normal canine prostate of a novel palladium-bacteriopheophorbide (WST09) photosensitizer for photodynamic therapy of prostate cancers. *Photochem Photobiol* 76:438–445.

Chen, Q., Z. Huang, D. Luck et al. 2002b. WST09 (TOOKAD) mediated photodynamic therapy as an alternative modality in treatment of prostate cancer. In *Optical Methods for Tumor Treatment and Detection: Mechanisms and Techniques in Photodynamic Therapy XI*. T. J. Dougherty, editor. SPIE, Bellingham, WA, 29–39.

Chen, Y., G. Li, and R. K. Pandey. 2004. Synthesis of bacteriochlorins and their potential utility in photodynamic therapy (PDT). *Curr Org Chem* 8:1105–1134.

Chevalier, S., M. Anidjar, E. Scarlata et al. 2011. Preclinical study of the novel vascular occluding agent, WST11, for photodynamic therapy of the canine prostate. *J Urol* 186:302–309.

ClinicalTrials.gov, National Library of Medicine (US). Identifier NCT01310894, Efficacy and Safety Study of TOOKAD Soluble for Localised Prostate Cancer Compared to Active Surveillance. (PCM301); 2011 March 1, [cited 2013 June 18];Available from: http://clinicaltrials.gov/show/NCT01310894.

Coleman, J. and A. Scherz. 2012. Focal Therapy of Localised Prostate Cancer by Vascular Targeted Photodynamic Therapy with WST-11 (TOOKAD Soluble). *EurUrol Rev* 72:106–108.

Cruess, A. F., G. Zlateva, A. M. Pleil, and B. Wirostko. 2009. Photodynamic therapy with verteporfin in age-related macular degeneration: A systematic review of efficacy, safety, treatment modifications and pharmacoeconomic properties. *Acta Ophthalmol* 87:118–132.

Davidson, S. R., R. A. Weersink, M. A. Haider et al. 2009. Treatment planning and dose analysis for interstitial photodynamic therapy of prostate cancer. *Phys Med Biol* 54:2293–2313.

Derycke, A. S., and P. A. de Witte. 2004. Liposomes for photodynamic therapy. *Adv Drug Deliv Rev* 56:17–30.

Diamond, I., S. G. Granelli, A. F. McDonagh et al. 1972. Photodynamic therapy of malignant tumours. *Lancet* 2:1175–1177.

Dole, K. C., Q. Chen, F. W. Hetzel et al. 2005. Effects of photodynamic therapy on peripheral nerve: *In situ* compound-action potentials study in a canine model. *Photomed Laser Surg* 23:172–176.

Donnelly, R. F., P. A. McCarron, D. I. Morrow, S. A. Sibani, and A. D. Woolfson. 2008. Photosensitiser delivery for photodynamic therapy. Part 1: Topical carrier platforms. *Expert Opin Drug Deliv* 5:757–766.

Dougherty, T. J., G. B. Grindey, R. Fiel, K. R. Weishaupt, and D. G. Boyle. 1975. Photoradiation therapy. II. Cure of animal tumors with hematoporphyrin and light. *J Natl Cancer Inst* 55:115–121.

Dragicevic-Curic, N., and A. Fahr. 2012. Liposomes in topical photodynamic therapy. *Expert Opin Drug Deliv* 9:1015–1032.

Eichwurzel, I., H. Stiel, K. Teuchner et al. 1998. Photochemistry and photophysics of the PDT-sensitizer bacteriochlorophyll-serine after short pulse excitation. *Photochem Photobiol* 62:77s.

Ethirajan, M., Y. Chen, P. Joshi, and R. K. Pandey. 2011. The role of porphyrin chemistry in tumor imaging and photodynamic therapy. *Chem Soc Rev* 40:340–362.

Fabre, M. A., E. Fuseau, and H. Ficheux. 2007. Selection of dosing regimen with WST11 by Monte Carlo simulations, using PK data collected after single IV administration in healthy subjects and population PK modeling. *J Pharm Sci* 96:3444–3456.

Feine, I., I. Pinkas, Y. Salomon, and A. Scherz. 2012. Local oxidative stress expansion through endothelial cells—A key role for gap junction intercellular communication. *PLoS One* 7:e41633.

Fiedor, L., A. A. Gorman, I. Hamblett et al. 1993. A pulsed laser and pulse radiolysis study of amphiphilic chlorophyll derivatives with PDT activity toward malignant melanoma. *Photochem Photobiol* 58:506–511.

Fiedor, L., V. Rosenbach-Belkin, and A. Scherz. 1992. The stereospecific interaction between chlorophylls and chlorophyllase. Possible implication for chlorophyll biosynthesis and degradation. *J Biol Chem* 267:22043–22047.

Fleshker, S., D. Preise, V. Kalchenko, A. Scherz, and Y. Salomon. 2008. Prompt assessment of WST11-VTP outcome using luciferase transfected tumors enables second treatment and increase in overall therapeutic rate. *Photochem Photobiol* 84:1231–1237.

Folkman, J. 1971. Tumor angiogenesis: Therapeutic implications. *N Engl J Med* 285:1182–1186.

Foote, C. S. 1991. Definition of type I and type II photosensitized oxidation. *Photochem Photobiol* 54:659.

Fu, Y., S. Li, Y. Zu et al. 2009. Medicinal chemistry of paclitaxel and its analogues. *Curr Med Chem* 16:3966–3985.

Gal, Y., N. Madar-Balakirski, R. Goldschmidt et al. 2012. Photogenerated oxido-nitrosative bursts in service of cancer therapy. *Submitted.*

Gerst, J. E., J. Sole, J. P. Mather, and Y. Salomon. 1986. Regulation of adenylate cyclase by beta-melanotropin in the M2R melanoma cell line. *Mol Cell Endocrinol* 46:137–147.

Glinert, I. S., E. Geva, C. Tempel-Brami et al. 2008. Photodynamic ablation of a selected rat embryo: A model for the treatment of extrauterine pregnancy. *Hum Reprod* 23:1491–1498.

Goldshaid, L., E. Rubinstein, A. Brandis et al. 2010. Novel design principles enable specific targeting of imaging and therapeutic agents to necrotic domains in breast tumors. *Breast Cancer Res* 12:R29.

Gross, S., A. Gilead, A. Scherz, M. Neeman, and Y. Salomon. 2003. Monitoring photodynamic therapy of solid tumors online by BOLD-contrast MRI. *Nat Med* 9:1327–1331.

Hanson, L. K. 1991. Molecular orbital theory of monomer pigments. In *Chlorophylls*. H. Scheer, editor. CRC Press, Boca Raton, FL. 993–1014.

Hartwich, G., L. Fiedor, I. Simonin et al. 1998. Metal-substituted bacteriochlorophylls. 1. Preparation and influence of metal and coordination on spectra. *J Am Chem Soc* 120:3675–3683.

Hetzel, F. W., Q. Chen, K. C. Dole et al. 2006. Evaluation of Tookad-mediated photodynamic effect on peripheral nerve in a canine model. In *Optical Methods for Tumor Treatment and Detection: Mechanisms and Techniques in Photodynamic Therapy XV*. D. Kessel, editor. SPIE, Bellingham, WA, 110–115.

Hetzel, F. W., Q. Chen, Z. Huang et al. 2002. Effect of WST09 mediated photodynamic therapy on normal canine prostate. *Int J Cancer*:367–367.

Huang, Z., Q. Chen, P. H. Brun et al. 2003. Studies of a novel photosensitizer palladium-bacteriopheophorbide (Tookad) for the treatment of prostate cancer. In *Optical Methods for Tumor Treatment and Detection: Mechanisms and Techniques in Photodynamic Therapy XII*. D. Kessel, editor. SPIE, Bellingham, WA, 104–114.

Huang, Z., Q. Chen, K. C. Dole et al. 2007. The effect of Tookad-mediated photodynamic ablation of the prostate gland on adjacent tissues—*In vivo* study in a canine model. *Photochem Photobiol Sci* 6:1318–1324.

Huang, Z., Q. Chen, D. Luck et al. 2005. Studies of a vascular-acting photosensitizer, Pd-bacteriopheophorbide (Tookad), in normal canine prostate and spontaneous canine prostate cancer. *Lasers Surg Med* 36:390–397.

Jain, R. K. 2005. Normalization of tumor vasculature: An emerging concept in antiangiogenic therapy. *Science* 307:58–62.

Jalili, A., M. Makowski, T. Switaj et al. 2004. Effective photoimmunotherapy of murine colon carcinoma induced by the combination of photodynamic therapy and dendritic cells. *Clin Cancer Res* 10:4498–4508.

Jemal, A., R. Siegel, E. Ward et al. 2009. Cancer statistics, 2009. *CA Cancer J Clin* 59:225–249.

Joppi, R., V. Bertele, and S. Garattini. 2006. Orphan drug development is progressing too slowly. *Br J Clin Pharmacol* 61:355–360.

Joppi, R., V. Bertele, and S. Garattini. 2009. Orphan drug development is not taking off. *Br J Clin Pharmacol* 67:494–502.

Jori, G., and E. Reddi. 1993. The role of lipoproteins in the delivery of tumour-targeting photosensitizers. *Int J Biochem* 25:1369–1375.

Katz, J. J., J. R. Norris, L. L. Shipman, M. C. Thurnauer and M. R. Wasielewski. 1978. Chlorophyll function in the photosynthetic reaction center. *Annu Rev Biophys Bioeng* 7:393–434.

Katz, S., Y. Vakrat, V. Brumfeld et al. 1998. Bacteriochlorophyll-serine generates only OH radicals under NIR illumination. In *Proceedings of the 7th Biennial Congress, The International Photodynamic Association*. T. Patrice, editor. ISPEN BIOTECH CD-ROM edition.

Kelleher, D. K., O. Thews, J. Rzeznik et al. 1999. Water-filtered infrared-A radiation: A novel technique for localized hyperthermia in combination with bacteriochlorophyll-based photodynamic therapy. *Int J Hyperthermia* 15:467–474.

Kelleher, D. K., O. Thews, A. Scherz, Y. Salomon, and P. Vaupel. 2003. Combined hyperthermia and chlorophyll-based photodynamic therapy: Tumour growth and metabolic microenvironment. *Br J Cancer* 89:2333–2339.

Kelleher, D. K., O. Thews, A. Scherz, Y. Salomon and P. Vaupel. 2004. Perfusion, oxygenation status and growth of experimental tumors upon photodynamic therapy with Pd-bacteriopheophorbide. *Int J Oncol* 24:1505–1511.

Kerbel, R., and J. Folkman. 2002. Clinical translation of angiogenesis inhibitors. *Nat Rev Cancer* 2:727–739.

Kolbasov, D., and A. Scherz. 2000. Asymmetric electron transfer in reaction centers of purple bacteria strongly depends on different electron matrix elements in the active and inactive branches. *J Phys Chem B* 104:1802–1809.

Koudinova, N. V., J. H. Pinthus, A. Brandis et al. 2003. Photodynamic therapy with Pd-Bacteriopheophorbide (TOOKAD): Successful *in vivo* treatment of human prostatic small cell carcinoma xenografts. *Int J Cancer* 104:782–789.

Liebmann, J., J. A. Cook, and J. B. Mitchell. 1993. Cremophor EL, solvent for paclitaxel, and toxicity. *Lancet* 342:1428.

Lindner, U., J. Trachtenberg, and N. Lawrentschuk. 2010. Focal therapy in prostate cancer: Modalities, findings and future considerations. *Nat Rev Urol* 7:562–571.

Lipson, R. L., and E. J. Baldes. 1960. The photodynamic properties of a particular hematoporphyrin derivative. *Arch Dermatol* 82:508–516.

Madar-Balakirski, N. 2011. The role of oxido-nitrosative bursts in tumor eradication induced by vascular targeted photodynamic therapy (VTP) with bacteriochlorophyll-derivatives. PhD thesis, Weizmann Institute of Science.

Madar-Balakirski, N., C. Tempel-Brami, V. Kalchenko et al. 2010. Permanent occlusion of feeding arteries and draining veins in solid mouse tumors by vascular targeted photodynamic therapy (VTP) with Tookad. *PLoS One* 5:e10282.

Marcovich, A. L., A. Brandis, O. Daphna et al. 2012. Stiffening of rabbit corneas by the bacteriochlorophyll derivative WST11 using near infrared light. *Invest Ophthalmol Vis Sci* 53:6378–6388.

Mazor, O., A. Brandis, V. Plaks et al. 2005. WST11, a novel water-soluble bacteriochlorophyll derivative; cellular uptake, pharmacokinetics, biodistribution and vascular-targeted photodynamic activity using melanoma tumors as a model. *Photochem Photobiol* 81:342–351.

Moan, J., and Q. Peng. 2003. An outline of the hundred-year history of PDT. *Anticancer Res* 23:3591–3600.

Moore, C. M., M. Emberton, and S. G. Bown. 2011. Photodynamic therapy for prostate cancer—an emerging approach for organ-confined disease. *Laser Surg Med* 43:768–775.

Moore, C. M., D. Pendse, and M. Emberton. 2009. Photodynamic therapy for prostate cancer—a review of current status and future promise. *Nat Clin Pract Urol* 6:18–30.

Moser, J. G., V. Rosenbach-Belkin, A. Brandis et al. 1998. Bacteriochlorophyllide-serine: Cellular uptake kinetics and transformation. In *Proceedings of the 7th Biennial Congress, The International Photodynamic Association*. T. Patrice, editor. ISPEN BIOTECH CD-ROM edition.

Mouraviev, V., and T. J. Polascik. 2006. Update on cryotherapy for prostate cancer in 2006. *Curr Opin Urol* 16:152–156.

Mroz, P., J. T. Hashmi, Y. Y. Huang, N. Lange, and M. R. Hamblin. 2011. Stimulation of anti-tumor immunity by photodynamic therapy. *Expert Rev Clin Immunol* 7:75–91.

Nagy, J. A., and H. F. Dvorak. 2012. Heterogeneity of the tumor vasculature: The need for new tumor blood vessel type-specific targets. *Clin Exp Metastasis* 29:657–662.

Nathan, T. R., D. E. Whitelaw, S. C. Chang, et al. 2002. Photodynamic therapy for prostate cancer recurrence after radiotherapy: a phase I study. J Urol 1684 Pt 1:1427–1432.

Nguyen, C. T., and J. S. Jones. 2011. Focal therapy in the management of localized prostate cancer. *BJU Int* 107:1362–1368.

Nomura, T., and H. Mimata. 2012. Focal therapy in the management of prostate cancer: An emerging approach for localized prostate cancer. *Adv Urol* 2012:391–437.

Noy, D., L. Fiedor, G. Hartwich, H. Scheer, and A. Scherz. 1998. Metal-substituted bacteriochlorophylls. 2. Changes in redox potentials and electronic transition energies are dominated by intramolecular electrostatic interactions. *J Am Chem Soc* 120:3684–3693.

Pandey, R. K., D. A. Bellnier, K. M. Smith, and T. J. Dougherty. 1991. Chlorin and porphyrin derivatives as potential photosensitizers in photodynamic therapy. *Photochem Photobiol* 53:65–72.

Plaks, V., Y. Posen, O. Mazor et al. 2004. Homologous adaptation to oxidative stress induced by the photosensitized Pd-bacteriochlorophyll derivative (WST11) in cultured endothelial cells. *J Biol Chem* 279:45713–45720.

Posen, Y., V. Kalchenko, R. Seger et al. 2005. Manipulation of redox signaling in mammalian cells enabled by controlled photogeneration of reactive oxygen species. *J Cell Sci* 118:1957–1969.

Preise, D., O. Mazor, N. Koudinova et al. 2003. Bypass of tumor drug resistance by antivascular therapy. *Neoplasia* 5:475–480.

Preise, D., R. Oren, I. Glinert et al. 2009. Systemic antitumor protection by vascular-targeted photodynamic therapy involves cellular and humoral immunity. *Cancer Immunol Immunother* 58:71–84.

Preise, D., A. Scherz, and Y. Salomon. 2011. Antitumor immunity promoted by vascular occluding therapy: Lessons from vascular-targeted photodynamic therapy (VTP). *Photochem Photobiol Sci* 10:681–688.

Rak, J., and R. S. Kerbel. 1996. Treating cancer by inhibiting angiogenesis: New hopes and potential pitfalls. *Cancer Metastasis Rev* 15:231–236.

Rosenbach-Belkin, V., L. Chen, L. Fiedor, Y. Salomon, and A. Scherz. 1998. Chlorophyll and bacteriochlorophyll derivatives as photodynamic agents. In *Photodynamic Tumor Therapy 2nd and 3rd Generation Photosensitizers*. J. G. Moser, editor. Harwood Academic Publishers, Amsterdam, 117–126.

Rosenbach-Belkin, V., L. Chen, L. Fiedor et al. 1996. Serine conjugates of chlorophyll and bacteriochlorophyll: Photocytotoxicity *in vitro* and tissue distribution in mice bearing melanoma tumors. *Photochem Photobiol* 64:174–181.

Salomon, Y., M. Zohar, J. O. Dejordy et al. 1993. Signaling mechanisms controlled by melanocortins in melanoma, lacrimal, and brain astroglial cells. *Ann NY Acad Sci* 680:364–380.

Scardino, P. T. 2009. Focal therapy for prostate cancer. *Nat Rev Urol* 6:175.

Scheer, H. 1991. Structure and occurence of chlorophylls. In *Chlorophylls*. H. Scheer, editor. CRC Press, Boca Raton, FL, 3–30.

Scheer, H. 2003. Chemistry and spectroscopy of chlorophylls. In *CRC Handbook of Organic Photochemistry and Photobiology*. F. Lenci and W. Horspool, editors. CRC Press, Boca Raton, FL.

Schellhammer, P. F., D. A. Kuban, and A. M. el-Mahdi. 1993. Treatment of clinical local failure after radiation therapy for prostate carcinoma. *J Urol* 150:1851–1855.

Scherz, A., A. Brandis, O. Mazor, Y. Salomon, and H. Scheer. 2004. Water-soluble anionic bacteriochlorophyll derivatives and their uses. U.S. Patent WO 2004045492.

Scherz, A., A. Brandis, Y. Salomon, D. Eren, and A. Cohen. 2005. Cationic bacteriochlorophyll derivatives and uses thereof. U.S. Patent WO 2005120573.

Scherz, A., L. Goldshaid, and Y. Salomon. 2009. RGD-(bacterio)chlorophyll conjugates for photodynamic therapy and imaging of necrotic tumors. U.S. Patent WO 2009107139.

Scherz, A., S. Katz, Y. Vakrat et al. 1998. Bacteriochlorophyll-serine based photochemotherapy; type III PDT? In *Photosynthesis: Mechanisms and Effects*. G. Garab, editor. Kluwer Academic Publishers, Dordrecht, 4207–4212.

Scherz, A., Y. Salomon, A. Brandis, and H. Scheer. 2003. Palladium-substituted bacteriochlorophyll derivatives and use thereof. U.S. Patent 6569846.

Scherz, A., Y. Salomon, L. Fiedor, and A. Brandis. 2000. Chlorophyll and bacteriochlorophyll derivatives, their preparation and pharmaceutical compositions comprising them. U.S. Patent 6147195.

Scherz, A., Y. Salomon, E. Rubinstein et al. 2008. Conjugates of RGD peptides and porphyrin or (bacterio)chlorophyll photosensitizer and their uses. U.S. Patent WO 2008023378.

Schreiber, S., S. Gross, A. Brandis et al. 2002. Local photodynamic therapy (PDT) of rat C6 glioma xenografts with Pd-bacteriopheophorbide leads to decreased metastases and increase of animal cure compared with surgery. *Int J Cancer* 99:279–285.

Schwartz, S. K., K. Abolon, and H. Vermund. 1955. Some relationships of porphyrins, x-rays and tumors. *Univ Minn Med Bull* 27:7–8.

Senge, M. O., and J. C. Brandt. 2011. Temoporfin (Foscan®, 5,10,15,20-tetra(m-hydroxyphenyl)chlorin)—a second-generation photosensitizer. *Photochem Photobiol* 87:1240–1296.

Shlyk-Kerner, O., I. Samish, D. Kaftan et al. 2006. Protein flexibility acclimatizes photosynthetic energy conversion to the ambient temperature. *Nature* 442:827–830.

Sibani, S. A., P. A. McCarron, A. D. Woolfson, and R. F. Donnelly. 2008. Photosensitiser delivery for photodynamic therapy. Part 2: Systemic carrier platforms. *Expert Opin Drug Deliv* 5:1241–1254.

Tarin, T., S. Kimm, P. Zhao et al. 2012. 1505 Photothermal effect in vascular targeted photodynamic therapy. *J Urology* 187:e609.

Tempel-Brami, C., I. Pinkas, A. Scherz, and Y. Salomon. 2007. Detection of light images by simple tissues as visualized by photosensitized magnetic resonance imaging. *PLoS One* 2:e1191.

Teuchner, K., H. Stiel, D. Leupold et al. 1997. Fluorescence and excited state absorption in modified pigments of bacterial photosynthesis a comparative study of metal-substituted bacteriochlorophylls a. *J Luminescence* 72–74:612–614.

Trachtenberg, J., A. Bogaards, R. A. Weersink et al. 2007. Vascular targeted photodynamic therapy with palladium-bacteriopheophorbide photosensitizer for recurrent prostate cancer following definitive radiation therapy: Assessment of safety and treatment response. *J Urol* 178:1974–1979; discussion 1979.

Trachtenberg, J., R. A. Weersink, S. R. Davidson et al. 2008. Vascular-targeted photodynamic therapy (padoporfin, WST09) for recurrent prostate cancer after failure of external beam radiotherapy: A study of escalating light doses. *BJU Int* 1025:556–562.

Tregub, I., J. Schmidt-Sole, J. O. Dejordy et al. 1992. Application of chlorophyll and bacteriochlorophyll derivatives to PDT of malignant melanoma. In *Proceedings of the Second International Congress on Interaction of Light with Biological Systems*. M. Holick and A. Kligman, editors. de Gruyter, New York, 354–361.

Vakrat-Haglili, Y., L. Weiner, V. Brumfeld et al. 2005. The microenvironment effect on the generation of reactive oxygen species by Pd-bacteriopheophorbide. *J Am Chem Soc* 127:6487–6497.

Vass, I., and K. Cser. 2009. Janus-faced charge recombinations in photosystem II photoinhibition. *Trends Plant Sci* 14:200–205.

Verma, S., G. M. Watt, Z. Mai, and T. Hasan. 2007. Strategies for enhanced photodynamic therapy effects. *Photochem Photobiol* 83:996–1005.

Vilensky, J., N. V. Koudinova, A. Harmelin, A. Scherz, and Y. Salomon. 2005. Vascular-targeted photodynamic therapy (VTP) of a canine-transmissible venereal tumour in a murine model with Pd-bacteriopheophorbide (WST09). *Vet Comp Oncol* 3:182–193.

Voet, D., and J. G. Voet. 2011. *Biochemistry.* John Wiley & Sons, Hoboken, NJ.

von Tappeiner, H., and A. Jodlbauer. 1904. Über die Wirkung der photodynamischen (fluorescierenden) Stoffe auf Protozoen und Enzyme. *Arch Klin Med* 80:427–487.

Vrouenraets, M. B., G. W. Visser, G. B. Snow, and G. A. van Dongen. 2003. Basic principles, applications in oncology and improved selectivity of photodynamic therapy. *Anticancer Res* 23:505–522.

Watanabe, T., and M. Kobayashi. 1991. Electrochemistry of chlorophylls. In *Chlorophylls*. H. Scheer, editor. CRC Press, Boca Raton, FL, 287–316.

Weersink, R. A., A. Bogaards, M. Gertner et al. 2005a. Techniques for delivery and monitoring of TOOKAD (WST09)-mediated photodynamic therapy of the prostate: Clinical experience and practicalities. *J Photochem Photobiol B* 79:211–222.

Weersink, R. A., J. Forbes, S. Bisland et al. 2005b. Assessment of cutaneous photosensitivity of TOOKAD (WST09) in preclinical animal models and in patients. *Photochem Photobiol* 81:106–113.

Wilt, T. J., M. K. Brawer, K. M. Jones et al. 2012. Radical prostatectomy versus observation for localized prostate cancer. *N Engl J Med* 367:203–213.

Zhu, T. C., and J. C. Finlay. 2008. The role of photodynamic therapy (PDT) physics. *Med Phys* 35:3127–3136.

Zilberstein, J., A. Bromberg, A. Frantz et al. 1997. Light-dependent oxygen consumption in bacteriochlorophyll-serine-treated melanoma tumors: On-line determination using a tissue-inserted oxygen microsensor. *Photochem Photobiol* 65:1012–1019.

Zilberstein, J., S. Schreiber, M. C. Bloemers et al. 2001. Antivascular treatment of solid melanoma tumors with bacteriochlorophyll-serine-based photodynamic therapy. *Photochem Photobiol* 73:257–266.

39

Photodynamic Therapy in Ophthalmology

Patrycja
Nowak-Sliwinska
*Swiss Federal Institute of
Technology (EPFL)*

Andrea Weiss
*Swiss Federal Institute of
Technology (EPFL)*

Michel Sickenberg
*Swiss Federal Institute of
Technology (EPFL)*

Hubert van den Bergh
*Swiss Federal Institute of
Technology (EPFL)*

39.1 Introduction

Ocular photodynamic therapy (PDT) was introduced as effective treatment for neovascular forms of age-related macular degeneration (AMD) and choroidal neovascularization (CNV) secondary to pathologic myopia in the late 1990s. It is worth mentioning that verteporfin PDT (v-PDT) was the first approved therapy for the treatment of subfoveal lesions and, at that time, was the only solution to avoid progressive vision loss and blindness. Verteporfin, or benzoporphyrin derivative monoacid ring A (BPD-MA), is a second-generation lipophilic/amphiphilic photosensitizer (PS) with an absorption peak centered at 689 nm. It is administrated as a liposomal formulation called Visudyne (Novartis Pharmaceuticals), which is a mixture of the two regioisomers, each a racemic mixture of two enantiomers. Other PSs, such as hematoporphyrin derivative, were used in preclinical and clinical studies for ocular disorders with encouraging outcomes. These agents were later abandoned, however, due to their associated side effects and poor tissue penetration, and the search for new photosensitizers continued. Several other photosensitizers were tested in phase I/II clinical ophthalmological trials, including tin etiopurpurin (SnET2) and Lutetium texaphyrin (Lu-tex); however, they also failed to achieve market approval (Koh and Haimovici 2004).

In the following sections, the clinical efficiency of v-PDT will be discussed in both nonmalignant and cancerous ocular disorders. Moreover, this chapter reviews the recent combination strategies aiming to improve PDT outcome in the clinic and the perspectives in ophthalmic PDT.

39.2 Nonmalignant Disorders

39.2.1 Choroidal Neovascularization

Choroidal neovascularization (CNV) is encountered most often in patients with age-related macular degeneration (AMD) but also occurs as a consequence of other disorders such as pathologic myopia, inflammation, angioid streaks, trauma, or choroidal rupture. Figure 39.1a depicts the retinograph of a healthy human eye with the macula in the center. CNV can develop in the central part of the macula, called the fovea. The optical coherence tomography (OCT) scan presented in Figure 39.1b shows different layers of the healthy retina. The maintenance of healthy macular function is dependent on the integrity of the microstructure of the different layers in this tissue, which include the choroid, Bruch's membrane, retinal pigmented epithelium (RPE), and the neural retina containing the photoreceptors. A schematic cross section of the retina's "layered" structure within the macula and pathologic CNV is shown in Figure 39.1d. CNVs are due to the process of neoangiogenesis and vascular sprouting from the choroid into Bruch's membrane. Neovessels are frequently restricted to the area below the RPE and near Bruch's membrane but are sometimes even capable of transgressing the RPE (Au Eong 2006; Coppens et al. 2011; Koh et al. 2011) (see Figure 39.1c). This process is thought to be stimulated by the production of inflammatory and proangiogenic factors by the RPE cells as a response to inflammation and oxidative stress in the RPE and Bruch's membrane (Qazi et al. 2009).

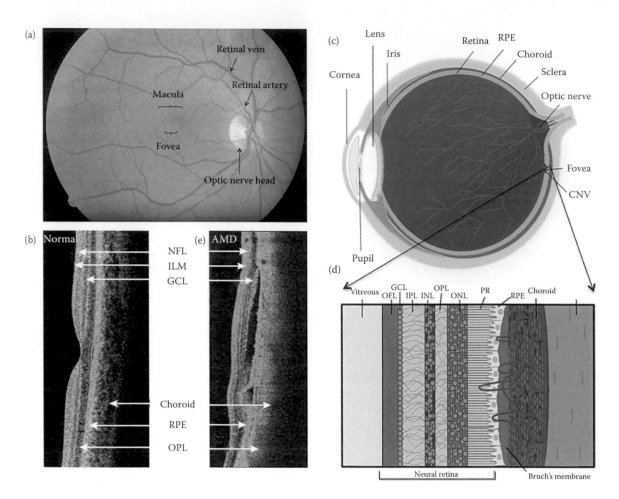

FIGURE 39.1 **(See color insert.)** (a) Retinograph of a healthy human eye. (b) Optical coherence tomography (OCT) scan with different layers of the healthy retina. (c) Schematic drawing of a cross section of the human eye. (d) Schematic cross section of the macula. (e) OCT image of the AMD-diagnosed retina visualizing the components of the neovascular membrane. GCL = ganglion cell layer, ILM = integral limiting membrane, IPL = inner plexiform layer, NFL = nerve fiber layer, OFL = optic fiber layer, ONL = outer nuclear layer, OPL = outer plexiform layer, RPE = retinal pigment epithelium.

39.2.1.1 Age-Related Macular Degeneration

The history of treatment of CNV associated with age-related macular degeneration (AMD) has been extensively reviewed before (Nowak-Sliwinska 2012; van den Bergh 2001). Exudative or wet AMD associated with choroidal neovascularization is the leading cause of vision loss in the elderly populations of developed countries (Klein et al. 1992). This disease results in the gradual destruction of the high-resolution central field of vision, a process that is strongly correlated with aging. The wet form of AMD is the most severe form, frequently leading to rapid vision loss. It is characterized by abnormal leaky CNV, which results in retinal edema. The newly formed blood vessels are pathological and functionally abnormal, leaking lipids, fluid, and blood into the retina, causing the edema and retinal thickening that is often also associated with vision impairment (see Figure 39.1c through e). If this leakage extends to longer times, it leads to scar tissue in the retina and a scotoma (a dark region in our visual image). Patients with wet AMD can lose as much as one letter a month on the Early Treatment Diabetic Retinopathy Study (ETDRS) vision chart without treatment.

In our current understanding of the etiology of wet AMD, the cells of the RPE play a crucial role in the pathological processes (Ting et al. 2009). RPE cells play an important role in maintaining the structural integrity and healthy function of the endothelium of the neighboring choriocapillaries and the photoreceptors through the expression of a variety of growth factors (Strauss 2005), including proangiogenic vascular endothelial growth factor A (VEGF-A) and antiangiogenic pigment epithelium-derived factor (PEDF) (Barnstable and Tombran-Tink 2004). Bruch's membrane is located between the RPE and choroid and is composed mainly of a complex of collagen and elastin layers. Lipophilic material that is shed by photoreceptors is deposited in and under the RPE and accumulates to contribute to the formation of structures referred to as drusen. This is a process that frequently increases with age. The drusen can easily be seen through examination of the retina with a standard ophthalmoscope. The size and number of drusen are often correlated with disease progression and have been used as successful predictors for patient outcome (Seddon et al. 2011). The drusen can be further characterized into two subgroups:

hard drusen are roundish, are yellow, and have clearly defined boundaries, while soft drusen are pale, are yellow, and have a poorly defined border. Additionally, the soft drusen are known to cause basal laminal deposits that accumulate between the RPE cell plasma membrane and the choroidal capillaries.

In the macular region of the eye, the metabolic requirements of the retina are provided by the choroidal capillary bed as there are no retinal vessels for this purpose. There are two types of photoreceptors. The cones, which number between 6 and 7 million, are mainly located in the central region of the macula, called the fovea, and serve for high resolution and color vision. The rods, approximately 120 million, provide vision in dim light and allow for peripheral vision. Vertebrate photoreceptors are composed of a photosensitive outer segment, an inner segment containing the cell's metabolic center, and a synaptic terminal that is connected to the second-order neurons of the retina. The photosensitive outer segment is composed of a series of discrete membranous discs and is connected to the inner segment by a modified, nonmotile cilium (Besharse and Pfenninger 1980). The membranous discs of the outer segment utilize photon-absorbing retinal molecules in order to initiate the visual process. Retinal can become embedded in the protein opsin in different environments. This can lead to various spectral sensitivities in the cones. New discs are created near the cilium, while the older discs are moved outward toward the ends of the cones and close to the RPE. To allow for the growth of new discs, aged (oxidized) discs are shed at the distal tip of the cone and undergo phagocytosis in the neighboring RPE cells. The RPE is also responsible for controlling the transport of fluids between the choroid and the neural retina. This transport includes, among other things, oxygen for the macular region, in particular for the fovea. The impaired function of the RPE due to age causes the accumulation of lipophilic material (i.e., the formation of drusen) in the RPE and Bruch's membrane (Green 1999; Schmidt-Erfurth et al. 1998). These deposits negatively affect the transport of molecules, as well as ions and oxygen, through the RPE and Bruch's membrane.

The resulting restricted oxygen transport causes hypoxia and an increase in hypoxia-inducible factor-1 (HIF-1), which subsequently leads to increases in the expression of VEGF-A. Other hypoxia-inducible genes, which are believed to play a role in this process and possibly contribute to the following angiogenesis, include glycolytic enzymes, erythropoietin, or inducible nitric oxide synthase. The process of angiogenesis begins with the development of increased permeability in the choriocapillaries, which are exposed to VEGF, allowing for the extravastation of plasma proteins (i.e., fibrinogen). Fibrinogen then clots to generate fibrin, forming a growth support for the newly growing vessels. Endothelial cells also begin to multiply faster, have increased mobility, and form buds, which express integrins (Connolly et al. 1989). These activated endothelial cells also produce matrix metalloproteinases, which can degrade the extracellular matrix and allow for the migration of proliferating endothelial cells toward the angiogenic stimuli. Through this process, new vessel sprouts are formed, which can then become integrated in the vascular network by forming capillary loops through the anastomotic connections of neighboring sprouts (see Figure 39.1d).

The brief etiology of wet AMD as described above is still uncertain. Nevertheless, for the purposes of this chapter, we support the hypothesis that the presence of lipophilic deposits in and around Bruch's membrane in the form of drusen may not only affect the diffusion of factors but also limit oxygen diffusion (van den Bergh 2001). The blood supply to the choroid is very high, and the partial pressure of oxygen (pO_2) at the choriocapillaries is more elevated than in most other perfused tissues in the human body. This is related to the high local oxygen consumption in the foveal region. Thickening and edema (Figure 39.1e) of the area around Bruch's membrane may result in an additional decrease in the pO_2 in the neural retina and RPE due to the increased distance from the choriocapillaries.

Exudative AMD can by chance be clinically diagnosed at an early stage, when visual acuity (VA) may still be close to normal. At these early stages of disease, however, drusen may already be seen in Bruch's membrane with varying number and size distributions.

The dry form of advanced AMD is also frequently referred to as central geographic atrophy and accounts for approximately 90% of all AMD cases (Biarnes et al. 2011). This dry form of AMD is associated with some cell death and retinal thinning but is generally not associated with rapid vision loss as in the case of the wet form of the disease.

v-PDT was approved by the Food and Drug Administration (FDA) in 2000 as an effective treatment strategy for exudative AMD. This treatment involves the intravenous administration of Visudyne, followed by the local activation of this photosensitizer using light with a wavelength of 689 nm. Most often, PDT is applied with what is referred to as "standard-fluence" light administration (i.e., a light dose of 50 J/cm^2 applied at an intensity of 600 mW/cm^2 for 83 s). Standard-fluence PDT is used in clinical trials, such as, the Treatment of Age-Related Macular Degeneration with Photodynamic Therapy (TAP) trial, which was composed of two simultaneous double-masked placebo-controlled randomized arms with patients suffering subfoveal CNV secondary to AMD (Treatment of Age-Related Macular Degeneration with Photodynamic Therapy (TAP) Study Group 1999). The 609 patients enrolled in this study were re-treated every 3 months, if judged necessary based on fluorescein angiograms showing any recurrence or persistence of leakage. Twelve months after the first treatment, PDT-treated patients showed an improvement in visual acuity of 1.3 lines on the ETDRS chart, as compared to the treatment group who were administered a sham treatment (Bressler 2001). An open-label extension of this study for an additional 3 years showed that visual acuity could be maintained at the level observed at the 1-year checkpoint, resulting in the recommendation that v-PDT treatment be continued beyond 24 months if fluorescein leakage from CNV was still observed at this time point (Kaiser 2006).

Due to retinal inflammation observed after PDT in the above studies and the potential 42 J/cm^2 for increased treatment selectivity, reduced-fluence v-PDT (300 mW/cm^2) protocols were later proposed (Michels et al. 2006). The verteporfin in subfoveal minimally classic CNV (VIM) trial employed both standard and reduced-fluence v-PDT protocols and showed stabilization of vision in both treatment groups versus the placebo group. This study, however, did not demonstrate a treatment benefit in terms of visual outcome for the reduced-fluence v-PDT protocol. It was also found in this study that re-treatment with v-PDT was necessary due to the reperfusion of PDT-occluded vessels (Costa et al. 2003), as well as due to the induction of angiogenesis (Geliskeni et al. 2004; Rudolf et al. 2004; Weiss et al. 2012). Both the reperfusion of blood vessels and the activation of angiogenesis are likely to be results of a combination of factors, including PDT-induced inflammation (Gollnick et al. 2003; Schmidt-Erfurth et al. 2005), PDT-induced enhanced expression of VEGF-A (Rudolf et al. 2004; Zhang et al. 2005) or other growth factors, and PDT-induced tissue hypoxia (Schmidt-Erfurth et al. 2003). Even prior to v-PDT, CNV associated with AMD is characterized by elevated levels of certain angiogenic cytokines, including VEGFs, fibroblast growth factors (bFGFs), and angiopoietins (Das and McGuire 2003; Grossniklaus and Green 2004; Hera et al. 2005). Findings indicating that PDT also results in the increased expression of some of these cytokines has led to clinical trials for AMD that exclude v-PDT treatment, instead targeting these cytokines with topically applied intravitreously injected compounds that block some of these cytokines (Das and McGuire 2003).

39.2.1.2 Pathologic Myopia

Myopic CNV is often associated with socioeconomic setbacks and varies among different countries and continents. The Verteporfin in Photodynamic Therapy (VIP) study was a multi-center, double-masked, placebo-controlled, randomized clinical trial in 28 ophthalmology centers in Northern America and Europe. Seventy-seven of 81 patients (95%) in the verteporfin group, compared with 36 of 39 patients (92%) in the placebo group, completed this 24-month examination. The change in VA at the 24-month examination favored a benefit in the verteporfin treatment group. In the v-PDT group, an average VA improvement of at least five letters (equivalent to at least one line) was observed in 32 verteporfin-treated cases (40%) versus only 5 placebo-treated cases (13%). Additionally, an improvement of at least 15 letters (at least three lines) was seen in 10 verteporfin-treated cases (12%) versus zero placebo-treated cases. The mean re-treatment number was 5.1, and no additional photosensitivity adverse effects associated with v-PDT were observed (Blinder et al. 2003). Similarly, a prospective, interventional study was conducted in Asian patients by Lam et al. (2004). After 24 months of follow-up, the median VA improvement was 1.7 lines; however, the mean number of PDT re-treatments required in the first 2 years was 2.3, which was significantly lower than in the VIP study at 2-year follow up the visual benefit was maintained vs placebo, but primary outcome was not statistically significant (Blinder et al. 2003).

39.2.1.3 Idiopathic CNV

Idiopathic CNV differs from CNV due to AMD. In this condition, lesions are smaller, are less aggressive in growth rate, and usually occur at a younger age. The v-PDT protocol used in treating classic CNV in AMD was also found to be effective in idiopathic CNV in phase I and II clinical studies (Sickenberg et al. 2000). No deterioration in visual acuity was observed, and the majority of patients gained at least one line of vision. Reduction in the size of the leakage area from classic CNV was already noted in all patients 1 week after v-PDT, and a complete absence of leakage from classic CNV was observed in almost half of the patients. Other studies have also reported retrospective analysis of improved VA. Another study in Asian patients (17 cases) showed that stable or improved vision was achieved in 94% of the patients after 12 months of follow-up, with a statistically significant improvement in mean VA compared to baseline. The mean number of re-treatments over 12 months was 1.8, and clinically important complications related to the use of v-PDT were encountered (Chan et al. 2003b).

39.2.1.4 Polypoidal Choroidal Vasculopathy

A unique form of CNV is polypoidal choroidal vasculopathy (PCV), which is characterized by a branching vascular network that terminates in polypoidal lesions. It may account for up to 50% of cases of wet AMD in African and Asian countries. Similarly to AMD, the high expression of VEGF-A was histologically observed in PCV. However, anti-VEGF treatment in PCV has been shown to have a limited efficacy (Tsujikawa et al. 2010). Although exudative changes regressed with concomitant improvement of visual acuity, complete regression of polypoidal lesions was only observed in 25% of treated eyes (Hikichi et al. 2010). However, a number of studies showed very encouraging effects of PDT on PCV: complete regression of the polypoidal lesions in 95% of eyes and stable or even improved VA 1 year after treatment were observed (Chan et al. 2004). Moreover, visual recovery after PDT was reported to be more favorable in eyes with PCV than in those with AMD: 7.0 letters with AMD compared to 8.0 letters in PCV (Gomi et al. 2008). However, even though all polypoidal lesions regress with PDT, its effect on branching does not seem to be permanent, and thus, recurrence of polypoidal lesions may occur in the long run (Tsuchiya et al. 2009). As recently reported, 64% of eyes successfully treated with PDT showed a recurrence of the polypoidal lesions during the follow-up at 24 months (Akaza et al. 2008).

39.2.1.5 Inflammation

CNV can also occur as a complication of punctate inner choroidopathy (PIC), uveitis, multifocal choroiditis (MC), and panuveitis. PIC is a relatively uncommon inflammatory multifocal chorioretinopathy that affects predominantly young myopic women. It is characterized by the presence of multiple, small, well-defined, yellow-white fundus lesions limited to the posterior pole of the eye where there is an absence of flare, and inflammatory cells in the anterior chamber or vitreous cavity (Amer and Lois 2011). MC is a chronic inflammatory disease also affecting myopic women, that is characterized by multiple punched-out chorioretinal lesions at the posterior pole and midperiphery, as well as abnormalities associated with vitreitis.

Visual deterioration can be a result of a number of complications, including cystoid macular edema, CNV, foveal scarring, or epiretinal membrane. The prognosis in PIC and MC patients is generally good; however, about 30% of cases may develop subfoveal CNV leading to vision loss. There is a strong indication for the success of v-PDT or its combination with anti-inflammatory or antiangiogenic compounds. Prospective and retrospective studies on patients with CNV due to PIC treated with v-PDT showed improved VA in most cases, as elegantly reviewed by Chan et al. (2010). Combining PDT with intravitreal triamcinolone (IVTA, 4 mg) seems to be a promising strategy in the treatment of idiopathic CNV and CNV secondary to PIC, as it resulted in fewer treatment sessions and superior visual improvement compared to other treatment options (Chan et al. 2008).

39.2.1.6 Angioid Streaks

Angioid streaks are small breaks in a weakened Bruch's membrane. They occur in patients with systemic diseases such as pseudoxanthoma elasticum, Paget's disease of bone, or sickle hemoglobinopathy. The rationale for v-PDT in CNV due to angioid streaks is similar to that for CNV due to AMD or pathologic myopia: to occlude new vessels and reduce leakage or lesion growth, thereby reducing the risk of vision loss. There are some case reports of patients with CNV due to angioid streaks who were treated with v-PDT with results depending on the baseline VA. In a placebo-controlled study on 17 patients, a smaller decrease in mean VA was observed in 10 v-PDT-treated patients (from 20/126 to 20/500), as compared to untreated patients (mean decrease from 20/160 to 20/640), over an average period of 18 months (Arias et al. 2006). Other studies reporting patients with a better baseline VA showed better v-PDT treatment outcomes (less than three lines of VA lost at 12 months) (Menchini et al. 2004). Also, other case series suggest that v-PDT is generally well tolerated and might limit, or slow, vision loss compared with the expected natural course of CNV due to angioid streaks (Menchini et al. 2004). Further studies, however, are needed to assess long-term safety and efficacy profiles.

39.2.2 Central Serous Chorioretinopathy

Central serous chorioretinopathy (CSCR) is a serous detachment of the neurosensory retina over an area of leakage from the choriocapillaries through the RPE lesions that occurs mostly in middle-aged men. Visual loss or permanent symptoms may appear in cases with persistent focal leakage or chronic diffuse leakage (Chan et al. 2003a). CSCR occurs most commonly in Asians and Caucasians. Some patients can develop CSCR as a complication of CNV. PDT can be an effective treatment in this case, as an angioocclusive treatment may lead to hypoperfusion of the choriocapillaries and narrowing of choroidal vessels, thereby reducing choroidal exudation and inducing vascular remodeling. The standard regimen of v-PDT was evaluated in two prospective case series. In the first of these studies, 10 eyes were identified as having subfoveal or juxtafoveal CNV secondary to CSCR and received v-PDT during a mean of 12.6 months, with six eyes (60%) with two lines of VA and no eye losing two lines or more (Chan et al. 2003a). As recently reported, the success of PDT for CSCR depends on the degree of hyperfluorescence seen in indocyanine green angiography (ICGA). In eyes without hyperfluorescence, another treatment should be explored in order to absorb the subretinal fluid (SRF) (Inoue et al. 2010). This recent study also showed that half-fluence (25 J/cm^2) v-PDT was effective in the treatment of an acute symptomatic central serous chorioretinopathy (Smretschnig et al. 2012). Therefore, from the studies discussed above, v-PDT has been demonstrated to be a promising treatment modality for patients with chronic CSCR by facilitating resolution of the serous macular detachment. Longer follow-up, however, is needed to determine the safest PDT-treatment parameters.

39.2.3 Symptomatic Choroidal Nevus

Choroidal nevus is a benign freckle in the eye. Treatment of choroidal nevus together with SRF using v-PDT was recently reported (Garcia-Arumi et al. 2012). All tumors of 17 patients presented at least two risk factors for the disease (i.e., peripapillary location, orange pigment, symptoms, SRF). SRF was reduced in all eyes and completely disappeared in nine eyes (53%) after PDT. Tumor thickness increased in 3 eyes (18%) and remained unchanged in 13 eyes (76%), and one lesion (6%) shrank down to a flat chorioretinal scar. Longer follow-up is necessary to determine the long-term PDT efficacy in these patients.

39.3 Malignant Disorders

Eye tumors can be of benign or malignant origin. Vascular tumors of the retina and choroid, while considered to be benign, are associated with significant visual impairment in many cases. Ophthalmic cancers, although uncommon, may arise from various eye and orbital structures. Melanoma, retinoblastoma, and squamous cell carcinoma constitute the main types of malignant eye tumors. Treatment options for eye tumors generally include a combination of laser photocoagulation, cryosurgery, transpupillary thermotherapy, radiation therapy, local resection, and adjuvant chemotherapy. PDT provides two major clinical advantages in the field of photochemotherapy of intraocular tumors. First, it treats malignancies selectively with a substantial increase in eye tolerance, and second, the most important clinical advantage, v-PDT provides the potential ability to treat posterior neoformations on the macula and close to critical structures such as the optic disk while preserving the adjacent anatomic structures, providing efficient tumor control with desirable cosmetic results. Moreover, this treatment modality might be a treatment option in patients who require ocular treatment only. The major vascular tumors of the retina include retinal capillary hemangioma, cavernous hemangioma of the retina, retinal vasoproliferative tumor, and racemose hemangiomatosis of the retina or Wyburn–Mason syndrome. Some of these are discussed below.

39.3.1 Retinal Hemangioma

Retinal capillary hemangioma (RCH, retinal hemangioblastoma) is a benign angiomatous hamartoma of the retina. Retinal capillary hemangiomas occurring as isolated tumors have also been referred to as angiomatosis retinae, retinal angiomatosis, retinal hemangioblastoma, and von Hippel disease. If these tumors occur with tumors in other systems, especially central nervous system hemangioblastomas and renal cell carcinomas, then they are named von Hippel–Lindau (VHL) disease (Maher et al. 2011). Although retinal hemangiomas are benign vascular tumors appearing as an orange mass in the posterior pole of the eye, they can progress and result in visual loss by complicated retinal exudates, exudative retinal detachment, an epiretinal membrane, and vitreous hemorrhage. Various treatment modalities have been reported, including observation, laser photocoagulation, plaque radiotherapy, or vitreoretinal surgery with limited success (Sachdeva et al. 2010; Singh et al. 2002). Moreover, intravitreal injection of anti-VEGF did not decrease the tumor size or the recurrence rate. Since PDT ablates aberrant choroidal vasculature while sparing the retinal vasculature, it is reasonable to use this treatment option in patients with retinal hemangiomas. Treatment of retinal capillary hemangioma is based on tumor size, location, the presence of SRF or retinal traction, and visual acuity. For larger tumors and those cases in which visual acuity is affected, other treatment modalities including laser photocoagulation, cryotherapy, PDT, radiotherapy, and various other vitreoretinal procedures can be employed.

39.3.2 Retinal Vasoproliferative Tumor

Retinal vasoproliferative tumors of the retina (VPTR) are very rare, and when their occurrence is secondary, the underlying cause has been attributed to inflammatory, vascular, traumatic, dystrophic, and degenerative retinal disease. They present as a pinkish-yellow nodular vascular mass in the pre-equatorial retina, often located inferior to the retina. Multiple treatment options exist for retinal vasoproliferative tumors including cryotherapy and plaque brachytherapy. Photocoagulation and PDT have been used in a limited number of cases with some success. Three patients with VPTRs who presented with macular exudative changes were treated with one session of v-PDT with the following specifications: 6 mg/m^2 body surface of verteporfin and a light dose of 100 J/cm^2 at 689 nm delivered in 166 s (Blasi et al. 2006). At the 1-year follow-up, all tumors responded with a reduction in size, and an improvement in VA (mean 4.7) was observed with no additional re-treatment. A larger number of cases and longer follow-up times, however, are required to further assess these treatment modalities.

39.3.3 Choroidal Hemangioma

Choroidal vascular tumors include circumscribed choroidal hemangioma and diffuse choroidal hemangioma (Singh et al. 2005).

For these tumors, treatment can often be challenging because visual symptoms secondary to exudative retinal detachment as well as a variety of other mechanisms are common and are a major source of long-term visual disability.

Circumscribed choroidal hemangiomas (CCH) are benign hamartomas appearing as an orange choroidal mass with indistinct margins that blend with the surrounding choroid. Their treatment is based on the location of the tumor, the presence of SRF, and the extent of symptoms. Current treatment options include laser photocoagulation, radiotherapy, transpupillary thermotherapy (TTT), and PDT. PDT is the preferred method of treatment in subfoveal or juxtafoveal circumscribed choroidal hemangiomas because other treatments are associated with significant treatment morbidity for tumors at this location. In general, tumor regression is most dramatic following the first session of PDT and is usually evident within 3 months (Madreperla 2001). There are case studies reporting on the effectiveness of v-PDT in diffuse choroidal hemangioma in patients with Sturge–Weber syndrome (SWS) (Tsipursky et al. 2011). There are important differences between circumscribed hemangiomas and SWS-associated diffuse choroidal hemangiomas. In SWS patients, hemangiomas predominately manifest as diffuse bodies in the leptomeninges and skin as well as in the choroid. These hemangiomas are usually unilateral and ipsilateral to the angiomatous malformation of the skin. The diffuse choroidal hemangiomas of SWS may have localized areas of excessive thickening simulating circumscribed choroidal hemangiomas. These patients are most likely to develop secondary retinal detachment with shifting of the SRF. PDT turned out to be an effective treatment option for visual deterioration from exudative retinal detachment in patients with diffuse choroidal hemangiomas.

Diffuse choroidal hemangiomas appear on ophthalmic examination as orange, diffuse choroidal thickening. As with circumscribed hemangiomas, there may be an associated exudative retinal detachment that often does not manifest until adolescence. Short-term treatment success using PDT has also been reported in these conditions (Anand 2003). PDT offers potential advantages over radiotherapy including avoidance of radiation and associated complications, ease of delivery, and fewer side effects. Larger studies with long-term follow-up are required to fully assess the efficacy of PDT in the treatment of diffuse choroidal hemangioma. In addition to visual loss from exudative retinal detachment, congenital glaucoma is seen in approximately 70% of cases of Sturge–Weber syndrome.

39.3.4 Melanoma

Uveal melanomas are the most common primary malignant tumor inside the eye in adults. They arise from uveal melanocytes or from other pigmented cells. They can be divided into three types based on anatomic location: iris melanoma, ciliary body melanoma, and choroidal melanoma. Treatment results, based mostly on the case reports, indicate that small and preferably low-pigmented or nonpigmented melanomas respond favorably to PDT, whereas pigmented melanomas were less or not at all responsive due to light absorption by melanin and hemoglobin. For example, Canal-Fontcuberta et al. (2012) reported a case study where pigmented choroidal melanomas receiving PDT as a primary treatment were followed by progressive tumor growth that led to enucleation years after, suggesting that a single session of v-PDT was not effective as a primary treatment for pigmented small choroidal melanomas.

39.3.5 Squamous Cell Carcinoma

The standard treatment options for squamous cell carcinoma (SCC) of the conjunctiva are often complicated by the potential to damage adjacent ocular structures as well as by high recurrence rates of treated lesions. Currently available therapeutic options include surgical excision, cryotherapy, and radiation, as well as investigational approaches with topical chemotherapy, interferon, and antiviral drugs. In an interventional case study, Barbazetto et al. (2004) noted SCC tumor regression in all patients 1 month after standard-dose v-PDT. Two patients had complete regression (clinical and angiographic observation) after one or two treatments for the entire follow-up time. One particular tumor involved large areas of the conjunctiva and cornea. In this case, only the treated areas showed tumor regression. PDT caused minimal temporary local irritation in two patients and small conjunctival hemorrhages and mild transient chemosis in the eyes directly after treatment. There are case reports on successful v-PDT of conjunctival ocular-surface squamous neoplasia extending into the cornea (Çekiç et al. 2011). Half of the mass regressed after first PDT. A second treatment of PDT resulted in a near-complete cure of the remaining lesion within 2 weeks. The patient remained stable for 13 months.

39.3.6 Choroidal and Uveal Metastasis

Metastases in the eye from tumors in distant organs are the most common form of intraocular cancer and can lead to significant vision loss. The uvea is the most common site for ocular metastasis. Within the uvea, 88% of metastases are to the choroid, followed by metastases to the iris (9%) and ciliary body (2%). Metastases that appear in and around the eye are usually from breast (in women) or lung carcinoma (in men). Other less common sites of origin include the prostate, kidney, thyroid, and gastrointestinal tract. Treatment options of choroidal metastasis include laser photocoagulation, cryotherapy, chemotherapy, radiotherapy, or PDT. v-PDT has been reported to have clinical benefits not only for benign tumors such as choroidal or retinal hemangiomas but also for malignancies such as choroidal melanoma and uveal metastasis. The advantage of PDT lies in intraluminal photothrombosis in endothelial structures.

Nine choroidal metastases, all associated with shallow SRF, were treated with v-PDT (Kaliki et al. 2012). After PDT, complete control with resolution of SRF was achieved in seven tumors (78%). These tumors had a mean tumor thickness reduction of 39%. Two tumors failed to respond to PDT and were further treated with plaque radiotherapy. Improvement or stabilization of vision was achieved in seven eyes. PDT-related complications included intraretinal hemorrhage in only one eye. This showed that PDT could effectively destroy a malignant tissue and induce an antitumor activity.

39.4 PDT in Combination with Other Treatment Modalities

PDT for the treatment of CNV demonstrated positive results despite some observed side effects on the RPE cells and the necessity of re-treatments. These factors of v-PDT prompted the PDT field to search for efficient strategies to improve the efficiency and selectivity of v-PDT and reduce the treatment burdens. PDT in combination with other treatment modalities has been extensively investigated in in vitro and in vivo models by our group and by others (Weiss et al. 2012). These include the combination with antiangiogenic strategies or angiostatic steroids, reported most frequently for CNV due to AMD (see Table 39.1 and Figure 39.2). The VEGF Inhibition Study in Ocular Neovascularization (VISION) demonstrated that in the group given pegaptanib sodium at 0.3 mg, 70% of patients lost fewer than 15 letters of visual acuity, as compared with 55% in the control group. Thus pegaptanib sodium turned out to be an effective therapy for AMD. However, it does not lead to any improvement in mean visual acuity (Chakravarthy et al. 2006). The Anti-VEGF Antibody for the Treatment of Predominantly Classic Choroidal Neovascularization in AMD (ANCHOR) study was a multicenter, double-masked, phase III trial comparing patient-reported visual function in patients with neovascular AMD treated with either two doses of monthly ranibizumab or with v-PDT. Patients were enrolled 1:1:1 into either v-PDT or monthly intravitreal ranibizumab (0.3 or 0.5 mg) injections. The need for v-PDT (active or sham) re-treatment was evaluated every 3 months by fluorescein angiography. This study revealed that ranibizumab treatment provided greater clinical benefit after 2 years than v-PDT in patients with AMD with predominantly classic CNV (Brown et al. 2006).

Unique characteristics in the pathogenesis of AMD such as oxidative process, inflammation, and cytokine expression contribute to disease development and progression and can be potential targets for intervention. Pharmacologic inhibition of VEGF-A decreases the proliferation of endothelial cells and recruitment of other cell types, such as leukocytes, which can express the cytokines and proteases necessary to develop and maintain neovessels. Once neovascularization is stabilized, it will not respond to anti-VEGF treatment. PDT destroys existing neovessels, effectively acting as a "CNV eraser," and could be combined with an anti-VEGF strategy aimed at inhibiting further neovessel growth and leakage and corticosteroids, which will have an anti-inflammatory and antifibrosis effect (Figure 39.3). Thus, by affecting separate pathways, one may presume synergies that might lead to a better treatment outcome.

The pathogenesis of AMD is in partly due to oxidative stress and inflammation leading to angiogenesis, which includes the increased expression of various cytokines. All of these processes in the disease development can be targeted for therapy.

TABLE 39.1 Summary of Clinical Trials with v-PDT for Choroidal Neovascularization

Study	Compound(s)	Treatment Regimen	Follow-Up Period (Months)	Result	Ref.
TAP	v-PDT	v-PDT	24	v-PDT reduced the risk of moderate or severe vision loss in patients with CNV due to AMD.	Treatment of Age-Related Macular Degeneration with Photodynamic Therapy (TAP) Study Group (1999)
VIP		v-PDT (Northern America, Europe)	24	In the v-PDT group, VA improvement by at least five letters was observed in patients with CNV due to pathologic myopia.	Blinder et al. (2003)
Lam et al.		v-PDT (Asian patients)	24	Median VA improvement was 1.7 lines, but the mean number of PDT re-treatments required in the first 2 years was 2.3—lower than in the VIP study.	Lam et al. (2004)
VISION	Pegaptanib sodium	Pegaptanib sodium (0.3 mg)	12	70% of patients lost fewer than 15 letters of visual acuity, as compared with 55% among the controls. Improvement in mean visual acuity was not observed.	Chakravarthy et al. (2006)
ANCHOR	Ranibizumab v-PDT	Ranibizumab (0.3 mg or 0.5 mg) v-PDT	12	Ranibizumab was clearly superior to PDT with respect to both VA and anatomic (lesion size and CNV leakage) efficacy outcomes.	Brown et al. (2006)
FOCUS	v-PDT Ranibizumab	Ranibizumab 0.5 mg monthly or sham injections monthly followed by SF v-PDT at day zero	24	Combination therapy was more effective than v-PDT alone and had a low rate of adverse events.	Antoszyk et al. (2008)
PROTECT	v-PDT Ranibizumab	Same-day SF v-PDT and ranibizumab 0.5 mg in CNV AMD	12	Improved VA; lesions were stabilized with minimal treatment required after month 3.	Schmidt-Erfurth and Wolf (2008)
MONT BLANC	v-PDT Ranibizumab	SF v-PDT, ranibizumab 0.5 mg in CNV AMD	12	VA improvements in the combination group are noninferior to a ranibizumab alone with three ranibizumab doses followed by injections on a monthly regimen.	Spitzer et al. (2008)
DENALI		SF v-PDT, ranibizumab 0.5 mg RF v-PDT, ranibizumab 0.5 mg Ranibizumab 0.5 mg in CNV AMD	12	DENALI did not demonstrate noninferior visual acuity gain for v-PDT combination therapy compared with ranibizumab monthly monotherapy.	http://www.qltinc.com/newsCenter/2010/100615.htm
EVEREST		SF v-PDT, ranibizumab 0.5 mg in PCV	12	Complete regression of polypoidal lesions in combination therapy group.	Sagong et al. (2012)
RADICAL	v-PDT Ranibizumab Dexamethasone	QF v-PDT + within 2 h by ranibizumab (0.5 mg) + dexamethasone (0.5 mg) HF v-PDT + ranibizumab (0.5 mg) + dexamethasone (0.5 mg) Ranibizumab only (0.5 mg)	24	Significantly fewer re-treatment visits required with combination therapies than with ranibizumab monotherapy. Mean VA change from baseline was not statistically different among the treatment groups.	
Triple therapy	v-PDT Bevacizumab Dexamethasone	DF v-PDT, 16 h later, dexamethasone (800 μg) and bevacizumab (1.5 mg)	9	Less than one-fourth of the patients treated with this regimen required additional treatment.	Augustin (2009)
	v-PDT Bevacizumab TA	SF v-PDT + immediate injection of bevacizumab (1.25 mg) + TA (4 mg) + bevacizumab (1.25 mg) every 3 months	6	Short-term results of this study (at 6 months) showed low rate of re-treatments, sustained CNV closure efficacy, and visual acuity improvement.	Yip et al. (2009)

Note: QF = quarter fluence (15 J/cm², 180 mW/cm²), SF = standard fluence (50 J/cm², 600 mW/cm²), VA = visual acuity.

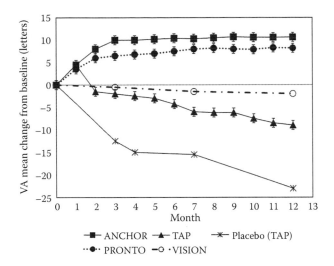

Furthermore, it seems reasonable to presume that more effective therapy can be achieved, with a more long-term improvement in vision as a goal, by targeting more than one of these processes at the same time. The closure of unwanted neovasculature by the combination of PDT with angiostatic compounds to prevent neovascularization as well as corticosteroids to prevent inflammation may provide an effective multilevel treatment for AMD. This combination therapy is also expected to be more effective due to the fact that PDT tends to result in the increased production VEGF-A (Fingar 1996; Nowak-Sliwinska et al. 2010) and thus the induction of angiogenesis (Nowak-Sliwinska et al. 2011, 2012; van Beijnum et al. 2011), as well as some degree of inflammation.

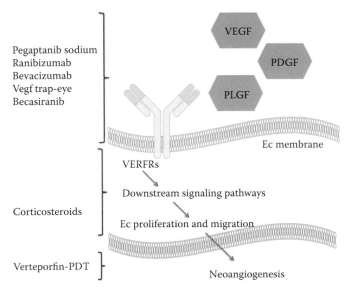

The search for long-lasting treatment for AMD is still underway, and the current status of combination trials is summarized below.

The RhuFab V2 Ocular Treatment Combining the Use of Visudyne° to Evaluate Safety (FOCUS) trial was a 2-year phase I/II study designed to compare the safety, tolerability, and efficacy of ranibizumab treatment in combination with v-PDT versus v-PDT alone. In this study, patients with subfoveal, predominantly classic CNV due to AMD received monthly intravitreal injections of ranibizumab (0.5 mg) or sham injections. All patients received PDT on day zero, followed by quarterly re-treatments as needed. The results of this study showed that the combination of ranibizumab and v-PDT was more effective than v-PDT alone and resulted in a low rate of associated adverse effects during the 2-year study period (Antoszyk et al. 2008).

In another study, the PROTECT trial (ranibizumab administrated in conjugation with PDT with verteporfin in patients with occult or predominantly classic subfoveal CNV due to AMD), a phase II open-label multicenter trial, patients received a same-day standard dose of v-PDT and an intravitreal injection of ranibizumab, followed by three monthly injections of ranibizumab (0.5 mg). An average visual acuity improvement of 6.9 letters at month 4 and 2.4 letters at 9 months was seen as compared to visual acuity measured at 2 months. Additionally, at the 9-month time point, all lesions were inactive with no recurrent leakage. Same-day v-PDT and ranibizumab was shown to be safe and was not associated with severe vision loss or inflammation. Lesions were stabilized, with minimal treatment being required after 3 months (Schmidt-Erfurth and Wolf 2008).

The SUMMIT trial was a multicenter, 12-month, double-blind noninferiority randomized phase III clinical program including a European (MONT BLANC), Northern American (DENALI), and Asian (EVEREST) study. This trial aimed at determining if the combination of ranibizumab and standard-dose v-PDT is better than ranibizumab monotherapy in patients with subfoveal CNV secondary to AMD (Spitzer et al. 2008). At the 12-month time point of the MONT BLANC trial, an average visual acuity improvement of 2.5 was seen in the combination therapy treatment group as compared to 4.4 letters in the monotherapy arm. This confirmed the idea that combining standard-fluence v-PDT with ranibizumab (0.5 mg) can result in visual acuity improvements that are equally effective, if not more, than a ranibizumab monotherapy regimen with three ranibizumab loading doses followed by injections on a monthly as-needed basis.

In another study, the DENALI study, patients received standard-fluence v-PDT and intravitreal ranibizumab (0.5 mg) combination therapy, reduced-fluence v-PDT and intravitreal ranibizumab (0.5 mg) combination therapy, or monthly ranibizumab monotherapy (0.5 mg). Patients receiving v-PDT combination therapy received ranibizumab treatment on day 1 and months 1 and 2. At the 12-month checkpoint, patients in the standard-fluence combination group gained an average of 5.3 letters from baseline, and patients in the reduced-fluence combination group gained an average of 4.4 letters. By comparison, patients receiving the monthly monotherapy of ranibizumab gained on average 8.1 letters at the 12-month time point. This study did not show that v-PDT combination therapy was less effective in terms of gain in visual acuity as compared to ranibizumab monthly monotherapy.

Many studies have also confirmed the importance of inflammation in the progression of AMD (Cook and Figg 2010; Couch and Bakri 2011). Patients with AMD have elevated systemic inflammatory biomarkers such as C-reactive protein (CRP), interleukin-6 (IL-6), and homocysteine. Localized inflammation and the activation of the complement cascade have been observed in the formation of drusen, a characteristic of dry AMD (Haines et al. 2005). These findings support the use of corticosteroids, a class of anti-inflammatory agents, in the treatment of AMD. Additionally, PDT is also known to induce inflammatory processes, and its combination with corticosteroids has already been investigated. Patients treated with v-PDT in combination with anti-inflammatory triamcinolone acetonide showed a reduced need for re-treatment as compared to v-PDT alone. However, many of these patients did not show an improvement in visual acuity (Couch and Bakri 2011), and this combination has also been associated with some adverse side effects, including cataract progression and ocular hypertension (Frimpong-Boateng et al. 2009), likely due to the corticosteroids. Retrospective analysis of PCV patients who underwent v-PDT with or without IVTA with a follow-up of 2 or more years was also reported (Lai et al. 2010). Twenty-seven eyes of 27 patients were analyzed, with 12 eyes treated by PDT monotherapy and 15 eyes treated by combined PDT with IVTA. PDT reduced the risks of vision loss in patients with symptomatic PCV in the short term, but the effect might not be sustained after 1 year. The adjunctive use of IVTA during PDT did not appear to result in additional benefit for treating PCV.

Recently, the randomized prospective study (EVEREST) showed a 6-month efficacy of v-PDT in combination with ranibizumab for PCV. Complete regression of polyploidal lesions was achieved at 6 months in 77.8% after combination therapy as compared to 71.4% after PDT alone and 28.6% after intravitreal injection of ranibizumab. The VA outcome in this study was +10.9, +7.5, and +9.2, respectively. To date, there is limited information on combination therapy for PCV that is refractory to anti-VEGF therapy, but due to limited effectiveness of the anti-VEGF treatment itself in PCV, combined therapy might be an option when persistent or recurrent exudative change is seen post–anti-VEGF treatments. Reduced-fluence-rate v-PDT combined with intravitreal bevacizumab was reported recently by Sagong et al. (2012). At 12 months, the BCVA improved in 56% of eyes by three lines or more, was stable in 37% of eyes, and decreased in one eye because of recurrence of polyps. Reduced-fluence PDT combined with bevacizumab for PCV seemed to be effective for improving vision and reducing complications.

The RADICAL trial was a phase II, multicenter, randomized, single-masked study, aimed to compare the effects of reduced-fluence v-PDT and ranibizumab combination therapies (with or without the anti-inflammatory agent dexamethasone) to ranibizumab monotherapy in subjects with CNV secondary to AMD.

This study included four treatment groups: quarter-fluence v-PDT (180 mW/cm^2, 15 J/cm^2) followed within 2 h by intravitreal ranibizumab (0.5 mg) and then intravitreal dexamethasone (0.5 mg); half-fluence v-PDT (300 mW/cm^2, 25 J/cm^2) followed within 2 h by intravitreal ranibizumab (0.5 mg) and then intravitreal dexamethasone (0.5 mg); half-fluence v-PDT (300 mW/cm^2, 25 J/cm^2) followed within 2 h by intravitreal ranibizumab (0.5 mg); and ranibizumab monotherapy (0.5 mg). At the 24-month checkpoint, significantly fewer re-treatment visits were required in the group of patients receiving combination therapy as compared to the ranibizumab monotherapy groups. The average improvement in visual acuity was not statistically significant between the different treatment groups as sample sizes were insufficient to draw definitive conclusions regarding visual acuity outcomes.

The use of triple therapy of v-PDT, anti-VEGF and anti-inflammation agents has also been investigated by Augustin (2009) in a noncomparative, interventional case study of 104 patients. In this study, v-PDT was delivered using the reduced-fluence protocol (42 J/cm^2 delivered over 70 s), followed after 16 h by a vitrectomy and the administration of dexamethasone (800 µg) and bevacizumab (1.5 mg) into the center of the eye. The mean follow-up of 40 weeks showed a significant increase of visual acuity (1.8 lines) and a decrease in retinal thickness (182 µm). Additionally, no serious adverse effects were observed. Less than 25% of the patients treated with this regimen required additional treatment, which, when administered, was either a repeat of the triple therapy or anti-VEGF monotherapy during the follow-up period.

Another study investigating triple therapy was the study by Yip et al. (2009) that included patients (36 eyes) with CNV secondary to AMD. Patients were treated with a standard-fluence v-PDT (600 mW/cm^2; 50 J/cm^2) followed by immediate intravitreal injection of bevacizumab (1.25 mg) and the corticosteroid, triamcinolone acetonide (4 mg), followed by bevacizumab (1.25 mg) every 3 months. Six-month results of this study showed a reduced necessity for re-treatments, sustained vasculature closure, and visual acuity improvement. Finally, Bakri et al. (2009) investigated the use of same-day triple therapy with reduced-fluence PDT (300 mW/cm^2; 25 J/cm^2), intravitreal dexamethasone (4 mg), and bevacizumab (1.25 mg) in patients with AMD (Bakri et al. 2009). In this study, visual acuity was maintained, and decreased macular thickness was observed in patients with and without previous anti-VEGF therapy. The triple therapy appeared to reduce the number of anti-VEGF injections needed in some patients and stabilized vision in some patients not responding to anti-VEGF therapy (Bakri et al. 2009).

39.5 Recent Advances and Perspectives

In recent years, a revival of the use of PDT has occurred both in ophthalmology and oncology. This is, in part, due to the availability of new techniques for more selective light and/or photosensitizer delivery. A good example is the recently published oscillatory PDT (OPDT) using verteporfin (Peyman et al. 2011).

OPDT was effective in treating CNV lesions and central serous retinopathy (CSR). It is expected that OPDT is an improvement of standard PDT due to reduced side effects. Advances in ophthalmic PDT could be achieved by using newly developed PSs with more rapid clearance and better selectivity. A question that still remains to be answered is whether more selective targeting of choroidal neovascular membranes would improve the therapeutic outcome of PDT, leading to reduced re-treatment rates (Madar-Balakirski et al. 2010). Moreover, PDT is being used to treat new vascular-based disorders, like retinoblastoma. Retinoblastoma is the most common malignant intraocular tumor in children. It occurs in one or both eyes, visible as a whitish or yellowish spots through the pupil. It results in decreasing vision or loss of vision. The efficacy of PDT for this indication in xenografted nude mouse models has been studied recently with two PSs: mTHPC (administrated intraperitoneally) and verteporfin (injected intravenously) (Aerts et al. 2010).

Another advancement in the field of PDT is the development of new drug delivery systems. One such system includes a small semipermeable polymer capsule with a device called encapsulated cell technology (ECT), being tested by Neurotech SA with support from the National Eye Institute (NEI, Bethesda, MD). This device is being investigated to release a protein, ciliary neurotrophic factor (CNTF) into the vitreous of the patient's eyes. CNTF is produced by genetically modified cells contained in the ECT. The results from phase III trials for RP and dry AMD were presented at the annual 2011 ARVO meeting in Fort Lauderdale, FL. The results showed promise for this device, with the release of CNTF resulting in stabilization of vision in patients who started with good visual acuity (at least 20/63) and a reduction in the progression of lesion size in geographic atrophy.

The current use of PDT is driven mainly by its combination with other treatment strategies affecting distinctly separate pathways. Very fast development of new, more selective compounds occurs in the antiangiogenesis field. Over the last decades, many new mechanisms in the process of angiogenesis have been identified, leading to the development of angiostatic compounds with novel targets and mechanisms. The FDA approved Aflibercept/VEGF Trap (Regeneron, Tarrytown, NY) in 2011 for wet AMD based on phase III clinical trial (http://www.ncbi.nlm.nih.gov/pubmed/23503202), and small interfering RNA targeted against VEGF is also being evaluated (Campa and Harding 2011). Tubulin-binding agents like combretastatin (Nambu et al. 2003) or OC-10X are in clinical trials. A major achievement in the field of cancer is the development of many receptor tyrosine kinase inhibitors. Due to their small size, these molecules present the advantage of delivery through oral or topical administration, reducing the chances of complications associated with intravitreal injection. We have recently shown that the clinically used small molecule kinase inhibitors Nexavar (sorafenib), Tarceva (erlotinib), and Sutent (sunitinib) have therapeutic potential for application in combination with PDT (Nowak-Sliwinska et al. 2010). Another tyrosine kinase inhibitor, Vatalanib (Novartis, East Hanover, NJ), is being investigated in the ADVANCE clinical study. Vatalanib (PTK787/ZK 222584; Bayer Schering Pharma AG, Berlin, and Novartis, East Hanover, NJ) is an oral tyrosine kinase inhibitor (TKI) that inhibits the two key kinases—KIT and the platelet-derived growth factor receptor-a (PDGFRa)—and all three isoforms of the VEGF receptors (VEGFR-1, VEGFR-2, and VEGFR-3).

Recent developments in the treatment of AMD also involve the genetic factors influencing AMD. β-amyloid (A-beta) accumulates in the drusen of dry AMD patients but not in those without AMD, indicating that A-beta is a viable target for AMD treatment. A study by Ding et al. (2011) investigated the administration of antibodies against A-beta in a mouse model of AMD. This study showed a significant therapeutic affect and an improvement in visual acuity. The monoclonal antibody ponezumab (PF-04360365) targeted to A-beta is being tested by Pfizer and is currently undergoing phase II tests for Alzheimer's disease.

Another potential treatment for AMD is replacing damaged and unhealthy RPE by healthy progenitor cells. In a study series of 10 patients (among them 4 with AMD), human neural retinal progenitor cell layers and RPE were transplanted. All four patients with AMD had vision of 20/200 or worse and experienced improved visual acuity. Moreover, there was no graft rejection during a 6-year follow-up (Yang et al. 2008). In conclusion, PDT represents an important treatment modality in a wide spectrum of ocular disorders. Knowing its limitations will help to improve PDT and to achieve the preservation and long-term improvement of vision.

References

http://www.qltinc.com/newsCenter/2010/100615.htm.

Aerts, I., P. Leuraud, J. Blais et al. 2010. In vivo efficacy of photodynamic therapy in three new xenograft models of human retinoblastoma. *Photodiagnosis Photodyn Ther* 7:275–283.

Akaza, E., R. Mori, and M. Yuzawa. 2008. Long-term results of photodynamic therapy of polypoidal choroidal vasculopathy. *Retina* 28:717–722.

Amer, R., and N. Lois. 2011. Punctate inner choroidopathy. *Surv Ophthalmol* 56:36–53.

Anand, R. 2003. Photodynamic therapy for diffuse choroidal hemangioma associated with Sturge Weber syndrome. *Am J Ophthalmol* 136:758–760.

Antoszyk, A. N., L. Tuomi, C. Y. Chung, and A. Singh. 2008. Ranibizumab combined with verteporfin photodynamic therapy in neovascular age-related macular degeneration (FOCUS): Year 2 results. *Am J Ophthalmol* 145:862–874.

Arias, L., O. Pujol, M. Rubio, and J. Caminal. 2006. Long-term results of photodynamic therapy for the treatment of choroidal neovascularization secondary to angioid streaks. *Graefes Arch Clin Exp Ophthalmol* 244:753–757.

Au Eong, K. G. 2006. Age-related macular degeneration: An emerging challenge for eye care and public health professionals in the Asia Pacific region. *Ann Acad Med Singapore* 35:133–135.

Augustin, A. 2009. Triple therapy for age-related macular degeneration. *Retina* 29:S8–S11.

Bakri, S. J., S. M. Couch, C. A. McCannel, and A. O. Edwards. 2009. Same-day triple therapy with photodynamic therapy, intravitreal dexamethasone, and bevacizumab in wet age-related macular degeneration. *Retina* 29:573–578.

Barbazetto, I. A., T. C. Lee, and D. H. Abramson. 2004. Treatment of conjunctival squamous cell carcinoma with photodynamic therapy. *Am J Ophthalmol* 138:183–189.

Barnstable, C. J. and J. Tombran-Tink. 2004. Neuroprotective and antiangiogenic actions of PEDF in the eye: Molecular targets and therapeutic potential. *Prog Retin Eye Res* 23:561–577.

Besharse, J. C., and K. H. Pfenninger. 1980. Membrane assembly in retinal photoreceptors I. Freeze-fracture analysis of cytoplasmic vesicles in relationship to disc assembly. *J Cell Biol* 87:451–463.

Biarnes, M., J. Mones, J. Alonso, and L. Arias. 2011. Update on geographic atrophy in age-related macular degeneration. *Optom Vis Sci* 88:881–889.

Blasi, M. A., A. Scupola, A. C. Tiberti, P. Sasso, and E. Balestrazzi. 2006. Photodynamic therapy for vasoproliferative retinal tumors. *Retina* 26:404–409.

Blasi, M. A., A. C. Tiberti, A. Scupola et al. 2010. Photodynamic therapy with verteporfin for symptomatic circumscribed choroidal hemangioma: Five-year outcomes. *Ophthalmology* 117:1630–1637.

Blinder, K. J., M. S. Blumenkranz, N. M. Bressler et al. 2003. Verteporfin therapy of subfoveal choroidal neovascularization in pathologic myopia: 2-year results of a randomized clinical trial—VIP report no. 3. *Ophthalmology* 110:667–673.

Bressler, N. M. 2001. Photodynamic therapy of subfoveal choroidal neovascularization in age-related macular degeneration with verteporfin: Two-year results of 2 randomized clinical trials-tap report 2. *Arch Ophthalmol* 119:198–207.

Brown, D. M., P. K. Kaiser, M. Michels et al. 2006. Ranibizumab versus verteporfin for neovascular age-related macular degeneration. *N Engl J Med* 355:1432–1444.

Campa, C., and S. P. Harding. 2011. Anti-VEGF compounds in the treatment of neovascular age related macular degeneration. *Curr Drug Targets* 12:173–181.

Canal-Fontcuberta, I., D. R. Salomao, D. Robertson et al. 2012. Clinical and histopathologic findings after photodynamic therapy of choroidal melanoma. *Retina* 32:942–948.

Çekiç, O., Y. Bardak, and N. Kapucuoğlu. 2011. Photodynamic therapy for conjunctival ocular surface squamous neoplasia. *J Ocul Pharmacol Ther* 27:205–207.

Chakravarthy, U., A. P. Adamis, E. T. Cunningham, Jr. et al. 2006. Year 2 efficacy results of 2 randomized controlled clinical trials of pegaptanib for neovascular age-related macular degeneration. *Ophthalmology* 113:1508.e1–25.

Chan, W. M., T. Y. Lai, T. T. Lau et al. 2008. Combined photodynamic therapy and intravitreal triamcinolone for choroidal neovascularization secondary to punctate inner choroidopathy or of idiopathic origin: One-year results of a prospective series. *Retina* 28:71–80.

Chan, W. M., D. S. Lam, T. Y. Lai et al. 2003a. Choroidal vascular remodelling in central serous chorioretinopathy after indocyanine green guided photodynamic therapy with verteporfin: A novel treatment at the primary disease level. *Br J Ophthalmol* 87:1453–1458.

Chan, W. M., D. S. Lam, T. Y. Lai et al. 2004. Photodynamic therapy with verteporfin for symptomatic polypoidal choroidal vasculopathy: One-year results of a prospective case series. *Ophthalmology* 111:1576–1584.

Chan, W. M., D. S. Lam, T. H. Wong et al. 2003b. Photodynamic therapy with verteporfin for subfoveal idiopathic choroidal neovascularization: One-year results from a prospective case series. *Ophthalmology* 110:2395–2402.

Chan, W. M., T. H. Lim, A. Pece, R. Silva, and N. Yoshimura. 2010. Verteporfin PDT for non-standard indications—A review of current literature. *Graefes Arch Clin Exp Ophthalmol* 248:613–626.

Connolly, D. T., D. M. Heuvelman, R. Nelson et al. 1989. Tumor vascular permeability factor stimulates endothelial cell growth and angiogenesis. *J Clin Invest* 84:1470–1478.

Cook, K. M., and W. D. Figg. 2010. Angiogenesis inhibitors: Current strategies and future prospects. *CA Cancer J Clin* 60:222–243.

Coppens, G., L. Spielberg, and A. Leys. 2011. Polypoidal choroidal vasculopathy, diagnosis and management. *Bull Soc Belge Ophtalmol* 39–44.

Costa, R. A., M. E. Farah, J. A. Cardillo, D. Calucci, and G. A. Williams. 2003. Immediate indocyanine green angiography and optical coherence tomography evaluation after photodynamic therapy for subfoveal choroidal neovascularization. *Retina* 23:159–165.

Couch, S. M., and S. J. Bakri. 2011. Review of combination therapies for neovascular age-related macular degeneration. *Semin Ophthalmol* 26:114–120.

Das, A., and P. G. McGuire. 2003. Retinal and choroidal angiogenesis: Pathophysiology and strategies for inhibition. *Prog Retin Eye Res* 22:721–748.

Ding, J. D., L. V. Johnson, R. Herrmann et al. 2011. Anti-amyloid therapy protects against retinal pigmented epithelium damage and vision loss in a model of age-related macular degeneration. *Proc Natl Acad Sci USA* 108:E279–287.

Fingar, V. H. 1996. Vascular effects of photodynamic therapy. *J Clin Laser Med Surg* 14:323–328.

Frimpong-Boateng, A., A. Bunse, F. Rufer, and J. Roider. 2009. Photodynamic therapy with intravitreal application of triamcinolone acetonide in age-related macular degeneration: Functional results in 54 patients. *Acta Ophthalmol* 87:183–187.

Garcia-Arumi, J., L. Amselem, K. Gunduz et al. 2012. Photodynamic therapy for symptomatic subretinal fluid related to choroidal nevus. *Retina* 32:936–941.

Gelisken, F., B. A. Lafaut, W. Inhoffen et al. 2004. Clinicopathological findings of choroidal neovascularisation following verteporfin photodynamic therapy. *Br J Ophthalmol* 88:207–211.

Gollnick, S. O., S. S. Evans, H. Baumann et al. 2003. Role of cytokines in photodynamic therapy-induced local and systemic inflammation. *Br J Cancer* 88:1772–1779.

Gomi, F., M. Ohji, K. Sayanagi et al. 2008. One-year outcomes of photodynamic therapy in age-related macular degeneration and polypoidal choroidal vasculopathy in Japanese patients. *Ophthalmology* 115:141–146.

Green, W. R. 1999. Histopathology of age-related macular degeneration. *Mol Vis* 5:27.

Grossniklaus, H. E., and W. R. Green. 2004. Choroidal neovascularization. *Am J Ophthalmol* 137:496–503.

Haines, J. L., M. A. Hauser, S. Schmidt et al. 2005. Complement factor H variant increases the risk of age-related macular degeneration. *Science* 308:419–421.

Heier, J. S., D. Boyer, Q. D. Nguyen et al. 2011. The 1-year results of CLEAR-IT 2, a phase 2 study of vascular endothelial growth factor trap-eye dosed as-needed after 12-week fixed dosing. *Ophthalmology* 118:1098–1106.

Hera, R., M. Keramidas, M. Peoc'h et al. 2005. Expression of VEGF and angiopoietins in subfoveal membranes from patients with age-related macular degeneration. *Am J Ophthalmol* 139:589–596.

Hikichi, T., H. Ohtsuka, M. Higuchi et al. 2010. Improvement of angiographic findings of polypoidal choroidal vasculopathy after intravitreal injection of ranibizumab monthly for 3 months. *Am J Ophthalmol* 150:674–682.

Inoue, R., M. Sawa, M. Tsujikawa, and F. Gomi. 2010. Association between the efficacy of photodynamic therapy and indocyanine green angiography findings for central serous chorioretinopathy. *Am J Ophthalmol* 149:441–446.

Jurklies, B., and N. Bornfeld. 2005. The role of photodynamic therapy in the treatment of symptomatic choroidal hemangioma. *Graefes Arch Clin Exp Ophthalmol* 243:393–396.

Kaiser, P. K. 2006. Verteporfin therapy of subfoveal choroidal neovascularization in age-related macular degeneration: 5-year results of two randomized clinical trials with an open-label extension: TAP report no. 8. *Graefes Arch Clin Exp Ophthalmol* 244:1132–1142.

Kaliki, S., C. L. Shields, S. A. Al-Dahmash, A. Mashayekhi and J. A. Shields. 2012. Photodynamic therapy for choroidal metastasis in 8 cases. *Ophthalmology* 119:1218–1222.

Klein, R., B. E. Klein, and K. L. Linton. 1992. Prevalence of age-related maculopathy. The Beaver Dam Eye Study. *Ophthalmology* 99:933–943.

Koh, A., T. H. Lim, K. G. Au Eong et al. 2011. Optimising the management of choroidal neovascularisation in Asian patients: Consensus on treatment recommendations for anti-VEGF therapy. *Singapore Med J* 52:232–240.

Koh, S., and R. Haimovici. 2004. *Photodynamic Therapy of Ocular Diseases.* Lippincott Williams & Wilkins, Philadelphia, PA.

Lai, T. Y., C. P. Lam, F. O. Luk et al. 2010. Photodynamic therapy with or without intravitreal triamcinolone acetonide for symptomatic polypoidal choroidal vasculopathy. *J Ocul Pharmacol Ther* 26:91–95.

Lam, D. S., W. M. Chan, D. T. Liu et al. 2004. Photodynamic therapy with verteporfin for subfoveal choroidal neovascularisation of pathologic myopia in Chinese eyes: A prospective series of 1 and 2 year follow up. *Br J Ophthalmol* 88:1315–1319.

Madar-Balakirski, N., C. Tempel-Brami, V. Kalchenko et al. 2010. Permanent occlusion of feeding arteries and draining veins in solid mouse tumors by vascular targeted photodynamic therapy (VTP) with Tookad. *PLoS One* 5:e10282.

Madreperla, S. A. 2001. Choroidal hemangioma treated with photodynamic therapy using verteporfin. *Arch Ophthalmol* 119:1606–1610.

Maher, E. R., H. P. Neumann, von Hippel-Lindau, R. S. 2011. Disease: a clinical and scientific review. *Eur J Hum Genet* 19(6):617–623.

Menchini, U., G. Virgili, U. Introini et al. 2004. Outcome of choroidal neovascularization in angioid streaks after photodynamic therapy. *Retina* 24:763–771.

Michels, S., F. Hansmann, W. Geitzenauer, and U. Schmidt-Erfurth. 2006. Influence of treatment parameters on selectivity of verteporfin therapy. *Invest Ophthalmol Vis Sci* 47:371–376.

Nambu, H., R. Nambu, M. Melia, and P. A. Campochiaro. 2003. Combretastatin A-4 phosphate suppresses development and induces regression of choroidal neovascularization. *Invest Ophthalmol Vis Sci* 44:3650–3655.

Nowak-Sliwinska, P. 2012. Anti-angiogenic treatment for exudative age-related macular degeneration: New strategies are underway. *Curr Angiogen* 1:318–334.

Nowak-Sliwinska, P., M. Storto, T. Cataudella et al. 2012. Angiogenesis inhibition by the maleimide-based small molecule GNX-686. *Microvasc Res* 83:105–110.

Nowak-Sliwinska, P., J. van Beijnum, M. van Berkel, H. van den Bergh, and A. Griffioen. 2010. Vascular regrowth following photodynamic therapy in the chicken embryo chorioallantoic membrane. *Angiogenesis* 13:281–292.

Nowak-Sliwinska, P., J. R. van Beijnum, A. Casini et al. 2011. Organometallic ruthenium(II) arene compounds with anti-angiogenic activity. *J Med Chem* 54:3895–3902.

Peyman, G. A., M. Tsipursky, N. Nassiri, and M. Conway. 2011. Oscillatory photodynamic therapy for choroidal neovascularization and central serous retinopathy; a pilot study. *J Ophthalmic Vis Res* 6:166–176.

Qazi, Y., S. Maddula, and B. K. Ambati. 2009. Mediators of ocular angiogenesis. *J Genet* 88:495–515.

Rudolf, M., S. Michels, U. Schlotzer-Schrehardt, and U. Schmidt-Erfurth. 2004. [Expression of angiogenic factors by photodynamic therapy]. *Klin Monatsbl Augenheilkd* 221:1026–1032.

Sachdeva, R., H. Dadgostar, P. K. Kaiser, J. E. Sears, and A. D. Singh. 2010. Verteporfin photodynamic therapy of six eyes with retinal capillary haemangioma. *Acta Ophthalmol* 88:e334–e340.

Sagong, M., S. Lim, and W. Chang. 2012. Reduced-fluence photodynamic therapy combined with intravitreal bevacizumab for polypoidal choroidal vasculopathy. *Am J Ophthalmol* 153:873–882.

Schmidt-Erfurth, U. M., S. Michels, C. Kusserow, B. Jurklies, and A. J. Augustin. 2002. Photodynamic therapy for symptomatic choroidal hemangioma: Visual and anatomic results. *Ophthalmology* 109:2284–2294.

Schmidt-Erfurth, U., J. Miller, M. Sickenberg et al. 1998. Photodynamic therapy of subfoveal choroidal neovascularization: Clinical and angiographic examples. *Graefes Arch Clin Exp Ophthalmol* 236:365–374.

Schmidt-Erfurth, U., M. Niemeyer, W. Geitzenauer, and S. Michels. 2005. Time course and morphology of vascular effects associated with photodynamic therapy. *Ophthalmology* 112:2061–2069.

Schmidt-Erfurth, U., U. Schlotzer-Schrehard, C. Cursiefen et al. 2003. Influence of photodynamic therapy on expression of vascular endothelial growth factor (VEGF), VEGF receptor 3, and pigment epithelium-derived factor. *Invest Ophthalmol Vis Sci* 44:4473–4480.

Schmidt-Erfurth, U., and S. Wolf. 2008. Same-day administration of verteporfin and ranibizumab 0.5 mg in patients with choroidal neovascularisation due to age-related macular degeneration. *Br J Ophthalmol* 92:1628–1635.

Seddon, J. M., R. Reynolds, Y. Yu, M. J. Daly, and B. Rosner. 2011. Risk models for progression to advanced age-related macular degeneration using demographic, environmental, genetic, and ocular factors. *Ophthalmology* 118:2203–2211.

Sickenberg, M., U. Schmidt-Erfurth, J. W. Miller et al. 2000. A preliminary study of photodynamic therapy using verteporfin for choroidal neovascularization in pathologic myopia, ocular histoplasmosis syndrome, angioid streaks, and idiopathic causes. *Arch Ophthalmol* 118:327–336.

Singh, A. D., P. K. Kaiser, and J. E. Sears. 2005. Choroidal hemangioma. *Ophthalmol Clin North Am* 18:151–161, ix.

Singh, A. D., M. Nouri, C. L. Shields, J. A. Shields, and N. Perez. 2002. Treatment of retinal capillary hemangioma. *Ophthalmology* 109:1799–1806.

Smretschnig, E., S. Ansari-Shahrezaei, S. Moussa et al. 2012. Half-fluence photodynamic therapy in acute central serous chorioretinopathy. *Retina* 32:2014–2019.

Spitzer, M. S., F. Ziemssen, K. U. Bartz-Schmidt, F. Gelisken, and P. Szurman. 2008. Treatment of age-related macular degeneration: Focus on ranibizumab. *Clin Ophthalmol* 2:1–14.

Strauss, O. 2005. The retinal pigment epithelium in visual function. *Physiol Rev* 85:845–881.

Ting, A. Y., T. K. Lee, and I. M. MacDonald. 2009. Genetics of age-related macular degeneration. *Curr Opin Ophthalmol* 20:369–376.

Treatment of Age-Related Macular Degeneration with Photodynamic Therapy (TAP) Study Group. 1999. Photodynamic therapy of subfoveal choroidal neovascularization in age-related macular degeneration with verteporfin: One-year results of 2 randomized clinical trials—TAP report. *Arch Ophthalmol* 117:1329–1345.

Tsipursky, M. S., P. R. Golchet, and L. M. Jampol. 2011. Photodynamic therapy of choroidal hemangioma in Sturge-Weber syndrome, with a review of treatments for diffuse and circumscribed choroidal hemangiomas. *Surv Ophthalmol* 56:68–85.

Tsuchiya, D., T. Yamamoto, R. Kawasaki, and H. Yamashita. 2009. Two-year visual outcomes after photodynamic therapy in age-related macular degeneration patients with or without polypoidal choroidal vasculopathy lesions. *Retina* 29:960–965.

Tsujikawa, A., S. Ooto, K. Yamashiro et al. 2010. Treatment of polypoidal choroidal vasculopathy by intravitreal injection of bevacizumab. *Jpn J Ophthalmol* 54:310–319.

van Beijnum, J. R., P. Nowak-Sliwinska, E. Van den Boezem, P. Hautvast, W. A. Buurman, A. W. Griffioen. 2013. Tumor angiogenesis is enforced by autocrine regulation of high-mobility group box 1. *Oncogene* 17;32:363–374.

van den Bergh, H. 2001. Photodynamic therapy of age-related macular degeneration: History and principles. *Semin Ophthalmol* 16:181–200.

Weiss, A., H. van den Bergh, A. W. Griffioen, and P. Nowak-Sliwinska. 2012. Angiogenesis inhibition for the improvement of photodynamic therapy: The revival of a promising idea. *Biochim Biophys Acta* 1826:53–70.

Yang, Z., C. Stratton, P. J. Francis et al. 2008. Toll-like receptor 3 and geographic atrophy in age-related macular degeneration. *N Engl J Med* 359:1456–1463.

Yip, P. P., C. F. Woo, H. H. Tang, and C. K. Ho. 2009. Triple therapy for neovascular age-related macular degeneration using single-session photodynamic therapy combined with intravitreal bevacizumab and triamcinolone. *Br J Ophthalmol* 93:754–758.

Zhang, X., F. Jiang, Z. G. Zhang et al. 2005. Low-dose photodynamic therapy increases endothelial cell proliferation and VEGF expression in nude mice brain. *Lasers Med Sci* 20:74–79.

Zhang, Y., W. Liu, Y. Fang et al. 2010. Photodynamic therapy for symptomatic circumscribed macular choroidal hemangioma in Chinese patients. *Am J Ophthalmol* 150:710–715.

40

Photodynamic Therapy in Dermatology

Alexis Sidoroff
Medical University of Innsbruck

40.1 Introduction

Already at the time of the discovery of its mechanism and coining the term "photodynamic" in the early 20th century, dermatologic disorders had been treated with photodynamic therapy (PDT) by von Tappeiner in Germany and Finsen in Norway. After this, it took about 70 years—until the works of Tom Dougherty—for the idea of using PDT to be picked up again (Dougherty 1984). The main drawback of early PDT was the use of systemic photosensitizers like Photofrin and hematoporphyrin derivative (HPD), which led to a very long photosensitivity of the treated patients. This made systemic PDT not very attractive for the treatment of small and localized cutaneous lesions.

The breakthrough for PDT in dermatology came with the idea to use a sensitizer prodrug—namely 5-aminolevulinic acid (5-ALA)—that was small enough to penetrate the skin barrier and thus could be applied topically (Kennedy, Pottier, and Pross 1990). This not only limited photosensitization to the treated area but also drastically reduced the time of PDT procedures. Two additional factors made this procedure extremely appealing: First, the transformation of the prodrug to the active sensitizer—protoporphyrin IX (PPIX)—occurs much faster in different kinds of diseased tissue as compared to normal skin, making the treatment selective, and second, PPIX is used up during the illumination process (photobleaching effect), which makes overtreatment virtually impossible.

To understand the current status of PDT in dermatology, a brief look at its historical development over the last decades is necessary. As ALA was not a registered drug but a commercially available chemical substance, and the legal regulations were different in the 1980s, the idea of a topical PDT spread to a few university-based centers worldwide. Concurrently, treatment of patients started quite quickly, coupled with extensive basic research on the mechanisms going on in PDT. In spite of a large number of promising results in an academic setting and a good understanding of the mechanisms involved in PDT, the problem of turning it into a licensed drug/procedure remained. Only with the interest of pharmaceutical companies were necessary clinical trials performed, making PDT an approved treatment modality in many countries. Because of the evidence from many publications and guidelines (Braathen et al. 2007; Morton, McKenna, and Rhodes 2008), PDT has now positioned itself as a well-accepted therapy in the dermatological field.

40.2 The Standard Procedure of Topical PDT in Dermatology

As described elsewhere in this book, PDT basically needs three things to work: a diseased "target tissue" with sufficient amounts of oxygen, a photosensitizer, and light. A combination of the three leads to tissue destruction by formation of reactive oxygen species (ROS). In dermatology, an agent that can be topically applied to the diseased tissue is favorable (see above), and sensitization should be selective for diseased tissue.

A standard procedure in topical PDT could be described as follows (Christensen et al. 2010): After some sort of lesion preparation (mainly to remove crusts and scales that could impair penetration), a sensitizer prodrug (usually ALA or its methyl ester) is applied on the diseased area (often under an occlusive dressing). After the precursor has penetrated the surface and reached the diseased tissue, it is being transformed predominantly to PPIX. The optimal time point for illumination is when the ratio of PPIX between the diseased tissue and the healthy surrounding is highest, making the treatment as selective as possible. After removal of an eventual dressing and/or remnants of the sensitizer preparation, fluorescence diagnosis can be performed. Then PDT irradiation with visible light takes place. In terms of illumination specifications, wavelength should clearly match one of the absorption peaks of PPIX, fluence should be sufficient to obtain the desired therapeutic result, but the energy of the light sources should not reach the level of thermal effects.

PDT has to be regarded as a procedure where all three components itemized above are subject to a wide range of given or intended variations. This explains why so-called standards may vary over time and should be taken into account when comparing data from the literature.

40.3 Lesion Preparation

In the early days of dermatological PDT, the importance of lesion preparation was greatly underestimated. But with increasing experience and efforts to improve PDT, it became clear that scales and crusts can have negative effects on both the penetration of the topical sensitizer precursor and light. Pretreatment with emollients or keratolytic agents prior to PDT can be helpful. Now it is generally recommended that scales and crusts are removed with a curette before application of the sensitizer (Christensen et al. 2010). Bleeding should be avoided and, if it occurs, stopped before PDT. In superficial pigmented basal cell carcinomas (BCCs), pigment can often easily be removed by curettage. In nodular BCCs (nBCCs), deeper tumor debulking within the tumor borders can make thicker lesions more responsive to PDT. This debulking usually takes place several days before the PDT session. New approaches to enhance sensitizer penetration include pretreatment with an ablative or fractional carbon dioxide (CO_2) laser (Togsverd-Bo et al. 2012) or the use of microneedles (Clementoni, B-Roscher, and Munavalli 2010).

40.4 Photosensitizers

5-ALA was introduced to the dermatological world by the works of Kennedy and Pottier in the early 1990s (Kennedy, Pottier, and Pross 1990). This prodrug penetrates the skin and is metabolized to PPIX, which generates ROS after illumination, leading to apoptosis and necrosis of targeted tissue. Many centers used or still use "homemade" 5-ALA preparations, prepared by hospital pharmacies, most commonly at a concentration of 20% in a cream base, but many variations have been tried (gel formulations; addition of penetration enhancers like glycolic acid, oleic acid, or dimethyl sulfoxide; different concentrations depending on the indication; etc.). Other attempts to improve 5-ALA activity included raising the temperature during incubation (Gerritsen et al. 2009) or adding the ferrochelatase inhibitor desferrioxamine to increase PPIX production (Fijan, Honigsmann, and Ortel 1995).

In the United States, an ALA solution (Levulan, DUSA Pharmaceuticals, Wilmington, MA) in a special application device (Kerastick) has been approved for the treatment of non-hyperkeratotic actinic keratosis (AK) in conjunction with a blue light source (Blue-U). Although incubation time for the licensed Levulan/Kerastick application was specified to be between 18 and 24 h, protocols with incubation times of 1 or 3 h have been described with response rates in AKs of 76% or 85%, respectively.

European industry has modified 5-ALA by methyl esterification. A range of clinical trials has been performed with the methylated ester of ALA, usually referred to as MAL (Metvix or Metvixia, developed by Photocure S.A., Norway, and now owned by Galderma Pharmaceuticals, Paris, France). These trials with methyl aminolevulinate (160 mg/g) have led to the approval of MAL by regulatory authorities in many countries for the treatment of AK, Bowen's disease (BD), superficial BCCs (sBCCs), and partially nBCCs up to a thickness of 2 mm. Incubation time for MAL (i.e., the time after which the highest PPIX concentration in the tumor as compared with healthy surrounding tissue could be demonstrated) is about 3 h. Apart from this application time—which usually should be between 5 and 6 h—the treatment procedures with ALA creams and the commercial MAL preparation are very comparable. In spite of some chemical differences (MAL being more lipophilic), the few trials comparing ALA to MAL did not show any significant difference in response rates. Recently, a self-adhesive patch containing ALA gained approval for the treatment of mild to moderate AK, its advantages being that pretreatment is not necessary and a very standardized single treatment cycle with red light is considered sufficient (Alacare, Spirig AG, Egerkingen, Switzerland) (Szeimies et al. 2010b). In addition, BF-200, an ALA nanoemulsion gel to be used with red light, has been compared to MAL and placebo in the treatment of AK (Ameluz, Biofrontera AG, Leverkusen, Germany) and showed favorable results (Szeimies et al. 2010a).

40.5 Light Sources and Illumination Parameters

Ideal therapeutic light sources must emit a wavelength that corresponds to one of the absorption peaks of PPIX (main peak at

410 nm; Q bands at 505, 540, 580, and 630 nm). Moreover, tissue optics demand that longer wavelengths penetrate deeper into the skin (which is why most of the light sources target the 630-nm peak, i.e., is in the range of visible red light). The type of light sources thus varies from lasers (one wavelength, small illumination field) over light-emitting diode (LED) arrays (narrow wavelength spectrum, medium illumination field) to filtered xenon arc and metal halide lamps (broader wavelength spectrum, large illumination field), fluorescent lamps, and recently, daylight (very wide spectrum, very large illumination field). In addition, intense pulsed light (IPL) sources have been used. Theoretically, they would be quite suitable, but unfortunately, the cutoff filter notations provided by the IPL manufacturers are not reflected in the phototoxic effect (Maisch et al. 2011). Thus, without detailed measurement of the emitted spectrum and intensity distribution, the biologic effect described or experienced for one light source cannot be transferred to another IPL. The lack of comparability also holds true for all other types of light sources. This has to be taken into account when interpreting data in the literature. Today LED arrays [e.g., the Aktilite 16 and 128 (Galderma, France) or the Omnilux (Phototherapeutics Ltd., United Kingdom)] are very commonly used because of their easy handling and the shorter illumination times needed. Recently, portable LED devices have been investigated (Attili et al. 2009). Their main drawback is the small illumination field. As well, lasers have the disadvantage that, even with diffuser lenses, only small fields can be illuminated, and they are comparatively expensive, which leads to usage mainly where they are already available. Lamps [like the Waldmann PDT 1200L (Waldmann, Germany)] have the advantage of larger illumination fields but are quite bulky devices that require longer illumination times. For light sources that are marketed for use in PDT, dosimetry is standardized for the treatment of nonmelanoma skin cancer (NMSC). In inflammatory skin diseases, lower fluences and multiple treatments are usually used.

One approach to improve topical PDT in dermatology is to fraction illumination time (i.e., to interrupt light exposure, allowing reoxygenation and additional formation of PPIX in the "dark" intervals). This has shown to be beneficial in a couple of trials (de Haas et al. 2008).

Another strategy that evolved over the last few years is to use natural daylight as a light source in the treatment of AK (Wiegell et al. 2011). Clinical studies in Scandinavia have done so with application of MAL for 30 min and subsequent exposure to daylight (2.5 and 1.5 h). Although this procedure is very suitable especially for the treatment of multiple AKs on the head because of its ease of use and the fact that it is nearly painless, it is still an off-label method. Issues like the influence of weather conditions and the need for application of an appropriate sunscreen need further clarification.

40.6 Fluorescence Diagnosis

PPIX fluorescence can be demonstrated either with a simple handheld Wood's light [emitting long-wave ultraviolet (UVA) radiation] or sophisticated computer-assisted systems using a charge-coupled device (CCD) camera and digital image analysis. While the interpretation of what can be seen with the simple routine techniques, the computer-based systems allow for a semiquantitative measurement of fluorescence. It has to be clear, though, that only superficial fluorescence can be detected. Fluorescence diagnosis thus can be of some additional help in determining tumor borders; detecting subclinical lesions, residual lesions, and recurrences; or discriminating them from scars (Neus et al. 2009).

Rather than initial fluorescence intensity, the amount of photobleaching seems to be correlated to clinical response (Tyrrell, Campbell, and Curnow 2010), but fluorescence intensity seems to be proportional to the level of pain that patients might experience during therapy (Wiegell et al. 2008).

Detection of fluorescence during Mohs micrographic surgery to determine tumor borders has so far not provided consistent evidence of improvement of the clinical outcome (Lee, Kim, and Kim 2010).

40.7 Clinical Effects and Pain Management

Although a few cases of intolerance to the sensitizer (prodrug) are known (Korshoj et al. 2009), application of the agent itself does not, in general, cause any problems. After the time it takes to metabolize ALA or MAL to PPIX, the tissue is photosensitized. Exposure to light of the appropriate wavelength(s) leads to apoptosis and necrosis of tissue due to ROS. Localized photosensitivity can last for up to 2 days (ALA half-life of 24 h, MAL-induced PPIX clearing within 24 to 48 h from normal skin).

The most relevant side effect is a stinging, burning sensation during illumination. This perception of pain is very variable from patient to patient and even from lesion to lesion. The detailed mechanisms of this perception are not totally clear, but the current thinking is that it is caused by a direct depolarization of nerve C- and Aδ fibers (type II pain, not mediated by prostaglandins and bradykinine) and/or tissue damage by ROS. Most patients tolerate PDT without analgesic or anesthetic intervention. A recent study tried to describe the patients that are at high risk of suffering from pain to an extent that an intervention is necessary (Miller et al. 2011). Risk factors are mainly the size and site of the area to be treated (large fields, well-innervated areas, sensitive skin types, preexisting inflammation). The second treatment session within a treatment cycle (usually 7 days after the first session) also seems to be more painful than the first. Quite a large number of pain management strategies have been tried. Topical anesthetics (tetracain gel, lignocaine/prilocain mixture, morphine gel) have not shown to be helpful; neither was pretreatment with capsaicin. Systemic nonsteroidal antiphlogistic drugs (NSAR) are of little help, as the block mediators are only secondarily involved in PDT pain.

In many cases, the use of a simple cold-air fan or cold-water spray or packs can reduce discomfort to a tolerable level. Cold-airflow devices can significantly cool down the treated area (Pagliaro et al. 2004), although there are some concerns that this might reduce the efficacy of the treatment by reducing oxygenation due to vasoconstriction or slowing down biochemical reactions (Tyrrell, Campbell, and Curnow 2011).

Nerve blockers, at sensitive sites such as the forehead or the entire scalp, seem to relieve pain of PDT quite well (Halldin et al. 2009; Paoli et al. 2008). A side-to-side comparison in PDT of field cancerization resulted in a mean visual analog score (VAS) of 1.3 on the anesthetized side versus 7.5 on the other. Another study where extensive AKs on the face were treated showed VAS on the nerve block side of 1.0 versus 6.4 on the unanesthetized side.

Differences in pain between ALA and MAL have been of particular interest. It was first reported that MAL-PDT was less painful than ALA-PDT in treating AK (Moloney and Collins 2007). In these comparative trials, ALA application time was 5–6 h in contrast to 3 h for MAL. In a comparative acne trial where both substances were applied for the same time (3 h), this difference in pain perception was not present (Moloney and Collins 2007). In tape-stripped healthy skin, again, ALA-PDT was more painful than MAL-PDT. Explanations given were that ALA is taken up by delta-aminobutyric acid receptors of nerve fibers while MAL is not and/or that ALA is less selective for diseased tissue (Wiegell et al. 2003). High accumulation of PPIX in circumscribed diseased tissue occurs in both ALA- and MAL-PDT. Differences seem to be relevant only when larger areas of normal skin are within the treatment field.

In addition, it has to be mentioned that the effect of such simple measures as distraction of the patient during illumination should not be underestimated.

After illumination, erythema and edema have to be expected, followed by crust formation and erosions that take 2 to 6 weeks to heal. Heavy urticating inflammatory responses have been described (Kerr, Ferguson, and Ibbotson 2007). Postinflammatory pigment changes are rather the exception, and the risk of hyperpigmentation can probably be reduced by avoiding UV exposure in the days after treatment (Moseley et al. 2006). If scars occur after PDT, they are most probably the consequence of tissue damage by the disease itself (e.g., thicker tumors) rather than a side effect of the treatment. The same is probably true for hair loss in treated areas, although there is a sensitization of the pilosebaceous unit. In general, the cosmetic outcome after PDT is excellent and superior to other treatment modalities (especially surgery and cryotherapy). Although PDT can trigger antitumorigenic and protumorigenic mechanisms, the risk of promoting carcinogenesis or tumor progression seems to be very low (Morton, McKenna, and Rhodes 2008).

40.8 Oncological Indications

NMSCs have been primary targets for PDT in dermatology since its very beginning (Sidoroff 2007; Sidoroff and Thaler 2010; Szeimies et al. 2005). In the last few years, the concept of field cancerization and the fact that AK has been recognized to be a noninvasive (i.e., in situ) cancer rather than a "precancerous" lesion fortified the need for field treatment modalities and thus—among others—PDT (Braathen et al. 2012).

40.8.1 PDT of AK

AKs share the same molecular changes and risk factors as invasive squamous cell carcinoma (SCC). Although the progression of a single AK to invasive cancer is a rare event, the numbers are growing because of demographic reasons and lifestyle (the population is getting older, and total UV exposure is increasing). AKs respond very well to PDT, especially when located on the face and scalp region. On the other hand, one must be aware that there is a plethora of treatment options for AK, from the easy-to-apply cryotherapy to topical treatments that can be applied at home (imiquimod, diclofenac sodium, 5-fluorouracil) to minor invasive surgical procedures (curettage, CO_2 laser ablation). The wide variability of clinical constellations demands an individualized treatment decision. Especially where multiple, ill-demarcated AKs prevail, where the cosmetic result is important, or where correct self-treatment by the patient cannot be guaranteed, PDT will play out its strengths. Another aspect is that management of a field disease should include both treatment and secondary prevention. Evidence for the preventive potential of PDT derives from a split-face placebo controlled study, where the time to develop new AKs in sun-damaged skin took much longer on the treated side, and two studies showing the decrease of p53 expression (Szeimies et al. 2012).

In therapeutic trials, thin or moderately thick lesions show clearance rates up to 92% 3 months after treatment. These numbers drop to 63–69% if patients are followed for a year. However, numbers like these have to be interpreted with care. Especially in the context of AKs as manifestations of a field disease, it is hard to draw a line between recurrences and new lesions.

The basic conclusion of therapeutic trials is that there is enough evidence to support PDT as a very suitable treatment for AKs (evidence level 1, strength of recommendation A) (Piacquadio et al. 2004; Szeimies et al. 2002).

A randomized intraindividual trial (1501 AKs in 119 patients) compared PDT with methyl aminolevulinate [application time 3 h followed by illumination with red LED light (634 +/−3 nm)] with cryotherapy (Morton et al. 2006). After the first treatment cycles, PDT showed an initial cure rate of 87% versus 76% in the cryotherapy group. After nonresponders were re-treated, both groups showed approximately the same response (89% vs. 86%). With the same PDT protocol, a clearance rate of 93% in thin and 70% in moderately thick lesions could be shown after two treatments 1 week apart. The recommendation for the licensed use of MAL-PDT is to treat AKs once (with the protocol described above) and to re-treat lesions after 3 months if necessary.

AKs on acral sites do not respond as well to PDT as face and/or scalp lesions (off-label indication). But this also holds true for other treatment modalities and is probably due to a larger number of thick lesions at these sites. MAL-PDT has been compared to cryotherapy, with an inferior clearance rate of 78% at 6 months (cryotherapy 88%), and imiquimod, which was significantly more effective in moderately thick lesions (58% vs. 37%) and comparably effective in thin lesions (72%).

PDT has also been investigated to treat actinic cheilitis. Although a response can be clearly seen, the rather low long-term histological clearance rates suggest that the treatment procedure has to be optimized. One promising way could be to combine PDT with other topical treatments. MAL-PDT followed by a treatment with 5% imiquimod cream achieved a clinical cure rate of 80% and a histological cure rate of 73%.

As mentioned before, there is a variety of modifications to the standard protocol that is being used in dermatology: pretreatment of lesions, sensitizer application with an adhesive patch, ambulatory light delivery (by portable light sources or exposure to daylight), fractionated illumination regimens, and many more. Most of these can be applied to the treatment of AKs. One of these strategies recently led to the approval of ALA-PDT with the BF-200 nanoemulsion (clearance rate 90% vs. 83% with MAL-PDT) (Szeimies et al. 2010a).

40.8.2 PDT in BD

Although BD [squamous cell carcinoma (SCC) in situ] is by far less frequent than AK, PDT has established itself as a standard BD treatment. Especially in lesions where surgery would result in disturbing scars (large lesions) or at sites of poor healing (e.g., lower legs, hands and feet, skin predamaged by irradiation dermatitis or epidermolysis bullosa), guidelines recommend PDT as the treatment of choice (evidence level 1, strength of recommendation A) (Cox, Eedy, and Morton 2007). Studies report response rates between 86% and 93% after 3 months and of 68% to 71% after 2 years with excellent cosmetic results. One treatment cycle consists of two treatments 7 days apart and can be repeated as necessary. In another study, the cure rate was 76% after 16 months after two MAL-PDT sessions.

Although published numbers and sustained response rates are quite low, PDT can also be considered in the treatment of erythroplasia of Queyrat (penile intraepithelial neoplasia) (Paoli et al. 2006). With respect to the alternative of relatively invasive and mutilating radical surgical removal of the tumor, repeated PDT sessions under thorough follow-up can be a therapeutic option in affected patients.

40.8.3 PDT of Invasive SCC

As depth is a limiting factor in topical PDT, thicker lesions have a high risk of not being sufficiently treated. In addition, the less differentiated SCC cells are, the less they seem to be sensitive to PDT. This is reflected in cure rates of 57% for microinvasive and 26% for nodular invasive SCC so that PDT in its actual form cannot be recommended in the treatment of SCC where a curative approach due to the risk of metastasis is essential (Calzavara-Pinton et al. 2008).

40.8.4 PDT of sBCC

Clearance rates of up to 97% have been reported in primary sBCCs. A review of 12 studies resulted in a weighted initial clearance rate of 87% (Peng et al. 1997). A long-term follow-up

observation showed a recurrence rate of 22% after 3 years with no further increase at 5 years. When compared with cryotherapy, ALA-PDT and MAL-PDT were as efficacious but resulted in better cosmetic outcome with shorter healing times. As well, surgery did not result in statistically better cure rates in sBCC (92% initial clearance for PDT and 99% for surgery, 9% and 0% recurrences at 1 year). Again, cosmetic results were better in PDT. PDT is recommended as the treatment of choice for sBCC, especially in extensive low-risk lesions where excision would lead to large scars (evidence level 1, strength of recommendation A). The recommended standard protocol is two sessions of PDT 7 days apart with a repeat cycle 3 months later, if necessary. Fractionated illumination protocols have been tried (de Haas et al. 2008) and shown to be superior to continuous illumination (97% vs. 89% after 12 months).

40.8.5 PDT of nBCC

Interpretation of published results is more complex for nBCCs. Although evidence is strong enough to recommend PDT as fair treatment for nBCCs (evidence level 1, strength of recommendation B), many aspects have to be considered. First, PDT will be less effective the thicker the lesion is. Then, lesions in the so-called H-zone do not respond as well. In addition, a semantic problem arises where lesion preparation is concerned. The curettage that is recommended prior to PDT of AKs or sBCCs to remove hyperkeratotic material, scales, and crusts can be extended in nBCCs to debulk the tumor. This obviously reduces the thickness of tissue treated with PDT. While some might consider this lesion preparation within the tumor borders as a self-evident part of the treatment procedure, others interpret it as a combination of a surgical intervention with PDT. Unfortunately, details on lesion preparation/debulking are often not given in publications, making it hard to compare the resulting response rates.

In general, the response rates for nBCC are inferior to surgical excision (initial clearance of 98%, recurrences of 4% for surgery vs. clearance of 91%, recurrences of 14% for PDT at 5 years). The comparison of 12 studies showed a weighted initial clearance rate of 53% (Peng et al. 1997). One study reports a clearance rate of 91% in primary nBCC. This rate dropped to 76% after a 3-year follow-up, but no additional recurrences occurred at 5 years. Two trials using the standard protocol (i.e., two sessions of PDT 7 days apart, second treatment cycle after 3 months if necessary) showed quite different response rates, between 33% and 73% (up to 89% in facial lesions). The difference might be due to different usage of prior tumor debulking. As in sBCC, comparison between ALA- and MAL-PDT with cryotherapy resulted in equivalent efficacy (76% overall clearance rate after 5 years) but superior cosmetic outcome with PDT. A fractionated protocol with 20 and 80 J/cm^2 clearing was maintained at 2 years in 80% of the lesions. A similar result (81% clearance rate) could be achieved at 5 years' follow-up in a trial using lesion curettage and ALA-PDT with dimethylsulfoxide as a penetration enhancer.

In the special clinical constellation of patients with Gorlin–Goltz syndrome (naevoid basal cell carcinoma syndrome), PDT can help to reduce the need for surgical intervention and is thus often much appreciated by affected patients (Morton, McKenna, and Rhodes 2008).

Morpheic BCCs respond very poorly to PDT, so this treatment modality is not recommended in these lesions. In conclusion, the higher risk of recurrence has to be considered when using PDT in the treatment of nBCCs. Still, it can be a treatment option when taking into account relative contraindications for surgery, patients' preferences, cosmetic concerns, or comorbidities. This risk of recurrence has to be reflected in the follow-up of patients treated with PDT.

40.8.6 PDT of NMSC in Organ Transplant Recipients

NMSC in organ transplant recipients (OTRs) is not only much more frequent but also behaves much more aggressively than in nonimmunosuppressed patients. In addition, the lesions can be widely spread over the body. PDT is being investigated and used in the management of such patients not only to treat lesions but also with the idea of prevention in mind (Basset-Seguin et al. 2011). It is clear that OTRs have to be monitored very strictly for NMSC and for their response to treatment due to their early metastatic potential. As well, efficacy of PDT (and other nonsurgical treatments) might be impaired by differences in immune response factors.

PDT has shown to be effective, with an overall 90% clearance rate (56 of 62) of AK in a randomized controlled trial comparing MAL-PDT to placebo PDT in two 2 × 2 cm areas. A 71% clearance was found in another trial with lower response on acral sites (40%). Another study targeted 32 facial lesions: 21 BCCs, 8 AKs, and 1 keratoacanthoma could be cured, whereas 2 invasive SCCs did not respond. In a comparative trial, MAL-PDT was shown to be more effective than 5-FU, but the total numbers of patients was very low. Eight of nine lesion areas could be cleared with PDT compared to one of nine with 5-FU (lesional area reduction was reported to be 100% in PDT vs. 79% in 5-FU). The difference between OTRs and immunocompetent patients was assessed in another trial: At 4 weeks, response rates were comparable (86% in OTRs, 94% in immunocompetent patients), but 2 months later, recurrences appeared, especially on OTRs, so that their response rate dropped to only 48%, while it still was 72% in the control group.

Other trials tried to assess the preventive potential of PDT in OTR with rather inconsistent results. An open intraindividual randomized trial showed a significant delay in the occurrence of new lesions in 27 renal OTRs after surface debridement and one single MAL-PDT: 9.6 versus 6.8 months. At month 12, 62% of patients were free from new lesions on the PDT side versus 35% on the untreated contralateral control side. Treating 81 OTRs with MAL-PDT at baseline and day 7 and then at months 3, 9, and 15 also resulted in a significant reduction of the occurrence of new lesions when compared to an untreated

control site in the same patient. The finding that this beneficial effect was lost after 27 months suggests that repeated treatments (at intervals yet to be defined) probably have to follow a long-term strategy.

All the studies mentioned above have used red light, following the European approach where deep penetration of light is desired in the treatment of NMSC. One trial with ALA and blue light did not show a reduced number of SCCs in the treated area after 2 years in 40 OTRs, whereas in another trial that used ALA treatments with blue light every 4 to 8 weeks for 2 years in 12 OTRs, a reduction of new SCCs compared with the year before PDT could be demonstrated (95% at 24 months).

40.8.7 PDT of Extramammary Paget's Disease

Some case reports and small case series suggest improvement of extramammary Paget's disease (EMPD) after PDT. This can have a positive impact on the quality of life for affected patients, but due to the nature of the disease, clinical improvement at the surface does not mean clearance from tumor. Histology (if necessary at multiple sites) is absolutely crucial to assess response to treatment and has not been performed in all published reports. In one study with five patients, ALA-PDT cleared 8 of 16 lesions at 6 months, with 3 three more recurring after an additional 3–4 months (Shieh et al. 2002). Another small series described four responders to MAL-PDT in seven patients with recurrent EMPD, and two other patients responding to ALA-PDT have been published. Histologically confirmed clearance has been described in four patients where ALA has been applied with adhesive patches.

40.8.8 PDT of Cutaneous Lymphoma

That PDT is effective in cutaneous T-cell lymphoma has been reported in a small case series and single case reports. Mostly, the same treatment regimens (with ALA or MAL) as for BCC are being used, but multiple treatments have to be performed. PDT was used in the treatment of early-stage disease or in an adjuvant setting with extensive erosive mycosis fungoides (MF) (Zane et al. 2006). A few reports also exist showing that cutaneous B-cell lymphoma responded to PDT (Mori et al. 2006). As availability of well-documented data is sparse, further studies are needed to assess the real potential of PDT in the treatment of cutaneous lymphoma.

40.9 Nononcological Indications

PDT has been tried in a large number of benign cutaneous diseases of the pilosebaceous unit, infections, inflammatory skin diseases, genetic disorders, and sclerotic skin diseases with very variable success (Karrer and Szeimies 2007; Kohl and Karrer 2011). PDT would be especially helpful in two situations: (1) very common diseases with a high target population and (2) (rare) diseases for which a good treatment option is still lacking.

40.9.1 PDT in Acne

Given the number of studies, the use of PDT in acne seems to be of great interest. The huge amount of protocol variations in these studies concerning pretreatment, sensitizer concentration, illumination parameters (low-dose vs. high-dose PDT), number of sessions, and others reflect the still-ongoing search for an optimal protocol. PDT usually leads to pain and a severe inflammatory reaction, and sustained effects and destruction of sebaceous glands are more likely to be achieved by high-dose PDT with red light (Sakamoto, Torezan, and Anderson 2010). The conclusion so far is that PDT can lead to long-term acne remission, but the optimal treatment parameters to put this benefit into relation with side effects and practicability have yet to be found.

40.9.2 PDT in Psoriasis

Psoriasis is another common disease with a high impact on public health, and initial research showed that PDT could be helpful. Unfortunately, further studies could not fulfill these expectations. PDT with the actual modalities is painful, cumbersome, and not reliable. One study in particular showed the clear superiority of narrowband UVB treatment (311 nm) (Beattie et al. 2004).

40.9.3 PDT of Viral Warts

Genital warts also belong to the group of diseases where high recurrence rates with standard treatments are not satisfactory. Thus, PDT has been tried either as monotherapy or as add-on therapy to CO_2 laser vaporization. Smaller studies showed quite promising results, but in the largest prospective, randomized trial where ALA-PDT in addition to CO_2 laser ablation vs. CO_2 laser ablation alone was used, a recurrence rate of about 50% in both groups was reported, stating that PDT was of no additional benefit (Szeimies et al. 2009).

Viral hand and foot warts are a very common problem in dermatology, especially as they can be very refractory to standard treatment modalities. Study results are quite promising. A clearance rate of 100% could be demonstrated in a trial when ALA-PDT was combined with a pulsed dye laser as a light source (Smucler and Jatsova 2005). PDT was quite painful in some patients, so the current status is that an optimized treatment protocol has to be established before PDT can become a therapy of wider use in recalcitrant warts.

Successful treatments with ALA-PDT have also been reported for plane facial warts and periungual warts. Data on MAL-PDT in this indication are very limited so far.

40.9.4 PDT in Cutaneous Leishmaniasis

A couple of observations have reported good results for ALA- or MAL-PDT in the treatment of cutaneous leishmaniasis. A review of six studies could summarize 77 lesions in 39 patients with healing of lesions in 94% to 100% of cases (van der Snoek et al. 2008). It is not totally clear if the therapeutic effect of PDT is only due to nonspecific tissue destruction or if the parasites are selectively affected (as has been shown in an *in vitro* experiment).

40.9.5 PDT in Localized Scleroderma

The finding that topical ALA-PDT was effective in five patients with localized scleroderma (Karrer et al. 2000) is of particular interest, as pathology in this disease is found in layers of the dermis where a direct effect of PDT is not expected. Thorough analysis of the mechanism demonstrated that PDT could induce the production of collagen-degrading matrix metalloproteinases (MMP1 and MMP3) by fibroblasts.

40.9.6 PDT in Other Diseases

There is a long list of other diseases where PDT has been tried and where mostly single case reports or small case series have reported beneficial effects. Within these studies, affections as diverse as molluscum contagiosum, Darier's disease, lichen sclerosus, perioral dermatitis, chondrodermatitis nodularis helicis, radiodermitis, hypertrophic scars, and others can be found. Only further studies will tell not only if these results are reproducible but also if treatment is practicable and cost-effective.

40.9.7 Antimicrobial PDT in Dermatology

The concept of antimicrobial PDT (described elsewhere in this book) could also have some indications in dermatology. So far, treatment of mycoses has not been very successful with ALA-PDT, but this might change with the development of new sensitizers that directly and selectively affect unwanted microorganisms. It is also important to note that PDT for the removal of acne is in part based on the destruction of the bacterium *Propionobacterium acnes*.

40.9.8 Photodynamic Rejuvenation (Photorejuvenation)

Clinical observations have shown that patients that received PDT for the treatment of AK also showed a general improvement of the signs of chronic sun damage in the treated areas. This led to a range of (mostly split-face) studies assessing the effect of PDT on signs of photoaging like fine wrinkles, mottled pigmentation, sallowness, roughness, and telangiectasias. Different light sources were used (light emitting diodes, intense pulse lightsource, pulsed dye laser), but in summary, PDT can be used to effectively improve the cosmetic appearance of sun-damaged skin. Upregulation of collagen production and increased epidermal proliferation have been observed as mechanisms (Kohl et al. 2010).

40.10 Conclusion and Outlook

PDT has proven to be a very effective treatment in superficial NMSCs, sometimes being recommended as first-line therapy by published guidelines (Braathen et al. 2007; Morton, McKenna, and Rhodes 2008). Approval of drugs and light sources in many countries makes this therapy more and more available. Cost-effectiveness has been demonstrated and published (Aguilar et al. 2010; Annemans et al. 2008), but reimbursement policies vary greatly from country to country. It is to be hoped that the benefits patients derive from PDT will soon be more widely accepted. In addition to NMSC, there is a wide spectrum of diseases where PDT has already shown promising results. But here, more randomized controlled trials are necessary to deliver the evidence needed to make PDT a standard in the therapeutic spectrum.

References

Aguilar, M., M. de Troya, L. Martin, N. Benitez, and M. Gonzalez. 2010. A cost analysis of photodynamic therapy with methyl aminolevulinate and imiquimod compared with conventional surgery for the treatment of superficial basal cell carcinoma and Bowen's disease of the lower extremities. *J Eur Acad Dermatol Venereol* 24:1431–1436.

Annemans, L., K. Caekelbergh, R. Roelandts et al. 2008. Real-life practice study of the clinical outcome and cost-effectiveness of photodynamic therapy using methyl aminolevulinate (MAL-PDT) in the management of actinic keratosis and basal cell carcinoma. *Eur J Dermatol* 18:539–546.

Attili, S. K., A. Lesar, A. McNeill et al. 2009. An open pilot study of ambulatory photodynamic therapy using a wearable low-irradiance organic light-emitting diode light source in the treatment of nonmelanoma skin cancer. *Br J Dermatol* 161:170–173.

Basset-Seguin, N., K. Baumann Conzett, M. J. Gerritsen et al. 2011. Photodynamic therapy for actinic keratosis in organ transplant patients. *J Eur Acad Dermatol Venereol* 27:57–66.

Beattie, P. E., R. S. Dawe, J. Ferguson, and S. H. Ibbotson. 2004. Lack of efficacy and tolerability of topical PDT for psoriasis in comparison with narrowband UVB phototherapy. *Clin Exp Dermatol* 29:560–562.

Braathen, L. R., C. A. Morton, N. Basset-Seguin et al. 2012. Photodynamic therapy for skin field cancerization: An international consensus. International Society for Photodynamic Therapy in Dermatology. *J Eur Acad Dermatol Venereol* 26:1063–1066.

Braathen, L. R., R. M. Szeimies, N. Basset-Seguin et al. 2007. Guidelines on the use of photodynamic therapy for nonmelanoma skin cancer: An international consensus. International Society for Photodynamic Therapy in Dermatology, 2005. *J Am Acad Dermatol* 56:125–143.

Calzavara-Pinton, P. G., M. Venturini, R. Sala et al. 2008. Methylaminolaevulinate-based photodynamic therapy of Bowen's disease and squamous cell carcinoma. *Br J Dermatol* 159:137–144.

Christensen, E., T. Warloe, S. Kroon et al. 2010. Guidelines for practical use of MAL-PDT in non-melanoma skin cancer. *J Eur Acad Dermatol Venereol* 24:505–512.

Clementoni, M. T., M. B-Roscher, and G. S. Munavalli. 2010. Photodynamic photorejuvenation of the face with a combination of microneedling, red light, and broadband pulsed light. *Lasers Surg Med* 42:150–159.

Cox, N. H., D. J. Eedy, and C. A. Morton. 2007. Guidelines for management of Bowen's disease: 2006 update. *Br J Dermatol* 156:11–21.

de Haas, E. R., H. S. de Bruijn, H. J. Sterenborg, H. A. Neumann, and D. J. Robinson. 2008. Microscopic distribution of protoporphyrin (PpIX) fluorescence in superficial basal cell carcinoma during light-fractionated aminolaevulinic acid photodynamic therapy. *Acta Derm Venereol* 88:547–554.

Dougherty, T. J. 1984. Photodynamic therapy (PDT) of malignant tumors. *Crit Rev Oncol Hematol* 2:83–116.

Fijan, S., H. Honigsmann, and B. Ortel. 1995. Photodynamic therapy of epithelial skin tumours using delta-aminolaevulinic acid and desferrioxamine. *Br J Dermatol* 133:282–288.

Gerritsen, M. J., T. Smits, M. M. Kleinpenning, P. C. van de Kerkhof, and P. E. van Erp. 2009. Pretreatment to enhance protoporphyrin IX accumulation in photodynamic therapy. *Dermatology* 218:193–202.

Halldin, C. B., J. Paoli, C. Sandberg, H. Gonzalez, and A. M. Wennberg. 2009. Nerve blocks enable adequate pain relief during topical photodynamic therapy of field cancerization on the forehead and scalp. *Br J Dermatol* 160:795–800.

Karrer, S., C. Abels, M. Landthaler, and R. M. Szeimies. 2000. Topical photodynamic therapy for localized scleroderma. *Acta Derm Venereol* 80:26–27.

Karrer, S., and R. M. Szeimies. 2007. [Photodynamic therapy: Non-oncologic indications]. *Hautarzt* 58:585–596.

Kennedy, J. C., R. H. Pottier, and D. C. Pross. 1990. Photodynamic therapy with endogenous protoporphyrin IX: Basic principles and present clinical experience. *J Photochem Photobiol B* 6:143–148.

Kerr, A. C., J. Ferguson, and S. H. Ibbotson. 2007. Acute phototoxicity with urticarial features during topical 5-aminolaevulinic acid photodynamic therapy. *Clin Exp Dermatol* 32:201–202.

Kohl, E., and S. Karrer. 2011. Photodynamic therapy for photorejuvenation and non-oncologic indications: Overview and update. *G Ital Dermatol Venereol* 146:473–485.

Kohl, E., L. A. Torezan, M. Landthaler, and R. M. Szeimies. 2010. Aesthetic effects of topical photodynamic therapy. *J Eur Acad Dermatol Venereol* 24:1261–1269.

Korshoj, S., H. Solvsten, M. Erlandsen, and M. Sommerlund. 2009. Frequency of sensitization to methyl aminolaevulinate after photodynamic therapy. *Contact Dermatitis* 60:320–324.

Lee, C. Y., K. H. Kim, and Y. H. Kim. 2010. The efficacy of photodynamic diagnosis in defining the lateral border between a tumor and a tumor-free area during Mohs micrographic surgery. *Dermatol Surg* 36:1704–1710.

Maisch, T., A. C. Moor, J. Regensburger et al. 2011. Intense pulse light and 5-ALA PDT: Phototoxic effects *in vitro* depend on the spectral overlap with protoporphyrine IX but do not match cut-off filter notations. *Lasers Surg Med* 43:176–182.

Miller, I. M., J. S. Nielsen, S. Lophaven, and G. B. Jemec. 2011. Factors related to pain during routine photodynamic therapy: A descriptive study of 301 patients. *J Eur Acad Dermatol Venereol* 25:1275–1281.

Moloney, F. J., and P. Collins. 2007. Randomized, double-blind, prospective study to compare topical 5-aminolaevulinic acid methylester with topical 5-aminolaevulinic acid photodynamic therapy for extensive scalp actinic keratosis. *Br J Dermatol* 157:87–91.

Mori, M., P. Campolmi, L. Mavilia et al. 2006. Topical photodynamic therapy for primary cutaneous B-cell lymphoma: A pilot study. *J Am Acad Dermatol* 54:524–526.

Morton, C., S. Campbell, G. Gupta et al. 2006. Intraindividual, right-left comparison of topical methyl aminolaevulinate-photodynamic therapy and cryotherapy in subjects with actinic keratoses: A multicentre, randomized controlled study. *Br J Dermatol* 155:1029–1036.

Morton, C. A., K. E. McKenna, and L. E. Rhodes. 2008. Guidelines for topical photodynamic therapy: Update. *Br J Dermatol* 159:1245–1266.

Moseley, H., S. Ibbotson, J. Woods et al. 2006. Clinical and research applications of photodynamic therapy in dermatology: Experience of the Scottish PDT Centre. *Lasers Surg Med* 38:403–416.

Neus, S., T. Gambichler, F. G. Bechara, S. Wohl, and P. Lehmann. 2009. Preoperative assessment of basal cell carcinoma using conventional fluorescence diagnosis. *Arch Dermatol Res* 301:289–294.

Pagliaro, J., T. Elliott, M. Bulsara, C. King, and C. Vinciullo. 2004. Cold air analgesia in photodynamic therapy of basal cell carcinomas and Bowen's disease: An effective addition to treatment: A pilot study. *Dermatol Surg* 30:63–66.

Paoli, J., C. Halldin, M. B. Ericson, and A. M. Wennberg. 2008. Nerve blocks provide effective pain relief during topical photodynamic therapy for extensive facial actinic keratoses. *Clin Exp Dermatol* 33:559–564.

Paoli, J., A. Ternesten Bratel, G. B. Lowhagen et al. 2006. Penile intraepithelial neoplasia: Results of photodynamic therapy. *Acta Derm Venereol* 86:418–421.

Peng, Q., T. Warloe, K. Berg et al. 1997. 5-Aminolevulinic acid-based photodynamic therapy. Clinical research and future challenges. *Cancer* 79:2282–2308.

Piacquadio, D. J., D. M. Chen, H. F. Farber et al. 2004. Photodynamic therapy with aminolevulinic acid topical solution and visible blue light in the treatment of multiple actinic keratoses of the face and scalp: Investigator-blinded, phase 3, multicenter trials. *Arch Dermatol* 140:41–46.

Sakamoto, F. H., L. Torezan, and R. R. Anderson. 2010. Photodynamic therapy for acne vulgaris: A critical review from basics to clinical practice: Part II. Understanding parameters for acne treatment with photodynamic therapy. *J Am Acad Dermatol* 63:195–211; quiz 211–212.

Shieh, S., A. S. Dee, R. T. Cheney et al. 2002. Photodynamic therapy for the treatment of extramammary Paget's disease. *Br J Dermatol* 146:1000–1005.

Sidoroff, A. 2007. [Photodynamic therapy of cutaneous epithelial malignancies. An evidence-based review]. *Hautarzt* 58:577–584.

Sidoroff, A., and P. Thaler. 2010. Taking treatment decisions in non-melanoma skin cancer—The place for topical photodynamic therapy (PDT). *Photodiagnosis Photodyn Ther* 7:24–32.

Smucler, R., and E. Jatsova. 2005. Comparative study of aminolevulic acid photodynamic therapy plus pulsed dye laser versus pulsed dye laser alone in treatment of viral warts. *Photomed Laser Surg* 23:202–205.

Szeimies, R. M., S. Karrer, S. Radakovic-Fijan et al. 2002. Photodynamic therapy using topical methyl 5-aminolevulinate compared with cryotherapy for actinic keratosis: A prospective, randomized study. *J Am Acad Dermatol* 47:258–262.

Szeimies, R. M., C. A. Morton, A. Sidoroff, and L. R. Braathen. 2005. Photodynamic therapy for non-melanoma skin cancer. *Acta Derm Venereol* 85:483–490.

Szeimies, R. M., P. Radny, M. Sebastian et al. 2010a. Photodynamic therapy with BF-200 ALA for the treatment of actinic keratosis: Results of a prospective, randomized, double-blind, placebo-controlled phase III study. *Br J Dermatol* 163:386–394.

Szeimies, R. M., V. Schleyer, I. Moll et al. 2009. Adjuvant photodynamic therapy does not prevent recurrence of condylomata acuminata after carbon dioxide laser ablation-a phase III, prospective, randomized, bicentric, double-blind study. *Dermatol Surg* 35:757–764.

Szeimies, R. M., E. Stockfleth, G. Popp et al. 2010b. Long-term follow-up of photodynamic therapy with a self adhesive 5-aminolaevulinic acid patch: 12 months data. *Br J Dermatol* 162:410–414.

Szeimies, R. M., L. Torezan, A. Niwa et al. 2012. Clinical, histopathological and immunohistochemical assessment of human skin field cancerization before and after photodynamic therapy. *Br J Dermatol* 167:150–159.

Togsverd-Bo, K., C. S. Haak, D. Thaysen-Petersen et al. 2012. Intensified photodynamic therapy of actinic keratoses with fractional CO$_2$ laser: A randomized clinical trial. *Br J Dermatol* 166:1262–1269.

Tyrrell, J. S., S. M. Campbell, and A. Curnow. 2010. The relationship between protoporphyrin IX photobleaching during real-time dermatological methyl-aminolevulinate photodynamic therapy (MAL-PDT) and subsequent clinical outcome. *Lasers Surg Med* 42:613–619.

Tyrrell, J., S. M. Campbell, and A. Curnow. 2011. The effect of air cooling pain relief on protoporphyrin IX photobleaching and clinical efficacy during dermatological photodynamic therapy. *J Photochem Photobiol B* 103:1–7.

van der Snoek, E. M., D. J. Robinson, J. J. van Hellemond, and H. A. Neumann. 2008. A review of photodynamic therapy in cutaneous leishmaniasis. *J Eur Acad Dermatol Venereol* 22:918–922.

Wiegell, S. R., J. Skiveren, P. A. Philipsen, and H. C. Wulf. 2008. Pain during photodynamic therapy is associated with protoporphyrin IX fluorescence and fluence rate. *Br J Dermatol* 158:727–733.

Wiegell, S. R., I. M. Stender, R. Na, and H. C. Wulf. 2003. Pain associated with photodynamic therapy using 5-aminolevulinic acid or 5-aminolevulinic acid methylester on tape-stripped normal skin. *Arch Dermatol* 139:1173–1177.

Wiegell, S. R., H. C. Wulf, R. M. Szeimies et al. 2011. Daylight photodynamic therapy for actinic keratosis: An international consensus: International Society for Photodynamic Therapy in Dermatology. *J Eur Acad Dermatol Venereol* 26:673–679.

Zane, C., M. Venturini, R. Sala, and P. Calzavara-Pinton. 2006. Photodynamic therapy with methylaminolevulinate as a valuable treatment option for unilesional cutaneous T-cell lymphoma. *Photodermatol Photoimmunol Photomed* 22:254–258.

Photodynamic Therapy in the Gastrointestinal Tract

Drew Schembre
Swedish Gastroenterology

41.1 Introduction

In 2010, approximately 275,000 new gastrointestinal (GI) malignancies were reported in the United States, representing about 18% of all new cancers, but accounting for nearly a quarter of the cancer deaths (Jemal et al. 2011). Surgery remains the best chance of cure for most of these cancers; however, at the time of diagnosis, most patients with GI cancers will not be candidates for curative surgery. For these patients, palliation of cancer-related symptoms and prolongation of life become the focus. Along with chemotherapy and radiation therapy, photodynamic therapy (PDT) has been used primarily as a palliative tool in advanced GI cancers. Since the 1980s, PDT has been used to treat obstructing cancers of the esophagus and other organs as well as to destroy early cancers and precancerous tissue. Over the past 30 years, the use of PDT for esophageal cancers and precancerous Barrett's esophagus has declined; however, interest has risen for its use in the biliary tract and other GI cancers.

41.2 Photosensitizers

The biology of PDT and a list of numerous photosensitizers are outlined in other chapters in this book. PDT for GI disease emerged in 1995 after the US Food and Drug Administration (FDA) approval of partially purified hematoporphyrin derivative (HpD) porfimer sodium or Photofrin for palliation of obstructing esophageal cancers. Photofrin has been produced and marketed by a number of different companies and is now distributed by Pinnacle Biologics, Inc. (Bannockburn, Illinois). Porfimer sodium is activated by red light in the 630 nm range as this provides deeper tissue penetration. The drug provides excellent tissue photosensitivity and may even foster selective uptake and retention by neoplastic tissues. It must be administered intravenously and has few direct side effects at therapeutic doses but has a half-life of about 3 weeks. This leads to cutaneous photosensitivity that can last for a month or more.

Porfimer sodium and 5-aminolevulinic acid (5-ALA) Levulan (DUSA Pharmaceuticals, Inc., Wilmington, Massachusetts) have been used to treat precancerous Barrett's esophagus and precancerous esophageal metaplasia of squamous epithelium. 5-ALA has the advantage of a significantly shorter half-life than other photosensitizers and therefore has reduced cutaneous photosensitivity. It can also be given intravenously as well as orally and even topically. PDT with 5-ALA utilizes laser light illumination in the blue or green light spectrum, instead of red light with porfimer sodium, which limits the depth of treatment. There is currently no FDA-approved indication for porfimer sodium PDT beyond the esophagus or upper airway in the United States or for any use of 5-ALA in the GI tract.

Another potent photosensitizer, meta-tetrahydroxy-phenyl chlorin (mTHPC, or temoporfin) Foscan (Biolitec AG, Jena, Germany), has shown promise for use in the treatment of GI disease. It produces an effect similar to porfimer sodium but has a shorter half-life which results in cutaneous photosensitivity for only about 2–3 weeks. Other photosensitizers such as 2-[1-hexyloxyethyl]-2-devinyl-pyrophenophorbide-a (HPPH) have undergone some limited study. Neither drug carries an FDA indication for treatment within the GI tract.

41.3 Light Sources and Delivery Devices

Since virtually the entire GI tract is inaccessible to direct light sources, light must be transmitted through endoscopes that are inserted into the body through the mouth or anus. Flexible endoscopy has been the foundation of nonsurgical interventions in the GI tract for the last half-century and utilizes instruments with long, controllable shafts. Older instruments relied on fiber-optic bundles to both illuminate the target and present the image to the operator. Newer devices incorporate small charge-coupled device (CCD) video chips at the endoscope tip to transmit high-resolution images to video screens. Endoscopes can easily access and image the esophagus, stomach, duodenum, and colon. More sophisticated tools can reach bile and pancreatic ducts as well as the deeper reaches of the small intestine. In theory, virtually every centimeter of the luminal GI tract is accessible to PDT via endoscopy.

A variety of fiber-optic catheters have been designed to pass through instrument channels of endoscopes. The most commonly used devices have been conical diffusing probes of different lengths that can selectively illuminate a fixed length of lumen. Calculating light dose is a function of the total output of the laser divided by the length of the probe, multiplied by the duration of tissue exposure. For example, to treat a 5 cm esophageal tumor, the endoscopist may choose a 2.5 cm diffusing probe and treat in two cycles, being careful not to overlap the segments and overtreat an area of tissue. Because peristaltic contractions occur throughout the GI tract, some movement is expected during the treatment. Close attention to scope and probe position is therefore essential. The bright treatment light often overwhelms the CCD processors and causes a whiteout of the video screen. Pausing the treatment periodically is important to ensure accurate targeting and avoiding overlapping treatment areas. Centering balloons have also been developed to reduce shadowing behind tissue folds, to ensure a constant distance between fiber and target, and to reduce movement during treatment. However, variable light absorption from the balloons themselves and the expense and difficulty of handling balloons have resulted in only sporadic use of these devices. All probes need to be calibrated before use, since a portion of the energy transmitted from the light source is lost within the fiber itself. Optical glass fibers generally do not bend, so placing probes in the bile duct for PDT of cholangiocarcinomas has been a challenge. While flexible probes have been available in Europe, they have not been approved for use in the United States. Consequently, short, 1 cm, or at most, 2 cm fibers have been used for even long bile duct tumors, requiring multiple careful placements under fluoroscopic guidance for bile duct tumors.

Lasers have been the light sources of choice for GI PDT because of their ability to deliver high-energy, single-frequency light consistently through a diffusing fiber. Large and expensive light sources such as the Ar dye and metal vapor lasers have been used for GI PDT; however, the potassium titanyl phosphate (KTP) laser combined with a 630 nm converting dye module was the most popular laser in the 1990s. Less expensive and much more portable diode lasers about the size of a desktop computer have become the laser of choice for most GI PDT cases using porfimer sodium. More information about lasers and delivery devices can be found in other chapters in this book.

41.4 Applications

41.4.1 Esophageal Cancer

Over 17,000 new cases of esophageal cancer occur each year in the United States and lead to almost 15,000 deaths (http://seer.cancer.gov/statfacts). Esophageal cancer has a high mortality in part because of its insidious onset: by the time patients experience symptoms such as difficulty swallowing, tumors are usually bulky and may have already metastasized. Around the world, esophageal cancer represents the sixth leading cause of cancer death, especially from squamous cell cancers of the mid- and proximal esophagus. Environmental factors such as tobacco and alcohol and certain nutritional deficiencies contribute significantly to the development of squamous cell cancers. Squamous cell cancers are much more common in an area extending from the Middle East to China. In Japan, squamous cell cancers are common enough that endoscopic screening programs have been implemented to identify and treat these types of tumors. In the United States, men of African descent suffer a significantly higher rate of esophageal cancer, especially squamous cell cancer, while men of European descent have a higher incidence of adenocarcinoma. The lifetime risk of esophageal cancer in the United States is less than 1% for men and about 0.3% for women, increasing with age and peaking in the mid-70s. Adenocarcinoma usually develops in the distal esophagus and esophagogastric junction arising from glandular tissue, often as a result of prolonged acid exposure due to gastroesophageal reflux. Cigarette smoking and chewing tobacco also increase the risk of developing esophageal adenocarcinoma.

Curing an established esophageal cancer usually requires surgery. Unfortunately, about half of patients harbor unresectable disease at the time of presentation and many others with technically operable disease will be unfit for or decide against an attempt at surgical cure. Many more will recur, even after a successful surgery. Chemotherapy and radiation therapy may eradicate some smaller tumors but is more often used as an adjunct to surgery or to reduce symptoms such as dysphagia (difficulty swallowing). This may help maintain nutrition, improve quality of life, and lengthen survival. Conventional chemotherapy and radiotherapy are not without their own side effects, including bone marrow suppression, esophageal stricture, and even fistula formation. Preoperative chemo/radiation therapy in advanced disease probably yields better results than surgery alone or surgery followed by chemo/radiation therapy.

Reducing dysphagia has become a major focus of esophageal cancer treatment because so many esophageal cancer patients do not undergo surgery. Chemotherapy and radiotherapy are often effective at shrinking both esophageal squamous and adenocarcinomas initially, but as tumors recur, different types of chemotherapy may become less effective and radiation therapy cannot usually be repeated. A variety of techniques have been developed to either push aside or "core out" obstructing tumors in an attempt to keep the esophagus open. These have included serial dilation, alcohol injection, laser vaporization, cryotherapy, placement of expandable stents, and ablation with PDT. PDT was greeted with enthusiasm as a palliative modality for esophageal cancer after it was shown to be as effective for relieving dysphagia with a few complications (Lightdale et al. 1995). Two hundred and eighteen patients with advanced esophageal cancer were randomized to PDT with porfimer sodium versus Nd:YAG laser tissue ablation therapy. While dysphagia improved in both groups, relief persisted longer in the PDT group. Moreover, perforations occurred in 7% of the Nd:YAG laser group but only in 1% of the PDT patients. Other studies confirmed this. Litle et al. (2003) showed that PDT improved dysphagia scores in 85% of patients with advanced adeno and squamous cell cancers with a median duration of 66 days. Complications included perforation (2%), stricture (2%), pleural effusions (4%), and cutaneous photosensitivity (6%). The procedure-related mortality rate was 1.8%, and median survival was 4.8 months (Litle et al. 2003).

Use of PDT for palliation of advance esophageal cancers has decreased over the last decade for several reasons. Despite effectiveness, the need to avoid sunlight for a month or more has deterred some patients. Cost for medication has increased in recent years and centers performing PDT remain limited. Other ablative treatments such as cryotherapy have experienced renewed interest. But the biggest reason for the decrease in PDT use has been the evolution of expandable metal stents. Stents are widely available, are easy to place, and relieve dysphagia effectively and rapidly. Although no formal head-to-head trials with PDT have been published, an abstract from Canto et al. (2002) described 56 patients with inoperable esophageal cancer who were randomized to receive either PDT or an expandable stent. Although both groups experienced significant improvement in swallowing and experienced a similar number of side effects, relief of dysphagia lasted longer among stent patients than PDT patients. Moreover, reintervention was more common and cost was over three times higher in the PDT group. Ultimately, there was no improvement in survival in either group (Canto et al. 2002).

PDT still has a place in the treatment of esophageal cancer. PDT would be appropriate for patients with short strictures high in the esophagus where stenting can compromise the airway, or at the esophagogastric junction where stenting can lead to significant reflux. Cryotherapy has proven to be a useful alternative to PDT in these settings as well, although dosimetry can be more difficult to estimate and results have been highly variable (Greenwald et al. 2010). Renewed interest has appeared for treating large-area, early squamous cell cancers, especially in Asia. In a recent study from Japan, 38 patients with early (T1) squamous cell cancers without clear lymph node involvement but that were too large to remove endoscopically were treated with PDT with porfimer sodium and resulted in a complete remission in 87% with a 5 year survival rate of 76% (Tanaka et al. 2011). This suggests that PDT may have real utility in this population of patients with early cancer. In fact, much of the use of PDT in the esophagus has focused on the treatment of early cancer rather than palliation of late cancers. The most successful applications of PDT have been for the prevention of the

development of esophageal cancer in individuals with precancerous changes.

The precursor to adenocarcinoma is a condition known as Barrett's esophagus (BE), named after a British surgeon who incorrectly described the condition as a congenital rather than acquired anomaly in 1950. Barrett's esophagus can develop in some people in the setting of chronic esophageal acid exposure. In these individuals, the acid-injured esophagus heals inflamed squamous mucosa by replacing it with glandular cells similar to stomach or intestinal lining. Pathologists describe this new lining as "intestinal metaplasia." BE secretes a mucus layer from specialized goblet cells that protects it from acid injury. Paradoxically, this tissue is much more unstable than the original squamous tissue and can undergo malignant transformation at a rate of up to 0.5% per year. Barrett's esophagus does not turn to cancer all at once. Instead, it usually progresses through a series of cellular and genetic changes referred to as dysplasia. Dysplasia can be further stratified as low or high grade based on the degree of morphologic change. Pathologists review tissue biopsies obtained during surveillance endoscopy, usually performed at 3 to 5 year intervals among patients with known BE. Low-grade dysplasia (LGD) suggests the beginning of the accumulation of genetic damage that leads to cancer, and high-grade dysplasia (HGD) represents the final step before cells turn frankly cancerous. The rate of progression from HGD to cancer ranges from 16% to 80% per year (Wolfsen 2005). Increased surveillance is recommended for individuals found to harbor LGD, and intervention, either with ablative therapies such as PDT or esophagectomy, remains the standard of care for BE with HGD.

BE occurs in a small fraction of individuals who suffer from acid reflux, perhaps 5%–8%, but accounts for the vast majority of adenocarcinoma of the esophagus. Research in the early 1990s showed that destroying the mucosal layer of BE with heat or laser light and subsequently eliminating acid exposure in the esophagus can allow the esophagus to heal with normal squamous mucosa instead of intestinal metaplasia. This led to great interest in using PDT to reverse BE in hope of reducing the risk of progression to esophageal cancer.

PDT appealed to researchers because instead of providing a series of small burns the way a laser or contact heat device might, it could provide controlled, uniform, limited-depth tissue destruction over a large area. Since BE can extend upward from the esophagogastric junction several centimeters or more, the ability to treat a large area at one time was very important. In a pilot study, Overholt, Panjehpour, and Haydek (1999) treated 84 patients with dysplasia or early cancer with porfimer sodium PDT and eliminated HGD in 88% and LGD in 92% of patients. In addition, 43% of patients showed complete reversal of all intestinal metaplasia. Additional studies showed similar results. Wolfsen, Woodward, Raimondo (2002), using porfimer sodium or another hematoporphyrin derivative, combined initial treatment with PDT with follow-up focal ablation with argon plasma coagulation and cleared HGD in 88% of patients. Significant side effects were demonstrated in all of these studies and included chest pain, difficulty swallowing, weight loss, and cutaneous photosensitivity. In addition, esophageal stricturing occurred in up to a third of patients requiring subsequent endoscopic dilation. Attempts at reducing stricture formation through a variety of techniques including concurrent use of oral steroids have been unsuccessful.

In an effort to reduce side effects associated with porfimer sodium PDT, investigators in Europe used ALA as a photosensitizer at 30 mg/kg and illuminated the esophagus with green light at 514 nm. In a double-blind, randomized, placebo-controlled trial of 36 patients with LGD, English investigators saw dysplasia disappear after treatment with ALA PDT in 98% of patients at 12 months with a reduction of overall Barrett's area by about 30% (Ackroyd et al. 2000). Other smaller trials have shown similar results; however, the use of ALA PDT has been limited by the consistent inability of this modality to eliminate Barrett's and promote uniform squamous reepithelialization. Additional concerns have arisen over the persistence of subsquamous dysplastic glandular tissue after ALA PDT.

The success of early studies of porfimer sodium PDT led to the pivotal, international, multicenter randomized controlled study of PDT versus acid suppression alone in 208 patients at 30 sites, known commonly as the PHO-BAR study (Overholt et al. 2005). In this highly structured study, PDT eradicated HGD in 77% of patients compared to resolution with acid suppression alone in 39% ($p < 0.001$), and at follow-up resulted in fewer cases of progression to cancer (13% vs. 20%, $p < 0.006$). This study was hampered by the rigid protocol that prevented follow-up treatment with other modalities including argon plasma coagulation (APC) and mucosal resection, but nevertheless, clearly demonstrated the efficacy of PDT for ablation of HGD and its ability to reduce progression to esophageal cancer. Side effects included some degree of photosensitivity in 69% of patients, mostly mild, and strictures requiring dilation in 36%. Detractors cited the high rate of complications and the inability to prevent progression to esophageal cancer as reasons to continue to insist on esophagectomy in the setting of HGD, but the tide had already begun to turn against surgery in favor of endoscopic treatments.

Ultimately, successful use of PDT for BE dysplasia paved the way for development of radiofrequency ablation (RFA) devices that can uniformly destroy the superficial layer of the esophagus without the same level of side effects. The Halo 360 from Barrx Medical Inc. (Covidien, Sunnyvale, California) uses an inflatable balloon ringed with a 3 cm coil of electrodes that delivers a precise, short pulse of radiofrequency heat energy that creates a superficial burn within the esophagus. In a large trial of 127 patients, RFA led to eradication of HGD in 81% of participants and complete reversal of BE in 77% (Shaheen et al. 2009). Side effects were minimal, consisting of some temporary substernal pain, one case of bleeding, and stricture formation requiring dilation in 6% of patients. Because of the relative ease of use of RFA, its lower side effect profile, and lower cost with at least equivalent results, RFA has virtually replaced PDT for treating BE dysplasia in the United States.

41.4.2 Stomach

Each year almost a million people worldwide develop gastric cancer and it results in approximately 650,000 deaths. Although decreasing in incidence in the United States, gastric cancer still affects over 20,000 individuals each year and just over half eventually die from it (http://seer.cancer.gov/statfacts). Survival has improved slightly over the last four decades as a result of advances in surgical therapy and adjuvant care but remains at about 22% at 5 years. Surgery still offers the only reasonable chance for long-term survival for patients with localized, invasive disease. Unfortunately, at the time of diagnosis, two-thirds of patients will have advanced disease. Just like with esophageal cancer, endoscopic screening programs in endemic areas have resulted in higher rates of detection of gastric cancer at earlier and more treatable stages.

Detection of disease limited to the mucosa is rarely associated with lymph node or distant metastases. Current therapies for early-stage disease rely less on radical surgery and more on endoscopic therapy. Early enthusiasm for PDT in this setting has been replaced by the rapid application of endoscopic mucosal resection (EMR) and endoscopic submucosal dissection (ESD) techniques. Nevertheless, interest remains high for PDT to augment EMR/ESD and for use in areas of the stomach such as the cardia and proximal lesser curve not easily approachable by other endoscopic techniques.

Early clinical studies of PDT in the stomach appeared positive but were marred by small numbers, limited follow-up, and variable outcome measures. For instance, one early study suggested a 100% response rate for gastric cancers treated with hematoporphyrin derivative (HPD) or dihematoporphyrin ethers and esters (DHE) but follow-up was only 2–19 months (Mimura 1994). Nakamura et al. (2001) also reported using DHE to treat seven patients with early gastric cancer, resulting in a complete response in all patients. In a large study, 120 patients in Japan were treated with a variety of protocols at seven hospitals, and led to a response in all participants, although cancer recurred in 23% of patients within several months (Kato et al. 1990).

Many patients in these series may have harbored invasive or metastatic disease at the time of treatment. In addition, early photosensitizers and light sources may have resulted in undertreatment of a number of patients. Mimura et al. (1996) reported two groups of patients with early gastric cancer sensitized with either HPD or DHE but treated with different light sources. Those treated with an Ar dye laser (ADL, Spectra-Physics, Mountain View, California) achieved a cure in 13 of 23 (57%) of mucosal cancers, 10 of 19 (53%) submucosal cancers, and 0 of 2 cancers involving the muscularis propria. Among the group that was treated with an excimer dye laser (EDL, Hamamatsu Photonics, Hamamatsu, Japan), 15 of 15 (100%) superficial cancers, 9 of 12 (75%) submucosal cancers, and 1 of 5 (20%) T_2 lesions were cured (Mimura et al. 1996).

Cylindrical diffusing fibers are ideal in the esophagus, but the stomach with its greater size and variable shape is more suited to treatment with a linear projecting microlens fiber that targets lesions much like a flashlight. These fibers have a fixed focal length, usually about 10 mm, and have a treatment diameter of about the same size. Because most lesions appropriate for PDT are relatively small, two or more overlapping treatment areas can encompass the treatment area.

5-ALA appears to be less useful in the stomach. In one study of treatment of precancerous lesions in the stomach, a complete response was achieved in only two of seven cases (Smolka 2006). This may have been in part due to inadequate light exposure, but the study highlights the difficulty of using PDT in the stomach with its many contours and undulations, especially with a mild photosensitizer. Ell et al. (1998) treated 22 patients with early gastric cancers with a much more potent photosensitizer, mTHPC, at 0.075 mg/kg followed 96 h later with illumination at 652 nm to provide 20 J/cm². This resulted in a complete remission in 16 of 22 patients (73%) that persisted at follow-up of 12–20 months. No severe reactions were reported, although skin photosensitivity was noted in seven patients and 12 reported transient abdominal pain (Ell et al. 1998).

Palliation of gastric cancer with PDT has not been well studied. Limited proximal gastric cancers causing dysphagia can be treated much like esophageal cancers, with circumferential PDT via a cylindrical diffusing fiber. However, because established cancers of the stomach tend to infiltrate deeply and broadly, palliation of advanced stomach cancers with PDT is unlikely to succeed. The exception may be among the subgroup of gastric cancers that bleed. Yanai et al. (2002) reported using a combination of PDT plus an infusion of activated autologous T lymphocytes to control aggressive oozing in two elderly patients with unresectable gastric cancers.

41.4.3 Duodenum and Ampulla

New cases of small bowel cancer occur in fewer than 8000 patients in the United States each year and lead to death in about 20% of these cases. Despite the length of the small intestine, malignancies of the duodenum, jejunum, or ileum account for only 2% of all GI cancers (http://seer.cancer.gov/statfacts). Individuals with conditions such as Peutz–Jeghers and familial polyposis syndromes, celiac disease, immunodeficiency states, and Crohn's disease may be at increased risk of developing small bowel malignancies. Environmental risk factors for small bowel cancers are poorly characterized, but probably include dietary factors, cigarette smoking, alcohol intake, and obesity.

Unfortunately, most duodenal cancers present late, with bleeding, jaundice, or gastric outlet obstruction. For all but the earliest cases, surgery represents the only chance at cure. Endoscopic stenting may help palliate obstructive problems associated with late-stage disease and surgical bypass may occasionally be used for palliation. Palliative PDT, except for bleeding, has not been described in this setting and is unlikely to be successful. PDT for early duodenal and ampullary cancers and advanced adenomas has been described in a handful of case reports. Regula et al. (1995) reported using ALA PDT to treat three duodenal adenomas and three ampullary carcinomas at 50–100 J/cm². While

ALA allowed identification of all lesions via photoporphyrin IX fluorescence, only minimal necrosis was seen after treatment. In another study, six patients with duodenal polyps treated with DHE, 2.5 mg/kg, followed by illumination at 200 J/cm² led to the eradication of small (<3 mm) polyps, over half of flat polyps ranging from 4 to 10 mm but less than half of sessile polyps between 4 and 10 mm. Side effects included one case of pancreatitis and at transient elevation of liver function tests in five of six individuals (Saurin, Chayvialle, and Ponchon 1999).

About 60% of individuals with familial adenomatous polyposis (FAP) syndromes develop duodenal and ampullary polyps, and when they are left untreated, many will progress to cancer. Duodenal and other small-bowel cancers represent a leading cause of premature death in FAP patients who have undergone prophylactic colectomy. Mlkvy et al. (1995) reported treating six patients with FAP and duodenal and rectal stump polyps with ALA or DHE PDT. Follow-up was short; however, lesions treated with DHE PDT appeared to respond better than those treated with ALA (Mlkvy et al. 1995).

Abulafi et al. (1995) treated 10 nonsurgical patients with ampullary cancers with HPD at the higher dose of 4 mg/kg followed by illumination at 630 nm to 50–200 J/cm² and repeated this at 3–6 month intervals, up to five times. Three patients sustained a complete response while four patients had decreased tumor bulk and three showed no change. The only side effect noted was skin photosensitivity in three patients.

As in the stomach, endoscopic mucosal resection and endoscopic ampullectomy have been used increasingly to treat early-stage duodenal cancers and dysplastic adenomas. PDT is unlikely to provide a dependable alternative to these minimally invasive therapies. Similarly, in cases of established cancer, surgery or endoscopic stenting remain the mainstays of treatment, although further research into this area is welcome.

41.4.4 Biliary System

With an incidence of about 1/100,000, cholangiocarcinoma (CCa) qualifies as a rare disease (http://seer.cancer.gov/statfacts). Intrahepatic lesions may present with nonspecific symptoms such as weight loss, abdominal pain, or malaise, while those with extrahepatic involvement typically present with painless jaundice. Tumors may spread widely in the biliary system before diagnosis and only about 20%–30% of patients will be candidates for potentially curative surgery at presentation (Nekeeb et al. 1996). Median survival without treatment is only about 3–6 months with death usually resulting from liver failure or cholangitis due to biliary obstruction. Chemotherapy has been marginally effective at reducing symptoms and extending survival (Singh and Patel 2006). The liver is quite sensitive to radiation, which limits its use. Intraductal brachytherapy may be beneficial in select cases but is technically difficult and not widely available. Liver transplantation has been curative in small series.

Risk factors for CCa include primary sclerosing cholangitis (PSC), choledochal cysts, chronic intraductal stones, parasitic infections, and exposure to certain industrial chemicals. Chronic viral hepatitis, excessive alcohol use, and obesity may also increase risk (Chapman 1999). Staging is described by the tumor/node/metastasis (TNM) system, but the older Bismuth–Corlette classification system is still widely used largely because it describes potential resectability.

Biliary drainage remains the cornerstone of CCa treatment. Good drainage reduces symptoms, preserves liver function, helps avoid cholangitis, and improves survival. Surgical drainage can often be quite difficult and has been associated with significant morbidity and mortality with survival no better than minimally invasive techniques. Expandable metal stents probably provide superior palliation than do plastic stents because they tend to stay patent longer, migrate less frequently, and require fewer repeat interventions (Chahal and Baron 2006).

Case reports of PDT for CCa began emerging in the 1990s, but it was not until Ortner et al. (1998) published their first experience with PDT in 1998 that the therapy received widespread attention. In this uncontrolled, observational pilot study, nine patients with inoperable, Bismuth type III/IV CCa were treated with PDT followed by stenting and were compared to three patients treated by stenting alone. Study end points were survival, resolution of jaundice, and quality of life. Photosensitization was achieved with porfimer sodium 2 mg/kg. Red light at 630 nm was applied via an endoscopically placed cylindrical diffusing fiber to provide 180 J/cm² of energy. At 2 months, serum bilirubin in the PDT group fell by 67% and the Karnofsky performance index increased from 32% to 69% ($p = 0.008$). Median survival was 439 days in the PDT group compared with only 74 days in the control group ($p < 0.05$). The PDT group demonstrated no mortality at 30 days and 1 year survival was 78%. Side effects in the PDT group were limited to minor skin hyperpigmentation in all patients and fever and abdominal pain in one patient.

The same authors published the results of a multicenter, randomized controlled study of PDT plus stenting versus stenting alone for unresectable cholangiocarcinoma 5 years later (Ortner et al. 2003). Seventy patients were enrolled in this study; 20 were randomized to PDT via a treatment protocol identical to their initial study and 19 were randomized to bilateral plastic stenting alone. Thirty-one patients who either refused consent to randomization or were otherwise excluded were followed in an open arm, consisting of PDT plus stenting. Patients underwent a mean of 1.5 PDT treatments (range: 1–4) and had bilateral, 15 cm plastic stents changed every 3 months. Median survival in the PDT + stenting group was 493 days compared to only 98 days in the stenting-alone group ($p < 0.0001$). The PDT + stenting group also experienced an increased quality of life based on Karnofsky performance scores. Further, survival in the open-label group was similar to the PDT + stenting arm with median survival of 426 days. The study was terminated early due to the quickly apparent survival difference between the groups. There were no treatment-related deaths but cholangitis and skin photosensitivity were more frequent in the PDT group.

Other studies have shown similar, if less dramatic results. Witzigmann et al. (2006) treated 68 patients with sodium porfimer PDT followed by stenting and compared them to 56 patients treated with stenting alone. Median survival in the PDT group was 12 months, compared to 6.4 months in the stent group ($p < 0.01$). Biliary drainage and Karnofsky performance scores were also significantly better among the PDT group. Zoepf et al. (2005) conducted a randomized trial of 32 patients with cholangiocarcinoma, half of whom were treated with a newer hematoporphyrin derivative, Photosan-3 (SeeLab, Wesselburenerkoog, Germany), 2 mg/kg, followed by irradiation at 633 nm via diode laser at 200 J/cm², followed by plastic stenting versus stenting alone. Stents were generally changed at 3 month intervals in all patients. Median survival for the PDT group was 21 months compared to 7 months for the control group ($p = 0.01$). Cholangitis was more common in the PDT group than in the control group but was successfully treated in all cases.

In the most recent large study from the United States, Kahaleh et al. (2008) showed a significant survival benefit among 19 patients treated with porfimer sodium PDT plus plastic stents versus 29 who were treated with stents alone. Mean survival was 16.2 months in the PDT group compared to 7.4 in the control group. Cholangitis occurred in over a third of patients in both arms, resulting in two deaths in the control arm.

Despite the demonstrated efficacy of PDT for CCa, it is still considered experimental by the FDA in the United States. Nevertheless, interest in PDT for CCa persists and studies continue. CCa treatment appears to be the most promising application for PDT in the GI tract.

41.4.5 Pancreas

Pancreatic adenocarcinoma is the second most common GI malignancy in the United States, after colon cancer, with about 44,000 new cases in 2012 (http://seer.cancer.gov/statfacts). Surgical resection affords the only real chance at cure; however, only about 20% of newly diagnosed pancreas cancer patients can undergo definitive surgery. About 40% will have metastatic disease at presentation, and another 40% will have locally advanced unresectable tumors. Median survival is 8 to 12 months for those with locally advanced disease and less for those with metastatic disease. Even among those who undergo successful surgery, 5 year survival is only 25% to 30% among patients with node negative disease and about 10% in those with positive nodes (Trede, Schwall, and Saeger 1990). Only recently have positive results emerged from trials of chemotherapy and chemo/radiotherapy, and these have been modest, extending average survival by perhaps 6 months, often in exchange for the risk of significant toxicity.

Early *ex vivo* studies suggested that PDT can have a toxic effect of pancreatic cancer cells. In addition, PDT may initiate an immune response by the host to surviving cancer cells. Regula et al. showed that hamsters with transplanted human pancreas tumors survived longer after ALA PDT than untreated hamsters. Further, the photosensitizer HPD and others appear to accumulate preferentially in pancreatic cancer cells compared to the healthy pancreas (Schroder et al. 1998). The photosensitizer mTHPC when used in animals created large areas of necrosis but also appeared to affect viscera, leading to duodenal perforations (Mlkvy et al. 1996).

Pancreas tumors pose a particularly difficult challenge for treatment with PDT as they are often not approachable via the usual endoscopic means, even pancreatoscopy. Bown et al. (2002) worked around this problem in the first phase I trial of PDT for pancreatic cancer in humans by approaching the tumors percutaneously. Sixteen patients sensitized with 0.15 mg/kg of mTHPC underwent treatment through a series of 19-gauge needles implanted under computed tomography (CT) guidance into unresectable pancreatic tumors. Thin diffusing fibers were passed 3 mm beyond the needles and illuminated with 652 nm red light for a median of 240 J. All patients demonstrated substantial tumor necrosis by CT scan at 2 weeks. Two patients experienced significant bleeding from the gastroduodenal artery as a result of local necrosis. No patient demonstrated clinical pancreatitis. Median survival was 9.5 months with 7 of 16 surviving over 1 year.

Endoscopic delivery of PDT for pancreas tumors has not been described in humans. However, Chan et al. (2004) demonstrated successful PDT to the pancreas as well as liver, kidney, and spleen via light probes passed through endoscopic ultrasound-guided 19-gauge needles in pigs.

New photosensitizers will likely be developed with further specificity for pancreatic tumors and new targeting and delivery systems for light will probably follow. Wide prevalence and poor survival with other therapies underscore the need for additional studies of PDT for pancreas cancer.

41.4.6 Colon and Rectal Cancers

Despite the success of screening programs and public awareness that colon cancer and pre-cancerous polyps and pre-cancerous polyps can be identified by colonoscopy, colon cancer still accounts for 10% of all cancer deaths in the United States. About 140,000 new cases of colon cancer are diagnosed each year in the United States, of which about two-thirds arise in the colon and the remainder in the rectum. In 2012, more than 51,000 individuals were expected to die of the disease, making it the second leading cause of cancer death after lung cancer (http://seer.cancer.gov/statfacts). Risk factors include family history, especially polyposis syndromes such as FAP, Lynch syndrome, and others; medical conditions such as inflammatory bowel disease, obesity, prior cholecystectomy, and diabetes; and likely certain environmental influences such as alcohol, cigarette smoking, prior irradiation, and low-fiber and high-fat diets. However, the majority of colon cancers occur in people with no obvious risk factors. Protective factors might include a low-fat, high-fiber diet, the use of aspirin or nonsteroidal anti-inflammatory drugs (NSAIDs), as well as the use of statins and folic acid.

The idea of using PDT for curing early colon and rectal cancers and palliating advanced ones has been around for many years. Animal studies have shown that colon cancers can be readily sensitized for PDT and that the colon wall can withstand high levels of energy without perforation. But the clinical application of PDT in the colon has been limited, largely due to the relative ease and reliability of polypectomy for large polyps, surgical resection for localized cancer, and endoluminal stenting for palliating advanced disease.

Numerous case reports and small studies, many from 20 years ago or more, have looked at using PDT to eradicate colon polyps as well as to palliate bulky colon and rectal cancers. These studies suffer from a lack of consistent treatment and standard outcomes, small numbers, and short follow-up.

In one of the first studies, McCaughan (1992) treated eight patients with rectal cancers with dihematoporphyrin ether and reported an average survival of 16 months after treatment. In another early study, HPD was used to treat nine villous adenomas in eight patients previously incompletely treated with Nd:YAG laser ablation (Loh et al. 1994). Seven polyps were completely eradicated and there were no significant complications. In the United Kingdom, 10 nonsurgical patients with colorectal cancers were treated with HPD. Two patients with small cancers had complete responses while patients with larger cancers demonstrated persistent symptoms such as pain and urgency and one patient experienced posttreatment bleeding (Barr et al. 1990). In a French report of 21 patients with rectosigmoid cancers treated with HPD, 50% demonstrated an initial complete response (Patrice et al. 1990). Spinelli, Mancini, and Dal Fante (1995) reported technical success treating 27 patients with advanced or obstructing distal colon cancers, although follow-up was short and end points were not clearly defined.

Other studies have shown less positive results. Of 14 patients with residual or recurrent colorectal cancer in the pelvis treated with PDT with DHP, only two showed a prolonged complete response, although several others reported decreased pain and urgency (Herrera-Ornelas et al. 1986). In a large study of 93 patients with inoperable colon cancers, including 18 with large polyps, PDT using porfimer sodium was found to be less effective than Nd:YAG laser ablation for debulking large tumors (Krasner 1989).

Novel therapies have been designed to treat colon cancer metastatic to the liver. In a report by Vogel et al. (2004), nine patients with liver metastases were treated with HPBC followed by percutaneous laser light illumination. This resulted in a complete response in five patients. The only side effects noted were two cases of cutaneous photosensitivity (Vogl et al. 2004).

PDT has also been used to treat uncontrolled bleeding associated with rectal cancers. Four patients who had failed other therapies responded to dihematoporphyrin ether PDT (McCaughan et al. 1996). HPD PDT has also been used to treat bleeding associated with radiation proctitis.

As endoscopic techniques continue to improve and stent technology advances, PDT for colon and rectal cancer will likely be reserved for a few indications such as low rectal tumors associated with bleeding and urgency and perhaps for large field defects such as FAP in the retained rectum. This may change if newer, more potent, and more specific photosensitizers are developed. Animal studies have shown that photosensitizers can be linked to specific molecules such as carcinoembryonic antigen (CEA) and granulocyte stimulating factor in an attempt to further focus the phototoxic effects of treatment (Carcenac et al.

1999; Golab et al. 2000). Research in these areas as well as in new therapies for metastatic lesions may trigger renewed interest in PDT for the lower GI tract.

References

Abulafi, A. M., J. T. Allardice, N. S. Wiliams et al. 1995. Photodynamic therapy for malignant tumours of the ampulla of Vater. *Gut* 36:853–856.

Ackroyd, R., N. J. Brown, M. F. Davis et al. 2000. Photodynamic therapy for dysplastic Barrett's oesophagus: A prospective, double blind, randomized placebo controlled trial. *Gut* 47:612–617.

Barr, H., N. Krasner, P. B. Boulos et al. 1990. Photodynamic therapy for colorectal cancer: A quantitative pilot study. *Br J Surg* 77:93–96.

Bown, S. G., A. Z. Rogowska, D. E. Whitelaw et al. 2002. Photodynamic therapy for cancer of the pancreas. *Gut* 50:549–557.

Canto, M. I., C. Smith, L. McClelland et al. 2002. Randomized trial of PDT vs. stent for palliation of malignant dysphagia: Cost-effectiveness and quality of life. *Gastrointest Endosc* 55:AB100.

Carcenac, M., C. Larroque, R. Langlois et al. 1999. Preparation, phototoxicity and biodistribution studies of anti-carcino-embryonic antigen monoclonal antibody-phthalocyanine conjugates. *Photochem Photobiol* 70:930–936.

Chahal, P., and T. H. Baron. 2006. Endoscopic palliation of cholangiocarcinoma. *Curr Opin Gastroenterol* 22:551–560.

Chan, H. H., M. S. Nishioka, M. Mino et al. 2004. EUS-guided photodynamic therapy of the pancreas: A pilot study. *Gastrointest Endosc* 59:95–99.

Chapman, R. W. 1999. Risk factors for biliary tract carcinogenesis. *Ann Oncol* 10(Suppl):308–311.

Ell, C., L. Gossner, A. May et al. 1998. Photodynamic ablation of early cancers of the stomach by means of mTHPC and laser irradiation: Preliminary clinical experience. *Gut* 43:345–349.

Golab, J., G. Wilcznski, R. Zagozdzon et al. 2000. Potentiation of the anti-tumour effects of Photofrin-based photodynamic therapy by localized treatment with G-CSF. *Br J Cancer* 82:1485–1491.

Greenwald, B. D., J. A. Dumont, J. A. Abrams et al. 2010. Endoscopic spray cryotherapy for esophageal cancer: Safety and efficacy. *Gastrointest Endosc* 71:686–693.

Herrera-Ornelas, L., N. J. Petrelli, A. Mittelman et al. 1986. Photodynamic therapy in patients with colorectal cancer. *Cancer* 57:677–684.

Jemal, A., F. Bray, M. M. Center et al. 2011. Global cancer statistics. *CA Cancer J Clin* 61:69–90.

Kahaleh, M., R. Misra, V. M. Shami et al. 2008. Unresectable cholangiocarcinoma: comparison of survival in biliary stenting versus stenting with photodynamic therapy. *Clin Gastro Hep* 6:290–297.

Kato, H., T. Kito, K. Furuse et al. 1990. Photodynamic therapy in the early treatment of cancer. *Gan To Kagaku Ryoho* 17:1833–1838.

Krasner, N. 1989. Laser therapy in the management of benign and malignant tumours in the colon and rectum. *Int J Colorectal Dis* 4:2–5.

Lightdale, C. J., S. K. Heier, N. E. Marcom et al. 1995. Photodynamic therapy with porfimer sodium versus thermal ablation therapy with Nd:YAG laser for palliation of esophageal cancer a multicenter randomized trial. *Gastrointest Endosc* 42:507–512.

Litle, V. R., J. D. Luketich, N. A. Christie et al. 2003. Photodynamic therapy as palliation for esophageal cancer: Experience in 215 patients. *Ann Thor Surg* 76:1687–1693.

Loh, C. S., P. Bliss, S. G. Bown et al. 1994. Photodynamic therapy for villous adenomas of the colon and rectum. *Endoscopy* 26:243–246.

McCaughan, J. S., Jr. 1992. Miscellaneous treatments. In *Photodynamic Therapy for Malignancies: Clinical Manual*. J. S. McCaughan Jr., editor. Landes, Austin, TX, 135–144.

McCaughan, J. S., P. C. Hawley, J. C. LaRosa et al. 1996. Photodynamic therapy to control life-threatening hemorrhage from hereditary hemorrhagic telangiectasia. *Lasers Surg Med* 19:492–494.

Mimura, S. 1994. Photodynamic therapy for early gastric cancer. 5th International Photodynamic Association Biennial Meeting, Amelia Island, FL.

Mimura, S., H. Narahara, H. Uehara et al. 1996. Photodynamic therapy for gastric cancer. *Gan To Kagaku Ryoho* 23:41–46.

Mlkvy, P., H. Messmann, H. Debinski et al. 1995. Photodynamic therapy for polyps in familial adenomatous polyposis—A pilot study. *Eur J Cancer* 31A:1160–1165.

Mlkvy, P., H. Messmann, M. Pauer et al. 1996. Distribution and photodynamic effects of meso-tetrahydroxyphenylchlorin (mTHPC) in the pancreas and adjacent tissues in the Syrian golden hamster. *Br J Cancer* 73:1473–1479.

Nakamura, H., H. Yanai, and J. Nishikawa et al. 2001. Experience with photodynamic therapy for the treatment of early gastric cancer. *Hepato-Gastroenterol* 48:1599–1603.

Nekeeb, A., H. A. Pitt, T. A. Sohn et al. 1996. Cholangiocarcinoma: A spectrum of intrahepatic, perihilar, and distal tumors. *Ann Surg* 224:463–473.

Ortner, M. E., K. Caca, F. Berr et al. 2003. Successful photodynamic therapy for nonresectable cholangiocarcinoma: A randomized prospective study. *Gastroenterol* 125:1355–1363.

Ortner, M. E., J. Liebetruth, S. Schreiber et al. 1998. Photodynamic therapy of nonresectable cholangiocarcinoma. *Gastroenterol* 114:536–542.

Overholt, B. F., C. J. Lightdale, K. K. Wang et al. 2005. Photodynamic therapy with porfimer sodium for ablation of high-grade dysplasia in Barrett's esophagus: International, partially blinded, randomized phase III trial. *Gastrointest Endosc* 62:488–498.

Overholt, B. F., M. Panjehpour, and J. M. Haydek. 1999. Photodynamic therapy for Barrett's esophagus: Follow-up in 100 patients. *Gastrointest Endosc* 49:1–7.

Patrice, T., M. T. Foultier, S. Yactayo et al. 1990. Endoscopic photodynamic therapy with haematoporphyrin derivative in gastroenterology. *J Photochem Photobiol B* 6:157–165.

Regula, J., A. J. MacRobert, A. Gorchein et al. 1995. Photosensitisation and photodynamic therapy of oesophageal, duodenal and colorectal tumours using 5 aminolaevulinic acid induced protoporphyrin IX—A pilot study. *Gut* 36:67–75.

Saurin, J. C., J. A. Chayvialle, and T. Ponchon. 1999. Management of duodenal adenomas in familial adenomatous polyposis. *Endoscopy* 31:472–478.

Schroder, T., I. W. Chen, M. Sperling et al. 1988. Hematoporphyrin derivative uptake and photodynamic therapy in pancreatic carcinoma. *J Surg Oncol* 38:4–9.

SEER Cancer Stat Fact Sheet, Available at http://seer.cancer.gov/statfacts, retrieved August 2012.

Shaheen, N. J., P. Sharma, B. F. Overholt et al. 2009. Radio-frequency ablation in Barrett's esophagus with dysplasia. *NEJM* 360:2277–2288.

Singh, P., and T. Patel. 2006. Advances in the diagnosis, evaluation and management of cholangiocarcinoma. *Curr Op Gastroenterol* 22:294–299.

Smolka, J. 2006. In vivo fluorescence diagnostics and photodynamic therapy of gastrointestinal superficial polyps with aminolevulinic acid. A clinical and spectroscopic study. *Neoplasma* 53:418–423.

Spinelli, P., A. Mancini, and M. Dal Fante. 1995. Endoscopic treatment of gastrointestinal tumors: Indications and results of laser photocoagulation and photodynamic therapy. *Sem Surg Oncol* 11:307–318.

Tanaka, T., S. Matono, T. Nagano et al. 2011. Photodynamic therapy for large superficial squamous cell carcinoma of the esophagus *Gastrointest Endosc* 73:1–6.

Trede, M., G. Schwall, and H. D. Saeger. 1990. Survival after pancreatoduodenectomy. 118 consecutive resections without an operative mortality. *Ann Surg* 211:447–458.

Vogl, T., K. Eichler, M. G. Mack et al. 2004. Interstitial photodynamic laser therapy in interventional oncology. *Eur Radiol* 14:1063–1073.

Witzigmann, J., F. Berr, U. Ringel et al. 2006. Surgical and palliative management and outcome in 184 patients with hilar cholangiocarcinoma. *Ann Surg* 244:230–239.

Wolfsen, H. C. 2005. Present status of photodynamic therapy for high-grade dysplasia in Barrett's esophagus. *J Clin Gastroenterol* 39:198–202.

Wolfsen, H. C., T. A. Woodward, and M. Raimondo. 2002. Photodynamic therapy for Barrett's esophagus and early esophageal adenocarcinoma. *Mayo Clin Proc* 77:1176–1181.

Yanai, H., Y. Kuroiwa, N. Shimizu et al. 2002. The pilot experience of immunotherapy-combined photodynamic therapy for advanced gastric cancer in elderly patients. *Int J Gastrointest Cancer* 32:139–142.

Zoepf, T., R. Jakobs, J. C. Arnold et al. 2005. Palliation of nonresectable bile duct cancer: Improved survival after photodynamic therapy. *Am J Gastro* 100:2426–2430.

Photodynamic Applications in Brain Tumors

Sam Eljamel
Ninewells Hospital and Medical School

42.1 Introduction

Malignant brain tumors (MBTs) encompass primary malignant brain tumors (PMBTs) and secondary brain tumors (SBTs). MBTs affect 5–10 out of 100,000 people annually and are responsible for 3% of all cancer deaths worldwide. These tumors are the second most common cause of cancer death in young people and the sixth most common cause of productive-years-loss in the community. MBTs carry very dismal prognosis and their diagnosis almost always means a death sentence deferred by merely 36 weeks or less (Obwegeser et al. 1995; Eljamel 2004). PMBT (intracranial gliomas) represent 38%–40% of primary brain tumors and glioblastoma multiforme (GBM) is the most common PMBT (Stupp 2007). GBMs are divided into two: primary and secondary GBMs. Primary GBMs arise *de novo*, represent 60% of GBMs, and are diagnosed mainly in people over 50 years of age. Secondary GBMs, on the other hand, represent 40% of GBMs and are most common in people under 45 years of age. Several researchers have studied GBMs extensively to assess their prognosis and factors associated with better survival. Three risk groups were identified (Lamborn, Chang, and Prados 2004): a low-risk group consisting of young patients (under 40 years of age) with tumors located in the frontal lobe; a moderate-risk group consisting of patients between 40 and 65 years of age and Karnofsky performance scores (KPSs) >70 who had surgical resection; and a high-risk group of patients >65 years of age and patients between 40 and 65 years who had KPS <80 or had tumor biopsy only. Several recent studies assessed the impact of genetic mutations on GBMs' outcome (Rich et al. 2005; Kleihues, Burger, and Cavenee 1997). Loss of heterozygosity (LOH) on chromosome group 10q, the most frequent gene abnormality for both primary and secondary GBMs, is associated with poor survival. It occurs in 60%–90% of GBMs. This mutation appears to be specific for GBMs and is found rarely in other tumor grades. Mutations of p53, a tumor suppressor gene, were among the first genetic alterations identified in astrocytic brain tumors and appear to be deleted or altered in approximately 25%–40% of all GBMs and particularly in secondary GBMs (Watanabe et al. 1997). Despite extensive clinical trials, the median survival remained about 12 months with fewer than 25% surviving for 2 years and fewer than 10% surviving for 5 years. In a series of 279 patients, only five (1.8%) survived for 3 years (Scott et al. 1998). Clearly, newer approaches for the management of GBMs are necessary and multimodality therapeutic approaches would be required to improve the outcome of these tumors. The joint tumor section of the American Association of Neurological Surgeons (AANS) and the Congress of Neurological Surgeons (CNS) in 2008 produced guidelines for the management of newly diagnosed GBMs (Olsen et al. 2008). The authors recommended maximum safe surgical resection of newly diagnosed GBMs (evidence level II), followed by 60 Gy of postoperative radiotherapy to the enhancing lesion (evidence level I) and for radiotherapy to include a 2 cm cuff around the enhancing lesion (evidence level II). The guidelines also recommended concurrent and postoperative temozolomide in newly diagnosed GBM (evidence level I) and BCNU (carmustine wafers) in those who undergo craniotomy (evidence level II). Temozolomide offers a mere 26% 2 year survival compared to 10% in placebo controls, a 14.6 month median survival compared to 12.1 months in placebo group, and a 7.2 month tumor-progression-free (TPF) survival compared to 5 months among placebo controls (Olsrn et al. 2008). Carmustine implants, on the other hand, prolong survival to 13.9 months compared to 11.9

months in the placebo group (Vecht et al. 1990). The poor outcome of these tumors is due to local invasion and local relapse. The vast majority of these tumors recur locally within 2 cm of the resection margin and patients often succumb to and die from local recurrence, indicating that a more aggressive local therapy is required to eradicate these tumors. However, complete radical surgical excision is hindered by the elusive nature of these tumors: a significant amount of tumor cells are invisible to the naked eye even with the aid of the surgical microscope. The ability of these cancers to disguise themselves makes their identification at surgery almost impossible. Most of these tumors have invaded the brain widely by the time they manifest clinically, making wider excision margin out of the question in most cases par polar lesions in noneloquent brain.

SBTs are much more common than PMBTs (Klos et al. 2004). Brain metastases have an incidence of 12 per 100,000 per year and 15% to 40% of systemic cancers metastasize to the brain (Scott et al. 1998). The most frequent sources of brain metastases are lung cancer (40%–50%), breast cancer (15%–25%), and malignant melanomas (5%–20%) (Klos et al. 2004). The frequency of metastatic brain tumors appears to be rising as a result of improved brain imaging and more effective treatment of these primary cancers leading to improved survival. Common clinical features similar to PMBT include headache, neurological deficit, and seizures. Detailed neuropsychological testing demonstrated cognitive impairment in 65% of patients, usually across multiple domains.

If SBTs are left untreated, they may rapidly become fatal. The local control rate of the primary tumor is paramount to patients' survival. Surgery is indicated in cases of a solitary metastasis, particularly when the patient is young (less than 65 years of age), with good KPS (>70), and when the systemic cancer is under control (Guillano et al. 2008). However, eradication of brain metastasis is not always possible because the basic oncological principle of resecting the tumor en mass with a cuff of normal brain tissue cannot be applied. Therefore, adjuvant therapies are often used such as whole-brain radiotherapy (WBRT), stereotactic radiosurgery (SRS), and chemotherapy (ChT) (Komblith 1995).

Because of the dismal prognosis of PMBT and SBT, the following questions are worth asking: What if we could discover a way to enhance our vision to see what is a tumor and what is not? What if we could find a way to selectively kill residual tumor cells without harming surrounding brain tissue? Photodiagnosis (PD) and photodynamic therapy (PDT) technologies hold an important key to unlock this mystery and answer these questions.

42.2 PD in MBT

The basic principle of PD is selectivity of photosensitizers in MBT, and the fluorescence of cells containing the photosensitizers is detected because they absorb blue light (373–404 nm) and emit red light (Aziz et al. 2009; Eljamel 2004, 2008; Zilidis et al. 2008). Hence, using a long-pass filter in the light path allows visualization of photosensitized tumor cells. Selective fluorescence occurs because hyperproliferating (malignant) tissue preferentially accumulates photosensitizers, therefore allowing

surgeons to spot malignant tissue from normal, healthy tissue, maximizing tumor resection and minimizing brain damage. This can also be used to spot pockets of malignant transformation in low-grade gliomas.

Clinical PD application in MBT was first reported in a prospective observational study of 52 consecutive high-grade gliomas in 2000 (Stummer et al. 2000). The authors found PD to be selective (99.6%, only one fluorescent biopsy out of 223 fluorescent biopsies did not contain tumor cells) and highly sensitive (81.6%, 222 biopsies of 272 fluorescent biopsies contained tumor), and PD resulted in complete resection of enhancing lesions on magnetic resonance imaging (MRI) in 63% of patients when 5-aminolevulinic acid was used (5-ALA) (Stummer et al. 2000). Furthermore, this study demonstrated that leaving more than 2 cm^3 tumor was an adverse prognostic factor for survival. The 5-ALA study group conducted a randomized controlled trial (RCT) in which 349 patients were randomized between white light and blue light groups (Stummer et al. 2006), and 176 patients were randomized with blue light exposure. There were no significant differences between the study and control groups in terms of age, gender, KPS, or location of tumor. Forty patients were excluded from the analysis because the underlying pathology was not high-grade glioma (of which 15 were metastases, 7 low-grade gliomas, and 14 benign lesions). Complete surgical excision of the enhancing lesion on MRI was achieved in 64% (112/176) under blue light compared to 38% (65/173) under white light ($p < 0.0001$). Tumor-free survival (TFS) was longer in patients who had blue light resection (5.1 months vs. 3.6 months, $p = 0.022$). Multivariate analysis of patients in this study and other similar studies had identified that a KPS score of 70 or more, age less than 65 years, and extent of surgical resection were independent good prognostic factors. Excision of 98% of the enhancing lesion extended life expectancy by at least 4.2 months compared to resections leaving more than 2% residual tumor. The median survival of GBM when complete resection was achieved was 16.7 months compared to 11.8 months in controls (Pichlmeier et al. 2008). PD had the potential to excise much more than the enhancing lesion on MRI. High-grade gliomas extend beyond gadolinium enhancement. Figure 42.1 demonstrates the size difference of a GBM using MRI enhancement, surgical microscopy, and fluorescence. The GBM was located in the right anterior temporal lobe and was excised by right temporal lobectomy. The resected right temporal lobe was sectioned, and the maximum diameter of the tumor was measured under white light (35 mm) and blue light (42 mm). The maximum diameter of the tumor measured 17 mm on MRI (Eljamel 2010). My study group explored fluorescence in 114 consecutive intracranial lesions and found variable sensitivity; GBM had 87%; primary cerebral lymphoma, ependymoma, myeloma, meningiomas, and metastatic colon, breast, and lung carcinomas had high sensitivity as well (Eljamel 2009). Most false-negative biopsies observed in studies using PD technology were the result of tumor necrosis rather than active invasive tumor edge. There was also good correlation between the amount of viable tumor cells and the intensity of fluorescence during PD: red fluorescence can only be seen in

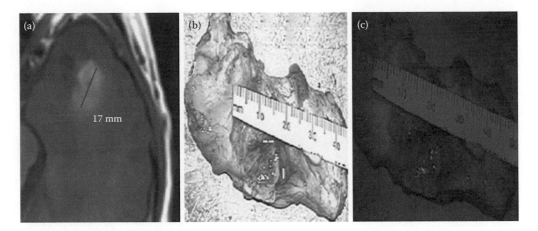

FIGURE 42.1 A temporal lobe GBM as seen on (a) MRI, (b) white light inspection, and (c) blue light, demonstrating that enhanced MRI underestimates the actual size of the tumor.

solid-packed viable malignant tumor, while pink fluorescence was observed where viable tumor cells were mixed with normal brain or radionecrosis in patients who had previous radiation therapy.

42.2.1 Fluorescence Image-Guided Surgery

Most of the evidence base for PD came about through fluorescence image-guided surgery (FIGS). The evidence was compelling in favor of its use in GBM and grade III astrocytoma [anaplastic astrocytoma (AA)] that the European Medicines Agency–Europa (EMAE) granted a license for the use of 5-ALA-induced fluorescence (Gliolan, Medac, Germany). Stummer et al. achieved 64% complete resection using 5-ALA-induced FIGS in a randomized controlled trial compared to 36% in the control white light group (Eljamel et al. 2008). Stummer et al. (2006) using Photofrin (Porfimer sodium, Axcan Pharma, Canada) and 5-ALA-induced FIGS in a randomized controlled trial was able to completely excise the lesion in 77%, while Kostron and coworkers using meta-tetrahydroxyphenylchlorin (mTHPC) (Foscan, Biolitec, Ireland) was able to achieve 75% complete excision in a matched controlled study (Kostron et al. 2006). Another study using mTHPC reported 87.9% sensitivity and 95.7% specificity in 22 patients with GBM (Zimmermann et al. 2001). FIGS had significantly ($p < 0.001$) prolonged time to tumor progression (Eljamel et al. 2008; Kostron et al. 2006; Stummer et al. 2006); it did not prolong survival in the first study (Stummer et al. 2008) because the number of patients who did not achieve total excision in the study group (36%) was sufficient to nullify this effect and the study was not powered to answer this important question. Furthermore, when tumors recurred in the two study groups, there was no control for rescue treatments. When the survival of patients who had total excision of their tumor was compared to survival of those who had not (122 patients vs. 121 patients), patients who had total excision had a median survival of 16.7 months compared to 11.8 months ($p = 0.0001$) (Stummer et al. 2006). Other FIGS users had obtained similar results obtaining maximum surgical resection (Grutsch et al. 2009; Diez et al.

2011); the latter group demonstrated that by using FIGS, they achieved excision of 98% or more of the tumors in their patients.

A typical 5-ALA FIGS procedure is outlined below. Please note that the procedure below may be adapted to other photosensitizers (Eljamel 2008).

1. Patients are counseled and consent is obtained. If the patient suffers from chronic liver failure, 5-ALA-based procedures must be avoided.
2. Three hours before surgery, the patient is given 2 mg/kg body weight ALA orally mixed in 30 mL of water (to use other photosensitizers, administer these according to the recommended times; e.g., Foscan is given IV 96 h before surgery and Photofrin is given IV 48 h before surgery).
3. The surgical procedure is planned using image-guided technology for localization of the craniotomy (Figure 42.2).

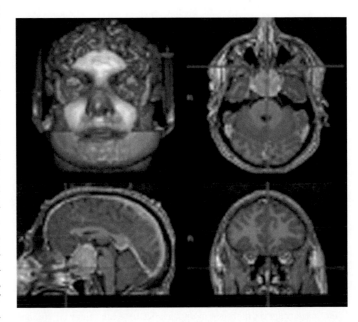

FIGURE 42.2 Screen photograph of neurosurgical plan using image-guided surgery.

4. Once the craniotomy is performed and the dura is opened, surface of the brain is inspected.
   a. If the tumor is visible on brain surface, proceed with surgical resection of easily recognizable tumor to debulk it.
   b. If the tumor is not visible on the surface of the brain, use the image guidance system to locate the tumor and proceed to remove the easily recognized bulk of the tumor.
5. The remainder of the tumor is then removed using FIGS as follows:
   a. Ambient operating procedure lights are shut off.
   b. Blue light is turned on.
   c. Observe the fluorescent tumor (bright red) shining through the blue surgical field (Figures 42.3 and 42.4).
   d. Remove any fluorescent (tumor) tissue.
   e. If the field became unclear switch back to standard light.
   f. Once the field is dry, the FIGS procedure is continued as described above until all remarkable fluorescence tissue is excised.
6. Once the resection is completed, secure hemostasis.
7. Give any additional adjunct intraoperative therapy planned at this point [e.g., PDT, gliadel, or intraoperative radiotherapy (IORT)].
8. Close the dura, craniotomy, and scalp in the normal fashion.
9. Continue postoperative care as usual after recovery of the patient from anesthesia. Be aware that skin and retina of patients are photosensitive for about 24 h in the case of 5-ALA and much longer in other photosensitizers, and the anesthesiologist and nurses need to be aware of this to avoid continuous pulse oxymetry and fundal retinal examinations during systemic photosensitivity.

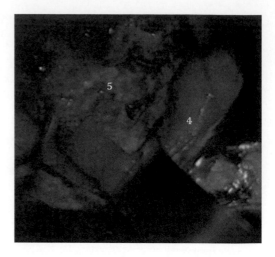

FIGURE 42.4 Intraoperative photograph of the same surgical field under blue light, demonstrating red fluorescence of residual tumor (5) while normal brain is blue (4).

The effect of FIGS is to achieve total resection of the enhancing lesion or at least excise more than 98% of the enhancing lesion.

42.2.2 Fluorescence Image-Guided Biopsy

Forty percent of GBMs transform from lower grades. The treatment of mixed gliomas should be based on the higher-grade component. However, during surgery, it would not be possible to distinguish low- from high-grade components using white light and basic imaging. Since 5-ALA-induced fluorescence occurs in GBM and AA while grade I and grade II astrocytomas do not, adding 5-ALA and using blue light to identify the higher-grade component of transformed glioma is the best way forward to guarantee taking representative biopsy samples to allow adequate treatment. A study published in *Cancer* 2010 (Widhalm et al. 2010) demonstrated that 100% of grade II astrocytoma was negative for fluorescence and 89% of AA was positive for fluorescence in mixed lesions.

Frame-based stereotactic biopsy still plays an important part to diagnose deep-seated lesions in the brain that are not suitable for cytoreductive surgery. However, false-negative results of the biopsy are not uncommon and could be as high as 11% (Feiden et al. 1991). For this reason, some authors suggest taking multiple biopsies, increasing the risk of the procedure. Using 5-ALA-induced fluorescence and inspecting the biopsy specimen under blue light once it has been extracted has the potential of increasing the sensitivity of stereo biopsy (Figure 42.5) and reducing the risk by negating the need for multiple biopsies.

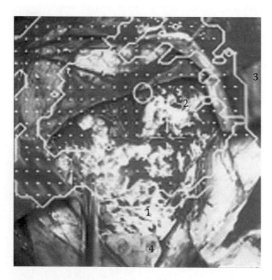

FIGURE 42.3 Intraoperative photograph of a tumor under white light, demonstrating no difference in appearance between tumor (1) and normal brain (4); the outline projected in the surgical field from the image-guided surgery (2) represents the head up display contour of the tumor.

FIGURE 42.5 Stereotactic biopsy block visualized under blue light.

42.3 PDT in MBT

PDT is based on the principles of photosensitizer administration followed by laser light exposure of the field to specific wavelength that is capable of converting the harmless photosensitizer into a killer substance leading to tumor cell death. Photosensitizers such as 5-ALA, Photofrin, and temoporfin absorb light energy and produce reactive oxygen species within the cells (Eljamel 2004, 2005; Feiden et al. 1991). In tumor cell experiments, glioma spheroids, GBM tumor models, and in clinical practice, PDT had been shown to induce tumor cell death, apoptosis, and immune responses (Eljamel 2004, 2005; Feiden et al. 1991). A study comparing the sensitivity of cancer cells and neurons reported that at 15 μg/mL of temoporfin, PDT killed almost all of tumor cells compared to only 10% of neurons in the absence of blood brain barrier in 3D cell cultures (Phillips 2008). Another clinical study that examined the use of stereotactic PDT in 31 deep-seated malignant gliomas with no additional treatment revealed complete resolution of the lesion in 64% on postoperative MRI, reduction of tumor volume by 50%–90% in 26%, and no change in 10% (Kaneko 2008a). Therefore, the combination of photosensitizer selectivity in tumor cells, high concentration of photosensitizer in tumor cells, and specificity of laser light to each photosensitizer makes PDT a selective local anticancer treatment that kills any residual tumor cells after maximum safe surgical resection. The limitation of PDT in the brain is the extent of light penetration within brain tissue. Because of this limitation, PDT may not impart the desired benefit if the thickness of residual tumor was beyond laser light reach. It is therefore essential that maximum safe surgical resection is performed first (using FIGS) before PDT (Aziz et al. 2009; Eljamel et al. 2008).

PDT in the brain is a local selective therapy without systemic side effects compared to chemotherapy or the secondary effects on brain and delayed cancers compared to radiotherapy. It can be repeated as many times as the patient and surgeon wish. It can be combined with surgery, FIGS, and intraoperative identification of the tumor to maximize safe resection at the same sitting without delay or fear of wound breakdown. PDT has a very low side effects profile compared to conventional radiotherapy and chemotherapy. The concentration of the photosensitizers is high in GBM cells compared to normal brain (normal brain to GBM is 1 to 3 of 400) (Eljamel 2003). Early studies of PDT in high-grade gliomas demonstrated increased median survival and increased 2-year survival from 18% to 28% (Figure 42.6) (Eljamel 2005; Kaneko 2008a; Perria et al. 1980; Pitchlmeier et al. 2008).

One of the largest experiences of PDT in MBT was from the Royal Melbourne Hospital, Australia. A total of 350 patients were treated with hematoporphyrin derivative (HPD), and 230 J/cm² was used without balloon diffuser. One hundred thirty-eight patients (of which before 78 GBMs and 58 AAs) were available for analysis with minimum follow-up of 3 years. HPD 5 mg/kg was given IV 24 h before surgery and 70–240 J/cm² KTP laser was used. Twenty-nine percent of these patients

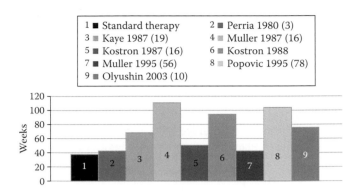

FIGURE 42.6 A bar graph representing the average survival of GBMs treated with PDT in phase I/II trials in the twentieth century compared to standard treatment. References given in the artwork are adopted from Eljamel (2005). Numbers in parentheses are the number of patients treated.

received chemotherapy. The median survival of newly diagnosed GBM was 14.3 months, for recurrent GBM 13.5 months, and for 2-year survival 28%. The median survival of newly diagnosed AA was 76.5 months, for recurrent AA 66.6 years, and for 2-year survival 37%. Older patients typically had worse prognoses than their younger counterparts [hazard ratio 1.25, 95% confidence interval (CI) 1.05–1.49, $p = 0.010$]. This was independent of tumor grade and whether newly diagnosed or recurrent GBMs. Among patients with newly diagnosed GBMs, light dose ≥ 230 J/cm² was associated with better prognosis (hazard ratio 0.502, 95% CI 0.27–0.94; $p = 0.033$). For recurrent GBMs, there was no statistically significant association between survival and light dose. Tumor location (frontal or other) was not associated with better survival ($p = 0.540$), nor was there any significant association with survival with regard to gender or use of concomitant chemotherapy (Kaye et al. 1987; Popovic et al. 1995; Stylli et al. 2005). The cumulative experience from Innsbruck University, Austria, included 25 patients with newly diagnosed GBM treated with HPD-PDT with a median survival of 18 months and 67 patients with recurrent GBM with a median survival of 7 months (Kostron 2009). It also included 22 patients with recurrent GBM treated with temoporin-PDT with a median survival of 9 months. The Hokkaido University (Japan) experience included 290 patients treated with 5-ALA-FIGS and 35 patients treated with HPD-PDT (Kaneko 2006). The median survival of GBM was 20.5 months and for AA 36 months. The North American experience (Toronto, Canada), using Photofrin-PDT included 96 high-grade glioma patients (total number reported 112 patients, 11 metastases, and one malignant meningioma). Photofrin 2 mg/kg was given IV 13–36 h before surgery and KTP laser intracavity, and patients survived for 1 year and 2% survived for 2 years (Muller and Wilson 2006b). Meta-analysis of observational studies of PDT in high-grade gliomas included more than 1000 patients with median survival of 16.1 months for newly diagnosed GBMs and 10.3 months for recurrent GBMs. These survival outcomes were better than standard therapy. However, advocates of evidence-based medicine criticized these studies because they were observational in nature and were not double-blind randomized placebo-controlled crossover trials (DBRPCCTs). DBRPCCT could not be applied to every intervention particularly in surgery because the surgical technique plays a bigger role and neither the patient nor the surgeon could be blinded. DBRPCCTs are not applicable to craft-based procedures such as neurosurgery (Smith et al. 2003). However, PDT in high-grade gliomas had been subjected to several prospective trials: a phase III multicenter randomized trial in North America using Photofrin 2 mg/kg PDT adjuvant to surgical resection compared to standard surgical resection (Muller and Wilson 2006b), 5-ALA-PDT RCT in Germany (Stepp et al. 2007), temoporfin-PDT with matched controls in Austria (Kostron et al. 2006), and Photofrin 2 mg/kg PDT in conjunction with 5-ALA-FIGS in Scotland (Eljamel et al. 2008). The first RCT included 43 patients in the PDT group and 34 patients in the control group (Muller and Wilson 2006b). The median survival of the PDT group was 11 months (95% CI, 6–14 months) compared to 8 months (95% CI, 3–10 months) in the control group, an increase of 38% in median survival, which was statistically significant. However, the life curves crossed over at 15 months. There were multiple reasons of the late failures in 5-ALA-FIGS procedure: no information about extent of surgical resection in the two groups, which is very important in such surgical trials; no matching of postrecurrence therapies such as chemotherapy; and further surgery, as many patients in the control group had PDT on recurrence. Secondary treatments were likely to have canceled any earlier PDT effect, and residual tumor left after resection in each group is likely to have reduced the impact of PDT. The second RCT combined ALA-guided FIGS and interstitial PDT (Stepp et al. 2007). The 5-ALA dose was 20 mg/kg given orally 3 h before anesthesia and 200 J/cm^2 was used for PDT with a 635 nm laser. The 6 month progression-free survival was 41% in the experimental group compared to 21% in the control group ($p < 0.001$). The Austrian trial recruited 26 patients and was matched by a similar group in the same institution (Kostron et al. 2006). This trial was not randomized but had a control group; it combined temoporfin-guided FIGS and PDT at 0.15 mg/kg IV temoporfin and 20 J/cm^2 at 652 nm laser. The sensitivity and selectivity of FIGS in this study was 87.9% and 95.7%, respectively (172 biopsy samples), and had a true prediction rate of 90.7%. Complete resection of enhancing lesion was achieved in 75% of the study group compared to 52% in matched controls. The median survival was 9 months compared to 3.5 months in matched controls. The final trial recruited 42 patients (Eljamel et al. 2008); 27 patients were eligible for analysis (14 were metastatic tumors on histology and one who died of unrelated causes was excluded), 13 were randomized to ALA+Photofrin FIGS and repetitive PDT, and 14 were randomized to standard therapy. The two groups were well matched for other prognostic factors such as age and location. They were also matched for secondary therapies on relapse such as chemotherapy and radiotherapy. The dose of PDT was 500 J/cm^2 fractionated into five sessions over 5 days. The relapse-free survival of the study group was 8.6 months compared to 4.8 months in the control group ($p < 0.01$). In the intent-to-treat analysis, the median survival was 52.8 (95%

CI, 40–65) weeks in the study group compared to 24.2 (95% CI, 18–30) weeks in controls ($p < 0.001$). The median KPS scores were 70 and 80 before and after PDT in the study group compared to 80 and 70 before and after treatment in the control group. Although patients in the study group were worse off before surgery, their KPS scores improved to much better scores postoperatively compared to controls (an absolute difference of 20 points).

42.3.1 PDT Brain Safety

An adverse event (AE) is any adverse change in health or side effect that occurs in a person who participates in a clinical trial while the patient is receiving the treatment (study medication, application of the study device, etc.) or within a prespecified period of time after their treatment has been completed. AEs are classified as serious (SAE) or minor (MAE); expected (EAE) or unexpected (UAE); and study-related (SRAE), possibly study-related (PSRAE), or not study-related (NSRAE). In PDT studies, skin and retinal photosensitivity are expected AEs. In our experience at the Scottish PDT center, we have carried out more than 400 treatments using 5-ALA and Photofrin intracavity PDT with balloon diffuser and 630 nm diode laser in 150 patients in brain and encountered AE in seven patients: three (2%) patients developed deep venous thrombosis (DVT) (0.7% per Photofrin-PDT treatment, none in ALA-mediated PDT), two (1.3%) patients developed skin photosensitivity because of nonadherence to light protection during summer months (0.18% per Photofrin-PDT treatment), two (1.3%) patients developed post-PDT brain swelling requiring treatment (0.18% per Photofrin-PDT treatment; no brain swelling was noted in newly diagnosed lesions after five doses of PDT), one (0.1%) balloon diffuser fracture due to poor catheter fixation (0.03% per Photofrin-PDT treatment), and one (0.1%) patient developed skin necrosis and CSF leak of previously irradiated skin flap (0.03% per Photofrin-PDT treatment). These side effects are not different from those encountered during standard therapy. Photofrin-PDT in 20 patients with recurrent GBM using escalating doses of Photofrin from 0.75 to 2 mg/kg and three methods of light delivery was reported (Whelan et al. 2009; Schmidt et al. 2004). The authors reported no neurotoxicity by escalating the Photofrin dose and no neurotoxicity in intracavity 630 nm [light-emitting diode (LED) or laser] with balloon diffuser. The authors reported neurotoxicity in two patients treated with interstitial PDT; one developed ataxia and one facial weakness. These SAEs were expected with any treatment of intraparenchymal brain tumors without cytoreductive therapy. In fact, this expected SAE of the biopsy of such tumors and inserting multiple laser fibers in a brain tumor inevitably leads to SAEs in a significant number of patients. In my opinion, these two SAEs were the result of the interstitial insertion of the fibers rather than the PDT treatment. The authors concluded that the rate of AEs in this study was within the acceptable range of their practice in this area of neuro-oncology. The authors had encountered no side effects in their patients who had gliomas in the posterior cranial fossa and close to the brain stem. Another earlier study of 20 patients treated with HPD and 630 nm light reported five expected skin photosensitivity, none of which were SAEs (Stylli et al. 2005). Two of 136 HPD-PDT-treated patients in Australia (2%) reported excessive sunburn related to skin photosensitization (EAE) (Kostron et al. 1988). In both cases, they had failed to adhere to written instructions regarding sun exposure, similar to our own experience. The Canadian early series (Muller and Wilson 2006a) reported 2.7% mortality, 0.9% intracavity hemorrhage, 3.6% DVT, 3.6% wound infection, 0.9% CSF leak, 1.8% skin photosensitivity, and 5.7% worsening neurological deficit. All these AEs are not specific to PDT and expected AEs of cranial tumor surgery with the exception of skin photosensitivity. Skin photosensitivity is specific to PDT and is an expected AE in patients who do not adhere to written and verbal light protection instructions. These rates of complications were unique to this study and were not replicated in any subsequent observational studies or any prospective RCT of PDT. These complications encountered by these authors simply reflect the influence of four factors: interstitial insertion of KTP laser fibers in some patients, varied light dose, varied time to treatment, and the steep learning curve. The same team had not reported any high complication rates in the RCT they carried out after their earlier experience (Muller and Wilson 2006b). Furthermore, our patients tolerated 500 J/cm² divided into five sessions. On the other hand, interstitial and postradiation PDT may be associated with a higher complication rate. A study of 18 patients with interstitial PDT reported no side effects with a total dose of 1500–3700 J compared to two out of six patients who received a total dose of 3700–4400 J and three out of six patients who received a total dose of 4400–5900 J/cm² (Krishnamurthy et al. 2000). This study is not powered enough or randomized to answer this question. It was superseded by RCTs that did not demonstrate SAE (Stummer et al. 2000; Stylli et al. 2005; Kostron 2009). In FIGS, the AE profile of the blue light group (overall 42.8%) did not significantly differ from that of the white light group (44.5%) (Stummer et al. 2006). Dysphasia (AE) was 3.5% versus 0.6% ($p = 0.07$, not significant), seizures were 6% versus 2.9% ($p = 21$, not significant), and hemiparesis was 4% versus 2.3% ($p = 0.4$, not significant).

42.3.2 Method of Intracavity PDT of MBT

1. The chosen photosensitizer is administered according to recommendation of its use (see Section 42.3).
2. Wait the required time between photosensitizer administration to photoirradiation.
3. The surgery is planned and executed as described under FIGS (Eljamel 2008).
4. A balloon diffuser is prepared as follows:
   a. The appropriate laser fiber is inserted into a sterile fiber sheet.
   b. The laser fiber and sterile sheet complex is passed into the catheter of the balloon diffuser (Figure 42.7A).
   c. The catheter, sterile sheet, and laser fiber complex is then passed to the surgical site via a tiny stab incision at the side of the skin incision. Make sure that it passes via one of the craniotomy burr holes to avoid kinks in

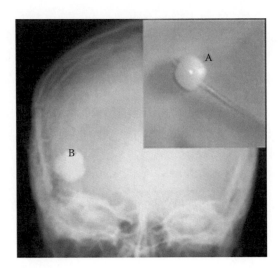

FIGURE 42.7 Photograph of the balloon diffuser. A = outside before insertion, B = on x-ray.

the PDT fiber complex and for easy extraction of the PDT fiber complex once PDT is finished.
   d. The balloon is inflated to fill the surgical resection cavity with a mixture of 0.8% intralipid, Hartman's solution, and radio-opaque solution (0.4 mL intralipid + 12.5 mL Hartman's solution + 12.5 mL contrast) (Eljamel 2008).
5. Measure the diameter of the balloon diffuser and work out the laser dose according to Table 42.1.
6. After dural closure, the craniotomy is closed in the usual fashion, avoiding any kinks in the PDT-fiber complex.
7. Provide the first treatment according to Table 42.1.
8. Further treatments can be provided at the bedside (Figure 42.8). I have given five treatments to patients who had not had radiotherapy and did not encounter complications; in patients who had radiotherapy, the maximum number of safe treatments is three.
9. Once the final PDT dose is finished, aspirate the balloon diffuser to deflate it. If you are in doubt, perform x-rays to make sure the balloon is fully deflated before extracting it (Figure 42.7B).

TABLE 42.1 Guidelines of Laser Fiber Therapy Using Photofrin Brain Intracavity PDT

Lesion/Balloon Size (cm)	0.5	0.75	1	1.25	1.50
Laser Fiber Diffuser (cm)	1	1	1	1	1.5
Laser Power (mW)	400	400	400	400	600
Dose (J/cm²)	100	100	100	100	100
Irradiance (mW/cm²)	509	226	127	81	85
PDT Duration (s)	196	442	785	1227	1178
Lesion/Balloon Size (cm)	1.75	2	2.25	2.5	3
Laser Fiber Diffuser (cm)	1.5	1.5	1.5	2.5	2.5
Laser Power (mW)	600	600	600	1000	1000
Dose (J/cm²)	100	100	100	100	100
Irradiance (mW/cm²)	62	48	38	51	35
PDT Duration (s)	1604	2094	2651	1963	2827

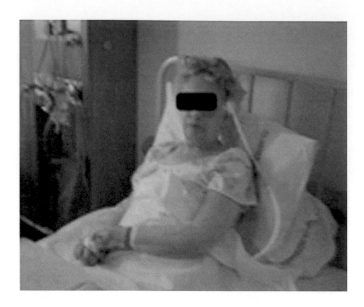

FIGURE 42.8 Photograph of a patient receiving PDT at the bedside.

42.3.3 Photosensitizers Used in PD and PDT

Several photosensitizers have been used in brain cancers. These include HPD (Kaneko 2008b; Kaye et al. 1987; Perria et al. 1980; Popovic et al. 1995; Stylli et al. 2005), porfimer sodium (Photofrin) (Eljamel et al. 2008; Muller and Wilson 2006a,b), 5-ALA (Eljamel et al. 2008; Stepp et al. 2007; Stummer et al. 2000, 2006), and mTHPC (Foscan) (Kaneko 2008b). However, the most commonly used photosensitizers in MBT today are 5-ALA, Photofrin, and Foscan. These photosensitizers are not licensed at the moment for treating brain tumors by PDT in most countries, and their use in brain tumors is off-label on a named-patient basis. However, ALA (Gliolan) is licensed in Europe for FIGS.

42.3.3.1 ALA

ALA has been used in many fields to detect and treat precancer and cancerous lesions. It is a natural precursor for the heme synthesis in all living mammalian cells. Each cell metabolizes ALA along a set pathway to heme, producing protoporphyrin-IX (PpIX) along the way. However, conversion of PpIX to heme is blocked in cancer cells, leading to accumulation of PpIX in the tumor (Figure 42.9). It can be applied topically or taken orally. In PD and PDT of brain tumors, it is given orally 3–4 h before induction of anesthesia in a mixture of nonfizzy orange juice at 15–20 mg/kg body weight. The active compound in the cells is PpIX, which can be activated by violet-blue light (375–440 nm) for PD (Figure 42.10) and diode laser light (635 nm) for PDT, tissue depth penetration is 7–15 mm, the irradiation dose required for PDT is 100 J/cm², and skin photosensitivity continues for 24–36 h after ingestion.

42.3.3.2 Photofrin (Porfimer Sodium)

Photofrin is given intravenously at 2 mg/kg body weight 48 h before PDT. It is activated by diode laser light at a 630 nm

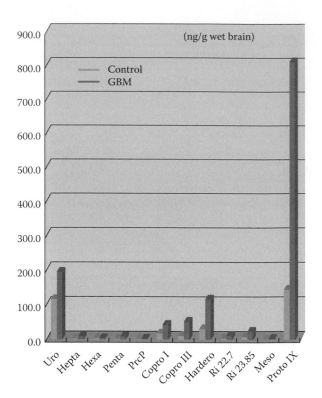

FIGURE 42.9 Concentration of ALA in GBM and brain. (Courtesy of Dr. Kaneko.)

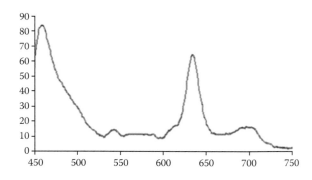

FIGURE 42.10 PpIX spectrum at 632 nm.

wavelength. The depth of tissue penetration is from 7 to 12 mm, and the irradiation dose is 100 J/cm². Skin photosensitivity carries on for 30 days, after which it reduces significantly and by 90 days it is usually gone.

42.3.3.3 Foscan (Temoporfin)

Foscan is given intravenously at a dose of 0.15 mg/kg body weight about 96 h before PDT. It is activated at a 652 nm wavelength and its tissue penetration is 10–15 mm. The irradiation dose is 20 J/cm² and its skin photosensitivity continues for about 2 weeks.

Therefore, the photosensitizer is administered either orally (ALA) or intravenously (Photofrin and Foscan) 3 h (ALA), 48 h (Photofrin), or 96 h (Foscan) before surgery to allow sufficient concentration of the photosensitizer in tumor cells. Patients are counseled to take adequate protection from bright light and the sun to prevent skin and retinal damage.

42.4 Specific Brain PD and PDT Equipment

PD-assisted brain tumor resection can be performed using commercially available systems such as the Olympus QTV55 and CLV-520 PD system attached to an endoscope with a long-pass observation filter or the Zeiss OPMI Pentero or NC4 microscope with a PD system, or other surgical microscopes or endoscopes with the correct modifications available from Lieca, Muller, or Storz. PDT, on the other hand, requires an inflatable diffuser balance filled and a diode laser that is capable of illuminating the postresection tumor cavity with the appropriate laser wavelength for the photosensitizer, such as the InGaAIP laser diode CW (630 ± 3 nm, Diomed, Cambridge, UK) for Photofrin PDT.

42.5 Conclusions

1. PD and PDT of MBT are safe, selective, and sensitive and did not increase the AE profile of surgical resection of these tumors.
2. FIGS is associated with significant improvement of tumor-free survival and significant increase in complete excision of enhancing PMBT on MRI.
3. PDT of PMBT is associated with improved survival in both newly diagnosed and recurrent PMBT.
4. PDT after maximum safe surgical excision using FIGS technology significantly improved tumor-free survival, median survival, and quality of life in patients with newly diagnosed PMBT.
5. PDT for SBT improved the local control rate of these tumors.
6. Fluorescence image-guided biopsy (FIGB) has the potential to improve the sensitivity of stereotactic biopsy and reduce its complications.

These conclusions are based on the presented evidence discussed in this chapter, including large observational studies and RCTs. Due to aforementioned compelling reasons, the only primary outcomes that can be used in future studies of MBT is tumor-free survival, completeness of resection, and quality of life postoperatively. These primary outcomes have already been proved in two RCT using FGIS and PDT, and it would be unethical to deny such treatment if it was available to patients. Future efforts should be channeled to study the interaction of FIGS, PDT, and other treatment modalities.

References

Aziz, F., S. Telara, H. Moseley et al. 2009. Photodynamic therapy adjuvant to surgery in metastatic carcinoma in brain. *Photodiagn Photodyn* 6:227–230.

Díez Valle, R., S. Tejada Solis, M. A. Idoate Gastearena et al. 2011. Surgery guided by 5-aminolevulinic fluorescence in glioblastoma: Volumetric analysis of extent of resection in single-center experience. *J Neurooncol* 102(1):105–113.

Eljamel M. S. 2003. New light on the brain: The role of photo-sensitising agents and laser light in the management of invasive intracranial tumours. *Technol Cancer Res Treat* 2(4):303–309.

Eljamel, M. S. 2004. Photodynamic assisted surgical resection and treatment of malignant brain tumors; technique, technology and clinical application. *Photodiag Photodyn* 1:93–98.

Eljamel, M. S. 2005. Brain PDD and PDT unlocking the mystery of malignant gliomas. *Photodiagn Photodyn* 1:303–310.

Eljamel, M. S. 2008. Fluorescence image guided surgery of brain tumors: Explained step-by-step. *Photodiag Photodyn* 5:260–263.

Eljamel, M. S. 2009. Which intracranial lesions would be suitable for 5-aminolevulenic acid-induced fluorescence-guided identification, localization, or resection? A prospective study of 114 consecutive intracranial lesions. *Clin Neurosurg* 56:93–97.

Eljamel, M. S. 2010. Photodynamic applications in brain tumors: A comprehensive review of the literature. *Photodiagn Photodyn* 7(2):76–85.

Eljamel, M. S., C. Goodman, and H. Moseley. 2008. ALA and Photofrin(R) fluorescence-guided resection and repetitive PDT in glioblastoma multiforme: A single centre phase III randomised controlled trial. *Lasers Med Sci* 23:361–367.

Feiden, W., U. Steude, K. Bise, and O. Gündisch. 1991. Accuracy of stereotactic brain tumor biopsy: Comparison of the histologic findings in biopsy cylinders and resected tumor tissue. *Neurosurg Rev* 14:51–56.

Gautschi, O. P., K. van Leyen, D. Cadosch, G. Hildebrandt, and J. Y. Fournier. 2009. Fluorescence guided resection of malignant brain tumors—Breakthrough in the surgery of brain tumors. *Praxis* (Bern 1994) 98(12):643–647.

Kaneko, S. 2006. Photodynamic therapy in high grade gliomas. Proceedings of the 7th International PDT Symposium, Brixen, Italy.

Kaneko, S. 2008a. Recent advances in PDD and PDT for malignant brain tumors. APLS Proceedings 1–4.

Kaneko, S. 2008b. A current overview: Photodynamic diagnosis and photodynamic therapy using ALA in neurosurgery. *JJSLSM* 29:135–146.

Kaye, A. H., G. Morstyn, and D. Brownbill. 1987. Adjuvant high-dose photoradiation therapy in the treatment of cerebral glioma, a phase I/II study. *J Neurosurg* 67:500–505.

Kleihues, P., P. C. Burger, and W. K. Cavenee. 1997. Glioblastoma. In *WHO Classification: Pathology & Genetics of Tumours of the Nervous System*. P. Kleihues and W. K. Cavenee, editors. International Agency for Research on Cancers, Lyon, France, 16–24.

Klos, K. J., and B. P. O'Neill. 2004. Brain metastases. *Neurologist* 10:31–46.

Komblith, P. 1995. The role of cytotoxic chemotherapy in the treatment of malignant brain tumors. *Surg Neurol* 44:551–552.

Kostron, H. 2009. Photodynamic applications in neurosurgery. Proceedings of the International Photodynamic Therapy (IPA), Seattle, WA.

Kostron, H., T. Fiegele, and E. Akatuna. 2006. Combination of FOSCAN® mediated fluorescence guided resection and photodynamic treatment as new therapeutic concept for malignant brain tumors. *Med Laser Appl* 21:285–290.

Kostron, H., E. Fritsch, and V. Grunert. 1988. Photodynamic therapy of malignant brain tumors; a phase I/II trial. *Br J Neurosurg* 2:241–248.

Krishnamurthy, S., S. K. Powers, P. Witmer, and T. Brown. 2000. Optimal light dose for interstitial photodynamic therapy in treatment for malignant brain tumors. *Lasers Surg Med.* 27:224–234.

Lamborn, K. R., S. M. Chang, and M. D. Prados. 2004. Prognostic factors for survival of patients with glioblastoma: Recursive partitioning analysis. *Neurooncol* 6(3):227–235.

Muller, P., and B. Wilson. 2006a. Photodynamic therapy of brain tumors—A work in progress. *Lasers Surg Med* 38:384–389.

Muller, P., and B. Wilson. 2006b. A randomized two arm clinical trial of Photophrin PDT and standard therapy in high grade gliomas, phase III trial. Proceedings of the 6th International PDT Symposium, Brixen, Italy.

Obwegeser, A., M. Ortler, M. Seiwald, H. Ulme, and H. Kostron. 1995. Therapy of glioblastoma multiforme, a cumulative experience of 10 years. *Acta Neurochirur (Wein)* 137:29–33.

Olsen, J. J., and T. Ryken. 2008. Guidelines for the treatment of newly diagnosed glioblastoma: Introduction. *J Neurooncol* 89:255–258.

Perria, C., T. Capuzzo, G. Cavagnaro, R. Datti, N. Francaviglia, C. Rivano, and V. E. Tercero. 1980. Fast attempts at the photodynamic treatment of human gliomas. *J Neurosurg Sciences* 1980; 24:119–129.

Phillips, J. B. 2008. Differences in sensitivity to mTHPC-mediated photodynamic therapy of neurons, glial cells and MCF7 cells in a 3-dimensional cell culture model. Proceedings of the 7th International PDT Symposium, Brixen, Italy.

Pichlmeier, U., A. Bink, G. Schackert, and W. Stummer. 2008. Resection and survival in glioblastoma multiforme: An RTOG recursive partitioning analysis of ALA study patients. *Neurooncol* 10:1025–1034.

Popovic, E. A., A. H. Kaye, and J. S. Hill. 1995. Photodynamic therapy of brain tumors. *Semin Surg Oncol* 11(5):335–345.

Rich, J. N., C. Hans, B. Jones et al. 2005. Gene expression profiling and genetic markers in glioblastoma survival. *Cancer Res* 65:4051–4058.

Scott, J. N., N. B. Rewcastle, P. M. Brasher et al. 1998. Long-term glioblastoma multiforme survivors: A population-based study. *Can J Neurol Sci* 25:197–201.

Schmidt, M. H., G. A. Meyer, K. W. Reichert et al. 2004. Evaluation of photodynamic therapy near functional brain tissue in patients with recurrent brain tumors. *J Neurooncol* 67:201–207.

Smith, G. C. S., and J. P. Pell. 2003. Hazardous journey parachute use to prevent death and major trauma related to gravitational challenge: Systematic review of randomised controlled trials. *BMJ* 327:1459–1461.

Stepp, H., T. Beck, T. Pongratz et al. 2007. ALA and malignant glioma: Fluorescence-guided resection and photodynamic treatment. *J Environ Pathol Toxicol Oncol* 26:157–164.

Stummer, W., A., Novotny, H. Stepp et al. 2000. Fluorescence guided resection of glioblastoma multiforme by using 5-aminolevulenic acid induced porphyrins; a prospective study in 52 consecutive patients. *J Neurosurg* 93:1003–1013.

Stummer, W., U. Pitchimeier, T. Meinel et al. 2006. Fluorescence-guided surgery with 5-aminolevulinic acid for resection of malignant glioma: A randomized controlled multicentre phase III trial. *Lancet Oncol* 7:392–401.

Stummer, W., H. J. Reulen, T. Meinel et al. 2008. Extent of resection and survival in glioblastoma multiforme: Identification of and adjustment for bias. *Neurosurgery* 62(3):564–576.

Stupp, R. 2007. Malignant gliomas: ESMO clinical recommendations for diagnosis, treatment and follow up. *Ann Oncol* 18 (Suppl 2):60–70.

Stylli, S. S., A. H. Kaye, L. MacGregor, M. Howes, and P. Rajendra. 2005. Photodynamic therapy of high grade glioma—Long term survival. *J Clin Neurosci* 12:389–398.

Vecht, C. J., C. J. Avezaat, W. L. van Putten, W. M. Eijkenboom, and S. Z. Stefanko. 1990. The influence of the extent of surgery on the neurological function and survival in malignant glioma. A retrospective analysis in 243 patients. *J Neurol Neurosurg Psychiatry* 53:466–471.

Watanabe, K., K. Sato, W. Biernat et al. 1997. Incidence and timing of p53 mutations during astrocytoma progression in patients with multiple biopsies. *Clin Cancer Res* 3:523–530.

Whelan, H., and S. Strivatsal. 2009. The proverbial light at the end of the tunnel in brain-tumor treatment. *SPIE* 10.1117/2.1200 907.1670, http://spie.org/documents/Newsroom/Imported/ 1670/1670_3646_1_2009-07-08.pdf.

Widhalm, G., S. Wolfsberger, G. Minchev et al. 2010. 5-aminolevulinic acid is a promising marker for detection of anaplastic foci in diffusely infiltrating gliomas with nonsignificant contrast enhancement. *Cancer* 116(6):1545–1552.

Zilidis, G., F. Aziz, S. Telara, and M. S. Eljamel. 2008. Fluorescence image-guided surgery and repetitive photodynamic therapy in brain metastatic malignant melanoma. *Photodiagn Photodyn Ther* 5:264–266.

Zimmermann, A., M. Ritsch-Marte, and H. Kostron. 2001. mTHPC-mediated photodynamic diagnosis of malignant brain tumors. *Photochem Photobiol* 74(4):611–616.

43

Photodynamic Therapy for Malignant Pleural Disease

Charles B. Simone
University of Pennsylvania

Melissa Culligan
University of Pennsylvania

Joseph S. Friedberg
University of Pennsylvania

Keith A. Cengel
University of Pennsylvania

43.1 Introduction

The pleural space exists between two layers of mesothelium that envelop the chest wall (parietal pleura) and lung (visceral pleura) (Figure 43.1). Normally this space contains an extremely thin layer of serous fluid that provides for lubrication and hydrostatic coupling between the visceral and parietal surfaces. Due to its contiguous nature, the pleural space allows malignancies involving pleural tissues the ability to rapidly spread to cover the entire surface of the chest wall, lung, mediastinum, and pericardium. The large surface area and the close proximity of the pleura to physiologically critical structures make cancers that involve the pleura among the most lethal and difficult-to-treat malignancies with which a clinician is faced.

Pleural cancers can be primary, arising from the pleura itself, or secondary, arising from metastatic spread of tumors originating from outside of the pleura. Secondary cancers of the pleura are much more common than primary cancers and result from direct extension of thoracic tumors (e.g., cancers of the lung or thymus) or hematologic spread of extrathoracic tumors (e.g., extremity sarcoma, ovarian cancer). Generally, the presence of a malignant effusion and/or bulky pleural disease renders a cancer stage IV. Primary pleural malignancies are relatively rare, and of these cancers, malignant pleural mesothelioma (MPM) is the most common, accounting for approximately 3000 new cancer diagnoses in the United States each year.

The treatment of pleural malignancies is typically palliative in nature. Pleural malignancies commonly present as an effusion, with the chief symptom being dyspnea. Other common symptoms include pain and anorexia. Palliative therapies include physiologically directed therapies such as drainage of pleural effusion, medically directed therapies such as chemotherapy, and symptom-directed therapies such as narcotic analgesics. Median survival times with palliative therapies vary considerably with the stage of disease at presentation but generally range from 6 to 14 months. As a general rule, definitive treatments for pleural malignancies must attempt to address the combined difficulties of widespread local disease and the high risk of systemic disease and remain investigational. Because of the diffuse nature of pleural cancers, which coat and contaminate all surfaces and structures in the chest cavity, surgery is not a stand-alone treatment option. Thus, the role of surgery is to achieve a macroscopically complete resection (MCR) as a part of a multimodal treatment program where other modalities are employed to attack the residual microscopic disease that will be present after surgery.

The appeal of photodynamic therapy (PDT) in the treatment of pleural malignancies is that it has the potential to combine selective destruction of cancerous tissue compared to normal tissue with the ability to treat and conform to relatively large surface areas. Moreover, the intrinsic, physical limitation in the depth of visible light penetration through tissue limits PDT damage to deeper structures, thereby providing additional potential for tumor cell selectivity. In combination with MCR and systemic chemotherapy to address the local and systemic elements of the cancer, PDT is currently being investigated as an attractive intraoperative adjuvant therapy option.

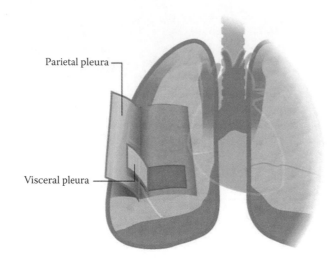

Parietal pleura

Visceral pleura

FIGURE 43.1 The pleural space between two layers of mesothelium enveloping the chest wall (parietal pleura) and lung (visceral pleura).

43.2 Surgery for Pleural Malignancies

Due to the difficulties noted above, even the most aggressive surgical resections are unlikely to result in curing a patient with a pleural malignancy. Thus, the goal of surgery in the definitive setting is to provide MCR and allow for treatments that can address the local (and systemic) microscopic disease that will almost certainly remain after surgery. Many surgery-based treatments employ extrapleural pneumonectomy (EPP), which achieves MCR through en bloc resection of the parietal pleura, diaphragm, pericardium, and lung. EPP can be combined with intracavitary or systemic chemotherapy and/or hemithoracic ionizing radiation therapy (RT). The alternative to EPP is radical pleurectomy/decortication (P/D) that is performed with the goal of achieving MCR while also attempting to preserve the lung and other structures such as the diaphragm, pericardium, and phrenic nerve. While lung-preserving options have an intuitive attraction, the retention of normal lung can significantly alter the therapeutic index of adjuvant therapies such as chemotherapy or RT. There is no surgical procedure accepted as the standard of care for surgical cytoreduction of pleural malignancies, and the selection of EPP versus P/D depends on a number of factors, including the expertise of the thoracic surgeon, plans for adjuvant/neoadjuvant treatment, pulmonary involvement by the malignancy, and patient comorbidities.

43.2.1 Criteria for Surgical Cytoreduction

Several important criteria exist for patients being considered for surgical cytoreduction. First and foremost, the patient must be medically fit for a surgery of this magnitude, and medical clearance includes multisystem evaluation of renal, hepatic, and cardiorespiratory function as well as evaluation of nutrition/metabolism. The second is that the disease must be considered surgically resectable to MCR, and in general, disease should be confined to one hemithorax so that MCR could leave the patient with no evidence of disease. Finally, the patient must be able to give true informed consent and understand that surgery for their disease is of an investigational nature. In addition to radiographic imaging, many groups who treat pleural malignancies will employ invasive staging, such as bronchoscopy, mediastinoscopy, or other lymph node biopsy techniques; contralateral video thoracoscopic pleural biopsies; and/or laparoscopy with peritoneal lavage and biopsies. Any evidence of radiographically occult metastases serves as a contraindication for surgery.

43.2.2 Local Recurrence after Definitive Surgery

Local recurrence rates following surgical resection of pleural malignancies range from 30% to 70% (Forster et al. 2003; Krug et al. 2009; Tilleman et al. 2009; Zellos et al. 2009). In patients with malignant pleural mesothelioma, the ipsilateral thoracic cavity has been reported as the first site of recurrence in approximately 1 of 3 of patients following EPP and 2 of 3 of patients following P/D. It is important to remember that these results and recurrence rates reflect a number of treatment plans, but all reflect a multimodal approach to the disease, with surgery alone resulting in nearly 100% local recurrence. In patients with non-small-cell lung cancer (NSCLC), even an isolated finding of positive postresection pleural cytology decreases the median recurrence-free survival to 1 year in early-stage NSCLC patients as opposed to similar patients without positive cytology who experience a 70% 5-year progression-free survival (Aokage et al. 2009). These results outline the need for adjuvant therapies to surgery that can reduce these high local recurrence rates in order to make surgery a more viable therapeutic option for patients with pleural malignancies. Moreover, as newer systemic agents increase rates of systemic disease control, local control of pleural malignancies will become more critical in potentially curative multimodality therapy regimens.

43.2.3 Overall Survival in Multimodality Protocols

The MPM patient populations treated in multimodality management protocols that include surgery are heterogeneous, and results can be difficult to compare between trials. While many studies include patients with diverse stages, survival reports are often not prepared for patients with specific stages. In addition, in protocols involving neoadjuvant chemotherapy, survival is frequently reported in the proportion of patients who complete all phases of therapy without a clear accounting for how the staging makeup of the population changes for patients who progress through all portions of the multimodality treatment (i.e., how enriched is the final population for early-stage patients). Finally, since MPM is a relatively rare disease, individual institutional series tend to be relatively small, and all of the caveats for comparison of small, potentially heterogeneous populations apply. In an attempt to increase the number of patients analyzed, Flores and colleagues (2008) collaboratively pooled the treatment results for patients at multiple institutions and reported on the overall survival of 663 patients with MPM treated using surgically based multimodality therapy. The overall survival for patients with early-stage (I/II) MPM ranges from 18 to 46 months, while the overall survival for patients with locally advanced (stage III/IV) MPM ranges from 4 to 13 months. A recent multicenter trial of neoadjuvant chemotherapy, EPP, and postoperative hemithoracic radiotherapy reported a median overall survival of 17.3 months in 39 patients with stage I/II and 16.8 months in 36 patients with stage III/IV MPM (Krug et al. 2009). These results appear to be fairly well representative of many studies using this trimodality approach. There is a recent report on 35 patients with MPM treated with radical pleurectomy (RP) followed by pemetrexed-based chemotherapy and radiotherapy to drain sites. In this series, 30 of 35 patients had negative hilar/mediastinal (N1/N2) lymph nodes, and 16 stage I/II patients and 19 stage III/IV patients demonstrated median overall survivals of approximately 14 and 30 months, respectively (Bolukbas et al. 2011). In the majority of surgically based series reported, patients with spread of MPM to mediastinal lymph nodes experience dramatically decreased overall survivals, typically ranging from 8 to 12 months.

43.3 Initial Preclinical Development of PDT for Pleural Malignancies

The appeal of PDT for treatment of malignancies involving serosal surfaces such as those of the peritoneum or the pleura is that PDT has the potential to combine selective destruction of cancerous tissue compared to normal tissues with the ability to treat and conform to relatively large surface areas. However, in patients with serosal malignancies, the uptake ratios, when compared to clinically relevant normal tissues such as bowel or lung, are not nearly as dramatic as the preclinical models predicted for the first- or second-generation photosensitizers (Cengel et al. 2007). Nevertheless, restricting the application of light to the region of the malignancy, thereby avoiding some normal tissues and minimizing damage, can gain some tumor cell selectivity. The intrinsic, physical limitation in the depth of visible light penetration through tissue limits PDT damage to deeper structures, thereby providing additional potential for tumor cell selectivity and also allowing the treatment of relatively large surface areas with acceptable toxicity.

The strong theoretical rational for serosal PDT was first tested in animal models of peritoneal carcinomatosis by Douglass and colleagues (1983). In these experiments, rabbits with Brown-Pierce epithelioma implants in the serosa of the bowel, liver, pancreas, or bladder were treated with hematoporphyrin derivative (HPD)-mediated PDT (5 mg/kg HPD and 631 nm light). On days 5–7 following HPD-PDT, extensive tumor necrosis was reported. However, these experiments used a single focal spot of light at a very high fluence (300 J/cm²) that most likely resulted in a combination of thermal and PDT-mediated tumor cell killing. Tochner and colleagues (1985) built upon this work

with a series of experiments using a mouse model of ovarian peritoneal carcinomatosis. In these experiments, they evaluated HPD-mediated PDT using 50 mg/kg HPD and 514 nm light. Mice were injected intraperitoneally with ovarian embryonal cancer cells and then randomly assigned on day 9 following tumor inoculation to receive no treatment, treatment with HPD alone, treatment with light alone, or treatment with HPD + light (HPD-PDT). The HPD and light (9.6 J delivered over 16 min) were both delivered intraperitoneally, and the intraperitoneal tumor burden at the time of treatment was 2–4 g. All of the untreated control animals, as well as the HPD-only and light-only treated animals, died of progressive disease between days 20 and 23 following tumor inoculation. The mice treated with a single treatment with HPD-PDT on day 9 showed prolonged survival, but only 1 animal (out of 16 total) survived past day 34. This animal survived >50 days and was presumably cured. However, a second group of mice that received two treatments with HPD-PDT on days 9 and 15 showed apparent cure of disease in 6 out of 16 animals. It should be noted that in this tumor model, eradication of the tumor is difficult to achieve with the administration of intraperitoneal chemotherapy. A 70% cure rate is observed with intraperitoneal doxorubicin but only if the agent is administered 2 days after tumor inoculation, when the tumor burden is low. If doxorubicin is administered on the same day as the PDT was given (day 9), a cure rate of less than 20% is observed, presumably because of a higher tumor burden (Ozols et al. 1979; Tochner et al. 1985).

At this time, PDT was being developed clinically for treatment of endobronchial lung cancer. The early work with intraperitoneal cancers and the similarity of natural history of pleural and peritoneal malignancies stimulated preclinical efforts to determine the potential applicability of this new therapy for pleural malignancies. Initial work demonstrated that cell lines derived from multiple thoracic malignancies, including malignant mesothelioma, were highly sensitive to PDT-induced cytotoxicity *in vitro* (Pass and Pogrebniak 1992). Additional work by several groups demonstrated that subcutaneous human mesothelioma tumor xenografts could be treated effectively by porfimer sodium-mediated PDT.

One major difficulty with PDT in the peritoneal and pleural cavities was that the geometry and anatomy could preclude the delivery of light through "shadowing" effects. One solution for this is to use advanced light dosimetry systems, and these continue to be developed in ongoing clinical trials. Another opportunity for improving the light dose distribution is in the design of delivery procedures and devices. Using preclinical modeling, two fairly simple yet elegant solutions were developed for this problem that are still in use today in the operating room. The first solution is to use a light-scattering medium in the cavity to allow more light to reach potentially shadowed areas. A number of different biocompatible substances were tested, but the best of these was found to be a combination of phosphate buffered saline and intralipid, which is the biocompatible fat component used in total parenteral nutrition (Perry et al. 1989). This is prewarmed to body temperature (37°C) and used to fill the

FIGURE 43.2 A modified endotracheal tube filled with 0.1% intralipid in normal saline is used to disperse the light from a flat-cut optical fiber.

entire body cavity and then changed over repeatedly during the procedure to prevent the small amount of blood that leaks from separate capillary beds during the tumor resection from opacifying the medium. The second development was in the design of the light delivery device (DeLaney et al. 1991). This consists of a modified endotracheal tube with a flat-cut fiber-optic diffuser inside the center channel. When sealed and filled with diluted intralipid, this device becomes a flexible, spherical light source that is well suited to manual delivery of light into all areas of the chest and peritoneal cavities (Figure 43.2).

43.4 Clinical Results with PDT for MPM

Trials using porfimer sodium pleural PDT make up the majority of the clinical experience with pleural PDT. In the early 1990s, phase I and II studies were performed at Roswell Park Cancer Center and the U.S. National Cancer Institute (NCI). Pass and colleagues (1994) performed a phase I trial of pleural PDT in patients with pleural malignancies (40 MPM, 8 NSCLC, 6 sacroma/fibrous tumor of the pleura) using 2 mg/kg porfimer sodium and a drug light interval of 24–48 h. The protocol required that the tumor had to be surgically resectable to leave any residual disease with a thickness of 5 mm or less. Fifty-four patients were enrolled, and of these, 42 went on to have intraoperative PDT; the remaining 12 patients were unable to complete PDT due to intraoperative complications (3 patients) or inability to resect to within 5 mm (9 patients). An initial 33 patients received a photosensitizer with a 48 h drug light interval, and the final 9 patients received the photosensitizer with a 24 h drug light interval. Light (630 nm) was delivered using the combined intralipid plus spherical light diffuser as described above, and light dose was measured using a real-time, four-position, non-isotropic dosimetry system. Patients enrolled at the 48 h drug light interval dose cohorts received light doses up to 35 J/cm^2 with one patient out of six at 17.5 and 35 J/cm^2 dose levels,

experiencing toxicities that were possibly related to protocol therapy. Nine additional patients were treated at the 24 h drug light interval, with two out of three patients treated at a 32.5 J/cm^2 dose level, experiencing esophageal perforations that were considered dose-limiting toxicities. Six patients were treated at the 30 J/cm^2 dose level with no patients experiencing dose-limiting toxicity, and therefore, this was declared the maximally tolerated dose (MTD). While not designed to measure overall survival, the median survival for the overall cohort was 13 months, which was considered promising and worthy of further study. Takita and colleagues treated 23 patients with MPM on a phase II trial of surgical resection and intraoperative porfimer sodium-mediated PDT using a drug light interval of 48 h and 20–25 J/cm^2 of 630 nm light (Moskal et al. 1998; Takita et al. 1994). In this trial, light was delivered using four percutaneously placed, spherically diffusing light fibers, and dose was determined according to measurements obtained preoperatively using an optical phantom. Six patients had disease resection via an EPP and 15 by RP, and 2 patients were considered unresectable and therefore did not receive PDT. In the survival analysis of patients who received resection and PDT, patients with stage III and stage IV disease had a median survival of 7 months, which was not a superior outcome as compared with historical controls, and five patients with stage I and stage II disease were alive and had no evidence of disease (NED) up to 33 months postoperatively. In the 1998 update of this trial, 40 patients were included, with 24 stage III/IV patients having a median survival of 10 months but stage I/II patients having a median survival of 36 months with a 2-year survival rate of 61%. In this experience, it was concluded that pleural PDT using this particular delivery method was most appropriate for patients with early-stage disease.

Due to the promising phase I data at the National Cancer Institute (NCI), a randomized phase III study of surgery versus surgery + PDT was performed in patients with MPM (Pass et al. 1997). In this trial, patients with histologically proven, surgically resectable MPM confined to one hemithorax were randomly assigned to receive intraoperative PDT using porfimer sodium and a drug light interval of 24 h versus no intraoperative adjuvant therapy. All patients received postoperative, adjuvant chemoimmunotherapy using tamoxifen, cisplatin, and subcutaneous interferon alpha. Of the 63 total patients who were enrolled and underwent randomization, 15 were not included due to a variety of factors including inability to debulk the tumor to within 5 mm, disease progression prior to surgery, or patient refusal of surgery. Of the remaining patients, 25 received PDT, and 23 did not. Of note, 26 patients had tumor resected via EPP (14 PDT, 11 non-PDT), and 23 patients had tumor resected via RP (11 PDT, 12 non-PDT). With a median potential follow-up among all patients (regardless of survival) of 23 months from surgery, there was no difference in the median overall survival in the PDT group versus the non-PDT group (14.1 vs. 14.4 months, respectively). Similarly, no difference was detected in median recurrence-free survival (8.5 vs. 7.7 months, respectively). In a follow-up publication determining prognostic factors for survival, preoperative tumor volume was found to be a highly significant predictor of survival. Interestingly, in this analysis, the median survival of patients receiving EPP was 11 months, while the survival of patients receiving RP was 22 months ($P = .07$).

Continued evolution of the pleural PDT technique at the University of Pennsylvania led to the initiation of a pilot trial of porfimer sodium-mediated PDT for patients with malignant pleural disease using increasingly sophisticated surgical and PDT dosimetry techniques. The results of treatment of a total of 52 patients with MPM have been reported by this group to date. In one study, the survival was compared between 14 patients receiving EPP with intraoperative PDT and 14 receiving RP with intraoperative PDT (Friedberg et al. 2011). All patients received PDT using porfimer sodium, a drug light interval of 24 h, and 60 J/cm² light delivered using a real-time, seven-position, isotropic dosimetry system (determined to be the equivalent of 30 J/cm² using the older nonisotropic detector system). In these patients, the majority of whom had locally advanced (stage III/IV) disease, RP-PDT was found to have a superior overall survival (median survival not reached with 2.1 year median follow-up) as compared to EPP (median survival of 8.4 months). A follow-up study including 38 patients with MPM (37/38 stage III/IV) treated using RP-PDT as described above demonstrated a remarkable median overall survival of 31.7 months (Friedberg et al. 2012). In 31 patients with epithelial MPM, the median overall survival was 41.2 months, with 27 mediastinal lymph-node-positive (N2) patients experiencing 31.7-month median overall survival and 11 mediastinal lymph-node-negative (N0/N1) patients experiencing a median overall survival of 57.1 months. Interestingly, there was a clear separation between the progression-free survival and overall survival curves for patients with epithelial MPM (Figure 43.3), which is unusual when compared to other surgical series. The reasons for the differences between this experience and prior porfimer sodium-mediated PDT trials remain an area of active investigation but may be related to improvements in both PDT and surgical techniques. In the prior NCI experience, patients receiving RP fared better than those receiving EPP, although any difference in PDT outcome has not been analyzed or reported for these two treatments.

PDT that uses the second-generation photosensitizer meta-tetrahydroxy-phenyl chlorine (mTHPC) has been evaluated in mesothelioma patients in several clinical studies. There was a pilot study of eight patients who received EPP-PDT using 0.3 mg/kg of mTHPC with a 48 h drug light interval and 10 J/cm² of 652 nm light (Ris et al. 1991, 1996). One patient died of contralateral pneumonia 6 days after the operation. The other seven patients had local control of their disease by computed tomography (CT) scan but developed distant metastases or contralateral disease 4 to 18 months after the procedure. A phase I study of EPP-PDT was performed by Bass and colleagues in patients with MPM (Schouwink et al. 2001). In this study, patients received mTHPC with doses ranging from 0.075 to 0.15 mg/kg and a drug light interval of 4 or 6 days. Real-time light dosimetry was provided using a nonisotropic,

four-detector system and a novel, saline-filled applicator that filled and conformed to the walls of the entire chest cavity. Dose-limiting toxicity including PDT-related death was reached at a dose level including 0.15 mg/kg mTHPC, a drug light interval of 4 or 6 days, and light fluence of 10 J/cm². Therefore, the maximally tolerated dose in this trial was 0.1 mg/kg mTHPC with a 4-day drug light interval and 10 J/cm² of 652 nm light. In a parallel trial conducted with a similar design at the University of Pennsylvania, 26 patients were treated with surgical resection and intraoperative mTHPC-mediated PDT. In this trial, 7 patients received EPP, and 19 patients received RP. Light was initially delivered using the saline-filled applicator; however, it was noted that blood clots frequently formed between the applicator and the chest wall, and the majority of patients were then treated using intralipid and the modified endotracheal tube applicator. In this trial, the maximally tolerated dose was slightly lower at 0.1 mg/kg mTHPC with a 6-day drug light interval and 10 J/cm² of 652 nm light. Two out of three patients treated at the 0.1 mg/kg mTHPC/10 J/cm² dose level with a 4-day drug light interval experienced a cytokine-related capillary leak syndrome that was felt to be definitely related to protocol therapy and eventually lead to patient demise. Despite the use of the second-generation photosensitizer, the median overall survival in both of these trials was 10–12.4 months. Thus, while second-generation sensitizers may have pharmacologically and biophysically superior properties, initial attempts showed no demonstrable clinical advantage as compared to initial trials with porfimer sodium. Currently, a phase I trial is currently underway at the University of Pennsylvania and Roswell Park Cancer Institute using RP and the isotropic, real-time light detector system.

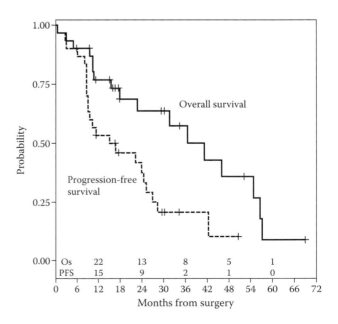

FIGURE 43.3 Progression-free and overall survival in patients with epithelial mesothelioma treated with RP-PDT followed by pemetrexed-based chemotherapy.

43.5 Clinical Results with PDT for NSCLC

NSCLC with pleural spread is incurable, with median survival rates ranging from 6 to 9 months and surgery alone failing to locally control this disease or extend survival beyond the accepted treatment of palliative chemotherapy. Based on promising phase I results, a pilot phase II trial of porfimer sodium-mediated PDT was performed to investigate the efficacy of combined surgery and PDT for either recurrent or primary NSCLC with pleural spread (Friedberg et al. 2004). Twenty-two patients were enrolled in this study, and five did not receive light delivery due to unresectability (three patients received Photofrin but were deemed unresectable intraoperatively). The 17 remaining patients received porfimer sodium (2 mg/kg) 24 h prior to gross tumor resection, and the hemithorax was illuminated with 630 nm light to a measured dose of 30 J/cm^2. The majority of this population had poor prognostic features, including N2 nodes and bulky pleural disease. Local control of pleural disease at 6 months was achieved in 11 of 15 (73%) of evaluable patients, and median overall survival for all 22 patients was 21.7 months. Toxicities that were deemed either probably or possibly related to PDT included postoperative fever, peripheral edema, and cutaneous photosensitivity. Two patients died in the immediate postoperative period, one of which was from adult respiratory distress syndrome that could have been related to PDT, surgery, or both. The other patient died of pneumonia in the contralateral lung following a difficult, 2-month-long postoperative course. These results are highly encouraging in this population of patients and suggest that additional investigation in this area is warranted.

43.6 Effect of Drug Light Interval and the Importance of Lung-Sparing Surgery

While preclinical modeling suggested that mTHPC would have superior PDT results using a 48 h drug light interval, clinical studies with the photosensitizer have demonstrated that the drug light interval has a critical impact on toxicity of therapy that includes a capillary leak syndrome presumably related to cytokines released during or in response to therapy. Of note, the maximally tolerated dose for PDT occurred with a drug light interval of 6 days in the University of Pennsylvania experience, where most patients received a lung-sparing surgery, but with a drug light interval of 4 days in the Netherlands Cancer Institute experience, where all patients received EPP. It seems likely that the PDT of normal lung tissue contributed to cytokine release and was an important factor in the lower maximally tolerated dose in the patients receiving lung-sparing surgery (RP). Indeed, in the ongoing 2-[1-hexyloxyethyl]-2-devinyl pyropheophorbide-a-mediated PDT trial, one patient treated with a 24 h drug light interval experienced a very similar capillary leak syndrome that was presumably mediated by cytokines. One possibility is that increased vascular damage produced by PDT with the shortened drug light interval may explain the combined observations of increased toxicity

with lung-sparing procedures and shorter drug light interval. This hypothesis is currently being tested in preclinical models, and all remaining patients (29 to date) in this trial have been treated with a 48 h drug light interval and have not experienced a similar type of complication. Thus, lung-sparing procedures may yield an increased rate of inflammatory complications when combined with PDT. However, when comparing the results of EPP to RP, it seems clear that the retention of normal lung is a critical factor in promoting extended overall survival following tumor resection and pleural PDT. In addition, patients treated with RP-PDT appear to experience prolonged survival recurrence that contributes to the superior overall survival. It is possible that the presence of additional normal lung tissue may contribute directly to this increased overall survival by allowing patients to receive more aggressive salvage therapy or better tolerate the metabolic burden of recurrent disease. If this effect were independent of PDT, then one would expect patients receiving radical correctly followed by multiagent chemotherapy to experience a similar prolongation of overall survival as compared to progression-free survival. In the limited published experience for these patients thus far, this does not appear to be the case. An additional possibility is that PDT of the normal lung tissue, along with microscopic residual cancer cells, not only promotes increased inflammatory complications but also provides the opportunity for increased antitumor immune response. These exciting possibilities are currently being investigated in both preclinical and clinical studies.

43.7 Conclusions

For more than 25 years, PDT has been investigated as a treatment for malignant pleural disease. In patients with epithelial MPM, RP-PDT consisting of a macroscopically complete, lung-sparing surgical resection followed by intraoperative porfimer sodium-mediated PDT performed with a real-time, isotropic light detector system appears to provide remarkable overall survival results as compared to other multimodality treatments. In patients with NSCLC and pleural spread, a select subset may experience prolonged overall survival with intraoperative PDT, although frequently, the nature of this disease makes lung-sparing surgery less feasible. Ongoing clinical trials are needed to define the most appropriate patient population for this treatment. In addition, more study, including possibly a randomized trial of RP with or without PDT followed by adjuvant pemetrexed-based chemotherapy in patients with epithelial MPM, is needed to definitively establish the role of PDT in the management of pleural disease and simultaneously identify the key mechanisms by which PDT might promote increased overall survival in these patients.

References

Aokage, K., J. Yoshida, G. Ishii et al. 2010. The impact on survival of positive intraoperative pleural lavage cytology in patients with non-small-cell lung cancer. *J Thorac Cardiov Surg.* 139:1246–1252.

Bolukbas, S., C. Manegold, M. Eberlein et al. 2011. Survival after trimodality therapy for malignant pleural mesothelioma: Radical pleurectomy, chemotherapy with cisplatin/pemetrexed and radiotherapy. *Lung Cancer* 71:75–81.

Cengel, K. A., E. Glatstein, and S. M. Hahn. 2007. Intraperitoneal photodynamic therapy. *Cancer Treat Res* 134:493–514.

DeLaney, T. F., P. D. Smith, G. F. Thomas et al. 1991. A light-diffusing device for intraoperative photodynamic therapy in the peritoneal or pleural cavity. *J Clin Laser Med Surg* 9:361–366.

Douglass, H. O., Jr., H. R. Nava, K. R. Weishaupt et al. 1983. Intra-abdominal applications of hematoporphyrin photoradiation therapy. *Adv Exp Med Biol* 160:15–21.

Flores, R. M., H. I. Pass, V. E. Seshan et al. 2008. Extrapleural pneumonectomy versus pleurectomy/decortication in the surgical management of malignant pleural mesothelioma: Results in 663 patients. *J Thorac Cardiov Sur* 135:620–626.

Forster, K. M., W. R. Smythe, G. Starkschall et al. 2003. Intensity-modulated radiotherapy following extrapleural pneumonectomy for the treatment of malignant mesothelioma: Clinical implementation. *Int J Radiat Oncol Biol Phys* 55:606–616.

Friedberg, J. S., M. J. Culligan, R. Mick et al. 2012. Radical pleurectomy and intraoperative photodynamic therapy for malignant pleural mesothelioma. *Ann Thorac Surg* 93:1658–1667.

Friedberg, J. S., R. Mick, M. Culligan et al. 2011. Photodynamic therapy and the evolution of a lung-sparing surgical treatment for mesothelioma. *Ann Thorac Surg* 91:1738–1745.

Friedberg, J. S., R. Mick, J. P. Stevenson et al. 2004. Phase II trial of pleural photodynamic therapy and surgery for patients with non-small-cell lung cancer with pleural spread. *J Clin Oncol* 22:2192–2201.

Krug, L. M., H. I. Pass, V. W. Rusch et al. 2009. Multicenter phase II trial of neoadjuvant pemetrexed plus cisplatin followed by extrapleural pneumonectomy and radiation for malignant pleural mesothelioma. *J Clin Oncol* 27:3007–3013.

Moskal, T. L., T. J. Dougherty, J. D. Urschel et al. 1998. Operation and photodynamic therapy for pleural mesothelioma: 6-year follow-up. *Ann Thorac Surg* 66:1128–1133.

Ozols, R. F., G. Y. Locker, J. H. Doroshow et al. 1979. Chemotherapy for murine ovarian cancer: A rationale for ip therapy with adriamycin. *Cancer Treat Rep* 63:269–273.

Pass, H. I., T. F. DeLaney, Z. Tochner et al. 1994. Intrapleural photodynamic therapy: Results of a phase I trial. *Ann Surg Oncol* 1:28–37.

Pass, H. I., and H. Pogrebniak. 1992. Photodynamic therapy for thoracic malignancies. *Semin Surg Oncol* 8:217–225.

Pass, H. I., B. K. Temeck, K. Kranda et al. 1997. Phase III randomized trial of surgery with or without intraoperative photodynamic therapy and postoperative immunochemotherapy for malignant pleural mesothelioma. *Ann Surg Oncol* 4:628–633.

Perry, R. R., S. Evans, W. Matthews et al. 1989. Potentiation of phototherapy cytotoxicity with light scattering media. *J Surg Res* 46:386–390.

Ris, H., H. Altermatt, B. Nachbur et al. 1996. Intraoperative photodynamic therapy with m-tetrahydroxyphenylchlorin for chest malignancies. *Lasers Surg Med* 18:39–45.

Ris, H. B., H. J. Altermatt, R. Inderbitzi et al. 1991. Photodynamic therapy with chlorins for diffuse malignant mesothelioma: Initial clinical results. *Br J Cancer* 64:1116–1120.

Schouwink, H., M. Ruevekamp, H. Oppelaar et al. 2001. Photodynamic therapy for malignant mesothelioma: Preclinical studies for optimization of treatment protocols. *Photochem Photobiol* 73:410–417.

Takita, H., T. S. Mang, G. M. Loewen et al. 1994. Operation and intracavitary photodynamic therapy for malignant pleural mesothelioma: A phase II study. *Ann Thorac Surg* 58:995–998.

Tilleman, T. R., W. G. Richards, L. Zellos et al. 2009. Extrapleural pneumonectomy followed by intracavitary intraoperative hyperthermic cisplatin with pharmacologic cytoprotection for treatment of malignant pleural mesothelioma: A phase II prospective study. *J Thorac Cardiov Surg* 138:405–411.

Tochner, Z., J. B. Mitchell, F. S. Harrington et al. 1985. Treatment of murine intraperitoneal ovarian ascitic tumor with hematoporphyrin derivative and laser light. *Cancer Res* 45:2983–2987.

Zellos, L., W. G. Richards, L. Capalbo et al. 2009. A phase I study of extrapleural pneumonectomy and intracavitary intraoperative hyperthermic cisplatin with amifostine cytoprotection for malignant pleural mesothelioma. *J Thorac Cardiov Surg* 137:453–458.

Clinical Photodynamic Therapy in the Chinese Region

Yih-Chih Hsu
Chung Yuan Christian University

Zheng Huang
University of Colorado

44.1 Introduction

After more than 30 years of preclinical study and clinical practice, photodynamic therapy (PDT) has now become an accepted drug-device therapeutic modality and is an established component of modern photomedicine in the Chinese region.

In the 1980s and 1990s, PDT, also known as photodynamic treatment, was often described in Chinese literature as laser- or photochemotherapy. Due to the high expectation of tumor selectivity of hematoporphyrin derivative (HPD), HPD-based PDT was also often called hematoporphyrin photoradiation therapy or hematoporphyrin (derivative) laser antitumor therapy. The first PDT case was performed on an eyelid tumor by Dr. Jin Zhou at Beijing Tong-Ren Hospital in July 1981. The novelty of this modality and its great therapeutic potential for the treatment of malignant diseases immediately drew attention and excitement in China and Taiwan. As a part of the government-organized campaign, the Chinese began to make HPD in the early 1980s after Thomas Dougherty (Roswell Park Cancer Institute) rediscovered Schwartz's HPD in the 1970s. The first domestic HPD was successfully prepared from hematoporphyrin isolated from animal blood in 1980. Over the past two decades, the Chinese continued their substantial efforts trying to create better photosensitizers and light sources. Several dozen photosensitizers have been synthesized and their photodynamic effects have been evaluated in a variety of *in vitro* and *in vivo* tests. Some promising drugs and PDT light sources have undergone clinical investigational studies and a few have obtained regulatory approvals. In addition, some Chinese studies and PDT protocols have certainly reached advanced levels and could be valuable to the international PDT communities. This chapter will highlight some unique accomplishments and experiences of PDT in the Chinese region.

44.2 Clinical PDT Progress

44.2.1 Regulatory Status of PDT

44.2.1.1 Photosensitizers

HiPorfin. This HPD product is marketed under the brand name of HiPorfin (hematoporphyrin injection) by Chongqing Huading Modern Biopharmaceutics Co. Ltd. It was approved for oncological indications by the State Food and Drug Administration (SFDA) in 2001.

Aila. Alia is a topical formulation containing predrug aminolevulinic acid (ALA). It is developed by Shanghai Fudan Zhangjiang BioPharmaceutical Co. Ltd. and was approved for the treatment of genital warts in early 2007. Fudan Zhangjiang is currently organizing multicenter clinical trials to seek SFDA approval to use Aila for the treatment of moderate to severe acne vulgaris.

Hemoporfin. Hematoporphyrin monomethyl ether (HMME) is a synthetic mixture of the two positional isomers of 7(12)-(1-methoxyethyl)-12(7)-(1-hydroxyethyl)-3,8,13, 17-tetramethyl-21H,23H-porphin-2,18-dipropionic acid. HMME-mediated vascular-targeting PDT has been primarily used to treat port-wine stain (PWS) birthmarks in China since the early 1990s. HMME has been licensed to Fudan Zhangjiang BioPharmaceutical Co. Ltd., and its new drug application for treating PWS was filed under the brand name of Hemoporfin (Hemoporfin Injection), which was approved for clinical trial by the SFDA in 2005. It is expected that it will obtain its regulatory approval in 2012.

Deuteporfin. Deuteporfin is an HPD preparation that mainly consists of 3 (or 8)-(1-methoxyethyl)-8(or 3)-(1-hydroxyethyl)-deuteroporphyrin IX (MHD), 3,

8-di(1-methoxyethyl)-deuteroporphyrin IX (DMD), 3(or 8)-(1-methoxyethyl)-8(or 3)-vinyl-deuteroporphyrin IX (MVD), 3(or 8)-(1-hydroxyethyl)-8(or 3)-vinyl-deuteroporphyrin IX (HVD), and a small portion of protoporphyrin IX (Pp). MHD and DMD are the two main active components (>85%) of Deuteporfin. It has been licensed to Fudan Zhangjiang BioPharmaceutical Co. Ltd., and its new drug application for treating oncological indications was filed under the brand name of Deuteporfin Injection.

Suftalan Zinc. Suftalan Zinc is a disulfonated di-phthalimidomethyl phthalocyanine zinc ($ZnPcS_2P_2$) formula that contains Cremophor EL 2% (V/V), propylene glycol 20% (V/V), and NaCl 0.9% (W/W). It was developed by the Institute of Research on Functional Materials of Fujian University and licensed to Longhua Pharmaceutical Co. Ltd (Liu 2007). The new drug application for esophageal cancer was filed under the brand name of Suftalan Zinc (Photocyanine Injection), which was approved for clinical trials by the SFDA in early 2008.

Radion. Radion is a topical formulation containing the prodrug ALA and is currently under research and development by Pharmapower Biotechnology Co. The indication of Radion is for the treatment of oral precancerous lesions. A preclinical hamster buccal pouch PDT animal study was completed in November 2011 at Chung Yuan Christian University (Chiang et al. 2012; Hsu et al. 2012). Formal human clinical studies will start in December 2013 through a multicenter clinical trial to seek drug approval to use Radion for the treatment of oral precancerous lesions.

In addition to new photosensitizer development, several PDT photosensitizers with regulatory approvals from Western countries have been imported to the China and Taiwan regions via channels after receiving either official SFDA approvals (e.g., Visudyne) or Taiwan Food and Drug Administration (TFDA) approvals (e.g., Visudyne, Photofrin). Others are under bridge study trials after institutional review board (IRB) approvals in order to obtain approvals from the TFDA in Taiwan (e.g., Foscan) or under special import permissions to fill hospital and patient special needs in China (e.g., Photofrin) and in Taiwan (e.g., Photosan).

44.2.1.2 Light Sources

As in the United States and Europe, currently the Chinese SFDA also requires the filing of a new PDT photosensitizer along with the particular light source and specified clinical indications. Nevertheless, multiple different light sources might be approved and used in different clinics. Various lasers and noncoherent light sources have been developed and used in China and Taiwan (Table 44.1); however, some high-power diode lasers and light applicators (e.g., microlens and diffuser fibers) still need to be imported.

TABLE 44.1 Light Sources Used for PDT Procedures in the Chinese Region

Light Sources	Wavelength (nm)	Indications
Argon ion laser	514.5	PWS
Argon-pumped dye laser	630	Bladder, GI, oral, and skin cancers
LED	635	Oral precancer and cancer
LED	600–700	Skin cancer
Copper bromide laser	511, 578	Bladder cancer
Copper vapor laser	511, 578	Bladder, GI, and skin cancer, PWS
Diode laser	630	GI, lung, and oral cancers, ocular PDT, brain tumor
Doubled frequency YAG-pumped-pulsed-dye laser	630	Brain tumor
Gold vapor laser	630	GI cancer
He-Ne laser	630	GI, oral cancer, and skin cancers
Krypton laser	413	PWS
KTP laser	532	PWS

Note: GI = gastrointestinal, He–Ne = helium–neon, KTP = potassium titanyl phosphate, LED = light-emitting diode, PDT = photodynamic therapy, PWS = port-wine stain, YAG = yttrium aluminum garnet.

In Taiwan, light-emitting diode (LED) light sources have been developed and used in PDT of oral precancerous and cancer lesions for 10 years. A portable device with sophisticated interface called LED Wonderlight (Figure 44.1) has been well designed and developed for PDT by MedEx Healthcare Inc., Taoyuan, Taiwan. The well-recognized animal model of hamster buccal pouch precancerous and cancerous lesions was treated with PDT with successful therapeutic results (Chiang et al. 2012; Hsu et al. 2012). The same findings were reported in human oral precancers and cancers subjected to clinical PDT using LED light also with successful outcomes (Tsai et al. 2004; Yu et al. 2009). Most importantly, previous studies showed no significant difference in clinical outcomes of oral erythroleukoplakia and oral verrucous hyperplasia when either LED or laser light

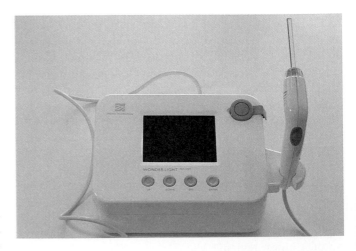

FIGURE 44.1 Innovative LED PDT Wonderlight by MedEx Healthcare.

was used for topical ALA-PDT (Lin et al. 2010; Yu et al. 2009). Regulatory approvals of Wonderlight in different countries are also in process at the same time.

44.2.2 Clinical Status of PDT

44.2.2.1 Precancerous and Cancerous Diseases

Over the past 30 years, Chinese clinicians have treated several thousand cancer patients and published a great number of clinical reports. The majority of these reports can be classified as case reports or describe single hospital experiences that were generally focused on assessing feasibility and/or efficacy. In terms of the study design, those reports can be described as pilot studies, comparison studies, or nonrandomized phase I/II/III trials. A comprehensive reference list has been published by the authors based on a literature survey of both Chinese and English scientific journals (Huang 2006a,b, 2008; Xu 2007).

An assay of clinical data reported by Chinese physicians between 1990 and 2001 shows that the vast majority of 3878 oncological cases were gastrointestinal (GI) cancers (esophageal cancer and stomach cancer) that were followed by bladder cancer and nasopharyngeal cancer (NPC) (Ding et al. 2004). Because of the geographic significance of certain precancerous and cancerous diseases in the Chinese region, local clinicians have developed unique PDT protocols to treat precancerous lesions in NPC and oral cavity and liver cancer.

> *Oral precancer and cancer.* Oral cancer is the fifth most common cancer in the world (Jemal 2005; Lingen, Sturgis, and Kies 2001; Parkin 2005), and it is an extremely important disease in Taiwan. It has become the fourth leading cancer in the male population and constitutes a significant portion (>6%) of all malignancies (Department of Health 2006). It is one of the leading causes of morbidity and mortality each year. The main etiologies that cause oral squamous cell carcinoma in Taiwan are areca quid chewing, cigarette smoking, and alcohol consumption (Ko et al. 1995; Kwan 1976). The first time clinical PDT was conducted on oral topical ALA-PDT using 635 nm LED red light to treat oral precancerous lesions took place in Taiwan. More than 130 cases of oral precancers of various types were treated with topical ALA-PDT using 635 nm LED red light once or twice a week. Successful treatment of human oral verrucous hyperplasia, oral leukoplakia, and extensive verrucous carcinoma with topical 5-ALA-mediated PDT were treated with modified protocols of ALA-PDT composed of multiple 3 min irradiations with a LED red (635 ± 5 nm) light. Topical ALA-PDT with fractionated irradiations by a LED red light at 635 ± 5 nm is an effective and successful treatment modality for the above-mentioned precancers and cancer diseases (Chen et al. 2004, 2005a,b).
>
> *NPC.* In some areas of China, the morbidity and the mortality of NPC are still high. Chinese physicians started

to explore the feasibility of HPD-based PDT for the treatment of newly diagnosed and recurrent NPC since the later 1980s. Complete remission (CR) was reported between 33% and 55%. Three- and five-year survival rates were reported as 44.6% and 25.4%, respectively. Since Photofrin and Diomed 630 laser were introduced into China in the early 2000s, they have been used in several newly established PDT centers for treating various malignant tumors including advanced nasopharyngeal carcinoma. Although Chinese data represent different subsets of patients, PDT protocols, and clinical outcomes, nevertheless, the data suggest that for patients with advanced NPC who have exhausted all standard treatment options, PDT might offer the possibility of an improvement in quality of life and good tumor control.

> *Primary liver cancer.* Primary liver cancer is still one of the most significant health problems in China, even though more patients are being diagnosed early and tend to undergo resection while receiving a better conservative treatment, which ensures a better prognosis. The Cancer Research Centre of Xiamen University pioneered an interventional PDT procedure for treating advanced liver cancer in the early 1990s. Interstitial PDT was performed under local anesthesia 48 h after administering HPD. First, positioning probes (18G) were inserted into the tumor through percutaneous punctures under ultrasound guidance. Light delivery fibers (400 μm core diameter, 1 cm diffuser tip) were then placed into the tumor. An argon laser-pumped dye laser (630 nm) was coupled into three diffuser fibers for simultaneous irradiation of three treatment spots at a time. A light dose of 220 J/cm^2 was delivered at fluence rates of 300–350 mW/cm for each spot.

Post-PDT examination showed various degrees of decrease in alpha-fetoprotein (AFP) levels after a single treatment or multiple treatments. Long-term follow-up of 70 patients (up to 5 years) showed that multiple treatments might prolong survival. Among them, 1 year survival rate was 10% in the one-session group ($n = 30$), 50% in the two-session group ($n = 12$), 75% in the three-session group ($n = 12$), and 92% in the four-session (or more) group ($n = 16$), respectively.

44.2.2.2 Noncancerous Diseases

> *PWS birthmarks.* PWSs are congenital vascular malformations characterized by ectatic capillaries in the papillary layer of the dermis. Pulsed dye laser (PDL)-mediated selective photothermalysis is the treatment of choice in North America and Europe. However, its role in the treatment of PWSs has been challenged by vascular-targeted PDT in China. The first clinical study using HPD was carried out in Chinese People's Liberation Army (PLA) General Hospital (known as Beijing 301 Hospital) in the early 1990s. Since then, various

photosensitizers and light sources have been used for the treatment of PWSs of various subtypes and severities in patients of all ages. Although no formal randomized controlled trial (RCT) has been conducted in China, a retrospective analysis shows that PDT was as effective as PDL for pink flat lesions and is more effective than PDL for purple flat lesions (Yuan et al. 2007). In order to avoid prolonged skin photosensitization associated with HPD, in recent times, the second-generation photosensitizer Hemoporfin has been used. It is expected that Hemoporfin will obtain its regulatory approval in 2012.

Genital warts. Genital condylomata acuminata (CA) or genital warts are the most prevalent sexually transmitted disease and are closely associated with human papillomavirus (HPV) infections. Currently, there is no completely satisfactory treatment option available for managing CA. Although *anti-HPV vaccination may reduce the* future incidence *of HPV infection*, there is still an urgent need to develop noninvasive therapies that can remove warty lesions and meanwhile eliminate HPV infections. The feasibility of ALA-mediated PDT for the treatment of CA in the urethra and external genital areas of male and female patients has been explored in China since the later 1990s. Results suggested that ALA-PDT is associated with a low recurrence rate possibly owing to the eradication of HPV infections (Chen et al. 2007; Wang et al. 2004). In addition, ALA-mediated photodiagnosis and PDT were also useful for subclinical and latent HPV infections. Aila, a topical formulation containing ALA, was approved for the treatment of genital warts in early 2007.

Acne. The main etiologic factors of acne vulgaris include excessive sebum production, ductal hypercornification, and bacterial colonization associated with *Propionibacterium acnes (P. acnes).* Therefore, potential therapeutic targets are the infundibulum, sebaceous gland, *P. acnes* bacteria, and any of the components of the sebaceous follicle that might modulate the inflammatory response. Early studies suggest that topically applied ALA can be converted into PpIX in acne lesions, and ALA-based PDT is a potentially useful method for managing localized persistent acne and for patients who were unable to tolerate isotretinoin or antibiotics. Several laser- or LED-based light sources of various wavelengths have been successfully used in topical PDT for the treatment of moderate to severe acne in Chinese patients since 2005 (Wang et al. 2010). Currently, ALA-based PDT treatment is marketed as a cosmetic product in China.

References

Chen, H. M., C. T. Chen, H. Yang et al. 2004. Successful treatment of oral verrucous hyperplasia with topical 5-aminolevulinic acid-mediated photodynamic therapy. *Oral Oncol* 40:630–637.

Chen, H. M., C. T. Chen, H. Yang et al. 2005a. Successful treatment of an extensive verrucous carcinoma with topical 5-aminolevulinic acid-mediated photodynamic therapy. *J Oral Pathol Med* 34:253–256.

Chen, H. M., C. H. Yu, P. C. Tu et al. 2005b. Successful treatment of oral verrucous hyperplasia and oral leukoplakia with topical 5-aminolevulinic acid-mediated photodynamic therapy. *Lasers Surg Med* 37:114–122.

Chen, K., B. Z. Chang, M. Ju, X. H. Zhang, and H. Gu. 2007. Comparative study of photodynamic therapy vs CO_2 laser vaporization in treatment of condylomata acuminata: A randomized clinical trial. *Br J Dermatol* 156:516–520.

Chiang, C.-P., W.-T. Huang, J.-W. Lee, and Y.-C. Hsu. 2012. Effective treatment of 7,12-dimethylbenz(a)anthracene-induced hamster buccal pouch precancerous lesions by topical photosan-mediated photodynamic therapy. *Head Neck* 34(4):505–512.

Department of Health, The Executive Yuan, Taiwan. 2009. Taiwan area main causes of death, 2008. R.O.C. Cancer registry annual report in Taiwan area.

Ding, X. M., Y. Gu, F. G. Liu, J. Zeng. 2004. Review of photodynamic therapy of neoplasms in the past 12 years in China—Analysis of 3878 cases. *Chin J Clin Rehab* 8:2014–2017.

Hsu, Y.-C., D.-F. Yang, C.-P. Chiang, J.-W. Lee, and M.-K. Tseng. 2012. Successful treatment of 7,12-dimethylbenz(a)anthracene-induced hamster buccal pouch precancerous lesions by topical 5-aminolevulinic acid-mediated photodynamic therapy. *Photodiag Photodyn Ther* 9:310–318.

Huang, Z. 2006a. Photodynamic therapy in China: Over 25 years of unique clinical experience. Part one—History and domestic photosensitizers. *Photodiag Photodyn Ther* 3:3–10.

Huang, Z. 2006b. Photodynamic therapy in China: Over 25 years of unique clinical experience. Part two—Clinical experience. *Photodiag Photodyn Ther* 3:71–84.

Huang, Z. 2008. An update on the regulatory status of PDT photosensitizers in China. *Photodiag Photodyn Ther* 5:285–287.

Jemal, A., T. Murray, E. Ward et al. 2005. Cancer statistics, 2005. *CA Cancer J Clin* 55:10–30.

Ko, Y. C., Y. L. Huang, C. H. Lee et al. 1995. Betel quid chewing, cigarette smoking and alcohol consumption related to oral cancer in Taiwan. *J Oral Pathol Med* 24:450–453.

Kwan, H. W. 1976. A statistical study on oral carcinomas in Taiwan with emphasis on the relationship with betel nut chewing: A preliminary report. *J Formos Med Assoc* 75:497–505.

Lin, H. P., H. M. Chen, C. H. Yu et al. 2010. Topical photodynamic therapy is very effective for oral verrucous hyperplasia and oral erythroleukoplakia. *J Oral Pathol Med* 39:624–630.

Lingen, M., E. M. Sturgis, and M. S. Kies. 2001. Squamous cell carcinoma of the head and neck in nonsmokers: Clinical and biologic characteristics and implications for management. *Curr Opin Oncol* 13:176–182.

Liu, W., N. Chen, H. Jin et al. 2007. Intravenous repeated-dose toxicity study of $ZnPcS_2P_2$-based-photodynamic therapy in beagle dogs. *Regul Toxicol Pharmacol* 47:221–231.

Parkin, D. M., F. Bray, J. Ferlay, and P. Pisani. 2005. Global cancer statistics. *CA Cancer J Clin* 55:74–108.

Tsai, J.-C., C.-P. Chiang, H.-M. Chen et al. 2004. Photodynamic therapy of oral dysplasia with topical 5-aminolevulinic and light-emitting diode array. *Laser Surg Med* 34:18–24.

Wang, X. L., H. W. Wang, H. S. Wang et al. 2004. Topical 5-aminolaevulinic acid-photodynamic therapy for the treatment of urethral condylomata acuminata. *Br J Dermatol* 151:880–885.

Wang, X. L., H. W. Wang, L. L. Zhang, M. X. Guo, and Z. Huang. 2010. Topical ALA PDT for the treatment of severe acne vulgaris. *Photodiag Photodyn Ther* 7:33–38.

Xu, D. Y. 2007. Research and development of photodynamic therapy photosensitizers in China. *Photodiag Photodyn Ther* 4:13–25.

Yu, C. H., H. P. Lin, H. M. Chen et al. 2009. Comparison of clinical outcomes of oral erythroleukoplakia treated with photodynamic therapy using either light-emitting diode or laser light. *Laser Surg Med* 41:628–633.

Yuan, K. H., Q. Li, W. L. Yu, C. Zhang, and Z. Huang. 2007. Retrospective analysis of treatment of port wine stain birthmarks—Photodynamic therapy vs pulsed dye laser. *Photodiag Photodyn Ther* 4:100–105.

Photodynamic Therapy and Fluorescent Diagnostics in the Russian Federation

Elena G. Vakulovskaya
Russian Cancer Research Center
of the Russian Academy of
Medical Sciences (RAMS)

45.1 Introduction

The development of photodynamic therapy (PDT) in the Russian Federation (RF) is based on domestic photosensitizers (PSs) and domestic lasers. The investigation and implementation of PDT took place at well-known scientific schools of chemistry and laser physics. Experimental and clinical trials were done in the RF, and the results were published mostly in Russian scientific journals. Experimental investigation in the PDT field began in the RF more than 20 years ago in the early 1990s with the domestic first-generation PS known as Photogem (PG) analogous to Photofrin 2 in the West. PG is a mixture of monomeric and oligomeric hematoporphyrin compounds with absorption peaks at 405, 505, 580, and 630 nm (Mironov, Seyanov, and Pizhik 1992). Clinical trials were carried out at N. N. Blochin Russian Cancer Research Center of RAMS (RCRC), P. A. Hertzen Moscow Research Oncological Institute (MROI), and the Laser Medicine State Research Center in patients with gastrointestinal cancer (GIC), lung cancer (LC), oral cancer (OC), and skin cancer (SC) (Kuvshinov et al. 1996; Sokolov et al. 1996; Stranadko et al. 1996; Vakulovskaya et al. 1996a). As a result of these trials, PG was approved for clinical use in the RF in 1999, but was not widely used because the more potent second-generation PS, Photosens (PTS), appeared and clinical trials using PTS began in 1993.

Up to the present time, six PSs have received federal approval in the RF for clinical application: PG, Photosens, and Alasens (AS) (FSUE SSC, Niopic, Moscow, RF), Radachlorin (RC, Rada-Pharma, Moscow, RF), Photodithazin (PD, Veta-Grand, Moscow, RF), and Photolon (PL, RUE Belmedpreparary, Belarus). Clinical trials for PL were done in Belarus and the approval in the RF was achieved in 2006 after a positive clinical study in patients with basal cell cancer (BCC).

From 1995 onward, experimental and clinical studies of PTS and AS, together with the investigation of new PS and lasers, have been done in concert with Moscow government programs for development of new methods of diagnostics and treatment of oncological and nononcological diseases. Since 1996, official clinical trials of second-generation PSs have been performed at well-known clinical institutions in accordance with the scientific and ethical principles of the Helsinki Declaration of the World Medical Association, reflected in OCT 42-511-99 "Regulations on Good Clinical Trials in the Russian Federation," the rules of the International Conference on Harmonisation of Technical Requirements for Registration of Pharmaceuticals for Human Use (ICH GCP), and protocols approved by the Pharmacology Committee of the RF and the Ethics Committee of the RF as multicenter open prospective studies. Complete clinical examination was provided to all patients, including biopsy proven tumor type in accordance with the recommendations of the World Health Organization (WHO) and the International Cancer Union. All patients signed informed consents stating the investigational nature of treatment, alternate treatments available, and the side effects to be expected. We will discuss the results of these well-controlled trials because of the high level of the PDT procedures and good follow-up of patients in them. Currently, when PDT is an approved treatment modality, it is performed in many different clinics, often for investigation in other pathology based on local institutional protocols, or without it; usually these trials are less controlled. The tumor response (WHO) 2 months after PDT was evaluated as follows:

complete response (CR), no evidence of tumor determined by two observations not less than 4 weeks apart; partial response (PR), more than a 50% decrease in total tumor load, stabilization (St), reduction less than 50% in total tumor size; and progression (P), new tumor sites appearing or increase in tumor mass more than 25%. Fluorescent diagnosis (FD) diagnostic capability was evaluated as: sensitivity (S), specificity (Sp), and accuracy (A).

45.2 Lasers and Light Sources for FD and PDT

FD was carried out with spectranalyzers LESA-6, LESA–01-Biospec [helium-neon (He–Ne) laser, λ = 633 nm, Biospec, Moscow, RF]; the analysis distinguished reflected laser signal and autofluorescence of tissues. Fluorescence contrast (FC) was estimated by the ratio of the levels of PS in tumor and surrounding tissue. We have used a spectral-fluorescent video complex with a sensitive charge-coupled device (CCD) camera to obtain black-and-white fluorescent images and a video-fluorescence diode device (630-675-01-Biospec). The following were used for FD with AS: sources of optical radiation (Biospec, Moscow, RF), a fluorescence bronchoscope, and cystoscope D-Light/AF System (Karl Storz GmbH, Germany) with a wavelength of 380 to 442 nm for colored fluorescent images.

At the beginning of our investigation, we tested a variety of therapeutic lasers: krypton laser (λ = 647, 675 nm), scanning electron-pumped semiconductor laser (λ = 670–674 nm, P 10 W), solid laser with doubled frequency (λ = 669 ± 1 nm, P 1,8 W (SPC Polus), dye lasers (copper vapor (λ = 640–690 nm, Yakhroma-2, SPC Istok), gold (λ = 620–650 nm, SPC Mechatron) (Kuvshinov et al. 1996; Sokolov et al. 1996; Vakulovskaya et al. 1996a,b). Some of these large lasers required a stationary position, a fixed connection to the water supply, and long-term cooling after operation. In the late 1990s compact semiconductor lasers appeared and are now widely used: LFT-670 (λ = 670 nm), LFT-675 (λ = 675 nm) (JSC Biospec) for PDT with PS, Milon (λ = 662 nm ± 3 nm, P 1.5–2.5 W) (Milon laser), and Atkus-2 (λ = 661 nm) (Poluprovodnikovye Pribory) for PDT with RC and PD. For the delivery of laser radiation, JSC Biospec in cooperation with RCRC developed quartz fiber catheters with various designs at the distal tip (with microlens with spherical and cylindrical diffusers of various length, lateral irradiation, and metalized), allowing the creation of different irradiation fields for both surface and interstitial irradiation. To decrease the adverse effects to vision, RCRC together with Institute of Astrophysics developed goggles with special filters for the physician in FD and PDT versions. All these lasers received federal approval for clinical use in the RF.

45.3 Clinical Studies with Photosens

PTS, a 0.2% aqueous solution of a mixture of disulfonated, trisulfonated, and tetrasulfonated aluminum phthalocyanines with an intensive absorption peak at 670 nm, was characterized with respect to chemical stability and high photodynamic activity (Lukiyanets 1998). It also provides intense fluorescence

emission. It was approved by the Ministry of Health and Pharmacology Committee of the RF in 2001–2008 after phase I–III multicenter clinical trials for skin cancer [BCC, squamous cell carcinoma (SCC), recurrences of melanoma], oropharyngeal cancer, cancer of the lip, cancer of the esophagus, GIC, recurrences of breast cancer (BC), skin metastases, mesothelioma and pleural metastases, and recurrences of cancer of larynx (CL). Presently waiting for approval are combined PDT + radiotherapy (RT) for skin metastases of BC and combined PDT + chemotherapy with 5-fluorouracil (5-FU) for skin cancer and gastric cancer. Clinical trials are now in progress for bladder cancer, prostate cancer, and early cancer of the cervix.

In the phase I clinical trials, PTS dissolved in 0.9% sodium chloride (NaCl) was injected intravenously at a dose of 2.0–2.5 mg/kg body weight, followed by a single laser irradiation 24 h after administration. The light dose ranged from 50 to 150 J/cm². We have observed a long-term increase in skin sensitivity to direct sunlight (8 weeks and more), leading to changes in the usual social activity of the patients because of the restricted light regimen. In many cases complications have developed (Sokolov et al. 1996; Vakulovskaya et al. 1996a). Close cooperation between RCRC and Institute of General Physics of Russian Academy of Sciences allowed the dynamic investigation of PTS pharmacokinetics by measuring the fluorescence spectra of the patients to estimate the accumulation of PS in tumor and surrounding skin and mucous membranes and the duration of drug retention in tumor and normal tissues (Stratonnikov et al. 1996). A dual-chamber model of PTS pharmacokinetics and biodistribution was utilized, where blood was the first chamber and tissues were the second, to predict possible residual quantities of PS in the patient's tissues. Our findings were subsequently confirmed by direct measuring of PS concentration in blood, obtained by invasive methods. Our spectrometric measurements revealed that the decrease in PTS dose down to 1.5 mg/kg of body weight did not affect the intensity of fluorescence, but FC in patients with tumors of different histological structure at doses down to 0.8 mg/kg significantly reduced it, while the content of PTS in tissues was still sufficient for effective PDT. During the phase II study, we conducted a randomized clinical study in BCC patients with two doses of PTS: 0.8 and 0.5 mg/kg (156 patients in the multicenter study). The efficacy of PDT in these groups, considering the distribution by sex, age, and tumor spread, was the same—CR in 91.6% and 90.4%, respectively, but skin phototoxicity with administration of a higher dose of PS was longer-lasting and more severe (Vakulovskaya 2006a; Vakulovskaya and Chental 1999). Taking into consideration these data, the dose was reduced to 0.5 mg/kg of body weight for phase III. In recent years, we have tested the dose of PS 0.3 and 0.4 mg/kg, and during repeated courses of the PDT, a dose of PS was chosen individually (0.2–0.4 mg/kg), depending on its residual concentration in tissues. Relatively high levels of fluorescence were observed in the tumor in the first hour after administration of PS, and a light regimen with irradiation after 1 h was tested. However, in this period, the PTS is still in circulation, binding to the endothelium of blood vessels, and there is no FC, so PDT provided at 1 h after administration of PTS leads to vessel destruction, rapid decrease in tissue oxygenation in the PDT area (Stratonnikov and Loschenov 2001; Stratonnikov, Meerovich, and Loschenov 2000; Stratonnikov et al. 2002), and impaired treatment efficacy and increases the risk of tumor regrowth.

We have found during dynamic fluorescence investigation that significant residual concentration of PTS remained in the tumor after the first laser irradiation, and therefore, we developed fundamentally new PDT regimens comprising a single injection of PS followed by multiple (2–6) laser irradiations within 24–72 h interval and comprising light dose fractionation (single light dose 80–100 J/cm², total light dose up to 600 J/cm²) during a PDT session. The selection of the interval between sessions was based on the estimation of recovery of blood flow and oxygenation in the PDT area. Because the cell damage during PDT has a threshold character, multiple laser irradiations enhanced tumor damage and allowed each PDT to take place under optimal tissue oxygenation, therefore relatively reducing the damage to adjacent normal tissue. The evaluation of the dynamics of blood oxygenation in the PDT area (on the basis of absorption spectra of PS) carried out in BCC and CL patients allowed us to develop multifractionation modes of PDT, when during the PDT procedure after delivering every 25–30 J/cm² of light dose in accordance with changes in oxygenation, we were using 3–5 min intervals with no light exposure in between for the recovery of tissue oxygenation in the PDT area (Stratonnikov et al. 2002; Vakulovskaya 2006a). Surface and interstitial irradiation or the combination of both was studied for PDT.

We have found fluorescence in all tumor sites after injection of PTS, with the exception of pigmented tumors (melanoma, nevus, and papillomas). Nonhomogeneity of drug distribution was seen in tumors more than 2 cm in diameter. PTS photobleaching was seen in all patients during the PDT session (Stratonnikov, Meerovich, and Loschenov 2000). Visible areas of tumor necrosis developed after PDT, and changes of emission curves were seen at spectroscopy indicating necrotic damage. The intensity of fluorescence and FC in patients with skin metastases of BC and primary multiple BCC differed in areas of former RT and beyond. In 64.1%–82.6% of patients with different cancers, fluorescent borders exceeded clinically defined ones. In 4.4%–11.2% of solid cancers and in 61.2% of patients with BC metastases, additional fluorescence zones have been detected and morphological verification of cancer was obtained in 93.8%–100% of cases. The FC of benign skin tumors and papillomas of the larynx was low (1.1–1.4; *P* < 0.01). Overall, FD had high S, A (100.0%), and SP (95.8%–100.0%) in all groups of patients, but because of long-term skin phototoxicity, it is unlikely to be used alone and is therefore likely to be performed for PDT planning and control in order to improve treatment results by more precise tumor delineation and detection of subclinical lesions (Vakulovskaya 2006a; Vakulovskaya et al. 1996a).

The efficacy of PDT was high in all groups of patients with the exception of melanoma. In 89 patients with T1-4N0M0 skin cancer and its recurrences, we achieved CR in 82.0%, PR in 16.9%, and St in 1.1. With respect to tumor morphological structure, better response was achieved in BCC patients: 86.2% CR and 13.8% PR; in T1-2 cases, the efficacy was higher: 93.8% CR. The

efficacy of PDT in patients with SCC was also high (88.9%); however, CR was much lower: 44.4%. The differences were statistically significant (*P* < 0.01) (Figure 45.1a and b). We achieved only poor responses in 18 cases of melanoma, although other authors reported significant responses in pigmented melanoma (Sokolov et al. 1996; Stranadko et al. 1996). In patients with locally advanced tumors, we have provided combined treatment PDT + cryosurgery of residual tumor. This combination has reduced the time needed for necrosis formation and increased the efficacy up to 97.8% CR (Chental et al. 1997). All these patients showed no recurrence during 6 years of follow-up (survival: 2-year survival, 100%; 3-year survival, 97.8%; disease-free survival: 1-year survival, 94.4%; 3-year survival, 93.0%) (Vakulovskaya 2006a).

In patients with recurrent head and neck cancer, we achieved CR in 66.7%, PR in 29.8%, and St in 3.5. With respect to tumor localization, higher efficacy was reached in cancer of the larynx (Figure 45.2) and cancer of the lower lip T1-3N0M0—CR 82.4 and 76.5 compared to 52.5 in OC (*P* < 0.01 and *P* < 0.05, respectively). In the entire group, 1-year survival was 98.3%, 2-year survival was 93.0%, and 3-year survival was 87.7%. Disease-free survival rates in the cohort of patients with CR after PDT were as follows 1-year survival, 63.2% and 94.7%; 2-year survival, 59.7% and 89.5%, respectively (Vakulovskaya 2006a; Vakulovskaya and Chental 1999).

In patients with primary T1-2N0-xM0 and recurrent GIC (histology: adenocarcinoma, nondifferentiated cancer, SCC), CR was observed in 26.1%, PR in 65.2%, and St in 8.7%. No recurrence was seen in 60% of primary gastric cancer patients with CR in 3–5 years of follow-up (Figure 45.3A and B). In cases of PR, multiple PDT courses (2–6) with intervals of 2–10 months in between have been carried out. Long-term stabilization (mean time to progression 21.2 ± 3.8 months) and improvement of quality of life have been achieved. The survival of patients was as follows: 1 year, 95.7%; 2 years, 86.9%; and 3 years, 65.2% (Vakulovskaya 2006a). MROI reported a higher rate of CR—up to 74% in early GIC and LC (Sokolov et al. 2000).

Combined treatment modalities were studied during recent years to study the possibilities of including PDT as a part of conventional anticancer therapy. This is of special interest with PTS taking into account its long-term retention in patients. Combined PDT + chemotherapy (CT) with 5-FU in patients with skin cancer and recurrences of gastric cancer after surgery was performed as a phase III open prospective study. PDT in standard mode (a single injection of PTS at a dose 0.3 mg/kg, three irradiations with intervals of 24–72 h) with intravenous injection of 5-FU in a dose of 375 mg/m² per 1–5 days of treatment has been provided. No significant increase in the toxicity of treatment was seen. The direct efficacy 2 months after PDT was 100% overall response rate with a prevalence of PR. In a locally approved protocol, we are now studying the comparison of combined PDT + endoscopic resection of mucosa versus endoscopic resection of mucosa alone and PDT alone for early gastric cancer (26 patients). Preliminary results have demonstrated high efficacy and safety in all three arms of treatment (Vakulovskaya et al. 2008b).

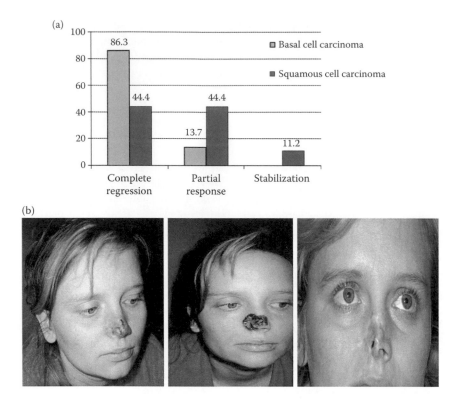

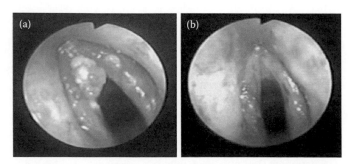

FIGURE 45.1 (a) With respect to tumor morphological structure, better response was achieved in BCC patients: 86.2% CR, 13.8% PR. In T1-2 cases, the efficacy was higher: 93.8% CR. The efficacy of PDT in patients with SCC was also high (88.9%); however, CR was much lower: 44.4%. The differences were statistically significant (*P* < 0.01). (b) Patient H, 26 years old. DS: Locally advanced recurrence of BCC (tumor infiltrating nose, right and left parts of nasal cavity, nasal septum, and destroying right cartilage) after chemoradiotherapy, laser destructions. Patient before PDT (left), after PDT session (middle), and 3 years after PDT (CR) (right).

PDT was carried out in 110 women with BC recurrences (skin metastases) after conventional treatment failed; the majority of patients had previously received multiple courses of chemotherapy and radiotherapy (40–80 Gy) with an interval of at least 2 months prior to PDT. Thirty-four patients had solitary metastases, and 69 had multiple metastases (histology: infiltrating ductal carcinoma or adenocarcinoma). In all patients, skin and subcutaneous metastases were the only sign of the disease. The interval between combined treatment and recurrence ranged from 6 months to 8 years. In seven patients with widespread recurrences and multiple drug resistance, PDT was provided as a salvage therapy. Two months after PDT, we

FIGURE 45.2 Recurrence of cancer of the larynx. Endophotographs before (a) and after (b) PDT. Complete response.

achieved CR in 51.5%, PR in 35.9%, St in 3.9, and progression in 8.7%. With respect to the number of metastases, better response was achieved in women with solitary metastases (91.2% CR and 8.8% PR). In patients with multiple metastases, we only found 31.9% CR, 49.3% PR, 5.8% St, and 13.0% progression with local or distant metastasis. In 1 year after PDT, 53 women with CR in 37.8% demonstrated no sign of progression, while new sites of skin metastases appeared in 18.9%, recurrence in the PDT fields was found in 3.7%, and distant metastasis to lung and bones in 39.6%. Overall survival was 1 year, 82.5%; 2 years, 70.8%; and 3 years, 54.4%. Disease-free survival in the whole cohort and in patients with PR, respectively, was 1 year, 19.4% and 37.8%; 2 years, 15.5% and 30.2%; and 3-years, 11.7%, and 22.6%. PDT was repeated in 12 patients, CR was achieved in 9 patients, and PR in 3 patients. In seven cases of salvage therapy, short-term stabilization (4–6 weeks) was obtained (Vakulovskaya 2006a; Vakulovskaya et al. 2002).

In 22 patients with T1-T2N0M0 BC, single or multiple interstitial PDT was performed as a preoperative treatment. Metalized catheters were injected into the tumor through standard needles under ultrasound control in one or two dimensions. Modified radical mastectomy or breast-conserving surgery had been performed 7–10 days after PDT with subsequent histological examination. Visual areas of necrosis of tumor were observed, and PR was found in 17 patients (77.3%). Histologic examination

(A)

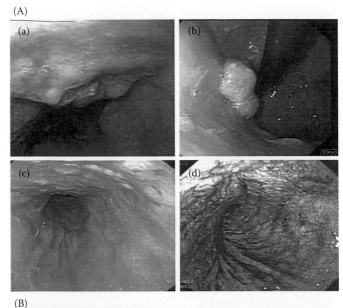

(B)

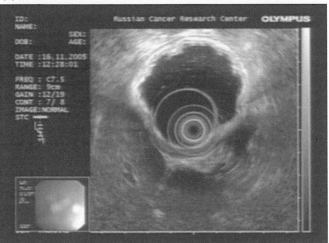

FIGURE 45.3 (A) Patient R, 79 years old. DS. Gastric cancer T2N0-xM0. Endophotographs before PDT (a, b) and 9 months after PDT (c, d). Complete response, no recurrence for 6 years. (B) Endoscopic ultrasound examination of stomach 9 months after PDT. Complete response.

showed infiltrating ductal carcinoma or adenocarcinoma with areas of necrosis and expressed hemorrhage, and pathomorphosis of grade 2–4. Pathomorphosis of grade 1 or no change with reduction of tumor mass less than 50% was seen in five patients (23.7%). No adverse events in the postoperative period were found in patients. They were followed up without signs of recurrence for 24 months.

In a group of 17 inoperable BC patients, we provided palliative PDT with multiple interstitial irradiation of primary tumor and metastatic lymph nodes. As a direct result of the treatment, reduction of tumor mass and grade 4 pathomorphosis was achieved. The direct efficacy of treatment (1 month after PDT) was PR with grade 4 pathomorphosis, significant

fibrosis, sclerosis, and rarely seen altered tumor cell complexes. Stabilization was reached for 3–6 months. Progression (without local progression but distant lung or liver metastases) was found 6 months after PDT in all patients (Vakulovskaya et al. 2002).

Combined PDT + RT in women with multiple skin metastases of BC has been performed as a phase III study. PDT in standard mode with simultaneous RT (2 Gy × 5 days/week) up to 60–65 Gy (1 arm) and 40–45 Gy (2 arms) has been performed. Previous experimental studies had showed the radiosensitizing effect of PTS *in vitro* on cell lines and additive antitumor effect of PDT and RT on *in vivo* models of transplanted tumors (Kubasova 2012). Direct clinical results demonstrated high efficacy and safety in both arms of the treatment; long-term results will be analyzed in the future. The safety and efficacy of the combination of PDT with laser hyperthermia was investigated in BCC and multiple skin metastases of BC.

Traditional side effects of PDT are pain during laser irradiation and skin phototoxicity. Different types of anesthesia have been used during laser irradiation in patients with SC and OC. Because of the increased solar sensitivity of the skin, initial administration of PTS in doses of 1.5–2.0 mg/kg caused hyperpigmentation of exposed parts of the skin in all patients, sunburns in 63.3%, and dermatitis in 14.7% of cases. Investigation of oxidative status showed significant reduction of the natural antioxidant (beta-carotene and alpha-tocopherol) content in plasma after PDT, depending mostly on its initial level. Administration of high doses of antioxidants significantly increased the plasma levels, thus decreasing the risk of development of oxidative stress at the biochemical level (Bukin et al. 1997). Reducing the dose of PTS to 0.8 mg/kg (phase II) and later to 0.4–0.5 mg/kg (phase III) with administration of a complex of antioxidants, we have significantly reduced the frequency of side effects to 6.3% hyperpigmentation, 5.6% sunburns, and 0.7% dermatitis ($P < 0.01$), and decrease the duration of solar sensitivity of skin to 4 weeks (Vakulovskaya 2006a; Vakulovskaya and Chental 1999).

Thus, PDT with PTS could be used with high efficiency as a radical treatment providing long-term local and systemic control alone or in combination with other anticancer modalities in skin cancer (except for melanoma), head and neck tumors (HNTs), early GIC, and LC, as well as palliative therapy in BC, recurrence of GIC, LC, and HNT. PDT resulted in good cosmetic and high functional outcome and could be carried out in elderly patients with severe comorbidities.

Nononcological PDT with Photosens has been investigated in several indications. For psoriasis, a 4 h application of 1% water solution of PTS was combined with 8–12 irradiations. More than 100 patients were enrolled, and the main effect of treatment was stabilization of disease. Nonhealing wounds and ulcers (up to 60 cm²) were treated with a 4 h application of 0.2% water solution of PTS and four to five multiple irradiations, resulting in shortening of healing time, and the antibacterial effect of PDT was confirmed. Choroidal neovascularization and ophthalmic tumors have also been treated.

45.4 Clinical Studies with Radachlorin

Radachlorin (Rada-Pharma) is a 0.35% aqueous solution of three chlorins extracted from microalgae of the *Spirulina* genus, including sodium chlorin e6 (88%–90%), sodium chlorin p6 (5%–7%), and purpurins (1%–5%), with intense absorption peaks at 400 ± 2, 504 ± 2, 534 ± 2, 608 ± 2, and 662 ± 2 nm. RC has high photodynamic activity and provides intense fluorescence. RC has been approved after phase I–II multicenter clinical trials (RCRC, MROI, Chelyabinsk clinical hospital) for skin cancer (BCC). Eighty-four patients were included in the trial at three institutions and 42 patients were involved at RCRC. Phase I clinical trials were performed as an open study in 14 patients with T1–T4N0M0 BCC, and recurrences of BCCBS, and phase II clinical trials were performed as a randomized, double-blind study in 28 patients with T2N0M0 BCC and recurrences of BCC. Primary T1–T4N0M0 BCC was in 28.6% patients and recurrent BCC in 71.4%. Primary multiple cancer was diagnosed in 21 cases (50%). RC at doses of 0.6, 1.2, or 2.4 mg/kg of body weight was injected intravenously, and PDT was done 3 h after injection. Surface irradiation with a power density of 100–300 mW/cm² and light doses of 200 or 300 J/cm², depending on the RC dose, were delivered. Spectral-fluorescent examination was conducted before RC administration, and each hour after it, during and for 6 days after PDT. The intensity of the tumor fluorescence reached a maximum 3 h after injection of RC. In 27 patients (64.3%) with 1.2 mg/kg RC, the average FC was 4.1 ± 1.1; in cases with previous RT, the FC was lower (3.1 ± 0.6); however, these differences were not statistically significant ($P > 0.05$). No increase in fluorescence intensity was noted in patients who received the higher dose (2.4 mg/kg RC). The intensity of fluorescence in nine patients (21.4%) with 0.6 mg/kg RC was lower, with an FC of 2.6 ± 0.4; the difference was statistically significant ($P < 0.05$). In 78.6% of cases, fluorescent borders exceeded clinically defined borders, while in 21.4% of cases, the borders coincided. Thirty-one additional fluorescence zones were detected in 12 patients with primary multiple BCC (28.6% of the total patients, 40.0% of patients with multiple primary BCC), and morphological verification of BCC was achieved in 30 cases (96.8%). FD demonstrated high sensitivity (95.8%), specificity (97.6%), and accuracy (97.4%). During the PDT session, significant photobleaching of RC was seen in all cases, reaching autofluorescence levels at the end of it. Fluorescent control can serve as an additional criterion for light dose selection; in the absence of PS in the tumor, further increase in a light dose would lead only to more significant damage of surrounding tissues. Rapid elimination of RC was found by means of spectroscopy in all patients; fluorescence was detected in healthy skin and mucosa of patients for 4–6 days after administration. Two months after PDT, CR was achieved in 73.8% of cases, PR in 23.8%, and St in 2.4% (Figure 45.4). The efficacy of treatment depended on tumor spread: CR in patients with T1–2N0M0 was 90% with no recurrence during 3 years of follow-up. In patients with recurrent BCC, CR was obtained only in 70.0% of cases and PR in 30.0%, significantly lower when comparing the results in previous groups ($P < 0.05$). The 1-year

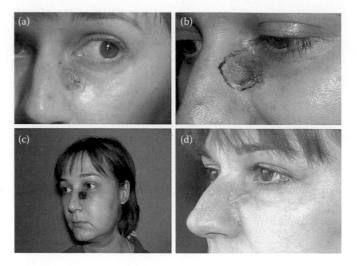

FIGURE 45.4 Patient Y, 32 years old. DS: Recurrence of BCC after radiotherapy and cryosurgery. Photos before (a, b) and after PDT with RC (c, d). Complete response.

disease-free survival was 94.9% and 2-year disease-free survival was 88.9% (Vakulovskaya 2006a; Vakulovskaya et al. 2004; Vakulovskaya, Zalevskaya, and Zalevskaya 2010). The efficacy of PDT in the cohort of 84 patients was higher because of the prevalence of T1-2 cases.

Results of pilot studies with RC in HNT, GIC, cholangiocarcinoma, and advanced LC in limited groups of patients as well as studies in inflammatory diseases of oral cavity have been published.

45.5 Clinical Studies with Photoditazin

Photoditazin (Veta Grand, PD), the di(N-methylglycamine salt of chlorine e6), is an aqueous solution of three chlorins, including sodium chlorin e6 (92.5%–94.5%), sodium chlorin p6 (3.0%–3.5%), and purpurins (less than 4.0%) with absorption peaks at 400 ± 2, 504 ± 2, 534 ± 2, 608 ± 2, and 662 ± 2 nm. It was approved in 2006 after multicenter clinical trials for skin cancer (BCC) and palliative treatment for recurrences of LC. Trials took place at RCRC, Medical Radiology Research Center (Obninsk), and N. N. Petrov Oncological Institute (Saint Petersburg). Thirty patients were included in the BCC trial at three institutions and 10 patients were enrolled at RCRC (four cases of T1–T2N0M0 and six cases of recurrences of BCC). Primary multiple cancer was diagnosed in six cases. PD at a dose of 0.7 mg/kg of body weight was injected intravenously, and PDT was carried out 2 h later. The intensity of the fluorescence in tumors reached a maximum 2–3 h after PD administration. FC ranged from 2.4:1 to 4.9:1 in patients with previous RT; FC was lower. Ten additional fluorescence zones were detected in five patients with primary multiple BCC, and morphological verification of BCC was obtained in nine cases. Two months after PDT, we achieved CR in nine patients and PR in one patient. In the multicenter study, CR was obtained in 92.8% of the cases and PR in 7.2% of the cases. In 30 patients with advanced inoperable lung cancer, the direct result of PDT was St of disease and clinical improvement (reduction in cough, hemoptysis, and dyspnea, and resolution of atelectasis) was noted in nearly 50% of patients (Ragulin et al. 2010). Studies in the recurrence of GIC, bladder cancer, brain tumors, and pleural metastases in limited groups of patients have been published.

Comparing the efficacy of PDT with PTS or RC and the tolerability of treatment in BCC patients, we could see similar results in primary T1-2 tumors. RC caused only short-term increase in solar sensitivity of the skin and was preferable in such cases. However, in locally advanced and recurrent BCC, the efficiency 2 months after PDT with PTS was significantly higher ($P < 0.005$): 85.1% compared to 70.0%, and the recurrence rate was lower. In these cases, PTS is therefore the PS of choice.

45.6 Clinical Studies with Alasens

The alternative possibility of creating effective concentrations of PS in tumors is stimulating the production of endogenous photoactive compounds by administration of their precursors. Among such chemical compounds is 5-aminolevulinic acid, produced under the name of Alasens (AS) by FSUE SSC Niopic. AS was approved by the Ministry of Health and Pharmacology Committee of the RF in 1999–2008 after phase I–II multicenter clinical trials for FD in skin cancer (BCC, SCC), skin metastases of BC, bladder cancer, OC, GIC, cancer of the larynx, LC, early cancer of the endometrium, and cancer of the cervix. Awaiting approval after trials are applications for FD of peritoneal metastasis in patients with ovarian cancer (OC) and GIC, and fluorescence-assisted surgery in patients with primary and metastatic brain tumors. Trials now in progress are PDT in primary and recurrent bladder cancer and adjuvant PDT in primary and metastatic brain tumors.

AS has been used in four different forms: 20% cream and biodegradable polymer film for application, 3% sterile solution for instillation, 3% solution for oral administration in a dose 20 or 30 mg/kg of body weight, and for inhalation. In our studies, we combined spectroscopy as a first step in the diagnostic procedure and acquired fluorescent images as the second step in the majority of patients. Spectroscopy allows us to get quantitative characteristics of fluorescence intensity, and therefore, it could serve as a criterion in differential diagnosis. We recorded the shape and amplitude of the signal and the integrated intensity of the fluorescence of protoporphyrin IX (PP-IX) in different parts of tumors in patients. Tumor borders were defined as well as the intensity of fluorescence of the normal skin of the hand, face, and mucosa of lower lip of patients. After oral administration of AS or using biodegradable polymer films, we examined the dynamics of accumulation of PP-IX in both tumor and normal skin and mucosa before 1, 2, 3, 4, and 24 h after AS administration. The intensity of fluorescence in tumor grew, reaching maximum 3–4 h after AS administration (Figure 45.5). No significant difference in fluorescence intensity was found in women with cancer of the cervix comparing local and oral AS administration. Short-term 24 h increase in the fluorescence intensity of the patient's normal skin and mucosa was observed after oral AS administration at doses of 20 and 30 mg/kg body weight. No increase in the sensitivity of skin to direct sunlight and short-term (a few hours) increase in PP-IX fluorescence and sensitivity of skin in places of local application of AS-containing cream were observed (Vakulovskaya 2006a,b; Vakulovskaya et al. 2002).

We observed increased fluorescence of all tumor sites with the exception of pigmented melanoma and pigmented nevus. In widespread tumors, significant heterogeneity of distribution of PP-IX was observed with marked increase in fluorescence intensity in ulcerated areas. FC was 2.2–19.0 and significantly differed between the different morphological forms of cancer. The intensity of fluorescence was positively higher in patients with SSC and metatypical cancer than in BCC patients. The FC level of benign tumors was less than 1.8, and the higher the FC, the higher the probability of an aggressive malignant tumor. Fluorescence borders of the tumor were close to clinical borders in cases of exophytic forms of tumor growth, but significantly exceeded clinically detected borders in cases of infiltrative

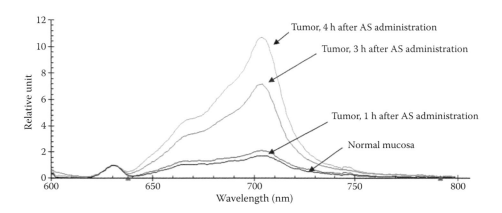

FIGURE 45.5 Fluorescence spectra of PP-9 accumulation in tumor during 4 h after oral administration of 20 mg/kg AS. Patient P. DS: Cancer of oral cavity.

tumor growth and were observed with frequencies from 39.6% (in peritoneal metastases of OC) up to 71.8% (in skin metastases of BC). Additional florescence zones were found in 13.8% in SC, 16.7% in OC, 77.8% in metastatic BC, and 3.8% in LC. Morphological verification was obtained in 89.5% of cases in SC, 87.5% in oropharyngeal cancer, 99.3% in metastatic BC, and 100% in LC. The sensitivity of FD in all groups of patients was from 98.7% up to 100%, the specificity from 78.3% to 98.1%, and the accuracy from 96.0% to 99.1% (Vakulovskaya 2005, 2006; Vakulovskaya et al. 2002).

PDT in 15 BCC patients was carried out with a single laser irradiation at a power density of 100–300 mW/cm^2 and a light dose of 100 J/cm^2. Superficial necrosis was formed in 12 patients (80%); in three cases (20%) CR was observed with no recurrence during 4-year follow-up. PDT with AS could therefore be recommended for treatment of superficial T1N0M0 BCC.

AS-induced fluorescent control of transurethral resection (TUR) of bladder cancer was carried out in 280 cases of primary and recurrent bladder cancer. In all patients, cystoscopy started with careful inspection under white light and removal of visual tumors of the bladder wall after brief illumination with blue light for estimation of fluorescence of already detected tumors and their borders that did not lead to significant photobleaching. After standard TUR, the endoscopy was continued under blue light and all fluorescent lesions were documented and resected, and biopsies from fluorescence zones and fluorescence negative zones were taken. In 13 patients, biopsies were collected in black boxes and spectroscopy in the dark was carried out. We were able to compare the efficacy of FD to conventional white light endoscopy in the same group of patients, as well as assess the advantage over TUR conducted only in blue light because in our conditions, photobleaching was less significant and the probability of detecting small lesions increased. We detected fluorescence in all visible tumors, and in 92% of cases, additional fluorescence zones were found, while morphological verification of tumor was obtained in 61.1%. Carcinoma in situ was detected in 48.5% of tumor-positive biopsies and superficial papillary tumors (Ta-1) in 51.5%. High-grade dysplasia was found in 14.8%, low to moderate dysplasia in 13.0%, and inflammation in 11.1% of cases. We found by means of spectroscopy that the intensity of fluorescence in cancer was 7.6-fold higher, in high-grade dysplasia 5.9-fold higher, and inflammation only 1.4-fold higher in comparison to normal urothelium. Spectroscopy could improve the specificity of FD. Sensitivity of FD was 100%, accuracy 86.4%, and specificity 70.0%, and when the results of biopsies with high-grade dysplasia were included, it increased up to 79.0%. The sensitivity of conventional cystoscopy in white light examination was 60.7% and accuracy 75.2%; these values are significantly lower compared to FD ($P < 0.005$) (Kudashev 2002; Vakulovskaya 2006a).

In a 3-year follow-up prospective study, the rate of recurrence in matched groups of patients with and without fluorescent control during TUR differed dramatically. The rate of recurrence in group of conventional treatment was significantly six- to eightfold higher than in the group with fluorescent control of

TUR up to the first 12 months after TUR, reaching maximum at 9 months. Later on, the difference in the rate of recurrences began to decrease and after 18 months of follow-up, it became minimal. The decrease in the rate of overlooked tumors for TUR with AS-induced fluorescence cystoscopy resulted in a relative decrease in the number of recurrences (Kudashev 2002).

During phase II clinical trials, we tested fluorescent laparoscopy (FL) in OC patients, evaluating the efficacy, toxicity, and safety of the procedure. As a first step, we carried out under general anesthesia conventional laparoscopy in white light with inspection and detection of metastatic disease in the peritoneum, pelvic, and abdominal cavity. We continued with spectroscopic investigation, which allowed us to evaluate the efficacy of FL in comparison with conventional laparoscopy for detecting peritoneal metastasis in the same group of patients (Vakulovskaya et al. 2005).

FL was carried out in 38 patients with T3-4 stage OC for evaluating the completeness of response after complex therapy (cytoreductive surgery including extirpation of uterus, ovariectomy, and omentectomy) + six to eight courses of CT, and was also used in cases of suspicion of disease recurrence. Histology in all cases was papillary cyst adenocarcinoma. Complete clinical and instrumental investigation (ultrasound, computed tomography) showed no signs of peritoneal metastasis or liquid in peritoneal cavity. No other clinical or instrumental signs of recurrence with the exception of more than a twofold increase in CA-125 were detected. The mean interval after finishing the combined treatment was 12.3 months.

Laparoscopy in white light and FL with detection of fluorescent zones, borders of dissemination, and intensity of accumulation of PP-IX in metastases of OC after adjusting for normal peritoneum was performed 4 h after AS oral administration at a dose 20 mg/kg of body weight in a 150 mL water solution. Metastases of OC in the peritoneum were found in 26 patients (68.4%) in white light. Intense fluorescence zones on the peritoneal surface were detected in 35 patients (92.1%), and the number of zones in patients was from 1 to 7, the sizes were 0.2–0.9 cm, and the FC between metastases and adjacent normal peritoneum was 2.5–8.0. Even in patients with visually detected metastases, additional fluorescent zones were observed. In four patients, no metastases were detected both during traditional laparoscopy in white light and during FL. Two hundred and five biopsies were taken from all fluorescent zones and 38 from nonfluorescent zones for cytological and histological examination. Morphological verification of OC metastases was obtained in 96.1% of biopsies from fluorescent zones. In one patient, biopsies from fluorescence zones revealed endometriosis (false-positive result). No tumor was found in nonfluorescent zones. FL showed significantly higher specificity (81.0%), sensitivity (100%), and accuracy (70.6%) compared to conventional laparoscopy in white light: S 61.5%, A 70.6% ($P < 0.005$) (Poddubny et al. 2005; Vakulovskaya 2006a; Vakulovskaya et al. 2005). We have also carried out FL in GIC patients and hepatic cancer, but in these cases, we found widespread metastases even in conventional laparoscopy, and thus the results of FL were less impressive. At

MROI, the same protocol was carried out mostly in gastric cancer patients with well-diagnosed carcinomatosis of the peritoneum with liquid in peritoneal cavity detected before laparoscopy.

In 2006, we began a phase III clinical trial of intraoperative fluorescence-guided resection (FGR) with AS in patients with malignant glioma and brain metastases, evaluating the efficacy, toxicity, and safety of the procedure. Intraoperative FGR was performed in 18 patients (22–67 years old) with recurrent malignant glioma (RMG) (six cases) and solitary metastatic brain tumors (12 cases) of different origin (LC, BC, kidney cancer, melanoma) for evaluating the completeness of resection. Traditional preoperative and postoperative clinical and instrumental investigation [CT, magnetic resonance tomography (MRT)] was done in all patients. Intraoperative FGR was carried out using an oral AS dose of 20 mg/kg of body weight in 150 mL of water solution given 3 h before induction of anesthesia. Intraoperative tumor fluorescence was visualized using the operating microscope Panther and D-light AF system (Karl Storz). Spectroscopy with detection of fluorescent zones, borders of tumor, or dissemination of metastases and intensity of accumulation of PP-IX before, during, and after surgical removal was carried out. Intense red fluorescent zones were found in all patients under blue light with the exception of zones of necrosis in two patients and one patient with solitary metastasis of melanoma. Different intensity of fluorescence was found in solid (high intensity) and infiltrative parts (low intensity) of RMG. Metastases provided less intense fluorescence. No fluorescence was found in normal brain. We obtained morphological verification of tumor in all biopsies from fluorescent zones. Complete surgical removal of fluorescent zones was done in 11 patients. The completeness of resection was confirmed by postoperative MR imaging. No side effects were found after AS administration or during operation. We demonstrated the safety and pronounced efficacy of intraoperative FGR with AS for completeness of tumor removal both in RMB patients and patients with brain metastases by highlighting residual solid and infiltrating tumor compared to traditional resection in white light (Vakulovskaya 2006, Karakhan, and Aleshin 2007; Vakulovskaya et al. 2008a). The high rate of radical resection demonstrated with FGR should give a new impetus to the study of whether complete resection influences survival in randomized trials on a firmer statistical basis.

Our experience show pronounced efficacy of FD with AS in patients with cancer of different localizations. FD provided diagnostically significant information about tumor advance and tumor borders and allowed identification of subclinical specific lesions with high sensitivity and specificity. It led to significant increases in sensitivity, specificity, and accuracy compared to traditional examination, laparoscopy, or endoscopy in white light due to early detection of metastases and carcinoma *in situ*.

Nononcological FD and PDT with AS has been studied in the following conditions: skin diseases (psoriasis with low effect, acne), inflammatory diseases in dentistry, virus-associated papillomas and precancerous lesions of the cervix, vulval condylomata acuminata, and vulvovaginal candidiasis in women.

Antiviral, antibacterial, and antimycotic effects of AS-PDT were also demonstrated.

After 20 years of investigation, six domestic photosensitizers and a variety of therapeutical lasers, spectroscopy apparatus, and diagnostic diode devices have federal approval for fluorescent diagnostics and PDT of cancer of different localizations in the RF. The Ministry of Health has included FD and PDT as high-technology treatment and diagnostic procedures in state oncological standards. However, in spite of further development and the increasing number of clinics using these technologies, PDT is still mostly carried out only at well-known institutions in Moscow and in a few big cities.

References

Bukin, Y., E. Vakulovskaya, V. Chental, and V. Draudin. 1997. Deficiency of beta-carotene and vitamin E and its correction after photodynamic therapy of cancer. 6th World Congress on Clinical Nutrition, Antioxidants and Disease, Alberta, Canada, 53.

Chental, V., E. Vakoulovskaia, N. Abdoullin et al. 1997. Combination of photodynamic therapy and cryosurgery in treatment of advanced skin cancer. *Acta Bio-Opt Info Med* 3:28.

Kubasova, I. Y. 2012. PDT possibilities in treatment of malignant tumors (experimental study). Moscow, doctoral thesis, 42 [in Russian].

Kudashev, B. V. 2002. The application of fluorescent diagnostics to enhance the radicalism of bladder transurethral resection. Moscow, Ph.D. thesis, 122.

Kuvshinov, Y. P., B. K. Poddybny, A. F. Mironov et al. 1996. Endoscopic photodynamic therapy of tumors using gold vapor laser. In Laser Use in Oncology—CIS Selected Papers, *Proc SPIE* 2728:206–209.

Lukiyanets, E. A. 1998. New sensitizers for photodynamic therapy. *Russ Chem J* 42(5):9–16.

Mironov, A. F., A. S. Seyanov, and V. M. Pizhik. 1992. Hematoporphyrin derivatives: Distribution in a living organism. *J Photochem Photobiol B Biol* 16:341–346.

Poddubny, B. K., A. N. Gubin, V. N. Sholokhov et al. 2005. Modern methods of laparoscopic diagnostics of malignant tumors of abdominal cavity [in Russian]. *Curr Oncol* 7(3):130–133.

Ragulin, Y. A., M. A. Kaplan, V. N. Medvedev, V. N. Kapinus, and V. V. Peters. 2010. Photodynamic therapy for treatment of endobronchial tumors. *Photodiagn Photodyn Ther* 7(Suppl 1):S.15.

Sokolov, V. V., V. I. Chissov, R. I. Yakubovskaya et al. 1996. Multicourse PDT of malignant tumors: The influence of primary tumor, metastatic spreading and homeostasis of cancer patients. *Proc SPIE* 2924:322–329.

Sokolov, V. V., N. N. Zharkova, E. V. Filonenko, L. V. Telegina, and E. S. Karpova. 2000. Present-day potentialities of endoscopic diagnostics and treatment of the early cancer in respiratory and digestive tracts. *Proc SPIE* 3909:2–12.

Stranadko, E. F., O. K. Skobelkin, G. S. Litwin, and T. V. Astrakhankina. 1996. Clinical photodynamic therapy of malignant neoplasms. *Proc SPIE* 2325:240–246.

Stratonnikov, A. A, N. E. Edinak, D. V. Klimov et al. 1996. The control of photosensitizer in tissue during photodynamic therapy by means of absorption spectroscopy. *Proc SPIE* 2924:49–56.

Stratonnikov, A. A., N. V. Ermishova, G. A. Meerovich et al. 2002. Simultaneous measurement of photosensitizer absorption and fluorescence in patients undergoing photodynamic therapy. *Proc SPIE* 4613:162–173.

Stratonnikov, A. A., and V. B Loschenov. 2001. Evaluation of blood oxygen saturation in vivo from diffuse reflectance spectra. *J Biomed Opt* 6(4):457–467.

Stratonnikov, A. A., G. A. Meerovich, and V. B. Loschenov. 2000. Photobleaching of photosensitizers applied for photodynamic therapy. *Proc SPIE* 3909:81–91.

Vakulovskaya, E. 2005. Fluorescent diagnostics with Alasense in oral cancer patients. *Oral Oncol* (Suppl)13(1):100.

Vakulovskaya, E. G. 2006a. Photodynamic therapy and fluorescent diagnostics of tumors [in Russian]. Moscow, 264.

Vakulovskaya, E. 2006b. Photodynamic therapy and fluorescent diagnostics of head and neck cancer with second-generation photosensitizers. In *Current Research on Laser Use in Oncology: 2000–2004, Proc SPIE* 5973:08-1–08-6.

Vakulovskaya, E. and V. Chental. 1999. New approaches to photodynamic therapy of tumors with Al phthalocyanine. In *Laser Use in Oncology II, Proc SPIE* 4059:32–38.

Vakulovskaya, E. G., V. V. Chental, N. A. Abdoullin et al. 1996a. Photodynamic therapy of head and neck tumors. *Proc SPIE* 2924:309–313.

Vakulovskaya, E., V. Chental, V. Letyagin et al. 2002. Photodynamic therapy and fluorescent diagnostics of breast cancer with Photosense and Alasense. *Proc SPIE* 4612:174–177.

Vakulovskaya, E., V. Chental, G. Meerovich, M. Ulasuyk, and E. Lukyanets. 1996b. Photodynamic therapy of spread skin malignancies with scanning electron-pumped semiconductive laser. In Laser Use in Oncology—CIS Selected Papers, *Proc SPIE* 2728:210–213.

Vakulovskaya, E., A. Gubin, E. Vakurova, and B. Poddubny. 2005. Laparoscopic fluorescent diagnostics of peritoneal dissemination of ovarian cancer with Alasense. *Eur J Cancer* (Suppl) 3(2):278.

Vakulovskaya, E. G., V. B. Karakhan, and V. A. Aleshin. 2007. Fluorescent diagnostics with Alasense in patients with primary and metastatic brain tumors [in Russian]. *Russ Biotherap J* 6(1):64.

Vakulovskaya, E. G., V. B. Karakhan, A. A. Stratonnikov, and V. A. Aleshin. 2008a. Fluorescent-guided resection with Alasense in patients with primary and metastatic brain tumors. *Photodiagn Photodyn Ther* 5(I.1):86.

Vakulovskaya, E. G., Y. P. Kemov, A. V. Reshentnikov, and I. D. Zalevsky. 2004. Photodynamic therapy and fluorescent diagnostics of skin cancer with radochlorin and Photosense: Comparing efficacy and toxicity. *Proc SPIE* 5315:148–151.

Vakulovskaya, E. G., Y. P. Kuvshinov, B. K. Poddubny, O. A. Malikhova, and I. S. Stilidi. 2008b. Photodynamic therapy of gastrointestinal cancer with Photosense. *Photodiagn Photodyn Ther* 5(I.1):73.

Vakulovskaya, E. G., L. I. Zalevskaya, and O. I. Zalevskaya. 2010. Photodynamic therapy and combined treatment modalities in head and neck cancer patients with Radochlorin. *Photodiagn Photodyn Ther* 7(Suppl 1):S.17.

V

Low-Level Laser (Light) Therapy (LLLT)

46

Chromophores (Photoacceptors) for Low-Level Laser Therapy

Tiina I. Karu
Russian Academy of Sciences

46.1 Introduction

The first publications about low-level laser therapy (LLLT) (then called laser biostimulation) appeared more than 40 years ago. Since then, more than 4000 studies have been published on this still-controversial topic (Tuner and Hode 2010). In the 1960s and 1970s, doctors in Eastern Europe, and especially in the Soviet Union and Hungary, actively developed laser biostimulation. However, scientists around the world harbored an open skepticism about the credibility of studies stating that low-intensity visible-laser radiation acts directly on an organism at the molecular level. Supporters in Western countries, such as Italy, France, and Spain, as well as in Japan and China, also adopted and developed this method, but the method was—and still remains—outside mainstream medicine. In the past several years, some excellent experimental work was performed in the United States (Anders 2009; Eells et al. 2003, 2004; Pal et al. 2007; Wong-Riley et al. 2001, 2005; Wu et al. 2009). The controversial points of laser biostimulation, which were topics of great interest at that time, were analyzed in reviews that appeared in the late 1980s (Karu 1987, 1989). Since then, medical treatment with coherent-light sources (lasers) or noncoherent light [light-emitting diodes (LEDs)] has passed through its childhood and adolescence. Most of the controversial points from the childhood period are no longer topical. Currently, low-power laser therapy—also called LLLT or photobiomodulation—is considered part of light therapy as well as physiotherapy (Karu 2003, 2007). In fact, light therapy is one of the oldest therapeutic methods used by humans [historically as sun therapy and later as color light therapy and

ultraviolet (UV) therapy]. A short history of experimental work with colored light on various kinds of biological subjects can be found elsewhere (Karu 1987, 1989). The use of lasers and LEDs as light sources was the next step in the technological development of light therapy.

It is clear now that laser therapy cannot be considered separately from physiotherapeutic methods that use such physical factors as low-frequency pulsed electromagnetic fields, microwaves, time-varying, static, and combined magnetic fields, focused ultrasound, and direct-current electricity. Some common features of biological responses to physical factors have been briefly analyzed (Karu 1998).

As this handbook makes abundantly clear, by the 21st century, a certain level of development of (laser) light use in therapy and diagnostics (e.g., photodynamic therapy, optical tomography) had been achieved. In low-power laser therapy, the question is no longer whether light has biological effects but rather how radiation from therapeutic lasers and LEDs works at the cellular and organism levels and what the optimal light parameters are for different uses of these light sources.

This chapter is organized as follows. First, Section 46.2 reviews very briefly one of the historically topical issues in LLLT: whether coherent and polarized light has additional benefits in comparison with noncoherent light at the same wavelength and intensity. This question was discussed in detail in the work of Karu (2003).

Second, direct activation of various types of cells via light absorption in mitochondrial respiratory chain is described in Section 46.3. Primary photoacceptors and mechanisms of light action on cells and mechanisms of cellular signaling are considered in detail. Section 46.4 describes briefly enhancement of cellular metabolism via activation of nonmitochondrial photoacceptors and possible indirect effects via secondary cellular messengers, which are produced by cells as a result of direct activation.

46.2 Role of Light Coherence and Polarization in LLLT

Low-power laser therapy is used by physiotherapists (to treat a wide variety of acute and chronic musculoskeletal aches and pains), by dentists (to treat inflamed oral tissues and to heal diverse ulcerations), by dermatologists (to treat edema, indolent ulcers, burns, and dermatitis), by rheumatologists (to relieve pain and treat chronic inflammations and autoimmune diseases), and by other specialists as well as general practitioners. Laser therapy is also widely used in veterinary medicine (especially in racehorse-training centers) and in sports-medicine and rehabilitation clinics (to reduce swelling and hematoma, relieve pain, improve mobility, and treat acute soft-tissue injuries). Lasers and LEDs are applied directly to the respective areas (e.g., wounds, sites of injuries) or to various points on the body (acupuncture points, muscle-trigger points). Several applications will be considered in Chapters 49 through 58.

One of the most topical and widely discussed issues in the low-power-laser-therapy clinical community has been whether the coherence and polarization of laser radiation have additional benefits as compared with monochromatic light from a conventional light source or LED with the same wavelength and intensity.

Two aspects of this problem must be distinguished: the *coherence of light* itself and the *coherence of the interaction* of light with matter (biomolecules, tissues). These problems are considered in detail in publications (Karu 2003, 2011).

The coherent properties of light are not manifested when the beam interacts with a biotissue on the molecular level. This problem was first considered years ago (Karu 1987). The question then arose of whether coherent light was needed for laser biostimulation or was it simply a photobiological phenomenon. The conclusion was that under physiological conditions, the absorption of low-intensity light by biological systems is of purely noncoherent (i.e., photobiological) nature because the rate of decoherence of excitation is many orders of magnitude higher than the rate of photoexcitation. The time of decoherence of photoexcitation determines the interaction with surrounding molecules (under normal conditions less than 10^{-12} s). The average excitation time depends on the light intensity (at an intensity of 1 mW/cm^2 this time is around 1 s). At 300 K in condensed matter for compounds absorbing monochromatic visible light, the light intensity at which the interactions between coherent light and matter start to occur was estimated to be above the gigawatts per square centimeter level (Karu 1987). Note that the light intensities used in clinical practice are not higher than tens or hundreds of milliwatts per square centimeter. Indeed, the

stimulative action of various bands of visible light at the level of organisms and cells was known long before the advent of the laser.

The spatial (lateral) coherence of the light source is unimportant due to strong scattering of light in biotissue when propagated to the depth $L \gg \ell_{sc}$, where ℓ_{sc} is the free pathway of light in relation to scattering. This is because every region in a scattering medium is illuminated by radiation with a wide angle ($\phi \sim 1$ rd). This means that $\ell_{coh} = \lambda$ (i.e., the size of spatial coherence ℓ_{coh} decreases to the light wavelength).

The length of longitudinal coherence, L_{coh}, is important when bulk tissue is irradiated because this parameter determines the volume of the irradiated tissue, V_{coh} (Karu 2003, 2011). In this volume, the random interference of scattered light waves and formation of random nonhomogeneities of intensity in space (speckles) occur. For noncoherent-light sources, the coherence length is small (tens to hundreds of microns). For laser sources, this parameter is much higher. Thus, the additional therapeutic effect of coherent radiation, if this indeed exists, depends not only on the length of L_{coh} but also, and even mainly, on the penetration depth into the tissue due to absorption and scattering (i.e., by the depth of attenuation). Table 1 in Karu (2003) summarizes qualitative characteristics of coherence of various light sources, as discussed above. The most recent experimental data allow us to suggest that the coherence length can play a role in laser phototherapy of gingival inflammation (Qadri et al. 2007). Considered within the framework of this qualitative picture is that some additional (i.e., additional to those effects caused by light absorption by photoacceptor molecules) manifestation of light coherence for deeper tissue is quite possible. This qualitative picture also explains why coherent and noncoherent light with the same parameters produce the same biological effects on cell monolayer (Karu 2011). It was established experimentally that elementary processes in cells after light absorption do not depend on the degree of beam polarization (Karu et al. 2008a). Some additional (therapeutic) effects from the coherent and polarized radiation can appear only in deeper layers of the bulk tissue. To date, no experimental work has been performed to qualitatively and quantitatively study these possible additional effects. In any case, the main therapeutic effects occur due to light absorption by cellular photoacceptors.

46.3 Enhancement of Cellular Metabolism via Activation of Respiratory Chain: A Universal Photobiological Action Mechanism

46.3.1 Cytochrome *c* Oxidase as the Photoacceptor in the Visible-to-Near-Infrared Spectral Range

Photobiological reactions involve the absorption of a specific wavelength of light by the functioning photoacceptor molecule.

The photobiological nature of low-power laser effects (Karu 1987, 1988, 1989, 2007) means that some molecule (photoacceptor) must first absorb the light used for the irradiation. After promotion of electronically excited states, primary molecular processes from these states can lead to a measurable biological effect at the cellular level. The main problem is knowing which molecule is the photoacceptor. When considering the cellular effects, this question can be answered by action spectra.

A graph representing photoresponse as a function of wavelength λ, wave number λ^{-1}, frequency v, or photon energy e is called an action spectrum. The action spectrum of a biological response resembles the absorption spectrum of the photoacceptor molecule. The existence of a structured action spectrum is strong evidence that the phenomenon under study is a photobiological one (i.e., primary photoacceptors and cellular signaling pathways exist; Hartman 1983).

The first action spectra in the UV-visible–near-infrared (NIR) region (from 313 to 860 nm) were recorded in the early 1980s for deoxyribonucleic acid (DNA) and ribonucleic acid (RNA) synthesis rate, growth stimulation of *Escherichia coli*, and protein synthesis by yeasts for the purpose of investigating the photobiological mechanisms of laser biostimulation (reviews: Karu 1987, 1988, 1989, 1990, 1998, 1999). In addition, other action spectra were recorded in various ranges of visible wavelengths: photostimulation of formation of E-rosettes by human lymphocytes, mitosis in L cells, exertion of DNA factor from lymphocytes in the violet-green range (Gamaleya et al. 1983), and oxidative phosphorylation by mitochondria in the violet-blue range (Vekshin 1991). These spectra were recorded for narrow ranges of the optical spectrum and with a limited number of wavelengths, which prevented identification of the photoacceptor molecule.

It was suggested in 1988 that the mechanism of low-power laser therapy at the cellular level was based on the absorption of monochromatic visible and NIR radiation by components of the cellular respiratory chain (Karu 1988). Absorption and promotion of electronically excited states cause changes in redox properties of these molecules and acceleration of electron transfer (primary reactions). Primary reactions in mitochondria of eukaryotic cells were supposed to be followed by a cascade of secondary reactions (photosignal transduction and amplification chain or cellular signaling) occurring in cell cytoplasm, membrane, and nucleus (Karu 1988; for reviews, see Karu 1999, 2007, 2008).

It is remarkable that the five action spectra that were analyzed in the work of Karu and Kolyakov (2005) had very close (within the confidence limits) peak positions in spite of the fact that these processes occurred in different parts of the cells (nucleus and plasma membrane). However, there were differences in peak intensities. Five of these action spectra only for the red-to-NIR range (wavelengths that are important in low-power laser therapy) are presented in Figure 46.1.

Two conclusions can be drawn from the action spectra recorded. First, the fact that the peak positions were found to be practically the same suggests that the primary photoacceptor is the same. Second, the existence of the action spectra implied the existence of cellular signaling pathways inside the cell between photoacceptor and the nucleus as well as between the photoacceptor and the cell membrane.

The bands of the action spectra were identified by analogy with the absorption spectra of the metal–ligand system characteristic of this spectral range (Karu and Afanasyeva 1995) (for reviews, see Karu 1998, 1999, 2007). It was concluded that the ranges 400–450 and 620–680 nm were characterized by the bands pertaining to a complex associated with charge transfer in a metal–ligand system, and within 760–830 nm, these were d-d transitions in metals, most probably in Cu(II). The range 400–420 nm was found to be typical of a $\pi-\pi^{\star}$ transition in a porphyrin ring. A comparative analysis of lines of possible d-d transitions and charge–transfer complexes of Cu with our action spectra suggested that the photoacceptor was the terminal enzyme of the mitochondrial respiratory chain cytochrome *c* oxidase. It was suggested that the main contribution to the 825 nm band was made by the oxidized Cu_A, to the 760 nm band by the reduced Cu_B, to the 680 nm band by the oxidized Cu_B, and to the 620 nm band by the reduced Cu_A. The 400–450 nm band was more likely the envelope of a few absorption bands in the 350–500 nm range (i.e., a superposition of several bands). Analysis of the band shapes in the action spectra and the line-intensity ratios also led to the conclusion that cytochrome *c* oxidase cannot be considered a primary photoacceptor when fully oxidized or fully reduced but only when it is in one of the intermediate forms (partially reduced or mixed-valence enzyme) (Karu and Afanasyeva 1995; for reviews see Karu 1998, 2007) that have not yet been identified. Figure 46.2 illustrates these conclusions for the red-to-NIR spectral region.

Taken together, the terminal respiratory-chain oxidases in eukaryotic cells (cytochrome *c* oxidase) and in prokaryotic cells of *E. coli* (cytochrome *bd* complex, Tiphlova and Karu 1991a) where believed to be photoacceptor molecules for red-to-NIR radiation. In the violet-to-blue spectral range, flavoproteins [e.g., NADH-dehydrogenase (Karu 1998) in the beginning of the respiratory chain] are also among the possible photoacceptors like the terminal oxidases.

One important step in identifying the photoacceptor molecule is to compare the absorption and action spectra. For recording the absorption of a cell monolayer and investigating the changes in absorption under irradiation at various wavelengths of monochromatic light, a sensitive multichannel registration method was developed (Karu et al. 1998, 2001a, 2005a). Recall that the absorption spectra of individual living cells were recorded at wavelengths of up to 650 nm years ago with the aim of identifying the respiratory chain enzymes. The absorption spectrum of whole cells in the visible region was found to be qualitatively similar to that of isolated mitochondria (Chance and Hess 1959). The extension of optical measurements from the visible spectral range to the far-red and NIR regions (650–1000 nm) was undertaken late in the 1970s for the purpose of monitoring the redox behavior of cytochrome *c* oxidase *in vivo*. These studies led to

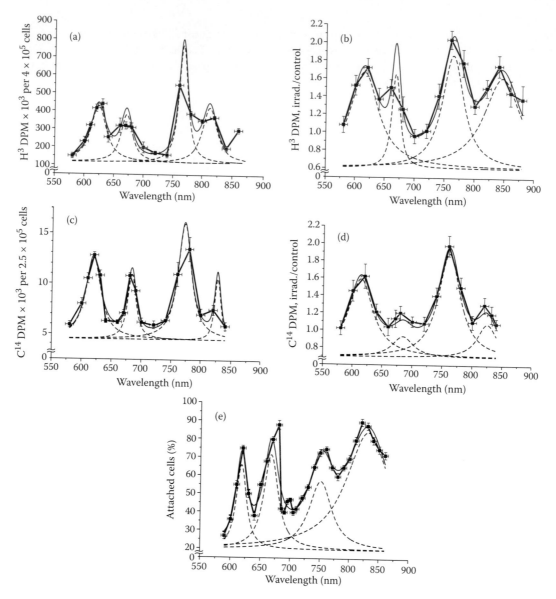

FIGURE 46.1 Action spectra in the region of 580–860 nm for (a) stimulation of DNA synthesis rate in log-phase and (b) plateau-phase HeLa cultures; (c) stimulation of RNA synthesis rate in log-phase and (d) plateau-phase cultures; and (e) increase of cell attachment to glass matrix. Experimental curves (-<-<-), curve fittings (–), and Lorentzian fittings (– –) are shown. Dose: 100 J/m² (a–d) or 52 J/m² (e). (Adapted from Karu, T. I. and S. F. Kolyakov, *Photomed Laser Surg* 23: 355–361, 2005.)

the discovery of an "NIR window" into the body and the development of NIR spectroscopy for monitoring tissue oxygenation (Jöbsis-vander Vliet 1999).

Figure 46.3 presents the intact absorption spectra of HeLa cell monolayer (A, B, C) and the same spectra after the irradiation at 830 nm (A₁, B₁, C₁) as well as two action spectra for the comparison (D, E). Table 46.1 shows the peak positions in all these absorption and action spectra as resolved by Lorentzian fitting. All experimental details can be found in Karu et al. (2005a).

For quantitative characterization purposes, as well as for comparison between the recorded absorption spectra, we decided to use intensity ratios between certain absorption bands. We used

the band present in all absorption spectra near 760 nm (exactly at 754, 756, 767, 765, and 762 nm) (Table 46.1) as a characteristic band for the relatively reduced photoacceptor. The band used by us to characterize the relatively oxidized photoacceptor was the one near 665 nm (exactly at 666, 661, and 665 nm) in spectra B, B₁, C, and C₁ (Table 46.1). This band is so weak it could not be resolved by the Lorentzian fitting method in spectra A and A₁ belonging to the most strongly reduced photoacceptor in our experiments. For this reason, we used in our intensity calculations for spectra A and A₁ absorption on the curve fitting level at 665 nm. The gray vertical lines in Figure 46.3 mark the bands chosen. The intensity ratio I_{760}/I_{665} was calculated to characterize every spectrum. In these simple calculations, we used only the

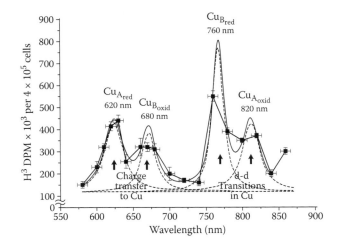

FIGURE 46.2 Action spectrum for stimulation of DNA synthesis rate on the cellular level. Suggested absorbing chromophores of the photoacceptor, cytochrome *c* oxidase, are marked. Original curve (-■-), curve fitting (━), and Lorentzian fitting (– –) are shown. (After Karu 2010.)

peak intensities (peak heights) and not the integral intensities (peak areas) that are certainly needed for further developments. In the case of equal concentrations of the reduced and oxidized forms of the photoacceptor molecule, the ratio I_{760}/I_{665} should be equal to unity. When the reduced forms prevail, the ratio I_{760}/I_{665} is greater than unity, and it is less than unity in cases where the oxidized forms dominate. Recall that the internal electron transfer within the cytochrome *c* oxidase molecule causes the reduction of the molecular oxygen via several transient intermediates of various redox states (Chance and Hess 1959; Jöbsis-vander Vliet 1999).

The magnitude of the I_{760}/I_{665} criterion is 9.5 for spectrum A, 1.0 for spectrum B, and 0.36 for spectrum C. By this criterion, irradiation of the cells, whose spectrum is marked by A (I_{760}/I_{665} = 9.5) causes the reduction of the absorbing molecule (I_{760}/I_{665} for spectrum A_1 is equal to 16). Irradiation of the cells characterized by spectrum B also causes the reduction of the photoacceptor, as evidenced by the increase of the I_{760}/I_{665} ratio from 1.0 to 2.5 in spectrum B_1. In the spectrum of the cells with initially more reduced photoacceptor (spectrum A), irradiation causes reduction to a lesser extent (16/9.5 = 1.7) than in that of the cells with initially less reduced photoacceptor (spectrum B). The intensity ratio in this case is 2.5/1 = 2.5.

Figure 46.3 also presents two action spectra, one for the stimulation of the DNA synthesis (D) and the other for the stimulation of the attachment of the cells to a glass matrix (E). Recall that under ideal conditions, the action spectrum should mimic the absorption spectrum of the light-absorbing molecule whose photochemical alteration causes the effect (Hartman 1983; Lipson 1995).

The two action spectra presented in Figure 46.3D and E are characterized by four bands whose peak positions are situated close to each other, namely, 624 and 618, 672 and 668, and 767 and 751 nm, respectively (Table 46.1). There is a significant

difference in the peak positions of spectra D and E at wavelengths over 800 nm (813 and 831 nm, Table 46.1). However, the peak positions at 813 and 831 nm are resolved by deconvolution in absorption spectra A, A_1, and C, C_1 (Table 46.1). Comparison between the absorption and action spectra presented in Figure 46.3 shows evidence that all the bands present in action spectra D and E are present in the absorption spectra as well (Table 46.1). There are more peaks resolved by the Lorentzian fitting method in the absorption spectra than in the action spectra. This controversy can be explained by the definition of the action spectrum that mirrors the absorption spectrum of the primary photoacceptor. This is an advantage and a specificity of the action-spectrum spectroscopy as compared to other types of spectroscopy. It is well known that the transient species of the cytochrome *c* oxidase turnover are extremely difficult to confidently identify by optical means in physiological conditions. The primary photoacceptor is believed to be one of the turnover intermediates of cytochrome *c* oxidase that has not as yet been identified (Karu et al. 1998, 2005a).

The I_{760}/I_{665} intensity ratio is 2.4 for spectrum D and 0.74 for spectrum E in Figure 46.3, which means that the redox state of the photoacceptor molecule differs between these two spectra, it being more reduced in spectrum D. As far as the I_{760}/I_{665} intensity ratio is concerned, spectrum D is close to absorption spectrum B_1. The two photoresponses whose action spectra are presented in Figure 46.3D and E belong to reactions occurring in different parts of the cell, namely, in the nucleus and in the plasma membrane, respectively. It means that the cellular signaling cascades from the photoacceptor (Karu 1988, 2008) can differ as well. It cannot also be ruled out that it is different intermediates of the cytochrome *c* oxidase turnover that play the role of the photoacceptor for these two cellular responses. Redox-absorbance changes after irradiation at 632.8 nm were also measured in *E. coli* cells (Dube et al. 1997).

Changes in the absorption of HeLa cells were accompanied by conformational changes in the molecule of cytochrome *c* oxidase [measured by circular dichroism (CD) spectra; Kolyakov et al. 2001; Karu et al. 2001b]. Distinct maxima in CD spectra (the spectra were recorded from 250 to 780 nm) of control cells were found at 566, 634, 680, 712, and 741 nm. After irradiation at 820 nm, the most remarkable changes in peak positions as well as in CD signals were recorded in the range 750 to 770 nm—an appearance of a new peak at 767 nm and its shift to 757 nm after the second irradiation. Also, the peaks at 712 and 741 nm disappeared, and a new peak at 601 nm appeared. It was suggested that the changes in degree of oxidation of the chromophores of cytochrome *c* oxidase caused by the irradiation were accompanied by conformational changes in their vicinity. It was further suggested that these changes occurred in the environment of Cu_B (Karu et al. 2001b). It is known that even small structural changes in the binuclear site of cytochrome *c* oxidase control both rates of the dioxygen reduction and rates of internal electron- and proton-transfer reactions (Chance and Hess 1959). Our suggestion that cytochrome *c* oxidase is the photoacceptor responsible for various cellular responses connected with light

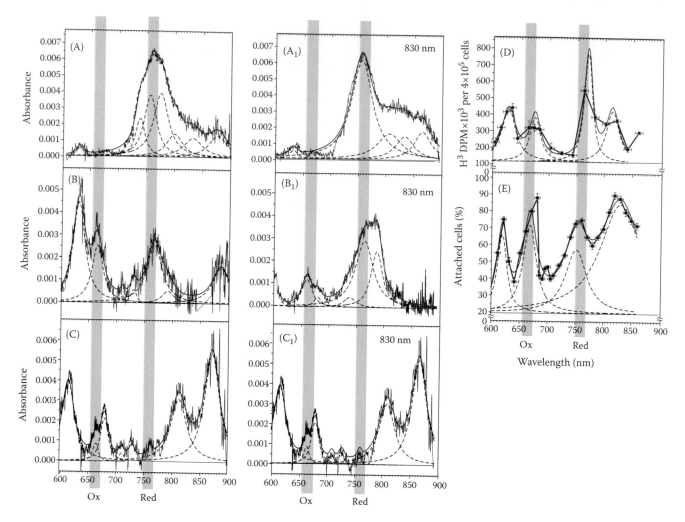

FIGURE 46.3 Absorption spectra of HeLa cell monolayer: (A–C) prior to and (A₁–C₁) after irradiation at 830 nm. A, A₁: enclosed cuvette, B, B₁: open cuvette, C, C₁: air-dry monolayer. Original spectrum, curve fitting (–) and Lorentzian fitting (– – –) are shown. (Adapted from Karu, T. I. et al., *J Photochem Photobiol B: Biol* 81:98–106, 2005.) Action spectra for (D) stimulation of DNA synthesis and (E) stimulation of cell adhesion to a glass matrix, measured respectively 1.5 h after irradiation of HeLa cell monolayer ($D = 100$ J/m², $t = 10$ s, $I = 10$ W/m²) and 30 min after irradiation of HeLa cell suspension ($D = 52$ J/m², $t = 40$ s, $I = 1.3$ W/m²). Experimental curves (-<-<-), curve fitting (–), and Lorentzian fitting (– – –) are shown as described in Karu and Kolyakov (2005). The gray lines mark the bands characteristic for relatively reduced photoacceptor near 770 nm and those for relatively oxidized photoacceptor near 675 nm (for details see Karu et al. 2005a).

therapy in the red-to-NIR region was later confirmed (Eells et al. 2003, 2004; Pastore et al. 2000; Wong-Riley et al. 2001, 2005).

Absorption measurements on cell monolayers after irradiation at 632.8 nm provided evidence that the shape of a dose dependence curve strongly depends on the initial redox state of irradiated cells (Karu et al. 2008b). The irradiation (three times at $\lambda = 632.8$ nm, dose $= 6.3 \times 10^3$ J/m², $\tau_{irrad.} = 10$ s, $\tau_{record.} = 600$ ms) of cells initially characterized by relatively oxidized cytochrome c oxidase caused first a reduction of the photoacceptor and then its oxidation (a bell-shaped curve) (Figure 46.4a). The irradiation by the same scheme of the cells with initially relatively reduced cytochrome c oxidase caused first oxidation and then a slight reduction of the enzyme (a curve opposite to the bell-shaped curve; Figure 46.4b). These experimental results demonstrate that the irradiation at 632.8 nm causes either a (transient) relative

reduction of the photoacceptor, putatively cytochrome c oxidase, or its (transient) relative oxidation depending on the initial redox status of the photoacceptor. The maximum in the bell-shaped dose dependence curve or the minimum of the reverse curve is the turning point between prevailing of oxidation or reduction processes. Our results show that bell-shaped dose dependences usually recorded for various cellular responses (for reviews, see Karu 1987, 1989, 1998) are characteristic also for redox changes in the photoacceptor, cytochrome c oxidase. Let us emphasize that another type of dose-response curve (Figure 46.4b) can be recorded very rear in studies of cellular responses (e.g., in special conditions of cultivation of cells).

The results of various studies (Karu et al. 1998, 2001a, 2005a) support the suggestion made earlier (Karu 1988) that the mechanism of low-power laser therapy at the cellular level is based on

TABLE 46.1 Peak Positions in Absorption and Action Spectra of HeLa Cells in Red-to-NIR Region as Resolved by Lorentzian Fitting

Absorption Spectra						Action Spectra		Characterization
A $R^2 = 0.99$	A_1 $R^2 = 0.98$	B $R^2 = 0.95$	B_1 $R^2 = 0.98$	C $R^2 = 0.95$	C_1 $R^2 = 0.95$	D DNA synthesis $R^2 = 0.97$	E Adhesion $R^2 = 0.91$	
–	–	–	–	616	616	–	618	Oxidized photoacceptor
(630)	(634)	633	–	–	–	624	–	Reduced photoacceptor
–	–	666	661	665	665	672	668	Gray line in Figure 46.3 — Oxidized photoacceptor
–	–	–	681	681	681	–	–	
–	–	(711)	–	(712)	(712)	–	–	Reduced photoacceptor
736	–	(730)	739	(730)	(730)	–	–	
754	756	767	765	(762)	(762)	767	751	Gray line in Figure 46.3
773	–	–	–	–	–	–	–	
797	–	(791)	788	–	–	–	–	
–	807	–	–	813	813	813	–	Oxidized photoacceptor
830	834	–	–	–	–	–	831	
874	867	880	–	872	872	Not measured	Not measured	

Source: Adapted from Karu, T. I. et al., *J Photochem Photobiol B: Biol* 81:98–106, 2005a.

Note: A, B, C absorption spectra before, and A_1, B_1, C_1 after irradiation at 830 nm. R^2: mean-square deviation of fitting. Weak bands are marked with brackets.

the increase in oxidative metabolism in mitochondria, which is caused by electronic excitation of components of the respiratory chain (in particular, cytochrome *c* oxidase). Our results also provide evidence that various wavelengths (670, 632.8, and 820 nm) can be used for increasing respiratory activity. The wavelengths that were used in these experiments (Karu et al. 1998, 2001a,b, 2005a) were chosen in accordance with the maxima in the action spectra (Figure 46.1). Note that 632.8 nm (He–Ne laser) and 820 nm (diode laser or LED) have been until now among the most common wavelengths used in therapeutic light sources (Tuner and Hode 2010).

It must be emphasized that when excitable cells (e.g., neurons, cardiomyocites) are irradiated with monochromatic visible light, photoacceptors are also believed to be the components of

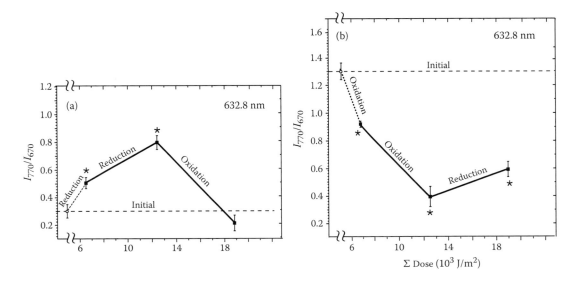

FIGURE 46.4 Dependences of the peak intensity ratio-criterion I770/I670 in absorption spectra of HeLa cell monolayer on the dose of radiation (λ = 632.8 nm). The asterisks indicate statistical significance ($p < 0.05$) from the initial values. The initial redox state of cells differs in two sets of experiments being relatively more oxidized by the same criterion (a) or relatively reduced by the same criterion (b). (Modified from Karu, T. I. et al., *Photomed Laser Surg* 26: 593–599, 2008.)

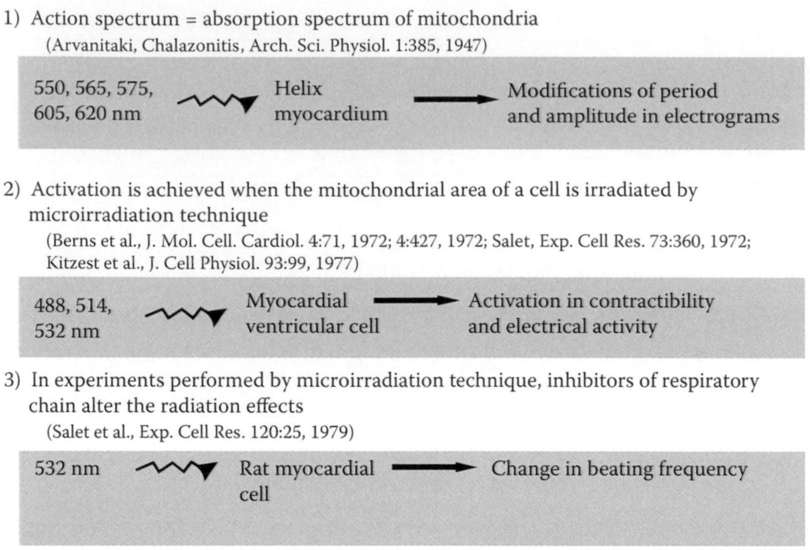

1) Action spectrum = absorption spectrum of mitochondria
   (Arvanitaki, Chalazonitis, Arch. Sci. Physiol. 1:385, 1947)

| 550, 565, 575, 605, 620 nm | ⟿ Helix myocardium | ⟶ Modifications of period and amplitude in electrograms |

2) Activation is achieved when the mitochondrial area of a cell is irradiated by microirradiation technique
   (Berns et al., J. Mol. Cell. Cardiol. 4:71, 1972; 4:427, 1972; Salet, Exp. Cell Res. 73:360, 1972; Kitzest et al., J. Cell Physiol. 93:99, 1977)

| 488, 514, 532 nm | ⟿ Myocardial ventricular cell | ⟶ Activation in contractibility and electrical activity |

3) In experiments performed by microirradiation technique, inhibitors of respiratory chain alter the radiation effects
   (Salet et al., Exp. Cell Res. 120:25, 1979)

| 532 nm | ⟿ Rat myocardial cell | ⟶ Change in beating frequency |

FIGURE 46.5 Experimental data obtained from irradiation of excitable cells and evidencing that photoacceptors are located in the mitochondria.

the respiratory chain. Since the publication in 1947 of a study by Arvanitaki and Chalazonitis (1947), it has been known that mitochondria of excitable cells have photosensitivity. Some of the experimental evidence concerning excitable cells is summarized briefly in Figure 46.5. These experiments were not performed in connection with light therapy. Experimental data (Berns et al. 1972; Berns and Salet 1972; Salet et al. 1979) made it clear that monochromatic visible radiation could cause (via absorption in mitochondria) physiological and morphological changes in nonpigmented excitable cells that do not contain specialized photoreceptors. Later, similar irradiation experiments were performed with neurons in connection with low-power laser therapy (Balaban et al. 1992). Clinical developments of these findings can be found in other publications (for a review, see Tuner and Hode 2010).

46.3.2 Primary Reactions after Light Absorption

Historically, the first mechanism, proposed in 1981 before recording of the action spectra, was the singlet-oxygen hypothesis (Karu et al. 1981). It is known that photoabsorbing molecules like porphyrins and flavoproteins (some respiratory-chain components belong to these classes of compounds) can be reversibly converted to photosensitizers (Giese 1980). Based on visible-laser-light action on RNA synthesis rates in HeLa cells and spectroscopic data for porphyrins and flavins, the hypothesis was put forward that the absorption of light quanta by these molecules was responsible for the generation of singlet oxygen $^1O_2$ and, therefore, for stimulation of the RNA synthesis rate (Karu et al. 1981) and the DNA synthesis rate (Karu et al. 1982). This possibility has been considered for some time as a predominant suppressive reaction when cells are irradiated at higher doses and

intensities (Karu 1989, 1998). This suggestion was made before the photoacceptor cytochrome *c* oxidase was found.

The primary mechanisms of light action after absorption of light quanta by cytochrome *c* oxidase and the promotion of electronically excited states have been studied only partly. The suggestions made to date are summarized in Figure 46.6; for simplicity, only singlet states (S_0 and S_1) are shown. However, triplet states are also involved.

Historically, the first mechanism proposed after understanding that cytochrome *c* oxidase in the photoacceptor was the redox properties alteration hypothesis in 1988 (Karu 1988).

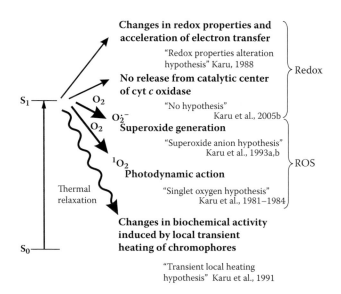

FIGURE 46.6 Possible primary reactions in the photoacceptor molecule (cytochrome *c* oxidase) after promotion of excited electronic states. ROS = reactive oxygen species.

Photoexcitation of certain chromophores in the cytochrome *c* oxidase molecule (like Cu_A and Cu_B or hemes *a* and a_3) influences the redox state of these centers, and consequently, the rate of electron flow in the molecule (Karu 1988, 1999).

The latest developments indicate that under physiological conditions, the activity of cytochrome *c* oxidase is also regulated by nitric oxide (NO) (Brown 1999). This regulation occurs via reversible inhibition of mitochondrial respiration. It was hypothesized (Karu et al. 2004a,b) that laser irradiation and activation of electron flow in the molecule of cytochrome *c* oxidase could reverse the partial inhibition of the catalytic center by NO and in this way increase the O_2-binding and respiration rate (NO hypothesis; Figure 46.6). This may be a factor in the increase in the concentration of the oxidized form of Cu_B (as can be integrated by results presented in Figure 46.6). Experimental results on the modification of irradiation effects with donors of NO did not exclude this hypothesis (Karu et al. 2004a, 2005b), as commented by Lane (2006). Note also that under pathological conditions, the concentration of NO is increased, mainly due to the activation of macrophages producing NO (Hothersall et al. 1997). This circumstance also increases the probability that the respiration activity of various cells will be inhibited by NO. Under these conditions, light activation of cell respiration may have a beneficial effect.

When electronic states of the photoabsorbing molecule are excited with light, a noticeable fraction of the excitation energy is inevitably converted to heat, which causes a local transient increase in the temperature of absorbing chromophores (transient local heating hypothesis; Figure 46.6) (Karu et al. 1991a). Any appreciable time- or space-averaged heating of the sample can be prevented by controlling the irradiation intensity and dose appropriately. The local transient rise in temperature of absorbing biomolecules may cause structural (e.g., conformational) changes and trigger biochemical activity (cellular signaling or secondary dark reactions) (Karu et al. 1991a).

In 1993, it was suggested (Karu et al. 1993a) that activation of the respiratory chain by irradiation would also increase production of superoxide anions (superoxide anion hypothesis; Figure 46.6). It has been shown that the production of $O_2^{\bullet-}$ depends primarily on the metabolic state of the mitochondria (Forman and Boveris 1982).

The belief that only one of the reactions discussed above occurs when a cell is irradiated and excited electronic states are produced is groundless. The question is, which mechanism is decisive? It is entirely possible that all the mechanisms discussed above lead to a similar result—a modulation of the redox state of the mitochondria (a shift in the direction of greater oxidation). However, depending on the light dose and intensity used, some of these mechanisms can prevail significantly. Experiments with *E. coli* provided evidence that, at different laser-light doses, different mechanisms were responsible—a photochemical one at low doses and a thermal one at higher doses (Karu et al. 1994).

46.3.3 Cellular Signaling (Secondary Reactions)

If photoacceptors are located in the mitochondria, how then are the primary reactions that occur under irradiation in the respiratory chain connected with DNA and RNA synthesis in the nucleus (the action spectra in Figure 46.1a through d) or with changes in the plasma membrane (Figure 46.1e)? The principal answer is that between these events are secondary (dark) reactions (cellular signaling cascades or photosignal transduction and amplification chain; Figure 46.7).

Figure 46.7 presents the latest version of a scheme of cellular signaling cascades (Karu 2008), which was proposed first in the paper of Karu (1988) to explain the increase in the DNA synthesis rate after the irradiation of HeLa cells with monochromatic visible light and later named as a mitochondrial signaling pathway (Karu et al. 2004a).

cDNA microarray technique has been used to analyze gene expression profiles in irradiated cells (McDaniel et al. 2010; Jaluria et al. 2007; Zhang et al. 2003). These experiments provided a more detailed answer on how the cellular signaling between mitochondria and the nucleus works. The gene expression profiles irradiated at 628 nm light evidenced that 111 genes of 10 categories were upregulated (Zhang et al. 2003). The activated genes were grouped into functional categories and were mostly genes that directly or indirectly play roles in the enhancement of cell proliferation and the suppression of apoptosis (Zhang et al. 2003). Gene expression in human skin fibroblasts was found to be altered after irradiation and it depended also on light intensity (McDaniel et al. 2010). What is more important is the result that two genes, siat7e and lama4, were found to be involved in the

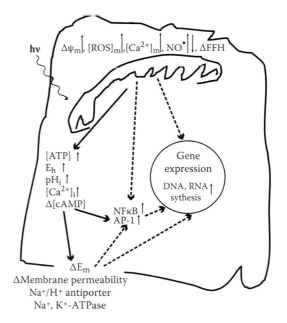

FIGURE 46.7 A schematic explaining putative mitochondrial retrograde signaling pathways after absorption visible and IR-A radiation (marked hv) by the photoacceptor, cytochrome *c* oxidase. Arrows ↑ and ↓ mark increase or decrease in the values, brackets [ ] mark concentration, ΔFFH: changes in mitochondrial fusion–fission homeostasis, AP-1: activator protein-1, and NF-κB: nuclear factor kappa B. Experimentally proved (→) and theoretically suggested (----➤) pathways are shown. (Adapted from Karu, T. I., *Photochem Photobiol* 84: 1091–1099, 2008.)

regulation of anchorage-dependent HeLa cell adhesion (Jaluria et al. 2007). Recall that HeLa cell attachment was a model used abundantly in our studies of modification of light effects with chemicals and these results are referred above (for a review, see Karu 2007).

Figure 46.7 suggests three regulation pathways. The first one is the control of the photoacceptor over the level of intracellular adenosine triphosphate (ATP). It is known that even small changes in ATP levels can significantly alter cellular metabolism (Brown 1992; Karu 2010a). However, in many cases, the regulative role of redox homeostasis has proved to be more important than that of ATP. For example, the susceptibility of cells to hypoxic injury depends more on the capacity of cells to maintain the redox homeostasis and less on their capacity to maintain the energy status (Chance and Hess 1959; Karu 2010b).

The second and third regulation pathways are mediated through the cellular redox state. This may involve redox-sensitive transcription factors (NF-κB and AP-1 in Figure 46.7) or cellular signaling homeostatic cascades from cytoplasm via cell membrane to nucleus (Figure 46.7). As a whole, the scheme in Figure 46.7 suggests a shift in overall cell redox potential in the direction of greater oxidation due to the irradiation. Details can be found in Karu (2008, 2010b).

It was suggested in the 1980s that activation of cellular metabolism by monochromatic visible light is a redox-regulated phenomenon (Karu 1988, 1989). Specificity of the light action is as follows: the radiation is absorbed by the components of the respiratory chain, and this is the starting point for redox regulation. The experimental data from following years have supported this suggestion. Let us recall here the results of the work of Alexandratou et al. (2002). They irradiated human fibroblasts at $\lambda = 647$ nm and observed, using confocal laser scanning microscopy at the single-cell level in real time, changes in the mitochondrial membrane potential ($\Delta\Psi$), intracellular pH (pH$_i$), intracellular calcium $\left(Ca_i^{2+}\right)$ alterations, and generation of reactive oxygen species (ROS) after the irradiation. After the laser irradiation, a gradual alkalization of pH$_i$ with its later normalization to the basal level occurred during 15 min. The maximal increase in mitochondrial membrane potential, $\Delta\Psi$ was observed 2 min after the 15 s irradiation and reached 30% of its basal value. $\Delta\Psi$ was back on the basal level approximately 4 min after irradiation. Recurrent spikes of Ca_i^{2+} were triggered and ROS were generated as a result of the irradiation (Alexandratou et al. 2002).

Dependencies of various biological responses (i.e., secondary reactions) on the irradiation dose, wavelength, pulsation mode, and intensity are available (for reviews, see Karu 1987, 1989, 1998, 2007). The main features are mentioned below. First, dose-biological response curves are usually bell-shaped, characterized by a threshold, a distinct maximum, and a decline phase like the curve in Figure 46.4a. However, at a very certain conditions (when the photoacceptor is rather reduced), the dose dependence curve can have the form shown in Figure 46.4b. Second, in most cases, the photobiological effects depend only on the radiation

dose and not on the radiation intensity and exposure time (the reciprocity rule holds true), but in other cases, the reciprocity rule proves invalid (the irradiation effects depend on light intensity). One example of dependence of the DNA synthesis rate on radiation intensity can be found in Figure 46.8.

Third, although the biological responses of various cells to the irradiation may be qualitatively similar, they may have essential quantitative differences. Fourth, the biological effects of irradiation depend on the wavelength (action spectra). The biological responses of the same cells to pulsed and continuous-wave (CW) light of the same wavelength, average intensity, and dose can vary (for a detailed review, see Karu 1998).

Figure 46.9 explains magnitudes of low-power laser effects as being dependent on the initial redox status of a cell. The main idea expressed in Figure 46.9 is that cellular response is weak or absent (the dashed arrows on the right side) when the overall redox potential of a cell is optimal or near optimal for the particular growth conditions. The cellular response is stronger when the redox potential of the target cell is initially shifted (the arrows on left side) to a more reduced state (and intracellular pH, pH$_i$, is lowered). This explains why the degrees of cellular responses can differ markedly in different experiments and why they are sometimes nonexistent. A jump in pH$_i$ due to irradiation has been measured experimentally [0.20 units in mammalian cells (Chopp et al. 1990) and 0.32 units in *E. coli* (Quickenden et al. 1995)].

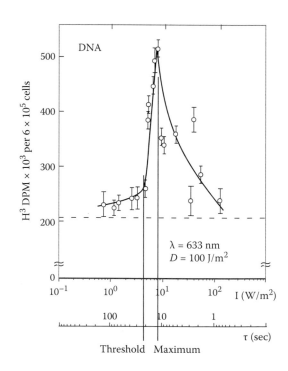

FIGURE 46.8 Dependence of stimulation of DNA synthesis rate on light intensity or irradiation time at a constant dose measured 1.5 h after irradiation of log-phase HeLa cells with a continuous-wave dye laser pumped by an argon laser ($\lambda = 633$ nm, $I_{max} = 80$ W/m²). Dashed line shows the control level. (Modified from Karu et al. 1984a,b.)

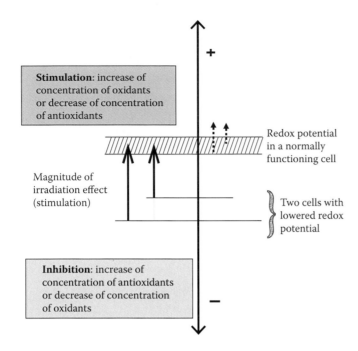

FIGURE 46.9 Schematic illustration of the action principle of monochromatic visible and NIR radiation on a cell. Irradiation shifts the cellular redox potential in a more oxidized direction. The magnitude of cellular response is determined by the cellular redox potential at the moment of irradiation.

Various magnitudes of low-power laser effects (strong effect, weak effect, or no effect at all) have always been one of the most criticized aspects of low-power laser therapy. An attempt was made to quantify the magnitude of irradiation effects as dependent on the metabolic status of *E. coli* cells (Tiphlova and Karu 1991b). The correlation was found between the amount of ATP in irradiated cells and the initial amount of ATP in control cells (Karu et al. 2001).

Thus, variations in the magnitude of low-power laser effects at the cellular level are explained by the overall redox state (and pH_i) at the moment of irradiation. Cells with a lowered pH_i (in which the redox state is shifted to the reduced side) respond stronger than cells with a normal or close-to-normal pH_i value.

46.4 Enhancement of Cellular Metabolism via Activation of Nonmitochondrial Photoacceptors: Indirect Activation/Suppression

The redox-regulation mechanism cannot occur solely via the respiratory chain (see Section 46.3). Redox chains containing molecules that absorb light in the visible spectral region are usually key structures that can regulate metabolic pathways. One such example is NADPH-oxidase of phagocytic cells, which is responsible for nonmitochondrial respiratory burst. This multicomponent enzyme system is a redox chain that generates ROS as a response to the microbicidal or other types of activation.

Irradiation with the He–Ne and semiconductor lasers and LEDs (for a review, see Karu 1998) can activate this chain. The features of radiation-induced nonmitochondrial respiratory burst, which was in our experiments quantitatively and qualitatively characterized by measurements of luminol-amplified chemiluminescence (CL) (Karu 1998), must be followed. First, nonmitochondrial respiratory burst can be induced both in homogeneous cell populations and cellular systems (blood, spleen cells, and bone marrow) by both CW and pulsed lasers and LEDs. Qualitatively, the kinetics of CL enhancement after irradiation is similar to that after treatment of cells with an object of phagocytosis, *Candida albicans*. Quantitatively, the intensity of CL induced by radiation is approximately one order of magnitude lower. This is true for He–Ne laser radiation (Karu et al. 1989) as well as for radiation of various pulsed LEDs (Karu et al. 1993a,b). Second, irradiation effects (stimulation or inhibition of CL) on phagocytic cells strongly depend on the health status of the host organism (Karu et al. 1993a, 1995, 1996a,b,c, 1997). This circumstance can be used for diagnostic purposes. Third, there are complex dependencies on irradiation parameters; irradiation can suppress or activate the nonmitochondrial respiratory burst (Karu et al. 1993b, 1995, 1996c, 1997). These problems have been reviewed in detail elsewhere (Karu 1998).

Finally, ROS, the burst of which is induced by direct irradiation of phagocytes, can activate or deactivate other cells that were not directly irradiated. In this way, indirect activation (or suppression) of metabolic pathways in nonirradiated cells occurs. Cooperative action among various cells via secondary messengers [ROS, lymphokines, cytokines (Funk et al. 1992), and NO (Naim et al. 1996)] requires much more attention when the mechanisms of low-power laser therapy are considered at the organism level.

This chapter did not consider systemic effects of low-power laser therapy at the organism level. The mechanisms of these effects have not yet been established. Perhaps NO plays a role as a secondary messenger for systemic effects of laser irradiation. A possible mechanism connected with the NO-cytochrome *c* oxidase complex was considered earlier (Section 46.3.2). In addition, mechanisms of the analgesic effects of laser radiation (Mrowiec et al. 1997) and systemic therapeutic effects that occur via blood irradiation (Vladimirov et al. 2000) could be connected with NO.

Recent studies have demonstrated that a number of nonphagocytic cell types, including fibroblasts, osteoblasts, endothelial cells, chondrocytes, kidney mesangial cells, and others, generate ROS (mainly superoxide anion) in low concentrations in response to stimuli (Sbarra and Strauss 1988). The function of this ROS production is not yet known. It is believed that an NADPH-oxidase (probably different from that in phagocytic cells) is present in nonphagocytic cells as well (Sbarra and Strauss 1988). To date, the effects of irradiation on this enzyme have not yet been studied.

Another example of important redox chains is NO-synthases, a group of redox-active P450-like flavocytochromes that are responsible for NO generation under physiological conditions

(Sharp and Shapman 1999). So far, the irradiation effects on these systems have not been investigated.

46.5 Conclusion

This chapter considered three principal ways of activating individual cells by monochromatic (laser) light. The photobiological action mechanism via activation of the respiratory chain is a universal mechanism. Primary photoacceptors are terminal oxidases (cytochrome *c* oxidase in eukaryotic cells, and, for example, cytochrome *bd* complex in the prokaryotic cell of *E. coli*) as well as NADH-dehydrogenase (for the blue-to-red spectral range). Primary reactions in or with a photoacceptor molecule lead to photobiological responses at the cellular level through cascades of biochemical homeostatic reactions (cellular signaling or photosignal transduction and amplification chain). Crucial events of this type of cellular metabolism activation occur due to a shift of cellular redox potential in the direction of greater oxidation. Metabolism activation via the respiratory chain occurs in all cells susceptible to light irradiation. Susceptibility to irradiation and capability for activation depend on the physiological status of irradiated cells; cells whose overall redox potential is shifted to a more reduced state (e.g., certain pathological conditions) are more sensitive to irradiation. The specificity of final photobiological response is determined not at the level of primary reactions in the respiratory chain but at the transcription level during cellular signaling cascades. In some cases, only partial activation of cell metabolism happens (e.g., priming of lymphocytes). All light-induced biological effects depend on the parameters of the irradiation (wavelength, dose, intensity, radiation time, CW or pulsed mode, and pulse parameters).

Second possibility to influence cellular metabolism by irradiation is the following. Other redox chains in cells except the respiratory chain can also be activated by irradiation. In phagocytic cells, the irradiation initiates a nonmitochondrial respiratory burst (production of ROS, especially superoxide anion) through activation of NADPH-oxidase located in the plasma membrane of these cells. The irradiation effects on phagocyting cells depend on the physiological status of the host organism as well as on radiation parameters.

Third, direct activation of cells can lead to the indirect activation of other cells. This occurs via secondary messengers released by directly activated cells: ROS produced by phagocytes, lymphokines and cytokines produced by various subpopulations of lymphocytes, or NO produced by macrophages or as a result of NO-hemoglobin photolysis of red blood cells.

Coherent properties of laser light are not manifested at the molecular level by light interaction with biotissue. The absorption of low-intensity laser light by biological systems is of a purely noncoherent (i.e., photobiological) nature. At the cellular level, biological responses are determined by absorption of light with photoacceptor molecules. Coherent properties of laser light are unimportant when the cellular monolayer, the thin layer of cell suspension, and the thin layer of tissue surface are irradiated. In these cases, the coherent and noncoherent light with the same wavelength, intensity, and dose provide the same biological response. Some additional (therapeutic) effects from coherent and polarized radiation can occur only in deeper layers of bulk tissue.

References

Alexandratou, E., D. Yova, P. Handris, D. Kletsas, and S. Loukas. 2002. Human fibroblast alterations induced by low power laser irradiation at the single cell level using confocal microscopy. *Photochem Photobiol Sci* 1:547–552.

Anders, J. J. 2009. The potential of light therapy for central nervous system injury and disease. *Photomed Laser Surg* 27:379–380.

Arvanitaki, A., and N. Chalazonitis. 1947. Reactiones bioelectriques a la photoactivation des cytochromes. *Arch Sci Physiol* 1:385–405.

Balaban, P., R. Esenaliev, T. Karu et al. 1992. He-Ne laser irradiation of single identified neurons. *Lasers Surg Med* 12:329–337.

Berns, M. W., D. C. L. Gross, W. K. Cheng, and D. Woodring. 1972. Argon laser microirradiation of mitochondria in rat myocardial cell in tissue culture. II. Correlation of morphology and function in single irradiated cells. *J Mol Cell Cardiol* 4:71–83.

Berns, M. W., and C. Salet. 1972. Laser microbeam for partial cell irradiation. *Int Rev Cytol* 33:131–155.

Brown, G. C. 1992. Control of respiration and ATP synthesis in mammalian mitochondria and cells. *Biochem J* 284:1–213.

Brown, G. C. 1999. Nitric oxide and mitochondrial respiration. *Biochem Biophys Acta* 1411:351–363.

Chance, B., and B. Hess. 1959. Spectroscopic evidence of metabolic control. *Science* 129:700–708.

Chopp, H., Q. Chen, M. O. Dereski, and F. W. Hetzel. 1990. Chronic metabolic measurement of normal brain tissue response to photodynamic therapy. *Photochem Photobiol* 52:1033–1038.

Dube, A., P. K. Gupta, and S. Bharti. 1997. Redox absorbance changes of the respiratory chain components of *E. coli* following He–Ne laser irradiation. *Lasers Life Sci* 7:173–178.

Eells, J. T., M. M. Henry, P. Summerfelt et al. 2003. Therapeutic photobiomodulation for methanol-induced retinal toxicity. *Proc Natl Acad Sci USA* 100:3439–3444.

Eells, J., M. T. Wong-Riley, J. VerHoeve et al. 2004. Mitochondrial signal introduction in accelerated wound and retinal healing by near-infrared light therapy. *Mitochondrion* 4:559–567.

Forman, N. J., and A. Boveris. 1982. Superoxide radical and hydrogen peroxide in mitochondria. In *Free Radicals in Biology*, Vol. 5. A. Pryor, editor. Academic Press, New York, 65–90.

Funk, J. O., A. Kruse, and H. Kirchner. 1992. Cytokine production in cultures of human peripheral blood mononuclear cells. *J Photochem Photobiol B: Biol* 16:347–355.

Gamaleya, N. F., E. D. Shishko, and G. B. Yanish. 1983. New data about mammalian cells photosensitivity and laser biostimulation. *Dokl Akad Nauk SSSR (Moscow)* 273:224–227.

Giese, A. C. 1980. Photosensitization of organisms with special reference to natural photosensitizers. In *Lasers in Biology and Medicine*. E. Hillenkampf, R. Pratesi, and C. Sacchi, editors. Plenum Press, New York, 299.

Hartman, K. M. 1983. Action spectroscopy. In *Biophysics*. W. Hoppe, W. Lohmann, H. Marke, and H. Ziegler, editors. Springer-Verlag, Heidelberg, 115–134.

Hothersall, J. S., F. Q. Cunha, G. H. Neild, and A. Norohna-Dutra. 1997. Induction of nitric oxide synthesis in J774 cell lowers intracellular glutathione: Effect of oxide modulated glutathione redox status on nitric oxide synthase induction. *Biochem J* 322:477–486.

Jaluria, P., M. Betenbaugh, K. Kontatopoulos, B. Frank, and J. Shiloah. 2007. Application of microarrays to identify and characterize genes involved in attachment dependence in He-La cells. *Metab Eng* 9:241–248.

Jöbsis-vander Vliet, F. F. 1999. Discovery of the near-infrared window in the body and the early development of near-infrared spectroscopy. *J Biomed Opt* 4:392–396.

Karu, T. I. 1987. Photobiological fundamentals of low-power laser therapy. *IEEE J Quantum Electron* 23:1703–1717.

Karu, T. I. 1988. Molecular mechanism of the therapeutic effect of low-intensity laser radiation. *Lasers Life Sci* 2:53–74.

Karu, T. I. 1989. Photobiology of low-power laser effects. *Health Phys* 56:691–704.

Karu, T. 1990. Effects of visible radiation on cultured cells. *Photochem Photobiol* 52:1089–1098.

Karu, T. 1998. *The Science of Low Power Laser Therapy*. Gordon & Breach, London.

Karu, T. 1999. Primary and secondary mechanisms of action of visible-to-near IR radiation on cells. *J Photochem Photobiol B Biol* 49:1–17.

Karu, T. I. 2003. Low power laser therapy. In *Biomedical Photonics Handbook*. T. Vo-Dinh, editor. CRC Press, Boca Raton, FL, 48-1–48-25.

Karu, T. 2007. *Ten Lectures on Basic Science of Laser Phototherapy*. Prima Books AB, Grängesberg.

Karu, T. I. 2008. Mitochondrial signaling in mammalian cells activated by red and near IR radiation. *Photochem Photobiol* 84:1091–1099.

Karu, T. I. 2010a. Mitochondrial mechanisms of photobiomodulation in context of new data about multiple roles of ATP. *Photomed Laser Surg* 28:159–160.

Karu, T. I. 2010b. Multiple roles of cytochrome c oxidase in mammalian cells under action of red and IR-A radiation. *IUBMB Life* 62:607–610.

Karu, T. I. 2011. Light coherence. Is this property important for photomedicine? Photobiological Sciences Online. K. C. Smith, editor. American Society for Photobiology. Retrieved from http://www.photobiology.info/Coherence.html.

Karu, T. I., and Afanasyeva, N. I. 1995. Cytochrome oxidase as primary photoacceptor for cultured cells in visible and near IR regions. *Dokl Akad Nauk (Moscow)* 342:693–695.

Karu, T. I., N. I. Afanasyeva, S. F. Kolyakov, and L. V. Pyatibrat. 1998. Changes in absorption spectra of monolayer of living cells after irradiation with low intensity laser light. *Dokl Akad Nauk (Moscow)* 360:267–270.

Karu, T. I., N. I. Afanasyeva, S. F. Kolyakov, L. V. Pyatibrat, and L. Welser. 2001a. Changes in absorbance of monolayer of living cells induced by laser radiation at 633, 670, and 820 nm. *IEEE J Sel Top Quantum Electron* 7:982–988.

Karu, T., T. Andreichuk, and T. Ryabykh. 1993a. Changes in oxidative metabolism of murine spleen following diode laser (660–950 nm) irradiation: Effect of cellular composition and radiation parameters. *Lasers Surg Medicine* 13:453–462.

Karu, T., T. Andreichuk, and T. Ryabykh. 1993b. Suppression of human blood chemiluminescence by diode laser radiation at wavelengths 660, 820, 880 or 950 nm. *Laser Therapy* 5:103–109.

Karu, T. I., T. N. Andreichuk, and T. P. Ryabykh. 1995. On the action of semiconductor laser radiation ($\lambda = 820$ nm) on the chemiluminescence of blood of clinically healthy humans. *Lasers Life Sci* 6:277–282.

Karu, T. I., G. S. Kalendo, and V. S. Letokhov. 1981. Control of RNA synthesis rate in tumor cells HeLa by action of low-intensity visible light of copper laser. *Lett Nuov Cim* 32:55–59.

Karu, T. I., G. S. Kalendo, V. S. Letokhov, and V. V. Lobko. 1982. Biostimulation of HeLa cells by low intensity visible light. *Nuov Cim D* 1:828–840.

Karu, T. I., and S. F. Kolyakov. 2005. Exact action spectra for cellular responses relevant to phototherapy. *Photomed Laser Surg* 23:355–361.

Karu, T. I., S. F. Kolyakov, L. V. Pyatibrat, E. L. Mikhailov, and O. N. Kompanets. 2001b. Irradiation with a diode 820 nm induces changes in circular dichroism spectra (250–780 nm) of living cells. *IEEE J Sel Top Quantum Electron* 7:976–981.

Karu, T. I., L. V. Pyatibrat, and N. I. Afanasyeva. 2004a. A novel mitochondrial signaling pathway activated by visible-to-near infrared radiation. *Photochem Photobiol* 80:366–372.

Karu, T. I., L. V. Pyatibrat, and N. I. Afanasyeva. 2005b. Cellular effects of low power laser therapy can be mediated by nitric oxide. *Lasers Surg Med* 36:307–314.

Karu, T. I., L. V. Pyatibrat, and G. S. Kalendo. 2001. Studies into the action specifics of a pulsed GaAlAs laser ($\lambda = 820$ nm) on a cell culture. I. Reduction of the intracellular ATP concentration: Dependence on initial ATP amount. *Lasers Life Sci* 9:203–210.

Karu, T. I., L. V. Pyatibrat, and G. S. Kalendo. 2004b. Photobiological modulation of cell attachment via cytochrome c oxidase. *Photochem Photobiol Sci* 3:211–216.

Karu, T. I., L. V. Pyatibrat, S. F. Kolyakov, and N. I. Afanasyeva. 2005a. Absorption measurements of a cell monolayer relevant to phototherapy: Reduction of cytochrome c oxidase under near IR radiation. *J Photochem Photobiol B: Biol* 81:98–106.

Karu, T. I., L. V. Pyatibrat, S. Kolyakov, and N. I. Afanasyeva. 2008b. Absorption measurements of cell monolayers relevant to mechanisms of laser phototherapy: Reduction or oxidation of cytochrome c oxidase under laser radiation at 632.8 nm. *Photomed Laser Surg* 26:593–599.

Karu, T. I., L. V. Pyatibrat, S. V. Moskvin, S. Andreev, and V. S. Letokhov. 2008a. Elementary processes in cells after light absorption do not depend on the degree of polarization: Implications for the mechanisms of laser phototherapy. *Photomed Laser Surg* 26:77–82.

Karu, T. I., T. P. Ryabykh, T. A. Sidorova, and Ya. V. Dobrynin. 1996a. The use of chemiluminescence test to evaluate the sensitivity of blast cells in patients with hemoblastoses to antitumor agents and low-intensity laser radiation. *Lasers Life Sci* 7:1–10.

Karu, T. I., T. P. Ryabykh, and S. N. Antonov. 1996b. Different sensitivity of cells from tumor-bearing organisms to countinuous-wave and pulsed laser radiation (λ = 632.8 nm) evaluated by chemiluminescence test. I. Comparison of responses of murine splenocytes: Intact mice and mice with transplanted leukemia EL-4. *Lasers Life Sci* 7:91–98.

Karu, T. I., T. P. Ryabykh, and S. N. Antonov. 1996c. Different sensitivity of cells from tumor-bearing organisms to countinuous-wave and pulsed laser radiation (λ = 632.8 nm) evaluated by chemiluminescence test. II. Comparison of responses of human blood: Healthy persons and patients with colon cancer. *Lasers Life Sci* 7:99–106.

Karu, T. I., T. P. Ryabykh, and V. S. Letokhov. 1997. Different sensitivity of cells from tumour-bearing organisms to countinuous-wave and pulsed laser radiation (λ = 632.8 nm) evaluated by chemiluminescence test. III. Effect of dark period between pulses. *Lasers Life Sci* 7:141–156.

Karu, T. I., T. R. Ryabykh, G. E. Fedoseyeva, and N. I. Puchkova. 1989. Induced by He–Ne laser radiation respiratory burst on phagocytic cells. *Lasers Surg Med* 9:585–588.

Karu, T., O. Tiphlova, R. Esenaliev, and V. Letokhov. 1994. Two different mechanisms of low-intensity laser photobiological effects on *Escherichia coli*. *J Photochem Photobiol B Biol* 24:155–161.

Karu, T. I., O. A. Tiphlova, Yu. A. Matveyets et al. 1991. Comparison of the effects of visible femtosecond laser pulses and continuous wave laser radiation of low average intensity on the clonogenicity of *Escherichia coli*. *J Photochem Photobiol B Biol* 10:339–345.

Karu, T. I., S. F. Kolyakov, L. V. Pyatibrat, E. L. Mikhailov, and O. N. Kompanets. 2001. Irradiation with a diode 820 nm induces changes in circular dichroism spectra (250–780 nm) of living cells. *IEEE Journal of Selected Topics in Quantum Electronics* 7:976–981.

Lane, N. 2006. Power games. *Nature* 443:901–903.

Lipson, E. D. 1995. Action spectroscopy: Methodology. In *CRC Handbook of Organic Chemistry and Photobiology*. W. H. Horspool and P.-S. Song, editors. CRC Press, Boca Raton, FL, 1257.

McDaniel, D. H., R. A. Weiss, R. G. Geronemus et al. 2010. Varying rates of wavelengths in dual wavelength LED photomodulation alters gene expression profiles in human skin fibroblasts. *Laser Surg Med* 42:540–546.

Mrowiec, J., A. Sieron, A. Plech, G. Cieslar, T. Biniszkiewicz, and R. Brus. 1997. Analgesic effect of low-power infrared laser radiation in rats. *Proc. SPIE* 3198:83–87.

Naim, J. O., W. Yu, K. M. L. Ippolito, M. Gowan, and R. J. Lanzafame. 1996. The effect of low level laser irradiation on nitric oxide production by mouse macrophages. *Lasers Surg Med Suppl* 8:7.

Pal, G., A. Dutta, K. Mitra et al. 2007. Effect of low intensity laser interaction with human skin fibroblast cells using fiber-optic nano-probes. *J Photochem Photobiol B* 86:252–261.

Pastore, D., M. Greco, and S. Passarella. 2000. Specific helium-neon laser sensitivity of the purified cytochrome c oxidase. *Int J Rad Biol* 76:863–870.

Qadri, T., P. Bohdanecka, J. Tunér et al. 2007. The importance of coherence length in laser phototherapy of gingival inflammation—A pilot study. *Lasers Med Sci* 22:245–251.

Quickenden, T. R., L. L. Daniels, and L. T. Byrne. 1995. Does low-intensity He–Ne radiation affect the intracellular pH of intact *E. coli*? *Proc SPIE* 2391:535–538.

Salet, C. 1971. Acceleration par micro-irradiation laser du rhythme de contraction de cellular cardiaques en culture. *CR Acad Sci Paris* 272:2584–2592.

Salet, C, G. Moreno, and F. Vinzens. 1979. A study of beating frequency of a single myocardial cell. III. Laser microirradiation of mitochondria in the presence of KCN or ATP. *Exp Cell Res* 120:25–32.

Sbarra, A. J. and R. R. Strauss, editors. 1988. *The Respiratory Burst and its Photobiological Significance*. Plenum Press, New York.

Sharp, R. E. and S. K. Chapman. 1999. Mechanisms for regulating electron transfer in multicentre redox proteins. *Biochem Biophys Acta* 1432:143–151.

Tiphlova, O., and T. Karu. 1991a. Action of low-intensity laser radiation on *Escherichia coli*. *CritRev Biomed Eng* 18:387–412.

Tiphlova, O., and T. Karu. 1991b. Dependence of *Escherichia coli* growth rate on irradiation with He–Ne laser and growth substrates. *Lasers Life Sci* 4:161–166.

Tuner, J., and L. Hode. 2010. *New Laser Therapy Handbook*. Prima Books, Grängesberg.

Vekshin, N. A. 1991. Light-dependent ATP synthesis in mitochondria. *Mol Biol (Moscow)* 25:54–58.

Vladimirov, Y., G. Borisenko, N. Boriskina, K. Kazarinov, and A. Osipov. 2000. NO-hemoglobin may be a light-sensitive source of nitric oxide both in solution and in red blood cells. *J Photochem Photobiol B Biol* 59:115–121.

Wong-Riley, M. T., X. Bai, E. Buchman, and H. T. Whelan. 2001. Light-emitting diode treatment reverses the effect of TTX on cytochrome c oxidase in neurons. *Neuroreport* 12:3033–3037.

Wong-Riley, M. T., H. L. Liang, J. T. Eells et al. 2005. Photobiomodulation directly benefits primary neurons functionally inactivated by toxins: Role of cytochrome c oxidase. *J Biol Chem* 280:4761–4771.

Wu, X., A. E. Dmitriev, M. J. Cardoso et al. 2009. 810-nm wavelength light: an effective therapy for transected or contused rat spinal cord. *Lasers Surg Med* 41:36–41.

Zhang, Y., S. Song, C.-C. Fong et al. 2003. cDNA microarray analysis of gene expression profiles in human fibroblast cells irradiated with red light. *J Invest Dermatol* 120:849–857.

47

Low-Level Laser Therapy Signaling Pathways

Da Xing
South China Normal University

Shengnan Wu
South China Normal University

47.1 Background

After the invention of the laser in 1960, its medical applications interested Hungarian physician Endre Mester. He used a low-level red (632.8 nm) He–Ne laser to irradiate animals and, eventually, people. His early work showed that wound healing was accelerated by laser irradiation (Mester et al. 1968, 1971), from which interest in low-level laser therapy (LLLT), also known as low-power laser therapy (LPLT), biostimulation, low-energy laser irradiation (LELI), cold laser, and photobiomodulation of both humans and animals spread from Europe to most of the world. These methods were applied to numerous ailments (Tata and Waynant 2011). Since 1960, more and more researchers have been involved with LLLT, which has become one of the many techniques utilized in modern phototherapy (Tata and Waynant 2011). Specialists and general practitioners use LLLT to treat a broad range of conditions, but the most common indications are local pain relief (Gur et al. 2004), anti-inflammation (Antunes et al. 2007), and wound healing (Posten et al. 2005). Initially, how LLLT worked for the treatment of different conditions was nothing but a mystery, and without information about its mechanism of action, efficient manipulation of LLLT was difficult.

The most important question about LLLT is no longer whether light has any biological effects, but rather how it works at cellular and organism levels and what treatment protocols are optimal for each specific indication (Karu 2003). LLLT's potential mechanisms of action have been uncovered by numerous research teams that contributed to the field of laser therapy. One of the most influential in elucidating potential mechanisms of LLLT treatment are Tiina Karu and colleagues whose over 30 years of research, over 164 scientific journals, and four books (Karu 1989, 1998, 2007) advance our understanding, particularly of the primary and secondary mechanisms of action of red to near-infrared (NIR, 760–1440 nm) light on cells (Tata and Waynant 2011). Scientists have mostly studied various cellular signaling pathways that appear induced by LLLT at red to NIR range Orna Halevy (Shefer et al. 2001, 2003), Mitsuhiro Ohshima (Miyata et al. 2006), and Da Xing (Gao et al. 2009; Feng, Zhang, and Xing 2012) have furthered understanding of pivotal mitogen-activated protein kinase (MAPK)/extracellular signal-regulated protein kinase (ERK) pathways in LLLT-induced skeletal muscle regeneration and wound healing. Hsin-Su Yu et al. have shown the significance of MAPK/c-Jun NH$_2$-terminal kinase (JNK) pathways in LLLT-induced melanoma cell proliferation, which opened new options for improving for vitiligo treatment (Yu et al. 2003; Hu et al. 2007). LLLT effects in some cells can be dependent on phosphoinositide 3-kinase (PI3K)/protein kinase B (Akt) pathways that were discovered by Da Xing (Zhang et al. 2009) and Shan-hui Hsu (Chen, Hung, and Hsu 2008). Timon Cheng-Yi Liu (Duan et al. 2001) and Da Xing (Gao et al. 2006) studied the effects of LLLT on the protein kinase C (PKC) pathway, which has a role in increased cell viability and neutrophils' respiratory burst. Da Xing also reported findings on the activation of Src kinase following LLLT treatment (Zhang, Xing, and

Gao 2008). Eijiro Jimi focused on bone morphogenetic protein (BMP)/Smad pathways [the latter transduce extracellular signals to the nucleus where they activate transforming growth factor beta (TGF-β)], which are involved in LLLT-induced bone formation (Hirata et al. 2010). Margherita Greco (Yamamoto et al. 2001), Orna Halevy (Ben-Dov et al. 1999), Chen (Zhang et al. 2011), Lilach Gavish (Gavish et al. 2004), Yoshimitsu Abiko (Yamamoto et al. 2001), and Tetsuro Takamatsu (Taniguchi et al. 2009) studied the regulation of cell cycle proteins during LLLT-induced cell proliferation and differentiation. Da Xing (Zhang et al. 2008; Zhang, Zhang, and Xing 2010; Zhang, Wu, and Xing 2012), Orna Halevy (Shefer et al. 2002), Mason C. P. Leung (Yip et al. 2011), Alessandro Giuliani (Giuliani et al. 2009), and OkJoon Kim (Lim et al. 2009) have studied the antiapoptotic pathway induced by LLLT for promoting the future potential use in neuroprotection, or slowing the loss of functional neurons. Da Xing also investigated high-fluence LLLT-triggered proapoptotic signal cascades, but this time for potential phototherapy modality for cancer (Wang et al. 2005a; Chu, Wu, and Xing 2010; Sun, Wu, and Xing 2010; Huang, Wu, and Xing 2011; Wu et al. 2007, 2009, 2011.)

The purpose of the current work is to review the available literature on molecular signaling pathways involved in LLLT stimulation for the enhancement of cell proliferation, differentiation, and apoptosis.

47.2 Cell Proliferation and Differentiation by LLLT

Cell proliferation is one of the most important biostimulatory effects of LLLT that provides wide areas of application in clinical practice. Following LLLT irradiation, increased cell proliferation has been observed in many cell types in vitro, including myoblasts (Ben-Dov et al. 1999; Shefer et al. 2001, 2003; Zhang et al. 2011) for skeletal muscle regeneration, fibroblasts (Miyata et al. 2006; Taniguchi et al. 2009; Zungu, Hawkins, and Abrahamse 2009; Chen et al. 2011), keratinocytes (Yu et al. 1996; Gavish et al. 2004), and endothelial cells for regeneration of skin, veins, and arteries (Chen, Hung, and Hsu 2008; Feng, Zhang, and Xing 2012) and also to accelerate the process of wound healing, melanocytes (Yu et al. 2003; Hu et al. 2007), and for treatment of vitiligo, a condition of loss of the pigment of the skin, hair, or eyes, human osteoblasts (Yamamoto et al. 2001; Hirata et al. 2010; Kiyosaki et al. 2010; Saygun et al. 2012) for bone formation, Schwann cells (SCs), primary neural cells, pheochromocytoma cells, human neuroblastoma cells, human-derived glioblastoma cells (Zhang et al. 2008; Giuliani et al. 2009; Lim et al. 2009; Murayama et al. 2012; Yazdani et al. 2012; Zhang, Wu, and Xing 2012) for neuroprotection, mesenchymal stem cells (MSCs) (Peng et al. 2012) for regenerative medicine, African green monkey fibroblast cells (COS-7), human lung adenocarcinoma cells (ASTC-a-1), and HeLa cells are slowly growing epithelial adenocarcinoma commonly used for in vitro tests of human cell lines (Gao et al. 2006, 2009; Zhang, Xing, and Gao 2008; Zhang et al. 2009) have also increased cell proliferation in response to LLLT. Nevertheless, cellular proliferation mechanisms induced by LLLT are not well understood. Various mechanisms for the mitogenic effects of LLLT have been proposed, such as ligand-free dimerization and activation of specific receptors that are in the correct energetic state to accept the laser energy (Shefer et al. 2001). These mechanisms lead to autophosphorylation and downstream effects, as well as activation of calcium (Ca^{2+}) channels, resulting in increased intracellular Ca^{2+} concentration and cell proliferation (Gao and Xing 2009). Red to NIR light is believed to be absorbed by cytochrome c oxidase (CcO), one of the mitochondrial respiratory chain components, resulting in the increase in reactive oxygen species (ROS), adenosine triphosphate (ATP), and cyclic adenosine monophosphate (cAMP), as well as initiating a signaling cascade (via proteins such as cytokines) that promotes cellular proliferation and cytoprotection. Light irradiation can also alter cellular homeostasis parameters (such as pH), cell redox state, and expression of redox-sensitive factors (such as NF-κB) (Gao et al. 2009).

47.2.1 Increased Mitochondrial Activity by LLLT

CcO is the terminal enzyme of the respiratory chain in eukaryotic cells, mediating the transfer of electrons from cytochrome c to molecular oxygen. CcO is the most understood photoacceptor and photosignal transducer in the red to NIR range of light (Karu 1999; Pastore, Greco, and Passarella 2000; Eells et al. 2004; Karu, Pyatibrat, and Kalendo 2004; Karu et al. 2005; Liang et al. 2006), and recently, Tiina I. Karu provided a critical review to highlight the role of CcO as an initial photoacceptor of LLLT and ATP as a crucial signaling molecule in biomodulation by LLLT (Karu 2010) (Figure 47.1). Many researchers in past decades have successfully documented the extracellular synthesis of ATP or ATP extrasynthesis in isolated mitochondria and various intact cells under monochromatic light of distinct wavelengths (Karu, Pyatibrat, and Kalendo 1995; Karu 2007). Using ATP as an intercellular signaling molecule provides critical understanding of the complicated mechanisms of photobiomodulation. CcO absorption of light can increase mitochondrial transmembrane potential ($\Delta\Psi m$), ATP, cAMP, and ROS, leading to increased energy availability of the cell and modulation of signal transduction (Karu, Pyatibrat, and Kalendo 1995; Karu, Pyatibrat, and Afanasyeva 2005; Hu et al. 2007; Tafur and Mills 2008). These biochemical and cellular changes in turn lead to macroscopic effects such as increased cell proliferation or accelerated wound healing (Eells et al. 2004; Maiya, Kumar, and Rao 2005; Hu et al. 2007).

Increased mitochondrial energy signals by LLLT were detected in many wavelengths and cell types. Greco et al. (2001) reported that 632.8 nm He–Ne laser irradiation at 12 mW/cm² at 0.24 J/cm² fluence increased the $\Delta\Psi m$ in isolated hepatocytes and the uptake of Ca^{2+} by mitochondria. The uptake of Ca^{2+} in mitochondrial membranes stimulates metabolic energy by generating ATP (Greco et al. 2001). The increase in Ca^{2+} level by LLLT was assumed to induce expression of c-Fos because

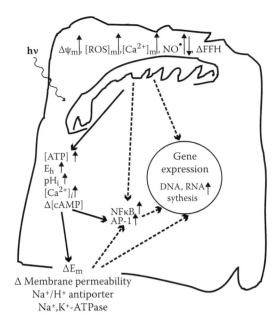

FIGURE 47.1 A schematic explanation of the putative mitochondrial retrograde signaling pathways after absorption of visible and IR-A radiation (marked as hν) by the photoacceptor, CcO. Arrows $\uparrow$ and $\downarrow$ indicate an increase or decrease in value, brackets [ ] mark concentration, ΔFFH indicates changes in mitochondrial fusion–fission homeostasis, and experimentally proved ($\rightarrow$) and theoretically suggested ($\rightarrow$) pathways are shown. (Adapted from Karu, T. I., *IUBMB Life* 62: 607–610, 2010.)

c-Fos upregulation was totally abolished in the absence of Ca^{2+} (Greco et al. 2001). It is reported that c-Fos can dimerize with c-Jun to form the activator protein 1 (AP-1) transcription factor, which upregulates the transcription of various genes involved in cell proliferation and differentiation (Ameyar, Wisniewska, and Weitzman 2003; Hess, Angel, and Schorpp-Kistner 2004). Thus, these findings demonstrate the connection between components of mitochondrial energy signaling and secondary cellular gene expression upon LLLT irradiation. Gavish et al. (2004) demonstrated that HaCaT human keratinocytes irradiated with 780 nm Ti–Sa laser (1.85 mW/cm^2 at 2 J/cm^2) increased $\Delta\Psi$m immediately as indicated by MitoTracker, a fluorescent dye that distributes itself across the mitochondrial inner membrane according to $\Delta\Psi$m. Yu et al. (2003) and Hu et al. (2007) reported that in A2058 human melanoma cells, an immediate increase in $\Delta\Psi$m, ATP, and cAMP via enhanced CcO activity could be observed during cell proliferation induced by 632.8 nm He–Ne laser irradiation (1.0 and 2.0 J/cm^2). Silveira et al. (2009) evaluated the effect of 904 nm As–Ga laser (5 J/cm^2) on mitochondrial respiratory chain complexes I, II, III, and IV (CcO), as well as on succinate dehydrogenase activities after traumatic muscular injury. Compared with the sham group of injured muscle without LLLT treatment, the four complexes of succinate dehydrogenase significantly increased activities after LLLT. Increased ATP synthesis may thus be involved in the muscle healing process accelerated by LLLT (Silveira et al. 2009). Zungu, Hawkins, and Abrahamse (2009) demonstrated that 632.8 nm He–Ne laser irradiation at 3 mW/cm^2 at 5 J/cm^2 fluence increased intracellular

Ca^{2+}, which then increased $\Delta\Psi$m, ATP, and cAMP in human skin fibroblastoid cells. Ultimately, the irradiation caused photobiomodulation to restore homeostasis of injured cells (Zungu, Hawkins, and Abrahamse 2009). Dias et al. (2011) reported that 780 nm Ga–Al–As laser (125 mW/cm^2 at 20 J/cm^2) stimulated the oxidative metabolism and expression of matrix metalloproteinase (MMP) in the masseter muscle, which may indicate a matrix remodeling process. Chen et al. (2011) observed intracellular ROS production and significant activation of NF-κB in primary murine embryonic fibroblasts (MEFs) following an 810 nm diode laser irradiation at the fluence of 0.3, 3, and 30 J/cm^2. As antioxidants inhibit NF-κB activation after LLLT, ROS may play an important role in laser-induced NF-κB signaling pathways through production of ROS (Chen et al. 2011). Moreover, LLLT likewise induced increased cellular ATP levels, indicating upregulation of mitochondrial respiration (Chen et al. 2011), and the expression of antiapoptosis and prosurvival genes responsive to NF-κB may explain various clinical effects of LLLT (Chen et al. 2011).

Cellular mechanisms involved in LLLT neuroprotection were reported to be associated with increased ATP levels. Transcranial LLLT improves behavioral disturbances following embolic strokes in embolized rabbits, and clinical rating scores in acute ischemic stroke patients. Lapchak and Taboada (2010) evaluated the effect of 808 nm LLLT on cortical ATP levels using rabbit small clot embolic stroke model. Cortical ATP content in the ischemic cortex decreased in embolized compared to that in healthy rabbit. This decrease was attenuated by LLLT at either continuous wavelength (CW) or pulse wavelength (PW), an important demonstration for insight into the molecular mechanisms associated with clinical improvement following LLLT (Lapchak and Taboada 2010). Sommer et al. (2012) showed that LLLT-induced clearance of amyloid-β-peptide 42 (Aβ_{42}) aggregates, a root cause of Alzheimer's disease (AD), occur at the expense of cellular ATP. When compared with those of nonirradiated cells, ATP levels significantly increased in Aβ_{42}-free human neuroblastoma cells irradiated with 670 nm In–Ga–Al–P diode laser (100 mW/cm^2 at 1 J/cm^2) (Sommer et al. 2012). More importantly, amounts of Aβ_{42} aggregates were significantly lower in irradiated cells than in nonirradiated cells (Sommer et al. 2012).

47.2.2 Signaling Pathways Involved in LLLT-Induced Cell Proliferation and Differentiation

47.2.2.1 MAPK Pathway

MAPK signal transduction pathways are among the most widespread mechanisms of eukaryotic cell regulation and osmosensing (Johnson and Lapadat 2002; Boutros, Chevet, and Metrakos 2008; Krishna and Narang 2008). All eukaryotic cells possess multiple MAPK pathways activated by a distinct set of stimuli, allowing cells to respond accordingly to multiple and divergent inputs. These stimuli may act through different receptor families coupled with MAPK pathways, such as receptor tyrosine kinases (RTKs), G protein-coupled receptors (GPCRs), cytokine receptors, and Ser/Thr kinase receptors. Activation of MAPK pathways coordinate diverse cellular activities, including gene expression, cell cycle machinery, cellular metabolism, motility, survival, apoptosis, and differentiation. In mammals, six distinct groups of MAPKs have been characterized: ERK1/2, JNK1/2/3, p38 (p38 a/b/g/d), ERK7/8, ERK3/4, and ERK5, the first three of which are the most studied. Several of the MAPK proteins have splice variants that increase the cascade diversity. The ERK pathway seems to play an important role in hormone and growth factor–induced cell proliferation, cell differentiation, and cellular transformation. The stress-activated protein kinase (SAPK)/JNK and p38 MAPK can be activated by stress signals such as high osmolarity, heat shock, ultraviolet (UV) irradiation, and proinflammatory cytokines [tumor necrosis factor (TNF-α)] (Boutros, Chevet, and Metrakos 2008; Johnson and Lapadat 2002; Krishna and Narang 2008).

47.2.2.1.1 ERK Pathway

Shefer et al. (2001) firstly reported that LLLT (He–Ne laser, 632.8 nm, 1.06 J/cm²) could induce the phosphorylation of c-Met [receptor of hepatocyte growth factor (HGF)] (Figure 47.2). The c-Met is a kind of RTKs, previously shown to activate the ERK pathway and promote proliferation of i28 mouse skeletal muscle myoblasts (Shefer et al. 2001). However, LLLT could not induce the phosphorylation of the TNF receptor, which activates the p38 MAPK and JNK pathways (Shefer et al. 2001) (Figure 47.2). Fifteen minutes after LLLT stimulation, phospho-specific antibodies detected the activation of ERK, but not JNK and p38 MAPK kinases (Shefer et al. 2001) (Figure 47.3). ERK pathway has biphasic regulation and has a dual function during myogenesis. A decline in ERK activity is concomitant with permanent withdrawal from the cell cycle. Stimulation increases

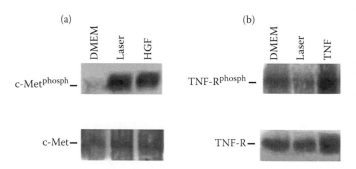

FIGURE 47.2 Differential receptor phosphorylation by LLLT. (a) Starved i28 cells were either irradiated for 3 s or treated with recombinant human HGF (100 units/mL). At 10 min postirradiation, cells were lysed and then immunoprecipitated with anti-c-Met, followed by immunodetection using antiphosphotyrosine. (b) Cells were irradiated as in (a) or exposed to TNF-α (50 ng/mL) for 10 min. Cell lysates were immunoprecipitated with anti-TNF-R followed by an immunoblot with antiphosphoserine. Parallel samples were immunoblotted with anti-c-Met or anti-TNF-R as a gel loading control. (Adapted from Shefer, G. et al., *J Cell Physiol* 187: 73–80, 2001.)

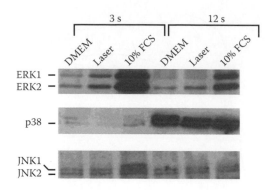

FIGURE 47.3 Effect of LLLT on signal transduction pathways at various times. Starved i28 cells were irradiated for 3 and 12 s or refed with DMEM-10% FCS. Western blot analysis was performed with antibodies to the phosphorylated forms of ERK1/2, JNK, and p38. (Adapted from Shefer, G. et al., *J Cell Physiol* 187: 73–80, 2001.)

transcription in myoblasts and inhibits myoblast differentiation at early stages, yet is induced at later stages of differentiation, causing myotubes to elongate, fuse, and survive, suggesting a dual role during myogenesis (Bennet and Tonks 1997; Gredinger et al. 1998). In contrast, the p38 MAPK pathway is specifically involved in muscle cell differentiation to form myotubes (Cuenda and Cohen 1999; Zetser, Gredinger, and Bengal 1999; Puri et al. 2000). There is a biphasic dose response to LLLT exemplified by the work of Halevy and coworkers, where low levels of myosin heavy chain (MHC) protein expression (an indicator of myoblast differentiation) followed LLLT, and their former results, demonstrating that a single 3-s LLLT inhibits the differentiation of primary satellite cells yet stimulates proliferation (Ben-Dov et al. 1999). Therefore, the authors concluded that in activating ERK but not JNK and p38 MAPK, LLLT induces the activation and proliferation of quiescent satellite cells and delays their differentiation (Shefer et al. 2001). They suggest that the delay may be due to a balance between active ERK1/2 and inactive p38 MAPK. Studies in other cell types support the assumption that LLLT stimulates activation of ERK but not JNK and p38 MAPK. Miyata et al. (2006) reported that ERK1/2 was phosphorylated 5–30 min following LLLT (Ga–Al–As laser, 810 nm, 20.79 J/cm²); however, p38 MAPK or JNK phosphorylation did not occur in human dental pulp-derived fibroblast-like cells. Kiyosaki et al. (2010) reported that in mouse osteoblastic cells (MC3T3-E1), irradiation with 830 nm Ga–Al–As diode laser at 1.91 J/cm² fluence significantly increased the expression of insulin-like growth factor-I (IGF-I) and Runx2, as well as the phosphorylation of ERK, which all contributed to bone formation. These studies firmly indicate that LLLT does not work via a stress signal pathway, but rather through a mitogenic pathway that leads to cell proliferation.

Protein kinases form a substantial subset of ERK1/2 targets. Mnk1/2 are serine/threonine kinases with putative MAPK phosphorylation sites within the activation loops of their catalytic and conserved C-terminal ERK-interacting domains. Mnk1/2 serve as common substrates for growth factor-stimulated ERK1/2 and stress-activated p38, thereby possibly integrating signals from multiple cellular stimuli. Once activated, Mnk1/2 phosphorylate the eukaryotic initiation factor 4E (eIF4E) on Ser209 in vitro. As a result, protein-synthesizing ribosomes and additional protein synthesis initiation factors are recruited to mRNA (Pearson et al. 2001). In brief, three mechanisms mainly regulate the eIF4E protein (Gao and Xing 2009): regulation of eIF4E expression, eIF4E phosphorylation, and phosphorylation-dependent dissociation of the translational-repressor protein PHAS-I (phosphorylated heat- and acid-stable protein regulated by insulin 1) from eIF4E. The partial or nonphosphorylation form of PHAS-I inhibits the activity of eIF4E through strong interaction. Shefer et al. (2003) reported that 15 min after LLLT stimulation (He–Ne laser, 632.8 nm, 1.06 J/cm^2), the phosphorylation of eIF4E increased in i28 mouse skeletal muscle myoblasts. Moreover, LLLT may activate eIF4E by elevating the phosphorylation of PHAS-I as well as the expression of cyclin D1 (Shefer et al. 2003). In addition, the presence of the MEK inhibitor, PD98059, abolished the LLLT-induced ERK1/2 activation and eIF4E phosphorylation, as well as dramatically reduced the expression of cyclin D1 (Shefer et al. 2003), which is further evidence that the involvement of the ERK/eIF4E pathway was involved in LLLT-induced protein translation (Figure 47.4).

ERK is also important in angiogenesis, the formation of new blood vessels from preexisting ones, and represents an excellent therapeutic target for the treatment of wounds and cardiovascular disease. The first mammalian transcription factor to be cloned (Black, Black, and Azizkhan-Clifford 2001), specificity protein 1 (Sp1) is a member of the Sp and Krüppel-like factor (Sp/KLF) family and regulates the expression of thousands of genes involved in an array of cellular processes, such as cell growth, proliferation, and angiogenesis, in response to physiologic and pathological stimuli (Santiago et al. 2007; Li et al. 2011; Gong et al. 2012). Sp1 contains a prototypic Cys2/His2-type zinc finger motif, through which it bonds directly to the

GC box element of deoxyribonucleic acid (DNA) and activates or represses gene transcription (Briggs et al. 1986; Kadonaga and Tjian 1986; Kadonaga et al. 1987). One of the most important angiogenic switch molecules is vascular endothelial growth factor (VEGF) because it significantly impacts the behavior of endothelial cells, including their migration, proliferation, and differentiation (Leppanen et al. 2010). High GC-rich motifs in the proximal regions of VEGF promoter are regulated by Sp1 (Schafer et al. 2003). Feng, Zhang, and Xing (2012) reported that He–Ne laser (632.8 nm, 12.74 mW/cm^2 at 1.8 J/cm^2) activated the ERK/Sp1 pathway to promote VEGF expression and vascular endothelial cell proliferation. For the first time, LLLT was shown to enhance DNA-binding and Sp1 transactivation activity of VEGF promoter in human umbilical vein endothelial cells (HUVEC-CS) (Feng, Zhang, and Xing 2012). Moreover, Sp1-regulated transcription was in an ERK-dependent manner. The LLLT-activated ERK translocated from the cytoplasm to nucleus and led to increasing interaction with Sp1. This triggered phosphorylation of Sp1 on Thr453 and Thr739 (Figure 47.5), which resulted in the upregulation of VEGF expression (Feng, Zhang, and Xing 2012). Furthermore, mithramycin-A or Sp1 short hairpin RNA (shRNA) suppressed the promotive effect of LLLT on cell cycle progression and proliferation. Similarly, this LLLT effect was also significantly abolished by the inhibition of ERK activity (Feng, Zhang, and Xing 2012). These findings highlight the importance of ERK/Sp1 pathway in angiogenesis, providing potential strategy for LLLT treatment where angiogenesis will have beneficiary effects (Figure 47.6).

EGF stimulation causes sequential activation of the small GTPase Ras protein, Raf-1, MAPK kinase (MEK1/2), and two ERK isoforms (ERK1/2) (Boutros, Chevet, and Metrakos 2008;

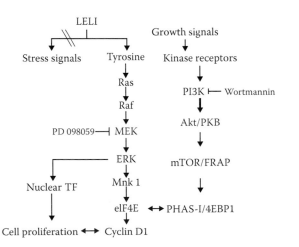

FIGURE 47.4 A proposed model for LLLT signaling in skeletal muscle satellite cells. (Adapted from Shefer, G. et al., *Biochim Biophys Acta* 1593: 131–139, 2003.)

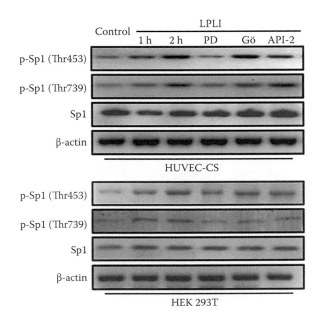

FIGURE 47.5 Representative Western blot assay for detecting the levels of Sp1 phosphorylation following different treatments. (Adapted from Feng, J. et al., *Cell Signal* 24: 1116–1125, 2012.)

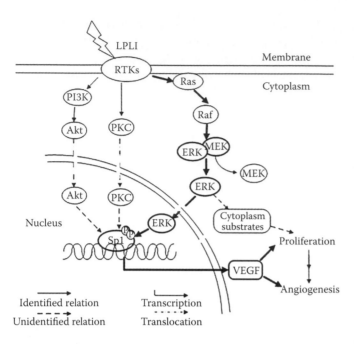

FIGURE 47.6 The model of signaling pathways activated by LLLT promoting vascular endothelial cell proliferation and angiogenesis. (Adapted from Feng, J. et al., *Cell Signal* 24: 1116–1125, 2012.)

Krishna and Narang 2008). Gao et al. (2009) reported that LLLT (He–Ne laser, 632.8 nm, 16 mW/cm² at 0.8 J/cm²) induced the formation of H-Ras-based circular ruffles in COS-7 cells (Figure 47.7). While LLLT activated Ras, the expression of YFP-H-Ras (N17), a dominant negative H-Ras, blocked the generation of

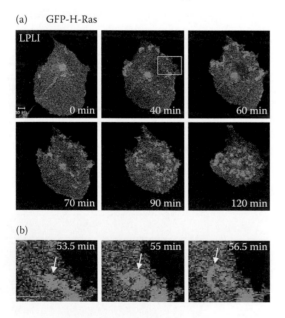

FIGURE 47.7 The formation of circular ruffles induced by LLLT. (a) Time-series images after LLLT. COS-7 cells were transfected with GFP-H-Ras and starved for 24 h. Then, the cells were treated with He–Ne laser (0.8 J/cm²) and recorded by LSM microscope. (b) Selected image series of ROI indicated by the white rectangle in (a). (Adapted from Gao, X. et al., *J Cell Physiol* 219: 535–543, 2009.)

LLLT-induced circular ruffles (Gao et al. 2009). Similarly, the PI3K inhibitor, wortmannin, potently suppressed the formation of LLLT-induced circular ruffles in a dose-dependent manner, demonstrating the process dependency on PI3K (Gao et al. 2009). Upon growth factor stimulation, the circular dorsal ruffles facilitate static cells for subsequent motility (Krueger et al. 2003). Thus, this work provides new information on the mechanisms of the biological effects of LLLT.

47.2.2.1.2 JNK Pathway

Hu et al. (2007) reported that He–Ne laser irradiation (632.8 nm, 1 J/cm²) immediately increased $\Delta\Psi$m, ATP, and cAMP levels via enhanced CcO activity as well as promoting JNK phosphorylation that in turn activated the transcription factor AP-1 in A2058 human melanoma cells. Hu et al.'s study was the first to demonstrate the activation of JNK following LLLT irradiation. These results were inconsistent with the aforementioned study indicating that LLLT activated ERK, but not JNK and p38 MAPK during cell proliferation (Shefer et al. 2001; Miyata et al. 2006; Kiyosaki et al. 2010). The discrepancy may be because melanoma cells initiate different signaling pathways compared with other cell lines during the proliferation induced by LLLT. cAMP is reported to be an important intracellular messenger that can regulate differentiation and proliferation. Karu and Tiphlova (1987) found that light irradiation caused cAMP elevation that in turn stimulated both DNA and RNA synthesis. In an investigation by Hu et al. (2007), cAMP analog (8-bromo-cAMP) significantly enhanced JNK phosphorylation induced by LLLT. It has been reported that cAMP elevating agents can induce activity of the transcription factor AP-1 (Pomerance et al. 2000). Taken together, these findings suggest that CcO/$\Delta\Psi$m/ATP/cAMP/JNK/AP-1 is involved in the regulation of melanoma cell proliferation induced by LLLT (Figure 47.8).

In molecular biology, AP-1 is a heterodimeric transcription factor composed of proteins belonging to the c-Fos, c-Jun, activating transcription factor (ATF), and Jun dimerization protein (JDP) families. AP-1 regulates gene expression in response to various stimuli, including cytokines, growth factors, and stress, as well as bacterial and viral infections (Hess, Angel, and

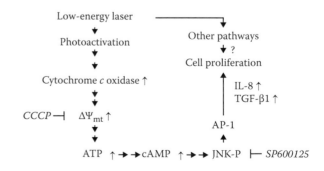

FIGURE 47.8 A proposed mechanism for He–Ne laser-induced cell proliferation in the A2058 melanoma cell. (Adapted from Hu, W. et al., *J Invest Dermatol* 127: 2048–2057, 2007.)

Schorpp-Kistner 2004). In turn, AP-1 controls several cellular processes that include differentiation, proliferation, and apoptosis (Ameyar, Wisniewska, and Weitzman 2003). The nerve growth factor (NGF) promoter region, which includes the AP-1 element, is important in basal and modulated NGF gene expression in cells within neuronal targets (D'Melo and Heinrich 1991). Yazdani et al. (2012) investigated the effect of 810 nm diode laser irradiation (1 and 4 J/cm^2) on human SC proliferation and neurotrophic factor gene expression in vitro. Two irradiated groups showed significant increase in proliferation in comparison with the control group (Yazdani et al. 2012). Meanwhile, SCs treated with laser highly increased the NGF gene expression (Yazdani et al. 2012). However, it is worthwhile to note that the treated and control groups showed no significant difference in brain-derived neurotrophic factor (BDNF) and glial cell-derived neurotrophic factor (GDNF) gene expression (Yazdani et al. 2012). On the other hand, these results are inconsistent with those from laser irradiation of olfactory ensheathing cell (OEC) proliferation, which is known to be highly similar to SCs in neural regeneration. In one study testing the effects of LLLT on OEC cells, it was found that LLLT increased the gene expression of BDNF and GDNF (Byrnes et al. 2005). This may indicate a dependency on cell type, as SCs and OECs differently respond to laser irradiation.

AP-1 induces factor/cytokine expression, such as those of transforming growth factor-β1 (TGF-β1) (Rui et al. 2012) and interleukin-8 (IL-8) (Fong et al. 2008). Melanoma cells secrete a variety of growth factors either constitutively or subsequently for induction by other cytokines (Matte et al. 1994). These growth factors/cytokines may act in autocrine or paracrine fashion on the host environment to stimulate growth. As the most abundant isoform of the TGF-β family, TGF-β1 rises in the plasma of melanoma patients, especially those with metastatic lesions (Krasagakis et al. 1998). IL-8 is a multifunctional cytokine that can stimulate proliferation of melanoma cells and keratinocytes in both an autocrine and a paracrine fashion (Schadendorf et al. 1993). Hu et al. (2007) reported that 632.8 nm He–Ne laser irradiation enhanced IL-8 (0.5, 1, and 2 J/cm^2) and TGF-β1 (1 and 2 J/cm^2) release from A2058 human melanoma cells. Their study also showed that the JNK inhibitor SP600125 significantly reduced IL-8 and TGF-β1 release induced by LLLT at the fluence of 1.0 J/cm^2 (Hu et al. 2007).

47.2.2.2 PI3K/Akt Pathway

PI3K constitutes a lipid kinase family characterized by their ability to phosphorylate the inositol ring 3′-OH group in inositol phospholipids to generate the second messenger phosphatidylinositol-3, 4, 5-trisphosphate (PIP3). RTK activation induces P13K to produce PIP3 and PIP2 at the inner side of the plasma membrane. These phospholipids interact with Akt, causing its translocation to the inner membrane where it is phosphorylated and activated by 3-phosphoinositide-dependent protein kinase-1 (PDK1) and PDK2. Activated Akt modulates the function of numerous substrates involved in

the regulation of cell survival, cycle progression, and growth (Fresno et al. 2004). Shefer et al. (2003) reported that LLLT (He–Ne laser, 632.8 nm, 1.06 J/cm^2)-induced PHAS-I phosphorylation in i28 mouse skeletal muscle myoblasts was abolished after preincubation with the PI3K inhibitor, wortmannin, suggesting that LLLT regulates PHAS-I phosphorylation in a PI3K-dependent pathway. Concomitantly, LLLT enhanced Akt phosphorylation, which was attenuated in the presence of wortmannin (Shefer et al. 2003). Thus, the PI3K/Akt pathways were involved in LLLT-induced protein translation because the phosphorylation form of PHAS-I promoted the activity of eIF4E through disassociation. LLLT has biostimulatory effects on various cell types by enhancing the production of several cytokines and growth factors. Saygun et al. (2012) reported that the irradiated osteoblasts (diode laser, 685 nm, 14.3 mW/cm^2 at 2 J/cm^2) revealed higher proliferation, viability, basic fibroblast growth factor (bFGF), IGF-I, and receptor of IGF-I (IGFBP3) expression compared with the nonirradiated control group. The MAPK/ERK and/or PI3K/Akt pathway participated in the signal transduction downstream of these growth factors in bone formation (Ling et al. 2010).

Zhang et al. (2009) investigated the activity of Akt and its effects on cell proliferation induced by LLLT in COS-7 cells. LLLT (He–Ne laser, 632.8 nm, 12.74 mW/cm^2 at 1.2 J/cm^2) induced a gradual and continuous activation of Akt (Zhang et al. 2009). Moreover, wortmannin completely abolished the activation of Akt, suggesting that this LLLT-induced activation is a PI3K-dependent event. The Src family was involved in Akt activation as demonstrated by the partial inhibition of Akt activity in samples treated with PP1 (an inhibitor of Src family) (Zhang et al. 2009). Haynes et al. (2003) reported that Src kinase mediates PI3K/Akt-dependent rapid endothelial nitric oxide (NO) synthase (eNOS) activation by estrogen. GFP-Akt fluorescence imaging and Western blot analysis demonstrated the activation of Akt as a multistep process in response to LLLT, involving membrane recruitment, phosphorylation, and membrane detachment (Zhang et al. 2009). More importantly, LLLT promoted cell proliferation through PI3K/Akt activation because cellular viability was significantly inhibited by PI3K inhibitor (Zhang et al. 2009). All these studies prove that the PI3K/Akt signaling pathway has a significant role in LLLT-triggered cell proliferation (Figure 47.9).

Studies in vitro have shown that Akt can directly phosphorylate eNOS and activate the enzyme, thereby producing NO (Iwakiri et al. 2002). Thus, the PI3K/Akt/eNOS signaling pathway is critical for the maintenance of endothelial vascular tone and integrity. NO was shown to promote angiogenesis and vasculogenesis, which are indispensable processes for tissue growth (Duda, Fukumura, and Jain 2004). When NO is constitutively produced at low levels, eNOS can be transiently stimulated by hormones or various extracellular stimuli to elevated NO levels (Janssens et al. 1992; Simoncini et al. 2000). In angiogenesis, the proliferation and migration of endothelial cells play critical roles (Chen et al. 2004). Many growth factors modulate endothelial

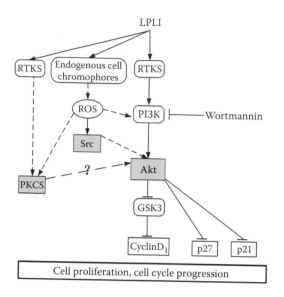

FIGURE 47.9 A model of the signaling pathways of Akt activation induced by LLLT. Dashed lines indicate that the downstream changes are yet to be confirmed by further studies. Well-studied downstream effects are indicated by solid lines. (Adapted from Zhang, L. et al., *J Cell Physiol* 219: 553–562, 2009.)

cell proliferation, migration, and angiogenesis via PI3K/Akt/eNOS signaling pathway. Chen, Hung, and Hsu (2008) reported that 632.5 nm He–Ne laser irradiation increased HUVEC proliferation, migration, and NO secretion, and also promoted angiogenesis. LLLT (<0.26 J/cm²) increased eNOS protein expression and gene expression in endothelial cells (Chen, Hung, and Hsu 2008) (Figure 47.10). Another PI3K inhibitor, LY294002, inhibited the increase in eNOS expression, indicating that the activation of PI3K/Akt pathway is critical for increased expression of eNOS upon LLLT irradiation (Chen, Hung, and Hsu 2008) (Figure 47.10). Thus, these findings suggest that the PI3K/Akt/eNOS pathway is involved in the endothelial cell proliferation induced by LLLT.

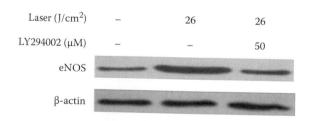

FIGURE 47.10 HUVECs were pretreated with the indicated concentration of LY294002 (50 mM), treated with or without laser irradiation (0.26 J/cm²), and incubated at standard conditions for 24 h. Whole cell lysates were prepared and probed with anti-eNOS antibody to detect eNOS protein expressions. Cells treated with DMSO (<0.01%) were used as the control group for each experiment and β-actin levels were used as a loading control. (Adapted from Chen, C. H. et al. *Lasers Surg Med* 40: 46–54, 2008.)

47.2.2.3 PKC Pathway

PKC is a family of serine/threonine kinase, composed of at least 13 isozymes that have an impact on cell proliferation, differentiation, angiogenesis, and apoptosis. The 13 members of the PKC family can be divided into three subfamilies based on their second messenger requirements: (1) conventional (or classical) PKCs, requiring Ca^{2+} and diacylglycerol (DAG) for activation, (2) novel PKCs, requiring DAG but not requiring Ca^{2+} for activation, and (3) atypical PKCs, requiring neither Ca^{2+} nor DAG for activation. Such activation of PKC isozymes depends on RTKs and GPCRs (Ali et al. 2009). Activated RTKs can induce the catalytic activity of phospholipase C (PLC)-gamma, which catalyzed the hydrolysis of some phospholipids, and then increased the concentrations of DAG and inositol triphosphate (IP3) in the cytoplasm. IP3 causes the endoplasmic reticulum (ER) to release Ca^{2+}, which works with DAG to activate PKCs (Newton 2009).

Numerous cell modes exhibited increased Ca^{2+} levels upon LLLT. Cohen et al. (1998) reported that irradiation of mouse spermatozoa by 630 nm He–Ne laser enhanced the intracellular Ca^{2+} levels and cellular fertilizing potential. Lavi et al. (2003) reported that LLLT of 40 mW/cm² at 3.6 J/cm² induced a transient increase in the Ca^{2+} level in rat cardiomyocytes without any cellular damage. However, LLLT at 40 mW/cm² at 12 J/cm² induced a linear increase in the Ca^{2+} level and damaged the cells (Lavi et al. 2003). This increase in the Ca^{2+} level due to high energy doses of light could be attenuated into a nonlinear small rise through the presence of extracellular catalase during illumination (Lavi et al. 2003). The different kinetics of Ca^{2+} level elevation following various light irradiations correspondingly represents different adaptation levels to oxidative stress. The adaptive response of cells to LLLT represented by the transient increase in the Ca^{2+} level can explain beneficial effects of LLLT.

Using specific inhibitors, Duan et al. (2001) reported that the RTK/PLC/PKC/nicotinamide adenine dinucleotide phosphate oxidase (NADPH) signal transduction pathway participated in respiratory burst of neutrophils induced by LLLT (He–Ne laser, 632.8 nm, 71 mW/cm² at 300 J/cm²). Gao et al. (2006) investigated the involvement of PKCs in ASTC-a-1 cell proliferation induced following LLLT (He–Ne laser, 632.8 nm, 16 mW/cm² at 0.8 J/cm²). They constructed an ASTC-a-1 cell line that stably expressed PKC activity and used fluorescence resonance energy transfer (FRET) reporter (CKAR) for real-time monitoring of PKC activity. The increasing dynamics was monitored during cell proliferation using FRET imaging on laser scanning confocal microscope and spectrofluorometric analysis on a luminescence spectrometer (Gao et al. 2006). PKC activation was also demonstrated in rat pheochromocytoma cells (PC12) following LLLT irradiation (Zhang et al. 2008). On the other hand, the decreasing dynamics of PKC activity was monitored in cells treated with high fluence LLLT (60 J/cm²) (Gao et al. 2006), which was previously reported to induce cell apoptosis (Wang et al. 2005a). PKCs therefore play an important role in the LLLT-induced biological effects (Figure 47.11).

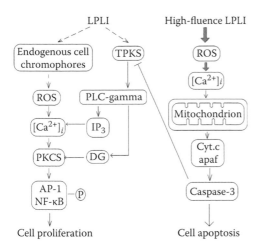

FIGURE 47.11 A model of the signal pathways of PKC activation and inactivation induced by laser irradiation. Broken lines indicate that the downstream changes produced by the upstream are not proven clearly. Solid lines indicate that the downstream effects induced by the upstream have already been proven. The thick arrowheads mean the corresponding effects are much stronger than the thin ones. (Adapted from Gao, X. et al., *J Cell Physiol* 206: 441–448, 2006.)

47.2.2.4 Src Kinase Pathway

Src is a member of the SFKs and has a critical role in cell adhesion, invasion, proliferation, survival, and angiogenesis during tumor development (Kim, Song, and Haura 2009). SFKs are composed of nine family members that share similar structure and function. Overexpression or high activation of SFKs frequently occurs in tumor tissues, making them one of the key components in multiple signaling pathways that are important in oncogenesis. SFKs can interact with RTKs, such as epidermal growth factor receptor (EGFR) and VEGF receptor. SFKs can affect cell proliferation via the Ras/ERK pathway and regulate gene expression via transcription factors such as signal transducer and activator of transcription (Stat) molecules. Sun, Wu, and Xing 2010 reported that high-fluence LLLT (He–Ne laser, 632.8 nm, 200 mW/cm² at 80 and 120 J/cm²) activated the ROS/Src/Stat3 pathway and thus promoted cell survival against proapoptotic signals. Zhang et al. (2009) reported that upon LLLT (He–Ne laser, 632.8 nm, 12.74 mW/cm² at 1.2 J/cm²), the SFKs became involved in Akt activation, as demonstrated by the partial inhibition of Akt activity in COS-7 cells treated with PP1 (an inhibitor of the SFKs).

ROS are considered to be the key secondary messengers produced by LLLT. Tyrosine kinases are well-known targets of ROS and can be activated by oxidative events. Zhang et al. studied the signaling pathways mediated by ROS upon the stimulation of 632.8 nm He–Ne laser irradiation. Using a Src FRET reporter (Wang et al. 2005b) and confocal laser scanning microscope, dynamic Src activation was observed in HeLa cells immediately after LLLT treatment (Zhang, Xing, and Gao 2008). Moreover, the activation occurred in a dose-dependent manner (Zhang et al. 2008) (Figure 47.12). The increase in Src phosphorylation at

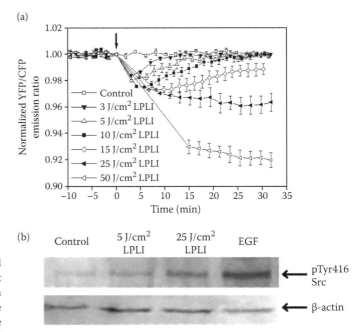

FIGURE 47.12 Dynamics of Src activity induced by LLLT in HeLa cells. (a) The time courses of YFP/CFP emission ratio of Src reporter in response to different doses of LLLT. (b) HeLa cells were treated with 5 J/cm² LLLT, 25 J/cm² LLLT, or EGF or kept as control. Five minutes after irradiation, cells were subjected to Western blotting assay. Cell lysates from various samples were probed for phospho-Src (Tyr416) (upper) and Src (to show comparable protein loading among various samples; lower). (Adapted from Zhang, L. et al., *Cell Physiol Biochem* 22: 215–222, 2008.)

Tyr416 was detected by Western blotting (Zhang et al. 2008). The presence of dehydroascorbic acid (DHA), catalase alone (CAT), or the combination of CAT and superoxide dismutase significantly abolished the activation of Src by LLLT (Zhang et al. 2008). In contrast, the broad spectrum PKC inhibitor GÖ6983 loading did not exhibit this response (Zhang et al. 2008). Treatment of HeLa cells with exogenous hydrogen peroxide (H_2O_2) also resulted in a concentration-dependent Src activation (Zhang et al. 2008). Therefore, LLLT induces ROS-mediated Src activation that may play an important role in the biostimulatory effect of LLLT (Figure 47.13).

47.2.2.5 BMP/Smad Pathway

LLLT induced osteoblast differentiation and bone formation both in vivo and in vitro. Numerous researchers tried to denote the molecular mechanism by which LLLT stimulates osteoblast differentiation and bone formation. Hirata et al. (2010) reported that 805 nm Ga–Al–As laser irradiation enhanced BMP2-induced alkaline phosphatase (ALP) activity in C2C12 mouse myoblast cells (Figure 47.14a). In addition, LLLT stimulated BMP2-induced phosphorylation of Smad1/5/8 and BMP2 expression but had no effect on the expression of the inhibitory Smads 6 and 7, BMP4, or IGF1 (Hirata et al. 2010) (Figure 47.14b). LLLT also enhanced the Smad-induced Id1

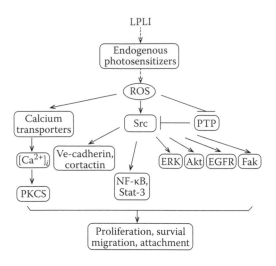

LPLI

Endogenous photosensitizers

ROS

Calcium transporters

Src

PTP

$[Ca^{2+}]_i$

Ve-cadherin, cortactin

ERK Akt EGFR Fak

PKCS

NF-κB, Stat-3

Proliferation, survial migration, attachment

FIGURE 47.13 A model of the signaling pathways of Src activation. Broken lines indicate that the downstream changes produced by the upstream are not proven clearly. Solid lines mean that the downstream effects induced by the upstream have been proved already. (Adapted from Zhang, J. et al., *J Cell Physiol* 217: 518–528, 2008.)

reporter activity as well as the expression of BMPs-induced transcription factors, such as Id1, Osterix, and Runx2 (Hirata et al. 2010). LLLT also stimulated BMPs-induced expression of type I collagen, osteonectin, and osteocalcin mRNA (Hirata et al. 2010). This enhancement of BMP2-induced ALP activity and Smad phosphorylation with laser irradiation was also observed in primary osteoblast differentiation (Hirata et al. 2010). The authors concluded that LLLT accelerates differentiation of BMP-induced osteoblasts by stimulating the BMP/Smad signaling pathway (Hirata et al. 2010). Fukuhara et al. (2006) also reported that LLLT (Ga–Al–As diode laser, 3.75 J/cm²) stimulated differentiation in osteoblastic cells, as

demonstrated with the increased Runx2 expression and ALP-positive colonies. Peng et al. (2012) demonstrated that noncoherent red light [light-emitting diode (LED), 620 nm, 1, 2, and 4 J/cm²] promoted MSC proliferation but did not induce osteogenic differentiation of MSCs in normal media. On the other hand, red light enhanced osteogenic differentiation and decreased proliferation of MSCs in media with osteogenic supplements (Peng et al. 2012).

Fávaro-Pípi et al. (2011) measured the temporal pattern of osteogenic gene expression after LLLT during the bone healing process. Histological results revealed intense new bone formation surrounded by highly vascularized connective tissue indicating a slight increase in osteogenic activity exposed to laser (Fávaro-Pípi et al. 2011). The quantitative real-time polymerase chain reaction (RT-PCR) showed that laser irradiation upregulated BMP4, ALP, and Runx2 after surgery (Fávaro-Pípi et al. 2011). These studies indicate that laser therapy improves bone repair in rats, as depicted with differential histopathological and osteogenic gene expression. It is reported that FGF and IGF signals regulated the expression of Runx2 through PI3K/Akt and/or MAPK/ERK pathway (Ling et al. 2010). Kiyosaki et al. (2010) reported that 830 nm Ga–Al–As diode laser at 1.91 J/cm² fluence stimulated in vitro mineralization through increased IGF-I and BMP production induced with Runx2 expression and ERK phosphorylation in mouse osteoblastic MC3T3-E1 cells. In another study, Saygun et al. reported that LLLT (685 nm, 14.3 mW/cm² at 2 J/cm²) increased the proliferation of osteoblast cells and stimulated the release of bFGF, IGF-I, and IGFBP3.

47.2.3 Cell Cycle-Specific Proteins Involved in Cell Cycle Progression by LLLT

Cell cycle is a crucial event in cell biology that consists of a series of repeated events allowing the cell to grow and

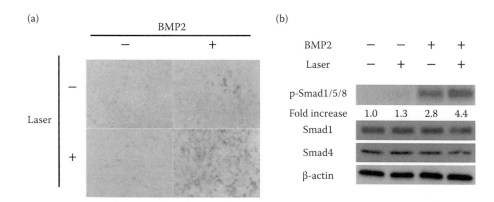

(a)

BMP2

−   +

Laser   −

+

(b)

BMP2   −   −   +   +
Laser   −   +   −   +

p-Smad1/5/8

Fold increase   1.0   1.3   2.8   4.4

Smad1

Smad4

β-actin

FIGURE 47.14 C2C12 cells were irradiated with a Ga–Al–As laser at 2.5 W, CW for 2 min or left unirradiated, then cultured for an additional 3 days in the presence or absence of BMP2 (100 ng/mL). (a) Cells were stained for ALP activity. (b) Total cell lysates were immunoblotted with antiphosphorylated Smad1/5/8, Smad1, or Smad4 antibodies, and anti-β-actin was used as a loading control. Numbers below the gels represent *n*-fold increases in the intensity of phosphorylated Smad1/5/8 relative to the corresponding Smad1 signals. (Adapted from Hirata, S. et al., *J Cell Biochem* 111: 1445–1452, 2010.)

duplicate in a correct manner (Massagué 2004). A normal eukaryotic cell cycle consists of four discrete phases: G1 (the period after cytokinesis and before the S phase), S (DNA synthesis), G2 (the period between S and M phases), and M (cell mitosis) (Nurse 1994). The normal progression of the four phases of cell cycle is very important for normal cell growth. The researchers proposed that the G1 phase of the mammalian cell could be subdivided into two phases: the early G1 phase and the late G1 phase. Cells in the early G1 phase could enter the G0 phase, unlike those in the late G1 phase. The G0 phase is the dormancy phase when the cell is not performing any function at all. The point of discrimination between G0 and G1 is associated with the restriction point of the cell cycle that is strictly controlled by multiple molecules.

47.2.3.1 c-Fos

c-Fos is a cellular proto-oncogene belonging to the immediate early gene (IEG) family of transcription factors. c-Fos has a leucine zipper DNA-binding domain and a transactivation domain at the C-terminus. c-Fos transcription is upregulated in response to many extracellular signals, such as growth factors. In addition, phosphorylation by MAPK, PKA, PKC, or cdc2 alters the activity and stability of c-Fos. c-Fos can dimerize with c-jun to form the AP-1 that upregulates the transcription of a diverse range of genes involved in all cellular activities, from proliferation and differentiation to defense against invasion and cell damage. Ca^{2+} can modulate the expression of a number of early-response eukaryotic genes in isolated cells, especially the expression of c-Fos gene (Li et al. 1996, Maturana et al. 2002). In the investigation by Greco et al. (2001), 632.8 nm He–Ne laser irradiation (0.24 J/cm²) triggered the activation of Ca^{2+}-dependent c-Fos gene and elevated the expression of c-Fos in isolated hepatocytes. The expression of c-Fos has been reported to give rise to a G0/G1 transition of the cell cycle in isolated hepatocytes (Loyer et al. 1996). Thus, the elevated expression of c-Fos induced by LLLT may cause quiescent cells to leave the G0 phase and enter the G1 phase.

47.2.3.2 Cyclin D1, Cyclin E, Cyclin A, PCNA, and PML

Cyclin D1 is a critical target for proliferative signals and is required for cell cycle progression in the G1 phase (Baldin et al. 1993). Cyclin E is induced at the G1/S boundary, and cyclin A is induced at a later phase of the cell cycle and is required for the cell to progress through the S phase (Girard et al. 1991). Proliferating cell nuclear antigen (PCNA) is a 36 kD protein located in the nucleus and usually found in proliferating cells and tumor cells. PCNA is closely related to DNA synthesis and plays an important role at the start of the cell proliferative stage. Ben-Dov et al. (1999) evaluated the effect of 632.8 nm He–Ne laser irradiation on the proliferation and differentiation of muscle satellite cells in vitro. They found that LLLT affected thymidine incorporation in primary rat satellite cells in a bell-shaped manner, with a peak at 3 s of irradiation (Ben-Dov et al. 1999). Three seconds of irradiation caused

an induction of cell cycle regulatory proteins: cyclin D1 (2 h), cyclin E (6 h), and cyclin A (9 h) in an established line of pmi 28 mouse satellite cells, and PCNA (19 h) in primary rat satellite cells (Ben-Dov et al. 1999). The MHC protein levels were found to be twofold lower in the irradiated cells than in the control cells, whereas the proliferation of the irradiated cells was twofold higher (Ben-Dov et al. 1999). Moreover, the fusion percentage was lower in the irradiated cells compared with the nonirradiated cells (Ben-Dov et al. 1999). In light of these results, it may be assumed that the promoting effects of laser irradiation on skeletal muscle regeneration in vivo may be due to its effect on the activation of early cell cycle regulatory genes in satellite cells, leading to increased proliferation and delay in cell differentiation. The following study by Shefer et al. (2003) showed that the induction of cyclin D1 expression was detected as early as 6 min after 632.8 nm He–Ne laser irradiation in i28 mouse myogenic cells, suggesting the entrance and progression through the G1 phase of the cell cycle. The sustained activation of ERK1 as a requirement for the continuous expression of cyclin D1 in the G1 phase and the detection of sustained activation of ERK1 during LLLT-induced cell proliferation have already been reported (Shefer et al. 2001). Zhang et al. (2011) also detected the increase in the expression of PCNA in primary rat muscle myoblasts. They found that 24 h after 632.8 nm He–Ne laser irradiation (6 mW/cm² at 1.08 J/cm²), 180 s PCNA expression was significantly higher than the non-irradiated control group in rat muscle myoblasts and was the same as 10% of the fetal bovine serum (FBS) refeeding group but less than that of 20% of the FBS group (Zhang et al. 2011). These findings imply that laser irradiation aids the cells to pass through the G1 phase and enter the S phase by affecting early cell cycle regulatory genes, and consequently stimulates cell proliferation.

Checkpoint regulatory proteins control the entrance of the cell cycle from a quiescent state, as well as the progression in the cell cycle. One such protein, the promyelocytic leukemia protein (PML), blocks the entry of the cells to the S phase of the cell cycle (de Thé et al. 1991). PML protein is typically concentrated in subnuclear PML oncogenic domains (PODs), together with other important cell cycle regulatory proteins such as p53. The localization and distribution of PML in the cells are related to the cell cycle progression. The redistribution of PML from PODs to the nucleoplasm indicates that the cell is leaving the quiescent stage and entering the S phase. Gavish et al. (2004) showed that upon 780 nm Ti–Sa laser irradiation (1.85 mW/cm² at 2 J/cm²) of HaCaT human keratinocytes, the $\Delta\Psi$m was increased immediately and the subnuclear distribution of PML was altered from discrete domains to its dispersed form within less than 1 h. Three hours after irradiation, the intensity of PML fluorescence was markedly reduced compared with control cells, suggesting that the PML protein is largely degraded (Gavish et al. 2004). Considering that $\Delta\Psi$m is a sensitive indicator of the energetic state of the mitochondria and of the whole cells, these results show the connection between the element of mitochondrial energy signals and secondary cellular reactions of the alteration of PML distribution in response to LLLT. In conclusion, LLLT may induce the cell cycle progression from G1 to the S phase through the redistribution and degradation of PML protein, thereby enabling cell proliferation.

47.2.3.3 MCM3

Minichromosome maintenance deficient 3 (MCM3) is a replication licensing factor that is a prereplication complex protein involved in the initiation of DNA replication in eukaryotic cells. Yamamoto et al. (2001) showed that upon LLLT (Ga–Al–As diode laser, 830 nm, 6.37 mW/cm² at 7.64 J/cm²), the mRNA levels of MCM3 were elevated in laser-irradiated MC3T3-E1 cells, when compared with the levels in the control cells. Radiolabeled thymidine incorporation increased with laser irradiation (Yamamoto et al. 2001), implying that LLLT enhances DNA replication via enhancement of MCM3 gene expression and plays an important role in osteoblast proliferation (Yamamoto et al. 2001).

47.2.3.4 p15

The cell proliferation and progression of the cell cycle are regulated by the sequential activities of various cyclin-dependent kinases (CDKs). CDK activity is dependent on the physical interaction with one of the cyclin proteins, which are the regulatory subunits of these complexes. In addition, this activity is negatively regulated by a group of proteins that are cell cycle regulators, collectively called CDK inhibitors (CKIs) and consisting of two groups, the INK4 family and the CIP/KIP family, classified according to structure and function (Hannon and Beach 1994; Sherr 2001). p15, an INK4 family member, regulates the cell cycle arrest at G1 by inhibiting CDK4 and CDK6 to inactivate the retinoblastoma family of tumor suppressor proteins (Hannon and Beach 1994; Sherr 2001). cAMP has a variety of effects on cell proliferation, including both inhibition (Kato et al. 1994; Haddad et al. 1999; Rao et al. 1999; Kim et al. 2001) and stimulation (Starzec et al. 1994; Iacovelli et al. 2007) cell growth. Taniguchi et al. (2009) reported that the proliferative effect on synovial fibroblasts by LLLT was abolished by cAMP. Their findings suggest that cAMP may be involved in the effect of LLLT on synovial fibroblast proliferation by modulating p15 subcellular localization. For details, they found that 660 nm LLLT (40 mW/cm² at 4.8 J/cm²) promoted HIG-82 rabbit synovial fibroblast proliferation and induced cytoplasmic localization of p15 (Taniguchi et al. 2009). Moreover, the proliferation of synovial fibroblasts was reduced by cAMP, whereas cAMP inhibitor SQ22536 induced p15 cytoplasmic localization, and as a result, elevated synovial fibroblast proliferation was observed (Taniguchi et al. 2009). Synovial fibroblast proliferation induced by 660 nm LLLT may contribute to further investigation on the biological effects and application of LLLT in regenerative medicine since synovial fibroblasts are important in maintaining the homeostasis of articular joints and have strong chondrogenetic capacity (Taniguchi et al. 2009).

47.3 LLLT Inhibits Cell Apoptosis

47.3.1 PI3K/Akt/GSK3β/Bax Pathway

Zhang, Zhang, and Xing (2010) demonstrated for the first time that LLLT inhibits mitochondrial proapoptotic cascades in human cancer cells by inactivating the glycogen synthase kinase 3 beta (GSK-3β)/Bax pathway. They have shown that LLLT (He–Ne laser, 632.8 nm, 12.74 mW/cm² at 1.2 J/cm²) could inhibit the activation of GSK-3β, Bax, and caspase-3 induced by staurosporine (STS) (Zhang, Zhang, and Xing 2010) (Figure 47.15). During the search for the mechanism, they found that LLLT could activate Akt, which was consistent with their previous findings (Zhang et al. 2009), even in the presence of STS. In this antiapoptotic process, the interaction between Akt and GSK-3β increased gradually, indicating that Akt interacts with and inactivates GSK-3β directly. Conversely, LLLT decreased the interaction between GSK-3β and Bax, with the suppression of Bax translocation to mitochondria, suggesting that LLLT inhibits Bax translocation by inactivating GSK-3β (Zhang, Zhang, and Xing 2010). Wortmannin, an inhibitor of PI3K, potently suppressed the activation of Akt and subsequent antiapoptotic processes induced by LLLT. Based on these studies, they have concluded that LLLT protects against STS-induced apoptosis upstream of Bax translocation via the PI3K/Akt/GSK-3β pathway (Figure 47.16). These findings raise the possibility of LLLT as a promising therapy for neurodegenerative disease induced by GSK-3β.

47.3.2 PI3K/Akt/YAP/p73/Bax Pathway

Previous studies show that p73 is vital for the mediation of the pathogenic process of AD, and Yes-associated protein (YAP) has been shown to regulate p73 positively in promoting apoptosis induced by anticancer agents (Basu et al. 2003; Strano et al. 2005; Matallanas et al. 2007). Zhang, Wu, and Xing (2011) first explored the functional role of YAP and its potential relationship with p73 in AD. They demonstrated that YAP accelerated apoptosis in response to Aβ₂₅₋₃₅ and that the nuclear translocation of YAP was involved in cellular signals regulating the apoptotic pathway. Aβ₂₅₋₃₅ induced YAP translocation from the cytoplasm to the nucleus, and accompanied increased phosphorylation on Tyr357 resulted in enhancement of interaction between YAP and p73 (Zhang, Wu, and Xing 2011). More importantly, the p73-mediated induction of Bax expression and activation were in a YAP-dependent manner (Zhang, Wu, and Xing 2011). YAP overexpression accelerated Bax translocation, upregulated Bax expression, and promoted caspase-3 activation (Zhang, Wu, and Xing 2011). Their results provide a potential therapeutic strategy for the treatment of AD through YAP/p73/Bax pathway inhibition. Akt is reported to phosphorylate YAP on Ser127 and thus inhibit its activity of interaction with p73. Furthermore, LLLT was shown to activate the PI3K/Akt pathway in their former study (Zhang et al. 2009). Based on these findings,

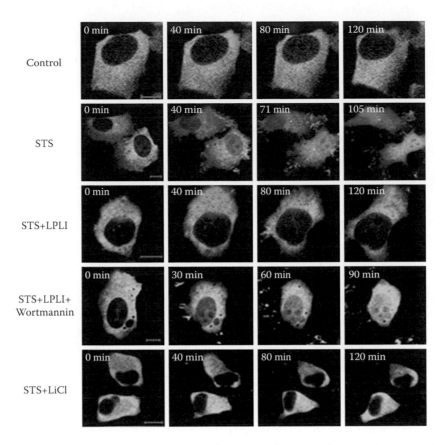

FIGURE 47.15 Dynamics of YFP-GSK-3β nuclear translocation in different conditions. LiCl was used as a negative control. (Adapted from Zhang, L. et al., *J Cell Physiol* **224**: 218–228, 2010.)

Zhang, Wu, and Xing (2012) further demonstrated that LLLT could inhibit Aβ$_{25-35}$-induced cell apoptosis through the activation of the Akt/YAP/p73 signaling pathway. They found that LLLT promotes YAP cytoplasmic translocation and inhibits Aβ$_{25-35}$-induced YAP nuclear translocation (Figure

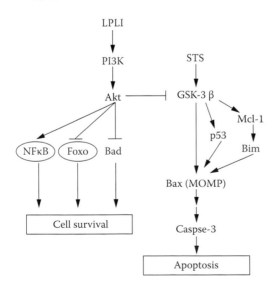

FIGURE 47.16 A model of the signaling pathways for LLLT inhibition on STS-induced cell apoptosis. (Adapted from Zhang, L. et al., *J Cell Physiol* 224: 218–228, 2010.)

47.17). Moreover, the cytoplasmic translocation occurred in an Akt-dependent manner (Figure 47.17). Activated Akt following LLLT irradiation phosphorylated YAP on Ser127 and resulted in decreased interaction between YAP and p73, thus suppressing the proapoptotic gene Bax expression following the Aβ$_{25-35}$ treatment (Zhang, Wu, and Xing 2012). On the other hand, the inhibition of Akt expression by siRNA significantly abolished the effect of LLLT (Zhang, Wu, and Xing 2012). Therefore, their findings directly point to a potential phototherapeutic strategy for the treatment of AD through the Akt/YAP/p73 signaling pathway (Zhang, Wu, and Xing 2012) (Figure 47.18).

47.3.3 Regulation on Bcl-2 Family Members

The Bcl-2 family members are major regulators of the apoptotic process (Reed 2008). This family consists of proapoptotic (e.g., Bax and Bak) and antiapoptotic (e.g., Bcl-2 and Bcl-X$_L$) molecules. Activated p53 mediates growth arrest and apoptosis by activating the expression of a number of cellular genes such as p21 and Bax (El-Deiry et al. 1993; Miyashita et al. 1994). Shefer et al. (2002) demonstrated that under serum deprivation conditions that normally lead to apoptosis, LLLT (He–Ne laser, 632.8 nm, 1.06 J/cm^2) stimulation was shown to promote the survival of fibers and their adjacent satellite cells, as well as cultured i28 mouse myogenic cells. More importantly, upon LLLT

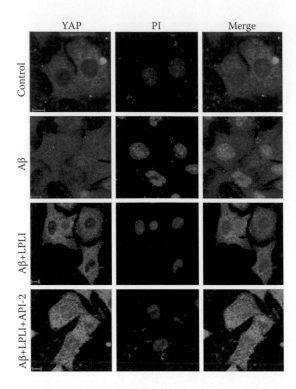

FIGURE 47.17 **(See color insert.)** LLLT inhibits $A\beta_{25-35}$-induced YAP translocation from cytoplasm to nucleus through Akt activation. PC12 cells were treated with $A\beta_{25-35}$ or/and API-2, an inhibitor of Akt, and then nucleus was stained with PI to differentiate from the cytoplasm. Representative images are shown. (Adapted from Zhang, H. et al., *Cell Signal* 24: 224–232, 2012.)

stimulation, the expression of the antiapoptotic protein Bcl-2 is markedly increased, whereas that of the p53, CDK inhibitor p21, and proapoptotic protein Bax is reduced (Shefer et al. 2002). The protective effect of elevating Bcl-2 expression in response to LLLT could be mediated by the suppression of p53 expression, or LLLT directly induces the expression of Bcl-2 at the post-transcriptional level. These findings imply that the protective

effects of LLLT against apoptosis and promoting survival are partly mediated with regulation of these factors. Upon He–Ne laser treatment at the fluence rates of 0, 0.5, 1.0, and 2.0 J/cm^2 on A2058 human melanoma cells, no obvious change was observed in the levels of p53 and Bcl-2 expression (Hu et al. 2007).

Zhang et al. (2008) investigated the antiapoptotic mechanism of 632.8 nm He–Ne laser via FRET and RT-PCR. The PC12 cells were treated with $A\beta_{25-35}$ for induction of apoptosis before LLLT treatment. The cell viability assays and morphological examinations showed that LLLT (0.52 mW/cm^2 from 0.156 to 0.624 J/cm^2) could inhibit the cell apoptosis (Zhang et al. 2008). An increase in PKC activation was dynamically monitored in the cells treated with LLLT only or $A\beta_{25-35}$ followed by LLLT treatment (Zhang, Xing, and Gao 2008). Furthermore, LLLT caused an increase in cell survival member Bcl-X_L mRNA and a decrease in the upregulation of cell death member Bax mRNA, which was caused by $A\beta_{25-35}$ (Zhang, Xing, and Gao 2008). Further data indicate that LLLT could reverse the increased level of Bax/Bcl-X_L mRNA ratio following $A\beta_{25-35}$ (Zhang, Xing, and Gao 2008). In addition, GÖ6983, a PKC inhibitor, could inhibit the decreased level of Bax/Bcl-X_L mRNA ratio (Zhang, Xing, and Gao 2008). Together, these data clearly indicate that LLLT inhibits $A\beta_{25-35}$-induced PC12 cell apoptosis via PKC-mediated regulation of the Bax/Bcl-X_L mRNA ratio.

Apoptosis, or programmed cell death, resulting from cerebral ischemia may be related to decreased levels of antiapoptotic factors, such as Akt and Bcl-2, and increased levels of proapoptotic factors, such as Bad, caspase-9, and caspase-3 activities. Yip et al. (2011) investigated the effects of 660 nm LLLT on the levels and activity of various antiapoptotic and proapoptotic factors following ischemia. The Akt, pAkt, Bcl-2, and pBad levels were significantly increased, whereas those of caspase-9 and activated caspase-3 were significantly decreased following laser irradiation (44 mW/cm^2 at 2.64, 13.2, and 24.6 J/cm^2) (Yip et al. 2011). Thus, LLLT may protect the brain by upregulating Akt, pAkt, pBad, and Bcl-2, as well as downregulating caspase-9 and caspase-3 following transient cerebral ischemia. This modality is a promising protective therapeutic intervention after strokes or other ischemic events.

47.3.4 Other Pathways

Giuliani et al. (2009) explored the effect of coherent red light irradiation with extremely low energy transfer on PC12 cells. They focused on the effect of pulsed laser irradiation (diode laser, 670 nm, 3 mW/cm^2 at 0.45 mJ/cm^2) on two distinct biological effects: neurite elongation under NGF stimulus on laminin-collagen substrate and cell viability during oxidative stress. They found that laser irradiation stimulated NGF-induced neurite elongation on a laminin-collagen-coated substrate and protects PC12 cells against oxidative stress (H_2O_2), as indicated by $\Delta\Psi m$ measurement and MTT assay (Giuliani et al. 2009). These effects could have positive implications for axonal protection.

NO is a major factor that contributes to the loss of neurons in ischemic stroke, demyelinating diseases, and other

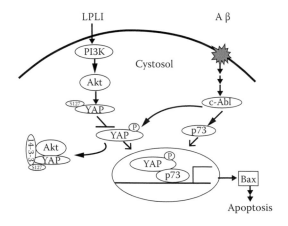

FIGURE 47.18 A model of the signaling pathways for LLLT inhibiting $A\beta_{25-35}$-induced cell apoptosis. (Adapted from Zhang, H. et al., *Cell Signal* 24: 224–232, 2012.)

neurodegenerative disorders (Patel et al. 1999). NO does not only function as a direct neurotoxin but also combines with superoxide (O_2^-) via a diffusion-controlled reaction to form peroxynitrite ($ONOO^-$), a species that contributes to oxidative signaling and cellular apoptosis (Virag et al. 2003; Soneja, Drews, and Malinski 2005; Szabo, Ischiropoulos, and Radi 2007). Lim et al. (2009) investigated the triggers of the apoptotic pathway using O_2^- scavenging with light irradiation (LED, 635 nm, 5 mW/cm² at 18 J/cm²) to block $ONOO^-$ production in SH-SY5Y human neuroblastoma cells. Cell survival was reduced to approximately 40% of control levels by sodium nitroprusside (SNP) treatment, and this reduction was increased to 60% with LLLT (Lim et al. 2009). Apoptotic cells were observed in the SNP-treated group, but the frequency of these was reduced in the irradiation group (Lim et al. 2009). NO, O_2^-, total ROS, and $ONOO^-$ levels were increased after SNP treatment, but O_2^-, total ROS, and $ONOO^-$ levels were decreased after irradiation despite the high NO concentration induced with the SNP treatment (Lim et al. 2009). Cytochrome *c* was released from mitochondria of SNP-treated SH-SY5Y cells, but not of irradiated cells, resulting in a decrease in caspase-3 and -9 activities in SNP-treated cells (Lim et al. 2009). Finally, these results show that LLLT promoting the scavenging of O_2^- protected against neuronal death by blocking the mitochondrial apoptotic pathway induced by $ONOO^-$ synthesis.

47.4 Inhibition of Cell Viability and Induction of Cell Apoptosis by High-Fluence LLLT

47.4.1 Inhibition of Cell Viability by High-Fluence LLLT

Wang et al. (2005a) measured the changes in cell viability in ASTC-a-1 cells induced by high-fluence LLLT treatment (He–Ne laser, 632.8 nm). Cell viability assay using the Cell Counting Kit-8 showed that when light irradiation fluence exceeded 126 mW/cm² at 60 J/cm², cell viability decreased sharply in a fluence-dependent manner, compared with the unirradiated group (Wang et al. 2005a). Zhang, Xing, and Gao (2008) investigated the effect of 632.8 nm He–Ne laser irradiation on cell viability in HeLa cells. They showed that LLLT could increase cell viability in a dose-dependent manner between 3 and 15 J/cm² (64.6 mW/cm²) (Zhang, Xing, and Gao 2008). However, this tendency of increasing cell viability ceased and changed to a decrease in cell viability when the LLLT dose was up to 64.6 mW/cm² at 25 J/cm². Besides, 64.6 mW/cm² at 50 J/cm² LLLT significantly inhibited cell viability compared with the control cells (Zhang, Xing, and Gao 2008). This opposite behavior of low and high doses of LLLT may be attributed to the increased amount of ROS generation with laser treatment, because they observed a similar result in H_2O_2-treated cells that a lower dose (≤100 µM) promotes cell viability, whereas a higher dose inhibits it (200 µM) (Zhang, Xing, and Gao 2008). LLLT has been suggested to induce photochemical reaction and activate several intracellular signaling pathways. ROS are considered

key secondary messengers produced with LLLT. ROS generated in the cellular level is known to act as a rapier because low levels can stimulate signal pathways that promote cell proliferation (Zhang et al. 2009), whereas high levels can result in oxidative damage (Huang, Wu, and Xing 2011). Zhang, Xing, and Gao (2008) observed that ROS generation increased as the laser dose increased from 3 to 50 J/cm² (64.6 mW/cm²).

In the following years, the inhibitive effect of high-fluence LLLT has been reported sequentially. Zungu, Hawkins, and Abrahamse (2009) reported that wounded, hypoxic, and acidotic cells irradiated using a 632.8 nm He–Ne laser at 3 mW/cm² at 5 J/cm² showed an increase in mitochondrial responses compared with nonirradiated cells, whereas 3 mW/cm² at 16 J/cm² showed a significant decrease. Murayama et al. (2012) reported that cell counts at 24 and 48 h after irradiation showed that LLLT (diode laser, 808 nm, 15 mW/cm² at 18, 36, and 54 J/cm²) suppressed the proliferation of A-172 human-derived glioblastoma cells in a fluence-dependent manner. A reduction in the number of viable cells was also demonstrated using a fluorescent marker for viable cells, calcein acetoxymethyl ester (Murayama et al. 2012). The reduction in cell viability was not associated with the morphological changes in the cells or necrotic cell death as demonstrated via propidium iodide staining (Murayama et al. 2012). LLLT also had a minimal effect on cell proliferation, as shown via 5-bromo-2′-deoxyuridine staining (Murayama et al. 2012).

47.4.2 Signaling Pathways Involved in High-Fluence LLLT-Induced Cell Apoptosis

47.4.2.1 Mitochondrial Apoptosis Pathway

Wang et al. (2005a) first reported that high-fluence 632.8 nm He–Ne laser irradiation could trigger cell apoptosis in human cancer cells. Using the FRET reporter SCAT3 (Kiwamu et al. 2003), activation of caspase-3, which marks the point of no return in apoptosis signaling (Hotchkiss et al. 2009), was observed in ASTC-a-1 cells 1 h after high-fluence LLLT (126 mW/cm² at 60 J/cm²) treatment (Wang et al. 2005a). This study provides a potential phototherapy modality using high-fluence LLLT for human cancers, with apoptosis being regarded as the major mode of cell death in cancer therapy (Kroemer, Galluzzi, and Brenner 2007).

Two distinct but convergent pathways exist: the death receptor pathway and the mitochondrial pathway (Hotchkiss et al. 2009). The death receptor pathway is activated when members of the TNF superfamily bind to the cell surface "death receptors," members of the TNF-receptor family. Ligation of these receptors initiates the formation of the multiprotein death-inducing signaling complex. The aggregation of this complex causes conformational changes in its components that trigger the catalytic activity of caspase-8, a central mediator of apoptosis. The interplay between proapoptotic and antiapoptotic members of the Bcl-2 family controls the mitochondrial apoptotic pathway. Caspase-9 regulates this pathway, which comes into play after intracellular sensors indicate overwhelming cell damage. The initiators of the pathway include increased intracellular ROS, DNA damage, unfolded protein response, and deprivation of growth factors. These initiators ultimately lead to increased mitochondrial permeability, thus promoting the release of proapoptotic proteins (e.g., cytochrome *c*) from the intermitochondrial membrane space into the cytosol. Activated caspase-8 (death receptor pathway) and caspase-9 (mitochondrial pathway) in turn mobilize caspases-3, 6, and 7, proteases that herald demolition of the cell by cleaving numerous proteins and activating DNases. Wu et al. (2007) explored the mechanisms involved in high-fluence LLLT-induced cell apoptosis in ASTC-a-1 and COS-7 cells. The following temporal sequence of cellular events was observed during the apoptotic process by high-fluence LLLT (He–Ne laser, 632.8 nm, 200 mW/cm^2 at 120 J/cm^2): (1) immediate generation of mitochondrial ROS, determined by measuring changes in fluorescence resulting from oxidation of intracellular dichlorodihydrofluorescein diacetate to dichlorodihydrofluorescein following laser irradiation, reaching a maximum level 60 min after irradiation; (2) onset of mitochondrial depolarization by measuring the reduction of cellular fluorescence of rhodamine 123 via confocal laser scanning microscopy, decreased 15 min after laser irradiation, reaching a minimum level 50 min after irradiation; and (3) activation of caspase-3, observed using FRET reporter SCAT3, between 30 and 180 min after laser irradiation. They also found that high-fluence LLLT does not activate caspase-8 by monitoring the cellular distribution of Bid-CFP reporter, indicating that the induced apoptosis is initiated directly from mitochondrial ROS generation and $\Delta\Psi$m decrease, independent of the caspase-8 activation (Wu et al. 2007). In addition, cytochrome *c* release and caspase-9 activation were subsequently observed in the following study (Chu, Wu, and Xing 2010; Wu et al. 2009). Thus, the authors conclude that 632.8 nm high-fluence LLLT induces cell apoptosis through the mitochondrial pathway.

47.4.2.2 ROS/MPT Pathway

Under physiological conditions, mitochondria exhibit a high $\Delta\Psi$m, intermembrane space (IMS) proteins are retained in IMS, proapoptotic members of the Bcl-2 family are in their inactive state [either soluble in the cytoplasm as Bax and Bid, or anchored to the mitochondrial outer membrane (MOM) as Bak], and the mitochondrial permeability transition (MPT) pore complex (MPTPC) ensures the exchange of metabolites between the cytosol and the matrix by virtue of its "flickering" activity (Kroemer, Galluzzi, and Brenner 2007). Under these circumstances, the interactions of hexokinase (HK) and cyclophilin D (CypD) with the scaffold structure of the MPTPC are likely to inhibit MOM permeabilization. One of the mechanisms associated with MOM permeabilization, which leads to the cytosol release of IMS proteins and eventually cell death, may occur through the long-lasting opening of the MPTPC, which, when associated with the loss of antiapoptotic interactions with HK and CypD, may lead to the dissipation of $\Delta\Psi$m,

followed by an osmotic imbalance that induces swelling of the mitochondrial matrix. Swelling may culminate in the physical rupture of the MOM because of the surface area of the mitochondrial inner membrane largely exceeding that of the MOM. Previous studies indicate that high-fluence LLLT (He–Ne laser, 632.8 nm) can induce cell apoptosis via the mitochondria pathway. Wu et al. (2009) further investigated the mitochondrial apoptotic process in ASTC-a-1 cells at a fluence of 200 mW/cm^2 at 120 J/cm^2. Cytochrome *c* release was ascribed to MPT because the release was prevented via cyclosporine (CsA), a specific inhibitor of MPT (Wu et al. 2009). Furthermore, mitochondrial permeability for calcein (~620 D) was another evidence for the MPT induction via high-fluence LLLT (Wu et al. 2009). CsA pretreatment also delayed mitochondrial depolarization upon high-fluence LLLT (Figure 47.19). A high-level intracellular ROS generation was observed after irradiation (Figure 47.20), which was observed previously (Wu et al. 2007). The photodynamically produced ROS caused the onset of MPT, as the ROS scavenger prevented MPT (Wu et al. 2009). Together, these results showed that high-fluence LLLT induced cell apoptosis via the CsA-sensitive MPT, which was ROS-dependent. The observed link between MPT and triggering ROS could be a fundamental phenomenon in high-fluence LLLT-induced cell apoptosis.

47.4.2.3 ROS/Akt/GSK3β Pathway

Wu et al. (2009) reported that CsA failed to prevent cell apoptosis induced with high-fluence LLLT, although MPT was demonstrated to be involved in the mitochondrial apoptotic process. This indicates the existence of other signaling pathways attributed to mitochondrial membrane permeabilization. Another mechanism for MOM permeabilization is discussed at length, which indicates that activated proapoptotic proteins of the Bcl-2 family (such as Bax) may assemble into large multimers, allowing for the release of IMS proteins (Kroemer, Galluzzi, and Brenner 2007). Following laser irradiation, Bax activation occurred after mitochondrial depolarization and cytochrome *c* release, indicating that Bax activation is a downstream event. In the presence of CsA, Bax was still activated at the end-stage apoptotic process following high-fluence LLLT, suggesting that Bax is involved in an alternative signaling pathway that was independent of MPT (Wu et al. 2009). Under high-fluence LLLT treatment, cell toxicity detection by pretreatment with DHA, CsA, or Bax RNAi demonstrated that the MPT signaling pathway was dominant, whereas Bax signaling pathway was secondary, and more important, ROS mediated both pathways (Wu et al. 2009).

The activation of GSK3β is proved to be involved in intrinsic apoptotic pathways under various stimuli (Grimes and Jope 2001). Huang, Wu, and Xing (2011) investigated the activity of proapoptotic factor GSK3β in high-fluence LLLT-induced apoptosis. Upon 633 nm He–Ne laser treatment at 200 mW/cm^2 at 120 J/cm^2, they found that GSK3β activation could promote high-fluence LLLT-induced apoptosis, which could be prevented

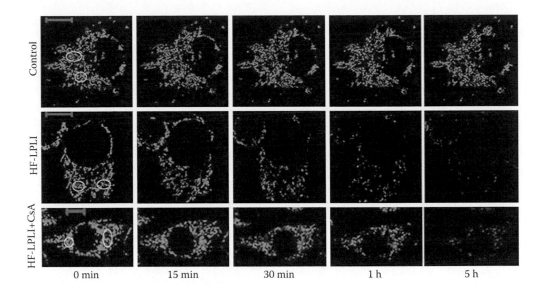

FIGURE 47.19 Time sequence of ΔΨm disappearance by high-fluence LLLT. Rhodamine 123 (green emission) was localized in mitochondria in ASTC-a-1 cells in response to the ΔΨm. Cells were treated with high-fluence LLLT or high-fluence LLLT in the presence of CsA. Fluorescence images were acquired by confocal microscopy. The time after the irradiation was indicated in each part. Cells with no treatment were control. Scale bars, 10 μm. (Adapted from Wu, S. et al., *J Cell Physiol* 218: 603–611, 2009.)

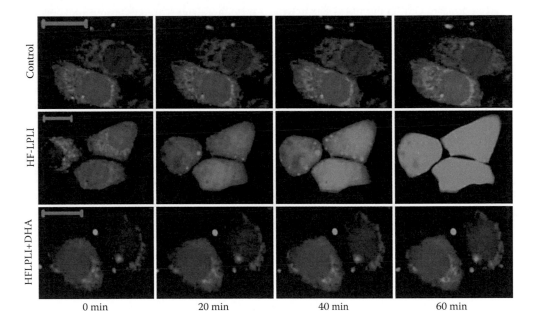

FIGURE 47.20 Time sequence of ROS generation induced by high-fluence LLLT. ASTC-a-1 cells were incubated with H_2DCFDA in serum-free medium for 30 min and then treated by high-fluence LLLT or high-fluence LLLT in the presence of DHA. The fluorescence images (green emission) were acquired by confocal microscopy. The time after the irradiation was indicated in each part. Cells with no treatment were control. Scale bars, 10 μm. (Adapted from Wu, S. et al., *J Cell Physiol* 218: 603–611, 2009.)

with lithium chloride (LiCl, a selective inhibitor of GSK3β) or with GSK3β-KD (a dominant-negative GSK3β). They also demonstrated that the activation of GSK3β with high-fluence LLLT was due to the inactivation of Akt (Figure 47.21), a widely reported and important upstream negative regulator of GSK3β, thus indicating the existence and inactivation of Akt/GSK3β signaling pathway (Huang, Wu, and Xing 2011). Furthermore, vitamin C, a ROS scavenger, completely prevented the inactivation of Akt/GSK3β pathway. This indicates that ROS generation is crucial for the inactivation of Akt/GSK3β pathway (Huang,

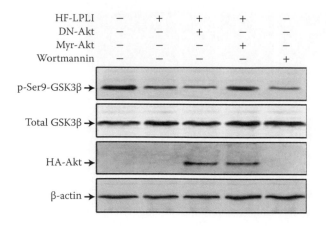

FIGURE 47.21 High-fluence LLLT induces activation of GSK3β through inactivation of Akt. ASTC-a-1 cells expressing Myr-Akt or DN-Akt were treated with high-fluence LLLT (120 J/cm²). Cells without any treatment were set as control. Wortmannin was used as a positive control. The graph was a representative Western blot analysis of phospho-Ser9-GSK3β in cells that received different treatments. (Adapted from Huang, L. et al., *J Cell Physiol* 226: 588–601, 2011.)

Wu, and Xing 2011). In addition, GSK3β promoted Bax activation with downregulating Mcl-1 upon high-fluence LLLT treatment (Huang, Wu, and Xing 2011). Together, they identify a new and important proapoptotic signaling pathway that consists of Akt/GSK3β inactivation for high-fluence LLLT stimulation. Their studies are expected to extend knowledge on the biological mechanisms induced with LLLT (Figure 47.22).

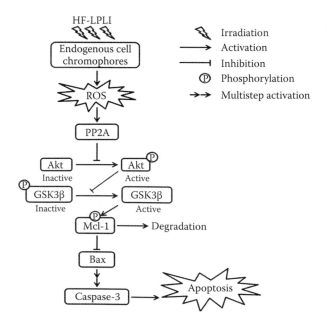

FIGURE 47.22 A model of inactivation of Akt/GSK3β signaling pathway is induced by high-fluence LLLT. (Adapted from Huang, L. et al., *J Cell Physiol* 226: 588–601, 2011.)

47.4.2.4 Mitochondrial Fission-Associated Apoptosis Pathway

Mitochondria are dynamic organelles that undergo continual fusion and fission to maintain their morphology and functions. However, the mechanisms involved in mitochondrial dynamics remain unclear (Chen and Chan 2005; Chan 2006; Liesa, Pacacin, and Zorzabo 2009). Wu et al. (2011) investigated the effect of mitochondrial ROS generation triggered with high-fluence LLLT (200 mW/cm² at 120 J/cm²) on mitochondrial dynamics in ASTC-a-1 and COS-7 cells. They found that upon high-fluence LLLT-triggered mitochondrial oxidative stress, mitochondria displayed a fragmented structure that was abolished with exposure to DHA, a ROS scavenger, thus indicating that oxidative stress can induce mitochondrial fragmentation (Wu et al. 2011). In addition, they found that high-fluence LLLT induced mitochondrial fragmentation via two processes: inhibiting fusion (Figure 47.23) and enhancing fission (Wu et al. 2011). Mitochondrial translocation of the profission protein dynamin-related protein 1 (Drp1) was observed following high-fluence LLLT, demonstrating apoptosis-related activation of Drp1 (Wu et al. 2011) (Figure 47.24). Notably, Drp1 overexpression increased mitochondrial fragmentation and promoted high-fluence LLLT-induced apoptosis by promoting cytochrome *c* release and caspase-9 activation. The opposite was observed with overexpression of mitofusin 2 (Mfn2), a profusion protein. Also, neither Drp1 overexpression nor Mfn2 overexpression affected mitochondrial ROS generation, mitochondrial depolarization, or Bax activation (Wu et al. 2011). They concluded that mitochondrial oxidative stress mediated through Drp1 and Mfn2 causes an imbalance in mitochondrial fission-fusion, resulting in mitochondrial fragmentation, which contributes to mitochondrial and cell dysfunction.

47.4.3 Signaling Pathways Involved in Self-Protection with High-Fluence LLLT

47.4.3.1 ROS/cdc25c/CDK1/Survivin Pathway

Survivin, a member of the inhibitor of apoptosis (IAP) family, can be upregulated by various proapoptotic stimuli, such as UV, photodynamic therapy, and cisplatin (Wall, O'Connor, and Plescia 2003; Ferrario, Rucker, and Wong 2007). Survivin is dramatically overexpressed in most human cancers and correlates with unfavorable prognosis, resistance to therapy, and accelerated rates of recurrence (Altieri 2001). Survivin is expressed in several subcellular compartments such as the cytosol, mitochondria, and nucleus (Altieri 2008). One of the critical requirements for survivin stability and function was recently identified in the phosphorylation on Thr34 by CDK1 (O'Connor, Grossman, and Plescia 2000), and a phosphorylation-mimetic survivin mutant strongly inhibited p53-induced apoptosis (Hoffman et al. 2002). This step has also been exploited for anticancer therapy, and inducible expression or adenoviral delivery of dominant negative mutant survivin

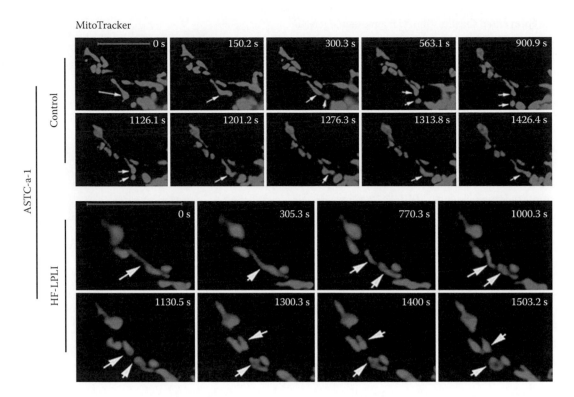

FIGURE 47.23 High-fluence LLLT inhibits mitochondrial fusion. ASTC-a-1 cells were stained with MitoTracker to localize mitochondria. Mitochondrial behavior in the cells was monitored for 25 min. Active fission and fusion (filled arrowhead) of individual mitochondria could be observed in the control cell (upper panel). Abnormal fission of individual mitochondria could be observed in high-fluence LLLT (120 J/cm^2)–treated cells (lower panel). Scale bars, 10 μm. (Adapted from Wu, S. et al., *FEBS J* 278: 941–954, 2011.)

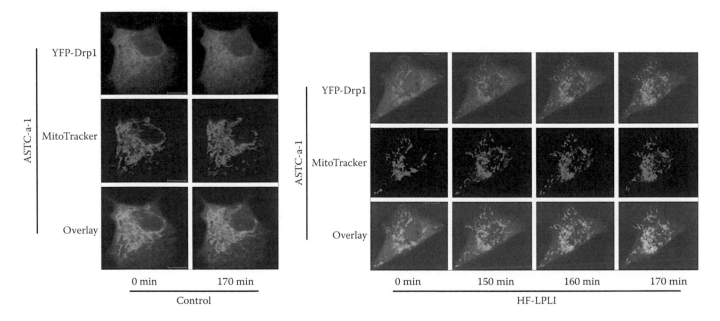

FIGURE 47.24 **(See color insert.)** ASTC-a-1 cells were transiently cotransfected with pYFP-Drp1 and pDsRed-mit. Forty-eight hours after transfection, cells coexpressing YFP-Drp1 and DsRed-mit were treated with high-fluence LLLT at 120 J/cm^2. The control group received no treatment. Representative confocal microscopic images reveal increased association of YFP-Drp1 with mitochondria in response to HF-LPLI. Scale bars, 10 μm. (Adapted from Wu, S. et al., *FEBS J* 278: 941–954, 2011.)

(T34A-survivin) prevented the phosphorylation of endogenous survivin, which resulted in caspase-9–dependent apoptosis (Grossman et al. 2001; Mesri et al. 2001). The activity of CDK1 has been known to be regulated by cdc25c phosphatase. The abrogation of Nox4-generated ROS resulted in the inhibition of cdc25c protein phosphatase activity (Yamaura, Mitsushita, and Furuta 2009), which leads to the speculation that the level of cdc25c activity is regulated by ROS. Chu, Wu, and Xing (2010) first studied the self-protection mechanism with high-fluence LLLT. They explored whether survivin is involved in the antitumor mechanism (Chu, Wu, and Xing 2010) (Figure 47.25). They reported high-fluence LLLT (632.8 nm He–Ne laser, 635 nm semiconductor laser) at 200 mW/cm² at 120 J/cm² upregulated survivin activity through the ROS/cdc25c/CDK1 signaling pathway in ASTC-a-1 cells (Chu, Wu, and Xing 2010). They found that the upregulation of survivin activity reduced laser-induced apoptosis, whereas the downregulation of the activity promoted apoptosis (Chu, Wu, and Xing 2010) (Figure 47.26). In addition, activated survivin delayed the mitochondrial depolarization, cytochrome *c* release, caspase-9, and Bax activation, all of which were typical proapoptotic events of cell apoptosis induced with high-fluence LLLT (Sun, Wu, and Xing 2010; Huang, Wu, and Xing 2011; Wu et al. 2007, 2009, 2011). The authors conclude that survivin can mediate self-protection during tumor cell apoptosis following high-fluence LLLT.

47.4.3.2 ROS/Src/Stat3 Pathway

Stat3 is an important transcription factor in the modulation of cell proliferation and apoptosis (Bromberg and Darnell Jr 2000;

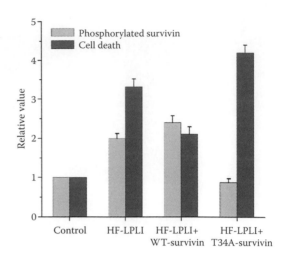

FIGURE 47.26 Analysis on correlation of the survivin phosphorylation level and the percentage of apoptotic cell death induced by high fluence LLLT. Values were normalized according to the control sample lanes that were arbitrarily set as 1. (Adapted from Chu, J. et al., *Cancer Lett* 297: 207–219, 2010.)

Levy and Lee 2002; Herrmann et al. 2007). Sun, Wu, and Xing (2010) investigated the changes in the activities of Stat3 in COS-7 cells upon 633 nm He–Ne laser (200 mW/cm² at 80 and 120 J/cm²) treatment and the underlying mechanisms via real-time single-cell analysis and Western blot analysis. They found that Stat3 was significantly activated with laser irradiation in a time- and dose-dependent manner (Sun, Wu, and Xing 2010). Stat3 activation attenuated laser-induced apoptosis, as shown by the enhancement of cellular apoptosis by both dominant negative Stat3 (DNStat3) overexpression and Stat3 RNAi (Sun, Wu, and Xing 2010). They also found that the major positive regulator of laser-induced Stat3 activation was Src kinase, the activation of which was reported in their former study (Zhang, Xing, and Gao 2008). ROS generation was essential for Src/Stat3 activation upon high-fluence LLLT, given that scavenging of ROS totally abrogated their activation (Zhang, Xing, and Gao 2008; Sun, Wu, and Xing 2010). This may explain the antiapoptosis function of Stat3, because the activated Stat3 molecules can dimerize and accumulate in the nucleus, where they induce transcription of many target genes, such as those encoding Bcl-2, Bcl-X_L, Mcl-1, survivin, cyclin D1, and c-Myc (Wang et al. 1999; Masuda et al. 2002; Yu and Jove 2004). Bcl-2, Bcl-X_L, and Mcl-1 are important members of antiapoptotic Bcl-2 family and have been reported to inhibit Bax activation under various apoptotic stimuli (Youle and Strasser 2008). Bax activation has been demonstrated to be an important step during high-fluence LLLT-induced apoptosis (Wu et al. 2009; Huang, Wu, and Xing 2011). Therefore, one of the cross-links between antiapoptotic and proapoptotic pathways under high-fluence LLLT is believed to be Stat3, via transcriptional upregulation of the antiapoptotic proteins to attenuate Bax activation. Their

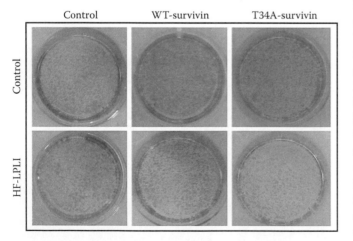

FIGURE 47.25 The effect of WT/T34A-survivin overexpression on high-fluence LLLT-induced reduction of colonogenic survival. ASTC-a-1 cells were transfected with WT-survivin or T34A-survivin for survivin overexpression and then were treated with high-fluence LLLT. Colonies formed within 1–2 weeks. After being stained with Giemsa dyes, colonies in the plates were observed on an inverted light microscope. (Adapted from Chu, J. et al., *Cancer Lett* 297: 207–219, 2010.)

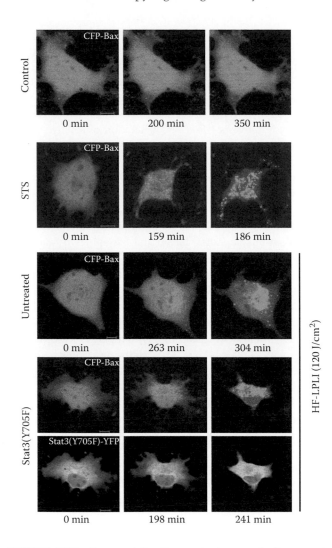

FIGURE 47.27 Representative time-series images of cells expressing CFP-Bax under high fluence LLLT with or without DN Stat3 (Y705F)-YFP overexpression. Scale bars, 10 μm. Cells treated with STS (1 μM) were used as a positive control. The data show that Stat3 inhibits Bax activation induced by high fluence LLLT. (Adapted from Sun, X. et al., *FEBS J* 277: 4789–4802, 2010.)

study supported this view, since the inhibition of Stat3 with DNStat3 obviously promoted Bax activation (Figure 47.27). On the other hand, the expression of survivin, an IAP, has also been found to be regulated via constitutively activated Stat3 in cancer cells (Aoki, Feldman, and Tosato 2003; Kanda et al. 2004), and thus, survivin may also participate in the negative feedback inhibition of apoptosis induced with high-fluence LLLT. However, the protein level of survivin was maintained in response to high-fluence LLLT, although obvious activation could be observed (Chu, Wu, and Xing 2010). These findings show that the ROS/Src/Stat3 pathway mediates a negative feedback inhibition of apoptosis induced with high-fluence LLLT in COS-7 cells (Figure 47.28). Their findings provide new insights into the mechanism of apoptosis following high-fluence LLLT and also extend the functional study of Stat3.

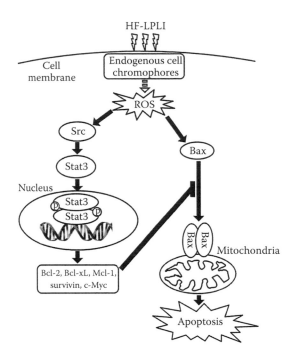

FIGURE 47.28 Model of the negative feedback inhibition induced by Stat3 activation on stimulation with high fluence LLLT. (Adapted from Sun, X. et al., *FEBS J* 277: 4789–4802, 2010.)

47.5 Perspectives

Clinical applications for LLLT include wound healing, pain attenuation, and various forms of inflammation regulation. Presently, a large number of clinical trials use LLLT, but a relatively small number of research studies on LLLT have also been reported. Currently, the wavelengths, dosage schedules, and appropriate conditions of laser irradiation are not well established. Thus, to facilitate the physician's capacity to optimally match the laser in clinical practices, the basic mechanisms of the biological effects of LLLT are examined. Recently, researchers have paid close attention to the signaling pathways involved in the biological effects of LLLT. Currently, extracellular stimuli are believed to trigger cellular responses such as proliferation, differentiation, and even apoptosis through the pathways of cellular signaling. Thus, we suppose that the investigation of the molecular events induced with LLLT can eventually reveal the mechanisms of LLLT. The close connection between mitochondrial retrograde signaling and cellular molecular events such as the activation or suppression of kinases in the cytoplasm and subsequent changes of downstream cascades is obvious. The elements of mitochondrial retrograde signaling can mediate the cellular molecular events, but at present, a few studies regarding the subject have been conducted. Further studies investigating the connection between mitochondrial retrograde signaling and cellular molecular events are needed to understand the basic mechanisms of LLLT. In this review, we summarized the studies on the molecular mechanisms of LLLT-induced proliferation, differentiation, and cell apoptosis. Among these studies, one of the most prominent trends is that advanced techniques extensively of cell biology are used less in the basic research of LLLT. Most results of the basic research studies are obtained using traditional techniques in cell biology. Thus, to reveal the molecular mechanisms of LLLT more clearly, traditional and innovative techniques should be combined. Eventually, these basic research studies on LLLT will lead to its application in clinical practice.

References

Ali, A. S., S. Ali, B. F. El-Rayes, P. A. Philip, and F. H. Sarkar. 2009. Exploitation of protein kinase C: A useful target for cancer therapy. *Cancer Treat Rev* 35:1–8.

Altieri, D. C. 2001. The molecular basis and potential role of survivin in cancer diagnosis and therapy. *Trends Mol Med* 7:542–547.

Altieri, D. C. 2008. Survivin, cancer networks and pathway-directed drug discovery. *Nat Rev Cancer* 8:61–70.

Ameyar, M., M. Wisniewska, and J. B. Weitzman. 2003. A role for AP-1 in apoptosis: The case for and against. *Biochimie* 85:747–752.

Antunes, H. S., A. M. Azevedo, L. F. S. Bouzas et al. 2007. Low power laser in the prevention of induced oral mucositis in bone marrow transplantation patients: A randomized trial. *Blood* 109:2250–2255.

Aoki, Y., G. M. Feldman, and G. Tosato. 2003. Inhibition of STAT3 signaling induces apoptosis and decreases survivin expression in primary effusion lymphoma. *Blood* 10:1535–1542.

Baldin, V., J. Lukas, M. J. Marcote, M. Pagano, and G. Draetta. 1993. Cyclin D1 is a nuclear protein required for cell cycle progression in G1. *Genes Dev* 7:812–821.

Basu, S., N. F. Totty, M. S. Irwin, M. Sudol, and J. Downward. 2003. Akt phosphorylates the yes-associated protein, YAP, to induce interaction with 14-3-3 and attenuation of p73-mediated apoptosis. *Mol Cell* 11:11–23.

Ben-Dov, N., G. Shefer, A. Irinitchev, A. Wernig, U. Oron, and O. Halevy. 1999. Low-energy laser irradiation affects satellite cell proliferation and differentiation in vitro. *Biochim Biophys Acta* 1448:372–380.

Bennet, A. M., and N. K. Tonks. 1997. Regulation of distinct stages of skeletal muscle differentiation by mitogen-activated protein kinases. *Science* 278:1288–1291.

Black, A. R., J. D. Black, and J. Azizkhan-Clifford. 2001. Sp1 and Krüppel-like factor family of transcription factors in cell growth regulation and cancer. *J Cell Physiol* 188:143–160.

Boutros, T., E. Chevet, and P. Metrakos. 2008. Mitogen-activated protein (MAP) kinase/MAP kinase phosphatase regulation: Roles in cell growth, death, and cancer. *Pharmacol Rev* 60:261–310.

Briggs, M. R., J. T. Kadonaga, S. P. Bell, and R. Tjian 1986. Purification and biochemical characterization of the promoter-specific transcription factor, Sp1. *Science* 234:47–52.

Bromberg, J., and J. E. Darnell Jr. 2000. The role of STATs in transcriptional control and their impact on cellular function. *Oncogene* 19:2468–2473.

Byrnes, K., X. Wu, R. Waynant, I. Ilev, and J. Anders. 2005. Low power laser irradiation alters gene expression of olfactory ensheathing cells in vitro. *Laser Surg Med* 37:161–171.

Chan, D. C. 2006. Mitochondria: Dynamic organelles in disease, aging, and development. *Cell* 125:1241–1252.

Chen, A. C., P. R. Arany, Y. Y. Huang et al. 2011. Low-level laser therapy activates NF-κB via generation of reactive oxygen species in mouse embryonic fibroblasts. *PLoS One* 6:e22453.

Chen, C. H., H. S. Hung, and S. H. Hsu. 2008. Low-energy laser irradiation increases endothelial cell proliferation, migration, and eNOS gene expression possibly via PI3K signal pathway. *Lasers Surg Med* 40:46–54.

Chen, H., and D. C. Chan. 2005. Emerging functions of mammalian mitochondrial fusion and fission. *Hum Mol Genet* 14:283–289.

Chen, J. X., M. L. Lawrence, G. Cunningham, B. W. Christman, and B. Meyrick. 2004. HSP90 and Akt modulate Ang-1-induced angiogenesis via NO in coronary artery endothelium. *J Appl Physiol* 96:612–620.

Chu, J., S. Wu, and D. Xing. 2010. Survivin mediates self-protection through ROS/cdc25c/CDK1 signaling pathway during tumor cell apoptosis induced by high fluence low-power laser irradiation. *Cancer Lett* 297:207–219.

Cohen, N., R. Lubart, S. Rubinstein, and H. Breitbart. 1998. Light irradiation of mouse spermatozoa: Stimulation of in vitro fertilization and calcium signals. *Photochem Photobiol* 68:407–413.

Cuenda, A., and P. Cohen. 1999. Stress activated protein kinase-2/p38 and arapamycin-sensitive pathway are required for C2C12 myogenesis. *J Biol Chem* 274:4341–4346.

de Thé, H., C. Lavau, A. Marchio et al. 1991. The PML-RAR alpha fusion mRNA generated by the t (15; 17) translocation in acute promyelocytic leukemia encodes a functionally altered RAR. *Cell* 66:675–684.

Dias, F. J., J. P. Issa, F. T. Vicentini et al. 2011. Effects of low-level laser therapy on the oxidative metabolism and matrix proteins in the rat masseter muscle. *Photomed Laser Surg* 29:677–684.

D'Melo, S. R., and G. Heinrich. 1991. Nerve growth factor gene expression: Involvement of a downstream AP-1 element in basal and modulated transcription. *Mol Cell Neurosci* 2:157–167.

Duan, R., T. C. Liu, Y. Li, H. Guo, and L. B. Yao. 2001. Signal transduction pathways involved in low intensity He-Ne laser-induced respiratory burst in bovine neutrophils: A potential mechanism of low intensity laser biostimulation. *Lasers Surg Med* 29:174–178.

Duda, D. G., D. Fukumura, and R. K. Jain. 2004. Role of eNOS in neovascularization: NO for endothelial progenitor cells. *Trends Mol Med.* 10:143–145.

Eells, J. T., M. T. Wong-Riley, J. VerHoeve et al. 2004. Mitochondrial signal transduction in accelerated wound and retinal healing by near-infrared light therapy. *Mitochondrion* 4:559–567.

El-Deiry, W. S., T. Tokino, V. E. Velculescu et al. 1993. WAF1, a potential mediator of p53 tumor suppression. *Cell* 75:817–825.

Fávaro-Pípi, E., D. A. Ribeiro, J. U. Ribeiro et al. 2011. Low-level laser therapy induces differential expression of osteogenic genes during bone repair in rats. *Photomed Laser Surg* 29:311–317.

Feng, J., Y. Zhang, and D. Xing. 2012. Low-power laser irradiation (LPLI) promotes VEGF expression and vascular endothelial cell proliferation through the activation of ERK/Sp1 pathway. *Cell Signal* 24:1116–1125.

Ferrario, A., N. Rucker, and S. Wong. 2007. Survivin, a member of the inhibitor of apoptosis family, is induced by photodynamic therapy and is a target for improving treatment response. *Cancer Res* 67:4989–4995.

Fong, Y. C., M. C. Maa, F. J. Tsai et al. 2008. Osteoblast-derived TGF-beta1 stimulates IL-8 release through AP-1 and NF-kappa B in human cancer cells. *J Bone Miner Res* 23:961–970.

Fresno, V. J. A., E. Casado, J. deCastro et al. 2004. PI3K/Akt signalling pathway and cancer. *Cancer Treat Rev* 30:193–204.

Fukuhara, E., T. Goto, T. Matayoshi, S. Kobayashi, and T. Takahashi. 2006. Optimal low-energy laser irradiation causes temporal G2/M arrest on rat calvarial osteoblasts. *Calcif Tissue Int* 79:443–450.

Gao, X., T. Chen, D. Xing, F. Wang, Y. Pei, and X. Wei. 2006. Single cell analysis of PKC activation during proliferation and apoptosis induced by laser irradiation. *J Cell Physiol* 206:441–448.

Gao, X., and D. Xing. 2009. Molecular mechanisms of cell proliferation induced by low power laser irradiation. *J Biomed Sci* 16:4.

Gao, X., D. Xing, L. Liu, and Y. Tang. 2009. H-Ras and PI3K are required for the formation of circular dorsal ruffles induced by low-power laser irradiation. *J Cell Physiol* 219:535–543.

Gavish, L., Y. Asher, Y. Becker, and Y. Kleinman. 2004. Low level laser irradiation stimulates mitochondrial membrane potential and disperses subnuclear promyelocytic leukemia protein. *Lasers Surg Med* 35:369–376.

Girard, F., U. Strausfeld, A. Fernandez, and N. J. Lamb. 1991. Cyclin A is required for the onset of DNA replication in mammalian fibroblasts. *Cell* 67:1169–1179.

Giuliani, A., L. Lorenzini, M. Gallamini et al. 2009. Low infrared laser light irradiation on cultured neural cells: Effects on mitochondria and cell viability after oxidative stress. *BMC Complement Altern Med* 9:8.

Gong, M., W. Yu, F. Pei et al. 2012. KLF6/Sp1 initiates transcription of the *tmsg*-1 gene in human prostate carcinoma cells: An exon involved mechanism. *J Cell Biochem* 113:329–339.

Greco, M., R. A. Vacca, L. Moro et al. 2001. Helium-neon laser irradiation of hepatocytes can trigger increase of the mitochondrial membrane potential and can stimulate c-fos expression in a Ca²⁺-dependent manner. *Lasers Surg Med* 9:433–441.

Gredinger, E., A. N. Gerber, Y. Tamir, S. J. Tapscott, and Bengal, E. 1998. Mitogen-activated protein kinase pathway is involved in the differentiation of muscle cells. *J Biol Chem* 273:10436–10444.

Grimes, C. A., and R. S. Jope. 2001. The multifaceted roles of glycogen synthase kinase 3beta in cellular signaling. *Prog Neurobiol* 65:391–426.

Grossman, D., P. J. Kim, J. S. Schechner, and D. C. Altieri. 2001. Inhibition of melanoma tumor growth in vivo by survivin targeting, *Proc Natl Acad Sci USA* 98:635–640.

Gur, A., A. J. Sarac, R. Cevik, O. Altindag, and S. Sarac. 2004. Efficacy of 904 nm gallium arsenide low level laser therapy in the management of chronic myofascial pain in the neck: A double-blind and randomize-controlled trial. *Lasers Surg Med* 35:229–235.

Haddad, M. M., W. Xu, D. J. Schwahn, F. Liao, and E. E. Medrano. 1999. Activation of a cAMP pathway and induction of melanogenesis correlate with association of p16(INK4) and p27(KIP1) to CDKs, loss of E2F-binding activity, and premature senescence of human melanocytes. *Exp Cell Res* 253:561–572.

Hannon, G. J., and D. Beach. 1994. p15INK4B is a potential effector of TGF-beta-induced cell cycle arrest. *Nature* 371:257–261.

Haynes, M. P., L. Li, D. Sinha et al. 2003. Src kinase mediates phosphatidylinositol 3-kinase/Akt-dependent rapid endothelial nitric-oxide synthase activation by estrogen. *J Biol Chem* 278:2118–2213.

Herrmann, A., M. Vogt, M. Monnigmann et al. 2007. Nucleocytoplasmic shuttling of persistently activated STAT3. *J Cell Sci* 120:3249–3261.

Hess, J., P. Angel, and M. Schorpp-Kistner. 2004. AP-1 subunits: Quarrel and harmony among siblings. *J Cell Sci* 117:5965–5973.

Hirata, S., C. Kitamura, H. Fukushima et al. 2010. Low-level laser irradiation enhances BMP-induced osteoblast differentiation by stimulating the BMP/Smad signaling pathway. *J Cell Biochem* 111:1445–1452.

Hoffman, W. H., S. Biade, J. T. Zilfou, J. Chen, and M. Murphy. 2002. Transcriptional repression of the anti-apoptotic survivin gene by wild type p53. *J Biol Chem* 277:3247–3257.

Hotchkiss, R. S., A. Strasser, J. E. McDunn, and P. E. Swanson. 2009. Cell death. *N Engl J Med* 361:1570–1583.

Hu, W., J. Wang, C. Yu, C. Lan, G. Chen, and H. Yu. 2007. Helium-neon laser irradiation stimulates cell proliferation through photostimulatory effects in mitochondria. *J Invest Dermatol* 127:2048–2057.

Huang, L., S. Wu, and D. Xing. 2011. High fluence low-power laser irradiation induces apoptosis via inactivation of Akt/GSK3β signaling pathway. *J Cell Physiol* 226:588–601.

Iacovelli, J., J. Lopera, M. Bott et al. 2007. Serum and forskolin cooperate to promote G1 progression in Schwann cells by differentially regulating cyclin D1, cyclin E1, and p27Kip expression. *Glia* 55:1638–1647.

Iwakiri, Y., M. H. Tsai, T. J. McCabe et al. 2002. Phosphorylation of eNOS initiates excessive NO production in early phases of portal hypertension. *Am J Physiol Heart Circ Physiol* 282:H2084–H2090.

Janssens, S. P., A. Shimouchi, T. Quertermous, D. B. Bloch, and K. D. Bloch. 1992. Cloning and expression of a cDNA encoding human endothelium-derived relaxing factor/nitric oxide synthase. *J Biol Chem* 267:14519–14522.

Reed, J. C. 2008. Bcl-2-family proteins and hematologic malignancies: History and future prospects. *Blood* 111:3322–3330.

Johnson, G. L., and R. Lapadat. 2002. Mitogen-activated protein kinase pathways mediated by ERK, JNK, and p38 protein kinases. *Science* 298:1911–1912.

Kadonaga, J. T., K. R. Carner, F. R. Masiarz, and R. Tjian. 1987. Isolation of cDNA encoding transcription factor sp1 and functional analysis of the DNA binding domain. *Cell* 51:1079–1090.

Kadonaga, J. T., and R. Tjian. 1986. Affinity purification of sequence-specific DNA binding proteins. *Proc Natl Acad Sci USA* 83:5889–5893.

Kanda, N., H. Seno, Y. Konda et al. 2004. STAT3 is constitutively activated and supports cell survival in association with survivin expression in gastric cancer cells. *Oncogene* 23:4921–4929.

Karu, T. I. 1989. *Photobiology of Low-Power Laser Therapy*. Routledge, New York.

Karu, T. I. 1998. *The Science of Low-Power Laser Therapy*. Gordon and Breach, London.

Karu, T. 1999. Primary and secondary mechanisms of action of visible to near-IR radiation on cells. *J Photochem Photobiol B* 49:1–17.

Karu, T. I. 2003. Low power laser therapy. In *Biomedical Photonics Handbook*. V. D. Tuan, editor. CRC Press, Boca Raton, FL, Chapter 48, 1–25.

Karu, T. 2007. *Ten Lectures on Basic Science of Laser Phototherapy*. Prima Books AB, Grängesberg.

Karu, T. I. 2008. Mitochondrial mechanisms of laser phototherapy. In *Proceedings of Light-Activated Tissue Regeneration and Therapy Conference*, Lecture Notes in Electrical Engineering, Vol. 12. R. Waynant and D. Tata, editors. Springer, Berlin, xxvii–xxxiv.

Karu, T. I. 2010. Multiple roles of cytochrome c oxidase in mammalian cells under action of red and IR-A radiation. *IUBMB Life* 62:607–610.

Karu, T. I., L. V. Pyatibrat, and N. I. Afanasyeva. 2005. Cellular effects of low power laser therapy can be mediated by nitric oxide. *Lasers Surg Med* 36:307–314.

Karu, T. I., L. V. Pyatibrat, S. F. Kolyakov, and N. I. Afanasyeva. 2005. Absorption measurements of a cell monolayer relevant to phototherapy: Reduction of cytochrome c oxidase under near IR radiation. *J Photochem Photobiol B* 81:98–106.

Karu, T. I., L. V. Pyatibrat, and G. S. Kalendo. 2004. Photobiological modulation of cell attachment via cytochrome c oxidase. *Photochem Photobiol Sci.* 3:211–216.

Karu, T., L. Pyatibrat, and G. Kalendo. 1995. Irradiation with He–Ne laser increases ATP level in cells cultivated in vitro. *J Photochem Photobiol B* 27:219–223.

Karu, T. I., and O. A. Tiphlova. 1987. Effect of irradiation with monochromatic visible light on cAMP content in Chinese hamster fibroblasts. *Il Nuovo Cimento* 9:1245–1251.

Kato, J. Y., M. Matsuoka, K. Polyak, J. Massague, and C. J. Sherr. 1994. Cyclic AMP-induced G1 phase arrest mediated by an inhibitor (p27Kip1) of cyclin-dependent kinase 4 activation. *Cell* 79:487–496.

Kim, L. C., L. Song, and E. B. Haura. 2009. Src kinases as therapeutic targets for cancer. *Nat Rev Clin Oncol* 6:587–595.

Kim, T. Y., W. I. Kim, R. E. Smith and E. D. Kay. 2001. Role of p27(Kip1) in cAMP- and TGF-beta2-mediated antiproliferation in rabbit corneal endothelial cells. *Invest Ophthalmol Vis Sci* 42:3142–3149.

Kiwamu, T., N. Takeharu, M. Atsushi, and M. Masayuki. 2003. Spatiotemporal activation of caspase revealed by indicator that is insensitive to environmental effects. *J Cell Biol* 160:235–243.

Kiyosaki, T., N. Mitsui, N. Suzuki, and N. Shimizu. 2010. Low-level laser therapy stimulates mineralization via increased runx2 expression and ERK phosphorylation in osteoblasts. *Photomed Laser Surg* 28:1.

Krasagakis, K., D. Tholke, B. Farthmann et al. 1998. Elevated plasma levels of transforming growth factor (TGF)-beta1 and TGF-beta2 in patient with disseminated malignant melanoma. *Br J Cancer* 77:1492–1494.

Krishna, M., and H. Narang. 2008. The complexity of mitogen-activated protein kinases (MAPKs) made simple. *Cell Mol Life Sci* 65:3525–3544.

Kroemer, G., L. Galluzzi, and C. Brenner. 2007. Mitochondrial membrane permeabilization in cell death. *Physiol Rev* 87:99–163.

Krueger, E. W., J. D. Orth, H. Cao, and M. A. McNiven. 2003. A dynamin-cortactin-Arp2/3 complex mediates actin reorganization in growth factor-stimulated cells. *Mol Biol Cell* 14:1085–1096.

Lapchak, P. A., and L. D. Taboada. 2010. Transcranial near infrared laser treatment (NILT) increases cortical adenosine-5′-triphosphate (ATP) content following embolic strokes in rabbits. *Brain Res* 1306:100–105.

Lavi, R., A. Shainberg, H. Friedmann et al. 2003. Low energy visible light induces reactive oxygen species generation and stimulates an increase of intracellular calcium concentration in cardiac cells. *J Biol Chem* 278:40917–40922.

Leppanen, V. M., A. E. Prota, M. Jeltsch et al. 2010. Structural determinants of growth factor binding and specificity by VEGF receptor 2. *Proc Natl Acad Sci USA* 107:2425–2430.

Levy, D. E., and C. Lee. 2002. What does Stat3 do? *J Clin Invest* 109:1143–1148.

Li, D. Q., S. B. Pakala, S. D. Reddy et al. 2011. Bidirectional autoregulatory mechanism of metastasis-associated protein 1-alternative reading frame pathway in oncogenesis. *Proc Natl Acad Sci USA* 108:8791–8796.

Li, S. L., N. Cougnon, L. Bresson-Bépoldin, S. J. Zhao, and W. Schlegel. 1996. c-fos mRNA and FOS protein expression is induced by Ca^{2+} influx in GH3B6 pituitary cells. *J Mol Endocrinol* 16:229–238.

Liang, H. L., H. T. Whelan, J. T. Eells et al. 2006. Photobiomodulation partially rescues visual cortical neurons from cyanide-induced apoptosis. *Neuroscience* 139:639–649.

Liesa, M., M. Palacin, and A. Zorzano. 2009. Mitochondrial dynamics in mammalian health and disease. *Physiol Rev* 89:799–845.

Lim, W. B., J. H. Kim, E. B. Gook et al. 2009. Inhibition of mitochondria-dependent apoptosis by 635-nm irradiation in sodium nitroprusside-treated SH-SY5Y cells. *Free Radical Bio Med* 47:850–857.

Ling, L., C. Dombrowski, K. M. Foong et al. 2010. Synergism between Wnt3a and heparin enhances osteogenesis via a phosphoinositide 3-kinase/Akt/RUNX2 pathway. *J Biol Chem* 285:26233–26244.

Loyer, P., S. Cariou, D. Glaise et al. 1996. Growth factor dependence of progression through G1 and S phases of adult rat hepatocytes in vitro. *J Biol Chem* 271:11484–11492.

Maiya, G. A., P. Kumar, and L. Rao. 2005. Effect of low intensity helium-neon (He–Ne) laser irradiation on diabetic wound healing dynamics. *Photomed Laser Surg* 23:187–90.

Massagué, J. 2004. G1 cell-cycle control and cancer. *Nature* 432:298–306.

Masuda, M., M. Suzui, R. Yasumatu et al. 2002. Constitutive activation of signal transducers and activators of transcription 3 correlates with cyclin D1 overexpression and may provide a novel prognostic marker in head and neck squamous cell carcinoma. *Cancer Res* 62:3351–3355.

Matallanas, D., D. Romano, K. Yee et al. 2007. RASSF1A elicits apoptosis through an MST2 pathway directing proapoptotic transcription by the p73 tumor suppressor protein. *Mol Cell* 27:962–975.

Matte, S., M. P. Colombo, C. Melani et al. 1994. Expression of cytokine/growth factors and their receptors in human melanoma and melanocytes. *Int J Cancer* 56:853–857.

Maturana, A., G. V. Haasteren, I. Piuz et al. 2002. Spontaneous calcium oscillations control c-fos transcription via the serum response element in neuroendocrine cells. *J Biol Chem* 277:39713–39721.

Mesri, M., N. R. Wall, J. Li, R. W. Kim, and D. C. Altieri. 2001. Cancer gene therapy using a survivin mutant adenovirus. *J Clin Invest* 108:981–990.

Mester, E., G. Lunday, M. Sellyei, and G. Gycnes. 1968. Untersuchungen üeber die hemmende bzw. foerdernde Wirkung der Laserstrahlen. *Arch Klin Chir* 322:1022.

Mester, E., T. Spiry, B. Szende, and J. G. Tota. 1971. Effects of laser rays on wound healing. *Am J Surg* 22:532–535.

Miyashita, T., S. Krajewski, M. Krajewska et al. 1994. Tumor-suppressor p53 is a regulator of Bcl-2 and BAX gene-expression in-vitro and in-vivo. *Oncogene* 9:1799–1805.

Miyata, H., T. Genma, M. Ohshima et al. 2006. Mitogen-activated protein kinase/extracellular signal-regulated protein kinase activation of cultured human dental pulp cells by low-power gallium–aluminium–arsenic laser irradiation. *Int Endod J* 39:238–244.

Murayama, H., K. Sadakane, B. Yamanoha, and S. Kogure. 2012. Low-power 808-nm laser irradiation inhibits cell proliferation of a human-derived glioblastoma cell line in vitro. *Lasers Med Sci* 27:87–93.

Newton, A. C. 2009. Lipid activation of protein kinases. *J Lipid Res* 50:S266–S271.

Nurse, P. 1994. Ordering S phase and M phase in the cell cycle. *Cell* 79:547–550.

O'Connor, D. S., D. Grossman, and J. Plescia. 2000. Regulation of apoptosis at cell division by p34cdc2 phosphorylation of surviving. *Proc Natl Acad Sci USA* 97:13103–1317.

Pastore, D., M. Greco, and S. Passarella. 2000. The specific helium-neon laser sensitivity of the purified cytochrome c oxidase. *Int J Rad Biol* 76:863–870.

Patel, R. P., J. McAndrew, H. Sellak et al. 1999. Biological aspects of reactive nitrogen species. *Biochim Biophys Acta* 1411:385–400.

Pearson, G., F. Robinson, T. B. Gibson et al. 2001. Mitogen-activated protein (MAP) kinase pathways: Regulation and physiological functions. *Endocr Rev* 22:153–183.

Peng, F., H. Wu, Y. Zheng, X. Xu, and J. Yu. 2012. The effect of noncoherent red light irradiation on proliferation and osteogenic differentiation of bone marrow mesenchymal stem cells. *Lasers Med Sci* 27:645–653.

Pomerance, M., H. B. Abdullah, S. Kamerji, C. Corrèze, and J. P. Blondeau. 2000. Thyroid-stimulating hormone and cyclic AMP activate p38 mitogen-activated protein kinase cascade. *J Biol Chem* 275:40539–40546.

Posten, W., D. A. Wrone, J. S. Dover et al. 2005. Low-level laser therapy for wound healing: Mechanism and efficacy. *Dermatol Surg* 31:334–340.

Puri, P. L., Z. Wu, P. Zhang et al. 2000. Induction of terminal differentiation by constitutive activation of p38 MAP kinase in human rhabdomyosarcoma cells. *Genes Dev* 14:574–584.

Rao, S., J. Gray-Bablin, T. W. Herliczek, and K. Keyomarsi. 1999. The biphasic induction of p21 and p27 in breast cancer cells by modulators of cAMP is posttranscriptionally regulated and independent of the PKA pathway. *Exp Cell Res* 252:211–223.

Reed, J. C. 2008. Bcl-2-family proteins and hematologic malignancies: History and future prospects. *Blood* 111:3322–3330.

Rui, H. L., Y. Y. Wang, H. Cheng, and Y. P. Chen. 2012. JNK-dependent AP-1 activation is required for aristolochic acid-induced TGF-β1 synthesis in human renal proximal epithelial cells. *Am J Physiol Renal Physiol* 302:F1569–F1575.

Santiago, F. S., H. Ishii, S. Shafi et al. 2007. Yin Yang-1 inhibits vascular smooth muscle cell growth and intimal thickening by repressing p21WAF1/Cip1 transcription and p21WAF1/Cip1-Cdk4-cyclin D1 assembly. *Circ Res* 101:146–155.

Saygun, I., N. Nizam, A. U. Ural et al. 2012. Low-level laser irradiation affects the release of basic fibroblast growth factor (bFGF), insulin-like growth factor-I (IGF-I), and receptor of IGF-I (IGFBP3) from osteoblasts. *Photomed Laser Surg* 30:149–154.

Schadendorf, D., A. Moller, B. Algermissen, M. Worm, M. Sticherling, and B. M. Czarbetzki. 1993. IL-8 produced by human malignant melanoma cells in vitro is an essential autocrine growth factor. *J Immunol* 151:2667–2675.

Schafer, G., T. Cramer, G. Suske et al. 2003. Oxidative stress regulates vascular endothelial growth factor-A gene transcription through Sp1- and Sp3-dependent activation of two proximal GC-rich promoter elements. *J Biol Chem* 278:8190–8198.

Shefer, G., I. Barash, U. Oron, and O. Halevy. 2003. Low-energy laser irradiation enhances de novo protein synthesis via its effects on translation-regulatory proteins in skeletal muscle myoblasts. *Biochim Biophys Acta* 1593:131–139.

Shefer, G., U. Oron, A. Irintchev, A. Wernig, and O. Halevy. 2001. Skeletal muscle cell activation by low-energy laser irradiation: A role for the MAPK/ERK pathway. *J Cell Physiol* 187:73–80.

Shefer, G., T. A. Partridge, L. Heslop et al. 2002. Low-energy laser irradiation promotes the survival and cell cycle entry of skeletal muscle satellite cells. *J Cell Sci* 115:1461–1469.

Sherr, C. J. 2001. The INK4a/ARF network in tumour suppression. *Nat Rev Mol Cell Biol* 2:731–737.

Silveira, P. C. L., L. A. da Silva, D. B. Fraga et al. 2009. Evaluation of mitochondrial respiratory chain activity in muscle healing by low-level laser therapy. *J Photochem Photobiol B* 95:89–92.

Simoncini, T., A. Hafezi-Moghadam, D. P. Brazil et al. 2000. Interaction of oestrogen receptor with the regulatory subunit of phosphatidylinositol-3-OH kinase. *Nature* 407:538–541.

Sommer, A. P., J. Bieschke, R. P. Friedrich et al. 2012. 670 nm laser light and EGCG complementarily reduce amyloid-β aggregates in human neuroblastoma cells: Basis for treatment of Alzheimer's disease? *Photomed Laser Surg* 30:54–60.

Soneja, A., M. Drews, and T. Malinski. 2005. Role of nitric oxide, nitroxidative and oxidative stress in wound healing. *Pharmacol Rep* 57:108–191.

Starzec, A. B., E. Spanakis, A. Nehme et al. 1994. Proliferative responses of epithelial cells to 8-bromo-cyclic AMP and to a phorbol ester change during breast pathogenesis. *J Cell Physiol* 161:31–38.

Strano, S., O. Monti, N. Pediconi et al. 2005. The transcriptional coactivator yes-associated protein drives p73 gene-target specificity in response to DNA damage. *Mol Cell* 18:447–459.

Sun, X., S. Wu, and D. Xing. 2010. The reactive oxygen species-Src-Stat3 pathway provokes negative feedback inhibition of apoptosis induced by high-fluence low-power laser irradiation. *FEBS J* 277:4789–4802.

Szabo, C., H. Ischiropoulos, and R. Radi. 2007. Peroxynitrite: Biochemistry, pathophysiology and development of therapeutics. *Nat Rev Drug Discov* 6:662–680.

Tafur, J., and P. J. Mills. 2008. Low-intensity light therapy: Exploring the role of redox mechanisms. *Photomed Laser Surg* 26:321–326.

Taniguchi, D., P. Dai, T. Hojo et al. 2009. Low-energy laser irradiation promotes synovial fibroblast proliferation by modulating p15 subcellular localization. *Lasers Surg Med* 241:232–239.

Tata, D. B., and R. W. Waynant. 2011. Laser therapy: A review of its mechanism of action and potential medical applications. *Laser Photonics Rev* 5:1–12.

Virag, L., E. Szabo, P. Gergely, and C. Szabo. 2003. Peroxynitrite-induced cytotoxicity: Mechanism and opportunities for intervention. *Toxicol Lett* 140-1:113–124.

Wall, N. R., D. S. O'Connor, and J. Plescia. 2003. Suppression of survivin phosphorylation on Thr34 by flavopiridol enhances tumor cell apoptosis. *Cancer Res* 63:230–235.

Wang, F., T. S. Chen, D. Xing, J. J. Wang, and Y. X. Wu. 2005a. Measuring dynamics of caspase-3 activity in living cells using FRET technique during apoptosis induced by high fluence low-power laser irradiation. *Lasers Surg Med* 36:2–7.

Wang, J. M., J. R. Chao, W. Chen et al. 1999. The antiapoptotic gene mcl-1 is up-regulated by the phosphatidylinositol 3-kinase/Akt signaling pathway through a transcription factor complex containing CREB. *Mol Cell Biol* 19:6195–2066.

Wang, Y. X., E. L. Botvinick, Y. H. Zhao et al. 2005b. Visualizing the mechanical activation of Src. *Nature* 434:1040–1045.

Wu, S., D. Xing, X. Gao, and W. R. Chen. 2009. High fluence low-power laser irradiation induces mitochondrial permeability transition mediated by reactive oxygen species. *J Cell Physiol* 218:603–611.

Wu, S., D. Xing, F. Wang, T. Chen, and W. R. Chen. 2007. Mechanistic study of apoptosis induced by high-fluence low-power laser irradiation using fluorescence imaging techniques. *J Biomed Opt* 12:064015.

Wu, S., F. Zhou, Z. Zhang, and D. Xing. 2011. Mitochondrial oxidative stress causes mitochondrial fragmentation via differential modulation of mitochondrial fission–fusion proteins. *FEBS J* 278:941–954.

Yamamoto, M., K. Tamura, K. Hiratsuka, and Y. Abiko. 2001. Stimulation of MCM3 gene expression in osteoblast by low level laser irradiation. *Lasers Med Sci* 16:213–217.

Yamaura, M., J. Mitsushita, and S. Furuta. 2009. NADPH oxidase 4 contributes to transformation phenotype of melanoma cells by regulating G2-M cell cycle progression. *Cancer Res* 69:2647–2654.

Yazdani, S. O., A. F. Golestaneh, A. Shafiee et al. 2012. Effects of low level laser therapy on proliferation and neurotrophic factor gene expression of human Schwann cells in vitro. *J. Photochem Photobiol B* 107:9–13.

Yip, K. K., S. C. Lo, M. C. Leung et al. 2011. The effect of low-energy laser irradiation on apoptotic factors following experimentally induced transient cerebral ischemia. *Neuroscience* 190:301–306.

Youle, R. J., and A. Strasser. 2008. The BCL-2 protein family: Opposing activities that mediate cell death. *Nat Rev Mol Cell Biol* 9:47–59.

Yu, H. S., K. L. Chang, C. L. Yu et al. 1996. Low-energy helium–neon laser irradiation stimulates interleukin-1 alpha and interleukin-8 release from cultured human keratinocytes. *J Invest Dermatol* 107:593–596.

Yu, H., and R. Jove. 2004. The STATs of cancer-new molecular targets come of age. *Nat Rev Cancer* 4:97–105.

Yu, H. S., C. S. Wu, C. L. Yu, Y. H. Kao, and M. H. Chiou. 2003. Helium-neon laser irradiation stimulates migration and proliferation in melanocytes and induces repigmentation in segmental-type vitiligo. *J Invest Dermatol* 120:56–64.

Zetser, A., E. Gredinger, and E. Bengal. 1999. p38 mitogen-activated protein kinase pathway promotes skeletal muscle differentiation. *J Biol Chem* 274:5193–5200.

Zhang, C. P., T. L. Hao, P. Chen et al. 2011. Effect of low level laser irradiation on the proliferation of myoblasts—The skeletal muscle precursor cells: An experimental in vitro study. *Laser Phys* 21:2122–21227.

Zhang, H., S. Wu, and D. Xing. 2011. YAP accelerates $A\beta_{25-35}$-induced apoptosis through upregulation of Bax expression by interaction with p73. *Apoptosis* 16:808–821.

Zhang, H., S. Wu, and D. Xing. 2012. Inhibition of $A\beta_{25-35}$-induced cell apoptosis by low-power-laser-irradiation (LPLI) through promoting Akt-dependent YAP cytoplasmic translocation. *Cell Signal* 24:224–232.

Zhang, J., D. Xing, and X. Gao. 2008. Low-power laser irradiation activates Src tyrosine kinase through reactive oxygen species-mediated signaling pathway. *J Cell Physiol* 217:518–528.

Zhang, L., D. Xing, X. Gao, and S. Wu. 2009. Low-power laser irradiation promotes cell proliferation by activating PI3K/Akt pathway. *J Cell Physiol* 219:553–562.

Zhang, L., D. Xing, D. Zhu, and Q. Chen. 2008. Low-power laser irradiation inhibiting $A\beta_{25-35}$-induced PC12 cell apoptosis via PKC activation. *Cell Physiol Biochem* 22:215–222.

Zhang, L., Y. Zhang, and D. Xing. 2010. LPLI inhibits apoptosis upstream of Bax translocation via a GSK-3β-inactivation mechanism. *J Cell Physiol* 224:218–228.

Zungu, I. L., E. D. Hawkins, and H. Abrahamse. 2009. Mitochondrial responses of normal and injured human skin fibroblasts following low level laser irradiation—An in vitro study. *Photochem Photobiol* 85:987–996.

48

Irradiation Parameters, Dose Response, and Devices

James D. Carroll
THOR Photomedicine Ltd

This chapter reviews the irradiation parameters, dose response, treatment intervals, and devices used in low-level laser therapy (LLLT).

48.1 Overview

For LLLT to be effective, the various irradiation parameters (wavelength, fluence, power, irradiance, pulse parameters, and some would argue coherence and polarization as well) need to lie within certain ranges and be applied for a suitable amount of time (usually seconds or minutes). LLLT is typically applied several times (1–10 treatment sessions) and at intervals ranging from twice a week to twice a day.

48.2 Irradiation Parameters

If the wrong irradiation parameters are used or applied for the wrong irradiation time (dose), then treatment will be ineffective. If the irradiance is too low and/or the time is too short, then there is no significant effect; alternatively, if the irradiance is too high and/or the treatment time is too long, then the benefit is lost and sometimes inhibitory effects are seen (Huang et al. 2009, 2011). Unfortunately, many authors of research papers fail to accurately measure or even report some of these parameters. This is due part to a poor appreciation of the relevance of these parameters by authors and reviewers, and also because some of these measurements require expensive instruments that need to be operated by trained engineers/physicists (Jenkins and Carroll 2011). The irradiation parameters are listed and described in Table 48.1.

48.3 Dose Parameters and Dose Response

Having established suitable irradiation parameters, they must be applied for an adequate amount of time. If the wrong irradiation parameters are used or applied for the wrong irradiation time, then treatment will be ineffective. Energy or fluence is often referred to as dose. These are different parameters and both are potentially flawed methods of recording the dose. Table 48.2 shows the formulas and discusses the limitations.

48.4 Dose Response and Dose Rate Effects

It has been shown many times that if insufficient energy is applied, then there is no effect, or if too much is applied, then there may be inhibitory effects, and that these results are also dependent on the irradiance of the beam (Huang et al. 2009, 2011) (the underlying mechanisms for this phenomenon is discussed elsewhere in Chapter 5). Typically irradiances between 10 and 100 mW/cm^2 applied for 30–150 s are effective at stimulating tissue repair and reducing inflammation in superficial tendinopathies, joint pain, and wounds. The World Association for Laser Therapy web site (www.walt.nu) maintains updated treatment guidelines for tendinopathies and joint conditions. Irradiance at the skin surface needs to be considerably greater for deeper targets such as low in back pain because light is highly scattered in tissue and absorbed by various tissues on the way down such that about 0.1% of incident irradiance on the skin surface might reach a target of 5 cm deep. There are times when inhibition is desirable, such as for analgesic

TABLE 48.1   Irradiation Parameters for LLLT

Irradiation Parameter	Unit of Measurement	
Wavelength	nm	Light consists of packets of electromagnetic energy that also have a wavelike property. Wavelength is expressed in nanometers and is visible from approximately the 400- to 750-nm range (du Nouy 1921; Graham and Hartline 1935; Hecht and Williams 1922; Kolb, Fernandez, and Nelson 1995); light from approximately 750 to 1500 nm is invisible and defined as near infrared. It is the structure of chromophores and their redox state that determines which wavelengths will be absorbed (Karu et al. 2008). LLLT devices are typically within the range of 600–1000 nm as there are many absorption peaks for cytochrome c oxidase in that range (Karu 2010; Sommer et al. 2001), they penetrate tissues reasonably well, and many clinical trials have been successful with them (though not in the 700–750 nm range) (Liang et al. 2008). There is some contention about wavelengths above 900 nm as they are probably absorbed by water and not by cytochrome c oxidase. However, excitation of CCO seems unlikely to be the mechanism, but these wavelengths do stimulate ATP production (Benedicenti et al. 2008) and they work clinically (Bjordal et al. 2008; Gur et al. 2004), so at this time, we speculate that absorption by water in the phospholipid bilayers may cause molecular vibration sufficient to perturb ion channels and alter cellular function. If deep penetration is required (>1 up to 5 cm), then wavelengths in the 690–850 nm range penetrate best (Smith 1991).
Power	W	Peak and average.
Beam area	cm²	Beam area is required for calculating irradiance, but is difficult to measure and frequently misreported (Jenkins and Carroll 2011). Diode laser beams are typically not round (more often they are like an ellipse), and the beams are usually brighter in the middle and gradually weaker toward the edge (Gaussian distribution). This has been poorly understood by many researchers and errors are frequently made when reporting the beam area. For example, many assume that the aperture of a device defines the beam size. It rarely does; the correct way to measure the beam area is with a beam profiler and report the $1/e^2$ area (Dickey and Holswade 2000). This is a job for a laser engineer/physicist and not a doctor/therapist.
Irradiance	W/cm²	Often called power density (technically incorrect), irradiance is the product of power (W)/beam area (cm²). This parameter is frequently misreported because of the difficulty with measuring the beam area (Dickey and Holswade 2000; Jenkins and Carroll 2011). Read more about measurement of the beam area above. Assuming we trust the parameters reported, then how much power density is required? Studies that have taken the trouble to measure beam irradiance carefully and take measurements at the target depth report successful tissue repair and anti-inflammatory effects in the range of 5–55 mW/cm² at the target tissue depth (Castano et al 2007; Lanzafame et al. 2007; Oron et al. 2001). Analgesia is a different matter; a systematic review of laboratory studies found higher irradiances of between 300 and 1730 mW/cm² necessary to inhibit nerve conduction in C fibers and A-delta fibers (Chow et al. 2011).
Pulse structure	Peak power (W), pulse frequency (Hz), pulse width (s), duty cycle (%)	If the beam is pulsed, then the power reported should be the average power and calculated as follows: Peak power (W) × pulse width (s) × pulse frequency (Hz) = average power (W). A review (Hashmi et al. 2010) concluded that pulsing in LLLT can be significantly better than CW; however, the optimal frequencies and pulse duration (or pulse intervals) are unknown at this time. It has been established that frequencies over 100 Hz have no different effects from a CW (not pulsed) beam and that much lower frequencies (around 10 Hz) may be more effective in reducing oxidative stress or increasing tissue regeneration. It is unknown at this time if this is frequency-, pulse-length-, or rest-period-dependent. Many laser systems produce a continuous beam only, but some produce a fixed pulse width and fixed pulse frequency. More common is a variable pulse frequency with a fixed duty cycle, e.g., a 10 Hz pulse with a 50% duty cycle has a pulse width of 1/10 × 0.5 = 0.05 s (50 ms) on and a 50 ms off period. If the pulse frequency is increased, then the average power remains constant so a 100 mW peak power laser will deliver 50 mW average power. Other duty cycles are sometimes used, e.g., 90:10 so 90% on, 10% off, so a 100 mW peak power laser will deliver 90 mW average. Another common format is a fixed pulse width and variable pulse frequency. In this format, changing the pulse frequency affects the average power. For example, if the pulse is 1 ms and the frequency is 10 Hz, then there are 10 pulses 1 ms wide. If the peak power were 100 mW, then the average power would be 0.1 W × 1 ms × 10 Hz = 1 mW average power. If the pulse frequency was increased to 20 Hz, the average power would increase to 2 mW. There are LLLT laser devices marketed as "superpulsed" because they have a very high peak power in the range 1–100 W and very narrow fixed pulse width (typically 200 µs), but the average power is usually limited to 10–100 mW. Superpulsed lasers are usually 904-/905 nm devices.

(continued)

TABLE 48.1 Irradiation Parameters for LLLT (Continued)

Irradiation Parameter	Unit of Measurement	
Coherence	Coherence length depends on spectral bandwidth	Coherent light produces laser speckle, which has been postulated to play a role in the photobiomodulation interaction with cells and subcellular organelles. The dimensions of speckle patterns coincide with the dimensions of organelles such as mitochondria, and the intensity gradients produced by these speckles may help improve clinical effects particularly in deep tissues where irradiance is low. No definitive trials have been published to date to confirm or refute this claim (Corazza et al. 2007; Zalevsky and Belkin 2011).
Polarization	Linear polarized or circular polarized	Polarized light may have different effects than otherwise identical nonpolarized light (or even 90° rotated polarized light). It is known that polarized light is rapidly scrambled in highly scattering media such as tissue (probably in the first few hundred micrometers).
		However, for birefringent protein structures such as collagen, the transmission of plane polarized light will depend on orientation. Several authors have demonstrated effects on wound healing and burns with polarized light (Durovic et al. 2008; Iordanou et al. 2002; Karadag et al. 2007; Oliveira et al. 2010).

ATP = adenosine triphosphate, CC = convergent close coupling, CW: continuous wave.

TABLE 48.2 Dose Parameters of Time/Energy/Fluence

Dose Parameter	Unit of Measurement	
Energy (joules)	J	Calculated as power (W) × time (s) = energy (joules).
		Using Joules as an expression of dose is potentially unreliable as it assumes an inverse relationship between power and time and it ignores irradiance. If a 100 mW laser is applied over two points on an Achilles tendon injury for 80 s, then 8 J has been delivered per point. What this does not tell you is the irradiance of the beam (see Table 48.1), which if too high could cause the treatment to fail. Systematic reviews have established that superficial tendon injuries should have a beam irradiance of less than 100 mW/cm² (Bjordal et al. 2008; Tumilty et al. 2010). Unfortunately, many authors have failed to report irradiance, so for that reason, the treatment effect is hard to replicate. A second problem is that of reciprocity. If the power is doubled and the time halved, the correct energy may be applied but the results may be different (Lanzafame et al. 2007; Schindl, Rosado-Schlosser, and Trautinger 2001). To be sure of replicating a successful treatment, ideally the same power, beam area, and time should be used. Using more powerful lasers as a way of reducing treatment time is not a reliable strategy.
Fluence (energy density)	J/cm²	Calculated as power (W) × time (s)/beam area = energy density (J/cm²).
		Using energy density as an expression of dose is also potentially unreliable as it assumes an inverse relationship between power, time, and irradiance. Again there is no reciprocity. If the power is doubled and the time halved, the correct energy may be applied but the results may be different (Lanzafame et al. 2007; Schindl, Rosado-Schlosser, and Trautinger 2001). If the beam area is halved, the irradiance may remain correct but the total energy applied will be halved and may not cover the whole pathology. To be sure of replicating a successful treatment, ideally the same power, beam area, and time should be used. Using more powerful lasers as a way of reducing treatment time is not a reliable strategy.
Irradiation time (seconds)	s	Given the lack of reciprocity described above, the safest way to record and prescribe LLLT is to define the irradiation parameters and then define the irradiation time and not rely on energy or fluence parameters alone. Treatment times vary significantly from a few seconds to many minutes but more often in the range 30–150 s (Bjordal et al. 2008, 2011; Chow et al. 2009).
Treatment interval	Hours, days, or weeks	The effects of different treatment interval is underexplored at this time, although there is sufficient evidence to suggest that this is an important parameter (Brondon, Stadler, and Lanzafame 2005; Lanzafame et al. 2007).With the exception of some early treatment of acute injuries, LLLT typically requires two or more treatments a week for several weeks to achieve clinical significance (Bjordal et al. 2003).

effects. Irradiances of at least 300 mW/cm² over nerves have been shown to slow conduction velocity and reduce amplitude in A-delta and C fibers in A-delta and C fibers (Chow et al. 2011).

48.5 Devices

There is a bewildering array of LLLT devices ranging in price from $100 to $100,000, some emitting a beam of just 1 mW and others as much as 10 W in a single beam. There are cluster arrays comprising just a few lasers or light-emitting diodes (LEDs) and others containing hundreds of emitters and delivering a combined output of 150 W of light from a 12,000 cm² canopy that treats an entire human body. Most devices are handheld and applied in contact by a doctor, therapist, or nurse; others may be on a pedestal and project the light onto the patient (see Figure 48.1).

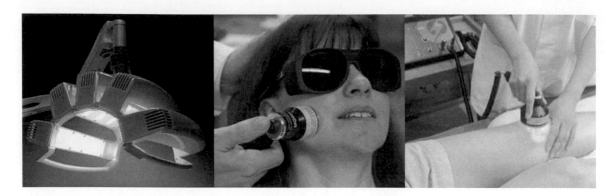

FIGURE 48.1 Examples of LLLT devices.

These devices typically offer wavelengths in the red and near-infrared, although other visible wavelengths and long infrared wavelengths are available (also see Table 48.1). Some devices are lasers while others are LEDs, some may be a mix of laser and LEDs, and some may be broad spectrum polarized lamps (400–2000 nm).

This intimidating range of devices makes a decision hard for the purchaser, but here are some suggestions.

1. Ask the manufacturer for a copy of their regulatory compliance certificate for your country [e.g., Food and Drug Administration (FDA) in the United States, Conformite Europeanne (CE) mark in Europe, Therapeutic Goods Administration (TGA) in Australia, Health Canada in Canada] and ask what the approved indications are.

2. When choosing wavelength, power, power density, and pulses, what are you intending to treat? First check the latest guidelines from the World Association for Laser Therapy (www.walt.nu) to see if they have issued guidelines. At the time of writing, there is advice on treating tendinopathies and arthritic joints.

3. Wounds are probably best treated with a large LED array emitting red light, but musculoskeletal pathologies need stronger near-infrared lasers (780–905 nm) to deactivate trigger points and reach deep anatomical targets such as in low back pain.

4. Dental applications need autoclavable or sterilizable intra-oral light guides, equine therapists need portability, and nerve regeneration devices need to be capable of several hours of sustained treatment with active cooling. Brain injuries, stroke, and Alzheimer's disease need devices specifically intended for the head as dosimetry is complex and overtreatment could have serious negative consequences.

5. Ignorance, exaggeration, and misinformation are frequently disseminated by Web sites and distributor representatives. Claims of treating through clothing, magic therapeutic pulse frequencies for your aura, or the benefits of special coherence are frequent stories to lure the LLLT novice. Shop around several suppliers, attend conferences, and work out who is trustworthy before making an expensive (but worthwhile) investment.

References

Benedicenti, S., I. M. Pepe, F. Angiero, and A. Benedicenti. 2008. Intracellular ATP level increases in lymphocytes irradiated with infrared laser light of wavelength 904 nm. *Photomed Laser Surg* 26(5):451–453.

Bjordal, J. M., R. J. Bensadoun, J. Tunèr et al. 2011. A systematic review with meta-analysis of the effect of low-level laser therapy (LLLT) in cancer therapy-induced oral mucositis. *Support Care Cancer* 19(8):1969–1977.

Bjordal, J. M., C. Couppé, R. T. Chow, J. Tunér, and E. A. Ljunggren. 2003. A systematic review of low level laser therapy with location-specific doses for pain from chronic joint disorders. *Aust J Physiother* 49(2):107–116.

Bjordal, J. M., R. A. Lopes-Martins, J. Joensen et al. 2008. A systematic review with procedural assessments and meta-analysis of low level laser therapy in lateral elbow tendinopathy (tennis elbow). *BMC Musculoskelet Disord* 9:75.

Brondon, P., I. Stadler, and R. J. Lanzafame. 2005. A study of the effects of phototherapy dose interval on photobiomodulation of cell cultures. *Laser Surg Med* 36(5):409–413.

Castano, A. P., T. Dai, I. Yaroslavsky et al. 2007. Low-level laser therapy for zymosan-induced arthritis in rats: Importance of illumination time. *Laser Surg Med* 39(6):543–550.

Chow, R., P. Armati, E. L. Laakso, J. M. Bjordal, and G. D. Baxter. 2011. Inhibitory effects of laser irradiation on peripheral mammalian nerves and relevance to analgesic effects: A systematic review. *Photomed Laser Surg* 29(6):365–381.

Chow, R. T., M. I. Johnson, R. A. Lopes-Martins, and J. M. Bjordal. 2009. Efficacy of low-level laser therapy in the management of neck pain: A systematic review and meta-analysis of randomised placebo or active-treatment controlled trials. *Lancet* 374(9705):897–908.

Corazza, A. V., J. Jorge, C. Kurachi, V. S. Bagnato. 2007. Photobiomodulation on the angiogenesis of skin wounds in rats using different light sources. *Photomed Laser Surg* 25(2):102–106.

Dickey, F. M. and S. C. Holswade, editors. 2000. *Laser Beam Shaping: Theory and Techniques.* Marcel Dekker, New York.

du Nouy, P. L. 1921. Energy and vision. *J Gen Physiol* 3(6): 743–764.

Durovic, A., D. Marić, Z. Brdareski, M. Jevtić, and S. Durdević. 2008. The effects of polarized light therapy in pressure ulcer healing. *Vojnosanit Pregl* 65(12):906–912.

Graham, C. H., and H. K. Hartline. 1935. The response of single visual sense cells to lights of different wave lengths. *J Gen Physiol* 18(6):917–931.

Gur, A., A. J. Sarac, R. Cevik, O. Altindag, and S. Sarac. 2004. Efficacy of 904 nm gallium arsenide low level laser therapy in the management of chronic myofascial pain in the neck: A double-blind and randomize-controlled trial. *Laser Surg Med* 35(3):229–235.

Hashmi, J. T., Y.-Y. Huang, S. K. Sharma et al. 2010. Effect of pulsing in low-level light therapy. *Laser Surg Med* 42(6): 450–466.

Hecht, S., and R. E. Williams. 1922. The visibility of monochromatic radiation and the absorption spectrum of visual purple. *J Gen Physiol* 5(1):1–33.

Huang, Y.-Y., A. C.-H. Chen, J. D. Caroll, and M. R. Hamblin. 2009. Biphasic dose response in low level light therapy. *Dose–Response* 7:358–383.

Huang, Y. Y., S. K. Sharma, J. Caroll, and M. R. Hamblin. 2011. Biphasic dose response in low level light therapy—An update. *Dose–Response* 9(4):602–618.

Iordanou, P., G. Baltopoulos, M. Giannakopoulou, P. Bellou, and E. Ktenas. 2002. Effect of polarized light in the healing process of pressure ulcers. *Int J Nurs Pract* 8(1):49–55.

Jenkins, P. A., and J. D. Carroll. 2011. How to report low-level laser therapy (LLLT)/photomedicine dose and beam parameters in clinical and laboratory studies. *Photomed Laser Surg* 29(12):785–787.

Karadag, C. A., M. Birtane, A. C. Aygit, K. Uzunca, and L. Doganay. 2007. The efficacy of linear polarized polychromatic light on burn wound healing: An experimental study on rats. *J Burn Care Res* 28(2):291–298.

Karu, T. I. 2010. Multiple roles of cytochrome c oxidase in mammalian cells under action of red and IR—A radiation. *IUBMB Life* 62(8):607–610.

Karu, T. I., L. V. Pyatibrat, S. F. Kolyakov, and N. I. Afanasyeva. 2008. Absorption measurements of cell monolayers relevant to mechanisms of laser phototherapy: Reduction or oxidation of cytochrome c oxidase under laser radiation at 632.8 nm. *Photomed Laser Surg* 26(6):593–599.

Kolb, H., E. Fernandez, and R. Nelson, editors. 1995. *The Organization of the Retina and Visual System.* University of Utah Health Sciences Center, Salt Lake City, UT.

Lanzafame, R. J., I. Stadler, A. F. Kurtz et al. 2007. Reciprocity of exposure time and irradiance on energy density during photoradiation on wound healing in a murine pressure ulcer model. *Laser Surg Med* 39(6):534–542.

Liang, H. L., H. T. Whelan, J. T. Eells, and M. T. Wong-Riley. 2008. Near-infrared light via light-emitting diode treatment is therapeutic against rotenone- and 1-methyl-4-phenylpyridinium ion-induced neurotoxicity. *Neuroscience* 153(4):963–974.

Oliveira, P. C., A. L. Pinheiro, J. A. Reis Jr. et al. 2010. Polarized light (lambda 400–2000 nm) on third-degree burns in diabetic rats: Immunohistochemical study. *Photomed Laser Surg* 28(5):613–619.

Oron, U., T. Yaakobi, A. Oron et al. 2001. Attenuation of infarct size in rats and dogs after myocardial infarction by low-energy laser irradiation. *Laser Surg Med* 28(3):204–211.

Schindl, A., B. Rosado-Schlosser, and F. Trautinger. 2001. Reciprocity regulation in photobiology. An overview (in German). *Hautarzt* 52(9):779–785.

Smith, K. 1991. The photobiological basis of low level laser radiation therapy. *Laser Therapy* 3:19–24.

Sommer, A. P., A. L. Pinheiro, A. R. Mester, R. P. Franke, and H. T. Whelan. 2001. Biostimulatory windows in low-intensity laser activation: Lasers, scanners, and NASA's light-emitting diode array system. *J Clin Laser Med Surg* 19(1):29–33.

Tumilty, S., J. Munn, S. McDonough et al. 2010. Low level laser treatment of tendinopathy: A systematic review with meta-analysis. *Photomed Laser Surg* 28(1):3–16.

Zalevsky, Z., and M. Belkin. 2011. Coherence and speckle in photomedicine and photobiology. *Photomed Laser Surg* 29(10):655–656.

49

Low-Level Laser Therapy: Clearly a New Paradigm in the Management of Cancer Therapy-Induced Mucositis*

Rene-Jean Bensadoun
*University Hospital and Faculty
of Medicine of Poitiers*

49.1 Introduction

Considerable oral toxicity of radiotherapy and/or chemotherapy in patients treated for cancer disease can cause patients to become discouraged and can alter their quality of life (Bensadoun et al. 2001; Elting et al. 2003; Sonis et al. 2004). In addition, such toxicity often necessitates alterations of treatment planning, with grave consequences in terms of tumor response and even survival (concept of dose intensity). With 5-fluorouracil (5FU) and head and neck radiotherapy, for example (Figures 49.1 and 49.2), acute mucosal toxic effect is the main limiting factor for which no clinically appropriate prophylaxis or efficacious antidote has been found to date. Management of oral mucositis (OM) is currently primarily directed at palliation of the symptoms and prevention of infections.

Mucositis is recognized as one of the principal dose-limiting factors during 5FU-based chemotherapy and is also one of the main intensity-limiting acute toxicity during radiotherapy and radiochemotherapy for head and neck cancer and during hematopoietic cell transplant conditioning. The frequency of its appearance varies from 12% in patients receiving adjuvant chemotherapy to 100% in patients submitted to radiotherapy of the oral cavity when the total dose exceeds 50 Gy (Dreizen 1990).

Pathologic evaluation of mucositis reveals mucosal thinning leading to a shallow ulcer thought to be caused by inflammation and depletion of the epithelial basal layer with subsequent denudation and bacterial infection. The wound healing response to this injury is characterized by inflammatory cell infiltration, interstitial exudate, fibrin, and cell debris producing a pseudomembrane analogous to the eschar of a superficial skin wound (Bensadoun et al. 2001; Sonis 2002).

The evaluation and scoring of mucositis and pain is a key point in this type of studies. Criteria for evaluation are the standard World Health Organization (WHO) scale for mucositis in the oral cavity and the oropharynx (subjective assessment), the National Cancer Institute Chemical Biology Consortium (NCI-CBC) scoring system also for mucositis (objective assessment), and a segmented visual analogic scale for pain (patient self-evaluation) (Spijkervet et al. 1989).

Management of OM is currently directed at palliation of the symptoms and prevention of infections (Bensadoun et al. 2006; Elting et al. 2008; Epstein and Schubert 2003; Raber-Durlacher et al. 1989; Worthington, Clarkson, and Eden 2007). Numerous agents and methods have been tested in an attempt to prevent or modulate cancer therapy-induced mucositis. Investigated strategies of mucositis prophylaxis include (Bensadoun et al. 2001):

---

* 2012 State of the Art and Results of a Literature Review with Meta-Analysis.

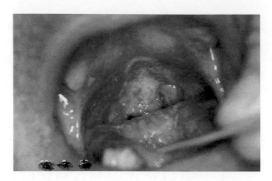

FIGURE 49.1   WHO grade 3 radiation-induced mucositis.

(1) administration of direct cytoprotectants such as amifostine (Hwang et al. 2004), prostaglandin E_2, silver nitrate, and beta-carotene; (2) pharmacologic manipulation of cytotoxic drug metabolism such as modulation of 5-FU metabolism with allopurinol or TGF-B3 (Sonis, Van Vugt, and Brien 1997); (3) modulation of 5FU by a pharmacokinetically based adaptation of dose (Thyss et al. 1986); (4) infection prophylaxis with topical antimicrobials like chlorhexidine or benzydamine; and (5) nonpharmacologic methods including oral cryotherapy (Lilleby et al. 2006). Clinical trials with these modalities have yielded inconsistent results; thus, none of them has become a standard adjunct with proven efficacy in modern cancer therapy. Some 12 different interventions have yielded partly positive results in controlled trials with varying degrees of scientific support, but no single intervention has emerged as the gold standard in cancer therapy-induced OM (Treister and Sonis 2007).

Irradiation by low-level laser therapy (LLLT or low-energy laser) (output power ranged from 5 to 200 mW), with He–Ne laser (wavelength 632.8 nm) or diode lasers of various wavelengths (630–680, 700–830, and 900 nm), have been reported to be a simple atraumatic technique (with no known toxicity in clinical setting), useful in the treatment of mucositis of various origins (Ciais et al. 1992; Migliorati et al. 2001; Pourreau-Schneider et al. 1992).

Irradiation by LLLT corresponds to a local application of a high photon density monochromatic light source. LLLT effects have been confirmed by numerous *in vitro* and *in vivo* studies

(Braverman and McCarthy 1989; Kreisler et al. 2003; Pereira et al. 2002; Pourreau-Schneider et al. 1990; Qadri et al. 2005) and are influenced by cell type, laser wavelength, and energy dose. Three main effects are suggested for this type of radiation (with adequate energy rate or fluence on the target): (1) analgesic effect (λ = 630–650 nm, λ = 780–900 nm), (2) anti-inflammatory effect (same wavelengths), which is the main effect when dealing with the prevention of toxicity (e.g., prevention of OM), and (3) wound healing effect (proved for He–Ne laser: λ = 632.8 nm and suggested for λ = 780–805 nm), all assessed by physical, biological, and experimental studies (Braverman and McCarthy 1989). The mechanism of action of the healing effect at a molecular and enzymatic level consists mainly of the activation of energy production in mitochondria [adenosine triphosphate (ATP)]. During oncological treatments, detoxification of free radicals and/or reduction of free radical formation, induced by chemotherapy and radiotherapy, are complementary effects (currently being studied by several teams). The preventive effect of LLLT raises a lot of interest but needs more experimental data to be confirmed (Figure 49.3).

Early reports first indicated that LLLT may have a beneficial effect on wound healing in humans (Mester and Mester 1987), but LLLT has remained controversial for this indication for decades. Several reviewers have questioned whether LLLT can induce any beneficial effects in cell cultures and animals (Braverman and McCarthy 1989; Lucas et al. 2002) and whether LLLT has any positive effect on the healing of human skin wounds (Hopkins et al. 2004; Kopera et al. 2005). Indeed, the literature for trials of skin wound healing is diverse with seemingly contradictory findings for LLLT of both positive (Hopkins et al. 2004) and nonsignificant (Kopera et al. 2005) values in humans. There may be several reasons for this, such as causal differences in wound pathology, presence of different bacteria, differences in laser irradiation procedures, and the laser parameters used. Due to the limited number of randomized placebo-controlled trials, no one has so far been able to identify dose-response patterns for LLLT in the healing of wounds in human skin. However, recent advances in other areas have identified fairly distinct dose-response patterns for LLLT in clinical studies of osteoarthritis (Bjordal et al. 2003; Brosseau et al. 2005; Chow et al. 2009) and tendinopathies (Bjordal, Couppé, and Ljunggren 2001; Tumilty

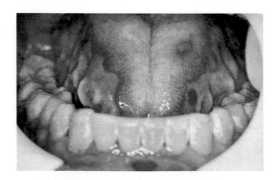

FIGURE 49.2   WHO grade 3 chemo-induced mucositis.

FIGURE 49.3   LLLT oral application.

et al. 2010). This development has partly originated from a better understanding of underlying mechanisms of LLLT actions. From an increasing number of laboratory trials, therapeutic windows have been determined for an analgesic effect (Bjordal et al. 2006; Gam, Thorsen, and Lonnberg 1993; Kreisler et al. 2004; Nes and Posso 2005), an anti-inflammatory LLLT effect (Arora et al. 2008; Basford 1995; Bensadoun and Ciais 2002; Genot and Klastersky 2005), and a dose-dependent, biostimulating LLLT effect on oral mucosal cells in terms of increased collagen production (Kreisler et al. 2003; Pereira et al. 2002) and fibroblast cell proliferation (Kreisler et al. 2004; Loevschall and Arenholt-Bindslev 1994; Pourreau-Scneider et al. 1990). There are some indications that healing of lesions inside the oral cavity may respond better to LLLT than skin wounds (Neiburger 1995). Results from clinical LLLT studies in minor oral surgery (Markovic and Todorovic 2007; Neckel et al. 2001; Posten et al. 2005; Qadri et al. 2005) also suggest that LLLT has a beneficial but a dose-dependent effect on pain and tissue healing.

These mechanisms do not seem to be wavelength-specific in the red and near infrared spectra. But optimal doses seem to deviate slightly between these wavelengths.

LLLT was initially reported to be effective in reducing the severity of OM lesions in a nonrandomized trial initiated in Nice, France (Ciais et al. 1992; Pourreau-Schneider et al. 1992). During recent years, there has been considerable interest into performing clinical LLLT trials in cancer therapy-induced OM, beginning with pilot studies (Schubert et al. 1994; Simoes et al. 2009; Whelan et al. 2002; Wong and Wilder-Smith 2002) and then randomized controlled studies (Abramoff et al. 2008; Antunes et al. 2008; Bensadoun et al. 1999; Chor et al. 2010; Cowen et al. 1997; Cruz et al. 2007; Genot-Klastersky et al. 2008; Kuhn et al. 2007, 2009; Maiya Arun, Sagar, and Fernandes 2006; Schubert et al. 2007). All but one of these studies confirmed the efficacy of LLLT in the prevention of cancer therapy-induced mucositis, especially for the reduction of maximal grade of mucositis, reduction of the duration of mucositis, and delay in the onset of mucositis.

The *2000–2004 Guidelines for Cancer Management (Mucositis)* mention that LLLT was a possible option, but also pointed out that expensive equipment and specialized training were needed because of variations in procedures, doses, and laser devices (Keefe et al. 2007; Rubenstein et al. 2004). In 2007, the actualization of the Multinational Association of Supportive Care in Cancer/International Society of Oral Oncology (MASCC/ISOO) mucositis guidelines put LLLT as a "recommended" method for the prevention of mucositis during bone-marrow transplantation (Keefe et al. 2007).

In this perspective, we initiated a systematic review of the clinical evidence with meta-analyses of the use of LLLT to prevent and treat OM for cancer patients and to identify possible success factors and optimal doses and procedures (Bjordal et al. 2011). This approach has successfully been applied for LLLT in other areas (Bjordal, Couppé, and Ljunggren 2001; Bjordal et al. 2003, 2006; Chow et al. 2009; Tumilty et al. 2010). The development of LLLT guidelines by the World Association for Laser Therapy (WALT) has been based on this approach. Recently, trial compliance with WALT guidelines for tendinopathies was shown to predict a positive treatment outcome in 92% of the available studies (Bjordal, Couppé, and Ljunggren 2001; Shea et al. 2002; Tumilty et al. 2010).

49.2 Literature Search and Exclusion Procedure

A literature search revealed 33 potentially relevant papers. Of these, nine studies were reviews and six studies were case studies, while another three were animal studies. Three controlled studies were excluded for lack of randomization, while one study lacked a placebo-control group (Bjordal et al. 2011).

The final sample consisted of 11 randomized placebo-controlled trials published from 1997 to 2009 with a total of 415 patients (Abramoff et al. 2008; Antunes et al. 2007; Bensadoun et al. 1999; Chor et al. 2009; Cowen et al. 1997; Cruz et al. 2007; Genot-Klastersky et al. 2008; Kuhn et al. 2007, 2009; Maiya Arun, Sagar, and Fernandes 2006; Schubert et al. 2007).

49.3 Methodological Quality

Methodological assessments were made independently according to the Jadad five-point scale. The assessors gave similar gradings for all the included studies, and a consensus meeting was not needed. Methodological quality was high for the included studies with a mean score of 4.10 (SD $\pm$ 0.74).

49.3.1 Relative Risk for the Development of OM after LLLT

Eight studies presented categorical data for the risk of developing OM during or after cancer therapy. There was a significant effect in favor of LLLT with a relative risk at 2.45 (95% CI: 1.85 to 3.18) for avoiding OM to occur in conjunction with cancer therapy. However, the analysis revealed significant heterogeneity ($I^2 = 54\%$, $p = 0.03$) between trials. An analysis of irradiation parameters revealed that one study deviated from the others by giving a considerably lower dose (0.18 J) and shorter irradiation time (3 s) than the other studies (Cruz et al. 2007). After subgrouping this trial in a separate category, heterogeneity was no longer present ($I^2 = 16\%$, $p = 0.31$) and the relative risk improved to 2.86 with a narrow confidence interval (95% CI: 2.15 to 3.82).

49.3.2 Subgroup Analysis of LLLT Wavelength Effects on the Relative Risk for Developing OM

The subgroup analysis revealed no heterogeneity between trials for the red (630–670 nm) and infrared (780–830 nm) subgroups, respectively ($p > 0.21$ and $I^2 < 32\%$), and there were no significant wavelength differences in relative risks between red at 2.72 (95% CI: 1.98 to 3.74) and infrared at 3.48 (95% CI: 1.79 to 6.75).

49.3.3 Effect of LLLT on the Duration of OM Grade 2 or Higher During Cancer Therapy

Six studies presented data for this outcome, and LLLT reduced significantly the number of days with OM grade 2 or worse with 4.38 (95% CI: 3.35 to 5.40).

49.3.4 Effects of LLLT on Mucositis Severity

Six trials presented seven different comparisons of continuous data for mucositis severity. As the trials used different mucositis index scales, the combined results were only calculated as the standardized mean difference (SMD). The combined SMD effect size was 1.33 (95% CI: 0.68 to 1.98) and corresponded to a very good effect.

However, heterogeneity was present and the reasons for heterogeneity were explored in a separate subgroup analysis of wavelengths without resolving the heterogeneity.

A further analysis of wavelength-specific doses revealed that a dose of 2 J with an infrared wavelength was ineffective with an SMD at 0.38 (95% CI: −0.19 to 0.96) in reducing mucositis severity, whereas a dose of 6 J was highly effective with an SMD at 2.17 (95% CI: 1.48 to 2.86) and without signs of heterogeneity between trials ($I^2 = 0\%$ and $p = 0.89$).

49.3.5 Pain-Relieving Effect of LLLT in OM

Four trials reported continuous data on pain intensity from different scales. The combined analysis revealed a significant effect of LLLT with an SMD at 1.22 (95% CI: 0.19 to 2.25), but also significant heterogeneity caused by one trial (Maiya Arun, Sagar, and Fernandes 2006).

This trial differed clinically from other trials by a considerably longer treatment period of 6 weeks (other studies lasted 2–3 weeks). Removal of this study restored homogeneity ($I^2 = 0\%$ and $p = 0.58$) and reduced the effect size to 0.61 but now with narrow confidence intervals (95% CI: 0.29 to 0.94).

49.3.6 Side Effects of LLLT

All the studies reported possible side effects, but none found side effects or adverse effects beyond those reported for placebo LLLT. On the contrary, several trials reported that LLLT was well tolerated among patients.

49.4 Discussion

This systematic review has revealed moderate to strong evidence for the efficacy of LLLT in cancer therapy-induced OM. In guidelines from the American Cancer Society, the evidence behind LLLT is characterized as promising but with conflicting evidence with large operator variability and cost variability. Our analyses show that fairly inexpensive diode lasers (from $2500) with low optical outputs (10–100 mW) can be used with similar success as the more expensive gas lasers used in the early trials. However, diode lasers have longer coherence lengths, which seem to require slightly higher doses for optimal effects in other oral inflammatory disorders like gingivitis. After exploring the apparent discrepancies of the material, our subgroup analyses revealed plausible causes for the few conflicting results. The misunderstandings caused by reporting LLLT doses in joules per square centimeter with small laser spot sizes led to underdosing in one trial (Cruz et al. 2007).

Our analyses showed that there is scientific evidence with meta-analyses and narrow confidence intervals from high-quality randomized placebo-controlled trials.

From this evidence, we gather that a fairly simple treatment procedure can be synthesized:

- LLLT should be performed with diode laser outputs of 10–100 mW in a stationary manner (not scanning) with a minimum irradiation time of 30 s per point and a dose of 2–3 J for red wavelengths or 6 J for infrared wavelengths
- Between 6 and 20 points should be covered depending on the severity and distribution of mucositis in the oral cavity
- Lesions and inflammatory areas should be specifically targeted for the irradiation, and treatment can be applied daily or every other day

Our findings relate well to the emerging LLLT evidence in other inflammatory conditions such as rheumatoid arthritis and acute postoperative pain (Kreisler et al. 2004). The optimal clinical doses found in the current review are in the same range as those previously found for rheumatoid arthritis and postoperative pain. It is also interesting to note that the variety of different cancer therapies involved in the included trials did not seem to seriously interfere with the beneficial effects of LLLT. How LLLT efficacy compares to the efficacy of various pharmacological agents in controlling OM is outside the scope of this review.

In terms of side effects, LLLT was very well tolerated with hardly any withdrawals due to adverse events, and no serious incidents were reported.

Finally, we focus on the necessity of checking all laser parameters and treatment characteristics before initiating such laser administration. Laser parameters must include wavelength (nm), power (mW), joules per point (or "dose"), energy density, spot size, power density (mW/cm²), and laser machine calibration. Treatment characteristics should include the total number of joules in any single laser session, the total number of sessions, the frequency of sessions (treatment protection), the site(s) of treatment, and precise instructions regarding laser administration (contact pressure treatment or scanning application, preparation of the skin, etc.).

49.5 Conclusion

We conclude that there is moderate to strong evidence in favor of clinically relevant effects when LLLT is applied with optimal doses in cancer therapy–induced OM. Based on our dose-finding subgroup analyses, LLLT procedures can be made easy and inexpensively with the use of diode laser technology.

In 2009, a new national French cooperative study (six cancer centers) began, for the prevention of mucositis in patients treated with chemoradiation for head and neck cancer. This large study should confirm preliminary results, which will increase the level of evidence for the efficacy of LLLT for these types of patients.

References

Abramoff, M. M., N. N. Lopes, L. A. Lopes et al. 2008. Low-level laser therapy in the prevention and treatment of chemotherapy-induced oral mucositis in young patients. *Photomed Laser Surg* 26:393–400.

Antunes, H. S., E. M. Ferreira, V. D. de Matos, C. T. Pinheiro, and C. G. Ferreira. 2008. The impact of low power laser in the treatment of conditioning-induced oral mucositis: A report of 11 clinical cases and their review. *Med Oral Patol Oral Cir Bucal* 13:E189–E192.

Arora, H., K. M. Pai, A. Maiya, M. S. Vidyasagar, and A. Rajeev. 2008. Efficacy of He–Ne laser in the prevention and treatment of radiotherapy-induced oral mucositis in oral cancer patients. *Oral Surg Oral Med O* 105(2):180–186.

Basford, J. R. 1995 Low intensity laser therapy: Still not an established clinical tool. *Laser Surg Med* 16:331–342.

Bensadoun, R. J., and G. Ciais. 2002. Radiation and chemotherapy-induced mucositis in oncology: Results of multicenter phase III studies testing low energy laser. *J Oral Laser Appl* 2:115–120.

Bensadoun, R. J., J. C. Franquin, G. Ciais et al. 1999. Low-energy He/Ne laser in the prevention of radiation-induced mucositis. A multicenter phase III randomized study in patients with head and neck cancer. *Support Care Cancer* 7:244–252.

Bensadoun, R. J., F. Le Page, V. Darcourt et al. MASCC/ISOO mucositis group. 2006. Radiation-induced mucositis of the aerodigestive tract: Prevention and treatment. MASCC/ISOO mucositis group's recommendations. *Bull Cancer* 93(2):201–211.

Bensadoun, R. J., N. Magné, P. Y. Marcy, and F. Demard. 2001. Chemotherapy- and radiotherapy-induced mucositis in head and neck cancer patients: New trends in pathophysiology, prevention and treatment. *Eur Arch Otorhinolaryngol* 258(9):481–487.

Bjordal, J. M., R. J. Bensadoun, J. Tunèr et al. 2011. A systematic review with meta-analysis of the effect of low-level laser therapy (LLLT) in cancer therapy-induced oral mucositis. *Support Care Cancer* 19(8):1069–1077.

Bjordal, J. M., C. Couppé, R. T. Chow, J. Tuner, and E. A. Ljunggren. 2003. A systematic review of low level laser therapy with location-specific doses for pain from chronic joint disorders. *Aust J Physiother* 49:107–116.

Bjordal, J. M., C. Couppé, and A. E. Ljunggren. 2001. Low level laser therapy for tendinopathy. Evidence of a dose–response pattern. *Phys Ther Rev* 6:91–99.

Bjordal, J. M., Johnson, M. I., Iversen, V., Aimbire, F., and Lopes-Martins, R. A. 2006. Photoradiation in acute pain: A systematic review of possible mechanisms of action and clinical effects in randomized placebo-controlled trials. *Photomed Laser Surg* 24:158–168.

Braverman, B., and McCarthy, R. 1989. Effect of HeNe and infrared laser irradiation on wound healing in rabbits. *Laser Surg Med* 9:50–58.

Brosseau, L., V. Robinson, G. Wells et al. 2005. Low level laser therapy (classes I, II and III) for treating rheumatoid arthritis. *Cochrane Database Syst Rev* CD002049.

Chor, A., S. R. Torres, A. Maiolino, and M. Nucci. 2010. Low-power laser to prevent oral mucositis in autologous hematopoietic stem cell transplantation. *Eur J Haematol* 84(2):178–179.

Chow, R. T., M. I. Johnson, R. A. B. Lopes-Martins, and J. M. Bjordal. 2009. Efficacy of low-level laser therapy in the management of neck pain: A systematic review and meta-analysis of randomised, placebo or active-treatment controlled trials. *Lancet* 374(9705):1897–1908.

Ciais, G., M. Namer, M. Schneider et al. 1992. La laserthérapie dans la prévention et le traitement des mucites liées à la chimiothérapie anticancéreuse. *Bull Cancer* 79:183–191.

Cowen, D., C. Tardieu, M. M. Schubert et al. 1997. Low energy helium-neon laser in the prevention of oral mucositis in patients undergoing bone marrow transplant: Results of a double blind randomized trial. *Int J Radiat Oncol Biol Phys* 38(4):697–703.

Cruz, L. B., A. S. Ribeiro, A. Rech et al. 2007. Influence of low-energy laser in the prevention of oral mucositis in children with cancer receiving chemotherapy. *Pediatr Blood Cancer* 48:435–440.

Dreizen, S. 1990. Oral complications of cancer therapies. Description and incidence of oral complications. *NCI Monogr* 9.11–15.

Elting, L. S., C. Cooksley, M. Chambers et al. 2003. The burdens of cancer therapy. Clinical and economic outcomes of chemotherapy-induced mucositis. *Cancer* 98:1531–1539.

Elting, L. S., D. M. Keefe, S. T. Sonis et al. 2008. Patient-reported measurements of oral mucositis in head and neck cancer patients treated with radiotherapy with or without chemotherapy: Demonstration of increased frequency, severity, resistance to palliation, and impact on quality of life. *Cancer* 113:2704–2713.

Epstein, J. B., and Schubert, M. M. 2003. Oropharyngeal mucositis in cancer therapy. Review of pathogenesis, diagnosis, and management. *Oncology* 17:1767–1779, discussion 1779–1782, 1791–1792.

Gam, A. N., H. Thorsen, and F. Lonnberg. 1993. The effects of low level laser therapy on musculoskeletal pain: A meta-analysis. *Pain* 52:63–66.

Genot, M. T., and J. Klastersky. 2005. Low-level laser for prevention and therapy of oral mucositis induced by chemotherapy or radiotherapy. *Curr Opinion Oncol* 17:236–240.

Genot-Klastersky, M. T., J. Klastersky, F. Awada, and M. Paesmans. 2008. The use of low-energy laser (LEL) for the prevention of chemotherapy- and/or radiotherapy-induced oral mucositis in cancer patients: Results from two prospective studies. *Support Care Cancer* 16:1381–1387.

Hopkins, J. T., T. A. McLoda, J. G. Seegmiller, and G. David Baxter. 2004. Low-level laser therapy facilitates superficial wound healing in humans: A triple-blind, sham-controlled study. *J Athl Train* 39:223–229.

Hwang, W. Y., L. P. Koh, H. J. Ng et al. 2004. A randomized trial of amifostine as a cytoprotectant for patients receiving myeloablative therapy for allogeneic hematopoietic stem cell transplantation. *Bone Marrow Transplant* 34:51–56.

Keefe, D. M., M. M. Schubert, L. S. Elting et al. 2007. Updated clinical practice guidelines for the prevention and treatment of mucositis. *Cancer* 109:820–831.

Kopera, D., R. Kokol, C. Berger, and J. Haas. 2005. Does the use of low-level laser influence wound healing in chronic venous leg ulcers? *J Wound Care* 14:391–394.

Kreisler, M., A. B. Christoffers, B. Willershausen, and B. D'Hoedt. 2003. Effect of low-level GaAlAs laser irradiation on the proliferation rate of human periodontal ligament fibroblasts: An *in vitro* study. *J Clin Periodontol* 30:353–358.

Kreisler, M. B., H. A. Haj, N. Noroozi, and B. Willershausen. 2004. Efficacy of low level laser therapy in reducing postoperative pain after endodontic surgery—A randomized double blind clinical study. *Int J Oral Maxillofac Surg* 33:38–41.

Kuhn, A., F. A. Porto, P. Miraglia, and A. L. Brunetto. 2009. Low-level infrared laser therapy in chemotherapy-induced oral mucositis: A randomized placebo-controlled trial in children. *J Pediatr Hematol Oncol* 31:33–37.

Kuhn, A., G. Vacaro, D. Almeida et al. 2007. Low-level infrared laser therapy for chemo- or radiation-induced oral mucositis: A randomized placebo-controlled study. *J Oral Laser Appl* 7:175–181.

Lilleby, K., P. Garcia, T. Gooley et al. 2006. A prospective, randomized study of cryotherapy during administration of high-dose melphalan to decrease the severity and duration of oral mucositis in patients with multiple myeloma undergoing autologous peripheral blood stem cell transplantation. *Bone Marrow Transplant* 37:1031–1035.

Loevschall, H., and D. Arenholt-Bindslev. 1994. Effect of low level diode laser irradiation of human oral mucosa fibroblasts *in vitro*. *Laser Surg Med* 14:347–354.

Lucas, C., L. J. Criens-Poublon, C. T. Cockrell, and R. J. de Haan. 2002. Wound healing in cell studies and animal model experiments by low level laser therapy; were clinical studies justified? A systematic review. *Laser Med Sci* 17:110–134.

Maiya Arun, G., M. S. Sagar, and D. Fernandes. 2006. Effect of low level helium–neon (He–Ne) laser therapy in the prevention and treatment of radiation induced mucositis in head and neck cancer patients. *Indian J Med Res* 124:399–402.

Markovic, A., and L. Todorovic. 2007. Effectiveness of dexamethasone and low-power laser in minimizing oedema after third molar surgery: A clinical trial. *Int J Oral Maxillofac Surg* 36:226–229.

Mester, A. F., and Mester, A. 1987. Clinical data of laser biostimulation in wound healing. *Laser Surg Med* 7:78.

Migliorati, C. A., C. Massumoto, F. P. Eduardo et al. 2001. Low-energy laser therapy in oral mucositis. *J Oral Laser Appl* 1:97–101.

Neckel, C., and Å. Kukiz. 2001. Biostimulation: A comparative study into the postoperative outcome of patients after third molar extraction. *J Oral Laser Appl* 1:215–219.

Neiburger, E. J. 1995. The effect of low-power lasers on intraoral wound healing. *NY State Dent J* 61:40–43.

Nes, A. G., and M. B. Posso. 2005. Patients with moderate chemotherapy-induced mucositis: Pain therapy using low intensity lasers. *Int Nurs Rev* 52:68–72.

Pereira, A. N., C. P. Eduardo, E. Matson, and M. M. Marques. 2002. Effect of low-power laser irradiation on cell growth and procollagen synthesis of cultured fibroblasts. *Laser Surg Med* 31:263–267.

Posten, W., D. A. Wrone, J. S. Dover et al. 2005. Low-level laser therapy for wound healing: Mechanism and efficacy. *Dermatol Surg* 31:334–340.

Pourreau-Schneider, N., A. Ahmed, M. Soudry et al. 1990. Helium–neon laser treatment transforms fibroblasts into myofibroblasts. *Am J Pathol* 137(1):171–178.

Pourreau-Schneider, N., M. Soudry, J. C. Franquin et al. 1992. Soft-laser therapy for iatrogenic mucositis in cancer patients receiving high-dose fluorouracil: A preliminary report. *J Natl Cancer Inst* 84(5):358–359.

Qadri, T., L. Miranda, J. Tuner, and A. Gustafsson. 2005. The short-term effects of low-level lasers as adjunct therapy in the treatment of periodontal inflammation. *J Clin Periodontol* 32:714–719.

Raber-Durlacher, J. E., L. Abraham-Inpijn, E. F. van Leeuwen, K. H. Lustig, and A. J. van Winkelhoff. 1989. The prevention of oral complications in bone-marrow transplantations by means of oral hygiene and dental intervention. *Neth J Med* 34:98–108.

Rubenstein, E. B., D. E. Peterson, M. Schubert et al. 2004. Clinical practice guidelines for the prevention and treatment of cancer therapy-induced oral and gastrointestinal mucositis. *Cancer* 100(Suppl 9):2026–2046.

Schubert, M. M., F. P. Eduardo, K. A. Guthrie et al. 2007. A phase III randomized double-blind placebo-controlled clinical trial to determine the efficacy of low level laser therapy for the prevention of oral mucositis in patients undergoing hematopoietic cell transplantation. *Support Care Cancer* 15(10):1145–1154.

Schubert, M. M., J. C. Franquin, F. Niccoli-Filho et al. 1994. Effects of low-energy laser on oral mucositis: A phase I/II pilot study. *Cancer Researcher Weekly* 7:14.

Shea, B., D. Moher, I. Graham, B. Pham, and P. Tugwell. 2002. A comparison of the quality of Cochrane reviews and systematic reviews published in paper-based journals. *Eval Health Prof* 25:116–129.

Simoes, A., F. P. Eduardo, A. C. Luiz et al. 2009. Laser phototherapy as topical prophylaxis against head and neck cancer radiotherapy-induced oral mucositis: Comparison between low and high/low power lasers. *Laser Surg Med* 41:264–270.

Sonis, S. T. 2002. The biologic role for nuclear factor-kappa B in disease and its potential involvement in mucosal injury associated with anti-neoplastic therapy. *Crit Rev Oral Biol Med* 13(5):380–389.

Sonis, S. T., L. S. Elting, D. Keefe et al. 2004. Perspectives on cancer therapy-induced mucosal injury: Pathogenesis, measurement, epidemiology, and consequences for patients. *Cancer* 100 (Suppl 9):1995–2025.

Sonis, S. T., J. Scherer, S. Phelan et al. 2002. The gene expression sequence of radiated mucosa in an animal mucositis model. *Cell Prolif* 35(Suppl 1):93–102.

Sonis, S. T., A. G. Van Vugt, and J. P. Brien. 1997. Transforming growth factor-beta 3 mediated modulation of cell cycling and attenuation of 5-fluorouracil induced oral mucositis. *Oral Oncol* 33:47–54.

Spijkervet, F. K., H. K. van Saene, A. K. Panders, A. Vermey, and D. M. Mehta. 1989. Scoring irradiation mucositis in head and neck cancer patients. *J Oral Pathol Med* 18(3):167–171.

Thyss, A., G. Milano, N. Renée et al. 1986. Clinical pharmacokinetic study of 5-FU in continuous 5-day infusions for head and neck cancer. *Cancer Chemother Pharmacol* 16(1):64–66.

Treister, N., and S. T. Sonis. 2007. Mucositis: Biology and management. *Curr Opin Otolaryngol Head Neck Surg* 15:123–129.

Tumilty, S., J. Munn, S. McDonough et al. 2010. Low level laser treatment of tendinopathy: A systematic review with meta-analysis. *Photomed Laser Surg* 28(1):3–16.

Whelan, H. T., J. F. Connelly, B. D. Hodgson et al. 2002. NASA light-emitting diodes for the prevention of oral mucositis in pediatric bone marrow transplant patients. *J Clin Laser Med Surg* 20:319–324.

Wong, S. F., and P. Wilder-Smith. 2002. Pilot study of laser effects on oral mucositis in patients receiving chemotherapy. *Cancer J* 8(3):247–254.

Worthington, H. V., J. E. Clarkson, and O. B. Eden. 2007. Interventions for preventing oral mucositis for patients with cancer receiving treatment. *Cochrane Database Syst Rev* CD000978.

50

Low-Level Laser Therapy for Wound Healing

Lilach Gavish
The Hebrew University
Hadassah Medical School

50.1 Introduction

In the early 1970s, Endre Mester reported that ruby laser treatment accelerated healing of burn wounds in mice (Mester, Ludany, and Seller 1968). Since then, a growing body of evidence has shown that low-level laser therapy (LLLT) is potentially beneficial for wounds of a variety of etiologies. *In vitro* and *in vivo* studies have shown that LLLT affects almost every molecular aspect of wound healing including increasing adenosine triphosphate (ATP) levels (Karu 1999; Pastore et al. 1996); promotion of proliferation and migration of keratinocytes (Fushimi et al. 2012; Grossman et al. 1998), endothelial cells (Chen, Hung, and Hsu 2008; Kipshidze et al. 2001; Schindl et al. 2003), and fibroblasts (Hawkins and Abrahamse 2006; Houreld and Abrahamse 2007); increasing collagen synthesis (Labbe et al. 1990; Prabhu et al. 2012; Saperia et al. 1986); enhancing phagocytic and bactericidal activities of inflammatory cells (Duan et al. 2001; Hemvani, Chitnis, and Bhagwanani 2005; Kupin et al. 1982); and modulation of expression and secretion of relevant chemokines and cytokines (Peplow et al. 2011a). LLLT has been shown to positively affect the three principal phases of healing—inflammation, proliferation, and maturation—in acute wounds (Carvalho et al. 2010; Gal et al. 2006) as well as chronic wounds, and improves tensile strength (Yasukawa et al. 2007). Although the experimental evidence is sound, the efficacy of LLLT in the clinical setting is still controversial, showing mixed results. Nevertheless, since LLLT has a few contraindications and no reported side effects, the lure of this potentially useful treatment option has been hard to resist. Moreover, the marketing of a variety of user-friendly devices has paralleled the great increase in numbers and scope of published clinical studies.

This chapter presents an overview of the molecular and cellular mechanisms underlying the effect of LLLT on wound healing (Section 50.2). This is followed by a detailed review of preclinical and clinical studies of the effect of LLLT on specific clinical entities (Section 50.3), including acute surgical wounds (incisions, excisions, and avulsions), burn wounds, chronic ulcers (diabetes, venous, and pressure ulcers, as well as radiation/drug impaired), other wounds associated with certain specific skin diseases, and scars. Finally, after presenting evidence for and against the use of LLLT for wound healing, we will discuss current dogma concerning its accepted indications.

Although there are a number of similarities between the effect of LLLT on skin wounds and on other organs of the body, this chapter focuses only on the former, which is defined as any break in the skin that results in losing its integrity. Reviewed in this chapter are studies using different types of light sources that induce photobiostimulation, including lasers, light-emitting diodes (LEDs), and polychromatic lamps; and different wavelengths including visible red, near-infrared (NIR), as well as green and yellow visible light. Finally, if not specifically stated, the light source referred to in this chapter is LLLT.

50.2 Mechanism Underlying the Effect of LLLT on Wound Healing

Skin wound healing represents a series of carefully regulated interrelated processes and events that involve the participation of many different tissues and cell lineages. The ultimate goal of skin wound healing is to restore its structural contiguity and functionality. In order to understand how LLLT affects each of these processes, a brief overview of the wound healing process is necessary. Selected aspects of wound healing will be discussed in each section. A comprehensive description of the wound healing process can be found in the literature (Martin 1997; Stadelmann, Digenis, and Tobin 1998).

Wound healing begins with temporary repair achieved in the form of a fibrin clot that plugs the damage to the tissue and serves as a provisional matrix over and through which cells can migrate during the repair process. Inflammatory cells, and then fibroblasts and capillaries, invade the clot to form granulation tissue to fill the gap. Macrophages secrete growth factors and signaling elements to direct the process. Meanwhile, the reepithelialization processes begin at the damaged epidermal edges, and epithelial cells migrate forward to cover the denuded wound surface. Myofibroblasts position themselves at the wound edges to draw the wound margins together. Collagen is secreted by fibroblasts and remodels in the process of forming scar tissue. The end result of uncomplicated healing is a fine scar with little fibrosis, minimal if any wound contraction, and a return to near-normal tissue architecture and organ function.

Conceptually, the wound healing process is divided into three overlapping phases: the inflammatory phase, in which immune cells migrate to the wound immediately after the injury and prepare the wound environment for healing; the proliferative phase, in which new granulation tissue is produced to fill the wound space with increased production of collagen along with reepithelialization to close the wound in parallel with wound contraction; and the remodeling phase, in which continued remodeling of scar tissue matrix results in restoration of wound strength.

Experimental studies have shown that LLLT affects all three phases of wound healing.

50.2.1 Inflammatory Phase

The disruption of the epidermal barrier results in the release of prestored interleukin (IL)-1α and tumor necrosis factor (TNF)-α from keratinocytes, which attract other surrounding cells to the injury location. In the vicinity of the small blood vessels of the dermis, mast cells react to the stimulus by degranulating and release of prestored histamine and chemotactic factors (Ng 2010). Histamine induces regional vasodilation and increases capillary permeability, thereby enabling the recruited leukocytes to reach the injury site, remove the debris, and protect the tissue from pathogens. The leukocytes are replaced by macrophages that direct the process by secreting mitogens and fibroblast chemoattractants while simultaneously clearing the wound of old neutrophils. The increased capillary permeability also allows serum fluid rich in proteins to enter the interstitial space. Fibronectin deposition then creates scaffolding on which fibroblasts can migrate into the wound.

50.2.1.1 Sounding the Horn and Opening the Ways: Amplifying the Initial Chemoattractive Signal and Increasing Vasodilation

LLLT was shown to increase keratinocyte IL-1α secretion (Gavish et al. 2004; Yu et al. 1996) and stimulate histamine release (Wu et al. 2010) and mast cell degranulation (el Sayed and Dyson 1990), thereby amplifying the initial chemoattractive signal. Increased vasodilation following LLLT was demonstrated as early as 2 min after irradiation in a clinical study (Samoilova et al. 2008). This increase was shown to result from modulation of nitric oxide, a free radical synthesized by nitric oxide synthase (NOS) that functions as a neurotransmitter on vascular smooth muscle cells as demonstrated by abolishment of the effect after injection of the NOS inhibitor L-NMMA. Indeed, stimulation of NO after LLLT was demonstrated in macrophages (Gavish et al. 2008) and, to a lesser degree, in human endothelial cells (Chen, Hung, and Hsu 2008).

50.2.1.2 The First Line of Defense—Neutrophils and Monocytes: Resolving the Inflammatory Phase Quicker

Arrival of neutrophils at the wound site was accelerated (Gal et al. 2006), and phagocytic (Hemvani, Chitnis, and Bhagwanani 2005; Kupin et al. 1982) and bactericidal (Duan et al. 2001) activities of inflammatory cells were shown to be enhanced following LLLT. This resulted in earlier resolution of the inflammatory phase.

50.2.1.3 Edema: Removing Stagnant Tissue Fluid

Reduction of regional edema was shown following LLLT *in vivo* as early as 24 h after injury (Medrado et al. 2003) and in patients with lymphedema after mastectomy (Lau and Cheing 2009), reflecting acceleration of removal of stagnant tissue fluid. Lievens (1991) found that LLLT restored injured lymph vessels to their original morphologic pattern within a few days without increase in permeability. In contrast, the injured lymph vessels in the control, nonirradiated group regenerated as a network of small lymph vessels that were abnormally permeable.

50.2.1.4 Preparing the Tissue for the Next Phase

LLLT-conditioned medium from macrophages (Young et al. 1989) and T-lymphocytes (Agaiby et al. 2000) was shown to accelerate fibroblast and endothelial cell proliferation.

50.2.2 Proliferative Phase

50.2.2.1 Reepithelialization: Stimulation of Keratinocyte Migration and Proliferation

The proliferative phase begins 3–4 days after injury and lasts up to 3 weeks in an acute wound. This process consists initially of epithelial cell migration and proliferation. When wound closure is complete, keratinocytes undergo stratification and differentiation to restore the barrier. Inadequate reepithelialization is characteristic of most chronic wounds.

Keratinocyte migration and proliferation was shown to be stimulated by LLLT *in vitro* (Fushimi et al. 2012; Grossman et al. 1998). *In vivo* migration of epithelial cells was seen by 48 h, and complete regeneration of the epidermis was found to precede that of controls by 24 h (Gal et al. 2006) (see also Section 50.3).

50.2.2.2 Blood Supply: Stimulation of Endothelial Cell Proliferation and Promotion of Angiogenesis and Neovascularization

For healing to occur, wounds must have adequate blood flow to transport the necessary nutrients and to remove the resulting waste, local toxins, bacteria, and other debris. Angiogenesis and neovascularization are critical for the formation of the new granulation tissue. Vascular endothelial growth factor (VEGF) and basic fibroblast growth factor (bFGF) are potent angiogenic growth factors.

LLLT was found to stimulate endothelial cell proliferation directly (Chen, Hung, and Hsu 2008; Schindl et al. 2003) or indirectly by upregulating the expression and secretion of VEGF from arterial smooth muscle cells (Kipshidze et al. 2001) and T-lymphocytes (Agaiby et al. 2000). The ability of LLLT to promote angiogenesis *in vivo* by quantifying vascular density has been reported in a variety of experimental and clinical circumstances (see Section 50.3.1.2 and 50.3.1.3).

50.2.2.3 Fibroplasia: Increased Expression and Secretion of Fibroblast Growth Factors (bFGF, PDGF, and TGF-β), Stimulation of Fibroblast Proliferation, and Collagen Secretion

The fibroblast synthesizes and secretes a variety of substances essential for wound repair including collagen and other components of the ground substance. During the first 2–3 days after wounding, fibroblast activity is principally confined to cellular replication and migration rather than collagen synthesis. Therefore, the time of observation might explain discrepancies found between different studies focusing on these parameters.

LLLT was found to increase expression and secretion of bFGF (Byrnes et al. 2004; Yu et al. 2003), platelet-derived growth factor (PDGF), and transforming growth factor beta (TGF-β) (Safavi et al. 2008). These growth factors have been shown to induce fibroblast proliferation and collagen deposition. Others have also shown direct effects of LLLT on fibroblast proliferation (Hawkins and Abrahamse 2006; Houreld and Abrahamse 2007) as well as increased collagen expression and secretion (Labbe et al. 1990; Prabhu et al. 2012; Saperia et al. 1986).

50.2.2.4 Wound Contraction: Increasing Myofibroblast Proliferation

Wound contraction is most prominent in large surface wounds with extensive loss of cells and tissue. Myofibroblasts, which share morphological features in common with fibroblasts and smooth muscle cells, are positioned along the wound edges. Contraction of these cells decreases the gap between the dermal edges of the wound, thereby facilitating wound closure.

Increased myofibroblast proliferation was shown 3 days after excision injury followed by LLLT (Medrado et al. 2003).

50.2.3 Maturation and Remodeling: Modulation of Metalloproteinase Activity, Collagen Distribution, and Increased Tensile Strength

Tissue remodeling involves transition from granulation tissue to scar and is dependent on continued synthesis and catabolism of collagen. This process is regulated by matrix metalloproteinases (MMPs) and their inhibitors that are secreted by macrophages but also by stromal cells such as fibroblasts and smooth muscle cells. Wound tensile strength correlates with collagen content as well as distribution pattern.

Modulation of metalloproteinase activity was shown by gene expression analysis. Tissue inhibitor of MMP-2 (TIMP-2) was upregulated 12 h following LLLT of primary aortic smooth muscle cells. This was accompanied by reduced collagen degradation. However, 24 h after irradiation, the level of MMP-2 secretion and collagenase (MMP-1) gene expression were increased (Gavish, Perez, and Gertz 2006). Collagen distribution (orientation, thickness) was shown to be more matured (Medrado et al. 2008), and tensile strength was found to be increased (Yasukawa et al. 2007) following LLLT.

50.2.4 Additional Mechanisms

50.2.4.1 Attenuation of Apoptosis

Apoptosis is the process of programmed cell death that can be triggered both by internal and external signals. This process is mediated by caspase 3/7 and has distinct morphological characteristics. In relation to wound healing, apoptosis is responsible for the removal of inflammatory cells and granulation tissue. However, in some types of wounds, such as burns, abnormally increased apoptotic signals may result in severe tissue damage. LLLT was reported to decrease levels of caspase 3/7 activity *in vitro* 24 h after irradiation compared to nonirradiated controls. This occurred concomitantly with increased cell proliferation (Hawkins and Abrahamse 2007).

50.2.4.2 Shifting the Cytokine Microenvironment from Chronic to Acute

The composition of cytokines and growth factors in chronic wounds differs from the situation in normal tissue. Reduced blood supply and inadequate levels of nutrients and oxygen decrease the ability of the cells to produce sufficient ATP. Moreover, repeated trauma and infection results in continuous recruitment of inflammatory infiltrates to the site of injury, thereby increasing proinflammatory cytokines and changing the balance between the various growth factors as well as MMPs and TIMPs. This chronic wound microenvironment prevents proper reparative processes.

LLLT was shown to reduce the levels of proinflammatory cytokines IL-1β, monocyte chemoattractant protein-1 (MCP-1),

TNFα, tissue plasminogen activator (tPA), cyclooxygenase-2 (COX-2), and prostaglandin E2 (PGE2) (partial list) in a variety of *in vitro* experimental models. LLLT was shown to increase the ATP levels (Karu 1999; Pastore et al. 1996) in the cell by stimulating cytochrome c oxidase enzyme in the respiratory chain, thereby stimulating the mitochondrial membrane potential. This pathway leads to a buildup of proton gradient that propels ATP synthase to produce additional ATP. A detailed discussion of the anti-inflammatory properties of LLLT as well as the mechanism underlying ATP production can be found in Chapters 46 and 52 of this book.

The effect of LLLT appears to be dependent on the extent of the inflammatory response. In a study of lipopolysaccharide (LPS)-stimulated macrophages, it was found that at low levels of inflammatory response, LLLT did not appear to inhibit the inflammatory cytokines (Gavish et al. 2008). However, when the inflammatory response was robust, a significant inhibitory effect of LLLT on cytokine gene expression was found. The finding that LLLT reduces cytokines only above a certain threshold of LPS stimulation and response suggested that this suppressive effect may be minimal in noninflammatory conditions, which would appear to be of significant clinical importance (see Section 50.3.3).

50.3 Specific Clinical Applications

The following section reviews results from preclinical and clinical studies with emphasis on the underlying mechanism exploited for each specific indication.

50.3.1 Acute Surgical Wounds: Incisions, Excisions, and Avulsions

A skin incision is a lined surgical cut and is considered the least problematic type of wound because of the small volume. In a skin excision, a portion of the skin is removed, which results in a missing volume of tissue. The most common skin excision procedure is a punch biopsy. Avulsions, commonly known as skin flaps, consist of skin and subcutaneous tissue that survive based on anatomic preservation of the intrinsic blood supply. Skin flaps are widely used in plastic surgery for functional and cosmetic wound coverage. However, if the blood supply is insufficient, flap necrosis can occur.

50.3.1.1 Incisions: LLLT Accelerates Incision Healing by Early Resolution of the Inflammatory Phase

In a detailed histological study, Gal et al. (2006) showed that LLLT reduces the inflammatory phase of incision healing by 24 h. Yasukawa et al. (2007) suggested that irradiation every other day, beginning 1 day after injury, was the preferable irradiation protocol as evaluated by measurement of tensile strength. Using this irradiation protocol in a clinical randomized study, Carvalho et al. (2010) showed that LLLT improved the macroscopic appearance of the surgical scar and reduced scar thickness 6 months after surgery.

By histopathological analysis, healing of incision wounds created on the back of rats was accelerated by 24 h after treatment with 670 nm Ga–Al–As diode laser (25 mW/cm^2, 12 J/cm^2 per section) daily for 7 days (Gal et al. 2006). In the latter study, there were more polymorphonuclear leukocytes (PMNs) in the irradiated wound as compared to nonirradiated 24 h after injury. However, this situation was reversed by 48 h showing that LLLT accelerated the inflammatory phase. This acceleration continued through day 7 and included reepithelialization, fibroplasia, angiogenesis, and remodeling phases. These studies show that the wound should be irradiated as early as possible in order to take maximum advantage of the LLLT effects on the inflammatory phase.

Similar beneficial results were found in a clinical study of 28 patients who underwent surgery for inguinal hernias (Carvalho et al. 2010). The experimental group was irradiated with 830 nm, 13 J/cm^2 every other day up to 7 days postsurgery. Six months after surgery, the LLLT group showed improvement in macroscopic appearance (using the Vancouver Scar Scale), less pain (by Visual Analog Scale), better malleability, and reduction in scar thickness.

However, in a study on episiotomy incisions using a different protocol, no improvement was found in the rate of wound healing after LLLT (Santos et al. 2011). In the latter study, 52 postpartum women who had mediolateral episiotomies during their first normal delivery were randomly divided into two groups of 26—experimental and control. Four sessions of 660 nm laser, 3.8 J/cm^2 were applied directly to the episiotomy after suture, within 2 h postpartum, and 1 and 2 days later. No significant difference was found in healing score between the two groups up to 20 days after the procedure.

50.3.1.2 Excisions: LLLT Improves Excision Healing by Accelerating Fibroplasia, Angiogenesis, and Wound Contraction

Medrado et al. (2003) used 670 nm Ga–Al–As diode laser 4 J/cm^2 to treat excision wounds created on backs of rats. Measurement of the wound area up to 14 days showed that the irradiated group achieved 90% wound closure by 3 days, whereas the nonirradiated group reached the same degree of closure in over 8 days. Edema was less at 24 h, and numbers of PMNs at 48 h reduced in the irradiated compared to nonirradiated group showing a more rapid resolution of the inflammatory phase. Myofibroblasts were also significantly increased in the irradiated group at 72 h compared to controls.

Corazza et al. (2007) studied the effects of LLLT on angiogenesis. Histomorphometry of excision wounds treated with 660 nm laser or 635 nm LED 5 J/cm^2 showed accelerated angiogenesis on days 3, 7, and 14 compared to control, thereby improving perfusion to the healing tissue. In both studies, higher energies were less beneficial.

Nipple trauma from breast-feeding is characterized by wounds of various depths and widths. In a preliminary randomized controlled clinical study (Chaves et al. 2012), 10 lactating women with nipple trauma were treated with 860 nm pulsed

LED, 4 J/cm^2 twice a week for 4 weeks, or sham-irradiated. Lesion area was measured from digital photos. The experimental group showed significantly faster nipple healing (2 weeks) than the sham-irradiated control (4 weeks).

50.3.1.3 Skin Flaps: LLLT Improves Skin Flap Viability by Increasing Blood Supply and Angiogenesis as Well as Decreasing the Inflammatory Reaction, as Shown by Reduced Levels of COX-2 and ROS

Costa et al. (2010) used the rat skin flap (4 × 10 cm) model in which ischemia develops on the distal portion because of insufficient blood supply. Irradiation with He–Ne 3 J/cm^2 using the punctual contact technique was shown to prevent the development of necrosis and increased vascular density by more than 50% (Costa et al. 2010). It was also shown that irradiation at 660 nm (10.36 J/cm^2) reduced COX-2 expression in the skin-flap pedicle (Esteves Jr. et al. 2012), and 670 or 830 nm 36 J/cm^2 also resulted in lower levels of ROS and ROS activity as measured by malondialdehyde concentration reflecting reduced ischemia (Prado et al. 2010).

Regarding the importance of the irradiation intensity, no beneficial effects were seen when the energy was too low as in the study by Smith et al. (1992) who used He–Ne (0.082 J/cm^2) in a sweeping motion technique, or when the energy was too high as in the study by Esteves Jr. et al. (2012) who used 660 nm but achieved 260.7 J/cm^2.

Coronally advanced flap (CAF) is a surgical technique used in periodontology to cover roots exposed by gingival recession. The success rate of CAF varies widely. Since LLLT was shown to have beneficial effects on skin flaps, it was hypothesized that it can also improve the success rates of CAF (Ozturan et al. 2011). In a clinical study of patients who underwent CAF surgery for symmetrical gingival recessions, 70% of patients treated with 588 nm (4 J/cm^2) before and after suturing, but only 30% of nonirradiated, had full root coverage. Moreover, the irradiated test sites achieved significantly better scoring for gingival recess depth and width and keratinized tissue.

50.3.2 Burn Wounds

Burn injury is a complex traumatic event with a variety of local and systemic effects. The severity of the burn wound, classified as degrees I–III according to depth, may change over time. This is a result of potential necrosis at the zone of stasis, which is the intermediate zone between the peripheral zone and the central focus of the injury that contains devitalized tissue. The latter is characterized by vascular stasis and ischemia and can be salvaged only if revascularization is achieved within a few days. Topical agents containing NO, such as nitrofurazone, to improve blood supply are commonly used to reach that goal. However, reperfusion may also lead to additional injury by introducing increased inflammatory infiltrates, factors, and ROS to the wounded tissue, resulting in further damage and/or apoptosis. Infections by opportunistic pathogens are a major concern in burn wounds. It was found that macrophages and natural killer (NK) cells have reduced ability to ingest and eliminate pathogens in burn wounds. Skin grafts are used in order to stimulate reepithelization and accelerate wound closure. Burn scars are known to be difficult to treat because of their tendency to worsen with hypertrophy and contracture (Evers, Bhavsar, and Mailander 2010).

LLLT may improve blood supply to the zone of stasis through immediate vasodilation by stimulation of NO secretion and inducible NOS (iNOS) induction as well as accelerated angiogenesis. Apoptosis may be limited through the anti-inflammatory and antiapoptotic properties of LLLT. Reepithelialization might be accelerated either directly by stimulating keratinocyte migration and proliferation or indirectly after improving incorporation of skin grafts.

Early studies have shown a beneficial effect of LLLT on burn wounds. Mester et al. (1971) reported acceleration of healing and stimulation of epidermal regeneration of third-degree burns on the backs of mice that were irradiated twice a week for 3 weeks by ruby laser (694.3 nm 1 J/cm^2 × 2). Rochkind et al. (1989) showed that daily irradiation with He–Ne (10 J/cm^2) for 21 days accelerated healing of burn wounds. However, they showed that this regimen also had a systemic effect since irradiation of one side of bilaterally distributed burns enhanced recovery also of the contralateral, nonirradiated side, albeit to a lesser degree (Rochkind et al. 1989). This systemic effect in burn wounds most probably explains the lack of effect of LLLT in experiments that used models in which the control wounds were anatomically adjacent to the experimental in the same animal (Schlager et al. 2000).

It is also apparent that second-degree burns respond better to LLLT than third-degree burns. Ezzati et al. (2009) failed to show accelerated healing with 890 nm pulsed at 11.7 J/cm^2 in third-degree burns as compared to sham-irradiated controls. However, when subjecting second-degree burns to these energy settings, a significant acceleration in closure rate was demonstrated by 2 and 3 weeks compared to control (Ezzati, Bayat, and Khoshvaghti 2010).

Regarding improved blood supply, Renno et al. (2011) showed that 660 nm improved vascularization in second-degree burns by increasing VEGF. Increased vascularization was also demonstrated by Dantas et al. (2011). In the latter study, daily irradiation with a high-powered (1 W/cm^2) 780 nm laser with 5 J/cm^2 per point × 4 points resulted in a significant increase in the number of blood vessels as shown by increased CD31 antibody staining. In addition, the irradiated animals showed increased epithelization and collagen deposition and organization at 8 and 14 days compared to controls.

LLLT was not found effective in reducing infections of burn wounds. Microbiological samples from burn wounds at different days after injury did not show significant differences between laser treated and controls (Bayat et al. 2005; Ezzati et al. 2009).

50.3.3 Chronic Ulcers and Impaired Healing

A wound that is not continuously progressing toward healing may be described as chronic (Keast and Orsted 1998). Chronic wounds involve a variety of pathological processes including ischemia, extended inflammatory response that results in an imbalance between proteases and its inhibitors, nerve dysfunction, and infection. Reepithelialization is the stage where most chronic wounds fail in the healing process. The most common etiologies of chronic ulcers are diabetes, venous stasis, and pressure ulcers. Iatrogenic causes such as drugs or radiation can impair wound healing processes. Standard treatment regimens include maintenance of a moist wound bed, compression therapy (but offloading in pressure ulcers), debridement, infection control, and appropriate dressing.

LLLT is postulated to have more prominent effects in a stressed wound environment. LLLT was shown to shift the chemokine/cytokine microenvironment of the wound from chronic to acute, thereby reducing the inflammatory infiltrate and stimulating ATP synthesis in energy-depleted cells, enabling the wound to progress toward healing.

50.3.3.1 Diabetic and Venous Ulcers

Houreld and Abrahamse (2007) showed that LLLT stimulated proliferation, migration, and IL-6 secretion in an *in vitro* diabetic wounded fibroblast model. They found increased NO secretion but decreased TNFα secretion and apoptosis (Houreld, Sekhejane, and Abrahamse 2010). A variety of *in vivo* studies using genetic or chemically induced models of diabetes in rodents showed that LLLT accelerated wound healing and increased tensile strength of associated scar tissue (Al-Watban, Zhang, and Andres 2007; Peplow et al. 2011b; Reddy et al. 2001; Stadler et al. 2001). As in other models, enhanced reepithelialization, angiogenesis, and collagen deposition were prominent underlying mechanisms for this effect (Byrnes et al. 2004; Peplow et al. 2011b). Dosimetry studies conducted in diabetic models showed optimal wound closure rate with red light (630–660 nm), but accelerated healing was also reported with other wavelengths (Al-Watban, Zhang, and Andres 2007; Byrnes et al. 2004; Peplow et al. 2011b).

Problematic blood supply to the wound area occurs in diabetic as well as venous ulcers. Schindl et al. (2002) evaluated microcirculatory blood flow in a double-blinded study of 30 patients with diabetic microangiopathy by measuring skin temperature. They found a significant temperature increase after LLLT. However, the dose (30 J/cm²) was higher than that used in most studies and time of the treatment (50 min) was unusually long. Kleinman, Simmer, and Braksma (1996) recommended phototherapy as an effective additional treatment modality for diabetic ulcers. However, small RCTs evaluating the effects of phototherapy as an adjuvant to conventional treatment on chronic diabetes/venous leg ulcers reported mixed results. A comprehensive summary of clinical studies evaluating the effect of LLLT on chronic wounds can be found in a review by Whinfield and Aitkenhead (2009). Figure 50.1 is an example of LLLT on an infected varicose vein.

Landau et al. (2011) evaluated the effect of phototherapy on 16 subjects with diabetic or venous foot ulcers in a double-blind randomized controlled trial (RCT) using broadband visible light twice a day on the experimental group (*n* = 10 with 19 ulcers)

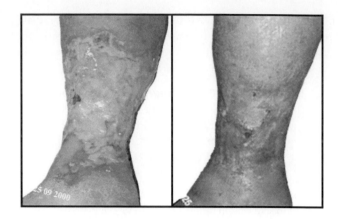

FIGURE 50.1 **(See color insert.)** Ulcus cruris as a result of infected varicose vein that failed conventional treatment in a 55-year-old incapacitated male. Nearly complete cure was achieved after 5 months of LLLT. The wound remained closed and pain-free at 9 months' follow-up. The patient returned to part-time work. (Courtesy of Dr. Yosef Kleinman, Wound Care Center, Bikur Cholim Hospital, Jerusalem, Israel.)

and nonhealing light projections for the placebo group (*n* = 6 with 6 ulcers). At 12 week follow-up, the wounds were closed in 9 out of 10 patients (90%) from the treatment group, whereas in the placebo group, only 2 out of 6 patients had closed wounds (33%) (Landau et al. 2011).

Kaviani et al. (2011) reported a beneficial effect of phototherapy on chronic diabetic foot wound healing in a placebo-controlled RCT of 23 subjects. Patients were randomized to receive placebo treatment (*n* = 10) or LLLT (*n* = 13) (685 nm, energy density 10 J/cm²) in addition to conventional therapy. At 4 weeks, the size of ulcers decreased significantly in the LLLT group (*p* = 0.04). After 20 weeks, complete wound healing was achieved in eight patients in the LLLT group but in only three patients in the placebo group. At week 4, the mean time of complete healing in LLLT patients (11 weeks) was less than that in placebo patients (14 weeks), although the difference was not statistically significant (Kaviani et al. 2011).

Favorable results of phototherapy were also reported by Minatel et al. (2009) in a placebo-controlled RCT of 14 subjects with chronic diabetic leg ulcers that failed to respond to other therapies. The treatment group received conventional treatment +1% silver sulfadiazine as well as 3 J/cm² of combined 660 and 890 LEDs twice a week. The placebo group received the same regimen, but the irradiation consisted of nonhealing phototherapy (<1 J/cm²). The number of completely healed wounds or wounds that achieved higher than 90% closure was significantly larger in the phototherapy group. The authors concluded that this irradiation regimen promotes rapid granulation and healing (Minatel et al. 2009).

Using a similar placebo-controlled RCT study design with an added control group, Caetano et al. (2009) evaluated the effect of these irradiation parameters on healing of 20 subjects (32 chronic venous ulcers) that failed to respond to other forms of treatment. The authors reported that medium- and large-sized ulcers healed significantly faster with phototherapy (Caetano et al. 2009). Lagan et al. (2002), in a small placebo-controlled study of 15 subjects with chronic venous leg ulcers, found no effect with a multisource pulsed diode array (660–950 nm; 12 J/cm²) after weekly treatments over 4 weeks. Kopera et al. (2005) also reported no effect on chronic venous ulcers of 4 J/cm², 685 nm phototherapy used as an adjunct to conventional therapy in a placebo-controlled study with 44 subjects. Likewise, Lundeberg and Malm (1991) reported no effect on the percentage of ulcer healed area, of HeNe 4 J/cm² or 904 nm 2 J/cm² compared to sham, in 46 and 42 venous leg ulcers, respectively. From the overview of clinical studies presented above, it is apparent that further, well-designed clinical studies with adequate numbers of subjects are required to determine whether LLLT, and which irradiation protocol, improves wound healing outcomes in diabetic and venous ulcers.

50.3.3.2 Pressure Ulcers

Pressure ulcers are a serious clinical problem associated with significant rates of morbidity and mortality particularly in the geriatric, ventilated, and spinal cord-injured population. Several clinical studies examined the effect of LLLT, but the results have been nonconsistent. In an intrapatient controlled study (patients with two pressure ulcers, one of which received polarized light therapy and the other acting as control), Iordanou et al. (2002) reported significant beneficial effects on wound closure and characteristics after 2 weeks compared to control wounds. In a prospective randomized controlled study, Schubert (2001) examined the effect of a combination of infrared and red pulsed LLLT on 72 trauma patients with stage II or III pressure ulcers and reported accelerated wound closure rates in the irradiated group. Dehlin, Elmstahl, and Gottrup (2007) reported a beneficial effect of LLLT in a randomized, double-blinded placebo-controlled study of 163 elderly patients with grade II pressure ulcers but not with grade III pressure ulcers (Dehlin, Elmstahl, and Gottrup 2003). On the other hand, Lucas, Van Gemert, and de Haan (2003) found no beneficial effects of 904 nm LLLT as adjuvant therapy five times a week over a period of 6 weeks in a prospective observer-blinded multicenter RCT on 86 patients with stage III pressure ulcers. In a double-blinded RCT, Taly et al. (2004) found no beneficial effects of multiwavelength light therapy on 35 spinal cord injury patients with pressure sores of stage II, III, and IV.

It can be concluded that using a variety of irradiation protocols on pressure ulcers, LLLT is effective mainly in pressure ulcers grade II but less so in grade III or higher.

50.3.3.3 Drug- and Radiation-Induced Ulcers

Wounds from animals treated with steroids or x-irradiation show delayed healing. Using a wound model of steroid treated animals, Pessoa et al. (2004) found that LLLT pulsed Ga–Al–As (904 nm) accelerated wound healing rate and had larger collagen deposition and decreased inflammatory infiltrates. In contrast, Lacjakova et al. (2010) found no effect of LLLT (670 nm) on wound healing in similar animals. Lowe et al. (1998) found that 890 nm had no stimulative effect on wound closure in wounds with impaired healing from x-ray. In fact, certain energies appeared to inhibit healing (Lowe et al. 1998). On the other

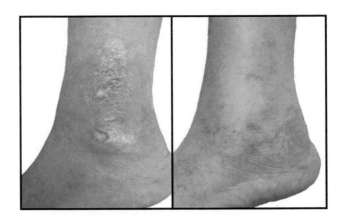

FIGURE 50.2 (**See color insert.**) Recalcitrant leg ulcers caused by hydroxyurea in a 67-year-old woman with essential thrombocytosis. LLLT (see regimen in Figure 50.1) resulted in complete resolution within 3 months. The patient remained ulcer-free for 1 year of follow-up. (Courtesy of Dr. Yosef Kleinman, Wound Care Center, Bikur Cholim Hospital, Jerusalem, Israel.)

hand, Schindl et al. (1999) reported that He–Ne 30 J/cm² twice weekly for 4 weeks stimulated significant dermal angiogenesis in a case of a chronic radiation ulcer compared to the pretreatment status. Figure 50.2 is an example of LLLT on Recalcitrant leg ulcers.

50.3.4 Wounds from Selected Skin Diseases and Pathogens

Acne vulgaris is one of the most common skin diseases experienced by 70%–90% of teenagers. The acne lesion is caused by sebaceous follicle plugging with entrapped *Propionibacterium acnes* bacteria. Herpes simplex virus type 1 often causes perioral and labial blisters (cold sores) that heal with a characteristic scab usually within less than 2 weeks. The infection itself is usually permanent with the virus residing in its latent form in nerve cells usually of the trigeminal ganglion, evading the immune system by interfering with major histocompatibility complex (MHC) class I presentation of the antigen on the cell surface. As a result of a variety of triggers—known and unknown—reactivation occurs with the characteristic lesions presenting on the skin of the dermatome whose innervation is supplied by nerve cells that host the virus. This reactivation can occur with varying frequencies. Radiation-induced dermatitis develops as a side effect in 80% of treated breast cancer patients. It is characterized by redness, dryness, itching, and peeling of the skin and may be severe enough to interrupt the treatment protocol. Psoriasis is a T cell-mediated autoimmune disease involving, among other phenomena, increased keratinocyte proliferation along with inflammation and angiogenesis. The psoriatic lesion is characterized by itchy hyperkeratotic plaques.

50.3.4.1 Acne Vulgaris

Several clinical studies addressed the effect of phototherapy on acne vulgaris. Phototherapy was shown to significantly reduce the number of inflammatory acne lesions and to accelerate the healing process, reduce occurrence of relapses, and prevent or reduce scars. It is worth noting that LED systems, designed for home use, are currently commercially available that facilitate daily treatment protocols. Sequential application of visible blue (415 nm) and visible red (633 or 660 nm) LED-based arrays was reported to reduce inflammatory acne lesions by 80% at 8 and 12 weeks after final treatment (Goldberg and Russell 2006; Sadick 2008) being at least as good as alternative medications such as 5% benzoil peroxidase (Papageorgiou, Katsambas, and Chu 2000) and photodynamic therapy (PDT) with methyl aminolevulinate (MAL) (Horfelt et al. 2009) without some of the related side effects. A combination of visible blue light and NIR (830 nm) with the same protocol was less effective. It showed only 44% improvement in inflammatory lesions (Sadick 2009). Blue light induces selective oxidative stress by excitation of endogenous photosensitizers produced by the *P. acnes* bacteria as part of its normal metabolism (Choi et al. 2011).

50.3.4.2 Herpes Simplex Virus-1 (Cold Sores)

LLLT, during herpes simplex virus-1 (HSV-1) outbreak or during the prodromal stage (symptom-free), accelerated HSV-1 related wound healing and reduced recurrence rate (Schindl and Neumann 1999). It is hypothesized that the prolonged relapse time between attacks is not a direct inhibitory effect on the virus but rather a result of stimulation of the immune system by the LLLT.

In a double-blinded RCT, Schindl and Neuman (1999) reported that a total of 10 irradiations with 690 nm, energy density 48 J/cm^2 during the symptom-free period significantly lowered the frequency of local recurrence in patients with recurrent herpes labialis infections. Fifty patients were monitored for 52 weeks. The mean recurrence-free interval in the laser group was 37.5 weeks (range, 2–52 weeks), and the interval in the placebo sham-irradiated group was 3 weeks (range: 1–20 weeks). In an RCT, de Carvalho et al. (2010) compared 10 weekly LLLT irradiations (n = 41) with topical acyclovir [5%, five times/day (n = 30)] and found that LLLT-treated patients had a greater decrease in lesion size and inflammatory edema related to recurrent herpes lesions. Nevertheless, the monthly recurrence rate was not statistically significant between the two groups, showing that LLLT was as good as topical acyclovir. In a 5-year follow-up study, Munoz Sanchez et al. (2012) reported that 322 patients receiving LLLT in the blister and the prodromal stage had obvious reduction in HSV-1 recurrent rate as compared to the pre-treatment period.

50.3.4.3 Radiation-Induced Dermatitis

DeLand et al. (2007) evaluated the potential of LED in reducing acute radiation-induced dermatitis after lumpectomy. Daily LED treatments were given to 19 patients immediately after x-ray treatments during the 5.5 weeks of standard treatment course. These were compared to 28 retrospectively analyzed age-matched controls. It was concluded that LED significantly reduces the frequency and degree of radiation-induced skin reactions and the frequency of treatment interruption due to

severe skin reaction. However, Fife et al. (2010), in a randomized, controlled double-blind study of 35 patients, using a similar device but a slightly different protocol, found no difference in incidence of radiation-induced skin reactions or therapy interruptions. Despite the differences in outcome, both studies agreed on the safety of LED treatments, showing no evidence of harm to the patients.

50.3.4.4 Psoriasis

Ablon (2010) studied the effects of combination of red and NIR LEDs (biweekly for up to 5 weeks) on chronic recalcitrant psoriatic lesions in nine patients. Follow-ups at 3–8 months showed high clearance rates of the lesions without side effects and high level of patient satisfaction.

50.3.5 Scars

Hypertrophic scars and keloids result from an abnormal, fibrous, wound healing process. These scars are characterized by an overabundance of collagen and are often associated with raised, pruritic lesions in the wound margins, with keloids often extending beyond the wound margins, or contracture where elastic fibers may be replaced by collagen. A scar is a dynamic structure with ongoing synthesis and degradation of collagen. LLLT was shown to influence the remodeling process by increasing TGF-b1, collagen deposition and synthesis, fibroblast proliferation, and tensile strength. On the other hand, LLLT was shown to modulate MMPs, and in certain cases, decrease TGF-b1 expression. Therefore, although LLLT might worsen a scar by stimulating collagen formation, it might alternatively stimulate collagen degradation, thereby favorably influencing scar remodeling, or, under the appropriate local microenvironmental circumstances, attenuate its formation.

Several studies evaluated the potential of phototherapy in improving scars. Gaida et al. (2004) used a 670 nm laser in a group of 19 patients with 19 burn scars twice a week for 8 weeks. Part of the scar of each patient was not irradiated and served as a control. Seventeen out of 19 lesions showed macroscopic improvement after the treatment as measured by the Vancouver Scar Scale (VSS) (Sullivan et al. 1990). Newer scars reacted better than older scars. All but one of the patients who had reported pain or pruritus before the treatment experienced relief of symptoms. No negative effects were reported. Barolet and Boucher (2010) presented a case series of three patients with bilateral scars that were removed surgically. After scar removal, patients self-treated daily with LED in the NIR spectrum (805 nm, 30 mW/cm^2) on one side of the affected area, while the other side served as nonirradiated control. Significant improvements were reported at the irradiated site versus the control as assessed macroscopically by VSS. Measurements of skin surface microtopography revealed that scar height was significantly reduced in the irradiated scars compared to control. Sasaki et al. (2009) compared the effect of 830 Ga–Al–As laser alone with laser plus steroidal topical cream in 20 patients with hypertrophic scars or keloids. Significant improvement was found in both groups by the fifth treatment in the itching, thickness, pliability, erythema, and hardness. The topical steroid plus LLLT group achieved better improvement than LLLT alone. In an RCT by Carvalho et al. (2010), surgical incisions were irradiated prophylactically. The LLLT group showed improvement in macroscopic appearance of the scars (by VSS), better malleability, and reduction in thickness. In all of the above studies, no significant treatment-related adverse effects were reported.

50.4 Conclusions

Reviews in the last decade present conflicting views regarding the effectiveness of LLLT on wound healing. In a review published in 2002 on cell and animal studies related to wound healing, Lucas et al. (2002) stated that although *in vivo* studies show a pooled effect in favor of LLLT, the evidence is inconclusive, and that LLLT should not be considered a valuable treatment (adjuvant) for patients for this purpose. Posten et al. (2005) reached the same conclusion in a review covering the years 1965–2003 and added that although favorable results were shown in rodent models that have loose skin, the results were not duplicated in larger animal models such as pigs that have tight skin similar to humans. Whinfield and Aitkenhead (2009) reviewed the literature and stated that although *in vitro* and *in vivo* investigations suggest that phototherapy may stimulate cell activity and promote tissue repair, reports on clinical trials do not provide sufficient evidence to establish the usefulness of phototherapy as an effective tool in wound care. In the same year, Fulop et al. (2009) published a meta-analysis of studies from 2000 to 2007 and concluded that phototherapy is a highly effective form of treatment for tissue repair, with stronger supporting evidence resulting from experimental animal studies than human studies. A year later, Peplow, Chung, and Baxter (2010) published a review covering studies from 2003 to 2008 of experimental wound healing in rodent models and demonstrated the ability of laser or monochromatic light to photobiomodulate wound healing processes. They concluded that the evidence strongly supports the case for further controlled research in humans.

The gradual shift with time toward a more favorable opinion is mainly due to improvements in comprehensiveness of reported details and optimization of irradiation parameters for specific indications. Nevertheless, although more and more placebo-controlled RCTs are published, many of them are still underpowered. Moreover, the immense range of possibilities regarding device configurations, irradiation settings, and treatment modes and schedules have confounded efforts to replicate successful experiments.

In conclusion, LLLT has been shown to accelerate the healing of wounds of a variety of etiologies while reducing the risk of infection and other complications. From this review, the state of the art regarding the effect of LLLT on wound healing can be characterized as rapidly developing with considerable potential. Vast information has been accumulated regarding irradiation settings for specific indications, and medical centers dedicated

to wound healing report high rates of successful resolution of complicated wounds. However, the outcomes of certain treatment regimens have, in some cases, been based on studies lacking appropriate control groups. One of the ramifications of insufficient clinical evidence is that while LLLT is approved by US regulatory authorities for the treatment of pain and inflammation, the Food and Drug Administration (FDA) has not approved LLLT for treatment of wounds. This has also prevented formal medical associations from publishing official guidelines for the use of LLLT for these entities, which is critical for maintaining appropriate standards of care. The recent proliferation of well-designed clinical trials appears to be rapidly overcoming these shortcomings, and in so doing, creating a well-based standard of care to the great benefit of those suffering from otherwise refractory wounds.

References

Ablon, G. 2010. Combination 830-nm and 633-nm light-emitting diode phototherapy shows promise in the treatment of recalcitrant psoriasis: Preliminary findings. *Photomed Laser Surg* 28:141–146.

Agaiby, A. D., L. R. Ghali, R. Wilson, and M. Dyson. 2000. Laser modulation of angiogenic factor production by T-lymphocytes. *Lasers Surg Med* 26:357–363.

Al-Watban, F. A., X. Y. Zhang, and B. L. Andres. 2007. Low-level laser therapy enhances wound healing in diabetic rats: a comparison of different lasers. *Photomed Laser Surg* 25:72–77.

Barolet, D., and A. Boucher. 2010. Prophylactic low-level light therapy for the treatment of hypertrophic scars and keloids: a case series. *Lasers Surg Med* 42:597–601.

Bayat, M., M. M. Vasheghani, N. Razavi, S. Taheri, and M. Rakhshan. 2005. Effect of low-level laser therapy on the healing of second-degree burns in rats: A histological and microbiological study. *J Photochem Photobiol B* 78:171–177.

Byrnes, K. R., L. Barna, V. M. Chenault et al. 2004. Photobiomodulation improves cutaneous wound healing in an animal model of type II diabetes. *Photomed Laser Surg* 22:281–290.

Caetano, K. S., M. A. Frade, D. G. Minatel, L. A. Santana, and C. S. Enwemeka. 2009. Phototherapy improves healing of chronic venous ulcers. *Photomed Laser Surg* 27:111–118.

Carvalho, R. L., P. S. Alcantara, F. Kamamoto, M. D. Cressoni, and R. A. Casarotto. 2010. Effects of low-level laser therapy on pain and scar formation after inguinal herniation surgery: a randomized controlled single-blind study. *Photomed Laser Surg* 28:417–422.

Chaves, M. E., A. R. Araujo, S. F. Santos, M. Pinotti, and L. S. Oliveira. 2012. LED phototherapy improves healing of nipple trauma: a pilot study. *Photomed Laser Surg* 30:172–178.

Chen, C. H., H. S. Hung, and S. H. Hsu. 2008. Low-energy laser irradiation increases endothelial cell proliferation, migration, and eNOS gene expression possibly via PI3K signal pathway. *Lasers Surg Med* 40:46–54.

Choi, M. S., S. J. Yun, H. J. Beom, H. R. Park, and J. B. Lee. 2011. Comparative study of the bactericidal effects of 5-aminolevulinic acid with blue and red light on Propionibacterium acnes. *J Dermatol* 38:661–666.

Corazza, A. V., J. Jorge, C. Kurachi, and V. S. Bagnato. 2007. Photobiomodulation on the angiogenesis of skin wounds in rats using different light sources. *Photomed Laser Surg* 25:102–106.

Costa, M. S., C. E. Pinfildi, H. C. Gomes et al. 2010. Effect of low-level laser therapy with output power of 30 mW and 60 mW in the viability of a random skin flap. *Photomed Laser Surg* 28:57–61.

Dantas, M. D., D. R. Cavalcante, F. E. Araujo et al. 2011. Improvement of dermal burn healing by combining sodium alginate/chitosan-based films and low level laser therapy. *J Photochem Photobiol B* 105:51–59.

de Carvalho, R. R., E. de Paula., F. Ramalho et al. 2010. Effect of laser phototherapy on recurring herpes labialis prevention: an *in vivo* study. *Lasers Med Sci* 25:397–402.

Dehlin, O., S. Elmstahl, and F. Gottrup. 2003. Monochromatic phototherapy in elderly patients: a new way of treating chronic pressure ulcers? *Aging Clin Exp Res* 15:259–263.

Dehlin, O., S. Elmstahl, and F. Gottrup. 2007. Monochromatic phototherapy: effective treatment for grade II chronic pressure ulcers in elderly patients. *Aging Clin Exp Res* 19:478–483.

DeLand, M. M., R. A. Weiss, D. H. Mcdaniel, and R. G. Geronemus. 2007. Treatment of radiation-induced dermatitis with light-emitting diode (LED) photomodulation. *Lasers Surg Med* 39:164–168.

Duan, R., T. C. Liu, Y. Li, H. Guo, and L. B. Yao. 2001. Signal transduction pathways involved in low intensity He–Ne laser-induced respiratory burst in bovine neutrophils: A potential mechanism of low intensity laser biostimulation. *Lasers Surg Med* 29:174–178.

El Sayed, S. O., and M. Dyson. 1990. Comparison of the effect of multi-wavelength light produced by a cluster of semiconductor diodes and of each individual diode on mast cell number and degranulation in intact and injured skin. *Lasers Surg Med* 10:559–568.

Esteves Jr., I., I. B. Masson, C. T. Oshima et al. 2012. Low-level laser irradiation, cyclooxygenase-2 (COX-2) expression and necrosis of random skin flaps in rats. *Lasers Med Sci* 27:655–660.

Evers, L. H., D. Bhavsar, and P. Mailander. 2010. The biology of burn injury. *Exp Dermatol* 19:777–783.

Ezzati, A., M. Bayat, and A. Khoshvaghti. 2010. Low-level laser therapy with a pulsed infrared laser accelerates second-degree burn healing in rat: A clinical and microbiologic study. *Photomed Laser Surg* 28:603–611.

Ezzati, A., M. Bayat, S. Taheri, and Z. Mohsenifar. 2009. Low-level laser therapy with pulsed infrared laser accelerates third-degree burn healing process in rats. *J Rehabil Res Dev* 46:543–554.

Fife, D., D. J. Rayhan, S. Behnam et al. 2010. A randomized, controlled, double-blind study of light emitting diode photomodulation for the prevention of radiation dermatitis in patients with breast cancer. *Dermatol Surg* 36:1921–1927.

Fulop, A. M., S. Dhimmer, J. R. Deluca et al. 2009. A meta-analysis of the efficacy of phototherapy in tissue repair. *Photomed Laser Surg* 27:695–702.

Fushimi, T., S. Inui, T. Nakajima et al. 2012. Green light emitting diodes accelerate wound healing: Characterization of the effect and its molecular basis *in vitro* and *in vivo*. *Wound Repair Regen* 20:226–235.

Gaida, K., R. Koller, C. Isler et al. 2004. Low level laser therapy—A conservative approach to the burn scar? *Burns* 30:362–367.

Gal, P., B. Vidinsky, T. Toporcer et al. 2006. Histological assessment of the effect of laser irradiation on skin wound healing in rats. *Photomed Laser Surg* 24:480–488.

Gavish, L., Y. Asher, Y. Becker, and Y. Kleinman. 2004. Low level laser irradiation stimulates mitochondrial membrane potential and disperses subnuclear promyelocytic leukemia protein. *Lasers Surg Med* 35:369–376.

Gavish, L., L. Perez, and S. D. Gertz. 2006. Low-level laser irradiation modulates matrix metalloproteinase activity and gene expression in porcine aortic smooth muscle cells. *Lasers Surg Med* 38:779–786.

Gavish, L., L. S. Perez, P. Reissman, and S. D. Gertz. 2008. Irradiation with 780 nm diode laser attenuates inflammatory cytokines but upregulates nitric oxide in lipopolysaccharide-stimulated macrophages: Implications for the prevention of aneurysm progression. *Lasers Surg Med* 40:371–378.

Goldberg, D. J., and B. A. Russell. 2006. Combination blue (415 nm) and red (633 nm) LED phototherapy in the treatment of mild to severe acne vulgaris. *J Cosmet Laser Ther* 8: 71–75.

Grossman, N., N. Schneid, H. Reuveni, S. Halevy, and R. Lubart. 1998. 780 nm low power diode laser irradiation stimulates proliferation of keratinocyte cultures: Involvement of reactive oxygen species. *Lasers Surg Med* 22:212–218.

Hawkins, D., and H. Abrahamse. 2006. Effect of multiple exposures of low-level laser therapy on the cellular responses of wounded human skin fibroblasts. *Photomed Laser Surg* 24:705–714.

Hawkins, D. H., and H. Abrahamse. 2007. Time-dependent responses of wounded human skin fibroblasts following phototherapy. *J Photochem Photobiol B* 88:147–155.

Hemvani, N., D. S. Chitnis, and N. S. Bhagwanani. 2005. Helium-neon and nitrogen laser irradiation accelerates the phagocytic activity of human monocytes. *Photomed Laser Surg* 23:571–574.

Horfelt, C., B. Stenquist, C. B. Halldin, M. B. Ericson, and A. M. Wennberg. 2009. Single low-dose red light is as efficacious as methyl-aminolevulinate—Photodynamic therapy for treatment of acne: Clinical assessment and fluorescence monitoring. *Acta Derm Venereol* 89:372–378.

Houreld, N., and H. Abrahamse. 2007. Irradiation with a 632.8 nm helium–neon laser with 5 J/cm² stimulates proliferation and expression of interleukin-6 in diabetic wounded fibroblast cells. *Diabetes Technol Ther* 9:451–459.

Houreld, N. N., P. R. Sekhejane, and H. Abrahamse. 2010. Irradiation at 830 nm stimulates nitric oxide production and inhibits pro-inflammatory cytokines in diabetic wounded fibroblast cells. *Lasers Surg Med* 42:494–502.

Iordanou, P., G. Baltopoulos, M. Giannakopoulou, P. Bellou, and E. Ktenas. 2002. Effect of polarized light in the healing process of pressure ulcers. *Int J Nurs Pract* 8:49–55.

Karu, T. 1999. Primary and secondary mechanisms of action of visible to near-IR radiation on cells. *J Photochem Photobiol B* 49:1–17.

Kaviani, A., G. E. Djavid, L. Ataie-Fashtami et al. 2011. A randomized clinical trial on the effect of low-level laser therapy on chronic diabetic foot wound healing: A preliminary report. *Photomed Laser Surg* 29:109–114.

Keast, D. H., and H. Orsted. 1998. The basic principles of wound care. *Ostomy Wound Manage* 44:24–28, 30–1.

Kipshidze, N., V. Nikolaychik, M. H. Keelan et al. 2001. Low-power helium: Neon laser irradiation enhances production of vascular endothelial growth factor and promotes growth of endothelial cells *in vitro*. *Lasers Surg Med* 28:355–364.

Kleinman, Y., S. Simmer, and Y. Braksma. 1996. Low power laser therapy in patients with diabetic foot ulcers: Early and long term outcome. *Laser Therapy* 8:205–208.

Kopera, D., R. Kokol, C. Berger, and J. Haas. 2005. Does the use of low-level laser influence wound healing in chronic venous leg ulcers? *J Wound Care* 14:391–394.

Kupin, V. I., V. S. Bykov, A. V. Ivanov, and V. Larichev. 1982. Potentiating effects of laser radiation on some immunological traits. *Neoplasma* 29:403–406.

Labbe, R. F., K. J. Skogerboe, H. A. Davis, and R. L. Rettmer. 1990. Laser photobioactivation mechanisms: *In vitro* studies using ascorbic acid uptake and hydroxyproline formation as biochemical markers of irradiation response. *Lasers Surg Med* 10:201–207.

Lacjakova, K., N. Bobrov, M. Polakova et al. 2010. Effects of equal daily doses delivered by different power densities of low-level laser therapy at 670 nm on open skin wound healing in normal and corticosteroid-treated rats: a brief report. *Lasers Med Sci* 25:761–766.

Lagan, K. M., T. Mckenna, A. Witherow et al. 2002. Low-intensity laser therapy/combined phototherapy in the management of chronic venous ulceration: A placebo-controlled study. *J Clin Laser Med Surg* 20:109–116.

Landau, Z., M. Migdal, A. Lipovsky, and R. Lubart. 2011. Visible light-induced healing of diabetic or venous foot ulcers: A placebo-controlled double-blind study. *Photomed Laser Surg* 29:399–404.

Lau, R. W., and G. L. Cheing. 2009. Managing postmastectomy lymphedema with low-level laser therapy. *Photomed Laser Surg* 27:763–769.

Lievens, P. C. 1991. The effect of a combined HeNe and I.R. laser treatment on the regeneration of the lymphatic system during the process of wound healing. *Lasers Med Sci* 6:193–199.

Lowe, A. S., M. D. Walker, M. O'Byrne, G. D. Baxter, and D. G. Hirst. 1998. Effect of low intensity monochromatic light therapy (890 nm) on a radiation-impaired, wound-healing model in murine skin. *Lasers Surg Med* 23:291–298.

Lucas, C., L. J. Criens-Poublon, C. T. Cockrell, and R. J. de Haan. 2002. Wound healing in cell studies and animal model experiments by low level laser therapy; were clinical studies justified? A systematic review. *Lasers Med Sci* 17:110–134.

Lucas, C., M. J. Van Gemert, and R. J de Haan. 2003. Efficacy of low-level laser therapy in the management of stage III decubitus ulcers: a prospective, observer-blinded multicentre randomised clinical trial. *Lasers Med Sci* 18:72–77.

Lundeberg, T., and M. Malm. 1991. Low-power HeNe laser treatment of venous leg ulcers. *Ann Plast Surg* 27:537–539.

Malm, M., and T. Lundeberg. 1991. Effect of low power gallium arsenide laser on healing of venous ulcers. *Scand J Plast Reconstr Surg Hand Surg* 25:249–251.

Martin, P. 1997. Wound healing—aiming for perfect skin regeneration. *Science* 276:75–81.

Medrado, A. P., A. P. Soares, E. T. Santos, S. R. Reis, and Z. A. Andrade. 2008. Influence of laser photobiomodulation upon connective tissue remodeling during wound healing. *J Photochem Photobiol B* 92:144–152.

Medrado, A. R., L. S. Pugliese, S. R. Reis, and Z. A. Andrade. 2003. Influence of low level laser therapy on wound healing and its biological action upon myofibroblasts. *Lasers Surg Med* 32:239–244.

Mester, E., M. Ludany, and M. Seller. 1968. The simulating effect of low power laser ray on biological systems. *Laser Rev* 1:3.

Mester, E., T. Spiry, B. Szende, and J. G. Tota. 1971. Effect of laser rays on wound healing. *Am J Surg* 122:532–535.

Minatel, D. G., M. A. Frade, S. C. Franca, and C. S. Enwemeka. 2009. Phototherapy promotes healing of chronic diabetic leg ulcers that failed to respond to other therapies. *Lasers Surg Med* 41:433–441.

Munoz Sanchez, P. J., J. L. Capote Femenias, A. Diaz Tejeda, and J. Tuner. 2012. The effect of 670-nm low laser therapy on herpes simplex type 1. *Photomed Laser Surg* 30:37–40.

Ng, M. F. 2010. The role of mast cells in wound healing. *Int Wound J* 7:55–61.

Ozturan, S., S. A. Durukan, O. Ozcelik, G. Seydaoglu, and M. C. Haytac. 2011. Coronally advanced flap adjunct with low intensity laser therapy: a randomized controlled clinical pilot study. *J Clin Periodontol* 38:1055–1062.

Papageorgiou, P., A. Katsambas, and A. Chu. 2000. Phototherapy with blue (415 nm) and red (660 nm) light in the treatment of acne vulgaris. *Br J Dermatol* 142:973–978.

Pastore, D., C. Di Martino, G. Bosco, and S. Passarella. 1996. Stimulation of ATP synthesis via oxidative phosphorylation in wheat mitochondria irradiated with helium–neon laser. *Biochem Mol Biol Int* 39:149–157.

Peplow, P. V., T. Y. Chung, and G. D. Baxter. 2010. Laser photo-biomodulation of wound healing: a review of experimental studies in mouse and rat animal models. *Photomed Laser Surg* 28:291–325.

Peplow, P. V., T. Y. Chung, B. Ryan, and G. D. Baxter. 2011a. Laser photobiomodulation of gene expression and release of growth factors and cytokines from cells in culture: a review of human and animal studies. *Photomed Laser Surg* 29:285–304.

Peplow, P. V., T. Y. Chung, B. Ryan, and G. D. Baxter. 2011b. Laser photobiostimulation of wound healing: reciprocity of irradiance and exposure time on energy density for splinted wounds in diabetic mice. *Lasers Surg Med* 43:843–850.

Pessoa, E. S., R. M. Melhado, L. H. Theodoro, and V. G. Garcia. 2004. A histologic assessment of the influence of low-intensity laser therapy on wound healing in steroid-treated animals. *Photomed Laser Surg* 22:199–204.

Posten, W., D. A. Wrone, J. S. Dover, K. A. Arndt, S. Silapunt, and M. Alam. 2005. Low-level laser therapy for wound healing: mechanism and efficacy. *Dermatol Surg* 31:334–340.

Prabhu, V., S. B. Rao, S. Chandra et al. 2012. Spectroscopic and histological evaluation of wound healing progression following low level laser therapy (LLLT). *J Biophotonics* 5:168–184.

Prado, R., L. Neves, A. Marcolino et al. 2010. Effect of low-level laser therapy on malondialdehyde concentration in random cutaneous flap viability. *Photomed Laser Surg* 28:379–384.

Reddy, G. K., L. Stehno-Bittel, and C. S. Enwemeka. 2001. Laser photostimulation accelerates wound healing in diabetic rats. *Wound Repair Regen* 9:248–255.

Renno, A. C., A. M. Iwama, P. Shima et al. 2011. Effect of low-level laser therapy (660 nm) on the healing of second-degree skin burns in rats. *J Cosmet Laser Ther* 13:237–242.

Rochkind, S., M. Rousso, M. Nissan et al. 1989. Systemic effects of low-power laser irradiation on the peripheral and central nervous system, cutaneous wounds, and burns. *Lasers Surg Med* 9:174–182.

Sadick, N. S. 2008. Handheld LED array device in the treatment of acne vulgaris. *J Drugs Dermatol* 7:347–350.

Sadick, N. S. 2009. A study to determine the effect of combination blue (415 nm) and near-infrared (830 nm) light-emitting diode (LED) therapy for moderate acne vulgaris. *J Cosmet Laser Ther* 11:125–128.

Safavi, S. M., B. Kazemi, M. Esmaeili et al. 2008. Effects of low-level He–Ne laser irradiation on the gene expression of IL-1beta, TNF-alpha, IFN-gamma, TGF-beta, bFGF, and PDGF in rat's gingiva. *Lasers Med Sci* 23:331–335.

Samoilova, K. A., N. A. Zhevago, N. N. Petrishchev, and A. A. Zimin. 2008. Role of nitric oxide in the visible light-induced rapid increase of human skin microcirculation at the local and systemic levels: II. Healthy volunteers. *Photomed Laser Surg* 26:443–449.

Santos, J. O., S. M. Oliveira, M. R. Nobre, A. C. Aranha, and M. B. Alvarenga. 2011. A randomised clinical trial of the effect of low-level laser therapy for perineal pain and healing after episiotomy: A pilot study. *Midwifery* 28:e653–e659

Saperia, D., E. Glassberg, R. F. Lyons et al. 1986. Demonstration of elevated type I and type III procollagen mRNA levels in cutaneous wounds treated with helium–neon laser. Proposed mechanism for enhanced wound healing. *Biochem Biophys Res Commun* 138:1123–1128.

Sasaki, K., T. Ohshiro, T. Ohshiro, and Y. Taniguchi. 2009. A prospective study of the influence that topical steroid exerts in low reactive level laser therapy (LLLT) for the treatment of hypertrophic scars and keloids. *Laser Therapy* 18: 137–141.

Schindl, A., G. Heinze, M. Schindl, H. Pernerstorfer-Schon, and L. Schindl. 2002. Systemic effects of low-intensity laser irradiation on skin microcirculation in patients with diabetic microangiopathy. *Microvasc Res* 64:240–246.

Schindl, A., H. Merwald, L. Schindl, C. Kaun, and J. Wojta. 2003. Direct stimulatory effect of low-intensity 670 nm laser irradiation on human endothelial cell proliferation. *Br J Dermatol* 148:334–336.

Schindl, A., and R. Neumann. 1999. Low-intensity laser therapy is an effective treatment for recurrent herpes simplex infection. Results from a randomized double-blind placebo-controlled study. *J Invest Dermatol* 113:221–223.

Schindl, A., M. Schindl, L. Schindl et al. 1999. Increased dermal angiogenesis after low-intensity laser therapy for a chronic radiation ulcer determined by a video measuring system. *J Am Acad Dermatol* 40:481–484.

Schlager, A., K. Oehler, K. U. Huebner, M. Schmuth, and L. Spoetl. 2000. Healing of burns after treatment with 670-nanometer low-power laser light. *Plast Reconstr Surg* 105:1635–1639.

Schubert, V. 2001. Effects of phototherapy on pressure ulcer healing in elderly patients after a falling trauma. A prospective, randomized, controlled study. *Photodermatol Photoimmunol Photomed* 17:32–38.

Smith, R. J., M. Birndorf, G. Gluck, D. Hammond, and W. D. Moore. 1992. The effect of low-energy laser on skin-flap survival in the rat and porcine animal models. *Plast Reconstr Surg* 89:306–310.

Stadelmann, W. K., A. G. Digenis, and G. R. Tobin. 1998. Physiology and healing dynamics of chronic cutaneous wounds. *Am J Surg* 176:26S–38S.

Stadler, I., R. J. Lanzafame, R. Evans et al. 2001. 830-nm irradiation increases the wound tensile strength in a diabetic murine model. *Lasers Surg Med* 28:220–226.

Sullivan, T., J. Smith, J. Kermode, E. Mciver, and D. J. Courtemanche. 1990. Rating the burn scar. *J Burn Care Rehabil* 11:256–260.

Taly, A. B., K. P. Sivaraman Nair, T. Murali, and A. John. 2004. Efficacy of multiwavelength light therapy in the treatment of pressure ulcers in subjects with disorders of the spinal cord: a randomized double-blind controlled trial. *Arch Phys Med Rehabil* 85:1657–1661.

Whinfield, A. L., and I. Aitkenhead. 2009. The light revival: does phototherapy promote wound healing? A review. *Foot (Edinb)* 19:117–124.

Wu, Z. H., Y. Zhou, J. Y. Chen, and L. W. Zhou. 2010. Mitochondrial signaling for histamine releases in laser-irradiated RBL-2H3 mast cells. *Lasers Surg Med* 42:503–509.

Yasukawa, A., H. Hrui, Y. Koyama, M. Nagai, and K. Takakuda. 2007. The effect of low reactive-level laser therapy (LLLT) with helium–neon laser on operative wound healing in a rat model. *J Vet Med Sci* 69:799–806.

Young, S., P. Bolton, M. Dyson, W. Harvey, and C. Diamantopoulos. 1989. Macrophage responsiveness to light therapy. *Lasers Surg Med* 9:497–505.

Yu, H. S., K. L. Chang, C. L. Yu, J. W. Chen, and G. S. Chen. 1996. Low-energy helium–neon laser irradiation stimulates interleukin-1 alpha and interleukin-8 release from cultured human keratinocytes. *J Invest Dermatol* 107:593–596.

Yu, H. S., C. S. Wu, C. L. Yu, Y. H. Kao, and M. H. Chiou. 2003. Helium–neon laser irradiation stimulates migration and proliferation in melanocytes and induces repigmentation in segmental-type vitiligo. *J Invest Dermatol* 120:56–64.

51

Low-Level Laser Therapy in the Treatment of Pain

Roberta Chow
The University of Sydney

51.1 LLLT and Pain

51.1.1 Early History

One of the most important applications of low-level laser therapy (LLLT) is in the treatment of pain with a history commencing within a few years of the production of the first ruby laser in 1960 (Maiman 1960). Early clinical use related to laser acupuncture (LA) (Bischko 1980), where laser irradiation (LI) was applied instead of needles to single points identified by the principles of Traditional Chinese medicine (TCM) (Deng et al. 1987). Helium–Neon (He–Ne) laser devices were used almost exclusively for LA and were characterized by power outputs of less than 10 mW and energy densities (EDs) less than 4 J/cm². As diode technology developed power outputs of laser devices used for pain increased from the original 1 mW devices to 1 W pulsed, defocused, ablative lasers such as Nd:YAG.

Over time, clinical use of laser devices diverged from the single-point application of LA to LLLT where lasers were applied to multiple local points at sites of pathology. The pain-modulating effects of LLLT were postulated to be mediated by direct, cumulative, and multiple photochemical effects in local tissue (Mester et al. 1968), rather than "stimulation" of acupuncture points where the therapeutic effect is said to be mediated by quantum oscillations on the system of meridians ("Qi") moving along acupuncture meridians (Kroetlinger 1980). Research into local tissue effects and mechanisms of pain relief in both modes of application has been the subject of continuing research. This has resulted in an exponential increase in the number of randomized controlled trials for LLLT in painful conditions in the last decade.

51.2 Painful Conditions Treated

LLLT is used in a range of painful conditions (Table 51.1), from acute pain following tissue injury within minutes to the treatment of persistent (chronic) pain that has been present for years (Table 51.2). The wide spectrum of conditions treated suggests that multiple mechanisms are operational in achieving pain relief.

TABLE 51.1 Acute Conditions Treated with LLLT

Sprains, strains (all joints and ligaments)
Whiplash injury
Postsurgical pain
Dental pain
Herpes zoster
Herpes labialis
Migraine
Cervicogenic headaches
Tension (muscle contraction) headaches
Back pain—acute flare of chronic pain
Sciatica
Gout—acute flare
Arthritis—acute flare
Exacerbation of chronic conditions
Tendinitis—Achilles, supraspinatus
Radiculopathies—lumbar, cervical
TMJ disorders

TABLE 51.2 Chronic Painful Conditions Treated with LLLT

Osteoarthritis (all joints)
Rheumatoid arthritis
Other arthridities
Whiplash associated disorder
Frozen shoulder
Post-herpetic neuralgia
Trigeminal neuralgia
Neck pain
Back pain
Lateral and medial epicondylitis
Tendinopathies
Myofascial pain syndrome
Trigger points
Chronic regional pain syndrome
Fibromyalgia

Over the last three decades, many authors have proposed that the direct effect of lasers on nerves is one such mechanism (Baxter et al. 1994; Kasai et al. 1996; Kono et al. 1993; Shimoyama et al. 1992b). Multiple effects of lasers on nerves, predominantly inhibitory, have been described and there is a large body of evidence to support a neural hypothesis for laser-induced pain relief (Chow et al. 2011). This chapter explores these mechanisms. In order to understand how LLLT modulates pain, however, it is essential to understand the underlying neurophysiological basis of pain and how LI effects changes in nerves.

51.3 The Neurophysiology of Pain and the Pain Matrix

The International Association for the Study of Pain (IASP) defines pain as "an unpleasant sensory and emotional experience associated with actual or potential tissue damage, or described in terms of such damage" (Merskey and Bogduk 1994).

This definition acknowledges the complexity of the phenomenon of pain beyond nociception, which is best understood in the biopsychosocial model (Engel 1977). In this model, the intensity of the noxious stimulus together with the "meaning" of the perceived pain for the individual as to how damaging it may be, as well as the social context in which it occurs, are all integrated into the patient's pain experience. Activation of nociceptors constitutes the "bio" component of the biopsychosocial model and in the clinical application of LLLT inhibition of nociceptors constitutes a major mechanism for pain relief.

Peripheral nerve endings of nociceptors, the thinly myelinated Aδ and unmyelinated, slow-conducting C fibers, lie within the epidermis forming a complex neural network (Kennedy et al. 2005; Lauria 1999), which transduce noxious stimuli, such as heat, mechanical force, and the chemical stimuli of inflammatory neuropeptides into action potentials (Siddall and Cousins 1998). The location of these endings is very superficial and is within the penetration depths of all laser wavelengths (Figure 51.1). Laser effects on these nerves rather than deeper nerves or structures may underlie effects of very-low-power lasers. These afferent fibers penetrate the basilar laminar and aggregate to form the larger nerve trunks, which also contain fibroblasts, Schwann cells, and capillaries. The anatomy of the individual neurons making up the larger nerve fibers is very different from that of compact cells such as fibroblasts, whose response to lasers has been well documented. Cell bodies of neurons lie within the dorsal root ganglia (DRG), which lie in proximity to the spinal cord at each spinal level. The trigeminal ganglion is equivalent to the DRG at lower spinal levels. The elongated cytoplasm of the neuron extends distally from the cell body up to 1 m in length, as in the sciatic nerve, to bare nerve endings in skin, while the proximal, short portion of the neuron enters the dorsal horn.

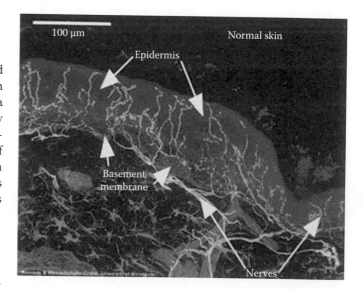

FIGURE 51.1 Skin section showing superficial peripheral nerve endings. (From Kennedy, W. et al., Pathology and quantitation of cutaneous innervation. In *Peripheral Neuropathy*, P. Dyck and P. Thomas, editors. W.B. Saunders, Philadelphia, PA, 873, 2005. With permission.)

Afferent signals from these sensory neurons synapse directly with second-order neurons or indirectly via interneurons within laminae 1 to 5 of the dorsal horn. Action potentials generated in the second-order neurons ascend to a complex network of multiple centers within the midbrain to the cortex, known as the pain matrix, identified in functional magnetic resonance imaging (fMRI) studies (Tracey and Johns 2010). Nerve tracts from higher centers descend to the spinal cord synapsing in the dorsal horn, modulating ascending and descending activity to and from the subcortical and cortical areas to the spinal cord. It is the unique structure of nerves and the complex feedback system from peripheral nerves in the skin to the central nervous system (i.e., skin to the brain) (Backonja and Lauria 2010) that is driven by neuroplasticity (i.e., the capacity of the nervous system to respond to changing input) (Ji and Woolf 2001; Woolf and Salter 2000), which underlie the capacity of lasers to reduce pain both in the short and long term.

The direct effects of lasers in nerves are operational at the level of the epidermal neural network, nerves to blood vessels (nervi vasorum) in the skin and subcutaneous tissues, sympathetic ganglia, and the neuromuscular junctions within muscles and nerve trunks.

51.4 Using Lasers to Block Nerve Conduction

The dominant findings of a review of laser effects on the electrophysiological activity of nerves demonstrated largely inhibitory effects (Chow et al. 2011). In human studies, continuous wave (CW), infrared (IR) LI reduced conduction velocity (CV) in median (Baxter et al. 1994), superficial radial (Kramer and Sandrin 1993), and sural nerves (Cambier et al. 2000; Hadian and Moghagdam 2003). Pulsed IR laser or laser not applied directly over the course of the nerve did not alter nerve conduction (Baxter et al. 1992). Conduction block in both motor nerves [as measured by the compound muscle action potential (CMAP)] and sensory nerves (somatosensory-evoked potentials) was also seen in a series of experiments with 30 s transcutaneous 808 and 650 nm LI to four points overlying rat sciatic nerve (Yan, Chow, and Armati 2011), seen in Figure 51.2a through d (CMAPS not shown). Neural

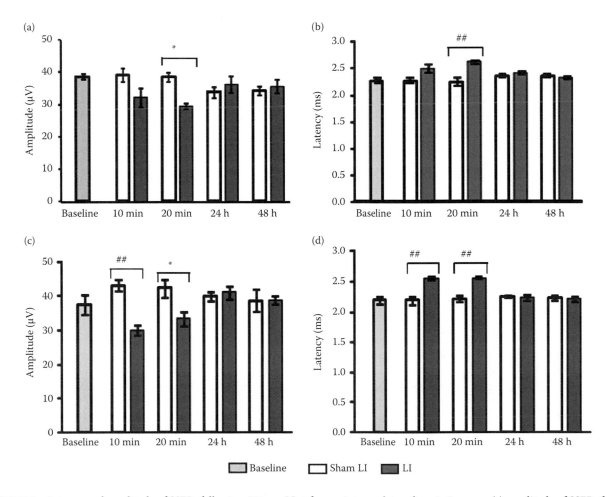

FIGURE 51.2 Latency and amplitude of SSEPs following 650 nm LI at four points overlying the sciatic nerve: (a) amplitude of SSEPs following 650 nm LI or sham LI (control), (b) latency of SSEPs following 650 nm LI or sham LI (control), (c) amplitude of SSEPs following 808 nm LI or sham LI (control), and (d) latency of SSEPs following 808 nm LI or sham LI (control). $^*p < 0.5$; $^{**}p < 0.01$.

blockade, up to 30%, occurred within 10 to 20 min of laser application and reversed by 24 h. Although it is not possible to extrapolate these findings directly to the pain-relieving effects of lasers, it provides unequivocal evidence that transcutaneous lasers can inhibit nerve conduction in underlying nerves.

Animal studies, which used a large number of different experimental models, were more informative with regard to dissecting out the relationship between nerve conduction and pain relief. Not only did these studies establish that both visible and IR lasers could cause conduction block in nerve trunks but also a series of studies demonstrated that lasers specifically inhibited action potentials generated by noxious and proinflammatory stimuli in nociceptors (i.e., lasers were nociceptor-specific) (Kasai et al. 1994; Mezawa et al. 1988). Blockade of action potential amplitudes initiated by mechanical and thermal noxious stimuli occurred with both visible and IR LI. Demonstrating this effect was a study of 830 nm (40 mW) CW laser that blocked pinch, cold, and heat applied to rat paw but not brush stimulation (Tsuchiya, Kawatani, and Takeshige 1994). As light touch (brush) is conveyed by Aβ fibers, which were not affected and only nociceptors were blocked, this illustrates the specificity of lasers of this wavelength for Aδ and C fibers. Furthermore, a parallel study showed that rats without Aδ and C fibers, which were destroyed at birth by capsaicin, could not respond to painful stimuli or LI, although brush sensitivity as in the previous experiment was unaffected. Other wavelengths, such as 1064 nm (Nd:YAG) laser, which is used widely in dentistry, have also demonstrated C fiber specificity in similar models (Wesselmann, Kerns, and Rymer 1992; Wesselmann, Lin, and Rymer 1991).

A similar set of experiments also using an *in vivo* rat nerve model demonstrated that action potentials generated by proinflammatory stimuli, such as formalin or turpentine in the skin of rat paw, were blocked by 830 nm laser (Sato et al. 1994; Shimoyama et al. 1992b; Tsuchiya, Kawatani, and Takeshige 1994). These studies are also important in unequivocally demonstrating that laser could block the afferent nerve response generated by proinflammatory agents applied to peripheral nerve.

51.5 Blockade in Peripheral Nerves to Reduce Central Effects

The concept that lasers can block noxious peripheral stimuli within skin is clearly supported, the fact that lasers applied to peripheral nerves has an effect in the proximal nerve and at the dorsal horn is important in understanding how lasers can modulate the pain matrix and have long-term effects in pain reduction. More importantly in understanding how lasers can modulate the pain matrix and have long term effects in pain reduction, is the fact that lasers applied to peripheral nerves have an effect in the proximal nerve and at the dorsal horn. A number of studies show this effect. In a single-neuron experiment, bradykinin applied to the axon generated an action potential in the cell body in a separate chamber (Jimbo et al. 1998). Application of laser (830 nm, 16.2 mW) for 1 min following bradykinin blocked the action potentials in the

cell body. In an *in vivo* model, action potentials generated by electrical stimulation to the sciatic nerve were measured at the L5 dorsal roots (Kono et al. 1993). When 632.8 nm, 1 mW, 100 Hz laser was applied proximal to the site of stimulation in the peripheral nerve, the action potentials generated at the dorsal root were significantly reduced. Extending the model, injection of bradykinin to rat face resulted in increase in mitochondrial proliferation in the trigeminal nucleus, while 830 nm, 60 mW, CW laser applied daily for 12 days to the site of injection reduced the extent of proliferation to control levels (Maeda 1989). Similarly, electrical activity in the trigeminal nucleus from noxious stimulation of root pulp was blocked by 830 nm laser to the root (Wakabayashi et al. 1993).

These series of experiments demonstrate that lasers applied to skin have a local inhibitory effect on the peripheral nerve endings at the site of injury or inflammation that blocks proximal effects within the nerve root. The cascade effect leads to suppressed synaptic activity in second-order neurons so that cortical areas of the pain matrix would not be activated.

51.6 How Does LI Inhibit Action Potentials?

The elongated structure of the neuron makes it uniquely vulnerable to the effects of LI. Adenosine triphosphate (ATP), which is the source of energy for all cells including the neuron, is transported via fast axonal flow along the cytoskeleton by molecular motors in ATP-rich mitochondria from the cell body where it is synthesized. The cytoskeleton is made up of microtubules of α- and β tubulin, which form the "monorail" system for transport along which other neurotransmitter-rich organelles move. When the cytoskeleton is disrupted, fast axonal flow and hence transport of mitochondria and other organelles is blocked, leading to the reduction in available ATP for nerve function. This leads to a disruption not only of cytoskeletal function but also of any ATP-dependent activities.

Immunohistochemistry demonstrated that 830 nm (300 mW), 808 nm (450 mW), and 650 nm (35 mW) lasers to cultured rat DRG neurons, shown in Figure 51.3, caused the formation of reversible varicosities, or beading, along axons, in which mitochondria pile up where the cytoskeleton is disrupted (Chen et al. 1993; Chow, David, and Armati 2007). This is also seen with 1064 nm Nd:YAG laser in tibial nerves (Wesselmann, Kerns, and Rymer 1994). The density of varicosities increased with increasing EDs of LI. The relationship between varicosity formation and slowing of CV has been demonstrated with local anesthetic agents, which also cause varicosities in DRG neurons (Nicolson, Smith, and Poste 1976; Poste, Papahadjopoulos, and Nicholson 1975). The authors developed a mathematic model showing how varicosity formation slowed CV. This suggests that varicosity formation may be a common mechanism for analgesic effects of laser and local anesthetic agents.

Further to the immunohistochemical studies above and live imaging studies in DRG neurons, 830 nm lasers also inhibited fast-axonal flow and decreased mitochondrial membrane

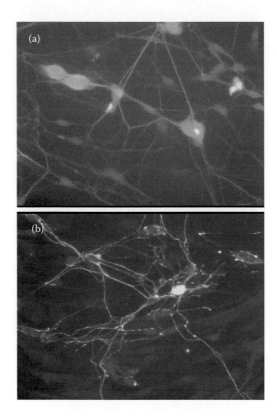

FIGURE 51.3 Representative micrographs of (a) control cultured rat dorsal root ganglion neurons, and (b) 30 s 830 nm laser with varicosities shown by beta tubulin aggregates along axons 1 h postexposure.

potential (MMP) in DRG neurons. As MMP is a surrogate measure of ATP, a reduction of ATP will lead to decreased activity in all ATPases. As an example, 830 nm LIs increased Na+K+ATPase at low EDs and inhibited levels by high dose LI, the latter consistent with slowing of fast axonal flow (Kudoh et al. 1989).

As Na+K+ATPase maintains the resting potential of the nerve, increased or decreased activity would have profound effects on the electrophysiology of the nerve. Depolarization blockade consistent with the above findings has been identified following 632.8 nm laser to nerves (Shimoyama et al. 1992a). Tetrodotoxin, a depolarizing sodium channel blocker, abolishes the effects of lasers on nerves, which suggests that lasers act via sodium channel activity (Jimbo et al. 1998; Miura and Kawatani 1996). Paradoxically, hyperpolarization of individual nerve cells has also been described by intracellular recording of cervical sympathetic ganglion neurons, although extracellular recording of the nerve showed decreased amplitude (Shimoyama et al. 1992a).

51.7 Consequences of LI-Induced Nociceptor Blockade

Laser-induced blockade of action potentials in nociceptors has important consequences, as seen in Figure 51.4. The most immediate effect of nociceptor blockade is pain relief. This occurs within a few minutes of LLLT application, which is consistent with the experimental findings of the timing of onset of conduction blockade in somatosensory-evoked potentials (SSEPs). The causal effect between nerve block and analgesia is well illustrated in a study in normal human teeth where 240 s of Nd:YAG laser caused analgesia equivalent to that of a standard topical anesthetic agent, EMLA, with concurrent decreased electrical activity in the pulpo-dental nerve (Chan and Armati 2012). The important clinical consequence of this for the patient is reduced drug intake, improved mobility, and reduced anxiety. A second consequence of neural blockade is inhibition of peripheral nerve sensitization in which increased sensitivity occurs (i.e., lowered threshold of activation in response to the release of local

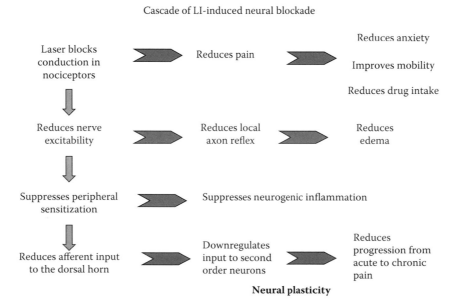

FIGURE 51.4 Laser-induced blockade of action potentials in nociceptors.

proinflammatory neuropeptides such as bradykinin and substance P). Peripheral sensitization results not only in lowering of the activation threshold of nerves but also in the release of proinflammatory neuropeptides such as substance P from the nerves themselves, which causes neurogenic inflammation. Inhibition of peripheral sensitization reduces the intensity of local inflammatory response and the local axon reflex limiting local vasodilation and swelling in acute injury and inflammation. Reducing afferent input to the dorsal horn leads to downregulation of synaptic transmission at the spinal cord level, which can limit the progression of acute to chronic pain. In persistent pain, reducing the tonic input of upregulated nociceptors and their synaptic connections can lead to downregulation (i.e., long-term depression) of spinal and cortical centers associated with pain, reducing the burden of persistent pain (Klein et al. 2004).

51.8 Lasers and Neurotransmitters

Modulation of neurotransmitters has also been postulated as a mechanism for pain relief (Navratil and Dylevsky 1997). Alteration in neurotransmitters associated with pain relief seen in both animal and human experiment mechanisms in pain relief is less defined. For example, serotonin levels are increased with treatment of myofascial pain with laser therapy (Ceylan, Hizmetli, and Silig 2004; Walker 1983), β-endorphin and precursors increase with trigger point treatment (Laakso et al. 1994), and endorphins in skin are also increased with locally applied laser (Peres e Serra and Ashmawi 2010).

51.9 LLLT and Pain Relief

LLLT has both short- and long-term effects in ameliorating pain. The immediate pain-relieving effects, which can occur within minutes of application, are affected by neural blockade in peripheral and sympathetic nerves, particularly nociceptors, by reduction of muscle spasm, and by reduction in local edema, especially in acute injury. Longer-term effects occurring within days to weeks and lasting months to years are affected by modulation of the inflammatory response and stimulation of tissue healing. These latter effects are the basis of the long-term benefits of LLLT, including the preventative effects of LLLT in reducing the progression of acute to chronic pain.

51.10 Clinical Implications in Different Types of Pain

The broad IASP definition of pain does not identify the different manifestations of pain that are important clinically and relevant to LLLT. There are, in fact, three different types of pain, in which the neurophysiology is substantially different; namely, nociceptive pain, neuropathic pain, and central pain. Clinically, LLLT can treat both nociceptive and neuropathic pain, which can coexist in conditions such as back pain and neck pain. Currently, central pain is not amenable to LLLT.

51.10.1 Nociceptive Pain

Nociceptive pain is the most common type of pain treated by LLLT. This includes acute pain conditions or acute exacerbations of chronic pain such as those listed in Table 51.1. The critical feature in nociceptive pain is the initiation of action potentials by noxious stimuli, the blockade of which will relieve pain. In nociceptive pain, pain is proportional to the stimulus (i.e., the stronger the stimulus, the greater the pain). In acute injury, whether accidental, such as trauma, or deliberate, such as surgery, tissue and cellular disruption result in the formation of an "inflammatory soup" that is made up of chemical irritants released from disrupted cells and blood. Nerves in the region of local injury become sensitized by these chemicals and proinflammatory neuropeptides, the threshold for activation of these receptors decreases, and patients experience pain. Tissue repair processes are also initiated by this process so that over hours and days, cells such as macrophages and neutrophils migrate to the site of injury to clear cellular debris. At the same time, fibroblasts become active and begin to lay down new tissue to repair the defect. These latter effects also play an important part in long-term modulation of pain as tissues heal.

Neck pain is representative of many musculoskeletal conditions where pain arises from injury to or arthritis of facet joints, enthesitis of ligaments, spasm of muscles with trigger-point formation, as well as nerve impingement from intervertebral disc prolapse (Chow, Barnsley, and Heller 2006). Pathology from each or all can contribute to the clinical presentation of neck pain and back pain, often described as multicompartment disease in spinal pathology. LLLT treatment of the neck is supported by strong evidence from a systematic review, making it one of the most evidence-based treatments for neck pain (Chow et al. 2009). Similarly, nociceptors around knee (Bjordal et al. 2007) and other joints affected by osteoarthritis, crystal arthropathies such as gout (Soriano et al. 2006), enthesitis such as lateral epicondylitis (Bjordal et al. 2008), and tendinopathies (Bjordal, Couppe, and Ljunggren 2001) will be activated by mechanical and proinflammatory stimuli and are all responsive to LLLT.

51.10.2 Neuropathic Pain

Neuropathic pain is pain arising from damage to nerves, and in contrast to nociceptive pain, the pain experienced is out of proportion to the stimulus: light touch may cause severe pain (hyperalgesia) and normal movements are painful (allodynia). Neuropathic pain results in changes to the receptors that increase the sensitivity of the nerve and clinically is of a burning, sickening quality, different from that of nociceptive pain. Damage can occur due to infection, such as postherpetic neuralgia from herpes zoster (Moore 1996), injury, such as compression of the sciatic nerve by a prolapsed disc (Konstantinovic et al. 2010), or the median nerve in carpal tunnel syndrome (Naeser 2006) or trigeminal neuralgia (Eckerdal and Bastian 1996; Walker et al. 1988). There is evidence of benefit in these conditions by direct application of laser to peripheral nerve endings in skin or overlying nerve trunk.

51.10.3 Myofascial Pain

Myofascial pain is defined as chronic pain arising from muscles, fascia, and ligaments with the term classified within the International Classification of Diseases (ICD.9.CM: 729.1). It is associated with muscle pain often as localized areas of tenderness as trigger points with tenderness and stiffness. Trigger points are discrete, focal, hyperirritable spots located in a taut band of skeletal muscle that produce local pain and demonstrate reproducible patterns of referred pain when palpated. Myofascial pain and trigger points can occur as a result of microtrauma or chronic postural strain, resulting in chronic spasm and pain. It is proposed that local tenderness in acute muscle pain is caused by peripheral sensitization of local muscle nociceptors (Mense 2003), leading to release and accumulation of inflammatory mediators at this site (Shah et al. 2008). Central sensitization of dorsal horn neurons from ongoing nociception leads to perpetuation of the pain, spasm cycle, and the persistence of trigger points and myofascial pain.

Several lines of evidence provide support for neurally mediated inhibitory effects, which reduce muscle pain and spasm. Direct laser-induced blockade of amplitude and latencies of CMAPs with 808 nm (450 mW CW) and 650 nm (35 mW CW) indicates that lasers can directly inhibit motor nerves. Studies of lasers on the neuromuscular junction end-plate potentials show reduced amplitude of evoked end-plate potentials with release of acetylcholine with 830 nm (86 mW CW, ED: 12 J/cm^2) but not with 808 nm laser at 4 J/cm^2 (Nicolau et al. 2004a) or 655 nm laser at the same EDs (Nicolau et al. 2004b). Clinical studies have demonstrated the reduction of tenderness and pain by treatment of trigger points (Carrasco et al. 2009; Laakso, Richardson, and Cramond 1997; Snyder-Mackler, Bork, and Bourbon 1986; Snyder-Mackler et al. 1989; Waylonis et al. 1988). A study by Airaksinen et al. (1989) is of particular interest as not only did 904 nm laser significantly reduced tenderness in neck and shoulder girdle trigger points but also increased pain threshold in the contralateral untreated trigger points 15 min later. Of particular note is the treatment response, which occurs within the same time frame in which animal studies show conduction block of CMAPs within 10 to 20 min. Moreover, the response on the contralateral side of the body suggests modulation through crossover of second-order neurons in the spinal cord affecting the same segment. It also suggests that care must be taken when using the contralateral side of the body as a control in randomized controlled studies. The IASP fact sheet on myofascial pain now includes lasers as an evidence-based treatment for trigger points (IASP 2009).

51.10.4 Persistent (Chronic) Pain

Persistent (chronic) pain is defined as pain of more than 3 months' duration. The neurophysiology of persistent pain involves a process of central sensitization in which changes occur within the spinal cord that lower the threshold of afferent stimulation. These changes occur acutely, within 24 h of injury, and begin to resolve within days of recovery. In some circumstances, the changes persist rather than resolve, resulting in perpetuation of the pain. This process of windup underpins the changes of long-term activation leading the pain persisting in the long term.

51.11 Neurogenic Inflammation

The anti-inflammatory effects of LLLT constitute one of the most important of the pain relieving mechanisms of LLLT. However, neurogenic inflammation constitutes a specific inflammatory process with special relevance to nerve effects of LLLT and pain. Neurogenic inflammation is sterile inflammation caused by the release of proinflammatory neuropeptides, such as substance P, bradykinin PGE$_2$ from nerve endings, fibroblasts, Schwann cells and mast cells, that are released locally. This can occur in acute injury or following activation of chronic injury such as Achilles tendinitis.

Reduction in PGE$_2$ and other inflammatory markers have been demonstrated in cellular, tissue, and *in vivo* animal and human studies. Cells such as neutrophils and macrophages associated with inflammation also show activation with LLLT, so that LLLT promotes resolution of the inflammatory process necessary for tissue repair. This latter effect occurs over weeks and is the last step in the process of long-term pain relief.

51.12 Sympathetically Maintained Pain

Chronic regional pain syndrome (CRPS) types I and II are one of the most challenging types of pain to treat. The former is related to chronic pain in which the sympathetic nervous system becomes recruited in the perpetuation of the pain. The pain is characterized by its onset with minimal trauma, as well as the intensity of the pain and changes in color and swelling of the limb (usually) related to autonomic dysfunction. Patients are often exquisitely tender to touch and many patients cannot tolerate even light touch. In this condition, central sensitization must be downregulated and requires long-term treatment with small doses of laser. Specialized laser systems have been used to treat CRPS, but it remains a long and difficult condition to treat. Treatment can be directed to the stellate ganglion to induce blockade, and this has been successfully used to treat other conditions of the head and neck (Basford et al. 2003; Hashimoto et al. 1997; Kemmotsu 1997; Takeyoshi et al. 1996).

51.13 Preventative Aspects of Pain Relief

Preventative effects of laser in reducing the progression of acute to chronic pain is an important but less well known consequence of early treatment. This can be preemptive treatment prior to surgery or immediately after surgery or injury. Such an effect is seen in different clinical scenarios. Significant reduction in postoperative pain occurred following open cholecystectomy when 830 nm, 60 mW CW laser for 6–8 min was administered before the dressing was applied and the patient was still anesthetized (Moore et al. 1992). In dentistry, several studies show significant reduction in postoperative pain when a laser was delivered immediately following surgical removal of third molar teeth using a red, 632.8 LI nm laser (Clokie, Bentley, and Head 1991; Markovic and Todorovic 2006) and endodontic surgery, with 809 nm (Kreisler et al. 2004). Patients treated with 904 nm lasers for acute neck pain had lower recurrence of their pain at 6-month follow-up (Soriano et al. 1996). Preexercise IR LI 904 nm, 15 mW (average power), with 1 or 3 J enhanced skeletal muscle performance and decreased postexercise skeletal muscle damage and inflammation in rats (de Almeida et al. 2011).

Mechanisms for preventative effects of laser may relate to stabilization of mast cells with reduced degranulation (Trelles et al. 1989) and release of proinflammatory mediators that sensitize peripheral nerves. Reduction of peripheral sensitization in all the above experimental and clinical scenarios would limit the development of inflammation and reduce nociceptor sensitization, thereby reducing neurogenic inflammation and pain (Niissalo et al. 2002).

51.14 Effects of LLLT on Lymphatic Effects

Not all pain-relieving effects of lasers relate specifically to effects on nerves. Modulation of lymphatic activity is an important component of pain relief. Localized swelling limits mobility and causes impingement of tendons, such as in supraspinatus tendinopathy or around inflamed nerve roots. Swelling is recognized as one of the pathophysiological signs of injury and inflammation, and small reductions in interstitial fluid or swelling in tendons may be sufficient to reduce pain and mobility.

More generalized swelling as in lymphedema following postmastectomy and other postcancer surgery is a cause of pain in these clinical situations. Swelling of a limb is associated with pain and heaviness, and taut bands in axillae or groin are painful and restrict mobility. Studies show the efficacy of 904 nm lasers in postmastectomy lymphedema; however, not only is there an immediate effect in reducing swelling but also a delayed benefit occurring several weeks after the initial treatment (Carati et al. 2003; Pillar and Thelander 1995).

Mechanisms for these effects relate to increasing the motoricity of the endothelial cells that make up the lymphagion, which are pulsatile and take up interstitial fluid by opening up small intercellular spaces. In addition, lymphogenesis, in which new lymphatics are formed, increase the capacity to reduce swelling in the long term.

51.15 Increased Pain after Treatment

Although the evidence for treatment benefits is clear and side effects are reported as minimal or absent, anecdotal evidence suggests that patients may experience increased pain within the first 24 to 48 h following treatment. A search of the literature failed to reveal consistent reporting of side effects. In a systematic review of neck pain, as an example, only 50% of the studies evaluated side effects, and of those reporting side effects, there was no significant reporting of adverse events such as increased pain. Possible mechanisms for this may be related to the production of growth factors, which can be locally painful.

51.16 Future Directions for Lasers in the Treatment of Pain

The evidence for laser treatment of pain is gaining strength, although use in mainstream medicine remains limited. There are various reasons for slow uptake, not the least of which is a lack of awareness of LLLT, even among users of lasers. Mechanisms for pain-relieving effects are now supported by strong evidence in laboratory-based experiments as well as high levels of evidence for clinical conditions from which treatment of a wide spectrum of illnesses can be extrapolated. An epidemic of persistent pain is predicted over the next decade in developed countries with aging populations, leading to a serious demand on health budgets, costing billions of dollars, and resulting in a dramatic reduction in quality of life for individuals. Cost of drugs with limited efficacy, as well as their increasingly recognized serious side effects, creates an imperative to identify nondrug therapies, such as LLLT, for pain. Future research should continue to examine optimal parameters for pain relief; however, cost-effectiveness should be included as one of the outcomes in clinical trials. There is an imperative to find new therapies for pain, and LLLT has the capacity if utilized to its full extent to improve quality of life and significantly reduce the costs of health budgets.

References

Airaksinen, O., P. Rantanen, K. Pertti, and P. Pontinen. 1989. Effects of the infrared laser therapy at treated and nontreated trigger points. *Acupuncture Electro* 14:9–14.

Backonja, M. M., and G. Lauria. 2010. Taking a peek into pain, from skin to brain with ENFD and QST. *Pain* 151:559–560.

Basford, J. R., P. Sandroni, P. A. Low et al. 2003. Effects of linearly polarized 0.6–1.6 M irradiation on stellate ganglion function in normal subjects and people with complex regional pain (CRPS I). *Laser Surg Med* 32:417–423.

Baxter, G. D., J. Allen, D. Walsh, J. Bell, and J. Ravey. 1992. Localisation of the effect of low energy laser irradiation upon conduction latencies in the human median nerve *in vivo*. *J Physiol Lond* 446:445P.

Baxter, G. D., D. M. Walsh, J. M. Allen, A. S. Lowe, and A. J. Bell. 1994. Effects of low intensity infrared laser irradiation upon conduction in the human median nerve *in vivo*. *Exp Physiol* 79:227–234.

Bischko, J. 1980. Examples of the clinical use of acupuncture. *J Belge Med Phys Rehabil* 3:209–214.

Bjordal, J., C. Couppe, and A. Ljunggren. 2001. Low-level laser therapy for tendinopathy: Evidence of a dose-response pattern. *Phys Ther Rev* 6:91–99.

Bjordal, J., M. I. Johnson, R. A. Lopes-Martins et al. 2007. Short-term efficacy of physical interventions in osteoarthritic knee pain. A systematic review and meta-analysis of randomised placebo-controlled trials. *BMC Musculoskel Dis* 8:51.

Bjordal, J. M., R. A. Lopes-Martins, J. Joensen et al. 2008. A systematic review with procedural assessments and meta-analysis of low-level laser therapy in lateral elbow tendinopathy (tennis elbow). *BMC Musculoskel Dis* 9:75.

Cambier, D., K. Blom, E. Witvrouw et al. 2000. The influence of low intensity infrared laser irradiation on conduction characteristics of peripheral nerve: A randomised, controlled, double blind study on the sural nerve. *Laser Med Sci* 15:195–200.

Carati, C. J., S. N. Anderson, B. J. Gannon, and N. B. Piller. 2003. Treatment of postmastectomy lymphedema with low-level laser therapy: A double-blind, placebo-controlled trial. *Cancer* 98:1114–1122.

Carrasco, T. G., L. D. Guerisoli, D. M. Guerisoli, and M. O. Mazzetto. 2009. Evaluation of low intensity laser therapy in myofascial pain syndrome. *Cranio* 27:243–247.

Ceylan, Y., S. Hizmetli, and Y. Silig. 2004. The effects of infrared laser and medical treatments on pain and serotonin degradation products in patients with myofascial pain syndrome. A controlled trial. *Rheumatol Int* 24:260–263.

Chan, A., and P. Armati. 2012. Pulsed Nd:YAG laser induces pulpal analgesia: A randomized clinical trial. *J Dent Res* 91(Suppl 1):S79–S84.

Chen, M., K. Shimada, K. Fujita et al. 1993. Neurite elongation from cultured dorsal root ganglia is inhibited by Ga-Al-As diode laser irradiation. *Laser Life Sci* 5:237–242.

Chow, R., P. Armati, E. L. Laakso, J. M. Bjordal, and G. D. Baxter. 2011. Inhibitory effects of laser irradiation on peripheral mammalian nerves and relevance to analgesic effects: A systematic review. *Photomed Laser Surg* 29:365–381.

Chow, R., M. David, and P. Armati. 2007. 830-nm laser irradiation induces varicosity formation, reduces mitochondrial membrane potential and blocks fast axonal flow in small and medium diameter rat dorsal root ganglion neurons: Implications for the analgesic effects of 830-nm laser. *J Peripher Nerv Syst* 12:28–39.

Chow, R. T., L. B. Barnsley, and G. Z. Heller. 2006. The effect of 300 mW, 830 nm laser on chronic neck pain: A double-blind, randomized, placebo-controlled study. *Pain* 124:201–210.

Chow, R. T., R. Lopes-Martins, M. Johnson, and J. M. Bjordal. 2009. Efficacy of low-level laser therapy in the management of neck pain: A systematic review and meta-analysis of randomised, placebo and active treatment controlled trials. *Lancet* 374:1897–1908.

Clokie, C., K. C. Bentley, and T. W. Head. 1991. The effects of the helium-neon laser on postsurgical discomfort: A pilot study. *J Can Dent Assoc* 57:584–586.

de Almeida, P., R. A. Lopes-Martins, S. S. Tomazoni et al. 2011. Low-level laser therapy improves skeletal muscle performance, decreases skeletal muscle damage and modulates mRNA expression of COX-1 and COX-2 in a dose-dependent manner. *Photochem Photobiol* 87:1159–1163.

Deng, L., Y. Gan, S. He et al. 1987. *Chinese Acupuncture and Moxibustion*. Foreign Languages Press, Beijing.

Eckerdal, A., and H. Bastian. 1996. Can low reactive level laser therapy be used in the treatment of neurogenic facial pain? A double blind placebo controlled investigation of patients with trigeminal neuralgia. *Laser Ther* 8:247–251.

Engel, G. 1977. The need for a new medical model: A challenge for biomedicine. *Science* 196:129–136.

Hadian, M., and B. Moghagdam. 2003. The effects of low power laser on electrophysiological parameters of sural nerve in normal subjects: A comparison between 670 and 780 nm wavelengths. *Acta Med Iran* 41:138–142.

Hashimoto, T., O. Kemmotsu, H. Otsuka, R. Numazawa, and Y. Ohta. 1997. Efficacy of laser irradiation on the area near the stellate ganglion is dose-dependent: A double-blind cross-over placebo-controlled study. *Laser Ther* 9:7–12.

International Association for the Study of Pain (IASP). 2009. Global Year against Musculoskeletal Pain October 2009–October 2010 Myofascial Pain. IASP.

Ji, R. R., and C. J. Woolf. 2001. Neuronal plasticity and signal transduction in nociceptive neurons: Implications for the initiation and maintenance of pathological pain. *Neurobiol Dis* 8:1–10.

Jimbo, K., K. Noda, H. Suzuki, and K. Yoda. 1998. Suppressive effects of low-power laser irradiation on bradykinin evoked action potentials in cultured murine dorsal root ganglia cells. *Neurosci Lett* 240:93–96.

Kasai, S., T. Kono, T. Sakamoto, and M. Mito. 1994. Effects of low-power laser irradiation on multiple unit discharges induced by noxious stimuli in the anesthetized rabbit. *J Clin Laser Med Surg* 12:221–224.

Kasai, S., T. Kono, Y. Yasuhiro et al. 1996. Effect of low-power laser irradiation on impulse conduction in anaesthetized rabbits. *J Clin Laser Med Surg* 14:107–109.

Kemmotsu, O. 1997. Editorial: Laser irradiation in the area near the stellate ganglion. *Laser Ther* 9:5–6.

Kennedy, W., G. Wendelschafter-Crabb, M. Polydefikis, and J. McArthur. 2005. Pathology and quantitation of cutaneous innervation. In *Peripheral Neuropathy*, P. Dyck and P. Thomas, editors. W.B. Saunders, Philadelphia, PA, 873.

Klein, T., W. Magerl, H.-C. Hopf, J. Sandkühler, and R.-D. Treede. 2004. Perceptual correlates of nociceptive long-term potentiation and long-term depression in humans. *J Neurosci* 24:964–971.

Kono, T., S. Kasai, T. Sakamoto, and M. Mito. 1993. Cord dorsum potentials suppressed by low power laser irradiation on a peripheral nerve in the cat. *J Clin Laser Med Surg* 11:115–118.

Konstantinovic, L. M., Z. M. Kanjuh, A. N. Milovanovic et al. 2010. Acute low back pain with radiculopathy: A double-blind, randomized, placebo-controlled study. *Photomed Laser Surg* 28:553–560.

Kramer, J. F., and M. Sandrin. 1993. Effect of low-power laser and white light on sensory conduction rate of the superficial radial nerve. *Physiother Can* 45:165–170.

Kreisler, M., H. Haj, N. Noroozi, and B. Willershausen. 2004. Efficacy of low level laser therapy in reducing postoperative pain after endodontic surgery—A randomized double blind clinical study. *Int J Oral Max Surg* 33:38–41.

Kroetlinger, M. 1980. On the use of laser in acupuncture. *Acupuncture Electro* 5:297–311.

Kudoh, C., K. Inomata, K. Okajima, M. Motegi, and T. Ohshiro. 1989. Effects of 830nm gallium aluminium garsenide diode laser radiation on rat saphenous nerve sodium-potassium-adenosine triphosphatase activity: A possible pain attenuation mechanism examined. *Laser Ther* 1:63–67.

Laakso, E., T. Cramond, C. Richardson, and J. Galligan. 1994. Plasma ACTH and beta-endorphin levels in response to low-level laser therapy (LLLT) for myofascial trigger points. *Laser Ther* 6:133–142.

Laakso, E., C. Richardson, and T. Cramond. 1997. Pain scores and side effects in response to low level laser therapy (LLLT) for myofascial trigger points. *Laser Ther* 9:67–72.

Lauria, G. 1999. Innervation of the human epidermis. A historical review. *Ital J Neurol Sci* 20:63–70.

Maeda, T. 1989. Morphological demonstration of low reactive laser therapeutic pain attenuation effect of the gallium aluminium arsenide diode laser. *Laser Ther* 1:23–30.

Maiman, T. 1960. Stimulated optical radiation in ruby. *Nature* 187:493.

Markovic, A. B., and L. Todorovic. 2006. Postoperative analgesia after lower third molar surgery: Contribution of the use of long-acting local anesthetics, low-power laser, and diclofenac. *Oral Surg Oral Med O* 102:e4–e8.

Mense, S. 2003. The pathogenesis of muscle pain. *Curr Pain Headache Rep* 7:419–425.

Merskey, H., and N. E. Bogduk. 1994. *Classification of Chronic Pain. Descriptions of Chronic Pain Syndromes and Definitions of Pain Terms.* IASP Press, Seattle, WA.

Mester, E., G. Ludany, M. Selyei, B. Szende, and G. Tota. 1968. The stimulating effects of low power laser-rays on biological systems. *Laser Rev* 1:3.

Mezawa, S., K. Iwata, K. Naito, and H. Kamogawa. 1988. The possible analgesic effect of soft-laser irradiation on heat nociceptors in the cat tongue. *Archs Oral Biol* 33:693–694.

Miura, A., and M. Kawatani. 1996. Effects of diode laser irradiation on sensory ganglion cells from the rat. *Pain Res* 11:175–183.

Moore, K. 1996. Laser therapy in post-herpetic neuralgia and lasers and pain treatment. The Laser Exchange, abstract from the MALC conference, November 2006 (online).

Moore, K., N. Hira, I. Broome, and J. Cruikshank. 1992. The effect of infra-red diode laser irradiation on the duration and severity of post-operative pain: A double-blind trial. *Laser Ther* 4:145–148.

Naeser, M. A. 2006. Photobiomodulation of pain in carpal tunnel syndrome: Review of seven laser therapy studies. *Photomed Laser Surg* 24:101–110.

Navratil, L., and I. Dylevsky. 1997. Mechanisms of the analgesic effect of therapeutic lasers *in vivo*. *Laser Ther* 9:33–39.

Nicolau, R., M. Martinez, J. Rigau, and J. Tomas. 2004a. Neurotransmitter release changes induced by low power 830 nm diode laser irradiation on the neuromuscular junction. *Lasers Surg Med* 35:236–241.

Nicolau, R. A., M. S. Martinez, J. Rigau, and J. Tomas. 2004b. Effect of low power 655 nm diode laser irradiation on the neuromuscular junctions of the mouse diaphragm. *Lasers Surg Med* 34:277–284.

Nicolson, G. L., J. R. Smith, and G. Poste. 1976. Effects of local anesthetics on cell morphology and membrane-associated cytoskeletal organization in BALB/3T3. *J Cell Biol* 68:395–402.

Niissalo, S., M. Hukkanen, S. Imai, J. Tornwall, and Y. T. Konttinen. 2002. Neuropeptides in experimental and degenerative arthritis. *Ann NY Acad Sci* 966:384–399.

Peres e Serra, A., and H. A. Ashmawi. 2010. Influence of naloxone and methysergide on the analgesic effects of low-level laser in an experimental pain model. *Rev Bras Anestesiol* 60:302–310.

Pillar, N. B., and A. Thelander. 1995. Treating chronic post-mastectomy lymphoedema with low level laser therapy: A cost effective strategy to reduce severity and improve the quality of survival. *Laser Ther* 7:163–168.

Poste, G., D. Papahadjopoulos, and G. Nicholson. 1975. Local anesthetics affect transmembrane cytoskeletal control of mobility and distribution of cell surface receptors. *Proc Natl Acad Sci USA* 72:4430–4434.

Sato, T., M. Kawatani, C. Takeshige, and I. Matsumoto. 1994. Ga-Al-As laser irradiation inhibits neuronal activity associated with inflammation. *Acupunct Electrother Res* 19:141–151.

Shah, J. P., J. V. Danoff, M. J. Desai et al. 2008. Biochemicals associated with pain and inflammation are elevated in sites near to and remote from active myofascial trigger points. *Arch Phys Med Rehabil* 89:16–23.

Shimoyama, M., Y. Fukuda, N. Shimoyama, K. IIjima, and T. Mizuguchi. 1992a. Effect of HeNe laser irradiation on synaptic transmission of the superior cervical ganglion in the rat. *J Clin Laser Med Surg* 10:337–342.

Shimoyama, N., K. Iijima, M. Shimoyama, and T. Mizuguchi. 1992b. The effects of helium-neon laser on formalin-induced activity of dorsal horn neurons in the rat. *J Clin Laser Med Surg* 10:91–94.

Siddall, P. J., and M. J. Cousins. 1998. Introduction to pain mechanisms—Implications for neural blockade. In *Neural Blockade in Clinical Anesthesia*, M. Cousins and P. Bridenbaugh, editors. Lippincott-Raven, Philadelphia, PA, 675–713.

Snyder-Mackler, L., A. Barry, A. Perkins, and M. Soucek. 1989. Effects of helium-neon laser irradiation on skin resistance and pain with trigger points in the neck and back. *Phy Ther* 69:336–341.

Snyder-Mackler, L., C. Bork, and B. Bourbon. 1986. Effects of helium-neon laser on musculoskeletal trigger points. *Phys Ther* 68:223–225.

Soriano, F., V. Campana, M. Moya et al. 2006. Photomodulation of pain and inflammation in microcrystalline arthropathies: Experimental and clinical results. *Photomed Laser Surg* 24:140–150.

Soriano, F., R. Rios, M. Pedrola et al. 1996. Acute cervical pain is relieved with gallium arsenide (GaAs) laser radiation. A double blind preliminary study. *Laser Ther* 8:149–154.

Takeyoshi, S., R. Takiyama, S. Tsuno et al. 1996. Low reactive-level infrared diode laser irradiation of the area over the stellate ganglion and stellate ganglion block in treatment of allergic rhinitis: A preliminary study. *Laser Ther* 8:159–164.

Tracey, I., and E. Johns. 2010. The pain matrix: Reloaded or reborn as we image tonic pain using arterial spin labelling. *Pain* 148:359–360.

Trelles, M., E. Mayayo, L. Miro et al. 1989. The action of low reactive laser therapy (LLLT) on mast cells: A possible pain relief mechanism examined. *Laser Ther* 1:27–30.

Tsuchiya, D., M. Kawatani, and C. Takeshige. 1994. Laser irradiation abates neuronal responses to nociceptive stimulation of rat-paw skin. *Brain Res Bull* 34:369–374.

Wakabayashi, H., M. Hamba, K. Matsumoto, and H. Tachibana. 1993. Effect of irradiation by semiconductor laser on responses evoked in trigeminal caudal neurons by tooth pulp stimulation. *Laser Surg Med* 13:605–610.

Walker, J. 1983. Relief from chronic pain by low power laser irradiation. *Neurosci Lett* 43:339–344.

Walker, J., L. Akhanjee, M. Cooney et al. 1988. Laser therapy for pain of trigeminal neuralgia. *Clin J Pain* 3:183–187.

Waylonis, G., S. Wilke, D. O'Toole, D. Waylonis, and D. Waylonis. 1988. Chronic myofascial pain: Management by low-output helium-neon laser therapy. *Arch Phys Med Rehab* 69:1017–1020.

Wesselmann, U., J. Kerns, and W. Rymer. 1992. Laser effects on myelinated and non-myelinated axons in rat peroneal nerve. *Soc Neurosci Abstr* 18:134.

Wesselmann, U., J. Kerns, and W. Rymer. 1994. Laser effects in myelinated and nonmyelinated fibres in the rat peroneal nerve: A quantitative ultrastructural analysis. *Exp Neurol* 129:257–265.

Wesselmann, U., S. Lin, and W. Rymer. 1991. Effects of Q-switched Nd:YAG laser irradiation on neural impulse propagation: II. Dorsal roots and peripheral nerves. *Physiol Chem Phys Med NMR* 23:81–100.

Woolf, C. J., and M. W. Salter. 2000. Neuronal plasticity: Increasing the gain in pain. *Science* 288:1765–1767.

Yan, W., R. Chow, and P. J. Armati. 2011. Inhibitory effects of visible 650-nm and infrared 808-nm laser irradiation on somatosensory and compound muscle action potentials in rat sciatic nerve: Implications for laser-induced analgesia. *J Peripher Nerv Syst* 16:130–135.

52

Low-Level Laser Therapy in Arthritis and Tendinopathies

Jan Magnus Bjordal
University of Bergen

Rodrigo A. B.
Lopes-Martins
Universidade de São Paulo

52.1 Arthritis and Tendinopathies: Prevalence and Pathology

Musculoskeletal disorders are burdens to modern society, and they account for a considerable amount of sick-listing and permanent work disability. Disorders of the spinal column are common, while arthritis and tendinitis are more common in the extremities. Arthritis may have a systemic rheumatic inflammatory origin and cause inflammation in several joints across the life span. Both rheumatoid arthritis and osteoarthritis exhibit inflammatory and degenerative signs. Rheumatoid arthritis may be aggressive and result in severe inflammation and eventually joint erosions, but the development of pharmaceutical agents like tumor necrosis factor-alpha (TNF-α) inhibitors (infliximab, etanercept, etc.) has brought new possibilities for inhibiting inflammatory processes, pain, and disability. In this chapter, we will focus more on osteoarthritis and tendinitis, which seem to exhibit more similarities than differences.

52.2 Osteoarthritis Pathology

Osteoarthritis has a strong relation to older age and degeneration of tissue cartilage and other joint structures. Progression of cartilage degeneration is predominantly slow and almost never seen in uninjured joints before the fourth decade of the human life span. Inflammatory signs like synovitis, subchondral bone marrow lesions, and edema seem to be associated with cartilage degeneration and a poorer prognosis in the clinical setting (Dore et al. 2010; Hunter et al. 2011). Sustained inflammation in experimental animal models has been shown to induce infiltration of mononuclear cells and considerable degenerative changes in cartilage, joint fluid, and synovia. Moderate to severe osteoarthritis has limited, if any, possibility for cartilage repair. In particular, moderate to severe osteoarthritis usually involves irreparable destruction of the osteochondral endplate, resulting in exposure of subchondral bone to inflammatory cytokines from the synovial fluid, altered biomechanical conditions, and impaired subchondral vascular architecture (Suri and Walsh 2012).

For tendinitis, on the other hand, inflammation is prevalent at the onset and early stages of this diagnosis and tendinitis may be seen across the whole life span. It is often seen as a result of repetitive overload, and while sports-induced tendinitis occurs in all age groups, it is often seen in sporting adolescents (Cook et al. 2000). The wider term *tendinopathy* has been introduced because the inflammatory component (-nitis) seems less dominant than degeneration in long-term tendon symptoms (Khan et al. 2002). Broadly under the tendinopathy term, tendinitis is characterized as an intratendinous and peritendinous inflammatory condition with edema and largely intact tendon morphology (Andres and Murrell 2008). Partial or complete tendon ruptures are more common in older age (Murrell and Walton 2001) and will exhibit inflammatory signs in the early stages of repair.

52.3 Tendon Pathology

Tendinopathies are unspecific tendon disorders where inflammation occurs at least in the early stages. Morphological changes may be present in terms of partial ruptures and degenerative changes. Tendinopathies are frequent and difficult to treat, disabling recreational and professional athletes as well as ordinary people in their workplaces. Its prevalence has been reported between 2% of active workers and 55% of athletes (Lian, Engebretsen, and Bahr 2005). However, the etiology and pathogenesis of tendinopathy are not yet fully explained. The high prevalence, with the fact that it often becomes chronic, means that these diseases are a major socioeconomic problem where medical interventions and therapies for rehabilitation are limited (Silverstein et al. 2006).

The most common cause of tendinitis is overuse by heavy load extension of the tendons, accounting for around 30% of lesions (Sharma and Maffulli 2006). This may trigger a relaxation of collagen fibers, which do not support the mechanical traction, eventually presenting partial ruptures developing intense and painful local inflammatory reaction. This local inflammatory reaction may contribute to tissue degradation and failure, as well as the promotion of a chronic state of the frame (Battery and Maffulli 2011).

The inflammation in these tissues has generated controversy due to its atypical relationship to low vascularization of the tendons. From the central body of the tendon, smaller vascular plexuses are reaching the synovial sheaths and providing blood supply to the epitendon, mesotendon, and endotendon (Leversedge et al. 2002).

In chronic inflammation, remodeling of the tendon sheath and synovial tissue fill the space of synovial inflammatory cells. These changes in the tendon tissue lead to a massive disturbance of the tendon, reducing its functionality. Morphological changes with decreased fibril size and increased density of tenocyte cells as well as increased activation of substance P and glutamate receptors are typical in chronic tendinopathies (Kongsgaard et al. 2010; Schizas et al. 2012). The inflammatory response may be associated with the development of tendinopathy, such as increased expression of cyclooxygenase-2 and changes in the levels of certain inflammatory mediators, including prostaglandin E_2, TNF-α, and transforming growth factor (TGF)-β1 detected in tendinitis (Marcos et al. 2011; Fredberg and Stengaard-Pedersen 2008). Interleukin (IL)-1b is a major inflammatory mediator responsible for the induction of certain matrix metalloproteinases (MMPs) such as MMP-1, MMP-3, and MMP-13. In tendon injuries the specific levels of MMP-13 also increased. Other *in vitro* work has shown that MMP-13 is in part dependent on the increase in IL-1b and that certain strategies for inhibiting MMPs could be an option to reduce the degeneration of the

extracellular matrix on tendinopathy (Sun et al. 2008). MMPs can alter the biological activity of cytokines by proteolytic processes, increasing the levels of both IL-1 and its receptor IL-1RII (Bellehumeur et al. 2005). Even in the absence of inflammatory cells in the tendon lesion, inflammatory mediators can be present in tendinopathies. The level of prostaglandin E_2, thromboxane, bradykinin, and interleukin (IL-6) appears to be increased during tendinitis with pain signals.

Cyclooxygenase (COX)-2 and the growth factor TGF-β also appear increased in patellar tendinopathy, and therefore tendon pain processes may be partly due to an increase of substance P. Some MMPs (MMP-2, MMP-3, and MMP-9) may stimulate the release of latent TGF-β to reduce its output. One study showed that both IL-1b and TNF-α can increase the expression of MMP-13 models of tendinitis. However, it also appears that metalloproteinase is expressed in healthy tendon cells, suggesting the involvement of MMP-13 in maintaining the homeostasis of the extracellular matrix endotenon (Nomura et al. 2007).

52.4 Pharmacological Inflammatory-Modulating Treatment in Arthritis and Tendinopathies

Nonsteroidal anti-inflammatory drugs (NSAIDs) are the most frequently recommended or administered therapy for traumatic injuries, soccer injuries (Tscholl et al. 2010), postoperative pain (Derry et al. 2009), neck and low back disorders (Chou and Huffman 2007), osteoarthritis (Jordan et al. 2003), and rheumatoid arthritis (Greenberg et al. 2009). These surveys suggest that more than half of these indications are treated with NSAIDs. Although inflammation is mostly associated with acute conditions, NSAIDs are also the most prevalent treatment for chronic pain in Europe (Breivik et al. 2006). Another common anti-inflammatory therapy is glucocorticoid injections, which are recommended for osteoarthritis (Godwin and Dawes 2004) and subacute and chronic tendinopathies (Gaujoux-Viala, Dougados, and Gossec 2009). Although anti-inflammatory drug treatments dominate arthritic and tendinopathy pain management, they are also a cause for concern due to their adverse effects (Johnsen et al. 2005).

52.5 Pharmacological Treatment for Prevention of Degenerative Changes

A second pharmacological treatment option is to target the degenerative component in arthritis. This avenue has to be considered as less successful than inflammatory-modulating strategies, with unconvincing clinical evidence of effects. In osteoarthritis, oral intake of chondroitin sulfate and glucosamine compounds has been viewed as promising for two decades. But independently funded large-scale studies and systematic reviews have failed to detect clinically relevant effects of either compound over placebo (Clegg et al. 2006). Hyaluronic acid compounds have also been injected intra-articularly in osteoarthritic joints in a large number of trials, but the effects are small and seem to diminish in recent meta-analyses (Zhang et al. 2010). Recently, there has been a similar upsurge in interest for platelet-rich plasma injections for promoting tendon repair in tendinopathies. But although promising results have appeared in the laboratory (Zhai et al. 2012), convincing clinical results are still missing (Abate et al. 2012).

52.6 Low-Level Laser Therapy Effects on Tendon and Arthritis Pathology

Low-level laser (or light) therapy (LLLT) has been a controversial modality with conflicting results over the past three decades (Devor 1990). The scattered positive clinical results in the early stages were unexplained, and unsupported speculation about mechanism was prevalent during the 1980s. For any therapy entering mainstream medicine, its acceptance was inevitably coupled to credible mechanisms of biological action. The research activity followed initially the track of biostimulation, and the first signs of LLLT biostimulation of biological activity by LLLT were reported by Endre Mester in 1973 (Mester and Jaszsagi Nagy 1973). During the 1980s, the track of inflammatory modulation or reduction was next to come, and the first animal trial where LLLT showed effects in an experimental inflammation model was published in 1985 (Bagnasco 1985).

These two tracks have more recently been broadened to include increased collagen synthesis (Reddy, Stehno-Bittel, and Enwemeka 1998), bone formation (Pires-Oliveira et al. 2010), cartilage synthesis (Cressoni et al. 2010), and reduced apoptosis (Carnevalli et al. 2003) for biostimulation, while inflammatory modulation has been extended to altering cytokine gene expression (Marcos et al. 2011), COX and COX-2 levels (Marcos et al. 2011), and the expression and secretion of the following inflammatory cytokines: prostaglandin (PG) E_2, TNF-α, COX-2, and IL-1 in addition to reducing oxidative stress (Lubart et al. 2005).

Both tracks have been extensively investigated by reviews of nearly 200 experimental LLLT studies (Gao and Xing 2009; Bjordal et al. 2006, 2010). For both tracks, the vast majority of experimental studies show consistently positive results for one or more LLLT doses and fairly systematic dose-response patterns. There is still some uncertainty how these findings can best be translated into clinical settings (Basford 1995), but several reviews have managed to identify the optimal dose in tendinopathies (Bjordal, Couppé, and Ljunggren 2001) and osteoarthritis (Bjordal et al. 2003).

52.6.1 Effect Size Comparisons of Optimal LLLT versus Pharmaceuticals Like Optimal NSAIDs and Glucocorticoid Steroids in Laboratory Trials

In four out of five head-to-head comparisons between optimal doses of LLLT and NSAIDs, human equipotent doses of NSAIDs did not exert significantly different effects from LLLT. Observations were measured within the first 3 days after experimental injury, and studies included the NSAIDs indomethacin (Honmura et al. 1992), meloxicam (Campana et al. 1999), celecoxib (Aimbire et al. 2005), and diclofenac (Albertini et al. 2004). The only study showing significantly less effect of LLLT (Viegas et al. 2007) did not report irradiation time. However, this may have been as short as 4 s if the reported parameters are used for calculation of irradiation time. There is mixed evidence for the comparison between the anti-inflammatory effect of glucocorticoid steroids (dexamethasone) and LLLT. The glucocorticoid steroid dexamethasone exhibited similar anti-inflammatory effects as LLLT in two studies (Castano et al. 2007; Reis et al. 2008), but it was slightly superior to LLLT in three other studies (de Morais et al. 2010; Ma et al. 2012; Pessoa et al. 2004) in reducing inflammation.

52.6.2 Inhibitory Drug Interactions with Anti-Inflammatory Effects of LLLT in Animal Studies

Five animal studies reported inhibitory interactions with glucocorticoid steroids that reduced the anti-inflammatory effects from LLLT (de Morais et al. 2010; Pessoa et al. 2004).

52.6.3 Dose-Response Patterns for LLLT in Animal Studies and Effectiveness of LLLT versus LED

The investigated wavelengths between 632–680, 810–830, and 904 nm induced significant anti-inflammatory effects in cell and animal trials. We have previously estimated that the median value for laser mean optical output was 25 mW in the animal studies where anti-inflammatory LLLT effects have been demonstrated (Bjordal et al. 2010). Likewise, the lower limits for achieving anti-inflammatory effects were found to be 0.6 J/cm² in animal studies, and the lower limit for irradiation time was 16 s. Median irradiation time was 80 s in animal studies, but an arthritis study reported significantly better effect when 3 J/cm² was delivered with low-power density (5 mW/cm²) over a longer period of time (600 s) than high-power density (50 mW/cm²) over a shorter period of time (60 s) (Castano et al. 2007). The exact upper limits for anti-inflammatory effects could not be identified, but evidence for positive effects from power densities above 135 mW/cm² and doses above 15 J/cm² is lacking. There is scattered evidence that 15 J/cm² may be the upper limit for effectiveness as nonsignificant effects have been found for cellular activity with doses of 20 J/cm² (Dias et al. 2011) in rat muscle and 21 J/cm² in fibroblast cells (Frigo et al. 2010). This upper dose limit of effectiveness at around 15 J seems to be present in human studies as well (Dundar et al. 2007; Krynicka et al. 2010). Power density is another critical LLLT parameter, particularly in the treatment of tendinopathies. Too high power density (>200 mW/cm²) seems to have a negative impact on collagen synthesis (Loevschall and Arenholt-Bindslev 1994) in the laboratory and lacking effects on clinical signs in tendinopathy (Papadopoulos et al. 1996; Tumilty et al. 2008). The last critical factor for optimal phototherapy is that both infrared and red lasers seemed significantly better than LED with red wavelength in the same experimental model of zymosan-induced arthritis (de Morais et al. 2010).

52.6.4 Do the Positive Effects Translate from the Laboratory to Clinical Settings with Human Patients?

There are indices that the anti-inflammatory effect of LLLT can (Carrinho et al. 2006) be translated to humans. LLLT significantly reduced the inflammatory cytokine PGE_2 in the synovia of rheumatoid arthritis joints (Amano et al. 1994), and female musculoskeletal pain patients with treatment success showed significantly decreased PGE_2 blood serum levels when compared to patients with no improvement (Mizutani et al. 2004). Our research group has shown that LLLT reduced PGE_2 levels as measured by microdialysis in acute episodes of aggravation in subacute unilateral Achilles tendinopathies. In heavy-muscle strength training, we found that levels of creatine kinase activity, a cytokine involved in the early stage of muscle damage, and c-reactive protein, a marker of systemic inflammation, were reduced after LLLT in animals and humans (Leal Junior et al. 2009, 2010). In our previous review of acute pain in humans (Bjordal et al. 2006), we found that nearly all studies with nonsignificant effects were underdosed below 0.5 J.

52.6.5 How Do We Successfully Translate the Therapeutic Window from Cell and Small Animal Studies to Clinical Use in Humans: J/cm² versus Energy Delivered in J

In the laboratory, cell studies are usually performed with irradiation through a diffuser lens distributing the laser dose evenly over the whole surface of the Petri dish. Dose reporting in these studies should be using J/cm² as the whole area of cells is covered by laser irradiation. Most animal studies are performed with small animals like mice and rats with very small volumes (mm³) of pathology. Broadly speaking, most of the pathology will then be covered by laser irradiation because the treated area is less than 1 cm² in size. However, in humans, the pathology may extend to large areas, typically some 10–100 cm² and the typical laser aperture is <1 cm² in size. The short-circuiting occurs when clinicians and researchers calculate the dose in J/cm² from the spot size of the laser aperture and not from the size of the pathology. With the very small apertures (<0.05 cm²) in some commercially available lasers, even small mean optical outputs of 20–50 mW will result in very short and ineffective irradiation times of a few seconds to deliver 2–3 J. There is a long line of otherwise well-designed LLLT trials that have made the mistake of using too short and ineffective irradiation times of less than

10 s (Brosseau et al. 2005; Cruz et al. 2007; Meireles et al. 2010). Likewise, there is a long line of LLLT trials (de Bie et al. 1998; Yurtkuran et al. 2007) that have made the mistake of covering only a single point (<1 cm²) of large pathological areas (>100 cm²) in ankle and knee synovitis. We can illustrate the error by taking a laser with mean optical output of 100 mW and change the lens/aperture spot size from 0.1 to 1 cm². With the smallest aperture, it will take 3 s to deliver 3 J, whereas it will take 30 s to deliver 3 J with the 1 cm² spot size. Most readers will recognize that the energy delivered with the smallest spot size in 3 s is only 0.3 J, while the energy delivered with the large spot size is 10 times higher at 3 J.

How do we solve this problem of misusing the spot size for dose calculations in J/cm²? There is little reason to believe that spot size between 0.1 and 1 cm² in itself should have any significant impact on treatment effects. But there is massive evidence suggesting that the energy delivered is one of the critical factors for achieving effects. We do not at the current time have exact data on the size of the different categories of musculoskeletal pathology. The pathological area will vary considerably and individually with body size, degree of edema or joint effusion, and anatomical location. The best evidence solution to minimize the dose reporting problem has been forwarded by World Association for Laser Therapy (WALT) (WALT 2006): it is simply to leave out the area size in cm² and instead report the dose in energy delivered (in J) per irradiated point together with the number of irradiated points and other laser parameters (wavelength in nm, power density in mW/cm², and irradiation time in s) in clinical LLLT studies.

52.6.6 What the Clinical LLLT Studies Tell Us

Today, the PedRo database of clinical studies holds 186 randomized controlled clinical LLLT trials, and 98 of these have been performed with acceptable method scores (60% or more of methods criteria met) but mixed efficacy results. We have at the current stage identified some 26 LLLT trials in tendon disorders and 22 trials in arthritis or joint disorders. It is typical for the LLLT literature that results are mixed, but the overall combined effect size regardless of dose or application procedure is still on the positive side (Bjordal et al. 2007, 2008; Chow et al. 2009; Enwemeka et al. 2004). The optimal dose ranges in arthritis and tendon disorders have been adopted from these meta-analyses and implemented into dosage recommendations by WALT (www.walt.nu). The validity of the WALT recommendations have recently been verified by two systematic reviews finding that trials adhering to WALT recommendations showed consistently and significantly positive results as opposed to trials using doses outside WALTS recommendations (Jang and Lee 2012; Tumilty et al. 2010). If further descriptions of how to perform LLLT studies in these conditions are needed, we suggest three of our own papers for descriptions of procedure and dosing (Alfredo et al. 2012; Bjordal et al. 2008; Stergioulas et al. 2008).

52.6.7 Practical Hints for Clinical LLLT Use in Arthritis, Synovitis, and Tendon Disorders

The available LLLT literature touches a wide range of disorders, and the literature is not always informative. In most clinical settings, LLLT can be used in combination with active exercises to control possible inflammatory responses to loading from joints and tendons (Alfredo et al. 2012; Stergioulas et al. 2008). LLLT also exhibits effects on active trigger points when irradiated, and this has been demonstrated in animal and human studies (Chen et al. 2008; Ceccherelli et al. 1989). The positive effect on active myofascial trigger points also seems to work very well in neck pain patients (Chow et al. 2009). It should be emphasized that active trigger points may be masking underlying pathology in joints and tendons, and such joint or tendon pathology should always be the main target of LLLT irradiation.

Glucocorticoid steroids are likely to inhibit the anti-inflammatory effect of LLLT. Several clinical LLLT trials on tendinopathy have recruited more than 40% of their patients from failures after steroid injection therapy (Haker and Lundeberg 1991). For tendinopathies, it is important to remember that the negative clinical effects on connective tissue repair from steroids are present from 3 to 6 months after injections. Bear in mind that many other conditions like rheumatic, allergic, and pulmonary disorders may be treated with glucocorticoid steroids. LLLT may reduce some of the negative effects from glucocorticoids on tissue repair and may be used if the therapist informs that this is the purpose of LLLT treatment. If not, the LLLT therapist will most likely be blamed for lacking or negative effects.

References

Abate, M., P. Di Gregorio, C. Schiavone et al. 2012. Platelet rich plasma in tendinopathies: How to explain the failure. *Int J Immunopathol Pharmacol* 25:325–334.

Aimbire, F., R. Albertine, R. G. Magalhaes et al. 2005. Effect of LLLT Ga-Al-As (685 nm) on LPS-induced inflammation of the airway and lung in the rat. *Lasers Med Sci* 20:11–20.

Albertini, R., F. S. Aimbire, F. I. Correa et al. 2004. Effects of different protocol doses of low power gallium–aluminum–arsenate (Ga–Al–As) laser radiation (650 nm) on carrageenan induced rat paw ooedema. *J Photochem Photobiol B* 74:101–107.

Alfredo, P. P., J. M. Bjordal, S. H. Dreyer et al. 2012. Efficacy of low level laser therapy associated with exercises in knee osteoarthritis: A randomized double-blind study. *Clin Rehabil* 26:523–533.

Amano, A., K. Miyagi, T. Azuma et al. 1994. Histological studies on the rheumatoid synovial membrane irradiated with a low energy laser. *Lasers Surg Med* 15:290–294.

Andres, B. M., and G. A. Murrell. 2008. Treatment of tendinopathy: What works, what does not, and what is on the horizon. *Clin Orthop Relat Res* 466:1539–1554.

Bagnasco, G. 1985. Mid-laser treatment of inflammation experimentally induced with formalhyde. *Med Laser Report* 3:19–22.

Basford, J. R. 1995. Low intensity laser therapy: Still not an established clinical tool. *Lasers Surg Med* 16:331–342.

Battery, L., and N. Maffulli. 2011. Inflammation in overuse tendon injuries. *Sports Med Arthrosc* 19:213–217.

Bellehumeur, C., T. Collette, R. Maheux et al. 2005. Increased soluble interleukin-1 receptor type II proteolysis in the endometrium of women with endometriosis. *Hum Reprod* 20:1177–1184.

Bjordal, J. M., C. Couppé, and A. E. Ljunggren. 2001. Low level laser therapy for tendinopathy. Evidence of a dose–response pattern. *Phys Ther Rev* 6:91–99.

Bjordal, J. M., C. Couppe, R. T. Chow, J. Tuner, and E. A. Ljunggren. 2003. A systematic review of low level laser therapy with location-specific doses for pain from chronic joint disorders. *Aust J Physiother* 49:107–116.

Bjordal, J. M., M. I. Johnson, V. Iversen, F. Aimbire, and R. A. Lopes-Martins. 2006. Low level laser therapy in acute pain: A systematic review of possible mechanisms of action and clinical effects in randomized placebo-controlled trials. *Photomed Laser Surg* 24:158–168.

Bjordal, J. M., M. I. Johnson, R. A. Lopes-Martins et al. 2007. Short-term efficacy of physical interventions in osteoarthritic knee pain. A systematic review and meta-analysis of randomised placebo-controlled trials. *BMC Musculoskelet Disord* 8:51.

Bjordal, J. M., R. A. B. Lopes-Martins, J. Joensen, and V. V. Iversen. 2010. The anti-inflammatory mechanism of low level laser therapy and its relevance for clinical use in physiotherapy. *Phys Ther Rev* 15:286–293.

Bjordal, J. M., R. A. Lopes-Martins, J. Joensen et al. 2008. A systematic review with procedural assessments and meta-analysis of low level laser therapy in lateral elbow tendinopathy (tennis elbow). *BMC Musculoskelet Disord* 9:75.

Breivik, H., B. Collett, V. Ventafridda, R. Cohen, and D. Gallacher. 2006. Survey of chronic pain in Europe: Prevalence, impact on daily life, and treatment. *Eur J Pain* 10:287–333.

Brosseau, L., G. Wells, S. Marchand et al. 2005. Randomized controlled trial on low level laser therapy (LLLT) in the treatment of osteoarthritis (OA) of the hand. *Lasers Surg Med* 36:210–219.

Campana, V. R., M. Moya, A. Gavotto et al. 1999. The relative effects of HeNe laser and meloxicam on experimentally induced inflammation. *Laser Therapy* 11:36–41.

Carnevalli, C. M., C. P. Soares, R. A. Zangaro, A. L. Pinheiro, and N. S. Silva. 2003. Laser light prevents apoptosis in Cho K-1 cell line. *J Clin Laser Med Surg* 21:193–196.

Carrinho P. M., A. C. Renno, P. Koeke et al. Comparative study using 685-nm and 830-nm lasers in the tissue repair of tenotomized tendons in the mouse. 2006. *Photomed Laser Surg* 24:754–758.

Castano, A. P., T. Dai, I. Yaroslavsky et al. Low-level laser therapy for zymosan-induced arthritis in rats: Importance of illumination time. *Lasers Surg Med* 39:543–550.

Ceccherelli, F., L. Altafini, G. Lo Castro et al. 1989. Diode laser in cervical myofascial pain: A double-blind study versus placebo. *Clin J Pain* 5:301–304.

Chen, K. H., C. Z. Hong, F. C. Kuo, H. C. Hsu, and Y. L. Hsieh. 2008. Electrophysiologic effects of a therapeutic laser on myofascial trigger spots of rabbit skeletal muscles. *Am J Phys Med Rehabil* 87:1006–1014.

Chou, R., and L. H. Huffman. 2007. Medications for acute and chronic low back pain: A review of the evidence for an American Pain Society/American College of Physicians clinical practice guideline. *Ann Intern Med* 147:505–514.

Chow, R. T., M. I. Johnson, R. A. Lopes-Martins, and J. M. Bjordal. 2009. Efficacy of low-level laser therapy in the management of neck pain: A systematic review and meta-analysis of randomised placebo or active-treatment controlled trials. *Lancet* 374:1897–1908.

Clegg, D. O., D. J. Reda, C. L. Harris et al. 2006. Glucosamine, chondroitin sulfate, and the two in combination for painful knee osteoarthritis. *N Engl J Med* 354:795–808.

Cook, J. L., K. M. Khan, Z. S. Kiss, and L. Griffiths. 2000. Patellar tendinopathy in junior basketball players: A controlled clinical and ultrasonographic study of 268 patellar tendons in players aged 14–18 years. *Scand J Med Sci Sports* 10:216–220.

Cressoni, M. D., H. H. Giusti, A. C. Piao et al. 2010. Effect of GaAlAs laser irradiation on the epiphyseal cartilage of rats. *Photomed Laser Surg* 108:1083–1088.

Cruz, L. B., A. S. Ribeiro, A. Rech et al. 2007. Influence of low-energy laser in the prevention of oral mucositis in children with cancer receiving chemotherapy. *Pediatr Blood Cancer* 48:435–440.

de Bie, R. A., A. P. Verhagen, A. F. Lenssen et al. 1998. Efficacy of 904 nm laser therapy in musculoskeletal disorders. *Phys Ther Rev* 3:1–14.

de Morais, N. C., A. M. Barbosa, M. L. Vale et al. 2010. Anti-inflammatory effect of low-level laser and light-emitting diode in zymosan-induced arthritis. *Photomed Laser Surg* 28:227–232.

Derry, P., S. Derry, R. A. Moore, and H. J. McQuay. 2009. Single dose oral diclofenac for acute postoperative pain in adults. *Cochrane Database Syst Rev* CD004768.

Devor, M. 1990. What's in a laser beam for pain therapy? *Pain* 43:139.

Dias, F. J., J. P. Issa, F. T. Vicentini et al. 2011. Effects of low-level laser therapy on the oxidative metabolism and matrix proteins in the rat masseter muscle. *Photomed Laser Surg* 29:677–684.

Dore, D., A. Martens, S. Quinn et al. 2010. Bone marrow lesions predict site-specific cartilage defect development and volume loss: A prospective study in older adults. *Arthritis Res Ther* 12:R222.

Dundar, U., D. Evcik, F. Samli, H. Pusak, and V. Kavuncu. 2007. The effect of gallium arsenide aluminum laser therapy in the management of cervical myofascial pain syndrome: A double blind, placebo-controlled study. *Clin Rheumatol* 26:930–934.

Enwemeka, C. S., J. C. Parker, D. S. Dowdy et al. 2004. The efficacy of low-power lasers in tissue repair and pain control: A meta-analysis study. *Photomed Laser Surg* 22:323–329.

Fredberg, U., and K. Stengaard-Pedersen. 2008. Chronic tendinopathy tissue pathology, pain mechanisms, and etiology with a special focus on inflammation. *Scand J Med Sci Sports* 18:3–15.

Frigo, L., G. M. Favero, H. J. Lima et al. 2010. Low-level laser irradiation (InGaAlP-660 nm) increases fibroblast cell proliferation and reduces cell death in a dose-dependent manner. *Photomed Laser Surg* 28(Suppl 1):S151–S156.

Gao, X., and D. Xing. 2009. Molecular mechanisms of cell proliferation induced by low power laser irradiation. *J Biomed Sci* 16:4.

Gaujoux-Viala, C., M. Dougados, and L. Gossec. 2009. Efficacy and safety of steroid injections for shoulder and elbow tendonitis: A meta-analysis of randomised controlled trials. *Ann Rheum Dis* 68:1843–1849.

Godwin, M., and M. Dawes. 2004. Intra-articular steroid injections for painful knees. Systematic review with meta-analysis. *Can Fam Physician* 50:241–248.

Greenberg, J. D., M. C. Fisher, J. Kremer et al. 2009. The COX-2 inhibitor market withdrawals and prescribing patterns by rheumatologists in patients with gastrointestinal and cardiovascular risk. *Clin Exp Rheumatol* 27:395–401.

Haker, E., and T. Lundeberg. 1991. Is low-energy laser treatment effective in lateral epicondylalgia? *J Pain Symptom Manag* 6:241–246.

Honmura, A., M. Yanase, J. Obata, and E. Haruki. 1992. Therapeutic effect of Ga–Al–As diode laser irradiation on experimentally induced inflammation in rats. *Lasers Surg Med* 12:441–449.

Hunter, D. J., W. Zhang, P. G. Conaghan et al. 2011. Systematic review of the concurrent and predictive validity of MRI biomarkers in OA. *Osteoarthr Cartilage* 19:557–588.

Jang, H., and H. Lee. 2012. Meta-analysis of pain relief effects by laser irradiation on joint areas. *Photomed Laser Surg* 30:405–417.

Johnsen, S. P., H. Larsson, R. E. Tarone et al. 2005. Risk of hospitalization for myocardial infarction among users of rofecoxib, celecoxib, and other NSAIDs: A population-based case-control study. *Arch Intern Med* 165:978–984.

Jordan, K. M., N. K. Arden, M. Doherty et al. 2003. EULAR Recommendations 2003: An evidence based approach to the management of knee osteoarthritis: Report of a Task Force of the Standing Committee for International Clinical Studies Including Therapeutic Trials (ESCISIT). *Ann Rheum Dis* 62:1145–1155.

Khan, K. M., J. L. Cook, P. Kannus, N. Maffulli, and S. F. Bonar. 2002. Time to abandon the "tendinitis" myth. *Brit Med J* 324:626–627.

Kongsgaard, M., K. Qvortrup, J. Larsen et al. 2010. Fibril morphology and tendon mechanical properties in patellar tendinopathy: Effects of heavy slow resistance training. *Am J Sports Med* 38:749–756.

Krynicka, I., R. Rutowski, J. Staniszewska-Kus, J. Fugiel, and A. Zaleski. 2010. The role of laser biostimulation in early post-surgery rehabilitation and its effect on wound healing. *Ortop Traumatol Rehabil* 12:67–79.

Leal Junior, E. C., R. A. Lopes-Martins, P. de Almeida et al. 2010. Effect of low-level laser therapy (GaAs 904 nm) in skeletal muscle fatigue and biochemical markers of muscle damage in rats. *Eur J Appl Physiol* 108:1083–1088.

Leal Junior, E. C., R. A. Lopes-Martins, A. A. Vanin et al. 2009. Effect of 830 nm low-level laser therapy in exercise-induced skeletal muscle fatigue in humans. *Lasers Med Sci* 24:425–431.

Leversedge, F. J., K. Ditsios, C. A. Goldfarb et al. 2002. Vascular anatomy of the human flexor digitorum profundus tendon insertion. *J Hand Surg Am* 27:806–812.

Lian, O. B., L. Engebretsen, and R. Bahr. 2005. Prevalence of jumper's knee among elite athletes from different sports: A cross-sectional study. *Am J Sports Med* 33:561–567.

Loevschall, H., and D. Arenholt-Bindslev. 1994. Effect of low level diode laser irradiation of human oral mucosa fibroblasts *in vitro*. *Lasers Surg Med* 14:347–354.

Lubart, R., M. Eichler, R. Lavi, H. Friedman, and A. Shainberg. 2005. Low-energy laser irradiation promotes cellular redox activity. *Photomed Laser Surg* 23:3–9.

Ma, W. J., X. R. Li, Y. X. Li et al. 2012. Antiinflammatory effect of low-level laser therapy on Staphylococcus epidermidis endophthalmitis in rabbits. *Lasers Med Sci* 27:585–591.

Marcos, R. L., E. C. Leal Junior, F. de Moura Messias et al. 2011. Infrared (810 nm) low-level laser therapy in rat Achilles tendinitis: A consistent alternative to drugs. *Photochem Photobiol* 87:1447–1452.

Meireles, S. M., A. Jones, F. Jennings et al. 2010. Assessment of the effectiveness of low-level laser therapy on the hands of patients with rheumatoid arthritis: A randomized double-blind controlled trial. *Clin Rheumatol* 29:501–509.

Mester, E., and E. Jaszsagi Nagy. 1973. The effect of laser radiation on wound healing and collagen synthesis. *Studia Biophys* 35:227–230.

Mizutani, K., Y. Musya, K. Wakae et al. 2004. A clinical study on serum prostaglandin E$_2$ with low-level laser therapy. *Photomed Laser Surg* 22:537–539.

Murrell, G. A., and J. R. Walton. 2001. Diagnosis of rotator cuff tears. *Lancet* 357:769–770.

Nomura, M., Y. Hosaka, Y. Kasashima et al. 2007. Active expression of matrix metalloproteinase-13 mRNA in the granulation tissue of equine superficial digital flexor tendinitis. *J Vet Med Sci* 69:637–639.

Papadopoulos, E. S., R. W. Smith, M. I. D. Cawley, and R. Mani. 1996. Low-level laser therapy does not aid the management of tennis elbow. *Clin Rehabil* 10:9–11.

Pessoa, E. S., R. M. Melhado, L. H. Theodoro, and V. G. Garcia. 2004. A histologic assessment of the influence of low-intensity laser therapy on wound healing in steroid-treated animals. *Photomed Laser Surg* 22:199–204.

Pires-Oliveira, D. A., R. F. Oliveira, S. U. Amadei, C. Pacheco-Soares, and R. F. Rocha. 2010. Laser 904 nm action on bone repair in rats with osteoporosis. *Osteoporos Int* 21:2109–2114.

Reddy, G. K., L. Stehno-Bittel, and C. S. Enwemeka. 1998. Laser photostimulation of collagen production in healing rabbit Achilles tendons. *Lasers Surg Med* 22:281–287.

Reis, S. R., A. P. Medrado, A. M. Marchionni et al. 2008. Effect of 670-nm laser therapy and dexamethasone on tissue repair: A histological and ultrastructural study. *Photomed Laser Surg* 26:307–313.

Schizas, N., R. Weiss, O. Lian et al. 2012. Glutamate receptors in tendinopathic patients. *J Orthop Res* 30:1447–1452.

Sharma, P., and N. Maffulli. 2006. Biology of tendon injury: Healing, modeling and remodeling. *J Musculoskelet Neuronal Interact* 6:181–190.

Silverstein, B. A., E. Viikari-Juntura, Z. J. Fan et al. 2006. Natural course of nontraumatic rotator cuff tendinitis and shoulder symptoms in a working population. *Scand J Work Environ Health* 32:99–108.

Stergioulas, A., M. Stergioula, R. Aarskog, R. A. Lopes-Martins, and J. M. Bjordal. 2008. Effects of low-level laser therapy and eccentric exercises in the treatment of recreational athletes with chronic Achilles tendinopathy. *Am J Sports Med* 36:881–887.

Sun H. B., Y. Li, D. T. Fung et al. 2008. Coordinate regulation of IL-1beta and MMP-13 in rat tendons following subrupture fatigue damage. *Clin Orthop Relat Res* 466:1555–1561.

Suri, S., and D. A. Walsh. 2012. Osteochondral alterations in osteoarthritis. *Bone* 51:204–211.

Tscholl, P., J. M. Alonso, G. Dolle, A. Junge, and J. Dvorak. 2010. The use of drugs and nutritional supplements in top-level track and field athletes. *Am J Sports Med* 38:133–140.

Tumilty, S., J. Munn, J. H. Abbott et al. 2008. Laser therapy in the treatment of Achilles tendinopathy: A pilot study. *Photomed Laser Surg* 26:25–30.

Tumilty, S., J. Munn, S. McDonough et al. 2010. Low level laser treatment of tendinopathy: A systematic review with meta-analysis. *Photomed Laser Surg* 28:3–16.

Viegas, V. N., M. E. Abreu, C. Viezzer et al. 2007. Effect of low-level laser therapy on inflammatory reactions during wound healing: Comparison with meloxicam. *Photomed Laser Surg* 25:467–473.

WALT. 2006. Consensus agreement on the design and conduct of clinical studies with low-level laser therapy and light therapy for musculoskeletal pain and disorders. *Photomed Laser Surg* 24:761–762.

Yurtkuran, M., A. Alp, S. Konur, S. Ozcakir, and U. Bingol. 2007. Laser acupuncture in knee osteoarthritis: A double-blind, randomized controlled study. *Photomed Laser Surg* 25:14–20.

Zhai, W., N. Wang, Z. Qi, Q. Gao, and L. Yi. 2012. Platelet-rich plasma reverses the inhibition of tenocytes and osteoblasts in tendon-bone healing. *Orthopedics* 35:e520–e525.

Zhang, W., G. Nuki, R. W. Moskowitz et al. 2010. OARSI recommendations for the management of hip and knee osteoarthritis: Part III: Changes in evidence following systematic cumulative update of research published through January 2009. *Osteoarthr Cartilage* 18:476–499.

Low-Level Laser Therapy and Light-Emitting Diode Therapy on Muscle Tissue: Performance, Fatigue, and Repair

Cleber Ferraresi
Federal University of São Carlos

Nivaldo A. Parizotto
Federal University of São Carlos

53.1 Introduction to Muscle Fatigue

The intense use of muscles during high-intensity exercise or during repeated muscle contractions leads to a decrease in muscle performance and appearance of peripheral muscle fatigue (Allen, Lamb, and Westerblad 2008; Westerblad, Bruton, and Katz 2010). Muscle fatigue is a complex phenomenon with many theories and scientific evidence to explain its process of appearance. Among the scientific evidence, we highlight the depletion of energy sources such as phosphocreatine, glycogen, increased amounts of phosphate inorganic (Pi), adenosine diphosphate (ADP), Ca^{2+}, Mg^{2+}, H^+, and lactate, and decreased sensitivity of myofibrils to Ca^{2+} and higher production or accumulation of reactive oxygen species (ROS) and reactive nitrogen species (RNS) during exercises (Allen, Lamb, and Westerblad 2008; Westerblad and Allen 2011; Westerblad, Bruton, and Katz 2010).

Peripheral muscle fatigue can affect one or more of the following events during muscle contraction (Allen, Lamb, and Westerblad 2008; Westerblad, Bruton, and Katz 2010): (1) action potential generation at the neuromuscular junction, (2) propagation of action potential along the sarcolemma and also through the T-tubule system, (3) activation of voltage-dependent sensors at the walls of T-tubules for opening Ca^{2+} channels in the sarcoplasmic reticulum, (4) Ca^{2+} release from the sarcoplasmic reticulum into the sarcoplasm, (5) binding of Ca^{2+} to troponin C (TnC) and movement of tropomyosin that exposes the binding site of actin with myosin, (6) formation of cross-bridges and beginning of muscle contraction, (7) constant pumping of Ca^{2+} into the sarcoplasmic reticulum and decreased concentration of Ca^{2+} in the sarcoplasm, and (8) muscle relaxation.

The energy sources for adenosine triphosphate (ATP) synthesis used during muscle contraction may be predominantly aerobic (oxidative) or anaerobic metabolism (lactic and alactic) (Allen, Lamb, and Westerblad 2008; Westerblad, Bruton, and Katz 2010). The aerobic metabolism obtains energy from the Krebs cycle through the oxidation of acetyl coenzyme A (acetyl-CoA) and reduction of cofactors as nicotinamide dinucleotide (NAD) and flavin dinucleotide (FAD) to provide protons and electrons to the electron transport chain (ETC) in the mitochondrial cristae. In the ETC, oxygen (O_2) is the final acceptor of protons and electrons to give syntheses of ATP and to produce metabolic water. Considering anaerobic metabolism, O_2 does not participate in ATP synthesis. ATP can be produced from hydrolysis of phosphocreatine (anaerobic alactic metabolism) and/or from the oxidation of NADH+H by pyruvate from glycolysis. In addition, during the second process, there is production of lactate (anaerobic lactic metabolism) (Allen, Lamb, and Westerblad 2008; Westerblad, Bruton, and Katz 2010).

There are different types of physical exercises that promote specific adaptations in muscle tissue and biochemical adjustments (energy metabolism) as well as structural changes that lead to better physical performance (Aagaard 2004; Fry 2004; Liu et al. 2003; Tonkonogi and Sahlin 2002; Tonkonogi et al. 2000; Westerblad, Bruton, and Katz 2010).

Strength training or high-intensity exercises (1) promote a greater energy recruitment from anaerobic metabolism (metabolic change), (2) increase the cross-sectional area of skeletal muscle (hypertrophy) through microlesions, (3) modify the contractile characteristics of the muscle fibers (transition between type I, IIx, and IIb fibers to type IIa—structural change), and (4) increase the recruitment, timing, and frequency of firing of muscle motor units in activity (neural change) and lead to the development of muscular strength (Aagaard 2004; Allen, Lamb, and Westerblad 2008; Fry 2004; Liu et al. 2003; Westerblad, Bruton, and Katz 2010).

In contrast, endurance or low-intensity exercises (1) promote a greater energy recruitment from aerobic metabolism (metabolic change), (2) stimulate muscle fibers to develop more mitochondria, increase the size of those already in existence, and provide a greater mitochondrial density and oxidative enzymes in the muscle fibers (predominantly type I fibers—structural change), and (3) increase ATP synthesis by mitochondrial pathway and increase resistance to muscle fatigue in exercises (Coffey and Hawley 2007; Hawley 2009; Sahlin et al. 2007; Tonkonogi and Sahlin 2002; Tonkonogi et al. 2000, Westerblad, Bruton, and Katz 2010).

Some authors have used low-level laser therapy (LLLT) to accelerate these metabolic and structural changes in muscle with the goal of preventing or reducing muscle fatigue (Bakeeva et al. 1993; Chow et al. 2009; Enwemeka 2009; Fulop et al. 2009; Karu 1999; Lopes-Martins et al. 2005). Studies on this issue have used experimental models to identify possible effects of LLLT on muscle tissue subjected to mechanical and metabolic stress from exercise. In particular, resistance to fatigue and improved muscle energy metabolism have been the focus of these pioneering researches (Lopes-Martins et al. 2006; Vieira et al. 2006).

53.2 LLLT and Light-Emitting Diode Therapy for Improvement of Muscle Performance

53.2.1 Experimental Models

Lopes-Martins et al. (2006) reported the effects of LLLT (655 nm) on muscle fatigue in rats. Tibialis anterior muscle fatigue was induced by neuromuscular electrical stimulation and measured the reduction of torque and increased muscle damage from blood levels of creatine kinase (CK). LLLT was applied at a single point of the tibialis anterior before fatigue induction. The results showed a reduced fatigue at a dose of 0.5 J/cm^2 and decreased muscle damage at doses of 1.0 and 2.5 J/cm^2.

Vieira et al. (2006) verified the effects of LLLT (780 nm) on energy metabolism related to muscle fatigue in rats trained on a treadmill with a load corresponding to the anaerobic threshold for 30 consecutive days. After each workout, rats were irradiated on a single point on the femoral quadriceps, tibialis anterior, soleus, and gluteus maximus. The results showed a greater inhibition of enzymatic activity of lactate dehydrogenase (LDH), especially the LDHA isoform (pyruvate reductase) in the muscles of trained and irradiated rats, also including heart muscle (not irradiated), suggesting there were systemic effects of LLLT.

The results of these previous studies (Lopes-Martins et al. 2006; Vieira et al. 2006) encouraged other researchers to develop more experimental studies in order to identify other interactions between LLLT and muscle tissue subjected to different physical exercises as well as the mechanisms of action of LLLT to reduce damage and muscle fatigue (de Almeida et al. 2011; Leal Junior et al. 2010a; Liu et al. 2009; Sussai et al. 2010; Xu et al. 2008).

Liu et al. (2009) trained rats on a treadmill in the declined plane (–16 slope) at a speed of 16 m/min until the animals' exhaustion. These authors applied LLLT (632.8) on a single point on gastrocnemius muscle, and the results showed inhibited inflammation, reduced CK activity in blood serum, reduced muscle levels of malondialdehyde (MDA) 24 and 48 h after exercise, and increased activity of superoxide dismutase (SOD), which is an antioxidant enzyme.

Other studies also verified the effects of LLLT in rats that underwent exercise (Sussai et al. 2010) and induced fatigue by neuromuscular electrical stimulation (de Almeida et al. 2012; Leal Junior et al. 2010a). Sussai et al. (2010) investigated the effects of LLLT (660 nm) on CK levels in blood plasma and muscle cell apoptosis of rats after a swimming protocol for induction of muscle fatigue. The LLLT was applied for 40 s on a single point of the gastrocnemius muscle immediately after fatigue protocol. Compared to the control group, the LLLT group had lower levels of CK and apoptosis 24 and 48 h after induction of muscle fatigue.

Leal Junior et al. (2010a) developed an experimental model very similar to the study of Lopes-Martins et al. (2006). These authors investigated the effects of LLLT (904 nm) in rat muscle fatigue induced by neuromuscular electrical stimulation. The LLLT was applied on a single point on the tibialis anterior with different treatment times and total energies before muscle fatigue induction. Groups 1 and 3 J had the highest force peak compared to the control group (without LLLT) and groups with 0.1 and 0.3 J. The blood lactate levels were lower in all groups irradiated with LLLT. The CK level in blood was also lower in irradiated groups except for group 3 J.

Using a similar protocol, de Almeida et al. (2011) reproduced the work of Leal Junior et al. (2010a) and identified that LLLT (904 nm) and energy of 1 J significantly decreased the CK level in blood, reduced levels of mRNA expression for cyclooxygenase-2 (COX-2), and increased the expression of COX-1 compared to other groups.

53.2.2 Clinical Trials: Acute Responses

The vast majority of papers involving LLLT and exercise in humans investigated the acute effects of this therapy on muscle performance in high-intensity exercises (Baroni et al. 2010a; de Almeida et al. 2012; De Marchi et al. 2012; Gorgey, Wadee, and Sobhi 2008; Leal Junior et al. 2008, 2009c, 2010b). Only three studies have investigated the chronic effects of this therapy (de Brito Vieira et al. 2012; Ferraresi et al. 2011, 2012).

The first published work was a clinical pilot study conducted by Gorgey, Wadee, and Sobhi (2008). The authors applied the LLLT (808 nm) in pulsed mode (pulse repetition of 1 to 10,000 per second) for 5 min (low energy) and 10 min (high energy) on the femoral quadriceps before muscle fatigue induction by neuromuscular electrical stimulation. The results showed that LLLT groups had a lower percentage of muscle fatigue compared to the control group that had statistical significance. There was no significant difference between groups irradiated with LLLT.

Leal Junior et al. (2008, 2009c, 2010b) and de Almeida et al. (2012) investigated the effects of LLLT on the biceps brachii performance in a double-blind placebo-controlled trial with very similar methodologies. Leal Junior et al. (2008) investigated the effects of LLLT (655 nm) applied on the biceps brachii before Scott bench exercise. The exercise was performed with 75% of the load corresponding to maximum voluntary contraction (MVC) until exhaustion. Five points were irradiated for 100 s on each point. The results showed a significant increase in the number of repetitions of the LLLT group compared to the placebo group. However, there was no increase in time of exercise and there was no decrease in blood lactate levels.

With the same experimental design, these authors used other LLLT (830 nm) applied on biceps brachii at four points (Leal Junior et al. 2009c). The radiation time was 50 s at each point before starting exercise in Scott bench. The results were the same as previously reported: the LLLT group increased the number of repetitions compared to placebo.

de Almeida et al. (2012) attempted to identify in a single study which wavelength would be better to increase the biceps brachii performance in high-intensity exercise. The LLLT (660 or 830 nm) was applied on biceps brachii at four points for 100 s on each point before starting the Scott bench exercise. The LLLT groups (660 or 830 nm) had a greater force and force peak compared to

placebo. However, there was no significant difference between irradiated groups.

Another work reporting a double-blind placebo-controlled and randomized trial also involved biceps brachii fatigue (Leal Junior et al. 2010b). This study investigated the effects of LLLT (cluster with five diodes of 810 nm) applied before Scott bench exercise with 75% of MVC until exhaustion. Two points of the biceps brachii were irradiated for 30 s on each point. There were increases in the number of repetitions and time of exercise, and decreasing levels of lactate, CK, and c-reactive protein after exercise.

Baroni et al. (2010a) investigated the effects of LLLT (cluster with five diodes of 810 nm) on energetic metabolism, muscle damage, and delayed onset muscle soreness (DOMS) in young males after five sets of 15 eccentric contractions of the femoral quadriceps on the isokinetic dynamometer. There were six points of LLLT radiation on the femoral quadriceps during 30 s on each point. The LLLT applied before exercise promoted a smaller increase in LDH activity at 48 h after exercise, a smaller increase in CK in blood at 24 and 48 h after exercise, and less loss of MVC immediately and 24 h after exercise. However, the DOMS measured by visual analog scale (VAS) was equal to the LLLT and placebo groups.

De Marchi et al. (2012) verified the effects of LLLT (cluster with five diodes of 810 nm) on fatigue, oxidative stress, muscle damage, and human physical performance on a treadmill. Six points of the femoral quadriceps were irradiated as performed by Baroni et al. (2010a); four points on the hamstrings and two points on gastrocnemius muscles were irradiated before progressive exercise performed on a treadmill until exhaustion. The time of radiation was 30 s on each point. The LLLT group increased the absolute and relative maximal oxygen uptake as well as the time of exercise compared to placebo. The activities of LDH, muscle damage (CK), and lipid damage thiobarbituric acid reactive substances (TBARS) were all significantly higher only in the placebo group after exercise. The activity of SOD, an antioxidant enzyme, was also decreased only in the placebo group after exercise.

53.2.3 Clinical Trials: Chronic Responses

Unlike previous researches, Ferraresi et al. (2011, 2012) and de Brito Vieira et al. (2012) verified the effects of LLLT on physical training programs in randomized and controlled clinical trials.

Ferraresi et al. (2011) investigated the effects of LLLT (cluster with six diodes of 808 nm) on a physical training program with load corresponding to 80% of one repetition maximum: 1 RM. The training program was performed twice a week for 12 weeks and LLLT was applied immediately after each workout. Seven regions of the femoral quadriceps were irradiated for 10 s per region. The LLLT group had a greater percentage gain of 1 RM compared to the trained group without LLLT and the control group after the training program. Only the LLLT group significantly increased the mean of peak torque and peak torque of knee extensor muscles in isokinetic dynamometry.

The second study was reported by de Brito Vieira et al. (2012). This study verified the effects of LLLT (cluster with six diodes of 808 nm) on a physical training program of low intensity (load relative to the anaerobic ventilatory threshold) on a stationary bicycle. The physical training was conducted three times a week for nine consecutive weeks. Five regions of the femoral quadriceps were irradiated with LLLT for 10 s per region. The LLLT group was the only group to decrease the fatigue index of the knee extensor muscles in isokinetic dynamometry.

Ferraresi et al. (2012) conducted a study showing modulations in gene expression of human muscle by LLLT (cluster with six diodes of 808 nm). With the same experimental design and parameters of LLLT used in a previous study (Ferraresi et al. 2011), the authors investigated changes of the load at 1 RM test and modulation of gene expressions in muscle tissue of the 10 young males allocated into two equal groups: LLLT and without LLLT. Biopsies from the vastus lateralis muscle were performed before and after the training program and analyzed by microarrays to identify modulation of gene expression throughout the whole human genome. Preliminary analysis identified that the LLLT group had more significant increase in load at 1 RM test, and the genes PPARGC1-α (mitochondrial biogenesis), mTOR (protein synthesis and muscle hypertrophy), and VEGF (angiogenesis) were significantly upregulated only in the LLLT group compared to the non-LLLT control group. In addition, the genes MuRF1 (protein degradation and muscle atrophy) and IL-1β (inflammation) were downregulated (Ferraresi et al. 2012) only in the LLLT group.

53.3 LEDT and Exercise

53.3.1 Clinical Trials: Acute Responses

Recently, light-emitting diode therapy (LEDT) has begun to be investigated for the same previous purpose as LLLT—increased performance in human physical exercises (Baroni et al. 2010b; Leal Junior et al. 2009a,c; Vinck et al. 2006). LEDT uses more accessible light sources and has a larger area compared to the light-emitting laser diode.

The first study using LEDT for muscle performance was conducted by Vinck et al. (2006). These authors investigated the effects of LEDT (cluster with 32 LEDs of 950 nm) on DOMS after fatigue induction of the biceps brachii using an isokinetic dynamometer. The authors found no significant differences between the LEDT and placebo groups for the peak torque and the level of pain measured by algometer and VAS.

Leal Junior et al. (2009a) compared the effects of a cluster with 69 LEDs (34 LEDs of 660 nm and 35 LEDs of 850 nm) with the single LLLT diode (810 nm) on the physical performance of volleyball players. Three tests of high-intensity effort (Wingate test) were conducted on nonconsecutive days, with each test lasting 30 s. The LEDT (cluster with 69 LEDs), LLLT, or placebo treatments were administered on two points on the belly of the rectus femoris muscle for 30 s of radiation on each point before starting the tests. The LEDT group had a significant reduction of CK levels in blood after exercise compared to the LLLT and placebo groups. However, there was no increased performance and significant reduction of lactate levels in the LEDT group compared to the other groups.

Using the same cluster LED employed in their previous study (Leal Junior et al. 2009a), Leal Junior et al. (2009b) applied the LEDT for 30 s of radiation on a single point on the belly of biceps brachii muscle before exercise at 75% MVC at Scott bench. The LEDT group increased the number of repetitions and time of contraction in the proposed exercise, and also reduced CK levels in blood, lactate, and protein c-reactive after exercise compared to placebo.

Baroni et al. (2010b) applied LEDT on three points of the femoral quadriceps muscle before starting a protocol of isokinetic dynamometry for induction of muscle fatigue. The radiation time was 30 s on each point. The results showed that LEDT was effective in preventing the decrease in knee extensor peak torque.

Comparing the effects of LEDT with other modalities of treatment, Leal Junior et al. (2011) tested the effects of cryotherapy and LEDT (cluster with 69 LEDs) on muscle recovery of six athletes in a randomized double-blind placebo-controlled trial. Athletes performed three Wingate tests on three different days. After each test, the athletes received placebo, LEDT, or cryotherapy (5 min of body immersion in water at 5°C). The LEDT was applied on two points of the femoral quadriceps, two points on hamstrings, and two points on gastrocnemius. LEDT did not increase the muscle work at Wingate test and did not modulate c-reactive protein, but significantly decreased CK and lactate levels in blood.

Tables 53.1 through 53.3 show all the parameters of LLLT and LEDT, muscle group irradiated, type of exercise performed, and the mode of application in experimental models (Table 53.1), LLLT in humans (Table 53.2), and LEDT in humans (Table 53.3).

LLLT and LEDT have been used in different ways among researchers. These differences comprise the number of radiation points on the muscle group, the parameters used, and the timing of radiation (before or after exercise).

The number of irradiation points appears to be an important parameter to effectively cover the muscle group (Baroni et al. 2010b; de Brito Vieira et al. 2012; De Marchi et al. 2012; Ferraresi et al. 2011; Gorgey, Wadee, and Sobhi 2008; Leal Junior et al. 2009a,b, 2011; Vinck et al. 2006); the irradiation points should be designed to cover the largest area and better distribute the energy applied on muscles (de Brito Vieira et al. 2012; Ferraresi et al. 2011). Results of studies using clusters of LLLT or LEDT appear to be better for reducing muscle fatigue and increasing muscle performance compared to the use of a single diode (de Brito Vieira et al. 2012; Ferraresi et al. 2011; Leal Junior et al. 2008, 2009a, 2010b). Figure 53.1 illustrates some examples of the number and distribution of the irradiation points using LLLT or LEDT on femoral quadriceps.

The parameters of LLLT and LEDT (particularly wavelength) used in experimental models and in clinical trials vary. However, near-infrared radiation seems to be the most common wavelength used in clinical trials (Baroni et al. 2010a; de Brito Vieira et al.

TABLE 53.1 LLLT and Exercise: Experimental Models

Reference	LLLT Parameters	Muscle/Exercise	Application Mode
(Vieira et al. 2006)	Laser 780 nm Diode area 0.039 cm², 15 mW, 37.5 mW/cm² 0.15 J per diode (10 s); 3.8 J/cm² 4 application points: total energy of 0.6 J (1 point on femoral quadriceps, 1 point on gluteus maximum, 1 point on tibialis anterior, 1 point on soleus)	Femoral quadriceps, gluteus maximum, tibialis anterior, soleus Exercise: running on treadmill	Contact After exercise
(Lopes-Martins et al. 2006)	Laser 655 nm Diode area 0.08 cm², 2.5 mW, 31.25 mW/cm² Groups: 0.5 J/cm² (32 s)/1 J/cm² (80 s)/2.5 J/cm² (160 s) 1 application point	Tibialis anterior Exercise: neuromuscular electric stimulation	Contact Before exercise
(Leal Junior et al. 2010a)	Laser 904 nm Diode area 0.2 cm², 15 mW, 75 mW/cm² Groups: 0.1 J (7 s)/0.3 J (20 s)/1 J (67 s)/3 J (200 s) 1 application point	Tibialis anterior Exercise: neuromuscular electric stimulation	Contact Before exercise

(Liu et al. 2009)	Laser 632 nm
	Diode area 0.2 cm²
	Groups:
	   12 J/cm², 4 mW, 20 mW/cm², 10 min
	   28 J/cm², 9 mW, 46 mW/cm², 10 min
	   43 J/cm², 14 mW, 71 mW/cm², 10 min
	1 application point
		Gastrocnemius
		Exercise: downhill running on treadmill
			Contact
			After exercise

(Sussai et al. 2010)	Laser 660 nm Diode area 0.03 cm², 100 mW, 3.3 mW/cm² 4 J per diode (40 s), 133.3 J/cm² 1 application point	Gastrocnemius Exercise: swimming with load	Contact After exercise
(de Almeida et al. 2011)	Laser 904 nm Diode area 0.2 cm², 15 mW, 75 mW/cm² Groups: 0.1 J (7 s)/0.3 J (20 s)/1 J (67 s)/3 J (200 s) 1 application point	Tibialis anterior Exercise: neuromuscular electric stimulation	Contact Before exercise

2012; De Marchi et al. 2012; Ferraresi et al. 2011, 2012; Gorgey, Wadee, and Sobhi 2008; Leal Junior et al. 2009a,d, 2010b; Vinck et al. 2006), although some studies have used clusters of LEDs with mixed wavelengths (red and near-infrared) in the same device (Baroni et al. 2010b; Leal Junior et al. 2009b, 2011).

The dosimetry and time of radiation of LEDT or LLLT on the muscles are still under investigation in clinical trials (Baroni et al. 2010a,b; de Brito Vieira et al. 2012; De Marchi et al. 2012; Ferraresi et al. 2011, 2012; Gorgey, Wadee, and Sobhi 2008; Leal Junior et al. 2009a,b,c, 2010b, 2011; Vinck et al. 2006). There is a wide variation in energy, power, power density, and irradiation time, but there is a consensus on how to apply (contact). The clinical trials that targeted the acute effects of LLLT on muscle fatigue increased the total energy (J), power (W), and power density (W/cm²) used in each study (Baroni et al. 2010a; De Marchi et al. 2012; Leal Junior et al. 2008, 2009a,d, 2010b), as reported in Table 53.2. However, the total energy and power were lower in clinical trials that target chronic effects (de Brito Vieira et al. 2012; Ferraresi et al. 2011, 2012).

Clinical trials that used LEDT before exercise used total energies much higher to reduce muscle fatigue than those in studies using LLLT (Baroni et al. 2010b; Leal Junior et al. 2009a,c). However, power and power density were smaller, as might be expected from the larger spot size (Baroni et al. 2010b; Leal Junior et al. 2009a,c; see Table 53.3). There is no consensus on the best parameters to be used in LLLT or LEDT, and therefore, more studies are needed to investigate the dose response of these therapies to reduce muscle fatigue and to repair muscle damage (Huang et al. 2009).

Clinical trials using LLLT and LEDT before exercise reported a preventive effect against mitochondrial dysfunction and muscle damage mediated by ROS and RNS, as well as modulation of energetic metabolism (Baroni et al. 2010a,b; de Almeida et al. 2012; De Marchi et al. 2012; Gorgey, Wadee, and Sobhi 2008; Leal Junior et al. 2008, 2009a,b,c,d, 2010b; Lopes-Martins et al. 2006). LLLT and LEDT radiation after physical exercise aims not only to remedy the mitochondrial and metabolic dysfunction but also to repair microlesions produced from mechanical and metabolic stress resulting from muscle contraction and ROS/RNS (de Brito Vieira et al. 2012; Ferraresi et al. 2011, 2012; Leal Junior et al. 2011; Liu et al. 2009; Sussai et al. 2010; Vieira et al. 2006). LLLT or LEDT radiation before or after physical exercises has distinct features as well as common mechanisms of action. The third section of this chapter will describe the mechanisms of action of LLLT and LEDT to prevent muscle fatigue and to repair muscle damage.

TABLE 53.2 LLLT and Exercise in Humans

Reference	LLLT Parameters	Muscle/Exercise	Application Mode
(Gorgey, Wadee, and Sobhi 2008)	Cluster with 4 diodes laser 808 nm 500 mW, 8.3 mW/cm^2 3 J (5 min) 7 J (10 min)	Femoral quadriceps Exercise: isokinetic dynamometer	Scanning Before exercise
(Leal Junior et al. 2008)	Laser 655 nm Diode area 0.01 cm^2, 50 mW, 5 W/cm^2 5 J per diode (100 s), 500 J/cm^2; 4 application points: total energy 20 J	Biceps brachii Exercise: Scott bench	Contact Before exercise
(Leal Junior et al. 2009c)	Laser 830 nm; Diode area 0.0028 cm^2, 100 mW, 35.7 W/cm^2 5 J (50 s), 1785 J/cm^2 4 application points: total energy 20 J	Biceps brachii Exercise: Scott bench	Contact Before exercise
(Baroni et al. 2010a)	Cluster with 5 diodes laser 810 nm Diode area 0.029 cm^2, 200 mW, 6.89 W/cm^2 6 J per diode (30 s), 206.89 J/cm^2 30 J per application point (5 × 6 J) 6 application points: total energy 180 J (2 points on vastus medialis, 2 points on vastus lateralis, 2 points on rectus femoris)	Femoral quadriceps Exercise: isokinetic dynamometer	Contact Before exercise
(Leal Junior et al. 2010b)	Cluster with 5 diodes laser 810 nm Diode area 0.0364 cm^2, 200 mW, 5.495 W/cm^2 6 J per diode (30 s), 164.85 J/cm^2 30 J per application point (5 × 6 J) 2 application points: total energy 60 J	Biceps brachii Exercise: Scott bench	Contact Before exercise
(de Almeida et al. 2012)	Laser 660 nm Diode area 0.0028 cm^2, 50 mW, 17.85 W/cm^2 5 J (100 s), 1785 J/cm^2 4 application points: total energy 20 J versus Laser 830 nm Diode area 0.0028 cm^2, 50 mW, 17.85 W/cm^2 5 J (100 s), 1785 J/cm^2 4 application points: total energy 20 J	Biceps brachii Exercise: MVC per 60 s in Scott bench	Contact Before exercise
(De Marchi et al. 2012)	Cluster with 5 diodes laser 810 nm Diode area 0.0364 cm^2, 200 mW, 5.495 W/cm^2 6 J per diode (30 s), 164.85 J/cm^2 30 J per application point (5 × 6 J) 12 application points: total energy 360 J (2 points on rectus femoris, 2 points on vastus medialis, 2 points on vastus lateralis, 4 points on hamstrings, 2 points on gastrocnemius)	Femoral quadriceps, hamstrings, gastrocnemius Exercise: running on treadmill until exhaustion	Contact Before exercise
(Ferraresi et al. 2011)	Cluster with 6 diodes laser 808 nm Diode area 0.0028 cm^2, 60 mW, 21.42 W/cm^2 0.6 J per diode (10 s), 214.28 J/cm^2 3.6 J per application region (0.6 J × 6) 7 regions distributed on femoral quadriceps: total energy 25 J	Femoral quadriceps Exercise: leg press and isokinetic dynamometer	Contact After exercise
(de Brito Vieira et al. 2012)	Cluster with 6 diodes laser 808 nm Diode area 0.0028 cm^2, 60 mW, 21.42 W/cm^2 0.6 J per diode (10 s), 214.28 J/cm^2 3.6 J per application region (0.6 J × 6) 5 regions distributed on femoral quadriceps: total energy 18 J	Femoral quadriceps Exercise: cycle ergometer and isokinetic dynamometer	Contact After exercise
(Ferraresi et al. 2012)	Cluster with 6 diodes laser 808 nm Diode area 0.0028 cm^2, 60 mW, 21.42 W/cm^2 0.6 J per diode (10 s), 214.28 J/cm^2 3.6 J per application region (0.6 J × 6) 7 regions distributed on femoral quadriceps: total energy 25 J	Femoral quadriceps Exercise: leg press and isokinetic dynamometer	Contact After exercise

TABLE 53.3   LEDT and Exercise in Humans

Reference	LLLT Parameters	Muscle/Exercise	Application Mode
(Vinck et al. 2006)	Cluster with 32 LEDs, 850 nm Cluster total area 18 cm^2, 160 mW 3.2 J/cm^2 per application point (360 s) 1 application point	Biceps brachii Exercise: isokinetic dynamometer	Contact After exercise
(Leal Junior et al. 2009a)	Laser 810 nm Diode area 0.036 cm^2, 200 mW, 5.50 W/cm^2 6 J per diode (30 s), 164.84 J/cm^2 2 application points: total energy 12 J versus Cluster with 69 LEDs Diode area 0.2 cm^2 34 LEDs 660 nm, 10 mW, 1.5 J/cm^2, 0.05 W/cm^2 35 LEDs 850 nm, 30 mW, 0.015 W/cm^2, 4.5 J/cm^2 0.3 J LED 660 nm (30 s) 0.9 J LED 850 nm (30 s) 41.7 J per application point (30 s) 2 application points: total energy 83.4 J	Rectus femoris Exercise: Wingate test	Contact Before exercise
(Leal Junior et al. 2009b)	Cluster with 69 LEDs Diode area 0.2 cm^2 34 LEDs 660 nm; 10 mW, 1.5 J/cm^2, 0.05 W/cm^2 35 LEDs 850 nm; 30 mW; 0.015 W/cm^2, 4.5 J/cm^2 0.3 J LED 660 nm (30 s) 0.9 J LED 850 nm (30 s) 41.7 J per application point (30 s) 1 application point	Biceps brachii Exercise: Scott bench	Contact Before exercise
(Baroni et al. 2010b)	Cluster with 69 LEDs: Diode area 0.2 cm^2; 34 LEDs 660 nm; 10 mW; 1.5 J/cm^2; 0.05 W/cm^2 35 LEDs 850 nm; 30 mW; 0.015 W/cm^2; 4.5 J/cm^2 0.3 J LED 660 nm (30 s) 0.9 J LED 850 nm (30 s) 41.7 J per application point (30 s) 3 application points: total energy 125.1 J (2 points on rectus femoris, 2 points on vastus medialis, 2 points on vastus lateralis)	Femoral quadriceps Exercise: isokinetic dynamometer	Contact Before exercise
(Leal Junior et al. 2011)	Cluster with 69 LEDs: Diode area 0.2 cm^2; 34 LEDs 660 nm; 10 mW; 1.5 J/cm^2; 0.05 W/cm^2 35 LEDs 850 nm; 30 mW; 0.015 W/cm^2; 4.5 J/cm^2 0.3 J LED 660 nm (30 s) 0.9 J LED 850 nm (30 s) 41.7 J per application point (30 s) 5 application points: total energy 208.5 J (2 points on femoral quadriceps, 2 points on hamstrings, 1 point on gastrocnemius) versus Cryotherapy (5°C) Lower limb immersion at 5°C for 5 min	Femoral quadriceps, hamstrings, and gastrocnemius Exercise: Wingate test	Contact After exercise

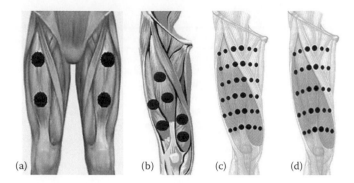

FIGURE 53.1 Number of LLLT or LEDT radiation points applied on femoral quadriceps muscle. (a: From Leal Junior, E. C. et al., *Photomed Laser Surg* 27:617–623, 2009; b: From De Marchi, T. et al., *Lasers Med Sci* 27:231–236, 2012; c: From Ferraresi, C. et al., *Lasers Med Sci* 26:349–358, 2011; d: From de Brito Vieira, W. H. et al., *Lasers Med Sci* 27:497–504, 2012.)

53.4 Mechanisms of Action of LLLT and LEDT for Performance, Repair, and to Prevent Muscle Fatigue

53.4.1 Energetic Metabolism

53.4.1.1 Mitochondrial Pathway

The infrared radiation emitted by LLLT and LEDT radiation seems to act on cellular energy metabolism, stimulating photochemical and photophysical events in the cell mitochondria (Bakeeva et al. 1993; de Brito Vieira et al. 2012; Ferraresi et al. 2011; Hayworth et al. 2010; Huang et al. 2009; Karu 2010; Manteifel and Karu 2005; Silveira et al. 2009). Photochemical and photophysical changes may result in increased mitochondrial membrane potential (Passarella et al. 1988) and higher enzyme activity in the respiratory chain (Hayworth et al. 2010; Huang et al. 2009; Karu 2010; Silveira et al. 2009). The structural change includes the formation of giant mitochondria through the merging of membranes of smaller and neighboring mitochondria (Manteifel, Bakeeva, and Karu 1997). These changes enable mitochondria to provide higher levels of respiration and ATP to cells (de Brito Vieira et al. 2012; Ferraresi et al. 2011; Hayworth et al. 2010; Silveira et al. 2009). Some studies reported an improvement of enzyme activity of the complex IV [cytochrome c oxidase (CCO)] (Karu 2010) in the mitochondrial ETC in skeletal muscle (Hayworth et al. 2010). Silveira et al. (2009) showed that all complexes of the mitochondrial ETC (complexes I, II, III, and IV) had their enzyme activity improved after near-infrared radiation.

Considering the effects of LLLT and LEDT on mitochondria, some hypotheses have been proposed to take advantage of the beneficial effects of LLLT and LEDT on endurance and also on strength exercises.

Endurance or low-intensity exercises. The oxidative capacity of muscle fibers is proportional to its mitochondrial density, since these cellular organelles can completely oxidize energy substrates (glucose, fatty acids, and proteins) for ATP synthesis during muscle contraction (Coffey and Hawley 2007; Hawley 2009; Sahlin et

al. 2007; Tonkonogi and Sahlin 2002; Tonkonogi et al. 2000; Westerblad, Bruton, and Katz 2010). Endurance or low-intensity exercise is a powerful stimulus to promote mitochondrial biogenesis in its own right, favoring aerobic metabolism and reducing muscle fatigue from metabolic origin, as the accumulation of Pi, ADP, H^+, and lactate in the sarcoplasm (Allen, Lamb, and Westerblad 2008; Coffey and Hawley 2007; Hawley 2009; Westerblad, Bruton, and Katz 2010). However, when LLLT and/or LEDT are added to the effects of endurance exercise on the mitochondria, the adaptive process can be increased. Giant and more functional mitochondria (higher enzyme activity) can provide high levels of cellular respiration and ATP synthesis (Hayworth et al. 2010; Huang et al. 2009; Silveira et al. 2009) during these exercises, which gives increased oxygen consumption (De Marchi et al. 2012) and reduced muscle fatigue (de Brito Vieira et al. 2012).

Strength or high-intensity exercises. These types of exercises have anaerobic metabolism, which can be lactic and alactic (Allen, Lamb, and Westerblad 2008; Westerblad and Allen 2011), and cover exercise performed in most studies involving LLLT (Baroni et al. 2010a; de Almeida et al. 2012; De Marchi et al. 2012; Ferraresi et al. 2011; Gorgey, Wadee, and Sobhi 2008; Leal Junior et al. 2008, 2009c, 2010b) and LEDT in humans (Baroni et al. 2010b; Leal Junior et al. 2009a,c; Vinck et al. 2006). LLLT and LEDT may increase muscle performance and reduce fatigue by three mechanisms suggested by Ferraresi et al. (2011) and de Brito Vieira et al. (2012):

1. *Mitochondrial ATP*: LLLT and LEDT seem to increase mitochondrial activity, providing higher levels of cell respiration and ATP synthesis (Hayworth et al. 2010; Huang et al. 2009; Silveira et al. 2009). As the muscle fiber recruitment is hierarchical and depends on fiber type (type I, IIa, IIb, and IIx, respectively) (Hodson-Tole and Wakeling 2009), larger amounts of ATP can be provided by those fibers with oxidative potential (type I and IIa fibers) during strength training or high-intensity exercises (Ferraresi et al. 2011).

2. *Phosphocreatine resynthesis*: Strength or high-intensity exercise consumes large amounts of ATP from hydrolysis of phosphocreatine (PCr), which is catalyzed by muscle CK enzyme in the sarcoplasm: PCr + ADP + Cr ↔ ATP. This consumption of ATP is faster than the rate of PCr resynthesis, producing an excess of creatine (Cr), ADP, and Pi in the sarcoplasm of muscle fibers (IIa, IIb, and IIx, especially), contributing to the process of fatigue (Allen, Lamb, and Westerblad 2008; Westerblad, Bruton, and Katz 2010). However, high concentrations of Cr and ADP stimulate respiration and mitochondrial ATP synthesis in muscle fibers of type I and IIa (high and mean mitochondrial density, respectively) (Tonkonogi and Sahlin 2002), integrating aerobic and anaerobic alactic metabolisms in this

type of exercise (Figure 53.2a through c) (Ferraresi et al. 2011). This metabolic integration is caused by PCr resynthesis by the mitochondrial Cr shuttle (Bakeeva et al. 1993; Harridge 2007; Hawke and Garry 2001). This shuttle system captures the creatine, ADP, and Pi from ATP consumed during muscle contraction and transports it to the mitochondrial matrix, crossing the inner mitochondrial membrane through an adenine nucleotide translocase. The mitochondrial ATP is delivered to muscle using the same pathway but in the opposite direction, providing energy for the PCr resynthesis, which is catalyzed by CK near the site of muscle contraction. Together with the use of PCr energy to resynthesize ATP for muscle contraction, there is also renewed production of Cr, ADP, and Pi. As both ADP and Pi follow the pathway previously described, the Cr is transported to the intermembrane space of muscle mitochondria where the mitochondrial CK catalyzes the reaction of PCr resynthesis also using mitochondrial ATP. Finally, the PCr is transported to the site of muscle contraction and provides the energy necessary for contraction and increasing the ratio of ATP/ADP (Figure 53.2b). Thus, the PCr resynthesis that occurs during intervals of exercises (Ferraresi et al. 2011) as well as during high-intensity exercises (de Brito Vieira et al. 2012) could provide more ATP resynthesis and could become the predominant energy source for maximal tests of muscle strength or during short high-intensity exercises (Ferraresi et al. 2011).

3. *Lactate oxidation by mitochondria*: Blood lactate has been used to measure muscle fatigue during strength training or high-intensity exercise (Allen, Lamb, and Westerblad 2008; Westerblad, Bruton, and Katz 2010). When the oxygen supply is insufficient and/or delayed

for optimal mitochondrial ATP synthesis, pyruvate is reduced to lactate by NADH+H in the glycolysis process. This reaction is catalyzed by cytosolic LDH with production of lactate (Brooks et al. 1999). Lactate is transported to the mitochondrial matrix by monocarboxylate transporters and by means of oxidized nicotinamide adenine dinucleotide (NAD$^+$) and the enzyme mitochondrial lactate dehydrogenase; lactate is oxidized to pyruvate. Next, the reduced NAD (NADH) is oxidized in the ETC (respiratory chain) and provides electrons and protons required for mitochondrial ATP synthesis. Pyruvate is oxidized to acetyl-CoA in the Krebs cycle and continues to be oxidized for ATP production in the ETC (Figure 53.2c). This mechanism of lactate oxidation in mitochondria was first proposed by Brooks et al. (1999) and was recently discussed by Ferraresi et al. (2011).

53.4.2 Enzyme Modulation

LDH (EC 1.1.1.27) is the enzyme responsible for the reduction of pyruvate to lactate (pyruvate + NADH + H $\leftrightarrow$ H-lactate + NAD$^+$) for ATP synthesis in anaerobic lactic metabolism. The LDH activity is often combined with blood lactate measurements to infer the magnitude of this energy metabolism, as well as the intensity of exercise and how efficient is the buffering of lactic acid to prevent metabolic acidosis (H-lactate + NaHCO$_3$ $\rightarrow$ Na-CO$_2$ + H$_2$O + lactate) and to prevent fatigue (Allen, Lamb, and Westerblad 2008; de Brito Vieira et al. 2012; Westerblad, Bruton, and Katz 2010).

LLLT seems to modulate LDH activity in physical exercise, as demonstrated and suggested in previous studies (Baroni et al. 2010a; de Brito Vieira et al. 2012; De Marchi et al. 2012; Vieira et al. 2006). These studies demonstrated that LDH activity is inhibited by LLLT even in the period in which the supply of O$_2$ is slow or inadequate for mitochondrial ATP synthesis to enhance

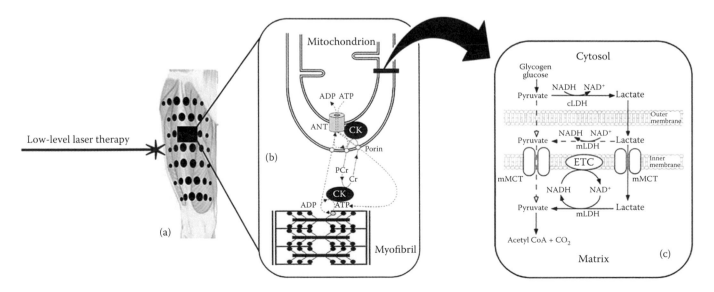

FIGURE 53.2 (a) LLLT radiation points on femoral quadriceps muscle, (b) mitochondrial creatine shuttle mechanism, and (c) lactate oxidation by mitochondrial pathway. (Based on Ferraresi, C. et al., *Lasers Med Sci* 26:349–358, 2011.)

muscle performance (Baroni et al. 2010a; de Brito Vieira et al. 2012; De Marchi et al. 2012; Vieira et al. 2006).

The enzymes of aerobic metabolism are also modulated by LLLT and LEDT such as complexes I, II, III, and IV of the mitochondrial ETC (Hayworth et al. 2010; Huang et al. 2009; Karu 2010; Silveira et al. 2009) stimulating aerobic metabolism. Other enzymes of aerobic metabolism are possibly modulated by LLLT and LEDT such as citrate synthase and other enzymes of the Krebs cycle because the Krebs cycle provides part of the protons and electrons necessary for ATP synthesis in the ETC.

53.5 ROS and RNS

Superoxide anion $\left(O_2^{\cdot-}\right)$ is the primary free radical ROS produced mainly by mitochondria and nicotinamide adenine dinucleotide phosphate (NADPH) oxidase in skeletal muscle. In the mitochondria, $O_2^{\cdot-}$ is produced as a byproduct of oxidative phosphorylation and therefore its formation is always associated with mitochondrial activity, especially during physical exercise (Allen, Lamb, and Westerblad 2008; Powers and Jackson 2008; Westerblad and Allen 2011). $O_2^{\cdot-}$ is produced mainly at complexes I and III of the ETC, which release $O_2^{\cdot-}$ in the mitochondrial matrix and at the intermembrane space, respectively. The production rate of $O_2^{\cdot-}$ is approximately 0.15% of oxygen consumption in mitochondria together with the enzymatic activities of NADPH oxidase, xanthine oxidase, and lipoxygenase that also produce $O_2^{\cdot-}$ (Allen, Lamb, and Westerblad 2008; Powers and Jackson 2008; Westerblad and Allen 2011). Beyond $O_2^{\cdot-}$, hydrogen peroxide (H_2O_2) and hydroxyl radicals $(OH^{\cdot})$ are also important ROS (Allen, Lamb, and Westerblad 2008; Powers and Jackson 2008; Westerblad and Allen 2011).

Nitric oxide $(NO^{\cdot})$ is the primary nitrogen free radical. It is synthesized from L-arginine amino acid by nitric oxide synthase (NOS), can be formed from the inorganic anions, nitrate $\left(NO_3^-\right)$ and nitrite $\left(NO_2^-\right)$, and can interact with $O_2^{\cdot-}$ to form peroxynitrite $(ONOO^-)$ (Allen, Lamb, and Westerblad 2008; Powers and Jackson 2008; Westerblad and Allen 2011). In adult skeletal muscle, there is expression of the three isoforms of NOS: neuronal (nNOS or type 1), inducible (iNOS or type 2), and endothelial (eNOS or type 3). However, the predominant isoform is nNOS (Allen, Lamb, and Westerblad 2008; Powers and Jackson 2008; Westerblad and Allen 2011).

ROS and RNS are produced by mitochondria, by NADPH oxidase activity in the sarcoplasmic reticulum and transverse tubules, and also by phospholipase A_2 and xanthine oxidase (Powers and Jackson 2008). However, there are defenses against ROS and RNS activity in cells and tissues. These defenses are mainly the antioxidant activity of SOD, glutathione peroxidase (GPX), and catalase (CAT) (Powers and Jackson 2008). There are three isoforms of SOD (SOD1-3) and five isoforms of GPX (GPX1-5), each with its specific location in the cell and extracellular places. For detailed information, see the review by Powers and Jackson (Powers and Jackson 2008).

Superoxide anion is rapidly broken down into hydrogen peroxide by antioxidant enzymes (reaction 1) such as mitochondrial

superoxide dismutase (SOD2). Hydrogen peroxide (H_2O_2) is broken down by catalase (reaction 2) and glutathione peroxidase with conversion of reduced glutathione (GSH) to oxidized glutathione (GSSG) (reaction 3) (Allen, Lamb, and Westerblad 2008; Powers and Jackson 2008; Westerblad and Allen 2011). Moreover, hydrogen peroxide can react with metals such as Fe^{2+} (reaction 4: Fenton reaction), which produces hydroxyl radicals ($OH^{\cdot}$ and OH^-), which are extremely reactive (Allen, Lamb, and Westerblad 2008; Powers and Jackson 2008; Westerblad and Allen 2011). Fe^{3+} is present in myoglobin, hemoglobin, and CCO and can be reduced to Fe^{2+} by superoxide radicals and then produce hydroxyl radicals (reaction 5: Haber–Weiss reaction). Hydroxyl radicals have a short half-life because they react with any organic molecules and can produce damage of proteins, deoxyribonucleic acid (DNA), and lipids that are present in the cell membrane and therefore lead to tissue damage (Allen, Lamb, and Westerblad 2008; Powers and Jackson 2008; Westerblad and Allen 2011).

Reaction 1: $O_2^{\cdot-} + O_2^{\cdot-} + 2H^+ \rightarrow H_2O_2 + O_2$

Reaction 2: $2H_2O_2 \rightarrow 2H_2O + O_2$

Reaction 3: $2GSH + H_2O_2 \rightarrow GSSG + 2H_2O$

Reaction 4: (Fenton reaction): $H_2O_2 + Fe^{2+} \rightarrow Fe^{3+} + OH^{\cdot} + OH^-$

Reaction 5: (Haber − Weiss reaction): $O_2^{\cdot-} + H_2O_2 \rightarrow OH^{\cdot} + OH^- + O_2$

During physical activity, the contraction of skeletal muscle is the main source of ROS and RNS production, and it produces deleterious effects on muscle fibers, such as cellular and tissue damage, loss of contractile function, and exercise performance (Allen, Lamb, and Westerblad 2008; Lamb and Westerblad 2011; Powers and Jackson 2008; Westerblad and Allen 2011). ROS and RNS can induce early onset of muscle fatigue in exercise through mechanisms involved in muscle contraction (Allen, Lamb, and Westerblad 2008; Powers and Jackson 2008), such as the following:

Decreased ATP synthesis. $NO^{\cdot}$ production in the mitochondria can reduce ATP synthesis by inhibition of CCO, muscle CK, and glyceraldehyde-3-phosphate dehydrogenase (glycolytic pathway), thereby reducing ATP production (Powers and Jackson 2008).

Regulation of sarcoplasmic reticulum and $Ca^{2\pm}$ release. ROS increases Ca^{2+} content in the sarcoplasm and promotes a slow Ca^{2+} reuptake. It also inhibits the activity of the ATPase enzyme of the Ca^{2+} pump in the sarcoplasmic reticulum (SERCA) leading to impairment of hydrolysis of ATP used for operating the Ca^{2+} pump (Allen, Lamb, and Westerblad 2008; Powers and Jackson 2008). $NO^{\cdot}$ also inhibits the activity of SERCA and results in decreased reuptake of Ca^{2+} from the sarcoplasm to the sarcoplasmic reticulum (Powers and Jackson 2008). Prolonged exposure to ROS and RNS can induce a lasting release of Ca^{2+} through changes of the receptor sensitivity of channels for Ca^{2+} of the sarcoplasmic reticulum (Westerblad and Allen 2011).

Contractile proteins. ROS and RNS can change the structure of contractile proteins, decrease the sensitivity of myofibrils to Ca^{2+} and oxidize actin, myosin, and troponin C, impairing the formation of cross bridges, muscle contraction, and production of muscle strength (Allen, Lamb, and Westerblad 2008; Lamb and Westerblad 2011; Powers and Jackson 2008; Westerblad and Allen 2011).

Potential action. ROS seems to reduce the Na^+-K^+ pump activity and increase amounts of extracellular K^+ that reduces the action potential for muscle fiber depolarization during exercise, contributing to the onset of early muscle fatigue (Westerblad and Allen 2011).

ROS and RNS after muscle fatigue. ROS decreases the sensitivity of myofibrils to Ca^{2+} by accumulation of H_2O_2 (via SOD2) and/or $OH^{\bullet}$ (via Fenton reaction), impairing contractile function (Lamb and Westerblad 2011; Westerblad and Allen 2011). High levels of $O_2^{\bullet-}$ can also inhibit the release of Ca^{2+} from sarcoplasmic reticulum and/or react with $NO^{\bullet}$ to produce $ONOO^-$, which also affects Ca^{2+} release from the sarcoplasmic reticulum (Lamb and Westerblad 2011; Westerblad and Allen 2011). Furthermore, $NO^{\bullet}$ can also bind to transition metals (Powers and Jackson 2008), such as Fe, and impair mitochondrial function (Huang et al. 2009; Vladimirov, Osipov, and Klebanov 2004).

Muscle damage. The lipids that make up the muscle cell membrane (sarcolemma) are attacked by ROS and RNS during the process of lipid peroxidation. Disruption of the sarcolemma promotes muscle cell death and all the contents are released into the extracellular environment, causing inflammation (degradation of cellular content), edema, pain, and loss of contractile function (Allen, Lamb, and Westerblad 2008; Powers and Jackson 2008). In this process, blood levels of muscle CK are increased, making it a useful parameter to measure muscle damage (Markert et al. 2011).

LLLT and LEDT have been used to combat ROS and RNS produced during physical exercise (Baroni et al. 2010a,b; de Almeida et al. 2012; De Marchi et al. 2012; Leal Junior et al. 2009a,c, 2010a,b, 2011; Liu et al. 2009; Sussai et al. 2010; Xu et al. 2008) for improvement in mitochondrial function that contributes to the reduction of muscle fatigue and the increase in muscle performance. LLLT and LEDT use CCO as the primary photoacceptor, and the main effects of this interaction are increased ATP synthesis and increased mitochondrial function (Hayworth et al. 2010; Huang et al. 2009; Karu 2010; Silveira et al. 2009; Vladimirov, Osipov, and Klebanov 2004; Xu et al. 2008). The relationship between light, mitochondria, ROS, and RNS involves the reduction of ROS and the photodissociation of NO-CCO, contributing to the restoration of oxygen consumption and ATP synthesis in mitochondria (Huang et al. 2009; Vladimirov, Osipov, and Klebanov 2004).

CCO-NO photodissociation is based on the hypothesis that NO may compete with oxygen to bind to the iron–sulfur complex (complex I) and to centers of iron and copper (complex IV) in the respiratory chain and inhibit the mitochondrial ATP synthesis (Huang et al. 2009; Vladimirov, Osipov, and Klebanov 2004) mainly in metabolically stressed cells (Huang et al. 2009), as occurs after muscle contraction (Xu et al. 2008). However, the binding between NO-CCO can be broken by visible and near-infrared light energy (Karu, Pyatibrat, and Afanasyeva 2005; Vladimirov, Osipov, and Klebanov 2004), restoring mitochondrial function to ATP synthesis (Hayworth et al. 2010; Huang et al. 2009; Silveira et al. 2009; Vladimirov, Osipov, and Klebanov 2004; Xu et al. 2008). Thus, the use of LLLT or LEDT after physical exercise may be more effective compared to its use before exercises.

It is known that physical exercise may decrease intracellular pH (Allen, Lamb, and Westerblad 2008). An accumulation of H^+ can inactivate the Cu–Zn–SOD (SOD1 and 3) through a protonation of the residue of histidine 61 (Hys61) in the active center of these enzymes (Vladimirov, Osipov, and Klebanov 2004). However, LLLT can reverse this process and reactivate SOD1 and 3 (Vladimirov, Osipov, and Klebanov 2004) (Figure 53.3).

De Marchi et al. (2012) reported that SOD activity did not change even after intense exercise in the LLLT group, thereby increasing physical performance compared to the placebo-LLLT group that had decreased SOD activity and had the worst performance. Liu et al. (2009) found increases of 44% and 58% in SOD activity 24 and 48 h after eccentric exercise and irradiation with LLLT.

Possibly LLLT and LEDT improve mitochondrial function and dismutation of superoxide anion $\left(O_2^{\bullet-}\right)$ via SOD, and decrease formation of $ONOO^-$ (Vladimirov, Osipov, and Klebanov 2004). In addition, LLLT can reduce H_2O_2 via CAT and GPX and can reduce the formation of hydroxyl radicals that contribute to the lower damage of the muscle cell membrane, as evidenced by lower lipid peroxidation and lower blood levels of muscle CK reported in previous studies (DeMarchi et al. 2012; Liu et al. 2009). Also, the activation of calpain and caspase mediated by ROS (Powers and Jackson 2008) can be inhibited by LLLT and LEDT, since these therapies can modulate oxidative stress (DeMarchi et al.

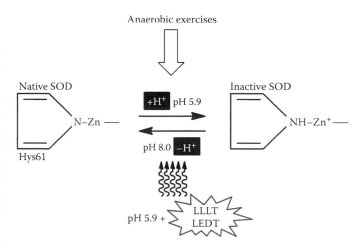

FIGURE 53.3 Deprotonation of hystidine and formation of the N–Zn bond to restore the active center structure and activity of SOD by LLLT or LEDT. (Based on Vladimirov, Y. A. et al., *Biochemistry (Mosc)* 69:81–90, 2004.)

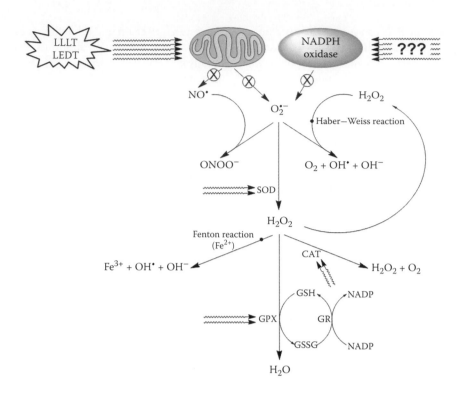

FIGURE 53.4 Effects of LLLT or LEDT on ROS and RNS production and antioxidant enzymes. ⊗ = decreased or inhibited production. (Based on Allen, D. G. et al., *Physiol Rev* 88:287–332, 2008.)

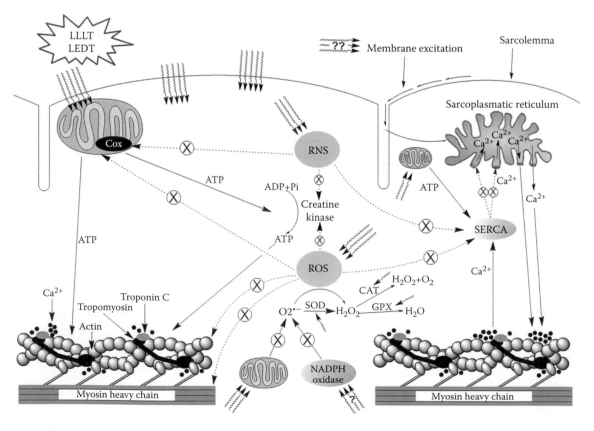

FIGURE 53.5 Effects of LLLT or LEDT on ROS, RNS, mitochondria, and muscle contraction. SERCA = sarcoplasmatic reticulum Ca^{2+} pump. Dashed line = inhibited activity and ⊗ = decreased or inhibited function. (Based on Allen, D. G. et al., *Physiol Rev* 88:287–332, 2008; Ferraresi, C. et al., *Lasers Med Sci* 26:349–358, 2011; Huang, Y. Y. et al., *Dose Response* 7:358–383, 2009; Powers, S. K. and M. J. Jackson, *Physiol Rev* 88:1243–1276, 2008.)

2012; Vladimirov, Osipov, and Klebanov 2004). Figures 53.4 and 53.5 illustrate the action of LLLT and LEDT on antioxidant enzymes and ROS and RNS in muscle contraction, respectively.

However, besides increased mitochondrial function and activity of antioxidant enzymes such as SOD, CAT, and GPX, other mechanisms of action of LLLT and LEDT may be involved in the reduction of muscle fatigue and the increase in muscle performance (Baroni et al. 2010a,b; de Almeida et al. 2012; de Brito Vieira et al. 2012; De Marchi et al. 2012; Ferraresi et al. 2011; Leal Junior et al. 2009a,c, 2010a,b, 2011; Liu et al. 2009; Sussai et al. 2010; Xu et al. 2008). These may be (1) improvement in the sensitivity of myofibrils and Ca^{2+} channels to the Ca^{2+} ion, (2) increase in Ca^{2+} uptake from sarcoplasm to sarcoplasmic reticulum via Ca^{2+} pump (ATP-dependent), (3) improvement in the formation of cross bridges and production of contractile force, (4) increase in the activity of Na^+–K^+ pump (ATP-dependent) to reduce the excess of extracellular K^+ and ensure the depolarization of muscle fibers to continue the exercise, and (5) lower muscle damage and leakage of muscle contents into the blood, such as muscle CK and myoglobin.

53.6 Repair of Muscle Damage

Muscle damage is a disruption in the myofibrils and is a process accompanied by inflammation at the site of injury and contractile function losses (Jarvinen et al. 2005). The regeneration of muscle tissue after different injuries is a process that usually occurs in six phases taking 21 days overall (Jarvinen et al. 2005): phase 1 occurs 2 days after injury with necrosis of injured tissue muscle, acute inflammation with swelling, increased collagen deposition, increased connective tissue and scarring formation; phase 2 occurs 3 days after injury with satellite cell activation; phase 3 occurs 5 days after injury with fusion of myoblasts at the site of injury and increased density of connective tissue; phase 4 occurs 7 days after injury with regenerating muscle cells extending outside of the former cylinders of the basal lamina to the injured site and then invading the scar tissue; phase 5 occurs 14 days after injury with the scar in the injured site reducing in size and new myofibrils filling the gap at the injured site; and phase 6 occurs 21 days after injury with interweaving and fusion of the ends of myofibrils and with further reduction in connective and scar tissue.

The regeneration of muscle tissue is limited but does allow activation, proliferation, and differentiation of muscle satellite cells into new myonuclei and/or myofibrils (Charge and Rudnicki 2004; Hawke and Garry 2001; Jarvinen et al. 2005; Kuang and Rudnicki 2008). Satellite cells are located at the basal layer of myofibrils and may be quiescent, proliferating, or differentiating according to the harmful stimuli (Charge and Rudnicki 2004; Hawke and Garry 2001; Kuang and Rudnicki 2008). These cells are considered reserve cells, because although they are not stem cells such as in embryonic development, these cells are able to renew the myogenesis program in response to a muscle injury (Mauro 1961) such as microlesions from strength training or high-intensity exercises (Charge and Rudnicki 2004; Chen and Goldhamer 2003; Hawke and Garry 2001; Holterman and Rudnicki 2005; Kuang and Rudnicki 2008; Wilborn et al. 2009).

Muscle injuries promote inflammation and higher concentrations of neutrophils and macrophages at the site of injury that produce chemoattractants to attract the satellites cells to the region (Charge and Rudnicki 2004; Hawke and Garry 2001). Satellite cells are activated and proliferate and differentiate by molecular pathways such as modulation of gene expression related to quiescence (Pax7, c-Met, Myf5), proliferation/activation (MyoD1, Myf5, M-cadherin), and differentiation (desmin, MRF4, and myogenin) (Charge and Rudnicki 2004; Hawke and Garry 2001).

Satellite cells that have already differentiated can be fused to the damaged muscle fibers as new myonuclei, providing gene expression that encodes new contractile proteins or improve the myonuclear domain (Harridge 2007; Petrella et al. 2008; Vierck et al. 2000). In addition, these cells can be activated and form new myoblasts that fuse to form myotubes that will produce new myofibrils (Charge and Rudnicki 2004; Hawke and Garry 2001).

Once the importance of satellite cells in the process of muscle repair became known, some authors investigated the effects of LLLT on these cells (Bibikova and Oron 1994; Roth and Oron 1985; Weiss and Oron 1992). Studies in experimental models have shown that LLLT increases the formation of new myofibrils that fill the gap in muscle injury and contractile characteristics can return to the site of injury (Roth and Oron 1985; Weiss and Oron 1992). Bibikova and Oron (1994) observed that LLLT (632.8 nm) promoted a greater maturation of young myofibrils compared to those not irradiated. Shefer et al. (2002) found an increase in the number and activation of satellite cells around the myofibrils irradiated with LLLT (632.8 nm). They also observed that LLLT (632.8 nm) was effective in increasing the levels of Bcl-2 (cell viability) and decreasing BAX (apoptosis). Nakano et al. (2009) reported that LLLT (830 nm) increased satellite cells (by BrdU labeling) and angiogenesis and maintained the diameter of rat myofibrils submitted to the process of muscle atrophy.

Currently, there is no study that investigated LLLT or LEDT and satellite cell activity in humans. Only *in vitro* studies and experimental models are reported in scientific literature, which show a positive influence of LLLT on the cell cycle, proliferation, and activation of these cells (Ben-Dov et al. 1999; Shefer et al. 2002, 2003; Weiss and Oron 1992).

In addition to satellite cells, the control of inflammation, proteolysis, ROS and RNS, and synthesis and remodeling of collagen and ATP levels are fundamental to the success of muscle damage repair (de Souza et al. 2011; Dourado et al. 2011; Liu et al. 2009; Luo et al. 2013; Mesquita-Ferrari et al. 2011; Renno et al. 2011; Rizzi et al. 2006; Silveira et al. 2009, 2013). Studies in experimental models have shown that LLLT can modulate amounts of collagen at the site of injury (de Souza et al. 2011; Rizzi et al. 2006; Silveira et al. 2013), inhibit inflammation by lower expression of TNF-α and COX-2 (Liu et al. 2009; Mesquita-Ferrari et al. 2011; Renno et al. 2011), decrease ROS by smaller lipid peroxidation (TBARS and lower levels of MDA) (Liu et al. 2009; Luo et al. 2013; Rizzi et al. 2006; Silveira et al. 2013), reduce RNS by inhibiting synthesis of iNOS (Rizzi et al. 2006), decrease the expression of NF-κβ and CK (related to proteolysis) (Liu et al. 2009; Rizzi et al. 2006; Silveira et al. 2013), increase activity of SOD (Liu et al.

2009; Luo et al. 2013), increase the expression of VEGF receptor (VEGFR-1) in the capillaries and in satellite cells (Dourado et al. 2011), and increase activity of respiratory chain for ATP synthesis and repair muscle damage (Silveira et al. 2009).

53.7 Gene Expression Effects

Several studies report specific molecular pathway signaling for each type of physical training where the expression and/or suppression of specific genes are essential for better physical performance (Adhihetty et al. 2003; Bodine 2006; Coffey and Hawley 2007; Favier, Benoit, and Freyssenet 2008; Fluck 2006; Hawley 2009; Stepto et al. 2009).

Strength training or high-intensity exercise have well-defined signaling pathways that involve specific genes related to muscle satellite cells at the quiescent state (c-Met genes, Myf5, and Pax7), activated (genes MyoD1, Myf5, M-cadherin) and differentiation to form new myonuclei and/or myofibrils (genes desmin, myogenin, and MRF4) in response to microlesions (Charge and Rudnicki 2004; Chen and Goldhamer 2003; Hawke and Garry 2001; Holterman and Rudnicki 2005; Kuang and Rudnicki 2008; Wilborn et al. 2009). Also, strength training or high-intensity exercise is influenced by genes related to protein synthesis (muscle hypertrophy) as IGF-1, AKT, mTOR, and p70^{S6K} (Bodine 2006; Coffey and Hawley 2007), genes regulating protein degradation (atrophy) as TNF-α, FOXO, FBXO32, TRIM63, or MuRF1, MSTN, E3 ubiquitin ligases, and enzymes involved in the proteolytic process by a ubiquitin–proteasome system (Bodine 2006; Coffey and Hawley 2007; Favier, Benoit, and Freyssenet 2008; Glass 2005; Hawley 2009).

Gene expression changes related to endurance training or low-intensity exercise include mitochondrial biogenesis involving upregulation of CCO, NRF 1/2, TFAM, and PPARGC-1α (Bodine 2006; Coffey and Hawley 2007; Favier, Benoit, and Freyssenet 2008; Glass 2005; Hawley 2009). When PPARGC-1α is upregulated, there is an indication of mitochondrial biogenesis and more fatty acids oxidation in the muscles also relating to the transition between the fiber types as type II to type I, which has greater oxidative potential (Adhihetty et al. 2003; Coffey and Hawley 2007; Fluck 2006). This transition provides biomechanical adaptations such as resistance to fatigue in strenuous exercise at low to medium intensity (Adhihetty et al. 2003; Coffey and Hawley 2007; Fluck 2006).

The identification of the molecular mechanisms modulated by LLLT and LEDT can aid the understanding of the effects of these therapies on the gene expression changes related to different types of exercises (Coffey and Hawley 2007; Hawley 2009; Stepto et al. 2009), as previously suggested (de Brito Vieira et al. 2012; Ferraresi et al. 2011). However, the literature has a lack of studies involving gene expression in humans, because it involves invasive procedures like muscle biopsy (Stepto et al. 2009).

With this perspective in mind, Ferraresi et al. (2012) investigated the effects of LLLT (cluster with six diodes of 808 nm) on the physical performance of 10 young males undergoing physical strength training. Preliminary results showed that the LLLT group had upregulated the genes PPARGC1-α (mitochondrial biogenesis), mTOR (protein synthesis and muscle hypertrophy), and VEGF (angiogenesis). Furthermore, only the LLLT group downregulated MuRF1 (protein degradation and muscle atrophy) and IL-1β (inflammation) (Figure 53.6).

53.8 Possible Mechanisms of Action

53.8.1 Changes in Relationship between ADP, Pi, Mg²⁺, Ca²⁺, and pH

The accumulation of Pi, NAD, and Mg²⁺ in the sarcoplasm may decrease Ca²⁺ release from the sarcoplasmic reticulum, decrease

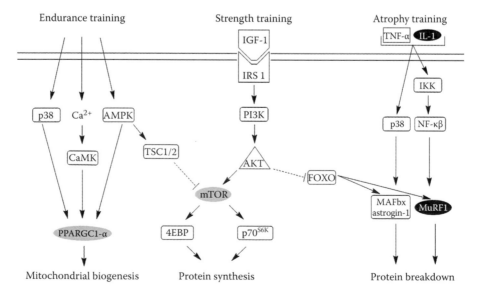

FIGURE 53.6 Mitochondrial biogenesis, protein synthesis, and protein breakdown signaling modulated by LLLT associated with exercise. Gray boxes = upregulated by LLLT and black boxes = downregulated by LLLT. (Based on Coffey, V. G. and J. A. Hawley, *Sports Med* 37:737–763, 2007; Glass, D. J., *Int J Biochem Cell Biol* 37:1974–1984, 2005; Hawley, J. A., *Appl Physiol Nutr Metab* 34:355–361, 2009.)

Ca^{2+} reuptake by Ca^{2+} pump, promote a loss of myofibril sensitivity to Ca^{2+}, and cause a decay in contractile force conducting to muscle fatigue (Allen, Lamb, and Westerblad 2008). During tetanic contractions for induction of early fatigue, Pi may reduce the bond between Ca^{2+} and troponin C (TnC), inhibiting or even reversing the Ca^{2+} pump in the sarcoplasmic reticulum, contributing to the higher concentration of Ca^{2+} into the sarcoplasm. In contrast, in the stages of late fatigue, Pi can inhibit the channels of Ca^{2+} release, or bind to Ca^{2+} (Pi-Ca^{2+}) in the sarcoplasmic reticulum, decreasing the available amount of Ca^{2+} to initiate muscle contraction (Allen, Lamb, and Westerblad 2008).

The decreased pH in the cell cytosol reduces the affinity of Ca^{2+} in muscle fibers, probably due to competition between the H^+ and Ca^{2+} to bind to TnC (Allen, Lamb, and Westerblad 2008). The LLLT and LEDT may have some modulating effect on the Ca^{2+} release from sarcoplasmic reticulum and the binding of Ca^{2+} to TnC through modulation of cellular energy metabolism via mitochondria, PCr resynthesis, and decreased enzyme activity of LDH.

The Ca^{2+} pump, in particular, may have its function improved by LLLT and LEDT, since it is dependent on ATP (Tonkonogi and Sahlin 2002) and/or LLLT and LEDT may facilitate the dissociation of Pi-Ca^{2+} and increase the availability of Ca^{2+} for muscle contraction. These mechanisms involving the Ca^{2+} ion may perhaps explain the action of LLLT in reducing muscle fatigue during tetanic contractions induced by neuromuscular electrical stimulation (de Almeida et al. 2011; Leal Junior et al. 2010a; Lopes-Martins et al. 2006) and the action of LLLT and LEDT on high-intensity exercise (Baroni et al. 2010a,b; de Almeida et al. 2012; de Brito Vieira et al. 2012; De Marchi et al. 2012; Ferraresi et al. 2011; Leal Junior et al. 2008, 2009b,c, 2010b).

53.8.2 Excitability of Muscle Fiber and Electromyography

Muscle contraction depends on the electrical excitation of the muscle fibers. When the muscle fiber is excited, the transmission of excitation begins with generation of an action potential that propagates rapidly through the sarcolemma and T-tubules. The rapid spread of the action potential ensures a more uniform muscle contraction. This contraction involves Ca^{2+} release that is directly linked to Na^+ channels (voltage-dependent) and K^+ (ATP-dependent) in the sarcolemma and T-tubules (Allen, Lamb, and Westerblad 2008).

Repeated muscle contraction has been associated with induction of fatigue because it promotes an increased Na^+ influx and K^+ efflux from muscle cells. This K^+ efflux occurs at every action potential, impairing the depolarization of the muscle fiber by an accumulation of this ion in the T-tubules and their cisterns. Consequently, there is a decrease in the force of muscle contraction. To avoid this imbalance between intracellular and extracellular Na^+ and K^+, the Na^+–K^+ pump works by pumping K^+ into the muscle cell (capturing the K^+ also in T-tubules) and pumping Na^+ to the outside, allowing new action potentials to propagate through muscle fibers and T-tubules to release Ca^{2+} and promote muscle contraction (Allen, Lamb, and Westerblad 2008).

LLLT and LEDT can possibly influence the excitation of the muscle fibers and reduce muscle fatigue by indirectly modulating the Na^+–K^+ pump, which is ATP-dependent. Mitochondria surrounding the T-tubules and Ca^{2+} cisterns can possibly increase their ATP synthesis as a result of LLLT and LEDT. Thus, a greater availability of ATP may improve the function of this pump and prevent an accumulation of extracellular K^+ that reduces muscle fatigue (Allen, Lamb, and Westerblad 2008; Nielsen and de Paoli 2007). Also, a higher conductance of the Cl^- channels in the T-tubules is important to reduce the accumulation of K^+ in these tubules and restore the excitability and membrane potential of the muscle fibers (Allen, Lamb, and Westerblad 2008). Perhaps the Cl^- channels are also modulated by LLLT and LEDT.

Action potentials that reach muscle fibers and cause their depolarization can be identified by electromyography (EMG) during muscle contraction in exercises (Allen, Lamb, and Westerblad 2008; Enoka 2012). This assessment tool may have an important role to elucidate possible effects of LLLT and LEDT on the neuromuscular junction, excitability of muscle fibers, and rate of activation and recruitment of motor units for inferring the magnitude of muscle fatigue.

53.9 New Perspectives

53.9.1 Duchenne Muscular Dystrophy

Duchenne muscular dystrophy (DMD) affects about 1 in 3500 male live births. The most common symptoms are abnormal gait, calf hypertrophy, and difficulty in getting up from the ground between 2 and 5 years of age (Manzur and Muntoni 2009). The clinical progression consists of general muscle weaknesses and the child becomes dependent on a wheelchair and experiences respiratory failure, heart disease, and death in early adulthood (Manzur and Muntoni 2009).

DMD is a lethal disease characterized by mutation in the dystrophin gene located on the Xp21 chromosome that leads to severe reduction or loss of the dystrophin protein, which is responsible for protecting the sarcolemma against mechanical stress of repeated muscle contractions (Manzur and Muntoni 2009; Markert et al. 2011). This protection consists of the indirect connection between the cell cytoskeleton (actin and intermediate myofilaments) with extracellular matrix, thus avoiding damage and degeneration of muscle fibers and allowing the transmission of the generated tension in the contractile process (Manzur and Muntoni 2009). The diagnosis is usually done through measurements of CK levels in blood serum that are 10 to 100 times higher in DMD children compared to children without DMD (Manzur and Muntoni 2009).

The clinical manifestation of DMD consists basically of destruction of the muscle cells and consequently results in muscle weakness and its consequent disabilities (Markert et al. 2011). In this process, there are five mechanisms that influence and/or worsen the clinical status of DMD patients (Markert et al. 2011): (1) mechanical weakness of the sarcolemma, (2) inappropriate Ca^{2+} influx, (3) aberrant molecular signaling, (4) increased oxidative stress, and (5) recurrent muscle ischemia. These five mechanisms are outlined below:

1. *Mechanical weakness of the sarcolemma.* The dystrophin protein is linked to glycoprotein to form a complex called dystrophin–glycoprotein. This complex has the function of stabilizing mechanically the sarcolemma during cycles of shortening and lengthening of muscle contractions as well as during isometric contractions (Markert et al. 2011).

2. *Inappropriate Ca²⁺ influx.* Ca²⁺ influx must be controlled in all cells in order to maintain homeostasis. An excess of intracellular Ca²⁺ can increase the activity of Ca²⁺-dependent proteases such as calpain and thus generate more damage and death of muscle cells (Markert et al. 2011). Also, high levels of intracellular Ca²⁺ stimulate production of ROS in the mitochondria, contributing to increased lipid peroxidation of the cell membranes as well as damage and death of muscle cells (Markert et al. 2011). This process creates more inflammation and ROS in the site of injury and adjacent tissues that contribute to increasing the area of the primary lesion, becoming a vicious cycle. Therapies that promote an appropriate Ca²⁺ influx and better mitochondrial function are fundamental to decreasing the deleterious effects of DMD (Markert et al. 2011).

3. *Aberrant molecular signaling.*

   (a) Interactions between muscle and immune system. The nuclear factor-kappa-β (NF-κβ) gene regulates the expression of proinflammatory genes and muscular atrophy. Upregulation of this gene promotes an increased inflammatory response with increased levels of cytokines and chemokines in the dystrophic area and an increased immune response mediated by leukocytes (Markert et al. 2011). The modulation of this gene seems to be important for DMD patients, controlling the degeneration and regeneration of muscle fibers (Markert et al. 2011).

   (b) Satellite cells. An absence or reduction of satellite cells can affect muscle regeneration (Markert et al. 2011). In the consequence of numerous cycles of lesion and muscle regeneration, DMD patients have a significant decrease in the satellite cell population over time (Abdel, Abdel-Meguid, and Korraa 2007; Markert et al. 2011). In this process, satellite cells of these patients have an abnormal level of proliferation and differentiation that leads to a fewer spare cells and viable cells for muscle regeneration (Markert et al. 2011). Consequently, there is a replacement of muscle tissue by fibrous or fat tissues that leads to losses of muscle function (Markert et al. 2011). Genes such as Wnt (Wingles int) and MyoD may be targets for therapies that modulate the expression of its mRNAs because Wnt inhibits muscle regeneration pathway and MyoD is responsible for the differentiation of satellite cells into myoblasts.

4. *Increased oxidative stress.* ROS and RNS can damage the cell sarcolemma by the lipid peroxidation process. It also promotes cell death and muscle inflammation, which generates higher amounts of ROS and RNS in the dystrophic muscle tissue (Markert et al. 2011).

5. *Recurrent muscle ischemia.* DMD patients do not have a normal level of nNOS gene expression. During physical activity, these patients do not have a good blood influx into the muscles by an imbalance between the adrenergic signs of local vasoconstriction and vasodilation mediated by NO, leading to ischemia, sarcolemma instability, and tissue damage (Markert et al. 2011).

Only a few studies have investigated the effects of LLLT or LEDT in DMD patients (Abdel, Abdel-Meguid, and Korraa 2007). Abdel, Abdel-Meguid, and Korraa (2007) evaluated the oxidative stress, lipid peroxidation, and apoptosis of 30 DMD patients compared to 20 patients without DMD (control). The results showed that DMD patients had greater MDA and protein carbonyls concentration in blood plasma (greater oxidative stress), greater cell apoptosis (higher gene expression for Bax), and minor amounts of nitric oxide also in blood plasma (Abdel, Abdel-Meguid, and Korraa 2007). After these comparisons, the authors irradiated with LLLT (632.8 nm, 2.5 J/cm², 10 mW) a portion of whole blood of DMD patients. The results showed that LLLT reduced oxidative stress (MDA and protein carbonyls), reduced apoptosis

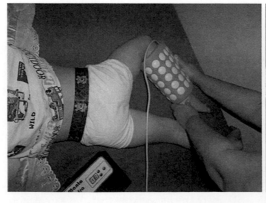 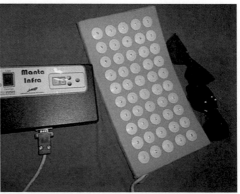

FIGURE 53.7 LEDT on a DMD patient. (Photos from authors' collection.)

(lower expression of BAX), and increased amounts of nitric oxide (Abdel, Abdel-Meguid, and Korraa 2007).

Considering this report, Ferraresi and Parizotto are conducting a case study report on the use of LEDT in DMD patients. These authors have observed clinical improvement of a DMD patient with three irradiations weekly of LEDT (array of 50 LEDs, 850 nm, total energy of 60 J, and total power of 5 W) on the trunk (ventral and dorsal), arms and forearms, and thighs and legs (Figure 53.7). There was an increase in daily activities, a decrease in blood levels of muscle CK, and lower incidence of pain and muscle cramps (unpublished data).

The mechanisms of action of LLLT and LEDT on muscular dystrophy are still uncertain and we have little scientific evidence. Possibly these therapies could help the treatment of this genetic disease by mechanisms reported previously: (1) modulations of oxidative stress (De Marchi et al. 2012), (2) Ca^{2+} influx, (3) inflammation (Ferraresi et al. 2012), and (4) molecular signaling. LLLT has been effective to upregulate genes related to protein synthesis (mTOR), mitochondrial biogenesis (PPARGC-1α), angiogenesis (VEGF), and genes related to proliferation and differentiation of satellite cells, and can downregulate genes related to inflammation (IL1-β and COX-2) and muscular atrophy (NF-κβ and MuRF1) (Ferraresi et al. 2011, 2012; Renno et al. 2011; Rizzi et al. 2006).

References

Aagaard, P. 2004. Making muscles "stronger": Exercise, nutrition, drugs. *J Musculoskelet Neuronal Interact* 4:165–174.

Abdel, S. E., I. Abdel-Meguid, and S. Korraa. 2007. Markers of oxidative stress and aging in Duchene muscular dystrophy patients and the possible ameliorating effect of He:Ne laser. *Acta Myol* 26:14–21.

Adhihetty, P. J., I. Irrcher, A. M. Joseph, V. Ljubicic, and D. A. Hood. 2003. Plasticity of skeletal muscle mitochondria in response to contractile activity. *Exp Physiol* 88:99–107.

Allen, D. G., G. D. Lamb, and H. Westerblad. 2008. Skeletal muscle fatigue: Cellular mechanisms. *Physiol Rev* 88:287–332.

Bakeeva, L. E., V. M. Manteifel, E. B. Rodichev, and T. I. Karu. 1993. [Formation of gigantic mitochondria in human blood lymphocytes under the effect of an He–Ne laser]. *Mol Biol (Mosk)* 27:608–617.

Baroni, B. M., E. C. Leal Junior, T. De Marchi et al. 2010a. Low level laser therapy before eccentric exercise reduces muscle damage markers in humans. *Eur J Appl Physiol* 110:789–796.

Baroni, B. M., E. C. Leal Junior, J. M. Geremia, F. Diefenthaeler, and M. A. Vaz. 2010b. Effect of light-emitting diodes therapy (LEDT) on knee extensor muscle fatigue. *Photomed Laser Surg* 28:653–658.

Ben-Dov, N., G. Shefer, A. Irintchev et al. 1999. Low-energy laser irradiation affects satellite cell proliferation and differentiation in vitro. *Biochim Biophys Acta* 1448:372–380.

Bibikova, A., and U. Oron. 1994. Attenuation of the process of muscle regeneration in the toad gastrocnemius muscle by low energy laser irradiation. *Lasers Surg Med* 14:355–361.

Bodine, S. C. 2006. mTOR signaling and the molecular adaptation to resistance exercise. *Med Sci Sports Exerc* 38:1950–1957.

Brooks, G. A., H. Dubouchaud, M. Brown, J. P. Sicurello, and C. E. Butz. 1999. Role of mitochondrial lactate dehydrogenase and lactate oxidation in the intracellular lactate shuttle. *Proc Natl Acad Sci U S A* 96:1129–1134.

Charge, S. B., and M. A. Rudnicki. 2004. Cellular and molecular regulation of muscle regeneration. *Physiol Rev* 84:209–238.

Chen, J. C., and D. J. Goldhamer. 2003. Skeletal muscle stem cells. *Reprod Biol Endocrinol* 1:101.

Chow, R. T., M. I. Johnson, R. A. Lopes-Martins, and J. M. Bjordal. 2009. Efficacy of low-level laser therapy in the management of neck pain: A systematic review and meta-analysis of randomised placebo or active-treatment controlled trials. *Lancet* 374:1897–1908.

Coffey, V. G., and J. A. Hawley. 2007. The molecular bases of training adaptation. *Sports Med* 37:737–763.

de Almeida, P., R. A. Lopes-Martins, T. De Marchi et al. 2012. Red (660 nm) and infrared (830 nm) low-level laser therapy in skeletal muscle fatigue in humans: What is better? *Lasers Med Sci* 27:453–458.

de Almeida, P., R. A. Lopes-Martins, S. S. Tomazoni et al. 2011. Low-level laser therapy improves skeletal muscle performance, decreases skeletal muscle damage and modulates mRNA expression of COX-1 and COX-2 in a dose-dependent manner. *Photochem Photobiol* 87:1159–1163.

de Brito Vieira, W. H., C. Ferraresi, S. E. de Andrade Perez, V. Baldissera, and N. A. Parizotto. 2012. Effects of low-level laser therapy (808 nm) on isokinetic muscle performance of young women submitted to endurance training: A randomized controlled clinical trial. *Lasers Med Sci* 27:497–504.

De Marchi, T., E. C. Leal Junior, C. Bortoli et al. 2012. Low-level laser therapy (LLLT) in human progressive-intensity running: Effects on exercise performance, skeletal muscle status and oxidative stress. *Lasers Med Sci* 27:231–236.

de Souza, T. O., D. A. Mesquita, R. A. Ferrari et al. 2011. Phototherapy with low-level laser affects the remodeling of types I and III collagen in skeletal muscle repair. *Lasers Med Sci* 26:803–814.

Dourado, D. M., S. Favero, R. Matias, T. Carvalho Pde, and M. A. da Cruz-Hofling. 2011. Low-level laser therapy promotes vascular endothelial growth factor receptor-1 expression in endothelial and nonendothelial cells of mice gastrocnemius exposed to snake venom. *Photochem Photobiol* 87:418–426.

Enoka, R. M. 2012. Muscle fatigue—From motor units to clinical symptoms. *J Biomech* 45:427–433.

Enwemeka, C. S. 2009. Intricacies of dose in laser phototherapy for tissue repair and pain relief. *Photomed Laser Surg* 27:387–393.

Favier, F. B., H. Benoit, and D. Freyssenet. 2008. Cellular and molecular events controlling skeletal muscle mass in response to altered use. *Pflugers Arch* 456:587–600.

Ferraresi, C., T. de Brito Oliveira, L. de Oliveira Zafalon et al. 2011. Effects of low level laser therapy (808 nm) on physical strength training in humans. *Lasers Med Sci* 26:349–358.

Ferraresi, C., R. Panepucci, R. Reiff et al. 2012. Molecular effects of low-level laser therapy (808 nm) on human muscle performance. *Phys Ther Sport* 13:e5.

Fluck, M. 2006. Functional, structural and molecular plasticity of mammalian skeletal muscle in response to exercise stimuli. *J Exp Biol* 209:2239–2248.

Fry, A. C. 2004. The role of resistance exercise intensity on muscle fibre adaptations. *Sports Med* 34:663–679.

Fulop, A. M., S. Dhimmer, J. R. Deluca et al. 2009. A meta-analysis of the efficacy of phototherapy in tissue repair. *Photomed Laser Surg* 27:695–702.

Glass, D. J. 2005. Skeletal muscle hypertrophy and atrophy signaling pathways. *Int J Biochem Cell Biol* 37:1974–1984.

Gorgey, A. S., A. N. Wadee, and N. N. Sobhi. 2008. The effect of low-level laser therapy on electrically induced muscle fatigue: A pilot study. *Photomed Laser Surg* 26:501–506.

Harridge, S. D. 2007. Plasticity of human skeletal muscle: Gene expression to in vivo function. *Exp Physiol* 92:783–797.

Hawke, T. J., and D. J. Garry. 2001. Myogenic satellite cells: Physiology to molecular biology. *J Appl Physiol* 91:534–551.

Hawley, J. A. 2009. Molecular responses to strength and endurance training: Are they incompatible? *Appl Physiol Nutr Metab* 34:355–361.

Hayworth, C. R., J. C. Rojas, E. Padilla et al. 2010. In vivo low-level light therapy increases cytochrome oxidase in skeletal muscle. *Photochem Photobiol* 86:673–680.

Hodson-Tole, E. F., and J. M. Wakeling. 2009. Motor unit recruitment for dynamic tasks: Current understanding and future directions. *J Comp Physiol B* 179:57–66.

Holterman, C. E., and M. A. Rudnicki. 2005. Molecular regulation of satellite cell function. *Semin Cell Dev Biol* 16: 575–584.

Huang, Y. Y., A. C. Chen, J. D. Carroll, and M. R. Hamblin. 2009. Biphasic dose response in low level light therapy. *Dose Response* 7:358–383.

Jarvinen, T. A., T. L. Jarvinen, M. Kaariainen, H. Kalimo, and M. Jarvinen. 2005. Muscle injuries: Biology and treatment. *Am J Sports Med* 33:745–764.

Karu, T. 1999. Primary and secondary mechanisms of action of visible to near-IR radiation on cells. *J Photochem Photobiol B* 49:1–17.

Karu, T. I. 2010. Multiple roles of cytochrome c oxidase in mammalian cells under action of red and IR-A radiation. *IUBMB Life* 62:607–610.

Karu, T. I., L. V. Pyatibrat, and N. I. Afanasyeva. 2005. Cellular effects of low power laser therapy can be mediated by nitric oxide. *Lasers Surg Med* 36:307–314.

Kuang, S., and M. A. Rudnicki. 2008. The emerging biology of satellite cells and their therapeutic potential. *Trends Mol Med* 14:82–91.

Lamb, G. D., and H. Westerblad. 2011. Acute effects of reactive oxygen and nitrogen species on the contractile function of skeletal muscle. *J Physiol* 589:2119–2127.

Leal Junior, E. C., V. de Godoi, J. L. Mancalossi et al. 2011. Comparison between cold water immersion therapy (CWIT) and light emitting diode therapy (LEDT) in short-term skeletal muscle recovery after high-intensity exercise in athletes-preliminary results. *Lasers Med Sci* 26:493–501.

Leal Junior, E. C., R. A. Lopes-Martins, B. M. Baroni et al. 2009a. Comparison between single-diode low-level laser therapy (LLLT) and LED multi-diode (cluster) therapy (LEDT) applications before high-intensity exercise. *Photomed Laser Surg* 27:617–623.

Leal Junior, E. C., R. A. Lopes-Martins, F. Dalan et al. 2008. Effect of 655-nm low-level laser therapy on exercise-induced skeletal muscle fatigue in humans. *Photomed Laser Surg* 26:419–424.

Leal Junior, E. C., R. A. Lopes-Martins, P. de Almeida et al. 2010a. Effect of low-level laser therapy (GaAs 904 nm) in skeletal muscle fatigue and biochemical markers of muscle damage in rats. *Eur J Appl Physiol* 108:1083–1088.

Leal Junior, E. C., R. A. Lopes-Martins, L. Frigo et al. 2010b. Effects of low-level laser therapy (LLLT) in the development of exercise-induced skeletal muscle fatigue and changes in biochemical markers related to postexercise recovery. *J Orthop Sports Phys Ther* 40:524–532.

Leal Junior, E. C., R. A. Lopes-Martins, R. P. Rossi et al. 2009b. Effect of cluster multi-diode light emitting diode therapy (LEDT) on exercise-induced skeletal muscle fatigue and skeletal muscle recovery in humans. *Lasers Surg Med* 41:572–577.

Leal Junior, E. C., R. A. Lopes-Martins, A. A. Vanin et al. 2009c. Effect of 830 nm low-level laser therapy in exercise-induced skeletal muscle fatigue in humans. *Lasers Med Sci* 24:425–431.

Liu, X. G., Y. J. Zhou, T. C. Liu, and J. Q. Yuan. 2009. Effects of low-level laser irradiation on rat skeletal muscle injury after eccentric exercise. *Photomed Laser Surg* 27:863–869.

Liu, Y., A. Schlumberger, K. Wirth, D. Schmidtbleicher, and J. M. Steinacker. 2003. Different effects on human skeletal myosin heavy chain isoform expression: Strength vs. combination training. *J Appl Physiol* 94:2282–2288.

Lopes-Martins, R. A., R. Albertini, P. S. Martins, J. M. Bjordal, and H. C. Faria Neto. 2005. Spontaneous effects of low-level laser therapy (650 nm) in acute inflammatory mouse pleurisy induced by carrageenan. *Photomed Laser Surg* 23:377–381.

Lopes-Martins, R. A., R. L. Marcos, P. S. Leonardo et al. 2006. Effect of low-level laser (Ga-Al-As 655 nm) on skeletal muscle fatigue induced by electrical stimulation in rats. *J Appl Physiol* 101:283–288.

Luo, L., Z. Sun, L. Zhang et al. 2013. Effects of low-level laser therapy on ROS homeostasis and expression of IGF-1 and TGF-beta1 in skeletal muscle during the repair process. *Lasers Med Sci* 28:725–734.

Manteifel, V., L. Bakeeva, and T. Karu. 1997. Ultrastructural changes in chondriome of human lymphocytes after irradiation with He–Ne laser: Appearance of giant mitochondria. *J Photochem Photobiol B* 38:25–30.

Manteifel, V. M., and T. I. Karu. 2005. Structure of mitochondria and activity of their respiratory chain in subsequent generations of yeast cells exposed to He–Ne laser light. *Izv Akad Nauk Ser Biol* 672–683.

Manzur, A. Y., and F. Muntoni. 2009. Diagnosis and new treatments in muscular dystrophies. *J Neurol Neurosurg Psychiatry* 80:706–714.

Markert, C. D., F. Ambrosio, J. A. Call, and R. W. Grange. 2011. Exercise and Duchenne muscular dystrophy: Toward evidence-based exercise prescription. *Muscle Nerve* 43:464–478.

Mauro, A. 1961. Satellite cell of skeletal muscle fibers. *J Biophys Biochem Cytol* 9:493–495.

Mesquita-Ferrari, R. A., M. D. Martins, J. A. Silva, Jr. et al. 2011. Effects of low-level laser therapy on expression of TNF-alpha and TGF-beta in skeletal muscle during the repair process. *Lasers Med Sci* 26:335–340.

Nakano, J., H. Kataoka, J. Sakamoto et al. 2009. Low-level laser irradiation promotes the recovery of atrophied gastrocnemius skeletal muscle in rats. *Exp Physiol* 94:1005–1015.

Nielsen, O. B., and F. V. de Paoli. 2007. Regulation of Na+-K+ homeostasis and excitability in contracting muscles: Implications for fatigue. *Appl Physiol Nutr Metab* 32:974–984.

Passarella, S., A. Ostuni, A. Atlante, and E. Quagliariello. 1988. Increase in the ADP/ATP exchange in rat liver mitochondria irradiated in vitro by helium-neon laser. *Biochem Biophys Res Commun* 156:978–986.

Petrella, J. K., J. S. Kim, D. L. Mayhew, J. M. Cross, and M. M. Bamman. 2008. Potent myofiber hypertrophy during resistance training in humans is associated with satellite cell-mediated myonuclear addition: A cluster analysis. *J Appl Physiol* 104:1736–1742.

Powers, S. K., and M. J. Jackson. 2008. Exercise-induced oxidative stress: Cellular mechanisms and impact on muscle force production. *Physiol Rev* 88:1243–1276.

Renno, A. C., R. L. Toma, S. M. Feitosa et al. 2011. Comparative effects of low-intensity pulsed ultrasound and low-level laser therapy on injured skeletal muscle. *Photomed Laser Surg* 29:5–10.

Rizzi, C. F., J. L. Mauriz, D. S. Freitas Correa et al. 2006. Effects of low-level laser therapy (LLLT) on the nuclear factor (NF)-kappaB signaling pathway in traumatized muscle. *Lasers Surg Med* 38:704–713.

Roth, D., and U. Oron. 1985. Repair mechanisms involved in muscle regeneration following partial excision of the rat gastrocnemius muscle. *Exp Cell Biol* 53:107–114.

Sahlin, K., M. Mogensen, M. Bagger, M. Fernstrom, and P. K. Pedersen. 2007. The potential for mitochondrial fat oxidation in human skeletal muscle influences whole body fat oxidation during low-intensity exercise. *Am J Physiol Endocrinol Metab* 292:E223–230.

Shefer, G., I. Barash, U. Oron, and O. Halevy. 2003. Low-energy laser irradiation enhances de novo protein synthesis via its effects on translation-regulatory proteins in skeletal muscle myoblasts. *Biochim Biophys Acta* 1593:131–139.

Shefer, G., T. A. Partridge, L. Heslop et al. 2002. Low-energy laser irradiation promotes the survival and cell cycle entry of skeletal muscle satellite cells. *J Cell Sci* 115:1461–1469.

Silveira, P. C., L. A. da Silva, D. B. Fraga et al. 2009. Evaluation of mitochondrial respiratory chain activity in muscle healing by low-level laser therapy. *J Photochem Photobiol B* 95:89–92.

Silveira, P. C., L. A. da Silva, C. A. Pinho et al. 2013. Effects of low-level laser therapy (GaAs) in an animal model of muscular damage induced by trauma. *Lasers Med Sci* 28:431–436.

Stepto, N. K., V. G. Coffey, A. L. Carey et al. 2009. Global gene expression in skeletal muscle from well-trained strength and endurance athletes. *Med Sci Sports Exerc* 41:546–565.

Sussai, D. A., T. Carvalho Pde, D. M. Dourado et al. 2010. Low-level laser therapy attenuates creatine kinase levels and apoptosis during forced swimming in rats. *Lasers Med Sci* 25:115–120.

Tonkonogi, M., and K. Sahlin. 2002. Physical exercise and mitochondrial function in human skeletal muscle. *Exerc Sport Sci Rev* 30:129–137.

Tonkonogi, M., B. Walsh, M. Svensson, and K. Sahlin. 2000. Mitochondrial function and antioxidative defence in human muscle: Effects of endurance training and oxidative stress. *J Physiol* 528 Pt 2:379–388.

Vieira, W., R. Goes, F. Costa et al. 2006. Adaptação enzimática da LDH em ratos submetidos a treinamento aeróbio em esteira e laser de baixa intensidade. *Revista Brasileira de Fisioterapia* 10:205–211.

Vierck, J., B. O'Reilly, K. Hossner et al. 2000. Satellite cell regulation following myotrauma caused by resistance exercise. *Cell Biol Int* 24:263–272.

Vinck, E., B. Cagnie, P. Coorevits, G. Vanderstraeten, and D. Cambier. 2006. Pain reduction by infrared light-emitting diode irradiation: A pilot study on experimentally induced delayed-onset muscle soreness in humans. *Lasers Med Sci* 21:11–18.

Vladimirov, Y. A., A. N. Osipov, and G. I. Klebanov. 2004. Photobiological principles of therapeutic applications of laser radiation. *Biochemistry (Mosc)* 69:81–90.

Weiss, N., and U. Oron. 1992. Enhancement of muscle regeneration in the rat gastrocnemius muscle by low energy laser irradiation. *Anat Embryol (Berl)* 186:497–503.

Westerblad, H., and D. G. Allen. 2011. Emerging roles of ROS/RNS in muscle function and fatigue. *Antioxid Redox Signal* 15:2487–2499.

Westerblad, H., J. D. Bruton, and A. Katz. 2010. Skeletal muscle: Energy metabolism, fiber types, fatigue and adaptability. *Exp Cell Res* 316:3093–3099.

Wilborn, C. D., L. W. Taylor, M. Greenwood, R. B. Kreider, and D. S. Willoughby. 2009. Effects of different intensities of resistance exercise on regulators of myogenesis. *J Strength Cond Res* 23:2179–2187.

Xu, X., X. Zhao, T. C. Liu, and H. Pan. 2008. Low-intensity laser irradiation improves the mitochondrial dysfunction of C2C12 induced by electrical stimulation. *Photomed Laser Surg* 26:197–202.

54

Low-Level Laser Therapy for Stroke and Brain Disease

Takahiro Ando
Massachusetts General Hospital
Keio University

Ying-Ying Huang
Massachusetts General Hospital
Harvard Medical School
Guangxi Medical University

Michael R. Hamblin
Massachusetts General Hospital
Harvard Medical School

54.1 Transmittance of Near-Infrared Laser for Brain Tissue

In the near-infrared (NIR) wavelength range (600–1000 nm), light absorption by the main chromophores such as water and hemoglobin in tissue is relatively low. For brain tissue, the highly transmitting and scattering characteristics of photons are often used in NIR spectroscopy (NIRS) and optical topography to gain information about the physiological processes (Kato et al. 1993; Maki et al. 1995; Villringer et al. 1993). Thus, the optical properties of human brain tissue for NIR light were well investigated in many studies (Bevilacqua et al. 1999; Choi et al. 2004; Stolik et al. 2007), which demonstrated that light can easily penetrate a few centimeters into the tissue even by transcutaneous irradiation. Figure 54.1 shows the structure of the brain. Lychagov et al. (2006) investigated the transmittance of 810 nm laser irradiation through *ex vivo* human skull and scalp. The value of transmittance varies from 0.5% to 5% in the case of a sample with scalp and from 1% to 16% in the case of a skull alone. Although the degree of photon absorption needed for low-level laser therapy (LLLT) is quite little as compared to that by the primal photoacceptors such as hemoglobin, melanin, and water, the NIR light in a brain tissue must be absorbed for photobiomodulation to occur. The laser fluence from 1 to 20 J/cm² is commonly used in LLLT for brain diseases, while the intensity generally varies from 5 to 50 mW/cm² depending on the actual light parameters and the therapeutic objects. On the assumption of the expected therapeutic power density at the dura mater, NIR laser irradiation with a few hundreds of milliwatts per square centimeter would be necessary for transcutaneous application of LLLT in a clinical situation. Meanwhile, the intensity should be controlled so that the irradiance on skin is well below the American National Standards Institute (ANSI) maximum permissible exposure (MPE) level. According to ANSI Z136.1, the skin exposure MPE for wavelengths ranging from 600 to 1000 nm is 200–800 mW/cm² (exposure duration, 10 to 3 × 10⁴ s). Figure 54.2 shows various brain diseases that have been treated with LLLT.

54.2 LLLT for Stroke

54.2.1 Stroke

Stroke is a cerebrovascular accident (CVA) in which blood flow to the brain is interrupted. Strokes can be classified into two main categories: ischemic and hemorrhagic. Ischemic strokes account for 85% of all strokes (Mayer 2003) and are commonly caused by the blockage of an artery in the brain due to a clot, thrombosis, embolism, or stenosis. Hemorrhagic strokes are the result of rupture of a cerebral artery, which is associated with spasm of the artery and intracerebral hemorrhage.

Stroke is the second most common cause of death worldwide and a major cause of disability (Donnan et al. 2011). In the United States, the disease is the third leading cause of death after heart disease and cancer, and is a major source of disability in persons older than age 60 years. Each year, approximately 795,000 people experience a new or recurrent stroke (Roger et al. 2011). Approximately 610,000 of these are first attacks and 185,000 are recurrent attacks (Roger et al. 2011). The mean expenses per

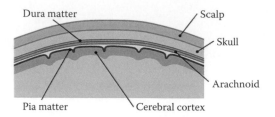

FIGURE 54.1 Schematic sagittal diagram of the vicinity of the brain surface. The brain is surrounded by scalp, skull, and three layers of protective membranes (meninges): dura mater, arachnoid mater, and pia mater.

person for stroke care in 2007 was estimated at $7657 (Roger et al. 2011). Internationally, millions of people have a new or recurrent stroke each year, and nearly a quarter of these people die.

For the treatment of stroke, there have been an enormous variety of agents and strategies that were justified by a pathophysiological rationale and clinically approved to date (Broussalis et al. 2012). Of the approximately 160 clinical trials of neuroprotection for ischemic stroke conducted as of late 2007 (Ginsberg 2008, 2009), only around 40 represent larger-phase completed trials and fully one-half of the latter utilized a window of treatment within 6 h, despite strong preclinical evidences that this delay exceeds the likely therapeutic window of efficacy in acute stroke. The conventional clinical trials of pharmacological and physical therapies for stroke are summarized in Figure 54.3.

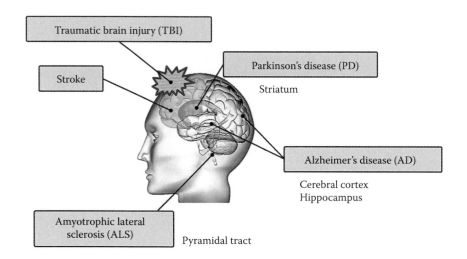

FIGURE 54.2 LLLT for various brain diseases. The illumination is applied for the targeted areas of the brain affected by a specific nervous system disease.

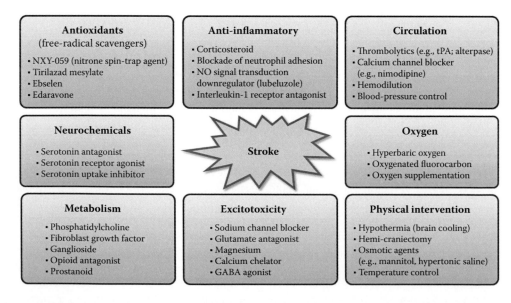

FIGURE 54.3 Conventional therapeutic strategies for stroke. The methodologies aim to restore cerebral flow and/or minimize the deleterious effects of ischemia on neurons (i.e., neuroprotection).

54.2.2 Animal Studies

Table 54.1 summarizes some of the published animal studies in a stroke model treated with LLLT. LLLT has been reported to induce angiogenesis (Kipshidze et al. 2001; Mirsky et al. 2002), modify growth factor signaling pathways (Zhang et al. 2003), and enhance protein synthesis (Hu et al. 2007; Karu 2008). In 2002, on the basis of the previous findings of LLLT, Leung et al. (2002) used a rat model of transient cerebral ischemia and showed that the direct irradiation of a 660 nm laser (spot area, 20 mm²; average power, 8.8 mW; pulse frequency, 10 kHz) for the cerebrum leads to a suppression of nitric oxide synthase (NOS) activity and an upregulation of transforming growth factor (TGF)-β1, which are considered neurotoxic and neuroprotective, respectively. In 2004, Lapchak, Wei, and Zivin (2004) showed that a transcranial irradiation of an 808 nm laser can treat the behavioral function deficits associated with embolic stroke in rabbits. In the study, LLLT was applied to the center of the parietal region of the skull at a power density of 7.5 or 25 mW/cm². The NIR laser irradiation within 6 h, but not 24 h, after the embolization significantly improved neurologic function at 3 weeks after stroke without unwanted side effects such as an extreme increase in both skin and brain temperature. Based on this study, a laser intensity of 7.5 mW/cm² and a wavelength of 808 nm were used in the following studies; the long-term safety of the transcranial application was confirmed by McCarthy et al. (2010). Moreover, Lapchak et al. (2007) reported in their further research that LLLT improved the behavioral performance following embolic strokes in rabbits when administered either as a continuous wave (CW) or

a pulsed wave (PW) (Lampl 2007). In 2006, De Taboada et al. (2006) demonstrated that 808 nm LLLT transcranially applied at 24 h poststroke effectively improves neurological function in rats following acute stroke. A remarkable improvement in neurological deficits was evident at 14, 21, and 28 days after stroke as compared to the nontreated rats. Another finding of this study was that there were no major differences in therapeutic effects regarding the location of laser irradiation: ipsilateral, contralateral, or bilateral sides of the stroke. Meanwhile, Oron et al. (2006) reported that transcranial application of LLLT at a delay of 24 h, but not at 4 h, after induction of acute stroke in rats significantly improves the neurological function. The number of newly formed neuronal cells was significantly elevated in the subventricular zone (SVZ), and thus the authors proposed that an underlying mechanism for the functional benefit after LLLT would be a possible induction of neurogenesis within 4 to 24 h poststroke. As the neuroprotective combination therapy of LLLT and therapeutic agents for stroke, Lapchak et al. (Lapchak et al. 2008) showed in 2008 that transcranial LLLT for embolic stroke model in rabbits did not affect increase in hemorrhage incidence induced by tissue plasminogen activator (tPA; Alteplase), and demonstrated that LLLT can be used either alone or in combination with tPA for curing patients with stroke.

Increased mitochondrial function in neural cells irradiated with an NIR laser is considered as one of the major mechanisms involved with the beneficial behavioral effects in LLLT for stroke. In 2010, Lapchak and De Taboarda (2010) evaluated the effect of 808 nm LLLT on cortical adenosine-5′-triphosphate (ATP) content using a rabbit small clot embolic stroke model.

TABLE 54.1 Animal Studies on LLLT for Stroke

Subject	Condition	Parameters	Effects	References
Rats	Transient cerebral ischemia	660 nm; 8.8 mW; 20 mm² at 5 mm distance; 44 mW/cm²; PW 10 kHz; 1, 5, or 10 min	Suppression of NOS activity and upregulation of TGF-β1 were observed.	Leung et al. 2002
Rabbits	Small clot embolic stroke	808 ± 5 nm; 7.5 or 25 mW/cm²; CW; 2 cm diameter laser probe; 10 min	LLLT within 6 h, but not 24 h, after the embolization significantly improved the neurologic function at 3 weeks poststroke.	Lapchak, Wei, and Zivin 2004
Rats	Acute stroke [middle cerebral artery occlusion (MCAO)]	808 nm; 7.5 mW/cm²; 0.9 J/cm² at cortical surface; 4 mm diameter fiber optic; 2 min	Significant improvement in the neurological deficits at 14, 21, and 28 days.	De Taboada et al. 2006
Rabbits	Small clot embolic stroke	808 ± 5nm; 7.5 mW/cm²; CW; 300 µs pulse at 1 kHz; 2.2 ms at 100 Hz; 0.9–1.2 J delivered to brain; 2 cm diameter laser probe; 2 min	LLLT administered 6 h following stroke in PW mode resulted in significant improvement of motor function.	Lapchak et al. 2007
Rats	Acute stroke (MCAO or filament insertion)	808 nm; 7.5 mW/cm²; 0.9 or 3.6 J/cm² at cortical surface; CW; PW 70 Hz; 4 mm diameter fiber optic	Improvement of neurological function when applied 24 h poststroke. The number of newly formed neuronal cells was significantly elevated in the SVZ.	Oron et al. 2006
Rabbits	Small clot embolic stroke	808 ± 5 nm; 7.5, 37.5 and 262.5 mW/cm²; CW; PW 100 Hz; 0.9, 4.5 and 31.5 J/cm² at cortical surface; 2 cm diameter laser probe; 2 min	CW and 100 Hz PW resulted in 41% and 157% increase in cortical ATP above sham-treated rabbits, respectively.	Lapchak and De Taboada 2010
Mice	Transient bilateral common carotid artery occlusion (BCCAO)	808 nm; 0.8, 1.6 and 3.2 W/cm²; CW; 3 mm diameter exposure; 45 min	Targeted increase of CBF without extreme increase in tissue temperature.	Uozumi et al. 2010

Note: Laser parameters are given in the following order: wavelength (nm), power (mW), power density (mW/cm²), energy (J), energy density (J/cm²), mode (CW or PW), spot size (cm²), irradiation time (s). The parameters are partially unavailable in several cases.

At 5 min after embolization, the rabbits were exposed to 2 min of transcranial LLLT with a CW or PW at 100 Hz on the skin surface posterior to bregma at midline. Measurement of the ATP content in the cerebral cortex was performed at 3 h after embolization. LLLT using CW (laser intensity, 7.5 mW/cm^2; total fluence, 0.9 J/cm^2) resulted in a 41% increase in cortical ATP, while use of 100 Hz PW (laser intensity, 37.5 mW/cm^2; total fluence, 4.5 J/cm^2) resulted in a 157% increase in cortical ATP above sham-treated embolized rabbits. The authors mentioned that greater improvement in functional scores might be achieved by optimizing the length of treatment and mode of laser irradiation for a clinical application (Lapchak 2012).

In 2010, Uozumi et al. (2010) examined the effect of 808 nm transcranial LLLT on cerebral blood flow (CBF) and directly measured nitric oxide (NO) in brain tissue during the NIR laser irradiation using transient bilateral common carotid artery occlusion (BCCAO) in mice. The results suggest that targeted increase in CBF is achieved by a transcranial laser irradiation (1.6 W/cm^2 for 45 min) without extreme increase of tissue temperature and is concerned with NOS activity and NO concentration, demonstrating that NIR LLLT can protect brain tissue from transient ischemia.

54.2.3 Clinical Studies

54.2.3.1 Acute Stroke

Transcranial LLLT has been shown to significantly improve the outcome in acute human stroke patients in several studies (Lampl et al. 2007; Stemer, Huisa, and Zivin 2010; Zivin et al. 2009). In 2007, Lampl et al. (2007) examined the safety and effectiveness of NIR laser therapy for treatment of patients within 24 h of ischemic stroke onset. The 120 enrolled patients involved in the trials were between ages 40 and 85 and were diagnosed with acute ischemic stroke within 24 h of the onset and whose measurable neurological severities measured by National Institutes of Health Stroke Scale (NIHSS) scores ranged from 7 to 22. After removing hair from the patient's scalp, transcranial irradiation of an 808 nm laser was applied to 20 predetermined locations on the scalp for 2 min at each site. The irradiation system was designed to deliver approximately 1 J/cm^2 of laser energy over the entire surface of the cortex regardless of stroke location. Seventy-nine patients received LLLT at a median of 16 h from stroke onset, resulting in 70% of the patients achieving a positive NIHSS outcome as compared with 51% of the control group. This study first showed the safety and effectiveness of NIR LLLT for the treatment of ischemic stroke in humans when initiated within 24 h of acute ischemic stroke, and thus the second larger trial was warranted.

The second clinical trial was nearly identical to the first trial as above but was larger and included patients 40 to 90 years of age (Zivin et al. 2009). The enrolled 660 acute stroke patients were randomly assigned to receive real or sham treatments of transcranial LLLT; two people were excluded by the declination before the following tests. Similar significant beneficial results were observed for the patients who had a moderate to a moderately severe ischemic stroke ($n = 434$) and received LLLT, but not for the patients who had a severe stroke. Therefore, the therapeutic effect of this trial was inadequate to meet conventional levels of formal statistical significance for the efficacy. However, when the data of the 778 patients in the two aforementioned studies were pooled, a much greater significant beneficial effect was seen for the transcranial LLLT group ($P = 0.003$) as compared to those who received the sham laser treatment (Stemer, Huisa and Zivin 2010). The investigators stated in their reports that a third definitive trial of transcranial LLLT for acute stroke with refined baseline NIHSS exclusion criteria was now planned (Lapchak 2010). Simultaneously, they planned to test the combination of LLLT and tPA (tissue plasminogen activator) for stroke as a separate trial on the basis of their previous animal study (Lapchak et al. 2008).

54.2.3.2 Chronic Stroke

LLLT may also promote chronic neuronal function restoration via trophic factor-mediated plasticity changes or possibly neurogenesis (Lapchak 2012). However, no studies today have been conducted to examine transcranial LLLT treatment of chronic stroke patients. Naeser et al. (1995) studied the application of NIR laser acupuncture (Whittaker 2004) (instead of needles) to stimulate acupuncture points on the body in chronic stroke patients with paralysis. Seven stroke patients ranging in age from 48 to 71 years, five of whom had a single left hemisphere stroke and two of whom had a single right hemisphere stroke, received laser acupuncture treatments beginning at 10 months to 6.5 years poststroke ($n = 6$); or at 1 month poststroke ($n = 1$). A 20 mW gallium–aluminum–arsenide (Ga–Al–As) CW laser (780 nm) was used directly on acupuncture points on the arm, leg, hand, and/or face for 20 or 40 seconds per point. All patients with improvement had lesion on computed tomography (CT) scan in <50% of the motor pathway areas (mild-moderate paralysis), while those with no improvement had lesion in >50% of the motor pathway areas (severe paralysis). Therefore, chronic stroke patients with paralysis improved when the paralysis was not severe, although a reduction in spasticity has been observed in severe cases (Naeser and Hamblin 2011), which were similar to the results of the studies with needle acupuncture (Naeser et al. 1994a,b). These findings are intriguing and suggest that some recovery of motor function can occur with needle or laser acupuncture applied to body acupuncture points in chronic stroke patients as well as conventional transcutaneous electrical nerve stimulation (TENS) for stroke (Ng and Hui-Chan 2007).

54.3 LLLT for Traumatic Brain Injury

54.3.1 Traumatic Brain Injury

Traumatic brain injury (TBI) occurs when an outside force impacts the head strongly enough to cause the brain to move within the skull, or if the force causes the skull to break and directly hurts the brain. Neuronal degeneration following TBI is believed to evolve in a biphasic manner consisting of the primary mechanical insult (primary injury) and a progressive secondary necrosis (secondary injury) (Siesjö et al. 1995). The cellular and physiologic disturbances include inflammation, oxidative stress, ionic imbalance, increased vascular permeability, mitochondrial dysfunction, and excitotoxic damage, which are shown in Figure 54.4. These processes result in brain edema, increased intracranial pressure (ICP), and impaired cerebral perfusion (Nortje and Menon 2004).

Each year in the United States, TBI and its sequelae affect at least 1.4 million people with an associated mortality of 50,000 and permanent disability of 90,000 (Langlois, Rutland-Brown, and Thomas 2005; Thurman et al. 1999). The long-term cognitive, psychosocial, and physical deficits that follow TBI lead to an annual economic burden of about 60 billion dollars (Faul et al. 2007; Finkelstein, Corso, and Miller 2006; Walker et al. 2009). However, these numbers markedly underestimate the incidence and costs. According to the U.S. Center for Disease Control and Prevention data (Faul et al. 2010), sports-related injuries are not listed among the top categories, but some have estimated the incidence of such head injuries at 1.6–3.8 million per year (Langlois, Rutland-Brown, and Wald 2006). Globally, the incidence of TBI is also increasing, particularly in developing countries (Maas, Stocchetti, and Bullock 2008).

The most frequent psychiatric manifestation seen following TBI is depression, which may affect one-third or more of patients (Fann, Hart, and Schomer 2009), and these patients usually need a variety of long-term services, including rehabilitative therapies, assistive technologies, and expensive medical equipment. Although a growing number of studies have been carried out for therapy of TBI, there are no standard, empirically validated interventions and clinical trials of pharmaceuticals have yet to be established.

54.3.2 Animal Studies

Since the pathophysiology of cerebral ischemia and trauma is rather similar (Leker and Shohami 2002), the success of transcranial LLLT for stroke encouraged researchers to investigate the therapeutic effects in animal models of TBI. Table 54.2 summarizes some of the published animal studies in the TBI model treated with NIR LLLT. In 2007, Oron et al. (2007) first evaluated the effects of LLLT for acute TBI in mice. Mice were subjected to closed-head injury (CHI) by using a weight-drop procedure, and at 4 h after CHI either sham or 808 nm LLLT was transcranially administered (1.2–2.4 J/cm² over 2 min irradiation with 10–20 mW/cm²). Laser was applied by placing a distal tip of a fiber optics on a point in the midline of the skull (sagittal suture) located 4 mm caudal to the coronal suture of the skull after skin had been removed in that region by small longitudinal incision. Neurobehavioral function was assessed by neurological severity score (NSS). At 5 days after TBI, the motor behavior was significantly better in the laser-treated group. At 28 days post-CHI, the brain tissue volume was examined. The mean lesion size of the laser-treated group (1.4%) was significantly smaller than that of the control group (12.1%). The researchers suggested various possible mechanisms, including an increase in ATP, total antioxidants, angiogenesis, neurogenesis, heat shock proteins content, and an antiapoptotic effect, which were similar observations reported after LLLT for ischemic heart skeletal muscles (Avni et al. 2005; Oron et al. 2001a,b; Shefer et al. 2002; Yaakobi et al. 2001).

In 2009, Moreira et al. (2009) analyzed the effect of NIR laser irradiation on local and systemic immunomodulation following cryogenic brain injury in rats. The cryogenic trauma was made on the top of the dura mater, and the injured site was irradiated

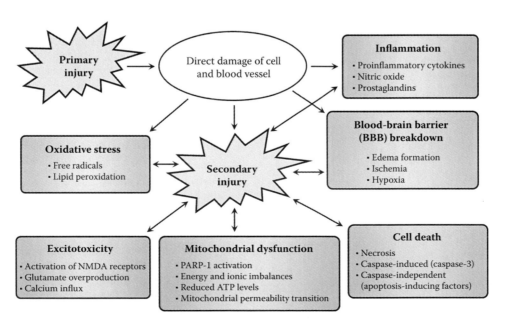

FIGURE 54.4 Processes caused in TBI. The cellular and physiologic disturbances include inflammation, oxidative stress, ionic imbalance, increased vascular permeability, mitochondrial dysfunction, and excitotoxic damage.

TABLE 54.2   Animal Studies on LLLT for TBI

Subject	Condition	Parameters	Effects	References
Mice	CHI	808 nm; 200 mW; 10–20 mW/cm²; CW; 1.2–2.4 J/cm²; 1.2 cm beam diameter; 2 min	Motor behavior was significantly better in the LLLT group at 5 days post-CHI, and the mean lesion size of the LLLT group (1.4%) was significantly smaller than that of the control group (12.1%) at 28 days after TBI.	Oron et al. 2007
Rats	Cryogenic brain injury	780 or 660 nm; 40 mW; CW; 3 or 5 J/cm²; 0.24 or 0.40 J in total; 0.042 cm²; 3 or 5 s (twice in 3 h interval)	LLLT can affect TNF-α, IL-1β, and IL-6 levels in the brain and in circulation within the first 24 h following cryogenic trauma.	Moreira et al. 2009
Rats	Cryogenic brain injury	780 nm; 40 mW; CW; 3 J/cm²; 0.24 J (0.12 J × 2 points); 0.04 cm²; 3 s each point	LLLT can prevent neuron death and severe astrogliosis.	Moreira et al. 2011
Mice	CHI	665, 810, or 980 nm; CW; 36 J/cm²; 150 mW/cm²; 4 min	Both 665 and 810 nm LLLT were highly effective for improving the neurologic performances of the mice during the succeeding 4 weeks, while 980 nm LLLT did not produce the same positive effects.	Wu et al. 2010
Mice	CHI	665, 730, 810, or 980 nm; CW; 36 J/cm²; 150 mW/cm²; 1 cm diameter spot; 4 min	TBI mice treated with 665 and 810 nm laser (but not with 730 or 980 nm) had a significant neurological outcome.	Wu et al. 2012
Mice	Controlled cortical impact (CCI)	810 nm; 50 mW/cm² at cortical surface; CW, PW 10 or 100 Hz; 36 J/cm²; 1 cm diameter spot; 12 min	LLLT with PW 10 Hz was the most effective for neurological improvement, although the other laser regimens (CW and PW 100 Hz) were also effective as compared to the sham-treated group.	Ando et al. 2011
Mice	CHI	808 nm; 10 mW/cm²; CW, PW 100 or 600 Hz; 1.2 J/cm² at cortical surface; 1.2 cm beam diameter; 2 min	The mice received LLLT with PW 100 Hz at 4 h posttrauma had the highest percentage (63%) of full recovery at 56 days post-TBI.	Oron et al. 2012
Mice	CCI	800 nm; 250, 500, or 1000 mW/cm²; CW; 0, 30, 60, 105, 120, or 210 J/cm²; 1.32 cm²; 2 or 7 min	The mice with CCI treated with 60 J/cm² (500 mW/cm² × 2 min) had significant improvement of the cognitive deficits and inhibition of microglial activation, either via an open craniotomy or transcranially.	Khuman et al. 2012
Mice	CCI	810 nm; 25 mW/cm² at cortical surface; CW; 18 J/cm²; 1 cm diameter spot; 12 min (either once, 3 daily, or 14 daily treatments)	Single or three times daily LLLT remarkably improved the neurological outcomes of TBI mice, although 14 daily treatments provided no benefit.	Xuan et al. 2012

Note: Laser parameters are given in the following order: wavelength (nm), power (mW), power density (mW/cm²), energy (J), energy density (J/cm²), mode (CW or PW), spot size (cm²), and irradiation time (s). The parameters are partially unavailable in several cases.

immediately and at 3 h after injury with CW diode lasers (wavelengths, 780 or 660 nm; total fluences, 3 or 5 J/cm²). The rats were sacrificed either at 6 or 24 h postlesion, and then the brain and blood samples were analyzed by an enzyme immunoassay technique (ELISA) test. The study concluded that LLLT can affect TNF-α, interleukin (IL)-1β, and IL-6 levels in the brain and in circulation within the first 24 h following cryogenic brain injury and thereby prevent cell death after TBI. Moreover, Moreira et al. (2011) reported in their further research in 2011 that 780 nm LLLT exerts its effect in wound healing following cryogenic injury to prevent neuron death and severe astrogliosis.

In 2010, Wu et al. (2010) investigated the effect of different wavelengths of transcranial LLLT for closed-head TBI in mice. TBI was produced by using a calibrated weight-drop apparatus. A neurologic functional score for each mouse was determined based on 10 standardized performance tests (involving beam balancing and maze exiting) administered at specified times. Mice were given a single treatment to the top of the head with 36 J/cm² (150 mW/cm² over 4 min) of a 665, 810, or 980 nm laser at 4 h after trauma. Both 665 and 810 nm lasers were highly effective for improving the neurologic performances of the mice during the succeeding 4 weeks, while 980 nm laser did not produce the same positive effects. Moreover, Wu et al. (2012)

reported in their subsequent research that TBI mice treated with 665 and 810 nm laser (but not with 730 or 980 nm) had a significant neurological improvement as compared to sham-treated controls. Morphometry of brain sections showed a reduction of deficits in 665 and 810 nm laser treated mouse brains at 28 days. The authors stated that the differences in the outcome can be explained by the absorption spectrum of cytochrome *c* oxidase (CCO), which is one of the candidate mitochondrial chromophores in transcranial LLLT and has distinct absorption bands around the wavelengths of 670 and 830 nm (Karu et al. 2005; Wong-Riley et al. 2005).

To examine the effects of various NIR laser modes (CW and PW) on LLLT outcome for TBI, Ando et al. (2011) compared the therapeutic effects using an 810 nm laser with a continuous mode and either a pulsed mode at 10 or at 100 Hz with 50% duty cycle. Mice were subjected to a single left lateral controlled cortical impact (CCI) with craniotomy. At 4 h after contusion, LLLT was applied to the top of the head using a 1 cm diameter spot with an average power density of 50 mW/cm² for 12 min, which corresponded to a total fluence of 36 J/cm². The laser irradiation pulsed at 10 Hz was the most effective, which was judged by improvement in NSS and body weight, although the other laser regimens (CW and PW 100 Hz) were also effective as compared to the sham-treated group. The brain lesion volume of mice treated with 10 Hz pulsed-laser irradiation was significantly lower than control group at 15 days and 4 weeks post-TBI. Additionally, they found an antidepressant effect of LLLT at 4 weeks as shown by forced swim and tail suspension tests, which have been used for the examination of depression and anxiety. The authors hypothesized that a possible reason why laser irradiation at 10 Hz PW mode was most effective for improving neurological outcome is that the frequency may affect the whole brain. Since the brain has waves with specific frequencies, such as alpha waves, beta waves, and theta waves, there exists the possibility of resonance occurring between the frequency of the pulsed light and that of the brain waves. Particularly relevant is the fact that oscillation of theta waves that have a prominent 4–10 Hz rhythm in the hippocampal region of all mammals (Green and Arduini 1954; Vanderwolf 1969; Winson 1972) is responsible for behavioral inhibition and attention, spatial memory, and navigation.

The study conducted by Oron et al. (2012) examined the long-term effect of LLLT with various laser modes (CW, PW 100, and 600 Hz) and at different treatment time points in mild to moderate CHI mice induced using a weight-drop device. An 808 nm laser irradiation was delivered for 2 min at a power density of 10 mW/cm^2, so that the fluence at the cortical surface of the cerebrum was 1.2 J/cm^2. The transcranial LLLT was employed at 4, 6, or 8 h posttrauma. The differences in NSS of the laser treated group at 6 and 8 h after TBI were 3.3- and 1.8-fold bigger than those of control nontreated group at day 56, respectively. The mice that received transcranial LLLT with PW 100 Hz at 4 h posttrauma had the highest percentage (63%) of full recovery (NSS = 0) at day 56. These data suggest that LLLT for TBI mice provides a significant long-term functional neurological benefit, and that the pulsed laser mode at 100 Hz is the preferred mode for such treatment.

In 2012, Khuman et al. (2012) proved that NIR LLLT improved cognitive deficits in CCI mice. Using an 800 nm laser device mounted 1 cm above the head of the mice, LLLT was applied directly to the contused parenchyma or transcranially in mice either at 4 h postcontusion or beginning 60–80 min after TBI. The total fluences were determined at either 0, 30, 60, 105, 120, or 210 J/cm^2. Cognitive function by Morris water maze (MWM), motor function by wire grip test, brain edema, lesion volume, and nitrosative stress by nitrotyrosine ELISA were assessed. The results showed that mice with CCI treated with 60 J/cm^2 (500 mW/cm^2 × 2 min) had significant improvement of the latency to the hidden platform and probe trails either via an open craniotomy or transcranially. An anti-inflammatory effect was noted via a significant reduction of microgliosis at 48 h with 60 J/cm^2. There was no significant difference on motor function (days 1 to 7), brain edema (24 h), nitrosative stress (24 h), or lesion volume (14 days) between the LLLT and control groups. They concluded that, although further dose optimization and mechanism studies are needed, the data suggest that LLLT might be a therapeutic option to improve cognitive recovery and limit inflammation after TBI.

More recently, Xuan et al. (2012) studied the effects of different treatment repetitions of 810 nm LLLT on neurobehavioral

and vestibulomotor functioning, histomorphological analysis, and histological evidence of neuroprotection and neurogenesis in CCI mice. The animals of the TBI treatment groups received transcranial LLLT (CW laser, intensity 25 mW/cm^2; total fluence 18 J/cm^2) either once at 4 h post-TBI, 3 daily treatments, or 14 daily treatments. They found that LLLT has beneficial effects in the acute treatment of TBI and demonstrated that TBI mice treated with single laser irradiation or three daily irradiations had significant improvements in NSS and wire-grip and motion test (WGMT). However, 14 daily laser treatments provided no benefit. Furthermore, the study reported LLLT for TBI in mice could significantly improve neural function, decrease lesion volume, augment cell proliferation, and even protect the brain against neuronal damage to some degree.

LLLT for TBI in mice remarkably improves neurological function and decreases a volume of trauma, where the therapeutic efficacy highly depends on the irradiation parameters. The augmented cell proliferation and protective effects for a brain tissue against neuronal damage were also observed in recent studies (Hashmi et al. 2010). To clarify the mechanisms of the therapeutic processes in LLLT for TBI, further studies will concentrate on the effects of LLLT on the hippocampus and SVZ in a TBI model, as these are considered to be the principal source of the neuroprogenitor cells. In addition, various examinations using another rodent will be performed in the near future.

54.3.3 Clinical Studies

In 2011, Naeser et al. (2011) for the first time reported the improved cognitive function after transcranial LLLT in two chronic TBI patients. The first patient (female) sustained closed-head TBI in a motor vehicle accident at the age of 59. At 7 years postinjury, she received her first transcranial LLLT using a red/NIR light-emitting diode (LED) device that had three square-shaped LED cluster heads, each with a dimension of 4.4 × 4.4 cm. The treatment area was 19.39 cm^2; each cluster head contained 49 diodes (40 NIR 870 nm diodes, 12.25 mW each, and 9 red 633 nm diodes, 1 mW each). Total power was 500 mW (±20%). The treatment was applied to her left and right forehead, and the treatment parameters were gradually increased from 8 J/cm^2 (310 s) per area to 20 J/cm^2 (774 s). After 8 weeks, she was able to work on a computer from 20 min to 3 h at a time. The patient continues to perform the nightly LED treatment herself at home (around 5 years); however, if she stops the treatment for more than 2 weeks, she regresses. The second patient (52-year-old woman) was a military officer who had a history of closed-head trauma (sports, military, and recent fall), and the magnetic resonance imaging (MRI) of her head showed frontoparietal atrophy. She was medically disabled for 5 months before transcranial LLLT. The laser treatment was applied bilaterally and to midline sagittal areas using a similar LED device that had one (or three), circular-shaped LED cluster heads, each with a diameter of 5.35 cm. The treatment area was 22.48 cm^2; each cluster head contained 61 diodes (52 NIR 870 nm diodes, and 9 red 633 nm diodes, 12–15 mW each). Total power was 500 mW (±20%). The duration of treatment for each area was gradually increased over 4 weeks from 7 min (9.3 J/cm^2 at scalp) to 10 min (13.3 J/cm^2 at scalp). After 4 months of nightly LLLT at home, her medical disability was discontinued, and then she returned to work full-time as an executive consultant at an international technology consulting firm. Neuropsychological testing performed after 9 months of LLLT showed significant improvement in executive function (inhibition itself, inhibition accuracy) and memory, as well as reduction in posttraumatic stress disorder (PTSD).

Schiffer et al. (2009) also reported a reduction in PTSD symptoms by treatment with transcranial LED. They gave one 8 min treatment (4 min at each of two sites on the forehead) with an 810 nm LED array to 10 patients with major depression, including seven with a history of substance abuse (six with a past history of opiate abuse and one with a past history of alcoholism), and nine with an anxiety disorder, including three with PTSD. The irradiance was 250 mW/cm^2 at 4 mm above the skin. The psychological state of patients was evaluated by various standard diagnostic tests. This study suggests that NIR LLLT may have utility for the treatment of depression and other psychiatric disorders.

In 2012, another case study using single-photon emission CT with N-isopropyl-[123I] p-iodoamphetamine (IMP-SPECT) to detect ischemic areas was reported by Nawashiro et al. (2012). A 40-year-old patient (man) in a persistent vegetative state following severe head injury was treated with an LED device from 228 days posttrauma. The LED device had a 78 × 72 mm square-shaped head with 23 diodes (peak wavelength 850 nm) and the total power was 299 mW. The device was applied at a distance of 5 mm away from the skin surface. The estimated beam area per diode on the skin was 1.14 cm^2 and the power density was 11.4 mW/cm^2 (energy density, 20.5 J/cm^2). The patient received transcranial LED treatments with placement of the LED device on the left and right forehead areas superior to the eyebrows for 30 min, two times daily, for 73 days. IMP-SPECT showed unilateral, left anterior frontal lobe focal increase of 20%, compared to the pretreatment value for regional CBF (rCBF) for this area. The patient showed some improvement in his neurological condition by moving his left arm/hand to reach the tracheostomy tube after LLLT. At 5 days after beginning the LED treatments, the patient began to move his left arm and hand to reach the site of tracheostomy when suctioning was performed, although he had never shown any spontaneous movement prior to LLLT. After LED treatment for 2 months, which ended 300 days postinjury, he always showed spontaneous movement, but did not obey commands. The authors observed no further improvement in his neurological condition after following LLLT.

Larger clinical studies are needed to gather a much larger cohort of patients to determine the factors that influence TBI treatment response based on level of severity, area of brain affected, and altered function, in order to better quantify how transcranial LLLT affects persons with TBI. Quantitative measures may include MRI, IMP-SPECT, and neuropsychiatric testing before and after treatment. Consistency among authors for quantifying the extent and location of brain injury or changes in CBF will be needed.

54.4 LLLT for Parkinson's Disease

Parkinson's disease (PD) is a progressive neurological disorder that results from degeneration of neurons and affects the control of body movements. Noninvasive transcranial LLLT has received much attention as one of the most promising methods to treat such neurodegeneration in the brain (Lapchak 2012).

Komel'kova et al. (2004) and Vitreshchak et al. (2003) reported the effects of He–Ne laser irradiation (wavelength, 632.8 nm; power density, 500 mW/cm^2) on enzyme activity of monoamine oxidase B (MAO B), Cu/Zn-superoxide dismutase (Cu/Zn-SOD), Mn-SOD, and catalase in blood from PD patients. In *in vitro* study, they used the blood samples of 10 PD patients with high ($n = 6$) or low ($n = 4$) MAO B and Cu/Zn-SOD activities. The effect of *in vivo* LLLT on neurological status and activities of the above enzymes was examined in 70 PD patients. LLLT consisted of five daily 20 min sessions of blood irradiation (intravenous irradiation *in vivo*), and the blood was analyzed on day 3 after the last irradiation session. LLLT significantly improved neurological status of PD patients assessed by Unified Parkinson's Disease Rating Scale (UPDRS). The improvement was accompanied by evident normalization of MAO B and Cu/Zn-SOD activities. The effects of LLLT on blood enzymes were much stronger after *in vivo* application of laser therapy than after *in vitro* blood irradiation.

Reduced or compromised axonal transport could underlie the progressive, relentless loss of dopaminergic nerve terminals in sporadic PD. Mitochondria supply the ATP needed to support axonal transport and contribute many other cellular functions essential for the survival of neuronal cells. Furthermore, mitochondria in PD tissues are metabolically and functionally compromised. In 2009, Trimmer et al. (2009) measured the velocity of mitochondrial movement in human transmitochondrial cybrid (cytoplasmic hybrid) neuronal cells bearing mitochondrial DNA from patients with sporadic PD and disease-free age-matched volunteer controls (CNT). PD and CNT cybrid neuronal cells were exposed to an 810 nm CW laser (50 mW/cm^2 for 40 s), and axonal transport of labeled mitochondria was measured. The velocity of mitochondrial movement in PD cybrid neuronal cells was significantly reduced as compared to mitochondrial movement in disease-free CNT cybrid neuronal cells. For at least 2 h after exposure to LLLT, the average velocity of mitochondrial movement in PD cybrid neurites was significantly increased and restored to CNT levels. Mitochondrial movement in CNT cybrid neurites was unaltered by LLLT. These findings demonstrate that NIR LLLT can increase axonal transport in model human dopaminergic neuronal cells, suggesting that LLLT could be developed as a novel treatment to improve neuronal function in patients with PD.

54.5 LLLT for Alzheimer's Disease

Alzheimer's disease (AD) is a devastating neurologic disease of the brain that leads to the irreversible loss of neurons and dementia. Although the cause of AD is still unknown, impaired oxidative balance, mitochondrial dysfunction, and disordered CCO play important roles in the pathogenesis of AD (Smith et al. 2000; Sullivan and Brown 2005). The disorder affects more than 37 million people worldwide, and the estimated direct and indirect annual cost of patient care in the United States alone is at least $100 billion (Rafii and Aisen 2009). Current Food and Drug Administration (FDA) approved drugs for AD do not prevent or reverse the disease, and provide only modest symptomatic benefits.

54.5.1 *In Vitro* Studies

One of the characteristic neuropathologic lesions associated with AD is the presence in the brain of senile plaques containing amyloid-β (Aβ) aggregated proteins, and there is a large body of research that has examined the relationship between Aβ peptides, Aβ-containing senile plaques, and AD (Selkoe 1991, 1994, 2001). Duan et al. (2001) have reported that within 24 h, cell apoptosis induced by $A\beta_{25-35}$ was significantly diminished with light irradiation. It is well established that $A\beta_{25-35}$ is a biologically active region that retains the toxicity of a full-length peptide associated with AD (Forloni et al. 1993; Pike, Walencewicz-Wasserman, and Kosmoski 1995), but LLLT diminishing $A\beta_{25-35}$-induced apoptosis can be another means for neuroprotection. In 2008, Zhang et al. (2008) investigated the inhibition of $A\beta_{25-35}$-induced pheochromocytoma (PC12) cell apoptosis by NIR laser irradiation (wavelength 632.8 nm). The monolayered cells were irradiated at a fluence rate of 0.52 mW/cm² for a fluence of 0.156–1.248 J/cm². By monitoring the dynamics of protein kinase C (PKC) activation and the change in the *bax* to *bcl-xl* ratio after Gö 6983 (specific inhibitor of PKC), $A\beta_{25-35}$, and laser treatment, the researchers concluded that $A\beta_{25-35}$ can induce cell apoptosis by increasing the level of *bax* mRNA, and the antiapoptotic effect of LLLT is dependent on the upregulation of *bcl-xl* and downregulation of *bax* via the PKC activation pathway. Therefore, they stated that the PKC-mediated regulation of *bax* and *bcl-xl* mRNA plays an important role in inhibiting $A\beta_{25-35}$-induced PC12 cell apoptosis.

Sommer et al. (2012) also presented the results of *in vitro* experiments to suppress Aβ deposition for treatment of AD. After internalization of $A\beta_{42}$ into human neuroblastoma (SH-EP) cells, they were irradiated with a moderately intense 670 nm laser (100 mW/cm²) and/or treated with epigallocatechin gallate (EGCG), which was shown to prevent the formation of extracellular fibrillary Aβ *in vitro* (Ehrnhoefer et al. 2008). EGCG reduced intracellular Aβ by 50% in a concentration-dependent manner. One-minute exposure to the laser light was already sufficient to reduce intracellular Aβ by 20%, a totally unexpected result. On the other hand, the dual treatment produced a reduction of more than 60%. Previously, the authors used EGCG with 670 nm LED to rejuvenate facial skin and exploited the capacity of EGCG to act as a scavenger of mitochondrial reactive oxygen species (ROS), activated by LED irradiation (Sommer and Zhu 2009). Thus, the results in SH-EP cells showed therapeutic relevance from two sides: identification of EGCG in the brains of animals after oral administration, indicating the penetration of EGCG across the blood–brain barrier (BBB), and the deep tissue penetration (several centimeters) of nonthermal red to NIR light, including human cranial and costal bone, delivering therapeutically relevant light doses into the brain and heart.

In 2010, Yang et al. (2010) tested *in vitro* the effects of LLLT at 632.8 nm on Aβ-induced ROS production through the activation of nicotinamide adenine dinucleotide phosphate (NADPH) oxidase and its downstream pathways involving phosphorylation of cytosolic phospholipase A_2 (cPLA$_2$), and expression of inflammatory factors including IL-1β and inducible nitric-oxide synthase (iNOS). Primary rat astrocytes were used to measure Aβ-induced production of superoxide anions. The data showed that 632.8 nm laser irradiation suppressed Aβ-induced superoxide production, colocalization between NADPH oxidase gp91[phox] and p47[phox] subunits, phosphorylation of cPLA$_2$, and the expressions of IL-1β and iNOS. The study demonstrated that LLLT was capable of suppressing cellular pathways of oxidative stress and inflammatory responses critical in the pathogenesis in AD.

54.5.2 *In Vivo* Studies

In 2011, De Taboada et al. (2011) and McCarthy et al. (2011) for the first time attempted to determine the effect of 808 nm LLLT (either CW or PW 100 Hz) in an amyloid-β protein precursor (AβPP) transgenic mouse model of AD. LLTT was administered 3 times/week at various doses, starting at 3 months of age, and was compared to a control group (no laser treatment). Treatment was continued for a total of 6 months. The numbers of Aβ plaques were significantly reduced and the levels of $A\beta_{1-40}$ and $A\beta_{1-42}$ were likewise decreased in the brain with administration of LLLT in a dose-dependent fashion. In addition, increased ATP levels and oxygen utilization were observed in the transgenic mice treated with LLLT, suggesting improved mitochondrial function.

Michalikova et al. (2008) treated middle-aged (12-month-old) female CD-1 mice with a daily 6 min exposure to 1072 nm LED light for 10 days and found that LLLT yielded a number of significant behavioral effects on testing in a three-dimensional maze. Middle-aged mice showed significant deficits in a working memory test, and LLLT reversed this deficit. Middle-aged mice treated with LED light were more considerate in their decision making, which resulted in an overall improved cognitive performance comparable with that of young (3-month-old) CD-1 mice. These observations suggest that LLLT could be applied in cases of general cognitive impairment in elderly persons.

54.6 LLLT for Amyotrophic Lateral Sclerosis

Amyotrophic lateral sclerosis (ALS) is a progressive neurodegenerative disorder involving primarily motor neurons in the cerebral cortex, brainstem, and spinal cord. The disorder leads to gradual muscle paralysis resulting in respiratory failure and death within 1–5 years after onset of clinical signs. Accumulating evidence suggests that mitochondrial dysfunction is involved in the pathogenesis of ALS. Moreover, oxidative stress is also thought to play an important role in the disease pathogenesis. Postmortem examination of spinal cord and brain tissues from ALS patients showed evidence of oxidative damage.

Moges et al. (2009) examined the synergistic effect of LLLT and vitamin B2 (riboflavin) on the survival of motor neurons in a mouse model of ALS. G93A SOD1 transgenic mice (ALS model) were divided into four groups: control, riboflavin alone, LLLT alone, and riboflavin with LLLT (combination). An 810 nm diode laser was utilized with the following parameters: 140 mW output power, a spot size of 1.4 cm², treatment duration of 120 s, and an energy density of 12 J/cm² per treatment site per day. The mice receiving LLLT were irradiated at three different sites: the primary motor cortex, the cervical enlargement of the spinal cord, and the lumbar enlargement of the spinal cord, for three consecutive days every week. Motor function was significantly improved only in the LLLT alone group at the early stage of the disease. Immunohistochemical expression of the astrocyte marker glial fibrillary acidic protein (GFAP) was significantly reduced in the cervical and lumbar enlargements of the spinal cord in the LLLT alone group. The lack of significant improvement in survival and motor performance indicates that the interventions were ineffective in altering disease progression in the G93A SOD1 mice. However, the findings have potential implications for the conceptual use of NIR light to treat other neurodegenerative diseases that have been linked to mitochondrial dysfunction.

54.7 Conclusion

In this contribution, we have reviewed the recent efforts of LLLT for stroke, TBI, and neurodegenerative diseases including AD, PD, and ALS. Evidence that transcranial LLLT is beneficial for the brain diseases has been rapidly accumulating. The almost complete lack of reports about side effects or adverse events associated with LLLT satisfies acceptance by both the medical profession and the general public. Meanwhile, the possible beneficial mechanisms and therapeutic processes in transcranial LLLT will be thoroughly investigated in further studies.

The clinical trials for stroke and TBI described above provide additional credibility to the use of LLLT for brain tissue. LLLT is steadily moving into mainstream medical practice in neurology as well as other departments. As the Western populations continue to age, the incidence of the degenerative diseases of old age will continue to increase and produce much more severe financial and societal burden. If LLLT can make a contribution to mitigating the loss of life, suffering, disability, and financial burden caused by brain disorders, the increasing efforts will be required for researchers in the field. Moreover, current ongoing works using nonlaser irradiation sources are encouraging and provide support for growth in the manufacture and marketing of affordable home-use LLLT devices.

References

Ando, T., W. Xuan, T. Xu et al. 2011. Comparison of therapeutic effects between pulsed and continuous wave 810-nm wavelength laser irradiation for traumatic brain injury in mice. *PLoS One* 6(10):e26212-1–e26212-9.

Avni, D., S. Levkovitz, L. Maltz et al. 2005. Protection of skeletal muscles from ischemic injury: Low-level laser therapy increases antioxidant activity. *Photomed Laser Surg* 23(3):273–277.

Bevilacqua, F., D. Piguet, P. Marquet et al. 1999. *In vivo* local determination of tissue optical properties applications to human brain. *Appl Opt* 38(22):4939–4950.

Broussalis, E., M. Killer, M. McCoy et al. 2012. Current therapies in ischemic stroke. Part A. Recent developments in acute stroke treatment and in stroke prevention. *Drug Discov Today* 17(7–8):296–309.

Choi, J., M. Wolf, V. Toronov et al. 2004. Noninvasive determination of the optical properties of adult brain: Near-infrared spectroscopy approach. *J Biomed Opt* 9(1):221–229.

De Taboada, L., S. Ilic, S. Leichliter-Martha et al. 2006. Transcranial application of low-energy laser irradiation improves neurological deficits in rats following acute stroke. *Lasers Surg Med* 38(1):70–73.

De Taboada, L., J. Yu, S. El-Amouri et al. 2011. Transcranial laser therapy attenuates amyloid-β peptide neuropathology in amyloid-β protein precursor transgenic mice. *J Alzheimers Dis* 23(3):521–535.

Donnan, G. A., S. M. Davis, M. W. Parsons et al. 2011. How to make better use of thrombolytic therapy in acute ischemic stroke. *Nat Rev Neurol* 7(7):400–409.

Duan, R., T. C. Liu, Y. Li et al. 2001. Signal transduction pathways involved in low intensity He-Ne laser-induced respiratory burst in bovine neutrophils: A potential mechanism of low intensity laser biostimulation. *Lasers Surg Med* 29(2):174–178.

Ehrnhoefer, D. E., J. Bieschke, A. Boeddrich et al. 2008. EGCG redirects amyloidogenic polypeptides into unstructured, off-pathway oligomers. *Nat Struct Mol Biol* 15(6):558–566.

Fann, J. R., T. Hart, and K. G. Schomer. 2009. Treatment for depression after traumatic brain injury: A systematic review. *J Neurotrauma* 26(12):2383–2402.

Faul, M., M. M. Wald, W. Rutland-Brown et al. 2007. Using a cost-benefit analysis to estimate outcomes of a clinical treatment guideline: Testing the Brain Trauma Foundation guidelines for the treatment of severe traumatic brain injury. *J Trauma* 63(6):1271–1278.

Faul, M., L. Xu, M. M. Wald et al. 2010. *Traumatic Brain Injury in the United States: Emergency Department Visits, Hospitalizations and Deaths, 2002–2006*. Centers for Disease Control and Prevention, National Center for Injury Prevention and Control.

Finkelstein, E. A., P. S. Corso, and T. R. Miller. 2006. *The Incidence and Economic Burden of Injuries in the United States*. Oxford University Press, New York.

Forloni, G., R. Chiesa, S. Smiroldo et al. 1993. Apoptosis mediated neurotoxicity induced by chronic application of beta amyloid fragment 25–35. *Neuroreport* 4(5):523–526.

Ginsberg, M. D. 2008. Neuroprotection for ischemic stroke: Past, present and future. *Neuropharmacology* 55(3):363–389.

Ginsberg, M. D. 2009. Current status of neuroprotection for cerebral ischemia: Synoptic overview. *Stroke* 40(3 Suppl):S111–S114.

Green, J. D., and A. A. Arduini. 1954. Hippocampal electrical activity in arousal. *J Neurophysiol* 17(6):533–557.

Hashmi, J. T., Y.-Y. Huang, B. Z. Osmani et al. 2010. Role of low-level laser therapy in neurorehabilitation. *Phys Med Rehabil* 2(12 Suppl 2):S292–S305.

Hu, W. P., J. J. Wang, C. L. Yu et al. 2007. Helium–neon laser irradiation stimulates cell proliferation through photostimulatory effects in mitochondria. *J Invest Dermatol* 127(8):2048–2057.

Karu, T. I. 2008. Mitochondrial signaling in mammalian cells activated by red and near-IR radiation. *Photochem Photobiol* 84(5):1091–1099.

Karu, T. I., L. V. Pyatibrat, S. F. Kolyakov et al. 2005. Absorption measurements of a cell monolayer relevant to phototherapy: Reduction of cytochrome c oxidase under near IR radiation. *J Photochem Photobiol B* 81(2):98–106.

Kato, T., A. Kamei, S. Takashima et al. 1993. Human visual cortical function during photic stimulation monitoring by means of near-infrared spectroscopy. *J Cereb Blood Flow Metab* 13(3):516–520.

Khuman, J., J. Zhang, J. Park et al. 2012. Low-level laser light therapy improves cognitive deficits and inhibits microglial activation after controlled cortical impact in mice. *J Neurotrauma* 29(2):408–417.

Kipshidze, N., V. Nikolaychik, M. H. Keelan et al. 2001. Low-power helium: Neon laser irradiation enhances production of vascular endothelial growth factor and promotes growth of endothelial cells *in vitro*. *Lasers Surg Med* 28(4):355–364.

Komel'kova, L. V., T. V. Vitreshchak, I. G. Zhirnova et al. 2004. Biochemical and immunological induces of the blood in Parkinson's disease and their correction with the help of laser therapy. *Patol Fiziol Eksp Ter* 1:15–18.

Lampl, Y. 2007. Laser treatment for stroke. *Expert Rev Neurother* 7(8):961–965.

Lampl, Y., J. A. Zivin, M. Fisher et al. 2007. Infrared laser therapy for ischemic stroke: A new treatment strategy: Results of the NeuroThera Effectiveness and Safety Trial-1 (NEST-1). *Stroke* 38(6):1843–1849.

Langlois, J. A., W. Rutland-Brown, and K. E. Thomas. 2005. The incidence of traumatic brain injury among children in the United States: Differences by race. *J Head Trauma Rehabil* 20(3):229–238.

Langlois, J. A., W. Rutland-Brown, and M. M. Wald. 2006. The epidemiology and impact of traumatic brain injury: A brief overview. *J Head Trauma Rehabil* 21(5):375–378.

Lapchak, P. A. 2010. Taking a light approach to treating acute ischemic stroke patients: Transcranial near-infrared laser therapy translational science. *Ann Med* 42(8):576–586.

Lapchak, P. A. 2012. Transcranial near-infrared laser therapy applied to promote clinical recovery in acute and chronic neurodegenerative diseases. *Expert Rev Med Devices* 9(1):71–83.

Lapchak, P. A., and L. De Taboada. 2010. Transcranial near infrared laser treatment (NILT) increases cortical adenosine-5′-triphosphate (ATP) content following embolic strokes in rabbits. *Brain Res* 1306:100–105.

Lapchak, P. A., M. K. Han, K. F. Salgado et al. 2008. Safety profile of transcranial near-infrared laser therapy administered in combination with thrombolytic therapy to embolized rabbits. *Stroke* 39(11):3073–3078.

Lapchak, P. A., K. F. Salgado, C. H. Chao et al. 2007. Transcranial near-infrared light therapy improves motor function following embolic strokes in rabbits: An extended therapeutic window study using continuous and pulse frequency delivery modes. *Neuroscience* 148(4):907–914.

Lapchak, P. A., J. Wei, and J. A. Zivin. 2004. Transcranial infrared laser therapy improves clinical rating scores after embolic strokes in rabbits. *Stroke* 35(8):1985–1988.

Leker, R. R. and E. Shohami. 2002. Cerebral ischemia and trauma different etiologies yet similar mechanisms: Neuroprotective opportunities. *Brain Res Rev* 39(1):55–73.

Leung, M. C., S. C. Lo, F. K. Siu et al. 2002. Treatment of experimentally induced transient cerebral ischemia with low energy laser inhibits nitric oxide synthase activity and up-regulates the expression of transforming growth factor-beta 1. *Lasers Surg Med* 31(4):283–288.

Lychagov, V. V., V. V. Tuchin, M. A. Vilensky et al. 2006. Experimental study of NIR transmittance of the human skull. *Proc SPIE* 6085:60850T-1–60850T-5.

Maas, A. I., N. Stocchetti, and R. Bullock. 2008. Moderate and severe traumatic brain injury in adults. *Lancet Neurol* 7(8):728–741.

Maki, A., Y. Yamashita, Y. Ito et al. 1995. Spatial and temporal analysis of human motor activity using noninvasive NIR topography. *Med Phys* 22(12):1997–2005.

Mayer, S. A. 2003. Ultra-early hemostatic therapy for intracerebral hemorrhage. *Stroke* 34(1):224–229.

McCarthy, T. J., L. De Taboada, P. K. Hildebrandt et al. 2010. Long-term safety of single and multiple infrared transcranial laser treatments in Sprague–Dawley rats. *Photomed Laser Surg* 28(5):663–667.

McCarthy, T., J. Yu, S. El-Amouri et al. 2011. Transcranial laser therapy alters amyloid precursor protein processing and improves mitochondrial function in a mouse model of Alzheimer's disease. *Proc SPIE* 7887:78870K-1–78870K-13.

Michalikova, S., A. Ennaceur, R. van Rensburg et al. 2008. Emotional responses and memory performance of middle-aged CD1 mice in a 3D maze: Effects of low infrared light. *Neurobiol Learn Mem* 89(4):480–488.

Mirsky, N., Y. Krispel, Y. Shoshany et al. 2002. Promotion of angiogenesis by low energy laser irradiation. *Antioxid Redox Signal* 4(5):785–790.

Moges, H., O. M. Vasconcelos, W. W. Campbell et al. 2009. Light therapy and supplementary riboflavin in the SOD1 transgenic mouse model of familial amyotrophic lateral sclerosis (FALS). *Lasers Surg Med* 41(1):52–59.

Moreira, M. S., I. T. Velasco, L. S. Ferreira et al. 2009. Effect of phototherapy with low intensity laser on local and systemic immunomodulation following focal brain damage in rat. *J Photochem Photobiol B* 97(3):145–151.

Moreira, M. S., I. T. Velasco, L. S. Ferreira et al. 2011. Effect of laser phototherapy on wound healing following cerebral ischemia by cryogenic injury. *J Photochem Photobiol B* 105(3):207–215.

Naeser, M. A., M. P. Alexander, D. Stiassny-Eder et al. 1994a. Acupuncture in the treatment of paralysis in chronic and acute stroke patients—Improvement correlated with specific CT scan lesion sites. *Acupunct Electrother Res* 19(4):227–249.

Naeser, M. A., M. P. Alexander, D. Stiassny-Eder et al. 1994b. Acupuncture in the treatment of hand paresis in chronic and acute stroke patients: Improvement observed in all cases. *Clin Rehab* 8(2):127–141.

Naeser, M. A., M. P. Alexander, D. Stiassny-Eder et al. 1995. Laser acupuncture in the treatment of paralysis in stroke patients: A CT scan lesion site study. *Am J Acupuncture* 23(1):13–28.

Naeser, M. A., and M. R. Hamblin. 2011. Potential for transcranial laser or LED therapy to treat stroke, traumatic brain injury, and neurodegenerative disease. *Photomed Laser Surg* 29(7):443–446.

Naeser, M. A., A. Saltmarche, M. H. Krengel et al. 2011. Improved cognitive function after-transcranial, light-emitting diode treatments in chronic, traumatic brain injury: Two case reports. *Photomed Laser Surg* 29(5):351–358.

Nawashiro, H., K. Wada, K. Nakai et al. 2012. Focal increase in cerebral blood flow after treatment with near-infrared light to the forehead in a patient in a persistent vegetative state. *Photomed Laser Surg* 30(4):231–233.

Ng, S. S., and C. W. Hui-Chan. 2007. Transcutaneous electrical nerve stimulation combined with task-related training improves lower limb functions in subjects with chronic stroke. *Stroke* 38(11):2953–2959.

Nortje, J., and D. K. Menon. 2004. Traumatic brain injury: Physiology, mechanisms, and outcome. *Curr Opin Neurol* 17(6):711–718.

Oron, A., U. Oron, J. Chen et al. 2006. Low-level laser therapy applied transcranially to rats after induction of stroke significantly reduces long-term neurological deficits. *Stroke* 37(10):2620–2624.

Oron, A., U. Oron, J. Streeter et al. 2007. Low-level laser therapy applied transcranially to mice following traumatic brain injury significantly reduces long-term neurological deficits. *J Neurotrauma* 24(4):651–656.

Oron, A., U. Oron, J. Streeter et al. 2012. Near infrared transcranial laser therapy applied at various modes to mice following traumatic brain injury significantly reduces long-term neurological deficits. *J Neurotrauma* 29(2):401–407.

Oron, U., T. Yaakobi, A. Oron et al. 2001a. Attenuation of infarct size in rats and dogs after myocardial infarction by low-energy laser irradiation. *Lasers Surg Med* 28(3):204–211.

Oron, U., T. Yaakobi, A. Oron et al. 2001b. Low-energy laser irradiation reduces formation of scar tissue after myocardial infarction in rats and dogs. *Circulation* 103(2):296–301.

Pike, C. J., A. J. Walencewicz-Wasserman, and J. Kosmoski. 1995. Structure-activity analyses of beta-amyloid peptides: Contributions of the beta 25–35 region to aggregation and neurotoxicity. *J Neurochem* 64(1):253–265.

Rafii, M. S., and P. S. Aisen. 2009. Recent developments in Alzheimer's disease therapeutics. *BMC Med* 7:7.

Roger, V. L., A. S. Go, D. M. Lloyd-Jones et al. 2011. Heart disease and stroke—2011 update: A report from the American Heart Association. *Circulation* 123(4):e18–e209.

Schiffer, F., A. L. Johnston, C. Ravichandran et al. 2009. Psychological benefits 2 and 4 weeks after a single treatment with near infrared light to the forehead: A pilot study of 10 patients with major depression and anxiety. *Behav Brain Funct* 5:46–55.

Selkoe, D. J. 1991. The molecular pathology of Alzheimer's disease. *Neuron* 6(4):487–498.

Selkoe, D. J. 1994. Normal and abnormal biology of the beta-amyloid precursor protein. *Annu Rev Neurosci* 17:489–517.

Selkoe, D. J. 2001. Presenilin, Notch, and the genesis and treatment of Alzheimer's disease. *Proc Natl Acad Sci USA* 98(20):11039–11041.

Shefer, G., T. A. Partridge, L. Heslop et al. 2002. Low-energy laser irradiation promotes the survival and cell cycle entry of skeletal muscle satellite cells. *J Cell Sci* 115(Pt 7):1461–1469.

Siesjö, B. K., K. Katsura, Q. Zhao et al. 1995. Mechanisms of secondary brain damage in global and focal ischemia: A speculative synthesis. *J Neurotrauma* 12(5):943–956.

Smith, M. A., C. A. Rottkamp, A. Nunomura et al. 2000. Oxidative stress in Alzheimer's disease. *Biochim Biophys Acta* 1502(1):139–144.

Sommer, A. P., J. Bieschke, R. P. Friedrich et al. 2012. 670 nm laser light and EGCG complementarily reduce amyloid-β aggregates in human neuroblastoma cells: Basis for treatment of Alzheimer's disease? *Photomed Laser Surg* 30(1):54–60.

Sommer, A. P., and D. Zhu. 2009. Facial rejuvenation in the triangle of ROS. *Cryst Growth Des* 9(10):4250–4254.

Stemer, A. B., B. N. Huisa, and J. A. Zivin. 2010. The evolution of transcranial laser therapy for acute ischemic stroke, including a pooled analysis of NEST-1 and NEST-2. *Curr Cardiol Rep* 12(1):29–33.

Stolik, S., J. A. Delgado, A. Pérez et al. 2000. Measurement of the penetration depths of red and near infrared light in human "ex vivo" tissues. *J Photochem Photobiol B* 57(2–3):90–93.

Sullivan, P. G., and M. R. Brown. 2005. Mitochondrial aging and dysfunction in Alzheimer's disease. *Prog Neuro-Psychoph* 29(3):407–410.

Thurman, D. J., C. Alverson, K. A. Dunn et al. 1999. Traumatic brain injury in the United States: A public health perspective. *J Head Trauma Rehabil* 14(6):602–615.

Trimmer, P. A., K. M. Schwartz, M. K. Borland et al. 2009. Reduced axonal transport in Parkinson's disease cybrid neurites is restored by light therapy. *Mol Neurodegener* 4:26.

Uozumi, Y., H. Nawashiro, S. Sato et al. 2010. Targeted increase in cerebral blood flow by transcranial near-infrared laser irradiation. *Lasers Surg Med* 42(6):566–576.

Vanderwolf, C. H. 1969. Hippocampal electrical activity and voluntary movement in the rat. *Electroencephalogr Clin Neurophysiol* 26(4):407–418.

Villringer, A., J. Planck, C. Hock et al. 1993. Near infrared spectroscopy (NIRS): A new tool to study hemodynamic changes during activation of brain function in human adults. *Neurosci Lett* 154(1–2):101–104.

Vitreshchak, T. V., V. V. Mikhailov, M. A. Piradov et al. 2003. Laser modification of the blood *in vitro* and *in vivo* in patients with Parkinson's disease. *Bull Exp Biol Med* 135(5):430–432.

Walker, P. A., S. K. Shah, M. T. Harting et al. 2009. Progenitor cell therapies for traumatic brain injury: Barriers and opportunities in translation. *Dis Model Mech* 2(1–2):23–38.

Whittaker, P. 2004. Laser acupuncture: Past, present, and future. *Lasers Med Sci* 19(2):69–80.

Winson, J. 1972. Interspecies differences in the occurrence of theta. *Behav Biol* 7(4):479–487.

Wong-Riley, M. T., H. L. Liang, J. T. Eells et al. 2005. Photobiomodulation directly benefits primary neurons functionally inactivated by toxins: Role of cytochrome c oxidase. *J Biol Chem* 280(6):4761–4771.

Wu, Q., Y.-Y. Huang, S. Dhital et al. 2010. Low level laser therapy for traumatic brain injury. *Proc SPIE* 7552:755206-1–755206-8.

Wu, Q., W. Xuan, T. Ando et al. 2012. Low-level laser therapy for closed-head traumatic brain injury in mice: Effect of different wavelengths. *Lasers Surg Med* 44(3):218–226.

Xuan, W., Q. Wu, Y.-Y. Huang et al. 2012. *In vivo* studies of low level laser (light) therapy for traumatic brain injury. *Proc SPIE* 8211:82110A-1–82110A-10.

Yaakobi, T., Y. Shoshany, S. Levkovitz et al. 2001. Long-term effect of low energy laser irradiation on infarction and reperfusion injury in the rat heart. *J Appl Physiol* 90(6):2411–2419.

Yang, X., S. Askarova, W. Sheng et al. 2010. Low energy laser light (632.8 nm) suppresses amyloid-β peptide-induced oxidative and inflammatory responses in astrocytes. *Neuroscience* 171(3):859–868.

Zhang, L., D. Xing, D. Zhu et al. 2008. Low-power laser irradiation inhibiting $A\beta_{25-35}$-induced PC12 cell apoptosis via PKC activation. *Cell Physiol Biochem* 22(1–4):215–222.

Zhang, Y., S. Song, C. C. Fong et al. 2003. cDNA microarray analysis of gene expression profiles in human fibroblast cells irradiated with red light. *J Invest Dermatol* 120(5):849–857.

Zivin, J. A., G. W. Albers, N. Bornstein et al. 2009. Effectiveness and safety of transcranial laser therapy for acute ischemic stroke. *Stroke* 40(4):1359–1364.

55

Low-Level Light Therapy for Nerve and Spinal Cord Regeneration

Nivaldo A. Parizotto
Federal University of Sao Carlos

55.1 Introduction

Posttraumatic nerve lesion and repair is still a challenge for rehabilitation. If the nerve is damaged, it can result in motor and sensory disabilities. In the clinical set, it represents serious problems and it is recognized that there are about 50,000 peripheral nerve repair procedures in the United States annually (Moges et al. 2011). For this reason, it is particularly important to develop clinical protocols to enhance nerve regeneration. Many papers have been published trying to explain the mechanisms of the low-intensity laser or other kinds of light therapy that improve the peripheral nerve repair in different models in animals and humans, including *in vitro* studies. It is very important to provide early movements to the muscles and create a way to allow improvement in the life conditions of patients submitted to peripheral nerve lesion (Gigo-Benato, Geuna, and Rochkind 2005; Moges et al. 2011; Zhang et al. 2010). The specific purpose of photostimulation in the nerves is to facilitate the physiological reconnection of those nerves to respective muscles. It is not easy. Many factors interfere in this process, including the size of lesion, the distance from proximal stub to distal part of the nerve, expression of the growth factors, morphology adaptations in the cell body of the neurons, and extracellular matrix reorganization to open this pathway in the periaxonal environment (Rochkind, Geuna, and Shainberg 2009).

Currently, the technical aspect of surgical procedures is limited to the new materials (Curtis et al. 2011), dexterity and competence of trained people (e.g., microsurgeons) in order to perform a good job (Nectoux, Taleb, and Liverneaux 2009), who can do autografts and allografts with different materials, end-to-end suture, and side-to-end suture, and use different kinds of glues.

Light is a very handy tool to use in nerve lesions, and its use has been increasing in the last several years. Both clinical and experimental results are encouraging the use of lasers and other light sources to stimulate the tissue repair process. This procedure can improve the function of the muscle when its nerve was damaged and treated by light (Gigo Benato et al. 2004). Many tests are used to measure the functional effect of light in the animal model, such as functional sciatic index (Santos et al. 2012) and grasping test (Moges et al. 2011; Santos et al. 2012). The function of muscles can be accessed by electrophysiological tests and by asking a person to perform a certain movement where the muscle is governed from that nerve, as in the Medical Research Council Grading System for injured patients (Rochkind, Geuna, and Shainberg 2009).

Many researchers have done studies about the different factors interfering on the interaction between light and tissue, like wavelength (Barbosa et al. 2010), power, frequency (Chen et al. 2005), the regiment of irradiation (Gigo-Benato et al. 2004), the placement of irradiation (Rochkind et al. 2001), and so on. There are papers that show that the association of different wavelengths can improve the outcome in the nerve and muscle (Gigo-Benato et al. 2004).

There is also some agreement about the best wavelengths of light and a range of acceptable dosages to be used (irradiance and fluence), but there is no agreement on whether continuous wave or pulsed light is best and on what factors govern the pulse parameters to be chosen. The molecular and cellular mechanisms of low-level light therapy (LLLT) are outlined. The type of pulsed light sources available and the parameters that govern their pulse structure are outlined. In this chapter, studies that have compared continuous wave and pulsed light in both animals and patients are reviewed, and frequencies used in other pulsed modalities used in physical therapy and biomedicine are compared to those used in

LLLT. There is some evidence that pulsed light does have effects that are different from those of continuous wave light. However, further work is needed to define the best results for different disease conditions and pulse structures (Hashmi et al. 2010).

55.2 LLLT for Nerve Regeneration

A recently published review (Rochkind, Geuna, and Shainberg 2009) about phototherapy on the nerve repair process shows good results using lasers, light-emitting diodes (LEDs) (Serafim et al. 2012), and some kinds of light devices (Anders, Geuna, and Rochkind 2004). In the review, the authors explain the mechanisms underlying LLLT using many previous results (Rochkind, Geuna, and Shainberg 2009).

Our group (Gigo-Benato et al. 2010) performed a study investigating the effects of 660 and 780 nm LLLT using different energy densities (10, 60, and 120 J/cm^2) on neuromuscular and functional recovery as well as on matrix metalloproteinase (MMP) activity after crush injury in the rat sciatic nerve. This study was done in rats, which received transcutaneous LLLT irradiation at the lesion site for 10 consecutive days postinjury and were sacrificed 28 days after injury. Both the sciatic nerve and tibialis anterior muscles were analyzed. Nerve analyses consisted of histology (light microscopy) and measurements of myelin, axon, and nerve fiber cross-sectional area (CSA). S-100 labeling was used to identify myelin sheath and Schwann

cells (see Figure 55.1). Muscle fiber CSA and zymography were carried out to assess the degree of muscle atrophy and MMP activity, respectively. LLLT 660 nm using both 10 and 60 J/cm^2 restored muscle fiber, myelin, and nerve fiber CSA compared to the normal group (N). Furthermore, it increased MMP-2 activity in the nerve and decreased MMP-2 activity in the muscle and MMP-9 activity in the nerve. In contrast, 780 nm LLLT using 10 J/cm^2 decreased MMP-9 activity in the nerve compared to the crush group (CR) and N; it also restored normal levels of myelin and nerve fiber CSA. Both 60 and 120 J/cm^2 decreased MMP-2 activity in the muscle compared to CR and N. Laser with 780 nm did not prevent muscle fiber atrophy. Functional recovery in the irradiated groups did not differ from the nonirradiated CR. These data suggest that 660 nm LLLT with low (10 J/cm^2) or moderate (60 J/cm^2) energy densities is able to accelerate neuromuscular recovery after nerve crush injury in rats.

It is very interesting because the best results of Rochkind's group (Anders, Geuna, and Rochkind 2004; Gigo-Benato, Geuna, and Rochkind 2005; Rochkind et al. 2001, 2007; Rochkind, Geuna, and Shainberg 2009) were obtained with wavelengths in the near-infrared band (780 nm), which is in disagreement with the findings in our group. However, there are several factors involved in these reactions that may explain differences in the results with various types of tissues, including peripheral nerves. In earlier papers (Rochkind et al. 1987a,b, 1989, 2001), this same group had good outcomes with laser in

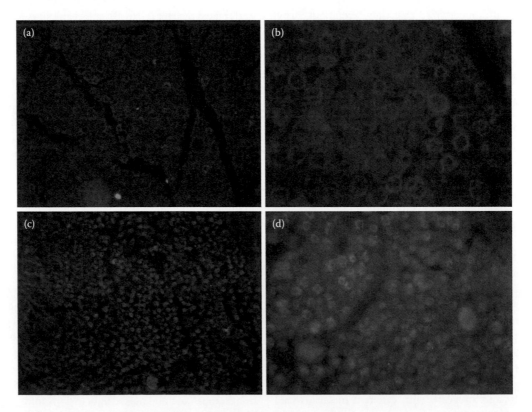

FIGURE 55.1   (See color insert.) Micrograph of sciatic nerve submitted to LLLT treatment and control evaluable at 28 days after the lesion. The immunomarker is S-100, which stains the Schwann cells. More staining of the Schwann cells can be seen in two magnifications (20× in the left and 100× in the right) of the laser-treated (3 times a week for 4 weeks, c and d) than the control (sham, a and b).

the red band (633 nm He–Ne laser), confirming the possibility of achieving good results in other spectrum bands.

It can be seen in a recent study that laser therapy with CW at 660 nm produced partial gait recovery in a complete resection of the rat sciatic nerve, and a morphometric analysis revealed an increase in axon and fiber diameter in the injured animals that received light. The intensity of the effect was correlated positively to the fluence used (10 and 50 J/cm^2). Compared to the animals receiving LLLT at 808 nm, the wavelength in the red band was better than in the infrared, and the latter was similar to controls (Medalha et al. 2012). This paper confirms the findings of our published paper (Gigo-Benato et al. 2010). The functional recovery was observed earlier by some authors (Barbosa et al. 2010; Belchior et al. 2009). It is interesting to see papers showing negative results using laser in the infrared band (Bagis et al. 2003). Many times the irradiation on the other side of the same animal can hide the real effects of the light.

Interesting screening was done by Anders et al. (1993) with regenerating facial nerves in rats using laser irradiation. They did an initial analysis of the primary wavelength and penetration of laser in rat's skin. Later, using the best results for wavelength, they applied the laser irradiation on different schedules, starting soon after the lesion (crush injury) on consecutive or alternate days. Using the retrograde transport of horseradish peroxidase (HRP) by the neurons labeled in the facial motor nucleus, it was determined that a maximal increase occurs in the following parameters: daily treatment with He–Ne laser, 633 nm, 8.5 mW, 90 min. These data indicated that transcutaneous low energy laser increases the rate of regeneration on the rat facial nerve following crush injury (Anders et al. 1993).

According to the study of Rosner et al. (1993), the laser irradiation (He–Ne 632.8 nm) on the moderately crushed optic nerves of rats could delay the degeneration after 2 weeks of treatment through the retina. The compound action potential (CAP) of the optic nerves was measured and higher electrical activity was observed when compared to the injured nonirradiated nerves. The better irradiation conditions were 10 to 35 mW of output 2 or 3 min per day during 2 weeks. These authors commented that if they used more than 3 min per day or repeated the irradiation twice a day for 4 to 7 days, it could cause damage in the optic nerve and did not delay that electrical activity (Rosner et al. 1993). On the other hand, if we try what can happen in the normal nerve (median) in healthy human beings subjected to infrared 830 nm laser irradiation, we can see a reduction in motor and sensory latencies in the treated limbs relative to the control group, but these effects are quite limited and appear to be limited to the distal portion of the nerve (Basford et al. 1993).

A double-blind randomized study evaluated the therapeutic effect of low-level laser irradiation (LLLI) on peripheral nerve regeneration after complete transection and direct anastomosis of the rat sciatic nerve. After this procedure, 13 of 24 rats received postoperative LLLI, with a wavelength of 780 nm laser applied transcutaneously 30 min daily for 21 consecutive days to corresponding segments of the spinal cord and the injured sciatic nerve. Positive somatosensory-evoked responses were found

in 69.2% of the irradiated rats compared to 18.2% of the nonirradiated rats. Immunohistochemical staining in the laser-treated group showed an increase in the total number of axons and better quality of the regeneration process due to an increased number of large-diameter axons compared to the nonirradiated control group. The study suggests that postoperative LLLI enhances the regenerative processes of peripheral nerves when using anastomosis after compete transection (Shamir et al. 2001).

As is known, growth factors are involved in a variety of responses in the process of nerve regeneration. One study investigated the effects of diode (Ga–Al–As) laser irradiation at an effective energy density of 5 or 20 J/cm^2 on cell growth factor-induced differentiation and proliferation in pheochromocytoma cells (PC12 cells), and whether those effects were related to activation of the p38 pathway. Laser irradiation at 20 J/cm^2 significantly decreased the number of PC12 cells, while no difference was seen between the 5 J/cm^2 group and the control group. Western blotting revealed marked expression of neurofilament and beta-tubulin, indicating greater neurite differentiation in the irradiation groups than in the control group at 48 h. Irradiation also enhanced the expression of phospho-p38. The decrease in the number of cells after laser irradiation was accelerated by p38 inhibitor, while neurite differentiation was upregulated by laser irradiation even when the p38 pathway was blocked. This result suggests that laser irradiation upregulated neurite differentiation in PC12 cells involving p38 and another pathway (Saito et al. 2011).

The therapeutic effect on neuronal regeneration was verified by finding elevated immunoreactivities (IRs) of growth-associated protein-43 (GAP-43), which is upregulated during neuronal regeneration. The animals received a standardized crush injury of the sciatic nerve, mimicking the clinical situations accompanying partial axonotmesis. The injured nerve received calculated LLLI therapy immediately after injury and for four consecutive days thereafter. The walking movements of the animals were scored using the sciatic functional index (SFI). The SFI level was higher in the laser-treated rats at 3–4 weeks while the SFIs of the laser-treated and untreated rats reached normal levels at 5 weeks after surgery. In immunocytochemical study, although GAP-43 IRs increased both in the untreated control and the LLLI-treated groups after injury, the number of GAP-43 IR nerve fibers was much more increased in the LLLI group than of those in the control group. The elevated numbers of GAP-43 IR nerve fibers reached a peak 3 weeks after injury and then declined in both the untreated control and the LLLI groups at 5 weeks, with no differences in the numbers of GAP-43 IR nerve fibers of the two groups at this stage. This immunocytochemical study using GAP-43 antibody study shows for the first time that LLLI has an effect on the early stages of the nerve recovery process following sciatic nerve injury (Shin et al. 2003).

Previous studies have proposed that proliferation and release of certain growth factors by different types of cells can be modulated by LLLT. A study was formulated to demonstrate the effect of laser irradiation on human Schwann cell (SC) proliferation and neurotrophic factor gene expression *in vitro*. Human SCs were harvested from sural nerve that was obtained from an organ donor followed by treatment with an 810 nm, 50 mW diode laser (two different energies: 1 and 4 J/cm²) on three consecutive days. SC proliferation was measured after first irradiation on days 1, 4, and 7 by the MTT assay. Real-time polymerase chain reaction (PCR) analysis was utilized on days 5 and 20 to evaluate the expression of key genes involved in nerve regeneration consisting of nerve growth factor (NGF), brain-derived neurotrophic factor (BDNF), and glial-derived neutrophic factor (GDNF). Evaluation of cellular proliferation following 1 day after laser treatment revealed significant decrease in cell proliferation compared to the control group. However, on day 7, significant increase in proliferation was found in both the irradiated groups in comparison with the control group. No significant difference was found between the laser-treated groups. Treatment of SCs with laser resulted in significant increase in NGF gene expression on day 20. The difference between the two treated groups and the control group was not significant for BDNF and GDNF gene expression. These results demonstrate that LLLT stimulate human SC proliferation and NGF gene expression *in vitro* (Yazdani et al. 2012).

Clinicians have turned the focus of their attention to more effective methods of promoting nerve regeneration, target-organ reinnervation, and functional restoration after peripheral nerve injury. Surgeons use different guide or conduits to increase the speed of the nerve regeneration. Neurotrophic factors are tested to improve the nerve repair process. A study was carried out using lasers on the nerve lesion covered by tubulization with biodegradable nerve conduit. It is a composite containing genipin-cross-linked gelatin annexed with β-tricalcium phosphate (TCP) ceramic particles served as a nerve guide. They used a multi-cluster with 20 lasers at 660 nm (large area of irradiation, 314 cm²), 3.84 J/cm², 5 min daily during 21 days irradiated transcutaneously on the sciatic nerve area. After 4 or 6 weeks, morphometric, immunohistochemical, and functional changes could be seen in the laser-treated rats compared to the sham group. The results support the proposition that the laser can promote both morphological and functional improvements in the process of nerve repair (Shen, Yang, and Liu 2011).

The autograft is currently the gold standard treatment for nerve repair. In a study using acellular allograft rat sciatic nerve, laser, autograft, and normal control group, it was found that the myelinated fiber number of the regenerated nerve in the laser group increased compared to the allograft group, but did not reach the level of the autograft group. The difference in axon diameter among the three groups was not significant. Although myelin sheath thickness of the regenerated nerve in the autograft group was greater than that of laser and allograft groups, the difference was not significant. Under an electron microscope, myelin sheath degeneration was obvious and the thickness was uneven, and myelin sheath degeneration in the laser and autograft groups was minimal and the thickness was even. These results indicate that Ga–Al–As laser is similar to autograft with the exception of axon medullization (Zhang et al. 2010).

The location of the application of low-intensity lasers appears to be quite important. Regularly we use applications directly on

the site of the lesion so not invasive mode (transcutaneous). Yet distance applications may be helpful in some situations. Positive results were shown in an interesting paper by Rochkind et al. (2001), who reported that by applying to corresponding segments over the spinal cord root nerve they were able to achieve similar results under the electrophysiologic view point.

55.3 Central Nervous System Injury (Spinal Cord Injury)

Rochkind's group was the first to begin studies of the use of LLLI in the treatment of severely injured peripheral nervous systems (PNSs) and central nervous systems (CNSs). The irradiation method was proposed by Rochkind and has been modified over the last few years. LLLI in specific wavelengths and energy density is known to maintain the electrophysiological activity of severely injured peripheral nerves in rats, preventing scar formation (at injury site) as well as degenerative changes in the corresponding motor neurons of the spinal cord, thus accelerating regeneration of the injured nerve. Laser irradiation applied to the spinal cord of dogs following severe spinal cord injury and implantation of a segment of the peripheral nerve into the injured area diminished glial scar formation, induced axonal sprouting in the lesion, and restoration of locomotor function. The use of laser irradiation in mammalian CNS transplantation shows that laser therapy prevents extensive glial scar formation (a limiting factor in CNS regeneration) between a neural transplant and the host brain or spinal cord. Abundant capillaries were seen to develop in the laser-irradiated transplants and were of crucial importance in their survival. Intraoperative clinical use of laser therapy following surgical treatment of the tethered spinal cord (resulting from myelomeningocele, lipomyelomeningocele, thickened filum terminale, or fibrous scar) increases functional activity of the irradiated spinal cord. In a previous experimental work, we showed that direct laser treatment on nerve tissue promotes restoration of the electrophysiological activity of the severely injured peripheral nerve, prevents degenerative changes in neurons of the spinal cord, and induces proliferation of astrocytes and oligodendrocytes. This suggested a higher metabolism in neurons and improved ability for myelin production under the influence of laser treatment. The tethering of the spinal cord causes mechanical damage to neuronal cell membranes leading to metabolic disturbances in the neurons. For this reason, Rochkind's group believes that using LLLI may improve neuronal metabolism, prevent neuronal degeneration, and promote improved spinal cord function and repair (Rochkind and Ouaknine 1992). The possible mechanism of LLLI is currently under investigation. Using electron paramagnetic resonance in cell culture models, Rochkind's group found that at low radiation doses, singlet oxygen is produced by energy transfer from porphyrin (not cytochrome as commonly assumed), which is known to be present in the cell. At low concentrations, singlet oxygen can modulate biochemical processes taking place in the cell and trigger accelerated cell division. On the other hand, at high concentrations, singlet oxygen damages the cell (Lavi et al. 2003; Lubart et al. 2005; Rochkind et al. 1988).

If we are looking at immune response, we can find an excellent paper showing interesting results about this subject. Byrnes et al. (2005a) demonstrated that 810 nm light (150 mW, 1.589 J/cm^2) could modulate the immune response in the spinal cord and also improve the axonal regeneration and functional recovery of hemisection spinal cord adult rats. They accessed the function by locomotor task tests like footfalls, time to cross the ladder, base of support, stride length, and angle of rotation in the walking gait. Significant improvements were found in the functions comparing photobiomodulation (PBM) groups to controls. The immune response was quantified at 48 h, 14 days, and 16 days after the spinal cord injury by invasion/activation of different cell types on the lesion site. Only the group that received light had double immunolabeled cells after 10 weeks. It represents the axons transected during the initial lesion and had regrown to vertebral level L3. The expression of iNOS, transforming growth factor b (TGFβ), four proinflammatory cytokines: interleukin (IL) 1β (IL1β), tumor necrosis factor α (TNFα), IL6, and granulocyte-macrophage colony stimulating factor (GM-CSF), and two chemokines (MIP1α and MCP-1) was assessed at 6 h. PBM resulted in a significant suppression of IL6 expression postinjury with a 171-fold decrease in MCP-1 at 6 h, while the control group increased 66% for the same chemokine. A fivefold suppression of iNOS transcription at this time point was found in the PBM group in comparison to the control group. Interestingly, IL1β, TNFα, and MIP1α, which have maximum expression at 3 h postinjury or earlier, were not found to be altered by PBM at 6 h postinjury. Another finding in this study determined that PBM significantly altered the invasion of a number of cell types that play a substantial role after SCI. Immunolabeling for macrophages/activated microglia, T lymphocytes, and astrocytes was significantly decreased postinjury; these cell types are involved in secondary damage to the spinal cord after injury.

Other groups introduced light therapy as a method to treat central and peripheral nervous tissue lesions. SCI is a severe central nervous system trauma with no effective restorative therapies. The effectiveness of light therapy on SCI caused by different types of trauma was studied. Two SCI models were used: a contusion model and a dorsal hemisection model. Infrared light (810 nm) was applied transcutaneously at the lesion site immediately after injury and daily for 14 consecutive days. A laser diode with an output power of 150 mW was used for the treatment. The daily dosage at the surface of the skin overlying the lesion site was 1589 J/cm^2 (0.3 cm^2 spot area, 2997 s). Mini-ruby was used to label corticospinal tract axons, which were counted and measured from the lesion site distally. Functional recovery was assessed by footprint test for the hemisection model and open-field test for the contusion model. Rats were euthanized 3 weeks after injury. The average length of axonal regrowth in the rats in the light treatment (LLLT) groups with the hemisection (6.89 ± 0.96 mm) and contusion (7.04 ± 0.76 mm) injuries was significantly longer than that in the comparable untreated control groups (3.66 ± 0.26 mm, hemisection; 2.89 ± 0.84 mm,

contusion). The total axon number in the LLLT groups was significantly higher compared to the untreated groups for both injury models. For the hemisection model, the LLLT group had a statistically significant lower angle of rotation compared to the controls. For the contusion model, there was a statistically significant functional recovery in the LLLT group compared to untreated control. Light therapy applied noninvasively promotes axonal regeneration and functional recovery in acute SCI caused by different types of trauma. These results suggest that light is a promising therapy for human SCI (Wu et al. 2009).

55.4 Clinical Results

Many studies were done investigating the effects of phototherapy on nerve regeneration in humans. Such research requires large investments of time and money to achieve the necessary data. The nerve repair process is too long and too difficult to collect data and compare to regular clinical and surgical treatments.

Figure 55.2 shows an example of a good clinical result for facial nerve treatment with LLLT after postsurgical removal of a gemistocystic astrocytoma in the temporobasal region to approach the cavernous sinus. The laser treatment was done with the following parameters: λ = 830 nm, energy density = 120 J/cm^2, power density of 120 mW/cm^2, 5 times per week. The recovery of movement was evident and the patient was happy with the result.

It is known that injury of a major nerve trunk frequently results in considerable disability associated with loss of sensory and motor functions. Spontaneous recovery of long-term severe incomplete peripheral nerve injury is often unsatisfactory.

A good pilot study was conducted to investigate prospectively the effectiveness of low power laser irradiation (780 nm) in the treatment of patients suffering from incomplete peripheral nerve and brachial plexus injuries for 6 months up to several years. A randomized, double-blind, placebo-controlled trial was performed on 18 patients who were randomly assigned placebo (nonactive light: diffused LED lamp) or low-power laser irradiation (wavelength, 780 nm; power, 250 mW). Twenty-one consecutive daily sessions of laser or placebo irradiation were applied transcutaneously for 3 h to the injured peripheral nerve (energy density, 450 J/mm^2) and for 2 h to the corresponding segments of the spinal cord (energy density, 300 J/mm^2). Clinical and electrophysiological assessments were done at baseline, at the end of the 21 days of treatment, and 3 and 6 months thereafter. The laser-irradiated and placebo groups were in clinically similar conditions at baseline. The analysis of motor function during the 6-month follow-up period compared to baseline showed statistically significant improvement in the laser-treated group compared to the placebo group. No statistically significant difference was found in sensory function. Electrophysiological analysis also showed statistically significant improvement in recruitment of voluntary muscle activity in the laser-irradiated group compared to the placebo group. This pilot study suggests that in patients with long-term peripheral nerve injury, noninvasive 780 nm laser phototherapy can progressively improve

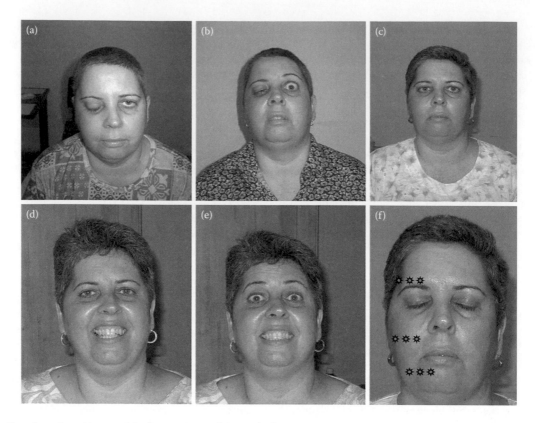

FIGURE 55.2 Female patient, 43 years old, after postsurgical removal of a gemistocystic astrocytoma in the temporobasal region to approach the cavernous sinus. The patient had ptosis, deviation from the central axis of the orbit to adduction, hypoesthesia, and palsy of the left facial muscles (a) that could be seen at the initial aspect (first day). (b) The patient after 11 days of starting the laser treatment (infrared laser $\lambda = 830$ nm, energy density = 120 J/cm^2, power density of 120 mW/cm^2, 5 times per week). (c) The patient after 55 days of the treatment. (d, e) The outcome of the treatment after 4 months. (f) We can see the points used to treat the patient, addressing three points on each branch of the facial nerve. (Image courtesy of Prof. Luiz Ferreira Monteiro.)

nerve function, which can lead to significant functional recovery (Anders, Geuna, and Rochkind 2004).

55.5 Laser Dosage

There is a broad range of dosages used in the reviewed papers. In animals, the doses used depend on the experimental model, the animal species, the goal of the tests, and so on. In human beings, we have suggestions of dosage for spinal cord: the laser irradiation should be transcutaneously directly above the projection of the corresponding segments of the spinal cord, at the base of the nerve root, bilaterally. Each side should be irradiated for 60 min a day (150 J/mm^2), totaling 120 min a day (300 J/mm^2) (Rochkind, Geuna, and Shainberg 2009; Wu et al. 2009).

In the peripheral nerve area, laser irradiation should be performed noninvasively and transcutaneously directly above the projection of the injured nerve, which should be divided into three parts: proximal, injured area, and distal. Each section should be irradiated for 60 min a day (150 J/mm^2), totaling 180 min a day (450 J/mm^2) (Anders, Geuna, and Rochkind 2004; Rochkind, Geuna, and Shainberg 2009).

If we consider the spot size of the lasers or LED sources, there are many options, but as we know, the nerve dimensions are small but scattered in to different areas. It is interesting to reach the largest possible area of the nerve supply. The wavelength can be chosen according to need, but for humans, good penetration is important. This is accomplished with lasers in the infrared range of the spectrum (780–830 nm), which can penetrate approximately 4 cm from the skin (Gigo-Benato, Geuna, and Rochkind 2005).

55.6 Concluding Remarks

There is still much to be learned about LLLT and its many proposed clinical applications, including those injuries and diseases of peripheral nerves and the spinal cord. However, with the current data available, we can still benefit patients with this alternative treatment that is not expensive and whose growth in applications is being felt and observed in clinical practice.

The underlying mechanisms are still not fully clear on how the laser works on the repair of nerves and the spinal cord. Clinical applications are already underway and some are already published (Rochkind 2009).

Some suggestions are being put forth about the possible mechanisms involved in photostimulated nerve repair. Stimulation of SCs to enhance neural tube formation, activation of neurotrophic factors to accelerate axoplasm production, and incremental formation of new blood vessels to supply sufficient nutrients are being proposed.

As a mechanism involved in the nervous system repair by phototherapy, we can find the upregulation of a number of neurotrophic growth factors (like BDNF and GDNF) and extracellular matrix proteins known to support neurite outgrowth (Byrnes et al. 2005b). The combination of the application of drugs, growth factors, procedures of transplantation, and autografts and allografts with phototherapy can be a good option to treat some diseases and traumatic lesions on the nerves and spinal cord.

Many researchers around the world study the influences of light (especially lasers and LEDs) on nerve repair, as well as other processes like tissue repair, in the modulation of inflammation and pain control. All these results show that light therapy is safe and has no side effects. This modality can become a good therapeutic option to treat different peripheral nerve diseases and injuries. It will be possible, with a good background of knowledge in this field supported by data from different researchers, to pave the way for acceptance and standardization of this innovative therapy in the clinic (Moges et al. 2011).

References

Anders, J. J., R. C. Borke, S. K. Woolery, and W. P. Van de Merwe. 1993. Low power laser irradiation alters the rate of regeneration of the rat facial nerve. *Lasers Surg Med* 13:72–82.

Anders, J. J., S. Geuna, and S. Rochkind. 2004. Phototherapy promotes regeneration and functional recovery of injured peripheral nerve. *Neurol Res* 26:233–239.

Bagis, S., U. Comelekoglu, B. Coskun et al. 2003. No effect of GA-AS (904 nm) laser irradiation on the intact skin of the injured rat sciatic nerve. *Lasers Med Sci* 18:83–88.

Barbosa, R. I., A. M. Marcolino, R. R. de Jesus Guirro et al. 2010. Comparative effects of wavelengths of low-power laser in regeneration of sciatic nerve in rats following crushing lesion. *Lasers Med Sci* 25:423–430.

Basford, J. R., H. O. Hallman, J. Y. Matsumoto et al. 1993. Effects of 830 nm continuous wave laser diode irradiation on median nerve function in normal subjects. *Lasers Surg Med* 13:597–604.

Belchior, A. C., F. A. dos Reis, R. A. Nicolau et al. 2009. Influence of laser (660 nm) on functional recovery of the sciatic nerve in rats following crushing lesion. *Lasers Med Sci* 24:893–899.

Byrnes, K. R., R. W. Waynant, I. K. Ilev et al. 2005a. Light promotes regeneration and functional recovery and alters the immune response after spinal cord injury. *Lasers Surg Med* 36:171–185.

Byrnes, K. R., X. Wu, R. W. Waynant, I. K. Ilev, and J. J. Anders. 2005b. Low power laser irradiation alters gene expression of olfactory ensheathing cells *in vitro*. *Lasers Surg Med* 37:161–171.

Chen, Y. S., S. F. Hsu, C. W. Chiu et al. 2005. Effect of low-power pulsed laser on peripheral nerve regeneration in rats. *Microsurgery* 25:83–89.

Curtis, N. J., E. Owen, D. M. Walker, and H. Zoellner. 2011. Comparison of microsuture, interpositional nerve graft, and laser solder weld repair of the rat inferior alveolar nerve. *J Oral Maxillofac Surg* 69:e246–e255.

Gigo-Benato, D., S. Geuna, A. de Castro Rodrigues et al. 2004. Low-power laser biostimulation enhances nerve repair after end-to-side neurorrhaphy: A double-blind randomized study in the rat median nerve model. *Lasers Med Sci* 19:57–65.

Gigo-Benato, D., S. Geuna, and S. Rochkind. 2005. Phototherapy for enhancing peripheral nerve repair: A review of the literature. *Muscle Nerve* 31:694–701.

Gigo-Benato, D., T. L. Russo, E. H. Tanaka et al. 2010. Effects of 660 and 780 nm low-level laser therapy on neuromuscular recovery after crush injury in rat sciatic nerve. *Lasers Surg Med* 42:673–682.

Hashmi, J. T., Y. Y. Huang, S. K. Sharma et al. 2010. Effect of pulsing in low-level light therapy. *Lasers Surg Med* 42:450–466.

Lavi, R., A. Shainberg, H. Friedmann et al. 2003. Low energy visible light induces reactive oxygen species generation and stimulates an increase of intracellular calcium concentration in cardiac cells. *J Biol Chem* 278:40917–40922.

Lubart, R., M. Eichler, R. Lavi, H. Friedman, and A. Shainberg. 2005. Low-energy laser irradiation promotes cellular redox activity. *Photomed Laser Surg* 23:3–9.

Medalha, C. C., G. C. Di Gangi, C. B. Barbosa et al. 2012. Low-level laser therapy improves repair following complete resection of the sciatic nerve in rats. *Lasers Med Sci* 27:629–635.

Moges, H., X. Wu, J. McCoy et al. 2011. Effect of 810 nm light on nerve regeneration after autograft repair of severely injured rat median nerve. *Lasers Surg Med* 43:901–906.

Nectoux, E., C. Taleb, and P. Liverneaux. 2009. Nerve repair in telemicrosurgery: An experimental study. *J Reconstr Microsurg* 25:261–265.

Rochkind, S. 2009. Phototherapy in peripheral nerve regeneration: From basic science to clinical study. *Neurosurg Focus* 26:E8.

Rochkind, S., L. Barr-Nea, A. Bartal et al. 1988. New methods of treatment of severely injured sciatic nerve and spinal cord. An experimental study. *Acta Neurochir Suppl (Wien)* 43:91–93.

Rochkind, S., L. Barr-Nea, N. Razon, A. Bartal, and M. Schwartz. 1987a. Stimulatory effect of He-Ne low dose laser on injured sciatic nerves of rats. *Neurosurgery* 20:843–847.

Rochkind, S., S. Geuna, and A. Shainberg. 2009. Chapter 25: Phototherapy in peripheral nerve injury: Effects on muscle preservation and nerve regeneration. *Int Rev Neurobiol* 87:445–464.

Rochkind, S., L. Leider-Trejo, M. Nissan et al. 2007. Efficacy of 780-nm laser phototherapy on peripheral nerve regeneration after neurotube reconstruction procedure (double-blind randomized study). *Photomed Laser Surg* 25:137–143.

Rochkind, S., M. Nissan, M. Alon, M. Shamir, and K. Salame. 2001. Effects of laser irradiation on the spinal cord for the regeneration of crushed peripheral nerve in rats. *Lasers Surg Med* 28:216–219.

Rochkind, S., M. Nissan, L. Barr-Nea et al. 1987b. Response of peripheral nerve to He–Ne laser: Experimental studies. *Lasers Surg Med* 7:441–443.

Rochkind, S., and G. E. Ouaknine. 1992. New trend in neuroscience: Low-power laser effect on peripheral and central nervous system (basic science, preclinical and clinical studies). *Neurol Res* 14:2–11.

Rochkind, S., M. Rousso, M. Nissan et al. 1989. Systemic effects of low-power laser irradiation on the peripheral and central nervous system, cutaneous wounds, and burns. *Lasers Surg Med* 9:174–182.

Rosner M., M. Caplan, S. Cohen et al. 1993. Dose and temporal parameters in delaying injured optic nerve degeneration by low-energy laser irradiation. *Lasers Med Sci* 13:611.

Saito, K., S. Hashimoto, H. S. Jung, M. Shimono, and K. Nakagawa. 2011. Effect of diode laser on proliferation and differentiation of PC12 cells. *Bull Tokyo Dent Coll* 52: 95–102.

Santos, A. P., C. A. Suaid, M. Xavier, and F. Yamane. 2012. Functional and morphometric differences between the early and delayed use of phototherapy in crushed median nerves of rats. *Lasers Med Sci* 27:479–486.

Serafim, K. G., P. Ramos Sde, F. M. de Lima et al. 2012. Effects of 940 nm light-emitting diode (led) on sciatic nerve regeneration in rats. *Lasers Med Sci* 27:113–119.

Shamir, M. H., S. Rochkind, J. Sandbank, and M. Alon. 2001. Double-blind randomized study evaluating regeneration of the rat transected sciatic nerve after suturing and postoperative low-power laser treatment. *J Reconstr Microsurg* 17:133–137; discussion 138.

Shen, C. C., Y. C. Yang, and B. S. Liu. 2011. Large-area irradiated low-level laser effect in a biodegradable nerve guide conduit on neural regeneration of peripheral nerve injury in rats. *Injury* 42:803–813.

Shin, D. H., E. Lee, J. K. Hyun et al. 2003. Growth-associated protein-43 is elevated in the injured rat sciatic nerve after low power laser irradiation. *Neurosci Lett* 344:71–74.

Wu, X., A. E. Dmitriev, M. J. Cardoso et al. 2009. 810 nm wavelength light: An effective therapy for transected or contused rat spinal cord. *Lasers Surg Med* 41:36–41.

Yazdani, S. O., A. F. Golestaneh, A. Shafiee et al. 2012. Effects of low level laser therapy on proliferation and neurotrophic factor gene expression of human Schwann cells *in vitro*. *J Photochem Photobiol B* 107:9–13.

Zhang, L. X., X. J. Tong, X. H. Yuan, X. H. Sun, and H. Jia. 2010. Effects of 660-nm gallium–aluminum–arsenide low-energy laser on nerve regeneration after acellular nerve allograft in rats. *Synapse* 64:152–160.

56

Low-Level Laser Therapy in Dentistry

Daiane Thais Meneguzzo
São Leopoldo Mandic

Leila Soares Ferreira
University of São Paulo

56.1 Introduction

Dentistry is the field of medicine that involves the study, diagnosis, prevention, and treatment of diseases, disorders, and conditions of the oral cavity, maxillofacial area, and the adjacent and associated structures and their impact on the human body. Muscle, nervous, connective (blood and bone), and epithelial tissues make up the oral cavity, and many treatments need to involve these different tissues simultaneously. Inflammatory processes result from all kinds of tooth interventions in which pulp tissue is stimulated such as caries lesions, cavity preparation, restorative procedures, tooth bleaching, orthodontic movement, and masticatory dysfunction. Furthermore, different types of surgical procedures such as tooth extraction, implant placement, and grafting produce inflammation and need considerable time to heal.

Low-level laser therapy (LLLT) is known to modulate the inflammatory process, promote analgesia, and accelerate wound healing (see Chapters 50–52) (Bjordal et al. 2006; Chow, David, and Armati 2007; Demidova-Rice et al. 2007). Moreover, LLLT can also perform antimicrobial activity using photodynamic therapy (PDT) when combined with a photosensitizer (Rajesh et al. 2011) (see Chapter 34).

LLLT is indicated to treat or act as an adjuvant treatment of several lesions encountered in dentistry (see Table 56.1). The benefits provided by the application of LLLT in all types of dental treatments include promoting comfort, providing analgesia, speeding wound healing, and improving the quality of all oral tissue regeneration in patients, all with no side effects (see Table 56.1). Furthermore, dentists can expect to obtain better clinical results in their treatments or at least less clinical failure when LLLT is added to the conventional treatment by improving patient capability of wound healing. Use of LLLT provides better clinical outcomes, better acceptance by patients, and better professional recognition.

56.2 Dentistry Clinical Protocols

Clinical experience has shown that many key characteristics considered important to achieve optimal LLLT results widely differ among patients. In the literature, the existence of a great variety of protocols proposed for the same injury raises doubts and controversies. Also, the development of randomized clinical trials to determine LLLT protocols is very challenging, mainly due to the difficulty in obtaining a large number of patients with similar injuries, as well as the need to choose an equally effective protocol for everyone. As with many other kinds of health treatment, the individual optimization of the dose seems to be essential to achieve optimum results. Therefore, the greatest challenge for the clinician is to develop individualized protocols.

The light absorption may vary depending on the patient's skin phototype (a classification system based on a person's sensitivity to sunlight), immune system condition, metabolism, general systemic condition, presence of endogenous chromophores, and the depth, color, and stage of the lesion. Moreover, the laser device (power, wavelength, spot area, energy) and therapeutic management selected by the professional as the treatment regimen (number of LLLT sessions per week) and mode of irradiation (punctual or scanning mode, distance from tissue) will affect LLLT results. The benefits of LLLT are directly related to the correct selection of the protocol. The following 10 steps will help professionals to develop individualized clinical protocols to achieve better clinical results with LLLT.

Step 1: Make the correct diagnosis. This first step is critical to initiate protocol development. Examinations or reports conducted by other health professionals may be required. In case of nerve damage, (e.g., neurological), diagnosis should exclude the presence of tumors or other lesions in the affected region to be irradiated.

TABLE 56.1   Main Applications of LLLT in Dentistry and Expected Effects

Sections of Dentistry	LLLT Indications in Dentistry	Expected LLLT Effects
Endodontics	Pain after endodontic instrumentation (Tunér and Hode 2004) Pulp inflammation and diagnostics (Mohammadi 2009)	Pain relief, bone stimulation, modulation of inflammatory response
Oral pathology	Mucositis (Gautam et al. 2012; Carvalho et al. 2011b) Aphthous ulcers (De Souza et al. 2010) Herpes (Carvalho et al. 2010; Muñoz-Sanchez et al. 2012) Lichen planus (Jajarm, Falaki, and Mahdavi 2011) Burning mouth syndrome (Carvalho et al. 2011a) Xerostomy (Lončar et al. 2011) Paralysis (Meneguzzo et al. 2010) Neuralgia (Hsieh et al. 2012)	Acceleration of tissue healing, modulation of inflammatory response, pain relief; salivary gland biostimulation, nerve regeneration

Oral surgery	Tissue healing (Ozcelik et al. 2008)
	Postoperative treatment of
	  • Soft tissue surgeries (Almeida et al. 2009)
	  • Tooth extraction (Aras and Güngörmüş 2010; Marković et al. 2006; Ribeiro et al. 2011)
	  • Implants (Campanha et al. 2010; Khadra 2005; Lopes et al. 2005; Maluf et al. 2010)
	  • Grafting (da Silva and Camilli 2006; Torres et al. 2008)
	Paresthesia (Rochkind et al. 2007; Khullar et al. 1996)
	Trismus (Aras and Güngörmüş 2010)
	Alveolitis (Tunér and Hode 2010)
	Bisphosphonate-related osteonecrosis of the jaws (Scoletta et al. 2010)
		Acceleration of wound healing, control of pain and edema, optimization of implant osseointegration, enhancement of local microcirculation, bone stimulation, improvement of mouth opening, nerve regeneration, muscle relaxation

Orthodontics	Acceleration of tooth movement (Sousa et al. 2011; Doshi-Mehta and Bhad-Patil 2012) Reduction of tooth pain (Xiaoting, Yin, and Yangxi 2010)	Bone remodeling, modulation of inflammatory response, pain relief
Pediatric	Primary herpetic gingivostomatitis (Navarro et al. 2007) Oral mucositis in children (Cauwels and Martens 2011) Trauma (soft and hard tissue) (Olivi, Genovese, and Caprioglio 2009)	Acceleration of tissue healing, pain relief, modulation of inflammatory response
Periodontics	Gingival grafting (Almeida et al. 2009) Gingival inflammation (Pesevska et al. 2012)	Modulation of inflammatory response, pain relief, tissue healing
Prosthodontics	TMJD and myalgias (temporomandibular disorders) (Aras and Güngörmüş 2010; Mazzetto et al. 2010; Salmos-Brito et al. 2013) Myalgias (Venezian et al. 2010)	Pain relief, modulation of inflammatory response, muscle relaxation
Restorative dentistry	Dentinal hypersensitivity (Yilmaz et al. 2011) Anesthesia (Tunér and Hode 2010)	Modulation of inflammatory response, pain relief, tertiary dentin formation

Step 2: Characterization of the lesion. This step comprises the observation of phase, size, and location of injury, quality of the tissues involved in the lesion and adjacent tissues, local microcirculation, the presence or absence of acute inflammation (pain, swelling, heat, redness, loss of function) or chronic inflammation (lesion length), the accessibility of the lesion to irradiation, causal factors, and conventional techniques that have been already used and their results, among other features.

Step 3: Characterization of the patient. This step includes phototype, systemic condition, immune system response, metabolism, age, body weight, medication used, general emotional state, and availability to perform the treatment (frequency, the need of home care sessions).

Step 4: Recognize the main LLLT desired effect. Based on the knowledge that LLLT has three main clinical effects—modulation of inflammatory response, production of analgesia, and acceleration of wound

healing—the clinician has to determine what is the most important desired effect in order to decide the best energy to achieve it. For example, in aphthous ulcer treatment, even though wound healing is necessary, the pain relief effect must be prioritized. The summary in Table 56.2 suggests an energy therapeutic

TABLE 56.2   Energy Therapeutic Window: Minimum and Maximum of Energy per Point According to the Clinical Desired Effect of LLLT and Depth of Target Tissue

Type of Tissue	Healing Effect	Modulation of Inflammatory Process	Analgesic Effect
Superficial tissues (skin and oral mucosa)	0.2–1 J per point	1–3 J per point	>3 J per point
Intermediate tissues (muscle, gland, lymph nodes)	1–2 J per point	2–4 J per point	>4 J per point
Deep tissues (bone, nerve)	2–3 J per point	3–5 J per point	>5 J per point

window based on the LLLT benefits desired and the depth of the target tissue.

Step 5: Choose the LLLT protocol. The patient (step 3) and lesion (step 2) characteristics as well as the laser device should be considered in choosing the LLLT protocol. Wavelength and energy comprise the basic laser parameters. In general, lasers with red wavelength are indicated for superficial tissues and infrared lasers for deep tissues (Chung et al. 2011). In dentistry, the most common laser device used in LLLT is the diode laser with wavelengths between 660 and 830 nm. Energy protocols already published or suggested may work as a starting point and can be modified or adjusted when necessary according to the data in Table 56.3.

Step 6: Determine the number of irradiation points and distance between them. The number of points will depend on the size of the lesion, and as general rule, it is suggested that one point of irradiation should cover an area of 1 cm². Thereafter, this number of points can increase or decrease depending on the desired effect previously determined: healing/biostimulation, anal-

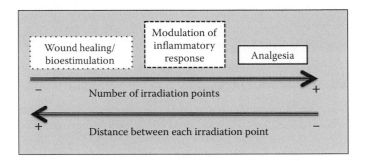

FIGURE 56.1 Number of irradiation points and distance between them can influence LLLT effects (healing/biostimulation, analgesia, or modulation of inflammation).

gesia, or modulation of inflammation, as shown in Figure 56.1.

Step 7: Evaluation method of LLLT effectiveness. There are several methods to evaluate the patient's clinical outcome, such as the visual analogue scale (VAS) pain scale, patient daily reports, photographs and/or videos, and detailed description of the signs and symptoms, as well as the evolution of the lesion, palpation of the area, determination of affected area, and mouth opening, among other forms of assessment depending on the lesion.

Step 8: Establishment of LLLT session frequency and irradiation mode. The frequency of LLLT will depend on the cellular metabolism of the tissue. Daily irradiation is recommended for superficial soft tissue healing, every 24/48 h for muscle and blood, every 48/72 h for bone tissue, and every 72 h or once a week for nervous tissue. The irradiation mode can be punctual (high-power density) or scanning (lower power density), perpendicular to the target tissue when possible (to avoid energy loss), and in contact mode (high-energy) or distant from tissue (less energy).

Step 9: Perform laser irradiation. Before starting the laser irradiation, check: dose, punctual and contact mode when possible, disposable protective barrier for laser equipment and goggles.

Step 10: Monitoring patient clinical outcome. All the information collected regarding the patient's response to LLLT must be recorded and compared every session in order to observe and determine whether the protocol is effective or not. The follow-up can be done every three sessions according to the clinical evolution method of the evaluation chosen. In cases with slow or close to no response, the LLLT protocol must be modified according to Table 56.3.

56.3 Key Clinical Indications and Protocols

In this section, the main clinical indications for LLLT in dentistry followed by the corresponding basic laser protocol are

TABLE 56.3 Clinical Features Relevant for the Development of Protocols

According to the clinical features observed, it is necessary to increase (⇑) or decrease (⇓) the energy. Scanning irradiation is also sometimes recommended (S).	
Acute inflammation	⇓
Chronic inflammation	⇑
Prodromal phase (infection)	⇑
Large lesions	⇑
Small lesions	⇓
Superficial tissue	⇓
Deep tissue	⇑
Sensitive target tissue	⇓/S
Local ischemia	⇑
Local pain	⇑
Local edema	⇑
Local paresthesia	⇑
Local hyperesthesia	⇓
Inability to do perpendicular irradiation over target tissue	⇑
Impossibility of irradiation in contact with the target tissue due to local sensitivity, severe pain, or difficult access	⇑/S
Patient skin phototype I–II	⇑
Patient skin phototype > III	⇓
Severe systemic involvement	⇓
Immunologically compromised patients	⇓
Accelerated metabolism (babies)	⇓
Slow metabolism (elderly or systemically compromised patients)	⇓
Pronounced fat tissue at the site of irradiation	⇑
Local lesions caused by systemic medication (e.g., medications that cause xerostomia)	⇑
Emotionally debilitated patient	⇑
Patient with limited availability for LLLT sessions	⇑

presented. It is noteworthy that all protocols can be individualized when necessary according to clinical evolution (see the 10-step protocol development above).

56.3.1 Wound Healing

Many studies have been performed demonstrating the clinical effects of LLLT on the acceleration of wound healing in different types of lesions (see Table 56.1). The modulation of the inflammatory response combined with the increase in fibroblast cell proliferation (Peplow, Chung, and Baxter 2010), collagen synthesis (Almeida-Lopes et al. 2001), and secretion of growth factors (Toyokawa et. al. 2003) leads to a faster tissue repair and allows chronic processes to resolve (Angelov et al. 2009). Thus, LLLT represents an outstanding adjuvant therapy for soft tissue healing, or in some situations is the treatment itself.

56.3.1.1 Oral Mucosa Healing

LLLT stimulates mucosal healing after soft or hard tissue surgeries (Ozcelik et al. 2008) and can be an adjunct to nonsurgical periodontal treatment (Aykol et al. 2011; Quadri et al. 2005). Suggested laser protocol: daily red laser irradiation, 0.5–1.0 J per point (Figure 56.2).

56.3.1.2 Aphthous Ulcers

LLLT should be applied daily using an analgesic protocol to treat pain, the patient's major complaint, followed by a biostimulatory protocol to accelerate healing (De Souza et al. 2010). Suggested laser protocol: infrared laser, 2.0–3.0 J per point (analgesic); red laser, 0.5–1.0 J per point (healing) (Figures 56.3 and 56.4).

56.3.1.3 Herpes Labialis

This viral disease is characterized by painful vesicular and ulcerative lesions often present in the labial and perilabial skin region or oral mucosa. The manifestation of herpes has three phases—prodromal, vesicular, and crust—with different symptomatology. Thus, different LLLT treatment approaches are suggested for each phase.

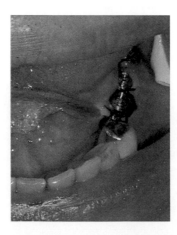

FIGURE 56.2 Red diode laser irradiation for soft tissue healing in postoperative of implant placement (0.7 J per point, 660 nm).

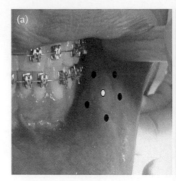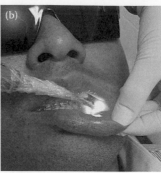

FIGURE 56.3 Aphthous ulcer irradiation points (a) and red laser light during irradiation with 660 nm, 100 mW, 1 J per point (b).

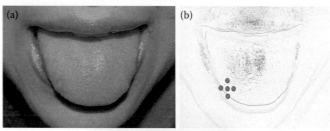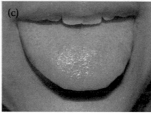

FIGURE 56.4 **(See color insert.)** Treatment of an aphthous ulcer lesion with one session of LLLT. (a) Initial lesion, (b) schematic drawing of irradiation points performed with a diode laser, 660 nm, 3 J per point, and (c) clinical result after 3 days of LLLT treatment.

Prodromal phase: The initial stage of the lesion, with erythema, itching, or burning before the characteristic rash of vesicles. LLLT treatment promotes stimulation of local immunological system and can inhibit the viral infection progress. Suggested laser protocol: red or infrared laser, 4.0–6.0 J per point, with a minimum of 20 J total energy delivered.

Vesicular phase: Presence of one or more vesicles with pain and local swelling. The use of LLLT in this phase with biostimulatory energy is controversial because of the risk of exacerbation of the manifestation. Thus, LLLT should be performed with high energy to promote pain and swelling reduction, as well as the inhibition of virus replication, and consequently allowing faster wound healing. Suggested laser protocol: red or infrared laser, 4.0–6.0 J per point, with a minimum of 20 J total energy delivered.

Crust phase: Stage of tissue repair. Healing process is usually delayed due to the fact that the patient is constantly

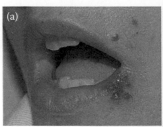
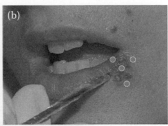
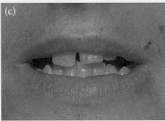

FIGURE 56.5 Treatment of the crust phase of herpes lesion. (a) Initial lesion, (b) red diode laser irradiation and schematic irradiation points of 0.7 J each, and (c) clinical result after 7 days of one session of LLLT treatment.

wetting the injury while brushing teeth or taking a bath. LLLT accelerates wound healing and promotes stimulation of local immunological system allowing frequency and intensity reduction of further herpes manifestation (Muñoz Sanchez et al. 2012). Suggested laser protocol: daily irradiation, red laser, 0.5–1.0 J per point (Figure 56.5).

56.3.1.3.1 Preventative Treatment

In most cases, the herpes manifestations are associated with the decline of the patient's immune response. Preventative treatment with LLLT will stimulate the local immunological system (labial and perilabial area), decreasing the intensity and frequency of manifestations (De Carvalho et al. 2010). Suggested laser protocol: red or infrared laser, 2.0–3.0 J per point in all labial and perilabial areas, 10 sessions, once a week.

56.3.1.3.2 Primary Herpetic Gingivostomatitis

Primary herpetic gingivostomatitis is the most common manifestation of primary herpes simplex virus (HSV) infection during childhood. The painful sores in oral cavity and labial and perilabial areas cause difficulty in feeding, requiring LLLT treatment as soon as possible (Navarro et al. 2007). Suggested laser protocol: daily red laser, 2.0–3.0 J per point or in a scanning mode over oral mucosa and perilabial area.

56.3.1.4 Oral Mucositis

Oral mucositis (OM) is characterized by inflammation and ulceration of oral mucosa with severe pain caused by the toxicity of chemotherapy and/or oral radiotherapy (for details see Chapter 49). LLLT has been demonstrated to prevent and reduce the lesion severity and pain and hasten OM healing (Carvalho et al.

2011b; Gautam et al. 2012; Schubert et al. 2007). Currently, LLLT is the only treatment able to heal and prevent OM. Suggested laser preventive protocol: weekly red laser irradiation, 2.0–3.0 J per point all over the mouth (minimum of four points in each site: tongue, hard and soft palate, buccal mucosa, and lips). Suggested laser treatment protocol: daily red laser, 3.0–4.0 J per point all over the lesions in the mouth. In severe pain areas, infrared laser irradiation can be also performed with 4.0–6.0 J per point.

56.3.1.5 Lichen Planus

Lichen planus is a chronic inflammatory disease with unknown etiology that affects the oral mucosa and can be very painful. LLLT is an alternative treatment to stimulate mucosal healing and reduce pain. Suggested laser protocol: red (Jajarm, Falaki, and Mahdavi 2011) or infrared laser (Cafaro et al. 2010), 1.0–2.0 J per point.

56.3.2 Control of Pain and Edema

56.3.2.1 Immediate Postoperative Control of Pain and Edema

After a surgical procedure, the inflammatory process is immediately activated and its severity will depend on the type of surgery and the patient's systemic condition. Therefore, it is important to perform LLLT immediately after the surgery in order to control the initial events of inflammation and prevent edema commencement (Tunér and Hode 2010). The protocol to promote pain and edema control can be used for several types of surgical procedures such as implants, tooth extraction (Marković and Todorović 2006), grafting, and periodontal and orthognathic surgeries. Since LLLT promotes the increase in local microcirculation, it is recommended to wait at least 15 min after the procedure in order to avoid extra bleeding. Suggested laser protocol: infrared laser around suture, 2 J per point; red laser over the suture, 0.5–1 J per point (Figure 56.6).

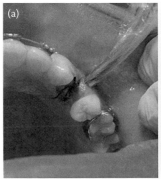
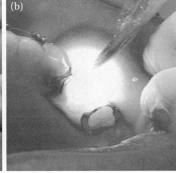

FIGURE 56.6 LLLT irradiation immediately after the surgery to control the initial events of inflammation and prevent edema (a), and LLLT laser irradiation with 660 nm, 100 mW, 0.5 J per point (b).

56.3.2.2 Treatment of Installed Edema

LLLT anti-inflammatory and biostimulatory effects over the local microcirculation and lymph nodes allow reduction of edema (Aras and Güngörmüş 2010; Tunér and Hode 2010). Suggested laser protocol: the irradiation can be applied over related lymph nodes and swollen area; infrared laser, 2.0–3.0 J per point (Figure 56.7).

56.3.3 Bone Biostimulation

56.3.3.1 General Stimulation of Bone Repair

Unlike soft tissue healing, bone repair is difficult to be clinically visualized. Nevertheless, the benefits of laser irradiation on bone tissue include a stimulatory effect on osteoblasts (proliferation and production of bone matrix), an increase in the activity of alkaline phosphatase, tissue NO, and calcium levels, and promotion of pronounced angiogenesis (Ihsan 2005; Khadra 2005). Although there is an increase in cellular activity (bone resorption and formation), no structural alterations are observed in the irradiated bone (Khadra 2005; da Silva and Camilli 2006). Suggested laser protocol: infrared laser, 2.0–4.0 J per point.

56.3.3.2 Implant Osseointegration

LLLT can assist in the osseointegration of implants. It improves the initial contact between bone and implant, increases the proliferation and adhesion of osteoblasts to the implant surface, enhances the content of calcium, and promotes angiogenesis in bone tissue (Campanha et al. 2010; Lopes et al. 2005; Maluf et al. 2010). A more pronounced response is achieved when LLLT is performed in the first 14 days when cells are proliferating and

secreting matrix that will be mineralized later. Suggested laser protocol: infrared laser, 2.0–4.0 J per point.

56.3.3.3 Bone Grafting

LLLT increases local microcirculation and has a positive biomodulatory effect on bone tissue repair. It is able to improve collagen fiber deposition and the amount of well-organized bone trabeculae at early stages (da Silva and Camilli 2006; Torres et al. 2008). Suggested laser protocol: start irradiation together with soft tissue (some points could be the same), red laser, 0.5–1.0 J per point for 2 weeks, followed by bone biostimulation protocol.

56.3.3.4 Tooth Movement (Orthodontics)

There is increasing evidence that LLLT is able to accelerate tooth movement and reduce dental pain after orthodontic archwire is applied (Tortamano et al. 2009; Doshi-Mehta and Bhad-Patil 2012; Sousa et al. 2011). Suggested laser tooth movement protocol: after application of orthodontic archwire, infrared laser, 2.0–3.0 J per point. Suggested laser protocol for orthodontic pain: infrared laser, 4.0–6.0 J per point at the apex of each tooth.

56.3.3.5 Bisphosphonate-Related Osteonecrosis of the Jaw

Bisphosphonates (BPs) are widely used for the treatment of osteoporosis, malignant hypercalcemia, bone metastasis of solid cancers, and multiple myeloma bone diseases. However, BPs may have serious side effects including a rare but painful condition called bisphosphonate-related osteonecrosis of the jaw (BRONJ). Prevention and treatment approaches have not been established to date, but antibiotics are recommended to control the infection and avoid surgical intervention after uncontrolled bone loss. LLLT is suggested in order to increase local microcirculation and stimulate bone tissue repair as well as to reduce the pain caused by BRONJ (Scoletta et al. 2010). Suggested laser protocol: red laser 1.0–2.0 J per point (initial bone necrosis); red and infrared laser 2.0–4.0 J per point (advanced bone necrosis and pain).

56.3.4 Nerve Tissue Healing

56.3.4.1 Paresthesia

In dentistry, paresthesia mainly affects the lingual, mentual, and inferior alveolar nerves. The main causes are implant, orthognathic or third molar extraction surgeries, mandibular fractures, compressive lesions, impacted teeth, or infections. LLLT has the potential to promote nerve biostimulation and increase local microcirculation that is essential for nerve regeneration. The literature states that LLLT accelerates the sensory perception recovery in time and magnitude (Khullar et al. 1996), and better results are achieved when early laser irradiation is performed (Rochkind et al. 2007). Suggested laser protocol: red or infrared laser, 1.0–4.0 J per point. The energy and wavelength protocol must vary each session to prevent the nerves from getting used to the stimulus (Figure 56.8).

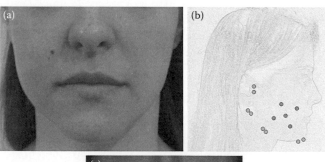
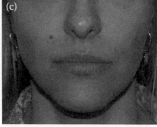

FIGURE 56.7 Treatment of edema with one session of LLLT. (a) Initial lesion, (b) schematic drawing of irradiation points performed with a diode laser, 880 nm, 2 J per point, and (c) clinical result after 3 days of LLLT treatment.

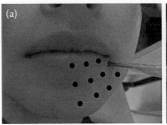

FIGURE 56.8 LLLT irradiation for paresthesia due to third molar extraction. (a) Laser irradiation and extra oral schematic irradiation points, and (b) intraoral laser irradiation over mentual nerve.

56.3.4.2 Facial Paralysis

LLLT can be an alternative treatment if the paralysis is peripheral (facial nerve damage) and has been diagnosed by a neurologist. It recovers the function of the nerve tissue affected, normalizing the facial muscle contractions and facial aesthetics (Meneguzzo et al. 2010). Suggested laser protocol: red or infrared laser, from 1.0 to 4.0 J per point. The energy and wavelength protocol must vary each session to prevent the nerves from getting used to the stimulus.

56.3.4.3 Neuralgia

Trigeminal neuralgia involves a dysfunction of the trigeminal nerve (cranial nerve V). The cause of the dysfunction of the trigeminal nerve is usually unknown and produces episodes of severe pain that can last from seconds to minutes. The diagnostic is clinical and is made by a neurologist. Diseases of the jaw, teeth, or sinuses or compression of the trigeminal nerve by a tumor or an aneurysm should be discarded. LLLT has been used as an alternative pain control treatment mainly when a patient does not respond to drug treatment or when the condition is recurrent despite surgery (Hsieh et al. 2012). The duration of the LLLT effects may vary depending on a patient's age (younger patients have a better prognosis), number of affected nerve branches, and time elapsed since the onset of disease. Suggested laser protocol: red or infrared laser, from 4.0 to 20.0 J per point. Low energy is used in the presence of hypersensitivity to light touch, and high energy is used in severe pain episodes.

56.3.5 General Dentistry Applications

56.3.5.1 Dentin Sensitivity

Dental sensitivity is one of the most common patient complaints in dental clinics. LLLT is a powerful anti-inflammatory and local analgesic, and in the long term accelerates tertiary dentin formation. Thus, LLLT is indicated to treat dental sensitivity after cavity preparation or bleaching or due to trauma. In cases of dentinal hypersensitivity, a proper diagnosis and removal of the etiological factor is essential for treatment success. Unlike the products and techniques designed to treat dentinal hypersensitivity by sealing dentinal tubules, LLLT acts directly on pulp tissue (Yilmaz et al. 2011). Suggested laser protocol: infrared laser, 2.0–4.0 J per point (Figure 56.9).

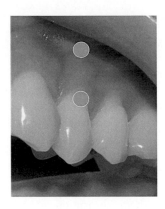

FIGURE 56.9 LLLT irradiation points (cervical and apical) for pain reduction in dentinal hypersensitivity.

56.3.5.2 Temporomandibular Joint Disorders

Temporomandibular joint disorder (TMJD) involves the inflammation of temporomandibular joint (TMJ), masticatory muscles, and nerves. LLLT promotes modulation of inflammatory response, pain relief, muscle relaxing, and nerve regeneration indicated in cases of myalgia, capsulitis, and trismus (Aras and Güngörmüş 2010; Mazzetto et al. 2010; Salmos-Brito et al. 2013). All LLLT benefits can contribute to a better outcome of TMJD conventional treatment. Suggested laser protocol on TMJD: infrared laser, 2.5–4.0 J per point over TMJ and 4.0–6.0 J per point over the muscles that refer pain or present trismus (Figure 56.10).

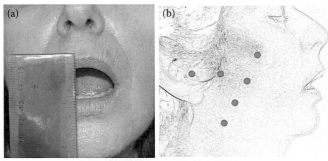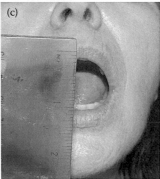

FIGURE 56.10 Patient presenting trismus due to TMJD. (a) Initial mouth opening, (b) image with laser irradiation points (IR laser, 100 mW, 4 J per point), and (c) final mouth opening.

56.3.5.3 Alveolitis

LLLT has analgesic and healing effect in the treatment of alveolitis acting on pain control, modulation of inflammatory reaction, and increase in microcirculation (Tunér and Hode 2010). Suggested laser protocol: infrared laser, 4.0–6.0 J per point (analgesic); red laser, 0.5–1.0 J per point (mucosal healing); red laser, 1.0–2.0 J inside the alveolar space (to increase microcirculation).

56.3.5.4 Anesthesia

LLLT improves local microcirculation and therefore can improve local anesthetic absorption when local laser irradiation is performed prior to the liquid injection. It can also reduce pain caused by introduction of the needle and numbness in the lip after anesthesia, which is particularly important in pediatric dental practice (Tunér and Hode 2010). Suggested laser protocol: red laser, 2.0–3.0 J per point to increase microcirculation; 4.0–8.0 J per point to reduce injection pain.

56.3.5.5 Xerostomia

Xerostomia is a common patient complaint characterized by oral dryness within the elderly population. It is caused by a reduction in normal salivary secretion of the major salivary glands (parotid, submandibular, and sublingual) usually due to drugs, head and neck chemotherapy, and the presence of a syndrome. LLLT effects on salivary glands have been demonstrated to be not only stimulating but also regenerative (Lončar et al. 2011). Suggested laser protocol: red and infrared laser over the salivary glands, 2.0–3.0 J per point.

References

Almeida, A. L., L. A. Esper, M. C. Sbrana et al. 2009. Utilization of low-intensity laser during healing of free gingival grafts. *Photomed Laser Surg* 27(4):561–564.

Almeida-Lopes, L., J. Rigau, R. A. Zângaro et al. 2001. Comparison of the low level laser therapy effects on cultured human gingival fibroblasts proliferation using different irradiance and same fluence. *Lasers Surg Med* 29(2):179–184.

Angelov, N., S. Pesevska, M. Nakova et al. 2009. Periodontal treatment with a low-level diode laser: Clinical findings. *Gen Dent* 57(5):510–513.

Aras, M. H., and M. Güngörmüş. 2010. Placebo-controlled randomized clinical trial of the effect two different low-level laser therapies (LLLT)—intraoral and extraoral—on trismus and facial swelling following surgical extraction of the lower third molar. *Laser Med Sci* 25(5):641–645.

Aykol, G., U. Baser, I. Maden et al. 2011. The effect of low-level laser therapy as an adjunct to non-surgical periodontal treatment. *J Periodontol* 82(3):481–488.

Bjordal, J. M., M. I. Johnson, V. Iversen et al. 2006. Low-level laser therapy in acute pain: A systematic review of possible mechanisms of action and clinical effects in randomized placebo-controlled trials. *Photomed Laser Surg* 24(2):158–168.

Cafaro, A., G. Albanese, P. G. Arduino et al. 2010. Effect of low-level laser irradiation on unresponsive oral lichen planus: Early preliminary results in 13 patients. *Photomed Laser Surg* 28(2):S99–S103.

Campanha, B. P., C. Gallina, T. Geremia et al. 2010. Low-level laser therapy for implants without initial stability. *Photomed Laser Surg* 28(3):365–369.

Carvalho, P. A., G. C. Jaguar, A. C. Pellizzon et al. 2011a. Effect of low-level laser therapy in the treatment of burning mouth syndrome: A case series. *Photomed Laser Surg* 29(12):793–796.

Carvalho, P. A., G. C. Jaguar, A. C. Pellizzon et al. 2011b. Evaluation of low-level laser therapy in the prevention and treatment of radiation-induced mucositis: A double-blind randomized study in head and neck cancer patients. *Oral Oncol* 47(12):1176–1181.

Cauwels, R. G., and L. C. Martens 2011. Low level laser therapy in oral mucositis: A pilot study. *Eur Arch Paediatr Dent* 12(2):118–123.

Chow, R. T., M. A. David, and P. J. Armati. 2007. 830 nm laser irradiation induces varicosity formation, reduces mitochondrial membrane potential and blocks fast axonal flow in small and medium diameter rat dorsal root ganglion neurons: Implications for the analgesic effects of 830 nm laser. *J Peripher Nerv Syst* 12(1):28–39.

Chung, H., T. Dai, S. K. Sharma et al. 2012. The nuts and bolts of low-level laser (light) therapy. *Ann Biomed Eng* 40(2):516–533.

da Silva, R. V., and J. A. Camilli. 2006. Repair of bone defects treated with autogenous bone graft and low-power laser. *J Craniofac Surg* 17(2):297–301.

De Carvalho, R. R., F. de Paula Eduardo, K. M. Ramalho et al. 2010. Effect of laser phototherapy on recurring herpes labialis prevention: An *in vivo* study. *Laser Med Sci* 25(3):397–402.

De Souza, T. O., M. A. Martins, S. K. Bussadori et al. 2010. Clinical evaluation of low-level laser treatment for recurring aphthous stomatitis. *Photomed Laser Surg* 28(2):S85–S88.

Demidova-Rice, T. N., E. V. Salomatina, A. N. Yaroslavsky et al. 2007. Low-level light stimulates excisional wound healing in mice. *Lasers Surg Med* 39(9):706–715.

Doshi-Mehta, G., and W. A. Bhad-Patil. 2012. Efficacy of low-intensity laser therapy in reducing treatment time and orthodontic pain: A clinical investigation. *Am J Orthod Dentofacial Orthop* 141(3):289–297.

Gautam, A. P., D. J. Fernandes, M. S. Vidyasagar et al. 2012. Low level helium neon laser therapy for chemoradiotherapy induced oral mucositis in oral cancer patients—A randomized controlled trial. *Oral Oncol* 48(9):893–897.

Hsieh, Y. L., L. W. Chou, P. L. Chang et al. 2012. Low-level laser therapy alleviates neuropathic pain and promotes function recovery in rats with chronic constriction injury-possible involvements in hypoxia-inducible factor 1α (HIF-1α). *J Comp Neurol* 520(13):2903–2916.

Ihsan, F. R. 2005. Low-level laser therapy accelerates collateral circulation and enhances microcirculation. *Photomed Laser Surg* 23(3):289–294.

Jajarm, H. H., F. Falaki, and O. Mahdavi. 2011. A comparative pilot study of low intensity laser versus topical corticosteroids in the treatment of erosive-atrophic oral lichen planus. *Photomed Laser Surg* 29(6):421–425.

Khadra, M. 2005. The effect of low level laser irradiation on implant-tissue interaction. *In vivo* and *in vitro* studies. *Swed Dent J* 172:1–63.

Khullar, S. M., B. Emami, and A. Westermark et al. 1996. Effect of low-level laser treatment on neurosensory deficits subsequent to sagittal split ramus osteotomy. *Oral Surg Oral Med O* 82(2):132–138.

Lončar, B., M. M. Stipetić, M. Baričević et al. 2011. The effect of low-level laser therapy on salivary glands in patients with xerostomia. *Photomed Laser Surg* 29(3):171–175.

Lopes, C. B., A. L. Pinheiro, S. Sathaiah et al. 2005. Infrared laser light reduces loading time of dental implants: A Raman spectroscopic study. *Photomed Laser Surg* 23(1):27–31.

Maluf, A. P., R. P. Maluf, C. R. Brito et al. 2010. Mechanical evaluation of the influence of low-level laser therapy in secondary stability of implants in mice shinbones. *Laser Med Sci* 25(5):693–698.

Marković, A. B., and L. Todorović. 2006. Postoperative analgesia after lower third molar surgery: Contribution of the use of long-acting local anesthetics, low-power laser, and diclofenac. *Oral Surg Oral Med O* 102(5):c4–c8.

Mazzetto, M. O., T. H. Hotta, and R. C. Pizzo. 2010. Measurements of jaw movements and TMJ pain intensity in patients treated with GaAlAs laser. *Braz Dent J* 21(4):356–360.

Meneguzzo, D. T., L. S. Ferreira, and F. Cunha et al. 2010. Treatment of peripheral facial paralysis with low intensity laser therapy—Case report. *J Bras Laser* 2(11):20–24.

Mohammadi, Z. 2009. Laser applications in endodontics: An update review. *Int Dent J* 59(1):35–46.

Muñoz Sanchez, P. J., J. L. Capote Femenías, A. Díaz Tejeda et al. 2012. The effect of 670-nm low laser therapy on herpes simplex type 1. *Photomed Laser Surg* 30(1):37–40.

Navarro, R., M. Marquezan, D. F. Cerqueira et al. 2007. Low-level-laser therapy as an alternative treatment for primary herpes simplex infection: A case report. *J Clin Pediatr Dent* 31(4):225–228.

Olivi, G., M. D. Genovese, and C. Caprioglio. 2009. Evidence-based dentistry on laser paediatric dentistry: Review and outlook. *Eur J Paediatr Dent* 10(1):29–40.

Ozcelik, O., M. Cenk Haytac, A. Kunin et al. 2008. Improved wound healing by low-level laser irradiation after gingivectomy operations: A controlled clinical pilot study. *J Clin Periodontol* 35(3):250–254.

Peplow, P. V., T. Y. Chung, and G. D. Baxter. 2010. Laser photobiomodulation of proliferation of cells in culture: A review of human and animal studies. *Photomed Laser Surg* 28(1):S3–S40.

Pesevska, S., M. Nakova, I. Gjorgoski, N. Angelov et al. 2012. Effect of laser on TNF-alpha expression in inflamed human gingival tissue. *Laser Med Sci* 27(2):377–381.

Quadri, T., L. Miranda, and J. Tunér et al. 2005. The short-term effects of low-level lasers as adjunct therapy in the treatment of periodontal inflammation. *J Clin Periodontol* 32(7):714–719.

Rajesh, S., E. Koshi, K. Philip et al. 2011. Antimicrobial photodynamic therapy: An overview. *J Indian Soc Periodontol* 15(4):323–327.

Ribeiro, A. S., M. C. de Aguiar, M. A. do Carmo et al. 2011. 660 AsGaAl laser to alleviate pain caused by cryosurgical treatment of oral leukoplakia: A preliminary study. *Photomed Laser Surg* 29(5):345–350.

Rochkind, S., V. Drory, M. Alon et al. 2007. Laser photo-therapy (780 nm), a new modality in treatment of long-term incomplete peripheral nerve injury: A randomized double-blind placebo-controlled study. *Photomed Laser Surg* 25(5):436–442.

Salmos-Brito, J. A., R. F. de Menezes, C. E. Teixeira et al. 2013. Evaluation of low-level laser therapy in patients with acute and chronic temporomandibular disorders. *Laser Med Sci* 28(1):57–64.

Schubert, M. M., F. P. Eduardo, K. A. Guthrie et al. 2007. A phase III randomized double-blind placebo-controlled clinical trial to determine the efficacy of low level laser therapy for the prevention of oral mucositis in patients undergoing hematopoietic cell transplantation. *Support Care Cancer* 15(10):1145–1154.

Scoletta, M., P. G. Arduino, L. Reggio et al. 2010. Effect of low-level laser irradiation on bisphosphonate-induced osteonecrosis of the jaws: Preliminary results of a prospective study. *Photomed Laser Surg* 28(2):179–184.

Sousa, M. V., M. A. Scanavini, E. K. Sannomiya et al. 2011. Influence of low-level laser on the speed of orthodontic movement. *Photomed Laser Surg* 29(3):191–196.

Torres, C. S., J. N. dos Santos, J. S. Monteiro et al. 2008. Does the use of laser photobiomodulation, bone morphogenetic proteins, and guided bone regeneration improve the outcome of autologous bone grafts? An *in vivo* study in a rodent model. *Photomed Laser Surg* 26(4):371–377.

Tortamano, A., D. C. Lenzi, A. C. Haddad et al. 2009. Low-level laser therapy for pain caused by placement of the first orthodontic archwire: A randomized clinical trial. *Am J Orthod Dentofacial Orthop* 136(5):662–667.

Toyokawa, H., Y. Matsui, J. Uhara et al. 2003. Promotive effects of far-infrared ray on full-thickness skin wound healing in rats. *Exp Biol Med* 228(6):724–729.

Tunér, J., and L. Hode. 2010. *The New Laser Therapy Handbook.* Prima Books, Grangesberg.

Venezian, G. C., M. A. da Silva, R. G. Mazzetto et al. 2010. Low level laser effects on pain to palpation and electromyographic activity in TMD patients: A double-blind, randomized, placebo-controlled study. *Cranio* 28(2):84–91.

Xiaoting, L., T. Yin, and C. Yangxi. 2010. Interventions for pain during fixed orthodontic appliance therapy. A systematic review. *Angle Orthod* 80(5):925–932.

Yilmaz, H. G., S. Kurtulmus-Yilmaz, E. Cengiz et al. 2011. Clinical evaluation of Er,Cr:YSGG and GaAlAs laser therapy for treating dentine hypersensitivity: A randomized controlled clinical trial. *J Dent* 39(3):249–254.

Low-Level Laser Therapy and Stem Cells

Heidi Abrahamse
University of Johannesburg

57.1 Introduction

Developments in adult stem cell (ASC) potentiation have contributed to excitement in the field of stem cell-based therapy. Not only the use of ASCs increases therapeutic treatment possibilities but also the successful use of multipotent cells for gene therapy has been demonstrated in animal models (Rizvi et al. 2006). The concurrent ability of stem cells (SCs) to either contribute to disease development, as identified in cancer SCs (CSCs), or to replace diseased tissue by induced differentiation using selected growth factors has highlighted the intricate molecular and cellular mechanisms. Adipose-derived SCs (ADSCs) are capable of self-renewal and respond well to induced differentiation (Tarnok, Ulrich, and Bocsi 2010). Autoimmunity and transplant rejection may become limitations when selective induction of immunological nonresponsiveness to specific antigens or tissues becomes possible using autologous cell sources (Ichim et al. 2010). CSCs can initiate tumorigenesis, generate differentiated daughter cells, or undergo self-renewal and are thought to instigate tumor recurrence posttreatment. Therapy targeting CSCs have failed to provide feasible alternatives to conventional cancer treatment. Low-intensity laser irradiation (LILI) induces a biostimulatory response in several tissue types in addition to a dose-response effect to the detriment of cellular degeneration. LILI has been applied clinically for the treatment of a variety of disorders including, pain, inflammation, cancer, skin diseases, soft tissue injuries, and many more. It includes the use of low-intensity light devices delivering light in the visible to near infrared wavelength of approximately 400 to 1000 nm. LILI of different intensities can either inhibit or stimulate cellular processes, activating signaling cascades that ultimately lead to cellular modulation. Light energy is absorbed by light-absorbing molecules called chromophores in the cells, which direct and convert the light energy to be harvested in the form of chemical energy through the photochemical synthesis of adenosine triphosphate (ATP) (Gao and Xing 2009). The mechanism whereby this occurs is not well understood but is thought to be facilitated by the mitochondrial respiratory complexes resulting in production of reactive oxygen species, cyclic adenosine monophosphate (cAMP) synthesis, and influx of intracellular calcium. A significant increase in ATP production has been identified using a range of different wavelengths including 632.8, 830, and 904 nm and LILI and in a number of different cell types such as fibroblasts, keratinocytes, osteoblasts, lymphocytes, and endothelial cells (Drochioiu 2010; Gao and Xing 2009). The augmentation of SC-based therapies to potentially modulate regenerative processes using noninvasive methods such as LILI holds great potential (Lin et al. 2010). The study and development of SC therapy coincide with the development of other highly investigative and therapeutic disciplines such as tissue and genetic engineering, molecular biology, and biocompatible polymer synthesis leading to significant advances in the field of regenerative medicine. Peptide-based biopolymers are emerging as a new class of biomaterials due to their unique chemical, physical, and biological properties. Applications of these engineered biomolecules include tissue engineering (TE) where they serve as injectable scaffolds that form gels *in vivo* via physical or chemical means and provide a minimally invasive route to deliver tissue scaffolds.

57.2 Stem Cells

SCs are regarded as undifferentiated cells that are capable of self-renewal, proliferation, production of a great number of differentiated progeny cells, and regeneration of tissues (Blau, Brazelton, and Weimann 2001). There is a great therapeutic potential of multilineage SCs for TE applications. Two general types of SCs are potentially useful for this application: embryonic SCs (ESCs) and adult (autologous) SCs (Zuk et al. 2001). Traditionally, ESCs are isolated from the inner cell mass (ICM) of blastocysts; however, the harvesting of these cells results in the death of the embryo, which has led to ethical, religious, and political issues (Moore 2007). In contrast, ADCs, by virtue of their nature, are immune-compatible and have no ethical issues associated with their use (Zuk et al. 2001). Subcutaneous adipose tissue is an active and highly complex tissue composed of several different cell types and is derived from the mesodermal germ layer and contains a supportive stromal vascular fraction (SVF) that can be easily isolated. This SVF contains a heterogeneous mixture of cells including preadipocytes (Jurgens et al. 2008; Raposio et al. 2007; Schäffler and Büchler 2007). The preadipocytes are considered as the multipotent SCs (ADSCs) that have similar properties to bone marrow mesenchymal SCs (BM-MSCs) (Fraser et al. 2006). ADSCs are ideal for cellular therapy applications due to various factors: they can be harvested, multiplied, and handled easily, efficiently, and noninvasively, they are pluripotent and have a proliferative capacity comparable to BM-MSCs, and the morbidity to donors is considerably less, requiring only local anesthesia and a short wound healing time. Human ADSCs (hADSCs) can be expanded in an undifferentiated state and have multipotent differentiation capacity along the classical mesenchymal lineages of adipogenesis, osteogenesis, chondrogenesis, and myogenesis (de Villiers, Houreld, and Abrahamse 2009).

Differentiation is characterized by the acquirement of cell type-specific phenotypic, morphologic, and functional features. SCs repopulate in a vigorous fashion in a specified tissue *in vivo*, requiring that the SCs are attracted to a specific tissue where they differentiate into that tissue cell type that can continue the function/s of the tissue (Verfaillie 2006). SCs can be isolated from a variety of mammalian tissues and organs throughout development and into adulthood. These cells arise from the pluripotent cells found in the ICM of blastocysts, the ESCs.

Around the time of gastrulation, these cells divide into the three germ layers and gradually mature into committed organ- and tissue-specific SCs that display varying degrees of proliferation and self-renewal. During the development of the embryo, SCs proliferate and their progeny undergoes a process of progressive lineage restriction to ultimately produce the terminally differentiated cells that will form mature tissues. While diversification of cell types is complete at or soon after birth, numerous

tissues in the adult organism undergo physiological turnover and repair and consequently must contain a population of relatively flexible SCs. ADCs are fairly quiescent or slowly proliferating cells but have the capacity for resuming proliferation to replace injured and/or dead cells. Thus, these SCs are assumed to be responsible for the growth and development of tissues and organs during development and for tissue homeostasis and repair throughout life (Gritti, Vescovi, and Galli 2002).

57.2.1 Characterization

Traditionally, SCs are classified as those that are derived either from the embryo or from adult tissues. ESCs, embryonic germ cells (EGCs), and embryonic carcinomal cells (ECCs) are derived from a preimplantation embryo (i.e., the ICM of blastocysts, morula, and single blastomeres, primordial germ cells, and teratocarcinomas, respectively) (Baharvand et al. 2007). Numerous studies have shown that a population of ADCs are present and can be isolated from within specific areas called niches in most adult mammalian tissues/organs such as bone marrow (BM), kidneys, lungs, brain, skin, eyes, gastrointestinal tract (GIT), liver, ovaries, prostate, testis, breast (Mimeault and Batra 2006), trabecular bone, adipose tissue, periosteum, synovium, heart valves, deciduous teeth, and skeletal muscles (Batten, Rosenthal, and Yacoub 2007).

57.2.2 Embryonic Stem Cells

ESCs are pluripotent SCs harvested from the ICM of a blastocyst, the liquid-filled sphere of the early embryo produced by cleavage of the ovum approximately 5–8 days after fertilization (Barberi et al. 2005; Gomillion and Burg 2006; Moore 2007; Noguchi 2007). The origin of hESCs from the preimplantation embryo distinguishes ESCs from other pluripotent SC lines, namely, EGCs and ECCs. Cleavage-stage human embryos are grown to the blastocyst stage and the ICM is isolated and plated onto mitotically inactivated murine embryonic fibroblast (MEF) feeder layers in tissue culture (Figure 57.1). The ICM cell outgrowths are propagated on feeder layers in medium containing serum alone or serum replacement media and basic fibroblast growth factor (bFGF) (Odorico, Kaufman, and Thompson 2001).

Presently, several factors have been identified as required for hESC growth. For hESC self-renewal, bFGF has been shown to be crucial. Three other requirements are as follows: (1) feeder cells, conditioned media, or cytokines, such as Wnt3a or transforming growth factor beta (TGFβ), (2) matrix (e.g., fibronectin, collagen, or laminin), and (3) fetal bovine serum (FBS) or serum replacement (Lu et al. 2006). Several mammalian pluripotent ESC lines derived from the ICM of blastocysts have been established (Mimeault and Batra 2006). Pluripotent ESCs have indefinite replicative potential and the capacity to differentiate into all three germ layers that occur in the development of mammals: ectoderm, endoderm, and mesoderm (Amit et al. 2000; Thompson et al. 1998; Young and Carpenter 2006). These three germ layers might subsequently generate a variety of organized

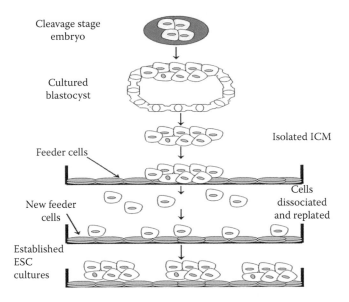

FIGURE 57.1 Cultured human blastocysts are grown from cleavage stage embryos and the ICM is isolated and plated onto MEF feeder layers, where the colonies are sequentially expanded and cloned to produce an established human ESC culture. (Adapted from Odorico, J. S. et al., *Stem Cells* 19:193–204, 2001.)

tissue structures involving complex epithelial–mesenchymal interactions. The endoderm layer may produce cells to create lung, pancreas (insulin-producing β cells, somatostatin-producing α cells, and polypeptide-producing γ cells), liver (hepatocytes), and GIT and urogenital tract cells. The mesodermal layer may produce cells to create kidney, bone (leukocytes, erythrocytes, thrombocytes, osteoblasts, chondrocytes, adipocytes, myoblasts, endothelial cells), and heart (cardiomyocytes). The ectodermal layer may produce cells to create the eye, nervous system (neurons, astrocytes, and oligodendrocytes), and skin (Mimeault and Batra 2006).

57.2.3 Adult Stem Cells

Most tissue-specific ADCs are thought to be multipotent but no longer pluripotent like ESCs. ADCs are generated during development past the stage of gastrulation. During gastrulation, pluripotent cells are fated to become mesoderm, endoderm, and ectoderm, and subsequently tissue-specific fate decisions are made (Figure 57.2) (Verfaillie 2006).

Thus, ADCs have lost their pluripotency and have obtained restricted, tissue-specific differentiation capabilities (Verfaillie 2006). ADCs might provide solutions to regenerative medicine that avoid the ethical and legal problems associated with cloning and the use of ESCs. Until recently, SCs from adult tissues

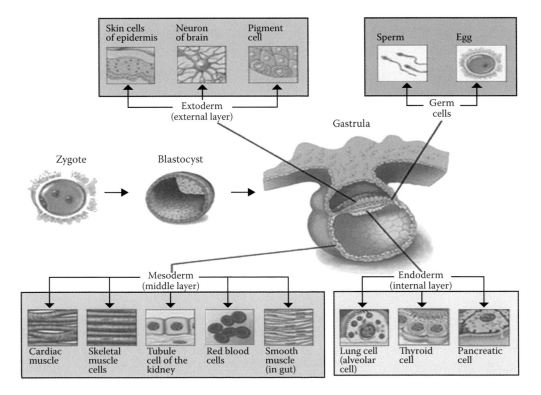

FIGURE 57.2 ADCs are fated to become mesoderm, ectoderm, and endoderm, and then follow a tissue-specific fate. (From National Center for Biotechnology Information (NCBI), What is a cell? Available at http://www.ncbi.nlm.nih.gov/About/primer/geneticscell.html, 2009. Accessed April 15, 2012.)

were believed to be restricted in their capacity to produce tissues other than those from which they originated from. A number of studies have contested this view and suggested that ADCs are plastic (i.e., they can differentiate not only into their original source tissue but also into cells of unrelated tissues) (Turksen 2004). ADCs are responsible for regenerating damaged tissue and maintaining tissue homeostasis (e.g., physiological replenishment of blood and skin cells) (Blau, Brazelton, and Weimann 2001).

57.2.3.1 ADCs from Endodermal Origin

ADCs of endodermal origin may produce lung, pancreatic, GIT, liver, and urogenital cells (Mimeault and Batra 2006). Typically, the stem/progenitor cells of the pulmonary epithelium are considered to form the mucus and basal membranes of the proximal airways, type II pneumocytes in the alveoli, and the Clara cells in the bronchioles. In the proximal airways (trachea and bronchi), the basal and mucus secretory cells are generally believed to be SCs. The basal and parabasal cells form a multipotent cell reserve that survives injury, unlike the surrounding epithelium. *In vitro* procedures that involve denuding the trachea have demonstrated the ability of basal cells to produce all major cell phenotypes found in the trachea, including ciliated, goblet, basal, and granular secretory cells. Clara cells have been shown to be progenitors of ciliated cells and progeny Clara cells in the bronchioles. Type II pneumocytes proliferate following injury, indicating that they are SCs. They restore the alveolar epithelium following generalized injury by oxidants by giving rise to either squamous type I pneumocytes or new type II pneumocytes, the former being destroyed following most types of lung injury (Bishop 2004).

The mammalian adult pancreas has three tissue types: the ductal tree, the endocrine islets of Langerhans (composed of insulin-producing β-cells, glucagon-producing α-cells, somatostatin-producing δ-cells, and the pancreatic polypeptide-producing γ-cells), and the exocrine acini (that produces digestive enzymes). A great deal of evidence has suggested the presence of pancreatic SCs (PSCs) in the ductal and/or islets regions of the mammalian pancreas (Mimeault and Batra 2006). PSCs have been shown to exhibit a remarkable potential for self-renewal and display an immense versatility in their differentiation potential (Ciba et al. 2009).

Epithelial cell lineages within the GIT are replenished frequently, approximately every 2–7 days under physiological conditions, contributing to the generation of new cell progenitors that repopulate damaged tissues during pathological injury such as ulceration and inflammation. This process is regulated by multipotent SCs localized within the niches in the intestinal crypts and gastric glands. These SCs can give rise to all the cell types in the crypt, including goblet cells, absorptive cells, Paneth cells, and enteroendocrine cells (Mimeault and Batra 2006). The adult mammalian liver has been shown to display a high regenerative capability. The liver is composed of mainly two epithelial cell types, bile ductular cells and hepatocytes. Hepatic oval cells are recognized as participating in liver

regeneration under certain conditions, and they have also been implicated in hepatic carcinogenesis (Oh, Hatch, and Petersen 2002). Hepatic oval cells appear to reside in the smallest units of the bile duct epithelium within the periportal region (termed the canals of Hering and/or periductular region) (Mimeault and Batra 2006). When liver damage is so severe that large numbers of hepatocytes are lost, hepatic oval cells appear in the peripheral regions of the liver. These cells are thought to have the capacity to proliferate in a clonogenic fashion, and possess oligopotent capability to differentiate into both hepatocytes and bile ductular cells (Oh, Hatch, and Petersen 2002). The continuation of vertebrate species is dependent upon successful reproduction. Oogenesis and spermatogenesis are analogous in that the end product is a haploid cell that is capable of forming a zygote after successful fertilization. Conversely, the cellular differentiation leading to the formation of an oocyte or spermatozoon is rather different. The fundamental difference between these two processes is the presence of an adult population of SCs, the spermatogonial SCs in the testis, which provide a source of undifferentiated spermatogonia throughout the lifetime of the male. In contrast, the traditional understanding of oogenesis in mammals is that primordial germ cells in the fetus differentiate into oogonia that proliferate, enter mitosis, and then arrest in prophase I of meiosis before birth. Hence, a female animal is born with a complete, finite population of oocytes that either are used in ovulation or are lost by follicular atresia (McLean 2006).

57.2.3.2 ADCs from Mesodermal Origin

SCs of mesodermal origin give rise to bone (leukocytes, erythrocytes, thrombocytes, osteoblasts, chondrocytes, adipocytes, myoblasts, and endothelial cells), kidney, and heart (Mimeault and Batra 2006). BM is a well-organized tissue composed of fundamental elements from the stroma and hematopoietic system and is located at the center of larger bones. BM contains at least two discernible SC populations, namely, hematopoietic SCs (HSCs) and stromal (or mesenchymal) SCs (Mimeault and Batra 2006; Prentice 2003). The differentiation of HSCs in BM ensures the continuous replenishment of all types of 12 hematopoietic cell lineages by giving rise to erythrocytes, leukocytes, and thrombocytes, which show a limited lifespan. The BM stroma is a highly vascularized and complex structure containing MSCs and extracellular matrix (ECM) elements supporting for hematopoiesis (Mimeault and Batra 2006). Human MSCs are characterized by their capacity to differentiate into cell types of the mesodermal lineage (Lee and Kemp 2006).

The most characterized and abundant source of MSCs is BM; however, recent reports have indicated that MSCs also exist in a variety of tissues including skeletal muscle and skin (Tuan, Boland, and Tuli 2002; Wagner et al. 2005), dental pulp, adipose tissue (Chamberlain, Ashton, and Middleton 2007; Lee and Kemp 2006; Tuan, Boland, and Tuli 2002; Wagner et al. 2005), amniotic fluid (Chamberlain, Ashton, and Middleton 2007; Kadivar et al. 2006), periosteum (Chamberlain, Ashton, and Middleton 2007; Tuan, Boland, and Tuli 2002), umbilical cord blood (Wagner et al. 2005), and peripheral blood (Tuan, Boland, and Tuli 2002; Wagner et al. 2005. Phenotypically, MSCs express a number of markers, none of which are specific to MSCs. It is generally agreed that human adult MSCs do not express costimulatory molecules CD80, CD86, or CD40 or the adhesion molecules CD31 [platelet/endothelial cell adhesion molecule (PECAM)-1], CD56 (neuronal cell adhesion molecule-1), or CD18 [leukocyte function-associated antigen-1 (LFA-1)]. They also do not express the hematopoietic markers CD45, CD14, CD34, or CD11, but they can express CD44, CD90, CD73, CD105, CD71, and Stro-1 as well as the adhesion molecules CD106 [vascular adhesion molecule (VCAM)-1], CD166 [activated leukocyte adhesion molecule (ALCAM)], intracellular adhesion molecule (ICAM)-1, and CD29 (Chamberlain, Ashton, and Middleton 2007).

57.2.3.3 ADCs from Ectodermal Origin

SCs originating from the ectoderm have the capacity to give rise to cells that form tissues of neuronal, ocular, and dermal cells (Mimeault and Batra 2006). Mature nervous tissue has for a long time been thought to be incapable of cell renewal and structural regeneration, especially in mammals. However, studies performed over the last few years have illustrated the presence of SCs in the mammalian central nervous system (CNS). The mammalian brain develops as a tube containing a cerebrospinal fluid-filled ventricular cavity. The uncommitted proliferating neural precursors reside initially in the ventricular zone (VZ), and then, as development continues, in the subventricular zone (SVZ) that forms beneath it (Gritti, Vescovi, and Galli 2002). Neuronal SCs (NSCs) from the SVZ can differentiate into glia and neurons, while other NSCs, located in the subgranular cell layer of the hippocampus in a region termed the dentate gyrus, may give rise to granule cell projection neurons (Mimeault and Batra 2006).

The interfollicular epidermis of human skin is a multilayered stratified squamous epithelium that is continuously renewed throughout life and provides the first line of defense against damaging environmental agents and thus constitutes a vital barrier to prevent destructive elements from disturbing tissue homeostasis and to keep indispensable fluids inside the body. The interfollicular epidermis maintains homeostasis by proliferation of keratinocytes in the basal layer attached to the underlying basement membrane. As basal cells detach from the basement membrane, they withdraw from the cell cycle and initiate a program of terminal differentiation to gradually move upward to the skin surface. Replenishment of the differentiation compartment depends on the proliferation of an SC population in the basal layer (Zouboulis et al. 2008).

The human ocular surface epithelium includes the limbal, corneal, and conjunctival stratified epithelia (Mimeault and Batra 2006). The cornea is a complex multilayered structure that can fulfill its role of transparency, refraction, photoprotection, and protection of the internal ocular structures from the external environment. The primary source of corneal epithelium is considered to be a population of SCs residing in the limbal region, which give rise to transient amplifying cells (TACs). These, in turn, divide to create the progressively more differentiated, nondividing cells of the anterior epithelium (Boulton and Albon 2004).

57.2.4 SC Properties

A number of properties are ascribed to SCs, including their ability to undergo self-renewal and differentiation, and their properties of plasticity and potency (Melton and Cowen 2006; Morrison, Shah, and Anderson 1997).

57.2.4.1 Self-Renewal and Differentiation

The capacity to undertake self-renewal and differentiation are two characteristic properties of SCs. Self-renewal can be defined as creating an entire phenocopy of SCs through mitosis, where at least one progeny cell maintains the SC capacity to self-renew and differentiate (Niwa 2006). At the single-cell level, self-renewal is defined as the ability of progenitor cells to divide while giving rise to at least one cell that remains identical to the progenitor cell. The conflict with the established SC model is the fact that the properties of proliferative potential and self-renewal are lost during differentiation (Zipori 2005). For example, it is not yet clear whether human HSCs self-renew for an entire lifespan, but rather that successive subsets of SC clones may become activated with increasing age (Morrison, Shah, and Anderson 1997). To put this into perspective, SCs are functionally multipotent self-renewing cells in the highest level of the lineage hierarchy that proliferate to make differentiated cells of a given tissue *in vivo*. However, it is important to restrict this definition to single cells that, once developed, self-renew for the lifetime of the organism in order to distinguish between SCs and the many types of transient progenitor cells with restricted self-renewal. Rather than considering SCs as undifferentiated cells, it may be better to think of them as appropriately differentiated for their specific tissue niches, with the prospect of displaying more potential phenotypes in alternative niches (van der Kooy and Weiss 2000).

57.2.4.2 Asymmetric versus Symmetric Cell Division

The most dynamic and regenerative SCs are defined by their capacity to eternally reconstitute complete tissues from a single cell. The ability to replenish a tissue requires that the SCs undergo one or more of three types of mitotic divisions: (1) replicating/symmetric division, where both daughter cells retain SC properties, (2) differentiating division, where through replication both daughter cells commit further down the lineage, and (3) asymmetric or self-renewal division, where one daughter cell differentiates and one maintains SC properties (Figure 57.3) (Naveiras and Daley 2006). SCs are thought to undergo repeated, fundamentally determined asymmetric cell divisions producing one SC and one differentiated cell. Symmetric cell divisions allow for regulation of the size of the SC pool by factors that control the probability of self-renewing as opposed to differentiating divisions (Morrison, Shah, and Anderson 1997).

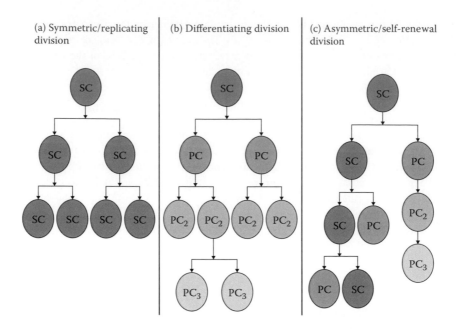

FIGURE 57.3 Possible patterns of cell division in SCs. (a) Symmetric/replicating cell division where both progenies remain SCs, (b) the progenies of SCs that undergo differentiating division both differentiate through replication down a particular lineage, and (c) asymmetric/self-renewal division where one progeny differentiates and the other retains SC properties. SC = stem cell, PC = progenitor cell, PC2 and PC3 = committed progenitor cells undergoing further differentiation through a cell lineage. (Adapted from Naveiras, O., and Daley, G. Q. *Cell Mol Life Sci* 63:760–766, 2006.)

57.2.4.3 SC Niches

The capacity of SCs to both self-renew and differentiate into various cell types enables these versatile cells to produce and repair organs and tissues, thereby maintaining tissue homeostasis and providing a continuous supply of new cells to replace highly differentiated, yet short-lived, cell types such as sperm, skin, and blood. The critical decision between self-renewal and differentiation must be tightly controlled. If SCs underwent unchecked self-renewal, the population could abnormally increase the number of partially differentiated proliferative cells in which secondary mutations could arise, leading to tumorigenesis. On the other hand, if too many progeny cells were produced, the SC pool would become depleted. Studies have revealed that the inherent features of SC fate are tightly regulated by the cells and proteins that constitute the extracellular environment or "niche" that the SCs inhabit (Jones and Fuller 2006; Wurmser, Palmer, and Gage 2004). Two classical mechanisms of development can give rise to progeny cells that follow different fates. Asymmetric cell division can produce daughter cells that, even though they reside in the same microenvironment, follow different pathways. Second, the orientation of the plane of division can place two daughter cells in different microenvironments, which may then specify individual cell fates by intercellular signaling. Therefore, SC self-renewal, number, division, and differentiation are likely to be regulated through integration of intrinsic factors, as well as extrinsic factors provided by the immediate microenvironment or niche (Jones and Fuller 2006). The SC niche is a dynamic multicellular structural unit consisting of an SC and its adjacent cells (Rajasekhar and Vemuri 2005). Histologically, an SC niche can be defined as the immediate interaction between the SC, the surrounding supporting mesenchymal cells, and the basement membrane that separates them (Naveiras and Daley 2006) (Figure 57.4).

The SC niche involves a complex interplay of short- and long-range signals between SCs, their progeny, and their neighboring cells. A broad range of secreted factors regulate SC proliferation and fate. The families of the TGF-βs and the Wnts (family of signaling proteins) show functional conservation between species and between tissues that self-renew through population asymmetry or asymmetric divisions. Of the TGF-β signaling protein family, at least two members are important in the differentiation of neural crest SCs in *Drosophila*. Wnts activate transcription via a complex pathway involving β-catenin. Cell–cell interactions mediated by integral membrane proteins is another external control, because although secreted factors can potentially act over many cell diameters, other signals that control SC fate require cell–cell interactions. Adhesion to the ECM is mediated most extensively by the family of integrins, proteins that hold the cells in the correct place in a tissue. Loss or alteration of integrin expression ensures exit from the SC niche through differentiation or apoptosis. Integrins are also signaling receptors and can directly activate growth factor receptors. Finally, the ECM can potentially appropriate and modulate the local

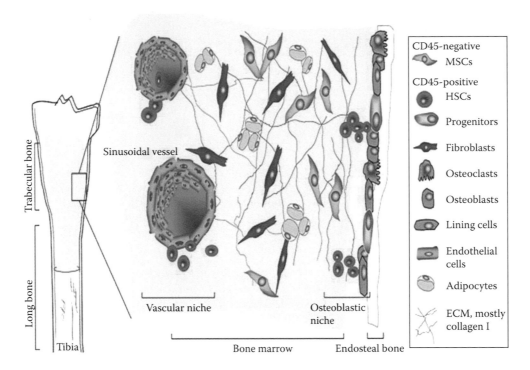

FIGURE 57.4 SC niches in BM. In BM, HSCs and their progeny populate the vascular niche, which is surrounded by stromal cells that are derived from MSCs. MSCs with true SC attributes are part of the stroma, while MSCs, which are committed osteoprogenitor cells, reside in the osteoblastic niche. (From Grassel, S. and Ahmed, N., *Front Biosci* 1(12):4946–4956, 2007.)

concentration of secreted factors available in the SC niche (Watt and Hogan 2000).

57.2.5 Regulation of SC Function

SC self-renewal, viability, proliferation, commitment, and differentiation depend on both intrinsic and extrinsic elements. The former includes an assortment of regulatory molecules present in a cell according to the specific tissue or lineage to which the cell belongs; the latter includes all the different cell types and cell products that form part of the microenvironment in which the cell develops (niche). In other words, SC function ultimately depends on intrinsic cell regulators modulated by exterior factors (Mayani 2003).

57.2.5.1 Intrinsic Elements

Intrinsic regulators of SC function include proteins responsible for setting up symmetric and asymmetric cell divisions, molecular regulators of the cell cycle, molecules that act as the mitotic clock (sets up the number of rounds of division within transit-amplifying populations), and nuclear transcription factors controlling gene expression. SC fate may be influenced by certain cytoplasmic or plasma membrane proteins inherited by each progeny cell during mitotic division. Symmetric or asymmetric divisions that an SC undergoes may depend on equal or unequal distribution of such proteins (e.g., Notch 1). Abundant evidence

also suggests that transcriptional factors control SC fate (Mayani 2003).

57.2.5.2 Extrinsic Elements

SCs develop within certain microenvironments consisting of different cell types and their products. Together, these external elements provide signals that control SC behavior by modulating expression and activity of the intrinsic elements present in a cell. For example, in postnatal life, blood cell formation takes place primarily in the BM. The SCs here are surrounded by different cell types including the stromal (macrophages, fibroblasts, endothelial cells, and adipocytes) and accessory (e.g., lymphocytes) cells. These cells produce and secrete a variety of proteins including ECM and cytokines that influence the SC physiology (Mayani 2003).

57.2.6 Plasticity

Traditionally, ADCs were thought to be committed to a particular cell fate to produce cells from the tissue of origin but not cells from unrelated tissues (Lakshmipathy and Verfaillie 2005); that is, HSCs would give rise to blood cells only, neural SCs would produce solely neurons, astrocytes, and oligodendrocytes, and satellite cells of muscles would produce muscle only. However, during the last few years, evidence from various studies has proved otherwise (Mayani 2003). Many examples

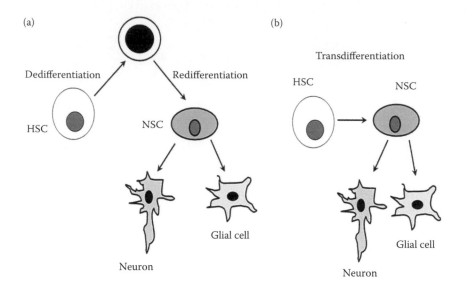

FIGURE 57.5 Models of SC plasticity. There are at least two possible mechanisms by which an SC committed to a specific tissue generates other cells of unrelated tissues. In the first model (a), an SC would dedifferentiate into a more primitive and plastic cell, and then redifferentiate into an SC of another tissue. The second model (b) shows transdifferentiation, where the SC directly takes on another differentiation path without the use of intermediaries. (Adapted from Mayani, H., *Arch Med Res* 34:3–15, 2003.)

have been reported in which there has been conversion of SCs from the tissue they pertain to into tissues that share the same embryonic origin (intragerm layer conversion). Transgerm layer conversion has also been observed, where certain SCs give rise to developmentally unrelated cell types. For example, bone marrow SCs (BMSCs) (mesoderm) may contribute to the regeneration of extrahematopoietic organs such as the liver that is of endodermal origin (Gritti, Vescovi, and Galli 2002). Although the exact mechanism of SC plasticity is unknown, one possibility is that an SC dedifferentiates into a more primitive and plastic state and then redifferentiates (Figure 57.5a). A second possibility is that an SC transdifferentiates, namely, an SC from a particular tissue directly takes on another differentiation path without proceeding into intermediate stages. Any of these processes may or may not involve cell division (Figure 57.5b) (Mayani 2003).

57.2.6.1 Dedifferentiation

Extensive evidence suggests that urodele (tailed) amphibians can regenerate an amputated limb, eye, tail, jaw, or heart structure through a process that appears to involve dedifferentiation of mature cells near the edge of the wound, forming a cluster of progenitor cells referred to as the blastema, and subsequent activation of a regenerative process reminiscent of the developmental program that originally functioned to specify limb formation (Tsai, Kittappa, and McKay 2002; Wagers and Weissman 2004). In many respects, these cells resemble mammalian SCs: they proliferate in response to injury and are multipotent (Tsai, Kittappa, and McKay 2002). Dedifferentiation of cells in adult mammals has not been clearly and unequivocally documented,

and at present, no evidence directly supports dedifferentiation or transdifferentiation events as an explanation for SC plasticity (Wagers and Weissman 2004).

57.2.6.2 Transdifferentiation

Two main requirements are needed for defining a cell conversion as transdifferentiation: (1) expression of specific biochemical and molecular markers should be different before and after the process of transdifferentiation, and (2) the establishment of a direct ancestor-descendent relationship at the cell level (Gritti, Vescovi, and Galli 2002). A final explanation for the plasticity of ADCs is that cell fusion occurs. This process may underlie at least some of the observations of SC plasticity and was first demonstrated by coculture of genetically marked ESCs with genetically marked NSCs or BM cells wherein the fused tetraploid cells retained the genetic marker expressed in the neural or BM cells as well as the pluripotency of the ESCs (Lakshmipathy and Verfaillie 2005).

57.2.7 Potency

SCs typically fall into a number of groups according to their developmental potential: totipotent, pluripotent, multipotent, oligopotent, and unipotent SCs. Totipotent cells (fertilized egg or zygote) may differentiate into all constituent cell types of an embryo (i.e., totipotent SCs can be transplanted into the uterus and can give rise to a full, viable organism) (Moore 2007; Ringe et al. 2002). These cells are capable of producing all the different cell types of an animal, including those that do not form part of the embryo, such as placental cells (Mayani 2003). Pluripotent cells (e.g., ESCs) have the ability to develop into every cell of the organism except the trophoblasts of the placenta (the supporting tissue of the embryo). Multipotent cells (ADCs) have the potential to differentiate into several different cell types that are all within a physiological system, a particular organ, or a particular tissue (e.g., HSCs that generate blood cell restricted progenitors) (Moore 2007; Ringe et al. 2002). Oligopotent SCs produce two or more lineages of cells within a particular tissue (e.g., neural SCs give rise to subsets of neurons in the brain). Finally, unipotent cells self-renew and they give rise to a single mature cell type (e.g., spermatological SCs that generate spermatozoa) (Moore 2007; Wagers and Weissman 2004).

57.2.8 Adipose-Derived SCs

In the past few years, it has been shown that MSCs have an innate ability for proliferation, self-renewal, and differentiation toward mature tissues, depending on the surrounding microenvironment (Sanz-Ruiz et al. 2008). Originally it was thought that these cells were found exclusively in the BM, but adipose tissue is also a mesodermally derived tissue that has since been shown to contain a stromal structure that is similar to BMSCs (Sanz-Ruiz et al. 2008; Strem et al. 2005). In mammals, there are three functionally different types of adipose tissue: brown adipose tissue (BAT), white adipose tissue (WAT), and BM adipose tissue (BMAT). BAT and WAT tissues both participate in energy balance but assume opposite functions: BAT functions as an energy-dissipating organ, whereas WAT is the principal energy storage system for the organism (Casteilla and Dani 2006; Sanz-Ruiz et al. 2008). Brown and white adipocytes display both lipogenic and lipolytic activities, but the main role of brown adipocytes is to dissipate energy by heat, whereas white adipocytes specialize in storing and mobilizing energy as triglycerides (Casteilla and Dani 2006). Inguinal WAT cells have a larger resident population of SCs and greater plasticity; thus they are considered an optimal choice for cellular therapy (Sanz-Ruiz et al. 2008). WAT has a remarkable capacity to dynamically shrink and expand throughout the lifespan of an individual (Fraser et al. 2006; Strem et al. 2005). This capacity is mediated by the existence of vascular and nonvascular cells providing a pool of stem and progenitor cells with distinctive regenerative capacity (Strem et al. 2005). Small increases in volume can be accommodated by changes in lipid concentration stored in individual adipocytes (hypertrophy), but larger changes are mediated by the generation of new adipocytes (hyperplasia) accompanied by expansion and remodeling of the adipose vasculature. These larger changes are due to populations of stem and progenitor cells within the adipose tissue (Fraser et al. 2006). Subcutaneous adipose tissue is an active and highly complex tissue composed of several different cell types, is derived from the mesodermal germ layer, and contains a supportive SVF that can be easily isolated. This SVF contains a heterogeneous mixture of cells including mature adipocytes, preadipocytes, fibroblasts, SMCs, endothelial cells, and resident

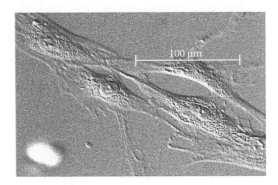

FIGURE 57.6 Differential interference contrast (DIC) microgram of isolated ADSCs. ADSCs have an average length of between 100 and 200 μm.

monocytes/macrophages and lymphocytes (Jurgens et al. 2008; Raposio et al. 2007; Schäffler and Büchler 2007). The preadipocytes are considered as the multipotent SCs termed ADSCs that have similar properties to BMSCs (Figure 57.6) (Fraser et al. 2006). The SC population in adipose tissue is responsible for the replacement of mature adipocytes within the tissue throughout the lifetime of an adult. ADCs can be isolated from adipose tissue and lipo-aspirates in considerable numbers and exhibit stable growth and proliferation kinetics *in vitro* (Mvula, Moore, and Abrahamse 2010).

ADSCs are ideal for cellular therapy applications for several reasons: they can be harvested, multiplied, and handled easily, efficiently, and noninvasively, they are pluripotent and have a proliferative capacity comparable to BM-MSCs, and the morbidity to donors is considerably less, requiring only local anesthesia and a short wound healing time (Ogawa 2006). The most significant feature of adipose tissue as a cell source is that the relative expandability of this tissue allows for large quantities of SCs to be obtained without difficulty and at minimal risk (Fraser et al. 2006). hADSCs can be expanded in an undifferentiated state and have multipotent differentiation capacity along the classical mesenchymal lineages of adipogenesis, osteogenesis, chondrogenesis, and myogenesis. Nonmesenchymal lineages have also been investigated and the transdifferentiation abilities of ADSCs confirmed, demonstrating that these cells can differentiate into bone, cartilage, fat, heart, nerve, liver, and smooth muscle (Fraser et al. 2006; Schäffler and Büchler 2007; Strem et al. 2005; Zuk et al. 2002) (Figure 57.7).

57.3 Low-Intensity Laser Irradiation

LILI refers to the use of photons to alter biological activity. Nonthermal laser light, from the red to the near-infrared region of the spectrum, is used in this type of therapy. Since the effects of LILI are biochemical and not thermal, there is no damage caused to living tissue at a cellular level (Hawkins and Abrahamse 2006, 2007). Laser irradiation at various intensities

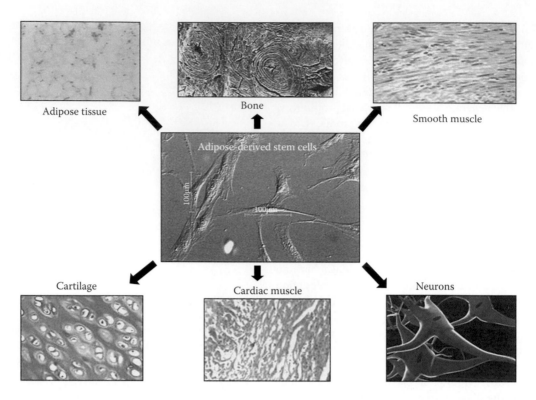

FIGURE 57.7 Multipotent differentiation capacity of ADSCs into bone, cartilage, fat (adipose tissue), heart (cardiac muscle), nerve (neurons), and smooth muscle. (From de Villiers, J. A. et al., *Stem Cell Rev Rep* 5(3):256–265, 2009.)

has been found to be able to stimulate or inhibit an assortment of cellular processes. LILI of a particular wavelength can initiate cascades, such as proliferation at a cellular level (Moore et al. 2005). LILI has been shown to increase viability, protein expression, and migration of SCs *in vitro*, and to stimulate proliferation of various types of SCs (Gasparyan, Brill, and Makela 2004; Mvula et al. 2008).

57.3.1 Lasers

The word *laser* is an acronym for *light amplification by stimulated emission of radiation* (Karu 1998; Pöntein 1992; Tunér and Hode 2002). From a practical standpoint, a laser can be considered as a source of a narrow beam of monochromatic coherent light in the visible, infrared (IR), or ultraviolet (UV) parts of the spectrum. The basic differences between lasers and conventional light sources are the following: the output beam is narrow and polarized, the emission is coherent, the light is highly monochromatic, and the intensity can be very high. However, it must be remembered that not every laser produces a narrow beam of coherent, monochromatic light (Karu 1998). A laser always includes the following parts: an energy source (power supply), lasing medium (solid, liquid, or gas), and a resonating cavity (Figure 57.8) (Tunér and Hode 2002).

The energy source can be an electric charger, flashlights, a chemical reaction chamber, a radio frequency, and so forth. The energy from the generating unit is fed into a lasing medium,

which could have the following compositions: a solid state electrically nonconducting crystal (solid laser); a gaseous state, a mixture of gases in a closed vessel (gas laser); or a liquid state where a liquid is in a cuvette (liquid laser, dye laser). The atoms of the lasing medium absorb the applied energy. Their electrons are excited to a higher orbit, where they have higher energy content, and it is through this process that the excited electrons

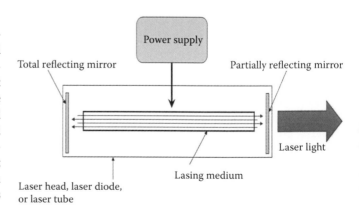

FIGURE 57.8 Principle design of a laser. All lasers have a power supply or energy source, lasing medium (solid, liquid, or gas), and a resonating cavity. (Adapted from Tunér, J., and L. Hode, *Laser Therapy Clinical Practice and Scientific Background*, 1st ed. Prima Books, Grängesberg, 8–19, 2002.)

store excess energy. The shape of the lasing medium is vital for effective amplification of the emitted light; this is the purpose of the resonance cavity, which normally gives the medium a long, thin shape. At both ends of the lasing medium, there is a mirror. One mirror reflects normally while the other reflects 80%–99%. The remaining 1%–20% is released through an aperture as a laser beam (Pöntein 1992).

57.3.1.1 Monochromaticity

Laser light is concentrated in a narrow range of wavelengths and therefore produces the purest, or most monochromatic, light available (Watson 1997).

57.3.1.2 Coherence

Coherence usually means order or synchronicity. Waves of coherent light stay in phase in long trains of waves (Figure 57.9) (Tunér and Hode 2002). When all the emitted photons bear a constant relationship with one another in both time and phase, the light is said to be coherent (Watson 1997). The length of these trains of waves (coherence length) may vary from one light source to another. When a rough surface is illuminated with visible laser light, a grainy quality can be seen in the light (called laser speckles) that occurs as a result of interference between the different light beams. This is because light with an adequately high coherence can be "combined" or "destroyed" in the same way that waves of water combine to give higher peaks or destroy each other when they meet (Tunér and Hode 2002).

57.3.1.3 Output Power

Power is measured in watts. The power output or strength of a laser is therefore measured in watts or milliwatts (Tunér and Hode 2002; Watson 1997). A higher power output means that reaching a desired dose or fluence (input energy, measured in joules per square centimeter) is much more quickly achieved because the energy is the same as power multiplied by time (Tunér and Hode 2002).

LED: one color (monochromatic) and waves not in phase (noncoherent)

LASER: one color (monochromatic) and waves in phase (coherent)

FIGURE 57.9 The waves of coherent light stay in phase with each other, as opposed to noncoherent light where the individual waves are not in phase with one another. (Adapted from Appleton Healthcare. *Frequently Asked Questions.* Available at http://www.appletonhc.com/laserfaq.html. Accessed June 24, 2009.)

57.3.2 Continuous and Pulsed Lasers

A light source usually emits light at a constant intensity, which is known as continuous wave (CW) emission. A pulsed light source can be pulsed by interrupting the light (e.g., mechanically by using a chopper) or electrically by switching its electrical power on and off (Tunér and Hode 2002).

57.3.2.1 Power Density

Power density can be described as "light intensity" or "light concentration" and is measured in watts per square centimeter (Tunér and Hode 2002). To calculate the power density of a laser, the surface area of light applied is divided by the output power of the laser (Watson 1997).

57.3.2.2 Energy Density

Energy density is measured in joules per square centimeter. As light produced from a laser is at a constant time, energy distribution is directly proportional to the power density (Watson 1997).

57.3.2.3 Collimation

The light from a gas laser is usually rather parallel and therefore nondivergent, while the light from a diode laser that does not have collimating optics is usually more divergent, with an angle of "spread" of approximately 30°–90° (Tunér and Hode 2002).

57.4 Lasers and SCs

Laser irradiation at various intensities has been found to be able to stimulate or inhibit a range of cellular processes. LILI of a particular wavelength can initiate signaling cascades leading to proliferation at a cellular level (Moore et al. 2005). Laser irradiation has been found to promote the proliferation of rat mesenchymal and cardiac SCs *in vitro* at 1 and 3 J/cm² (Tuby, Maltz, and Oron 2007) as well as promote proliferation and induce differentiation of human osteoblasts in culture at 632 nm with a power output of 10 mW (Stein et al. 2005). Red light at 647 nm enhances osteogenic differentiation in MSCs shown by an increase in alkaline phosphatase (ALP) activity, Alizarin red staining, higher mRNA expression of osteocalcin and collagen type I, and increased immunofluorescence (IF) staining against CD44 (Kim et al. 2009). LILI can change the metabolism of SCs and increase ATP activity, and in so doing increase migration in SCs, indicating the possibility that SC reaction to light irradiation could be one of the factors of light therapy (Gasparyan, and Brill, and Makela 2004). LILI at 685 nm has been found to stimulate the SC proliferation rate in *Dugesia tigrina* during regeneration (De Souza et al. 2005). It has been shown that 5 J/cm² of laser irradiation at a wavelength of 635 nm positively affects hADSCs by increasing cellular proliferation, viability, and expression of β1-integrin (Mvula et al. 2008) and Thy-1, and LILI in combination with EGF enhances the proliferation of ADSCs (Mvula, Moore, and Abrahamse 2010).

Recently, Wu et al. (2011) demonstrated the activation and expression of 119 genes affected by LILI after irradiation at a fluence of 0.5 J/cm² using microarray analysis. Most of these genes were involved in cell proliferation, apoptosis, and the cell cycle. Furthermore, five of the genes were confirmed using real-time *polymerase chain reaction* (PCR) (Akt1, Ptpn6, Stk17b, Ccnd1, and Pik3ca), all of which were identified as playing an important role in the PI3K/Akt/mTOR/eIF4E signaling pathway mediating the effects of LILI on the proliferation of MSCs (Wu et al. 2011).

57.4.1 ADSC Applications

ADSCs have emerged as a new and promising type of SC with two distinct advantages over previously used ADCs: (1) easy and repeatable access allows harvesting of high amounts of adipose tissue, and (2) ADSCs have an increased potential to proliferate and expand in culture (Sanz-Ruiz et al. 2008). As mentioned previously, ADSCs can be differentiated into the classical mesenchymal lineages and can also extend to certain nonmesenchymal lineages such as neuronal cells (ectodermal) and liver cells (endodermal).

57.4.1.1 Osteogenesis

Bone diseases, such as osteoporosis and osteopenia, affect several millions of people throughout the world. Studies for repairing bone defects are performed in gene and cellular therapies and in pharmacology. Suitable cells needed for TE should exhibit immunocompatible and self-regenerative potential (Elabd et al. 2007). The ability of MSCs to undergo osteogenesis is well known (Strem et al. 2005). Many reports have demonstrated that adipocytes and osteocytes share a common cell precursor. Osteoblastic gene expression is detected in stromal cell lines capable of undergoing adipocyte differentiation (Casteilla and Dani 2006). Observations of patients with progressive osseous heteroplasia, a rare disorder in which calcified nodules form in subcutaneous adipose depots, provided physiological evidence that similar cells exist in adipose tissue (Strem et al. 2005). For osteogenic differentiation to be induced in ADSCs, the cells are cultured in basal culture medium supplemented with 1 nM dexamethasone, 2 mM β-glycerolphosphate, and 50 μM ascorbate-2-phosphate for approximately 14 days (Bunnell et al. 2008).

Under osteogenic conditions, similar to those of MSCs, ADSCs are observed to express genes and proteins associated with an osteoblast phenotype, such as alkaline phosphatase (Fraser et al. 2006; Schäffler and Büchler 2007; Strem et al. 2005), collagen type I (Fraser et al. 2006; Schäffler and Büchler 2007; Strem et al. 2005), osteopontin (Fraser et al. 2006; Schäffler and Büchler 2007; Strem et al. 2005), osteonectin (Strem et al. 2005), osteocalcin (Elabd et al. 2007; Fraser et al. 2006; Strem et al. 2005), bone sialo protein, RunX-1, bone morphogenic protein (BMP)-2, BMP-4, BMP receptors I and II, parathyroid (PTH)-receptor (Strem et al. 2005), dentin matrix protein 1 (DMP1), phosphate-regulating gene with homology to endopeptidase genes present on the X chromosome (PHEX), core-binding factor alpha subunit 1 (CBFA-1) (Elabd et al. 2007; Fraser et al. 2006), and popdoplanin (Ppm) (Elabd et al. 2007). ADSCs are

also capable of forming mineralized matrix *in vitro* in both long-term two-dimensional or three-dimensional osteogenic cultures (Elabd et al. 2007; Schäffler and Büchler 2007; Strem et al. 2005). These results indicate that ADSCs have the potential to differentiate into real mechanosensitive bonelike cells and might therefore provide a promising tool for bone TE (Schäffler and Büchler 2007). BMSC differentiation into osteoblasts is prone to a dose-dependent effect following LILI since both energy densities of 2 and 4 J/cm² increase the proliferation and its differentiation to osteoblasts (Soleimani et al. 2011).

57.4.1.2 Chondrogenesis

Articular cartilage has limited potential for self-repair; therefore, cartilage damage eventually results in osteoarthritis. Various therapies exist, and one in particular, autologous chondrocyte transplantation (ACT), is well established in orthopedic surgery to treat cartilage defects. The long-term results of ACT are promising. To perform ACT, a cartilage biopsy is taken from the articular joint, which is associated with donor site morbidity. Consequently, adult MSCs from BM or adipose tissue have been contemplated as an alternative cell type for transplantation (Mehlhorn et al. 2009).

hADSCs have been shown to differentiate into chondrocyte-like cells in the presence of growth factors from the TGF-β1 superfamily (Fraser et al. 2006; Mehlhorn et al. 2009). High-density cultures of MSCs and ADSCs generate cellular nodules that produce large amounts of cartilage-related ECM molecules, including collagen II and IV, PRELP, sulfated proteoglycans, and aggrecan. ADSCs seeded onto alginate discs and implanted into immunodeficient mice exhibited prolonged (12 weeks) synthesis of cartilage matrix molecules, including collagen II and IV and aggrecan (Strem et al. 2005). Mehlhorn et al. (2009) seeded ADSCs onto poly(lactide-*co*-glycolide) (PLGA) scaffolds, treated them in TGF-β1-containing media for 3 weeks *in vitro*, and then implanted them in a subcutaneous pocket of nude mice. Chondro-specific marker molecules such as cartilage oligomeric matrix protein, collagen type II and X, and aggrecan substantially increased during the 3 weeks *in vitro*. After 8 weeks *in vivo*, samples pretreated with TGF-β1 continued expressing collagen type II and aggrecan, and collagen type II was found in the ECM. These results signify that ADSCs have the potential to differentiate into chondrocyte-like cells and as a result could represent a suitable instrument for TE of cartilage.

57.4.1.3 Adipogenesis

A soft tissue defect is commonly defined as a large tissue void within the subcutaneous fat layer of the skin that often results in a change of the normal tissue contour (Gomillion and Burg 2006). Conventional soft tissue-grafting procedures have had a few clinical successes for soft tissue reconstruction and augmentation. However, the need for secondary surgical procedures to harvest autologous tissues and an average 40%–60% reduction in graft volume over time are considerable disadvantages of current autologous fat transplantation procedures. For adipose TE, the use of ADSCs may be more useful than differentiated adipocytes because mature adipocytes, although readily available, have low expandability and poor ability for volume retention following *in vivo* soft tissue reconstruction. ADSCs can be isolated from the patient via a minimally invasive procedure, and they are greatly expandable in culture and can readily be differentiated into adipose tissue-forming cells by exposure to an adipogenic medium (Choi et al. 2006). ADSC-derived adipocytes develop important features of mature adipocytes such as lipolytic capacity upon catecholamine stimulation, secretion of typical adipokines such as leptin and adiponectin, and antilipolytic activity mediated by α2-adrenoreceptors. ADSCs retain their adipocyte differentiation capacity through multiple passages (Schäffler and Büchler 2007).

Choi et al. (2006) conducted a study to evaluate the use of a combination of ADSCs and injectable PLGA spheres for adipose TE. Adipogenesis was examined in nude mice injected subcutaneously with ADSCs (group I), PLGA spheres (group II), or ADSCs attached to PLGA spheres (group III). The ADSCs were attached to PLGA spheres and cultured in adipogenic medium containing 3-isobutyl-1-methylxanthine, dexamethasone, insulin, and indomethacin for 7 days (group III) and injected into the neck of nude mice. Groups I and II were used as controls. The mice were sacrificed after 4 and 8 weeks. Newly formed adipose tissue was observed in groups II and III, but at no time in group I. The largest volume of newly formed tissue was seen in group III. *In vivo* adipogenesis was evaluated by oil red O staining, and in group III, the injected cells and PLGA spheres aggregated and regenerated adipose tissue. Newly formed adipose tissue at 8 weeks showed that the extent of staining were more similar to native adipose tissue as compared to the newly formed tissue at 4 weeks. RT-PCR data showed that peroxisome proliferator-activated receptor gamma (PPARγ) and enhancer binding protein factor beta (C/EBPα) were expressed strongly in group III at 8 weeks. These results show that after 8 weeks, the adipogenic ADSCs attached to the PLGA spheres differentiated into adipocytes and regenerated adipose tissue. This study suggests that injectable spheres and ADSCs are useful materials for adipose TE. The injectable spheres used in the study leave only minimal postoperation scars and maintain their shape. Therefore, the combination of ADSCs and PLGA spheres can be used in a clinical setting as an adipose TE technique that results in noninvasive three-dimensional soft tissue filler.

57.4.1.4 Myogenesis

57.4.1.4.1 Cardiomyogenic Potential

Cardiovascular diseases like heart valve disease, myocardial infarction, and subsequent heart failure are chief causes of morbidity and mortality. Currently, the valves, conduits, and patches for replacement therapy or repair are flawed and subject patients to one or more ongoing risks such as limited durability, thrombosis, increased susceptibility to infection, and need for reoperations due to lack of growth in pediatric patients. TE offers a promising solution for the above problems (Wu et al. 2006).

CD29+ murine ADSCs from BAT can differentiate into cardiomyocytes with high efficiency, as well as endothelial cells and SMCs, and consequently repair damaged myocardium (Yamada et al. 2006). These findings were based on morphology, electrophysiological parameters, and molecular and protein expression. The group transplanted CD29+ BAT-derived cells into the infarct border zone of an acute myocardial infarction model in rats. Immunohistochemistry revealed that implanted cells expressed markers found in SMCs, endothelial cells, and cardiomyocytes. RT-PCR displayed expression of cardiomyocyte-specific genes in the BAT-derived cells such as α-MHC, β-MHC, α-skeletal actin, α-cardiac actin, brain natriuretic peptide (BNP), GATA-4, and NKx2.5. A cardiac function test, measured by echocardiology, revealed that there was improved ventricular function and reduction of the infarction area due to the transplantation of the CD29+ BAT-derived cells. These findings suggest that BAT-derived cells could potentially be a useful resource for cardiomyocyte regeneration.

One group presented data suggesting that porcine BMSCs as well as ADSCs engrafted in the infarct region of female pigs improve cardiac function and perfusion after intracoronary cell transplantation (Valina et al. 2007). Myocardial infarction was induced and 2 million cultured autologous cells were intracoronary injected. Relative and absolute perfusion significantly decreased after 28 ± 3 days after BMSC and ADSC administration compared to carrier administration. There was a significant increase in left ventricular function, the thickness of the ventricular wall in the infarction area, and improved vascular density of the border zone after the administration of SCs as compared to the carrier administration. Colocalization of the grafted cells with SMα-a and von Willebrand factor was observed, with the incorporation into recently formed blood vessels.

57.4.1.4.2 Myogenic Potential

Duchene muscular dystrophy (DMD) is an X-linked genetic disorder that is characterized by progressive muscle weakness and degeneration. Cell therapy is being pursued as a possible treatment modality for the repair of defective muscle in DMD. Vieira et al. (2008) showed that ADSCs participate in myotube formation resulting in the restoration of dystrophin. To differentiate hADSCs into myogenic lineage, SCs were grown in growth media (Dulbecco's Modified Eagle Medium–High Glucose DMEM-HG, 10% FBS) supplemented with 0.1 μM dexamethasone, 50 μM hydrocortisone, and 5% horse serum for 45 days. IF studies showed expression of alpha-actin in the differentiated culture and RT-PCR revealed expression of myogenic determination (MyoD), telethonin, and dystrophin. Western blot confirmed the presence of dystrophin in early passage, high-cell-density cultures.

Two different types of cocultures were tested. In the first, equal amounts of green fluorescent protein (GFP) negative myoblasts stained with 4-6-diamidino-2-phenylindole (DAPI) and GFP-positive ADSCs were cocultured. In the second culture, the DMD myotubes were stained with DAPI and GFP-positive ADSCs were added at a ratio of 3:1. The cultures were then subjected to fusion media (FM) that induces myoblasts to coalesce and form multinucleated structures. The conclusion for the coculture experiment was that ADSCs participate in the generation of human myotubes through cellular fusion as all GFP+ syncytia presented at least one DAPI-stained nucleus and expressed dystrophin. ADSCs were plated with DMD myotubes, and it was revealed that ADSCs are able to fuse with DMD myotubes and restore dystrophin. Dystrophin expression was evaluated by RT-PCR, and it was found that ADSCs, when differentiated into muscle cells, can express dystrophin at a level equal to that of normal myoblasts. ADSCs also have the potential to differentiate into SMC lineage (Rodriguez et al. 2006), which will be discussed in detail in the following section.

57.4.1.5 Hepatic Differentiation

Most liver diseases lead to hepatocyte dysfunction with the ultimate possibility of organ failure. The replacement of diseased hepatocytes by SCs and the stimulation of endogenous or exogenous regeneration by SCs are the main aims of liver-directed cell therapy (Taléns-Visconti et al. 2007).

A recent study suggested that under certain specific inducing conditions, ADSCs can be induced to differentiate into hepatic lineage *in vitro* (Taléns-Visconti et al. 2007). ADSCs were pretreated by serum deprivation for 2 days and cultured in DMEM supplemented with 20 ng/mL epidermal growth factor (EGF) and 10 ng/mL bFGF. The cells were then incubated with step-one differentiation medium consisting of DMEM supplemented with 20 ng/mL hepatocyte growth factor (HGF), 10 ng/mL bFGF, and 4.9-mM nicotinamide, for 7 days, followed by treatment with step-two maturation medium consisting of the same media (protocol A) or DMEM supplemented with 20 ng/mL oncostatin M (OMS), 1 μM dexamethasone, and 10 μL/mL ITS + premix (100 μM insulin, 6.25 μg.mL transferrin, 3.6 μM selenous acid, 1.25 mg/mL BSA, and 190 μM linoleic acid) (protocol B). Morphologically the cells changed from the fibroblast-like morphology of ADSCs to a round cuboidal shape when cultured in two-step differentiation media, using either protocol A or B. The expression of hepatic genes was assessed by RT-PCR. Cytokeratin 18 (CK-18) and 19 (CK-19) expression was not significantly changed by induction of differentiation. Expression of albumin (ALB), transthyretin (TTR), cytochrome 2E1 (CYP 2E1), and C/EBPβ increased with the time of differentiation in ADSCs when protocol A was employed. When protocol B was used, a minor increase was observed, and although there was a lower increase than with protocol A, the differences were still significant when compared to the control. This study indicates that ADSCs may become a valuable and promising *in vitro* alternative tool as compared to adult hepatocytes for artificial liver devices and for preclinical drug testing.

57.4.1.6 Neurogenic Differentiation

Unlike many other tissues, the nervous system has a restricted capacity for self-repair, with mature nerve cells lacking the ability to regenerate, and NSCs have a limited capability to generate new functional neurons in response to injury (Kang et al. 2003).

ADSCs can be differentiated into a neural cell lineage called olfactory ensheathing cells (OECs), using a coculture of the

SCs with the OECs on three-dimensional scaffolds (Wang et al. 2007). The differentiated cells had similar morphology and antigenic phenotypes [p75NTR+/Nestin+/glial fibrillary acidic protein (GFAP)] of OECs. The group's results indicated that ADSCs had the potential to differentiate into OEC-like cells on three-dimensional scaffolds *in vitro*. These cells are believed to be important for the repair of damaged CNS, and as a result, the transplantation of OECs is a promising potential therapy for spinal cord injury. ADSCs are known to secrete multiple growth factors and therefore have cytoprotective effects in a variety of injury models. A study investigated the neuroprotective effects of ADSCs in a rat intracerebral hemorrhage (ICH) model and found that ADSC transplantation prompted functional recovery, reduced apoptosis and cerebral inflammation, and reduced brain atrophy and glial proliferation. ICH was induced via stereotaxic infusion of collagenase, and then human ADSCs were intravenously administered 24 h after ICH induction. Three days after ICH induction, acute inflammation markers such as terminal transferase Deoxyuridine Triphosphate (dUTP) nick end labeling (TUNEL) and myeloperoxidase (MPO) levels as well as brain water content had significantly reduced in the ADSC-transplanted rats. ADSC transplantation attenuated neurological defects from 4 to 5 weeks post-ICH and reduced both glial proliferation and brain atrophy. Transplanted ADSCs were found to densely populate perihematomal areas at 6 weeks and express endothelial markers but not neuronal or glial markers. In brief, the transplantation of ADSCs in the IHC model reduced chronic brain degradation and acute cerebral inflammation, and promoted long-term functional recovery (Kim et al. 2007).

Using LILI at 810 nm wavelength and a fluence of 3 J/cm^2, Soleimani and coinvestigators demonstrated the enhancement of BMSC proliferation and differentiation into neurons in a dose-dependent manner. While lower dosages seem to induce proliferation and differentiation (3 J/cm^2), higher dosages are required to induce differentiation but reduce proliferation (6 J/cm^2) (Soleimani et al. 2011). Not only do fluence and wavelength affect these two different properties of BMSCs, but they also affect the outcome of differentiation.

57.4.1.7 Smooth Muscle Cell Applications

Smooth muscle cells (SMCs) play a role in a large range of physiological and pathological conditions because they constitute the principal layer of all SMC tissues. They are also known to play a vital role in a large number of major human diseases, such as arteriosclerosis, asthma, hypertension, and cancer (Sinha et al. 2006).

57.4.1.7.1 Arteriosclerosis

Arteriosclerosis encompasses spontaneous arteriosclerosis, autologous arterial or venous graft arteriosclerosis, restenosis after percutaneous transluminal coronary angioplasty, and transplant arteriosclerosis. In all types of arteriosclerosis, a key event is an accumulation of vascular SMC in the intima (Mayr and Xu 2001). The SMCs of the arterial media play a principal role in structural and functional alterations of the arterial wall. The switch from the contractile to the synthetic phenotype appears to be an early crucial event in arteriosclerosis. This phenotypic modulation permits SMCs to migrate into the intima, proliferate, and secrete ECM components (Tukaj, Bohdanowicz, and Kubasik-Juraniec 2004).

57.4.1.7.1.1 Cardiovascular Applications of SMCs

Cardiovascular disease is the leading cause of mortality around the world, increasing the demand for small diameter blood vessels. Current treatments include stents, vascular grafts (currently the gold standard), and angioplasty (Liu et al. 2007; Wong et al. 2003). Venous grafts suffer from several major disadvantages: pain and discomfort at the donor site, limited availability, especially for repeat grafting procedures, limited replicative capacity of cells from older donors, and high 10-year failure rate (Wong et al. 2003). The aim of small-diameter blood vessel TE is to create autologous replacement vessels for use in vascular replacement surgery, among others (Krenning et al. 2008). These conduits need to closely resemble the native vessel, not only in dimension but also in physiological responsiveness, cellular constituents, and mechanical properties, as well as in its relaxing and antithrombotic factors released at the cellular level (Campbell et al. 2004).

A novel method for isolating SMCs from ovine BM using a tissue-specific promoter and fluorescence-activated cell sorting was developed by Liu et al. (2007) for potential use in cardiovascular tissue regeneration applications. When compared to vascular SMCs, BM-derived smooth muscle progenitor cells (BM-SMPCs) showed higher proliferation, exhibited similar morphology, and expressed several SMC markers such as Myosin Heavy Chain (MHC), smoothelin, calponin, α-actin, SM22, and caldesmon. When embedded in fibrin hydrogels, the BM-SMPCs contracted the matrix and displayed receptor- and nonreceptor-mediated contractility, which indicated that these cells can generate force in response to vasoreactive agonists. Tissue-engineered blood vessels were prepared from BM-SMPC and BM-derived endothelial cells and implanted into the jugular vein of lambs. Five weeks postimplantation, the grafted tissues showed a confluent endothelial layer over the medial layer in which BM-SMPCs were aligned circumferentially and synthesized significant amounts of collagen, as well as high amounts of elastin organized in a fibrillar network very similar to native vessels. These results suggest that BM-SMPC may be of use in studying SMC differentiation and have potential for development of cell therapies for the treatment of cardiovascular disease.

The application of LLLT to autologous BM of rats postmyocardial infarction (-MI) offers a novel approach to induce BM-derived MSCs, which are consequently recruited from the circulation to the infarcted heart and markedly attenuate the scarring process postmyocardial infarction. In a recent study, this was demonstrated, opening the possibility that this approach may also be applied to other injured organs or organs undergoing degenerative processes such as neurodegenerative diseases (Tuby, Maltz, and Oron 2011).

57.4.1.7.2 Asthma

Chronic persistent asthma is characterized by poorly reversible airway obstruction and shows marked inflammatory and architectural changes associated with airway wall thickening. Increased airway smooth muscle content is believed to be one of the principle contributors to airway wall thickening and occurs as a result of hyperplastic and/or hypertrophic growth (Hirst 1996).

57.4.1.7.2.1 Respiratory System Applications of SMCs

Airway smooth muscle is present in the bronchial tree of most vertebrates. It encircles the entire airway beneath the level of the main bronchus in a generally circular orientation, except at high lung volumes. These cells are present in the central and peripheral airways, are more transverse in central airways, and are somewhat more prominent in the peripheral airways. Airway SMCs are arranged in a helical or geodesic pattern that is more apparent when the lung is fully inflated and in the peripheral airways (James and Carroll 2000).

One group evaluated the feasibility of an engineered human bronchial mucosa as a model to study cellular interactions in asthma (Chakir et al. 2001). Human bronchial fibroblasts from normal and asthmatic donors were incorporated into collagen gel, and bronchial epithelial cells were seeded over this gel and cultured in an air–liquid interface in the presence or absence of T lymphocytes. Biopsy specimens from this engineered mucosa were taken for ultrastructural and structural analysis, T lymphocytes were harvested, and interleukin 5 (IL-5) was localized. The histologic analysis demonstrated that engineered mucosa with normal bronchial cells presented a pseudostratified ciliated epithelium with the presence of mucus secretory cells. These features were confirmed with electron microscopy and were comparable with those observed in normal bronchial tissues. However, in engineered mucosa from asthmatic subjects, the tissue structure was disorganized, principally the epithelial cell arrangement. The percentage of IL-5+ lymphocytes was significantly higher in engineered bronchial mucosa from asthmatic subjects compared with mucosa from normal volunteers. The authors claim that using TE, they produced an *in vitro* model of bronchial mucosa from normal and asthmatic subjects, which could be used in the future to better comprehend the key mechanisms involved in the inflammation and repair of respiratory tissue.

57.4.1.7.3 Bladder Disease

A variety of injuries can lead to the damage or loss of the bladder, requiring the eventual replacement or repair of the organ (Atala et al. 2006). During the last decade, reconstructive urology with the use of specialized surgical equipment and allogenic or autologous tissue has improved surgical outcomes. TE aims to regenerate urological organs and structures to a point where full physiological functions are restored (Sievert, Amend, and Stenzl 2007).

Atala et al. (2006) explored an alternative approach to cystoplasty as a treatment of patients with end-stage bladder disease using autologous engineered bladder tissues. Seven patients with myelomeningocele with high-pressure or poorly compliant bladders were identified as candidates for cystoplasty and bladder biopsies were obtained. Muscle and urothelial cells were grown in culture, and then seeded on a biodegradable bladder-shaped scaffold (collagen or collagen-polyglycolic acid). Seven weeks after the biopsy, the autologous engineered bladder constructs were used for reconstruction and implanted with or without an omental wrap. Cytograms, ultrasounds, bladder biopsies, serial urodynamics, and serum analyses were then done. Follow-up analyses were done in a mean range of 46 months. Postoperatively, the mean bladder leak point pressure decreased with capacity and the compliance and volume increase was greatest in the composite engineered bladders with an omental wrap. After surgery, bowel function returned promptly. Metabolic consequences were not noted, no urinary calculi formed, renal function was preserved, and mucus production was normal. An adequate structural architecture and phenotype was displayed by the engineered bladder biopsies. This research showed that engineered bladder tissues created with autologous cells seeded on collagen-polyglycolic acid scaffolds and wrapped in omentum after implantation can be used as an alternative to cystoplasty in patients with end-stage bladder disease.

57.4.2 ADSCs Differentiated into SMCs

Smooth muscle is an active component of the cardiovascular, reproductive, urinary, and intestinal systems and has been the subject of intense research in the field of regenerative medicine. A restriction to date has been a reliable source of healthy SMCs, as biopsies normally lead to low cell harvest that needs to be expanded at length before therapeutic use (Rodriguez et al. 2006). In addition, previous research has shown that SMCs that are obtained from diseased tissue can differ phenotypically and functionally from normal healthy SMCs, which consequently restricts their use (Dozmorov et al. 2007; Lin et al. 2004). There is a variety of diseases associated with SMC pathology. The use of SCs for cell-based TE provides a promising possible alternate to current treatment strategies for smooth muscle repair. A limitation to date has been a reliable source of SMCs for these applications (Rodriguez et al. 2006).

Rodriguez et al. (2006) differentiated processed lipoaspirate (PLA) cells (ADSCs) into functional (therefore contractile) SMCs. To induce differentiation, PLA cells were cultured in smooth muscle induction media [consisting of MCDB131 media (Sigma) supplemented with 1% FBS and 100 units/mL of heparin], which brought about genetic expression, at a transcriptional and translation level, of SMC-specific α-actin, SM22, caldesmon, calponin, smoothelin, and MHC. Upon induction, PLA cells assumed typical SMC morphology. To examine functionality of differentiated SMCs, cells were subjected to carbachol and atrophine. After 1 min exposure to 10^{-5} M, carbachol cells began to contract, remaining in this state for up to 10 min, which shows similar contractile function as that of SMCs *in vivo*. The muscarinic antagonist atrophine (10^{-4} M) was shown to block the effects mediated by carbachol. The cells exhibited SMC characteristics both phenotypically and functionally.

Hence, ADSCs have the potential to differentiate into functional SMCs that could be used in therapeutic applications as a source of healthy SMCs. LILI can positively affect hADSCs by increasing cellular proliferation, viability, and protein expression (Mvula et al. 2008; Mvula, Moore, and Abrahamse 2010). This highlights yet another potential advantage for the use of these cells in TE applications, as the initial number of SCs can be increased before beginning differentiation, which would lead to a greater amount of differentiated SMCs for use in therapeutic applications. Recently, de Villiers et al. (2011) presented evidence that LILI at a wavelength of 636 nm and a fluence of 5 J/cm² (8.59 mW/cm²) positively affects isolated hADSCs *in vitro* by increasing cellular proliferation and viability and by maintaining SC morphology and protein expression for up to 72 h. Laser irradiation at this fluence and wavelength does not appear to induce differentiation into another cell type even after an extended period of time. However, SCs isolated from human adipose tissue can be differentiated into SMCs using retinoic acid (RA). Since low yields of SCs are isolated and their proliferation rate is low, LILI could be a useful tool in relation to TE and SC therapies as the initial number of isolated SCs could be significantly expanded before commencing differentiation, which would lead to a higher yield of differentiated cells. Cell therapies that involve SMC differentiation may offer an alternative treatment for diseases that involve SMC pathology such as gastrointestinal diseases, bladder dysfunction, urinary incontinence, and cardiovascular diseases (de Villiers et al. 2011).

57.5 SCs and TE

ADCs offer scientists and medical researchers the opportunity to produce autologous *de novo* tissues *ex vivo* with fewer ethical dilemmas. SCs could be isolated from a patient's own adipose tissue and then cultured on tissue-specific scaffolds or matrices that in combination with lineage specific growth factors would allow the SCs to differentiate and develop into the required tissue, such as bone, skin, or muscle. This tissue could then be grafted back into the same patient. Being autologous, there would be no need for patients to be relegated to severe regimes of immunosuppressant drugs and less chance of graft versus host disease in which the tissue is usually rejected (Moore 2007). Nanotechnology and TE are promising scientific fields in the development of advanced materials useful to human health. The preparation of a nanocarrier for the controlled release of a photosensitizer compound associated with low-level laser therapy (LLLT) for skin wound-healing treatment has been indicated, which may also be applicable to other skin diseases. A biological model can be used as an *in vitro* skin equivalent based on a three-dimensional culture of fibroblasts and MSCs and denominated by a dermal equivalent (DE). Results show that it is possible to use the photomodulation process to control the wound healing in a scratching process and to induce the biomolecule release, both of which are related to the inflammatory wound-healing process. A dependence on enzymatic activity relating to LLLT applications indicates a potential application in wound-healing processes based on phototherapy and nanotechnology (Primo et al. 2011).

57.6 SCs and Cancer

Normal stem and cancer cells share the ability to self-renew, and therefore SCs may now also prove to be useful in the study of cancer as it has been shown that many of the classical cancer signal transduction pathways, such as Wnt, Shh, and Notch may also regulate SC development, self-renewal, and function. It has now been shown that tumors, including those of blood (Bonnet and Dick 1997), breast (Al Hajj et al. 2003; Dontu et al. 2003), testis (Rorth et al. 2000), and brain (Hemmati et al. 2003), originate from a subset of tumor cells that have SC characteristics. It appears that these subsets or CSCs are SCs that normally have highly regulated self-renewal signaling pathways, but due to transformation, the usually well-regulated pathways become dysregulated and result in uncontrolled self-renewal and the formation of a tumor. This could be a reason why tumors reappear after traditional cancer therapies such as surgery, radiation, and chemotherapy. These treatments rely on removing solid tumors and targeting rapidly dividing cells. SC division is not a "rapid" process and is therefore immune to these treatments. Self-renewal is a hallmark property of both stem and neoplastic cells. In understanding the properties and biology of normal and cancer SCs, it may also be possible to elucidate treatments for curing cancer (Moore 2007).

Normal ASCs undergo an initial genetic mutation to become founder cells of solid tumors, allowing these tumors to contain a subpopulation of self-renewing and expanding SCs known as CSCs. It has even become evident that cancer can originate from normal SCs in a variety of different cancers, and since cancer progression is an evolutionary process that generates multiple clones with a fresh identity, eradication of cancer can be futile if only one type of SC is targeted (Alison, Lim, and Nicholson 2011). Postulated to regenerate tumors posttreatment, CSCs have been the basis of controversy for several years. Having shown to initiate tumorigenesis, CSCs have provided a novel model of cancer biology that holds great promise for future therapeutic developments (Tomasson 2009). The cellular heterogeneity of cancer cells has been appreciated for decades, and yet the idea of developing more efficient and sensitive treatment regimes based on the concept of treatment-resistant CSCs was met with extreme enthusiasm with the identification of the first CSCs in 1977 (Hamburger and Salmon 1977). CSCs, like normal SCs, can, during cell division, either generate more differentiated daughter cells or undergo self-renewal, showing the significance of the cellular pathways governing self-renewal (Tomasson 2009). Therapeutic intervention should target key signaling pathways that are active in CSC self-renewal. These pathways include the Wnt pathway (important for self-renewal and maintenance of SCs), the Notch pathway (overactivated expression of receptors and ligands leading to transformation), and the Hedgehog pathway (maintenance of SCs) (Abrahamse 2011).

57.7 Ethical Considerations

Traditionally, SCs have been isolated from the ICM of blastocysts—the liquid-filled sphere of the early embryo produced by cleavage of an ovum about 8 days after fertilization. Cells from the ICM are essentially pluripotent; however, harvesting these cells results in the death of the embryo. Understandably this has led to political, religious, and ethical issues, especially in the development of human ESC lines (Moore 2007). Most SC biologists agree that hESCs have a vast potential for treating human injury and disease based on the observation that these cells have the potential to differentiate into most, if not all, cell types of the adult human body. These cells could also be used as a source of essential growth factors or signaling molecules (Gruen and Grabel 2006). Possible alternatives to traditional ESC lines that would appease this central objection to ESC research are as follows: (1) an ESC from a single blastomere isolated from an eight-cell-stage embryo by coculturing it with a preexisting ESC line and then isolating the blastomere-derived cells after growth adaptation in culture (commonly used in preimplantation genetic analysis in assisted reproduction clinics), (2) ESC lines can also be isolated from a blastocyst produced by using somatic cell nuclear transfer (SCNT), where the nucleus is removed from a donor somatic cell and injected into a denucleated donor egg cell, and then grown to the blastocyst stage *in vitro* and the ICM isolated (Moore 2007; Gruen and Grabel 2006), and (3) altered nuclear transfer (ANT) is a method that employs a nucleus carrying a mutation, so that the cell is deficient and cannot support development, so although ESCs are isolated, they cannot be implanted *in utero* (Gruen and Grabel 2006; Scott and Reijo Pera 2008). Another concern with ESC-based therapy is possible tumor formation from undifferentiated transplanted cells (Cha and Falanga 2007; Gruen and Grabel 2006). The ability of ESCs to proliferate readily is seen *in vivo* where they can form teratocarcinoma-like tumors in adult mice if injected subcutaneously, into the testis, or intramuscularly. However, studies using mouse ESCs have suggested a means to inhibit tumorigenicity following transplantation by using a more differentiated derivative of ESCs for grafting, and these cells are less likely to generate a tumor (Gruen and Grabel 2006).

57.8 Concluding Remarks

It is of the utmost importance that SC research is supported. Most studies that have assayed the differentiation potential of tissue SCs have used adult cells. If we can indeed maintain and propagate ADCs *in vitro*, the implications and applications of such a tool are quite possibly endless. The demonstration that SCs from adult tissue have a broad differentiation potential is particularly interesting from a therapeutic point of view. One important application is TE in which a patient's own somatic SCs from a particular tissue might be used in autologous cell therapy (Clark and Frisen 2001) where SCs can be used in the production of autografts (Bianco and Robey 2001). It is of great consequence that studies are carried out that would be beneficial in further assessing the long-term health effects and safety of SCs. Elucidation of the proliferative pathways in SCs could be beneficial in further understanding cellular differentiation and oncogenesis. Although there are definite political, religious, and moral ethical issues associated with SC research, there are also the ethical implications of *not* proceeding with SC research (Moore 2007). Consider the ethical aspects of not pursuing this line of research: Would it be ethical to prohibit the discovery of possible cures for cancers and debilitating diseases and disorders? If good moral and ethical common sense are applied to the regulation of SC research, it should proceed to benefit and expand current therapeutic modalities.

References

Abrahamse, H. 2011. Inducing stem cell differentiation using low intensity laser irradiation: A possible novel therapeutic intervention. *Eur J Biol* 6(5):695–698.

Al Hajj, M., M. S. Wicha, A. Benito-Hernandez, S. J. Morrison, and M. F. Clarke. 2003. Retrospective identification of tumorigenic breast cancer cells. *Proc Natl Acad Sci USA* 100:3983–3988.

Alison, M. R., S. M. L. Lim, and L. J. Nicholson. 2011. Cancer stem cells: Problems for therapy? *J Path* 223:147–161.

Amit, M., M. K. Carpenter, M. S. Inokuma et al. 2000. Clonally derived human embryonic stem cell lines maintain pluripotency and proliferative potential for prolonged periods of culture. *Dev Biol* 227:271–278.

Appleton Healthcare. *Frequently Asked Questions.* Retrieved 24 June, 2009 from http://www.appletonhc.com/laserfaq.html.

Atala, A., S. Bauer, S. Soker, J. J. Yoo, and A. B. Retik. 2006. Tissue-engineered autologous bladders for patients needing cytoplasty. *Lancet* 367:1241–1246.

Baharvand, H., A. Fathi, D. van Hoof, and G. H. Salekdeh. 2007. Concise review: Trends in stem cell proteomics. *Stem Cells* 25:1888–1903.

Barberi, T., L. M. Willis, N. D. Socci, and L. Studer. 2005. Derivation of multipotent mesenchymal precursors from human embryonic stem cells. *PLoS Med* 2(6):0554–0560.

Batten, P., N. A. Rosenthal, and M. H. Yacoub. 2007. Immune response to stem cells and strategies to induce tolerance. *Phil Trans R Soc B* 362:1343–1356.

Bianco, P., and P. G. Robey. 2001. Stem cells in tissue engineering. *Nature* 414:118–121.

Bishop, A. E. 2004. Pulmonary epithelial stem cells. *Cell Prolif* 37:89–96.

Blau, H. M., T. R. Brazelton, and J. M. Weimann. 2001. The evolving concept of a stem cell: Entity or function? *Cell* 105:829–841.

Bonnet, D., and J. E. Dick. 1997. Human acute myeloid leukemia is organized as a hierarchy that originates from a primitive hematopoietic cell. *Nat Med* 3:730–737.

Boulton, M., and J. Albon. 2004. Stem cells in the eye. *Int J Biochem Cell Biol* 36:643–657.

Bunnell, B. A., M. Flaat, C. Gagliardi, B. Patel, and C. Ripoll. 2008. Adipose-derived stem cells: Isolation, expansion and differentiation. *Methods* 45:115–120.

Campbell, J. H., P. Walker, W. Chue et al. 2004. Body cavities as bioreactors to grow arteries. *Int Congr Ser* 1262:118–121.

Casteilla, L., and C. Dani. 2006. Adipose tissue-derived cells: From physiology to regenerative medicine. *Diabetes Metab* 32:393–401.

Cha, J., and V. Falanga. 2007. Stem cells in cutaneous wound healing. *Clin Dermatol* 25(1):73–78.

Chakir, J., N. Pagé, Q. Hamid et al. 2001. Bronchial mucosa produced by tissue engineering: A new tool to study cellular interactions in asthma. *J Allergy Clin Immunol* 107(1):36–40.

Chamberlain, G., J. Fox, B. Ashton, and J. Middleton. 2007. Concise review: Mesenchymal stem cells: Their phenotype. differentiation capacity, immunological features, and potential for homing. *Stem Cells* 25:2739–2749.

Choi, Y. S., S. M. Cha, Y. Y. Lee et al. 2006. Adipogenic differentiation of adipose tissue derived stem cells in nude mouse. *Biochem Biophys Res Commun* 345(2):631–637.

Clarke, D., and J. Frisen. 2001. Differentiation potential of adult stem cells. *Curr Opin Genet Dev* 11:575–580.

De Souza, S. C., E. Munin, L. Procopio Alves, M. A. Castillo Salgado, and M. T. T. Pacheco. 2005. Low power laser radiation at 685 nm stimulates stem-cell proliferation rate in *Dugesia tigrina* during regeneration. *J Photochem Photobiol B Biol* 80:203–207.

de Villiers, J. A., N. N. Houreld, and H. Abrahamse. 2009. Adipose derived stem cells and smooth muscle cells: Implications for regenerative medicine. *Stem Cell Rev Rep* 5(3):256–265.

de Villiers, J. A., N. N. Houreld, and H. Abrahamse. 2011. Influence of low intensity laser irradiation on isolated human adipose derived stem cells over 72 hours and their differentiation potential into smooth muscle cells using retinoic acid. *Stem Cell Rev Rep* 7(4):869–882.

Dontu, G., M. Al Hajj, W. M. Abdallah, M. F. Clarke, and M. S. Wicha. 2003. Stem cells in normal breast development and breast cancer. *Cell Prolif* 36(Suppl 1):59–72.

Dozmorov, M. G., B. P. Kropp, R. E. Hurst, E. Y. Cheng, and H. Lin. 2007. Differentially expressed gene networks in cultures smooth muscle cells from normal and neuropathic bladder. *J Smooth Muscle Res* 43(2):55–72.

Drochioiu, G. 2010. Laser induced ATP formation: Mechanism and consequences. *Photomed Laser Surg* 28(4):573–574.

Elabd, C., C. Chiellini, A. Massoudi et al. 2007. Human adipose tissue-derived multipotent stem cells differentiate *in vitro* and *in vivo* into osteocyte-like cells. *Biochem Biophys Res Commun* 361:342–348.

Fraser, J. K., I. Wulur, Z. Alfonso, and M. H. Hedrick. 2006. Fat tissue: An underappreciated source of stem cells for biotechnology. *Trends Biotechnol* 24(4):150–154.

Gao, X., and D. Xing. 2009. Molecular mechanisms of cell proliferation induced by low power laser irradiation. *J Biomed Sci* 16:4–30.

Gasparyan, L., G. Brill, and A. Makela. 2004. Influence of low level laser radiation on migration of stem cells. *Laser Florence* 5968:58–63.

Gomillion, C. T., and K. J. L. Burg. 2006. Stem cells and adipose tissue engineering. *Biomaterials* 27:6052–6063.

Grassel, S., and N. Ahmed. 2007. Influence of cellular microenvironment and paracrine signals on chondrogenic differentiation. *Front Biosci* 1(12):4946–4956.

Gritti, A., A. L. Vescovi, and R. Galli. 2002. Adult neural stem cells: Plasticity and developmental potential. *J Physiol–Paris* 96:81–90.

Gruen, L., and L. Grabel. 2006. Concise review: Scientific and ethical roadblocks to human embryonic stem cell therapy. *Stem Cells* 24:2162–2169.

Hamburger A. W., and S. E. Salmon. 1977. Primary bioassay of human tumor stem cells. *Science* 197:461–463.

Hawkins, D. H., and H. Abrahamse. 2006. Effect of multiple exposures of low-level laser therapy on the cellular responses of wounded human skin fibroblasts. *Photomed Laser Surg* 24:705–714.

Hawkins, D., and H. Abrahamse. 2007. Changes in cell viability of wounded fibroblasts following laser irradiation in broad-spectrum or infrared light. *Laser Chem* 2007:71039.

Hemmati, H. D., I. Nakano, J. A. Lazareff et al. 2003. Cancerous stem cells can arise from pediatric brain tumors. *Proc Natl Acad Sci USA* 100:15178–15183.

Hirst, S. L. 1996. Airway smooth muscle cell culture: Application to studies of airway wall remodelling and phenotype plasticity in asthma. *Eur Respir J* 9:808–820.

Ichim, T. E., R. J. Harman, W. Min et al. 2010. Autologous stromal vascular fraction cells: A tool for facilitating tolerance in rheumatic disease. *Cell Immunol* 264:7–17.

James, A., and N. Carroll. 2000. Airway smooth muscle in health and disease; methods of measurement and relation to function. *Eur Respir J* 15:782–789.

Jones, D. L., and M. T. Fuller. 2006. Stem cell niches. In *Essentials of Stem Cell Biology*, 1st ed. R. Lanza, editor. Academic Press, London, 43.

Jurgens, W. J. F. M., M. J. Oedayrajsingh-Varma, M. N. Helder et al. 2008. Effect of tissue-harvesting site on the yield of stem cells derived from adipose tissue: Implications for cell-based therapies. *Cell Tissue Res* 332:415–426.

Kadivar, M., S. Khatami, Y. Mortazavi, M. Taghikhani, and M. A. Shokrgozar. 2006. Multilineage differentiation activity by the human umbilical vein-derived mesenchymal stem cells. *Iran Biomed J* 10(4):175–184.

Kang, S. K., D. H. Lee, Y. C. Bae et al. 2003. Improvement of neurological deficits by intracerebral transplantation of human adipose tissue-derived stromal cells after cerebral ischemia in rats. *Exp Neur* 183:355–366.

Karu, T. 1998. *The Science of Low-Power Laser Therapy*, 1st ed. Gordon and Breach, Amsterdam, 43.

Kim, H. K., J. H. Kim, A. A. Abbas et al. 2009. Red light of 647 nm enhances osteogenic differentiation in mesenchymal stem cells. *Lasers Med Sci* 24:214–222.

Kim, J. M., S. Lee, K. Chu et al. 2007. Systemic transplantation of human adipose stem cells attenuated cerebral inflammation and degeneration in a hemorrhagic stroke model. *Brain Res* 1183:43–50.

Krenning, G., J. R. A. J. Moonen, M. J. A. van Luyn, and M. C. Harmsen. 2008. Vascular smooth muscle cells for use in vascular tissue engineering obtained by endothelial-to-mesenchymal transdifferentiation (EnMT) on collagen matrices. *Biomaterials* 29:3703–3711.

Lakshmipathy, U., and C. Verfaillie. 2005. Stem cell plasticity. *Blood Rev* 19:29–38.

Lee, J., and D. M. Kemp. 2006. Human adipose-derived stem cells display myogenic potential and perturbed function in hypoxic conditions. *Biochem Biophys Res Commun* 341:882–888.

Lin, F., S. F. Josephs, D. T. Alexandrescu et al. 2010. Lasers, stem cells and COPD. *J Transl Med* 8(16):1–10.

Lin, H. K., R. Cowan, P. Moore et al. 2004. Characterization of neuropathic bladder smooth muscle cells in culture. *J Urol* 171:1348–1352.

Liu, J. Y., D. D. Swartz, H. F. Peng et al. 2007. Functional tissue-engineered blood vessels from bone marrow progenitor cells. *Circ Res* 75:618–628.

Lu, J., R. Hou, C. J. Booth, S. Yang, and M. Snyder. 2006. Defined culture conditions of human embryonic stem cells. *Proc Natl Acad Sci USA* 103(15):5688–5693.

Mayani, H. 2003. A glance into somatic stem cells biology: Basic principles, new concepts, and clinical relevance. *Arch Med Res* 34.3–15.

Mayr, M., and Q. Xu. 2001. Smooth muscle cell apoptosis in arteriosclerosis. *Exp Geron* 36:969–987.

McLean, D. J. 2006. Vertebrate reproductive stem cells: Recent insights and technological advances. *Semin Cell Dev Biol* 17:534–539.

Mehlhorn, A. T., J. Zwingmann, G. Finkenzellar et al. 2009. Chondrogenesis of adipose-derived adult stem cells in a poly-lactide-*co*-glycolide scaffold. *Tissue Eng Part A* 15(00):1–9.

Melton, D. A., and C. Cowen. 2006. Stemness: Definitions, criteria, and standards. In *Essentials of Stem Cell Biology*, 1st ed. R. Lanza, editor. Academic Press, London, xxv, xxvi.

Mimeault, M., and S. K. Batra. 2006. Concise review: Recent advances on the significance of stem cells in tissue regeneration and cancer therapies. *Stem Cells* 24:2319–2345.

Moore, P., T. D. Ridgeway, R. G. Higbee, E. W. Howard, and M. D. Lucroy. 2005. Effect of wavelength on low-intensity laser irradiation-stimulated cell proliferation *in vitro*. *Lasers Surg Med* 36:8–12.

Moore, T. J. 2007. Stem cell Q and A—An introduction to stem cells and their role in scientific and medical research. *Med Tech SA* 21(1):3–6.

Morrison, S. J., N. M. Shah, and D. J. Anderson. 1997. Regulatory mechanisms in stem cell biology. *Cell* 88:287–298.

Mvula, B., T. Mathope, T. Moore, and H. Abrahamse. 2008. The effect of low level laser therapy on adipose derived stem cells. *Lasers Med Sci* 23(3):277–282.

Mvula, B., T. J. Moore, and H. Abrahamse. 2010. Effect of low-level laser irradiation and epidermal growth factor on adult human adipose-derived stem cells. *Lasers Med Sci* 25:33–39.

National Center for Biotechnology Information (NCBI). 2009. What is a cell? Retrieved 15 April 2012 from http://www.ncbi.nlm.nih.gov/About/primer/geneticscell.html.

Naveiras, O., and G. Q. Daley. 2006. Stem cells and their niche: A matter of fate. *Cell Mol Life Sci* 63:760–766.

Niwa, H. 2006. Mechanisms of stem cell self-renewal. In *Essentials of Stem Cell Biology*, 1st ed. R. Lanza, editor. Academic Press, London, 55.

Noguchi, H. 2007. Stem cells for the treatment of diabetes. *Endocrinology* 54(1):7–16.

Odorico, J. S., D. S. Kaufman, and J. A. Thompson. 2001. Multilineage differentiation from human embryonic stem cell lines. *Stem Cells* 19:193–204.

Ogawa, R. 2006. The importance of adipose-derived stem cells and vascularised tissue regeneration in the field of tissue transplantation. *Curr Stem Cell Res Ther* 1:13–20.

Oh, S., H. M. Hatch, and B. E. Petersen. 2002. Hepatic oval 'stem' cell in liver regeneration. *Semin Cell Dev Biol* 13:405–409.

Pöntein, P. J. 1992. *Low Level Laser Therapy as a Medical Treatment Modality*, 1st ed. Art Urpo Ltd., Tampere, 13, 17, 18.

Prentice, D. A. 2003. *Adult Stem Cells*. Retrieved 2 May 2012 from http://www.stemcellresearch.org/facts/prentice.htm.

Primo, F. L., M. B. da Costa Reis, M. A. Porcionatto, and A. C. Tedesco. 2011. *In vitro* evaluation of chloroaluminum phthalocyanine nanoemulsion and low-level laser therapy on human skin dermal equivalents and bone marrow mesenchymal stem cells. *Curr Med Chem* 18(22):3376–3381.

Rajasekhar, V. K., and M. C. Vemuri. 2005. Molecular insights into the function and prospects of stem cells. *Stem Cells* 23:1212–1220.

Raposio, E., I. Baldelli, F. Benvenuto et al. 2007. Characterisation and induction of human pre-adipocytes. *Toxicol In Vitro* 21:330–334.

Ringe, J., C. Kaps, G. Burmester, and M. Sittinger. 2002. Stem cells for regenerative medicine: Advances in the engineering of tissues and organs. *Naturwissenschaften* 89:338–351.

Rizvi, A. Z., J. R. Swain, P. S. Davies et al. 2006. Bone marrow-derived cells fuse with normal and transformed intestinal stem cells. *Proc Natl Acad Sci USA* 103(16):6321–6325.

Rodriguez, L. V., Z. Alfonso, R. Zhang et al. 2006. Clonogenic multipotent stem cells in human adipose tissue differentiate into functional smooth muscle cells. *Proc Natl Acad Sci USA* 103(32):12167–12172.

Rorth, M., E. Rajpert-De Meyts, L. Andersson et al. 2000. Carcinoma in situ in the testis. *Scand J Urol Nephrol Suppl* 166–186.

Sanz-Ruiz, R., M. E. Fernandez Santos, M. Dominguez Munoa et al. 2008. Adipose tissue-derived stem cells: The friendly side of a classic cardiovascular foe. *J Cardiovasc Trans Res* 1(1):55–63.

Schäffler, A., and C. Büchler. 2007. Concise review: Adipose tissue-derived stromal cells—Basic and clinical implications for novel cell-based therapies. *Stem Cells* 25:818–827.

Scott, C. T., and R. A. Reijo Pera. 2008. The road to pluripotence: The research response to the embryonic stem cell debate. *Hum Mol Genet* 17(1):R3–R9.

Sievert, K., B. Amend, and A. Stenzl. 2007. Tissue engineering for the lower urinary tract: A review of a state of the art approach. *Eur Urol* 52:1580–1589.

Sinha, S., B. R. Wamhoff, M. H. Hoofnagle et al. 2006. Assessment of contractility of purified smooth muscle cells derived from embryonic stem cells. *Stem Cells* 24:1678–1688.

Soleimani, M., E. Abbasnia, M. Fathi et al. 2012. The effects of low-level laser irradiation on differentiation and proliferation of human bone marrow mesenchymal stem cells into neurons and osteoblasts—An *in vitro* study. *Lasers Med Sci* 27:423–430.

Stein, A., D. Benayahu, L. Maltz, and U. Oron. 2005. Low-level laser irradiation promotes proliferation and differentiation of human osteoblasts *in vitro*. *Photomed Laser Surg* 23(2):161–166.

Strem, B. M., K. C. Hicok, M. Zhu et al. 2005. Multipotential differentiation of adipose tissue-derived stem cells. *Keio J Med* 54(3):132–141.

Taléns-Visconti, R., A. Bonora, R. Jover et al. 2007. Human mesenchymal stem cells from adipose tissue: Differentiation into hepatic lineage. *Toxicol In Vitro* 21:324–329.

Tarnok, A., H. Ulrich, and J. Bocsi. 2010. Phenotypes of stem cells from diverse origin. *Cytometry A* 77A:6–10.

Thompson, J. A., J. Itskovitz-Eldor, S. S. Shapiro et al. 1998. Embryonic stem cell lines derived from human blastocysts. *Science* 282:1145–1147.

Tomasson, M. H. 2009. Cancer stem cells: A guide for skeptics. *J Cell Biochem* 106:745–749.

Tsai, R. Y. L., R. Kittappa, and R. D. G. McKay. 2002. Plasticity, niches, and the use of stem cells. *Dev Cell* 2:707–712.

Tuan, R., G. Boland, and R. Tuli. 2002. Adult mesenchymal stem cells and cell based tissue engineering. *Arthritis Res Ther* 5(1):32–45.

Tuby, H., L. Maltz, and U. Oron. 2007. Low-level laser irradiation (LLLI) promotes proliferation of mesenchymal and cardiac stem cells in culture. *Lasers Surg Med* 39:373–378.

Tuby, H., L. Maltz, and U. Oron. 2011. Induction of autologous mesenchymal stem cells in the bone marrow by low-level laser therapy has profound beneficial effects on the infarcted rat heart. *Lasers Surg Med* 43:401–409.

Tukaj, C., J. Bohdanowicz, and J. Kubasik-Juraniec. 2004. The growth and differentiation of aortal smooth muscle cells with microtubule reorganisation—An *in vitro* study. *Folia Morphol* 63:51–57.

Tunér, J., and L. Hode. 2002. *Laser Therapy Clinical Practice and Scientific Background*, 1st ed. Prima Books, Grängesberg, Sweden, 8–19.

Turksen, K., editor. 2004. *Adult Stem Cells*, 1st ed. Humana Press, Clifton, NJ, 1.

Valina, C., K. Pinkernell, Y. Song et al. 2007. Intracoronary administration of autologous adipose tissue-derived stem cells improves left ventricular function, perfusion, and remodelling after acute myocardial infarction. *Eur Heart J* 28:2667–2677.

Van der Kooy, D., and S. Weiss. 2000. Why stem cells? *Science* 287:1439–1441.

Verfaillie, C. 2006. Adult stem cells: Tissue specific or not? In *Essentials of Stem Cell Biology*, 1st ed. R. Lanza, editor. Academic Press, London, 23, 24.

Vieira, N. M., V. Brandalise, E. Zucconi et al. 2008. Human multipotent adipose derived stem cells restore dystrophin expression of Duchenne skeletal muscle cells *in vitro*. *Biol Cell* 100:231–241.

Wagers, A. J., and I. L. Weissman. 2004. Plasticity of adult stem cells. *Cell* 116:639–648.

Wagner, W., F. Wein, A. Seckinger et al. 2005. Comparative characteristics of mesenchymal stem cells from human bone marrow, adipose tissue, and umbilical cord blood. *Exp Hemat* 33:1402–1416.

Wang, B., J. Han, Y. Gao et al. 2007. The differentiation of rat adipose-derived stem cells into OEC-like cells on collagen scaffolds by co-culturing with OECs. *Neurosci Lett* 421:191–196.

Watson, J. 1997. *Lasers in Engineering*. Retrieved August 25, 2008 from http://vcs.abdn.ac.uk/ENGINEERING/lasers/lasers.html.

Watt, F. M., and B. L. M. Hogan. 2000. Out of Eden: Stem cells and their niches. *Science* 287:1427–1430.

Wong, J. W., A. Velasco, P. Rajagopalan, and Q. Pham. 2003. Directed movement of vascular smooth muscle cells on gradient-compliant hydrogels. *Langmuir* 19:1908–1913.

Wu, K., Y. L. Liu, B. Cui, and Z. Han. 2006. Application of stem cells for cardiovascular grafts tissue engineering. *Transpl Immunol* 16:1–7.

Wu, Y., J. Wang, D. Gong et al. 2012. Effects of low-level laser irradiation on mesenchymal stem cell proliferation: A microarray analysis. *Lasers Med Sci* 27:509–519.

Wurmser, A. E., T. D. Palmer, and F. H. Gage. 2004. Cellular interactions in the stem cell niche. *Science* 304:1253–1254.

Yamada, Y., X. Wang, S. Yokayama, N. Fukuda, and N. Takakura. 2006. Cardiac progenitor cells in brown adipose tissue repaired damaged myocardium. *Biochem Biophys Res Commun* 342:662–670.

Young, H., and M. K. Carpenter. 2006. Characterisation of human embryonic stem cells. In *Essentials of Stem Cell Biology*, 1st ed. R. Lanza, editor. Academic Press, London, 265.

Zipori, D. 2005. The stem state: Plasticity is essential, whereas self-renewal and hierarchy are optional. *Stem Cells* 23:719–726.

Zouboulis, C. C., J. Adjaye, H. Akamatsu, G. Moe-Behrens, and C. Niemann. 2008. Human skin stem cells and the aging process. *Exp Geron* 43:986–997.

Zuk, P. A., M. Zhu, P. Ashjian et al. 2002. Human adipose tissue is a source of multipotent stem cells. *Mol Biol Cell* 13:4279–4295.

Zuk, P. A., M. Zhu, H. Mizuno et al. 2001. Multilineage cells from human adipose tissue: Implications for cell-based therapies. *Tissue Eng* 7:211–228.

Low-Level Light Therapy for Cosmetics and Dermatology

Pinar Avci
Massachusetts General Hospital

Theodore Nyame
Harvard Medical School

Michael R. Hamblin
Massachusetts General Hospital

58.1 Introduction

Low-level laser (or light) therapy (LLLT), phototherapy, or photobiomodulation refers to the use of photons to alter biological activity. Nonthermal, coherent light sources (lasers) or noncoherent light sources consisting of filtered lamps or light-emitting diodes (LEDs) are used in this type of therapy for reducing pain and inflammation, augmenting tissue repair and regeneration, and preventing tissue damage (Chung et al. 2012; Gupta et al. 2012). In the last few decades, nonablative laser therapies have been used increasingly for the aesthetic treatment of fine wrinkles, photoaged skin, and scars, a process known as photorejuvenation. More recently, they have also been used for inflammatory acne (Seaton et al. 2006). Their potential use for other dermatological conditions and cosmetics such as vitiligo, psoriasis, photoprotection, hair regrowth, and fat reduction has been shown by several studies. In this chapter, we will briefly mention about these cosmetic and dermatological applications of LLLT, starting with its current and potential use in cosmetic dermatology and various skin conditions, hair loss treatment, and lastly fat reduction procedures and cellulite treatment.

58.2 LLLT in Dermatology

58.2.1 LLLT for Skin Rejuvenation

Aging skin starts showing its first signs between late 20s and early 30s, and it usually presents with wrinkles, dyspigmentation, telangiectasia, and loss of elasticity. Several modalities have been developed in order to reverse the dermal and epidermal signs of aging skin. Retinoic acid, a vitamin A derivative, dermabrasion,

chemical peels, and ablative laser resurfacing with carbon dioxide (CO_2) or erbium:yttrium–aluminum–garnet (Er:YAG) lasers or a combination of these wavelengths are just a few examples (Branham and Thomas 1996; Airan and Hruza 2005; Paasch and Haedersdal 2011). LED, a novel light source for nonthermal, nonablative skin rejuvenation, has been suggested to be effective for improving wrinkles and skin laxity through its regenerative effects such as increasing collagen production, inducing proliferation of fibroblasts, as well as enhancing microcirculation and vascular perfusion in the skin (Abergel et al. 1987; Yu, Naim, and Lanzafame 1994; Weiss et al. 2004; Bhat et al. 2005; Dierickx and Anderson 2005; Russell, Kellett, and Reilly 2005; Weiss et al. 2005a,c; Barolet et al. 2009). One study in accordance with this hypothesis is a split-face, single-blinded clinical study that was carried out to evaluate the effect of LLLT on skin texture and appearance of individuals with aged/photoaged skin (Barolet et al. 2009). Profilometry quantification demonstrated that more than 90% of individuals had a reduction in rhytid depth and surface roughness, and 87% of the individuals reported that they have experienced a reduction in the Fitzpatrick wrinkling severity score following 12 LED treatments (Barolet et al. 2009).

Using different pulse sequence parameters, a multicenter clinical trial was also conducted, with 90 patients receiving eight LED treatments over 4 weeks (Geronemus et al. 2003; McDaniel et al. 2003; Weiss, McDaniel, and Geronemus 2004, 2005a). According to this study, over 90% of patients improved by at least one Fitzpatrick photoaging category, and 65% of patients demonstrated global improvement in facial texture, fine lines, erythema, and pigmentation. Best results were obtained at 4 to 6 months following completion of eight treatments, and prominent increase in collagen in the papillary dermis and reduced Matrix metalloproteinase-1 (MMP-1) were common findings.

58.2.2 LLLT for Acne

Acne vulgaris is a common skin disorder, with a reported prevalence varying from 35% to over 90% among adolescents (Stathakis, Kilkenny, and Marks 1997). Pathogenesis of acne has not yet been identified; however, current consensus is that it involves follicular hyperconification, increased sebum secretion, colonization of *Propionibacterium acnes* (*P. acnes*), and inflammation (Lee, You, and Park 2007). Current treatments for acne vulgaris include topical and oral medications such as topical antibiotics, topical retinoids, benzoyl peroxide, salicylic acid, or azaleic acid. In severe cases, systemic antibiotics such as tetracycline and doxycycline, oral retinoids, and some hormones may also be indicated (Aziz-Jalali, Tabaie, and Djavid 2012). Despite many options that are available for treatment of acne vulgaris, many patients still respond inadequately to treatment or experience some adverse effects.

Phototherapy (light, lasers, and photodynamic therapy) was presented as an alternative therapeutic modality for treatment of acne vulgaris and was proposed to have fewer side effects when compared to other treatment options (Rotunda, Bhupathy, and Rohrer 2004). One mechanism of action for phototherapy is through absorption of light (specifically blue light) by porphyrins that are produced by *P. acnes* that in turn cause a photochemical reaction and form reactive free radicals and singlet oxygen species (Ross 2005; Lee, You, and Park 2007). On the other hand, red light is assumed to exert its effects through its anti-inflammatory properties (Rotunda, Bhupathy, and Rohrer 2004; Sadick 2008).

Among several studies demonstrating effects of LLLT for treatment of acne vulgaris, one study used 630 nm red spectrum LLLT with fluence 12 J/cm² twice a week for 12 sessions in conjunction with 2% topical clindamycin, and significant reduction in active acne lesions after 12 sessions of treatment has been observed (Aziz-Jalali, Tabaie, and Djavid 2012). However, the same study showed no significant effects when the wavelength was changed to 890 nm (Aziz-Jalali, Tabaie, and Djavid 2012). Furthermore, combination of blue light (antibacterial effect) and red light (anti-inflammatory effect) was proposed to have synergistic effects in acne treatment, which later on have been supported by several studies (Papageorgiou, Katsambas, and Chu 2000; Goldberg and Russell 2006; Lee, You, and Park 2007; Sadick 2008).

58.2.3 LLLT for Herpes Virus

Herpes simplex virus (HSV) infection is one of the most common viral infections that is chronic and lifelong. It has two types: HSV-1 and HSV-2; while HSV-1 causes primarily mouth, throat, face, eye, and central nervous system infections, HSV-2 causes primarily anogenital infections. However, each may cause infections in all areas as well. Following primary infection

and resolution of the lesions, the virus moves through the nerve endings and establishes a latent state at sensory ganglia, most commonly the trigeminal ganglion (de Paula Eduardo et al. 2011). The exposition of the host to various types of physical or emotional stress such as fever, immunosuppression, or exposure to UV light causes virus reactivation and migration through sensory nerves to skin and mucosa, localizing especially on lips and the perioral area (de Paula Eduardo et al. 2011). While in immunocompetent individuals it presents mainly as cold sores, in immunocompromised patients, HSV-1 infection can lead to potentially harmful complications. Among both immunocompetent and immunocompromised patients, up to 60% will experience a prodromic stage, after which the outbreaks develop through the stages of erythema, papule, vesicle, ulcer, and crust, until healing is achieved and is accompanied by pain, burning, itching, or tingling at the site of blister.

While several antiviral drugs such as acyclovir and valacyclovir are being used for controlling the recurrent herpes outbreaks, since their mechanism of action is through interfering with viral multiplication, there is a very narrow window in which oral intake may bring benefit to any patient; only limited reduction in the lesion healing time has been observed. Moreover, development of drug-resistant HSV strains is increasing especially among immunocompromised patients (Whitley, Kimberlin, and Roizman 1998). Therefore, new alternatives to current therapeutic modalities are in demand. Recently, LLLT has been investigated for accelerated healing, reducing symptoms and influencing length of the recurrence period (Bello-Silva et al. 2010; de Paula Eduardo et al. 2011; Muñoz Sanchez et al. 2012). Among 50 patients with recurrent perioral herpes simplex infection, when LLLT was applied daily for 2 weeks during recurrence-free periods, it led to a decrease in the frequency of herpes labialis episodes (Schindl and Neumann 1999). In a separate study with similar irradiation parameters, investigators achieved a significant prolongation of remission intervals from 30 to 73 days in patients with recurrent HSV infection (Landthaler, Haina, and Waidelich 1983).

Even though mechanism of action is still not clear, indirect effect of LLLT on cellular and humoral components of the immune system being involved in antiviral responses rather than a direct virus-inactivating effect could be considered (Korner et al. 1989).

58.2.4 LLLT for Hypertrophic Scars and Keloid

Hypertrophic scars and keloids are benign skin tumors that usually form following surgery, trauma, or acne and are difficult to eradicate. Epidemiology as well as the optimal treatment of these lesions are yet to be discovered (Bouzari, Davis, and Nouri 2007; Louw 2007; Wolfram et al. 2009). LLLT's use as a prophylactic method to alter the wound healing process in order to avoid or attenuate the formation of hypertrophic scars or keloids has been investigated by Barolet and Boucher (2010). Following scar revision by surgery or CO_2 laser ablation on bilateral areas, single scar was treated daily by the patients at home with near infrared

(NIR)-LED 805 nm at 30 mW/cm² and 27 J/cm². One patient had pre-auricular linear keloids bilaterally that have occurred following a face-lift procedure, and surgical scar revision/excision has been performed. Another patient had hypertrophic scars on chest bilaterally following severe acne, and CO_2 laser was used for resurfacing. The last patient had hypertrophic scars on the back bilaterally post-excision, and again CO_2 laser was used for resurfacing (Barolet and Boucher 2010). As a result, significant improvements on the NIR-LED treated versus the control scar have been reported by the authors, while no significant treatment-related adverse effects have been observed (Barolet and Boucher 2010).

58.2.5 LLLT for Burns

Burn, one of the main complications of laser treatments, is many times devastating for a patient. In order to test whether LLLT may facilitate healing in such cases, nine patients with a variety of second-degree burns from nonablative devices were given LED therapy once a day for 1 week. According to both the patient and the physician, healing occurred 50% faster (Weiss et al. 2005c). However, when Schlager et al. (2000) tested the 670 nm, 250 mW, 2 J/cm² laser light on the healing of 30 rats, which were burned on both flanks, and compared them with control wounds, which received no LLLT, neither macroscopic nor histological evaluation demonstrated accelerated wound healing.

Burn scars are difficult to treat because of their tendency to worsen with hypertrophy and contracture. Since there is no ideal treatment, LLLT was tested as a possible treatment option. Nineteen patients with 19 burn scars were treated with a 400 mW, 670 nm, 4 J/cm² LLLT for twice a week over 8 weeks, and following treatment, scars were softer and more pliable (Gaida et al. 2004). Relief from pruritus and pain and sometimes improved pattern of scars within the mesh grafts were also reported. However, these effects were sometimes limited, and it is difficult to expect a complete disappearance of the scars. It is also important to note that patients with burn scars not older than 12 months achieved greater results following treatment (Gaida et al. 2004).

58.2.6 LLLT for Psoriasis

Psoriasis is a chronic and recurrent inflammation skin disease, affecting 1%–3% of the population (Stern et al. 2004; Gelfand et al. 2005). Although the etiology is not completely known, it results from interactions between systemic, genetic, immunological, and environmental factors (Zhang 2012). Psoriasis is characterized by well-demarcated plaques due to hyperproliferation of keratinocytes, mediated by T cell lymphocytes that are attacking rather than defending normal skin (Griffiths and Barker 2007).

The main affected regions are knee, elbows, scalp, nails, lower back, or sacrum, but the plaques can be present on whole body and the severity is measured by total body surface area involvement or plaque severity. Different types of psoriasis exist such as chronic plaque psoriasis or psoriasis vulgaris (Griffiths and Barker 2007), flexural (inverse) psoriasis (van de Kerkhof et al. 2007; Laws and Young 2010), guttate psoriasis (Krishnamurthy et al. 2010), erythrodermic psoriasis (Laws and Young 2010), palmar-plantar psoriasis, facial psoriasis, and scalp psoriasis, and almost all types cause significant morbidity and poor quality of life (Finlay et al. 1990). Several modalities have been evaluated for therapeutic management of psoriasis including application of topical agents, systemic administration of drugs, photodynamic therapy, ultraviolet radiation, and laser therapy. The introduction of UVB and later of UVA treatment for psoriasis revolutionized the therapeutic modality to treat this disease, although some studies showed that repeated and excessive exposures to UVB increased the risk of skin carcinoma and also the PUVA reduced but did not eliminate the risk to induce skin carcinoma. First studies with laser treatment for psoriasis have been investigated since the 1980s with the carbon dioxide ablative laser (Bekassy and Astedt 1985), helium–neon laser (Colver, Cherry, and Ryan 1984), and red light photodynamic therapy (Berns et al. 1984). Laser can offer some advantages over other treatments since it can be applied solely on the lesion without affecting the surrounding skin and it has no or only limited systemic effects. It can be also used in combination with other treatment modalities in order to improve the treatment of resistant lesions. Many studies were carried out using excimer laser, which selectively emits light at a 308 nm wavelength (Asawanonda et al. 2000; Trehan and Taylor 2002; Gattu, Rashid, and Wu 2009). The results observed in these studies were similar to those obtained using UVB treatment. Laser prevented replication of epidermal cells and induced localized suppression of immune function helping to decrease the inflammation observed in psoriasis (Railan and Alster 2008). However, whether a long-term use of excimer laser might cause induction of carcinogenesis is still uncertain. Pulsed dye laser (PDL) at 585 nm, which is commonly used for treatment of vascular disorders, was suggested as an alternative. Increased vascularity has been observed in psoriasis, and several studies proved the efficacy of PDL in treatment of psoriasis (Ilknur et al. 2006; De Leeuw et al. 2009). Recently, LLLT has been investigated as an alternative treatment option for psoriasis plaques. A recent preliminary study investigated the efficacy of combination of two different wavelengths (830 and 630 nm) to treat recalcitrant psoriasis using a LED irradiation. All patients with psoriasis, resistant to conventional therapy, were enrolled and were treated sequentially with 830 and 630 nm wavelengths in two 20 min sessions with 2 days between sessions for 4 or 5 weeks. At the end of this study, while no adverse side effects have been observed, resolution of psoriasis has been reported (Ablon 2010). The study was limited to a small number of patients; however, the results observed encourage future investigations for use of LLLT in psoriasis.

58.3 LLLT for Hair Loss Treatment

In 2007, LLLT was approved by the FDA as a safe treatment for hair loss (Wikramanayake et al. 2012). Even though the mechanism of action is yet to be determined, laser phototherapy is assumed to stimulate anagen reentry into telogen hair follicles, increase rates of proliferation in active anagen hair follicles, prevent premature catagen development, and prolong duration of anagen phase (Leavitt et al. 2009; Wikramanayake et al. 2012). Miura et al. (1999) observed that following irradiation of the backs of Sprague–Dawley rats with linear polarized infrared laser, there was an upregulation of hepatocyte growth factor (HGF) and HGF activator expression. Another study reported increased skin temperature and blood flow following linear polarized irradiation around the stellate ganglion area (Wajima et al. 1996). Weiss et al. demonstrated that LLLT modulates 5-α reductase, which converts testosterone into DHT, and also alters vascular endothelial growth factor (VEGF) gene expression, which has a significant role in hair follicle growth, and thus stimulates hair growth (Yano, Brown, and Detmar 2001; Castex-Rizzi et al. 2002; Weiss et al. 2005b). LLLT's effects in modulating inflammatory processes and immunological responses suggest that these responses may also have a role in hair regrowth (Meneguzzo et al. 2013). A study conducted by Wikramanayake et al. (2012) on C3H/HeJ mouse model of alopecia areata (AA) was consistent with this assumption where in mice treated with laser comb, decreased inflammatory infiltrates as well as increased number of hair follicles with majority in anagen phase were noted.

58.3.1 LLLT for AA

As previously mentioned, modulatory effects of LLLT on inflammation might have a significant role in treatment of AA (Wikramanayake et al. 2012). In order to test the effect of linear polarized infrared irradiation in treatment of AA, a study has been conducted among six men and nine women using a medical instrument operating on polarized linear light with a high output (1.8 W) of infrared radiation (600–1600 nm) (Yamazaki et al. 2003). Scalp was irradiated for 3 min once every week or every 2 weeks, and additionally, carpronium chloride 5% was applied twice daily to all the lesions (Yamazaki et al. 2003). Furthermore, oral antihistamines, cepharanthin, and glycyrrhizin (extracts of Chinese medicine herbs) were prescribed (Yamazaki et al. 2003). The authors of this study have reported that in 47% of the patients, hair growth occurred 1.6 months earlier in irradiated areas than in nonirradiated areas (Yamazaki et al. 2003). Another study by Wikramanayake et al. (2012) tested the hair growth effects of LLLT on C3H/HeJ mouse model of AA using a 655 nm HairMax Laser Comb, which emits 9 beams. Mice were irradiated 20 s daily three times per week for a total of 6 weeks (Wikramanayake et al. 2012). At the end of the treatment, hair regrowth was observed in all the laser-treated mice, while no difference has been observed in sham-treated group (Wikramanayake et al. 2012). Moreover, histology results demonstrated increased number of anagen hair follicles in laser-treated mice (Wikramanayake et al. 2012).

58.3.2 LLLT for Androgenetic Alopecia

Shukla et al. (2010) investigated the effect of helium–neon (He-Ne) laser (632 nm, at doses of 1 and 5 J/cm^2 at 24-h interval for 5 days) on the hair follicle growth cycle of testosterone-treated and untreated Swiss albino mice skin. The results showed that exposure to the He–Ne laser at a dose of 1 J/cm^2 led to significant increase in percentage of hair follicles at anagen phase compared to that of the control group, which received no testosterone treatment nor He–Ne laser irradiation. Exposure to the He–Ne laser at a dose of 5 J/cm^2 led to a significant decrease compared to the control group, which might be expected considering the biphasic effect of LLLT (Shukla et al. 2010; Chung et al. 2012). Moreover, testosterone treatment led to the inhibition of hair growth with respect to the control group, which was characterized by a significant reduction in the percentage of catagen follicles (Shukla et al. 2010). However, in the testosterone-treated mice that received He–Ne laser at a dose of 1 J/cm^2, a significant increase in the percentage of hair follicles with respect to only testosterone-treated mice has been observed. When the testosterone-treated mice received He–Ne laser at 5 J/cm^2, relative percentage of hair follicles at the telogen phase doubled (Shukla et al. 2010). Since the hair growth promoting effect of He–Ne laser (1 J/cm^2) was much higher for the testosterone-treated mice than the control, it can be suggested that cells growing at a slower rate or under stress conditions respond better to the stimulatory effects of LLLT. Another important observation in this study that is worth mentioning is that in He–Ne laser (1 J/cm^2) irradiated skin, some of the anagen follicles appeared from a higher depth and possessed a different orientation (Shukla et al. 2010). Such follicles are known to represent the late anagen stage in the hair growth cycle, so the presence of these follicles may suggest that laser irradiation prolongs the anagen phase (Muller-Rover et al. 2001; Philp et al. 2004). Furthermore, in testosterone-treated and He-Ne (1 J/cm^2) irradiated skin, hair follicles were seen to originate from the middle of the dermis, and these follicles represent early anagen phase (Shukla et al. 2010). These two observations taken together conclude that majority of catagen and telogen follicles reenter into anagen phase as a result of low-level laser irradiation at 1 J/cm^2.

Using 655 nm red light and 780 nm infrared light once a day for 10 min, 24 AGA male patients were evaluated by a group of investigators (Kim, Park, and Lee 2007). Evaluation has been performed via global photography and phototrichogram (Kim, Park, and Lee 2007). Following 14 weeks of treatment, density of hairs and anogen/telogen ratio on both the vertex and occiput was significantly increased, and 83% of the patients reported to be satisfied with the treatment (Kim, Park, and Lee 2007).

Satino and Markou (2003) conducted a clinical study in order to test the efficacy of LLLT on hair growth and tensile strength on 28 male and 7 female AGA patients. Each patient received a 655 nm HairMax LaserComb, and they were asked to use it for 5 to 10 min every other day for a period of 6 months (Satino and Markou 2003). In terms of hair tensile strength, the results revealed greater improvement in the vertex area for males and temporal area for females; however, both sexes benefited in all areas significantly (Satino and Markou 2003). In terms of hair count, both sexes and all areas had significant improvement, but males and the vertex area had the best results (Satino and Markou 2003). Same device was tested by Leavitt et al. (2009) in a double-blind, sham device-controlled, multicenter, 26 week trial randomized study among 110 male AGA patients. Subjects used the device three times per week for 15 min for a total of 26 weeks (Leavitt et al. 2009). Significantly greater increase in mean terminal hair density as well as significant improvements in overall hair regrowth, slowing of hair loss, thicker feeling hair, better scalp health, and hair shine were reported (Leavitt et al. 2009).

58.3.3 LLLT for Chemotherapy-Induced Alopecia

The incidence of cancer-induced alopecia is close to 65% among the patients receiving chemotherapy, and it has significant negative psychological effects (Trueb 2009). LLLT may promote hair regrowth in chemotherapy-induced alopecia; however, as of today, the number of studies is still limited. One study investigated this possible effect of LLLT using a rat model. Different regiments of chemotherapy were given to each rat in conjunction with an LLLT device, and hair regrowth was observed to occur much earlier in all laser-treated rats than in sham-treated rats. Furthermore, LLLT treatment did not compromise the efficacy of chemotherapy by causing localized protection of the cancer cells (Wikramanayake et al. 2013).

58.4 LLLT for Fat Reduction and Cellulite Treatment

58.4.1 Lipoplasty and Liposuction

The modern concept of suction-assisted lipectomy was developed through many innovations popularized as far back as the 1920s. The first clinical introduction attempted to remove fat from the calves of a dancer. This endeavor ended in eventual death (Thorek 1939). Lipoplasty was reintroduced in 1974 by Fischer (1990), who utilized oscillating blades within a cannula to remove subcutaneous tissue. In 1983, Illouz reported his 5-year experience with a new lipoplasty technique utilizing large cannulas and suction tubing to safely remove fat from different regions in the body (Illouz 1983). This ushered in the era of modern lipoplasty. Over the ensuing decades, the concept of tumescent technique decreased blood loss and subsequent morbidity associated with liposuction and improved results.

58.4.2 LLLT for Fat Reduction and Cellulite Treatment

LLLT is one of the newest innovations to lipoplasty. The concept introduced by Niera et al. in 2000 utilizes a low-level laser with a dose rate that causes little detectable temperature rise in the tissue and no macroscopic changes in the tissue structure (Niera, Solarte, and Reyes 2000; Neira et al. 2002). The advances of LLLT in lipoplasty are based on determining the optimum wavelength and power necessary to augment lipoplasty without altering the macroscopic structure of the tissues (Oschmann 2000). The application of LLLT was derived from prior investigations of LLLT in wound healing, pain relief, and edema prevention (King 1989; Baxter et al. 1991). The literature supports that wavelengths between 630 and 640 nm are optimum for biomodulation (Fröhlich 1968, 1970, 1975; van Breugel and Bar 1992; Al-Watban and Zang 1996; Sroka et al. 1997). Utilizing low-level external laser therapy with a diode laser at 635 nm and a maximal power of 10 mW at energy values of 1.2 to 3.6 J/cm^2, Neira et al. (2002) made several astonishing observations regarding the effects of LLLT on adipocytes. Investigations using scanning electron microscopy (SEM) and transmission electron microscopy (TEM) demonstrated transient pore formation within the plasma membranes of adipocytes. This was hypothesized to enable the release of intracellular lipids from adipocytes. As an adjunct to liposuction, it was expected to decrease the time spent in the operating room, increase the volume of fat extracted, and enable the surgeon to expend less energy to aspirate fat.

Although the initial data were met with much enthusiasm, a rigorous study by Brown et al. (2004) called into question the observations surrounding LLLT. These studies used a 635 nm LLLT device of 1 J/cm^2 to determine whether alteration in adipocyte structure or function was modulated following LLLT. In their experiment, cultured human preadipocytes after 60 min of irradiation did not change appearance compared with non-irradiated control cells (Brown et al. 2004). Furthermore, histological assessment of lipoaspirates in a porcine model exposed to LLLT (30 min) and human lipoaspirates failed to demonstrate transitory pores using SEM (Brown et al. 2004). Additional data called into question the ability of red light (635 nm) to penetrate effectively below the skin surface and into the subdermal tissues (Kolari and Airaksinen 1993). In a supportive commentary, Peter Fodor wrote, "One could postulate that the presence of the black dots on scanning electron microscopy images on the surface of fat cells reported by Neira et al. could represent an artifact" (Brown et al. 2004). Since the presentation of Brown et al.'s data in 2004, a barrage of new literature regarding the efficacy of LLLT has been published.

58.4.2.1 LLLT's Mechanism of Action on Fat Reduction

In Neira's original paper, the effects of LLLT on adipocytes were attributed to formation of transitory micropores, which were visualized on SEM (Neira et al. 2002). This was theorized to enable the release of intracellular lipids from adipocytes. Based on these data, it was suggested that up to 99% of fat could be released from the adipocytes via application of 635 nm, 10 W intensity LLLT for a period of 6 min (Neira et al. 2002). When irradiated adipocytes were recultured, they demonstrated the ability to recover their native cellular structure. Caruso-Davis et al. (2011) further confirmed the viability of these adipocytes following irradiation using a live-dead assay. Increases in levels of reactive oxidant species (ROS) levels following LLLT have been proposed to induce lipid peroxidation within the cell membrane. This was thought to cause temporary damage that presents as transient pores. (Geiger, Korytowski, and Girotti 1995; Karu 2008; Tafur and Mills 2008; Chen et al. 2011). However, in an attempt to replicate Neira et al.'s (2002) data, Brown et al. (2004) failed to visualize any transitory micropores. As of today, no additional SEM studies have documented these pores, but many papers have indirectly supported this theory. Another proposed mechanism of action for release of lipids was through activation of the complement cascade, which could cause induction of adipocyte apoptosis and subsequent release of lipids (Caruso-Davis et al. 2011). To investigate the complement activation theory, Caruso-Davis et al. (2011) exposed differentiated human adipocytes to plasma. With or without irradiation, there was no difference in complement-induced lysis of adipocytes (Caruso-Davis et al. 2011). Although no enzymatic assays were done to determine levels of complement within the plasma, the group concluded that laser does not activate complement.

Further evidence suggests that LLLT increases levels of cytoplasmic cyclic adenosine monophosphate (cAMP) (Karu et al. 1985; Karu 1999). cAMP is known to activate protein kinase, which further activates enzymatic cytoplasmic lipase. This enzyme in turn breaks down triglycerides into fatty acids and glycerol, which can both penetrate the adipocyte membrane (Honnor, Dhillon, and Londos 1985; Nestor, Zarraga, and Park 2012). However, the findings of Caruso-Davis et al. (2011) from *in vitro* studies on human fat cells obtained from subcutaneous fat irradiated with 635–680 nm LLLT for 10 min demonstrated no increase in glycerol and fatty acids suggesting that fat loss from the adipocytes in response to laser treatment is not due to a stimulation of lipolysis. Furthermore, increased triglyceride levels within the supernatant lend support to the formation of pores in adipocytes. Although these mechanisms have been worked out independently, the mechanism by which triglycerides would traverse the adipocyte lipid membrane remains the most enigmatic.

Following the initial results, Neira et al. (2002) took samples of adipose tissue from lipectomy samples removed from patients both with and without tumescent technique and irradiated them with a 635 nm, 10 mW diode laser with total energy values of 1.2–3.6 J/cm^2 for 0 to 6 min and found out that the tumescent technique facilitated laser light penetration and intensity and thus improved fat liquefaction. A similar experiment was repeated *in vitro* on 12 healthy women undergoing lipectomy, but this time tumescent technique was applied followed by external laser therapy using LLLT; both superficial and deep fat samples were extracted from the infraumbilical area of all patients studied (Neira et al. 2002). Results again showed that LLLT had a synergistic effect when applied together with tumescent solution (Neira et al. 2002). Without laser exposure, the adipose tissue remained intact and adipocytes maintained their original spherical structure. The enhancing effect of tumescent solution was postulated to be caused by increased epinephrine-induced stimulation of cAMP production by adenyl cyclase and/or improved light penetration and intensity facilitated by tumescent solution (Neira et al. 2002).

58.4.2.2 LLLT for Cellulite Treatment

Cellulite is a major cosmetic concern found in approximately 85% of postpubertal women (Gold et al. 2011). Women with this condition present with irregular skin dimpling, most frequently on the tight and buttocks, that resembles orange peels. Even though pathophysiology of cellulite is still not known, enlarged adipocytes, weakened connective tissue, and reduced microcirculation are possibly the triggering factors underlying this condition (Gold et al. 2011). Various devices and topical agents are available for treatment of cellulite; however, the results are only temporary. Based on the previously mentioned studies demonstrating LLLT's stimulatory effects on circulation, collagen formation, and fat reduction, it can be proposed as an alternative method for cellulite treatment. A multicenter open-label study was conducted among 83 subjects with mild to moderate cellulite to test the efficacy of a low-level, dual wavelength (650 and 915 nm) laser in combination with a massage device that has a vacuum and contoured rollers (Gold et al. 2011). The results of the study demonstrated not only improvement in appearance of cellulite but also the fact that 71% of the treatment thighs lost circumference compared to 53% of control thighs (Gold et al. 2011). Sasaki et al. (2007) used topical phosphatidylcholine-based anticellulite gel with LED array at wavelengths of 660 and 950 nm to test the effects of LED on cellulite reduction. The results of this study were interesting since LLLT alone failed to show improvement; however, when combined with a topically applied phosphatidylcholine-based anticellulite gel, patients showed improvement in cellulite reduction (Sasaki et al. 2007). Eight out of nine subjects experienced significant improvement in the thighs treated with phosphatidylcholine-based anticellulite gel and LED treatments, and it was further confirmed by clinical examination, measurements, and ultrasound evaluations, which showed a significant reduction in hypodermal thickness (Sasaki et al. 2007). However, 18 months following treatment, five of the improved thighs reverted back to their original cellulite grade, and only three improved to maintain their status.

LLLT seems to be a promising treatment modality as an alternative to other current treatment options, especially when used in combination with other methods. Current literature shows that it has almost no side effects. However, more studies with larger number of subjects are necessary.

References

Abergel, R. P., R. F. Lyons, J. C. Castel, R. M. Dwyer, and J. Uitto. 1987. Biostimulation of wound healing by lasers: Experimental approaches in animal models and in fibroblast cultures. *J Dermatol Surg Oncol* 132:127–133.

Ablon, G. 2010. Combination 830-nm and 633-nm light-emitting diode phototherapy shows promise in the treatment of recalcitrant psoriasis: Preliminary findings. *Photomed Laser Surg* 281:141–146.

Airan, L. E., and G. Hruza. 2005. Current lasers in skin resurfacing. *Facial Plast Surg Clin North Am* 131:127–139.

Al-Watban, F., and X. Y. Zang. 1996. Comparison of the effects of laser therapy on wound healing using different laser wavelengths. *Laser Ther* 19968:127–135.

Asawanonda, P., R. R. Anderson, Y. Chang, and C. R. Taylor. 2000. 308-nm excimer laser for the treatment of psoriasis: A dose-response study. *Arch Dermatol* 1365:619–624.

Aziz-Jalali, M. H., S. M. Tabaie, and G. E. Djavid. 2012. Comparison of red and infrared low-level laser therapy in the treatment of acne vulgaris. *Indian J Dermatol* 572:128–130.

Barolet, D., and A. Boucher. 2010. Prophylactic low-level light therapy for the treatment of hypertrophic scars and keloids: A case series. *Lasers Surg Med* 426:597–601.

Barolet, D., C. J. Roberge, F. A. Auger, A. Boucher, and L. Germain. 2009. Regulation of skin collagen metabolism *in vitro* using a pulsed 660 nm LED light source: Clinical correlation with a single-blinded study. *J Invest Dermatol* 12912:2751–2759.

Baxter, G. D., A. J. Bell, J. M. Allen, and J. Ravey. 1991. Low level laser therapy: Current clinical practice in Northern Ireland. *Physiotherapy* 773:171–178.

Bekassy, Z., and B. Astedt. 1985. Laser surgery for psoriasis. *Lancet* 28457:725.

Bello-Silva, M. S., P. M. de Freitas, A. C. Aranha et al. 2010. Low- and high-intensity lasers in the treatment of herpes simplex virus 1 infection. *Photomed Laser Surg* 281:135–139.

Berns, M. W., M. Rettenmaier, J. McCullough et al. 1984. Response of psoriasis to red laser light (630 nm) following systemic injection of hematoporphyrin derivative. *Lasers Surg Med* 41:73–77.

Bhat, J., J. Birch, C. Whitehurst, and S. W. Lanigan. 2005. A single-blinded randomised controlled study to determine the efficacy of Omnilux Revive facial treatment in skin rejuvenation. *Lasers Med Sci* 201:6–10.

Bouzari, N., S. C. Davis, and K. Nouri. 2007. Laser treatment of keloids and hypertrophic scars. *Int J Dermatol* 461:80–88.

Branham, G. H., and J. R. Thomas. 1996. Rejuvenation of the skin surface: Chemical peel and dermabrasion. *Facial Plast Surg* 122:125–133.

Brown, S. A., R. J. Rohrich, J. Kenkel et al. 2004. Effect of low-level laser therapy on abdominal adipocytes before lipoplasty procedures. *Plast Reconstr Surg* 1136:1796–1804; discussion 1805–1806.

Caruso-Davis, M. K., T. S. Guillot, V. K. Podichetty et al. 2011. Efficacy of low-level laser therapy for body contouring and spot fat reduction. *Obes Surg* 216:722–729.

Castex-Rizzi, N., S. Lachgar, M. Charveron, and Y. Gall. 2002. [Implication of VEGF, steroid hormones and neuropeptides in hair follicle cell responses]. *Ann Dermatol Venereol* 1295 Pt 2:783–786.

Chen, A. C., P. R. Arany, Y. Y. Huang et al. 2011. Low-level laser therapy activates NF-kB via generation of reactive oxygen species in mouse embryonic fibroblasts. *PLoS One* 67:e22453.

Chung, H., T. Dai, S. K. Sharma et al. 2012. The nuts and bolts of low-level laser (light) therapy. *Ann Biomed Eng* 402:516–533.

Colver, G. B., G. W. Cherry, and T. J. Ryan. 1984. Lasers, psoriasis and the public. *Br J Dermatol* 1112:243–244.

De Leeuw, J., R. G. Van Lingen, H. Both et al. 2009. A comparative study on the efficacy of treatment with 585 nm pulsed dye laser and ultraviolet B-TL01 in plaque type psoriasis. *Dermatol Surg* 351:80–91.

de Paula Eduardo, C., L. M. Bezinelli, F. de Paula Eduardo et al. 2011. Prevention of recurrent herpes labialis outbreaks through low-intensity laser therapy: A clinical protocol with 3-year follow-up. *Lasers Med Sci* 27:1077–1083.

Dierickx, C. C., and R. R. Anderson. 2005. Visible light treatment of photoaging. *Dermatol Ther* 183:191–208.

Finlay, A. Y., G. K. Khan, D. K. Luscombe, and M. S. Salek. 1990. Validation of sickness impact profile and psoriasis disability index in psoriasis. *Br J Dermatol* 1236:751–756.

Fischer, G. 1990. Liposculpture: The "correct" history of liposuction. Part I. *J Dermatol Surg Oncol* 1612:1087–1089.

Fröhlich, H. 1968. Long-range coherence and energy storage in biological systems. *Int J Quantum Chem* 25:641–649.

Fröhlich, H. 1970. Long range coherence and the action of enzymes. *Nature* 2285276:1093.

Fröhlich, H. 1975. The extraordinary dielectric properties of biological materials and the action of enzymes. *Proc Natl Acad Sci USA* 7211:4211–4215.

Gaida, K., R. Koller, C. Isler et al. 2004. Low level laser therapy—A conservative approach to the burn scar? *Burns* 304:362–367.

Gattu, S., R. M. Rashid, and J. J. Wu. 2009. 308-nm excimer laser in psoriasis vulgaris, scalp psoriasis, and palmoplantar psoriasis. *J Eur Acad Dermatol Venereol* 231:36–41.

Geiger, P. G., W. Korytowski, and A. W. Girotti. 1995. Photodynamically generated 3-beta-hydroxy-5 alpha-cholest-6-ene-5- hydroperoxide: Toxic reactivity in membranes and susceptibility to enzymatic detoxification. *Photochem Photobiol* 623:580–587.

Gelfand, J. M., R. Weinstein, S. B. Porter et al. 2005. Prevalence and treatment of psoriasis in the United Kingdom: A population-based study. *Arch Dermatol* 14112:1537–1541.

Geronemus, R. G., Weiss R. A., Weiss, M. A. et al. 2003. Non-ablative LED photomodulation light activated fibroblast stimulation clinical trial. *Lasers Surg Med* 25:22.

Gold, M. H., K. A. Khatri, K. Hails, R. A. Weiss and N. Fournier. 2011. Reduction in thigh circumference and improvement in the appearance of cellulite with dual-wavelength, low-level laser energy and massage. *J Cosmet Laser Ther* 131:13–20.

Goldberg, D. J., and B. A. Russell. 2006. Combination blue (415 nm) and red (633 nm) LED phototherapy in the treatment of mild to severe acne vulgaris. *J Cosmet Laser Ther* 82:71–75.

Griffiths, C. E., and J. N. Barker. 2007. Pathogenesis and clinical features of psoriasis. *Lancet* 3709583:263–271.

Gupta, A., P. Avci, M. Sadasivam et al. 2012. Shining light on nanotechnology to help repair and regeneration. *Biotechnol Adv.* doi: 10.1016/j. biotechadv.2012.08.003.

Honnor, R. C., G. S. Dhillon, and C. Londos. 1985. cAMP-dependent protein kinase and lipolysis in rat adipocytes.

II. Definition of steady-state relationship with lipolytic and antilipolytic modulators. *J Biol Chem* 26028:15130–15138.

Ilknur, T., S. Akarsu, S. Aktan, and S. Ozkan. 2006. Comparison of the effects of pulsed dye laser, pulsed dye laser + salicylic acid, and clobetasole propionate + salicylic acid on psoriatic plaques. *Dermatol Surg* 321:49–55.

Illouz, Y. G. 1983. Body contouring by lipolysis: A 5-year experience with over 3000 cases. *Plast Reconstr Surg* 725:591–597.

Karu, T. 1999. Primary and secondary mechanisms of action of visible to near-IR radiation on cells. *J Photochem Photobiol B* 491:1–17.

Karu, T. I. 2008. Mitochondrial signaling in mammalian cells activated by red and near-IR radiation. *Photochem Photobiol* 845:1091–1099.

Karu, T. I., V. V. Lobko, G. G. Lukpanova, I. M. Parkhomenko and I. Chirkov. 1985. Effect of irradiation with monochromatic visible light on the cAMP content in mammalian cells. *Dokl Akad Nauk SSSR* 2815:1242–1244.

Kim, S. S., M. W. Park, and C. J. Lee. 2007. Phototherapy of androgenetic alopecia with low level narrow band 655-nm red light and 780-nm infrared light. *J Am Acad Dermatolog.* American Academy of Dermatology 65th Annual Meeting. AB112.

King, P. R. 1989. Low level laser therapy: A review. *Lasers Med Sci* 43:141–150.

Kolari, P. J., and O. Airaksinen. 1993. Poor penetration of infrared and helium neon low power laser light into the dermal tissue. *Acupunct Electrother Res* 181:17–21.

Korner, R., F. Bahmer, and R. Wigand. 1989. The effect of infrared laser rays on herpes simplex virus and the functions of human immunocompetent cells. *Hautarzt* 406:350–354.

Krishnamurthy, K., A. Walker, C. A. Gropper, and C. Hoffman. 2010. To treat or not to treat? Management of guttate psoriasis and pityriasis rosea in patients with evidence of group A Streptococcal infection. *J Drugs Dermatol* 93:241–250.

Landthaler, M., D. Haina, and W. Waidelich. 1983. Treatment of zoster, post-zoster pain and herpes simplex recidivans in loco with laser light. *Fortschr Med* 10122:1039–1041.

Laws, P. M., and H. S. Young. 2010. Topical treatment of psoriasis. *Expert Opin Pharmacother* 1112:1999–2009.

Leavitt, M., G. Charles, E. Heyman, and D. Michaels. 2009. HairMax LaserComb laser phototherapy device in the treatment of male androgenetic alopecia: A randomized, double-blind, sham device-controlled, multicentre trial. *Clin Drug Investig* 295:283–292.

Lee, S. Y., C. E. You, and M. Y. Park. 2007. Blue and red light combination LED phototherapy for acne vulgaris in patients with skin phototype IV. *Lasers Surg Med* 392:180–188.

Louw, L. 2007. The keloid phenomenon: progress toward a solution. *Clin Anat* 201:3–14.

McDaniel, D. H., J. Newman, R. Geronemus et al. 2003. Non-ablative non-thermal LED photomodulation—A multi-center clinical photoaging trial. *Lasers Surg Med* 15:22.

Meneguzzo, D. T., L. A. Lopes, R. Pallota et al. 2013. Prevention and treatment of mice paw edema by near-infrared low-level laser therapy on lymph nodes. *Lasers Med Sci* 28:973–980.

Miura, Y., M. Yamazaki, R. Tsuboi, and H. Ogawa. 1999. Promotion of rat hair growth by irradiation using Super Lizer™. *Jpn J Dermatol* 109:2149–2152.

Muller-Rover, S., B. Handjiski, C. van der Veen et al. 2001. A comprehensive guide for the accurate classification of murine hair follicles in distinct hair cycle stages. *J Invest Dermatol* 1171:3–15.

Muñoz Sanchez, P. J., J. L. Capote Femenías, A. Díaz Tejeda, and J. Tunér. 2012. The effect of 670-nm low laser therapy on herpes simplex type 1. *Photomed Laser Surg* 301:37–40.

Neira, R., J. Arroyave, H. Ramirez et al. 2002. Fat liquefaction: Effect of low-level laser energy on adipose tissue. *Plast Reconstr Surg* 1103:912–922; discussion 923–925.

Nestor, M. S., M. B. Zarraga, and H. Park. 2012. Effect of 635 nm low-level laser therapy on upper arm circumference reduction: A double-blind, randomized, sham-controlled trial. *J Clin Aesthet Dermatol* 52:42–48.

Niera, R., E. Solarte, and M. A. e. a. Reyes. 2000. Low level assisted lipoplasty: A new technique. In Proceedings of the World Congress on Liposuction, Dearborn, Michigan.

Oschmann, J. I. 2000. *Energy Medicine: The Scientific Basis.* Churchill Livingstone, Edinburgh.

Paasch, U., and M. Haedersdal. 2011. Laser systems for ablative fractional resurfacing. *Expert Rev Med Dev* 81:67–83.

Papageorgiou, P., A. Katsambas, and A. Chu. 2000. Phototherapy with blue (415 nm) and red (660 nm) light in the treatment of acne vulgaris. *Br J Dermatol* 1425:973–978.

Philp, D., M. Nguyen, B. Scheremeta et al. 2004. Thymosin beta4 increases hair growth by activation of hair follicle stem cells. *FASEB J* 182:385–387.

Railan, D., and T. S. Alster. 2008. Laser treatment of acne, psoriasis, leukoderma, and scars. *Semin Cutan Med Surg* 274:285–291.

Ross, E. V. 2005. Optical treatments for acne. *Dermatol Ther* 183:253–266.

Rotunda, A. M., A. R. Bhupathy, and T. E. Rohrer. 2004. The new age of acne therapy: Light, lasers, and radiofrequency. *J Cosmet Laser Ther* 64:191–200.

Russell, B. A., N. Kellett, and L. R. Reilly. 2005. A study to determine the efficacy of combination LED light therapy (633 nm and 830 nm) in facial skin rejuvenation. *J Cosmet Laser Ther* 73–4:196–200.

Sadick, N. S. 2008. Handheld LED array device in the treatment of acne vulgaris. *J Drugs Dermatol* 74:347–350.

Sasaki, G. H., K. Oberg, B. Tucker, and M. Gaston. 2007. The effectiveness and safety of topical PhotoActif phosphatidylcholine-based anti-cellulite gel and LED (red and near-infrared) light on Grade II-III thigh cellulite: A randomized, double-blinded study. *J Cosmet Laser Ther* 92:87–96.

Satino, J. L., and M. Markou. 2003. Hair regrowth and increased hair tensile strength using the hairmax lasercomb for low-level laser therapy. *Int J Cos Surg Aest Dermatol* 5:113–117.

Schindl, A., and R. Neumann. 1999. Low-intensity laser therapy is an effective treatment for recurrent herpes simplex infection. Results from a randomized double-blind placebo-controlled study. *J Invest Dermatol* 1132:221–223.

Schlager, A., K. Oehler, K. U. Huebner, M. Schmuth, and L. Spoetl. 2000. Healing of burns after treatment with 670-nanometer low-power laser light. *Plast Reconstr Surg* 1055:1635–1639.

Seaton, E. D., P. E. Mouser, A. Charakida et al. 2006. Investigation of the mechanism of action of nonablative pulsed-dye laser therapy in photorejuvenation and inflammatory acne vulgaris. *Br J Dermatol* 1554:748–755.

Shukla, S., K. Sahu, Y. Verma et al. 2010. Effect of helium-neon laser irradiation on hair follicle growth cycle of Swiss albino mice. *Skin Pharmacol Physiol* 232:79–85.

Sroka, R., C. Fuchs, M. Schaffer et al. 1997. Biomodulation effects on cell mitosis after laser irradiation using different wavelengths. *Laser Surg. Med.* Supplement 9:6.

Stathakis, V., M. Kilkenny, and R. Marks. 1997. Descriptive epidemiology of acne vulgaris in the community. *Australas J Dermatol* 383:115–123.

Stern, R. S., T. Nijsten, S. R. Feldman, D. J. Margolis, and T. Rolstad. 2004. Psoriasis is common, carries a substantial burden even when not extensive, and is associated with widespread treatment dissatisfaction. *J Investig Dermatol Symp Proc* 92:136–139.

Tafur, J., and P. J. Mills. 2008. Low-intensity light therapy: Exploring the role of redox mechanisms. *Photomed Laser Surg* 264:323–328.

Thorek, M. 1939. Plastic reconstruction of the female breasts and abdomen. *Am J Surg* 432:268–278.

Trehan, M., and C. R. Taylor. 2002. Medium-dose 308-nm excimer laser for the treatment of psoriasis. *J Am Acad Dermatol* 475:701–708.

Trueb, R. M. 2009. Chemotherapy-induced alopecia. *Semin Cutan Med Surg* 281:11–14.

van Breugel, H. H., and P. R. Bar. 1992. Power density and exposure time of He–Ne laser irradiation are more important than total energy dose in photo-biomodulation of human fibroblasts *in vitro*. *Lasers Surg Med* 125:528–537.

van de Kerkhof, P. C., G. M. Murphy, J. Austad et al. 2007. Psoriasis of the face and flexures. *J Dermatolog Treat* 186:351–360.

Wajima, Z., T. Shitara, T. Inoue, and R. Ogawa. 1996. Linear polarized light irradiation around the stellate ganglion area increases skin temperature and blood flow. *Masui* 454:433–438.

Weiss, R. A., D. H. McDaniel, and R. G. Geronemus. 2004. Non-ablative, non- thermal light emitting diode (LED) phototherapy of photoaged skin. *Laser Surg Med* 16:31.

Weiss, R. A., D. H. McDaniel, R. G. Geronemus, and M. A. Weiss. 2005a. Clinical trial of a novel non-thermal LED array for reversal of photoaging: Clinical, histologic, and surface profilometric results. *Lasers Surg Med* 362:85–91.

Weiss, R. A., D. H. McDaniel, R. G. Geronemus, and M. A. Weiss. 2005b. LED photomodulation induced hair growth stimulation. *Ann Meet Am Soc Laser Med Surg*, Orlando.

Weiss, R. A., D. H. McDaniel, R. G. Geronemus et al. 2005c. Clinical experience with light-emitting diode (LED) photomodulation. *Dermatol Surg* 319 Pt 2:1199–1205.

Weiss, R. A., M. A. Weiss, R. G. Geronemus, and D. H. McDaniel. 2004. A novel non-thermal non-ablative full panel LED photomodulation device for reversal of photoaging: Digital microscopic and clinical results in various skin types. *J Drugs Dermatol* 36:605–610.

Whitley, R. J., D. W. Kimberlin, and B. Roizman. 1998. Herpes simplex viruses. *Clin Infect Dis* 263:541–553; quiz 554–555.

Wikramanayake, T. C., R. Rodriguez, S. Choudhary et al. 2012. Effects of the Lexington LaserComb on hair regrowth in the C3H/HeJ mouse model of alopecia areata. *Lasers Med Sci* 272:431–436.

Wikramanayake, T. C., A. C. Villasante, L. M. Mauro et al. 2013. Low-level laser treatment accelerated hair regrowth in a rat model of chemotherapy-induced alopecia (CIA). *Lasers Med Sci* 28:701–706.

Wolfram, D., A. Tzankov, P. Pulzl, and H. Piza-Katzer. 2009. Hypertrophic scars and keloids—A review of their pathophysiology, risk factors, and therapeutic management. *Dermatol Surg* 352:171–181.

Yamazaki, M., Y. Miura, R. Tsuboi, and H. Ogawa. 2003. Linear polarized infrared irradiation using Super Lizer is an effective treatment for multiple-type alopecia areata. *Int J Dermatol* 429:738–740.

Yano, K., L. F. Brown, and M. Detmar. 2001. Control of hair growth and follicle size by VEGF-mediated angiogenesis. *J Clin Invest* 107:409–417.

Yu, W., J. O. Naim, and R. J. Lanzafame. 1994. The effect of laser irradiation on the release of bFGF from 3T3 fibroblasts. *Photochem Photobiol* 592:167–170.

Zhang, X. 2012. Genome-wide association study of skin complex diseases. *J Dermatol Sci* 662:89–97.

VI

Surgical Laser Therapy

Laser and Intense Pulsed Light Treatment of Skin

Rui Yin
Third Military Medical University
Massachusetts General Hospital

Garuna Kositratna
Massachusetts General Hospital

R. Rox Anderson
Massachusetts General Hospital

59.1 Introduction

Light has long been used as a tool for the restoration of health. The healing powers of sunlight became one of the earliest recorded treatments in Western medicine (Bettman 1979; Kelly 2009), used for a wide variety of medical conditions, such as smallpox and tuberculosis (Bettman 1979). In 1963, Goldman et al. (1963a) first clearly described ruby laser-induced injury to pigmented skin, including hair follicles. In the following years, the ruby laser was explored as a treatment for many conditions, with little regard for absorption of light energy by various tissues. A historical case reported in 1983 was that of a young boy treated for a vascular malformation that resulted in severe epidermal damage. In the same year, Anderson and Parrish (1983) published the theory of selective photothermolysis (SP), which describes a strategy for using optical pulses to selectively affect pigmented "targets." A wavelength(s) is chosen that is preferentially absorbed by a particular chromophore associated with the target (melanin, hemoglobins, tattoo inks, or other pigments), and a pulse duration is chosen that is sufficiently short to limit heat transfer from the target structure during the laser pulse. By providing a microscopic target-selective treatment, SP enabled new laser treatments to be developed with minimal risk of gross tissue injury, for example, treatment of microvascular skin malformations and removal of pigmented hair follicles, pigmented lesions, and tattoos (Anderson and Parrish 1983; Grossman et al. 1996). With this concept came an explosion in the number of lasers and light sources in dermatology to accommodate a spectrum of aesthetic procedures with minimal pain (Bashkatov et al. 2005; Tanzi, Lupton, and Alster 2003; Tseng et al. 2009). In 2008, nearly 75 million aesthetic light procedures were performed, and the number is expected to double because of a growing and demanding young consumer market. Herein we introduce the applications of lasers and intense pulsed light in dermatology, including the potential for further development.

59.2 Pioneering Studies and the Development of Laser Treatment Theory

Based on an aspect of Einstein's quantum theory of light, the first operating laser was made by Maiman in 1960 (Maiman 1960). It consisted of a ruby (chromium-doped sapphire) crystal rod excited by a xenon flashlamp to produce a bright pulse of 693.7 nm, deep red light of about 1 ms duration. Ophthalmologists first described retinal photocoagulation using the laser; this was the first report of a potential biomedical laser application (Zaret et al. 1961), one that ultimately failed using ruby laser but led to many other laser applications in ophthalmology. In 1963, the first study of laser effects on skin was presented by Goldman et al., who noted selective injury of pigmented structures by the laser. They quickly realized the potential for selective treatment of colored skin lesions and conducted an impressive series of pioneering studies during the 1960s and early 1970s. Their 0.5 ms pulse duration, 5 J energy, 694 nm wavelength, normal-mode ruby laser was the first laser to be used for treating port-wine stains (PWSs), tattoos (Goldman et al. 1967), and pigmented lesions including metastatic melanoma. Goldman's clinical observations were important for the development of dermatologic laser use and at times surprising. However, epidermal damage was frequent with a risk of hyperpigmentation and/or scarring. These disconcerting side effects led to an era in which pulsed lasers were rarely used in medicine. From about 1970 to 1985, surgeons adopted continuous-wave (CW) lasers for tissue cutting and coagulation. This era led to classical surgical uses of CO_2, argon-ion, and neodymium:yttrium–aluminum–garnet (Nd:YAG) lasers. CW lasers remain in widespread use for open and endoscopic surgeries.

59.2.1 Selective Photothermolysis

SP (Anderson and Parrish 1983) offers a rationale for developing different tissue-selective lasers. The aim of SP is to provide permanent thermal damage of targeted structures with the surrounding tissue held intact. An essential requirement of SP is that the target structures or a nearby associated structure has stronger optical absorption relative to the surrounding tissue. Choice of wavelength is generally related to the spectral absorption of target chromophores (hemoglobins, melanin, lipids, tattoo ink, or external dyes), to the spectral penetration of light to the target structures, and to avoidance of competing chromophores. For example, the strongest optical absorption of hemoglobin species occurs at blue wavelengths near 420 nm, but these wavelengths penetrate poorly into skin and suffer strong absorption by melanin in the epidermis. Therefore, the first laser specifically designed and built around SP theory was a yellow organic dye laser operating near the 577 nm absorption maximum of oxyhemoglobin. This is still the gold standard for treatment of children with PWSs, a microvascular malformation lesion. The second central requirement of SP is that a pulse of light be delivered that is short enough to confine heat to the target structures during the laser pulse. The concept of a thermal relaxation time (TRT) of the intended target structure (Anderson 2003) is useful, which can be thought of as the time necessary for significant cooling of the targets. TRT varies with the square of the target structure diameter. For most biological objects, the TRT in seconds is approximately equal to the target size in millimeters squared. In classical SP theory, the pulse duration should be approximately equal to or less than TRT. Since small objects cool more rapidly, the pulse duration of lasers used for SP in dermatology covers a range from nanoseconds (for small targets such as subcellular organelles, tattoo ink nanoparticles, and individual cells) to milliseconds (for multicellular targets such as blood vessels and hair follicles). However, in many applications, the target absorption is nonuniform over its area and a part of the target exhibits weak or no absorption, but the other part exhibits significant absorption. Furthermore, many targets for SP are actually at some distance from the associated target chromophore (i.e., the site where light is actually absorbed and converted to heat). If this is the case, the weakly absorbing part of the target has to be damaged by heat diffusion from the highly pigmented/strongly absorbing part. For example, the biological target to damage a blood vessel is its endothelium and vessel wall, not the erythrocytes that contain the chromophore molecule hemoglobin. For effective thermal injury to vessels, heat must propagate from erythrocytes into the surrounding vessel wall. The same seems to be true for permanent laser hair removal: the targets are epithelial stem cells, which lie at the outermost part of the outer root sheath some distance from the pigmented hair shaft (Altshuler et al. 2001). Based on the spatial separation of the chromophore and desired target, an extended theory of SP was proposed that requires diffusion of heat from the chromophore to the desired target for destruction. In essence, the extended theory leads to choosing pulse durations that more nearly match the TRT of the entire target structure (e.g., the entire hair follicle).

59.2.2 Laser Skin Resurfacing

Thermal confinement by use of short laser pulses was also applied to tissue vaporization, starting with CO_2 surgical lasers and then including erbium, thulium, and holmium lasers. These infrared (IR) lasers emit wavelengths that are absorbed by water. For skin resurfacing, CO_2 or erbium lasers are used, corresponding to optical penetration depths of about 20 and 2 μm, respectively. Thus, it is possible to heat, vaporize, and remove a thin layer of tissue on the micrometer scale when pulses in the microsecond or a few milliseconds time domain are used. Classical skin resurfacing with these devices is a highly effective treatment for photoaged skin; the entire epidermis and a portion of superficial dermis is vaporized, which after healing restores the epidermis and superficial dermis. Classical laser resurfacing is much less popular at present, due to risks of infection, scarring, and permanent hypopigmentation in a fraction of patients. Still, the procedure remains a valuable one for patients with severe photoaging changes in the epidermis and dermis, and for other indications including the off-label one of limiting the rate of appearance of skin cancers in patients with field cancerization. Laser skin resurfacing works by producing a controlled superficial skin burn that stimulates the skin to entirely replace its epithelium (epidermis) from stem cells in the hair follicles and to repair and remodel the underlying dermis.

59.2.3 Fractional Photothermolysis

To overcome some of the problems of laser skin resurfacing, the concept of fractional photothermolysis was developed (Manstein et al. 2004). A focused laser beam is used to create an array of microscopic, cylindrical treatment zones (MTZs) of controlled width (typically, 0.15–0.4 mm) and depth (typically 0.1–2 mm deep). A single MTZ would probably go unnoticed, but when up to 2 million MTZs are created on a human face, the skin is potently stimulated to repair and remodel. Unlike conventional resurfacing, each MTZ is surrounded by unexposed tissue that initiates a rapid healing response. Facial skin will tolerate fractional laser treatment that kills or removes up to about 40% of the skin. "Density" of treatment is adjusted by controlling the size and number of MTZs per unit of skin area, with typical treatment for photoaging at a density of 10%–30%. The depth of each MTZ is precisely controlled by varying the

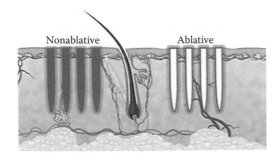

FIGURE 59.1 Fractional laser interactions with skin cause deep, narrow columns of thermal injury. Ablative fractional lasers (CO_2, erbium) vaporize a channel that is surrounded by residual thermal damage.

energy delivered with each laser microbeam exposure. As with classical laser resurfacing, there are both ablative (CO_2 laser and erbium laser, at 10.6 and 2.9 μm wavelength, respectively) and nonablative (mid-IR lasers, at 1.3–2 μm wavelengths) versions of fractional laser treatment. Epidermal and dermal disruption occurs in each MTZ, stimulating rapid epidermal repair and remodeling of the dermal extracellular matrix. Each MTZ leads to shedding of a small amount of necrotic skin tissue (Figure 59.1). Hantash and colleagues demonstrated that the degenerated dermal material is incorporated into columns of microscopic epidermal necrotic debris (MEND), which is shuttled through the epidermis and exfoliated. Dermal remodeling stimulated by fractional laser treatment proceeds for several months. A fundamental aspect of fractional laser treatment is that scarring is avoided. Even though each MTZ corresponds to a deep skin "burn," the diameter of MTZ is limited to a few hundred micrometers so that the adjacent unexposed tissue is able to heal by remodeling. There appears to be a size limit for physical wounds to initiate scarring. Rather than making an array of very small scars, fractional laser treatment stimulates the process of dermal remodeling without stimulating the process of scarring. The precise mechanisms responsible for lack of scarring when wounds are less than about 0.3 mm diameter remain unknown.

59.3 Clinical Applications

59.3.1 Vascular Lesions

59.3.1.1 Classification of Vascular Lesions

Vascular lesions of the skin are classified primarily as being either malformations or proliferations. Congenital vascular malformations include PWSs, arteriovenous malformations (AVMs), venous malformations (VMs), and some lymphatic lesions. Acquired vascular malformations include telangiectases, spider angiomas, cherry angiomas, venous lakes, and leg vein varicosities. Infantile hemangiomas are the most common proliferative vascular lesions, which usually appear within the first month of life, subsequently grow rapidly, and thereafter involute (Kern et al. 2000; Mulliken and Glowacki 1982).

59.3.1.2 Laser Treatment of Vascular Lesions

The argon laser and other CW visible light lasers were used for patients with PWSs in the 1970s and early 1980s (Cosman 1980; Goldman 1980). The laser emits two major lines at 488 and 514 nm, which are preferentially absorbed by hemoglobin. However, argon lasers are unable to generate high-energy pulses that are necessary for SP. Despite emitting wavelengths that are selectively absorbed by blood, argon laser treatment caused a high incidence of scarring and permanent hypopigmentation because the exposure times greatly exceed the TRT of PWS target vessels (Astner and Anderson 2005). This laser is now nearly obsolete for medical applications. At present, based on the concept of SP (Anderson and Parrish 1981, 1983), a variety of different lasers and light sources are useful in the treatment of vascular lesions. The most selective and versatile of these are millisecond-domain pulsed lasers at 532 nm [green high-energy pulsed potassium–titanyl–phosphate (KTP) laser], 585–595 nm [yellow pulsed dye lasers (PDL)], 755 nm (near-IR alexandrite laser), 810 nm (near-IR diode laser), and 1064 nm (near-IR Nd:YAG laser) (Tan, Murray, and Kurban 1989; Tan, Sherwood, and Gilchrest 1989). Broadband xenon-flashlamp sources [intense pulse light (IPL)] have increasingly been employed for the treatment of acquired vascular lesions in adults.

59.3.1.2.1 Telangiectases

The treatment of telangiectatic blood vessels on the face is the most commonly cosmetic requested procedure. The vessel diameter in these lesions is between 0.2 and 0.5 mm. For many years, the treatment of choice was PDL, using fluences between 8 and 15 J/cm², a 5–10 mm spot size, and pulse durations of about 6–10 ms (Astner and Anderson 2005). The clinical end point of PDL, KTP, or IPL treatment is immediate target vessel coagulation or disappearance. Earlier versions of PDL were unable to produce long pulses. With these devices, a significant downtime of 7–10 days needs to be taken into consideration when treating large cosmetic areas as a result of purpura after PDL. Makeup can be applied immediately after treatment. The recent generation of PDLs is capable of purpura-free treatment, achieved by using pulse durations longer than about 10 ms. Long-pulse KTP lasers are perhaps the ideal source for treatment of telangiectasia, and IPLs are also particularly useful for telangiectasia treatment. All of these devices should be used in combination with some form of skin cooling, particularly in patients with tanning or constitutive skin pigmentation. IPLs are versatile sources but in general are not as powerful or as target-selective as lasers, showing inferior results at similar fluence for treatment of PWSs (Bjerring, Christiansen, and Troilius 2003; Raulin, Hellwig, and Schonermark 1997) and a higher incidence of side effects. Typical IPLs use green or green-yellow filters and fluences between 32 and 40 J/cm² in a 20 ms pulse width to treat telangiectasias. Rosacea is a common condition prevalent in fair-skinned and especially Celtic individuals, with associated telangiectasia, sebaceous gland hypertrophy, and vascular flushing. Matted telangiectasias and facial ruddiness associated with rosacea may better be treated with IPL devices than with traditional PDL devices, which can cause purpura formation despite somewhat better final results when a PDL is used. The vascular flushing of rosacea often responds well to IPL, PDL, or KTP laser treatment when performed with large spot sizes. Due to optical scattering losses, exposure spot sizes smaller than about 7 mm often fail to treat the deep dermis, where arteriolar blood flow control is likely to be associated with flushing.

59.3.1.2.2 Spider Angioma

These acquired lesions consist of a central arteriole that supplies numerous superficial branches, leading to a red spiderlike appearance. The target vessel diameter is typically 0.1–0.5 mm, corresponding to TRTs of about 10–250 ms. The purpose of laser treatment is closure of the feeding arteriole and the subsequent treatment of the superficial branches. Diascopy is used to arrest flow in the central arteriole, followed by one or two single pulses to the area upon release of diascopy and refill of superficial branches, which then become amenable to laser treatment. It is important to treat all branches to avoid recurrence. Focal lesions can easily be treated using traditional PDL devices, with the side effect of temporary posttreatment pupura limited to the treatment site. PDL treatment at fluences between 8 and 10 J/cm² and 5 or 7 mm spot size with dynamic cryogen cooling device (DCD) settings of 30 ms spray duration and 30 ms delay time between the cryogen and laser pulse will generally yield very good results. PDLs, KTP, or IPL devices without DCD can also be used if excellent skin cooling is provided by cold air or contact cooling devices. A second treatment may be required, and retreatment evaluation should occur at 4–6 weeks after the initial treatment (Astner and Anderson 2005).

59.3.1.2.3   Cherry Angioma

Cherry angiomas are benign cutaneous microvascular aneurisms commonly appearing after the third decade of life. The diameter of individual vessels ranges from 0.1 to 1 mm, with either focal or generalized distribution and predilection of the trunk and proximal extremities (West and Alster 1998a). Unlike the spider angioma, these are low-flow lesions easily eradicated by KTP, PDL, or IPL treatment. Thin papular lesions can easily be treated with single pulses; more hypertrophic lesions may need two single pulses—one with and one without diascopy—to allow complete clearance of the angioma. On occasion, cherry angiomas may require a second treatment 4–6 weeks after initial PDL treatment. Angiokeratoma is a similar-appearing lesion, in which the microvascular malformation is associated with epidermal and dermal growth forming a small, hard, red papule. Angiokeratomas are more difficult to eradicate, but good results are often achieved by applying multiple and/or higher fluence pulses.

59.3.1.2.4 Pyogenic Granuloma

Pyogenic granulomas (PGs) are a proliferating microvascular and tissue mass stimulated by cutaneous injury (e.g., sharp

trauma, arthropod bites, and other penetrating injuries), which is also increased in patients with PWSs. It has an inflammatory component, with significant hyperproliferation of superficial cutaneous vasculature, and always presents with a central feeding arteriole comparable to the anatomy of a spider angioma. Because of varying thickness of these lesions, they are not always amenable to laser surgery or only after the hypertrophic papular aspect of the lesion has been removed surgically. Similarly, treatment may be more successful if lesions are treated early and if diascopy is employed to arrest flow in the feeding arteriole, release, and subsequent treatment of the superficial component. If laser treatment is considered, PDL remains the treatment of choice. Fluences, pulse width, spot size, and cooling should be set at standard parameters as previously outlined; however, multiple pulses may be required to allow successful closure of the arteriole. The nonvascular residues of PG have been treated with CO₂ laser with good cosmetic results.

59.3.1.2.5 Venous Lake

Venous lakes are commonly acquired VMs often presenting between the fourth and fifth decade of life. Common locations include the lip and oral mucosa and facial or neck skin. Venous lakes typically respond well to laser treatment. In superficial lesions, PDL with the aid of diascopy may be sufficient for a successful treatment and remains the first-line treatment in superficial lesions. However, thicker and more nodular lesions require wavelengths with increased penetration. Nd:YAG, alexandrite, or diode lasers (800–810 nm) with contact cooling and adjustable pulse durations allow for safe and effective treatment of these lesions. The procedure is painful; hence, the use of local anesthesia, such as a mental nerve block for lower lip lesions, is usually needed. A local anesthetic without added epinephrine should be used, since epinephrine sometimes constricts the target vessels to the extent that there is insufficient blood for absorption of the optical energy. Nd:YAG, alexandrite, or diode lasers operating near 800 nm wavelength with exposure spot sizes greater than 6 mm and delivered with contact skin cooling work very well for these lesions and other VMs. A typical set of treatment parameters using diode laser is fluence from 45 to 60 J/cm² and pulse duration of at least 30 ms using a spot size of 12 mm delivered through a cold sapphire window held with gentle pressure against the skin. Similarly, long-pulse Nd:YAG lasers can be used for hypertrophic and nodular lesions extending deeper into the tissue; pulse widths between 30 and 100 ms and fluences of about 100–150 J/cm² can achieve vessel occlusion, yielding the clinical end point of immediate vessel disappearance. With all treatment modalities, two or more treatment sessions may be required, especially with hypertrophic and nodular lesions extending deeper into the tissue, and the combination of different wavelengths may yield better clinical outcomes.

59.3.1.2.6 Leg Vein Anomalies

Leg vein anomalies usually arise from reflux of blood due to incompetent venous valves (Astner and Anderson 2005). Laser or IPL treatment of leg veins is difficult because of a wide range in size and depth, a wide variety in flow, and the many different types of leg vein dilatation. Sclerotherapy is the preferred treatment for superficial, isolated leg veins. However, before treatment with sclerotherapy or with a laser, all patients should be assessed for the presence of deep vein disease. This is easily done using Doppler ultrasound to detect reflux of blood into the deep vein system when firm pressure is applied to the abdomen. If saphenous vein reflux is present, a procedure called endovenous laser treatment (EVLT) is performed, in which direct laser destruction of saphenous vein near the junction with femoral vein eliminates reflux into the saphenous vein system. This is highly effective in more than 80% of patients and often leads to the spontaneous disappearance of the superficial, visible dilated veins. Lasers useful for treating superficial spider-leg veins include Nd:YAG, diode, alexandrite, and PDL. Long-pulse durations up to 100 ms and excellent skin cooling should be used. These pulse durations allow theoretically to selectively heat larger vessels with relative sparing of smaller vessels, as they cool faster than larger veins. PDL, KTP, and IPL sources have shown some efficacy for the treatment of fine matted telangiectases; however, treatments are commonly associated with side effects of postinflammatory hyperpigmentation (PIH) that can persist for extended periods of time. KTP (Alam et al. 2003) and PDL with adjustable pulse durations (Braverman 1989) have shown increased clinical efficacy and improved side effect profile.

The role of laser treatment of superficial leg vein ectasias remains controversial because of the high incidence of treatment-related side effects and pigmentary changes. The technique is painful, multiple treatment sessions are required, and recurrence is common. This does not compare favorably with the lower cost, higher efficacy, and lower incidence of side effects from sclerotherapy. However, in some needle-phobic patients, the light sources are a useful alternative to sclerotherapy.

59.3.1.2.7 *Hemangioma*

Hemangiomas are common benign tumors of infancy. They have a characteristic clinical course marked by early proliferation and followed by spontaneous involution. Spontaneous involution in many instances produces better cosmetic outcome than active intervention (Loo and Lanigan 2002). The classical approach to infantile hemangiomas was "benign neglect," meaning to monitor the child for possible harm as the hemangioma grew and then slowly regressed. However, this classical approach is also regressing, since the presence of an obvious hemangioma persisting until school age can strongly influence the child's social development, and because more than one-fourth of these lesions leave a permanent atrophic scarlike defect that is eventually treated by surgical or laser correction. Furthermore, the lesions are most common on the head and neck, occurring in about 5% of all girls. About 10% of hemangiomas pose a threat to functions such as normal vision, feeding, or breathing. A similar percentage of these lesions ulcerate during their growth phase, inducing strong pain followed by a high risk of infection and scarring. The recent discovery that most (about 70%) infantile hemangiomas

are growth-arrested and then involute when the child is treated with the nonselective β-adrenergic blocking agent propranolol has led to a major shift in medical management of these tumors, as well as a shift in opinion among many pediatricians about preventive treatment. Medical management of threatening hemangiomas is primarily with oral propranolol, which entails a risk of hypoglycemia and hypotension, and/or with oral or locally injected corticosteroids that also induce growth arrest. The mechanisms involved remain largely unknown. A topical β-blocker eye drop medication for glaucoma (timolol) has also been shown to be effective for inducing growth arrest and involution of early stage infantile hemagiomas, which poses far less risk than systemic medications. With these changes in medical management, the role of PDL for treatment of infantile hemangioma has decreased. Current recommendations suggest the treatment of a very early, macular precursor lesion in an effort to diminish or even prevent growth (Astner and Anderson 2005). Treatment parameters are generally with the shortest available pulse duration of 0.4 ms, a fluence of 5–7 J/cm², and cooling with 30 ms of dynamic cryogen spray. Low-dose treatment may trigger early involution of the superficial component and prevent growth of the deep component in combined lesions. The treatment of proliferating lesions remains controversial despite good evidence (Batta et al. 2002) that laser treatment can accelerate involution. The most accepted role of PDL is for treatment of ulcerated hemangiomas, which usually reepithelialize rapidly after PDL treatment. Ulcerated, bleeding, and/or painful tumors respond effectively to PDL (Barlow, Walker, and Markey 1996; Kolde 2003; Lacour et al. 1996). However, the treatment parameters are generally set at low fluence (6 J/cm²) to avoid additional ulceration yet sufficient to induce superficial vessel closure and regression.

Because of limited penetration of current laser technologies, laser treatment appears to be ineffective in reducing the bulk of deeper and larger hemangiomas (Scheepers and Quaba 1995). A nonrandomized trial looked at the efficacy of PDL in treating 225 hemangiomas (Poetke, Philipp, and Berlien 2000). This study showed that superficial lesions responded well to PDL, but none of the mixed hemangiomas involuted completely and early laser treatment may not prevent proliferative growth of the deeper component as alluded to by previous studies (Ashinoff and Geronemus 1991; Garden, Bakus, and Paller 1992; Haywood, Monk, and Mahaffey 2000). Apart from PDL, both the Nd:YAG and the KTP lasers can also be employed for intralesional treatment of bulky tumors (Achauer et al. 1999; Burstein et al. 2000). Although hampered by hypertrophic scarring, the Nd:YAG has been used successfully to debulk large hemangiomas. PDL and KTP are useful lasers for treatment of the residual telangiectasia that persists commonly after involution of hemangioma. Fractional laser treatment appears to offer good to excellent results as a nonsurgical means to improve thickness, texture, and color of the underlying fibrofatty residue, but there are only several reports of this new procedure. In a small study, 12 patients with atrophic scar or fibrofatty tissue secondary to residual hemangiomas were treated by ablative fractional erbium:yttrium–scandium–gallium–garnet (Er:YSGG) laser, operating at 2790 nm wavelength or with combined sequential 595 nm PDL and 1064 nm Nd:YAG. Half of the 12 patients improved substantially (Alcantara Gonzalez et al. 2012). Surgical resection of involuted hemangioma is also effective but should be reserved for sites in which a surgical excision scar is acceptable.

59.3.1.2.8 Vascular Malformations

The classification of vascular malformation includes capillary malformation (PWS), VM, arterial malformation (AM), AVM, and lymphatic malformation (LM). Among these diseases, capillary malformation (e.g., PWS or nevus flammeus) is very common (0.3% of newborns) (Alper and Holmes 1983). PWSs are irregularly shaped and red or violaceous that are present at birth and never disappear. The standard treatment for PWS is laser therapy. Laser treatment achieves observable lightening of PWS by reducing the number and size of abnormal target vessels in the dermis by means of SP (Anderson and Parrish 1983) based on hemoglobin as the target chromophore (Lanigan and Taibjee 2004). Over the past two decades, PDL with a wavelength of 585 nm and a pulse duration of 0.45 ms was initially the method of choice for the treatment of PWS (Taieb et al. 1994; Wlotzke et al. 1996). The advent of cryogen spray cooling about 15 years ago allowed higher fluences to be used without causing epidermal damage, which increased efficacy somewhat (Zenzie et al. 2000). Other advances, including the use of longer wavelengths of 585–600 nm and availability of longer pulse duration, have made relatively little difference in PWS treatment. Multiple treatments are necessary to achieve maximal lightening; however, complete clearing of PWS when treatment is started after the age of 6 months is rare (Garden, Polla, and Tan 1988; Scherer et al. 2001). In contrast, early treatment is associated with a higher rate of complete clearance, and for children less than 1 year of age can generally be performed with analgesia but without anesthesia, at a time when the PWS has not grown substantially as the child grows and before the child forms a physical self-image. It is published that up to about 20%–30% of PWSs are resistant to PDL treatment (Renfro and Geronemus 1993), but the definition of "resistant" is questionable. Only rarely does a PWS fail to lighten at all in response to PDL treatment. Also controversial are reports of PWS "recurrence" years after laser treatment, because the natural history of PWS whether treated or untreated is to darken over time due to progressive dilation of vessels. These lesions are predominantly a malformation of venules with a paucity of sympathetic innervation in the anatomic distribution of nerves (dermatomes). The primary defect is likely to be neural rather than vascular.

PDL devices are the most frequently and widely accepted treatment of PWS (Fitzpatrick et al. 1994b; Garden, Polla, and Tan 1988). In general, PDL parameters for childhood PWS are 585–600 nm wavelength, 5–10 J/cm² fluence, 0.45–3 ms pulse duration, and a spot size of 5–12 mm, delivered with dynamic cryogen spray cooling. KTP lasers capable of generating 532 nm pulses with similar fluence, spot size, and pulse duration are also useful for PWS treatment, but older, less powerful versions of KTP laser are to be avoided because of the high risk of scarring. In general, IPLs pose a higher risk and are less effective for PWS compared with PDLs, but nonetheless, some IPLs are useful for treatment of fair-skinned adults with PWSs.

Although alexandrite lasers are primarily used for hair removal, they also have been used for treating vascular lesions. In a retrospective case review, Izikson, Avram, and Tannous (2008) evaluated 20 patients with either hypertrophic or PDL-resistant PWS. Patients were treated with a 755 nm alexandrite laser only or in combination with other lasers, including PDL. The number of treatment sessions with the alexandrite laser varied from 3 to 10. All three patients with hypertrophic PWS showed significant lightening after treatment with alexandrite laser, either in combination with PDL or alternating with PDL. Twelve patients with resistant PWS showed moderate lightening, three had mild lightening, and one patient showed no response. Two patients developed isolated and small hypertrophic scars after blistering. In a direct comparison study, McGill et al. (2007) compared PDL, alexandrite, KTP, Nd:YAG lasers, and IPL in patients with therapy-resistant PWS (average of 20 previous PDL treatments). They found that the alexandrite laser was the most effective treatment device, lightening PWS in 10 out of 16 patients. However, this laser resulted in hyperpigmentation and scarring in four patients. In conclusion, the alexandrite laser is worth considering for the treatment of patients with therapy-resistant PWS; however, care needs to be taken because of the possible side effects.

Diode lasers with wavelengths of 800, 810, and 930 nm were introduced to treat deeper, larger-caliber vessels. Small vessels < 0.4 mm were not very responsive (Dover and Arndt 2000). Whang, Byun, and Kim (2009) investigated in a small case series a dual-wavelength approach with an 800 nm diode laser (17–30 J/cm², 30 ms, spot size not noted) followed by PDL treatment (585 nm, 6–9 J/cm², 0.5 ms, spot size not noted) in the same treatment session. Eight patients showed moderate to excellent responses, and a ninth patient had a fair response. Patients received 1–3 treatments that were well tolerated. In conclusion, randomized controlled studies confirming the efficacy and safety of diode lasers for the treatment of PWS are lacking.

The 1064 nm Nd:YAG laser pulses are characterized by a deeper penetration depth than KTP lasers, PDL, Alexandrite laser, diode laser, and IPL devices. Theoretically, this parameter results in a higher impact on deep-lying and thicker blood vessels, which is particularly advantageous in matured or hypertrophic PWSs, respectively. Yang et al. (2005) compared the PDL (595 nm, 8 J/cm², 1.5 ms, 7 mm) with the pulsed Nd:YAG laser in 17 patients. For pulsed Nd:YAG laser treatment, the following parameters were applied: 1064 nm, 40–130 J/cm², 4–10 ms pulse duration, spot size 5–7 mm. A clearing of 50%–70% could be obtained with both lasers after three treatment sessions, but patients preferred the Nd:YAG laser because purpura was less pronounced. There was one case of severe scarring in this study, which was the first to use millisecond-pulsed Nd:YAG laser for

PWS, and the authors cautioned that scarring after Nd:YAG laser treatment may be frequent. This has turned out to be true, and a combined PDL-Nd:YAG laser pulse strategy was developed. Alster and Tanzi (2009a) treated 25 patients with recalcitrant or hypertrophic PWSs with a dual 595 nm PDL and 1064 nm Nd:YAG laser. The time delay between the delivery of the two wavelengths was 0.5–1 s. Laser treatment was delivered at 6 to 8 week intervals with a mean number of treatments of 3.8 on the face and 4.9 on the extremities. Moderate clinical improvement (25%–50%) was seen in 48% of patients and mild improvement (1%–25%) in 52% of patients. Side effects were limited to mild purpura and vesicle formation in one patient that resolved without sequelae in 6 days. The synergies between PDL and Nd:YAG 1064 nm lasers (hybrid lasers) have also been evaluated by other authors and proved to be effective (Borges da Costa et al. 2009). Better outcomes with good to excellent responses in 63.2% of 19 patients after an average of 2.9 Nd:YAG treatments (1064 nm, 60–210 J/cm^2, 10–30 ms, 5–7 mm) were reported by Civas et al. (2009); however, they did not treat recalcitrant PWS as in the aforementioned studies. The 1064 nm Nd:YAG laser also showed to be safe and effective for treating hypertrophic PWS of the lip (Kono et al. 2009). In conclusion, the 1064 nm Nd:YAG laser is a somewhat risky but important device in the armamentarium of treatment options for PWS, particularly in case of nodularity and hypertrophy.

Klein et al. (2011) evaluated the efficacy and side effects of IPL treatment of PWSs in a direct comparison to two versions of PDLs in untreated ($n = 11$) and previously treated ($n = 14$) PWSs. Lesion clearance and the incidence of side effects were evaluated. In previously untreated PWSs as well as in pretreated PWSs, IPL treatments were rated significantly ($p < 0.05$) better than treatments with a conventional 585 nm, 0.4 ms PDL, and no different than treatment with a longer-pulse, longer-wavelength PDL. However, neither source was tested over a full range of fluences to assess the fluence–response relationship. In general, such comparison studies are difficult to interpret. Side effects were few in all settings. Although PDL devices differ from one another, the output of IPL devices covers a much broader range of all parameters—spectral power distribution (wavelengths), pulse energy, pulse duration, and temporal pulse structure. Many IPLs produce wavelengths inappropriate for PWS treatment and/or have inadequate output. Many IPLs are also not calibrated during use, such that large variations exist between devices even from the same manufacturer or the same device over time. The evidence is clear that some well-filtered, high-energy IPLs capable of sufficient fluence at short pulses offer a useful treatment for PWSs in fair-skinned adults (Goldman et al. 1963b, 1967). In contrast, in an intraindividual randomized clinical trial, Faurschou et al. (2009) compared the efficacy and adverse events of a recent generation PDL and a recent generation IPL. These authors found significantly better PWS lightening with PDL treatment compared to IPL treatment, and almost all of these patients preferred treatment with PDL. In summary, some IPLs are promising alternatives to lasers for some patients.

59.3.2 Tattoo Removal

Tattooing has been around for at least 5000 years in Western culture, and probably dates back much further. The prevalence of tattoos in the United States is now high, with about half of the adult population between the ages of 18 and 35 "sporting" at least one tattoo. Tattoos consist of insoluble light-absorbing nanoparticles, which upon introduction into the dermis are phagocytosed, where they remain inside fibroblasts and other resident cells for a lifetime. A large amount of the tattoo ink is also transported and retained in lymph nodes, and some ink is transported to lung, liver, and other organs. Removing tattoos began with abrasive and destructive measures to destroy the tattoo and, unfortunately, the skin it was contained in. Using the principles of SP, most tattoos can be "removed" without destroying the surrounding skin (i.e., without leaving a scar). The TRT for tattoo ink particles is in the nanosecond (1 ns = 10^{-9} s) range. Q-switched lasers emitting high-energy pulses from about 10 to 100 ns duration at visible and near-IR wavelengths absorbed by various tattoo ink colors are used for tattoo removal. Laser exposure selectively kills ink-containing cells, releasing the ink particles, which are then partially eliminated from skin by shedding or by transport to lymph nodes over a period of several weeks. After each treatment, residual ink remains. Therefore, a typical course is to treat once per month. The number of treatments necessary for adequate ink removal varies widely, from about 4 to more than 20 treatments. Unfortunately, about 25% of tattoos are resistant to treatment by lasers, depending mainly on tattoo ink color, chemical composition, and body site. The duration, cost, and uncertainty of laser tattoo removal are largely unsolved problems that deter many individuals from choosing to remove an unwanted tattoo. There have been misguided attempts to remove tattoos with IPL devices, which are physically incapable of generating high-energy nanosecond optical pulses. The risk of scarring is very high with IPL or long-pulse laser treatment of tattoos due to excessive thermal damage.

In general, the color of laser output must be complementary to that of the tattoo for successful tattoo removal, for example, using green laser pulses to remove red tattoo ink and vice versa. Black carbon tattoos are the most reliable to remove, especially homemade or radiotherapy tattoos made with India ink. White tattoos containing titanium dioxide and most yellow tattoo inks are the most difficult to remove. Goldman et al. (1963b, 1965) first reported tattoo removal using the Q-switched ruby laser in several cases without scarring and also noted that millisecond pulses from the normal-mode ruby laser resulted in thermal damage to the treated area. Independent efforts by Reid and colleagues in Scotland and by Taylor and colleagues in the United States found that Q-switched ruby laser could remove black tattoos with minimal risk of scarring (Bernstein 2006). Several other Q-switched lasers have since become, and remain, the mainstay of modern tattoo removal. New paradigms for tattoo removal are currently being explored, with the Q-switched lasers still remaining the preferred method. There are presently three Q-switched lasers available for tattoo removal: ruby (red,

694 nm), alexandrite (near-IR, 755 nm), and Nd:YAG (near-IR 1064 nm and green 532 nm).

The Q-switched ruby laser was the first to be used for treating tattoos and pigmented lesions (Goldman et al. 1963b, 1965). Typically, the Q-switched ruby laser has historically been extremely effective at removing black, blue, and green tattoo pigments (Scheibner et al. 1990; Taylor et al. 1990). However, even some amateur tattoos have proven quite resistant to removal with any of the Q-switched lasers (Bernstein 2006), and there is moderately high incidence of hypopigmentation following treatment (Leuenberger et al. 1999). Q-switched Nd:YAG laser is presently the most commonly used device for tattoo removal (Figure 59.2). At its fundamental wavelength of 1064 nm, black tattoo inks absorb strongly and there is only weak absorption by the competing natural chromophores melanin, hemoglobins, and water. Using a KTP and other crystals, the frequency of this laser can be doubled, thus halving the wavelength to 532 nm in the green range. This enables effective treatment of dark tattoo pigments such as black and dark blue using the 1064 nm wavelength, as well as removal of red and orange pigments using the 532 nm wavelength. The skin response strongly associated with tattoo removal is immediate whitening. An ash-white, raised plaque occurs immediately in the laser exposure spot. If there is epidermal disruption or immediate whitening when adjacent nontattooed skin is exposed, the fluence should be reduced. Immediate whitening of the tattoo is caused by residual gas bubbles remaining after the violent process of cavitation, in which steam bubbles form, expand, and collapse around the laser-heated tattoo ink particles. When treating patients with darkly pigmented skin and/or in whom there is a significant risk for keloid scarring or destruction of natural pigment as a consequence of tattoo removal, the Q-switched Nd:YAG laser appears

to be an excellent choice because of minimal absorption by melanin. The Q-switched alexandrite laser emits light at 755 nm, an intermediate wavelength between the ruby and Nd:YAG lasers. Alexandrite lasers offer advantages over earlier ruby lasers, often with greater reliability, faster repetition rates, and lower production costs, although modern ruby lasers seem to have alleviated some of these early deficiencies (Bernstein 2006). Alexandrite lasers are capable of removing black, blue, and some green ink. Green, yellow, or white inks are often the only colors left behind after treatment with Nd:YAG lasers. Zelickson et al. (1994) studied the clinical, histopathologic, and ultrastructural effects of all three types of Q-switched lasers in an animal model. They found that red brown, dark brown, and orange pigment responded best to the Nd:YAG laser, the alexandrite laser was most effective for removing blue and green pigment, and the Q-switched ruby laser was most effective for removing purple and violet pigment. The 532 nm wavelength of the Nd:YAG laser was best for removing red pigment. All lasers were found to be equivalent for removing black tattoo pigment. In practice, a combination of Q-switched Nd:YAG laser with either ruby or alexandrite Q-switched lasers is usually needed for treatment of a multicolored tattoo.

The technique of combining ablative fractional resurfacing (AFR), which creates channels through the skin, or nonablative fractional resurfacing and Q-switched lasers for tattoo removal, appears to be promising. A small pilot controlled clinical trial investigated this combination technique on three patients, all of whom received fractionated carbon dioxide (ablative) or 1550 nm laser (nonablative) treatment to half of the tattoo in addition to Q-switched ruby laser (QSRL) treatment given to the entire tattoo. The addition of fractional laser treatment to conventional Q-switched laser treatment enhanced tattoo removal, eliminated blistering, shortened recovery time, and diminished treatment-induced hypopigmentation. No scarring was observed in this small pilot study. Other strategies for potentially improving laser tattoo removal include the use of picosecond (10^{-12} s) domain pulses and the use of repetitive laser treatments given on the same day. Most tattoo ink particles are 40 to 300 nm in size (Baumler et al. 2000; Ross et al. 1998). While the TRT of such particles is in the low-nanosecond domain, it has been shown that inertial confinement produces additional efficiency and efficacy for tattoo removal. Inertial confinement is defined by a pulse duration shorter than the time for a pressure wave to traverse the particle diameter. Under this condition, extreme mechanical stress can be created in the tattoo ink nanoparticles, leading in theory to particle fracture or explosion. Despite the violence of such processes, they are occurring on the nanometer scale such that gross skin injury is still unlikely to occur. There have been several independent comparisons of nanosecond versus picosecond laser pulse treatment of tattoos in animals and humans, all of which found more effective and somewhat less color-dependent tattoo removal using picosecond laser pulses. The picosecond lasers yielded greater pigment lightening than QS Nd:YAG laser and mean depth of pigment alteration was greater for picosecond pulses (Izikson et al. 2010).

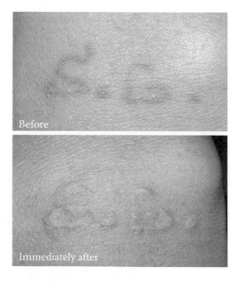

FIGURE 59.2 The immediate whitening response of this black carbon tattoo treated with a Q-switched Nd:YAG laser is due to temporary gas bubbles in the dermis where ink-containing cells were vaporized. A series of treatments are needed for substantial or complete, permanent lightening of most tattoos.

In a recent human study, Kossida et al. compared removal of tattoos using Q-switched ruby laser delivered in a conventional single pass with that of the same laser delivered several times on the same day, separated by 20 min. These investigators found greatly increased effectiveness for tattoo removal when multiple passes were used. The 20 min interval between passes was based on fading of the immediate whitening response (as the dermal gas bubbles gradually dissolve), and the same study showed that each pass produced subsequently deeper interactions within the dermis.

59.3.3 Laser Treatment of Facial Photoaging

Facial skin "rejuvenation" (Tanzi, Lupton, and Alster 2003) is probably a misnomer, since only some aspects of the appearance of youth can be produced. Despite this, the term is not entirely wrong because the mechanisms of improvement after laser treatment of aged and photoaged skin include formation of new skin tissue in response to a well-controlled wound. Ablative skin resurfacing lasers are carbon dioxide (CO_2; 10,600 nm) and Er:YAG (2930 to 2940 nm) lasers, which are strongly absorbed by water. CO_2 laser resurfacing is still considered a gold standard treatment for facial photoaging despite a high risk of side effects, because of strong efficacy (Apfelberg and Smoller 1997; Rabe et al. 2006; Tanzi, Lupton, and Alster 2003; West and Alster 1998b). The Er:YAG laser, developed to reduce the morbidity associated with CO_2 laser resurfacing (Tanzi, Lupton, and Alster 2003), has demonstrated comparable results with fewer side effects (Munker 2001; Weiss et al. 1999). Undesired effects of ablative systems include possibility of infections, scarring, and pigmentary alterations. While temporary hyperpigmentation is common in pigmented skin following laser resurfacing, the most prevalent problem is permanent hypopigmentation occurring in almost one-third of fair-skinned people treated for facial photoaging. There is also significant morbidity until reepithelialization occurs, which typically occurs between 4 and 7 days after treatment. Full recovery may require a month or more. Therefore, the long downtime period has limited its use (Tajirian and Goldberg 2011). Fractional ablative technology, a relatively recent development that uses these same lasers to produce an array of many microscopic, deep laser wounds, is a significant advance over classical laser resurfacing. Efficacy is generally acceptable, while there is less downtime, lower risk of infection or scarring, and very close to zero risk of permanent hypopigmentation common in conventional laser resurfacing. Recently, fractional laser therapy (FLT) has become a widely accepted modality for skin rejuvenation and has also been used for various other skin diseases, including pigment disorders, scars, stretch marks, and morphea (Alexiades-Armenaka et al. 2011; Tajirian and Goldberg 2011). A 2790 nm wavelength Er:YSGG laser was also introduced for fractional treatments (Smith and Schachter 2011). Acne scarring also responds well to fractional laser treatment with these devices (Berne, Nilsson, and Vahlquist 1984).

Nonablative fractional lasers use mid-IR lasers to create columns of thermal injury without removal of tissue. There is preservation of the stratum corneum, minimizing some risks and downtime. Other nonablative lasers that improve facial photoaging include pulsed dye, diode, and Nd:YAG. One of the most popularly used modalities for rejuvenation is IPL. With all of these laser and IPL treatments, focal thermal damage is created in a pattern that leads to reactive synthesis of new dermal matrix, including collagen (Stam-Posthuma et al. 1998). The IPLs are very useful for improvement of the pigmentation, telangiectasia, and ruddiness seen with photoaging. However, significant changes in skin texture and wrinkles have proved less achievable with the IPL alone (Fodor et al. 2009). Each of these treatment modalities is performed multiple times, usually several weeks apart. They are popular among patients who are unwilling or unable to undergo a postoperative recovery period associated with ablative procedures. In general, the nonablative treatments are much less effective than ablative lasers for the treatment of photoaging (Paquette, Badiavas, and Falanga 2001), but can provide many cosmetic benefits, including improvement of skin texture, discoloration, and preexistent scarring with minimal patient downtime (Doherty et al. 2009). A controlled half-face study of the 1450 nm diode laser demonstrated significant clinical improvement in periorbital rhytides as well as increases in dermal collagen assessed histologically. Recently, a clinical study described the safety and efficacy results of treatment with a fractional nonablative 1540 nm erbium:glass laser in 51 patients with Fitzpatrick skin types II to IV for striae (Otley et al. 1999). However, in a separate study, 25 dermatologists clinically evaluated patients after 1450 nm diode treatment, and although all patients reported mild to moderate improvement, only 2 of the 25 dermatologists recorded a significant positive treatment effect, suggesting that modest changes induced by the laser may not be clinically meaningful (Humphreys et al. 1996). This is not the case for ablative laser procedures.

59.3.4 Hair Removal

Permanent reduction of pigmented hair by lasers or IPLs is still a fast-growing trend 15 years after its invention (Blume-Peytavi and Hahn 2008). This is another application of the extended theory of SP (Altshuler et al. 2001). Devices for hair removal include normal-mode ruby laser (694 nm; the device originally used by Grossman et al. to establish that lasers can permanently remove pigmented hair), normal-mode alexandrite laser (755 nm), PDLs (800, 810 nm), long-pulsed Nd:YAG laser (1064 nm), and IPL sources (590–1200 nm) (Bjerring et al. 2000; Drosner and Adatto 2005; Gorgu et al. 2000; Lou et al. 2000). Today, the alexandrite laser, diode laser, and IPL devices cover the majority of hair removal treatments, whereas the ruby laser has a minor role owing mainly to its cost and limited reliability. The long-pulsed Nd:YAG laser is particularly useful for hair removal and for treating pseudofolliculitis barbae, a beard hair disorder, in darkly pigmented individuals.

Hair removal with lasers and IPL sources is generally regarded as safe treatment when performed with proper choice of operating parameters and proper technique (Figure 59.3). There are published studies describing the evidence supportive of laser

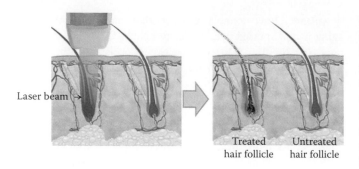

Laser beam →

Treated hair follicle Untreated hair follicle

FIGURE 59.3 Laser hair removal is an example of SP. Light absorbed by melanin in the hair shaft causes local heating and destruction of pigmented hair follicles.

hair removal (LHR) efficacy (Haedersdal and Gotzsche 2006; Haedersdal and Wulf 2006), including randomized controlled trials and nonrandomized controlled trials. Treatment parameters in terms of wavelength, pulse duration, spot size, and fluence must be individually selected and adjusted to the clinical situation before commencing treatment sessions. Pulse width, or pulse duration, is an important consideration that in professional settings may be controlled. Longer pulse widths and longer wavelengths are generally safer for darker skin, but some efficacy may be sacrificed. Spot size, or the width of the laser beam at the skin surface, also affects treatment outcome, as larger spot sizes suffer less attenuation from lateral scattering as the radiation penetrates the dermis. Larger spot sizes also allow faster treatment. An ideal patient for LHR has dark, thick terminal hairs, light skin color, and a normal hormonal status. People with very thin, blonde, red, or white hairs usually do not respond sufficiently well to currently available lasers and IPL systems.

The typical adverse events observed after LHR include discomfort/pain, redness, and swelling around the treatment area. These acute effects are transient. Beyond these common adverse events, there are more serious skin effects such as pigment changes (e.g., hypopigmentation or hyperpigmentation, burns, blistering, and crusting) that in rare cases may result in scar formation. Paradoxical hair stimulation has been noted on the face after laser and IPL treatments in approximately 5% of women of Mediterranean and Middle Eastern descent (Haedersdal, Beerwerth, and Nash 2011).

Recently, home-use lasers and IPLs for hair removal have been developed and sold at much lower prices than for professional devices. Some of these home-use devices provide impressive results, with 40%–60% hair reduction when measured 3 months after treatment. Professional treatments are somewhat more effective; as with hair styling and haircuts, the choice of whether to seek professional care or to try it at home is now available (Alster and Tanzi 2009b; Elm et al. 2010; Gold 2010).

59.3.5 Pigmented Lesions

A major application of SP is laser treatment of pigmented lesions. Pigmented lesions are classified anatomically as epidermal,

dermal, or mixed, and as being either melanocytic (due to abnormal melanocytes) or melanotic (due to abnormal amount or location of melanin pigment). Melanin is the target chromophore for pigmented lesions, which occurs as an intracellular organelle about the same size of a tattoo ink particle. Melanin absorption covers the entire visible spectrum and much of the near-IR spectrum, such that a range of lasers and IPLs are useful. For melanocytic lesions, Q-switched (QS) lasers are often used, for the same rationale that they are used to remove tattoos—the target is a cell containing cytoplasmic pigment particles. QS lasers used in pigmented lesions include the QS ruby (QSR, 694 nm), the QS Nd:YAG (532 and 1064 nm), and the QS alexandrite (QSA, 755 nm) lasers. When scattered, individual melanocytes are the target (as in nevus of Ota, a classic melanocytic lesion), QS lasers should be used. The long-pulsed ruby, alexandrite, diode, Nd:YAG lasers or IPLs have also been used for the treatment of pigmented lesions. These long-pulse devices are appropriate when the target is a multicellular structure such as the entire pigmented epidermis. Some melanocytic lesions such as congenital nevi should not be treated with a laser unless there is gross disfigurement without a surgical treatment option. Other melanocytic lesions that are more strongly associated with melanoma, such as dysplastic nevi, should not be treated at all with lasers or IPLs. The proliferation of IPLs into the hands of estheticians who have not been trained to recognize premalignant or malignant pigmented lesions is a disturbing situation. No data are available regarding the frequency with which melanoma or melanoma precursor lesions are treated as if these are cosmetic lesions.

59.3.5.1 Epidermal Lesions

The pigment of epidermal lesions is located in the epidermis and usually predominantly near the dermoepidermal junction, such as lentigines, café au lait patches, the epidermal variant of melasma, and so on. Lasers that have been used to treat lentigines include the CO_2, argon, 532 nm QS Nd:YAG, QS ruby, QS alexandrite, and long-pulsed alexandrite laser. Regarding laser ablation of tissue, pulsed CO_2 (10,600 nm) or Er:YAG lasers (2940 nm) can be used in the treatment of epidermal pigment by precisely removing the epidermis, a superficial form of laser skin resurfacing (Dover et al. 1988; Fitzpatrick, Goldman, and Ruiz-Esparza 1994a). QS ruby laser was the first to be successfully used to treat dermal pigmentary lesions without scarring, and it is also excellent for the treatment of many epidermal pigmented lesions, such as solar lentigines, ephelides, labial lentigo, and some cases of PIH. Kilmer et al. (1994) reported the efficacy of one treatment with a frequency-doubled QS Nd:YAG (532 nm) in 37 patients with solar lentigines. The results were superior at higher fluences, and they found 75% clearance of 60% of the lesions treated (Kilmer et al. 1994). The outcome and treatment of lentigines vary according to skin type. Light-skinned patients have better results and less complication, whereas dark-skinned patients have a higher risk of hyperpigmentation. Chan et al. (2000) found that the long-pulsed 532 nm Nd:YAG laser (2 ms pulse duration, 6.5–8 J/cm² fluence, 2 mm spot size) can result in a decreased risk of PIH in the treatment of lentigines in Asian patients.

The 532 nm wavelength is useful at low fluences for epidermal pigment such as in lentigines and café au lait patches. However, the recurrence rates of café au lait patches are inconclusive after laser treatment. Chan (2004) compared the normal-mode ruby laser (NMRL) to the QSR in the clearing of café au lait patches in 33 patients. Preliminary data indicated that the risk of recurrence was lower using the NMRL, 42.4% recurrence versus 81.8% with QSRL 3 months after a single treatment. Recently, a retrospective study evaluated the efficacy and safety of the Q-switched 755 nm alexandrite laser for the treatment of café au lait patches in 48 Chinese patients. They received an average of 3.2 treatments and the mean follow-up time was 16.1 months. Good to excellent responses were seen in 54.1% of patients and poor responses were seen in 16.7% of patients. Recurrence was noted in five patients (10.4%) and transient hypopigmentation was observed in one patient. Therefore, the recurrence rate of café au lait patches varies according to the laser used, but there are variable rates within the same laser. More than one treatment is sometimes necessary to achieve full clearing of the lesion. Among epidermal pigmented lesions, simple lentigines (due to sun exposure and aging) are the most commonly treated with these lasers and with IPLs. In general, lentigines respond very well, with permanent and selective removal after one or several treatments.

59.3.5.1.1 Becker's Nevus

Becker's nevus is a hamartoma of hair follicles and muscle characterized by hypertrichosis and brown hyperpigmentation. Histologically, the pigmentation of Becker's nevus is related to increased melanin in basal cells. Traditional laser treatment of Becker's nevus may involve the use of ablative and Q-switched lasers. The pigmented portion of Becker's nevus has been shown to improve with QS lasers, including the QSRL, QS Nd:YAG, Er:YAG laser, and the 1550 nm fractional erbium-doped fiber laser (Nanni and Alster 1998; Trelles et al. 2004; Tse et al. 1994). The hypertrichosis portion has been improved with long-pulsed ruby laser and long-pulsed alexandrite laser.

The typical adverse events observed after ablative lasers include posttreatment edema, erythema, burning, and crusting, pigmentary changes, acne flares, herpes simplex infection, scars, and milia formation. Erythema and pigmentary changes can last several months, whereas Q-switched lasers can selectively damage epidermal and dermal melanin without removing the entire epidermis, which decreases the adverse effects associated with treatment. Therefore, complete clearing of the lesions is rarely achieved, and many treatment sessions are necessary with Q-switched lasers.

Fractional photothermolysis has also been used to reduce the pigmentation associated with Becker's nevus. Glaich et al. (2007) reported two patients treated by using the 1550 nm fractional erbium-doped fiber laser. Both patients underwent five to six treatment sessions at 4-week intervals and achieved greater than 75% pigment reduction at 1 month of follow-up. No reduction in terminal hairs was noted. However, Becker's nevus is one of the most difficult pigmented lesions to achieve a satisfactory result.

There is great variability between patients, and within the lesion, to laser treatment. Usually, residual and resistant areas are present that yield a patchy appearance after laser treatment that are often less appealing than the original appearance. For these reasons, Becker's nevi should be treated in selected patients who are willing to consider surgical excision in the event of a poor outcome to laser treatment.

59.3.5.1.2 Postinflammatory Hyperpigmentation

PIH can result from hemosiderin or melanin deposition with limited options for treatment. Taylor and Anderson (1994) first reported the laser treatment by using QS ruby laser (694 nm, 40 ns). The results showed that the QSRL does not provide an effective treatment for PIH. After that, others confirmed that laser treatment of PIH is difficult (Kim and Cho 2010a,b; Rokhsar and Ciocon 2009). Kim et al. described three cases with improvement of PIH after treatment using the 1064 nm QS Nd:YAG laser with low fluence. With the application of the principle fractional photothermolysis (FP), so far there have been some case reports that show the efficacy of FPL. Rokhsar and Ciocon (2009) reported a patient with CO_2 laser-induced hyperpigmentation refractory to topical treatment who improved significantly after five sessions of 1550-nm FPL. Katz et al. (2009) evaluated a 1550 nm fractionated erbium-doped laser for the treatment of PIH in one patient who received three treatments and achieved greater than 95% improvement. In contrast, Kroon reported that nonablative 1550 nm FLT was not effective for the treatment of PIH.

59.3.5.1.3 Melasma

Melasma is a common pigmentary condition characterized by brown patches on the sun-exposed areas of the face and most commonly occurs in women, especially in Asian women. Although many treatment modalities have been studied, the efficacy is limited, there is no cure, and recurrences are common. Laser therapy has been alternatively used to improve melasma, but caution must be exercised when treating this condition because worsening of the disease or PIH may occur after treatment (Jones and Nouri 2006). There are a lot of laser modalities used to treat the melasma, including QS laser and FP. The 1064 nm QS Nd:YAG with a big beam spot and low fluences can increase the lightness of melasma (Cho, Kim, and Kim 2009a; Jeong et al. 2010; Suh et al. 2011; Wang and Liu 2009). The combinational treatment, using intense pulsed light (IPL) and low fluence QS Nd:YAG laser, also can elicit a clinical resolution in the mixed type melasma with long-term benefits (Na, Cho, and Lee 2012). Nonablative fractional laser (Kroon et al. 2011) and ablative fractional laser (Trelles, Velez, and Gold 2010) therapy are safe and comparable in efficacy and recurrence rate. It may be a useful alternative treatment option for melasma when topical bleaching is ineffective or not tolerated. Different laser settings and long-term maintenance treatment should be tested in future studies. The common side effect of treatment by different laser modalities is dyschromia, including hyperpigmentation and hypopigmentation. The risk of recurrence and refractory after treatment are also mentioned. Therefore, the optimal treatment parameters for melasma should be selected in individual therapy. The combinational treatment, such as using laser, chemical peeling, and brightening agents, is also an improvement approaching and can reduce the side effects of laser therapy.

59.3.5.2 Dermal Melanocytic Lesions

Dermal melanocytic lesions such as nevus of Ota, nevus of Ito, some melanocytic nevi, and blue nevi contain melanocytes at various levels of differentiation deep in the dermis. As a result, devices with longer wavelengths provide superior clearance because the wavelength is able to penetrate to the appropriate depth.

59.3.5.2.1 Nevus of Ota and Nevus of Ito

Nevus of Ota and Nevus of Ito are blue-gray lesions most common among Asians, with similar clinical and histological character, except in different locations. The response of these disfiguring, life-long lesions to QS lasers is usually excellent, with permanent and selective removal. Lasers used for nevus of Ota can successfully improve the lesions, including QSRL (Goldberg and Nychay 1992), QS Nd:YAG laser (Omprakash 2002; Sharma, Jha, and Mallik 2011), and *Q-switched alexandrite laser* (QSAL) (Moreno-Arias and Camps-Fresneda 2001). More recently, a fractionated laser has been used to treat nevus of Ota effectively. In 2008, Kouba, Fincher, and Moy (2008) reported a case of complete clearance of nevus of Ota using a fractionated 1440 nm Nd:YAG laser. It has also been suggested that treatment of nevus of Ota in childhood leads to better results with fewer treatment sessions and fewer complications (Watanabe and Takahashi 1994), although in many cases, the lesion becomes first apparent at puberty. Hypopigmentation was the most common complication, then hyperpigmentation. Scarring formation caused by high fluences and inappropriate posttreatment care rarely occurs.

59.3.5.2.2 Melanocytic Nevi

This is a common lesion that is dermal in nature. When possible, congenital or acquired melanocytic nevi should be removed surgically, which is the treatment of first choice. A combination of QS and long-pulsed laser appears to be most effective for treatment of cases in which excision is not a good option. Another combined approach uses the ablative laser, such as CO_2 laser, NMRL (Chan 2004), to remove the epidermis followed by QSRL or QSAL for deeper penetration. The method can remove epithelial cells so that the QSRL or QSAL can more effectively penetrate to the deep nests of cells. This technique is thought to create microscopic scarring that covers underlying nest cells leading to cosmetic improvement.

59.3.5.2.3 Congenital Nevi

Treatment of congenital nevi is controversial because of the potential for malignant transformation. Laser treatment is an alternative choice for small (diameter <1 cm) lesions with uniform color and smooth regular borders, excluding atypical-looking lesion, melanoma, or other malignant pigmented lesions.

Any suspicious, atypical-looking lesion should be biopsied before beginning treatment (Jones and Nouri 2006). Surgical resection is usually the first choice of congenital nevi. In other cases, congenital nevi are enormous, covering a large fraction of the body surface. A combination of surgical and laser treatment is sometimes useful in these cases.

In summary, Q-switched lasers and IPLs are effective in treating many pigmented lesions. In general, the lasers with shorter wavelengths (QSN 532 nm, IPL 515 nm) are more effective on epidermal/superficial pigmented lesions, and longer wavelengths (QSR 694 nm, QSN 1064 nm, QSA 755 nm, and IPL 1200 nm) work better on deeper dermal pigment. The QSR and QSN 532 nm are excellent for solar lentigines, ephelides, nevus of Ota, and melanocytic nevi with a low incidence of scarring and purpura. The QSA 1064 nm has similar efficacy but has a lower incidence of hypopigmentation than QSR and is safer than QSR in dark skin types (Fitzpatrick types IV–VI).

59.3.6 Laser Management of Acne and Acne Scarring

Acne is considered the most common skin disorder and affects millions of people every year. There are many treatments that are currently used for acne, including topical and systemic medications. Although medication is the first line of treatment for acne, many current therapies have drawbacks involving patient compliance, systemic toxicities, and bacterial resistance (Nouri and Ballard 2006). Lasers are now established options in the physical modalities to treat acne. Laser treatment of acne has the benefit of a faster response compared with the 6 to 8 weeks needed to see the beneficial effects of traditional therapies such as oral antibiotics and retinoic acid (Seaton et al. 2003). Lasers and light sources used to treat inflammatory acne include PDLs, KTP, IPL, 1450 nm diode laser and long-pulsed Nd:YAG laser. Ablative and nonablative fractional lasers have shown efficacy for treatment of acne scarring in several small studies. The number of treatments will vary according to several factors including patient response after each laser session, laser technology used, and parameters of treatment.

PDL is useful for treating erythema (redness) associated with acne, and acne scarring (Alster and McMeekin 1996). Treatments may be safely performed on all skin types and over hair-bearing areas without fear of follicular destruction. However, patients with darker skin types (IV–VI) have a higher risk for postoperative skin dyspigmentation, especially hyperpigmentation. Purpura is the main complication, lasting for 7–10 days. Purpura has been advocated as a clinical end point and resolves without sequelae; however, purpura is not necessary to obtain therapeutic response (Kwok and Rao 2012). Some studies showed that PDL appears to be a well-tolerated and effective treatment of inflammatory acne lesions (Seaton et al. 2003). However, some trial results showed that the PDL did not prove very efficacious in acne lesions. The 532 nm KTP laser has also been reported to have a beneficial effect on vascular lesions as well as stimulating collagen production (Nataloni 2003). It can be used to treat the erythema and red scars of acne with a low side effect profile (Yilmaz et al. 2011). The output wavelength of 532 nm corresponds with the first peak of the oxyhemoglobin absorption curve (Silver and Livshots 1996). The penetration depth of the KTP laser is confined to the superficial dermis of the skin and is less suitable for deeper vessels, especially in comparison with alexandrite or Nd:YAG lasers, which penetrate the entire dermis. A number of green and yellow output lasers and IPLs are able to activate porphyrin for photodynamic therapy using aminolevulinic acid (ALA) or its esters. In an open-label, split-face study, the combined use of topical ALA and a 532 nm laser PDT showed promising results for acne treatment (Sadick 2010). IPL devices have also shown benefit for inflammatory acne treatment, for example, reducing red macules of acne in one split-face controlled trial (Chang et al. 2007). At present, IPL combined with ALA-PDT can significantly improve inflammatory acne lesions and provide greater, longer-lasting, and more consistent improvement than IPL alone in the treatment of moderate to severe acne vulgaris (de Leeuw et al. 2010; Oh et al. 2009; Taub 2007) with minimal side effects. However, there is strong evidence that ALA-PDT is most effective when performed with long contact times (2–4 h between application and light exposure) and activation with 630–640 nm red light at high fluences (30–200 J/cm^2) rather than with IPL or pulsed lasers.

Near- and mid-IR lasers have also been shown to improve inflammatory acne. Nd:YAG 1064 nm laser pulses in the microsecond domain may be useful for treating inflammatory acne, and Q-switched Nd:YAG delivered at low fluence has been shown to significantly improve acne scarring (Keller et al. 2007; Kwok and Rao 2012). A 1450 nm diode laser in the IR spectrum is approved by the US Food and Drug Administration (FDA) for acne treatment. This laser uses cryogen spray cooling to protect the epidermis while nonselectively heating the upper dermis. A decrease in inflammatory acne lesions and acne scars was noted after 1450 nm laser sessions for facial rhytides, followed by a prospective study showing improvement of acne on the back (Lewis and Benedetto 2002). It has been theorized that this laser can cause thermal damage to the sebaceous glands, which may temporarily arrest their sebaceous output (Paithankar et al. 2002). Paithankar et al. (2002) showed significant clearance of inflammatory acne that lasted at least 6 months. Erythema was the most common side effect, and transient hyperpigmentation occurred in a minority of the subjects.

Several clinical studies have presented the efficacy and safety of FP lasers for improving acne scars (Chapas et al. 2008; Cho et al. 2009b, 2010). A retrospective study found that ablative FP had an improvement range of 26%–83%, whereas nonablative FP had an improvement range of 26%–50% for acne scar (Ong and Bashir 2012). The main complications of FP are erythema, edema, crusting and scaling, and PIH (Chapas et al. 2008; Cho et al. 2009b, 2010). Ablative FP laser caused erythema for 3–14 days that resolved by 12 weeks, whereas nonablative FP laser generated erythema between 1 and 3 days and it was resolved within a week. A higher proportion of patients (up to 92.3%) (Chan et al. 2010; Cho et al. 2009b; Jung et al. 2010) who underwent ablative

FP experienced PIH than those who had nonablative FP (up to 13%) (Chan et al. 2007; Cho et al. 2010; Mahmoud et al. 2010). The maximum duration of PIH in ablative FP was up to 6 months, whereas in nonablative FP, it lasted for up to 1 week. The procedure with ablative FP was relatively uncomfortable compared with nonablative FP. The pain score with ablative FP ranged from 5.90 to 8.10 (scale 1–10) (Manuskiatti et al. 2010) and with nonablative FP from 3.90 to 5.66 (scale 1–10) (Mahmoud et al. 2010). Fractional photothermolysis appears to be one of the most effective treatments for acne scarring, with fewer side effects and complications than laser resurfacing or dermabrasion.

59.4 Laser-Assisted Drug Delivery

Transdermal drug delivery systems have been widely accepted for administration of systemic drugs and topical drugs and are attractive for many reasons, including increased patient acceptability, avoidance of gastrointestinal disturbances, and bypassing of hepatic first-pass metabolism and increasing topical efficacy. However, drug transport through the skin is fundamentally limited by cutaneous barrier function (Lee et al. 2001). The skin barrier can be partially overcome by removal of the stratum corneum (SC) such as with tape-stripping, which cannot be precisely controlled (Bronaugh and Stewart 1985). Laser-assisted drug delivery has the potential for facilitating topical drug uptake with reduced onset time. A previous study shows that the photomechanical wave generated by a ruby laser could enhance the skin transport of 5-ALA by *in vivo* topical application. Lee et al. (2001) studied the influence of an Er:YAG laser on the transdermal delivery of lipophilic and hydrophilic drugs across skin *in vitro*. The results indicated a significant increase in the permeation of the drugs across skin pretreated with an Er:YAG laser, and hydrophilic molecules could more easily permeate the skin barrier than lipophilic drugs. The laser intensity and its spot size were found to play an important role in controlling transdermal delivery of drugs. A randomized, double-blind, crossover study compared the efficacy and adverse event profile of laser-assisted topical anesthesia before venipuncture, finding more rapid onset of anesthesia that was independent of laser fluence (Koh et al. 2007). Recently, AFR has been shown to greatly facilitate penetration and distribution of topically applied drugs, since the ablated laser holes extend into dermis, thereby possibly acting as channels for drug uptake. Haedersdal et al. (2010) used methyl 5-aminolevulinate (MAL) to evaluate the efficacy of drug delivery by CO_2 laser AFR. AFR before or after MAL application enhanced drug delivery in dermis versus MAL alone. Therefore, AFR appears to be a clinically practical means for enhancing uptake of many other topical skin medications.

59.5 Laser Diagnostic Microscopy

The need to improve the diagnostic accuracy and sensitivity for skin tumors has led to the development of new noninvasive, *in vivo* techniques including ultrasound, dermoscopy, digital photography, confocal scanning laser microscopy (CSLM), high-resolution magnetic resonance imaging, and optical coherence tomography (OCT). Among these, optical coherence tomography and CSLM use lasers.

59.5.1 Optical Coherence Tomography

OCT is an emergent *in vivo* imaging technique based on the interferometry measurement of IR light backscattered from various depths inside the tissue. OCT allows high-resolution, high-speed, two- or three-dimensional, cross-sectional visualization of microstructural morphology of tissues. OCT provides depth-resolved images with resolution ranging from 2 to 10 μm and depth up to several millimeters depending on tissue type (Gambichler, Jaedicke, and Terras 2011) (Figure 59.4). OCT is widely used in ophthalmology and is being developed in many domains of medicine: cardiology, gastroenterology, urology, surgery, neurology, rheumatology, pneumology, gynecology, and dentistry for both research and clinical practice. In 1997, OCT was introduced in dermatology to assess skin structure in clinical settings (Welzel et al. 1997) and has improved substantially since. It is now being increasingly employed in clinical skin research (Gambichler, Jaedicke, and Terras 2011). A recent study by Morsy et al. suggests that OCT can measure epidermal thickness in psoriasis and that these measurements correlate with several other parameters of disease severity. Inflammatory skin diseases such as contact dermatitis have already been extensively studied *in vivo* using OCT (Coulman et al. 2011; Welzel 2001; Welzel et al. 1997; Welzel, Bruhns, and Wolff 2003). OCT is also a promising method for objectively monitoring the activity of cutaneous lupus and the effects of treatment over time *in vivo* (Gambichler et al. 2007). Intradermal or subepidermal blistering can easily be detected on OCT images. Intraepidermal blisters can be distinguished from subepidermal blisters by looking for the location of the cleft at the lateral border of the blister. Besides

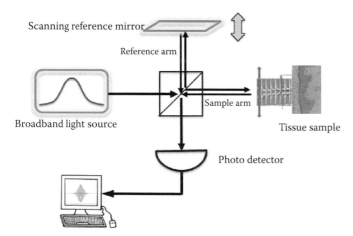

FIGURE 59.4 OCT uses an interferometer to detect reflected light returning from different depths inside the tissue. When the path length of the sample arm is equal to that of the reference arm, the detector measures interference of light waves. As the reference mirror is scanned (double arrow), different depths in the tissue are detected.

the level of blisters, its content can also be determined by means of OCT *in vivo*, and it can be used to detect the blistering disease (Welzel 2001; Welzel et al. 1997). However, it is not possible to differentiate between pemphigoid lesions and pemphigus-like diseases such as subcorneal pustular dermatosis and Darier's disease because the variation is too subtle (Mogensen et al. 2008). OCT images can clearly distinguish dilated blood vessels or vascular abnormalities from normal tissue (Chen et al. 1997; Izatt et al. 1997; Nelson et al. 2001; Salvini et al. 2008; Zhao et al. 2010). With OCT, structural parameters of vascular lesions such as epidermal thickness and depth of the dilated blood vessels can be visualized. This information can be helpful in the selection of appropriate treatments (Salvini et al. 2008). The skin and finger- and toenails can also be assessed using OCT, which helps to distinguish onychomycosis from other nail diseases such as psoriasis, lichen planus, and dystrophic nail changes (Abuzahra et al. 2010). In previous studies, OCT also demonstrated potential for imaging of skin infestations such as larva migrans and scabies (Mogensen et al. 2009; Welzel et al. 1998). The application of OCT on detection of skin cancer, including nonmelanoma and melanoma skin cancer, is promising. Commercial OCT devices for skin have appeared on the market recently. For nodular basal cell carcinomas, OCT appears to be capable of visualizing tumor cell aggregates in superficial dermis. It remains to be determined whether OCT is clinically useful for defining surgical excision margins, detecting morpheaform (infiltrating) basal cell carcinomas, or differentiating between squamous cell, basal cell, and other tumors. Hamdoon et al. (2011) demonstrated that OCT-guided photodynamic therapy is a promising approach to efficiently discriminate between tumor-involved and noninvolved margins. It reduced the untoward healthy tissue necrosis and provided an encouraging monitoring of the healing process (Hamdoon et al. 2011). Gambichler, Jaedicke, and Terras (2011) detected significant differences between benign nevi and malignant melanoma in regard to micromorphological features visualized by OCT. However, the resolution of OCT in that study was not high enough to reveal the morphology of single cells, although it is possible to evaluate the architecture of a lesion (de Giorgi et al. 2005). Systematic studies for the diagnosis of pigmented lesions with OCT with regard to sensitivity and specificity are still lacking (Gambichler, Jaedicke, and Terras 2011). Recent advances employing analysis of the phase component of spectral-domain OCT have led to much higher resolution of about 0.5 μm, enabling single-cell image resolution similar to that of CSLM. Another technical advance is Doppler OCT, in which the frequency shift of light returning from skin due to the motion of blood can be imaged. Doppler OCT yields an unprecedented, detailed view of cutaneous microvasculature.

59.5.2 Confocal Scanning Laser Microscopy

CSLM has become an invaluable tool for a wide range of investigations in the biological and medical sciences for imaging thin optical sections in living and fixed specimens ranging in thickness up to 200 μm. CSLM allows for the examination of the epidermis and papillary dermis at a resolution approaching histological detail (Branzan, Landthaler, and Szeimies 2007). Limited by depth of penetration, CSLM mainly applies to the epidermis and superficial dermis. Limited also by the absence of specific stains for *in vivo* microscopy, CSLM relies on endogenous sources of contrast and on the user's ability to recognize microscopic morphologies (Figure 59.5). Nonetheless, besides its noninvasiveness, avoiding artifacts of tissue processing and staining and the speed of imaging are its advantages when compared with histology. Dynamic processes like inflammatory cell migration, blood flow, wound healing and changes in melanin content and in epidermal morphometry in pathologic conditions can be visualized in real time (Huzaira et al. 2001). Unlike biopsy-based microscopy, CSLM does not destroy a skin site such that interval changes, early response to therapy, and growth or regression of lesions can be monitored and in some cases quantified. CSLM has been used to describe pigmented skin lesions, including benign melanocyte nevus (Marghoob et al. 2005; Pellacani et al. 2004), atypical nevi (Busam et al. 2002; Pellacani et al. 2005), and melanomas (Busam et al. 2002; Curiel-Lewandrowski et al. 2004; Pellacani et al. 2005), epithelial tumors, including actinic keratoses and squamous cell carcinoma (Chung et al. 2004), and basal cell carcinoma (Gonzalez and Tannous 2002). Other cutaneous diseases can be analyzed by CSLM, including psoriasis, sebaceous gland hyperplasia (Gonzalez et al. 1999), cherry angiomas, allergic and irritant contact dermatitis (Nyren, Kuzmina, and Emtestam 2003), tinea pedis and onychomycosis, folliculitis, and so on.

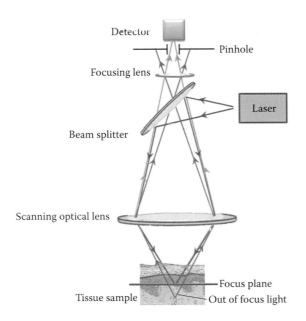

FIGURE 59.5 Confocal microscopes project light returning from a focal point inside the tissue to a pinhole placed in a conjugate focal plane. The pinhole rejects light returning from above, below, or adjacent to the focal point. The focused laser beam or the sample is then scanned to create an image of the tissue at the plane of the focus.

References

Abuzahra, F., F. Spoler, M. Forst et al. 2010. Pilot study: Optical coherence tomography as a non-invasive diagnostic perspective for real time visualisation of onychomycosis. *Mycoses* 53:334–339.

Achauer, B. M., C. Chang, V. M. Vanderkam et al. 1999. Intralesional photocoagulation of periorbital haemangiomas. *Plast Reconstr Surg* 103:11–16.

Alam, M., N. E. Omura, J. S. Dover, and K. A. Arndt. 2003. Clinically significant facial edema after extensive treatment with purpura-free pulsed-dye laser. *Dermatol Surg* 29:920–924.

Alcantara Gonzalez, J., P. Boixeda, M. T. Truchuelo Diez, J. C. Lopez Gutierrez, and P. J. Olasolo. 2012. Ablative fractional yttrium–scandium–gallium–garnet laser for scarring residual haemangiomas and scars secondary to their surgical treatment. *J Eur Acad Dermatol Venereol* 26:477–482.

Alexiades-Armenaka, M., D. Sarnoff, R. Gotkin, and N. Sadick. 2011. Multi-center clinical study and review of fractional ablative CO_2 laser resurfacing for the treatment of rhytides, photoaging, scars and striae. *J Drugs Dermatol* 10:352–362.

Alper, J. C., and L. B. Holmes. 1983. The incidence and significance of birthmarks in a cohort of 4,641 newborns. *Pediatr Dermatol* 1:58–68.

Alster, T. S., and T. O. McMeekin. 1996. Improvement of facial acne scars by the 585 nm flashlamp-pumped pulsed dye laser. *J Am Acad Dermatol* 35:79–81.

Alster, T. S., and E. L. Tanzi. 2009a. Combined 595-nm and 1,064-nm laser irradiation of recalcitrant and hypertrophic port-wine stains in children and adults. *Dermatol Surg* 35:914–918; discussion 918–919.

Alster, T. S., and E. L. Tanzi 2009b. Effect of a novel low-energy pulsed light device for home-use hair removal. *Dermatol Surg* 35:483–489.

Altshuler, G. B., R. R. Anderson, D. Manstein, H. H. Zenzie, and M. Z. Smirnov. 2001. Extended theory of selective photothermolysis. *Lasers Surg Med* 29:416–432.

Anderson, R. R. 2003. Dermatologic history of the ruby laser: The long story of short pulses. *Arch Dermatol* 139:70–74.

Anderson, R. R., and J. A. Parrish. 1981. Microvasculature can be selectively damaged using dye lasers: A basic theory and experimental evidence in human skin. *Lasers Surg Med* 1:263–276.

Anderson, R. R., and J. A. Parrish. 1983. Selective photothermolysis: Precise microsurgery by selective absorption of pulsed radiation. *Science* 220:524–527.

Apfelberg, D. B., and B. Smoller. 1997. UltraPulse carbon dioxide laser with CPG scanner for deepithelialization: Clinical and histologic study. *Plast Reconstr Surg* 99:2089–2094.

Ashinoff, R., and R. G. Geronemus. 1991. Capillary haemangiomas and treatment with the flashlamp-pumped pulsed dye laser. *Arch Dermatol* 127:202–205.

Astner, S., and R. R. Anderson. 2005. Treating vascular lesions. *Dermatol Ther* 18:267–281.

Barlow, R. J., N. P. Walker, and A. C. Markey. 1996. Treatment of proliferative haemangiomas with the 585 nm pulsed dye laser. *Br J Dermatol* 134:700–704.

Bashkatov, A. N., E. A. Genina, V. I. Kochubey, and V. V. Tuchin. 2005. Optical properties of human skin, subcutaneous and mucous tissues in the wavelength range from 400 to 2000 nm. *J Phys D: Appl Phys* 38:2543–2555.

Batta, K., H. M. Goodyear, C. Moss et al. 2002. Randomised controlled study of early pulsed dye laser treatment of uncomplicated childhood haemangiomas: Results of a 1-year analysis. *Lancet* 360:521–527.

Baumler, W., E. T. Eibler, U. Hohenleutner et al. 2000. Q-switch laser and tattoo pigments: First results of the chemical and photophysical analysis of 41 compounds. *Lasers Surg Med* 26:13–21.

Berne, B., M. Nilsson, and A. Vahlquist. 1984. UV irradiation and cutaneous vitamin A: An experimental study in rabbit and human skin. *J Invest Dermatol* 83:401–404.

Bernstein, E. F. 2006. Laser treatment of tattoos. *Clin Dermatol* 24:43–55.

Bettman, O. H. 1979. *A Pictorial History of Medicine*. Charles Thomas Publisher, New York.

Bjerring, P., K. Christiansen, and A. Troilius. 2003. Intense pulsed light source for the treatment of dye laser resistant port-wine stains. *J Cosmet Laser Ther* 5:7–13.

Bjerring, P., P. M. Cramers, H. Egekvist et al. 2000. Hair reduction using a new intense pulsed light irradiator and a normal mode ruby laser. *J Cutan Laser Ther* 2:63–71.

Blume-Peytavi, U., and S. Hahn. 2008. Medical treatment of hirsutism. *Dermatol Ther* 21:329–339.

Borges da Costa, J., P. Boixeda, C. Moreno, and J. Santiago. 2009. Treatment of resistant port-wine stains with a pulsed dual wavelength 595 and 1064 nm laser: A histochemical evaluation of the vessel wall destruction and selectivity. *Photomed Laser Surg* 27:599–605.

Branzan, A. L., M. Landthaler, and R. M. Szeimies. 2007. *In vivo* confocal scanning laser microscopy in dermatology. *Lasers Med Sci* 22:73–82.

Braverman, I. M. 1989. Ultrastructure and organization of the cutaneous microvasculature in normal and pathologic states. *J Invest Dermatol* 93:2S–9S.

Bronaugh, R. L., and R. F. Stewart. 1985. Methods for *in vitro* percutaneous absorption studies V: Permeation through damaged skin. *J Pharm Sci* 74:1062–1066.

Burstein, F. D., C. Simms, S. R. Cohen et al. 2000. Intralesional laser therapy of extensive haemangiomas in 100 consecutive paediatric patients. *Ann Plas Surg* 44:188–194.

Busam, K. J., C. Charles, C. M. Lohmann et al. 2002. Detection of intraepidermal malignant melanoma *in vivo* by confocal scanning laser microscopy. *Melanoma Res* 12:349–355.

Chan, H. H., W. K. K. Fung, S. Y. Ying et al. 2000. An *in vivo* trial comparing the use of different types of 532 nm Nd:YAG lasers in the treatment of facial lentigines in oriental patients. *Dermatol Surg* 26:743–749.

Chan, H. H., D. Manstein, C. S. Yu et al. 2007. The prevalence and risk factors of post-inflammatory hyperpigmentation after fractional resurfacing in Asians. *Lasers Surg Med* 39:381–385.

Chan, H. H. L., and T. Kono. 2004. The use of lasers and intense pulsed light sources for the treatment of pigmented lesions. *Skin Ther Lett* 9:1–5.

Chan, N. P., S. G. Ho, C. K. Yeung, S. Y. Shek, and H. H. Chan. 2010. Fractional ablative carbon dioxide laser resurfacing for skin rejuvenation and acne scars in Asians. *Lasers Surg Med* 42:615–623.

Chang, S. E., S. J. Ahn, D. Y. Rhee et al. 2007. Treatment of facial acne papules and pustules in Korean patients using an intense pulsed light device equipped with a 530- to 750-nm filter. *Dermatol Surg* 33:676–679.

Chapas, A. M., L. Brightman, S. Sukal et al. 2008. Successful treatment of acneiform scarring with CO_2 ablative fractional resurfacing. *Lasers Surg Med* 40:381–386.

Chen, Z., T. E. Milner, S. Srinivas et al. 1997. Noninvasive imaging of *in vivo* blood flow velocity using optical Doppler tomography. *Opt Lett* 22:1119–1121.

Cho, S. B., J. S. Kim, and M. J. Kim. 2009a. Melasma treatment in Korean women using a 1064-nm Q-switched Nd:YAG laser with low pulse energy. *Clin Exp Dermatol* 34:e847–850.

Cho, S. B., S. J. Lee, S. Cho et al. 2010. Non-ablative 1550-nm erbium-glass and ablative 10 600-nm carbon dioxide fractional lasers for acne scars: A randomized split-face study with blinded response evaluation. *J Eur Acad Dermatol Venereol* 24:921–925.

Cho, S. B., S. J. Lee, J. M. Kang et al. 2009b. The efficacy and safety of 10,600-nm carbon dioxide fractional laser for acne scars in Asian patients. *Dermatol Surg* 35:1955–1961.

Chung, V. Q., P. J. Dwyer, K. S. Nehal et al. 2004. Use of ex vivo confocal scanning laser microscopy during Mohs surgery for nonmelanoma skin cancers. *Dermatol Surg* 30:1470–1478.

Civas, E., E. Koc, B. Aksoy, and H. M. Aksoy. 2009. Clinical experience in the treatment of different vascular lesions using a neodymium-doped yttrium aluminum garnet laser. *Dermatol Surg* 35:1933–1941.

Cosman, B. 1980. Experience in the argon laser therapy of port wine stains. *Plast Reconstr Surg* 65:119–129.

Coulman, S. A., J. C. Birchall, A. Alex et al. 2011. *In vivo, in situ* imaging of microneedle insertion into the skin of human volunteers using optical coherence tomography. *Pharm Res* 28:66–81.

Curiel-Lewandrowski, C., C. M. Williams, K. J. Swindells et al. 2004. Use of *in vivo* confocal microscopy in malignant melanoma: An aid in diagnosis and assessment of surgical and nonsurgical therapeutic approaches. *Arch Dermatol* 140:1127–1132.

de Giorgi, V., M. Stante, D. Massi et al. 2005. Possible histopathologic correlates of dermoscopic features in pigmented melanocytic lesions identified by means of optical coherence tomography. *Exp Dermatol* 14:56–59.

de Leeuw, J., N. van der Beek, P. Bjerring, and H. A. Neumann. 2010. Photodynamic therapy of acne vulgaris using 5-aminolevulinic acid 0.5% liposomal spray and intense pulsed light in combination with topical keratolytic agents. *J Eur Acad Dermatol Venereol* 24:460–469.

Doherty, S. D., C. B. Doherty, J. S. Markus, and R. F. Markus. 2009. A paradigm for facial skin rejuvenation. *Facial Plast Surg* 25:245–251.

Dover, J. S., and K. A. Arndt. 2000. New approaches to the treatment of vascular lesions. *Lasers Surg Med* 26:158–163.

Dover, J. S., B. R. Smoller, R. S. Stern, S. Rosen, and K. A. Arndt. 1988. Low-fluence carbon dioxide laser irradiation of lentigines. *Arch Dermatol* 124:1219–1224.

Drosner, M., and M. Adatto. 2005. Photoepilation: Guidelines for care from the European Society for Laser Dermatology (ESLD). *J Cosmet Laser Ther* 7:33–38.

Elm, C. M., I. D. Wallander, S. E. Walgrave, and B. D. Zelickson. 2010. Clinical study to determine the safety and efficacy of a low-energy, pulsed light device for home use hair removal. *Lasers Surg Med* 42:287–291.

Faurschou, A., K. Togsverd-Bo, C. Zachariae, and M. Haedersdal. 2009. Pulsed dye laser vs. intense pulsed light for port-wine stains: A randomized side-by-side trial with blinded response evaluation. *Br J Dermatol* 160:359–364.

Fitzpatrick, R. E., M. P. Goldman, and J. Ruiz-Esparza. 1994a. Clinical advantage of the CO2 laser superpulsed mode. Treatment of verruca vulgaris, seborrheic keratoses, lentigines, and actinic cheilitis. *J Dermatol Surg Oncol* 20:449–456.

Fitzpatrick, R. E., N. J. Lowe, M. P. Goldman et al. 1994b. Flashlamp-pumped pulsed dye laser treatment of port-wine stains. *J Dermatol Surg Oncol* 20:743–748.

Fodor, L., N. Carmi, A. Fodor, Y. Ramon, and Y. Ullmann. 2009. Intense pulsed light for skin rejuvenation, hair removal, and vascular lesions: A patient satisfaction study and review of the literature. *Annals of plastic surgery* 62:345–349.

Gambichler, T., J. Hyun, G. Moussa et al. 2007. Optical coherence tomography of cutaneous lupus erythematosus correlates with histopathology. *Lupus* 16:35–38.

Gambichler, T., V. Jaedicke, and S. Terras. 2011. Optical coherence tomography in dermatology: Technical and clinical aspects. *Arch Dermatol Res* 303:457–473.

Garden, J. M., A. D. Bakus, and A. S. Paller. 1992. Treatment of cutaneous haemangiomas by the flashlamp-pumped pulsed dye laser: Prospective analysis. *J Paediatr* 120:555–560.

Garden, J. M., L. L. Polla, and O. T. Tan. 1988. The treatment of port-wine stains by the pulsed dye laser. Analysis of pulse duration and long-term therapy. *Arch Dermatol* 124:889–896.

Glaich, A. S., L. H. Goldberg, T. Dai, J. H. Kunishige, and P. M. Friedman. 2007. Fractional resurfacing: A new therapeutic modality for Becker's nevus. *Arch Dermatol* 143:1488–1490.

Gold, M. H., A. Foster, and J. A. Biron. 2010. Low-energy intense pulsed light for hair removal at home. *J Clin Aesthet Dermatol* 3:48–53.

Goldberg, D. J., and S. G. Nychay. 1992. Q-switched ruby laser treatment of nevus of Ota. *J Dermatol Surg Oncol* 18:817–821.

Goldman, L. 1980. The argon laser and the port wine stain. *Plast Reconstr Surg* 65:137–139.

Goldman, L., D. J. Blaney, D. J. Kindel, Jr., and E. K. Franke. 1963a. Effect of the laser beam on the skin. Preliminary report. *J Invest Dermatol* 40:121–122.

Goldman, L., D. J. Blaney, D. J. Kindel, Jr., D. Richfield, and E. K. Franke. 1963b. Pathology of the effect of the laser beam on the skin. *Nature* 197:912–914.

Goldman, L., R. J. Rockwell, R. Meyer et al. 1967. Laser treatment of tattoos. A preliminary survey of three year's clinical experience. *JAMA* 201:841–844.

Goldman, L., R. G. Wilson, P. Hornby, and R. G. Meyer. 1965. Radiation from a Q-switched ruby laser. Effect of repeated impacts of power output of 10 megawatts on a tattoo of man. *J Invest Dermatol* 44:69–71.

Gonzalez, S., and Z. Tannous. 2002. Real-time, *in vivo* confocal reflectance microscopy of basal cell carcinoma. *J Am Acad Dermatol* 47:869–874.

Gonzalez, S., W. M. White, M. Rajadhyaksha, R. R. Anderson, and E. Gonzalez. 1999. Confocal imaging of sebaceous gland hyperplasia *in vivo* to assess efficacy and mechanism of pulsed dye laser treatment. *Lasers Surg Med* 25:8–12.

Gorgu, M., G. Aslan, T. Akoz, and B. Erdogan. 2000. Comparison of alexandrite laser and electrolysis for hair removal. *Dermatol Surg* 26:37–41.

Grossman, M. C., C. Dierickx, W. Farinelli, T. Flotte, and R. R. Anderson. 1996. Damage to hair follicles by normal-mode ruby laser pulses. *J Am Acad Dermatol* 35:889–894.

Haedersdal, M., F. Beerwerth, and J. F. Nash. 2011. Laser and intense pulsed light hair removal technologies: From professional to home use. *Br J Dermatol* 165 Suppl 3:31–36.

Haedersdal, M., and P. C. Gotzsche. 2006. Laser and photoepilation for unwanted hair growth. *Cochrane Database Syst Rev*:CD004684.

Haedersdal, M., F. H. Sakamoto, W. A. Farinelli et al. 2010. Fractional CO(2) laser-assisted drug delivery. *Lasers Surg Med* 42:113–122.

Haedersdal, M., and H. C. Wulf. 2006. Evidence-based review of hair removal using lasers and light sources. *J Eur Acad Dermatol Verereol* 20:9–20.

Hamdoon, Z., W. Jerjes, T. Upile, and C. Hopper. 2011. Optical coherence tomography-guided photodynamic therapy for skin cancer: Case study. *Photodiagnosis Photodyn Ther* 8:49–52.

Haywood, R. M., B. E. Monk, and P. J. Mahaffey. 2000. The treatment of early cutaneous capillary haemangiomata (strawberry naevi) with the tunable dye laser. *Br J Plast Surg* 53:302–307.

Humphreys, T. R., V. Werth, L. Dzubow, and A. Kligman. 1996. Treatment of photodamaged skin with trichloroacetic acid and topical tretinoin. *J Am Acad Dermatol* 34:638–644.

Huzaira, M., F. Rius, M. Rajadhyaksha, R. R. Anderson, and S. Gonzalez. 2001. Topographic variations in normal skin, as viewed by *in vivo* reflectance confocal microscopy. *J Invest Dermatol* 116:846–852.

Izatt, J. A., M. D. Kulkarni, S. Yazdanfar, J. K. Barton, and A. J. Welch. 1997. *In vivo* bidirectional color Doppler flow imaging of picoliter blood volumes using optical coherence tomography. *Opt Lett* 22:1439–1441.

Izikson, L., M. Avram, and Z. Tannous. 2008. Treatment of port wine stains with pulsed dye laser in patients with systemic lupus erythematosus: Practical considerations and complications. *J Cosmet Laser Ther* 10:223–225.

Izikson, L., W. Farinelli, F. Sakamoto, Z. Tannous, and R. R. Anderson. 2010. Safety and effectiveness of black tattoo clearance in a pig model after a single treatment with a novel 758 nm 500 picosecond laser: A pilot study. *Lasers Surg Med* 42:640–646.

Jeong, S. Y., J. B. Shin, U. C. Yeo, W. S. Kim, and I. H. Kim. 2010. Low-fluence Q-switched neodymium-doped yttrium aluminum garnet laser for melasma with pre- or post-treatment triple combination cream. *Dermatol Surg* 36:909–918.

Jones, C. E., and K. Nouri. 2006. Laser treatment for pigmented lesions: A review. *J Cosmet Dermatol* 5:9–13.

Jung, J. Y., J. H. Lee, D. J. Ryu et al. 2010. Lower-fluence, higher-density versus higher-fluence, lower-density treatment with a 10,600-nm carbon dioxide fractional laser system: A split-face, evaluator-blinded study. *Dermatologic surgery* 36:2022–2029.

Katz, T. M., L. H. Goldberg, B. F. Firoz, and P. M. Friedman. 2009. Fractional photothermolysis for the treatment of postinflammatory hyperpigmentation. *Dermatol Surg* 35:1844–1848.

Keller, R., W. Belda Junior, N. Y. Valente, and C. J. Rodrigues. 2007. Nonablative 1,064-nm Nd:YAG laser for treating atrophic facial acne scars: Histologic and clinical analysis. *Dermatol Surg* 33:1470–1476.

Kelly, K., editor. 2009. *The History of Medicine: The Middle Ages.* Infobase Publishing, New York.

Kern, S., C. Niemeyer, K. Darge et al. 2000. Differentiation of vascular birthmarks by MR imaging. An investigation of hemangiomas, venous and lymphatic malformations. *Acta Radiol* 41:453–457.

Kilmer, S. L., R. G. Wheeland, D. J. Goldberg, and R. R. Anderson. 1994. Treatment of epidermal pigmented lesions with the frequency-doubled Q-switched Nd:YAG laser. A controlled, single-impact, dose–response, multicenter trial. *Arch Dermatol* 130:1515–1519.

Kim, S., and K. H. Cho. 2010a. Treatment of facial postinflammatory hyperpigmentation with facial acne in Asian patients using a Q-switched neodymium-doped yttrium aluminum garnet laser. *Dermatol Surg* 36:1374–1380.

Kim, S., and K. H. Cho. 2010b. Treatment of procedure-related postinflammatory hyperpigmentation using 1064-nm Q-switched Nd:YAG laser with low fluence in Asian patients: Report of five cases. *J Cosmet Dermatol* 9:302–306.

Klein, A., W. Baumler, M. Landthaler, and P. Babilas. 2011. Laser and IPL treatment of port-wine stains: Therapy options, limitations, and practical aspects. *Lasers Med Sci* 26:845–859.

Koh, J. L., D. Harrison, V. Swanson, D. C. Norvell, and D. C. Coomber. 2007. A comparison of laser-assisted drug delivery at two output energies for enhancing the delivery of topically applied LMX-4 cream prior to venipuncture. *Anesth Analg* 104:847–849.

Kolde, G. 2003. Early pulsed-dye laser treatment of childhood haemangiomas. *Lancet* 361:348–349; author reply 349.

Kono, T., W. Frederick Groff, H. H. Chan, H. Sakurai, and T. Yamaki. 2009. Long-pulsed neodymium:yttrium-aluminum–garnet laser treatment for hypertrophic port-wine stains on the lips. *J Cosmet Laser Ther* 11:11–13.

Kouba, D. J., E. F. Fincher, and R. L. Moy. 2008. Nevus of Ota successfully treated by fractional photothermolysis using a fractionated 1440-nm Nd:YAG laser. *Arch Dermatol* 144:156–158.

Kroon, M. W., B. S. Wind, J. F. Beek et al. 2011. Nonablative 1550-nm fractional laser therapy versus triple topical therapy for the treatment of melasma: A randomized controlled pilot study. *J Am Acad Dermatol* 64:516–523.

Kwok, T., and J. Rao. 2012. Laser management of acne scarring. *Skin Therapy Lett* 17:4–6.

Lacour, M., S. Syed, J. Linward, and J. I. Harper. 1996. Role of the pulsed dye laser in the management of ulcerated capillary haemangiomas. *Arch Dis Child* 74:161–163.

Lanigan, S. W., and S. M. Taibjee. 2004. Recent advances in laser treatment of port-wine stains. *Br J Dermatol* 151:527–533.

Lee, W. R., S. C. Shen, H. H. Lai, C. H. Hu, and J. Y. Fang. 2001. Transdermal drug delivery enhanced and controlled by erbium:YAG laser: A comparative study of lipophilic and hydrophilic drugs. *J Control Release* 75:155–166.

Leuenberger, M. L., M. W. Mulas, T. R. Hata et al. 1999. Comparison of the Q-switched alexandrite, Nd:YAG, and ruby lasers in treating blue-black tattoos. *Dermatol Surg* 25:10–14.

Lewis, A. T., and A. V. Benedetto. 2002. Lasers in dermatology: Unapproved treatments. *Clin Dermatol* 20:700–714.

Loo, W. J., and S. W. Lanigan. 2002. Recent advances in laser therapy for the treatment of cutaneous vascular disorders. *Lasers Med Sci* 17:9–12.

Lou, W. W., A. T. Quintana, R. G. Geronemus, and M. C. Grossman. 2000. Prospective study of hair reduction by diode laser (800 nm) with long-term follow-up. *Dermatol Surg* 26:428–432.

Mahmoud, B. H., D. Srivastava, J. J. Janiga et al. 2010. Safety and efficacy of erbium-doped yttrium aluminum garnet fractionated laser for treatment of acne scars in type IV to VI skin. *Dermatol Surg* 36:602–609.

Maiman, T. 1960. Stimulated optical radiation in ruby. *Nature* 187:493–494.

Manstein, D., G. S. Herron, R. K. Sink, H. Tanner, and R. R. Anderson. 2004. Fractional photothermolysis: A new concept for cutaneous remodeling using microscopic patterns of thermal injury. *Lasers Surg Med* 34:426–438.

Manuskiatti, W., D. Triwongwaranat, S. Varothai, S. Eimpunth, and R. Wanitphakdeedecha. 2010. Efficacy and safety of a carbon-dioxide ablative fractional resurfacing device for treatment of atrophic acne scars in Asians. *J Am Acad Dermatol* 63:274–283.

Marghoob, A. A., C. A. Charles, K. J. Busam et al. 2005. *In vivo* confocal scanning laser microscopy of a series of congenital melanocytic nevi suggestive of having developed malignant melanoma. *Arch Dermatol* 141:1401–1412.

McGill, D. J., C. Hutchison, E. McKenzie, E. McSherry, and I. R. Mackay. 2007. A randomised, split-face comparison of facial hair removal with the alexandrite laser and intense pulsed light system. *Lasers Surg Med* 39:767–772.

Mogensen, M., H. A. Morsy, B. M. Nurnberg, and G. B. Jemec. 2008. Optical coherence tomography imaging of bullous diseases. *J Eur Acad Dermatol Venereol* 22:1458–1464.

Mogensen, M., L. Thrane, T. M. Jorgensen, P. E. Andersen, and G. B. Jemec. 2009. OCT imaging of skin cancer and other dermatological diseases. *J Biophotonics* 2:442–451.

Moreno-Arias, G. A., and A. Camps-Fresneda. 2001. Treatment of nevus of Ota with the Q-switched alexandrite laser. *Lasers Surg Med* 28:451–455.

Mulliken, J. B., and J. Glowacki. 1982. Hemangiomas and vascular malformations in infants and children: A classification based on endothelial characteristics. *Plast Reconstr Surg* 69:412–422.

Munker, R. 2001. Laser blepharoplasty and periorbital laser skin resurfacing. *Facial Plast Surg* 17:209–217.

Na, S. Y., S. Cho, and J. H. Lee. 2012. Intense pulsed light and low fluence Q-switched Nd:YAG laser treatment in melasma patients. *Ann Dermatol* 24:267–273.

Nanni, C. A., and T. S. Alster. 1998. Treatment of a Becker's nevus using a 694-nm long-pulsed ruby laser. *Dermatol Surg* 24:1032–1034.

Nataloni, R. 2003. Laser treatment comparable to oral antibiotics. 532 nm laser addresses multiple acne pathogens. *Dermatol Times* S32, S36.

Nelson, J. S., K. M. Kelly, Y. Zhao, and Z. Chen. 2001. Imaging blood flow in human port-wine stain in situ and in real time using optical Doppler tomography. *Arch Dermatol* 137:741–744.

Nouri, K., and C. J. Ballard. 2006. Laser therapy for acne. *Clin Dermatol* 24:26–32.

Nyren, M., N. Kuzmina, and L. Emtestam. 2003. Electrical impedance as a potential tool to distinguish between allergic and irritant contact dermatitis. *J Am Acad Dermatol* 48:394–400.

Oh, S. H., D. J. Ryu, E. C. Han, K. H. Lee, and J. H. Lee. 2009. A comparative study of topical 5-aminolevulinic acid incubation times in photodynamic therapy with intense pulsed light for the treatment of inflammatory acne. *Dermatol Surg* 35:1918–1926.

Omprakash, H. M. 2002. Treatment of nevus of OTA by Q-switched, frequency doubled, ND:YAG laser. *Indian J Dermatol Venereol Leprol* 68:94–95.

Ong, M. W., and S. J. Bashir. 2012. Fractional laser resurfacing for acne scars: A review. *Br J Dermatol* 166:1160–1169.

Otley, C. C., S. M. Gayner, I. Ahmed et al. 1999. Preoperative and postoperative topical tretinoin on high-tension excisional wounds and full-thickness skin grafts in a porcine model: A pilot study. *Dermatol Surg* 25:716–721.

Paithankar, D. Y., E. V. Ross, B. A. Saleh, M. A. Blair, and B. S. Graham. 2002. Acne treatment with a 1,450 nm wavelength laser and cryogen spray cooling. *Lasers Surg Med* 31:106–114.

Paquette, D., E. Badiavas, and V. Falanga. 2001. Short-contact topical tretinoin therapy to stimulate granulation tissue in chronic wounds. *J Am Acad Dermatol* 45:382–386.

Pellacani, G., A. M. Cesinaro, C. Grana, and S. Seidenari. 2004. *In vivo* confocal scanning laser microscopy of pigmented Spitz nevi: Comparison of *in vivo* confocal images with dermoscopy and routine histopathology. *J Am Acad Dermatol* 51:371–376.

Pellacani, G., A. M. Cesinaro, C. Longo, C. Grana, and S. Seidenari. 2005. Microscopic *in vivo* description of cellular architecture of dermoscopic pigment network in nevi and melanomas. *Arch Dermatol* 141:147–154.

Poetke, M., C. Philipp, and H. P. Berlien. 2000. Flashlamp-pumped pulsed dye laser for haemangiomas in infancy: Treatment of superficial vs mixed haemangiomas. *Arch Dermatol* 136:628–632.

Rabe, J. H., A. J. Mamelak, P. J. McElgunn, W. L. Morison, and D. N. Sauder. 2006. Photoaging: Mechanisms and repair. *J Am Acad Dermatol* 55:1–19.

Raulin, C., S. Hellwig, and M. P. Schonermark. 1997. Treatment of a nonresponding port-wine stain with a new pulsed light source (PhotoDerm VL). *Lasers Surg Med* 21:203–208.

Renfro, L., and R. G. Geronemus. 1993. Anatomical differences of port-wine stains in response to treatment with the pulsed dye laser. *Arch Dermatol* 129:182–188.

Rokhsar, C. K., and D. H. Ciocon. 2009. Fractional photothermolysis for the treatment of postinflammatory hyperpigmentation after carbon dioxide laser resurfacing. *Dermatol Surg* 35:535–537.

Ross, V., G. Naseef, G. Lin et al. 1998. Comparison of responses of tattoos to picosecond and nanosecond Q-switched neodymium: YAG lasers. *Arch Dermatol* 134:167–171.

Sadick, N. 2010. An open-label, split-face study comparing the safety and efficacy of levulan kerastick (aminolevulonic acid) plus a 532 nm KTP laser to a 532 nm KTP laser alone for the treatment of moderate facial acne. *J Drugs Dermatol* 9:229–233.

Salvini, C., D. Massi, A. Cappetti et al. 2008. Application of optical coherence tomography in non-invasive characterization of skin vascular lesions. *Skin Res Technol* 14:89–92.

Scheepers, J. H., and A. A. Quaba. 1995. Does the pulsed tunable dye laser have a role in the management of infantile hemangiomas? Observations based on 3 years' experience. *Plast Reconstr Surg* 95:305–312.

Scheibner, A., G. Kenny, W. White, and R. G. Wheeland. 1990. A superior method of tattoo removal using the Q-switched ruby laser. *J Dermatol Surg Oncol* 16:1091–1098.

Scherer, K., S. Lorenz, M. Wimmershoff, M. Landthaler, and U. Hohenleutner. 2001. Both the flashlamp-pumped dye laser and the long-pulsed tunable dye laser can improve results in port-wine stain therapy. *Br J Dermatol* 145:79–84.

Seaton, E. D., A. Charakida, P. E. Mouser et al. 2003. Pulsed-dye laser treatment for inflammatory acne vulgaris: Randomised controlled trial. *Lancet* 362:1347–1352.

Sharma, S., A. K. Jha, and S. K. Mallik. 2011. Role of q-switched Nd:YAG laser in nevus of ota: A study of 25 cases. *Indian J Dermatol* 56:663–665.

Silver, B. E., and Y. L. Livshots 1996. Preliminary experience with the KTP/532 nm laser in the treatment of facial telangiectasias. *Cosmet Dermatol* 9:61–64.

Smith, K. C., and G. D. Schachter. 2011. YSGG 2790-nm superficial ablative and fractional ablative laser treatment. *Facial Plast Surg Clin North Am* 19:253–260.

Stam-Posthuma, J. J., J. Vink, S. le Cessie et al. 1998. Effect of topical tretinoin under occlusion on atypical naevi. *Melanoma Res* 8:539–548.

Suh, K. S., J. Y. Sung, H. J. Roh et al. 2011. Efficacy of the 1064-nm Q-switched Nd:YAG laser in melasma. *J Dermatolog Treat* 22:233–238.

Taieb, A., L. Touti, M. Cony et al. 1994. Treatment of port-wine stains with the 585-nm flashlamp-pulsed tunable dye laser: A study of 74 patients. *Dermatology* 188:276–281.

Tajirian, A. L., and D. J. Goldberg. 2011. Fractional ablative laser skin resurfacing: A review. *J Cosmet Laser Ther* 13:262–264.

Tan, O. T., S. Murray, and A. K. Kurban. 1989. Action spectrum of vascular specific injury using pulsed irradiation. *J Invest Dermatol* 92:868–871.

Tan, O. T., K. Sherwood, and B. A. Gilchrest. 1989. Treatment of children with port-wine stains using the flashlamp-pulsed tunable dye laser. *N Engl J Med* 320:416–421.

Tanzi, E. L., J. R. Lupton, and T. S. Alster. 2003. Lasers in dermatology: Four decades of progress. *J Am Acad Dermatol* 49:1–31; quiz 31–34.

Taub, A. F. 2007. A comparison of intense pulsed light, combination radiofrequency and intense pulsed light, and blue light in photodynamic therapy for acne vulgaris. *J Drugs Dermatol* 6:1010–1016.

Taylor, C. R., and R. R. Anderson. 1994. Ineffective treatment of refractory melasma and postinflammatory hyperpigmentation by Q-switched ruby laser. *J Dermatol Surg Oncol* 20:592–597.

Taylor, C. R., R. W. Gange, J. S. Dover et al. 1990. Treatment of tattoos by Q-switched ruby laser. A dose–response study. *Arch Dermatol* 126:893–899.

Trelles, M. A., I. Allones, M. Velez, and G. A. Moreno-Arias. 2004. Becker's nevus: Erbium:YAG versus Q-switched neodimium:YAG? *Lasers Surg Med* 34:295–297.

Trelles, M. A., M. Velez, and M. H. Gold. 2010. The treatment of melasma with topical creams alone, CO2 fractional ablative resurfacing alone, or a combination of the two: A comparative study. *J Drugs Dermatol* 9:315–322.

Tse, Y., V. J. Levine, S. A. McClain, and R. Ashinoff. 1994. The removal of cutaneous pigmented lesions with the Q-switched ruby laser and the Q-switched neodymium: yttrium–aluminum–garnet laser. A comparative study. *J Dermatol Surg Oncol* 20:795–800.

Tseng, S. H., P. Bargo, A. Durkin, and N. Kollias. 2009. Chromophore concentrations, absorption and scattering properties of human skin *in-vivo*. *Opt Express* 17:14599–14617.

Wang, H. W., and K. Y. Liu. 2009. [Efficacy and safety of low-energy QS Nd:YAG and QS alexandrite laser for melasma]. *Zhongguo Yi Xue Ke Xue Yuan Xue Bao* 31:45–47.

Watanabe, S., and H. Takahashi. 1994. Treatment of nevus of Ota with the Q-switched ruby laser. *N Engl J Med* 331:1745–1750.

Weiss, R. A., A. C. Harrington, R. C. Pfau, M. A. Weiss, and S. Marwaha. 1999. Periorbital skin resurfacing using high energy erbium:YAG laser: Results in 50 patients. *Lasers Surg Med* 24:81–86.

Welzel, J. 2001. Optical coherence tomography in dermatology: A review. *Skin Res Technol* 7:1–9.

Welzel, J., M. Bruhns, and H. H. Wolff. 2003. Optical coherence tomography in contact dermatitis and psoriasis. *Arch Dermatol Res* 295:50–55.

Welzel, J., E. Lankenau, R. Birngruber, and R. Engelhardt. 1997. Optical coherence tomography of the human skin. *J Am Acad Dermatol* 37:958–963.

Welzel, J., E. Lankenau, R. Birngruber, and R. Engelhardt. 1998. Optical coherence tomography of the skin. *Curr Probl Dermatol* 26:27–37.

West, T. B., and T. S. Alster. 1998a. Comparison of the long-pulse dye (590–595 nm) and KTP (532 nm) lasers in the treatment of facial and leg telangiectasias. *Dermatol Surg* 24:221–226.

West, T. B., and T. S. Alster. 1998b. Improvement of infraorbital hyperpigmentation following carbon dioxide laser resurfacing. *Dermatol Surg* 24:615–616.

Whang, K. K., J. Y. Byun, and S. H. Kim. 2009. A dual-wavelength approach with 585-nm pulsed-dye laser and 800-nm diode laser for treatment-resistant port-wine stains. *Clin Exp Dermatol* 34:e436–e437.

Wlotzke, U., U. Hohenleutner, T. A. Abd-El-Raheem, W. Baumler, and M. Landthaler. 1996. Side-effects and complications of flashlamp-pumped pulsed dye laser therapy of port-wine stains. A prospective study. *Br J Dermatol* 134:475–480.

Yang, M. U., A. N. Yaroslavsky, W. A. Farinelli et al. 2005. Long-pulsed neodymium:yttrium-aluminum-garnet laser treatment for port-wine stains. *J Am Acad Dermatol* 52:480–490.

Yilmaz, O., N. Senturk, E. P. Yuksel et al. 2011. Evaluation of 532-nm KTP laser treatment efficacy on acne vulgaris with once and twice weekly applications. *J Cosmet Laser Ther* 13:303–307.

Zaret, M. M., G. M. Breinin, H. Schmidt et al. 1961. Ocular lesions produced by an optical maser (laser). *Science* 134:1525–1526.

Zelickson, B. D., D. A. Mehregan, A. A. Zarrin et al. 1994. Clinical, histologic and ultrastructural evaluation of tattoos treated with three laser systems. *Lasers Surg Med* 15:364–372.

Zenzie, H. H., G. B. Altshuler, M. Z. Smirnov, and R. R. Anderson. 2000. Evaluation of cooling methods for laser dermatology. *Lasers Surg Med* 26:130–144.

Zhao, S., Y. Gu, P. Xue et al. 2010. Imaging port wine stains by fiber optical coherence tomography. *J Biomed Opt* 15:036020.

60

Therapeutic Uses of Lasers in Eye Care

60.1 Introduction

There has been a revolution in ophthalmic diagnostics and therapeutics since the invention of lasers in 1960. This chapter will summarize the most common therapeutic uses of ophthalmic lasers, starting at the front of the eye (anterior segment) with corneal refractive uses, lenticular refractive and cataract uses, and glaucoma treatments, and then moving to the back of the eye (posterior segment) to look at the treatment of retinal disorders.

Gas, liquid, and solid state lasers have been used to treat ocular conditions. Laser eye surgery employs the three laser damage mechanisms: photothermal, photochemical, and photomechanical (photodisruptive) damage (Wormington 2003).

The majority of ophthalmic lasers use a slit lamp biomicroscope to deliver laser energy to the eye. The laser energy is directed from the laser via a fiber optic cable or mirrors in an articulated arm to the slit lamp biomicroscope and then into the patient's eye. If the output of the laser is not visible (e.g., 1064 nm in an Nd:YAG laser), then a different laser with a visible output is used to focus the invisible laser beam on the targeted tissue. This additional aiming laser is either a helium–neon (He–Ne) laser (632.8 nm) or more often a red diode laser (e.g., 635, 640, or 670 nm). Some of the slit lamp delivery systems utilize contact lenses with mirrors or focusing lenses to reach the target tissue. Other delivery systems include binocular indirect ophthalmoscopes, scanning systems, external contact probes, or internal endoscopic probes.

The variable controls on ophthalmic lasers can include laser spot size (e.g., µm), pulse duration (e.g., s, ms, ns, or fs), energy (e.g., mJ or J) on pulsed lasers, and power (e.g., mW or W) on continuous wave (CW) lasers.

Table 60.1 gives a summary of the primary characteristics and applications of the more common ophthalmic therapeutic lasers.

60.2 Anterior Segment

60.2.1 Corneal Refractive Procedures

Laser refractive eye surgery is used to change and improve the refractive state of the eye and to decrease dependency on glasses or contact lenses. Thus, it means that a laser can be used to treat myopia, hyperopia, and astigmatism. This can be accomplished by using the laser to remodel the cornea or perform cataract surgery. Today the most common lasers used in these procedures include the excimer and the femtosecond lasers (FSLs).

There has been a steady evolution of laser refractive surgery since the 1980s. New techniques of laser surgery began with photorefractive keratectomy (PRK), followed by the advent of laser *in situ* keratomileusis (LASIK), laser epithelial (or subepithelial) keratomileusis (LASEK), and epipolis- (epi-) LASIK. This evolution has also involved the use of advanced software, larger ablation zones, improved anterior segment imaging, wavefront and topography technology, and accurate eye-tracking. Because of these changes, there has been an increase in efficacy, safety, and predictability of the surgical outcome.

The primary use of lasers in the anterior segment is for corneal refractive or glaucoma procedures. The corneal refractive procedures can be broken down into surface treatment techniques and lamellar (flap) treatment techniques.

60.2.1.1 Surface Treatment Techniques

By using the ArF excimer laser with a wavelength of 193 nm and a pulse duration of 10–20 ns, it was discovered that the corneal surface could be ablated, removing about 0.25 µm of corneal stroma per laser pulse (Preferred Practice Pattern Guidelines, Refractive Errors & Refractive Surgery 2007; Taneri, Weisberg, and Azar 2011). The damage mechanism is photochemical and

TABLE 60.1 Summary of Common Therapeutic Ophthalmic Lasers: Primary Characteristics and Applications

Laser Type	Wavelength (nm)	Active Medium	Temporal Mode	Primary Damage Mechanism	Possible Applications
Helium–neon (He–Ne)	543.5 (green) 632.8 (red)	Gas: Helium and Neon	CW	Low-power	Aiming lasers Laser pointers
Argon (Ar)	488 (blue) 514.5 (green)	Gas: Argon	CW	Photothermal Photochemical	Laser iridotomy Laser iridoplasty Laser trabeculoplasty Laser suture lysis Laser sclerostomy Cyclophotocoagulation Retinal photocoagulation PDT Oculoplastic surgery
Krypton (Kr)	531 (green) 568 (yellow) 647 (red)	Gas: Krypton	CW	Photothermal	Laser iridotomy Laser trabeculoplasty Retinal photocoagulation PDT Oculoplastic surgery
Carbon dioxide (CO_2)	10,600 (far-IR)	Gas: CO_2	CW	Photothermal	Laser phacolysis Laser sclerostomy Oculoplastic surgery
Excimer (ArF)	193 (UV)	Gas: Argon fluoride	Pulsed	Photochemical	LASIK PRK PTK LASEK Laser trabeculotomy Laser sclerostomy
Nd:YAG	1064 (near-IR) Frequency-doubled, 532 (green)	Solid state: Neodymium ions in yttrium, aluminum, garnet matrix	CW Pulsed: Q-switched	Photodisruption (pulsed) Photothermal (CW)	Posterior capsulotomy Laser iridotomy Laser trabeculoplasty Laser goniopuncture Laser phacolysis Laser sclerostomy Cyclophotocoagulation Retinal photocoagulation Oculoplastic surgery

Laser	Lasing medium	Wavelength (nm)	Mode	Interaction	Applications
Various solid state and diode	Solid state (e.g., Nd:YVO₄) Diode (e.g., gallium aluminum arsenide)	532 (green) 577 (yellow) 689 (red) 810 (near-IR) 1030 (near-IR) 1040 (near-IR) 1043 (near-IR) 1045 (near-IR) 1053 (near-IR) 1064 (near-IR) Other visible and near-IR wavelengths	CW Pulsed (e.g., fs)	Photodisruption (pulsed) Photothermal (CW)	Laser iridotomy; Laser iridoplasty; Laser sclerostomy; Laser suture lysis; Laser trabeculoplasty; Cyclophotocoagulation; LASIK flap; Penetrating keratoplasty; Lamellar keratoplasty; Endothelial keratoplasty; Tunnels for ICRS; Astigmatic keratotomy; Intrastromal presbyopia correction; Lenticule extraction; Anterior capsulorhexis; Lens fragmentation; Posterior capsulotomy; Clear corneal incisions; Retinal photocoagulation; PDT; Oculoplastic surgery; Aiming lasers; Laser pointers
Dye	Fluorescent dyes	310–1200 (UV, visible, IR)	CW pulsed	Photothermal Photochemical	Laser iridotomy; Laser sclerostomy; Laser suture lysis retinal photocoagulation; PDT; Oculoplastic surgery

Source: Updated and adapted from Wormington, C. M., *Ophthalmic Lasers*, Butterworth-Heinemann, Philadelphia, 2003.

Note: CW = continuous wave, ICRS = intracorneal ring segments, IR = infrared, LASEK = laser epithelial keratomileusis, LASIK = laser in situ keratomileusis, PDT = photodynamic therapy, PRK = photorefractive keratectomy, PTK = phototherapeutic keratectomy, UV = ultraviolet.

photoablative. This allowed fairly precise corneal reshaping to decrease the radius of corneal curvature and thus correct low and moderate levels of myopia. This first technique, PRK, was approved by the FDA in 1995. Later developments allowed the correction of hyperopia and astigmatism.

In PRK, the central corneal epithelium is removed chemically (e.g., with alcohol), mechanically (e.g., with a blade or brush), or by laser (Taneri, Weisberg, and Azar 2011). The excimer laser is then used to ablate the corneal stroma. For example, by flattening the curved stroma to a given depth, a myopic correction can be accomplished. The cornea is then allowed to heal with new epithelial cells migrating to cover the corneal surface. Postprocedure problems include photophobia, tearing, halos, and significant pain in the early stages of healing. Possible complications include corneal scarring, subepithelial haze formation, induced irregular or regular astigmatism, corneal ectasia, development or exacerbation of dry eye symptoms, decreased corneal sensitivity, recurrent corneal erosion, symptomatic overcorrection/undercorrection, partial regression of effect, visual aberrations (e.g., glare), decreased contrast sensitivity, and corneal infiltrates (Preferred Practice Pattern Guidelines, Refractive Errors & Refractive Surgery 2007). Some of these complications can lead to a loss of two or more lines of best spectacle-corrected visual acuity (BSCVA) in up to about 1% of patients at 1 year following the procedure (Shortt and Allan 2006).

60.2.1.2 Lamellar (Flap) Treatment Techniques

LASIK was developed to minimize pain and delayed corneal healing, as well as to allow treatment of higher myopic levels than could be achieved with PRK (Barsam and Allan 2012). It has become the most common laser refractive technique. In LASIK, a microkeratome or an FSL is used to create a flap of corneal tissue (Kymionis et al. 2012). Like a hinged door, the flap is lifted, the excimer laser is used to ablate and recontour the corneal stroma, and then the flap is replaced. The final uncorrected visual acuity after the LASIK and PRK procedures is comparable at 12 months postoperatively (Shortt and Allan 2006). The loss of two or more lines of BSCVA in LASIK and PRK is also comparable (Chen et al. 2012).

LASEK combines some features of PRK and LASIK (Preferred Practice Pattern Guidelines, Refractive Errors & Refractive Surgery 2007). It attempts to preserve the epithelium by using alcohol to loosen it, and then an epithelial trephine and spatula are used to roll up the epithelium, leaving it attached with a superior or nasal hinge. After the photoablation of the stroma, the epithelium is unrolled back over the ablated corneal stroma, and a bandage contact lens is placed on the cornea. Clinical outcomes of PRK and LASEK are similar. However, postoperative pain and the time for reepithelialization are reduced in LASEK compared to PRK.

Epi-LASIK is an alternative procedure developed to minimize some of the problems with PRK and LASIK (Preferred Practice Pattern Guidelines, Refractive Errors & Refractive Surgery 2007). Instead of using alcohol to loosen the epithelium as in LASEK, an epikeratome with a blunt blade is used to lift the epithelial layer away from Bowman's membrane. After using the

excimer laser to ablate Bowman's membrane and part of the corneal stroma, the epithelial layer is replaced.

60.2.1.3 Customized Laser Vision Correction

Wavefront technology has impacted the delivery of laser refractive procedures (Smadja et al. 2012). The total aberration of the eye can be represented by the shape of the wavefront that in turn can be described mathematically with a series of polynomials (Zernike polynomials). The standard clinical refraction deals with the sphere, cylinder, and axis, and these are called lower-order aberrations. There are also other optical aberrations that can contribute to a decrease in the patient's vision or an increase in visual symptoms. These other aberrations are termed higher-order aberrations (HOAs), and they consist of terms that describe conditions like spherical aberration and coma.

All of the laser vision correction procedures attempt to correct the lower-order aberrations. The conventional or standard LASIK procedure attempts to minimize just the patient's spherocylindrical error (i.e., the lower-order aberrations). Wavefront-optimized LASIK uses the wavefront data to minimize the spherical aberration (one of the HOAs) induced by the standard or conventional LASIK procedure (El Awady, Ghanem, and Saleh 2011; Smadja et al. 2012, 2013). Wavefront-guided LASIK goes further by incorporating more HOAs and attempts to use the wavefront aberrometer measurements to determine the laser ablation profile (Smadja et al. 2012). Topography-guided LASIK uses data from the spherocylindrical correction and the shape of the cornea (obtained by a corneal topography instrument) to determine the laser ablation profile (El Awady, Ghanem, and Saleh 2011). A new approach, optical ray-tracing guided LASIK, involves development of an individualized eye model using an optical ray-tracing algorithm (Smadja et al. 2012). This eye model takes into account not only wavefront data on the whole eye but also the curvature and shape of the anterior and posterior corneal surfaces, corneal and lens thicknesses, axial ocular length, anterior chamber depth, and data associated with the lens surfaces. Initial findings look promising.

60.2.1.4 Intraoperative Complications

Most of the intraoperative complications are associated with the use of a mechanical microkeratome (Preferred Practice Pattern Guidelines, Refractive Errors & Refractive Surgery 2007; Shah, Shah, and Vogelsang 2011; Zhang, Chen, and Xia 2013). In those procedures that require the creation of a corneal flap, there is a risk of malfunction of the microkeratome and the production of a buttonhole flap, a flap without a hinge (free-cap), or an irregular flap (e.g., a flap that is too thick or wrinkled). Other problems include epithelial defect formation and a decentered ablation. Most of these complications are minimized when an FSL is used instead of a microkeratome. This is largely because the FSL is more precise and accurate than a microkeratome. Another possible complication with the FSL is the creation of an opaque bubble layer (OBL) (Kymionis et al. 2012; Soong and Malta 2009). This involves the accumulation of gas bubbles above and below the resection plane in the intralamellar space. This may result in problems using the laser system eye tracker and when lifting the flap. The OBL dissipates in about 30 min.

60.2.1.5 Postoperative Complications

The most frequent complication of laser vision correction is dry eye, especially after LASIK where the corneal nerves are transected (Shtein 2011). This is often transient and usually improves in the weeks and months after the procedure. Other complications that can occur include pain, epithelial ingrowth, scarring, haze, undercorrection, overcorrection, partial regression of effect, induced irregular astigmatism, halos, glare, corneal ectasia, dislocation of the flap, flap stria, infectious keratitis, pressure-induced stromal keratopathy, central toxic keratopathy and diffuse lamellar keratitis (DLK; this is more common when the FSL is used to create a flap) (Binder 2010; Chen et al. 2012; Preferred Practice Pattern Guidelines, Refractive Errors & Refractive Surgery 2007; Randleman and Shah 2012).

With use of the FSL, there is a possibility of a late-onset transient photophobia (transient light-sensitivity syndrome) (Kymionis et al. 2012). This consists of mild pain and photophobia, even though there is decent uncorrected distance visual acuity and an unremarkable slit lamp examination (Preferred Practice Pattern Guidelines, Refractive Errors & Refractive Surgery 2007). This syndrome usually disappears in a few weeks after surgery.

Fortunately, serious complications from laser vision correction are fairly rare. One of the most serious complications would be loss of two or more lines of best spectacle-corrected vision, and this occurs in from 0% to 3% of LASIK procedures (Preferred Practice Pattern Guidelines, Refractive Errors & Refractive Surgery 2007).

One of the reasons for dissatisfaction with laser vision correction is residual refractive error. In other words, in some patients, the refractive error was not completely eliminated, and the patient may not see 20/20 or better without glasses. Even with successful vision correction, some patients are dissatisfied due to dry eyes or night vision problems (e.g., halos, starbursts, and/or glare).

60.2.1.6 FSL-Assisted Corneal Surgery

Ophthalmic FSLs (fs = 10^{-15} s) are diode-pumped solid state lasers. Depending on the manufacturer, they have a pulse width of from 200 to ~800 fs. Their output is in the near-infrared in the region of 1020–1060 nm with a pulse energy of a few hundred nanojoules to a few microjoules (Binder 2010). Also depending on the manufacturer, their repetition rate is from 12 kHz to ~21 MHz.

The tissue interaction and damage mechanism of FSLs is photomechanical (Soong and Malta 2009). They produce photodisruption because the energy of the laser is delivered in a few hundred femtoseconds with a laser beam spot size of ~1 μm. This results in tremendous power delivery with many photons hitting the tissue at the same time and thus producing optical breakdown. The plasma that is created vaporizes a small amount of tissue and produces a shockwave and then a cavitation bubble.

That bubble expands at supersonic speed, then it slows down and contracts. This process leaves a very small gas bubble. These cavitation bubbles create a tissue cleavage plane, and hence the ability to create a precise corneal flap or a corneal incision. Because the energy is delivered in an ultrashort time, there is minimal damage to collateral tissue, significantly less damage than occurs with the nanosecond pulses of a Q-switched Nd:YAG laser. The laser pulse duration is much shorter than the tissue thermal relaxation time. So there is essentially no thermal damage to adjacent tissue. Also because of the extremely short duration of the laser pulse, the amount of laser energy delivered can be decreased in comparison with the amount of energy delivered in a nanosecond pulse of the Nd:YAG laser. This is because the threshold for optical breakdown is proportional to the square root of the pulse duration (Juhasz et al. 1996; Sun et al. 2007).

A major milestone in the evolution of corneal refractive surgery has been the introduction of FSLs. Over the last decade, the use of FSLs to create a corneal flap in LASIK surgery has almost replaced the use of a mechanical microkeratome to perform that task (Kymionis et al. 2012). The FSLs provide improved versatility, reproducibility, precision, biomechanical flap stability, and safety (Chen et al. 2012; Kim, Sutton, and Rootman 2011). Using the FSLs allows more precise specification of the diameter, depth, and hinge of the flap. A variation of this technique that involves creation of a very thin flap (e.g., 90 μm flap) is called sub-Bowman keratomileusis (SBK) (Prakash et al. 2011).

The characteristics of ophthalmic FSLs also allow them to be used for anterior and posterior lamellar keratoplasty, Descemet-stripping endothelial keratoplasty (DSEK), customized-trephination penetrating keratoplasty, astigmatic keratectomy/keratotomy, tunnel creation for intrastromal ring segments, implantation of a keratoprosthesis, and lenticule extraction (He, Sheehy, and Culbertson 2011; Farid and Steinert 2010).

In penetrating keratoplasty (PKP, "corneal transplant"), the FSL may be used for trephination of both donor and patient corneas generating better centered and regular cuts than the traditional mechanical trephination. FSLs can be used for creating reproducible, interlocking graft configurations that may result in self-adhesive, sutureless penetrating keratoplasty. A number of biomechanically more stable wound configurations can be generated with the FSL leading to faster wound healing. Astigmatic keratectomy/keratotomy can be used to correct high astigmatism resulting from penetrating keratoplasty, as well as from deep anterior lamellar keratoplasty (DALK) and DSEK (Kim, Sutton, and Rootman 2011; Kymionis et al. 2012).

Intracorneal ring segments (ICRS) are clear, curved polymethylmethacrylate segments that can be implanted in peripheral corneal tunnels made with an FSL. The ICRS can be used to correct mild to moderate myopia and have been used in the treatment of post-LASIK corneal ectasia, keratoconus, and pellucid marginal degeneration (Kim, Sutton, and Rootman 2011; Kymionis et al. 2012). A variation on this technique involved ICRS implantation followed later by same-day topography-guided PRK and corneal collagen cross-linking (CXL) to treat selected keratoconus patients (Iovieno et al. 2011). There is also a report of just

PRK and CXL (Kymionis et al. 2011). The PRK improves vision by treating the irregular astigmatism. The use of CXL involves the application of riboflavin molecules to the cornea with subsequent exposure to Ultraviolet A (UVA) radiation. This process strengthens the cornea by inducing interfibrillar cross-links of the stromal collagen fibers. Hence, this inhibits or stops the progression of the thinning of the cornea in keratoconus.

A promising new form of corneal refractive surgery employs only the FSL without the use of the excimer laser to ablate stromal tissue (Kymionis et al. 2012; Sekundo, Kunert, and Blum 2011; Shah, Shah, and Vogelsang 2011). This new technique involves removal of corneal tissue (a lenticule) with or without the making of a flap. The FSL can make precise incisions in three dimensions under computer control. Thus, the anterior and posterior surfaces of the lenticule can be made with curves that correct for a given amount of refractive error. Femtosecond lenticule extraction (FLEx) utilizes an FSL to carve out an intrastromal corneal lenticule and make a hinged flap. The flap is then raised, the lenticule is removed, and the flap is laid back down. Small incision lenticule extraction (SMILE) involves the creation of a corneal lenticule without a flap. The FSL can be used to carve out the intrastromal corneal lenticule and make a small incision in the cornea through which the lenticule can be extracted. This method reduces postoperative irritation, due to the small corneal incision, and it lessens the adverse effect on corneal sensibility and tear production because the procedure transects fewer corneal nerves.

60.2.1.7 Procedures for Correction of Presbyopia

New options for the treatment of presbyopia with lasers are emerging. Besides monovision LASIK where one eye is corrected for near vision and the other eye is corrected for distance vision, the excimer laser can also be used to make multifocal corneal ablations (Uthoff et al. 2012). A new option using an FSL is called INTRACOR (Holzer et al. 2012). The FSL is used to deliver laser photodisruptive pulses entirely within the corneal stroma in a series of cylindrical rings. These rings start in the posterior stroma and end in the anterior stroma. They result in a multifocal, hyperprolate corneal shape change that increases the depth of focus and hence improves near vision with a minimal effect on distance vision. The procedure is fast, is painless, and has few complications.

Different types of corneal inlays have also been used to treat presbyopia (Limnopoulou et al. 2013). One promising technique involves insertion of a corneal inlay in the nondominant eye under the flap created for simultaneous LASIK (Tomita et al. 2012). The inlay can be an opaque material with a central small aperture (e.g., 1.6 mm diameter). Better near vision results from the increase in the depth of focus produced by the small aperture, like the pinhole effect obtained with a small aperture for an f-stop camera. A variation of that technique involves insertion of the corneal inlay in a small intrastromal pocket. Some advantages of the procedure are that the technique is reversible and minimally invasive, and the small pocket minimizes damage to corneal nerves and alterations in the biomechanical properties of the cornea.

Another technique for treating presbyopia is to use the FSL to increase the elasticity of aging crystalline lenses, hence partially restoring accommodation (Reggiani Mello and Krueger 2011). The laser is used to create microincisions that increase the sliding of the lens fibers inside the aging lens. The process creates bubbles inside the lens that gradually fade away after a few days. The early results are promising.

60.2.1.8 Phototherapeutic Keratectomy

The excimer laser can be used to treat pathologies that occur in the anterior portion of the cornea (Rapuano 2011). These include epithelial basement membrane dystrophy, superficial corneal dystrophies, and anterior corneal opacities and scars. By ablation of thin layers of corneal tissue, phototherapeutic keratectomy (PTK) can eliminate elevated corneal lesions and remove anterior stromal opacities while leaving a smooth corneal surface. Some of the complications that can occur with PTK are delayed healing of the epithelium, pain, infection, haze, corneal scarring, induced refractive errors (e.g., hyperopia), glare, and recurrence of the primary corneal disease (Rapuano 2011).

60.2.2 Cataract Procedures

Cataracts are the leading cause of blindness in the world and the leading cause of visual impairment in Americans of European, African, and Hispanic/Latino descent (American Academy of Ophthalmology Cataract and Anterior Segment Panel 2011). Cataract surgery is the most frequently performed surgery in the United States, and it is undergoing a significant evolution because of FSLs (Roberts et al. 2013a,b). FSLs are very promising as a tool for assisting or performing cataract surgery. Currently, cataract surgery involves the manual formation of a circular hole in the anterior lens capsule (capsulorhexis, a continuous curvilinear capsulotomy), the breaking up (phacoemulsification) of the cataractous lens using ultrasound energy, aspiration of the fragments, and then insertion of a synthetic intraocular lens (IOL). The use of an FSL automates some of the steps in cataract surgery and makes them easier and more precise. These steps include the making of the anterior capsulorhexis, the sectioning and softening of the crystalline lens (laser lens fragmentation), the cutting of a clear corneal incision to access the anterior chamber, and the cutting of corneal- or limbal-relaxing incisions to treat astigmatism. Besides enhancing accuracy, the FSL may enhance safety and refractive outcomes as well.

The FSL capsulorhexis is more centered, reproducible, and precisely sized than when doing a manual capsulorhexis with freehand tearing and pulling of the capsule. Current FSL systems are optically guided, using either optical coherence tomography (OCT) or Scheimpflug technology to image the anterior segment of the eye (He, Sheehy, and Culbertson 2011). This allows the determination of the position of the cornea and its thickness, the iris boundaries, the anterior chamber depth, the iridocorneal angle, and the position and thickness of the lens. This information increases the precision and reproducibility of the laser incisions, resulting in a more centered capsulorhexis and better IOL centration. This advantage is especially important for successful implantation of the premium advanced-technology IOLs (e.g., multifocal, accommodative, toric, or aspheric IOLs).

The FSL can also be used for lens fragmentation and laser-assisted phacofragmentation. The ultrasound energy used to break up the cataract in conventional cataract surgery can cause injury to the corneal endothelial cells. The use of the FSL can supplant the use of ultrasound energy in softer cataracts or reduce the amount of ultrasound energy necessary to break up harder, denser cataracts (Conrad-Hengerer et al. 2012). This can result in fewer complications.

The FSL capsulorhexis and lens fragmentation can be done before opening the eye. Then the FSL can be used to make a self-sealing, clear corneal incision to introduce the tools for phacoemulsification and/or aspiration. The laser lens fragmentation minimizes the risk of corneal endothelial injury and capsule complications partly because it minimizes the amount of ultrasound energy necessary for the surgery.

There is also an interesting report of the use of the FSL to photobleach the age-associated yellowing of human donor lenses (Kessel et al. 2010). The laser intensity used for this photolysis was significantly below the threshold for photomechanical effects. Instead, the laser used a photochemical mechanism to bleach the lens. This resulted in an increase in light transmission through the lens. Further work needs to be done to explore the potential of this method for noninvasive cataract treatment.

60.2.2.1 Laser Posterior Capsulotomy

In cataract surgery, the posterior capsule and part of the anterior capsule that surrounds the crystalline lens are usually left intact. Then an IOL is implanted in this capsular bag after the cataract has been extracted. Following the surgery, the most common complication is opacification of the posterior capsule by proliferation of residual epithelial cells. This occurs in up to 50% of patients during the first 5 years after surgery and can be even higher in children (American Academy of Ophthalmology Cataract and Anterior Segment Panel 2011; Wormington 2003). Newer IOL designs can result in a significant decrease in this percentage.

This opacification can cause decreased visual acuity and functional impairment. The indication for surgery is interference with the patient's visual needs but with the potential for functional benefit. A laser posterior capsulotomy is the usual treatment, carried out with the Q-switched Nd:YAG laser (American Academy of Ophthalmology Cataract and Anterior Segment Panel 2011; Wormington 2003). The laser output is a near-infrared (1064 nm wavelength) beam with a pulse duration of about 5 ns, a spot size of 7–20 μm diameter, and an energy of about 1 mJ per pulse. The beam is focused just behind the opacified posterior capsule. When the pulse is fired, optical breakdown occurs due to multiphoton absorption, and this creates a plasma just behind the posterior capsule. The rapidly expanding plasma generates a shock wave, and this shock wave creates a hole in the opacified capsule. This typically restores the patient's vision.

Some of the complications of posterior capsulotomy include IOL pits and cracks, a spike in IOP, inflammation/uveitis, cystoid macular edema, vitreous prolapse, and retinal detachment (American Academy of Ophthalmology Cataract and Anterior Segment Panel 2011; Wormington 2003). Most of these problems can be treated with medications or, if necessary, surgery.

60.2.3 Glaucoma Procedures

Glaucoma is the second leading cause of blindness in the world after cataract (Quigley and Broman 2006). Glaucoma is a group of diseases that can be defined as "an optic neuropathy associated with characteristic structural damage to the optic nerve and associated visual dysfunction that may be caused by various pathological processes" (Foster et al. 2002). Glaucoma is often associated with increased IOP, but it can also occur when IOP is statistically normal. Primary open-angle glaucoma is found in patients with open anterior chamber drainage angles and no identifiable secondary cause. Primary angle-closure glaucoma is found in patients who have narrow occludable drainage angles and no identifiable secondary cause. To slow the progression of glaucoma, the current main treatment is to reduce IOP whether it is high or within statistically normal levels. In open-angle glaucoma, the initial treatment is usually medical (topical and/or oral drugs). If that fails to reduce IOP or if the patient cannot or will not use medications, then laser trabeculoplasty is often used. If that fails, surgical treatment can be performed (e.g., trabeculectomy). As a last resort, cyclodestructive treatment can be used for refractory glaucoma. In closed-angle glaucoma, the main treatment is a laser peripheral iridotomy (LPI). We will review a few of the more common laser methods of reducing IOP in patients with glaucoma.

60.2.3.1 Argon and Diode Laser Trabeculoplasty

Argon laser trabeculoplasty (ALT) uses an argon laser to place a series of laser burns in the trabecular meshwork (the peripheral drainage area of the eye in front of the iris and behind the cornea). This area of the anterior segment of the eye is termed "the angle," formed in the periphery where the cornea intersects the iris. The indication for ALT is open-angle glaucoma where IOP is not low enough to prevent progression of damage. ALT can be used to improve aqueous outflow and hence decrease IOP. ALT is more effective in some types of open-angle glaucoma than it is in others (Rolim de Moura, Paranhos, Jr., and Wormald 2007; Samples et al. 2011; Wormington 2003). It can also be used as initial therapy in selected patients.

Commonly, ALT employs a 50 μm spot size with a 0.1 s duration starting at a power level of 300 or 500 mW and increasing the power until the desired end point is reached (usually a blanching of the trabecular meshwork or a small bubble). Then that setting is used for the remainder of the burns. One method is to treat 180° of the angle with 40 to 50 evenly spaced burns, and if that does not reduce IOP enough, the remaining 180° can be treated. Another method is to treat 360° with about 100 burns initially (Wormington 2003).

In more than 75% of ALT treatments in previously untreated eyes, there is a clinically significant decrease in IOP (American Academy of Ophthalmology Glaucoma Panel, Primary Open-Angle Glaucoma 2010). However, within 5 years in 30% to more than 50% of treated eyes, additional medical or surgical treatments are needed. With this subsequent rise of IOP, repeated ALT treatments are not very effective.

Of the potentially serious complications of ALT, a transient IOP elevation in the first few hours after the procedure is the most common. Topical medications can be used to prevent or treat the transient IOP increase. Some of the other complications of ALT include the formation of peripheral anterior synechiae, anterior uveitis, hemorrhage/hyphema, corneal edema, and mild ocular pain (Rolim de Moura, Paranhos, Jr., and Wormald 2007; Wormington 2003).

Diode laser trabeculoplasty (DLT) with an 810 nm diode laser was similar in effect and complications to ALT (Samples et al. 2011; Rolim de Moura, Paranhos, Jr., and Wormald 2007). Micropulse DLT (MDLT) with a wavelength of 810 nm, a 300 μm spot size, 100 ms pulse duration, and an energy of 0.6 mJ/pulse has also been investigated (Meyer and Lawrence 2012; Samples et al. 2011). Another method, titanium sapphire laser trabeculoplasty (TLT) (790 nm, 200 μm spot size, 7 ms pulse duration, and an energy of 30 to 50 mJ) is being evaluated (Meyer and Lawrence 2012; Samples et al. 2011). It exhibited similar safety and efficacy as was found with ALT.

Another variation is called patterned laser trabeculoplasty using a frequency-doubled (532 nm) Nd:YAG laser (Turati et al. 2010). Initially the pulse duration is 10 ms, and the power is increased until blanching of the trabecular meshwork occurs with a single spot. That power is then used with a reduction to a 5 ms pulse duration that does not produce an ophthalmoscopically visible lesion. A computer-guided pattern scanning algorithm is then used to deliver multiple 100 μm laser spots in several arcs to cover the trabecular meshwork with more than 1000 spots.

60.2.3.2 Selective Laser Trabeculoplasty

Since 1998, there has been an increase in the use of selective laser trabeculoplasty (SLT) to treat open-angle glaucoma. This uses a Q-switched, frequency-doubled (532 nm) Nd:YAG laser to selectively target pigmented trabecular meshwork cells. SLT uses a 400 μm spot size with a 3 ns duration starting with an energy of 0.6 mJ and increasing the energy until fine bubbles appear and then reducing the energy until they stop appearing (Barkana and Belkin 2007). Decreased thermal damage to the trabecular meshwork occurs because the fluence level delivered by SLT is a couple of orders of magnitude below that of ALT and it is delivered in a few nanoseconds (which is less than the thermal diffusion time of the tissue) (Murthy and Latina 2009). Both 180° and 360° applications have been used, and both are reasonable not only for treatment after failure of medical therapy but also as initial therapy.

The IOP-lowering effect of SLT after 5 years is similar to the effect of ALT with both showing a gradual decrease in effectiveness (Avery et al. 2013; Samples et al. 2011). Unlike ALT, a repeat

SLT appears to be safe and effective after an initially successful 360° SLT (Samples et al. 2011).

Similar complications have been reported after SLT and ALT. However, there is a report of fewer complications with low-energy SLT (using half of the energy level in the conventional SLT control group) compared with conventional SLT (Tang et al. 2011).

60.2.3.3 LPI and Iridoplasty

Angle-closure glaucoma accounts for almost half of the cases of glaucoma worldwide (Quigley and Broman 2006). Aqueous fluid is released continuously by the ciliary body in the posterior chamber just behind the iris. The aqueous then flows from the posterior chamber to the anterior chamber by moving between the anterior surface of the lens and the posterior surface of the iris through the pupil. From the anterior chamber, the aqueous then flows into the anterior chamber angle, through the trabecular meshwork, into Schlemm's canal, and into collector channels. Direct contact between the lens and the iris can lead to a relative or absolute blockage of the aqueous flow, resulting in an increase in pressure in the posterior chamber. This pressure can push the iris forward closing the angle and preventing the aqueous from draining out of the eye. This decrease in outflow with no change in inflow can result in even higher IOP. The increased pressure in the eye can lead to glaucomatous optic neuropathy and loss of vision.

LPI is the main treatment for primary angle-closure and primary angle-closure glaucoma that is due to absolute or relative pupillary block (American Academy of Ophthalmology Glaucoma Panel, Primary Angle Closure 2010; Meyer and Lawrence 2012; Wormington 2003). An LPI creates a new opening in the iris peripherally in either the 11 or 1 o'clock position usually, and this allows the aqueous to bypass the pupillary block. Thus, the indications for LPI include acute angle-closure glaucoma, chronic angle-closure glaucoma, and various miscellaneous conditions that include pupillary block. Eyes that have narrow, occludable angles may be an indication for performing a prophylactic LPI.

Some of the more common possible complications from LPI include IOP spikes, transient iritis, closure of the iridotomy site, pupillary distortion, corneal damage, lenticular damage, retinal damage, iris hemorrhage/hyphema, monocular diplopia, glare, and formation of posterior synechiae (American Academy of Ophthalmology Glaucoma Panel, Primary Angle Closure 2010; Meyer and Lawrence 2012; Wormington 2003). Pretreatment with apraclonidine can be used to blunt the IOP spike, and corticosteroids can be used to treat the iritis.

LPI can be performed with a Q-switched Nd:YAG laser, an argon laser, or a combination of the two (Meyer and Lawrence 2012; Wormington 2003). The most commonly used laser for LPI is the Nd:YAG. The argon laser requires a pigmented iris to absorb the visible radiation in order to burn a hole in the iris, but the Nd:YAG does not. LPI by Nd:YAG laser alone can result in an iris hemorrhage from the disrupted vessels. Applying pressure to the globe can tamponade a hemorrhage. Alternatively, in patients with darkly pigmented, thick irides, the argon laser can be used to thin the iris and cauterize any disrupted blood vessels, followed by the use of the Nd:YAG laser to create a hole in the iris. Iridotomies made with the Nd:YAG laser rarely close, but argon LPIs tend to close in 16% to 34% of patients (Wormington 2003).

To accomplish argon laser iridotomy, a number of different types of thermal delivery can be used. These include contraction burns (500 μm spot size, 0.5 s duration, 200–400 mW power), stretch burns (200 μm spot size, 0.2 s duration, 200 mW power), penetration burns (50 μm spot size, 0.01 to 0.02 s or 0.1 to 0.2 s duration, 600–1500 mW power), or cleanup burns (50 μm spot size, 0.01 to 0.02 s, 400–600 mW power or 0.1 to 0.2 s duration, 200–600 mW power) (Wormington 2003).

To accomplish Nd:YAG laser iridotomy, the energy used to produce the photodisruption can range from 1 to 15 mJ per pulse (Wormington 2003). The minimum energy necessary and the minimum number of shots to produce the effect are employed.

In cases where the angle is appositionally closed and a laser iridotomy cannot be performed or medical treatment has failed, laser peripheral iridoplasty can be utilized as well as paracentesis and incisional iridectomy (American Academy of Ophthalmology Glaucoma Panel, Primary Angle Closure 2010; Ng, Ang, and Azuara-Blanco 2012). Laser iridoplasty can also be employed when mechanical and structural factors in the iris and in the angle (e.g., angle-closure due to the position of the size of the lens or nanophthalmos) contribute to an increase in IOP, and LPI has no effect on the pressure. In the laser iridoplasty procedure, laser burns are applied to the peripheral iris resulting in a contraction of the iris, pulling the iris away from the trabecular meshwork. This can open the angle, allow aqueous outflow, and decrease IOP. An argon or diode laser can be used to perform the procedure typically with a spot size of 500 μm, a 0.5 s duration, and an initial power of about 50 to 200 mW (Ng, Ang, and Azuara-Blanco 2012). The power is gradually increased until iris shrinkage is seen.

Possible complications from laser iridoplasty include a transient IOP spike, intraocular inflammation, pain or irritation, corneal burns, a need for retreatment, pupil distortion, or failure to open the angle (Wormington 2003). Angle closure may persist after a patent LPI due to factors like a thicker iris, plateau iris syndrome, ciliary block, or smaller anterior chamber angle dimensions (Ng and Morgan 2012).

60.2.3.4 Cyclophotocoagulation

When the usual methods of treating glaucoma (e.g., medication, LTP, LPI, and trabeculectomy) are not adequate, cyclodestructive procedures can be used to decrease the production of aqueous fluid, and hence lower the pressure in the eye. They can also be used to treat blind, painful eyes with elevated IOP. Because of its higher complication rate, laser cyclophotocoagulation (CPC) is a procedure of last resort. The laser energy can be delivered via transpupillary CPC, via transscleral CPC, or by endocyclophotocoagulation (ECP) (Ishida, 2013; Meyer and Lawrence 2012; Wormington 2003).

The argon laser (488 nm, 50 to 200 μm spot size, 0.1 to 0.2 s pulse duration, and 300 to 1000 mW power) can be used for direct transpupillary exposure of the ciliary processes that produce the aqueous fluid. But this procedure is not used much because of the difficulty in visualizing the ciliary processes through the pupil. Transscleral CPC can be accomplished with a noncontact Nd:YAG laser (free-running mode, 20 ms, and 4–8 J; or CW mode, 1.5 s, 7 W), a contact Nd:YAG laser (CW mode, 0.7 to 1.0 s, 7 W and up, titrated to avoid an audible "pop") with the probe applied against the sclera posterior to the limbus, a noncontact diode laser (CW, 810 nm, 100 to 400 μm spot size, 900 ms or 1 s pulse duration, 1200 to 1500 mW), or a contact diode laser (CW, 670 nm, 10 s, 430 mW; or CW 810 nm, 0.3 to 0.5 s at 1.3 to 2.0 W, 1.3 s at 3 W, 1.5 to 2 s at 1.5 W, or 2 s at 3 W, with the power set at a level just below where an audible "pop" is heard) (Lin 2008; Wormington 2003). One of the main difficulties with transscleral CPC is the inability to visualize the target. However, it appears to be simple, safe, and effective and may be useful in patients with good vision (Lin 2008; Meyer and Lawrence 2012).

ECP is the most common laser cyclodestructive procedure and utilizes an intraocular laser probe to treat the ciliary processes. The endoscope allows simultaneous viewing of the target, and hence fewer complications. Argon and diode lasers have been used. Diode laser ECP (CW, 810 nm, 0.3 to 0.9 W) is safe and effective at lowering IOP in refractory glaucoma patients (Lin 2008). The laser spot is applied until shrinkage and whitening of the ciliary processes occurs (180° to 360° are treated). ECP is often used in combination with cataract surgery of a glaucoma patient.

Potential serious complications of CPC include hypotony, phthisis bulbi, significant vision loss, choroidal detachment, and retinal detachment (Ishida 2013; Lin 2008). Pain, conjunctival burns, anterior segment uveitis, fibrin exudate, cystoid macular edema, pupil distortion, IOP spike, cataract progression, sympathetic ophthalmia, and hyphema are some of the other complications that can occur (Ishida 2013; Lin 2008). Complication occurrence rates depend on factors such as the type and severity of the glaucoma, the treatment protocol, and the type of laser. The potential complications of ECP include all of the above except for conjunctival burns. ECP also involves the potential for risk to the crystalline lens, rupture of zonules, and the risks of an intraocular procedure (e.g., endophthalmitis).

60.2.3.5 Nd:YAG Laser Goniopuncture

Deep sclerectomy is a nonpenetrating glaucoma filtration procedure that involves a conjunctival filtration bleb. When deep sclerectomy does not decrease the IOP enough, laser goniopuncture (LGP) may be used to lower IOP further (Anand and Pilling 2010). LGP involves using the Nd:YAG laser (free-running mode, 2 to 6 mJ energy, and between 1 and 20 shots) to puncture the trabeculo-Descemet's membrane (TDM) to increase the flow of aqueous through the TDM. Potentially serious complications include late acute IOP rise, hypotony, peripheral anterior synechiae, blebitis, and late bleb leak, choroidal detachment, iris incarceration, and hyphema (Anand and Pilling 2010).

60.2.3.6 Excimer Laser Trabeculotomy

To accomplish an *ab interno* excimer laser trabeculotomy (ELT), a XeCl excimer laser has been used (308 nm, 200 μm spot size, 80 ns pulse duration, and 1.2 mJ energy). The quartz fiber-optic laser probe was inserted into the eye through a 1.2 mm incision, and while under visual control, the probe tip was placed in contact with the trabecular meshwork. The laser energy was delivered in eight spots, 500 μm from each other in the anterior trabeculum. Each laser application creates a small hole in the trabecular meshwork and the inner wall of Schlemm's canal to allow the aqueous to leave the anterior chamber, move into the canal, and then into the outflow collector channels. This increase in outflow reduces IOP. In a 2 year, randomized controlled clinical trial comparing ELT to SLT, there were no statistically significant differences between the results of the two techniques (Babighian et al. 2010).

60.3 Posterior Segment (Retina)

60.3.1 Photocoagulation

When lasers are utilized to treat retinal disorders (American Academy of Ophthalmology Retina Panel 2008a,b,c), the most common damage mechanism employed is photothermal (Wormington 2003), and most of this damage is due to photocoagulation. This occurs when laser photons are absorbed by molecules of the ocular tissue, and this increases the average kinetic energy of the molecules. The temperature of the tissue thus rises, leading to denaturation of the proteins and coagulation. When this happens, there is visible whitening of the tissue.

The most common lasers used for treating the retina include the argon laser (514.5 nm, with previous use of 488 nm), the diode laser (variable within a range of ~530–850 nm), and the frequency-doubled Nd:YAG laser (532 nm). A new treatment option is the use of a micropulsed 577 nm diode laser (Sivaprasad et al. 2010). The shorter pulse duration (μs) of the micropulsed lasers limits the thermal spread and hence decreases collateral damage. This means the laser treatment can be subthreshold without a visible burn end point. Other lasers that have been used in the past include the krypton laser (647.1 nm, with some use of 568.2 and 530.8 nm) and tunable dye lasers (variable from 310–1200 nm).

The main pigments in the retina that absorb laser radiation are melanin, hemoglobin, and macular xanthophyll (Wormington 2003). Melanin absorbs strongly in the UV and blue region of the spectrum, and then absorption decreases monotonically as the wavelength increases. Retinal photocoagulation is mediated mostly by absorption of melanin in the retinal pigment epithelium cells and the choroidal melanocytes. Hemoglobin has absorption peaks at 555 nm for reduced hemoglobin and two peaks at 542 and 577 nm for oxyhemoglobin. Thus, hemoglobin absorbs strongly the argon 514.5 nm, krypton 568.2 nm, and diode 577 nm lines.

Absorption drops off dramatically in the red (e.g., Kr 647.1 nm) and infrared (e.g., Nd:YAG 1064 nm) regions of the spectrum. The xanthophyll pigment absorbs strongly in the blue (400–500 nm) region of the spectrum. Thus, using the blue Argon 488 nm line will result in damage to the macular retina and should not be used in that area.

The deepest penetration of tissue is in the near infrared region of the spectrum (Wormington 2003). For wavelengths less than 1400 nm, the longer the wavelength, the greater the transmission through the tissue, and thus the deeper the penetration. For wavelengths more than 1400 nm, the depth of penetration decreases because of an increase in water absorption. Hence, the Nd:YAG laser 1064 nm line can penetrate the sclera and treat the ciliary body, whereas the ArF excimer laser 193 nm line can only penetrate the first ~2 μm of corneal tissue.

The most common method of delivering laser energy to the retina is via slit lamp biomicroscopy with or without the assistance of contact lenses. Other delivery methods include binocu-

TABLE 60.2 Some Indications for Retinal Laser Photocoagulation

Proliferative diabetic retinopathy (pan-retinal photocoagulation)

Clinically significant diabetic macular edema (targeted focal laser coagulation)

Macular edema and neovascularization secondary to branch retinal vein occlusion

CNV (e.g., the wet form of age-related macular degeneration)

Peripheral neovascularization

Central serous chorioretinopathy

Fundus vascular anomalies

Retinopexy of retinal tears for prevention of RD

Far peripheral retinal tears

Threshold retinopathy of prematurity

Small ocular tumors of the choroid and retina

Note: CNV = choroidal neovascularization, RD = retinal detachment.

TABLE 60.3 Complications of Retinal Photocoagulation

Inadvertent laser burn of the fovea, cornea, iris, or lens

Choroidal effusion

Secondary CNV

Retinal pigment epithelium tears

Macular edema exacerbation

Macular pucker

Immediate and late visual acuity loss

Restriction of visual field

Nyctalopia (decreased night vision)

Color vision changes

Preretinal and subretinal fibrosis

Laser scar expansion

Tractional retinal detachment

Pain

Vitreous hemorrhage

Persistent or recurrent CNV after photocoagulation

Note: CNV = choroidal neovascularization.

lar indirect ophthalmoscopes with a condensing lens, external probes, and internal endoscopic probes.

Table 60.2 gives some of the indications for laser retinal photocoagulation (Yanoff and Duker 2009). Table 60.3 shows possible complications of retinal photocoagulation (Yanoff and Duker 2009).

References

American Academy of Ophthalmology Cataract and Anterior Segment Panel. 2011. Preferred Practice Pattern® Guidelines. Cataract in the Adult Eye American Academy of Ophthalmology. San Francisco, CA. Retrieved from http://www.aao.org/ppp.

American Academy of Ophthalmology Retina Panel. 2008a. Preferred Practice Pattern® Guidelines. Age-Related Macular Degeneration. American Academy of Ophthalmology, San Francisco, CA; (2nd printing 2011). Retrieved from http://www.aao.org/ppp.

American Academy of Ophthalmology Retina Panel. 2008b. Preferred Practice Pattern® Guidelines. Diabetic Retinopathy. American Academy of Ophthalmology, San Francisco, CA; (4th printing 2012). Retrieved from http://www.aao.org/ppp.

American Academy of Ophthalmology Retina Panel. 2008c. Preferred Practice Pattern® Guidelines. Posterior Vitreous Detachment, Retinal Breaks, and Lattice Degeneration. American Academy of Ophthalmology, San Francisco, CA; Retrieved from http://www.aao.org/ppp.

American Academy of Ophthalmology Glaucoma Panel. 2010. Preferred Practice Pattern® Guidelines. Primary Angle Closure. American Academy of Ophthalmology, San Francisco, CA. Retrieved from http://www.aao.org/ppp.

American Academy of Ophthalmology Glaucoma Panel. 2010. Preferred Practice Pattern® Guidelines. Primary Open-Angle Glaucoma. American Academy of Ophthalmology, San Francisco, CA. Retrieved from http://www.aao.org/ppp.

American Academy of Ophthalmology Refractive Management/Intervention Panel. 2012. Preferred Practice Pattern® Guidelines. Refractive Errors & Refractive Surgery. American Academy of Ophthalmology, San Francisco, CA. Retrieved from http://www.aao.org/ppp.

Anand, N., and R. Pilling. 2010. Nd:YAG laser goniopuncture after deep sclerectomy: Outcomes. *Acta Ophthalmol* 88:110–115.

Avery, N., G. S. Ang, S. Nicholas, A. Wells. 2013. Repeatability of primary selective laser trabeculoplasty in patients with primary open-angle glaucoma. *Int Ophthalmol.* [Epub ahead of print].

Babighian, S., L. Caretti, M. Tavolato, R. Cian, and A. Galan. 2010. Excimer laser trabeculotomy vs 180 degrees selective laser trabeculoplasty in primary open-angle glaucoma. A 2-year randomized, controlled trial. *Eye* 24:632–638.

Barkana, Y., and M. Belkin. 2007. Selective laser trabeculoplasty. *Surv Ophthalmol* 52:634–654.

Barsam, A., and B. D. Allan. 2012. Excimer laser refractive surgery versus phakic intraocular lenses for the correction of moderate to high myopia. *Cochrane Database Syst Rev* 1:CD007679.

Binder, P. S. 2010. Femtosecond applications for anterior segment surgery. *Eye Contact Lens* 36:282–285.

Chen, S., Y. Feng, A. Stojanovic, M. R. Jankov, 2nd, and Q. Wang. 2012. IntraLase femtosecond laser vs mechanical microkeratomes in LASIK for myopia: A systematic review and meta-analysis. *J Refract Surg* 28:15–24.

Conrad-Hengerer, I., F. H. Hengerer, T. Schultz, H. B. Dick. 2012. Effect of femtosecond laser fragmentation on effective phacoemulsification time in cataract surgery. *J Refract Surg* 28(12):879–883.

El Awady, H. E., A. A. Ghanem, and S. M. Saleh. 2011. Wavefront-optimized ablation versus topography-guided customized ablation in myopic LASIK: Comparative study of higher order aberrations. *Ophthalmic Surg Las Im* 42:314–320.

Farid, M., and R. F. Steinert. 2010. Femtosecond laser-assisted corneal surgery. *Curr Opin Ophthalmol* 21:288–292.

Foster, P. J., R. Buhrmann, H. A. Quigley, and G. J. Johnson. 2002. The definition and classification of glaucoma in prevalence surveys. *Br J Ophthalmol* 86:238–242.

He, L., K. Sheehy, and W. Culbertson. 2011. Femtosecond laser-assisted cataract surgery. *Curr Opin Ophthalmol* 22:43–52.

Holzer, M. P., M. C. Knorz, M. Tomalla, T. M. Neuhann, and G. U. Auffarth. 2012. Intrastromal femtosecond laser presbyopia correction: 1-year results of a multicenter study. *J Refract Surg* 28:182–188.

Iovieno, A., M. E. Legare, D. B. Rootman et al. 2011. Intracorneal ring segments implantation followed by same-day photorefractive keratectomy and corneal collagen cross-linking in keratoconus. *J Refract Surg* 27:915–918.

Ishida, K. 2013. Update on results and complications of cyclophotocoagulation. *Curr Opin Ophthalmol* 24(2):102–110.

Juhasz, T., G. A. Kastis, C. Suarez, Z. Bor, and W. E. Bron. 1996. Time-resolved observations of shock waves and cavitation bubbles generated by femtosecond laser pulses in corneal tissue and water. *Lasers Surg Med* 19:23–31.

Kessel, L., L. Eskildsen, M. van der Poel, and M. Larsen. 2010. Non-invasive bleaching of the human lens by femtosecond laser photolysis. *PLoS One* 5:e9711.

Kim, P., G. L. Sutton, and D. S. Rootman. 2011. Applications of the femtosecond laser in corneal refractive surgery. *Curr Opin Ophthalmol* 22:238–244.

Kymionis, G. D., D. M. Portaliou, G. A. Kounis et al. 2011. Simultaneous topography-guided photorefractive keratectomy followed by corneal collagen cross-linking for keratoconus. *Am J Ophthalmol* 152:748–755.

Kymionis, G. D., V. P. Kankariya, A. D. Plaka, D. Z. Reinstein. 2012. Femtosecond laser technology in corneal refractive surgery: A review. *J Refract Surg* 28(12):912–920.

Limnopoulou, A. N., D. I. Bouzoukis, G. D. Kymionis et al. 2013. Visual outcomes and safety of a refractive corneal inlay for presbyopia using femtosecond laser. *J Refract Surg* 29(1):12–18.

Lin, S. C. 2008. Endoscopic and transscleral cyclophotocoagulation for the treatment of refractory glaucoma. *J Glaucoma* 17:238–247.

Meyer, J. J., and S. D. Lawrence. 2012. What's new in laser treatment for glaucoma? *Curr Opin Ophthalmol* 23:111–117.

Murthy, S., and M. A. Latina. 2009. Pathophysiology of selective laser trabeculoplasty. *Int Ophthalmol Clin* 49:89–98.

Ng, W. S., G. S. Ang, and A. Azuara-Blanco. 2012. Laser peripheral iridoplasty for angle-closure. *Cochrane Database Syst Rev* 2:CD006746.

Ng, W. T., and W. Morgan. 2012. Mechanisms and treatment of primary angle closure: A review. *Clin Experiment Ophthalmol* 40:e218–e228.

Prakash, G., A. Agarwal, D. A. Kumar et al. 2011. Femtosecond sub-Bowman keratomileusis: A prospective, long-term, intereye comparison of safety and outcomes of 90- versus 100-μm flaps. *Am J Ophthalmol* 152:582–590.

Quigley, H. A., and A. T. Broman. 2006. The number of people with glaucoma worldwide in 2010 and 2020. *Br J Ophthalmol* 90:262–267.

Rapuano, C. J. 2011. Excimer laser phototherapeutic keratectomy. In *Cornea*. J. Krachmer, M. Mannis, and E. Holland, editors. Mosby, Philadelphia, PA.

Reggiani Mello, G. H., and R. R. Krueger. 2011. Femtosecond laser photodisruption of the crystalline lens for restoring accommodation. *Int Ophthalmol Clin* 51:87–95.

Roberts, T. V., M. Lawless, S. J. Bali, C. Hodge, G. Sutton. 2013a. Surgical outcomes and safety of femtosecond laser cataract surgery: A prospective study of 1500 consecutive cases. *Ophthalmology* 120(2):227–233.

Roberts, T. V., M. Lawless, C. C. Chan et al. 2013b. Femtosecond laser cataract surgery: technology and clinical practice. *Clin Experiment Ophthalmol* 41(2):180–6.

Rolim de Moura, C., A. Paranhos, Jr., and R. Wormald. 2007. Laser trabeculoplasty for open angle glaucoma. *Cochrane Database Syst Rev* CD003919.

Samples, J. R., K. Singh, S. C. Lin et al. 2011. Laser trabeculoplasty for open-angle glaucoma. A report by the American Academy of Ophthalmology. *Ophthalmology* 118:2296–2302.

Sekundo, W., K. S. Kunert, and M. Blum. 2011. Small incision corneal refractive surgery using the small incision lenticule extraction (SMILE) procedure for the correction of myopia and myopic astigmatism: Results of a 6 month prospective study. *Br J Ophthalmol* 95:335–339.

Shah, R., S. Shah, and H. Vogelsang. 2011. All-in-one femtosecond laser refractive surgery. *Tech Ophthalmol* 9:114–121.

Shortt, A. J., and B. D. Allan. 2006. Photorefractive keratectomy (PRK) versus laser-assisted in-situ keratomileusis (LASIK) for myopia. *Cochrane Database Syst Rev* CD005135.

Shtein, R. M. 2011. Post-LASIK dry eye. *Expert Rev Ophthalmol* 6:575–582.

Sivaprasad, S., M. Elagouz, D. McHugh, O. Shona, and G. Dorin. 2010. Micropulsed diode laser therapy: Evolution and clinical applications. *Surv Ophthalmol* 55:516–530.

Smadja, D., G. Reggiani-Mello, M. R. Santhiago, and R. R. Krueger. 2012. Wavefront ablation profiles in refractive surgery: Description, results, and limitations. *J Refract Surg* 28:224–232.

Smadja, D., M. R. Santhiago, G. R. Mello et al. 2013. Corneal higher order aberrations after myopic wavefront-optimized ablation. *J Refract Surg* 29(1):42–48.

Soong, H. K., and J. B. Malta. 2009. Femtosecond lasers in ophthalmology. *Am J Ophthalmol* 147:189–197.

Sun, H., M. Han, M. H. Niemz, and J. F. Bille. 2007. Femtosecond laser corneal ablation threshold: Dependence on tissue depth and laser pulse width. *Lasers Surg Med* 39:654–658.

Taneri, S., M. Weisberg, and D. T. Azar. 2011. Surface ablation techniques. *J Cataract Refract Surg* 37:392–408.

Tang, M., Y. Fu, M. S. Fu et al. 2011. The efficacy of low-energy selective laser trabeculoplasty. *Ophthalmic Surg Las Im* 42:59–63.

Tomita, M., T. Kanamori, G. O. T. Waring et al. 2012. Simultaneous corneal inlay implantation and laser in situ keratomileusis for presbyopia in patients with hyperopia, myopia, or emmetropia: Six-month results. *J Cataract Refract Surg* 38:495–506.

Turati, M., F. Gil-Carrasco, A. Morales et al. 2010. Patterned laser trabeculoplasty. *Ophthalmic Surg Las Im* 41:538–545.

Uthoff, D., M. Polzl, D. Hepper, and D. Holland. 2012. A new method of cornea modulation with excimer laser for simultaneous correction of presbyopia and ametropia. *Graefes Arch Clin Exp Ophthalmol* 250:1649–1661.

Wormington, C. M. 2003. *Ophthalmic Lasers.* Butterworth-Heinemann, Philadelphia, PA.

Yanoff, M., and J. S. Duker, editors. 2009. *Ophthalmology.* Expert Consult Premium Edition: Enhanced Online Features and Print. Mosby, St. Louis, MO.

Zhang, Y., Y. G. Chen, Y. J. Xia. 2013. Comparison of Corneal Flap Morphology Using AS-OCT in LASIK With the WaveLight FS200 Femtosecond Laser Versus a Mechanical Microkeratome. *J Refract Surg* 29(5):320–324.

61

Lasers Used in Dentistry

Kirpa Johar
Johars Laser Dental Clinic

Vida de Arce
Massachusetts General Hospital

61.1 Introduction

Discovery of novel drugs, treatments, and testing of consumer products in the field of dentistry is a multibillion dollar business. The use of lasers for dentistry has increased at an exponential rate in recent years. This review will cover the invention of lasers and strategies for targeted tissue disruption, incision, bonding, and so forth. Lasers in dentistry that have been reported are then discussed under six broad headings: lasers used in dentistry, hypersensitivity treatment, lasers in preventative dentistry, lasers in restorative dentistry, lasers in pediatric dentistry, lasers in implant dentistry, laser-assisted endodontics, laser-assisted surgery, laser-assisted periodontics, laser-assisted cosmetic dentistry, and lasers as an adjunct to orthodontics. Not only are lasers widely employed in dentistry but also have been employed using a wide variety of lasers for multiple tissue types to treat an ever-growing number of dental conditions and diseases.

We recapitulate laser treatment of dentin hypersensitivity, as well as prevention, detection, and treatment of caries, pits, and fissures, pulp and its uses in restorative dentistry and oral surgery (e.g., ablation, composite adhesion, soft tissue procedures, and implants), and cosmetic surgery (crown lengthening, teeth whitening, and gingival depigmentation). The combination of the aforementioned soft tissue intervention and mucogingival surgery with lasers is used to supplement orthodontic procedures as well. There has been a remarkable increase in the number and variety of laser therapies in dentistry in recent years, and the basic strategy for constructing procedures is outlined.

61.1.1 Lasers Used in Dentistry

61.1.1.1 History of Lasers

As early as 1917, Albert Einstein formulated the theory of stimulated emission, whereby a photon of the right wavelength can stimulate an atom in a high-energy state to create another photon of the same wavelength and direction of travel. Stimulated emission becomes the basis for researching harnessing photons to increase the energy of light (National Academy of Engineering 2012). This led to developing a microwave amplification by stimulated emission of radiation (MASER), an acronym coined by Charles Townes James Gordon and Herbert Zeiger in 1958, that uses high-frequency molecular oscillation to excited molecules of ammonia gas that amplify and generate short-wavelength radio

waves (National Academy of Engineering 2012). Further study by Theodore Maiman led to the development of the first laser device, a ruby laser, in 1960 (National Academy of Engineering 2012). A dental practitioner, Dr. Leon Goldman, first used the ruby laser on the tooth in 1965 (Convissar 2010). Two decades later, Dr. Terry D. Myers heeded Dr. Leon Goldman's call to develop what became the first laser designed specifically for general dentistry (Convissar 2010). Currently, a number of laser wavelengths are used in dentistry (Convissar 2010).

61.1.2 Different Lasers Used for Intraoral Applications

A brief outline of the literature indicates lasers used for dental applications include lasers in the following spectrums of light:

- Visible: Argon (400–500 nm), Potassium titanyl phosphate (KTP) (532 nm) and DIAGNOdent (655 nm) (Parker 1998) (see Figure 61.1a)

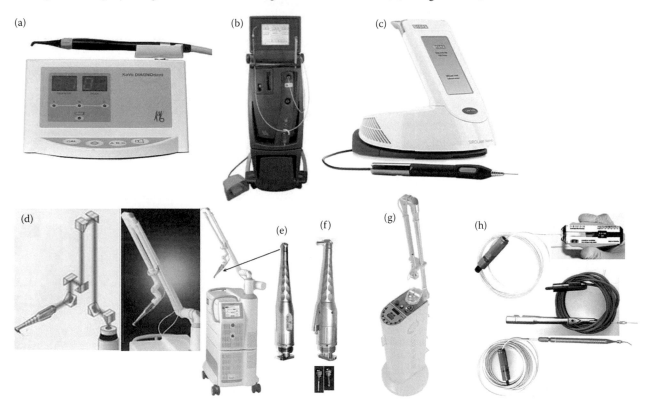

FIGURE 61.1 Laser delivery systems for intraoral applications. (a) DIAGNOdent (Diagnodent Kavo dental GmBH, Bismarcking 39, 88400 Biberach). (b) Nd:YAG laser, the preferred laser for periodontal procedures (Nd:YAG lasers, Millennium Dental Technologies, 10945 South Street, Suite #109-A, Cerritos, California 90703). (c) Diode lasers are the most commonly used lasers for oral soft tissue surgical and periodontal procedures (diode lasers, Sirona Dental GmBH, Sirona Straße 1, A-5071 Wals BEI Salzburg, Austria). (d) Articulated arm of Er:YAG laser. (e) Er:YAG noncontact handpiece (Fatona, Stegne 7, 1000 Ljubljana, Slovenia). (f) Er:YAG contact handpiece with sapphire tips. (g) Carbon dioxide laser, the preferred laser for surgical procedures (CO_2 lasers, Union Medical Co. Ltd., 522-6, Yongheon-Dong, Uljeongbu-Si, Gyeonggi-do, Korea). (h) Optical fiber. Diagram showing 200, 400, and 600 μm fiber optic cables. Top: a 200 μm fiber being inserted into a fiber stripper. Middle: a stripped 600 μm fiber can fit into a handpiece that can then guide the fiber into a cannula, such as the straight cannula, which is for the anterior teeth. Bottom: a diode handpiece through which the 400 μm fiber-optic cable is passed. It is shown with a contrangled cannula for the posterior teeth. The cannula is the terminal end of the handpiece through which the fiber-optic cable reaches the operating field. Ceramic scissors are used to cut the fiber-optic cable after every use (not shown).

- Near-infrared: Nd-YAG lasers (1064 nm) (Aoki et al. 2004; Yukna, Carr, and Evans 2007) (see Figure 61.1b) and diode lasers (800–830 nm, 980 nm) (Hilgers and Tracey 2004) (see Figure 61.1c)
- Mid-infrared: Erbium, namely, Er:YAG (2940 nm) (Aoki et al. 2004; Hibst and Keller 1996) (see Figure 61.10d through f) and Er,Cr:YSGG (2790 nm) (Aoki et al. 2004; Chen 2002)
- Far-infrared: CO_2 laser (10,600 nm) (Aoki et al. 2004; Moritz et al. 1998) (see Figure 61.1g)

61.1.3 Delivery Systems of Lasers for Intraoral Applications

There are several ways to conduct the laser beam of various wavelengths to the target tissue (Wigdor 2008).

61.1.3.1 Optical Fibers

Diode and Nd:YAG lasers (803–1064 nm) and some of the visible laser wavelengths (532 nm) are transmitted through optic fibers that run from the terminal angled tip of the handpiece directly to the tissue of the diode handpiece through which the fiber optic cable is passed. The straight cannula is used for the anterior teeth and the contrangled cannula is for the posterior teeth. The cannula is the terminal end of the handpiece through which the fiber-optic cable reaches the operating field. Ceramic scissors are used to cut the fiber-optic cable after every use (see Figure 61.1h).

61.1.3.2 Hollow Fibers

Hollow fibers have mirrored inner walls that reflect the photons along their inner axes and transmit the energy. This laser transmission technology is used in some Er:YAG and CO_2 lasers (Goharkhay et al. 2009).

61.1.3.3 Articulated Arm

The articulated arm method of laser beam transmission uses rigid and interlocked systems of mirrors to transport energy (Figure 61.1d, left) and is the most efficient method of transmission. This system of transmission is used by some Er:YAG (Figure 61.1d, right) and CO_2 lasers (Figure 61.1g).

61.1.4 Laser Effects on Tissue

Laser light has four different interactions with the target tissue: (1) reflection, (2) transmission, (3) scattering, or (4) absorption (Figure 61.2).

- *Reflection* is when the beam is redirected off the surface, with no effect on the target tissue.
- *Transmission* is laser energy going directly through the tissue, with no effect on the target tissue.
- *Scattering* laser light weakens the intended energy.
- *Absorption* of the laser light energy into the intended target tissue allows the laser energy to interact with the tissue components, such as pigment and water content. Depending on the tissue characteristics and on the laser

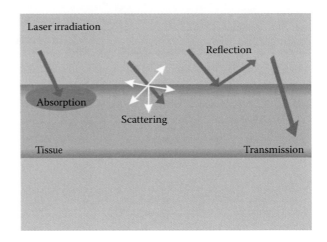

FIGURE 61.2 Diagram of the effects of laser on tissue.

wavelength, absorption of laser energy results in heating to the point of vaporization or stimulating a tissue response (such as tissue fluorescence or a healing response) (Benjamin 2011a,b). The term *chromophore* is applied to a material or tissue component that attracts a specific laser wavelength (1989). These different interactions are based on how the light is absorbed by the principle components or chromophores of the irradiated tissue (Goldman 1981).

61.1.5 Spot Size and Emission Mode

Spot size. With a hollow wave guide or an articulated-arm delivery system, there is a precise spot at the point where the energy is the greatest. This focal point is used for incision (Hall 1971) and excision surgery. For fiber-optic delivery systems, the focal point is at or near the tip of the fiber, which again has the greatest energy.

When the handpiece is moved away from the tissue and away from the focal point, the beam is defocused (out of focus), becomes more divergent, and therefore delivers less energy to the surgical site. At a small divergent distance, the beam can cover a wider area, which would be useful in achieving hemostasis (see Figure 61.3).

Emission mode. Dental laser devices can emit light energy in two modalities as a function of time: (1) constant on and (2) pulsed on and off.

The pulsed lasers can be further divided into *gated* and *free-running modes* in delivering energy to the target tissue. There are three different emission modes:

1. *Continuous wave mode*: the beam is emitted at only one power level for as long as the operator depresses the foot switch.
2. *Gated-pulse mode*: there are periodic alterations of the laser energy, similar to a blinking light. This mode is achieved by the opening and closing of a mechanical shutter in front of the beam path of a continuous wave emission.

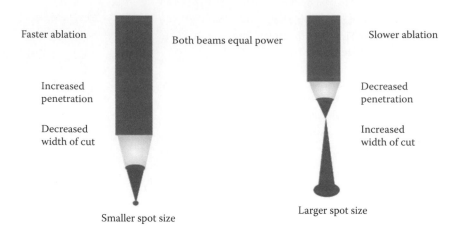

FIGURE 61.3   Diagram of the effect of spot size on ablation.

3. *Free-running pulsed* (FRP) *mode*: This is referred to as a *true-pulsed mode*. This emission is unique in that large peak energies of laser light are emitted for usually microseconds, followed by a relatively long time in which the laser is off.

61.2 Hypersensitivity Treatment

The use of lasers has opened up new dimensions in the treatment of dentin hypersensitivity. The lasers used for the treatment of dentin hypersensitivity are divided into two groups:

- Low output power lasers (He–Ne and Ga–Al–As lasers)
- Middle output power lasers [Nd:YAG, Er:YAG, and carbon dioxide lasers (Hall 1971)]

61.3 Lasers in Preventive Dentistry

61.3.1 Laser-Assisted Caries Prevention

Laser energy is utilized to improve the resistance of dental enamel to acid attack (Kwon et al. 2003), such as that involved during the caries process (Clarkson and Rafter 2001).

The mechanism by which laser irradiation is proposed to improve caries resistance ranges from changes in the enamel composition to a physical seal achieved by meeting the surface through partial fusion and recrystallization of enamel prisms only (Parker et al. 2007). The enamel surface is sealed by the laser and is less permeable to the subsequent diffusion of ions into and from the enamel.

61.3.2 Laser-Assisted Caries Detection

DIAGNOdent is a handheld laser fluorescent caries detection device (Huth et al. 2008). It has two versions that can be aimed precisely at the affected area: DIAGNOdent classic and DIAGNOdent pen. It is used to detect incipient and subsurface carious lesions, respectively. Reflected light indicates the carious lesion through fluorescence. DIAGNOdent offers unparalleled

diagnosis when compared with an explorer or a radiograph (Parker 2007) (Figure 61.1a).

61.3.2.1 Advantages

DIAGNOdent is a flexible, battery-operated mobile unit. It can be used at a precise location, which aids in minimal cavity preparation and restoration. It provides a simple, fast, and painless detection of caries, thereby making it an extremely reliable diagnostic aid. It helps in early detection of pathologic changes, demineralization, and changes in enamel composition.

61.3.3 Laser-Assisted Sealants

Erbium laser in combination with acid etching has been proposed as a noninvasive technique for pretreatment of pits and fissures. Sealants are one of the primary methods of caries prevention (Simonsen 2005).

When using sealants on pits and fissures of newly erupted permanent teeth, the following aspects must be considered:

- Age of the tooth eruption, anatomy of pits and fissures, patient history in relation to the presence of occlusal caries on the primary and permanent teeth, and oral hygiene habits and conditions.
- Pretreatment techniques used on enamel prior to the application of sealants vary.

Noninvasive techniques include etching with 37% orthophosphoric acid or air abrasion followed by acid etching. Invasive techniques involve the use of a bur mounted on a high-speed handpiece to assist in opening the deep and narrow fissures.

The erbium laser in combination with acid etching has been proposed as a noninvasive technique for pretreatment of pits and fissures. Advantages of using an Er:YAG laser for pit and fissure sealants are decontamination and improved sealant adhesion, decontamination and cleansing of fissures, improved adhesive strength of the sealant to enamel, improved quality of the margins in order to avoid secondary decay, and longer duration of the sealant. Particularly for deep fissures, laser pretreatment of pits and fissures followed by application of orthophosphoric acid is a treatment of choice because of its decontaminating and cleansing properties and its minimally invasive micropreparations (Hicks et al. 1993).

61.4 Lasers in Restorative Dentistry

61.4.1 Laser Hard Tissue Interaction

When incident laser energy directed onto dental hard tissue is absorbed by the prime chromophores (water or carbonated hydroxyapatite), the following effect occurs: For both Er:YAG and Er,Cr:YSGG wavelengths (Visuri, Walsh, Jr., and Wigdor 1996), (1) this energy is absorbed primarily by water and is converted to heat, which causes superheating and (2) a phase transfer in the subsurface water, resulting in a disruptive expansion in the tissue (Armengol, Jean, and Marion 2000). A common term for this effect is called *spallation* (Cozean et al. 1997).

61.4.2 Factors Affecting Ablation Efficiency

Factors that affect ablation efficiency are discussed in detail below and include

- Operator laser techniques such as correct focal length, angulation, and tip condition (Niemz 2007)
- Water content
- Laser parameters such as power, pulse duration, and pulse frequency (Hz) (Levy, Koubi, and Miserendino 1998)

Laser Techniques

The operator's use of laser techniques influences laser to tissue interaction (1989). It affects *fluence,* or radiant exposure, through the correct focalization of the laser beam and through the speed of movement of the handpiece, with deposition of more or less energy in terms of time and surface area.

Fluence. The density of energy, or fluence, irradiated into the tissue is dependent on the diameter of the fiber or tip used. A smaller diameter increases the density of energy delivered into the tissue. Fluence also depends on the technique used. Focusing or defocusing the laser beam increases or decreases, respectively, the density of the energy. It is also affected by the amount of irradiated surface covered per unit of time, which is closely related to the speed of one's hand moving the laser.

Focus and beam divergence. The laser beam can work in contact and noncontact mode. The cutting efficiency is best when the tip is close to the target tissue (1 mm). Once the energy exits from the tip, it diverges rapidly in the direction of 13.2°. As the distance exceeds 1 mm, both energy density and absorption are decreased. Continuous contact should be avoided during procedures on hard tissue because this causes the tips to become more likely to break.

Angulation. The ideal impact angle for laser ablation of enamel is perpendicular to the orientation of the prisms, and thus angled to the tooth surface.

Tip condition. The condition of the tip surface, which is made of sapphire or quartz, is crucial for the effective delivery of energy to the tissue. Accidental contact with the tooth surface, impact from ablation debris, wear from high-energy contact, and effects of long-time use lead to deterioration of the tip surface. The condition of the tip affects the beam profile and can increase heating and thus dehydration of the tissue.

Water

Water spray. During erbium laser hard tissue ablation, irrigation water is used to cool the tooth structure and clean the surface. Water spraying avoids thermal structural damage and simultaneously removes debris from the ablation site. In the absence of an extrinsic water spray, the tooth would overheat (Hibst and Keller 1996).

Water film. When laser energy strikes the tooth surface covered by a film of water, rapid absorption and vaporization occur, thus creating a shockwave of significant magnitude (Kang, Rizoiu, and Welch 2007). This shockwave displaces water at the point of contact, creating an open channel to the underlying tooth surface. The channel formation starts at approximately 5 μs during the radiation process.

Energy. The laser energy absorbed into the tissues in the form of thermal energy is modulated by water spray. This energy will ablate hard tissues upon reaching a specific threshold value (Benjamin 2011a).

Ablation threshold. A significant amount of energy is needed to bring water in the tissue to boiling and to provide the energy for phase transformation from liquid to gas. The ablation threshold is currently known to be approximately 12 to 20 J/cm^2 for enamel and 8 to 14 J/cm^2 for dentin (Dela Rosa et al. 2004) for Er:YAG and Er,Cr:YSGG lasers (Parker et al. 2007).

Power

Pulse frequency. The emission mode of mid-infrared wavelength lasers used on hard tissue is defined as FRP mode and is designed to elicit the emission of continuous pulses that repeat over time with a steady frequency. The FRP mode influences (1) the power and thus the ablative effectiveness and (2) the thermal diffusion between pulses that enables the tissue to cool.

Power density. The power applied per surface unit (cm^2) in a unit of time (seconds) is the power density. It depends on the choice of tips and their diameters. The smaller the diameter of the tip, the higher the density applied, given the same energy emitted, which allows for more efficient ablation.

Pulse duration. The duration of the pulse is important in determining the peak power and the thermal effect. The shorter the pulse, the higher the peak power, and the longer the pulse, the greater the thermal effect on the tissue. The pulse duration of erbium lasers is very close to the *thermal relaxation time* of enamel and dentin. Therefore, longer pulse duration with the erbium can induce potential thermal damage to the tissues.

Pulse shape. This is also known as profile of the laser pulse. The ideal laser beam is emitted in fundamental transverse mode, represented by a Gaussian profile with a typical symmetrical bell-shaped curve. The steeper the rise of the energy pulse, the faster the rise in temperature, which will allow a more efficient ablation of the hard tissue with reduced energy dispersion in heat.

61.4.3 Surface Characteristics of Laser-Prepared Cavities

The results of cavities prepared with Er:YAG and Er:YSGG lasers are very similar. The cavity ground is observed to be rough and the ablation edge is uneven. The roughness of the site of fracture extends deep into the microstructure, so a large adhesive surface is available for bonding with filling materials. The ablation edge in dentin is less irregular than in enamel because dentin is less brittle due to higher organic content. The dentinal tubules are opened. These open tubules offer a key for bonding between the restorative material and dentin. *Structural fragmentation* with an erbium laser is produced by a single burst of energy. Laser-assisted cavity preparation draws on a minimalistic approach for removal of only diseased tissue and is compatible with composite-resin restorative materials (Evans et al. 2000) (Figure 61.4).

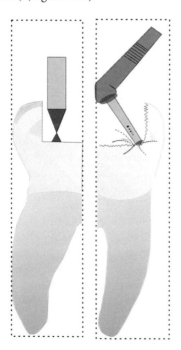

FIGURE 61.4 Diagram of surface characteristics of erbium laser prepared cavity showing irregular margins compared to the 90° cavo-surface margin with the AIRotor. Lasers provide minimally invasive restorations ideally suited for present-day composites.

61.4.4 Erbium Handpiece and Tips

The mid-infrared lasers such as Er:YAG and Er:YSGG as well as the far infrared wavelengths, including the CO_2 lasers, are equipped with a handpiece with a terminal mirror and tips to transfer the photon energy to the targeted tissues (Figure 61.1d). Some handpieces are designed to work at far contact (Figure 61.1e), while the others have terminal tips and work at close contact (Figure 61.1f). Laser transmission devices have simple handpieces that are either angled or straight. The ideal handpiece must be light, small, and ergonomic.

61.4.4.1 Far Contact Handpiece

This handpiece uses a special lens placed in the terminal part of the handpiece instead of the tip. The lens focuses the laser beam at a distance of 5 to 15 mm from the target tissue [e.g., Fontona Fidelis laser (Slovenia), Er:YAG laser, of 2940 nm wavelength at 250 mJ and 20 Hz (short pulse) with high-speed evacuation] (Figure 61.1e). Any uncontrolled movement of the operator or the patient could move the focal spot from the target tissue, resulting in increased error for angulation and the direction of the laser beam focal point.

61.4.4.2 Close Contact Handpiece

In order to interact with various oral tissues, close contact handpieces (Figure 61.1f) have tips of different shapes (conical, cylindrical, and chisel), diameters, and lengths (Figure 61.4). The tips are placed closer to the target tissue without making physical contact, resulting in better precision of laser intervention. Disadvantages include fragility, wear of the tips, and energy loss through the tip. Tips made of sapphire offer marginal improvements in energy delivery but are more expensive and less able to be reconditioned, and their rigidity places these tips at a greater risk of fracture during use.

61.4.5 Laser-Assisted Composite Adhesion

The laser energy absorbed by enamel causes the enamel surface to be heated to a high temperature, generating microcracks in the surface, which aids in the enhancing adhesion of composite to the tooth structure. The surfaces appear similar to the acid-etched surfaces. This results in significant improvement in shear bond strength of resin composites to surfaces irradiated by the laser. Etching within dentin results in carbonization (char) due to its high organic content. Moreover, there is localized melting on the dentinal surfaces causing sealing of the dentinal tubules and thereby reducing microleakage and enhancing the bond of the final composite restoration.

Some studies have concluded that primary dentin treated with Er,Cr:YSGG laser at lower wattage (0.5 W, 50 mJ) did not require etching for bonding (Sung et al. 2005). As the energy levels increase, it is beneficial to add etching as a part of the conditioning to provide adequate bonding. Studies have also shown that when the dye penetration method is used, composite restorations using glass ionomer cement exhibit less microleakage

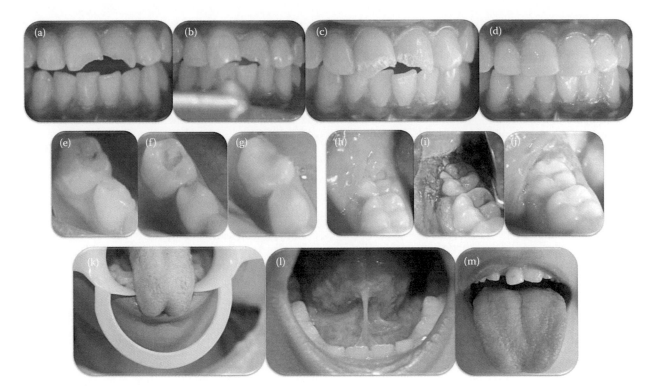

FIGURE 61.5 **(See color insert.)** Clinical case of laser-assisted restoration of tooth 11 and 21. This hard tissue procedure was performed using a Fotona Fidelis laser (Slovenia). The procedure was performed with an Er:YAG laser of 2940 nm wavelength at 250 millijoule (mJ) and 20 Hz (short pulse) with high-speed evacuation and with water on a 25-year-old female patient (a–d) and a 9-year-old child (e–g). (a) Ellis type 2 fracture on tooth 11 and 21. (b) Tooth surface prepared with Er:YAG (R02 handpiece, 250 millijoule (mJ), 20 Hz with water spray). (c) Lased tooth surface postablation. (d) Definitive restoration after Ellis type 2 fracture. (e) Occlusal caries on the first permanent molar. (f) Laser-assisted cavity preparation. (g) After composite restoration. (h, i, j) Laser-assisted operculectomy. This soft tissue procedure was performed using a 980 nm diode laser manufactured by Microscientific Instruments, New Delhi, India. The procedure was performed using a diode laser with a 400 µm optical fiber at 2.5 W power and (continuous pulse) with high-speed evacuation and with no water or air, on a 23-year-old male patient. (h) Pericoronitis with respect to tooth 48. (i) Immediately postoperative. (j) One week postoperative. (k) Preoperative view of lingual frenum showing the inability to extend the tongue with observed cleft of the tip of the tongue. (l) Er:YAG-assisted frenectomy. (m) After lingual frenectomy. Immediately postoperative, showing the ability to extend the tongue. This pediatric soft tissue procedure was performed using a Fotona Fidelis laser, an Erbium:YAG laser, with a wavelength of 2940 nm at 100 mJ and 20 Hz (very long pulse) with high-speed evacuation, and with no water or air, on a 7-year-old boy.

in laser-prepared cavities than in the conventional types. Better marginal adaptation between the prepared tooth surface and the composite resin was obtained when Er:YAG laser preparation was followed by total acid etching.

Fluence, power density, pulse length, laser angulation, focus mode, and amount of air–water spray are all factors that can cause substructural damage to the overall dentin. A final conditioning at low power is advisable on enamel and dentin (Clarkson and Rafter 2001) (Figure 61.5a through g).

61.5 Lasers in Pediatric Dentistry

61.5.1 Introducing the Pediatric Patient to the Laser

Several factors contribute to making lasers the preferred mode of treatment in pediatric dentistry:

- They are minimally invasive.
- There is no direct contact with the tooth surface.

- They reduce the need for local anesthetics.
- They create a favorable psychological impact on the child (Parkins 2000) in that they are seen as tools that use "magical" light and water to clean the teeth.

"Tell, show, do" is one technique used prior to the actual surgical aspect of pediatric dental therapy, during which the dentist explains what will be done, demonstrates before the therapy how it will be done, and then performs what was explained and demonstrated (e.g., shining the laser light at low power on the child's hand) (2005).

61.5.2 Pediatric Restorative Dentistry

Laser-assisted dental care has changed the way dentists prepare diseased teeth, ablate bone, and treat soft tissue abnormalities and diseases. An entire new standard of care is becoming a reality. The erbium lasers have helped create a positive treatment atmosphere, with most pediatric patients undergoing dental caries treatment without fear.

Using the concept of minimally invasive restorative procedures, the erbium laser allows the operator to remove only the diseased tissue, thus preserving much more of the healthy, unaffected tooth structure. Lasers also prevent the small microfractures that occur in enamel when using conventional dental handpieces (Clarkson and Rafter 2001; Levy, Koubi, and Miserendino 1998).

The erbium laser creates its ablation effect by being absorbed by water within the hydroxyapatite of the tooth structure. This heats up the water within the mineral, creating microexplosions (Venugopalan, Nishioka, and Mikic 1996) of the hydroxyapatite out of the tooth. Erbium lasers can be more specific for decay (Featherstone and Nelson 1987) than conventional instruments because the erbium laser is absorbed by water, and hard tissue contains more water when decayed than when healthy (Figure 61.1d through f).

61.5.3 Indirect and Direct Pulp Capping

Laser-assisted pulp capping (Clement, Willemsen, and Bronkhorst 2000) uses the positive effects of infrared radiation to create a biological base (decontaminated area and coagulation of exposed pulp) for the development of the dentinal bridge. The laser-assisted indirect pulp capping technique makes pulp capping much more predictable as the erbium laser can detoxify the surface up to 300 μm in depth (Schoop et al. 2004). Cavity preparation with the lasers is generally done without anesthesia (Parkins 2000). At the time of exposure, during direct pulp capping, laser-induced analgesia leaves the patient refractory to sensory perception. As a result, accidental pulp exposure is often asymptomatic.

61.5.4 Pulpotomy and Pulpectomy

Lasers can be quite effective, chemical-free, alternative for treating pulp tissue (Clement, Willemsen, and Bronkhorst 2000). Pulpotomy using lasers provide a safe, effective, and nonchemical alternative, that is especially valuable for children. Lasers provide a safe, effective, nonchemical alternative for pulpotomy (Kotlow 2004).

61.5.4.1 Laser-Assisted Pulpotomy

Pulpotomy is defined as amputation of the pulp chamber while preserving the vitality of the radicular pulp tissue. The tooth cavity is prepared with erbium family lasers at the appropriate parameters. Once the caries is removed, the pulp chamber is opened at the exposed site. When preparing the cavity with traditional techniques, CO_2 (Moritz 2006) and Nd:YAG (Liu 2006) lasers can be used for the vaporization and coagulation of the coronal pulp.

61.5.4.2 Laser-Assisted Pulpectomy

Pulpectomy is a procedure performed on primary teeth having irreversible pulpitis or necrosis caused by pulp infection following trauma or penetration of caries (Liu, Chen, and Chao 1999). After administration of local anesthesia on vital teeth

and isolation of the field, the pulp chamber can be opened using a high-speed handpiece or a laser at 2.5 to 3 W at 20 Hz. The erbium laser with curved tips of different diameters allows for a fast cleansing of the pulp chamber (600 μm, 80° curved tip) and easy access to the root canal (400 μm, 80° curved tip). The entire procedure takes approximately 10 min. Alternately, the laser radiation is used between the saline solution and 10% hydrogen peroxide irrigation alone.

61.5.5 Pediatric Soft Tissue Procedures

The young and naive oral mucosal tissue needs to be treated with a lot of finesse. A wide array of the soft tissue procedures may be performed with lasers in the pediatric dental office.

Erbium lasers may be used to accomplish many soft tissue procedures with little or no bleeding when used at lower energy settings than for hard tissue procedures and without water spray.

The most commonly encountered soft tissue abnormalities in children, such as ankyloglossia (Figure 61.5k through m), abnormal lingual frenum attachments, and growths associated with natal and neonatal teeth, may interfere with nursing or normal development and hence call for an early but minimally invasive intervention. In older children, gingival hypertrophy (Sarver and Yanosky 2005), pericoronal infections, exposure of unerupted teeth, or aphthous ulcer pain may all require early intervention.

When lasers are used, several problems are almost nonexistent: postoperative bleeding, suturing, need to inject pre- and post-operative, local anesthesia, or post-operative antibiotics. Another advantage of lasers is scar-free healing with minimal postoperative complications.

61.6 Lasers in Implant Dentistry

Lasers are used in various stages of implant dentistry. From surgical placement to prosthetic delivery, to treating infected peri-implant tissues, lasers have proved to be beneficial in many ways.

61.6.1 Preparation of the Surgical Site

The first step of implant surgery is to prepare the surgical site, such as reducing bacterial load with laser treatment. Lasers are bactericidal. The bactericidal effect is profound and almost instantaneous. The clinician needs to expose the surgical site to the laser energy for a few seconds. Before osteotomy development, the soft tissue can be sterilized much more effectively with a laser than with rinsing or swabbing.

61.6.2 Decontamination and Implant Placement

A CO_2 laser has a distinct advantage over contact lasers because CO_2 lasers are not in direct contact. Thus, it is a simple procedure to place a wide-angle handpiece on the CO_2 laser and use the laser out of focus, which would increase the spot size on the tissue even more. Sterilization of a large osteotomy site with a CO_2 laser takes a few seconds. As the surgery progresses, the clinician and assistant should be diligent in keeping the surgical site free of saliva. In cases where infected tissue is present, the clinician uses a spoon curette to remove gross amounts of soft tissue easily and quickly, and then uses a laser to remove any tissue tags. The entire inner surface of the extraction socket can then be decontaminated with a laser.

61.6.3 Soft Tissue Osteotomy

The next objective in laser implant surgery is preparation of the osteotomy through soft and then hard tissues.

The soft tissue is removed in a 3 to 4 mm diameter down to the crest of bone. This can be of a thickness of 1 to 2 mm or 3 to 4 mm, depending on the biotype. If the tissue is thicker, using a diode or Nd:YAG laser would be more time-consuming compared to erbium (Levy, Koubi, and Miserendino 1998) and CO_2 lasers. The unobstructed vision, excellent hemostasis, and efficiency of cutting through all tissue biotypes and thickness make the CO_2 laser most suitable for these procedures.

A sterile cut is less likely to become infected. Lasers incise tissue without creating the cascade of events that leads to swelling and inflammation. Since lasers seal off the lymphatic and blood vessels, there is a clinically measurable reduction in pain, swelling, and postoperative complications. No suturing is required. Fewer and milder analgesics and antibiotics are required as patients have a less traumatic postoperative course. Lasers have excellent hemostatic properties that lead to decreased bleeding, which facilitates intraoperative hemorrhage control.

61.6.4 Hard Tissue Osteotomy

Lasers cannot be used for osteotomy procedures as of yet. Research is ongoing in hopes of using erbium drills for osteotomies of the future. An advantage of a laser osteotomy would be that it is a noncontact procedure; thus, there is no friction between the laser tip and bone. The temperature increase in osseous tissue using an erbium laser is minimal provided proper parameters are maintained and adequate water spray is used. Studies show better healing and faster new bone formation with erbium lasers.

61.6.5 Uncovering Implants

When it is necessary to uncover an integrated implant after healing is complete, occasionally the implant body is covered not only by soft tissue but also by newly formed bone up to 2 to 3 mm thick. After locating the implant radiographically, the soft tissue must be removed. If bone has formed over the top of the implant, an erbium laser can efficiently and safely accomplish the uncovering process. For thick tissue, the CO_2 wavelength is most efficient to remove significant tissue quickly and maintain excellent visualization of the surgical site.

61.6.6 Peri-Implantitis

Peri-implantitis can create pockets around implants, and therefore, probing around the implants becomes part of the examination and diagnosis. Before any laser surgery is attempted, the initial surgery and healing process should be evaluated (Goldman 1981). A rigidly placed implant with no crestal bone loss and adequate zones of attached gingival should be present.

61.6.7 Future of Lasers in Implant Dentistry

The future use of lasers in implant dentistry has the promise of replacing bone drills if they can have the ability to control depth of cutting. This would allow the use of erbium lasers in osteotomy site preparations. A nonsterile drill can contaminate the surgical site while an erbium laser could make the same cut in the bone without mechanical trauma. Lasers sterilize as they cut, so using a laser in the osteotomy site would reduce the postoperative risk of infection and promote a successful outcome. Studies have shown bone ablation with Er:YAG promotes ingrowth of new bone around inserted titanium metal implants and that osseointegration can occur. Diode, CO_2, and erbium lasers have the potential to improve the clinician's ability to deliver the highest quality of care while providing a more comfortable experience for the patient with fewer postoperative problems.

61.7 Laser-Assisted Endodontics

61.7.1 Laser Doppler Flowmetry

A false diagnosis of pulp vitality may lead to an unnecessary endodontic procedure. Laser Doppler flowmetry (LDF) was developed to assess blood flow in microvascular systems by sending a signal to the tissue and detecting how long it takes to be detected by a photodetector (Figure 61.6). LDF is a diagnostic aid to assess pulp vitality by measuring blood flow within

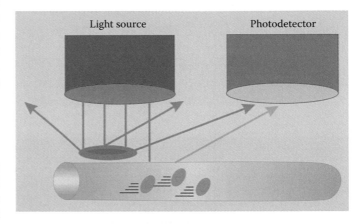

FIGURE 61.6 Laser Doppler flowmetry. When the light source hits a moving cell within the blood vessel, there is a shift in the light spectrum detected by a photodetector. When the same source hits a stationary cell or tissue, the light beam gets scattered off and there is no shift in the light spectrum.

the pulp, and a dental dam is used to prevent other sources of a Doppler signal from interfering with testing (Polat et al. 2004). This technique may be a sensitive and accurate means for pulp vitality testing because it reflects vascular rather than neural responsiveness used in other methods.

61.7.2 Limitations of Conventional Endodontics

Bacteria can still persist in the complex network of dentinal tubules and microcanals that are unreachable by conventional techniques. Despite mechanical removal, irrigation, and disinfection of the canals. In addition, a smear layer is formed on the instrumented walls of the root canal during this treatment (Gutmann 1984).

61.7.3 Laser-Assisted Root Canal Sterilization

Laser therapy following traditional procedures can improve results when different wavelengths with different modes of action are used to facilitate cleaning and debridement of the root canals and three-dimensional decontamination of the endodontic system (Figure 61.5e through g). In various laser systems used in dentistry, the emitted energy can be delivered into the root canal system via a thin optical fiber (Nd:YAG, KTP-Nd:YAG, Er:YSGG, argon, diode) or a hollow tube [CO_2, Er:YAG (Kimura et al. 2002)]. The potential bactericidal effect of laser irradiation can be effectively used for additional cleaning and disinfecting of the root canal system after biomechanical instrument usage.

61.7.3.1 Cleaning and Disinfecting the Root Canal System

When the cavity is prepared with traditional techniques, both the near-infrared (diode and Nd:YAG) and far-infrared CO_2 lasers can be used for decontamination of the cavity and coagulation of the exposed pulp (Moritz et al. 1998). If an erbium laser

is available, the entire procedure—access cavity preparation, decontamination of the cavity, and coagulation of the exposed pulp—can be done with the same laser.

The emission of laser energy from the tip of the optical fiber or the laser tip is directed along the root canal and not necessarily laterally along the root canal walls (Figure 61.7).

A new endodontic side-firing spiral tip has been designed to fit the shape and the volume of the root canals (Stabholz et al. 2003) prepared by NiTi rotary instrumentation. This emits the erbium laser irradiation laterally to the walls of the root canal through spiral slits located all along the length of the tip (Stabholz et al. 2003). The tip is sealed at its far end, preventing the transmission of irradiation to or through the apical foramen of the tooth (Stabholz et al. 2003).

61.7.3.2 Obturation of the Root Canal System

The rationale for introducing laser technology to assist in obturating the root canal system is based on the following three assumptions about laser utility (Stabholz et al. 2003): (1) using the laser irradiation as a heat source for softening gutta-percha (root canal sealer), which is employed as the obturating material, (2) using the laser as a means to condition the dentinal walls before placing an obturation bonding material, and (3) laser irradiation at 5 W and 20 Hz helped reduce the apical microleakage irrespective of the sealer or the technique used (Blum, Parahy, and Machtou 1997). Currently, procedural simplification seems to be the only proven advantage of using lasers in obturation.

61.8 Laser-Assisted Surgery

61.8.1 Advantages of Laser-Assisted Oral Surgery

Lasers are gradually becoming the standard of care for many surgical procedures, given the advantages of improved visualization, hemostasis, and reduced discomfort. Patients typically experience reduced postoperative swelling and pain.

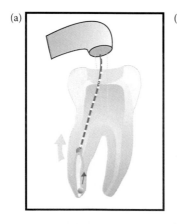
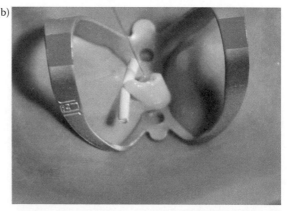

FIGURE 61.7 (a) Endodontic sterilization diagram showing a 200 μm cable inserted in an apicocoronal direction. (b) The 200 μm cable used for endodontic sterilization. This procedure was performed using a 980 nm diode laser manufactured by Microscientific instruments, New Delhi, India. The procedure was performed with a diode laser using a 200 μm optical fiber of at 2.5 W power and (continuous pulse) with high-speed evacuation, and with no water or air, on a 35-year-old female patient.

Tissue healing and scarring are also improved with the use of the laser due to a combination of decreased lateral tissue damage, less traumatic surgery, and more precise control of depth of tissue damage. Laser wounds heal with minimal scar formation and intraoral laser wounds often can be left unsutured.

Lasers help maintain sterile conditions during surgery. The hemostatic nature of laser allows surgery to be performed more precisely and accurately because of increased visibility of the surgical site. A CO_2 laser can seal blood vessels of approximately 500 µm or less (Figure 61.1g).

61.8.2 Techniques Used for Oral Surgery

The three fundamental photothermal techniques generally using lasers to perform various intraoral surgical procedures are (1) incision and excision, (2) ablation and vaporization, and (3) hemostasis and coagulation. These are described below.

61.8.2.1 Incision and Excision

The incision technique removes only a representative portion of a lesion as well as the adjacent normal tissue (Hall 1971). Lasers are commonly used in surgery as a light scalpel, to make relatively deep, thin cuts. Excision technique requires the removal of the entire lesion with at least 2 to 3 mm of peripheral margin. This technique is preferred for oral lesions that are 1 cm or less and for minor, solid, and exophytic lesions.

61.8.2.1.1 Procedure

The outlining can be done on most machines by using an intermittent, pulsed, or gated mode with a rate of 10 to 20 pulses per second and low fluence per pulse to allow superficial mark on the surface of the target without deep penetration. After this, the laser can be put into continuous mode and the dots can be connected to create the desired incision. Once the appropriate depth has been reached, grasp the tissues with forceps. Apply slight traction, excised by horizontally undercutting the tissue with the laser in focused mode.

61.8.2.2 Ablation and Vaporization

Tissue ablation is used when the surgeon wants to remove only the surface of the target tissue or perform superficial removal of the tissue over a larger area. The most common examples for intraoral use are mucosal lesions such as leukoplakia, dysplasia, and papillary hyperplasia.

Vaporization is accomplished by using larger spot sizes, which decreases significantly the power density and subsequent depth of effect. Increasing the width of laser allows larger areas of tissue to be covered in a relatively short period. This technique allows the surgeon to demarcate clearly the extent of vaporization in a controlled slow fashion using intermittent pulsing mode. At this point, the laser is defocused by pulling it away from the target, allowing the beam to widen (Figure 61.7a and b). The defocused beam is transversed along the lesion in a series of strokes (Figure 61.8).

61.8.2.3 Hemostasis Technique

The use of a CO_2 laser generally results in a bloodless surgical field. The hemostatic effect is not caused by coagulation of blood but by the contraction of vascular wall collagen. This contraction results in constriction of the vessel opening and hemostasis.

61.8.2.3.1 Procedure

The hemostasis technique uses a small spot size (somewhere between the focused beam used for incision and the defocused beam used for ablation). The laser is passed over the tissue until the bleeding ceases. If bleeding continues, it indicates that a vessel larger than the possible lateral thermal zone of the laser is involved and this requires standard hemostatic techniques to arrest the bleeding.

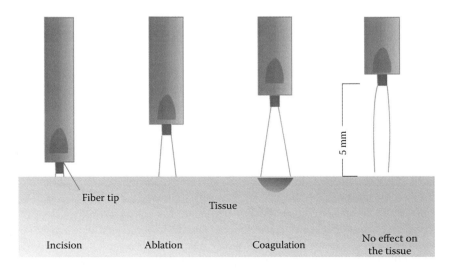

FIGURE 61.8 Diagrammatic representation of the effect of focused and defocused laser beam on the target tissue. A focused beam results in a well-defined incision whereas a defocused beam causes ablation.

61.8.3 Laser Soft Tissue Interaction

The laser tissue interaction varies in relation to both the different laser wavelengths used and the oral tissue treated. The interaction is primarily determined by the wavelength's affinity for specific chromophores of different tissues. The laser energy absorbed and/or diffused into the tissue is transformed, causing mainly photochemical and photothermic effects.

61.8.3.1 Wavelength

Laser wavelengths absorbed by melanin, hemoglobin, and water match the wavelengths of the main chromophores of the oral soft tissues such as the gingiva and mucosa (Benjamin 2011a).

Visible laser wavelengths such as argon (514 nm) and KTP (532 nm) lasers have a very good hemostatic effect in treating vascular lesions, which require a superficial penetration from 100 μm to 1 mm. Near-infrared wavelengths (803 to 1340 nm) and the diode, Nd:YAG, and Nd:YAP lasers are commonly used for cutting, vaporization, and decontamination of soft tissue (Moritz 2006).

61.8.3.2 Type of Tissue Lased

The oral soft tissue contains a variety of healthy and pathologic tissue types, such as keratinized and nonkeratinized gingival tissue and mucosa (Sarver and Yanosky 2005). Inflamed tissue that contains more blood and therefore more pigment and hemoglobin will react favorably with wavelengths in the visible and near-infrared regions (Figure 61.7b).

61.9 Laser-Assisted Periodontics

61.9.1 Lasers in Periodontal Treatment

Lasers are used as a viable alternative in the treatment of periodontal disease when the goal of laser treatment, through decontamination and debridement (Sculean et al. 2004), is to reduce the total bacterial load in the gingival pockets as well as the inflammation index (Cobb 2006).

Thin, flexible light conductor systems lead the laser beam to almost any desired location, facilitating easy use in the area of periodontal therapy (Figure 61.9).

61.9.2 Laser-Assisted Pocket Elimination

Lasers are excellent tools in pocket elimination due to the bactericidal effect and their capability to remove infected sulcular epithelium.

Most laser therapies have antimicrobial properties. Nd:YAG and diode lasers are absorbed by bacteria, especially those with pigmentation, thereby reducing colonization. Decreasing bacterial load in the soft tissue wound site can enhance wound healing with less postoperative discomfort. CO_2 and erbium laser wavelengths are absorbed by the water content in the cells, causing

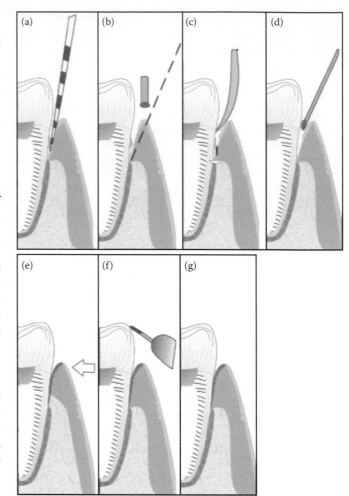

FIGURE 61.9 Diagrammatic representation of laser-assisted periodontal procedure of pocket elimination. (a) Determination of pocket depth using periodontal probe. (b) Pulsed laser irradiation selectively dissects epithelium and denatures diseased tissues and pathological proteins. (c) Removal of concrement and tartar using ultrasonic scalers and curettes. (d) Laser furnishes pocket debridement and establishes coagulation. (e) Compression of pocket wall against the root surface after surgery to facilitate the formation of static fibrin clot at the gingival crest. (f) Occlusal adjustments done with diamond points to prevent trauma from occlusion. (g) Reattached gingiva postsurgery.

cell vaporization when the temperature of the intracellular water exceeds 100°C (Seka et al. 1995) (Figure 61.1g).

61.9.2.1 Procedure

Initial therapy includes tooth cleaning and curettage to remove plaque and calculus to which laser radiation can exert optimal effect at the target destination (Coluzzi 2002). The light conductor is a fiber between 200 and 400 μm in diameter within a dental probelike "hook" that can easily maneuver within the periodontal pocket (Moritz 2006). After the laser activation, the fiber is swept using sinusoidal movements from the base to the outside of the pocket within 5 s (using 15 Hz, 1.5 W Nd:YAG laser; 2.5 W diode laser) (Moritz 2006). This is necessary to irradiate the tissues and to simultaneously prevent overheating (Moritz 2006).

61.9.3 Laser-Assisted Mucogingival Surgeries

Lasers are used to treat a spectrum of mucogingival conditions such as high frenal attachments, gingival recessions, inadequate width of attached gingiva, and shallow vestibule.

61.9.3.1 Gingivectomy

Lasers are used to treat acute, hypertrophic, and hyperplastic gingivitis, and inflammation of the gingival (Kravitz and Kusnoto 2008).

61.9.3.2 Frenectomy

Frenectomy is the complete removal of the frenum, including its attachment to the underlying bone.

61.9.3.3 Laser Surgical Approach

The laser allows for precise surgical incisions and excellent hemostatic control (Hall 1971). When analyzing which mode to use for each type of laser, the principle concept for all wavelengths is that minimum effective energy must be used. The lower the energy applied, the less damage to the targeted tissue and the faster the healing process.

61.9.3.3.1 Procedure

The incision begins at the coronal attachment and moves the laser tip unidirectionally, pulling on the lip for tension. The incision is made at the sagittal plane first, along the axis of the frenum, from the point of gingival insertion toward the lip, with the power gradually increased from 1.5 to 1.75 W and the frequency at 25 or 30 Hz. After the first incision, the intervention continues with a "V" design cut, using a shorter pulse mode. The collagen fibers are vaporized and a light incision of the periosteum is made along the mucogingival junction using a higher percentage of air–water spray to better control thermal damage (Olivi et al. 2010). Hemostasis is achieved with low power (0.25 W, 10 Hz), in long pulse mode.

61.9.3.4 Vestibuloplasty

Vestibuloplasty is also known as the lip switch technique or transpositional flap. It is done when the soft tissues from the inner aspect of the lip are shifted to a favorable zone on the alveolar bone so that the increase in the denture bearing area is achieved. It is indicated when the patient has a bone height of 15 mm or more in the anterior region. The mucosa must be healthy and exhibit no fibrosis, scarring, or hyperplasia.

61.9.3.4.1 Procedure

The laser incision is made at the junction of keratinized and nonkeratinized tissue. The surface of the nonkeratinized tissue is gently ablated, which allows for a new band of keratinized tissue to form as it heals.

61.9.3.5 Pericoronitis

Pericoronitis refers to inflammation of the gingiva in relation to the crown of an incompletely erupted tooth. Laser operculectomy removes the small portion of inflamed soft tissue covering the crown of the erupting tooth (Figure 61.5h through j).

61.10 Laser-Assisted Cosmetic Dentistry

61.10.1 Laser-Assisted Teeth Whitening

The objective of laser power bleaching is to excite the bleaching agents using the very efficient source of light energy of lasers (1998). By placing photons of one specific wavelength that approximates the absorption spectrum of the bleaching agent rather than using a light source that emits multiple wavelengths, the chemical reaction proceeds at a faster rate, thereby decreasing exposure time of the bleaching agent to the tooth (Goharkhay et al. 2009).

KTP, argon, and diode lasers (Figure 61.1c) are commonly used for in-office bleaching treatment (Verheyen et al. 2006).

61.10.1.1 Procedure

Prior to bleaching, oral prophylaxis needs to be carried out. Prophylactic pastes should be avoided as they contain fluoride and oils that will interfere with the bleaching process. Pumice is used for polishing before the bleaching procedure. Teeth are isolated with a rubber dam or light-cured dam. The bleaching agent is prepared and carefully placed on the enamel. The laser is activated following the manufacturer's instructions for exposure time. The used bleaching agent should be wiped off with a wet gauze or cotton and fresh material reapplied, repeating application and removal three times. Teeth are irrigated and isolation material removed. The new shade should be confirmed with a shade guide (Figure 61.10).

61.10.2 Laser-Assisted Crown Lengthening

Laser-assisted crown lengthening involves excising an appropriate amount of gingival tissue around the tooth and predictably removing a given height of only gingival tissue (soft tissue crown lengthening) or both gingival tissue and alveolar bone (osseous crown lengthening) (Pang 2008).

Crown lengthening procedures that are indicated within the aesthetic zone require special consideration to achieve predictable aesthetic results (Adams and Pang 2004). Whether they are performed for the purpose of exposing sound tooth structure or to enhance the appearance of definitive restorations, these procedures must be planned to satisfy biologic requirements while simultaneously avoiding deleterious aesthetic effect.

Soft tissue. The crown lengthening procedure is basically an excision of gingival soft tissue. Recommended placement of the excision is at least 2 mm coronal to the bottom of the gingival attachment to reduce the risk of root exposure and violation of biologic width rules (Figure 61.11).

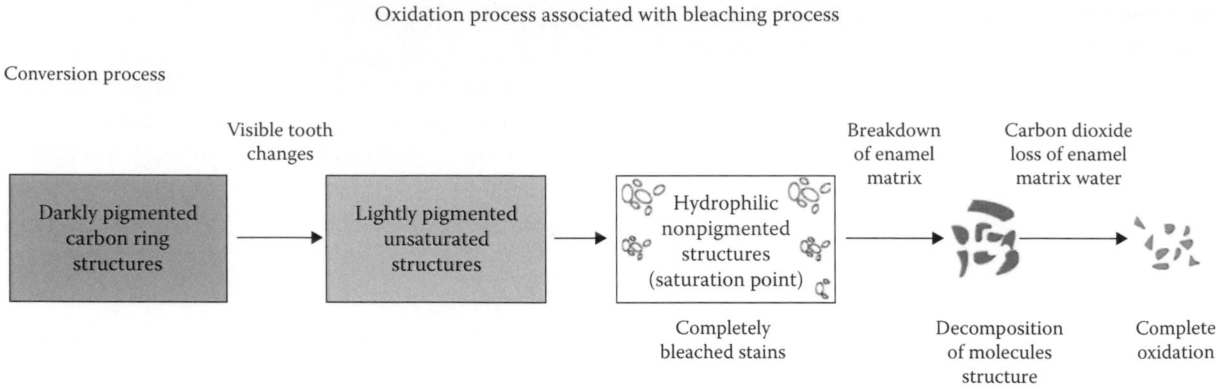

Oxidation process associated with bleaching process

Conversion process

| | Visible tooth changes | | | Breakdown of enamel matrix | Carbon dioxide loss of enamel matrix water |

Darkly pigmented carbon ring structures → Lightly pigmented unsaturated structures → Hydrophilic nonpigmented structures (saturation point)

Completely bleached stains — Decomposition of molecules structure — Complete oxidation

FIGURE 61.10 Schematic representation of oxidation process during laser-assisted bleaching. The optimal amount of bleaching occurs at the saturation point.

Hard tissue. To provide adequate biologic width from the margin of the restoration, a minimum of 3 mm of attached gingiva must remain over the underlying bone to create a healthy periodontal environment.

Erbium lasers have end-cutting tips with the advantage that the cooling water spray prevents the surgical site from overheating (Armengol, Jean, and Marion 2000; Fried et al. 2001), in contrast to the rotary friction heat released by conventional burs. Collateral thermal tissue damage (Benjamin 2011b) when using erbium lasers is less than with conventional techniques.

61.10.2.1 Procedure

Intraoral periapical radiographs are taken and probed for soundness with a periodontal probe. Gingival width is marked with a tissue-marking pencil at the estimated and desired position from canine to canine, taking into consideration the maintenance of the biological width. The gingival tissue above the marking is cut with the help of a laser tip, generally without anesthesia. The

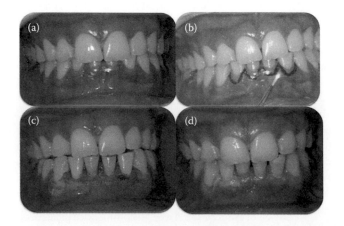

FIGURE 61.11 (a) Preoperative view of lower anteriors (central and lateral incisors) before crown lengthening. (b) During the procedure. (c) Immediate postoperative. (d) One week postoperative.

procedure parameters are set as power settings at 1.25 W, 7% water, and 11% air.

61.10.3 Laser-Assisted Gingival Depigmentation

Melanin, carotene, and hemoglobin are the most common natural pigments contributing to the normal color of the gums. In the past, gingival depigmentation has been carried out successfully several times using nonsurgical and surgical procedures. Recently, laser ablation has been recognized as the most effective, pleasant, and reliable method (Stabholz et al. 2003).

61.10.3.1 Laser Depigmentation

Diode, Nd:YAG, and erbium and CO_2 lasers work on the same principle with regard to laser depigmentation. Because the technique's objective is not to cut the tissue but rather to have the laser energy absorbed by the pigment deep in the epithelium, a low power is used. Nd:YAG and diode lasers may be used without contact with an inert tip such that their light will not be absorbed by the superficial layers of the tissue but will penetrate until absorbed by melanin. As the laser energy is absorbed by the melanin, the tissue lightens in color. Alternatively, the diode and Nd:YAG can be used in light contact with very light, brushstroke-like movements through the tissue. Erbium lasers can also be used in contact with the tissue, gently removing the tissue layer by layer until the epithelial layer containing the melanocytes is removed (Figure 61.12). Low power is used with a large spot size, this minimizes the power density and enables the dentist to cover a large area more quickly (see Figure 61.3).

61.11 Laser as an Adjunct to Orthodontics

There are many clinical situations that require soft tissue intervention to resolve mucogingival conditions that interfere with or complicate the ongoing orthodontic therapy.

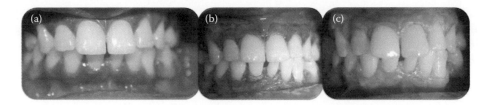

FIGURE 61.12 Laser-assisted depigmentation. This soft tissue procedure was performed using a 980 nm diode laser manufactured by Microscientific Instruments soft tissue laser, New Delhi, India. The procedure was performed with a diode laser using a 400 μm optical fiber of 980 nm wavelength at 1.5 W power and defocused mode with high-speed evacuation and with no water or air, on a 27-year-old male patient. (a) Preoperative, (b) immediately postoperative of depigmentation, and (c) 1 week postoperative.

Minimally invasive mucogingival surgery using laser techniques can be performed quickly, easily, and safely by orthodontists. Many procedures can be performed with a soft tissue laser, which can be divided into two categories: access gingivectomies and aesthetic procedures.

Access gingivectomies involve gaining or improving access to a tooth or allowing earlier or more ideal bracket or band position, thus improving the quality and reducing the treatment time (Kravitz and Kusnoto 2008). Aesthetic procedures involve improving gingival aesthetics, primarily of the anterior teeth, which can result in a marked improvement in the appearance of the completed treatment. Understanding the importance of keratinized tissue and the concept of biologic width is critical before performing any laser procedure.

Some of the procedures that benefit by the use of soft tissue lasers are access to unerupted or partially erupted teeth, labial or lingual frenum removal, and access for ideal bracket placement (Kravitz and Kusnoto 2008).

61.12 Conclusion

Patients benefit tremendously from laser dentistry due to the noninvasive nature of treatment, faster recovery time, and lesser postoperative pain. Lasers have the ability to selectively and precisely target the diseased tissue. Bactericidal effect, good hemostasis, reduced anesthesia, and elimination of sutures are some of the main benefits of laser-assisted dentistry. Advantages for the dentist include lesser operator fatigue, greater patient compliance to various dental procedures, and more satisfied patients. For these reasons, laser dentistry should be the gold standard for providing optimal dental care.

References

1989. Laser-tissue interaction. *IEEE Trans Biomed Eng* 36:1145–1243.

1998. Laser-assisted bleaching: an update. ADA Council on Scientific Affairs. *J Am Dent Assoc* 129:1484–1487.

2005. American Academy of Pediatric Dentistry reference manual 2005–2006. *Pediatr Dent* 27:1–211.

Adams, T. C., and P. K. Pang. 2004. Lasers in aesthetic dentistry. *Dent Clin North Am* 48:833–860, vi.

Aoki, A., K. M. Sasaki, H. Watanabe, and I. Ishikawa. 2004. Lasers in nonsurgical periodontal therapy. *Periodontology 2000* 36:59–97.

Armengol, V., A. Jean, and D. Marion. 2000. Temperature rise during Er:YAG and Nd:YAP laser ablation of dentin. *J Endod* 26:138–141.

Benjamin, S. D. 2011a. All lasers are not the same: Success requires knowledge and training. *Compend Contin Educ Dent* 32:66–68.

Benjamin, S. D. 2011b. Laser dentistry: Mainstream after 25 years. *Dentalaegis Tech* 32:1.

Blum, J. Y., E. Parahy, and P. Machtou. 1997. Warm vertical compaction sequences in relation to gutta-percha temperature. *J Endodont* 23:307–311.

Chen, W. H. 2002. Laser root canal therapy. *J Indiana Dent Assoc* 81:20–23.

Clarkson, B. H., and M. E. Rafter. 2001. Emerging methods used in the prevention and repair of carious tissues. *J Dent Educ* 65:1114–1120.

Clement, A. W., W. L. Willemsen, and E. M. Bronkhorst. 2000. [Success of direct pulp capping after caries excavations]. *Ned Tijdschr Tandheelkd* 107:230–232.

Cobb, C. M. 2006. Lasers in periodontics: A review of the literature. *J Periodontol* 77:545–564.

Coluzzi, D. J. 2002. Lasers and soft tissue curettage: An update. *Compend Contin Educ Dent* 23:1104–1111.

Convissar, R. A. 2010. *Principles and Practice of Laser Dentistry.* Mosby Elsevier, St. Louis, MO.

Cozean, C., C. J. Arcoria, J. Pelagalli and G. L. Powell. 1997. Dentistry for the 21st century? Erbium:YAG laser for teeth. *J Am Dent Assoc* 128:1080–1087.

Dela Rosa, A., A. V. Sarma, C. Q. Le, R. S. Jones, and D. Fried. 2004. Peripheral thermal and mechanical damage to dentin with microsecond and sub-microsecond 9.6 microm, 2.79 microm, and 0.355 microm laser pulses. *Laser Surg Med* 35:214–228.

Evans, D. J., S. Matthews, N. B. Pitts, C. Longbottom, and Z. J. Nugent. 2000. A clinical evaluation of an Erbium:YAG laser for dental cavity preparation. *Br Dent J* 188:677–679.

Featherstone, J. D., and D. G. Nelson. 1987. Laser effects on dental hard tissues. *Adv Dent Res* 1:21–26.

Fried, D., J. Ragadio, M. Akrivou et al. 2001. Dental hard tissue modification and removal using sealed transverse excited atmospheric-pressure lasers operating at lambda 9.6 and 10.6 microm. *J Biomed Opt* 6:231–238.

Goharkhay, K., U. Schoop, J. Wernisch et al. 2009. Frequency doubled neodymium:yttrium-aluminum-garnet and diode laser-activated power bleaching—pH, environmental scanning electron microscopy, and colorimetric *in vitro* evaluations. *Laser Med Sci* 24:339–346.

Goldman, L. 1981. Laser medicine in America: An overview. *Lasers Surg Med* 1:285–288.

Gutmann, J. L. 1984. Principles of endodontic surgery for the general practitioner. *Dent Clin North Am* 28:895–908.

Hall, R. R. 1971. The healing of tissues incised by a carbon-dioxide laser. *Br J Surg* 58:222–225.

Hibst, R., and U. Keller. 1996. Effects of water spray and repetition rate on the temperature elevation during Er:YAG laser ablation of dentine. *Proc SPIE* 2623:1.

Hicks, M. J., C. M. Flaitz, G. H. Westerman et al. 1993. Caries-like lesion initiation and progression around laser-cured sealants. *Am J Dent* 6:176–180.

Hilgers, J. J., and S. G. Tracey. 2004. Clinical uses of diode lasers in orthodontics. *J Clin Orthod* 38:266–273.

Huth, K. C., K. W. Neuhaus, M. Gygax et al. 2008. Clinical performance of a new laser fluorescence device for detection of occlusal caries lesions in permanent molars. *J Dent* 36:1033–1040.

Kang, H. W., I. Rizoiu, and A. J. Welch. 2007. Hard tissue ablation with a spray-assisted mid-IR laser. *Phys Med Biol* 52:7243–7259.

Kimura, Y., K. Yonaga, K. Yokoyama et al. 2002. Root surface temperature increase during Er:YAG laser irradiation of root canals. *J Endod* 28:76–78.

Kotlow, L. A. 2004. Lasers in pediatric dentistry. *Dent Clin North Am* 48:889–922, vii.

Kravitz, N. D., and B. Kusnoto. 2008. Soft-tissue lasers in orthodontics: An overview. *Am J Orthod Dentofacial Orthop* 133:S110–114.

Kwon, Y. H., O. W. Kwon, H. I. Kim, and K. H. Kim. 2003. Nd:YAG laser ablation and acid resistance of enamel. *Dent Mater J* 22:404–411.

Levy, G., G. F. Koubi, and L. J. Miserendino. 1998. Cutting efficiency of a mid-infrared laser on human enamel. *J Endod* 24:97–101.

Liu, J. F. 2006. Effects of Nd:YAG laser pulpotomy on human primary molars. *J Endodont* 32:404–407.

Liu, J. F., L. R. Chen, and S. Y. Chao. 1999. Laser pulpotomy of primary teeth. *Pediatr Dent* 21:128–129.

Moritz, A., editor. 2006. *Oral Laser Application.* Quintessence Publishing, Chicago, IL.

Moritz, A., U. Schoop, K. Goharkhay, and W. Sperr. 1998. The CO_2 laser as an aid in direct pulp capping. *J Endod* 24:248–251.

National Academy of Engineering. 2012. *Lasers and Fiber Optics Timeline.*

Niemz, M. H. 2007. *Laser-Tissue Interactions, Fundamental Applications.* Springer, Heidelberg.

Olivi, G., F. Angiero, S. Benedicenti et al. 2010. Use of the erbium, chromium:yttrium-scandium-gallium-garnet laser on human enamel tissues. Influence of the air-water spray on the laser-tissue interaction: Scanning electron microscope evaluations. *Lasers Med Sci* 25:793–797.

Pang, P. 2008. Lasers in cosmetic dentistry. *Gen Dent* 56:663–670; quiz 671–672, 767.

Parker, S. 1998. Lasers in dentistry. *Br Dent Nurs J* 57:7–9.

Parker, S. 2007. Low-level laser use in dentistry. *Br Dent J* 202:131–138.

Parker, S. P. A., A. A. Darbar, J. D. B. Featherstone et al. 2007. The use of laser energy for therapeutic ablation of intraoral hard tissues. *J Laser Dent* 15:78–86.

Parkins, F. 2000. Lasers in pediatric and adolescent dentistry. *Dent Clin North Am* 44:821–830.

Polat, S., K. Er, K. E. Akpinar, and N. T. Polat. 2004. The sources of laser Doppler blood-flow signals recorded from vital and root canal treated teeth. *Arch Oral Biol* 49:53–57.

Sarver, D. M., and M. Yanosky. 2005. Principles of cosmetic dentistry in orthodontics: Part 2. Soft tissue laser technology and cosmetic gingival contouring. *Am J Orthod Dentofacial Orthop* 127:85–90.

Schoop, U., W. Kluger, A. Moritz et al. 2004. Bactericidal effect of different laser systems in the deep layers of dentin. *Laser Surg Med* 35:111–116.

Sculean, A., F. Schwarz, M. Berakdar et al. 2004. Periodontal treatment with an Er:YAG laser compared to ultrasonic instrumentation: A pilot study. *J Periodontol* 75:966–973.

Seka, W., D. Fried, J. D. Featherstone, and S. F. Borzillary. 1995. Light deposition in dental hard tissue and simulated thermal response. *J Dent Res* 74:1086–1092.

Simonsen, R. J. 2005. Preventive resin restorations and sealants in light of current evidence. *Dent Clin North Am* 49:815–823, vii.

Stabholz, A., R. Zeltser, M. Sela et al. 2003. The use of lasers in dentistry: Principles of operation and clinical applications. *Compend Contin Educ Dent* 24:935–948; quiz 949.

Sung, E. C., T. Chenard, A. A. Caputo et al. 2005. Composite resin bond strength to primary dentin prepared with Er, Cr:YSSG laser. *J Clin Pediatr Dent* 30:45–49.

Venugopalan, V., N. S. Nishioka, and B. B. Mikic. 1996. Thermodynamic response of soft biological tissues to pulsed infrared-laser irradiation. *Biophys J* 70:2981–2993.

Verheyen, P., L. J. Walsh, J. Wernisch, U. Schoop, and A. Moritz. 2006. Laser-Assisted Bleaching. *Oral Laser Application* Chapter 10:407–448.

Visuri, S. R., J. T. Walsh, Jr., and H. A. Wigdor. 1996. Erbium laser ablation of dental hard tissue: Effect of water cooling. *Lasers Surg Med* 18:294–300.

Wigdor, H. 2008. Basic physics of laser interaction with vital tissue. *Alpha Omegan* 101(3):127–132.

Yukna, R. A., R. L. Carr, and G. H. Evans. 2007. Histologic evaluation of an Nd:YAG laser-assisted new attachment procedure in humans. *Int J Periodontics Restorative Dent* 27:577–587.

62

Lasers Used in Urology

Naeem Bhojani
*Indiana University
School of Medicine*

James Lingeman
*Indiana University
School of Medicine*

62.1 Introduction

Lasers have become an integral part of and have revolutionized the field of urological surgery. Theodore Maiman, an American physicist, created the first working laser using a ruby crystal in 1960 (Maiman 1960). The first description of a laser being used in the field of urology was by Parsons in 1966 with the application of a laser to a canine bladder (Parsons et al. 1966).

The word "laser" is an acronym for light amplification by stimulated emission of radiation. A laser is formed when energy passes through a special medium, and in doing so stimulates the atoms within that medium to emit light in exactly the same direction and at exactly the same wavelength. This light or photon is created from the relaxation of excited electrons. A typical laser includes a pumping source that provides energy, an amplifying medium, and an optical cavity or mirror that reflects the beam of light. Lasers are characterized as being monochromatic, collimated, and coherent beams of light. Lasers are monochromatic with a very narrow wavelength bandwidth, in contrast to light, which has a very wide bandwidth. Every ray has exactly the same direction (collimated) and every beam oscillates in phase with each other (coherent).

One of the most important aspects of lasers is their interaction with tissue. This interaction is responsible for the conversion of light to thermal energy. This conversion and transfer of thermal energy results in the elevation of the target tissue temperature and can lead to either coagulation or vaporization. When the target tissue is heated to below the boiling/vaporization temperature, it results in coagulative necrosis. However, when the target tissue is heated above the boiling/vaporization temperature, ablation or removal of tissue occurs.

Before the early 1980s, lasers were used sparingly in the field of urology. However, in the early 1980s, lasers became a vital part of the field and can now be considered standard of care.

Lasers in urology have been used to treat many common urological problems including benign prostatic hyperplasia (BPH), urogenital malignancies, urolithiasis, ureteropelvic junction (UPJ) obstruction, stricture disease, and tissue lesions. There are several different types of lasers with different specifications that are being employed in the field of urology today (Table 62.1). The goal of this chapter is to summarize the types of lasers and their clinical applications in urology.

62.2 Holmium:Yttrium– Aluminum–Garnet

The holmium:yttrium–aluminum–garnet (Hol:YAG) laser, which can be either an end-firing or side-firing laser, emits light at a wavelength of 2140 nm. Unlike other lasers, this laser emits energy in a series of pulses with a pulse duration of 350 ms. This allows the laser to emit light energy in a very short period of time, which makes it capable of very high peak powers. However, as this laser is pulsed, the energy that is transferred to the tissue is quickly absorbed, resulting in rapid dispersion of heat. Due to the wavelength, this laser is highly absorbed by water, which leads to vaporization without deep coagulative necrosis. The depth of tissue penetration is 0.4 mm. The absorption coefficient is very high, which leads to a high transfer of energy to tissue leading to vaporization. Coagulation is excellent and coagulation necrosis is minimal. The Hol:YAG laser, using different fiber sizes, is employed in a wide variety of clinical scenarios (Table 62.2). The clinical applications for the Hol:YAG laser include the treatment of kidney stones, BPH, UPJ obstruction, stricture disease, and urogenital malignancy. More recently, this laser is being used in a novel technique known as natural orifice transluminal endoscopic radical prostatectomy (NOTES RP) for the treatment of prostate cancer (Humphreys et al. 2011).

TABLE 62.1   Laser Types and Their Description

Laser Type	Wavelength (nm)	Depth of Penetration (mm)	Type of Laser	Clinical Capacity	Clinical Applications
Hol:YAG	2100	0.4	Pulsed	Vaporization Resection Enucleation	BPH UPJ obstruction Lithotripsy Stricture disease Urogenital cancer Prostate cancer
Nd:YAG	1064	10	Pulsed or continuous	Coagulation	Superficial bladder and penile cancer Bladder hemangioma
KTP	532	0.8	Continuous	Vaporization	BPH
Diode	940 980 1470	5	Pulsed or continuous	Vaporization	BPH
Tm:YAG	2000	0.25	Continuous	Vaporization Resection Enucleation	BPH Prostate cancer

TABLE 62.2   Hol:YAG Laser Fiber Selection According to Clinical Application

Application	Laser Fiber (u)	Settings: Energy (J)	Settings: Frequency
Bladder stones	1000	0.6–1	20–40
NOTES RP	550	2	20–40
BPH	550	2	20–50
Urethral stricture disease	550	0.6–1	6–15
UPJ obstruction	200 or 365	0.6–1	6–15
Renal stones or tumors (antegrade)	200, 365, or 550	0.6–1	6–15
Renal stones or tumors (retrograde)	200	0.6–1	6–15
Distal and midureteral stones or tumors	200 or 365	0.6-1	6-15

Stone disease affects about 10% of the population. Many of these patients will require some form of lithotripsy. The Hol:YAG laser has become the standard of care for these patients. It can be employed both antegradely as well as retrogradely, and it can fragment all types of stones in any location. A review of the literature demonstrates the effectiveness of the holmium laser for the treatment of renal stones with a fragmentation rate above 95% and a stricture and perforation rate below 1% (Biyani, Cornford, and Powell 1998; Devarajan et al. 1998; Grasso and Chalik 1998; Gould 1998; Scarpa et al. 1999; Shroff et al. 1996; Sofer and Denstedt 2000; Yip, Lee, and Tam 1998).

An important application for the Hol:YAG laser is for the treatment of BPH, also known as holmium laser enucleation of the prostate (HoLEP). HoLEP has been confirmed, with level 1 data including both short- and long-term results, to be the most effective laser for the treatment of BPH for glands of all sizes. This is due to the short length of hospital stay (LOS), the short length of catheterization (LOC), and the minimal bleeding associated with this laser (Kuntz et al. 2004; Wilson et al. 2006). As well, HoLEP significantly improves the patient's urinary peak flow rates as well as their American Urological Association (AUA) symptom scores (Ahyai et al. 2010). Unlike other ablative treatments for BPH, HoLEP enucleates the prostate tissue, which can then be subjected to pathological analysis. Rigorous studies have determined that HoLEP is better than open prostatectomy

for the treatment of large prostate glands in terms of both perioperative and postoperative outcomes (Kuntz, Lehrich, and Ahyai 2008). For treatment of smaller prostate glands, HoLEP is superior compared to transurethral resection of the prostate (TURP) in terms of perioperative outcomes and at least equivalent for postoperative outcomes. There are also data to suggest that it is superior to TURP even for postoperative outcomes of smaller prostate glands (Ahyai et al. 2010). Finally, and most importantly, the beneficial outcomes of HoLEP are long term, durable, and reproducible and can be applied to prostates of any size (Humphreys et al. 2008; Krambeck, Handa, and Lingeman 2010).

Similar to HoLEP, the Hol:YAG laser can also be used to treat BPH by ablating prostate tissue with a technique called holmium laser ablation of the prostate (HoLAP). HoLAP was first reported in 1994 using a 60 W laser. Using the newer 80 and now 100 W lasers, HoLAP has been shown through two randomized control trials (RCTs) to be comparable to both TURP and KTP (Elzayat et al. 2009; Mottet et al. 1999). In one RCT, it was as effective as TURP in the improvement of patient's International Prostate Symptom Score (IPSS) and Qmax (Mottet et al. 1999). Hospital stay and LOC were shorter in the HoLAP group in this study. In another study comparing HoLAP and KTP, it was shown that functional outcomes improved equally and significantly in both groups (Elzayat et al. 2009).

Patients with narrowed UPJs or those with UPJ crossing vessels have been known to not only suffer pain but also be at risk of developing renal insufficiency. Classically, this condition, which is usually congenital, is treated with a pyeloplasty (Brooks et al. 1995). However, an alternative procedure called endopyelotomy, using a Hol:YAG laser to incise the narrowed region, can also be performed. In a recent review, 78% of 153 renal units were successfully treated with Hol:YAG endopyelotomy (Phillips and Landman 2007; Ost et al. 2005). Overall, regardless of the approach (antegrade or retrograde), the success rate for endopyelotomy is around 80% (Kletscher et al. 1995; Knudsen et al. 2004; Kunkel and Korth 1990; Ost et al. 2005; Mendez-Torres, Urena, and Thomas 2004; Motola, Badlani, and Smith 1993; Shalhav et al. 1998).

Similar to UPJ obstructions, patients with ureteral strictures can be treated with an endoscopic Hol:YAG laser treatment. These strictures, unlike UPJ obstructions, are usually iatrogenic and not congenital. A recent review demonstrated that 64% of 86 ureteral and/or ureteroenteric strictures were treated successfully (Phillips and Landman 2007).

The entire genitourinary system can be affected with transitional cell carcinoma (TCC). While the gold standard for the treatment of upper urinary tract TCC is nephroureterectomy, some patients can avoid this treatment using Hol:YAG laser endoscopic techniques. These techniques can be done both retrogradely or antegradely. The rate of recurrence depends on the tumor grade and stage. One study found recurrence rates of 18%, 33%, and 50% for tumor grades 1, 2, and 3, respectively, after treatment with the Hol:YAG laser (Jarrett et al. 1995).

Along with the already extensive list of Hol:YAG laser clinical applications, a new and revolutionary technique has recently been reported. This technique, also known as NOTES RP, uses similar settings to HoLEP, but instead of removing most of the prostate, it removes the prostate in its entirety. Therefore, NOTES RP is being used as a definitive treatment for prostate cancer. Although the NOTES RP technique is in its early stages, the achievable benefits are evident. Not only do patients avoid painful incisions, but there is minimal bleeding, early return to work, and possibly better postoperative erectile function (Humphreys et al. 2011).

62.3 Neodymium:Yttrium–Aluminum–Garnet

The neodymium:yttrium–aluminum–garnet (Nd:YAG) laser emits light at a wavelength of 1064 nm (Natalin et al. 2008). At this frequency, this laser is poorly absorbed by water, and therefore, it leads to deep coagulative necrosis and more extensive thermal injury. This laser has a depth of penetration of ≥ 1 cm (Marks and Teichman 2007). Its past clinical indications included surgery for BPH and lithotripsy. However, due to the depth of penetration resulting in significant postoperative irritation symptoms and the laser's inability to fragment all stone types, this laser is no longer used in these clinical scenarios and other more effective lasers have replaced it. The Nd:YAG

is, however, still being used to treat superficial bladder tumors, superficial penile lesions and bladder hemangiomas (Natalin et al. 2008). The maximum power output of this laser is 60 W, but it should be limited to 35 W to minimize the risk of bladder perforation. The advantage of this laser is that it can be used in the outpatient setting with local anesthesia. As well, within the bladder, there is no risk of an obturator reflex, which can lead to bladder perforation. The disadvantage of this laser is that there is no tissue retrieval, the laser has a deep level of tissue penetration, and there is a rare but important risk of hollow viscus organ injury during bladder tumor ablation.

62.4 Potassium–Titanyl–Phosphate

The potassium–titanyl–phosphate (KTP) laser is generated by passing a Nd:YAG laser through a KTP crystal to double the frequency and produce a 532 nm wavelength. The laser energy is delivered by a side-firing glass fiber. Initial prostate vaporization used 60 W KTP lasers, but due to slower vaporization times, newer 80 W KTP and 120 W lithium triborate laser systems have been developed. Even more recently, a 180 W system has been introduced. Due to the 532 nm laser wavelength, which is very close to the peak absorption of hemoglobin, this laser produces excellent hemostasis and transmission through water (Marks and Teichman 2007). As well, KTP lasers have a depth of penetration of 0.8 mm and produce only 1–2 mm of coagulation, making it safe for vaporization of the prostate. In the field of urology, this laser is used exclusively for the treatment of BPH.

The use of the KTP laser to treat BPH has attracted much attention due to the fact that the short-term outcomes are excellent and the learning curve is short. In a randomized controlled trial, KTP was shown to be equivalent to TURP in terms of the improvement in the patient's maximum peak flow rates and in the patient's BPH symptoms (evaluated using the International Prostate Symptom score) (Bouchier-Hayes et al. 2006). In another RCT, KTP was shown to have a shorter LOC and a shorter LOS compared to TURP (Horasanli et al. 2008). However, there is very limited long-term data to support these results. As well, KTP cannot be used for all prostate sizes and is typically used to treat prostate glands under 60 g. Furthermore, KTP ablates tissue as opposed to HoLEP, which enucleates it, and therefore, there is no tissue available for pathological analysis. As well, after KTP, there is minimal to no change in the patient's level of prostate specific antigen (PSA). In comparison, after HoLEP, there is a substantial decrease in the patient's level of PSA. PSA in this situation can be used as a surrogate to represent the amount of tissue that has been removed.

62.5 Diode Lasers

The diode laser uses special diodes to generate energy. Various wavelengths are available, including 940, 980, and 1470 nm. At these wavelengths, this laser is absorbed very well by both water and hemoglobin and therefore has very good hemostatic as well as vaporization properties (Gravas et al. 2011). The depth of penetration is 5 mm. The main use of this laser is for the treatment of BPH. The main advantage of this laser is that it is an inexpensive laser that does not require a cooling device. The main disadvantage of this laser is that very little and only short-term data are available at this time.

62.6 Thulium:Yttrium–Aluminum–Garnet

The thulium:yttrium–aluminum–garnet (Tm:YAG) laser emits light at a wavelength of 2000 nm. Like the Hol:YAG laser, Tm:YAG is highly absorbed by water (depth of penetration is 0.25 mm) and therefore offers excellent and rapid vaporization of tissue. Unlike the Hol:YAG laser, Tm:YAG is continuous, and therefore, it produces a high mean energy output resulting in a significant charring effect on tissue. Since these lasers are diode pumped, they are small and portable, without the need for water-cooling systems (Marks and Teichman 2007). The main clinical use for the Tm:YAG laser is for the treatment of BPH. Recent reports have demonstrated good perioperative and postoperative outcomes including improvements in maximal peak flow rates and in the patient's urinary symptoms. However, these results are all short term (Bach et al. 2011; Netsch et al. 2012a,b). Similar to the Hol:YAG laser, this laser has also recently been used for the treatment of prostate cancer (Nagele et al. 2012).

62.7 Conclusion

Over the past decade, the evolution of the laser in the field of urology has been dramatic and evolutionary. From coagulation to vaporization, resection, and enucleation, lasers have become an integral part of the field of urology. Much research has been done in this domain, but it is evident that better quality data with long-term durable results need to be performed for the majority of lasers in urology. To date, it is clear that with its multipurpose applications and the available short- and long-term data, the Hol:YAG laser should be considered the current standard of care. Additionally, until and unless more level 1 evidence is reported on other laser alternatives, Hol:YAG will and should remain the gold standard.

References

Ahyai, S. A., P. Gilling, S. A. Kaplan et al. 2010. Meta-analysis of functional outcomes and complications following transurethral procedures for lower urinary tract symptoms resulting from benign prostatic enlargement. *Eur Urol* 58:384–397.

Bach, T., C. Netsch, L. Pohlmann, T. R. Herrmann, and A. J. Gross. 2011. Thulium:YAG vapoenucleation in large volume prostates. *J Urol* 186:2323–2327.

Biyani, C. S., P. A. Cornford, and C. S. Powell. 1998. Ureteroscopic holmium lasertripsy for ureteric stones. Initial experience. *Scand J Urol Nephrol* 32:92–93.

Bouchier-Hayes, D. M., P. Anderson, S. Van Appledorn, P. Bugeja, and A. J. Costello. 2006. KTP laser versus transurethral resection: Early results of a randomized trial. *J Endourol* 20:580–585.

Brooks, J. D., L. R. Kavoussi, G. M. Preminger, W. W. Schuessler, and R. G. Moore. 1995. Comparison of open and endourologic approaches to the obstructed ureteropelvic junction. *Urology* 46:791–795.

Devarajan, R., M. Ashraf, R. O. Beck, R. J. Lemberger, and M. C. Taylor. 1998. Holmium: YAG lasertripsy for ureteric calculi: An experience of 300 procedures. *Br J Urol* 82:342–347.

Elzayat, E. A., M. S. Al-Mandil, I. Khalaf, and M. M. Elhilali. 2009. Holmium laser ablation of the prostate versus photoselective vaporization of prostate 60 cc or less: Short-term results of a prospective randomized trial. *J Urol* 182:133–138.

Gould, D. L. 1998. Holmium:YAG laser and its use in the treatment of urolithiasis: Our first 160 cases. *J Endourol* 12:23–26.

Grasso, M., and Y. Chalik. 1998. Principles and applications of laser lithotripsy: Experience with the holmium laser lithotrite. *J Clin Laser Med Surg* 16:3–7.

Gravas, S., A. Bachmann, O. Reich et al. 2011. Critical review of lasers in benign prostatic hyperplasia (BPH). *BJU Int* 107:1030–1043.

Horasanli, K., M. S. Silay, B. Altay et al. 2008. Photoselective potassium titanyl phosphate (KTP) laser vaporization versus transurethral resection of the prostate for prostates larger than 70 mL: A short-term prospective randomized trial. *Urology* 71:247–251.

Humphreys, M. R., N. L. Miller, S. E. Handa et al. 2008. Holmium laser enucleation of the prostate—outcomes independent of prostate size? *J Urol* 180:2431–2435; discussion 2435.

Humphreys, M. R., J. S. Sauer, A. R. Ryan et al. 2011. Natural orifice transluminal endoscopic radical prostatectomy: Initial perioperative and pathologic results. *Urology* 78:1211–1217.

Jarrett, T. W., P. M. Sweetser, G. H. Weiss, and A. D. Smith. 1995. Percutaneous management of transitional cell carcinoma of the renal collecting system: 9-year experience. *J Urol* 154:1629–1635.

Kletscher, B. A., J. W. Segura, A. J. LeRoy, and D. E. Patterson. 1995. Percutaneous antegrade endopyelotomy: Review of 50 consecutive cases. *J Urol* 153:701–703.

Knudsen, B. E., A. J. Cook, J. D. Watterson et al. 2004. Percutaneous antegrade endopyelotomy: Long-term results from one institution. *Urology* 63:230–234.

Krambeck, A. E., S. E. Handa, and J. E. Lingeman. 2010. Holmium laser enucleation of the prostate for prostates larger than 175 grams. *J Endourol* 24:433–437.

Kunkel, M., and K. Korth. 1990. [Long-term results following percutaneous pyeloplasty]. *Urologe A* 29:325–329.

Kuntz, R. M., S. Ahyai, K. Lehrich, and A. Fayad. 2004. Transurethral holmium laser enucleation of the prostate versus transurethral electrocautery resection of the prostate: A randomized prospective trial in 200 patients. *J Urol* 172:1012–1016.

Kuntz, R. M., K. Lehrich, and S. A. Ahyai. 2008. Holmium laser enucleation of the prostate versus open prostatectomy for prostates greater than 100 grams: 5-year follow-up results of a randomised clinical trial. *Eur Urol* 53:160–166.

Maiman, T. 1960. Stimulated optical radiation in ruby. *Nature* 187:493–494.

Marks, A. J., and J. M. Teichman. 2007. Lasers in clinical urology: State of the art and new horizons. *World J Urol* 25:227–233.

Mendez-Torres, F. R., R. Urena, and R. Thomas. 2004. Retrograde ureteroscopic endopyelotomy. *Urol Clin North Am* 31:99–106.

Motola, J. A., G. H. Badlani, and A. D. Smith. 1993. Results of 212 consecutive endopyelotomies: An 8-year followup. *J Urol* 149:453–456.

Mottet, N., M. Anidjar, O. Bourdon et al. 1999. Randomized comparison of transurethral electroresection and holmium: YAG laser vaporization for symptomatic benign prostatic hyperplasia. *J Endourol* 13:127–130.

Nagele, U., A. G. Anastasiadis, U. Walcher et al. 2012. Natural orifice (NOTES) transurethral sutureless radical prostatectomy with thulium laser support: First patient report. *World J Urol* 30:625–631.

Natalin, R. A., C. K. Phillips, R. V. Clayman and J. Landman. 2008. Urologic laser types and instrumentation. *Arch Esp Urol* 61:971–977.

Netsch, C., T. Bach, T. R. Herrmann and A. J. Gross. 2012a. Thulium:YAG VapoEnucleation of the prostate in large glands: A prospective comparison using 70- and 120-W 2-μm lasers. *Asian J Androl* 14:325–329.

Netsch, C., T. Bach, L. Pohlmann, T. Herrmann and A. J. Gross. 2012b. Comparison of 120–200 W 2 μm thulium:yttrium-aluminum-garnet vapoenucleation of the prostate. *J Endourol* 26:224–229.

Ost, M. C., J. D. Kaye, M. J. Guttman, B. R. Lee, and A. D. Smith. 2005. Laparoscopic pyeloplasty versus antegrade endopyelotomy: Comparison in 100 patients and a new algorithm for the minimally invasive treatment of ureteropelvic junction obstruction. *Urology* 66:47–51.

Parsons, R. L., J. L. Campbell, M. W. Thomley, C. G. Butt, and T. E. Gordon, Jr. 1966. The effect of the laser of dog bladders: A preliminary report. *J Urol* 95:716–717.

Phillips, C. K., and J. Landman. 2007. Lasers in the upper urinary tract for non-stone disease. *World J Urol* 25:249–256.

Scarpa, R. M., A. De Lisa, D. Porru, and E. Usai. 1999. Holmium:YAG laser ureterolithotripsy. *Eur Urol* 35:233–238.

Shalhav, A. L., G. Giusti, A. M. Elbahnasy et al. 1998. Adult endopyelotomy: Impact of etiology and antegrade versus retrograde approach on outcome. *J Urol* 160:685–689.

Shroff, S., G. M. Watson, A. Parikh et al. 1996. The holmium: YAG laser for ureteric stones. *Br J Urol* 78:836–839.

Sofer, M., and J. Denstedt. 2000. Flexible ureteroscopy and lithotripsy with the Holmium: YAG laser. *Can J Urol* 7: 952–956.

Wilson, L. C., P. J. Gilling, A. Williams et al. 2006. A randomised trial comparing holmium laser enucleation versus transurethral resection in the treatment of prostates larger than 40 grams: Results at 2 years. *Eur Urol* 50:569–573.

Yip, K. H., C. W. Lee, and P. C. Tam. 1998. Holmium laser lithotripsy for ureteral calculi: An outpatient procedure. *J Endourol* 12:241–246.

Lasers Used in Otolaryngology

James A. Burns
Massachusetts General Hospital

Anca M. Barbu
Massachusetts General Hospital

63.1 Historical Review

Throughout the 20th century, seminal innovations such as the surgical microscope and general endotracheal anesthesia greatly enhanced the precision and success of endolaryngeal surgery. This culminated in the introduction of the carbon dioxide (CO_2) laser to surgery in the 1970s by Polanyi, Bredermeier, and Davis (1970), Jako (1972), Strong (1975), Strong et al. (1976), and Vaughan (1978). They coupled the CO_2 laser to the surgical microscope, thereby creating a new means of precise hemostatic dissection. Clinically, the CO_2 laser is a fundamental tool for the endolaryngeal surgeon; however, its use has been limited to the operating room because the energy cannot be delivered through a fiber.

Anderson and Parrish (1981) proposed concepts of selective photothermolysis that allowed for specific targeted damage to cells by "suitably brief pulses" of optical radiation based on properties of the target tissue. Anderson (Anderson and Parrish 1983) applied selective photothermolysis to the development of yellow light (585–600 nm) pulsed-dye lasers (PDLs) for treatment of vascular malformations of the skin by targeting oxyhemoglobin. This concept eventually evolved into two angiolytic lasers: the 585 nm PDL and the 532 nm pulsed potassium–titanyl–phosphate (KTP) laser. These wavelengths are precisely selected to target absorbance peaks of oxyhemoglobin (~571 and ~541 nm) and can fully penetrate intralumenal blood and thereby deposit heat uniformly into the vessel, thereby causing intravascular coagulation and "photoangiolysis" of the subepithelial microcirculation. The short pulse width is precisely selected to contain the heat to the vessel without causing collateral damage to the extravascular soft tissue from heat conduction. The output of these lasers is transmitted through a thin flexible glass fiber (≤0.6 mm). Therefore, PDL and KTP lasers are well suited for use through the channel of a flexible laryngoscope in the office as well as the speculum of a direct laryngoscope in the operating room. Most recently, a 2 µm continuous wave thulium laser, which retains some of the key cutting and ablative characteristics of the CO_2 laser but it is delivered through glass fibers, has been introduced into clinical practice.

While laryngeal applications have driven the use of lasers in otolaryngology, clinical efficacy has been shown for various lasers in a variety of treatment strategies within the broad spectrum of otolaryngologic diseases. The CO_2 laser has been used for facial skin resurfacing and in the treatment of rhinophyma. The neodymium:yttrium–aluminum–garnet (Nd:YAG) laser has been used for hereditary hemorrhagic telangiectasia (HHT) and treatment of chronic nasal obstruction due to turbinate hypertrophy. The argon laser has clinical utility in middle ear surgery during stapedectomy and lysis of adhesions due to chronic ear disease and in some skin resurfacing surgery during treatment of rhinophyma.

Lasers discussed in this chapter can lead to the enhanced treatment of a number of benign and malignant otolaryngologic disorders. There is an expanding group of applications for which fiber-based technologies have shifted the treatment paradigms from management in the operating room with general endotracheal anesthesia to the office setting with local anesthesia. Chronic epithelial proliferative diseases of the vocal fold surface such as papillomatosis and dysplasia can now be aggressively treated using photoangiolytic lasers with maximum preservation and/or restoration of voice.

63.2 Carbon Dioxide Laser

63.2.1 Laryngeal Disease

The CO_2 laser functions primarily as a hemostatic scalpel when the beam is focused. When the beam is defocused, the CO_2 laser can also be used effectively to ablate and cytoreduce epithelial disease such as diffuse papillomatosis. Operating at a wavelength 10.6 μm in the infrared region, CO_2 lasers deliver nonionizing electromagnetic radiation that is well absorbed by water, which is ubiquitous in the laryngeal soft tissues. Care must be taken near phonatory membranes because thermal energy can result in fibrosis of the delicate superficial lamina propria (SLP), which is the primary oscillator responsible for voice production. It is reasonable to use the CO_2 on the vibratory membranes when (1) there is no functional SLP present as may be encountered in patients who have had previous surgery, and (2) cancer has already invaded and replaced the SLP (Vaughan 1978).

The aiming beam (line-of-sight) delivery system using a joystick and foot pedal offers surgeons greater ease in performing precise bimanual surgery. However, the enhanced manual dexterity of the joystick is offset by the vaporization and ablation of varied amounts of the phonatory membranes. The CO_2 laser is also valuable in treating selected posterior glottal disorders that require arytenoidectomy or dissection of subepithelial stenosis. The microspot CO_2 laser is ill suited to treat benign subepithelial masses of the phonatory vocal fold such as nodules, polyps, and cysts. These lesions are optimally resected by cold-instrument tangential dissection with maximal preservation of underlying SLP and complete preservation of overlying epithelium.

63.2.2 Facial Skin Resurfacing

Resurfacing of photoaged skin with laser is a well-established tool for treatment of facial rhytids. The use of the CO_2 laser in skin resurfacing is advantageous because the thermal damage to the surrounding tissue causes an insult that prompts fibroblasts to increase collage production and results in the observed clinical effects (Ward and Baker 2008). There are three treatment modalities for facial skin resurfacing: ablative skin resurfacing (ASR), nonablative dermal remodeling (NDR), and fractional photothermolysis (FP) (Manstein et al. 2004).

The most effective laser treatment option for repair of most photodamaged skin has been ASR with the pulsed CO_2 laser; however, the side effects are significant (Manstein et al. 2004). In the first week following treatment, there is oozing, edema, crusting, and a burning discomfort. Adverse effects include acneiform eruption, herpes simplex outbreak, bacterial infections, and hyperpigmentation (Shamsaldeen, Peterson, and Goldman 2011). Therefore, ASR with Er:YAG laser was introduced because at 2940 nm, its wavelength is much closer to an absorption maximum of water (3000 nm), and therefore, there is shallower absorption depth, leading to less residual thermal damage (Manstein et al. 2004).

In 2004, Manstein et al. introduced the concept of FP for cutaneous remodeling in order to overcome the problems with ASR and NDR. FP creates microscopic thermal wound zones (MTZs) and specifically spares tissue surrounding each wound. Each MTZ allows fast epidermal repair and short migratory paths for keratinocytes (Manstein et al. 2004). The density of MTZs and the amount of space between them can be varied for a given energy level. The FP modality is currently used for treatment of photoaging, acne scarring, and skin laxity, among others. Few complications have been reported with FP and the neck-banding and ectropion seen in treatment of patients treated by fractional CO_2 laser are much less than what is seen with traditional CO_2 lasers and dermabrasion (Fife, Fitzpatrick, and Zachary 2009). Therefore, fractional resurfacing thru FP has resulted in a dependable skin rejuvenation system with minimal downtime and predictable results (Collawn 2007).

63.3 Photoangiolytic Lasers

63.3.1 Pulsed-Dye Laser (585 nm)

Anderson and Parrish's (1981) development of the concept of selective photothermolysis for the treatment of dermatologic vascular malformations led to the development of the 585 nm PDL because its wavelength is precisely selected to target an absorbance peak of oxyhemoglobin (approximately 571 nm) and to fully penetrate the intralumenal blood, thereby depositing heat uniformly into the vessel. The laser pulse width (0.5 ms) is precisely selected to contain the heat to the vessel without causing collateral damage to the extravascular soft tissue from heat conduction. Pilot studies were performed using the 585 nm PDL for laryngeal papillomatosis by Bower, Flock, and Waner (1998) and McMillan, Shapshay, and McGilligan (1998). Shortly thereafter, large-scale investigations in the treatment of a spectrum of laryngeal lesions, including vocal fold dysplasia (Franco et al. 2003), papillomatosis (Franco et al. 2002), and ectasias/varices, were reported (Zeitels 2002). The aberrant and/or abundant microvasculature present in each of the aforementioned lesions is a key feature when considering photoangiolysis for surgical management. The microcirculation could be targeted to involute laryngeal lesions (dysplasia, cancer, papilloma, and varices) or facilitate cold-instrument resection (ectasias and polyps) while minimizing thermal trauma to the surrounding soft tissue, SLP, and epithelium. In theory, this would be ideal for maintaining the pliability of the vocal folds' layered microstructure (SLP and epithelium) and glottal sound production.

PDLs have been effective in treating papillomatosis and dysplasia without the associated clinically observed soft-tissue complications associated with the CO_2 laser (thermal damage, tissue necrosis, SLP scarring, and anterior commissure web formation) (Zeitels 2004). The presumed mechanism of disease regression is the selective destruction of the subepithelial microvasculature and separation of the epithelium from the underlying SLP by denaturing the basement membrane zone-linking proteins. This results in ischemia to the diseased mucosa, albeit not permanently. This microvascular angiolysis approach restricts survival and growth of neoplastic epithelium while minimizing cytotoxicity to the delicate layered microstructure (SLP) of the vocal fold.

A potential disadvantage of PDL treatment is that it can be difficult to accurately quantify the energy delivery and real-time tissue effects, despite the fact that this laser is unlikely to cause substantial soft-tissue injury to the vocal folds. Furthermore, given the extremely short pulse width (approximately 0.5 ms), it is not unusual for the vessel walls of the microcirculation to rupture, resulting in extravasation of blood into the surrounding tissue. In laryngeal lesions such as papillomatosis, the extravasated blood diverts the laser energy in the form of a heat sink, which diminishes the effectiveness and reduces the selectivity of the laser.

63.3.2 Potassium–Titanyl–Phosphate Laser (KTP, 532 nm)

The KTP laser is a green-light laser with a wavelength of 532 nm, which coincides with one of the absorbance peaks of oxyhemoglobin. Like the PDL, this laser has a fiber-based delivery system. It has been used to treat papilloma, dysplasia, and vascular lesions within the larynx (Zeitels et al. 2006), and as a new surgical strategy in the management of early vocal fold cancer (Zeitels et al. 2008).

Comparative experiments between KTP and PDL lasers using the chick chorioallantoic membrane to simulate vocal fold microvasculature reveal several advantages of the KTP laser over the PDL (Broadhurst et al. 2007; Burns et al. 2008). The longer pulse width of the KTP laser (15 ms) as compared to the PDL (0.5 ms) created better coagulation and diminished vessel rupture, which had been experienced clinically with the PDL. The longer pulse width of the KTP laser also takes advantage of the fact that the energy delivery time is less than the thermal relaxation time of the tissue. Consequently, there is minimal collateral extravascular thermal soft tissue trauma compared with using the same laser in a continuous mode. The KTP laser has been used clinically to treat papilloma and dysplasia (Burns, Zeitels, and Anderson 2007). There has been less blood extravasation into the surrounding tissue, and the cytology of overlying diseased (papilloma and dysplasia) epithelium has been virtually unaltered by the subepithelial photoangiolysis.

63.3.3 Office-Based Applications of the KTP Laser

Photoangiolytic lasers are now used routinely in office-based laryngeal surgery to involute premalignant laryngeal disease and papilloma without resection (Zeitels 2004; Zeitels et al. 2006; Rees et al. 2006). The selectivity of photoangiolytic lasers leads to improved vocal outcomes by allowing for aggressive treatment of dysplasia and papilloma with maximum preservation of the layered microstructure of the vocal fold including the SLP. Our current treatment strategies utilizing office-based laryngeal

laser surgery are limited to treatment of dysplasia and papilloma, and we do not favor the use of office-based techniques in the management of malignancy, microvascular angiomata, benign phonotraumatic lesions such as polyps or chronic inflammatory conditions such as polypoid corditis unless the patient's medical comorbidities prohibit general anesthesia (Zeitels 2004). Office-based laryngeal laser surgery sacrifices a certain degree of precision due to the loss of binocular visualization, high-powered magnification, and an immobile and insensate operative field that exist when surgery is done to patients who are under general anesthesia. Due to the avoidance of multiple general anesthetics and the ability to treat regrowth of disease more often with less recovery time, office-based laryngeal laser surgery is advantageous in cases of recurrent dysplasia and papilloma.

Dysplasia and papillomatosis are the two most common indications for photoangiolysis using the pulsed KTP laser both in the operating room and in the office (Burns et al. 2010). Current strategy involves treating patients initially in the operating room where the extent of epithelial disease and the prior surgically induced soft-tissue changes can be adequately assessed. The KTP laser has proven its utility in ablating disease with maximum preservation of the underlying SLP, and this laser is used in almost every patient with dysplasia or papilloma.

63.4 Nd:YAG Laser

63.4.1 Rhinology: Turbinate Hypertrophy

The Nd:YAG laser emits light at a 1064 nm wavelength, making it the longest penetration depth (4 mm) of any of the surgical lasers. Within the subspecialty of rhinology in the head and neck, the Nd:YAG has been mainly described as in the role of reducing hyperplastic inferior nasal turbinates. A review article looking at over 2000 cases of laser treatment of hyperplastic inferior nasal turbinates concludes that irrespective of laser type, laser treatment can be considered a useful, cost-effective modality to treat turbinates, and it had comparable or better results than most traditional conventional techniques, such as electrocautery, cryotherapy, or conchotomy (Janda et al. 2001). Although various lasers have been used in this treatment, the solid-state Nd:YAG laser can be described as an effective tool for the reduction of hyperplastic inferior nasal turbinates. By penetrating into the soft tissues to coagulate zones of the turbinate's venous plexus with relative preservation of superficial epithelial layers, the Nd:YAG laser can cause less postoperative nasal crusting compared to other lasers (Janda et al. 2001).

63.4.2 Rhinology: HHT

The vascular malformations of inferior turbinates and nasal septum that characterize hereditary HHT and lead to intractable nosebleeds can be effectively managed with the Nd:YAG laser. The Nd:YAG laser can coagulate these large vessels readily and be delivered on a fiber that can bend within the nasal cavity for optimal delivery. However, again one must consider the significant depth of penetration with this laser, which may be problematic if one is delivering treatment to areas near the lamina papyracea, as the medial rectus muscle adjacent to it is dark in color and can readily absorb the laser's thermal effect. Thereby, the Nd:YAG should only be used on the septum and inferior turbinates where there is less risk of injury given a further location to and from any vital adjacent structures.

63.5 Argon Laser

The argon laser operates in the visible range (488–514 nm, blue-green light) in a continuous mode. Its small spot size is ideal during otologic surgery in the small confines of the middle ear. In the head and neck, the argon laser has most often been used in otologic surgery for stapedotomy (Perkins 1980; Strunk and Quinn 1993), or middle ear adhesions (DiBartolomeo and Ellis 1980), but has also been used in treatment of cutaneous lesions such as rhinophyma (Sadick et al. 2008) and in eustachian tuboplasty (Poe, Metson, and Kujawski 2003). During stapedectomy, the small spot size of the argon laser facilitates precise placement of a control hole prior to removal of the small stapes footplate. The argon laser has a shallow depth of penetration, thereby minimizing the thermal effect to the facial nerve or the saccule of the inner ear (both of which lie in close proximity to the stapes footplate).

As the argon laser is strongly absorbed by hemoglobin and melanin, it can also be used to treat vascular cutaneous lesions of the head and neck. Rhinophyma, a benign inflammatory growth of the caudal one-third of the nose, is a disease process most commonly affecting men in their fifth to seventh decades that can be both cosmetically and functionally impairing. The increased vascularity at the caudal aspect of the nose from the proliferation of sebaceous glands allows for multiple treatment modalities, including the use of either argon or CO_2 laser. The argon laser's advantage is that there is selective coagulation of blood capillaries. Both the argon and CO_2 laser provide effective hemostasis during sculpting of the nasal tip, but can cause dermal necrosis due to uncontrolled depth of tissue destruction and scar contraction (Sadick et al. 2008).

63.6 Thulium—2 μm Continuous Wave Laser

63.6.1 Laryngeal Disease

The thulium laser is a diode-pumped solid-state laser that has a thulium-doped YAG laser rod, which produces a continuous wave beam with a wavelength of 2.013 μm. This wavelength has a target chromophore of water and therefore simulates the hemostatic cutting properties of the CO_2 laser. Its fiber-based delivery system offers a distinct advantage in laryngeal surgery during three-dimensional and tangential cutting needed during endoscopic partial laryngectomy. The present authors have used this laser to perform a number of endoscopic partial laryngectomy procedures in both the glottis and the supraglottis. The most remarkable observation during a reported series of endoscopic

partial laryngectomies involving the glottis and supraglottis was that the procedure was never halted to stop bleeding from laser dissection during any case (Zeitels et al. 2006). Although preliminary observations suggest that there is increased thermal damage on the soft tissues at the margin of the cancerous section compared with the CO_2 laser, *ex vivo* studies indicate that the increased thermal effect is not excessive (Burns et al. 2007).

The thulium laser can also be used through the side-port working channel of a flexible laryngoscope to perform ablation of a variety of benign and malignant epithelial lesions of the larynx in an office-based setting with topical anesthesia. The clinical indication for use of the thulium laser in this way can damage the delicate layered microstructure of phonatory membranes, so care must be taken to properly select patients who require less selective debulking of their disease. The optimal clinical scenarios are still evolving for the thulium laser, but the overall positive preliminary clinical experience thus far (Zeitels et al. 2006) warrants further prospective investigations to determine its optimal application in laryngeal surgery.

63.6.2 Future Applications of Robotic Surgery Using Thulium Laser

The current trend toward minimally invasive surgery for upper aerodigestive tract cancer utilizes transoral laser resection along with transoral robot-assisted surgery. Experienced head and neck cancer surgeons report excellent survival and swallow function results with transoral laser microsurgery (TLM) as a primary treatment for advanced oropharyngeal malignancy (Haughey, Hinni, and Salassa 2011). Further, technological advances in robotic-assisted surgery continue to enhance the scope of this type of surgery (Hartl et al. 2011). With its fiber-based delivery system, the thulium laser is ideally suited for use with robotic surgical systems and provides effective hemostatic cutting during complex transoral removal of large upper aerodigestive tract tumors. Preliminary investigations are underway to determine the optimal parameters for use of the thulium laser during robotic-assisted surgery.

References

Anderson, R. R., and J. A. Parrish. 1981. Microvasculature can be selectively damaged using dye lasers: A basic theory and experimental evidence in human skin. *Lasers Surg Med.* 1:263–276.

Anderson, R. R., and J. A. Parrish. 1983. Selective photothermolysis: Precise microsurgery by selective absorption of pulsed radiation. *Science* 220:524–527.

Bower, C. M., S. Flock, and M. Waner. 1998. Flash pump dye laser treatment of laryngeal papillomas. *Ann Otol Rhinol Laryngol* 107:1001–1005.

Broadhurst, M. S., J. B. Kobler, L. M. Akst et al. 2007. Effects of pulsed KTP laser parameters on vessel ablation in the avian chorioallantoic membrane (CAM): Implications for vocal fold mucosa photoangiolysis. *Laryngoscope* 117(2):220–225.

Burns, J. A., A. D. Friedman, M. J. Lutch, R. E. Hillman, and S. M. Zeitels. 2010. Value and utility of 532nm pulsed potassium-titanyl-phosphate laser in endoscopic laryngeal laser surgery. *J Laryngol Otol* 124(4):407–411.

Burns, J. A., J. B. Kobler, J. T. Heaton, R. R. Anderson, and S. M. Zeitels. 2008. Predicting clinical efficacy of photoangiolytic and cutting/ablating lasers using the chick chorioallantoic membrane model: Implications for endoscopic voice surgery. *Laryngoscope* 118(6):1109–1124.

Burns, J. A., J. B. Kobler, J. T. Heaton et al. 2007. Thermal damage during thulium laser dissection of laryngeal soft tissue is reduced with air-cooling: An *ex-vivo* calf model study. *Ann Otol Rhinol Laryngol* 116(11):853–857.

Burns, J. A., S. M. Zeitels, and R. R. Anderson. 2007. 523 nm Pulsed KTP laser treatment of papillomatosis under general anesthesia. *Laryngoscope* 117(8):1500–1504.

Collawn, S. S. 2007. Fraxel skin resurfacing. *Ann Plastic Surg* 58(3):237–240.

DiBartolomeo, J. R., and M. Ellis. 1980. The argon laser in otology. *Laryngoscope* 90(11):1786–1796.

Fife, D. J., R. E. Fitzpatrick, and C. B. Zachary. 2009. Complications of fractional CO_2 laser resurfacing: Four cases. *Lasers Surg Med* 41(3):179–184.

Franco, R. A., S. M. Zeitels, W. A. Farinelli, and R. R. Anderson. 2002. 585-nm pulsed dye laser treatment of glottal papillomatosis. *Ann Otol Rhinol Laryngol* 111:486–492.

Franco, R. A., S. M. Zeitels, W. A. Farinelli, W. Faquin, and R. R. Anderson. 2003. 585-nm pulsed dye laser treatment of glottal dysplasia. *Ann Otol Rhinol Laryngol* 112(9) Part 1:751–758.

Hartl, D. M., A. Ferlito, C. E. Silver et al. 2011. Minimally invasive techniques for head and neck malignancies: current indications, outcomes and future directions. *Eur Arch Otorhinolaryngol* 268(9):1249–1257.

Haughey, B. H., M. L. Hinni, and J. R. Salassa. 2011. Transoral laser microsurgery as primary treatment for advanced-stage oropharyngeal cancer: A United States multicenter study. *Head Neck* 33(12):1683–1694.

Jako, G. J. 1972. Laser surgery of the vocal cords. *Cope* 82:2204–2215.

Janda, P., R. Sroka, R. Baumgartner, G. Grevers, and A. Leunig. 2001. Laser treatment of hyperplastic inferior nasal turbinates: A review. *Lasers Surg Med* 28(5):404–413.

Manstein, D., G. S. Herron, R. K. Sink, H. Tanner, and R. R. Anderson. 2004. Fractional photothermolysis: A new concept for cutaneous remodeling using microscopic patterns of thermal injury. *Lasers Surg Med* 34(5):426–438.

McMillan, K., S. M. Shapshay, and J. A. McGilligan. 1998. A 585-nanometer pulsed dye laser treatment of laryngeal papillomas: Preliminary report. *Laryngoscope* 108:968–972.

Perkins, R. C. 1980. Laser stepedotomy for otosclerosis. *Laryngoscope* 90(2):228–240.

Poe, D. S., R. B. Metson, and O. Kujawski. 2003. Laser eustachian tuboplasty: A preliminary report. *Laryngoscope* 113(4):583–591.

Polanyi, T., H. C. Bredermeier, and T. W. Davis. 1970. CO_2 laser for surgical research. *Med Biol Eng Comput* 8:548–858.

Rees, C. J., S. L. Halum, R. C. Wijewickrama et al. 2006. Patient tolerance of in-office pulsed dye laser treatments to the upper aerodigestive tract. *Otolaryngol Head Neck Surg* 134:1023–1027.

Sadick, H., B. Goepel, C. Bersch et al. 2008. Rhinophyma: Diagnosis and treatment options for a disfiguring tumor of the nose. *Ann Plastic Surg* 61(1):114–120.

Shamsaldeen, O., J. D. Peterson, and M. P. Goldman. 2011. The adverse events of deep fractional CO_2: A retrospective study of 490 treatments in 374 patients. *Lasers Surg Med* 43(6):453–456.

Strong, M. S. 1975. Laser excision of carcinoma of the larynx. *Laryngoscope* 85:1286–1289.

Strong, M. S., C. W. Vaughan, S. R. Cooperband et al. 1976. Recurrent respiratory papillomatosis: Management with the CO_2 laser. *Ann Otol Rhinol Laryngol* 85:508–516.

Strunk, C. L., Jr., and F. B. Quinn, Jr. 1993. Stapedectomy surgery in residency: KTP-532 laser versus argon laser. *American J Otol.* 14(2):113–117.

Vaughan, C. W. 1978. Transoral laryngeal surgery using the CO_2 laser: Laboratory experiments and clinical experience. *Laryngoscope* 88:399–420.

Ward, P. D., and S. R. Baker. 2008. Long-term results of carbon dioxide laser resurfacing of the face. *Arch Facial Plastic Surg* 10(4):238–243.

Zeitels, S. M., L. M. Akst, J. A. Burns et al. 2006. Pulsed angiolytic laser treatment of ectasias and varices in singers. *Ann Otol Rhinol Laryngol* 115:571–580.

Zeitels, S. M., J. A. Burns, G. Lopez-Guerra, R. R. Anderson, and R. E. Hillman. 2008. Photoangiolytic laser treatment of early glottic cancer: A new management strategy. *Ann Otol Rhinol Laryngol* 117(7), part 2:1–24.

Zeitels, S. M., R. A. Franco, S. H. Dailey et al. 2004. Office-based treatment of glottal dysplasia and papillomatosis with the 585-nm pulsed dye laser and local anesthesia. *Ann Otol Rhinol Laryngol* 113(4): 265–276.

64

Laser Targeting to Nanoparticles for Theranostic Applications

James Ramos
Arizona State University

Huang-Chiao Huang
Massachusetts General Hospital

Kaushal Rege
Arizona State University

64.1 Introduction

In several diseases, including cancer, there is an urgent need for formulating platforms that are capable of (1) detection, (2) treatment, and (3) determination of therapeutic effect (therapeutic response) of the diseased tissue (Young, Figueroa, and Drezek 2012). Detection involves the specific labeling of malignant tissue so that its localization can be differentiated from that of healthy tissue. Following labeling and localization of the malignant tissue, a wide variety of options exist for treating the disease depending on the extent of the disease. One such option is the targeted delivery of a therapeutic payload, including chemotherapeutic drugs or biologicals, for treatment of the disease. In addition to targeted delivery, control of the payload release via an external stimulus, as opposed to passive release, may be desirable in many cases. This doubly ensures that the payload reaches the desired targeted tissue and will only be released in its specific location. Following treatment, determination of the therapeutic effect is also desirable and may be accomplished in a similar way as the initial labeling. The challenge to encompass these needs in a single modality has resulted in a push toward the formulation of theranostic treatment options, wherein a single platform can address multiple or all of these needs. Recent investigations in nanotechnology have made it possible to develop single-platform theranostics options. More specifically, photoresponsive nanoparticles have demonstrated potential as viable theranostic platforms. Light absorption and surface modification properties of these nanoparticles can be exploited for targeting and labeling diseased tissues via multiple imaging modalities. In addition, their ability to absorb specific wavelengths of light has been exploited for photothermal ablation of malignant tissue, controlled release of treatment payloads, and combination treatments resulting in synergistic therapeutic options. In this chapter, we will discuss the utility of photoresponsive nanoparticles, which possess the ability to absorb laser light, as novel theranostic options for the diagnosis and treatment of disease.

64.2 Imaging and Diagnostics

Noninvasive and minimally invasive imaging methods for bioimaging are invaluable tools for clinical diagnostics of diseased tissue. Current bioimaging techniques range from being able to image whole organisms to specific molecular imaging. Emerging imaging techniques have the potential to aid early diagnostics and treatment regimens with potential applications in detecting early-stage cancer, guides for drug and biological therapies, and even for guiding surgery, among several others. The ability to use nanoparticles as contrast agents has attracted increased attention, as they can be formulated to overcome drawbacks such as chemical and photostability, detection limits, and compatibility, associated with conventional contrast agents (Hahn et al. 2011). The use of imaging modalities using near-infrared (NIR) lasers have been studied, since NIR light penetrates tissue at greater depths compared to visible wavelengths of light, demonstrates minimal tissue scattering, and possesses reduced phototoxicity. Thus, nanoparticles synthesized and tuned to absorb NIR light have been widely explored for use in multiple imaging modalities (Durr et al. 2007; Li and Gu 2010; Wang et al. 2005). This section will discuss current methods to exploit NIR light absorbing nanoparticles for use in imaging.

64.2.1 Two-Photon Imaging

Two-photon imaging is a technique in which femtosecond to picosecond pulsed, high-intensity NIR Ti:S lasers are used to deliver a high photon flux to a specimen of interest (Perry, Burke, and Brown 2012). Due to their surface plasmon resonance property, noble metal nanoparticles have emerged as viable targets for two-photon luminescence (TPL). TPL is a serial process in which photons are sequentially absorbed and emission results from the recombination of electrons in the sp-band and holes in the d-band. Using the same methods, two-photon fluorescence (TPF) with fluorophores can also be achieved by excitation due to the absorption of two photons nearly instantaneously. Gold nanorods (GNRs) have emerged as particularly appealing due to ease of synthesis methods, controllable tuning of their longitudinal plasmon absorbance in the NIR region, and reduced plasmon dampening leading to increased TPL signal (Durr et al. 2007).

In a study performed by the Prasad group (Zhu et al. 2010), GNRs were synthesized and modified with a polyelectrolyte multilayer coating resulting in shielding of the toxic cetyltrimethylammonium bromide (CTAB) layer but maintaining a net positive charge. The GNR assemblies were then modified with transferrin that was deposited onto the surface via electrostatic interactions with the polyelectrolyte coating. These GNR-based assemblies were used to treat human pancreatic carcinoma (Panc-1) cells that overexpress the transferrin receptor *in vitro* and were subsequently imaged using TPL. Cells treated with transferrin-GNR assemblies demonstrated a significantly higher TPL signal compared to those treated with unmodified GNR assemblies and those presaturated with transferrin and then treated with transferrin-GNR assemblies (competitive inhibition). The Ben-Yakur group (Durr et al. 2007) demonstrated targeted delivery of anti-epidermal growth factor receptor (EGFR) antibody-conjugated GNRs for labeling of EGFR-overexpressing A431 skin cancer cells. A 60-fold reduction in laser power was capable of giving a GNR TPL signal compared to the autofluorescence of the cells; this also corresponded to an increase in emitted signal for equal excitation intensity by three orders of magnitude. In a study carried out by the Chang group (Wang et al. 2005), the TPL signal of a single GNR demonstrated an intensity that was 58-fold times higher than the signal from a single rhodamine dye molecule. They also demonstrated GNR TPL *in vivo* by injecting the GNRs into the tail vein in a mouse model. GNR TPL was employed to monitor the flow of GNRs in the blood vessels of the mouse earlobe, and it was found that the GNR signal was threefold greater than the autofluorescence of the blood and surrounding tissue. These studies demonstrate the potential for using nanoparticle systems that absorb NIR light with TPL for detection and labeling of cancer. In principle, localization can be achieved by targeting nanoparticles to specific receptors overexpressed on cancer cells. This could further be explored for targeted and guided delivery or relevant drugs for therapeutic effect.

The Gu group (Li and Gu 2010) carried out a study in which transferrin was conjugated to PEGylated GNRs and used for targeted treatment of the HeLa human cervical cancer cell line. Autofluorescence of HeLa cells was observed at laser powers above 35 mW, while laser powers as high as 20 mW were needed in order to obtain a sufficiently strong signal in case of

the fluorescein isothiocyanate (FITC) dye. However, in the case of the transferrin-conjugated PEGylated GNRs as probes, it was found that laser powers as low as 0.3 mW were sufficient for a strong luminescence signal. Photothermal induction of apoptosis was also possible with significantly lesser laser power for the transferrin-conjugated PEGylated GNR-treated cells than in untreated cells. Thus, in addition to targeted laser-based labeling of cancer, this study demonstrates the ability to combine TPL imaging and cancer photothermal therapy using a single nanoparticle platform. These studies are indicative of photoresponsive nanoparticle-based multimodal platforms.

64.2.2 Photoacoustic Imaging

Photoacoustic imaging (PAI) is emerging as a powerful noninvasive imaging technique. In photoacoustic tomography (PAT), short-pulsed laser beams are used to elicit thermoexpansion of an absorber. This then creates pressure, or photoacoustic (PA) waves, that are detected by a wideband ultrasonic detector and allows for three-dimensional reconstruction of the optical absorption properties of the internal tissue structure. Due to their unique optical absorption properties, gold nanoparticles have proven to serve as useful contrast agents in PAT (Mallidi et al. 2009; Yang et al. 2009).

In a study from the Emelianov group (Bayer et al. 2011), silica-coated GNRs (SiO_2-GNRs) were employed with PAI to differentiate between different cell types. This was accomplished by tuning the NIR absorption peak of one set of SiO_2-GNRs to 780 nm and bioconjugating them with human epidermal growth factor receptor 2 (HER2). The second set was set to NIR absorption at 830 nm and was bioconjugated with EGFR. The different SiO_2-GNRs were then used to treat A431 cells, which overexpress EGFR, and MCF7 cells, which overexpress HER2. Following treatment and upon analysis of the signal intensity as a function of laser wavelength, the PA signal was correlated to optical absorption spectra of the different SiO_2-GNRs. It was possible to differentiate between malignant cell types by using nanoparticles that possess different NIR light absorption properties. In principle, this can allow for multiplexing, leading to the detection of different disease states or phenotypes. Thus, it would be possible to formulate different therapy regimes on these nanoparticle platforms that can be employed and targeted for each respective disease and can be guided via PAT.

In a rat model study performed by the Xia group (Cal et al. 2011), PAT was used to quantitatively evaluate gold nanocage transport in the lymphatic system and uptake in the lymph nodes. Concentration, size, and surface characteristics of the gold nanocages were optimized for imaging at a depth of approximately 12 mm. It was found that 50 nm gold nanocages demonstrated a stronger PA signal compared to 30 nm nanocages due to their larger optical absorption cross section. However, the larger nanocages exhibited slower transport and accumulation in the sentinel lymph node (SLN). Additionally, it was found that gold nanocages with neutral surface charge had the fastest transport to SLNs, while positively charged gold nanocages exhibited the slowest transport. It was also found that the gold nanocages were transported to the second and third axillary lymph nodes, which can have potential implications for use in metastatic lymph node mapping. A study from the Wang group (Kim et al. 2010) demonstrated the use of bioconjugated gold nanocages for active targeting to subcutaneously inoculated melanomas in mouse models, resulting in 300% higher signal enhancement compared to PEG-gold nanocages that demonstrate passive targeting due to the EPR effect. These studies demonstrate the ability to couple different targeted nanoparticles with PAT for *in vivo* imaging. Additionally, the importance of modifying the nanoparticles with targeting moieties is underlined, since the PAT signal greatly decreases in the absence of targeting abilities.

The Li group (Lu et al. 2011) demonstrated the use of hollow gold nanospheres (HAuNs) as a single platform for both PAT as well as photothermal therapy. Intravenously injected HAuNs were targeted to integrins overexpressed in both glioma and angiogenic blood vessels in orthotopically inoculated U87 glioma in nude mice. Following injection, PAT was used to image HAuNs for determining the localization of the tumor in the brain. Subsequently, they were able to irradiate the tumor location with NIR laser and observed a tumor-ablation effect. Although tumor reoccurrence was observed approximately 10 days following ablation, a significant increase in median survival time was observed in cases of mice treated with NIR light. It was mentioned that additional or different treatments can be employed using the same system to further reduce the instance of tumor recurrence. This study demonstrated the use of a single platform for PAT imaging as well as guidance of photothermal therapy. This work serves as an example of a single nanoparticle platform developed as a theranostic option for both tumor imaging and photothermal therapy.

64.3 Light-Triggered Drug Delivery Systems

Nanotechnology-based drug delivery systems have progressed significantly in the recent past. Nanoparticles are ideal vectors for drug loading because of their high surface area-to-volume ratio, enhanced site-specific targeting, capability of cell membrane penetration, and high biocompatibility after proper surface modification. Several US Food and Drug Administration (FDA)-approved lipid- or polymer-based nanoparticles have shown great promise for altering pharmacodynamics, reducing systemic drug toxicity, and increasing drug concentration at desired locations for human clinical use. Several inorganic agents are in preclinical studies and clinical trials (Huang et al. 2011a). The ability of the vectors to control the release of drug molecules at desired sites has been demonstrated to further improve treatment efficacy in numerous preclinical studies. Several controlled release mechanisms, including laser irradiation, diffusion, microenvironmental pH shift, and enzymatic activity, have been proposed (Cheng et al. 2010; Park et al. 2010; Sershen et al. 2000; West et al. 2002; Wu et al. 2010). Employing light as an external stimulus to trigger cargo release from vectors is an attractive option when precise spatial and temporal controlled release is required. This section reviews different light-triggered drug delivery systems.

64.3.1 Gold Nanoparticles

Light-induced plasmonic heating for cargo release from nanovectors has been extensively investigated in numerous preclinical studies but has not yet progressed to clinical trials. In general, plasmonic heating of gold nanoparticles can induce the following mechanisms to trigger cargo release: photothermal-induced surface coating collapse/shrinkage, cleavage of the covalent gold–thiol bonds between nanoparticle and payload, and nanoparticle shape transformation leading to surface area reduction. Different shapes of gold nanoparticles, including nanoshells, nanorods, and nanocubes, demonstrate a response to NIR light and are promising platforms for controllable temporal and spatial drug release.

NIR-triggered drug release using polymer-gold nanoshell composites was first introduced by Dr. Naomi Halas, Dr. Jennifer West, and others in 2000 (Sershen et al. 2000; West et al. 2002). In this study, heating of plasmonic nanoshells upon pulsed laser irradiation resulted in the collapse of thermoresponsive copolymers (N-isopropylacrylamide and acrylamide), resulting in triggered release of albumin. Gold nanoshells (also known as AuroShells) are currently in clinical trials for laser treatment of head and neck tumors (http://www.clinicaltrials.gov). In a related study, Yavuz et al. (2009) demonstrated light-triggered release of doxorubicin (DOX) drugs from copolymer-covered gold nanocages to induce death of breast cancer cells *in vitro*. Shiotani et al. (2007) employed NIR light to trigger the rapid shrinkage of the poly (N-isopropylacrylamide) hydrogels coated around GNRs and release rhodamine-labeled dextrans. Recently, we demonstrated the formation of GNR-elastin-like polypeptide composites that, upon NIR irradiation, simultaneously resulted in photothermally induced moderate hyperthermia and release of chemotherapeutic drugs, resulting in synergistic ablation of human prostate cancer cells *in vitro* (Huang et al. 2011b).

A number of photoreaction studies using gold nanoparticles and pulsed laser irradiation have demonstrated successful release of bound biomolecules via linkage cleavage. The Rotello group (Agasti et al. 2009) investigated the use of gold nanospheres as a transporter for the anticancer drug 5-fluorouracil (5-FU). The 5-FU moieties were attached to the photocleavable and zwitterionic thiol ligands already functionalized on the gold surface. Upon ultraviolet (UV) radiation at 365 nm, photolytic cleavage was observed and allowed for a controlled release of the attached 5-FU moieties. It was also found that human breast adenocarcinoma cells (MCF-7) treated with 5-FU-conjugated gold nanospheres and exposure to UV irradiation resulted in significant decrease in cell viability compared to cells treated with only light or only 5-FU-conjugated gold nanospheres. In addition to release of drug molecules, triggered release of DNA by light and mediated by photocleavable cationic gold nanoparticles was reported by Han et al. (2006). They synthesized positively charged gold nanospheres with a photoactive o-nitrobenzyl ester linkage associated with deoxyribonucleic acid (DNA) by means of electrostatic interactions. Near-UV irradiation allowed cleavage of the nitrobenzyl linkage, which resulted in the release of positively charged alkyl amines from residual negatively charged carboxylate groups. This charge reversal further allowed for efficient release of the bound DNA, presumably due to electrostatic repulsion. Delivery of fluorescein amidite (FAM)-labeled 37-mer DNA and release into mouse embryonic fibroblast cells was observed with significant nuclear localization following UV irradiation.

Light-triggered cargo release from GNRs is attributed not only to the cleavage gold–thiol bonds but also to shape transformation (from cylindrical rods to spheres), leading to a reduction in nanoparticle surface area. Shuji Yamashita, Katayama, and Niidome (2009) reported the release of more than 60% of conjugated PEG molecules from GNRs following pulsed laser irradiation. Chen et al. (2006) self-assembled enhanced green fluorescent protein (EGFP) encoded DNA onto GNRs via the gold–thiol bond and studied gene expression using HeLa cells upon femtosecond NIR laser irradiation. NIR laser irradiation triggered the release of DNA, which led to subsequent EGFP expression in localized regions after internalization; free EGFP DNA with and without laser and EGFP DNA-GNR conjugates alone did not result in transgene expression. Wijaya et al. (2009) demonstrated the selective controlled release of two independent DNA oligonucleotides from two different GNR populations by matching their characteristic peak absorption wavelength (λ_{max}) in the NIR region. Specifically, 6-carboxyfluorescein-labeled DNA conjugated on shorter nanorods (λ_{max} at 800 nm) and tetramethylrhodamine-labeled DNA conjugated on longer nanorods (λ_{max} at 1100 nm) were separately released by using 800 and 1100 nm wavelength lasers, respectively.

Gold nanoparticles have also been employed to deliver photosensitizers for combined photothermal therapy (PTT) and photodynamic therapy (PDT). PDT involves light activation of photosensitizers leading to the formation of reactive oxygen species at desired locations. Zaruba et al. (2010) demonstrated that intravenously injected porphyrin-modified gold nanospheres (14.7 nm) were more effective in reducing basaloid squamous carcinoma volume in nude mice compared to porphyrin alone. Kah et al. (2008) reported simultaneous excitation of hypericin and gold nanoshells using a wide band illumination above 585 nm. Combination treatment resulted in a pronounced death of nasopharyngeal carcinoma (CNE2) cells compared to PDT and PTT alone *in vitro*. Recently, Jang et al. (2011) self-assembled Al(III) phthalocyanine chloride tetrasulfonic acid (AlPcS4) onto GNRs via electrostatic interactions and intravenously injected GNR–photosensitizer complexes into squamous cell carcinoma (SCC7) bearing mice. An additive effect in the reduction of tumor growth was observed upon sequential PTT/PDT treatment.

64.3.2 Lipid-Based Nanoparticles

Liposomes, composed of phospholipid bilayer membranes, are one of the most widely investigated nanoscale carriers (Torchilin 2005).

However, the inability of liposomes to readily release encapsulated drug molecules can potentially compromise therapeutic efficacy. Payload release from liposomes can be light-triggered remotely by the introduction of light-responsive inorganic nanoparticle (Babincova et al. 1999; Paasonen et al. 2007; Volodkin, Skirtach, and Mohwald 2009). In general, three types of liposome–nanoparticle complexes have been investigated: (1) nanoparticles in the inner aqueous core, (2) nanoparticles in the lipid membrane, and (3) nanoparticles in the vicinity of/on the liposome surface (Kojima et al. 2008; Park et al. 2006). Upon laser irradiation, photothermal effect (heat generation) and/or photoacoustic effect (microbubble formation and collapse, cavitation) from inorganic nanoparticles can cause liposomal membrane disruption and facilitate release of liposome contents.

In 1999, Babincova et al. (1999) synthesized thermosensitive dipalmitoylphosphatidylcholine (DPPC) liposomes that encapsulated dextran-magnetite (Fe_3O_4) nanoparticles and dye molecules. In the presence of pulsed laser (10 ns duration, 50 mJ/pulse, and 380 nm), heat transfer from Fe_3O_4 nanoparticles resulted in a temperature increase above the liposomal gel-liquid crystal transition temperature (~41°C) and led to significant drug escape. In a subsequent study (Paasonen et al. 2010), hydrophobic gold nanospheres were embedded in the lipid bilayer of distearoylphosphatidylcholine (DSPC)/DPPC liposomes. UV light-triggered calcein dye release from liposomes was only observed in the presence of gold nanoparticles using human retinal pigment epithelial (ARPE-19) cells. For *in vivo* applications, the use of UV-visible light (200–750 nm) to trigger cargo release from liposome–nanoparticle complexes is less efficient due to absorption of UV-visible light by tissues. Thus, choosing a wavelength in the near NIR (750–1000 nm) window to minimize absorption by tissue is preferred (Qin and Bischof 2012). In 2008, Wu et al. (2008) triggered 6-carboxyfluorescein release from DPPC liposome-hollow gold nanoshell complexes by pulsed NIR laser (130 fs duration, 670 µJ/pulse, and 800 nm). The release of fluorescent dye was suggested to occur due to lipid bilayer perforation, which was a result of the nanoshell photoacoustic effect and not due to permanent liposomal damage induced by thermally triggered phase transition. Moreover, nanoshells tethered to liposomes were found to provide the highest cargo release efficacy upon pulsed laser irradiation compared to liposomes with nanoshells encapsulated within or suspended freely outside. Park et al. (2010) demonstrated that GNR-mediated photothermal heating can result in release of DOX, an antitumor drug, from thermally sensitive liposomes (Gaber et al. 1995). In this *in vitro* study, MDA-MB-435 human melanoma cells were incubated with DOX-liposome and GNRs and then subjected to continuous wave (CW) laser exposure (810 nm, 0.75 W/cm² for 15 min). DOX-related toxicity toward cancer cells was found to be most significant when DOX-liposome, GNR, and laser irradiation were all present.

In addition to liposome–gold nanoparticle complexes, hollow gold nanospheres were employed to modulate the release of paclitaxel (PTX), an anticancer drug, from biodegradable poly(lactic-*co*-glycolic acid) (PLGA) microspheres (Chen et al. 2006). In the study, human U87 gliomas and MDA-MB-231 mammary-tumor-bearing nude mice were intratumorally injected with PLGA microspheres (loaded with hollow gold nanosphere and PTX drugs) and subjected to NIR irradiation (~800 nm). Light irradiation triggered PTX release from PLGA microspheres, resulting in significant decrease in tumor volume compared to mice treated with PTX-loaded microspheres alone and gold nanosphere-loaded microspheres plus laser.

64.4 Conclusions

The field of nanotechnology and the use of nanoparticles with the ability to absorb laser light have tremendous potential as theranostics in medicine. Through functionalization for malignant tissue targeting, laser-targeted nanoparticles have been used for diseased tissue labeling and imaging, therapeutic payload delivery with controlled release, and PTT. Future investigations are needed to gather more preclinical and clinical safety and efficacy information on these promising systems in order to facilitate their transition to clinical use.

References

Agasti, S., A. Chompoosor, C. You et al. 2009. Photoregulated release of caged anticancer drugs from gold nanoparticles. *J Am Chem Soc* 131:5728–5729.

Babincova, M., P. Sourivong, D. Chorvat, and P. Babinec. 1999. Laser triggered drug release from magnetoliposomes. *J Magn Magn Mater* 194:163–166.

Bayer, C. L., Y.-S. Chen, S. Kim et al. 2011. Multiplex photoacoustic molecular imaging using targeted silica-coated gold nanorods. *Biomed Opt Exp* 2:1828–1835.

Cal, X., W. Li, C.-H. Kim et al. 2011. *In vivo* quantitative evaluation of the transport kinetics of gold nanocages in a lymphatic system by noninvasive photoacoustic tomography. *ACS Nano* 5:9658–9667.

Chen, C., Y. Lin, C. Wang et al. 2006. DNA-gold nanorod conjugates for remote control of localized gene expression by near infrared irradiation. *J Am Chem Soc* 128:3709–3715.

Cheng, Y., A. C. Samia, J. Li et al. 2010. Delivery and efficacy of a cancer drug as a function of the bond to the gold nanoparticle surface. *Langmuir* 26:2248–2255.

Durr, N. J., T. Larson, D. K. Smith et al. 2007. Two-photon luminescence imaging of cancer cells using molecularly targeted gold nanorods. *Nano Lett* 7:941–945.

Gaber, M. H., K. L. Hong, S. K. Huang, and D. Papahadjopoulos. 1995. Thermosensitive sterically stabilized liposomes—formulation and *in-vitro* studies on mechanism of doxorubicin release by bovine serum and human plasma. *Pharm Res* 12:1407–1416.

Hahn, M. A., A. K. Singh, P. Sharma, S. C. Brown, and B. M. Moudgil. 2011. Nanoparticles as contrast agents for *in-vivo* bioimaging: Current status and future perspectives. *Anal Bioanal Chem* 399:3–27.

Han, G., C.-C. You, B.-J. Kim et al. 2006. Light-regulated release of DNA and its delivery to nuclei by means of photolabile gold nanoparticles. *Angew Chem Int Ed Engl* 45: 3165–3169.

Huang, H. C., S. Barua, G. Sharma, S. K. Dey, and K. Rege. 2011a. Inorganic nanoparticles for cancer imaging and therapy. *J Control Release* 155:344–357.

Huang, H.-C., Y. Yang, A. Nanda, P. Koria, and K. Rege. 2011b. Synergistic administration of photothermal therapy and chemotherapy to cancer cells using polypeptide-based degradable plasmonic matrices. *Nanomedicine* 6:459–473.

Jang, B., J. Y. Park, C. H. Tung, I. H. Kim, and Y. Choi. 2011. Gold nanorod-photosensitizer complex for near-infrared fluorescence imaging and photodynamic/photothermal therapy *in vivo*. *ACS Nano* 5:1086–1094.

Kah, J. C. Y., R. C. Y. Wan, K. Y. Wong et al. 2008. Combinatorial treatment of photothermal therapy using gold nanoshells with conventional photodynamic therapy to improve treatment efficacy: An *in vitro* study. *Lasers Surg Med* 40:584–589.

Kim, C., E. C. Cho, J. Chen et al. 2010. *In vivo* molecular photoacoustic tomography of melanomas targeted by bioconjugated gold nanocages. *ACS Nano* 4:4559–4564.

Kojima, C., Y. Hirano, E. Yuba, A. Harada, and K. Kono. 2008. Preparation and characterization of complexes of liposomes with gold nanoparticles. *Colloids Surf B Biointerfaces* 66:246–252.

Li, J.-L., and M. Gu. 2010. Surface plasmonic gold nanorods for enhanced two-photon microscopic imaging and apoptosis induction of cancer cells. *Biomaterials* 31:9492–9498.

Lu, W., M. P. Melancon, C. Xiong et al. 2011. Effects of photoacoustic imaging and photothermal ablation therapy mediated by targeted hollow gold nanospheres in an orthotopic mouse xenograft model of glioma. *Cancer Res* 71:6116–6121.

Mallidi, S., T. Larson, J. Tam et al. 2009. Multiwavelength photoacoustic imaging and plasmon resonance coupling of gold nanoparticles for selective detection of cancer. *Nano Letters* 9:2825–2831.

Paasonen, L., T. Laaksonen, C. Johans et al. 2007. Gold nanoparticles enable selective light-induced contents release from liposomes. *J Control Release* 122:86–93.

Paasonen, L., T. Sipila, A. Subrizi et al. 2010. Gold-embedded photosensitive liposomes for drug delivery: Triggering mechanism and intracellular release. *J Control Release* 147:136–143.

Park, J.-H., G. von Maltzahn, L. L. Ong et al. 2010. Cooperative nanoparticles for tumor detection and photothermally triggered drug delivery. *Adv Mater* 22:880–885.

Park, S. H., S. G. Oh, J. Y. Mun, and S. S. Han. 2006. Loading of gold nanoparticles inside the DPPC bilayers of liposome and their effects on membrane fluidities. *Colloids Surf B Biointerfaces* 48:112–118.

Perry, S. W., R. M. Burke, and E. B. Brown. 2012. Two-photon and second harmonic microscopy in clinical and translational cancer research. *Ann Biomed Eng* 40:277–291.

Qin, Z. P., and J. C. Bischof. 2012. Thermophysical and biological responses of gold nanoparticle laser heating. *Chem Soc Rev* 41:1191–1217.

Sershen, S. R., S. L. Westcott, N. J. Halas, and J. L. West. 2000. Temperature-sensitive polymer-nanoshell composites for photothermally modulated drug delivery. *J Biomed Mater Res A* 51:293–298.

Shiotani, A., T. Mori, T. Niidome, Y. Niidome, and Y. Katayama. 2007. Stable incorporation of gold nanorods into N-isopropylacrylamide hydrogels and their rapid shrinkage induced by near-infrared laser irradiation. *Langmuir* 23:4012–4018.

Shuji Yamashita, Y. N., Y. Katayama, and T. Niidome. 2009. Photochemical Reaction of poly(ethylene glycol) on gold nanorods induced by near infrared pulsed-laser irradiation. *Chem Lett* 3:226–227.

Torchilin, V. P. 2005. Recent advances with liposomes as pharmaceutical carriers. *Nat Rev Drug Discov* 4:145–160.

Volodkin, D. V., A. G. Skirtach, and H. Mohwald. 2009. Near-IR remote release from assemblies of liposomes and nanoparticles. *Angew Chem Int Ed Engl* 48:1807–1809.

Wang, H. F., T. B. Huff, D. A. Zweifel et al. 2005. *In vitro* and *in vivo* two-photon luminescence imaging of single gold nanorods. *Proc Natl Acad Sci USA* 102:15752–15756.

West, J. L., S. R. Sershen, N. J. Halas, S. J. Oldenburg, and R. D. Averitt. 2002. Temperature-sensitive polymer/nanoshell composites for photothermally modulated drug delivery.

WM. Marsh Rice University. US Patent 6428811 B1, filed July 14, 2000, and issued February 5, 2002.

Wijaya, A., S. Schaffer, I. Pallares, and K. Hamad-Schifferli. 2009. Selective release of multiple DNA oligonucleotides from gold nanorods. *ACS Nano* 3:80–86.

Wu, G., A. Milkhailovsky, H. A. Khant et al. 2008. Remotely triggered liposome release by near-infrared light absorption via hollow gold nanoshells. *J Am Chem Soc* 130:8175–8177.

Wu, W. T., T. Zhou, A. Berliner, P. Banerjee, and S. Q. Zhou. 2010. Smart core-shell hybrid nanogels with Ag nanoparticle core for cancer cell imaging and gel shell for pH-regulated drug delivery. *Chem Mater* 22:1966–1976.

Yang, X., E. W. Stein, S. Ashkenazi, and L. V. Wang. 2009. Nanoparticles for photoacoustic imaging. *Wiley Interdiscip Rev Nanomed Nanobiotechnol* 1:360–368.

Yavuz, M. S., Y. Cheng, J. Chen et al. 2009. Gold nanocages covered by smart polymers for controlled release with near-infrared light. *Nat Mater* 8:935–939.

Young, J. K., E. R. Figueroa, and R. A. Drezek. 2012. Tunable nanostructures as photothermal theranostic agents. *Ann Biomed Eng* 40:438–459.

Zaruba, K., J. Kralova, P. Rezanka et al. 2010. Modified porphyrin-brucine conjugated to gold nanoparticles and their application in photodynamic therapy. *Org Biomol Chem* 8:3202–3206.

Zhu, J., K.-T. Yong, I. Roy et al. 2010. Additive controlled synthesis of gold nanorods (GNRs) for two-photon luminescence imaging of cancer cells. *Nanotechnology* 21:285106.

Tomas Hode
Immunophotonics Inc.

Xiaosong Li
*The First Affiliated Hospital of
Chinese PLA General Hospital*

Mark Naylor
*Dermatology Associates
of San Antonio*

Lars Hode
Swedish Laser-Medical Society

Peter Jenkins
Irradia LLC

Gabriela Ferrel
*Hospital Nacional Edgardo
Rebagliati Martins*

Robert E. Nordquist
Immunophotonics Inc.

Orn Adalsteinsson
*International Strategic
Cancer Alliance*

John Lunn
*Commonwealth Medical
Research Institute*

Michael R. Hamblin
Massachusetts General Hospital

Luciano Alleruzzo
Immunophotonics Inc.

Wei R. Chen
University of Central Oklahoma

65

Laser Immunotherapy

65.1 Introduction

The ultimate control of cancer lies in the host's immune surveillance and defense system. Many immunotherapy strategies have been proposed to target cancer, including cytokine therapy, dendritic cell (DC)-based vaccines, and immune-activating antibodies, either alone or in various combinations with other therapies. However, thus far, these immunotherapy strategies have yielded low response rates, usually in the range of 5%–10% among treated patients. Novel approaches are therefore needed to increase the efficacy of immunotherapies. Ideally, such novel approaches should achieve a potent systemic, tumor-specific immunological response through a minimally invasive, local intervention to suppress local tumors and at the same time eradicate metastases at distant sites, while providing antitumor immunity to the host with minimal adverse side effects.

Laser-assisted immunotherapy [also known as laser immunotherapy (LIT)], first proposed by Chen et al. (1997), is a drug/device combination therapy that utilizes a local intervention to induce a systemic antitumor immunity. The two principles underlying this therapy are (1) the local destruction of tumor cells resulting from direct delivery of laser energy into the tumor,

which liberates tumor antigens and in itself induces a local immune response, and (2) the local administration of an immunoadjuvant to elicit a much stronger systemic immune response. The fundamental mechanism behind LIT is the activation of antigen-presenting cells (APCs), such as DCs, and subsequent exposure of the activated APCs to tumor antigens *in vivo* so that a tumor-specific T cell response is induced. LIT may therefore be considered an *in situ* autologous cancer vaccine (trademarked as in CVAX) that utilizes whole tumor cells as the source of tumor antigens without the need for *ex vivo* preparations.

65.2 Photothermal Effects

65.2.1 Hyperthermia and Immune Responses

Hyperthermia has been extensively studied in the context of cancer therapy, and it has been demonstrated that different temperature intervals give rise to different responses (e.g., Zhang et al. 2008, and references therein). For example, at fever-range temperatures (39°C–41°C), immune-cell activity such as APCs and T cells can be modulated, whereas at heat-shock temperatures (41°C–43°C), an increase in tumor antigen expression has been observed. At the cytotoxic temperature range (>43°C), tumor cells will undergo coagulative necrosis. This, in turn, can lead to the release of damage-associated molecular pattern molecules (DAMPs), as well as the release of tumor antigens (e.g., den Brok et al. 2004; Mukhopadhaya et al. 2007), which then can be harvested by APCs to induce a systemic antitumor immune response.

DAMPs are intracellular molecules that are normally hidden within living cells, which acquire immunostimulatory properties upon exposure or secretion by damaged/dying cells (Garg et al. 2010). These molecules have the ability to exert various effects on APCs, such as maturation, activation, and antigen processing/presentation. DAMPs vary greatly depending on the type of cell (epithelial or mesenchymal) and injured tissue. Protein DAMPs include intracellular proteins, such as heat-shock proteins or high-mobility group box 1 (HMGB1), and proteins derived from the extracellular matrix that are generated following tissue injury. Tumor deoxyribonucleic acid (DNA) may also be released into the extranuclear space and/or extracellular microenvironment, functioning as a DAMP. Many of these DAMPs have been reported to activate DC via toll-like receptor (TLR) ligation and complement activation, which in itself is consistent with the observation that hyperthermia activates APCs (Zhang et al. 2008).

65.2.2 Laser Ablation of Tumors

Shortly after the first functioning laser was built (Maiman 1960), reports were published on laser hyperthermia (ablation) of tumors using animal models (McGuff et al. 1963, 1965). The authors noted a "delayed" response to the treatment, with additional tumor shrinkage during the weeks following tumor ablation, but the effects were not understood. Laser ablation of

tumors started to gain popularity as lasers became more widespread in clinical settings (e.g., Amin et al. 1993; Brown 1983; Harries et al. 1994; Steger et al. 1989). More recent studies have pointed to immune responses associated with laser ablation (e.g., Haraldsdóttir et al. 2011; Möller et al. 1998).

An important aspect of hyperthermia through laser ablation is that a thermal gradient is generated inside the tissue, which means that the entire immunostimulatory temperature range may be covered (Figure 65.1). Heating of the tissue can either be generated directly through an interstitial optical fiber (e.g., Haraldsdóttir et al. 2008; Le et al. 2011) or by means of first injecting a light-absorbing agent into the tumor, such as indocyanine green (e.g., Chen et al. 1995), gold nanoparticles (Hirsch et al. 2003), or single-walled carbon nanotubes (Zhou et al. 2012), and then applying the laser irradiation noninvasively. It should be noted that at temperatures above 105°C, carbonization and evaporation of tissue are induced, which is generally considered undesirable in laser ablation because it changes the optical properties of the tissue and reduces the light penetration into the surrounding tissue (e.g., Iizuka et al. 2000). In addition, such a high temperature results in complete denaturing of the cell proteins, which diminishes the role of thermal treatment in releasing viable tumor antigens to stimulate targeted immunological responses in the host.

A unique feature of laser ablation (as opposed to other thermal ablation techniques) is that nonthermal photobiological effects (photobiomodulation) are likely to occur in the tumor and the surrounding tissue (Figure 65.1). Photobiomodulation (e.g., Chung et al. 2012; Tunér and Hode 2004) is induced through photochemical and photophysical reactions in cellular photoreceptor molecules (Karu 2007). Redox-active heme and Cu metal centers in cytochrome *c* oxidase have been identified as possible photoacceptors for red and near-infrared (IR) light, whereas

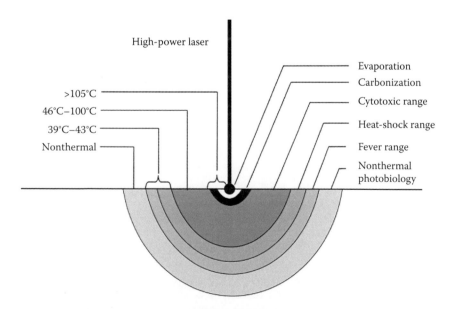

FIGURE 65.1 Diagram of idealized temperature profile following interstitial laser ablation.

flavoproteins such as NADH-dehydrogenase or glutamate dehydrogenase have been suggested as possible photoacceptors for blue and red light. The biological responses include increased adenosine triphosphate (ATP) production (Karu 2007), modulation of reactive oxygen species (ROS) (Alexandratou et al. 2002; Lubart et al. 2005; Stadler et al. 2000), and activation of transcription factors such as NF-κB (Aihara, Yamaguchi, and Kasai 2006; Rizzi et al. 2006). These initial reactions, in turn, lead to a cascade of processes that include increased cell proliferation and migration, modulation of cytokine levels, growth factors and inflammatory mediators, and increased tissue oxygenation. In the case of cancer, these reactions may also lead to tumor cell apoptosis (Wu et al. 2009).

Many cellular signaling pathways are regulated by the intracellular redox state, and it is believed that extracellular stimuli induce cellular responses such as proliferation, differentiation, and apoptosis (Karu 2007). Even though some similar or even identical steps in the cellular signaling occur, the final cellular responses to stimuli can differ due to the existence of different modes of regulation of transcription factors and induction of protein phosphorylation. This circumstance could explain why a limited number of photoacceptors could be responsible for inducing such a diverse cellular response. Multiple levels of regulation exist not only in signal transduction but also in transcription, translation, and posttranslational modification (Karu 2008). However, the effects of photobiomodulation in the tumor microenvironment are poorly understood, and further studies are needed to delineate positive or negative effects.

The observation that thermal ablation may induce a local immune-stimulating effect as well as generate tumor antigen release led to the development of ablation-based autologous vaccines (e.g., Möller et al. 1998; Sanchez-Ortiz et al. 2003). However, as the immune systems of cancer patients are compromised and the level of their regulatory T cells is often increased, tumor antigen release and immune stimulation from the destruction of tumor tissue alone are likely not sufficient to induce a potent antitumor response (Chen et al. 1996; den Brok et al. 2004; Sinkovics and Horvath 2000). Therefore, additional immunological interventions are likely required to potentiate the immune system to achieve a sustained immune response against residual and/or metastatic tumor deposits.

65.3 Combination of Laser Ablation and Immunotherapy

The limited efficacy of current treatment modalities on cancer, particularly late-stage, metastatic cancers, has led investigators to design and implement new approaches, such as targeted therapies in combination with immunotherapy (Vanneman and Dranoff 2012). Chen et al. (1997) proposed the combination of laser ablation and the application of a novel immunoadjuvant, glycated chitosan (GC), later referring to the therapy as "laser immunotherapy" (LIT) (Chen et al. 2001). Several studies on LIT were performed with a consistent demonstration

of an immediate photothermal destruction of neoplastic cells and a long-term response of the immunological defense system against residual primary and metastatic tumor cells (Chen et al. 1997, 1999, 2001, 2003a). Eradication of primary and metastatic tumor burden, along with increases in survival, was generally observed using an aggressive, poorly immunogenic rat mammary DMBA-4 tumor model.

The observed successes in these experiments would be less significant if the tumor model was highly immunogenic; however, the lack of tumor resistance observed in rats following freeze–thaw lysate immunization and surgical resection revealed a low level of immunogenicity in the DMBA-4 tumor model (Chen et al. 2003a). This implies that the immune responses in this tumor model must be augmented by additional factors, such as direct tumor ablation and active immunologic stimulation using an immunoadjuvant.

Overall, the results clearly indicate that the immunoadjuvant plays an essential role in inducing antitumor immunity with LIT, leading to long-term survival of tumor-bearing rats. However, the choice of immunoadjuvant in LIT also appears to be important; Chen et al. (2005) evaluated the effects of four different immunoadjuvants used in LIT in the treatment of rat tumors and found a significant difference in survival. Female Wistar Furth rats (6 to 8 weeks of age, 150 to 200 g) were subcutaneously inoculated with DMBA-4 metastatic mammary tumor (10^5 viable tumor cells) in the inguinal fat pad 7 to 10 days before treatment.

The primary tumor generally became palpable after 5 to 7 days, and the remote inguinal and axillary metastases appeared 15 to 20 days after inoculation. LIT treatment was initiated when the primary tumor reached 0.2 to 0.5 cm^3 (usually on day 10). The immunoadjuvants included 1% GC aqueous solution (0.2 mL/rat; $n = 48$ rats in two experiments), 50% complete Freund's adjuvant (0.2 mL/rat; Sigma, St. Louis, Missouri; $n = 33$ rats), 50% incomplete Freund's adjuvant (0.2 mL/rat; Sigma; $n = 30$ rats), and *Corynebacterium parvum* (35 μg/rat; East Coast Biologics, North Berwick, Maine; $n = 32$ rats).

Laser ablation was augmented by an intratumoral injection of 0.25% indocyanine green 2 h before irradiation with an 805 nm diode laser. Animals were observed daily, and tumor burden was measured twice a week. All immunoadjuvants resulted in a statistically significant increase in animal survival rate compared to that of control animals ($p < 0.05$). The 1% GC appeared to be the most effective immunoadjuvant with a 29% long-term survival rate (Table 65.1).

65.3.1 Mechanisms of Laser Immunotherapy with GC

It is not fully understood why GC appears to be such an effective immunoadjuvant, but it is possible that the carbohydrate structure (Chen et al. 2003b) plays a role: Genomic approaches to defining DC and Langerhans cells have identified a series of genes that encode lectin or lectin-like receptors (Figdor, van Kooyk, and Adema 2002). Many of these lectins are

TABLE 65.1   Long Term Survival Rates Following LIT Treatment with Four Different Immunoadjuvants

Treatment	No. of Rats	Long Term Survival (%)
Control	38[a]	0
Laser ablation + Glycated Chitosan	48[b]	29
Laser ablation + Complete Freund's Adjuvant	33	18
Laser ablation + Incomplete Freund's Adjuvant	30	7
Laser ablation + *C-parvum*	32	9

Source: Chen, W. R. et al., *Photochem Photobiol* 81:190–195, 2005.

Note: Statistical significance was observed when the glycated chitosan adjuvant was compared to the *C-parvum* ($p = 0.009$) and incomplete Freund's adjuvant ($p = 0.03$). Although not significant, a noticeable improvement in survival was observed when compared to complete Freund's adjuvant that had a comparable cure rate (18%). A relative weak survival rate was observed following treatment with the incomplete Freund's adjuvant and *C-parvum*.

[a] Tumor-bearing control rat data was collected from several control groups in different studies.

[b] Data collected from two separate experiments.

members of the calcium-dependent C-type lectin family, and there is evidence that this type of lectins can preferentially bind to carbohydrate-bearing antigens (Sallusto et al. 1995). These C-type lectins have been associated with antigen uptake and may be important in the migration of DC and their interactions with lymphocytes. This may explain our experimental observations that GC activates DC (Zhou et al. 2012, supplemental).

In contrast to TLR ligands whose signaling pathways are mediated via Myd88 and NF-κB activation (Kawai and Akira 2006), signaling mediated by C-type lectins is much less understood (Mascanfroni et al. 2011). The brisk cytokine response caused by TLR agonists is responsible for the marked adverse side effects when these substances are used in tumor immunotherapy protocols (Ahonen et al. 2008). The notable lack of pronounced side effects in the clinical use of GC in LIT could be explained by the much reduced production of inflammatory cytokines with C-type lectin ligands (van Vliet et al. 2006). The galactose/N-acetyl-galactosamine (GalAc) specific lectin (MGL), also known as CD301 or CLEC10A, is a good candidate for the receptor for GC because it is the only C-type lectin known to date that recognizes galactose residues (Denda-Nagai et al. 2010).

In a study by Zhou et al. (2011), the direct immunostimulation properties of GC on macrophages were evaluated using nitric oxide and tumor necrosis factor-alpha (TNFα) secretion analysis, which are indicative of macrophage activation. In addition, the immunopotentiation properties of GC were evaluated during tumor treatment by analyzing TNFα secretion of macrophages *in vitro* stimulated by laser-treated tumor cells, and by analyzing cell cytotoxicity of spleen cells *in vivo* treated by laser and GC. The murine mammary tumor, EMT6, and murine macrophage, RAW264.7, cell lines were utilized in these experiments. Overall, the experiments determined that GC could enter

into macrophages to stimulate nitric oxide generation and TNFα secretion, and that GC could further enhance TNFα secretion in macrophages stimulated by laser immunotherapy in treated tumor cells.

To confirm the antitumor immunity induced by LIT, a tumor rechallenge experiment was performed (Chen et al. 2001). The protective ability of induced immunity was evaluated in several groups of successfully treated tumor-bearing rats that were challenged repeatedly with increased inoculation dose of viable tumor cells. In addition, resistance to tumor challenges after laser immunotherapy and the inhibition of tumor growth was evaluated in naive rats.

In these experiments, 15 rats that had been successfully treated with LIT were rechallenged with 10^6 viable tumor cells 120 days after initial inoculation. Eighteen naive age-matched rats (25 weeks of age) were inoculated with 10^6 viable tumor cells for comparative purposes. All of the successfully treated rats showed total resistance to the challenge, with neither primary tumors nor metastases observed; however, the age-matched control rats developed primary and metastatic tumors and died within 30 days after inoculation. A separate group of young rats (approximately 8 weeks of age) were inoculated with 10^5 viable tumor cells. Survival appeared to be dependent on the tumor dose, with control rats inoculated with 10^5 and 10^6 viable tumor cells surviving on average 33 and 28 days, respectively.

Successfully treated rats usually experienced a gradual regression in both treated primary tumor and untreated metastases. After the first rechallenge, the rats from several experimental groups were followed by two subsequent challenges in a time interval from 1 to 5 months, with 10^6 viable tumor cells. The rats successfully treated by laser immunotherapy were totally refractory to all three tumor challenges (Table 65.2).

Following the rechallenge experiments, an adoptive immunity transfer experiment was performed (Chen et al. 2001) to further study the induced long-term immunological memory in the successfully treated rats. Four groups of naive female Wistar Furth rats were inoculated with tumor cells as follows: Group A consisted of tumor-bearing control rats inoculated with 10^5 viable tumor cells without treatment, group B consisted of rats inoculated with the tumor cells admixed with the spleen cells from a control tumor-bearing rat, group C consisted of rats inoculated with the tumor cells admixed with the spleen cells from a tumor-bearing rat successfully treated by laser immunotherapy 28 days after tumor rechallenge, and group D consisted of rats inoculated with the tumor cells admixed with the spleen cells from a naive rat without prior tumor exposure.

The experiment was performed in duplicate, and the survival of rats from both experiments was combined. No primary or metastatic tumors were observed in rats in group C, indicating that the spleen cells from successfully LIT-treated rats provided 100% protection to the recipients. Multiple metastases and death within 35 days of tumor inoculation were observed in all the tumor-bearing control rats in group A. There was no protection

TABLE 65.2 Tumor Rechallenge in Rats That Were Previously Cured with LIT

Group	No. of Rats	No. of Tumor Cells	Tumor Rate	Death Rate 40 Days	Survival (Days)
	\multicolumn{5}{Tumor Rechallenge in Rats Previously Cured with Laser Immunotherapy}				
Cured rats[a]	15	10^6	0%	0%	>120
Cured rats[b]	15	10^6	0%	0%	>120
Cured rats[c]	15	10^6	0%	0%	>120
Age-matched tumor control rats[d]	18	10^6	100%	100%	28.2 ± 2.8
Young tumor control rats[e]	20	10^5	100%	100%	32.7 ± 3.5

Note: All successfully treated rats showed complete resistance to three subsequent tumor rechallenges.

[a] First challenge. Tumor-bearing rats cured by laser immunotherapy, rechallenged with viable tumor cells 120 days after the initial inoculation.

[b] Second challenge. Tumor-bearing rats cured by laser immunotherapy, rechallenged with viable tumor cells a second time after the first challenge.

[c] Third challenge. Tumor-bearing rats cured by laser immunotherapy, rechallenged with viable tumor cells a third time after the second challenge.

[d] Untreated rats the same age as the cured rats at time of inoculation (no previous tumor exposure).

[e] Untreated rats that were 8 weeks of age at the time of inoculation (no previous tumor exposure).

provided by the spleen cells from a healthy rat in group D. A single rat out of 10 rats in group B survived; however, this rat developed both primary tumor and metastases. All rats in group C were rechallenged 60 days after the adoptive immunity transfer, and all withstood the challenge. The immune spleen cells of the rats in group C were collected and admixed with tumor cells in the same ratio as in the first adoptive transfer to evaluate the ability of these animal spleen cells in protecting a subsequent cohort of normal recipient rats (*n* = 6) that were inoculated with this admixture. The immune cells from the group C rats protected 5 of 6 naive rats, as neither primary or metastatic tumors were observed. The rat that died had a prolonged survival time (60 vs. 30 days) and a delayed tumor emergence after inoculation (37 vs. 7 to 10 days), in comparison with the control group. The tumor resistance by the successfully treated rats when rechallenged strongly suggests that the tumor-selective immunity has a long-lasting effect.

In another experiment, sera was obtained from successfully treated tumor-bearing rats after rechallenge and was analyzed for tumor-selective antibodies using the tumor tissue immuno-fluorescence and immunoperoxidase assays (Chen et al. 1999). The histochemical assays using both live and preserved tumor cells revealed strong antibody binding using sera from successfully treated tumor-bearing rats when compared to sera from untreated control tumor-bearing rats. Furthermore, a Western blot assay was conducted using sera from naive, control tumor-bearing, and successfully treated tumor-bearing rats. There was a lack of staining in the assay using naive rat sera, indicating that it did not contain any tumor-selective antibodies. Two distinct bands were observed at approximately 45 and 35 kDa using sera from successfully treated tumor-bearing rats. The 45 kDa band was observed with a much weaker intensity after probing the blot containing tumor proteins with serum from the control tumor-bearing rat; however, the 35 kDa band was absent in the control tumor-bearing rat. These results indicate that laser immunotherapy resulted in the production of certain rats' antibodies that bind or intensify the binding of specific tumor proteins.

65.4 Preliminary Clinical Trials

65.4.1 Advanced Breast Cancer

With the promising preclinical outcomes of LIT, two investigator-driven breast cancer trials were performed at two sites, one of which is a major hospital in Lima, Peru (Li et al. 2011). The investigators acquired institutional review board (IRB) and government approvals prior to the trials. Nineteen patients with advanced breast cancer (12 stage IV, 7 stage III, life expectancy 3–6 months) were treated by laser immunotherapy (received at least one LIT treatment), most of whom had responded poorly or not at all to conventional modalities such as chemotherapy, radiation, surgery, and hormonal therapy. Three patients withdrew prematurely due to unrelated reasons, leaving 16 evaluable patients. Biopsies and medical imaging [computed tomography (CT) scans, dual-phase positron emission tomography (PET) scans, etc.] were used for the evaluation of the primary lesions and metastasis.

The primary efficacy parameter was the best overall response by the investigators' assessments using Response Evaluation Criteria in Solid Tumors (RECIST). Complete response (CR) was defined as disappearance or lack of qualifying metabolic activity of all target lesions. Partial response (PR) was defined as a ≥30% decrease from baseline in activity or in the sum of the longest diameter of target lesions. Progressive disease (PD) is defined as a ≥20% increase in the sum of the longest diameter of target lesions or the appearance of one or more new lesions. Stable disease (SD) was defined as neither sufficient reduction to qualify for PR nor sufficient increase to qualify for PD.

Of the 16 patients available for evaluation in LIT clinical trials, 63% were progression-free (90% if triple-negative patients are excluded), 50% displayed PR (73% excluding triple-negative patients), and 19% showed CR within 12 months (27% excluding triple-negative patients).

Case report 1. The patient was a 70-year-old female diagnosed with stage 4 breast cancer. She received left breast partial mastectomy with no chemotherapy or radiation. A second occurrence in the primary location was found 2 years after the surgery. The patient received two LIT treatments in total. The size of the tumor in the left breast was 1.1 cm prior to LIT treatment. Two pulmonary metastatic nodules were observed in the left lung. Three months after the first LIT treatment, both of the left lung metastases were no longer viable. However, a new pulmonary metastatic nodule was observed in the right lung. Seven months after the first LIT treatment, all lung metastases were no longer viable (Figure 65.2). CT scans showed only residual opacities within the lung, and the primary tumor demonstrated a quantitative partial metabolic response. The tumor attached to the chest wall later resolved, and two years after the first treatment the patient was still tumor free.

Case report 2. The patient was a 47-year-old female diagnosed with stage 4 breast cancer. She had previously received AC (doxorubicin/cyclophosphamide) for four cycles, paclitaxel for three cycles, capecitabine plus ixabepilone for three cycles, and hormonal therapy with tamoxifen. However, she was resistant to chemotherapy and hormonal therapy. This patient received four LIT treatments in total. Clinical observations before LIT showed that the sizes of the two tumors in the right breast of the patient were 6 × 4.5 and 2 × 2 cm. Pulmonary metastatic nodules were observed in bilateral lungs. Two-and-a-half months after the first LIT treatment, one of the lung metastases had disappeared, but the remaining three were still at the same level. Nine months after the first LIT treatment, CT scans showed that the lung was normal without any metastasis (Figure 65.3). The 1-year evaluation confirmed the absence of the lung metastases. This patient is completely tumor-free at the time of writing this report.

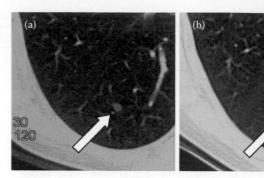

FIGURE 65.3 CT scan of one of the four lung metastases. (a) Arrow indicates a small tumor in the left lung prior to LIT treatment. (b) Image taken 12 months after first LIT treatment, and the tumor has dissolved completely (arrow). Additional figures are presented in Li et al. (2011). The patient is still in remission 2.5 years after the first LIT treatment.

65.4.2 Advanced Melanoma

Using the same principle of laser immunotherapy, 11 late-stage (IIIB, IIIC, and IV) melanoma patients were treated in the United States using imiquimod, a U. S. Food and Drug Administration (FDA)-approved drug as the immune stimulant (Li et al. 2010; Naylor et al. 2006). Imiquimod is a TLR agonist and is only approved for topical application in the United States, which makes it suitable only for treating cutaneous melanoma metastases. Complete responses were observed in six patients, although eventual relapse was the rule (no durable complete responses longer than 2 years). However, cutaneous relapses could be re-treated successfully. Relapses in other organs generally led to fatal outcomes because they could not be treated with surface technology. As discussed in Chen et al. (2005), the choice of immunoadjuvant may be an important factor in the overall success of LIT, and further studies are needed to better optimize the clinical outcome and increase the long-term survival of patients with advanced melanoma.

65.5 Summary

LIT represents a new and significantly improved approach to other autologous cancer vaccines, both by eliminating the need of *ex vivo* preparations and by using a new and highly potent nontoxic immunoadjuvant GC. The fundamental mechanism behind LIT is believed to be based on the activation of APCs (such as DCs) *in vivo* and subsequent exposure of the activated APCs to tumor antigens *in situ* through a practical, local procedure. This is achieved by (1) laser ablation of any one accessible tumor with a laser to devitalize the tumor, induce a heat-shock response, and liberate tumor antigens, and (2) local application of an immune-stimulating drug, which further boosts the immune system to process tumor antigens and induce a systemic antitumor immune response. LIT is thus an *in situ* autologous cancer vaccine that uses whole tumor cells as the sources of tumor antigens from each individual patient without preselection of tumor antigens or *ex vivo* preparation.

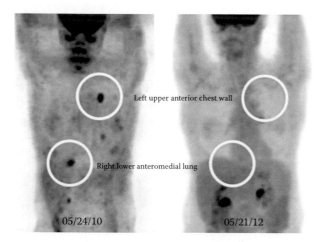

FIGURE 65.2 PET scans before treatment (left) and two years after first treatment (right) of a stage 4 patient in trial site 1. Seven months after first treatment there was a complete resolution of all lung metastases, whereas the tumor attached to the chest showed a quantitative partial metabolic response (which later resolved). The PET image to the right (taken two years after the first treatment) demonstrates complete metabolic resolution of all neoplastic sites.

Previous research has shown that LIT with GC as an immunoadjuvant activates macrophages, induces maturation of DCs, enhances T cell proliferation, and increases interferon-gamma (IFN-γ) secretion and HSP70 expression. Furthermore, it was shown that LIT induces tumor-specific immunity, with a demonstrated infiltration of tumor-specific CD4 and CD8 cells into the tumors following the treatment. Studies also revealed that sera from LIT-cured tumor-bearing rats contained antibodies that bound to the plasma membrane of both living and preserved tumor cells and bound to certain tumor cell proteins, indicating induction of tumor-selective antibodies. Successfully treated rats could acquire long-term resistance to tumor rechallenge, which would indicate the presence of tumor-specific memory T cells, and adoptive immunity could be transferred using spleen cells from successfully treated rats, which would indicate tumor-specific cytotoxic T cells.

Finally, preliminary human trials on advanced breast cancer and melanoma have yielded very promising results, indicating that the combination of laser ablation and an immune stimulation drug, particularly GC, may represent a step toward a systemic control of metastatic cancer.

References

Ahonen, C. L., A. Wasiuk, S. Fuse et al. 2008. Enhanced efficacy and reduced toxicity of multifactorial adjuvants compared with unitary adjuvants as cancer vaccines. *Blood* 111:3116–3125.

Aihara, N., M. Yamaguchi, and K. Kasai. 2006. Low-energy irradiation stimulates formation of osteoplast-like cells via RANK expression *in vitro*. *Lasers Med Sci* 21:24–33.

Alexandratou, E., D. Yova, P. Handris, D. Kletsas, and S. Loukas. 2002. Human fibroblast alterations induced by low power laser irradiation at the single cell level using confocal microscopy. *Photochem Photobiol Sci* 1:547–552.

Amin, Z., J. J. Donald, A. Masters et al. 1993. Hepatic metastases: Interstitial laser photocoagulation with real-time US monitoring and dynamic CT evaluation of treatment. *Radiology* 187:339–347.

den Brok, M. H., R. P. Sutmuller, R. van der Voort et al. 2004. *In situ* tumor ablation creates an antigen source for the generation of antitumor immunity. *Cancer Res* 64:4024–4029.

Brown, S. G. 1983. Phototherapy of tumors. *World J Surg* 7:700–709.

Chen, W. R., R. L. Adams, K. E. Bartels, and R. E. Nordquist. 1995. Chromophore-enhanced *in vivo* tumor cell destruction using an 808-nm diode laser. *Cancer Lett* 94:125–131.

Chen, W. R., R. L. Adams, A. K. Higgins, K. E. Bartels, and R. E. Nordquist. 1996. Photothermal effects on murine mammary tumors using indocyanine green and an 808-nm diode laser: An *in vivo* efficacy study. *Cancer Lett* 98:169–173.

Chen, W. R., R. L. Adams, R. Carubelli, and R. E. Nordquist. 1997. Laser-photosensitizer assisted immunotherapy: A novel modality in cancer treatment. *Cancer Lett* 115:25–30.

Chen, W. R., W.-G. Zhu, J. R. Dynlacht, H. Liu, and R. E. Nordquist. 1999. Long-term tumor resistance induced by laser photo-immunotherapy. *Int J Cancer* 81:808–812.

Chen, W. R., A. K. Singhal, H. Liu, and R. E. Nordquist. 2001. Laser immunotherapy induced antitumor immunity and its adoptive transfer. *Cancer Res* 61:459–461.

Chen, W. R., S. W. Jeong, M. D. Lucroy et al. 2003a. Induced anti-tumor immunity against DMBA-4 metastatic mammary tumors in rats using a novel approach. *Int J Cancer* 107:1053–1057.

Chen, W. R., R. Carubelli, H. Liu, and R. E. Nordquist. 2003b. Laser immunotherapy: A novel treatment modality for metastatic tumors. *Mol Biotechnol* 25:37–43.

Chen, W. R., M. Korbelik, K. E. Bartels et al. 2005. Enhancement of laser cancer treatment by a chitosan-derived immunoadjuvant. *Photochem Photobiol* 81:190–195.

Chung. H., T. Dai, S. K. Sharma et al. 2012. The nuts and bolts of low-level laser (light) therapy. *Ann Biomed Eng* 40:516–533.

Denda-Nagai, K., S. Aida, K. Saba et al. 2010. Distribution and function of macrophage galactose-type C-type lectin 2 (MGL2/CD301b): Efficient uptake and presentation of glycosylated antigens by dendritic cells. *J Biol Chem* 285:19193–19204.

Figdor, C. G., Y. van Kooyk, and G. J. Adema. 2002. C-type lectin receptors on dendritic cells and Langerhan cells. *Nat Rev Immunol* 2:77–84.

Garg, A. D., D. Nowis, J. Golab et al. 2010. Immunogenic cell death, DAMPs and anticancer therapeutics: An emerging amalgamation. *Biochim Biophys Acta* 1805:53–71.

Haraldsdóttir, K. H., K. Ivarsson, S. Gotberg et al. 2008. Interstitial laser thermotherapy (ILT) of breast cancer. *Eur J Surg Oncol* 34:739–745.

Haraldsdóttir, K. H., K. Ivarsson, K. Jansner, U. Stenram, and K.-G. Tranberg. 2011. Changes in immunocompetent cells after interstitial laser thermotherapy of breast cancer. *Cancer Immunol Immunother* 60:847–856.

Harries, S. A., Z. Amin, M. E. Smith et al. 1994. Interstitial laser photocoagulation as a treatment for breast cancer. *Br J Surg* 81:1617–1619.

Hirsch, L. R., R. J. Stafford, J. A. Bankson et al. 2003. Nanoshell-mediated near-infrared thermal therapy of tumors under magnetic resonance guidance. *Proc Natl Acad Sci USA* 100:13549–13554.

Iizuka, M. N., I. A. Vitkin, M. C. Kolios, and M. D. Sherar. 2000. The effects of dynamic optical properties during interstitial laser photocoagulation. *Phys Med Biol* 45:1335–1357.

Karu, T. I. 2007. *Ten Lectures on Basic Science of Laser Phototherapy*. Prima Books, Grängesberg.

Karu, T. I. 2008. Mitochondrial signaling in mammalian cells activated by red and near-IR radiation. *Photochem Photobiol* 84:1091–1099.

Kawai, T., and S. Akira. 2006. TLR signaling. *Cell Death Differ* 13:816–825.

Le, K., X. Li, D. Figueroa et al. 2011. Assessment of thermal effects of interstitial laser phototherapy on mammary tumors using proton resonance frequency method. *J Biomed Opt* 16:128001.

Li, X., G. L. Ferrel, M. C. Guerra et al. 2011. Preliminary safety and efficacy of laser immunotherapy for the treatment of metastatic breast cancer patients. *Photochem Photobiol Sci* 10:817–821.

Li, X., M. F. Naylor, H. Le et al. 2010. Clinical effects of *in situ* photoimmunotherapy on late-stage melanoma patients—A preliminary study. *Cancer Biol Ther* 10:1081–1087.

Lubart, R., M. Eichler, R. Lavi, H. Friedman, and A. Shainberg. 2005. Low-energy laser irradiation promotes cellular redox activity. *Photomed Laser Surg* 23:3–9.

Maiman, T. H. 1960. Stimulated optical radiation in ruby. *Nature* 187:493–494.

Mascanfroni, I. D., J. P. Cerliani, S. Dergan-Dylon et al. 2011. Endogenous lectins shape the function of dendritic cells and tailor adaptive immunity: Mechanisms and biomedical applications. *Int Immunopharmacol* 11:833–841.

McGuff, P. E., D. Bushnell, H. S. Soroff, and R. A. Deterling, Jr. 1963. Studies of the surgical applications of laser (light amplification by stimulated emission of radiation). *Surg Forum* 14:143–145.

McGuff, P. E., R. A. Deterling, Jr., L. S. Gottlieb et al. 1965. Laser surgery of malignant tumors. *Chest* 48:130–139.

Möller, P. H., K. Ivarsson, U. Stenram, M. Radnell, and K.-G. Tranberg. 1998. Comparison between interstitial laser thermotherapy and excision of an adenocarcinoma transplanted into rat liver. *Br J Cancer* 77:1884–1892.

Mukhopadhaya, A., J. Mendecki, X. Dong et al. 2007. Localized hyperthermia combined with intratumoral dendritic cells induces systemic antitumor immunity. *Cancer Res* 67:7798–7806.

Naylor, M. F., W. R. Chen, T. K. Teague, L. Perry, and R. E. Nordquist. 2006. *In situ* photoimmunotherapy: A tumor-directed treatment modality for melanoma. *Br J Dermatol* 155:1287–1292.

Rizzi, C. F., J. L. Mauriz, D. S. F. Correa et al. 2006. Effects of low-level laser therapy (LLLT) on the nuclear factor (NF)-signaling pathway in traumatized muscle. *Lasers Surg Med* 38:704–713.

Sallusto, F., M. Cella, C. Danieli, and A. Lanzavecchia. 1995. Dendritic cells use macropinocytosis and the mannose receptor to concentrate macromolecules in the major histocompatibility complex class II compartment: Downregulation by cytokines and bacterial products. *J Exp Med* 182:389–400.

Sanchez-Ortiz, R. F., N. Tannir, K. Ahrar, and C. G. Wood. 2003. Spontaneous regression of pulmonary metastases from renal cell carcinoma after radio frequency ablation of primary tumor: An *in situ* tumor vaccine? *J Urol* 170:178–179.

Sinkovics, J. G., and J. C. Horvath. 2000. Vaccination against human cancers (review). *Int J Oncol* 16:81–96.

Stadler, I., R. Evans, B. Kolb et al. 2000. *In vitro* effects of low-level laser irradiation at 660 nm on peripheral blood lymphocytes. *Lasers Surg Med* 27:255–261.

Steger, A. C., W. R. Lees, K. Walmsley, and S. G. Brown. 1989. Interstitial laser hyperthermia: A new approach to local destruction of tumors. *BMJ* 299:362–365.

Tunér, J., and L. Hode. 2004. *The Laser Therapy Handbook*. Prima Books, Grängesberg.

Vanneman, M., and G. Dranoff. 2012. Combining immunotherapy and targeted therapies in cancer treatment. *Nat Rev Cancer* 12:237–251.

van Vliet, S. J., E. van Liempt, T. B. Geijtenbeek, and Y. van Kooyk. 2006. Differential regulation of C-type lectin expression on tolerogenic dendritic cell subsets. *Immunobiology* 211:577–585.

Wu, S., D. Xing, X. Gao, and W. R. Chen. 2009. High fluence low-power laser irradiation induces mitochondrial permeability transition mediated by reactive oxygen species. *J Cell Physiol* 218:603–611.

Zhang, H.-G., K. Mehta, P. Cohen, and C. Guha. 2008. Hyperthermia on immune regulation: A temperature's story. *Cancer Lett* 271:191–204.

Zhou, F., S. Song, W. R. Chen, and D. Xing. 2011. Immunostimulatory properties of glycated chitosan. *J X-ray Sci Technol* 19:285–292.

Zhou, F., S. Wu, S. Song et al. 2012. Antitumor immunologically modified carbon nanotubes for photothermal therapy. *Biomaterials* 33:3235–3242.

66

Tissue Repair by Photochemical Cross-Linking

Irene E. Kochevar
Massachusetts General Hospital

Robert W. Redmond
Massachusetts General Hospital

66.1 Introduction

Sutureless joining of tissue surfaces has been a long-sought goal because use of sutures induces inflammation and subsequent fibrosis and may leave gaps between suture points. Light-initiated techniques to bond tissues together were introduced over 30 years ago. Ideally, these approaches produce a strong, immediate, watertight bond without damage to adjacent tissue that maintains its strength over a long period. Such techniques might be especially appropriate for microsurgery since they circumvent the need for skilled placement of hair-fine sutures in small delicate structures. Two approaches, differing in mechanism, have been developed for light-initiated tissue bonding. The first method developed operates by a thermal mechanism involving rapid absorption of laser energy at the junction of the tissues and a temperature rise above the denaturation temperature of collagen. The partially denatured collagen molecules are believed to interact and intertwine so that, upon cooling, a continuous seal forms. The second mechanism involves photochemical reactions that initiate formation of covalent cross-links between collagen molecules on the tissue surfaces. Photochemical cross-linking can occur without a temperature increase or protein denaturation, and it produces a continuous molecular level seal between the two surfaces. This review focuses on tissue bonding by photochemical cross-linking.

66.2 Photochemical Linking

Three general types of photoactivated cross-linking have been investigated:

1. Zero-length cross-linking, where the covalent bond is formed directly between protein molecules
2. Bridging cross-links, where the light-activated agent is integral to the cross-link itself
3. Photochemical glues incorporating an extrinsic protein in the bond

The earliest studies of protein cross-linking for tissue repair were performed by Judy et al. (1993a,b, 1994, 1996a,b) using bifunctional 1,8-naphthalimide dyes connected by a flexible bridge. Naphthalimides absorb at short visible wavelengths and react with nucleophilic amino acid residues (e.g., tryptophan, cysteine, and methionine). Initial studies showed oxygen-independent consumption of dye and isolated amino acids in solution, and an electron transfer mechanism was proposed for covalent attachment of nucleophilic amino acids to the terminal naphthalimides (Judy et al. 1993b). Illumination of hydrophobic and hydrophilic bifunctional naphthalimides with proteins such as F-actin monomer and collagen gave dimer, trimer, and higher covalently bound aggregates (Judy et al. 1993b, 1994). A preliminary study of tissue bonding using porcine dura sections *ex vivo* gave bond shear strengths of up to 425 g/cm^2 (1.14 × 10^4 N/m^2), but large fluences (>1000 J/cm^2) were required (Judy et al. 1993a). A hydrophilic analog was used for bonding thin sections of cadaveric human meniscus and articular cartilage, producing bond sheer strengths of up to 1.8 and 1.2 kg/cm^2 for meniscal and articular cartilage tissues, respectively, with a fluence of 3902 J/cm^2 delivered at an irradiance of 200 mW/cm^2 at 458 nm (Judy et al. 1996b). It does not appear that these compounds proceeded to *in vivo* preclinical or clinical studies for tissue repair but sparked interest in photochemical tissue bonding.

Givens and coworkers used a photoactivated bifunctional diazopyruvoyl cross-linking agent (Givens et al. 2003; Timberlake et al. 2005). Ultraviolet A (UVA) light induced loss of N$_2$ and formation of reactive ketene intermediates that are susceptible to nucleophilic attack by amines to form amide bonds. Photolysis of the diazo group at each end of the molecule allows reaction with lysine or hydroxylysine on adjacent collagen molecules to form cross-links. Initial experiments with a hydrophobic compound showed high-strength bonding between model gelatin strips (~100 N/cm^2) illuminated for 400 s with 320–540 nm light at 328 mW (Givens et al. 2003). When applied to *ex vivo* rabbit Achilles tendon strips, bonding was observed, but at lower strength (~3.8 N/cm^2), presumably due to reduced optical penetration in the tendon (Givens et al. 2003). In feasibility tests for corneal bonding, it was possible to bond the cornea to itself or to glass, but this was only efficient under dehydrated conditions where water did not interfere with the cross-linking reactions (Timberlake et al. 2005). A multifunctional cross-linking agent based on poly(amidoamine) (PAMAM) dendrimers terminated with diazopyruvoyl groups improved water solubility but behaved in a similar manner to the bifunctional analog (Givens et al. 2008).

Khadem, Truong, and Ernest (1994) introduced photodynamic tissue glues in ophthalmology. Dye/protein mixtures were light-activated to produce heat-free bonding. Using riboflavin-5-phosphate (RF-5P) as a blue light-activated agent with 18% fibrinogen solder as a resorbable glue, full-thickness corneal incisions were sealed using 488 nm light (irradiance = 19.1 W/cm^2). Burst pressures of 219 mm Hg were obtained. Mechanistic studies identified both singlet oxygen and other reactive oxygen species as potential reactive intermediates (Khadem, Truong, and Ernest 1994). When sealing partial thickness corneal wounds in rabbits, the glue was rapidly resorbed and the wounds demonstrated less dehiscence, a more rapid epithelial healing, and less collagen disorganization that control eyes (Goins et al. 1997). An improved photodynamic agent, chlorin e6 (Ce6), covalently linked to a bovine serum albumin (BSA) carrier protein, was used to close scleral wounds in human cadaveric eyes (Khadem et al. 1999). Immediate burst strengths proved rather weak (78 mm Hg) but improved (207 mm Hg) when unconjugated BSA was included in the mixture. Using red light Janus green (JG) was more efficient than Ce6 when conjugated to BSA for sealing of full-thickness corneal incisions (Khadem et al. 2004). Day 1 leak pressures were 430, 357, and 190 mm Hg for JG, Ce6, and unglued wounds, respectively. However, the leak

pressure decreased from day 1 to day 7 due to biodegradation of the glue.

Although the agents developed by Givens and coworkers operated by clearly defined mechanisms (Givens et al. 2003, 2008; Timberlake et al. 2005), the processes proposed for photo-cross-linking by other agents are based on simple model systems in solution. Photosensitized cross-linking of proteins in solution was actively studied in the 1970s. Dubbelman et al. showed that photo-cross-linking of spectrin in red blood cell membranes was a two-step process involving dye-sensitized formation of singlet oxygen that oxidized histidine followed by nucleophilic attack by amino-containing moieties on oxidized histidine to form the cross-link (Dubbelman, de Goeij, and van Steveninck 1978; Verweij, Dubbelman, and Van Steveninck 1981). Spikes and coworkers used N-(2-hydroxypropyl) methacrylamide polymer models with pendant histidine and lysine groups and showed singlet oxygen involvement in rose bengal (RB)-sensitized oxidation of histidine and then cross-linking between His-Lys or His-His units (Shen et al. 1996a). Spikes then demonstrated that RB-sensitized photo-cross-linking of ribonuclease A proceeded via singlet oxygen formation and oxidation of histidine residues, which can then react with other His or Lys residues (Shen et al. 1996b).

66.3 RB and Green-Light Photo-Cross-Linking

RB is a halogenated xanthene dye and a well-known photosensitizing agent activated by visible light (Allen et al. 1991; Dahl, Midden, and Neckers 1988). It has an absorption maximum around 550 nm with an absorption coefficient of ~7.5 × 10^4 M^{-1} cm^{-1} in aqueous solution. The lowest excited singlet state is short-lived and relaxes primarily by efficient intersystem crossing to the lowest excited triplet state (Rodgers 1983). The excited triplet state can generate singlet oxygen in high yields by energy transfer to oxygen (Redmond and Gamlin 1999) and can also undergo electron transfer reactions to and from other substrates to generate radical ions and free radicals (Lambert and Kochevar 1997) [see Section 66.11 for discussion of the photochemical tissue bonding (PTB) mechanism].

RB has several advantages as a tissue photo-cross-linking agent. It has a history of clinical use (Balkissoon and Weld 1965) and is already Food and Drug Administration (FDA) approved as a diagnostic agent (Kim 2000). RB strongly binds to collagen in tissue, which limits its penetration into tissue and localizes the subsequent photoactivation to near the surface of tissue (Yao et al. 2011). In addition, RB phototoxicity to cells in tissue is very limited under the irradiation conditions used for bonding (Yao et al. 2011).

The term PTB has been used to describe the sealing of tissues with RB and green light, although this technology includes a wide range of dyes and wavelength ranges. PTB has shown advantages over conventional suture repair of wounds in many different tissue and organ systems in preclinical studies. Published studies include those showing the effectiveness of

PTB for sealing full-thickness corneal wounds (Mulroy et al. 2000; Proano et al. 2004a,b; Verter et al. 2011), bonding amniotic membrane to cornea for limbal cell transplantation and repairing corneal surface defects (Gu et al. 2011; Wang et al. 2011), closing skin wounds and attaching skin grafts (Chan, Kochevar, and Redmond 2002; Kamegaya et al. 2005; Yang et al. 2012), reattaching severed peripheral nerve and tendons, (Chan et al. 2005; Henry et al. 2009a,b; Johnson et al. 2007; O'Neill et al. 2009a,b) anastomosing blood vessels (O'Neill et al. 2007), and vocal fold flap repair (Franco et al. 2011). A clinical study of PTB for sealing skin excisions has also been completed (Tsao et al. 2012). The results of these studies, as well as results of unpublished studies, are described in the following sections. PTB has also been taken up by other researchers, notably by Lauto and coworkers, who adapted a prior laser welding approach to bond tissue to chitosan film adhesives by adding RB to the chitosan formulation and photochemically cross-linking the chitosan adhesive to tissue. These RB+chitosan films bonded strongly to calf intestine with a force of 15 Pa when illuminated with 110 J/cm² at 532 nm (Lauto et al. 2010).

66.4 Corneal Surgery

The transparency of the cornea, essential for its function as the major refracting element of the eye, can be compromised by sutures that can produce inflammation and infection leading to opaque scars with visual acuity loss and astigmatism. Closing full-thickness cornea incisions, such as those made for cataract removal, with PTB was demonstrated *ex vivo* and *in vivo* in rabbit eyes (Mulroy et al. 2000; Proano et al. 2004b). RB was applied to the wound edges and irradiated at 514 nm. The strength of the seal was judged from the intraocular pressure causing leakage (IOP_L). Watertight seals ($IOP_L > 300$ mm Hg; normal IOP = 15–20 mm Hg) were achieved in *ex vivo* eyes. Neither dye nor light was effective independently. The IOP_L increased with

increasing fluence, but decreased at high irradiances due to thermal damage. *In vivo*, the IOP_L reached >500 mm Hg at 192 J/cm². No reduction in the IOP_L was observed up to 14 days after surgery and long-term adverse responses were not observed.

Corneal transplants (penetrating keratoplasty) require excising a central corneal disc, and then placing 16 to 24 running sutures to seal the new cornea into the recipient eye. In a rabbit model for penetrating keratoplasty, PTB sealed the incision after placement of 16 interrupted sutures (Proano et al. 2004a). The IOP_L achieved (410 mm Hg) was significantly greater than that for sutures alone (250 mm Hg), suggesting that PTB is useful as an adjunct to sutures for penetrating keratoplasty.

PTB enabled a novel approach for sealing irregularly shaped full-thickness corneal wounds, such as formed by trauma (Verter et al. 2011). Amniotic membrane, a thin, translucent collagenous tissue, was sealed with PTB over corneal wounds (Figure 66.1A). A 532 nm fluence of 150 J/cm² produced IOP_L of 261 and 448 mm Hg *ex vivo* and *in vivo*, respectively. The IOP_L increased with increasing fluence and was significantly stronger than using sutures or fibrin sealant (Figure 66.1B). The temperature increased to only 30°C during the irradiation. This approach to sealing eye wounds is rapid and requires less skill than sutured closure.

Bonding amnion to cornea with PTB has also been used for amniotic membrane transplantation and limbal stem cell (LSC) transplantation. Currently, amnion is sutured to the cornea for treating injuries. In a rabbit model for ocular surface defect, amnion was secured using sutures or PTB (Wang et al. 2011). The inflammatory cells and the level of tissue necrosis factor alpha (TNF-α) after PTB were significantly lower than after sutures on day 3 and many fewer neovessels were present after PTB at day 28. The collagen fibers in the PTB group were well organized and oriented as assessed by second harmonic generation microscopy, suggesting that PTB treatment led to less corneal scarring. Suturing an amniotic membrane/LSC graft to the

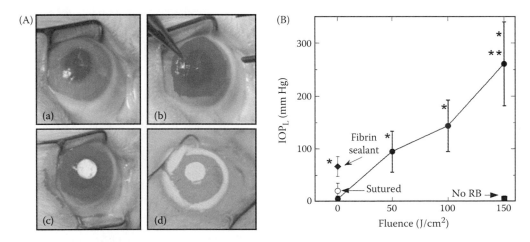

FIGURE 66.1 (**See color insert.**) Sealing amniotic membrane over corneal wounds. (A) Steps in the procedure: (a) full-thickness incision in cornea, (b) RB-stained amnion placed on cornea, (c) opaque disk placed to block light from entering through pupil, and (d) irradiation with 532 nm laser. (B) The IOP_L was measured immediately after bonding treatment. *p < 0.05 versus control, ** *p* < 0.05 versus 100 J/cm². (Adapted from Verter, E. E. et al., *Invest Ophthalmol Vis Sci* 52:9470–9477, 2011. Used with permission from the Association for Research in Vision and Ophthalmology.)

cornea has recently become a major treatment option for LSC deficiency, a severe eye condition that, untreated, causes loss of vision. Securing the LSC graft with PTB on amnion in a rabbit model of LSC deficiency provided better outcomes than sutured attachment (Gu et al. 2011). At 28 days, the corneal opacity scores, neovascularization scores, neutrophil infiltration, and formation of new blood vessels were significantly lower after securing the graft with PTB compared with suture attachment.

66.5 Skin Surgery

The minimal inflammatory response to PTB compared to sutures suggested that surgical scars would be less obvious. PTB was evaluated for epidermal closure in layered skin wound repair along with standard suture closure and octyl cyanoacrylate tissue adhesive in porcine skin (Kamegaya et al. 2005). Incisions and excisions were treated with 75–150 J/cm² 532 nm radiation at 0.56 W/cm². Wounds that were evaluated at 2, 4, and 6 weeks for cosmetic outcomes and histological scar width did not differ among the treatment groups, suggesting that PTB did not initiate adverse responses.

In a clinical study, the effectiveness of PTB was compared to standard interrupted epidermal sutures for closure of skin excisions in a split-lesion, paired comparison study of 31 skin excisions (Tsao et al. 2012). Following deep closure with absorbable sutures, one-half of each wound was sealed with nonabsorbable sutures or treated with PTB (100 J/cm², 0.5 W/cm²). After 2 weeks, the PTB-treated side showed less erythema and better overall appearance than the sutured sides (Figure 66.2). The

scars produced after PTB at 6 months were rated better than the suture scar in overall appearance and had a narrower width than after sutured closure. These results indicated that PTB produces effective wound sealing and less scarring than closure with conventional epidermal sutures.

The ability of PTB to close wounds in very thin eyelid skin was evaluated using the dorsal skin (~0.5 mm thick) of the SKH-1 hairless mouse (Yang et al. 2012). The immediate seal strength was equivalent for all PTB fluences (25, 50, or 100 J/cm²) and for sutures. The ultimate strength after PTB was greater than the controls on days 1 and 3 and less inflammatory infiltrate was observed. Significantly, the average procedure time for sutured closure was about twice as long as for PTB (25 J/cm²). Thus, PTB seals wounds in thin, delicate skin more rapidly than suturing and does not require painful suture removal.

Because RB is phototoxic to cultured cells, a study was carried out to determine whether PTB killed skin cells during closure of skin incisions (Yao et al. 2011). RB was retained in an ~100 μm wide band on the wound wall in porcine skin (Figure 66.3a). Using 100 J/cm² 532 nm radiation, the percentage of dead cells

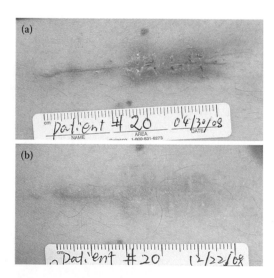

FIGURE 66.2 Epidermal closure of a skin excision wound after removal of a nevus. Absorbable sutures were used to close deep dermal layers before sealing with interrupted sutures or PTB. (a) Two weeks immediately after suture removal. (b) Six months postsurgery. (Tsao, S., M. Yao, H. Tsao, F. P. Henry, Y. Zhao, J. J. Kochevar, R. W. Redmond and I. E. Kochevar: Light-activated tissue bonding for excisional wound closure: a split-lesion clinical trial. *Br J Dermatol.* 2012. 166. 555–563. Copyright Wiley-VCH Verlag GmbH & Co. KGaA. Reproduced with permission.)

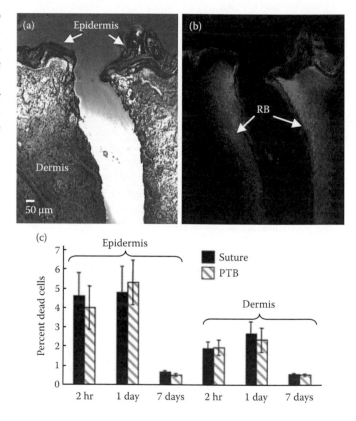

FIGURE 66.3 Lateral penetration of RB into wound surface of incision made in porcine skin viewed by (a) reflectance, (b) fluorescence confocal microscopy, (c) assessment of PTB phototoxicity. Incisions in rabbit skin *in vivo* were closed with sutures or PTB. Percent dead cells within 100 μm of the dermal incision surface. (Yao, M., A. Yaroslavsky, F. P. Henry, R. W. Redmond and I. E. Kochevar: Phototoxicity is not associated with photochemical tissue bonding of skin. *Lasers Surg Med.* 2011. 42. 123–131. Copyright Wiley-VCH Verlag GmbH & Co. KGaA. Reproduced with permission.)

did not differ between PTB-treated and control incisions in porcine skin *ex vivo* at 24 h or in rabbit skin at 2 h and 3 and 7 days (Figure 66.3b). Modeling of light distribution in skin indicated that the fluence rate was ~0.5 W/cm^2 in the mid-dermis (~350 mm). Thus, although sufficient RB and light were present, PTB did not cause phototoxicity when used to repair skin wounds.

Attaching skin grafts with PTB has the potential advantage of sealing the graft over the entire wound, thus increasing the ability of the graft to resist shear stress. In an *ex vivo* study, the dermal surface of two porcine split thickness skin grafts was treated with PTB (Chan, Kochevar, and Redmond 2002). Immediate adherence increased with increasing RB concentration and with increasing laser fluence. The maximal surface temperature during irradiation was less than 40°C and the skin grafts retained cell viability and collagen organization after bonding, suggesting that PTB might improve graft adherence. Preliminary results suggest that PTB can seal split-thickness skin grafts to wound beds in porcine skin.

66.6 Peripheral Nerve Repair

The rationale for using PTB in nerve repair is the formation of a watertight seal over the repair site of transected nerves and nerve grafts. Intimate sealing of the collagenous epineurium prevents leakage of the neurotrophic and neurotropic factors that are essential for axonal regrowth, prevents axonal escape from the endoneural architecture, and reduces inflammation and scarring caused by suture trauma. By optimizing the internal neural environment for regeneration, we hypothesized an improvement in nerve regrowth.

66.6.1 Direct Neurorrhaphy

Initial studies were performed in a rat model following transection of the sciatic nerve (Johnson et al. 2007). A small segment of endoneurium was removed from the distal stump before applying RB to the stumps, forming a proximal-in-distal cuff and illuminating with 532 nm light. The functional recovery with the PTB approach was equivalent to standard microsurgical neurorrhaphy, but the cuff shortened the nerve and an improved approach was devised that did not induce tension across the repair.

66.6.2 Photosealing with Amniotic Membrane

In the same model, the nerve stumps were approximated and reattached using sutures with and without sealing of the

coaptation site with a human amniotic membrane (HAM) wrap (O'Neill et al. 2009b) (Figure 66.4). HAM is a thin membrane (~40 μm), avascular with antiangiogenic and anti-inflammatory properties, and negligible immune response, and contains growth factors that promote wound healing. The sciatic function index (SFI) was used to quantitate functional recovery (0 = complete recovery, –100 = complete loss of function). Following standard neurorrhaphy (Group 1), the repair was wrapped with HAM and either sutured in place (Group 2) or sealed by RB/PTB cross-linking (Group 3). Group 3 showed a statistically significant improvement in SFI (–55.7) compared to Group 2 (–68.6) or Group 1 (–70.8) at 12 weeks postrepair. Histomorphometry also highlighted superior regeneration in Group 3 than in Group 1. These results suggest that watertight sealing of the repair site is important for optimal outcomes in nerve repair.

Repair of transection of rabbit common peroneal nerve was evaluated using electrophysiological testing (Henry et al. 2009b). Group 1 received neurorrhaphy (N) alone; Group 2 received an additional HAM wrap around the repair site, secured to the epineurium with a 10/0 nylon suture. In Group 3, the HAM was stained with RB, wrapped around the repair site, and sealed to the nerve by 1 min of 532 nm (0.5 W/cm^2) light. A significant improvement in distal compound muscle action potential was observed by 90 days in Group 3 compared to Group 1 and by 120 days compared to Group 2. Histology and histomorphometry also demonstrated significant increases in myelin thickness, fiber and axon diameter, and G value distal to the repair site for Group 3. Axonal outgrowth through the epineurium occurred in Groups 1 and 2 but not Group 3, where HAM wrap was photochemically sealed over the repair site (Figure 66.5).

66.6.3 Repairs Involving Nerve Deficit

A nerve deficit requires a conduit or nerve graft to bring the nerve into continuity. A HAM tube, photo-cross-linked with RB and green light to improve rigidity and degradation resistance, was used as a nerve conduit (Figure 66.6). The stiffer central area bridged the nerve gap and the uncross-linked ends were then stained with RB, slipped over the nerve stumps, and illuminated to seal the coaptation sites (O'Neill et al. 2009a).

The PTB-sealed cross-linked HAM conduit bridged a 1 cm deficit with equivalent functional recovery and nerve regrowth as autologous nerve graft in a rat sciatic nerve model (O'Neill et al. 2009a). At 12 weeks postrepair, the regenerating nerve had regrown completely through the conduit lumen. The autologous

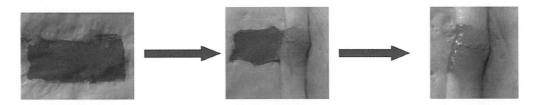

FIGURE 66.4 Left: segment of HAM stained with 0.1% RB. Center: HAM is wrapped around the neurorrhaphy site. Right: HAM wrap following PTB sealing over the repair site.

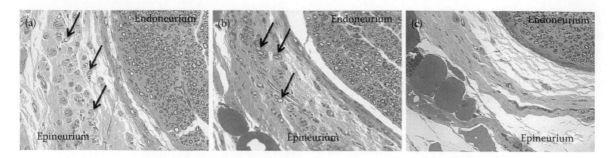

FIGURE 66.5 200 × toluidine blue sections, 5 mm distal to repair with axonal sprouting with regeneration into and through the epineurium (arrows). (a) Group 1. (b) Group 2. (c) Group 3. PTB sealing prevented axonal sprouting into epineurium with normal neural microarchitecture.

FIGURE 66.6 Schematic showing synthesis of HAM conduit with central region stained with RB and internally cross-linked with green light to form a tube.

nerve graft and PTB-sealed cross-linked HAM conduit also showed improved histomorphometry results compared to the sutured cross-linked HAM conduit. Sections taken within the regenerating nerve at 12 weeks postrepair demonstrated statistically significant increases in fiber count, fiber diameter, and myelin thickness.

66.7 Blood Vessel Anastomosis

Elimination of sutures and associated collateral tissue damage is an unmet goal for microsurgical vascular repair. Suture placement is inevitably associated with trauma to the vessel, foreign body reaction, inflammation, and scarring. Sutures also create irregularities of the endothelium with a consequent prothrombotic effect. Light-activated vascular repair is an attractive concept to achieve a watertight seal and avoid suture complications. In a preliminary *ex vivo* study, porcine brachial arteries were transected and repaired by sutures or PTB. Repair strength was measured by clamping the ends of the arterial segment and measuring the leak pressure (LP). PTB produced significantly higher LPs (1105 mm Hg) than suture repair (352 mm Hg). In an *in vivo* study, the left femoral artery of rats was clamped and sharply transected 1 cm distal to its exit from the inguinal ligament (O'Neill et al. 2007). One group was repaired with six 10–0 monofilament nylon sutures. In the PTB group, a small arteriotomy was made in the tibial artery and a 24G angiocatheter inserted into the femoral artery to support the lumen. Following

transection, the proximal segment of the artery was drawn into the distal end to create a 2 mm tissue overlap. RB was applied and the overlapped area was exposed to 532 nm light (0.5 W/cm² for 30 s). All arteries in both suture and PTB repair groups were patent at 6 h postrepair, and at 8 weeks, the patency rate was 80% for both groups with no evidence of aneurysm formation. Bleeding was absent from the repair site in the PTB-treated vessels, in contrast to the suture repair group.

These results showed that PTB is a feasible sutureless vascular repair technique that produces a strong and immediate seal with short- and long-term patency rates equal to those for standard suture repair. Current studies in this area are focused on the production of permanent and temporary stent devices to support the arterial lumen during repair.

66.8 Tendon Repair

Total rupture of a tendon or ligament is almost always an indication for surgery. Suture repair provides mechanical strength to the repair during healing. However, sutures do not produce bonding of ruptured ends at the molecular level. Moreover, infection, inflammation, adhesions, and scarring frequently result. A prolonged period of joint immobilization is typically used to protect from rerupture, but earlier mobilization is now thought to be advantageous in reducing adhesions and calf muscle wastage and minimizing deep venous thrombosis. PTB could cross-link the torn tendon ends, strengthen the repair, and enhance wound healing.

In early experiments, a rat Achilles tendon rupture was repaired by direct application of RB to the surfaces of the torn tendon that were then bought into contact and illuminated to reattach the tendon stumps (Chan et al. 2005). Although this approach did not provide sufficient mechanical strength as a primary repair method, it did enhance the healing process at early times postoperatively. A limitation to direct PTB reattachment of tendon stumps was the penetration depth of green light in tendon of only 680 μm.

In recent studies, PTB was used to bond a strong supporting biocompatible sleeve around the repair site. A new, strong, biocompatible nanofiber biomaterial, electrospun silk (ES), when bonded over the tendon repair site, was tested for its ability to reduce the tensile load across the healing tendon and enhance repair.

A mat of randomly aligned ES fibers was created with approximately 50 μm thickness, and 700 nm fiber diameter was used in PTB augmentation of surgical repairs of ruptured rabbit Achilles tendon. The ES mat, stained with RB, was wrapped twice around the repair site and illuminated for 6 min at 0.3 W/cm². Experimental groups were (1) suture repair (SR), (2) SR + ES/PTB, and (3) ES/PTB with only a stay suture used to approximate the tendon stumps. The tendon complex was harvested at days 7, 14, and 28 for biomechanical testing of repair strength. The ES/PTB treatment on its own did not generate a repair approaching the strength of sutures at early times, but as an augmentation of standard suture technique, it produced significant strengthening of the repair site. Additionally, in both groups where ES/PTB was applied, there was a significant reduction in adhesions. Thus, PTB augmentation of standard repair has potential for improved tendon repair and better functional outcomes.

66.9 Vocal Fold Repair

PTB has the potential to bond together delicate tissues such as vocal fold that are difficult to suture (i.e., sutures tear the tissues). RB was applied to microflaps created in the canine vocal fold and the tissue was irradiated with 100 J/cm² 532 nm light (Franco et al. 2011). Compared to microflaps that were not sealed with PTB, the flap could not be dislodged with a stream of air. The PTB and control wounds healed with the same subepithelial reaction (i.e., the same scarring), suggesting that PTB may allow for more predictable healing after microflap resection and may reduce the need for postoperative voice rest.

66.10 Colon Anastomosis

Penetrating bowel wounds can have devastating outcomes and require immediate surgical attention to restore bowel integrity. Resection and subsequent bowel anastomosis is difficult, and leakage occurs up to 30% following traumatic injury possibly leading to infection, peritonitis, and sepsis. Standard suture or staple closure techniques generate inflammation that weakens closure strength and induces adhesion formation and painful bowel obstruction.

We evaluated whether augmentation of a standard sutured anastomosis by light-activated sealing of a wrap of amniotic membrane over the anastomotic line could increase the strength of anastomosis in rat bowel. Following surgical transection of the colon, groups underwent anastomosis by standard suture with or without additional photochemical sealing using 532 nm light (80 s, 0.5 W/cm²). After 3 and 7 days, a colon section was removed and strength of anastomosis evaluated by burst pressure (BP) measurement. A fourfold increase in burst pressure at day 3 was observed in the photochemically augmented repair (BP = 94 mm Hg) compared to suture repair alone (BP = 25 mm Hg). Additionally, a significant decrease in intraabdominal, postsurgical adhesions was seen in the photochemically augmented repair. Thus, augmenting standard suture technique for colon anastomosis with a light-activated amnion wrap increases repair strength at early times postrepair and reduces problematic abdominal adhesions.

66.11 PTB Mechanism

The photoprocesses of RB in solution have been well studied. In tissue, however, these processes are likely to differ because RB associates strongly with tissue proteins (largely collagen in connective tissue), thus limiting its diffusion and its nearest neighbor reaction partners. The RB/protein association is apparent from the red shift of the absorption maximum in cornea and amniotic membrane to 562 nm from 550 nm in PBS (Verter et al. 2011). Association of RB with collagen in solution also produces a 562 nm maximum and decreases the RB fluorescence yield (Y. Tang, unpublished). RB binds strongly to collagen since RB is not extracted from dermis and cornea by phosphate buffered saline (PBS) (unpublished) and only partially extracted from amniotic membrane even after 24 h incubation (Verter et al. 2011). Tight binding to tissue proteins is presumably the reason that RB penetration into dermis is limited to ~100 μm (Figure 66.3).

The reactions initiated by RB after light absorption are likely to occur from the long-lived excited triplet state (^{3}RB) (Equation 66.1), which forms with a quantum yield of ~0.90 (Gandin, Lion, and Van de Vorst 1983). Electron transfer from electron-rich amino acids such as lysine, arginine, or histidine to ^{3}RB would produce an amino acid cation radical (AA$^{•+}$) and RB anion radical (RB$^{•-}$) (Equation 66.2). Highly reactive carbon-centered radicals (AA$^•$), which may initiate cross-link formation, can be formed from AA$^{•+}$ by deprotonation (Equations 66.4 and 66.5). Formation of cross-links by this mechanism may involve oxygen after the initial electron transfer. Energy transfer from ^{3}RB produces singlet oxygen, which initiates protein–protein cross-linking by oxidizing amino acid side chains, especially histidine (Equation 66.6) (Shen et al. 1996a; Tsao et al. 2012). Oxidized histidine, possibly an imidazolone, reacts with certain nucleophilic amino acids to form protein–protein cross-links (Equation 66.7) (Au and Madison 2000).

$$RB \xrightarrow{\;532\,nm\;} {}^1RB \rightarrow {}^3RB \quad \text{excitation} \tag{66.1}$$

$$^3RB + AA \rightarrow RB^{•-} + AA^{•+} \quad \text{electron transfer} \tag{66.2}$$

$$^3RB + O_2 \rightarrow RB + {}^1O_2 \quad \text{energy transfer} \tag{66.3}$$

$$AA^{•+} \xrightarrow{\;-H^+\;} AA^{•} \quad \text{free radical formation} \tag{66.4}$$

$$AA^{•}, AA^{•} \rightarrow AA - AA \quad \text{protein–protein cross-link} \tag{66.5}$$

$$^1O_2 + AA \rightarrow \text{oxidized AA (Ox} - AA) \quad \text{AA oxidation} \tag{66.6}$$

$$Ox - AA + \text{nucleophilic AA} \rightarrow Ox - AA - AA \quad \text{protein–protein cross-link} \tag{66.7}$$

The relative contributions of initial electron transfer and energy transfer/singlet oxygen mechanisms will be highly influenced by the level of oxygen available. For some PTB applications, the oxygen level is expected to be low because the tissue is not vascularized, whereas for others, the oxygen level may be higher because oxygen can diffuse to the RB-stained surface during the irradiation. Also, the oxygen level may not remain constant during a PTB treatment because reactions of $^1O_2$ with tissue components consume oxygen, leading to oxygen depletion. In fact, based on results of studies of photodynamic therapy, oxygen depletion might be expected at the high irradiances used in PTB.

Experimentally, both electron transfer and energy transfer initiation mechanisms appear to occur. When RB was applied to the cornea surface and incisions in cornea and then irradiated to cross-link collagen, the absorption maximum decreased and blue-shifted, indicating oxygen-independent reactions (unpublished results). In these cases, it appears that the rate of oxygen diffusion into the corneal stroma was not sufficient to support an oxygen-dependent mechanism and cross-linking was initiated by an electron transfer process. In contrast, oxygen was required for photobonding amniotic membrane to cornea; the bond strength was much greater when the irradiation was carried out in oxygen than under nitrogen (Verter et al. 2011). Singlet oxygen was involved in the cross-linking mechanism since greater bond strength was produced in D_2O compared to H_2O, (since $^1O_2$ has an inherently longer lifetime in D_2O than in H_2O). In addition, RB photobleached without the blue shift in absorption spectrum expected in an oxygen-poor environment. Since amnion is <50 μm thick, it appears that sufficient oxygen diffused through the membrane to support formation of $^1O_2$. The relative efficiency for formation of covalent protein–protein cross-links initiated by electron transfer versus energy transfer mechanisms has not been investigated in detail, but the stronger bonding of amnion to cornea observed in the presence of oxygen suggests that energy transfer/singlet oxygen initiation is more efficient (Verter et al. 2011).

66.12 Summary

PTB that is initiated by photochemical mechanisms has been shown to attain the goals of sutureless joining of tissue surfaces in preclinical studies as well as in an initial clinical study. PTB produced less inflammation, fibrosis, and scarring than sutures and formed an immediate watertight seal. This technology appears to be especially appropriate for microsurgical applications and for closure of delicate, difficult-to-suture tissues. Further clinical studies are warranted.

References

Allen, M. T., M. Lynch, A. Lagos, R. W. Redmond, and I. E. Kochevar. 1991. A wavelength dependent mechanism for rose bengal-sensitized photoinhibition of red cell acetylcholinesterase. *Biochim Biophys Acta* 1075:42–49.

Au, V., and S. A. Madison. 2000. Effects of singlet oxygen on the extracellular matrix protein collagen: Oxidation of the collagen crosslink histidinohydroxylysinonorleucine and histidine. *Arch Biochem Biophys* 384:133–142.

Balkissoon, B., and R. Weld. 1965. A rapid, simple and inexpensive method of measurement of blood volume and hepatic function utilizing rose bengal I. *Ann Surg* 162:881–885.

Chan, B. P., C. Amann, A. N. Yaroslavsky et al. 2005. Photochemical repair of Achilles tendon rupture in a rat model. *J Surg Res* 124:274–279.

Chan, B. P., I. E. Kochevar, and R. W. Redmond. 2002. Enhancement of porcine skin graft adherence using a light-activated process. *J Surg Res* 108:77–84.

Dahl, T. A., W. R. Midden, and D. C. Neckers. 1988. Comparison of photodynamic action by Rose Bengal in Gram-positive and Gram-negative bacteria. *Photochem Photobiol* 48:607–612.

Dubbelman, T. M., A. F. de Goeij, and J. van Steveninck. 1978. Photodynamic effects of protoporphyrin on human erythrocytes. Nature of the cross-linking of membrane proteins. *Biochim Biophys Acta* 511:141–151.

Franco, R. A., J. R. Dowdall, K. Bujold et al. 2011. Photochemical repair of vocal fold microflap defects. *Laryngoscope* 121: 1244–1251.

Gandin, E., Y. Lion, and A. Van de Vorst. 1983. Quantum yield of singlet oxygen production by xanthene derivatives. *Photochem Photobiol* 37:271–278.

Givens, R. S., G. T. Timberlake, P. G. Conrad et al. 2003. A photoactivated diazopyruvoyl cross-linking agent for bonding tissue containing type-I collagen. *Photochem Photobiol* 78:23–29.

Givens, R. S., A. L. Yousef, S. Yang, and G. T. Timberlake. 2008. Collagen cross linking agents: Design and development of a multifunctional cross linker. *Photochem Photobiol* 84:185–192.

Goins, K. M., J. Khadem, P. A. Majmudar, and J. T. Ernest. 1997. Photodynamic biologic tissue glue to enhance corneal wound healing after radial keratotomy. *J Cataract Refract Surg* 23:1331–1338.

Gu, C., T. Ni, E. E. Verter et al. 2011 Photochemical tissue bonding: A potential strategy for treating limbal stem cell deficiency. *Lasers Surg Med* 43:433–442.

Henry, F. P., D. Cote, M. A. Randolph et al. 2009a. Real-time *in vivo* assessment of the nerve microenvironment with coherent anti-Stokes Raman scattering microscopy. *Plast Reconstr Surg* 123:123S–130S.

Henry, F. P., N. A. Goyal, W. S. David et al. 2009b. Improving electrophysiologic and histologic outcomes by photochemically sealing amnion to the peripheral nerve repair site. *Surgery* 145:313–321.

Johnson, T. S., A. C. O'Neill, P. M. Motarjem et al. 2007. Photochemical tissue bonding: A promising technique for peripheral nerve repair. *J Surg Res* 143:224–229.

Judy, M. M., L. Chen, L. Fuh et al. 1996a. Photochemical cross-linking of type I collagen with hydrophobic and hydrophilic 1,8 naphthalimide dyes. *Proc SPIE* 2681:53–55.

Judy, M. M., L. Fuh, J. L. Matthews, D. E. Lewis, and R. Utecht. 1994. Gel electrophoretic studies of photochemical cross-linking of type I collagen with brominated 1,8-naphthalimide dyes and visible light. *Proc SPIE* 2118:506–509.

Judy, M. M., J. L. Matthews, R. L. Boriack et al. 1993a. Heat-free photochemical tissue welding with 1,8-naphthalimide dyes using visible (420 nm) light. *Proc SPIE* 1876:175–179.

Judy, M. M., J. L. Matthews, R. L. Boriack et al. 1993b. Photochemical cross-linking of proteins with visible-light-absorbing 1,8-naphthalimides. *Proc SPIE* 1882:305–308.

Judy, M. M., J. L. Matthews, H. R. Nosir et al. 1996b. Bonding of human meniscal and articular cartilage with photoactive 1,8-naphthalimide dyes. *Proc SPIE* 2671:251.

Kamegaya, Y., W. A. Farinelli, A. V. Vila Echague et al. 2005. Evaluation of photochemical tissue bonding for closure of skin incisions and excisions. *Lasers Surg Med* 37:264–270.

Khadem, J., M. Martino, F. Anatelli, M. R. Dana, and M. R. Hamblin. 2004. Healing of perforating rat corneal incisions closed with photodynamic laser-activated tissue glue. *Lasers Surg Med* 35:304–311.

Khadem, J., T. Truong, and J. T. Ernest. 1994. Photodynamic biologic tissue glue. *Cornea* 13:406–410.

Khadem, J., A. A. Veloso, Jr., F. Tolentino, T. Hasan, and M. R. Hamblin. 1999. Photodynamic tissue adhesion with chlorin(e6) protein conjugates. *Invest Ophthalmol Vis Sci* 40:3132–3137.

Kim, J. 2000. The use of vital dyes in corneal disease. *Curr Opin Ophthalmol* 11:241–247.

Lambert, C. R., and I. E. Kochevar. 1997. Electron transfer quenching of the rose bengal triplet state. *Photochem Photobiol* 66:15–25.

Lauto, A., D. Mawad, M. Barton et al. 2010. Photochemical tissue bonding with chitosan adhesive films. *Biomed Eng Online* 9:47.

Mulroy, L., J. Kim, I. Wu, P. Scharper et al. 2000. Photochemical keratodesmos for repair of lamellar corneal incisions. *Invest Ophthalmol Vis Sci* 41:3335–3340.

O'Neill, A. C., M. A. Randolph, K. E. Bujold et al. 2009a. Preparation and integration of human amnion nerve conduits using a light-activated technique. *Plast Reconstr Surg* 124:428–437.

O'Neill, A. C., M. A. Randolph, K. E. Bujold et al. 2009b. Photochemical sealing improves outcome following peripheral neurorrhaphy. *J Surg Res* 151:33–39.

O'Neill, A. C., J. M. Winograd, J. L. Zeballos et al. 2007. Microvascular anastomosis using a photochemical tissue bonding technique. *Lasers Surg Med* 39:716–722.

Proano, C. E., D. T. Azar, M. C. Mocan, R. W. Redmond, and I. E. Kochevar. 2004a. Photochemical keratodesmos as an adjunct to sutures for bonding penetrating keratoplasty corneal incisions. *J Cataract Refract Surg* 30:2420–2424.

Proano, C. E., L. Mulroy, E. Jones et al. 2004b. Photochemical keratodesmos for bonding corneal incisions. *Invest Ophthalmol Vis Sci* 45:2177–2181.

Redmond, R. W., and J. N. Gamlin. 1999. A compilation of singlet oxygen yields from biologically relevant molecules. *Photochem Photobiol* 70:391–475.

Rodgers, M. A. J. 1983. Picosecond fluorescence studies of rose bengal in aqueous micellar dispersions. *Chem Phys Lett* 78:509–514.

Shen, H. R., J. D. Spikes, P. Kopecekova, and J. Kopecek. 1996a. Photodynamic crosslinking of proteins. I. Model studies using histidine- and lysine-containing N-(2-hydroxypropyl) methacrylamide copolymers. *J Photochem Photobiol B* 34:203–210.

Shen, H. R., J. D. Spikes, P. Kopeckova, and J. Kopecek. 1996b. Photodynamic crosslinking of proteins. II. Photocrosslinking of a model protein-ribonuclease A. *J Photochem Photobiol B* 35:213–219.

Timberlake, G. T., A. L. Yousef, S. R. Chiles, R. A. Moses, and R. S. Givens. 2005. Bonding corneal tissue: Applications of photoactivated diazopyruvoyl cross-linking agent. *Photochem Photobiol* 81:1180–1185.

Tsao, S., M. Yao, H. Tsao et al. 2012. Light-activated tissue bonding for excisional wound closure: A split-lesion clinical trial *Br J Dermatol* 166:555–563.

Verter, E. E., T. E. Gisel, P. Yang et al. 2011. Light-initiated bonding of amniotic membrane to cornea. *Invest Ophthalmol Vis Sci* 52:9470–9477.

Verweij, H., T. M. Dubbelman, and J. Van Steveninck. 1981. Photodynamic protein cross-linking. *Biochim Biophys Acta* 647:87–94.

Wang, Y., I. E. Kochevar, R. W. Redmond, and M. Yao. 2011. A light-activated method for repair of corneal surface defects. *Lasers Surg Med* 43:481–489.

Yang, P., M. Yao, S. L. Demartelaere, R. W. Redmond, and I. E. Kochevar. 2012. Light-activated sutureless closure of wounds in thin skin. *Lasers Surg Med* 44:163–167.

Yao, M., A. Yaroslavsky, F. P. Henry, R. W. Redmond, and I. E. Kochevar. 2011. Phototoxicity is not associated with photochemical tissue bonding of skin. *Lasers Surg Med* 42:123–131.

VII

Other Phototherapies and Future Outlook

Optical Guidance for Cancer Interventions

Brian C. Wilson
University of Toronto
University Health Network

Robert Weersink
University Health Network

67.1 Introduction

The purpose of this chapter is to provide, mainly through specific examples, an overview of current and emerging optics-based techniques and technologies for active guidance of therapeutic interventions. The primary focus is on applications in oncology and specifically in the treatment of patients with solid tumors, but it is not difficult to see how some of the approaches could be used in other major therapeutic areas such as cardiovascular and neurological diseases and even in management of patients with localized infections. That is, the concepts apply to any situation where the physician intervenes by a treatment that is locally targeted to diseased tissue and where additional information can help guide the treatment and so improve the outcome.

Clearly, there are a variety of standard radiological imaging modalities, which to different degrees are already used to guide cancer treatments: x-ray computed tomography (CT), magnetic resonance imaging (MRI), and positron emission tomography (PET). This includes some very advanced approaches; for example, intraoperative MRI for real-time surgical guidance in brain tumor resection (Busse et al. 2006; Kubben et al. 2011), volumetric CT (Ahn et al. 2011), or MRI-based (Poetter et al. 2011; Roels et al. 2009) planning of radiation treatments and monitoring of tumor shrinkage during a course of therapy [so-called adaptive radiotherapy (Ling et al. 2000)], and serial PET scanning to monitor the response of tumors during a course of chemotherapy so that the drug regime can be modified according to whether the tumor appears to be responding (Terasawa, Dahabreh, and Nihashi 2010; Wang et al. 2012).

In the optical-guidance domain, therapeutic endoscopy is a well-established approach for active intervention under real-time visual guidance. Thus, in the management of patients with Barrett's esophagus (premalignant transformation of the esophageal mucosa that is associated with high risk of developing esophageal dysplasia with progression to invasive cancer), endoscopic mucosal resection is a standard and minimally invasive procedure in which the mucosa is stripped using an argon plasma under direct endoscopic visualization (Dulai et al. 2005; Wang and Sampliner 2008). While this is clearly an example of optical guidance of a cancer intervention—in this case, with preventative rather than curative intent—it does not involve any optical visualization beyond standard endoscopy. On the other hand, fluorescence imaging to identify dysplastic lesions within the field of transformed mucosa with subsequent destruction of the dysplasia tissue [including by laser ablation or photodynamic therapy (PDT)] would be more representative or the type of advances with which this chapter is concerned.

67.2 Current Cancer Therapies and Unmet Needs

The three main pillars of cancer treatment are surgery, radiation therapy, and chemotherapy. Each is used across a broad range of tumor types and stages, and they are often used in combination; for example, surgical resection followed by radiation and drugs targeting residual tumor or distant metastases. In addition, there are a variety of other modalities that are niche in the sense of being used for only a limited range of tumor types/stages, are not yet widely adopted into clinical practice, and/or are still at the developmental (preclinical or clinical trials) stage. These include: PDT (Agostinis et al. 2011; Dolmans, Fukumura, and Jain 2003; Pinthus et al. 2006), the use of compounds (photosensitizers) that are taken up by tumor cells (and/or tumor microvasculature) and are subsequently activated by light of an appropriate wavelength, generating cytotoxic species, most commonly singlet oxygen (see Chapters 23–47); various forms of biophysical interventions to destroy tumor tissue by heating [thermal therapy using high-power near-infrared lasers (Carpentier et al. 2011; Lindner et al. 2009), ultrasound (Berge et al. 2011; Keller et al. 2010), microwaves (Sherar et al. 2004; Wolf et al. 2012), radiofrequency radiation (Buscarini et al. 2001; Salas, Castle, and Leveillee 2011)] or freezing [cryotherapy (Clyne 2011; Foley et al. 2011)]; biologic therapies, such as immunotherapy (Mellman, Coukos, and Dranoff 2011; Sharma et al. 2011), that aim to use the body's own immune system to kill the tumor cells; and treatments based on targeting the putative tumor stem cells that may be responsible for sustaining tumor growth.

Some of the unmet needs in which optical techniques may be valuable for interventional guidance with each of these modalities are summarized in Table 67.1. These are divided into two categories:

- *Pretreatment*, which is treatment planning to help identify and accurately localize in 3-D the therapeutic target and/ or to characterize the tumor in order to determine the optimum treatment approach

- *During treatment*, by providing active guidance to enable the planned treatment to be delivered, including feedback to control the treatment parameters and/or, in the case of treatments comprising multiple "fractions" (e.g., of radiation or drugs), to measure the response to treatment so that adjustments can be made dynamically

In the first category, we do not include the use of optical techniques for cancer diagnostics (i.e., for screening and detection of lesions), which is an important and rapidly evolving field in itself. That is, we assume that the diagnosis has been made and that the tumor has been at least grossly localized. Also, this chapter does not cover the plethora of laboratory-based methods for cancer discovery in which optical approaches play increasingly important roles (e.g., in flow cytometry, microscopy, bioanalytics, small animal imaging, genomic profiling, etc.).

In addition to applications pre-intervention and during intervention, optical guidance can contribute yet another category:

- *Posttreatment*, determining the response of the tumor and possibly also any effects on normal, nontarget tissues, for the purpose of correlating these to the delivered treatment in order to refine the delivery for subsequent patients

Although there are some common needs between modalities, such as accurately defining the real tumor margins (as opposed to simply the morphologic margin), some of the requirements are very treatment-modality specific. The specific needs may also be different between acute treatments that are delivered in a

TABLE 67.1 Categories of Unmet Needs in the Delivery of Personalized Cancer Treatments in Which Existing Optical Techniques May Contribute to Optimizing Outcome

Treatment Modality and Primary Objectives of Optical Guidance	Unmet Clinical Needs Amenable to Optical Guidance	
	Pre-Intervention	During Intervention
Surgery (to reduce re-excision rates)	Define tumor margins and coregister with radiological (3-D) volume Localize critical adjacent structures Determine optimum surgical approach Guide biopsy for tumor confirmation/staging	Identify and localize residual (mesoscopic) tumor tissue following bulk resection Identify and quantify tumor margin depth in resection specimens
Radiation therapy (to minimize risk of geographic miss and reduce collateral damage to normal tissues)	Define tumor margins (contours) and coregister with radiological (3-D) volume to generate accurate treatment plan	Track changes in tumor extent between fractions to modify radiation fields as tumor shrinks
Chemotherapy (to minimize toxicity and cost of using ineffective drugs)	Determine baseline tumor metabolic status	Track changes in status to modify drug regimen Monitor drug uptake in tumor
PDT (to optimize drug–light distribution to match the target area or volume)	Define tumor target area or volume to plan placement of light sources Measure photosensitizer concentration in target tissue	Monitor dynamic changes in tissue optical properties to adapt the light delivery Monitor photosensitizer, photobleaching, and/or photoproducts as indirect means of online dosimetry Monitor singlet oxygen generated during treatment as a direct dose metric Monitor vascular or cellular changes in target tumor to assess immediate response Monitor light and/or photosensitizer levels in adjacent tissues to minimize collateral damage
Biophysical therapies (to monitor outer boundary of ablation zone)		Monitor vascular or cellular changes in target tumor to assess immediate response

single session (such as surgical resection or most of the biophysical therapies) and fractionated/extended treatments in which increments are delivered over a period of days or weeks (as in radiation, drug, and biologic therapies). In some applications, the optical technique may be used as a stand-alone method, for example, in monitoring photobleaching in PDT. However, in most cases, the optical information is complementary to other, often image-based, information. Thus, for example, in surgical guidance, the optical data (e.g., fluorescence images or spectra) are best used with, and indeed may need to be spatially coregistered with, x-ray CT or MRI images taken before or during surgery. Similarly, the information on tumor metabolic response to chemotherapeutic drugs by diffuse optical tomography may be used together with longitudinal MRI or PET imaging. Indeed, one of the huge strengths of optical techniques is that they are highly compatible with and complementary to other assessment and guidance methods. In addition, in many instances it is also possible to integrate optical technologies with the actual treatment technology, for example, in robotic surgery or thermal therapies. In all cases, the primary rationale in adding the photonics tools is to provide the optimum treatment for individual patients: that is, *to deliver personalized cancer treatment.*

67.3 Photonics Technologies Relevant to Interventional Guidance

The successful convergence of optical technologies and interventional guidance depends on both the clinical requirements and the (present or emerging) capabilities of the photonic techniques. This is a rapidly changing landscape, since the drive to personalized cancer medicine (Desmond-Hellmann 2012), of which a key feature is the ability to tailor treatments to individual patient needs and tumor characteristics, is accelerating, while biophotonic imaging and spectroscopy methods and instruments

TABLE 67.3 Performance Requirements for Interventional Guidance

Requirement	Rationale
Minimally invasive	Avoids increasing risk of intervention
Quantitative	Measurement standardization for predictive accuracy
Low cost	Facilitates adoption into clinical practice
Real time	Provides online feedback to modify the treatment
Compatible with/ registration to other imaging modalities	Facilitates integration with existing practice and provides complementary information
Compact	Enables point-of-care applications

and their applications are proliferating. Personalized medicine requires precision in treatment delivery and in the measurements used to adapt or alter the individual treatment. Precision implies quantifiable metrics, not simply statistical information across a patient population.

There are also a number of performance requirements for any technology to be useful for interventional guidance, as summarized in Table 67.3. Many of these requirements are met or could be achieved with existing or emerging optics-based or optics-enabled techniques, and several are found in multiple optical techniques. Thus, for example, since the light is being used here to probe rather than modify tissue, and it is nonionizing, optical techniques are of low risk. Compared to, for example, most radiological imaging techniques, they are also low cost, are compact, and can be used in many clinical environments without the need for specialized infrastructure or services. Of course, the key requirement is that they provide useful information that can substantively alter the interventional procedure. This will be illustrated by the examples in the next section.

TABLE 67.2 Types of Information That Can Be Utilized for Interventional Guidance and Examples of Current Optical Techniques That Provide or Potentially Could Provide This Information Based on Techniques That Have Been Reported in Clinical Trials for These Purposes

	Optical Technique				
Type of Information	Fluorescence Spectroscopy/Imaging	Raman Spectroscopy	OCT	DOT	DRS/EES
Tumor margin	Identification of mesoscopic/microscopic disease	Endogenous point measurements (Haka et al. 2005)	Tissue microstructure (Hamdoon et al. 2010; Nguyen et al. 2009)		Tissue microstructure, cell size and density (Keshtgar et al. 2010; Perelman et al. 1998; Suh et al. 2011)
Completeness of resection	Identification of mesoscopic/microscopic disease	Endogenous point measurements	Depth of tumor margin (Nguyen et al. 2009)		
Tumor functional status	Measured with exogenous agents; antibodies, proteins, etc.	Contrast-enhanced SERs based on concentration of agent	Vascular flow	Blood concentration and oxygen saturation; water and lipid concentration	

Note: DOT, diffuse optical tomography; DRS, diffuse reflectance spectroscopy; EES, elastic scattering spectroscopy; OCT, optical coherence tomography; SERS, surface enhanced Raman scattering.

67.4 Specific Examples

In this section, a number of specific examples will be presented briefly to illustrate some of the concepts underlying interventional guidance, the current state of the art in each case, some of the remaining challenges, and opportunities for future development.

67.4.1 Fluorescence Guidance in Tumor Resection

For many patients with solid tumors, the first line of therapy is surgery to achieve as complete a tumor resection as possible. However, the success rate can be highly variable, both for different tumor size and stages and between surgeons. This either places the patient at risk for less effective second-line therapy (e.g., radiation or chemotherapy) with concomitant side effects or necessitates a second surgery. The other important surgical need is to guide lymph node dissection, since many tumors spread first through the local lymphatic system, so that removing tumor-involved nodes has major impact on the risk of subsequent metastases. The causes of incomplete tumor resection are that it is not technically possible or safe to remove all the known tumor tissue, the true tumor margin is not known precisely beforehand, or the surgeon is unable to visualize the residual tumor tissue. Optical techniques have potential to address the last two deficiencies, and a wide variety of techniques are being investigated for these purposes, both imaging and point spectroscopy based. At this time, fluorescence-image-guided resection (FGR) is probably the most extensively studied and at the most advanced stage of clinical investigation and utilization. This has recently been reviewed in detail (Valdes et al. 2012) in the particular case of brain tumors, but many of the concepts apply across tumor sites.

As with any fluorescence-based application in medicine, a fundamental issue is whether an exogenous fluorophore (targeted or nontargeted) should be used: The pros and cons of each are summarized in Table 67.4.

An example of the endogenous contrast is the use of tissue autofluorescence imaging in guiding surgery of patients with oral cancer using violet/blue-light excitation. Thus, for squamous cell oral carcinoma, there is a high rate of local recurrence (up to 30%), but simply taking larger surgical margins can lead

to significant functional deficits and poor quality of life. Initial studies have shown that using autofluorescence imaging intraoperatively to define the surgical margin more precisely and reliably significantly reduces the local recurrence rate compared to standard clinical/radiological identification (from 7/22 to 0/38 at 3 years in one study cohort). An example of current instrumentation and a clinical case is shown in Figure 67.1 that demonstrates (1) how difficult it can be to identify the true tumor boundary on a white-light image; (2) how, as a result, a wide margin has to be used to reduce the risk of missing tumor tissue;

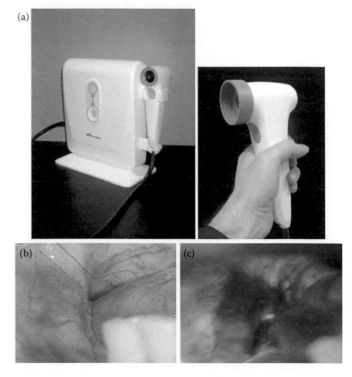

FIGURE 67.1 Example of autofluorescence imaging for surgical guidance in oral cancer. (a) Imaging system with handpiece (Velscope, Burnaby, BC, Canada), (b) white-light image of a lesion (carcinoma in situ) with the tumor margin outlined (dashed line) and the target resection margin (solid line) as assessed by a surgical oncologist, and (c) corresponding autofluorescence image. Note that there are regions of positive fluorescence lying outside the tumor margin identified on the white-light image. (Courtesy of Dr. Catherine Poh and Dr. Calum Macaulay, BC Cancer Center, Vancouver, British Columbia, Canada.)

TABLE 67.4 Comparison of Autofluorescence and the Use of Exogenous Fluorescent Contrast Agents for Surgical Guidance

	Advantages	Limitations
Autofluorescence	Reduced risk and regulatory barriers No time delay or optimization required	Spectral signature not known a priori Fluorescence emission can be weak Highest information content is with UV excitation
Exogenous fluorescence	Known spectral signature Usually strong signals Can use tumor-specific biomarker targeting to increase specificity Biomarker multiplexing is feasible With nanoparticles, can combine multiple imaging and therapeutic functions	Potential toxicity Cost Need to optimize dose and timing

(3) the complementary information provided by the autofluorescence image in defining the true tumor margin; and (4) the technical challenge of accurately coregistering these images because of variations in magnification and orientation. A multicenter randomized trial of this technique is now in progress in Canada (Poh et al. 2011).

In the case of exogenous contrast agents to delineate tumors, there are many options using either untargeted molecules or tumor-targeted fluorophores. For the former, indocyanine green (ICG), which has a distinctive green fluorescence, is already in clinical use with wide-field and endoscopic imaging systems in a variety of surgical-guidance applications, including cancer resection (Tobis et al. 2011; Winer et al. 2010). Several other agents developed primarily for PDT of tumors have also been used in surgical guidance. The approach that has been most fully developed is with aminolevulinic acid (ALA)-induced protoporphyrin IX (PpIX) (see Chapter 23). ALA is an essential precursor in heme biosynthesis, in which the penultimate step is the fluorescent (and photodynamically active) compound PpIX. Through several metabolic and biochemical mechanisms, this usually reaches significantly higher levels in tumor than normal host tissues without depending on specific tumor biomarker targeting. The fluorescence spectrum of PpIX is shown in Figure 67.2a. Like all porphyrins, PpIX can be excited across a wide part of the visible spectrum: Most studies have used violet/blue excitation, since this gives the strongest red fluorescence signal from the tumor surface.

ALA-PpIX fluorescence imaging and point spectroscopy have been investigated for a range of different tumors and to guide endoscopic resection. A major focus has been in brain tumors, particularly malignant gliomas, driven by the grim survival prospects of these patients (typically ~1 year in the case of high-grade tumor) and the fact that virtually all succumb from local recurrence of disease. Figure 67.2b shows a white-light image of the brain after "complete" standard resection and a corresponding ALA-PpIX fluorescence image that reveals areas of residual tumor. This approach has now been incorporated into neurosurgical microscopes, and multicenter clinical trials (Stummer et al. 2006) have shown that FGR significantly improves completeness of resection. However, survival is extended only by about 4 months, so that there is clearly room for further development to increase the sensitivity and specificity of the technique. Avenues being explored involve a range of alternative exogenous fluorescence contrast agents (including agents based on nanoparticles for targeted fluorophore delivery or intrinsically fluorescent quantum dots), as well as the use of other light–tissue interactions such as fluorescence lifetime spectroscopy (Lloyd et al. 2010), optical coherence tomography (Hamdoon et al. 2010; Nguyen et al. 2009), and Raman spectroscopy (Haka et al. 2005; Keller et al. 2011; Shim et al. 1999). Each has potential strengths and weaknesses in terms of sensitivity, specificity, and ease/cost of implementation.

With respect to fluorescence quantification, the fundamental challenge, as illustrated in Figure 67.3a, is that the detected fluorescence signal at the tissue surface depends not only on the concentration of the fluorophore (e.g., PpIX) but also on the depth-dependent attenuation of the excitation and emitted light by the tissue. There are additional factors affecting the measurement, particularly when imaging deep within a (brain) tumor resection cavity, such as the limited field of view, the variable distance of the surface from the imaging system, and the curvature of the cavity. The effects of light attenuation can be corrected for if the optical absorption coefficient

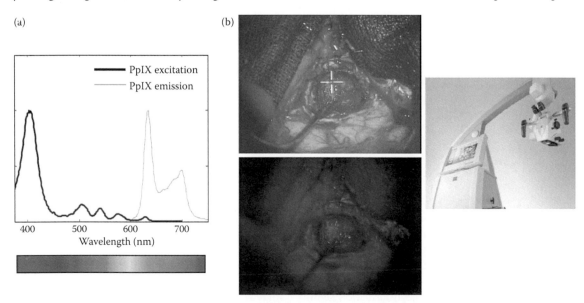

FIGURE 67.2 **(See color insert.)** ALA-PpIX fluorescence imaging for guiding resection of malignant brain tumors. (a) PpIX fluorescence excitation and emission spectra and (b) white-light and PpIX fluorescence images at the end of white-light resection of high-grade glioma. (Courtesy of Dr. David Roberts and colleagues, Dartmouth-Hitchcock Medical Center, Hanover, NH, USA.) The residual tumor is clearly visible as red fluorescence on the blue scattered light background. Also shown is the working head of a fluorescence-enabled neurosurgical microscope (Pentero, Zeiss, Germany) used to take these images.

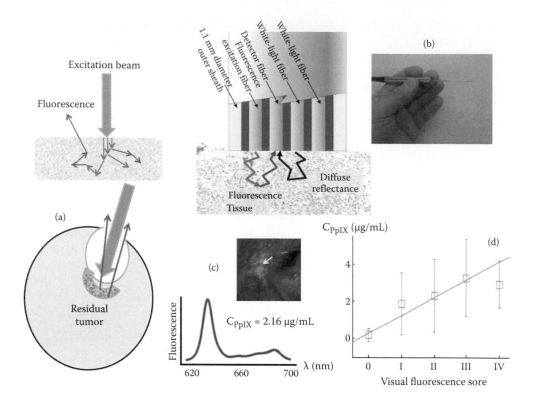

FIGURE 67.3　Fluorophore quantification for surgical guidance. (a) Schematic of the tissue attenuation (loss of excitation and emitted photons through absorption and scattering) and geometric effects in making accurate quantitative measurements of fluorophore concentration in tissue. (b) Example of a fiber-optic point probe in which diffuse reflectance spectroscopy is used to determine the tissue optical properties and thereby to correct the measured fluorescence spectrum and calculate the absolute fluorophore concentration. (From Kim et al. 2010., *J Biomed Opt* 15(6), 067006) (c) Example of a PpIX fluorescence image at the end of white-light resection and the corresponding PpIX fluorescence spectrum and derived concentration (C_{PpIX}) value. (Courtesy of Dr. David Roberts.) (d) Plot of CPpIX versus the nonparametric score of the surgeon's perception of the fluorescence brightness in high-grade glioma cases. (Courtesy of Marco Brantsch.)

and (transport) scattering coefficient are known at both the fluorescence excitation and detection wavelengths. These can be determined by measuring the diffuse reflectance spectrum at two or more source–detector distances on the tissue surface (Figure 67.3b). This has recently been implemented in a point-probe configuration and applied as a complementary technique to ALA-PpIX–based fluorescence image-guided resection in brain tumor patients, by measuring the *absolute* PpIX concentration, C_{PpIX}, at the surface of the resection cavity (Figure 67.3c).

Several points have emerged from the initial clinical trials of this technology:

- As expected, there is very high variability in C_{PpIX} from point to point and patient to patient, even for the same tumor type and grade.
- Nevertheless, C_{PpIX} does correlate strongly with histopathological markers of malignancy (Valdes et al. 2011a).
- Use of the quantitative fluorophore information significantly increases the sensitivity and specificity of residual tumor detection (Valdes et al. 2011b).
- Residual tumor can be detected whose PpIX fluorescence is not visible using the current fluorescence-enabled neurosurgical microscope.

While these results are highly positive, it would clearly be desirable to make quantitative fluorescence in full-imaging mode, but this is far from trivial. For instance, while full-spectrum (hyperspectral) reflectance imaging is possible using, for example, a fast tunable filter in front of the fluorescence camera (Valdes et al. 2012), the point algorithm for deriving the tissue optical coefficients cannot simply be applied pixel by pixel to the images, since there is significant cross talk between adjacent pixels. Various approaches to this problem are being explored, including the use of spatial light modulation techniques (Konecky et al. 2012).

67.4.2 Radiation Therapy Planning: Registration of Endoscopy to Volumetric Imaging

Radiation treatment planning has, of necessity, become increasingly sophisticated in the past decade with the introduction of 3-D conformal and intensity-modulated techniques. The ability to tailor radiation delivery to precisely match the tumor target decreases the toxicity to normal tissue and so enables dose escalation for improved cure rates. However, this requires increased accuracy in defining (contouring) the target volume. Several forms of 3-D radiological imaging can be used for this purpose: CT provides spatially accurate anatomical information, MRI provides high soft tissue contrast, and PET can provide functional information (Gregoire and Mackie 2011).

In endoluminal cancers such as in the head and neck, esophagus, and bronchus, optical endoscopy is often used in staging, since it reveals superficial areas of visible disease beyond that seen on radiological volumetric imaging. No effort has been made to date to relate these findings quantitatively to the volumetric imaging used for treatment planning. Rather, the clinician "connects" them subjectively using anatomical landmarks. A method of registering 2-D images to 3-D anatomical imaging has recently been developed, so that tumor margins localized endoscopically can be accurately transferred into the 3-D treatment planning space (Weersink et al. 2011). Coregistration of 2-D and 3-D image information has also been previously explored for surgical and biopsy guidance (Caversaccio et al. 2008; Fried, Parikh, and Sadoughi

2008) and applied in bronchial (Higgins et al. 2008) and head and neck tumors (Caversaccio et al. 2008; Lapeer et al. 2008). In general, accurate endoscopic (2-D) to radiological (3-D) registration includes optical or electromagnetic tracking devices attached to the endoscope tip to monitor its position and orientation; registration of patient, image, and tracking coordinate systems; and various forms of virtual visualization. After registration, a virtual endoscopy image can be generated from the 3-D volumetric images to match that of the real (i.e., optical) endoscopic image.

Figure 67.4 shows an example of endoscopic contouring in a patient with early-stage glottic cancer. The primary nodule, barely visible on the CT images, is clearly seen on endoscopy, which also reveals a rim of suspicious tissue along the glottis. This region was contoured on the real endoscope image and projected onto the 3-D surface of the virtual endoscopic view (Figure 67.4b). It was then imported into radiotherapy treatment planning software to aid in contouring the true tumor target. By

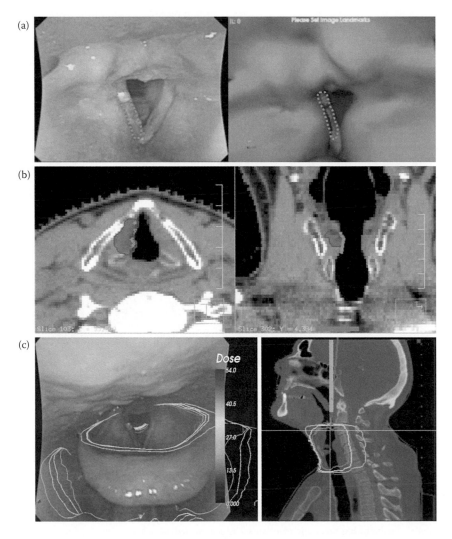

FIGURE 67.4 **(See color insert.)** Example of combining real (white-light) and virtual (CT-based) endoscopy in planning radiation treatment of a glottis tumor. (a) Real (left) and virtual (right) images with the tumor margins outlined. (b) Coregistered images showing different slices of the real endoscopy-defined tumor superimposed on the virtual endoscopic mass. (c) Planned radiation dose contours superimposed on the real endoscopic image (left) and a CT volumetric image slice (right).

reversing the projection process, the radiation dose can also be displayed on the endoscopic images (Figure 67.4c). This is valuable to assess changes in the tumor (i.e., response) or normal tissues (i.e., potential toxicity) between radiation dose fractions, allowing the treatment to be adaptively optimized.

While this example is solely anatomy based, other types of optical information could clearly be incorporated, such as fluorescence images (autofluorescence or biomarker-targeted exogenous fluorophores) or Raman spectral measurements, which could add functional/molecular information to identify the true tumor margins and monitor the tumor response, while optical coherence tomography or confocal endomicroscopy could provide structural images at a resolution greatly exceeding that of radiological volume imaging.

67.4.3 Monitoring Chemotherapy Response of Breast Cancer Using Diffuse Optical Tomography and Diffuse Optical Spectroscopy

As with most cancers, breast tumors have significantly higher levels of hemoglobin and water relative to normal breast tissue

and higher light scattering. Near-infrared (NIR) optical spectroscopy is well suited to quantifying these physiological and structural parameters. Hence, diffuse NIR measurements should allow monitoring of tumor responses to chemotherapy where this involves, for example, antiangiogenic responses that decrease total hemoglobin (Hb_t), oxygenation and water content in tumor. For many deep-seated tumors, the limited light penetration in tissue negates the noninvasive use of optical measurements, but the size, shape, and accessibility of the breast make diffuse optical measurements feasible. In this case, the poor spatial resolution of the optical images (~5–10 mm at depth) is not of major concern, since the critical information is the overall changes in an already localized tumor mass. Diffuse optical tomography/spectroscopy (DOT/DOS) measurements have been implemented in several forms (Culver et al. 2003; Intes 2005; Pogue et al. 2001; Tromberg et al. 2008). DOS is typically performed using simple surface-based probes with multiple sources and detectors. In DOT, the patient lies prone with the uncompressed breast positioned in a liquid that matches typical tissue optical properties. The measurement geometry can be planar using multiple sources and a charge-coupled device (CCD) array detector to make direct (photon time-of-flight) diffuse

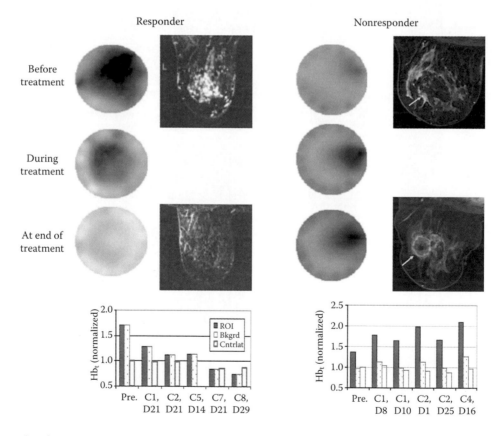

FIGURE 67.5 Examples of DOT monitoring of the response of breast cancer to neoadjuvant chemotherapy, showing the images of total hemoglobin concentration (Hb_t) and corresponding magnetic resonance images immediately before, during, and at the end of treatment. The graphs show the relative Hb_t at different treatment cycles and days in each cycle, measured over the tumor region of interest (ROI), the background tissue in the same breast, and the contralateral breast. The reduction in tumor Hb_t is apparent in the patient, who showed radiological and pathological response to chemotherapy, while little change (or even an increase) was seen in the nonresponding patient. (Courtesy of Dr. Brian Pogue and Dr. Shudong Jiang, Dartmouth College, Hanover, NH, USA.)

transmittance images (Choe et al. 2005), or multiple point sources and detectors can be placed around the breast (Pogue et al. 2001) to provide input to a 3-D reconstruction algorithm. Multiple wavelengths matching major tissue components allow spectral decomposition into physiological parameters.

The first attempts to measure chemotherapy response employed surface-based DOS (Jakubowski et al. 2004; Tromberg et al. 2005). In the subset of patients who responded, total hemoglobin and water content decreased, as did tissue scattering, with most change observed early in the first 5 days of treatment. This allowed the treatment to be modified in nonresponders. Recent DOT studies have used either prototype (Cerussi et al. 2011; Pakalniskis et al. 2011) or commercially available systems (Soliman et al. 2009) (SoftScan ART, Montreal, Canada). Example results are shown in Figure 67.5. As illustrated by these two cases, in patients responding to treatment, consistent reductions have been reported in total hemoglobin (Hb_t), oxyhemoglobin (HBO_2) and water and scattering within 1–2 weeks following initiation of treatment, and there is also recent evidence of correlation between Hb_t changes and biomarkers of tumor-induced angiogenesis (Pakalniskis et al. 2011).

67.4.4 Monitoring Vascular and Cellular Responses to PDT

PDT is a potential treatment modality for a wide spectrum of diseases, including destruction of solid tumors and premalignant (dysplasic) lesions (Dolmans, Fukumura, and Jain 2003; Hamblin and Mroz 2008). As a light-based therapy, it is clearly amenable to the use of optical interrogation, both for guiding treatment and monitoring treatment responses (see Chapter 25). The former includes, for example, online dosimetry of the light, photosensitizer, and cytotoxic singlet oxygen that is generated (Wilson and Patterson 2008). For the latter, since the cellular and tissue responses to PDT can be very rapid, monitoring the biological effects of PDT during the actual treatment delivery is feasible. This is particularly the case for vascularly targeted PDT, where the intent is to shut down the abnormal neovasculature that is associated with tumor growth. (In a major nononcological application of PDT for treatment of wet-form age-related macular degeneration, the shutdown of the choroidal neovasculature in the eye can be observed directly on the fundus camera used to localize the abnormal tissue and direct the treatment light with fluorescein fluorescence contrast to highlight the vessels; see Chapter 39.)

Technically, the most advanced use of optical guidance for PDT treatment delivery and dosimetry has been in PDT of prostate cancer, in which multiple interstitial optical fibers are used to deliver light to the whole prostate gland in patients who have local recurrence following radiation treatment. Several groups have developed somewhat different implementations of treatment planning and of integrated light delivery and online monitoring of the light, photosensitizer, and/or tissue oxygenation using fiber-optic probes (Davidson et al. 2009; Swartling et al. 2010; Weersink et al. 2005; Yu et al. 2006). These measurements

can be used to adjust the light delivery online during treatment in order to deliver the most complete, precise, and uniformly effective PDT dose to the target as possible, while minimizing collateral damage to the local normal tissue structure and function. Some of the complexity of simultaneously monitoring the (changes in) light, photosensitizer, and oxygen during treatment can be circumvented by monitoring the photobleaching of the photosensitizer as a surrogate measure of the effective PDT dose, and this has been used, for example, in individualizing treatments of basal cell skin cancer (Cottrell, Oseroff, and Foster 2006).

Reports of monitoring the biological response of the tissue during treatment delivery in patients include laser Doppler and Doppler optical coherence tomography imaging of the vascular responses in skin (Wang et al. 1997) and esophagus (Standish et al. 2007), respectively, and NIR diffuse correlation spectroscopy to measure local or tumor volume averaged changes in the tissue hemodynamics (Yu 2012). Examples of such responses are shown in Figure 67.6. To our knowledge, such information has not been used to date to modify treatment delivery, but this is likely in the future as confidence is gained in the robustness of the measurements through retrospective analysis to determine their ability to predict clinical outcome.

Other techniques for monitoring PDT responses that have been used only in preclinical animal models to date include bioluminescence imaging, both constitutive to show surviving cells and under specific gene promotion to show stress responses (Moriyama et al. 2004), and speckle-variance optical coherence tomography (OCT) to show blood flow stasis (Mariampillai et al. 2008). An additional emerging technique is photoacoustic imaging, in which an image is generated

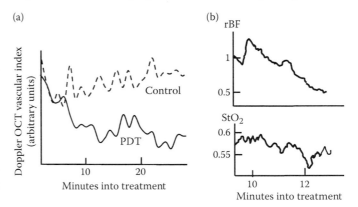

FIGURE 67.6 Examples of monitoring of the vascular responses to Photofrin-mediated PDT. (a) Barrett's esophagus: The dashed line shows the baseline variations in untreated tissue (after Standish et al. 2007). (b) Diffuse correlation spectroscopy measurements of local relative blood flow (top) and blood oxygen saturation (bottom) changes in prostate at a specific location relative to the treatment fiber during m-tetrahydrophenylchlorin (mTHPC)-mediated (mTHPC)-mediated PDT: Note the marked change in relative blood flow (rBF) with relatively small change in StO_2 in this case. (Courtesy of Dr. G. Yu, Dr. A. Yodh, and colleagues at the University of Pennsylvania, Pennsylvania, Philadelphia, USA.)

through a series of biophysical effects following a short (~ns) laser pulse; local absorption leads to localized heating that causes thermoelastic tissue expansion that generates an acoustic wave detectable by an external transducer to form an image of the light absorption pattern (Wang and Sampliner 2008). The absorber can be either endogenous chromophores (particularly Hb and HbO₂) or exogenous dyes (including the PDT photosensitizer itself) or nanoparticles. This technique combines the molecular specificity of light with the deep tissue penetration of ultrasound and so should be of considerable value in PDT response monitoring (Hirao et al. 2009; Xiang et al. 2007).

67.4.5 Monitoring Coagulation in Thermal Therapies

The last decade has seen rapid clinical development of several thermally ablative therapies for localized cancers: high-power (~5–10 W) NIR lasers (Carpentier et al. 2011; Lindner et al. 2009), high-intensity focused ultrasound (HIFU) (Berge et al. 2011; Keller et al. 2010), and radiofrequency (RF) and microwave radiation (Wolf et al. 2012; Sherar et al. 2004). For example, RF is approved for hepatic and renal cancer, while HIFU has shown promise in prostate cancer and uterine fibroids. Laser thermal therapy has also been used clinically in the liver (Sequeiros et al. 2010) and prostate (Lindner et al. 2009) and for brain metastasis (Carpentier et al. 2011). In most cases, these treatments aim to destroy relatively small localized tumors inside larger organs.

Treatment monitoring must be online to achieve the objectives of complete tumor ablation and protection of critical adjacent tissues. The relative importance of each objective depends on the clinical situation; for example, treatment margins in RF ablation of hepatic cancer can be generous, since complete ablation is important and some damage localized damage to normal liver is acceptable, while in prostate cancer, small volumes of rectal damage can have a large negative impact.

Optical monitoring of tissue coagulation is primarily based on changes in tissue (elastic) scattering resulting from protein denaturation, which is usually much larger than changes in absorption (Skinner et al. 2000; Yaroslavsky et al. 2002). Other techniques have been tested on ex vivo tissues, including fluorescence (Buttemere et al. 2004) and photoacoustics (Larin, Larina, and Esenaliev 2005; Moriyama et al. 2004). In both cases, the scattering changes also play a role in the detected signal. For direct in vivo monitoring of coagulation-induced scattering changes, fiber-optic probes can be inserted at the outmost limit of the tumor mass to detect changes in the diffuse optical signal (Balbierz and Johnson 2006). This only interrogates a small volume of tissue around the fiber tip and gives little indication of treatment progression until the coagulation front actually crosses the tip. In the case of laser thermal therapy, the treatment light itself (typically NIR) can also serve to probe the tissue, measuring either the fluence rate or radiance with a fiber-optic probe. The change in signal then relates to change in the tissue between the treatment fiber tip and the probe fiber, as illustrated in Figure 67.7a. Measurement of the radiance (i.e., the

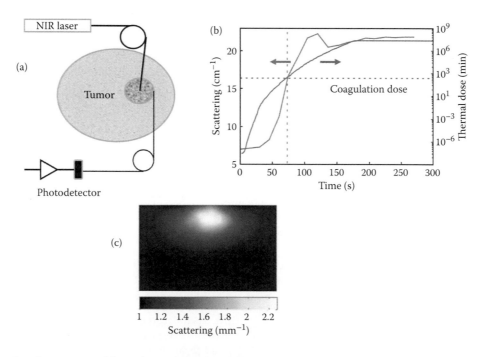

FIGURE 67.7 Examples of monitoring of thermal coagulation for solid tumor destruction. (a) Schematic of interstitial fiber-optic probe monitoring of the local light fluence rate or radiance. (b) Scattering coefficient (black) measured interstitially in prostate using the probe shown in Figure 67.3b during laser thermal coagulation of focal cancer, along with calculated thermal dose (gray). (c) Simulated DOT image of a photocoagulation zone in tissue based on the reconstructed transport scattering coefficient. (Courtesy of Jie Hie, University of Toronto, Toronto, Canada.)

directionally dependent light flux), which requires a specially designed fiber tip that picks up light only within a limited cone, has been found to be more sensitive than the local fluence rate, especially as the coagulation front traverses the probe tip (Chin et al. 2004).

While such measurements are simple to perform, they are not very quantitative and lack volumetric and spatial information on the extent of coagulation. More quantitative local measurements can be achieved with the same type of probe as used to measure the local absorption and scattering coefficients for correcting fluorescence monitoring (Figure 67.3b). Figure 67.7b illustrates one such measurement, where the derived (transport) scattering coefficient (μ'_s) is tracked over time during laser thermal therapy of focal prostate cancer treatment, and shows a several-fold increase upon coagulation. These changes can, in principle, be imaged using DOT. To date, this has been investigated by simulating the growth of a coagulation volume, calculating the changes in the diffuse optical signals that would result from this, and reconstructing the corresponding 3-D coagulation volume (see Figure 67.7c). There is potential to apply transrectal DOT, which has been demonstrated for prostate tumor imaging (Jiang et al. 2011), to monitor the exact position of the photocoagulation boundary in photothermal therapy of focal prostate cancer in near real time, so that treatment could be stopped at a time that gives maximum tumor destruction with minimal risk of rectal wall damage, which has limited this treatment option to date (Lindner et al. 2009; Raz et al. 2010).

67.5 Summary and Future Directions

It is clear that we have barely scratched the surface of potential applications of optical techniques—both stand-alone and in combination with nonoptical methods such as the various forms of radiological imaging—to guide cancer interventions. The need for this will grow as the efforts increase to achieve a higher degree of personalized cancer treatment, in which the goal is to achieve high precision of treatment delivery based on the specific characteristics of individual patients and their cancers. While, to date, most of the approaches to optical guidance of interventions have used optical techniques and instruments that were developed initially for cancer detection and diagnosis, the above examples illustrate the need for and challenges in developing intervention-specific methods. Guiding interventions may require different biological/biophysical information from that used in tumor detection and in many cases require a significantly different technological implementation, especially to integrate the optical method into the normal interventional workflow.

One of the major generic challenges in applying optical techniques to interventional monitoring is the need for standardization in the metrics, technology, and data analysis/presentation so that multicenter clinical trials can be made reliable and reproducible. This problem may be even more pressing than in diagnostic applications, since the impact on patient outcome of technical error or uncertainty in interventional guidance is so immediate and direct. Thus, for example, if the absolute concentration of a fluorophore like PpIX is used to guide the surgeon's decision as to whether to continue resection (based on tumor probability vs. C_{PpIX} relationship established through clinical trials), then the measurement had better be accurate and consistent between surgical users. This will require setting up rigorous standard operating procedures with the appropriate technical underpinning (optical phantoms, calibration techniques, etc.). The creation by the National Institutes of Health (NIH) of the Network for Translational Research in Optical Imaging is a promising initiative that should help meet this challenge, as is recent focus on establishing proper calibration protocols for optical imaging (Tromberg et al. 2008).

Acknowledgments

Examples shown in the figure were financially supported as follows: Figures 67.2 and 67.3, National Institutes of Health grant no. R01NS052274-01A; Figure 67.4, Ontario Institute for Cancer Research and the Kevin and Sandra Sullivan Chair in Surgical Oncology of the Princess Margaret Hospital Foundation; Figure 67.7, CHRP grant no. 777607408.

References

Agostinis, P., K. Berg, K. A. Cengel et al. 2011. Photodynamic therapy of cancer: An update. *CA Cancer J Clin* 61: 250–281.

Ahn, P. H., C.-C. Chen, A. I. Ahn et al. 2011. Adaptive planning in intensity-modulated radiation therapy for head and neck cancers: Single-institution experience and clinical implications. *Int J Radiat Oncol Biol Phys* 80:677–685.

Balbierz, D. J., and T. Johnson. 2006. Tissue Biopsy and Treatment Apparatus and Method. *US Patent#* 7,025,765.

Berge, V., E. Baco, A. A. Dahl, and S. J. Karlsen. 2011. Health-related quality of life after salvage high-intensity focused ultrasound (HIFU) treatment for locally radiorecurrent prostate cancer. *Int J Urol* 18:646–651.

Buscarini, L., E. Buscarini, M. Di Stasi et al. 2001. Percutaneous radiofrequency ablation of small hepatocellular carcinoma: Long-term results. *Eur Radiol* 11:914–921.

Busse, H., A. Schmitgen, C. Trantakis et al. 2006. Advanced approach for intraoperative MRI guidance and potential benefit for neurosurgical applications. *J Magn Reson Imaging* 24:140–151.

Buttemere, C. R., R. S. Chari, C. D. Anderson et al. 2004. *In vivo* assessment of thermal damage in the liver using optical spectroscopy. *J Biomed Opt* 9:1018–1027.

Carpentier, A., R. J. McNichols, R. J. Stafford et al. 2011. Laser thermal therapy: Real-time MRI-guided and computer-controlled procedures for metastatic brain tumors. *Lasers Surg Med* 43:943–950.

Caversaccio, M., J. G. Giraldez, R. Thoranaghatte et al. 2008. Augmented reality endoscopic system (ARES): Preliminary results. *Rhinology* 46:156–158.

Cerussi, A. E., V. W. Tanamai, D. Hsiang et al. 2011. Diffuse optical spectroscopic imaging correlates with final pathological response in breast cancer neoadjuvant chemotherapy. *Philos Trans A Math Phys Eng Sci* 369:4512–4530.

Chin, L. C. L., B. C. Wilson, W. M. Whelan, and I. A. Vitkin. 2004. Radiance-based monitoring of the extent of tissue coagulation during laser interstitial thermal therapy. *Opt Lett* 29:959–961.

Choe, R., A. Corlu, K. Lee et al. 2005. Diffuse optical tomography of breast cancer during neoadjuvant chemotherapy: A case study with comparison to MRI. *Med Phys* 32:1128–1139.

Clyne, M. 2011. Prostate cancer: Focal cryotherapy—Results of a COLD search. *Nat Rev Urol* 8:648.

Cottrell, W. J., A. R. Oseroff, and T. H. Foster. 2006. Portable instrument that integrates irradiation with fluorescence and reflectance spectroscopies during clinical photodynamic therapy of cutaneous disease. *Rev Sci Instrum* 77:064302.

Culver, J. P., R. Choe, M. J. Holboke et al. 2003. Three-dimensional diffuse optical tomography in the parallel plane transmission geometry: Evaluation of a hybrid frequency domain/continuous wave clinical system for breast imaging. *Med Phys* 30:235–247.

Davidson, S. R. H., R. A. Weersink, M. A. Haider et al. 2009. Treatment planning and dose analysis for interstitial photodynamic therapy of prostate cancer. *Phys Med Biol* 54:2293–2313.

Desmond-Hellmann, S. 2012. Toward precision medicine: A new social contract? *Sci Transl Med* 4:(129)ed3.

Dolmans, D., D. Fukumura, and R. K. Jain. 2003. Photodynamic therapy for cancer. *Nat Rev Cancer* 3:380–387.

Dulai, G. S., D. M. Jensen, G. Cortina, L. Fontana, and A. Ippoliti. 2005. Randomized trial of argon plasma coagulation vs. multipolar electrocoagulation for ablation of Barrett's esophagus. *Gastrointest Endosc* 61:232–240.

Foley, P., K. Merlin, S. Cumming et al. 2011. A comparison of cryotherapy and imiquimod for treatment of actinic keratoses: Lesion clearance, safety and skin quality outcomes. *J Drugs Dermatol* 10:1432–1438.

Fried, M. P., S. R. Parikh, and B. Sadoughi. 2008. Image-guidance for endoscopic sinus surgery. *Laryngoscope* 118:1287–1292.

Gregoire, V., and T. R. Mackie. 2011. State of the art on dose prescription, reporting and recording in intensity-modulated radiation therapy (ICRU Report No. 83). *Cancer Radiother* 15:555–559.

Haka, A. S., K. E. Shafer-Peltier, M. Fitzmaurice et al. 2005. Diagnosing breast cancer by using Raman spectroscopy. *Proc Natl Acad Sci USA* 102:12371–12376.

Hamblin, M. R., and P. Mroz, editors. 2008. *Advances in Photodynamic Therapy: Basic, Translational, and Clinical.* Artech House, Norwood, MA.

Hamdoon, Z., W. Jerjes, G. McKenzie, A. Jay, and C. Hopper. 2010. Assessment of tumour resection margins using optical coherence tomography. *Head Neck Oncol* 2:O7.

Higgins, W. E., J. P. Helferty, K. K. Lu et al. 2008. 3D CT-video fusion for image-guided bronchoscopy. *Comput Med Imaging Graph* 32:159–173.

Hirao, A., S. Sato, D. Saitoh et al. 2009. *In vivo* photoacoustic monitoring of photosensitizer in skin: Application to dosimetry for antibacterial photodynamic treatment. Proc. SPIE 7177, Photons Plus Ultrasound: Imaging and Sensing, February 12, 2009, doi:10.1117/12.808480.

Intes, X. 2005. Time-domain optical mammography SoftScan: Initial results. *Acad Radiol* 12:934–947.

Jakubowski, D. B., A. E. Cerussi, F. Bevilacqua et al. 2004. Monitoring neoadjuvant chemotherapy in breast cancer using quantitative diffuse optical spectroscopy: A case study. *J Biomed Opt* 9:230–238.

Jiang, Z., D. Piao, G. R. Holyoak et al. 2011. Trans-rectal ultrasound-coupled spectral optical tomography of total hemoglobin concentration enhances assessment of the laterality and progression of a transmissible venereal tumor in canine prostate. *Urology* 77:237–242.

Keller, M. D., S. K. Majumder, M. C. Kelley et al. 2010. Autofluorescence and diffuse reflectance spectroscopy and spectral imaging for breast surgical margin analysis. *Lasers Surg Med* 42:15–23.

Keller, M. D., E. Vargis, G. N. de Matos et al. 2011. Development of a spatially offset Raman spectroscopy probe for breast tumor surgical margin evaluation. *J Biomed Opt* 16:077006.

Keshtgar, M. R. S., D. W. Chicken, M. R. Austwick et al. 2010. Optical scanning for rapid intraoperative diagnosis of sentinel node metastases in breast cancer. *Br J Surg* 97:1232–1239.

Kim, A., M. Khurana, Y. Moriyama, and B. C. Wilson. 2010. Quantification of *in vivo* fluorescence decoupled from the effects of tissue optical properties using fiber-optic spectroscopy measurements. *J Biomed Opt* 15:067006.

Konecky, S. D., C. M. Owen, T. Rice et al. 2012. Spatial frequency domain tomography of protoporphyrin IX fluorescence in preclinical glioma models. *J Biomed Opt* 17:056008.

Kubben, P. L., K. J. ter Meulen, O. E. M. G. Schijns et al. 2011. Intraoperative MRI-guided resection of glioblastoma multiforme: A systematic review. *Lancet Oncol* 12:1062–1070.

Lapeer, R., M. S. Chen, G. Gonzalez, A. Linney, and G. Alusi. 2008. Image-enhanced surgical navigation for endoscopic sinus surgery: Evaluating calibration, registration and tracking. *Int J Med Robot Comput Assist Surg* 4:32–45.

Larin, K. V., I. V. Larina, and R. O. Esenaliev. 2005. Monitoring of tissue coagulation during thermotherapy using optoacoustic technique. *J Phys D Appl Phys* 38:2645–2653.

Lindner, U., R. A. Weersink, M. A. Haider et al. 2009. Image guided photothermal focal therapy for localized prostate cancer: Phase I trial. *J Urol* 182:1371–1377.

Ling, C. C., J. Humm, S. Larson et al. 2000. Towards multidimensional radiotherapy (MD-CRT): Biological imaging and biological conformality. *Int J Radiat Oncol Biol Phys* 47:551–560.

Lloyd, W. R., R. H. Wilson, C.-W. Chang, G. D. Gillispie, and M.-A. Mycek. 2010. Instrumentation to rapidly acquire fluorescence wavelength-time matrices of biological tissues. *Biomed Opt Express* 1:574–586.

Mariampillai, A., B. A. Standish, E. H. Moriyama et al. 2008. Speckle variance detection of microvasculature using swept-source optical coherence tomography. *Opt Lett* 33:1530–1532.

Mellman, I., G. Coukos, and G. Dranoff. 2011. Cancer immunotherapy comes of age. *Nature* 480:480–489.

Moriyama, E. H., S. K. Bisland, L. Lilge, and B. C. Wilson. 2004. Bioluminescence imaging of the response of rat gliosarcoma to ALA-PpIX-mediated photodynamic therapy. *Photochem Photobiol* 80:242–249.

Nguyen, F. T., A. M. Zysk, E. J. Chaney et al. 2009. Intraoperative evaluation of breast tumor margins with optical coherence tomography. *Cancer Res* 69:8790–8796.

Pakalniskis, M. G., W. A. Wells, M. C. Schwab et al. 2011. Tumor angiogenesis change estimated by using diffuse optical spectroscopic tomography: Demonstrated correlation in women undergoing neoadjuvant chemotherapy for invasive breast cancer? *Radiology* 259:365–374.

Perelman, L. T., V. Backman, M. Wallace et al. 1998. Observation of periodic fine structure in reflectance from biological tissue: A new technique for measuring nuclear size distribution. *Phys Rev Lett* 80:627–630.

Pinthus, J. H., A. Bogaards, R. Weersink, B. C. Wilson, and J. Trachtenberg. 2006. Photodynamic therapy for urological malignancies: Past to current approaches. *J Urol* 175:1201–1207.

Poetter, R., P. Georg, J. C. A. Dimopoulos et al. 2011. Clinical outcome of protocol based image (MRI) guided adaptive brachytherapy combined with 3D conformal radiotherapy with or without chemotherapy in patients with locally advanced cervical cancer. *Radiother Oncol* 100:116–123.

Pogue, B. W., S. P. Poplack, T. O. McBride et al. 2001. Quantitative hemoglobin tomography with diffuse near-infrared spectroscopy: Pilot results in the breast. *Radiology* 218:261–266.

Poh, C. F., J. S. Durham, P. M. Brasher et al. 2011. Canadian Optically-Guided Approach for Oral Lesions Surgical (COOLS) trial: Study protocol for a randomized controlled trial. *BMC Cancer* 11:462.

Raz, O., M. A. Haider, S. R. H. Davidson et al. 2010. Real-time magnetic resonance imaging-guided focal laser therapy in patients with low-risk prostate cancer. *Eur Urol* 58:173–177.

Roberts, D. W., P. A. Valdes, B. T. Harris et al. 2012. Glioblastoma multiforme treatments with clinical trials for surgical resection (aminolevulinic acid). *Neurosurg Clin N Am* 23:371–377.

Roels, S., P. Slagmolen, J. Nuyts et al. 2009. Biological image-guided radiotherapy in rectal cancer: Challenges and pitfalls. *Int J Radiat Oncol Biol Phys* 75:782–790.

Salas, N., S. M. Castle, and R. J. Leveillee. 2011. Radiofrequency ablation for treatment of renal tumors: Technological principles and outcomes. *Expert Rev Med Devices* 8:695–707.

Sequeiros, R. B., J. Kariniemi, R. Ojala et al. 2010. Liver tumor laser ablation—Increase in the subacute ablation lesion volume detected with post procedural MRI. *Acta Radiol* 51:505–511.

Sharma, P., K. Wagner, J. D. Wolchok, and J. P. Allison. 2011. Novel cancer immunotherapy agents with survival benefit: Recent successes and next steps. *Nat Rev Cancer* 11:805–812.

Sherar, M. D., J. Trachtenberg, S. R. H. Davidson, and M. R. Gertner. 2004. Interstitial microwave thermal therapy and its application to the treatment of recurrent prostate cancer. *Int J Hyperthermia* 20:757–768.

Shim, M. G., B. C. Wilson, E. Marple, and M. Wach. 1999. Study of fiber-optic probes for *in vivo* medical Raman spectroscopy. *Appl Spectrosc* 53:619–627.

Skinner, M. G., S. Everts, A. D. Reid et al. 2000. Changes in optical properties of ex vivo rat prostate due to heating. *Phys Med Biol* 45:1375–1386.

Soliman, H., A. Gunasekara, M. Rycroft et al. 2009. Functional imaging of neoadjuvant chemotherapy response in women with locally advanced breast cancer using diffuse optical spectroscopy. *Clin Cancer Res* 16:2504–2516.

Standish, B. A., V. X. D. Yang, N. R. Munce et al. 2007. Doppler optical coherence tomography monitoring of microvascular tissue response during photodynamic therapy in an animal model of Barrett's esophagus. *Gastrointest Endosc* 66:326–333.

Stummer, W., U. Pichlmeier, T. Meinel et al. 2006. Fluorescence-guided surgery with 5-aminolevulinic acid for resection of malignant glioma: A randomised controlled multicentre phase III trial. *Lancet Oncol* 7:392–401.

Suh, H., O. A'Amar, E. Rodriguez-Diaz et al. 2011. Elastic light-scattering spectroscopy for discrimination of benign from malignant disease in thyroid nodules. *Ann Surg Oncol* 18:1300–1305.

Swartling, J., J. Axelsson, G. Ahlgren et al. 2010. System for interstitial photodynamic therapy with online dosimetry: First clinical experiences of prostate cancer. *J Biomed Opt* 15(5):058003.

Terasawa, T., I. J. Dahabreh, and T. Nihashi. 2010. Fluorine-18-fluorodeoxyglucose positron emission tomography in response assessment before high-dose chemotherapy for lymphoma: A systematic review and meta-analysis. *Oncologist* 15:750–759.

Tobis, S., J. Knopf, C. Silvers et al. 2011. Near infrared fluorescence imaging with robotic assisted laparoscopic partial nephrectomy: Initial clinical experience for renal cortical tumors. *J Urol* 186:47–52.

Tromberg, B. J., A. Cerussi, N. Shah et al. 2005. Imaging in breast cancer: Diffuse optics in breast cancer: Detecting tumors in pre-menopausal women and monitoring neoadjuvant chemotherapy. *Breast Cancer Res* 7:279–285.

Tromberg, B. J., B. W. Pogue, K. D. Paulsen et al. 2008. Assessing the future of diffuse optical imaging technologies for breast cancer management. *Med Phys* 35:2443–2451.

Valdes, P. A., A. Kim, F. Leblond et al. 2011a. Combined fluorescence and reflectance spectroscopy for *in vivo* quantification of cancer biomarkers in low- and high-grade glioma surgery. *J Biomed Opt* 16:116007.

Valdes, P. A., F. Leblond, A. Kim et al. 2011b. Quantitative fluorescence in intracranial tumor: Implications for ALA-induced PpIX as an intraoperative biomarker. *J Neurosurg* 115:11–17.

Valdes, P. A., F. Leblond, V. L. Jacobs, K. D. Paulsen, and D. W. Roberts. 2012. *In vivo* fluorescence detection in surgery: A review of principles, methods and clinical applications. *Curr Med Imag Rev* 8:211–232.

Wang, I., S. Andersson Engels, G. E. Nilsson, K. Wardell, and K. Svanberg. 1997. Superficial blood flow following photodynamic therapy of malignant non-melanoma skin tumours measured by laser Doppler perfusion imaging. *Br J Dermatol* 136:184–189.

Wang, K. K., and R. E. Sampliner. 2008. Updated guidelines 2008 for the diagnosis, surveillance and therapy of Barrett's esophagus. *Am J Gastroenterol* 103:788–797.

Wang, Y., C. Zhang, J. Liu, and G. Huang. 2012. Is 18F-FDG PET accurate to predict neoadjuvant therapy response in breast cancer? A meta-analysis. *Breast Cancer Res Treat* 131:357–369.

Weersink, R. A., A. Bogaards, M. Gertner et al. 2005. Techniques for delivery and monitoring of TOOKAD (WST09)-mediated photodynamic therapy of the prostate: Clinical experience and practicalities. *J Photochem Photobiol B* 79:211–222.

Weersink, R. A., J. Qiu, A. J. Hope et al. 2011. Improving superficial target delineation in radiation therapy with endoscopic tracking and registration. *Med Phys* 38:6458–6468.

Wilson, B. C., and M. S. Patterson. 2008. The physics, biophysics and technology of photodynamic therapy. *Phys Med Biol* 53:R61–R109.

Winer, J. H., H. S. Choi, S. L. Gibbs-Strauss et al. 2010. Intraoperative localization of insulinoma and normal pancreas using invisible near-infrared fluorescent light. *Ann Surg Oncol* 17:1094–1100.

Wolf, F. J., B. Aswad, T. Ng, and D. E. Dupuy. 2012. Intraoperative microwave ablation of pulmonary malignancies with tumor permittivity feedback control: Ablation and resection study in 10 consecutive patients. *Radiology* 262:353–360.

Xiang, L., D. Xing, H. Gu et al. 2007. Pulse laser integrated photodynamic therapy and photoacoustic imaging. *Proc. SPIE* 6437:6437B. Photons Plus Ultrasound: Imaging and Sensing 2007: The Eighth Conference on Biomedical Thermoacoustics, Optoacoustics, and Acousto-optics, doi:10.1117/12.698771.

Yaroslavsky, A. N., P. C. Schulze, I. V. Yaroslavsky et al. 2002. Optical properties of selected native and coagulated human brain tissues *in vitro* in the visible and near infrared spectral range. *Phys Med Biol* 47:2059–2073.

Yu, G. 2012. Near-infrared diffuse correlation spectroscopy in cancer diagnosis and therapy monitoring. *J Biomed Opt* 17:010901.

Yu, G., T. Durduran, C. Zhou et al. 2006. Real-time in situ monitoring of human prostate photodynamic therapy with diffuse light. *Photochem Photobiol* 82:1279–1284.

68

Phototherapy for Newborn Jaundice

Brendan K. Huang
Yale University

Michael A. Choma
Yale University

It is common for newborns that are a few days old to have a yellowish tinge on their skin. This physical exam finding is known as jaundice. The yellowing is due to accumulation of a chemical known as bilirubin in the skin. For the vast majority of babies, a form of bilirubin called *unconjugated bilirubin* causes the jaundice. Likewise, for the vast majority of babies, the jaundice will resolve without harming the newborn. However, in a small subset of infants, levels of unconjugated bilirubin can rise to dangerous levels and lead to a devastating neurological condition known as kernicterus. For these newborns, phototherapy with light in the blue wavelength range can lower unconjugated bilirubin levels and prevent kernicterus. The story of how millions of newborns came to be prescribed blue-light phototherapy for unconjugated hyperbilirubinemia started with a simple observation about premature babies who were put in the sunlight. It is a story of how researchers took this observation from bedside to bench with the elucidation of the photochemistry of bilirubin, and back to the bedside with new phototherapeutic interventions based on their findings.

In this chapter, we will describe the underlying physiological mechanisms that lead to newborn jaundice and its potential neurological sequelae. We will then discuss the original clinical insights that led to using blue light as a therapy for neonatal jaundice, as well as the photochemistry of this therapy that was subsequently elucidated. We will discuss the clinical evidence for using blue-light therapy and finally discuss recent developments in phototherapy for neonatal jaundice.

68.1 Physiology and Biochemistry of Newborn Jaundice

68.1.1 Heme Metabolism and Regulation

In order to understand the pathophysiology of newborn jaundice caused by unconjugated hyperbilirubinemia, one must first understand the physiology of *heme*. Heme is an iron-containing prosthetic group that allows hemoglobin to carry oxygen (Suchy 2011). Because of this critical duty in the body, hemoglobin is present in large quantities in red blood cells (erythrocytes). At the same time, heme is toxic to tissues when released from erythrocytes, and its degradation is highly regulated.

Structurally, heme consists of a *porphyrin* ring, a highly conjugated aromatic ring coordinating an iron molecule (Suchy 2011). Typical of π-delocalized molecules, porphyrin rings have large absorptions in the ultraviolet (UV) and visible range and exhibit bright colors, the specific wavelength of which depends on their exact structure. As an example, heme gives blood its red color, bilirubin gives bile a greenish-yellowish tinge, and the breakdown products of bilirubin give urine its yellow color and stool its brown color (McDonagh and Lightner 1985). The fact that these molecules absorb photons in the visible spectrum means that their excited electronic structures have the opportunity to undergo photochemical reactions (Patrice et al. 2003). Specifically, unconjugated bilirubin has a chemical transition induced by this absorption that underlies the therapeutic mechanism of blue-light therapy for newborn jaundice.

When erythrocytes are destroyed, the resulting hemoglobin must be broken down. Regulated destruction of senescent cells occurs by phagocytic cells in the reticuloendothelial system, including the spleen (Knutson and Wessling-Resnick 2003; Maisels 2005). When erythrocytes are destroyed in the bloodstream, a protein known as haptoglobin binds hemoglobin and transports it to the liver for processing to bilirubin (Knutson and Wessling-Resnick 2003). This haptoglobin-mediated process is particularly important during *intravascular hemolysis*, where erythrocytes are destroyed at pathologic levels in the blood (e.g., immune-mediated erythrocyte destruction). As suggested by the name, hemoglobin consists of a globin protein structure combined with a heme moiety, and the globin and heme portion are separately metabolized. In the context of hyperbilirubinemia, the pathway of interest is the metabolism of this porphyrin ring, specifically the metabolism of heme to bilirubin and the subsequent elimination of bilirubin.

Heme is degraded to unconjugated bilirubin by a pathway with two main enzymes: heme oxygenase and biliverdin reductase. Heme contains reduced iron, and unbound heme can lead to the production of reactive oxygen species that are toxic to tissues and can itself can cause further hemolysis (Kumar and Bandyopadhyay 2005). Consequently, heme oxygenase is found in the reticuloendothelial system, where senescent erythrocytes are routinely destroyed. Heme oxygenase is responsible for opening the porphyrin ring and catabolizing heme to biliverdin, a step that concomitantly releases carbon monoxide and iron (Knutson and Wessling-Resnick 2003; Kumar and Bandyopadhyay 2005; Maisels 2005). Biliverdin reductase, using nicotinamide adenine dinucleotide phosphate (NADPH) as a substrate, reduces the terminal carbon on biliverdin to produce unconjugated bilirubin (Knutson and Wessling-Resnick 2003; Kumar and Bandyopadhyay 2005; Maisels 2005). In all, these biochemical pathways in the reticuloendothelial system minimize the time that free heme is active and quickly converts it to unconjugated bilirubin.

68.1.2 Bilirubin Metabolism and Regulation

Unconjugated bilirubin, also known as *indirect bilirubin*, is a main metabolite of heme. While not as immediately reactive as heme, it is neurotoxic at high levels and needs to be further metabolized and subsequently excreted by the gastrointestinal system. The enzymes responsible for bilirubin metabolism and excretion are localized primarily in the hepatocytes of the liver (Maisels 2005).

Unconjugated bilirubin has several important chemical properties consistent with its hydrophobic nature: It is insoluble in water, it partitions into lipid bilayers, and it has significant hydrophobic interactions with proteins (McDonagh and Lightner 1985). Indeed, a significant fraction of circulating unconjugated bilirubin is albumin bound. Because of these chemical properties, unconjugated bilirubin needs to be made more water soluble before removal is possible (Lightner and McDonagh 1984). The liver primarily accomplishes this step by conjugating bilirubin to glucuronic acid; the result is *conjugated* (or *direct*) *bilirubin*, which contains several polar groups. Shown in Figure 68.1 is the structure of conjugated bilirubin, in which glucuronic acid can be added to unconjugated bilirubin at several sites (McDonagh and Lightner 1985). After conjugation, bilirubin is then excreted in the bile and finally the stool. Based on this process, bilirubin has historically been divided into two main pools: conjugated and unconjugated, where the sum of the two is *total bilirubin*.

Circulating unconjugated bilirubin is processed and excreted from the liver (Suchy 2011). First, it is transported across hepatocyte membranes. Albumin-bound unconjugated bilirubin can be transported across the membrane by three different transporters expressed on the sinusoidal surface of hepatocytes. Next, bilirubin is conjugated to glucuronic acid by uridine diphosphate (UDP)-glucuronosyltransferases in hepatocytes. Finally, conjugated bilirubin is transported into the bile by the transporter MRP2.

After bile is transported from the liver into the gallbladder, it is then secreted into the small intestine for elimination in stool. Stooling is the major means of eliminating unconjugated

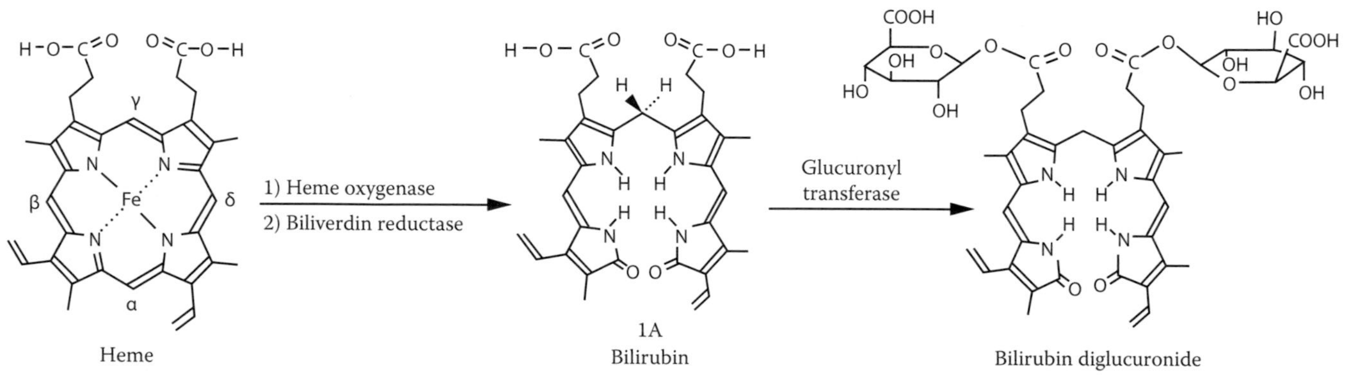

FIGURE 68.1 Structures of heme, bilirubin (unconjugated bilirubin), and bilirubin diglucuronide (conjugated bilirubin). Key enzymes are depicted as well. All three share the same base porphyrin ring. (Reproduced from McDonagh, A.F. and D.A. Lightner, *Pediatrics*, 75, 443–455, 1985. With permission.)

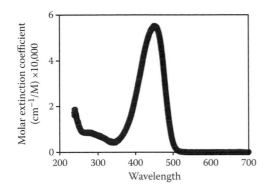

FIGURE 68.2 Bilirubin absorption spectrum shown as wavelength-dependent molar extinction coefficient for bilirubin dissolved in carbon tetrachloride. (Data are from Prahl, S. Bilirubin. Oregon Medical Laser Center Optical Properties Spectra, 2012. http://omlc.ogi.edu/spectra/PhotochemCAD/html/119.html, accessed May 1, 2012.)

bilirubin both at baseline and during blue-light phototherapy (Suchy 2011). Additionally, further modification of conjugated bilirubin can occur by intestinal flora. These microbes chemically modify conjugated bilirubin, including modification to urobilinogen. This modified pool can cross the intestinal wall and reenter systemic circulation in what is known as enterohepatic circulation. In the process of enterohepatic circulation, the reabsorbed bilirubin metabolites, which are still water soluble, can be reexcreted by the kidney, a very minor means of unconjugated bilirubin elimination (Suchy 2011). Note that urobilin, generated by oxidizing urobilinogen, is excreted by the kidney and gives urine its characteristic yellow color (Figure 68.2).

68.1.3 Unconjugated versus Conjugated Hyperbilirubinemia as Causes of Neonatal Jaundice

Neonatal jaundice is the yellowing of skin as a result of increased levels of total bilirubin in infants. From the different stages of bilirubin metabolism and excretion described above, critical steps can be identified where dysfunction can lead to increased bilirubin levels in the body.

There are a number of less common conditions that can either increase the production of unconjugated bilirubin or decrease the metabolism and excretion of unconjugated bilirubin, all of which can lead to jaundice (Maisels 2005). Immune-mediated hemolysis, glucose-6-phosphate dehydrogenase deficiency [G6PD deficiency, a metabolic disease that makes erythrocytes more susceptible oxidative damage (Glader 2012)], and sepsis can all increase erythrocyte destruction and unconjugated bilirubin levels. In Crigler–Najjar syndrome, deficiency in UDP-glucuronosyltransferase causes a backup of unconjugated bilirubin, leading to severe jaundice and neurological defects. Also note that cholestasis due to biliary atresia, a congenital malformation of the biliary tract that requires surgical management, can cause a conjugated hyperbilirubinemia. Biliary atresia is rare but can present with jaundice in the newborn period. Moreover,

liver failure due to any number of causes can also present with jaundice. Taken together, while blue-light phototherapy is the treatment of choice for an otherwise healthy newborn with elevated unconjugated bilirubin, the presence of jaundice may also indicate more severe underlying disease. As such, the use of blue-light phototherapy for newborn jaundice is not reflexive but rather prescribed by a pediatrician after careful clinical and laboratory evaluation of the patient.

Although many of the less common conditions are noteworthy owing to their clinical severity (e.g., sepsis, immune-mediated hemolysis), the most common causes of unconjugated hyperbilirubinemia in infants relate to the otherwise normal transition from fetal to neonatal life [e.g., breastfeeding failure jaundice and breast milk jaundice (Wong and Bhutani 2012a)]. Breastfeeding failure jaundice is characterized by reduced fluid intake by a newborn secondary to inadequate intake of breast milk. Inadequate fluid intake can lead to hypovolemia, which predisposes a newborn to elevated levels of unconjugated bilirubin. The diagnosis of breastfeeding failure jaundice implies three important factors (Wong and Bhutani 2012a). First, hypovolemia, not the breast milk itself, is believed to be the cause of the jaundice. Second, breastfed newborns are at an increased risk of jaundice and kernicterus compared to formula-fed newborns. Third, since a breast milk-fed newborn is dependent on milk for maintaining adequate hydration, inadequate intake and resultant jaundice typically occur in the first week of life. In contrast, breast milk jaundice occurs when unconjugated bilirubin levels are elevated beyond the first week of life, presumably secondary to a still-unidentified factor in breast milk (Wong and Bhutani 2012a). Other factors also predispose to unconjugated hyperbilirubinemia in the newborn (Maisels 2005). Infant UDP-glucuronosyltransferase activity is lower than that of adults. Additionally, compared to adults, newborns have higher hematocrit levels (fraction of blood occupied by erythrocytes) and shorter erythrocyte lifetimes. Due to all these factors, many infants develop an unconjugated hyperbilirubinemia that is typically transient, is typically not harmful, and generally resolves within a week (Maisels 2006). Nevertheless, at high levels, bilirubin can be neurotoxic, and a small subset of jaundiced infants would go on to develop bilirubin neurotoxicity if not treated with blue-light phototherapy.

68.1.4 Kernicterus as a Devastating Consequence of Unconjugated Hyperbilirubinemia

Unconjugated bilirubin is a neurotoxin and can have devastating effects on the infant brain at high circulating levels. Kernicterus classically has been used to refer to the accumulation of bilirubin in the brain but now also denotes the classic long-term neurological effects of hyperbilirubinemia. *Bilirubin-induced neurologic dysfunction* (BIND) more generally refers to both the long- and short-term neurological effects of bilirubin.

Acute bilirubin encephalopathy (ABE) refers to the short-term effects of BIND. ABE is classically divided into three stages

(Dennery, Seidman, and Stevenson 2001; Maisels 2005; Shapiro 2003). In the first stage, the infant becomes lethargic, becomes less responsive, and may exhibit hypotonia. In the second stage, the infant becomes irritable, even less responsive, and has a hypertonia with a typical arched back and hyperextended neck, known as retrocollis–opisthotonos. In the advanced phase, the infant may develop a shrill cry, refuse to feed, be apneic, have a fever, and may even go into a coma.

Long-term BIND, or kernicterus, has several classic features (Dennery, Seidman, and Stevenson 2001; Maisels 2005; Shapiro 2003). Choreoathetoid cerebral palsy can be present with tremors, sudden jerking movements, or involuntary motions. Decreased muscle tone also can be present. Hearing loss, thought to be secondary to auditory nerve damage, also occurs, with loss concentrated at high frequencies. Children can also develop gaze abnormalities, with difficulties especially in upward gaze. Finally, 75% of children develop dental dysplasia, most commonly with enamel hypoplasia and potentially some green discoloration of the teeth.

The clinical manifestations of kernicterus can appear with variable time frame (Dennery, Seidman, and Stevenson 2001; Maisels 2005; Shapiro 2003). Hypotonia, active deep tendon reflexes, and delayed motor skill generally are noticed within the first year. Choreoathetosis, upward gaze, and hearing loss, on the other hand, may take longer to manifest. These late manifestations can make epidemiological tracking of kernicterus more difficult.

68.2 Initial Phototherapy Trials and the Origins of Blue-Light Therapy

In order to prevent kernicterus, unconjugated bilirubin must be removed from the body. One invasive way to remove bilirubin is with an exchange transfusion, in which a volume of blood is removed and then replaced with donor blood that does not contain unconjugated bilirubin. In this procedure, a central catheter is inserted in order to remove the patient's blood and replace it with that of a donor. Exchange transfusion is costly, is labor intensive, and carries a number of risks including cardiac, gastrointestinal, and infectious complications. Its main use is in rapidly reducing very high unconjugated bilirubin levels (Keenan et al. 1985; Maisels 2005).

Exchange transfusion was the only available therapy for neonatal jaundice for many years, until an insight in the 1950s led to the discovery of phototherapy as an alternative treatment. A nurse at Essex Hospital routinely put premature babies in the sunlight, believing that the fresh air would be beneficial to them. It was noted that one baby had unjaundiced skin except for a triangle of yellow that had been shielded from the sun by a blanket. Later, when bilirubin levels were being drawn on other infants, it also was found that leaving blood samples in sunlight led to lower-than-expected laboratory readings of bilirubin (Dobbs and Cremer 1975).

Based on these observations, a team at Essex began initial trials in using phototherapy to treat unconjugated hyperbilirubinemia.

The team started by first putting infants in sunlight for short periods of time and then measuring bilirubin levels, from which they discovered that sunlight exposure for as few as 2 h could lower bilirubin levels (Dobbs and Cremer 1975). Soon afterward, they started searching for an artificial light source to recreate the effects of sunlight. It was noted that serum from jaundiced patients had a distinct absorbance peak at 420 nm and that after exposure to sunlight, this peak shifted to 550–650 nm, coincident with a decrease in serum bilirubin. These data supported the use of blue light for phototherapy. Additionally, the team also demonstrated that the most effective wavelengths were visible, not UV or infrared (IR). In 1958, the results of the first phototherapy trials were published. These trials demonstrated that exposing infants to blue fluorescent light could significantly reduce serum bilirubin levels (Cremer, Perryman, and Richards 1958).

This initial trail was an important landmark both in pediatric care and in the medical use of light, as it showed that phototherapy could achieve important clinical end points and, in some situations, replace a very invasive procedure (i.e., exchange transfusions). In particular, light phototherapy avoided the complications of central line placement as well as the risks of blood transfusion (e.g., infection and transfusion reactions).

Subsequent trials continued to demonstrate the efficacy of phototherapy using blue light to reduce serum bilirubin and reinforced its advantages over exchange transfusion (John 1975). At this point, however, there were still two unanswered questions. First, physicians knew that phototherapy could empirically lower bilirubin levels, but they did not understand the specific photochemistry. Second, physicians knew that phototherapy could reduce serum bilirubin levels, but rational, evidence-based criteria for patient selection did not exist.

68.3 Elucidation of the Blue-Light Photoproducts of Unconjugated Bilirubin

Beginning in the 1970s, scientists worked from bedside to bench to elucidate the photochemistry of unconjugated bilirubin degradation. A jaundiced rat known as the Gunn rat was well known and enabled much of the photochemistry research to be conducted (Gunn 1944). It was determined that bilirubin photochemistry in the blue wavelength range involves three distinct mechanisms: configurational isomerization, structural isomerization, and photooxidation (McDonagh and Lightner 1985), summarized in Figure 68.3. Understanding the photochemistry was central to later studies that revealed the role of these photoproducts in lowering unconjugated hyperbilirubinemia levels in jaundiced newborns.

In the first two of these mechanisms, photoisomerization, photon absorption provides the energy needed for reorganizing the geometric configuration or chemical bonding structure of a molecule while leaving its chemical formula unchanged. A

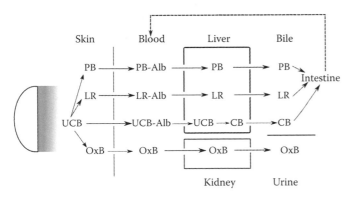

FIGURE 68.3 Schematic showing major metabolic pathways of unconjugated bilirubin (UCB) both physiologically and in phototherapy. Physiological pathways include binding to albumin (UCB-Alb), conversion to conjugated bilirubin (CB), and excretion through intestine. Metabolism via blue-light therapy includes three different pathways: conversion to configurational isomers [photobilirubin (PB)], structural isomers [lumirubin (LR)], and photooxidation products [oxidated bilirubin (OxB)]. Excretion of PB and LR are biliary, while excretion of OxB is renal. (Reprinted with permission from Lightner, D.A. and A.F. McDonagh, *Acc. Chem. Res.* 17, 417–424. Copyright 1984 American Chemical Society.)

classic example of photoisomerization involves the retinal molecule rhodopsin, in which light is absorbed by 11-*cis*-retinal, causing a *cis–trans* isomerization, leading to all-*trans*-retinal, which in turn activates G protein-mediated signaling (Pollard 2008). In the case of bilirubin metabolism, photoisomerization can involve a reaction that rotates around a π-double bond (configurational isomerization) or a reaction that changes the covalent bonding structure of the molecule (structural isomerization).

In configurational isomerization of bilirubin, a change in three-dimensional structure occurs by rotation around one of two double bonds, located at carbons 4 and 15. The normal configuration is *4Z,15Z*, but photoisomerization leads to *4E,15Z*, *4Z,15E*, and *4E,15E*, with the dominant species being *4Z,15E* (McDonagh, Palma, and Lightner 1980; McDonagh and Ramonas 1978). In the structural isomerization reaction, photoexcitation can drive a cyclization in a portion of the molecule, leading to a molecule known as lumirubin. Lumirubin exhibits two diastereomers, each with a rotated methyl group around C2.

One natural question to ask is how photoisomerization facilitates physiological excretion without changing the chemical formula of the molecule. As we described earlier, unconjugated bilirubin is a water-insoluble molecule, and the conjugation process allows the molecule to become water soluble and thus be excreted. That unconjugated bilirubin is water insoluble, however, is not immediately obvious from its chemical structure. Unconjugated bilirubin contains a number of polar groups, including carbonyl, carboxylic acids, and amine groups that can freely participate in hydrogen bonding. In its natural three-dimensional structure, however, the molecule

is folded up in such a way as to bury these hydrogen-bonding groups internally (McDonagh and Lightner 1985). In both cases of photoisomerization, the three-dimensional structure changes to allow hydrogen bonding to occur with the outside solvent, allowing excretion in the bile (McDonagh et al. 1980). In lumirubin, a large proprionic acid group at C12 sterically hinders the molecule from folding in the same manner, forcing hydrophilic residues to be exposed to the solvent (McDonagh and Lightner 1985).

The third process occurring in blue-light therapy is photooxidation (Lightner, Linnane, and Ahlfors 1984). In its electronically excited state, unconjugated bilirubin can react with oxygen and be broken down to colorless dipyrroles. The resulting products are fragments of the original bilirubin molecule with significantly lower molecular weight and greater water solubility. These products are excreted by the kidney and appear in the urine.

68.3.1 The Central Role of Lumirubin

After research into the blue wavelength photochemistry of unconjugated bilirubin, the next issue was to determine which photoproducts were the most physiologically relevant in terms of reducing levels of unconjugated bilirubin. When phototherapy was first used, scientists had assumed that photooxidation was the major pathway. However, photochemical studies in Gunn rats showed that immediately after light therapy was started on a rat, configurational isomers of bilirubin appeared in the serum and bile (McDonagh and Ramonas 1978; Ostrow 1971). Only after the configurational isomers appeared did lumirubin levels eventually rise, and they rose to lower serum levels compared to configurational isomers. These data argued against the physiological relevance of photooxidative products and argued for the general relevance of photoisomers and the specific relevance of configurational isomers. This perspective also was supported by the fact that configurational isomerization is more quantum efficient than structural isomerization is (Ennever 1988).

Subsequent research in newborns with unconjugated hyperbilirubinemia put the Gunn rat data in a more clinical context. Surprisingly, this research found that lumirubin, the more slowly accumulating photoisomer, is likely the physiologically relevant photoproduct (Ennever et al. 1985, 1987; Ennever 1988). It was found that although configurational isomers have serum concentrations about fivefold higher than that of lumirubin, lumirubin was the dominant photoisomer excreted in bile as measured from duodenal bilirubin photoproducts (Ennever et al. 1985, 1987). One presumed underlying mechanism is that although configurational isomers appear more rapidly, they also can rapidly convert back to unconjugated bilirubin, thereby limiting the net rate at which they are excreted into bile (Ennever et al. 1985, 1987; Ennever 1988). Additionally, configurational isomers can be absorbed by the intestines and enter enterohepatic circulation, potentially limiting the efficiency of configurational isomer elimination (Lightner and McDonagh 1984). Although the evidence is not completely conclusive, the available data are consistent with lumirubin having a higher net rate of excretion, suggesting that structural isomers are the most important photoproduct in the blue-light therapy of newborns with unconjugated hyperbilirubinemia.

In summary, there are several properties of unconjugated bilirubin that render blue-light therapy effective. First, it is photoabsorptive, a property that comes from having electronic transitions with $E = h\nu$ in the visible range. Specifically, delocalized π systems are responsible for these visible spectral absorptions. Second, the unconjugated bilirubin is photoreactive. Photon absorption can facilitate E/Z transisomerization or ring cyclization. Further, these photoproducts have properties that allow them to be more readily eliminated in the bile compared to unconjugated bilirubin. Finally, blue-light photochemistry must be selective for unconjugated bilirubin. For example, heme and bilirubin both exhibit photoabsorption at blue wavelengths, but only unconjugated bilirubin exhibits photoreactivity. Photoreactivity with other molecules is not a routine concern. However, clinical guidelines list a family history of porphyria as a contraindication to blue-light phototherapy. Porphyrias are a very rare class of conditions in which heme synthesis is compromised and photosensitive precursor molecules can potentially aggregate in the skin, leading to photodamage (Soylu, Kavukcu, and Turkmen 1999). These porphyrin precursors are both photoabsorptive and photoreactive, and their presence in the case of this rare condition causes increased toxicity from blue light (Fritsch et al. 1997).

68.4 Clinical Aspects of Blue-Light Therapy for Newborn Unconjugated Hyperbilirubinemia

68.4.1 Evidence and Guidelines

It was clear from the first trials that blue-light therapy reduced unconjugated bilirubin levels. However, given the fact that jaundice is very common in newborn infants, and given that very few of these infants will develop kernicterus, the question naturally arises as to which patients should be treated. In the 1990s, researchers addressed the question of risk stratification in newborns with unconjugated hyperbilirubinemia.

Since the introduction of phototherapy for neonatal jaundice, kernicterus has been very rare. Because of the low incidence of kernicterus, it is difficult to directly address the question of which infants were at highest risk for kernicterus (Johnson, Bhutani, and Brown 2002). Rather, in a landmark study, a proxy end point was used: which infants, based on bilirubin levels drawn prior to hospital discharge, would go on to have highly elevated unconjugated hyperbilirubin at a later point in time (Bhutani, Johnson, and Sivieri 1999). That is, it was assumed that a high degree of elevation after an initial in-hospital level was a risk factor for kernicterus. In this study, a large number of healthy newborns born at or near term ($\geq$36 weeks gestational age) had bilirubin levels drawn prior to and after hospital discharge. The results gave a risk stratification of infants based on their bilirubin levels and allowed researchers to predict with some certainty those infants who were more likely to become hyperbilirubinemic at a later time.

Several previous observations were confirmed through this study. The data showed that nearly all infants have some rise in their serum bilirubin over the first 5 days of life and explicitly showed what the magnitude of this rise was. The data also showed that levels of unconjugated bilirubin were normally distributed and quantitated how that normal distribution evolved over the first week of life. Most importantly, data were generated to guide selection of those infants who were more likely to benefit from phototherapy, based on a proxy end point of hyperbilirubinemia. In order to prevent the rare but devastating outcome of kernicterus, it is important to generate selection criteria with a low rate of false-negatives; at the same time, false-positives can be tolerated because the intervention is relatively safe. Indeed, using the 40th percentile as a cutoff, none of the infants in the low-risk group went on to develop subsequent hyperbilirubinemia, making the sensitivity essentially 100%.

Current American Academy of Pediatrics (AAP) guidelines for phototherapy combine these data with other important risk stratification criteria in order to decide which infants to treat (Subcommittee on Hyperbilirubinemia 2004). As discussed in Section 68.1.4, there are a number of conditions that may preclude an infant to overproducing heme or undermetabolizing unconjugated bilirubin, and these conditions were identified by the AAP as risk factors for developing hyperbilirubinemia. These factors include isoimmune hemolytic disease, G6PD deficiency, asphyxia, or sepsis. Other risk factors are early signs of physiologic stress or kernicterus including significant lethargy, temperature instability, acidosis, or albumin <3.0 g/dL. Additionally, the gestational age of the infant is a major risk factor, with babies above 38 weeks being at lower risk than those at 35–37 weeks.

The overall guidelines of when to initiate phototherapy are summarized graphically in Figure 68.4. If a baby is above 38 weeks gestational age and well, then he or she can have a relatively high serum bilirubin without needing the intervention. If a baby is 38 weeks with risk factors or 35–37 weeks and well, then the serum bilirubin cutoff is slightly lower to begin phototherapy. If the baby is 35–37 weeks with any of the above-listed risk factors, then the serum bilirubin cutoff is the most stringent for intervention.

68.4.2 Dosing

The total delivered dose in phototherapy is a function of the intensity of the light, the exposed body surface area, the penetration into the skin, and the duration of the exposure (Vreman, Wong, and Stevenson 2004). In practice, since the total body surface area and exposure are maximized with whatever practical constraints exist and light penetration in to the skin cannot be controlled for, that leaves light intensity as the only variable that is modulated. In terms of temporal duration, although

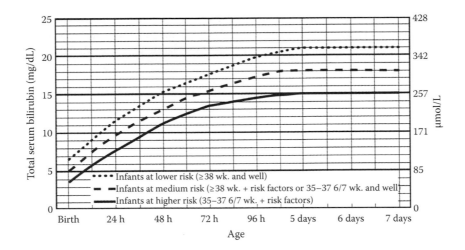

FIGURE 68.4 American Association of Pediatrics guidelines for initiation of phototherapy in infants at 35 or more weeks' gestation. Curves give phototherapy initiation cutoffs based on total serum bilirubin, gestational age, and risk factors that include isoimmune hemolytic disease, G6PD deficiency, asphyxia, significant lethargy, temperature instability, sepsis, acidosis, or albumin <3.0 g/dL. (Reproduced with permission from Subcommittee on Hyperbilirubinemia, *Pediatrics,* 114, 297–316, 2004. With permission.)

the original trials in the 1950s had intermittent exposure, current guidelines state that the infant should be as continuously exposed to light as possible, with breaks only for necessary care of the infant, including time with the parents and feeding (Cremer, Perryman, and Richards 1958; Subcommittee on Hyperbilirubinemia 2004).

In order to better understand dosing, researchers studied the fractional decreases in serum bilirubin over 24 h as a function of light intensity, shown in Figure 68.5. This curve plateaus between 30 and 60 µW/cm²/nm over a 425–475 nm range, at which point increasing dosing will not increase further bilirubin clearance (Maisels 1996; Tan 1982). Guidelines for standard phototherapy are based in part on these data. The AAP recommends 30 µW/cm²/nm over approximately the 460–490 nm range for phototherapy (Bhutani and Newborn 2011). These recommendations include confirming the delivered dose using a radiometer.

68.4.3 Treatment Side Effects

Phototherapy is generally safe, but as with any treatment, it has side effects (Maisels 2005; Wong and Bhutani 2012b). Transient red rashes can be noted, as well as loose stools. Infants are mostly naked to increase body surface area exposure to light, and insensible water losses and heat losses can occur. Indeed, several studies have shown increased transepidermal water losses in phototherapy (Maayan-Metzger et al. 2001). As such, hydration status should be closely monitored. Additionally, in order to prevent any potential phototoxicity to the retina, the eyes are covered. Burns from the light source are a very rare adverse event. There is conflicting evidence that nevi are increased in children who have undergone phototherapy (Mahe et al. 2009; Matichard et al. 2006).

Other toxicities are specific to certain populations of patients. As mentioned in Section 68.3.1, patients who have cutaneous porphyria and are sensitive to sunlight may develop severe blistering, and consequently, a family history of congenital porphyria is a contraindication to therapy (Subcommittee on Hyperbilirubinemia 2004). Phototherapy in patients receiving photosensitizers (e.g., fluorescein) can result in skin damage (Maisels 2005). Additionally, as discussed in Section 68.1.3, cholestatic jaundice is a condition in which bile cannot be excreted into the small intestine. As a result, infants with cholestatic jaundice undergoing phototherapy may develop what is known as "bronze baby syndrome," a bronze discoloration of the skin, urine, and secretion that may be due to copper porphyrins that result from photodegradation (Rubaltelli, Jori, and Reddi 1983; Subcommittee on Hyperbilirubinemia 2004).

Perhaps one of the largest but most incalculable side effects of phototherapy is the fact that parents cannot hold and bond with their infant. It is unclear to what extent this barrier to bonding

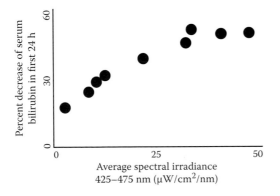

FIGURE 68.5 Dose–response curve for blue-light phototherapy. Decrease in serum bilirubin concentration is shown as a function of average spectral irradiance over 425–275 nm. (Adapted from Maisels, M.J., *Pediatrics,* 98, 283–287, 1996. With permission.)

has negative health effects, but it is one of the most apparent costs for parents of children undergoing therapy.

68.5 Light Sources

The initial light sources during the 1950s were fluorescent lamps with spectra matched as best as possible to the absorption of unconjugated bilirubin. Since then, a number of different light sources have been used to produce a similar spectrum, most recently with the introduction of light-emitting diode (LED) sources.

One must consider several practical factors when selecting a light source (Bhutani and Newborn 2011; Vreman, Wong, and Stevenson 2004). The most important is the emission spectrum: current recommendations from the AAP recommend light in approximately the 460–490 nm range (Bhutani and Newborn 2011). At the same time, emitting outside this bandwidth can have negative side effects. Emitting in the IR may cause unnecessary heating, and emitting in the UV can cause deoxyribonucleic acid (DNA) damage. Excessive heating means that the light cannot be placed in close proximity to the infant, decreasing the effective irradiance reaching the child. Practically speaking, traditional setups involving bulbs require that the baby be in the phototherapy bassinette while he or she receives therapy; other setups like fiber-optic "blankets" allow the child to be picked up and nursed while receiving treatment. We will consider four commonly used light sources and setups, focusing on the aforementioned qualities of each.

68.5.1 Fluorescent Sources

Fluorescent tubes were the first used light source and are still in use today (Ennever et al. 1984). Fluorescent lamps work by generating a short arc of electrons that excite mercury, argon, or another gas in the tube. Based on the characteristic of their electronic structure, these gases then emit photons in narrowly defined bands. The incident photons from these bands, which are often in the UV, then collide with the phosphor coating on the lamp, where they are converted into visible spectra. The spectrum of a fluorescent tube, then, tends to exhibit sharp peaks corresponding to the emission spectrum of the gas, with a wider wavelength range corresponding to the interaction of the UV light with the phosphorescent material (Grossweiner et al. 2005). Fluorescent lights used today are designed to emit at blue wavelengths, minimizing UV and IR emission.

68.5.2 Halogen Sources

Halogen sources, most commonly tungsten–halogen or quartz–iodine lamps, are incandescent sources that emit via blackbody radiation. The halogen in these lamps, in contrast to halogens present in fluorescent tubes, prolong the life of the tungsten filament and are not involved in any fluorescent process (Grossweiner et al. 2005). As a result of the underlying blackbody process, halogen spotlights emit significant amounts of near-IR and UV radiation. The UV radiation, which is generally greater than fluorescent tubes, can also be filtered by coatings. The IR radiation can generate heating and may cause thermal burns if placed too close (Maisels 2005).

68.5.3 Fiber-Optic Setups

Fiber-optic setups can use any source in principle but historically have been coupled with metal halide sources. Fiber-optic setups can guide light into a "blanket" that can be wrapped around the baby (Vreman, Wong, and Stevenson 2004). The major advantage is that the baby can be held and nursed while still receiving phototherapy, and blindfolding is unnecessary. Fiber-optic setups tend to transmit much lower intensity overall than direct sources and do not utilize the full body surface area; indeed, currently available sources may have lower efficacy than conventional methods of phototherapy (Mills and Tudehope 2001).

68.5.4 LEDs

The use of LEDs is a more recent development in the treatment of neonatal jaundice. LEDs work on the principle of electrons and holes recombining in specifically designed semiconductor materials. When the recombination occurs, the photon emitted is equal to the bandgap energy of the LED, which, in turn, is a function of the specific doping of the semiconductor material. LEDs, like fluorescent lamps, have varying emission spectra but in general are characterized by narrow spectra overall (Grossweiner et al. 2005). Thus, they produce little excess heat from IR radiation and almost no UV, making them very efficient and safe. Because LEDs have very low heat output, they can be placed close to the infant, including below the infant, or even embedded into moldable silicon as a blanket. Such setups can maximize light delivery to the skin. More recently, LEDs have been combined with fiber-optic cables for convenient light delivery. LEDs were developed as a light source much later than fluorescent lights and have only been incorporated into phototherapy in the last decade (Maisels 2005), with evidence suggesting that LEDs are as effective as traditional phototherapy sources (Kumar, Chawla, and Deorari 2011). Most recently, low-cost LED systems have been developed for use in low-resource settings.

68.6 Future Directions

In this chapter, we have discussed neonatal jaundice, beginning with the underlying physiology and pathophysiology of the condition. We took a historical approach to describe how phototherapy evolved from an initial observation in premature infants to an understanding of photochemistry, and then to better elucidation of clinical guidelines. Phototherapy for neonatal jaundice is a well-understood topic in many regards, but there are still open questions for the future. One important question that remains is better defining which children need phototherapy. Because kernicterus is such a rare event (Johnson, Bhutani, and Brown 2002), risk–benefit calculations of whether to treat can be difficult to make. On the one hand, phototherapy decreases a severe yet rare event; on the other hand, treatment itself confers a small risk to a large number of infants. Moreover, the specific risk reduction afforded by phototherapy in reducing neurotoxicity is not yet firmly established (Bhutani and Newborn 2011). These issues complement the growing interest in at-home phototherapy (Bhutani and Newborn 2011) and in expanding the use of phototherapy on a global scale (Bhutani 2012).

Standard protocols of certain aspects of phototherapy have yet to be defined. It is not yet established from evidence, for example, when children should be taken off of phototherapy. The role of transcutaneous bilirubin measurements, a method of spectroscopically and noninvasively determining bilirubin levels, has yet to be definitively defined (Maisels 2005). Finally, most data regarding treatment have been taken from relatively healthy term and near-term infants. Premature babies, on the other hand, are highly susceptible to unconjugated hyperbilirubinemia, and it is not clear when these children should begin to receive phototherapy (Watchko and Maisels 2003).

Overall, phototherapy in neonatal jaundice nicely illustrates the potential therapeutic implications of light, and the story of blue-light therapy nicely illustrates the interplay between basic physiology, photochemistry, and photonic sources. These in turn illustrate the interplay between clinical observations and basic sciences. As with many therapeutic interventions, the understanding of phototherapy has come a long way since its initial introduction but still has multiple questions to be answered in the future.

References

Bhutani, V. 2012. Editorial: Building evidence to manage newborn jaundice worldwide. *Indian J Pediatr* 79:253–255.

Bhutani, V. K., L. Johnson, and E. M. Sivieri. 1999. Predictive ability of a predischarge hour-specific serum bilirubin for subsequent significant hyperbilirubinemia in healthy term and near-term newborns. *Pediatrics* 103:6–14.

Bhutani, V. K., and C. F. Newborn. 2011. Phototherapy to prevent severe neonatal hyperbilirubinemia in the newborn infant 35 or more weeks of gestation. *Pediatrics* 128:E1046–E1052.

Cremer, R. J., P. W. Perryman, and D. H. Richards. 1958. Influence of light on the hyperbilirubinaemia of infants. *Lancet* 1:1094–1097.

Dennery, P. A., D. S. Seidman, and D. K. Stevenson. 2001. Neonatal hyperbilirubinemia. *New Engl J Med* 344:581–590.

Dobbs, R. H., and R. J. Cremer. 1975. Phototherapy. *Arch Dis Child* 50:833–836.

Ennever, J. F. 1988. Phototherapy for neonatal jaundice. *Photochem Photobiol* 47:871–876.

Ennever, J. F., A. T. Costarino, R. A. Polin, and W. T. Speck. 1987. Rapid clearance of a structural isomer of bilirubin during phototherapy. *J Clin Invest* 79:1674–1678.

Ennever, J. F., I. Knox, S. C. Denne, and W. T. Speck. 1985. Phototherapy for neonatal jaundice: *In vivo* clearance of bilirubin photoproducts. *Pediatr Res* 19:205–208.

Ennever, J. F., M. Sobel, A. F. McDonagh, and W. T. Speck. 1984. Phototherapy for neonatal jaundice: *In vitro* comparison of light sources. *Pediatr Res* 18:667–670.

Fritsch, C., K. Bolsen, T. Ruzicka, and G. Goerz. 1997. Congenital erythropoietic porphyria. *J Am Acad Dermatol* 36:594–610.

Glader, B. 2012. Genetics and pathophysiology of glucose-6-phosphate dehydrogenase deficiency. In *UpToDate*. D. Basow, editor. UpToDate, Waltham, MA.

Grossweiner, L. I., L. R. Jones, J. B. Grossweiner, and B. H. G. Rogers. 2005. *The Science of Phototherapy: An Introduction*. Springer, Dordrecht.

Gunn, C. K. 1944. Hereditary acholuric jaundice in the rat. *Can Med Assoc J* 50:230–237.

John, E. 1975. Phototherapy in neonatal hyperbilirubinaemia. *Austr Paediatr J* 11:49–52.

Johnson, L. H., V. K. Bhutani, and A. K. Brown. 2002. System-based approach to management of neonatal jaundice and prevention of kernicterus. *J Pediatr* 140:396–403.

Keenan, W. J., K. K. Novak, J. M. Sutherland, D. A. Bryla, and K. L. Fetterly. 1985. Morbidity and mortality associated with exchange transfusion. *Pediatrics* 75:417–421.

Knutson, M., and M. Wessling-Resnick. 2003. Iron metabolism in the reticuloendothelial system. *Crit Rev Biochem Mol* 38:61–88.

Kumar, P., D. Chawla, and A. Deorari. 2011. Light-emitting diode phototherapy for unconjugated hyperbilirubinaemia in neonates. *Cochrane Database Syst Rev* CD007969.

Kumar, S., and U. Bandyopadhyay. 2005. Free heme toxicity and its detoxification systems in human. *Toxicol Lett* 157:175–188.

Lightner, D. A., W. P. Linnane, 3rd, and C. E. Ahlfors. 1984. Bilirubin photooxidation products in the urine of jaundiced neonates receiving phototherapy. *Pediatr Res* 18:696–700.

Lightner, D. A., and A. F. McDonagh. 1984. Molecular mechanisms of phototherapy for neonatal jaundice. *Acc Chem Res* 17:417–424.

Maayan-Metzger, A., G. Yosipovitch, E. Hadad, and L. Sirota. 2001. Transepidermal water loss and skin hydration in preterm infants during phototherapy. *Am J Perinatol* 18:393–396.

Mahe, E., A. Beauchet, P. Aegerter, and P. Saiag. 2009. Neonatal blue-light phototherapy does not increase nevus count in 9-year-old children. *Pediatrics* 123:E896–E900.

Maisels, M. J. 1996. Why use homeopathic doses of phototherapy? *Pediatrics* 98:283–287.

Maisels, M. J. 2005. Jaundice. In *Avery's Neonatology: Pathophysiology & Management of the Newborn*. M. G. MacDonald, M. M. K. Seshia, and M. D. Mullett, editors. Lippincott Williams & Wilkins, Philadelphia, PA, 768–846.

Maisels, M. J. 2006. Neonatal jaundice. *Pediatr Rev* 27:443–454.

Matichard, E., A. Le Henanff, A. Sanders et al. 2006. Effect of neonatal phototherapy on melanocytic nevus count in children. *Arch Dermatol* 142:1599–1604.

McDonagh, A. F., and D. A. Lightner. 1985. 'Like a shrivelled blood orange'—Bilirubin, jaundice, and phototherapy. *Pediatrics* 75:443–455.

McDonagh, A. F., L. A. Palma, and D. A. Lightner. 1980. Blue light and bilirubin excretion. *Science* 208:145–151.

McDonagh, A. F., and L. M. Ramonas. 1978. Jaundice phototherapy: Micro flow-cell photometry reveals rapid biliary response of Gunn rats to light. *Science* 201:829–831.

Mills, J., and D. Tudehope. 2001. Fibreoptic phototherapy for neonatal jaundice (Review). *Cochrane Database Syst Rev.* 1:CD002060.

Ostrow, J. D. 1971. Photocatabolism of labeled bilirubin in the congenitally jaundiced (Gunn) rat. *J Clin Invest* 50:707–718.

Patrice, T., A. C. E. Moor, B. Ortel, and T. Hasan. 2003. Mechanisms of photodynamic therapy. In *Photodynamic Therapy*. T. Patrice, editor. The Royal Society of Chemistry, Cambridge.

Pollard, T. D. 2008. *Cell Biology,* Saunders, Philadelphia, PA.

Prahl, S. Bilirubin. Oregon Medical Laser Center Optical Properties Spectra, 2012. http://omlc.ogi.edu/spectra/PhotochemCAD/html/119.html (accessed May 1, 2012).

Rubaltelli, F. F., G. Jori, and E. Reddi. 1983. Bronze baby syndrome: A new porphyrin-related disorder. *Pediatr Res* 17:327–330.

Shapiro, S. M. 2003. Bilirubin toxicity in the developing nervous system. *Pediatr Neurol* 29:410–421.

Soylu, A., S. Kavukcu, and M. Turkmen. 1999. Phototherapy sequela in a child with congenital erythropoietic porphyria. *Eur J Pediatr* 158:526–527.

Subcommittee on Hyperbilirubinemia. 2004. Management of hyperbilirubinemia in the newborn infant 35 or more weeks of gestation. *Pediatrics* 114:297–316.

Suchy, F. J. 2011. Hepatobiliary function. In *Medical Physiology.* W. F. Boron and E. L. Boulpaep, editors. Saunders, Philadelphia, PA.

Tan, K. L. 1982. The pattern of bilirubin response to phototherapy for neonatal hyperbilirubinaemia. *Pediatr Res* 16:670–674.

Vreman, H. J., R. J. Wong, and D. K. Stevenson. 2004. Phototherapy: Current methods and future directions. *Semin Perinatol* 28:326–333.

Watchko, J. F., and M. J. Maisels. 2003. Jaundice in low birth-weight infants: Pathobiology and outcome. *Arch Dis Child Fetal Neonatal Ed* 88:F455–F458.

Wong, R. J., and V. K. Bhutani. 2012a. Pathogenesis and etiology of unconjugated hyperbilirubinemia in the newborn. In *UpToDate.* D. Basow, editor. UpToDate, Waltham, MA.

Wong, R. J., and V. K. Bhutani. 2012b. Treatment of unconjugated hyperbilirubinemia in term and late preterm infants. In *UpToDate.* D. Basow, editor. UpToDate, Waltham, MA.

Biological Evidence of the Efficacy of Light Therapy in Psychiatric Disorders

Anne-Marie Gagné
*Centre de Recherche de
l'Institut Universitaire en
Santé Mentale de Québec*

Marc Hébert
*Centre de Recherche de
l'Institut Universitaire en
Santé Mentale de Québec*

69.1 Introduction

Bright-light therapy (BLT) was first introduced in 1893 by Niels Ryberg Finsen to boost the immune system and was part of the work for which he later received the Nobel Prize in 1903. It was, however, only in the early 1980s that light therapy began to be used in psychiatry by Dr. Norman Rosenthal and colleagues in order to treat seasonal depression (Rosenthal et al. 1984). This simple treatment, which consists of exposing the patient to a very bright light source for about 30 min every day, provides a success rate similar to common antidepressants (Partonen and Lonnqvist 1996). At first, BLT was almost exclusively used for the treatment of seasonal affective disorder (SAD), which is characterized by a mood lowering during fall and winter followed by a spontaneous remission usually occurring in spring and summer, when the photoperiod increases (Rosenthal et al. 1984). Even today, BLT is strongly associated with SAD. The efficacy and efficiency of BLT to alleviate the symptoms of this pathology is widely accepted and very well documented (Golden et al. 2005; Lam, Terman, and Wirz-Justice 1997). It is noteworthy, however, that most of the reviews supporting the efficiency of BLT rely on significant decreased depression scores on various psychometric questionnaires and inventories. Indeed, most authors use a response criterion first proposed by Terman, Terman, and Rafferty (1990), which is a 50% reduction of the Structured Interview Guide for the Hamilton Rating Scale for Depression – Seasonal Affective Disorder version (SIGH-SAD) or Hamilton Depression Rating Scale with Atypical Depression Supplement (SIGH-ADS). There is, however, a strong placebo effect associated with light therapy. As reported in a study by Eastman et al. (1998), it may take up to 3 weeks to distinguish the light treatment effect against a placebo effect when treating SAD. This is why in the present review we wish to focus on the biological evidence for BLT efficacy, which is less likely influenced by a placebo effect. In SAD, BLT is recognized to impact several biological parameters such as neurotransmitters and hormonal levels, retinal physiology, and circadian entrainment, to name a few. BLT may also be beneficial for patients affected with other psychiatric pathologies less commonly associated with this therapy such as eating disorders, antepartum and postpartum depression, Parkinson's disease, and sleep disturbances. In addition, we will briefly describe how light and more particularly natural sunlight may change the neurochemical state of healthy individuals.

69.2 Light, Seasons, and Neurotransmitters

69.2.1 Serotonin

Variations in the levels of serotonin are found all year long and seem to correlate at least partially with changes in the photoperiod. For instance, when measured in the human cerebrospinal fluid (CSF) of healthy individuals, the metabolite of serotonin, 5-hydroxyindole acetic acid (5-HIAA), will peak during April and reach a nadir in October (Luykx et al. 2012). These equinoxal peak and nadir are in accordance (in Northern latitudes) with the onset and remission of SAD symptoms, when days become longer than nights and vice versa. Variations of metabolite content are thought to be at least partially due to the photoperiod changes and consequently the likelihood of being exposed to light, since Lambert et al. (2002) demonstrated that the human brain secretes more serotonin on sunny days, irrespective of the month of the year, as measured in the cerebral blood. They also reported that the serotonin level was not correlated with any other weather parameter such as atmospheric pressure, humidity, and so forth. Another study using positron emission tomography (PET) showed that a daily amount of sunshine and global radiation from the sky positively influenced postsynaptic serotonin-1A receptor binding in limbic brain regions (e.g., amygdala, posterior cingulate cortex, and parahippocampal cortex) of normal controls (Spindelegger et al. 2011). In contrast to Lambert et al. (2002), who did not observe any impact of the weather in the previous days of the experiment, Spindelegger et al. (2011) noticed a cumulative effect on the binding over a 5-day period. The fact that these authors obtained their results on limbic areas is of particular interest since those anatomical structures are thought to be involved in depression disorders (Purves et al. 2004). Of interest, one of those light-responsive structures (i.e., the amygdala), has been recently found to respond differently in symptomatic SAD participants compared to normal controls when exposed to green and blue light and then challenged with an auditory emotional stimulus. In fact, blue light was shown to enhance responses to an emotional stimulus in SAD-afflicted people but not in normal controls, whereas green light decreases these responses. Unfortunately, SAD patients were not tested during the remission phase so it is not clear if these responses represent a state marker or a biomarker of SAD (Vandewalle et al. 2011). Lastly, it would be interesting to verify if those discrepancies in amygdala activity between depressed SAD patients and controls could be related to a disturbance of the postsynaptic serotonin-1A binding during the light adaptation phenomenon that was reported in the brain structure by Spindelegger's team. It has also been demonstrated that tetrahydrobiopterin (BH4) levels increase following 5 days of light therapy (Hoekstra et al. 2003). This cofactor is a key player in the hydroxylation of tryptophan, which in turn promotes serotonin production. Even if BH4 levels could be an indicator of serotonin activity, it cannot be concluded that BLT was responsible of the increase in serotonin levels as it was not measured in this study. To conclude, evidence supporting an action of light on the serotoninergic function and its relevant structures (i.e., amygdala and limbic areas) are numerous. Advances in *in vivo* imaging technologies will probably continue to corroborate the positive impact of light on this neurotransmitter.

69.2.2 Dopamine

One of the first pieces of evidence that BLT could have an impact on dopamine in patients with SAD was found in the late 1980s.

Those studies used the eye blink rate as a dopamine measure, which is considered to be a good marker of central dopaminergic function (Kaminer et al. 2011). At first, Depue et al. (1988) noticed in a small sample that depressed SAD participants ($N = 4$) had an increased eye blink rate compared to their matched healthy controls. The anomaly disappeared after remission following BLT. Subsequently, the authors replicated their study with a larger sample, but this time after the summer spontaneous remission of SAD patients (Depue et al. 1990). They did replicate their finding of increased blink rate compared to controls during the depressive state, but also when remitted in summer, suggesting that the measure represented a marker of the disease. Another team found that BLT decreased the blink rate in premenopausal depressed SAD female subjects, albeit at baseline, there was no difference between groups (Barbato et al. 1993). Because the eye blink rate has been linked to dopaminergic function (Kaminer et al. 2011), it is believed that BLT normalized an abnormal dopaminergic state in those SAD female subjects.

Another piece of evidence of the impact of light on dopaminergic function is that BLT seems to restore the thermoregulatory function in patients with SAD during the depressive state (Arbisi et al. 1989). Indeed, during the depressive state, patients needed more time to come back to their initial body temperature after a sustained exercise challenge on a treadmill. After 2 weeks of BLT, the heat loss of SAD patients was not different from that observed in either the control group matched for VO2 max values or SAD patients during summer in the euthymic state.

In addition, natural light seems to enhance the dopaminergic function, at least in healthy individuals, as demonstrated using single photon emission computed tomography (SPECT). In fact, sunshine duration calculated 30 days prior to the SPECT scan was positively correlated with striatal dopamine D_2/D_3 receptor availability (Tsai et al. 2011).

69.2.3 Melatonin

The ability of the so-called "darkness hormone" to be suppressed by bright light (Lewy et al. 1980), the fact that its secretion duration is affected by the photoperiod (in laboratory conditions) (Wehr 1991), as well as the annual fluctuations featuring higher melatonin levels during the winter–spring seasons (Martikainen et al. 1985) probably explain why melatonin has generated so much interest in SAD research. The link between melatonin and SAD has been strongly associated to the phase shift hypothesis proposed by Lewy et al. (1988) suggesting that SAD at least partially originates from a mismatch between circadian rhythms. This asynchrony between endogenous cycles, namely, sleep/wake pattern and melatonin onset/off, would trigger depression (Lewy et al. 1987). Nevertheless, melatonin assessments as a marker of circadian disturbances have produced conflicting results with anomalies reported in some (Danilenko et al. 1994; Wehr et al. 2001), but not all, studies (Checkley et al. 1993; Partonen et al. 1996). Various factors could account for those disparities, but it is generally agreed that the therapeutic effect of BLT on SAD symptoms is unlikely to be due to the suppression of melatonin per se, but rather to the internal resynchronization of biological rhythms including the melatonin rhythm. For more details, the reader is invited to consult more exhaustive reviews on this particular hypothesis (Lewy et al. 2007; Srinivasan et al. 2006).

It seems, however, that the strong impact of light on melatonin could be central in other pathologies. For instance, BLT has been successfully used to alleviate some Parkinson's symptoms (Willis and Turner 2007). Within 2 weeks, Willis and Turner (2007) noticed not only an improvement of the motor impairments, namely, bradykinesia, rigidity, and dyskinesia, but also a decreased need for dopaminergic medication (L-dopa, bromocriptine). This effect might be mediated through a light-related inhibition of melatonin secretion, which in turn would increase dopamine activity. Indeed, melatonin is known to exacerbate Parkinson's symptoms, probably because of its inhibiting effect on dopamine production (Zisapel 2001). It was concluded that BLT should be seriously considered as an adjunctive treatment for Parkinson's, especially because it has very few side effects.

A similar trend is observed in restless leg syndrome (RLS), another pathology linked to dopamine (Michaud et al. 2002). Indeed, while administration of melatonin increases the severity of the symptoms, BLT slightly but significantly reduced the leg discomfort even though it had no effect on the leg movements per se (Whittom et al. 2010). Once again, this improvement following BLT is thought to be a result of melatonin suppression by light. The efficacy of BLT to alleviate some symptoms of RLS has been corroborated by other studies (Mitchell 2010), with improvement still visible 4 weeks after the end of the treatment (Mitchell et al. 2011). Of interest, the prevalence of RLS increases with the latitude where the photoperiod is marked by great variation throughout the year (Koo 2011).

69.3 SAD, Light, and Retina

The origin of the pathophysiology of SAD remains elusive, but ocular sensitivity hypotheses have been the focus of many investigations with mixed results. The link between SAD and the retina arose from the fact that the effect of BLT appears to be mediated by eyes (Wehr et al. 1987). There are, however, various means to investigate retinal or ocular sensitivity, and depending on the technique used, contrasting results have been reported.

69.3.1 Psychophysical Light Detection

In the early 1990s, Oren, Joseph-Vanderpool, and Rosenthal (1991) were the first to postulate that SAD patients may present a retinal defect that could account for depression when days become shorter and darker. Using the dark adaptation threshold (DAT) test, they noticed that the rod photoreceptors (night vision) in the SAD group were adapting faster to darkness than normal controls (Oren, Joseph-Vanderpool, and Rosenthal 1991), which does not support the retinal hyposensitivity hypothesis. Those results were corroborated by another team that challenged SAD people to the DAT test not only for rod adaptation but also

for cones (day vision), during both winter and summer (Terman and Terman 1999a). Whereas SAD patients demonstrated an increased retinal sensitivity in summer during the euthymic state when compared to winter, depressed normal controls did not show any seasonal fluctuations. Compared to controls, they also demonstrated a light hypersensitivity during winter. Overall, these DAT studies suggest that patients appear to have better light responsiveness than controls all year around, which does not support the hyposensitivity hypothesis to explain the origin of SAD.

69.3.2 Electrooculogram

In contrast to studies using subjective psychophysical light detection, performing the electrooculogram (EOG) technique reveals a light hyposensitivity in SAD patients (Ozaki et al. 1993). This trend was observed even after remission following BLT in winter. In a follow-up study, it was also reported that while normal controls increased their light sensitivity in winter when compared to summer, SAD people did not show this seasonal variation (Ozaki et al. 1995). The authors concluded that the increase in sensitivity during the winter season represented an adaptation mechanism to compensate for light reduction occurring at that time of the year. It was proposed that this compensatory defect in SAD people could therefore trigger the depressive symptoms.

It is noteworthy that although authors of these studies are presenting the EOG as a measure of light sensitivity, we have to be cautious with such interpretation since the real meaning of a higher or lower EOG Arden ratio and its cellular origin are still unclear. In fact, the Arden ratio is composed of a dark trough over a light peak obtained by performing saccadic movement during a period of dark adaptation followed by a period of light adaptation. The change in potential is due to the resting potential of the retinal pigment epithelium, and although healthy photoreceptors are necessary to produce the light peak that composes the Arden ratio, it cannot be said what is the origin of a variation within the normal range of such a ratio. In fact, a higher ratio could be linked to an increased response observed in the dark trough or a decreased response in the light peak and vice versa.

69.3.3 Electroretinogram

Investigation using the electroretinogram (ERG) technique also demonstrated retinal anomalies in SAD. This technique is the only one that allows the direct assessment of selective functioning of rods and cones. The first to report ERG anomalies in SAD were Lam et al. (1992). The authors reported that SAD's rod response was diminished compared to normal healthy controls (Lam et al. 1992). However, this difference was found only in the female group. In contrast, male patients showed an increased rod response.

Using the same technique, Hebert, Dumont, and Lachapelle (2002) replicated a decreased rod response in a group of subsyndromal SAD (S-SAD) people. S-SAD is a subtype of the pathology that affects as much as 15% of the population (North America), in which symptoms although bothersome do not reach clinical significance (Kasper et al. 1989). The authors observed reduced rod sensitivity only in winter during the depressive episode. On a subsequent study with SAD patients, they were able to replicate their finding (i.e., a decreased rod sensitivity in winter in SAD people compared to healthy individuals) (Hebert et al. 2004). These results were corroborated by the same group who noticed not only a decreased rod sensitivity in SAD patients compared to the control group but also that this impaired retinal function could be restored following remission either after 4 weeks of BLT or in summertime (Lavoie et al. 2009). This research represented the first demonstration of the therapeutic effect of light on retinal functions in SAD patients.

Lastly, it has been demonstrated that the retinas of SAD patients do not react the same way as healthy individuals following recent light history (Gagne and Hebert 2011). This team demonstrated that SAD patients decrease their rod response following 60 min of BLT when compared to the baseline light exposure or to a very dim light exposure (5 lx). This phenomenon was not observed in the control group who instead increase their rod response following the dim light exposure compared to the baseline and the BLT condition. This was observed in both the winter and summer seasons. Though this trend was assessed after only a single BLT session and therefore could not represent any cumulative effect, this result suggests that SAD people might not react the same way as controls when photoperiod changes naturally occur through the year.

Overall, ocular sensitivity studies appear to demonstrate some anomalies in S-SAD and SAD patients, although the exact nature of such anomalies remains elusive. But the ERG appears to represent a good tool to investigate such disorders including treatment efficacy. It should be pointed out that the retina, which is part of the central nervous system, has been identified as a relevant organ for investigation for psychiatric illnesses.

69.4 Nonseasonal Depression

69.4.1 Major Depression

As a treatment for major depression, BLT appears to be beneficial as an adjunct remedy along with antidepressant. For instance, the combination of BLT and sertraline produces a larger decrease in depression scores than when sertraline is given alone (Martiny et al. 2005). However, a subsequent study showed that although BLT enhanced the antidepressant effect of sertraline in depressed patients, the therapeutic response was not sustained 4 weeks after the ending BLT administration (Martiny et al. 2006). A similar potentiation effect was found with citalopram, but in this trial, the antidepressant response was also hastened in patients receiving BLT compared to the placebo group (Benedetti et al. 2003). Along with the same antidepressant, a double-blind study showed that BLT had a dose response effect in a group of poststroke patients diagnosed with major depression. Indeed, a high regimen of BLT (10,000 lx) yielded a better improvement of depressive symptoms than a medium regimen (4000 lx)

(Sondergaard et al. 2006). In a group of 28 depressed teenagers, BLT combined with fluoxetine produced a larger decrease in depression scores but also improved the evening melatonin secretion that seemed deficient when those same subjects were evaluated at baseline or after receiving fluoxetine combined with the placebo treatment (Niederhofer and von Klitzing 2011). An augmentation of the antidepressant effect of selective serotonin reuptake inhibitors (SSRIs) has also been documented in a small group ($N = 13$) of drug-resistant depressed women with a comorbid borderline personality diagnosis following 6 weeks of BLT (Prasko et al. 2010). Finally, BLT can also be added to enhance the effect of wakeful therapy (sleep deprivation) sometimes used in major depression (Loving, Kripke, and Shuchter 2002).

BLT may be beneficial also as a stand-alone treatment for depression. This is particularly true in the elderly since the daily amount of medication is sometimes already high and diverse. In those cases, interpharmacological interactions become more probable and the use of BLT might represent a good alternative. A randomized double-blind placebo-controlled trial performed on 90 depressed elderly patients demonstrated that BLT administered in the morning significantly reduced depression scores (Lieverse et al. 2011). It also had an impact on some endocrinological markers such as an increased steepness of the melatonin rise and a reduced cortisol level of 37% compared to the placebo group. A sleep improvement was also observed during the trial.

69.4.2 Antepartum and Postpartum Depression

Antepartum and postpartum depression could affect as many as 20% of expectant mothers and new mothers (Gavin et al. 2005). Because pregnancy and lactation represent situations during which medication is often unsuitable, light therapy has been successfully used in some studies. In 2000, Corral, Kuan, and Kostaras (2000) were the first to publish a case study in which they reported improvements in depression scores of two women diagnosed with postpartum depression treated with BLT over a 4-week period. A follow-up study (Corral et al. 2007), however, failed to detect a significant difference between BLT and placebo with both conditions yielding to a reduction of the depressive symptoms. It is noteworthy that the dim light used for the placebo condition was uncommonly bright, with an intensity of 600 lx. This could account for the fact that many women responded to the placebo, and that although the bright-light group appeared to show a more pronounced improvement trend, it was not statistically significant.

Antepartum depression has also generated some interest in the field of light therapy. Oren et al. (2002) tested a group of 16 pregnant women diagnosed with major depression. Three weeks of treatment were enough to diminish depression scores (Hamilton Depression Rating Scale) by about 50%, without any side effect on the pregnancy (Oren et al. 2002) except nausea in two women. This is supported by a recent trial with a larger sample ($N = 27$) that demonstrated that BLT was more effective than a placebo to treat antepartum depression (Wirz-Justice et al. 2011). In contrast to Corral et al. (2007), the placebo used in this study was a much dimmer light (i.e., 70 lx instead of 600 lx). Another trial on antepartum showed that 10 weeks of BLT improved depressive symptoms with the same magnitude as antidepressant medication (Epperson et al. 2004).

The effect of BLT in these females is not likely due to a lack of light exposure as there was no difference in the illumination pattern of postpartum depressed mothers compared to normal women. Similarly, the illumination levels were not correlated with mood (Wang et al. 2003).

69.5 Eating Disorders

Some studies documented the impact of BLT on eating disorders, namely, bulimia, anorexia, and binge eating. Recently, it was demonstrated that BLT can restore temperature circadian rhythms in bulimic and anorexic patients. This suggests that this therapy might improve disturbed eating pattern because of the physiological links between light and food intake circadian oscillators (Yamamotova, Papezova, and Vevera 2008).

69.5.1 Bulimia Nervosa

Bulimia nervosa is known to worsen during winter (Ghadirian et al. 1999), and this fact may explain why some clinicians proposed the idea of adding BLT as an adjunct treatment for this eating disorder. It is in 1989 that Lam (1989) and coworkers first reported a case study of a young bulimic woman who had showed a decrease in binge/purge episode following BLT administrated during winter. In a follow-up placebo-controlled study, they replicated their findings in a larger group of patients ($N = 17$). They noticed, however, that BLT had a greater effect on bulimic patients with concurrent diagnosis of SAD than in the non-SAD group (Lam et al. 1994). Later, similar results were obtained by Braun et al. (1999) in a double-blind placebo-controlled study. Interestingly, in this study, it was found that BLT could not improve the depression scores more than the dim light condition, meaning that this result was independent of the depression level. This contrasts with the study performed by Blouin et al. (1996), who reported exactly the opposite results following BLT: a decrease in depression scores when compared to placebo and no effect on the frequency, size, or content of binge-eating episodes. Braun's team explained this discrepancy by the fact that depression levels at the beginning of the studies were not comparable (i.e., they were less severe in their sample and therefore harder to significantly improve).

69.5.2 Anorexia

As with bulimia, the first evidence of the positive impact of BLT on anorexia was a case report about a young woman who showed an exacerbation of her symptoms during winter months (Ash, Piazza, and Anderson 1998). At her second admission, a BLT treatment was offered. She progressively increased her food intake and showed an improvement of the depression level. Later, a short trial of BLT administrated over a 5-day period

showed clinical improvement on a small sample of chronic anorexic women (Daansen and Haffmans 2010). Indeed, these women demonstrated a better core eating pathology (e.g., drive for thinness, body dissatisfaction, current body image, and body ideal image), less depression symptoms, as well as a decrease in global distress. Those effects were, however, of short duration as the follow-up assessment 3 months after the end of the treatment revealed a partial loss of the improvement. It was concluded that BLT should be administered again after a few months. It is noteworthy that only 5 days of BLT represents a very short treatment regimen for any pathology, and therefore the clinical improvement loss after 3 months could have been predicted. In fact, a subsequent study concluded that 6 weeks of daily BLT might not be enough to draw conclusions about its efficacy in anorexia nervosa (Janas-Kozik et al. 2011). In that research, though the girls receiving BLT did not increase their body mass index more than the girls who received only cognitive therapy at the end of study, this increase started earlier in the BLT group. The BLT group also demonstrated less severe degree of depression at the end of the treatment.

69.5.3 Night Eating Syndrome

Night eating syndrome (NES) is a disorder not included yet in *Diagnostic and Statistical Manual of Mental Disorders* (DSM-IV-R) but is often defined as follows: an eating pattern involving morning anorexia, evening hyperphagia, and insomnia or sleeplessness at least three to four times a week (Striegel-Moore et al. 2006). Friedman et al. (2002) were the first to document the impact of BLT on NES. This was a case-study of a 51-year-old overweight woman presenting depression symptoms despite paroxetine treatment for the past 2 years. Two weeks of BLT not only treated complete NES symptoms but also improved depression score to a normal level (Friedman et al. 2002). NES symptoms, however, returned 1 month after the end of BLT. A similar case was reported 2 years later (Friedman et al. 2004). Though evidence for the usefulness of BLT for NES is scant, it is noteworthy that this disorder is a relatively new topic of interest in psychiatry. Another study has suggested that BLT should be considered for NES, particularly because of circadian rhythm profile disturbances found in that population (Goel et al. 2009). Indeed, a group of 14 patients demonstrated decreased amplitudes in various daily rhythms such as food intake, ghrelin, insulin, and cortisol.

69.6 Side Effects of BLT and Potential Hazards

The fact that BLT has very few side effects probably explains why this treatment is often used in order to alleviate symptoms of seasonal-, dopamine-, melatonin-, or circadian-related pathologies. Because of the nonpharmacological nature of BLT, side effects tend to be underestimated. Users, researchers, and care providers should know (1) that side effects do exist even if they are generally considered as mild and (2) how to adjust the administration in order to maintain benefits from the treatment and avoid treatment discontinuation. Reported side effects include headaches, dry eyes, eye strain, dizziness, nausea, agitation, excitation, jumpiness/jitteriness, elation, overactivity, and feeling "weird" (Labbate et al. 1994; Terman and Terman 1999b). Fortunately, most of the side effects resolve spontaneously within a few days or weeks of the treatment (Terman and Terman 1999b). Also, using BLT as an adjunct treatment has been reported to sometimes worsen the side effects of antidepressant medication such as trimipramine (Muller et al. 1997).

Lastly, it should be pointed out that BLT is an over-the-counter treatment that is not regulated by the Food and Drug Administration (FDA) or any equivalent organization devoted to protect the public health. Information about eventual hazards consequent to abuse and/or long-term effects is very scant. To our knowledge, only one study revealed no cumulative damage after years of light therapy, but this relies on observations made on a rather small sample of 17 users (Gallin et al. 1995). It has been demonstrated, however, that BLT, in particular, blue-light therapy, even in a safety standard, affects significantly the retinal physiology in normal controls after a single exposure (Gagne and Hebert 2011; Gagne et al. 2007, 2011). This raises some important concerns with regard to the growing popularity of those light devices used to treat SAD symptoms and various other disorders. Reme et al. (1996) already suggested developing specific exposure standards for BLT, as the actual norms provided by the American Conference of Industrial Hygienists (ACGIH) are based on a single administration and not from chronic use of bright-light sources, but manufacturers are not obliged to follow any advice. To summarize, this is not to say that any actual apparatus available on the market is dangerous. But as BLT is mostly used as a self-administered treatment, adoption of specific regulation regarding the development of new devices would be recommended in order to safeguard the general public.

69.7 Conclusion

To conclude, the use of BLT seems to be becoming more widespread in the psychiatric therapeutic sphere. While some people are still tempted to think that the symptomatological relief following BLT represents only a placebo effect, more and more objective and biological pieces of evidence support the direct action of light on brain chemistry, as demonstrated by advanced imaging techniques, blood analysis, and retinal response. The way BLT acts on the nervous system remains unclear but has undeniable effects, both subjective and biological. As a stand-alone treatment in SAD, major or perinatal depression, it allows the remarkable advantage of being an alternative to medication. As an adjunct treatment, BLT can help to avoid pharmacological interactions with the possibility of even potentiating the positive action of the medication already administered. Compared to most antidepressant compounds, BLT probably has the best risk/benefit ratio, and research trying to integrate this treatment to other pathologies than SAD should be strongly encouraged.

References

Arbisi, P. A., R. A. Depue, M. R. Spoont, A. Leon, and B. Ainsworth. 1989. Thermoregulatory response to thermal challenge in seasonal affective disorder: A preliminary report. *Psychiatry Res* 28:323–334.

Ash, J. B., E. Piazza, and J. L. Anderson. 1998. Light therapy in the clinical management of an eating-disordered adolescent with winter exacerbation. *Int J Eat Disord* 23:93–97.

Barbato, G., D. E. Moul, P. Schwartz, N. E. Rosenthal, and D. A. Oren. 1993. Spontaneous eye blink rate in winter seasonal affective disorder. *Psychiatry Res* 47:79–85.

Benedetti, F., C. Colombo, A. Pontiggia et al. 2003. Morning light treatment hastens the antidepressant effect of citalopram: A placebo-controlled trial. *J Clin Psychiatry* 64:648–653.

Blouin, A. G., J. H. Blouin, H. Iversen et al. 1996. Light therapy in bulimia nervosa: A double-blind, placebo-controlled study. *Psychiatry Res* 60:1–9.

Braun, D. L., S. R. Sunday, V. M. Fornari, and K. A. Halmi. 1999. Bright light therapy decreases winter binge frequency in women with bulimia nervosa: A double-blind, placebo-controlled study. *Compr Psychiatry* 40:442–448.

Checkley, S. A., D. G. Murphy, M. Abbas et al. 1993. Melatonin rhythms in seasonal affective disorder. *Br J Psychiatry* 163:332–337.

Corral, M., A. Kuan, and D. Kostaras. 2000. Bright light therapy's effect on postpartum depression. *Am J Psychiatry* 157:303–304.

Corral, M., A. A. Wardrop, H. Zhang, A. K. Grewal, and S. Patton. 2007. Morning light therapy for postpartum depression. *Arch Womens Ment Health* 10:221–224.

Daansen, P. J., and J. Haffmans. 2010. Reducing symptoms in women with chronic anorexia nervosa. A pilot study on the effects of bright light therapy. *Neuro Endocrinol Lett* 31:290–296.

Danilenko, K. V., A. A. Putilov, G. S. Russkikh, L. K. Duffy, and S. O. Ebbesson. 1994. Diurnal and seasonal variations of melatonin and serotonin in women with seasonal affective disorder. *Arctic Med Res* 53:137–145.

Depue, R. A., P. Arbisi, S. Krauss et al. 1990. Seasonal independence of low prolactin concentration and high spontaneous eye blink rates in unipolar and bipolar II seasonal affective disorder. *Arch Gen Psychiatry* 47:356–364.

Depue, R. A., W. G. Iacono, R. Muir, and P. Arbisi. 1988. Effect of phototherapy on spontaneous eye blink rate in subjects with seasonal affective disorder. *Am J Psychiatry* 145:1457–1459.

Eastman, C. I., M. A. Young, L. F. Fogg, L. Liu, and P. M. Meaden. 1998. Bright light treatment of winter depression: A placebo-controlled trial. *Arch Gen Psychiatry* 55:883–889.

Epperson, C. N., M. Terman, J. S. Terman et al. 2004. Randomized clinical trial of bright light therapy for antepartum depression: Preliminary findings. *J Clin Psychiatry* 65:421–425.

Friedman, S., C. Even, R. Dardennes, and J. D. Guelfi. 2002. Light therapy, obesity, and night-eating syndrome. *Am J Psychiatry* 159:875–876.

Friedman, S., C. Even, R. Dardennes, and J. D. Guelfi. 2004. Light therapy, nonseasonal depression, and night eating syndrome. *Can J Psychiatry* 49:790.

Gagne, A. M., P. Gagne, and M. Hebert. 2007. Impact of light therapy on rod and cone functions in healthy subjects. *Psychiatry Res* 151:259–263.

Gagne, A. M., and M. Hebert. 2011. Atypical pattern of rod electroretinogram modulation by recent light history: A possible biomarker of seasonal affective disorder. *Psychiatry Res* 187:370–374.

Gagne, A. M., F. Levesque, P. Gagne, and M. Hebert. 2011. Impact of blue vs red light on retinal response of patients with seasonal affective disorder and healthy controls. *Prog Neuropsychopharmacol Biol Psychiatry* 35:227–231.

Gallin, P. F., M. Terman, C. E. Reme et al. 1995. Ophthalmologic examination of patients with seasonal affective disorder, before and after bright light therapy. *Am J Ophthalmol* 119:202–210.

Gavin, N. I., B. N. Gaynes, K. N. Lohr et al. 2005. Perinatal depression: A systematic review of prevalence and incidence. *Obstet Gynecol* 106:1071–1083.

Ghadirian, A. M., N. Marini, S. Jabalpurwala, and H. Steiger. 1999. Seasonal mood patterns in eating disorders. *Gen Hosp Psychiatry* 21:354–359.

Goel, N., A. J. Stunkard, N. L. Rogers et al. 2009. Circadian rhythm profiles in women with night eating syndrome. *J Biol Rhythms* 24:85–94.

Golden, R. N., B. N. Gaynes, R. D. Ekstrom et al. 2005. The efficacy of light therapy in the treatment of mood disorders: A review and meta-analysis of the evidence. *Am J Psychiatry* 162:656–662.

Hebert, M., C. W. Beattie, E. M. Tam, L. N. Yatham, and R. W. Lam. 2004. Electroretinography in patients with winter seasonal affective disorder. *Psychiatry Res* 127:27–34.

Hebert, M., M. Dumont, and P. Lachapelle. 2002. Electrophysiological evidence suggesting a seasonal modulation of retinal sensitivity in subsyndromal winter depression. *J Affect Disord* 68:191–202.

Hoekstra, R., D. Fekkes, B. J. van de Wetering, L. Pepplinkhuizen, and W. M. Verhoeven. 2003. Effect of light therapy on biopterin, neopterin and tryptophan in patients with seasonal affective disorder. *Psychiatry Res* 120:37–42.

Janas-Kozik, M., M. Krzystanek, M. Stachowicz et al. 2011. Bright light treatment of depressive symptoms in patients with restrictive type of anorexia nervosa. *J Affect Disord* 130:462–465.

Kaminer, J., A. S. Powers, K. G. Horn, C. Hui, and C. Evinger. 2011. Characterizing the spontaneous blink generator: An animal model. *J Neurosci* 31:11256–11267.

Kasper, S., T. A. Wehr, J. J. Bartko, P. A. Gaist, and N. E. Rosenthal. 1989. Epidemiological findings of seasonal changes in mood and behavior. A telephone survey of Montgomery County, Maryland. *Arch Gen Psychiatry* 46:823–833.

Koo, B. B. 2011. Restless legs syndrome: Relationship between prevalence and latitude. *Sleep Breath* 16:1237–1245.

Labbate, L. A., B. Lafer, A. Thibault, and G. S. Sachs. 1994. Side effects induced by bright light treatment for seasonal affective disorder. *J Clin Psychiatry* 55:189–191.

Lam, R. W. 1989. Light therapy for seasonal bulimia. *Am J Psychiatry* 146:1640–1641.

Lam, R. W., C. W. Beattie, A. Buchanan, and J. A. Mador. 1992. Electroretinography in seasonal affective disorder. *Psychiatry Res* 43:55–63.

Lam, R. W., E. M. Goldner, L. Solyom, and R. A. Remick. 1994. A controlled study of light therapy for bulimia nervosa. *Am J Psychiatry* 151:744–750.

Lam, R. W., M. Terman, and A. Wirz-Justice. 1997. Light therapy for depressive disorders: Indications and efficacy. *Mod Probl Pharmacopsychiatr* 25:215–234.

Lambert, G. W., C. Reid, D. M. Kaye, G. L. Jennings, and M. D. Esler. 2002. Effect of sunlight and season on serotonin turnover in the brain. *Lancet* 360:1840–1842.

Lavoie, M. P., R. W. Lam, G. Bouchard et al. 2009. Evidence of a biological effect of light therapy on the retina of patients with seasonal affective disorder. *Biol Psychiatry* 66:253–258.

Lewy, A. J., J. N. Rough, J. B. Songer et al. 2007. The phase shift hypothesis for the circadian component of winter depression. *Dialogues Clin Neurosci* 9:291–300.

Lewy, A. J., R. L. Sack, C. M. Singer, and D. M. White. 1987. The phase shift hypothesis for bright light's therapeutic mechanism of action: Theoretical considerations and experimental evidence. *Psychopharmacol Bull* 23:349–353.

Lewy, A. J. and R. L. Sack. 1988. The phase-shift hypothesis of seasonal affective disorder. *Am J Psychiatry* 145:1041–1043.

Lewy, A. J., T. A. Wehr, F. K. Goodwin, D. A. Newsome, and S. P. Markey. 1980. Light suppresses melatonin secretion in humans. *Science* 210:1267–1269.

Lieverse, R., E. J. Van Someren, M. M. Nielen et al. 2011. Bright light treatment in elderly patients with nonseasonal major depressive disorder: A randomized placebo-controlled trial. *Arch Gen Psychiatry* 68:61–70.

Loving, R. T., D. F. Kripke, and S. R. Shuchter. 2002. Bright light augments antidepressant effects of medication and wake therapy. *Depress Anxiety* 16:1–3.

Luykx, J. J., S. C. Bakker, E. Lentjes et al. 2012. Season of sampling and season of birth influence serotonin metabolite levels in human cerebrospinal fluid. *PLoS One* 7:e30497.

Martikainen, H., J. Tapanainen, O. Vakkuri, J. Leppaluoto, and I. Huhtaniemi. 1985. Circannual concentrations of melatonin, gonadotrophins, prolactin and gonadal steroids in males in a geographical area with a large annual variation in daylight. *Acta Endocrinol (Copenh)* 109:446–450.

Martiny, K., M. Lunde, M. Unden, H. Dam, and P. Bech. 2005. Adjunctive bright light in non-seasonal major depression: Results from clinician-rated depression scales. *Acta Psychiatr Scand* 112:117–125.

Martiny, K., M. Lunde, M. Unden, H. Dam, and P. Bech. 2006. The lack of sustained effect of bright light, after discontinuation, in non-seasonal major depression. *Psychol Med* 36:1247–1252.

Michaud, M., J. P. Soucy, A. Chabli, G. Lavigne, and J. Montplaisir. 2002. SPECT imaging of striatal pre- and postsynaptic dopaminergic status in restless legs syndrome with periodic leg movements in sleep. *J Neurol* 249:164–170.

Mitchell, U. H. 2010. Use of near-infrared light to reduce symptoms associated with restless legs syndrome in a woman: A case report. *J Med Case Reports* 4:286.

Mitchell, U. H., J. W. Myrer, A. W. Johnson, and S. C. Hilton. 2011. Restless legs syndrome and near-infrared light: An alternative treatment option. *Physiother Theory Pract* 27:345–351.

Muller, M. J., E. Seifritz, M. Hatzinger, U. Hemmeter, and E. Holsboer-Trachsler. 1997. Side effects of adjunct light therapy in patients with major depression. *Eur Arch Psychiatry Clin Neurosci* 247:252–258.

Niederhofer, H., and K. von Klitzing. 2011. Bright light treatment as add-on therapy for depression in 28 adolescents: A randomized trial. *Prim Care Companion CNS Disord* 13:01194.

Oren, D. A., J. R. Joseph-Vanderpool, and N. E. Rosenthal. 1991. Adaptation to dim light in depressed patients with seasonal affective disorder. *Psychiatry Res* 36:187–193.

Oren, D. A., K. L. Wisner, M. Spinelli et al. 2002. An open trial of morning light therapy for treatment of antepartum depression. *Am J Psychiatry* 159:666–669.

Ozaki, N., N. E. Rosenthal, D. E. Moul, P. J. Schwartz, and D. A. Oren. 1993. Effects of phototherapy on electrooculographic ratio in winter seasonal affective disorder. *Psychiatry Res* 49:99–107.

Ozaki, N., N. E. Rosenthal, F. Myers, P. J. Schwartz, and D. A. Oren. 1995. Effects of season on electro-oculographic ratio in winter seasonal affective disorder. *Psychiatry Res* 59:151–155.

Partonen, T., and J. Lonnqvist. 1996. Moclobemide and fluoxetine in treatment of seasonal affective disorder. *J Affect Disord* 41:93–99.

Partonen, T., O. Vakkuri, C. Lamberg-Allardt, and J. Lonnqvist. 1996. Effects of bright light on sleepiness, melatonin, and 25-hydroxyvitamin D(3) in winter seasonal affective disorder. *Biol Psychiatry* 39:865–872.

Prasko, J., M. Brunovsky, K. Latalova et al. 2010. Augmentation of antidepressants with bright light therapy in patients with comorbid depression and borderline personality disorder. *Biomed Pap Med Fac Univ Palacky Olomouc Czech Repub* 154:355–361.

Purves, D., G. J. Augustine, D. Fitzpatrick et al., editors. 2004. *Neuroscience*, 3rd ed. Sinauer Associates, Sunderland, MA.

Reme, C. E., P. Rol, K. Grothmann, H. Kaase, and M. Terman. 1996. Bright light therapy in focus: Lamp emission spectra and ocular safety. *Technol Health Care* 4:403–413.

Rosenthal, N. E., D. A. Sack, J. C. Gillin et al. 1984. Seasonal affective disorder. A description of the syndrome and preliminary findings with light therapy. *Arch Gen Psychiatry* 41:72–80.

Sondergaard, M. P., J. O. Jarden, K. Martiny, G. Andersen, and P. Bech. 2006. Dose response to adjunctive light therapy in citalopram-treated patients with post-stroke depression. A randomised, double-blind pilot study. *Psychother Psychosom* 75:244–248.

Spindelegger, C., P. Stein, W. Wadsak et al. 2011. Light-dependent alteration of serotonin-1A receptor binding in cortical and subcortical limbic regions in the human brain. *World J Biol Psychiatry* 13:413–422.

Srinivasan, V., M. Smits, W. Spence et al. 2006. Melatonin in mood disorders. *World J Biol Psychiatry* 7:138–151.

Striegel-Moore, R. H., D. L. Franko, A. May et al. 2006. Should night eating syndrome be included in the DSM? *Int J Eat Disord* 39:544–549.

Terman, J. S., and M. Terman. 1999a. Photopic and scotopic light detection in patients with seasonal affective disorder and control subjects. *Biol Psychiatry* 46:1642–1648.

Terman, M., and J. S. Terman. 1999b. Bright light therapy: Side effects and benefits across the symptom spectrum. *J Clin Psychiatry* 60:799–808; quiz 809.

Terman, M., J. S. Terman, and B. Rafferty. 1990. Experimental design and measures of success in the treatment of winter depression by bright light. *Psychopharmacol Bull* 26:505–510.

Tsai, H. Y., K. C. Chen, Y. K. Yang et al. 2011. Sunshine-exposure variation of human striatal dopamine D(2)/D(3) receptor availability in healthy volunteers. *Prog Neuropsychopharmacol Biol Psychiatry* 35:107–110.

Vandewalle, G., M. Hebert, C. Beaulieu et al. 2011. Abnormal hypothalamic response to light in seasonal affective disorder. *Biol Psychiatry* 70:954–961.

Wang, E. J., D. F. Kripke, M. T. Stein, and B. L. Parry. 2003. Measurement of illumination exposure in postpartum women. *BMC Psychiatry* 3:5.

Wehr, T. A. 1991. The durations of human melatonin secretion and sleep respond to changes in daylength (photoperiod). *J Clin Endocrinol Metab* 73:1276–1280.

Wehr, T. A., W. C. Duncan, Jr., L. Sher et al. 2001. A circadian signal of change of season in patients with seasonal affective disorder. *Arch Gen Psychiatry* 58:1108–1114.

Whittom, S., M. Dumont, D. Petit et al. 2010. Effects of melatonin and bright light administration on motor and sensory symptoms of RLS. *Sleep Med* 11:351–355.

Willis, G. L., and E. J. Turner. 2007. Primary and secondary features of Parkinson's disease improve with strategic exposure to bright light: A case series study. *Chronobiol Int* 24:521–537.

Wirz-Justice, A., A. Bader, U. Frisch et al. 2011. A randomized, double-blind, placebo-controlled study of light therapy for antepartum depression. *J Clin Psychiatry* 72:986–993.

Yamamotova, A., H. Papezova, and J. Vevera. 2008. Normalizing effect of bright light therapy on temperature circadian rhythm in patients with eating disorders. *Neuro Endocrinol Lett* 29:168–172.

Zisapel, N. 2001. Melatonin-dopamine interactions: From basic neurochemistry to a clinical setting. *Cell Mol Neurobiol* 21:605–616.

70

Future Developments in Photomedicine and Photodynamic Therapy

Thierry J. Patrice
Laboratoire de Photobiologie des Cancers

Trying to predict the future developments of a relatively recent medical field is a real challenge, particularly at the conclusion of a comprehensive textbook on photomedicine. I am grateful to the editors for having thought of me for this task! but it may appear to be a poisoned gift made even more poisonous since I have been very honored to accept this task. Having participated in the photodynamic therapy (PDT) adventure as a student since its early stages, I will share my vision of the future. Photomedicine, although as old as the sun in being a freely available source of energy, has found its modern applications only very recently. Some of these applications will disappear or have already disappeared, while some others, still in the laboratory stage, will progressively emerge. There are two driving forces guiding this emergence: chemistry and technologies (particularly nanotechnology) on one hand, which are developing so quickly and make apparently impossible things easily feasible, and economic realism on the other hand, which will at the same time promote new fields but also limit their expansion.

PDT actually belongs to the field of destructive tools that can destroy cells and tissues among a powerful arsenal of knives, x-rays, and chemicals not only in oncology but also against bacteria or damaged skin. In the future, other applications that are not merely directly destructive will appear less "primitive" and more sophisticated. Since the beginning of the PDT story, we have thus been told that new powerful photosensitizer chemicals were needed. It led to a competition between companies that unfortunately has temporarily ended because of a lack of original concepts. At the same time, photosensitizers had to be selective toward their attributed target (i.e., cancer cells or bacteria) and do as little damage as possible to normal tissues. This dream of a magic bullet has not yet been achieved and research will first have to address this point. We think that restarting PDT with new chemicals is a point of a critical importance, although some specialists have claimed that the search for new drugs is useless. As a result of this decline in chemistry, the actual chemical research is aimed at including existing photosensitizers

into more complex structures as nanoparticles having specific capabilities aimed at compensating to a certain extent the lack of efficacy or of selectivity of raw photosensitizers. However, if this activity is of interest for the future, because more results will still keep being reported, it will make photosensitizers more expensive, and improvement may be proportionally minimal. The more the number of steps taken to get the final drug, the more expensive it is, and as a consequence, the more patients are needed to get a proper revenue return. This is highly problematic because the problem of PDT is precisely the paradox of a great efficacy but on a very limited number of patients within very narrow clinical indications. However, some approaches are of interest, particularly those that use cells' peculiar enzymatic equipment to provoke the release of an increased amount of photosensitizer in a given targeted area. Techniques involving sugars as targeting vehicles are simple to handle and thus could be cost-effective in the future. Finally, we have to keep in mind that a big-pharma company must find one new compound with a high return on investment (ROI) every 5 years per given field of activity. If not, it is difficult to justify new investments to shareholders.

Tomorrow as yesterday I thus would bet on new and much more powerful photosensitizers being developed. Actual indications will further develop and new indications will arise from this research of critical importance. What could these new photosensitizers be, or rather as I am not a chemist, where could we look for them? Again the past offers us possible answers. For years I heard during congresses people proposing theories about how PDT works. Now these presentations are still there, but only as a kind of ritual update at the beginning of each conference essentially for those who would newly join the PDT community. But how PDT works really is still unsolved. We still do not even know how to predict the fluorescence quantum yield of a molecule in a cuvette. How do we know what the physical mechanisms *in vivo* are? How much do these mechanisms vary from one sensitizer to another? How relevant are they to any of them?

We recently screened a series of indirubin derivatives. The main conclusion was that efficacy was simply unpredictable not only on cells but also the yield of singlet oxygen ($^1O_2$) production *in vitro* (Olivier et al. 2008). A strong and intense research effort is needed to clearly establish the physical laws governing efficacy *in vivo*. This research in turn would help in modifying existing or next-generation sensitizers not randomly, but logically, and lead to new and appropriate sensitizers. A compilation of already performed studies could be a starting point, although it would consume much time and money, but it seems to me difficult and meaningless to start from scratch. Some data probably exist that an expert insight would reanimate and consider from a new angle thanks to the new analytic tools in chemistry.

A well-established intellectual approach in science is called an analogical process. A phenomenon is noted in nature and reproducing it leads to a new technological progress. Light is deleterious through induced photoreactions in the presence of oxygen for most living species. It is rather the exception than the rule that molecules and complex systems can survive at the earth's surface. Looking for photoreactive models of chemicals among those that are present at the surface of earth in plants or animals is an additional complication, particularly when we do not know clearly what makes a chemical photoreactive. To identify a photosensitizer from our on-earth position, we actually have first to imagine how to make the molecule photoreactive and then test it. Several drugs within those already in clinical assays are of marine origin (i.e., Tookad) (US patent 08071648-1993). For 4 years we have been looking in a cooperative project with IFREMER (an oceanographic institution in France) for photosensitizers in marine animals (i.e., bacteria or microalgae). When we submitted our project to get funding, we estimated that 5% of extracts could be photoreactive. Reviewers claimed that a maximum of 2% would be more realistic. In fact, nearly 50% of the 140 algae extracts tested are photosensitive, some of them being 30 times more photoreactive than m-THPC that had been used as our gold standard for a long time (Figure 70.1) (Morlet et al. 1995).

Some of the photoreactive substances discovered within this project are similar to already known chemical structures, but something different makes them much more reactive than what would have been expected from the simple basic chemical structure of the core of the molecule. This smart detail is not yet completely identified, but it is likely that new classes of powerful and biocompatible photosensitizers will be discovered.

If many photosensitizers absorb in the red, there are much less producing $^1O_2$ with a high quantum yield, and even less working *in vivo* with a reasonable efficacy. In the absence of development of innovative sensitizers, an alternative approach could be the use of two-photon excitation (2PA). Two photons hitting their target in an interval of time much shorter than the relaxation time of the molecule (i.e., a femtosecond) would produce an effect comparable to a photon of half the wavelength of 2PA photons. Two photons at 800 nm would thus have the high penetration of 800 nm of light into tissues but the photochemical effect of a 400 nm photon. We have plenty of chemicals absorbing at 400 nm likely to support such an excitation. This approach had been considered as early as 1997 and is applicable to already designed sensitizers

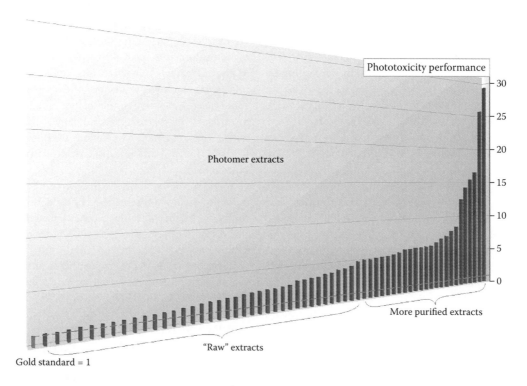

FIGURE 70.1 Nearly 50% of the 140 algae extracts tested are photosensitive, some being 30 times more photoreactive than m-THPC, which has been used as our long-time gold standard. (From Morlet, L. et al., *J Photochem Photobiol B* 28:25–32, 1995.)

(Beck et al. 2007; Fisher et al. 1997). The limit of this highly exciting (*stricto sensu*) field is that there is no device available likely to be used in clinics because 2PA requires a monomodal transmission, something possible only in short and extremely thin optic fibers that in addition can resist this high power without destruction. Some experiments using new types of hollow-core crystal fibers have been performed. Another limit is represented by the fact that sensitizers used in the past as single-photon absorbers have low 2PA cross-section values and resulted in a low PDT effect, which means that 2PA PDT requires new drugs based on existing ones but modified in order to increase the 2PA cross section (Ogawa and Kobuke 2008). When (and if) this kind of technology will be ready to enter the clinical area, it certainly would change the whole PDT process itself and all applications would have to reconsider their parameters.

All the above innovations are likely to occur in the near future and could in addition be mixed together and combined into completely new procedures. Each could completely renew the PDT world and new indications could arise. The reason for this is that PDT is a very young procedure by comparison to others such as radiotherapy. Working and fighting during a lifelong attempt to push PDT forward can sometimes lead us to forget that fact. The main effect of what has been done until now is to leave us standing at a starting line, and most of the progress has yet to be made. This youthfulness can be considered a major advantage for PDT compared to the situation that exists in most other technical fields of medicine. The possible future progression is much more, and therefore, objective reasons to invest in that field are also much more compelling. It is also true when we consider PDT not globally but in particular medical fields. PDT was first used for the palliative treatment of recurrent breast cancers. This application disappeared almost completely with the exception of some compassionate applications, and then PDT applications appeared for early bronchial cancers (Edell and Cortese 1987; Hayata et al. 1984). In parallel, PDT was applied to corresponding early cancers in gastroenterology (Patrice et al. 1990) and in the head and neck field. The reasons for the disappearance of these applications are numerous, and it is useless to spend too much time detailing them, but the main one is simply that there are too few patients for a treatment that is expensive and does not efficiently challenge the surgeon. The palliative treatment of cancers had also been proposed, but the PDT efficacy is in question when the mass is huge, metastases widespread, and the expected survival time is short. Of these old applications, only skin PDT remains as a routine application, because PDT applications in dermatology can perform better than the surgeon's knife with the use of 5-aminolevulinic acid (ALA) or its derivatives as prodrugs.

Oncology applications that I can see developing in the future are in the skin area and in neurosurgery. Skin, because the healing effect after PDT is beautiful, and the access to light delivery is simple. PDT is thus easy to manage and results are good, especially when lesions are numerous in a single patient. Depending on how fast new sensitizers will arrive on the market, if any, endoscopic applications for early cancers could be reevaluated in combination with other more systemic treatment modalities. Neurosurgery raises specific problems that PDT could also solve. Malignant brain tumors are often impossible to be surgically removed, and since the ratio of sensitizer concentration in tumor/healthy tissues is high (Kostron 2010), it offers an additional possibility to better visualize the tumor during surgery. However, despite much hard work in the field, results were and are still not yet at the stage of a significant improvement (Kostron 2010). One reason for that, besides the extreme malignancy of glioblastoma, is that even with a high tumor/normal ratio, the concentration within tumor tissues remains low due to the existence of the blood–brain barrier, a natural filter isolating the brain matter from the rest of the circulation. As well, tumor cell uptake is heterogeneous, with some cells being strongly loaded and some others not. Improving the power of photosensitizers would dramatically change things and could make PDT the standard treatment for these awful tumors. Good treatment efficacy could be achieved even with a low sensitizer concentration and a heterogeneous distribution within the cells. All tumor cells would thus be efficiently treated—some more than others—but with an overall favorable outcome.

If PDT could become more efficient, thus providing a good result in such a short duration that it would not necessitate specific procedures for the patient (i.e., anesthesia), it would find more indications and thus market and investments would grow in parallel. Among unsolved problems that could be addressed by PDT in oncology is Barrett's esophagus. This is the transformation caused by acid aggression during chronic gastroesophageal reflux of the normal esophageal mucosa into a new abnormal form, initially attributed to Norman Barrett (1957) but described earlier by others, with a high risk of malignant transformation into an adenocarcinoma of the lower part of the esophagus. This abnormal mucosa cannot regress spontaneously, it is spread over variable distances in the esophagus, and it occupies a variable portion of its circumference. Within this area, there could be deep areas of tissues already transformed into cancers. The speed of the transformation is unpredictable and obliges a systematic long-term follow-up to detect cancers as early as possible, leading to significant costs (around $50,000 per patient for a prevalence of 1% over 60 years old) to which the question of the ratio cost–benefit over risk is open. Preventive surgical resection is impossible due to a morbidity higher than the one induced by the cancer. Many endoscopic treatments have been proposed but all are lengthy, requiring many sessions, and provide a heterogeneous response over the Barrett's surface. PDT could be the only modality able to destroy the whole surface homogeneously, deep enough to prevent any recurrence but superficial enough to respect the esophageal wall and avoid any risk of perforation. The purpose of this surface treatment would be to remove totally the Barrett's mucosa and thus limit further follow-up to a simple and limited endoscopy. Again, it implies a fast and reliable treatment, with an adjustable depth that is possible in PDT by simply tuning light delivery. Recent analysis of posttreatment aneuploidy could additionally improve the follow-up efficacy (Dunn et al. 2010).

Surprisingly, PDT that had been proposed in previous times to be a smart treatment, recognizing its target at the cellular level, now finds its applications for tomorrow in wide surface treatments or large volumes. Tomorrow I would also bet on antimicrobial PDT becoming widespread simply because $^1O_2$ is a reactive species already involved naturally in our human defenses against microorganisms (Hampton, Kettle, and Winterbourn 1998). I do not think that this would someday totally replace antibiotics, but it could limit developing resistance to antibiotics. Rather, in a number of septic indications, it could be much more efficient than classical approaches and as another benefit reduce doses of antibiotics (Dai, Huang, and Hamblin 2009). There are several clinical indications where bioavailability of antibiotics is poor, particularly for anatomical reasons, something that can only be partially bypassed by increasing drug doses. Infectious arthritis or periodontitis are among these indications for which classical treatments can be challenging. One main problem that remains is that susceptibility of bacteria to photosensitizers needs to be largely improved, particularly for Gram-negative bacteria. Related to such applications is the problem of water purification in developing countries. Years ago, we suggested that PDT could be a treatment of choice in countries with a low per-capita income for cost reasons (Patrice 1999). More recently, advances in PDT applications suggested that PDT could be efficient for destroying bacteria and parasites using sunlight as the source of energy (Abu Samra et al. 2011). If it were proved to be realistic on a large scale, using more efficient sensitizers than the ones already available could represent a major PDT application affecting potentially millions of people.

Besides humans, PDT could also find applications in pets thanks to its cost-effectiveness. Fibrosarcoma is a terrible cancer arising in cats with a multifocal form and can be vaccine vehicle-induced. Surgery associated with radiotherapy is the most often cited treatment approach. The detection by fluorescence of all diseased areas that would allow individual treatment of each tumor site may be an alternative to the suggested but perhaps unrealistic 7 cm safety excision zone around a tumor. Sterilization of the tumor bed after a surgical resection could represent a reasonable option if PDT could be performed fast enough, thus using high-performance sensitizers for these poorly vascularized tumors. Skin epithelioma of the nose in cats may also benefit from PDT as they are frequently relapsing and surgery is limited by its collateral damage (Bexfield et al. 2008).

Among its other capabilities, PDT had been demonstrated by several authors (Gollnick and Brackett 2010; Luna et al. 2000) to stimulate antitumor immunity. Although the mechanism is unclear, it may be related to the indirect effects that have been documented by Chakraborty et al. (2009) and Olivier et al. (2009). PDT could be considered not only as a local treatment *per se* but also as a source of $^1O_2$, inducing a series of secondary oxidative species having various capabilities. In such a respect, PDT would not be a therapy but a simple tool allowing to explore oxidation either in physiological or in pathological situations. Singlet oxygen has the unsurpassable advantages of being a natural species, generated *in situ* without inducing any artifact and able to directly react with organic compounds, thus

bypassing the spin interdiction, making O_2 in its triplet ground state nonreactive. Any study exploring oxidation processes from energy production to inflammatory reactions could benefit from $^1O_2$ used as a source of oxidation. Photoreactions would thus be of major importance and would find applications within most basic research fields.

In addition, $^1O_2$ is by itself a mysterious species of a short lifetime, varying with solvents from nanoseconds to microseconds (Bensasson, Truscott, and Land 1993; Snyder et al. 2006), and is difficult to detect and quantify. However, its role during evolution started probably at the same time that oxygen began to be used for the metabolism of archeobacteria. Exploring the way it is produced and deactivated is also a way to study how life learned to manage oxidative processes and how it adapted or eventually governed oxidation-induced changes during evolution. If the very first cells were submitted to a Darwinist pressure, why should it not be considered that it is still the case, particularly for cancer cells? A cancer cell is a part of its host and the question is to know what made it diverge from the regular cell pathways. Mutations or aberrant genetic expression are involved, but applying Darwinism, one could also consider that these are secondary to an external pressure, among which could be oxidative stress. Cancer cells would therefore be, from such a point of view, normal cells responding abnormally to an external pressure and adapting. Identifying how cells of today changed step by step from those of yesterday, exhibiting the evidence of these changes in the nonexpressed part of the genome, could provide by analogy new strategies for the control of cancer cell growth. PDT, which is a unique source of $^1O_2$, would consequently play a major role within this strategy.

What makes PDT a promising medical procedure is its cost-effectiveness, which has been documented in various medical fields. However, the structure of medical expenses in our developed countries, whatever the level of analysis—for instance big-pharma companies, hospitals, doctors, or insurance companies—is not in favor of cheap treatment modalities. Each group with the exception of patients has a direct interest in using expensive methods. Therefore, PDT that has a comparable efficacy but is cheaper than other techniques could be significantly attractive either in countries with low revenues or in the medical fields where there will be no or little overlap with existing methods. The development in developing countries is slow as they generally lack sufficient cash income. When there is no real standard PDT treatment, development is also slow as everything has to be demonstrated from the concept to the results in clinics. It is thus of critical importance to analyze precisely the markets targeted. Thanks to the debt crisis, in the future, one can expect a change in the reimbursement philosophy of health expenditures in a way that would reinforce PDT.

We started from the recent past at the beginning of this chapter to end with a future located in the deep past. This may in some ways be a kind of summary of the whole of PDT history: a modern technology using a brilliant light but finding its roots in the dark of our origins to someday lighten our future. It is impossible to imagine that PDT, using light, oxygen, and molecules—most of which with a tetrapyrrolic ring as a core are agents of critical importance for life on earth—would simply vanish from the arsenal of our treatment tools without providing any result. I am ready to bet on a bright PDT future, despite some present difficulties related to the way we use it.

References

Abu Samra, N., F. Jori, A. Samie, and P. Thompson. 2011. The prevalence of *Cryptosporidium* spp. oocysts in wild mammals in the Kruger National Park, South Africa. *Vet Parasitol* 175:155–159.

Barrett, N. R. 1957. The lower esophagus lined by columnar epithelium. *Surgery* 41:881–894.

Beck, T. J., M. Burkanas, S. Bagdonas et al. 2007. Two-photon photodynamic therapy of C6 cells by means of 5-aminolevulinic acid induced protoporphyrin IX. *J Photochem Photobiol B* 87:174–182.

Bensasson, R. V., T. G. Truscott, and E. J. Land. 1993. *Excited States and Free Radicals in Biology and Medicine: Contributions from Flash Photolysis and Pulse Radiolysis.* Oxford University Press, New York.

Bexfield, N. H., A. J. Stell, R. N. Gear, and J. M. Dobson. 2008. Photodynamic therapy of superficial nasal planum squamous cell carcinomas in cats: 55 cases. *J Vet Intern Med* 22:1385–1389.

Chakraborty, A., K. D. Held, K. M. Prise et al. 2009. Bystander effects induced by diffusing mediators after photodynamic stress. *Radiat Res* 172:74–81.

Dai, T., Y. Y. Huang, and M. R. Hamblin. 2009. Photodynamic therapy for localized infections—State of the art. *Photodiagnosis Photodyn Ther* 6:170–188.

Dunn, J. M., G. D. Mackenzie, D. Oukrif et al. 2010. Image cytometry accurately detects DNA ploidy abnormalities and predicts late relapse to high-grade dysplasia and adenocarcinoma in Barrett's oesophagus following photodynamic therapy. *Br J Cancer* 102:1608–1617.

Edell, E. S., and D. A. Cortese. 1987. Bronchoscopic phototherapy with hematoporphyrin derivative for treatment of localized bronchogenic carcinoma: A 5-year experience. *Mayo Clin Proc* 62:8–14.

Fisher, W. G., W. P. Partridge, Jr., C. Dees, and E. A. Wachter. 1997. Simultaneous two-photon activation of type-I photodynamic therapy agents. *Photochem Photobiol* 66:141–155.

Gollnick, S. O., and C. M. Brackett. 2010. Enhancement of antitumor immunity by photodynamic therapy. *Immunol Res* 46:216–226.

Hampton, M. B., A. J. Kettle, and C. C. Winterbourn. 1998. Inside the neutrophil phagosome: Oxidants, myeloperoxidase and bacterial killing. *Blood* 92:3007–3017.

Hayata, Y., H. Kato, C. Konaka et al. 1984. Photoradiation therapy with hematoporphyrin derivative in early and stage 1 lung cancer. *Chest* 86:169–177.

Kostron, H. 2010. Photodynamic diagnosis and therapy and the brain. *Methods Mol Biol* 635:261–280.

Luna, M. C., A. Ferrario, S. Wong, A. M. Fisher, and C. J. Gomer. 2000. Photodynamic therapy-mediated oxidative stress as a molecular switch for the temporal expression of genes ligated to the human heat shock promoter. *Cancer Res* 60:1637–1644.

Morlet, L., V. Vonarx-Coinsmann, P. Lenz et al. 1995. Correlation between meta(tetrahydroxyphenyl)chlorin (m-THPC) biodistribution and photodynamic effects in mice. *J Photochem Photobiol B* 28:25–32.

Ogawa, K., and Y. Kobuke. 2008. Recent advances in two-photon photodynamic therapy. *Anticancer Agents Med Chem* 8:269–279.

Olivier, D., S. Douillard, I. Lhommeau, and T. Patrice. 2009. Photodynamic treatment of culture medium containing serum induces long-lasting toxicity in vitro. *Radiat Res* 172:451–462.

Olivier, D., M. A. Poincelot, S. Douillard et al. 2008. Photoreactivity of indirubin derivatives. *Photochem Photobiol Sci* 7:328–336.

Patrice, T. 1999. Photodynamic therapy in developing countries. *Rev Contemp Pharmacol* 10:75–78.

Patrice, T., M. T. Foultier, S. Yactayo et al. 1990. Endoscopic photodynamic therapy with hematoporphyrin derivative for primary treatment of gastrointestinal neoplasms in inoperable patients. *Dig Dis Sci* 35:545–552.

Snyder, J. W., E. Skovsen, J. D. Lambert, L. Poulsen, and P. R. Ogilby. 2006. Optical detection of singlet oxygen from single cells. *Phys Chem Chem Phys* 8:4280–4293.

Index

Page numbers followed by f and t indicate figures and tables, respectively.

T - #0195 - 111024 - C886 - 280/210/41 - PB - 9780367576295 - Gloss Lamination